13th Americas Conference on Wind Engineering 2017

Gainesville, Florida, USA
21 - 24 May 2017

Volume 1 of 2

ISBN: 978-1-5108-5482-6

Printed from e-media with permission by:

Curran Associates, Inc.
57 Morehouse Lane
Red Hook, NY 12571

proceedings.com

Some format issues inherent in the e-media version may also appear in this print version.

Printed by Curran Associates, Inc. (2018)

American Association for Wind Engineering (AAWE)
c/o Cermak Peterka Peterson, Inc.
1415 Blue Spruce Drive, #3
Fort Collins, Colorado 80524
USA

Phone: (970) 498-2334
Fax: (970) 221-3124

info@aawe.org

Additional copies of this publication are available from:

Curran Associates, Inc.
57 Morehouse Lane
Red Hook, NY 12571 USA
Phone: 845-758-0400
Fax: 845-758-2633
Email: curran@proceedings.com
Web: www.proceedings.com

TABLE OF CONTENTS

VOLUME 1

KEYNOTE SPEAKERS

BLUFF BODY AERODYNAMICS I

BLUFF BODY AERODYNAMICS II

BRIDGES

CATASTROPHE MODELING I

CATASTROPHE MODELING II

CATASTROPHE MODELING III

CATASTROPHE MODELING IV

CODES AND STANDARDS I

CODES AND STANDARDS II

CODES AND STANDARDS III

CODES AND STANDARDS IV

COMPUTATIONAL WIND ENGINEERING AND CYBERPHYSICAL MODELING I

COMPUTATIONAL WIND ENGINEERING AND CYBERPHYSICAL MODELING II

HIGH-RISE AND SUPER-TALL BUILDINGS III

LOW-RISE BUILDINGS I

LOW-RISE BUILDINGS II

LOW-RISE BUILDINGS III

VOLUME 2

NHERI II

OTHER STRUCTURES AND WIND ENGINEERING APPLICATIONS I

OTHER STRUCTURES AND WIND ENGINEERING APPLICATIONS II

POSTER SESSION

RESILIENCE

THUNDERSTORMS AND DOWNBURSTS

TORNADOES I

TORNADOES II

TORNADOES III

TORNADOES IV

TORNADOES V

TORNADOES VI

UTILITIES

WIND DAMAGE ASSESSMENT

WIND ENERGY

WIND-DRIVEN RAIN, DEBRIS AND EMBERS

The 13th Americas Conference on Wind Engineering (13ACWE)
Gainesville, Florida USA, May 21-24, 2017

A framework for the aerodynamics of low-rise buildings in tornadoes: Can boundary layer wind tunnels give us everything we need?

Gregory A. Kopp, Chieh-Hsun Wu

Boundary Layer Wind Tunnel Laboratory, Faculty of Engineering,
University of Western Ontario, London, ON, Canada

ABSTRACT: A comprehensive understanding of wind loads in tornadoes is currently lacking. In the paper, we examine the effects of turbulence and flow structure on the wind loads of roofs of low-rise buildings in tornadoes. To provide a framework for assessing these wind loads, quasi-steady theory models obtained from simultaneous velocity and pressure measurements are used. This allows the differences between the tornado and atmospheric boundary layer flow structures to be examined in some detail. The hypothesis that tornadic wind loads are not substantially different than those in ABLs is examined within the framework.

1 INTRODUCTION

1.1 *Background*

There has been increased recent interest in the wind loads on buildings in tornadoes. In fact, papers on tornadoes at wind engineering conferences is one of the most rapidly growing areas. This is due, in part, to the many new tornado-like vortex simulators that are emerging around the world including those at Iowa State University (Haan et al., 2010; Hu et al. 2011), Texas Tech University (Mishra et al., 2008a,b), Tokyo Polytechnic University (Sabaheesh et al., 2012), the University of Western Ontario (Refan and Hangan, 2016), amongst others. There has also been growing interest in considering design methods for tornadoes, perhaps to combat the significant economic losses that are incurred annually, which are of a similar order to those caused by hurricanes in the United States (Simmons et al., 2015). Quantifying wind loads in tornadoes is critical for a range of purposes, including the development of loss/risk models for the insurance industry, for engineering design considerations, and even for public interest in severe weather. The recently-formed ASCE Standards Committee on Tornado Wind Speed Estimation is re-visiting the Enhanced Fujita (EF) Scale and the methods for determining tornado wind speeds, while it is expected that ASCE 7-22 will have a chapter on design wind loads for tornadoes.

What is currently lacking is a comprehensive understanding of the nature of building aerodynamics in tornadoes. As such, wind loads in tornadoes are often assessed using boundary layer wind tunnel results directly (e.g., Kopp et al., 2017; Gavanski and Kopp, 2017) or via boundary layer wind tunnel results with multiplication factors to account for tornadic wind effects (e.g., Amini and van de Lindt, 2013; Van de Lindt et al., 2013). Gavanski and Kopp (2017) argue that use of boundary layer wind tunnel results should give upper-bound estimates of failure-inducing wind speeds, with any tornadic effects reducing the speeds by increasing the aerodynamic coefficients, as caused by, for example, the vertical component of the wind or the effects of the static pressure drop in the core of tornadoes. The tornado-vortex-simulator results of Haan et al. (2010) suggest that the multiplication factor could be in the range from 1.8 to 3.2, although this assumes a perfectly-sealed building, which is not the case for almost all buildings. Such increases, if accurate, would decrease the failure-inducing wind speeds by 25 – 45%. In terms of wind loads, there is a need to establish the external aerodynamics of buildings in tornadoes as well as the details of the internal pressures. The objective of this paper is to address the question of the external building aerodynamics, with a focus on low-slope roofs on low-rise buildings. Internal pressure dynamics will not be examined herein, nor the myriad of issues with respect to determining tornado wind speeds via damage.

1.2 Approach

To resolve the question of external building aerodynamics in tornado wind fields, experimental work with direct comparisons to results from boundary layer wind tunnels is required. In particular, we need (i) a common framework for assessing the wind loads, (ii) a quantitative understanding of the role of small-scale and large-scale turbulence on wind loads, (iii) a quantitative understanding of the role of wind direction, particularly the vertical components of the wind, to provide a basis for (iv) comparative experiments.

1.2.1 Quasi-steady theory

Quasi-Steady Theory (QST) assumes that pressure fluctuations can be estimated solely from simultaneous velocity fluctuations. In other words, the pressure is a function solely of the velocity, $\vec{V}$, for any location on any building, and not on the rate of change of velocity, $d\vec{V}/dt$. This is a significant simplification compared to the Navier-Stokes equations, which indicate that the surface pressure fluctuations are an integral over the flow domain of a function of the velocity gradient tensor. Banks and Meroney (2001) examined QST detail and have confirmed, for corner vortices on low-rise buildings, that point pressures do, indeed, depend on $d\vec{V}/dt$. However, Wu and Kopp (2016), using a similar approach, indicated that QST is able to estimate area-averaged pressures with good accuracy. In this paper, the hypothesis being investigated is that QST is a viable framework for determining area-averaged tornadic wind loads on low-rise buildings.

Several caveats are required. There is some significant complexity related to variations of the tornado wind field caused by changes in the basic vortex structure as a function of swirl ratio, along with the (mostly unknown) effects of ground roughness on the mean and turbulent wind fields close to the ground. Turbulence levels in tornadoes are also unknown, as is the energy content at various scales (i.e., velocity spectra). For our purposes here, we will assume that the tornado is large compared to the size of the building such that the instantaneous velocity and static-pressure gradients are relatively small across the building, although these will be assessed. Related to this, the swirl ratios considered will be such that the flow structure is (at least nominally) a single core vortex. Of course, the turbulence structure and transient effects may be different between tornadoes and atmospheric boundary layers, and this may alter the wind loads. In fact, this is primarily what is being assessed with the current approach.

1.2.2 Effects of turbulence

There are still open issues regarding the effects of turbulence for low-rise-building wind loads. For example, the roles of small-scale and large-scale turbulence are not fully resolved such that concepts like Peter Irwin's Partial Turbulence Simulation (Irwin, 2008; Asghari-Mooneghi et al., 2016) remain controversial. Concepts such as how the energy levels at various scales alter the wind loads remain unresolved, as well as even whether using such scale cut-offs can be fully corrected.

In general, wind tunnel testing approaches for high-rise buildings avoid this issue by scaling the entire boundary layer, including the full spectrum of turbulence, and directly measuring the resulting buffeting loads. (No discussion of resonant loads will be examined herein, since the focus is on low-rise buildings and short duration of loading.) Since the mean wind tunnel speed is controlled, results are initially related to mean speeds, with gust speeds estimated from velocity profiles measured independently of the wind loading experiments. In this sense, the traditional testing approach treats the wind field as a black box that acts as a transfer function to convert the input mean wind speed into fluctuating loads or peak loads used in design. Knowledge of the roles of the various scales of turbulence is not required, as long as they are fully resolved in the wind tunnel boundary layer.

For low-rise buildings, things are not as simple, as pointed out so elegantly by Tieleman (2003). Usually, it is assumed that an integral scale within a factor of two of the full-scale value is considered sufficient, with the smaller-scales matched precisely (Surry, 1982; Irwin, 2008). While it is known that corrections to wind loads for too-small-values of the integral scale are needed, these are not generally conducted in commercial, or even research, studies. This will have a more profound effect when large model scales, such as 1/20 in testing of small components like roof-mounted solar arrays, are used in boundary layer wind tunnels designed for typical length scales an order of magnitude smaller, such as 1/200. There have been approaches to address this issue, such as Irwin's correction of the statistically-estimated peak loads using QST (Asghari-Mooneghi et al., 2016; after Irwin, 2008), although there is not yet a general consensus on how to deal with these issues, as noted above.

Returning to tornadoes, the above discussion raises many questions. Several of the research groups examining tornado wind loads are using the traditional approach of having a fully-scaled tornado vortex, which they translate over a building model, and which is assumed to be at the same scale. Aerodynamic coefficients are formed from the instantaneously-measured, dimensional forces (or pressures) using separately-measured wind speeds and static pressures (which are usually measured outside of the vortex flow). This will work as long as one can be assured that the tornado-like vortices are correct in every respect including the turbulent structure and scales. However, we suspect that one runs into the same problems that currently exist for low-rise building aerodynamics. Questions arise as to what to do to correct the test methods if we need to (or actually have) a scale mismatch or want tornadoes of a slightly different size, or ground roughness, or translation speed, etc. For determination of design wind loads, one would need to create many scenarios to account for the range of possibilities, or develop a consensus on what the design tornado should be. Testing parameters could be endless! To address this, we posit that methods that quantify the effects of turbulent scales and energy levels on wind loads are required. The quasi-steady theory provides the necessary framework to do so.

1.3 Hypotheses for Building Aerodynamics in Tornadoes

Two hypotheses are examined in this paper. The first, as outlined above, is that a quasi-steady model can account for tornado-induced wind loads as a function of the magnitude and direction of the wind. To accomplish this, we use the approach of Wu and Kopp (2016), as applied to area-averaged roof pressures.

The second hypothesis, which will be more controversial, is that we can develop the quasi-steady model using boundary layer wind tunnel data for the same building model. The importance of this hypothesis is that it provides a framework for assessing how similar or different tornado wind loads are as compared to atmospheric boundary layer wind loads on low-rise buildings. With such a model, if successful, we could conceivably account for changes in wind loads analytically for a variety of tornado wind fields. In other words, such a model would allow for changes in the translation speed of the vortex, the tornado size, the strength of rotation, etc., within certain, well-defined bounds. The explicit assumptions are made that the tornado vortex has a single core (i.e., is not multi-core) and that the building dimensions are small compared to the size of the vortex core. Thus, the objective of the paper, stated more precisely than in Section 1.1, is to develop and assess a quasi-steady model for low-rise building aerodynamics in tornado wind fields using boundary layer wind tunnel data as the basis for the model. This provides a framework to determine how tornado-like vortices may alter the building aerodynamics.

1.4 Structure of the Paper

In Section 2, the Quasi-Steady (QS) Model is developed, based on boundary layer wind tunnel test data in an open terrain. In Section 3, the QS Model is assessed by examining its performance in predicting wind load statistics in both smoother and rougher terrains. Since the model will be shown to be reasonably effective, Section 4 will provide a brief overview of the aerodynamic mechanisms such that appropriate limits can be set. Section 5 will discuss the tornado-vortex experimental set-up and approach in the WindEEE Dome, while Section 6 provides preliminary results and comparisons with BLWT results. Section 7 will provide final comments, conclusions and future research needs.

2 QUASI-STEADY MODEL FOR UPLIFT ON THE TTU-WERFL BUILDING

2.1 Boundary Layer Wind Tunnel Experimental Set-up

In the current work, the measurement configurations are identical to the experimental set-up of Wu and Kopp (2016), but additional terrain roughnesses are considered. The six terrains, measured in Boundary Layer Wind Tunnel 2 (BLWT2) at the University of Western Ontario (UWO), are identical to those used by Akon and Kopp (2016), with the important turbulence characteristics provided in Table 1 and Figure 1. Table 1 provides the roof height turbulence intensities, I_u, the ratio of the streamwise integral scale to building height, L_x/H, and the Jensen numbers, H/z_o. The terrain labels are for (approximately) open water (1), open country (2), and suburban (3) terrains. The label 'S' indicates a smaller integral scale, while the label 'L' indicates a larger integral scale. For further details, see Akon and Kopp (2016). Figure 1 provides the streamwise velocity spectra at the roof height for the six terrains, from which the differences in turbulence energy at scales relative to the building height can be seen. The results for terrains 1S, 2S and 3L will the primary ones presented herein, due to space limitations.

Table 1: Upstream turbulence characteristics of six conditions simulated in BLWT2

Terrain	I_u (%)	L_x/H	H/z_o
1S	13	6	540
1L	14	13	600
2S	17	8	290
2L	17	11	600
3S	26	7	56
3L	27	12	71

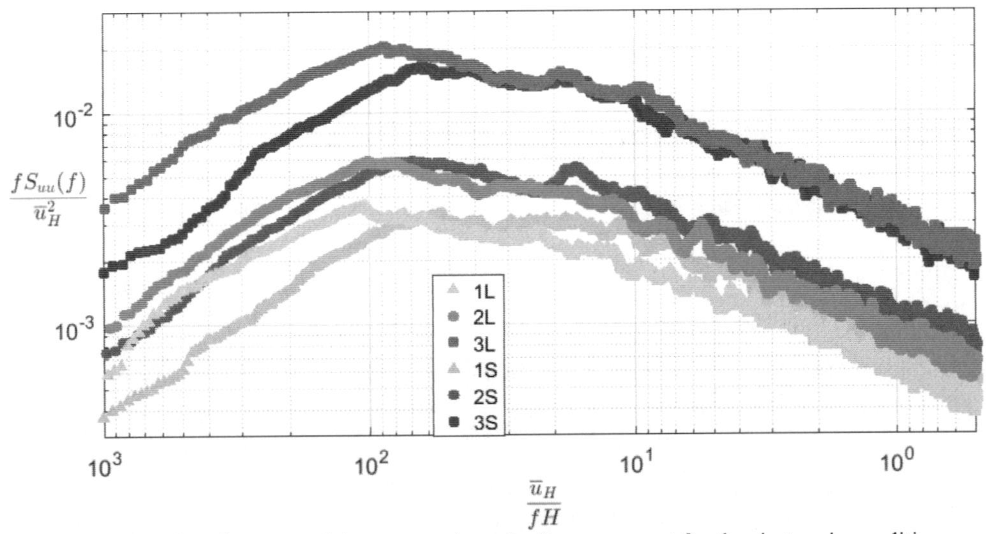

Figure 1. Reduced spectra of the streamwise velocity component for the six terrain conditions.

Surface pressure measurements were made on a 1/50 geometrically-scaled model of Texas Tech University (TTU) 'WERFL' Building (Levitan and Mehta, 1992a,b; Smith, 2010). The modelled building has plan dimensions of 18.3 cm × 27.5 cm and an eave height of 7.8 cm. Figure 2(a) shows the 204 uniformly-distributed taps. The Cartesian coordinate system used to define the space and velocity components is also included in the figure. Based on this definition, wind direction normal to the shorter wall is 0° in azimuth, whereas the wind direction normal to the longer wall is 90° in azimuth. The building model was placed in the high-speed test section of BLWT2, where the six upstream terrain roughness conditions were applied. Detailed tubing system and frequency responses for the pressure measurement can be found in Ho et al. (2005). The Reynolds number for these experiments was $Re = \overline{U_H}H/\upsilon$ = 53,000, where $\overline{U_H}$ is the mean wind speed at the eave height, H is the eave height, and ν is the kinematic viscosity.

Simultaneous measurements of wind velocity at a point location were also conducted, as described by Wu and Kopp (2016). A Cobra probe (TFI Corp., Model 900) was placed one building height above the leading edge of the roof surface at the mid-plane of the long wall. This point location of the velocity measurements, which is denoted as m, and shown in Figure 2(a), was selected to obtain the velocities representative of the flow at the building location, while minimizing the effects on the building pressures. The particle image velocimetry (PIV) measurements of Akon and Kopp (2016) and Akon (2017) indicate that this location is outside of the region where there are significant distortion effects caused by the building. The synchronized velocity and pressure measurements were conducted for mean wind azimuth angles varied from 0° to -90°, in increments of 5° (see Figure 2a). Furthermore, the measurement point, m, was fixed with respect to the building for each mean wind azimuth angle in order to analyze the velocity-pressure data using QST. The synchronized pressure and velocity time series were sampled at 625 Hz for 200 seconds for each (mean) wind direction. Further details pertaining to these experiments can be found in Wu and Kopp (2016).

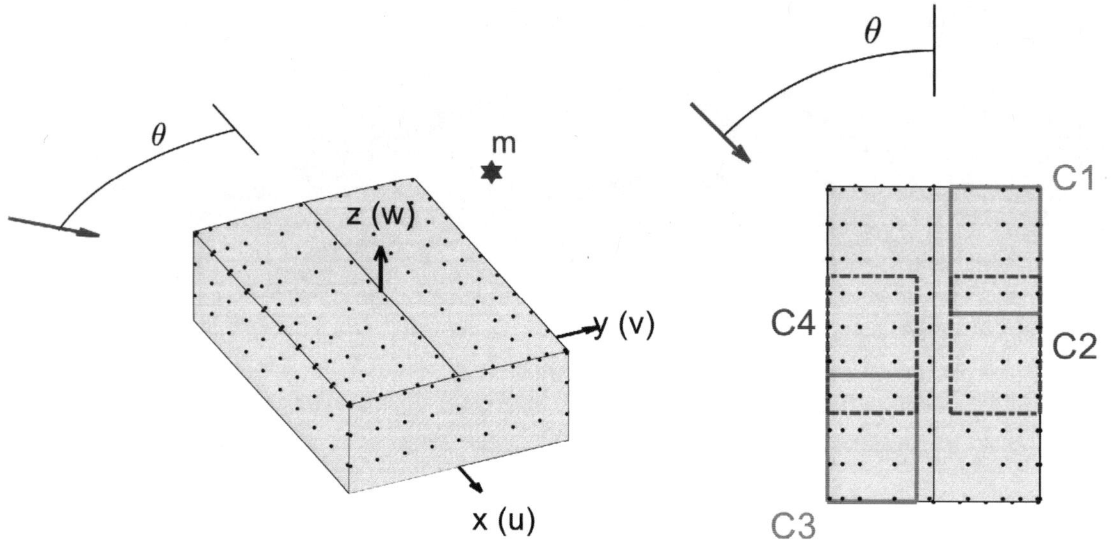

(a) (b)

Figure 2. (a) Pressure tap locations on the 1/50-scale TTU-WERFL building model, along with the cobra probe measurement location at point *m*. (b) Panel regions designated for area-averaged surface pressure analyses. Note that areas C1 and C2 overlap, as do C3 and C4.

2.2 Quasi-Steady Model

The quasi-steady theory has been a common approach in wind engineering for determination of fluctuating wind loads, either on low-rise buildings (e.g., Letchford et al., 1993) or tall buildings (e.g., Kawai, 1983). This method estimates the instantaneous roof surface pressure coefficients, p, from the instantaneous velocity vector, $\mathbf{u}_m$, at a point, m:

$$p - p_\infty = 0.5\,\rho\left|\mathbf{u}_m\right|^2 Cp_{inst} \tag{1}$$

By normalizing the instantaneous pressure to a predetermined dynamic pressure, i.e., $0.5\,\rho u_{ref}^2$, the pressure coefficient, Cp, can be defined as

$$Cp = \frac{p - p_\infty}{0.5\,\rho u_{ref}^2} \tag{2}$$

Hence, the QS estimation in Eq. (1) can alternatively be written in the form of a pressure coefficient, i.e.,

$$Cp = \frac{\left|\mathbf{u}_m\right|^2}{u_{ref}^2} Cp_{inst} \tag{3}$$

Note that the velocity vector used here is composed of the three components in Cartesian coordinate so that the magnitude of $\mathbf{u}$ satisfies

$$\left|\mathbf{u}\right|^2 = u^2 + v^2 + w^2 . \tag{4}$$

The corresponding azimuth, θ, and elevation, β, angles of the wind velocity vector are defined, respectively, as

$$\theta = \tan^{-1}\left(\frac{v}{u}\right) \quad \text{and} \quad \beta = \tan^{-1}\left(\frac{w}{\sqrt{u^2 + v^2}}\right) \tag{5}$$

On the right-hand side of Eq. (3), the ratio $|\mathbf{u}_m|^2/u_{ref}^2$ accounts for the contribution of instantaneous dynamic pressure, whereas Cp_{inst} denotes the 'instantaneous function,' which depends on the variations of the pressure that are not captured by the wind speed and direction variations. Thus, one of the advantages in considering the QS formulation in Eq. (3) is that the instantaneous building surface pressure is separated into the contribution from the instantaneous dynamic pressure and the contribution from other aerodynamic effects, such as the wind directions, temporal gradients, and body-generated turbulence embedded in Cp_{inst}.

The QS vector model includes both wind azimuth and elevation angles in order to relate the instantaneous wind vector to instantaneous surface pressures. The effects of wind azimuth variations are treated as non-linear functions and handled in the manner suggested by Richards et al. (1995). By assuming that the QS model holds for the measurements at each mean wind azimuth angle, the instantaneous function can be obtained from the information of mean building surface pressure coefficients and the probability of the wind azimuth fluctuation (see Richards et al., 1995; Banks and Meroney, 2001; Wu and Kopp, 2016). The final form of the instantaneous function can be conveniently represented by a Fourier series, i.e.,

$$Cp_{inst} = \sum_{k=0}^{N1} a_{1k}\cos(k\theta) + b_{1k}\sin(k\theta) \tag{6}$$

where a_{1k} and a_{2k} are the fitted coefficients for a k-th order model.

The wind elevation angle has been found to affect the variation of instantaneous pressure coefficients for regions on the roof (Letchford and Marwood, 1997; Sharma and Richards, 1999; Richards and Hoxey, 2004; Wu and Kopp, 2016). If a specific value of Θ is selected and, thus, $E[\beta|\Theta]$ is fixed and denoted as $\overline{\beta}$, the variation of Cp_{inst} due to changing β has been found to be approximately linear by various researchers (Letchford and Marwood, 1997; Sharma and Richards, 1999; Richards and Hoxey, 2004). Therefore, it can be written as

$$Cp_{inst}(\theta,\beta) = Cp_{inst}(\theta,\overline{\beta}) + B(\theta)\Delta\beta \tag{7}$$

where $B(\Theta)$ denotes the gradient, $dCp_{inst}/d\beta$, at specific Θ while $\Delta\beta$ represents the fluctuating elevation angle, $\beta - \overline{\beta}$. Note that the $Cp_{inst}(\theta,\overline{\beta})$ in Eq. (8) is represented by the Fourier series shown in Eq. (6). In the current work, the changes of Cp_{inst} due to the changes of β are found by subtracting $Cp_{inst}(\theta,\overline{\beta})$, defined in Eq. (6), from the conditionally-averaged values of $Cp_{inst}(\Theta, \beta)$ (Banks and Meroney, 2001), i.e.,

$$Cp_{inst}(\theta,\beta) = E[Cp_{inst}|\theta,\beta] \tag{8}$$

where $E[\bullet]$ represents the expected value, in this case conditioned on the instantaneous values of both θ and β. Because the gradient, $dCp_{inst}/d\beta$, may also vary with respect to Θ, $B(\Theta)$ can also be fit to a Fourier series, i.e.,

$$B(\theta) = \sum_{k=0}^{N_2} a_{2k}\cos(k\theta) + b_{2k}\sin(k\theta) \tag{9}$$

The QSM was developed for the two areas, C1 and C2, indicated in Figure 2(a) using the data from each of the six terrains. Each of these areas includes 16 pressure taps, with a total plan area (model scale) of 90cm^2, compared to the total plan area of the entire roof of 503 cm^2. In Wu and Kopp (2016), the QS models were developed using the instantaneous wind speed. In the current work, we consider temporally-averaged wind speeds that have equivalent spatial durations of L_c/H = 10, where L_c is the equivalent spatial length, $\Delta t \cdot \overline{u_H}$, and H is the height of the building model.

This moving-average approach improves the performance of the quasi-steady model by "filtering" uncorrelated small scale velocity fluctuations at the measurement location, as discussed below.

In addition, the pressure coefficients are defined such that later comparisons with the situation for tornado-like vortices can be made. Since tornadoes have substantial static pressure drops in the core, these effects must be removed (to the extent they can). To this end, the spatially-averaged pressure over the four walls of the building was removed from the external pressure. In other words, the instantaneous, wall-averaged pressure replaces p_∞ in Eq. (1) or (2) and the resulting pressure coefficients can be viewed as being approximately equal to net roof pressures for the case of a building with background wall leakage. In this way, the primary effects of the static pressure drop can be removed so that differences in the external aerodynamic loads can be assessed.

3 APPLICATION OF QUASI-STEADY MODEL IN BOUNDARY LAYER TURBULENCE

Figure 3(a) shows the $Cp_{inst}(\theta)$ results for area, C1, in the roof corner as a function of azimuth, θ, obtained via Eqn. (6). Details of the method can be found in Wu and Kopp (2016). The dashed lines represent the QS model obtained in each terrain, while the solid line is the model obtained in the terrain, 2S (i.e., open, smaller integral scale). Figure 3(b) shows the same, but for area, C2, along the edge of the roof near the middle of the longer side. One can see that there are some differences in the QS models, although the overall trends are similar. These figures were chosen to highlight the largest differences, which are about $\Delta Cp_{inst} \sim 0.1$ for terrain 3L.

Also included in Figure 3 are the conditional averages of $Cp_{inst}(\theta,\beta)$ for area C1 and $\bar{\theta} = -90^{\circ}$ and area C2 and $\bar{\theta} = -60^{\circ}$. Note that the strength of the colour in the plotted symbols relates to the number of points included in the conditional average, with "faded" points using many fewer data points – further details can be found in Wu and Kopp (2016). The range of wind directions in the conditional averages is a function of the velocity fluctuations in the terrain, where a much larger range of angles occurs for terrain 3L compared to 1S. For $-2.5^{\circ} < \beta < +2.5^{\circ}$, the conditional averages of $Cp_{inst}(\theta,\beta)$ closely follow the values for $Cp_{inst}(\theta)$, as would be expected. For positive elevation angles, i.e., upwardly-directed, the magnitude of Cp_{inst} is increased, while for negative (downwardly-directed) elevation angles, the magnitude of the loads decreases (as discussed in detail by Wu and Kopp, 2016).

Figure 4 provides example time series of the azimuth and elevation angles of the instantaneous wind vector, along with the Cp_{inst} and Cp, for panel areas C1 and C2 under the 2S terrain condition. Four seconds of the measured data obtained from $\bar{\theta} = -90^{\circ}$ are used for C1, and data obtained from $\bar{\theta} = -60^{\circ}$ is used for the panel area C2. For the azimuth and elevation angles, the instantaneous signals are plotted with dots. These signals are smoothed by applying the moving-average technique using an equivalent duration (or cut-off length) of $L_c = 0.8$m ($10H$).

The time series of Cp_{inst} shown in Figure 4 are obtained directly by dividing the instantaneous roof pressure by the smoothed instantaneous dynamic pressure (measured at location m). Note that the Cp_{inst} signal is highly smoothed because the dynamic pressure fluctuations are taken out in the calculation of Cp_{inst}. These measured Cp_{inst} are also compared to the QS model including smoothed azimuth angle (i.e., QS-θ) and both smoothed azimuth and elevation angles (i.e., QS-θ-β). The general match between the measured Cp_{inst} and both QS models can be observed in this plot, with clearly improved performance for the QS-θ-β model. However, it is also clear that the variations associated with rapidly peaks are not captured. However, this is better represented via the measured and QS-θ-β predicted Cp time series (see definition in Eqs. (2) and (3)) shown in Figure 4(d). Note that the θ and β used in the QS model for Cp prediction are the moving-averaged signals. Good agreement between measured and QS-θ-β predicted Cp can be observed, although the high frequency components are clearly absent. The accuracy is further assessed below, although the QS models can be seen to smooth the data.

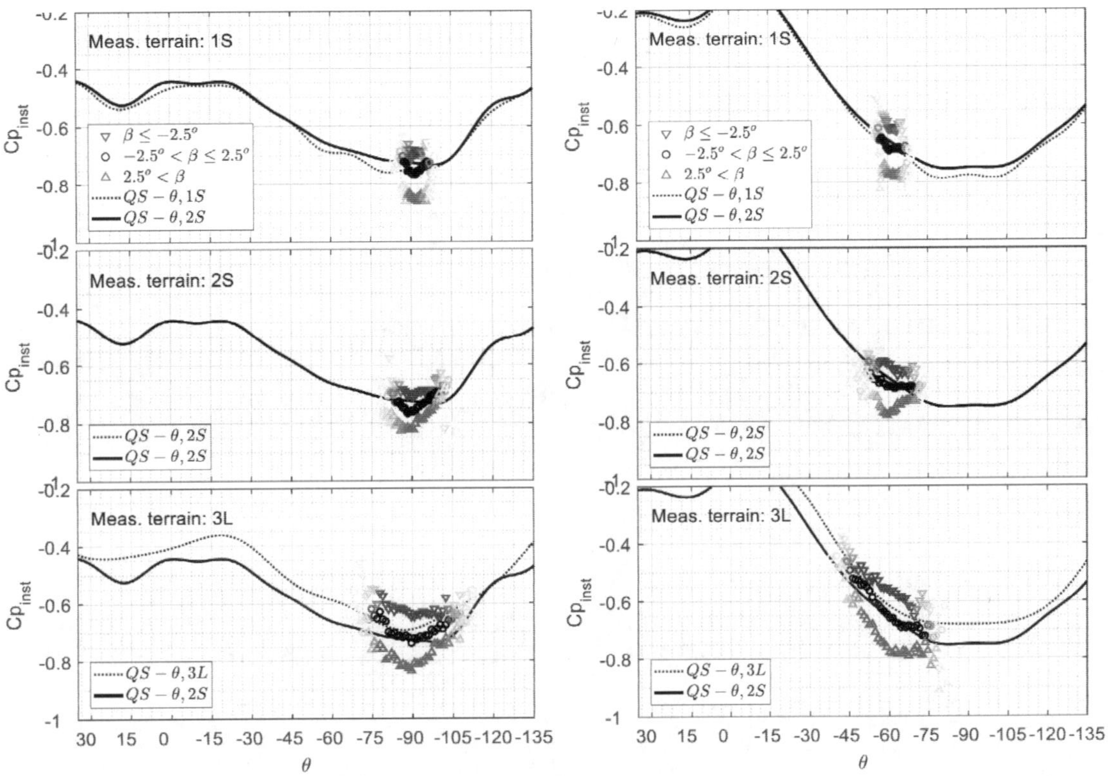

(a) (b)

Figure 3. Comparison of $Cp_{inst}(\Theta, \beta)$ established in the QS models to the $E[Cp_{inst} | \Theta, \beta]$ obtained from terrains 1S, 2S and 3L. Note that $L_c = 10H$ is used to filter the velocity signal: Results for (a) area C1, with $\bar{\theta}$ = -90° and (b) area C2 with $\bar{\theta}$ = -60°.

Figure 5 provides further information on the performance of the QS models developed in terrain 2S. The ratio of the power spectral density obtain via the QS model, $S_{pp,QS}$, to that obtained from the measurements, $S_{pp,M}$, is shown. The same conditions for Figures 3-4 are also considered in Figure 5. As can be seen, the ratio for wave numbers is close to 1 for the large scales ($\overline{u_m}/f >$ ~3m). The value of 3m is related to the low-pass-filtering effects of the moving average, which is for an equivalent length of $L_c = 10H = 0.8$m. For $\overline{u_m}/f < 0.8$m, all of the velocity fluctuations have been suppressed by the moving average. Thus, for the large-scale fluctuations, relative to the size of the building, the terrain-2S QS model is able to capture the pressure fluctuations. Small-scale fluctuations, associated with the body-generated turbulence are not captured when the moving-average is used because these are "filtered" by the area-averaging process. One can see that when the unfiltered velocity is used as an input, the high-frequency (small-scale) fluctuations are over-estimated using the QS model. It is an open question as to whether these should be included; however, herein we have generally chosen to remove them (for the tornado wind field where we use a time series approach), but examine the reasons for this in Section 4 below.

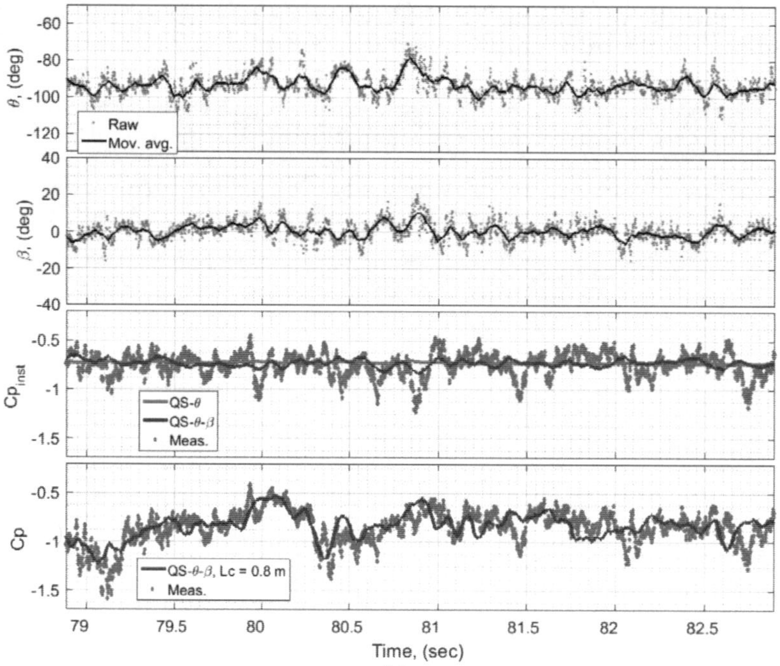

(a)

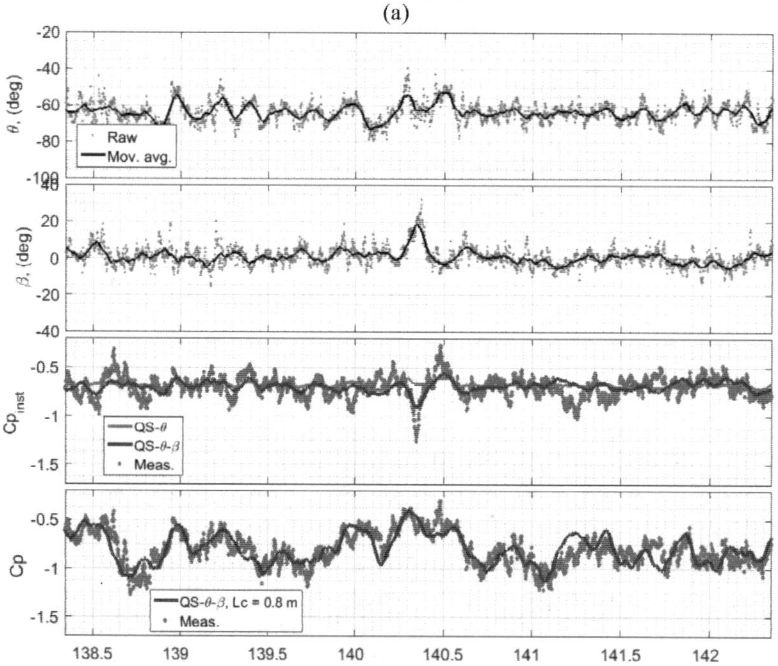

(b)

Figure 4. Raw and smoothed (moving averaged signals with $L_c = 10H$) wind directions and time series of Cp_{inst} and Cp at location, m, with terrain 2S. (a) Panel area C1, $\overline{\theta} = -90^{\circ}$; (b) area C2, $\overline{\theta} = -60^{\circ}$.

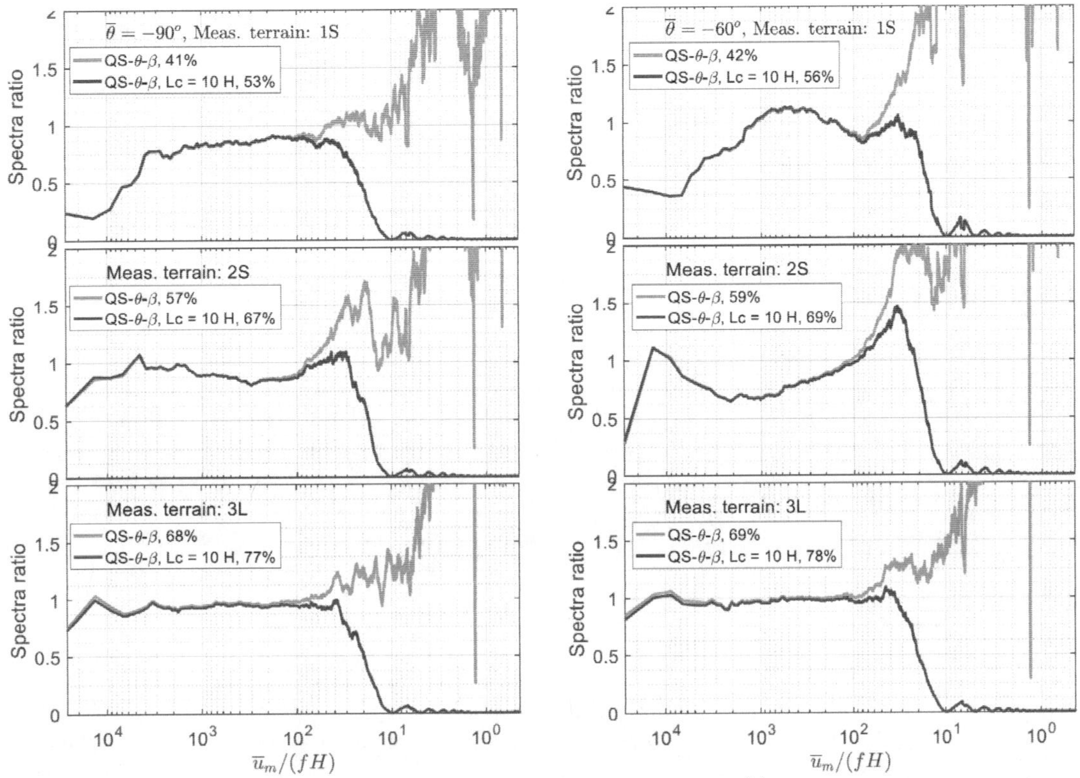

(a) (b)

Figure 5. Ratio of power spectral density of pressure obtained from the QS model to the measured values. The zero-time lag correlation coefficient between the two signals is shown in the legend. The QS model is established using terrain 2S and $L_c = 10H$ is used to filter the velocity signal: Results for (a) area C1 with $\overline{\theta}$ = -90° and (b) area C2 with $\overline{\theta}$ = -60°.

Figure 6 shows the probability density functions of the pressure fluctuations obtained via the QS model with the wind speed time histories and the directly measured values using both the instantaneous and moving-average filtered wind speeds. The figures indicate that the moving-average results are able to capture the bulk of the distribution, but not the tails, which are governed to a significant extent by the local velocity fluctuations of the smaller scales. Leaving the small-scale velocity fluctuations in place, rather than using a moving average improves the performance of the QS model in the tails of the distribution. In fact, because the ratios are higher than for the small scales in Figure 5, the peak pressures in the tails are over-estimated when the instantaneous speeds are used. However, because the small-scale fluctuations at location m are uncorrelated with the body-generated turbulence in the separated shear layer, the correlation coefficients between the estimated and actual pressure time histories (indicated by the percent values in the legends for both Figures 5 and 6) are reduced when the instantaneous velocity time series is used, in contrast to those using the moving-average speed.

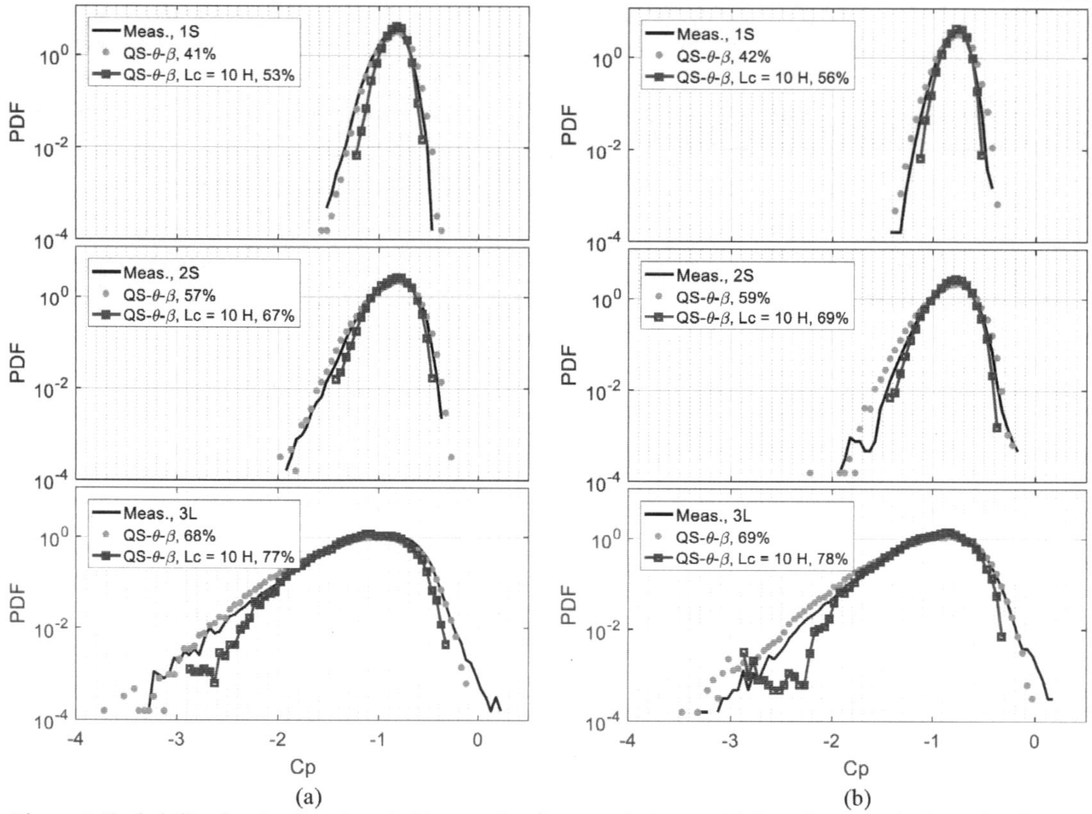

Figure 6. Probability density functions (with zero-time lag correlation coefficient shown in the legend) of Cp. QS model is established using terrain 2S and $L_c = 10H$ is used to smooth the velocity signal: Results for (a) area C1 with $\bar{\theta}$ = -90° and (b) area C2 with $\bar{\theta}$ = -60°.

4 DISCUSSION OF AERODYNAMIC MECHANISMS

The results presented in Section 3 suggest that there are not significant changes in the aerodynamic mechanisms associated with these six terrains, for this building shape and size. This will be examined in more detail in this section (within the space constraints), both with respect to the large scales and small scales. With respect to the large scales, it is fair to say that the QS aerodynamic coefficients developed in one terrain (2S in this case) can be used to accurately estimate the pressure variations in other terrains, both rougher and smoother. This is remarkable given that the differences in energy level in the streamwise velocity fluctuations at the cut-off scale are an order of magnitude different between terrain 1S (I_u = 13%) and 3L (I_u = 27%), i.e., at the values of
$\frac{fS_{uu}(f)}{\overline{u_H}^2}$ at $\frac{\overline{u_H}}{fH} = 10$.

As discussed by Saathoff and Melbourne (1997), and first shown by Gartshore (1973), it is the small-scale turbulence on the stagnation streamline which alters the character of the separation bubble and the separated shear layer. Thus, the effects of the order of magnitude differences in the values of $fS_{uu}(f)/\overline{u_H}^2$ for scales smaller than the cut-off also need to be understood. Of course, area-averages may mask subtle but important differences, so here we discuss these issues with respect to the point pressure distributions within the separated flow at the roof edge.

In uniform smooth flow for 2D bluff bodies, the largest pressure fluctuations occur near the mean reattachment point due to Kelvin-Helmholtz vortices which form and pair and grow along the separated shear layer (Cherry et al., 1984). Since the inflow is smooth, this is all body-generated turbulence. This is illustrated in Figure 7(a), with data from Saathoff and Melbourne (1997) and Cherry et al. (1984). When small-scale turbulence is added to the free stream, the separated shear layer becomes turbulent earlier because of a by-pass transition mechanism (Lander et al., 2016). This turbulence moves the reattachment point closer to the leading edge (Saathoff and Melbourne, 1997; Akon, 2017; Morrison and Kopp, 2017) and, more importantly, moves the peak value of pressures even closer to the leading edge. In other words, the largest magnitude fluctuations move closer to the leading edge relative to the mean reattachment point, even though the mean reattachment point also moves closer to the leading edge. This is illustrated in Figure 7(a), with data again from Saathoff and Melbourne (1997) for $I_u = 8\%$ and $L_x/D = 0.4$ and 2.1. For turbulence intensities and scales in the ABL, the fluctuating pressures are largest even closer to the leading edge. This can be seen in the full-scale field data from the TTU-WERFL building (Smith, 2010), shown in Figure 7(a), where $I_u = 17\%$ and $L_x/H = 35$ (based on a fit to the von Karman spectrum). Note that the method of Akon and Kopp (2016) is used to determine the reattachment point from the TTU data.

Based on these results, it can be seen that it is the turbulence intensity that is most important to this change in the fluctuating pressures. Saathoff and Melbourne (1997) discuss the mechanisms, which are related to the proximity of the shear layer vortices to the building surface. Integral sales apparently have a limited influence on this, although they clearly impact the magnitude of Cp_σ in the separation bubble with larger scales leading to larger fluctuation levels. There is a subtlety here though, because the full-scale experiments of Morrison and Kopp (2017) in the IBHS Research Center wind tunnel indicate that it is the scales in the range of

$$1 < \frac{\overline{u_H}}{fH} < 10$$

which are critical. So, it is the small-scale turbulence which is critical to setting the aerodynamic mechanisms, and which lead to the peak pressures in the tails of the distribution, while the large-scale turbulence sets the overall magnitude of the pressure (or load) fluctuations, as discussed by Bearman and Morel (1983).

Figure 7(b) shows the standard deviations of the pressure coefficients, Cp_σ, from the current experiments, as a function of the distance from the leading edge, x, normalized by the mean reattachment length, X_r. Several observations can be made. First, both the turbulence intensity and integral scales affect the level of pressure fluctuations, with larger values of both increasing the pressure fluctuations. Second, inspite of these changes in the magnitude of Cp_σ, the location of $(Cp_\sigma)_{max}$ is the same for five of the six terrains, relative to X_r. The smoothest terrain, 1S, appears to be close to a boundary of the aerodynamic mechanism since the peak in Cp_σ is slightly further from the leading edge. This suggests that the dynamic processes in the separated shear layers are no longer changing and the surface pressure fluctuations are responding passively to changes in the upstream turbulence. If this is indeed the case, one would then expect quasi-steady theory to be able to predict the changes caused by the turbulence levels and scales. The QS model would be unable to predict the different distributions seen in Figure 7(a). As in section 3, quasi-steady models were developed for the point pressures in terrain 2S and used to predict the point pressures in the other five terrains. The results are presented in Figure 7(b).

Figure 7(b) shows the predicted Cp_σ values using the QS-θ–β models for each location based on the 2S terrain model. The similarity between the measured and QS model is significant, with the largest changes picked up by the models. Further from the leading edge, the quality of the prediction diminishes, particularly beyond the reattachment point. This is where the separated shear layer has broken down (see Akon, 2017, for details) and the local wind speed fluctuations play a much greater role.

(a)

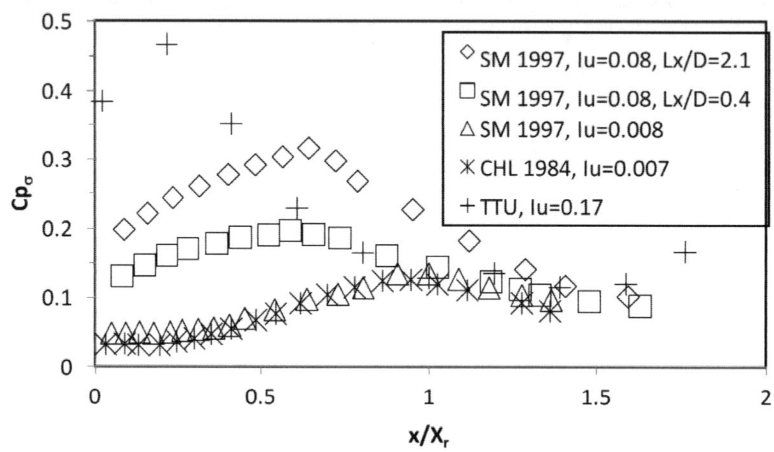

(b)

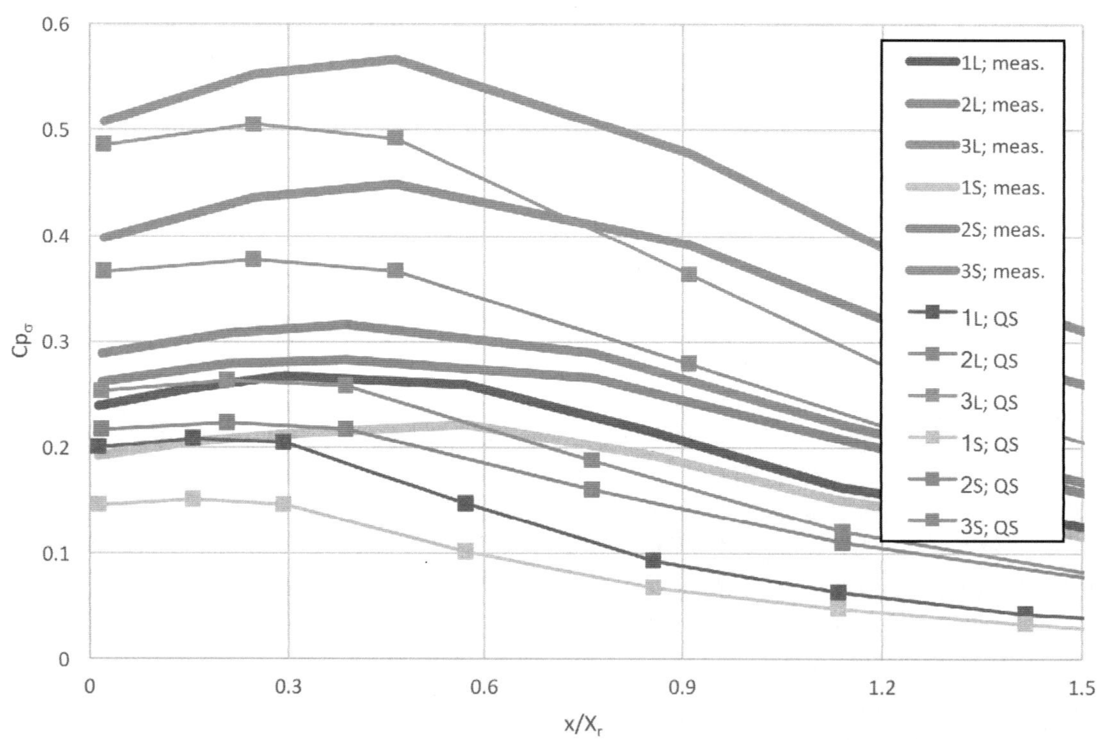

Figure 7. Standard deviations of the pressure coefficients, Cp_σ, as a function of distance from the edge roof along the mid-plane of body, for (a) various published experiments, (b) the current experiments and the QS-θ-β model results (developed in terrain 2S).

In the current experiments, the reference location for the wind speed is 1H above the roof surface, which is well above the separated shear layer (above 5 times the maximum thickness). The turbulent velocity fluctuations at this height are uncorrelated with those in the separated shear layer (SSL), which are of much larger magnitude, being 5 to 10 times higher in the SSL than in the free stream above the building. Thus, placing the reference probe in this location will lead to more significant distortions, which are not likely to improve the QS model. Nevertheless, Akon (2017) observed that the highest correlation between wind speed and pressure was at the leading edge, just outside the shear layer. This suggests that, to develop a model for the pressure fluctuations caused by the body-generated turbulence, using a second probe, close to the body may work. However, Akon (2017) found the highest correlation was with the instantaneous swirling strength within the mean separation bubble for the body-generated component.

The differences between the measured and modeled values in Figure 7(b) indicate that the body-generated fluctuations are a small proportion of the total energy near the leading edge, but only up to half the total beyond the mean reattachment length. It seems probable that some sort of stochastic correction may be possible for these small-scale motions. The work of Akon (2017) indicates that the shear layer turbulence is randomly aligned with the ABL turbulence. However, the basic vortex structure at the moments of peak pressure are similar across these terrains. This suggests that a simple kinematic model may suffice to capture the body-generated effects which are so important to the estimation of the peak pressures at the extremes of the pdfs shown in Figure 6. If a time series approach is not required, Figure 5 (and also the results in Wu and Kopp, 2016) indicates the fluctuations outside the shear layer will capture the reasonably accurate statistics for area-averages.

So, in summary, the QS model is able to capture the bulk of the fluctuation energy within the roof separation bubble. The missing energy from the body-generated turbulence at scales smaller than the cut-off at $L_c = 10H$ is only a small proportion of the total. While the turbulence in the separation bubble is not self-similar, with different levels of production of turbulence energy (Akon, 2017), the basic vortex structures in the separation bubbles have similar intensity levels (for this range of terrains with this building geometry). Thus, the overall dynamics of the separation bubble flows are similar. Given this robust behaviour across such different turbulence levels, it is worthwhile to examine what happens in a tornado-like vortex. The use of the same QS-θ–β model, developed in the 2S terrain, is applied as a basis for assessing the building aerodynamics in a tornado-like vortex in the next sections.

5 EXPERIMENTAL SET-UP IN THE WINDEEE DOME

The same building model used in the boundary layer wind tunnel (BLWT) experiments is used in the WindEEE Dome experiments. Figure 8 shows the basic layout, although the coordinate system provided in Figure 2 holds for these experiments. The primary difference between the BLWT and WindEEE experiments (notwithstanding the wind field, of course) is in the position of the Cobra probes for the tornado experiments. The building is placed on one side of the translating vortex, near the location of the radius of maximum winds. The radius to maximum winds was chosen to give the greatest elevation angles (see, e.g., the wind fields in Haan et al., 2017). The translation speed of the vortex was about 0.057 m/s. The diameter of the vortex core was about 0.8m such that the core diameter was about 3 times the largest plan dimension of the building model. As such, the building size is relatively large and one may expect significant velocity gradients across the model. In addition, the relatively slow translation (i.e., the ratio of the maximum rotational speed to forward translation speed is about 400) provides a significant test for QS models developed in ABL simulations because of the potentially different turbulence structure, even though the swirl ratio was small (about 0.5), such that the vortex structure was (nominally) of a single core. The Reynolds number for the data reported herein is about 100,000 although the complete range in the experiments was from about 20,000 up to about 120,000.

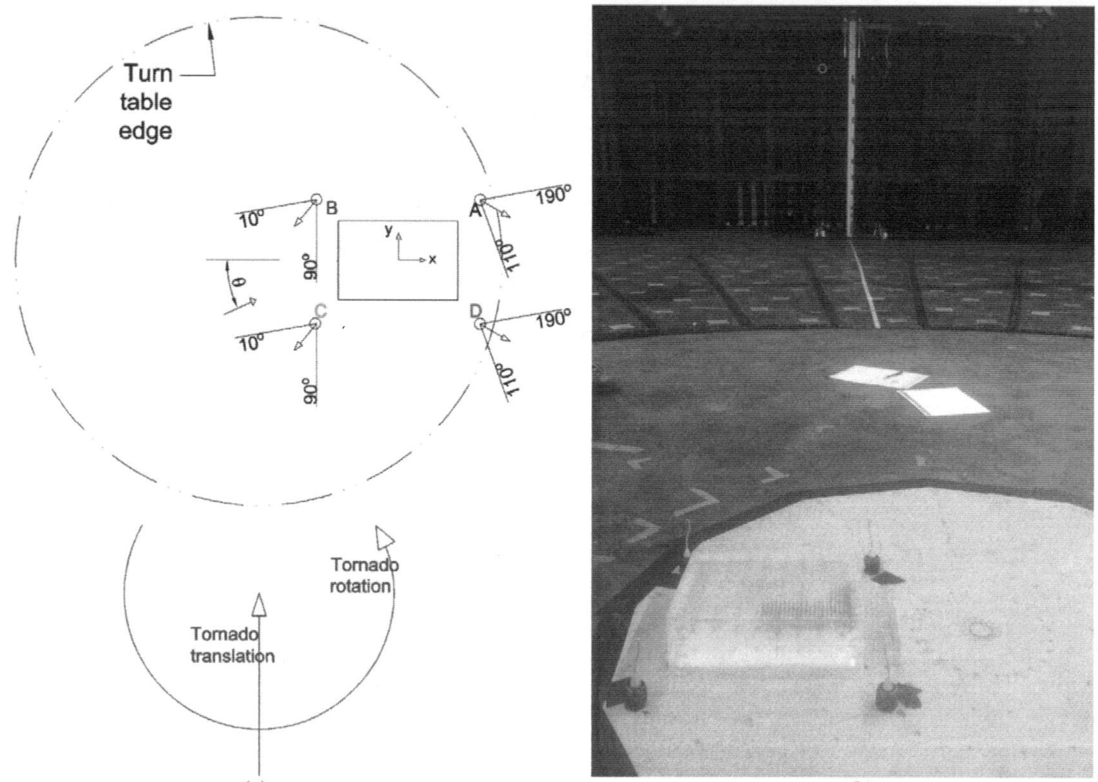

(a) (b)

Figure 8. (a) Schematic drawing and (b) a photograph of the experimental setup in WindEEE Dome (colour-coded for the time histories shown in later figures).

The floor of the WindEEE Dome experiments was set as nominally smooth, which is physically smoother than the BLWT experiments. However, the near-ground flow field is generally different, and it is not clear what role ground roughness plays in the mean field or turbulence structure of the tornado-like vortex flow. In any case, the nominal distance from inlet to the center of the turntable is about 13m. Thus, there is a significant fetch for the development of the near-surface flow (to the extent that it may be important, since the primary mean velocity gradients are horizontal, and not vertical, in the tornado vortex). Turbulence characteristics will be discussed further below.

The pressure coefficients were obtained relative to the static pressure measured in the bellmouth chamber above the turntable. The pressure coefficients were, then, re-referenced to the average pressure over the four walls. In this way, all coefficients can be thought of as being net coefficients. The data obtained in the BLWT experiments were handled in the same way, as discussed earlier.

Four Cobra probes were placed around the building model, one near each corner. These are labelled as A, B, C, and D in Figure 8. They were placed 0.9H from each corner, with a measurement height of 1.2H above ground. The probes were aligned so they could make measurements when they were approximately windward. They were adjusted for various vortex passages, with Figure 8 indicating the ranges of wind angles for which velocities could be measured. This choice of positions was made as a compromise since the position used in the BLWT experiments could not be easily replicated in WindEEE. This is discussed in more detail later, but positioning probes within the non-stationary flow field offers certain challenges. In the present analysis, the pressure coefficients are formed using the velocities measured at point D. It is noted that we do see hints of wake effects from the probes on the wall pressures, which is something which still requires analysis.

Figure 9 shows the wind speeds (magnitudes), azimuth and elevation angles for three passages of the tornado vortex, each with a slightly different alignment of the probes. Each of the four colours in the figures represents the data for one of the probes. The complete range of what was measured in each run is presented such that, where there is no data, the actual values or angles were outside of the acceptable range for the probe in that position. As can be seen, wind speeds ranged from 2 – 25 m/s, azimuth angles from about 10° to 200°, and elevation angles from about -10° to +40°. Nominal turbulence intensity values (which could be obtained because of the slow translation speed) were found to range from about 8% to more than 20% at the probe height (with potentially higher in the missing segments near the radius of maximum winds).

The focus of the current analysis is when probe D is windward for areas C3 and C4 in Figure 2. By symmetry, these are related to areas C1 and C2. Figure 10 provides some examples of the simultaneously-measured wind speeds and directions along with the surface pressures. The pressure patterns in these images are clearly similar to those one would expect for the given wind directions. (Also shown in Figure 10 are ground-plane pressures, which are also referenced to the average wall pressure.)

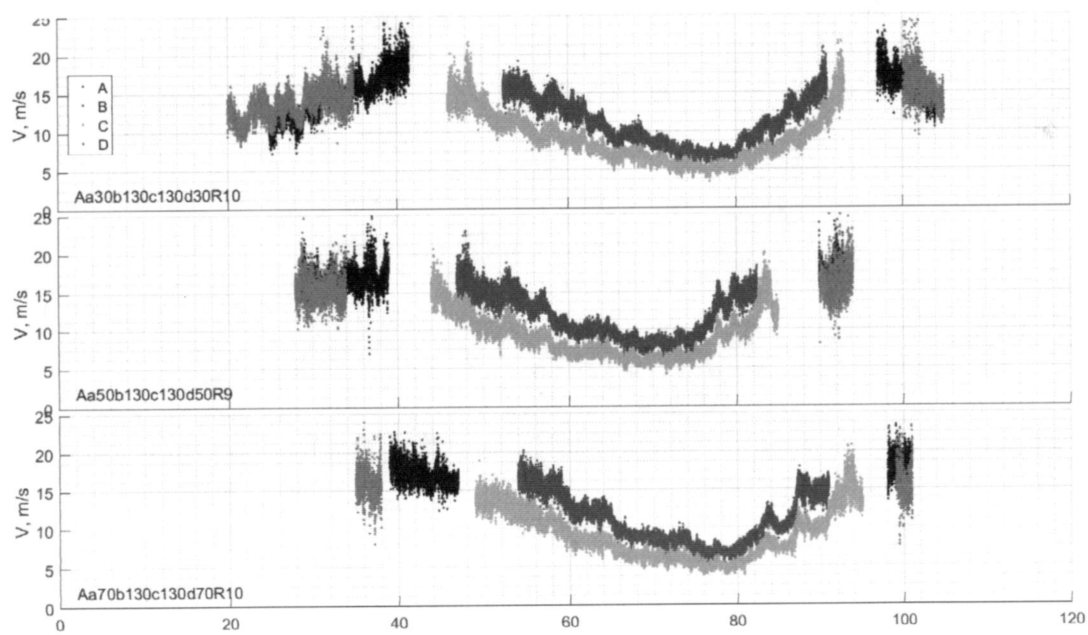

(a)

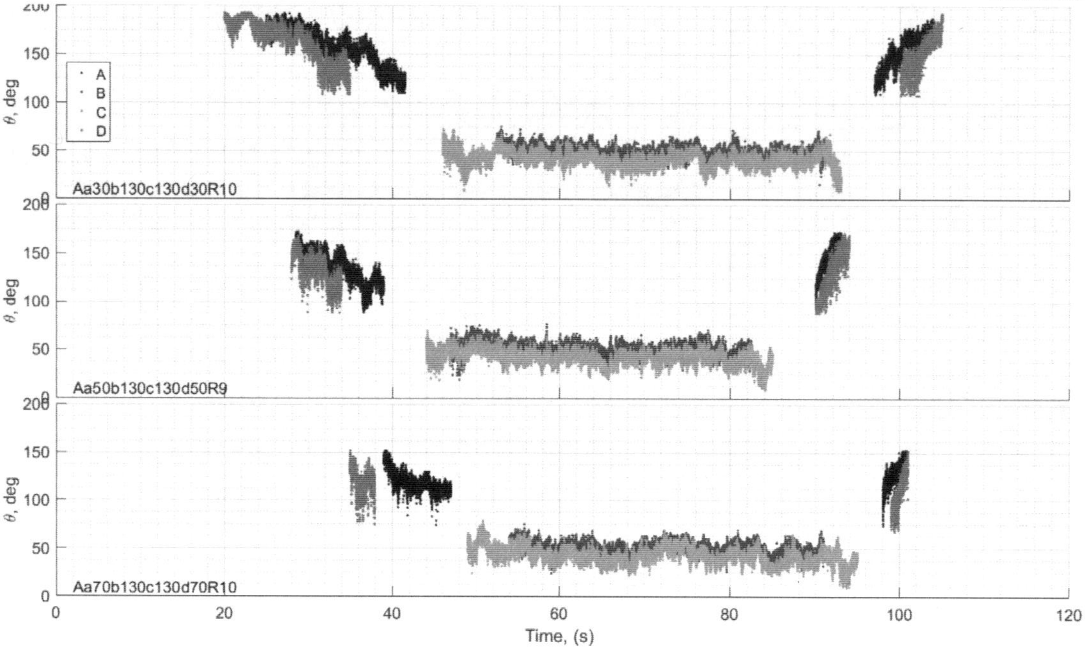

(b)

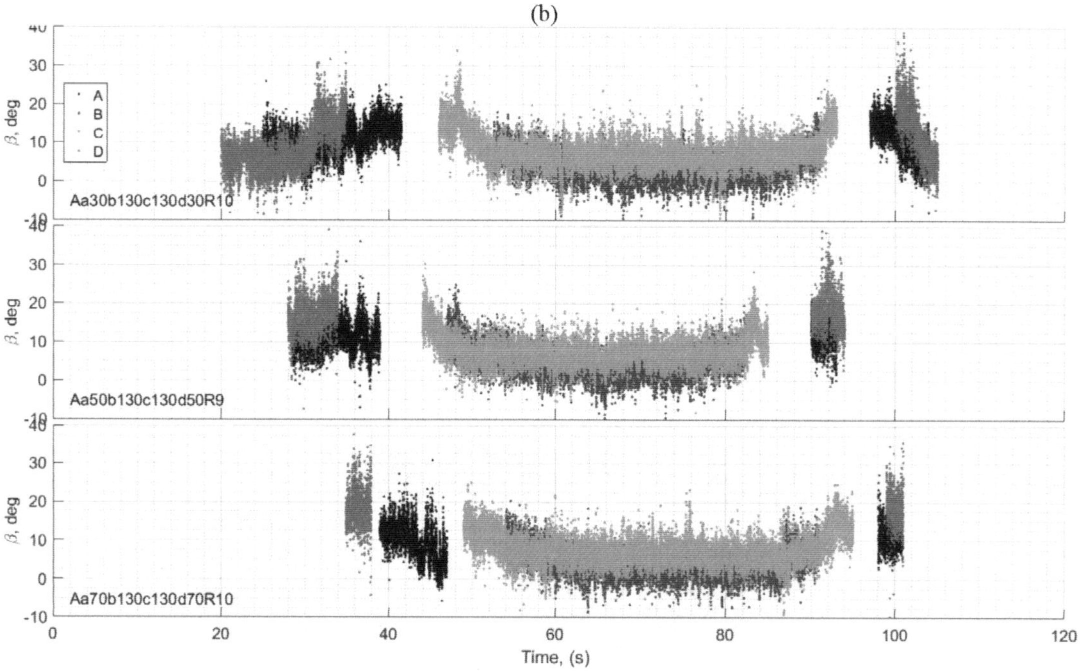

(c)

Figure 9. Velocity measured near the building model due to the translating tornado-like vortex: (a) wind speed, (b) azimuth angle, and (c) elevation angle.

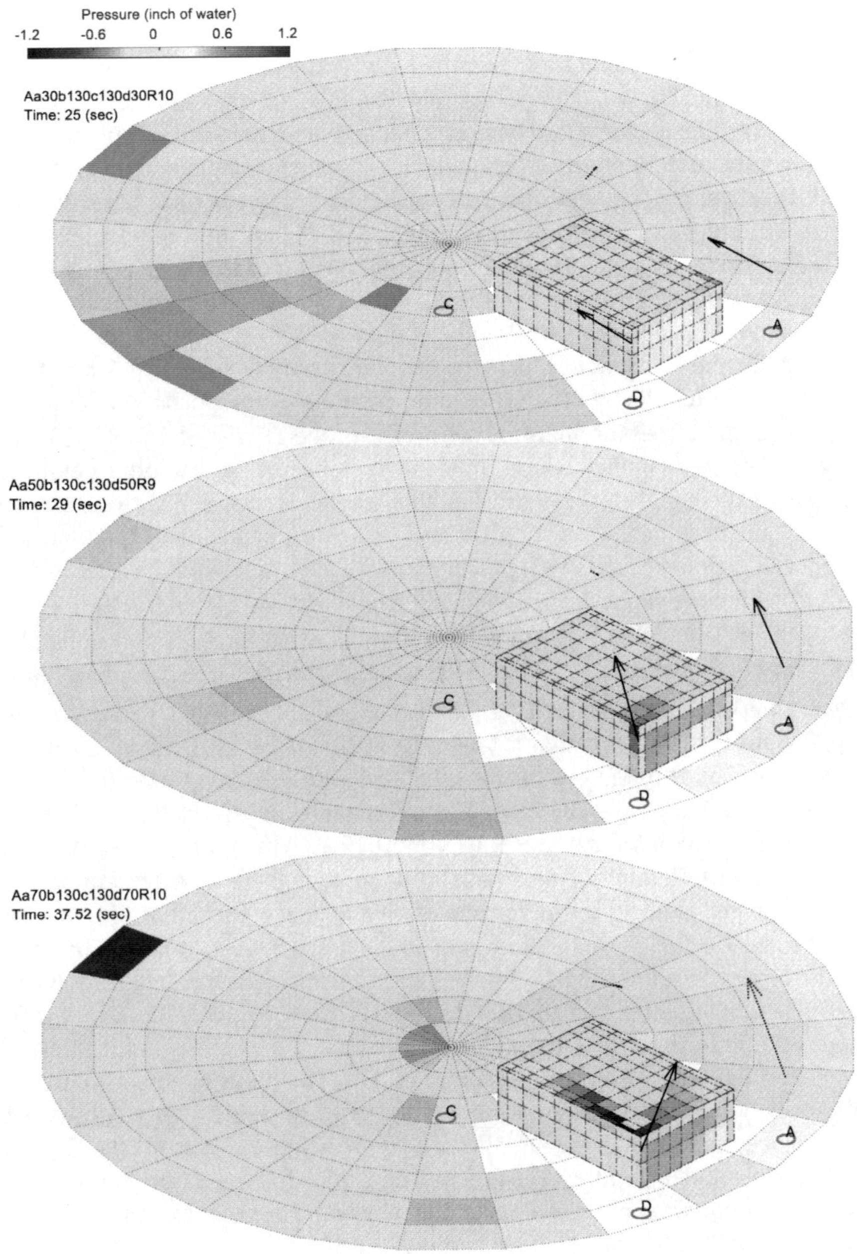

Figure 10. Snapshots of ground and building surface pressures (reference to the average building wall pressure) along with velocity vectors measured near building.

6 ANALYSIS OF THE TORNADO-VORTEX RESULTS

The QS model established using BLWT terrain 2S is used to estimate the surface pressures in the tornadic wind field. However, the instantaneous function from the QS model had to re-referenced to that for the probe at 1.2H height, upstream from the building. Figure 11 shows the time series of Cp_{inst} obtained from the direct measurements, along with those from the QS model, for panel areas C3 and C4. Note that probe D is located upstream with respect to these panels, which is used in the analysis. Like for the BLWT data in Figure 4, the QS-θ-β model is clearly an improvement, although it does not capture all the aerodynamic effects, with a mean offset occurring in places.

Figure 12 shows the conditionally-averaged Cp_{inst} values obtained from the measurements along with the QS model coefficients obtained in the BLWT for terrain 2S, with the black line representing the case neglecting the elevation angle, and the red line representing $\beta = 30^{o}$. The conditionally-averaged results were obtained from five runs for each of three probe orientation set-ups. As with the data presented in Figure 3, the darker the symbol, the greater the quantity of data included in the average. Faded symbols should be viewed with caution as they may not be fully converged statistically, but we include them to provide some perspective since there are some interesting trends. It should also be mentioned that, because of the position of the probes and the building, the most accurate measurement direction for probe D will be for $\theta \sim 135^{o}$, although its magnitude may be distorted (likely reduced) by the presence of the building behind it. Thus, we will focus on the results in the range around this angle. Because only one vortex structure, translation speed, and building position (distance to vortex centre and orientation angle) was considered, the mean elevation angle will be set by these parameters (and is fairly high).

At $\theta \sim 135^{o}$, $\bar{\beta}$ is is in the range from 15° to 20° (upwards), as can be seen in Figure 11. In the range from $\theta = 120^{o}$ to 150°, the Cp_{inst} values for $\beta = 15^{o}$ to 25° are within the band provided by the QS model. However, there are some differences, with the conditional-average having a different slope (i.e., $dCp_{inst}/d\theta$) in Figure 12(a) for area C3. For area C4, the match is better, although $E[Cp_{inst}|\Theta, \beta]$ for $\beta = 15^{o}$ to 25° is closer to the QS model for $\beta = 0^{o}$.

Other observations can be made. As discussed above, angles in the range $\beta = 15^{o}$ to 25° are the most common for $\theta = 120^{o}$ to 150°. The other elevation angles are much less common (Figure 11), as indicated by the faded symbols used in Figure 12, which are, therefore, less reliable. Nevertheless, there is a potentially important point in that there is a tendency for the largest magnitude coefficients to be for the smaller elevation angles. In particular, the largest magnitude coefficients tend to be for $\beta = 5^{o}$ to 15° with the smallest for $\beta = 25^{o}$ to 45°. This is precisely the opposite trend as observed for the ABL. While there is not sufficient data to determine whether such a conclusion is robust; if it is indeed true, it suggests that the turbulence structure of the vortex may be altering the aerodynamics. This may be related to a large expansion of the separated flow, much like the phenomenon of stall on aircraft wings. This is a point that needs careful study.

The other aspect which needs examination is the wall pressures, as there may be different azimuthal dependence than for ABL winds. Figure 13 shows the contour of the $E[Cp_{inst}|\Theta, \beta]$ and the corresponding QS model. Figures 13(a,b) indicate that, for the same wind azimuth, the QS model predicts the high-suction region near the leading edge of longer wall well, while it underestimates the suction on the shorter leading edge corner. However, based on the wall pressure pattern in Figure 13(c), it appears that a better performance of the QS model may be obtained from a wind azimuth angle being shifted by 30°. However, in this case the suction near the longer leading edge is underestimated even though the wall pressures are a better match. Hence, the aerodynamic effects of tornadic wind on the roof surface pressures do not appear to be exactly the same as for boundary layer straight-line winds, even though the pressure patterns are somewhat similar for this building shape. Clearly, further, detailed investigations of the wall (and internal) pressures are required to assess these effects. QST provides a useful framework for this assessment.

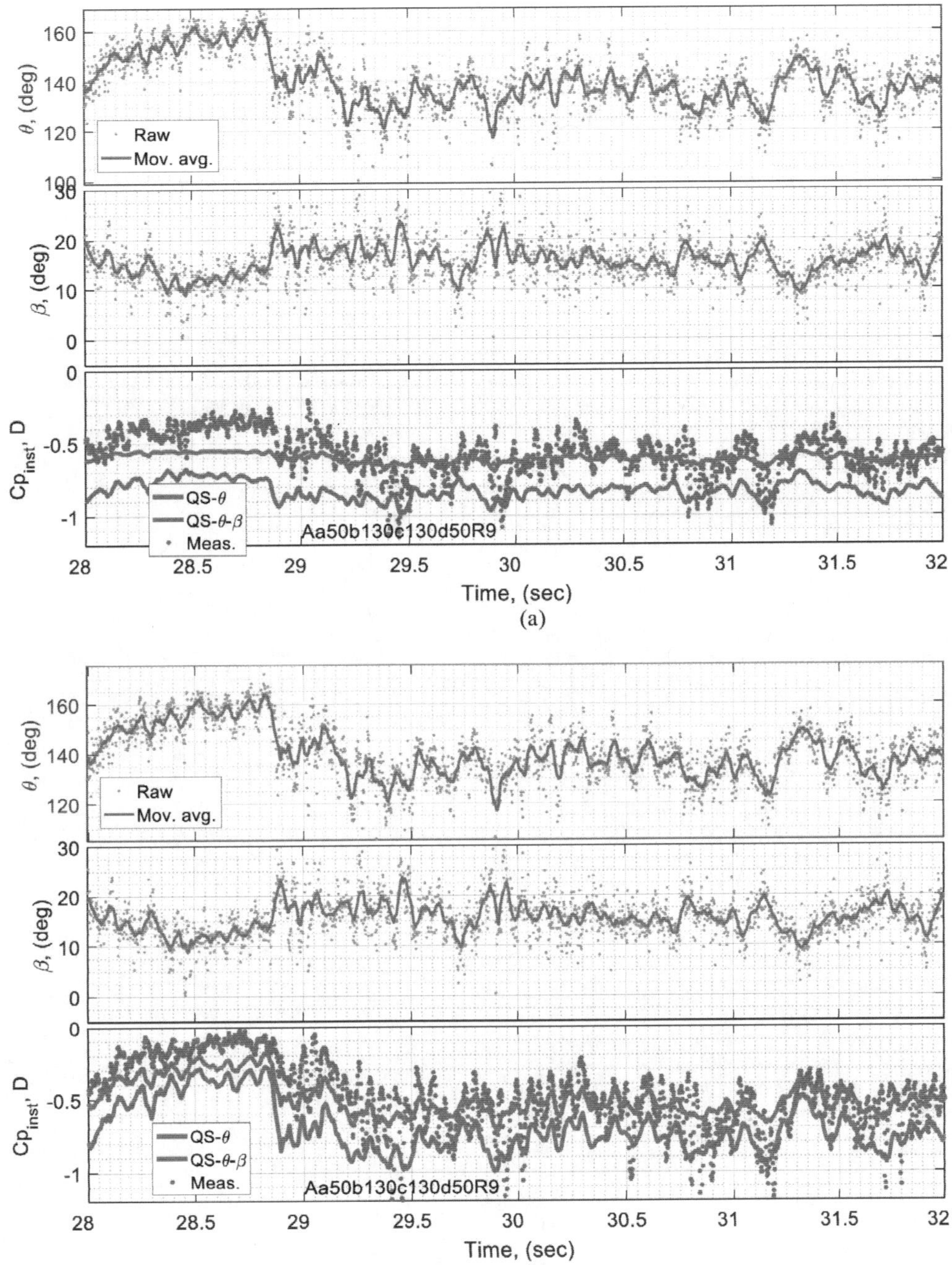

Figure 11. Raw and smoothed (moving averaged signals using window $\Delta t = 0.04$ sec) wind directions and Cp_{inst} time series obtained from probe D: (a) Panel area C3, and (b) area C4.

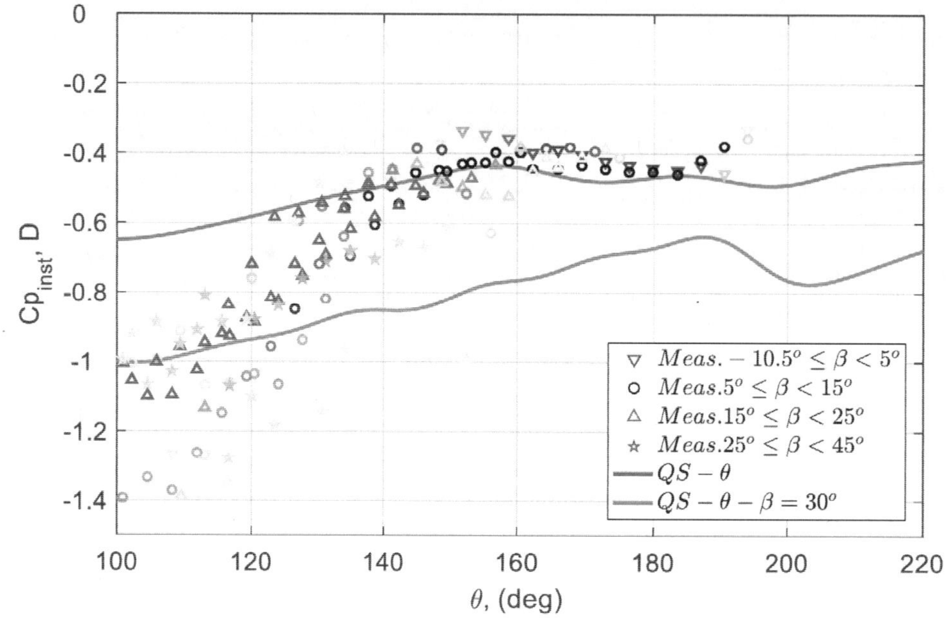

(a)

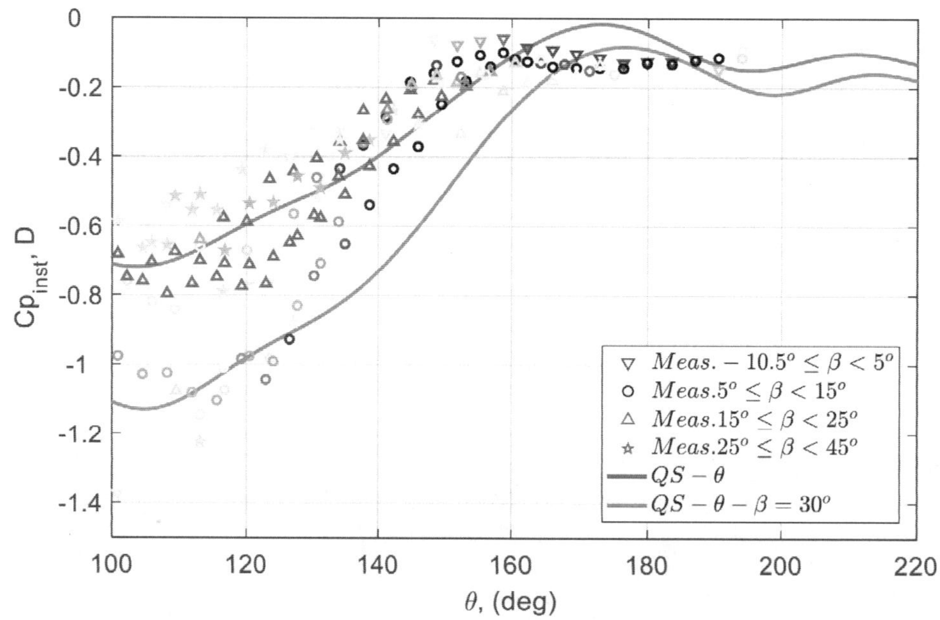

(b)

Figure 12. $E[Cp_{inst}|\Theta, \beta]$ and the instantaneous coefficients of the QS model: (a) Panel area C3, and (b) area C4.

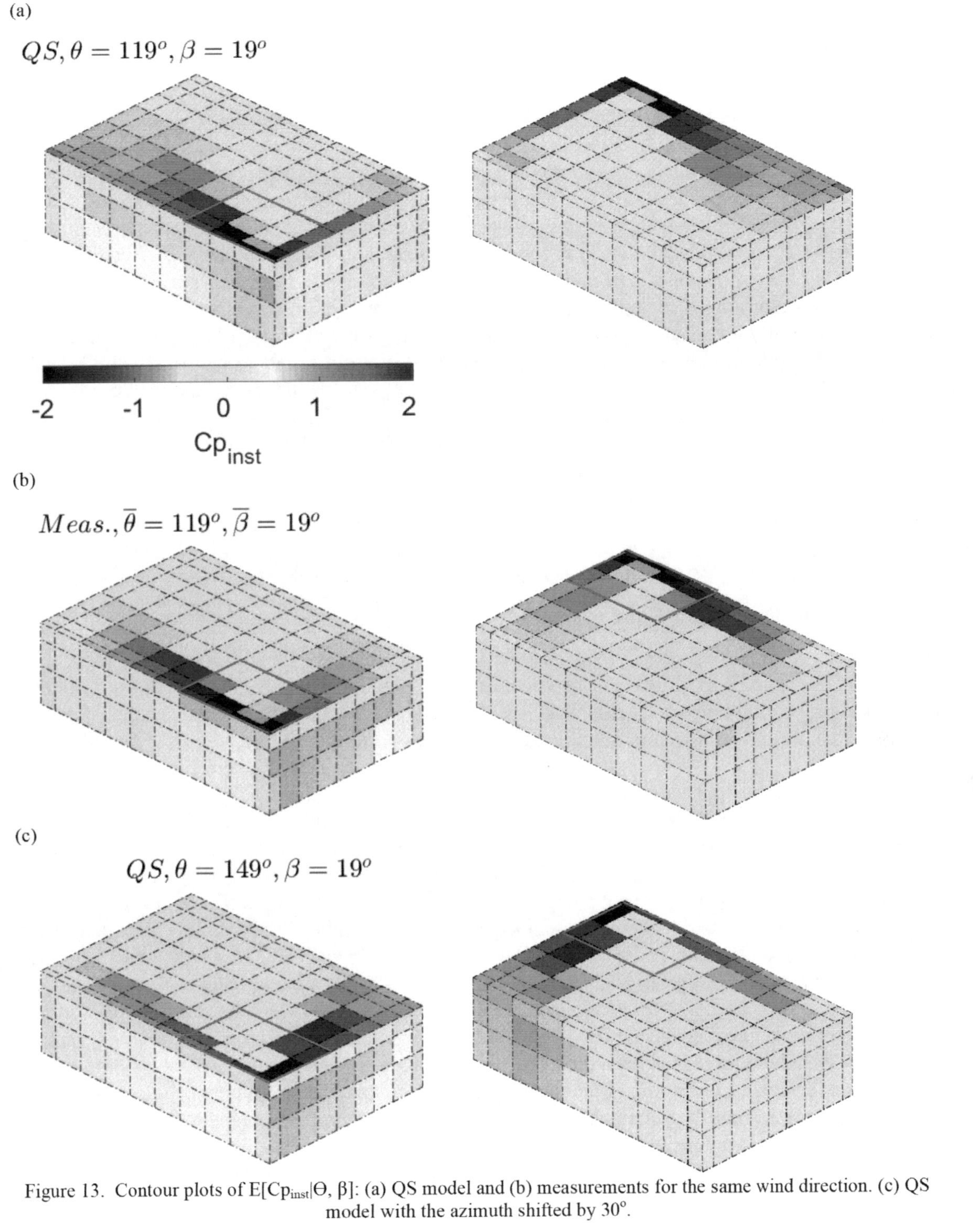

Figure 13. Contour plots of $E[Cp_{inst}|\Theta, \beta]$: (a) QS model and (b) measurements for the same wind direction. (c) QS model with the azimuth shifted by 30°.

7 FINAL DISCUSSION AND CONCLUSIONS

7.1 Concluding Remarks

Boundary Layer Wind Tunnel (BLWT) experiments were conducted with six distinct terrain simulations and a scale model of the TTU-WERFL building. Simultaneous pressure and velocity measurements were conducted. Quasi-Steady (QS) models were developed that considered the effects of azimuth (θ) and elevation (β) angles on area-averaged pressure coefficients. It was found that the use of a QS model developed in one terrain could be used to predict the pressure variations of the large-scale wind speed fluctuations in other terrains, using a time history approach. The small-scale fluctuations play a less significant role on the area-averages and were filtered out using a moving-averaged wind speed. It should be emphasized, however, that small-scale turbulence on the stagnation streamline is critical to setting the building aerodynamics, so must be present in the wind field. In addition, the body-generated turbulence in the separated shear layer, which is altered by the turbulence in boundary layer along the stagnation streamline, is critical to setting the tails of the probability density functions of the pressures, even for fairly-large area averages.

Simultaneous pressure and velocity measurements were conducted on the same building model in a tornado-like vortex flow. The QS model, developed from the BLWT experiments, was used to assess differences and similarities in the wind loads between BL and tornadic wind fields. It was found that there are some general similarities, although elevation angle and wall pressure effects exhibit some differences, which require further investigation. It appears that there may be differences in wind loads up to 40% higher in tornadoes, although this conclusion is still speculative. This form of QS model is clearly promising as the basis for determining such impacts.

7.2 Issues to be Resolved – Future Research

Based on the above conclusions, there are yet many questions to be addressed with respect to building aerodynamics in tornadoes. This work is attempting to address the effects of turbulence on the building aerodynamics. These are at least three significant questions which remain open: (i) What is the optimum approach to obtaining simultaneous wind speeds and pressures in translating vortex flows? (ii) What is it about the tornadic wind field that alters the aerodynamics? (iii) Are the wall (and internal) pressures altered to a significant extent?

It is well known that obtaining simultaneous wind speed and pressure data is challenging because the presence of the building distorts the wind field near the building (wind speed magnitudes, turbulence levels and wind directions), where the correlation between wind speeds and pressures are highest. The probes can also distort the flow on the building, particularly by altering the surface pressure fluctuations in the wake of the probe. Putting a probe further away causes challenges with the correlation dropping off rapidly with distance from the building. Analysis by Wu (2017) suggests that mounting probes above the leading edge of the roof may be best for optimizing the correlation with the surface pressures without significant flow distortion. The measurements presented here for the BLWT flows had the probe at a minimum of 5 mean separation bubble thicknesses above the roof edge. This is straightforward to accomplish in boundary layer wind tunnels, but is much more difficult in tornado simulators because there may not be positions to mount probes above the building models without some significant support structures, which may alter the swirling and upward flows. This may be where simultaneous PIV and pressure measurements are required, although this is a challenge to do routinely or automatically, since stereoscopic PIV would be required with the current analysis approach. This is a challenge for large-scale facilities where the cameras and laser optics may need to be in flow field (e.g., Pratt and Kopp, 2014; Akon and Kopp, 2016). Clearly, much work remains.

In BLWT simulations of the ABL, it has been found that the pressure coefficients vary linearly as a function of elevation angle, over the ranges that are observable. Our results suggest that this may not hold for tornado-like vortices. We suspect that the nature of the turbulence and gust structure lead to some different aerodynamic loading than seen in ABL winds. Our simulation is probably a worst-case scenario (for a relatively small building to tornado size ratio) because of the slow translation speed in these experiments. Some PIV experiments would be helpful in this regard, but this may also be a situation where LES-based experiments can aid our understanding. It is not clear whether there are subtle static pressure effects, such as enhancement of corner (and other) vortices due to a static pressure difference across the building, or net loads are altered because of the wall pressure differences, or changes in the strength of the separated flow caused by large-scale vertical flows such that separated shear layers move much further from the roof surface, or just that transient effects are different. These are all subtle changes that require flow field details together with the quasi-steady framework.

Finally, it is important to emphasize that the largest uncertainties are still associated with the tornado wind fields, where we really have limited data, limited understanding, and limited models. This is particularly true in the lower 10m where most of our buildings and infrastructure are.

ACKNOWLEDGEMENTS

The authors are grateful to many colleagues for interesting conversations pertaining to this topic, particularly, Drs. Fahad Akon, Fred Haan, Daniel Lander, Chris Letchford, and Murray Morrison. Funding for this research was provided by the National Science and Engineering Research Council (NSERC) of Canada through the Collaborative Research and Development program and the Institute of Catastrophic Loss Reduction (ICLR). The long-term and enthusiastic support of Mr. Paul Kovacs is gratefully acknowledged. The authors also acknowledge the kind support of Mr. Gerry Dafoe in conducting the WindEEE Dome experiments.

REFERENCES

Akon, A.F., 2017, Effects of turbulence on the separating-reattaching flow above surface-mounted, three-dimensional bluff bodies, PhD Thesis, University of Western Ontario, London, Canada.

Akon, A.F. & Kopp, G.A., 2016, Mean pressure distributions and reattachment lengths for roof-separation bubbles on low-rise buildings, J. Wind Eng. Ind. Aerodyn., 155, 115-125.

Amini, M., van de Lindt, J. (2013). "Quantitative insight into rational tornado design wind speeds for residential wood frame structures using a fragility approach." ASCE J. Struct. Eng., 140 (7), 04014033.

Asghari-Mooneghi, M., Irwin, P. & Chowdhury, A.G., 2016, Partial turbulence simulation method for predicting peak wind loads on small structure and building appurtenances, J. Wind Eng. Ind. Aerodyn., 157, 47-62.

Banks, D. & Meroney, R. N., 2001, The applicability of quasi-steady theory to pressure statistics beneath roof-top vortices. J. Wind Eng. Ind. Aerodyn., 89, 569 – 598.

Bearman, P.W. & Morel, T., 1983, Effect of free stream turbulence on the flow field around bluff bodies, Prog. Aerospace. Sci., 20, 97-123.

Cherry, N.J., Hillier, R. & Latour, M.E.M.P., 1984, Unsteady measurements in a separated and reattaching flow, J. Fluid Mech., 144, 13.

Gartshore, I.S., 1973, The effects of free stream turbulence on the drag of rectangular two-dimensional prisms, Boundary Layer Wind Tunnel Laboratory Report BLWT-4-73, London, Canada, 25 pp.

Gavanski, E. & Kopp, G.A., 2017, Fragility assessment of roof-to-wall connection failures for wood-frame houses in high winds, ASCE-ASME J. Risk Uncertainty Eng. Sys., Part A: Civil Eng. (to appear).

Haan, F.L., Balaramudu, V.K. & Sarkar, P.P., 2010, Tornado-induced wind loads on a low-rise building, J. Struct. Eng., 136(1), 106–116.

Haan Jr., F.L., Sarkar, P.P., Kopp, G.A. & Stedman, D.A., 2017, Critical wind speeds for tornado-induced vehicle movements, J. Wind Eng. Ind. Aerodyn. (to appear).

Hiller, R. & Cherry, N.J., 1981, The effects of stream turbulence on separation bubbles, J. Wind Eng. Ind. Aerodyn., 8, 49.

Ho, T.C.E., Surry, D., Morrish, D. & Kopp, G. A., 2005, The UWO contribution to the NIST aerodynamic database for wind loads on low buildings: Part 1. Archiving format and basic aerodynamic data, J. Wind Eng. Ind. Aerodyn., 93, 1 – 30.

Hu, H., Yang, Z., Sarkar, P. & Haan, F., 2011, Characterization of the wind loads and flow fields around a gable-roof building model in tornado-like winds, Experiments in Fluids, 51, 835-851.

Irwin, P.A., 2008, Bluff body aerodynamics in wind engineering, J. Wind Eng. Ind. Aerodyn., 96, 701.

Kawai, H., 1983, Pressure fluctuations on square prisms – applicability of strip and quasi-steady theories, J. Wind Eng. Ind. Aerodyn., 13, 197 – 208.

Kopp, G.A., Hong, E., Gavanski, E., Stedman, D. & Sills D.M.L., 2017, Assessment of wind speeds based on damage observations from the Angus (Ontario) Tornado of 17 June 2014, Can. J. Civil Eng., 44, 37-47.

Lander, D.C., Letchford, C.W., Amitay, M. & Kopp, G.A., 2016, The influence of the bluff body shear layers on the wake of a square prism in a turbulent flow, Physical Review Fluids, 1(4), doi:10.1103/PhysRevFluids.1.044406.

Letchford, C.W., Iverson, R.E. & McDonald, J.R., 1993, The application of the quasi-steady theory to full scale measurements on the Texas Tech Building, J. Wind Eng. Ind. Aerodyn. 48, 111-132.

Letchford. C.W. & Marwood, R., 1997, On the influence of v and w component turbulence on roof pressures beneath conical vortices, J. Wind Eng. Ind. Aerodyn., 69-71, 567-577.

Levitan M.L. & Mehta K.C., 1992a, Texas Tech field experiments for wind loads part I: Building and pressure measuring system, J. Wind Eng. Ind. Aerodyn., 43, 1565.

Levitan M.L. & Mehta K.C., 1992b, Texas Tech field experiments for wind loads part II: Meteorological instrumentation and terrain parameters, J. Wind Eng. Ind. Aerodyn., 43, 1577.

Mishra, A.R., James, D.L. & Letchford, C.W., 2008, Physical simulation of a single-celled tornado-like vortex, Part A: Flow field characterization, J. Wind Eng. Ind. Aerodyn., 96, 1243-1257.

Mishra, A.R., James, D.L. & Letchford, C.W., 2008, Physical simulation of a single-celled tornado-like vortex, Part B: Wind loading on a cubical model, J. Wind Eng. Ind. Aerodyn., 96, 1258-1273.

Pratt, R. N. & Kopp, G. A., 2014, Velocity field measurements above the roof of a low-rise building during peak suctions. J. Wind Eng. Ind. Aerodyn., 133, 234 – 241.

Refan, M. & Hangan, H., 2016, Characterization of tornado-like flow fields in a new model scale wind testing chamber, J. Wind Eng. Ind. Aerodyn., 151, 107-121.

Richards, P.J., Hoxey R.P. & Wanigaratne, B.S., 1995, The effect of directional variation on the observed mean and rms pressure coefficients, J. Wind Eng. Ind. Aerodyn., 54/55, 359-367.

Richards, P. J. & Hoxey, R. P., 2004, Quasi-steady theory and point pressures on a cubic building, J. Wind Eng. Ind. Aerodyn., 92, 1173 – 1190.

Richards, P. J. & Hoxey, R. P., 2012, Pressures on a cubic building – Part 2: Quasi-steady and other processes, J. Wind Eng. Ind. Aerodyn., 102, 87 – 96.

Saathoff, P.J. & Melbourne, W. H., 1997, Effects of free-stream turbulence on surface pressure fluctuations in a separation bubble, J. Fluid Mech., 337, 1-24.

Sabaheesh, G.R., Matsui, M. & Tamura, Y., 2012, Dependence of surface pressures on a cubic building in tornado like flow on building location and ground roughness, J. Wind. Eng. Ind. Aerodyn., 103, 50-59.

Simmons, K.M., Kovacs, P. & Kopp, G.A., 2015, Tornado damage mitigation: benefit/cost analysis of enhanced building codes in Oklahoma, Weather, Climate and Society, 7(2), 169-178.

Smith D., 2010, Validation of wind and wind-induced pressure data collected at the Institute for Building and Home Safety's state-of-the-art multi-peril applied research and training facility, Wind Science and Engineering Research Center, Lubbock, Texas.

Surry, D., 1982, Consequences of distortions in the flow including mismatching scales and intensities of turbulence, Wind Tunnel Modeling for Civil Eng. App., ed. T.A. Reinhold, Cambridge University Press, Cambridge, 137.

Tieleman, H.W., 2003, Wind tunnel simulation of wind loading on low-rise structures: a review, J. Wind Eng. Ind. Aerodyn., 91, 1627.

Van de Lindt, J.W., Pei, S., Dao, T., Graettinger, A., Prevatt, D.O., Gupta, R., Coulbourne, W., 2013, Dual-objective-based tornado design philosophy, J. Struct. Eng., 139(2), 251-263.

Wu, C.-H., 2017, PhD Thesis, University of Western Ontario, London, Canada.

Wu, C.-H. & Kopp, G.A., 2016, Estimation of wind-induced pressures on a low-rise building using quasi-steady theory, Frontiers Built Env., 2(5), doi: 10.3389/fbuil.2016.00005.

13th Americas Conference on Wind Engineering
Gainesville, Florida, USA
May 21 - 24, 2017

Tornado Damage Modeling: Understanding the Role of Atmospheric Pressure Change, Wind Directionality, and Transient Wind Field Intensity

Peter J. Vickery[a], Sudhan Banik[b], Lawrence A. Twisdale Jr.[c], and Long Phan[d]

[a]*Principal Engineer, IntraRisk Division, Applied Research Associates, Raleigh, NC, United States, pvickery@ara.com*
[b]*Senior Engineer, IntraRisk Division, Applied Research Associates, Raleigh, NC, United States, sbanik@ara.com*
[c]*Executive Vice-President, IntraRisk Division, Applied Research Associates, Raleigh, NC, United States, ltwisdale@ara.com*
[c]*Leader, Structures Group, National Institute of Standards and Technology, Gaithersburg, MD, United States, long.phan@nist.gov*

ABSTRACT

Modeling loads and damage to buildings in tornado winds has received increased attention of the last decade. A number of experimental studies performed in tornado simulators have suggested that wind pressures on buildings in tornadoes are much different than those produced in atmospheric boundary layer (ABL) flow conditions.Direct comparisons of exterior pressure coefficients derived in tornado simulators to those developed using ABL wind tunnels have been clouded through differences in normalization of wind speeds and though the contribution of the atmospheric pressure change (APC) to the exterior pressures.

The tornado load model is coupled with a time-stepping damage and windborne debris models. While the building is undamaged, the internal pressure is computed assuming uniform leakage over the exterior of the building and a spatially invariant internal pressure. Following envelope breach, a new value of the internal pressure is computed, taking into breach size and background leakage. The model considers failures to roof deck, roof trusses, fenestrations, and walls.

The damage model was validated through comparisons to observed damage of two big box stores during the Joplin tornado. Structural models of the two big stores were developed using design drawings provided by NIST. The damage model was able to reproduce observations that one of the buildings was destroyed and that the damage on the other building varied from complete at one end of the building and undamaged at the other end.

The study brought to light the need for a better understanding of the spatial correlation of the wind loads and the need to increase data to properly include the effect of vertical winds on flat roofed buildings. Information on background and/or concentrated leakage of buildings is also critical to properly model the internal pressures in large buildings.

KEYWORDS

Tornado, Damage, Model, APC, Joplin

Motion-induced Wind Forces of a Slender Prism

Zengshun Chen [a], K.T.Tse [a], K.C.S. Kwok [b]

[a]*Department of Civil and Environmental Engineering, The Hong Kong University of Science and Technology, Hong Kong, China*
[b]*Center for Infrastructure Engineering, Western Sydney University, Penrith, NSW, Australia*

ABSTRACT: The motion-induced wind forces of a slender prism were investigated using a forced vibration technique. The prism was driven to oscillate by an actuator in cross-wind direction and the unsteady distributed pressures, under different wind velocities and oscillations, were measured. The unsteady generalized and local aerodynamic force coefficients and the motion-correlated force coefficients of the prism, which are functions of reduced wind velocity and oscillation, were analyzed. It shows that the effects of structural motion on the unsteady force coefficients are significant in the cross-wind direction while the effects are slight in the along-wind direction. Furthermore, in the cross-wind direction, the coefficients tend to increase with oscillating amplitudes at low wind speeds while they are at a quasi-steady state at high wind speeds. The study advances the understanding of the effect of structural motion on three-dimensional prisms, which can be utilized to improve response predictions of the prisms.

KEYWORDS: Unsteady pressure, forced vibration, slender prism

1 INTRODUCTION

Architectural structures tend to be increasingly high and slender, and accurately obtaining the wind load of such structures is becoming more important than before. Generally, a high frequency base balance (HFBB) or a static synchronous multi-pressure sensing system (SMPSS) wind tunnel test is carried out to evaluate the wind load of structures. The HFBB and SMPSS techniques are effective. However, they are static measurements and the effect of structural motion is therefore excluded. As is known, the effect is usually small. But when it is in-phase with wind velocity, the effect acts as aerodynamic damping and cannot be neglected: at low wind speeds, motion-correlated aerodynamic damping is positive, and neglecting its effect can lead to overestimation when predicting structural response; on the other hand, at high wind speeds, motion-correlated aerodynamic damping is negative, and neglecting its effect can result in underestimated prediction in structural response. Also, it has already been confirmed that the differences of wind load measured by a static and a dynamic test are mainly ascribed to the effect of structural motion [1, 2]. Therefore, the unsteady wind loads of structures, which include the effect of structural motion should be acquired.

This study aims to investigate the distributed unsteady aerodynamic force of a slender square prism using a forced vibration technique. Wind tunnel tests of the prism were performed using a forced vibration apparatus. In the tests, the model was driven to oscillate under different oscillating amplitudes, and the distributed unsteady pressure and responses were measured simultaneously. Then, the unsteady force coefficient and the motion-correlated force coefficient (aerodynamic damping and stiffness terms) of the prism, under different oscillations and wind velocities, were determined and analyzed. This study advances our understanding of the effect of motion on three-dimensional prisms, which can be utilized to evaluate the aerodynamic damping of structures and improve the response predictions of the structures.

2 EXPERIMENTAL SETUPS

2.1 *Wind flow field*

A forced vibration test was carried out in the high-speed section of the wind tunnel at the Hong Kong University of Science and Technology with dimensions 29.2 m (length) × 3 m (width) × 2 m (height). The terrain category II defined in the AS/NZS 1170.2:2002 was simulated in the wind tunnel by adjusting roughness elements and spires in the upstream of the test section. The mean wind speed and flow turbulence intensity were measured from a hot-wire anemometer.

2.2 *Forced vibration technique*

The dimensions of the test model in the present study are 50.8 mm (*D*)× 50.8 mm (*D*)× 915 mm (*H*) with an aspect ratio being around 18:1. Here, *D* donates the width and depth of the model and *H* is the height of the model. The blockage ratio of the test model is therefore 0.78%, which is much smaller than the critical value of 5% [3]. The prism was driven to oscillate sinusoidally in the cross-wind direction using a forced vibration system which consisted of an actuator, a signal generator and a power amplifier (Fig. 1). The oscillating frequency f was set to 7.8 Hz and the root-mean-square (RMS) response ratio of the prism σ_y / D ranged from 6% to 20%. Here, σ_y is the RMS tip oscillating amplitude of the test model. The response was measured using a *LED* sensor installed at the bottom of the test rig (Fig. 1). Before the test, the relationship between the response measured at the tip and at the bottom by using *LED* sensors was well calibrated. The reduced wind speed (V_{ref}/fD, V_{ref} is reference wind speed) at the tip of the prism ranged from 7 to 20. Correspondingly, the Reynolds number *Re* varied from 9.89×10^3 to 2.86×10^4 with the characteristic length being 50.8 mm. The sampling frequency was 500 Hz and the duration was 100 s.

2.3 *Pressure measurements*

The unsteady distributed pressure of the prism was measured using a high-speed scanning pressure equipment. The reference of static pressure was selected on the upstream side of the prism at a height of 1.1 m. The sampling frequency and duration were set the same with that in the previous section. There were 162 pressure taps in total distributed on 9 levels of all 4 faces of the prism. The distribution of the pressure taps is presented in Fig. 1 (d). It should be noted that the pressure should be observed simultaneously with the oscillations of the prism.

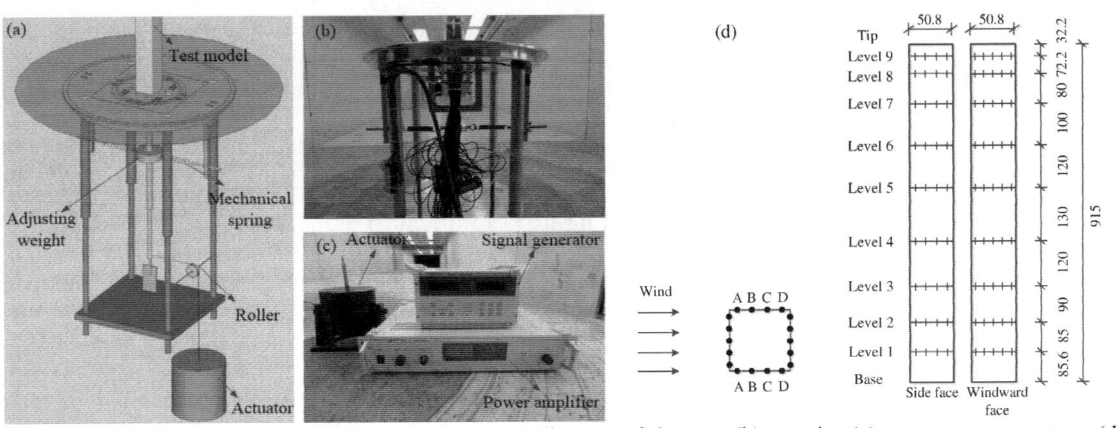

Figure 1. The forced vibration device and system: (a) diagram of the test; (b) test rig; (c) measurement system; (d) pressure distribution.

3 RESULTS AND DISCUSSIONS

3.1 *Unsteady force coefficients*

The observed force coefficients of the prism may be affected by turbulence, vortex shedding as well as fluid-structure interaction [4]. The coefficients can be used to predict responses of structures, in which the RMS coefficient is significant in studying the background and resonant components [5]. In the cross-wind direction, the mean lift force coefficients are close to zero. Therefore, only the RMS lift force coefficients of the prism are concerned. The mean and RMS drag force coefficients, and the RMS lift force coefficients ($\bar{C}_D$, $\tilde{C}_D$, $\tilde{C}_L$) under different wind speeds and oscillating amplitudes are presented in Fig. 2.

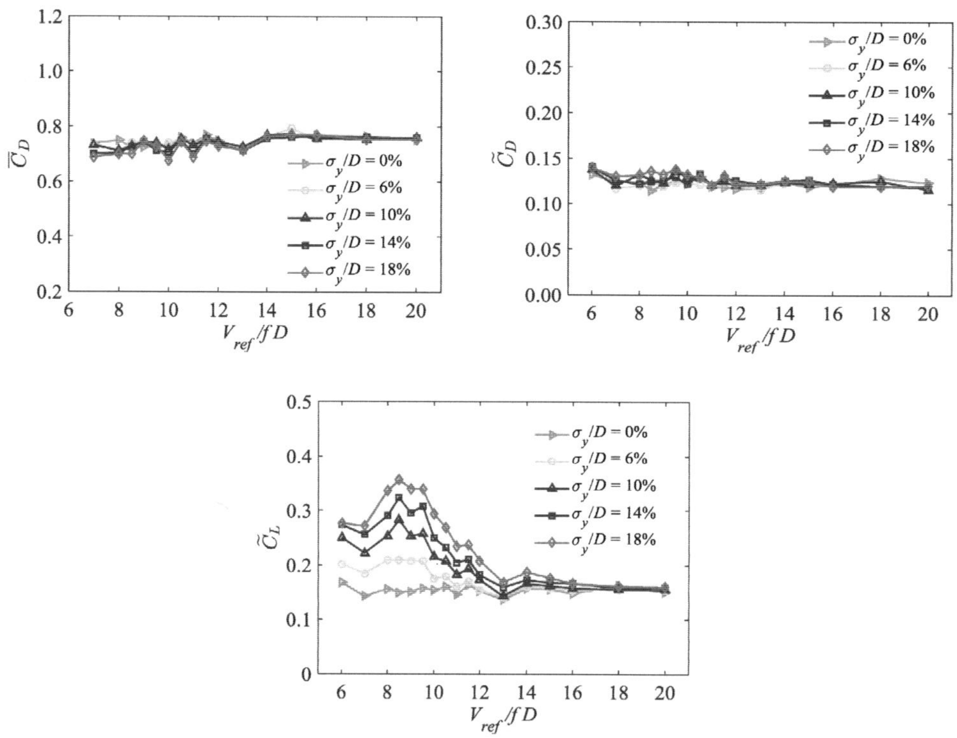

Figure 2. The generalized unsteady aerodynamic force of the prism

In Fig. 2, the mean and RMS drag force coefficients at high wind speeds are close to those at low wind speeds and tend to be constant with increasing reduced wind speeds. Furthermore, the coefficients do not change with increasing oscillating amplitudes and the values are in close agreement with those of the stationary case. In the cross-wind direction, the RMS lift force coefficients decrease rapidly at wind speed of 10 which is in the lock-in range. It is interesting to find that, in the low wind speed range, the RMS lift coefficients increase with increasing oscillating amplitudes, while the coefficients are close to those of the stationary case and are constant in the high wind speed range. The above results at low reduced wind speeds may be partially ascribed to the changes in unsteady flow separation due to the motion-induced flow angles of the model [1]. The separation would have significant effect on the lift force coefficients while having slight effect on the drag force coefficients. The above results also suggest that the flow is in a quasi-steady state at high wind speeds even under large oscillations. In addition, the quasi-steady theory is applicable in the along-wind direction at all wind velocities and in the cross-wind direction at high wind velocities (away from the lock-in range).

3.2 *Motion-correlated force coefficients*

3.2.1 *Generalized motion-correlated force coefficients*

The motion-correlated force is divided into in-phase with oscillating displacement component and in-phase with oscillating velocity component [1, 6], which correspond to motion-correlated aerodynamic stiffness term λ and motion-correlated aerodynamic damping term μ, respectively. The details of the definitions can be found in a previous study [1]. The generalized motion-correlated force coefficients are obtained from integrating the local force coefficients and weighted by modal shape over the height of the prism. The coefficients of the prism in the cross-wind directions under different oscillating amplitudes and wind speeds are presented in Fig. 3. It should be noted that, since motion has slight effect on wind forces in the along-wind direction, only the coefficients in the cross-wind direction are concerned hereafter.

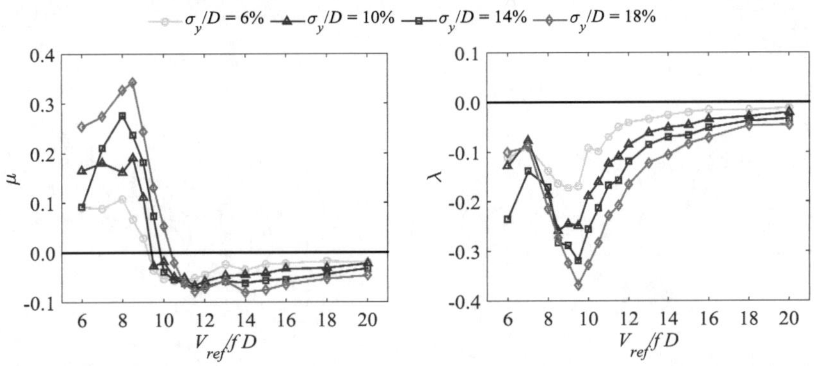

Figure 3. The generalized motion-correlated force coefficients in the cross-wind direction: (a) aerodynamic damping force coefficients; (b) aerodynamic stiffness force coefficients

The motion-correlated force coefficients in the cross-wind direction (Fig. 3) are interesting: in Fig. 3 (a), the motion-correlated aerodynamic damping coefficients are positive at low wind speeds and change from positive to negative at wind speeds around 10, which is in the lock-in range. At high wind speeds, the coefficients are negative and tend to be constant. Furthermore, the coefficients are largely affected by oscillating magnitude and increase with increasing amplitudes, especially in the low wind speed range. In addition, the crossing points (zero points from the positive to the negative) tend to be delayed with increasing oscillating amplitudes. In Fig. 3 (b), the motion-correlated aerodynamic stiffness coefficients are negative and increase rapidly in the lock-in range. The magnitudes tend to increase with oscillating amplitudes. This may be attributed to the change of vortex shedding and unsteady flow separation, which are affected by the model motion.

3.2.2 *Local motion-correlated force coefficients*

The only local motion-correlated force coefficients (aerodynamic stiffness term λ_z, aerodynamic damping term μ_z) the prism in the cross-wind direction under different wind speeds and oscillating amplitudes are presented in Fig. 4.

In Fig. 4, the trends of the local motion-correlated aerodynamic force coefficients are in close agreement with those of the generalized coefficients: at low wind speeds, the damping force coefficients are positive and cross the zero points within the lock-in range, while at high wind speeds, the coefficients tend to be constant; the stiffness force coefficients are negative and peaks exist within the lock-in range. Furthermore, the force coefficients tend to be constant at low levels (i.e. level 2) and motion has little effect on these low levels. Along the height of the prism, the magnitudes of the force coefficients tend to increase. This is ascribed to the different oscillating amplitudes along the height of the prism (tip amplitudes being larger than those at lower levels). However, it is found that the maximum magnitudes occur at level 8 instead of the tip level, and this is induced by the end effect. This is to say, for a three-dimensional prism, the separation of the unsteady flow filed is affected by the end of the prism and therefore the aerodynamic characteristic is changed at the tip end. In addition, at levels 6 to 9, the zero crossing points tend to increase with the height of the prism and also different from those at low levels. It may be caused by the different lock-in effects along the height of the prism and this will be further discussed in the following sections. Besides, Fig. 4 also shows that oscillations would enlarge the magnitudes of the force coefficients.

All the above results indicate that neglecting the effect in the cross-wind direction may overestimate or underestimate the aerodynamic forces and the corresponding predicted responses of structures (overestimate at low wind speeds and underestimated at high wind speeds). This study advances the understanding of the motion-correlated aerodynamic forces of structures, which can help improve response predictions of the structures.

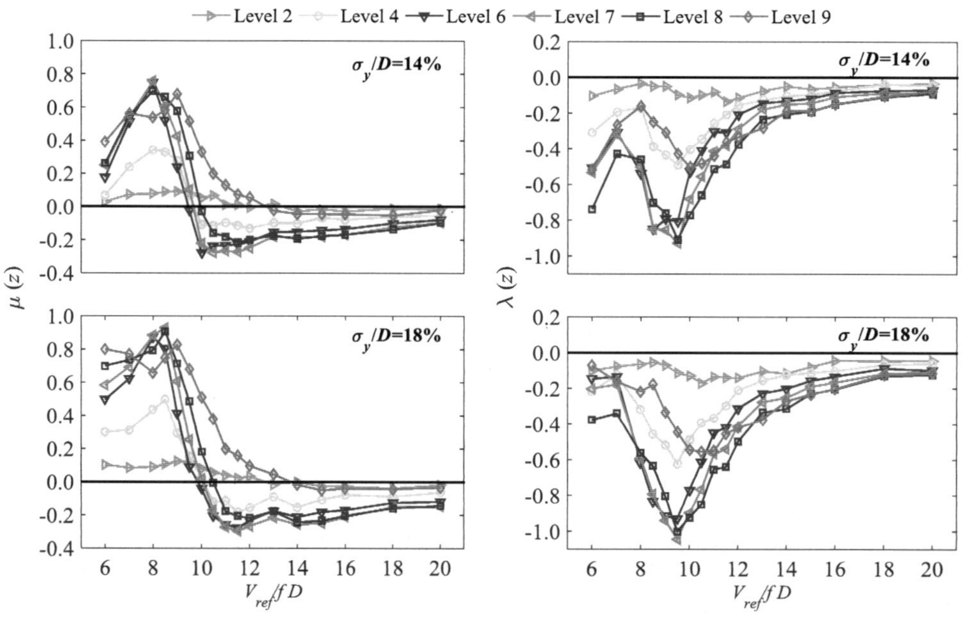

Figure 4. The local motion-correlated force coefficients in the cross-wind direction: (a) aerodynamic damping force coefficients; (b) aerodynamic stiffness force coefficients

4 CONCLUDING REMARKS

The investigation of the unsteady aerodynamic forces of the prism regarding the effect of structural motion using a forced vibration technique leads to several conclusions. The major findings in brief are as follows:

(a) The drag force coefficients tend to be constant and are in quasi-steady state under different wind velocities and oscillations. The lift force coefficients increase with oscillating amplitudes at low wind speeds while the coefficients are close to those of the stationary case and are constant at high wind speeds. The RMS lift force coefficients decrease rapidly at wind speeds around 10, which is in the lock-in range. This suggests that motion has great effect on aerodynamic forces of a structure in the cross-wind direction while has little effect on those in the along-wind direction.

(b) In the cross-wind direction, the motion-correlated aerodynamic damping coefficients are positive at low wind speeds and change from positive to negative at wind speeds around 10, which is within the lock-in range. At high wind speeds, the coefficients are negative and tend to be constant. The coefficients are largely affected by and increase with oscillating amplitude, especially at low wind speeds. The motion-correlated aerodynamic stiffness coefficients are negative and rapidly increase in the lock-in range. The magnitudes tend to increase with oscillating amplitudes. The maximum magnitudes occur at level 8 instead of the tip level 9, which is due to the end effect.

5 REFERENCES

1 Cooper, K., et al., Unsteady aerodynamic force measurements on a super-tall building with a tapered cross section, Journal of wind engineering and industrial aerodynamics, 72 (1997), 199-212.

2 Bearman, P. and E. Obasaju, An experimental study of pressure fluctuations on fixed and oscillating square-section cylinders, Journal of Fluid Mechanics, 119 (1982), 297-321.

3 Holmes, J.D., Wind loading of structures. CRC Press,2015.

4 Kwok, K. and W.H. Melbourne, Wind-induced lock-in excitation of tall structures, Journal of the Structural Division, 107 (1981), 57-72.

5 Lin, N., et al., Characteristics of wind forces acting on tall buildings, Journal of Wind Engineering and Industrial Aerodynamics, 93 (2005), 217-242.

6 Steckley, A., Motion-induced wind forces on chimneys and tall buildings, PhD dissertation, University of Western Ontario: London, Canada,1989.

A new framework for vehicle safety assessment under hazardous driving conditions

Guangyang Hou[a], Suren Chen[b]

[a] *Department of Civil & Environmental Engineering, Colorado State University, Fort Collins, CO 80523, United States, guangyang.hou@colostate.edu*
[b] *Department of Civil & Environmental Engineering, Colorado State University, Fort Collins, CO 80523, United States, suren.chen@colostate.edu*

ABSTRACT

Under some adverse driving condition (i.e. strong wind, icy or snowy road surface), vehicles often experience increased risks if single-vehicle accidents. Most existent studies of vehicle safety focused on a single vehicle moving at a constant speed. In reality, a vehicle accelerates, decelerates, and switches lanes along with other vehicles and it becomes more realistic to evaluate the safety of a vehicle in the context of realistic traffic flow. Vehicle safety performance is therefore affected by the specific traffic environment, including traffic density, speed limit, lane numbers, vehicle types and the driving behaviors. A new framework of traffic safety assessment of traffic flow is developed with three parts. Firstly, stochastic traffic flow is simulated with an improved cellular automaton (CA)-based traffic flow simulation model is developed, which considers limited acceleration/deceleration capabilities, realistic vehicle properties and different driving behaviors among all the drivers; secondly, with the single-vehicle accident (SVA) simulation model, the whole process of a rollover or sideslip accidents can be simulated under hazardous driving conditions, including wind gust, snowy or icy road surface, and road curvature. With detailed time-dependent vehicle information from the CA traffic flow simulation, the likelihood of accident occurrence of each vehicle in the traffic flow at any time is predicted with the SVA model; finally, for a specific traffic flow, even under the same adverse driving condition, accidents of all vehicles do not occur at the same time due to different trajectories (i.e. velocity, position). In this study, vehicle accident vulnerability of the traffic flow is defined as the ratio of the number of vehicles that are vulnerable to accidents to the number of all vehicles in the traffic flow for a certain period of time. Vehicle accident vulnerability of the whole traffic flow can then be assessed for a specific hazardous scenario. Following the proposed framework, traffic safety of vehicles in the stochastic traffic of a "roadway–bridge–roadway" system is investigated for the demonstration purposes.

KEYWORDS

Vehicle safety, Accident, Cellular automaton, Hazardous driving conditions

Unsteady wind forces of a slender prism

Zengshun Chen-1[a], K. T. Tse-2[a], K. C. S. Kwok-3[b]

[a] *Institution, Location, Country, email*
[a]*Department of Civil and Environmental Engineering, The Hong Kong University of Science and Technology, Hong Kong, China,* zchenba@connect.ust.hk, timkttse@ust.hk
[b]*Center for Infrastructure Engineering, Western Sydney University, Penrith, NSW, 2751, Australia,* K.kwok@westernsydney.edu.au

ABSTRACT

Architectural structures tend to be high and slender, and accurately obtaining the wind load of such structures is becoming more important than before. Generally, a high frequency base balance (HFBB) or a static synchronous multi-pressure sensing system (SMPSS) wind tunnel test is carried out to evaluate the wind load of structures. The HFBB and SMPSS techniques are effective. However, they are static measurements and the effect of structural motion is therefore excluded. As is known, the effect is usually small. But when it is in-phase with wind velocity, the effect acts as aerodynamic damping and cannot be neglected: at low wind speeds, motion-correlated aerodynamic damping is positive, and neglecting its effect can lead to overestimation when predicting structural response; on the other hand, at high wind speeds, motion-correlated aerodynamic damping is negative, and neglecting its effect can result in underestimated prediction in structural response. Also, it has already been confirmed that the differences of wind load measured by a static and a dynamic test are mainly ascribed to the effect of structural motion. Therefore, the unsteady wind loads of structures, which include the effect of structural motion should be acquired.

The unsteady aerodynamic forces acting on a slender prism were investigated using a forced vibration technique. The prism was driven to oscillate by an actuator and the unsteady distributed pressures, under different wind velocities and oscillations, were measured. The measurement was calibrated with respect to driving oscillation, aerodynamic force coefficient, as well as motion-correlated force coefficient. Then, the unsteady generalized and local aerodynamic force coefficients and the motion-correlated force coefficients of the prism, which are functions of reduced wind velocity and oscillation, were analyzed. It shows that the effects of structural motion on the coefficients are significant in the cross-wind direction while the effects are slight in the along-wind direction. Furthermore, in the cross-wind direction, the coefficients tend to increase with oscillation amplitudes at low wind speed range while they are at a quasi-steady state at high wind speed range. These characteristics were analyzed from the perspectives of generalized and pointwise spectra and *Strouhal numbers* of the prism. The study advances the understanding of the effect of structural motion on three-dimensional prisms, which can be utilized to improve response predictions of the prisms.

KEYWORDS

Forced vibration technique; Unsteady pressure measurements; Aerodynamic damping; Aerodynamic stiffness

Characteristics of the instability of a shear layer formed on a square prism

Daniel Lander[a,b], Daniel Moore [a,b], Michael Amitay [b], Chris Letchford [a]

[a]Department of Civil and Environmental Engineering, Rensselaer Polytechnic Institute, Troy, NY, USA, landed@rpi.edu
[b]Center for Flow Physics and Control, Department of Mechanical, Aerospace and Nuclear Engineering, Rensselaer Polytechnic Institute, Troy, NY, USA.

ABSTRACT

The transition-to-turbulence in the shear layers of a square prism are one of the defining features of the flow. Indeed, the Reynolds number range $1{,}000 < Re_D < 200{,}000$ is known as the "shear layer transition regime" on account of the dramatic changes observed in the wake which arise from highly unstable shear layer, which contains the Kelvin-Helmholtz (KH) instability. These regions of concentrated gradients play a central role in many of the complex dynamics of the flow: the KH instability is responsible for the first signatures of turbulence in the flow; they act as conduits for the passage of vorticity, while their presence is indispensable to the formation of the von-Karman (VK) vortex street, arguably the most distinct instability in fluid mechanics. The presentation will focus on the scaling characteristics of the shear layer instability such as frequency, f_{KH}, and critical length, θ_c and the associated vortex dynamics of the transition process, as a function of Re_D. These flow features will be presented in the context of wind tunnel testing of bluff bodies relevant to Wind Engineering..

KEYWORDS

Bluff body shear layers

Experimental Investigation of Vortex-Induced Vibrations of a Rectangular Section

Mehedy Mashnad

Principal, Walter P. Moore & Associates, Dallas, TX, USA, mmashnad@walterpmoore.com

ABSTRACT

Vortex-induced vibrations of a spring-mounted rigid rectangular section model with two degrees of freedom were studied in a wind tunnel. The along-wind to across-wind aspect ratio (*B/D*) of the model section is 4. This paper presents and discusses a number of observations made during those experiments such as vertical amplitudes versus reduced wind speeds, rotational amplitudes versus reduced wind speeds, onset reduced wind speeds of lock-in response, normalized amplitudes versus Scruton number, and characteristics of transient and steady-state regimes of the response histories.

Experiments were performed in smooth and nominally two-dimensional flow conditions using an open-return type low-speed, low-turbulence wind tunnel at the University of Illinois, Urbana-Champaign. The working cross section of the tunnel is 0.91 m × 1.22 m and maximum attainable speed is 71 m/sec. A force-balance with vertical and torsional degrees of freedom was used. Strain gages mounted on the flexural support elements of the force-balance measured vertical displacement and rotation. A pitot tube placed in downstream region from the centerline of the model recorded dynamic pressure while a pitot tube in the upstream region recorded the free-stream wind speeds.

Vertical and torsional structural frequencies of the model are measured to be 3.45 Hz and 6.3 Hz, respectively, while vertical and torsional damping ratio are 0.26% and 0.18%, respectively. Strouhal number measured on a fixed rectangular model is 0.153. Reynolds number ranged between 0 and 2300. Two lock-in regions for vertical motion at reduced wind speeds ranging from 3.5 to 4.5 and 6 to 9, while two lock-in regions for torsional motion at reduced speeds ranging from 2.4 to 3 and 4.4 to 6 were observed. Normalized vertical amplitude (Amplitude/D) in first lock-in region is 0.006 and in second lock-in region 0.045. Rotational amplitude for first lock-in region is 0.002 radians and for second lock-in region 0.019 radians. During lock-in oscillations, vertical and rotational response signals exhibited nearly sinusoidal structure and the frequency of oscillation is essentially equal to the natural frequency of the model.

To understand the mechanism of lock-in vibration, transient time histories - both decaying and growing responses - were recorded. It is observed that decaying and growing responses of the model reached limit-cycle oscillations inside the lock-in regions leading to moderate-amplitude self-limiting vibration at a given wind speed. The steady-state amplitude of oscillation continues to grow with increasing wind speed until a threshold speed is reached after which the oscillation begins to diminish. Logarithmic forms of the envelopes of the transient displacement and rotational time histories obtained by Hilbert Transform indicate that the slopes of the envelopes are nonlinear. It implies net damping ratio of the oscillator is not constant but depends on displacement amplitudes. Frequency of transient time history is found independent of response amplitudes.

Experimental studies on the vortex-induced vibrations of non-circular bluff sections are scarce in the literature. This paper is intended to complement the existing knowledge base for rectangular sections representative of tall buildings, long-span bridges, and other structures.

KEYWORDS

Four or five keywords or phrases, comma delineated, same font and size as abstract

13th Americas Conference on Wind Engineering
Gainesville, Florida, USA
May 21 - 24, 2017

Dynamic coupling effects on long-span bridges under concurrent wind, traffic and earthquake

Yufen Zhou [a], Suren Chen [b]

[a] *Colorado State University, Fort Collins, USA, yufen.zhou@colostate.edu*
[b] *Colorado State University, Fort Collins, USA, suren.chen@colostate.edu*

ABSTRACT

A long-span bridge is often susceptible to the excitations from wind due to its slenderness and low structural damping. In addition, a long-span bridge usually supports considerable amount of traffic, which can significantly influence the dynamic performance of the bridge along with wind on a daily basis. Meanwhile, long-span bridges may experience low to moderate earthquakes of certain probability occurrence especially when they are located adjacent to faults. To appropriately assess the structural safety and serviceability risks, it is important to rationally predict the performance of a long-span bridge-traffic system under possible concurrent presence of dynamic loadings. However, this presents a challenging engineering problem because various coupling effects are involved among the bridge and multiple excitations, which are originated from wind-bridge interaction, traffic-bridge interaction, coupling between the bridge superstructure and substructure and their secondary coupling forces. Existing methodology is not able to consider comprehensive coupling effects properly for dynamic assessment of long-span bridges under wind, traffic and earthquake.

This study will introduce a new simulation methodology on long-span bridges under wind, traffic and earthquake to take into account the comprehensive coupling effects involved among the bridge structure and multiple dynamic excitations. The methodology starts with the equations of motion of the bridge superstructure, substructure and vehicles in the traffic flow under wind and earthquake. By deriving the coupled earthquake force on the bridge superstructure and vehicles, the constrained degrees of freedom on the bridge substructure can be eliminated from the equations of motion. Different from traditional formulation of earthquake forces, the derived earthquake force acting on the bridge superstructure has a coupling term in additional to traditional earthquake inertia force. The coupling term in the total earthquake force on the bridge superstructure is expressed as the function of vehicle response, aerodynamic matrices, bridge-vehicle coupling matrices, earthquake ground motion inputs and the influence matrix on the bridge superstructure. Because of the dynamic interaction between the bridge and vehicles, earthquake force will also be generated on the vehicles, which is expressed by bridge-vehicle coupling matrices, earthquake ground motion inputs and the influence matrix on the bridge superstructure. The final equations of motion of the bridge and vehicles in the traffic flow are obtained based on bridge superstructure and vehicles in the traffic flow. An iterative algorithm is utilized to obtain the coupled forces and the global response of the bridge and vehicles.

Numerical analysis is conducted on a prototype long-span bridge under excitations from wind, traffic and earthquake. The bridge dynamic performance with excitations of different intensities is investigated through the proposed coupled model. In typical dynamic scenarios, the contribution of various coupling effects among the bridge, vehicles, wind and earthquake on the bridge superstructure is quantified. The dynamic responses of the bridge from the proposed coupled model are then compared with those of the bridge without considering coupling effects. It is found that the coupling effects among the bridge, vehicles, wind and earthquake are significant and should be taken into account in predicting the dynamic performance of the long-span bridge under concurrent dynamic loads.

KEYWORDS

Long-span bridges, wind, traffic, earthquake, coupling

13th Americas Conference on Wind Engineering
Gainesville, Florida, USA
May 21 - 24, 2017

Spanwise Correlation of Vortex-induced Force on Rectangular 5:1 Cylinder

Yanguo Sun, Xiaonan Wang, Haili Liao, Mingshui Li

Research Center for Wind Engineering, Southwest Jiaotong University, Chengdu, China
Department of Bridge Engineering, Southwest Jiaotong Unviersity, Chengdu, China
syg222@163.com

ABSTRACT

Vortex-induced vibration (VIV) usually occurs on long and slender structures with bluff cross section such as rectangular, and wind tunnel tests for a section model is a general methodology to evaluate its VIV performance at present. The distributions of vortex-induced aerodynamic force of the whole body surface show spatial or three-dimensional characteristics, and the aerodynamic forces along the span are not fully correlated.

The vortex-induced aerodynamic forces acting on an oscillatory rectangular 5:1 cylinder are investigated in details by synchronized surface pressure measurement by using free and forced motion testing apparatus separately in current research. The distribution characteristics of surface pressure and the spanwise correlation of measure points in different parts of the bluff body are studied at various frequencies and amplitudes of vibration.

The testing results show that the attachment and loss of vortex exist in body's surface which cause the vibration. The aerodynamic forces along the span are not fully correlated, which means that if the VIV response of prototype structures is directly estimated and evaluated by the testing results from section model wind tunnel tests, consequently the overestimate results of VIV response may be made. It's also shown that the spanwise correlation of the rectangular section model is affected at a large extent by the change of vibration amplitudes and the spanwise spacing. The correlation coefficients will increase with the increase of amplitude and decrease with the increase of spanwise spacing, and both the free and forced motion tests methods show the similar results.

The spanwise correlations of measure points on upper surface and bottom which parallel to the direction of the wind show an increase tendency with the increase of non-dimensional frequency (the ratio between the frequency of section model oscillation and the frequency of vortex shedding) at first, and the maximum value of correlation coefficients occurs at non-dimensional frequency around 1.0, and the correlation decreases after that. However, for the measure points in the wake area, the non-dimensional frequency at which the maxima of the curves occurred is about 0.80 to 0.85 rather than 1.0, which is quite different with the points in upper surface and bottom. Furthermore, the correlation of lift force (whole cross section) is better than the correlation of measure point of wake. At last, a correlation function for this bluff body was fitted based on current testing results in the form of exponential which can be adopted by the there-dimensional VIV analysis of long and slender structures.

KEYWORDS

vortex-induced vibration, bluff-body aerodynamics, wind tunnel tests, spanwise correlation, rectangular cylinder

Numerical simulation of aeroelastic instability and behavior of a rectangular prism beyond critical

Jiawei Wan[a], Haili Liao[b], Ahsan Kareem[c]

[a] *Department of Civil Engineering, Southwest Jiaotong University, Chengdu, China, jiawei.wan@outlook.com*
[b] *Department of Civil Engineering, Southwest Jiaotong University, Chengdu, China, hlliao@home.swjtu.edu.cn*
[c] *Department of Civil Engineering and Geological Sciences, University of Notre Dame, Indiana, USA, Ahsan.Kareem@nd.edu*

ABSTRACT

The aeroelastic instability and post critical behavior of a rectangular prism modeled as a single-degree-of-freedom with aspect (chord to depth) ratio of 5 to 1 undergoing torsional flutter is numerically investigated using the large eddy simulation (LES) and fluid-structure interaction (FSI) schemes. The applicability of a numerical approach in reproducing limit cycle oscillations (LCO) of a bluff body, whose amplitude increases with the flow velocity and experiences nonlinear aerodynamics, is examined. The spatially-filtered incompressible Navier-Stokes equations are solved utilizing a Runge–Kutta (RK) based projection method. The *k*-equation LES model with Van Driest damping near solid wall is adopted for subgrid scale stress modeling. The dynamic response of rectangular prism is computed through the Newmark-beta method, and a partitioned semi-implicit strong coupling scheme is employed to account for coupling between the flow and the prism. The implementation of explicit RK based projection method for dynamic mesh together with the semi-implicit flow-structure coupling scheme in the OpenFOAM[@] library is introduced. A pre-defined mesh motion function is utilized to accommodate the relative motion between the oscillating prism and the stationary boundary. The grid quality at different oscillation amplitude is evaluated by computing the equivalent skew-angle of each cell. First, the flow field around the stationary and oscillating rectangular prism experiencing forced rotational motion at various amplitudes are simulated. Different levels of reduced velocities are considered by changing the oscillation frequency while keeping the incoming flow speed constant. Subsequently, the stationary rectangular prism or the one experiencing oscillations will be released from its zero angle of incidence position with the velocity proportional to the product of amplitude and frequency in a fully developed flow condition including FSI simulation to undergo free vibration. The aeroelastic response of the rectangular prism at various reduced velocity and initial motion status can be investigated. Finally, a single-degree-of-freedom nonlinear modeling scheme is introduced to predict the dynamic response of prism. The results are compared with those obtained from FSI simulation and wind tunnel experiments. The pressure fluctuations and unsteady loads acting on the rectangular prism are examined to calculate the energy transfer to the prism from the flow in different test cases. The flow field around the oscillating prism is also demonstrated to supplement understanding the mechanisms involving nonlinear aerodynamics of rectangular prisms.

KEYWORDS

Fluid structure interaction, Limit cycle oscillation, Projecting method, Coupling scheme, OpenFOAM[@]

13th Americas Conference on Wind Engineering
Gainesville, Florida, USA
May 21 - 24, 2017

Aerodynamic Pressures Measured at Multiple Taps on the Surfaces of a Square Cylinder in Smooth Flow with Various Angles of Attack

DongHun Yeo[a], Matthew Kovarek[b], Iosif I. Shinder[c], and Emil Simiu[d]

[a] *National Institute of Standards and Technology, Gaithersburg, MD. 20899, USA,* donghun.yeo@nist.gov
[b] *National Institute of Standards and Technology, Gaithersburg, MD. 20899, USA,* matthew.kovarek@nist.gov
[c] *National Institute of Standards and Technology, Gaithersburg, MD. 20899, USA,* iosif.shinder@nist.gov
[d] *National Institute of Standards and Technology, Gaithersburg, MD. 20899, USA,* emil.simiu@nist.gov

ABSTRACT

Aerodynamic characteristics of bluff bodies such as circular and square cylinders have been investigated for decades, as they help to understand the complex physics of flows around bluff bodies and provide a foundation for Computational Fluid Dynamics (CFD) simulations. Aerodynamic pressures on such structures as buildings and bridges are affected by signature turbulence due to flow separation from the structures, and by turbulence in the incoming atmospheric boundary layer flow. As a first step toward understanding the joint effects of the two types of turbulence it is useful to examine the effects of signature turbulence alone.

Modern pressure measurement techniques offer an opportunity to do so in ways that were not heretofore possible. A detailed fundamental experimental study conducted at NIST was devoted to the investigation of the effects of signature turbulence on the pressure distribution on the surfaces of a square cylinder under incoming smooth flow with various angles of attacks. The study used the NIST close-circuit wind tunnel testing facility, whose dimensions are 1.5 m in width × 1.2 m in height × 12 m in length. Pressures were synchronously measured at the sampling rate of 625 Hz using 90 pressure taps on the sides of the cylinder with dimensions 0.1 m × 0.1 m × 1.5 m. The tubing system connecting the multi-channel pressure scanner to the pressure taps consisted of (i) two tubes with 1.37 mm and 0.86 mm diameters, and (ii) a restrictor with 2.79 mm length and 0.22 mm diameter. The measured pressures were corrected using a dynamic pressure calibration program developed by NIST that accounts for amplitude alteration and phase shift of the measured fluctuating pressures due to the tubing system, as well as for the phase change due to the measuring sequence in the multi-channel scanners. The Reynolds number was 6.5×10^4 based on the incoming 10 m/s wind speed and 0.1 m cylinder sides. Tests were performed for angles of attack in 2-degree increments obtained by rotating the cylinder along its spanwise axis.

Results showed that the angle of attack significantly influences (i) the distribution of the pressure coefficients on the surface of the cylinder, (ii) the frequency of the fluctuating pressures, and (iii) the characteristics of their coherence. The results provide insight into the characteristics of aerodynamic pressures as affected by the signature turbulence and valuable fundamental information required for the detailed validation of CFD simulation results.

KEYWORDS

Aerodynamic pressure, spatial distribution, simultaneously measured pressures, wind engineering, wind tunnel testing.

This page intentionally left blank.

13th Americas Conference on Wind Engineering
Gainesville, Florida, USA
May 21 - 24, 2017

Computational aerodynamic shape optimization of long-span bridge decks –optimal fairing design for H-shaped decks

T. H. Birhane, M.Sc.[a], G. T. Bitsuamlak, Ph.D., P.Eng.[b] and J. P. King, Ph.D., P.Eng.[c]

[a] *Department of Civil and Environmental Engineering, The University of Western Ontario, IGAB-Room#3N60 - London, ON, Canada; email: tbirhane@uwo.ca*
[b] *Department of Civil and Environmental Engineering, The University of Western Ontario, IGAB-Room#3N60 - London, ON, Canada; email: gbitsuam@uwo.ca*
[c] *Boundary Layer Wind Tunnel Laboratory, The University of Western Ontario, London, ON, N6A 5B9, Canada; email: jpck@blwtl.uwo.ca*

ABSTRACT

In the wake of Tacoma Narrows Bridge failure, considerable advancements have been achieved in enhancing the aerodynamic performance of long-span bridges through deck shape modifications. Shape modifications to existing long-span bridges as in the case of White Bronx Bridge or development of aerodynamically efficient bridges as in the case of the proposed Messina Straight Bridge have been mostly carried out through the experimental based "cut and try" procedure. This procedure is believed to be cost effective if it is coupled with computational tools. Though the identification of aerodynamic mitigation measures in wind engineering applications has been guided by the series of wind tunnel tests, computational tools have been applied effectively to design optimum aerodynamic shapes in aeronautical and automobile industries. However a framework for applying such tools for long-span bridges needs to be developed and to be tested for practical implementation. In this study, a computational aerodynamic shape optimization (CASO) framework which involves integrated use of computational fluid dynamics (CFD), radial basis neural network and numerical optimization routines has been developed. Unlike the traditional cut-and-try approach, the method is self-contained in such way that the shape modification process, which aims to identify the best behaving deck profile within the bound of specified alternatives, is automatically handled by the optimization algorithms. This circumvents the direct involvement of design engineers in the decision making at each intermediate design iteration – try and error shape modification - cycles. The framework is applied to design fairings for generic elongated H-shaped decks with the objective of reducing the lateral wind load and improving its aerodynamic stability under smooth flow. Various forms of fairing profiles are tested including triangular and curved profiles defined by splines. Optimal fairing shapes are evaluated for each different deck forms of varying breadth-depth ratio, and deck eccentricities. Numerically evaluated optimal fairing shapes are compared with that of Bronx Whitestone Bridge and Deer Isle Bridge, which were designed based on wind tunnel tests. It is shown that sharper triangular fairings are effective to reduce wind induced drag, but shorter fairings with height around 60% to 70% of deck depth are effective to improve the aerodynamic stability of elongated H-shaped decks. Furthermore, asymmetric triangular fairings are found to be effective to improve the aerodynamic performance of asymmetric decks.

KEYWORDS

Long-span bridges, aerodynamics, computational shape optimization, computational fluid dynamics, neural network.

Identification of Flutter Derivatives of Truss Deck from Gust Response

Hoang Trong LAM[a], Hiroshi KATSUCHI[b], Hitoshi YAMADA[c]

[a]*Yokohama National University, Yokohama, Japan*
[b]*Yokohama National University, Yokohama, Japan*
[c]*Yokohama National University, Yokohama, Japan*

ABSTRACT: This study is to clarify the effects of oncoming turbulence on the self-excited force of a suspended long span bridge deck. In the study reported herein, the more challenging is the application of a system identification method to identify flutter derivatives from gust responses for the section model. The gust response is obtained by an experimental wind tunnel test for a trussed deck section. The output only time domain analysis stochastic system identifications: co-variance stochastic system (SSI-cov) and data driven stochastic system method (SSI-data) are proposed to extract simultaneously all flutter derivatives from two degrees of freedom system. The results are also compared with those from smooth flow as well as free decay response.

KEYWORDS: System identification, flutter derivatives, gust response, SSI-cov, SSI-data.

1 INTRODUCTION

Long-span cable supported bridges are highly susceptible to wind excitation and the aerodynamic instability such as flutter phenomenon play an important role in the design of these bridges. This phenomenon is well represented in term of flutter derivatives that are a function of the reduced frequency and body shape and can be identified from a section model test by using a system identification method. System identification techniques that have been applied to identify the flutter derivatives can be classified into two broad groups. A deterministic system identification technique is applied only for a free decaying vibration signal or impulse response. Therefore, in this system the buffeting force and their responses are considered as external noise, so this causes more difficulties at high wind velocity such as noise increase due to turbulence. The other is a stochastic system identification technique, which is applied for not only free decaying signal but also buffeting response. In latter system identification technique the deterministic knowledge of the input is replaced by the assumption that the input is a realization of a stochastic process.

The wind in the atmospheric boundary layer is always turbulent. Therefore any research of wind-induced vibration problems must consider this issue. Not many researchers have focused clearly on the effects of turbulence on aeroelastic forces, especially at high turbulence intensity. In addition, there is a shortcoming in the free vibration technique. At a high wind velocity the extraction of FDs cannot be obtained accurately because the aerodynamic damping of vertical mode is too high and vertical free-vibration data is too short for analysis. From these considerations, the idea for applying the stochastic system identification (SSI) to estimate FDs from gust responses of a bridge trussed deck section is brought.

This study concentrates on the investigation of turbulence effects on FDs of a trussed bridge deck by employing a section model wind tunnel test in a turbulent flow. The output only system identification SSI is proposed to extract FDs from buffeting response. Tests are also carried out with the free vibration method in order to compare with the proposed method.

2 WIND TUNNEL TEST

A wind tunnel test was conducted in a closed-circuit wind tunnel of Yokohama National University. The investigated profile is trussed deck section (Figure. 1). The width and depth of the section model are 363mm and 162.5mm, respectively. The unit mass is 8.095 kg/m and moment of inertia is 0.2281 kgm^2/m. The vertical frequency and damping ratio are 1.869Hz and 0.0051, respectively. The torsional frequency and damping ratio are 3.296Hz and 0.004186, respectively. The tests have been carried out in both smooth and different turbulent flows. The turbulent flows used in this study were generated with biplane wooden grids. The turbulent properties are controlled by changing the distance to the model. The flow conditions and turbulence properties are shown in Table 1.

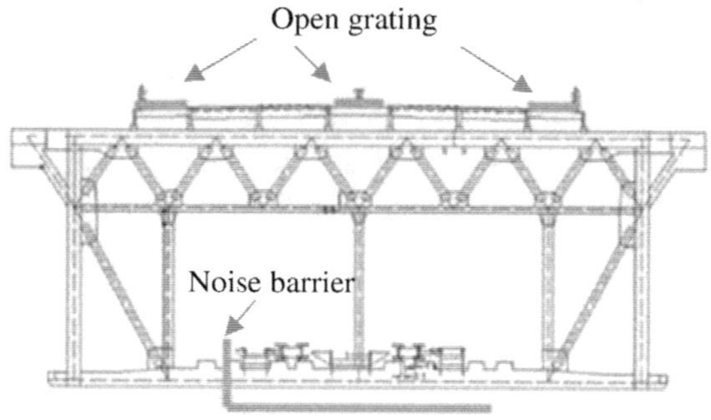

Figure 1 Truss deck section model

Table 1 Turbulence intensity and scale

	I_u (%)	I_w (%)	L_u (cm)	L_w (cm)
Case 1	6.2	4.6	11.3	9.1
Case 2	9.1	6.9	9.0	8.7
Case 3	15.6	13.2	6.8	6.4

3 IDENTIFICATION OF FLUTTER DERIVATIVES

3.1 *Stochastic discrete-time state-space system*

Considering a 2 DOF section model of bridge deck in turbulent flow, equation of motion is written by

$$\begin{aligned} m[\ddot{h} + 2\xi_h\omega_h\dot{h} + \omega_h^2 h] &= L_{se} + L_b \\ I[\ddot{\alpha} + 2\xi_\alpha\omega_\alpha\dot{\alpha} + \omega_\alpha^2\alpha] &= M_{se} + M_b \end{aligned} \tag{1}$$

where m and I are mass and mass moment of inertial per unit length, respectively; $\omega_h=2\pi f_h$ and $\omega_a=2\pi f_a$ are circular frequencies of heaving and pitching mode; ξ_h and ξ_a are damping ratio to critical; L_{se} and M_{se} are the aerodynamic self-excited lift and moment, respectively, given as

$$\begin{aligned} L_{se} &= \frac{1}{2}\rho U^2 B\left[K_h H_1^*(K_h)\frac{\dot{h}}{U} + K_\alpha H_2^*(K_\alpha)\frac{B\dot{\alpha}}{U} + K_\alpha^2 H_3^*(K_\alpha)\alpha + K_h^2 H_4^*(K_h)\frac{h}{B}\right] \\ M_{se} &= \frac{1}{2}\rho U^2 B^2\left[K_h A_1^*(K_h)\frac{\dot{h}}{U} + K_\alpha A_2^*(K_\alpha)\frac{B\dot{\alpha}}{U} + K_\alpha^2 A_3^*(K_\alpha)\alpha + K_h^2 A_4^*(K_h)\frac{h}{B}\right] \end{aligned} \tag{2}$$

where ρ is the air density; U is the mean wind velocity; B is the width of bridge deck; $K_i=\omega_i B/U$ is the reduced frequency ($i=h$, α); H^*_i and A^*_i (i=1,2,3,4) are the flutter derivatives. L_b and M_b (factorized into matrix B_2 and input vector $u(t)$) are the buffeting forces in the vertical and torsional directions, respectively. By substituting L_{se}, M_{se}, L_b and M_b into Eq. (1) and moving the aerodynamic damping and stiffness terms to the left hand side, Eq. (1) can be transformed into a first-order state equation by Eq. (3)

$$\dot{x}(t)=\begin{bmatrix} 0 & I \\ -[M]^{-1}[K^e] & -[M]^{-1}[C^e] \end{bmatrix}_{4\text{x}4} x(t)+\begin{bmatrix} 0 \\ [M]^{-1}B_2 \end{bmatrix} u(t) \tag{3}$$

where $x(t)$ is state vector; $[M]$ is mass matrix; $[C^e]$ is gross damping matrix including the structural damping and aerodynamic damping; $[K^e]$ is gross stiffness matrix including the structural stiffness and aerodynamic stiffness.

In the modal analysis, sometimes the input is unknown and measurements are mostly sampled at discrete-time. On the other hand, it is impossible to measure all DOFs and measurements always have disturbance effects. For all these reasons, the continuous deterministic system will be converted to suitable form: discrete-time stochastic state-space model as follow:

$$\begin{aligned} x_{k+1} &= Ax_k + w_k \\ y_k &= Cx_k + v_k \end{aligned} \tag{4}$$

where $x_k=x(k\Delta t)$ is the discrete-time state vector containing the discrete sample displacement and velocity; w_k is the process noise; v_k is the measurement noise due to sensor inaccuracy. Assuming w_k and v_k are zero mean and $\{x_k\}$, $\{w_k\}$ and $\{v_k\}$ are mutually independent, the output covariance R=E[$y_{k+i}\ y_k^T$] for any arbitrary time lags $i\Delta t$ can be considered as impulse response (Eq. (5)) of the deterministic linear time-invariant system A, C and G where G= E[$x_{k+1}\ y_k^T$] is the next state-output covariance matrix.

$$R_i = CA^{i-1}G \tag{5}$$

3.2 *Covariance driven stochastic system identification (SSI-cov)*

The heart of SSI-cov method is the ERA developed by Juang & Pappa (1985). It is the famous technique for modal parameter identification from free vibration response or impulse response. The key step in the SSI-cov system is the computation of output covariance which can be expressed assuming ergodicity process as:

$$R_i = E[y_{k+1}\ y_k^T] = \lim_{N\to\infty}\frac{1}{N}\sum_{k=0}^{N-1} y_{k+i}y_k^T \approx \frac{1}{N}\sum_{k=0}^{N-1} y_{k+i}y_k^T \tag{6}$$

where i is time lag; N is finite number of data
All output covariance R_i in Eq. (6) is stored in block Toeplitz matrix as:

$$T_{1|i} = \begin{bmatrix} R_i & R_{i-1}\cdots R_1 \\ R_{i+1} & R_i\ \cdots\ R_2 \\ \cdots & \cdots\ \ \cdots \\ R_{2i-1} & R_{2i-2}\cdots R_i \end{bmatrix}_{ixi} \tag{7}$$

Combining Eq. (5) and Eq. (7), then block Toeplitz matrix can be decomposed as follows

$$T_{1|i} = \begin{bmatrix} C \\ CA \\ \cdots \\ CA^{i-1} \end{bmatrix} [A^{i-1}G \ \ A^{i-2}G \cdots AG \ \ G] = O_i C_i \tag{8}$$

On the other hand, observability matrix O_i and controllability matrix C_i can be obtained from singular value decomposition (SVD) of the Toeplitz matrix:

$$T_{1|i} = \begin{bmatrix} U_1 & U_2 \end{bmatrix} \begin{bmatrix} S_1 & 0 \\ 0 & 0 \end{bmatrix} \begin{bmatrix} V_1^T \\ V_2^T \end{bmatrix} = U_1 S_1 V_1^T \tag{9}$$

where U and V are orthonormal matrices and S is a diagonal matrix containing singular values. The number of non-zero gives the rank of the decomposed matrix and coincides with the size n of the state-space matrix A. Comparing the Eq. (8) with Eq. (9), we can rewrite that

$$\begin{aligned} O_i &= U_1 S_1^{1/2} \\ C_i &= S_1^{1/2} V_1^T \end{aligned} \tag{10}$$

Now the realization of all system matrices A and C is achieved. The state matrix A can be obtained by decomposing a shift block Toeplitz matrix:

$$T_{2|i+1} = O_i A C_i \tag{11}$$

Combining the Eq. (10) and Eq. (11) gives:

$$A = O_i^+ T_{2|i+1} C_i = S_i^{-1/2} U_1^T T_{2|i+1} V_1 S_1^{-1/2} \tag{12}$$

where $(.)^+$ is the pseudo-inverse of a matrix. The output matrix C equals the first l rows of O_i where l number of outputs (in this study l=2 for 2 DOF system).

3.3 *Data driven stochastic system identification (SSI-data)*

SSI-data works directly with time-series of experimental data, without need to convert output data to correlation, covariance or spectra. The main step of SSI-data is a projection of the row space of the future outputs into the row of past outputs. The orthogonal projection P_i is defined as:

$$P_i = Y_f / \mathbf{Y_p} = Y_f Y_p (Y_p Y_p^T)^{-1} Y_p \tag{13}$$

where the matrix Y_f and Y_p are the under half part and upper half part of a bock Hankel matrix H, respectively.

$$H = \begin{bmatrix} Y_{(1:j-2i)} \\ Y_{(2:j-2i+1)} \\ \cdots \\ Y_{(2i:j)} \end{bmatrix} = \begin{bmatrix} Y_p \\ \hline Y_f \end{bmatrix} \begin{matrix} \updownarrow li \\ \updownarrow li \end{matrix} \tag{14}$$

The main theorem of stochastic subspace identification states is that the projection P_i can be factorized as the product of observability matrix O_i and the Kalman filter state sequence $\hat{X}_i$. The observability matrix O_i and the Kalman filter sequence $\hat{X}_i$ are obtained by applying SVD to the projection matrix P_i. The Kalman state sequences $\hat{X}_i, \hat{X}_{i+1}$ are calculated using only output data. The system matrices can now be recovered from over determined set of linear equations and obtained by extending Eq. (4)

$$\begin{pmatrix} \hat{X}_{i+1} \\ Y_{i/i} \end{pmatrix} = \begin{pmatrix} A \\ C \end{pmatrix} \begin{pmatrix} \hat{X}_i \end{pmatrix} + \begin{pmatrix} \rho_w \\ \rho_v \end{pmatrix} \tag{15}$$

where $Y_{i/i}$ is a Hankel matrix with only one block row. Since the Kalman state sequence and the outputs are known and the residuals $(\rho^T{}_w\ \rho^T{}_v)^T$ are uncorrelated with $\hat{X}_i$, the set of equation can be solved for A and C in a least-squares where (.)ᵘ is pseudo-invert of a matrix.

$$\begin{pmatrix} A \\ C \end{pmatrix} = \begin{pmatrix} \hat{X}_{i+1} \\ Y_{i/i} \end{pmatrix} \begin{pmatrix} \hat{X}_i \end{pmatrix}^{\dagger} \tag{16}$$

3.4 *Identification of flutter derivatives*

The modal parameters of system can be obtained by solving the eigenvalue problem state matrix A as

$$A = \Psi \Lambda \Psi^{-1}, \qquad \Phi = C\Psi \tag{18}$$

where Ψ is the complex eigenvector; Λ the complex eigenvalue is the diagonal matrix; Φ is the mode shape matrix. When the complex modal parameters are known, the gross damping C_e and gross stiffness K_e in Eq. (3) are determined by following:

$$\begin{bmatrix} K^e & C^e \end{bmatrix} = -M \begin{bmatrix} \Phi\Lambda^2 & \Phi^*(\Lambda^*)^2 \end{bmatrix} \begin{bmatrix} \Phi & \Phi^* \\ \Phi\Lambda & \Phi^*\Lambda^* \end{bmatrix}^{-1} \quad and \quad \begin{cases} \overline{C}^e = M^{-1}C^e; & \overline{K}^e = M^{-1}K^e \\ \overline{C} = M^{-1}C^0; & \overline{K} = M^{-1}K^0 \end{cases} \tag{19}$$

where C^0 and K^0 the mechanical damping and stiffness matrix of system under no-wind condition.

Thus, the flutter derivatives of 2 DOF can be defined as

$$\begin{aligned} &H_1^*(K_h) = -\frac{2m}{\rho B^2 \omega_h}(\overline{C}_{11}^e - \overline{C}_{11}),\ H_2^*(K_\alpha) = -\frac{2m}{\rho B^3 \omega_\alpha}(\overline{C}_{12}^e - \overline{C}_{12}),\ H_3^*(K_\alpha) = -\frac{2m}{\rho B^3 \omega_\alpha^2}(\overline{K}_{12}^e - \overline{K}_{12}) \\ &H_4^*(K_h) = -\frac{2m}{\rho B^3 \omega_h^2}(\overline{K}_{11}^e - \overline{K}_{11}),\ A_1^*(K_h) = -\frac{2I}{\rho B^3 \omega_h}(\overline{C}_{21}^e - \overline{C}_{21}),\ A_2^*(K_\alpha) = -\frac{2I}{\rho B^4 \omega_\alpha}(\overline{C}_{22}^e - \overline{C}_{22}) \\ &A_3^*(K_\alpha) = -\frac{2I}{\rho B^4 \omega_\alpha^2}(\overline{K}_{22}^e - \overline{K}_{22}),\ A_4^*(K_h) = -\frac{2I}{\rho B^4 \omega_h^2}(\overline{K}_{21}^e - \overline{K}_{21}) \end{aligned} \tag{20}$$

4 EXTRACTION FLUTTER DERIVATIVES FROM EXPERIMENT DATA

The decay and buffeting signals are acquired at a sampling frequency 100Hz and these samples are set to zero mean before operating with Matlab (Figure 2).

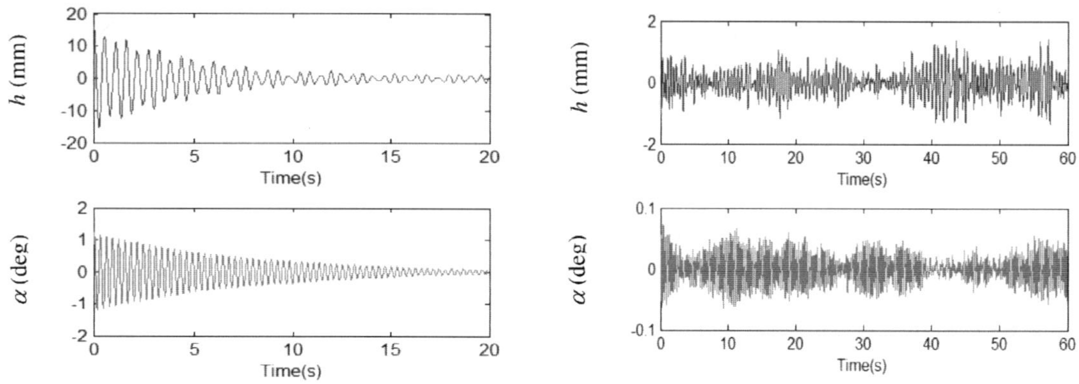

(a) Free decay response (U=2.91m/s) (b) Buffeting response (U=2.91m/s)

Figure 2 Response of the bridge deck section model (h-heaving; α–torsional) under turbulent flow (I_u=9.1%)

4.1 *Comparision flutter derivatives between SSI-cov and SSI-data*

Figures 3 and 4 show the flutter derivatives with respect to reduced wind speed. H_1^*, H_4^*, A_1^* and

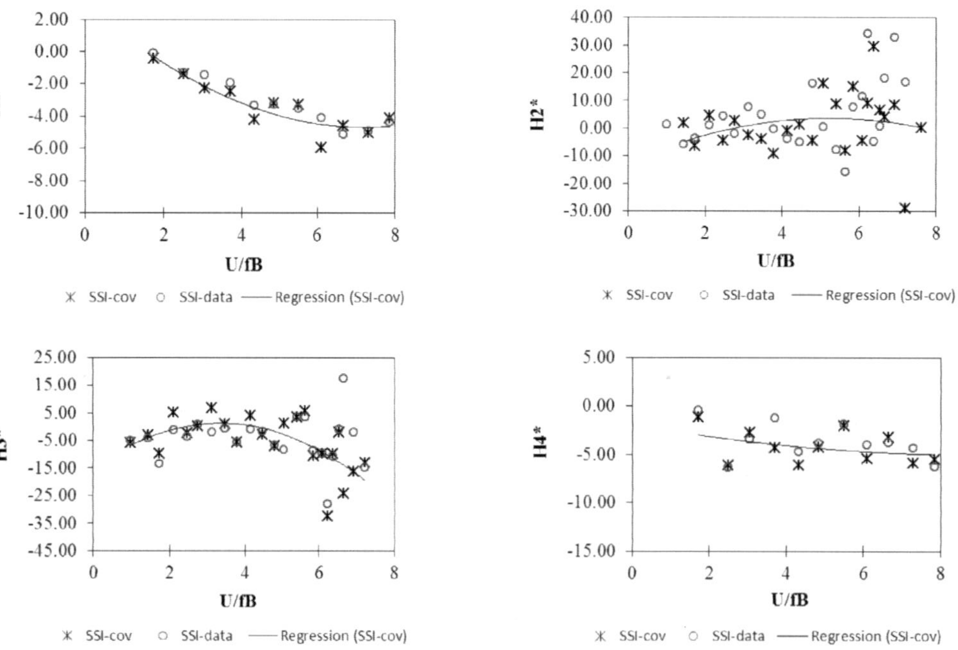

Figure 3 FDs (H_i) of the bridge section model under turbulent flow (I_u=9.1%) from free decay response

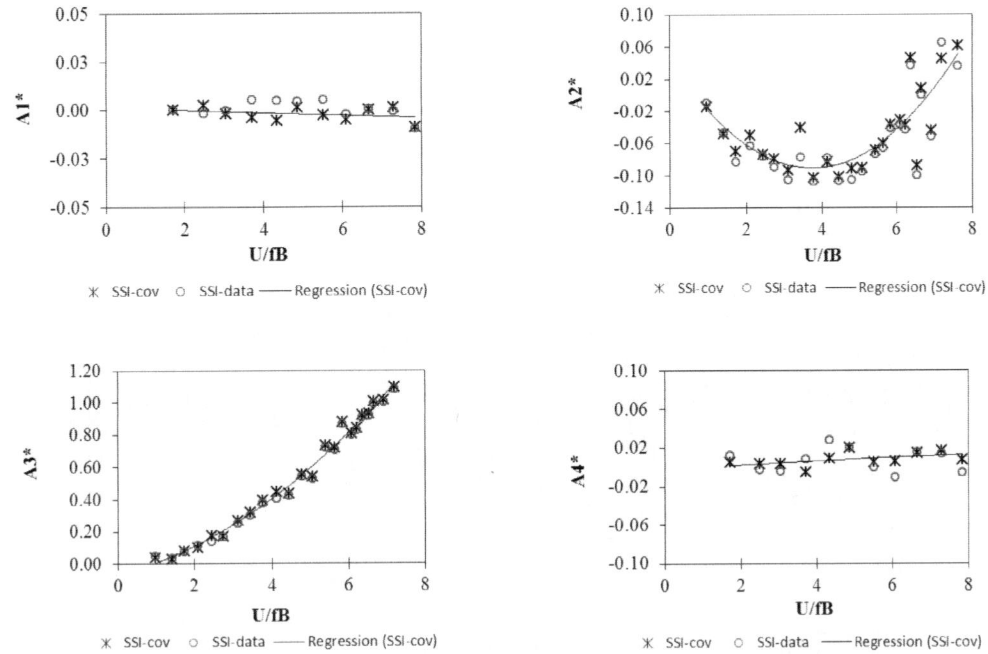

Figure 4 FDs (A_i) of the bridge section model under turbulent flow (I_u=9.1%) from free decay response

A_4^* associated with vertical oscillation were identified using the vertical frequency and H_2^*, H_3^*, A_2^* and A_3^* associated with torsional oscillation were calculated using torsional frequency. In general, the results from two methods show the same trend and both techniques are working well. The method SSI-cov shows the results (H^*_1 and A^*_2) somewhat scattering than those obtained by SSI-data. The SSI-data method is appreciably faster than the SSI-cov. Both methods start with data reduction step. In the SSI-cov, the reduction is due to take covariance of output of the Toeplitz matrix, while SSI-data algorithm is obtained by projecting the row space of the future outputs into the row of the past output. However SSI-data avoids the calculation of covariance between the outputs data.

4.2 *Extract of flutter derivatives from buffeting response*

In general, at high wind velocity, the aerodynamic damping of heaving mode is too high and vertical free response is too short. Therefore, the extraction of FDs cannot be accomplished accurately. In addition, free decay mechanism is not practical to describe real bridges behavior in field. On the other hand, the extraction of FDs from buffeting response is more closely reflected to a real bridge behavior under turbulent wind field. The bridge deck section model will vibrate under the excitation of turbulent flow even at small wind velocity. The method is simpler than free vibration technique because of no operator corrupts by exciting the section model. It is reasonable to extract FDs from buffeting response.

Figures 5 and 6 show the flutter derivatives of the bridge deck obtained by the SSI-data from both decaying response and buffeting response of 1DOF and 2DOF systems under turbulent

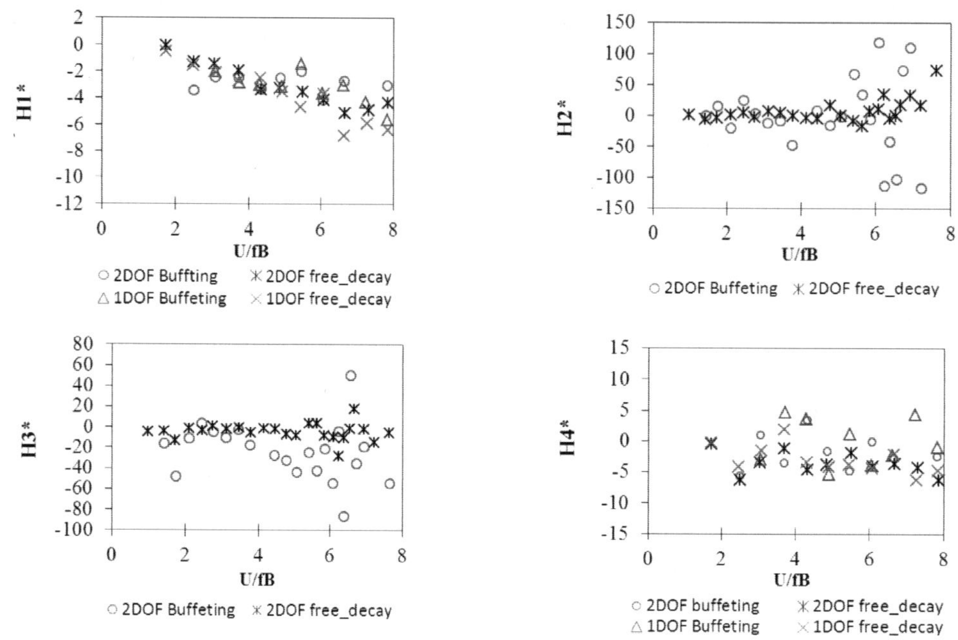

Figure 5 FDs (H_i) by single mode and coupled mode from free decay and buffeting responses (I_u=9.1%)

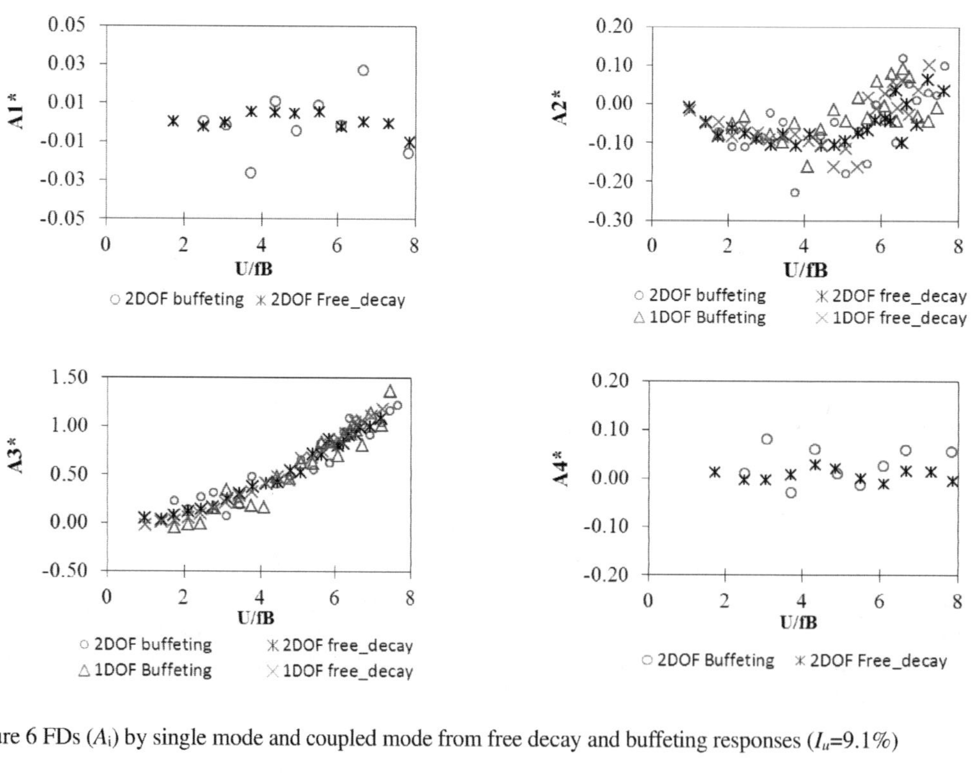

Figure 6 FDs (A_i) by single mode and coupled mode from free decay and buffeting responses (I_u=9.1%)

flow (I_u=9.1%). Generally, most FDs are in agreement with both free decay and buffeting response of 1DOF and 2DOF systems. The flutter derivative H_1^* extracted from buffeting response is slightly higher than that obtained from free decay response. Coupled terms (H_2^* and H_3^*) extracted from buffeting response are more scattering than those from free decay, particularly at high reduced wind velocity. This might be due to the fact that the sectional profile is a truss bridge deck section, which exhibits only pure torsional flutter. A_2^* extracted from buffeting response is scattering at high reduced wind velocities similarly to that obtained from free decay response.

4.3 *The effects of turbulence on flutter derivatives*

In order to clarify the effects of oncoming flow turbulence on FDs, the SSI-data method is applied to extract FDs from buffeting response with different turbulence intensity. Figures 7 and 8 show the flutter derivatives under smooth and turbulent flows with the different turbulence intensity versus reduced wind speed. The truss bridge deck section tends to exhibit torsional flutter and less coupled flutter. The torsional damping term A_2^* is important role on torsional flutter instability since its positive/negative value corresponds to the aerodynamic instability/stability of torsional fluter. In this experiment, the onset flutter is defined as A_2^* crossing the zero reduced velocity. As shown in Figure 8, under smooth flow, the positive value A_2^* can be seen at high reduced wind speed (V_r>5.5). Total torsional damping is almost controled by A_2^*. The effects of turbulent flow on FDs are aforementioned but significant effects can be seen.

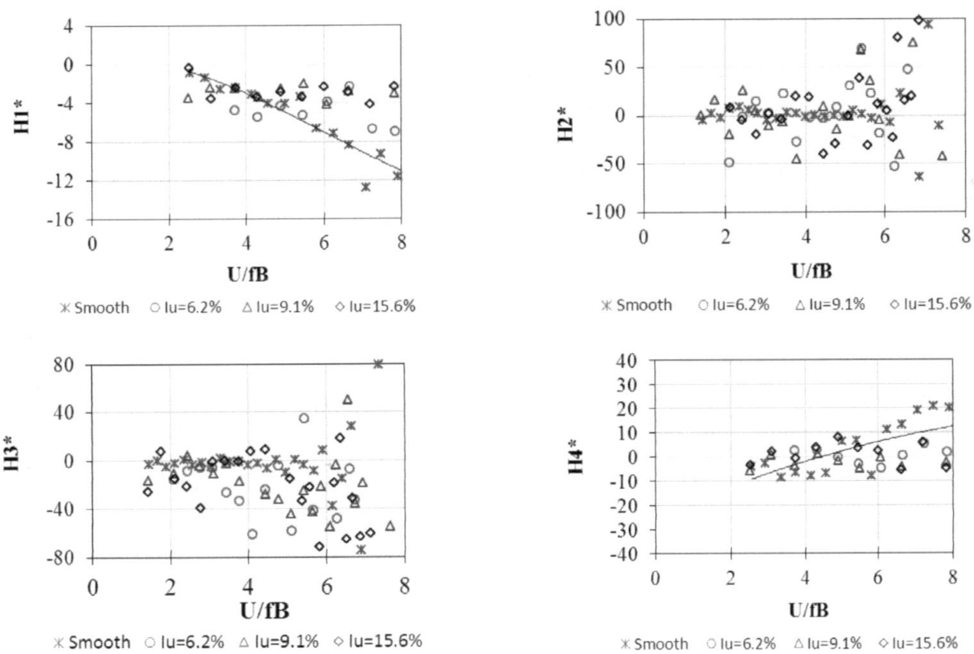

Figure 7 FDs (H_i) under smooth and turbulent flows by buffeting response (solid curve are fitted polynomial of smooth case)

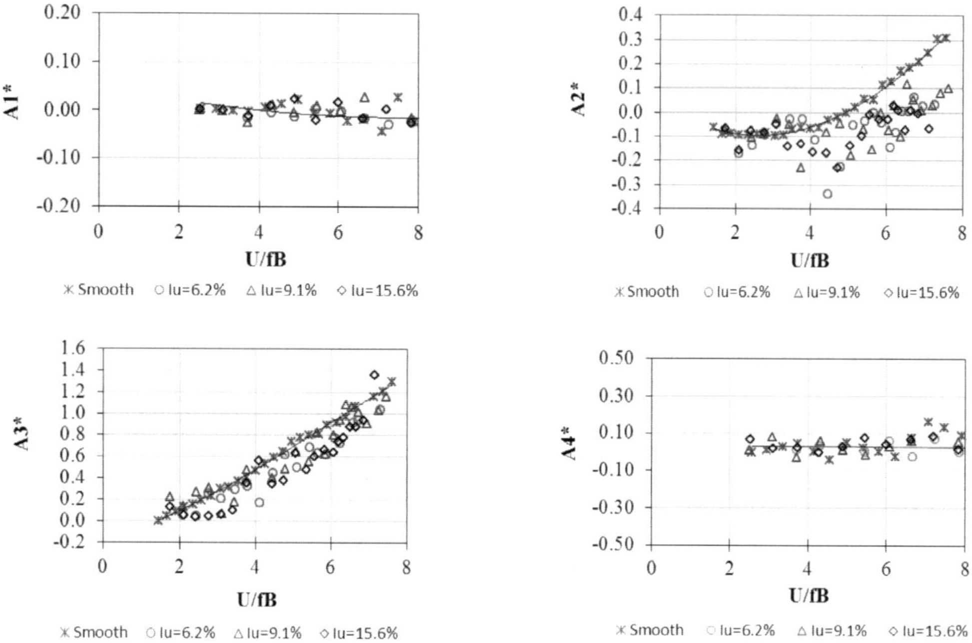

Figure 8 FDs (A_i) under smooth and turbulent flows by buffeting response (solid curve are fitted polynomial of smooth case)

A_2^* becomes positive at V_r of around 6.5 to 7 under I_u = 6.2% and 9.1%, whereas in case of I_u = 15.6%, A_2^* does not become positive up to V_r = 8. On the other hand, the effects of different turbulence intensity on FDs are fairly modest.

It can be found from figure 7 that in smooth flow H_1^* decreases more than that extracted from turbulent flows. This means that damping of heaving mode under smooth flow is higher than that in turbulent flow. Turbulence has very small effect on vertical and torsional frequency terms H_4^* and A_3^*. These values extracted from buffeting response are somewhat less than those in smooth flow.

5 CONCLUSIONS

This study investigated the effects of turbulence on flutter derivatives of a truss bridge deck section by using wind-tunnel test and output only state space stochastic system identification technique. Conclusions from this study are summarized as follows:

- FDs can be well obtained from buffeting response. An advantage of this technique is easier to obtain buffeting response and less time consuming. In particular, at high wind velocity, the length of time history data can be obtained long enough for extracting FDs, which more closely reflects a real bridge behavior
- Both SSI-cov and SSI-data methods show good results even under turbulent flows because an advantage of those methods are considered buffeting force and response as inputs instead of noise.

- Turbulent flow significantly affects on dynamic response and flutter derivatives of a truss bridge deck section. Turbulence induces buffeting response but increase flutter critical velocity, which is illustrated clearly in aredynamic torsional damping term A_2^*.

6 REFERENCES

1. Sarkar P.P., Jones N.P. and Scanlan R.H., System identification for estimation of flutter derivatives, *J. Wind Eng. Ind. Aerodyn*, **42**, pp.1243-1254, 1992.
2. Yamada H, Miyata T. and Ichikawa H., Measurement of aerodynamic coefficients by system identification method, *J. Wind Eng. Ind. Aerodyn*, **42**, pp.1255-1263, 1992.
3. Kirkegaar P.H. and Andersen P., State space identification of civil engineering structures from output measurement, *Proc. of the 15th International Modal Analysis Conference*, Orlando, Florida, USA, 1997.
4. Peeters B. and Roeck G.D., Reference based stochastic subspace identification for output-only modal analysis, *Mechanical Systems and Signal Processing*, **13(6)**, pp.855-878, 1999.
5. Juang, J.N., and Pappa, R.S. (1985), "An eigensystem realization algorithm (ERA) for modal parameter identification and model reduction", *J. Guid. Contr. Dyn.*, 299-318.
6. Simiu E. and Scanlan R.H., Wind Effects on Structures -3rd edition, John Wiley & Sons, New York, 1996.

13th Americas Conference on Wind Engineering
Gainesville, Florida, USA
May 21 - 24, 2017

Experiments to understand Dry Galloping of Stay Cables and Conductor Cables in Single and Grouped Configurations

Mohammad Jafari-1[a], Partha Sarkar-2[b]

[a] *Iowa State University, Ames, USA, mjafari@iastate.edu*
[b] *Iowa State University, Ames, USA, ppsarkar@iastate.edu*

ABSTRACT

There are multiple flow-induced vibration phenomena that cause instabilities in cables used in cable-supported structures and high-voltage power transmission lines. The aeroelastic phenomena "rain-wind induced vibration (RWIV)" and "Karman vortex-induced vibration" that cause large-amplitude motions in inclined and/or yawed stay-cables at relatively low wind speeds (<20m/s) have been studied widely. In contrast, "wake-induced vibration" and "galloping or dry-cable vibration" of cables in single- and bundled or grouped- cable configurations have not been adequately researched. Past investigations of these phenomena that were based on wind tunnel tests and a handful of CFD studies have been largely restricted to individual cables with a smooth surface (stay cables) in smooth, stationary and uniform flow. Although some researchers have pointed out that the main reason of galloping instability is axial flow behind the cable which causes negative aeroelastic damping, the mechanism of onset condition for this type of vibration is not fully clear.

In this paper, results from the first set of wind tunnel tests that were designed to understand the mechanism of dry cable vibration of smooth and grooved cables in single- and bundled or grouped- configuration are presented. The static wind tunnel tests of a smooth cable (L/D=12, D=127mm) and a helically grooved aluminum conductor (L/D=30, D=50mm) in yawed and smooth flow conditions were conducted in the Aerodynamic-Atmospheric Boundary Layer (AABL) Wind and Gust Tunnel (2.44m W x 1.83m H) located at Iowa State University. The measurements include velocity distribution in the wake of an individual cable, pressure distribution on the surface of the individual cable in single and grouped configurations and the model end forces. For grouped configurations, the measurements were done for distances of 3.5D and 6.5D between cables over a range of Reynolds numbers. The aerodynamic force coefficients were calculated for the cables tested at different yaw angles varying between 0° and 45° and Reynolds number. The preliminary results indicate that both drag and lift coefficients are functions of the yaw angle and that galloping instability will occur over a range of yaw angles between 30° and 40° which confirms previous criterion for yawed flow condition for onset of dry galloping.

KEYWORDS

Galloping Instability, Yawed Cylinder, Stay Cable, Aluminum Conductor, Wind-induced Vibration

13th Americas Conference on Wind Engineering (13 ACWE)
Gainesville, Florida, USA, May 21-24, 2017

13th Americas Conference on
WIND ENGINEERING

Effect of Hurricane Wind Direction on Loss Estimation

Bejoy P. Alduse[a], Prashant Rawal[b], Weichiang Pang[c]

[a] *Clemson University, South Carolina, USA, balduse@clemson.edu*
[b] *Clemson University, South Carolina, USA, prawal@clemson.edu*
[c] *Clemson University, South Carolina, USA, wpang@clemson.edu*

ABSTRACT

Importance of considering wind direction on event based risk assessment has been pointed out by several researchers. Sensitivity of various components of the building, to direction of extreme winds is investigated in this study using Probable Maximum Loss ratio as a measure. Probable maximum loss for a simulated event is the maximum loss that can occur to the building (or a set of buildings) during the event under study. Loss ratio functions are determined for eight different wind directions using an adaptable vulnerability framework. Wind speed and direction distributions for octants are obtained using data simulated from 100,000 year catalog. Results from considering only the vulnerability show that if the wind direction uncertainty is low there is a tendency of the probable maximum loss to deviate from the loss predicted assuming uncertain wind direction or when the loss is predicted using an average loss function. Loss ratio probabilities at two locations with different wind speed-direction characteristics are also studied using the same building. Preliminary results indicate that the influence of wind direction is well reflected in the probable loss. Location with predominant wind direction oriented along the weaker direction of the building will show a higher probability of loss, compared to one with direction of wind along other directions. These results can be used as a platform for further research and decision making under different scenarios of loss estimation. One example scenario would be before and after the hurricane path is accurately known.

KEYWORDS

Catastrophe modelling, wind vulnerability, wind direction, hurricane

1. Introduction

US faces challenges from managing catastrophe losses to buildings, especially the ones arising from hurricane, primarily due to uncertainty in the hazard and vulnerability. Although hurricane events are hazards with sufficient warning, the uncertainties associated with the perils make the post event effects difficult to manage and recover. Catastrophe models have been used in predicting the hurricane losses, managing the risks and adapting precautionary measures. Consider an example of predicting the hurricane loss for a season or a period of three to five years. Accuracy of predicting hurricane tracks have been increasing, however, a lesser noted point of concern is the abstract way the vulnerability modules of most of the popular risk models deal with important aspects of the vulnerability of the buildings. Some of those vulnerability modules were based mainly on the relevant codes of construction and building stock that was available at that point of time. It is noted that the human activity along the shores have increased recently in the past considerably, mainly due to increased economic freedom and preference for the sea-shore dwellings. Also, buildings are becoming more complex and many renovations had undergone recently in the previously existing buildings due to strict code enforcements and to accommodate the growing needs. It should be also noted that several buildings facing the ocean at ocean front, would face a low probability a threat from debris impact, the threat from direct winds pressures and water intrusion will however be higher. Hence it is important to consider all these aspects of the buildings, in developing its vulnerability model. Vulnerability frameworks and the vulnerability curves would become obsolete if they do not consider the latest code enforcements or if they did not incorporate the unsymmetrical nature of the building. The effects of the recent changes in construction, changing exposure around the building are important points to consider while obtaining the vulnerability curve.

A change of hurricane track can trigger a rapid change in the amount of loss that is predicted by these models. Buildings experience wind mainly from a few directions when the hurricane passes through. This could be an angle of attack within a sector of 90° or up to 180° (Powell and Houston, 1996), depending on the location of the building with respect to the path of eye. However, it is noted that this issue is not exclusively addressed in vulnerability models and vulnerability curves are hardly based on the direction of wind attacking the building. The change in hurricane track can result in change in wind direction too. It is also observed that if the buildings are large, or if there exists considerable unsymmetrical features in the building or if there are pre-existing weak locations in the building, or if the building is sufficiently shielded from other buildings or a part of it; will result in some regions of the building becoming weaker and other regions stronger. This will also result in loss associated with a particular direction(s) becoming more severe. However in order to understand this phenomenon and implement this in a loss prediction framework, the vulnerability module should be suitably tuned. A careful consideration of each of the items discussed above should be incorporated into the framework. This paper discusses how the vulnerability framework for the building were developed by carefully studying the factors that affect the losses. Detailed discussions will follow to describe the key changes observed. After that a loss estimation example will be presented using the newly developed vulnerability curves and the existing models. A comparison will also be made by using the curves that were developed by assuming equal probability for the wind directions.

It is observed from literature that the importance of wind direction is less significant for a long period, because the hurricane tracks tend to be random, resulting in neutralizing the concentration effect during a season or a short period. However, for short term predictions, wind direction will be very important, since many important financial decisions like post-event recovery, triggers of catastrophe bonds etc. are increasing becoming dependent on the short term hurricane loss.

2. Background Study

A vulnerability module plays a key role in the loss estimation framework of the catastrophe models. While a vast amount of research have been ongoing recently for the hazard modules, literature on wind vulnerability model from a hurricane loss prediction perspective is limited. It is also noted that the current models of vulnerability hardly represent the recent changes in the construction codes or it gives abstract approach to the exposure and construction details of the study building. Hazus database for example carries around 35,000 vulnerability curve for the wind loss prediction. However, the wind properties of the buildings are dealt with in an abstract sense. Also, it can be observed from Hazus database, that features of building like aspect ratio or floor area of the building do not influence the loss ratio prediction. However, evaluation of financial loss using our model shows that properties like floor area and aspect ratio are important. This could be a problem, since, the increase in the vulnerable envelope region and value of the building may not be a linear relation. Earlier we have observed similar trend in our analysis of wind vulnerability of buildings with different sizes. Hence it is likely that the loss ratio for larger buildings are usually lesser than the smaller buildings (Menun and Rahnama, 2012). Using similar argument based on wind load concepts, it is also possible that diagonal direction of wind could be more vulnerable to roof failure, exposing larger area to higher uplift forces under direct wind attack. A larger area of the roof will experience higher uplift pressure compared to wind blowing perpendicular to the short or long wall. Also it is observed from many reports that cornering winds have been more dangerous to the roofs, owing to high suctions developed at the corners and subsequent domino effect failure of the roof cover.

Site surveys after hurricane events suggest influence of wind direction noting the predominant damage scenarios observed in the affected areas. For example Wakamato and Black (1994) observes the damage of trees and buildings after hurricane Andrew in 1992, in different locations to be concentrated in a certain direction. Authors suggest the presence of severe winds in narrow range of the angle for one side of the hurricane track. The damage directions mentioned in this study, shows the presence of strong surface winds. While the locations far from the track may indicate tangential wind direction, the locations near the track will show strong winds directed towards the track, owing to the "pull" from the low pressure eye.

The dynamics of hurricane wind on land will be different from the hurricane wind on sea. Inflow angle in the land is generally reported to be close to 40° (Zhang et al. 2012). The inflow angle of surface wind is generally modelled as a function of central pressure and distance of location. The surface roughness of the location also influences the inflow angle. In this study we will be using the inflow angles simulated using the approach described in Zhang et al. (2012), with suitable modifications for land.

A detailed study is presented in FPHLM on the vulnerability of the low rise building. This document lays a strong foundation for future study of hazard and vulnerability issues. Although this document will ensure a detailed approach to the problem of determining vulnerability, a generalized treatment of the problem and scalability is not well addressed. Grayson et al (2013) discusses effects of three dimensional wind borne debris and subsequent envelope loss in the building in a residential community. Studies by Lin et al and others also suggest these important points. It should be noted that the building could have multiple exposure characteristics around it. The building may be completely guarded by another building next to it or it could be well exposed to debris impact from the nearby buildings.

It should be noted that hurricane losses are not complete if water intrusion is not considered. Several researchers and FEMA reports shows significant upswing of the loss curve beyond a certain wind speed, which primarily is associated with water intrusion due to main envelopes giving way to rain. A water intrusion model is presented in Pita et al., and many discussions on wind driven rain are well laid out in Blocken (2004) and Bitsuamlak et al. (2009). It should be also noted that water intrusion through the imperfections on the windows and doors are also dependent on the wind direction. In this study we limit the scope to water entering the building to roof cover, roof sheathing and fenestration loss.

Above discussions highlight the importance of wind direction as an important parameter as observed from recent literature. Hence wind direction needs to be an integral part of developing wind vulnerability curves. In order to effectively consider the influence of different parameters, and to differentiate the loss scenarios of buildings with different orientation, it is important to consider the wind direction as an input into the vulnerability framework. It is important to associate the loss curves with the wind direction that was used in determining it. Above discussion was mainly focused on the wind direction effects, while it should be noted that the vulnerability model needs to be flexible enough to carry out the above calculations, and many more parameters of the building to carry out a better approach to the problem. Hence it is a challenging task to carefully consider the effects of each parameter and incorporate that in the building vulnerability framework.

3. Theory and proposed approach

In this section we first discuss the theoretical concepts and the overall approach to vulnerability framework by focusing on the wind direction effects. A brief overview of the hazard module will be presented. Then we present the simulation framework for finding the probable maximum loss.

The core of the vulnerability model is a Monte Carlo simulation engine with multiple modules within it. The first module addresses the uncertainty in building, second addresses the directional effect and third carries out calculation for all wind speeds. This study focuses on researchers in insurance industry. Primary and secondary modifiers are key terms in the popular commercial vulnerability models. Within the first module, the uncertainty in primary modifiers and secondary modifiers are addressed. This include mainly, occupancy, year of construction, construction type, number of stories and location. It should be noted that the secondary modifiers are more sensitive to wind lads. This include the roof slope, type of roof construction, roof-to-wall connections, truss spacing, window protection and wind debris environment.

The directional effect of wind loads are incorporated into the analysis based on detailed study of Grayson et al. (2013), Gurley et al. (2004), Vickery et al. (2005) and ASCE 07-10. The following items are important in considering the influence of wind direction on loss prediction.

1. Building vulnerability

- Pressure contour on roofs and walls 2. Roof shape 3. Fenestration layout 4. Complex shape of the building,
- Environmental factors – 1. Potential Upstream debris inventory and distribution. 2. Surface roughness

2. Hazard
   - Event frequency
   - Wind speed and direction distribution

It is important to understand the scope of work in the current study. Although all of the above items are in consideration for future study, we limit our scope for this current study. From vulnerability we use a single building which shows different directional loss characteristics. From the hazard we use the raw data generated using 100,000 year catalog for wind and direction distribution. Locations of different wind characteristics will also be used to present the loss ratio calculations. In the next section we discuss the highlights of the building vulnerability model and how we get the loss functions.

3.1. Building vulnerability

The uncertainties in the building is handled in the vulnerability framework. In this section we discuss how to obtain the building vulnerability function.

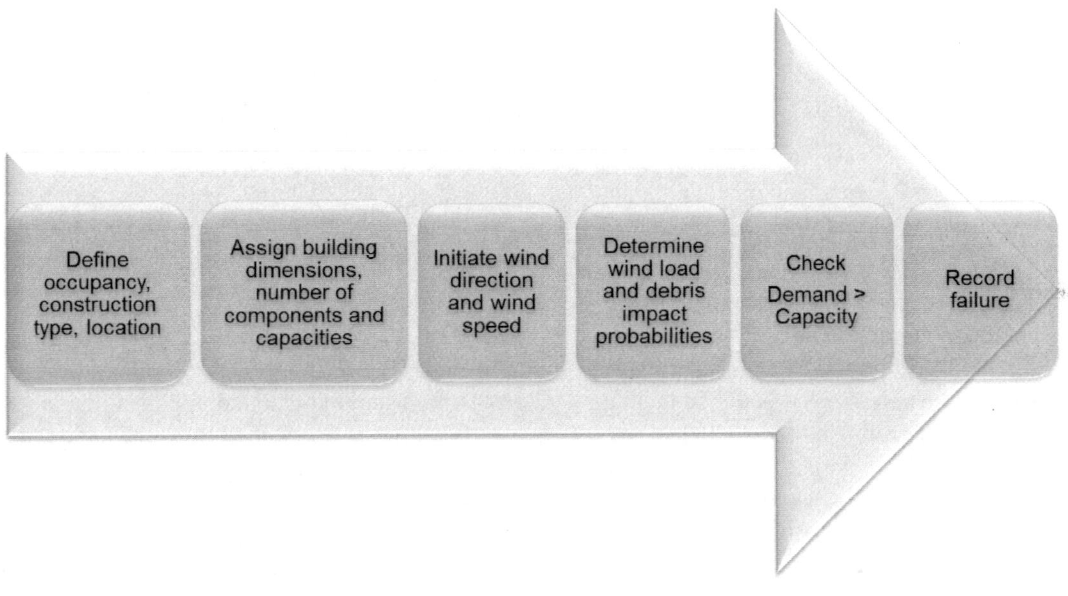

Figure 1 Conceptual Overview of the vulnerabilityf framework

The study building is described with its occupancy, construction type, stories, floor area and aspect ratio. The building is divided into the following components and each component is subdivided into certain number of units.

The components are assigned the capacity by sampling from a distribution unique to the component. For a component k, the capacities of each units are sampled from the distribution as shown below assuming Gaussian. (The capacity distribution does not have to be Gaussian)

$$P(C_k) = N(\mu_k, \sigma_k)$$

The wind load on the components are determined using the ASCE 7 – 10.

$$q = 0.00256k_z k_{zt}(GCp - GCpi)$$

The debris impact and consequent failure of window/doors are determined using the below formula

$$P = 1 - \exp(-N_{Debris} \times P_U \times P_B \times P_W \times P_F)$$

If there is a window failure the internal pressure is changed and wind loads are recalculated. A unit of the component is assigned as failed if the wind load on that unit of the component is exceeded. The interior loss due to wind driven rain, is determined by quantifying the rain water that can enter through breached openings.

The total repair cost for an individual building, at a given wind speed is determined using the percentage of units that are failed for each component and the Normal distribution of unit cost of repair for various percentage of damage. The cost of repair is determined using the formula below

$$Total\ repair\ cost = \sum_{k=1}^{K}\sum_{l=1}^{M_k}([(F_{kl})]/M_{kl}) \times N(\mu_{C_k}(F_{kl}/M_{kl}), \sigma_{C_k})$$

The Loss ratio is defined as

$$Loss\ ratio = (Total\ repair\ cost)/(Total\ value)$$

Total value of the building in this study considers the *envelope items* and *interior items* of the building. The *contents* (like furniture or electronic equipment) are not included, while items like dry wall partitions, water heater etc. are part of the *interior items*. In the MC simulation framework for the vulnerability, the simulations are run for all wind speeds and all wind directions. The loss ratio obtained from each wind direction is grouped separately to obtain loss functions for each wind direction. Once the loss ratio functions are obtained for each wind direction we need to obtain the wind speed intensities and the directions for the short term period using the hazard information described below.

3.2. Hazard – Obtaining wind speed and direction at different locations.

In the current scope of work, we consider the loss estimation using different levels of uncertainty in the maximum wind speed and direction at a location. The wind speed and angle distribution for a location is obtained using a simulation based approach (Georgiou, 1980; Vickery, 200; Liu 2014). The wind direction at the gradient level is currently used in the study. More accurate method to include knowledge of surface level wind direction and inflow angles based on recently published articles like Zhang et al. (2012) are going on.

Raw data from this simulations can be used for further analysis. The wind speed and direction is binned into the octants each sector with 45° azimuthal extension. The maximum speed and direction of the maximum wind speed for each bin is also recorded. Hence a probability distribution function associated (assuming Gaussian) with each octant is obtained. For i[th] octant

$$P_i(\theta_i < \theta \leq \theta_{i+1}) = N(\mu_{\theta i}, \sigma_{\theta i})$$

$$P_i(V|\theta_i < \theta \leq \theta_{i+1}) = N(\mu_{vi}, \sigma_{vi})$$

3.3. Proposed approach for estimation of distribution of probable maximum loss for a short period

Probable maximum loss is defined as the maximum loss that can be observed during the simulated event. The following steps describe the process to find the probable maximum loss using the loss ratio function obtained from the vulnerability framework and the raw data of maximum wind speeds and its direction from hazard module.

1. Raw data of wind speed and direction for a location is generated using a simulation based approach. Once the raw data is obtained the maximum wind speed and the wind direction distribution can be obtained by sampling for any period.
2. We choose a 5 year period to obtain the wind speed and direction distributions for each bin (or octant). Let us call this distribution D_m where say $m = 1{:}\,1000$
3. Once this distribution is obtained we run the MC simulation for a set of hypothetical storms for this location (say $n = 1{:}\,10{,}000$ samples below)
   a) Randomly generate number of storms $N_{i=1\ldots n}$ that can strike the location in a 5 year period
   b) For each of these events, sample a wind speed and direction from the distribution D_m obtained in 2.
   c) Choose the loss ratio function, from vulnerability model based on the (local coordinate) wind direction in b above.
   d) Sample a loss ratio value using the statistics of the loss ratio function at the wind speed and direction.
   e) Do this for all N_n events and record the losses for all the events. Sum up the losses from all the events. So there will be 10,000 loss ratio values. These are the maximum probable loss for the hypothetical events.
4. Once the 10,000 loss ratios are obtained we can get one curve describing the pdf of each loss ratio. (i.e.: Loss ratio vs pdf or Loss ratio vs Exceedance probability)
5. Repeat steps 2 to 4, 1000 times (each with unique D_m) to get 1000 pdf or exceedance probability plots.

The proposed approach is tested using a simple residential building example as shown below.

4. Example

This study is carried out in three steps.

- *Obtaining directional loss ratio functions* - In the first step we select a residential building and generate direction based loss functions for this building. Furthermore the percentage difference in loss ratio for the most vulnerable direction with respect to the average loss function will be explained.
- *Sensitivity of wind direction uncertainty on probable maximum loss* - In this second step we carry out a simple sensitivity study of wind direction uncertainty assuming a Gaussian distribution for the wind direction distribution. This study is carried out to understand the relative performance of an "average loss function" and a direction based loss function, for different levels of uncertainty in associated with the wind direction.
- *Hazard intensity variation with direction* - In the third step, two study locations are selected and the wind speed database at the locations are generated using the 100,000 year simulation catalog. The proposed MC simulation approach (section 3.3) described in previous section is carried out using the raw data for individual location.

4.1. Obtaining directional loss ratio functions

In order to demonstrate our hypothesis, an example residential building is selected based on examples presented in Grayson (2014) and Cope (2004). The loss functions for the building for eight different wind directions need to be first obtained. The properties of the components are as shown below.

Table 1Geometry of the building

Engineering property	**Value**
Floor area	2464 sqft.
Roof shape	Gable
Length	56'
Breadth	44'
Eave height	10'
Roof slope	5/12
Truss spacing	4'

Table 2 Properties used to model the components

Component	**Value or Description**	**Dimensions (ft. or sqft)**	**Mean Wind Resistance Capacity (lbs. or psf.)**	**COV of Wind Resistance Capacity**	**Distribution type for Wind pressure Resistance Capacity**
Roof cover	Shingle	3'x1'	70 psf.	0.4	Normal

Roof sheathing	NA	8' × 4'	150 psf.		Lognormal
Wall sheathing	NA	8' × 4'	126 psf.	0.4	Lognormal
Roof to wall connection*	NA	NA	3720 lb. (Interior) 3780 lb. (Gable)	0.2 (Population) 0.05 (Sample)	Normal
Window 1	Annealed	5'x3'	50 psf.	0.20	Normal
Window 2	Annealed	6'x5'			
Sliding Glass Door	Annealed	6'x8'			
Door	Wood	3'x8'			

Using the vulnerability framework the loss ratio functions are obtained as shown below. The simulations were run for 10mph increment of wind speeds and for each wind direction. For each wind direction 10,000 simulations were carried out. It is observed that in this particular example, the loss functions differ from each other at each wind speed. The most sensitive wind direction is found to be at direction 6 (at 225° to the building). The difference between the loss ratio functions occur mainly due to the fact that the pressure contour is different for different angles of attack and the fenestration layout is also

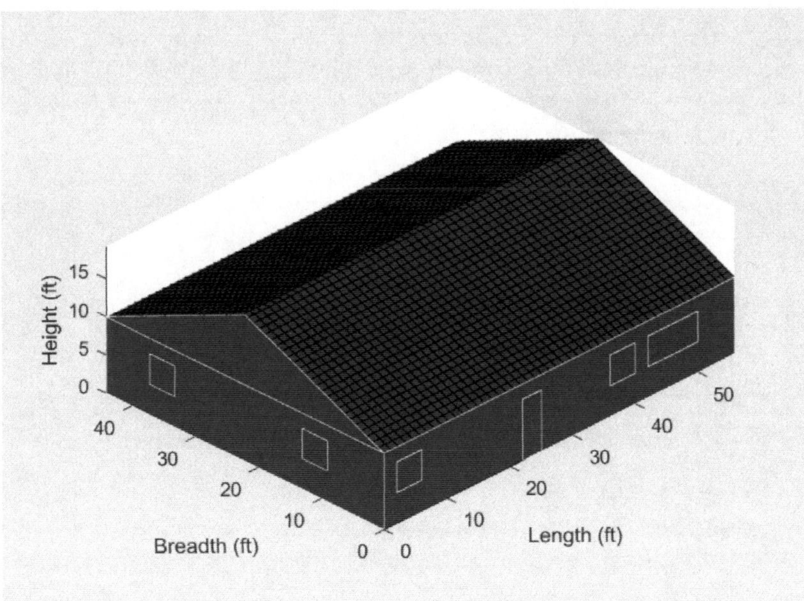

Figure 23D model of the residential building

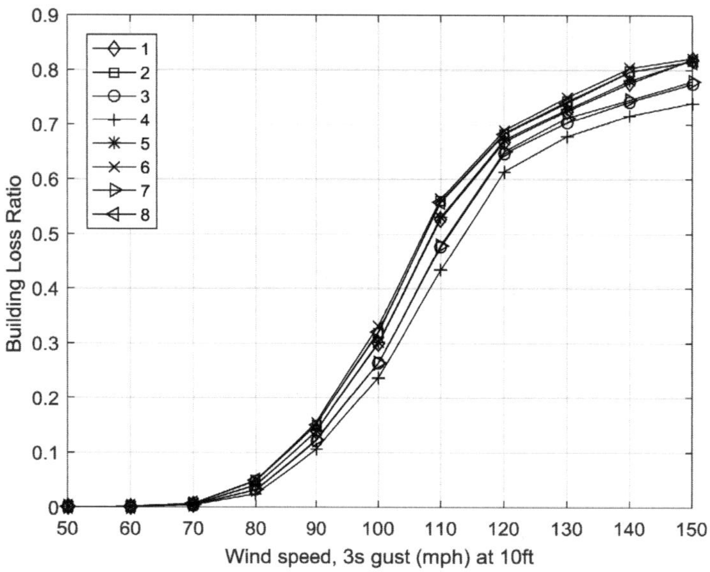

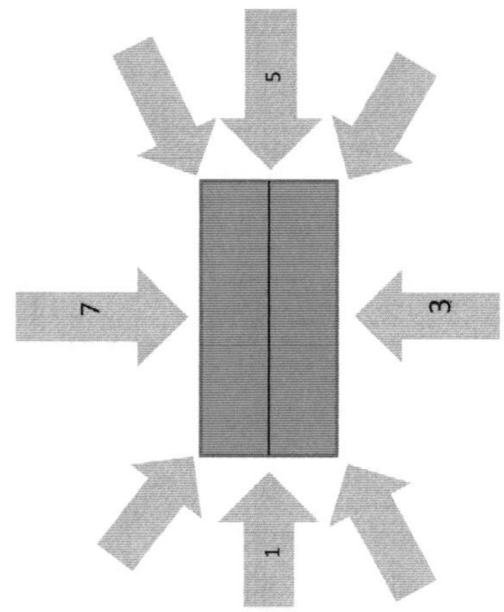

Wind direction w.r.t. building. Direction 1 is perpendicular to short wall

Figure 3 Loss functions for different wind directions

A comparison is given below. It is observed that the percentage difference of the most vulnerable directional loss function with respect to average loss function is changes from 30% to close to 20% in the wind speed range of 80mph to 100mph. The difference occurs mainly due to the fenestration layout and diagonal wind resulting in more roof areas under uplift.

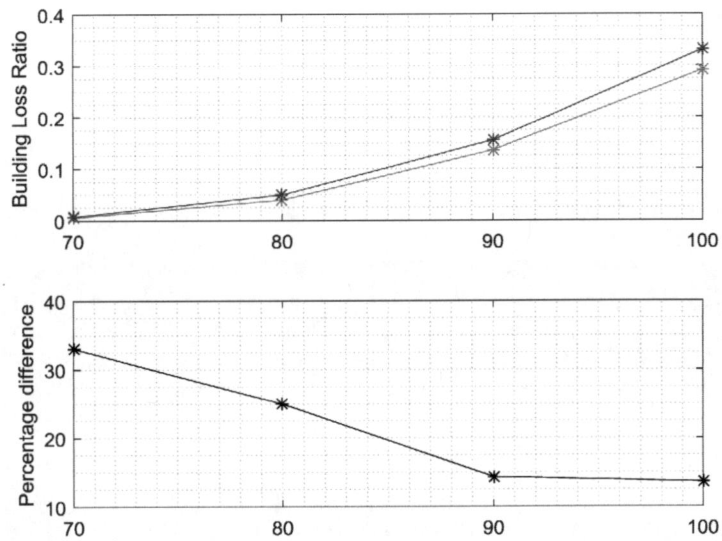

Figure 4 Comparison of most vulnerable direction with average loss function

4.2. Sensitivity of PML to wind direction uncertainty

In order to carry out this study we assume that the probability of wind speed and direction at a particular location follows a Gaussian distribution. Different levels of wind direction uncertainties are assumed. The mean angle of attack at direction 6 for different COV $COV_\theta = [0.05, 0.10, 0.20, 0.40]$

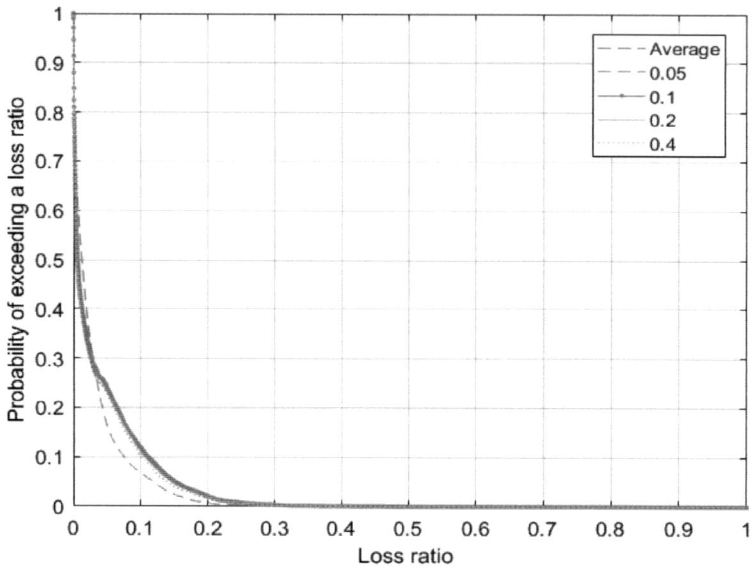

Figure 5 Probability of exceedance of loss ratio values for different levels of directional uncertainties

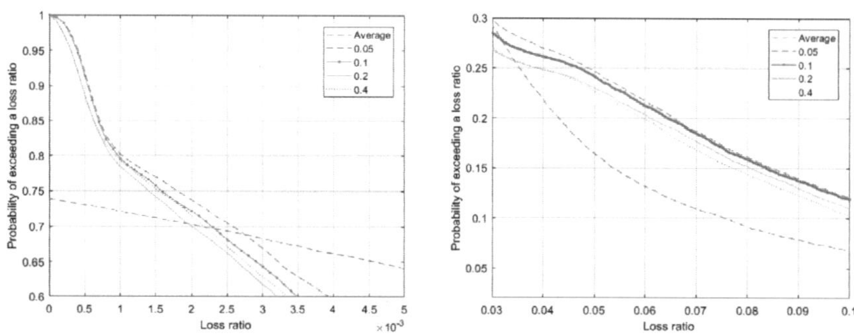

Figure 6 Zoom in for Figure 5. Comparison of PML ratio for different percentage levels of angle of attack uncertainty and PML ratio using an average loss function. It is possible that the average loss function can under predict the PML compared to loss from directional loss functions, when direction is more accurately known.

It is noted that for this study building, the exceedance probability from two different approaches can differ significantly. In the low loss ratio values (in this case 0 to approximately 0.0025), the average loss function under predicts the PML compared to that from direction based loss function. However, the observation at higher loss ratios is opposite. This topic hence needs to be further investigated and how the combination of hazard and vulnerability uncertainty will propagate into loss prediction.

4.3. Wind hazard

Two study locations are chosen to understand the influence of wind direction on the loss estimation. The wind speed database at the location is generated using 100,000 year simulation catalog. Only the storms that cross a 250km radius of the site (or location) are considered for the wind speed calculation. For each storm that pass through this radius, the maximum wind speed and the wind direction for each octant at the location are obtained. The following figure describes the measurement of wind direction. It should be noted that the wind hazard direction and vulnerability local coordinate direction are measured in opposite sense.

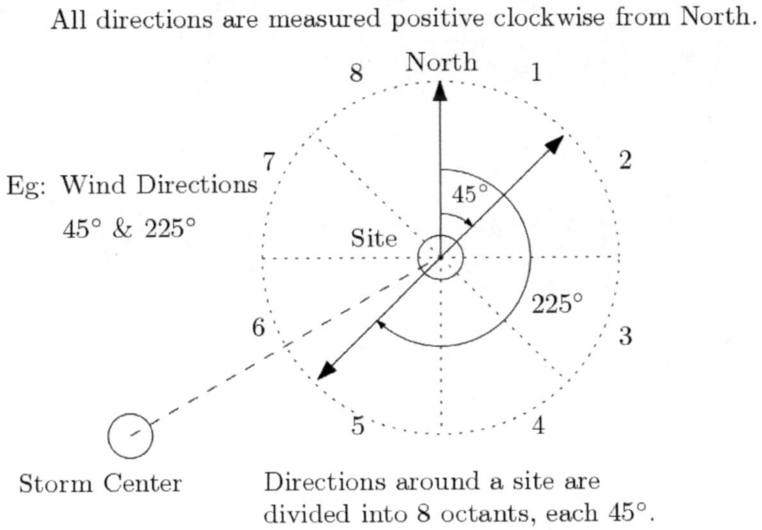

Figure 7Measurement of wind direction for Hazard module. The wind direction is measured from North clockwise. Region is divided into 8 octants.

The relative layout of the mileposts are given below.

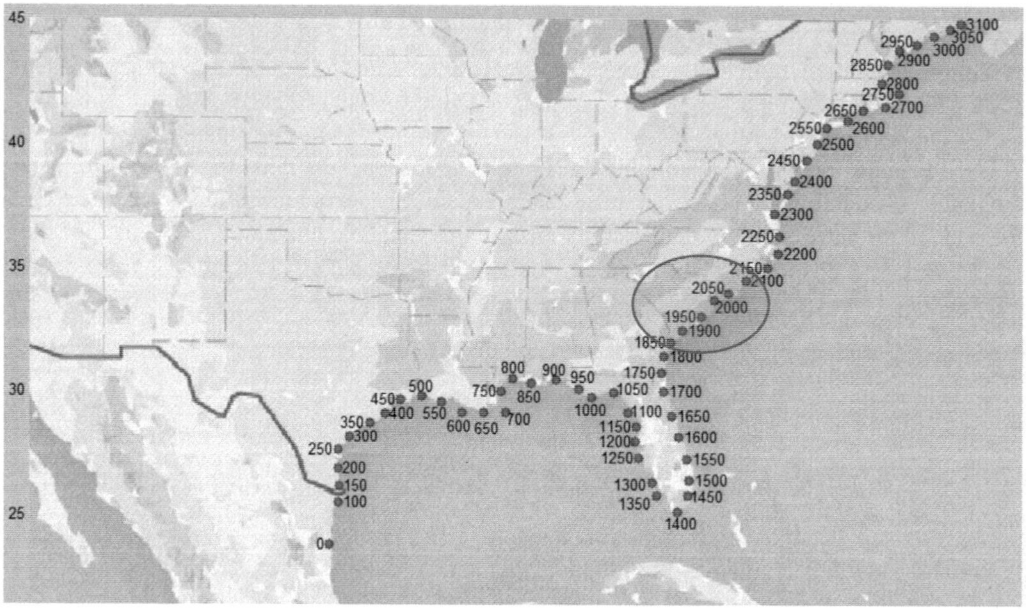

Figure 8 Layout of mileposts on the US coast

The figure below shows the polar plot of the mean wind speed at Location 1 and 2 (1900 milepost and 2000 milepost) under study.

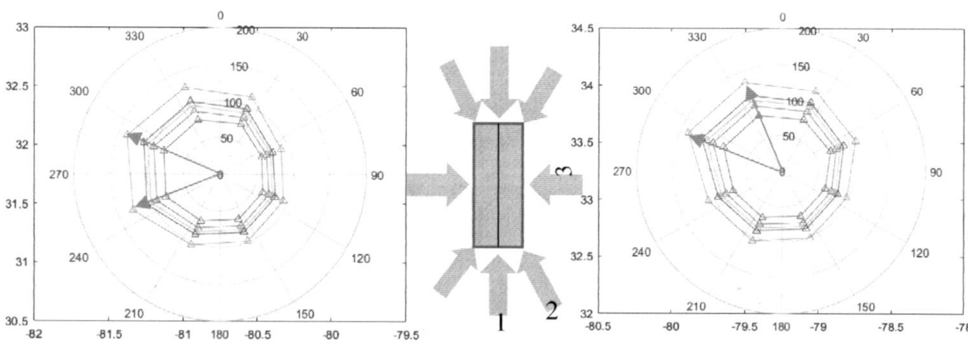

[50 100 150 200] MRI Wind speeds for each bin at 1900 mile post

[50 100 150 200] MRI Wind speeds for each bin at 2000 mile post

Figure 9 Change in wind MRI (50 through 200 year) wind speeds (mph) for each octants at different mile posts. It is noted that the direction of strongest wind changes with respect to location.

As described in the proposed approach (section 3.3), a MC based study was done to determine the PML for different loss ratio values using the raw data of wind speed obtained using 100,000 catalog of hurricanes. For current study we consider, MP1900 and MP 2000. Using a simple approach to determine average storm frequency and sampling of wind speeds and directions we obtain loss ratio distribution are then sampled for each storm. The probable loss ratio values are given in Figure 10. This is preliminary results based on simple but yet technically correct assumptions. We intend to carry out more rigorous analysis for each steps described in the MC simulation to determine the PML. The results are further investigated in the next section.

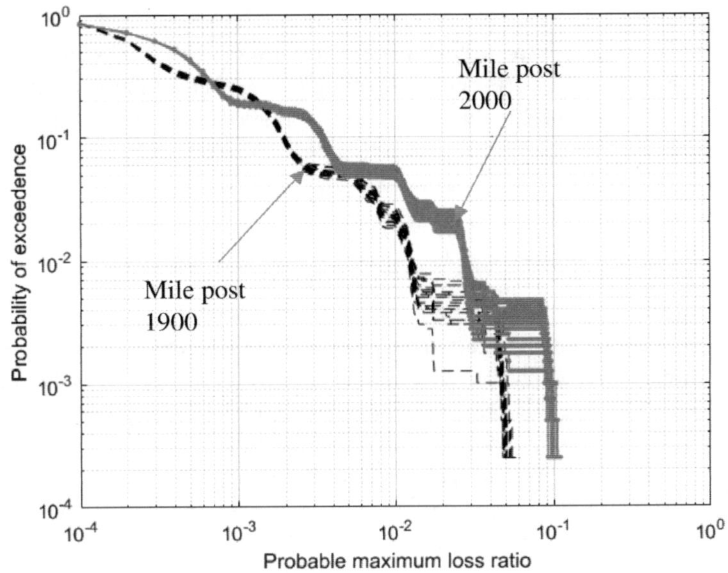

Figure 10 PML comparison for two different mileposts on the map for multiple simulations

5. Results and discussions

After a careful consideration of different factors affecting the loss estimation and using an adaptable vulnerability framework, the influence of wind direction on loss estimation has been studied. We considered multiple scenarios in this study. First we considered the situation in which the knowledge of wind direction changed from very certain to highly uncertain. In this case the vulnerability curve for individual direction, are found to be different for each angle of attack. The probable maximum loss ratio (PML) associated with maximum wind speed observed at the location during the event, is given in Figure 6. The PML value is found to be over predicted for the small loss ratio case, where as it is under predicted at relatively higher loss ratio values, using the average loss function compared to directional based loss function. Interestingly, as the uncertainty of the wind direction increases, it moves towards the PML curve obtained using the average loss function. This could be a representation of a more uncertain scenario about wind direction, (for example a long term risk prediction in the future, where wind direction is highly uncertain), and the use of average loss function will yield comparable results. In case, if the event has already occurred and the wind direction are more accurately known, the loss estimation should preferably be carried out using the direction based loss function.

A MC simulation algorithm is proposed to find the PML ratio for short term loss prediction. The number of storms is estimated using a simple technique and wind speed- direction pairs are sampled from the raw data. The loss ratio is then determined for a large number of trials. This approach is repeated for milepost 1900 and 2000, which differ in wind direction probability. The building is assumed to be oriented with local direction 1 along north. The loss functions for the building (Figure 3 and Figure 4) show that the wind direction 135° (global direction in the hazard approach) is weaker than 90° (global). The expected result is reflected in Figure 10 reasonably well, where the probable maximum loss ratio at 2000 mile post under study is influenced by the predominant wind direction oriented along the weaker direction of the building compared to that at 1900 milepost. This is preliminary result and further improvement of the overall approach still need to be made.

6. Conclusions

Although buildings are subjected to winds from different direction, during a hurricane event, the directions of maximum speed are generally concentrated in a sector of the angles. This results in significant influence of the directional effect of the hurricane wind on the dollar loss. This phenomenon is studied using the capabilities of the vulnerability framework, a wind database model and a simple MC approach to simulate Probable Maximum Loss ratio. It is observed that the average loss scenario and uncertain wind direction scenario are comparable. If the loss estimation is done once the hurricane track (and in effect the wind direction) are more accurately known, the probable maximum loss ratio would significantly deviate from the average loss function scenario (or when the wind direction is highly uncertain). Hence a wind direction based loss curve would be more accurate in estimating losses when the wind direction is more accurately known. The effect of wind direction at two different locations, with different wind speed-direction characteristics are compared by using the raw data obtained at the two locations. Preliminary results show that the effect of wind direction is propagated into the probable loss. Future study is aimed at considering the influence of various variables in the vulnerability framework and hazard module on the direction based loss prediction. For example influence of dimensions, aspect ratio and roof shape for each direction needs to be quantified. A more detailed study of the various variables that influence the frequency of the events at a location and wind directionality for short term period need to be carried out. Furthermore the study need to be conducted for locations away from the sea.

References

[1] Grayson JM, Pang W, Schiff S. Building envelope failure assessment framework for residential communities subjected to hurricanes. Eng. Struct. 2013; 51:245-58.

[2] FEMA P. 58-1 (2012) Seismic performance assessment of buildings (volume 1-Methodology). Federal Emergency Management Agency, Washington.

[3] Lin N, Vanmarcke E. Windborne debris risk analysis—Part I. Introduction and methodology. Wind and Structures 2010; 13:191.

[4] Cope AD. Predicting the vulnerability of typical residential buildings to hurricane damage. ProQuest Dissertations and Theses 2004.

[5] Vickery PJ, Skerlj PF, Lin J, Twisdale Jr LA, Young MA, Lavelle FM. HAZUS-MH hurricane model methodology. II: Damage and loss estimation. Nat Hazards Rev 2006;7:94-103.

[6] Vickery PJ, Skerlj PF, Steckley AC and Twisdale LA. Hurricane wind field model for use in hurricane simulations. Journal of Structural Engineering; 126(10):1203-1221

[7] Liu F. Projections of future US design wind speds due to climate change for estimating hurricane losses 2014 All Dissertations. Paper 1305.

[8] Sparks PR, Bhinderwala SA. Relationship between residential insurance losses and wind conditions in Hurricane Andrew 1994:111-24.

[9] Bitsuamlak GT, Chowdhury AG, Sambare D. Application of a full-scale testing facility for assessing wind-driven-rain intrusion. Build Environ 2009;44:2430-41.

[10] Blocken B, Carmeliet J. A review of wind-driven rain research in building science. J Wind Eng Ind Aerodyn 2004;92:1079-130.

[11] Pita G, Pinelli J, Cocke S, Gurley K, Mitrani-Reiser J, Weekes J, Hamid S. Assessment of hurricane-induced internal damage to low-rise buildings in the Florida Public Hurricane Loss Model. J Wind Eng Ind Aerodyn 2012;104:76-87.

[12] Wakimoto R M, Black P G. Damage survey of hurricane Andrew and its rleationship to the eyewall. Bulletin of the American Meteorological Society 1994;189-200.

[13] Zhang J A and Uhlorn E W. Hurricane sea surface inflow angle and an observation-based parametric model American Meteorological Society 2012; 3587-3604.

[14] Menun C, Rahnama M. The relationship between the wind damage sustained by a residential building and its floor area. Advances in Hurricane Engineering 2012; 295-304.

13th Americas Conference on Wind Engineering
Gainesville, Florida, USA
May 21 - 24, 2017

Hurricane Storm Surge Content Damage and Additional Living Expenses

Mohammad Baradaranshoraka [a], Jean-Paul Pinelli [b], Kurt Gurley [c]

[a] *Florida Institute of Technology, Melbourne, Florida, USA, mbaradaransh2014@my.fit.edu*
[b] *Florida Institute of Technology, Melbourne, Florida, USA, pinelli@fit.edu*
[c] *University of Florida, Gainesville, Florida, USA, kgurl@ce.ufl.edu*

ABSTRACT

Each year hurricanes cost millions of dollars of damage to the built environment in the United States. Per the National Oceanic and Atmospheric Administration (NOAA), in 2010, 123.3 million or 39 percent of the United States population lived in a coastal county. Most these people live in buildings prone to hurricane hazard, which could result in wind induced damage (due to high wind and/or wind driven rain infiltration) and water induced damage (due to storm surge and flooding). In addition to damages to the building, high winds and inundation can result in damages to the content as well. The approach of a hurricane followed by possible damages can cause the occupants to flee from their homes and not been able to return immediately. The resulting expenses associated with individuals living away from their homes are called additional living expenses.

This paper focuses on a new methodology for calculating the water induced damage to the content and the additional living expenses, due to storm surge, or fresh water inundation. It is implemented in the Florida Public Hurricane Loss Model. The National Flood Insurance Program (NFIP) covers water induced losses, with the exception of additional living expenses. The paper describes how the NFIP claim data is used in the development, validation and calibration of the water induced content damage model. The paper also describes the development of the additional living expenses model. Results will be presented and discussed.

KEYWORDS

Hurricane, storm surge, content damage, additional living expenses

Characteristic Events Based Typhoon Catastrophe Risk Assessment Model for Japan

Filmon Habte[a] and Nozar Kishi[b]

[a] *Karen Clark and Company (KCC), Boston, Massachusetts, USA (fhabte@karenclarkandco.com)*
[b] *Karen Clark and Company (KCC), Boston, Massachusetts, USA (nkishi@karenclarkandco.com)*

ABSTRACT

Natural hazards such as tropical and extratropical storms, earthquakes, floods, and convective storms are highly destructive and quite unpredictable. Japan is located in the North West Pacific Tropical Cyclone Basin, and every year on average around 10 tropical cyclones get within 300 km off its coast, and about 2 to 3 of them make landfall. We have developed a catastrophe model to address the risk associated with Typhoons in Japan for use by insurance, reinsurance as well as NGOs and governmental institutions. KCC's catastrophe model constitute of events, hazard, vulnerability and financial modules. The events module is comprised of events with different intensities and corresponding frequencies/probabilities, developed from historical tropical cyclone data. The hazard module models the wind intensity at each location; the vulnerability module then links the hazard intensity to expected damage, and the financial module reports expected loss. For development of the events catalogue, the coast of Japan has been divided into regions with similar typhoon climatology in a manner that also included bypassing tropical cyclones. For each region, a set of stochastic events are developed that results in events with different intensities, track path, transitional speed, and their corresponding probabilities of occurrence (ranging from 0.002 to 0.5). The probabilities are selected through a stratified sampling approach (CE; Characteristics Events) emphasizing on those which are commonly used in the insurance industry. In the characteristic events development process, Typhoon track data from several databases including the Regional Specialized Meteorological Center (RSMC), Joint Typhoon Warning Center (JTWC) and China Meteorological Agency for the North West Pacific basin were analyzed, cleaned, duplicated eliminated, missing data injected, and completed. Although different historical typhoon databases resulted in similar characteristic track path and translational speeds, in regards to the characteristic intensities we found that such similarity was strong only for the low probability events.

KEYWORDS

Natural Hazard, Typhoon, Japan, Characteristic Events, Catastrophe Modelling

13th Americas Conference on Wind Engineering
Gainesville, Florida, USA
May 21 - 24, 2017

Generalized wind loading chain: A time-frequency perspective

Ahsan Kareem[a], Yanlin Guo[b], Liang Hu[c]

[a]Univ. of Notre Dame, Notre Dame, Indiana, USA, kareem@nd.edu
[b]Colorado State Univ., Fort Collins, Colorado, USA, yanlin@colostate.edu
[c]Univ. of Notre Dame, Notre Dame, Indiana, USA
& Huazhong Univ. of Sci. & Tech., Wuhan, China, lhu1@nd.edu

ABSTRACT

This paper highlights the development of a generalized wind loading chain (GWLC) to facilitate seamless response analysis under stationary as well as non-stationary wind events. More importantly, it offers a time-frequency domain modeling framework that is attentive to the nature of non-stationary winds (e.g. tropical storms, tornadoes, downbursts and gust fronts) and makes a case for a shift in the current design paradigm. To account for non-stationary winds, the framework recasts the well-established stationary wind loading chain in a time-frequency format for non-stationary winds utilizing evolutionary power spectral density (EPSD) and wavelet transform (WT). The formulation of GWLC seamlessly reverts to the conventional stationary scheme due to Prof. Davenport widely used for stationary winds as the time-dependence is disregarded. In this framework, the non-stationary features of the five components of the wind loading chain, i.e. gusts/turbulence in wind field, aerodynamic transfer function, aerodynamic force, structural dynamic transfer function and response, are represented respectively by either EPSD or wavelet-based scalogram. Detailed closed-form formulation that relates these five components of the chain for the buffeting response analysis is introduced. The usefulness of the generalized wind loading chain has been demonstrated through two applications, including the estimation of the extreme response under non-stationary winds leading to time variant Gust Effects Factor (GEF) and the formulation of Gust Front Factor (GFF) for design loads in gust front winds based on this framework. In closing, the overall framework of the GWLC is demonstrated through a numerical example.

KEYWORDS

Generalized wind loading chain, non-stationary winds, evolutionary power spectral density, wavelet transform

Hurricane loss analysis of wood-frame structures in Florida

Grzegorz Kakareko[a], Sungmoon Jung[a], O. Arda Vanli[b], Amanuel Tecle[c], Omar Khemici[c], Mahmoud Khater[c]

[a]*Department of Civil and Environmental Engineering, Florida A&M University – Florida State University College of Engineering, Tallahassee, FL 32310, USA*
[b]*Department of Industrial and Manufacturing Engineering, Florida A&M University – Florida State University College of Engineering, Tallahassee, FL 32310, USA*
[c]*CoreLogic, Oakland, CA 94612-1938, USA*

ABSTRACT: This research presents a multi-scale loss analysis of wood frame residential structures located in Florida. In this study, we conducted two separate loss analyses. The first was loss of singular house structure. Based on different components capacities, we developed three different house models named the weak (houses built before hurricane Andrew), the average (current houses with roof subjected to roof uplift), and the strong (built from the strongest possible components). The second analysis covers three modules: the hazard module that utilizes wind occurrence in the region, the vulnerability module that uses developed vulnerability functions, and the financial module that quantifies different loss characteristics. The results of this paper may help homeowners make mitigation decisions. Moreover, regional analysis can help identify parts of Florida that are most subjected to hurricane loss.

KEYWORDS: Hurricane, Vulnerability, Damage Ratio, Regional Loss.

1. INTRODUCTION

This research presents a multi-scale loss analysis of wood frame residential structures located in Florida. Between the years 1851-2015, Florida was hit by 37 hurricanes (Jarrell et al. 2001; Landsea 2016). Moreover, among the 10 most costly hurricanes in United States history, 7 occurred in Florida. Those two facts contributed to rank Florida as the most susceptible state to damage caused by high winds.

In this study we conducted two separate loss analyses. The first was loss of singular house structure, which was described by vulnerability functions. Vulnerability is a function that takes wind speed as an input and produces a damage ratio defined as a ratio of the repair cost of damage caused by high wind, over the total replacement cost of the house. The second was Florida regional analysis, which used the vulnerability functions developed in the first analysis.

Many researchers contributed to the topic of hurricane loss estimation (Amirinia and Jung 2016; Amirinia and Jung 2017; Amirinia et al. ; Khanduri and Morrow 2003; Pinelli et al. 2004; Pita et al. 2012; Simmons et al. 2002; Watson Jr and Johnson 2004). State-of-art of vulnerability estimation (Pita et al. 2014; Walker 2011) can be divided into two main methods. The first is to deliver vulnerability functions based on data claims from prior events (Friedman 1984; Leicester and Reardon 1976; Sparks et al. 1994). A disadvantage of this method is generalization of different house structures without considering certain parameters among them. One advantage of this method is simplicity (simple function approximation methods). The second method is based on engineering principles to quantify the possible hurricane loss (Gurley et al. 2005; Pinelli et al. 2011; Pinelli et al. 2004; Pinelli et al. 2009; Pinelli et al. 2003; Vickery et al. 2006a; Vickery et al.

2006b). Based on house characteristics (geometry, localization of the house, capacities of components) and the wind load, engineering methods approximate the damage stage of the house elements for the certain wind speed.

In this research we used a Monte Carlo method to deliver vulnerability functions. Based on different components capacities, three types of structures were investigated: the weak, the average, and the strong. The weak structure represents houses that were built before Hurricane Andrew which occurred in 1992 and were not retrofitted yet. The average represents houses that represent the average condition for building and occupation standard, but the capacities of their roof-to-wall connections and roof sheathing were chosen in a way to subject the roof structure will be damaged by roof uplift. The strong house represents house structures with the highest possible components capacities, built from materials that should resist higher wind loads.

Florida regional analysis delivered in this study was divided in three separate parts. In first analysis, the hazard module calculates wind occurrence probability. This part was based on: site specific joint distributions, climatology database, track/pressure joint distribution, and other storm data. The probabilistic hurricane model used in this research simulates around 16,000 stochastic storms that may affect Florida. The second is the vulnerability model which depends on vulnerability functions conducted in this research. The third part is the financial module, where the aim was to estimate the average annual loss (mean annual loss) and different aggregate exceedance probabilities (calculated loss based on aggregate effect of all possible hurricane events that can occur in a year).

The findings of this study contribute into the studies developed by previous researchers and models develop by Florida Commission on Hurricane Loss Projection Methodology (AIR WORLDWIDE 2015; Applied Research Associates 2002b; CoreLogic EQECAT 2015; FPHLM 2015; Hamid et al. 2010; Pinelli et al. 2011; Pinelli et al. 2004; Pinelli et al. 2003; RiskLink 15.0 2015; Simmons et al. 2002; Watson Jr and Johnson 2004).

The possible outcomes of this paper may help private homeowners in mitigation decision making (Applied Research Associates 2002a, 2002b, 2008; Pinelli et al. 2009). The regional analysis quantifies regional loss, which helps local counties and insurance companies to quantify the potential regional loss and its mitigation.

2 METHODOLOGY OF MULTI-SCALE LOSS ESTIMATION

2.1. Vulnerability estimation

Vulnerability functions obtained in current research were developed based on Monte Carlo simulation method. A MATLAB program was created, pursuing the overall algorithm described in Figure 1. The input values of the program were house dimensions, type of roof, site location, surrounding environment and capacities of house components (roof-to-wall connection, roof sheathing, roof cover, windows, doors, garage door, and walls). The failure of a particular element was assumed when the load exceeded the capacity of the element. For conversion of wind speed into the loads we use the modified formulas from ASCE 7 (2010) (Cope 2004).

The first element checked was openings (windows, garage doors and doors). Windows have two failure modes: failure by pressure, and failure by debris impact (Gurley et al. 2005) (Equation 1).

$$p_D(V) = 1 - exp(-A \cdot N_A \cdot B \cdot C \cdot D) \quad (1)$$

$p_D(V)$ is the probability of impact, A is a fraction of missile objects, N_A is a total number of missile objects, B is the number of the objects that are currently in the air, C is a fraction of the unprotected windows in the wall, and D is the probability if described object destroys the window.

Failure of the openings may produce the additional internal pressure. Internal pressure was increased if Equation 2 was satisfied by changing the category of the building from enclosed to partially enclosed (ASCE 7 2010).

$$A_o > 1.10\ A_{oi} \tag{2}$$

In this equation, A_o is the total area of openings in the wall that receives positive external pressure, A_{oi} is the sum of the areas of openings in the building envelope excluding A_o.

The next house component checked was roof structure (roof cover, roof sheathing, and roof-to-wall connection). Roof cover is the main barrier preventing water penetration. Water penetration is the main reason for internal loss, which is a significant part of the total loss. For internal loss calculation, the approach from Holmes (1999) was utilized. Roof sheathing and roof cover have the same failure mode (failure by pull out). If the roof sheathing panel is damaged, the corresponding roof cover is also assumed as damaged. The last element of the roof structure is the roof-to-wall connection. Capacity of roof to wall connection correlates to roof uplift damage mode Equation 3. If the roof is uplifted, the whole roof structure (roof cover, roof sheathing, and roof-to-wall connection) is destroyed.

$$R > (W - D) \tag{3}$$

In this equation, R is the sum of the capacity of the roof-to-wall connections that have not failed; W is uplift wind load; and D is the dead load, which has beneficial effect on the overall roof capacity.

The next house component in our study were walls. We considered three failure modes. The first two are sheathing failure and lateral failure (Cope 2004). The third is foundation failure due to the shear force. The first mode is progressive whereas the second and third modes are abrupt. For each wind speed, we investigated eight wind direction, the final loss is a mean over the directions.

The damage of every element was assigned into the damage matrices. Based on the damage matrices and replacement costs (RS Means 2015), the vulnerability at a particular wind speed was calculated.

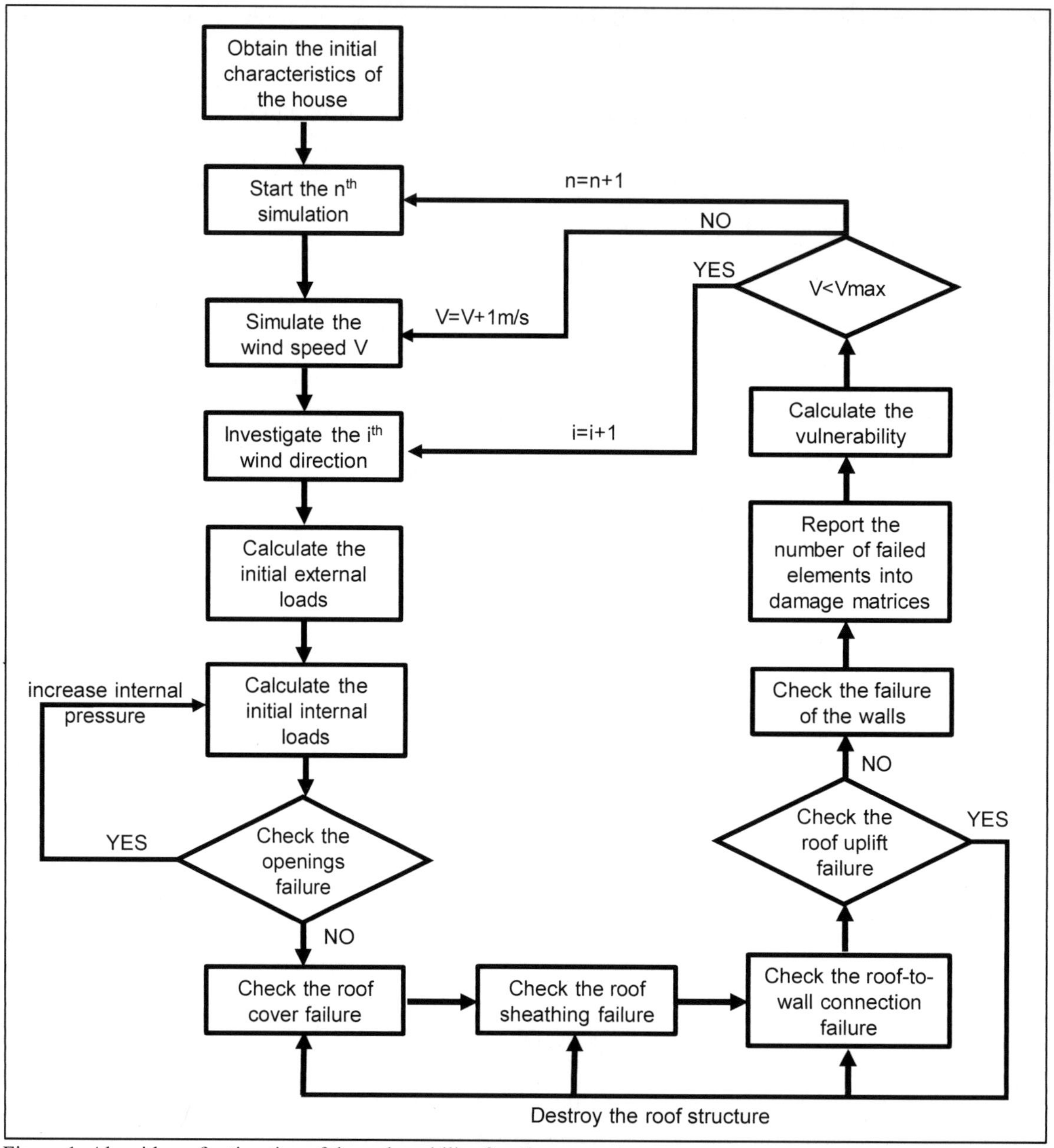

Figure 1. Algorithm of estimation of the vulnerability function.

2.1 *Florida regional hurricane loss analysis*

The loss analysis was performed in collaboration with CoreLogic company which developed a fully validated software for loss estimation. The regional analysis was conducted for each county and for the whole state of Florida.

The methodology used in this paper was similar to CoreLogic North Atlantic Hurricane Model (NAHM), which is a probabilistic model designed to estimate the wind and storm surge losses to properties due to the occurrence of hurricanes along the 3,100 miles of US coastline from Texas to Maine.

The loss estimation software calculates the Average Annual Loss (AAL) (Equation 4) or the mean annual loss, which is the sum from all possible events, the product of the mean loss and the annual frequency of occurrence of each event. Other loss metric includes the Aggregate Exceedance Probability (AEP) loss distribution. The AEP provides the probability of exceeding a loss level resulting from the aggregate effect of all possible events that could occur in a single year. The stochastic event set includes over 16,000 which affect Florida. Figure 2 describes the algorithm that was utilized for the regional loss estimation. The regional loss estimation can be divided into three parts, namely, the hazard module, the vulnerability module, and the financial module. The hazard module estimates the wind speed at all property locations, defined by their latitude and longitude coordinates or street addresses, using a proprietary wind speed model. The model computes wind speeds using the storm's maximum sustained wind speed, the filling rate, the radius to maximum winds, the storm track, the translation speed, the gust factor, the storm profile (attenuation of wind speed outward from the center), and the friction caused by local terrain and man-made structures. The vulnerability module provides damage estimates for each of the property assets being analyzed. This module uses a set of vulnerability functions (developed in this research) to calculate damage to the building structures. The financial module provides insurance loss estimates using insured values, limits, deductibles and facultative and/or treaty reinsurance information from probabilistic distributions of computed damage for each site obtained from the vulnerability module. This summary addresses the loss estimates for wind effects only. Storm surge losses follow a similar approach with the use of corresponding hazard and vulnerability but are outside the scope of this research.

$$AAL = \sum_{i=1}^{k} fe_i \cdot Le_i \tag{4}$$

In this equation, fe_i is an event frequency; Le_i is event loss; k is a number of events and i is a stochastic hurricane event simulated in the analysis.

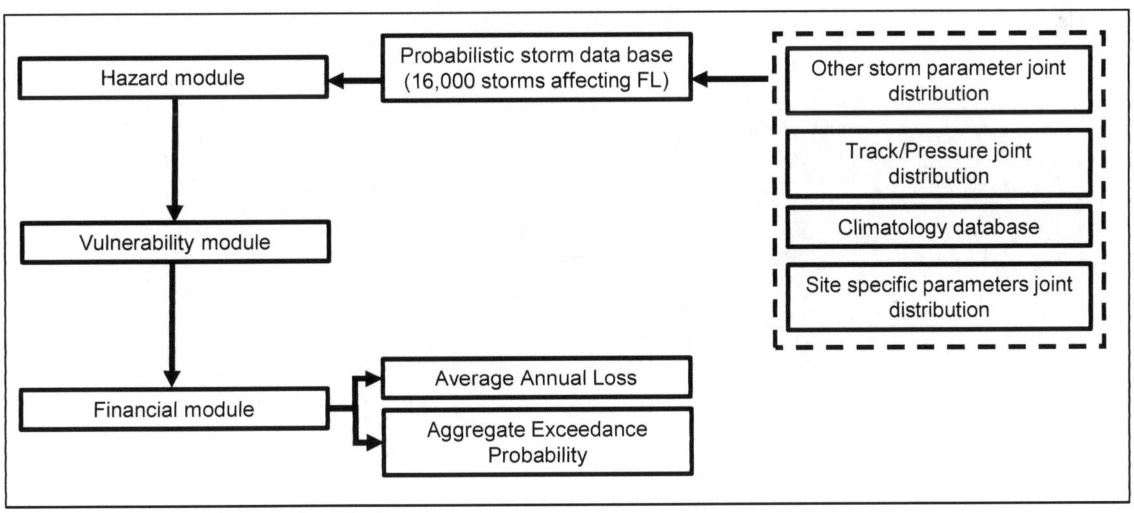

Figure 2. Algorithm of regional analysis.

3 RESULTS OF MULTI-SCALE LOSS ESTIMATION

3.1 *Vulnerability estimation*

For the geometry of the structure that has been investigated the following dimensions have been chosen: $overhang = 0.61m\ (2ft)$, $length = 18.29m\ (60\ ft)$, $width = 11.58m\ (38\ ft)$, $walls\ height = 3.05m\ (10\ ft)$, $roof\ height = 4.14m\ (13.6\ ft)$, $roof\ angle = 23^o$, $truss\ spacing = 0.61m\ (2ft)$. This chosen geometry was based from Pinelli et al. (2003), where a significant number of houses were investigated and this structure was identified as the most representative one.

Tables 1, 2, and 3 describe capacities used for vulnerability estimation. The weak structure represents houses built before Hurricane Andrew in 1992 (after hurricane Andrew, the building code has been changed to decrease the future hurricane loss). The average and strong types represent current, occupied houses. The capacities of the average structure were chosen to show the importance of roof structure in hurricane loss estimation. The strong house model has the strongest possible components capacities among wood residential structures.

Table 1. Capacity of house components- weak version

Component	Mean	COV	Source
Sheathing (6d 6/12)	2614 Pa (54.6 psf)	0.11	(Vickery et al. 2006b)
Roof cover (tiles)	2442 Pa (51 psf)	0.2	(Gurley et al. 2005)
R2W (toe nail)	1846 N (415 lb)	0.25	(Vickery et al. 2006b)
Windows (tall)	2500 Pa (52.2)	0.2	(Gurley et al. 2005)
Garage doors	1430 Pa (30 psf)	0.2	(Gurley et al. 2005)
Wall panels	2614 Pa (54,6 psf)	0.11	(Vickery et al. 2006b)
Doors	2394 Pa (50 psf)	0.2	(Vickery et al. 2006b)

Table 2. Capacity of house components- average version

Component	Mean	COV	Source
Sheathing (8d 6/6)	4932 Pa (103 psf)	0.11	(Vickery et al. 2006b)
Roof cover (shingles)	3352 Pa (70 psf)	0.2	(Gurley et al. 2005)
R2W (clip)	3069 N (690 lb)	0.2	(Gurley et al. 2005)
Windows (medium)	3330 Pa (69.6 psf)	0.2	(Gurley et al. 2005)
Garage doors	1430 Pa (30 psf)	0.2	(Gurley et al. 2005)
Wall panels	4932 Pa (103 psf)	0.11	(Vickery et al. 2006b)
Doors	4788 Pa (100 psf)	0.2	(Gurley et al. 2005)

Table 3. Capacity of house components- strong version

Component	Mean	COV	Source
Sheathing (8d 6/6)	6300 Pa (131 psf)	0.14	(Datin et al. 2010)
Roof cover (shingles)	3352 Pa (70 psf)	0.2	(Gurley et al. 2005)
R2W (wrap)	5840 N (1200 lb)	0.1	(Ellingwood et al. 2004)
Windows (small)	5000 Pa (52.2 psf)	0.2	(Gurley et al. 2005)
Garage doors	2490 Pa (52 psf)	0.2	(Gurley et al. 2005)
Wall panels	6300 Pa (131 psf)	0.14	(Datin et al, 2010)
Doors	4788 Pa (100 psf)	0.2	(Gurley et al. 2005)

Figure 3 shows the vulnerability curves of the three different house structures. As a validation of our vulnerability curves, Figure 3 includes curves from Hazus (HAZUS MHMR 2003). These curves are fully validated based on historical loss data. For the average model, the significant increase of damage can be observed after 150 mph. The increase is caused by roof uplift and will be discussed in more details later. Because of the stronger roof-to-wall connection, roof uplift did not occur in the strong model (see Figure 4), the main difference between the average and the strong models beyond 150 mph.

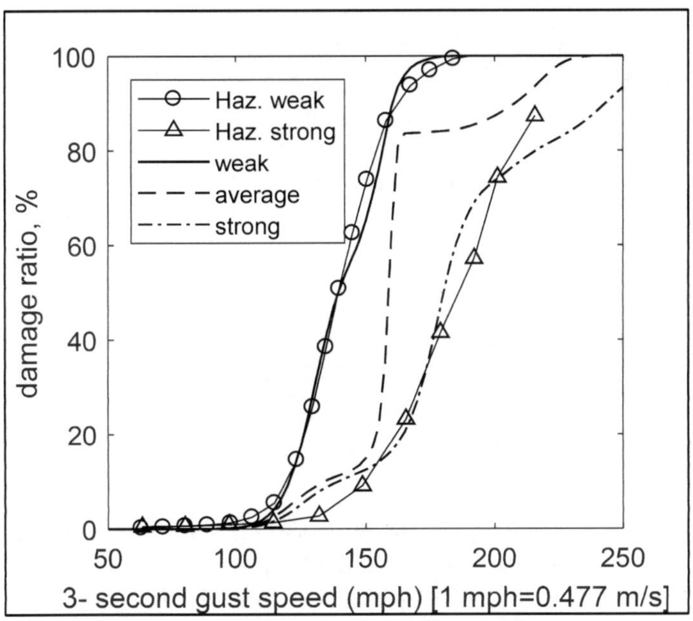

Figure 3. Comparison of different vulnerability models

Roof has an important role in residential structures in case of the hurricane loss. The first element of roof structure is the roof cover, which prevents the water from penetrating the interior of the house. The second element is the roof sheathing, which gives the additional stiffness to the house structure and is an additional barrier against water penetration. The third roof component is the roof-to-wall connection. Roof uplift is the worst failure scenario for the roof structure, on one particular wind speed the whole roof structure is destroyed, which often produces collapse of the house.

The important factor in this type of failure is ratio of the roof-to-wall connection capacity and the roof sheathing capacity. Houses with strong roof sheathing need appropriate roof to wall connection to allow roof sheathing to be destroyed progressively. Figure 4 illustrates how to prevent the roof uplift damage mode. The x-axis represents the different sheathing capacities, from 6d 6/12 .nailing as the weakest to 8d 6/6 nailing as the strongest. The y-axis represents roof-to-wall connection capacities starting from toe nail as the weakest and ending with strap as the strongest connection type. We analyzed combinations of roof-to-wall connections and roof sheathing and developed the safe zone area, which prevents from roof uplift damage mode. The investigated structures are marked on Figure 4. The average house falls inside the roof uplift zone. This difference causes an abrupt increase of the damage ratio after 150 mph (see Figure 3). Improving the roof-to-wall connection with a stronger type would prevent the roof uplift and significantly decrease the vulnerability function of this structure.

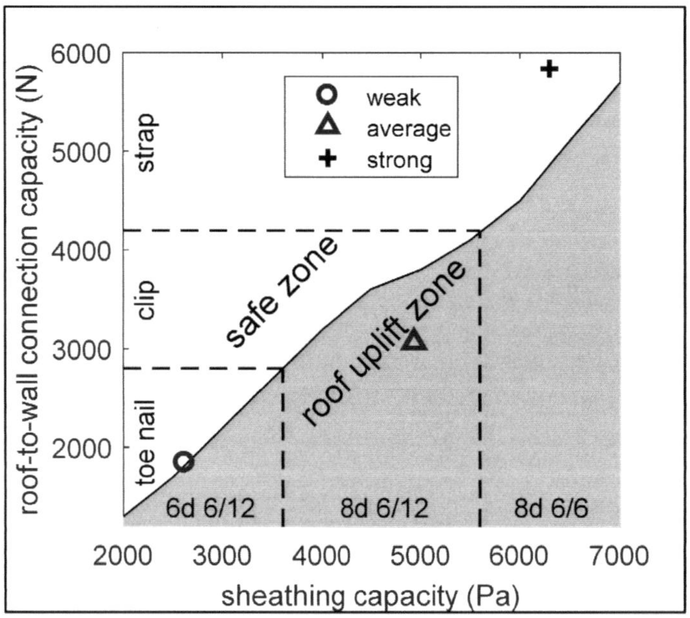

Figure 4. Roof uplift safe zone

3.2 *Florida regional analysis of wood frame structures*

This section presents the results for regional loss analysis with utilized vulnerability functions developed earlier. All results were normalized to loss costs. The dollar loss of the region was normalized to the total replacement value of all houses in the region (only the building value without the price of the lot). Moreover, the results shown are for policies with zero deductible and limit equal to the building replacement value. No contents and no additional living expenses coverages were assumed in the analysis performed in this research.

Figure 5 represents the AAL for the weak, the average, and the strong structure. Following the regional Florida classification (Gurley et al. (2005) and Pinelli et al. (2003)), the AAL decreased from South (Keys) to North Florida. The main factor of this order was the wind hazard speed, which is higher in the South and decreases as we move North.

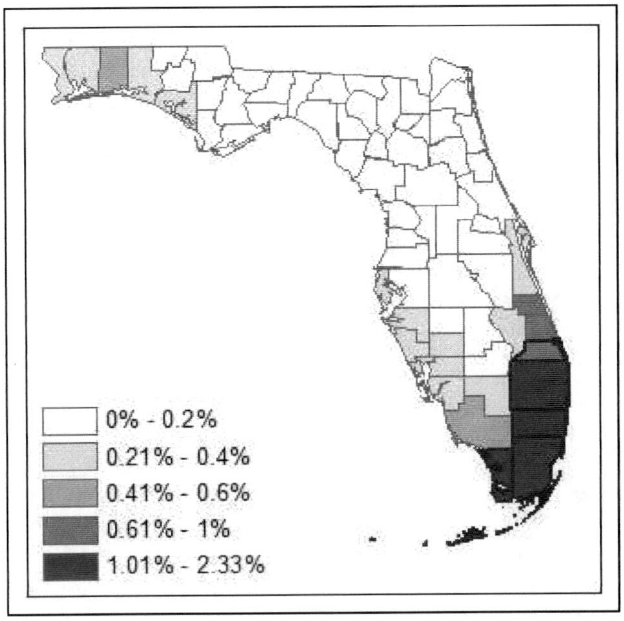

(a)

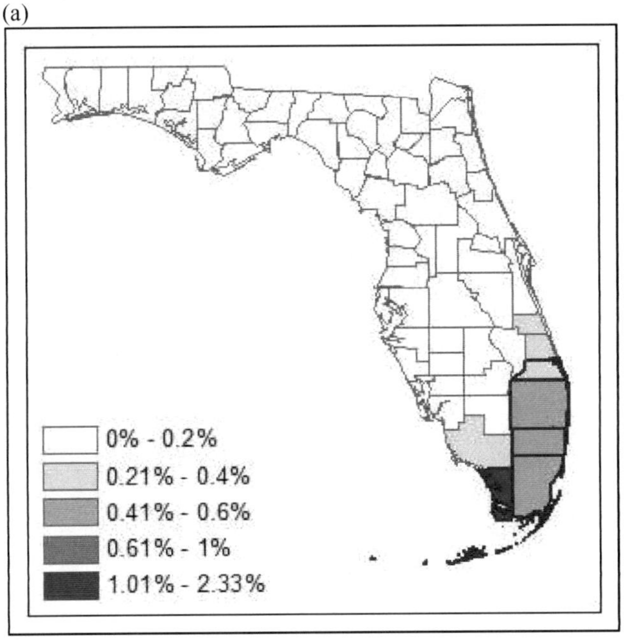

(b)

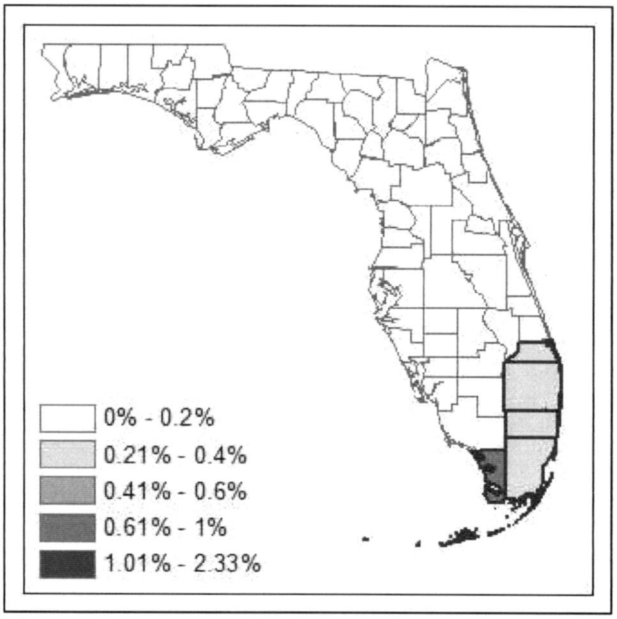

(c)
Figure 5. Average annual loss: (a) weak; (b) average; (c) strong

Based on our analysis, the counties with the higher possible hurricane loss are Monroe, Miami-Dade, Broward, Palm Beach, and Martin (bolded on Figure 5). Monroe was hit by 15 hurricanes between 1960-2008 time period. If a 250-Year AEP events were to hit Monroe county, the aggregate loss from multiple events per year for weak, average and strong model would exceed approximately 67.64%, 40.91%, and 22.54% of the existing property values. Only by improving house capacities of house components, the estimated loss decreased by about three times. This fact underlines the need to retrofit the existing structures that were not changed after hurricane Andrew. Moreover, the main differences between the average and the strong models is caused by different roof-to-wall connection. Differences among those two models show how much the regional loss can be decreased only by improving roof-to-wall connections.

Table 4. Percentage loss rate for weak model

County	AAL	AEP		
		50-Year	100-Year	250-Year
Monroe	2.33%	34.90%	50.26%	67.64%
Miami-Dade	1.46%	25.74%	48.77%	73.76%
Broward	1.27%	22.25%	44.09%	67.36%
Palm Beach	1.19%	20.43%	40.26%	61.31%
Martin	0.98%	15.41%	35.62%	58.18%

Table 5. Percentage loss rate for average model

County	AAL	AEP		
		50-Year	100-Year	250-Year
Monroe	1.08%	16.54%	27.20%	40.91%
Miami-Dade	0.58%	8.56%	16.64%	29.40%

Broward	0.46%	7.23%	12.88%	22.36%
Palm Beach	0.45%	6.54%	12.59%	23.28%
Martin	0.38%	5.19%	10.58%	20.28%

Table 6. Percentage loss rate for strong model

County	AAL	AEP		
		50-Year	100-Year	250-Year
Monroe	0.64%	8.78%	14.01%	22.54%
Miami-Dade	0.36%	6.42%	11.17%	16.74%
Broward	0.31%	5.63%	9.89%	14.57%
Palm Beach	0.29%	5.10%	9.05%	13.68%
Martin	0.24%	4.15%	7.99%	12.55%

Table 7 presents the summary of the damage for the state of Florida. The same trend is visible with the significant different in percentage loss improving the house components capacities. Only by improving the roof to wall connection in the average house the AAL will decrease by 0.08%.

Table 7. State results (without wind uncertainty)

Case ID	AAL	AEP		
		50-Year	100-Year	250-Year
Weak	0.53%	8.10%	12.41%	17.10%
Average	0.21%	2.85%	4.55%	6.82%
Strong	0.13%	1.94%	2.92%	3.95%

Some researchers have already investigated the importance of uncertainties in hurricane loss estimation (Li and Ellingwood 2006). Ellingwood and Tekie (1999) investigated the uncertainty in wind load in ASCE 07 formulas. They specified the wind load factors distribution. As an additional analysis, we utilized Ellingwood and Tekie (1999) results into our vulnerability estimations. The new vulnerabilities functions were implemented in the regional analysis. Table 8 represents the estimated state level results summary with included wind uncertainty. Table 8 should be compared to Table 7 (without uncertainty). It can be observed that the wind uncertainty makes the loss results more conservative. The utilization of wind load uncertainty did not change the state loss results significantly, but added an important element to the overall analysis.

Table 8. State results (with wind uncertainty)

Case ID	AAL	AEP		
		50-Year	100-Year	250-Year
Weak	0.69%	10.34%	15.29%	20.62%
Average	0.32%	4.49%	7.65%	12.03%
Strong	0.16%	2.28%	3.50%	4.90%

4 CONCLUSION

This paper investigated the multi-scale loss analysis of wood frame residential structures. The paper was divided into two separate parts. The first part was loss analysis of singular house structure. The loss of single house was described based on vulnerability functions. The authors delivered three vulnerability models (weak, average, strong) for one of the most popular Florida wood geometries. The second analysis was the Florida regional analysis, which was based on the vulnerability functions devolved earlier. This analysis was done in collaboration with CoreLogic company, which developed a fully validated software for loss estimation. The methodology used for regional loss estimation was similar to CoreLogic North Atlantic Hurricane Model (NAHM), which is a probabilistic model designed to estimate the wind and storm surge losses to properties due to the occurrence of hurricane.

Figure 3 presented vulnerability models developed in this research. The significant difference is visible between the model average and the strong model after 150 mph wind speed. This difference was caused by roof uplift that occurred in the average model. Only by improving the roof to wall connections in the average model, we are able to drastically decree the loss. Figure 4 described how to avoid this type of failure and specified the safe zone where the roof is not subjected to damage by uplift. Figure 8 quantified the distribution of AAL across all Florida counties for investigated house models. The top five counties with the highest hurricane loss costs were Monroe, Miami-Dade, Broward, Palm Beach, and Martin. We specified the loss results for three returns periods of 50, 100 and 250 years. The main driver, which decided about this classification, was the wind hazard, was higher in those counties. The main difference between the average and the strong model was type of roof-to-wall connection. Only by improving roof-to-wall connections we could significantly decrease the predicted loss (see Table 5,6, and 8). The stochastic event set used in this research is a set of likely occurring storms and projected from historical data and stochastic simulations. An additional discussion was included about the wind load uncertainty in the regional loss estimation. Table 8 presented the state-wide results with wind load uncertainties included. The additional uncertainties did not change the state-wide results drastically but added an important factor into overall analysis and should be included in the future research.

5 ACKNOWLEDGMENTS

This publication was supported by the National Sea Grant College Program of the U.S. Department of Commerce's National Oceanic and Atmospheric Administration (NOAA), Grant No. NA 14OAR4170108, Subcontract No UFDSP00010509 made to Florida State University. The views expressed are those of the authors and do not necessarily reflect the view of these organizations.

6 REFERENCES

1 AIR WORLDWIDE, 2015. The air hurricane model: Air atlantic tropical cyclone model v15.0.1. Florida Commission on Hurricane Loss Projection Methodology.

2 Amirinia, G., Jung, S., 2016. Time domain analysis of unsteady aerodynamic forces on a parked wind turbine tower subjected to high winds, 8th International colloquium on bluff body aerodynamics and applications.

3 Amirinia, G., Jung, S., 2017. Along-wind buffeting responses of wind turbines subjected to hurricanes considering unsteady aerodynamics of the tower. Engineering Structures 138, 337-350.

4 Amirinia, G., Jung, S., Alduse, B.P., Effect of different hurricane spectrums on wind turbine loads and responses.

5 Applied Research Associates, 2002a. Development of loss relativities for wind resistive features of residential buildings with five or more units.
6 Applied Research Associates, 2002b. Development of loss relativities for wind resistive features of residential structures.
7 Applied Research Associates, 2008. Florida residential wind loss mitigation study.
8 ASCE 7, 2010. Minimum design loads for buildings and other structures. American Society of Civil Engineers.
9 Cope, A.D., 2004. Predicting the vulnerability of typical residential buildings to hurricane damage. University of Florida.
10 CoreLogic EQECAT, 2015. A component of the eqecat north atlantic hurricane model in risk quantification and engineering. Florida Commission on Hurricane Loss Projection Methodology.
11 Datin, P.L., Prevatt, D.O., Pang, W., 2010. Wind-uplift capacity of residential wood roof-sheathing panels retrofitted with insulating foam adhesive. Journal of Architectural Engineering 17, 144-154.
12 Ellingwood, B.R., Rosowsky, D.V., Li, Y., Kim, J.H., 2004. Fragility assessment of light-frame wood construction subjected to wind and earthquake hazards. Journal of Structural Engineering 130, 1921-1930.
13 Ellingwood, B.R., Tekie, P.B., 1999. Wind load statistics for probability-based structural design. Journal of Structural Engineering 125, 453-463.
14 FPHLM, 2015. Florida public hurricane loss model 6.1. Florida Commission on Hurricane Loss Projection.
15 Friedman, D.G., 1984. Natural hazard risk assessment for an insurance program. Geneva Papers on Risk and Insurance, 57-128.
16 Gurley, K., Pinelli, J., Subramanian, C., Cope, A., Zhang, L., Murphree, J., 2005. Florida public hurricane loss projection model engineering team. Final report, International Hurricane Research Center, Florida International University, Miami.
17 Hamid, S., Kibria, B.G., Gulati, S., Powell, M., Annane, B., Cocke, S., Pinelli, J.-P., Gurley, K., Chen, S.-C., 2010. Predicting losses of residential structures in the state of florida by the public hurricane loss evaluation model. Statistical methodology 7, 552-573.
18 HAZUS MHMR, 2003. Multi-hazard loss estimation methodology. FEMA http://www.fema.gov.
19 Holmes, W., 1999. Multi-hazard loss estimation methodology: Technical manual. Washington, DC: Federal Emergency Management Agency.
20 Jarrell, J.D., Mayfield, M., Rappaport, E.N., Landsea, C.W., Jarrell, J., Mayfield, M., Rappaport, E., Landsea, C., 2001. The deadliest, costliest, and most intense united states hurricanes from 1900 to 2000 (and other frequently requested hurricane facts). NOAA, Technical Memorandum NWS-TPC-3.
21 Khanduri, A., Morrow, G., 2003. Vulnerability of buildings to windstorms and insurance loss estimation. Journal of wind engineering and industrial aerodynamics 91, 455-467.
22 Landsea, C., 2016. How many direct hits by hurricanes of various categories have affected each state? Atlantic Oceanographic and Meteorological Laboratory, http://www.aoml.noaa.gov/hrd/tcfaq/E19.html.
23 Leicester, R., Reardon, G., 1976. Case studies of wind damage to domestic and other low-rise buildings, Symposium on Natural Hazards in Australia. Australian Academy of Science, Canberra.
24 Li, Y., Ellingwood, B.R., 2006. Hurricane damage to residential construction in the us: Importance of uncertainty modeling in risk assessment. Engineering structures 28, 1009-1018.
25 Pinelli, J.-P., Pita, G., Gurley, K., Torkian, B., Hamid, S., Subramanian, C., 2011. Damage characterization: Application to florida public hurricane loss model. Natural Hazards Review.
26 Pinelli, J.-P., Simiu, E., Gurley, K., Subramanian, C., Zhang, L., Cope, A., Filliben, J.J., Hamid, S., 2004. Hurricane damage prediction model for residential structures. Journal of Structural Engineering 130, 1685-1691.
27 Pinelli, J.-P., Torkian, B.B., Gurley, K., Subramanian, C., Hamid, S., 2009. Cost effectiveness of hurricane mitigation measures for residential buildings, Proceedings, 11th America's Conference on Wind Engineering, American Association of wind Engineering.(www. aawe. org).
28 Pinelli, J.-P., Zhang, L., Subramanian, C., Cope, A., Gurley, K., Gulati, S., Hamid, S., 2003. Classification of structural models for wind damage predictions in florida, Proc., 11th Int. Conf. on Wind Engineering.
29 Pita, G., Pinelli, J.-P., Cocke, S., Gurley, K., Mitrani-Reiser, J., Weekes, J., Hamid, S., 2012. Assessment of hurricane-induced internal damage to low-rise buildings in the florida public hurricane loss model. Journal of Wind Engineering and Industrial Aerodynamics 104, 76-87.
30 Pita, G., Pinelli, J.-P., Gurley, K., Mitrani-Reiser, J., 2014. State of the art of hurricane vulnerability estimation methods: A review. Natural Hazards Review 16, 04014022.

31 RiskLink 15.0, 2015. North atlantic hurricane models. Florida Commission on Hurricane Loss Projection Methodology.
32 RS Means, 2015. Residential cost data. R.S. Means Company.
33 Simmons, K.M., Kruse, J.B., Smith, D.A., 2002. Valuing mitigation: Real estate market response to hurricane loss reduction measures. Southern Economic Journal, 660-671.
34 Sparks, P.R., Schiff, S.D., Reinhold, T., 1994. Wind damage to envelopes of houses and consequent insurance losses. Journal of Wind Engineering and Industrial Aerodynamics 53, 145-155.
35 Vickery, P.J., Lin, J., Skerlj, P.F., Twisdale Jr, L.A., Huang, K., 2006a. Hazus-mh hurricane model methodology. I: Hurricane hazard, terrain, and wind load modeling. Natural Hazards Review 7, 82-93.
36 Vickery, P.J., Skerlj, P.F., Lin, J., Twisdale Jr, L.A., Young, M.A., Lavelle, F.M., 2006b. Hazus-mh hurricane model methodology. Ii: Damage and loss estimation. Natural Hazards Review 7, 94-103.
37 Walker, G.R., 2011. Modelling the vulnerability of buildings to wind. Canadian Journal of Civil Engineering 38, 1031-1039.
38 Watson Jr, C.C., Johnson, M.E., 2004. Hurricane loss estimation models. Bulletin of the American Meteorological Society 85, 1713.

Hurricane Risk in Florida as it Relates to the Evolution of Building Codes

Bahareh Kordi [a], Karthik Ramanathan [b], Tim Johnson [c], Farid Moghim [d], Sarah Bobby [e]

[a] *AIR Worldwide, 131 Dartmouth Street, Boston, MA 02116, email: bkordi@air-worldwide.com*
[b] *AIR Worldwide, 131 Dartmouth Street, Boston, MA 02116, email: kramanathan@air-worldwide.com*
[c] *AIR Worldwide, 131 Dartmouth Street, Boston, MA 02116, email: tjohnson@air-worldwide.com*
[d] *AIR Worldwide, 131 Dartmouth Street, Boston, MA 02116, email: fmoghim@air-worldwide.com*
[e] *AIR Worldwide, 131 Dartmouth Street, Boston, MA 02116, email: sbobby@air-worldwide.com*

ABSTRACT: Building codes are often updated as warranted by improvements in science and engineering, as new data become available. Following Hurricane Andrew, significant changes were introduced to the South Florida Building Code. In 2001, the first statewide Florida Building Code (FBC) was introduced. FBC 2010 has introduced significant changes in its wind provisions. The changes were mainly due to the introduction of new design wind speed maps, which also triggered the redefinition of the wind borne debris region. Terrain exposure categories and their definitions have also changed over time in the FBC. As all these changes take effect continually, the vulnerability of at-risk building stock also changes. This study quantifies the impact of building code related changes on the hurricane risk of residential homes in Florida in terms of probabilistic loss metrics, such as average annual loss, using AIR Hurricane Model for the United States.

KEYWORDS: Florida Building Code, Hurricane Risk, Catastrophe Modeling, Expected Loss, ASCE 7.

1 INTRODUCTION

Florida is no stranger to hurricane threats; in fact, the average frequency of land falling hurricanes is almost 1.5 times that of the second highest risk hurricane prone state, Texas [9]. As a result, Florida has led the way in terms of developing and adopting building codes that address high expected hurricane risk. However, building codes are often reflections of the state of scientific knowledge at the time of their implementation. Occasionally, nature challenges the state of knowledge and uncovers its shortcomings in both the perceived risk and our understanding of the buildings' behavior in extreme natural events. For example, Hurricane Andrew in 1992 highlighted significant deficiencies in the built environment which culminated in the adoption of a very stringent South Florida Building Code (SFBC) for Miami-Dade and Broward counties to mitigate losses from future events [10]. However, statewide adoption of these measures did not happen until 2001 with the introduction of the first edition of the FBC, synonymously referred to as FBC 2001. Hurricane Andrew also stressed the importance of enforcement of the building codes. Poor enforcement practices led to a highly susceptible building stock which exacerbated the losses. Florida has since improved and currently is one of the most stringent states with respect to building code enforcement [11].

In addition to the evolution of building codes in Florida, the load standards on which the building codes are based (e.g. American Society of Civil Engineers (ASCE) ASCE-7 standard [6-8]) also evolve due to new research and technology. With respect to wind design, the ASCE 7 wind load standard dictates the design levels to meet the building code requirements. Significant changes occurred to the ASCE 7 wind load standards in 1998 and again in 2010. For instance, in 2010 ASCE 7 introduced risk category specific design wind speed maps. These changes to the wind maps also triggered changes to the wind borne debris region which was in turn redefined in FBC 2010. Terrain exposure categories and their definitions have also changed over time in FBC and ASCE 7.

While building code changes happen at discrete points in time, vulnerability changes to the built environment as a whole often vary smoothly due to effects of structural aging, evolution of building materials, and the development of new and better construction practices, to mention a few. The aforementioned factors usually fall outside the scope of building codes, but are noteworthy. As all these changes take effect continually, the at-risk building stock becomes a conglomeration of various building code standards depending on the year and code in place at the time of construction. Insurance claims data serve as an excellent means to not only validate catastrophe models and respective vulnerability functions, but also glean insight into the aforementioned aspects. Figure 1 shows the normalized mean damage ratio by year of construction from insurance claims data pertinent to the 2004-2005 hurricane seasons in Florida. Normalization is done by dividing the claims based average damage ratio for a given year of construction by the claims based average damage ratio for a building built in 1980. The following are some of the noteworthy observations.

- Actual insurance claims data shows continuous changes in vulnerability over the years post 1995. The behavior is steady until this year
- The gradual changes in vulnerability may be attributed to adoption and enforcement of subsequent building codes in Florida coupled with factors outside the scope of these building code updates. These include aging and deterioration of structures along with evolution of building materials, building technology and construction practices

Despite the availability of actual claims data, it is important to note that the data is not statistically significant across the whole range of local intensities that the damage functions strive to capture. Further, the data captures damageability across a family of wood frame single family homes which comprise significant variability among them in terms of size, shapes, roof geometries, roof cover materials etc. to mention a few, not to forget the variability in local intensity and hazard as these locations are spread out across the entire state. Catastrophe models strive to capture the temporal variation in vulnerability due to the aforementioned aspects at a given location. Extrapolating beyond that, they also attempt to capture the spatial variation in vulnerability across these time bands.

Evaluating the relative performance of these building code changes and their impact on the inherent hurricane risk to buildings is difficult to quantify without actual wind events. Catastrophe risk assessment models serve as excellent platforms to not only project future hurricane losses, but also help test the impact of regional/local building codes and their effectiveness to the exposed hurricane risk [20,19,12]. These models integrate a stochastic hazard coupled with detailed views of building vulnerability in order to estimate reliable loss distributions. The study presented in this paper assesses the impact of the evolution of building codes in Florida on the vulnerability of residential single family homes using AIR Hurricane Model for the United States [13]. This model has been validated using past hurricane event claims data from insurance companies and is used to assess the aforementioned impact measured in terms of probabilistic loss metrics, such as the expected loss in a given year commonly termed as the average annual loss (AAL). Subsequent sections of this paper provides extensive details about the evolution of building codes in Florida and its relationship to evolving wind load standards given in ASCE 7, including wind borne debris and terrain definitions. Vulnerability implications are then presented by modeling wood frame single family homes built at different time periods in various parts of the state to explore the time- and space- dependent variation in their hurricane risk.

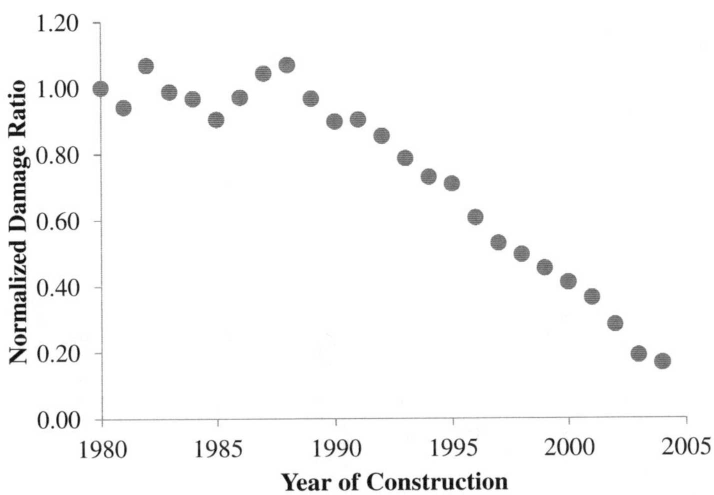

Figure 1. Normalized damage ratios from insurance claims data by year of construction for residential wood frame single family homes in Florida

2 EVOLUTION OF BUILDING CODES IN FLORIDA

Unlike many countries where a single national building code applies to all regions of the country, in the United States many different building codes have been in effect at different times and localities. In addition, to design and construct a safe structure in the U.S., engineers follow two sets of guidelines: building codes and building standards. Building codes are described by the Federal Emergency Management Agency (FEMA) as "… sets of regulations governing the design, construction, alteration and maintenance of structures;" in other words, building codes must be followed by law [14]. Building codes are legally adopted by state and/or local jurisdictions through legislative processes and are enforced by local governments. They provide minimum life safety requirements for buildings and their occupants against certain hazards. Building standards are generally adopted by reference into the building codes. Building standards are developed by professional organizations, such as ASCE, and offer in-depth provisions based on recent advancements in building, material, and construction research. They also furnish minimum design loads against certain atmospheric and geological natural hazards to protect the well-being of building occupants.

Prior to the statewide adoption of FBC 2001, two building codes were in effect in Florida: the Standard Building code (SBC) [21], which was used in most of Florida except in Miami-Dade and Broward counties, and the SFBC, which was in effect in Miami-Dade and Broward counties. FBC 2001 became effective on March 1, 2002 and superseded all local codes. The Code was based on the 1997 SBC with 1999 Revisions. All local governments in the state are required by law to adopt and enforce the FBC. Individual jurisdictions may amend the code to be more stringent when local conditions warrant such action. The 2004 version of FBC became effective in November of 2004; this code adopted the 2003 edition of the International Building

Code (IBC) and the International Residential Code (IRC) with state amendments. New editions of the FBC are published regularly; generally a new publication is released every three years. Comprehending the changes to the building code can be broken down into four main categories:

1. Understanding the changes to the corresponding wind load standards
2. Tracking the evolution of terrain exposure categories
3. Tracking the evolution of wind-borne debris region
4. Monitoring the changes in design requirements given in the building codes

While details regarding the inherent building vulnerabilities and changes in design requirements are presented in Section 3 of this paper, the next set of sub-sections provide extensive details about items 1-4.

2.1 *Changes to ASCE Wind Load Standards*

In parallel with building code evolution is the evolution of wind load standards that are referenced in the building codes. The ASCE-7 wind load standards provide guidance regarding the design wind speeds that should be used to develop the design wind loading conditions for the structure. The 1988 version of the wind standards referenced a fastest mile wind speed. The 1998 version of the wind load standard utilized a stochastic simulation that captured the frequency and intensity of simulated hurricane tracks which could then be translated into a wind contour map representing the 3-sec gust design wind speeds. In 2010, ASCE introduced new design wind speed maps, with changes due to:

1. A shift in the design methodology from Allowable Strength Design (ASD) to Load and Resistance Factor Design (LRFD) in order to make the wind and earthquake methodologies consistent
2. Introduction of risk category factors in the design wind maps, resulting in three design wind maps for three risk categories
3. New research findings and data which resulted in the re-simulation of the hurricane winds

Prior to FBC, the design wind speed map in the SBC was based on ASCE 7-88. The ASCE 7-98 wind map was used as the reference design wind speed map in FBC 2001 and has not changed throughout the newer versions of the FBC up until the 2010 version, in which the design wind speed is based on ASCE 7-10. The ASCE 7-10 design wind maps, which were developed using the most recent hurricane data, provides uniformity in Mean Recurrence Intervals (MRIs) to the hurricane and non-hurricane states. As stated in the Commentary of the standard, the previous basic wind speed maps yielded predictions of 50-100 year return period peak gust wind speeds along the coast [8]. Thus when translating the new maps to an ASD equivalent 50-year MRI design wind speed map, we will see that the changes generally result in lower design wind speeds (and therefore lower design wind pressures) along the hurricane prone region, while further inland the design wind pressures remain mostly unchanged. Furthermore, as the risk categories' "importance factors" were included in the design wind speed maps, ASCE 7-10 contains three wind speed maps: one each for risk categories I and II, and a single map for Risk Categories III and IV, which have the same importance factor. Figure 2 shows the design wind speed maps as per ASCE 7-88 (which is identical to the map used in ASCE 7-93), ASCE 7-98 (which is identical to the map used in ASCE 7-02 and 7-05), and the ASCE 7-10 design wind speed map for Risk Category II, which is shown for consistency with the previously mentioned maps. Tied to these changes are corresponding movements to the wind borne debris regions (WBDRs), which is discussed in the next section.

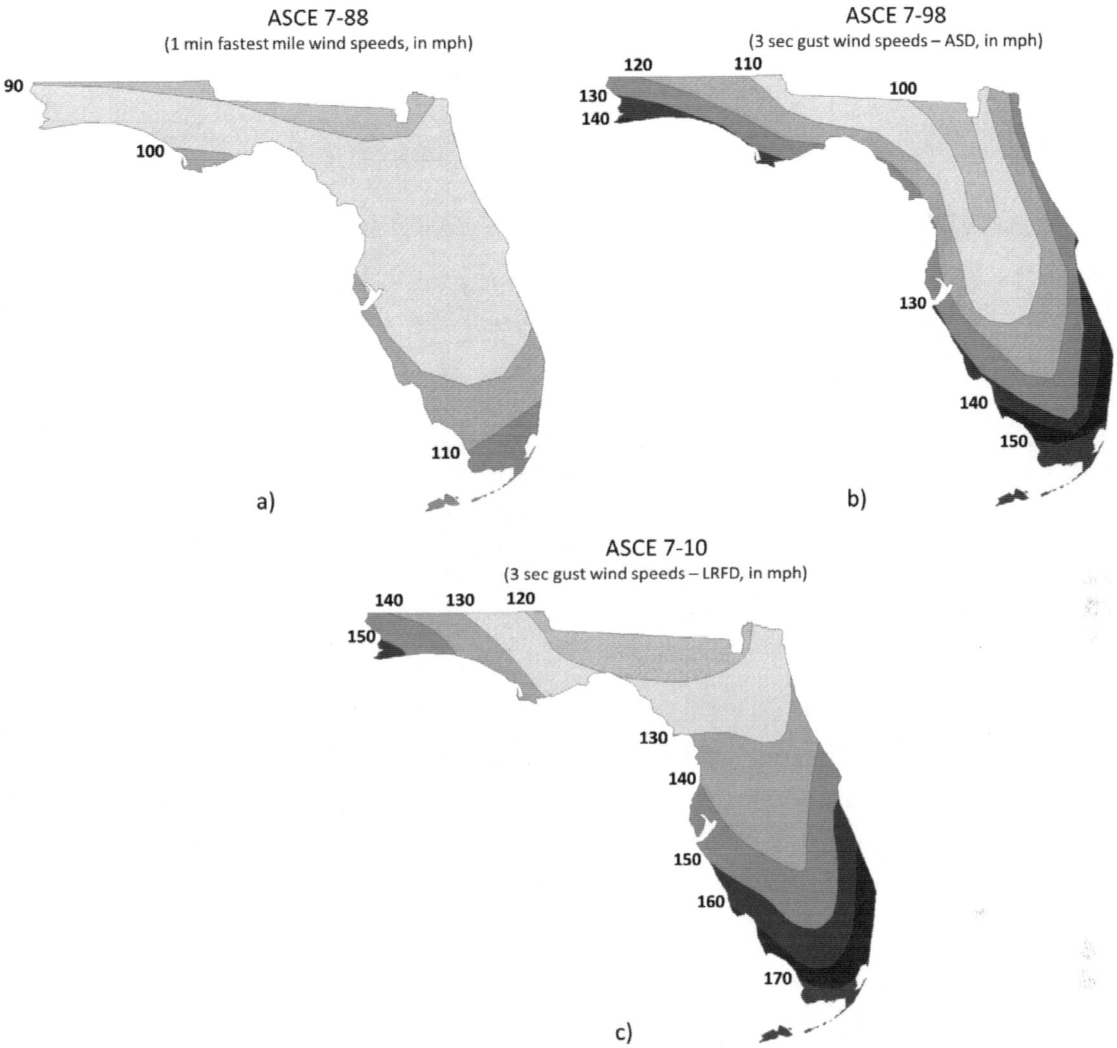

Figure 2. ASCE 7 Design Wind Speed Maps for the Following ASCE 7 Standards: a) ASCE 7-88; b) ASCE 7-98; c) ASCE 7-10 for Occupancy Category II Buildings

Table 1 shows the design wind speed in ASCE 7-93, ASCE 7-05, and ASCE 7-10 for occupancy category II buildings for select cities in Florida. It also shows the ratio between the design wind velocity pressure in ASCE 7-05 (FBC 2007) and ASCE 7-10 (FBC 2010) for those cities. The table indicates the ratio of the velocity pressures between ASCE 7-05 and ASCE 7-10 in two cases: the first is for a structure classified using the same exposure category in both versions of the code, and the second is for a structure located within the first 600 feet of the coast, which would be classified as Exposure C in ASCE 7-05 and Exposure D in ASCE 7-10. The comparison is done for a one-story (15 foot height) house (Risk Category II).

Table 1. The ratio between the design wind velocity pressure in ASCE 7-10 (FBC2010) and ASCE 7-05 (FBC2007) for 15 ft tall Risk Category II buildings.

	City	Design Wind Speed* (mph)			ASCE 7-10 Velocity Pressure / ASCE 7-05 Velocity Pressure	
		ASCE 7-93 (fastest mile)	ASCE 7-05 (3-sec gust, ASD)	ASCE 7-10 (3-sec gust, LRFD)	Same Exposure	Exp.C (ASCE 7-05) *to* Exp.D (ASCE 7-10)
FL Panhandle	Tallahassee	92	108	118	0.72	NA
	Panama City	99	128	134	0.66	0.80
	Fort Walton Beach	97	135	144	0.68	0.83
	Pensacola	97	137	151	0.73	0.88
	Milton	95	130	146	0.76	NA
North FL	Jacksonville	92	115	125	0.71	0.86
	St. Augustine	95	121	130	0.69	0.84
	Daytona Beach	97	121	137	0.77	0.93
	Ocala	95	100	130	1.01	NA
	Gainesville	93	100	125	0.94	NA
Central FL	Titusville	97	122	143	0.82	1.00
	Melbourne	98	127	151	0.85	1.03
	Fort Pierce	99	140	159	0.77	0.94
	Bradenton	100	127	147	0.80	0.97
	Tampa	99	116	141	0.89	1.07
	St. Petersburg	100	124	145	0.82	0.99
	Orlando	96	106	136	0.99	NA
	Lakeland	98	106	137	1.00	NA
South FL and Keys	West Palm Beach	105	144	169	0.83	1.00
	Boca Raton	108	142	167	0.83	1.01
	Key Largo	110	150	178	0.84	1.02
	Key West	110	150	180	0.86	1.05
	Naples	103	133	166	0.93	1.13
	Fort Myers	101	124	154	0.93	1.12
	Port Charlotte	100	122	150	0.91	1.10
	Sarasota	100	128	149	0.81	0.99

*Design wind speeds are taken from http://windspeed.atcouncil.org/

As expected, due to the uniform representation of hazard in the new design wind maps, the design velocity pressure for hurricane prone regions, which includes Florida, with no change in exposure assignment has become smaller in ASCE 7-10 than ASCE 7-05. However, the differences decrease in areas within 600 feet of the coast, where the exposure D is assigned.

2.2 *Changes to the Terrain Exposure Categories*

From a design perspective, engineers need to consider the effects of the local terrain around the location of a proposed building site. Depending on the site built environment, the wind field may be exposed to different local turbulent effects or surface drag effects. ASCE 7 defined four broad categories referred to as the "exposure categories" which are used to adjust the design pressures based on the local terrain, namely, Exposures A, B, C and D. The definitions from ASCE 7-98 (2000) are provided as follows:

Exposure A: Represents large city centers with at least 50% of buildings having a height in excess of 70 feet.

Exposure B: Represents a large spectrum of urban environments but generally is defined as urban and suburban areas, wooded areas or other terrain with numerous closely spaced obstructions having the size of single-family dwellings or larger. Exposure B is assumed the default unless the site meets the definition of another exposure type.

Exposure C: Open terrain with scattered obstructions having heights generally less than 30 feet including open country, grasslands, and shorelines in hurricane prone regions.

Exposure D: Flat, unobstructed areas exposed to flowing over open water (Excluding Hurricane prone regions) for a distance of 1 mile. Exposure D extends inland from the shoreline a distance of 1500 feet or 10 times the height of the building structure.

The incorporation of these exposure categories into the FBC added some geographical reference points and removed the Exposure D portion since Florida is a hurricane-prone state. Exposure A was rarely applicable and was entirely removed in later versions of the FBC. Exposure B was considered the default exposure unless the location met the requirements of Exposure C. Exposure C was geographically referenced to the coastline and represented the first 1500 feet from the "coastal construction control line" or within 1500 feet of the mean high tide line, whichever was less. Exposure C was also applicable to barrier islands in the coastal building zone set forth in the Florida statues 161.55(5) as well as the entire jurisdiction of Miami-Dade and Broward counties.

In ASCE 7-02 and subsequent versions, more clarity was provided and the exposure categories were supplemented with "surface roughness categories," which were intended to remove ambiguity. This did not significantly alter the application of the exposure category, and as such there were not significant regional changes in assigning one exposure over the other.

In 2010, ASCE 7-10 re-introduced Exposure D as an applicable exposure category for hurricane-prone regions along the East Coast and Gulf of Mexico. The previous exclusion from hurricane prone regions was due to the idea that during hurricanes, winds and surf conditions created significant roughness on the sea surface that was more in line with Exposure C than Exposure D. The improvement in aircraft reconnaissance and reanalysis indicated that this was not the case and the surface friction was more in line with Exposure D during hurricanes. So, while the wind pressures generally decreased along the coastline, the reintroduction of Exposure D compensated the reduction where Exposure D was applicable. In ASCE 7-10, Exposure D is applicable to locations at least within the first 600 feet from the coastline, as long as the water surface prevails 5000 feet in the upwind direction. The prevalent terrain exposure categories including the default categories as per FBC 2001 and FBC 2010 are shown in Figure 3 below.

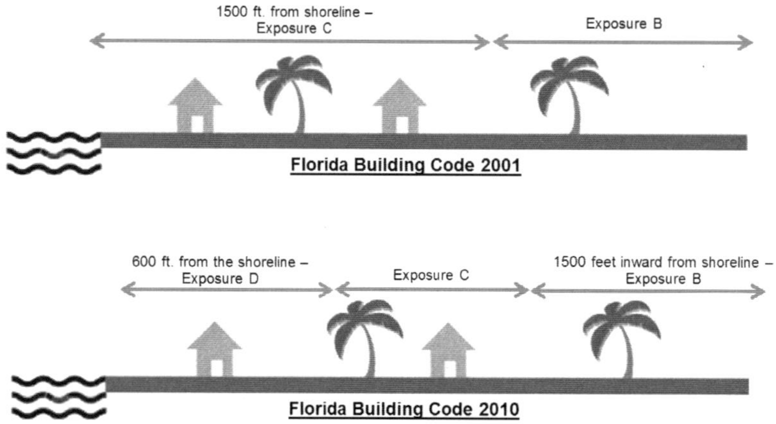

Figure 3. The prevalent terrain exposure categories including the default categories as per FBC 2001 and FBC 2010

2.3 *Changes to the Wind-borne Debris Region*

Changes to the wind maps also trigger changes to the WBDR, as the WBDR definition is based on the design wind maps. These are areas where the potential for damage due to flying debris is high. WBDR in FBC 2001 was defined as any area where the design wind speed is in excess of 120 mph, with the exception of Panhandle region, as well as the areas within one-mile of the coast where the design wind speed is above 110 mph. In FBC 2010, however, with the introduction of new design wind maps, the wind borne debris region is defined as the areas where the design wind speed is in excess of 140 mph as well as the areas within one-mile of the coast where the design wind speed is above 130 mph. This change significantly alters the wind-borne debris region in some areas of Florida, namely in the panhandle and southern part of the state (see Figure 4). FBC 2010 requires that all glazed opening be protected from the wind borne impact in the wind borne debris region. However, this was not a requirement until FBC 2004 Revision 2007, as long as the building was designed as a partially open structure.

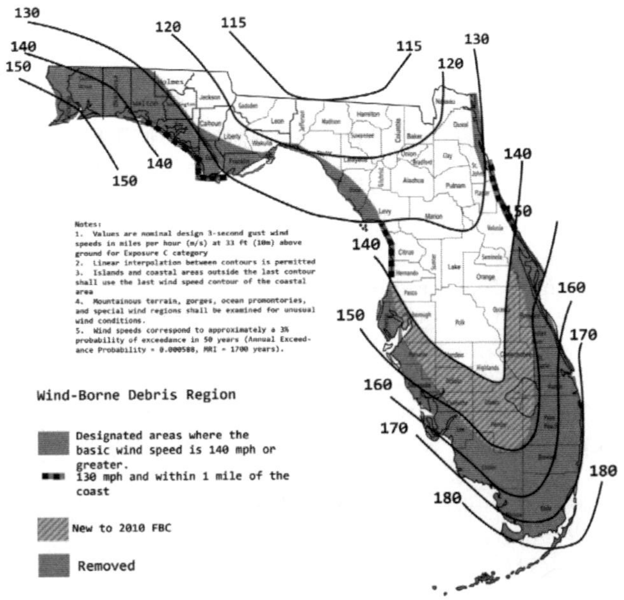

Figure 4. Changes introduced to the WBDR in FBC 2010 for Category II buildings [15].

2.4 *Changes to the other Wind Provisions*

In addition to the evolution of wind load provisions in the building codes, the wind resistant features also play a huge role in the vulnerability of the structures. Not only do these feature change over time, they are also location dependent. Below are some examples of these:

- In Miami-Dade and Broward counties, the minimum of six fasteners per strip shingle is the minimum requirement, while in the rest of Florida this is reduced to four
- For buildings located in the Miami-Dade and Broward counties, the whole building envelope should be impact resistant, while in the rest of Florida only the glazed openings should be protected in the wind borne debris region. The provision has become more stringent in FBC 2010, requiring all ventilation openings in an exterior wall into an attic space be protected against wind borne debris impact if the building is located in the wind borne debris region
- Miami-Dade and Broward counties have their own product testing standards

3 BUILDING VULNERABILITIES IN FLORIDA'S RESIDENTIAL BUILDING STOCK

Insights gained from storm resistant design and construction in Florida is nothing new. The earliest documented evidence of building vulnerability in Florida dates back to the insights gleaned from the 1926 South Florida Hurricane. Several building vulnerabilities that were identified from this event and detailed below continued to exist in the local building inventory until the time Hurricane Andrew made landfall in southern Dade County and caused over $15 billion in insured losses in Florida. Building vulnerabilities identified during the 1926 South Florida Hurricane include [18]:

- Light wooden truss roof systems and roof to wall connections
- Lack of attention to small detail, anchors, ties, bracing, and reinforcing

The fundamental conclusion that was drawn was the need to strengthen and better enforce adoption of building codes and the necessity to equip building departments and code officials to inspect and supervise workmanship.

While there are no notable references to building vulnerabilities post the 1926 South Florida Hurricane, the occurrence of the 1992 Hurricane Andrew was a watershed moment for building vulnerabilities and renewed emphasis on wind/hurricane design in Florida and around the country at large. Several building vulnerabilities were identified [16] including:

- Buildings designed and constructed with well-defined load paths performed fine. When these critical load transfer paths were not in evidence, damage ranged from considerable to total, depending on the type of architecture and construction
- Significant failures due to failure of exterior wall openings including windows, garage doors, sliding glass doors, and French doors leading to internal pressurization of the building envelope
- Catastrophic failures of light wood-frame buildings were observed more frequently in comparison to other site built construction types such as masonry. These total failures were primarily due to
    - Lack of bracing and load path continuity at wood-frame gable ends
    - Poor fastening and subsequent separation of roof sheathing from roof trusses
    - Inadequate roof truss bracing or bridging (see Figure 5 below)
    - Improper sill plate-to-foundation or sill plate-to-masonry connections

Figure 5. Roof failure due to inadequate bracing from the 1992 Hurricane Andrew [16]

- Failures in masonry wall buildings were usually attributable to one or more of the following:
    - Lack of or inadequate vertical wall reinforcing
    - Poor mortar joints between masonry walls and monolithic slab pours
    - Lack of or inadequate tie beams, horizontal reinforcement, tie columns, and tie anchors
    - Missing or misplaced hurricane straps between the walls and roof structure
- Loss of roof sheathing and consequent rainfall penetration through the roof magnified damage by a factor of five over that suffered by buildings whose roofs remained intact or suffered only minor damage [1]

The 2004 and 2005 hurricane seasons [20] served as a good natural laboratory to test the efficacy of building codes, primarily the FBC 2001 and the IBC and IRC. Building structural capacities appeared to have improved since Hurricane Andrew in 1992 because of stronger building codes and better enforcement, resulting in less structural damage overall even from intense hurricanes such as 2004's Hurricane Charley. Buildings designed to codes and standards that were revised after Hurricane Andrew in 1992 performed better than the older building stock. New wood frame houses built in accordance with FBC 2001 and the 2000 IRC performed well structurally, including those located in areas that experienced winds of up to 150 mph (3-second gust) in Charley. For each structure, load path was accounted for throughout the structure, including the connection of the roof deck to supporting trusses and rafters. Loss of roof decking on newer homes was rare. This is not to say that buildings were undamaged during the 2004-2005 hurricane seasons [17]. Winds primarily damaged building envelope components and accessory structures, as shown in Figure 6. Building envelope failure led to widespread damage to building interiors throughout the paths of the hurricanes.

During Hurricane Frances, which had wind speeds well below design levels, the long duration of the winds caused fatigue and subsequent damage to some building envelope systems. The majority of building damage was caused by:

- Intrusion of wind driven rain leading to significant content losses due to insufficient wind resistance of the building envelope
- Impact of wind borne debris

Both these observations lead to the conclusion that requirements and guidance related to load path for non-structural components and cladding elements still required improvement and have been addressed in subsequent versions of the FBC.

Figure 6. Damage to a post 2001 built structure's building envelope due to wind borne debris without any major structural damage during 2004 Hurricane Charley

4 CAT MODELING FRAMEWORK

The catastrophe modeling framework for the AIR Hurricane Model for the United States is shown in Figure 7. The hazard module includes both the event generation and local intensity calculation. The event generation element evaluates event location, severity, and frequency based on a number of uncertain source parameters with the goal of simulating a large number of possible yet realistic future scenarios. Once the possible events are generated, the local intensity element propagates the event across the affected area and evaluates the local intensity, in terms of, e.g., surface wind speed or storm surge depth.

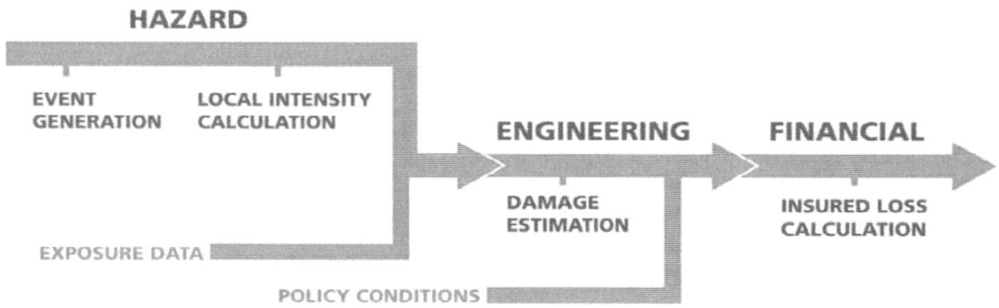

Figure 7. Components of AIR Hurricane Model for the United States [13].

Once the local hazard intensity is known the intensities of each event are superimposed onto a database of exposed properties and the engineering module is used to calculate the resulting monetary damage of the structures. This damage is calculated using construction and occupancy-dependent damage functions that relate the hazard intensity to a loss metric. Damage functions may also be referred to as vulnerability functions by structural engineers and loss functions by actuaries. In the present context, these functions illustrate the relationship between the intensity of the hazard - measured in terms of, e.g. the wind speed or storm surge depth - and the damage ratio, which is defined as the ratio of the building's repair cost to its total replacement value. Since the objective of catastrophe models is the estimation of future monetary losses that an insurer may pay in the aftermath of a catastrophe, it is necessary to account for not only the building type's general response in the damage functions, but also building code enforcement, construction practices, socio-economic effects, social preparedness, and claims adjustment and insurance practices, among other things. Needless to say, these resulting damage functions are region and time-specific in order to account for these additional aspects. They also provide an estimate of the mean damage ratio corresponding to each intensity as well as a complete probability distribution around the mean. This probability distribution around the mean is necessary to capture the so-called secondary uncertainty, or the variability in damage that is observed after catastrophic events when different structures experience different degrees of damage at the same hazard intensity.

While the focus of this study is entirely devoted to the engineering or damage estimation module shown in Figure 7, it is imperative to understand various techniques used to formulate damage functions [19-20]. These techniques can be broadly classified into three main types: The first one is rooted into an engineering based component framework, where the vulnerability of each component in the structure (such as the roofs, walls, connections etc.) is estimated using the load and resistance method in a simulation framework and then combined to yield the overall building vulnerability using combination rules that are either cost based or performance based. This method has some limitations, in the sense that it might not capture the effect of certain important factors on building vulnerability, such as the effect of the claim adjustment practices or losses due to trees falling on the buildings, as well as the complex interactions between component responses unless the underlying model explicitly accounts for progressive collapse. The second is to use insurance claims data and formulate damage functions statistically using the same after assigning a local intensity to each claim location. This method suffers from the fact that statistically significant and nominally identical data is often not being available to formulate damage functions for the range of various construction and occupancy types. Further, at higher intensities, the lack of data is even more pronounced, further leading to questionable estimates of vulnerability. The final method used to develop vulnerability functions is by combining the power of engineering based framework and statistical analysis of insurance claims data to develop damage functions to overcome the aforementioned shortcomings of both techniques. This last method is employed in AIR Hurricane Model for the United States, and therefore is the method utilized for developing the vulnerability functions used for this study.

The financial module is the last step in the catastrophe modeling framework and calculates insured losses by applying the policy conditions of the cedant, or the insurance company purchasing reinsurance, to the total damage estimates. Typically the catastrophe model reports the loss estimates in terms of their probability distribution or in terms of an exceedance probability curve. Other reported loss metrics include the expected annual loss, that is the mean of the annual loss distribution, also referred to as Average Annual Loss (AAL), which is a heavily used metric in the insurance industry.

5 CASE STUDY AND DISCUSSION

Although catastrophe models are widely used in the insurance industry to understand the hurricane risk from a regional perspective, this study explores their use for evaluating the temporal and spatial variability in building strength due to constantly evolving building codes and their corresponding enforcement practices, along with factors such as structural aging and building technology factors. This case study evaluates these aspects by comparing the AAL for individual buildings located across different regions and built in time periods in the state of Florida. Since, for a given location the hurricane hazard is the same, differences in AAL for buildings constructed at different times is used to evaluate the change in vulnerability of buildings due primarily to building codes, their corresponding wind load standards, and code enforcement. One story, wood frame, single family homes were modeled using AIR Hurricane Model for the United States in eleven locations throughout the state of Florida with year built values increasing in one year increments from 1994 to 2013. These eleven locations were strategically picked to encompass a range of design wind speeds, presence within the WBDR, and terrain categories (see Figure 8 and Table 2). The hurricane catastrophe model was then run through stochastic simulations that generate 50,000 realizations of the next hurricane season to obtain the wind only AAL for each building type. Although AIR Hurricane Model for the United States captures damage due to wind and storm surge, the wind only losses and the AAL estimates are considered in this study to correlate them with changes to building codes and other aspects discussed in the aforementioned sections of this paper. For a similar evaluation of several other coastal states please see ref. [12].

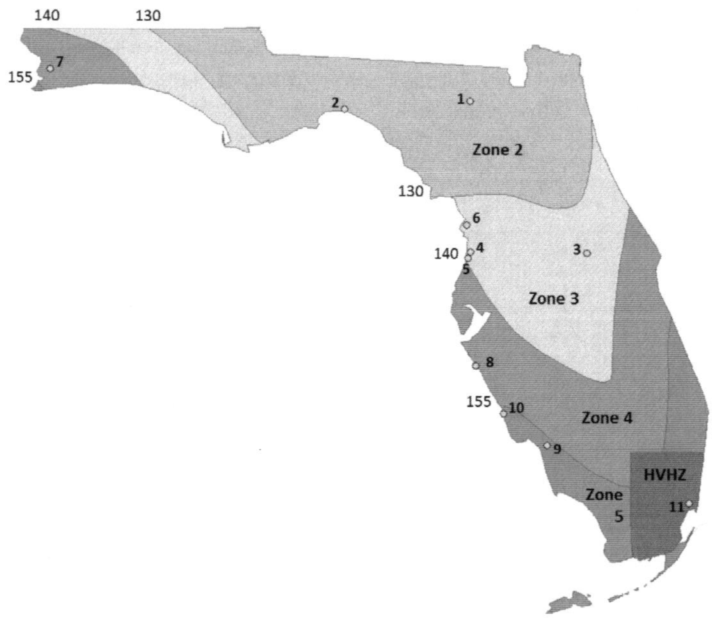

Figure 8. Eleven locations of structures used in case study with Post-2011 wind zone classifications

Table 2. Wind Zone Properties in Florida

Year-Built Band	Wind Zone	Wind Speed (mph)		WBDR	Terrain
		ASD	LRFD		
Pre-1995	3	80 - 90		N/A*	N/A*
	4	90 - 100		N/A	N/A
	5	>= 100		N/A	N/A
1995 – 2002	2	90 - 100		N/A	N/A
	3	100 - 110		N/A	N/A
	4	110 - 120		N/A	N/A
	5	>= 120		N/A	N/A
	HVHZ**	HVHZ		N/A	N/A
2002 – 2011	1	< 120		No	B
	2	>= 110		Yes	C
	3	>= 120		No	B
	4	>= 120		Yes	B
	5	>= 120		Yes	C
	HVHZ	HVHZ		Yes	C
Post-2011	2	89 – 101	115 – 130	No	B, C
	3	101 – 108	130 – 140	No	B, C
	3WP	101 – 108	130 – 140	Yes	B, C
	4	108 – 120	140 – 155	Yes	B, C
	5	>= 120	>= 155	Yes	B, C
	HVHZ	HVHZ	HVHZ	Yes	C

*N/A suggests that this parameter did not exist in the version of the building code that was implemented at a certain point in time.

**High Velocity Hurricane Zone (HVHZ) includes Miami-Dade and Broward Counties.

The percent change in AAL for the co-located properties designed to different levels as dictated by the building codes at the time of the selected year of construction is shown in Table 3. Negative percent changes indicate that the newer building is less vulnerable than the older building at the same location. It is obvious for all transitions at the given locations that the newer structures were built under more stringent and better enforced building codes and therefore are less vulnerable. Expected losses significantly decrease between the Pre-1995 (1) and 1995-2002 (2) year built bands for the locations chosen in this case study. This is due to more stringent building codes being adopted in Florida after the devastation caused by Hurricanes Andrew and Hugo.

Table 3. Wind zones and percent change in AAL for wood frame single family homes in Florida

Location	Zones By Year-Built Band				Percent Change in AAL Between Building Code Transition Years		
	Pre-1995 (1)	1995-2001 (2)	2002-2011 (3)	Post 2011 (4)	1994 to 1995	2001 to 2002	2011 to 2012
1	3	2	1	2 B	-17.74%	-33.18%	0.00%
2	4	4	2	2 C	-30.23%	-39.14%	0.00%
3	4	3	1	3 B	-17.44%	-34.35%	-8.27%
4	4	4	1	3 C	-29.60%	-21.34%	-27.49%
5	4	5	1	3 WP B	-29.58%	-21.37%	-22.55%
6	4	4	2	3 WP C	-30.44%	-39.05%	-23.11%
7	4	5	3	4 B	-32.32%	-26.82%	-28.98%
8	5	5	5	4 C	-28.18%	-42.24%	-27.20%
9	5	5	4	5 B	-31.73%	-27.50%	-45.17%
10	5	5	5	5 C	-28.69%	-42.86%	-44.83%
11	5	HVHZ	HVHZ	HVHZ	-60.92%	-40.43%	-32.94%

While the results presented in Table 3 elicit the temporal variation in vulnerability, Figure 9 helps look at the spatial variation in vulnerability across Florida for residential homes built post 2011. The figure shows vulnerability functions for wood frame single family homes built in accordance with FBC version 2010 located across Florida. In catastrophe modeling, damage functions—or, more precisely, mean damage functions—relate the hazard intensity to the mean damage ratio (MDR), which is the expected value of the ratio of the repair cost of the asset to its replacement value [22]. Increased design wind speeds along the coast makes these buildings less vulnerable in comparison to their inland counterparts within the same building code era. This is seen in Figure 9 when comparing damage functions for homes in HVHZ with respect to zones 5, 4, 3, 2, respectively, in increasing order of vulnerability. Similarly, buildings designed as per FBC 2010 have lower vulnerabilities when designed to terrain exposure C in comparison to exposure B. Changes in window protection requirements typically see significant changes in structural vulnerability. Significant decreases in vulnerability were observed in cases where window protection is required as per FBC 2010 (as in zones 4, 5, HVHZ) but not in other areas of the state (zones 2, 3).

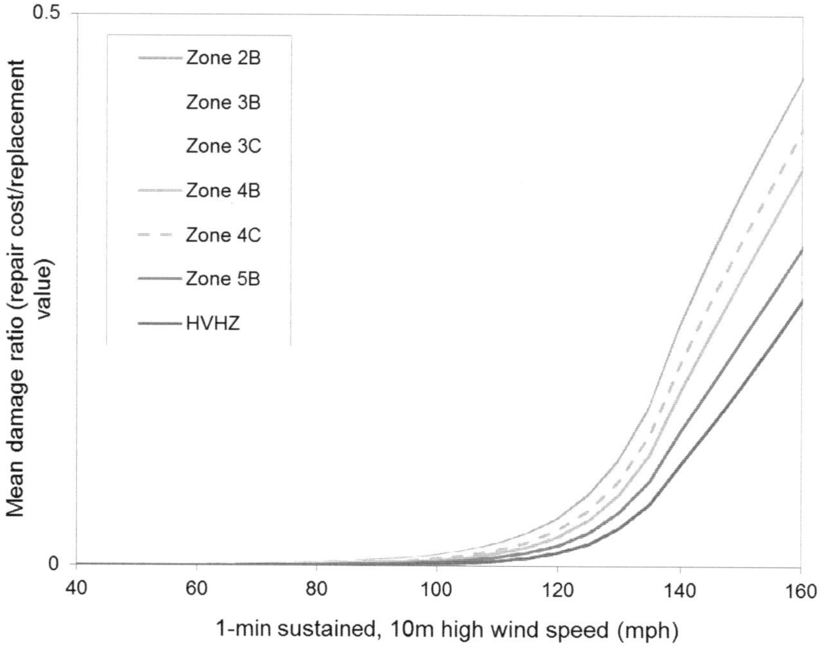

Figure 9. Damage functions for post 2011 built wood frame single family homes in Florida by zone (see Figure 8 and Table 2 for zonal descriptions)

Although a decrease in vulnerability is exhibited throughout Florida due to code changes, it should be noted that this observation should not necessarily be expected in other states. For example, a number of locations on the coast in Virginia actually saw a vulnerability increase with the adoption of the 2012 ICC codes due to significantly decreased wind speeds and a corresponding removal of the WBDR along the coast. As the focus of this study is on Florida, detailed information about other states is not provided here; however more information about vulnerability changes expected in other states may be found in [12].

It is also important to note that structural aging along with newer and better construction practices and building material based technologies will cause newer structures to perform better than older structures built under the same building code. This is clearly seen in Figure 10 below, which shows the vulnerability based on the normalized AAL for a structure located in Orlando with year-built values ranging between 1994 and 2013. Normalization is done by dividing the AAL for a given year of construction by the AAL for a building built in 1994. From 1995 to 2001, the Standard Building Code was used in Orlando and while the vulnerability of the structures built in this time period is similar, it is obvious that even within this unique year-built band the older structures are more vulnerable based on normalized AAL. This observation is also evident for the structures built in the 2002-2011 era. It is additionally clear that the vulnerability changes suddenly between 1994 and 1995, between 2001 and 2002, and between 2011 and 2012. This is due to the fact that new building codes with significant changes to wind design sections came into effect during the years of 1995, 2002, and 2012.

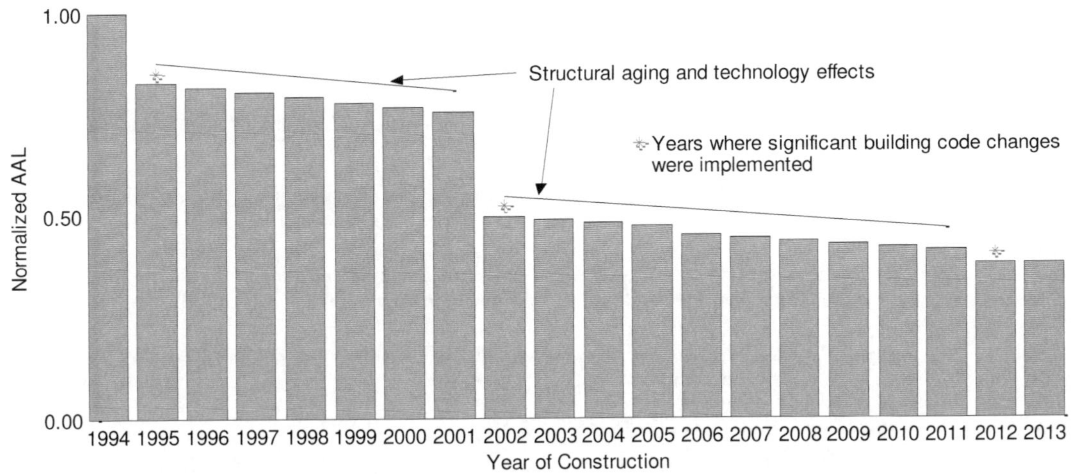

Figure 10. Normalized AAL for structure located in Orlando by Year-Built (Normalized to 1994 AAL)

6 CONCLUSIONS

Continually evolving building codes and their corresponding wind load standards in Florida lead to changes in the inherent wind vulnerability of the built environment. The recent updates to the wind hazard and design criteria in the 2010 Florida Building Code and referenced ASCE 7-10 standard lead to a decrease in design wind pressures in Florida. The design wind speed changes also affected related wind design requirements, including wind borne debris regions.

This paper discussed these changes across time and their potential impact on the hurricane wind vulnerability of structures in the state of Florida. To gain an understanding of these vulnerability impacts a case study was presented that compared a very commonly used actuarial metric called the average annual loss (AAL) calculated for one story, wood frame, single family homes built in different time periods and at different locations in the state. The AIR Hurricane Model for the United States, which incorporates vulnerability impacts due to different building codes, enforcement, and construction practices, was used to estimate these AALs. Moderate to larger changes in vulnerability in a number of locations were observed. In the locations tested, the newer codes specified stronger buildings and a decrease in building vulnerability was generally observed. Changes in window protection requirements due to modifications of the wind borne debris regions generally caused notable changes in structural vulnerability. Less drastic vulnerability changes also occurred in other locations, primarily due to less drastic design wind speed changes.

7 REFERENCES

Journals

1 P.R. Sparks, S.D. Schiff, T.A. Reinhold, 1994. Wind Damage to Envelopes of Houses and Consequent Insurance Losses, J. Ind. Aerodyn., 53(1994) 145-155.

Monographs, Multi-author volumes, Proceedings

2 2001 Florida Building Code, International Code Council (ICC), 2002.
3 2004 Florida Building Code, International Code Council (ICC), 2005.

4 2007 Florida Building Code, International Code Council (ICC), 2009.
5 2010 Florida Building Code, International Code Council (ICC), 2012.
6 ASCE 7-88, Minimum Design Loads for Buildings and Other Structures, American Society of Civil Engineers, Reston, VA, 1990.
7 ASCE 7-98, Minimum Design Loads for Buildings and Other Structures, American Society of Civil Engineers, Reston, VA, 2000.
8 ASCE 7-10, Minimum Design Loads for Buildings and Other Structures, American Society of Civil Engineers, Reston, VA, 2010.
9 National Hurricane Center (NHC), HURDAT2, accessed 2016 http://www.nhc.noaa.gov/data/#hurdat
10 Florida Building Commission Report to the 2008 Legislature, Florida Department of Community Affairs, Tallahassee, FL, 2008.
11 Rating the States 2015: An Assessment of Residential Building Code and Enforcement Systems for Life Safety and Property Protection in Hurricane-Prone Regions. Insurance Institute for Business & Home Safety (IBHS), 2015, Tampa, FL.
12 S. Bobby, T. Johnson, B. Kordi, F. Moghim, and K. Ramanathan, The Impact of Temporal and Spatial Variation in Building Codes on the Hurricane Wind Vulnerability of Residential Single Family Homes, Structures Congress 2017, Denver, CO, April 6-8 2017.
13 AIR Hurricane Model for the United States, AIR Worldwide, Boston, MA, 2016.
14 Building Codes. Federal Emergency Management Agency (FEMA), 2016, https://www.fema.gov/building-codes
15 Changes to the Wind Speed Maps and Wind Design – 2010 Florida Building Codes, Building a safer Florida, Inc., http://www.floridabuilding.org/fbc/Wind_2010/Flyer_Wind_January2012.pdf
16 Building Performance: Hurricane Andrew in Florida, Observations, Recommendations and Technical Guidance, FIA-22, Federal Emergency Management Agency (FEMA), 1993.
17 Mitigation Assessment Team Report: Hurricane Charley in Florida, FEMA 488. Federal Emergency Management Agency (FEMA), 2005.
18 T. Eefting, Structural Lessons of the South Florida Hurricane, Florida Engineer and Contractor, September 1927, pp. 162-170.
19 H. He, V. Jain, G. Leiva, and C. Kafali, Modeling Relative Spatial and Temporal Hurricane Wind Vulnerability of the Building Stock in the US, 11th Americas Conference on Wind Engineering, San Juan, Puerto Rico, June 22-26 2009.
20 V.K. Jain., J. Guin, and H. He, Statistical Analysis of 2004 and 2005 Hurricane Claims Data, 11th Americas Conference on Wind Engineering, San Juan, Puerto Rico, June 22-26 2009.
21 1999 Standard Building Code, Southern Building Code Congress International, Inc., 1999.
22 K. Ramanathan, Anatomy of a Damage Function, AIR Currents, April 2017.

The 13th Americas Conference on Wind Engineering (13ACWE)
Gainesville, Florida USA, May 21-24, 2017

Vulnerability analysis of steel roofing cladding considering wind directionality

Xiaowen Ji [a], Guoqing Huang [a], Xinxin Zhang [b], Gregory A. Kopp [c]

[a] *School of Civil Engineering, Southwest Jiaotong University, Chengdu, China*
[b] *Berkshire Hathaway Specialty Insurance, Boston, MA, USA*
[c] *Boundary Layer Wind Tunnel Laboratory, Faculty of Engineering, University of Western Ontario, London, London, ON, Canada*

ABSTRACT: Steel roofing is widely used for non-residential facilities, however, very vulnerable to high winds. The paper addresses a damage estimation framework considering wind loading correlation and wind directionality effects. In this framework, external pressures were measured from wind tunnel testing. A proper orthogonal decomposition (POD) method is introduced to interpolate external wind pressures. Internal pressures due to openings in the building envelope are taken into account by simulation. Then, the internal forces on fasteners distributed on the steel roof are evaluated, with corresponding peak values estimated by a Gumbel conversion approach. Furthermore, the failure probability of a single cladding element and the damage ratio for the whole roof are determined based on Monte Carlo simulation (MCS), where the correlation among internal forces of fasteners is incorporated by a Nataf transformation. Finally, wind directionality effects are integrated in order to provide a comprehensive damage assessment for the roofing.

KEYWORDS: Wind damage estimation; Steel roofing; Internal pressure; Correlation; Wind directionality

1 INTRODUCTION

Metal structures are widely used for low-rise buildings, especially for non-residential buildings. Based on the statistics from Metal Building Manufacturers Association (MBMA), approximately 65% of non-residential low-rise buildings are built with metal structures in USA (e.g., Ref. [1]). Among these metal structures, lightweight steel structures represent a significant proportion and are popular for warehouses, sheds, airplane hangars and industrial buildings, which are vulnerable during hurricanes (or typhoons or tropical cyclones), thunderstorms and tornadoes (e.g., Refs. [2,3]). For example, typhoon "Rananim" in 2004 devastated industrial buildings in Zhejiang Province, China, including collapsed area of 2.72 million m^2 and damaged area of 7.56 million m^2 [4].

Post-event damage surveys have indicated that the majority of damage to steel structures is related to the breach of the envelope instead of the collapse of the main frame [5]. The breach of the roofing not only introduces losses to the building itself, but also triggers further damage to interior contents due to secondary perils. Additionally, business interruption increases indirect losses of income, which is a common concern for the insurance sector. Therefore, it is important to analyze and predict wind-induced damage for steel roofing in order to conduct damage mitigation and risk management assessments.

Damage analysis of low-rise buildings, especially wood-frame structures, under high winds has received significant attentions from the engineering community. Lee & Rosowsky [6] assessed the wind-induced fragility of roof sheathing for light wood-frame structures. Li & Ellingwood [7] proposed probabilistic risk assessment methods to evaluate performance and reliability of low-rise light-frame wood residential constructions in hurricane-prone region of the United States, where the importance of uncertainties is highlighted. Recently, aerodynamic databases have been applied to wind damage assessments. Zhao & Gu [8] presented a data-based wind vulnerability assessment model for metal buildings. Huang et al. [9] introduced a data-based probabilistic damage estimation approach for asphalt shingle roofing. Konthesingha et al. [10] developed a vulnerability model for metal-clad industrial building in a tropical cyclone region. Huang et al. [11] developed a damage estimation method for roof panels where the wind loading correlation was taken into account.

Despite these achievements, there is a need to develop a data-based integrated approach to incorporate important factors such as wind loading correlation and wind directionality effects. Cope et al. [12] showed that the correlation of the surface pressures varied with direction and became strong under quartering winds and winds perpendicular to the roof gable. Huang et al. [11] had found that the wind loading correlation may significantly influence the standard deviation (STD) of the damage ratio for roof panels. Although directionality effects have been widely recognized and incorporated in structural and cladding design, it has not yet been well addressed in vulnerability studies. To develop a comprehensive understanding of structural vulnerability, and for the sake of wind-induced damage mitigation and risk management, one needs to integrate the vulnerabilities of all directions with the local wind speed distribution within a framework that also considers wind load correlations and other influential factors.

Based on an illustrative low-rise building model whose wind pressure data were measured in a wind tunnel study, a wind damage estimation method incorporating the wind loading correlation and wind directionality for steel roofing is addressed in this paper. First, descriptions of the steel roofing and wind pressure data are introduced. Second, POD is adopted to interpolate the external wind pressure for roof locations where there are no pressure data. Third, the internal pressure is determined by the simulation. Fourth, the internal forces on fasteners distributed on the steel roof are computed with the aid of the influence-surface approach, with the corresponding peak internal forces estimated by a Gumbel conversion method. Fifth, the failure probability of a single panel and the damage ratio of whole roof are determined based on MCS, where the correlation among internal forces on fasteners is considered by a Nataf transformation. Sixth, the influence of wind directionality is incorporated in the damage estimation. In the end, concluding remarks are given.

2 DESCRIPTIONS OF WIND PRESSURES AND STEEL ROOFING

The illustrative prototype industrial building used in this study has a full-scale size of 19.05 m × 12.2 m × 3.66 m, a roof slope of 1:12, and is assumed to be located in suburban terrain. The wind pressure data were obtained from wind tunnel tests conducted at the University of Western Ontario (UWO), as reported by Ho et al. [13]. The model scale was 1:100, with 335 taps distributed on the roof top. These are shown as dots in Figure 1. The sampling frequency was 500 Hz with a sampling time of 100 s. The tests were conducted in suburban terrain with roughness length of about 0.3 m, under a reference wind speed of 13.7 m/s at the equivalent of 10 m above the ground which corresponds to a wind speed of 6.1 m/s at the roof height (3.66 m). The tests were carried out at various wind angles of attack (AOAs) with intervals of 5° from 0° to 90° and from 270° to 360°.

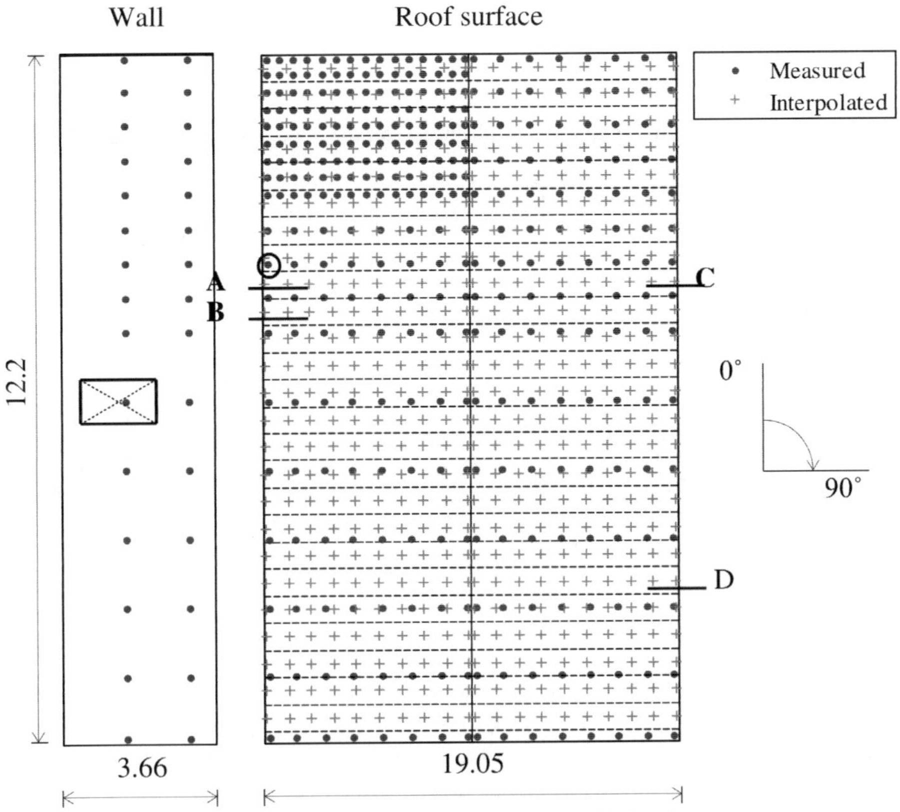

Figure 1. Tap locations and panel layout (Unit: m)

There are many types of steel cladding profiles that are commonly used in construction, such as trapezoidal and corrugated steel claddings. Due to the requirements of large spans, low price, and simplicity of construction, the high-strength trapezoidal steel cladding with closely spaced ribs, which is widely used as roofing cladding on low-rise, industrial buildings [14], is selected as the cladding type for this study. In the current study, the size of a single cladding panel is assumed to be 750 mm × 6096 mm with a thickness of 0.6 mm. The height of crest is 35 mm and ribs are closely spaced with an interval of 125 mm between two neighboring crests. The layout of the cladding on the roof is shown in Figure 1 where 50 (2 × 25) steel cladding panels are distributed on the roof. The cladding is made of high-strength steel G550 (yielding stress = 690 MPa). Self-tapping screws with head diameters of 11 mm are assumed as the connection fasteners. Further details can be found in Ref. [14].

For the cladding cross section, four screws are used to connect the roof cladding to the purlins at alternate crests. Along the rib, four screws are also uniformly distributed. The schematic description can be found in Figures 2a and b. The locations of screws on a cladding panel are represented by the coordinate system shown in Figure 2a. For example, x2y3 denotes the screw at the intersection of line x2 and line y3 on the panel. Two adjacent cladding panels overlap at marginal crests and share common screws, as shown in Figure 2c.

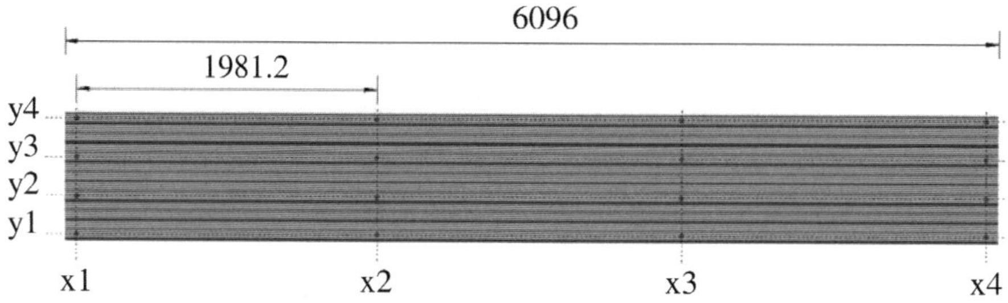

(a) Cladding configuration and screw distribution

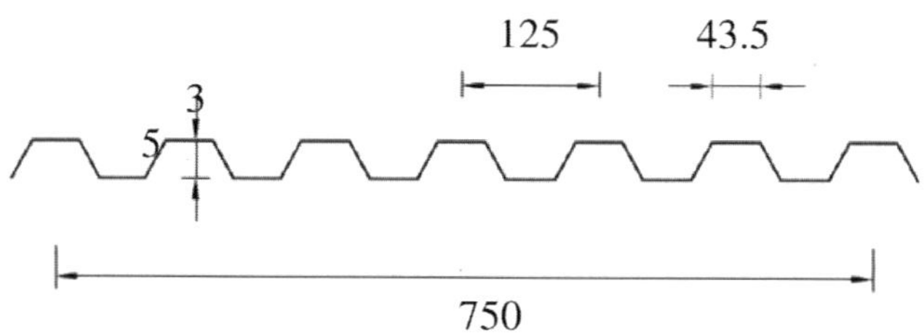

(b) Cross section

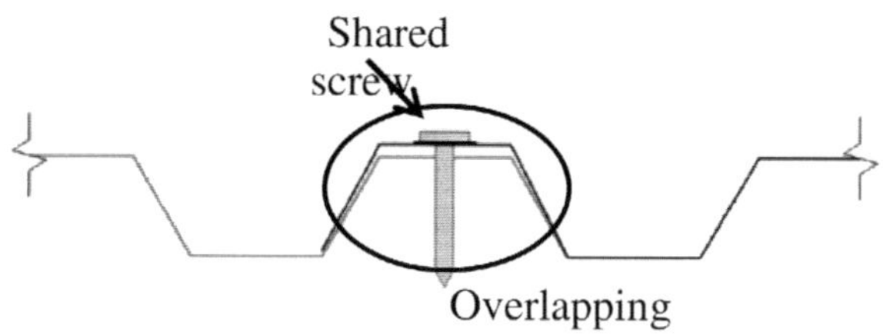

(c) Overlap

Figure 2. Configuration, dimension and joint type of cladding (Unit: mm)

3 POD-BASED WIND PRESSURE INTERPOLATIONS

To determine the wind-induced internal forces on a screw, ideally, the associated cladding panels should be assigned sufficient pressure taps in the wind tunnel experiment. However, not every panel satisfies this requirement, as shown in Figure 1. It can be seen that, for many panels, there no pressure taps. In order to evaluate the internal forces on the screws, ten proxy taps are evenly assigned along the central line of the each of the panels (see the "+" in Figure 1). The pressures at these locations are interpolated from existing taps. In the present study, POD will be used to reconstruct the wind pressure field on the low-rise building roof due to its adaptivity and easy implementation.

If one assumes that $\mathbf{C_P}(t) = \{C_{P_1}(t), C_{P_2}(t), \cdots, C_{P_N}(t)\}^T$ is a zero-mean N-variate fluctuating wind pressure coefficient vector, where N is the number of taps in the wind tunnel testing, POD can be used to find a set of optimal orthonormal basis vectors $\mathbf{\Theta} = [\Theta_1, \Theta_2, \cdots, \Theta_N]$ and $\mathbf{C_P}(t)$ can be expanded as

$$\mathbf{C_P}(t) = \mathbf{\Theta a}(t) = \sum_{i=1}^{N} \Theta_i a_i(t) \tag{1}$$

where the component $a_i(t)$ is the projection of $\mathbf{C_P}(t)$ on the basis vector Θ_i, $i = 1, 2, \cdots, N$. Note that the mean component has been removed as discussed by Tamura et al. [15]. The basis vector set $\mathbf{\Theta}$ can be determined from the eigenvalue equation (Bienkiewicz et al. [16])

$$\mathbf{R_p A \Theta} = \mathbf{\Theta \Lambda} \tag{2}$$

where $\mathbf{R_p}$ is the covariance matrix of $\mathbf{C_P}(t)$; $\mathbf{A} = diag(A_1, A_2, \cdots, A_N)$ and A_i is the tributary area of the i th pressure tap; $\mathbf{\Lambda} = diag(\lambda_1, \lambda_2, \cdots, \lambda_N)$ is diagonal matrix and λ_i is i th eigenvalue. If the pressure taps are distributed uniformly and tributary areas associated with different taps are identical, then Eq. (2) is rewritten as

$$\mathbf{R_p \Theta} = \mathbf{\Theta \Lambda}^{\#} \tag{3}$$

where $\mathbf{\Lambda}^{\#} = diag(\frac{\lambda_1}{A}, \frac{\lambda_2}{A}, \cdots, \frac{\lambda_N}{A})$ and A is the tributary area. If the pressure taps are distributed non-uniformly, Eq. (2) can be pre-multiplied by $\mathbf{A}^{1/2}$ and an equivalent form is given as (Jeong et al. [17]).

$$\mathbf{R_p^* \Theta^*} = \mathbf{\Theta^* \Lambda} \tag{4}$$

where $\mathbf{R_p^*} = \mathbf{A}^{1/2}\mathbf{R_p A}^{1/2}$ and $\mathbf{\Theta}^* = \mathbf{A}^{1/2}\mathbf{\Theta}$. Note that the modification guarantees that the transformed $\mathbf{R_p^*}$ is a real and symmetric matrix, and allows the Cholesky decomposition. The target orthogonal basis vector then can be obtained through the inverse transformation

$$\mathbf{\Theta} = \mathbf{A}^{-1/2}\mathbf{\Theta}^* \tag{5}$$

It is clear that the eigenvalue matrix Λ is unchanged in the transformation.

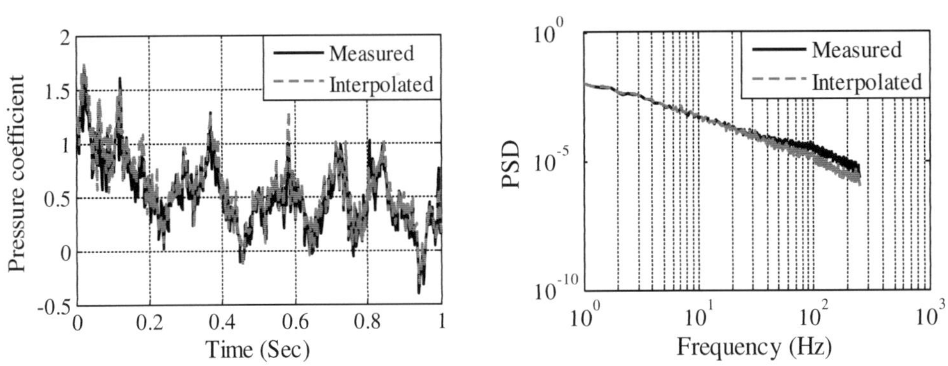

(a) Time histories of fluctuations (b) PSDs of fluctuations

Figure 3. Comparison of measured fluctuation and its interpolated counterpart at an existing tap (AOA of 315°)

Once the value of the basis vector for the proxy tap is estimated by the spatial interpolation, the fluctuation of the wind pressure at this tap can be interpolated from non-uniformly distributed taps. The interpolated fluctuation of the wind pressure coefficient at an existing tap (highlighted by a circle in Figure 2) is compared with the measured counterpart in Figure 3. Consistency can be found in time histories and lower frequency part of power spectral densities (PSDs). Similar results can be found for other taps. By the spatial interpolation, the mean component of the wind pressure coefficients at the proxy taps is obtained.

4 INCORPORATION OF INTERNAL PRESSURES

Internal pressure appears because of the leakage or openings of a building in response to external wind pressures. Because internal pressure has a significant influence on the structural cladding system, it has been investigated by many scholars (e.g., Refs. [18,19]).

Clearly, internal pressures are dependent on the damage status of the building envelope and cladding. In reality, the wind-induced damage on the building envelope and cladding is a progressive process and the determination of internal pressure is challenging in that additional wind tunnel experiments are usually required. To simplify the damage process, in this study, buildings under high winds are assumed to undergo the following three stages. Stage 1: Enclosed building. In the beginning, the building is nominally sealed with background leakage. In this case, the internal pressure is usually negative (i.e., suction) and increases the load on the windward wall and reduces the wind load on the roof. Stage 2: Partially-enclosed building. The wall, especially a door or window, may be broken due to wind-borne debris-impacts or larger net pressure loads. Loss of these components causes a dominant opening on the wall and makes the building partially enclosed. At this stage, the internal pressure usually becomes positive, which leads to an increase of the net wind loads on the roof and makes roof cladding more vulnerable to wind. Stage 3: Roof cladding loss or component failures. At this stage, portions of the roof panels may be removed, causing new openings on the roof. Accordingly, the internal pressure may decrease, to some extent. Because the roof cladding tends to be most vulnerable at the second stage, partially-enclosed buildings are studied herein. The related internal pressure can be obtained by either wind tunnel tests (such as the UWO data) or via simulation, with similar accuracy [19]. In this study, the latter will be adopted.

For a single opening in the wall, the governing equation for determining the internal pressure within the building volume is obtained from the unsteady form of the Bernoulli equation. This equation, which represents the conservation of energy of the flow through the opening, is (e.g., Refs. [18,20]):

$$\frac{\rho l_e V}{\gamma a P_0}\ddot{C}_{pi}+\left[\frac{\rho V \bar{U}_h}{2\kappa\gamma a P_0}\right]^2 \dot{C}_{pi}\left|\dot{C}_{pi}\right|+C_{pi}=C_{pe} \tag{6}$$

where ρ is the air density; l_e is the effective length of an "air slug"; V is the internal volume; γ is the ratio of specific heat capacities of air; a is the opening area; P_0 is the atmospheric pressure; $\bar{U}_h$ is the mean wind speed at reference height; κ is the discharge coefficient compensating for various energy losses; C_{pe} and C_{pi} denote the external and internal pressure coefficient, respectively. According to Vickery & Bloxham [21], in the present study, $l_e=\sqrt{a\pi}/2$

and $\kappa = 0.6$ will be adopted. It should be mentioned that the effect of envelope shear stresses is not incorporated here. However, it must be considered for narrow and long openings. Because Eq. (6) has the form of a spring-mass-damper system, the internal pressure has a natural frequency that is known as the Helmholtz frequency, $\frac{1}{2\pi}\sqrt{\frac{\kappa\gamma a P_0}{\rho l_e V}}$, which is the resonant frequency of the internal pressures (e.g., Refs. [18,19]).

Similar to Oh et al. [19], a rectangular dominant opening is assumed on one wall (see Figure 1). The ratio of the opening area to the wall area is assumed to be 3.2%, while the building porosity is considered to be negligibly small for the current study. While the current study is illustrative, neglecting the leakage should be reasonable for the damage estimation when one considers the other uncertainties. A 4th-order Runge-Kutta method can be used to solve the above nonlinear differential equation. The simulated internal pressure C_{pi} is presented together with a corresponding external pressure C_{pe} in Figure 4a. Note that the external pressure has been multiplied by -1 such that suction coefficients are presented. It can be seen internal pressures are almost coincident with the external pressures for this particular case with the amplitude of the internal pressures being slightly larger than that of the external pressure. PSD functions are also given in Figure 4b. The peak in PSD of C_{pi} is the Helmholtz frequency, which is 2.4 Hz in full scale. Together with (interpolated) external pressures on the roof, the simulated internal pressure will be used to estimate the net wind load on the cladding.

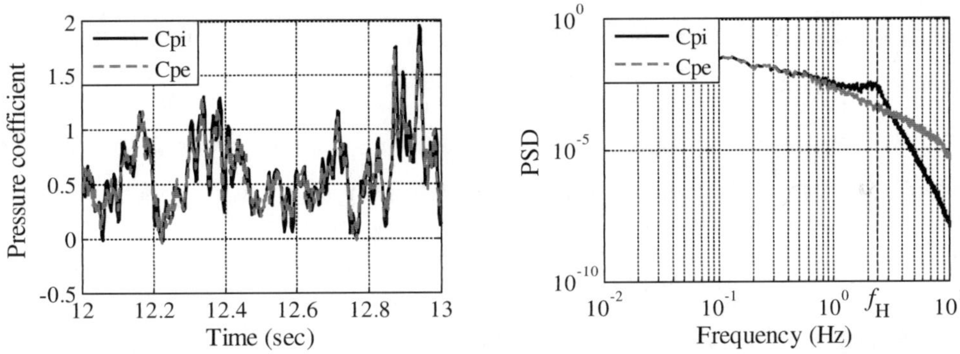

(a) Time series (b) PSD

Figure 4. Time series and spectra of internal and external pressure coefficients for 30 m/s and 315°

5 DETERMINATION OF INTERNAL FORCES AND THEIR PEAKS ON SCREWS

High wind speeds can cause large uplift forces on roof cladding, and these forces will be transferred to the self-tapping screw fasteners. As a consequence, dimpling around a screw arises initially and cracks appear. Eventually, the steel cladding will be pulled through the screw connection (i.e., pull-through failure). It should be noted that for cladding systems on metal buildings, failures due to low-cycle fatigue (e.g., Refs. [22,23,24]) are not considered in the current methodology and analysis; rather, failures mainly due to stress concentrations or internal forces in the screws are considered in the development of the method.

According to Mahaarachchi & Mahendran [14], the internal forces of screws on any cladding panel can be estimated individually on a single- or half-cladding-panel basis. In this study, to improve the calculation efficiency, the (linear) influence coefficient approach is used to estimate the internal forces on a screw (e.g., Refs. [24,25]). Accordingly, the internal force of a screw can be obtained as follows

$$X(t) = \iint q(x, y, t) I_c(x, y) dxdy \tag{7}$$

where $I_c(x, y)$ is the internal-force, influence coefficient at position (x, y); $q(x, y, t)$ is the corresponding net pressure (summation of the external and internal pressures). Because the layout of screws is the same for all cladding sheets and each cladding sheet is bi-symmetric, only the internal force influence coefficients for 4 screws are needed.

Usually, the duration of a pressure time history of wind pressure from a wind tunnel is 20 to 60 min in full scale. The same length is applied to the internal forces on the screw. A Gumbel conversion method, which was introduced by Cook & Mayne [26], can project Gumbel parameters for a given longer interval based on those determined from a shorter interval. Numerous experimental results have validated the appropriateness of the Gumbel distribution in modeling the peak wind pressure (e.g., Refs. [26,27]). The conversion method is appropriate to estimate PDFs of peak internal forces for hundreds of screws due to its relative simplicity, accuracy and efficiency.

Let W_{T_1} denote the peak value of the internal force process over T_1 min, say, 1 min. Suppose its cumulative distribution function (CDF) follows Gumbel distribution,

$$\Psi_{W_{T_1}}(w) = \exp\{-\exp[-\alpha_{w_{T_1}}(w - \hat{w}_{T_1})]\} \tag{8}$$

where the mode $\hat{w}_{T_1}$ is the most likely value and $1/\alpha_{w_{T_1}}$ is the dispersion. If the CDF of the peak value W over T min, say, 10 min or 1 h, is given as

$$\Psi_W(w) = \exp\{-\exp[-\alpha_w(w - \hat{w})]\} \tag{9}$$

If peak values from different subsamples over T_1 min are mutually independent, then the peak value distribution over T min can be computed as (e.g., Ref. [26])

$$\Psi_W(w) = [\Psi_{W_{T_1}}(w)]^{T/T_1} = \exp\{-\exp[-\alpha_{w_{T_1}}(w - \hat{w}_{T_1}) + \ln(T/T_1)]\} \tag{10}$$

The mode and dispersion parameters are given as

$$\hat{w} = \hat{w}_{T_1} + \ln(T/T_1)/\alpha_{w_{T_1}}; \quad \alpha_w = \alpha_{w_{T_1}} \tag{11}$$

The product $\Pi_w = \alpha_w \hat{w}$ is a dimensionless characteristic product.

In this paper, the internal force time history is divided into subsamples with intervals of 1 min, under various wind speeds. For instance, the number of subsamples is about 33 under 30 m/s. Then peak values from subsamples are fitted by Gumbel distribution. And the CDF of peak values over 1 min is converted to that over 10 min via Eqs. (10) and (11).

6 CONSIDERATION OF CORRELATION FOR PEAK INTERNAL FORCES

6.1 *Derivation the relation between correlations for peak forces over 1 and 10 min*

Cope et al. [12] illustrated that the high correlation of the surface pressures is induced by quartering winds and winds perpendicular to the roof gable. Subsequently, highly correlated wind pressures may lead to high correlations among internal forces on screws. Huang et al. [11] showed that highly correlated pressures could cause larger variation of the damage ratio, or higher risk for roof components. Hence, the correlation among the peak internal forces should be considered for estimates of the wind-induced damage on steel cladding. However, it is difficult to determine the correlation for peak values over 10 min directly using the current UWO data with such limited length (around 20-40 min in the full scale).

To make the proposed framework feasible, an alternative to approximately estimate the correlation for peak internal forces over 10 min will be introduced. In this alternative, the relationship for the correlations of peak internal forces over 1 min and 10 min is assumed to be equivalent to that for peak wind pressures.

The very long wind pressure data used to derive the relation between correlations for peak wind pressures over T_1 = 1 min and T = 10 min were obtained from Boundary Layer Wind Tunnel II at UWO. Two 1:50 scale house models, FL27 and FL30, located in the suburban terrain under different AOAs and surroundings were tested. The sampling frequency for model scale is 400 Hz and the sampling time for model scale is 3 h. Based on the assumption that a full-scale wind speed at 10 m height in suburban exposure is 31.7 m/s, equivalent to a wind speed = 24.2 m/s at the mean roof height of 4 m (The mean roof heights of both houses are roughly 4 m), the velocity scale is 1:5, the sampling frequency and time in full scale are 40 Hz and 30 h, respectively. More details can be found in Ref. [28].

For each tap, the very long data (30 h) are divided into 1800 segments of 1-min duration and 180 segments of 10-min duration in the following analysis. Correlation coefficients among peak pressure coefficients over 1 min and 10 min are estimated for all test conditions. Because the sampling frequency and time in full scale vary with the different wind speed, these variations may influence the correlation of peak values. Thus correlation coefficients under wind speeds of 12.1 and 36.3 m/s at the roof height are also investigated. Figure 5 shows the relationship between correlation coefficients over two time intervals for two test conditions (FL27, 120°, isolated and FL30, 130°, surrounded) under three different wind speeds, where ρ_T and ρ_{T_1} are correlation coefficients for peak values over T and T_1 min. It can be seen that the relationships are similar for different building models, AOAs, surroundings and wind speeds.

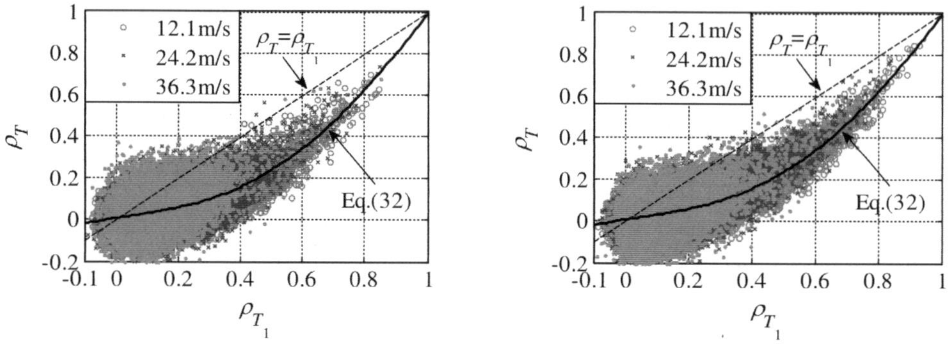

(a) FL27, 120°, isolated (b) FL30, 130°, surrounded

Figure 5. Correlation coefficients of peak pressure coefficients over T_1 and T

The relationship between correlation coefficients of peak values over 1 min and 10 min is fitted by the following polynomial

$$\rho_T = \sum_{i=0}^{4} a_i \rho_{T_1}^i \tag{12}$$

where the coefficients a_i for i = 0 to 4 are determined as 0.007, 0.241, -0.232, 1.716 and -0.741, respectively. Although the uncertainties in the lower correlation for peak values are large, they should not have the significant effect on the roof damage estimation and can be neglected for simplicity.

6.2 *Nataf transformation*

To incorporate the correlation among peak internal forces in the cladding damage estimation, the Nataf transformation is adopted (e.g., Ref. [29]), which relates correlated non-Gaussian variables to correlated Gaussian counterparts.

Assume a random variable vector with components made up of the peak internal forces on screws such that $\mathbf{W} = [W_1, W_2, \cdots, W_n]^T$ whose marginal CDF is $F_{W_j}(w_j)$, $j = 1, 2, \cdots, n$ and n is the number of screws. This vector can be transformed to its corresponding standard Gaussian random vector $\mathbf{Z} = [Z_1, Z_2, \cdots, Z_n]^T$ by

$$Z_j = \Phi^{-1}[F_{W_j}(w_j)], \quad j = 1, 2, \cdots, n \tag{13}$$

where $\Phi(\cdot)$ denotes the standard Gaussian CDF. The Nataf transformation can be derived as (e.g., Ref. [11])

$$f_{\mathbf{W}}(\boldsymbol{w}) = f_{W_1}(w_1) f_{W_2}(w_2) \cdots f_{W_n}(w_n) \frac{\varphi_n(\mathbf{z}, \mathbf{R}_{\mathbf{Z}})}{\varphi(z_1)\varphi(z_2)\cdots\varphi(z_n)} \tag{14}$$

where $f_{\mathbf{W}}(\boldsymbol{w})$ is the joint PDF of peak internal forces and $\varphi_n(\mathbf{z}, \mathbf{R}_{\mathbf{Z}})$ is the standard Gaussian n-variate joint PDF with the correlation matrix R_Z.

With known ρ_{jk}^{w}, the correlation coefficient between W_j and W_k, can be expressed as

$$\rho_{jk}^{w}=\int_{-\infty}^{\infty}\int_{-\infty}^{\infty}\left(\frac{w_j-\mu_j}{\sigma_j}\right)\left(\frac{w_k-\mu_k}{\sigma_k}\right)\varphi_2(z_j,z_k,\rho_{jk}^{z})dz_jdz_k \tag{15}$$

where μ_j and σ_j are the mean and STD of W_j, respectively; and ρ_{jk}^{z} (correlation coefficient between Z_j and Z_k) is an element in Rz. An important lemma is that ρ_{jk}^{w} is a strictly monotonic function with respect to ρ_{jk}^{z} [29]. This leads to one-to-one mapping between ρ_{jk}^{w} and ρ_{jk}^{z}. To avoid iteration in above equation, a series of empirical formulae were fitted by Liu & Der Kiureghian [29]. If W_j and W_k both follow Gumbel distribution, Eq. (15) can be approximated as

$$\rho_{jk}^{z}=1.064\rho_{jk}^{w}-0.069(\rho_{jk}^{w})^2+0.005(\rho_{jk}^{w})^3 \tag{16}$$

Obviously, both the marginal CDF and correlation of the non-Gaussian variables are maintained under the Nataf transformation.

7 DAMAGE ESTIMATION OF STEEL ROOFING

7.1 *Failure probability and damage ratio of a roof cladding*

Based on a comprehensive parametric study for the pull-through failures of the trapezoidal steel cladding at the screw connections, Mahaarachchi & Mahendran [14] developed strength formulae for trapezoidal steed claddings with closed space ribs. When G550 steel is used, the mean of the strength R is given as

$$\mu_R=0.04\times\left(4.7-\frac{20f_yd_h}{E}\right)^2\left(\frac{h_c}{h_p}\right)^{3/4}\left(\frac{W_t}{W_c}\right)^{1/5}\left(12+\frac{1500t^2}{Ld_h}\right)^{1/3}d_htf_y \tag{17}$$

where the steel yield stress $f_y=690$ MPa; the diameter of screw head $d_h=11$ mm; the Young's modulus $E=200$ GPa; the cladding crest height $h_c=35$ mm; the crest pitch $h_p=125$ mm; the cladding trough width $W_t=81.5$ mm; the cladding crest width $W_c=43.5$ mm; the cladding thickness $t=0.6$ mm and the span between purlins $L=1981.2$ mm. The coefficient of variation (the ratio of the STD to the mean) is 0.12. Furthermore, a capacity reduction factor ϕ is required to consider the possible variations in material, fabrication and loading effects. As suggested by Mahaarachchi & Mahendran [14], $\phi=0.6$ is adopted hereafter for illustration, although the value of this factor deserves further investigation. The cladding strength R is assumed to follow Gaussian distribution, in the current case, with mean of $\phi\mu_R$ and STD of $0.12\phi\mu_R$.

Once the peak internal force on the screw is larger than the cladding strength around the screw, the cladding may suffer from the pull-through failure. Due to the lack of the research and experimental data, the peak internal force and the cladding strength around a screw are regarded to be independent. The failure probability of cladding around a screw is given by

$$s = \iint_{r \le w} f_R(r) f_W(w) dr dw \quad (18)$$

where $f_R(r)$ is PDF of the cladding strength around the screw.

Previous investigations showed that 90% wind load would be redistributed to two adjacent screws on the same crest if a screw fails to take the load [10]. As a consequence, internal forces on these two screws will significantly increase, and the redistributed internal force may exceed the cladding strength and the corresponding cladding sheet will undergo the failure very likely. Hence, it is appropriate to assume that the failure of one of the screws on a cladding panel leads to the failure of the whole panel.

Apparently, the failure probability of a cladding panel is determined by the peak internal forces and resistances at the screws on that cladding. The failure probability can be expressed as

$$p = 1 - \iint_{w_1 < r_1} \iint_{w_2 < r_2} \cdots \iint_{w_{n_s} < r_{n_s}} f_{\mathbf{R}}(\boldsymbol{r}) f_{\mathbf{W}}(\boldsymbol{w}) d\boldsymbol{r} d\boldsymbol{w} \quad (19)$$

where $f_{\mathbf{R}}(\boldsymbol{r})$ is the joint Gaussian PDF of strengths and can be determined based on the assumption of independence. Note that $f_{\mathbf{W}}(\boldsymbol{w})$, which is defined in section 6, has incorporated the correlation among peak internal forces. In this study, the number of screws, n_s, is 8 due to the symmetric layout of screws.

Suppose the number of cladding panels over the entire roof is N_C. To estimate the overall damage of claddings over the roof, the damage ratio is used. It describes the extent of damage on the roof cladding and is defined as the percentage of total failed panels, i.e.,

$$D = M_C / N_C \quad (20)$$

where M_C is the failed cladding number, and M_C and D are random variables.

7.2 *Monte Carlo simulation*

MCS is more efficient than a numerical method in estimating the damage ratio over the entire roof as well as the failure probability for each single cladding. It should be noted that the internal forces on many pairs of screws will be fully correlated. For example, the internal forces on screws x2y2 and x2y3 on a cladding are identical. One of two fully correlated forces in the simulation is taken out and the simulated force can be adopted for both.

Assume MCS is repeated for n_t rounds. In the m th simulation, the correlated peak internal forces at all screws are simulated via the Nataf transformation, firstly. The peak internal forces at all screws can be simulated in the following way. The correlated Gaussian vector Z is simulated according to

$$\mathbf{R}_{\mathbf{Z}} = \mathbf{L}\mathbf{L}^T ; \ \mathbf{Z} = \mathbf{L}^{-1}\mathbf{U} \quad (21)$$

where U is the independent standard Gaussian vector and the lower triangular matrix L shall be obtained by Cholesky decomposition of R_Z. Once the sample of Gaussian vector Z is generated, that of the non-Gaussian vector W can be simulated accordingly. Further details can be found in Huang et al. [11].

In addition, Cholesky decomposition for matrix R_Z may be not applicable because negative eigenvalues may still exist. These negative values appear due to highly correlated forces on different screws and/or computational errors. To solve this difficulty, the R_Z matrix should be manipulated. First, this matrix can be written as

$$\mathbf{R}_Z = \mathbf{V}^T \mathbf{\Omega} \mathbf{V} \tag{22}$$

where V is the eigenvector matrix and Ω is the diagonal eigenvalue matrix. The negative eigenvalue in matrix Ω can be replaced by a small positive value such as 0.001 in order to make the Cholesky decomposition available. Results show that the simulation is not sensitive to the minor changes in the eigenvalues. After the correlated peak internal forces have simulated, the independent resistances associated with those internal forces are generated accordingly. Note that the correlation of resistances is neglected due to a lack of available data.

Let $f_{l,m}$ denote whether the l th cladding has failed in the m th simulation such that $f_{l,m} = 0$ or 1 for undamaged or damaged, respectively. The failure probability of the l th cladding is

$$p_l = \frac{1}{n_t} \sum_{m=1}^{n_t} f_{l,m} \tag{23}$$

where $\sum_{m=1}^{n_t} f_{l,m}$ is the number of the failure for the l th cladding in the simulations. The damage ratio for m th simulation is given by

$$d_m = \frac{1}{N_C} \sum_{l=1}^{N_C} f_{l,m} \tag{24}$$

where $\sum_{l=1}^{N_C} f_{l,m}$ is the number of failed cladding panels in the m th simulation. According to discussions by Huang et al. (2016), the random variable D will approach the Gaussian distribution approximately. The mean and STD of damage ratio are determined as

$$\mu_D = \frac{1}{n_t} \sum_{m=1}^{n_t} d_m \; ; \quad \sigma_D = \sqrt{\frac{1}{n_t} \sum_{m=1}^{n_t} (d_m - \mu_D)^2} \tag{25}$$

7.3 *Results and discussions*

In the following discussions, the correlation for peak values over 10 min will be denoted as C_{10min}. For comparison, another three types of correlations will be also used to investigate the influence of the correlation of peak internal forces: the correlation for the parent process (denoted by C_{par}), the correlation for peak values over 1 min (denoted by C_{1min}), and no correlation (denoted by C_0). Note that C_{par} and C_{1min} are estimated based on the limited data, and C_{10min} is approximately estimated from C_{1min}.

Table 1 shows the correlation coefficients for internal forces on selected screws under wind speed of 30 m/s at roof height and AOA of 315° (The corresponding cladding numbers are marked in Figure 1). It can be observed that the correlation coefficients decreases from C_{par} to C_{1min} and further C_{10min}, which is consistent with the results in Luo & Huang [30]. Also strong correlations can be found even for the distant screws. For example, the correlation coefficients between x2y2 on Cladding A and x2y3 on Cladding D are estimated as 0.917, 0.838 and 0.691 for C_{par}, C_{1min} and C_{10min}, respectively.

Table 1 Correlation coefficients among internal forces (wind speed = 30 m/s and AOA = 315°)

Screw	x2y2 (Cladding A)	x3y2 (Cladding A)	x4y1 (Cladding B)	x2y4 (Cladding C)	x2y3 (Cladding D)
x2y2 (Cladding A)	1				
x3y2 (Cladding A)	0.956 0.890 0.783	1			
x4y1 (Cladding B)	0.910 0.839 0.691	0.957 0.894 0.789	1		
x2y4 (Cladding C)	0.920 0.835 0.685	0.952 0.808 0.639	0.979 0.941 0.877	1	
x2y3 (Cladding D)	0.917 0.838 0.691	0.945 0.791 0.611	0.956 0.846 0.704	0.962 0.898 0.797	1

Note: Numbers represent C_{par}, C_{1min} and C_{10min} from top to bottom in a cell.

In total, 10,000 rounds of simulations are conducted based on C_{par}, C_{1min}, C_{10min} and C_0, respectively. Once the peak internal forces and resistances at all screws are simulated, the failures for all screws can be judged and corresponding failure probabilities can then be estimated by the simulation. Also failure probabilities can be evaluated numerically from Eq. (18). It should be pointed out that the correlation of peak internal force has no influence on screw failures. The contour map for the failure probabilities of all 608 screws under the wind speed of 30 m/s at roof height and AOA of 315° is shown in Figure 6. It can be seen that screws close to the leading edges are more vulnerable. Also it is seen that screws along the line x2 (e.g., x2y3) on a cladding panel have larger failure probabilities, which indicates cladding failure possibly starts from these positions. This is attributed to the facts that these screws have relatively larger tributary areas and suffer from relatively larger wind loading.

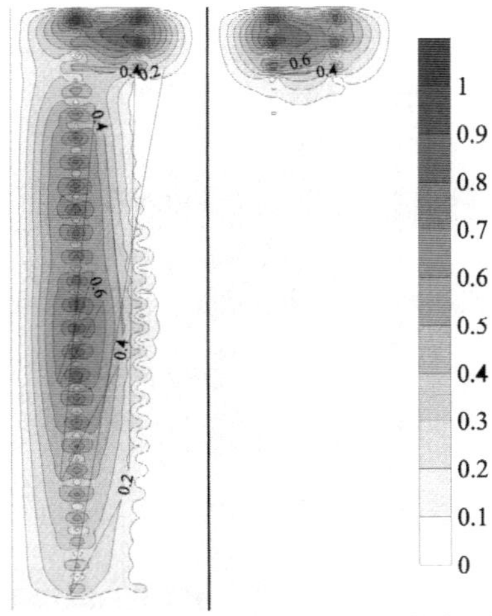

Figure 6. Failure probabilities for screws under speed of 30 m/s and AOA of 315°

Subsequently, failure probabilities of all cladding panels can be obtained. From Eq. (19), the correlation of peak internal forces will affect these probabilities. Results show that C_{par}, C_{1min}, C_{10min} and C_0 have little effect on the failure probability of a cladding. With C_{10min} for illustration, failure probabilities of all cladding under wind speeds of 27, 30 and 33 m/s and AOAs of 270°, 315° and 360° are shown in Figure 7. Here, the progressive damage can be observed. With the increase of the wind speed, the breach will spread from the windward side to leeward side under AOA of 270°, and start from the windward corner and spread radially to the leeward corner under 315°, and gradually extend from one edge to another one under 360°. It can also be seen that the roof faces the highest risk for 270°; not for 315°. This observation can be explained as follows. Typically, the roof damage associated with AOA of 315° is expected to be largest due to the existence of the conical vortex if only the external pressures are considered. However, when a breakage on a windward wall is assumed, the internal pressure has a significant influence on the roof damage. In current study, the internal pressure under AOA of 270° on Stage 2 makes the roof most vulnerable.

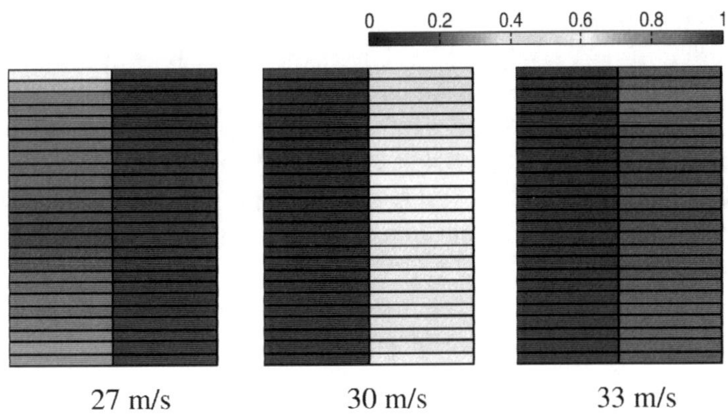

(a) AOA of 270°

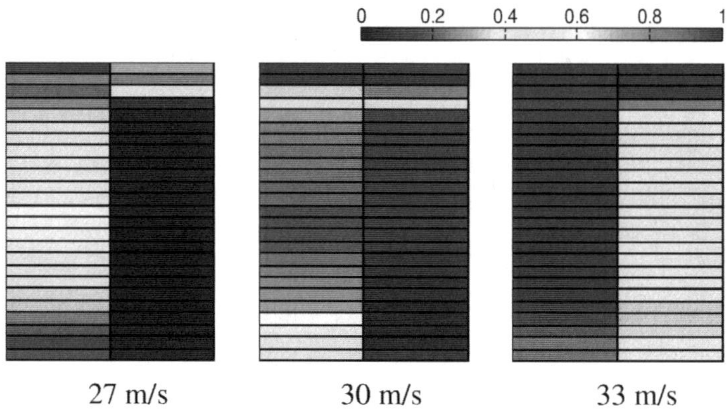

(b) AOA of 315°

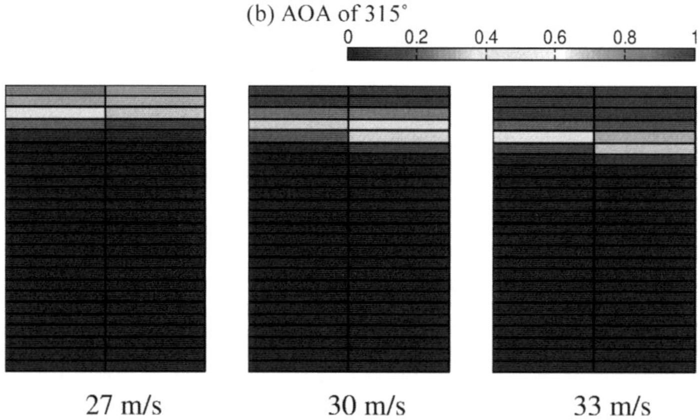

(c) AOA of 360°

Figure 7. Claddings failure probabilities under wind speeds of 27, 30 and 33 m/s, and AOAs of 270°, 315° and 360°

The means and STDs of the damage ratios for C_{par}, C_{1min}, C_{10min} and C_0 under various wind speeds and AOAs are plotted in Figure 8. Results show that the correlation has a negligible effect on the mean damage ratio. This is similar to results reported by Huang et al. [11], where the wind loading correlation has no effect on the mean damage ratio for the roof panels. However, the STD may be sensitive to the correlation, i.e., it becomes larger with the increase of the correlation. For example, under wind speed of 30 m/s and AOA of 315°, the STDs of the damage ratio are about 0.155, 0.145 and 0.135 for C_{par}, C_{1min} and C_{10min}, and that for C_0 drops to less than 0.05. This is attributed to the fact that higher correlations among internal forces on screws cause the failures at different cladding panels more likely. Especially, if the wind loading correlations approach unit, those claddings may either fail or survive together, which will greatly increase the variation of damage ratio.

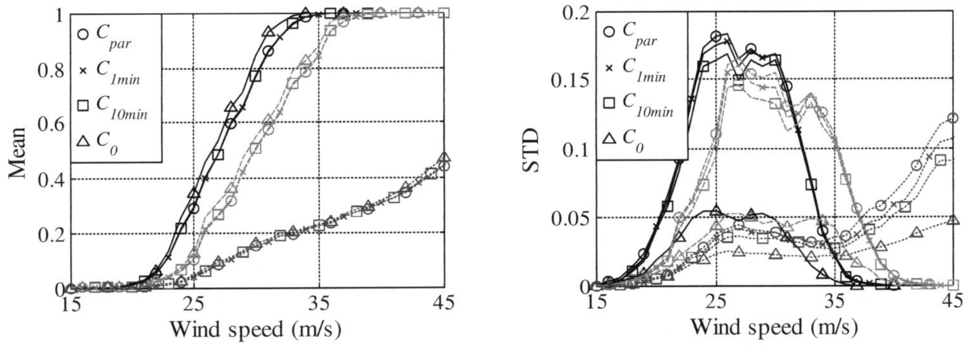

(a) Mean (b) STD

Figure 8. Damage ratios for various wind speeds and AOAs (solid line: 270°, dash line: 315°, dot line: 360°)

8 DIRECTIONALITY IN DAMAGE ESTIMATION

While the effect of wind directionality on probabilistic estimation of wind load effects of structures has drawn significant attentions for load design purpose (e.g., Refs. [31,32,33]), its study in vulnerability analysis has been limited in the literature. For example, ASCE 7-10 specifies a directionality reduction factor of 0.85 for the structural design of roofs and claddings [34]. However, the directionality effect on vulnerability, i.e., damage status of a building altered by the directionality effect of the actual building aerodynamics (determined via the wind tunnel study), has rarely been considered as a separate parameter in previous literatures.

The consideration of directionality effects is also important since the wind climate for a prescribed location is unique and will introduce directionality of extreme wind speed, i.e., there may be a dominant direction of wind speed in terms of both the frequency and magnitude. The overall vulnerability for a particular building at a particular location should be an integration of the vulnerabilities conditioned on prescribed directions, over all directions. Thus, the following discussions focus on a fully probabilistic method which considers the wind directionality, directional wind speed correlations as well as uncertainties in the damage ratio within a unified framework.

The meteorological data of wind speed records are from the Automated Surface Observation System (ASOS) operated by National Oceanic and Atmospheric Administration (NOAA) at Baltimore, MD, USA. 10-min mean wind speeds at 10-m height are derived from this dataset and then converted to those at the roof height of the building adopted in this paper. Wind speeds covering 360 degrees are categorized into eight representative sectors with each denoted by their central directions $\alpha_i = 45^\circ, 90^\circ, \cdots, 360^\circ$ ($i = 1, 2, \cdots, 8$ indicates directions NE, E, $\cdots$, N, respectively, as shown in Figure 12). Monthly maximum wind speed data are selected in each sector. In total, 156 monthly (from 2000 Jan to 2013 Dec) maxima are collected for every sector. In the following discussions, the damage analysis of steel cladding is performed with the consideration of the wind loading correlation among the screw peak forces over 10 min (i.e., C_{10min}).

8.1 *Without consideration of variation of damage ratio*

Denote the joint CDF of directional annual maximum wind speeds as $H(v_1, v_2, \cdots, v_{n_d})$ where n_d is the total number of directions being partitioned. Such joint CDF can be derived from the multivariate extreme value theory using a Gaussian Copula model, which is expressed as [33]

$$H(v_1, v_2, \cdots, v_{n_d}) = G_{n_d}(\Phi^{-1}[\Psi_{V_1}(v_1)], \Phi^{-1}[\Psi_{V_2}(v_2)], \cdots, \Phi^{-1}[\Psi_{V_{n_d}}(v_{n_d})]) \tag{26}$$

where Φ^{-1} is the inversed CDF of standard Gaussian distribution; G_{n_d} is the CDF of n_d-dimensional normal distribution with the zero mean and the covariance matrix Σ in which $\Sigma_{ii} = 1$ and $\Sigma_{ij} = \Sigma_{ji} = E\left[\Phi^{-1}[\Psi_{V_i}(v_i)], \Phi^{-1}[\Psi_{V_j}(v_j)]\right]$; and $\Psi_{V_i}(v_i)$ (i = 1, 2, $\cdots$, n_d) is the CDF of annual maximum wind speed in the i th direction. It is proper to adopt the Gumbel distribution in Eq. (9) to model the annual maximum wind speed in any direction. Hence, the CDF of the annual maximum can be determined from that of monthly maximum following Eq. (11). The extreme wind speeds for 50 and 500 -year return period in each direction are plotted in Figure 9. It can be observed that N and NW are two dominant directions in terms of the magnitude of the mean wind speed.

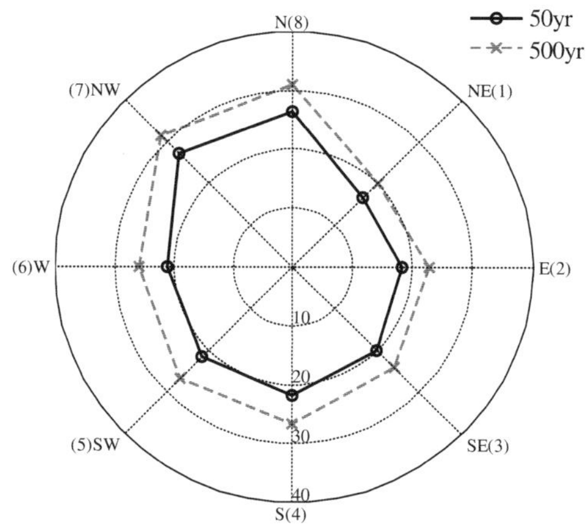

Figure 9. Extreme wind speeds for 50 and 500 –year return period (Unit: m/s)

In engineering practice and the insurance industry, the mean damage ratio is often of primary concern. The mean damage ratio for a component such as steel roofing can be treated as a function of the wind speed from the i th direction. Alternatively, the wind speed v_i^d, producing a given damage level d in that direction, can be obtained from Figure 8a. Therefore, for a component subjected to wind loads from all directions, the probability not exceeding the damage level d can then be calculated as

$$P(D \le d) = H(v_1^d, v_2^d, \cdots, v_{n_d}^d) \tag{27}$$

If the directional extreme wind speeds can be assumed to be mutually independent, the estimation by Eq. (27) can be replaced by

$$P(D \le d) = \prod_{i=1}^{n_d} \Psi_{V_i}(v_i^d) \tag{28}$$

The damage of the steel roofing for Y-year return period, d_Y, can then be determined from

$$Y = 1/[1 - P(D \le d_Y)] \tag{29}$$

The probabilities for various damage levels are illustrated in Figure 10, where the roofing damages considering wind directionality and those in 8 directions are included. It can be seen NW direction (7 th) retains the dominance among 8 directions due to the larger mean wind speed and the higher damage risk in this direction. In contrast, another wind speed dominant N direction (8 th) has less influence due to the lower damage risk. Besides, the consideration of wind directionality effect produces a larger damage estimation compared with those in 8 individual directions. This is attributed to the fact that $H(v_1^d, v_2^d, \cdots, v_{n_d}^d) \le \min[\Psi_{V_i}(v_i^d)]$. Furthermore, the directional extreme wind speeds are shown to be mutually independent, as seen in the almost identical estimations by Eqs. (27) and (28). This can be attributed to a dominant NW direction and low correlations among mean wind speeds across different directions.

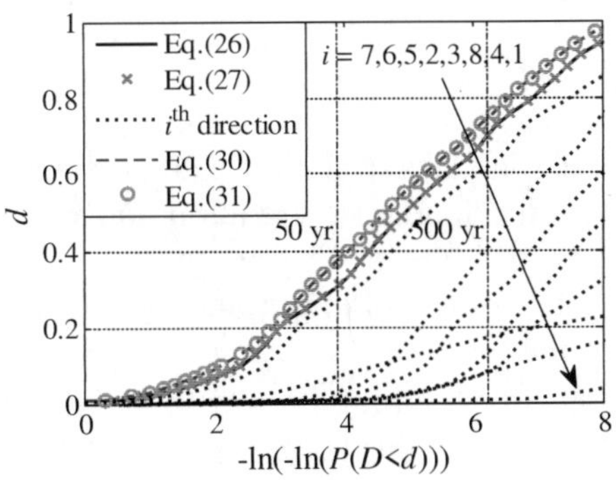

Figure 10. Estimations of roofing damage for various levels d

8.2 *With consideration of variation of damage ratio*

According to Huang et al. [11], the STD of the damage ratio is larger when the wind loading has a stronger correlation. This variation around the mean damage ratio causes additional risk, which is sometimes termed as "secondary uncertainty". Ignoring the influence of the secondary uncertainty may result in an inadequate assessment of the risk. The following discussions address the issue of quantifying the damage considering the variation of damage ratio.

The probability of D not exceeding damage level d in all directions can be expressed as

$$P(D \le d) = \int \cdots \int F_{D_1, D_2, \cdots, D_{n_d}}(d|v_1, d|v_2, \cdots, d|v_{n_d}) h(v_1, v_2, \cdots, v_{n_d}) dv_1 dv_2 \cdots dv_{n_d} \tag{30}$$

where $F_{D_1, D_2, \cdots, D_{n_d}}(d_1|v_1, d_2|v_2, \cdots, d_{n_d}|v_{n_d})$ is the joint CDF of damage ratios conditioned on wind speed v_i ($i = 1, 2, \cdots, n_d$). If the roofing damage is mutually independent across directions, Eq. (30) becomes

$$P(D \le d) = \int \cdots \int \prod_{i=1}^{n_d} F_{D_i}(d|v_i) h(v_1, v_2, \cdots, v_{n_d}) dv_1 dv_2 \cdots dv_{n_d} \tag{31}$$

where $F_{D_i}(d|v_i)$ is the CDF of damage ratio in i th direction under v_i, which is served as a complimentary fragility and obeys exceeding lognormal distribution [6]. If the directional extreme wind speeds are mutually independent, which is the particular case for present study, the above formula can be approximated by

$$P(D \le d) = \prod_{i=1}^{n_d} \int F_{D|i,v}(d) \psi_{V_i}(v) dv \tag{32}$$

where $\psi_{V_i}(v)$ is the PDF of extreme wind speed in the i th direction.

Numerical results show that damage ratios across directions are almost independent, which can be explained by the fact that both the peak force and resistance on a fastener are independent for different directions. Hence, Eq. (32) can be used to estimate the damage. The result considering the damage ratio variation is plotted in Figure 10. It is clearly seen that this yields an estimation of a higher damage level in a certain return period (e.g., 50-year). That is, the consideration of damage variation leads to a larger risk for roofing claddings.

9 CONCLUSIONS

The paper presents an integrated damage estimation method for steel roofing claddings in high winds. In this method, both the wind loading correlation and wind directionality effects are incorporated. For illustration, the widely used high-strength trapezoidal steel cladding is chosen as the roof sheathing which is connected to building frame by screws. The wind loading is derived from the wind pressure data from a wind tunnel. The method addressed in this paper provides a general framework applicable for all cladding, roof components, terrains and so on, where data is available. Some observations and conclusions are given as follows.

(1) The failure probability and damage ratio can be estimated by Nataf transformation - based MCS accurately when the wind loading correlation is considered. Results show that the wind loading correlation may noticeably increase the variation of the cladding damage. (2) The wind directionality is incorporated into damage estimations in this study. Methods for consideration of mean damage ratio and the variation of damage ratio are developed. Results show that the wind directionality may lead to a larger damage risk for claddings in high winds and its consideration will provide a reliability-consistent solution in risk management.

The presented damage estimation framework for the low-rise building roof can be a useful tool in making damage assessments for new or existing buildings. Furthermore, the method may benefit the performance-based design for the low-rise buildings (e.g., Ref. [35]).

10 ACREFERENCES

1 Dabral, A., and Ewing, B. T. (2009). Analysis of wind-induced economic losses resulting from roof damage to a metal building. J. Business Val. Econ. Loss Anal., 4(2).

2 Perry, D. C., Mcdonald, J. R., Saffir, H. S., and Asce, F. (1990). Performance of metal buildings in high winds. J. Wind Eng. Ind. Aerodyn., 36, 985-999.

3 Ginger, J. D., Henderson, D. J., Leitch, C. J., and Boughton, G. N. (2007). Tropical cyclone larry: estimation of wind field and assessment of building damage. Australian J. Struct. Eng., 7(3), 209-224.

4 Song F., and Ou J. (2009). Wind hazard damage estimation of industrial buildings. 7th Asia-Pacific Conf. on Wind Eng., Taipei, Taiwan, China.
5 NIST (2006). Performance of physical structures in Hurricane Katrina and Hurricane Rita: A reconnaissance report. National Institute of Standards and Technology, Gaithersburg, MD, USA.
6 Lee, K. H., and Rosowsky, D. V. (2005). Fragility assessment for roof sheathing failure in high wind regions. Eng. Struct., 27(6), 857-868.
7 Li, Y., and Ellingwood, B. R. (2006). Hurricane damage to residential construction in the us: importance of uncertainty modeling in risk assessment. Eng. Struct., 28(7), 1009-1018.
8 Zhao, M., and Gu M. (2011). Database-assisted wind vulnerability assessment for metal buildings. Proc. 13th Intl. Conf. on Wind Eng., Amsterdam, Netherlands.
9 Huang, G., He, H., Mehta, K. C., and Liu, X. (2015). Data-based probabilistic damage estimation for asphalt shingle roofing. J. Struct. Eng., 141(12), 04015065.
10 Konthesingha, K. M. C., Stewart, M. G., Ryan, P., Ginger, J., and Henderson, D. (2015). Reliability based vulnerability modelling of metal-clad industrial buildings to extreme wind loading for cyclonic regions. J. Wind Eng. Ind. Aerodyn., 147, 176-185.
11 Huang, G., Ji, X., Luo, Y., and Gurley, K. R. (2016). Damage estimation of roof panels considering wind loading correlation. J. Wind Eng. Ind. Aerodyn., 155, 141-148.
12 Cope, A. D., Gurley, K. R., Gioffre, M., and Reinhold, T. A. (2005). Low-rise gable roof wind loads: Characterization and stochastic simulation. J. Wind Eng. Ind. Aerodyn., 93(9), 719-738.
13 Ho, T. C. E., Surry, D., Morrish, D., and Kopp, G. A. (2005). The UWO contribution to the NIST aerodynamic database for wind loads on low buildings: Part 1. Archiving format and basic aerodynamic data. J. Wind Eng. Ind. Aerodyn., 93(1), 1-30.
14 Mahaarachchi, D., and Mahendran, M. (2009). Wind uplift strength of trapezoidal steel cladding with closely spaced ribs. J. Wind Eng. Ind. Aerodyn., 97(3), 140-150.
15 Tamura, Y., Ueda, H., Kikuchi, H., Hibi, K., Suganuma, S., and Bienkiewicz, B. (1997). Proper orthogonal decomposition study of approach wind-building pressure correlation. J. Wind Eng. Ind. Aerodyn., 72, 421-431.
16 Bienkiewicz, B., Ham, H. J., and Sun, Y. (1993). Proper orthogonal decomposition of roof pressure. J. Wind Eng. Ind. Aerodyn., 50, 193-202.
17 Jeong, S. H., Bienkiewicz, B., and Ham, H. J. (2000). Proper orthogonal decomposition of building wind pressure specified at non-uniformly distributed pressure taps. J. Wind Eng. Ind. Aerodyn., 87(1), 1-14.
18 Holmes, J. D., (1979). Mean and fluctuating internal pressures induced by wind. Proc. 5th Intl. Conf. on Wind Eng. 435-440., Fort Collins, Colorado, USA.
19 Oh, J. H., Kopp, G. A., and Inculet, D. R. (2007). The UWO contribution to the NIST aerodynamic database for wind loads on low buildings: Part 3. Internal pressures. J. Wind Eng. Ind. Aerodyn., 95(8), 755-779.
20 Holmes, J. D., and Ginger, J. D. (2012). Internal pressures – the dominant windward opening case – a review. J. Wind Eng. Ind. Aerodyn., 100(1), 70-76.
21 Vickery, B. J., and Bloxham, C. (1992). Internal pressure dynamics with a dominant opening. J. Wind Eng. Ind. Aerodyn., 41(1-3), 193-204.
22 Mahendran, M. (1990). Fatigue behaviour of corrugated roofing under cyclic wind loading. Civil Engineering Transaction, Institution of Engineers, Australia 32(4), 219-226.
23 Xu, Y. L. (1995). Fatigue performance of screw-fastened light-gauge-steel roofing sheets. J. Struct. Eng., 121(3), 389-398.
24 Henderson, D. J., and Ginger, J. D. (2011). Response of pierced fixed corrugated steel roofing systems subjected to wind loads. Eng. Struct., 33(12), 3290-3298.
25 Kopp, G.A. (2013). Wind Loads on Building Components and Cladding. Advanced Structural Wind Engineering (ed. Tamura, Y. & Kareem, A.), Springer, Tokyo, 177-195.
26 Cook, N. J., and Mayne, J. R. (1980). A refined working approach to the assessment of wind loads for equivalent static design. J. Wind Eng. Ind. Aerodyn., 6(1-2), 125-137.
27 Holmes, J. D., and Cochran, L. S. (2003). Probability distributions of extreme pressure coefficients. J. Wind Eng. Ind. Aerodyn., 91(7), 893-901.
28 Peng, X., Yang, L., Gavanski, E., Gurley, K., and Prevatt, D. (2014). A comparison of methods to estimate peak wind loads on buildings. J. Wind Eng. Ind. Aerodyn., 126, 11-23.
29 Liu, P. L., and Der Kiureghian, A. (1986). Multivariate distribution models with prescribed marginals and covariances. Prob. Eng. Mech., 1(2), 105-112.
30 Luo, Y., and Huang, G. (2016). Characterizing dependence of extreme wind pressures. J. Struct. Eng., 10.1061/(ASCE)ST.1943-541X.0001699, 04016208.
31 Simiu, E., and Scanlan, R. H. (1996). Wind effects on structures: fundamentals and applications to design.

32 Laboy-Rodríguez, S.T., Gurley, K. R., and Masters, F. J. (2014). Revisiting the directionality factor in ASCE 7. J. Wind Eng. Ind. Aerodyn., 133(4), 225–233.
33 Zhang, X., and Chen, X. (2015). Assessing probabilistic wind load effects via a multivariate extreme wind speed model: A unified framework to consider directionality and uncertainty. J. Wind Eng. Ind. Aerodyn., 147, 30-42.
34 ASCE. (2010). Minimum design loads for buildings and other structures. ASCE7-10, Reston, VA.
35 Ciampoli, M., Petrini, F., and Augusti, G. (2011). Performance-based wind engineering: towards a general procedure. Struct. Safety, 33(6), 367-378.

Impact of the Latest Florida Building Code on the Statewide Hurricane Risk

Bahareh Kordi[a], Karthik Ramanathan[b], Tim Johnson[c], Farid Moghim[d], Sarah Bobby[e]

[a] *AIR Worldwide, 131 Dartmouth Street, Boston, MA 02116, email: bkordi@air-worldwide.com*
[b] *AIR Worldwide, 131 Dartmouth Street, Boston, MA 02116, email: kramanathan@air-worldwide.com*
[c] *AIR Worldwide, 131 Dartmouth Street, Boston, MA 02116, email: tjohnson@air-worldwide.com*
[d] *AIR Worldwide, 131 Dartmouth Street, Boston, MA 02116, email: fmoghim@air-worldwide.com*
[e] *AIR Worldwide, 131 Dartmouth Street, Boston, MA 02116, email: sbobby@air-worldwide.com*

ABSTRACT

Building codes are often updated as warranted by improvements in the science that they are based on. Occasionally, nature challenges the state of knowledge and uncovers their shortcomings. Following 1992 Hurricane Andrew, significant changes were introduced to the South Florida Building Code for Miami-Dade and Broward Counties to mitigate losses for future events. However, statewide adoption of these measures did not happen until 2001 with the introduction of the Florida Building Code (FBC). Hurricane Andrew also stressed the importance of enforcement of the building codes. Poor enforcement practices lead to a highly susceptible building stock which exacerbated the losses. Florida has since improved and currently is one of the stringent states with respect to building code enforcement. Research has consistently showed that proper adoption and enforcement of stronger building codes often leads to reduction in the hurricane risk facing the built environment.

Since 2001, FBC as well as the wind load standard which it references (i.e., ASCE 7) has been updated. These wind standards dictate the design levels required to meet the building code requirements. Significant changes occurred to the ASCE 7 wind load standards in 1998 and again in 2010. In 2010, ASCE 7 introduced risk-category specific new design wind speed maps which replaced the building occupancy “importance factor”. The changes were due to (i) the shift in the design methodology from Allowable Strength Design to Load and Resistance Factor Design, (ii) introduction of risk category factors, and (iii) new research findings and data which resulted in the re-simulation of the hurricane winds. These changes to the wind maps also triggered changes to the wind borne debris region which was redefined in FBC 2010. Terrain exposure categories and their definitions have also changed over time in FBC and ASCE 7. As all these changes take effect continually, the at-risk building stock becomes a conglomeration of various building code standards depending on the year and code in place at the time of permitting. Evaluating the relative performance of these building code changes and their impact on the inherent hurricane risk to buildings is difficult to quantify without actual wind events. Nevertheless an attempt is made in the existing study to evaluate the changes introduced in the recent FBC updates and their impact on the statewide hurricane risk using AIR’s Hurricane Catastrophe Model for U.S.

Catastrophe risk assessment models serve as excellent platforms to not only project future hurricane losses, but also help test the impact of regional/local building codes and their effectiveness to the exposed hurricane risk. These models integrate a stochastic hazard, coupled with detailed views of building vulnerability to estimate reliable loss distributions. The aim of the current research is to investigate the impact of the evolution of FBCs since 2001 and wind load standards on the hurricane risk of residential homes in Florida. AIR’s Hurricane Catastrophe Model will be used to assess the aforementioned impact measured in terms of probabilistic loss metrics, such as expected loss in a given year commonly termed as the average annual loss.

KEYWORDS

Florida Building Code, Hurricane Risk, Catastrophe Modeling, Expected Loss, ASCE 7-10

Assessing Climate Change Impact on U.S. Atlantic Hurricane Hazards – Initial Conditions and Sea Surface Temperature Uncertainties

Thai-Hoa Le [a], Fangquian Liu [b], Weichiang Pang [c]

[a] *Clemson University, Clemson, SC 29634, USA*
[b] *Applied Research Associates/IntraRisk, Raleigh, NA 27615, USA*
[c] *Clemson University, Clemson, SC 29634, USA*

ABSTRACT: Hurricane is one the most devastating meteorological events that impacts the United States and Atlantic basins on a yearly basis. There are anecdotal evidences that show the hurricane hazards in terms of annual frequency and intensity are changing due to a changing climate patterns. This study examines the impact of climate change on U.S. Atlantic hurricane, with emphasis on uncertainty quantification of Sea Surface Temperature (SST). Projected deterministic future (up to year 2100) SST data obtained from the IPCC (The Intergovernmental Panel on Climate Change) study and probabilistic SSTs considering uncertainty for the Atlantic basins are utilized in a stochastic hurricane simulation program to investigate the influence of SST on hurricane intensity. Two models of the SST uncertainties, namely spatially uncorrelated and correlated SSTs, are considered in this study. Moreover, a randomization of fundamental model parameters, i.e. storm frequency, storm spawn vectors with an effect of storm seasonality are also investigated.

KEYWORDS: Atlantic Hurricane; Hurricane Wind; Climate Change; Parameter Randomization; Sea Surface Temperature; Uncertainty Quantification.

1 INTRODUCTION

1.1 *Background*

Hurricane is one of the most damaging meteorological events and natural hazards, which frequently impact the coastal regions of the United States (U.S.). Damages and losses from hurricanes are associated with both the direct and indirect hazards of hurricanes such as strong winds, storm surge and wind-borne debris. Recent hurricanes in the U.S. Atlantic basins have been observed to cause increasing losses in properties and human lives. The costliest hurricanes in ascending order, the 2005 Hurricane Katrina, 2012 Hurricane Sandy, 2008 Hurricane Ike and 2011 Hurricane Irene resulted in, respectively, $150.2 billion, $66.5 billion, $32.2 billion and $15.8 billion losses (Pielke Jr et al., 2008). Furthermore, the U.S. coastal population has increased by 39 % from 1970 to 2010 with more than 123 million people or 39 % of the U.S. population resides in the coastal areas. Coastal population growth and more high-value properties concentrated in the coastal regions are likely to result in future hurricane losses that greatly exceed the current levels of annual losses. Hurricane risk assessment and management are the keys to reduce the damages and losses due to hurricanes.

Hurricane simulation models for generating synthetic storm tracks were initially developed around 1990s for the purposes of hurricane hazards quantification, risk assessment and loss estimation. The current state-of-the-art hurricane simulation models can be broadly quantified into either empirical-based stochastic models (e.g. Vickery et al. 2000) or physical-based deterministic model (e.g. Emanuel et al. 2006). The former employs historical storm database and a stochastic simulation approach using Markov chain and autoregressive models to simulate tracks and hurricane parameters (Vickery et al., 2000, Vickery et al., 2009a), whereas a large-scale atmosphere and ocean environmental data, originated from both observations and global climate model, are used to generate synthetic hurricane tracks in the latter (Emanuel et al., 2006a-b). A physica-based model does not rely on the historical track data, but rather generates the synthetic storms having statistical agreement with observations (Lin et al., 2012). Therefore, one can apply the stochastic simulation method for historical tracks-rich regions, while the deterministic simulation method for historical tracks-sparse regions. The stochastic simulation model is the most commonly used model for wind engineering applications, and this method has been adopted to develop the design wind speed maps in the current U.S. building codes (i.e., ASCE7-10). In a stochastic hurricane simulation model, fundamental parameters, i.e., storm eye coordinate (longitude and latitude), translational speed (V_t) and heading angle (θ), radius to maximum winds (R_{max}), central pressure (P_c) and pressure profile parameter have been used to describe the complete path and state of conditions of a storm. Generally, the stochastic simulation framework consists of several computational modules: storm genesis, storm tracking, central pressure, decay, and hurricane wind field (Vickery et al., 1995, Vickery et al., 2000a-b, Vickery et al., 2005, Vickery et al., 2008).

In the recent years, the impact of non-stationary or changing climate on the U.S. Atlantic hurricane activities and their effects have gained increasing attention in the wind engineering community (Emanuel, 1987). The changing climate pattern is primarily caused by a growing concentration of heat-trapping gasses in the atmosphere, often referred to as greenhouse gasses (i.e., CO_2, NO_2, CH_4). A warmer climate is likely to result in more intense storms and possibly affect the annual frequency of storms as well (Emanuel, 1987). The impact of climate change on future hurricane hazard and risk may be quantified using projected future climate scenarios. Climate change directly affect the annual storm frequency and the SST. The Intergovernmental Panel on Climate Change (IPCC)'s Representative Concentration Pathway (RCP) scenarios have been developed recently for the climate change projections (IPCC-AR5). Four RCP scenarios, i.e. RCP2.6 (the best case scenario), RCP4.5, RCP6.0 and RCP8.5 (the worst case scenario) describe independent pathways of greenhouse gas emissions. Each RCP scenario is associated with a radiative forcing index in the year 2100. One can use IPCC's climate change scenarios to project atmospheric, environmental parameters, which may be employed in hurricane simulation models to project future hurricane activities. The impact of climate change on the hurricane activities using the RCP projections have been examined by some authors, i.e. Nishijima et al., 2012; Wang and Rosowsky, 2012; Mudd et al., 2014; Liu 2014; Liu and Pang, 2014; Cui and Caracoglia, 2016. Impact of the climate change on the hurricane intensity and amplification of hurricane winds have recently been discussed (Wang and Rosowsky, 2012; Emanuel, 2013; Mudd et al., 2014). More specifically, Wang and Rosowsky (2012) found that the 50-year maximum wind speeds for New York City may increase by 25 m/s to 35 m/s under the low greenhouse gas concentration scenario (i.e. RCP2.6). In a recent study by Emanuel (2013), a large number of simulated hurricane events were compared between those generated using the climate of 1950-2005 and that projected for 2006-2100 under the high greenhouse gas emission scenario (RCP 8.5). It was found that both the frequency and intensity of tropical cyclones increase during the 21st century. In addition, a 40% increase globally in the number of major events (Saffir-Simpson category 3 and higher events) can be expected.

1.2 *Limitations of Current Hurricane Simulation Models and Climate Change Quantification*

Some limitations and drawbacks in the current hurricane simulation models are as follows:

Annual storm frequency: In many current models, the statistical distribution of annual storm frequency is assumed to be stationary over years (e.g., Lee and Rosowsky, 2007; Wang and Rosowsky, 2012). A Poisson distribution and a negative binomial distribution are often employed for a random generation of the annual storm frequencies (e.g., Mudd et al., 2014; Pei et al., 2014; Wei and Caracoglia, 2016). A constant mean and a standard deviation of historical storm frequencies are utilized in these distributions. For instance, Pei et al. (2014) used the negative binomial distribution with a mean of 9.05 and a standard deviation of 3.98 storms per year in simulating the annual storm frequency. Utilization of the constant mean is a drawback, and more importantly, the probabilistic approach via using abovementioned distribution functions has a difficulty in generating the annual storm frequencies at the projection year. New annual frequency model was developed to consider the variation of storm frequency as a function of year, which enables the annual storm frequency to be projected at future years (Liu and Pang 2012; Liu 2014). Liu (2014) and Liu and Pang (2012) applied a continuous oscillatory function, which nonlinearly fits with the historical storm frequency data, to vary the annual storm frequency for future years. A probabilistic model for generating the annual storm frequency, accounting for both the time-varying mean and fluctuation is essential.

Effect of storm seasonality: The storm activities are affected by the season in a year. Based on historical data, the months with active hurricane activities range from July to October, with peak spawn rates around August and September. One of the reasons that the hurricane activities are most active during these months is because of the warm SST in those months. The effect of seasonality was considered (or not in an explicit way) in many existing simulation models. Previous studies used a unique and deterministic monthly SST map for every year of simulation to calculate the intensity of hurricanes (e.g., Liu, 2014; Liu and Pang, 2012; Mudd et al., 2014; Cui and Caracoglia 2016). To better capture the seasonality effect of hurricanes, explicit consideration of storm spawn probability and frequency for each month and the extremes of SST within each month may be required.

Initial conditions: In the current simulation models (Lee and Rosowsky, 2007; Wang and Rosowsky, 2012; Liu and Pang, 2012; Mudd et al., 2014; Cui and Caracoglia 2016), the initial conditions of a storm (spawn location, initial heading angle and initial forward speed) are randomly selected from the initial conditions of known historical storms (1814 historical spawn locations are available for data from 1851 to 2015). Sampling the initial conditions solely based on permutation of known observed conditions may result in underestimation of the true uncertainty of the initial storm conditions. Further improvement to the current stochastic hurricane simulation method could be made by producing a probabilistic model that specifically considers the uncertainty of the initial conditions.

Sea surface temperature uncertainty: Many existing models utilize the monthly average SST data to simulate the intensity of hurricanes. Extremes of SST are likely to occur during the formation of storms. Since deterministic average monthly SST data or maps are used in many existing hurricane simulation models, the influence of SST uncertainty on the storm intensity is not considered. This study explores the use of probabilistic SST data with spatial correlation to simulate hurricanes.

The aforementioned limitations inspire this study to improve the current hurricane simulation models by specifically consider the uncertainties of sea surface temperature and initial storm conditions. This study assesses the impact of climate change on the U.S. Atlantic hurricanes. The new stochastic hurricane simulation model developed in this study utilizes (1) a new seasonality-based storm frequency model that explicitly considers the month spawn probability, (2) a new initial storm conditions model (i.e. initial storm location, intensity and heading direction and translation speed), and (2) a probabilistic SST models that consider the spatial correlation and uncertainty of monthly SSTs.

In this study, the projected future annual storm frequency is modeled using two terms: (1) a time-varying mean storm frequency, and (2) a fluctuation term modeled using a Gaussian distribution with a zero mean. The storms spawned in each simulation year are assigned spawn date and month. The month for each storm is assigned based on the monthly spawn probability mass function (PMF) developed from past observations. The initial conditions of a storm are subsequently generated using 1-D, 2-D Gaussian kernel density functions developed based on historical storm spawn conditions. Furthermore, the influences of SST uncertainty for different IPCC RCP scenarios is examined by studying storms simulated: (1) with and without (i.e. deterministic mean monthly SSTs) uncertainty, and (2) with and without spatial correlation in the SST data. These improvements are incorporated into a new hurricane simulation program to account for the impact of climate change scenarios on the hurricane activities and wind speeds along the Gulf and Atlantic coasts.

2 CURRENT HURRICANE SIMULATION MODELS: BACKGROUND

2.1 *Storm Genesis*

Best tracks and fundamental environmental parameters of historical storms since 1857 have been recorded in HURDAT database for an usage (Jarvinen et al. 1984), the most updated hurricane database until 2015 is provided in HURDAT2 (Landsea et al., 2015). In each historical storm, every 6-hour storm locations and initial storm conditions, i.e. initial position in longitude and latitude coordinates, heading angle, forward speed, central pressure and radius to maximum wind speeds are provided (Jarvinen et al., 1984; Landsea et al., 2015). It is noteworthy to repeat that the historical initial storm positions have been used to depart every simulated storms in the current simulation models. Furthermore, the initial location for simulated storms can be randomly selected among entire historical storm spawns. Subsequently, movement of the storms in the next time steps are then generated using an empirical hurricane tracking model. Figure 1 indicates historical storm frequencies in observed years and all historical storm spawn vectors, e.g. initial storm locations, associated with translational speeds and heading angles in HURDAT2 database.

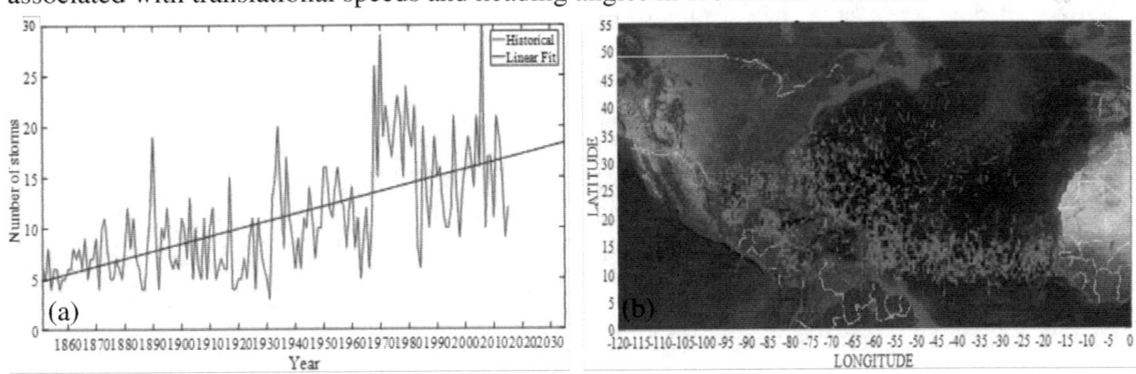

Figure 1. Historical annual storm frequencies from 1857-2015 (based on HURDAT2): (a) storm frequency versus year and a fitted linear trend, (b) storm spawn locations and initial heading directions presented as vectors.

2.2 *Storm Track Model*

The movement of a storm is modeled using the differences in logarithmic forward speeds and heading angles between two subsequent time steps (every 6 hours) following Vickery et al. (2000):

$$\ln V_{T_{i+1}} - \ln V_{T_i} = a_1 + a_2\psi + a_3\lambda + a_4 \ln c_i + a_5\theta_i + \varepsilon_{V_T}, \tag{1}$$

$$\theta_{i+1} - \theta_i = b_1 + b_2\psi + b_3\lambda + b_4 c_i + b_5\theta_i + b_6\theta_{i-1} + \varepsilon_\theta. \tag{2}$$

where, $V_{T_{i+1}}$ and V_{T_i} are the storm forward speeds at time steps *i+1* and *i*, respectively. θ_{i+1}, θ_i, and θ_{i-1} are the heading angles of a hurricane at time steps *i+1*, *i* and *i-1*, respectively. a_1 to a_5 and b_1 to b_6 are the grid specific regression constants developed using $5°\times5°$ grids over the entire Atlantic basins, including the Gulf of Mexico and the Caribbean Sea. ψ and λ are the latitude and longitude of the storm eye, respectively. ε_{V_T} and ε_θ are the residuals or random error terms.

2.3 *Central Pressure and Decay Models*

The intensity of a storm is typically quantified using the central pressure (P_c) at the storm eye, which is commonly expressed in terms of either Pascal (Pa) or millibar (mbar). The central pressure model can be expressed as a unitless and normalized quantity called the relative intensity (*I*). The relationship between the central pressure and the relative intensity is described in a dimensionless form by Darling (1991):

$$I = \frac{P_a - P_c + (1-RH)e_s}{(1-x)\left[P_a - (RH)e_s\right]}. \tag{3}$$

where, P_c is the central pressure of a hurricane (mbar); P_a is the standard atmospheric pressure (i.e., 1013 mbar); *RH* is the relative humidity of ambient air around the storm eye (taken as 0.75); *x* is the ratio of the minimum sustained central pressure and the partial pressure of ambient dry air (Darling 1991); and e_s is the saturation vapor pressure (mbar), given by the following equation:

$$e_s = 6.112\exp\left(\frac{17.67(T_s - 273)}{T_s - 29.5}\right), \tag{4}$$

where T_s denotes the Sea Surface Temperature or SST (Kelvin). The upper bound of the central pressure is set at 1013 mbar in the current simulation model, since the simulated central pressures are restricted by a physical constraint of the relative intensity, which ranges from 0 to 1.26 (Darling 1991).

In Vickery et al. (2000a), the relative intensity is described as a regressive function of the relative intensities and the SSTs of the previous three time steps:

$$ln(I_{i+1}) = d_0 + d_1\ln(I_i) + d_2\ln(I_{i-1}) + d_3\ln(I_{i-2}) + d_4 T_{s_{i+1}} + d_5(T_{s_{i+1}} - T_{s_i}) + \varepsilon_I, \tag{5}$$

where I_i is the relative intensity at time step *i*; d_0 to d_5 are the grid specific coefficients; T_{s_i} is the sea surface temperature at time step *i* (Kevin); and ε_I is the random error term. The central pressure deficit (or relative intensity) decays after the storm made landfall. It should be noted that, in many current simulation models, the SST at the storm eye is extracted from deterministic monthly average SST maps (12 month SST maps, one for each month).

Reduction of the storm intensity after the landfall is quantified by the decay model. The central pressure decay model (or filling rate model), developed by Vickery and Twisdale (1995), describes the decay of central pressure in terms of time after landfall, as follows:

$$\Delta P_c(t) = (P_a - P_{co})\exp(-at). \tag{6}$$

where, $\Delta P_c(t)$ is the central pressure deficit, computed as the difference between the ambient air pressure and the hurricane central pressure, at time *t* after landfall; P_{co} is the hurricane central pressure at landfall; and *a* is the decay coefficient. An empirical relationship for *a* is given by Vickery and Twisdale (1995):

$$a = f_0 + f_1(P_a - P_{co}) + \varepsilon_a, \tag{7}$$

where f_0 and f_1 are the regional varying filling rate constants; and ε_a is a normally distributed random error term. The decay parameter for each storm can be obtained by an analysis of the historical central pressures in the database.

2.4 *Radius to Maximum Winds and Wind field*

The size of a storm is measured by a distance between the storm eye and a location of the maximum wind speed, which is known as the radius to maximum winds ($R_{\max}$) (Vickery et al. 2000a). Vickery et al. (2000), Vickery and Wadhera (2008) suggested that the value of $R_{\max}$ can be described separately for the Atlantic basin and the Gulf of Mexico basin as a function of latitude (ψ in degree) and central pressure deficit ($\Delta P_c = P_a - P_c$ in mbar), as follows:

$$\ln\left(R_{\max\_Atlantic}\right) = 3.015 - 6.291 \times 10^{-5} \Delta P_c^2 + 0.0337\psi + \varepsilon_{Atlantic}, \tag{8a}$$

$$\ln\left(R_{\max\_Gulf}\right) = 3.859 - 7.700 \times 10^{-5} \Delta P_c^2 + \varepsilon_{Gulf}, \tag{8b}$$

where $\varepsilon_{Atlantic}$, ε_{Gulf} are the random error terms. In the simulation program, the coordinates of the storm eye (latitude and longitude), translational speed (V_T), heading angle (θ), central pressure (P_c) and radius to maximum winds ($R_{\max}$) are generated at a 6-hour time interval.

The wind field model, proposed by Georgiou (1985), was employed to compute the gradient hurricane wind speeds as:

$$V_g^2 = \frac{r}{\rho_a}\frac{dP_r}{dr} + V_g\left(V_T \sin\beta - fr\right). \tag{9}$$

where, V_g is the gradient wind speed and V_T is the forward speed of the hurricane; *r* is the radial distance from the hurricane eye to the location where the wind speed is being calculated; ρ_a is the air density; β is the angle measured from the hurricane forward direction to where the wind speed is being considered (clockwise is positive); *f* is the Coriolis parameter; and P_r is the air pressure at distance *r* from the hurricane eye. The air pressure field is provided as:

$$P_r = P_c + (P_a - P_c)\exp\left[-\left(\frac{R_{\max}}{r}\right)^B\right], \tag{10}$$

where *B* is the pressure profile parameter, also known as the *Holland B* parameter (Vickery et al. 2000), which are determined by the following equations:

$$B = 1.76 - 1.21\sqrt{A} + \varepsilon_B, \tag{11}$$

where ε_B is the random error term.

Coefficient *A* can be obtained as follows:

$$A = \frac{R_{\max} f}{\sqrt{2R_d T_s \ln\left(1+\frac{\Delta P}{P_c e}\right)}}, \tag{12}$$

where f is the Coriolis parameter; R_d is the gas constant of dry air; T_s is the sea surface temperature (Kelvin), P_c is the hurricane central pressure, and e is the base of natural logarithm.

The *Holland B* post landfall is expressed as:

$$B(t) = B_0 \exp(ct), \tag{13}$$

where $B(t)$ is the value of *B* at time *t* after landfall, B_0 is the value of *B* at landfall, and c is the coefficient empirically obtained as:

$$c = 0.0291 - 0.0429B_0 \le -0.005\,. \tag{14}$$

The gradient wind speed at location (r, β) is obtained by substituting Eq. (10) into Eq. (9) and solve for V_g:

$$V_g = \frac{1}{2}(V_T \sin\beta - fr) + \sqrt{\frac{1}{4}(V_T \sin\beta - fr)^2 + B\frac{P_a - P_c}{\rho_a}\left(\frac{R_{\max}}{r}\right)^B \exp\left[-\left(\frac{R_{\max}}{r}\right)^B\right]} \tag{15}$$

The gradient wind speed in Eq. (15) corresponds approximately to a 10-min sustained wind speed (Georgiou 1985). Subsequently, the 10-min sustained wind speed at the top of the boundary layer (i.e., 3000-m elevation) is converted to a 3-s gust design wind speed at 10-m elevation (Vickery et al., 2009a; ESDU, 1982; ESDU, 1983).

3 CLIMATE CHANGE SCENARIOS AND DETERMINISTIC SEA SURFACE TEMPERATURE

The future climate scenarios based on the Representative Concentration Pathways (RCPs) presented in the IPCC Assessment Report Number 5 (AR5) are used in this study. More specifically the RCP2.6, RCP4.5, RCP6.0 and RCP8.5 are employed for the climate change projections. It is noted that the RCPs are not accounting for fully-integrated indexes of socioeconomic, emissions and climate projections, as described in previous IPCC Special Report on Emissions (SRES) scenarios A1B, B1, and A1F1. The RCP scenarios are defined with a range of radiative forcing levels in the year 2100 in accordance with preindustrial values at +2.6, +4.5, +6.0 and +8.5 W/m^2. The time series of atmospheric, ocean, environmental parameters on air, sea and land, including the sea surface temperatures, have been digitally synthesized in the RCP scenarios by various global climate and atmospheric models. The time series data have been deposited into data portals for sharing with the greater research community. One such portal is GFDL-CMIP5 (GFDL).

Climate change influences the storm frequency and the SSTs. Therefore, one needs to consider the change in annual storm frequency and the SSTs for the future years of the particular climate change scenario (or RCP) of interest. For the historical baseline model, the monthly average SSTs of the HadISST (Hadley Centre sea Ice and Sea Surface Temperature) dataset from 1870 to 2013 (Met Office, Hadley Centre 2013) were used in the central pressure model. For the climate change scenarios, the SST projections were taken from the Geophysical Fluid Dynamics Laboratory (GFDL) of NOAA's Coupled Model Intercomparison Project Phase 5 (CMIP5) simulations under initial conditions of the various IPCC RCP scenarios (Taylor et al. 2012). It should be noted that the SSTs for various RCPs obtained from the GFDL-CMIP5 portal are deterministic (GFDL).

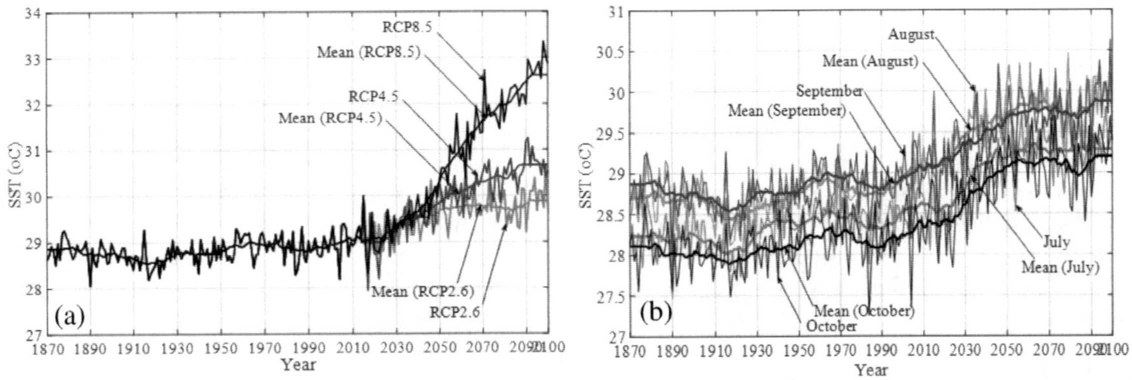

Figure 2. Samples deterministic SST time series of historical and simulated future values : (a) Temperature time series at longitude -70.5^0 and 24.5^0 latitude in August, projected for RCP2.6, RCP4.5 and RCP8.5 (with 15-year moving averaged means), (b) Temperature time series at longitude -70.5^0 and 24.5^0 latitude in July, August, September and October, projected for RCP2.6 (with 15-year moving averaged means).

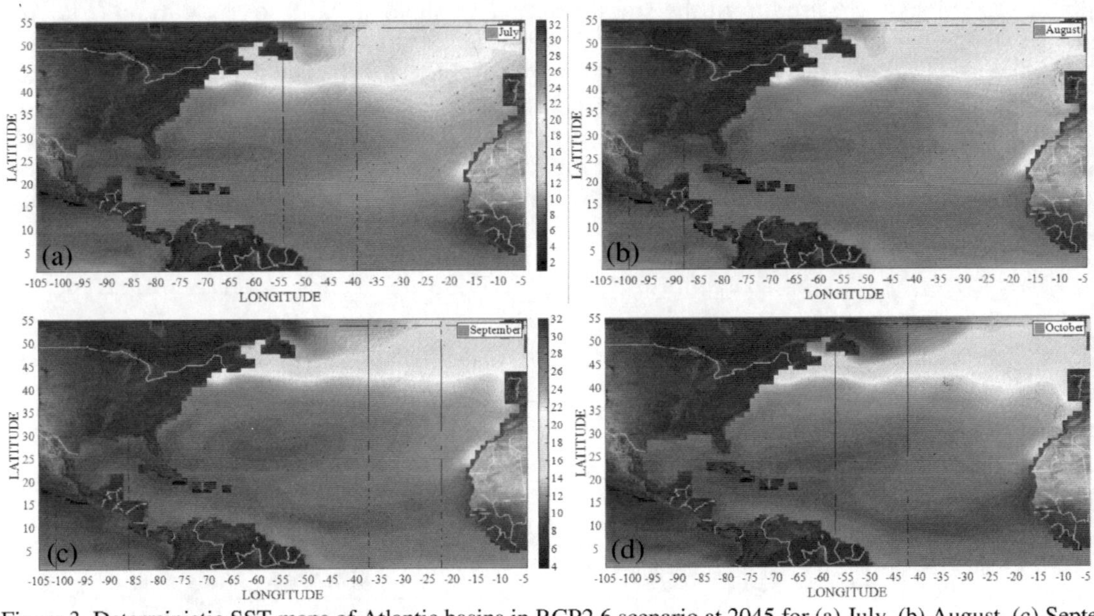

Figure 3. Deterministic SST maps of Atlantic basins in RCP2.6 scenario at 2045 for (a) July, (b) August, (c) September, and (d) October.

Figure 2 illustrates the time series of SSTs at a selected location (i.e. -70.5^0 longitude and 24.5^0 latitude) for three RCP scenarios. The time series at the selected location for the month of August for RCP2.6, RCP4.5 and RCP8.5 are shown in Figure 2(a), whereas those for July, August, September and October for RCP2.6 are shown in Figure 2(b). Time varying means or trends of the SST series can be obtained by using a 15-year moving average smoothing technique. Note that the moving average window (i.e. 15 years) was determined via an optimization approach as the most optimal window size for smoothing purpose. Figure 3 shows the deterministic SST maps projected for July to October in year 2035 for the RCP2.6 scenario.

4 PROBABILISTIC MODELS OF MODEL PARAMETERS

4.1 *Probabilistic Annual and Monthly Storm Frequencies*

In this study, the annual storm frequency at any projection year includes the effect of seasonality, specifically, the monthly storm probability mass function is utilized to determine the spawn month for each storm. A two-step simulation procedure is introduced to generate the numbers of storms for each simulation year and month. In step one, the annual storm frequency for a simulation year is generated based on the statistics of the historical storm frequencies and the projection of the trend into the future. Subsequently, simulated annual storm frequency at the simulation year from the first step is randomly grouped in months based on the monthly storm spawn probability mass function.

For the purpose of generating the annual storm frequencies for all the simulation years, historical record of the storm frequencies is decomposed into two terms of time-varying mean and dispersion. The mean of the annual storm frequency can be determined using a data smoothing method, whereas the fluctuation around the smoothed trend line (red curve shown in Figure 4(a)) is characterized using a zero-mean normal distribution. In this study, several continuous nonlinear oscillatory functions were used to fit the smoothed mean annual storm frequency trend (Liu and Pang, 2012). The decomposition of the historical storm frequencies into oscillatory mean function and the fluctuation around the smoothed mean is shown in Figure 4(a). The fitted oscillatory mean function takes on the form of $y = a_1 + a_2x + a_3xsin(a_4x + a_5)$, where $a_1 - a_5$ denote the fitted coefficients; *x* is the observation years; *y* is the annual storm frequency. Histogram of the dispersions, computed as the differences between the historical storm frequencies and the smoothed mean trend, is shown in Figure 4(b). Also shown in Figure 4(b) is the fitted normal probability density function (red curve) for the dispersion.

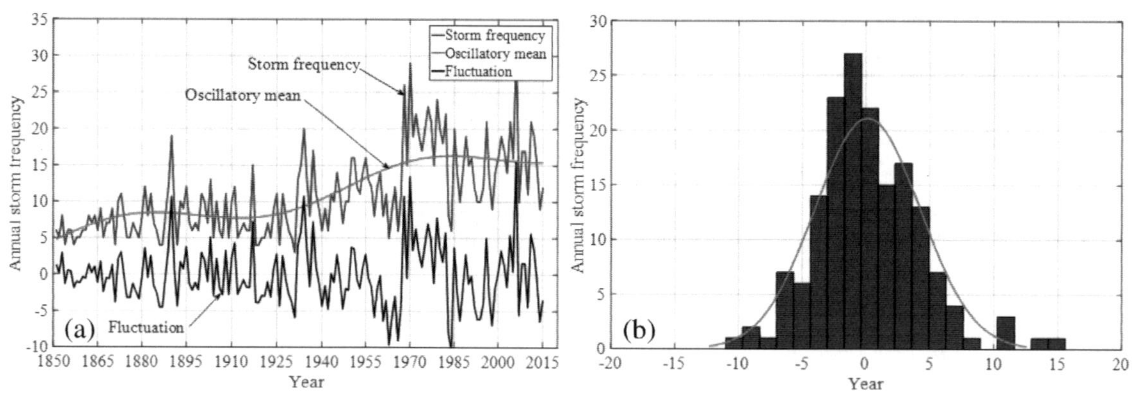

Figure 4. Estimate of oscillatory mean and fluctuation of historical storm frequencies: (a) Decomposition of oscillatory mean and fluctuation, (b) Histogram of fluctuation and fitted normal distribution

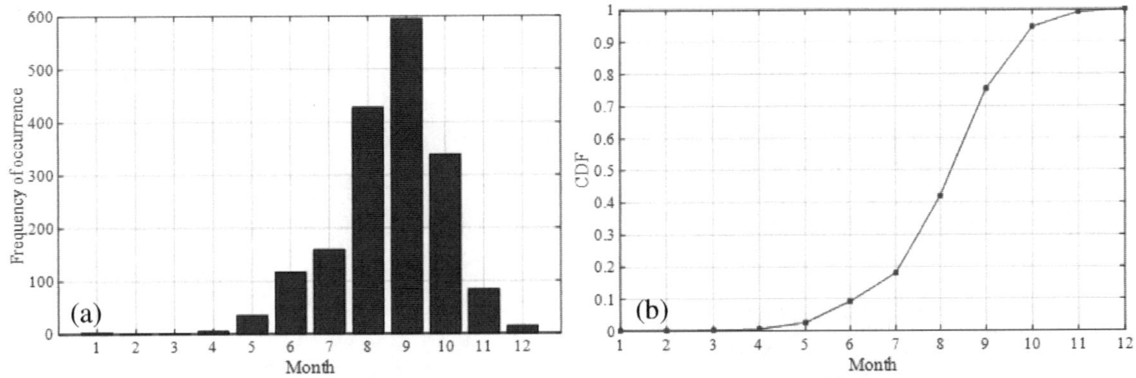

Figure 5. Probability of historical storm frequency in months: (a) Frequency of occurrence, (b) Empirical cumulative density function

Figure 5 depicts the monthly frequency of storm occurrences (i.e. the probability mass function of monthly spawn frequency) obtained from all historical storms. Subsequently, the probability of storm occurrences in the months and the cumulative density function were determined and used to simulate the spawn month for each storm.

4.2 *Initial Storm Conditions*

The initial spawn conditions of a storm consist of the spawn location in terms of the longitude and latitude coordinates, the translational speed and the heading angle. The simulation of the initial storm conditions is based on the non-parametric 1-D and 2-D Gaussian kernel distributions using the kernel density estimation (KDE) method. One of the advantages of the non-parametric KDE method is that the kernel density offers much greater flexibility in modeling a given dataset and it is not affected by specification bias. The Gaussian kernel density estimator used to model the storm spawn location is defined as:

$$\hat{f}(X) = \frac{1}{NH}\sum_{i=1}^{N} K\left(\frac{X - X_i}{H}\right), \tag{16}$$

where N is the total number of historical spawn locations; H is the bandwidth; X is the model spawn location; X_i is the historical spawn location, $K(.)$ is the Gaussian probability density function (PDF) kernel, defined as:

$$K(X) = \frac{1}{\sqrt{2\pi H}} e^{-(X-X_i)^2/(2H)}, \tag{17}$$

where $\sqrt{H}$ is the scale parameter. Note that the bandwidth, H, is a smoothing parameter which is an unknown to be determined.

The initial storm conditions, which include the storm spawn locations, the translational speeds and the heading angles, are simulated via four different steps. First, the 2-D KDE of spawn location probability density map for each month is determined. Second, the number of storms in each month is determined using the monthly spawn rate PMF (see Figure 5) and the actual location of each storm within a given month is determined using the KDE maps developed from the first step. Third, 1-D kernel densities of the translational speeds and the heading angles (see Figure 6(b) and 6(c)) are generated using data close to the spawn location determined in the second step .

Fourth, the initial translational speeds and heading angles are simulated using the 1-D kernel density functions determined in the previous step. Two issues may arise during the simulation of the initial storm conditions. The initial locations of a small fraction of the simulated storms may fall inside the land boundaries. For those storms that spawned on land, they are re-simulated until the spawn locations are all on the ocean. Some of the simulated storms may spawn in regions with sparse observations of historical events (i.e. not enough data points to reliably construct the 1-D KDE for translational speeds and heading angles). Flexible cell (FlexCell) treatment has been proposed, in which the dimension of the cell is expanded to capture enough historical data points for developing the 1-D kernel density functions for translational speeds and heading angles.

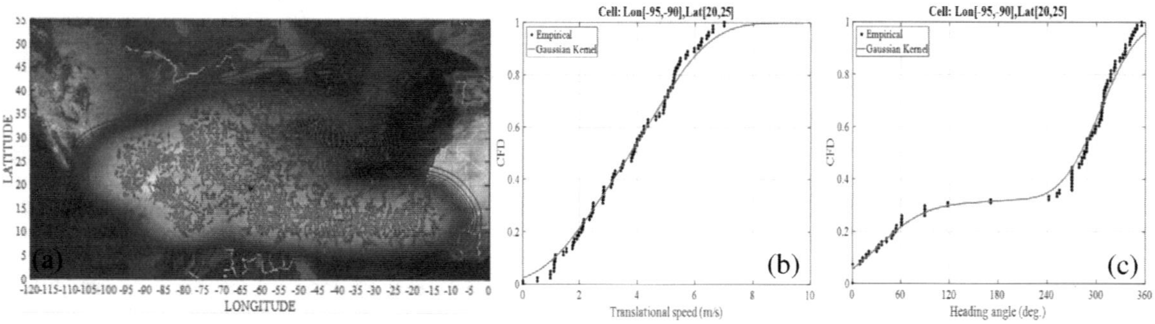

Figure 6. Example Gaussian kernel density estimations for (a) spawn locations, (b) translational speed, and (c) heading angles.

5 SEA SURFACE TEMPERATURE UNCERTAINTIES

The sea surface temperature (SST) models can be broadly characterized into two groups, namely, deterministic (without uncertainty) and probabilistic (with uncertainty) models. In the deterministic SST model, the monthly average SST maps are utilized to determine the SST for the simulated storm. The deterministic SST model has been used in many existing hurricane simulation models. This study explores the influence of SST uncertainty on the intensity of the simulated storms. Three probabilistic SST models with different assumptions for uncertainties are considered: (1) spatially uncorrelated uncertainties, (2) fully correlated uncertainties, and (3) spatially correlated uncertainties.

5.1 *Uncorrelated SST Simulation*

The simulation of uncorrelated SST is carried out as following. First, the deterministic SST time series for different RCP cases are decomposed into two parts, the 15-year moving average means and the corresponding fluctuation (uncertainty). Second, for each location (latitude and longitude), the SST uncertainties or errors around the moving average mean are subsequently fitted to normal distribution. Third, during simulation, the SST uncertainties for all locations are independently generated for each simulation year. Fourth, for each simulation year, the simulated SST uncertainties for all locations are combined with the SST means to generate twelve SST maps (one for each month). The simulated SST maps are used to compute the central pressure of a storm at time step.

5.2 *Correlated SST Simulation*

The use of spatially uncorrelated approach to simulate SSTs assume the SSTs are independents between locations of close proximity. This is not a realistic approach as SSTs of nearby locations are likely to be correlated. To simulate spatially correlated SST maps, a covariance matrix for the SST uncertainties, which describes the spatial correlation of SST uncertainties, is required. The theoretical background is briefly presented in the following equations.

$$[T'] = \{T_1(yr), T_2(yr), \dots, T_N(yr)\} = \begin{bmatrix} T_{11} & T_{12} & \cdots & T_{1N} \\ T_{21} & T_{22} & \cdots & T_{2N} \\ \vdots & \vdots & \vdots & \vdots \\ T_{M1} & T_{M2} & \cdots & T_{MN} \end{bmatrix}, \quad (18)$$

$$[M_U] = \{\mu_1, \mu_2, \dots, \mu_N\} = E[T_i(yr)], \ [C] = Cov\left(T_i(yr), T_j(yr)\right) = E\left[\left(T_i(yr) - \mu_i\right)\left(T_j(yr) - \mu_j\right)\right] \quad (19)$$

$$[C] = [R][R]' = Chol\left([C]\right), \quad (20)$$

$$[T']^* = [M_U] + [R]' \mathbb{N}(0, I, N). \quad (21)$$

where, $[T']$ and $[T']^*$, respectively, denote the uncorrelated SST matrix and the correlated SST matrix. $T_i(yr)$ is the time series of SST of all the locations in the Atlantic (i=1,N); N is the number of SST locations; M is the length of the projection years; $[M_u]$ is the mean SST matrix (the mean vector is organized in a matrix form, regarding to coordinates of SSTs); $[C]$ is the covariance matrix; $[R]$ is the lower triangular matrix; $Cov(.)$ is the covariance operator; $Chol(.)$ is the Cholesky decomposition operator; $\mathbb{N}(.)$ denotes the standard normal distribution.

The simulation procedure for the spatially correlated SST maps are implemented in the following steps. First, the SST time series, consisting of the SST data at every locations on the Atlantic basins for all projection years are organized in matrices. Second, the mean SST vector and the covariance matrix of the SST uncertainties are estimated. Third, the covariance matrix is Cholesky decomposed to determine the lower triangular matrix. Fourth, the probabilistic SST uncertainties are randomly generated based on Eq.(21). Finally, the simulated SST uncertainties and the mean SSTs are combined to generated the spatially correlated SST maps.

6 PRELIMINARY RESULTS AND DISCUSSIONS

A preliminary simulation was carried out using the previously discussed simulation methods to generate 10,000 realizations of hurricane events for year 2045 using the RCP 2.6 climate change scenario. A total of 169,537 storms were generated for 10,000 realizations of year 2045. The simulated storms were grouped into 12 months based on the month spawn rate PMF.

Figure 7 shows the KDE for the probability density of the spawn locations based on the historical data grouped by month. Note that for non-hurricane session or low hurricane activities months (i.e., January, February, March, April), there are insufficient numbers of the historical storm occurrences. Therefore, some months (February and March) were combined to create the spawn location probability density maps using the KDE method. Figure 8 shows the simulated storm spawn locations grouped by month. Good agreements between the simulation and historical spawn locations can be observed. Apparently, the simulated storm spawns are distributed more widely on entire Atlantic basins than the selection among the historical spawns.

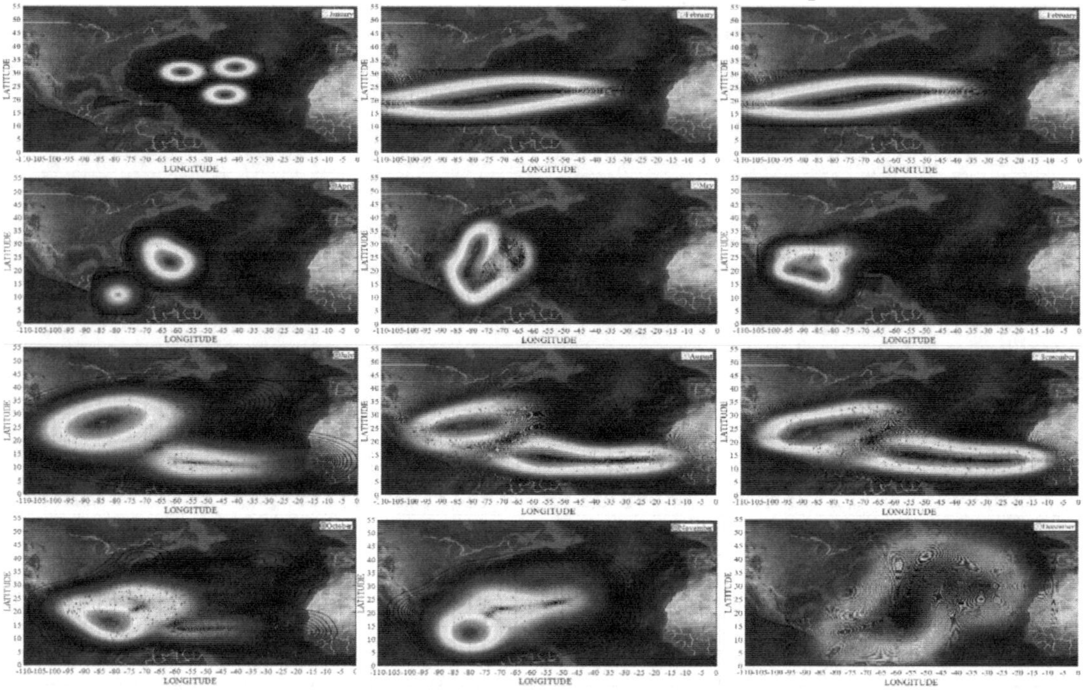

Figure 7. 2-D Gaussian kernel density estimation of historical storm spawn locations of the months of January to December.

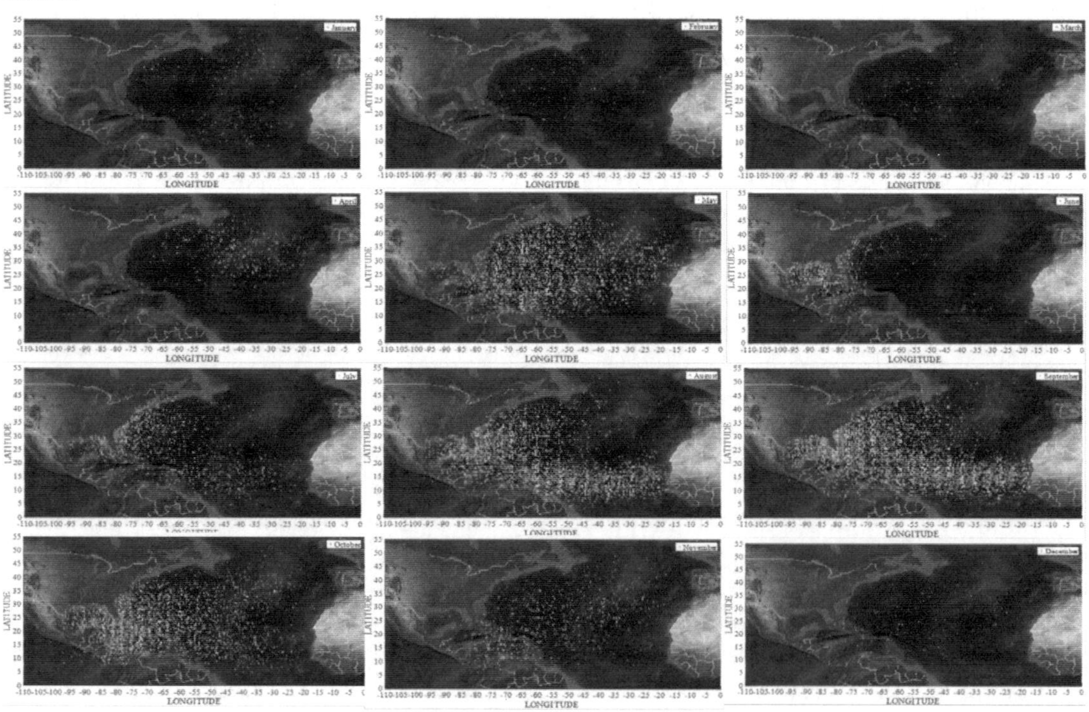

Figure 8. 10,000 realizations of simulated storm spawned in year 2045 for the months of January to December.

Kernel density estimations of the month-based simulated storm spawns have been validated to those of the historical storm spawns, in which there are good agreement in the kernel densities between the simulated and historical spawns observed.

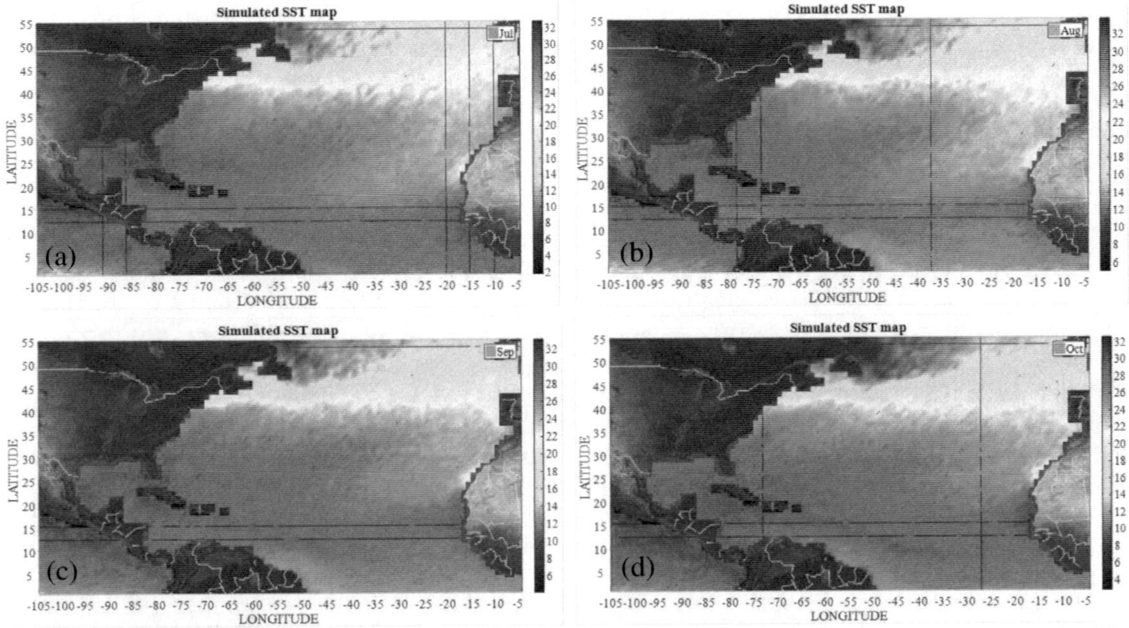

Figure 9. Simulated SST maps for the RCP 2.6 scenario for year 2045 assuming no spatial correlation in the SST uncertainties; (a) July, (b) August, (c) September, (d) October.

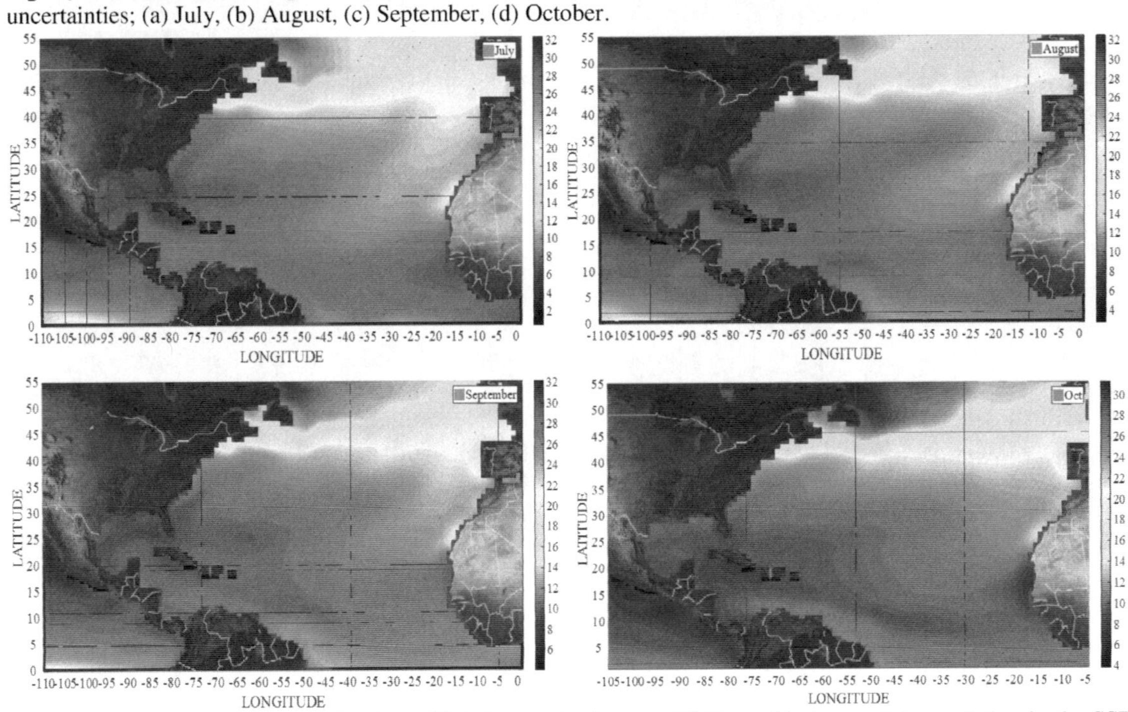

Figure 10. Simulated SST maps for the RCP 2.6 scenario for year 2045 considering spatial correlation in the SST uncertainties; (a) July, (b) August, (c) September, (d) October.

Figure 9 shows the simulated SST maps assumed no spatial correlation between the SST uncertainties for July, August, September and October for the RCP 2.6 scenario for year 2045. For comparison purpose, the spatially correlated SST maps are shown in Figure 10. As can be seen that the spatially correlated SST maps are smooth and more realistic when compared to the SST maps generated assuming no spatial correlation. The SST maps that do not consider spatial correlation show patches of heat spots or cool spots that are not realistic (Figure 9).

7 CONCLUSIONS

This study assesses the impact of climate change on the U.S. Atlantic hurricanes. A new stochastic hurricane simulation model has been developed to improve (1) the annual and monthly spawn frequencies by seasonality (i.e. change in monthly spawn rate), (2) the treatments of the initial conditions of storms (i.e. initial storm locations, magnitude and direction of translation speeds), and (3) the treatment of uncertainties in the SSTs considering various climate change scenarios. More specifically, the annual storm frequencies for future years is predicted are modeled with two terms: a time-varying mean (trend) and a fluctuation term using a zero-mean Gaussian distribution. The generated storms for each simulation year are grouped by month based on the monthly spawn rate probability mass function. The initial conditions for each simulated storm were generated using 1-D and 2-D Gaussian kernel density functions derived from the data of historical storms. Finally, a methodology to simulate monthly SST maps that consider spatial correlation in the SST uncertainties have been developed.

8 ACKNOWLEDGEMENTS

This research is funded in part by the American International Group (AIG), which its support is gratefully acknowledged.

9 REFERENCES

Cui, W., and Caracoglia, L., 2016. Exploring hurricane wind speed along US Atlantic coast in warming climate and effects on predictions of structural damage and intervention costs, Eng Struct 122, 2016, 209-225.

Darling, R.W.R., 1991. Estimating probabilities of hurricane wind speeds using a large-scale empirical model. J. Clim., 4(10), 1035-1046.

Emanuel, K., 1987. Dependence of hurricane intensity on climate, Nature, 326(6112), 483-485.

Emanuel, K., Ravela, S., Vivant, E., Risi, C., 2006a. A statistical– deterministic approach to hurricane risk assessment, American Meteorological Society, 299-314.

Emanuel, K., 2006b. Climate and tropical cyclone activity - A new model downscaling approach, American Meteorological Society, 1797-4802.

Engineering Sciences Data Unit (ESDU), 1982. Strong winds in the atmospheric boundary layer, Part 1: Mean hourly wind speed, Item 82026.

Engineering Sciences Data Unit (ESDU), 1983. Strong winds in the atmospheric boundary layer, Part 2: Discrete gust speeds, Item 83045.

Geophysical Fluid Dynamics Laboratory (GFDL) Data Portal. https://www.gfdl.noaa.gov/gfdl-cmip5-ipcc-ar5-database/

Georgiu, P.N., 1985. Design wind speeds in tropical cyclone-prone regions, Ph.D. dissertation, Univ. of Western Ontario, Canada.

Intergovernmental Panel on Climate Change (IPCC), Fifth Assessment Reports (AR5), Climate change 2014: Synthesis Report.

Jarvinen, B.R., Neumann, C.J., Davis, M.A.S., A tropical cyclone data tape for the North Atlantic Basin 1886–1983: Contents, limitations, and uses, NOAA Technical Memorandum No. NWSNHC-22, 1884, U.S. Dept. of Commerce, Washington, D.C.

Landsea, C., Franklin, J., Beven, J., 2015. The revised Atlantic hurricane database (HURDAT2).

Lee, K.H., Rosowsky, D.V., 2007. Synthetic hurricane wind speed records: Development of a database for hazard analyses and risk studies, Nat Hazards Rev., 8(2), 23-34.

Lin, N., Emanuel, K., Oppenheimer, M., Vanmarcke, E., 2012. Physically-based assessment of hurricane surge threat under climate change, J Nature Climate Change, 2, 462-467.

Liu, F.Q., 2014. Projections of future US design wind speeds due to climate change for estimating hurricane losses, Ph.D. Dissertation, Clemson Univ., USA.

Liu, F. Q., and Pang, W. C., 2012. Influence of climate change on the future hurricane wind hazards along the U.S. Eastern coast and Gulf of Mexico. ATC SEI: Advances in Hurricane Engineering Conf., ASCE, Reston, VA.

Mudd, L., Wang, Y., Letchford, C., Rosowsky, D., 2014. Assessing climate change impact on the U.S. East Coast hurricane hazard: Temperature, frequency, and track, Nat Hazards Rev., 15, 1-13.

Mudd, L., and Vickery, P.J., 2015. Advancements in Synthetic Hurricane Track Modeling in the Gulf of Mexico. Proceedings of 14th Int'l Conf. Wind Eng. (ICWE14), Porto Allegro, Brazil.

Nishijima, K., Maruyama, T., Graf, M., 2012. A preliminary impact assessment of typhoon wind risk of residential buildings in Japan under future climate change. Hydrol. Res. Lett., 6(1), 23–28.

Pei, B., Pang, W.C., Testik, F.Y., Ravichandran, N., Liu, F.Q., 2014. Mapping joint hurricane wind and surge hazards for Charleston, South Carolina, Nat Hazards, 74, 375-403.

Pielke Jr, R.A., Gratz, J., Landsea, C.W., Collins, D., Saunders, M.A., Musulin, R., 2008. Normalized hurricane damage in the United States: 1900–2005, Nat Hazards Rev, 9(1), 29–42.

Vickery, P.J., and Twisdale, L.A., 1995. Wind-field and filling models for hurricane wind-speed prediction. J Struct. Eng., 121(11), 1700–1709.

Vickery, P.J., Skerlj, P.F., Steckley, A.C., Twisdale, L.C., 2000a. Hurricane wind field model for use in hurricane simulations. J Struct. Eng., 126(10), 1203–1221.

Vickery, P.J., Skerlj, P.F., Twisdale, L.A., 2000b. Simulation of hurricane risk in the U.S. using empirical track model, J Struct. Eng., 126(10), 1222–1237.

Vickery, P.J., and Wadhera, D., 2008. Statistical models of Holland pressure profile parameter and radius to maximum winds of hurricanes from flight-level pressure and H* Wind data, J App Met & Climatology, 47(10), 2497-2517.

Vickery, P.J., Wadhera, D., Powell, M.D., Chen, Y., 2009. Hurricane wind field model for use in hurricane simulations. J Struct Eng, 126, 1222-1237.

Wang Y., and Rosowsky, D.V., 2012. Joint distribution model for prediction of hurricane wind speed and size, Struct. Saf., 35, 40–51.

The 13th Americas Conference on Wind Engineering (13ACWE)
Gainesville, Florida USA, May 21-24, 2017

An application of Cat Modeling for Hurricane Risk Mitigation Using Life Cycle Analysis

Ruilong Li[a], Karthik Ramanathan[b], Timothy Reinhold[c] Cagdas Kafali[d],

[a] *AIR Worldwide, 131 Dartmouth Street, Boston, MA 02116, U.S.A.* rli@air-worldwide.com
[b] *AIR Worldwide, 131 Dartmouth Street, Boston, MA 02116, U.S.A.* kramanathan@air-worldwide.com
[c] *Insurance Institute for Business and Home Safety, 5335 Richburg Road, Richburg, SC 29729, U.S.A.* treinhold@ibhs.org
[d] *AIR Worldwide, 131 Dartmouth Street, Boston, MA 02116, U.S.A.* ckafali@air-worldwide.com

ABSTRACT: Every year, there are significant losses from natural disasters, including hurricanes, earthquakes, hailstorms, high winds, and floods, to mention a few. Many mitigation measures have been developed to minimize future losses resulting from different natural disasters; however, it is very important to optimize mitigation guidelines to best suit an organization's budget and resource constraints, as well as their resiliency goals for various properties constructed under different building codes and regional norms. The IBHS FORTIFIED program is a set of engineering and building standards created to help minimize the damage from specific natural hazards and to create more resilient buildings in the United States. The program employs a systems-based approach to improving resilience and typically has three different levels of designations—Bronze, Silver, and Gold—which provide options that best fit an organization's resources and budget as well as the risk-levels.

In a mature economy such as the United States, the importance of costs associated with the installation of mitigation measures, their maintenance and management are being increasingly recognized. In this context, the introduction of life cycle benefit analysis concepts constitutes an important development in the choice of optimal mitigation measures. Although loss-cost benefit analysis is not a new concept, its application for hurricane risk assessment is beginning to gain widespread acceptance in the United States. The basic concept of life cycle benefit analysis is a simple one. For a particular structural design, the cost piece of the analysis consists of initial costs in mitigating an existing structure. The loss piece of the analysis examines the reduction in the expected loss to the structure due to the application of the specified mitigation measure for a specific time. Life cycle benefit analysis is then aimed at justifying the cost of mitigation against the reduction in losses over the life span of the structure.

The aim of the current study is to substantiate the choice of a specific level of designation in the IBHS FORTIFIED program (Bronze vs. Silver vs. Gold) towards enhancing the hurricane resistance of residential and commercial construction in the United States using loss-cost benefit analysis. The choice of mitigation measures is often location and year built dependent since both these attributes are influenced by the local building codes that are in place at a given location at the time of construction and strongly influence the resistance of the building being built. The study also investigates the optimality of the choice of specific levels of IBHS FORTIFIED designation in space and time using a cat model.

KEYWORDS: Catastrophe Models, IBHS FORTIFIED program, mitigation measures, optimality, hurricanes

1 INTRODUCTION

Natural hazards such as tropical and extratropical storms, severe thunderstorms and tornadoes are known for their destructive impacts on life, economy and environment. Although it is not possible to completely avoid damage due to such disasters, it may be possible to minimize their devastating effects by enhancing resilience in communities, that is, by reducing damageability of the built environment, consequences of these damages, and time to recovery. Damageability and recovery time can be reduced by increased system performance through retrofitting and rehabilitation. Most of builders intend to build a code compliant structure and generally do not consider making extra investments to reduce or further mitigate damage from natural disasters. Recognizing that code compliant construction represents minimum requirements, numerous mitigation measures, including the IBHS FORTIFIED program, have been developed to minimize future damage and increase the resilience of the built environment to natural disasters. To motivate building owners to retrofit existing buildings, it is very important to optimize the mitigation measures to best suit the limited budget as well as the resiliency goals for different structures built to different building codes and exposed to different dominant natural disasters. These decisions are based on target performance levels and cost estimates. Life-cycle risk analysis of systems is concerned with the performance of these systems subjected to natural and/or man-made hazards during their lifetime. The results from a life-cycle risk analysis can be used to identify an optimal retrofitting technique for structural/nonstructural systems from a collection of design alternatives.

In this paper, specific levels of designation in the IBHS FORTIFIED program (Bronze vs. Silver vs. Gold) are evaluated using the life cycle risk analysis and AIR's catastrophe modeling framework to guide the choice of the optimal measure given a structure of certain age and a certain location subjected to hurricane hazards. System performance is measured by the expected lifetime losses. The methodology is based on site hazard analysis, system fragility analysis and capacity and cost estimation. The methodology is probabilistic in nature since (1) the intensity and arrival time of different hurricane events (2) the loads acting on the system due to wind intensities, and (3) some of the system characteristics, are generally uncertain. Consequently, probabilistic models are used for characterizing natural hazards occurring at a given site. The input to the analysis consists of hurricane hazard, building properties, performance criteria, rehabilitation strategies, and a reference time. Lifecycle costs consist of initial cost related to the rehabilitation of the system and repair/replacement costs for bringing the damaged systems back to their original states. Section 2 briefly describes the catastrophe modeling framework which forms the basis of the analysis performed in this paper, section 3 briefly describes the hurricane risk model used in this study, section 4 presents a case study using the hurricane risk model and the IBHS FORTIFIED program and presents results from a lifecycle cost benefit analysis perspective, and section 5 provides concluding remarks.

2 CATRASTROPE MODELING FRAMEWORK

This study uses AIR's Hurricane Model for the United States [1] to assess the choice of optimal IBHS fortification in space and time to enhance the hurricane performance of single family homes within the U.S. hurricane prone states.

Figure 1 shows the framework for AIR's Hurricane Model for the United States. The event generation and local intensity calculation modules together comprise the hazard module within the framework. Event generation module deals with so-called source parameters and answers questions about where events are likely to occur, how large or severe are they likely to be, and how frequently they are likely to occur. The task of this module is to simulate all types of possible, yet realistic, future scenarios. Upon probabilistic generation of a potential future event, the local intensity module propagates the event across the affected area by estimating the local intensity (such as location specific surface wind speed or storm surge depth). In this component as well as in the event generation component, detailed scientific and meteorological data and algorithms are employed to model the local effects of each simulated event. Windstorm models, for example, use high-resolution digital land use/land cover data to calculate surface frictional effects. Estimates of surface roughness dictate, in part, the behavior of ground level wind speeds.

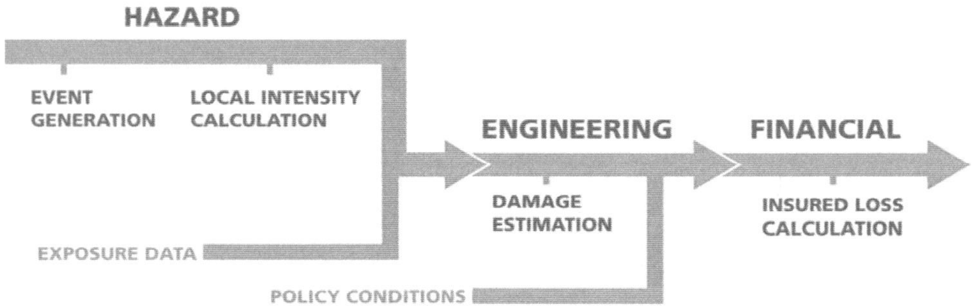

Figure 1: Components of AIR's Hurricane Model for the United States

The local intensities of each simulated event are superimposed onto a database of exposed properties. The engineering or damage estimation component then calculates the resulting monetary damage by applying mathematical relationships, called damage or vulnerability functions. Damage functions capture the relationship between the intensity of the event (measured in terms of wind speed or storm surge depth) to the damage ratio, which is the ratio of the repair cost of the building to its replacement value. While engineering is at the core of a catastrophe model's damage function development framework, the ultimate objective of a catastrophe model is not the estimation of physical damage states or idealized cost estimates; rather, it is the estimation of future monetary losses that an insurer will actually pay out in the aftermath of a catastrophe. To achieve this, factors other than a structure's engineering response must be accounted for such as building codes and their enforcement, claims adjustment and insurance practices, socio economic effects, societal preparedness and response, to mention a few. Damage functions are region-specific and time specific and reflect a thorough understanding of local building codes and construction practices. They provide not only estimates of the mean or expected damage ratio corresponding to each level of intensity, but, in addition, provide a complete probability distribution around the mean. Because different structures experience different degrees of damage for a given level of intensity, the damage functions need to capture this variability, or so-called secondary uncertainty.

Finally, insured losses are calculated by applying the cedant's specific policy conditions to the total damage estimates in the financial module of the catastrophe modeling framework. The model output is usually in the form of probability distributions of losses and their complement, exceedance probability curves, along with metrics such as average annual loss (AAL) which are used heavily in the insurance space.

3 AIR'S HURRICANE MODEL FOR THE UNITED STATES

AIR's Hurricane Model for the United States captures the complete profile of wind and storm surge risk leveraging the state of the art in terms of meteorology and engineering principles. The model is fully stochastic and incorporates a 100,000 year catalog of hurricanes signifying

100,000 realizations of the upcoming hurricane season. The relationship between wind intensity and building damage, described by damage or vulnerability functions varies depending on building characteristics, such as occupancy class, construction type and building height as well as regional and temporal variation in building standards as gleaned from the building's year of construction. In addition to these aforementioned primary characteristics, the model facilitates superior risk differentiation by considering the impact of several secondary risk characteristics associated with roofs, walls, connections etc. on the damageability. The list of model supported building and environmental features that impact a building's wind vulnerability is shown in Table 1 below. AIR's approach is a knowledge-based expert system designed to estimate the performance of residential and commercial buildings under wind loads. It is based on engineering principles and data regarding building performance during high winds.

The general, or base, damage functions used by the AIR Hurricane Model for the United States are individually developed for the "typical" building with certain "primary" risk characteristics, which include age, number of stories, construction type and occupancy class. The construction and occupancy classes are broadly defined, without reference to individual—or secondary—structural characteristics. AIR's general damage functions were validated and calibrated using extensive and detailed actual hurricane loss data. The relative abundance of hurricane loss data greatly facilitates the determination of average or "typical" building performance.

Modification functions are then applied to the general damage functions to reflect the performance enhancement or diminution of a wide variety of secondary structural and environmental characteristics listed in Table 1. These might include roof covering, roof pitch, type of window protection or proximity to trees (which are potential sources of wind-borne missiles). Detailed claims data from 2004 and 2005 storms have been used to validate the impact of these secondary risk characteristics on building vulnerability.

Table 1. List of primary and secondary building characteristics supported in AIR's Hurricane Model for the United States.

Primary Building Characteristics:

Occupancy
Construction
Number of stories
Year of construction/Age

Secondary Building Characteristics:

Category	Feature	Category	Feature
Location Roof Detail	Roof Geometry	Location Connection Detail	Roof Attached Structures
	Roof Pitch		Foundation Connection
	Roof Covering		Wall Attached Structures
	Roof Deck		Appurtenant Structures
	Roof Covering Attachment	Location Surrounding Detail	Tree Exposure
			Small Debris
	Roof Deck Attachment		Large Missile
	Roof Anchorage		Terrain Roughness
	Roof Year Built		Adjacent Building Height
Location Wall Detail	Glass Type	Location Building Detail	Floor of Interest
	Glass Percentage		Building Condition
	Window Protection		Seal of Approval
	Exterior Doors		Designated Structures (IBHS)

The hallmark of the model is the ability to capture spatial and temporal variation in vulnerability of the building stock by capturing the evolution of state and local building codes along with their enforcement. The model takes a comprehensive approach to developing damage functions for each region and year built in the United States, while drawing on many years of experience and research conducted at AIR. There are many sets of building codes released in the United States, with varying levels of adoption and enforcement. This, coupled with a lack of historical data, makes it somewhat challenging to assess the regional impact of building codes. AIR takes a top-down approach to incorporate information about building codes at a local level for all regions. First, the evolution over time of the national wind load standards (ASCE 7-98, ASCE 7-05, ASCE 7-10) and their impact was analyzed. Next, the wind design provisions of the model building codes (National Building Code (NBC), Standard Building Code (SBC), and International Council of Codes (ICC)) were evaluated. As state and local authorities may significantly amend codes, their adoption and amendment as related to wind design provisions was then examined. "Model" building—in terms of individual risk features such as roof covering, roof sheathing, roof-to-wall connection, and opening protection—that meets the building code requirements for that region and time period is defined and the degree of code enforcement in the region is analyzed. In some localities, enforcement of building codes is mandatory and changes to instituted codes are not allowed. In other localities, changes may be made to the building codes only if the modifications are more restrictive than the original codes. In some states, a state authority can declare certain codes to be mandatory statewide. Nonetheless, a locality can determine whether or not to adopt the "mandatory" code, and therefore a mandatory code is only practiced in those localities that choose to adopt it. The relative vulnerability of different model buildings is assessed based on the impact of secondary building features on the overall building vulnerability. By repeating this process for all hurricane prone regions in the United States, model buildings have been defined for all locations and time periods.

Catastrophe risk assessment models often get utilized to project potential future hurricane losses and price (re)insurance contracts. Given the explicit treatment of the evolution of building codes and their enforcement, they can be used to assess the impact of regional/local building codes and their effectiveness with respect to the exposed hurricane risk [2][3]. As in the present case, they can also be used to assess the optimality of potential retrofit strategies to increase the resilience of the exposed building stock to natural catastrophes.

4 CASE STUDY – LOSS-COST BENEFIT ANALYSIS FOR CHOICE OF OPTIMAL MITIGATION STRATEGY

Recognizing that code compliant construction represents minimum requirements, numerous mitigation measures have been developed to minimize future damage and increase the resilience of the built environment to natural disasters. To motivate building owners to retrofit existing buildings, it is very important to optimize the mitigation measures to best suit the limited budget as well as the resiliency goals for different structures built to different building codes and exposed to different dominant natural disasters by using catastrophe model based loss-cost benefit analysis. In this section, suites of mitigation measures included in the IBHS FORTIFIED program are evaluated using life cycle cost benefit analysis framework to guide the choice of the optimal measure given a structure of certain age and a certain location.

4.1 *IBHS Fortified Program*

IBHS FORTIFIED program is a set of engineering and building standards to strengthen the structures against natural disasters, primarily hurricanes and severe thunderstorms. The Fortified Standards are continuing to expand throughout the country. In fact, the Alabama Department of Insurance now offers wind mitigation discounts to the base insurance premiums for those risks mitigated using IBHS fortification standards. Three different levels of designation: Bronze, Silver and Gold, are introduced to minimize the damage from hurricanes, high wind and hail for existing or new residential and commercial structures. Each designation is focused on strengthening a particular component of the structure with higher levels of designation building on the prior levels. IBHS Bronze focuses its effort on the strengthening the roof, with specific provisions for roof covering, roof sheathing attachment, and anchorage of the roof deck, etc., to minimize the risk of water intrusion into the building. IBHS Silver designation requires satisfying all the requirements from Bronze as well as focusing mitigation on the wall envelope including requirements for gable end walls, windows and doors, etc. The Gold designation builds on the Silver designation requirements, and aims at the creation of a continuous load path from roof to foundation [4] [5][6][7][8][9]. Another superior set of standards with more stringent requirements than the Gold designations called Safer Living and Safer Business, are also available for design of new residential and commercial structures, respectively [10] [11]. In the AIR Hurricane Model for the United States, all four designations can be appropriately modeled for both residential and commercial structures. Given the variability in the hurricane hazard along the coastal United States coupled with the vast landscape of building code adoption and enforcement practices, the choice of an optimal hurricane mitigation strategy is not an easy one to make. What turns out to an optimal mitigation strategy at a certain location for a risk of a certain age might be sub-optimal for another location and a similar kind of building. This paper utilizes the loss cost benefit analysis platform to demonstrate its potential by investigating the optimality of the IBHS FORTIFIED Hurricane requirements. This case study assesses the benefit to cost ratio associated with increasing the resilience of wood frame single family homes of varying ages at select locations in the hurricane prone regions of the United States.

4.2 *Benefit-cost analysis*

Catastrophe risk assessment models often get utilized to project potential future hurricane losses and price (re)insurance contracts. Given the explicit treatment of the evolution of building codes and their enforcement, they can be used to assess the impact of regional/local building codes and their effectiveness with respect to the exposed hurricane risk [2][3]. The present study uses catastrophe models to guide the selection of optimal retrofit strategies in the form of IBHS fortification in a cost benefit analysis framework. It does not consider potential savings due to insurance or other types of credits that may help motivate building owners to take action.

One story wood frame single family homes of different ages and IBHS FORTIFIED Home Hurricane Designations are modeled in two locations, one each in Miami-Dade County, Florida and Baldwin County, Alabama, as shown in Figure 2. The building ages and IBHS FORTIFIED Home Hurricane Designation levels for these locations are described in Table 2. The choice of year of construction is made such that each value of year built falls within a particular building code era in the state. For instance in the case of Florida, a house built in the year 1992 at location 1 would conform to the requirements of the pre- Hurricane Andrew South Florida Building Code, while that built in the year 1998 would be in accordance with the provisions of the post- Hurricane Andrew South Florida Building Code that required shutters and used

ASCE 7- 88 to specify design wind loads. Houses built in 2005 would follow the requirements of the High Velocity Hurricane Zone in the 2001 Florida Building Code.

Figure 2. Case study locations

Table 2. Year of construction and IBHS Fortification options chosen for the test structures by location.

Location	Year of construction	IBHS Fortification				Total
		None	Bronze	Silver	Gold	
Location 1, Miami-Dade County, Florida	1992	✓	✓	✓	✓	4
	1998	✓	✓	✓	✓	4
	2005	✓	✓	✓	✓	4
Location 2, Baldwin County, Alabama	1992	✓	✓	✓	✓	4
	1998	✓	✓	✓	✓	4
	2015	✓	✓	✓	✓	4
						24

Cost–benefit analysis is a well-established tool for comparing alternative mitigation investments, though it is frequently used in scenario driven applications. This framework presents an opportunity to assess the cost effectiveness of IBHS fortifications through a risk-based framework. Accurate estimation of both cost and benefit for desired levels of mitigation plays a crucial role in the choice of an optimal mitigation strategy. The as-built costs and the costs to mitigate the case study structures to different levels of IBHS FORTIFIED Home standards is obtained by modeling the same in a commercially available replacement cost estimation product called 360Value [12]. The replacement cost estimates for the as-built homes in the Florida and Alabama locations are approximately \$220/sq.ft. and \$110/sq.ft., respectively, for a 2000 square-foot wood frame single family home.

Table 3 shows the cost of mitigating the modeled as-built structures of different ages at the two locations to different levels of IBHS fortification. The costs were also checked against data provided by the Insurance Institute for Business and Home Safety (IBHS). As shown in Table 3, a \$10,000 investment is required to upgrade a 1992 built home in Miami-Dade County to a bronze designation. This investment covers the cost to re-roof. However, the investments for the 1998 and 2005 built homes at the same location are based on the cost to strengthen the roof from inside the attic. The cost differential of \$9,500 between the bronze and silver designations in the Florida location for a 1992 built home arise from the need to provide opening protection, brace the gable ends and replace the garage door. However, this cost differential is only \$2,500 for the 1998 and 2005 built homes by virtue of the year built and the building codes that were in effect in Miami-Dade County, Florida at this time. Homes built to the post-Andrew South Florida Building Code and Florida Building Code version 2001, respectively, would already have opening protection as required by these codes thereby only requiring a nominal expenditure for vulnerabilities that may have been overlooked such as strengthening porches and carports etc. The \$10,000 additional investment for a 1992 built home to upgrade to gold from the silver designation is to tie the house together to ensure a continuous load path from the roof through the walls and on to the foundation. However, for the 1998 and 2005 built homes in Florida, the building codes ensure load path continuity and therefore fortification to silver standards automatically ensures a gold designation without any additional investment. Similar assessments apply to the Alabama homes built during different periods. The primary exception is that a 2015 home might well be fitted with code allowed wood structural panel shutters while the FORTIFIED program requires commercial products for opening protection in this area.

It is important to note that the cost estimates assume that the building owner decides to undertake the mitigation measures solely because they want to reduce their vulnerability. If the mitigation measures were undertaken as part of required maintenance, such as needing to re-roof, or as part of recovery from a damaging event, where insurance payments would be available to cover some of the costs, the mitigation related costs could be dramatically lower than the values assumed in this analysis. Lower costs would significantly increase the benefit-to-cost ratios. Improvements in building codes and building standards in the latter years with respect to wind design provisions is what leads to comparably smaller investments to mitigate to similar levels of IBHS fortification when compared with similar costs for older structures.

Table 3. Cost of mitigating the as-built structures by age to different IBHS Fortification levels

Location	IBHS Fortification	1992	1998	2005
Location 1, Miami-Dade County, Florida	Bronze	\$10,000	\$5,500	\$5,500
	Silver	\$19,500	\$8,000	\$8,000
	Gold	\$29,500	\$8,000	\$8,000
Location	**IBHS Fortification**	**1992**	**1998**	**2015**
Location 2, Baldwin County, Alabama	Bronze	\$7,500	\$7,500	\$5,500
	Silver	\$17,000	\$17,000	\$10,500
	Gold	\$27,000	\$27,000	\$10,500

In order to estimate the benefits side of the problem, the portfolio of 24 buildings described in Table 2 is modeled in AIR's Hurricane Model for the United States to estimate the hurricane risk to every building. Estimates of a commonly used actuarial metric, the average annual loss (AAL) is then determined for each of the case study structures by running the hurricane model through stochastic simulations that generate 10,000 realizations of the next hurricane season. Since the hazard is the same for a given location, changes in the AAL across different years and different levels of IBHS Fortification is a direct reflection of the change in vulnerability of the buildings due to different building codes and mitigation levels. Figure 3 shows the AAL estimates for the test buildings of different ages and mitigation levels at the two chosen locations. The following are some of the noteworthy observations:

- The AAL estimates are different for the two locations. This is expected given the differences in the hurricane hazard and also the building vulnerabilities dictated by the local building codes and their respective enforcements at these locations
- For a given location and mitigation level, AAL decreases with newer structures. This is due to the betterment of building codes and building standards along with their enforcement at the test locations.
- For a given year built, increase in the level of IBHS fortification usually leads to decreases in the AAL, which is along expected lines. While IBHS Bronze mitigation efforts focus on the roofs, IBHS Silver focuses mitigation efforts on both roofs and walls leading to better hurricane resistance and lower AALs. Likewise, IBHS Gold focuses on roofs, walls and foundation further leading to increases in hurricane resilience of the structure and further reductions in AALs.
- With Location 1 in Florida, especially in the latter years of construction i.e. 2005, the reduction in AALs with increased levels of IBHS Fortification is not well pronounced or is non-existent. This is because the test location falls within FBC's High Velocity Hurricane Zones where building design and construction prescriptions are very strict and are almost at the level of IBHS FORTIFIED Home Hurricane Gold Standards.

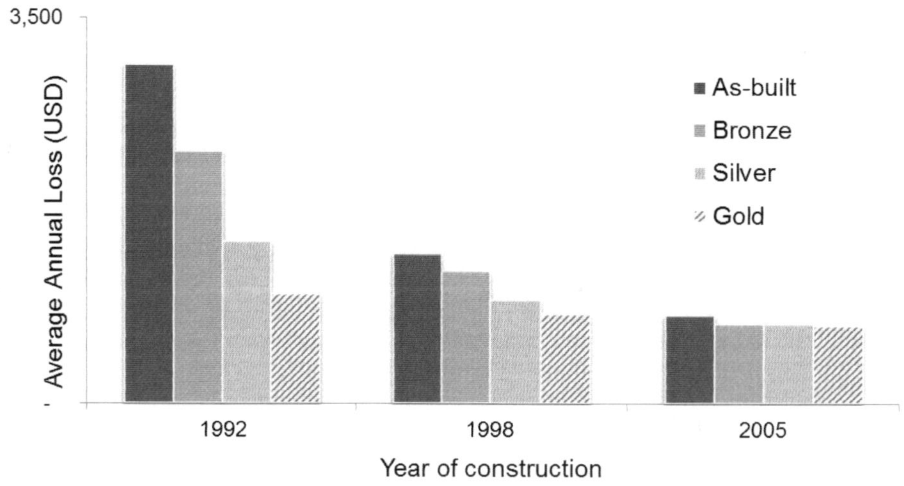

a) Location 1, Florida

b) Location 2, Alabama

Figure 4. Variability of AAL for test buildings of different ages and mitigation levels at the two case study locations

Having estimated the AAL from the Hurricane Model for various ages and mitigation levels, the benefit of a particular mitigation is evaluated as the difference between the present value of the AAL estimates without any mitigation, $AAL_{as\text{-}built}$, and the present value of the AAL estimate with a mitigation strategy employed, $AAL_{mitigated}$. This is expressed mathematically in equation (1). The present value of the AAL can be estimated using equation (2) where T is the service life of the structure and assumed as 50 years in this study.

$$Benefit_{mitigation} = AAL_{as\text{-}built} - AAL_{mitigated} \quad (1)$$

$$AAL_{present\ value} = \sum_1^T Model\ estimated\ AAL \quad (2)$$

Having estimated the cost and benefit, the benefit-cost ratio (BCR) is then assessed as the ratio between the benefit attained based on a specific mitigation strategy to the initial cost of employing the respective mitigation strategy, described mathematically in equation (3).

$$BCR_{mitigation} = \frac{Benefit_{mitigation}}{Cost_{mitigation}} \quad (3)$$

BCR may be perceived as a measure of the financial return for every dollar invested in hurricane mitigation measure under consideration. BCR greater than one indicates a positive return on investment and the mitigation measure with the largest BCR value has the largest expected savings in losses over the service life of the structure for every dollar of investment made in the mitigation strategy. Table 5 shows the BCR ratios for case study structures of varying ages and mitigation standards at the chosen locations. The BCR values shaded in green indicate the optimal mitigation strategy for the structure considered. Clearly, the results in Table 5 demonstrate that the optimal mitigation strategy for location 1 in Florida is not the same as location 2 in Alabama. Further, the choice of optimal mitigation strategy is also age dependent at a location. The analysis indicates IBHS bronze and silver are optimal mitigation strategies for the 1992 built home in the Florida location, while IBHS gold is the optimal strategy for the 1998 built home. In contrast, the analysis suggests the optimum mitigation strategy is bronze for both 1992 and 1998 built homes in the Alabama location for this case study. The stringent requirements for the high velocity hurricane zones in FBC 2001 and similar prescriptions for Baldwin County, Alabama already reduce the vulnerability of new homes thereby leading to no positive benefit-cost ratio based on the cost estimates considered in this case study. However, the costs are significantly reduced for achieving all levels of designation in these areas when bronze elements are included in new construction at the time the home is built. Experience with the FORTIFIED program in Alabama where thousands of homes have achieved a FORTIFIED designation reflect these benefits and costs. New construction typically achieves a gold designation while existing homes are generally mitigated so that they achieve a bronze designation – and that is typically done when the home needs a new roof or is part of a mitigation grant program.

Table 5. Cost benefit ratios for test structures of varying age and mitigation strategies.

Location	Year of construction	IBHS Fortification		
		Bronze	Silver	Gold
Location 1, Florida	1992	3.96	4.01	3.50
	1998	1.45	2.58	3.67
	2005	0.62	0.39	0.39
Location 2, Alabama	1992	2.57	1.74	1.21
	1998	2.40	1.63	1.14
	2015	0.27	0.36	0.54

5 CONCLUSION

The loss cost benefit analysis using IBHS FORTIFIED program in AIR US hurricane model confirmed the benefits of optimizing the different levels of mitigation measures. Different structures exposed to different severity of the hazard and constructed using different building code and construction practice will exhibit different vulnerabilities. Also, costs required to improve the structure to meet different levels of mitigation measure requirements vary significantly. Thus, with limited budget and resource constraints, as well as meeting their resiliency goals, it is important to clearly communicate the relative benefits and costs of implementing specific mitigation guidelines. It is also important to consider opportunities for incorporating mitigation measures when the costs will be minimized such as when a building is being re-roofed, when a major renovation or repair is undertaken, or when the building is being constructed. While this research was limited to two different locations in Alabama and Florida for hurricane only, it clearly shows risk models can be a valuable tool for assessing the potential benefits of mitigation measures and for assessing their relative benefit-cost ratios. It is clearly worthwhile to further study the optimality of the choice of specific levels of mitigation guidance such as those in the IBHS FORTIFIED program for different part of the country and also from a multi-hazard impact standpoint using catastrophe modeling.

6 REFERENCES

[1] AIR Worldwide (2015). AIR Hurricane Model for the United States. AIR Worldwide, Boston, MA.

[2] S. Bobby, T. Johnson, B. Kordi, F. Moghim, and K. Ramanathan, The Impact of Temporal and Spatial Variation in Building Codes on the Hurricane Wind Vulnerability of Residential Single Family Homes, Structures Congress 2017, Denver, CO, April 6-8, 2017.

[3] B. Kordi, K. Ramanathan, T. Johnson, F. Moghim, S. Bobby, Hurricane Risk in Florida as it Related to the Evolution of Building Codes, 13th American Conference on Wind Engineering (13ACWE), Gainesville, FL, May 21-24, 2017.

[4] Insurance Institute for Business & Home Safety, FORTIFED HOME, Technical Requirements Summary, New Residential, Single-Family Detached Homes, Hurricane.
[5] Insurance Institute for Business & Home Safety, FORTIFED HOME, Technical Requirements Summary, Existing Residential, Single-Family Detached Homes, Hurricane.
[6] Insurance Institute for Business & Home Safety, FORTIFED HOME, Technical Requirements Summary, New Residential, Single-Family Detached Homes, High Wind and Hail.
[7] Insurance Institute for Business & Home Safety, FORTIFED HOME, Technical Requirements Summary, Existing Residential, Single-Family Detached Homes, High Wind and Hail.
[8] Insurance Institute for Business & Home Safety, FORTIFED COMMERCIAL, Standards, Hurricane.
[9] Insurance Institute for Business & Home Safety, FORTIFED COMMERCIAL, Standards, High Wind and Hail.
[10] Insurance Institute for Business & Home Safety, FORTIFED for Safer Living, Builders Guide.
[11] Insurance Institute for Business & Home Safety, FORTIFED for Safer Business, Builders Guide.
[12] 360Value, 360Value Residential – Replacement Cost Estimates, Verisk Insurance Solutions, http://www.verisk.com/underwriting/property/360value-residential.html

13th Americas Conference on Wind Engineering
Gainesville, Florida, USA
May 21 - 24, 2017

Wind Damage Function and Exposure Database developed in the Framework of Hurricane Loss Prediction Model

Jeong Hee Oh[a], Piboon Apirakvorapinit[b], Mamta Gadhiya[c], Siamak Daneshvaran[d], Bin Pei[e]

[a] *Impact Forecasting, Aon Benfield, Chicago, IL, USA, jeong.oh@aonbenfield.com*
[b] *Impact Forecasting, Aon Benfield, Chicago, IL, USA, piboon.apirakvorapinit@aonbenfield.com*
[c] *Impact Forecasting, Aon Benfield, Chicago, IL, USA, mamta.gadhiya@aonbenfield.com*
[d] *Impact Forecasting, Aon Benfield, Chicago, IL, USA, siamak.daneshvaran@aonbenfield.com*
[e] *Impact Forecasting, Aon Benfield, Chicago, IL, USA, bin.pei@aonbenfield.com*

ABSTRACT

Economic or insured losses due to natural hazards such as hurricanes are estimated by the hazard model, the damage function model and the exposure data. This paper is concerned with the development and validation of the wind damage function for buildings and the exposure database in the framework of the hurricane loss prediction model.

The ultimate goal of the wind damage function for the loss prediction model is to represent vulnerability levels of an assets exposure (e.g., re/insurance portfolio). The portfolio damage model needs to be robust enough to cope with the large variability and inherent uncertainty in the characteristics of loads and resistance on the assets in an exposure. The wind tunnel testing data on NIST (National Institute of Standards and Technology) buildings conducted by University of Western Ontario was utilized to estimate the wind loads on the surface/cladding of the buildings. The spatial and directional distributions of peak pressures in various regions of building components (similar to zoning for components and claddings defined in ASCE 7-10) were sampled, and applied to different parts of the building envelope components (i.e., roof shingles, roof sheathing, roof frame, roof-wall connection, window protection, wall cladding and garage door) to mimic the failure mode of the typical buildings in an insurance portfolio. The wind resistance of the building components were also sampled using their distributions derived from testing reports, product literature and field observations of product performance. By calculating the failure wind speed for each building component, complying with the wind pressure equation defined in ASCE 7-10, large numbers of scenarios of physical damage to buildings subjected to hurricanes can be developed. The damage of each component was then converted into dollar loss that is normalized by the replacement cost of the building. Monte Carlo simulation is employed to develop probability distribution for physical damage in a series of wind speed intervals. The damage functions were tested and verified by comparing simulation results with claim data from the insurance industry. Hurricanes Wilma in 2005 and Matthew in 2016 were used as scenario events.

The U.S. exposure database was developed from State to census block levels for residential buildings for the year 2017. The building exposure values were derived at census block level using housing units data (i.e., number of units at a block estimated from data available from US Census Bureau), Unit floor area (sqft) data (available from HAZUS) and state-wise construction costs data (derived from recent data available from US Census Bureau, RS Means and HAZUS). The three dwelling types (i.e., single family dwellings, multi-family dwellings and mobile homes) were separately used to develop the residential exposure values. The exposure values for multi-family dwellings were derived separately for categories comprising 2-units, 3 to 4 units, 5 to 9 units, 10 to 19 units, 20 to 49 units and 50 & above units. The total exposure value of residential buildings in Florida at the end of 2017 is presented and validated by the U.S. Census Bureau's Net Stock of Residential Fixed assets.

KEYWORDS

Hurricanes, Damage, Exposure, Losses, Insurance

Fragility Model for Hurricane-induced Damage to Tall Building Clusters with Application to Urban Resilience Analysis

Yanlin Guo[a], John van de Lindt[b]

[a] *Colorado State University, Fort Collins, Colorado, USA, yanlin@colostate.edu*
[b] *Colorado State University, Fort Collins, Colorado, USA, jwv@engr.colostate.edu*

ABSTRACT

Tall buildings typically do not suffer significant structural damage from hurricane wind because they have engineered wind load resisting systems comprised of steel or concrete frames and/or shear walls contrary to non-engineered low-rise structures such as residential. The primary damage to tall buildings after hurricanes is mainly comprised of the damage to the cladding systems and façade, as well as the water intrusion through the building envelope resulting from that damage. Thus this paper presents the results of research that focused on the assessment of non-structural damage to tall buildings from hurricanes. The existing damage assessment models for mid/high-rise buildings often neglect the geometry of building clusters or simply assume a homogeneous configuration for simplification purposes. These simplified approaches likely introduce significant uncertainties for urban tall buildings whose geometries and distributions often vary. In this context, this research proposes a new fragility model for hurricane-induced damage to tall buildings, which explicitly considers both the individual building geometries and complex surrounding environment of urban areas, so that the uncertainties associated with geometries of individual buildings and building clusters can be propagated through the community or urban model to better provide risk-informed decision support. Moreover, to embrace the hurricane vulnerability of tall buildings into community or urban resilience assessment, the proposed fragility model is developed to assess the damage at both the individual building and community level. To develop such a fragility model, first, high fidelity physical models of debris impact, wind pressure damage and water intrusion for an individual building are established, considering the configuration of surround buildings in the complex urban area. Then the computationally efficient surrogate models are built based on physical models to facilitate Monte Carlo based probabilistic damage assessment. Damages of individual buildings are integrated to assess the total damage at the community level for an illustrative example.

KEYWORDS

Fragility model, tall building, non-structural damage, hurricane, urban resilience

The 13th Americas Conference on Wind Engineering (13ACWE)
Gainesville, Florida USA, May 21-24, 2017

Site-specific Building Vulnerability to Wind Perils – A Closed-form Framework

Jason Lin [a], Mohsen Rahnama[b]

[a] *Risk Management Solutions (Beijing), 6 Wudinghou St., Beijing, Beijing, CHINA*
[b] *Risk Management Solutions Inc., 7575 Gateway Blvd., Newark, California, USA*

ABSTRACT: The presented framework treats a building as a system comprising assemblies of component elements, for example, an assembly of roof shingles, or a collection of glass windows. Each element's physical attributes affect its probabilistic pressure/impact load and resistance, and lead to generally distinct failure probabilities among elements within an assembly for a given wind speed, primarily because loads vary spatially. A closed-form method is introduced to compute a probabilistic assembly damage ratio from these unequal element failure probabilities. A closed-form scheme is further introduced to obtain a probabilistic building damage ratio from multiple assembly damage ratios, accounting for nonlinear interactions among assemblies and multiple types of loads. This framework enables real-time computation of building vulnerability to wind perils, when provided with inputs on building shape and dimensions, component materials and qualities, exterior glazing and opening characteristics, interior partitions, surrounding topography, terrain roughness, tall tree and debris source distributions etc. along with certain other details affecting an individual building's aerodynamic and structural performances.

KEYWORDS: Insurance, Risk, Vulnerability, Damage, Wind, Site-specific

1 INTRODUCTION

Traditionally in the insurance industry, building vulnerability is primarily derived by statistically analyzing claim losses against estimated local wind speeds that cause the losses, sometimes with empirical adjustments based on simple engineering analyses and judgments. This approach has served the risk modeling community for several decades since the onset of catastrophe modeling in the 1980s. Nevertheless, relying solely on such an approach has inherently high uncertainties from four groups of sources:

- There are epistemic uncertainties in the *estimated* local wind speed that the damaged structure is purported to have experienced during a loss event, due to uncertainties in the wind field model, the geocoded location, and the residual effects of local topography, terrain surface roughness, and immediate surroundings interference, and so on.
- There are uncertainties associated with below-deductible and unclaimed losses estimated and added back in the analysis, and uncertainties in insurance-to-value ratios, inevitably present along with adjuster errors and other contaminations in the claim loss data. Together these undermine the accuracy of analyzed mean damage ratios (MDR) and loss distributions for given nominal wind speeds.

- There are uncertainties about the attributes of a building or a group of buildings that are insured and claimed losses for. Often a vulnerability curve is derived from losses for a collection of structures whose physical attributes are scarcely known. It is uncertain how representative a vulnerability curve so derived is for a given group of structures, which may have other attributes than those the curve is derived from.
- There is a general lack of claim data or loss experiences for some regions and/or classes of structures, which limits the development of vulnerability using such statistical approach.

As a relatively new breed, engineering-science-based vulnerability development relies first on understanding the physical mechanisms and process of wind damage to buildings as a basis, and then mathematically formulating the relationship of probabilistic damage level to given nominal wind speed for various parameters representing building attributes and site characteristics. A vulnerability model built as such is then examined and calibrated by comparing results with applicable claims data, with warranted adjustments to model assumptions.

Previous implementations of engineering-based hurricane vulnerability [1 & 2] use multi-layer, nested Monte Carlo simulations and detailed bookkeeping of the location and state of every element on a building, which is an extremely laborious and lengthy process. This approach limits the number of parametric variables and their variation ranges manageable with such models under constraints of computing time and storage. Such deficiency has effectively constrained the scope of the operational application of such models.

The framework and formulations presented herein theoretically and operationally advances vulnerability modeling for hurricanes, and potentially for other perils, in several ways.

First on the theoretic front, by delving into the detailed mechanism of component element failures, the framework introduces a physically and mathematically well-founded, three-parameter probability distribution model that represents the collective effects of element failures on assembly damage ratios and building damage ratio. The three parameters are closed-form functions of an aggregate of individual element failure probabilities that are derived from probabilistic wind loads on, and resistance capacities of, these individual elements using a conventional demand-supply model. Load and resistance data used are supported by engineering tests and research. As the model propagates the variability and uncertainty in load and resistance through the process into the assembly and building damage ratios, the three-parameter distribution inherently develops with a distributed density portion bounded between 0 and 1 (100% damage), along with probability masses at 0 and 1, respectively. Component assembly repair and replacement costs from regional construction and renovation data are incorporated in the process.

Operationally, the model engine developed runs million to billion times faster than Monte Carlo simulations to compute a mean damage curve or a probability distribution of damage ratios. This enables real-time assessment of vulnerability once building attribute inputs are provided and requires no data storage. The model also brings about exceptional connectivity with hazard and financial models and can be easily updated with new engineering data and research results. The mean damage curve and associated probabilistic damage ratio distributions are parametric, smooth, and continuous, compared to those pre-compiled from time-costly simulations that are discrete and require much larger data storage.

Details of the framework and key formulations are laid out in the following sections.

2 FRAMEWORK OUTLINE

Figure 1 illustrates a framework comprising essentially four computational steps, each of which involves one or more processes. Given information on site and building attributes as parametric variables, the framework is aimed to establish a functional relationship between the wind speed (hazard severity) as input and the building damage ratio as output. In catastrophe risk modeling, a basic functional relationship is the mean vulnerability function, or so-called damage curve where a given hazard severity corresponds with a mean damage ratio (MDR). A more comprehensive functional relationship is the *probabilistic vulnerability function*, or *damage surface* where a given hazard severity corresponds with a *probabilistic damage ratio distribution* defined in the domain [0, 1]. The framework illustrated in Figure 1 provides for the latter case where uncertainties in loads and resistances, including their inherent variabilities and those uncertainties resulting in Step 0 (due to lack of information or knowledge on site and building attributes), propagate through Steps 1, 2 and 3 all the way to the final building damage ratio output.

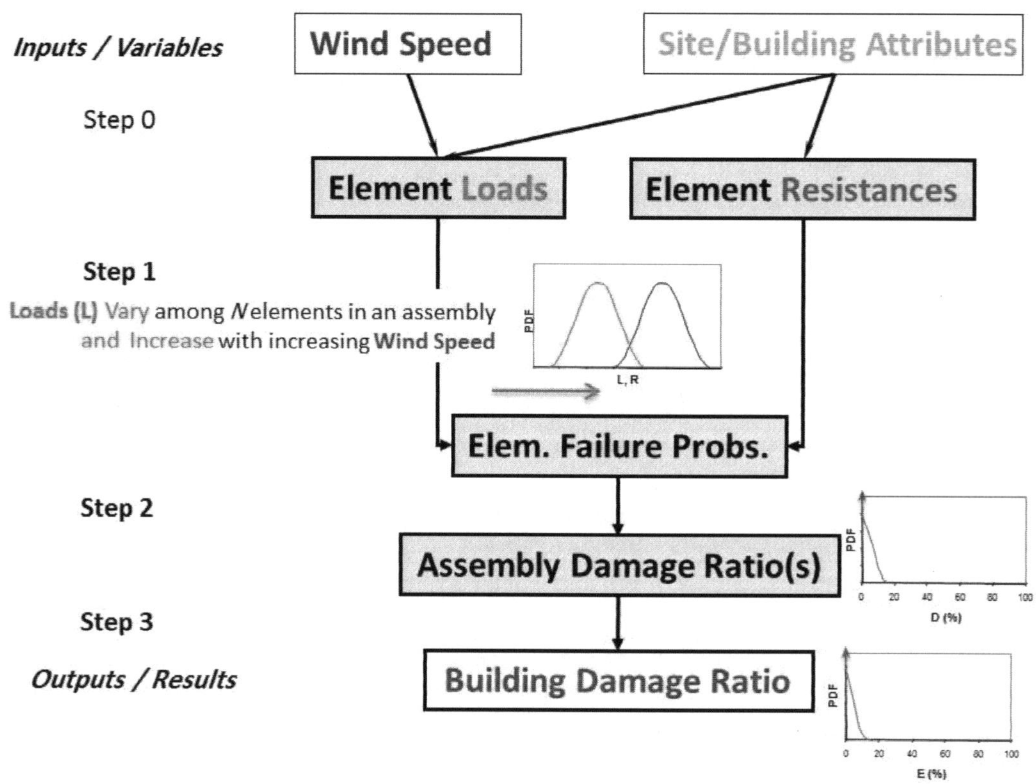

Figure 1. Flowchart illustrating the computational framework

In Figure 1, Step 0 involves following processes:

- Computations of load demands generated from wind-induced pressure, wind-borne debris impact and tree-fall impact on building exterior, for a given wind speed and a set of building and site attributes such as overall building height, footprint shape and dimensions, geometric setbacks, roof configurations, glazing area and unit size, surrounding topography and terrain roughness, major buildings nearby, tall tree and debris source distributions, etc. While wind tunnel test is a relatively reliable method to obtain wind pressure loads, it cannot cover all situations or scenarios encountered in vulnerability modeling. An automated wind pressure generalization method, so-called "virtual wind tunnel", used to estimate pressure loads for risk modeling will be discussed in separate papers, which generalizes directional pressure coefficient data obtained with wind tunnel and full-scale tests, primarily using aerodynamic similarity theory being applied to local flow structures around a building and around portions of a building such as corners and edges [3 & 4]. A wind-borne debris impact modeling methodology will also be presented in separate articles.

- Defining various component resistance capacities from building attribute information such as roof covering and wall cladding materials, their production grades, installation qualities and maintenance practices, glazing characteristics, etc. User inputs are utilized when provided, or otherwise regional building inventory statistic data are used as substitutes, which may yield higher uncertainty as reflected in the current model methodology. In either case, laboratory tests and field observations play an irreplaceable part in quantifying resistance capacities of various component materials and products, while building code requirements are sometimes utilized in vulnerability modeling as baseline references only.

The following subsections describe the key processes for Steps 1 through 3.

3 KEY FORMULATIONS

A building is a system constructed of component assemblies, each of which consists of equal or assumed equal elements, for example, a roofing assembly of asphalt shingles or clay tiles, or a fenestration assembly of windows of similar size and material, etc. As such, characterizing the building damage state by a set of defined assembly damage ratios is conducive to tractable modeling of building vulnerability. The framework and associated methodologies are centered on computing assembly damage ratios, derived from individual element failure probabilities and composed into building economic damage ratio.

3.1 *Element failure probabilities*

Given information on various types of load (pressure, impact etc.) and corresponding resistance, the failure probability (p_i) of an i-th element is determined by the probabilistic load demand versus the probabilistic resistance capacity through conventional reliability methodology as follows,

$$p_i = \int_0^\infty [1 - F_L(x)]_i [f_R(x)]_i dx \qquad (1)$$

where $F_L(x)$ is the CDF for load demand on an element, and $f_R(x)$ is the PDF for resistance capacity of the element. This formulation sums up the demand-capacity joint probability density where demand exceeds capacity, to arrive at the failure probability of an element. Note that the element failure probability generally varies among individual elements, since load often varies spatially and resistance capacity may vary among spatial zones. Also note that load magnitudes on all elements generally increase with wind speed.

3.2 *Assembly damage ratio distribution*

Consider an assembly a collection of equal-worth elements, and the probabilistic damage ratio of an assembly can be represented by the number of failures from a set of Bernoulli trials that have *unequal* failure probabilities, normalized by the number of trials, that is, the total number (N) of elements in an assembly. A closed-form approximation, upon mathematical manipulations, is obtained to compute the probability density function (PDF) of the assembly damage ratio, D (continualized random variable $D \in [0, 1]$, with functional variable $d \in [0, 1]$), as follows,

$$f_D(d) = N\frac{\Gamma(n+1)}{\Gamma(a+1)\Gamma(n-a+1)}p^a(1-p)^{n-a} \tag{2}$$

with intermediate variable a being related to the functional variable d by,

$$a = d \times N, \qquad d \in [0, 1] \text{ and } a \in [0, N]; \tag{3}$$

and the PDF parameters,

$$n = N\frac{\bar{p}^2}{\bar{p}^2 + \tilde{p}^2} \tag{4}$$

$$p = \frac{\bar{p}^2 + \tilde{p}^2}{\bar{p}} \tag{5}$$

where $\bar{p}$ is the average, and $\tilde{p}^2$ the mean squared deviations, of p_i by Eq. (1) ($i = 1, \ldots N$), the failure probabilities of N elements. The symbol $\Gamma(\ \ )$ represents the Gamma function.

To some degree, this PDF resembles the Binomial distribution's PMF, but is defined on a non-integer, continualized and normalized random variable $D \in [0, 1]$. It can also be defined on $A=D\times N \in [0, N]$ if not normalized by N, and Eq. (2) is to be re-scaled by $1/N$ so that N will drop from Eq. (2) to become a *two*-parameter PDF (n and p). When failure probabilities are all identical, it reduces to a Binomial, but continualized. It is also delighting that Eq. (2) autonomously includes a mass at 0 and another at 1, thus is an autonomous form of combined PDF-PMF, one that is much desired in loss modeling where loss is bounded at *0* and *1* (or max. exposure). Detailed derivation and Monte Carlo verification of Eqs. (2-5), evaluation, and other features and properties, will be presented in a separate paper along with real-world application examples beyond the current field.

The assembly mean damage ratio ($\bar{D}$) and variance (${\sigma_D}^2$) are given by,

$$\bar{D} = \bar{p} \tag{6}$$

and

$${\sigma_D}^2 = \frac{\bar{p}(1-\bar{p}) - \tilde{p}^2}{N} \tag{7}$$

They are related to the distribution parameters n, p and the normalizing factor N as,

$$\bar{D} = \frac{n}{N}p \tag{8}$$

and

$$\sigma_D{}^2 = \frac{n}{N^2}p(1-p) \tag{9}$$

Assembly damage ratios evaluated as above for a set of applicable building assemblies will in a large part define the physical damage state of a building system; however, before assembly damage ratios are aggregated into a building damage ratios three types of adjustment need to be considered:

a. There exist certain functional dependences of the damage ratio of one exterior assembly on that of another. For example, additional internal pressure due to wall fenestration breach adds to the total load alongside with exterior surface pressure, which in turn will change the damage ratios of roof deck and roof frame etc. from those without this consideration. Another example is that, when an underlying roof deck element (say a plywood panel) fails and dislodges, roof cover elements supported on it (say roof shingles) will be carried and fail. Similarly, if a section of a roof frame fails, the supported roof deck panels and roof shingles also fail. These are a sort of "*unidirectional*" dependences that can approximately modeled with functional dependence and accounted for either before or after applying the above formulation in Section 3.2 as appropriate.
b. Damage to some assemblies, such as interior finishes, equipment, furniture and electronics, depend firstly on damage or breach to such other assemblies as exterior wall fenestration and roof cover or deck, secondly on interior partition, and thirdly on their susceptibility to water and wind damage. Without damage to exterior assemblies that protect the interior assemblies, there is little "load" (wind and water infiltration) on these interior assemblies. As damage to exterior increases, damage to interior assemblies increases. As such, their mean damage ratios and standard deviations are estimated as functions of the damage ratios of exterior assemblies, interior partition and their susceptibility to water and wind damage. This concept and treatment also apply in the case of contents damage.
c. So far, we have discussed *physical* damage ratios of various assemblies. For them to be aggregated to a building damage ratio, the measurement of damage must be in an economic term or scale, that is repair or replacement cost, and *economic* damage is usually not linear to physical damage. A set of functional relationships between economic damage ratios and physical damage ratios for various assemblies need to be established to better quantify the overall building economic damage ratio. Normally, economic damage ratio of an assembly increases faster towards a total damage than its physical damage ratio. For example, if 5% of roof shingles is damaged, the repair cost, depending on how the damage is spatially distributed, is normally larger than 5% of the replacement cost of the entire roof shingle assembly; and if the physical damage reaches certain threshold, the entire assembly will likely be replaced by usual industrial practice.

In addition, the set of formulations in Section 3.2 is derived assuming independence among elements, and is verified to be a close approximation in scenarios where mutual dependence exists among elements. One such scenario is for those somehow interlocking roof shingles or roof tiles, or wall sidings, of which any one element's probabilistic outcome may alter the resistance and/or load of adjacent elements and thus their probabilistic outcomes. This *mutual* dependence would be better accounted for approximately as correlations among random variables than as functional dependence, for that the dependence here is *multidirectional* and no failure probability of any one of the elements in the relationship can be evaluated before the "trials" by relying on the knowledge of the outcomes of the others (unlike those scenarios described in paragraphs a. and b. above, where one or more are depended random variables while others are dependent random variables). Such correlations would alter the element failure probabilities (p_i) evaluated with Eq. (1) in Section 3.1 for element loads and resistances in their original state by transforming them into a new set of failure probabilities with an approximate correlation coefficient matrix, and this will thus alter the parameters n and p in this Section. In many cases, the results of the transformation are equivalent or very close to those obtained by properly modifying the distribution functions of the element load and/or resistance used as inputs in Eq. (1). This topic, one of the progressive failure problems, will be discussed in more detail with a separate paper.

3.3 *Building economic damage ratio*

A building is a system constructed of component assemblies, each of which shares a proportion in the total economic value or replacement cost of the building. Proportional assembly costs are available in such publications as RSMeans yearbooks, by building and assembly construction types for various geographic regions, which are helpful in evaluating building economic damage ratio from assembly damage ratios.

A cost-weighted sum of the assemblies' damage ratios is used to estimate the building economic damage ratio (denoted $E \in [0, 1]$), with its mean ($\bar{E}$) and variance (${\sigma_E}^2$) being,

$$\bar{E} = \{w\}^{\mathrm{T}}\{\bar{D}\} \tag{10}$$

$${\sigma_E}^2 = \{w\}^{\mathrm{T}}[\sigma_D][\rho][\sigma_D]\{w\} \tag{11}$$

where $\{w\}$ is the assembly cost weights vector; $\{\bar{D}\}$ is the assembly mean damage ratios vector and $[\sigma_D]$ is a diagonal matrix of assembly damage ratio standard deviations; and $[\rho]$ is a correlation matrix reflecting residual correlations between each pair of assembly damage ratios beyond those functional dependences discussed in Section 3.2.

It is verified that the PDF of the resulting building economic damage ratio can be described approximately by a reduced form of Eq. (2) as,

$$f_E(e) = m\frac{\Gamma(m+1)}{\Gamma(b+1)\Gamma(m-b+1)}q^b(1-q)^{m-b} \tag{12}$$

with intermediate variable b being related to the functional variable e by,

$$b = e \times m, \qquad e \in [0, 1] \text{ and } b \in [0, m]; \tag{13}$$

and the PDF parameters m and q are related to the distribution's moments (sample statistics) by,

$$q = \bar{E} \tag{14}$$

$$m = \frac{\bar{E}(1-\bar{E})}{{\sigma_E}^2} \tag{15}$$

4 APPLICATION EXAMPLE

A most recent and comprehensive use of the framework and formulations is upon a vulnerability model product developed in Risk Management Solutions Inc., which is named RiskAssessor.

Recall the processes illustrated in figure 1, a set of variables is first determined as shown in figure 2, which affect element loads and resistances and thus their failure probabilities, or other physical quantities, for various component assemblies as depicted in Figure 3. The four variables prefixed with "*" under category Building General in Figure 2 are the so-called primary variables for conventional vulnerability modeling.

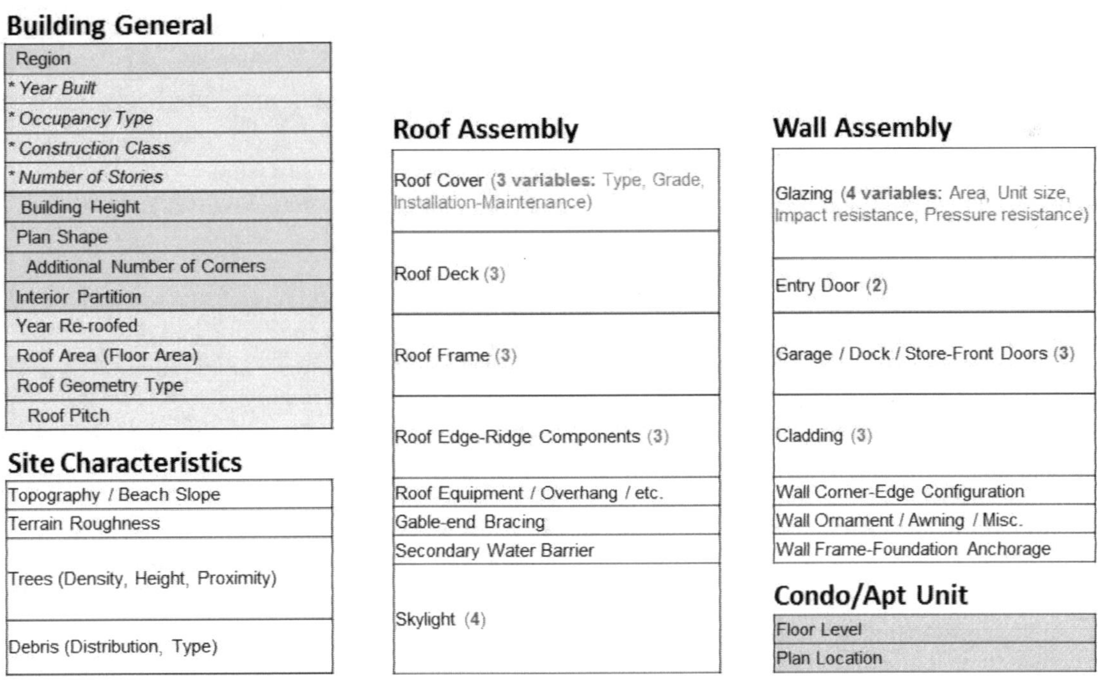

Figure 2. An example set of variables used

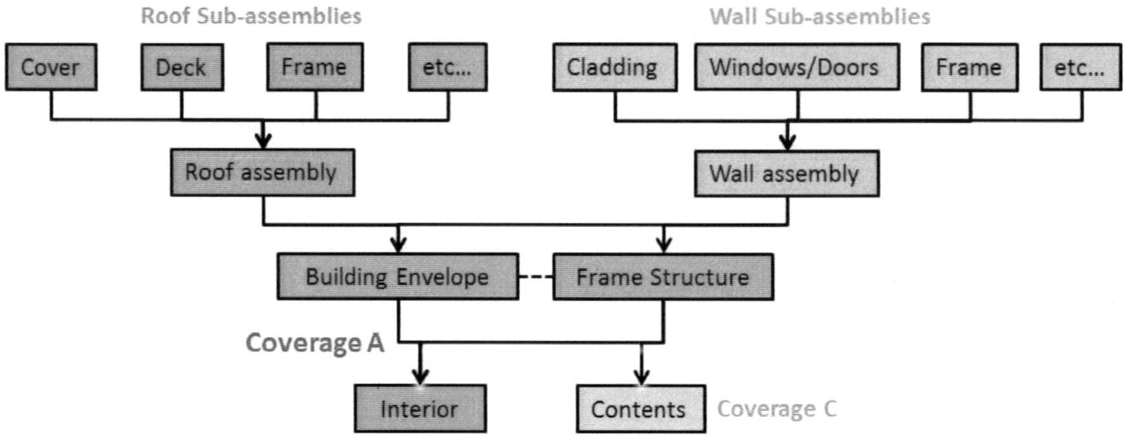

Figure 3. Building assemblies of elements

As mentioned in Section 2, once given the input values for the variables, probabilistic loads for wind-induced pressures, wind-borne debris and tree-fall impacts are computed real-time using built-in models that were developed based upon proper generalizations of results from wind tunnel tests, other field and laboratory tests, and upon analytical and numerical solutions of differential equations in some such cases as wind-borne debris generation, trajectory and terminal velocity (all to be presented and discussed in separate papers). On the other hand, probabilistic resistances to these loads are estimated based upon available data from engineering researches and product tests in laboratory or in field, with references to building code requirements or certified product grades in some cases. The above processes involve considerations of a variety of such factors as regional practice, code enforcement and compliance; structural forms and component materials, element unit size; construction quality and installation workmanship, building age and maintenance level; site terrain roughness and topography; forestation characteristics; building shape and dimensions, architectural or equipment fixtures, etc. For condo and renter polices, floor level and plan location of a unit are also accounted for, to differentiate the unit from the building average as a whole.

Using the resulting loads and resistances for the exterior components, element failure probabilities, assembly damage ratios, and finally the building and contents economic damage ratios, are then computed respectively with the set of formulations described in Sections 3.1, 3.2 and 3.3.

As an example, the computed mean building economic damage ratio (MDR) is shown in Figure 4 as a function of wind speed (3-sec maximum peak gust at 10m in local terrain), for a US structure with most of the variables being set to default values that are mostly drawn from regional building inventory statistics. The curve noted as WEB (stands for Wind Engineering Based vulnerability) is one derived from the methodology discussed herein, and the one noted as RL13 is excerpted from a historical vulnerability database that is developed using claim loss data. It is seen that, the two curves agree well for wind speeds below 150 mph, at which both curves indicate a MDR of about 23%. Within this lower wind speed range, tremendous claim loss data are available and used to develop the RL13 curve and thus lend it credibility. The two curves diverge somewhat beyond that point until they both reach a total loss, where loss data is scarce to none such that the RL13 curve is essentially an extrapolation from the lower part, mostly based on expert judgment.

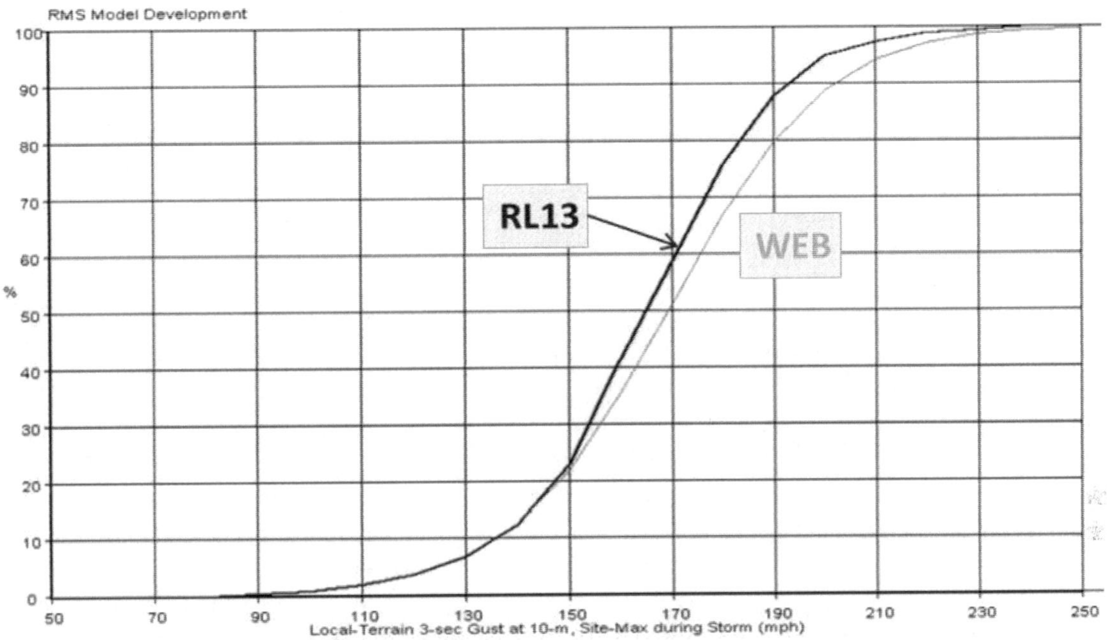

Figure 4. Example vulnerability curve comparison between current method and historical loss data based method (WEB – current; RL13 – historical)

Figure 5 shows a scatter plot of loss costs (annual average loss per $1000) computed for thousands of locations of various hurricane climates, indicating satisfactory agreement between WEB-derived results and RL13's.

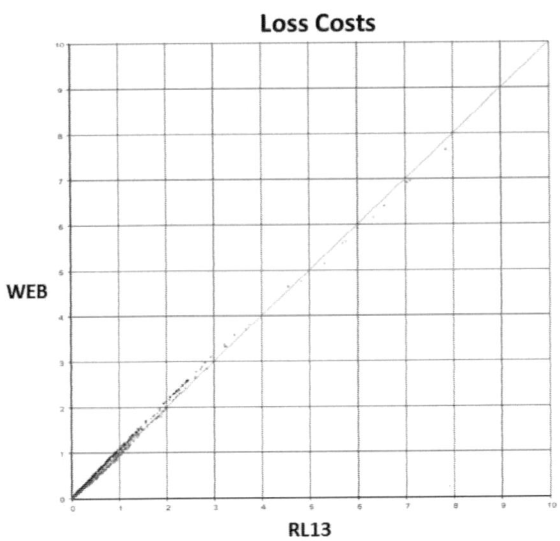

Figure 5. Loss cost catter plot comparison between current method and historical loss data based method

Figure 6 compares a set of vulnerability curves for various building box shapes and dimensions (only changing height and roof area) with all other variables held consistent. The smallest building yields the highest vulnerability (black curve) above all others shown, for two primary reasons. Firstly, the *proportion* of "hot zones" (roof corner and edge areas experiencing high loads) for a small roof area, is larger since the spatial extent of the "hot zones" on a roof is proportional to the building height instead of its planar dimensions as examined by numerous studies (see [3, 4 & 5], for example). Secondly, its interior space is small, such that once the building envelope is damaged or breached, say on the above roof hot zones, wall corners (wall "hot zones") or wall openings, larger *proportion* of its interior space can be reached by infiltrating rainwater and wind than those larger box buildings. Between the green and the purple buildings, the purple is 3 times the green in planar dimensions (9 times in roof area) and is of the same height as the green. It has lower vulnerability across most wind speeds than the green, for similar reasons described above. The fourth building shown (the dark brown box building) has the same planar dimensions as the green but 3 times its height. The dark brown box building has higher vulnerability at lower wind speeds than the green but crosses the green curve at mid-range of loss (at wind speed of about 170 mph) to become lower in vulnerability than the green. This is primarily because, at lower wind speeds, the dark brown has larger spatial extents of roof "hot zones" owing to its larger height than the green. At higher wind speeds, this scenario will maintain but it matters less in the sense that the potentially already damaged roof may affect the most the top floor but less the lower two floors, while at such higher wind speeds the green will potentially also has certain damage on the roof, starting from its own hot zones, and those damages will impact the entire interior floor space. The wall vulnerability tends to be less than the roof, and damage scenarios may be similar between the two buildings (the lower green building likely gets denser wind-borne missile hits than the higher dark brown building, which turns into a significant factor at high wind speeds). This phenomenon becomes even more pronounced in the case of the fifth building, the blue tall building, with its vulnerability curve crossing all buildings shown, at various wind speeds, to become the lowest at very high wind speeds.

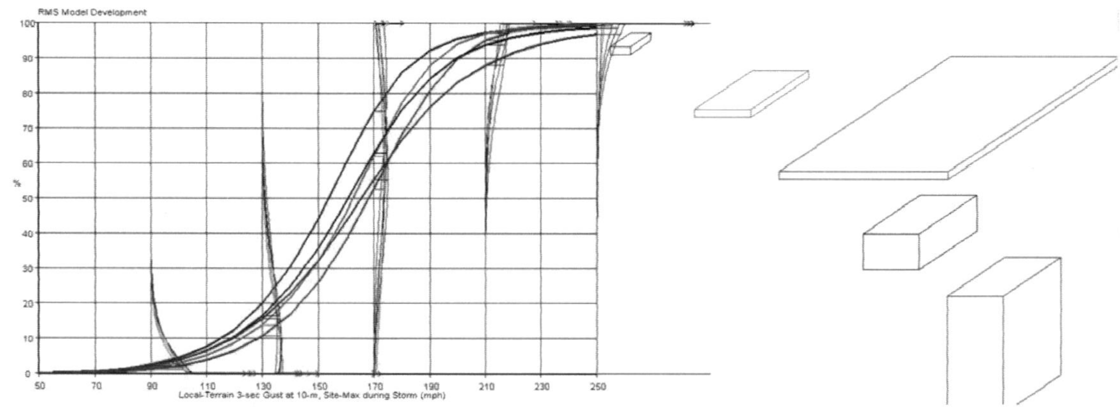

Figure 6. A set of vulnerability curves for various building shapes and dimensions (changing height and roof area only)

5 CONCLUDING REMARKS

This paper provides a closed-form computational framework and basic formulations that enable a site-specific, real-time estimation of individual building vulnerability to wind perils once the input values of a set of physical variables are provided. While the framework and formulations are herein presented for general use, some associated methodologies for determining required parameters, their values, and their quantitative effects in the processes, need to be discussed in a much greater length in separate papers. These will include those for loads and resistances, specific forms and evaluations of element and assembly correlations and dependences, and of economic damage as function of physical damage, and regional building inventory statistics, etc. There may be different ways or methods among researchers and practitioners in finding proper solutions for these topics. The application example provided is simply one realization of the framework and formulations on one practical problem based upon one set of solutions found for these topics, which are probably somewhat optimized for the current state of the art.

6 ACKNOWLEDGEMENTS

The authors are grateful to Risk Management Solutions Inc. for providing an opportunity to apply and test the framework, formulations and methodologies in so far the most comprehensive manner on an actual product for the insurance industry and risk management community. The first author also sincerely thanks his friends and colleagues for their moral and spiritual supports.

7 REFERENCES

1. P.J. Vickery, J. Lin, P.F. Skerlj, L.A. Twisdale Jr. and K. Huang, HAZUS-MH Hurricane Model Methodology. I: Hurricane Hazard, Terrain, and Wind Load Modeling, Natural Hazards Review, 7 (2, 2006) 82-93.
2. P.J. Vickery, P.F. Skerlj, J. Lin, L.A. Twisdale Jr. and M.A. Young, HAZUS-MH Hurricane Model Methodology. II: Damage and Loss Estimation, Natural Hazards Review, 7 (2, 2006) 94-103.
3. J.X. Lin, D. Surry and H.W. Tieleman, The distribution of pressure near roof corners of flat roof low buildings, J. Wind Eng. & Ind. Aerodyn., 56 (2, 1995) 235-265.
4. D. Surry and J.X. Lin, The effect of surroundings and roof corner geometric modifications on roof pressures on low-rise buildings, J. Wind Eng. & Ind. Aerodyn., 58 (1, 1995) 113-138.
5. J.X. Lin, and D. Surry, The variation of peak loads with tributary area near corners on flat low building roofs, J. Wind Eng. & Ind. Aerodyn., 77-78 (5, 1998) 185-196.

A New Framework for Modeling Tropical Cyclone Hazard

Bin Pei[a], Siamak Daneshvaran[b], Emmi Yonekura[c], Jeong Hee Oh[d], Mamta Gadhiya[e]

[a] *Impact Forecasting, Aon Benfield, Chicago, IL, USA, bin.pei@aonbenfield.com*
[b] *Impact Forecasting, Aon Benfield, Chicago, IL, USA, siamak.daneshvaran@aonbenfield.com*
[c] *RAND Corporation, Pittsburgh, PA, USA, yonekura@rand.org*
[d] *Impact Forecasting, Aon Benfield, Chicago, IL, USA, jeong.oh@aonbenfield.com*
[e] *Impact Forecasting, Aon Benfield, Chicago, IL, USA, mamta.gadhiya@aonbenfield.com*

ABSTRACT

Hurricane Matthew (2016) struck the US East Coast and ended the drought of major hurricanes ashore for over a decade. Hurricanes regain public attention and again become hot topics in insurance/reinsurance industry. To better serve the industry, a new framework is developed to model the hurricane hazard. This framework reflects the latest engineering and scientific development with a particular focus on implementing environmental variables to handle the changing background climate. The framework consists of four essential components: the genesis component, the track propagation component, the intensity component (relative intensity and filling models), and the wind speed component (wind field, boundary layer and terrain models). The modeling of the radius to maximum winds, the pressure-wind relationship, and the wind speed duration conversion are also considered. The datasets utilized in the modeling include the Atlantic hurricane database, the environmental reanalysis dataset, the sea surface temperature data, and the land cover land use database.

Using this framework, the full life cycle of tropical cyclones are modeled from their initiation to dissipation. The genesis model is first used to simulate the annual cyclone count and the initial cyclone parameters, including the date of the genesis, and the initial location, displacement and central pressure of each of the simulated tropical cyclones. The annual cyclone count and their initial locations are determined separately by random draws from Poisson distributions, whose rates are modeled as functions of the basin-wide and localized cyclone genesis indices. Once the cyclone initial parameters are determined, the track propagation model is then utilized to simulate the displacement field of the tropical cyclone. The displacement, which is described as the movements of the cyclone eye in latitudinal and longitudinal directions, is modeled as a function of the zonal and meridional winds at different pressure levels. Beta drift is considered by adding constants that vary with latitudes to the simulation equations. When a tropical cyclone is in the ocean, the relative intensity model is used to simulate the change of the cyclone central pressure. The relative intensity is modeled as a function of the sea surface temperature, as well as the Ventilation Index. Ventilation is reported to be effective in constraining tropical cyclone intensification. After a tropical cyclone makes landfall, the filling model is utilized to simulate the decay of the cyclone. An exponential decay function is applied and the decay rates are developed for different geographical regions. Once the full track of a tropical cyclone is simulated, surface wind speeds are calculated using a state-of-the-art wind field model coupled with a boundary layer model. A detailed terrain model is also developed based on the ESDU transition function taking into account a wind fetch of 10 km upstream roughness and the wind directionality. The results of applying the framework are validated using 100,000 years of tropical cyclone simulations by comparing not only the simulated cyclone parameters (e.g. displacement, central pressure, etc.) but also the calculated wind speeds with historical measurements. Both point-based time series and area-wise validations are performed and good agreements are observed.

KEYWORDS

Tropical cyclone simulation, environmental variables, cyclone genesis index, ventilation index, wind speed calculation

13th Americas Conference on Wind Engineering
Gainesville, Florida, USA
May 21 - 24, 2017

US Hurricane Wind Database with Hazard Curves and Applications

Prashant Rawal [a], Weichiang Pang [b], Bin Pei [c]

[a] *Clemson University, SC, USA, prawal@g.clemson.edu*
[b] *Clemson University, SC, USA, wpang@clemson.edu*
[c] *Aon Benfield, IL, USA, (formerly PhD Student at Clemson University) bin.pei@aonbenfield.com*

ABSTRACT

Conventional methods of structural design for wind loads currently use the wind maps presented in ASCE 7-10. In recent times though, various frameworks for performance-based wind engineering (PBWE), which requires the use of wind hazard curves at the sites of interest for analyses, have been proposed. A seismic hazard curve database for the US is already publicly available. However, a similar database for wind hazard is not yet publicly available. This study presents a database which addresses this need. A wind speed database with wind hazard curves was developed for a grid system covering the 30 Eastern US states using a 100,000-year simulated hurricane track database. This database is expected to assist further research and implementation of the PBWE methodologies. The usefulness of the wind hazard curves extends to hurricane risk assessment and insurance loss calculations as well. To perform such loss calculations, hurricane track data are needed. But the original track database covers a simulation period of 100,000 years, using which in its entirety can be computationally expensive. Therefore, to expedite the calculations, a method for reducing the track database to a desired size without significantly altering the wind hazards was introduced. In addition to the 30 Eastern US states, this wind speed database also covers the maritime exclusive economic zone (EEZ) along the Atlantic coast. This makes it useful for the design of off-shore structures. In order to make this database consistent with the ASCE 7-10 and therefore easier to be adapted by practitioners, the wind speeds were calculated using a methodology similar to that utilized for the generation of the ASCE 7-10 design wind speed maps. However, unlike ASCE 7-10, this database consists of only hurricane wind speeds without combination with non-hurricane winds. Finally, this paper discusses the suitability of various probability distributions in representing the simulated wind speed data and the hazard curves at every grid site. The distribution parameters are represented using equations which are functions of return periods. This enables the users to generate the hazard curve for any desired return periods. In addition to engineering design, the wind curves can also be used by insurance companies or catastrophe modelers to assessment the hurricane risk for a portfolio of structures.

KEYWORDS

Hurricane wind speed database, Performance based wind engineering, wind hazard curves, offshore structures design, hurricane risk assessment

Effects of spatial correlation of wind speed on hurricane wind induced losses

Bejoy P. Alduse[a], Weichiang Pang[b]

[a] *Clemson University, South Carolina, USA, balduse@clemson.edu*
[b] *Clemson University, South Carolina, USA, wpang@clemson.edu*

ABSTRACT

Catastrophe loss modelling for hurricane events are gaining popularity as it helps loss prediction and resource allocation. Hazus and Florida Public Hurricane Loss models are popular publicly available catastrophe loss models. It is observed that wind speed intensity data shows spatial correlation which is not well captured in most of the hazard models. Moreover, direction of wind plays a key role in vulnerability assessment, which is not well represented in public loss models.

Uncertainties of wind speed and direction should be well addressed for wind induced loss prediction during a hurricane event. Traditional approach uses the modelled peak wind speed at the location of the building to predict the dollar loss. However if wind speed correlation are not well addressed, the spatial loss prediction may be biased. In this study an approach to quantify the loss in a residential subdivision in South Florida focusing on the spatial correlation using multiple semi-variogram models is proposed. Regional study was conducted using ten selected hurricanes to obtain the peak wind speeds and directions at the subdivision. Unique wind vulnerability functions are assigned to the individual buildings. The uncertainties in the properties of the building and wind load are addressed in a separate wind vulnerability module. Concepts from FEMA P-58-1 framework are used to estimate the cost of repair. The proposed vulnerability functions also helps to quantify wind direction based risk for each building. A comparison is made with the dollar loss calculated using the conventional approach. Using the particular example, it was found that the conventional approach can under predict the dollar loss calculation by as much as 20%.

It is observed that spatial correlation of wind speed influences results of predicted damage. Hence it is important to consider wind speed correlation effects for loss modelling. It is proposed that studies should focus on damage states obtained by considering influence of wind speed correlation effects.

KEYWORDS

Catastrophe modelling, spatial correlation, wind vulnerability, wind direction, hurricane

Evaluating the uncertainty of combining wind and storm surge damage in the Florida Public Hurricane Loss Model

Mohammad Baradaranshoraka [a], Jean-Paul Pinelli [b], Kurt Gurley [c], Sneh Gulati [d]

[a] *Florida Institute of Technology, Melbourne, Florida, USA, mbaradaransh2014@my.fit.edu*
[b] *Florida Institute of Technology, Melbourne, Florida, USA, pinelli@fit.edu*
[c] *University of Florida, Gainesville, Florida, USA, kgurl@ce.ufl.edu*
[d] *Florida International University, Miami, Florida, USA, gulati@fiu.edu*

ABSTRACT

An uncertainty analysis that evaluates the reliability of the output of a model should include the variability of the inputs in addition to model assumptions. The probabilistic Florida Public Hurricane Loss Model (FPHLM) currently predicts separately loss from multiple hurricane related hazards (wind and rain, storm surge, and fresh water inundation). An engineering approach in conjunction with hazard information (intensity, timing) and statistical methods is being developed to calculate the combined damage and to assign damage causation from each hazard.

This paper addresses the uncertainty in the methodology which combines the results of wind-only and surge-only vulnerability models. The uncertainty is influenced by the inputs to this model, specifically the hazard specific damages calculated from the wind-only and surge-only vulnerability curves. The combined damage component of the FPHLM currently uses only the mean damage from the vulnerability curves to calculate the combined damage. This paper will explore the effect of including the uncertainties of each vulnerability curve and how they influence the uncertainty of the combined damage. Several case studies of hurricane damaged buildings will be used to validate this analysis.

In addition to the quantification of the uncertainty attached to the mean values of damage, a parametric uncertainty analysis identified the main contributors to the uncertainty. To take the uncertainty and sensitivity analysis a step forward, in addition to the analysis of the mean damages, a multi-criteria uncertainty analysis is performed to identify the uncertainty and sensitivity of the combined damage to variation of different components used to calculate the mean damages such as cost analysis.

KEYWORDS

Risk model, uncertainty analysis, hurricane, wind damage, storm surge damage, combined damage

Analytical Modeling of Container Stack Vulnerability

Derek Stedman [a], Rajkiran Vojjala[a]
[a]*Risk Management Solutions Inc., 7575 Gateway Blvd, Newark, California, USA*

ABSTRACT: Wind vulnerability of shipping containers at ports has not been widely studied in the past. However, wind damage to container stacks when exposed to high winds in typhoons and hurricanes is not an uncommon occurrence, resulting in significant losses to cargo stored within these containers. An analytical model was developed to quantify the degree to which stacked containers at ports are affected when exposed to high-winds in hurricanes. Monte Carlo simulation was performed considering 10,000 stack configurations. The results were used to obtain a distribution for impacted containers accounting for containers that slid or tipped off the stack configuration as well as ones that were penetrated by wind borne debris.

KEYWORDS: Marine Cargo, Wind damage, Vulnerability, Shipping Container, Ports, Monte Carlo Simulation.

1 INTRODUCTION

Vulnerability of cargo shipping containers when subject to hazardous perils such as high winds has not been widely studied. However, failure of container stacks in hurricanes is not an uncommon occurrence; 14% of ports surveyed by Tsai (1) experienced wind-related cargo damage over a five-year period. Natural disasters such as Hurricane Katrina (2005) and Superstorm Sandy (2012) caused significant damage to cargo in ports. While damage to port facilities is often researched, damage to cargo and more importantly, future potential of cargo losses has not been studied or quantified systematically. This paper describes an analytical model that was developed to quantify the container damage expected from future large natural catastrophes. The analytical model, based on Monte Carlo simulations, estimates the expected fraction of containers at port that are affected when exposed to high winds. The model does not quantify cargo damage itself when containers are affected. Cargo damage is highly dependent on the type of products shipped and the packaging methods used, which is a separate problem onto itself.

While in transit onboard a ship, train, or truck, containers in a stack are often locked together and lashed that restricts their movement. Port surveys and interviews with terminal operators revealed that there are a wide variety of practices to store and stack containers. Some containers are lashed together when stacked, some are stacked in stable configurations, while others are stacked without any lashing or tie downs or consideration for wind vulnerability at ports. In addition, the maximum stack height varies by port, often depending on the capabilities of the cranes at that port. Due to differences in port practices and relatively limited data, the model uses an average approach that is applicable to any generic port. Distributions used for the simulations are based on averages across many ports, increasing the uncertainty but also increasing the range of applicability. In this paper, a container stack is defined as a one container wide vertical array of containers. A container configuration is multiple stacks side by side, similar to how most containers in port are stored.

2 LITERATURE REVIEW

Most research on container vulnerability has pertained to containers in transit on a ship. This is because ships commonly experience high winds and motions while at sea that can lead to loss of cargo. Furthermore, the likelihood of a marine accident significantly increases if the increase in lateral wind loads on the ship due to loaded container weights are not adequately accounted for. Andersen (2) looked at the effect of shipboard container configurations on overall wind loads on a container ship and found uneven container stacks cause larger overall loads than even/tapered stacks. Extensive research performed by Fujiwara et al (3) used wind tunnel experiments to assess different wind load calculation methods, and proposed an empirical calculation methodology. Lindeen (4) leveraged empirical methodology from Fujiwara and applied it specifically to NYK Cool reefer ships. Computer modeling was used to calculate the increased wind load effects when the ships were fully loaded. Finally, Ueno et al. (5) estimated wind load coefficients for 76 different types of cargo ships and used them to generate a simplified method for calculating wind loading of a container ship based on area, size, and shape. All this research provides insights into shipboard wind loads due to cargo, but not necessarily on wind loads experienced on cargo container stacks when at port.

In general, minimal experimental and modeled information is available concerning container stack performance at ports. Most information is from surveys and onsite post-event research. Bos (6) researched the effect of winds on container handling and stacking, including calculation of wind speeds required to cause a basic container stack to slide or tip a stack of empty containers with a low (0.1) coefficient of friction. This research found that sliding can start at wind speeds as low as 22-45 mph (10-20 m/s). However, the paper did not state the measure of wind speed considered, i.e. if it was 10-minute, 1-minute, 3-second, or other wind gust. The measurement height or roughness length considered in the wind speed and load calculations were also not stated. Zhen et al. (7) performed wind tunnel pressure tests on model configurations of cargo containers at port providing critical overturning wind speeds for different rectangular container stack configurations. Only large and rectangular (minimum of 5 wide and 4 tall) container configurations were considered. In the real world, however, most container stacks are tiered and not perfect rectangles, often leaving highly exposed individual containers on the top (Figure 1).

Figure 1 Wind loads on this tiered container configuration will be quite different than those on a fully rectangular configuration

Tsai (1) sent a questionnaire to many different ports, with the goal of finding out how each port prepares for high wind scenarios. Some of the responses relevant to this paper are summarized below:

- While 80% of ports had stacking procedures designed to mitigate loss of empty containers, only 67% of ports had them for loaded containers.
- The definition of high winds varied across ports, ranging from 13.8 to 24 m/s (31 to 53 mph).
- 79% of ports used fixed anemometers supplemented with local weather forecasts for wind readings. Only 69% of these anemometers were in 'approved' locations.
- Only 39% of ports had specific high-wind procedures to mitigate losses.
- 78% of ports had experienced falling stacks of empty containers and 18% for loaded containers due to high winds.

These findings show that container movement due to high winds and resulting cargo damage at ports is common, and practices used to quantify high winds and mitigate damage to container stacks vary widely across ports all over the world.

3 CONTAINER STACK CONFIGURATION

Containers can be stacked in many configurations, generally up to eight containers high, which is the current limit of crane operations. Container stacks even at a single port or terminal are not identical; they vary in shape and configuration. A configuration which is five containers high but only one wide is more vulnerable to wind than a configuration with eight containers high but eight containers wide (Figure 2). Thus, the simulation-based model used a distribution of container configurations based on field observations at the Ports of Oakland, USA (8); Yokohama, Japan (9) and supplemented with an online survey of 25 different ports around the world (10).

Based on the information gathered, each container configuration was independently sampled as part of the simulation. Stack heights were sampled from a normal distribution constrained to minimum of 1 and maximum of 8-high. The Port of Oakland can only stack containers 5-high due to crane size restrictions. However, the distribution was adjusted to account for higher stacks to reflect practices at other ports around the world. The resulting distribution is shown in Figure 4, which compares reasonably with observed data at the Port of Oakland.

Figure 2 - The 4-wide by 4-tall configuration on the left will perform very differently from the single-wide, 4-tall configuration on the right

Configuration width at the base of the stack was also sampled from a normal distribution, again constrained between 1 and 8-wide. This was also based on data from reconnaissance at the Port of Oakland and then adjusted to better represent observations from ports worldwide. Above the bottom-most level, stack width is sampled uniformly based on the number of containers in the level below.

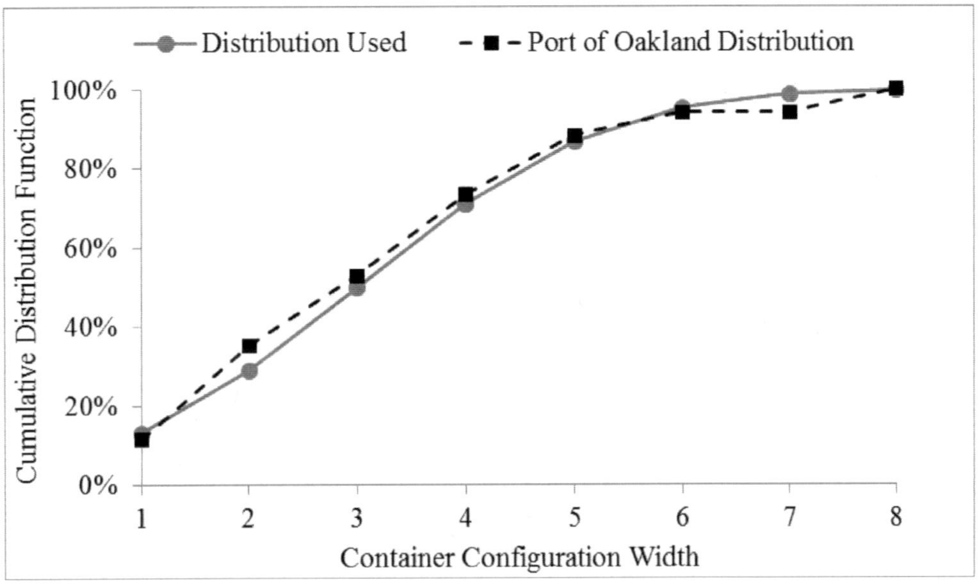

Figure 3 – Cumulative distribution for the container configuration base width.

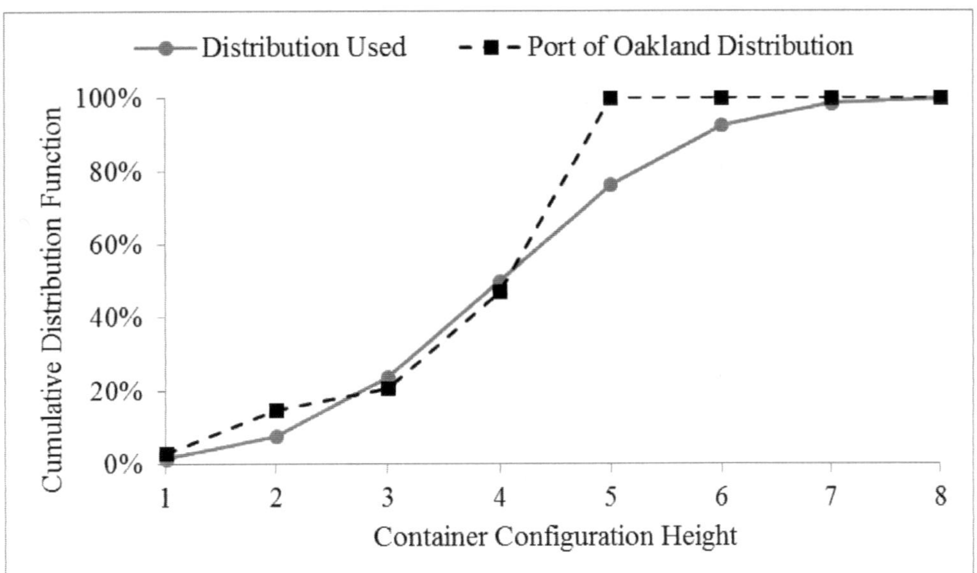

Figure 4 - Cumulative distribution for the container configuration height

Other parameters sampled as part of the container stack simulation were container weight, whether a container was fully sealed or not, and friction between containers.

Based on information gathered at the ports of Oakland and Yokohama, often no particular sorting of empty and loaded containers is done as part of normal operations. So, for the analytical simulations, all containers were assumed to be randomly mixed. Port authorities stated that at any given time 30% or more of all containers are empty. However, this varies by port. Ports that are export-heavy, such as those in China, will often see fewer empty containers stacked on premise than import-heavy ports, where loaded containers arrive and empty containers are ready to be shipped back. Thus, containers were considered as full, half full or empty using the weights in Table 1. 50% of the loaded containers were assumed to be at half their maximum weight due to variation in weight-to-volume ratio of different goods. For instance, a container filled completely with new television boxes would typically weigh less than half the maximum capacity. On the other hand, a container shipping heavy steel rods to its maximum weight capacity would only fill up 20% of the container volume.

Table 1 - Container load distributions assumed for simulation

Container Load	Mass (kg)	Probability
Empty	3,600	30%
Half Full	12,000	50%
Full	27,000	20%

Per port inspections and information from operators, most containers are not air-tight making them vulnerable to water penetration from rain and storm surge. Thus, each container was randomly sampled for its sealing condition with 95% probability of it being improperly sealed.

The final sampled parameter was the friction between the container under consideration and the container below it within a stack. Three different values for the coefficient of friction were used based on steel-steel static friction (Applied Industrial, (11)) as listed in Table 2. Static friction was considered to reflect the initial conditions of the containers. Once the container movement is initiated, kinetic friction that has a lower value than static friction governs i.e. the model assumes container movement continues unimpeded. Thus, the friction state between containers was sampled with varying likelihood for wind simulations. This was done to account for the higher probability of rain during windstorms, which reduces the friction between surfaces. However, even in these conditions not all containers will have wet surfaces due to shielding from other containers in the stack

Table 2 - Distribution of friction coefficients

State	Coefficient of Static Friction	Likelihood of friction state assumed
Lubricated	0.16	30%
Dry - min	0.5	60%
Dry - max	0.8	10%

10,000 container configurations were generated using this methodology, and the results were compared to a uniform distribution and observations from the Port of Oakland. The comparison was used as validation of stack height and width parameter assumptions and to verify their convergence.

4 LOADS

4.1 *Wind Loads*

Port locations are usually on the waterfront with no protection or shielding. Furthermore, there is not enough distance for a boundary layer transition to occur. Therefore, containers at the front of the lot will be directly exposed to the wind profile from the ocean. This allows for low roughness lengths to be used in the calculation of wind loading on container stacks at port. The wind load calculation only considered winds blowing normal to the broadside of the container configuration, which is the critical loading case. This was deemed acceptable since prevailing wind direction changes throughout the duration of the storm and most stacks experience high winds on their broadside for at least a short time in the event.

Wind loads were calculated using a basic drag equation [1], with a drag coefficient of 1.2 (Holmes (12)). The drag coefficient selected is for a 2-D wall, and includes the effects of negative pressure on the rear of the wall. While individual containers stacked with others may not experience negative pressures, since the entire configuration acts as one large wall, the leeward side of the overall stack would most likely experience some negative pressures. Thus, the use of this drag coefficient was deemed acceptable when considering the behavior of the entire stack.

$$F_D = \frac{1}{2}\rho C_D A V^2 \qquad [1]$$

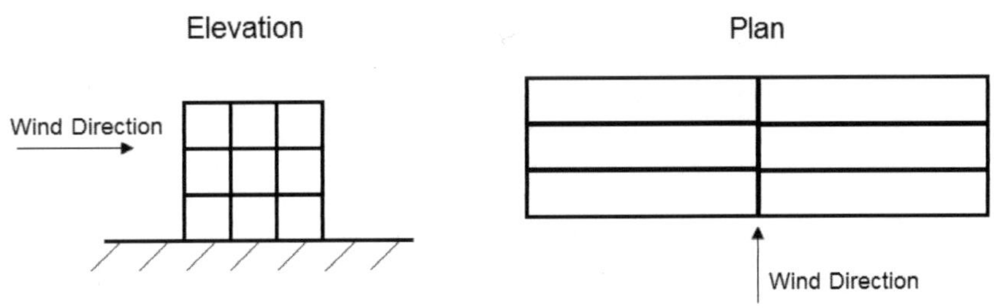

Figure 5 - Sketch of wind loading direction showing the length is far greater than the width

An open-water profile, with a zo value of 0.005m was used to calculate wind speeds. Due to the small height of a container relative to the boundary layer, windspeeds were considered uniform throughout the height of a single container, but varied across the container stack height that can be up to 20 m tall. The log-law [2] was used to determine windspeeds at the mid-height of each container. Normalized windspeed values, shown in Figure 8, were used as factors (kw) to scale wind loads for the input 3-second gust at 10 m reference height.

$$\bar{u}(z) = \frac{u_*}{k}\ln\left(\frac{z}{z_o}\right) \qquad [2]$$

The final step in determining wind loads affecting a container stack was the shielding coefficient (ks). As shown in Figure 9, containers are usually stacked end to end. This means while the outer row of containers will be fully exposed to high winds, stacks behind may experience reduced loading due to significant shielding effects. A digital survey of several ports revealed that as many as two-thirds of all container stacks at a port are likely to be shielded from high winds. On the other hand, there is also a possibility of wind speed-up when a lower stack is in front of a higher stack, increasing the wind load on the taller stack. To represent such local variations in wind loads due to shielding and speed-up effects, two factors were simulated for each container configuration. The first was a Boolean sample to determine if a given stack was shielded or not, with probability of shielding assumed to be 67%. The second factor was only sampled if a stack was shielded using a truncated normal distribution with a mean that varied based on stack height and values capped between 0.5 and 1.5. The shielding coefficient was assigned to either 1.0 (non-shielded), or the second sample (0.5-1.5) and then applied to the drag coefficient to scale the wind loads and an adjusted wind speed was calculated for each level in the container stack. Combining the basic wind force equation with the adjustment factors leads to a final wind load equation which can be used to calculate total wind loads on a container stack [3a and 3b].

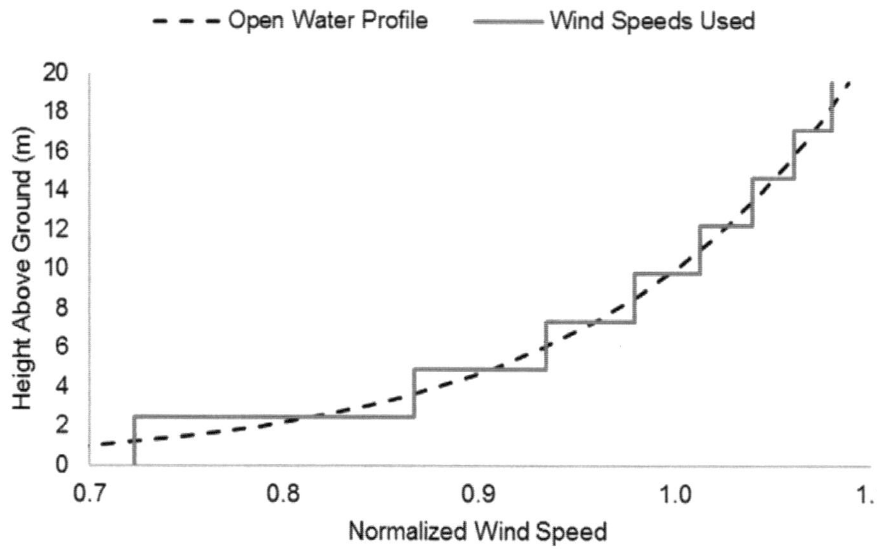

Figure 6 – Normalized open water and simplified wind profiles

Initial results revealed that it was improbable for an unlocked container stack to tip over without sliding first. Thus if a container stack was considered unlocked, it was tested for sliding failure only and if a container stack was locked together, it was tested for tipping failure.

Testing for sliding failure used sampled configuration weights and friction state information about each container. The failure condition [3e] was used to determine if and where failure occurred within a container stack. This was done by considering the total frictional forces resisting movement across the entire width of a stack at a given height [3c]. Working downwards from the top level, each stack was checked for sliding failure [3d]. If at any level, wind loads exceeded the frictional force, satisfying the failure condition [3e], the containers above that level were assumed to have fallen over and were removed from the stack. The above process was then repeated for the leftover stack after removing the 'failed' containers to capture any progressive failures till the bottom-most level was checked.

$$\text{Total wind force: } F = \sum_{j=1}^{Stack\ Height} F_j \quad [3a]$$

$$\text{Wind force at height } j\text{: } F_j = \frac{1}{2}\rho(V * k_w)^2 * (C_D k_S) \quad [3b]$$

$$\text{Wind force at height } j\text{: } F_j = \frac{1}{2}\rho(V * k_w)^2 * (C_D k_S) \quad [3b]$$

$$\text{Friction force at height } j\text{, row } k\text{: } F_F = \sum_{j=1}^{Stack\ Height} F_{Fkj} \quad [3d]$$

$$\text{Failure Condition: } 0 \geq F_{Fj} - F_j \quad [3e]$$

Each container stack was subject to increasing windspeeds and the above process repeated to yield the probability of stack failure as a function of windspeed.

For a stack with locked containers, an entire stack may tip if the aspect ratio is narrow enough. To calculate tipping, the moment-arm required to resist rotation about the base of the stack due to wind loads was calculated. If the arm (x in Figure 7) was longer than the width of a stack, the stack was considered to have tipped. Moment forces were calculated working through the width of each stack in the configuration from windward to leeward side. If a stack is not in equilibrium, the wind force will cause it to tip. However, the stack next to it (if the configuration is more than 1 container wide) is assumed to resist tipping. When moment equilibrium is calculated for the next container stack, an additional force, equal to that required to resist tipping of the previous stack (FR1 in Figure 11) is applied at 2/3 the previous stack's height, or the maximum height of the new stack, whichever is higher. This force, along with any additional wind forces due to exposed containers at the higher levels, is checked for tipping, and the process is repeated throughout the width of the configuration.

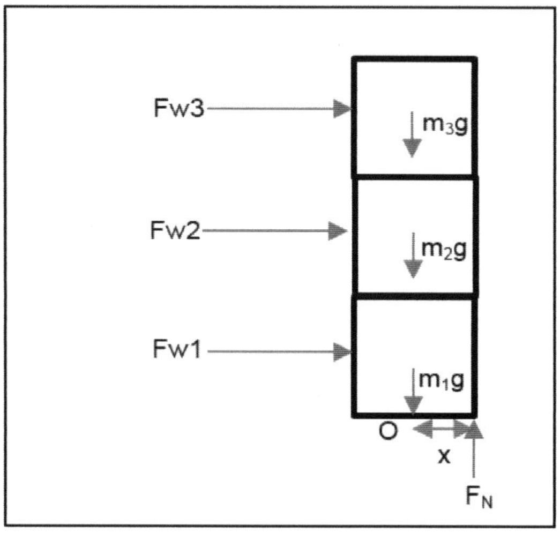

Figure 8 - A free body diagram of the forces on a 3-high, 1-wide container stack locked together, moments are calculated about "O".

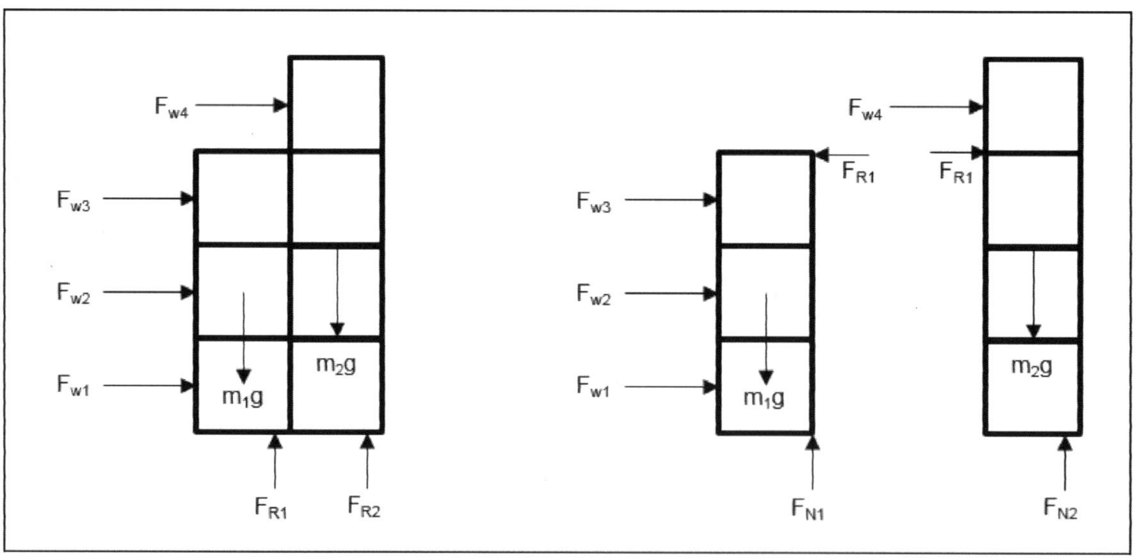

Figure 7 - A free body diagram showing how the forces are transferred through a 2-wide configuration.

4.2 *Wind-Borne Debris*

In addition to sliding or tipping over due to wind loading, container stacks have a significant risk of wind-borne debris impacting then, causing damage to the containers and cargo within. To account for this, wind-borne debris impacts were considered only to the outside containers of a configuration that may be directly exposed to debris. Containers on the inside of a stack are rarely exposed to debris unless the configuration is significantly disturbed by wind loads. In addition, not all exposed containers on the outside of a stack will be hit by debris, which was factored in the simulations. Given a wind-borne debris impact, the size of the debris was sampled as uniformly distributed. Due to the nearly infinite potential of debris shapes, sizes, and impact orientations, a single worst-case orientation was considered for the simulation. Impacts were all assumed to be from a wood 2x4 member (50x101 mm), commonly used in impact testing, normal to the container surface. The length of wood members was sampled from 20 equally-probable options ranging from 0.01m to 3m (0.03 feet to 9.8 feet) in length.

For each sampled debris length, the takeoff wind speed (Holmes (12)), the critical debris speed to puncture a container and the run-up distance required for the debris to reach this speed were calculated. The required windspeed for takeoff (Uf) is shown in equation [4], where l is the characteristic length, I (1.0), ρm is the density of the debris (500kg/m3), ρa is the air density (1.23 kg/m3) and Cf an aerodynamic force coefficient (0.5 – Holmes (12)).

The required debris run-up distance to reach critical speed was calculated for windspeeds between 20-100 m/s (50 and 250mph) [8]. Assuming the wood member to be a compact object in flight governed by equations [6] and [7], impact plausibility was determined based on a 50m run-up distance. If the debris would reach the critical wind speed in 50m or less, it was deemed a damaging hit. If the run-up was more than 50m, it was considered a non-damaging hit. 50m is significantly wider than the average gap between container configurations, but was chosen to account for wind directions not perfectly normal to container configuration faces which could still cause large debris impacts. If an exposed container sustained a damaging hit, it would be recorded as failed at that wind speed. All exposed containers were checked for damaging debris impacts at increasing windspeeds.

$$U_f = \sqrt{\frac{lI\rho_m g}{\frac{1}{2}\rho_a C_F}} \qquad [4]$$

$$V_d = \sqrt{\frac{Energy\ Required\ to\ Puncture\ Container}{2 * Mass\ of\ debris\ (m_d)}} \qquad [5]$$

$$Critical\ Velocity\ V_{cr} = \begin{bmatrix} V_d & if\ V_d > U_f \\ U_f & if\ V_d \leq U_f \end{bmatrix} \qquad [6]$$

$$Time\ to\ Critical\ Velocity{:}\ t_{cr} = \frac{V_{cr}}{k_d V_{wind}[V_{wind} - V_{cr}]} \qquad [7]$$

$$where\ k_d = \frac{\rho_{air}}{2\rho_{debris} l_{debris}} \qquad [7a]$$

$$Distance\ to\ Critical\ Velocity{:}\ d_{cr} = V_{wind}\left[t_d - \left(\frac{1}{k_d V_{wind}}\right) * \ln(1 + k_d V_{wind} t_d)\right] \qquad [8]$$

5 RESULTS

Convergence with 10,000 configurations was verified by comparing results from multiple different runs of the simulation, using an increasing number of configurations each time. The runs with only 1 configuration considered show a step-function as containers are affected. The configuration for "run 2" was particularly vulnerable and had been fully affected before the wind speed exceeded 60m/s. Conversely, "run 1" only ever sees 20% of containers affected due to a relatively resilient configuration. The range between the two runs shows the uncertainty in stack performance depending on the sampled configuration.

Even with just 10 configurations, results for the percentage of containers affected improves. However, the results still show discrete jumps. Results for 10,000 and 100,000 configurations showed less than 1% difference at any windspeed, and were repeatable. Thus, 10,000 configurations were deemed to be sufficient for this simulation exercise.

The wind simulation (Figure 23) results showed that at high wind speeds almost all containers would be affected to some extent, with the percentage of containers sliding or tipping gradually increasing with wind speed. Debris damage does not start occurring until ~40m/s which is required for lift off and enough acceleration of debris to cause container damage. The maximum number of containers affected by debris is much lower than 100% in the model as only container stacks on the outside of a configuration are at risk from debris impacts.

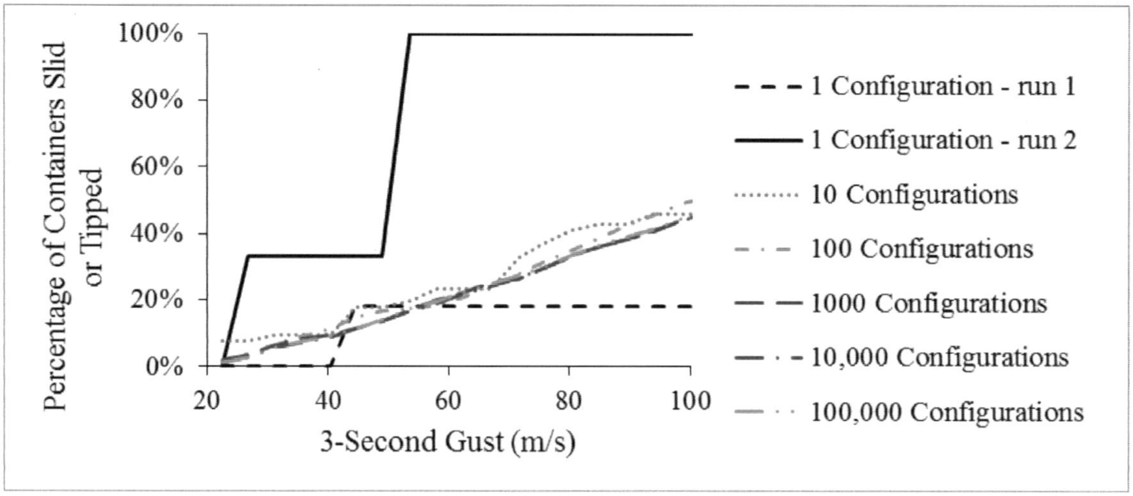

Figure 9 - Testing of various combinations showed convergence by 10,000 configurations.

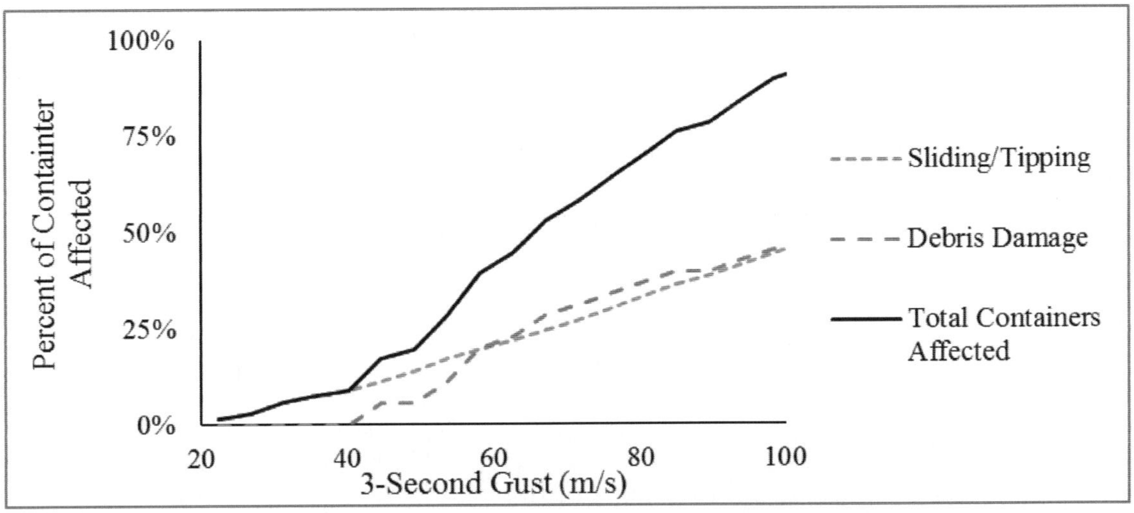

Figure 10 - Results of simulation in terms of total affected containers

6 CONCLUSIONS AND FUTURE WORK

A few key conclusions on the container susceptibility in hurricanes and windstorms.

- While not heavily studied, containers are significantly at risk of being affected by high-wind events. This is very likely to damage cargo inside the containers.
- Even low windspeeds can cause damage to vulnerable container configurations. On the other hand, even an extremely large hurricane is unlikely to affect all containers at a port due to some very large and wind-resistant configurations as well as shielding impacts due to normal stacking procedures.
- Ports and terminals could focus on changing the configuration of stacked containers to a more wind-resistant configuration as standard operating procedures at the onset of a hurricane or typhoon warning. This can have a significant impact on the potential number of containers sliding or tipped over in high-wind events, reducing the damage potential cargo within these containers.

The next steps of this research would involve performing similar analyses for other perils, such as storm surge and waves. In addition, variation of damage to different types of cargo contained within a container need to be studied to enable detailed damage estimations and the development of windspeed-based damage curves for different types of cargo being shipped in containers.

7 ACKNOWLEDGEMENTS

The authors would like to acknowledge the personnel at the ports of Oakland and Yokohama who generously provided facility tours and information.

8 REFERENCES

1. Tsai, k., Terminal Operations in High Wind Conditions, International Safety Panel, ICHCA International Ltd., Romford, United Kingdom, 2002.
2. Anderson, I.M.V., Wind Forces on Container Ships, Mercator, 2012.
3. Fujiwara, T., Tsukada, Y., Kitamura, F., Sawada, H. and Ohmatsu, S., Experimental Investigation and Estimation on Wind Forces for a Container Ship, Proceedings of the Nineteenth International Offshore and Polar Engineering Conference, Osaka, Japan, June 21-26, 2009.
4. Lindeen, C., Wind Resistance Generated by Containers on Reefer Vessels, KTH Center for Naval Architecture, Stockholm, 2008.
5. Ueno, M., Kitamura, F., Sogihnara, N. and Fujiwara, T., A Simple Method to Estimate Wind Loads on Ships, The 2012 World Congress on Advances in Civil, Environmental, and Materials Research, Seoul, Korea, August 26-30, 2012.
6. Bos, W. Wind influence on container handling, equipment and stacking, Port Technology International.
7. Zhen, W., Xingquan, P. and Chunhui, Z., Overturning Analysis of Harbor Containers Based on Wind Tunnel Test of Rigid Models, College of Civil Engineering, Huaqiao University, Quanzhou, China, 2011.
8. Patalano, A, SSQ Terminal, Port of Oakland, Oakland, USA, onsite interview May 5 2015.
9. Shibasaki, K., Port of Yokohama, Yokohama, Japan, onsite interview, Dec 8 2015.
10. Google, [Various maps/images of worldwide ports], Retrieved throughout 2015 and 2016 from maps.google.com
11. 55LE5400, LG Electronics USA Inc., 1000 Sylvan Avenue, Englewood Cliffs, New Jersey, 2010.
12. Coefficients of Friction "F" Applied Industrial Technologies, http://web.applied.com/site.c fm/CoefficientsofFriction.cfm, 2014.
13. Holmes, J.D., Wind Loading of Structures Second Edition, Taylor and Francis, New York, USA, 2007.

Probabilistic Model for Wind Pressure on Low-rise Buildings

Haoran Zhao[a], Mircea Grigoriu[b], Kurtis Gurley[c]
[a] *Cornell University, Ithaca, NY, USA, hz289@cornell.edu*
[b] *Cornell University, Ithaca, NY, USA, mdg12@cornell.edu*
[c] *University of Florida, Gainesville, FL, USA, kgurl@ce.ufl.edu*

ABSTRACT

A probabilistic model is developed and calibrated to the experimental wind-pressure records on the low-rise building. The data used in this paper are obtained in the boundary layer wind tunnel from University of Florida, where the surface pressures were measured on a low-rise building model at multiple wind speeds and directions. Because of the irregular shapes of the building model, the pressure records exhibit non-Gaussian features and complex spatial correlations. The proposed probabilistic model has the capability to capture these characteristics with relatively low computational effort. The construction of this model is based on 1) the singular value decomposition of time-space random functions, and 2) the translation model which matches the marginal distributions and the second-moment properties. Examples are provided to illustrate the accuracy and efficiency of this model.

KEYWORDS

Probabilistic model; Wind pressure; Low-rise building; Translation model

Thunderstorm Design Wind Speeds for Southwestern Ontario Applicable to Transmission Lines

Haitham Aboshosha, Thomas Mara, Peter Case

The Boundary Layer Wind Tunnel Laboratory, Western University, London, Ontario, Canada N6A 5B9

ABSTRACT

Recently, we have developed and validated a framework capable of predicting thunderstorm gust front speeds suitable for design. The framework is analogous to hurricane modeling where (i) a numerical model is used to describe thunderstorm gust front records seen at meteorological station as a function of a number model parameters (called thunderstorm parameters), then (ii) the model is coupled with Monte Carlo simulation to simulate thunderstorm gust front events happening in thousands of simulation years employing the probability distribution of the model parameters. In the current study, the framework is applied to study thunderstorms happening in Southwestern Ontario throughout analyzing records from a number of 26 meteorological stations. This is to obtain design speeds suitable for structure with small footprint (i.e. ~ in the order of 50 m width in plan view) and extended structures such as transmission lines. It was found that extended structure can experience speeds higher than those with small footprint up to 15 due to the localized nature of thunderstorm gust front. Thunderstorm design wind speeds were compared to design speeds based on treating all historical records as synoptic wind (i.e. a common practice in analyzing wind records) and design speeds specified in the NBCC. It was found out that speeds specified in the NBCC are higher than the thunderstorm speeds for structures with small footprint, but lower (down to 90%) than thunderstorm speeds for extended structures such as transmission lines. This explains the considerable number of transmission line failures in the region.

KEYWORDS

Thunderstorm, Gust front, Downdraft, Design Speed, Historical Records, Numerical Simulations, Localized Nature, Transmission Lines.

13th Americas Conference on Wind Engineering
Gainesville, Florida, USA
May 21 - 24, 2017

Wind Load Provisions for Roofs: Low-Rise Industrial and Institutional Buildings

Hatem Alrawashdeh[a], Ted Stathopoulos[b]

[a] *PhD student, Department of Building, Civil and Environmental Engineering, Concordia University, Canada, Corresponding Author: h_alraw@encs.concordia.ca*
[b] *Professor, Department of Building, Civil and Environmental Engineering, Concordia University, Canada, statho@bcee.concordia.ca*

ABSTRACT

Current literature on wind-induced pressures on roofs of low-rise buildings is generally limited to regular-shaped, mostly rectangular roofs with relatively small dimensions, say less than 60 m. Low-rise industrial and institutional buildings have commonly flat or nearly flat roofs with very large size and these geometries have not been considered in the formulation of wind code and standard provisions. On this basis, it is important to evaluate the applicability of the current national building code/standard provisions for very large roofs. This paper reports on an experimental study to further enrich the knowledge content of wind-induced pressures on flat roofs of low-rise buildings with large dimensions. Nine low-rise buildings (5, 7.5 and 10 m high) have been scaled down in an atmospheric boundary-layer wind tunnel to represent low-rise industrial and institutional buildings. The buildings have square roofs with full-scale dimensions of 60, 120 and 180 m, tested in open-country terrain exposure for different wind directions. Wind load provisions of different wind codes and standards of practice, namely: the American Society of Civil Engineers (ASCE 7, 2016), the National Building Code of Canada (NBCC, 2015), Australian/New Zealand Standard (AS/NZS 1170.2, 2011) and the European Standard (EN 1991-1-4, 2005), are critically reviewed and design wind pressures and roof zones derived from these wind codes and standards are compared with the experimental results. The comparisons show that that some of the current provisions may lead to excessively conservative designs by increasing the width of the edge and corner zones to unjustified sizes. A variety of approaches were considered in order to redefine the size of the edge and corner zones with respect to the magnitude of the pressure coefficients provided for all roofs in the different codes/standards when dealing with industrial or institutional low-rise buildings with large dimensions.

KEYWORDS

Industrial and institutional buildings, design large roofs, wind codes and standards, wind pressures, wind tunnel

Wind Loads on Roof-Mounted Solar Panels: Comparison of Codes and Standards

Hatem Alrawashdeh[a], Ted Stathopoulos[b]

[a] *PhD student, Department of Building, Civil and Environmental Engineering, Concordia University, Canada, Corresponding Author: h_alraw@encs.concordia.ca*

[b] *Professor, Department of Building, Civil and Environmental Engineering, Concordia University, Canada, statho@bcee.concordia.ca*

ABSTRACT

Several studies have been published recently regarding wind-induced pressures on solar panels and intended to generate wind standard and code provisions to assist practitioners with the design of solar panels installed on building roofs. This paper attempts to review the available literature and the current provisions of some national wind codes and standards of practice. This review includes but it is not limited to the results obtained by previous researchers in terms of their tests' methodologies, their findings on the parameters affecting the wind-induced pressures and wind loadings on solar panels and it aims to comprehend the findings and results obtained by means of atmospheric boundary layer wind tunnel studies. The paper also discusses and compares wind load provisions provided by different wind codes and standards, namely: American Society of Civil Engineers (ASCE 7, 2016), Australian/New Zealand Standard (AS/NZS 1170.2, 2011 with Amendment 2) and Japanese Industrial Standard (JISC 8955, 2011). Also, experimental values of extreme peak panel and area-averaged pressure coefficients are compared with the respective design values derived from wind codes and standards. The comparisons show several inconsistencies among the results of previous studies. The investigation of the current wind code and standard provisions for roof-mounted solar panels revealed difficulties of provision interpretations in evaluating the wind-induced loads on these elements. Also, some of these provisions may not be appropriate to evaluate the actual wind loads induced on single-array systems of solar panels and should not be used for such simple geometries.

KEYWORDS

Building roofs, codes and standards, solar panels, wind loads, wind tunnel

13th Americas Conference on Wind Engineering
Gainesville, Florida, USA
May 21 - 24, 2017

Experimental Investigation of Wind Effects on Canopies Attached to Mid-Rise Buildings

Amir Naeiji[a], Manuel Matus[b], Ioannis Zisis[c]

[a] *PhD Candidate, Department of Civil and Environmental Engineering, Florida International University, Miami, FL, USA, 33174,anaeiji@fiu.edu*

[b] *Graduate Student, Department of Civil and Environmental Engineering, Florida International University, Miami, FL, USA, 33174, mmatu016@fiu.edu*

[c] *Assistant Professor, Department of Civil and Environmental Engineering, Florida International University, Miami, FL, USA, 33174, izisis@fiu.edu*

ABSTRACT

Canopies attached to mid-rise buildings are often light-weight structures that function as shelter from sun and precipitation at the entrance of buildings. In this application, the canopy is not enclosed by walls except at one side where it is attached to the building. This type of enclosure makes the wind load that is applied on the canopy incomparable to the wind load applied on typical building roofs or free-standing canopies. The wind-induced load applied on both the top and bottom surfaces of the canopy is one of the principal design loads, however, building codes and wind standards do not provide sufficient wind design guidelines for such structures. It is apparent that the wind effect on these structures is not well known and the need for a codification study is of high importance.

The objective of the current study was to quantify the wind effects on canopies attached to mid-rise buildings. Several 1:20 scaled models were tested at Wall of Wind (WOW) Experimental Facility at Florida International University (FIU) in order to investigate the effect of various geometrical parameters such as canopy length, tilt angle, installation height and building height along with wind direction. The local pressure distribution at the upper and lower surfaces of the canopy, as well as the net pressure distribution, was obtained for each model configuration and the results were compared to evaluate the most critical loading scenarios. Finally, area averaged force coefficient versus effective area curves were obtained and enveloped, considering all geometrical configurations, to generate design force coefficients curves. These codified design curves can be incorporated into building codes and wind standards for the wind design of attached canopies to mid-rise buildings.

KEYWORDS

Attached canopy, Pressure distribution, Force coefficient, Are-averaged curves, Codes and standards

Aerodynamic Response of Roof-mounted Photovoltaic Panels

Amir Naeiji [a], Raphael Greenbaum [b], Ioannis Zisis [c], Arindam Chowdhury [d], Peter Irwin [e]

[a] *PhD Candidate, Department of Civil and Environmental Engineering, Florida International University, Miami, FL, USA, 33174, anaeiji@fiu.edu*

[b] *Research Scientist, Research Scientist, International Hurricane Research Center, Florida International University, Miami, FL, USA, 33174, rgreenba@fiu.edu*

[c] *Assistant Professor, Department of Civil and Environmental Engineering, Florida International University, Miami, FL, USA, 33174, izisis@fiu.edu*

[d] *Associate Professor, Department of Civil and Environmental Engineering, Florida International University, Miami, FL, USA, 33174, chowdhur@fiu.edu*

[e] *Professor of Practice, Department of Civil and Environmental Engineering, Florida International University, Miami, FL, USA, 33174, peairwi@fiu.edu*

ABSTRACT

Current design standard specifications do not address the effect of photovoltaic (PV) module vibrations under wind action although these vibrations have been found to be significant during full-scale tests. Preliminary studies showed that the effect of PV-system vibration should be considered in the design of mounting structure, and the <1 Hz fundamental natural frequency criterion in current standards to determine whether wind-induced dynamic effects should be considered or not, may not be applicable to PV systems. The objective of this research is to formulate a methodology that takes the effect of PV system vibration into account in the design practices of mounting structures.

Scaled rigid models are not suitable for investigating the dynamic effects, because such models do not simulate the dynamic properties of the systems and are unable to capture resonant response. For this study, a full-scale PV system was mounted on a low-slope roof and instrumented using load cells and accelerometers. In addition to the full-scale model, an identical large-scale pressure-tapped model was built. Both models were tested in the Wall of Wind (WOW) Experimental Facility at Florida International University (FIU). By comparing the force coefficients spectra of the flexible full-scale model and the identical rigid large-scale one, the mechanical admittance function (MAF) was developed. The MAF was then incorporated in the wind-induced response estimation of the scaled (rigid) model in order to generate the realistic dynamic effects of the PV system. This effort can lead to new code provisions to account for the wind-induced dynamic effects on small structures and ensure adequate systems design.

KEYWORDS

Full-scale test, Large-scale test, PV vibration, Mechanical admittance function

13th Americas Conference on Wind Engineering
Gainesville, Florida, USA
May 21 - 24, 2017

Historical Development of Wind Speeds for Structural Design in the Commonwealth Caribbean 1952-2008

Tony Gibbs FREng

Consulting Engineers Partnership Ltd, Barbados, tmgibbs@caribsurf.com

ABSTRACT

Little is known by the present generation of Commonwealth Caribbean engineers about wind analysis practices of the pre-2nd-World-War construction industry of the former British colonies in the Caribbean. After that war a degree of formality gradually took hold with respect to structural analysis for wind forces. The first identifiable guidance came in the form of the British Standard CP3:Chapter-V:Part-2:1952 – Wind Loads. That standard prescribed a wind speed of 75 miles per hour (averaging period unknown) for the severest British conditions. But that is what was used in the British colonies in the Caribbean. In the 1960s there was a general gravitation to the use of the South Florida Building Code with its higher wind speeds – typically 120 MPH fastest mile.

It was towards the end of the 1960s that Commonwealth Caribbean engineers embarked on the process of obtaining wind speed information specifically for their region. The first such exercise was undertaken by Harold C Shellard (formerly of the UK Meteorological Office) who worked in Barbados at the Caribbean Meteorological Institute for 3 years. His studies were based on the procedure developed by H C S Thom using limited historical data, mainly from airport meteorological stations in the islands. The results of his work were incorporated in the standard "Wind Loads for Structural Design – 1970". In 1981 Basil Rocheford, also of the Caribbean Meteorological Institute, revised Shellard's work with the benefit of another 12 years of wind-speed data. Rocheford further revised his own work in 1984.

As part of the development of the Caribbean Uniform Building Code (CUBiC) a new analysis of wind speeds was undertaken by Davenport, Georgiou and Surry of the University of Western Ontario in 1985. Those studies made use of the NHC HURDAT information, analysed to produce statistical distributions representing various tropical cyclone parameters. These were sampled sequentially at random. Tropical cyclones were generated and the computed wind fields were moved past the sites of interest. This simulation was run repeatedly. The UWO team also examined the effects of topography on wind speeds.

In 2007-08, in preparation for a major rewriting of CUBiC, a comprehensive study of wind speeds throughout the Caribbean Basin was undertaken by Peter J Vickery and D Wadhera of Applied Research Associates. The methodology used a hurricane track model that reproduced the frequency and characteristics of hurricane tracks and the hurricane wind field model to provide estimates of the wind speed.

In 2008 a World Bank study carried out by Georgia Tech, Peter Vickery and Tony Gibbs examined the effect of future climate change on wind speeds for structural design in the Eastern Caribbean. This study has led to specific guidance which has since been incorporated in the latest edition of the Organisation of Eastern Caribbean States Building Code.

KEYWORDS

History, limited data, wind speeds, standards

13th Americas Conference on Wind Engineering
Gainesville, Florida, USA
May 21 - 24, 2017

Development of New Wind Speed Maps for the ASCE 7-16 Standard

Marc Levitan[a], Frank Lombardo[b], Adam Pintar[c], Peter Vickery[d], Emil Simiu[e]

[a] National Institute of Standards and Technology, Gaithersburg, MD, USA, marc.levitan@nist.gov
[b] University of Illinois, Urbana IL, USA, lombaf@illinois.edu
[c] National Institute of Standards and Technology, Gaithersburg, MD, USA, adam.pintar@nist.gov
[d] Applied Research Associates, Inc. Raleigh, N. C., USA, pvickery@ara.com
[e] National Institute of Standards and Technology, Gaithersburg, MD, USA, emil.simiu@nist.gov

ABSTRACT

All wind speed maps in ASCE 7-16 have been updated, based on (i) a new analysis of non-hurricane wind data available through 2010 and (ii) improvements to the hurricane simulation model. Separate wind speed maps are now provided for Risk Category III and Risk Category IV buildings and structures, recognizing the higher reliabilities required for essential facilities and facilities whose failure could pose a substantial hazard to the community.

In order to determine appropriate mean recurrence intervals for each Risk Category, a reliability analysis was conducted consistent with the target reliabilities in the first row of ASCE 7-10 Table C1.3.1a. The resulting return periods for Risk Category I, II, and III structures remained unchanged, at 300, 700, and 1,700 years, respectively, and the return period for Risk Category IV structures was set at 3,000 years.

The non-hurricane wind speeds for the contiguous United States were estimated from peak gust speed data collected at 575 meteorological stations. The data at each station were classified by storm type, thunderstorm or non-thunderstorm, after removal of gusts associated with tropical cyclones. Recorded peak gusts from each station were corrected to standardize the observations to equivalent three-second peak gusts at 10 m height over open (Exposure C) terrain. At each station, there were at least 15 years of data, and there were sufficient numbers of both thunderstorm and non-thunderstorm observations to account for their potential differences when estimating wind speeds with specified mean recurrence intervals. The estimation was performed in two stages. In the first stage, a peaks-over-thresholds (POT) model was fitted to the data from each station. The POT model employed was a two-dimensional Poisson process, the parameters of which were allowed to vary with time to capture differences between thunderstorm and non-thunderstorm winds. The Poisson process has a tail length parameter that was set to zero, leading to Gumbel-like tails for the distribution of wind speeds (consistent with past practice in wind engineering). In the second stage, local regression was used to interpolate wind speeds at all points of a fine regular grid covering the contiguous United States, for all required mean recurrence intervals. This had the effect of spatially smoothing the noisy station estimates.

The hurricane wind speeds were based on the results of a Monte Carlo simulation using the same hurricane model utilized for ASCE 7-10 but with two updates, which better account for (i) the translation speed effects of fast moving storms, and (ii) the transition from hurricanes to extra-tropical storms in the northern latitudes (i.e., transition from warm core to cold core low pressure systems). The effects of the hurricane model revisions are to slightly reduce hurricane speeds in the northeast, extending from Maine to Virginia, while hurricane wind speeds are unchanged along the coast from Texas to the Carolinas.

The non-hurricane and hurricane winds were then combined as statistically independent events using the same general approach that has been used in previous editions of ASCE 7. These new maps better represent the regional variations in the non-hurricane extreme wind climate. Wind speeds generally decrease everywhere in the country, except in the northern Great Plains region where they remain approximately the same as in ASCE 7-10. This is reflective of the fact that the ASCE 7-10 (and predecessor) maps used a single uniform wind speed zone for most of the country (except the coasts), where the value for the zone set near the largest wind speed values anywhere in the zone, which occur in the northern Great Plains states.

KEYWORDS

Wind speed maps, ASCE 7, peaks-over-threshold, hurricane model, thunderstorm winds

Comparison of wind pressures in typical buildings computed according to current wind regulations of the Dominican Republic and ASCE 7

Norberto J. Rojas-Mercedes[a,c,d], Luis D. Aponte-Bermúdez[b],
Víctor Escalante-Cervera[c], and Douglas Carvajal-Sánchez[d]

[a] *Instituto Tecnológico de Santo Domingo INTEC, Santo Domingo, Dominican Republic,*
norberto.rojas@intec.edu.do
[b] *University of Puerto Rico, Mayagüez Campus, Puerto Rico,*
luisd.aponte@upr.edu
[c] *Epsa-Labco Consulting Engineers, Santo Domingo, Dominican Republic,*
victor.escalante@epsalabco.com
[d] *Universidad Iberoamericana, Santo Domingo, Dominican Republic,*
d.carvajal1@unibe.edu.do

ABSTRACT

The Dominican Republic, located in the middle of the Caribbean region, had been impacted by major hurricanes in the past few decades causing significant economic losses. Also, in the last 20 years, the vertical growth experienced in urban areas has led to the construction of buildings higher than 60 meters. This increase means that wind loads might control the design by lateral forces and serviceability. Despite being constantly threatened by hurricanes, the recommendations for wind design currently in use in the country were developed more than 35 years ago, and have not been updated ever since because wind engineering was not part of the curriculum in universities nationwide. Also, practitioner engineers give more attention to seismic design than wind design, as evidenced by a recent update to the seismic code in 2011. In 2015, the Civil Engineering School of the "Universidad Iberoamericana" (UNIBE) formally introduced Wind Engineering as a course subject on its bachelor degree program. In addition to extreme wind loads the Dominican Republic region had a high threat of seismic events since several active faults surround it. Even though historically the probability that a major hurricane strikes the island is greater than the occurrence of a devastating earthquake, the country does not have an updated reliable code to estimate wind loads. Here we present a comparison of the wind loads in two typical buildings obtained using the recommendations according to the current Dominican wind Code and ASCE 7 wind load provisions. The results would serve as a basis to motivate a complete revision of the current Dominican Wind Code and to promote research in wind engineering in the Dominican Republic.

KEYWORDS

ASCE 7, Wind Engineering, Dominican Republic, Code Comparison, Dominican wind code.

13th Americas Conference on Wind Engineering
Gainesville, Florida, USA
May 21 - 24, 2017

Geographic Based Parameters for Design Wind Speeds

Scott. D. Schiff, PhD[a]

[a] *Applied Technology Council, Atlanta, Georgia, USA, scott.schiff@atcouncil.org*

ABSTRACT

ASCE/SEI 7-16 now has three load hazards that are tied to design wind speeds. Most common are winds speeds associated with straight-line, thunderstorm, and hurricane winds. Over the past few editions of ASCE/SEI 7, there have been two significant changes to these maps. First was the transition in ASCE/SEI 7-05 from using a fastest mile wind speed reference to a 3-second gust. The motivation for making this change was measurement instrumentation for directly recording fastest mile wind speed was no longer being used and, more importantly, news agencies reporting wind speeds typically quoted 3-second gust wind speeds and a more direct comparison to design wind speed was desired. However, there were also subtle changes made to the wind speed contours based upon further analysis of actual and simulated wind speed data. The second significant change occurred in the release of ASCE/SEI 7-10 to directly provide wind speeds associated with a limit-state design event rather than a service-level design event. In addition, rather than using an adjustment factor to account for the importance of a building (typically related to occupancy), three maps were produced with return periods associated with Risk Category I, Risk Category II, and Risk Category III/IV buildings. For designs that also needed to consider service-level design events, maps were also provided for mean recurrence intervals of 10, 25, 50, and 100 years. Another change was the reintroduction of Exposure Category D and the associated reduction of design wind speeds along much of the East and Gulf Coasts. The wind speeds maps for ASCE/SEI 7-16 reflect minor adjustments in the wind speed contours. In general, the design winds for ASCE/SEI 7-10 and ASCE/SEI 7-16 are directly comparable.

One of the issues with using printed maps provided in ASCE/SEI 7 is the resolution and the ability of a user to accurately determine the design wind speed for a given site. The ATC Windspeed Website provides a convenient tool to obtain the design wind speed for multiple versions of ASCE/SEI 7. A user can input a street address or the latitude and longitude for a given site and acquire the design wind speed rounded up to a whole number. This website provides the same information as the maps and does not return design wind speeds for sites located in a special wind region. A user can request a report that echoes the input data, indicates the source map, and reports the output from the website.

Ongoing enhancements to the ATC Windspeed website include 1) determination as to whether a given site is considered to be in a windborne debris region, 2) the design wind speed for the consideration of a tornadic event, and 3) the design speed to be used in combination with the accumulation of ice. This paper presents a brief overview of the recent evolution of the ASCE/SEI 7 wind loads, and reports on development of the website, proper usage of the website or an application program interface (API) and proper interpretation and usage of provided information.

KEYWORDS

Wind Speed Maps, ASCE/SEI 7, Geographic Based Parameters Website

13th Americas Conference on Wind Engineering
Gainesville, Florida, USA
May 21 - 24, 2017

Directionality Factor Revisited

Peter J. Vickery[a]

[a]*Principal Engineer, IntraRisk Division, Applied Research Associates, Raleigh, NC, United States, pvickery@ara.com*

ABSTRACT

ASCE 7 defines K_d as the directionality factor. The factor takes into account the fact that both the wind climate and the building aerodynamics vary with direction, and it is unlikely that the direction having the highest wind load effect will align with the dominant wind direction. The research described herein presents values of K_d for both component and cladding loads and main wind force resisting loads. Probability distributions for K_d are given for both hurricane and non-hurricane prone areas of the United States.

K_d is defined as the ratio of the predicted N-year return period pressure (computed by integrating the full directional probability distribution of velocity pressure with the directional pressure coefficients, GCp) divided by the product of the maximum GCp over all wind directions and the *N*-year return period velocity pressure.

Distributions of K_d in hurricane prone regions were developed using pressure data obtained from twelve model lowrise buildings with varying geometries and roof slopes which were tested in the University of Western Ontario's Boundary Layer Wind Tunnel Laboratory (BLWTL), coupled with time series of wind speeds and direction from simulated hurricanes. The wind tunnel pressure data measured on the model buildings were a part of the NIST database assisted design (DAD) experiments (Main and Fritz, 2006). The hurricane wind speed and direction data were obtained from the hurricane model described in Vickery et al. (2009, 2010).

Probability distributions of K_d were developed for three roof zones, two wall zones, a range of return periods, four coastal locations and two inland locations. Example cumulative distribution functions for K_d are given in Figure 1 for roof corners (ASCE 7 Zone 3) for buildings located in Miami, FL (left) and Wilmington, NC (right).

Figure 2 presents summary plots showing the mean and coefficient of variation of K_d versus return period. The results presented in Figure 2 clearly show the increase in mean, and reduction in the coefficient of variation of K_d with increasing return period. In all cases the mean is well below the value of 0.85 given in ASCE 7.

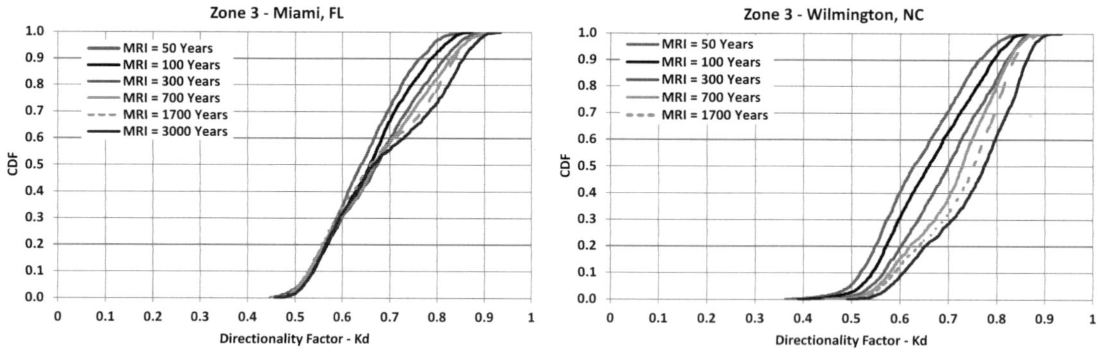

Figure 1. Cumulative distribution functions of K_d for various return periods negative roof pressures for buildings located in Miami, FL and Wilmington, NC.

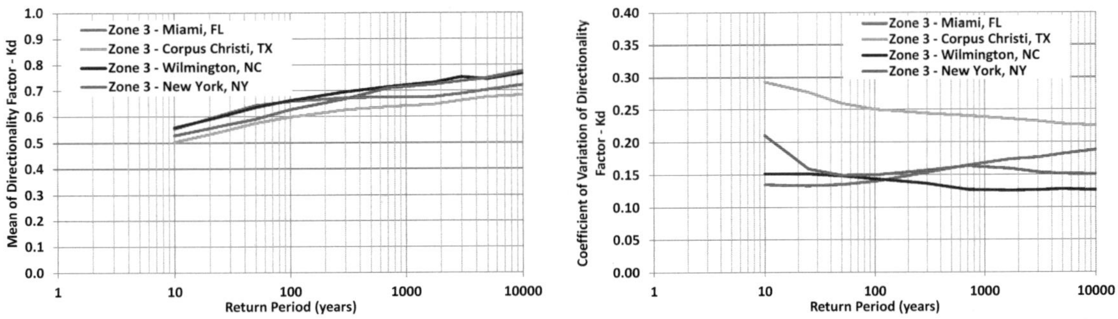

Figure 2. Variation of the mean and coefficient of variation of K_d as a function of return period.

KEYWORDS

ASCE 7, Hurricane, Directionality Factor

REFERENCES

Joseph A. Main and William P. Fritz (2006). "Database-Assisted Design for Wind: Concepts, Software, and Examples for Rigid and Flexible Buildings", NIST Building Science Series 180, March 2006.

Vickery, P.J., D. Wadhera, L.A. Twisdale Jr. and F. M. Lavelle (2009b). "United States hurricane wind speed risk and uncertainty," *J. Struct. Engrg.*. **135**, 301-320.

Vickery, P.J., D. Wadhera, J. Galsworthy, P.A. Irwin and L.A. Griffis (2010). "Ultimate wind load design gust speeds in the United States for use in ASCE 7," *J. Struct. Engrg.*. **136**, 613-625

The 13th Americas Conference on Wind Engineering (13ACWE)
Gainesville, Florida USA, May 21-24, 2017

13th Americas Conference on
WIND ENGINEERING

Comparison of Loads from IEC 61400-3 Extreme Conditions with Loads from Recently Observed Hurricane Models

Gholamreza Amirinia [a], Sungmoon Jung [b]

[a] *Department of Civil and Environmental Engineering, Florida A&M University – Florida State University, Tallahassee, United States, gamirinia@fsu.edu*
[b] *Department of Civil and Environmental Engineering, Florida A&M University – Florida State University, Tallahassee, United States, sjung@eng.fsu.edu*

ABSTRACT: This paper compares the buffeting responses of offshore wind turbines subjected to recently observed hurricane models with responses from IEC 61400-3 extreme load condition. For this purpose, first, the recent observations on hurricane turbulence models were discussed. Next, hurricane wind and wave fields were simulated based on the Saffir-Simpson hurricane wind scale. Later, a new formulation for addressing unsteady wind forces on the tower and modified NREL-FAST package was implemented to analyze structure-wind-wave-soil interaction of the NREL-5 MW wind turbine during hurricane. Finally, by using the short term structural responses, an extreme value analysis based on direct integration method with peak over threshold (POT) approach was carried out to compare the results with IEC 61400-3 recommendations. Results showed that for design of the tower and blades subjected to hurricane buffeting forces, the IEC 61400-3 recommendations for turbine class I are overestimated and underestimated respectively to be used for class S.

KEYWORDS: Offshore wind turbine, hurricane, extreme value analysis, IEC-61400-3.

1 INTRODUCTION

Many studies have been conducted to address different issues on wind turbines structural parts subjected to regular boundary layer winds (non-hurricane winds) [1-6]; however, hurricane winds and their effects on the structures still need more studies and observations. Thus in the last decade, much research has been conducted in order to clarify the nature and characteristics of hurricanes, their differences with regular high winds, and their effects on the structures [7-13].
Observations and analysis of hurricane surface winds revealed that turbulence spectrum of hurricane winds differs from that of non-hurricane high winds [7, 8, 12, 14-17]. Li et al. [16] and Caracoglia and Jones [14] showed that in the hurricane, the higher turbulence frequencies have higher level of energy; however, Schroeder and Smith [12], Yu et al. [17], and Jung and Masters [8] showed that hurricane spectrum has higher level of energy in low frequencies. Different turbulence energy models affect structures differently, while the mean wind speed and turbulence intensity are identical between models [18]. In this regard, mean responses of structures subjected to regular high winds and hurricane winds are comparable, whereas, the buffeting responses are different. Because of these differences in buffeting responses, conditions such as structural integrity and low cycle fatigue subjected to different hurricane turbulence models should be investigated.

IEC 61400-1 [19] and IEC 61400-3 [20] have recommendations for different wind turbine classes and load cases; however, for special events such as hurricane, they introduced wind turbine class S which the design variables should be defined by the designer. In order to compare the short term responses of the wind turbine subjected to hurricane with IEC 61400-1 [19] and IEC 61400-3 [20] recommendations, an extreme value analysis is necessary. Several studies has been conducted on extreme value analysis which considered various methods and approaches [4, 5, 21]. By considering the importance of new hurricane turbulence models on buffeting responses, an accurate extreme value analysis is necessary for comparing the short term results with IEC 61400-3 [20] recommendations. An accurate extreme value analysis and comparison with existing recommendations also assist to consider important issues in designing wind turbines class S.

This paper compares responses of offshore wind turbines subjected to recently observed hurricane turbulence models with IEC 61400-3 recommendations. For this purpose, first, the recent observations on hurricane turbulence models were discussed and compared with Kaimal spectrum model. At next step, according to importance of recent findings about hurricane winds, hurricane wind and wave fields were simulated based on the Saffir-Simpson hurricane wind scale. Next, the modified NREL-FAST [22, 23] package with a new formulation for addressing unsteady wind forces on the tower was implemented to analyze structure-wind-wave-soil interaction of the NREL-5 MW [24] wind turbine. Finally, by using the short term structural responses, an extreme value analysis based on direct integration method with peak over threshold (POT) approach was carried out to compare the results with IEC 61400-1 [19] recommendations.

2 HURRICANE BOUNDARY LAYER WINDS

The wind turbulence spectrum represents the energy distribution in turbulent wind [16]. Many studies investigated wind turbulence spectrums for non-hurricane winds [25-28]. Kaimal et al. [26], based on series of experiments, showed that all turbulence spectrums reduce to a limited family of curves and proposed a formula for wind spectrum as shown in eq. 1:

$$\frac{nS_{uu}}{\sigma_u^2} = \frac{21.6f}{(1+33f)^{5/3}} \tag{1}$$

where $f = nz/U(z)$ represents the normalized frequency, n is the frequency, S_{uu} is the spectral density of the longitudinal velocity fluctuation at height z, and σ is the standard deviation of longitudinal velocity fluctuation. On the other hand, hurricane field data observations during the last decade revealed that turbulence spectrum of the hurricane boundary layer winds is different from those of non-hurricane winds. According to the data observed by Schroeder and Smith [12], Yu et al. [17], Jung and Masters [8], Caracoglia and Jones [14], and Li et al. [16], the spectral distribution of hurricane winds shows a large variability. Schroeder and Smith [12], Yu et al. [17], and Jung and Masters [8] showed that compared to non-hurricane spectral models, hurricane spectrums had higher energy content in low frequencies. Eq. 2 shows a formula derived for hurricane wind turbulence spectrum [17] with high amount of energy in low frequencies:

$$\frac{nS_{uu}}{\sigma_u^2} = \frac{1}{\beta} \times \frac{p_1 f^2 + p_2 f + p_3}{f^3 + q_1 f^2 + q_2 f + q_3} \tag{2}$$

where $\sqrt{\beta} = \sigma/u^*$ is the turbulence ratio and u^* represents the friction velocity, and p_i and q_i are constants proposed by Yu et al. [17] as shown in Table 1.

Table 1. Coefficients of Yu et al. (2008) spectrum

Spectra	p_1	p_2	p_3	q_1	q_2	q_3
10m, Over Land	-0.9999	3.112	1.159×10^{-4}	18.64	1.188	3.35×10^{-3}
10m, Over Sea	-0.00598	0.1544	1.055×10^{-5}	0.4458	0.06486	9.754×10^{-5}

On the other hand, Li et al. [16] and Caracoglia and Jones [14] based on series of observations, presented an opposite results that higher frequencies contained larger amount of energy rather than low frequencies. Li et al. [16] provided spectrum models for their observation with high amount of energy in high frequencies as:

$$\frac{nS_u(n)}{\sigma_u^2} = \frac{16.66f}{1.72 + 237.24f^{5/3}} \tag{3}$$

Table 2 summarizes the spectrum models presented by mentioned researchers and used in this paper. In addition, Figure 1 shows the difference between recent observed models and the spectrum introduced by Kaimal et al. [26].

Table 2. Spectrum models summary

Model Name	Proposed by	Similar Observations
Model A	Yu et al. (2008)	Schroeder and Smith [12], Yu et al. [17], Jung and Masters [8]
Model B	Li et al. (2012)	Caracoglia and Jones [14], Li et al. [16]

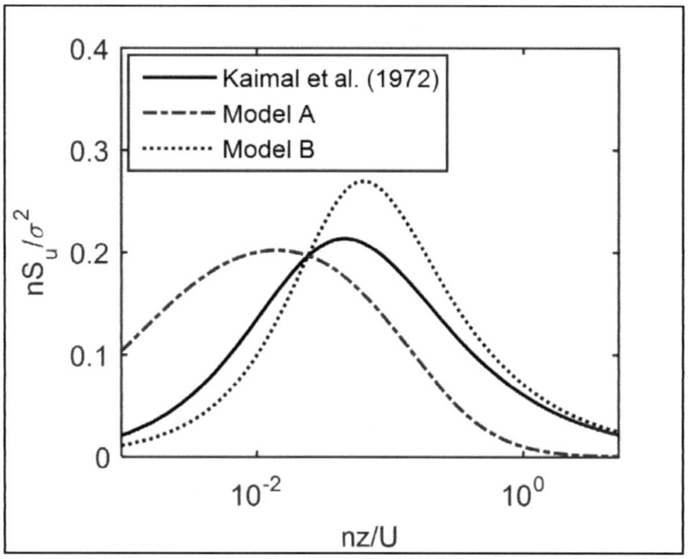

Figure 1. Comparison of hurricane spectrum models (over sea) and Kaimal spectrum

3 WIND BUFFETING FORCES ON WIND TURBINE

Previous studies showed that in parked condition the tower play an important role in total structural responses and unsteady analysis of the tower resulted in more realistic responses [22, 23]. Formulation of wind loading on the wind turbine tower is given in eq. 4 and eq. 5 as [29]:

$$\mathbf{F}_{\text{wind}}(z,t) = \begin{bmatrix} \bar{f}_D(z) \\ \bar{f}_L(z) \\ \bar{M}_\alpha(z) \end{bmatrix} + \begin{bmatrix} f_D(z,t) \\ f_L(z,t) \\ M_\alpha(z,t) \end{bmatrix} = \bar{\mathbf{f}} + \mathbf{B}_\text{f}\mathbf{v} + \mathbf{C}_{\text{ae}}\dot{\mathbf{r}} + \mathbf{K}_{\text{ae}}\mathbf{r} \tag{4}$$

$$\mathbf{v}(z,\text{t}) = [u/U \quad w/U]^{\text{T}} \tag{5}$$

where $\bar{\mathbf{f}} = [\bar{f}_\text{D}(z), \bar{f}_\text{L}(z), \bar{M}_\alpha(z)]^\text{T}$ are quasi-static forces, $\mathbf{B}_\text{f}\mathbf{v}$ are buffeting forces, $\mathbf{C}_{\text{ae}}\dot{\mathbf{r}}$ and $\mathbf{K}_{\text{ae}}\mathbf{r}$ are self-excited forces, u and w are longitudinal and lateral fluctuating parts of the wind, and $\mathbf{r}$ and $\dot{\mathbf{r}}$ are structural position and velocity vectors. In eq. 5 and eq. 6 it is assumed that the main flow direction is perpendicular to the rotor plane and wind field is homogenous [29].

Aerodynamic analysis and wind tunnel tests showed that in a line-like structure, the turbulence eddies hit the structure simultaneously [30, 31]; however, when the size of the structure increases, the turbulence eddies may not hit the structure at the same time which corresponds to aerodynamic unsteady forces. The aerodynamic unsteady forces can be addressed by aerodynamic admittance function [30]. In order to address unsteady forces on the tower, previous studies used rational functions for presenting aerodynamic admittance function [22, 23]. Figure 2 illustrates the aerodynamic admittance function presented by Davenport [30] compared with two terms rational function (RF) approximation ($m = 1$), measured rational function by Clobes and Peil [32], rectangular prism wind tunnel tests, and measured indicial functions by Chang et al. [33].

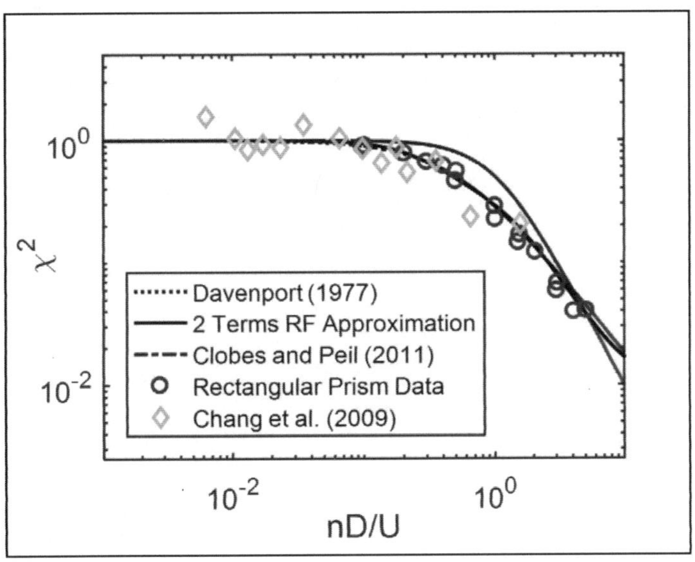

Figure 2. Aerodynamic admittance function

4 HURRICANE WIND AND WAVE MODELING

To investigate the contribution of blades and tower under hurricane condition, the NREL 5MW offshore monopile wind turbine [24] was selected for analysis. 60 wind field time series for each turbulence model (each one 600s) were generated by NREL-TurbSim [34] with frequency of 10 Hz. The mean wind speeds at 10 m height were equal to 30, 40, 50,60, 70, and 80 m/s (each speed 10 time series) in order to cover five Saffir-Simpson hurricane categories [35]. In addition, wave field was generated based on JONSWAP (Joint North Sea Wave Observation Project) spectrum

[36] by assuming 20 m water depth, 6 hours hurricane, and duration-limited wave. The wave length, $L = gT_P^2/2\pi$, for each peak wave period, T_P, ranges from 230 m to 613 m which represents shallow water in the depth of 20 m ($d/L > 0.5$). Hence the maximum wave height will be 0.9 times of the water depth [37] and the maximum wave height for the considered depth was limited to 18 m. The nonlinear wave forces due to large wind induced wave heights were computed by the HydroDyn package [38] through a hybrid model consist of potential flow theory and Morison's equation. The Saffir-Simpson wind characteristics and generated wave field is presented in Table 3.

Table 3. Saffir-Simpson hurricane categories and modeled wind and wave field

Saffir-Simpson Category	Wind Speed Range (m/s)	Modeled Wind Speed @10m (m/s)	H_s(m)	T_P (s)
1	33~42.5	30	8.4	12.1
2	42.5~49.2	40	12.2	15.1
3	49.2~57.8	50	16.6	17.2
4	57.8~69.7	60	18.0	19.8
5	69.7<	70	18.0	19.8
		80	18.0	19.8

5 EXTREME VALUE ANALYSIS

IEC 61400-3 [20] provides recommendations for extreme value analysis and long term load extrapolations for different wind turbine classes; however, for special conditions such as hurricane, it defines class S, where the values should be specified by the designer. In order to compare IEC 61400-3 [20] recommendations with recent hurricane observations, the wind turbine class I in a low turbulence offshore region (similar to recent observations [17]) was selected. In this case, the reference wind speed and turbulence intensity at hub height were 50 m/s and 12% respectively [20].

In order to conduct an extreme value analysis, there are several methods such as direct integration, first order reliability method (FORM), and inverse first order reliability method (IFORM) [1, 39]. The direct integration method which is the most straight forward method can be expressed as [39]:

$$P(M_{10\,min} > x) = \int_V (1 - F(x|V))f(V)dV \tag{6}$$

where $P(M_{10\,min} > x)$ is the probability that load M resulted from 10 minute simulation exceed the load value of x, $F(x|V)$ represents the cumulative distribution of load value x in the wind speed bin V, and $f(V)$ is the wind speed distribution. A common way for using the direct integration method is to use the method of peak-over threshold (POT). In the POT method, a threshold is selected and the local maxima are captured. Then the exceedance probability can be restated as:

$$P(M_{10\,min} > x) = \int_V (1 - F(x|V)^n)f(V)dV \tag{7}$$

where n is the expected number of peaks above the selected threshold. According to IEC 61400-1 [19] and IEC 61400-3 [20], the 10 minutes averaging duration is standard for wind turbine analysis. In this case, the probability to observe a load M larger than T-years load, M_T, can be expressed as:

$$P(M > M_T) = 1/(T \times 365.25 \times 24 \times 6) = 1.90 \times 10^{-5}/T \quad (8)$$

In order to compare IEC 61400-3 [20] recommendation and results from new hurricane turbulence models, the extreme wind speed model (EWM) with 50-years recurrence period (50 m/s @ hub height) was considered which reduced eq. 8 to:

$$P(M > M_{50}) = 3.80 \times 10^{-7} \quad (9)$$

6 RESULTS

Choosing an appropriate threshold and distribution for exceedance probability are important for extreme value analysis [1, 20, 39]. The distribution representing the exceedance probability of the local maxima in each wind speed should precisely represent the tail behavior. Figure 3 shows observed local maxima of tower buffeting base moment with different POTs where Weibull, lognormal, and generalized extreme value (GEV) distributions are fitted on the data (SD: Standard Deviation). By increasing the threshold values, the POT method leads to the method of global maxima.

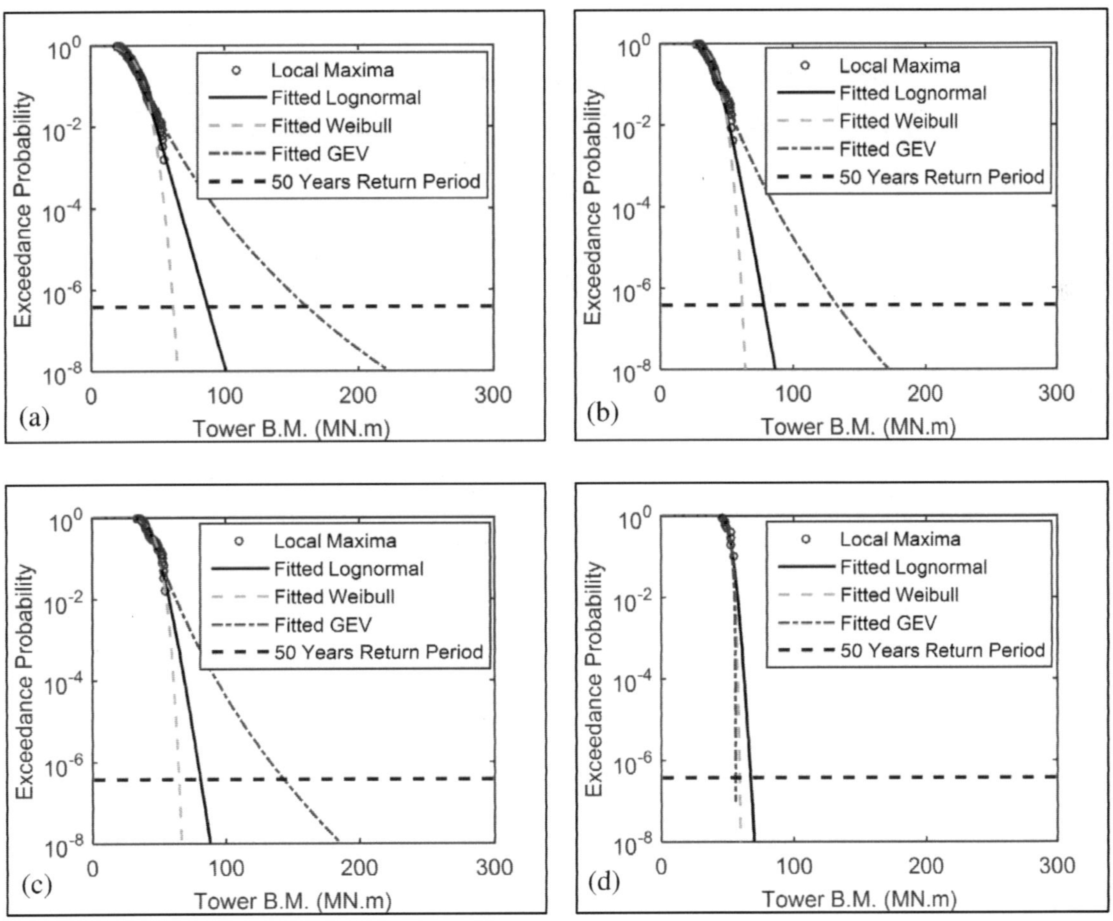

Figure 3. Fitted distributions on local maxima based on different thresholds as (a) 1.4SD, (b) 2.0SD, (c) 2.5SD, and (d) 3.0SD

The direct integration method with a POT of 1.4 ×SD [19] was applied for extreme value analysis of the tower fore-aft base and blade root edge-wise moments. Moreover, three mentioned distributions for fitting local maxima in this section were used for the analysis. Figure 4 shows the response extreme value analysis of the three turbulence models by assuming different distributions compared to IEC 61400-3 [20] recommendations. At 50-years return period, by assuming Weibull distribution for tower buffeting fore-aft moment, IEC 61400-3 [20] recommendations are larger than resulted extreme values from all turbulence models; however, IEC 61400-3 [20] recommendations are smaller than results of GEV assumption (Figure 4a, c, e). Hence, by assuming Weibull and GEV distributions respectively, IEC 61400-3 [20] recommendations for wind turbine class I are respectively conservative and non-conservative for designing the tower of wind turbine class S subjected to hurricane.

Blades response extreme value analysis showed that for Kaimal et al. [26] and Model B, all three distributions resulted in larger extreme values than IEC 61400-3 [20] recommendations (Figure 4b, f); however, for Model A, Weibull and GEV distributions resulted in smaller and larger values than IEC 61400-3 [20] recommendations respectively (Figure 4d).

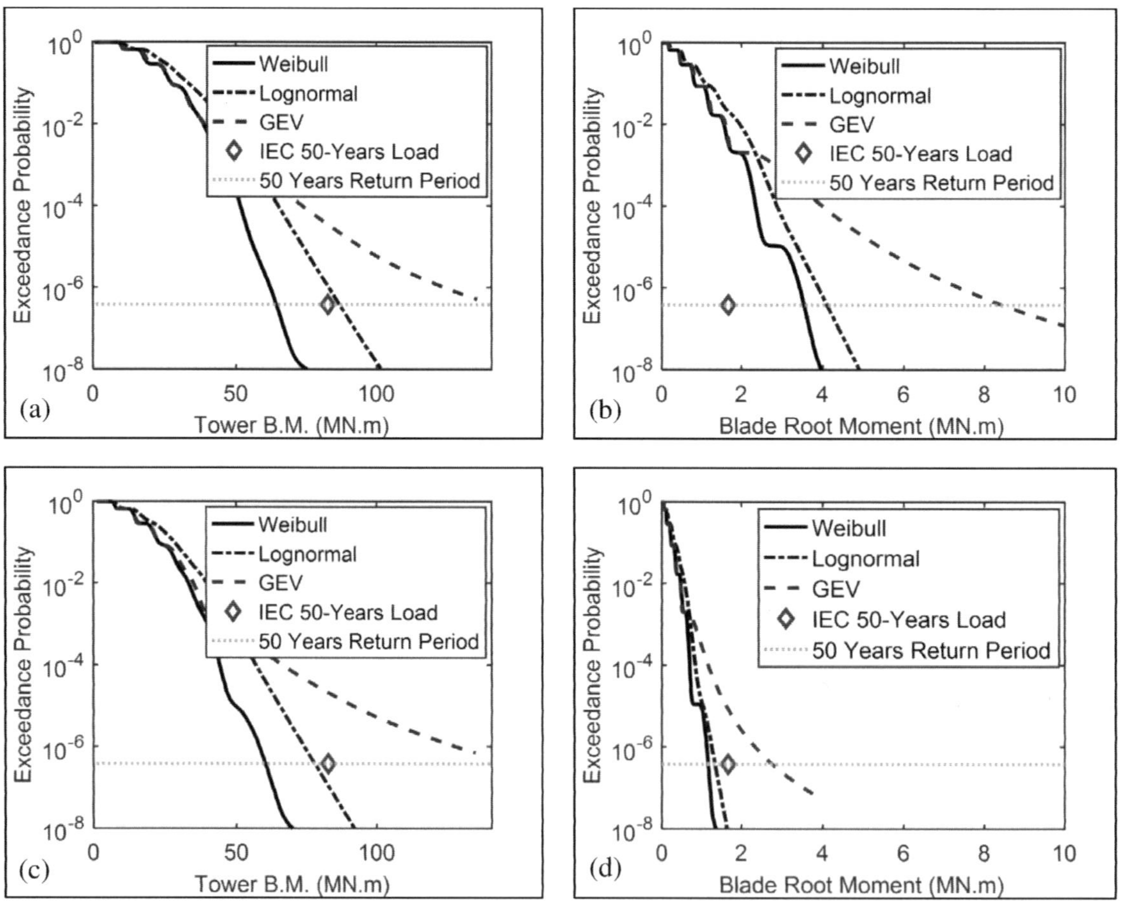

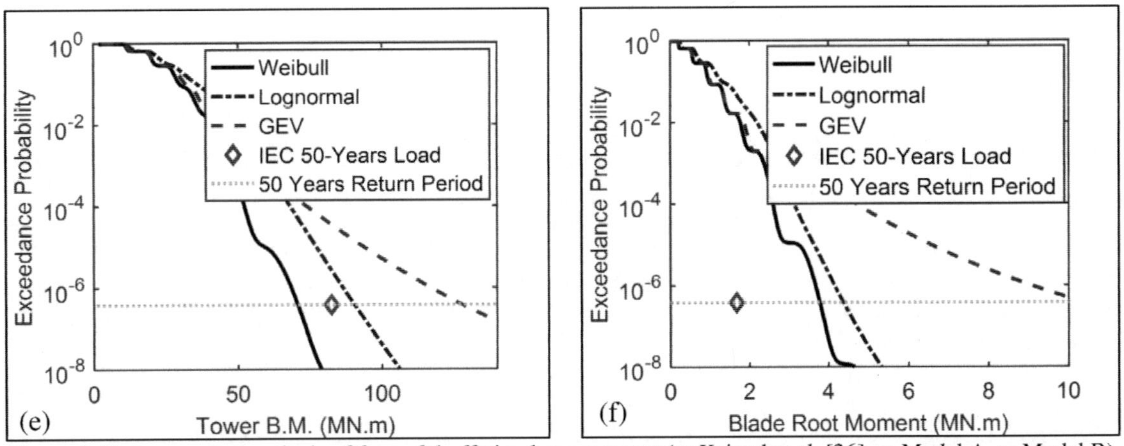

Figure 4. Extreme value analysis of fore-aft buffeting base moment (a: Kaimal et al. [26], c: Model A, e: Model B) and blade root edge-wise moment (b: Kaimal et al. [26], d: Model A, f: Model B) using different distributions

The comparison of the different distributions for extreme value analysis showed that, Weibull distributions has the smallest extreme values for the tower and blades. Hence the IEC 61400-3 [20] recommendations compared to Weibull distribution results, turns in the condition for proposing minimum changes in IEC 61400-3 [20]. By assuming Weibull distribution for local maxima, Figure 5a shows the extreme values of tower buffeting base moment for three different turbulence models. In this case the IEC 61400-3 [20] recommendations resulted in 30%, 37%, and 18% larger values than Kaimal et al. [26], Model A and Model B respectively. For blade root edge-wise moment, the IEC 61400-3 [20] recommendations resulted in underestimating the blade buffeting responses compared to conventional Kaimal et al. [26] model and spectrum Model B (Figure 5b).

Table 4 shows the values of tower buffeting fore-aft base moment and blade root edge-wise moment extrapolated for 50-years return period.

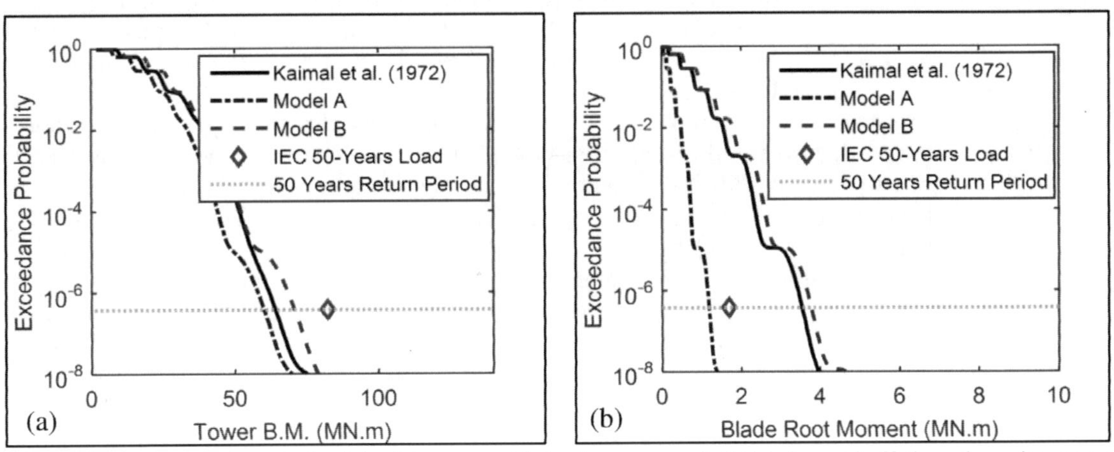

Figure 5. Extreme value analysis of the (a) tower buffeting base moment, (b) blade root buffeting edge-wise moment

Table 4. 50-years return period buffeting responses for different models

Response	Kaimal et al. [26]	Model A	Model B	IEC 61400-3 [20]
Tower Buffeting base moment (MNm)	64.07	60.12	70.68	83.38
Blade Root edge-wise moment (MNm)	3.54	1.17	3.78	1.67

The extreme value analysis of along-wind responses depicted that by assuming Weibull distribution for local maxima, for designing the tower, the IEC 61400-3 [20] recommendations for turbine class I can be used for class S; however, for designing the blades subjected to hurricane, IEC 61400-3 [20] should be reconsidered.

7 CONCLUSION

This paper studied the effects of newly observed hurricane turbulence models on offshore wind turbines by considering unsteady aerodynamic forces on the tower. The main goal was to compare the 61400-3 [10] recommendations for wind turbine class S with hurricane forces resulted from recent observations. For this purpose, first, the recent observations and presented turbulence models were discussed. The five Saffir-Simpson hurricane categories as well as wind induced wave field was generated. Then. The modified NREL-FAST package to address unsteady forces on the tower was implemented. At the end the followings were concluded:

- In order to compare the responses of short term analysis of the offshore wind turbine with IEC 61400-3 [20] recommendations for 50-years return period, an extreme value analysis was carried out. For the extreme value analysis, the direct integration method with peak over threshold (POT) approach was used. The extreme value analysis of along-wind responses depicted that by assuming Weibull distribution for local maxima, for designing the tower, the IEC 61400-3 [20] recommendations for turbine class I can be used for class S; however, for designing the blades subjected to hurricane, IEC 61400-3 [20] should be reconsidered. On the other hand, assuming GEV distribution showed that IEC 61400-3 [20] recommendation for turbine class I, should be revised for designing tower and blades of turbine class S subjected to hurricane.
- The comparison of three different distributions for fitting local maxima in extreme value analysis showed that, Weibull distribution has the smallest extreme values for tower and blades. Hence the IEC 61400-3 [20] recommendations compared to Weibull distribution results, will be the condition for proposing minimum changes in IEC 61400-3 [20]. By using the Weibull distributions, the results showed that for the tower fore-aft buffeting base moment, the IEC 61400-3 [20] recommendations resulted in 30%, 37%, and 18% larger values than Kaimal et al. [26], Model A and Model B respectively; however, the IEC 61400-3 [20] recommendations resulted in underestimating the blade edge-wise buffeting responses compared to conventional Kaimal et al. [26] model and spectrum Model B. Hence, for designing the tower for special events such as hurricane, the IEC 61400-3 [20] recommendations for turbine class I can be used for wind turbine class S; however, for designing the blades subjected to hurricane, IEC 61400-3 [20] should be reconsidered.

8 ACKNOWLEDGMENT

The research reported here is supported in part by the National Science Foundation CMMI Grant 1252736. Any opinions, findings, and conclusions expressed in this paper are those of the authors and do not necessarily reflect the views of the sponsor.

9 REFERENCES

[1] Agarwal P, Manuel L. Simulation of offshore wind turbine response for long-term extreme load prediction. Engineering structures. 2009; 31:2236-46.
[2] Hassanzadeh A, Hassanabad AH, Dadvand A. Aerodynamic shape optimization and analysis of small wind turbine blades employing the Viterna approach for post-stall region. Alexandria Engineering Journal. 2016; 55:2035-43.
[3] Hassanzadeh A, Naughton J. Design and analysis of small wind turbine blades with wakes similar to those of industrial scale turbines. APS Division of Fluid Dynamics Meeting Abstracts2016.
[4] Kim E, Manuel L. On the Extreme Rotor and Support Structure Response of an Offshore Wind Turbine in an Evolving Hurricane. ASME 2013 32nd International Conference on Ocean, Offshore and Arctic Engineering: American Society of Mechanical Engineers; 2013. p. V008T09A82-VT09A82.
[5] Kim E, Manuel L, Curcic M, Chen SS, Phillips C, Veers P. On the Use of Coupled Wind, Wave, and Current Fields in the Simulation of Loads on Bottom-Supported Offshore Wind Turbines during Hurricanes. NREL (National Renewable Energy Laboratory (NREL), Golden, CO (United States)); 2016.
[6] Sanaye S, Hassanzadeh A. Multi-objective optimization of airfoil shape for efficiency improvement and noise reduction in small wind turbines. Journal of Renewable and Sustainable Energy. 2014; 6:053105.
[7] Balderrama J, Masters F, Gurley K, Prevatt D, Aponte-Bermúdez L, Reinhold T, et al. The Florida coastal monitoring program (FCMP): A review. Journal of Wind Engineering and Industrial Aerodynamics. 2011; 99:979-95.
[8] Jung S, Masters F. Characterization of open and suburban boundary layer wind turbulence in 2008 Hurricane Ike. Wind and Structures. 2013; 17:135-62.
[9] Li Q, Xiao Y, Wu J, Fu J, Li Z. Typhoon effects on super-tall buildings. Journal of Sound and Vibration. 2008; 313:581-602.
[10] Powell MD, Houston SH, Reinhold TA. Hurricane Andrew's landfall in south Florida. Part I: Standardizing measurements for documentation of surface wind fields. Weather and Forecasting. 1996; 11:304-28.
[11] Powell MD, Vickery PJ, Reinhold TA. Reduced drag coefficient for high wind speeds in tropical cyclones. Nature. 2003; 422:279-83.
[12] Schroeder JL, Smith DA. Hurricane Bonnie wind flow characteristics as determined from WEMITE. Journal of Wind Engineering and Industrial Aerodynamics. 2003; 91:767-89.
[13] Vickery PJ, Wadhera D, Powell MD, Chen Y. A hurricane boundary layer and wind field model for use in engineering applications. Journal of Applied Meteorology and Climatology. 2009; 48:381-405.
[14] Caracoglia L, Jones NP. Analysis of full-scale wind and pressure measurements on a low-rise building. Journal of Wind Engineering and Industrial Aerodynamics. 2009; 97:157-73.
[15] Gong K, Chen X. Improved modeling of equivalent static loads on wind turbine towers. Wind and Structures. 2015; 20:609-22.
[16] Li L, Xiao Y, Kareem A, Song L, Qin P. Modeling typhoon wind power spectra near sea surface based on measurements in the South China sea. Journal of Wind Engineering and Industrial Aerodynamics. 2012; 104:565-76.
[17] Yu B, Chowdhury AG, Masters FJ. Hurricane wind power spectra, cospectra, and integral length scales. Boundary-layer meteorology. 2008; 129:411-30.
[18] Amirinia G, Jung S, Alduse BP. Effect of different hurricane spectrums on wind turbine loads and responses. AWEA Wind Power Conference. Orlando, US,2015.
[19] IEC 61400-1. IEC 61400-1: Wind turbines part 1: Design requirements. International Electrotechnical Commission. 2005.
[20] IEC 61400-3. Wind turbines – Part 3: Design requirements for offshore wind turbines. International Electrotechnical Commission. 2009; 3.

[21] Kim E, Manuel L. Hurricane-Induced Loads on Offshore Wind Turbines with Considerations for Nacelle Yaw and Blade Pitch Control. Wind Engineering. 2014; 38:413-24.
[22] Amirinia G, Jung S. Along-wind buffeting responses of wind turbines subjected to hurricanes considering unsteady aerodynamics of the tower. Engineering Structures. 2017; 138:337-50.
[23] Amirinia G, Jung S. Time domain analysis of unsteady aerodynamic forces on a parked wind turbine tower subjected to high winds. 8th International Colloquium on Bluff Body Aerodynamics and Applications. Boston, US. 2016.
[24] Jonkman JM, Butterfield S, Musial W, Scott G. Definition of a 5-MW reference wind turbine for offshore system development. National Renewable Energy Laboratory Golden, CO, USA; 2009.
[25] Davenport AG. The spectrum of horizontal gustiness near the ground in high winds. Quarterly Journal of the Royal Meteorological Society. 1961; 87:194-211.
[26] Kaimal J, Wyngaard J, Izumi Y, Coté O. Spectral characteristics of surface - layer turbulence. Quarterly Journal of the Royal Meteorological Society. 1972; 98:563-89.
[27] Tieleman HW. Universality of velocity spectra. Journal of Wind Engineering and Industrial Aerodynamics. 1995; 56:55-69.
[28] Von Karman T. Progress in the statistical theory of turbulence. Proceedings of the National Academy of Sciences of the United States of America. 1948; 34:530.
[29] Strømmen E. Theory of bridge aerodynamics: Springer Science & Business Media; 2010.
[30] Davenport A. The prediction of the response of structures to gusty wind. Safety of structures under dynamic loading. 1977; 1:257-84.
[31] Simiu E, Scanlan RH. Wind Effects on Structures: Fundamentals and Application to Design. New York: John Wiley & Sons; 1996.
[32] Clobes M, Peil U. Unsteady buffeting wind loads in the time domain and their effect on the life-cycle prediction of guyed masts. Structure and Infrastructure Engineering. 2011; 7:187-96.
[33] Chang B, Sarkar P, Phares B. Time-domain model for predicting aerodynamic loads on a slender support structure for fatigue design. Journal of Engineering Mechanics. 2009; 136:736-46.
[34] Jonkman BJ. TurbSim user's guide: version 1.50: National Renewable Energy Laboratory Golden, CO, USA; 2009.
[35] Simpson RH, Saffir H. The hurricane disaster potential scale. Weatherwise. 1974; 27:169.
[36] U.S. Army Coastal Engineering Research Center. Shore protection manual. 3rd ed. Washington, D.C.: Government Printing Office; 1977.
[37] Sorensen RM. Basic coastal engineering: Springer Science & Business Media; 2005.
[38] Jonkman J, Robertson A, Hayman G. HydroDyn user's guide and theory manual. National Renewable Energy Laboratory. 2014.
[39] Ragan P, Manuel L. Statistical extrapolation methods for estimating wind turbine extreme loads. Journal of Solar Energy Engineering. 2008; 130:031011.

A Comparison of Wind Tunnel, AS/NZS1170.2, and CFD Wind Speed-ups Over a Rounded Hill

AA Safaei Pirooz [a], Richard G.J. Flay [b]

[a]*University of Auckland, 24 Princes St, Auckland, New Zealand*
[b]*University of Auckland, 24 Princes St, Auckland, New Zealand*

ABSTRACT: Mean and gust wind speeds can be increased significantly by natural and man-made topography. Various multipliers provided in wind loading standards and codes can be used to predict these speed-ups, and accordingly, structures are designed to resist the effects of topographic wind speed-ups. The need to confirm the actual mean and gust wind speed-ups over simple two-dimensional hill shapes is a basic first step in the development of practical and simple prediction techniques suitable for more complex terrain. In this study, a 2D bell-shaped hill is modelled in ANSYS CFX, and results are compared with wind tunnel measurements, and the Standards AS/NZ1170.2 (2011) and AS 1170.2(1989). For the CFD study, three different turbulence models, viz the standard k-ε, k-ω and SST, are used and compared. Results show that k-ω and SST gave better predictions for separation and reattachment points and k-ε model estimates the mean speed-up ratio more accurately and is more comparable with the wind tunnel measurements.
KEYWORDS: CFD modelling, complex terrain, hills, wind tunnel, horizontally homogeneous boundary layer.

1 INTRODUCTION

The study of wind flow in the Atmospheric Boundary Layer (ABL) is important in many applications, such as wind loading on structures, prediction of local weather conditions, dispersion of pollutants, and designing wind farms. The increasing worldwide use of wind energy means that farms are being constructed in more complex terrain, especially in countries like New Zealand where almost all the available terrain is complex and hilly. Wind speeds above different topography and wind loads on structures are estimated by using standard codes. Therefore, it is of a great importance to make sure that values and methods given in standards are as accurate as possible. As the wind blows over hills, its velocity, direction and turbulence quantities change, and the mechanisms of these changes, due to the unpredictable behaviour of the wind, are not well understood. Estimating wind forces on structures, utilisation of wind power, meteorological stations, and many other phenomena are significantly affected by wind over hills. In practical many situations, these effects need to be predicted and this can only be done by better understanding and better data on wind flow over hills [1].

Computational fluid dynamics (CFD) methods, particularly commercial CFD packages, have become widespread and are used to model wind flow in the ABL. Therefore, reliable predictions of the wind's behaviour rely upon accurate modelling results. Despite the fact that the CFD packages are widely available, well-developed, and broadly verified and validated, several authors ([2], [3], [4], [5]) have demonstrated that these commercial CFD packages are not suitable for modelling the ABL flows using their standard settings. The standard settings mentioned include the turbulence models, sand grain wall-function models and neutrally stable atmospheric conditions. O'Sullivan, et al. [4] have shown that these methods are not able to provide sufficiently accurate results for modelling the wind flow in the ABL, particularly when more complex cases in the ABL are simulated.

Many computational wind engineering problems are sensitive to the approach flow velocity and turbulence property profiles. Therefore, it was strongly recommended by Richards, et al. [2] that the ABL should be modelled as *horizontally homogeneous*, which means that the statistical properties of the wind flow are independent of horizontal position. This assumption ensures that the velocity and turbulence property profiles are maintained throughout the domain in the absence of stream-wise gradients in the vertical profiles. Unintended differences, even minor changes, between inlet approach flow, and the outlet flows can affect the CFD simulations. Consequently, such results may not reflect correctly the real phenomena taking place in the ABL.

To date, the most successful CFD modelling of the ABL has been done by Richards, et al. [2]. They used the k-ε turbulence model and based on that, proposed a suitable set of boundary conditions to ensure a homogeneous boundary layer. The boundary conditions and profiles proposed by Richards and Hoxey have become a standard among researchers in computational wind engineering. However, it was later demonstrated by Hargreaves, et. al [3] that the profiles proposed by Richards and Hoxey are not sufficient by themselves to generate a horizontally homogeneous ABL in the CFD packages, and additional modifications to the wall and top boundary conditions are needed to produce a consistent ABL.

Many computational wind engineers use the Richards and Hoxey inlet profiles, which were derived for a shear driven flow with zero-gradient boundary at the top of the domain, inappropriately. Without the driving shear stress at the upper boundary, the problem becomes a pressure driven boundary layer. If a symmetry boundary is intended to be applied at the top boundary, the Richards and Hoxey profiles cannot be used, and instead the new inlet profiles recently proposed by Richards, et al. [6] based on the Deaves and Harris model, should be implemented.

In 2011, Richards, et al. [7] extended the initial work done in 1993 for the k-ε turbulence model, to provide similar results for other turbulence models, such as RNG k-ε, Wilcox k-ω, and the quasi-isotropic LRR Reynolds stress transport model. In their study, the new profiles for these turbulence models and their application within the commercial CFD code CFX 12.0 were described.

One of the first analytical solutions to describe the wind flow over low hills was presented by Jackson, et al. [1]. They showed how the size and shape of the hill and surface roughness can affect the wind speed and shear stress.

The first numerical and modelling studies regarding the effects of topography on wind flows were mostly concerned with the wind flow over shallow and gently sloping terrain (for example; [8], [1], [9]). Therefore, no analytical models have been developed to investigate the separation and significant changes in wind characteristics which occur in more complex and steeper terrain. To gain a better understanding of wind behaviour over such topography, accurate computational modelling and experimental tests are needed.

In a hilly country like New Zealand, the hill-shape multiplier is the most important factor to calculate the design wind speed in order to achieve a safe level of structural design for buildings. Recently research was carried out by Flay, et al. [10] to improve the method of analysing of design wind speed-ups over the rugged Belmont Hills in Wellington, New Zealand. Apart from the Australia/New Zealand wind loading standard AS/NZ 1170.2, six other international standards were also reviewed, and since all of them use similar approaches to estimate the wind speed-ups, so the results from the standards were all pretty much the same, and none of them was able to provide adequate factors to correctly predict the changes in wind speed in this complex terrain.

2 METHODOLOGY AND FORMULATION

2.1 *Wind Flow over a hill according to AS/NZ 1170.2 and AS 1170.2*

If the structure under consideration is located within a local topographic zone, the topographic multiplier is obtained from the appropriate standard, which in New Zealand is AS/NZ1170.2 [11]. In the latest version of the standard, the hill shape multiplier is based on gust wind speeds. Figs 2-3 show a schematic diagram of a bell-shaped hill from the standard, to which Eqs. (1) refer.

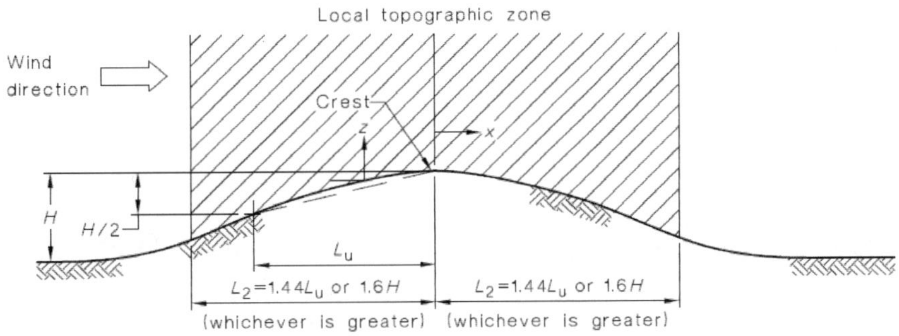

Fig 1. Hills and ridges [11]

(a) For $H/(2L_u) < 0.05$: $M_h = 1.0$

(b) For $0.05 \leq H/(2L_u) \leq 0.45$: $M_h = 1 + \left(\frac{H}{3.5(z+L_1)}\right)\left(1 - \frac{|x|}{L_2}\right)$

(c) For $H/(2L_u) > 0.45$; (1)

- Within the separation zone (Fig. 10): $M_h = 1 + 0.71\left(1 - \frac{|x|}{L_2}\right)$
- Elsewhere within the local topographic zone, M_h is given as 1(b)

H, L_u, x, and z are shown in Fig. 2. L_1 and L_2 are defined as follow:
L_1 = Length scale, to determine the vertical variation of M_h to be taken as the greater of $0.36L_u$ or $0.4H$.
L_2 = Length scale, to determine the horizontal variation of M_h, to be taken as $4L_1$ upwind for all types, and downwind for hills and ridges, or $10L_1$ downwind for escarpments [6].

Fig. 3 shows the expected separation zone for steep hills which starts at the crest, extending for a length of *H/4* downstream.

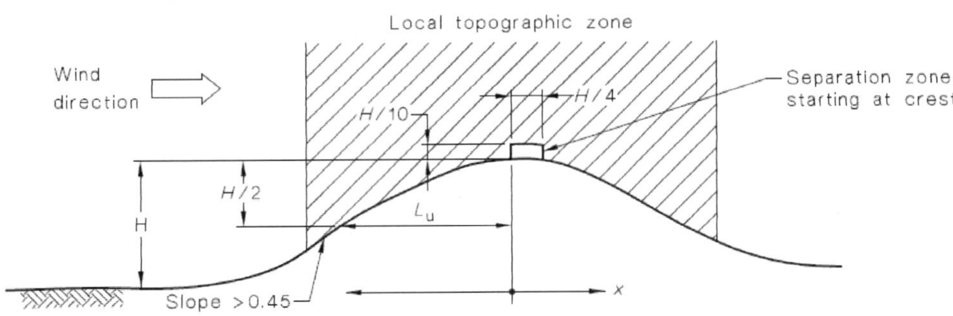

Fig 2. Separation zone for hills having upwind slopes greater than 0.45 [11]

In contrast, in a previous version of Australian standard [12], this multiplier (Eq. 2) was defined with respect to mean wind speeds.

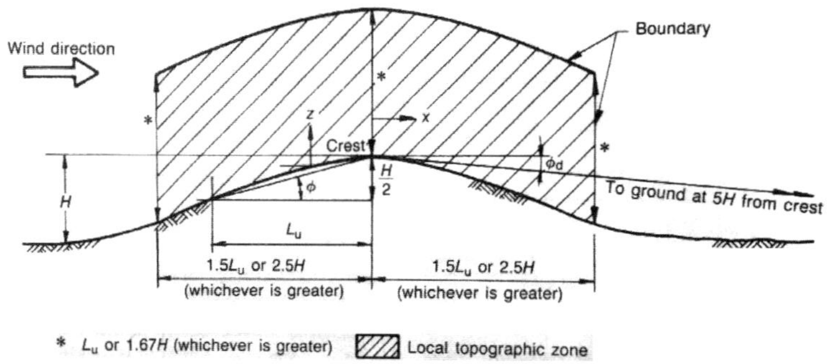

Fig 3. Hills and Ridges **[12]**

(a) For $\phi < 0.05$: $M_t = 1.0$

(b) $\bar{M}_t = 1 + (k_t s \phi')$ (2)

Where;

ϕ = The effective upwind slope of a hill = $H/(2L_u)$ (3a)

ϕ' = The lesser value of ϕ or 0.3 (3b)

k_t = a type factor for topography = $1.4 + 36(\phi_d - 0.05)$ with a maximum of 3.2 for hills and ridges (3b)

s = a position factor for topography effects

$$= \left(1 - \frac{|x|}{1.5L*}\right)\left(1 - \frac{0.6z}{H}\right) \quad for\ \phi \geq 0.3$$ (3b)

For negative values of x: $L* = 1.67H\ for\ \phi \geq 0.3$

For positive values of x: $L* = L_u\ or\ 1.67H\ whichever\ is\ greater$

ϕ_d = The average downward slope of a hill, ridge (3b)

2.2 *CFD modelling of the Atmospheric Boundary Layer*

There are two approaches commonly used for modelling the ABL: wall shear driven and pressure driven. In this study, the results from the CFD modelling are compared with wind tunnel results and predictions from the standard. In the wind tunnel and ABL the air flow is mainly generated by pressure difference. Therefore, to have a better comparison, it was decided to use the pressure driven CFD approach in the present study.

The details and derivation of equations for the pressure driven ABL have been elaborated by Richards, et al. [6]. In this approach the symmetry boundary condition is used at the top boundary. The inlet velocity and turbulence profiles for k-ε, k-ω, and SST turbulence models are as described in Eqs. (4 – 7) [6].

$$U(z) = \frac{u_\tau}{\kappa}\left(\ln\left(\frac{z}{z_0}\right) + C_{U1}\left(\frac{z}{H}\right) + C_{U2}\left(\frac{z}{H}\right)^2 + C_{U3}\left(\frac{z}{H}\right)^3 + C_{U4}\left(\frac{z}{H}\right)^4\right) \tag{4}$$

$$k(z) = u_\tau^{\,2}\left(C_{k1} + C_{k2}\left(1-\frac{z}{H}\right)^2 + C_{k3}\left(1-\frac{z}{H}\right)^4 + C_{k4}\left(1-\frac{z}{H}\right)^6\right) \quad (5)$$

$$\varepsilon(z) = \frac{C_\mu k(z)^2}{\kappa u_\tau z}\left(1 + (1+C_{U1})\left(\frac{z}{H}\right) + (1+C_{U1}+2C_{U2})\left(\frac{z}{H}\right)^2 + (1+C_{U1}+2C_{U2}+3C_{U3})\left(\frac{z}{H}\right)^3\right) \quad (6)$$

$$\omega(z) = \frac{k(z)}{\kappa u_\tau z}\left(1 + (1+C_{U1})\left(\frac{z}{H}\right) + (1+C_{U1}+2C_{U2})\left(\frac{z}{H}\right)^2 + (1+C_{U1}+2C_{U2}+3C_{U3})\left(\frac{z}{H}\right)^3\right) \quad (7)$$

The values of the coefficients used in Eqs. (4-7) are provided in [6] calculated by semi-analytical and CFD methods. By having values of the aerodynamic roughness and the reference velocity of 0.02 m (Category 2 [11]) and 10m/s (at height of 10m), respectively, the frictional velocity, and gradient height can be calculated for each turbulence model and the results are shown in Table 1.

Table 1 - Boundary layer parameters

	u_τ (m/s)	h (m)
$k-\varepsilon$	0.64105	1036
$k-\omega$	0.6415	1037
SST	0.6416	1037

It is essential to make sure that the inlet boundary conditions are combined with an appropriate rough wall treatment. Due to the fact that many CFD packages use an equivalent sand grain size, it is important to find the relationship between the aerodynamic roughness (z_0) and the sand grain roughness. This relationship for both CFX and Fluent is discussed by Blocken, et al. [5] who propose the relationship given in Eq. (8).

$$k_{s,ABL} = 29.6\, z_0 \qquad (CFX) \quad (8)$$

The inhomogeneity error relative to the inlet profile, for each of the variables U, k, ε, and ω as a function of downstream distance is calculated by Eq. (9).

$$e = 100\left|\frac{\varphi(x)-\varphi(x=0)}{\varphi(x=0)}\right| \quad (9)$$

The dimensions of the full scale hill modelled in CFX are shown in Fig. 4.

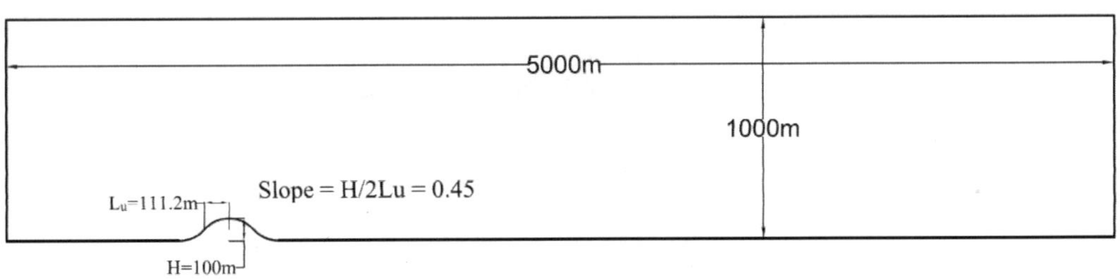

Fig 4. Schematic of the full scale hill

2.3 *Wind Tunnel setup*

Wind tunnel measurements were carried out in the boundary layer wind tunnel at the University of Auckland at a nominal scale of 1:500. The model tested was two-dimensional, extending across the full-width of the wind tunnel perpendicular to the flow. The wind tunnel test section is 20 m long, 2.5 m high, and 3.6 m wide. The reference wind tunnel speed at a height of 20 mm above the floor was 6.56 m/s. The standard terrain category 2 ($z_0 = 0.02\ m$, open grassland) [11] was replicated in the wind tunnel at a scale of 1:500.

Initially, to ensure that the replicated terrain was capable of producing the target profiles of velocity and turbulence intensity, the values were measured in the vertical direction and compared with recommended values from AS/NZ 1170.2 [11].

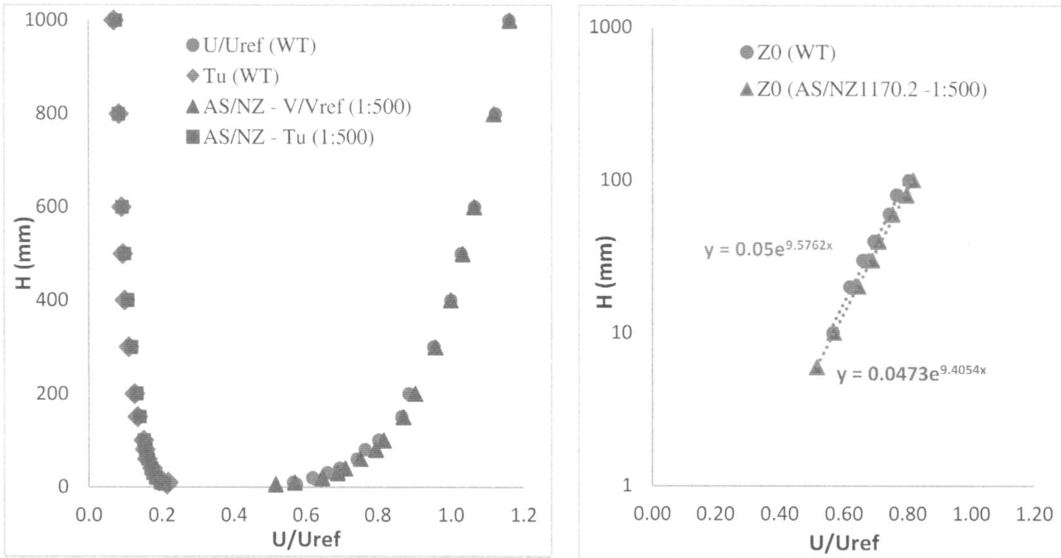

Fig 5. (a) Velocity and Turbulence intensity profiles; (b) Aerodynamic Roughness

The wind tunnel setup for measuring the vertical velocity profile is shown in Fig. 6(a). In addition, to estimate the flow separation and reattachment locations, some tufts were stuck to the leeward side of the hill, as shown in Fig. 6(b).

(a) (b)

Fig 6. Wind Tunnel Set up (a) Velocity profile measurement; (b) Estimating the separation and reattachment

The mean speed amplification factors presented in the results are simply the ratio of the mean speed divided by the mean speed for the reference profile at the same local height above the ground. In the following results and discussion, all dimensions are expressed as metres at nominal full scale, (e.g. the 20 mm model hill height is equivalent to a full-scale hill height of 10 m).

3 RESULTS AND DISCUSSION

3.1 *Homogeneous Atmospheric Boundary Layer*

By using the proposed inlet profiles in CFX a horizontally homogeneous ABL was produced as shown in Fig. 7. The peak that can be seen in the TKE profile at low levels is due to discretisation errors in the near wall region [7]. Also, it is noted in [3] that this spike occurs at the second cell from the wall and it is a feature of the $k - \varepsilon$ model.

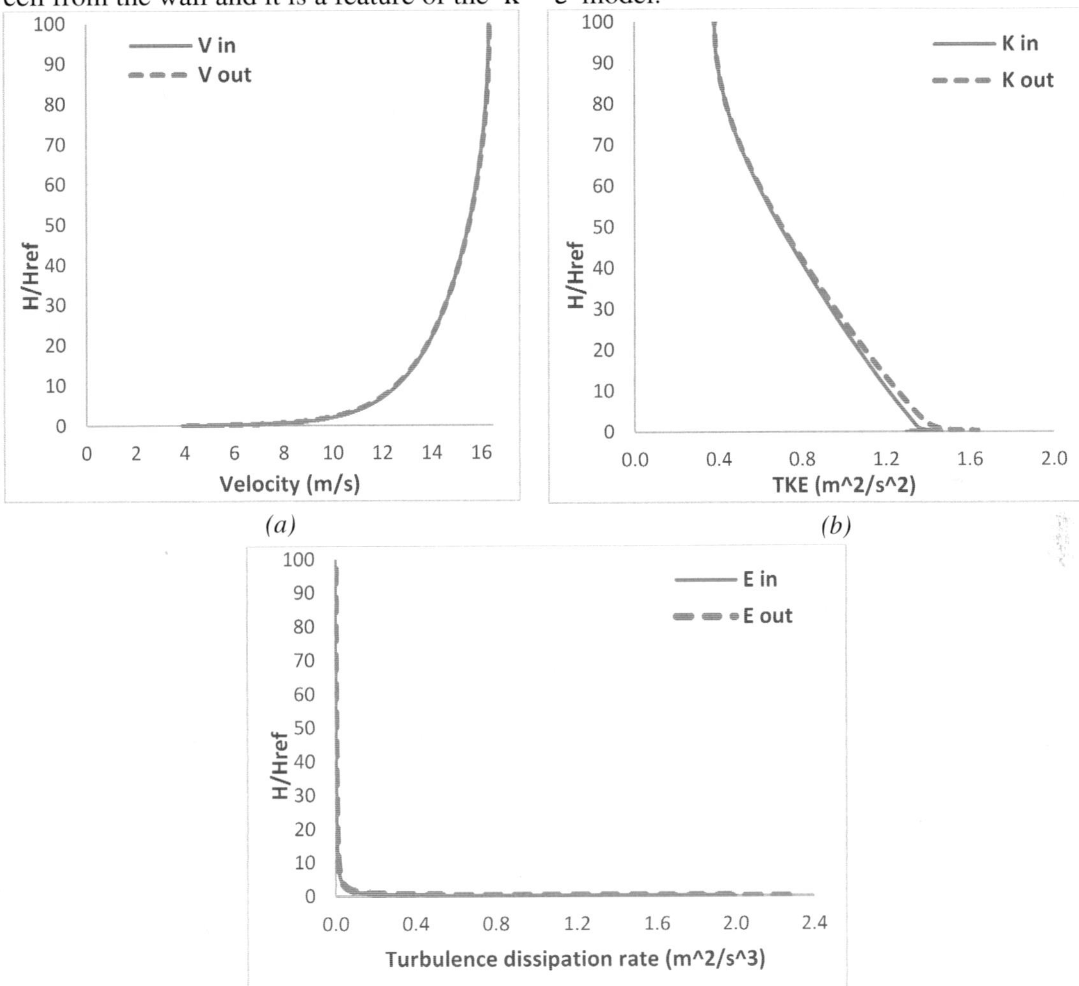

(a) *(b)* *(c)*

Fig 7. Horizontally homogeneous ABL modelled by using CFX for k-ε model: (a) Velocity; (b) Turbulence kinetic energy; (c) Turbulence dissipation rate

Three turbulence models, *k-ε*, *k-ω* and SST were used to model the ABL in CFX, then by using Eq. (9) the inhomogeneity errors were calculated (*Href* = 10 m) and are shown in Fig. 8.

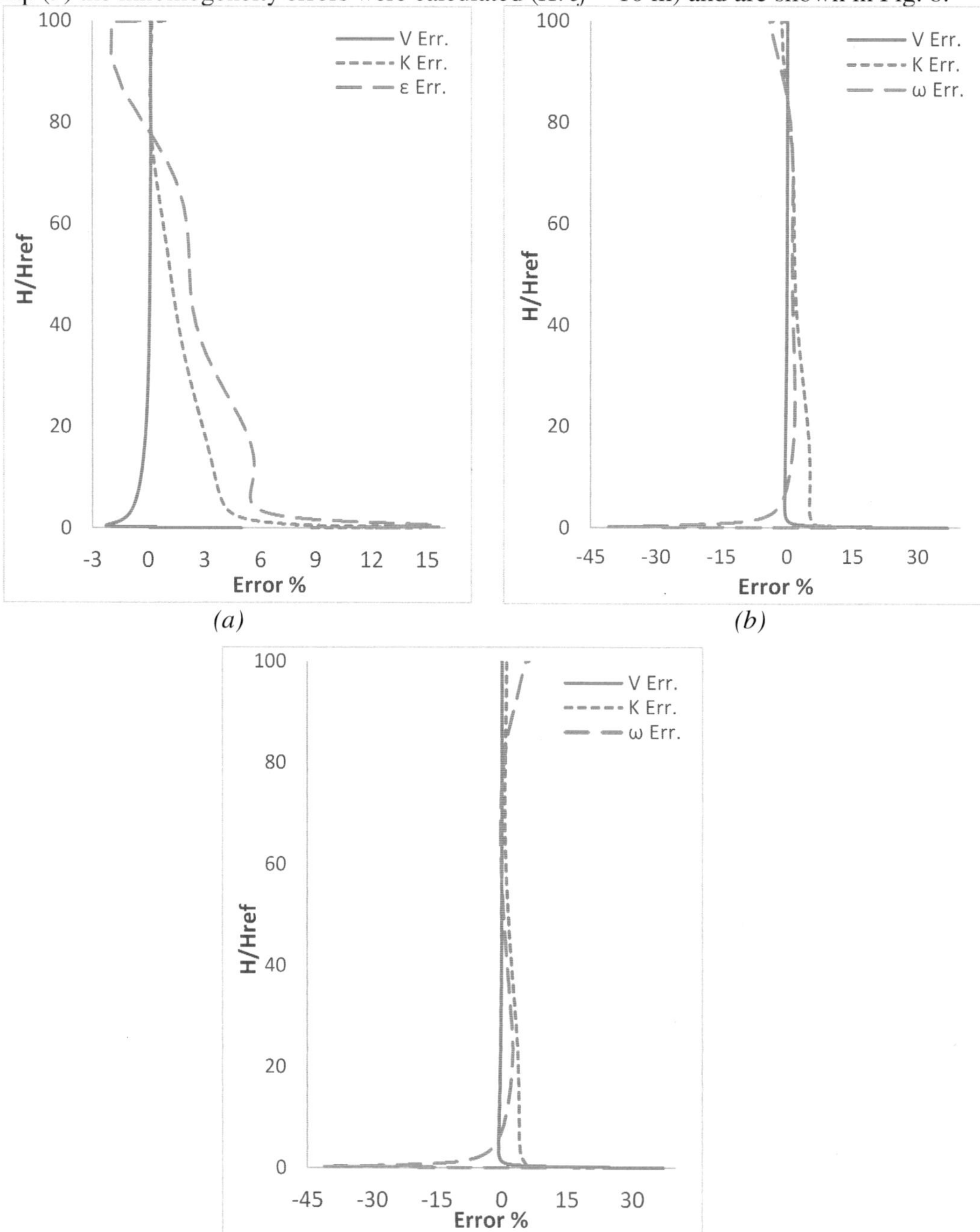

(a) *(b)* *(c)*

Fig 8. Inhomogeneity Error: (a) k-ε; (b) k-ω; (c) SST

As can been seen, the k-ω and SST models have less inhomogeneity error away from the surface. In contrast, k-ε shows less error near the wall.

3.2 *CFD Modelling*

The 2D full-scale hill was modelled in CFX with the same dimensions as shown in Fig. 4. In order to obtain accurate results and good visualisations, a fine mesh was generated near the surface (first layer thickness 1 m). In addition, the high resolution scheme was used for both the momentum and turbulence property equations to ensure reliability of the results.

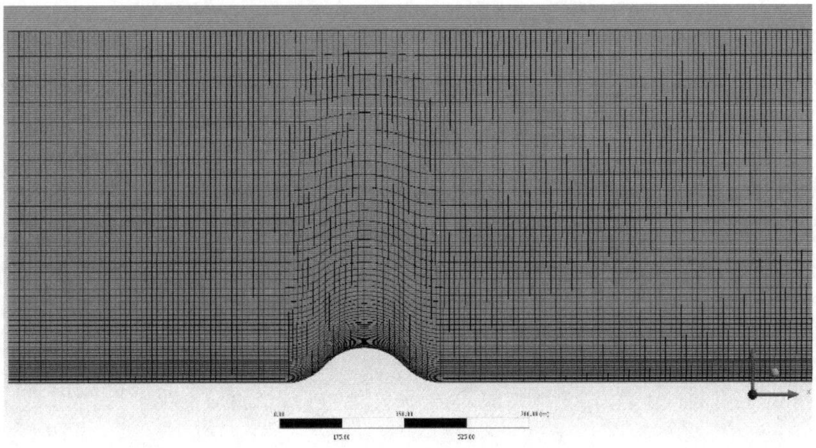

Fig 9. Mesh

The speed-up ratio is equal to the ratio of the mean velocity above the hill divided by the reference profile at the same local height above the ground. The mean speed-up ratios obtained from 2D CFD modelling are compared with AS 1170.2 – 1989, as discussed earlier.

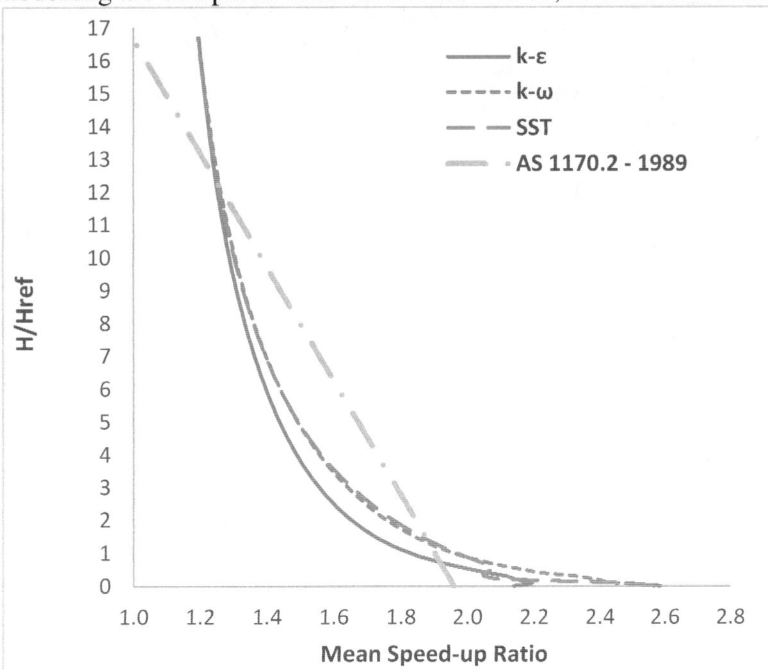

Fig 10. Mean Speed-up Ratio, 2D CFX results and AS 1170.2 - 1989

Fig. 10 illustrates that all the turbulence models predict higher mean speed-up ratios near the surface than that of the former standard. The other issue with the former standard is that it has a linear prediction of the speed-up ratio, unlike both the CFD modelling and the wind tunnel measurements. In addition, it can be seen that $k-\omega$ and SST have a similar predication for all heights, and have higher speed-up ratios near the surface compared to the $k-\varepsilon$ model.

Fig. 11 compares the separation and reattachment locations of the flow on the leeward side of the hill estimated by different turbulence models. The $k-\omega$ and SST models showed quite a similar wake region behind the hill. In contrast, the wake region produced in the $k-\varepsilon$ model is significantly larger, and the separation point in this model is more comparable with the separation zone shown in [11] (Fig. 2). Later in the paper these estimations are compared with wind tunnel results as well.

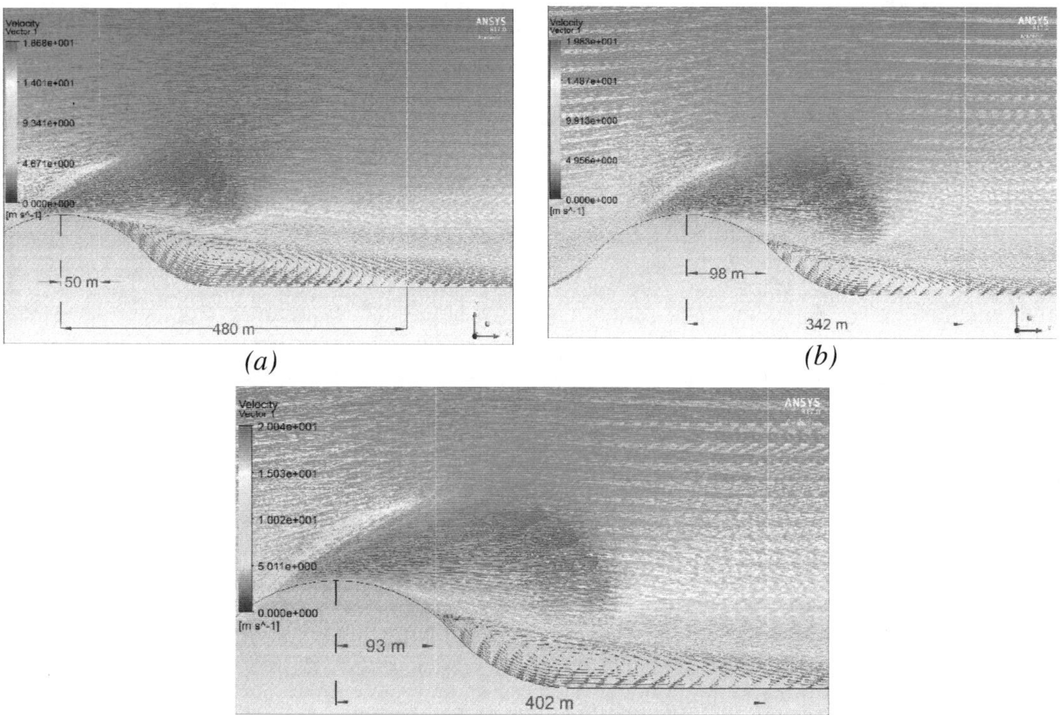

(a) (b)

(c)

Fig 11. 2D modelling prediction of Separation and reattachment point: (a) $\mathbf{k}-\boldsymbol{\varepsilon}$; (b) $\mathbf{k}-\boldsymbol{\omega}$; (c) SST

3.3 *Wind Tunnel Results*

In the wind tunnel experiment, both the mean and gust speeds were measured. In Fig. 12 the mean and gust speed-up ratios obtained from the wind tunnel measurements are compared with the former and current standards. As explained, the mean prediction in AS 1170.2 (1989) shows a linear trend which is not comparable with measurements and modelling results. However, the gust speed-up ratios recommended in AS/NZ 1170.2 are in a good agreement with ratios achieved from the wind tunnel.

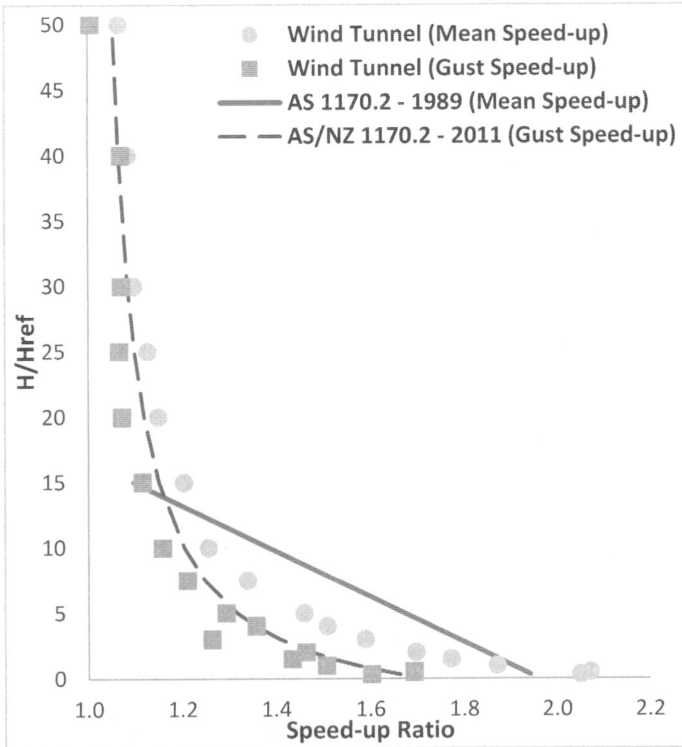

Fig 12. Wind Tunnel results; Mean and Gust speed-up ratios

Fig. 13 shows the approximate locations of the separation and reattachment points achieved in the wind tunnel. By comparing this estimation with 2D modelling results, it can be said that the k-ω and SST models provide better predictions of the separation and reattachment points, and the wake region.

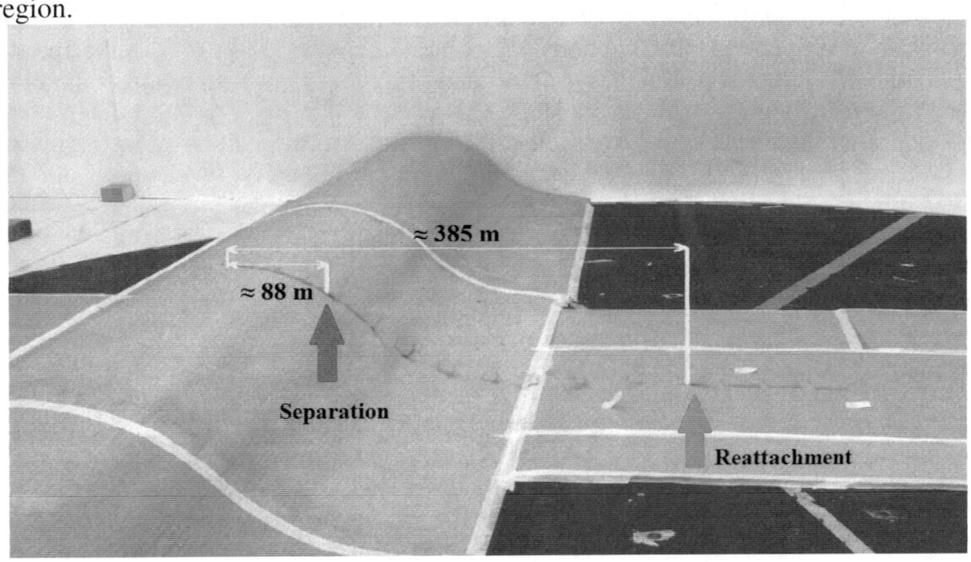

Fig 13. Separation and reattachment points in the wind tunnel

3.4 *Comparison of wind tunnel results and numerical results*

The mean speed-up ratios obtained from the wind tunnel measurements and 2D CFD modelling are compared in Fig. 14. At higher levels all the turbulence models predicted almost the same values. However, close to the surface the k-ε model provided better estimations compared to the wind tunnel results.

As mentioned earlier, the predictions of the separation zone from the k-ω and SST models were more comparable with the wind tunnel results. Thus, it can be said that both the k-ε turbulence model and the AS/NZ 1170.2 Standard overestimate the separation zone.

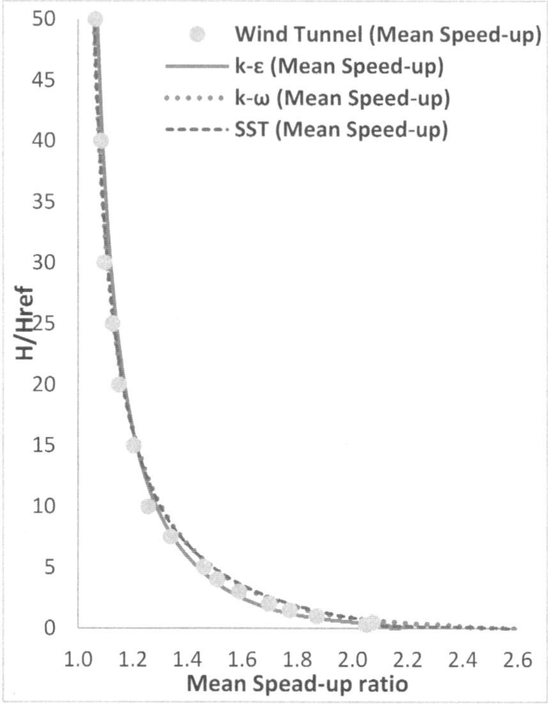

Fig 14. Comparison between wind tunnel and CFD results

4 CONCLUSIONS

This study was aimed at investigating the effects of a simple bell-shaped hill on wind flow in open terrain. For this purpose, CFD and wind tunnel studies were carried out at full-scale and at a nominal scale of 1:500, respectively. Three different turbulence models, namely k-ε, k-ω and SST, were used to simulate the homogenous ABL, and to compute and estimate the speed-up ratios and separation locations. A 2D hill was built and tested in the boundary layer wind tunnel at the University of Auckland.

The prediction of the former standard (from 1989) for the mean speed-up ratios is linear which shows poor agreement with both the CFD and wind tunnel results. The mean speed-up ratios obtained by using the k-ε turbulence model are more similar to the values calculated from the wind tunnel measurements than those of the k-ω and SST models. On the other hand, the separation and reattachment locations estimated by k-ω and SST were in a good agreement with the wind tunnel results. In addition, it was found out that the recommended separation zone length in the standard is an overestimate.

The wind tunnel gust speed-up ratios were compared with values from AS/NZ 1170.2. It was found that the values given in AS/NZ 1170.2 for the gust speed-up ratios are in good agreement with the 2D model tested in the wind tunnel. However, it must be noted that the 2D model does not represent the real situation, and the full-scale comparative study carried out for the complex Belmont Hill in Wellington showed that the standard is not capable of predicting accurate hill multipliers for real 3D complex terrain. Thus, a new and more accurate method is required in the standard in order to reflect the real phenomena taking place in complex terrain. Further work is underway to attempt to develop a more sophisticated prediction method for such hilly complex terrain.

5 REFERENCES

[1] P. S. Jackson and J. C. R. Hunt, "Turbulent wind flow over a low hill," *Quarterly Journal of the Royal Meteorological Society*, vol. 101, pp. 929-955, 1975.

[2] P. J. Richards and R. P. Hoxey, "Appropriate boundary conditions for computational wind engineering models using the k-Ɛ turbulence model," *WInd Engineering and Industrial Aerodynamics,* vol. 46 & 47, pp. 145-153, 1993.

[3] D. M. Hargreaves and N. G. Wright, "On the use of the k-Ɛ model in commercial CFD software to model the neutral atmospheric boundary layer," *Wind Engineering and Industrial Aerodynamics,* vol. 95, pp. 355-369, 2007.

[4] J. P. O'Sullivan, R. A. Archer and R. G. Flay, "Consistent boundary conditions for flows within the atmospheric boundary layer," *Wind Engineering and Industrial Aerodynamics*, vol. 99, pp. 65-77, 2011.

[5] B. Blocken, T. Stathopoulos and J. Carmeliet, "CFD simulation of the atmospheric boundary layer: wall function problems," *Atmospheric Environment,* vol. 41, pp. 238-252, 2007.

[6] P. J. Richards and S. E. Norris, "Appropriate boundary conditions for a pressure driven boundary layer," *Wind Engineering and Industrial Aerodynamics,* vol. 142, pp. 43-52, 2015.

[7] P. J. Richards and S. E. Norris, "Appropriate boundary conditions for computational wind engineering models revisited," *Wind Engineering and Industrial Aerodynamics*, vol. 99, pp. 257-266, 2011.

[8] A. J. Bowen, "The prediction of mean wind speeds above simple 2D hill shape," *Wind engineering and industrial aerodynamics,* vol. 15, pp. 259-270, 1983.

[9] H. W. Teunissen, "Wind tunnel and full scale comparisons of mean flow over an isolated low hill," *Wind engineering and Industrial aerodynamics,* vol. 15, pp. 271-286, 1983.

[10] R. G. Flay, R. Turner, M. Revell, P. Carpenter, P. Cenek and A. B. King, "Wind speed up over hills and complex terrain, and the risk to infrastructure," Launceston; Australia, 2012.

[11] AS/NZ1170.2:2011, Australia/New Zealand Standard, Structural design actions, Part 2: Wind actions, Standards Australia Limited/Standards New Zealand, 2011.

[12] Australian Standard, Australian Standard, SAA Loading Code, Part 2: Wind loads, North Sydney NSW: Standards Australia, 1989.

Toward a Codification of Thunderstorm/Downburst Winds through a Gust Front Factor: Model-based and Data-Driven Perspectives

Ahsan Kareem[a], Dae Kun Kwon[b]

[a] *NatHaz Modeling Laboratory, University of Notre Dame, Notre Dame, IN 46556,* kareem@nd.edu
[b] *NatHaz Modeling Laboratory, University of Notre Dame, Notre Dame, IN 46556,* dkwon@nd.edu

ABSTRACT

Unlike boundary layer winds, it is now well recognized that gust front winds resulting from a thunderstorm/downburst exhibit distinct nonstationarity involving wind shear, i.e., rapid changes in wind velocity for a short time interval. In addition to this nonstationary feature, the key distinguishing attributes are so-called "nose shape" profile, in which the maximum wind velocity generally occurs near the ground, potentially exposing structures to higher wind loads. Furthermore, such a change in the approach flow profile/kinematics would introduce a major change in the flow-structure interaction that may differ significantly from the corresponding boundary layer flow case. This is compounded by the inherent transient nature of energetic convective gusts that rapidly increase in amplitude and direction, raising serious questions regarding the applicability of conventional aerodynamic loading theories. For several decades, a number of studies have been focused on identifying the characteristics of such nonstationary gust front winds in a variety of manners such as experimental/numerical methods and rather limited full-scale measurements. Yet, a definitive design practice has rarely been suggested, partially due to little consensus on such characteristics among studies in conjunction with the lack of available data to corroborate and the difficulty to deal with nonstationarity. In an effort to establish a new design procedure for this type of wind load effect on structures, the gust front factor framework has been proposed by authors that encapsulates both the kinematic and dynamic features of gust front induced wind effects on structures which distinguish themselves from those experienced in conventional boundary layer flows, i.e., variation in the kinematics of the velocity profile and its effects on the associated aerodynamics; dynamic effects induced by the sudden rise in wind speed; non-stationarity of turbulence in gust front winds; transient aerodynamics. This is akin to the gust loading factor format used in codes and standards world-wide for the treatment of conventional boundary layer winds. The gust front factor expresses a generalized description of the genesis of the overall wind load effects on structures under both gust front and boundary layer winds and it reduces simply to the gust loading factor for the case of conventional boundary layer winds. This study revisits the gust front factor framework seeking to take the next step toward a codification of gust front winds from model-based and data-driven perspectives. In addition, a comparison of the model-based gust front factor with a recently introduced thunderstorm response spectrum technique to assess their relative performance.

KEYWORDS

Thunderstorm/Downburst; Codification; Gust front factor; Modeling; Design

The 13th Americas Conference on Wind Engineering (13ACWE)
Gainesville, Florida USA, May 21-24, 2017

Preliminary Design Guidelines for Temporary Securement of Wind-Damaged Low-Slope Roof Assemblies

Thomas L. Smith, AIA, RRC, F.SEI[a]
[a]*TLSmith Consulting Inc., Rockton, IL, USA, tlsmith@hughes.net*

ABSTRACT: When wind damages a roof assembly, a temporary roof is often needed to keep rain from leaking into the building. The damaged roof membrane or temporary roof needs to be adequately secured to avoid it from becoming damaged during subsequent winds. However, because of the lack of adequate guidelines for the design of temporary securement, many damaged roof membranes and temporary roofs have been inadequately secured. When inadequately secured, they are vulnerable to blow-off and present a risk to property and life-safety. For example, temporary ballast (e.g., concrete roof pavers and sandbags) have been blown from temporary roofs.

This paper addresses the waterproofing aspects of temporary roofs; however, the focus is on design considerations for temporary securement. The paper is based on the author's experience with temporary roofs over the past thirty years. A variety of temporary roof materials and securement options are discussed, along with guidance that designers can use to design cost-effective temporary roof securement that provides relatively low risk to property and life-safety until a permanent roof system is installed.

The paper also addresses documenting conditions prior to installation of temporary roofing, and it addresses roofing debris that may be on the roof and/or the ground.

The guidelines are intended for architects, engineers, roof consultants and roofing contractors who are involved with wind-damaged roof assemblies.

KEYWORDS: Low-slope[1] roofs, temporary roof system securement, wind blow-off, wind-damaged roofs, wind uplift.

1 INTRODUCTION

A wind-damaged roof should be investigated by a knowledgeable wind-damage investigator. The purpose of the investigation is to determine the cause(s) of the damage and obtain data for designing the roof repair/replacement. A quality investigation can help facilitate avoidance or resolution of disputes arising from the wind damage. Inadequate investigation and evaluation can result in a replacement roof assembly that possesses inadequate wind resistance, and it too may blow off. Reference 1 provides general investigation recommendations.

When a roof is damaged by wind, it is often obvious that the entire roof area needs to be replaced rather than just repairing the damaged area. However, in many instances, just repairing the damaged area is a viable and more economical approach. When determining whether it is appropriate to repair versus replace wind-damaged roofs, it is incumbent on the investigator to perform suitable field testing, and rigorous evaluation and analysis. With appropriate investigation, damaged or wind vulnerable areas outside the apparent damage periphery will likely be identified and corrected, thereby avoiding future wind damage. Reference 2 provides a protocol for evaluating wind-damaged roof assemblies.

---

[1] Low-slope is defined as roofs with a slope $\leq$ 1:3 (ASTM D1079).

Ideally, the investigation should be performed before temporary repairs are made. However, if water is leaking into the building or if a rainstorm is forecast before the investigation occurs, temporary repairs usually take precedence over preservation of the damage scene. If temporary repairs are performed or if wind-borne debris is removed from the roof before the investigation occurs, it is recommended that someone take photos of the roof before debris removal or temporary repair occurs. Initial photos may subsequently alert the investigator to roof areas that may have been damaged by wind-borne debris, and the photos may assist in the damage evaluation. It is also recommended that all roof system debris be retained for follow-up evaluation by the investigator.

Reference 1 also includes recommendations regarding notifications, safety precautions and temporary repairs. The recommendations are summarized and amplified below:

- Notifications: The building owner should notify the following parties that damage has occurred: Their insurance company, their attorney, the manufacturer of the roof system, the roof system designer, the roofing contractor that installed the system and the general contractor.
- Safety precautions: It generally is not advisable to attempt temporary repairs until after severe weather has subsided because of the danger of high winds and possibility of lightning. Caution should be exercised when investigating and making temporary repairs because the work may be conducted during poor lighting or weather conditions (Figure 1), and/or storm damage may have created an extremely hazardous condition (e.g., live electrical conductors, openings through the roof caused by blown-off roof deck or rooftop equipment, or broken gas lines).

Figure 1. Installation of sandbags on a detached roof membrane shortly before sundown.

- Temporary repairs: A professional roofing contractor should be retained to perform the temporary repairs. To the extent possible, perform the repair in a manner that will minimize changing the damage scene and facilitate a follow-up investigation. For example, if a roof membrane has blown away, consider loose laying and temporarily securing a single-ply membrane rather than mop-applying two plies of felt. With this approach, the temporary work can be removed easily during the investigation, allowing the investigator to inspect the membrane substrate. Emergency leak repairs made to a membrane should be marked on a roof plan to aid in locating them for later permanent repair. Temporary repairs should include removal or temporary reattachment of loose flashings and loose rooftop equipment. Remove debris from roof drains, scuppers and conductor heads. Ensure that the temporary roof has adequate drainage to avoid ponding-induced roof collapse. Remove debris from the roof so that is not blown off during a subsequent storm. Check lower-level roofs for damage by wind-borne debris from the damaged roof. Remove debris from the ground around the vicinity of the building.
- Water leakage: If water leaked into the building, the building owner should have someone control the spread of water and initiate drying of the interior.

References 3 – 6 provide emergency repair information (e.g., puncture repairs) on single-ply, spray polyurethane foam, built-up, modified bitumen and structural metal panel roofs. Reference 7 provides a guide for permanently repairing built-up, modified bitumen and single-ply membranes.

2 DAMAGE SCENARIOS

There are four primary damage scenarios: 1) Ballast blow-off and/or scour, 2) the roof membrane detaches, but does not blow off, 3) the roof membrane blows off, but can be temporarily repositioned and 4) the roof membrane blows off and can't be repositioned. These four scenarios are further discussed and shown below:

- Ballast (i.e., aggregate, concrete pavers or cementitious-coated extruded polystyrene panels) blow-off and/or scour: Reference 8 defines scour as the displacement of a few or several pieces of ballast, wherein the membrane is left exposed, but the displaced ballast is not blown off the roof. Ballast blow-off is like scour, except that some ballast is blown off the roof (Figure 2). If the roof is scoured, but blow-off does not occur, the consequences may be minimal. In some instances, there is no subsequent damage to the roof system. In other instances, the insulation boards shuffle below the membrane. Or, the membrane may be punctured by wind-borne debris or a seam may open due to ballooning. Ballooning is the phenomenon wherein the membrane lifts from the substrate and tends to rise in the shape of a balloon. Ballast may be displaced by the balloon, or ballooning may occur after the ballast has been displaced.

  With ballast scour or blow-off, the following actions are typically taken: The exposed membrane is checked for punctures, tears and ruptured seams – if damaged, either emergency or permanent membrane repairs are made. If insulation boards below the membrane shuffled, they are repositioned. Then the ballast is repositioned. If there is an insufficient amount of ballast to reposition because of ballast blow-off (Figure 3), temporary ballast is installed as recommended in Section 4 until replacement ballast can be procured and installed.

  Based on Reference 3, if the exposed area exceeds 4.6 square meters (50 square feet [sf]), or if ballast was blown off the roof, it is recommended that the ballast system be evaluated to determine if it complies with the current edition of ANSI/SPRI RP-4 (Reference 9). If ballasted with square-edged concrete pavers, and if the roof is outside of the RP-4 tables, it is recommended that the wind resistance of the pavers be evaluated in accordance with the recommendations given in Reference 10. Corrective action is recommended if the existing ballast design does not comply with RP-4 or Reference 10.

- The fully adhered or mechanically attached roof membrane detaches, but does not blow off (Figure 4): The membrane is checked for punctures, tears and ruptured seams – if damaged, emergency repairs are made. The loose membrane is temporarily secured with mechanical fasteners or ballast as recommended in Section 4.

- The fully adhered or mechanically attached roof membrane blows off, but can be temporarily repositioned (Figure 5): The same action as described at the previous bullet is taken.

- The roof membrane blows off and can't be reposition (Figure 6): Either a permanent or temporary roof is installed (see Section 4 for temporary roof securement recommendations).

Figure 2. Aggregate ballasted single-ply membrane. A tornado caused this unusual ballast scour pattern. The red arrow indicates where the photo on the right was taken. At some of the scour areas, there were rows of displaced ballast that could be repositioned to cover the roof membrane. At other areas, much of the scoured ballast was blown off the roof, thus necessitating installation of some type of temporary ballast or procurement of replacement aggregate.

Figure 3. Single-ply membrane ballasted with 0.6 x 1.2 meter (2'x4') interlocking cementitious-coated extruded polystyrene panels. Several of the panels were broken and several were blown off the roof. This is a view of the temporary securement that was installed by a roofing contractor. 200 x 400 millimeter (mm) (8" x 16") concrete pavers weighing 83 kilogram per square meter (kg/m^2) (17 pounds per square foot [psf]) were placed on some of the displaced panels (red arrow) and around the perimeter of the blow-off area (yellow arrows). Other displaced panels were not secured with pavers (blue arrow). A timber was also used for temporary securement. The author judged this temporary securement to be inadequate, and provided criteria for immediate installation of additional temporary pavers.

Figure 4. This fully adhered single-ply membrane detached from the insulation, but did not blow off the roof. It was punctured by wind-borne debris, and it tore at some of the roof penetrations. A membrane in this condition should be patched and temporarily secured as soon as possible to avoid more severe damage during a subsequent storm.

Figure 5. This is a mechanically attached single-ply membrane. Single-ply membranes can be repositioned relatively easily and secured to provide temporary watertight protection.

Figure 6. Most the aggregate surfaced built-up roof and insulation was blown off the roof. Some steel roof deck (red arrow) was also blown away. The yellow arrow indicates a deck opening where equipment was blown away. The damaged membrane can't be repositioned; a permanent or new temporary roof is needed to provide watertight protection. (Courtesy FEMA Mitigation Assessment Team [MAT], Hurricane Katrina 2005.)

3 TEMPORARY SECUREMENT OPTIONS

Damaged membranes that remain in place or that can be reposition (e.g., Figures 1 – 5), are typically temporarily secured with some type of ballast or mechanical fasteners. If the membrane is blown away (Figure 6), a single-ply membrane can be temporarily secured. Alternatively, a temporary or permanent membrane may be adhered (Figure 7). See Section 4 for securement recommendations.

Figure 7. This building was temporarily roofed with a two-ply built-up roof. (Courtesy FEMA MAT, Hurricane Charlie 2004.)

A variety of materials have been used for temporary ballast: Concrete pavers, sandbags, tires, water bags, wood boards and wood sheathing.

At the roof shown in Figure 8, a portion of the fully adhered single-ply membrane detached from the insulation. Concrete pavers were randomly placed to provide temporary securement. However, the paver layout was insufficient to prevent the roof membrane from lifting and progressively peeling beyond the pavers. The size of the detached area increased during subsequent modest winds. If a detached area subsequently spreads beyond the temporary ballast, the portion of the detached membrane that is not ballasted is very vulnerable. If a seam fails or the membrane tears, substantial leakage may occur (Figure 10). Also, if the size of the detached area increases because of ineffective temporary securement, the cost of repairing the damage is often increased.

Figure 8. The roof membrane was detached from the insulation on the left side of the red line. The area between the pavers and red line detached after the pavers were installed.

At the roof shown in Figure 9, most of the fully adhered single-ply membrane detached from the insulation. A roofing contractor placed rows of sandbags to provide temporary securement. Although attention was given to locating the sandbags, a subsequent moderate wind event caused the sandbags to shuffle out of position, which allowed the membrane to balloon and open (Figure 10). Sandbags near the roof edge had been tied together. Some of the sandbags fell off the roof, but the rope kept them from reaching the ground (Photo 10).

Figure 9. Temporary sandbag securement. This is the roof shown in Figure 1.

Figure 10. After the sandbags shown in Figure 10 shuffled, the membrane ballooned and opened.

Figure 11. Rope kept these sandbags from reaching the ground.

At the roof shown in Figure 12, most of the fully adhered single-ply membrane between the parapet and the equipment screen wall detached from the insulation. Tires were randomly placed to provide temporary securement. The ballast was inadequate to prevent propagation of the detachment beyond the ballast (yellow line). Ballast was not placed near the parapet in order to avoid overstressing the membrane's base securement at the parapet (red line). In several areas the ballast was too far apart (red oval).

Figure 12. Although effort was made to temporarily secure this detached membrane with tires, the tire placement was insufficient to avoid damage during subsequent modest winds.

At the roof shown in Figure 13, the fully adhered single-ply membrane detached from the insulation at a portion of the roof. Plastic bags partially filled with water (water bags) were placed to provide temporary securement. The blasting at the transition between the adhered and unadhered areas was inadequate to prevent propagation of the detachment beyond the ballast.

Figure 13. Temporary securement provided by water bags.

At the roof shown in Figure 14, temporary watertight protection was provided by a loose-laid tarp. The tarp was ballasted with wood sheathing. Most of the panels were warped (cupped), thus making them more susceptible to wind uplift. In several areas, the sheathing was too far apart, and there was lack of ballast in a corner region (red oval).

Figure 14. Temporary securement provided by wood sheathing. (Courtesy FEMA MAT, 2011 tornadoes.)

At the roof shown in Figure 15, the fully adhered single-ply membrane detached from the insulation at much of the roof, and some of the base flashing was detached and torn. Temporary securement was provided by rows of mechanical fasteners (screws and plates), which were stripped-in (yellow lines). Mechanical fasteners were also installed at the parapet (red arrows), but they were not stripped-in at the time the photo was taken.

Figure 15. Temporary securement provided by rows of mechanical fasteners.

4 TEMPORARY SECUREMENT RECOMMENDATIONS

The following temporary securement recommendations are based on experienced and informed engineering judgment. The recommended solutions offer an appropriate balance between the cost of the temporary securement and the level of protection that they provide. Recommendations are given for temporary securement provided by mechanical attachment, ballasting and fully adhering a membrane. Recommendations are also provided for monitoring the securement.

In most instances, a roofing contractor will initially install emergency securement, using materials that are readily available. This initial emergency securement is intended to quickly stabilize the roof. Soon after the initial emergency securement is installed, it should be assessed to determine if additional work should be performed to provide the desired level of temporary securement. Subsequent winds with sufficient speed to cause damage to the initial emergency securement could occur within a very short time (e.g. a day or a week). It is therefore important to quickly assess the adequacy of the emergency securement and upgrade it if deemed appropriate.

The goal of the temporary securement design is to keep material from blowing off the roof and to avoid propagation of roof membrane detachment.

4.1 *Temporary Mechanical Attachment*

Temporary securement with mechanical fasteners is recommended in lieu of temporary ballast, unless conditions preclude fastening (e.g., conduit embedded in a concrete deck, conduits attached directly to the underside of the deck or the deck type is not conducive to mechanical attachment). The following are recommended:

- Calculate design uplift loads for the corners, perimeter and field of the roof using the current edition of ASCE 7 and an appropriate serviceability mean recurrence interval (MRI). Figures CC-1 and CC-2 in ASCE 7-10 give 10-year and 25 year MRI speeds. For most of the continental U.S., the 10- and 25-year speeds are 34 and 38 meters/second (m/s) (76 and 84 miles per hour [mph]). These are "allowable stress design" speeds (i.e., load factor of 1.6). Therefore, the 0.6 W load combination in ASCE 7-10 should not be used. Table C6-7 in ASCE 7-05 provides a table for determining speeds associated with various MRIs, including a 5-year MRI. The 7-05 5-year MRI for most of the continental U.S. is 70 mph. In most instances, use of the 10-year MRI given in ASCE 7 is sufficiently conservative.

  After calculating the design uplift loads for the three roof zones, design the fastener row spacing and spacing of fasteners along the rows. For steel roof decks, specify that the fastener rows run perpendicular to the deck flanges (References 11 and 12). If there is a parapet, specify a row of fasteners near the roof/parapet interface (this is in addition to the existing base securement).
- If the membrane was adhered, and if portions of the roof are still adhered: At the interface between adhered and unadhered membrane install a row of fasteners over the adhered membrane, approximately 150 mm (6") inward from the adhered/unadhered boundary. Install an additional row of fasteners over the adhered membrane, approximately 300 mm (12") from the first row of fasteners. Space the fasteners along these rows at 300 mm (12") on center maximum.
- Specify a row of fasteners around the perimeter of penetrations (e.g., plumbing vents, roof drains and equipment curbs).
- If the base flashing is detached, and if the parapet height exceeds 1 m (3'), it is recommended that a row of fasteners be placed at intervals not exceeding 1 m (3') horizontally up the parapet. Add a row of fasteners near the coping unless the coping is believed to be capable of resisting as much load as the fastener rows below it.
- If the roof has a lightning protection system (LPS), typically it is not checked for continuity while the temporary roof is in place. However, if the building occupancy or contents are such that a functional LPS is desired by the building owner during the time that the temporary roof is in place, a qualified LPS contractor should be retained.
- If emergency ballast was installed prior to installation of the temporary mechanical securement, remove the ballast after the mechanical attachment work has been completed.

4.2 *Temporary Ballast*

There is lack of reliable design data to produce a truly engineered temporary ballast solution. Therefore, temporary ballasting is recommended only when mechanical attachment is not feasible. Use of emergency ballast prior to installation of mechanical attachment can be appropriate. For temporary ballast, sandbags are recommended. If sandbags are not available, 610 x 610 mm (24" x 24") concrete roof pavers are recommended. An advantage of sandbags is that they are less likely to damage the roof membrane. However, if they are in place for a prolonged time, they may be torn or the bags may deteriorate. Water bags can be inexpensive to deploy if a hose bibb is nearby, and they are inexpensive to demobilize. However, water bags are more susceptible than sandbags or pavers to shuffling caused by membrane ballooning. For emergency ballasting, use of other types of ballast can be appropriate when sandbags or pavers are not immediately available. If another type of ballast is used for the emergency work, it is recommended that the other type of ballast be replaced with sandbags or pavers when the temporary ballasting work is performed.

The following are recommended:

- For roof heights greater than 18 m (60'), temporary ballasting is not recommended because of the increased potential for ballast being tossed off the roof by membrane ballooning.
- Along the parapet, place a row of sandbags[2]. Along this row, leave a space of about 600 mm (24") between each bag. Place the long dimension of the bag parallel to the parapet.
- If there is no parapet, place the first row of bags about 1.2 m (4') from the roof edge.
- At the perimeter zone, place rows of sandbags at about 1.8 m (6') on center. Along these rows, leave a space of about 600 mm (24") between each bag.
- At the field zone, continue placement of rows of sandbags at about 1.8 m (6') on center, but leave a space of about 1.2 m (4') between the bags.
- If the membrane was adhered, and if portions of the roof are still adhered: At the interface between adhered and unadhered membrane, place a continuous row of sandbags (i.e., no space between the bags) on the unadhered membrane – place this row within about 150 mm (6") of the demarcation line. Place another row of continuous sandbags on the adhered membrane – place this row within about 150 mm (6") of the demarcation line. This will result in a gap of about 300 mm (12") between the two rows. Note: If roof drainage is such that these rows of sandbags will impede drainage, leave a space of about 25 mm (1") between the bags.
- At all roof penetrations (including pipes), place sandbags around the penetration (Figure 16). Where the penetration has a curb that is longer than 600 mm (24"), place a minimum of two sandbags at each side of the curb. At roof drains, keep the bags about 300 mm (12") from the drain so that water can flow to the drain.

Figure 16. Sandbags placed at a penetration to avoid membrane ballooning and subsequent tearing.

[2] These recommendations are based on sandbags that weigh approximately 23 kg (50 pounds).

- When applying emergency or temporary ballast, minimize stockpiling of ballast to avoid overloading the structure, and use caution in placing the ballast to avoid damaging the roof membrane.
- Prior to placing the temporary ballast, have a structural engineer verify that the structure can accommodate the proposed ballast load.

Figure 17 shows emergency ballast that was installed after a single-ply membrane detached from the insulation at a portion of the roof. The author recommended that additional ballast be installed, following the above recommendations (Figure 18).

Figure 17. View of placement of emergency ballast.

Figure 18. View of placement of temporary ballast. The double row of sandbags occurs at the interface between the adhered and unadhered areas.

If the wind damaged roof was ballasted, also see the first bullet in Section 2.

4.3 *Temporary Fully Adhered Membrane*

If all or most of the roof membrane is blown away (Figure 6), an alternative to temporarily mechanically attaching or ballasting a membrane is to fully adhere a membrane (Figure 7). If insulaion boards need to first be attached to the deck with fasteners or foam ribbon adhesive, the attachment should be designed to meet the uplift load prescribed in the building code if the membrane will be permanent. However, if the membrane will be torn off when the building is reroofed, or if it will remain in place and a new membrane mechanically attached to the deck, a more economical attachment of the temporary roof insulation may be used by following the recommendations in the first bullet of Section 4.1

4.4 Monitoring

It is recommended that the temporary roof be inspected to determine if the interface between adhered and unadhered membranes has moved (propagated), and determine whether temporary ballast has shuffled. As a conservative trigger for inspecting, it is recommended that the roof be inspected once a month (in some instances, inspections every other week may be appropriate). In addition to these periodic inspections, it is recommended that the roof be inspected if winds greater than 22 m/s (50 mph)[3] occur. In instances where there are special property or life-safety issues, a lower threshold may be prudent.

If the inspections reveal that relatively minor ballast shuffling occurred, repositioning of the ballast is not needed. If shuffling is significant, repositioning is recommended. Also, the ballast layout should be reevaluated to determine if additional ballast should be installed. If the interface between the adhered and unadhered membranes has moved, reposition the boundary ballast; or if the mechanical attachment was used at the interface, install additional fastener rows as recommended at bullet two in Section 4.1 Also check each drain to ensure that drainage is not impeded.

6 REFERENCES

1 T.L. Smith, "The roof system blew off – now what?," *Professional Roofing*, August 2000, p. 28.

2 T.L. Smith, "Evaluating Wind-Damaged Low-Slope Roof Assemblies: A Preliminary Protocol," *The 12th Americas Conference on Wind Engineering (12ACWE)*, June 2013.

3 *SPRI/NRCA Manual of Roof Inspection, Maintenance and Emergency Repair for Existing Single-ply Roofing Systems*, Single Ply Roofing Institute and National Roofing Contractors Association, 1992.

4 *Manual for Inspection and Maintenance of Spray Polyurethane Foam-based Roof Systems: A Guide for Building Owners*, National Roofing Contractors Association and The Society of Plastics Industry, Inc., Spray Polyurethane Foam Division, 1998.

5 *Manual for Inspection and Maintenance of Built-Up and Polymer-modified Bitumen Roof Systems: A Guide for Building Owners*, Asphalt Roofing Manufacturers Association and National Roofing Contractors Association, 2010.

6 *Manual for Inspection and Maintenance of Low-slope Structural Metal Panel Roof Assemblies: A Guide for Building Owners*, National Roofing Contractors Association, 2011.

7 *Repair Manual for Low-slope Membrane Roof Systems*, Asphalt Roofing Manufacturers Association, National Roofing Contractors Association and SPRI, 1997.

8 T.L. Smith, R.J. Kind and J.R. McDonald, "Hurricane Hugo: Evaluation of Wind Performance and Wind Design Guidelines for Aggregate Ballasted Single-ply Membrane Roof Systems," *Proceedings of the VII International Roofing and Waterproofing Congress*, 1992.

9 *ANSI/SPRI RP-4, Wind Design Standard For Ballasted Single-ply Roofing Systems*, 2013. Available at: https://www.spri.org

10 M. Asghari Mooneghi, T.L. Smith, P. Irwin, and A. Gan Chowdhury, "Concrete Roof Pavers: Wind Uplift Aerodynamic Mechanisms and Design Guidelines – A proposed Addition to ANSI/SPRI RP-4," *Proceedings of the 32nd RCI International Convention and Trade Show*, 2017.

11. M.S. Graham, "The Situation with Steel Decks," *Professional Roofing*, March 2017, p. 33.

12. J.M. Fisher and T. Sputo, "Are Your Roof Members Overstressed?", *Structure Magazine*, March 2017, p. 30.

[3] Gust, exposure C at 10 m [33'].

13th Americas Conference on Wind Engineering
Gainesville, Florida, USA
May 21 - 24, 2017

Development of Design Wind Speeds for Global Locations – Analysis of Extreme Straight Winds with Surface Wind Speed Observations

Fangqian Liu[a], Peter J. Vickery[b]

[a] *Staff Scientist , IntraRisk Division, Applied Research Associates, Raleigh, NC, United States, aliu@ara.com*
[b] *Principal Engineer, IntraRisk Division, Applied Research Associates, Raleigh, NC, United States, pvickery@ara.com*

ABSTRACT

Extreme wind hazards are usually associated with the occurrences of tropical cyclones, thunderstorms, extratropical cyclones and tornados. All these events, except for tornados, are considered in the quantification of the design wind speeds. The design wind speeds for tropical cyclone regions are typically developed using stochastic simulation models; while the design values for straight winds, which include thunderstorms, extratropical cyclones and etc., are quantified using statistical methods using surface wind observations. In this paper, the procedures for developing the design values for the straight winds outside the US, are demonstrated, using wind and weather observations from the global historical databases available in the National Climatic Data Center (NCDC) were utilized.

In the approach discussed herein, thunderstorms and non-thunderstorm events (e.g. extratropical cyclones) were treated as statistically independent events. At a particular location, the exceedance probability of thunderstorm winds was modeled with either a standard Type I extreme distribution of annual extremes or through a stochastic model developed from all thunderstorm wind measurements at that location. For non-thunderstorm winds, individual storm maxima method was employed to detect the wind speed peaks from which the exceedance probability was developed. The exceedance probability of straight winds is a combination of the exceedance probabilities from the thunderstorm and the non-thunderstorm winds.

The datasets utilized in the analysis include the Integrated Surface Database (ISD) and the Global Summary of the Day (GSOD) archived by NCDC. Both ISD and GSOD provide continuous surface observations for locations outside the US since as early as the beginning of the 20th century. Hourly measurements and daily summaries of sustained wind speeds, gust wind speeds, wind directions and information of the weather types are collected from global aviation and synoptic stations. Among these records, gust winds were not always available. Special efforts were made to investigate the quality of the data. It was found that restrictions and inconsistencies in data quality, even erroneous values, commonly exist in all the datasets. It was observed that the observations were gathered over eras of changing observation technologies and were operated by local meteorological authorities. Conversion factors were derived to accommodate the restrictions, such as the constrained measuring temporal resolution and the unknown duration of sustained winds. A data cleaning process was also introduced to identify the erroneous records. Those erroneous values, which mainly occurred when data were keyed in or during format transformation, stretched the wind speed distributions. It was shown that the unrealistically high values in the datasets could be identified by an "automatic" screening process along with a manual inspection on suspicious artifact types. The design wind speeds were reduced to reasonable levels using the data after cleaning.

Design wind speeds at locations in North America, Asian and Europe are given at the end of the paper. These locations were selected with various wind speed types (i.e. gust or sustained wind), different dominating wind hazard types and varying ocean basins. Design wind speeds from local building design codes were also collected and shown as a set of comparisons.

KEYWORDS

Design Wind Speed; Integrated Surface Database; Thunderstorm; Artifact in Archived Wind

13th Americas Conference on Wind Engineering
Gainesville, Florida, USA
May 21 - 24, 2017

Wind Performance Based Design – A Case Study

Ali Ashrafi[a], Antonio De Luca[b], Jason Garber[c], and Jon Galsworthy[d],

[a] *Thornton Tomasetti, New York, NY, USA, AAshrafi@ThorntonTomasetti.com*
[b] *Thornton Tomasetti, Fort Lauderdale, FL, USA, ADeLuca@ThorntonTomasetti.com*
[c] *RWDI, Guelph, ON, Canada, Jason.Garber@rwdi.com*
[d] *RWDI, Guelph, ON, Canada, Jon.Galsworthy@rwdi.com*

ABSTRACT

The concept of performance based design (PBD) in structural engineering consists of specifying acceptable levels of performance, including damage, that a building can undergo when subjected to a specific natural or human-made hazard. Service-level wind PBD has been widely adopted by practitioners: assessment of human comfort in tall buildings, and wind flow around buildings are an example. It is time to embrace this concept for strength-level PBD.

Wind forces are typically the governing lateral load when designing supertall buildings or tall buildings in non-seismic regions that are hurricane-prone. The design of tall buildings in hurricane-prone regions is typically performed using equivalent static forces as provided by the local building codes, or as obtained from wind tunnel testing. This approach is inherently conservative because of the assumptions needed to obtain equivalent static forces and load combinations. Research is ongoing to evaluate the benefits of using explicit dynamic analysis in designing structural systems against high-return period winds.

In this paper, the Authors will discuss the results of the dynamic analysis of a fifty-three story building subjected to wind forces and twisting moments obtained from boundary layer wind tunnel testing. The building, whose lateral system consists of reinforced concrete shear walls and coupling beams, was originally designed against the wind-tunnel based equivalent static forces. The effects of dynamic analysis on the design of the structure and how its performance is assessed are presented in this paper.

KEYWORDS

Performance Based Design, Dynamic Analysis, Nonlinear Analysis, Wind Tunnel Testin, Concrete Shearwalls,

Design Wind Pressure by Peaks-over-Threshold Method: Implementation and Comparative Performance

Dat Duthinh[a], Adam Pintar[b] and Emil Simiu[c]

[a] *National Institute of Standards and Technology, Gaithersburg, MD. 20899, USA, dduthinh@nist.gov*
[b] *National Institute of Standards and Technology, Gaithersburg, MD. 20899, USA, adam.pintar@nist.gov*
[c] *National Institute of Standards and Technology, Gaithersburg, MD. 20899, USA, emil.simiu@nist.gov*

ABSTRACT

Estimating properties of the distribution of the peak of a stationary process from a single finite realization is a problem that arises in a variety of science and engineering applications. Further, it is often the case that the realization is of duration T while the distribution of the peak is sought for a different duration, $T_1 > T$. For the particular case in which the marginal distribution of the process is Gaussian, closed-form expressions for the mean and standard deviation of the distribution of the peak are available. If the distribution is not Gaussian, a nonlinear mapping procedure, sometimes referred to in the literature as *translation*, has been developed, by which those statistics can be obtained. The translation approach depends upon the estimate of the marginal distribution of a non-Gaussian time series. Here, reliably estimating the tail of the distribution is crucial, but very difficult since in most cases $T_1 >> T$. Translation and other currently employed methods for estimating peaks of time series have drawbacks that motivated the development of a new procedure, based on the *peaks-over-threshold* (POT) method. One advantage of this approach is that it often results in an increased size of the relevant extreme value data set compared with *epochal* procedures.

This paper provides an improved and standardized method for analysis of peaks that can be used for incorporation of new wind tunnel results into the standard, to achieve consistency between proposals submitted by different organizations. The proposed procedure is based on a two-dimensional nonhomogeneous Poisson process model for the thresholded pressure coefficient. A criterion based on previous work by the authors is available, that allows an automated optimal choice of the threshold value to be made (according to a chosen metric) that influences strongly the estimated peak. This is the second advantage of the POT method over the epochal procedure for estimating peaks, which is often combined with Best Linear Unbiased Estimates *(BLUE)* of the Gumbel parameters, and is found to depend, in some cases very significantly, upon the number of partitions being used.

Two versions of the proposed new procedure called PotMax have been developed. One version, denoted by FpotMax, includes estimation of a tail length parameter with a similar interpretation to the generalized extreme value distribution tail length parameter. The second version, denoted by GpotMax, assumes that the tail length parameter vanishes, resulting in a tail of the Gumbel distribution type. Results show that PotMax estimates closely track observed peaks. Reliability theory, data and PotMax calculations show that the 80th percentile specified by ISO 4354 on BLUE estimates (or the 78th percentile used by many practitioners) is over-conservative.

KEYWORDS

Gumbel distribution, Monte-Carlo simulation, peaks-over-threshold, Poisson process, stationary time series.

History of Building Code Changes in Florida and its Effects on Wind Design Provisions.

Derek Stedman [a], Rajkiran Vojjala[a], Peter Datin[a]
[a]*Risk Management Solutions Inc., 7575 Gateway Blvd, Newark, California, USA*

ABSTRACT: Throughout history, building code creation and adoption has typically followed a large damaging event. This is evident in Florida, with the formation of the Florida Building Codes Study Commission in 1996, in response to Hurricane Andrew's devastating impacts on the building infrastructure. This review examines past and present building code adoptions in Florida; specifically addressing wind-load provisions. The review also attempts to quantify the impact of these changes by calculating wind loads using different versions of the code for a notional structure at seven locations across Florida.

KEYWORDS: Florida Building Code, ASCE 7, International Code Council

1 INTRODUCTION

Building code adoption, amendment and enforcement practices vary significantly between different US states. Since 2000, the International Code Council (ICC) has been releasing model building (IBC) and residential (IRC) codes for adoption. However, each state is responsible for determining it's building design standards. Thus, a lot of spatial and temporal variation exists, in terms of standards governing building performance in hurricanes and tropical storms.

Throughout history, building code creation and adoption has typically followed a large damaging event. This is evident in Florida, with the formation of the Florida Building Codes Study Commission in 1996, in response to Hurricane Andrew's devastating impacts on the building infrastructure. Areas that never experienced such a catastrophic event are often slower adopting new codes or are more likely to amend them in ways which weaken the wind resistance provisions. Due to the high amount of administrative procedure required to adopt and enforce a statewide building code, it often takes a long time for a new code to be adopted. In Florida, the commission formed in response to Hurricane Andrew ultimately resulted in the creation of the first Florida Building Code in 2001, nine years after the event. This review examines past and present building code adoptions in hurricane-prone regions of the US, and compares the differences in wind design requirements across the state of Florida due to building code changes.

2 THE INTERNATIONAL CODE COUNCIL

2.1 *Background*

The International Code Council (ICC) creates several model building codes, including the: International Building Code (IBC), International Residential Code (IRC) and International Plumbing Code (IPC), to name a few examples. The group was founded in 1994 by the joining of three code bodies, the Building Officials and Code Administrators International, Inc, (BOCA), the International Conference of Building Officials (ICBO) and the Southern Building Code Congress International, Inc. (SBCCI). These three organizations from different areas of the US recognized the need for a single set of building codes for the entire country. The first ICC code was the IBC 1st Edition, published in 1997. Since then, the ICC has published a new set of codes every three years, the most recent one in 2015. ICC codes are model codes, not codes designed by a governing entity that will be enforced, but as a model for a standardized building code that can be revised by an authority adopting the codes as that entity sees fit.

Wind loading provisions which appear in the IBC and IRC documents tend to follow the most recent edition of the American Society of Civil Engineers' Minimum Design Standards (ASCE 7). ASCE 7 is generally updated every three to six years, depending on the scale of the update and number of changes. The current set of ICC codes, the 2015 version (1), use the wind provisions from ASCE 7-10 (2), the most current version of ASCE 7 available to public.

2.2 *Adoption*

Different states have adopted the ICC -or other model codes at different times and with varying levels of enforcement. IBHS performed detailed studies (3,4) comparing the current status of building code adoptions in many states. This review generously refers to and expands on that work, adding research on previous building code adoptions in each state. The quality of available information was highly variable between states. While some states have detailed histories, others have extremely minimal information that is published. It is common for states to skip every other ICC release; part of the reasoning behind this is the legislative and resource requirements to learn and enforce a new code. Builders are often opposed to fast code update cycles because they must then change construction practices frequently, a time consuming and expensive process. States can also be generally grouped into one of three categories when adopting ICC codes: those adopting with positive amendments, those adopting with no or neutral amendments and those adopting with negative amendments. Positive amendments refer to those when the wind loading provisions were strengthened relative to the ICC base codes. Thus, negative amendments are those that weaken wind loading provisions.

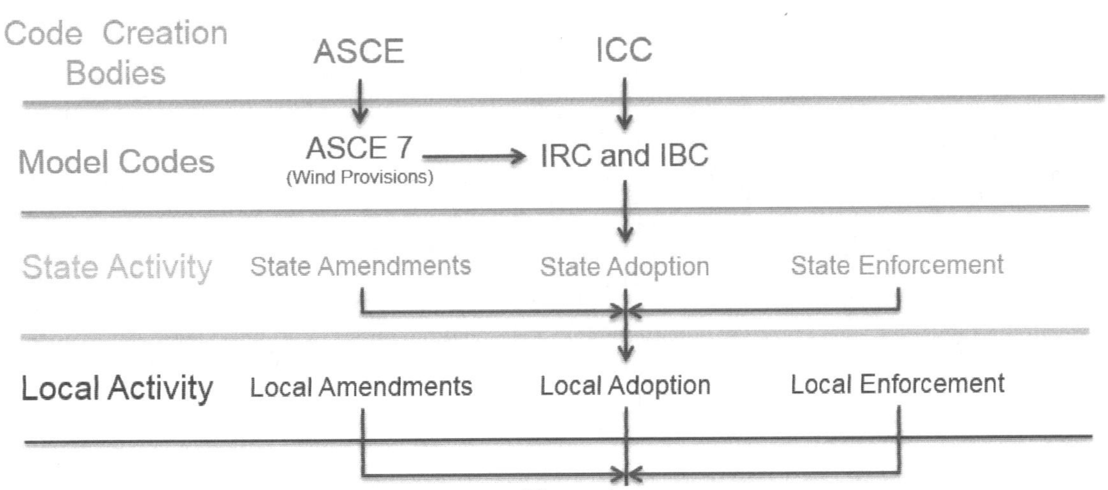

Figure 1 –
Structure of adoption and enforcement of model building codes and wind design standards

3 FLORIDA BUILDING CODE

3.1 *History*

The state of Florida has had a building code in place since the 1970s, when all municipalities were required to adopt and enforce one of four state-recognized model building codes. However, it has been noted that throughout the state, adoption was very inconsistent.

1992 saw Hurricane Andrew make landfall in Florida as a category 5 hurricane, causing widespread devastation. Damage from Andrew revealed that despite the state adoption requirement building code enforcement and adoption was poor. Consequently, the state government decided to act, resulting in the July 1996 creation of the Florida Building Codes Study Commission. The goal of the commission was to evaluate the existing building code system and recommend ways in which the system could be improved or reformed. The commission found no major weakness in the codes, but the inability for proper enforcement. Specifically, the large number of allowed model codes and differences between them made it difficult to properly enforce code-level construction. Thus, after reviewing the commission's findings, the Florida Legislature authorized the creation of the Florida Building Code to be the sole document for all building standards in Florida. This authorization also created the Florida Building Commission, responsible for development of the code and its implementation system.

The Florida Building Commission combined elements from the 1997 Southern Building Code (SBC 1997) and the 1997 South Florida Building Code (SFBC 1997) to create the first Florida Building Code (FBC) in 2001 (5). Wind loads for this code were determined using ASCE 7-98. However, buildings less than 60 feet tall could use simplified wind loading provisions, which were tables dictating pressures based on certain building characteristics. For coastal exposure, which in ASCE 7-98 (6) was changed from Exposure D to C, this exception only applied to buildings less than 30 feet tall. This simplified method gives the same design wind pressures as found in ASCE 7-02 (7), which also has a simplified method.

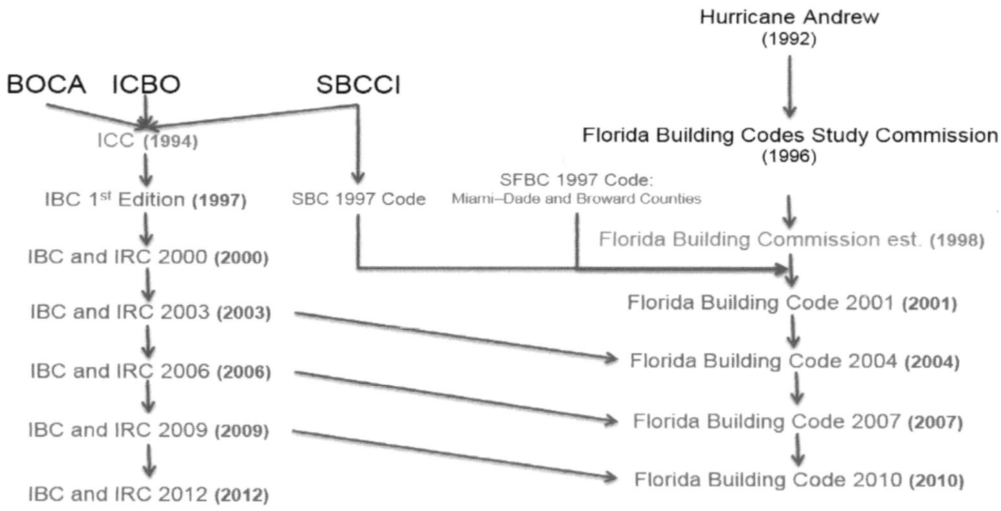

Figure 2 - Relationship between the ICC and FBC codes

A new version of the Florida building code has been released every three years. Starting with the 2004 FBC (8), the code provisions were mostly based off the previous year's ICC codes (as shown in Figure 2), but with some Florida-specific amendments. Unlike in many other states, code amendments in Florida tend to strengthen the wind design provisions of the code. However, like many other states, Florida allows local amendments to the building code only if the amendments are proven as more stringent than the base code and must be reported to the Florida Building Commission for approval.

3.2 *Summary of Major Changes*

FBC 2004 was released in October 2005, largely mirroring the 2003 IBC and IRC (9,10) releases, except for a few state-specific amendments. Many of these amendments were clarifications or better definitions to create slightly more stringent requirements. There were amendments that affected wind loads, specifically wind-borne debris protection. An example of this is wind-borne debris protection for garage doors. While the IBC and IRC codes do not require garage glazing to be protected from wind-borne debris, the FBC was amended to require similar opening protection on garage glazing as regular glazing. The FBC also increased the required nail size for fastening wind-borne debris protection (wood panels) from #6d to #8d nails. This small change in nail size causes a significant increase in the capacity of the nails. For example, calculating the relative withdrawal capacity, an 8d nail is expected to perform around 40% better than a 6d nail, due to the increase in length and diameter.

In 2006, Florida did not release a new code but only a supplement to the code (11). One of the significant changes in this supplement was the move from ASCE 7-98 to ASCE 7-02 for wind load calculations. However, wind loads in ASCE 7-02 were similar to ASCE 7-98. A small but important change in this supplement was the requirement to now use 8d ring shank nails on the entire building roof and not just in the edge zones. These nails have larger withdrawal capacity and provide increased roof strength and wind resistance.

The FBC 2007 (12) was not released until March 2009 and was based on the 2006 IBC and IRC codes (13,14). One of the important changes was an IRC change, which required additional design considerations for structures with certain basic windspeeds (R301.2.1.1). Previously, this only applied to locations with a basic windspeed greater than or equal to 110mph. In the 2007 code it became applied to locations with a basic windspeed greater than or equal to 100mph that are within hurricane-prone regions. This change increased the design strength requirements of buildings in areas such as Tallahassee and Orlando for instance. The 2006 IRC also added a new requirement that all exterior walls need a weather resistive barrier under the exterior material.

In 2012, the FBC 2010 (15) was released based on the 2009 IBC and IRC (16,17). One of the critical changes in this version for wind loading was the move to ASCE 7-10, which was the most significant update since ASCE 7-98. This version of ASCE 7-10 moved to limit-state design, removing the importance factor. Thus, instead of an external importance factor, basic windspeeds were re-calibrated to include it , providing a better indication of when major damage is expected to initiate. Figure 3 shows the basic windspeeds in Florida from ASCE 7-05 (18) and ASCE 7-10 (category II building), which reveals a clear increase in windspeeds. However, a direct comparison of windspeeds is not accurate for effects on wind loads because of the changes to importance factor and method of calculation. A more accurate comparison is discussed later.

Another significant change in ASCE 7-10 was the use of Exposure D (coastal) in hurricane-prone regions. Previously, the code assumed the surface roughness (zo) would continually increase with windspeed over water due to increasing size of waves. Research (Powell (19), Vickery et al. (20) and Powell et al. (21)) done more recently found this not to be true in most cases. So, Exposure D is required instead of Exposure C in coastal areas of hurricane-prone regions.

A final significant change which will be discussed in this paper (there are many others in the FBC 2010), is the change in designated wind-borne debris regions. FBC requirements for wind-speeds above which wind-borne debris protection is required were modified to account for the increase in basic windspeeds in ASCE 7-10. However, the wind contours were re-drawn in ASCE 7-10 based on research (Vickery (22)), which changed inland-filling models for hurricanes making landfall that resulted in lower inland windspeeds. The result of this change was lower decreases in basic windspeeds in central Florida, which consequently increased the extent of the wind-borne debris region. Figure 3 shows the extent of this change, where the shaded area represents the wind-borne debris region.

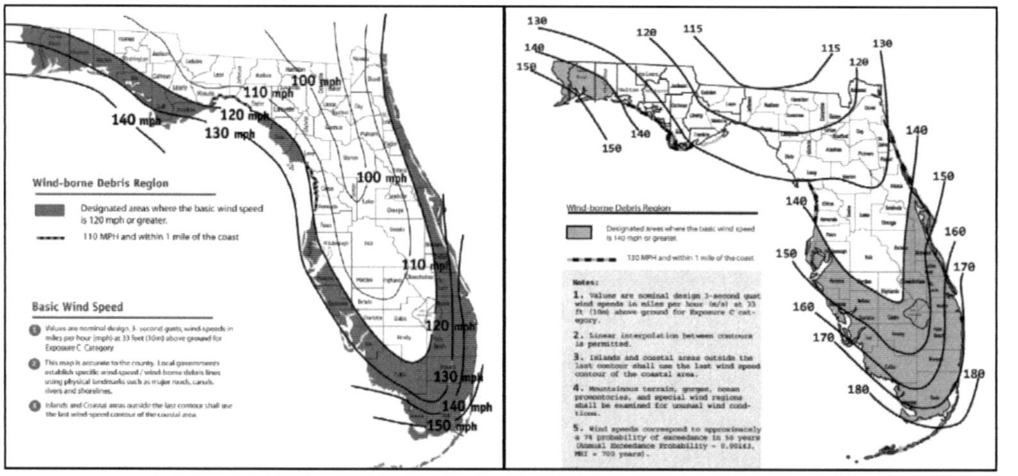

Figure 3 - Wind-borne debris regions (shaded). Left: ASCE 7-02 and 7-05. Right: ASCE 7-10

4 COMPARISON OF WIND LOADS ACROSS FLORIDA

To examine the regional variations and effect of the building code changes on wind loading, seven locations across Florida were examined (Figure 4, Table 1). At each location, design wind-speeds and pressures for the Main Wind Force Resisting System (MWFRS) and Components and Cladding were calculated for a notional structure. The notional structure was a small, two-story residential house, which typically is designed using the prescriptive Residential code instead of the building code. However, for the purposes of this exercise, wind loads were calculated based on the building code (which is also allowed for designing this type of structure). The assumed building specifications are listed below:

- Risk Category II
- Mean roof height of 30 feet
- No topographic effects – flat area
- Simple diaphragm structure

From 2001-2007 inclusive, all the coastal locations considered were within the wind-borne debris region and required the associated protection. In 2010, St Marks and Jacksonville were no longer considered to be in the wind-borne debris region. While this change is not reflected in the calculated wind pressures, it has significant implications for the wind resistance of glazed openings.

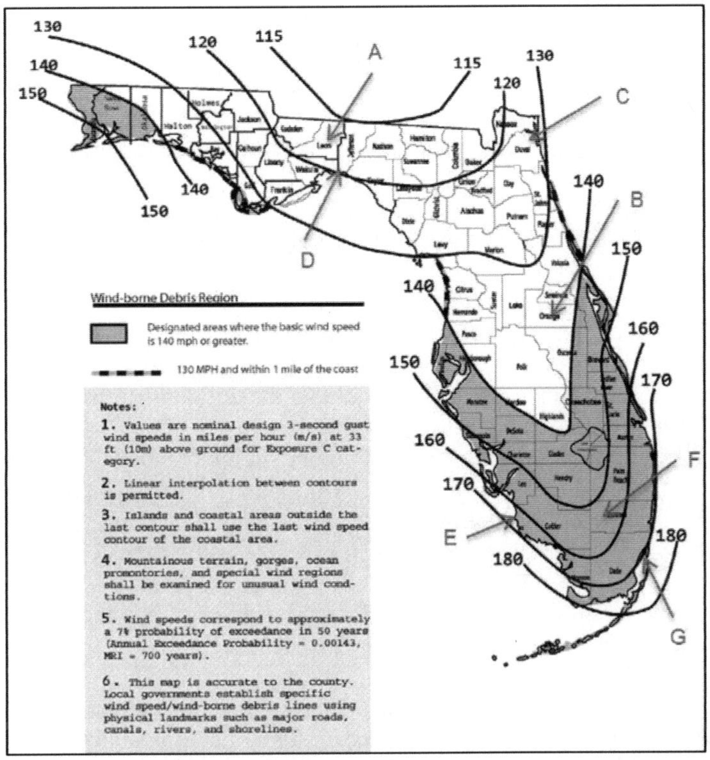

Figure 4 – Locations chosen for examination

Table 1 - List of locations chosen for examination

Location	City	Coastal or Inland	Exposure
A	Tallahassee	Inland	B
B	Orlando	Inland	B
C1	Jacksonville	Coastal	C or D
C2	Jacksonville	Inland	B
D1	St Marks	Coastal	C or D
D2	St Marks	Inland	B
E1	Naples	Coastal	C or D
E2	Naples	Inland	B
F	Broward County	Inland	B
G	Miami Beach	Coastal	C or D

To compare windspeeds from earlier versions of the code consistently with the 2010 code, a multiplication by the square root of 1.6 was performed, giving an accurate comparison between windspeeds across different years. This adjustment was performed to account for the change in windspeed methodology i.e. removal of importance factor in ASCE 7-10. Figure 5 shows adjusted basic windspeeds for all locations using various versions of the code since 2001. After the adjustment, the general trend noted was a decrease in the basic windspeeds in 2010. The most pronounced drop occurred for St Marks. The exception was Orlando, where the design windspeed increased due to changes in the basic windspeed map in the interior of Florida peninsula as discussed earlier.

In non-HVHZ counties, coastal location pressures were calculated in two ways: first assuming the structure was right on the coast (Exposure C or D depending on the version of the code, labeled Coastal) and second, assuming the structure was in fully suburban area (Exposure B, labeled Inland). This shows how the distance from the coastline affects design pressures. The next was to evaluate the difference in pressures throughout Florida in the years of interest when the code changed. To do this, the percentage difference between average MWFRS pressures at each location and the average at the Tallahassee location in 2001 was used (Figure 6) so that year to year changes could also be studied along with regional differences.

Figure 6 shows minimal changes between 2001 and 2007 pressures for all locations. However, significant changes in wind pressures are seen in 2010, corresponding with changes in the design windspeeds and the code methodology for considering the importance factor. The change in exposure classification of coastal locations in ASCE 7-10 to use Exposure D instead of Exposure C mitigates decrease in pressures for coastal areas.

In addition to comparing MWFRS pressures, components and cladding pressure differences were also compared at all the same locations (Figure 7) for the same notional structure. Components and cladding have higher pressure coefficients due to the smaller area – 100ft^2 in this case, causing larger differences across locations. In general, similar trends to MWFRS were observed except that the magnitude of difference in pressures is larger.

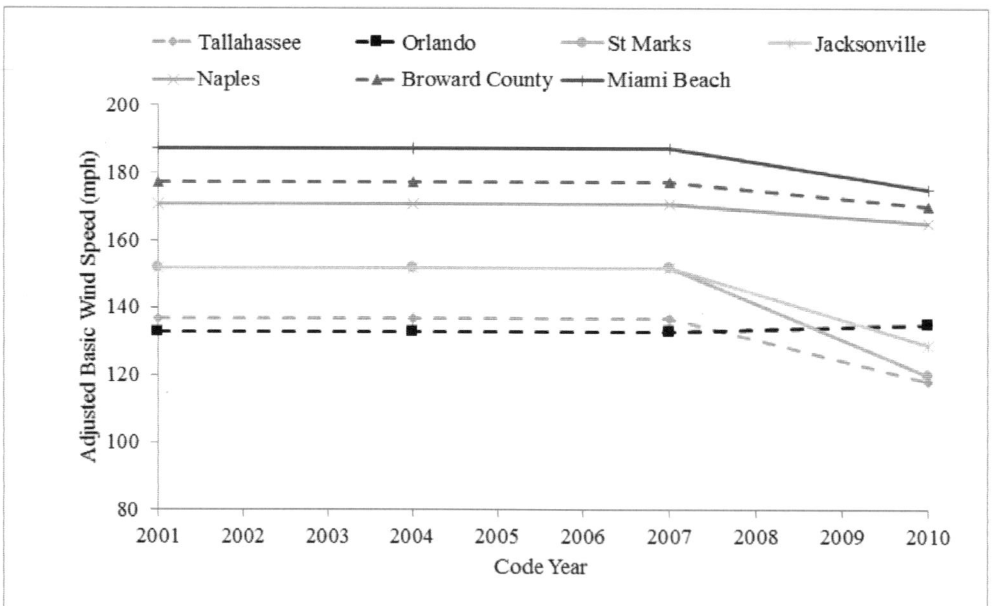

Figure 5 - Comparison of adjusted basic wind speeds.

Analysis of different pressures for the notional residential structure yields a few key insights into differences in design strengths across Florida. Comparing locations where both inland and coastal exposure was considered reveals coastal exposure clearly has higher wind loads than the corresponding inland locations. This is expected due to the higher wind exposure. The coastal exposure region extends 600 feet inland (or twenty times the building height). So, construction in Florida that is close to the coast is expected to be built to a higher standard. The exception is in the High Velocity Hurricane Zone (HVHZ), consisting of Miami-Dade and Broward counties, where all structures are considered Exposure C if inland or D if coastal. As a result, while there is still a difference between inland and coastal structures in these counties, it is less pronounced than in non-HVHZ regions as evident in Figure 6 and Figure 7.

A distinct pattern of grouping in wind load regionality can also be seen in the plots. Considering the pressures in the 2010 code, there are three distinct groups: the inland locations at Orlando, Jacksonville, St Marks and Tallahassee form one group with the lowest design wind pressures. Except for Orlando, these locations are all in Northern Florida. The second group is comprised of coastal locations at Jacksonville and St Marks along with inland location at Naples. The final group is coastal Naples, Broward, and Miami Beach. Broward and Miami Beach are in HVHZ and so have more stringent design criteria applied. Thus, in terms of expected building performance, Florida can be divided into distinct regions based on design pressures alone: North Florida, Central and Coastal Northern Florida, Coastal Central Florida, and the HVHZ counties. As such, buildings in each of these zones subjected to the same windspeeds in a hurricane would be expected to perform differently due to differences in design pressures and wind-borne debris requirements.

Code provisions are only as effective as their enforcement. The IBHS studies (3,4) examining state building code adoptions also addressed building code enforcement, providing a score out of 25. The enforcement score was based on a variety of weighted factors pertaining to certification requirements and training of code enforcement officials. Florida was assessed a score of 22/25 for enforcement (tied for 5th out of 18 states assessed), this generally high score indicates a high degree of adherence to the building code is expected – unlike some other states (Delaware, Mississippi), where state-level code enforcement is quite poor.

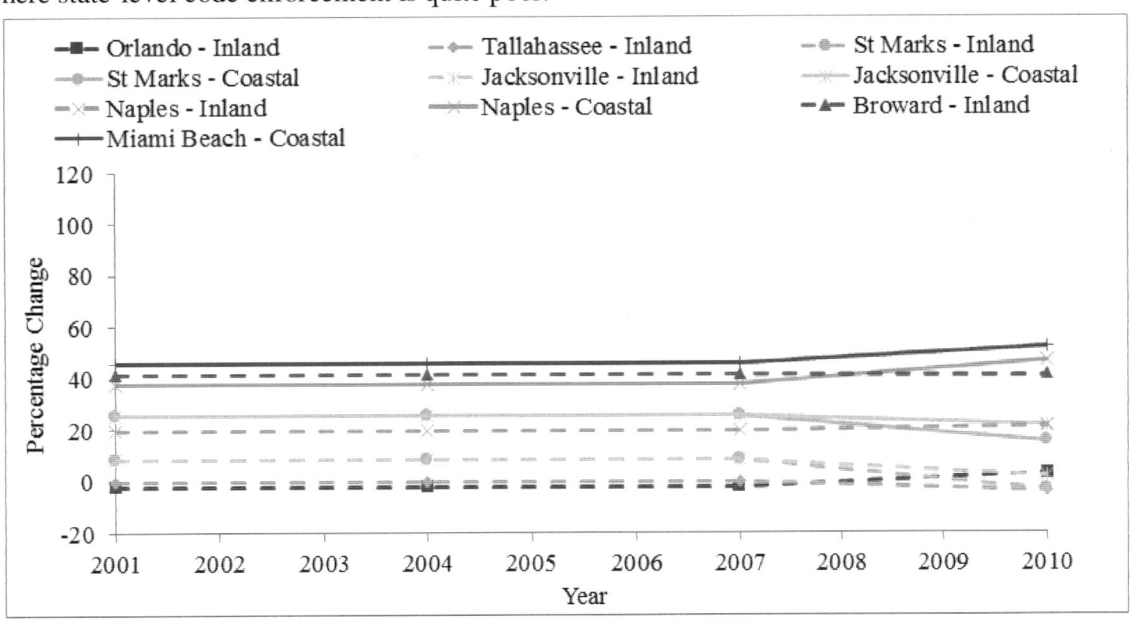

Figure 6 – Change in MWFRS pressures.

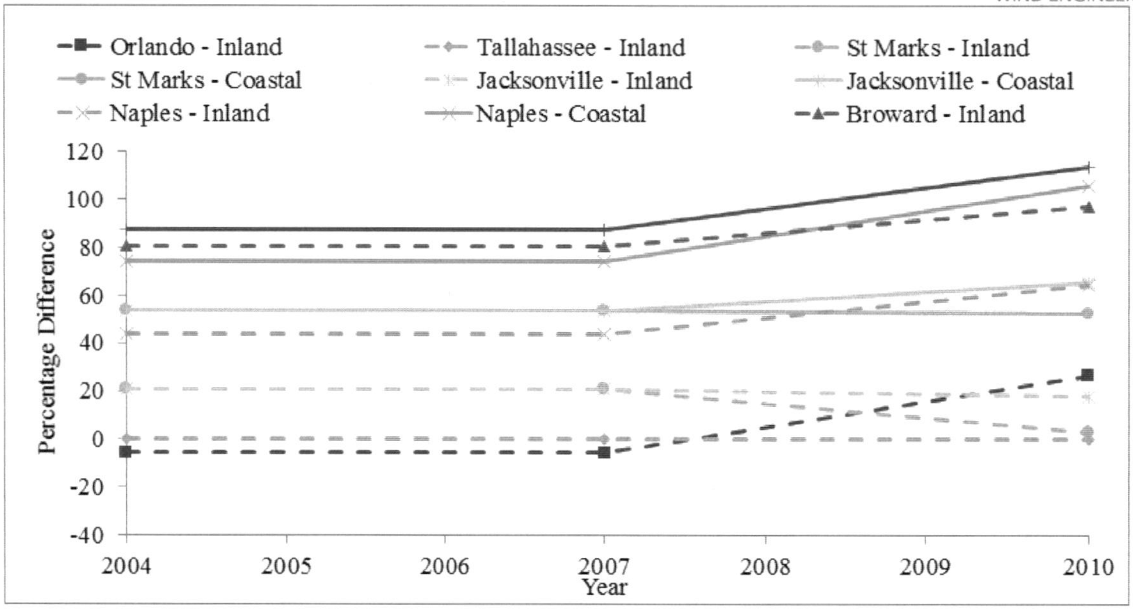

Figure 7 - Comparison of Components and Cladding Pressures

5 CONCLUSIONS

This paper presents a review of building code adoption and enforcement history in Florida. It shows that building code adoption and changes are often a result of a large and damaging event.

An examination of building design pressures for a notional structure in Florida was undertaken. The findings clearly show that building design requirements vary significantly within the state of Florida, which can be grouped into several distinct regions of expected building performance when subject to the same wind hazards.

Finally, the review revealed that small changes in the prescriptive portions of the building code, such as requiring ring-shank nails instead of regular nails, can have significant influence on the quality of building construction and resistance to wind loads. As previously mentioned, 8d nails can be expected to have a withdrawal capacity about 40% greater than a 6d nail, significantly increasing the resistance to uplift/negative pressures. Therefore, even seemingly trivial changes in the code require careful consideration when examining the influence of building codes on wind resistance of structures.

6 REFERENCES

1. ICC, International Building Code 2015 International Code Council, Washington, DC, 2015.
2. American Society of Civil Engineers-ASCE, ASCE 7-10: Minimum Design Loads for Buildings and Other Structures, American Society of Civil Engineers, Reston, VA., 2010.
3. IBHS, Rating the States An Assessment of Residential Building Code and Enforcement Systems for Life Safety and Property Protection in Hurricane Prone Regions, Institute for Business & Home Safety, Tampa, FL., 2011, www.DisasterSafety.org
4. IBHS, Rating the States An Assessment of Residential Building Code Enforcement Systems for Life Safety and Property Protection in Hurricane-Prone Regions Midterm Update, Institute for Business & Home Safety, Tampa, FL., 2013, www.DisasterSafety.org
5. FBC, 2001 Florida Building Code, State of Florida, Tallahassee, FL., 2001.
6. American Society of Civil Engineers-ASCE, ASCE 7-98: Minimum Design Loads for Buildings and Other Structures, American Society of Civil Engineers, Reston, VA., 1998.
7. American Society of Civil Engineers-ASCE, ASCE 7-02: Minimum Design Loads for Buildings and Other Structures, American Society of Civil Engineers, Reston, VA., 2002.
8. FBC, 2004 Florida Building Code, State of Florida, Tallahassee, FL., 2004.
9. ICC, International Building Code 2003, International Code Council, Washington, DC.,2003.
10. ICC, International Residential Code for One- and Two- Family Dwellings 2003, International Code Council, Washington, DC., 2003.
11. FBC, 2006 Supplement to the 2004 Florida Residential Code, State of Florida, Tallahassee, FL., 2006.
12. FBC, 2007 Florida Building Code, State of Florida, Tallahassee, FL.,2007.
13. ICC, International Building Code 2006, International Code Council, Washington, DC.,2006.
14. ICC, International Residential Code for One- and Two- Family Dwellings 2006, International Code Council, Washington, DC., 2006.
15. FBC, 2010 Florida Building Code, State of Florida, Tallahassee, FL.,2010.
16. ICC, International Building Code 2009, International Code Council, Washington, DC., 2009.
17. ICC, International Residential Code for One- and Two- Family Dwellings 2009, International Code Council, Washington, DC., 2009.
18. American Society of Civil Engineers-ASCE, ASCE 7-05: Minimum Design Loads for Buildings and Other Structures, American Society of Civil Engineers, Reston, VA., 2005.
19. Powell, M.D., High Wind Drag Coefficient & Sea Surface Roughness in Shallow Water, Report to Joint Hurricane Testbed, NOAA HRD-AOML, 2008.
20. Vickery, P.J., Wadhera, D., Powell, M.D. and Chen, Y., A Hurricane Boundary Layer and Wind Field Model for Use in Engineering Applications, Journal of Applied Meteorology and Climatology, Volume 48, 381-405, 2009.
21. Powell, M.D., Vickery,P.J., and Reinhold, T., Reduced Drag Coefficient for High Wind Speeds in Tropical Cyclones, Nature 422, 279-283, 2003.
22. Vickery, P.J., Simple Empirical Models for Estimating the Increase in Central Pressure or Tropical Cyclones after Landfall along the Coastline of the United States, Journal of Applied Meteorology, Volume 44, 1807-1826, 2005.

A review of Brazilian wind data

Matthew Bruce Vallis, Acir Mércio Loredo-Souza, Larissa Cassol Watrin

Laboratório de Aerodinâmica das Construções, Universidade Federal do Rio Grande do Sul, Porto Alegre, Brazil

ABSTRACT: Since 1988, *NBR6123,* Brazil's wind load code, has balanced the competing needs of safety and comfort against the construction costs of the nation's expanding skyline. The isopleth map of the regional basic wind speed, based on wind data from 1950-1974, is the starting point for all wind design load calculations in South America's largest nation. With the goal of optimizing wind loads, efforts are currently being made to update regional wind speeds across the country. An overview of both the civil and aeronautical surface weather observation networks is presented and a review of several data banks is conducted. An improved method of extracting data from raw weather observations is proposed. Using this approach, changes to wind data characteristics at observation stations are revealed with examples given, and the geographical locations of Brazil's most frequent strong wind regions are mapped.

KEYWORDS: Brazil, wind data, thunderstorm, NBR6123.

1 INTRODUCTION

As identified by Loredo-Souza (1), typical extreme wind events in Brazil have origins in thunderstorms (TS) and extended pressure systems (EPS). Both downbursts and tornadoes can be generated during TS events, with the number of reported tornados rising in recent years (2). A number of destructive downburst events during recent years have also been documented (3, 4). The first tropical cyclone (TC) to reach Brazil was recorded in 2005 (5). Design wind speeds as per NBR6123, determined by Vieira Filho (6) and Padaratz (7), were formulated without differentiation between these events. Furthermore, as with most national wind codes, NBR6123 provides only atmospheric boundary layer (ABL) models for the calculation of wind loads. It is proposed that future versions of NBR6123 will have separate design wind speeds and models for TS and ABL winds.

An examination of the fidelity of datasets is a critical step of any statistical analysis. However, this fundamental process may not be undertaken due to its time-consuming and tedious nature. The use of unverified data to generate statistical models could mislead those who make decisions. In the case of the estimation of design wind velocities, this could lead to the construction of structures unsuitable for their environments, or conversely, the overuse of materials.

2 OVERVIEW OF BRAZILIAN METEOROLOGICAL OBSERVATION SYSTEMS

There are two surface meteorological observation networks which operate on a nationwide basis and make observed data available to the public, both of which are controlled by federal government organizations.

DECEA (Departamento de Controle do Espaço Aéreo/Department of Airspace Control) is a military organization of the FAB (Força Aérea Brasileira/Brazilian Air Force), under the control of the Ministry of Defence, and is the organization responsible for the regulation and management of the AIS (Aeronautical Information Service). The observation and forecast of airspace weather conditions, particularly at Brazilian aerodromes, is part of this responsibility. Two further acronyms which play an integral part in the FAB's meteorological observation network are ICEA (Instituto de Controle do Espaço Aéreo/Institute of Airspace Control) and SISCEAB (Sistema de Controle do Espaço Aéreo Brasileiro/Brazilian Airspace Control System). ICEA is responsible for the management of data and associated products, and conducts research with the aim to develop and improve the SISCEAB. The operation of each individual weather station under the responsibility of the aerodrome operator, of which there are many (some of the most common are INFRAERO, SEIL, DAESP, FAB). However, the operational procedures to be followed at all stations are determined by FAB, and so for simplicity this network will herein be referred to as FAB.

A second system which runs parallel and independent is INMET (Instituto Nacional de Meteorologia/Brazilian Meteorological Service), and is under the control of the Ministry of Agriculture, Livestock and Supply. The role of INMET is to provide meteorological information through the collecting, monitoring, analysis and prevision of weather and climatic conditions. Both FAB and INMET adhere to procedures as outlined by ICAO (International Civil Aviation Organization) and WMO (World Meteorological Organization) respectively, and INMET is Brazil's representative to WMO. Figure 1 shows the location of surface observation stations of each network within Brazil.

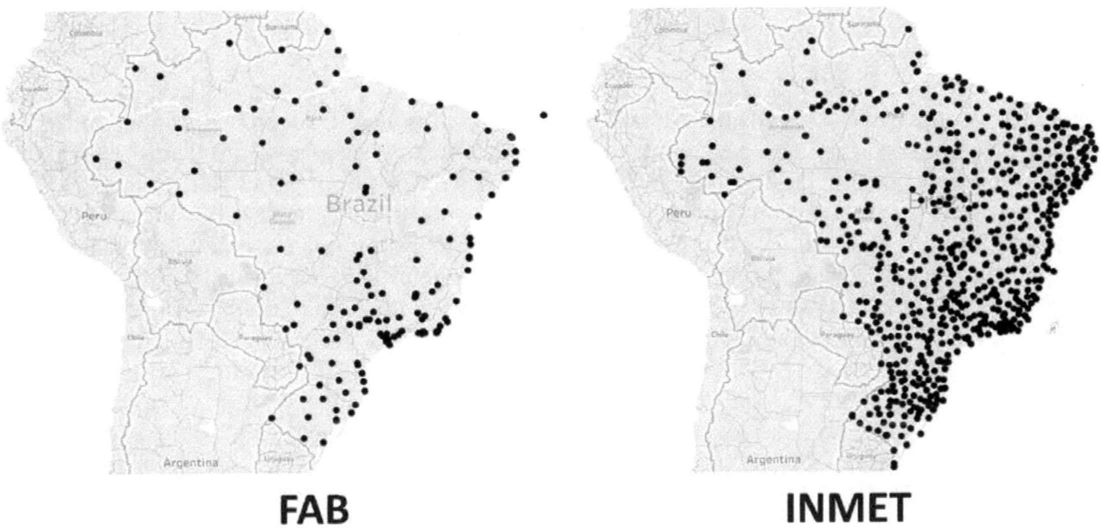

Figure 1. Land coverage of 147 FAB and 744 INMET surface observation stations.

2.1 *FAB*

A list of 147 Brazilian aerodromes (the name for airport within ICAO terminology) has been compiled for analysis. The hours of operation of each individual weather station are dependent of the opening hours of each particular aerodrome. At the start of 2017, only 52 of the 147 aerodromes were identified as operating 24 hours a day (H24), and 12 either no longer transmit wind observations or no meteorological data at all. It is common for aerodromes to change their operating hours over time, to cease transmitting weather observations or to move location. In addition, new airports are built and opened. Some examples of these changes include,

- SBRJ (Santos Dumont, Rio de Janeiro) operated as H24 up until December 1, 2009, when operations were reduced to H18. This changed again on July 1st, 2015, when the airport returned to a H24 operation.
- Several aerodromes have ceased transmitting weather observations in the last 10 years, including SBAM (Amapá, Amapá), SBCV (Caravelas, Bahia) and SBNM (Santo Angelo, Rio Grande do Sul).
- SBMG (Maringá, Paraná) shifted to the city's newer airport in 2001, approximately 12km away from the now unused aerodrome.
- SBAE (Arealva, São Paulo), SBDO (Dourados, Mato Grosso do Sul) and SBSG (São Gonçalo do Amarante, Rio Grande do Norte) have all been opened within the last 7 years.

Wind data is composed of a wind direction (*DIR*), a 10-minute mean speed (*V*) and occasionally a 3-second gust speed (*G*). Observations made at aerodromes are compiled with other meteorological data, such as temperature, and emitted as either a METAR/SPECI or SYNOP report. The observation and emission processes are not completely standardized across the entire network, and historical changes to these procedures are noticeable when examining time-series of wind data. Anemometer heights are to be installed at 10m in open terrain, however this is not always the case. Of the anemometer heights that have been confirmed by current investigations, approximately only 20% are installed at a height of 10m, with the remainder at heights of 9m, 8m and even 6m.

Prior to the turn of the millennium, wind observation was a manual operation, whereby the observer determined wind speeds by visually monitoring an indicator on an analog dial and mentally calculating the mean. Other meteorological parameters would also require the attention of the observer, and so it is highly unlikely the observer would monitor the wind readings for a complete 10-minute period, and even if it were the case, a precise calculation of mean and gust readings would be difficult to accomplish. The observations would be recorded by hand on sheets of paper with pre-prepared tables, with an exampled shown in Figure 2. The METAR/SPECI and SYNOP reports, which are required for immediate use by pilots, would be compiled manually and transmitted via radio, and electronically once the technology became available in later years. METAR reports are made on the hour, SPECI during the hour when a change of conditions occurs and SYNOP every 3-hours. METAR observations were made at 30-minute intervals for the period between November 1st, 1999 and October 31st, 2001.

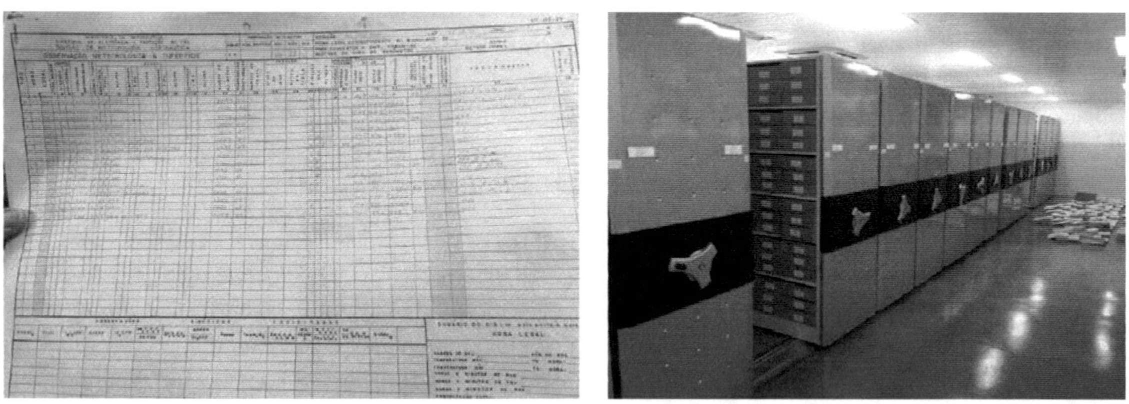

Figure 2. An example of meteorological observations from November 1st, 1988 (left), now archived at ICEA (right).

ICEA is responsible for the archiving, management and updating of the BDC (Banco de Dados Climatólogico do Comando da Aeronáutica/Climatologic Data Bank of the Air Force Command), which today is composed of three elements: the archive of physical historical observations, the electronic database stored on on-site servers, and the summary of historical meteorological data available to the public via internet (8).

Current observation protocol at airports with high traffic could be classified as semi-automated. The monitoring of wind speed and direction are digitalized and automated. This includes the calculation of 10-minute means and identification peak gust speeds. The data is compiled automatically into METAR/SPECI or SYNOP formats and the observer reviews both the individual data for each parameter, and the compiled observation reports. The observer is able to edit data and/or compiled weather reports as required. The WEBMET system transmits this data to the BDC, and observation reports to Banco OPMET (Banco Internacional de Dados Operacionais de Meteorologia/International Bank of Operational Meteorology Data) in Brasilia, which is then sent over the GTS (Global Telecommunication System). OPMET can be accessed via the website of REDEMET (Rede de Meteorologia do Comando da Aeronáutica/Meteorology Network of the Air Force Command) (9). Only one fully-automated surface weather station is currently part of the FAB's network. The ASOS (Automated Surface Observing System) was commissioned at SBSP (Congonhas, São Paulo) on March 3, 2015, and is the first of its kind in a Latin American aerodrome (10), and operates in AUTO mode from the hours of 0200 to 0700 UTC. The first ASOS operating in a South American aerodrome is located at SOCA (Cayenne, French Guiana), which first transmitted an AUTO METAR on February 15, 2009.

Identifying the exact positions of anemometers is an important step in order to correctly identify upstream roughness, or near-field obstacles, which would require a correction to the observed speed. The notion that a homogenous roughness can be assumed for all directions was discarded by Gatey (11). The AIP (Aeronautical Information Publication) contains the information of the number of anemometers in each aerodrome, identification of the primary, secondary and subsequent anemometers, their type and location in reference to the runways (12). Although the document is updated regularly, information is not available or out of date for many aerodromes. Confirmation of exact anemometer location, records of the dates of changes to position or heights of anemometers, and changes of the anemometers themselves, are difficult to obtain, however efforts are being made to contact each station individually to obtain information. Using information from the AIP, the four anemometers at SBGR (Guarulhos, São Paulo) were identified with their locations and operational rankings shown in Figure 3. Weather observations emitted over the GTS take wind data from the primary anemometer, with the remaining anemometers used to identify wind shear at the critical landing/takeoff positions.

Figure 3. The locations of the four anemometers at SBGR (1° denotes primary anemometer).

2.2 *INMET*

Under the scope of INMET are conventional and automated surface weather observation networks. As of the start of 2017, conventional stations total 220, and automated stations 524. Observations are made at conventional stations at the hours of 0000, 1200 and 1800 UTC, and include 10-minute mean wind speed and direction, but no gust observation. Analog anemographs are also in operation at conventional stations. Many of these stations have wind data available from 1961, however it is common for large periods to be missing from time-series. Automated weather stations are modern than their conventional counterparts, and new stations are being commissioned regularly. A total of 30 new stations were commissioned in 2016 alone. INMET's ASOS network is not contained to Brazilian territory, with 4 stations in Uruguay and a station in Antarctica. The first ASOS station was commissioned on May 7th, 2000 at the Universidade Federal Rural do Rio de Janeiro campus in Seropédica, and remains active as of the start of 2017.

Wind data observations are made hourly and include the 10-minute mean wind speed at the end of the hour, predominant wind direction during 10-minutes, and maximum 3-second gust speed of the last hour. Observational data made over the previous three months, as well as accurate localization of the anemometer and photos of the station, are available via the INMET website (13). Only INMET's ASOS network will be considered for this study.

2.3 *Wind observation protocols*

Both FAB and INMET surface observation networks report on *DIR*, *V* and *G*, however there are some differences in their approaches. Table 1 summarizes various aspects of wind observations for each organization, which differ with regards to measurement resolutions and units, as well as criteria for gust observations.

Table 1 Comparison of wind observation protocols

Feature	Organization	
	FAB	INMET
Observation	Manual or automated	Automated
Transmission	Manual or automated	Automated
Wind speed resolution (units)	1 (knot)	0.1 (m/s)
Time-average of V (s)	600	600
Time-average of G (s)	3	3
G observation condition	When $G \geq V + 10$kt	Always, hourly maximum
Wind direction resolution (°)	10	1
Height of anemometer (m)	6-10	10
Present weather ID	Yes	No

Whilst observations made at both FAB and INMET networks are hourly, FAB also reports observations made during the hour when conditions change suddenly. FAB observations also include qualitative present weather conditions, such as precipitation and thunderstorm activity. Exposure conditions of the anemometers also vary quite significantly. The open and flat nature of aerodromes means that anemometers are typically surrounded by few close proximity obstacles. INMET stations do not generally have this luxury, and anemometers are often located within meters of buildings and trees which influence wind observations. An example of the exposure of the primary anemometer at SBPA (Porto Alegre, Rio Grande do Sul) is compared to that of INMET's station A801, also in Porto Alegre, in Figure 4. It is clear that the anemometer at A801 is highly shielded by obstacles, while the anemometer at SBPA is in open terrain.

Figure 4. Aerial photographs of the primary anemometer at SBPA (left) and INMET station in Porto Alegre (right).

3 OVERVIEW OF DATA BANKS

3.1 *ICEA*

The BDC at ICEA is not available for public access, and time-series requested from the BDC cannot be issued due to Brazilian law. However, statistical data for a determined period of time can be requested (daily/monthly, maximums, minimums and means). The website makes available historical monthly maximum V for 105 stations (14). Frequency of occurrence per wind direction of V and G are also available, with wind speeds grouped in bands of 5kt up to 30kt, and a final group of +30kt for V, and bands of 10kt up to 60kt, and a final group of +60kt for G. The data represents a summary of all anemometers present at the aerodrome for the given period of time. Comparing the number of V observations made at SBGR during January of 1993 (855) with January of 1994 (3041) suggests an increase in either the number of anemometers in operation or frequency of observation, with the former being the most likely explanation. Without access to the time-series it is difficult to assess the quality of the data being represented, however questions are raised by the annual maximums at some stations. An example is V annual maximum series at SBGL (Galeão, Rio de Janeiro) as shown in Figure 5. A sharp increase in the number of +60kt observations is noted from 1998 onwards and is highly suspicious.

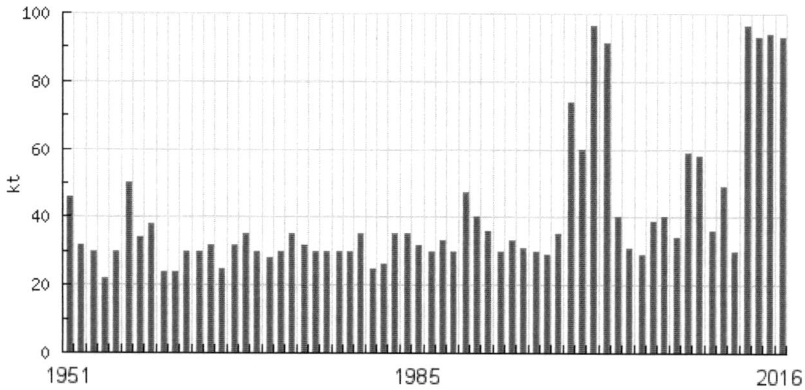

Figure 5. *V* annual maximums at SBGL (8).

3.2 *CPTEC-INPE*

CPTEC-INPE (Centro de Previsão de Tempo e Estudos Climáticos do Instituto Nacional de Pesquisas Espaciais/Center for Weather Forecasting and Climate Studies of the National Institute of Space Research) manages the BDM (Banco de Dados Meteorológicos/Meteorological Database) which is available for public access (14). The data is split into METAR and SYNOP observations, with stations being identified by their ICAO code and WMO number respectively. Historical data is available for many international stations and is not limited to Brazilian territory. METAR data is available from 1999 and SYNOP data is available from 1995, although data is very limited for the years prior to 2003. Data is presented in formatted tables, and is converted from the raw observation reports. The raw observation reports are not included. Wind observation data includes *DIR* and *V* (in m/s), but not *G*. Data from SPECI observations are not included in the database.

3.3 *NCEI/NCDC*

The NCEI (National Centers for Environmental Information), formerly NCDC (National Climatic Data Center), is a branch of NOAA (National Oceanic and Atmospheric Administration), which is overseen by the USA government's Department of Commerce. The NCEI global surface observation databases are likely the most extensive and comprehensive available to the public. The DS3505 database containing hourly surface observations (15) is typically the first stop for researchers and wind-tunnel laboratories when undertaking wind climate assessments. Examples of its utilization on a global scale are found in studies of Australia and Thailand by Burton and Allsop (16), Europe by Gatey (11), India by Kumar et al (17) and of the USA by Pryor et al (18). NCEI makes available data received over the GTS, which is from 1973 for many FAB stations, although there are a number of stations which have data available for earlier years, including SBBE (Val de Cans/Belém, Pará), SBBH (Pampulha/Belo Horizonte, Minas Gerais) and SBKP (Viracopos/Campinas, São Paulo). Data from INMET's conventional network is available for sustained periods of time, and data from INMET's ASOS is available as of June 2016.

Data appears in tabulated format and wind observations data include *DIR*, *V* and *G*, and are accompanied by qualifiers which indicate whether the data is suspicious or not. Indication of the origin of each observation is also given (i.e., SYNOP, METAR, SPECI or a combination of SYNOP/METAR), and each line of data is accompanied by the raw observation reports in the remarks section from September 1999 onwards. For the mixed SYNOP/METAR observations, the wind data is typically taken from the SYNOP observation which accompanies the data, and the wind data from the METAR report is omitted. A visual demonstration of the number of observations and types of the sources of data per year for station SBPA as provided by NCEI is shown in Figure 6, and is representative of many other stations located at aerodromes in Brazil. The NCEI's automated data extraction is problematic and is unable to deal with weather reports which are not formatted correctly.

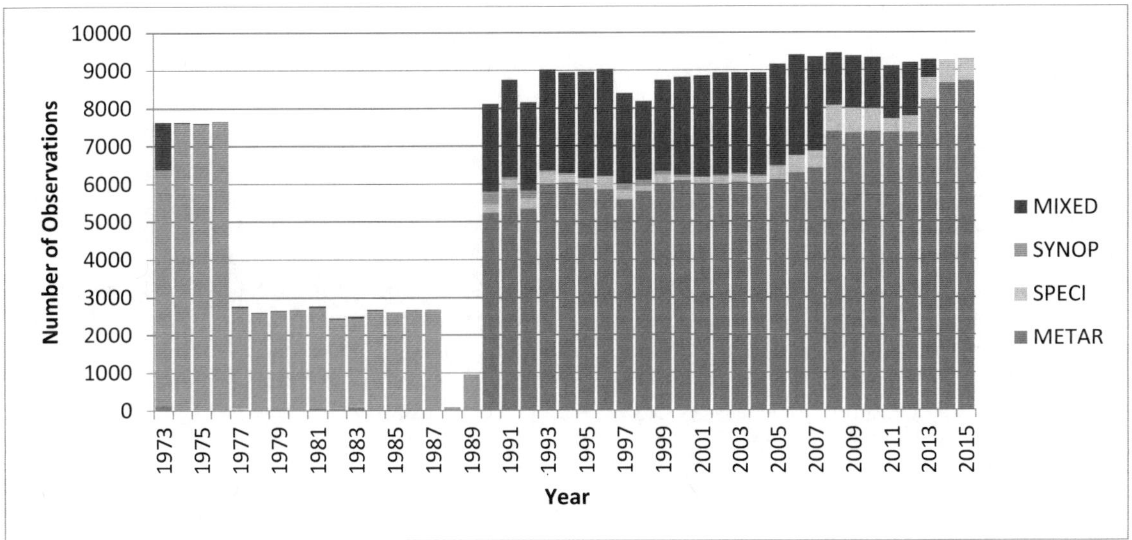

Figure 6. Composition of weather observations for SBPA as made available by NCEI (15).

Selection of the station is by a combination of its name and WMO number. Both should be checked and verified against another source of information prior to using data downloaded. In several cases it is found that NCEI merged data from an INMET conventional station with that of a FAB station. This is the case for SBCO (Canoas, Rio Grande do Sul) which is incorrectly assigned WMO 83967 by NCEI, a station which is located 10km to the south-west in the river port of Porto Alegre. In these cases, data originating from SYNOP reports should be omitted as these observations are recorded at different locations.

3.4 *Weather Underground*

Weather Underground, herein referred to as WU, is a privately owned digital resource which makes available weather forecasts and historical meteorological data on its website (19). For many countries outside of the USA, weather reports are available for surface observation stations from July, 1996. Wind observation parameters of *DIR*, *V* and *G* (in km/h) are extracted from, and accompanied by, original METAR/SPECI or SYNOP reports. In many cases, SYNOP and METAR/SPECI reports are present for the same time. The original source from which WU received raw observation reports is unknown, however due to the presence of Spanish within some weather reports, "MENSAJE REPARADO" (message repaired) or "MENSAJE RECHAZADO" (message rejected) are commonly found examples, it can be assumed the source is not Brazilian (Portuguese is spoken in Brazil and not Spanish). No messages with Spanish text are found after 2004.

3.5 *REDEMET*

REDEMET makes a number of meteorological products available to the public via its website (9). A visualization of current weather conditions is presented in map format, and weather observations (METAR, SPECI, SYNOP, PILOT, TEMP) are disseminated with forecasts and warnings (TAF, SIGWEX) for all of South America via Banco OPMET. METAR/SPECI observations are available from November 2002 onwards, however there are some discontinuities in the database. An example of a discontinuity is the period from November 1st, 2004 to December 31st, 2004, for which no data is available for any of the stations.

3.6 *INMET*

Although theoretically superior to the FAB network due to the higher number of stations and removal of human interference in the observation processes, data from INMET stations are affected by a number of different problems. Firstly, analysis of extreme wind speeds from the *V* time-series is unfeasible in most cases due to interference from nearby obstacles. Considering METAR/SPECI observations from SBPA (Porto Alegre, Rio Grande do Sul), the mean and maximum *V* wind speeds for 2016 were 3.5 m/s and 19m/s respectively. While for A801, INMET's ASOS in Porto Alegre, the mean and maximums for 2016 were 1.5 m/s and 6.7 m/s respectively. As both stations report the height of the anemometers to be 10m, the effect of shielding is clear. Analysis of the *V* time-series can be useful to determine changes in the characteristics of a station in order to qualify gust data or to assess the station's overall quality. Examples of gradual changes at three stations are seen in Figure 7.

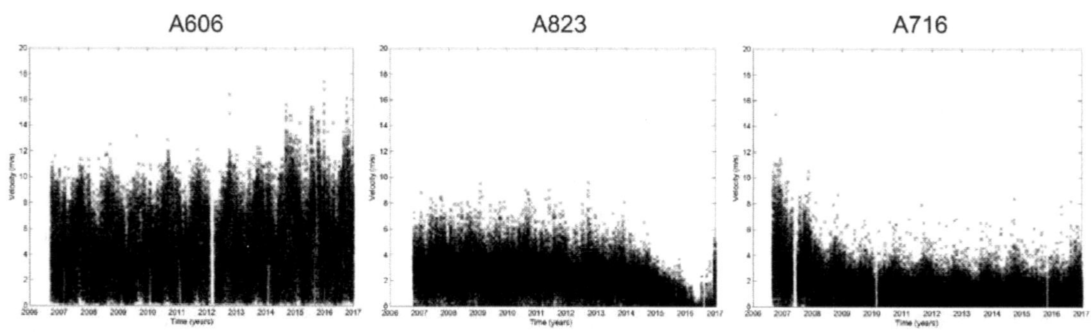

Figure 7. Time-series of hourly *V* data for A606 (Arial do Cabo, RJ), A823 (Inácio Martins, PR) and A716 (Ourinhos, SP).

Gust data from INMET stations appear to be useful for extreme wind analyses. A comparison of gust data for SBPA and A801 for the period between 2014-2016 shows maximum observed gusts of 36.0 m/s and 33.2 m/s respectively – much similar in magnitude than that of the 10-minute mean data. However, there are many stations which exhibit abrupt peaks in gust speed, which may or may not be the result of thunderstorm activity. The regular occurrence of gust speeds above 50 m/s suggests issues with the performance of the station, and there are cases in which these issues have been identified and resolved by INMET, such as A307 in Figure 8, and others in which the issues continue, as is the case for A044. Shifts in station characteristics can also be seen from the gust time-series. This type of behavior is shown for station A526 in Figure 8.

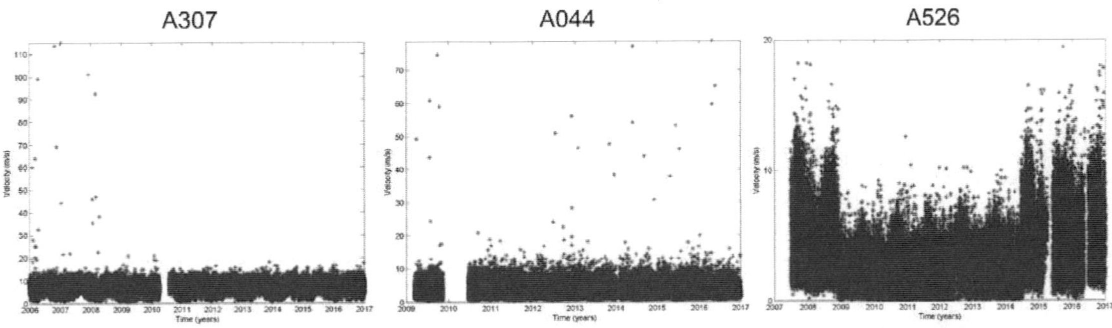

Figure 8. Time-series of hourly *G* data for A307 (Pertrolina, PE), A044 (Araguatins, TO) and A526 (Montalvania, MG).

In addition to these issues above, it is observed in Figures 7 and 8 that there are often long periods of time for which there is no wind data. It is also necessary to separate false peaks caused by equipment failure from real gust speeds measured during convective storms. There is no qualitative present weather information within the INMET ASOS dataset to assist with thunderstorm identification; however the use of other meteorological parameters could be used to identify such events. Using a combination of INMET observational data and satellite imagery, Ferreira (20) identified and analyzed over 9,000 downburst events in Brazil, finding that atmospheric pressure increased, while temperature and dew-point temperature decreased during the hour of strong wind gusts caused by a downburst event.

4 WEATHER OBSERVATION REPORTS

Weather observation reports are coded messages which contain observed meteorological data for a given location. As well as being of a different format, METAR/SPECI reports have a different purpose to SYNOP reports. The purely numeric SYNOP reports are made at simultaneous 3-hour intervals on a global scale in order to register synoptic events (i.e. the scale of approximately 1,000km). METAR/SPECI reports are alphanumeric with a main purpose of informing pilots of weather conditions during pre-flight preparation, but are also used by meteorologists to forecast weather.

Errors in the formation of the observation report are common, especially when compiled manually. Typographic errors within the observation can be identified, however it is difficult to identify an observation made from a misreading of a measurements by the observer. Automated programs utilized by the likes of CPTEC-INPE, NCEI and WU are able to extract, convert and tabulate large quantities of data from the raw observations, but can generate misleading data when the raw observations are not formatted correctly.

4.1 *METAR/SPECI*

Although highly standardized on an international level by ICAO, each country has slight variations in the reporting procedures of METAR/SPECI reports. For Brazil, reporting procedures are defined by FAB in two documents MCA 105-2 (21) and ICA 105-16 (22). According to these documents there are only 8 acceptable ways to report wind observations. The most commonly found 7 variations, with accompanying examples, are shown in Table 2. The 8th variation is used in the case of observed *V* or *G* above 100kt, which is represented by *P99KT* (*P* for 'plus'), however an example of its usage in real-life during this study was not encountered. It is of concern that this protocol does not permit for the accurate recording of wind speeds above 51m/s. Unless a gust is reported or wind direction varies, the wind observation field is strictly 5 digits in length, comprising of *DIR* (first three digits, multiples of 10, maximum 360, minimum of 0) and *V* (last two digits), and followed by *KT* denoting the units as knots. Directional convention is for wind from the North to be reported as 360, with 000 only used to report calm conditions, i.e. no wind.

Table 2 METAR/SPECI reports with correctly formatted wind observation fields

Type of wind observation	Date (UTC)	Example	Observed Data		
			V (kt)	*G* (kt)	*DIR* (°)
No wind	02/07/1996	SBLO 021100Z **00000KT** 9000 SCT100 …	0	-	-
No reading	02/08/2006	SBGO 022100Z **/////KT** 9999 FEW025 …	-	-	-
Standard reading	31/12/2015	SBRJ 312300Z **15006KT** 9999 FEW035 SCT100 ...	6	-	150
Standard with gust	17/12/2013	SBPA 171842Z **11014G24KT** CAVOK 29/18 …	14	24	110
Variable DIR (V<3kt)	29/12/2008	SBSP 291200Z **VRB02KT** CAVOK 25/18 …	2	-	Variable
Variable DIR (V≥3kt)	05/04/2011	SBSP 051800Z **19007KT 140V240** 9999 TS …	7	-	140-240
Gust with variable DIR	08/12/2003	SBPF 011600Z **27020G30KT 230V290** 5000 …	20	30	230-290

Current practice in Brazil is for METARs to be made on the hour. A wind-related SPECI report is issued when one of the following criteria is observed:

- *DIR* shifts by 60° or more in regards to the last observation made, and *V* is equal to or greater than 10kt before and/or after the change;
- *V* is 10kt greater or less than the last observation made;
- *G* is 10kt greater or less than the last observation made, and *V* is equal to or greater than 15kt before and/or after the change.

Errors in the preparation of METAR/SPECI observations are common, and from 2012 onwards the use of *COR* (representing "correction") denotes a METAR/SPECI report which corrects an erroneous observation previously emitted with the same time-stamp. Through the examination of raw METAR/SPECI reports it is found that there are an infinite number of ways to prepare an incorrectly formatted report made. However, there are several types of errors which are more common, and include the failure to use a space between fields, the joining of observations from different stations, report not in METAR/SPECI format (i.e. is SYNOP or TAF), too many or too few digits in the wind observation field, *DIR* not a multiple of 10 or more than 360, the letters *KT* missing after wind observation, incorrect format when reporting variable wind or gust observations, multiple wind observation fields, incorrect reporting of calm conditions, the use of the letter *O* instead of digit *0*, the use of the digit *9* instead of digit *0*, the use of the digit *7* instead of digit *1*, the use of letters instead of numbers in the wind observation field, no wind observation field and multiple reports for the same location and time-stamp. The automated routines used by CPTEC-INPE, NCEI and WU are developed to convert reports which are free of errors and often do not question the integrity of the report. In many cases this results in the extraction of incorrect data, or no data at all, with examples shown in Table 3. Values of *999.9* and *-9999.0* are understood to denote unreadable observations.

Table 3 Converted data from incorrectly prepared METAR/SPECI reports

Type of error	Date (UTC)	METAR/SPECI	Database V wind speeds (m/s)		
			NCEI	CPTEC	WU
Too many digits	25/03/2008	SBGL 252000Z **160045KT** 6000 BKN013 ...	23.1	0	23.1
Too few digits	13/09/2008	SBGL 130500Z **2705KT** 9000 FEW015 ...	999.9	2.6	-9999.0
"9" instead of "0"	14/11/1996	SBGR 142000Z **35993KT** 6000 SCT025 ...	1.5	N/A	47.8
Missing KT	19/01/2003	SBSP 190500Z **12003** CAVOK 23/19 ...	999.9	N/A	-9999.0
Incorrect KT	15/03/2009	SBGL 150500Z **33005KTKT** 9999 SCT010 ...	999.9	2.6	2.6

It is often the case that dates, times and locations do not match those within the METAR/SPECI report, particularly in the NCEI and WU databases for prior to 2006. Reports from REDEMET do not appear to have this problem. Examples of errors in station assignment are encountered in both NCEI and WU's databases for two strong wind observations at the same location. According to these two databases the following two SPECI observations were made at SBPA: 1) wind speeds of V = 30kt and G = 50kt observed at 1515 UTC on August 29th, 2005, 2) wind speeds of V = 22kt and G = 30kt observed at 2105 UTC on August 30th, 2005. No such records exist within the REDEMET database for SBPA at those times and dates. However, within the REDEMET database, the exact same observation as 1) was found for SBPF (Passo Fundo, Rio Grande do Sul) at 1515 UTC on August 29th, 2005, and observation 2) for SBCA (Cascavel, Paraná) at 2015 UTC on August 30th, 2005. After analyzing observations in the hours before and after the misplaced observations at SBPA, SBPF and SBCA, it was clear that these speeds were not observed at SBPA.

4.2 *SYNOP*

Reporting procedures for SYNOP observations are defined by FAB in document ICA 105-16 (22). The report is entirely numeric with the exception of the prefix *AAXX* which identifies the station as land based and not a ship. The number of groups of digits varies with each report, however the first five groups after the prefix are of most importance to surface wind observations. Examples of a correctly formatted standard observation and a report with no wind observation are shown in Table 4.

The first group of numbers is in the format $YYGGi_w$, where YY is the day of the month, GG is the hour of the day in 24hr format, and i_w is the units of the wind observation, which is defined as follows: i_w = 0 is estimated speed in m/s, i_w = 1 is measured speed in m/s, i_w = 3 is estimated speed in kt and i_w = 4 is measured speed in kt. The second group of numbers $IIiii$ identifies the location of the station as per its assigned WMO number. In the standard reading in Table 4, this would mean an observation at 18hrs on the 22nd day of the month, V reading in kt, for station 83612 (SBCG, Campo Grande). The third group contains information related to precipitation, station type, heights of cloud and visibility. The fourth group is of the format $Nddff$ where N is the total cloud cover, dd is direction of wind and is to be multiplied by 10 to convert to degrees (dd = 36 is North) and ff is observed speed in the units determined by i_w. Calm conditions are represented by $ddff$ = 0000. An optional group comes after in cases of velocities greater than *99*, and takes the format *00fff*, however no such use of this convention has been found while examining Brazilian data for this study. The fifth group contains observed temperature data in the format $1s_nTTT$, where *1* is intentionally used to identify the temperature grouping, s_n indicates the sign of the temperature reading, where s_n = 0 is positive and s_n = 1 is negative, and TTT represents the temperature, in 0.1C. In the first example of Table 4 the observed temperature is 19.6C.

Table 4 SYNOP reports with correctly formatted wind observation fields

Type of wind observation	Date (UTC)	Example	Observed Data		
			Units	V	DIR (°)
Standard reading	22/06/2012	AAXX 22184 83612 32470 61906 10196 ...	kt	6	190
No reading	22/06/2012	AAXX 22213 83612 42470 5//// 1//// ...	-	-	-

Errors in the preparation of SYNOP observations are common. Amongst the most commonly encountered types of errors are the incorrect reporting of calm conditions, wind direction greater than 360 degrees, incorrect reporting of date and time, groups joined together due to missing space, missing wind observation field and incorrect wind speed unit indicator. The last two types have a profound effect on the quality of NCEI, CPTEC-INPE and WU databases. The automated programs used by these organizations often fail to identify erroneous SYNOP observations and display incorrect wind speeds as a result.

The example in Table 5 compares the data extracted by these databases from SYNOP and METAR reports made at the same time and station. The V wind speed is much higher when extracted from the SYNOP when compared to the wind speed extracted from the METAR report. A comparison with METAR reports prior to and after the observation show the V = 5.2 m/s wind speed fit weather conditions for that period. Upon inspection of the SYNOP it is apparent that the fourth grouping, relating to the wind observation, is in fact the temperature data, and the wind data is omitted altogether from the report. Hence a V = 56kt reading is extracted from the report and converted to 28.8 m/s, when in fact the 56 is a partial reading of the temperature of 25.6C. The misreading of poorly formatted SYNOP observations is common in all three databases, with the tendency to generate false high wind speeds.

Table 5 SYNOP and METAR reports at 1800 UTC on Dec 6, 2003 for station SBPA/83971

Format	Observation	Converted V (m/s)		
		NCEI	CPTEC	WU
SYNOP	AAXX 06184 83971 32970 10256 20084 30129 40134 56012	28.8	28.8	-
METAR	SBPA 061800Z 29010KT CAVOK 26/08 Q1013	-	5.2	5.2

An examination of historical SYNOP observations in Brazil shows that i_w values infrequently indicate wind velocities in m/s and not kt. In these cases, the change in units is considered by the automated data extraction/conversion programs of NCEI and WU. However, when comparing these observations with METAR observations made before and after, it is apparent that the speed of these SYNOPs in m/s is double the magnitude of the others. This suggests that the infrequent indication of the use of m/s in SYNOP reports is due to erroneous i_w indicators, and the units should always be considered as knots unless a regular and frequent use of m/s is observed.

4.3 *PREPARATION OF WIND SPEED TIME–SERIES FROM WEATHER OBSERVATION REPORTS*

A method is proposed to extract wind observation data from METAR/SPECI and SYNOP reports with the aim is to reduce the number of errors in the time-series. It has been shown that automated extraction and conversion processes are problematic, particularly when the observations are prepared manually. A 5-stage semi-automated process is proposed and draws from the raw METAR/SPECI and SYNOP observations made available by REDEMET, WU and NCEI. The reports are processed based on the following ranking: 1) METAR/SPECI from REDEMET, 2) METAR/SPECI from WU, 3) METAR/SPECI from NCEI, 4) SYNOP from WU, 5) SYNOP from NCEI. The REDEMET database is composed of data received prior to being transmitted over the GTS, i.e. one process closer to the observation's origin, and is given the highest position of confidence. Observations from WU are given preference over NCEI due to less number of errors in the conversion of time, date and station information in the WU database. METAR/SPECI observations are given preference over SYNOP due to the reporting of gust data, as well as the problems created by SYNOP reports when the wind observation field is missing.

The first step involves the comparison of time, date and station information for each entry with that contained within the raw weather observation. Date information contained within the weather report is limited to only day of the month, and so the month and year of the observation can only be determined from the listing within the particular database. The process is semi-automated, and each observation is processed one at a time. An automated routine is employed to compare the day, hour and station information per entry with that within the raw weather observation. If all three are the same, the observation is classified as permissible. If not, the entry is manually compared to observations prior and after. If an understanding of the error can be determined, the date or time is changed manually, however if the date and time cannot be determined or the station is different, the observation is excluded. Global quality checks are also made on the time and date (e.g. time-stamp minute of observation cannot be greater than 59 minutes, and the date February 30 is not possible).

The second step is a fully-automated process which extracts wind observation data and any wind related present, recent and in-proximity weather indicators (thunderstorm, squall, funnel cloud, sandstorm, dust-storm). If the wind observation does not match the correct formatting options, as listed in Table 2 for METAR/SPECI reports and Table 4 for SYNOP reports, the wind observation data is not extracted and the entry is flagged. If there are multiple observations for the same time, date and station, all observations of the same combination of time, date and station are flagged, however, if only one of the observations has the *COR* identifier within the report, it is identified as being the correct report. In the case of multiple reports with the *COR* identifier, the observations with no *COR* identifier are excluded and the multiple *COR* observations are flagged. Observations are also flagged if $V \geq G$.

The third step is a manual operation which assesses all observations flagged during step 2. In the same way that we as humans are able to process and understand those who make grammatical or pronunciation errors in written or spoken language, it is possible to understand incorrectly formatted weather observations when placed in context. Flagged observations are examined one at a time and compared to observations in the hours before and after the time of observation. If the observation under examination can be understood, the *DIR*, *V* and *G* information is extracted manually; alternatively, if neither one of *DIR* and *V* can be determined, the observation is excluded from further analysis. Similarly, in the case of multiple observations flagged for the same date and time, only the most contextually appropriate observation is admitted to the time-series to be analyzed.

The fourth and fifth steps involve the identification of wind speeds deemed spurious. The fourth step is automated, and the *V* time-series is passed through a routine which identifies spurious data. The following set of criteria is similar to those as proposed by DeGaetano (23), Burton and Allsop (16) and Gatey (11), with the ith observation being assessed:

- $V_i \geq 30$kt,
- $V_i/V_{i+1} \geq 3$,
- $V_i/V_{i-1} \geq 3$,
- No accompanying G_i observation,
- No present, recent or in-proximity weather observation which is associated with strong wind (thunderstorm, sandstorm, dust-storm, squall).

The fifth and final step is the manual identification of spurious data. The *V* and *G* time-series are plotted and a visual inspection of each station is undertaken. Spurious data, mainly in the *V* time-series, is often easily identifiable. At this point, any spurious data is typically due to typing errors in two consecutive wind observations, a typographic error in an observation during an event which could produce strong winds (i.e. thunderstorm), or a typing error during a mild wind event (10-15kt). The dates and times of suspicious observations are noted and records of damages caused by strong winds are investigated using news websites and other online resources. Examples of the assessment of some questionable observations are shown in Table 6. No news reports were found regarding Case 1, however the *+TS* indicates the occurrence of a strong thunderstorm and so the observation was accepted. For Case 2, a news report of damaged transmission towers was encountered and coupled with the *+TS* observation the wind speed was accepted (24). No reports of damage caused by wind or storms were found for either Case 3 or 4 which were subsequently rejected from the time-series. The thunderstorm observation in Case 3 was prefixed *VC* (vicinity), meaning it was observed to be within 8-16km of the station. Observations in the hours before and after Case 4 were constantly between 10-15kt, making it highly likely that the observation was intended to be 14kt and not the reported 41kt.

Table 6 Investigation of suspicious weather observations

Case	Date	Observation	Decision
1	Nov. 2, 2007	SBMT 022300Z 27055KT 0100 +TSRA OVC020 //////CB 23/16 Q1010=	Accepted
2	Sep. 20, 2011	SBIH 202200Z 27050KT 1500 +TSRA FEW017 FEW027CB 23/23 Q1009=	Accepted
3	Nov. 26, 2015	SBEG 260900Z 24093KT 9999 VCTS SCT015 FEW025CB 24/21 Q1009=	Rejected
4	Dec. 20, 2015	SBMQ 201800Z 08041KT 9999 SCT023 24/23 Q1012=	Rejected

5 RESULTS

The composition of the time-series of a H24 station using the proposed method is reported in Figure 9 for SBPA. METAR/SPECI reports from REDEMET and WU are the most frequently used source, with each year after and including 1999 composed of multiple sources, showing that all sources have discontinuities. When compared with the NCEI data composition of SBPA in Figure 6, it is observed that NCEI data is lacking the METAR reports made at 30 minute intervals for the period between November, 1999 and October, 2001.

For this study, a strong wind observation is defined as $V \geq 30$kt (15m/s) and strong gust observation as $G \geq 40$kt (20m/s). A comparison of the number of strong wind and gust observations per year at SBPA is made in Figure 10 and Figure 11 respectively. Comparisons are made between NCEI wind data and the time-series created using the proposed method. In the case of V time-series, the NCEI data is also applied to the same filter as described in the fourth step of the proposed method, with the purpose of eliminating false peaks. It is noted that no manual extraction of spurious data, as described in the fifth step of the proposed method, was required for SBPA. In Figure 10, it is observed that all three sets of data are in agreeance from 2012 onwards, which is attributed to the reduced usage of mixed SYNOP/METAR reports by NCEI. Prior to this, the NCEI data regularly contains a higher number of strong V observations than the proposed method. The filtered NCEI data is in good agreeance with the proposed method from 2007 onwards. The proposed method gives greater confidence that false peaks are eliminated while data related to thunderstorms are preserved. With the substantial differences between the number of strong wind observations as reported by NCEI and the proposed method, the validity of NCEI data for years with no accompanying wind reports is questionable.

The number of strong gust observations, as reported by NCEI and the proposed method and shown in Figure 11, is closer in agreeance than that of the V observations. This is mainly due to the infrequency of reporting gusts and also reduced numbers of errors in the observation of strong gusts. The tendency is for the proposed method to have more than, or equal to, the number of strong gust observations when compared to the NCEI database. However, there are two years in which the NCEI database contains a single $G \geq 40$kt observation more than the proposed method. The extra observation in 2005 is explained by the incorrect reporting of station SBPF as SBPA, as described in Section 4.1, whereas the extra observation in 1997 is due to a 2-hour discontinuity of the WU data when the observation occurred. The discontinuity highlights the importance of drawing data from multiple sources for overlapping periods.

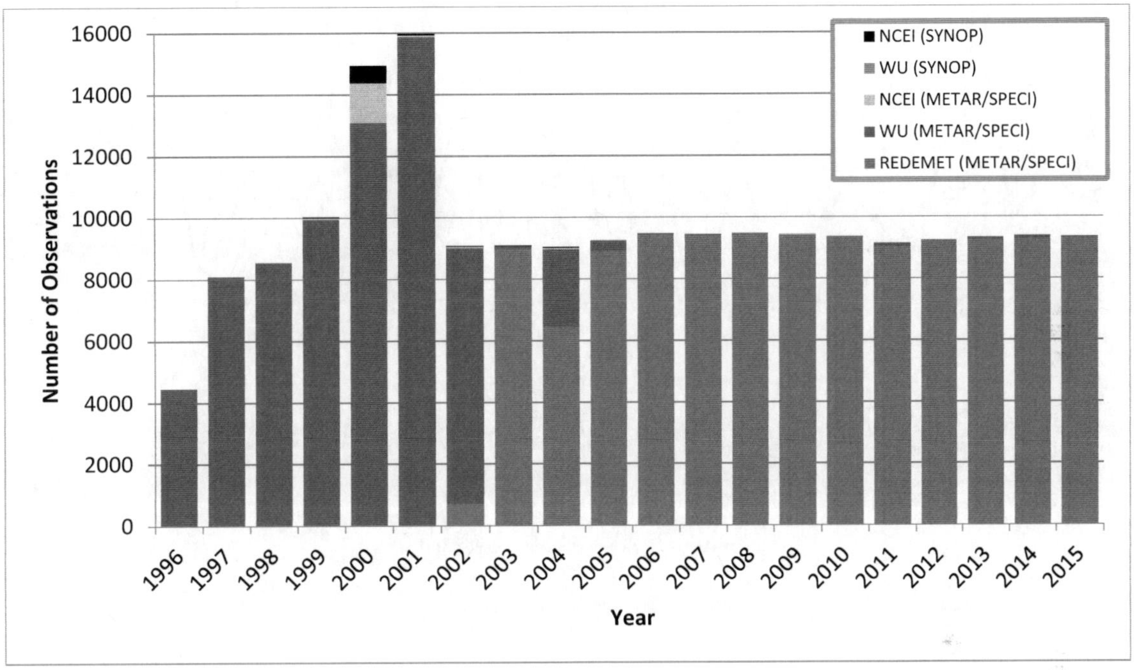

Figure 9. Composition of weather observation report types and sources for SBPA using the proposed method.

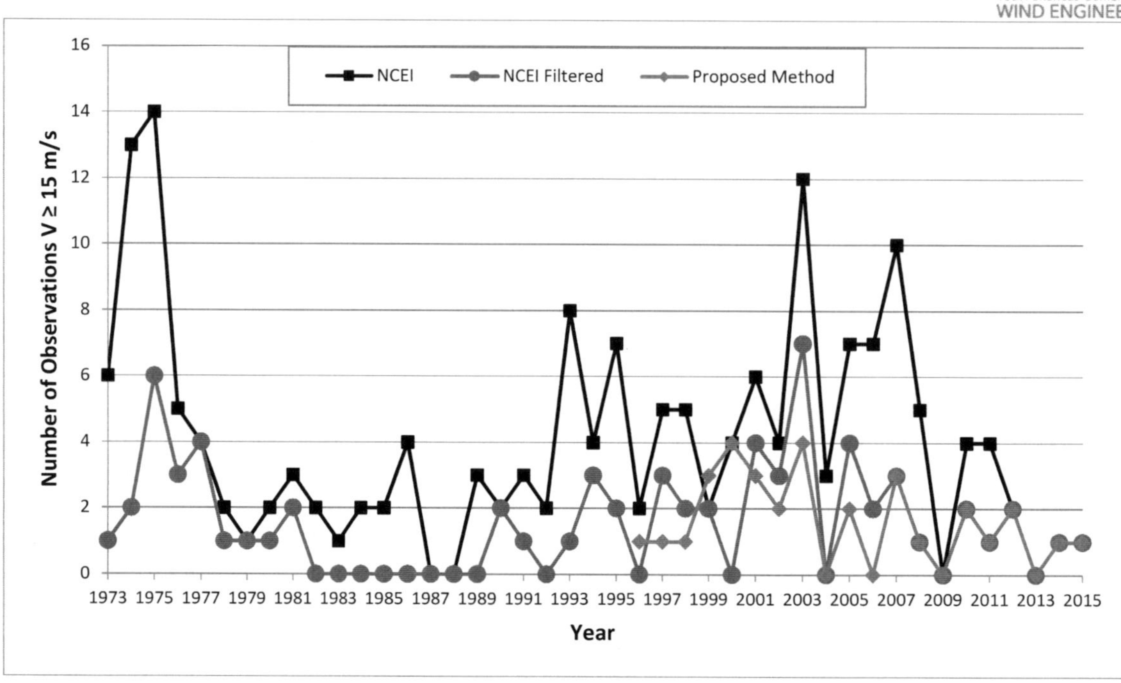

Figure 10. Number of strong wind observations per year for SBPA as per the proposed method, NCEI database and the NCEI database with a filter applied.

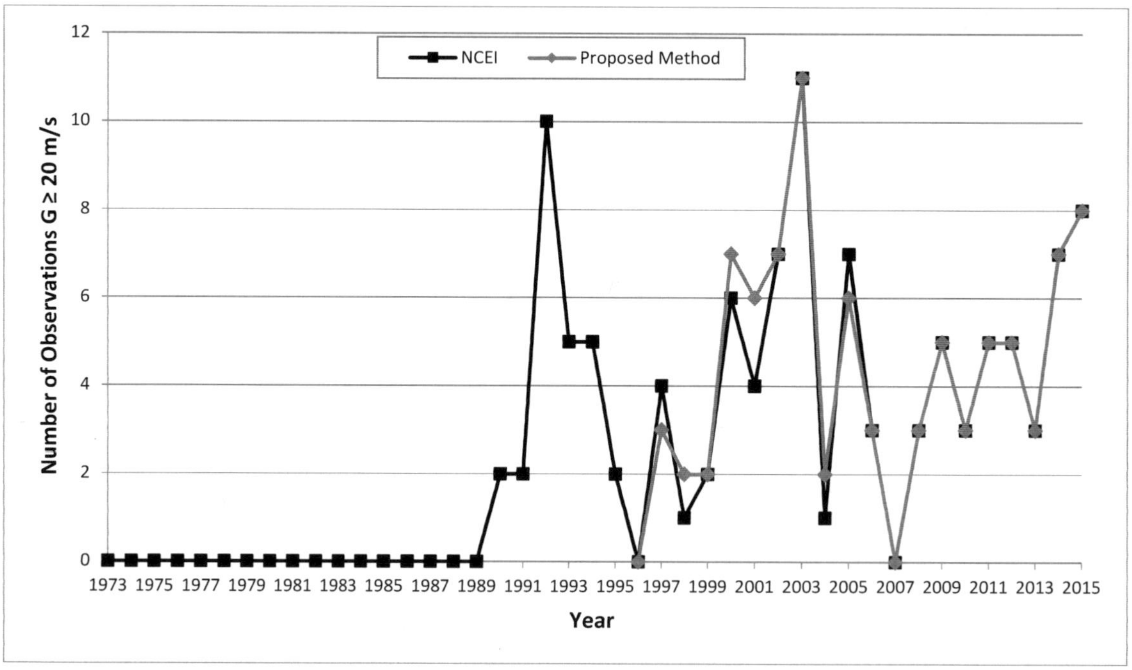

Figure 11. Number of strong gust observations per year for SBPA as per the proposed method and NCEI database.

The V and G time-series are then available for further analyses once spurious data has been removed from the time-series. It is found that the wind data measured at several FAB stations are often subject to abrupt or gradual changes unrelated to climate. Typical reasons include a change of anemometer, its height, position or exposure, or station operational procedures, including frequency of observation. Some examples are reported below.

A sudden decrease in the magnitudes of both V and G at SBKP are noted at the end of 2000 as observed in Figure 11, and would suggest either a decrease in the height of the anemometer, or the repositioning of the anemometer from atop a rooftop to an earth-based location. The installation of a new anemometer on December 13th, 2000 was confirmed by staff at SBKP. The previous height and location remains to be confirmed, however the newer anemometer was installed at a height of 8m, located approximately 300m South-West from the nearest group of structures which were single-story dwellings. The anemometer remained in operation until December 17th, 2002, when a new anemometer was reportedly commissioned at the same location and height. On November 8th, 2013, a new anemometer was reportedly commissioned at the same location but at a height of 10m. Observing the time-series in Figure 11, it is noted that the signal of the anemometer in operation between 2002 and 2013 is more similar to that of the anemometer installed at 10m, than that previously in operation from 2000 to 2002, raising the possibility that the anemometer installed in 2002 was actually at a height of 10m and not 8m.

The distribution of V observations is also substantially different for 1999 and 2001 as observed in Figure 12. In 1999, a strong bias for even numbered wind speeds is observed (i.e. 2,4,6,8…etc.), with 0kt, 10kt and 20kt being the most frequently made observations. This indicates the strong likelihood that manual wind observations procedures were in place at SBKP in 1999. The distribution for 2001 is substantially different to that of 1999, and the same level of bias for even numbered wind speeds is not observed. It is noted however, that there is a bias for 8kt, 10kt and 12kt observations as opposed to 7kt, 9kt and 11kt observations respectively. This implies the implementation of a more precise and disciplined manual observation process than that used in 1999, or a process in which the measurement is made automatically and discretion is used by the observer to alter the measurement on occasion.

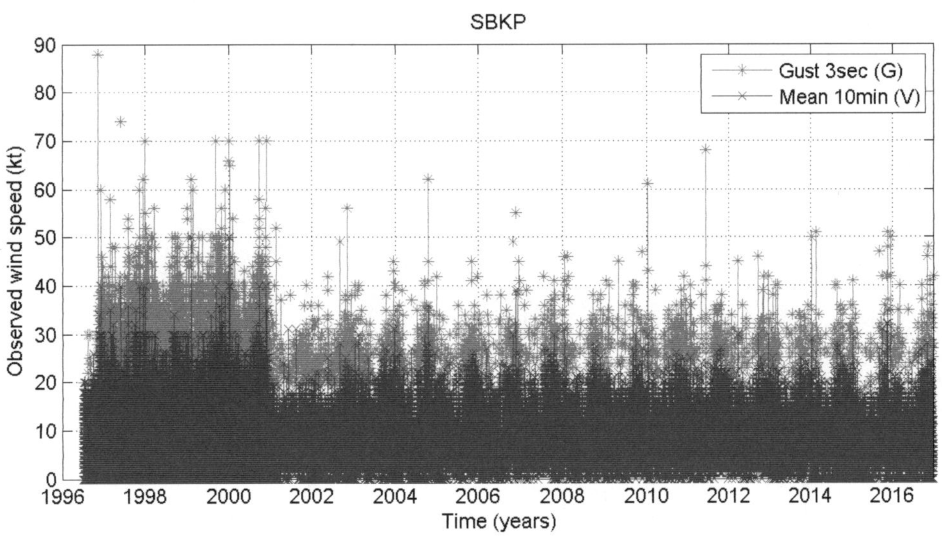

Figure 11. Observed V and G wind speed time-series for SBKP (Campinas, São Paulo).

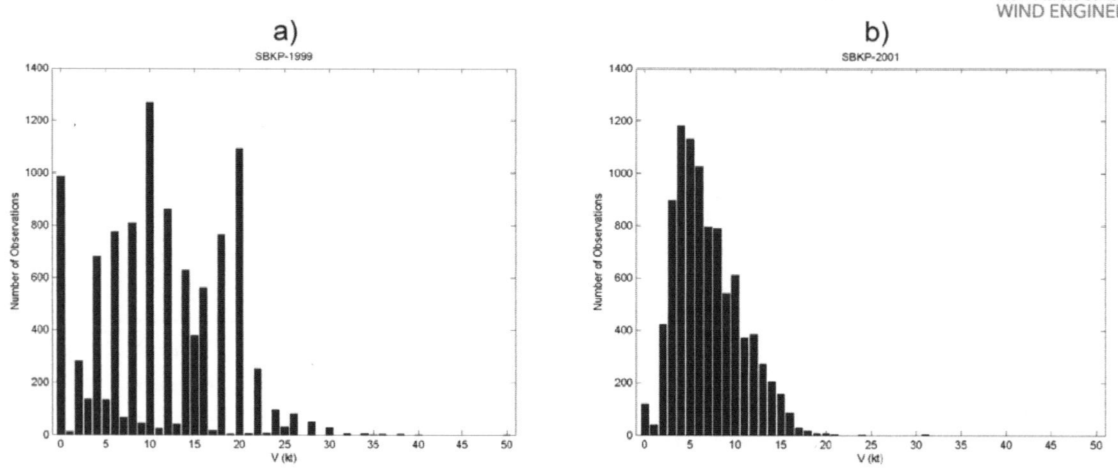

Figure 12. Frequency distribution of *V* observations at SBKP for a) 1999 and b) 2001.

The calculation of the ratio between even and odd numbered observations could be used to identify a change in operational procedures. Ignoring no-wind observations (i.e. $V = 0$kt), it is assumed this ratio should be around 50% for an automated process. This is the case for SBKP after December of 2000 as observed in Figure 13. Prior to this, the ratio of even to odd numbered observations was consistently above 80%. In the case of SBGR, the ratio is consistently around 50% from 1997 until the start of 2017. The decrease from over 80% to 50% occurs over a number years for SBLO (Londrina, Paraná), approximately from 2000 to 2008, and could indicate the gradual introduction of automated processes. Observations at SBBW (Barra do Garças, Mato Grosso) appear to have devolved from an automated process to a manual approach, as indicated by the abrupt increase from 50% to 80% in 2013.

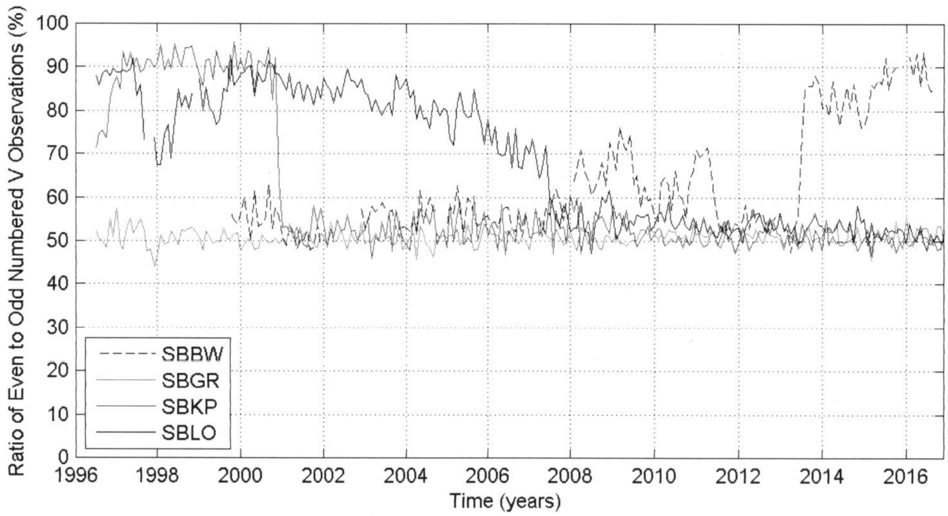

Figure 13. Monthly ratio of number of even to odd numbered *V* (in kt) observations for four different weather stations.

The changing of an anemometer can be problematic for the gust time-series. In the case of two anemometers with different response characteristics, wind speeds averaged over 10 minutes are expected to similar, however instantaneous or gust wind speeds may differ. Differences in gust measurement could also be attributed to different averaging intervals, i.e. 1 second or 3 seconds, or different types of anemometers, i.e. cup, vane, sonic.

Two examples of abrupt changes to the wind gust time-series are observed at SBGR and SBPB in Figures 14 and 16 respectively. For SBGR, an increase in the frequency of gust observations is noted occur at some point during 2005. Not a single gust equal to or over 50kt is observed from 1996 to 2005, however 8 such observations are made in the period from 2005 to 2016. The ability to determine peak gust wind speeds during TS events improved after 2005, as highlighted by data shown in Figure 15. Prior to 2005, only 1 or 2 days of the year had $G \geq 40$kt wind speeds during TS events. From 2006 onwards this number averaged at 3 days per year, with 8 days observed in 2010.

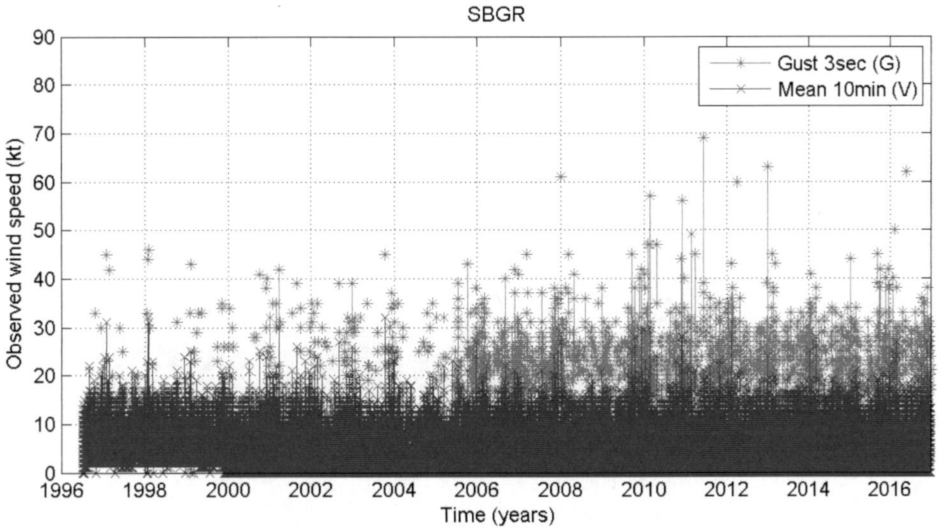

Figure 14. Observed V and G wind speed time-series at SBGR (Guarulhos, São Paulo).

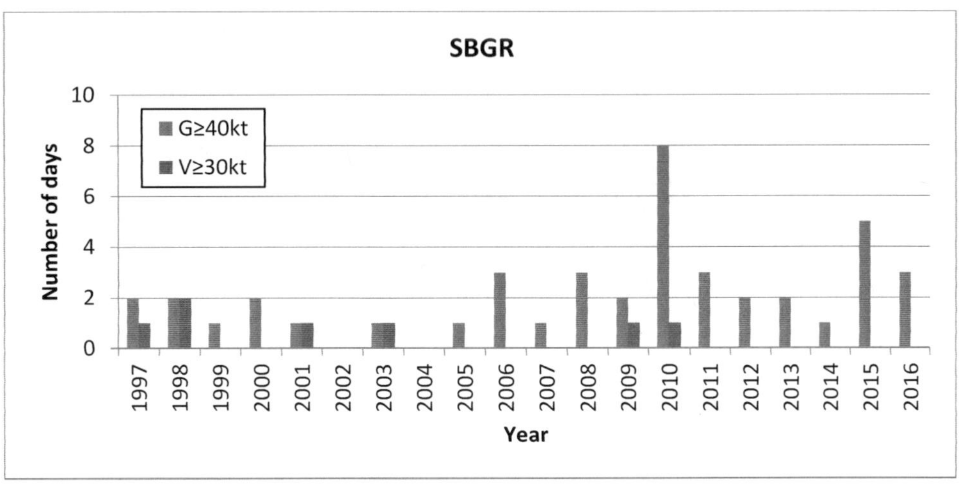

Figure 15. Number of days per year with strong wind, as identified by either $G \geq 40$kt or $V \geq 30$kt, during a thunderstorm event.

Three interesting observations are made for the time-series of SBPB in Figure 16. There is a notable lack of data for the period around 2002, as well as prior to 2000, due to the limitations of all three sources (REDEMET, WU and NCEI) in the provision of data for these periods. The wind-climate is highly seasonal due to the station's location in the ITCZ (Intertropical Convergence Zone). However, the most intriguing observation is the abrupt increase in magnitude of gust speed observations from 2015 onwards. The motivation for this increase is unknown and warrants further investigation.

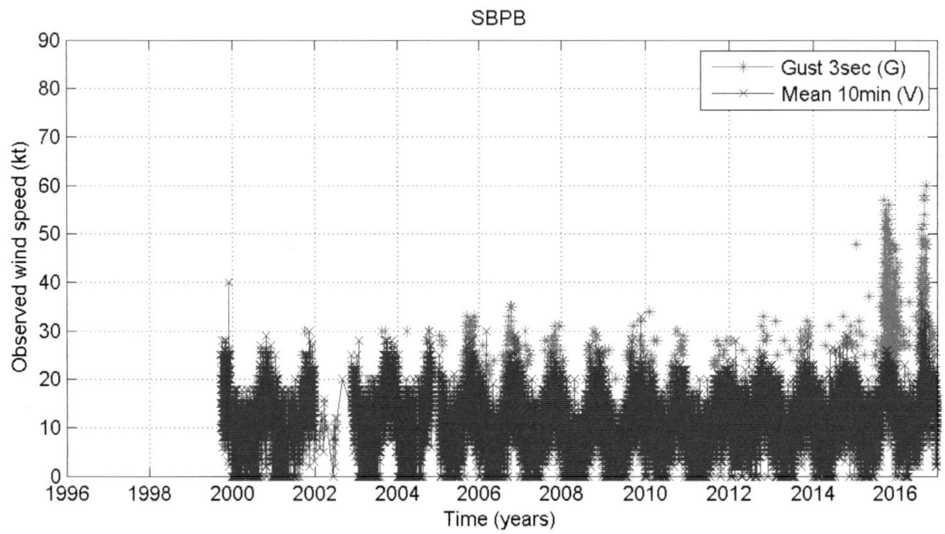

Figure 16. Observed V and G wind speed time-series at SBPB (Parnaíba, Piauí).

The locations of the most frequently observed strong winds are mapped in Figure 16. The average number of days per year that a strong wind is observed, N_D, is used as the comparative parameter. In order to reduce the effects caused by changes to anemometer heights and observation procedures, only data from 2007 to 2016 were analyzed. An attempt is made to present different types of wind mechanisms.

Figure 17a) shows the average number of days per year a strong wind observation is made during a thunderstorm event. It is assumed that these winds are mainly related to downbursts events, although there remains the possibility that some events were caused by tornadoes. The highest frequency of strong wind days related to thunderstorms is in the south and central areas of Brazil. The four locations with the highest frequency are SBCA (Cascavel, Paraná), SBCH (Chapecó, Santa Catarina), SBMG (Maringá, Paraná) and SBCG (Campo Grande, Mato Grosso do Sul). The north-eastern and northern extremities of the country are identified as having the lowest frequency of thunderstorm related strong wind events.

Figure 17b) shows the average number of days per year a strong wind observation is made outside of a thunderstorm. It is tempting to label these winds as a synoptic, however no examination of event duration was made and these events could include the intermediate gust-front wind as described by Kasperski (25). As such, the title of non-thunderstorm wind would be more appropriate. In addition to the central southern region of the country, the coastal regions of Rio de Janeiro and Espirito Santo are the locations with the highest frequency of non-thunderstorm winds. The high frequency of strong winds at SBBP in the north of Brazil is highly suspicious and due to the anomaly as presented in Figure 16.

Figure 17c) maps strong winds that are not observed during thunderstorms but that occur on the same day as an observed thunderstorm, likely to be the gust fronts which precede TS events. However a more thorough event based analysis, as opposed to a day based analysis, is required to confirm. Again, it is the southern region of Brazil which has the most number of events.

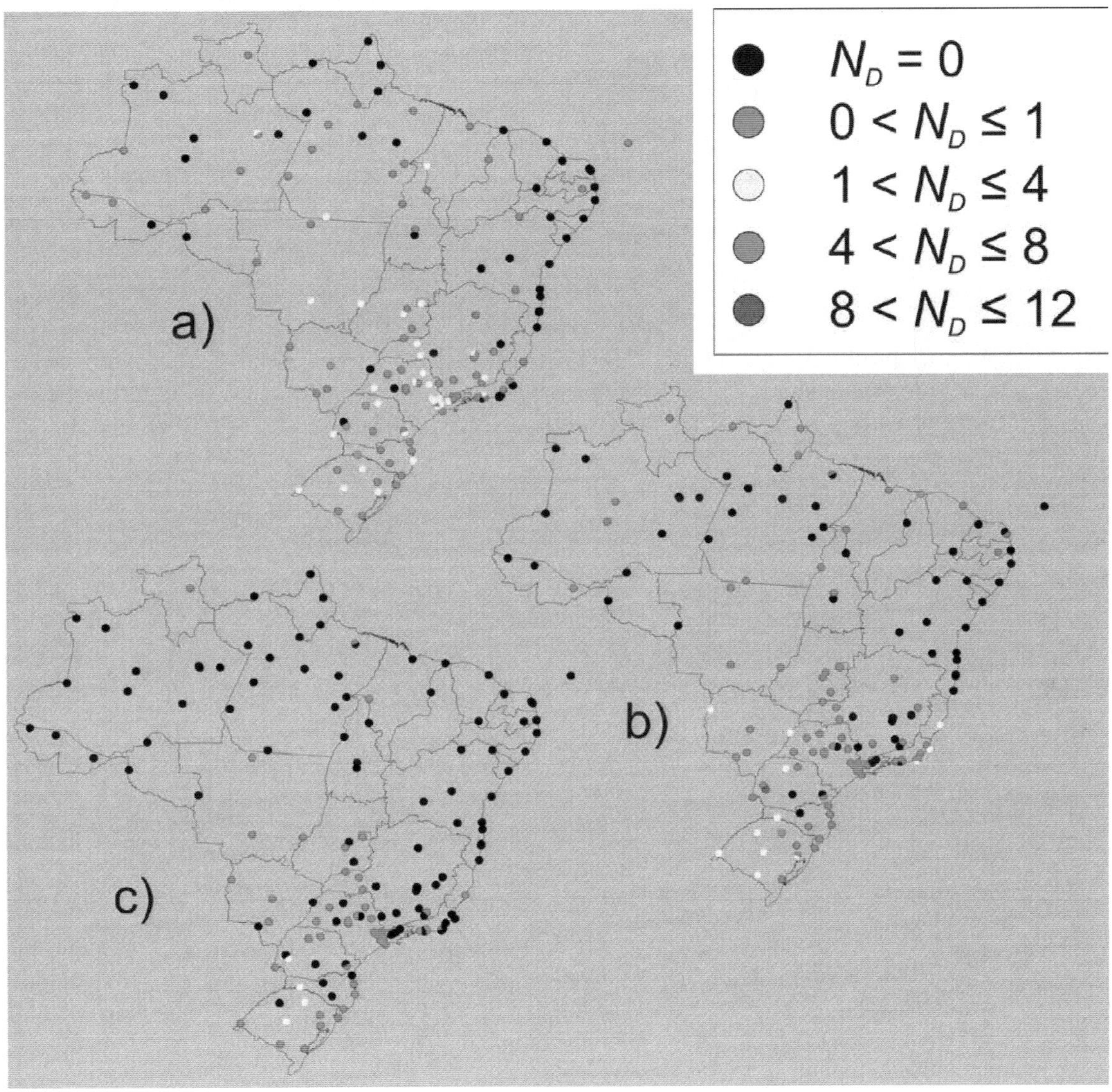

Figure 17. Average number of days per year (N_D), with a) strong wind speeds observed during TS events, b) strong wind speeds observed outside of TS events, c) strong winds speeds observed outside TS events, but on the same day as a TS event (Jan 1st 2007 to Dec 31st, 2016).

6 CONCLUSIONS

An overview of the surface weather observation networks of Brazil was conducted, with emphasis given to the FAB network. Problems associated with the automatic extraction of data within METAR/SPECI and SYNOP weather reports were exposed, and a proposed semi-automated method to extract wind observation data from weather reports was shown to eliminate false peaks in the 10minute mean wind speed with greater confidence than that obtained by automated filters, while also preserving winds observed during thunderstorms. Several issues associated with Brazilian wind data were demonstrated using examples. These issues include the change of anemometers, their heights and observation procedures. Further investigations are required before regional extreme wind models for the country can be produced, both in terms of individual station history and identification of the origins of extreme wind events. Southern Brazil was identified as the region of the country which has the highest frequency of both TS and non-TS strong winds.

7 ACKNOWLEDGEMENTS

Special thanks to Prof. Gilberto Fisch, Cel. Cleber Corrêa, Antonio Luis Carlos Neto, Prof. Ernani Nascimento and Vanessa Ferreira for assistance provided in their areas of expertese. This research is supported financially by the Brazilian National Council for Scientific and Technological Development (CNPq).

8 REFERENCES

1 A.M. Loredo-Souza, Meteorological events causing extreme winds in Brazil, Wind and Structures, 15 (No. 2) (2012), 177-188.
2 M.A.F. Silva Dias, An Increase in the Number of Tornado Reports in Brazil, Journal of Weather, Climate, and Society, 9 (No. 3) (2011), 209-217
3 A.M. Loredo-Souza, E.G Lima, M.B. Vallis, M.M Rocha, A.R Wittwer, M.G.K Oliveira, Full-scale downburst damage versus boundary layer wind tunnel pressures: a survey analysis, Proc. 8th Int. Coll. on Bluff Body Aerodyn. and Appl., Boston, MA, 2016.
4 A.P. Queiroz, C.S. Correa, Estudo de microexplosão e tornado no interior de São Paulo usando dados de radar meteorológico, Proc. XVII Simpósio de Aplicações Operacionais em Áreas de Defesa, São Jose dos Campos, Brazil, 2016.
5 A.M. Lored-Souza, M.J Paluch, Brazil storm Catarina: hurricane or extratropical cyclone?, Proc. 10th Am. Conf. on Wind Eng., Raton Rouge, LA, USA, 2005.
6 J.M.S. Vieira Filho, Velocidades máximas do vento no Brasil, Master's thesis, Universidade Federal do Rio Grande do Sul, 1975.
7 I.J. Padaratz, Velocidade básica do vento no Brasil, Master's thesis, Universidade Federal do Rio Grande do Sul, 1977.
8 BDC/ICEA, http://clima.icea.gov.br/clima/
9 REDEMET, http://www.redemet.aer.mil.br/
10 Ten Brig Ar R.R. Filho, Ordem do Dia, Dia Meteorológico Mundial 2015. http://www.redemet.aer.mil.br/?i=blog&y=2015&m=3, 19/03/2015.
11 D.A. Gatey, The analysis of extreme synoptic winds, Doctorate thesis, The University of Western Ontario, 2011.
12 DECEA, AIP: Publicação de Informação Aeronáutica, 2016.
13 INMET, http://www.inmet.gov.br/
14 CPTEC-INPE, Banco de dados meteorológicos, http://bancodedados.cptec.inpe.br/downloadBDM/
15 NCEI, DS3505, https://www7.ncdc.noaa.gov/CDO/dataproduct
16 M.D. Burton, A.C. Allsop, Predicting Design Wind Speeds from Anemometer Records: Some Interesting Findings, Proc. 11th Am. Conf. on Wind Eng., San Juan, Puerto Rico, 2009.

17 K.S. Kumar, C. Cini, V. Sifton, Assessment of design wind speeds for metro cities of India, Proc. 7th Int. Coll. on Bluff Body Aerodyn. and Appl., Shanghai, China, 2012.
18 S.C. Pryor, R.J. Barthelmie, D.T. Young, E.S. Takle, R.W. Arritt, D. Flory, W.J. Gutowski Jr., A. Nunes, J. Roads, Wind speed trends over the contiguous United States, J. of Geophysical Research, 114 (2009), D14105.
19 Weather Underground, https://www.wunderground.com/
20 V. Ferreria, Um estudo observacional de rajadas de vento geradas por tempestades severas no sul do Brasil, Master's thesis, Universidade Federal de Santa Maria, 2017.
21 DECEA, MCA 105-2, Manual de Estações Meteorológicas de Superfície, 2016.
22 DECEA, ICA 105-16, Códigos Meteorológicos, 2016.
23 A.T. DeGaetano, A Quality-Control Routine for Hourly Wind Observations, J. Atmos. Ocean. Tech., 14 (1997), 308-317
24 Cidade de Santarém, Forte tempestade causa estragos em Itaituba, Sep. 21, 2011, http://www.cidadedesantarem.com.br/2011/09/forte-tempestade-causa-estragos-em.html
25 M. Kasperski, A new wind zone map of Germany, J. Wind Eng. Ind. Aerodyn., 90 (2002), 1271-1287.

Wind load factors and design mean recurrence intervals of wind effects for use in the wind tunnel procedure

Emil Simiu[a], Adam Pintar[b], Dat Duthinh[c], and DongHun Yeo[d]

[a] National Institute of Standards and Technology, Gaithersburg, MD. 20899, USA, emil.simiu@nist.gov
[b] National Institute of Standards and Technology, Gaithersburg, MD. 20899, USA, adam.pintar@nist.gov
[c] National Institute of Standards and Technology, Gaithersburg, MD. 20899, USA, dduthinh@nist.gov
[d] National Institute of Standards and Technology, Gaithersburg, MD. 20899, USA, donghun.yeo@nist.gov

ABSTRACT

A 2004 Skidmore Owings and Merrill report notes that the ASCE 7 Standard is incomplete insofar as it provides no guidance on wind load factors appropriate for use with the Standard's wind tunnel procedure. The purpose of this paper is to contribute to such guidance. For a design mean recurrence interval of the wind effects N, the classical definition of wind load factor is

$$\gamma(N) \equiv \{1 + k\,\mathrm{COV}[p_{pk}(N)]\}$$

$$p_{pk}(N) \approx \gamma(N)\,\bar{p}_{pk}(N)$$

$$\mathrm{COV}[p_{pk}(N)] \approx \{\mathrm{COV}^2(E_z) + \mathrm{COV}^2(K_d) + \mathrm{COV}^2[G(\theta_m)] + \mathrm{COV}^2[C_{p,pk}(\theta_m)] + 2\,\mathrm{COV}^2[V(N)]\}^{1/2}$$

$$\bar{p}_{pk}(N) \approx a\,\bar{E}_z\,\bar{K}_d\,\bar{G}(\theta_m)\,\bar{C}_{p,pk}(\theta_m)\bar{V}^2(N)$$

where the covariance of the peak load p_{pk}, $\mathrm{COV}[p_{pk}(N)]$, accounts for uncertainties in the micrometeorological, wind directional, quasi-static and/or dynamics, aerodynamics and wind climatological elements that determine wind loads, the overbar denotes expectation, and k is established by consensus based on calibration against previous practice. A first observation based on those equations is that the mean of the peak aerodynamic coefficient $C_{p,pk}(\theta_m)$ in the last equation corresponds typically to a 57 percentile of is distribution (assumed to be of the Gumbel type), and that assuming that it corresponds to a 78 percentile, as is assumed by some practitioners, amounts to double-counting of its variability, which is attended to by its covariance. A second observation is that in estimating design wind loads, the various uncertainties should not be accounted for in isolation, for example by specifying peak pressure coefficients with percentage points higher than those corresponding to their expected values. Rather, to achieve risk-consistent designs, the uncertainties should be accounted for collectively, in terms of their joint effect on the design wind loading. For example, an uncertainty of 10 % in the aerodynamic coefficient C_{pk} will result in an uncertainty in the peak lower much lower than 10 %.

Based on the definition of the wind load factor, a simple, straightforward approach is presented that allows practitioners to use appropriate wind load factors applicable when those uncertainties are either the same as or different from those assumed in the development of the ASCE 7 Standard. Illustrations of the approach are presented for a variety of cases of practical interest. It was found that, in most cases, the wind load factor is not significantly different from the load factor explicitly or implicitly specified in the ASCE 7 Standard. Notably, the load factor is not affected significantly by errors associated with interpolations required in typical Database Assisted Design applications. However, if the available wind speed records are several times shorter than, say, 20 to 30 years, the wind load factor typically increases by amounts of the order of 15 %.

KEYWORDS

Aerodynamics, load factors, micrometeorology, safety, structural reliability, uncertainties, wind climatology, wind engineering.

Assessment of aerodynamic and aeroelastic performance improvement of computationally optimized bridge deck sections using LES and FSI simulations

T. H. Birhane, M.Sc.[a], G. T. Bitsuamlak, Ph.D., P.Eng.[b] and J. P. King, Ph.D., P.Eng.[c]

[a] *Department of Civil and Environmental Engineering, The University of Western Ontario, IGAB-Room#3N60 - London, ON, Canada; email: tbirhane@uwo.ca*
[b] *Department of Civil and Environmental Engineering, The University of Western Ontario, IGAB-Room#3N60 - London, ON, Canada; email: gbitsuam@uwo.ca*
[c] *Boundary Layer Wind Tunnel Laboratory, The University of Western Ontario, London, ON, N6A 5B9, Canada; email: jpck@blwtl.uwo.ca*

ABSTRACT

Long-span bridges are wind sensitive as a result of which their structural response developed from aerodynamic interactions with wind prevails over other environmental loads. Typical examples include but not limited to strong oscillations sustained by White Bronx Bridge in USA and Long Creek Bridge in Canada, and to a total collapse of the First Tacoma Narrows Bridge. To mitigate such unfavorable wind effects of the interaction, shape modifications of bridge decks such as partial streamlining, venting, and provision of appendages such as guide vanes have been implemented successfully. The advancement in computational power is enabling to simulate such interactions and to propose aerodynamic measures through automated procedure. This involves the use of CFD for aerodynamic analysis and numerical optimization tools for iterative selection of mitigation measures. One of the major challenges anticipated on application of numerical optimization to aerodynamic design of long-span bridge decks is the complexity associated with the selection and evaluation of objective functions. The evaluation of objective functions involves numerical simulation of turbulent wind flows and their interaction with bridge decks; which are governed by very complex equations "Navier-Stokes" (NSE) that are computationally expensive to solve. This is further complicated by the need to solve these equations repeatedly upon the modification of geometrical deck details as the optimization progresses. Regeneration of the CFD grid for every shape modification also adds to the complexity. To circumvent the anticipated high computational demand, we developed a computational aerodynamic shape optimization framework that adopts a two-tier approach. In the first tier reduced CFD models (simplified objective functions, and reduced flow models such as Reynolds Averaged NSE with the $k-\omega\ SST$ turbulence closure model) are used and hybrid optimization approaches are implemented to narrow-down the solution domain initially. In the second tier the CFD based aerodynamic analyses will progress to higher-accuracy, but computationally expensive simulations (such as three dimensional large eddy simulation, LES and fluid-structure interactions, FSI) for the narrowed-down solution domain. Optimal edge fairing shapes are designed for H-shaped bridge decks based on tier-1 approach, where the details are given in another paper by the same authors. This paper presents aerodynamic and aeroelastic factors that indicate aerodynamic performance improvement of H-shaped bridge decks due to the addition of edge fairings with numerically optimized shape. Static aerodynamic coefficients are evaluated using three-dimensional static section model and large eddy simulation for the flow. Aeroelastic responses, and flutter derivatives are evaluated using FSI analysis based on three-dimensional dynamic section model and unsteady $k-\omega\ SST$ turbulence model for the flow domain. The results of these refined analyses confirm the aerodynamic improvement as implied by tier-one approach of the proposed optimization framework.

KEYWORDS

Lon-span bridges, aerodynamics, computational fluid dynamics, optimization, fluid-structure interaction, flutter, aerodynamic derivatives.

13th Americas Conference on Wind Engineering
Gainesville, Florida, USA
May 21 - 24, 2017

CFD applications in Wind Engineering

Girma T. Bitsuamlak[a]

[a] *Department of Civil & Environmental Engineering /WindEEE Research Institute, Western University, ON, Canada, gbitsuam@uwo.ca*

ABSTRACT

Boundary layer wind tunnels are industry-wide accepted wind engineering tools. Recently, large scale wind testing facilities such as WindEEE Dome and Wall of Wind have been added to the list of tools for wind engineering. Steadily CFD is making important inroads augmenting these experimental approaches as well. For a wind engineer, the developments in CFD and the availability of a high performing computer are greatly beneficial. To explain these benefits, examples from different stages of Alan Davenport's wind loading chain (e.g. wind climate and wind load) are presented. In an application concerned with the characterization of incoming flows (wind climate). CFD can be as effective as wind tunnel tests for simulation of flows over terrain with topographical or orographic features, or inhomogeneous surface roughness in urban areas. This type of application is important for both structural engineering and wind energy applications. A second type of application is concerned with the estimation of preliminary global aerodynamic loads on structural systems and local loads on cladding. When augmented with numerical optimization, these preliminary load estimations create a new application area: aerodynamic optimization. Although these types of studies have been carried out in wind tunnels, CFD based procedures are unique because the extent of the search area for the targeted local or global modifications of the building shapes (or bridge sections) can lead to more optimal solutions. The final shape selection can still be tested in a wind tunnel to generate accurate design wind loads. Another wind load application example is the estimation of the level of pressure equalization in multi-layered air-permeable envelope systems such as roof pavers. CFD based applications can also be used in the estimation of wind speeds in pedestrian areas, where they have already been proven to complement wind tunnel tests. In addition, the integration of CFD with building energy simulations has proven to decrease the margin of error found in the use of energy simulation tools alone. This is primarily attributed to the ability for CFD to produce realistic site specific boundary conditions for energy simulators.

KEYWORDS

Computational fluid dynamics (CFD), wind load, aerodynamic optimization, pedestrian level wind, aerodynamic roughness, topography, wind engineering, building energy simulation

The 13th Americas Conference on Wind Engineering (13ACWE)
Gainesville, Florida USA, May 21-24, 2017

13th Americas Conference on
WIND ENGINEERING

Cyber-physical systems approach to optimization in wind engineering: parapet wall design

Michael L. Whiteman [a], Brian M. Phillips [b], Pedro L. Fernández Cabán [c], Forrest J. Masters [d], Jennifer A. Rice [e], Justin R. Davis [f]

[a] *University of Maryland, College Park, Maryland, USA, mlwii@umd.edu*
[b] *University of Maryland, College Park, Maryland, USA, bphilli@umd.edu*
[c] *University of Florida, Gainesville, Florida, USA, plfernandez@ufl.edu*
[d] *University of Florida, Gainesville, Florida, USA, masters@eng.ufl.edu*
[e] *University of Florida, Gainesville, Florida, USA, jrice@ce.ufl.edu*
[f] *University of Florida, Gainesville, Florida, USA, justin.r.davis@essie.ufl.edu*

ABSTRACT: This paper explores the use of cyber-physical systems (CPS) for optimal design in wind engineering. The approach combines the accuracy of physical wind tunnel testing with the ability to efficiently explore a solution space using numerical optimization algorithms. The approach is fully automated, with experiments executed in a boundary layer wind tunnel (BLWT), sensor feedback monitored by a high-performance computer, and actuators used to bring about physical changes in the BLWT. Because the model is undergoing physical change as it approaches the optimal solution, this approach is given the name "loop-in-the-model" testing.

The building selected for this study is a low-rise structure with a parapet wall of variable height. Parapet walls alter the location of the roof corner vortices, alleviating large suction loads on the windward facing roof corner and edges and setting up an interesting optimal design problem. In the BLWT, the model parapet height is adjusted using servo-motors to achieve a particular design. The model surface is instrumented with pressure taps to measure the envelope pressure loading. The taps are densely spaced on the roof to provide sufficient resolution to capture the change in roof corner vortex formation. Experiments are conducted using a boundary BLWT located at the University of Florida Natural Hazard Engineering Research Infrastructure (NHERI) Experimental Facility.

The proposed CPS approach enables the optimal solution to be found quicker than brute force methods, in particular for complex structures with many design variables. The parapet wall provides a proof-of-concept study with a single design variable that has a non-monotonic influence on a structure's wind load. This study focuses on envelope load effects, seeking the parapet height that minimizes roof and parapet wall suction loading. Implications are significant for more complex structures where the optimal solution may not be obvious and cannot be reasonably determined with traditional experimental or computational methods.

KEYWORDS: Cyber-physical systems, optimization, boundary-layer wind tunnel, parapet wall, NHERI

1 INTRODUCTION

Boundary layer wind tunnels (BLWT) remain a leading tool in wind engineering to characterize the pressure loading on wind-sensitive structures. In particular, BLWT testing is valuable when studying new structures for which the simplified provisions of ASCE 7-10 are inadequate or computational fluid dynamics approaches cannot be applied with confidence [1]. While BLWT testing has remained an industry standard for decades, there have been many recent advances in computationally-based optimization techniques for structural design. Meta-heuristic algorithms such as particle swarm and genetic algorithms are problem-independent algorithms that efficiently explore a complex solution space, providing new opportunities to study multi-variate and multi-objective optimization problems. While new optimization techniques have promise for delivering cost-effective design solutions for wind-sensitive structures, they must be combined with a method such as BLWT testing to evaluate the candidate solutions.

This study proposes the use of cyber-physical systems (CPS) for optimal design in wind engineering. The study will create a proof-of-concept that demonstrates the potential for replacing laborious trial-and-error design approaches with cyberinfrastructure-augmented BLWT modeling that produces optimal designs faster than purely experimental methods and with a higher degree of realism than purely computational methods. The approach is fully automated, with experiments executed in a BLWT, sensor feedback monitored by a high-performance computer, and optimization techniques used to bring about physical changes to the structural model in the BLWT. Because the model is undergoing physical change as it approaches the optimal solution, this approach is given the name "loop-in-the-model" testing. The CPS framework is illustrated in Figure 1.

The building selected for this study is a low-rise structure with a parapet wall of variable height. The sharp edges on bluff bodies, in particular windward roof edges on low-rise structures, cause a separation of the boundary layer and generate vortex flow with large suction loading. These loads are particularly severe for oblique approaching wind angles. Parapets are common on industrial and commercial buildings and help to alleviate extreme roof wind suction loads [2-5]. Parapet walls alter the location of the roof corner vortex and mitigate extreme corner and edge suction loads on the roof of the building. Conversely, parapet walls increase the downward roof loads which combine with other roof loads. ASCE 7-10 contains basic design guidance with regard to parapet loading, however it is not refined enough to infer what constitutes an optimal height or configuration.

In this study, the model parapet height is adjusted automatically using servo-motors to reach a particular candidate design. The building envelope is instrumented with pressure taps to measure the envelope pressure loading. The taps are densely spaced on the roof to provide sufficient resolution to capture the change in roof corner vortex formation. Experiments are conducted using a BLWT located at the University of Florida Natural Hazard Engineering Research Infrastructure (NHERI) Experimental Facility.

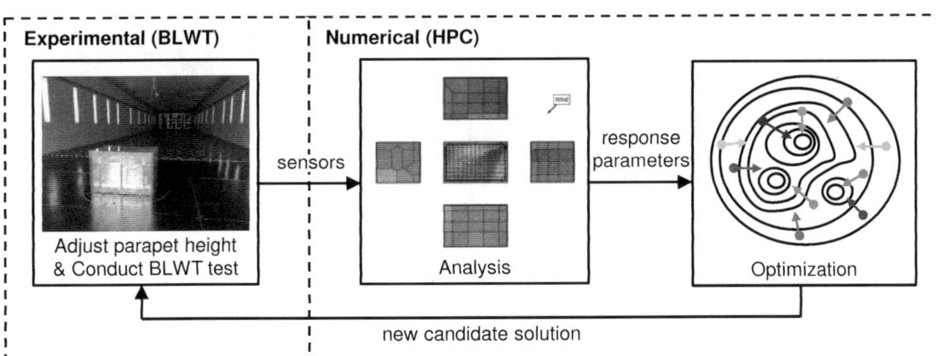

Figure 1. Diagram of CPS framework for optimal design under wind hazards

2 PARTICLE SWARM OPTIMIZATION

Particle swarm optimization (PSO) is a population-based stochastic optimization technique. PSO mimics social behavior where a population of individuals adapts to its environments by discovering and jointly exploring promising regions. This swarm intelligence method is based on the simulation of social interactions of members of a species, such as the movement of flocks of birds, schools of fish, and swarm of bees. Particle swarm optimization was inspired by evolutionary programming, genetic algorithms, and evolution strategies and shares similarities with genetic algorithms and evolutionary algorithms.

Particle swarm optimization is a non-gradient-based, meta-heuristic optimization method [6]. Non-gradient-based optimization techniques are especially useful in solving problems in structural engineering due to their versatility in handling multiple design variables. Particle swarm optimization efficiently explores a large number of candidate solutions over a large search space without prematurely converging, which can lead to non-intuitive solutions. The basic technique is easy to program because it is an inherently iterative process reliant on only a few formulas to govern the iterations. Also, the problem definition does not require continuity and is capable of handling nonlinear, nonconvex design spaces. Complexities can arise in the analysis of candidate solutions, the calculation of the objective function, application of constraints, and ensuring sufficient balance between exploration and exploitation of the design space. As a meta-heuristic method, there is no guarantee that a global optimal solution, or even bounded solution will be found [7]. Because the solution is not necessarily optimal, the solution from a PSO algorithm is more precisely termed a sub-optimal solution. The technique is relatively new with limited studies in structural engineering; however research is actively being conducted to improve this optimization approach with specific structural engineering considerations.

3 EFFECTS OF WIND ON LOW-RISE BUILDINGS WITH PARAPETS

Building detailing has a large influence on the distribution of pressures over the roof surface in both magnitude and direction. With regard to structures with parapet walls, most research focuses on local pressure distributions on the roof surface, as in components and cladding. Studies also include the use of parapets with non-uniform or modified geometries to reduce the extreme suction loads caused by the corner vortices [2]. Additionally, a few studies consider the effect of parapets on the underlying structural members [3, 9]. Recent studies have uncovered inconsistencies with previous research. These inconsistencies are likely the result of an insufficiently dense region of pressure taps in the upwind corner of the roof of the model, leading to the true peak suction pressure values not being measured [2, 3, 10]. Thus, it is essential to have a very high density of pressure taps in the upwind corner region to ensure that the peak suction pressures are captured.

Wind approaching at oblique angles to a building with a flat roof produces strong vortices near the upwind corner and edges of the roof [10]. These vortices are similar to the vortices that are produced at the leading edge of delta type wings and, as such, are also known as delta wing vortices. These vortices create an area of high suction on the surface of the roof near the corner [11]. Solid, continuous perimetric parapets taller than 1 m act to reduce both the mean and peak pressure coefficients in the corner region of these buildings. Most building codes, such as ASCE 7-10, allow for a pressure reduction over different regions of a roof in the presence of parapets; however there has not been extensive research conducted regarding accurate regions of reduction based upon the geometry of the building and parapet or on the optimal height of a parapet for a given low-rise building [1]. Additionally, research has primarily focused on the corner zones of roofs with limited research focusing on the edge and interior zones. The research regarding the edge and interior zones has mainly focused on mitigating local loading through the use of alternative geometries and not much regarding the effect of different heights of solid, perimetric parapets or on the optimal height of solid, perimetric parapets [5].

4 MODEL DEVELOPMENT

4.1 *Design variable*

Candidate design solutions must be physically created in the BLWT such that their envelope wind loads are accurately measured. For this proof-of-concept study, a single controllable design variable is sufficient to demonstrate the cyber-physical approach to optimization. Additionally, by limiting the study to a single design variable, unnecessary mechanical complexity is avoided and focus is instead placed on the optimization framework.

The design parameter selected is the parapet wall height of a low-rise building. The outer wall of the model is actuated by four stepper motors, one at each corner of the model. The inner core of the model remains stationary, maintaining a constant building height. As the outer wall rises above the inner model, a parapet wall is created. Strips made from Teflon PTFE are used between the inner model and outer wall to achieve smooth linear actuation. A foam gasket is used between the outer wall and the turntable to allow the outer wall to move while preventing air from leaking around the model. The model is shown in Figure 2, including the inner model (stationary) and outer wall (vertically movable).

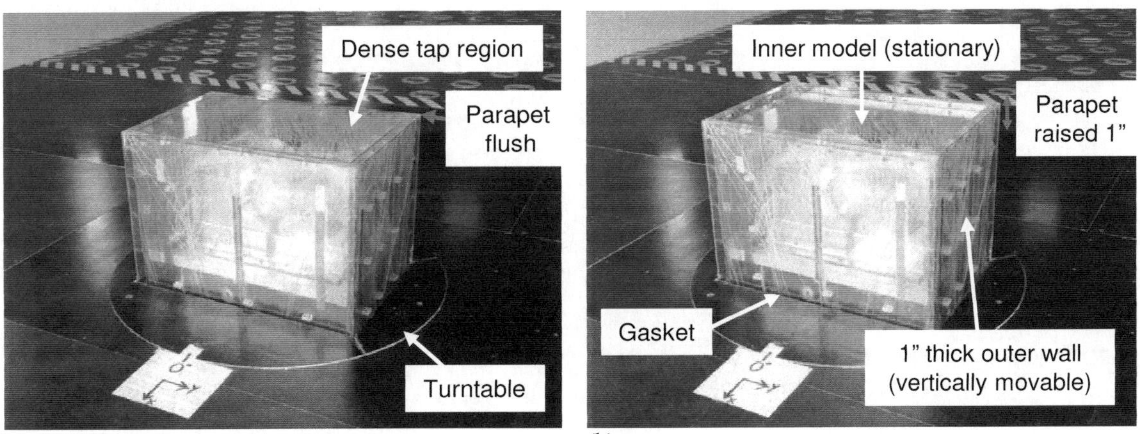

(a) (b)

Figure 2. (a) Building model with a 0 inch parapet wall and (b) a 1 inch parapet wall

The linear actuation of the outer wall around the inner core of the model is achieved using of Nanotec stepper motors. These motors are connected to the outer wall using polycarbonate triangular supports installed in the bottom corners. A PVC pipe is installed on the outside of the drive shaft of the stepper motor to protect the shaft from coming into contact with any urethane tubing. The stepper motor and its installation are shown in Figure 3. The setup for controlling the stepper motors is given in Figure 4. Data (i.e., commands from the coordinating computer) and power are passed through a slip ring on the BLWT turntable. A Raspberry Pi 3 within the turntable then sends commands to the stepper motor controllers, which in turn actuate the stepper motors. Encoders on the stepper motors provide feedback to ensure the desired displacement is reached in all four stepper motors.

(a) (b)

Figure 3. (a) Stepper motor and (b) stepper motor installed in corner of parapet wall with PVC shield

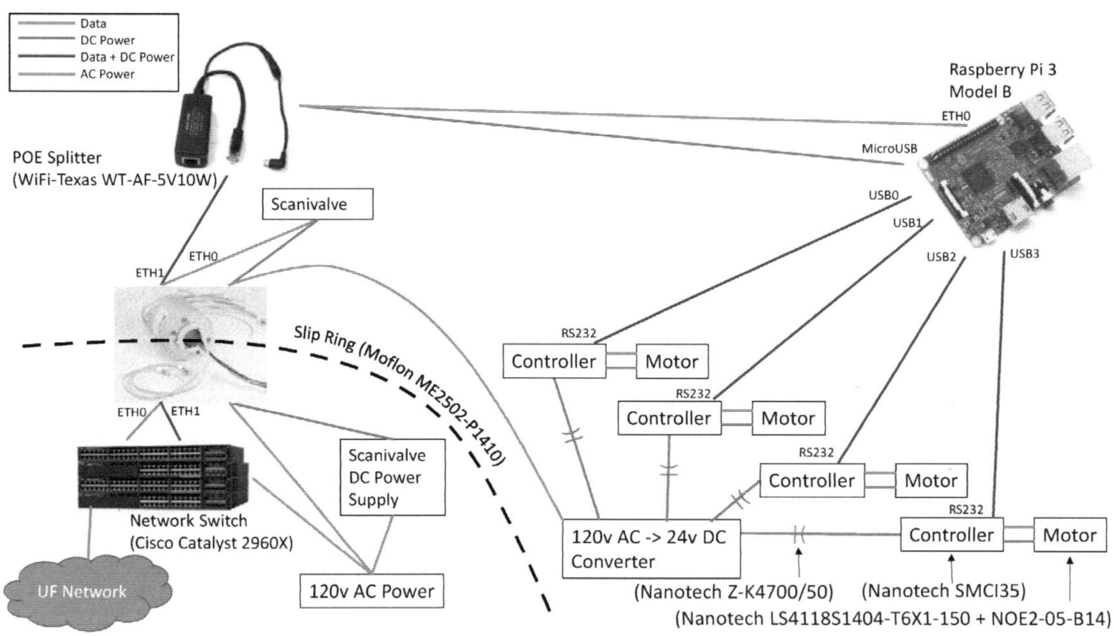

Figure 4. Wiring diagram for stepper motor control

4.2 *Model Geometry*

The low-rise building with a parapet wall is modeled after a two-story office building. The model and outer wall actuation system must fit within the 40 inch diameter turntable. Additionally, a length-to-width ratio of 1.5 is desired to create a rectangular model building. Model dimensions are selected as 29.25 inches × 19.50 inches in plan with a height of 20 inches.

By actuating the outer wall, a parapet wall of up to 5 inches model-scale is created. Urethane tubing and pressure taps are installed on the outer and inner sides of the parapet wall. This requires a total thickness of the model parapet wall (and thus outer wall) of at least 1 inch to accommodate the thickness of polycarbonate sheets, tubulation (metal ends of the urethane tubing), and minimum bend radius of 0.5 inches for the 0.063 inch diameter urethane tubing. The pressure taps on the outer and inner parapet walls are staggered to permit a thinner model parapet wall. The taps and tubing of the inner model and outer wall are shown in Figure 5.

Based on the physical model dimensions and the target two-story office building, a model to full-scale ratio of 1:18 is selected. This corresponds to a building with full-scale dimensions of 29.625 feet × 44.4375 feet in plan, 30 feet tall, and a parapet that is 1.5 feet thick. According to the Building Code Requirements for Masonry Structures, parapet walls should have a thickness of at least 8 inches [12]. The building model represents a realistic two-story full-scale building with a two by three bay steel frame.

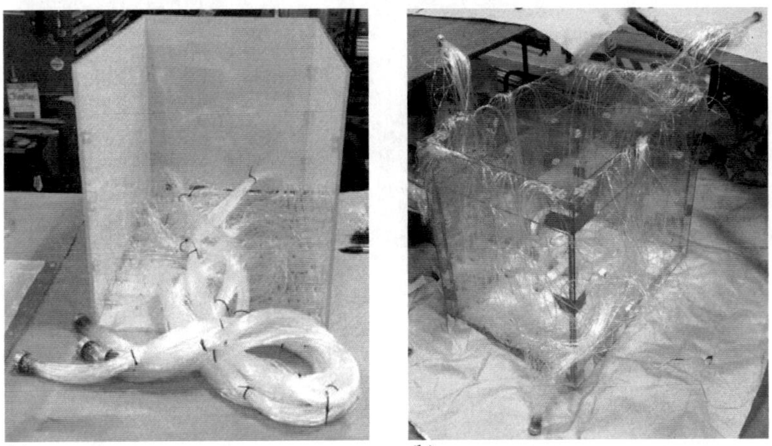

(a) (b)

Figure 5. (a) Taps and tubing of inner model (stationary) and (b) taps and tubing of outer wall (movable)

5 EXPERIMENTAL SETUP

5.1 *Experimental equipment*

The pressures on the model building surfaces are measured using Scanivalve ZOC33 pressure scanners. Experiments are conducted using a BLWT located at the University of Florida Natural Hazard Engineering Research Infrastructure (NHERI) Experimental Facility. The BLWT is 6.1 m wide with a 1 m turntable centered along the 6.1 m width 31.75 m downwind of 8 fans. The fans are kept at 1050 RPM for all testing, which corresponds to a reference height velocity of approximately 14 m/s. The the model building installed in the BLWT is shown in Figure 6.

Figure 6. Boundary layer wind tunnel with model low-rise building, upwind view

5.2 *Tap tributary areas*

The tributary area of a particular tap describes the area over which the tap's pressure is assumed to be acting. In this model, tap locations are variable due to the movable outer wall. Based on the parapet wall height, exposed tap locations and surface areas are first calculated. Then, the tap tributary areas are calculated using Voronoi diagrams derived from Delaney triangulation. This process is both reproducible and automated, which is particularly important because the geometry of the building changes with every candidate solution. The taps and tributary areas for the model with a parapet wall of 5 inches are depicted in the flattened view of Figure 7. The walls of the building are given by Surfaces 1 to 4. As the walls extend above the roof (from actuation), Surfaces 1 to 4 also form the outer parapet walls. The inner parapet walls are given by Surfaces 6 to 9. The edges that join the outer walls (Surfaces 1 to 4) and the inner parapet walls (Surfaces 6 to 9) in Figure 7 are at the same height in the model. I.e., the walls are flattened by rotating about the top of the parapet wall. Surfaces 5 and 10 are the top of the parapet wall and the roof, respectively. As the parapet height increases, the tributary areas for both the outer wall and inner parapet walls increase while the tributary areas for the top of the parapet wall and the roof remain constant.

Parapet: 5 in.

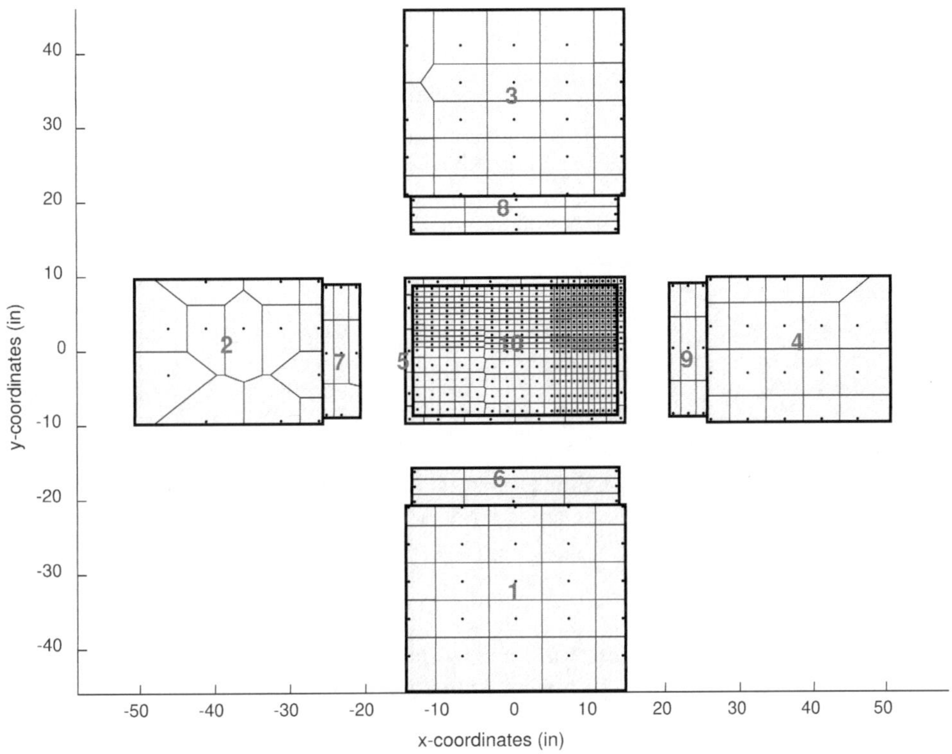

Figure 7. Tap locations and tributary areas on a flattened representation of the model with a parapet of 5 inches

5.3 *Wind simulation*

Simulation of upwind terrain roughness is performed via the Terraformer, an automated roughness element grid that rapidly reconfigures the height and orientation of 1116 roughness elements in a 62 × 18 grid to achieve desired upwind terrain conditions. The grid has a fetch length of 18.3 m. Dimensions of the elements are 5 cm by 10 cm, and they are spaced 30 cm apart in a staggered pattern. Height and orientation can be varied from 0-160 mm and 0-360 degrees, respectively. For this study, the Terraformer was configured to simulate open terrain for the given geometric scale (1:18).

Prior to placing the model in the tunnel, flow measurements were taken at the center of the test section using an automated gantry system instrumented with four Turbulent Flow Instrumentation Cobra pressure probes that measure u, v, and w velocity components and static pressure. For this study, roughness elements were raised to 20 mm and oriented with the wide edge perpendicular to the flow. Figure 8 includes the mean velocity profile and the measured longitudinal turbulence spectra at a height of 610 mm. The measured spectra was compared with the power spectra model in ESDU [13], and first derived by von Kármán for isotropic turbulence [14]. The mean velocity profile was normalized by the reference mean wind velocity U_{ref} measured at a height z_{ref} = 1.48 m. A roughness length estimate of 1.59 mm was obtained from a nonlinear least-squares fit of the log law in the inertial-sublayer (ISL) region (z ~ 150-900 mm), following the curve-fitting method in Karimpour et al. [15]. This results in an equivalent full-scale roughness length of 0.029 m, which is within the range of open terrain as defined in ASCE 7-10.

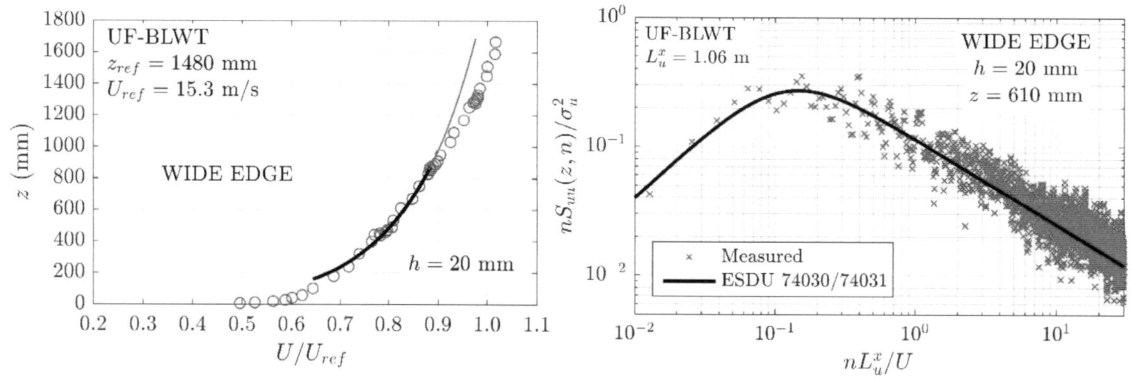

Figure 8. Mean velocity profile and longitudinal turbulence spectra (z = 610 mm) measured at the center of the test section for h = 20 mm and a wide edge windward element orientation.

5.4 *Pressure coefficients*

Differential pressures from 512 taps were measured simultaneously and sampled at 625 Hz. Data was collected for 120 seconds, corresponding to approximately 660 seconds full-scale assuming a basic wind speed of 40 m/s. Pressure coefficients are referenced to the velocity pressure at the model eave height. This velocity pressure was obtained indirectly by applying a reduction factor to pitot tube measurements at the freestream (z = 1.48 m). Maximum and minimum pressure coefficients were estimated from each tap pressure time history using a Fisher-Tippett Type I (Gumbel) distribution. The C_p time history was truncated into 50 segments of equal length. The peak maximum and minimum pressure coefficients from each segment were then taken, and the 78th percentile is then used to estimate the maximum and minimum C_p values.

6 OPTIMIZATION SETUP

The optimization problem is physically constrained by the model-scale minimum and maximum achievable parapet height of 0 and 4.5 inches, respectively. The problem-specific particle swarm optimization (PSO) parameters of $w, c_1,$ and $c_2$were selected to be equal in magnitude so that an equal weight would be placed on the particle's inertia, trust in itself, and trust in the swarm. After initial tests, a magnitude of 0.5 was selected for all problem-specific PSO parameters. Considering the time limits on experimental resources, a balance in needed between sufficient particles to create the PSO swarm effect and sufficient iterations to see convergence. Based on an estimated two minutes per BLWT run, one minute to set up the BLWT run, and a day of testing, five particles were selected.

The objective function is selected as a minimization of the suction on the roof, inner parapet walls, and top of the parapet considering all wind angles. As the parapet height increases, the suction decreases for the roof surface and top of the parapet wall and increases for the inner parapet wall surfaces. Critical minimum C_p values are observed for the roof, inner parapet wall, and top of parapet at approach wind angles of 45° and 90° (Figures 9 and 10). To minimize the number of BLWT runs, each candidate solution will only be evaluated at 45° and 90° with the dense roof taps upwind.

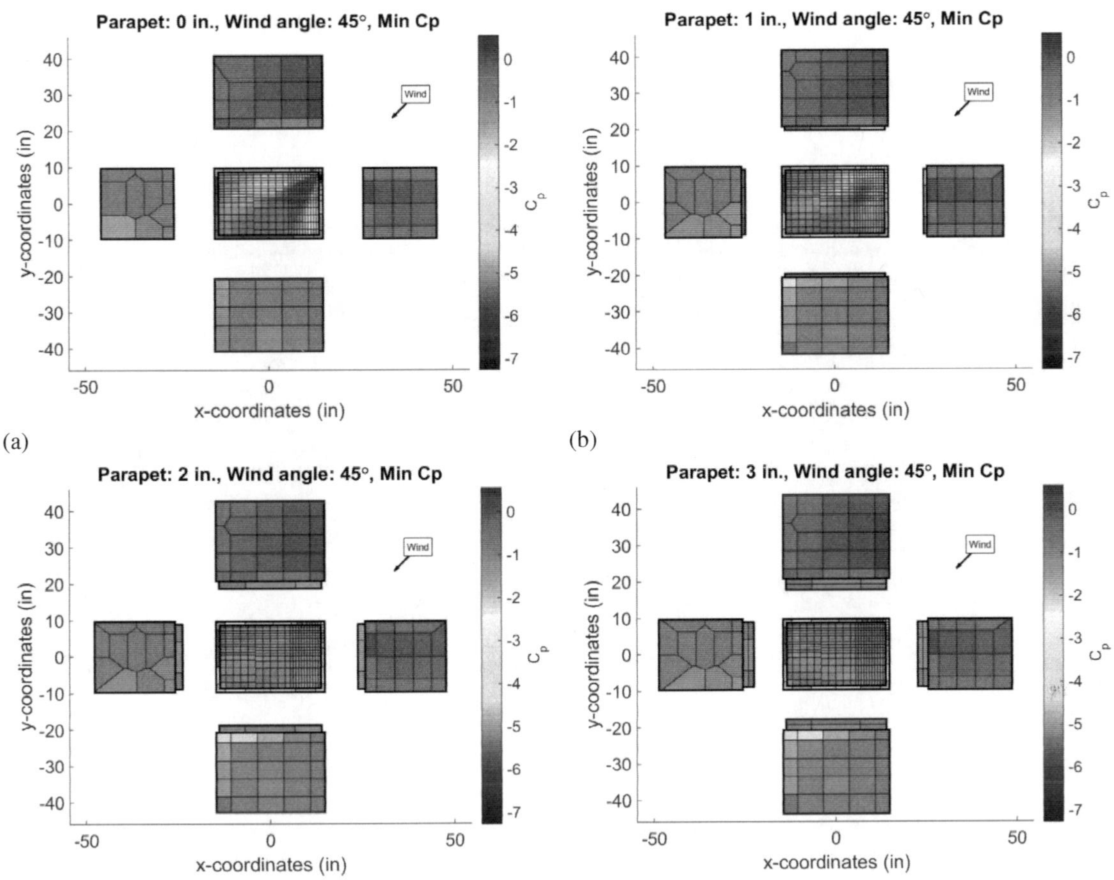

Figure 9. Minimum C_p for 45°, (a) 0 inch parapet, (b) 1 inch parapet, (c) 2 inch parapet, and (d) 3 inch parapet

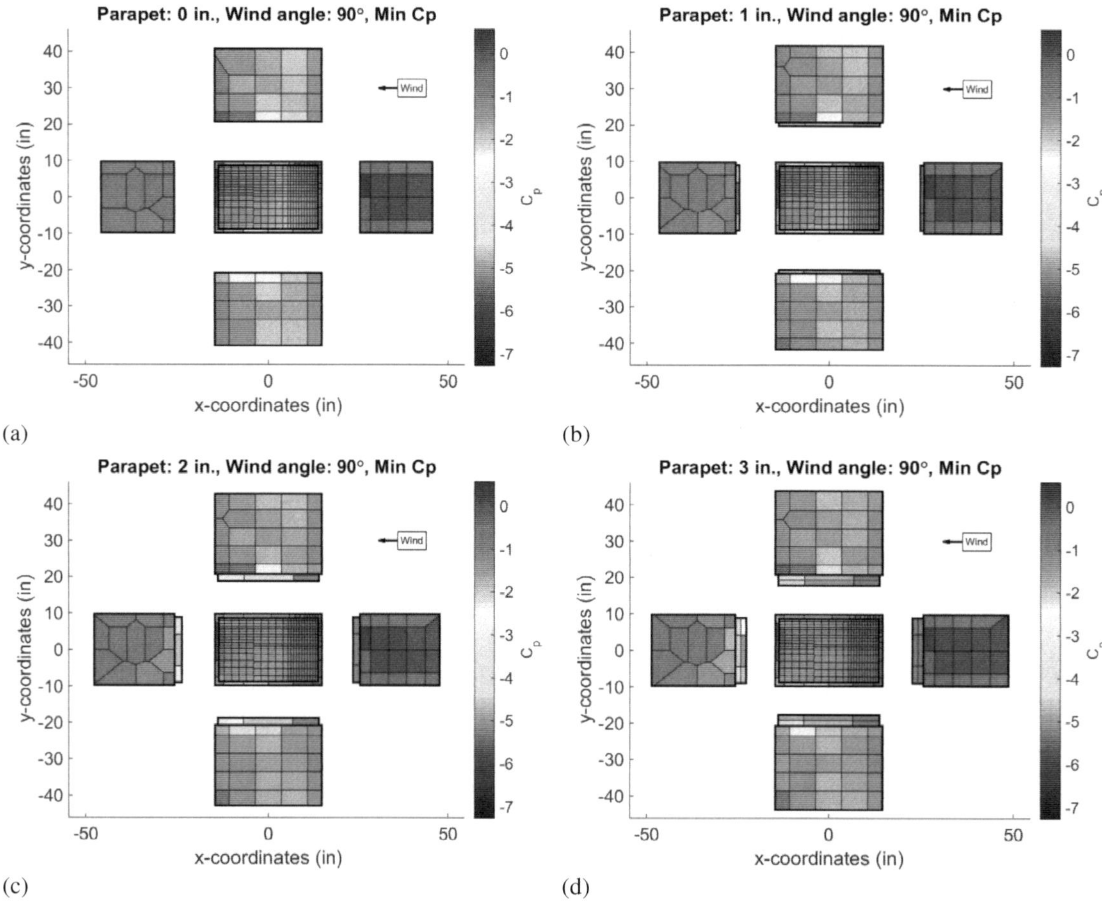

Figure 10. Minimum C_p for 90°, (a) 0 inch parapet, (b) 1 inch parapet, (c) 2 inch parapet, and (d) 3 inch parapet

7 OPTIMIZATION RESULTS AND ANALYSIS

A total of 13 design iterations were conducted for the 5 particles. The position of the particles was initially uniformly distributed across the range of positions. The convergence of the particles towards the optimum height of 2.69 inches is shown in Figure 11a. Four of the five particles converged to the global best cost. The one particle that does not converge is likely due to the particle being attracted to both its personal best cost (achieved at iteration 1) and the global best cost. The global best cost for each iteration are shown in Figure 11b. Points with both particle and cost identified represent an update to the global best cost. Figures 12 and 13 depict the envelope plot of the minimum C_p for the optimal parapet height at 45° and 90° respectively. This illustrates the balance in minimum C_p on the roof (Figure 12) and inner parapet wall surfaces (Figure 13). This balance is expected because the suction on the roof, inner parapet walls, and top of the parapet were given equal weight in the objective function.

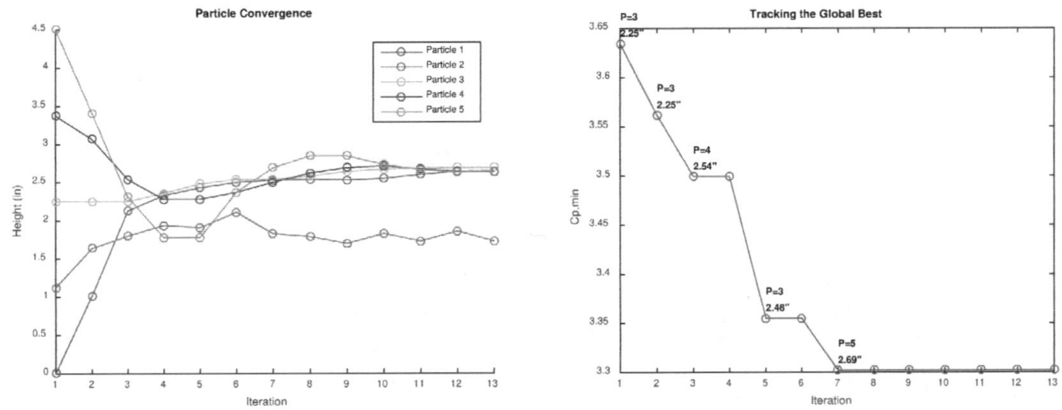

(a) (b)

Figure 11. (a) Particle convergence at each iteration and (b) global best cost at each iteration

The optimal result corresponds to a full-scale parapet height of 4.04 feet, an otherwise non-intuitive design. This parapet height simultaneously minimizes suction on the roof and inner parapet walls. According to the Building Code Requirements for Masonry Structures, the height of structural parapets should not exceed 3 times their thickness [12]. The optimal height found satisfies this limit of 4.5 feet as applied to the current building.

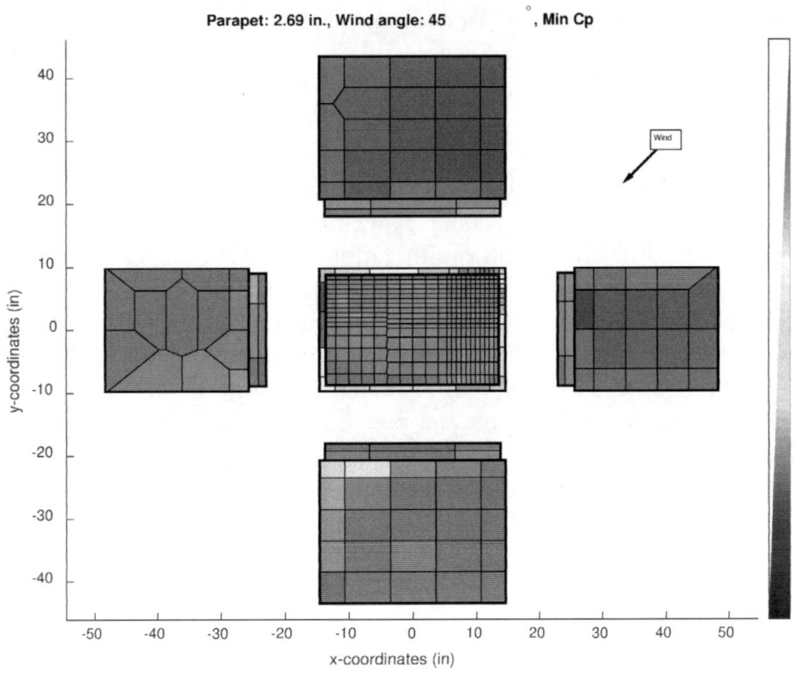

Figure 12. Minimum C_p for optimal parapet height, 45° wind angle shown

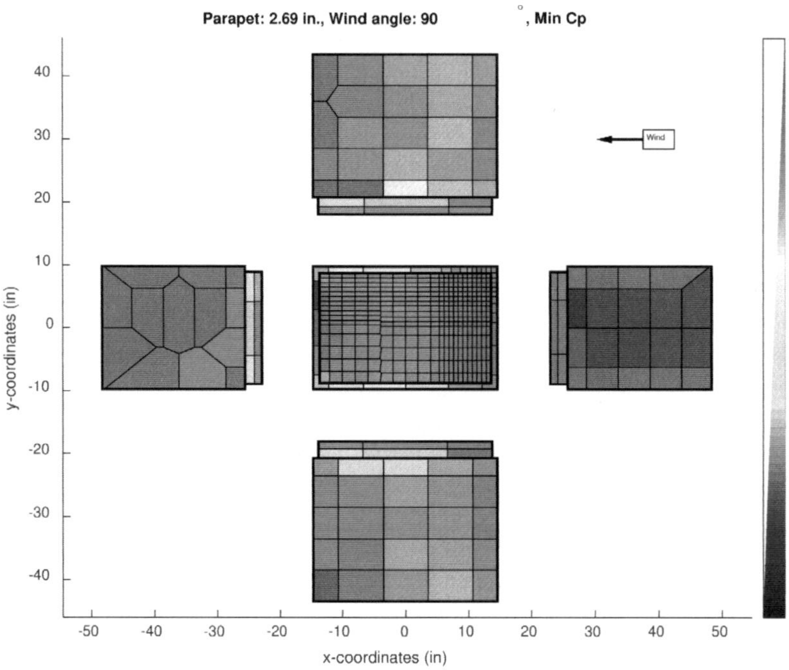

Figure 13. Minimum C_p for optimal parapet height, 90° wind angle shown

8 CONCLUSIONS

This study investigates the effect of wind on low-rise buildings with a solid parapet of variable height. The main goal is to develop a proof-of-concept cyber-physical approach to the optimal design of wind-sensitive structures. Particle swarm optimization (PSO) was used to explore the design space, producing of candidate solutions for evaluation in the BLWT. Particle swarm optimization algorithms are not problem specific, do not require gradients, and handle multiple design variables and constraints. In combination with BLWT, an accurate assessment of candidate solutions was achieved. The PSO-based approach was successful guiding the physical structure to an optimal state. Based on the objective function and constraints chosen, an optimal parapet height of 2.69 inches model-scale and 4.04 feet full-scale was found for the low-rise structure studied. Implications are significant for more complex structures where the optimal solution may not be obvious and cannot be reasonably determined with traditional experimental or computational methods.

9 ACKNOWLEDGEMENTS

This material is based upon work supported by the National Science Foundation (NSF) under Grant No. 1636039. Any opinions, findings, and conclusions or recommendations expressed in this material are those of the authors and do not necessarily reflect the views of NSF. The authors also acknowledge the NSF NHERI awardee that contributed to the research results reported within this paper under Grant No. 1520843: Natural Hazards Engineering Research Infrastructure: Experimental Facility with Boundary Layer Wind Tunnel, Wind Load and Dynamic Flow Simulators, and Pressure Loading Actuators (University of Florida).

10 REFERENCES

1 ASCE, *Minimum Design Loads for Buildings and Other Structures.* (2010). Reston, VA: American Society of Civil Engineers.

2 Kopp, Gregory A., et al. "Wind effects of parapets on low buildings: Part 1. Basic aerodynamics and local loads." *Journal of Wind Engineering and Industrial Aerodynamics*, vol. 93, no. 11, 10 Oct. 2005, pp. 817–841.

3 Kopp, Gregory A., et al. "Wind effects of parapets on low buildings: Part 2. Structural loads." *Journal of Wind Engineering and Industrial Aerodynamics*, vol. 93, no. 11, 10 Oct. 2005, pp. 843–855.

4 Kopp, Gregory A., et al. "Wind effects of parapets on low buildings: Part 3. Parapet loads." *Journal of Wind Engineering and Industrial Aerodynamics*, vol. 93, no. 11, 10 Oct. 2005, pp. 857–872.

5 Kopp, Gregory A., et al. "Wind effects of parapets on low buildings: Part 4. Mitigation of corner loads with alternative geometries." *Journal of Wind Engineering and Industrial Aerodynamics*, vol. 93, no. 11, 10 Oct. 2005, pp. 873–888.

6 Talbi, El-Ghazali. *Metaheuristics: From Design to Implementation.* Hoboken (NJ): John Wiley & Sons, 2009. Print.

7 Perez, R. E., & Behdinan, K. (2007). Particle Swarm Optimization in Structural Design. *Swarm Intelligence: Focus on Ant and Particle Swarm Optimization*, 373-377.

8 He, S., et al. "An improved particle swarm optimizer for mechanical design optimization problems." *Engineering Optimization*, vol. 36, no. 5, 2004, pp. 585–605.

9 Stathopoulos, T., et al. "Wind loads on parapets." *Journal of Wind Engineering and Industrial Aerodynamics*, vol. 90, no. 4-5, 2002, pp. 503–514.

10 Kind, R. J. "Worst suctions near edges of flat rooftops with parapets." *Journal of Wind Engineering and Industrial Aerodynamics*, vol. 31, no. 2-3, Dec. 1988, pp. 251–264.

11 Pindado, S., and J. Meseguer. "Wind tunnel study on the influence of different parapets on the roof pressure distribution of low-Rise buildings." *Journal of Wind Engineering and Industrial Aerodynamics*, vol. 91, no. 9, 2003, pp. 1133–1139.

12 ACI/ASCE/TMS, *Building code requirements and specification for masonry structures: containing Building code requirements for masonry structures (TMS 402-11/ACI 530-11 / ASCE 5-11), Specification for mason structures (TMS 602-11 / ACI 530.1-11 / ASCE 6-11) and companion commentaries / developed by the Masonry Standards Joint Committee (MSJC).* Boulder, Co., The Masonry Society, 2011.

13 ESDU. (1974). Characteristics of atmospheric turbulence near the ground. Part I: definitions and general information, *Engineering Sciences Data Unit*, Itm. No. 74030, 74031, London, UK.

14 Karimpour, A., Kaye, N. B., and Baratian-Ghorghi, Z. (2012). "Modeling the neutrally stable atmospheric boundary layer for laboratory scale studies of the built environment." *Building and Environment*, 49, 203-211

15 Von Karman, T. (1948). "Progress in the statistical theory of turbulence." *Proceedings of the National Academy of Sciences*, 34(11), 530-539.

13th Americas Conference on Wind Engineering
Gainesville, Florida, USA
May 21 - 24, 2017

13th Americas Conference on
WIND ENGINEERING

Real-Time Aerodynamics Hybrid Simulation: A Novel Aeroelastic Wind-Tunnel Model for Flexible Bridges

Teng Wu[a], Mettupalayam Sivaselvan[b]

[a] *University at Buffalo, Buffalo, NY 14260, USA,* tengwu@buffalo.edu
[b] *University at Buffalo, Buffalo, NY 14260, USA,* mvs@buffalo.edu

ABSTRACT

Regardless of the rapid development in computational fluid dynamics and semi-empirical analysis using parameters from sectional model tests, wind tunnel results from full aeroelastic models are still a final check for the wind design of long-span cable supported bridges. Although it is well known that the wind-induced bridge responses such as buffeting, flutter and vortex-induced vibration are typically multi-mode phenomena, full-bridge aeroelastic models can only accurately simulate the first few modes due to several inherent limitations. To advance aeroelastic modeling of flexible bridges in the wind tunnel, real-time hybrid simulation (RTHS) techniques, popularly used in the earthquake engineering community, are introduced here.

During a closed-loop RTHS test, the whole structure is partitioned into experimental and computational substructures, and the interactions among them are implemented by means of shake tables and dynamic actuators. A slightly different version of the RTHS method is developed here for the particular purpose of using it in aeroelastic modeling of flexible bridges. Specifically, the "skeleton" of the full-bridge model, characterizing the dynamic properties (e.g., mass, damping and stiffness of the structure), is numerically simulated based on computational structural dynamics, while the "skin" of the full-bridge model, characterizing the aerodynamic and aeroelastic properties, is physically modeled in the wind tunnel. At each time step, the aerodynamic inputs (gusts) and aeroelastic inputs (motions) to the bridge "skin" are obtained from the boundary-layer wind tunnel and simulation outputs of the bridge "skeleton", respectively. On the other hand, the dynamic inputs to the bridge "skeleton" are from the measured forces (and moments) on the bridge "skin". The interactions between the bridge "skeleton" and "skin" is accomplished based on a control system consisting of sensors, a network of electromagnetic and piezoelectric actuators, and controllers. With advancing of time steps, the time history of the wind-induced bridge responses could be obtained at the end of the proposed real-time aerodynamics hybrid simulation (RTAHS).

In this study, a long-span cable-stayed bridge example is utilized to demonstrate the simulation efficiency and fidelity of this new RTAHS methodology. The line-element-based 3-D finite element model with equivalent sectional properties, which could well capture the global dynamic behaviors of bridges, is employed for simulation of the "skeleton". Since the wind loadings on the towers and cables are relatively insignificant (or could be simply characterized using quasi-steady theory), only the bridge deck is considered in the modeling of the "skin" in the wind tunnel. The new RTAHS methodology is made possible by recent developments in control algorithms we have made at the University at Buffalo. The RTAHS implementation will utilize the same platform as the past work, based on LabVIEW RT. The system will operate in real time in the time scale of the wind tunnel model.

KEYWORDS

Wind tunnel, aeroelastic model, real-time hybrid simulation, bridges

Computational Fluid Dynamics Analysis on the Wake Effect of a Bridge Tower on High Side Vehicles

Nasi Zhang [a], Harold Bosch [b], Jerry Shen [c]

[a]*Genex/TFHRC, McLean, Virginia, USA*
[b]*FHWA/TFHRC, McLean, Virginia, USA*
[c]*Genex/TFHRC, McLean, Virginia, USA*

ABSTRACT: The aerodynamic force and moment change on a high side vehicle due to the wake effect of the bridge tower is studied through Computational Fluid Dynamics (CFD) modeling. Simple CFD models are firstly built and calibrated by using previous wind tunnel tests. The same modeling method is then used to simulate a 1:70 truck-trailer model passing through a bridge tower in the wind tunnel at the Aerodynamics Laboratory of the Turner-Fairbank Highway Research Center (TFHRC). The simulation is performed in both dynamic and stationary scenarios. Finally, a hybrid method based on the static aerodynamic force coefficients while considering the wind speed, wind angle and truck speed is presented. The comparisons show that the stationary simulation is not always an appropriate substitute for a dynamic one. The results of the hybrid method show better agreement with dynamic simulation.

KEYWORDS: Wake effect, High side vehicles, CFD simulation, Dynamic and stationary method, Quasi-static method (or hybrid method).

1 INTRODUCTION

Wind-induced vehicular accidents are reported all over the world, such as in Japan, China, UK, and USA[1-3]. Based on the Wyoming crash database, from January 1, 2001 to June 30, 2006, 3119 vehicles crash associated with high winds were entered. Seven hundred and seventy-six of them were selected to be inspected in detail based on their importance criteria (economic losses). Four hundred and three out of 776 were confirmed to be wind-related crashes, in which 141 accidents were blow-over crashes, 123 were push-over crashes, 47 were slide-over crashes, 39 were yaw instability crashes, 3 were blown-apart crashes, and 50 other crashes[4]. The wind-induced traffic safety problem might get worse when the high side vehicle (such as lorries, vans, trailers, etc) passes through a bridge tower or a narrow shielded area, because the forces and moments on the vehicle will undergo sudden change, which can result in vehicle veering from its lane, overturning, rotating and sliding. An appropriate geometric design of the bridge deck, tower and the wind barrier can mitigate the wake effect on vehicles and reduce the probability of accidents, especially for high side vehicles.

To investigate the wake effect of the bridge tower and the mitigation measures, a series of experimental and computational studies was initiated at the Turner-Fairbank Highway Research Center (TFHRC), including:

- Phase I: the computational fluid dynamics (CFD) analysis on the wake effect of bridge tower on vehicles;
- Phase II: wind tunnel tests on the vehicle passing through the wake of the bridge tower;
- Phase III: the systematic study on the influence of the geometry of bridge deck, tower, and wind barrier.

This paper mainly introduces the achievements in the Phase I analysis. CFD models are first built by using STAR-CCM+ (CD-adapco), and then calibrated by using data from a previous experimental study. Wind tunnel tests performed by Charuvisit *et al* [5] are used for the CFD model validation. The sliding mesh method is adopted here to carry out the movement of the truck in the calculation domain. A detailed description of the wind tunnel tests and the model validation were presented in reference [6], which is not duplicated herein. The comparisons were performed on the mean aerodynamic force coefficients of the truck, wind speed profiles along the truck path, as well as the dynamic force and moment on the moving truck at various wind speeds and wind angles. It was concluded that the proposed CFD model can simulate the wake effect of bridge tower very well.

The calibrated CFD model is then applied to simulate a truck-trailer passing through a rectangular bridge tower under different wind speeds and angles in both dynamic (moving truck) and stationary (static truck) modes. Furthermore, a hybrid (quasi-static) method is developed to calculate the force and moment change on the truck. The results from the dynamic simulation, stationary simulation and hybrid method are compared. The feasibility of performing stationary or hybrid tests instead of dynamic tests to analyze the wake effect is discussed. The data provided in Phase I study are taken as the guidance for the test design in Phase II.

2 HYBRID TEST METHOD

It is well known that the dynamic test is the most direct and accurate way of testing the bridge tower wake effect on a truck. Stationary test is taken as the alternative method due to the difficulties of carrying out the dynamic test in the laboratory. The hybrid method presented here is aimed at combining the static aerodynamic force coefficients of the truck and the resultant wind speed when there is a truck running on the path of the bridge deck to provide more realistic results than the stationary test method.

The idea of the hybrid test is to divide the truck into n parts (shown in Figure 1), and then sum the force and moment on each part to get the force and moment on the whole truck. Equations (1) and (2) are used to calculate the force and moment on each part, in which the local wind speed (u_i), wind angle (β_i) and the aerodynamic coefficient for parts ($C_{\cdot,i}$) are applied.

$$F_i = C_{force,i} \cdot \frac{1}{2} \rho u_i^2 A \tag{1}$$

$$M_i = C_{moment,i} \cdot \frac{1}{2} \rho u_i^2 A r \tag{2}$$

where F_i = the force on the ith part of the truck; M_i = the moment on the ith part of the truck; $C_{force,i}$ = the force coefficient of the ith part; $C_{moment,i}$ = the moment coefficient of the ith part; ρ = density of air; u_i = resultant wind on the ith part; A = frontal area of the truck (a constant value for each part); and r = specified length.

The wind effect on the whole truck is calculated by summing the local wind effect on each part as:

$$F = \sum_1^n F_i \tag{3}$$

$$M = \sum_1^n M_i \tag{4}$$

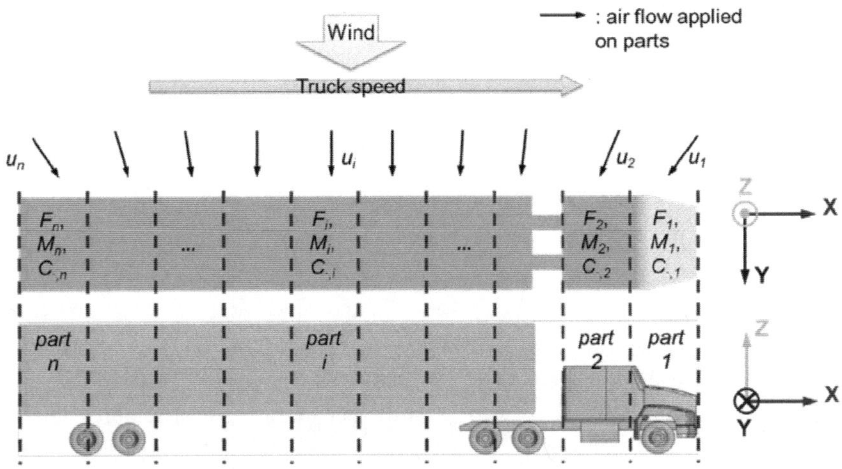

Figure 1 Illustration of divided parts on the truck

The following assumptions are applied in the proposed hybrid test method:

1. The force and moment on the whole truck can be calculated by summing of the force and moment applied on each part of the truck.
2. The force and moment on each part of the truck can be obtained by equations (1) and (2). The force and moment coefficients of each part, $C_{\cdot,i}$, are not influenced by the absolute value of mean wind speed, but only change with the horizontal wind angle (β).
3. The wind applied on each part of the truck when the truck moves on the bridge deck can be obtained by the vector addition of minus truck speed ($-\mathbf{u}_{truck,i}$) and the wind speed applied on this part ($\mathbf{u}_{wind,i}$). The local wind ($\mathbf{u}_i$) on each part of the truck when the truck moves to a certain location is:

$$\mathbf{u}_i = \mathbf{u}_{wind,i} - \mathbf{u}_{truck,i} \tag{4}$$

4. For simplification, the turbulence around the truck is not considered. Only the horizontal wind (wind speed in x and y direction in Figure 1) affects the force and moment on parts of the truck.

An example of calculating $\mathbf{u}_i$ is shown in Figure 2. In this example, x and y directions are the horizontal directions. The truck is assumed to move in x direction only. According to assumption 4, only wind speed in x and y direction, $\mathbf{u}_{wind,i}$(x,y), is considered, thus the resultant wind speed $\mathbf{u}_i$ will only have a component in x and y direction, $\mathbf{u}_i$(x,y).

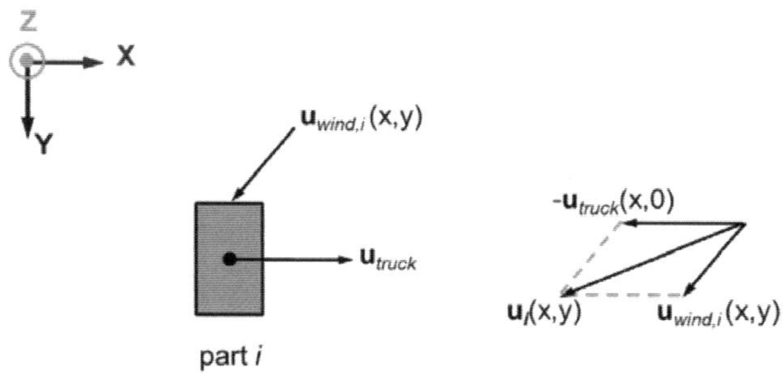

Figure 2 Illustration of calculating $\mathbf{u}_i$

The hybrid test is performed in the following steps:

1. The truck model is virtually divided into n parts.
2. The static wind tunnel tests are performed on truck model to determine the aerodynamic coefficient of the truck for the wind angle from -180° to +180°. The suggested maximum interval is 10°. Sufficient pressure taps are necessary to capture the pressure change on the truck.
3. The pressures obtained from Step 2 along the surface of the truck are used to determine the force and moment coefficients on each part of the truck, $C_{force,i}$ and $C_{moment,i}$. Note that the moment on each part is calculated about the geometry centroid of the truck (and not the geometry centroid of the part).
4. The static wind tunnel test is performed to measure the wind speed at all locations on the truck path (without the truck but with the bridge deck and tower).
5. Select a location for the truck. At this location calculate the resultant wind speed for each part of the truck from the measured wind speed ($\mathbf{u}_{wind,i}$) and the truck speed ($\mathbf{u}_{truck,i}$) by using equation (4).
6. The force and moment on each part of truck is calculated by using the resultant wind speed obtained in Step 4 and the force and moment coefficient obtained in Step 3 by using equation (1) and (2).
7. The force and moment on each part is added to obtain the force and moment on the whole truck by using equation (3) and (4).
8. The truck is moved to the next location. Steps 5 to 7 are repeated.

The flow chart of the proposed hybrid test method is shown in Figure 3. The advantage of this method is:

- It can consider the wind speed (horizontal) and truck speed at the same time;
- No dynamic wind tunnel test is needed. Only static wind tunnel tests to measure the aerodynamic coefficients of the truck and the wind speed along the truck path are required.

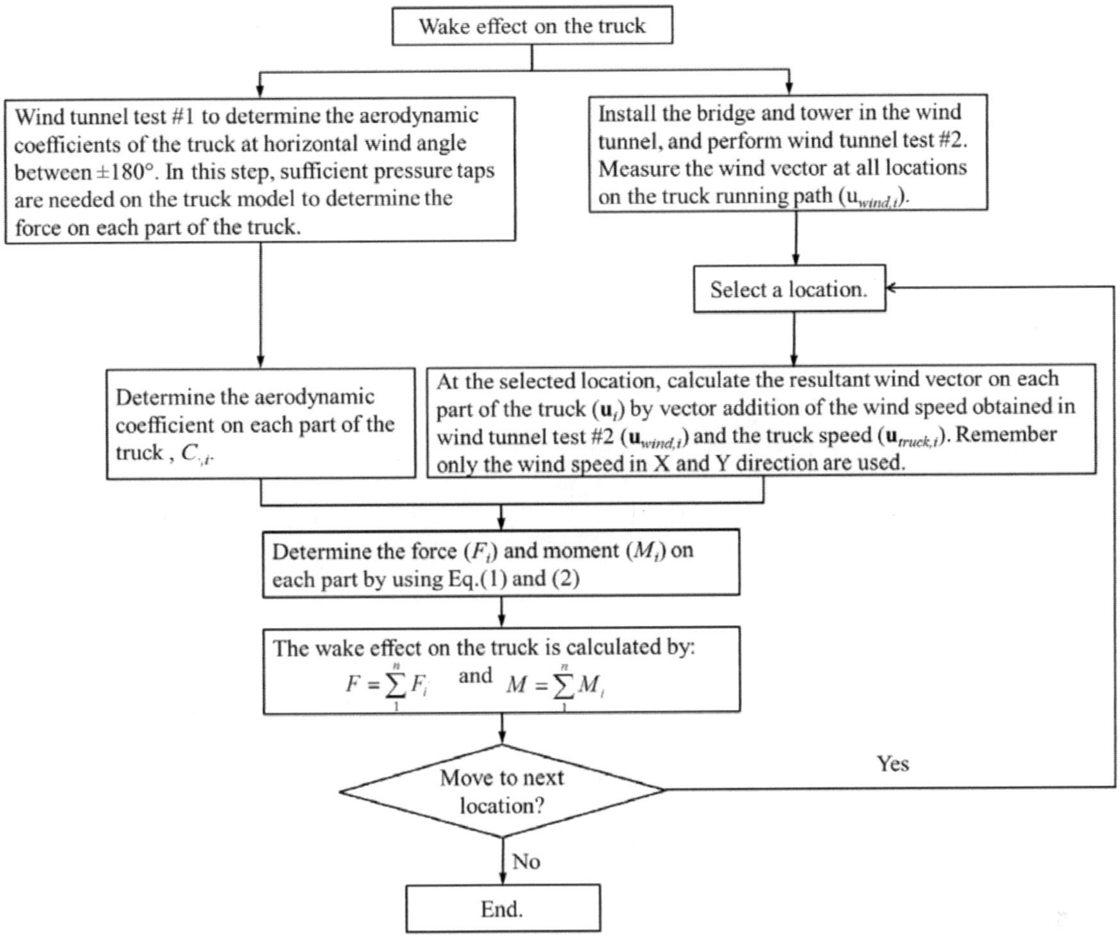

Figure 3 Flow chart of the proposed hybrid test

3 MODEL DESCRIPTION

The bridge model is built based on the Hale Boggs Memorial Bridge (also known as the Luling-Destrehan Bridge), and the cross section of the side span is used throughout the entire bridge as shown in Figure 4. The semi-trailer truck in Figure 5 is adopted as the studied vehicle, and its geometric centroid is calculated to be at 2.48 m and 11.33 m from the head and bottom of the truck for the prototype. The scale of 1:70 is used to fit the model into the wind tunnel at TFHRC. Figure 6 illustrates the CFD model, in which the bridge and truck are set up in the test section of the wind tunnel showing the coordinate system. Froude number similarity criterion is adopted to scale the relative physical quantities.

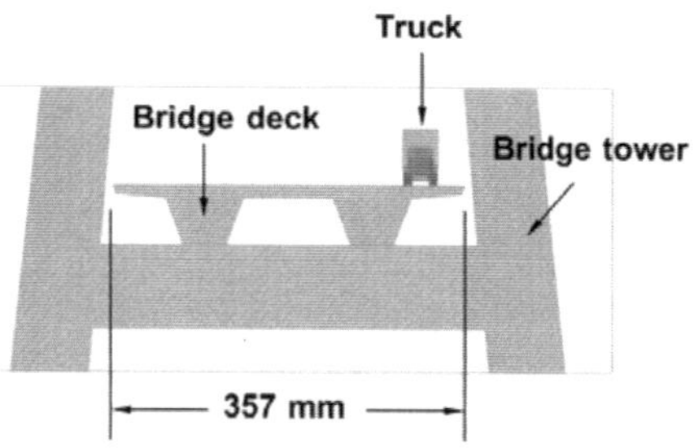

Figure 4. Bridge deck cross section (model)

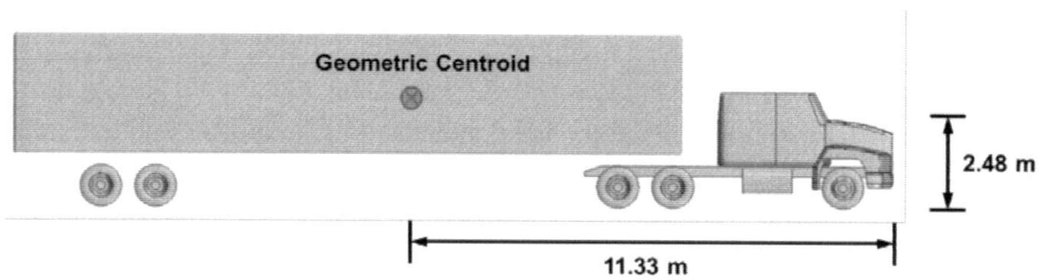

Figure 5. Truck-trailer (prototype)

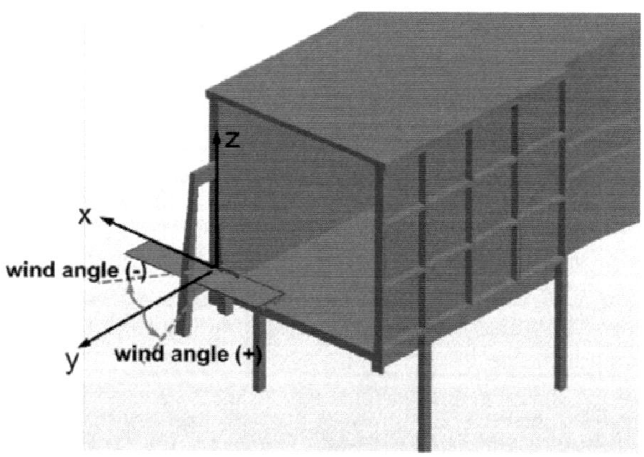

Figure 6. Illustration of the CFD model

Dynamic and stationary simulations are firstly performed using the CFD model. The combination of truck speed of 3 m/s, wind speed of 10 m/s, 3 m/s and 1 m/s, and wind angle of -30°, 0°, and 30° are studied. In the dynamic simulation, the truck model moves on the bridge deck during the simulation, and the instantaneous force and moment on the truck are obtained, while in the stationary simulation, the truck is placed at fixed positions on the bridge deck along the truck path. The force and moment on the truck are then measured at each location as shown in Figure 7.

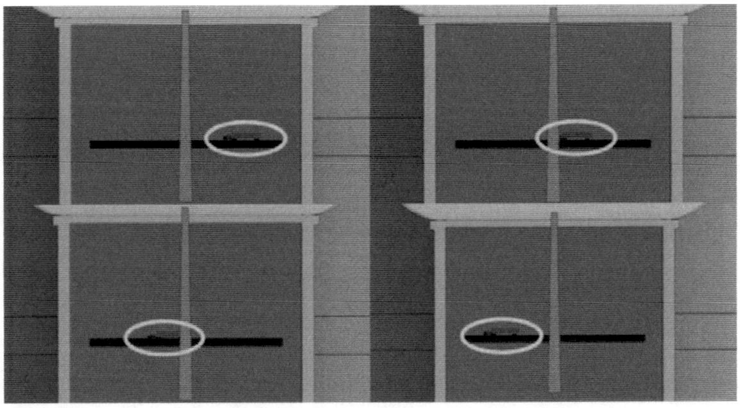

Figure 7. Illustration of the stationary simulation

4 COMPARISON ON SIMULATION RESULTS

A total of 9 cases are subjected to dynamic, stationary and hybrid CFD virtual testing. The parameters for each case are listed in Table 1.

Table 1. Parameters for case study

Case number	Wind speed	Wind angle	Truck speed
	m/s	°	m/s
Case 1	10	0	3
Case 2	10	30	3
Case 3	10	-30	3
Case 4	3	0	3
Case 5	3	30	3
Case 6	3	-30	3
Case 7	1	0	3
Case 8	1	30	3
Case 9	1	-30	3

In the hybrid testing, the truck is divided to 30 parts. Instead of wind tunnel tests, a series of CFD simulations is performed to obtain the force and moment on each part of the truck. As introduced in Section 1, it is believed that the CFD model can predict well the static aerodynamic force coefficients.

The simulation results for the nine cases in Table 1 are compared in Figure 8 to Figure 16, which include drag, side and lift force, as well as roll, pitch and yaw moment. It is found that:

1. When the ratio of wind speed to truck speed is high (10:3, Figure 8 to Figure 10), the stationary simulation and the hybrid simulation can represent the dynamic simulation results well in terms of the most forces and moments. Smaller difference is observed for side force, roll moment and yaw moment while larger error is found in drag force, lift force and pitch moment in many cases.
2. When the ratio of wind speed to truck speed is equal to 1 (3:3, Figure 11 to Figure 13), the results of the stationary simulation become somewhat inaccurate when compared with the dynamic results, especially for drag force, lift force and pitch moment. For side force, roll moment and yaw moment, although the stationary results still keep the same shape and trend as the dynamic results, the fit is not so good. In contrast, the results of the hybrid simulation still compare well with the dynamic results. More differences are

observed for the drag force, lift force and pitch moment in some cases (but not in all cases).

3. When the ratio of the wind speed to truck speed is low (1:3, Figure 14 to Figure 16), the stationary simulation is not able to accurately predict the force and moment on the truck induced by the wake effect of the bridge tower. However, the hybrid method still works well. It is also found that when the ratio of wind speed to the truck speed is low, the hybrid method can provide reasonable results in all types of force and moment, and is not limited to side force, roll moment and yaw moment.

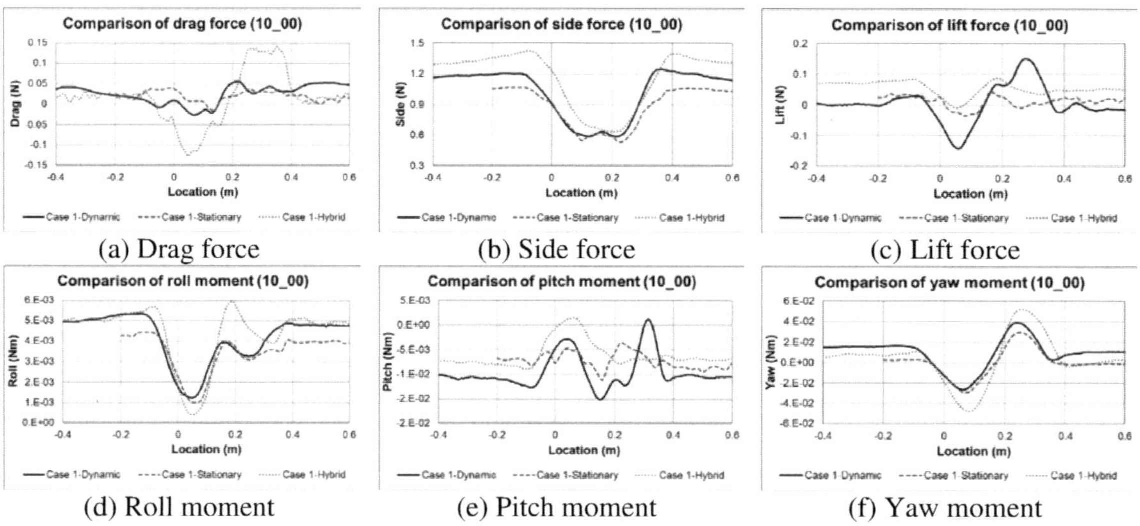

(a) Drag force (b) Side force (c) Lift force

(d) Roll moment (e) Pitch moment (f) Yaw moment

Figure 8 Comparison on Case 1 (wind speed: 10 m/s, truck speed: 3 m/s, wind angle: 0°)

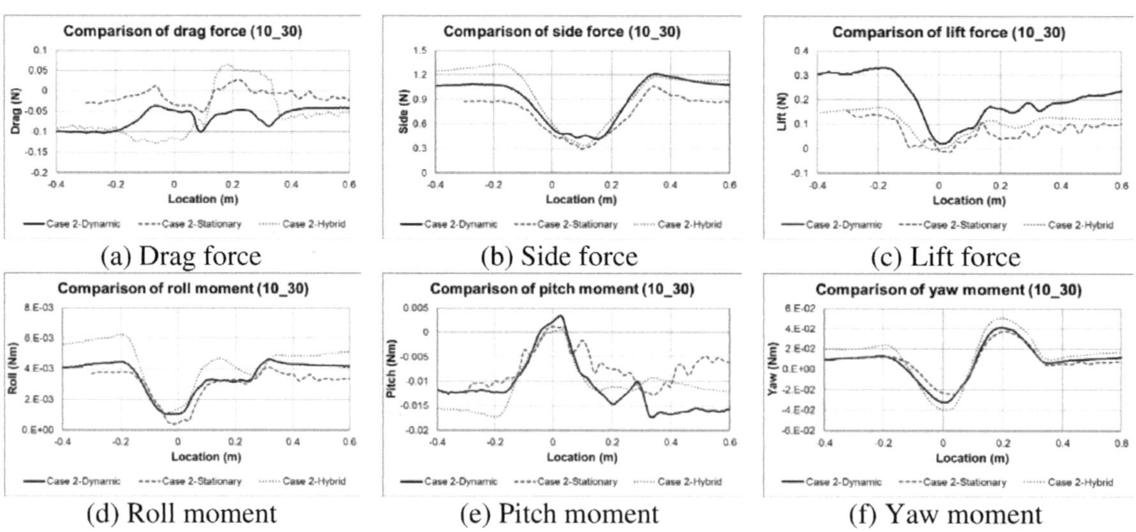

(a) Drag force (b) Side force (c) Lift force

(d) Roll moment (e) Pitch moment (f) Yaw moment

Figure 9 Comparison on Case 2 (wind speed: 10 m/s, truck speed: 3 m/s, wind angle: 30°)

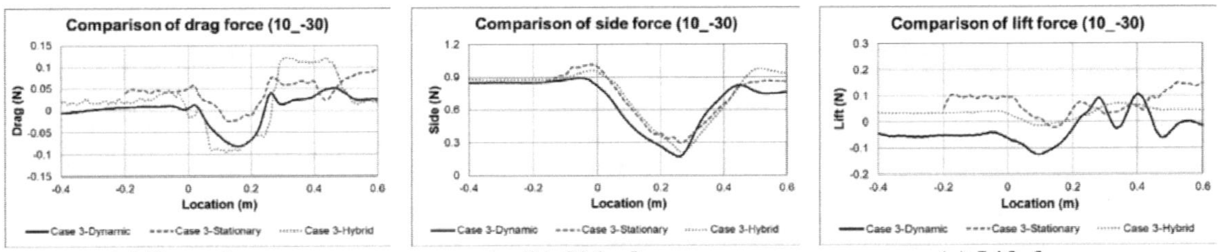

(a) Drag force (b) Side force (c) Lift force

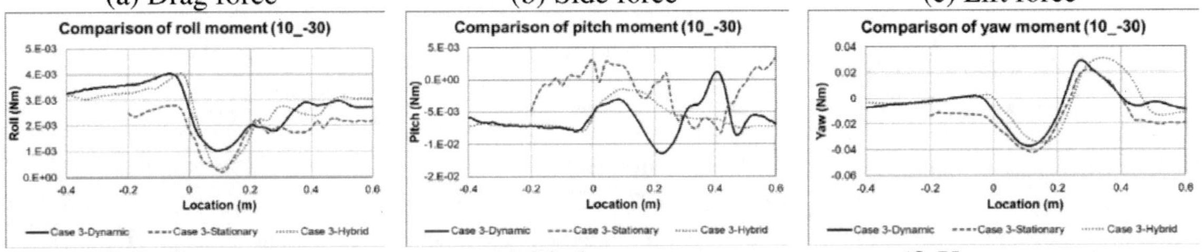

(d) Roll moment (e) Pitch moment (f) Yaw moment

Figure 10 Comparison on Case 3 (wind speed: 10 m/s, truck speed: 3 m/s, wind angle: -30°)

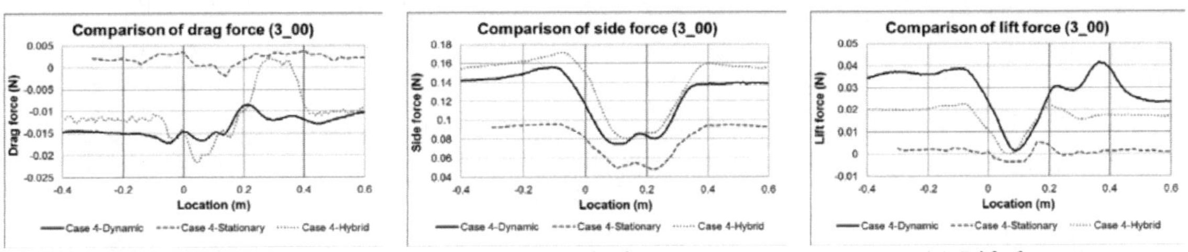

(a) Drag force (b) Side force (c) Lift force

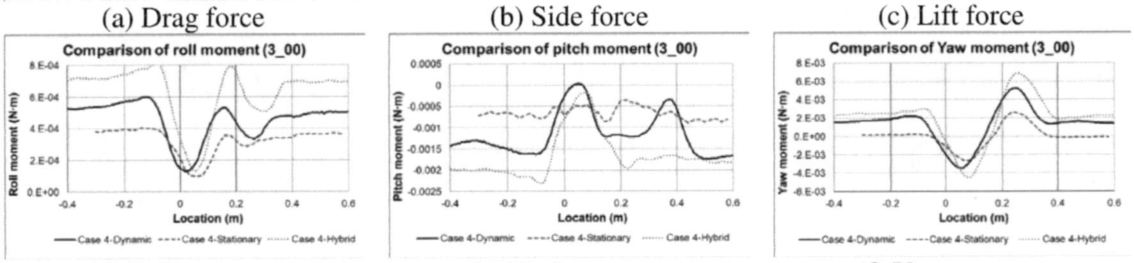

(d) Roll moment (e) Pitch moment (f) Yaw moment

Figure 11 Comparison on Case 4 (wind speed: 3 m/s, truck speed: 3 m/s, wind angle: 0°)

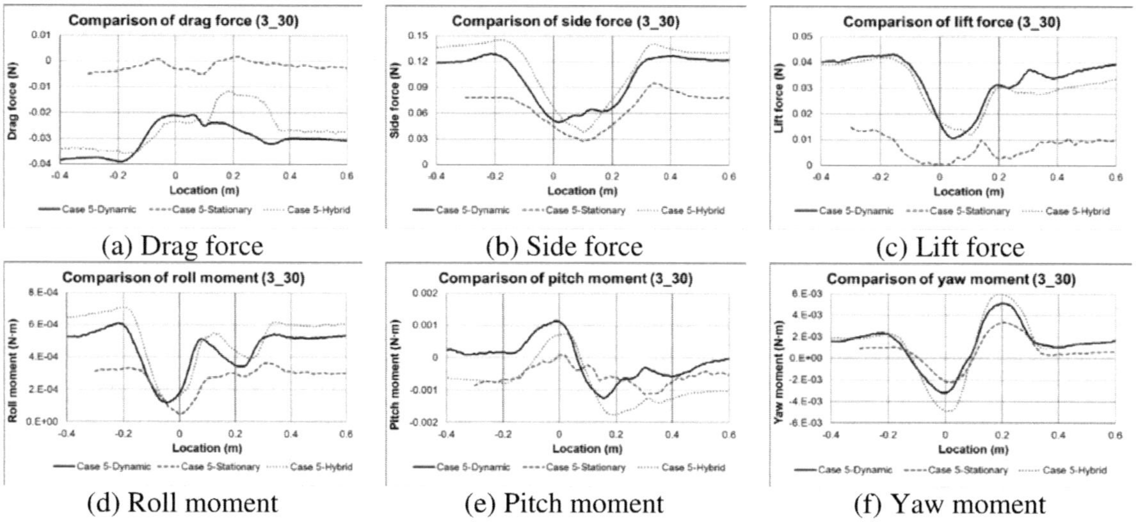

(a) Drag force (b) Side force (c) Lift force

(d) Roll moment (e) Pitch moment (f) Yaw moment

Figure 12 Comparison on Case 5 (wind speed: 3 m/s, truck speed: 3 m/s, wind angle: 30°)

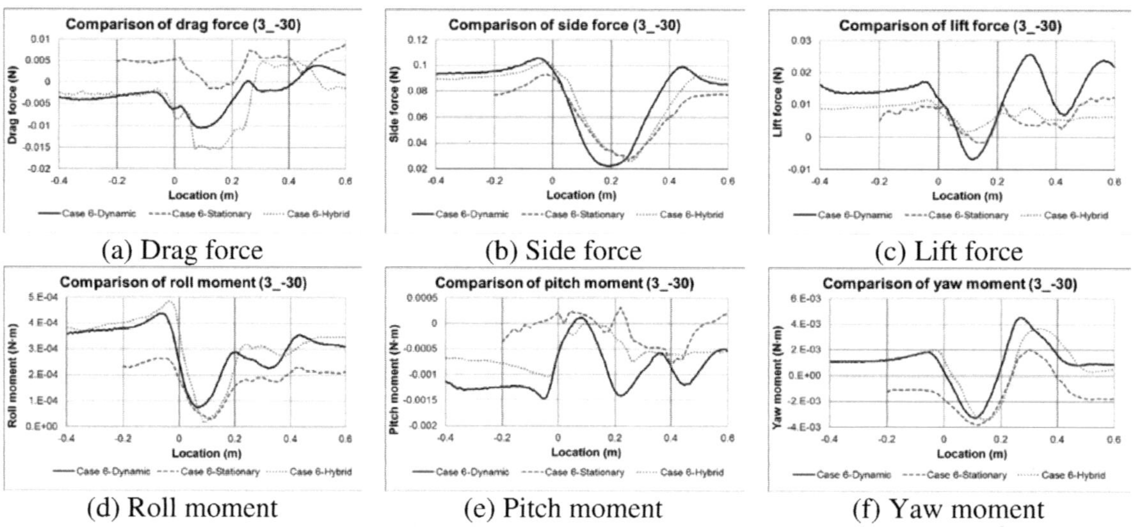

(a) Drag force (b) Side force (c) Lift force

(d) Roll moment (e) Pitch moment (f) Yaw moment

Figure 13 Comparison on Case 6 (wind speed: 3 m/s, truck speed: 3 m/s, wind angle: -30°)

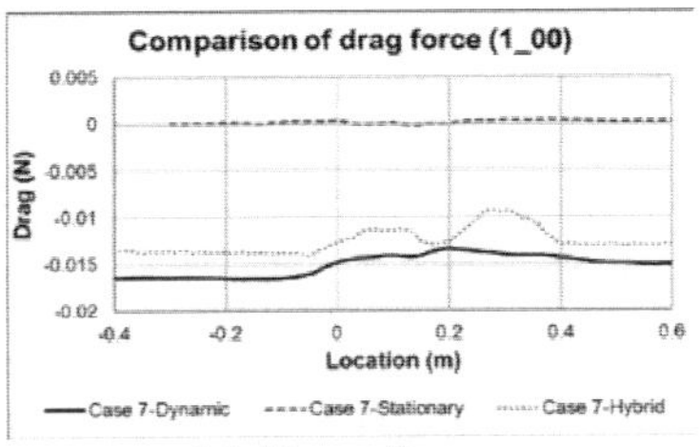

(a) Drag force

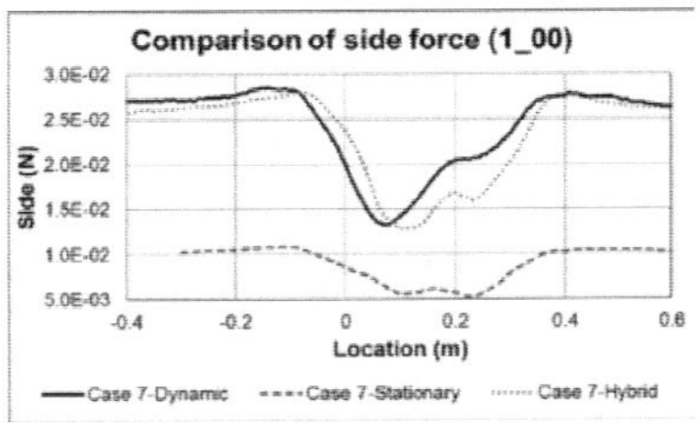

(b) Side force

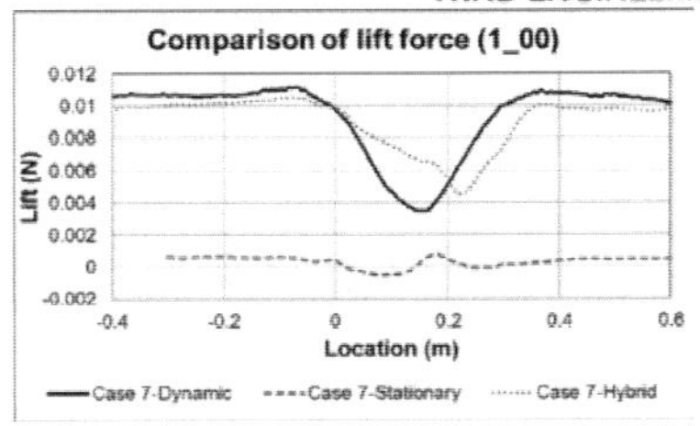

(c) Lift force

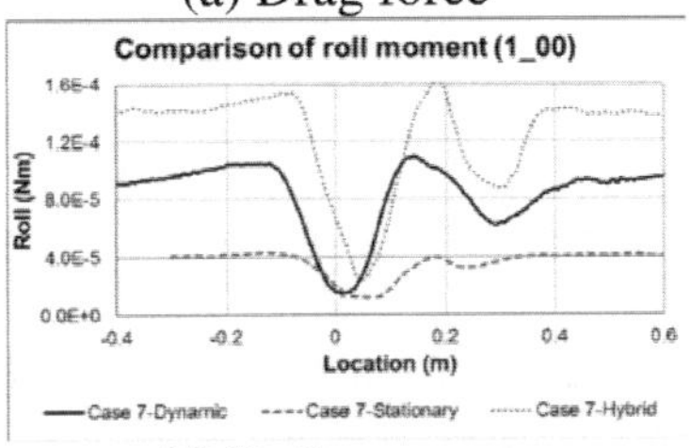

(d) Roll moment

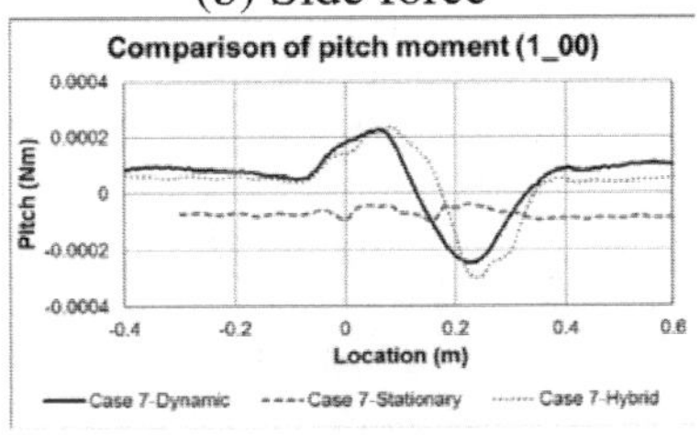

(e) Pitch moment

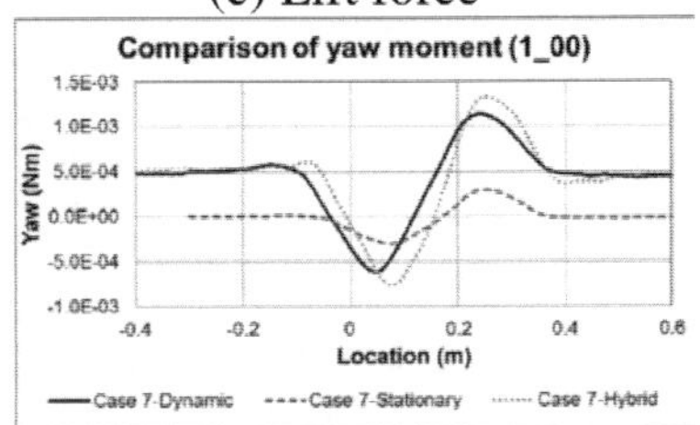

(f) Yaw moment

Figure 14 Comparison on Case 7 (wind speed: 1 m/s, truck speed: 3 m/s, wind angle: 0°)

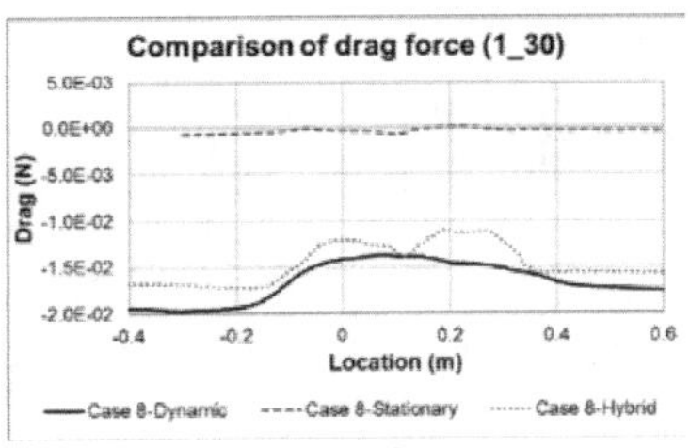

(a) Drag force

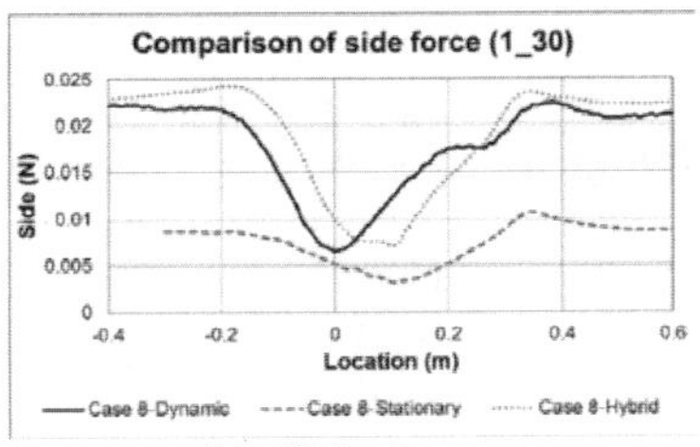

(b) Side force

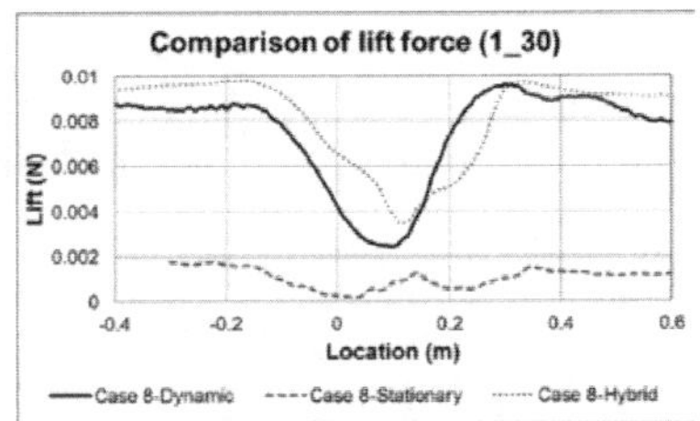

(c) Lift force

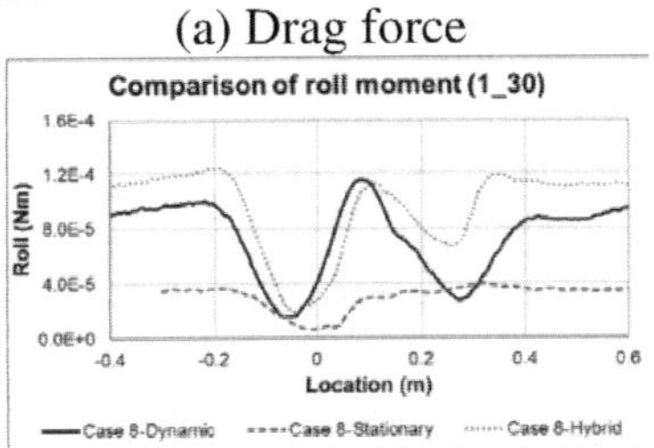

(d) Roll moment

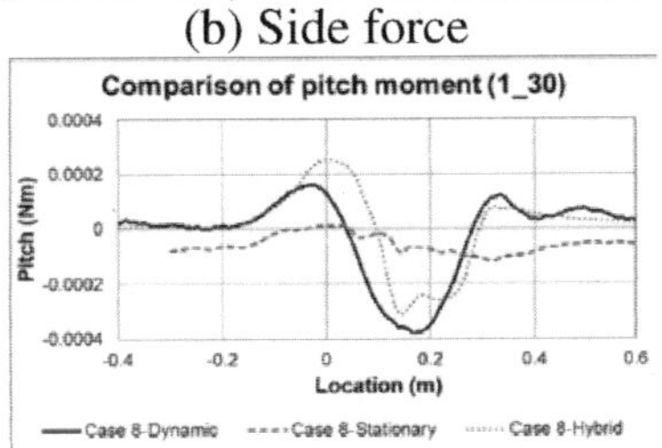

(e) Pitch moment

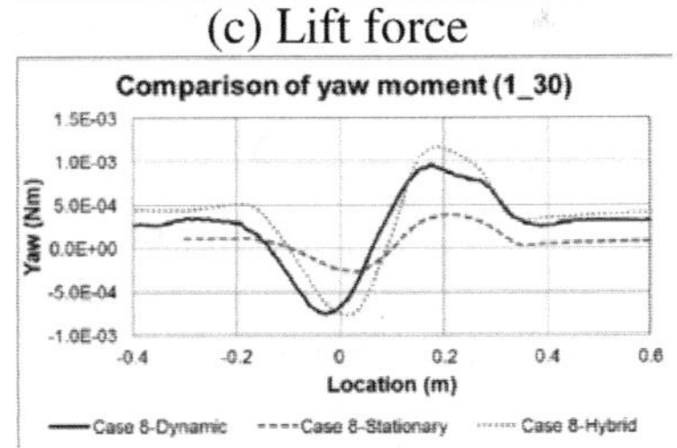

(f) Yaw moment

Figure 15 Comparison on Case 8 (wind speed: 1 m/s, truck speed: 3 m/s, wind angle: 30°)

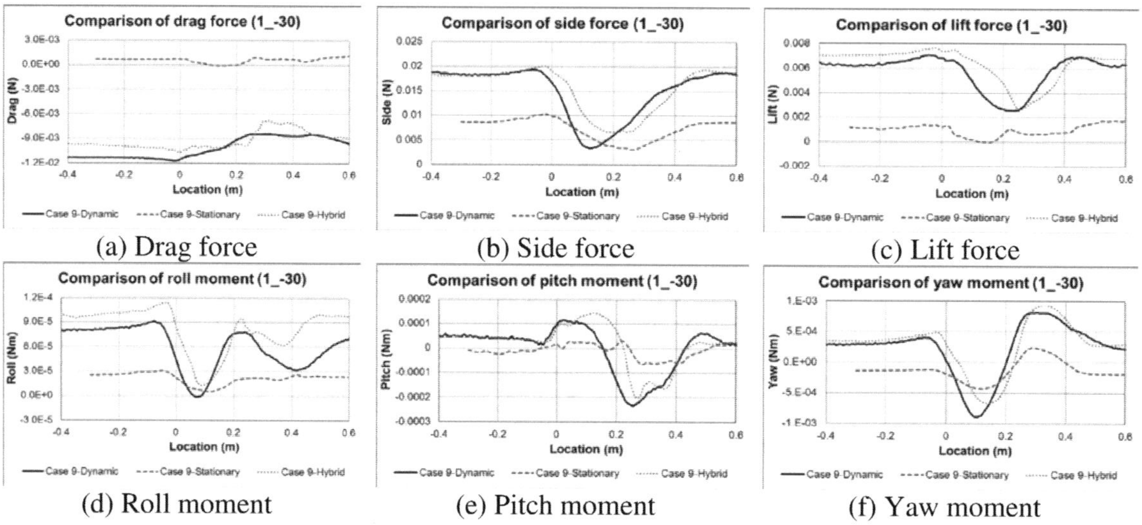

(a) Drag force (b) Side force (c) Lift force

(d) Roll moment (e) Pitch moment (f) Yaw moment

Figure 16 Comparison on Case 9 (wind speed: 1 m/s, truck speed: 3 m/s, wind angle: -30°)

5 CONCLUSIONS

In this paper, a hybrid test method is developed to evaluate the wake effect of a bridge tower on the moving truck to overcome difficulties in performing dynamic wind tunnel tests. The dynamic, stationary and hybrid tests are carried out through numerical means by using CFD simulations. The model wind speed of 1-10 m/s, model truck speed of 3 m/s, and wind angle of -30° to 30° are applied to a 1:70 truck passing through a bridge tower in the same scale.

Based upon the comparisons, it is concluded that:

1. The dynamic test is the most direct way to study the wake effect of a bridge tower. It is the preferred method in the test design and analysis.
2. The hybrid test method can be used as an alternative test method to analyze the wake effect of a bridge tower on the moving truck. The results of a hybrid test match the dynamic simulation results much better than the stationary test in a larger range of wind speed to truck speed ratio, because it can consider the wind speed and truck speed simultaneously.
3. The stationary test method can provide reasonable results when the ratio of wind speed to truck speed is high. However, it should be used with caution for low ratio of wind speed to truck speed. Although the threshold speed ratio for determining the appropriateness of stationary testing has not been determined yet, it seems to be larger than 1 according to the results.

6 ACKNOWLEDGEMENTS

The research reported herein was performed utilizing resources made available through the interagency agreement DTFH61-14-X-300002 between the U. S. Department of Transportation (DOT) and Department of Energy (DOE). The Transportation Research Analysis Computing Center (TRACC) at DOE's Argonne National Laboratory (ANL) provided high performance computing systems, CFD software, and invaluable technical advice, to assist researchers at the Federal Highway Administration's Aerodynamics Laboratory in performing these simulations.

REFERENCES

1. T. Fujii, T. Maeda, H. Ishida, T. Imai, K. Tanemoto and M. Suzuki, Wind-Induced Accidents of Train/Vehicles and Their Measures in Japan, Quarterly Report of RTRI (Railway Technical Research Institute) (Japan), 1999, 40(1): 50-55.
2. Y. Li, S. Qiang, H. Liao and Y.L. Xu, Dynamics of Wind-Rail Vehicle-Bridge Systems, Journal of Wind Engineering and Industrial Aerodynamics, 2005, 93(6): 483-507.
3. S.R. Chen and C.S. Cai, Accident Assessment of Vehicles on Long-Span Bridges in Windy Environments, Journal of Wind Engineering and Industrial Aerodynamics, 2004, 92(12): 991-1024.
4. J. McCarthy, Final Report: Evaluation of Intelligent Transportation System Alternatives for Reducing the Risks of Truck Rollover Crashes Due to High Winds, 2006.
5. S. Charuvisit, K. Kimura and Y. Fujino. Experimental and Semi-Analytical Studies on the Aerodynamic Forces Acting on a Vehicle Passing through the Wake of a Bridge Tower in Cross Wind, Journal of Wind Engineering and Industrial Aerodynamics, 2004, 92(9): 749-780.
6. N. Zhang, H.R. Bosch and J. Shen, Computational Fluid Dynamics Simulation and Parametric Study of Wake Effect of Bridge Tower on a Truck, Proceedings, Eighth Colloquium on Bluff Body Aerodynamics and Applications, Boston, Massachusetts, June 2016.

The 13th Americas Conference on Wind Engineering (13ACWE)
Gainesville, Florida USA, May 21-24, 2017

Influence of damping and force hypotheses for evaluating ductility demand of structures subjected to time history wind loading

Taleb Sabbek [a], Sébastien Langlois [a], Frédéric Légeron [b]

[a] *Departement of Civil Engineering, Université de Sherbrooke, 2500 boul. De l'Université, Sherbrooke, QC, Canada*
[b] *PMC – Bridge & Tunnel UAE District Manager, Parsons, Abu Dhabi PO5498, UAE*

ABSTRACT: Current wind design procedures require that structural behavior remains elastic under wind loading corresponding to a specific return period unlike seismic codes that set some specifications to account for the ductile behavior of structures. Based on both linear and nonlinear dynamic time history analyses of simplified shear frame structures, this study examines the sensitivity of the ductility demand to different analysis hypotheses, and gives recommendations for their adequate values to be used in extended parametric studies. In particular, the current results show that the use of admittance functions strongly influence the ductility demand of structures. It was also shown that the ductility demand of the structure varies significantly with the damping ratio. More specifically, it was found that a 1% damping ratio (including the aerodynamic damping) is a coherent and conservative value to be used for parametric studies on medium-to-high-rise structures in order to establish a relation between the ductility demand and the strength reduction factor for structures subjected to extreme wind loadings.

KEYWORDS: Time domain analysis, nonlinear analysis, extreme wind loading, damping, ductility.

1 INTRODUCTION

Unlike modern seismic codes that account for the ductile behavior of structures, current wind design procedures require that structural behavior remains elastic under wind loading corresponding to a specific return period. This assumption would naturally lead to an increase in the structural costs especially where the structures are exposed to extreme wind events (low probability of occurrence events) such as severe storms and hurricanes and/or where wind loadings are prevailing the overall structural response. In some situations, the structures are already designed with a specific value of ductility, and therefore they have more safety margin against failure compared to their brittle counterparts. Therefore, it is of both economic and safety interests to evaluate the ductility demand of structures under extreme wind events.

In the literature, several endeavors were attempted to understand the inelastic behavior of structures under wind loading. For example, Chen and Davenport [4] studied the vulnerability of tall buildings in hurricanes based on a simplified analytical approach developed earlier by Vickery [23] to estimate the nonlinear response of lightly damped structure under wind loading. Nonlinear time history analyses were conducted by Tschanz [21] and Gani and Légeron [8] to evaluate the effects of different analysis parameters on the ductility demand of structures under wind loading. They conducted parametric studies on the effect of both structural parameters (e.g., single degree of freedom(SDOF) or multi degree of freedom (MDOF), aerodynamic and structural damping) and wind force calculation hypotheses (e.g., real wind speed measurement, numerically generated wind speed and wind turbulence intensity) on the ductile behavior of structures. It is to be noted that most of these efforts have been performed on specified structures subjected to specific loading conditions so their results cannot be generalized to other cases. Although Gani and Légeron [8] have established a relation between the ductility demand of the structures and the force reduction factor based on a comprehensive parametric study, their work was limited to the nonlinear performance of SDOF structures. Yet, little work has been devoted to investigate the inelastic response of MDOF structures under extreme wind events. It is also of importance to pay a special attention to the analysis hypotheses to be used before being involved in extended parametric studies.

To this end, both linear and nonlinear dynamic time history analyses of simplified shear frame structures with one, two, and three degrees of freedom have been carried out. A computer program was developed in this study adopting Newmark method combined with Newton-Raphson error reduction algorithm to solve the equation of motion. The influence of using an admittance function for evaluating wind forces was studied for both real (hurricanes wind speed measurement) and numerically generated wind samples. The prime objectives of this study are: (1) to examine the sensitivity of structure response in term of its ductility demand to different analysis hypotheses (time-step analysis and wind samples durations) and (2) to propose some recommendations with respect to the appropriate values to be used in extended parametric studies.

After the description of the model and hypotheses used in this study, a brief presentation into the wind force calculation is given. This paper then presents the main part of the study with respect to the impact of both damping and force hypotheses in evaluating ductility demand of structures subjected to wind loading..

2 MODEL DESCRIPTION AND METHODOLOGY

This study investigates the performance of both single (SDOF) and multi (MDOF) (Figure 1a) degree of freedom structures under the action of wind loading applied at the corresponding structural masses. The dimensions (width × depth × height) of each story are selected, respectively at 25 × 25× 4 m^3, while variable stiffness, k_i and invariable concentrated mass, m_0 of 500 tons are assumed for each story. The structural ductility is defined as shown in Figure 1b by the ratio $\mu = u_m/u_y$ and the strength reduction factor by $\beta = f_y/f_0$; with u_m being the maximum displacement, u_y being the yield displacement, f_0 being the maximum elastic force and f_y being the yield strength. When β=1 the system remains elastic; otherwise ($0 < \beta < 1$), excursions in the plastic range can be occurred. The nonlinear behaviour of the structure is defined in this study by bilinear constitutive law with strain hardening of 5% and damping of Rayleigh type. Equation of motion was solved using Newmark (average acceleration) method in conjunction with the modified Newton-Raphson iterations.

An iterative design procedure was implemented to control the structural fundamental frequency, f_1. First, initial stiffness is assigned for all levels. Then a linear analysis is conducted; using the obtained maximum shear force at each level V_i, a reduction factor is applied to each stiffness level k_i as $k_i' = k_i(V_i/V_1)$. The structure with modified stiffness is then analyzed following the same steps until no significant change in stiffness has been achieved. The use of such procedure is required to insure a rational design for the structure with the same story drift at each level for the specified wind loading sample. The designed structure is then linearly (β=1) and non-linearly (β<1) analyzed according to specified reduction factors where $V_y = \beta V_0$.

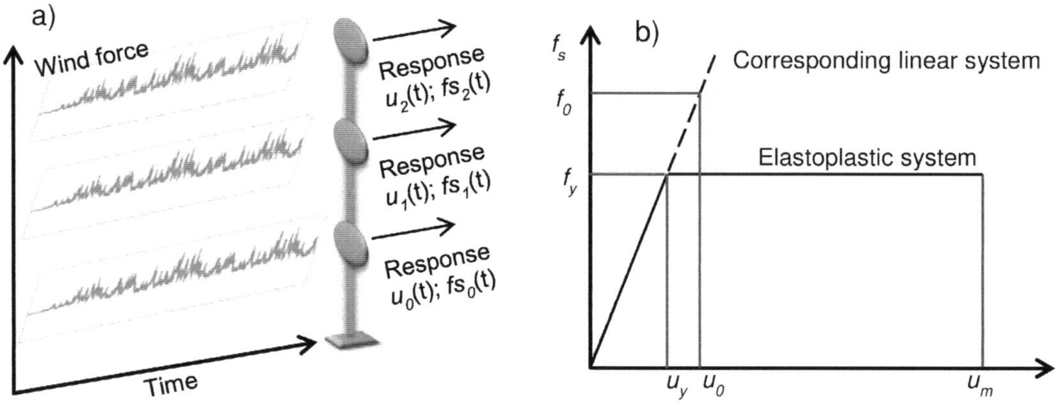

Figure 1. a) The MDOF system under investigation and b) linear and nonlinear constitutive force-deformation relationship (Chopra, 2011).

Table 1. Parameters studied in the analyses.

Parameter	Values
Analysis time-step, Δt(s)	0.1, 0.01, 0.001
Wind sample duration (min)	10, 20, 30, 60
Wind force calculation approach	quasi-static, admittance
Damping, ξ(%)	0, 0.1, 0.2, 0.5, 0.7, 1, 2, 3, 5

In this study, a total of more than 2000 have been analyzed using two different wind force calculation (quasi-static, admittance) approaches considering different values of analysis time-step, wind sample duration, and damping ratios. Table 1shows the different parameters to be evaluated in the context of this study.

3 GENERAL ANALYSIS HYPOTHESES

Although several studies have been documented in the literature pertaining to the nonlinear time history analysis of structures under extreme wind events, there are some important issues that were not explicitly discussed and therefore no conclusive recommendations have been reached. The selection of appropriate time step and wind loading duration are two of these issues that have a direct impact not only on the total analysis executing time but also on the precision of the obtained results. For example, it is customarily adopted that the use of a one hour wind sample in linear or nonlinear analyses of structures would be satisfactory [8,12,22,23,24]. However, there is no a clear justification for this assumption. In fact, it is of importance to pursue the previous assumptions with respect to the selection of the analysis time step and the used wind loading duration before starting an extended parametric study.

3.1 *Time-step analysis*

A sensitivity analysis of the appropriate time-step was carried out for both linear and nonlinear structural behavior. Table 2 shows the values of the maximum displacement obtained from linear analyses and ductility demand from nonlinear analyses of 2 DOF structure having f_1 =1Hz under the Katrina-2005 wind loading assuming damping ratio, ξ of 1%. Table 2 shows that the error in estimating the maximum displacement between a time-step loading, Δt of 0.1s and of 0.01s in the linear analysis is about 11%. This percentage reaches 30% in nonlinear analysis. The error in predicting the structural deformation if Δt is changed from 0.01s to 0.001s is within 2% for both linear and the nonlinear analyses. Therefore it can be concluded that Δt=0.01s is the optimal time-step to be used for linear and nonlinear analysis.

It should be noted that, because of nonlinear analysis is related to linear analysis (design procedure described above) the same time step must be used for both analysis cases otherwise error on ductility demand reaches 80%.

Table 2. Maximum displacement and ductility demand obtained from linear and nonlinear analyses.

Time-step	Linear		Nonlinear	
Δt (s)	Δ_{max} (mm)	Error (%)	Ductility (μ)	Error (%)
0.1	44.90	10.9	2.25	-31.2
0.01	40.54	0.1	3.32	1.6
0.001	40.50	0.0	3.27	0.0

3.2 *Wind sample duration*

Wind loading duration for numerically generated wind simulating winter storms may not have considerable effect because of the constant mean wind speed in opposition to hurricanes wind speed measurement where intervals covering maximum wind speed may be much longer because of mean wind speed variation (e.g. wind loading duration considered in the case of hurricane Isabel-2003 is 4 hours). A sensitivity study was carried out for generated wind speed comparing the maximum ductility demand for different wind loading durations (10, 20, 30, 60, 180 min). The obtained results showed that for almost all the cases analyzed (for wind loading with constant mean wind speed), no more yielding occurs when approaching 60 min; structures already reached its maximal displacement and behave linearly for the remainder analysis duration. Therefore maximal ductility demand can be obtained form 60 min wind loading sample for generated wind loading. Table 3 shows the error (with respect to the customarily used one-hour wind sample) in predicting the ductility demand obtained from nonlinear analyses of 3 DOF structure having f_1 =1Hz assuming damping ratio, ξ of 1% is around 15-25%.

Table 3. Maximum displacement and ductility demand obtained from linear and nonlinear analyses.

Duration (min)	Ductility (μ)	Error (%)
10	5.39	-15.1
20	4.76	-25.1
30	5.72	-9.9
60	6.35	0.0

However, due to stochastic nature of wind loading, structures subjected to generated wind loading will be analyzed under different wind loading samples with the same characteristics to obtain mean response. So the same sensitivity study was established for 10 wind loading samples and it was found that the maximal difference of ductility demand was reduced to 10% which can be considered satisfactory for this preliminary study. Therefore 10 min samples will be used for different parameter evaluation in this study; but for the extended parametric study 60 min wind loading samples will be used.

4 WIND FORCE CALCULATION

4.1 *Wind speed data*

In the context of this study, structures will be subjected to both real (hurricanes wind speed measurement) and numerically generated wind samples. Five hurricane wind data sets were provided by Florida Coastal Monitoring Program (FCMP) [2]: Isabel-2003; Ivan-2004; Katrina-2005; Rita-2005; and Wilma-2005. Details of these hurricane measurement projects are given in Masters [13]. Wind speed were measured at z = 5 m and z = 10 m simultaneously which permits analyzing single and 2 DOF structures. Numerically generated wind samples were obtained for different turbulence intensity *Iu*, using WindGen program developed by Hang et al. [10] based on Power Spectral Density (PSD) proposed by Simiu [17] for along wind component. It is to be noted that WindGen generates only turbulent part of wind speed *u(t)* which must be combined with mean wind speed $\overline{U}$ that is set constant at 40 m.s^{-1} throughout this study.

4.2 *Wind force calculation in the literature*

4.2.1 *Quasi-static wind force*

Different wind forces loading on building where used on previous studies. Vickery [23] used a one-hour-sample record obtained from wind tunnel measurements. The other studies [4,8,21,24] used a wind sample numerically generated according to Monte Carlo process. In their calculations, wind forces were calculated adopting the quasi-static approach in which the quasi-static total drag wind force at a given height z and a time t is given by Simiu and Scanlan [18]:

$$F(z,t) = \frac{1}{2}\rho_a C_d A[\overline{U}(z) + u(z,t)]^2 \tag{1}$$

where ρ_a is the air density, C_d is the drag coefficient and A is the exposed area to the wind loading. For prismatic structures C_d is between 1 to 3 depending on many parameters such as the shape of the structure, the inclination of the wind direction with respect to the surface of the structure, wind turbulence, and the structure dimensions [11]. The drag coefficient C_d is selected at 1 in this study.

4.2.2 *Aerodynamic admittance function*

Sears [16] theoretically derived the aerodynamic admittance function for lift on thin symmetrical airfoil [3]. Large structure response will be overestimated due to 'lack of correlation' [20] which leads to the introduction of the aerodynamic admittance concept by Vickery [22] and Davenport [6] who proposed an empirical aerodynamic admittance function based on experimental data for drag force on square plate. Aerodynamic admittance is used on the frequency domain with spectral stochastic approach for structural response estimation through gust factor method which was introduced in design codes [1,7]. The drag wind force on a plate by Vickery [22] can be given as :

$$\chi^2(f) = \left(\frac{1}{1+\left[\frac{2f\sqrt{A}}{\overline{U}}\right]^{\frac{4}{3}}} \right)^2 \tag{2}$$

where f is the frequency, $\overline{U}$ is the mean wind speed and A is the area of the plate.

4.2.3 *Indicial function*

Indicial admittance function is used in civil engineering applications to evaluate aeroelastic phenomena in the time domain for wind sensitive structures. A procedure presented in Chang et al. [3] based on the work of Scanlan [15] to determine indicial functions for slender support structures is used in the current study to evaluate the wind drag force corresponding to the specified aerodynamic admittance function to be used in time history analyses. The derivative of the indicial function is given as :

$$\phi'(s) = A_1 e^{-A_2 s} + A_3 e^{-A_4 s} \tag{3}$$

where s is non dimensional time, $s=\overline{U}t/c$ and A_i are constants.

The admittance function is related to the indicial function through:

$$\Phi(K) = \int_0^{\infty} \phi'(s) e^{-iK\sigma} d\sigma \tag{4}$$

Wind force calculation using the admittance function was given in Tamura and Kareem [20] as:

$$F(z,t) = \overline{F}(z) + f(t) \tag{5}$$

$$f(t) = \rho_a A C_d \overline{U} \int_0^s [u(s) - \phi'(s-\sigma)]\, d\sigma \tag{6}$$

$$\overline{F}(z) = \frac{1}{2}\rho_a C_d A \overline{U}^2(z) \tag{7}$$

where $s=\overline{U}t/\sqrt{A}$.

4.3 *Wind force calculation through indicial function*

First, the aerodynamic admittance functions given in ASCE7-10 [1] for different structural characteristics were calculated. Then using Matlab fitting tools, indicial function constants A_i were identified for each case. The obtained results are summarized in Table 4. Figure 2 shows an example of the calibration of the indicial function to fit aerodynamic admittance given in ASCE 7-10 [1]. It appears from Figure 2 that there is a certain divergence between the two curves after frequency of 1.0 Hz and admittance less than 0.1 where the wind effects on the structural response could be neglected.

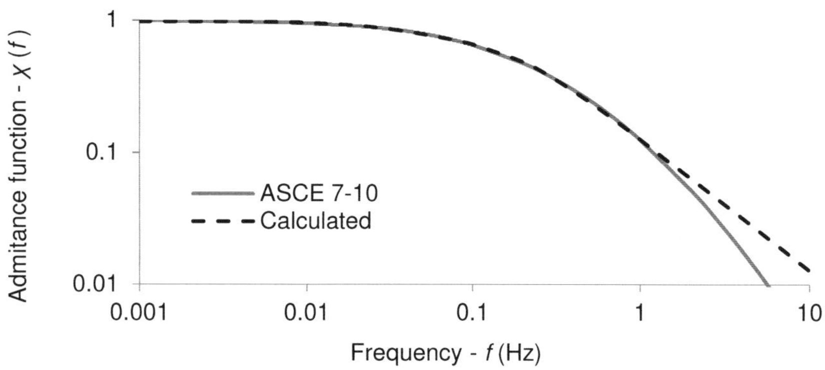

Figure 2. Admittance function from ASCE 7-10 compared to that calculated using the indicial function.

Table 4. Coefficients of indicial function corresponding to each structure geometry.

N° DOF	Height (m)	A_1	A_2	A_3	A_4
1	4	0.008261	0.0262	0.09176	0.138
2	8	0.008585	0.03096	0.1121	0.1594
3	12	0.00768	0.03161	0.1223	0.1665

4.4 *Structural response comparison*

Linear and nonlinear responses of structures subjected to wind forces calculated according to quasi-static and admittance approaches were compared in this section. Due to the limitation in space, only the results pertaining to the analyses of a structure (3 DOF; f_1=1 Hz; ξ=1%; β=0.8; I_u=30%) will be presented.

Figure 3 shows that the use of the indicial admittance function significantly reduced (by about 50%) the calculated wind force compared to the one estimated from the quasi-static approach. This reduction in the wind force, therefore, reduces the structural response. In linear analysis, Figure 4 shows that the displacement of the 1st level calculated adopting the indicial admittance function is 60% less the one computed with the quasi-static approach. Nonlinear response, on the other hand, is also noticeably affected by the wind force calculation methods as shown in Figure 5a, b even though the same reduction factor β=0.8 (20% reduction) was applied for both cases with reference to their linear counterparts. The hysteresis loop is larger in the case of aerodynamic admittance which leads to a higher ductility demand: $\mu_{(\text{Quasi-static})}$=1.35 and $\mu_{(\text{Admittance})}$=5.33 which is equivalent to a 75% underestimation of the ductility demand (Figure 5c). These results imply that, when performing time history analysis of structures under wind loading, ignoring aerodynamic admittance effect could lead to underestimation of the ductility demand. In this study it was decided to use the aerodynamic admittance approach to evaluate wind forces on structures.

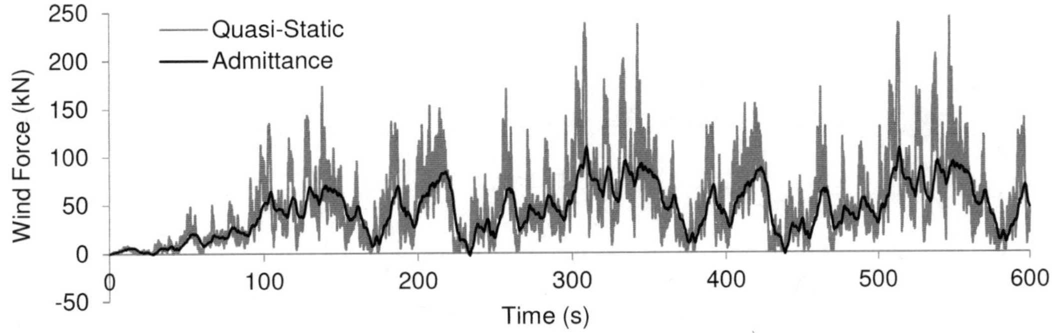

Figure 3. Wind force time history from quasi-static and admittance approach.

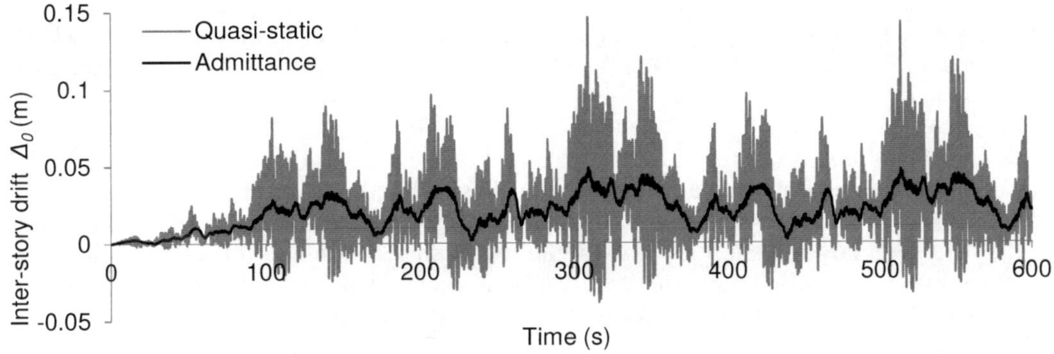

Figure 4. Linear response (displacement of 1st level) time history for quasi-static and admittance analysis case.

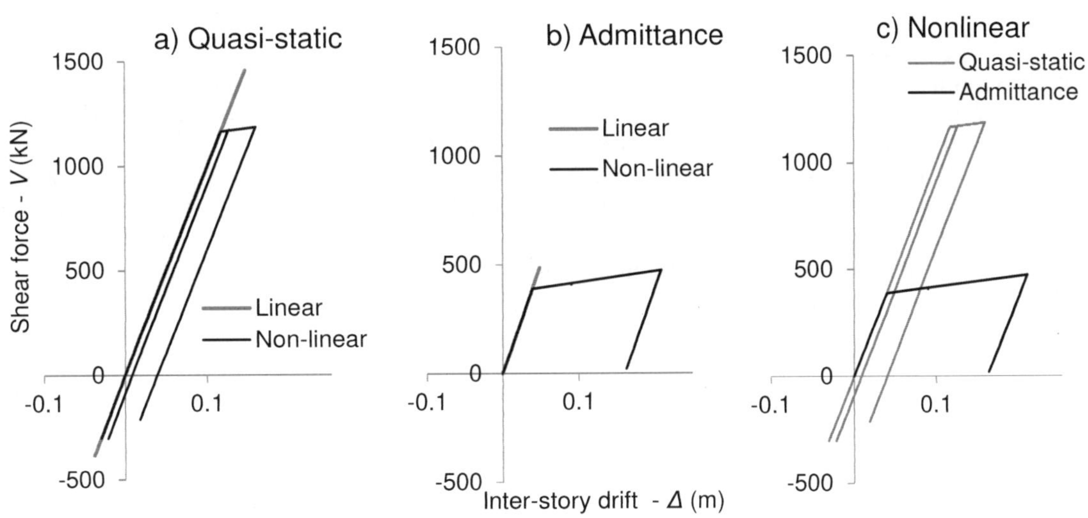

Figure 5. Force-displacement relationships for linear and nonlinear analyses from: a) quasi-static, b) admittance wind force calculation approach, and c) nonlinear analysis only.

5 DAMPING EVALUATION

5.1 *Recommended damping from literature*

Evaluation of damping is very important for vibrating systems. It allows the dissipation of energy and may have an important influence on the response of the structure. There is two main types of damping to be considered for buildings: aerodynamic and structural. They usually act in parallel to reduce structural response and total damping is given by:

$$\xi = \xi_a + \xi_s \tag{8}$$

where ξ, ξ_a and ξ_s are respectively the total, aerodynamic, and structural damping.

5.1.1 *Aerodynamic damping*

Aerodynamic damping is proportional to the relative speed between building and wind. Its effect appears mainly in flexible structures. Some empirical models are presented in literature to evaluate the aerodynamic damping according to several parameters (e.g., shape, dimensions, wind turbulence intensity). The aerodynamic damping is usually small for tall buildings (<1%) [21]. This was confirmed in Tamura and Kareem [20] who prescribed that ξ_a for typical tall building is about 0.5%.

5.1.2 *Structural damping*

Structural damping estimation for tall building under wind loading was investigated mainly by Tamura et al. [11,19]. An empirical equation giving the structural damping for steel tall building was established. For large displacement (ultimate limit states), Holmes [11] noted that much higher values may be used for the structural damping but without giving recommended values due to the lack of data measurement. This was confirmed by Tamura and Kareem [20] but they noted that "*damping ratio does not necessarily increase at higher amplitude level*". Recommended damping ratio to be used was proposed for different structures type and height based on Tamura et al. [19]: 1% for tall steel structure in the elastic range (higher 80 m) and 2% for structure less than 30 m.

5.1.3 *Rayleigh damping matrix*

Rayleigh damping matrix is commonly used for dynamic analyses [5,14]. Because of its simple mathematical formulation, Rayleigh damping is still the first choice to be used despite it has not a clear physical meaning. However it requires special attention for nonlinear seismic analysis as discussed in Hall [9] which explained through practical examples that damping force can be overestimated when stiffness decreases suddenly which leads to increase the structure velocity and the damping force may reaches high percentage compared to restoring force and even can be greater than it which is not realistic. Consequently it is important to evaluate real amount of damping introduced when performing nonlinear dynamic analysis.

5.2 *Sensitivity analysis on damping*

Since there is no explicit equations giving total damping for tall buildings under wind loading in the elastic and plastic range, it was pertinent to use a safe amount of damping. In the current analyses, a constant Rayleigh damping was adopted and the effect of aerodynamic damping will be evaluated later in this paper.

5.2.1 *Evaluation of the use of Rayleigh damping for nonlinear analysis*

In order to validate the use of Rayleigh damping for nonlinear analysis, the ratio of damping force f_d to restoring force f_s during nonlinear analysis was examined. For example, as shown in Figure 6 for a SDOF structure (f_1=0.5 Hz; ξ=1%; β=0.8) subjected to hurricane wind force from Isabel-2003, it was found that the ratio of damping force to restoring force was less than 0.6% for all the analysis duration. Figure 6 indicates also that the problem of the damping force over-estimation due to stiffness loss is not problematic in the considered cases, and Rayleigh damping can be used without problems even for nonlinear time-history analysis. This can be explained by the presence of mean wind speed which gives a considerable value for restoring force compared to damping force.

5.2.2 *Influence of damping on ductility demand*

Figure 7a shows the time histories of the inter-story drift from the linear analysis of a SDOF structure (f_1=0.5 Hz) subjected to Isabel-2003 wind loading with and without Rayleigh damping (ξ=1%). The displacement was significantly reduced by 40% when 1% of damping was introduced. For the nonlinear analysis, the results shown in Figure 7b indicate that the ductility demand is increased when 1% of damping was used. In fact, the reduction of maximal shear force in the linear analysis leads also to the reduction of yielding shear forces by the same amount which explain the higher ductility demand when damping amount is greater; due to design procedure, two structures with different yielding shear forces are analyzed. However when specifying the same yielding shear force and analyzing the same structure with different damping amount, ductility demand becomes greater for lower damping value. Therefore, it can be concluded that for nonlinear analysis conducted in this study, the introduction of higher damping amount leads to a higher plastic displacement which means higher ductility demand.

Figure 8a, b shows two different examples of the influence of damping amount on ductility demand for 2 analysis cases; a) 2 DOF structure (f_1=1 Hz) subjected to Katrina-2005 wind forces; and b) SDOF structure (f_1=0.5 Hz) subjected to Isabel-2003. Focusing on the cases where the aerodynamic damping is not accounted for, the damping has a different influence on the ductility demand. Unlike the former case where the damping has a significant effect on the ductility demand (Figure 8a), it appears from (Figure 8b), that the damping in the latter has no practical influence on the ductility demand. In Figure 8a, the ductility demand was reduced by 30% when ξ=0.5% is used and increased by 32% for ξ=2% with respect to the ductility demand at ξ=1%.

All of these results show the importance of the consideration of the right amount of damping when conducting nonlinear time history analysis; under estimating damping amount can lead to underestimation of nonlinear response and damage accumulation.

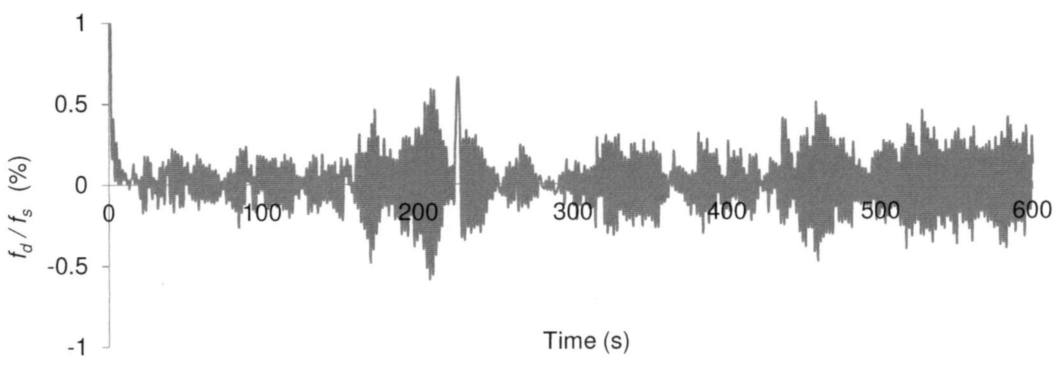

Figure 6. Ratio of damping force f_d to restoring force f_s time history

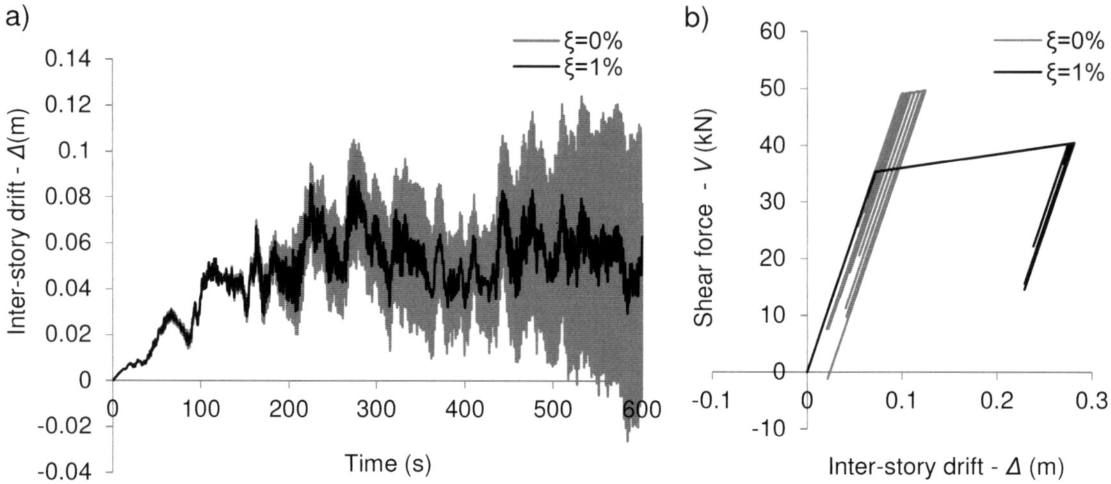

Figure 7. a) Linear time history response and b) nonlinear force displacement diagram for ξ = 0 % and 1%

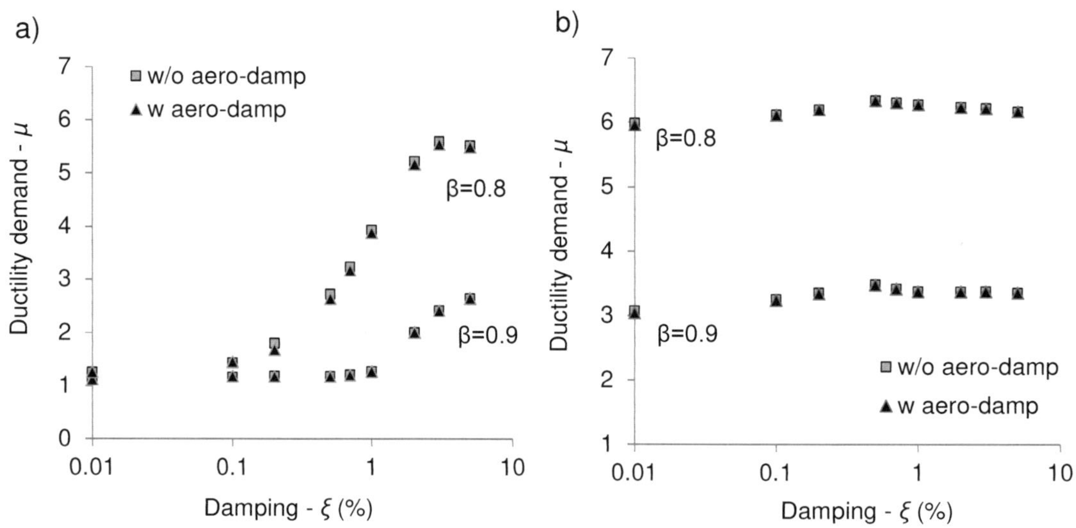

Figure 8. Ductility demand function of damping amount with consideration or not of aerodynamic damping.

5.2.3 *Aerodynamic damping evaluation*

Because of the sensitivity of nonlinear response of structures under wind events to the used amount of damping as previously demonstrated, the aerodynamic damping was evaluated. To do so, the wind forces acting on the building were calculated based on the relative speed between wind and structure. But due to the excessive calculation time, it was also pertinent to evaluate the amount of aerodynamic damping to be introduced as additional constant Rayleigh damping in addition to structural damping.

As it was shown in Figure 8a, b, for both situations where damping was highly influencing ductility demand or not, the aerodynamic damping effect is insignificant. This permits to neglect it and to only consider the recommended structural damping amount ξ=1% in the analyses of high rise buildings. However, it is important to note that the influence of aerodynamic damping may need to be re-evaluated for each analysis of high rise buildings.

6 CONCLUSION

The present study evaluated the influence of different hypotheses on ductility demand for structures subjected to time history wind loading and gives recommendations about adequate values to be used in extended parametric studies. The results presented in this paper lead to the following conclusions :

- Analysis time-step to be used on linear and nonlinear analysis must be the same otherwise error on ductility demand can reach 80%. The optimal recommended time-step to be used is $\Delta t=0.01s$ which yields an error of less than 2%.
- Wind loading duration influences considerably the ductility demand of structures. In the context of this type of study one hour wind loading samples are sufficient for generated wind loading samples whereas 3 or 4 hours wind loading samples is needed when using hurricane real wind measurement. Ten minute wind loading sample can be used when evaluating influence of different parameters with an error around 15% for only one wind loading sample and 10% for 10 wind loading sample.
- The application of indicial function related to the aerodynamic admittance when calculating wind forces on structures influences significantly the ductility demand compared to the quasi-static approach; the ductility demand may be underestimated by 75%. Therefore, it is recommended to use the aerodynamic admittance approach through indicial function when analyzing nonlinearly structures subjected to wind loading in the time domain.
- Constant Rayleigh damping can be used for nonlinear time history analysis of structures subjected to wind loading without producing excessive damping amount when yielding of structures occurs.
- Damping amount introduced in the analysis may influences significantly the ductility demand of structures whereas aerodynamic damping can be neglected in the context of this study; damping amount ξ=1% is the recommended value to be used.

7 ACKNOWLEDGEMENTS

The authors gratefully acknowledge the National Sciences and Engineering Research Council of Canada (NSERC) for this financial support of this research project and also the Fonds de recherche du Québec sur la nature et les technologies (FRQNT) for the financial support of the Centre d'études interuniversitaire des structures sous charges extrêmes (CEISCE). The authors would like to acknowledge the contribution of the Florida Costal Monitoring Program (FCMP) for providing hurricane wind data sets used in the present study.

8 REFERENCES

1 ASCE7-10, Minimum design loads for buildings and other structures. ASCE, Reston, VA, 2010.

2 J. Balderrama, F. Masters, K. Gurley,D. Prevatt, L. Aponte-Bermúdez, T. Reinhold, J.P. Pinelli, C. Subramanian, S. Schiff, A. Chowdhury, The Florida Coastal Monitoring Program (FCMP): a review, Journal of Wind Engineering and Industrial Aerodynamics, 99 (2011) 979–995.
3 B. Chang, P. Sarkar, B. Phares, Time-domain model for predicting aerodynamic loads on a slender support structure for fatigue design, Journal of Engineering Mechanics, 136 (6), 2010, 736-746.
4 D. Chen, A. Davenport, Vulnerability of tall buildings in typhoons, In: Advances in Structural Dynamics, 2000, vol. II, pp. 1455–1462.
5 A.K. Chopra, Dynamics of Structures: Theory and Applications to Earthquake Engineering, fourth edition Prentice Hall, 2011.
6 A.G. Davenport, Gust loading factors, ASCE J Struct Div 93 (1967) 11–34.
7 NF EN 1991-1-4/A1, Actions sur les structures : Actions générales-Actions du vent, AFNOR, France, 2010.
8 F. Gani, F. Légeron, Relationship between specified ductility and strength demand reduction for single degree-of-freedom systems under extreme wind events, Journal of Wind Engineering and Industrial Aerodynamics 109 (2012), 31-45.
9 J.F. Hall, Problems encountered from the use (or misuse) of Rayleigh damping, Earthquake Engineering and structural dynamics, 35 (2006) 525-545.
10 S. Hang, F. Gani, F. Légeron, Générateur de vent appliqué au génie civil, Manuel de logiciel, Université de Sherbrooke, 2005.
11 J.D. Holmes, Wind Loading of Structures, third edition CRC Press, 2015.
12 H. Hong, Accumulation of wind induced damage on bilinear SDOF systems, Wind and Structures, 7 (2004) 145–158.
13 F.J. Masters, Measurement, Modeling and Simulation of Ground-level Tropical Cyclone Winds, Ph.D, Thesis, University of Florida, 2004.
14 Paultre, P., 2005. Dynamique des structures, application aux ouvrages de génie civil, Hermes Science/Lavoisier.
15 R.H. Scanlan, Problematic on formulation of wind-force models for bridge decks, Journal of Engineering Mechanics, 119 (7), 1993, 1353-1375.
16 W.R. Sears, Some aspects of nonstationary airfoil theory and its practical applications, J Aeronaut. Sci., 8 (1941) 104-108.
17 E. Simiu, Wind spectra and dynamic alongwind response, ASCE Journal of the Structural Division, 100 (1974) 1897–1910.
18 E. Simiu, R. Scanlan, Wind Effects on Structures, third edition John Wiley & Sons Inc, 1996.
19 Y. Tamura, K. Suda, A. Sasaki, Damping in buildings for wind resistant design, In: Proceedings of the international symposium on wind and structures for the 21st century, 26–28 January 2000, Cheju, 2000, pp 115–130.
20 Y. Tamura, A. Kareem, Advanced Structural Wind Engineering, Springer Japan, 2013.
21 T. Tschanz, The Base Balance Measurement Technique and Application to Dynamic Wind Loading of Structure, Ph.D, Thesis, University of Western Ontario, London, Ontario, 1982.
22 B.J. Vickery, On the flow behind a coarse grid and its use as a model of atmospheric turbulence in studies related to wind loads on buildings, National Physical Laboratory Aero Rep. No. 1143, 1965.
23 B.J. Vickery, Wind action on simple yielding structures, Journal of Engineering Mechanics Division 96 (1970) 107–120.
24 T. Wyatt, H. May, The ultimate load behaviour of structures under wind loading, In: Proceedings of the Third International Conference on Wind Effects on Buildings and Structures, Tokyo, Japan, 1971, pp. 501–510.

The 13th Americas Conference on Wind Engineering (13ACWE)
Gainesville, Florida USA, May 21-24, 2017

Validation of CFD Models for a Virtual Aerodynamics Laboratory

M. Sitek[a], S. Lottes[b], C. Bojanowski[c], H. Bosch[d]

[a] *TRACC/ANL, Lemont, Illinois, U.S.A., msitek@anl.gov*
[b] *TRACC/ANL, Lemont, Illinois, U.S.A., slottes@anl.gov*
[c] *TRACC/ANL, Lemont, Illinois, U.S.A., cbojanowski@anl.gov*
[d] *FHWA/TFHRC, McLean, Virginia, U.S.A., harold.bosch@dot.gov*

ABSTRACT: Full scale Computational Fluid Dynamics (CFD) models of FHWA's Turner Fairbank Highway Research Center Aerodynamics Laboratory were developed, including the room, the furniture and the wind tunnel. The effort focused on identifying which models were most computationally efficient and accurate enough to serve as a virtual laboratory. The results were validated against measurements of velocity at various locations in the laboratory at different fan speeds. A comparison of the models was done in terms of the simulation time needed to obtain a converged solution. One of the models was chosen for further assessment and used to compute drag forces acting on specimens that had been tested in the laboratory. The results showed a good agreement between the computational and experimental wind flow in the laboratory room.

KEYWORDS: aerodynamics, CFD modeling, wind tunnel, virtual laboratory.

1 INTRODUCTION

The subsonic open-jet wind tunnel in the Aerodynamics Laboratory at the Turner-Fairbank Highway Research Center is used to study the responses to wind of a variety of bridge structures and roadside transportation hardware. Scaled models of the structures to be studied are built and then usually placed in a force balance just downstream of the wind tunnel exit. Figure 1 shows a specimen in the force balance. To assess the effects of the laboratory walls and furniture and other objects of significant size on the wind tunnel air jet in the test zone just downstream of the wind tunnel exit, a CFD model of the aerodynamics laboratory including all of these elements was built, and the results were provided in a technical report [1]. This CFD model was extensively revised using a new parts based meshing feature of the CFD software [2] that allows for easy movement and replacement of objects in the wind tunnel test zone as well as any other furniture and equipment in the room to match a current arrangement of interest in the physical laboratory. The original CFD model was reworked to provide a virtual wind tunnel laboratory in the form of several CFD models that could be used to extend the capabilities of the physical aerodynamics laboratory at the Turner-Fairbank facility.

Using the virtual model in conjunction with physical wind tunnel testing provides the following extended capabilities: (1) Much larger parametric series of tests can be performed than those that can be accomplished in the physical laboratory within available time and budget. Many more geometry variations in scale models of bridge structures and roadside hardware can be investigated in the virtual wind tunnel without the need to build a large number of physical scale models. (2) The responses of objects to wind speeds that are much higher than those that can be achieved in the physical laboratory due to the maximum rotation velocity of the physical fan can be studied using one of the CFD models. And, (3) tests that are very difficult or impossible to run in the physical laboratory can be done with one of the CFD models. Examples of such tests include objects moving with respect to each other, such as a truck moving on a bridge deck past bridge towers in high wind conditions, and temporal variations in wind speed that are not possible in the physical laboratory due to the inertia of the fan, and wind combined with rain.

The CFD models developed in the project included (1) full room and wind tunnel model with rotating fan, (2) full room and wind tunnel model with fan in a rotating reference frame that does not actually move with respect to the fan cage but imparts momentum to the air in the fan blade region that would be imparted by rotating blades, and (3) simplified CFD models that exclude fan and most of the interior of the wind tunnel and impose a wind velocity profile at the last screen in the wind tunnel as an inlet condition for flow into the room and in particular the test zone just downstream of the wind tunnel exit. The CFD models provide for variations of the wind tunnel physics including k-epsilon URANS and LES turbulence models.

2 CFD MODEL DEVELOPMENT

Three types of models were developed in this study: (1) full geometry of the room and the wind tunnel with rotating centrifugal fan, (2) full geometry of the room and the wind tunnel, but rotating reference frame used for fan with no grid motion, (3) full geometry of the room, with the internal components of the wind tunnel removed, the domain inlet at the last screen and outlets at the inflow surfaces of the fan cage with velocity inlet and pressure outlet boundary conditions defined at those locations.

(a)

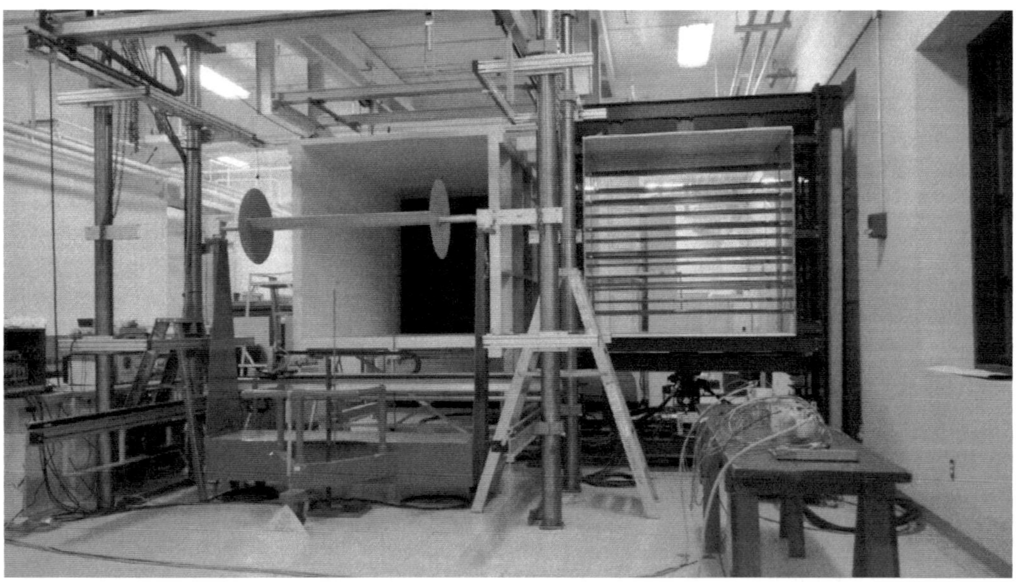

(b)

Figure 1. Views of the Aerodynamic Laboratory: (a) front of the extension of the wind tunnel and the turbulence generator, (b) a view of the room from the left side of the wind tunnel extension

In the first model type, the rotation of the fan was introduced explicitly. The rotating fan region was separated from the domain using a cylindrical geometry part with a sliding interface between the fan and the surroundings. The second type made use of a rotating reference frame feature available in the software. It imposed forces on the fluid that are induced by the rotating fan blades without changing the position of cell vertices. In the third model, all structural features from the inside of the wind tunnel up to the last downstream screen, were removed. The last screen was converted into a velocity inlet and the two interfaces between the fan inlet and the room were defined as pressure outlet boundaries.

Inside the full wind tunnel model there were nine mesh wire screens whose role was to make the flow more uniform in the downstream regions. They were introduced to the CFD model as porous baffle interfaces. The porosity, χ, of the screens was determined to be equal to 0.67, based on the specification sheets provided by TFHRC [3].

The pressure drop across a screen, modeled as a porous baffle, is computed with the following equation [2]:

$$\Delta p = -\rho(\alpha|v_n| + \beta)v_n, \tag{1}$$

$$\alpha = 0.5 \cdot 6.5 \left[\frac{1-\chi}{\chi^2}\right]\left[\frac{v_n d}{\chi v}\right]^{-1/3} \tag{2}$$

with ρ is the density of air, α is the porous inertial resistance, $v_n$is the velocity normal to the screen, d is the wire diameter, and the porous viscous resistance $\beta = 0$ [4].

Figure 2 illustrates a view of the model geometry of the laboratory with the test zone just downstream of the wind tunnel outlet. The room has dimensions of 27.4 m by 11.7 m by 4.9 m (90 ft by 38.4 ft by 16.1 ft). The wind tunnel is 16.8 m (55.1 ft) long, including the extension, and 3.8 m (12.5 ft) high in the highest point, which is at the screen closest to the outlet. The extension dimensions are 1.8 m by 1.8 m (6 ft by 6 ft). The differences in the wind tunnel geometry between the full and simplified models are illustrated in Figure 3. In the full model all parts of the structure are represented. The fan as well as 9 mesh screens, modeled as porous baffles, are visible inside the structure. The inside of the wind tunnel in the simplified model is empty and is not a part of the computational domain. The flow domain has a velocity inlet boundary condition surface at the screen nr 1 (see Figure 3(a)) and pressure outlet boundary condition at the fan air intake surfaces.

A series of measurements was performed in the aerodynamics laboratory. The velocity magnitude and direction were recorded across the laboratory room, in front of the wind tunnel extension and on the sides of the wind tunnel, 7 feet above the floor. The measurements were done at 437 rpm fan speed. Moreover, the velocity magnitude was measured inside the extension and in front of it at various fan speeds, from 100 rpm to 500 rpm. Figure 4 shows the locations of the points where the measurements were done in each case.

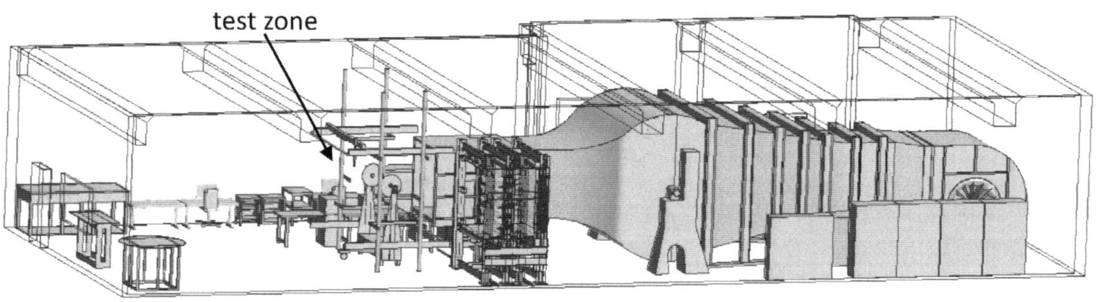

Figure 2. View of the domain geometry

(a)

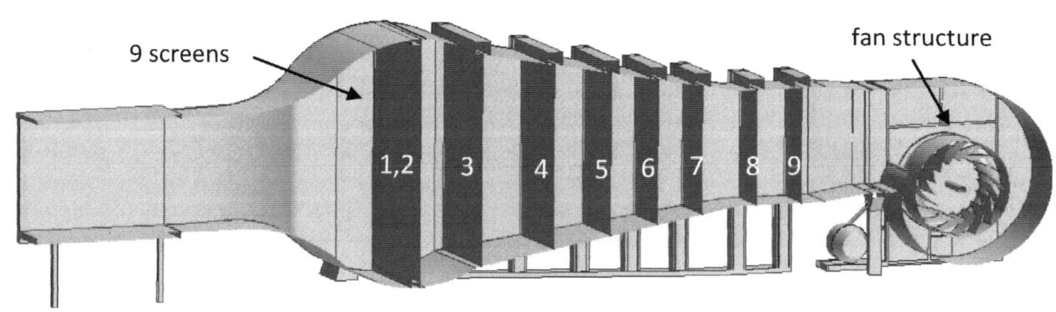

(b)

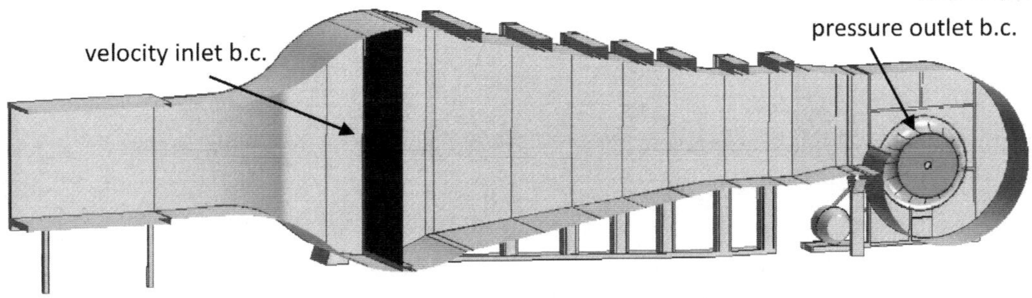

Figure 3. Views of the inside structure of the wind tunnel models: (a) full model with the fan and screens, (b) the simplified model with a velocity inlet b.c. at the last screen and pressure outlet b.c. at the fan air intake surfaces

(a)

(b)

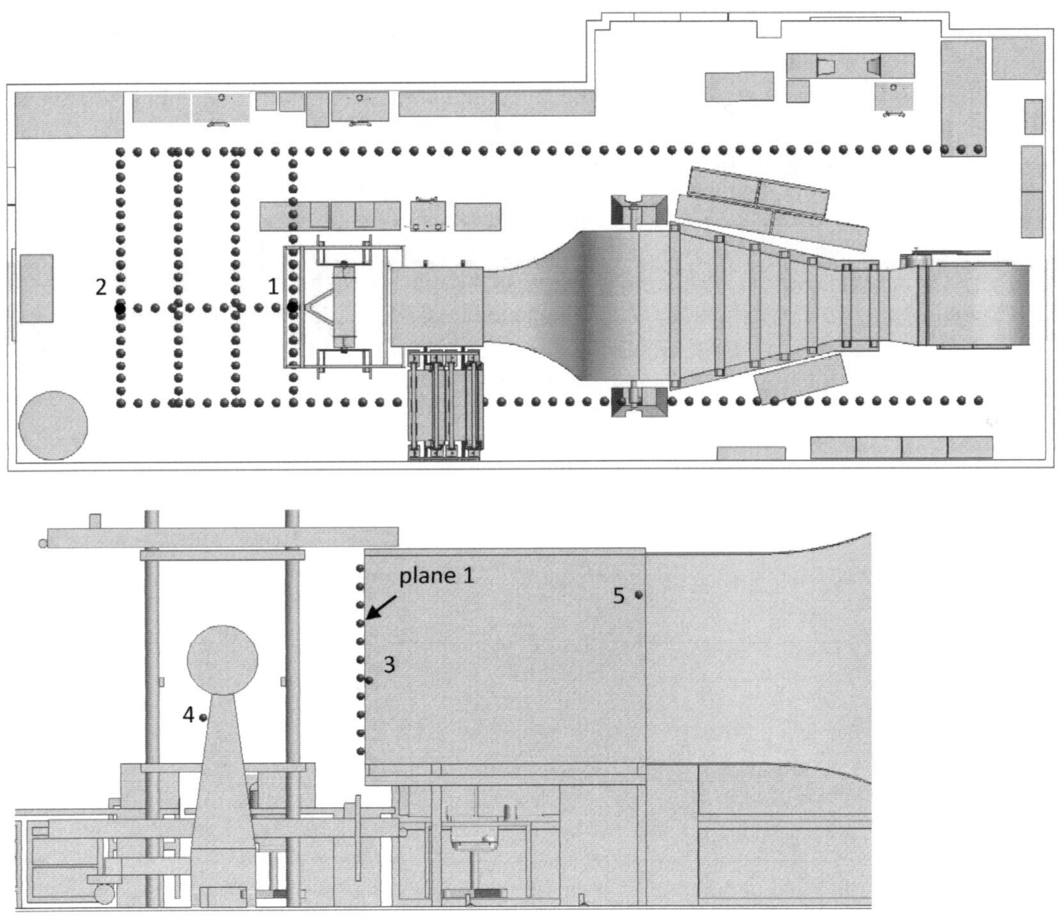

Figure 4 Locations of the measurement points: (a) across the laboratory room, and (b) inside, and in front of, the wind tunnel extension

3 DISCUSSION OF RESULTS

3.1 *Comparison of the solution computation clock time of the CFD models*

Several models with various solvers and modifications in geometry were prepared. Table 1 combines the information about the types of models, the solvers used, and the number of computational volume cells in the model. Also the run times and clock times are compared for runs on 64 cores, unless otherwise noted.

The model with rotating fan required the longest run time to obtain a converged solution. It took about 12 days on eight 16-core nodes to finish the simulation. This was the most complicated model in terms of the model setup. It used the segregated flow solver and a rigid body motion solver to model the rotation of the fan. The entire domain was meshed, including the inside of the wind tunnel between the mesh screens, and the fan cage. The cells had the smallest size in the region containing the fan in order to accurately resolve the fan blades and the flow induced by them. The rotation of the fan was modeled as a rigid body motion of the region around the fan axis with the specified rotational velocity. The sliding mesh model was used to couple the moving mesh region containing the fan to the stationary mesh of the region outside the fan. The time step used was small, 0.001 sec, which is a result of both the small mesh cell size and the need to temporally resolve the rotational motion of the fan. The finest mesh was assigned in the fan region, where the cells size is approximately 0.02 m and the highest flow velocity reaches about 34 m/s. This means that the flow coming off the fan blades passes through less than two cells in a time step.

A rotating reference frame feature made it possible to reduce the solution clock time, as the motion of the fan was not modeled explicitly. The moving reference frame does not change the position of cell vertices, but imposes the forces on the fluid that are induced by the rotation of solid surfaces. The simulation ran almost six times faster than the model with a rotating fan.

To reduce the computational time and resources needed, modifications in geometry were proposed. In the model presented here, most of the internal structure of the wind tunnel, including the fan, was removed from the flow domain. The mesh screen closest to the tunnel exit was transformed into a velocity inlet type of boundary with a uniform velocity distribution. The two surfaces where air enters the fan cage were changed into pressure outlet boundary condition surfaces in the simplified model. These modifications reduce the solution clock time significantly and the computations could be finished in under two hours.

Table 1 Types of model developed and their run times

Type	Solver	Number of cells [10^6]	Simulation time / clock time (four 16-core nodes, unless otherwise noted)	Clock time [days]
rotating fan	URANS, k-ε	6.3	1 sec/36 h (8 nodes)	12
rotating reference frame	URANS, k-ε	6.6	1 sec/40 min	2.2
	LES	7.7	1 sec/23 h (8 nodes)	7.7
simplified (uniform inlet velocity)	RANS, k-ε	2.4	1000 iter./40 min	0.06
	LES	4.4	1 sec (2000 iter.) /52 min	0.35

3.2 *Comparison of the experimental and computational results*

The velocity measurements at various locations in the laboratory room were used as a reference in the validation analysis of the model. The velocity values and direction are monitored on a horizontal plane, 7 feet above floor. Plots of the velocity magnitude along the probe lines were presented in [5]. In this paper the time averaged velocity magnitude values at two points, located in front of the wind tunnel exit (marked in Figure 4 (a) with dots) are given in Table 2. These two points were chosen because they were located in the air jet in line with the center of the wind tunnel exit, one closer to the wind tunnel and the other close to a wall. Both the laboratory measurements and the CFD computed velocities at these points oscillated about a mean value. The magnitude of the oscillations was much larger for the measurement point near the wall. The experimental and computational velocity vector fields are compared in Figure 5.

Table 2 Velocity magnitude at points 1 and 2, see Figure 4 (a)

Point number			1	2
Average Y velocity [m/s]	experiment		10.5	9.8
	model with rotating fan	URANS, k- ε	11.5	7.3
	model with rotating reference frame	URANS k-ε	11.1	7.6
		LES	10.4	9.5
	model with velocity inlet and pressure outlet boundary conditions	RANS, k- ε	10.7	10.7
		LES	10.8	7.8

The time-averaged velocity flow field measured in the lab is shown in Figure 5(a). The air jet was very uniform up to the front wall, with the maximum velocity reaching 10.5 m/s at point 1 and 9.8 m/s at point 2, at fan speed 437 rpm. The standard deviations of velocity magnitude and its direction at point 1 were equal 0.05 m/s and 1.46 degrees, respectively. The measurements at point 2 showed that the fluctuations had much higher amplitudes and resulted in standard deviations of velocity magnitude equal to 0.54 m/s and direction deviation equal to 7.38 degrees. Conditions at point 2 are very sensitive to the turning of the jet at the rear wall of the laboratory, and simulated points very close to the point 2 position matched the velocities to two significant figures. Therefore a very small change in turning angle of the jet may appear as a 30 percent under-prediction of velocity at that point. The raw data from laboratory measurements before averaging also showed large fluctuations with a fair amount of uncertainty in the values. The mismatch between experiment and simulation was not considered to be a significant problem in terms of using the model for studying wind effects on models of transportation infrastructure placed in the test zone.

The computational vector fields are shown in Figure 5 (b)-(f). The quantitative difference between the experimental and computational results was small. The difference in wind speed between experimental measurements and CFD results depended on the location in the room. The values were close to experimental in the wind tunnel test zone and its vicinity. Discrepancies appeared in the locations closer to the front wall due to variations in the position of the jet as it turned for the return flow. The models with recirculating air flow (Figure 5 (b)-(d)) showed oscillations of the velocity values across the room during the simulation, especially outside of the test zone and closer to the walls. Breaking the feedback loop of the recirculating air model in the simplified model (Figure 5 (e)-(f)) eliminated these oscillations as well as the decrease in velocity close to the walls. To assure the correctness of results, the mass flow through the wind tunnel was tracked. It was balanced and constant in all simulations, because even though the velocities in the jet varied during the simulation, the average velocities at the screens were constant. The results obtained with the LES solver (Figure 5 (d), (f)) showed a turbulent flow around the fan, with numerous eddies, which were smoothed out by the wind tunnel screens. The air flowing out of the wind tunnel exit was uniform. Eddies formed on the sides of the jet and also in the downwind part of the jet, as the air was deflected by the wall.

(a)

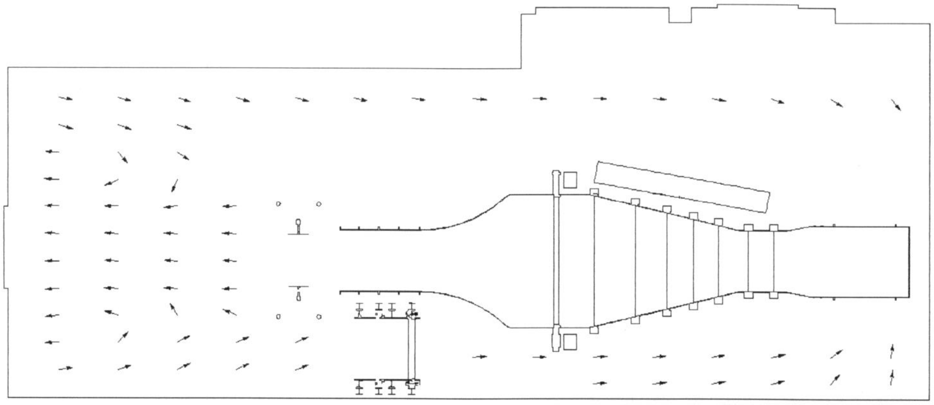

(b)

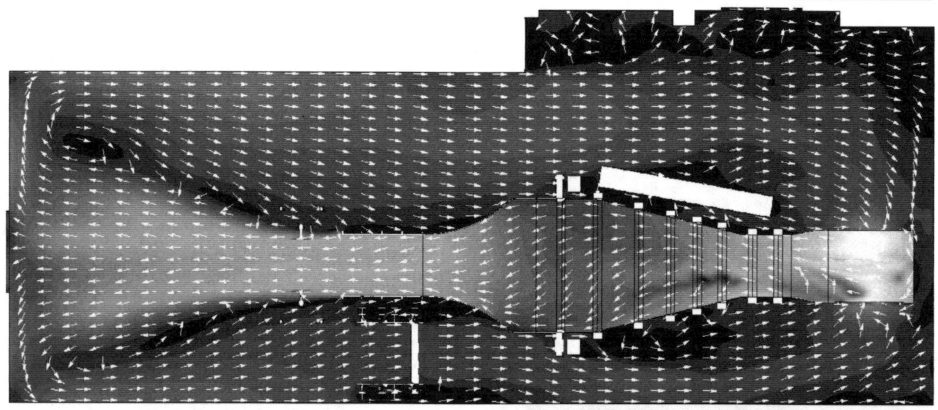

(c)

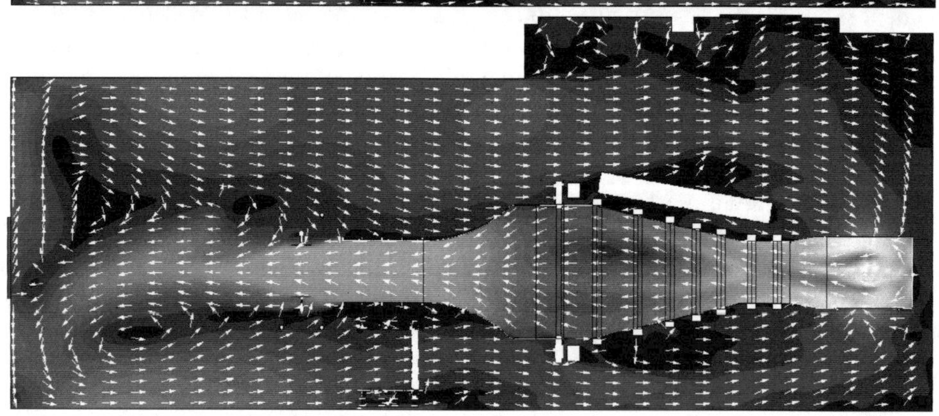

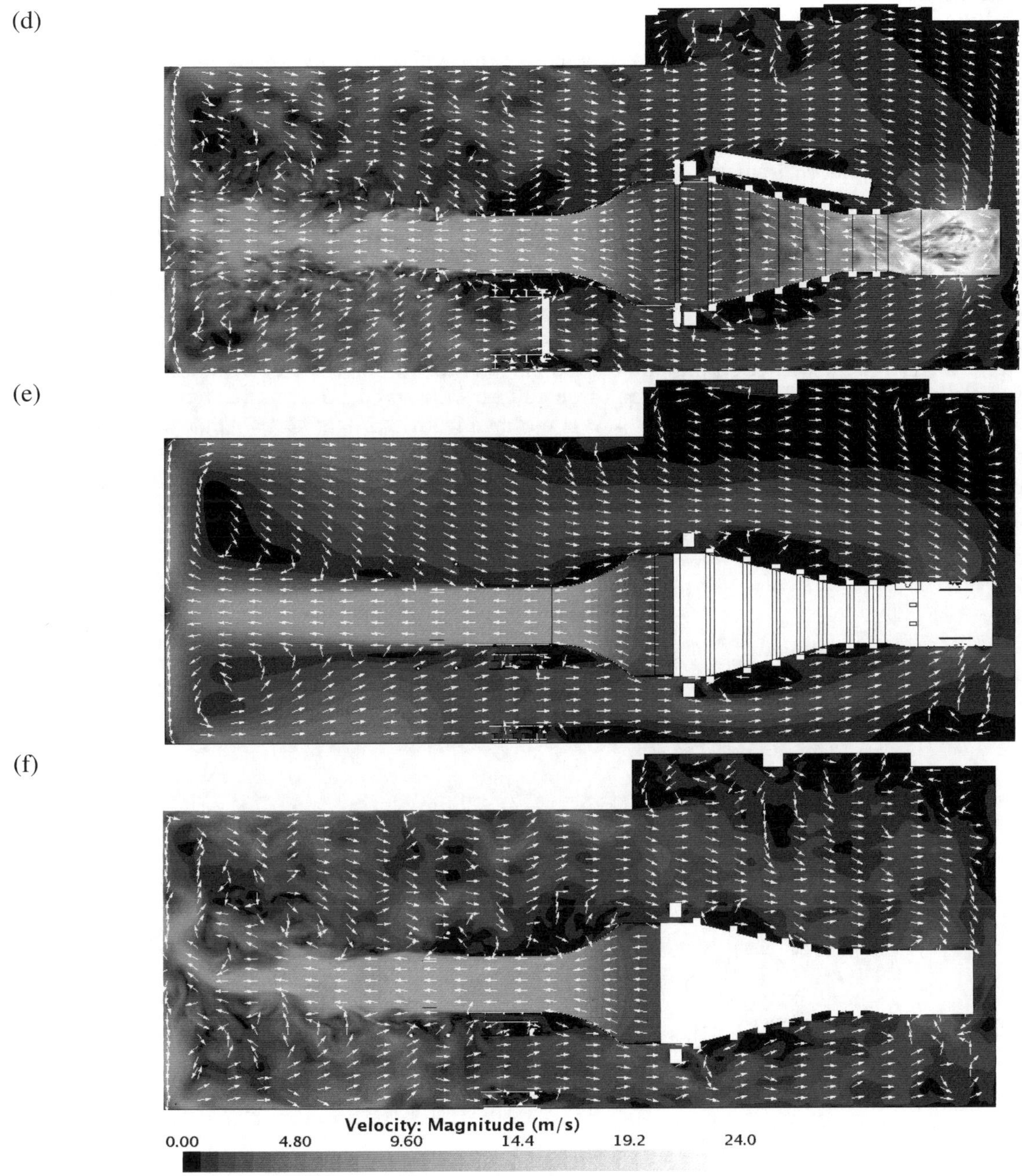

Figure 5 Velocity magnitude and vectors on a horizontal plane 7 feet above the floor, (a) experimental results mapped onto the CFD model, (b) model with rotating fan, (c) model with rotating reference frame, URANS k-e solver, (d) model with rotating reference frame, LES solver, (e) simplified model, URANS k-e solver, (f) simplified model, LES solver.

3.3 *Validation of the simplified model*

The full model with rotating reference frame was run with different fan speeds to calibrate the simplified model. At each fan speed the surface averaged velocities at the last screen were recorded. They were used to define the inlet velocity in the simplified model. Removing the inside structure of the wind tunnel reduced the number of volume cells in the model. Also, the air flow in the room was not recirculated anymore with consequent feedback effects and therefore the jet was more uniform and most of the oscillations in the jet flow velocity were eliminated. This model gave a converged solution much faster than the previous models, which allowed more cases to be run in a shorter time.

Simulations were run with inlet velocities corresponding to fan speeds in the range from 100 rpm to 500 rpm. Velocities were monitored at four locations inside and in front of the wind tunnel extension, as illustrated in Figure 4(b). Figure 6 presents a comparison of laboratory measurements with a linear function fitted to the points, and also computational results with 5% error bars are shown. A very good agreement between these values is observed. The relative deviation for point 3 varies from 4.7 % to 5.9%, for point 4, it is in range from -7.2% to -5.1%, and for point 5 from 2.8% to 4% deviation. The relative deviation increases with higher fan speed at points 3 and 5, and decreases for point 4. The deviation of velocities at plane 1 doesn't change significantly, with values between 3.6% and 3.8%.

(a)

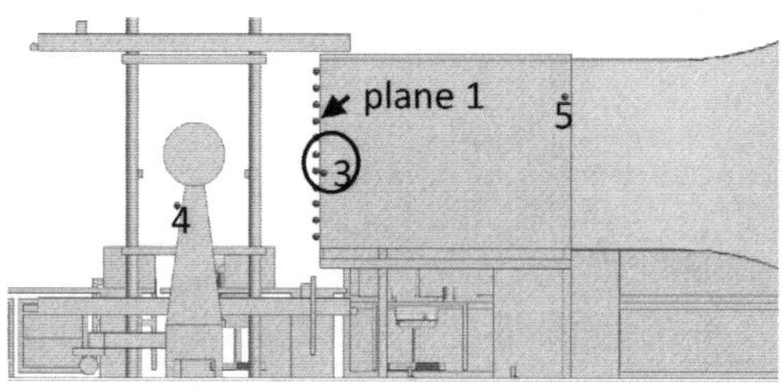

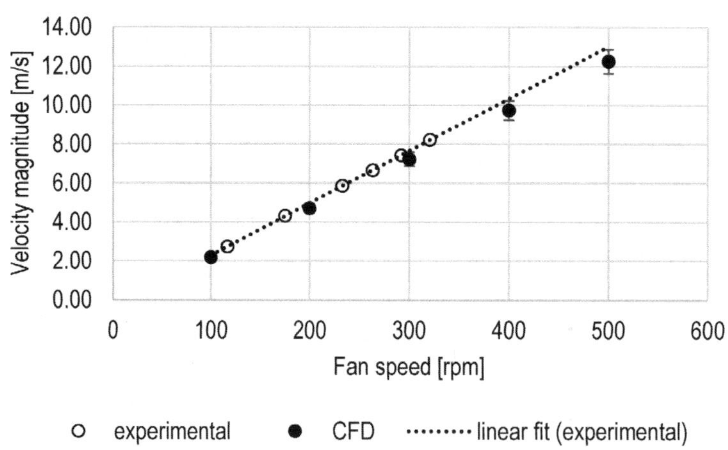

(b)

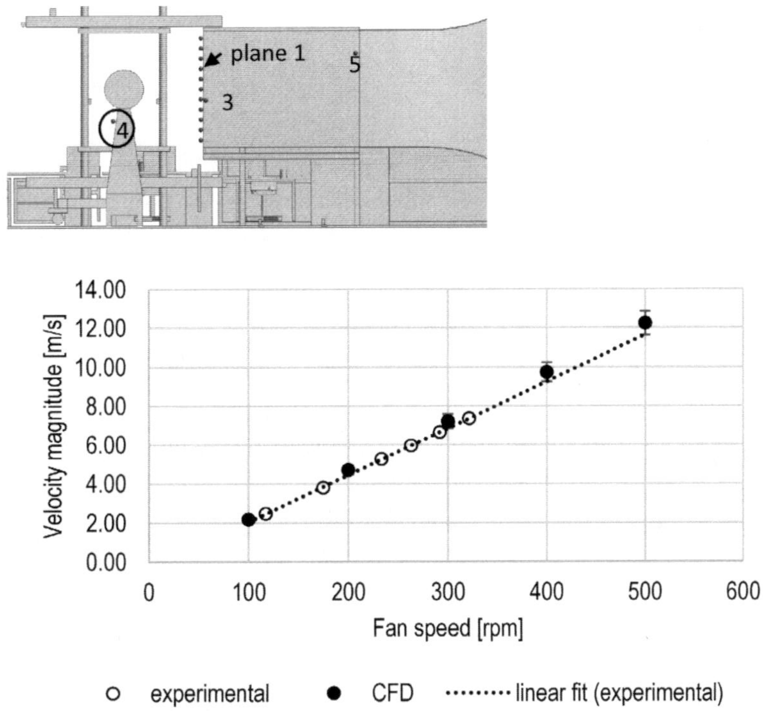
plane 1
3
4
5
Velocity magnitude [m/s]
14.00
12.00
10.00
8.00
6.00
4.00
2.00
0.00
0
100
200
300
400
500
600
Fan speed [rpm]
experimental
CFD
linear fit (experimental)

(c)

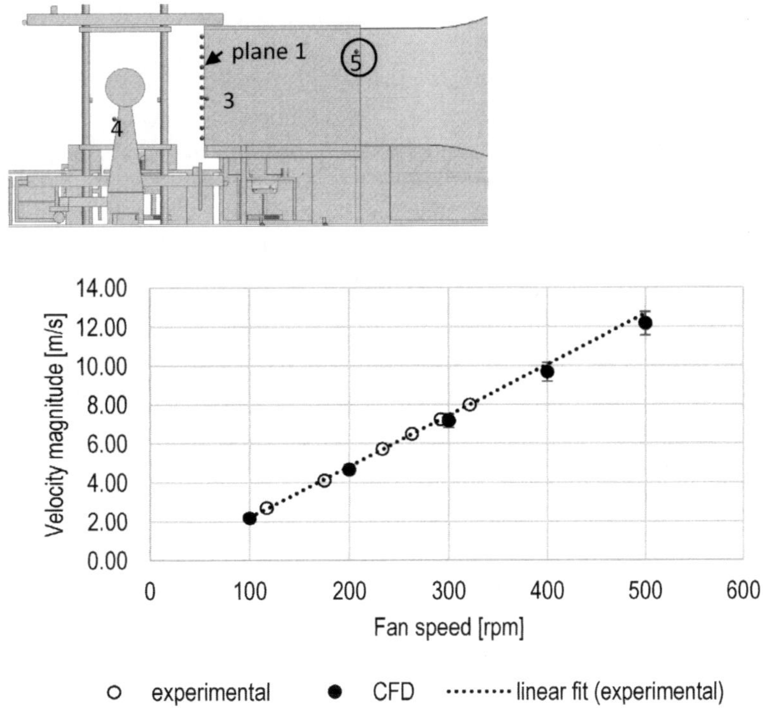
plane 1
3
4
5
Velocity magnitude [m/s]
14.00
12.00
10.00
8.00
6.00
4.00
2.00
0.00
0
100
200
300
400
500
600
Fan speed [rpm]
experimental
CFD
linear fit (experimental)

(d)

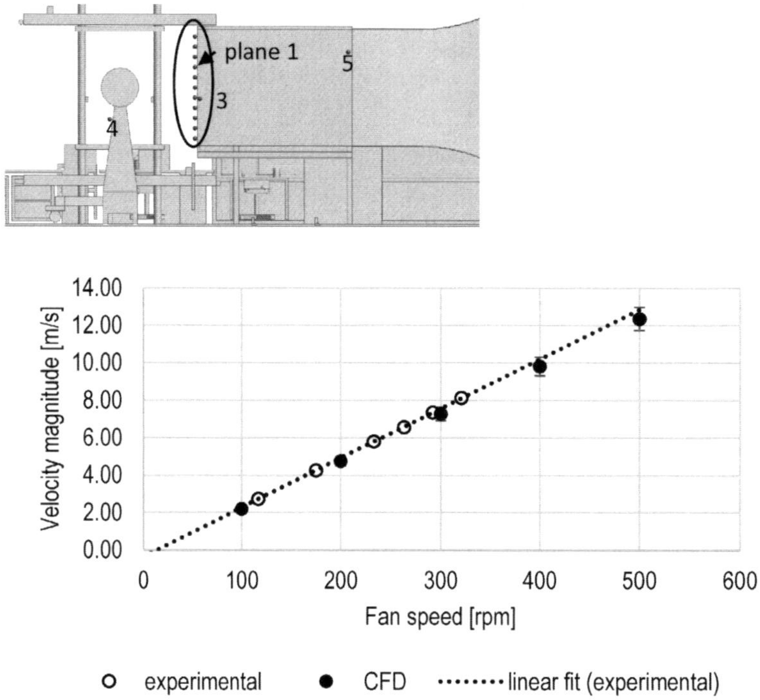

Figure 6 Velocity magnitude at: a) point 3, b) point 4, c) point 5, d) plane 1

3.4 *The virtual lab*

The main goal of developing the CFD model of the Aerodynamics laboratory room is to use it as a full size 'virtual lab' to expand the number of cases that can be tested beyond those that can be accomplished with experimental testing. The building of physical models and the testing of them in the wind tunnel can be very time consuming. When a very large number of parametric cases need to be tested, it may be more feasible to test a large subset of the models computationally. A small number of tests performed in the lab can be used as a check of the CFD model, which then can serve as a tool to conduct parametric sweeps on over a much bigger test matrix.

The simplified model described previously was used to simulate the airflow around a circular cylinder. Testing of uniform cylinders at various wind speeds was also performed in the wind tunnel, providing drag coefficient values. The sample was 1.5 m long with a uniform diameter. It was placed in the force balance in the test zone in front of the wind tunnel extension. At the ends, two circular plates were mounted, with a purpose to make the flow more uniform along the cylinder length. In this paper, one case of Reynolds number, Re=5 x 10^3, Re=ρUd/μ (where ρ is the air density and ρ=1.18 kg/m^3, U is the approach velocity and U=1.02 m/s, d is the cylinder diameter and d=0.075 m, μ is the dynamic viscosity and μ=18.55 x 10^6 kg/m/s) was considered. The measured drag coefficient was equal 0.91.

A URANS solver with two turbulence models, k-ε and k-ω, was used in the simulation. To get the required Reynolds number, the velocity at the inlet boundary (see Figure 3(b) for the inlet location) was adjusted accordingly and calculated to be equal to 0.23 m/s. At this inlet velocity, the value of approach velocity for the tested model is equal 1.02 m/s.

The obtained drag coefficients were 0.9 and 1.35 for k-ε and k-ω models respectively. The lift coefficient oscillates about zero for both turbulence models. The Strouhal number, calculated from the fluctuation frequency of the lift coefficient in time, S=fd/U (where f is the frequency), equals 0.21.

In the presented test case the simulation was fully converged in 10 seconds of simulation time. The simulation ran for approximately 200 hours on 8 computational nodes, each with 16 cores. The significant amount of time and resources is attributed to the high number of volume cells, about 7.7 million, and also to a small time step, adjusted to the smallest cell size, which is equal to 0.001 sec.

Because the CFD domain was three dimensional and the flow solution yields values at the cell centroids of every computational cell in the domain, it is possible to analyze the flow around the sample in greater detail. Figure 7 provides the velocity magnitude contour plots on: (a) the horizontal, and (b) the vertical symmetry planes of the sample. Note that the flow field results from CFD model solution provides much more detail than normally available with a limited number of measurements in the laboratory. Figure 8 illustrates streamlines of the velocity field around the cylinder. The flow is indeed three-dimensional. The distribution of the flow coming from the extension is not perfectly uniform. A small dip in the velocity values was registered in the central part of the jet, both in the computations and laboratory measurements. As a result, the wake behind the cylinder is not uniform along the length, even though Reynolds averaged Navier-Stokes equations are being solved.

The proposed model can be used to perform advanced aerodynamic analysis. Every simulation provides extensive information about the flow around the tested specimen, as well as outside of the testing area. It gives the capability to introduce modifications in the geometry of the room and analyze the influence on the airflow. The changes can consist of modifying the surroundings, e.g. by changing location of the surrounding furniture and equipment, or making changes in the wind tunnel structure itself, e.g. removing or adding screens, or lowering the porosity of the screens due to collected dust. It can therefore be used to test many more specimen geometries and conditions than can be fabricated and scheduled for testing in the physical wind tunnel laboratory. The model can also extend the range of wind speeds used for testing specimen responses to speeds greater than those that can be achieved in the physical laboratory.

Another advantage of the virtual wind tunnel is the ability to visualize the airflow in three dimensions, which is almost impossible to achieve in the experimental setup for a problem of this size. Thanks to the CFD visualization we can learn, among other things, about the pressure distribution on the surface of a specimen, the locations of the flow separation points, and also the characteristics of the wake behind the model. The ability to effectively see the three dimensional patterns in the flow field around a specimen of interest in visualizations of velocity, pressure, streamlines, high turbulence dissipation zones, and other values can allow an engineer to more easily identify the cause of problems in a design and determine changes that are most likely to yield improvements. Such changes can then be easily tested with a CFD computation. All of the information generated by CFD computation is relevant in the aerodynamic analysis and evaluation of existing bridge or roadside hardware designs and in developing new optimized structures. CFD visualization is also a very important tool used in communication of the results to others and often becomes an important element in the decision-making process.

(a)

(b)

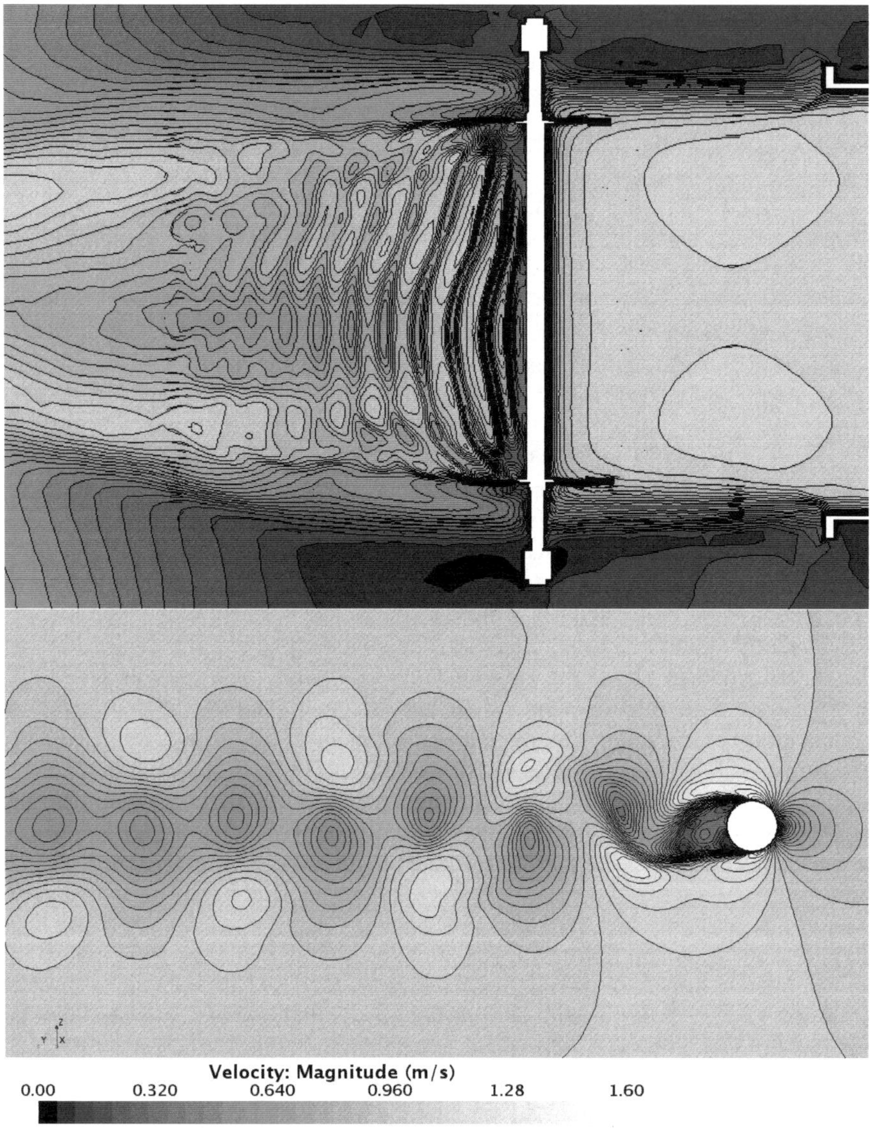

Figure 7 Velocity magnitude contour plots on: (a) the horizontal, and (b) the vertical symmetry planes of the sample

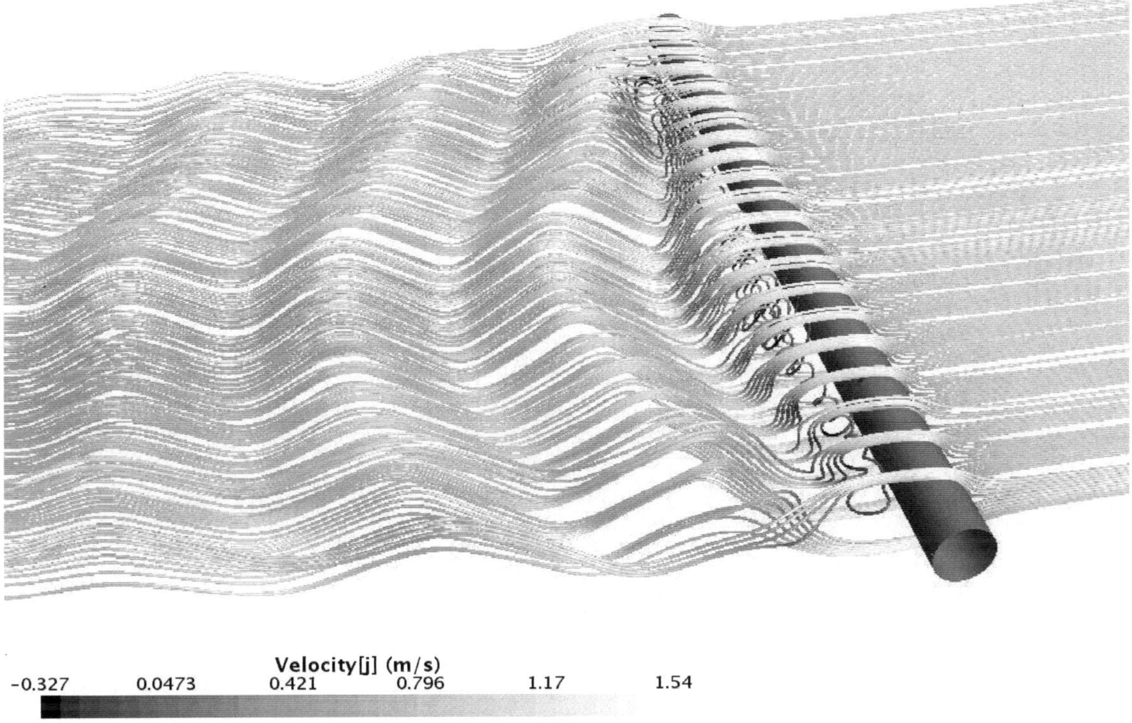

Figure 8 Streamlines of velocity around the cylinder

4 CONCLUSIONS

The goal of this paper was to present the work that has been done to create a virtual laboratory in CFD environment that could be used to run larger parametric tests or extend the experimental testing beyond the conditions available in the lab.

Models with different levels of complexity were tested. These models included (1) a full model with a rotating fan, (2) a full model of the fan in a rotating reference frame that did not require moving computational grids, and (3) simplified models without the fan and wind tunnel interior, but with appropriate inflow and outflow boundary conditions at the last screen near the wind tunnel exit and the fan cage inlets. The results were compared against experimental air velocity measurements at various locations around the room both inside and outside the testing area. CFD results for full models match well the experimental data in the air jet region and in particular in the test zone of the jet. Discrepancies in the flow characteristics appear in areas of less importance, in the proximity of walls and on the sides of the room. In the simulations with recirculating air flow, i.e. in the models with a rotating fan or rotating reference frame fan model, the jet slowed down and slightly changed direction.

In order to reduce the computational time and computer cluster resources needed to solve a problem, simplifications of the models were tested. The inside of the wind tunnel was removed from the domain, the last (furthermost from the fan) screen was changed into a velocity inlet boundary and the surfaces where the air enters the fan were modified into a pressure outlet boundary. Breaking the feedback loop of the recirculating flow field eliminated oscillations of the flow as well as improving the match with experimental velocities close to the walls. It also reduced the computational time to a few hours on the same number of cores as before.

The simplified model of the wind tunnel was used to test the effects of the air flow on a long circular cylinder with a uniform diameter. The recorded force coefficients were close to the laboratory measurements. The simplified models produce flow fields that are sufficiently close to those in the real aerodynamics laboratory to use the model in practical studies of transportation hardware, extending the number of geometry variations of specimens and flow conditions that are tested.

5 ACKNOWLEDGEMENTS

The funding for this project came from the Turner-Fairbank Highway Research Center, through Interagency Agreement Number DTFH61-14-X-300002 between DOT and DOE, and the work was performed under DOE's contract with UChicago Argonne, LLC, Contract No. DE-AC02-06-CH11357.

6 REFERENCES

1 S.A. Lottes, C. Bojanowski, Wind Tunnel Model of the Turner-Fairbank Highway Research Center Aerodynamics Laboratory, Technical Report ANL/ESD/13-3, Argonne National Laboratory, DOI: 10.2172/1111295, June, 2013.
2 CD-adapco, User Guide STAR-CCM+ version 11.02.010, 2016
3 McMaster-Carr, http://www.mcmaster.com/#wire-mesh/=emqkhc, last accessed July 2016
4 R. D. Mehta, Turbulent Boundary Layer Perturbed by a Screen, AIAA Journal, 2(9) (1985) 1335-1342
5 C. Bojanowski, S. Lottes, M. Sitek, N. Zhang, H. Bosch, Modeling of wind flow in a real-size wind tunnel at FHWA Turner-Fairbank Aerodynamics Laboratory, 4th AAWE Workshop, Miami, FL, USA, August 14-16, 2016

Numerical Design of a Field Measurement Campaign on Stanford's Campus

J. Sousa[a], C. Gorlé[b]
[a] *Civil and Environmental Engineering Department, Stanford University, US, jsousa@stanford.edu*
[b] *Civil and Environmental Engineering Department, Stanford University, US, gorle@stanford.edu*

ABSTRACT

The development of sustainable, healthy and comfortable urban environments motivates a continuous interest in using Computational Fluid Dynamic (CFD) solvers to model flow and transport processes in urban canopies. Commonly, numerical studies rely on wind tunnel measurements for model evaluation and validation. While these have significantly improved our understanding of this type of flows, it should also be acknowledged that a wind tunnel cannot fully represent the complexity of all flow phenomena involved in the atmospheric boundary layer. Considering that only a few full-scale experimental campaigns have been carried out in urban areas, there is a need for new and reliable measurement data to validate numerical models and to quantify the different sources of uncertainty in urban canopy flow simulations.

This paper reports on numerical simulations that were performed to set-up a pilot field measurement campaign on Stanford University's engineering quad. The objective is to inform strategic placement of several measurement sensors to enable a detailed study of the flow and transport processes in this realistic urban environment. Different sensors will be used to measure the mean and the unsteady signature of wind velocity, direction and temperature. In order to model the flow field on the engineering quad, the numerical geometry replicates all the buildings within an 800m radius from the measurement location. The top boundary is placed at a height of 5 H_{max}, where H_{max} (75m) corresponds to the highest building in the domain. The lateral boundaries are placed 300m away from the closest building to avoid unintended interactions. The inlet boundary conditions (wind velocity and direction) are estimated based on measurements from a local weather station. Reynolds-averaged Navier-Stokes simulations will be performed considering different scenarios. In the first phase it is assumed that the weather station data can be imposed directly on the boundaries of the computational domain. Considering the yearly variability in the inflow conditions, we will perform a set of simulations that enables constructing a polynomial chaos expansion (PCE) response surface. Sobol indices will be computed to identify locations on the engineering quad with a higher sensitivity to the inflow conditions. This knowledge will be essential to achieve an optimal spatial distribution of the sensors during the experimental campaign. Secondly, we will investigate any potential offsets between the wind conditions measured at the location of the weather station, and those at the locations of the boundaries of our computational domain. Using Bayesian inference we intend to update and improve the probability density functions for the inflow conditions in our simulations.

Once sufficient data is obtained from the pilot field measurement campaign it will be used to validate our computational results. Comparison with the measurements will inform future research on improving our current uncertainty quantification methodologies to enable the prediction of several quantities of interest such as wind speed, pollutant concentration, and temperature with quantified confidence intervals.

KEYWORDS

Field campaign, optimal sensor displacement, numerical uncertainty quantification, design of experiments

On the effects of minor changes in inlet wind direction on urban CFD simulations

Wise, D.J., Boppana, V.B.L., Poh, H.J.

Institute of High Performance Computing, Singapore

ABSTRACT

Computational Fluid Dynamics (CFD) studies are frequently used in the assessment of urban developments, enabling *a priori* analysis of, amongst other things, pedestrian wind and thermal comfort, and natural ventilation. With a wide variety of available software capable of performing such simulations and the ease at which they can be used, CFD is considered to be one of the main components of the design process in the construction of new buildings, where certain ventilation/performance criteria must be achieved. In spite of its prevalence, there are still significant gaps in the knowledge of how simulation parameters affect the results. As such, multiple guidelines have been developed serving to standardize and provide quality control for CFD. A few such recommendations are on the domain size, solution order, turbulence model, and inlet wind profile, alongside suggestions to perform simulations for a variety of wind directions. However, in complex built-environments, the effects of relatively small changes in wind directions on the wind corridors and thermal hot/cool spots are not thoroughly explored – while this is essential in land-scarce and densely populated places like Singapore and Hong Kong.

As part of the present study we perform CFD simulations of a typical residential development in Singapore and compare the results to field data. The study area is approximately 1 km x 1 km, and the mesh resolution ranges from 1.25 m on the wall surfaces to 40 m in the free stream. We choose to use a simulation methodology typical to the building industry, thereby accepting the well-known limitations of the steady RANS approach and the standard k-ε model. The simulations are performed for inlet wind directions within the range $220\pm15^\circ$ as this corresponds to the dominant driving wind conditions during the Southwest monsoon period of Singapore. Wind velocities are measured using sonic anemometers at 15 locations within the region of interest, ranging from building corridors to open parkland. By ensemble-averaging the field data according to the overlying mesoscale wind conditions – obtained from nearby meteorological stations – we show satisfactory agreement between simulation and ground truth data.

The comparison highlights the caution that must be taken when drawing conclusions from CFD results. Small changes in inlet direction can cause profound changes in the location and intensity of the simulated wind corridors. This has significant consequences for the building industry, where the results from the simulation of a single wind direction can be taken far too literally.

KEYWORDS

Steady RANS, wind direction, field measurements, Singapore.

Improvement of Tornado Simulation by Adjusting Boundary Conditions

Yi Zhao [a], Guirong Yan[b], and Ruoqiang Feng[c]

[a]*Department of Civil, Architectural and Environmental Engineering, Missouri University of Science and Technology, Rolla, MO 65409, US*
[b]*Department of Civil, Architectural and Environmental Engineering, Missouri University of Science and Technology, Rolla, MO 65409 (corresponding author), US*
[c]*The Key Laboratory of Concrete and Prestressed Concrete Structures of Ministry of Education, Southeast University, Nanjing, Jiangsu 210096, China.*

ABSTRACT: Tornadoes have resulted in significant fatalities and property damage each year. To mitigate this natural hazard, the research on numerically simulating tornadic wind field and investigating the wind effects on civil structures has been conducted. Considering that the setup of the boundary conditions has a significant influence on the generated wind flow when conducting CFD simulations, this study is to investigate how the sizes of the pressure outlet and velocity inlet affect the characteristics of the generated tornadic wind flow by taking the CFD simulation of a real-world tornado (Tornado Spencer) as an example. Based on the observations in this study, the way to improve the tornado simulation by adjusting physical boundary condition is proposed.

KEYWORDS: Tornado, CFD, Physical Boundary

1 INTRODUCTION

Tornadoes occur very frequently in the US, with an average of approximately 1200 reports per year according to NOAA statistics. They have resulted in significant fatalities and property damage each year. Due to the difficulties in obtaining the in situ radar-measured data and the high cost of experimental testing, numerical simulation of tornadic wind field and tornadic wind effects using CFD has been widely studied. The first numerical simulation of vortex structure was developed by Harlow and Stein (1974). Rotunno (1979) investigated the relationship between swirl ratio and vortex structure in a numerical model of a Ward-type vortex chamber. Nolan and Farrell (1999) explored the structure and dynamics of axisymmetric tornado-like vortices with a numerical model of axisymmetric incompressible flow.

When conducting CFD simulations, the setup of the boundary conditions has a significant influence on the generated wind flow. This study is to investigate how the sizes of the pressure outlet and velocity inlet affect the characteristics of the generated tornadic wind flow by taking the CFD simulation of a real-world tornado (Tornado Spencer) as an example.

2 SETUP of CFD SIMULATIONS

Tornado Spencer, rated at EF4, passed through Spencer, SD in 1998. To simulate this tornadic wind field, a transient, incompressible CFD simulation is conducted using ANSYS FLUENT. A cylindrical computational domain is developed to simulate the swirling wind, as shown in Figure. 1. The radius of the computational domain is $R = 800\,m$ and the total height is $(H + L) = 1100\,m$. The inflow is simulated by the velocity-inlet on the side wall with a height of H. A pressure-outlet is set up on the top wall as the outflow. Standard sea-level atmospheric conditions (pressure = 101,325 N/m^2, temperature = 298 K, and dynamic viscosity = 1.715×10^{-5} kg/(s.m) are adopted in the simulations. To simulate the practical situation, the roughness of the ground and translation of tornado are included in the simulation by properly setting up the boundary condition of the bottom wall. The height of roughness elements is 0.1 m and the roughness coefficient is set as 0.5. The translation speed assumes to be 15 m/s. The Detached Eddy Simulation (DES) is applied to simulate turbulence. The DES simulation combines the features of classical Reynolds Averaged Navier-Stokes (RANS) formulations with Large Eddy Simulations (LES) methods. The time step size is set as Δt=0.02s and the total simulation duration is 500 s (25000 time steps).

In the computational domain, except the velocity inlet, pressure outlet and the bottom wall, all other boundary surfaces are set up as no-slip walls. At the velocity inlet, the velocity data at low-level elevations collected by a Doppler on Wheels (DOW) mobile radar [1] was used to obtain the velocity profiles of the tangential, radial and vertical components using the data regression technique, as indicated by Equations (1)-(5) [1,7]. The radial, tangential and vertical velocities applied on the velocity inlet surface are input into FLUENT using User Define Function (UDF).

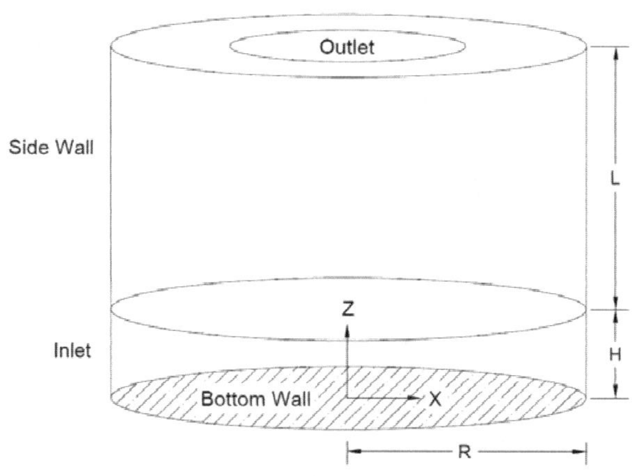

Figure 1. Cylindrical computational domain to simulate swirling wind field

$$T = 21.97\left(\frac{z}{20}\right)^{0.1189} \quad (1)$$

$$R = 11.61\left(\frac{z}{20}\right)^{0.4439} - 42.97,\ when\ z \geq 20 \quad (2)$$

$$R = -31.36\left(\frac{z}{20}\right)^{1.036},\ when\ z < 2 \quad (3)$$

$$w = 8.8373\left(\frac{z}{20}\right)^{-0.075} - 0.201\left(\frac{z}{20}\right)^{1.4439} + 0.07341z - 9.86439,\ when \geq 20 \quad (4)$$

$$w = -0.14496\left(\frac{z}{20}\right)^{1.8541} + 0.38507\left(\frac{z}{20}\right)^{2.036}, \; when < 20 \quad (5)$$

3 SIMULATION RESULTS

First, numerical simulations is run with 300 m as the height of the velocity inlet and 300 m as the diameter of the pressure outlet. The obtained three-dimensional (3D) streamlines and the velocity vector (resultant velocity) on a radial-vertical plane are presented in Figure 2 and Figure 3. As shown in Figure 2, the streamlines provide an overview of wind flow in the computational domain. It can be found that the flow initiates from the velocity inlet and forms tornado-like vortices at the central region of the computational domain, and then flows upward through the pressure outlet while rotating. Figure 3 presents the distribution of the velocity vectors on a radial-vertical plane. As shown in Figure 3, the air gets into the domain from the velocity inlet and flow towards the plane center. With flowing into the inner zone, the vertical component of the flow increases. In the central region of the computational domain, downdraft is formed and the wind speed is low. These two figures show that the tornado-like vortices have been successfully generated in the computational domain.

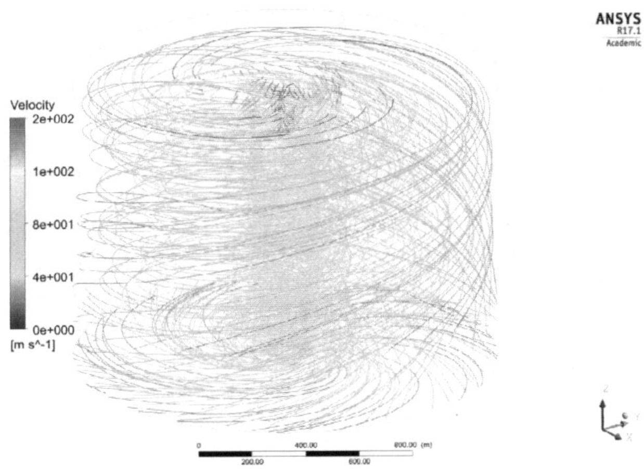

Figure 2. 3D streamlines of the generated tornadic wind flow

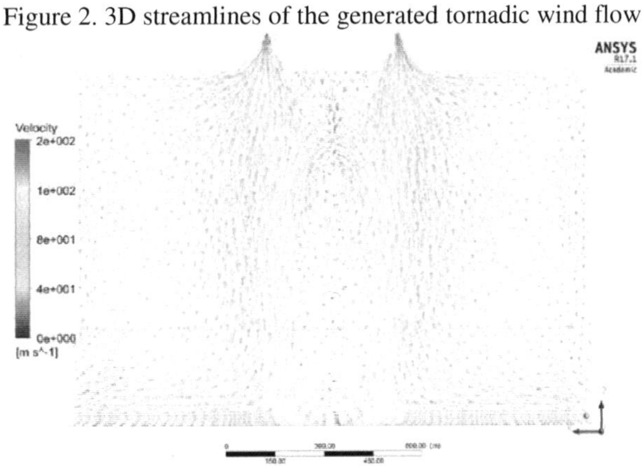

Figure 3. Velocity vector of the resultant velocity on a radial-vertical plane

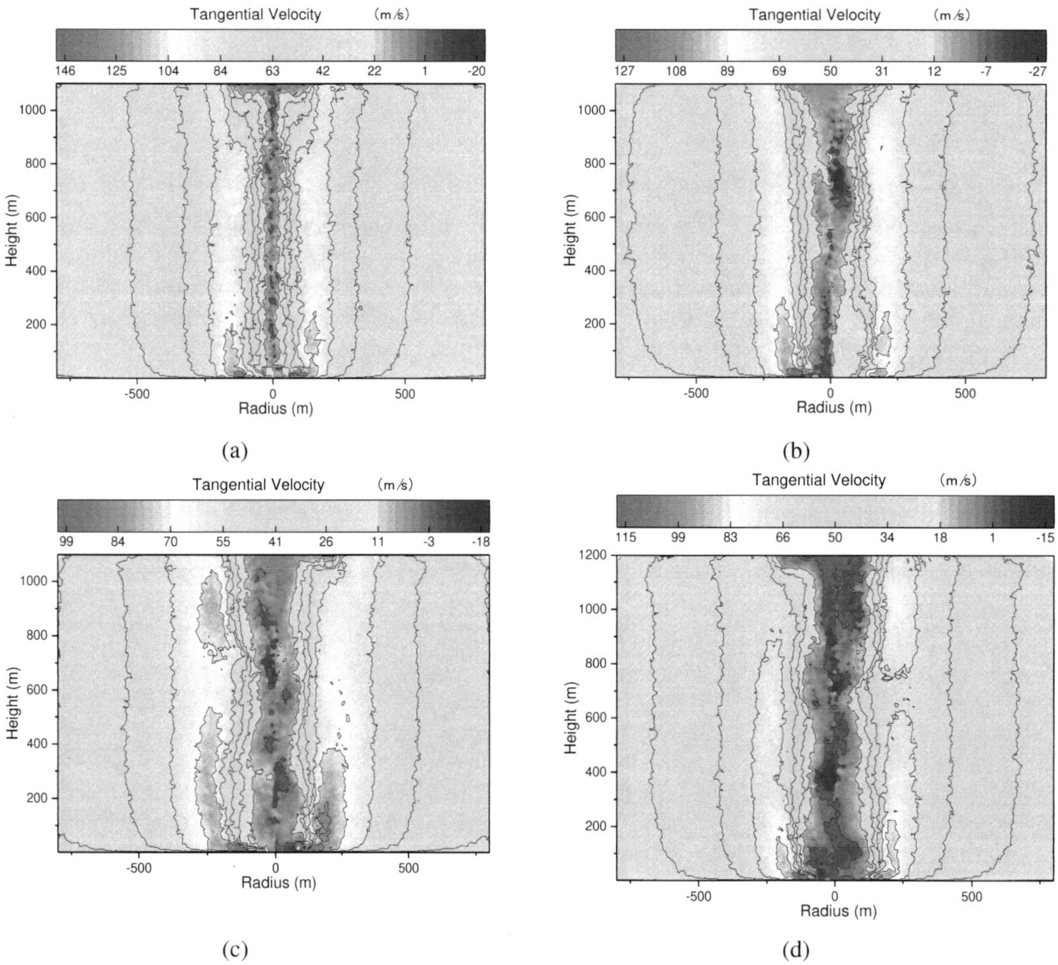

Figure 4. Contour of time-averaged tangential velocity on radial-vertical plane with different sizes of pressure-outlet and velocity-inlet. (a) Diameter of pressure outlet = 300 m; (b) Diameter of pressure outlet = 350 m; (c) Diameter of pressure outlet = 400m; (d) Diameter of pressure outlet = 400m and Height of velocity-inlet = 400m.

Second, the influence of the sizes of pressure outlet on the generated tornadic wind flow is investigated and discussed. Three diameters of pressure outlet are considered, as indicated in Fig. 4. Herein, the height of the velocity inlet is set as 400 m for all the three cases. Figures 4(a), 4(b) and 4(c) present the contour of instantaneous tangential velocity on a radial-vertical plane for each case. By comparing Figures 4(a), 4(b) and 4(c), as the size of the pressure outlet increases, the core radius increases and the tangential velocity decreases; the smaller the size of the pressure outlet, the more stable the flow

Third, the influence of the height of velocity inlet on the generated tornadic wind flow is investigated and discussed. By comparing Figure. 4(c) and 4(d), as the height of the velocity inlet is increased, the maximum velocity increases and the velocity at high elevations decreases.

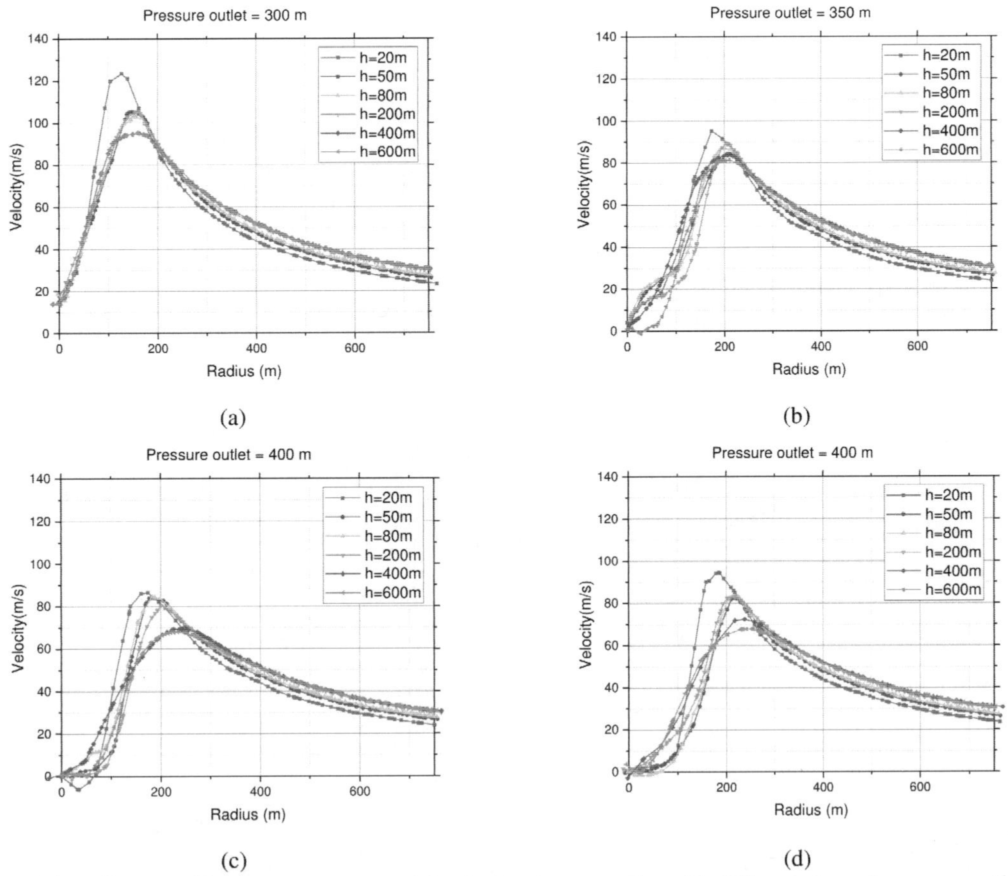

Figure 5. Distribution of instantaneous tangential velocity along a radius with different sizes of pressure-outlet and velocity-inlet. (a) Diameter of pressure outlet = 300 m; (b) Diameter of pressure outlet = 350 m; (c) Diameter of pressure outlet = 400m; (d) Diameter of pressure outlet = 400m and Height of velocity-inlet = 400m.

Figure 5 presents the instantaneous tangential velocity distribution from the models with different sizes of pressure-outlet and velocity-inlet. From Figure 5(a), 5(b) and 5(c), as the diameter of the pressure outlet increases, the maximum value of the tangential velocity decreases, while the core radius first increases and then remains almost constant. The maximum tangential velocities are 125 m/s, 115 m/s and 95 m/s associated with the 300m, 350m and 400m pressure-outlet. Meanwhile the core radius increases from 140m to 190 m. In addition, the maximum tangential velocity at the elevation of 20 m is much greater than the ones at other elevations when the size of the pressure outlet is small (300 m). With the size of pressure outlet increasing, this difference becomes smaller and smaller. From these observations, it can be concluded that the size of the pressure outlet has an influence on the generated wind flow. To be specific, with the size of pressure-outlet increasing, the maximum tangential velocity decreases and the core radius increases.

Figure 5(d) presents the tangential velocity profile for the model with 400 m-wide pressure-outlet and 400 m-high velocity-inlet, which enlarges the velocity-inlet and the total height of the domain comparing to the model in Figure 5(c). With the height of velocity-inlet increased, the maximum tangential velocity at 20 m increases significantly and the core radius also has a slight increase, while the maximum tangential velocities at other elevations almost remain the same between the two cases.

Figure 6 presents the distribution of tangential velocity along a radius based on the radar-measured data of Tornado Spencer (Kuai, L., 2008). By comparing Figure 5 with Figure 6, the maximum value of the tangential velocity of the model with 300 m-wide pressure-outlet is much larger than that based on radar-measured data. After increasing the diameter of the pressure-outlet, the maximum tangential velocity gets close to the radar-measured data. In the numerical simulation, at low elevations, the maximum velocity becomes similar to those at a high elevation, which is different from the radar-measured data. In Figure 5(d), the height of the velocity-inlet is increased to 400 m and it can be found that the maximum tangential velocity at a low elevation (20 m) are much greater than the values at high elevations, which matches the results based on radar-measured data better.

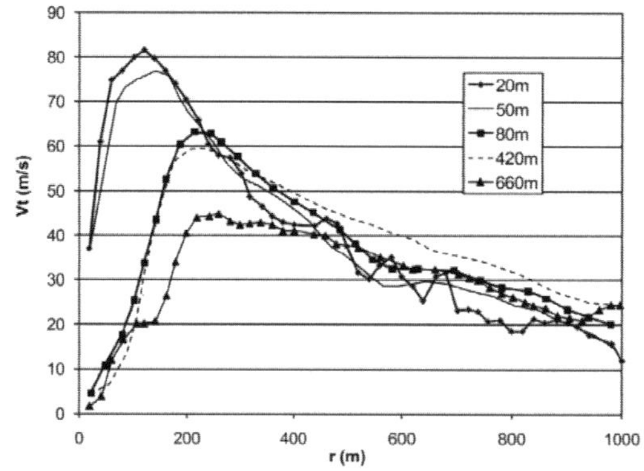

Figure 6. Distribution of tangential velocity, radar measured data of Tornado Spencer (Kuai, L., 2008)

4 CONCLUSION

In this study, a three-dimensional computational domain is developed to simulate tornadic wind flow. Several cases with different sizes of pressure-outlet and velocity inlet are studied to investigate the influence of different boundary parameters on the generated tornadic wind flow. The obtained results suggest that with the size of pressure outlet increasing, the maximum value of the tangential velocity decreases and the core radius increases. When the height of velocity-inlet increases, the maximum tangential velocity at a low elevation increases significantly, while it has little influence on the velocities at high elevations. These observations can help to adjust the CFD model using the radar-measured data as the input at the velocity inlet. For example, in order to get a lower velocity and larger core radius, it can be achieved by increasing the size of pressure-outlet. With the height of velocity-inlet increased, the difference of maximum tangential velocity between low elevation and high elevation increases.

5 ACKNOWLEDGEMENTS

The authors greatly appreciate the financial support from National Science Foundation, the Hazard Mitigation and Structural Engineering program, through the project of "Damage and Instability

Detection of Civil Large-scale Space Structures under Operational and Multi-hazard Environments" (Award No.: 1455709).

6 REFERENCE

1 J. Wurman, C.R. Alexander. The 30 May 1998 Spencer, South Dakota, storm. Part II: Comparison of observed damage and radar-derived winds in the tornadoes. Monthly weather review. 2005 Jan;133(1):97-119.
2 F.H. Harlow, L.R. Stein. Structural analysis of tornado-like vortices. Journal of the Atmospheric Sciences. 1974 Nov;31(8):2081-98.
3 D.C. Lewellen, W.S. Lewellen, J. Xia. The influence of a local swirl ratio on tornado intensification near the surface. Journal of the atmospheric sciences. 2000 Feb;57(4):527-44.
4 R. Rotunno. A study in tornado-like vortex dynamics. Journal of the Atmospheric Sciences. 1979 Jan;36(1):140-55.
5 Selvam RP, Millett PC. Computer modeling of tornado forces on buildings. Wind and Structures. 2003;6(3):209-20.
6 D.S. Nolan, B.F. Farrell. The structure and dynamics of tornado-like vortices. Journal of the Atmospheric Sciences. 1999 Aug;56(16):2908-36.
7 L. Kuai, F.L. Haan, W.A. Gallus Jr, P.P. Sarkar. CFD simulations of the flow field of a laboratory-simulated tornado for parameter sensitivity studies and comparison with field measurements. Wind and Structures. 2008 Mar 1;11(2):75-96.

Reproducing Tornado Spencer in CFD simulation based Doppler radar data

Yi Zhao [a], Guirong Yan [b], Kakkattukuzhy M. Isaac[c]

[a]*Department of Civil, Architectural and Environmental Engineering, Missouri University of Science and Technology, Rolla, MO, US, yzvxd@mst.edu*
[b]*Department of Civil, Architectural and Environmental Engineering, Missouri University of Science and Technology, Rolla, MO, US, yang@mst.edu*
[c]*Department of Mechanical and Aerospace Engineering, Missouri University of Science and Technology, Rolla, MO, US, isaac@mst.edu*

ABSTRACT

Tornadoes occur very frequently in the US, with an average of approximately 1200 reports per year according to NOAA statistics. While over 90 percent of tornadoes have an intensity of EF2 (Enhanced Fujita scale) or even lower, severe damage can be detected in civil structures after tornado striking. Experimental testing in the wind tunnel is a direct and traditional method to prevent the failure of the structures caused by tornadoes and to investigate the interaction between them. Several tornado simulators were developed in North America and used to study the wind effect on structures. Due to the small size of tornado simulator and limited power of fans, the experimental method also has some limitation in terms of the dimensional similarity and incapability of developing a cost-effective manner. However, CFD simulation is comparatively more effective and practical to simulate the tornado and investigate its wind effect on structures.

Benefited from the improvement of computing technique, CFD model can simulate wind flow in a precise manner. Setting up the boundary is a significant process to generate tornado in an accurate simulation. Using the data collected from natural tornado is one effective way for building the boundary condition. Tornado Spencer rated at EF4, passed through Spencer, SD in 1998 the velocity data from low-level elevation was collected by a Doppler on Wheels (DOW) mobile radar which can be used to simulate it. However, the radar data only provide a two-dimension velocity filed, which is a lack of vertical component.

The objective of this study is to investigate the vertical component of Tornado Spencer and develop the velocity inlet profile for CFD simulation. In this paper, two regression equations of the tangential and radial velocity are developed based on the radar data of Tornado Spencer. Although the vertical velocity profile is being excluded in original radar data, it's able to derive from the two regression equations based on the continuity equation. To validate this approach, a three-dimension computational domain is built in FLUENT to simulate Tornado Spencer. The regression equations are considered as the velocity inlet profile. The filtered time-dependent Navier-Stokes equations are selected as the governing equations and the turbulence is modeled by large eddy simulation (LES). The distribution of tangential and radial velocity is extracted to compare with the original radar data and the influence of vertical velocity component is investigated in the simulation.

KEYWORDS

Tornado Spencer, Doppler radar data, CFD simulation, vertical velocity, regression equation.

13th Americas Conference on Wind Engineering
Gainesville, Florida, USA
May 21 - 24, 2017

Generating a turbulent inflow for LES of wind loading

Giacomo Lamberti[a], Catherine Gorlè[b]

[a] *Stanford University, Stanford (CA), US, giacomol@stanford.edu*
[b] *Stanford University, Stanford (CA), US, giacomol@stanford.edu*

ABSTRACT

Computational fluid dynamics provides a unique tool to quantify pressure loads on buildings and assess wind hazards, but uncertainties in the simulations can compromise the accuracy of the results. One potential uncertainty is the inflow boundary condition, which should ultimately be defined by the local flow conditions around the building of interest. When using large-eddy-simulations (LES), both the mean flow and the turbulence statistics should be correctly reproduced. This study therefore aims to develop a method that would allow investigating the influence of uncertainties in LES inflow boundary conditions on the calculated wind loads.

A wide range of methods to generate a turbulent inflow in LES is available, and they can broadly be classified in two categories: precursor methods and synthetic turbulence generators. Precursor or recycling methods generate a turbulent ABL either from a separate simulation performed in advance, or calculated simultaneously in an upstream region of the flow domain. To generate a realistic ABL, appropriate roughness elements have to be represented in the domain to generate turbulence. The resulting boundary layer characteristics will depend on the chosen roughness configuration. Hence, the user only has indirect control over ABL statistics such as the roughness height, turbulence intensities and length scales. On the other hand, several synthetic turbulence generators that require direct specification of these ABL parameters have been developed, providing full control over the values near the inflow boundary. Their drawback is that the resulting turbulent velocity field is not a solution of the system of equations being solved, and the specified inflow conditions will develop towards an equilibrium solution downstream of the inlet. The development length will depends strongly on the specific boundary layer that one is modeling, and on the combination of subgrid and wall models used in the LES.

As a first step towards achieving our goal we performed LES simulations of an unperturbed ABL where the inflow conditions are defined using a previously developed divergence free digital filter method. It was found that the turbulence specified at the inflow tends to decay with downstream distance in the computational domain, resulting in turbulent intensities up to 50% lower at the actual building location. This decay can be explained by the fact that the turbulence generated at the inflow is not a solution of the system of equations being solved. In particular, the wall functions and subgrid models were not designed to compensate for the absence of roughness elements that would normally generate and sustain ABL turbulence in the computational domain. In ongoing work we are investigating whether this problem can be alleviated by implementing a method that automatically varies the turbulence parameters specified at the inflow, such that ultimately the desired turbulence characteristics at the building location are achieved. This method would then enable a detailed sensitivity study regarding the influence of ABL turbulence characteristics on pressure loads calculated using LES.

KEYWORDS

Wind loading, inflow condition, LES, atmospheric boundary layer

Synthetic inflow turbulence generation methods for LES study of tall building aerodynamics

Abiy Melaku[a], Girma Bitsuamlak[a], Ahmed Elshaer[a], Haitham Aboshosha[b]

[a]*Department of Civil & Environmental Engineering /WindEEE Research Institute, Western University, ON, Canada, amelaku@uwo.ca, gbitsuam@uwo.ca, aelshae@uwo.ca*
[b]*Boundary Layer Wind Tunnel Laboratory, Western University, ON, Canada, haboshos@uwo.ca*

ABSTRACT: Large-Eddy Simulation(LES) of ABL wind flows and their effect on buildings are highly sensitive to the transient inflow boundary conditions. Specifically, evaluation of aerodynamic load on tall buildings using LES requires appropriate specification of inlet flow field satisfying the target spacial and temporal correlations. In this paper, different synthetic inflow turbulence generation methods are statistically assessed in matching ABL flow statistics. The flow fields generated using these methods are utilized to study the aerodynamic response of Commonwealth Advisory Aeronautical Council (CAARC) building model using LES. Pressure distributions on different faces of the building and the base moment spectra resulted from LES are compared with previous wind tunnel experimental results. Finally, the challenges associated with inflow boundary condition specification for LES based study of tall building aerodynamics are highlighted.

KEYWORDS: ABL flow, Computational wind engineering(CWE), Tall building, CFD, LES, Inlet boundary conditions

1 INTRODUCTION

Recent developments in computer hardware and software technologies have greatly contributed in the application of Computational Fluid Dynamics (CFD) techniques to wide spectrum of engineering fields. Computational wind engineering is one of these areas, where CFD is used as a tool to study the effect of wind on the built environment. Majority of CWE applications are concerned in simulating the lower ABL flows, which are turbulent in nature. Among the available methods of modeling turbulence, LES gives more accurate result with reasonable computational cost (Dagne and Bitsuamlak, 2013). Due to this merit, LES has been used by many researchers in CWE applications, including, but not limited to, wind load evaluation on structures (Tamura et al., 2008; Tanaka et al., 2013; Dagnew and Bitsuamlak, 2013), pedestrian level wind (Blocken and Carmeliet, 2004; Stathopoulos, 2006; Mochida and Lun, 2008; Blocken et al. ,2012), pollution dispersion (Vardoulakis et al., 2003; Holmes and Morawska, 2006; Gousseau et al., 2012).

Specifying appropriate transient inlet velocity boundary condition is of an extreme importance in LES study of ABL flows (Tamura and Ono, 2003). In particular, for LES based study of tall building aerodynamics, as noted by Noda and Nakayama (2003) and Kareem (2008), the specified inlet boundary conditions should possess accurate representation of the incoming turbulence. This can be achieved by satisfying the target spacial and temporal correlations(Huang et al., 2010; Yan and Li, 2015). In addition, the generated velocity field must guarantee the divergence free condition for fast convergence of LES calculations.

There are several methods of defining inflow boundary condition in LES, and most literatures (Tabor and Baba-Ahmadi, 2010; Smirnov et al. 2001, Huang et al. (2010), Castro and Paz (2013) and Aboshosha et al. 2015) classified the inflow generation methods into three broad categories: *(i)* precursor database methods, *(ii)* recycling methods and *(iii)* synthetic inflow generation methods.

In synthetic inflow generation methods, the velocity time history at the inlet of the computational domain is numerically simulated as a stochastic stationary process. According to Tabor and Baba-Ahmadi (2010) and Huang et al. (2010), based on the method of generation used, the existing synthetic inflow generation methods can be sub-categorized into three main categories: *(i)* white noise approach, *(ii)* Fourier synthesis methods and *(iii)* proper orthogonal decomposition(POD) methods. In white noise approach the fluctuating velocity field is perturbed as a random component with the amplitude prescribed by the turbulent intensity (Aider et al, 2007). The generated velocity field using this method has no spacial and temporal correlation making it inaccurate for CWE application. In Fourier based methods, the turbulent velocity field is generated by superposing sinusoidal waves of different amplitude, frequency, and phase shift. The main advantage of the Fourier synthesis methods is that it is possible to impose the required spacial and temporal correlation in the generation procedure. Most inflow generation methods commonly used for specifying transient inlet boundary for LES, including inflow generation methods proposed by Kraichnan (1970), Kondo et al. (1997), Smirnov (2001), Huang et al. (2010), Castro and Paz (2013) and Aboshosha et al (2015) fall under this general category. In the third methods, proper orthogonal decomposition(POD) method is used to generate inflow condition based on existing experimental data at the inlet with low spacial or temporal resolution (Druault et al, 2004 and Perret et al., 2006). The inlet velocity field generated by POD method has high spacial and temporal correlation. Since most of the energy is contained only in the first few modes, only the first few modes are enough to capture most of the energy content of the fluctuation making this method computationally efficient.

After the first attempt of Kraichnan (1970) to numerically simulate divergence free velocity field, several types of synthetic inflow turbulence generation methods have been developed and used by many researchers. For example, Smirnov (2001) extended the method originally proposed by Kraichnan (1970) to generate anisotropic non-homogeneous velocity field by scaling and orthogonal transformation; Kondo et al. (1997) extended the spectral method used by Hoshiya (1972) to generate inflow with additional divergence free operation for satisfying flow continuity equation; Huang et al. (2010) developed a new general inflow generation method to fit any prescribed velocity power spectrum; Castro and Paz (2013) modified the method developed by Huang et al. (2010) to introduce spacial and temporal correlation in the original formulation; Aboshosha et al(2015) further modified the method developed by Huang et al. (2010) to satisfy the exponential coherency function and to better fit the ABL velocity spectra.

In this paper, existing synthetic inflow turbulence generation methods are reviewed briefly emphasizing their application for CWE. Specifically, three inflow generation methods, recently proposed by Huang et al. (2010), Castro and Paz (2013) and Aboshosha et al. (2015) are examined in detail by comparing with, both ABL wind flow statistics and experimental wind tunnel test. To assess the performance of each inflow generation technique, numerical evaluation of wind load on Commonwealth Advisory Aeronautical Council (Melbourne, 1980) is carried using LES and the results obtained are compared with High Frequency Pressure Integration(HFPI) test data. Furthermore, future perspective for improving the existing inflow generation for better statistical accuracy is also highlighted.

2 BRIEF REVIEW OF RANDOM FLOW GENERATION METHODS

2.1 *Kraichnan's Method*

The original work of Kraichnan (1970) was developed to generate stochastic multi-dimensional velocity field for particle dynamics. The generated flow field is multivariate, divergence free, statistically homogenous and isotropic. The value of the wave number vector, $\mathrm{k_n}$ is chosen to produce the desired isotropic energy distribution, $E(k)$. Since the velocity generated using this method is isotropic and homogeneous, application of the method to ABL flow simulation is limited due to nonhomogeneous and anisotropic nature of the flow.

2.2 *Smirnov's Method*

Modifying Kraichnan's technique, Smirnov (2001) proposed a new random flow generation method to produce inhomogeneous and anisotropic velocity fluctuations. The generation procedure involves scaling and orthogonal transformation operation applied to the continuous velocity field generated by superposing harmonic function. The fluctuations are scaled to match the time-averaged fluctuation prescribed by Reynolds stress tensor, and the length- and time-scales of the turbulence are found from previous RANS simulation or experimental data. The generated velocity field using this method fits the Gaussian spectral model expressed by Eq. (1).

$$E(k) = 16\left(\frac{2}{\pi}\right)^{\frac{1}{2}} k^4 \exp\left(2k^2\right) \tag{1}$$

The velocity field from Smirnov's method contains only eddies within the viscous subrange. As noted by Huang et a. (2010), Gaussian spectra given in Eq. (1) decays rapidly in the energy-containing subrange, ignoring the contribution of small size eddies in the inertial subrange. However, ABL turbulent flows are made up of all eddy sizes in the entire turbulence length scale. Hence, eddies in the inertial subrange play a vital role in transferring the energy from the large size eddies in the energy-containing range to smaller size eddies in the dissipation range. This makes Smirnov's method less applicable for LES based study of ABL flows. Commercial engineering packages like ANSYS Fluent implemented Smirnov's method as Spectral Synthesizer for unsteady simulations.

2.3 *DSRFG Method*

Huang et al. (2010) proposed a new general synthetic turbulence generation method called discretized and synthetized random flow generation (DSRFG) method. The major improvement of DSRFG over Smirnov's method is that its capability to generate inhomogeneous and anisotropic turbulence for any given target spectra. DSRFG was developed based on previous method proposed by Kraichnan's, which generates isotropic continuous flow field with energy spectra satisfying Dirac's function. Huang et al. (2010) has demonstrated that DSRFG can generate a fluctuating velocity field satisfying von Karman model. The study also demonstrated the performance of DSRFG over Smirnov's method by undertaking LES of tall building and reported a very good agreement with previous experimental wind tunnel result. The simulation equation for this method is given in Eq. (2) with corresponding parameters provided in Table 1.

$$u_i(\mathbf{x},t)=\sum_{m=1}^{M}\sum_{n=1}^{N} p_i^{m,n}\cos\left(\tilde{k}_j^{m,n}\tilde{x}_j+2\pi f_{m,n}\frac{t}{\tau_0}\right)+q_i^{mn}\sin\left(\tilde{k}_j^{m,n}\tilde{x}_j+2\pi f_{m,n}\frac{t}{\tau_0}\right) \quad (2)$$

$$p_i^{m,n}=sign\left(r_i^{m,n}\right)\sqrt{\frac{c}{N}S_{u_i}\left(f_m\right)\frac{\left(r_i^{m,n}\right)^2}{1+\left(r_i^{m,n}\right)^2}\Delta f}$$

$$q_i^{m,n}=sign\left(r_i^{m,n}\right)\sqrt{\frac{c}{N}S_{u_i}\left(f_m\right)\frac{1}{1+\left(r_i^{m,n}\right)^2}\Delta f}$$

$$\mathbf{k}^{m,n}\cdot\mathbf{p}^{m,n}=\mathbf{0}$$
$$\mathbf{k}^{m,n}\cdot\mathbf{q}^{m,n}=\mathbf{0}$$
$$\left|\mathbf{k}^{m,n}\right|=k_m$$

$$\overline{k_j^{m,n}}=\frac{k_j^{m,n}}{k_0},\tilde{x}_j=\frac{x_j}{L_s^j}$$

Table 1: Parameters used for different inflow generation methods

Parameters	Huang et al. (2010)	Castro and Paz (2013)	Aboshosha et al. (2015)
k_0	$f_{\min}/U$	$f_{\min}/U$	1
L_s^j	$1.5\sqrt{L_u^2+L_v^2+L_w^2}$	$1.7\sqrt{L_u^2+L_v^2+L_w^2}$	$1/\left(\gamma C_j\right)$
τ_0	1	0.75~1.5	1
$f_{m,n}$	$N\left(0,f_m\right)$	$N\left(0,f_m\right)$	$N\left(f_m,\Delta f\right)$
c	4	2	2
Δf	1	$f_{\max}/M$	$\left(f_{\max}-f_{\min}\right)/M$

The parameters $f_{\min}$ and k_0 are the minim frequency and wave number, respectively; $f_{\max}$ is the maximum frequency; $N\left(\mu,\sigma\right)$ represents normally distributed random number with μ mean and σ standard deviation; L_u, L_v and L_w are the integral length scale of turbulence in *x*-, *y*- and *z*-direction, respectively; C_j is the coherence decay constant in each direction(in the current study a constant value of 10 is used) and γ is a parameter used to tune the coherence function and its given by,

$$\gamma = \begin{cases} 3.7\beta^{0.3} & \beta < 6.0 \\ 2.1 & \beta \geq 6.0 \end{cases} \tag{3}$$

where $\beta = C\tilde{D}$; $\tilde{D}$ is taken as 0.5 times the maximum dimension of the object as recommended in Aboshosha et al. (2015), and C is the coherence decay constant in streamwise direction.

2.4 *MDSRFG Method*

Castro and Paz (2013) revised DSRFG originally proposed by Huang et al. (2010) by introducing time and space correlations parameters and called the modified method MDSRFG. The method has also showed very good convergence to the statistical target values as the number of discrete frequency segments (*M*) of Eq. (2) is increased. The time correlation is controlled by a scaling the parameter, τ_0 and the range of values commonly used are given in Table 1.

2.5 *CDRFG Method*

The common drawback DSRFG and MDSRFG is that the generated velocity fluctuations are not coherent. However, ABL flows are characterized by exponential correlation function defined as coherency function (Davenport, 1995). To alleviate this problem, Aboshosha et al. (2015) proposed a new inflow generation method modifying DSRFG. The equation used to simulate the velocity field using CDSRFG is given in Eq. (2) with its special parameters described in Table 1.

3 COMPARISON WITH ABL FLOW STATISTICS

In this study, the experimental data in Dagnew and Bitsuamlak (2013) is used to validate different inflow generation methods. The experiment was conducted at 1:400 scale on CAARC model. The CAARC building is a rectangular prismatic shaped building with 30.48m by 45.72m floor dimension and 182.88m height. The geometry of the building, streamwise mean velocity and turbulent intensity profiles are shown in Figure 1. The velocity profile is fitted to an open-exposure with a power law coefficient, $\alpha = 0.16$. The mean velocity at the building height in the model-scale is 12.84m/s. The statistical properties of the velocity fluctuations generated by DSRFG, MDSRFG and CDRFG are compared using von Karman spectrum as a target spectrum. The comparison is performed by generating velocity time history for a single point located at the building height. The lateral and vertical components of turbulent intensity, and all components of integral length scale of turbulence are determined using the ESDU by calculating the equivalent roughness height(z_o).

The fluctuating component of velocity in the stream-wise direction for each inflow generation method is shown in Figure 2. For the simulation, the maximum frequency, f_{max} is taken as 100Hz with a time step of 0.005s. The value of M and N in Eq. (2) are picked as 100 and 50, respectively, as recommended in Aboshosha et al. (2015). The power spectra for each components of velocity are depicted in Figure 3 together with the target von Karman spectrum. The standard deviations of the fluctuating component of velocities are also calculated and the results are summarized in Table 2 with the corresponding target values. As it can be seen from Figure 3 and Table 2, CDRFG produces better accuracy in matching the target three-dimensional wind spectra, as well as the standard deviation, compared to the other two methods.

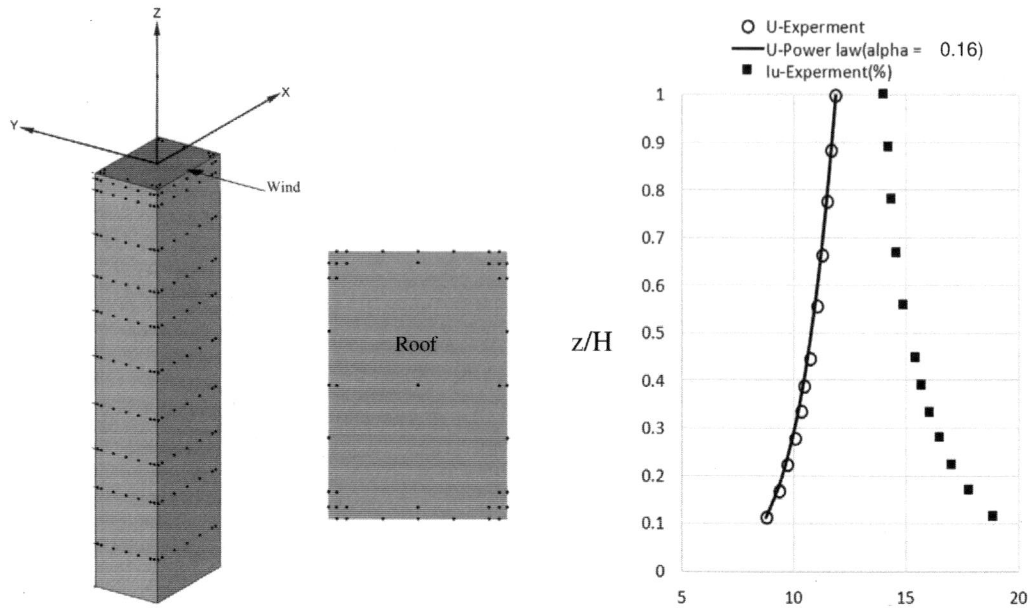

(a) (b) (c)

Figure 2: CAARC building geometry (a,b) and mean velocity and turbulence intensity profiles in model-scale(c)

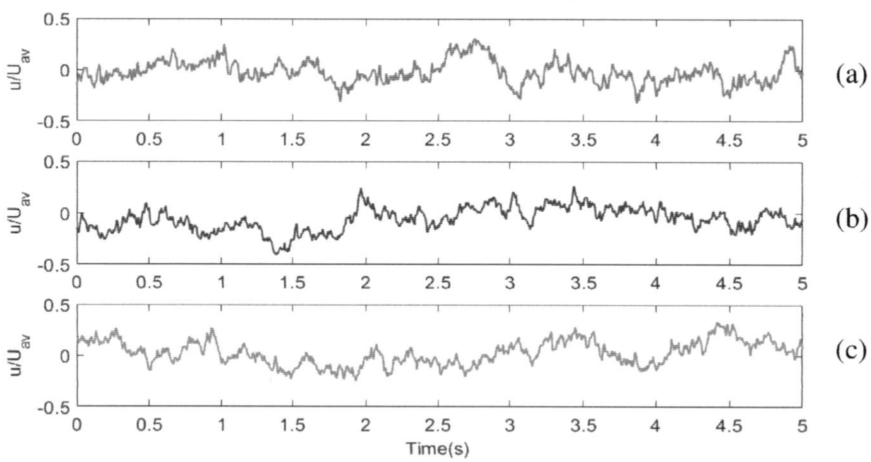

Figure 1: Horizontal component of velocity simulated using (a) CDRFG, (b) DSRFG and (c) MDSRFG [$U_{av} = 12.84 m/s, I_u = 11.9\%$]

Table 2: Comparison of standard deviation of velocity fluctuations from different methods

	σ_u	σ_v	σ_w
DSRFG (Huang et al., 2010)	1.521	1.150	0.796
MDSRFG (Castro and Paz, 2013)	1.660	1.084	0.776
CDSRFG (Aboshosha et al., 2015)	1.520	1.178	0.824
Target	1.529	1.195	0.846

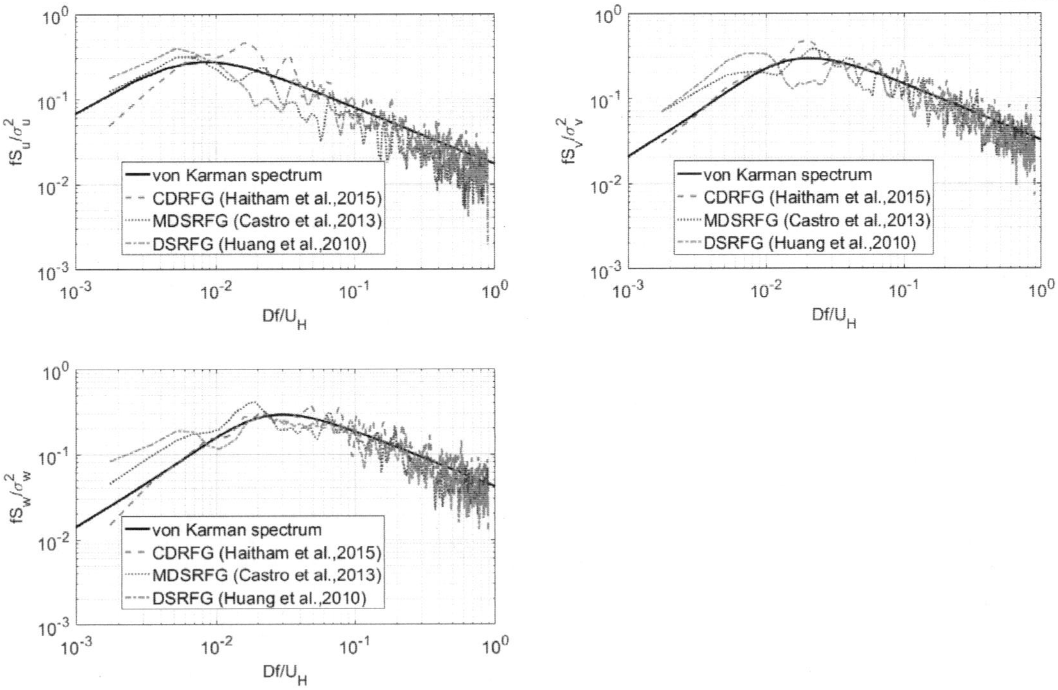

Figure 3: Comparison of velocity spectra with von Karman spectrum
(I_u=11.9%, I_v=9.3%, I_w=6.6%, L_u=1.97m, L_v=0.59m, L_w=0.40m)

4 LES APPLICATION AND COMPARISON WITH EXPERIMETAL DATA

In this section, inflow generation methods described above, namely; DSRFG, MDSRFG and CDRFG are used to generate inlet velocity boundary condition for evaluating wind induced load on a CAARC model. The inflow is generated using the ABL wind tunnel profile given in Figure 1. The wind tunnel simulation was conducted using HFPI model with 280 pressure taps recorded at 515 Hz sampling frequency. The experiment was conducted by Dagnew and Bitsuamlak (2013) at the wind tunnel testing facility of RWDI USA LLC.

4.1 *Discretization*

Proper discretization of the computational domain allows accurate capturing of the flow details and increases the cut-off frequency for the LES. Using appropriate grid size in different parts of the computational domain is crucial to provide the required accuracy at regions of interest and areas of high velocity gradients. Particularly, near walls (i.e. surface of the building and the ground), relatively dense mesh is needed with prism layer meshing. In addition, the region from the inlet face to the face of the building in the computational domain should be meshed finely to maintain small scale fluctuations up to the face of the building.

The dimension of the computational domain used and definition of coordinate system are shown in Figure 4. The computational domain spans *51B* (*B* is the width of the building) in the along wind direction(x),*21B* in transverse (y) direction and *3H* (*H* is the building height) in vertical (z) direction. The blockage ratio is 1.6% which is sufficiently less than 3%. The provision in AIJ (Architectural Institute of Japan) and COST (European Cooperation in the Field of Scientific and Technical Research) recommend *5H* for the height of the computational domain. In the current study, however, smaller value is used considering the relative uniformity of the streamwise velocity contours above a certain height as demonstrated by Huang et al. (2010) and Zhang et al. (2015) for a height of *2H*.

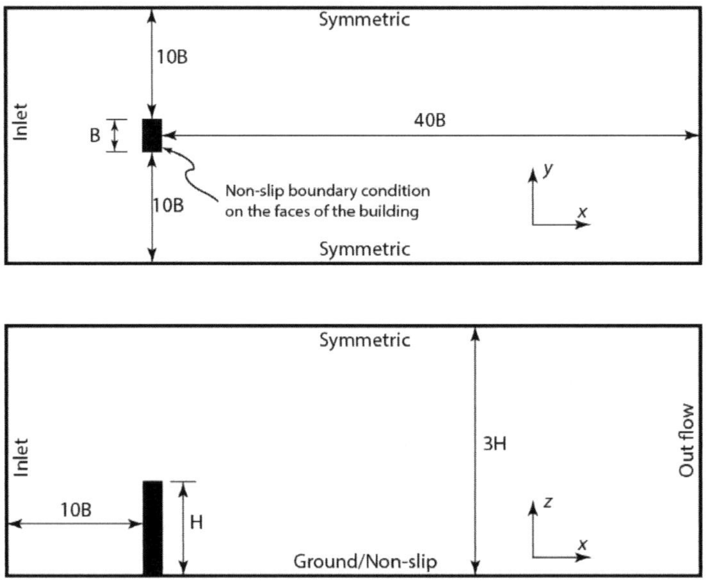

Figure 4: The computation domain and applied boundary conditions

The maximum Reynolds number at the building height is 8.64×10^4, which requires refined mesh close to the surface of the building. For better resolution near the walls, 10 prism layers were used with a stretching factor of 1.05. The first prism layer near the wall is located 0.002*B* away from wall keeping the corresponding wall coordinate y^+ bellow 5. The computational domain is meshed using polyhedral cells. The mesh is refined close to the building surface to account for the viscous effects and capture the fluctuations within the boundary layer region. Totally, ~3.1 million cells are used and different views of the computational grid used are shown in Figure 5.

For the current study, the LES simulation is conducted using Smagorinsky-Lilly sub-grid scale model with second order implicit time discretization. The time step used in the simulation is 0.0004s, which is sufficiently enough to keep the Courant–number below 1.0. The total physical time is fixed to be 5s for all cases. The simulations were conducted using high-performance computing facility of SHARCNET at the University of Western Ontario. The simulation took 72 hours on 64 processors for each case.

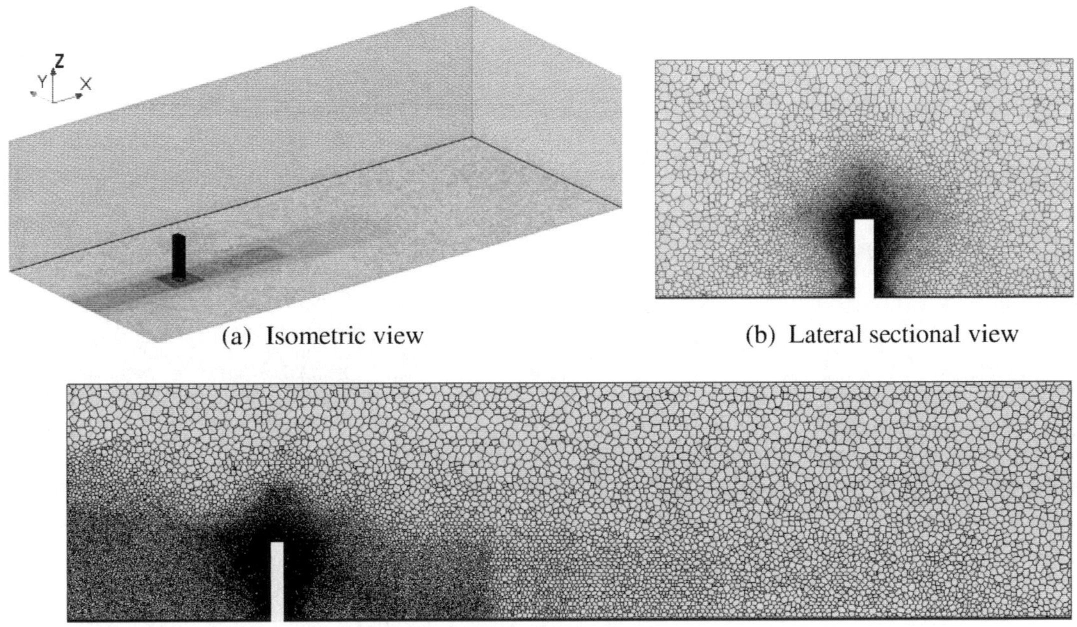

(a) Isometric view

(b) Lateral sectional view

(c) Along-wind sectional view

Figure 5: View of computational domain and grid

4.2 *Velocity and pressure distribution*

To demonstrate significance of quasi-static fluctuations, using inflows generated from CDRFG, DSRFG and MDSRFG, the instantaneous velocity field in stream-wise and transverse direction, are compared as shown in Figure 6. As it can be seen in Figure 6, there are large-scale and small-scale fluctuations that interact with the structure. This fluctuations in the velocity field play significant role in the distribution of pressure coefficient on the faces of the building, ultimately affecting the calculated wind induced load.

Figure 7 and Figure 8 show mean pressure coefficient distribution on wind-ward and lee-ward face of the building, respectively, utilizing different inflow generation techniques. The pressure coefficient on the building surface is calculated using Eq. (4).

$$C_p = \frac{p - p_0}{\frac{1}{2}\rho U_H^2} \quad (4)$$

where U_H is the reference wind velocity at the building height, $p - p_0$ is the dynamic pressure, and ρ is the air density which is taken to be $1.225 kg/m^3$ for the current study.

As it can be seen from Figure 7, for all inflow generation techniques used, the mean C_p values found from LES overlap with the experimental values in the wind-ward face to considerable accuracy. However, for lee-ward face, suction pressure coefficients found from CDRFG (Aboshosha et al., 2015) give good predication compared to DSRFG and MDSRFG (see Figure 8). The relative accuracy of CDRFG over DSRFG and MDSRFG can be attributed to the unique features of the method. CDRFG matches the velocity power spectrum to high degree as described in Section 3 and accounts the coherency function of ABL flow, which makes the method more suitable for ABL flow simulations.

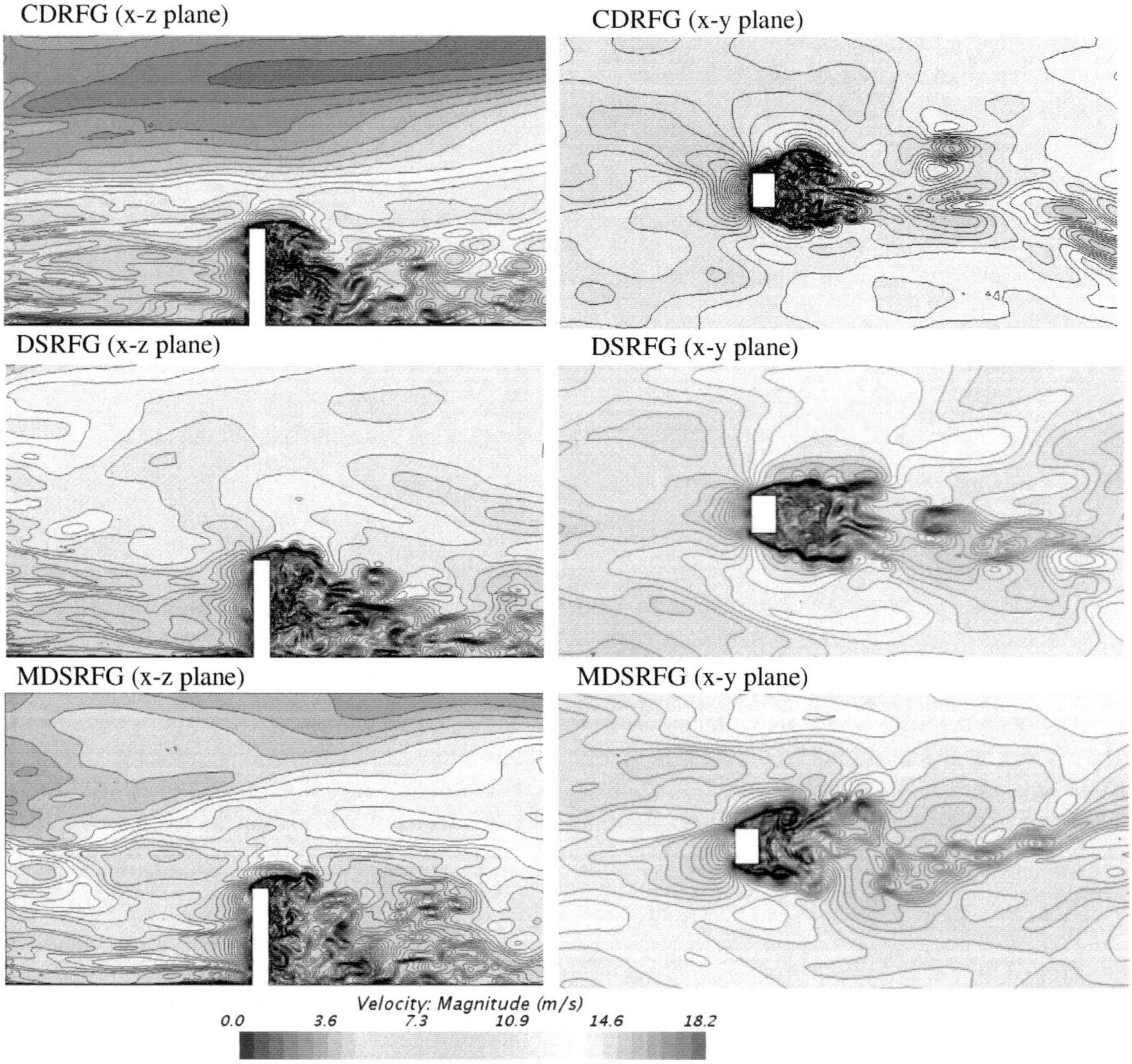

Figure 6: Instantaneous velocity field from different inflow generation methods (the x-y sections are taken at $z = 0.75H$)

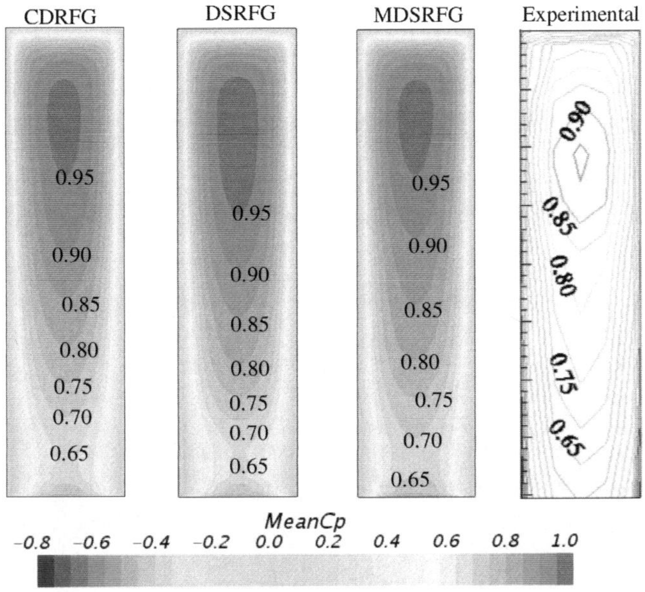

Figure 8: Comparison of mean pressure coefficient distribution in wind-ward face of the building

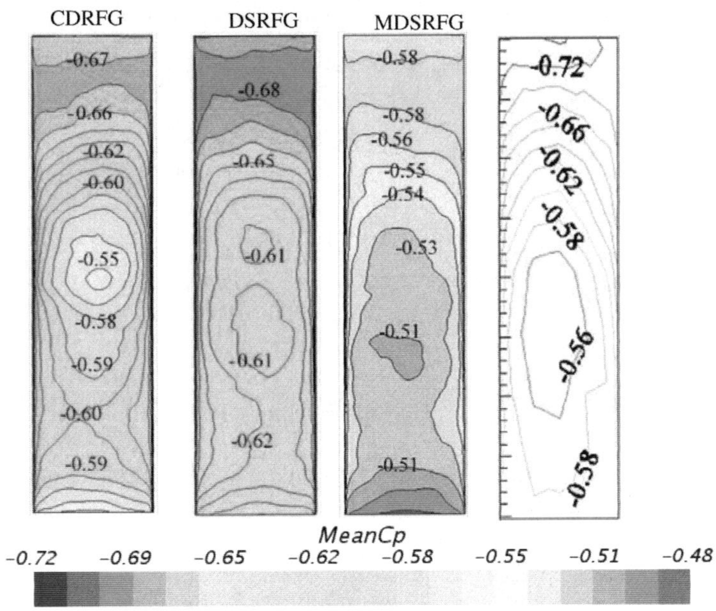

Figure 7: Comparison of mean pressure coefficient distribution in lee-ward face of the building

4.3 *Building response*

The along-wind, across-wind and torsional base moments of the building are calculated by integrating the pressure over the entire surface of the building. The base moments are normalized by their corresponding reference base moment values given in Eq. (5). The normalized base moment time histories are shown in Figure 9. To evaluate the frequency content of base moment response, the power spectral density (PSD) of the base moment in each direction are shown in Figure 10 for each inflow generation method, together with the experimental data.

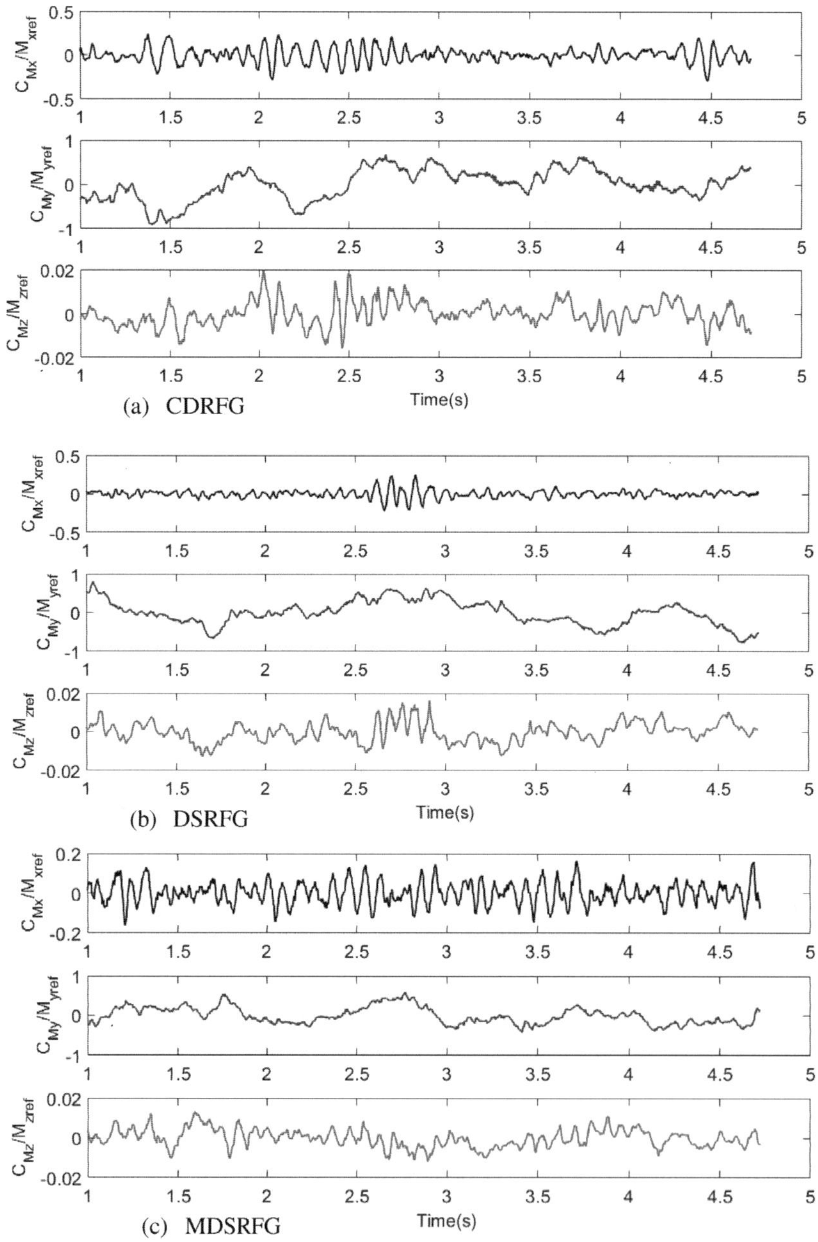

(a) CDRFG

(b) DSRFG

(c) MDSRFG

Figure 9: Normalized base moment time history

$$M_{xref} = \frac{1}{2}\rho V_h^2 BH^2, M_{yref} = \frac{1}{2}\rho V_h^2 DH^2, M_{zref} = \frac{1}{2}\rho V_h^2 DBH \qquad (5)$$

The result gained using each inflow generation techniques, as depicted in Figure 9, shows a very good agreement with the experimental result for across-wind response (M_x). However, for along-wind base moment spectrum (M_y), there is a slight deviation from the experimental values for higher frequency. This can be attributed to the number of frequency interval used in Eq. (2), smaller frequency intervals(Δf) are required to capture low frequency fluctuations. On the other hand, for the torsional moment spectra(M_z), for lower range of reduce frequency (Bf/V_H), the LES results of all inflow generation methods over predict the energy content of the response.

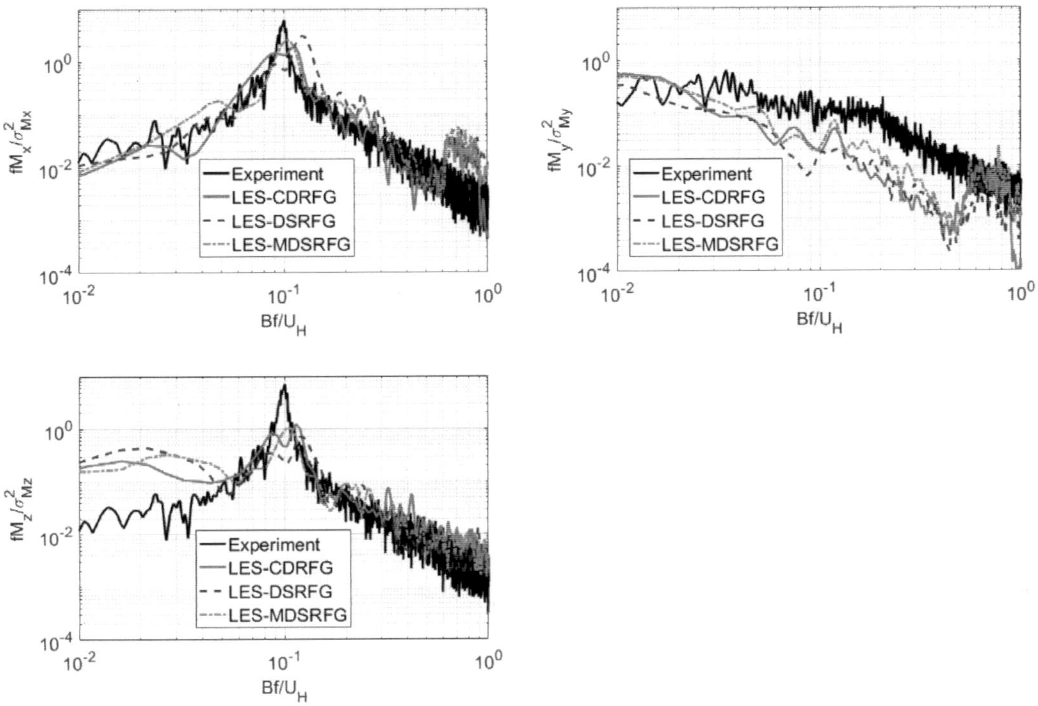

Figure 10: Base moment spectra

5 CONCLUSION

Existing synthetic inflow turbulence generation techniques were examined for their suitability in LES based study of tall building aerodynamics. Three inflow generation methods namely, DSRFG (Huang et al., 2010), MDSRFG (Castro and Paz (2013) and CDRFG (Aboshosha et al., 2015) were investigated in detail. Statistical comparison with ABL wind flow statistics showed that CDRFG achieved higher accuracy in matching the velocity spectra and the standard deviation the of the fluctuation. In addition, LES prediction of C_p values on CAARC model utilizing each inflow generation methods showed a very good agreement with wind tunnel experimental result. The same accuracy has been achieved by LES in predicting the across-wind base moment. However, for the along-wind, and torsion moment spectra, slight deviation from experimental result has been observed in certain range of frequencies. From the current study, it is noted that in the existing inflow generation methods there are different parameters which control the property of the generated velocity field. Therefore, detailed investigation of the pertinent parameters and the sensitivity of the generated flow field to these parameters need to be performed.

As a continuation of this work, a more compressive and new inflow generation method is being developed based on spectral method originally proposed by Shinozuka and Jan (1972). The new method will alleviate some of the drawbacks of the existing inflow generation methods discussed earlier.

6 ACKNOWLEDGEMENTS

The authors are thankful for the financial support from the Ontario Trillium Foundation (OTF). The authors would also like to acknowledge the support from SharcNet staffs at the University of Western Ontario.

7 REFERENCES

1. Aboshosha, Haitham, Ahmed Elshaer, Girma T. Bitsuamlak, and Ashraf El Damatty. "Consistent inflow turbulence generator for LES evaluation of wind-induced responses for tall buildings." *Journal of Wind Engineering and Industrial Aerodynamics* 142 (2015): 198-216.
2. Aider, Jean-Luc, and Alexandra Danet. "Large-eddy simulation study of upstream boundary conditions influence upon a backward-facing step flow." Comptes Rendus Mécanique 334.7 (2006): 447-453.
3. Blocken, Bert, and Jan Carmeliet. "Pedestrian wind environment around buildings: Literature review and practical examples." *Journal of Thermal Envelope and Building Science* 28.2 (2004): 107-159.
4. Blocken, Bert, W. D. Janssen, and Twan van Hooff. "CFD simulation for pedestrian wind comfort and wind safety in urban areas: General decision framework and case study for the Eindhoven University campus." *Environmental Modelling & Software* 30 (2012): 15-34.
5. Castro, Hugo G., and Rodrigo R. Paz. "A time and space correlated turbulence synthesis method for Large Eddy Simulations." Journal of Computational Physics 235 (2013): 742-763.
6. Dagnew, A., and Girma T. Bitsuamlak. "Computational evaluation of wind loads on buildings: a review." *Wind Struct* 16.6 (2013): 629-660.
7. Davenport, A. G. "How can we simplify and generalize wind loads?." Journal of Wind Engineering and Industrial Aerodynamics 54 (1995): 657-669.
8. Druault, Philippe, et al. "Generation of three-dimensional turbulent inlet conditions for large-eddy simulation." *AIAA journal* 42.3 (2004): 447-456.
9. Gousseau, P., Bert Blocken, and G. J. F. Van Heijst. "Large-Eddy Simulation of pollutant dispersion around a cubical building: Analysis of the turbulent mass transport mechanism by unsteady concentration and velocity statistics." *Environmental Pollution* 167 (2012): 47-57.
10. Holmes, Nicholas S., and Lidia Morawska. "A review of dispersion modelling and its application to the dispersion of particles: an overview of different dispersion models available." *Atmospheric environment* 40.30 (2006): 5902-5928.
11. Hoshiya, Masaru. "Simulation of multi-correlated random processes and application to structural vibration problems." 土木学会論文報告集 1972.204 (1972): 121-128.

12. Huang, S. H., Q. S. Li, and J. R. Wu. "A general inflow turbulence generator for large eddy simulation." Journal of Wind Engineering and Industrial Aerodynamics 98.10 (2010): 600-617.
13. Kareem, Ahsan. "Numerical simulation of wind effects: a probabilistic perspective." *Journal of Wind Engineering and Industrial Aerodynamics* 96.10 (2008): 1472-1497.
14. Kraichnan, Robert H. "Diffusion by a random velocity field." Physics of Fluids (1958-1988) 13.1 (1970): 22-31.
15. Kondo, K., S. Murakami, and A. Mochida. "Generation of velocity fluctuations for inflow boundary condition of LES." Journal of Wind Engineering and Industrial Aerodynamics 67 (1997): 51-64.
16. Melbourne, W. H. "Comparison of measurements on the CAARC standard tall building model in simulated model wind flows." *Journal of Wind Engineering and Industrial Aerodynamics* 6.1-2 (1980): 73-88.
17. Mochida, Akashi, and Isaac YF Lun. "Prediction of wind environment and thermal comfort at pedestrian level in urban area." *Journal of wind engineering and industrial aerodynamics* 96.10 (2008): 1498-1527.
18. Noda, Hiroshi, and Akihiko Nakayama. "Reproducibility of flow past two-dimensional rectangular cylinders in a homogeneous turbulent flow by LES." *Journal of wind engineering and industrial aerodynamics* 91.1 (2003): 265-278.
19. Perret, Laurent, et al. "Generation of turbulent inflow conditions for large eddy simulation from stereoscopic PIV measurements." International journal of heat and fluid flow 27.4 (2006): 576-584.
20. Shinozuka, Masanobu, and C-M. Jan. "Digital simulation of random processes and its applications." *Journal of sound and vibration* 25.1 (1972): 111-128.
21. Smirnov, A., S. Shi, and I. Celik. "Random flow generation technique for large eddy simulations and particle-dynamics modeling." Journal of Fluids Engineering 123.2 (2001): 359-371.
22. Stathopoulos, Ted. "Pedestrian level winds and outdoor human comfort." *Journal of Wind Engineering and Industrial Aerodynamics* 94.11 (2006): 769-780.
23. Tabor, G. R., and M. H. Baba-Ahmadi. "Inlet conditions for large eddy simulation: a review." Computers & Fluids 39.4 (2010): 553-567.
24. Tamura, Tetsuro, Kojiro Nozawa, and Koji Kondo. "AIJ guide for numerical prediction of wind loads on buildings." Journal of Wind Engineering and Industrial Aerodynamics 96.10 (2008): 1974-1984.
25. Tanaka, Hideyuki, et al. "Aerodynamic and flow characteristics of tall Buildings with various unconventional configurations." *International Journal of High-Rise Buildings* 2.3 (2013): 213-228.
26. Vardoulakis, Sotiris, et al. "Modelling air quality in street canyons: a review." *Atmospheric environment* 37.2 (2003): 155-182.
27. Yan, B. W., and Q. S. Li. "Inflow turbulence generation methods with large eddy simulation for wind effects on tall buildings." Computers & Fluids 116 (2015): 158-175.
28. Zhang, Yue, Wagdi G. Habashi, and Rooh A. Khurram. "Predicting wind-induced vibrations of high-rise buildings using unsteady CFD and modal analysis." *Journal of Wind Engineering and Industrial Aerodynamics* 136 (2015): 165-179.

Large eddy simulation of flow field over urban-like roughness using SGS model based on turbulent structures

Azusa ONO [a], Pham Van PHUC [b]

[a] *Institute of Technology, Shimizu Corporation, Tokyo, Japan, E-mail:ono.azusa@shimz.co.jp*
[b] *Institute of Technology, Shimizu Corporation, Tokyo, Japan, E-mail:p_phuc@shimz.co.jp*

ABSTRACT: In this study, Large Eddy Simulation (LES) is conducted to investigate the performance of a coherent structure Smagorinsky model (CSM), a subgrid-scale (SGS) model based on the coherent structures in grid-scale flow fields, for the urban canopy flow. The CSM is applied to the flow over staggered arrays of cubical obstacles, and the results are compared with those from standard Smagorinsky model (SM), windtunnel tests and direct numerical simulation (DNS). The mean velocity distribution given by both SGS models showed good agreement with the windtunnel tests and DNS. The spatial distribution of model parameter estimated by the CSM shows larger value above the urban canopy and smaller value behind the cube than that of the SM, given as Smagorinsky constant. This causes the difference in the magnitude of the turbulence intensity and Reynolds stress between the SM and the CSM.

KEYWORDS: LES, subgrid-scale model, coherent structure Smagorinsky model

1 INTRODUCTION

Flowfields within and above urban canopies with various configurations or densities have been studied to improve thermal comfort and air quality in urban areas. Large Eddy Simulation (LES) has been used to obtain detailed turbulence statistics. It solves large eddies in calculation and has been shown to implicitly account for small eddies by using a subgrid-scale (SGS) model. The best-known SGS model is the standard Smagorinsky model (SM). In the SM, one value of the Smagorinsky constant C_s, must be selected. However, it is difficult to determine an appropriate value of C_s for analyzing the entire flowfield involves various types of flow properties, such as flowfield around a bluff body. The dynamic SGS model based on the Smagorinsky model (DSM), which was proposed by Germano et al. (1991) and revised by Lillly (1992), successfully corrected this fault of the SM. In the DSM, the value of the model parameter C was determined as a variable of space and time following the properties of the flowfield. However, computation based on the DSM is not stable as the fluctuation of C is too large. In order to stabilize the fluctuation of C, various techniques and models have later been proposed. In recent study, a coherent structure Smagorinsky model (CSM), which is a SGS model based on coherent structures in the grid-scale, was proposed. The CSM is numerically stable. In addition, since the model reproduces the asymptotic behavior of the eddy viscosity to a wall, an explicit wall-damping function is not necessary. The CSM showed the same performance to the DSM with less computational time for turbulent channel flows (Kobayashi, 2005). However, the applicability of the CSM to an urban canopy flow including the flow separation at the building corners, the reattachment on the roof, the recirculation regions on the roof and walls and near wall flow with low Reynolds number is not clear.

In this study, the applicability of the CSM to flow field over urban-like roughness was investigated. LES of turbulent flow over staggered arrays of cubical obstacles have been carried out using the CSM and the SM. The results were compared with data from experiments and the direct numerical simulation (DNS).

2 NUMERICAL SIMULATION

2.1 *SGS model based on coherent structures*

In LES computation, by applying a filter operation with a filter width, the filtered momentum equation can be written as

$$\frac{\partial \overline{u_i}}{\partial t} + \overline{u_j}\frac{\partial \overline{u_i}}{\partial x_j} = -\frac{1}{\rho}\frac{\partial \overline{p}}{\partial x_i} + \frac{\partial}{\partial x_j}\left(\nu \frac{\partial \overline{u_i}}{\partial x_j}\right) - \frac{\partial \tau_{ij}}{\partial x_j} \tag{1}$$

$$\tau_{ij} = \overline{u_i u_j} - \overline{u_i}\,\overline{u_j} \tag{2}$$

where $\bar{f}$ is the filtered value of f and τ_{ij} is the traceless SGS stress tensor modeled in LES. The best-known model is the Smagorinsky model based on an eddy-viscosity,

$$\tau_{ij} = -2C\overline{\Delta}^2 \left|\overline{S}\right| \overline{S}_{ij} \tag{3}$$

$$\left|\overline{S}\right| = \sqrt{2\overline{S}_{ij}\overline{S}_{ij}} \tag{4}$$

where C is the model parameter, $\overline{\Delta}$ is the filter width and $|\overline{S}|$ is the magnitude of the velocity-strain tensor $\overline{S}_{ij}$. In the SM, the Smagorinsky constant C_s $(=\sqrt{C})$ must be changed depending on the flowfield. The model parameter C in Eq.(3) is determined as follows (Kobayashi, 2005):

$$C = C_2 \left|F_{CS}\right|^{3/2} F_{\Omega} \tag{5}$$

$$F_{\Omega} = 1 - F_{CS}\,,\; C_2 = 1/22 \tag{6}$$

$$F_{CS} = Q/E \tag{7}$$

$$Q = \frac{1}{2}\left(\overline{W}_{ij}\overline{W}_{ij} - \overline{S}_{ij}\overline{S}_{ij}\right) = -\frac{1}{2}\frac{\partial \overline{u_j}}{\partial x_i}\frac{\partial \overline{u_i}}{\partial x_j} \tag{8}$$

$$E = \frac{1}{2}\left(\overline{W}_{ij}\overline{W}_{ij} + \overline{S}_{ij}\overline{S}_{ij}\right) = \frac{1}{2}\left(\frac{\partial \overline{u_j}}{\partial x_i}\right)^2 \tag{9}$$

$$\overline{S}_{ij} = \frac{1}{2}\left(\frac{\partial \overline{u_j}}{\partial x_i} + \frac{\partial \overline{u_i}}{\partial x_j}\right),\quad \overline{W}_{ij} = \frac{1}{2}\left(\frac{\partial \overline{u_j}}{\partial x_i} - \frac{\partial \overline{u_i}}{\partial x_j}\right) \tag{10}$$

where F_{CS} in Eq.(7) is the coherent structure function defined as the second invariant normalized by the magnitude of a velocity gradient tensor E. As shown in Eq.(5), model parameter C is determined as a function of F_{CS}, which can reflect time varying characteristics of the turbulence structures on the grid-scale.

2.2 *Numerical conditions and methods*

Figure 1 shows a plan view of the model configuration. Cubes were aligned in a staggered array with a plan-area density of 0.25 in the computational domain, where the height of each cube (H) was 20 mm. This geometry is the same as that in the windtunnel measurements made by Castro et al. (2006) and DNS conducted by Coceal et al. (2007). Figure 2 shows the computational domain, which has streamwise, lateral and vertical dimensions 16H, 12H and 8H, respectively. An isotropic and uniform grid was used with grid size = H/16. Periodic boundary conditions were imposed in both streamwise and lateral directions to simulate an infinite array. An averaged pressure gradient was imposed in the streamwise direction as a driving force. The Reynolds number, defined by the mean velocity at the top of the domain and the building height H, was approximately 1.0×10^4, which is similar to that of measurement conducted by Castro et al.(2006).

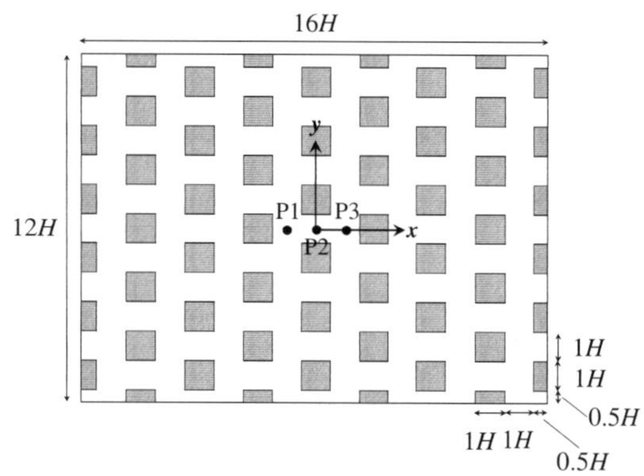

Figure 1. Plan view of the model configuration

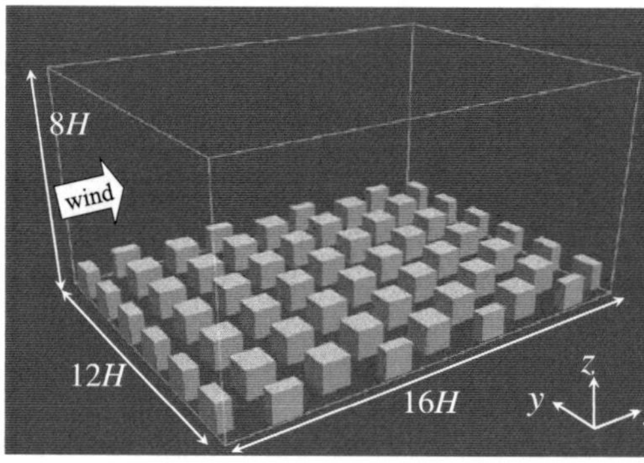

Figure 2. Computational domain

Table 1 lists the computational conditions. The numerical simulations in this study were performed using the open source toolbox OpenFOAM. The standard Smagorinsky model and the coherent structure Smagorinsky model were applied for the SGS model. In the case of the SM, the Smagorinsky constant C_s was set to 0.1. C_s is multiplied by a Van Driest type wall damping function in order to account for the near wall effect. Filtered-Linear scheme was used for the convection terms and second-order implicit scheme for unsteady terms were used. The PISO method was adopted for the pressure-velocity coupling of transient flow calculations. Turbulent statistics were collected and averaged over a duration of 400T as suggested by Coceal et al. (2007). T was defined as $T = H / u^*$, where u^* is the friction velocity derived from the total drag force acting on the cubes and ground surface.

3 RESULTS AND DISCUSSIONS

3.1 *Flowfield characteristics*

The results from the SM and the CSM were compared with those from the windtunnel measurement by Castro et al. (2006) and DNS by Coceal et al. (2007). Here, f' is the deviation from the time-averaged value $\langle f \rangle$ of f. Figure 3 compares the profiles of mean streamwise velocity normalized by friction velocity u^* at the three locations P1, P2 and P3 shown in Figure 1. The agreement is very good at all three locations. This indicates that the CSM showed a similar level of accuracy of the mean velocity to that obtained by using the SM. Furthermore, the CSM calculation cost half the computational time of the SM calculation. This is because the CSM does not need to calculate a wall damping function near the wall and ground.

Figures 4 and 5 compare the profiles of turbulence intensities and Reynolds stress normalized by u^* at the locations P1 and P3. The Reynolds stress and streamwise turbulence intensities of the SM and the CSM are in good agreement with that of the experiment and DNS both within and above the canopy. On the other hand, at the height of the cube at location P1, vertical turbulence intensities and the Reynolds stress of the SM and the CSM are somewhat lower than those of the experiment and DNS. In comparison between the results from two types of SGS models, the values of the streamwise turbulence intensity and the Reynolds stress from the CSM are smaller than that from the SM over the roughness. On the other hand, the CSM gives a large value of Reynolds stress around the height of the cube at location P1 (Figure 4). in comparison with the SM. There were no clear differences observed between the results of vertical turbulence intensities from two SGS models.

Table 1 Numerical settings

Software	OpenFOAM v2.1.0
SGS turbulence models	Standard Smagorinsky model ($C_s = 0.1$), Coherent structure Smagorinsky model
Domain size	$16H(x) \times 12H(y) \times 8H(z)$
Grid arrangement	$256(x) \times 192(y) \times 128(z)$
Unsteady terms	Second-order implicit scheme
Convective terms	Filtered-Linear
Algorism	PISO
Inlet / Outlet	Cyclic
Side	Cyclic
Top	Slip
Ground and building surfaces	Spalding's law

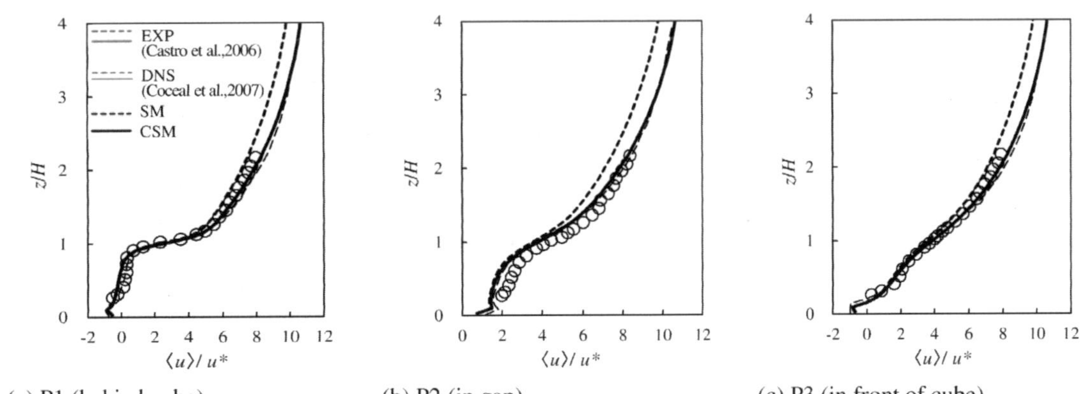

(a) P1 (behind cube) (b) P2 (in gap) (c) P3 (in front of cube)

Figure 3. Vertical profiles of mean streamwise velocity

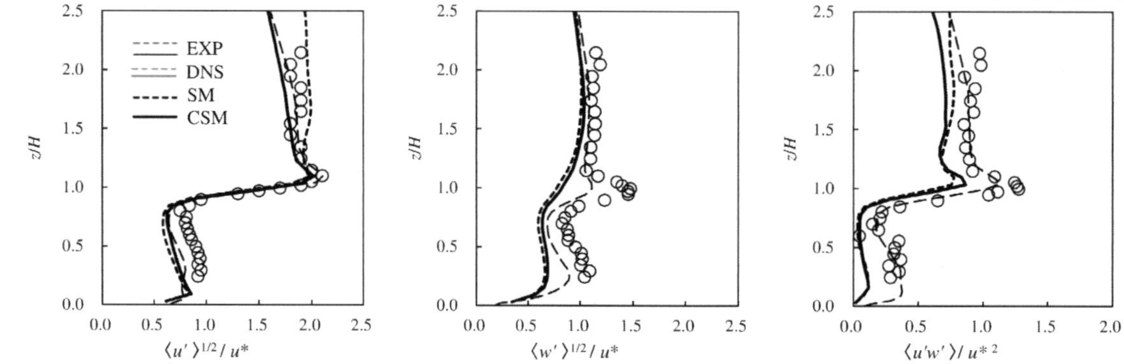

(a) Streamwise turbulence intensities (b) Vertical turbulence intensities (c) Reynolds stress

Figure 4. Vertical profiles of turbulence intensities and Reynolds stress at location P1 (behind cube)

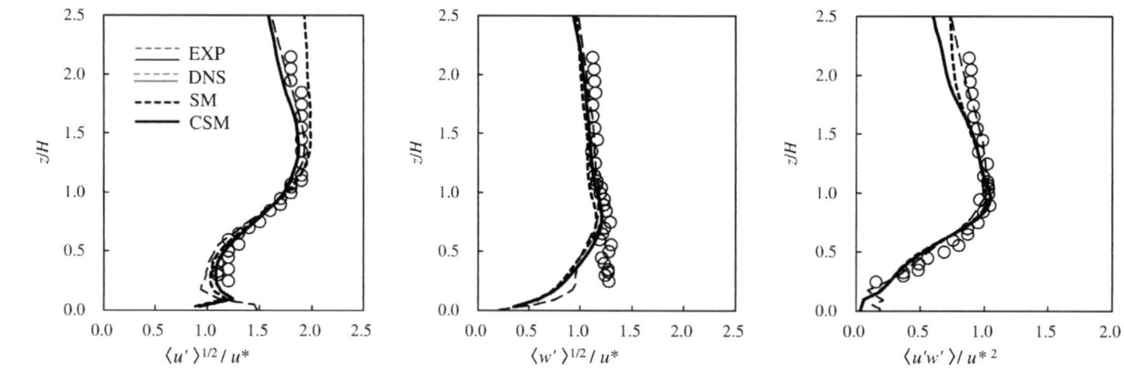

(a) Streamwise turbulence intensities (b) Vertical turbulence intensities (c) Reynolds stress

Figure 5. Vertical profiles of turbulence intensities and Reynolds stress at location P3 (in front of cube)

3.2 *Relation between the flowfield and the distribution of the model parameter*

As mentioned above, in the CSM, the model parameter is determined locally. In this section, the relation between the flowfield and the spatial distribution of the model parameter is discussed.

Figures 6 and 7 show the mean velocity vectors and the distribution of the time-averaged coherent structure function F_{CS} on the vertical plane at $y = 0$ and horizontal plane at $z = 0.5H$. In Figure 7, a positive value of $\langle F_{CS} \rangle$ is shown where the recirculation flow behind the cube and the reverse flow at the side corner caused by separation of the flow are observed (Figure 6). A negative value of $\langle F_{CS} \rangle$ appears at the windward wall, where impinging of the flow is observed. According to Eqs.(7)-(9), a positive value of F_{CS} means that the region has a strong rotation rather than a strain. The distribution of $\langle F_{CS} \rangle$ (Figure 7) is consistent with the flowfield shown in Figure 6.

Figure 8 illustrates the distribution of the time-averaged model parameter $\langle C \rangle$ (= C_s^2). The value of $\langle C \rangle$ is relatively small where a large positive value of $\langle F_{CS} \rangle$ is observed (Figure 7). $\langle C \rangle$ shows a small value in the area on the roof and around the rear corner of the cube in comparison to a given value as the Smagorinsky constant (C_s^2=0.01) in the SM computation. This is because the absolute value of $\langle F_{CS} \rangle$ is very small in this region (calculated by Eq.(5)).

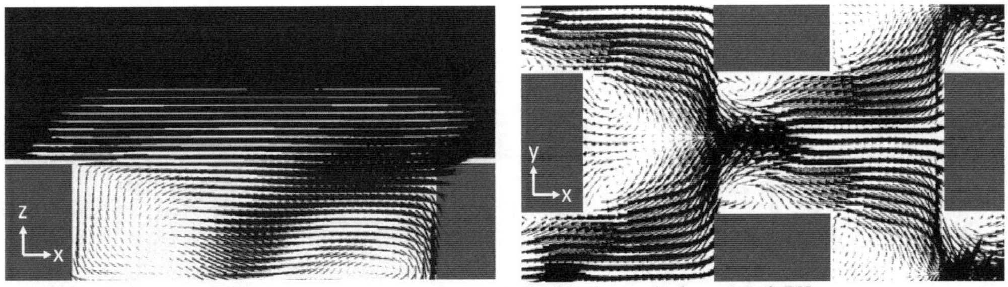

(a) Vertical plane ($y=0$) (b) Horizontal plane ($z=0.5H$)

Figure 6. Velocity vectors

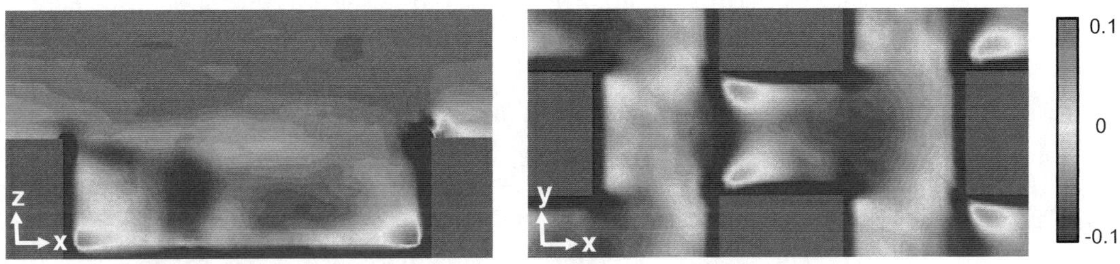

(a) Vertical plane ($y=0$) (b) Horizontal plane ($z=0.5H$)

Figure 7. Coherent structure function $\langle F_{CS} \rangle$

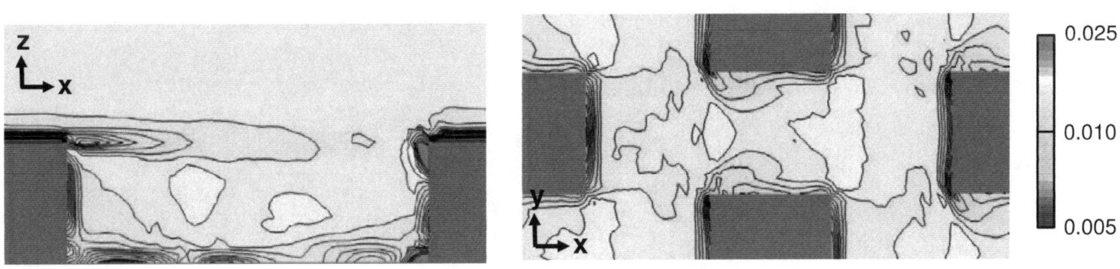

(a) Vertical plane ($y=0$) (b) Horizontal plane ($z=0.5H$)

Figure 8. Model parameter $\langle C \rangle$ ($=C_s^2$)

Figure 9 illustrates the vertical distribution of $\langle C \rangle$ at location P1, behind the cube. $\langle C \rangle$ shows a small value near the ground and wall of the cube. The damping effect in the narrow region near the wall, which is considered by using a damping function in the SM calculation, is reproduced only by modelling of SGS in the CSM calculation. $\langle C \rangle$ shows larger value than the given value in the SM computation, $C_s^2=0.01$, except for the region near the ground and wall.

Figures 10 and 11 illustrate the distributions of the time-averaged SGS eddy viscosity $\langle \nu_{SGS} \rangle$ ($= \langle C\bar{\Delta}^2|\bar{S}| \rangle$) calculated by the SM and the CSM, respectively. The results are normalized by the spatial and time averaged streamwise velocity at height of $2H$. $\langle \nu_{SGS} \rangle$ given by the CSM shows larger value in the area near the front corner and windward face, where a large $\langle C \rangle$ is observed (Figure 8), than that given by the SM. On the other hand, in the region around the rear corner, $\langle \nu_{SGS} \rangle$ in the SM shows a large

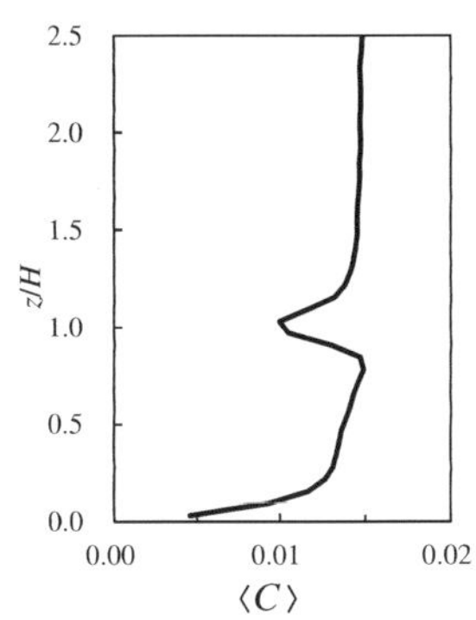

Figure 9. Vertical profile of model parameter $\langle C \rangle$

value. Such an increase of $\langle \nu_{SGS} \rangle$ does not appear in the results of the CSM, as a large value of $|\bar{S}|$ is canceled by a small $\langle C \rangle$ (Figure 8) in this region. The distribution of $\langle C \rangle$ of the CSM makes different distribution of $\langle \nu_{SGS} \rangle$ obtained from the SM.

Figure 12 compares the vertical profiles of $\langle \nu_{SGS} \rangle$ of the SM and the CSM at the locations P1 and P3. $\langle \nu_{SGS} \rangle$ of the CSM is larger than that of the SM at location P3. At location P1, $\langle \nu_{SGS} \rangle$ of the SM shows sharp maximum peak near the height of the cube. In the area around the rear corner of the cube, $\langle \nu_{SGS} \rangle$ of the CSM shows a small value in comparison with that of the SM, as shown in Figures 10 and 11. This causes a larger Reynolds stress of the CSM in this area in comparison to the SM (Figure 10) as the decrease of $\langle \nu_{SGS} \rangle$ means an increase of turbulent diffusion in the grid-scale. In the region over the roughness, $\langle C \rangle$ is larger than the given value in the SM, $C_s^2=0.01$ (Figure 9). This makes larger value of $\langle \nu_{SGS} \rangle$ in the CSM than that in the SM. Thus, Reynolds stress and turbulence intensity of the CSM are calculated smaller than those of the SM in this area. This indicates that the spatial distribution of model parameter C by the CSM causes the difference of the magnitude of the turbulence intensity and Reynolds stress between the SM and the CSM.

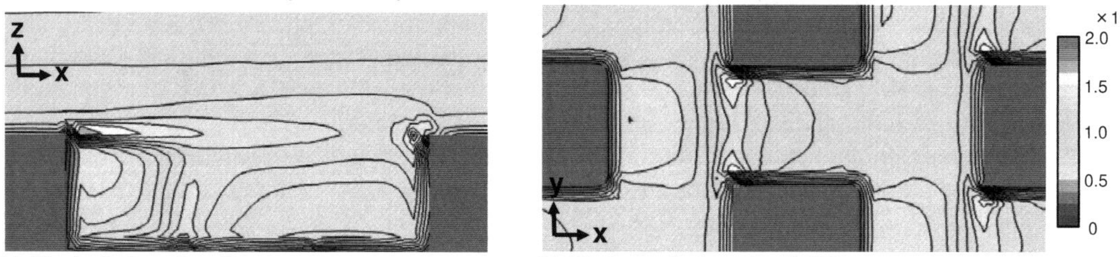

(a) Vertical plane ($y=0$) (b) Horizontal plane ($z=0.5H$)

Figure 10. SGS eddy viscosity $\langle \nu_{SGS} \rangle/(H \times \langle u \rangle_{(2H)})$ computed by the SM

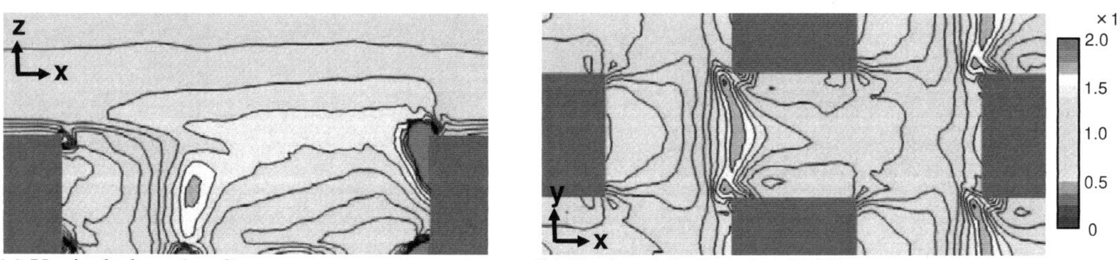

(a) Vertical plane ($y=0$) (b) Horizontal plane ($z=0.5H$)

Figure 11. SGS eddy viscosity $\langle \nu_{SGS} \rangle/(H \times \langle u \rangle_{(2H)})$ computed by the CSM

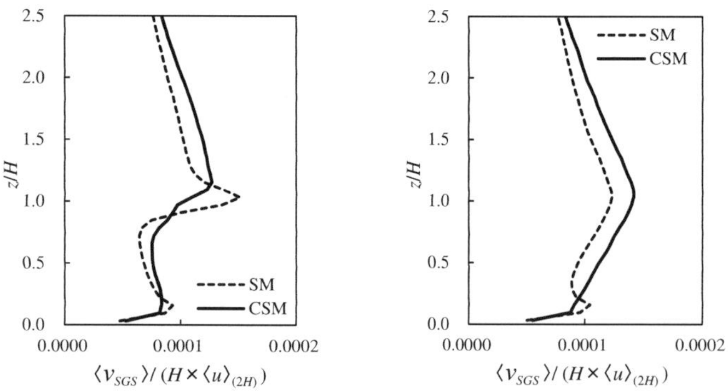

(a) P1 (behind cube) (b) P3 (in front of cube)

Figure 12. Vertical profiles of SGS eddy viscosity $\langle \nu_{SGS} \rangle/(H \times \langle u \rangle_{(2H)})$

4 CONCLUSIONS

In this study, the applicability of the CSM to a flowfield over an urban-like roughness was investigated. LES of turbulent flow over staggered arrays of cubical obstacles was carried out using the CSM and the SM. The CSM showed similar level of accuracy of the mean velocity to that obtained by using the SM, but it only required approximately half of the computational time. Although the CSM and the SM computation underestimated the magnitude of the turbulence intensity and Reynolds stress at the building corners, the CSM showed better agreement with experiments and the DNS than the SM. The spatial distribution of a time-averaged model coefficient computed by the CSM showed small values at the building height in comparison with the Smagorinsky constant given in the SM computation. This causes a difference in the magnitude of the turbulence intensity and Reynolds stress between the CSM and the SM.

References

Castro, I.P., Cheng,H. and Reynolds, R., Turbulence over urban-type roughness : deductions from wind-tunnel measurements, Boundary-Layer Meteorology, 118, pp.109-131, 2006

Coceal, C., Thomas, T.G., Castro, I.P. and Belcher, S.E., Mean flow and turbulence statistics over groups of urban-like cubical obstacles, Boundary-Layer Meteorology, 121, pp. 491-519, 2006

Coceal, O., Dobre, A., Thomas, T.G., Belcher, S.E., Structure of turbulent flow over regular arrays of cubical roughness. J Fluid Mech 589, 375–409, 2007

Germano, M., Piomelli, U., Moin, P. and Cabot, W.H., A dynamic subgrid scale eddy viscosity model, Pyhys. Fluids, A3(7), 1760-1765, 1991

Kobayashi, H., The subgrid-scale models based on coherent structures for turbulent channel flow, Phys. Fluids 17, 045104, 2005

Kobayashi, H., Ham, F. and Wu, X., Application of a local SGS model based on coherent structures to complex geometries, International Journal of Heat and Fluid Flow, 29, pp. 640-653, 2008

Lilly, D.K., A proposed modification of the Germano subgrid-scale closure method, Phys. Fluids, A4(3), 633-635, 1992

13th Americas Conference on Wind Engineering
Gainesville, Florida, USA
May 21 - 24, 2017

On the Validity of Equivalent Static Wind Loads for Tall Building Design

Sejun Park[a], DongHun Yeo[b] and Emil Simiu[c]
[a] *National Institute of Standards and Technology, Gaithersburg, MD, USA, sejun.park@nist.gov*
[b] *National Institute of Standards and Technology, Gaithersburg, MD, USA, donghun.yeo@nist.gov*
[c] *National Institute of Standards and Technology, Gaithersburg, MD, USA, emil.simiu@nist.gov*

ABSTRACT

Modern structural design for wind emerged in the 1960s as a synthesis of several developments, including the estimation of equivalent static wind loads; i.e., of time-invariant loads yielding peak horizontal building displacements equal to those due to fluctuating loads induced by flow normal to a face of isolated tall buildings with a rectangular shape in plan. For this case, fluctuating loads are assumed to be proportional to the fluctuations of the oncoming wind flow; the distribution of the wind loads over the building face is therefore known, and demand-to-capacity indexes (DCIs) used by structural engineers to size structural members, can be determined accordingly. However, the distribution of wind forces with height, and therefore internal forces induced by those forces, could not be determined from theoretical considerations or measurements, for (i) vortex-induced across-wind loading, (ii) wind speeds skewed with respect to a building face, and (iii) buildings with aerodynamics affected by the presence of nearby structures. For those cases, it was necessary to resort to subjective assessments – informed guesses -- by the wind engineer. The development of the high-frequency force balance (HFFB) in the late 1970s was an improvement over earlier approaches, although HFFB measurements of base moments and shears yield no information on the distribution of pressures on the facades. However, the development of the pressure scanner in the 1990s allows the simultaneous measurement of pressure time histories at large numbers of taps and, therefore, the determination of fluctuating wind loads at each floor level. In current practice, those fluctuating wind loads are converted into static wind loads at the floor levels, whose effects on the structure are purportedly equivalent to those induced by their fluctuating counterparts. The question arises whether those static loads produce effects -- other than the base moments and shears, e.g., internal forces and demand-to-capacity indexes, -- approximately equivalent to their counterparts induced by the fluctuating loads. To answer this question, consider a two-dimensional structure with fluctuating wind loads $f_1(t)$, $f_2(t),\ldots,f_n(t)$ at the first, second, …, n-th floor, and assume that all the floors have height h. Obtain the sum $h[f_1(t) + 2f_2(t) + \ldots + nf_n(t)] = M_b(t)$, where $M_b(t)$ is the time series of the base moment induced by those loads. We seek constant floor loads $F_1, F_2, \ldots, F_n$ such that $h(F_1 + 2F_2 + \ldots + nF_n) = \max_t[|M_b(t)|]$. This equation has an infinity of solutions, one of which is $F_j = \alpha_M \max_t [|f_j(t)|]$ $(j = 1, 2, \ldots, n)$ where, due to imperfect cross-correlations between functions $f_j(t)$, $\alpha_M < 1$, However, the equivalence of static and fluctuating forces must apply to all wind effects $R_m(t)$ of interest, where $R_m(t) = f_1(t)r_{m1} + f_2(t)r_{m2} + \ldots + f_n(t)r_{mn}$, and r_{mj} are influence coefficients; i.e., the unknowns F_j must satisfy the system of equations $F_1r_{m1} + F_2r_{m2} + \ldots + F_nr_{mn} = \max_t[R_m(t)]$ $(m = 1, 2, \ldots, m_{max})$. Since $m_{max} > n$, this is not possible. Using database-assisted design software, DCIs were calculated for 45 members of a 47-story, 160 m tall building with pressure data from the Tokyo Polytechnic University database. Calculations were performed for fluctuating loads and for static loads inducing base moments equal to the peak of the fluctuating base moments. The static loads induced DCIs that differed in 35 % of the cases by 20 % to 60 % from their peak fluctuating counterparts. In half of those cases the DCIs due to static loads were higher, and in half of the cases they were lower than those induced by the time-dependent loads. Possible alternative methods for defining equivalent static loads would similarly be unable to satisfy the requirements imposed by the fact that the number of equations m_{max} exceeds the number of unknowns n.

KEYWORDS

Database-assisted design, equivalent static wind loads, fluctuating wind loads, tall buildings.

13th Americas Conference on Wind Engineering
Gainesville, Florida, USA
May 21 - 24, 2017

Partial Atmospheric Boundary Layer Simulations Using Flow Databases Based on the Precursor Method

Liang Shi[a] and DongHun Yeo[b]

[a] *National Institute of Standards and Technology, Gaithersburg, MD. 20899, USA, liang.shi@nist.gov*
[b] *National Institute of Standards and Technology, Gaithersburg, MD. 20899, USA, donghun.yeo@nist.gov*

ABSTRACT

In partial atmospheric boundary layer (ABL) simulations the computational domain height is lower than the geostrophic height. Partial ABL simulations can significantly reduce the computational cost and can be used in the simulation of low-and mid-rise structures immersed in ABL flow. However, unlike full ABL flows, partial ABL flows experience at the top surface fluctuations of velocity and shear stresses, which require that turbulence effects be prescribed not only at the inlet but top boundary as well.

Turbulent ABL inflow simulations are generally performed by using either a synthetic turbulence generator or a precursor method. Synthetic turbulence is generated by using Fourier or other transformation techniques, under certain specified requirements on turbulence statistics, such as the energy spectrum, integral length scales, and the mean velocity profile. This method is computationally efficient, requiring only a one-stage simulation for wind engineering applications, but the quality of the synthetic turbulence is highly dependent on the way the generator is constructed. Such synthetic turbulence is typically too dissipative and disappears very fast in time because it does not possess the coherent structures for sustaining the turbulence generating cycle. Synthetic turbulence generators have not yet provided accurate representations of the turbulence inherent in the flow. In contrast, the precursor method realistically generates ABL turbulence because the flow structures are developed in the computational domain. However, this method requires a two-stage simulation: an ABL flow precursor simulation, followed by a main simulation with building(s) contained in the computational domain, using the flow field obtained from the first simulation, resulting in additional simulation time. A practical approach to mitigating the computational burden is to use an ABL flow database obtained from high-quality ABL precursor simulations. One of challenges in using CFD ABL flow databases could be due to a spatial and temporal resolution mismatch between the database and the user's main simulation.

To investigate the partial ABL flow in a simulation using ABL flow database as inflow condition, we performed partial ABL LES simulations with the precursor boundary conditions both at the inlet and at the top surface, where time-series of velocities were extracted from full ABL simulation results. Three different grid resolutions were employed to compare results with the full ABL simulation results. The simulations were performed using WindOF, an OpenFOAM package developed at NIST for computational wind engineering applications. Results show that the use of different grid resolutions could generate consistent flow statistics with those from the precursor full ABL simulations with periodic boundary conditions. This study supports the validity of the use of the CFD ABL flow databases provided by the precursor method.

KEYWORDS

CFD ABL flow database, Inflow boundary condition, Large-Eddy Simulation, OpenFOAM, Partial ABL simulation, Spatial and temporal resolution.

13th Americas Conference on Wind Engineering
Gainesville, Florida, USA
May 21 - 24, 2017

CFD-based Multi-objective Aerodynamic Shape Optimization of Twisted Tall Buildings

Fei Ding[a], Ahsan Kareem[b], Seymour MJ Spence[c]

[a] *NatHaz Modeling Laboratory, Department of Civil and Environmental Engineering and Earth Sciences, University of Notre Dame, Notre Dame, IN, USA, fding@nd.edu*
[b] *NatHaz Modeling Laboratory, Department of Civil and Environmental Engineering and Earth Sciences, University of Notre Dame, Notre Dame, IN, USA, kareem@nd.edu*
[c] *Department of Civil and Environmental Engineering, University of Michigan, Ann Arbor, MI, USA, smjs@umich.edu*

ABSTRACT

Current trend in tall building design is towards creating buildings with twisted forms through a gradual rotation of floor plans along their height. This trend mainly originates in the pursuit of timeless aesthetics by architects as a stunning variety of view makes twisted buildings recognizable and symbolic in modern cities. Nonetheless, tall buildings are highly sensitive and susceptible to wind-induced motion due to their slenderness. Specifically, turbulent eddies that are shed continuously from each side of the building are more coherent along the height of uniform cross-section buildings. As for twisted buildings, the floor plan is progressively rotated throughout the height, resulting in vortices being shed irregularly that help break up their structure. Therefore, a twisted building is considered to be more aerodynamically favorable since it effectively vitiates the coherence of vortex shedding.

However, relationship between the twist angle from the ground floor to the top floor plates and its corresponding aerodynamic loads on buildings has not been systematically and thoroughly investigated yet. Additional considerations of morphology and constructability demanded by the complexity of the aesthetic design space would compound difficulties in the design of twisted forms. To address above issues, an automated methodology for aerodynamic shape optimization of twisted tall buildings is proposed in this paper. A full 3D building model is used to carry out computational fluid dynamics (CFD) simulations in the atmospheric boundary layer to obtain its aerodynamic quantities. To achieve computational feasibility, surrogate-based modeling calibrated from a limited set of CFD simulations is introduced as a computationally inexpensive replacement for high-fidelity CFD. An automated design methodology proposed in this study allows the comprehensive search of optimal twisted forms in the design space guided by rigorous optimization algorithms. Objective functions are defined as minimizing aerodynamic loads on twisted buildings while constraints are associated with the desired architectural design and constructability considerations. This paper includes a case study that shows the capability of the proposed approach for identification of the optimal twisted forms of tall buildings. Comparisons with a selected set of wind tunnel tests are also presented.

KEYWORDS

Twisted tall building, vortex shedding, CFD, surrogate-based modeling, aerodynamic shape optimization

13th Americas Conference on Wind Engineering
Gainesville, Florida, USA
May 21 - 24, 2017

Tailoring of Long-span Bridge Decks to Improve their Aerodynamic Performance Using CFD Based Optimization Method

Yan Fang[a], Ahsan Kareem[b]

[a] *NatHaz Modeling Laboratory, Department of Civil and Environmental Engineering and Earth Sciences, University of Notre Dame, Notre Dame, IN, USA, yfang1@nd.edu*
[b] *NatHaz Modeling Laboratory, Department of Civil and Environmental Engineering and Earth Sciences, University of Notre Dame, Notre Dame, IN, USA, kareem@nd.edu*

ABSTRACT

Long-span bridges are very susceptible to wind actions. Their aerodynamic behaviors have been widely studied after the collapse of Tacoma Narrows Bridge among which the flutter instability is of the most concern. Tailoring the shape of long-span bridge decks can help modify the aerodynamics of bridge deck which has the promise of enhancing the performance of bridges, e.g., increasing the flutter speed. This paper focuses on shape optimization of long-span bridge decks utilizing computational fluid dynamics (CFD) based simulations. It employs multi-objective function in which flutter instability, drag, lift and torsional moment coefficients are taken into consideration to optimize the performance of bridges. A generic bridge deck is used as the initial shape and a surrogate-based approach that eliminates exhaustive number of simulations in conjunction with generic algorithms with a filter that constrains critical flutter speed is adopted to search for the optimal shape. An open source CFD software OpenFOAM with features of mesh morphing and dynamic meshing is utilized to simulate flow around 2D bridge decks using turbulence modeling scheme unsteady Reynolds-averaged Navier-Stokes (URANS) approach adopting Menter's SST k-ω turbulence model to calculate flutter derivatives and aerodynamic coefficients. The CFD code is validated through a comparison with experiments in a wind tunnel involving generic bridge decks.

KEYWORDS

Long-span bridges, computational fluid dynamics, shape optimization, flutter

Topology Optimization of Wind-Excited Uncertain Building Systems Subject to Performance Constraints: from Life-Safety to Habitability

Arthriya Suksuwan [a], Seymour M.J. Spence [a]

[a] *Department of Civil and Environmental Engineering, University of Michigan, Ann Arbor, MI, USA*

ABSTRACT: The need for civil engineers to design cost-effective structural systems to withstand natural disasters has led to the development of state-of-the-art frameworks that combine performance-based design concepts with topology optimization. To this end, there is still a lack of topology optimization frameworks that can handle performance constraints written in terms of first excursion probabilities over the duration of the event. This work proposes two approaches for the topology optimization of uncertain and dynamic wind-excited building systems subject to constraints on first excursion probabilities. The first approach derives solutions analytically while the second approach is based on solving a sequence of approximate sub-problems defined in terms of a probabilistic auxiliary variable. An example is presented in order to compare the two approaches.

KEYWORDS: Performance-based design, Topology optimization, Wind engineering, First excursion probability.

1 INTRODUCTION

This work aims to firstly validate the simulation-based framework developed in [1] for the topology optimization of uncertain and dynamic wind-excited tall building systems, and secondly discuss the extension of performance-based topology optimization frameworks to acceleration-based habitability constraints. In particular, the problems discussed in this work are formulated and solved in a setting consistent with the recently developed probabilistic wind PBD framework proposed in [2], therefore ensuring that the results are compatible with the latest advances in building design practices. For the convenient application of gradient-based optimizers, equivalent inverse constraints are formulated in terms of the second order response statistics and a reduced variate with target failure probability. Two approaches are proposed and evaluated, termed full approach and approximate approach. In the full approach, the insensitivity of the reduced variate to the change in design variables allows the sensitivities of the inverse constraints to be derived analytically in the time domain. The approximate approach, on the other hand, is based on solving a sequence of optimization sub-problems defined in terms of a probabilistic auxiliary variable vector [3-7], while the sensitivities of the constraints with respect to the design variables can be efficiently obtained using the standard adjoint method.

A case study consisting of a building system subject to stochastic wind loads is presented to demonstrate the potential of the full approach as well as the validity of the approximate approach. In particular, the lateral load-resisting system of a planar frame is optimized subject to a performance constraint defined in terms of the first excursion probability of the top floor displacement response. The exact solution is compared to the approximate solution. The advantages and disadvantages of both approaches are evaluated. The potential of extending the proposed approaches to account for occupant comfort criteria is also discussed.

2 PROBLEM STATEMENT

The primary goal of this work is to develop a framework for performance-based topology optimization for wind-excited structural systems. In particular, the specific aim is to identify the optimal topologies for the lateral load-resisting system that minimizes the material within a predefined design domain while ensuring that the optimized system meets a number of predetermined performance constraints written in terms of first excursion probabilities. To this end, it is convenient to write the performance constraints in their inverse form [8]. Under this condition, the topology optimization problem may be formally posed in the following form:

$$\begin{aligned} &\min_{\boldsymbol{\rho}} V(\boldsymbol{\rho}) = \sum_{e=1}^{n} \int_{\Omega_e} \rho_e \, \mathrm{d}\Omega_e \\ &\text{s.t. } dm_j^{(P_{0,j})}(\boldsymbol{\rho}) \leq \widetilde{dm}_j \, , j = 1, \dots, N_c \\ &0 \leq \rho_e \leq 1 \end{aligned} \tag{1}$$

where: $\boldsymbol{\rho} = \{\rho_1, \dots, \rho_n\}^T$ is a vector collecting the element-wise normalized material densities that are generally related to a vector of independent design variables, **x**, through a filter operator [9]; V represents the total material volume; e denotes the element number; n is the total number of elements in the design domain; Ω_e denotes the design domain of element e; $dm_j^{(P_{0,j})}$ is the damage measure associated with the predefined target failure probability for the jth performance constraint $P_{0,j}$; $\widetilde{dm}_j$ is the predefined threshold value indicating the damage state defined by damage measure dm_j; and N_c denotes the total number of performance constraints.

3 METHODOLOGY

To solve the topology optimization problem stated in Eq. (1), the performance of the structural system is described in this work through a damage measure defined as the ratio between the engineering demand parameter and the corresponding capacity as shown below (for the sake of clarity, subscript j is dropped):

$$dm^{(P_0)} = \frac{\hat{r}^{(P_0)}}{\bar{c}} \tag{2}$$

where $\hat{r}^{(P_0)}$ denotes the engineering demand parameter (e.g., absolute largest displacement response, absolute largest acceleration response, etc.) taken as the response threshold such that the first excursion probability over the event duration of the response process $r(t)$ coincides with P_0 while $\bar{c}$ is the mean/nominal capacity of the corresponding constrained component.

Two approaches, termed the full approach and the approximate approach, are presented in order to assess the performance of the structural systems. Sensitivity information is derived in order to enable the use of gradient-based optimization algorithms. In addition, the Solid Isotropic Material with Penalization (SIMP) method [10] is also adopted in order to obtain high quality black and white solutions.

3.1 *Full Approach*

The response threshold, $\hat{r}^{(P_0)}$, defined in Eq. (2) can be written in terms of second order statistics of the response process, $r(t)$, as:

$$\hat{r}^{(P_0)} = \mu_R + g^{(P_0)}\sigma_R \tag{3}$$

where μ_R and σ_R are the mean and the standard deviation of the response process $r(t)$ respectively while $g^{(P_0)}$ denotes the reduced variate associated with the first excursion probability of interest. Assuming a modal analysis framework, the response process may be written as:

$$r(t) = \sum_{m=1}^{N_m} B_m q_m(t) \tag{4}$$

$$\ddot{q}_m(t) + 2\zeta_m\omega_m\dot{q}_m(t) + \omega_m^2 q_m(t) = \frac{\boldsymbol{\phi}_m^T\mathbf{f}(t)}{M_m} \tag{5}$$

where: m indicates the mode number; N_m is the total number of modes considered; B_m are the modal participation coefficients; $q_m(t)$ is the modal response process; ζ_m represents the modal damping ratio; ω_m denotes the modal circular frequency; $\boldsymbol{\phi}_m$ represent the mth mode shape; M_m denotes the modal mass; and $\mathbf{f}(t)$ is the external aerodynamic loads.

To determine $\mathbf{f}(t)$, any wind simulation methods can be implemented. In particular, the stochastic wind model proposed in [11] is adopted in this work in which the site-specific largest mean hourly wind speeds over the building height is extrapolated using the relationship proposed in [12].

Having obtained the response process $r(t)$, the corresponding second order statistics can be determined through the following equations:

$$\mu_R = \sum_m B_m \int_0^t \boldsymbol{\phi}_m^T\boldsymbol{\mu}_\mathbf{f} h_m(t-\tau)\, \mathrm{d}\tau \tag{6}$$

$$\sigma_R = \sqrt{\sigma_R^2} = \sqrt{\sum_m \sum_n \iint_0^\infty B_m B_n R_{\mathrm{f}_m\mathrm{f}_n}(t_1 - t_2) h_m(t_1) h_n(t_2)\, \mathrm{d}t_1 \mathrm{d}t_2} \tag{7}$$

where $\boldsymbol{\mu}_\mathbf{f}$ is the mean of the external aerodynamic loads $\mathbf{f}(t)$, h_m represents the modal impulse response function, and $R_{\mathrm{f}_m\mathrm{f}_n}$ is the covariance function of the generalized forces.

Having defined the second order statistics of the response process through Eq. (6) and Eq. (7), if it is also assumed that the reduced variate $g^{(P_0)}$ is insensitive to changes in the design variables, the derivatives of the damage measure defined in Eq. (2) with respect to the design variables can be obtained analytically by taking the gradients of Eq. (6) and Eq. (7).

3.2 *Approximate Approach*

As mentioned earlier, the approximate approach is based on solving a sequence of optimization sub-problems. In particular, for each sub-problem, the engineering demand parameter is written in terms of the following auxiliary variable vector (AVV), $\psi^{(P_0)}$, [7]:

$$\boldsymbol{\psi}^{(P_0)} = \boldsymbol{\mu}_\mathbf{F} + g^{(P_0)}\frac{\mathbf{C}_\mathbf{F}\boldsymbol{\Gamma}_R}{\sigma_R} \tag{8}$$

where $\boldsymbol{\Gamma}_R$ is a vector of influence functions derived from the system stiffness matrix $\mathbf{K}$ while $\boldsymbol{\mu}_\mathbf{F}$ and $\mathbf{C}_\mathbf{F}$ are the mean and covariance matrix of the following vector valued stochastic process:

$$\mathbf{F}(t) = \mathbf{K}\boldsymbol{\Phi}_m\boldsymbol{q}_m(t) \tag{9}$$

where $\boldsymbol{\Phi}_m$ is the mode shape matrix of order m while $\boldsymbol{q}_m(t)$ is the modal response vector collecting the first m modal responses.

The interest in defining the AVV is that the following holds:

$$\hat{r}^{(P_0)} = \boldsymbol{\Gamma}_R^T \boldsymbol{\psi}^{(P_0)} \tag{10}$$

$$dm^{(P_0)} = \frac{\boldsymbol{\Gamma}_R^T \boldsymbol{\psi}^{(P_0)}}{\bar{c}} \tag{11}$$

In other words, the static application of the AVV to the structural system will generate a response level that is identical to the response threshold with target first excursion probability. In particular, by assuming the AVV to be independent of the design variables, an optimization sub-problem can be defined that takes the form of a standard static topology optimization problem which can be extremely efficiently solved through any gradient-based optimization algorithm after the sensitives have been estimated through the application of the standard adjoint method [7].

4 CASE STUDY

4.1 *Description*

The numerical example considered in this work consists in the topology optimization of a planar load-resisting system subject to stochastic wind loads. In particular, the objective is to minimize the material volume used while also ensuring that a performance constraint set on the first excursion probability of the top floor displacement is satisfied. The building has one bay and four stories. The story height is taken as 4 m while the bay width is taken as 5 m. A carried mass of 7000 kg/m^3 is considered. The design domain is discretized using Q4 elements while the secondary system (i.e. the external columns and horizontal floor beams) are modeled using frame elements. Each Q4 element has dimensions of 0.125×0.125 m with 0.1 m thickness and possesses concrete material properties with E = 21 GPa. The frame elements are steel wide-flange beams W8×21 with E = 199 GPa. The two types of elements are superimposed following the method presented in [13]. The total number of design variables in this problem is 5120. Symmetry is imposed about the vertical center line of the frame. The constraint is imposed on the displacement at the right external column line at the top floor. A mean/nominal capacity of 1/100 of the total building height is considered. The limiting value for the damage measure, $\widetilde{dm}_j$, is taken as 1. The reduced variate associated with predefined target failure probability P_0 is set as $g^{(P_0)} = 3.5$. Stochastic wind excitation is simulated for a 1-hour event using the model described in [11]. The reference wind speed at 10 m height was taken as 60 m/s. This reference wind speed was then translated into the site-specific wind intensity according to the method proposed in [12]. For the other parameters values: the air density was taken as 1.25 km/m^3; the pressure coefficient was taken as 1.3; the reference surface roughness was taken as 0.05 m; the site-specific surface roughness was taken as 0.5 m; and the influence width of the building was taken as 5 m. For each floor, the total number of random phase angle components was set as 4096 while the cut-off frequency was taken as 4.6 Hz. Information on the marginal distributions of the uncertain parameters associated with the wind simulation can be found in [1].

For the SIMP method, a penalty of three was considered. The Method of Moving Asymptotes (MMA) [14] was used in this work as the gradient-based optimizer.

4.2 *Results and Discussion*

The final topologies obtained from both approaches are illustrated in Fig. 1. In particular, from Fig. 1, it can be clearly seen that both approaches converge to very similar final designs. This strong agreement between the systems proves the validity of the approximate approach. The minor difference that can be seen between the systems is expected since the second order statistics were derived using different techniques. It is observed that the full approach leads to a slightly lower material volume than the approximate counterpart. In terms of the final performance of the systems, both final designs satisfied the first excursion constraint. In particular, it took a total of 262 iterations to reach convergence with the full approach while the approximate approach required about 100 more iterations for in order to reach final convergence. Notwithstanding this difference, the approximate approach converged within 5 minutes while the full approach took almost 40 hours to converge. The high computational cost of using the full approach is mainly due to the sensitivity analysis required at each iteration.

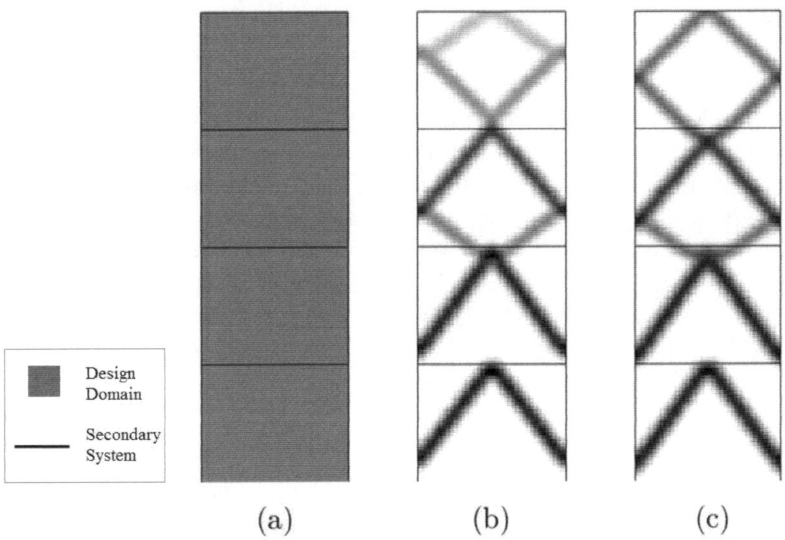

Figure 1: (a) Design Domain (b) Optimal topology derived from the full approach (c) Optimal topology derived from the approximate approach.

5 EXTENSION TO OCCUPANT COMFORT

The potential and the practicality of both the full and approximate approaches have been demonstrated through the case study in which only life-safety requirements defined in terms of the displacement response has been considered. In the case of wind-excited structural systems, especially tall buildings, the acceleration response occurring during moderate windstorms may have to be controlled in order to ensure occupant comfort. With this in mind, the next step is to extend the proposed framework to account for habitability requirements by adding performance constraints defined in terms of first excursion probabilities of acceleration responses. For the full approach, this is straightforward and simply requires the replacement of the displacement-based impulse response function with the acceleration-based impulse response function. The main drawback of the full approach is the numerical cost. To circumvent this, it may be useful to solve the problem in the frequency domain or state-space instead of the time domain. Another possible alternative would be to revisit the concept of the auxiliary variable vector and redefine the sub-problem in terms of an auxiliary spectrum that would enable the inclusion of acceleration-based performance constraints. If the latter is validated against the exact solution, its advantage with respect to efficiency is undoubtable. These aspects are currently under investigation.

6 CONCLUSIONS

This work proposed a framework for the performance-based topology optimization of structural systems subject to stochastic wind excitation. In particular, the aim was to define a procedure for identifying optimal topologies for the lateral load-resisting system of wind-excited systems while ensuring the satisfaction of performance constraints written in terms of target first excursion probabilities. To facilitate the use of gradient-based optimizers, equivalent inverse constraints were formulated in terms of the second order response statistics and a reduced variate associated with the target failure probability. Two approaches were evaluated, termed full approach and approximate approach. In the full approach, the sensitivity information was derived analytically in the time domain. The approximate approach centered on the definition of an approximate sub-problem defined in terms of a probabilistic auxiliary variable vector. A case study demonstrated the potential of the full approach as well as the validity of the approximate approach when dealing with displacement-based performance constraints. While the approximate approach required more iterations to achieve convergence, the computational time was a fraction of that required by the full approach. Finally, the potential of extending the approaches to occupant comfort criteria defined in terms of acceleration responses was discussed.

7 ACKNOWLEDGEMENTS

This research was supported by the NSF Grant No. CMMI-1462084. The authors also would like to thank Krister Svanberg for proving the MMA algorithm, which was used as the optimization algorithm in this work.

8 REFERENCES

1 S. Bobby, S. M. J. Spence, and A. Kareem, Data-driven performance-based topology optimization of uncertain wind-excited tall buildings, Struct. Multidisc. Optim., (2016) 1–24.
2 M. Ciampoli, F. Petrini, and G. Augusti, Performance-Based Wind Engineering: Towards a general procedure, Struct. Saf., 33 (2011) 367–378.
3 S. Bobby et al., Performance-based topology optimization for wind-excited tall buildings: A framework, Eng. Struct., 74 (2014) 242–255.
4 S. M. J. Spence and M. Gioffre, Efficient algorithms for the reliability optimization of tall buildings, J. Wind Eng. Ind. Aerodyn., 99 (2011) 691–699.
5 S. M. J. Spence and M. Gioffr`e, Large scale reliability-based design optimization of wind excited tall buildings, Prob. Eng. Mech., 28 (2012) 206–215.

6 S. M. J. Spence, M. Gioffre, and A. Kareem, An efficient framework for the reliability-based design optimization of large-scale uncertain and stochastic linear systems, Prob. Eng. Mech., 44 (2016) 174–182.

7 S. M. J. Spence and A. Kareem, Performance-based design and optimization of uncertain wind-excited dynamic building systems, Eng. Struct., 78 (2014) 133–144.

8 A. Der Kiureghian, Y. Zhang, and C.-C. Li, Inverse Reliability Problem, J. Eng. Mech., 120 (1994) 1154–1159.

9 B. Bourdin, Filters in topology optimization, Int. J. Numer. Meth. Engng., 50 (2001) 2143–2158.

10 M. P. Bendsøe, Optimal shape design as a material distribution problem, Struct. Optimization, 1.4 (1989) 193–202.

11 G. Deodatis, Simulation of Ergodic Multivariate Stochastic Processes, J. Eng. Mech., 122 (1996) 778–787.

12 F. Minciarelli et al., Estimates of extreme wind effects and wind load factors: Influence of knowledge uncertainties, Prob. Eng. Mech., 16 (2001) 331–340.

13 L. L. Stromberg et al., Topology optimization for braced frames: Combining continuum and beam/column elements, Eng. Struct., 37 (2012) 106–124.

14 K. Svanberg, The method of moving asymptotes - A new method for structural optimization, Int. J. Numer. Meth. Engng., 24 (1987) 359–373.

3D Performance-Based Topology Optimization of Uncertain Dynamic Tall Building Systems

Xihaier Luo[a], Seymour M.J. Spence[b], Ahsan Kareem[c]

[a]*Nathaz Modeling Laboratory, Dept. of Civil and Environmental Engineering and Earth Sciences, Univ. of Notre Dame IN, USA. E-mail: xluo1@nd.edu*
[b]*Dept. of Civil Engineering, Univ. of Michigan, Ann Arbor, MI, USA E-mail: smjs@umich.edu*
[c]*Nathaz Modeling Laboratory, Dept. of Civil and Environmental Engineering and Earth Sciences, Univ. of Notre Dame IN, USA. E-mail: kareem@nd.edu*

ABSTRACT

The artistry of high-rise building design is not only in satisfying life safety constraints but also probabilistic displacement-based serviceability constraints that must rigorously account for the dynamic nature of the system under moderate to severe wind excitation. Current structural assessment and optimization frameworks for serviceability criteria under wind-induced vibration is plagued by inevitable uncertainty in construction material characterization, building model idealization and stochastic wind loading conditions. It is therefore of fundamental importance to incorporate the state-of-the-art concept of performance-based design (PBD) into structural optimization algorithms in order to achieve truly efficient designs. Within the context of structural optimization, topology optimization was initially introduced as a computational material distribution method that is capable of providing innovative and efficient layouts for a preselected building material within a prescribed design space. Few works on 3D topology optimization are available due to the high computational cost of solving 3D problems notwithstanding how designs in 3D space are critical for directly achieving practical applications. In this regard, this paper introduces a framework for the 3D performance-based topology optimization of wind-excited tall building systems with respect to serviceability performance objectives. Performance-oriented objectives are identified through the modeling and simulation of dynamically-loaded building systems subject to various uncertainties. Meanwhile, serviceability constraints on non-structural performance criteria are seamlessly incorporated into the constraint functions of the 3D topology optimization problem through a set of explicit constraint functions. Based on solving a decoupled sequence of reliability analysis and structural optimization sub-problems, a robust sequential algorithm is defined for solving the original nested performance-based topology optimization problem. A case study is presented in order to demonstrate the validity and effectiveness of the proposed 3D topology optimization framework.

KEYWORDS

3D Topology optimization, Performance-based design, Uncertainty propagation, Tall buildings

13th Americas Conference on
WIND ENGINEERING

Wind-induced fatigue behavior of rooftop pinnacles and masts on tall buildings: A Machine Learning approach

Monica ArulJayachandran-1[a], Ahsan Kareem-2[b], Dae Kun Kwon-3[c]

[a] *NatHaz Modeling Laboratory, Univ. of Notre Dame, Notre Dame, IN 46556, USA, maruljay@nd.edu*
[b] *NatHaz Modeling Laboratory, Univ. of Notre Dame, Notre Dame, IN 46556, USA, kareem@nd.edu*
[c] *NatHaz Modeling Laboratory, Univ. of Notre Dame, Notre Dame, IN 46556, USA, dkwon@nd.edu*

ABSTRACT

With increasing height of buildings, the desire for architectural details such as pinnacles, spires and sign are also on the rise. Masts and towers carrying antenna are placed on rooftop of tall buildings which helps in transmission and broadcast. Compared to the building, these appendage structures are very critical in their functionality. Structurally they undergo asynchronous dynamic motion with respect to the tall building. Due to this, both the mean and fluctuating components of the stresses at the base varies at a significant level. The shape, weight, connection details and the building-top location render these structures fatigue sensitive that could be subjected to severe and sustained wind-induced vibration at different wind speeds and profiles.

Hence, the owners and stakeholders of several tall buildings have adopted for installation of real time monitoring programs to monitor the dynamic response of the building as well as to obtain sufficient data to understand the fatigue behavior of rooftop structural elements. Both wind and fatigue domains involve large amount of data. Thus continuous monitoring generates big data sets for wind and fatigue behavior. Conventional methods of data crunching have been reported to be inadequate in the literature. Hence it is evident that data science techniques such as machine learning approaches are ideal tools for the effective mining of the data. This approach is capable of revealing potential wind and fatigue life characteristics and response of the structural element under consideration. Published research work in this area is scarce.

In line with the observations presented above, the present study aims at analyzing and predicting the fatigue behavior of pinnacles and masts atop tall buildings (e.g. Burj Khalifa and Willis Tower) using the data obtained from accelerometers and ultrasonic wind anemometers installed on the buildings. The wind data is numerically processed to estimate the time dependent loading on the critical elements such as pinnacle/masts. The fatigue sensitive locations known as the hot spots are identified in the support bases of pinnacle/mast depending upon their detail category. The stress ranges at the hot spot are computed for both the along-wind and across-wind directions. The fatigue spectrum of stress ranges is computed using rain-flow cycle counting algorithm. The high-stress and low-stress cycles are computed to obtain the fatigue stress spectrum and fatigue evaluation is done using the Palmgren-Miner's rule. The Miner's rule is an excellent tool for residual life assessment of fatigue sensitive structures. Using data mining tools like Information Gain, the different parameters estimated are ranked in order of their potential for predicting fatigue strength. Predictive models are developed for the same using Random Forest (RF), Artificial Neural Networks(ANN) and a combination of other machine learning techniques. These tools provide accurate relationship between input parameters (e.g. maximum stress, stress ratio) and the number of cycles to failure. The present formulation is not only applicable for appended pinnacle/masts but also for the steel connections which are part of the tall buildings which undergo cyclic stresses throughout service life. The details of this computational frame work are reported in this paper.

KEYWORDS

Machine learning, fatigue evaluation, Miner's rule, Wind data

13th Americas Conference on Wind Engineering
Gainesville, Florida, USA
May 21 - 24, 2017

Wind Loading and Dynamic Performance of a Tall Timber Building

William A. Ashman[a], Lancelot Reres Reres[b], Hoa Thai Le[c], Weichiang Pang[d]

[a] *Clemson University, SC, USA, washman@clemson.edu*
[b] *Clemson University, SC, USA, lreres@clemson.edu*
[c] *Clemson University, SC, USA, thle@clemson.edu*
[d] *Clemson University, SC, USA, wpang@clemson.edu*

ABSTRACT

With the increased knowledge concerning and availability of timber products, namely cross-laminated timber (CLT), a great interest has been expressed in tall timber buildings. Motivation is greatly benefitted by the environmental advantages of timber construction. Since the allowable height of mass timber buildings is based on fire safety, and not limitations of strength or serviceability, the dynamic response of mass timber structures to wind loads, far-exceeding these allowable heights, is not well understood. Indeed, several buildings are under development now that exceed the scope of existing codes. For example, the University of British Columbia has an 18 story residential building underway, a 12 story construction is planned in Portland OR, and 10 story building scheduled for construction in New York City. Skidmore Owings and Merrill, LLP have recently produced a report outlining the limitations and strengths of a composite 'concrete jointed timber frame'. The report issued a call for greater research into this and similar new construction techniques. The research presents the most recent investigations of the wind loading and dynamic performance on a fully timber tall building. The scope of the research includes the base force measurements and the vibrational excitation of the tall timber building under wind loading. The base moment loading data is collected using a High Frequency Force Balance (HFFB) test. This test captures the spectra of modal forces. From the modal force data, base shear, torsion and moments can be computed. An aeroelastic scale model of 20-story timber building will be fabricated for the wind tunnel testing to determine the drift and acceleration of the tenable levels of the building, and to measure the vibrational excitation. Wind-induced vibration and aerodynamic behavior of the tall timber building will be subsequently evaluated. All experiments occur in the atmospheric boundary layer (ABL) wind tunnel at Clemson University. Results for both experiments will be disseminated in the full-paper. The results of the research will be applicable to developers and architects to advance the timber construction techniques in the high-rise timber building market, especially in high wind prone regions.

KEYWORDS

Tall building, vibration, serviceability, Cross Laminated Timber

13th Americas Conference on Wind Engineering
Gainesville, Florida, USA
May 21 - 24, 2017

A Comprehensive Framework for Performance-Based Wind Engineering of Tall Buildings Considering Wind Direction and Wind Speed

Wei Cui[a], Luca Caracoglia[b]

[a] *Northeastern University, Boston, Massachusetts, USA, cui.we@husky.neu.edu*
[b] *Northeastern University, Boston, Massachusetts, USA, lucac@coe.neu.edu*

ABSTRACT

Tall buildings are the land-mark and center of human activities. They greatly influence local culture, economics and other aspects of society. Therefore, accurate evaluation of structural performance under extreme hazards, such as earthquakes, high winds and man-made hazards, is necessary in modern building design.

Currently, most researchers use the Monte-Carlo sampling method as the core step to calculate the risk of structural failure of a tall building for different wind speed levels in order to simulate structural fragility curves. These curves can subsequently be used to examine structural integrity against wind hazards. Furthermore, they can be combined with the structure's local wind climate to evaluate the structural failure probability over structural lifetime.

However, the complexity of wind climate and its relationship with the building response still need to be investigated by researchers to more accurately assess the wind-induced structural response and to evaluate structural performance over lifetime more precisely. The state-of-the-art structural performance analysis against wind hazards is usually composed of three separated steps: (1) finding out the most unfavorable wind direction, which contributes the most to the extreme wind load either by wind tunnel test or through computational fluid dynamics; (2) constructing the structural fragility curves at different wind speeds; (3) incorporating the building's local wind climate data into the structural fragility curves to evaluate the structural failure probability. The major flaw in these steps is ignoring the possibility of multi-directionality in the mean wind direction, which may affect the structural vulnerability to winds, since the winds are not always coming from the worst possible direction. This paper will propose a simulation framework for tall buildings that combines a more comprehensive fragility analysis, accounting for variable wind direction, into the structural vulnerability evaluation. Furthermore, wind climate information will consider the correlation between wind speed and wind direction. As a result, the structural fragility curve (i.e. surface) will depend on two input variables: wind speed and wind direction. Consequently, the structural performance over the building lifetime will include the site wind speed and wind direction.

KEYWORDS

Performance-Based Wind Engineering, Tall Building aerodynamics, Wind Directionality, Fragility Curves

The 13th Americas Conference on Wind Engineering (13ACWE)
Gainesville, Florida USA, May 21-24, 2017

Estimation of Inelastic Crosswind Response of Tall Buildings using Equivalent Nonlinear Equation Approach

Changda Feng [a], Xinzhong Chen [b]

[a] *National Wind Institute, Texas Tech University, Lubbock, TX, USA, changda.feng@ttu.edu*
[b] *National Wind Institute, Texas Tech University, Lubbock, TX, USA, xinzhong.chen@ttu.edu*

ABSTRACT: This study addresses the analysis of stochastic crosswind response of tall buildings with bilinear hysteretic restoring force character. The building response is represented in fundamental modal response. Both wind speed regions with and without nonlinear aerodynamic damping effect are considered. The equivalent nonlinear equation (ENEL) approach is used for estimating response statistics, which include root-mean-square value, kurtosis, probability density distribution, extreme value distribution and fatigue damage. Response history analysis is also performed to demonstrate the accuracy of this analytical framework. The inelastic crosswind response characteristics are discussed though a parametric study to explore the potential benefit of inelastic building design.

KEYWORDS: Crosswind building response, nonlinear aerodynamic damping, hysteretic restoring force, extreme response, fatigue damage

1 INTRODUCTION

Crosswind response often controls the design of wind-excited tall buildings and other flexible structures such as chimney and towers. Extensive studies have been made to characterize the crosswind wind. loads and structural response. For very flexible buildings, the crosswind response at the vicinity of vortex lock-in wind speed needs to be carefully studied with consideration of nonlinear aerodynamic damping effect. Chen (2013 and 2014a) showed that the nonlinear aerodynamic damping effect drives the crosswind response to have hardening non-Gaussian distribution, which is responsible to the reduced peak factor. Chen (2013 and 2014b) presented the equivalent nonlinear equation (ENLE) approach, which is able to provide solutions of response statistics including RMS response, extreme value distribution and fatigue damage.

Current wind resistant structural design does not permit structural response beyond linear elastic range even at ultimate limit state. However, studies have shown that permitting controlled inelastic building response has a potential to result in a more economical structural design. In addition, tall buildings with hysteretic dampers and base isolation systems have inelastic restoring force characteristics. Therefore, there is a need to explore inelastic response analysis framework and to develop a better understanding of inelastic crosswind response of tall buildings. Feng and Chen (2017) introduced statistical linearization approaches for estimating inelastic crosswind response where the nonlinear aerodynamic damping effect was not accounted. The results demonstrated that the crosswind response can be reduced considerably due to hysteretic damping associated with inelastic response.

This study addresses the estimation of inelastic crosswind response of tall buildings using ENLE approach. Both wind speed regions with and without nonlinear aerodynamic damping effect are considered. Response history analysis is also performed to demonstrate the accuracy of this analytical framework. A parametric study is performed to better understand the inelastic response characteristics and to explore the potential benefit of inelastic building design over the current linear elastic design.

2 ANALYTICAL FRAMEWORK

2.1 *Equation of crosswind response*

The crosswind response of tall building is represented in terms of first modal response. The equation of motion in terms of non-dimensional displacement is represented as follows with bilinear hysteretic restoring force as shown in Fig. 1 (e.g., Lutes and Sarkani 2004):

$$\ddot{x} + 2\omega_s(\xi_s + \xi_a)\dot{x} + \alpha\omega_s^2 x + (1-\alpha)\omega_s^2 z = \frac{1}{2}\left(\frac{\rho B^2}{m_s}\right)\left(\frac{U^2}{B^2}\right)\eta_b \eta C_{Mb}(t) \quad (1)$$

$$\dot{z} = \dot{x}\{1 - u(z - x_y)u(\dot{x}) - u(-z - x_y)u(-\dot{x})\} \quad (2)$$

$$\xi_a = -\frac{1}{4}\left(\frac{\rho B^2}{m_s}\right)\eta_{se}\eta H_1^* \quad (3)$$

$$m_s = \frac{M_s}{\int_0^H \phi^2(y)dy}; \qquad \eta = \frac{H}{\int_0^H \phi^2(y)dy} \quad (4)$$

where M_s and ξ_s are generalized mass and damping ratio; ω_s is natural modal frequency calculated by using the initial generalized stiffness k, i.e., $\omega_s = 2\pi f_s = \sqrt{k/M_s}$; m_s is effective building mass per unit height; $x = x_1/B$ is generalized non-dimensional displacement, and is the building top displacement when mode shape $\phi(y)$ is normalized as $\phi(H) = 1$; ξ_a is aerodynamic damping ratio; α is ratio of the second stiffness to initial stiffness k; $z = z_1/B$ is non-dimensional hysteretic displacement; $u(\cdot)$ is unit step function; $x_y = x_{y1}/B$ is non-dimensional yield displacement; η_{se} and η_b are mode shape correction factors; η is a non-dimensional parameter related to mode shape; ρ is air density; U is wind speed at building top; B is building width; H is building height; and $C_{Mse}(t)$ and $C_{Mb}(t)$ are self-excited (motion-induced) and buffeting components of base bending moment coefficient, which are determined by high-frequency-force-balance (HFFB) measurement in wind tunnel.

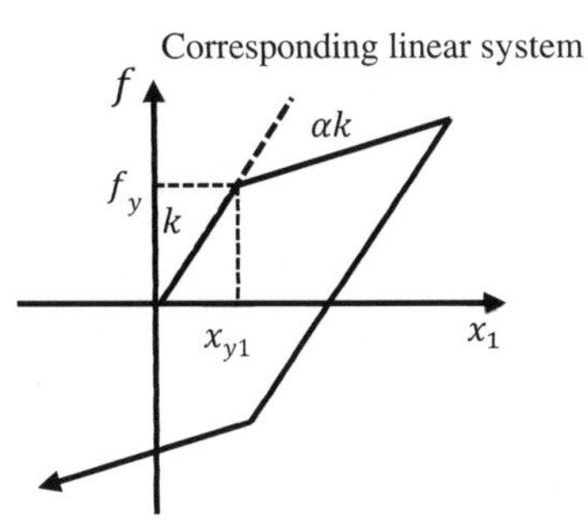

Fig. 1 Bilinear hysteretic force model

When wind speed is far lower than the vortex lock-in speed, the aerodynamic damping is low and positive thus can be neglected. However, it is not the case at the vicinity of vortex lock-in wind speed where the nonlinear aerodynamic damping effect is significant. The aerodynamic damping ratio ξ_a can be determined from forced-vibration test in wind tunnel with harmonic motion. At a given reduced frequency $K = fB/U$, ξ_a is a nonlinear function of vibration amplitude $A = x_{max}$, and can be expressed as follows for $\xi_{a1} = m_s\xi_a/\rho B^2$ (Chen 2013):

$$\xi_{a1} = a_1 + a_2 x_{max} + a_3 x_{max}^2 \tag{5}$$

where a_1, a_2 and a_3 are coefficients.

For stochastic response analysis, the aerodynamic damping needs to be expressed as a nonlinear function of time-varying velocity and/or displacement as (Chen 2013):

$$\xi_{a1} = A_1 + \frac{A_2}{K}\left(\frac{B|\dot{x}|}{U}\right) + \frac{A_3}{K^2}\left(\frac{B\dot{x}}{U}\right)^2 \tag{6}$$

where $A_1 = a_1$, $A_2 = 3\pi a_2/8$, and $A_3 = 4a_3/3$.

After modelling aerodynamic damping in term of velocity or displacement, the equation of motion can be solved in the time domain using a step-by-step integration method such as Newmark's method. It can also be solved by Runge-Kutta method based on the state-space equation. One of the objectives of this study is to develop an analytical approach which permits direct calculation of response statistics thus is computationally more effective.

2.2 *Determination of statistics of random response through ENLE approach*

For narrow-band inelastic response which does not significantly exceed the yield displacement, the system frequency can be regarded as the same of the corresponding linear system. The total system energy is expressed as:

$$E = \frac{1}{2}\omega_s^2 A^2; \quad A = \left(\frac{\dot{x}^2}{\omega_s^2} + x^2\right)^{1/2} \tag{7}$$

The influence of hysteretic force on the system can be represented in terms of equivalent hysteretic damping determined as follows from the energy dissertation per cycle of vibration (e.g., Caughey 1960; Roberts and Spanos 2003):

$$\xi_{heq} = \begin{cases} \dfrac{2x_y(1-\alpha)}{\pi A}\left(1 - \dfrac{x_y}{A}\right) & A > x_y \\ 0 & A \le x_y \end{cases} \tag{8}$$

which is represented in terms of system energy as

$$\xi_{heq} = \begin{cases} \dfrac{2(1-\alpha)}{\pi}\left(\dfrac{x_y\omega_s}{\sqrt{2E}} - \dfrac{x_y^2\omega_s^2}{2E}\right) & E > \dfrac{1}{2}\omega_s^2 x_y^2 \\ 0 & E \le \dfrac{1}{2}\omega_s^2 x_y^2 \end{cases} \tag{9}$$

The aerodynamic damping given in Eq. (5) is represented as functions of energy as

$$\xi_{aeq} = \left(\frac{\rho B^2}{m_s}\right)\left(A_1 + \frac{8}{3\pi}A_2\left(\frac{2E}{\omega_s^2}\right)^{\frac{1}{2}} + \frac{3}{4}A_3\left(\frac{2E}{\omega_s^2}\right)\right) \tag{10}$$

The equation of motion is represented as

$$\ddot{x} + 2(\xi_s + \xi_{aeq} + \xi_{heq})\omega_s\dot{x} + \omega_s^2 x = \frac{1}{2}\left(\frac{\rho B^2}{m_s}\right)\left(\frac{U^2}{B^2}\right)\eta_b\eta C_{Mb}(t) \tag{11}$$

Caughey (1964) showed that an exact solution for the probability density function (PDF) of amplitude can be derived as:

$$f_A(A) = \begin{cases} f_{A0}(A) \exp\left\{-\dfrac{4x_y^2\omega_s^3(1-\alpha)}{\pi^2 S_0}\left(-\ln\dfrac{A}{x_y}+\dfrac{A}{x_y}-1\right)\right\} & A > x_y \\ f_{A0}(A) & A \le x_y \end{cases} \quad (12a)$$

$$f_{A0}(A) = C_A A \exp\left(-\frac{\omega_s^3 A^2}{\pi S_0}\left\{\xi_s + \left(\frac{\rho B^2}{m_s}\right)\left(A_1 + \frac{16}{9\pi}A_2 A + \frac{3}{8}A_3 A^2\right)\right\}\right) \quad (12b)$$

where C_A is a normalization constant for the PDF function; and S_0 is the power spectrum of white noise loading excitation in terms of double-sided spectrum with respect to circular frequency. S_0 is given in terms of one-sided power spectrum of $C_{Mb}(t)$ at the structural frequency, $S_{CMb}(f_s)$, as

$$S_0 = \frac{1}{16\pi}\left(\frac{\rho B^2}{m_s}\right)^2\left(\frac{U^2}{B^2}\right)^2 \eta_b^2 \eta^2 S_{CMb}(f_s) \quad (13)$$

The PDF of displacement and the RMS value and kurtosis of displacement can be calculated as (Chen 2013):

$$f_x(x) = \frac{1}{\pi}\int_{|x|}^{\infty}\frac{f_A(A)}{\sqrt{A^2-x^2}}dA \quad (14)$$

$$\sigma_x^2 = \frac{1}{2}E[A^2]; \qquad \alpha_{4x} = \frac{3}{8}E[A^4]/\sigma_x^4 \quad (15)$$

The corresponding equivalent damping ratio can be re-calculated from σ_x:

$$\xi_{eq} = \xi_s + \xi_{aeq} + \xi_{heq} = \frac{\pi S_0}{2(2\pi f_s)^3\sigma_x^2} \quad (16a)$$

$$\xi_{aeq} = \left(\frac{\rho B^2}{m_s}\right)\left(A_1 + \frac{8}{3\pi}\frac{E[A^3]}{E[A^2]}A_2 + \frac{3}{4}\frac{E[A^4]}{E[A^2]}A_3\right) \quad (16b)$$

$$\xi_{heq} = \begin{cases} \dfrac{2x_y(1-\alpha)}{\pi}\dfrac{\int_{x_y}^{\infty}(A-x_y)f_A(A)dA}{E[A^2]} & A > x_y \\ 0 & A \le x_y \end{cases} \quad (16c)$$

It is obvious that in the case of deterministic response, Eqs. (16b) and (16c) are identical to Eqs. (5) and (8), respectively.

The CDF of extreme value of displacement within time duration T is calculated based on Poisson assumption of crossings:

$$F_{max}(x) = \exp\{-\nu(x)T\} \quad (17)$$

$$\nu(x) = \int_0^{\infty}\dot{x}f(x,\dot{x})d\dot{x} = f_s\int_x^{\infty}f_A(A)dA = f_s[1-F_A(x)] \quad (18)$$

where $\nu(x)$ is upcrossing rate at level x; and $F_A(x)$ is CDF of $A(t)$. The mean extreme value within duration T can be determined as 57%-fractile value, i.e., $F_{max}(x) = 0.57$.

The mean fatigue damage accumulated in time period T can be calculated from all individual cycles by adopting the Miner-Palmgren rule (e.g., Chen 2014b):

$$E[D] = \frac{2^m \nu_0 T \kappa_s^m}{K_0}E[A^m] \quad (19)$$

where K_0 and m are material constant and slope of the S-N curve $N = K_0 s^m$ (N is the cycle number to cause fatigue damage under stress level s); ν_0 is mean upcrossing rate at zero-mean level, and $\nu_0 = f_s$ for narrow-band process; κ_s is a constant and can be calculated as the stress due to inertial load with unit displacement.

3 RESULTS AND DISCUSSIONS

3.1 *Inelastic crosswind response without aerodynamic damping effect*

In this study, a square-shaped tall building in a very smooth terrain is considered as an example. The aspect ratio of building $H/B = 13.3$. The fundamental modal shape for both elastic and inelastic buildings is assumed as linear over the building height, i.e., $\phi(y) = y/H$. The aerodynamic damping ratio as function of vibration amplitude at different reduced wind speed is calculated using the model introduced by Watanabe et al. (1997) that was developed based on wind tunnel data (Vickery and Steckley 1993). The aerodynamic damping ratio at each reduced wind speed is fitted into second order polynomial function of amplitude (Chen 2013). For a parametric study, the bilinear hysteresis force model is defined with the parameter α=0.25, 0.75 and 0.95, and $x_y = \beta x_{ref}$, where x_{ref}=0.09 is pre-selected reference non-dimensional displacement of $x(t) = x_1(t)/B$, and β =0.2, 0.3, 0.4, 0.5, 0.7 and 0.9. The power spectrum of base bending moment coefficient is taken the value specified in the recommendations of Architectural Institute of Japan (AIJ 2004).

In the following, the inelastic crosswind response with building damping ratio of $\xi_s = 1\%$ and mass parameter $m_s/\rho B^2 = 172$ is investigated. To validate the ENLE approach, response history simulation is also performed using Runge-Kutta method based on state-space equation. The time step of 0.04 s and duration is 15 min for each sample, where the first 5-min response is disregarded to eliminate the transient dynamic effect. The time history of base bending moment coefficient is generated using spectral representation method. For a given wind speed, 20 response history samples are simulated to calculate ensemble averaged response statistics.

The accuracy of the ENLE approach is firstly examined for estimating inelastic crosswind response when the nonlinear aerodynamic damping effect is negligible. Feng and Chen (2017) has studied the performance of statistical linearization approach with Gaussian response assumption. Two linearization approaches have been compared: the first linearization approach is to linearize the hysteretic velocity as $\dot{z} \approx b_0' + b_1'x + b_2'\dot{x} + b_3'z$; and the second one is to linearize the hysteretic displacement as $z = b_0 + b_1 x + b_2 \dot{x}$. It has been demonstrated that both linearization approaches give good estimation of response RMS value in general, but fall short in capturing the hardening non-Gaussian distribution feature.

Fig. 2 shows the RMS value of inelastic response calculated by the ENLE approach and two statistical linearization approaches, which are compared with that from time history simulation. It is evident that the ENLE approach gives better estimation of response RMS value over the statistical linearization approaches, and the estimation is very agreeable to the result from response history simulation. Furthermore, it is shown in Fig. 3 that the ENEL approach can well capture the non-Gaussian feature in terms of the response kurtosis, thus leads to good estimation of peak factor, which is lower than that of Gaussian response.

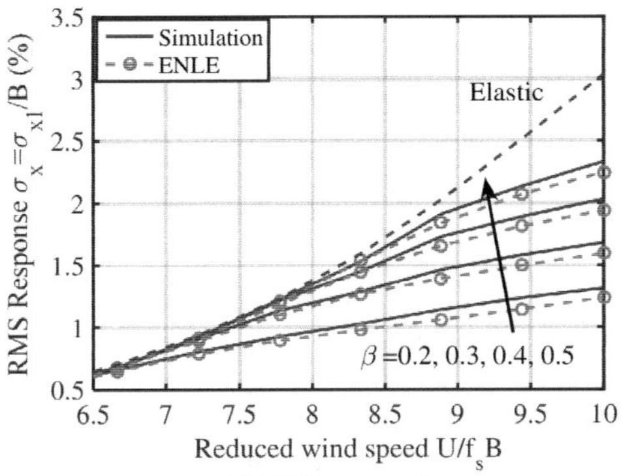

a) ENLE approach

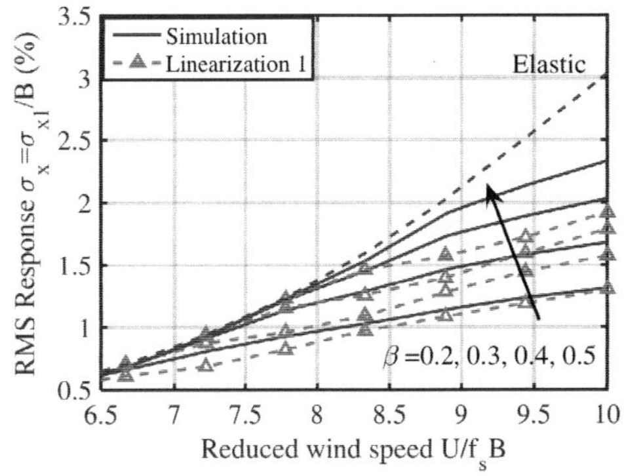

b) Linearization approach 1

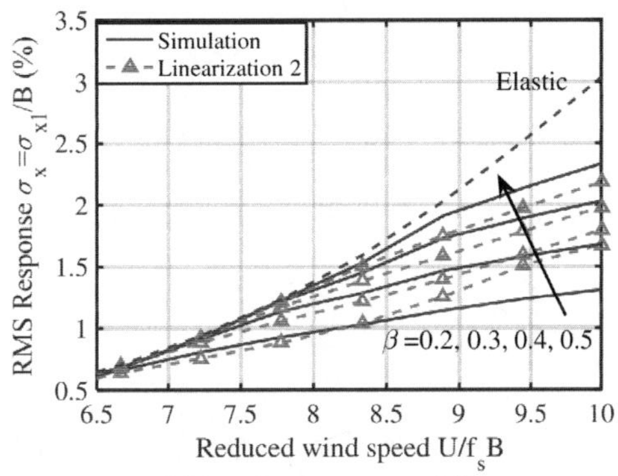

c) Linearization approach 2

Fig. 2 Inelastic response statistics without aerodynamic damping effect (α = 0.25)

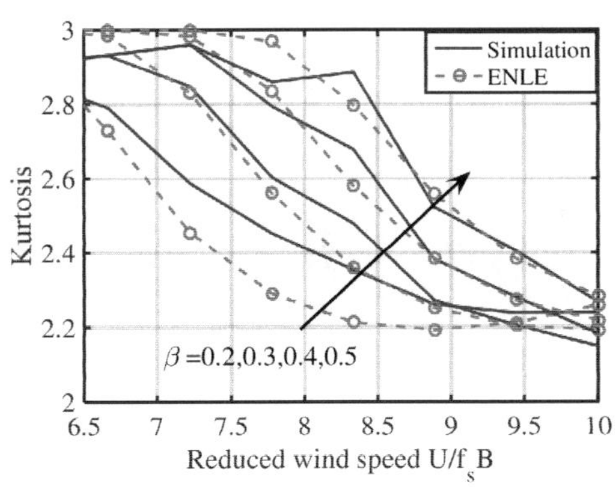

a) Response kurtosis

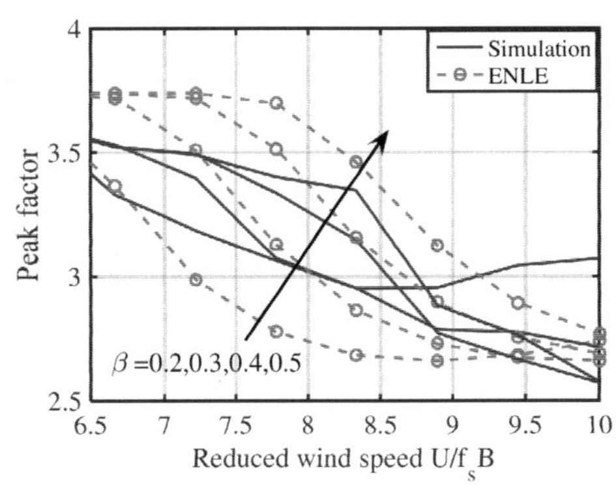

b) Peak factor

Fig. 3 Kurtosis and peak factor of inelastic response (α = 0.25)

3.2 *Inelastic crosswind response with aerodynamic damping effect*

In the case with the consideration nonlinear aerodynamic damping, the reference non-dimensional displacement x_{ref} is selected as 0.30. Fig. 4 displays time history samples of inelastic displacement ($\alpha = 0.25$, $\beta = 0.2$ and 0.5), and those of the corresponding linear elastic system at $U/U_{cr} = 1.3$, i.e., $U/f_sB = 14.4$. Fig. 5 shows the comparison of RMS value and kurtosis of displacement from response history simulation and the ENLE approach. It is evident that the RMS response and kurtosis can be accurately estimated by the ENLE approach. The RMS response calculated from the ENLE approach is very close but slightly lower than that of simulation. The slightly lower estimation seems to be attributed to disregard of a slight reduction in building frequency due to yielding. It is evident that the inelastic response is lower than the corresponding elastic response. The reduction in response is more significant for a low yield displacement, i.e., small value of β. This is due to the contribution of the hysteretic damping as shown in Fig. 6a) that leads to the increase in the total damping as shown in Fig. 6b). Both hysteretic damping and aerodynamic damping lead to hardening non-Gaussian effect if considered individually (Feng and Chen 2017; Chen 2013). The effect of hardening non-Gaussian introduced by nonlinear aerodynamic damping is reduced due to the addition of hysteretic damping. Therefore, the response kurtosis is not necessarily monotonous with yield displacement, i.e., β. As shown in Fig. 5b), at $U/f_sB = 22.2$, the kurtosis increases with the increasing β when $\beta = 0.3 - 0.9$, while that of $\beta = 0.2$ is between those of $\beta = 0.5$ and 0.7.

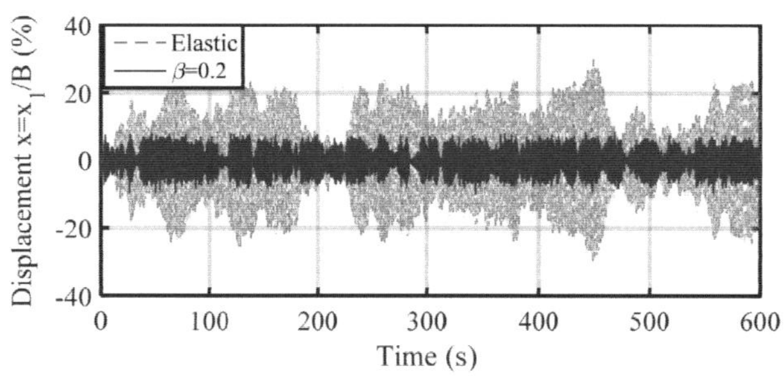

a) Elastic and inelastic with $\beta = 0.2$

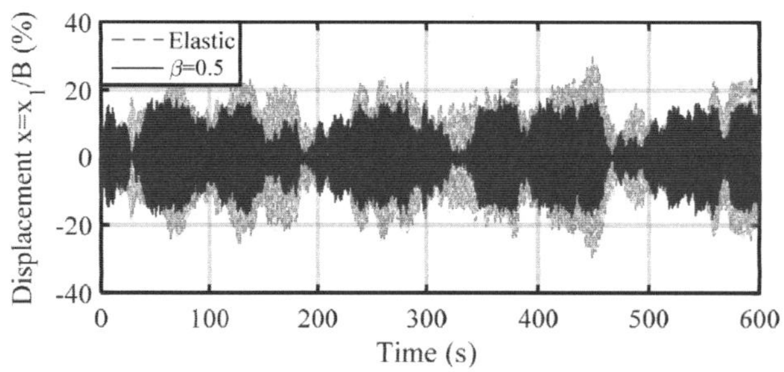

b) Elastic and inelastic with $\beta = 0.5$

Fig. 4 Time history sample of elastic and inelastic crosswind displacement ($U/f_sB = 14.4$, $\alpha = 0.25$)

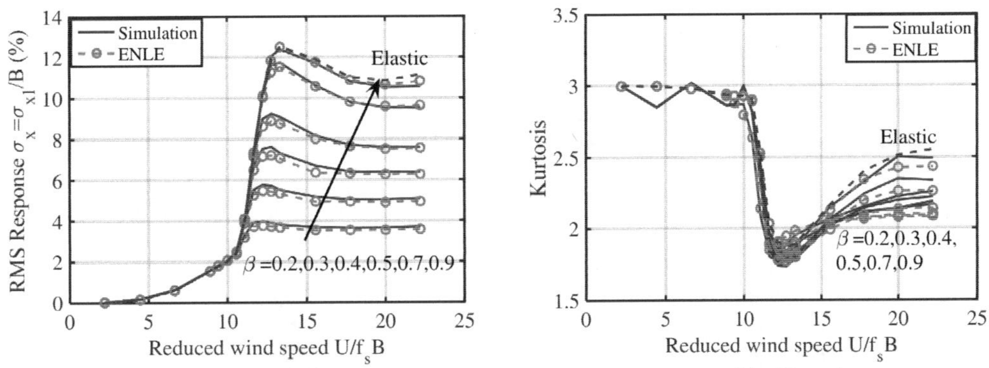

a) RMS b) Kurtosis

Fig. 5 Comparison of RMS response and kurtosis (α = 0.25)

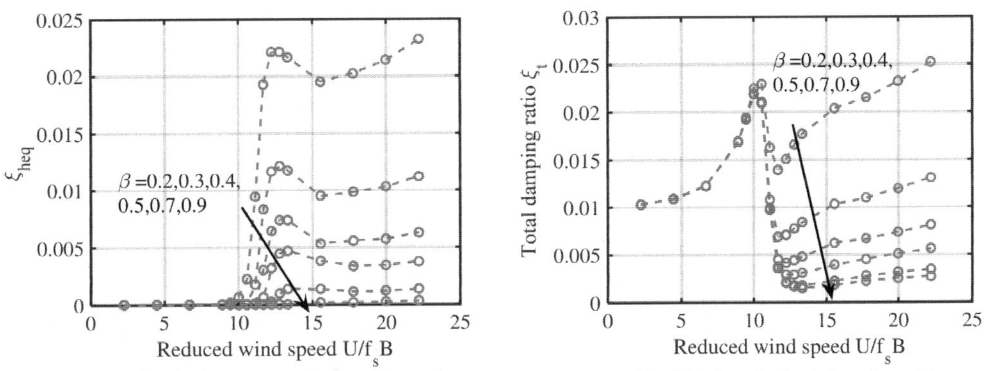

a) Equivalent hysteretic damping ratio b) Total equivalent damping ratio

Fig. 6 Equivalent damping ratio (α = 0.25)

The response peak factor, i.e., the ratio of mean extreme response to RMS response, is calculated using crossing rate analysis. Fig. 7 shows the comparison of estimations with that from response history simulations. The analytical approach gives accurate estimation of peak factor. Similar to the relationship between response kurtosis and yield displacement, the response peak factor is also not monotonous with yield displacement. Fig. 8 displays the comparison of ductility factor, i.e., the ratio of mean extreme displacement to yield displacement x_y. The ductility factor increases with decrease in yield displacement.

The mean fatigue damage of crosswind response is calculated through statistical moments of vibration amplitude. It is also computed from fatigue damage of underlying Gaussian response with a correction factor that is a function of response kurtosis (Chen 2014b). Fig. 9 compares the estimations with the simulation result estimated using rainflow cycle counting method. The slope of S-N curve $m = 3$, and mean fatigue damage is normalized by $\nu_0 T k_s^m / K_0$. The results re-confirm the accuracy of the analytical approaches. The fatigue damage decreases with decrease in yield displacement.

To gain better understanding of the inelastic crosswind response, response analysis with different second stiffness ratio α is also implemented and the results are shown in Fig. 10. The response is not sensitive to parameter α in general. A noticeable influence of parameter α can only be found when α is very close to one and the yield displacement is low.

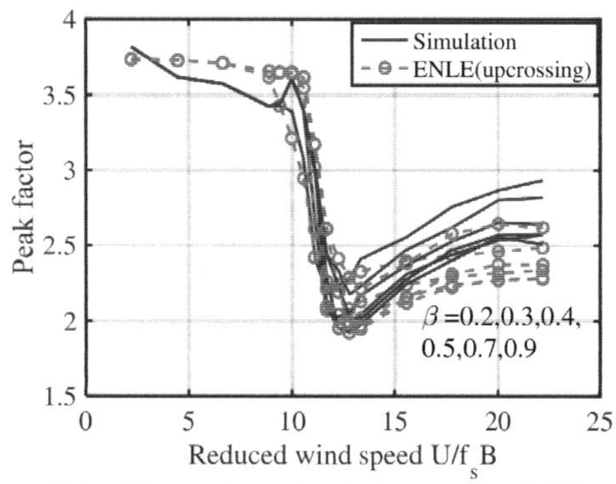

Fig. 7 Comparison of peak factors (α = 0.25)

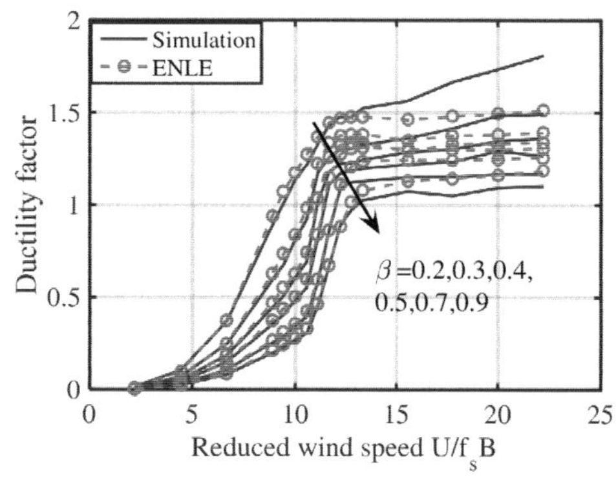

Fig. 8 Ductility factor (α = 0.25)

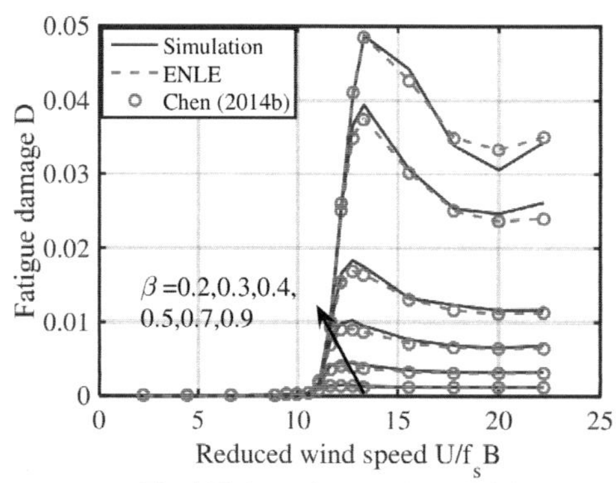

Fig. 9 Fatigue damage (α = 0.25)

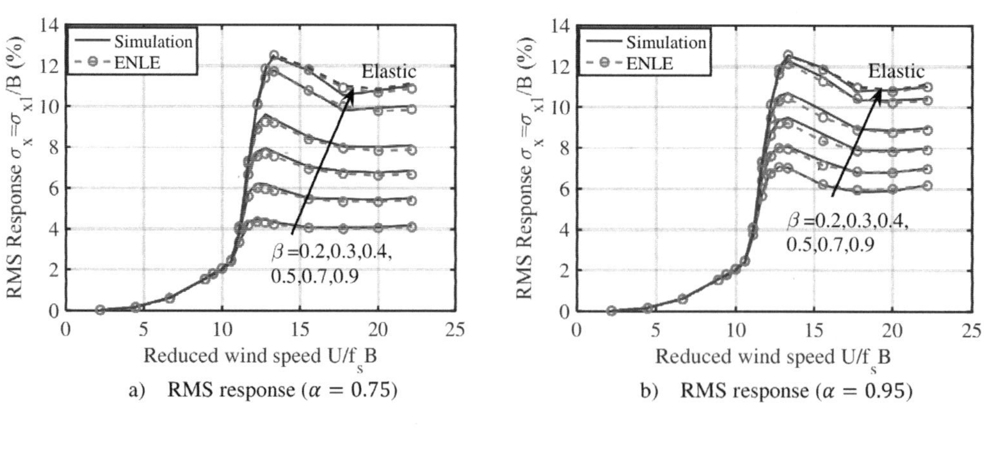

a) RMS response (α = 0.75)

b) RMS response (α = 0.95)

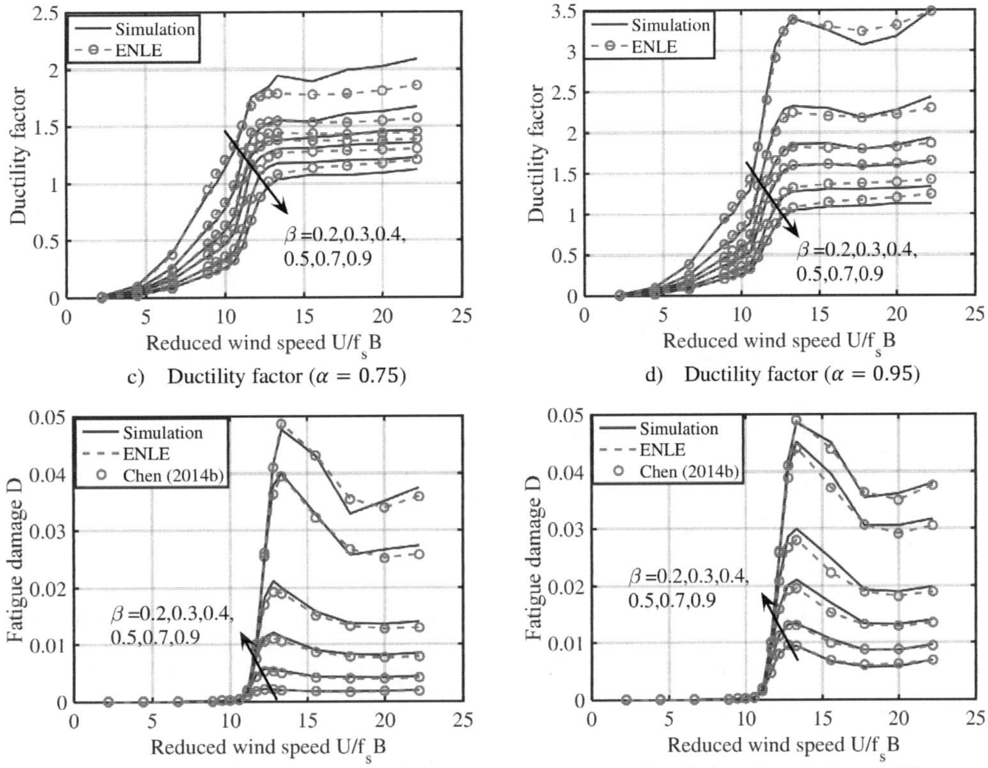

c) Ductility factor (α = 0.75)
d) Ductility factor (α = 0.95)
e) Fatigue damage (α = 0.75)
f) Fatigue damage (α = 0.95)

Fig. 10 Influence of second stiffness ratio

4 CONCLUSION

This study presented the ENLE approach for estimating inelastic crosswind response of tall buildings with bilinear hysteretic restoring force character. The comparison with the estimations from response history analysis illustrated the accuracy of the analytical approach. The study demonstrated the advantage of the ENLE approach over the statistical linearization approach for estimating narrow-band crosswind response when the non-Gaussian response character is noticeable. The non-Gaussian response distribution is resulted from hysteretic damping, and is more significant when nonlinear aerodynamic damping effect is further included.

The results demonstrated that the hysteretic damping resulted from yielding can considerably reduce the crosswind response due to added damping effect to the system. The fatigue damage also decreases but the ductility factor increases with decrease in yield displacement. The inelastic response is not sensitive to second stiffness ratio in general.

5 ACKNOWLEDGEMENTS

The support for this work provided in part by NSF grant No. CMMI-1400224 and CMMI-1536108 is greatly acknowledged.

6 REFERENCES

Architectural Institute of Japan (AIJ). (2004). *AIJ Recommendations for Load on Buildings*, AIJ, Tokyo.

Caughey, T. K. (1960). "Random excitation of a system with bilinear hysteresis." *J. Appl. Mech.*, 27(4), 649-652.

Caughey T. K. (1964). "On the response of a class of nonlinear oscillators to stochastic excitation." *In: Proc. Coll. Int. du Centre Nat. de la Research Scient.*, 148, Marseille, France, 393–402.

Chen, X. (2013). "Estimation of stochastic crosswind response of wind-excited tall buildings with nonlinear aerodynamic damping". *Eng. Struct.*, 56, 766-778.

Chen, X. (2014a). "Extreme value distribution and peak factor of crosswind response of flexible structures with nonlinear aeroelastic effect". *J. Struct. Eng.*, 140(12), 04014091.

Chen, X. (2014b). "Analysis of crosswind fatigue of wind-excited structures with nonlinear aerodynamic damping." *Eng. Struct*, 74, 145-156.

Feng, C., and Chen, X. (2017) "Wind-induced inelastic responses of tall buildings with bilinear hysteretic restoring force characteristics by statistical linearization approach." *J. Eng. Mech.* (Accepted)

Lutes, L.D., and Sarkani, S. (2004). *Random vibrations: analysis of structural and mechanical systems.* Elsevier Butterworth-Heinemann, Burlington, MA, USA.

Roberts, J. B., and Spanos, P. D. (2003). *Random vibration and statistical linearization.* Dover Publications Inc., Mineola, UY, USA.

Vickery, B. J., and Steckley, A. (1993). "Aerodynamic damping and vortex excitation on an oscillating prism in turbulent shear flow." *J. Wind Eng. Ind. Aerodyn.*, 49(1), 121-140.

Watanabe, Y., Isyumov, N., and Davenport, A. G. (1997). "Empirical aerodynamic damping function for tall buildings." *J. Wind Eng. Ind. Aerodyn.*, 72, 313-321.

13th Americas Conference on Wind Engineering
Gainesville, Florida, USA
May 21 - 24, 2017

13th Americas Conference on
WIND ENGINEERING

Combined Risk Prediction of Wind Loading with Changes Surrounding Buildings

Peter A. Irwin[a], Jason Garber[b]

[a] *RWDI, Guelph, Ontario, Canada, Peter.Irwin@rwdi.com*
[b] *RWDI, Guelph, Ontario, Canada, Jason.Garber@rwdi.com*

ABSTRACT

Wind tunnel testing is a well-established tool for the design of buildings for wind. A key benefit of site specific studies is that the influence of the immediate surroundings –favorable or otherwise – is inherent in the measurements. When future construction plans are known, it is a simple matter to expand the test programme to account for the impact. However, one issue that is receiving increasing attention is the extent to which beneficial impacts of shelter can be relied upon for reduced design loads. Related to this is the uncertainty as to the future state of the surroundings although experience suggests development in general is more likely to reduce loading than increase it. More recent editions of ASCE 7 have introduced guidelines limiting permissible reductions of wind tunnel derived loads below code analytical predictions, with provisions for additional testing of a so-called "unsheltered" configuration to justify further reductions.

The wording of these guidelines has left some room for interpretation. The authors have been involved with numerous studies where data from an unsheltered configuration have been collected. In some cases, the unsheltered configuration produced dramatic increases in the design load. A possible example is where a strong across-wind response is disrupted by buildings on adjacent sites.

A Combined Risk procedure has been developed to provide a rational means for incorporating the results of an unsheltered configuration into design load predictions. Examples of application of the approach are presented.

KEYWORDS

Wind loading, wind tunnel testing, risk, surrounding buildings, unsheltered

13th Americas Conference on Wind Engineering
Gainesville, Florida, USA
May 21 - 24, 2017

Analysis of aerodynamic configuration of high-rise buildings in Mexico

Hernández-Barrios Hugo[a], Huergo Ríos Iván[b]

[a] *Universidad Michoacana de San Nicolás de Hidalgo, Morelia, México, hugohernandezbarrios@yahoo.cm.mx*
[b] *Universidad Nacional Autónoma de México, FESA, UNAM, ing.huergo@hotmail.com*

ABSTRACT

In the last decade, a large number of medium-height buildings have been constructed in Mexico City, ranging from 90 m to 240 m. The design of high-rise buildings is mainly characterized by satisfy the requirements of seismic design without considering the requirements of the serviceability related to the dynamic effects of wind. The maximum permissible acceleration specified in the mexican wind design code is 0.04 g , however it does not consider the structure´s period of vibration or the the purpose for which it was designed. In this paper, the maximum acceleration due to the dynamic effects of wind is analized in mexican high-rise builidings along with its dynamic configuration.

The high-rise buildings are mathematically modeled as an equivalent continuous structure consisting of a combination of a flexural cantilever beam and a shear cantilever beam with the aim of identifying the modes shapes of the structure. The wind speed time-series are obtained through spectral simulation. The results show that a large number of buildings located in Mexico City do not satisfy the comfort requirements established in international wind design codes. It is demonstrated that the accelerations produced by the wind can be equal to or higher than those produced by a service earthquake. On the other hand, the aerodynamic configuration of more than 20 buildings constructed in Mexico City is reviewed and compared with the configuration recommended in the literature. The results indicate that the aerodynamic configuration in the great majority of the high-rise buildings is not considered in a suitable way to decrease the effects of the wind.

KEYWORDS

Comfort conditions, high-rise buildings in Mexico, aerodynamic configuration

13th Americas Conference on Wind Engineering
Gainesville, Florida, USA
May 21 - 24, 2017

Effect of the spatial correlation on the overpressure risk assessment for the mid/high-rise building envelop

Si Han Li-1[a], Jon Galsworthy-2[b], John Kilpatrick-3[c], Mark Chatten-4[d]

[a] *RWDI, Guelph, Canada, Sihan.Li@rwdi.com*
[b] *RWDI, Guelph, Canada, Jon.Galsworthy@rwdi.com*
[c] *RWDI, Guelph, Canada, John.Kilpatrick@rwdi.com*
[d] *RWDI, Guelph, Canada, Mark.Chatten@rwdi.com*

ABSTRACT

Mid- and high-rise buildings in tropical cyclone prone regions face wind induced risk, which includes building envelope failure, water ingress and associated loss of the interior of the building. The building envelope failure is often the first mode of failures causing the cascade failure of the other two failure modes. The failure of the building envelope, including the façade and secondary components, can cause direct and large financial loss. There are types of sources can cause building envelope failure, which are overpressure on the building façade and wind borne debris missiles. The overpressure failure happens when the structural design demand is less than the wind pressure/loads acting on the design components. To assess the overpressure risk of the building envelope, wind pressures experienced by the building envelope and the resistance of the building envelope must be obtained. The design wind pressure and pressure zones can be obtained from the building code. However, for the mid- or high-rise buildings, the wind pressures on the building surfaces are complex than values recommended by the building code due to the unique building geometry and building surroundings. Moreover, the spatial correlation of the wind pressures imposed on the building envelope is not provided by the building code. It appears as well that such correlation has not been considered in some of the public funded risk assessment model. The effect of neglecting such correlation on the assessed wind risk for the building envelope of mid/high rise building has not been yet studied. However, it is expected that the areas of a building experiencing wind loads that have strong correlation have higher probability of simultaneous failure than those without such correlation. Assuming the independence of the wind loads in these areas decreases the probability of such simultaneous failure, consequently underestimating the higher percentile and maximum value of the possible risk/loss. To investigate such effect, this study selects one high-rise building located in hurricane prone region. The extreme wind pressures on the surface of this building are obtained from the wind tunnel test. The resistance of the designed façade is determined based on estimated design wind pressures and considering the variability of the resistance by employing the probabilistic models. The wind induced risk for the part of selected building envelope is assessed by considering the wind pressures on the surface areas are independent or correlated. The assessed risks in two different scenarios are compared. The impact of the spatial correlation on the estimated wind induced overpressure risk for the building envelope is discussed.

KEYWORDS

façade, risk, high-rise building, wind pressure

13th Americas Conference on Wind Engineering
Gainesville, Florida, USA
May 21 - 24, 2017

A Performance-Based Design Frmework for Wind-Excited Multistory Buildings

Wei-Chu Chuang[a], Seymour M.J. Spence[b], Ahsan Kareem[c]

[a]Department of Civil and Environmental Engineering, University of Michigan, Ann Arbor, MI, United States, wechuang@umich.edu
[b]Department of Civil and Environmental Engineering, University of Michigan, Ann Arbor, MI, United States, smjs@umich.edu
[c]Nathaz Modeling Laboratory, Dept. of Civil and Environmental Engineering and Earth Sciences, Univ. of Notre Dame IN, USA. E-mail: kareem@nd.edu

ABSTRACT

Despite the fact that current building codes are prescriptive in nature, there is a strong interest in moving towards a performance-based design approach for identifying structures that meet sets of performance objectives defining target performance levels for systems subject to extreme winds. Extensive research on developing the principles of performance-based design (PBD) has facilitated its adoption in the field of seismic engineering. However, the same cannot be said for wind engineering where the need to define appropriate frameworks that embrace the principles of PBD still exists. In this paper, such a PBD framework is presented. In particular, a general simulation-based methodology is outlined that not only accounts for the inevitable uncertainties in predicting the response but also communicates performance through system-level measures, such as repair cost and downtime (i.e. the time necessary to restore the system to its original state), that are meaningful to decision-makers and/or stakeholders. Necessary steps for reaching this goal include the identification of appropriate models for describing the wind hazard, aerodynamic wind loads, structural response, as well as the definition of suitable fragility functions and consequence functions that rationally assess damage and losses respectively. In addition, in order to apply PBD to wind engineering, a full knowledge of the structural behavior, from elastic response to possible collapse, is essential for comprehensibly estimating the building's performance. In particular, the possibility of allowing wind excited structures to enter the inelastic regime needs to be taken into consideration for safer and more economical systems. This will also help avoid the possibility of undesirable collapse scenarios or unknown post-yield behavior. Considering the computational difficulties arising from the use of robust methods such as incremental dynamic analysis for modeling the non-linear response under long duration events, such as wind storms, an alternative approach based on the theory plasticity is adopted here to model the structure beyond its elastic limit. A complete picture of the post-yield behavior is therefore provided by the evaluation of the limit state separating plastic collapse from the safe state of dynamic shakedown. A case study consisting of a 40-story frame structure is presented to demonstrate the applicability of the proposed PBD framework. The strong potential of the probabilistic framework demonstrated by the efficiency of the procedure in systematically estimating both structural and non-structural damage and loss which are measures of immediate use to decision-makers.

KEYWORDS

Performance-based design, Aerodynamic loads, Monte Carlo simulation, Wind engineering, Dynamic shakedown

Gust Front Factor: Tropical Cyclone-Induced Wind Load Effects on Tall Buildings

Haifeng Wang [a], Teng Wu [a]

[a]*Department of Civil, Structural and Environmental Engineering, University at Buffalo, Buffalo, New York, USA*

ABSTRACT: Gust front factor (GFF) is a well-established analysis framework to consider the effects of unconventional wind loadings on structures. The existence of the supergradient wind region, where the tangential winds are larger than the gradient wind, has been frequently observed in tropical cyclones. This results in a vertical profile that cannot be well represented by conventional power or logarithmic law. Hence, the GFF framework is utilized in this study to investigate the tropical cyclone-induced wind load effects on tall buildings. More specifically, the kinematic loading effect of the tropical cyclone, which plays a dominant role under the GFF framework, is comprehensively examined based on the tropical cyclone wind profiles resulting from observations, empirical model and semi-empirical model. Buildings with heights in the range of 18-1000m are selected to calculate the kinematic effects sub-factor of GFF. The results suggest that it may be important to take the supergradient wind effects into account in the wind design of high-rise buildings in coastal areas to ensure the target safety and performance levels.

KEYWORDS: Gust Front Factor, Supergradient Wind, Tropical Cyclone, Tall Building

1 INTRODUCTION

The trend of tall building construction has been significantly enhanced by innovative structural systems and high strength materials in the past decades. More than half of the 100 highest buildings all over the world are completed since 2000, and the Burj Khalifa set the new world record in 2010 approaching 1000m [1]. A large number of researches have focused on investigating the wind-resistant performance of tall buildings under conventional boundary layer winds (e.g. [2–8]). The vertical mean wind profiles of the atmospheric boundary layer adopted in current building codes and standards follow the power or logarithmic law, indicating wind speed monotonically increases with height. On the other hand, several studies claimed that a supergradient wind region, where the tangential component of the mean winds are larger than the gradient wind, exists in the tropical cyclone wind profiles [9–12].

According to the numerous observations obtained from the widely implemented Global Positioning System (GPS) dropsondes, the supergradient flow is actually a common feature of tropical cyclone boundary layer [9, 13, 14]. In contrast to the conventional boundary layer winds, supergradient flows present a medium-level maximum wind speed. More specifically, the wind speed shows a logarithmic increase in the lower height, a maximum wind speed in the medium height and a gradual decrease to the top of the tropical cyclone boundary layer [9]. Hence, the utilization of power-law or logarithmic profile may result in underestimation of the wind load effects on tall buildings under the tropical cyclone without considering the medium-level maximum wind speed. In addition, the energetic convective gusts, existence of multiscale wind flows, variation of strength and translation of storm-relative positions may lead to strong nonstationarity of the tropical cyclone winds [15–21], raising serious questions regarding the applicability of the conventional aerodynamic loading theories in the estimation of structure response under tropical cyclone [21].

Early design methodology of the nineteen sixties shifted from vertically uniform approach flows to boundary-layer winds, which is viewed as the turning point in realistically capturing wind load effects on structures. During this time, Davenport proposed a gust loading factor (GLF) approach to consider wind-induced dynamic load effects under synoptic events [22]. To account for the unconventional spatiotemporal characteristics both in the mean and the fluctuation wind components of the thunderstorm downbursts, Kwon and Kareem [23] advanced the GLF to the gust front factor (GFF) framework for the analysis of structural response under the non-synoptic winds. The GFF framework was designed to take into account of wind characteristics that are different from the conventional boundary layer winds, i.e. the vertical mean wind profile, the time-varying wind speed, the nonstationary turbulence features and the transient aerodynamics [24]. The consideration of these unconventional wind characteristics embedded in the GFF framework makes it a powerful tool for the consideration of the wind load effects on the structures associated with the tropical cyclones and non-synoptic events.

In this study, the GFF framework is utilized to investigate the tropical cyclone-induced wind load effects on tall buildings. Previous researches (e.g.[23–25]) indicated that the kinematic effects typically play a dominant role in the GFF framework, while the dynamic effects either are relatively less significant or need further examination. Hence, only the kinematic loading effects of the tropical cyclone resulting from the vertical mean wind profile under the GFF framework are discussed here as the first step beyond conventional considerations. The tropical cyclone wind measurements are divided into three groups, namely the inner radius of maximum winds (RMW), the outer RMW and the outer core to obtain the corresponding vertical profiles. The vertical profile of gust fronts presents the "nose-shape" with its maxima near the ground, whereas the maximum mean wind speed in supergradient wind region locates in a relatively higher altitude and varies within a more extensive range. Therefore, buildings with heights in the range of 18m-1000m are selected for a comprehensive investigation on the kinematic loading effects of both tropical cyclones and thunderstorm downbursts. Results show that the downburst wind loadings present larger kinematic effects on buildings lower than 200m, while the response of higher buildings is dominated by the tropical cyclone wind loadings. This suggests that it may be important to take the supergradient wind region into account in the wind design of high-rise buildings in coastal areas to ensure the target safety and performance levels.

2 THEORETICAL BACKGROUND

To realistically capture the kinematics and dynamics characteristics of supergradient wind effects, the GFF framework proposed by Kwon and Kareem [23] is applied to tropical cyclones. In this section, previous researches focused on the GFF framework will be briefly reviewed for completeness.

The GFF framework was proposed to capture the unconventional characteristics of gust-front winds by introducing four separated sub-factors, namely the kinematic effects sub-factor, the pulse dynamics sub-factor, the structural dynamics sub-factor and the load modification sub-factor [23, 25]. By the combination of these four sub-factors and a velocity/pressure coefficient, the effects of wind characteristics that deviate from conventionally boundary layer theory can be considered in conjunction with the existing codes and standards, e.g. ASCE7. The design wind loads based on the GFF framework could be expressed as:

$$F_{design} = F_{ASCE7} K_{z,G-F} G_{G-F} \qquad (1)$$

where F_{design} is the design wind loads; F_{ASCE7} represents the recommendation of ASCE7; $K_{z,G-F}$ is the height-dependent modification factor, whereas G_{G-F} is a constant. The G_{G-F} can be best captured in terms of four underlying sub-factors of unconventional winds:

$$G_{G-F} = I_1 \cdot I_2 \cdot I_3 \cdot I_4 \tag{2}$$

where I_1 is the kinematic effects sub-factor characterizing the variation in the vertical profile; I_2 is the pulse dynamics sub-factor introduced by the sudden change in wind speed; I_3 is the structural dynamics sub-factor resulting from the nonstationarity of turbulence in unconventional winds and I_4 is the potential load modification sub-factor representing the transient aerodynamics.

Previous researches suggested that the kinematic effects sub-factor typically presents dominant contribution to the GFF, hence, only I_1 is discussed in this study. Following the GFF formulation defined by Kwon and Kareem [23], the kinematic effects sub-factor (I_1) for a building structure can be expressed as:

$$I_1 = \frac{\int_0^H V_{G-F}^2(z)\phi_1(z)dz}{\int_0^H V_{B-L}^2(z)\phi_1(z)dz} \tag{3}$$

where H is the height of the building; V_{G-F} and $V_{B-L}$are the unconventional and unconventional vertical profiles, respectively; ϕ_1 is the first mode shape of the building. Eq. 3 indicates that for a specific building, the kinematic effects sub-factor is determined by the vertical mean wind profiles.

3 VERTICAL MEAN WIND PROFILE

Both the conventional and unconventional vertical mean wind profiles to be employed in this study will be discussed in this section. Figure 1 shows a schematic comparison of vertical profiles of the conventional boundary layer, downburst and tropical cyclone boundary layer winds. The widely-used empirical/analytical profiles for conventional boundary layer and downburst winds will be first briefly reviewed, then the approaches to obtain the supergradient wind profile will be detailed.

3.1 *Conventional boundary layer profile*

While both the power-law and logarithmic wind profiles are popularly used for the conventional boundary layer, the former profile typically presents a better performance for heights larger than 200m [26]. Hence, the power-law wind profile is utilized here:

$$V_{B-L}(z) = \begin{cases} V_0 b\left(\frac{z}{10}\right)^{\alpha}, z \le z_g \\ V_0 b\left(\frac{z_g}{10}\right)^{\alpha}, z > z_g \end{cases} \tag{4}$$

where V_0 is the basic mean wind speed; z_g is the height of the boundary layer; b and α are two constants depending on the terrain type. It should be noted that the hourly averaged basic mean wind speed is utilized for all profiles used in this study without conversion to the 3-sec gust speed as defined in ASCE7.

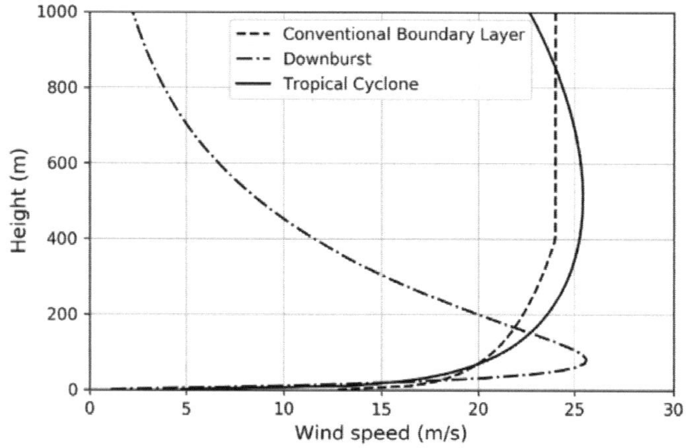

Figure 1. A schematic comparison of various wind profiles

3.2 *Downburst profile*

The downburst profile of gust-front winds has been well studied based on theoretical considerations, laboratory-scale experiments and computational fluid dynamics [27–31]. In this study, the empirical downburst profile proposed by Vicroy is utilized for the simplicity without loss of generality [32]:

$$V_{G-F}(z) = 1.354V_{\max}\left[\exp\left(\frac{b_1 z}{z_{\max}}\right) - \exp\left(\frac{b_2 z}{z_{\max}}\right)\right] \tag{5}$$

where V_{max} is the maximum horizontal wind speed; z_{max} is the height of the maximum horizontal wind speed; b_1 and b_2 are model constants set to be -0.22 and -2.75, respectively.

3.3 *Tropical cyclone*

Powell et al. (2003), based on the data from GPS dropsondes over the ocean, observed that the tropical cyclone vertical mean wind profile follows a logarithmic increase in the lower 200m, reaches its maximum, and then gradually decreases up to the launch altitude. Despite several studies based on theoretical considerations and numerical simulations, there is no well-established tropical cyclone profile for the overland situations due to a lack of observations. In order to realistically capture the kinematic loading effects of tropical cyclones on structures, a comprehensive analysis of the tropical cyclone profile is carried out based on Typhoon Fengshen measurements together with empirical and semi-empirical considerations in this study. More specifically, the wind profiles are discussed in terms of three tropical cyclone regions, namely inner RMW, outer RMW and outer core.

3.3.1 *Supergradient observation*

Typhoon Fengshen, which severely affected Hong Kong, was well observed by the meteorological stations. The closest distances from the track of Typhoon Fengshen to Shan Shui Po (SSP) station and Cheng Chau (CCH) station are 25 km and 45 km, respectively [33]. The hourly mean wind speed measurements of station CCH from 2008/06/24 to 2008/06/25 (totally 48 hours) are employed in this study [33].

The hourly averaged wind speeds measured at various distances to the center of Typhoon Fengshen can be utilized to obtain vertical profiles in different regions of a tropical cyclone. According to the track of the Typhoon Fengshen as shown in Figure 2(a), the full-scale measurements during three time periods (indicated using dark gray lines in the figure) with average distances from CCH site to the tropical cyclone center as 65km, 85km and 114km, respectively, are selected to obtain the wind profiles associated to the inner RMW, outer RMW and outer core regions as presented in Figure 2(b).

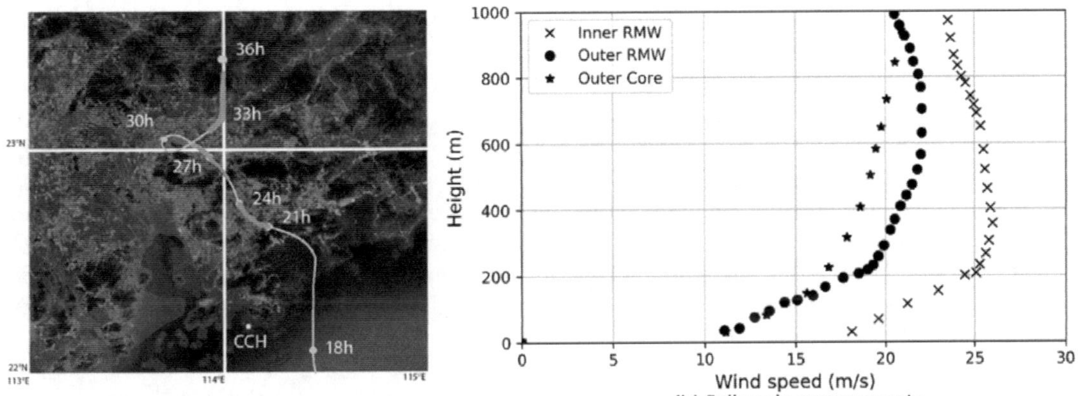

(a) Path of Typhoon Fengshen

(b) Full-scale measurments

Figure 2. Comparison of wind profiles for various locations.

3.3.2 *Empirical profile*

The observations from a single tropical cyclone may not represent the general features of super-gradient wind profile since it is significantly affected by the structure of tropical cyclones, complex terrain conditions and relative positions of the observation site [10, 11, 34, 35]. Based on the high horizontal- and vertical-resolution observations from the widely implemented GPS dropsondes, Vickery et al. [34] proposed an empirical, logarithmic-quadratic tropical cyclone model for the engineering applications expressed as:

$$V(z) = \frac{u_*}{k}\left[\ln\left(\frac{z}{z_0}\right) - a\left(\frac{z}{H^*}\right)^n\right] \tag{6}$$

where u_* is the friction velocity; k is the von Karman constant with a value of 0.4; z_0 is the surface roughness; H^* is the boundary layer height parameter; a and n are two constants that are suggested to be 0.4 and 2.0, respectively. Eq. 6 shows that the lower part of the empirical profile is governed by the logarithmic law while the upper part dominated by the quadratic part.

Although the empirical model was developed using the linear regression of dropsonde observations over ocean, it could be applied to the overland wind profile estimations by modifying the boundary layer height and the surface roughness. The transition of tropical cyclone from sea to land was validated by three landed tropical cyclones [34]. In this study, the observations of Typhoon Fengshen are fitted into the empirical model using least squares, which is then employed in the calculation of the kinematic effects sub-factor of GFF. Figure 3 shows the fitted empirical profiles for inner RMW, outer RMW, and outer core regions. While the monotonically increasing trend in the outer core region is well presented, it is noted that there is relatively large deviation for both inner and outer RMW empirical profiles from the full-scale observations.

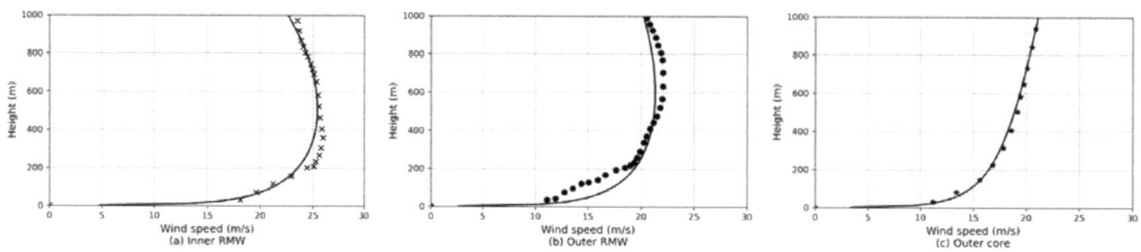

Figure 3. Fitted empirical wind profiles

3.3.3 *Semi-empirical profile*

Compared with the purely empirical model, a semi-empirical model derived from the theoretical considerations may provide a better representation of the tropical cyclone wind profiles. Snaiki and Wu [36] proposed a height-resolving, analytical wind field model for the tropical cyclone boundary layer using dimensional analysis of the Navier-Stokes equations, which was proved to be an excellent representation of tropical cyclone observations at heights beyond 200m. On the other hand, it is well known that the tropical cyclone profile below 200m follows the logarithmic law well [9, 33, 37]. As a result, a semi-empirical model used for the calculation of the kinematics sub-factor of GFF could be expressed as:

$$V(z) = \begin{cases} \frac{u_*}{k}\ln\left(\frac{z}{z_0}\right), & z \leq 200 \\ v_g\left\{1+\frac{a'H'}{km}\exp\left(-\frac{z}{H'}\right)\left[-(1+a')\cos\left(\frac{z}{h'}\right)+\sin\left(\frac{z}{H'}\right)\right]\right\}, & z>200 \end{cases} \tag{7}$$

where v_g is the gradient wind velocity at the top of the tropical cyclone boundary layer; a' is the profile parameter that determines the shape of the profile; H' is the boundary layer height parameter. The observations of Typhoon Fengshen are also fitted into the semi-empirical model using least squares. The comparisons of the fitted wind profiles and measurements are shown in Figure 4. The semi-empirical profile fits the observations of three regions well. Compared with the fitted empirical profiles in Figure 3, it is shown that the semi-empirical profiles better represent the full-scale observations, especially for the inner and outer RMW regions.

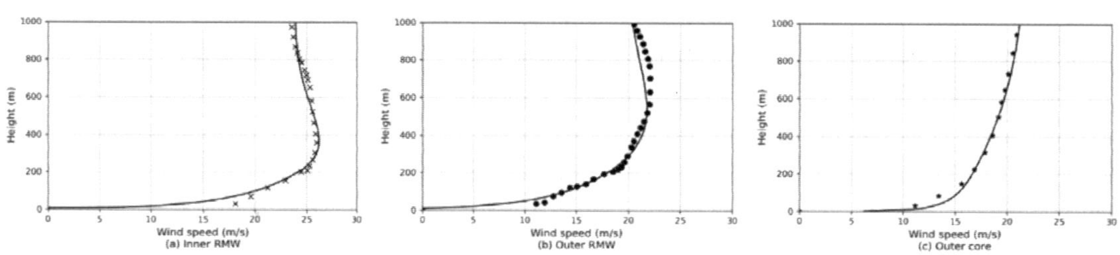

Figure 4. Fitted semi-empirical wind profiles

4 ESTIMATE OF KINEMATIC EFFECTS UNDER GUST FRONT FACTOR FRAMEWORK

With the vertical mean wind profiles discussed in Sect. 3, the kinematic effects sub-factor can be determined for buildings with various heights. In order to reasonably compare the kinematic effects of different wind profiles, a comparison criterion is firstly setup, then the comparison results based on the case study will be elaborated.

4.1 *Comparison criterion*

For reasonable estimation of the kinematic loading effects, it is necessary to set up a criterion to relate the vertical profiles of the conventional boundary layer, tropical cyclone and downburst. Unfortunately, the comparison criterion used in previous studies to set wind speeds of various vertical profiles equal to each other at 10m height is not easy to be applied to the tropical cyclone cases. This is partially due to the fact that the overland measurements of tropical cyclones near the surface level are usually inaccurate [38]. In this study, the employed comparison criterion is to set the gradient wind speed of tropical cyclone and conventional boundary layer profiles equal to each other and the maximum wind speed of tropical cyclone and downburst profiles equal to each other.

The boundary layer height has a significant effect on the comparison results and hence needs to be accurately determined. It is noted that the zero-wave-number depth scale of the analytical model proposed by Snaiki and Wu [36], involving the consideration of the radius, azimuthal angle, pressure profile and moving speed of the tropical cyclone, offers a good estimate of the boundary layer height. Accordingly, the boundary layer height utilized in this study is expressed as:

$$\delta = \frac{1}{(2K)^{\frac{1}{2}}(\xi_g \xi_{ag})^{\frac{1}{4}}} \tag{8}$$

where K stands for the turbulent diffusivity; ξ_g and ξ_{ag} are the absolute angular velocity and the vertical component of absolute vorticity of the gradient wind, respectively. The boundary layer heights for the inner RMW, outer RMW and outer core regions of Typhoon Fengshen calculated using Eq. 8 was determined as 952m, 1196m and 1573m, respectively.

4.2 *Case study*

For the sake of simplicity, the fundamental model shapes of the buildings used in the calculation of kinematic loading effects are modeled based on a continuous slender vertical cantilever beam [39, 40]. The kinematic effects sub-factor of the tropical cyclone inner RMW region for various building heights are determined and compared based on the measured, empirical and semi-empirical profiles. The obtained kinematic loading effects using the semi-empirical profiles of the inner RMW, outer RMW and outer core regions are then compared with those based on the downburst profiles for building with various heights.

4.2.1 *Comparison of kinematic effects sub-factor with measured, empirical and semi-empirical profiles*

Figure 5(a) shows the calculated kinematic effects sub-factors for the inner RMW region, and the corresponding wind speed profiles are given in Figure 5(b). As shown in Figure 5(a), the kinematic effects sub-factors of all three tropical cyclone profiles generally increase with height to maximum values and then decrease up to the height considered.

The major difference among these three profiles exists within the building height range of 18-200m. The measured profile presents fluctuations while the other two profiles monotonically increase with height. The fluctuations shown in Figure 5(a) can be explained by the fluctuating measurements of the wind speed in the corresponding height range as presented in Figure 5(b).

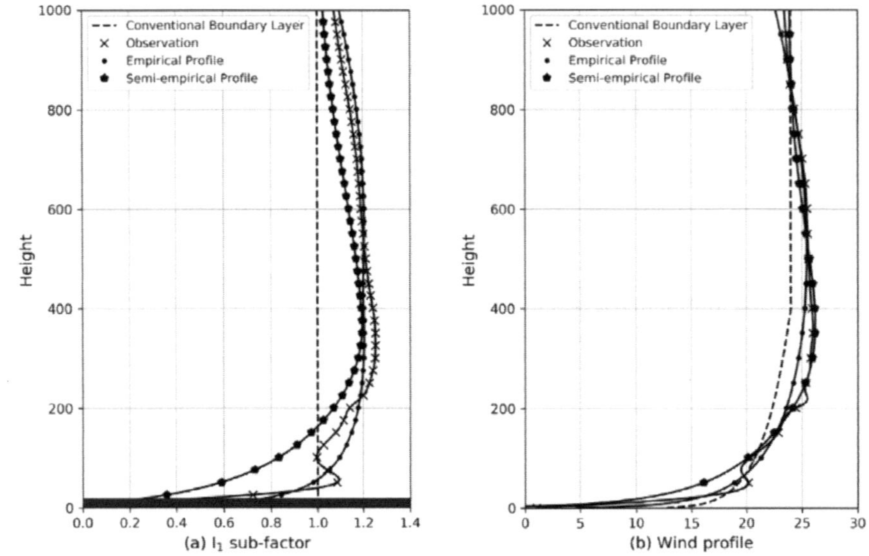

Figure 5. Comparison of I_1 sub-factor with measured, empirical and semi-empirical profiles

Compared with the empirical profile, the I_1 sub-factor of the semi-empirical profile is smaller in the range of 18-200m. This observation is reasonable as the wind speed from the fitted semi-empirical profile is smaller than that from the empirical profile, especially in the height range below 100m. Since the fitted semi-empirical profile presents a better representation of the tropical cyclones as mentioned in Sect. 3, it will be employed in the estimate of the kinematic effects sub-factors for the following discussion.

4.2.2 *Comparison of kinematic effects sub-factor with tropical cyclone and downburst profiles*

Figure 6 shows the kinematic effects sub-factors of the tropical cyclone in the inner RMW, outer RMW and outer core regions for building heights from 18m to 1000m. The I_1 sub-factors are larger than unity for building heights larger than 170m in the inner RMW region and larger than 270m in the outer RMW region. This indicates that the tropical cyclone profiles in the near RMW regions need to be considered for wind design of tall buildings. In the outer core region, the kinematic effects sub-factors of tropical cyclone profile are less than unity for the entire building height range.

In addition, the I_1 sub-factors of tropical cyclone winds in the inner RMW region show larger value for buildings higher than 200m compared to the downburst winds. In the outer RMW region, the I_1 sub-factors of tropical cyclone exceed those of downburst at a height beyond 230m. The comparison results further suggest the kinematic loading effects resulting from the tropical-cyclone supergradient winds need to be considered for buildings higher than 200m in coastal areas.

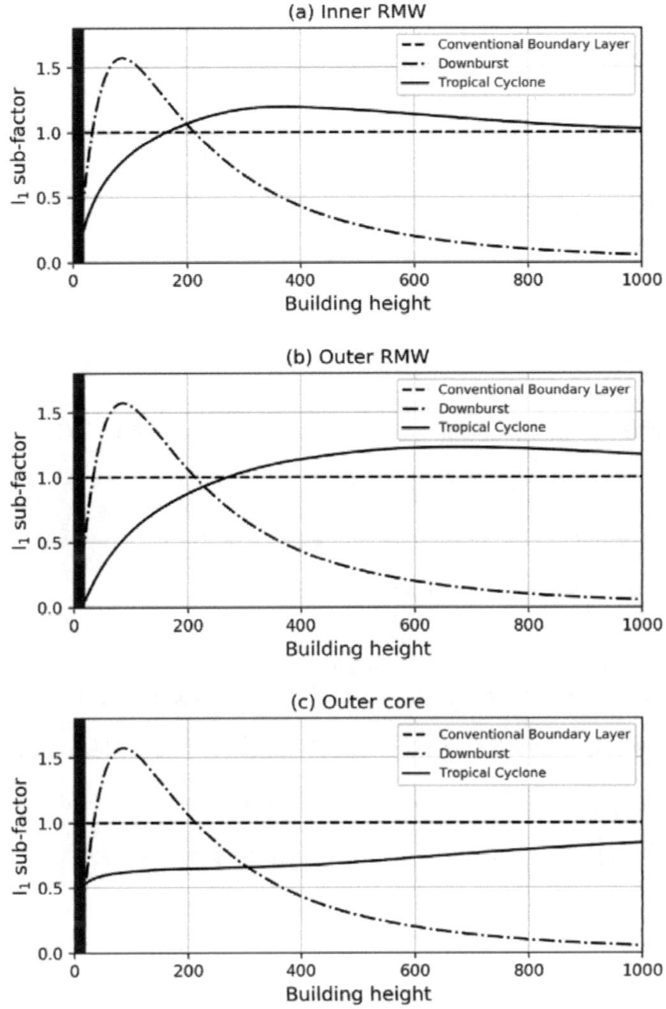

Figure 6. Comparison of I_1 sub-factor in three tropical cyclone regions

5 CONCLUDING REMARKS

The well-established gust front factor (GFF) analysis framework is utilized to investigate the kinematic loading effects of supergradient winds in the tropical cyclones as the first step beyond conventional considerations. The semi-empirical profile introduced in this study showed superior ability of representing the kinematic features of tropical cyclone to measured and empirical profiles. The existence of kinematic effects sub-factors that are larger than unity in the near RMW regions emphasizes the importance to consider the supergradient wind load effects of tropical cyclones. While the kinematic loadings of downbursts are generally significant for buildings lower than 200m, supergradient winds of tropical cyclones have a dominant mean load effects for structures that are taller than 200m.The investigation on the pulse dynamics, structural dynamics and potential load modification sub-factors due to the supergradient winds of the tropical cyclones is in progress. Also, it should be noted that several other factors such as the local maximum near surface winds resulting from the low-level jet and the modifications of wind-structure interaction due to the change of approach flow profile are not discussed here in the consideration of kinematic loading effects on tall buildings.

6 ACKNOWLEDGEMENT

The support for this project provided by the NSF Grant # CMMI 15-37431 is gratefully acknowledged.

7 REFERENCES

1 H. Tanaka, Y. Tamura, K. Ohtake, M. Nakai, and Y. Chul Kim, "Experimental investigation of aerodynamic forces and wind pressures acting on tall buildings with various unconventional configurations," J. Wind Eng. Ind. Aerodyn., 107–108 (2012), 179–191.

2 A. G. Davenport, "The Application of Statistical Concepts to the Wind Loading of Structures," Proc. Inst. Civ. Eng., vol. 19, no. 4, pp. 449–472, 1961.

3 Y. Tamura and S. Suganuma, "Evaluation of amplitude-dependent damping and natural frequency of buildings during strong winds," J. Wind Eng. Ind. Aerodyn., 59 (1996) 115–130.

4 G. Piccardo, G. and Solari, "Generalized equivalent spectrum technique." pp. 161–174, 1998.

5 Q. S. Li, J. Q. Fang, A. P. Jeary, and C. K. Wong, "Full scale measurements of wind effects on tall buildings," J. Wind Eng. Ind. Aerodyn., 74–76 (1998) 741–750.

6 Q. S. Li, K. Yang, N. Zhang, C. K. Wong, and A. P. Jeary, "Field measurements of amplitude-dependent damping in a 79-storey tall building and its efects on the structural dynamic responses," Struct. Des. Tall Build., 11 (2002) 129–153.

7 Q. S. Li, Y. Q. Xiao, J. Y. Fu, and Z. N. Li, "Full-scale measurements of wind effects on the Jin Mao building," J. Wind Eng. Ind. Aerodyn., 95 (2007) 445–466.

8 P. Mendis, T. Ngo, N. Haritos, a Hira, B. Samali, and J. Cheung, "Wind loading on tall buildings," EJSE Spec. Issue Load. Struct., 3 (2007) 41–54.

9 M. D. Powell, P. J. Vickery, and T. A. Reinhold, "Reduced drag coefficient for high wind speeds in tropical cyclones," Nature, 422 (2003) 279–283.

10 J. D. Kepert, "Observed Boundary Layer Wind Structure and Balance in the Hurricane Core. Part I: Hurricane Georges," J. Atmos. Sci., 63 (2006) 2169–2193.

11 J. D. Kepert, "Observed Boundary Layer Wind Structure and Balance in the Hurricane Core. Part II: Hurricane Mitch," J. Atmos. Sci., 63 (2006) 2194–2211.

12 P. J. Vickery, F. J. Masters, M. D. Powell, and D. Wadhera, "Hurricane hazard modeling: The past, present, and future," J. Wind Eng. Ind. Aerodyn., 97 (2009) 392–405.

13 J. Kepert, "The Dynamics of Boundary Layer Jets within the Tropical Cyclone Core. Part I: Linear Theory," J. Atmos. Sci., 58 (2001) 2485–2501.

14 J. Kepert and Y. Wang, "The Dynamics of Boundary Layer Jets within the Tropical Cyclone Core. Part II: Nonlinear Theory," J. Atmos. Sci., 58 (2001) 2485–2501.

15 J. Wurman and J. Winslow, "Intense Sub-Kilometer-Scale Boundary Layer Rolls Observed in Hurricane Fran," Science (80-.)., 280 (1998) 555–557.

16 K. Kosiba, J. Wurman, F. J. Masters, and P. Robinson, "Mapping of Near-Surface Winds in Hurricane Rita Using Finescale Radar, Anemometer, and Land-Use Data," Mon. Weather Rev., 141 (2013) 4337–4349.

17 Y. L. Xu and J. Chen, "Characterizing Nonstationary Wind Speed Using Empirical Mode Decomposition," 130 (2004) 912–920.

18 L. Wang and A. Kareem, "Modeling and Simulation of Transient Winds in Downbursts / Hurricanes," Proc. 10th Am. Conf. Wind Eng. Bat. Rouge, LA, no. Fujita, pp. 1–12, 2005.

19 X. Z. Chen, "Analysis of alongwind tall building response to transient nonstationary winds," J. Struct. Eng., 134 (2008) 782–791.

20 L. Song, Q. S. Li, W. Chen, P. Qin, H. Huang, and Y. C. He, "Wind characteristics of a strong typhoon in marine surface boundary layer," Wind Struct. An Int. J., 15 (2012) 1–15.

21 A. Kareem and T. Wu, "Wind-induced effects on bluff bodies in turbulent flows: Nonstationary, non-Gaussian and nonlinear features," J. Wind Eng. Ind. Aerodyn., 122 (2013) 21–37.

22 A. G. Davenport, "Gust Loading Factors," J. Struct. Div. ASCE, 93 (1967) 11–34, 1967.

23 D.-K. Kwon and A. Kareem, "Gust-Front Factor: New Framework for Wind Load Effects on Structures," J. Struct. Eng., 135 (2009) 717–732.

24 D. K. Kwon and A. Kareem, "Generalized Gust-front Factor : A Computational Framework for Wind Load Effects," pp. 1–44, 2013.

25 D. K. Kwon and A. Kareem, "Gust-front factor : A new framework for the analysis of wind load effects in gust-fronts," in 12th International Conference on Wind Engineering, 2007, pp. 767–774.

26 N. J. Cook, "The Deaves and Harris ABL model applied to heterogeneous terrain," J. Wind Eng. Ind. Aerodyn., 66 (1997) 197–214.

27 G. S. Wood, K. C. S. Kwok, N. A. Motteram, and D. F. Fletcher, "Physical and numerical modelling of thunderstorm downbursts," J. Wind Eng. Ind. Aerodyn., 89 (2001) 535–552.

28 M. T. Chay, F. Albermani, and R. Wilson, "Numerical and analytical simulation of downburst wind loads," Eng. Struct., 28 (2006) 240–254.

29 Y. Zhao, S. Cao, Y. Tamura, Z. Duan, and S. Ozono, "Simulation of downburst in a multiple fan wind tunnel and research on its load on high-rise structure by wind tunnel experiment," 2009 IEEE Int. Conf. Mechatronics Autom. ICMA 2009, no. 1, pp. 4506–4511, 2009.

30 M. S. Mason, D. F. Fletcher, and G. S. Wood, "Numerical simulation of idealised three-dimensional downburst wind fields," Eng. Struct., 32 (2010) 3558–3570.

31 D. Romanic, D. Parvu, and H. Hangan, "Downburst Reconstruction using Physical Simulation and Analytical Model with Application to Urban Environments," First Int. Conf. Urban Phys., no. September, pp. 1–12, 2016.

32 D. D. Vicroy, "A Simple, Analytical, Axisymmetric Microburst Model for Downdraft Estimation," 1991.

33 Y. C. He, P. W. Chan, and Q. S. Li, "Observations of vertical wind profiles of tropical cyclones at coastal areas," J. Wind Eng. Ind. Aerodyn., 152 (2016) 1–14.

34 P. J. Vickery, D. Wadhera, M. D. Powell, and Y. Chen, "A hurricane boundary layer and wind field model for use in engineering applications," J. Appl. Meteorol. Climatol., 48 (2009) 381–405.

35 W. F. Huang and Y. L. Xu, "Prediction of typhoon design wind speed and profile over complex terrain," Struct. Eng. Mech., 45 (2013) 1–18.

36 R. Snaiki and T. Wu, "Temperature and moisture effects on the tropical cyclone boundary layer: pressure and wind fields," 8th Int. Colloquium on Bluff Body Aerodynamics and Applications, 2016.

37 J. L. Franklin, M. L. Black, and K. Valde, "GPS Dropwindsonde Wind Profiles in Hurricanes and Their Operational Implications," Weather Forecast., 18 (2003) 32–44.

38 C.-K. Yu and L.-W. Cheng, "Dual-Doppler-Derived Profiles of the Southwesterly Flow Associated with Southwest and Ordinary Typhoons off the Southwestern Coast of Taiwan," J. Atmos. Sci., 71 (2014) 3202–3222.

39 L. Chen and C. W. Letchford, "A deterministic–stochastic hybrid model of downbursts and its impact on a cantilevered structure," Eng. Struct., 26 (2004) 619–629.

40 G. Solari and P. De Gaetano, "Thunderstorm response of structures by response spectrum technique," in ICWE14, 2015, pp. 1–12.

The 13th Americas Conference on Wind Engineering (13ACWE)
Gainesville, Florida USA, May 21-24, 2017

Interference Effects on Wind Loads on Tall Buildings

Bharat Singh Chauhan[a], Ashok Kumar Ahuja[b]

[a]*Indian Institute of Technology Roorkee, Roorkee, India, bharat32@gmail.com*
[b]*Indian Institute of Technology Roorkee, Roorkee, India, ahujashok@gmail.com*

ABSTRACT: This paper presents the effect on wind loads on rectangular plan tall buildings due to the presence of another adjacent building having a similar plan shape with equal variation in orientation of both the buildings with respect to a line along the width of the test section. Effect of height variation of the interfering building is also studied for all the arrangements. Using the 5 component load cell, mean components of the along-wind force, across-wind force, along-wind moment, across-wind moment and torsional moment are reported in the form of X-Y plots. Results obtained show that wind loads are highly affected due to the presence of another interfering building. Along wind force gets enhanced by 48.7 percent as compared to the isolated case. Across wind force is found to increase to 4.56 times and torsional moment is found to be as high as 15.82 times of that in the isolated case.

KEYWORDS: Wind Tunnel, Interference, Rectangular Cross-section, Orientation, Height Variation, Force Study.

1 INTRODUCTION

With the increase in human population, low-rise buildings are being replaced by tall buildings all over the world. The main aim of the construction of tall buildings is to accommodate many residential apartments and offices on a small plot area. Wind is one of the important loads considered for the design of tall buildings. Information available in the standards on wind loads of various countries [AS/NZS: 1170.2 (2002)[1], ASCE: 7-02 (2002)[2], EN: 1991-1-4 (2005)[3] and IS: 875 (Part-3) (2015)[4]] regarding wind pressure and wind force coefficients are limited to isolated condition only. When tall buildings are constructed close to one another, wind loads on them get changed. However, this information is not available in the standards. Although some of the researchers have carried out wind tunnel studies on the effects of interference on wind loads on tall buildings, yet the available information is not enough for the designers to make safe and economical design of tall buildings for wind loads under interference condition.

The significance of interference studies on wind loads on tall buildings has been shown by several research studies in the past. Khanduri et al.[5] analyzed the experimental data available on interference studies and concluded that the wind loads on the principal building are greatly affected if the interfering building is closely located. Xie & Gu[6] studied the effect of interference on square plan shaped principal building due to the presence of one and two interfering buildings also having square plan shape by varying the location, height and plan size of the interfering buildings. Amin[7] carried out wind tunnel experiments to study the effect of interference on rectangular plan shape buildings having aspect ratio of 1:2.5 and 1:1.5 due to the mutual presence of the buildings arranged in L - and T -shape patterns, concluding that the wind loads on the buildings are greatly affected by the arrangement, wind direction and relative dimensions of the buildings. Zhao & Lam[8] studied the effect of mutual presence on 5 square plan shaped buildings arranged in L- and T-shape patterns with a variation in spacing amongst the buildings and the wind direction. Kim et al.[9] carried out wind tunnel studies to understand the interference effects on local peak pressures on a square plan shape tall building due to the presence of an interfering building having a similar shape with a variation in wind direction, location and height of the interfering building. Kushal[10] had conducted wind tunnel experiments to study the effect of interference on rectangular plan shape tall building due to the presence of a square plan shape tall building by varying the spacing amongst the buildings arranged in L- and T- shape patterns. Pandey[11] studied the effect of presence of two square plan shape tall buildings on a similar plan shape principal building with a variation in wind direction and spacing of the buildings. Mara et al. [12] undertook wind tunnel experiments to study the aerodynamic and peak response interference factors on square plan shape buildings due to the presence of another identical building at different upstream positions. Yan & Li[13] carried out Wind tunnel experiments to understand the interference effects between identical tall buildings with aerodynamic modifications by varying the location of the interfering building. Going through the literature it can be observed that the wind loads are greatly affected by the relative dimensions and location of the buildings. However, it can be observed that no research is available for interference studies on commonly used rectangular plan shape tall buildings due to the presence of closely located another interfering tall building having a similar plan shape with a variation in the orientation of both the buildings.

Therefore, an attempt has been made here to study the effect of interference on rectangular plan shape tall buildings due to the presence of a closely located interfering building having a similar plan shape with change in orientation of the two buildings and relative variation in height of the interfering building. Models of the two rectangular plan buildings are arranged at equal orientation giving the arrangement a V-shape plan. The height of the interfering building model is also varied for all the different arrangements, as shown in Figure 1. Results of the force measurements are presented in the form of X-Y plots showing the variation of forces and moments with respect to the orientation angle for different heights of the interfering building.

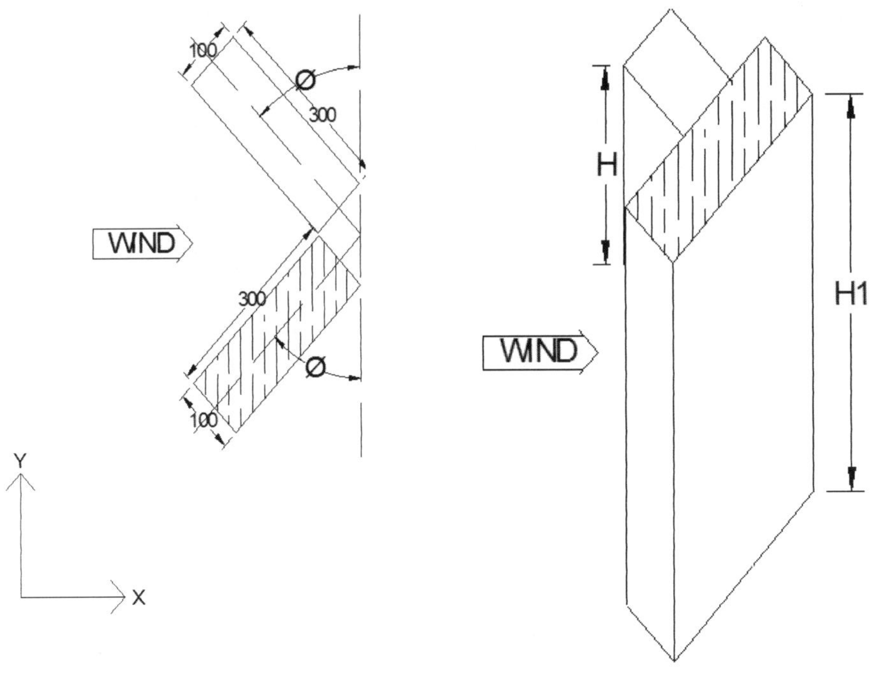

Figure 1. Plan and isometric view of different arrangements for experimentation. (All dimensions are in mm) [Ø=0^0, 15^0, 30^0, 45^0, 60^0, 75^0, 90^0; H =500 mm, 400 mm, 300 mm, 200 mm, 100 mm; H1 =500 mm]

2 MATERIALS AND METHODS

2.1 *Model Description*

Rigid models are made using plywood at a geometric scale of 1:200 to the corresponding prototype, having width to length ratio of 1:3 and width to height ratio of 1:5. The assumed prototype for the principal building considered in this study is of rectangular shape in plan having plan dimensions of 60m x 20m (i.e. 1200 m^2 area in plan) and having height of 100m. Similarly, the prototype considered for the interfering tall buildings in the study are also of rectangular shape in plan having plan dimensions of 60m x 20m (i.e. 1200 m^2 area in plan) but having variation in heights as 100m, 80m, 60m, 40m and 20m. Figure 1 shows the dimensions and arrangement of the models used for principal (hatched) and interfering buildings for wind tunnel tests.

2.2 *Wind Flow Characteristics*

An open circuit boundary layer wind tunnel having a cross-section of 2m x 2m and length of the test section as 15m is used for the testing of the models. Vortex generators for generating turbulence in horizontal plane, barrier wall for generating turbulence in vertical plane, and floor roughening cubical blocks of size 70mm, 50mm and 38mm are used on the upstream end of the test section to achieve the mean wind velocity profile of the approaching flow corresponding to power law exponent of 0.3. The wind velocity profile and the turbulence intensity profile used during the experimentation are shown in Figure 2.

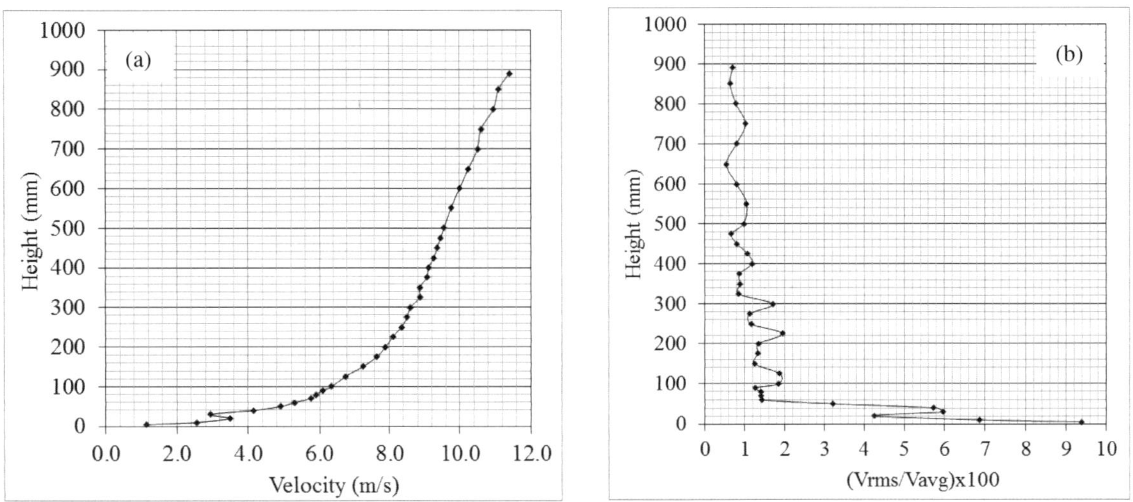

Figure 2. Wind characteristics used for experiments: (a) Mean wind speed profile; (b) Turbulence intensity profile.

2.3 *Measurement Technique*

Force measurements are conducted by placing the principal building plywood model of rectangular cross-section on top of the 5 component load cell and is tested under free stream mean wind velocity of 11.4 m/sec measured at 0.89 m height above the floor of the tunnel. Recording of observations are undertaken at an interval of 1 second for 60 seconds. The principal building model is tested for both isolated and interference conditions for 7 different variations in the reduction angle (Ø, Figure1) for the different heights of the interfering building models (Figure 1) making a total of 42 different arrangements. Photographs for some arrangements are added in Figure 3.

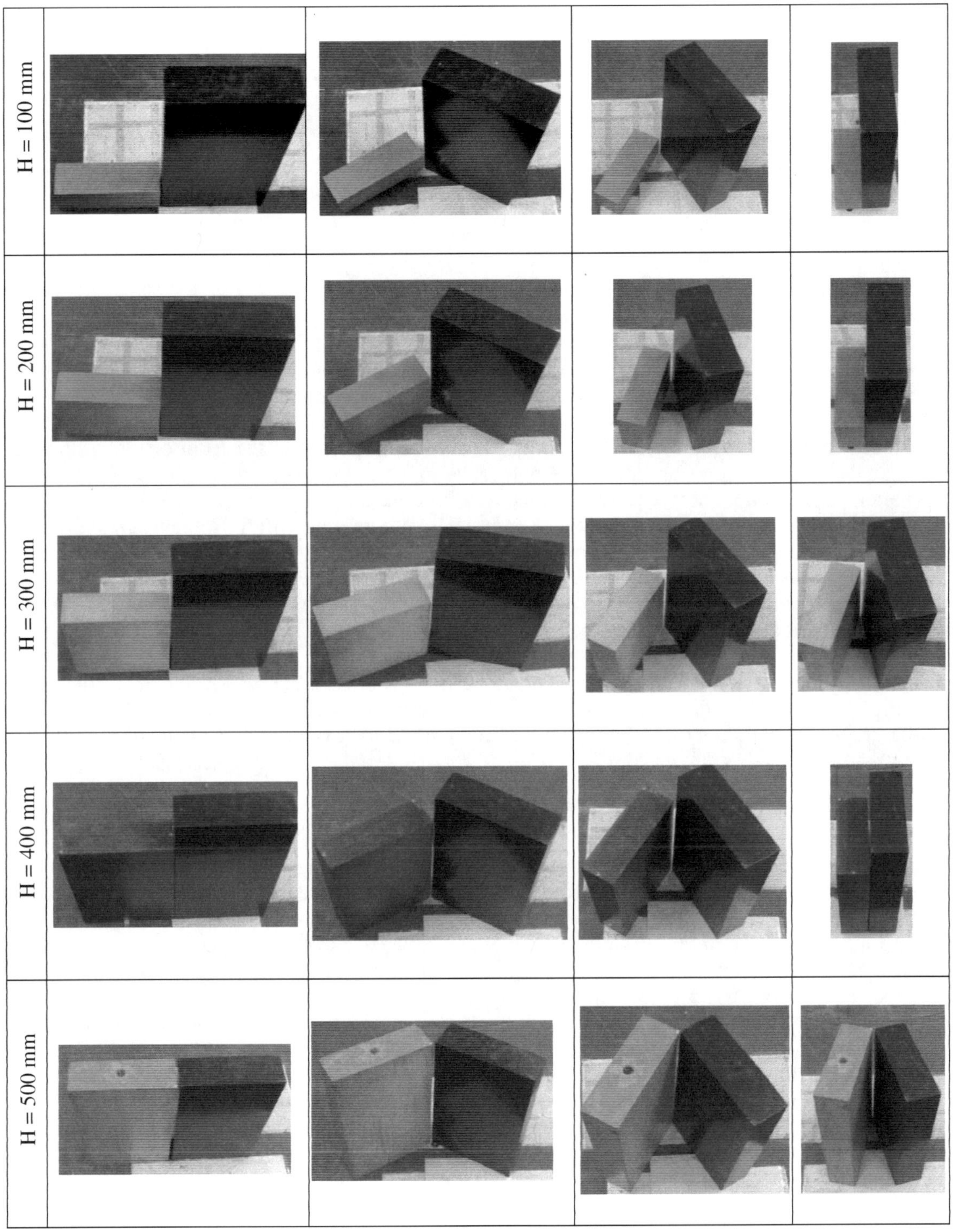

Figure 3. Photographs for some arrangements.

3 RESULTS AND DISCUSSION

Experimental results are reported in the form of X-Y plots with reduction angle (Ø, Fig. 1) on X axis and the corresponding value of interference factor on Y axis, where the interference factor being defined as:
Interference Factor (K_{FX} or K_{FY} or K_{MX} or K_{MY} or K_{MZ}) =

$$\frac{\text{Measurment Parameter (Fx or Fy or Mx or My or Mz) under interference condition}}{\text{Corresponding Parameter (Fx or Fy or Mx or My or Mz) under isolated condition}}$$

where,
Fx =Base shear force on the principal building along the wind flow;
Fy =Base shear force on the principal building across the wind flow;
Mx =Base overturning moment on the principal building about the X-axis;
My = Base overturning moment on the principal building about the Y-axis;
Mz = Twisting moment on the principal building about the Z-axis;

Interference factors for various measurement parameters are calculated using average values (for 60 readings) for the corresponding parameter and are reported in the form of line graphs in Figure 4. To gain an understanding of the numerical magnitude of the observations made during experimentation, numerical values are also reported in Figure 5 in form of line graphs.

Observing plots given in Figure 4 and Figure 5 , we note that with the increase in the reduction angle (Ø) the force in along wind direction (F_X, Fig. 5) is seen to decrease as expected due to the reducing exposed area, for both isolated as well as interference conditions. However, referring to the corresponding interference factor (K_{FX}, Fig. 4), it can be observed that an increase in the along wind force as high as 48.7 percent is noted due to entrapment of air for H=500 mm and Ø=75^0, compared to the isolated condition. This entrapment of air increases with the increase in the height of the interfering building which can be observed from the plots for K_{FX} for different heights of interfering building. K_{MY} is seen to follow a trend similar to K_{FX}. For both K_{FX} and K_{MY} similar graphs are obtained for different heights of the interfering building. From the plots for these two interference factors, it can also be observed that the effect of interference diminishes with decrease in the height of interfering building.

From the plot for across wind force (F_Y, Fig. 5), significant increase can be noted in interference conditions compared to the isolated case due to the change in the direction of wind streams towards the principal building caused by the presence of interfering building. From Figure 5, we can observe the increase in the across wind force (F_Y) with the increase in the reduction angle (Ø) up till Ø=75^0, after which there is a sudden drop in the magnitude of across wind force at Ø=90^0, because at this stage the two buildings become parallel to each other and the earlier reason for increase that is change in the direction of wind streams towards the principal building, is no longer valid. A value as high as 4.56 times the isolated value is observed (K_{FY}, Fig. 4) for H=500mm, and Ø=90^0, however this value is not due to the change in the direction of wind streams towards the principal building, but due to uneven suction on two parallel faces of the principal building oriented along the direction of wind, due to the presence of interfering building adjacent to one of the faces. Although, a very high value of K_{FY} is observed for Ø=90^0, but it can be seen that magnitude wise (Fig. 5) the value observed at Ø=75^0 is highest which is due to the change in the direction of wind streams towards the principal building. Plots for K_{MX} follow trend similar to K_{FY}. For different heights of the interfering building similar graphs can be seen for both K_{FY} and K_{MX}. Diminishing effect of interference with decrease in the height of interfering building can again be observed here for the two parameters.

Again observing the plot for across wind force (F_Y, Fig. 5) and across wind moment (M_X, Fig. 5), we find that for Ø between 0^0 and 15^0 the difference between the values for interference condition and isolated condition is very less, hence the small scattering shown by K_{FY} and K_{MX} has little significance. The value of K_{FY} and K_{MX} at Ø=15^0 is close to 1.0, thereafter the value of these two interference factors increase monotonically.

M_Z plot shows that for isolated condition (Fig. 5) the values are close to zero for Ø=0^0 and 90^0 due to the symmetry in flow of wind, and is highest at Ø=45^0 because at intermediate reduction angles the windward face is partially subjected to pressures and partially to suction. However, it can be seen from Figure 5 that for interference condition as the height of the interfering building is increased M_Z plots follow almost a similar trend to the isolated case but start shifting downwards. This might be because the presence of interfering building causes a suction close to the edge of the principal building at the apex of the V-shape plan, the suction caused at the edge of the principal building causes torsional moment in the principal building in a direction opposite to the torsional moment caused by the change in orientation of the principal building in isolated case, thus reducing the value of M_Z. This suction increases with the increase in the height of the interfering building, which is evident from the plots for M_Z (Fig. 5). Also, if we notice the plot for M_Z at Ø=0^0 it can be seen that torsional moment increase as the height of the interfering building is increased, this is because higher amount of pressure force acts on half of the principal building close to the interfering building, and this force increases as the height of the interfering building is increased. Now, at Ø=90^0 M_Z reverses its direction, this is because amongst the two parallel faces of the principal building along the direction of wind, one of the face is subjected to uniform suction due to the presence of interfering building adjacent to it, while the other face is subjected to higher suction at the leading edge compared to the lower suction at the trailing edge, thus producing a torsional moment in a direction opposite to that at Ø=0^0, hence changing the sign of the torsional moment in the plot.

Referring to K_{MZ} plot in Figure 4, it can be observed that largest deviation from the isolated case are seen for Ø=0^0 and Ø=90^0, due to the reasons discussed above. The highest value observed is as high as 15.82 times of the isolated condition at Ø=90^0 and H=500mm. For interference condition, at intermediate reduction angles the values for K_{MZ} are less than 1.0 due to the shifting of M_Z plots downward as discussed earlier. From the plot of K_{MZ}, noting the values at points having largest deviation from the values of isolated case (i.e. Ø=0^0 and Ø=90^0), it can again be deduced that as the height of the interfering building decreases the effect of interference decreases.

From Figure 4, it can be noted that an increase in the along wind force (K_{FX}) as high as 48.7 percent is noted for H= 500 mm and Ø=75^0, compared to the isolated condition. A value as high as 4.56 times the isolated value is observed for across wind force (K_{FY}, Fig. 4) for H=500mm, and Ø=90^0.Also, the highest value observed for torsional moment (K_{MZ}) is as high as 15.82 times of the isolated condition for Ø=90^0 and H= 500mm. The high values of interference factors, thus emphasize the importance of interference studies. From Figure 5, observing the plot for along wind force (F_X) for isolated condition, it can be seen that the magnitude of along wind force for Ø=90^0 is only 18.94% of the value observed for Ø=0^0, whereas the reduced exposed area for Ø=90^0 is 33.33% of the exposed area for Ø=0^0. Thus, showing that wind loads get modified with change in the relative dimensions of the building and hence corroborating importance of the undertaken research work.

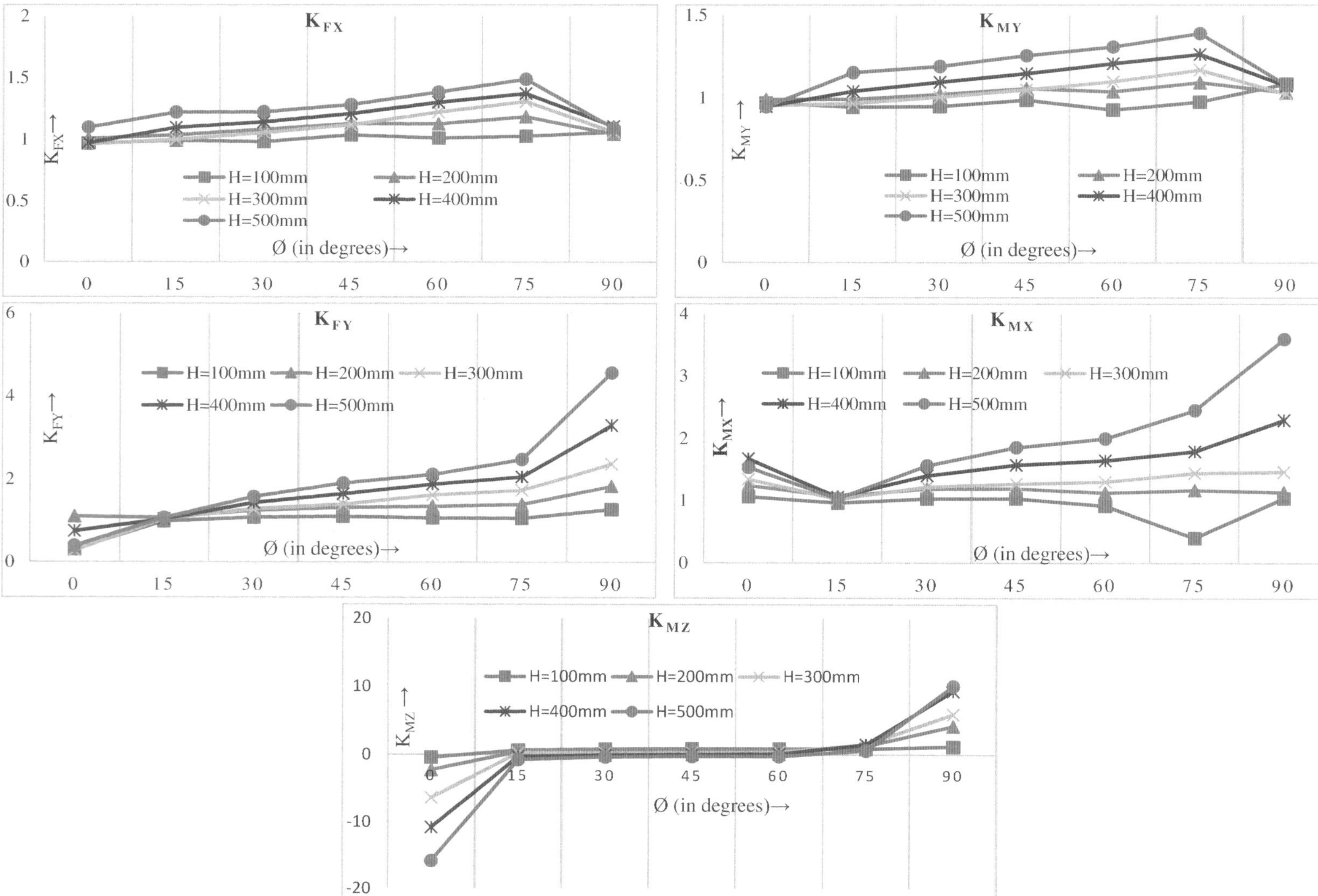

Figure 4. Line graphs representing the interference factors obtained for different arrangements.

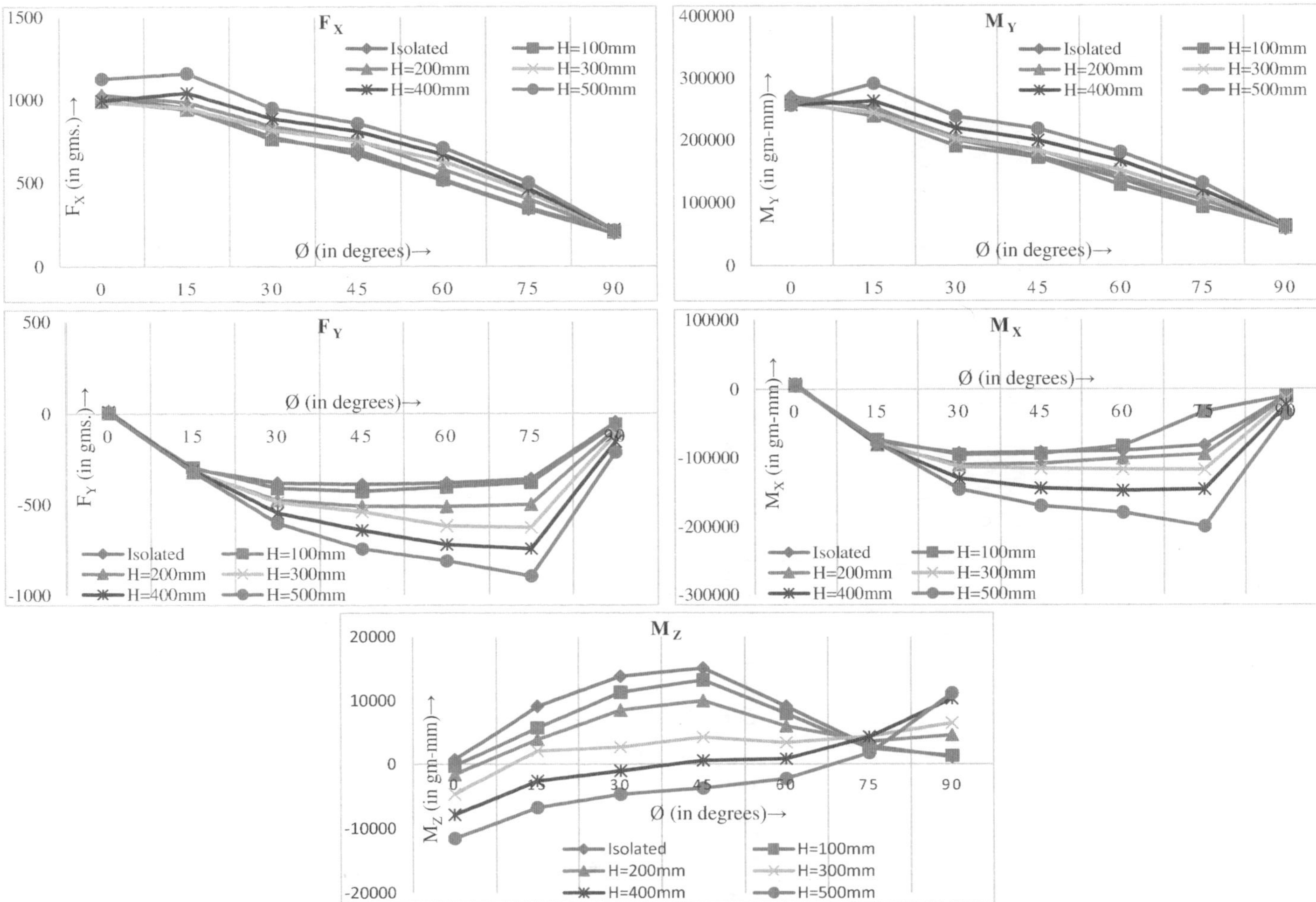

Figure 5. Line graphs representing the numerical values obtained for different parameters.

4 CONCLUSIONS

- The effect of interference on wind loads increases with increase in height of the interfering building.
- The force in along wind direction (F_X) decreases with increase in the reduction angle (Ø) due to the reducing exposed area, for both isolated as well as interference conditions.
- An increase in the along wind force as high as 48.7 percent is noted due to the entrapment of air, compared to the isolated condition.
- K_{MY} follows the trend similar to K_{FX}. For different heights of the interfering building similar graphs are obtained for both K_{FX} and K_{MY}.
- A value as high as 4.56 is observed for across wind force interference factor (K_{FY}), due to uneven suction on two parallel faces of the principal building.
- K_{MX} follows the trend similar to K_{FY}. Here also, for different heights of the interfering building similar graphs are obtained for both K_{FY} and K_{MX}.
- For interference condition as the height of the interfering building is increased M_Z plots follow almost a similar trend to the plot for isolated case, but start shifting downwards with increase in the height of interfering building.
- The highest value for K_{MZ} is as high as 15.82, which is due to the enhancement of wind pressure force near the interfering building.

5 ACKNOWLEDGEMENTS

The work presented in this paper is part of the research work being done by the first author for his Ph.D. degree under the supervision of second author.

6 REFERENCES

1 AS/NZS: 1170.2-2002, Australian/New Zealand standard: Structural design actions part-2: wind actions.
2 ASCE: 7-02-2002, American society of civil engineers: Minimum design loads for buildings and other structures.
3 EN: 1991-1-4-2005, Euro Code 1: Actions on structures – wind actions.
4 IS: 875 (Part-3) 2015, Code of practice for design loads (other than earthquake loads), for building and structures – Wind loads.
5 A.C. Khanduri, T. Stathopoulos and C. Bédard, Wind induced interference effects on buildings - a review of the state-of-the-art, Engineering Structures, 20 (1998), 617–630.
6 Z.N. Xie and M. Gu, Mean interference effects among tall buildings, Engineering Structures, 26 (2004), 1173–1183.
7 J. Amin, Effects of plan shape on wind induced response of tall buildings, Ph.D. Thesis, 2008, Department of Civil Engineering, Indian Institute of Technology Roorkee, India.
8 J.G. Zhao and K.M. Lam, Interference effects in a group of tall buildings closely arranged in an L- or T-shaped pattern, Wind and Structures, Vol. 11, No. 1 (2008), 1-18.
9 W. Kim, Y. Tamura and A. Yoshida, Interference effects on local peak pressures between two buildings, Journal of Wind Engineering and Industrial Aerodynamics, 99 (2011), 584–600.
10 T. Kushal, Effect of plan shapes on the response of tall buildings under wind loads, M.Tech. Thesis, 2013, Department of Civil Engineering, Indian Institute of Technology Roorkee, India.
11 S.C. Pandey, Influence of proximity on the response of tall buildings under wind loads, M.Tech. Thesis, 2013, Department of Civil Engineering, Indian Institute of Technology Roorkee, India.
12 T.G. Mara, B.K. Terry, T.C.E. Ho and N. Isyumov, Aerodynamic and peak response interference factors for an upstream square building of identical height, Journal of Wind Engineering and Industrial Aerodynamics, 133(2014), 200–210.
13 B. Yan and Q.S. Li, Wind tunnel study of interference effects between twin super-tall buildings with aerodynamic modifications, Journal of Wind Engineering and Industrial Aerodynamics, 156(2016), 129–145.

13th Americas Conference on Wind Engineering
Gainesville, Florida, USA
May 21 - 24, 2017

Experimental assessment of the effect of balconies on the wind load distribution on a high-rise building

L. Ludena[a], A.G. Chowdhury[b], P. Irwin[c], M. Moravej[d], B. Hajra[e], I. Zisis[f]

[a] *Florida International University, Miami, USA, email: llude001@fiu.edu*
[b] *Florida International University, Miami, USA, email: chowdhur@fiu.edu*
[c] *Florida International University, Miami, USA, email: peairwi@fiu.edu*
[d] *Florida International University, Miami, USA, email: mmora229@fiu.edu*
[e] *Florida International University, Miami, USA, email: bhajra@fiu.edu*
[f] *Florida International University, Miami, USA, email: izisis@fiu.edu*

ABSTRACT

Many residential buildings located close to coastal areas have balconies that are used by the building occupants for recreational purposes. However, during past hurricane events wind induced damage to buildings with balconies have caused losses worth millions of dollars. Due to limited information available in codes and standards on the safe design of buildings with balconies, additional research in this direction is needed. This paper presents results from an experimental study carried out at the Wall of Wind (WOW) experimental facility at Florida International University (FIU), USA, on a fifteen story tall residential building, for three different model scales: 1:25, 1:67 and 1:180 in an open terrain exposure. At each scale two models were built: a building without any balconies and a building with two types of balconies; namely continuous and discontinuous. Continuous balconies span from one end of the building to the other, while discontinuous balconies have partitions in between them. Mean and peak pressure coefficients were estimated on various locations along the building surface as well as the balcony for various cases. Results show that at 0° wind direction, the front wall experienced higher magnitudes of peak pressure coefficients for the case with balconies than the case without balconies. In general, the location of balcony influenced the pressure distribution across the building wall and had a considerable effect on the magnitude and distribution of peak pressures. Furthermore, results show that the wind loads on the balconies are quite sensitive to the model scale used for the experiments. Results from this large scale experimental study can greatly contribute to existing design standards. Future research must be directed towards understanding the wind load distribution on buildings for different balcony geometries and terrain exposure.

KEYWORDS

Wall of Wind, Wind load, Balconies, Residential building, Model scale.

The 13th Americas Conference on Wind Engineering (13ACWE)
Gainesville, Florida USA, May 21-24, 2017

The impact of time averaging of wind speed data on wind comfort and hazard analysis

Hung Nguyen Sinh[a], Stefano Cammelli[b] and Max Lee[c]

[a]*BMT Fluid Mechanics Ltd., New York City, USA,* hnguyensinh@bmtfm.com
[b]*BMT Fluid Mechanics Ltd., Teddington, UK,* scammelli@bmtfm.com
[c]*BMT Fluid Mechanics Ltd., New York City, USA,* mlee@bmtfm.com

ABSTRACT: Using readily accessible wind speed data from meteorological databases, such as the one made available by the National Oceanic and Atmospheric Administration (NOAA), has become a standard procedure in today's wind engineering consultancy and – when it comes to the assessment of the local wind microclimate – wind specialists would typically rely on mean-hourly wind speed records to define the parent distribution for a specific site. However, in the USA, for many of the available weather stations, "true hourly" mean wind speed information is not available and the "2-minute" mean wind speed prior to every top-of-hour is used as a replacement for the "true hourly" mean wind speed. Using Weibull parameters for San Francisco obtained both from a full and comprehensive set of wind records (truly continuous 2-min mean data) and from a reduced sub-set of it (discrete 2-min average at the end of each hour), this technical paper analyzes the impact of these on the appraisal of the pedestrian level wind microclimate within the context of the San Francisco Environmental Review Office criteria.

KEYWORDS: Wind microclimate, wind comfort, wind hazard, mean-hourly wind speed, boundary layer wind tunnel testing.

1 INTRODUCTION

One of the key challenges faced by sites being developed within the complex urban fabric of large North American cities is to ensure that the impact of such developments with regard to winds at pedestrian level is minimized. Tall buildings in particular have in fact the potential to cause downdrafts of fast upper-level winds, strong flow accelerations around building corners and funneling between neighbors. Planners would therefore typically expect to see due consideration from an early stage of the design, not only via a comprehensive and robust wind assessment but also by demonstrating in-built design measures that offset or even prevent unacceptable impacts within an existing 'present-day' context plus, more importantly, within a future context where other equally tall structures are expected to emerge. In this context, boundary layer wind tunnel testing plays a key role as one of the most reliable and cost-effective tools that can be used to quantify the impact of winds at street level both from a comfort and hazard conditions point of view.

Local measurements of mean and gust wind speeds within a boundary layer wind tunnel are typically normalized by the mean wind speed at a specific reference height (typically the highest point within the proposed development), and are known as 'speed-up' factors. By combining these 'speed-up' factors with site-specific statistics (e.g. probability distribution as a function of speed and direction) a detailed appraisal of the wind microclimate conditions can be obtained: these in turn will then need to be compared against relevant criteria in order to quantify the level of comfort and / or the presence of hazards within the site. The criteria that are used within this technical paper are the one presently specified in the San Francisco Environmental Review Office [1]. The principles of these criteria are summarized in Table 1.

Table 1. San Francisco Environmental Review Office criteria

Comfort criteria	*1-min equivalent wind speed (mph)*	*Probability of exceedance*
Pedestrian use areas	11	10%
Public seating areas	7	10%
Safety criteria	*1-hour equivalent wind speed (mph)*	*Probability of exceedance*
Wind hazard	26	0.0114%

It should be noted that the wind speed used in these criteria is an "equivalent wind" that combines the effects of both mean wind speed and turbulence intensity on people and it is described by the following Equation 1:

$$U_{equivalent} = U_{mean} \times \left(1 + 3 \times I_u\right) \tag{1}$$

where I_u is the local turbulence intensity and U_{mean} is the local mean wind speed.

The averaging period used in the San Francisco Environmental Review Office criteria [1] to assess comfort is 1 minute, a duration fully in line with the length of time over which the U.S. Weather Bureau observers make their hourly observations [2]. The San Francisco Environmental Review Office states in fact that "this interval is sufficiently close to the length of time that the wind actually takes to affect pedestrian's comfort that the Weather Bureau data can be used directly to satisfy those criteria". For wind hazard conditions, the averaging periods is instead 1 hour. Also, whilst wind hazard conditions are quantified throughout the entire 24 hours, wind comfort is only considered between 7am and 6pm [2].

Within the framework of the San Francisco Environmental Review Office criteria [1], this technical paper will analyze the impact of different time averaging of wind speed data on wind comfort analysis.

2 WIND DATA AND STATISTICAL ANALYSIS

The wind speed data used within the study presented in this technical paper were collected from the NOAA's [3] ISH database (ftp://ftp.ncdc.noaa.gov/pub/data). This database contains two sets of records: 2-min top-of-hour data and 2-min data every 2 minutes. From the latter dataset, the mean hourly wind data was derived by averaging the 30 values (every 2 minutes) for each hour. The treatment of directionality needed more attention: the most frequently appearing wind direction across the compass (16 sectors at 22.5° intervals) during each hour was selected as the representative one for that particular hour of data.

Figure 1 shows a sample extracted from this dataset with three types of wind speeds: continuous 2-min mean wind speed, 2-min average wind speed at the end of each hour and true mean- hourly wind speed. Statistical manipulation of this dataset allowed the directional probability distribution to be extracted (see Fig. 2). It is evident from Figure 2 that the wind sector between West and West-North-West is the most prevailing one for San Francisco: this is well in line with the requirements of the San Francisco Environmental Review Office criteria [1] and Arens et al. 's paper [2] which stipulate that only the WSW, W, WNW and NW winds should be accounted for in the wind microclimate analysis. Also, the reasonable match between the use of continuous 2-min mean and true mean-hourly wind speed data confirms that the wind directionality is well described by the latter dataset.

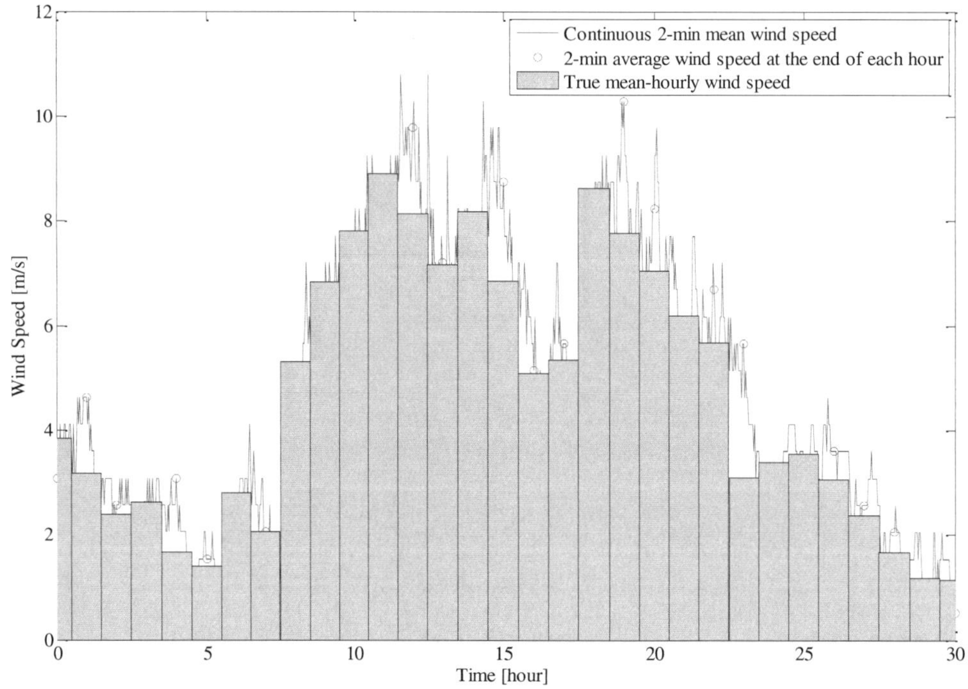

Figure 1. Sample extracted from the wind speed dataset.

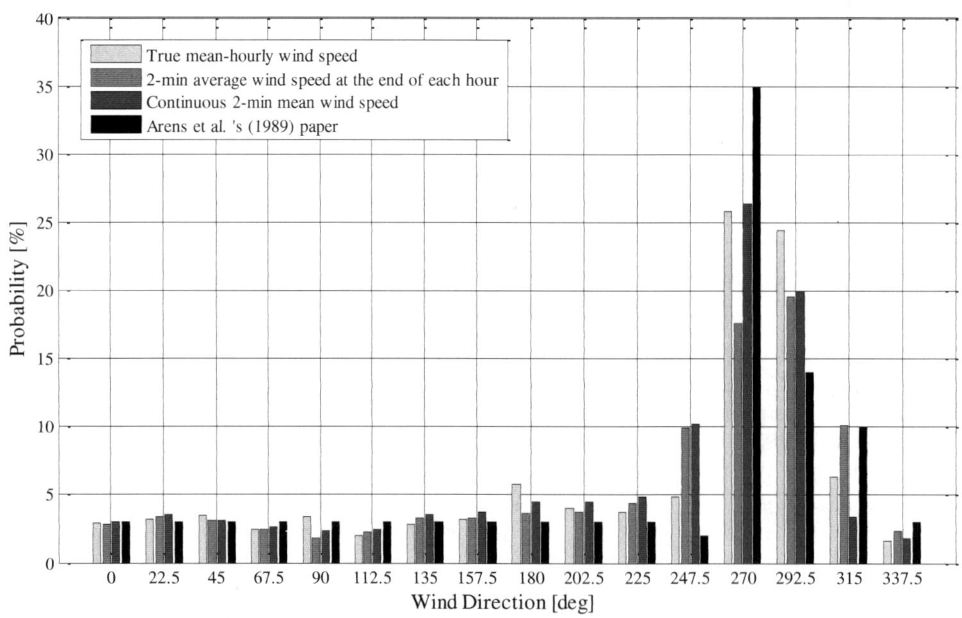

Figure 2. Directional probability distribution.

For wind microclimate analysis, the directional variation of the scale ("c") and shape ("k") parameters of the Weibull distribution are important. Figures 3 and 4, respectively. show a comparison of the scale and shape parameters obtained from 2-min average wind speeds at the end of each hour and true mean-hourly wind speeds for both all day(used for hazard analysis) and day-time hours (used for comfort analysis). From these graphs two features are clearly noticeable: firstly, the scale parameters obtained from the 2-min average wind speeds at the end of each hour are in general ~5% higher than those derived from the true mean-hourly records. Secondly, the strength of day-time winds is generally higher than the strength of night-time winds.

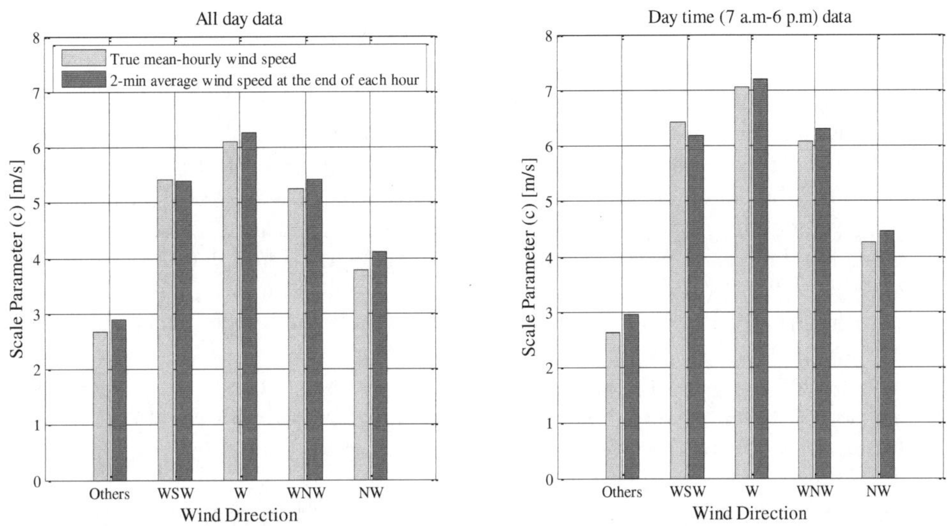

Figure 3. Directional variation of the "c" parameter of the Weibull distribution.

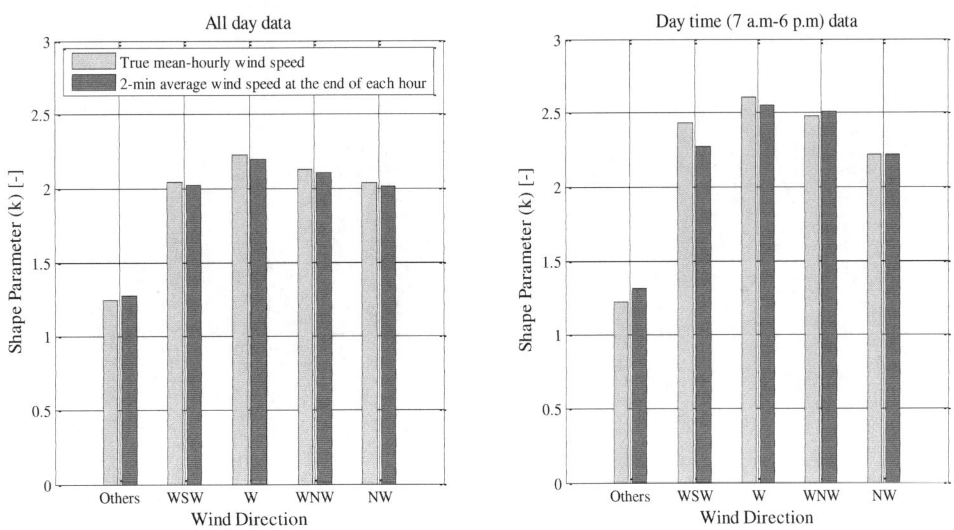

Figure 4. Directional variation of the "k" parameter of the Weibull distribution.

The parameters so far presented have been used on a commercial Project to fully examine the ultimate impact on the outcome of the wind comfort and hazard assessments.

3 A CASE STUDY

The proposed development that is used within this technical paper to test the impact of time averaging of wind speed data on wind comfort and hazard analysis within the context of the San Francisco Environmental Review Office criteria [1] comprises three towers of nominally rectangular plan form with a maximum height of approximately 150m.

The different steps that were followed in the assessment of this commercial Project were:

i. The assessment locations pertaining to the wind comfort and hazard study were identified and discussed in conjunction with the Planning Department. A total number of 78 pedestrian level test points were considered, including: side walk, building main entrances, car parks, bus stops and outdoor seating areas.
ii. A 1:300 scale wind tunnel model of the proposed development (see Figure 5) and of the area surrounding the site (extension of approximately 500m radius) was designed and constructed from digital three-dimensional information making use of a combination of computer numerically controlled (CNC) machining and robotic arms.
iii. Boundary layer wind tunnel testing was conducted for the four prevailing wind directions principally responsible for the driving wind conditions in San Francisco. All tests were conducted in BMT's Boundary Layer Wind Tunnel (www.bmtfm.com/about-us/wind-tunnel-facilities) which has a test section 4.8m wide, 2.4m high and 15m long with a 4.4m diameter turntable and an operating wind speed ranging between 0.2m/s and 45m/s.
iv. The measured 'speed-up' factors were combined with site-specific statistics to carry out a detailed assessment of the wind microclimate conditions within the site of the proposed development.

Figure 5. Close-up of the wind tunnel model.

Figure 6 shows the results of the pedestrian level wind comfort analysis where 'speed-up' factors measured in the wind tunnel have been combined with two different sets of wind statistics (one based on true mean-hourly wind speed and one based on the 2-min average wind speed at the end of each hour). Approximately 50 out of the total 78 locations exceed the wind comfort criteria (a 1-min wind speed of 11mph (4.9m/s) exceedance for more than 10% of the time) and, apart from location '67' – which changes from 0-7mph (0-3.1m/s) to 7-11mph (3.1-4.9m/s) – the two sets of results are largely identical.

Figure 6. Comfort analysis using the San Francisco Environmental Review Office criteria.

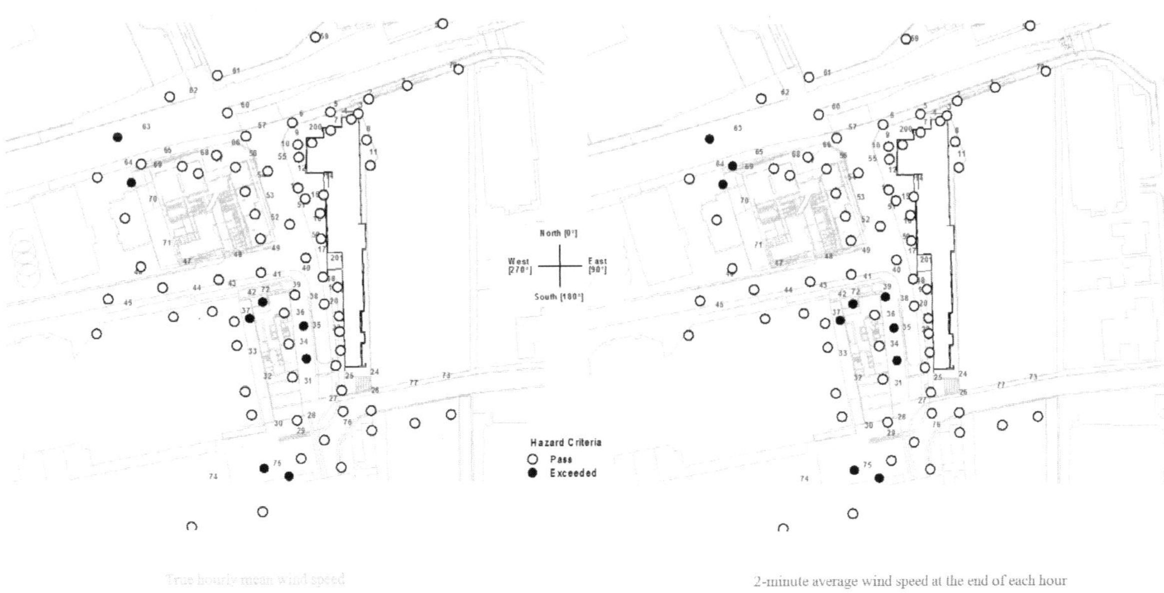

Figure 7. Hazard conditions using following the San Francisco Environmental Review Office criteria.

Figure 7 shows the results of the wind hazard conditions at pedestrian level where 'speed-up' factors measured in the wind tunnel have been combined with two different sets of wind statistics. Contrary to the wind comfort analysis, the results produced by the two datasets are very different. Moving, in fact, from the 2-minute average wind speed at the end of each hour to the true mean-hourly wind speed, the number of wind hazard exceedance locations reduces from 10 to 8 (i.e. a reduction of 20%). Also, the total number of wind hazard hours in one year when the 1-hour equivalent wind speed exceeds 26mph (11.6m/s) across all locations reduces from 81 to 54 (i.e. a reduction of ~34%) which is pretty substantial.

4 CONCLUSIONS

This paper analyzes the San Francisco wind climate starting from two different datasets: true mean-hourly wind speed and 2-minute average wind speed at the end of each hour. Directional Weibull distributions were initially derived from the two sets of data: this confirmed that the two prevailing wind directions in San Francisco are West and West-North-West, as indicated in [2]. The Weibull analysis also showed that the wind speeds obtained using the true mean-hourly data are approximately 5% lower than the ones obtained from the reduced set and that the wind speeds during daytime are stronger than the ones during nighttime. A wind microclimate case study featuring three towers with a maximum height of approximately 150m was then presented. By combining the wind statistics for San Francisco from the two datasets (duly transposed to the Project site) with the various "speed-up" factors measured in the wind tunnel, wind comfort and hazard conditions were assessed following the San Francisco Environmental Review Office criteria [1]. From a comfort point of view the two datasets provided very comparable results. However, the hazard conditions predicted by the true mean-hourly dataset showed a reduction in the number of hazard hours of approximately 34%: this clearly shows the level of inherent conservatism that is typically associated with the use of the more common and readily available 2-minute average wind speed at the end of each hour.

5 REFERENCES

1 San Francisco City Planning Department. San Francisco Planning Code, 1996.
2 E. Arens, D. Ballanti, C. Bennett, S. Guldman and B. White, Developing the San Francisco Wind Ordinance and its Guidelines for Compliance, Building and Environment, Vol. 24, No. 4, pp. 297-303, 1989.
3 The Integrated Surface Database (ISD), National Climatic Data Center (NCDC), U.S. Department of Commerce http://www.ncdc.noaa.gov/oa/climate/isd/index.php

The 13th Americas Conference on Wind Engineering (13ACWE)
Gainesville, Florida USA, May 21-24, 2017

PC-NARX Metamodeling of Non-Linear and Uncertain Wind Excited Dynamic Systems

Wei-Chu Chuang[a], Seymour M.J. Spence [a]

[a]*Department of Civil and Environmental Engineering, University of Michigan, Ann Arbor, MI, USA*

ABSTRACT: In order to analyze wind-excited systems within a probabilistic performance-based (PBD) setting, the estimation of the behavior of the system beyond the elastic limit is necessary. In the field of wind engineering, this can become computationally burdensome in light of the generally long duration of wind storms. This paper investigates the possibility of applying a computationally tractable approach using advanced non-intrusive metamodeling techniques, based on combining polynomial chaos expansions (PCEs) with non-linear autoregressive with exogenous input (NARX) models, for nonlinear systems subject to stochastic wind excitation of long duration. The application to a case study illustrates the practicality of such an approach.

KEYWORDS: Surrogate model, Nonlinear autoregressive with exogenous input models, Polynomial chaos expansion, Monte Carlo simulation, Wind excited dynamic systems.

1 INTRODUCTION

There is growing interest in applying the principles of probabilistic performance-based design (PBD) to the analysis of wind excited building systems [1-5]. In the application of such an approach, it is necessary to understand the response of the system over a full range of hazard levels, i.e. from serviceability to possible collapse scenarios. However, the extremely long duration of wind storms, typically in the order of hours, as well as the challenging issue of propagating the inevitable uncertainties through the nonlinear systems, which is required by modern probabilistic PBD [6], make the estimation of extreme responses computationally burdensome. To overcome this problem, a possible approach would be to use non-intrusive metamodeling techniques. Indeed, such an approach has recently been proposed to simulate non-linear dynamic systems subject to short duration stochastic excitation, such as seismic loads [7-9]. The key characteristic of this approach is the use of non-linear autoregressive with exogenous input (NARX) models for mimicking the dynamical behavior of the system while representing the uncertainties through polynomial chaos expansions (PCEs).

In this paper, the possibility of applying this approach for modeling nonlinear systems subject to stochastic wind excitation of long duration (i.e. in the order of hours) is investigated. A case study of a single-degree-of-freedom (SDOF) system is presented illustrating the practicality of applying the proposed PC-NARX model for simulating the dynamic response of wind excited systems.

2 THE METAMODEL

2.1 *The NARX model*

The nonlinear autoregressive model with exogenous inputs (NARX) is a commonly used model for representing a wide class of nonlinear systems in discrete time domain. In particular, the model is an expansion of past input and output defined as:

$$y(t) = G(\mathbf{z}(t)) + \varepsilon(t) = G(u(t), \cdots, u(t-n_u), y(t-1), \cdots, y(t-n_y)) + \varepsilon_t \quad (1)$$

where $y(t)$ and $u(t)$ are the time-dependent system output and input sequences; n_u and n_y are the maximum lags for the system output and input; $G(\square)$ is the underlying nonlinear mathematical model to be identified while ε_t is the residual error of the NARX model, assumed as a normal random variable with zero mean and standard deviation of $\sigma_\varepsilon(t)$. In a linear-in-parameters form, the above equation is commonly expressed as:

$$y(t) = \sum_{i=1}^{n_g} \vartheta_i g_i(\mathbf{z}(t)) + \varepsilon_t \quad (2)$$

in which n_g is the number of model terms $g_i(\mathbf{z}(t))$ that are functions of the regression vector while ϑ_i are the coefficients of the NARX model. The identification of an appropriate NARX model consists of two major steps, namely system detection and parameter estimation. System detection, i.e. the determination of which NARX terms, $g_i(\mathbf{z}(t))$, are in the model, is critically important in order to efficiently predict the output response and to avoid numerical and computational problems arising from the inclusion of spurious terms. Therefore, the simplest model that can appropriately represent the underlying dynamics of the system has to be identified [10,11]. The second step is to estimate the corresponding NARX coefficients of the identified model based on the purpose of the intended model. Then, the identified mathematical model, considered as a metamodel, can be used to predict future responses.

2.2 *The Polynomial Chaos Expansion*

Consider a deterministic mapping $Y = \mathcal{M}(\mathbf{X})$ where $\mathbf{X}=\{X_1, \ldots, X_M\}^T$ is a M-dimensional vector of the input random variables and Y is the scalar model output of interest. If the input random variables are considered independent, the aforementioned model can be represented through a generalized polynomial chaos expansion (PCE) as:

$$Y = \mathcal{M}(\mathbf{X}) = \sum_{\alpha \in \mathbb{N}^M} a_\alpha \psi_\alpha(\mathbf{X}) \quad (3)$$

where a_α are deterministic coefficients while $\psi_\alpha(\boldsymbol{X}) = \prod_{i=1}^{M} \pi^i_{\alpha_i}(X_i)$ are basis of multivariate orthonormal polynomials defined as the tensor product of the univariate polynomials, in which $\pi^i_{\alpha_i}$ represents a family of orthonormal polynomials. In practice, the full expansion of Equation 3 is computationally intractable. Thus, a sparse PCE based on hyperbolic truncation scheme [12] is employed in this work so that only those basis polynomials with total degree not greater than p and rank not greater than r are retained:

$$\mathcal{A}_q^{M,p} = \left\{\alpha \in \mathbb{N}^M : \|\alpha\|_q \le p, \|\alpha\|_0 \le r\right\} \quad (4)$$

where $0 < q \le 1$ is the parameter defining the hyperbolic truncation surface while the degree and rank of $\boldsymbol{\alpha}$ are respectively:

$$\|\alpha\|_q \equiv \left(\sum_{i=1}^{M} \alpha_i^q\right)^{1/q}, \quad \|\alpha\|_0 \equiv \sum_{i=1}^{M} \mathbf{1}_{\{\alpha_i > 0\}} \quad (5)$$

The truncated PCE may then be rewritten as

$$Y = \sum_{\alpha \in \mathcal{A}_q^{M,p}} a_\alpha \psi_\alpha(\mathbf{X}) + \varepsilon \equiv \boldsymbol{a}^T \psi(\mathbf{X}) + \varepsilon \quad (6)$$

where $\boldsymbol{a}$ and $\psi(\mathbf{X})$ are vectors of all coefficients and basis polynomials respectively and ε is the truncation induced error. In particular, the coefficients can be estimated by minimizing the mean square error of the model, which may be further reduced to a linear regression problem with least squares solution.

2.3 *A combined PCE and NARX model*

In order to efficiently simulate the extreme response of a wind-excited system, a metamodeling approach based on combining the NARX and PCE models is employed to simulate the dynamic response and propagate uncertainty through the nonlinear system. To this end, consider a numerical model $y(t,\mathbf{X}) = \mathrm{M}\,(u(t,\mathbf{X}))$ where $y(t, X)$ is the response time history of interest. The output response may be represented by the following NARX model with random variables:

$$y(t,\mathbf{X}) = \sum_{i=1}^{n_g} \vartheta_i(\mathbf{X}) g_i(\mathbf{z}(t)) + \varepsilon_g(t,\mathbf{X}) \tag{7}$$

where $\varepsilon_g(t, X)$ is the residual error assumed to be normally distributed with zero mean and variance of $\sigma_\varepsilon^2(t)$. The NARX coefficients can be expanded onto an appropriate PC basis through a truncated PCE, as follows:

$$\vartheta_i(\mathbf{X}) = \sum_{j=1}^{n_\psi} \vartheta_{i,j} \psi_j(\mathbf{X}) + \varepsilon_i \tag{8}$$

where $\psi_j(\mathbf{X})$ are multivariate orthonormal polynomials of $\mathbf{X}$; $\vartheta_{i,j}$ are associated deterministic PC coefficients and ε_i is the truncation error. Finally, the time dependent output time series can be represented by a PC-NARX model as:

$$y(t,\mathbf{X}) = \sum_{i=1}^{n_g} \sum_{j=1}^{n_\psi} \vartheta_{i,j} \psi_j(\mathbf{X}) g_i(\mathbf{z}(t)) + \varepsilon(t,\mathbf{X}) \tag{9}$$

where $\varepsilon(t, X)$ is the total error due to the truncation of both the NARX and PCE models.

In this work, a two-phase approach based on least angle regression (LARS) is applied to select the appropriate terms in the PC-NARX model [9]. First, relevant NARX regressors are identified by means of LARS [13]. In particular, the corresponding NARX coefficients are estimated by minimizing the one-step-ahead prediction error (PE). Considering the long duration characteristic of a typical wind storm, a NARX model optimized through a PE criterion may lose its accuracy due to error accumulation from the model output feedback. To avoid this, a different approach, calibrated in parallel configuration that yields the optimal simulation error (SE) of the entire time history, is applied here to estimate the corresponding NARX coefficients.

3 CASE STUDY

3.1 *Description*

In this section, the PC-NARX model is implemented to simulate the dynamic response of a wind excited system. The overall height of the structure, H, is 154.7 m (507.5 ft) with width W and length L respectively 36.6 m (120 ft) and 24.4 m (80 ft). The building density is 192 kg/m^3. In order to model the system as an equivalent SDOF oscillator, the first modal response, with linear mode shape over the building height $\varphi_1(z) = z / H$, of the building is considered. The dynamics of the system can therefore be described by the following equation of motion:

$$M\ddot{y}(t) + 2M\omega\xi\dot{y}(t) + f(y(t), \dot{y}(t)) = Q(t) \tag{10}$$

where $y(t)$, $\dot{y}(t)$ and $\ddot{y}(t)$ are respectively the first modal displacement, velocity and acceleration responses; $M = \varphi_1^{\mathrm{T}} m \varphi_1$ is the generalized mass in which m is the global mass matrix; $f(y(t), \dot{y}(t))$ is the generalized nonlinear restoring force while $Q(t) = \varphi_1^{\mathrm{T}} \boldsymbol{F}(t)$ is the generalized force with $\boldsymbol{F}(t)$ the external aerodynamic loads. The generalized damping ratio and natural modal frequency are considered deterministic with values $\xi = 0.02$ and $\omega = 1.44$ rad/s. The stochastic wind loads, $\boldsymbol{F}(t)$, acting on the system are estimated through a quasi-steady model [5,14]. The intensity of the extreme wind event is defined by the maximum wind speed to occur during the event at a height of interest z averaged over a fixed time interval T of one hour. In particular, the wind speed v used to generate the stochastic wind loads was defined by a Type II distribution with mean 30 m/s and standard deviation 3.5 m/s. The roughness length was taken as $z_{01} = 0.05$ while the meteorological height $H_{met} = 10$ m. Roughness length at the site of the structure, z_0, was assumed to be 0.02 m.

The nonlinearity of the SDOF system is modeled as a Duffing oscillator with restoring force defined as:

$$f(y(t), \dot{y}(t)) = M\omega^2 (y(t) + \beta y(t)^3) \tag{11}$$

where β is the parameter governing the nonlinear behavior, assumed as an uniform random variable between 9 and 11. To calibrate the metamodel, 200 simulations were carried out with input random variables generated by Latin hypercube sampling. A full NARX model was chosen as the following polynomial function:

$$g_i(t) = y^l (t - n_y) Q^m (t - n_u) \tag{12}$$

with $l + m \leq 3$, $0 \leq l \leq 3$, $0 \leq m \leq 1$, $n_y = 1, 2$, $n_u = 0, 1, 2$, leading to 10 candidate terms including constant term.

3.2 *Results*

The best NARX model selected by LARS to simulate the displacement response of the SDOF system Duffing oscillator contained 5 terms, namely, the constant term, $Q(t)$, $y(t\text{-}1)$, $y(t\text{-}2)$ and $y(t\text{-}1)^2$. The mean relative error over 200 simulations was $\bar{\varepsilon} = 0.0486$. In the second phase of the proposed approach, the NARX coefficients, estimated from the SE algorithm, were represented by sparse PCEs with maximum interaction rank $r = 2$ and truncation parameter $q = 1$.

Figure 1(a) reports a representative time history of the displacement response along with the errors associated with NARX and PC-NARX models. As can be seen, the PC-NARX results agree remarkably well with the reference responses. The model also possesses very good accuracy in simulation, even during the transient dynamics part of the response that occurs at the beginning of the time series. The extreme responses of interest, i.e. the maximum displacements, for all samples reproduced by the identified PC-NARX model are also compared with the actual values obtained from the numerical simulation, as shown in Figure 1(b). From Figure 1(b), it is seen that the PC-NARX model can estimate the extreme response of a nonlinear system with a high degree of accuracy.

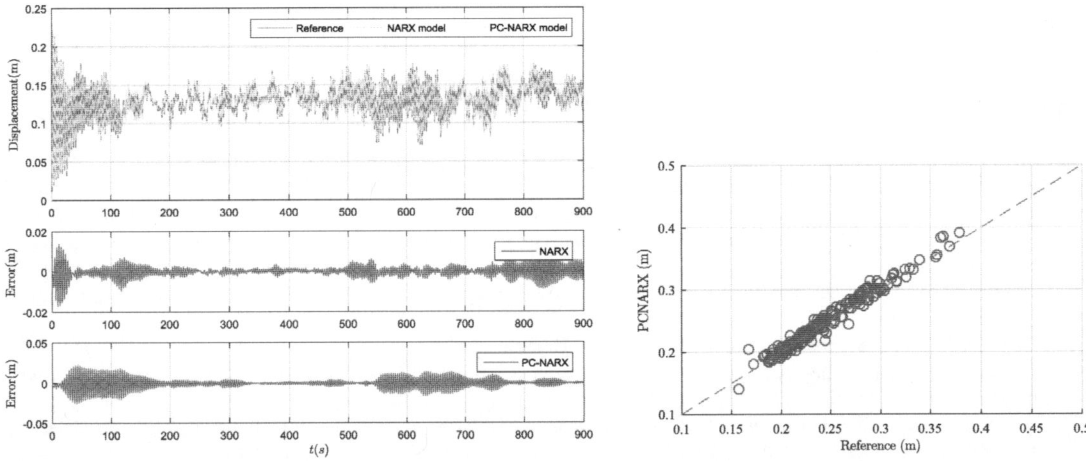

(a) (b)

Figure 1. Duffing oscillator example: (a) simulated displacement time history by means of the NARX and PC-NARX models with respective simulation errors; (b) simulated maximum displacement.

4 CONCLUDING REMARKS

This paper investigated the feasibility of using advanced non-intrusive metamodeling techniques to simulate the dynamic response of wind excited nonlinear systems. To achieve this goal, a recently proposed PC-NARX model is employed to capture the dynamics and propagate the uncertainties in the system. Considering the long duration characteristic of typical wind loads and the limits of applying such model, a refined model with parameters optimized through the simulation error approach is introduced. The case study of the SDOF Duffing oscillator demonstrates the applicability of such model. The efficiency of the metamodel and its ability to accurately simulate the nonlinear dynamic responses indicate the strong potential of the proposed approach wind engineering applications.

5 ACKNOWLEDGEMENTS

Support for this work was provided by the NSF Grant Number CMMI-1462084.

REFERENCES

1 S.M.J. Spence, A. Kareem, A Performance-based design and optimization of uncertain wind-excited dynamic building systems, Engineering Structures,78 (2014) 133-144.

2 S.M.J. Spence, W.C. Chuang, P. Tabbuso, E. Bernardini, A. Kareem, L. Palizzolo, A. Pirrotta, Performance-Based Engineering of Wind Excited Structures: A General Methodology, Structures Congress 2016, February 14-17, Phoenix, AR, USA.

3 S.M.J. Spence, E. Bernardini, A. Kareem, A First Step towards a General Methodology for the Performance-Based Design of Wind-Excited Structures, Structures Congress 2015, April 23-25, Portland, OR, USA.

4 E. Bernardini, S.M.J. Spence, D.K. Kwon, A. Kareem, Performance-based design of high-rise buildings for occupant comfort, Journal of Structural Engineering, 141 (2015).

5 P. Tabbuso, S.M.J. Spence, L. Palizzolo, A. Pirrotta, A. Kareem, An efficient framework for the elasto-plastic reliability assessment of uncertain wind excited systems, Structural Safety, 58 (2016) 69-78
6 Federal Emergency Management Agency (FEMA), Seismic performance assessment of buildings, Volume 1 Methodology (FEMA Publication P-58-1). Washington, DC, 2012.
7 M.D. Spiridonakos, E.N. Chatzi, Metamodeling of dynamic nonlinear structural systems through polynomial chaos narx models, Computers and Structures, 157 (2015) 99-113.
8 C.V. Mai, M.D. Spiridonakos, E.N. Chatzi, B. Sudret, LARS-based ARX PCE metamodel for computing seismic fragility curves, In Joint Engineering Mechanics Institute Conference 2016 (EMI 2016) &Probabilistic Mechanics & Reliability Conference 2016 (PMC 2016), 2016.
9 C.V. Mai, M.D. Spiridonakos, E.N. Chatzi, B. Sudret, Surrogate modelling for stochastic dynamical systems by combining NARX models and polynomial chaos expansions, International Journal for Uncertainty Quantification, 6 (2016) 313-339.
10 S.A. Billings, Nonlinear system identification: NARMAX methods in the time, frequency, and spatio-temporal domains, Wiley, 2013.
11 O. Nelles, Nonlinear system identification, Springer, Berlin, 2001.
12 G. Blatman, B. Sudret, Adaptive sparse polynomial chaos expansion based on least angle regression, Journal of Computational Physics, 230 (2011) 2345-2367.
13 B. Efron, T. Hastie, I. Johnstone, R. Tibshirani, Least angle regression, Annals of Statistics, 32 (2004) 407-499.
14 G. Deodatis, Simulation of ergodic multivariate stochastic processes, Journal of Engineering Mechanics, 122 (1996) 778-787.

13th Americas Conference on Wind Engineering
Gainesville, Florida, USA
May 21 - 24, 2017

On the Validity of Equivalent Static Wind Loads for Tall Building Design

Sejun Park[a], DongHun Yeo[b] and Emil Simiu[c]
[a] *National Institute of Standards and Technology, Gaithersburg, MD, USA,* sejun.park@nist.gov
[b] *National Institute of Standards and Technology, Gaithersburg, MD, USA,* donghun.yeo@nist.gov
[c] *National Institute of Standards and Technology, Gaithersburg, MD, USA,* emil.simiu@nist.gov

ABSTRACT

Modern structural design for wind emerged in the 1960s as a synthesis of several developments, including the estimation of equivalent static wind loads; i.e., of time-invariant loads yielding peak horizontal building displacements equal to those due to fluctuating loads induced by flow normal to a face of isolated tall buildings with a rectangular shape in plan. For this case, fluctuating loads are assumed to be proportional to the fluctuations of the oncoming wind flow; the distribution of the wind loads over the building face is therefore known, and demand-to-capacity indexes (DCIs) used by structural engineers to size structural members, can be determined accordingly. However, the distribution of wind forces with height, and therefore internal forces induced by those forces, could not be determined from theoretical considerations or measurements, for (i) vortex-induced across-wind loading, (ii) wind speeds skewed with respect to a building face, and (iii) buildings with aerodynamics affected by the presence of nearby structures. For those cases, it was necessary to resort to subjective assessments – informed guesses -- by the wind engineer. The development of the high-frequency force balance (HFFB) in the late 1970s was an improvement over earlier approaches, although HFFB measurements of base moments and shears yield no information on the distribution of pressures on the facades. However, the development of the pressure scanner in the 1990s allows the simultaneous measurement of pressure time histories at large numbers of taps and, therefore, the determination of fluctuating wind loads at each floor level. In current practice, those fluctuating wind loads are converted into static wind loads at the floor levels, whose effects on the structure are purportedly equivalent to those induced by their fluctuating counterparts. The question arises whether those static loads produce effects -- other than the base moments and shears, e.g., internal forces and demand-to-capacity indexes, -- approximately equivalent to their counterparts induced by the fluctuating loads. To answer this question, consider a two-dimensional structure with fluctuating wind loads $f_1(t)$, $f_2(t),\ldots,f_n(t)$ at the first, second, …, n-th floor, and assume that all the floors have height h. Obtain the sum $h[f_1(t) + 2f_2(t) + \ldots + nf_n(t)] = M_b(t)$, where $M_b(t)$ is the time series of the base moment induced by those loads. We seek constant floor loads $F_1, F_2, \ldots, F_n$ such that $h(F_1 + 2F_2 + \ldots + nF_n) = \max_t[|M_b(t)|]$. This equation has an infinity of solutions, one of which is $F_j = \alpha_M \max_t [|f_j(t)|]$ (j = 1, 2, …, n) where, due to imperfect cross-correlations between functions $f_j(t)$, $\alpha_M < 1$, However, the equivalence of static and fluctuating forces must apply to all wind effects $R_m(t)$ of interest, where $R_m(t) = f_1(t)r_{m1} + f_2(t)r_{m2} + \ldots + f_n(t)r_{mn}$, and r_{mj} are influence coefficients; i.e., the unknowns F_j must satisfy the system of equations $F_1r_{m1} + F_2r_{m2} + \ldots + F_nr_{mn} = \max_t[R_m(t)]$ (m = 1, 2,…, m_{max}). Since $m_{max} > n$, this is not possible. Using database-assisted design software, DCIs were calculated for 45 members of a 47-story, 160 m tall building with pressure data from the Tokyo Polytechnic University database. Calculations were performed for fluctuating loads and for static loads inducing base moments equal to the peak of the fluctuating base moments. The static loads induced DCIs that differed in 35 % of the cases by 20 % to 60 % from their peak fluctuating counterparts. In half of those cases the DCIs due to static loads were higher, and in half of the cases they were lower than those induced by the time-dependent loads. Possible alternative methods for defining equivalent static loads would similarly be unable to satisfy the requirements imposed by the fact that the number of equations m_{max} exceeds the number of unknowns n.

KEYWORDS

Database-assisted design, equivalent static wind loads, fluctuating wind loads, tall buildings.

Wind Loads on Flat Canopy Roofs

Neelam Rani[a], Ashok Kumar Ahuja[b]

[a]*Indian Institute of Technology Roorkee, Roorkee, India, neelam2310@gmail.com*
[b]*Indian Institute of Technology Roorkee, Roorkee, India, ahujashok@gmail.com*

ABSTRACT: Canopy roofs are used as shelters or sheds for bus and train passengers, entrance to buildings, for car and bicycle parking. Like other low-rise buildings, structural designers refer to relevant standards on wind loads during their design to arrive at correct values of wind forces which will be acting on them for their safety during wind storms. Standards on wind loads of various countries provide information about wind loads on certain types of canopy roofs. But this information is limited for few wind directions, roof slope and single span only. Although some of the researchers have also carried out wind tunnel studies on canopy roofs, yet the available information is not enough for the designers to make safe and economical design of multi-span canopy roofs for wind loads.An experimental study has, therefore, been carried out in an open circuit boundary layer wind tunnel on the models of single and two-span flat canopy roofs. Present paper describes the results of the study. Effects of wind incidence angle and spacing between the canopies are studied by changing their values. Values of wind pressure coefficients are calculated from the values of wind pressures measured at numerous pressure points on upper and lower surfaces of the roof and are reported in the form of pressure contours. Net upward or downward loads are also calculated from the values of wind pressures. It is observed that wind loads on flat canopy roofs are highly influenced by both wind direction as well as distance between them.

KEYWORDS: Flat Canopy Roof, Wind Incidence Angle, Wind Load, Pressure Coefficient.

1 INTRODUCTION

Canopies have different shapes like flat, pitched, mono slope, trough etc Asbestos sheet or galvanized sheet is generally used as roofing material for canopy roofs. Wind hits on both upper and lower roof surfaces of canopy causing their damage under wind storms if they are not properly designed. Codes of different countries [1, 2, 3, 4, 5] give limited information regarding wind pressure distribution on flat canopy roof. Roy[6]carried out research on canopy roof. Singh[7] did research on trough canopy roof with multi span. Netalini[8]studied the effects of different aspect ratios on vaulted canopy roofs is yet not enough. But the available information for flat canopy roofs. An attempt has therefore, been made to generate more information about wind pressure distribution on flat canopy roofs.

2 EXPERIMENTAL PROGRAM

2.1 *Details of model*

The prototype adopted for the study is a building with flat canopy roof and having a plan dimension of 12m×6m. The model is made of material known as perspex sheet at a scale of 1:40. Thickness of the perspex sheet is 3mm and it is fully transparent. The model has plan dimension 300mm×150mm and height of 75mm (Figure 1). Upper roof surface of the model has 35 pressure taps while lower roof surface of the model has 29 pressure taps (Figure 2). In the model, both upper and lower surfaces have same tapping location so that pressure difference can be calculated directly. Six columns are provided with 20mm×20mm cross sectional dimension. Four columns are placed at each corner and two are placed at the center of long edges. Wind direction during the experimental study is varied from 0° to 90° at an interval of 15° (Figure 3) in case of single-span canopy roof.

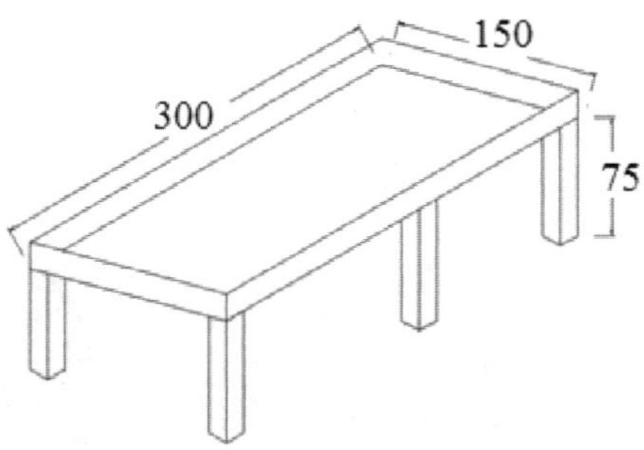

Figure 1. Isometric view of flat canopy roof (All dimension are in mm)

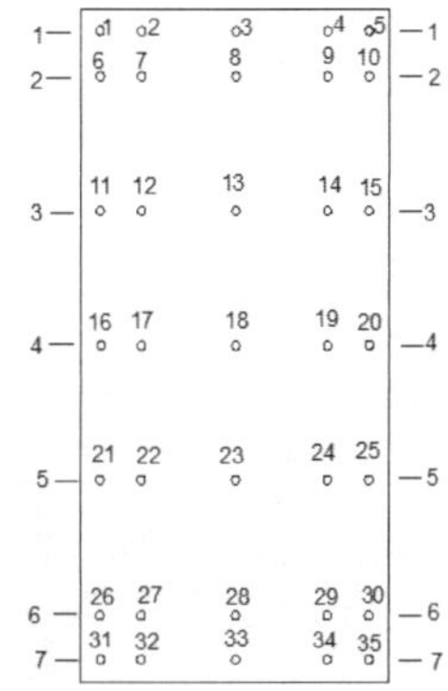

Figure 2.Pressure points on upper roof surface

2.2 *Wind flow characteristics*

Experiments for flat canopy are carried out in boundary layer wind tunnel at IIT Roorkee. The length of the wind tunnel is 15m with cross sectional dimension of 2m×2m. Surface rougness is generated with wooden blocks of 150mm×100mm×50mm size and vortex genrator fins. Models are tested at wind velocity of approximately 10 m/sec measured at 1m height from the floor level of the tunnel. Models are tested under wind simulation condtion corresponding to terrain catgoty 2. Figure 4 shows mean velocity profile inside the tunnel.

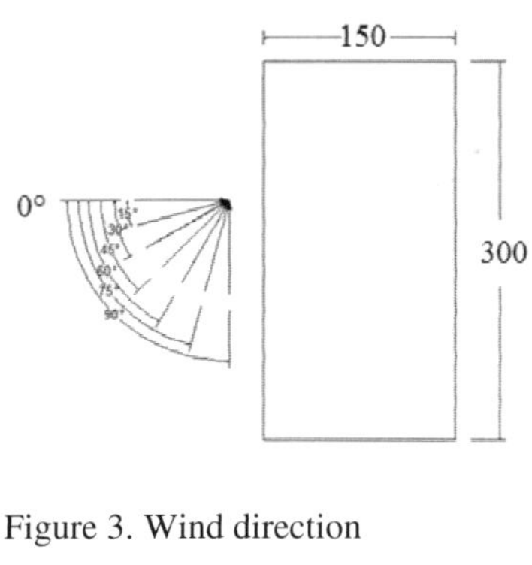

Figure 3. Wind direction

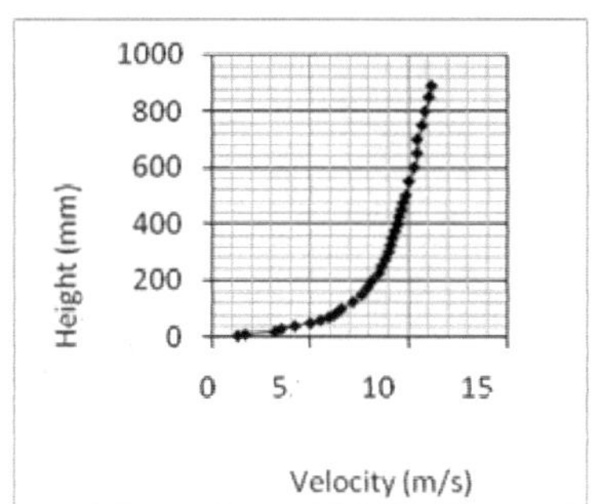

Figure 4. Velocity profile

2.3 *Measurment technique*

Perspex sheet model of the flat canopy roof with reactangular shape is placed at the centre of the turntable in such a way that points 1, 6, 11, 16, 21, 26 and 31 fall on windward edge at 0° wind incidence angle (Figure 5). Turntable is then rotated for 15°. At 0° wind incidence angle, wind flows perpendicular to the longitudinal axis of the model and at 90°, it comes parallel to the longitudinal axis. Wind pressure at each tapping point is mearsured with the help of pressure transduser for 60 seconds.

Plywood model of flat canopy roof without pressure points is placed close to perspex sheet model having pressure points on it (Figure 6) to measure the wind pressure distribution on two-span flat canopy roof. Spacing between two models (S) is varied as 0, 75, 150, 225, 300 mm (Figure 7) to study the effect of spacing between them. However, results of only two cases namely S=0 and 150 mm are reported in the paper due to paucity of space.

Figure 5. Single-span model inside the tunnel

Figure 6. Two span model inside the tunnel

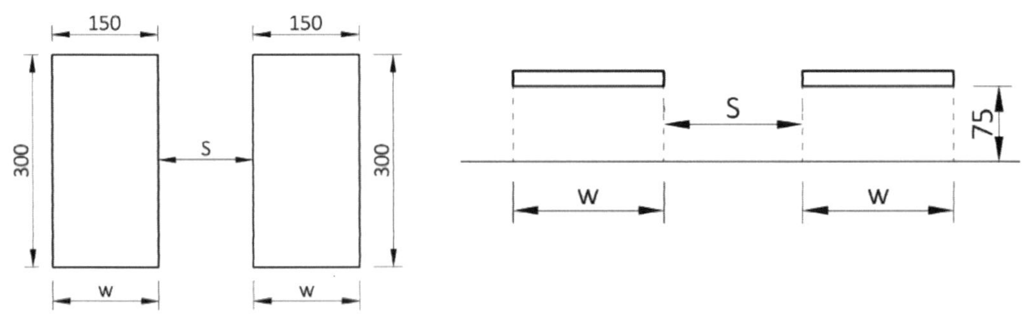

Figure 7. Models of two-span flat canopy roof (All dimensions are in mm)

3 RESULTS AND DISCUSSION

3.1 *Single-span flat canopy roof*

Figure 8 shows the variation of mean pressure coefficients over the upper roof surface in the form of contours. It is seen that pressure coefficients are having suction values and this suction decreases from windward to leeward side in each angle from 0° to 90°. On the upper roof surface, the variation of pressure coefficient lies between the value of -0.4 to -1.7.

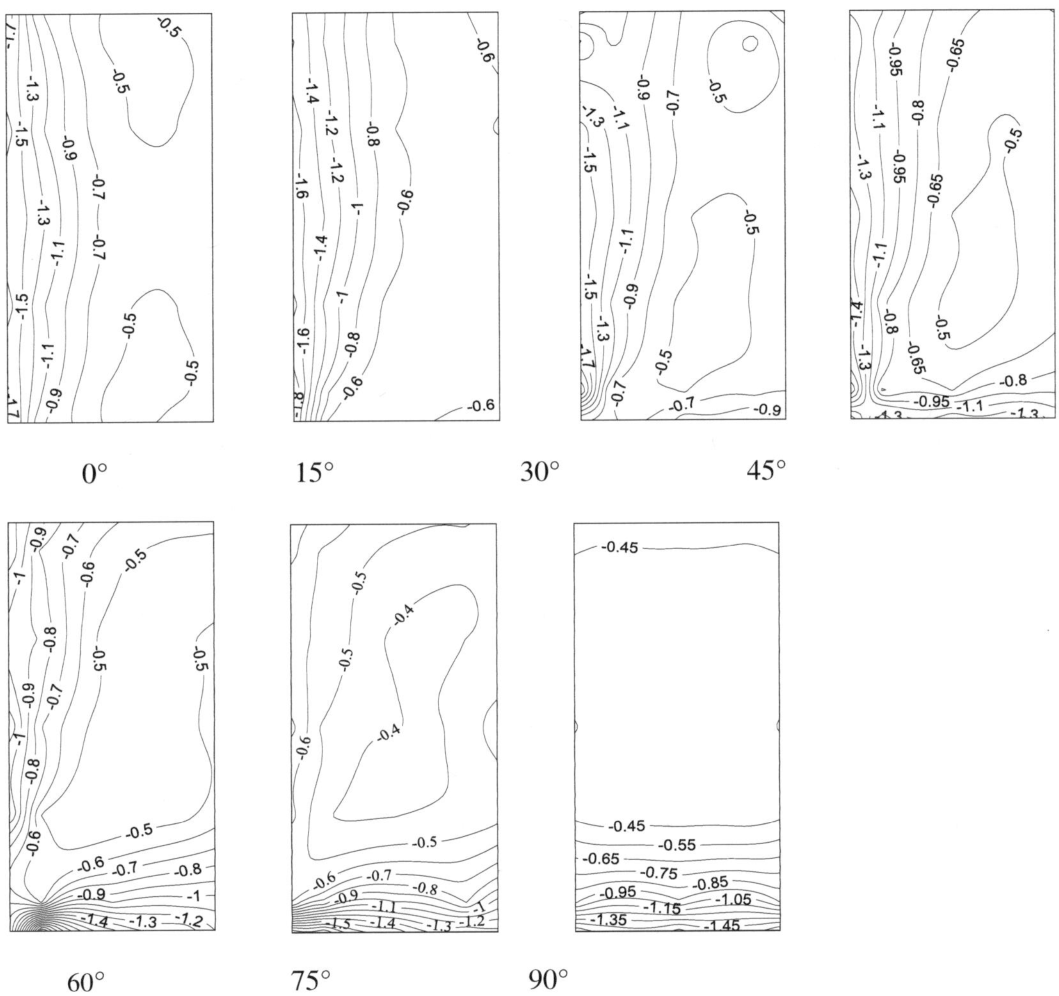

Figure 8. Contours of mean wind pressure coefficients on upper roof surface of single-span flat canopy roof

Figure 9 shows contours on lower roof surface of flat canopy roof. Variation of pressure coefficient for 0^0 and 15^0 wind incidence angle is almost similar. Suction value for lower roof surface is varying between -0.2 to -0.9. Windward edges have high suction values. Suction decreases from windward to leeward side. 90^0 wind incidence angle has higher suction values as compared to other wind incidence angles.

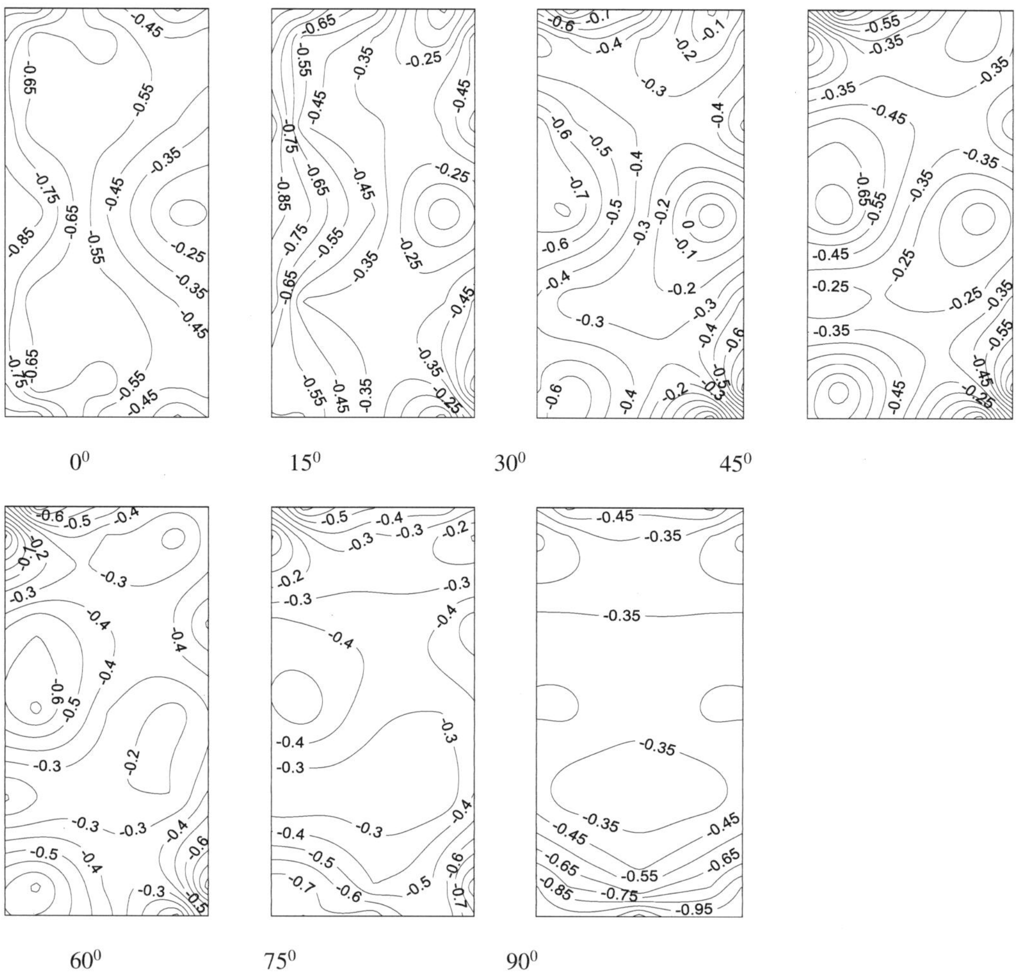

Figure 9. Contours of mean wind pressure coefficients on lower roof surface of single-span flat canopy roof

Figure 10 shows cross sectional variation of mean wind pressure coefficients at various sections on both upper and lower roof surfaces at 0° wind incidence angle. Sectional average values are evaluated from these values and shown in Figure 11. It is seen from Figure 8 and 11 that suction on upper roof surface is more as compared to lower roof surface. Resultant values of mean wind pressure coefficients are shown in Figure 12. It is noticed that entire roof is subjected to almost same value of upward suction.

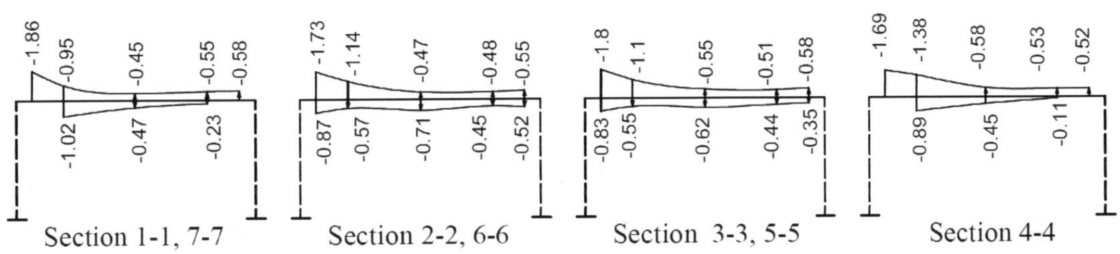

Figure 10. Cross sectional variation of mean wind pressure coefficients at 0° wind incidence angle

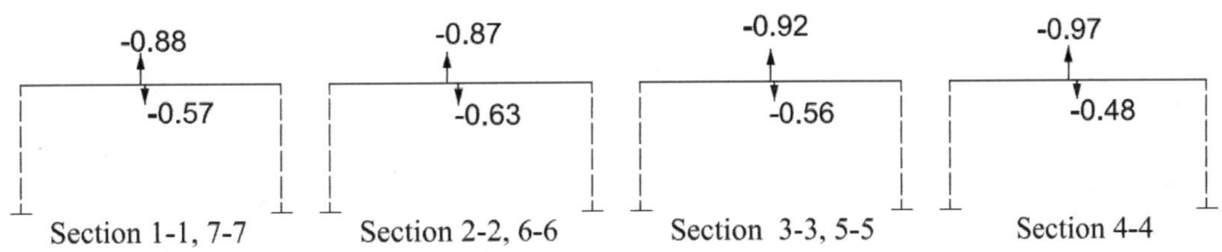

Figure 11. Sectional average values of mean wind pressure coefficients at 0° wind incidence angle

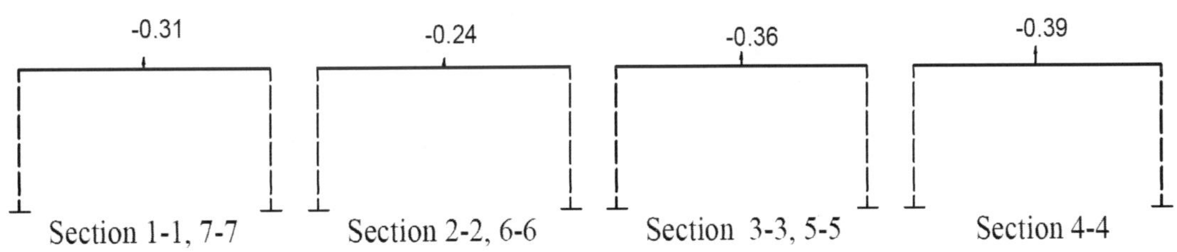

Figure 12. Resultant values of mean wind pressure coefficients at 0° wind incidence angle

3.2 *Two-span flat canopy roof*

3.2.1 *Without spacing*

Pressure coefficients on first span in two-span case without spacing is almost similar as compared to single-span case. On upper roof surface, pressure coefficient is -1.69 in single-span and -1.60 in two-span case on windward side. At lower roof surface again, the values are almost similar (Refer Figure 10 section 4-4 and Figure 15). But if we see second span, the variation of coefficient pressure coefficient is very less.

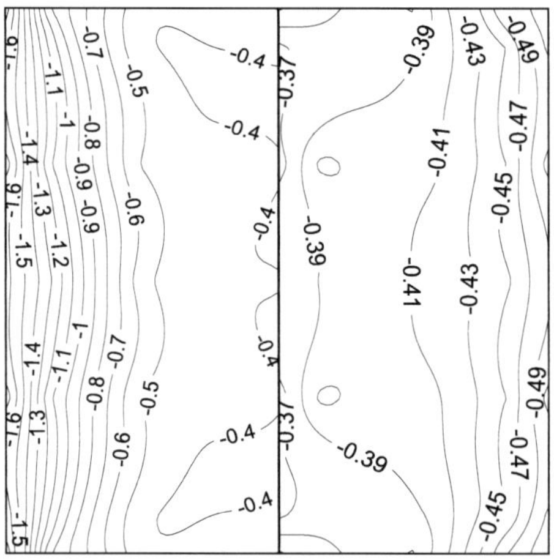

Figure 13. Contours of mean wind pressure coefficients on upper roof surface at 0° wind incidence angle

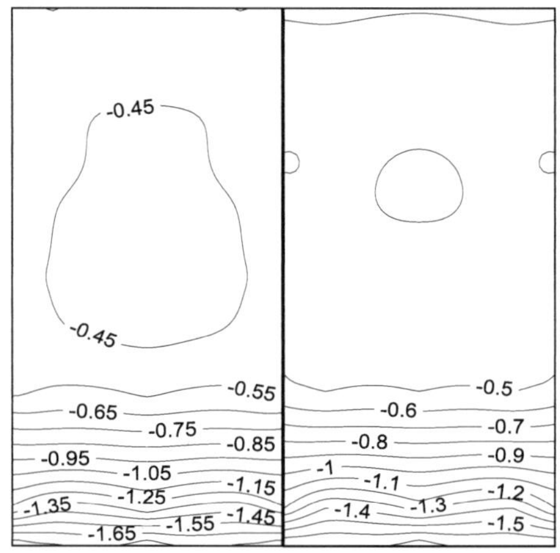

Figure 14. Contours of mean wind pressure coefficients on upper roof surface at 90° wind incidence angle

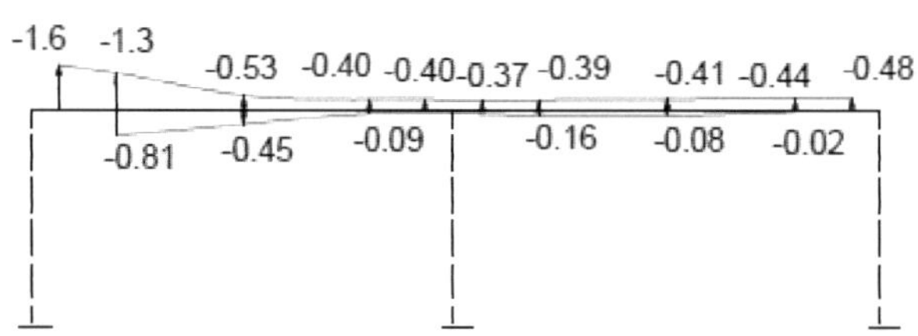

Figure 15. Cross sectional variation of mean wind pressure coefficients at 0° wind incidence angle for section 4-4

3.2.2 *Flat canopy roof with 150 mm spacing*

Figure 16 shows the variation of mean wind pressure coefficient on upper roof surface on the span with 150 mm spacing. Cross sectional variation of mean wind pressure coefficients are shown in Figure 17.

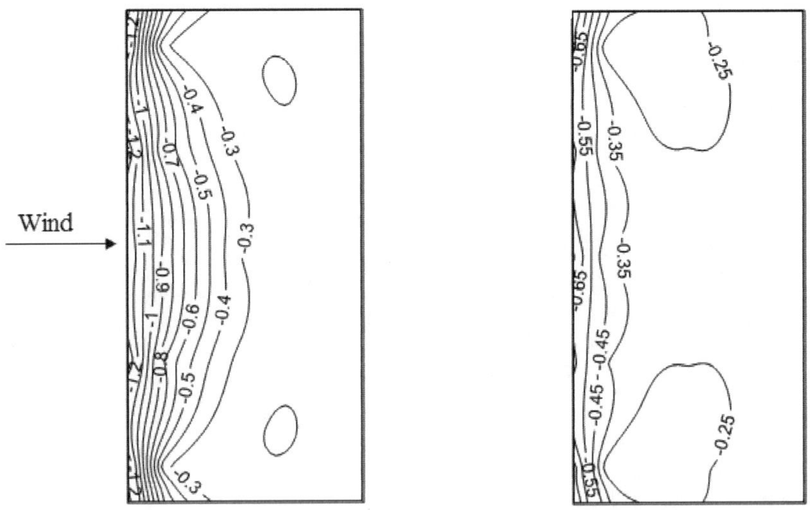

Figure 16. Contour of mean wind pressure coefficients on upper roof surface at 0° wind incidence angle

Figure 17. Cross sectional variation mean wind oressure coefficients at section 4-4 for 0° wind incidence angle

4 CONCLUSIONS

- Wind pressure distribution on flat canopy roof is highly affected by wind incidence angle
- Both upper and lower surfaces of flat canopy roof are subjected to suction at all wind incidence angles.
- Resultant wind load is upward when wind hits the roof perpendicular to its length.

5 ACKNOWLEDGEMENTS

The work presented in this paper is part of the research work being done by the first author for her Ph.D degree under the supervision of second author.

6 REFERENCES

1. AS/NZS: 1170.2, 2002, Structural Design Actions, Part-2: Wind Action.
2. ASCE, 2002, Minimum Design Loads for Building and Others Structures.
3. BS: 636999, 1995, Loading for Building: Part-2: Code of Practice for Wind Loads.
4. EN: 1991-1-4, 2005,
5. IS: 875-Part-3, 2015, Code of Practice for Design Loads (other than Earthquake Loads) for Buildings and Structures- Wind Loads.
6. A. K. Roy, 2009,Wind Loads on Canopy Roof Building, Ph.D Thesis, Dept. of Civil Engineering, Indian Institute of Technology, roorkee, Roorkee, India."
7. Singh, 2015, Wind Load on Canopy roof, M.Tech Thesis, Dept. of Civil Engineering, Indian Institute of Technology, Roorkee, Roorkee, India.
8. Netalini, Morel, Netalini, 2013,"Mean Loads on Vaulted Canopy Roofs", Journal of Wind Engineering and Industrial Aerodynamics, Vol. 119, 2013, pp 102-113.

The 13th Americas Conference on Wind Engineering (13ACWE)
Gainesville, Florida USA, May 21-24, 2017

Influence of terrain on the uplift loads on low-rise building roofs

Fakhruzzaman Arif[a], Gregory A. Kopp[b]

[a]*Boundary Layer Wind Tunnel Laboratory, University of Western Ontario, London, Ontario, Canada*
[b]*Boundary Layer Wind Tunnel Laboratory, University of Western Ontario, London, Ontario, Canada*

ABSTRACT: During extreme wind events, roof failure may lead to damage of the whole structure. In order to alleviate the effect, wind loads on roofs have been extensively investigated. This research aims to quantify the large values of uplift on the roofs of low-rise buildings with consideration of the effects of terrain. Pressure measurements as well as wind speed data were taken at the Boundary Layer Wind Tunnel of the University of Western Ontario to determine the most crucial scenario of the wind loading on the roof for various upstream boundary layer conditions. Besides, plan dimensions and eave heights of the building were also varied. Generally, corner vortices play a vital role in generating large uplift force on the roof surface in flat terrain. Moreover, separation bubble at the leading edge of low-rise buildings is also very significant to take into consideration. Our objective is to examine these points in terms of area-averages used in design. The results suggest that in case of flat terrain, corner vortices have the most dominant impact on the roofs among all angles of attack, whereas this effect reduces quickly in suburban terrain. In addition, peak pressure coefficients, GC_p, are dependent on the turbulence intensity and length scales, which may be important in some design scenarios.

KEYWORDS: Wind loads, Low-rise buildings, Terrains, Turbulence.

1 INTRODUCTION

Peak pressures acting on the roofs of low-rise buildings are important to study as these may cause failure of the roofs during windstorms. Numerous studies were conducted to understand the mechanism of mean pressures in the separated shear region. Most of these studies were performed for two-dimensional bluff bodies placed in uniform upstream flows, for example, Castro and Robins [1], Tieleman et al. [2] and Kim et al. [3]. Hillier and Cherry [4], Bienkiewicz and Sun [5] and, Saathoff and Melbourne [6] explained that fluctuating roof pressures depend on turbulence intensity and integral scale. However, only a few studies in literature are found to address peak pressures on roof surfaces where the buildings are exposed to various turbulent boundary layers. Melbourne [7] and, Saathoff and Melbourne [6], [8] investigated peak suctions over a blunt flat plate, while Pratt and Kopp [9] examined over the roof surface of a low-rise building model for open terrain. Case and Isyumov [10] found that larger wind loads occurred in open terrain than that of suburban one due to changes in the wind speeds. In order to conduct experiments in wind tunnel for different upstream boundary layer conditions, detailed explanations and requirements were described in Tieleman [11]. In addition, separation bubble is known to depend significantly on turbulence levels (i.e., the terrain condition), as shown by Akon and Kopp [12]. Moreover, as found by Richards and Hoxey [13] and Banks et al. [14], corner vortices have the most dominant impact on the roof surface. However, different approaches are used by various wind tunnel labs when there are scale mismatches in case of large model scales, which lead to different peak pressures. The aim of this study is to examine how magnitudes and critical wind directions of the point and area-averaged (e.g. Lin et al. [15] and Peterka et al. [16]) pressure coefficients change with terrain differences and building size.

2 EXPERIMENTAL APPROACH

2.1 *Building Models*

Wind tunnel tests were performed on several buildings, one of which was a scaled model of Texas Tech University “WERFL” building (TTU building). A detailed description of the TTU building can be found in Levitan and Mehta [17], [18]. On the roof surface, 950 pressure taps were uniformly distributed in 25 rows and 38 columns in a removable panel. Model-scale plan dimensions are shown in Figure 1. Figure 1(a) shows the arrangement of the high-resolution pressure tap layout. Building 2 has 3 times height (H2) of Building 1, which has similar height of TTU model (H1).

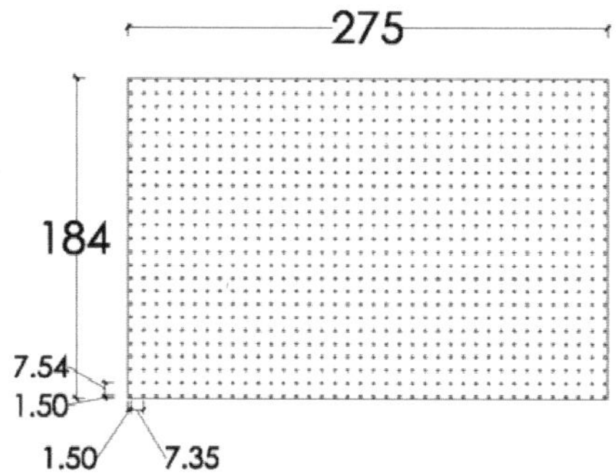

(a) Building 1 (B1)

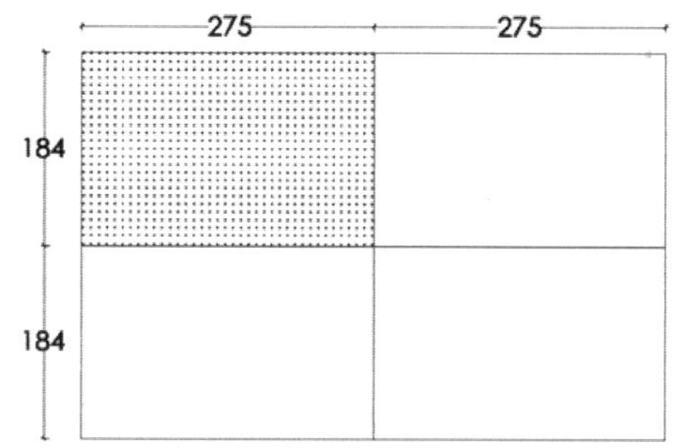

(b) Building 2 (B2)

Figure 1. Two different plan shapes of the model: (a), (b), where 950 pressure taps were uniformly spaced in one block and dummy blocks were attached for another.

2.2 *Terrain Simulation*

The experiments were carried out for six distinct upstream boundary layer conditions. There were three different surface roughness element arrangements indicated as 1, 2, and 3. These three configurations having 0.38 m tall barrier at the entrance of the upstream are denoted as 1L, 2L, and 3L, whereas the three having no barrier are denoted as 1S, 2S, and 3S. That means the number represents the roughness length of the terrain whereas L refers to the existence of the barrier with larger turbulence length scales and S refers to no barrier condition with smaller turbulence length scales. The different characteristics of six upstream boundary layer conditions, such as mean velocity profile, streamwise turbulence intensity, and integral scale, performed in Akon and Kopp [12], were similar in this experiment. Figure 2 represents the mean streamwise velocity profile, U/U_{H1} with respect to the distance from the ground, $y/H1$. It indicates that mean velocity profiles near the roof surface are almost same, irrespective of the terrain roughness. However, the turbulence levels and integral scales are very different.

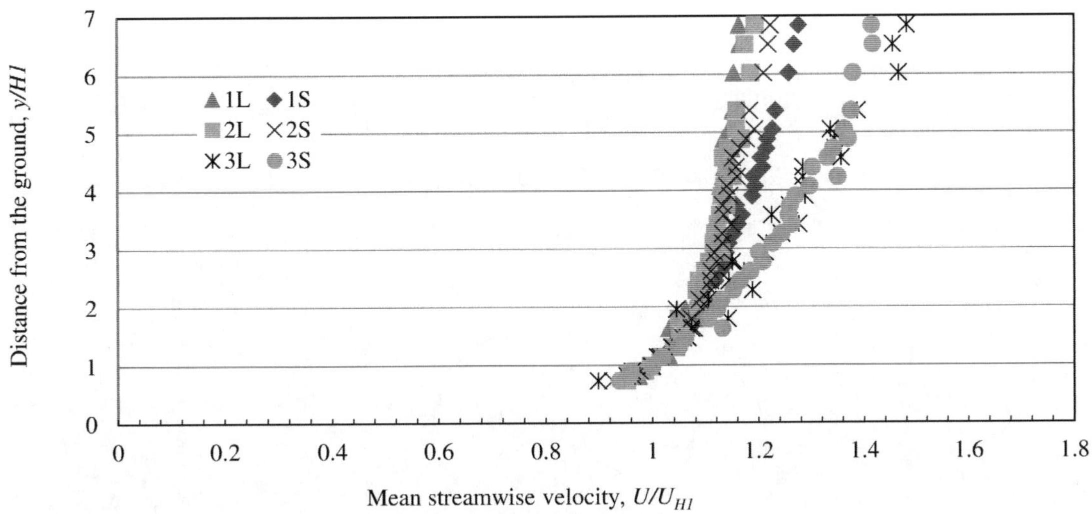

Figure 2. Mean velocity profile for various upstream conditions.

Table 1 refers to various features of six atmospheric upstream cases for two different heights, H1 and H2. To obtain desired upstream conditions, the roughness lengths (y_o) were varied in height from the wind tunnel floor surface, along with other turbulence-generating components. Here, turbulence intensity is the ratio of velocity fluctuation to mean velocity, whereas turbulence length scale, L_x, is a measure of the size of large energy containing eddies.

Table 1. Properties of the upstream boundary layer simulations

Terrain	Barrier (m)	Roughness length, y_o (m)	Turbulence intensity, I_u		Integral scale, L_x	
			$y= H1$	$y= H2$	$L_x/H1$	$L_x/H2$
1L	0.38	0.00013	14	10	13	4
1S	0	0.00014	13	9	6	2
2L	0.38	0.00014	17	13	11	5
2S	0	0.00027	17	13	8	2
3L	0.38	0.0011	27	25	12	3.5
3S	0	0.0014	26	22	7	3

2.3 *Testing Parameters*

Pressure data were obtained at 625 Hz for 120 seconds at each angle of wind in the experiment. As a consequence, for each pressure tap, time histories of 75000 pressure coefficients were taken at each angle of wind direction. Pitot-static tubes were fixed at a standard height of 147 cm from the wind tunnel floor where wind speed was measured 10.35 m/s approximately. Wind speed data were taken by two cobra probes at roof height (H1) and double of height (H1). The Reynolds number for the experiments, based on the mean wind speed at the mean roof, are in the range from 35500 to 117000. Synchronization of pressure and wind speed was done in the experiment. Figure

3 displays the photograph of the experimental setup of a model during the experiment in wind tunnel.

Figure 3. Photograph of the experimental setup of a model in wind tunnel.

2.4 *Analysis Process*

In this analysis, an area-averaging approach was implemented to investigate the pressure distribution on the roof surface of low-rise buildings using:

$$C_P = \sum_{i=1}^{m}\sum_{j=1}^{n} C_{pij}\frac{a_{ij}}{A} \tag{1}$$

In order to estimate reliable peak pressures from the wind tunnel data, extreme value analysis is required. In this analysis, the Lieblein BLUE method is applied to determine the extreme value distribution. A detailed explanation about this method was described in Lieblein [19]. The pressure coefficients, C_{P_H} which are referenced to the mean dynamic pressure at the roof height of the model, can be obtained from $C_{p_{measured}}$ as:

$$C_{P_H} = C_{p_{measured}}\left(\frac{\bar{V}_{pitot}}{\bar{V}_H}\right)^2 \tag{2}$$

where $\bar{V}_H$ is the mean velocity at the model height. Equation 2 provides least variability, which was discussed in Ho et al. [20]. Assuming that the 3-sec gust speed is approximately $\bar{V}_H + 3\sigma$, where σ is the standard deviation of the streamwise velocity fluctuations. Then the ASCE 7 – equivalent pressure coefficient is given by:

$$GC_p = C_{p_{measured}} \left(\frac{\overline{V}_{pitot}}{\overline{V}_H+3\sigma}\right)^2 \tag{3}$$

3 RESULTS

3.1 *Worst (enveloped) GC_p values*

Figure 4 indicates the worst suction GC_p values enveloped over all wind directions on the roof surfaces of B1H1 and B2H2 model for 1S and 1L upstream boundary layer conditions. One can see that the area-averaged pressure coefficients are dependent on turbulence length scales. The magnitude of suction GC_p is higher in case of 1L than 1S for both building models. It also depicts that peak pressure coefficients vary not only in the magnitude but also in the spatial patterns with respect to turbulence length scale.

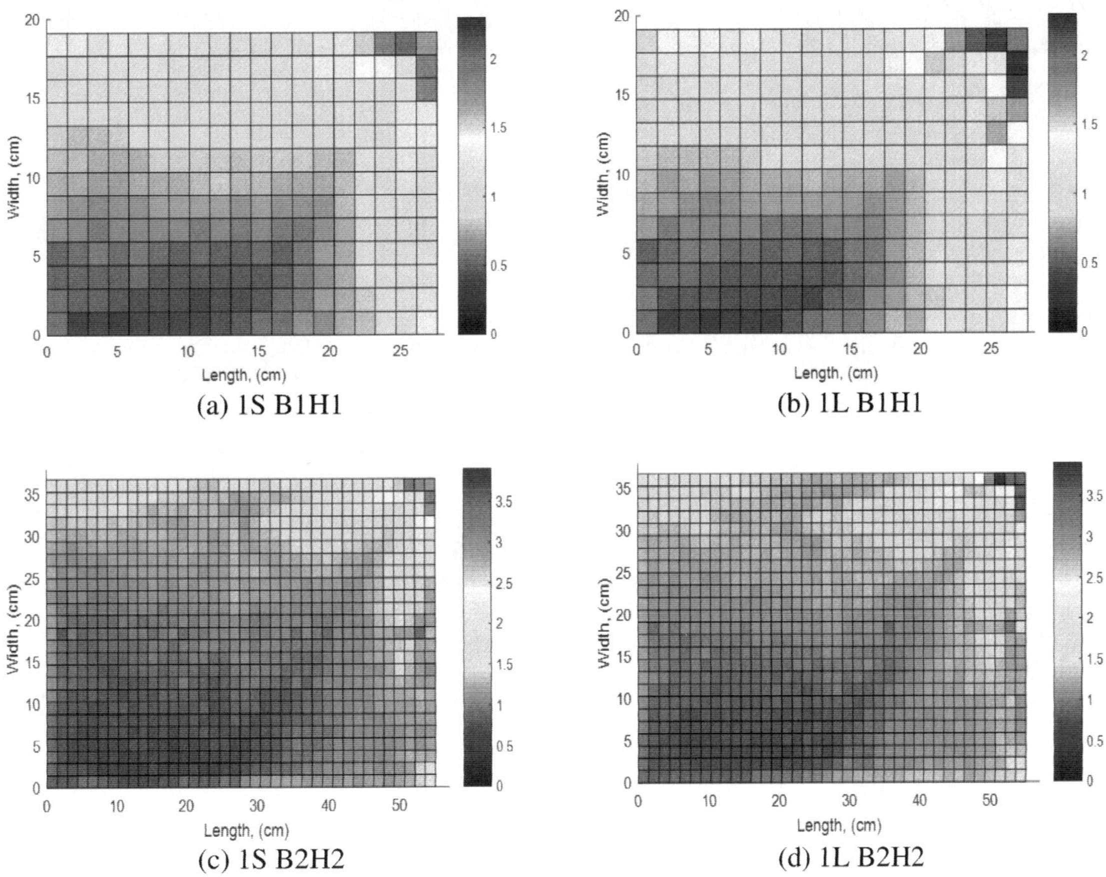

Figure 4. Comparison of worst suction GC_p over B1H1 and B2H2 models for different length scales.

Figure 5 shows the suction pressure coefficients, GC_p on the roof of B1H1 and B2H2 buildings for same suburban terrain. Here, the magnitude of suction pressures are compared from figure 4a, 4c and, 5. It can easily be concluded that the areas of the roof surface having more suction loads are larger in case of higher buildings than that of lower ones for both flat and suburban terrains.

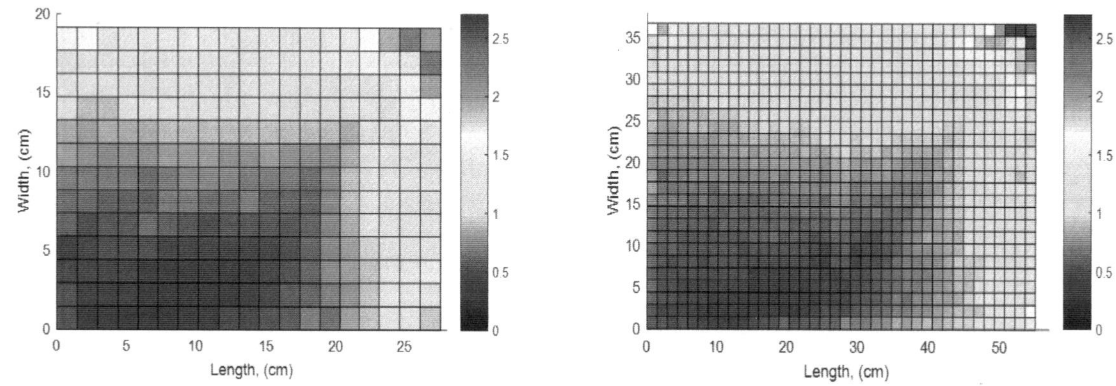

(a) 3L B1H1 (b) 3L B2H2

Figure 5. Comparison of worst suction GC_p over B1H1 and B2H2 models for different terrains.

3.2 *Critical Wind directions*

The critical wind conditions for the 1S and 3L upstream conditions over the roof surfaces of B1H1 and B2H2 models are plotted in Figure 6. It provides that corner vortex has an impact on the roofs for both cases. From the figure, it is clear that in case of flat terrain, corner vortices have the most crucial effect on the roof surface among all angles of attack. But this effect becomes weaker quickly in suburban terrain due to the effect of bubble separation. Besides, it depicts that excluding the area of corner vortices, worst suction GC_p occurs along the longer wall at windward direction, whereas it happens along the shorter wall at wind direction normal to the shorter one. In addition, the zone controlled by the corner vortices has the most dominant impact over the roof surface. However, normal winds are important over the bulk of the roof for the building model B2H2.

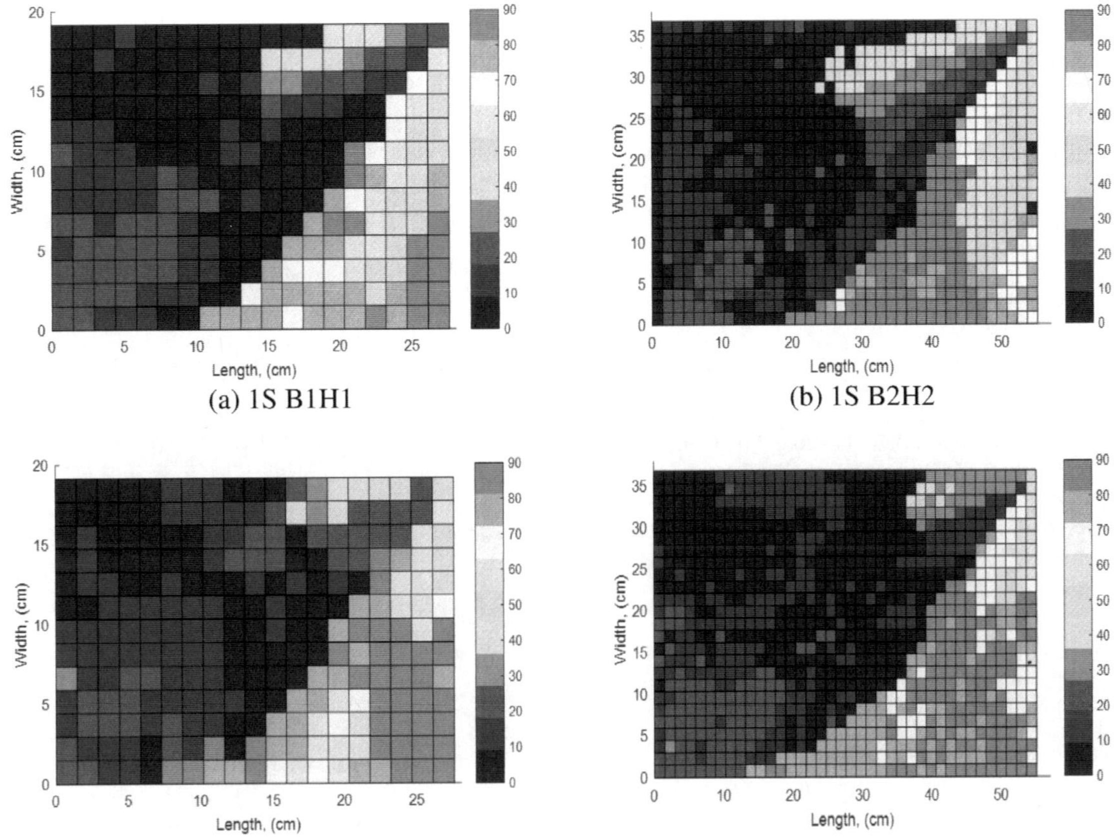

(a) 1S B1H1 (b) 1S B2H2

(c) 3L B1H1 (d) 3L B2H2

Figure 6. Critical wind direction over B1H1 and B2H2 models from 0 to 90 degree.

4 CONCLUSIONS

Surface pressure coefficients and wind speed data were taken to obtain the effect of terrain on the uplift loads on roof surface. Results indicate that area-averaged peak pressure coefficient, GC_p is sensitive to turbulence length scales and intensity. Here, L tends to larger values of peak pressure, GC_p than S. Besides, corner vortex has the most critical effect on the roof surface in flat terrain, while this impact decreases in suburban terrain for both building models. Moreover, larger peak suctions are observed for higher buildings compared to smaller ones.

5 ACKNOWLEDGEMENTS

This wok was funded by the Natural Sciences and Engineering Research Council (NSERC) of Canada under the Collaborative Research and Development program and by the Institute for Catastrophic Loss Reduction (ICLR). The ongoing support and encouragement provided by Mr. Paul Kovacs is gratefully acknowledged.

6 REFERENCES

1 I. P. Castro and A. G. Robins, The flow around a surface-mounted cube in uniform and turbulent streams, J. Fluid Mech., 79 (1977) 307–335.

2 H. W. Tieleman, Z. Ge, M. R. Hajj, and T. A. Reinhold, Pressures on a surface-mounted rectangular prism under varying incident turbulence, J. Wind Eng. Ind. Aerodyn., 91 (2003) 1095–1115.

3 K. C. Kim, H. S. Ji, and S. H. Seong, Flow structure around a 3-D rectangular prism in a turbulent boundary layer, J. Wind Eng. Ind. Aerodyn., 91 (2003) 653–669.

4 R. Hillier and N. J. Cherry, The effects of stream turbulence on separation bubbles, J. Wind Eng. Ind. Aerodyn., 8 (1981) 49–58.

5 B. Bienkiewicz and Y. Sun, Local wind loading on the roof of a low-rise building, J. Wind Eng. Ind. Aerodyn., 45 (1992) 11–24.

6 P. J. Saathoff and W H Melbourne, Effects of free-stream turbulence on surface pressure fluctuations in a separation bubble, J. Fluid Mech, 337 (1997) 1–24.

7 W. Melbourne, Turbulence effects on maximum surface pressures-a mechanism and possibility of reduction, Proc. 5th Int. Conf. on Wind Engineering, Fort Collins, CO, 1979, pp. 541-551.

8 P. J. Saathoff and W. H. Melbourne, The generation of peak pressures in separated/reattaching flows, J. Wind Eng. Ind. Aerodyn., 32 (1989) 121–134.

9 R. N. Pratt and G. A. Kopp, Velocity field measurements above the roof of a low-rise building during peak suctions, J. Wind Eng. Ind. Aerodyn., 133 (2014) 234–241.

10 P. C. Case and N. Isyumov, Wind loads on low buildings with 4 : 12 gable roofs in open country and suburban exposures, J. Wind Eng. Ind. Aerodyn., 77–78 (1998) 107–118.

11 H. W. Tieleman, Wind tunnel simulation of wind loading on low-rise structures: A review, J. Wind Eng. Ind. Aerodyn., 91 (2003) 1627–1649.

12 A. F. Akon and G. A. Kopp, Mean pressure distributions and reattachment lengths for roof-separation bubbles on low-rise buildings, J. Wind Eng. Ind. Aerodyn., 155 (2016) 115–125.

13 P. J. Richards and R. P. Hoxey, Wind loads on the roof of a 6 m cube, J. Wind Eng. Ind. Aerodyn., 96 (2008) 984–993.

14 D. Banks, R. N. Meroney, P. P. Sarkar, Z. Zhao, and F. Wu, Flow visualization of conical vortices on flat roofs with simultaneous surface pressure measurement, J. Wind Eng. Ind. Aerodyn., 84 (2000) 65–85.

15 J. X. Lin, D. Surry, and H. W. Tieleman, The distribution of pressure near roof corners of flat roof low buildings, J. Wind Eng. Ind. Aerodyn., 56 (1995) 235–265.

16 J. A. Peterka, N. Hosoya, S. Dodge, L. Cochran, and J. E. Cermak, Area-average peak pressures in a gable roof vortex region, J. Wind Eng. Ind. Aerodyn., 77–78 (1998) 205–215.

17 M. L. Levitan and K. C. Mehta, Texas Tech field experiments for wind loads part 1: building and pressure measuring system, J. Wind Eng. Ind. Aerodyn., 43 (1992) 1565–1576.

18 M. L. Levitan and K. C. Mehta, Texas tech field experiments for wind loads part II: meteorological instrumentation and terrain parameters, J. Wind Eng. Ind. Aerodyn., 43 (1992) 1577–1588.

19 J. Lieblein, Efficient Methods of Extreme-Value Methodology, NBSIR 74-602, Washington D.C., 1974, p. 1–24.

20 Ho, T.C.E., Surry, D., Morrish, D., Kopp, G.A., The UWO contribution to the NIST aerodynamic database for wind loads on low buildings: Part1. Archiving format and basic aerodynamic data, J. Wind Eng. Ind. Aerodyn., 93 (2005) 1–30.

Experimentally validated nonlinear modeling of low-rise building response under wind loading

Jing He [a], Fang Pan [b], and C.S. Cai [a]

Arindam Chowdhury [c] and Filmon Habte [c]

[a] Louisiana State University, Baton Rouge, Louisiana, USA (cscai@lsu.edu)

[b] Southwest Research Institute, San Antonio, Texas, USA (fang.pan@swri.org)

[c] Florida International University, Miami, (USA) (chowdhur@fiu.edu)

ABSTRACT

A nonlinear finite-element model is presented to study the performance of light-frame low-rise buildings under hurricane wind loads. This comprehensive model can accommodate various materials and structural connections with mechanics-based load-deformation characteristics such as framing-to-framing nail connections. The applied wind pressure data is measured from the Wall of Wind (WOW) Experimental Facility (EF) at Florida International University. The WOW experiments also helped in obtaining deflections and strains at key locations of the building system. The numerical modeling is used to assess the performance of the structure, specifically the most vulnerable parts such as roof sheathing and roof-to-wall as well as other framing-to-framing connections. The first failure conditions are obtained with respect to specific components such as the sheathing and nail. The predicted structural responses showed reasonable agreement when compared to the experimental measurements. The validated numerical model will be used for further progressive failure analysis to predict the nonlinear performance of low-rise building under extreme wind conditions.

KEYWORDS

low-rise building, wind, building, finite element, failure, nonlinear

Wind Pressure Characteristics on Gable Roofs

Xin Dong[a], Changke Jiao[b]

[a] *Tongji Architectural Design (Group) Co. Ltd, Room 533, No. 1230, Siping Road, Shanghai, China,* dongxinseu@163.com
b Engineering General Institute of Shanghai Construction Group, No. 700, Xinjun Road, Shanghai, China, jiaochk@126.com

ABSTRACT

Through Aerodynamic Database of Tokyo Polytechnic University, wind pressures on gable roofs with different pitch angles were investigated. Firstly, mean and fluctuating pressures were presented under normal and oblique flow. Secondly, peak factors and statistic peak suctions of each panel were calculated by Hermite Model. Thirdly, provisions of wind pressures on gable roof claddings from different country codes were discussed. Results indicated that gable roofs could be classified as low pitch angle ($0° \leqslant \theta \leqslant 15°$), medium pitch angle ($15° < \theta \leqslant 30°$) and high pitch angle ($30° < \theta \leqslant 45°$), according to the similar wind pressure distribution and statistic peaks. For most pitch angles, the unfavorable wind directions are oblique and the most susceptible panel is near the windward corner. Considering the difference of design wind pressures, gable roofs with low pitch angles can be divided into corner, longer edge, shorter edge and interior, while ridge and connection of shorter edge and ridge should be further divided as separated regions on gable roofs with medium and high pitch angles.

KEYWORDS

gable roofs, cladding, wind pressure, conical vortices, peak factor

Pressure coefficients on components and cladding from wind tunnel tests

Dat Duthinh [a], Matthew L. Gierson [b], and Brian M. Phillips [c]

[a] *National Institute of Standards and Technology, Gaithersburg, MD. USA,*
[b] *Dept. of Civil Eng., University of Maryland, College Park, MD. USA,*
[c] *Dept. of Civil Eng., University of Maryland, College Park, MD. USA,*

ABSTRACT: External pressure coefficients specified in the ASCE 7-10 Standard (2010, Reston, VA) used to determine design wind pressures for the components and cladding of buildings, are developed from wind tunnel test data that date back 30 to 50 years. In recent decades, advances in pressure measurement and computer technology have made it possible to obtain simultaneous pressure records, with high sampling rates, at many more wind tunnel pressure taps than was the case in the past. This paper presents two methods for analyzing wind pressure data on cladding and components of low-rise buildings. The aerodynamic force is obtained by aggregating values from pressure time series measured in recent wind tunnel tests by the University of Western Ontario (UWO, www.itl.nist.gov/div898/winds/homepage.htm) and Tokyo Polytechnic University (TPU, http://www.wind.arch.t-kougei.ac.jp/info_center/windpressure/lowrise/mainpage.html) at adjoining pressure taps by their respective tributary areas. For the regularly spaced taps of UWO, this operation is carried out for all sums of tributary areas that make up rectangles with aspect ratios not exceeding four. For the irregularly spaced taps of TPU, Voronoi diagrams are used to assign tributary areas, then user-defined grids of various size and shape are placed at various offsets over the building surface to perform area-averaging of the pressure time series. Considering all wind directions for which measurements are obtained in the wind tunnel, the peak pressures are determined assuming a Gumbel distribution, and are extrapolated to a standard storm duration. The external peak pressure coefficients are then plotted as functions of their corresponding area for various zones of the building enclosure to produce envelope plots similar to the ASCE 7-10 specifications on components and cladding. Results for four gable buildings analyzed show that the current ASCE 7-10 specifications severely underestimate the external pressure coefficients for components and cladding of low-rise buildings.

KEYWORDS: Components and cladding, gable roofs, low-rise buildings, walls, wind tunnel.

1 INTRODUCTION

Current ASCE 7-10 [1] (American Society of Civil Engineers 2010) specifications of wind pressures on low-rise buildings are based on data that are thirty to forty years old (see references in the next section). Advances in computer technology allow simultaneous recording of many more pressure taps (on the order of hundreds) than was possible a few decades ago. Furthermore, wind tunnel test measurements have become publicly available for considerably more building geometries. The most well-referenced publicly available sources of data are the National Institute of Standards and Technology [2] (NIST 2004) database, which covers tests conducted at the University of Western Ontario [3, 4, 5] (UWO, Ho et al. 2003, 2003, 2005), and the Tokyo Polytechnic University [6] (TPU, Tamura 2012) database.

The main purpose of this study is to establish clear and reproducible methodologies for using general wind tunnel test databases, namely the NIST-UWO database [2] (part A) and the TPU database [6] (part B), to calculate peaks of wind pressure over different size areas of building surfaces. The methodologies are essential components to updating wind pressure coefficients for the design of components and cladding, considering the deficiencies in current wind load specifications. Part C discusses uncertainties and compares the two methodologies explained in A and B.

2 BACKGROUND

Current ASCE 7-10 [1] wind load provisions on components and cladding of low-rise buildings are largely based on the work of Stathopoulos [7] (1979), supplemented by Meecham [8] (1988) for hip roofs, and updated by Stathopoulos et al. [9] (1999).

Wind pressure on building enclosures continues to be an active area of research. Gavanski et al. [10, 11] (2013 a, b) tested 1:50 scale models of houses with gable or hip roofs and concluded that roof slope and upstream terrain had the most significant effect on wind loads acting on roof sheathing, and that ASCE 7-10 standards substantially underestimated wind pressures, more so for gable roofs than for hip roofs. Pressure coefficients were averaged over the area of sheathing panels 4 ft × 8 ft (1.2 m × 2.4 m) and the Best Linear Unbiased Estimate (BLUE, Lieblein 1974) [12] method was used to determine the peak pressure coefficients. Vickery et al. (2013) [13] analyzed the same data and found that, for gable roof zone 1 (interior, Fig. 1), ASCE 7-10 wind pressure provisions were "woefully inadequate" for effective areas less than 50 ft^2 (5 m^2), but could be reduced for areas greater than 100 ft^2 (9 m^2); for zone 2 (edges), ASCE 7-10 provisions underestimated roof pressure for areas less than 100 ft^2 (9 m^2), but were adequate for larger areas. For zone 3 (corners), Vickery et al. (2013) [13] suggested that the ASCE 7-10 minimum area of 10 ft^2 (1 m^2) was too large because of the large pressure gradients at roof corners. Gavanski and Uematsu (2014) [14] analyzed wind pressure on walls of low-rise buildings and concluded that zoning for positive pressure is unnecessary as pressure was approximately uniform over the entire wall surface. For roof slopes less than 14°, positive wall pressures increased slightly as the roof slope decreased. For negative pressures, larger suctions were observed at the wall edges than in the interior, as expected, and the width of these higher suction zones increased with roof slope and eave height. ASCE 7-10 provisions underestimated positive pressures for almost all effective areas by a factor of 1.2 to 1.5. A similar tendency was found for negative pressures, though not for all building geometries. In fact, for several geometries, ASCE 7-10 provisions overestimated wall pressures on large effective areas.

PART A

Under the sponsorship of the NIST, the University of Western Ontario (UWO) [2, 3, 4, 5] conducted wind tunnel tests on 37 different buildings of one basic type: rectangular in plan; gable roofs with various slopes and no overhang; and open country and suburban terrain conditions (Ho et al. 2003, 2003, 2005) [3, 4, 5]. The roof slope ranged from ¼:12 (1.2°) to 6:12 (26.6°), the building width from 30 ft (9.1 m) to 160 ft (48.8 m), the building length from 45 ft (13.7 m) to 200 ft (61.0 m), and the eave height from 12 ft (3.7 m) to 40 ft (12.2 m). These test results underpinned the Database-Assisted Design (DAD) method for Main Wind Force Resisting Systems (Main and Fritz 2006, Habte et al. 2016) [2, 15, 16] and were used in the current study.

3 METHODOLOGY FOR REGULAR TAP DISTRIBUTION

Herein, the methodology is developed around the NIST-UWO database, and results for one building are presented for illustration. For more complete results, the reader is referred to Duthinh et al. (2015) [17].

3.1 *Select data*

Within NIST-UWO wind tunnel database [2], select the building, roof or wall zones to study. Zones currently defined in ASCE 7-10 (Fig. 1) are investigated. For wind loads, ASCE 7-10 divides a gable roof into four corner zones (zone 3), whose projections onto a horizontal plane are squares of side a, four edge zones (zone 2) of horizontally projected width a running along the edges, excluding the corners, and an interior zone (zone 1), which is the rest of the gable. a is defined as 10 % of the least horizontal dimension of the building, or 0.4 h, whichever is smaller, but not less than either 4 % of the least horizontal dimension or 3 ft (0.9 m). h is the eave height, or the mean roof height when the roof slope exceeds 10°. For walls, two edge zones (zone 5) of width a are defined, and the rest of the wall is called zone 4.

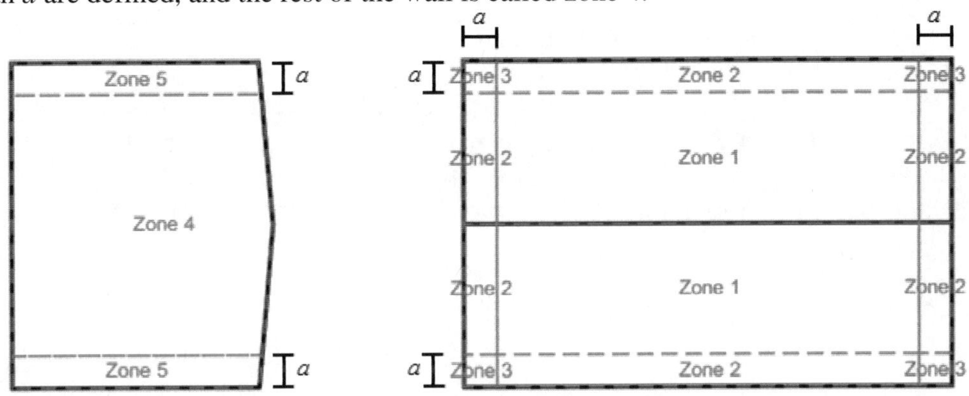

Figure 1. ASCE wall (left figure, elevation view with ground to the left) and roof (right figure, top view) zones for wind loads (adapted from ASCE 7-10 [1] Figs. 30.4-1 and 30.4-2A

3.2 *Combine areas*

The aerodynamic pressures induced on a structure by the wind field are random both in time and space. The randomness in space means that, at any given time, the pressures differ from point to point, that is, the spatial coherence of the pressures at any two points decreases as the distance between those points increases. For design purposes, the pressure on any given area A is, therefore, a function of A, and is defined as the aerodynamic force acting on area A divided by A. The time series of the aerodynamic force is obtained by summing up the product of pressure time series measured in wind tunnel tests at adjoining pressure taps by their respective tributary areas or cells.

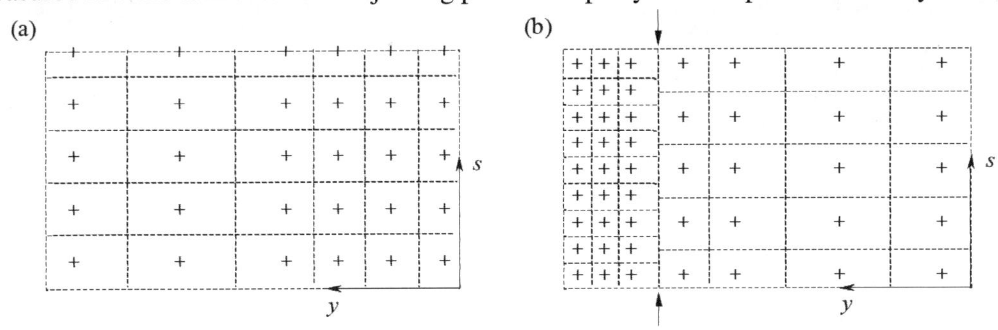

Figure 2. Definition of tributary areas or cells for taps on a particular face: a) simple tap array; b) more complicated tap array with varying tap density

Cell boundaries are straight lines equidistant to adjacent taps, but taps are at the center of cells only in a regular grid (Fig. 2a). The area *A* is the sum of those adjoining tributary areas. These operations are carried out for all combinations of tributary areas that make up rectangular areas *A*.

The summation of tributary areas is simplest when the grid of taps is regular, and is shown schematically in Figure 3. Special consideration must be given to the study zone edges and corners, which generally do not coincide with cell boundaries, and to places where grids of different densities merge (indicated by arrows in Fig. 2b). These situations are explained in Figures 4 and 5.

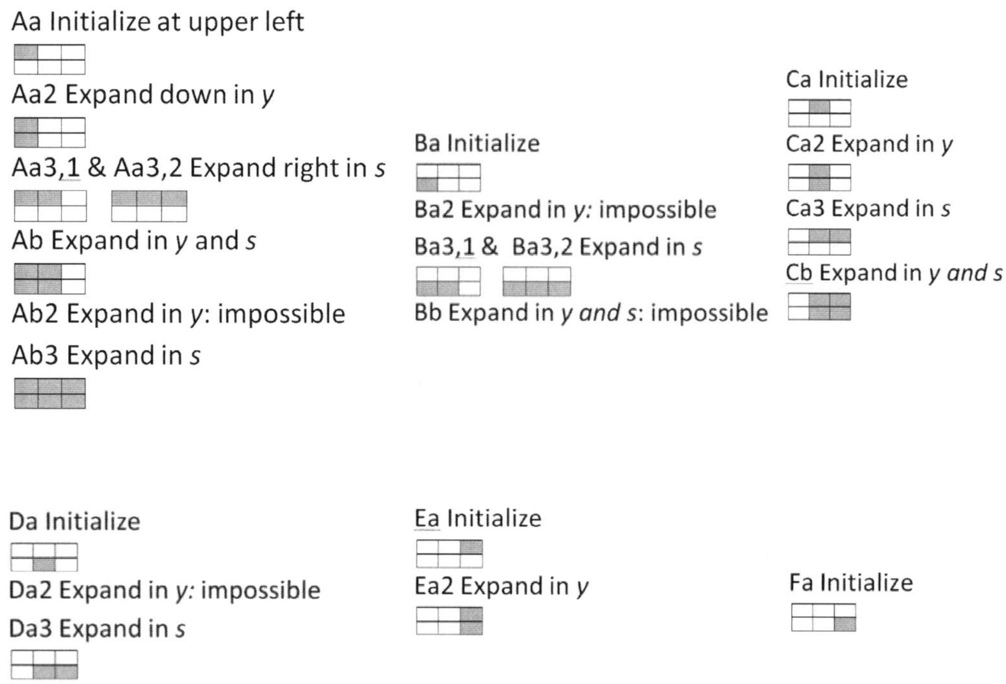

Figure 3. Cell area combinations

Figure 3 illustrates a case where the zone of interest only includes six cells, each with one pressure tap. The goal is to determine the pressure variation versus area *A* (shaded), for various sizes of *A* and regardless of the location of *A* in the zone of interest. The area summation procedure starts with the step denoted by Aa, which consists of the selection of one cell on the upper left corner. This is the first area *A*, where the number of cells is the same in both coordinate directions, denoted by *s* and *y*. The next step, denoted by Aa2, consists of adding a cell in the *y* direction downward. With this step, the lower boundary of the zone is reached; therefore, no additional cell can be added in direction *y*. To the cell selected in step Aa is added one cell in direction *s*, rightward, in step Aa3,1. In a next step, denoted by Aa3,2, an additional cell is added, again in direction *s*, rightward. Thus two additional areas *A* have been created in step Aa3. With step Aa3,2 the rightmost boundary of the zone has been reached, so further expansion in the direction *s* is not possible. Next, step Ab consists of adding to the cell selected in step Aa via expansion in both directions *y* downward and *s* rightward. Thus an area A consisting of four cells is created. Expansion in the direction *y* downward is attempted in step Ab2, but is not possible. Step Ab3 consists of expanding in the *s* direction rightward, which results in a six-cell area *A*. All possibilities of expansion from the single cell selected in step Aa being exhausted, one proceeds to the next initial cell (step Ba). The procedure is repeated until all possible initial cells have been used. Figure 3 shows six areas formed by one cell, seven areas formed by two cells, two areas formed by three cells, two areas formed by four cells, and one area formed by six cells, for a total of eighteen areas. In this method, all areas A are rectangular. If the cells are squares of unit area, for this zone of interest of area 6, there result 6 pressure time series for $A = 1$, 7 for $A = 2$, 2 for $A = 3$, 2 for $A = 4$ and 1 for $A = 6$. In this example of square cells, areas A have aspect ratios ranging from 1 to 3.

To limit the number of combinations for large zones of interest, and not to lose any data (as some of the cells along the long edges of roofs have an aspect ratio of 3.5), the aspect ratio of the rectangles formed by the aggregation of cells is limited not to exceed four. This aspect ratio covers many practical units of components and cladding, and allows consideration of long, narrow zones along the edges of roofs and walls. This choice also covers all "effective wind areas used to evaluate *(GCp)*", whose width, according to the Commentary of ASCE 7-10 [1], "need not be taken as less than one-third of the length of the area. This increase in effective wind area has the effect of reducing the average wind pressure acting on the component." How the aspect ratio of rectangular areas affect the peak area-averaged pressure will be discussed further in the results section. The number of area combinations increases very quickly with the size of the grid, e.g., a 19×8 grid produces 4290 areas, whereas a subset 16×7 of the same grid produces 2351 areas of aspect ratio ≤ 4.

Partial cells may be necessary and are accounted for at the edges and the corners of the study zone being considered, as the zone (of interest) boundaries may not coincide with cell boundaries. In Figures 4 and 5a the black lines represent the cell boundaries. In Figure 4, a corner of the zone being studied is shaded and includes one full cell and three partial cells. The pressures in the partial cell areas are the same as in the three full cells from which the partial cells are formed. The partial areas are considered when calculating area-weighted average pressures.

If the zone being studied overlaps areas of different tap density, the coarser density is used overall, and full and partial cell areas in the high density region are combined as needed. Figure 5a shows a portion of a study zone, with two grid densities. To conform with the coarser grid, the two rows of three cells at the bottom of the figure are transformed into two rectangles each (Fig. 5b).

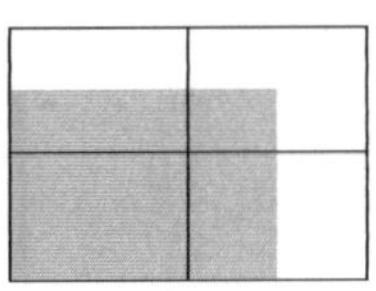

Figure 4. Partial cell areas

Figure 5a, b. Combination of areas with different tap density densities

3.3 *Average pressure over area*

Database-Assisted Design (DAD) software (Main and Fritz 2006) [15] was supplemented by software developed as part of this project to calculate the time series of wind pressure coefficients $C_{p3600}(A)$ on area $A = \Sigma\, a_i$ for each specified wind direction. The time-series $C_{p3600}(A)$ is the area-weighted average of pressure coefficients $C_{p,i}$ calculated from wind pressures p_i measured by taps i with their tributary areas a_i that overlap with the area A being studied. Typically, as explained in § 3.2, there are many combinations of a_i for the same value of A. As the peak of the sum is less than the sum of the peaks of several time series, the peak pressure decreases as the area increases. In physical terms, the spatial correlation of pressure peaks as measured by the taps decreases as the area increases, resulting in lower peak pressures over larger areas. p_i, $C_{p,i}$ and $C_{p3600}\,(A)$ are functions of time and wind direction. For simplicity, the subscript 3600 is omitted from $C_{p,i}$.

$$C_{p,i} = \frac{p_i - p_0}{\frac{1}{2}\,\rho\,V_{3600}^2}; \quad (1) \qquad C_{p3600}(A) = \frac{\sum(a_i\,C_{p,i})}{\sum a_i} \quad (2)$$

p_0 = static pressure at reference height (roof eave);
V_{3600} = mean hourly wind speed at reference height;
ρ = air density.

3.4 *Calculate peaks over time*

For each area combination $A = \Sigma\, a_i$, Rice's zero up-crossing method adapted to non-Gaussian time series (Sadek and Simiu 2002, Main 2011) [18, 19] was used to calculate minima and maxima over time of C_{p3600} for a 60-minute (full scale) windstorm. The Sadek-Simiu method uses the Gamma Distribution and the Normal Distribution to estimate the peaks corresponding, respectively, to the upper and lower tails of the time series' histograms. The peak distribution is represented by the Extreme Value Type I (Gumbel) distribution, and the values selected correspond to the mean of this distribution applied to the upper and lower tails. The extension of test results to a longer time than the test duration converted to full scale is discussed in the section on duration ratio.

The selection of peaks (in absolute value) of minima (negative pressure or suction) and maxima (positive pressure inward) of C_{p3600} for each area combination $A = \Sigma\, a_i$ results in two vectors of minima and maxima versus A for each wind direction. The gust effect factor G accounts for the variability of the pressure coefficient due to the randomness of the aerodynamic response and is introduced implicitly when the averaging process and the peak selection process are applied (Simiu, 2011) [20]. The peak pressures are functions of wind direction.

3.5 *Calculate peaks over wind direction*

Steps 3 and 4 are repeated for all wind directions tested. This results in two matrices of minima and maxima of GC_{p3600} for various areas and wind directions. From these two matrices, two vectors of peaks over all wind directions of minima and maxima of GC_{p3600} vs. A are selected.

3.6 *Normalize results*

The results are renormalized to 3-s wind gust speed V_3 as in ASCE 7-10, using the Durst (1960) [21] curve.

$$GC_p \equiv GC_{p3} = GC_{p3600}\left(\frac{V_{3600}}{V_3}\right)^2 = GC_{p3600}\left(\frac{1}{1.52}\right)^2 \qquad (3)$$

For a mean hourly wind speed of 80 mi/h (36 m/s), Durst (1960) [21] estimated the 5-s wind gust speed to have a mean of 118 mi/h (53 m/s) and a standard deviation of 12.7 mi/h (5.7 m/s). Since wind speed can be assumed to follow a Gaussian distribution, for 68 % of the data, the ratio of the 5-s wind gust speed to the hourly wind speed is (118 ± 12.7) / 80 = 1.475 ± 0.159. The uncertainty for the ratio of the 3-s gust speed to the hourly wind speed can be considered to be similar, 1.52 ± 0.16, for 68 % of the data.

3.7 *Repeat*

The procedure is repeated for other zones, terrains and buildings.

Note there are two consecutive steps in the selection of the peak wind pressures for design purposes: selection of peaks over time (§3.4), and selection of peaks over all wind directions (§3.5). The selection of the peaks over all wind directions is inherent in the envelope method, as defined in the ASCE 7 Standard. Finally, the extreme peaks corresponding to combinations of cells forming various areas A are chosen for the development of design specifications.

4 DURATION RATIO

The Sadek-Simiu (2002) [18] method uses a duration ratio that is explained by the following example. A wind tunnel tests a 1/100 model of size $D_m = 6.0$ in (0.152 m), by exposing it to mean wind speed $V_m = 30.0$ ft/s (9.144 m/s) in open country, for a duration $T_{ma} = 100$ s. The mean wind speed is calculated over the duration of the test, and is measured at a reference height ($h_{ref} = 6.8$ ft × 100 = 680 ft = 207.26 m full scale), then converted to eave height ($H = 40$ ft = 12.19 m) by the following power law (Ho, Surry and Morrish 2003) [3]. For a discussion of this power law and various values of the exponent proposed over the years, see Simiu and Scanlan 1996 p. 46 [22].

$$\frac{V_H}{V_{ref}} = \left(\frac{H}{h_{ref}}\right)^{1/7.446} \tag{4}$$

These data are used to design a building of size $D_p = 50$ ft (15.2 m) for a region with 3-s gust speed of 140 mi/h (225 km/h) measured at 33 ft (10 m) elevation. The hourly wind speed is 140 mi/h/1.52 = 92.11 mi/h (=135.09 ft/s = 41.17 m/s). The 33 ft (10 m) elevation is close enough to the eave height that no height correction is required. Scaling laws require (Simiu and Scanlan 1996 [22]):

$$\left(\frac{D}{VT}\right)_m = \left(\frac{D}{VT}\right)_p \tag{5}$$

where subscripts m and p stand for model and prototype respectively, and D, V and T refer to length, velocity and time. Thus:

$$T_m = T_p \frac{D_m}{D_p}\frac{V_p}{V_m} = 3600\frac{1}{100}\frac{135.09}{30.0} = 162 \text{ s} \tag{6}$$

The following duration ratio must be applied in the extreme value procedure to obtain the proper estimate for peak wind gusts:

$$\frac{T_m}{T_{ma}} = \frac{162 \text{ s}}{100 \text{ s}} = 1.62 \tag{7}$$

5 RESULTS FROM NIST-UWO DATABASE

Results are shown for one building in the UWO-NIST database, Building 7, tested in open country (data set jp1). The building was modelled at a scale of 1:100, and data were collected for 100 s at 500 Hz. Building 7 (Fig. 6c) is 40 ft (12 m) wide, 62.5 ft (19 m) long, 40 ft high (12 m, eave height) and has a roof slope of 1:12 (4.8°). It was tested for wind directions ranging from 0° to 90° and 270° to 360° every 5°. Due to building symmetry, and for similar terrain in all directions, only half of all possible wind directions need to be investigated. In Figure 6c, 0° and 90° are in the $+y$ and the $+x$ directions respectively. Selected results covering ASCE zones 1 to 5 are discussed in this section and summarized in Table 1.

Figure 6c shows an outline of Building 7, with dimensions in feet. Pressure sensors, or taps, are shown on half of the gable roof (called roof 3 here), and the interior zone (defined in ASCE 7-10 as zone 1) being investigated is delineated by blue lines. The area combination scheme of rectangles with aspect ratio less than or equal to 4 is used, and is also specialized to the case of squares (aspect ratio 1) that cover at least four taps. This last scheme has been used by other researchers (Morrison 2015) [23]. Each area combination results in a maximum and a minimum pressure coefficient (Figs. 6a, b). The current ASCE 7 specifications are plotted for comparison (+ means positive maxima, - means negative minima).

Figure 6a shows the current ASCE specifications underestimate negative pressures by a factor ranging from 1.3 for larger areas close to 1000 ft^2 (93 m^2) to 2.5 for smaller areas of 10 ft^2 (1 m^2) or less. The square area scheme also captures the same trend, albeit at fewer area combinations, as expected. Limiting the data to areas including at least four taps results in a lower limit of 37 ft^2 (3.4 m^2), compared with the area of 10 ft^2 (1 m^2) below which ASCE specifies the pressure coefficients to remain constant. ASCE underestimates positive pressures by a factor of 2.0 for areas between 5 ft^2 (0.5 m^2) and 30 ft^2 (3m^2), but the pressures are small, and the ASCE specifications are acceptable for larger areas. Figure 6b will be discussed later in a separate section on aspect ratio.

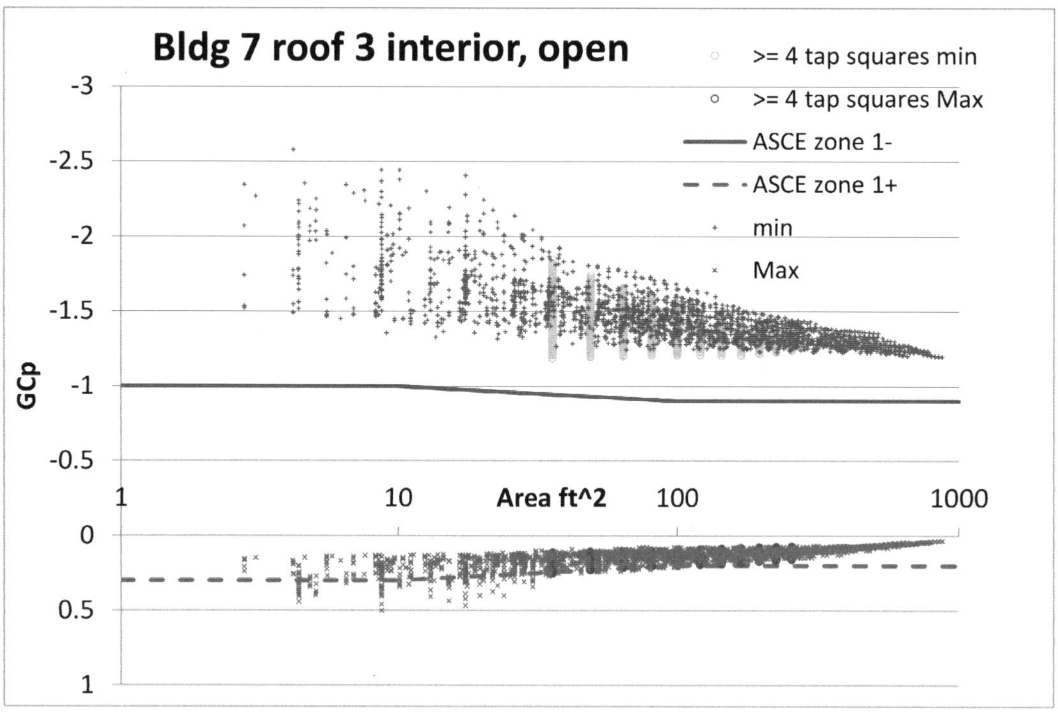

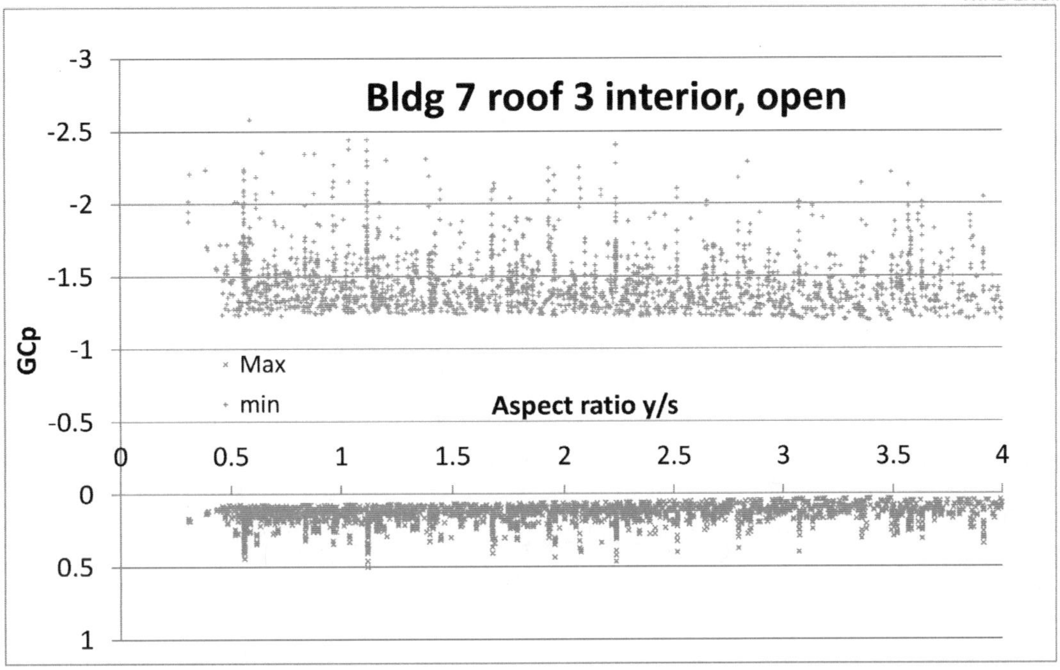

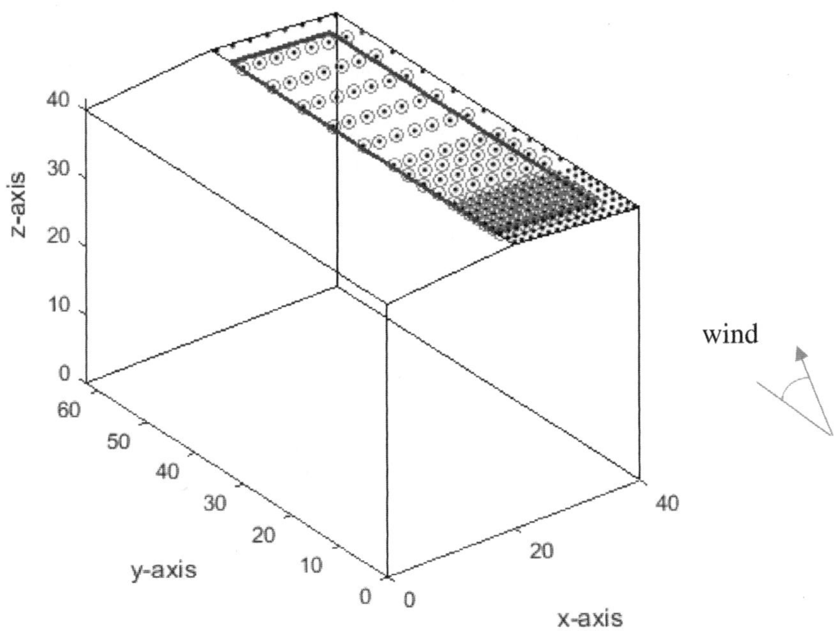

Figure 6a, b, c. Bldg 7 roof 3 ASCE zone 1 interior, with pressure taps (1 ft = 0.305 m, 1 ft^2 = 0.0929 m^2)

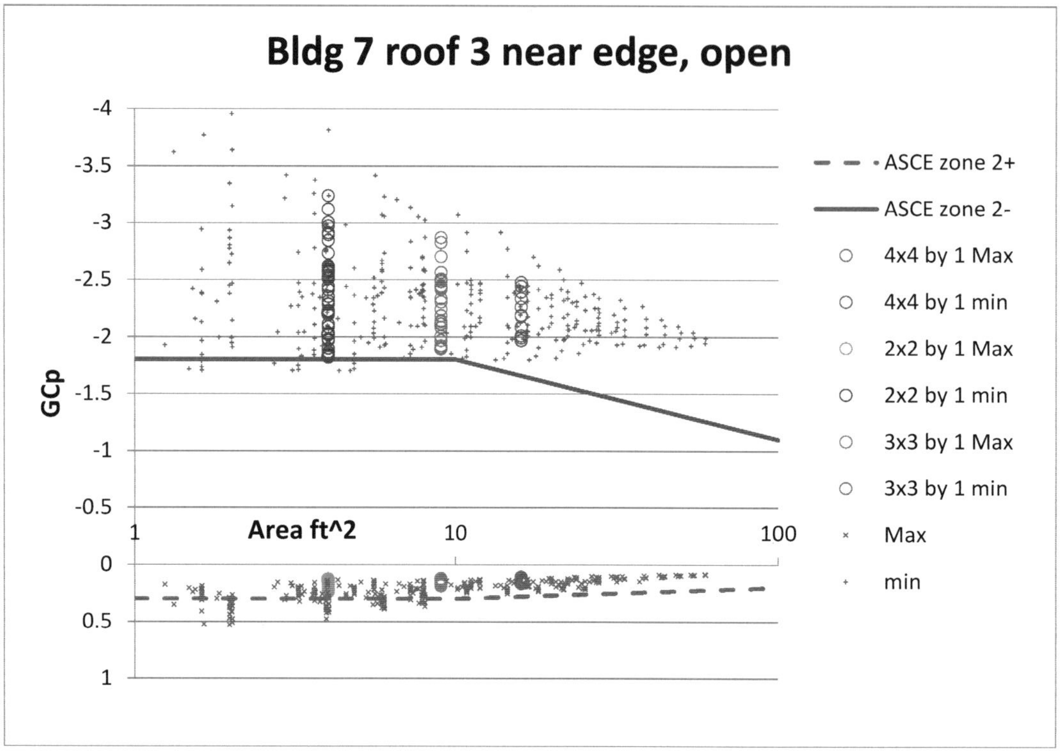
Bldg 7 roof 3 near edge, open
GCp
Area ft^2
-4
-3.5
-3
-2.5
-2
-1.5
-1
-0.5
0
0.5
1
1
10
100
ASCE zone 2+
ASCE zone 2-
4x4 by 1 Max
4x4 by 1 min
2x2 by 1 Max
2x2 by 1 min
3x3 by 1 Max
3x3 by 1 min
Max
min

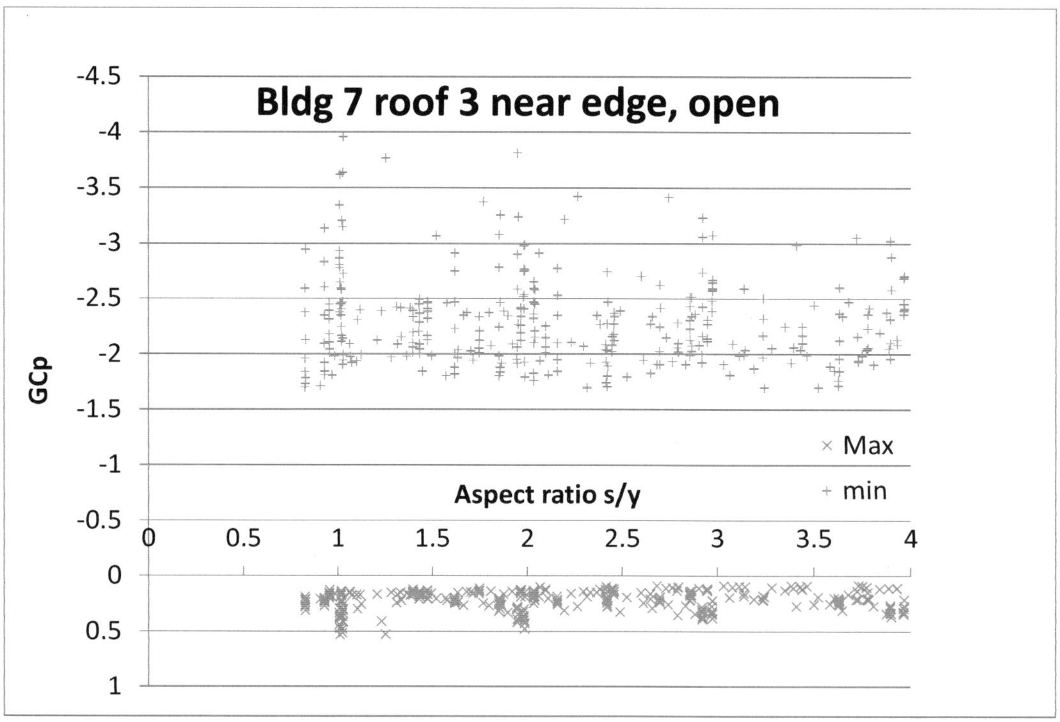
Bldg 7 roof 3 near edge, open
GCp
Aspect ratio s/y
-4.5
-4
-3.5
-3
-2.5
-2
-1.5
-1
-0.5
0
0.5
1
0
0.5
1
1.5
2
2.5
3
3.5
4
Max
min

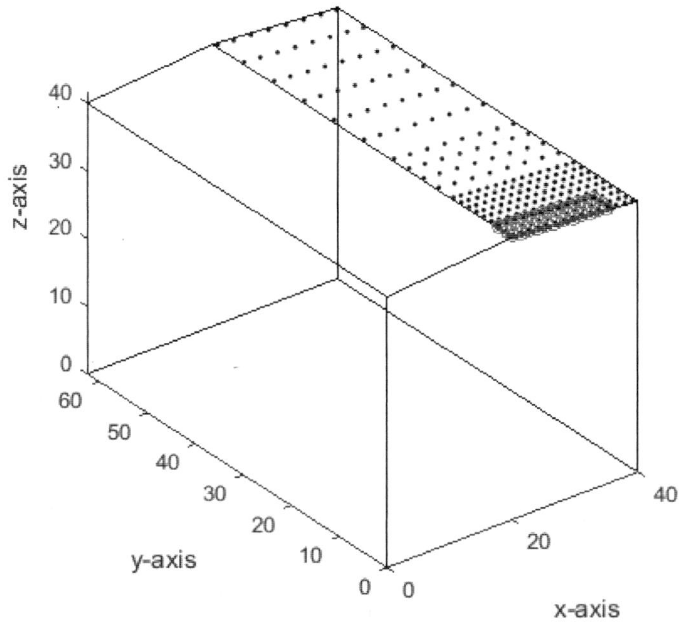

Figure 7a, b, c. Bldg 7 roof 3 ASCE zone 2 near edge (1 ft = 0.305 m, 1 ft^2 = 0.0929 m^2)

One particularly attractive feature of the NIST-UWO database is the high density of pressure taps on one part of the roof. This allows high resolution in edge and corner zones, shown in Figures 7 and 8 for Building 7, roof 3 (in the DAD software, the roof gables are numbered 2 and 3, and the side walls 1 and 4. The DAD software focuses on structural response in the transverse direction x and does not use the end wall pressure data). The far corner of roof 3 (x = 40, y = 62.5) does not have a sufficient number of taps for meaningful results.

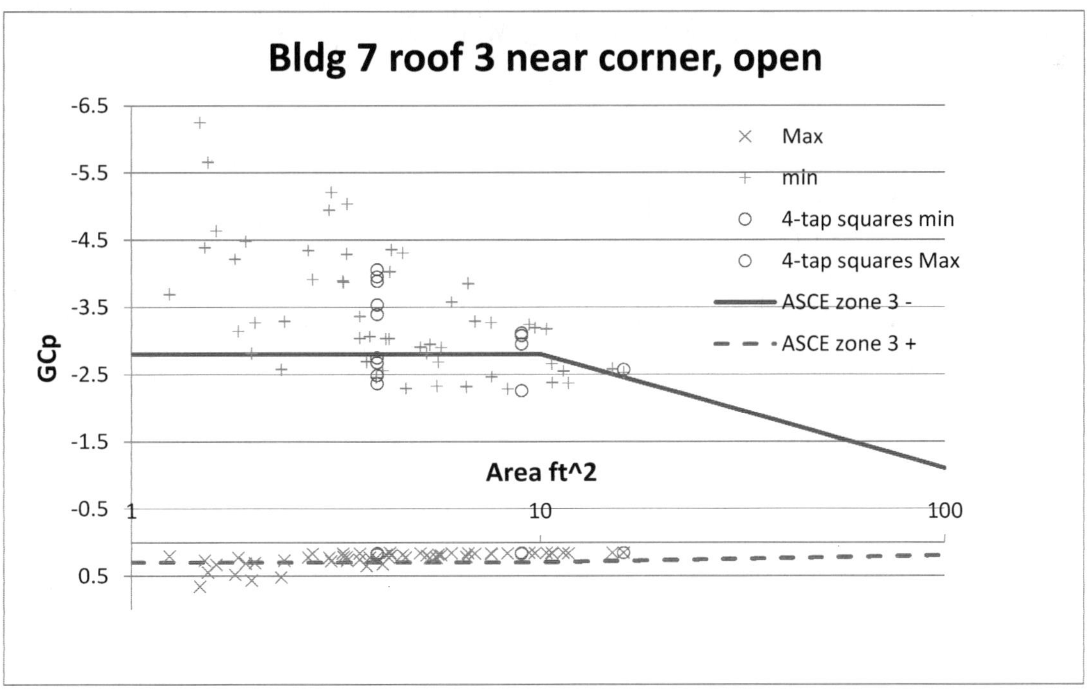

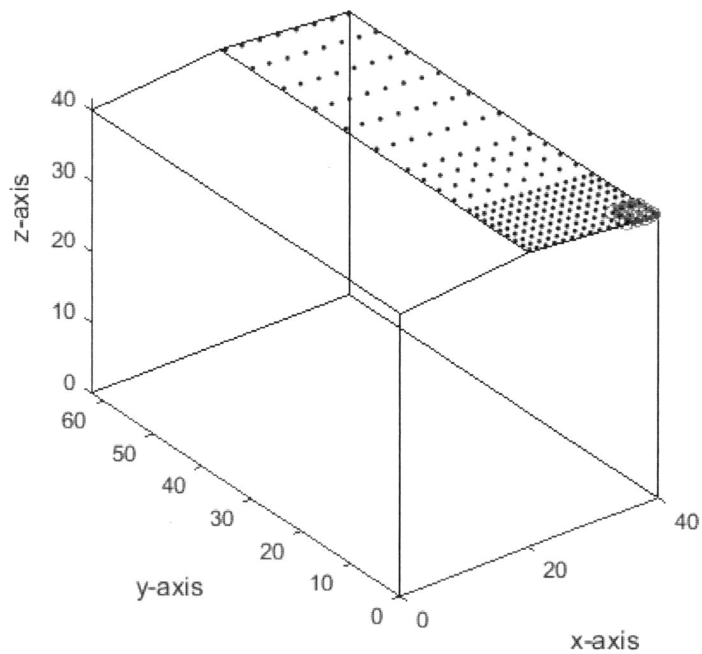

Figure 8a, b. Bldg 7 roof 3 ASCE zone 3 near corner, with pressure taps (1 ft = 0.305 m, 1 ft^2 = 0.0929 m^2)

Figure 7 shows the results for the roof edge on the windward side in the UWO tests. Observations are consistent with those regarding the interior zone just discussed. For negative pressures, current ASCE specifications underestimate the measurements by a factor ranging from 1.6 for the larger areas (60 ft^2 or 6 m^2) to 2.2 for the smaller areas (2 ft^2 or 0.2 m^2). ASCE underestimates positive pressures by a factor of 1.7 for areas smaller than 3 ft^2 (0.3 m^2), but the pressures are small, and the ASCE specifications are acceptable for larger areas. The notation "2×2 by 1 min" indicates negative pressures (or minima) estimated from square areas with a 2×2 arrangement of pressure taps, with the squares being shifted by 1 ft (0.30 m) in consecutive steps in both directions of the planar surface to cover the entire study zone. Similar notation holds for the maxima and other size squares.

Figure 8 shows that ASCE specifications underestimate negative pressures over virtually all of the areas within the corner zone, by up to a factor of 2.3. For areas that include at least four taps, the underestimation is by 1.5. On the other hand, positive pressure specifications are adequate. This corner is on the windward side in the UWO tests.

Figure 9 shows results for the far edge, part of ASCE zone 5, of Wall 1 of Building 7. ASCE underestimates negative pressures and positive pressures by a factor of 1.6. Note the tap density is not high in this zone, which is less than two-cell wide. The peaks of negative pressure occur for wind directions between 50° and 65°, and the peaks of positive pressures occur for wind directions of 10° (0° is $+y$ and 90° is $+x$). Since the tests were conducted only for wind directions between 0° and 90° and 270° and 360°, the corresponding peaks for both positive and negative pressure were missed for the right edge of Wall 1.

Figure 10 shows that for the interior, zone 4, of Wall 1 of Building 7, ASCE 7-10 specifications underestimate negative pressures by a factor ranging from 1.2 for areas of 1100 ft^2 (100 m^2) to 1.9 for areas of 60 ft^2 (6 m^2); and underestimate positive pressures by a factor of 1.8 over all areas. The rest of the results for Building 7 are summarized in Table 1. Overestimations and underestimations in Table 1 are reported in terms of GC_p.

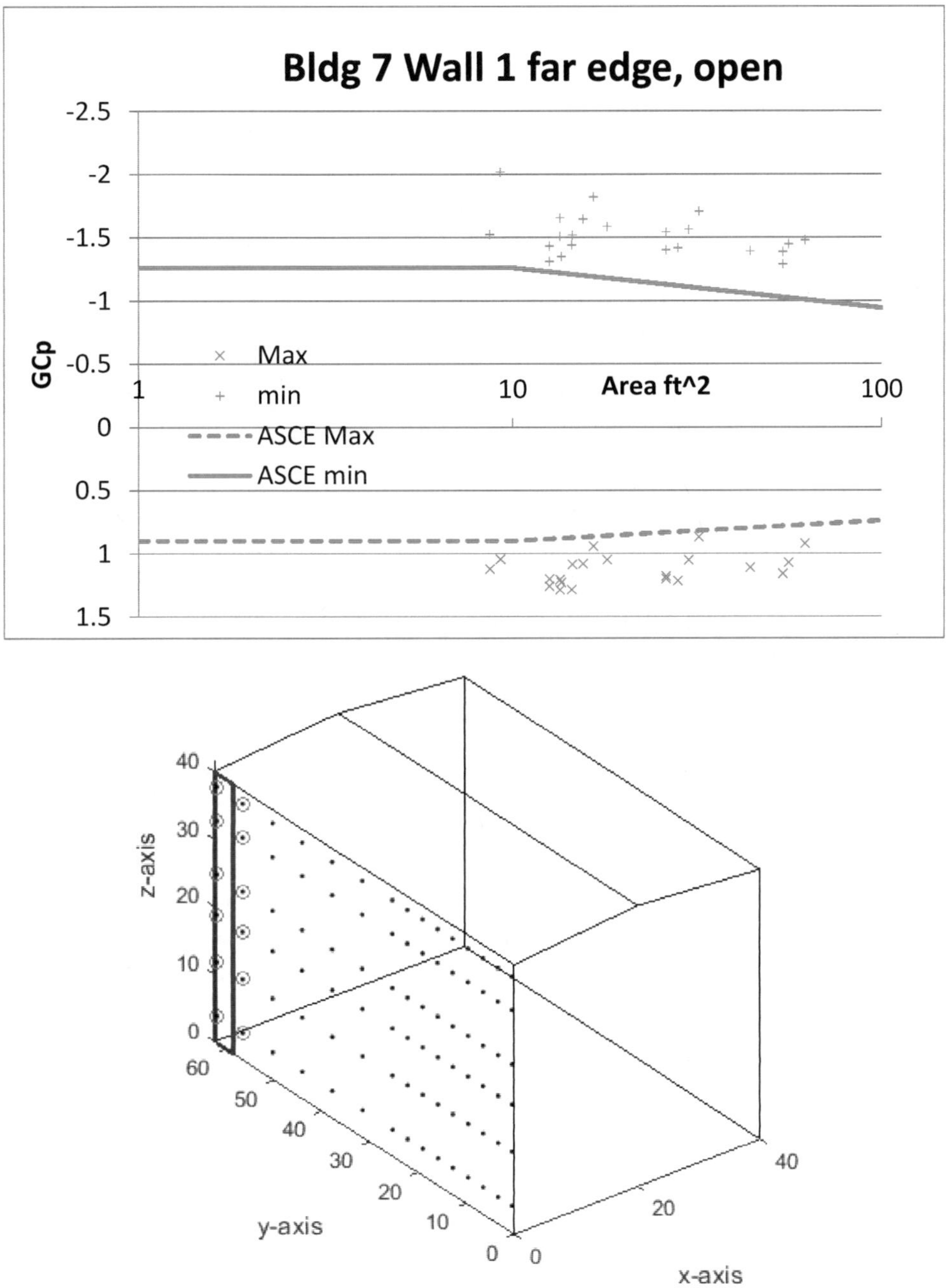

Figure 9a, b. Building 7 Wall 1 ASCE zone 5, far edge, with pressure taps (1 ft = 0.305 m, 1 ft^2 = 0.0929 m^2)

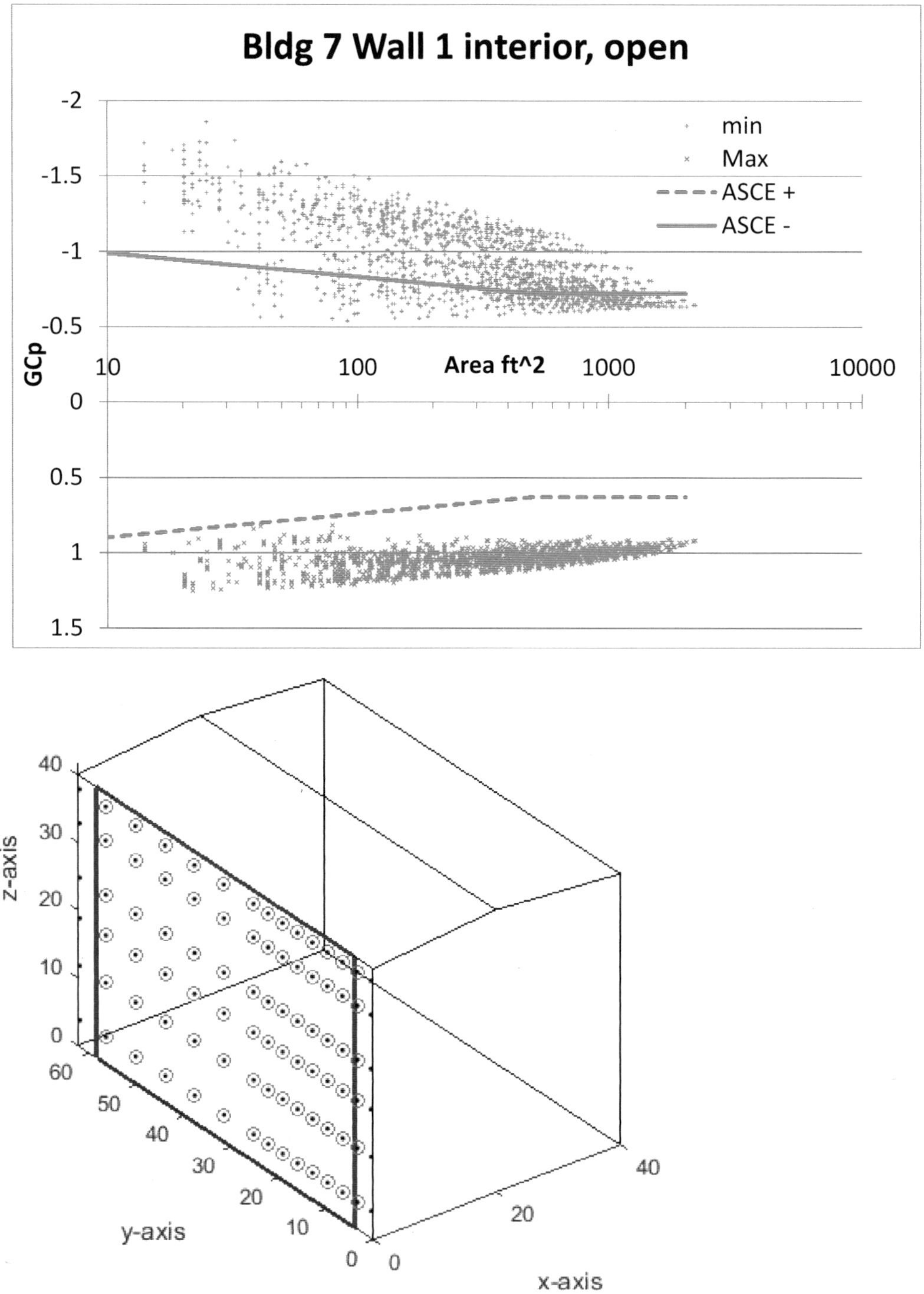

Figure 10a, b. Building 7 Wall 1 ASCE zone 4, interior, with pressure taps (1 ft = 0.305 m, 1 ft2 = 0.0929 m^2)

6 ASPECT RATIO

The choice of aspect ratios of four or less makes more effective use of the UWO data than, say, three or less. Results show a slight decreasing trend of GC_p as the aspect ratio increases (Figs. 6b, 7b). This is true for edge zones as well as interior zones, and is consistent with trends expected based on the ASCE 7-10 Commentary. The envelope of GC_p is not affected by extending the aspect ratio from three to four. (It had been expected that, along the edges of roofs and walls, and for certain wind directions, the spatial coherence of wind pressure may extend over longer shapes than those mentioned in the ASCE 7-10 Commentary.)

Table 1 Results for Building 7, open country
(ASCE underestimates by *n* means the ratio of the envelope of the data to the ASCE specification for that area is *n*.)

part	zone	- pressure	+ pressure
Roof 3	Near corner	ASCE underestimates by 1.5 for A > 4 cells, by 2.3 for smaller areas.	ASCE acceptable.
Roof 3	Long edge	ASCE underestimates by 1.8 for A = 70 ft^2 (7 m^2) to 2.0 for A = 7 ft^2 (0.7 m^2).	ASCE acceptable.
Roof 3	Far edge	ASCE underestimates by 1.1. This edge is on the leeward side.	ASCE acceptable
Roof 3	Near edge	ASCE underestimates by 1.6 for larger areas (60 ft^2 or 6 m^2) and 2.2 for smaller areas (2 ft^2 or 0.2 m^2).	ASCE underestimates by 1.7 for $A \leq 3$ ft^2 (0.3 m^2), acceptable for larger areas.
Roof 3	Interior	ASCE underestimates by 1.3 for A = 1000 ft^2 (93 m^2) to 2.5 for A = 10 ft^2 (1 m^2)	ASCE underestimates by 1.7 for 5 ft^2 $\leq A \leq$ 30 ft^2 (0.5 m^2 $\leq A \leq$ 3 m^2)
Roof 2	Long edge	ASCE underestimates by 1.8 for A = 70 ft^2 (7 m^2) to 2.0 for A = 7 ft^2 (0.7 m^2).	ASCE acceptable
Roof 2	Far edge	ASCE acceptable. This edge is on the leeward side.	ASCE acceptable
Roof 2	Near edge	ASCE underestimates by 1.5 for A = 70 ft^2 (7 m^2) to 2.0 for A = 7 ft^2 (0.7 m^2).	ASCE acceptable
Roof 2	Interior	ASCE underestimates by1.3 for A = 1000 ft^2 (93 m^2) to 2.5 for A = 8 ft^2 (0.8 m^2).	ASCE acceptable
Roof 2	Corners	ASCE underestimates by 1.6 for A = 5 ft^2 (0.5 m^2), acceptable for $A \geq$ 10 ft^2 (1 m^2).	ASCE acceptable
Wall 1	Interior	ASCE underestimates between 1.2 for A = 1100 ft^2 (100 m^2) and 1.9 for A = 60 ft^2 (6 m^2).	ASCE underestimates by 1.8
Wall 1	Near edge	ASCE underestimates by 1.2.	ASCE acceptable
Wall 1	Far edge	ASCE underestimates by 1.6. - peaks for wind between 50° and 65°.	ASCE underestimates by 1.6. + peaks for wind directions of 10°.
Wall 4	Interior	ASCE underestimates by 2.1 for A = 20 ft^2 (2 m^2), 1.4 for A = 1000 ft^2 (93 m^2). Linear trend of *log A* in between.	ASCE underestimates by 1.8 for A = 20 ft^2 (2 m^2), 2.0 for A = 500 ft^2 (46 m^2).
Wall 4	Near edge	ASCE underestimates by 1.6. - peaks for wind between 330° and 360°.	ASCE underestimates by 1.6. + peaks between 290° and 335°.
Wall 4	Far edge	ASCE underestimates by 1.2.	ASCE *overestimates* by 1.2.

PART B

7 METHODOLOGY FOR IRREGULAR TAP DISTRIBUTION

For more irregular tap distributions, such as those present in the TPU Aerodynamic Database for Low-Rise Buildings, a different methodology is needed to calculate area-weighted-average pressures. The methodology presented in Part B is based on a computational geometry approach to determining pressure tap tributary areas; this is in contrast to the rectangular tributary areas of Part A for more regular tap distributions. The methodology is presented using TPU's Aerodynamic Database for Low-Rise Buildings and the sub-database entitled Isolated Low-rise Buildings without Eave (Tamura, 2012) [6]. The following steps are used to calculate external wind pressure coefficients on low-rise buildings and compare them with ASCE 7-10 specifications:

Step 1: Select a wind tunnel database and building for analysis;

Step 2: Plot the building's unfolded surfaces and pressure taps;

Step 3: Determine pressure taps' tributary areas using Voronoi diagrams;

Step 4: Specify sets of rectangular grids for pressure-area evaluation;

Step 5: Overlay a selected rectangular grid over the building surface;

Step 6: For a rectangular grid and a wind direction θ, calculate the area-weighted average pressure coefficients using tributary areas and corresponding pressure tap time histories;

Step 7: Determine peak minimum and maximum pressure coefficients for a given windstorm duration using the same Sadek-Simiu (2002) [18] peak estimation method as above;

Step 8: Normalize results to a three-second gust speed for comparison with ASCE 7-10 charts;

Step 9: Repeat steps 6-8 to obtain an envelope of + and – peaks over all wind directions θ;

Step 10: Repeat steps 5-9 for the next rectangular grid;

Step 11: Separate the external pressure coefficient results by ASCE 7-10 zones.

Some of these steps are now explained in more detail. In Step 2, unfolding means that the six surfaces of the building (four walls and two halves of the gable roof) are considered in their own planes. In step 3, pressure taps are located within the boundaries of the building's unfolded surfaces and their tributary areas are determined by Voronoi (1908) [24] diagrams, which are well-suited for irregularly spaced taps. Voronoi diagrams are derived from Delaunay (1934) [25] triangulation, which connects a given set of point taps (Fig. 11a) to form triangles (Fig. 11b) that: (1) do not overlap, (2) cover the entire interior space formed by the taps, and (3) do not have any tap within the triangle's circumcircle. The corresponding Voronoi diagram is created by drawing perpendicular bisectors to the sides of the previously created triangles (Fig. 11c). Regions formed from these bisectors contain one tap each and bound an area that is closer to that tap than to any other tap (Fig. 11d). Voronoi diagrams are mathematically proven to be optimal for tributary area assignment and remove user arbitrariness from the process. MATLAB (2014) [26] has a function, Voronoi, that can generate both Delaunay triangulation and Voronoi diagrams.

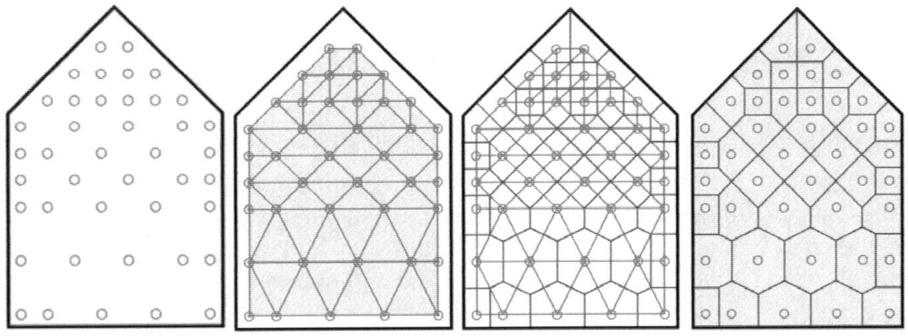

(a) (b) (c) (d)

Figure 11. Examples of pressure tap tributary area assignments

Step 4 introduces a grid for averaging pressure over area for comparison with ASCE's wind pressure charts. Rectangular grids of various grid cell size are superposed on the entire building flattened surface at various offsets in both orthogonal directions *x* and *y* of the planar surface. The minimum grid cell size is based on the smallest distance between individual taps, as there is no benefit in resolving a grid cell area smaller than the smallest tributary area. The largest grid cell size is selected based on the largest area of interest for specifying pressure vs. area in the design of components and cladding. The grid results in averages of pressure over the grid cell size, but these values depend on the position of the grid. To ensure that the highest area-averaged pressure is not missed, the grid is offset in the *x*- and *y*-directions at increments smaller than the grid cell *x*- and *y*-dimensions.

Step 5 applies a grid size and offset combination to the building surfaces. Figures 12(a, b) show a superposed 2 m × 2 m and a 2 m × 4 m grid respectively, with no offset. Figure 12(c) shows a 2 m × 4 m grid area with 1 m offset in the *x*-direction, and Figure 12(d) shows a 2 m × 2 m grid area with 1 m offset in the *x*- and *y*-directions. Where grid lines cross outside a building surface, partial or incomplete grid areas created by such lines are neglected.

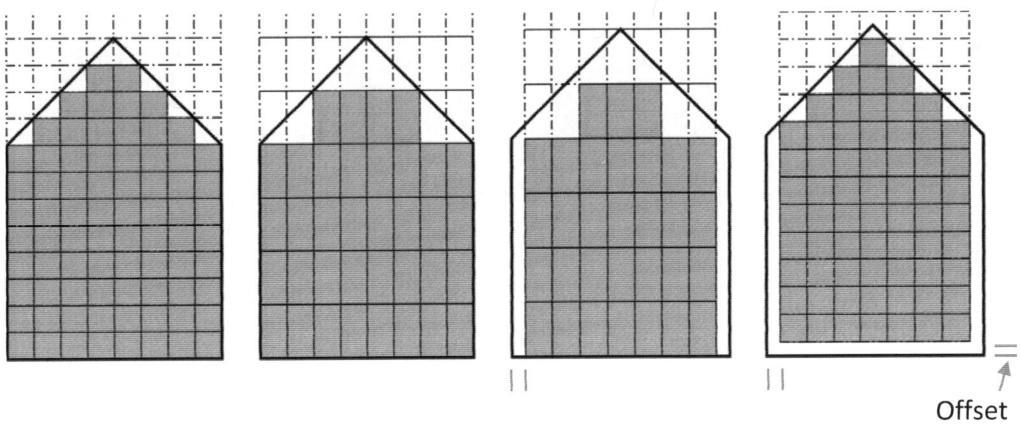

(a) (b) (c) (d)

Figure 12. Example of grid area sizes, increments and offsets

Step 6 calculates the pressure coefficients $GC_{p,t}$ for each area created by the rectangular grid for a particular wind direction θ, using Boolean algebra and a MATLAB function called Polybool to determine the intersection of the grid with the tributary area of each tap and area-average the pressure coefficient. Step 7 calculates peak pressure coefficients as described in part A, §3.4. A duration ratio of 6 was used, obtained by dividing the storm duration, 3600 s, by the full-scale duration of the TPU test, 600 s. Step 8 converts the pressure coefficients from an hourly wind speed to a three-second gust, as described in part A, §3.6.

Step 9 involves repeating steps 6 through 8 for all tested wind directions θ (0°, 15°, 30°, 45°, 60°, 75° and 90° with 0° parallel to the roof ridge) to obtain the worst-case envelope of pressure over all surfaces of the building. Step 10 repeats steps 6 through 8 for the next grid area and offset combination. In step 11, the worst-case peak minimum and maximum external pressure coefficients are plotted for the zones defined in ASCE 7-10 and compared with the standard's external pressure coefficients.

8 RESULTS FROM TPU DATABASE

The method outlined previously was programmed using MATLAB (2014) [26] to process the available gable buildings in TPU's low-rise building wind tunnel database, specifically case numbers 13 to 108 (Tamura, 2012) [6]. Three building models were analyzed (Table 2). The buildings selected for analysis match those analyzed by Hagos et al. (2014) [27].

Table 2. Buildings selected for analysis from TPU's Database

Test	Roof Type	Case Number	Breadth	Depth	Eave height	Roof Pitch
			B (m)	D (m)	H0 (m)	β (°)
TPU-1	Gable	61	16	24	12	4.8
TPU-2	Gable	93	16	40	12	4.8
TPU-3	Gable	95	16	40	12	14

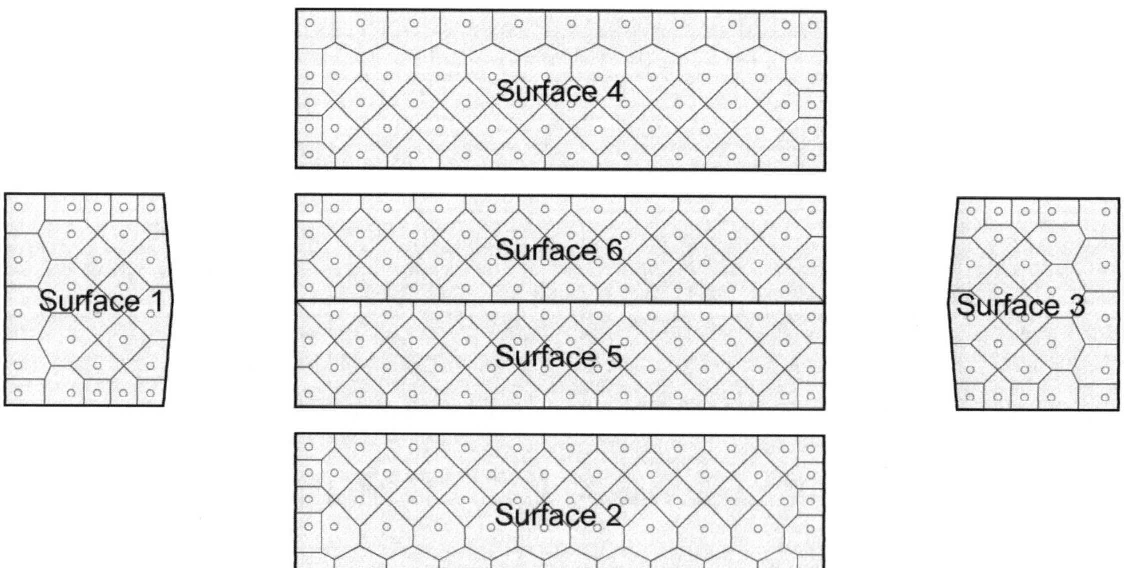

Figure 13. Tributary areas achieved using the Voronoi diagram for TPU-2

Having selected the database and buildings, the next steps are to plot the pressure tap locations and find their tributary areas with the aid of Voronoi diagrams. Figure 13 shows the Voronoi diagram applied to TPU-2 where the pressure taps are indicated by circles, bounded by the polygons which define the tributary areas.

With the tributary areas in place, the grid areas and offset combinations can then be specified. The smallest grid area was chosen as 2 m × 2 m and the largest as 7 m × 7 m. The smallest grid area was selected based on the minimum 2 m pressure tap spacing used in the TPU tests. The grid areas were incremented from 2 m × 2 m up to 7 m × 7 m, by increments of 0.5 m. Also, the grid was offset in increments of 0.5 m in the *x*- and *y*-directions. Based on these grid size and offset combinations, the total number of combinations is 9801, each grid with an aspect ratio of 3.5 or less, for consistency with part A.

The envelope plots of the peak minimum and maximum external wind pressure coefficients for a grid size of 2 m × 2 m with zero *x*- or *y*-offset are shown for TPU-1 in Figures 14 and 15, TPU-2 in Figures 16 and 17, and TPU-3 in Figures 18 and 19. Note that Figures 14-19 only show results for wind directions between 0° and 90° (i.e., wind blowing from bottom left quadrant). Extension of the results by symmetry are shown in Figures 23 and 24. The thin band of width $8(1/(\cos(14°)-1)) = 0.24$ m on either side of the roof ridge not covered by the grid is discernable in Figures 18 and 19. This determination of peak minimum and maximum wind pressure coefficients is repeated for the next grid area/offset combination, and so on.

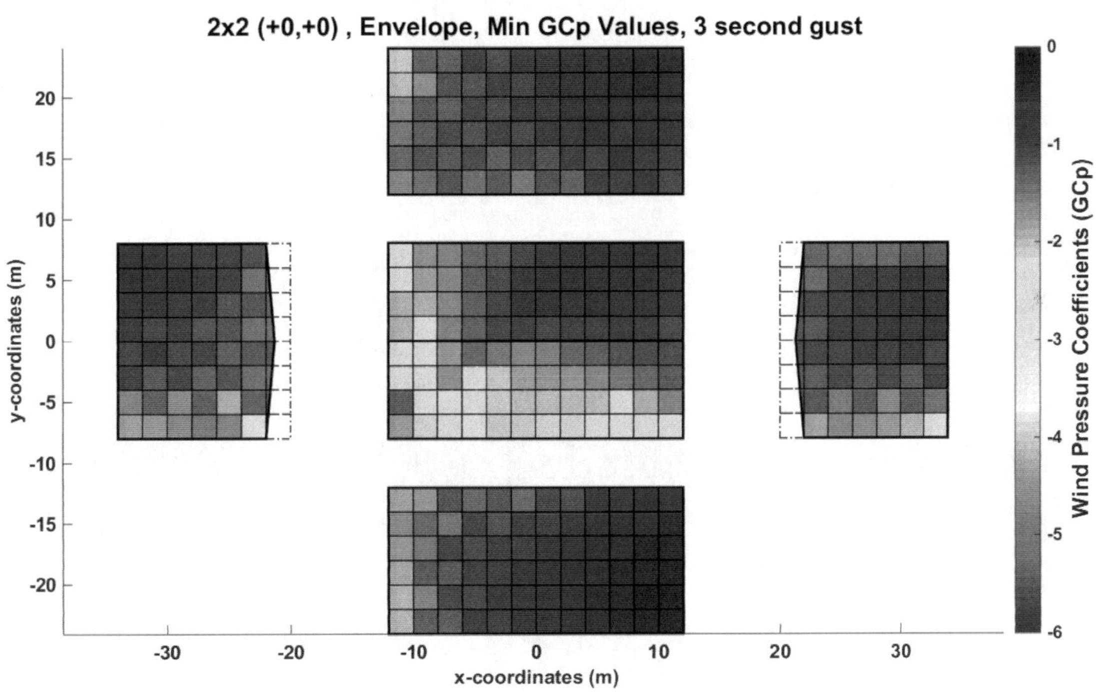

Figure 14. TPU-1 envelope of peak min. pressure, wind from lower left quadrant, grid 2 m × 2 m, offset 0

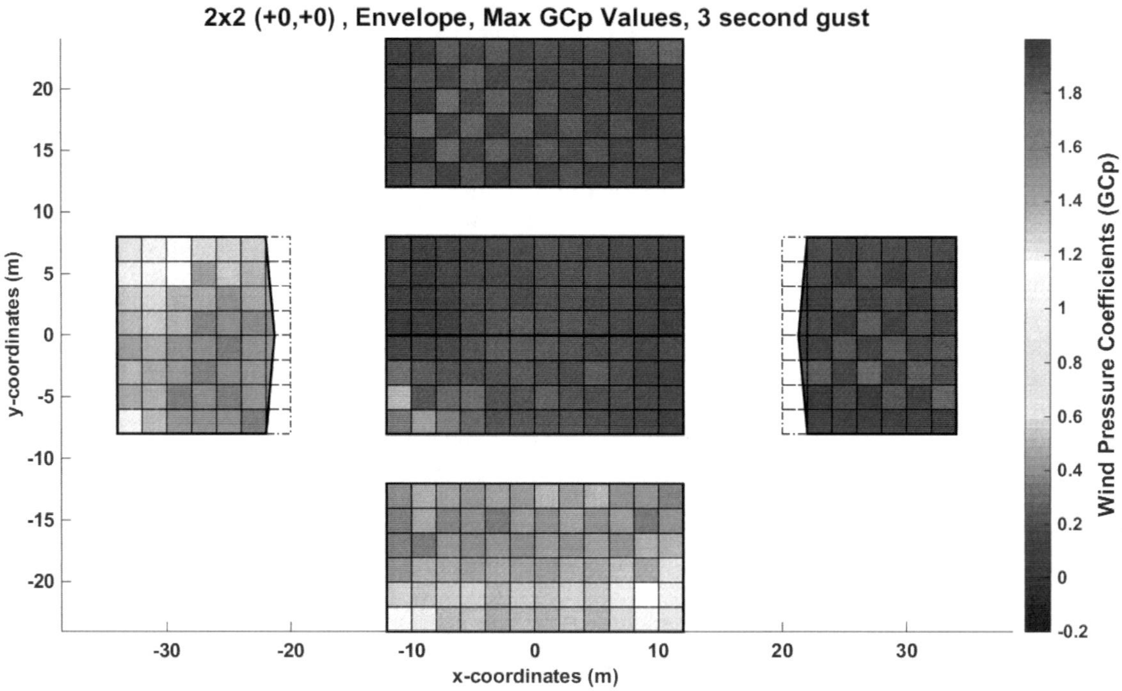

Figure 15. TPU-1 envelope of peak max. pressure, wind from lower left quadrant, grid 2 m × 2 m, offset 0

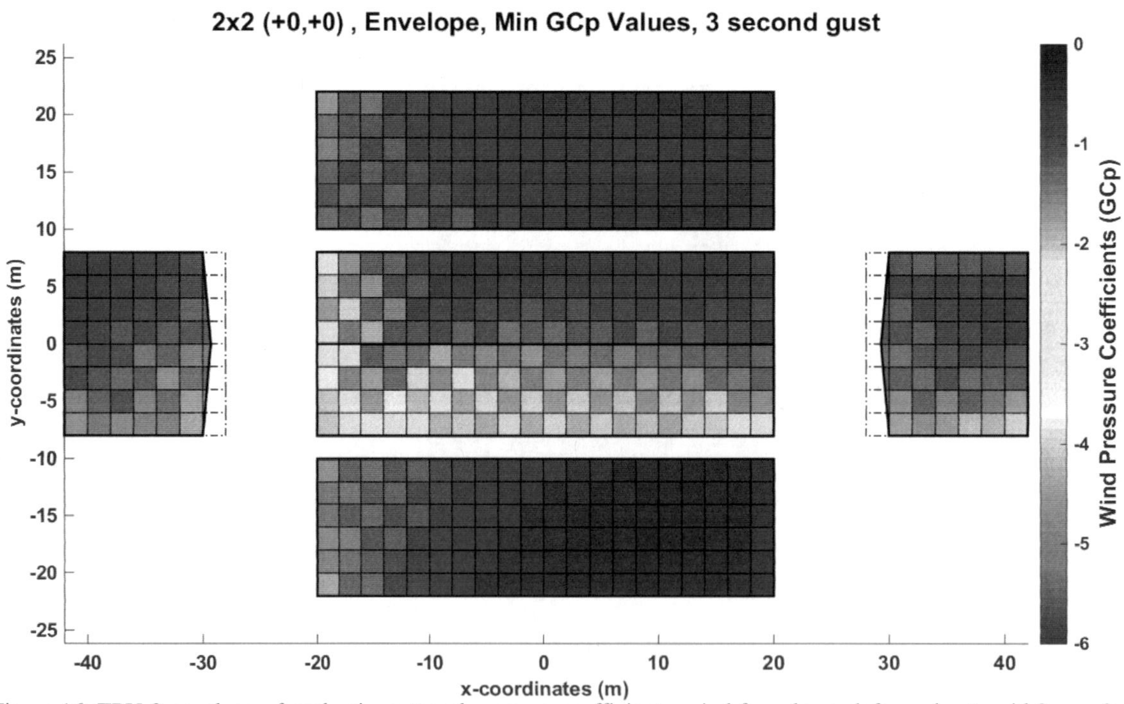

Figure 16. TPU-2 envelope of peak min. external pressure coefficients, wind from lower left quadrant, grid 2 m × 2 m, offset 0

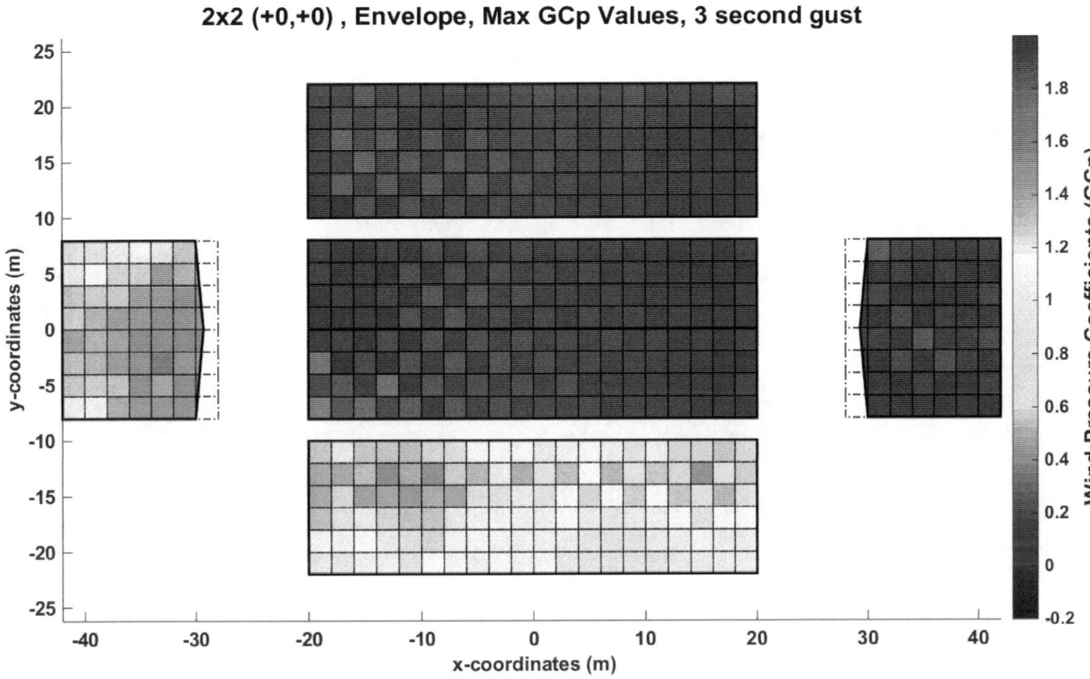

Figure 17. TPU-2 envelope of peak max. external pressure coefficients, wind from lower left quadrant, grid 2 m × 2 m, offset 0

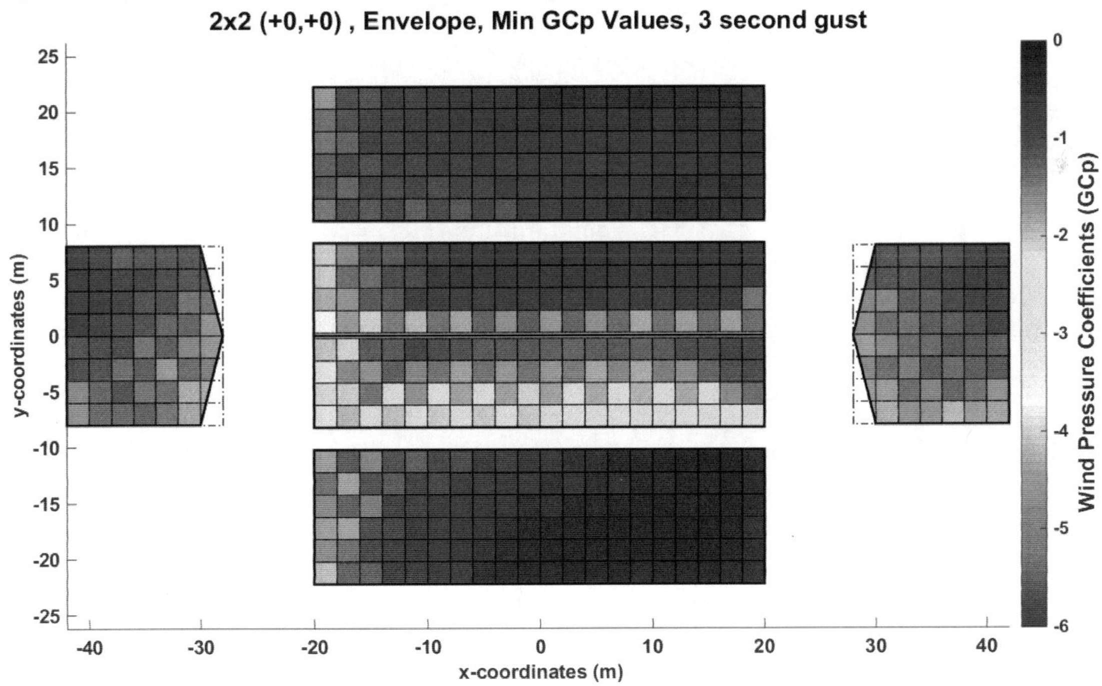

Figure 18. TPU-3 envelope of peak min. external pressure coefficients, wind from lower left quadrant, grid 2 m × 2 m, offset 0

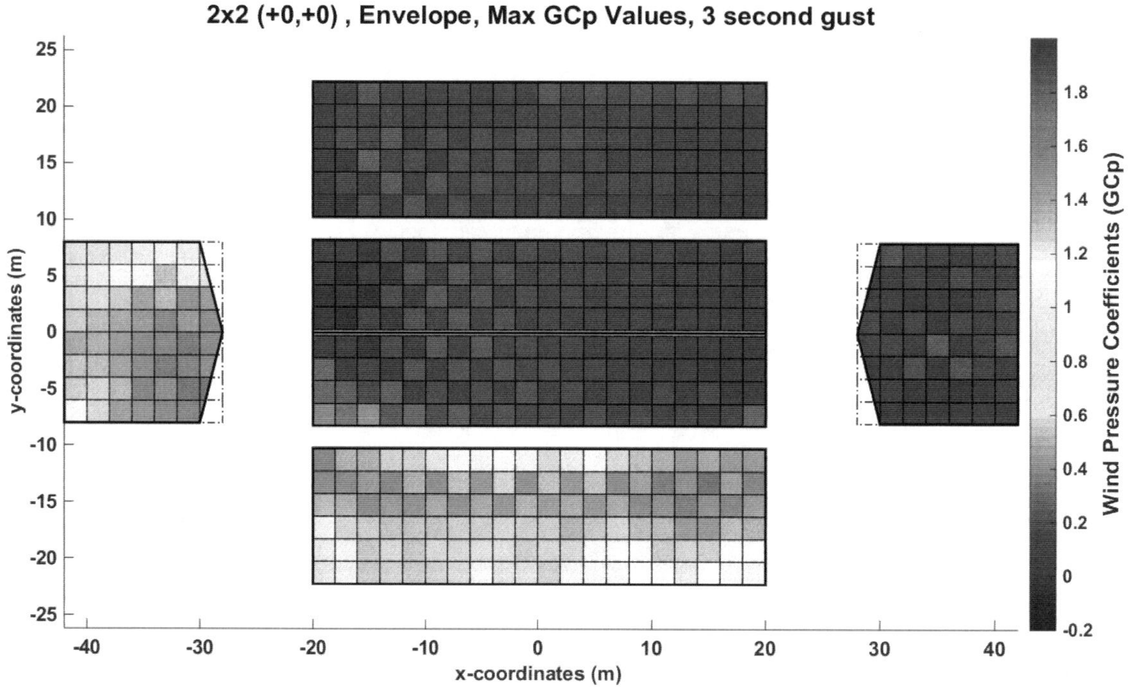

Figure 19. TPU-3 envelope of peak max. external pressure coefficient, wind from lower left quadrant, grid 2 m × 2 m, offset 0

Following the calculation of the worst-case peak external pressure coefficients for all grid area and offset combinations, the building is overlaid by ASCE 7-10's zones and the pressure results are distributed into their corresponding zones. This allows comparison of the results with ASCE 7-10's charts. The roof zones for TPU-1 and TPU-2 correspond to Figure 30.4-2A in ASCE 7-10, while TPU-3 corresponds to Figure 30.4-2B in ASCE 7-10. The wall zones for TPU 1 to 3 correspond to Figure 30.4-1 in ASCE 7-10. The width of the edge and roof corner zone is a = 1.61 m for TPU-1 and TPU-2, and a = 1.65 m for TPU-3. The zones applied to TPU-1 are shown in Figure 1. This comparison with ASCE7-10 is limited to the use and analysis of TPU wind tunnel pressure data, and how these data affect the ASCE design charts mentioned above. All other ASCE-7 specifications, e.g., temporal averaging, load factors, wind maps, etc. remain unchanged.

For the buildings considered, the width a of the edge zone (Fig. 1) is less than 2 m. Because TPU tap spacing is rather coarse, the smallest grid has dimensions 2 m × 2 m (e.g., see Fig. 14), wider than zones 2, 3, and 5 (Fig. 1). To populate these zones with pressure data, it was necessary to include partial grids: for zones 2, 3, and 5, if at least 50% of the grid area falls within the zone, pressure coefficients are calculated. In all other cases, the calculations only include grid areas that are completely contained within the zones.

The peak external pressure coefficients are assigned to their respective zones and the plots of the results are provided in Figures 20, 21 and 22 for TPU-1, -2 and -3 respectively.

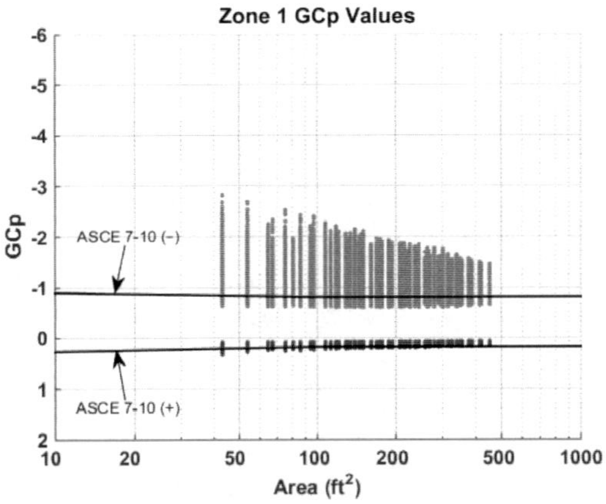

(a)

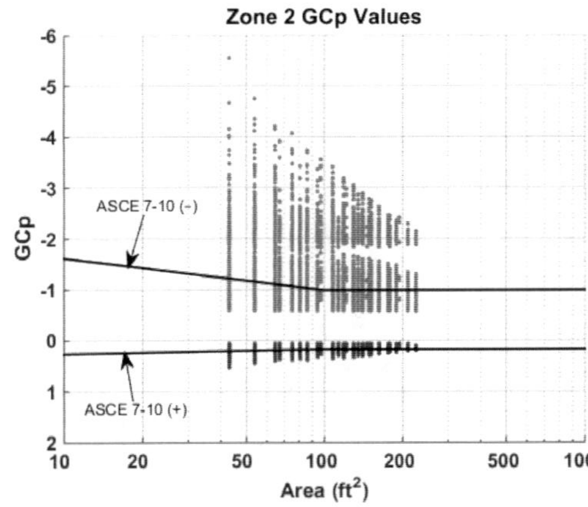

(b)

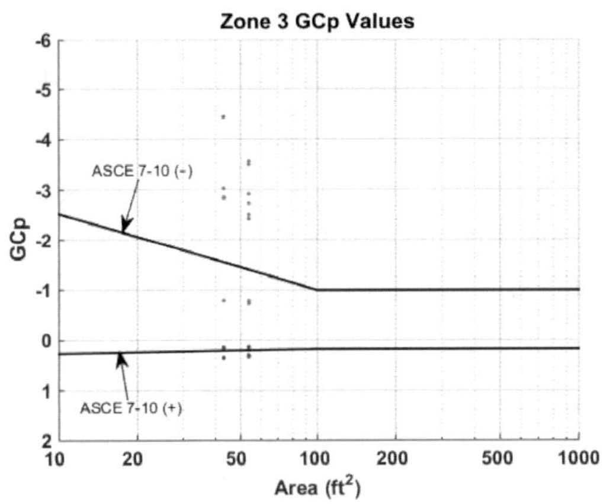

(c)

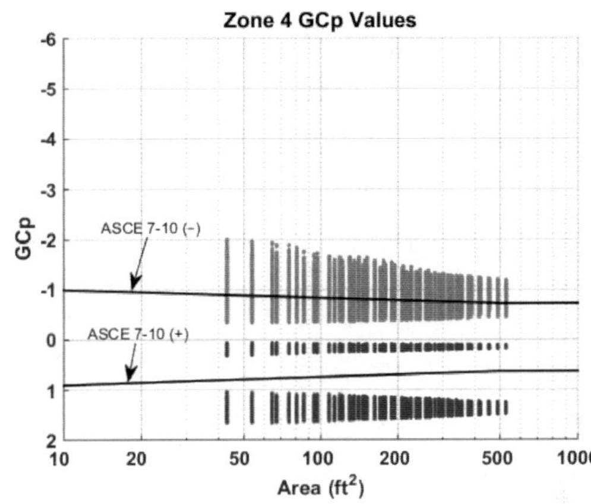

(d)

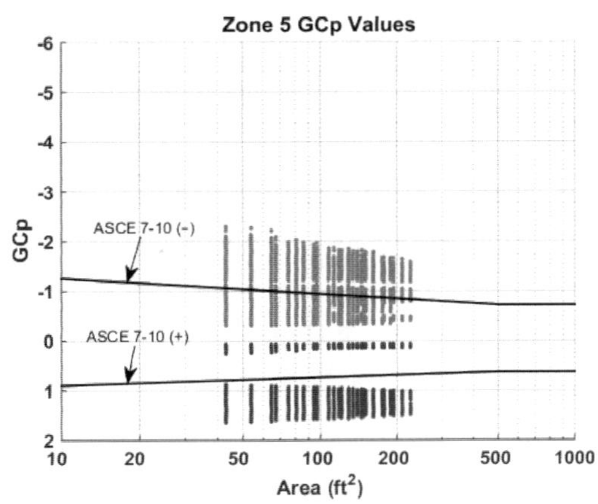

(e)

Figure 20. External pressure coefficients by zone for TPU-1 (1 $ft^2 = 0.093\ m^2$)

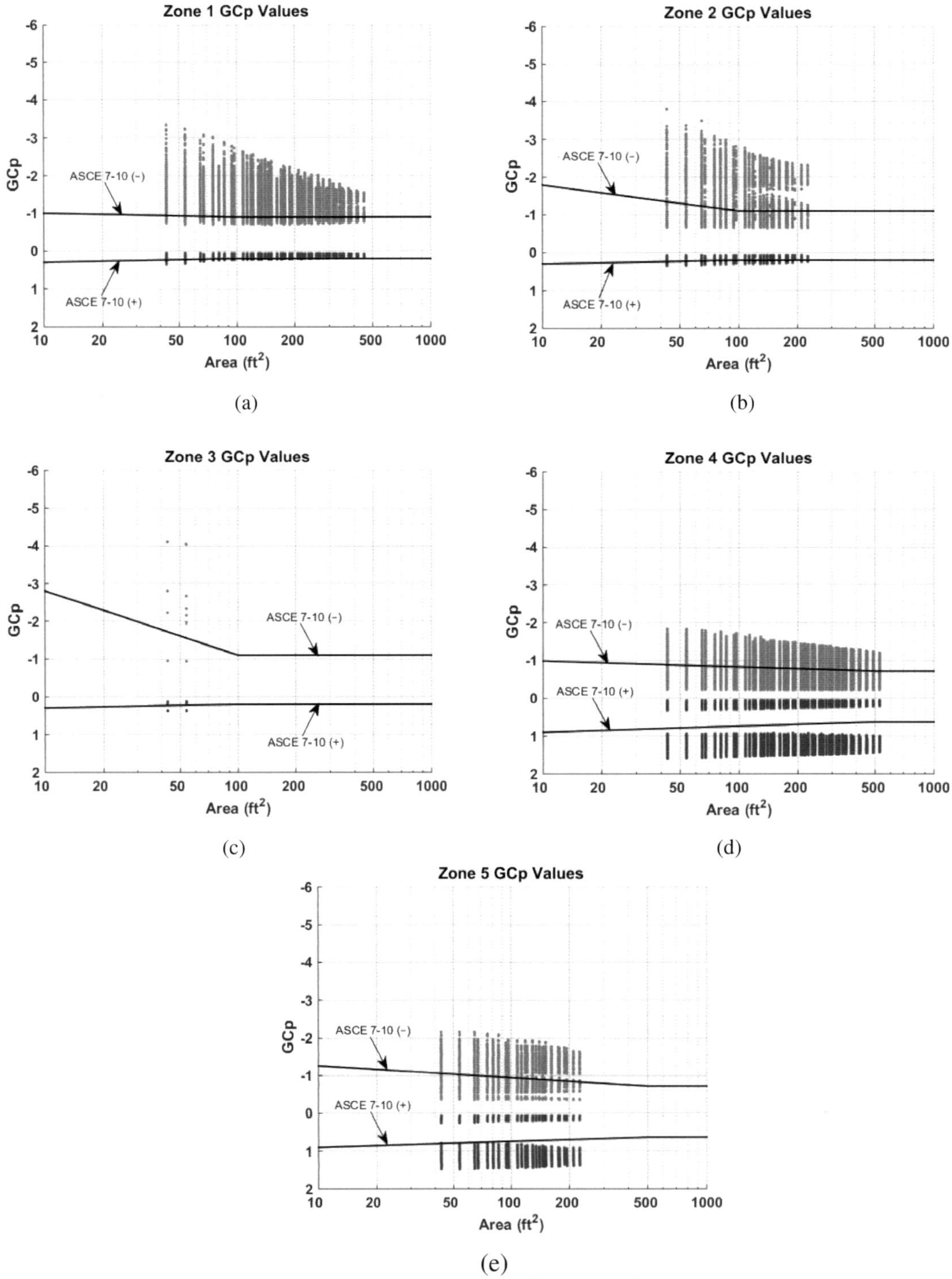

Figure 21. External pressure coefficients by zone for TPU-2 (1 ft^2 = 0.093 m^2)

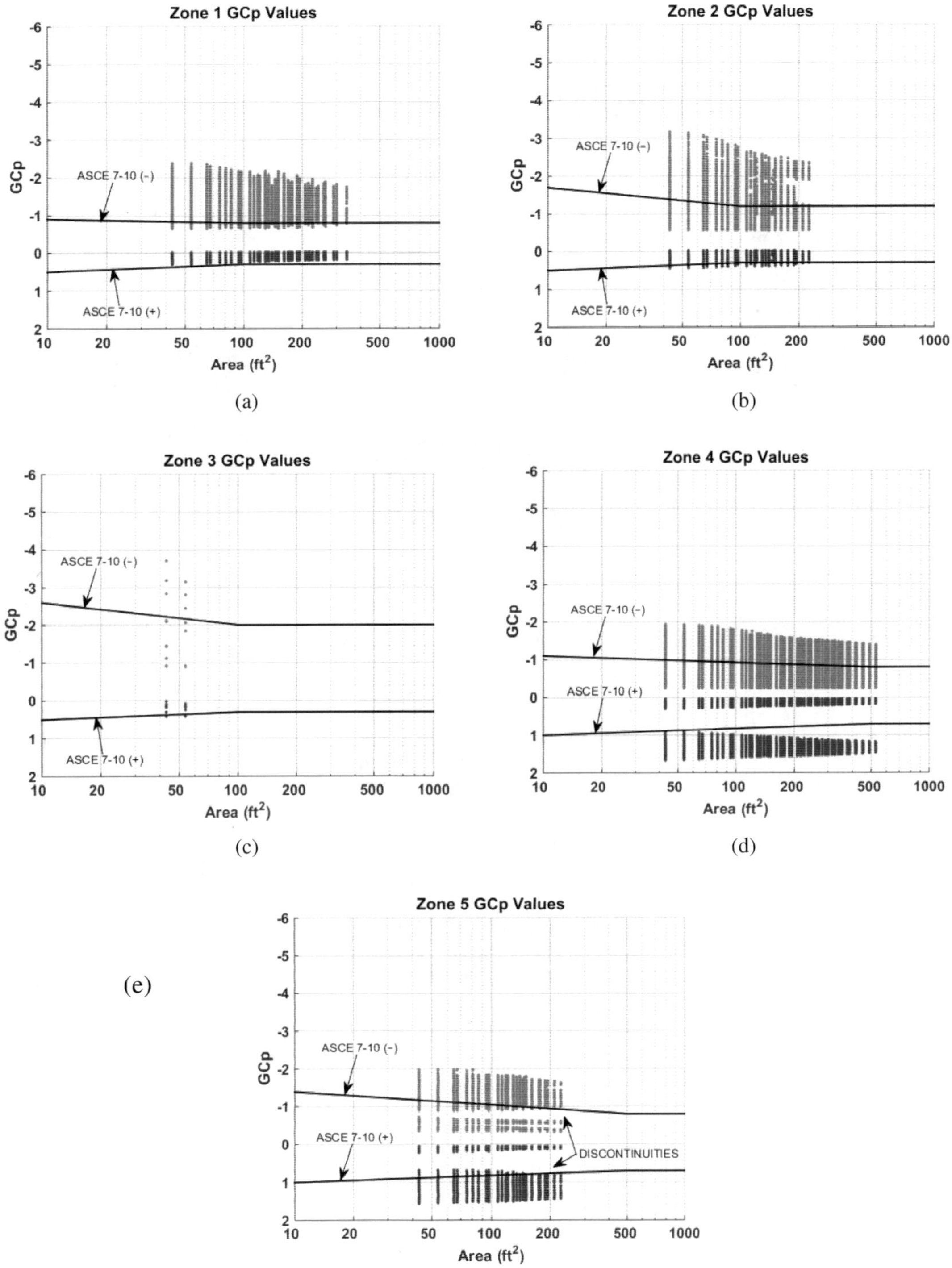

Figure 22. External pressure coefficients by zone for TPU-3 (1 ft^2 = 0.093 m^2)

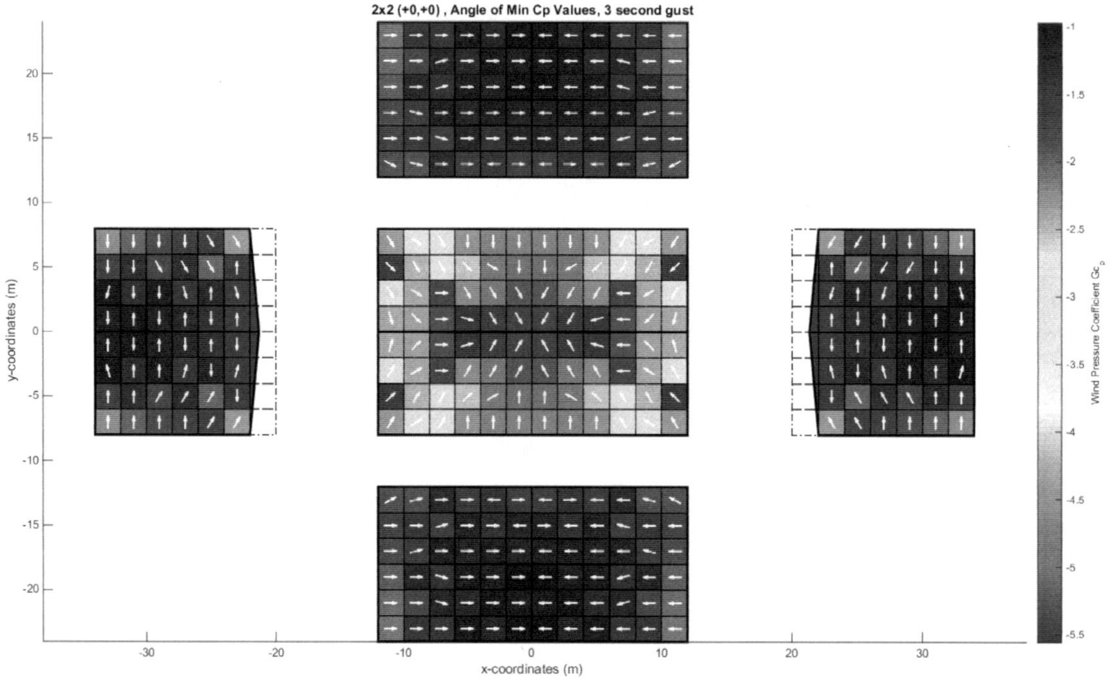

Figure 23. Worst-case wind directions for the negative (min.) pressures of TPU-1

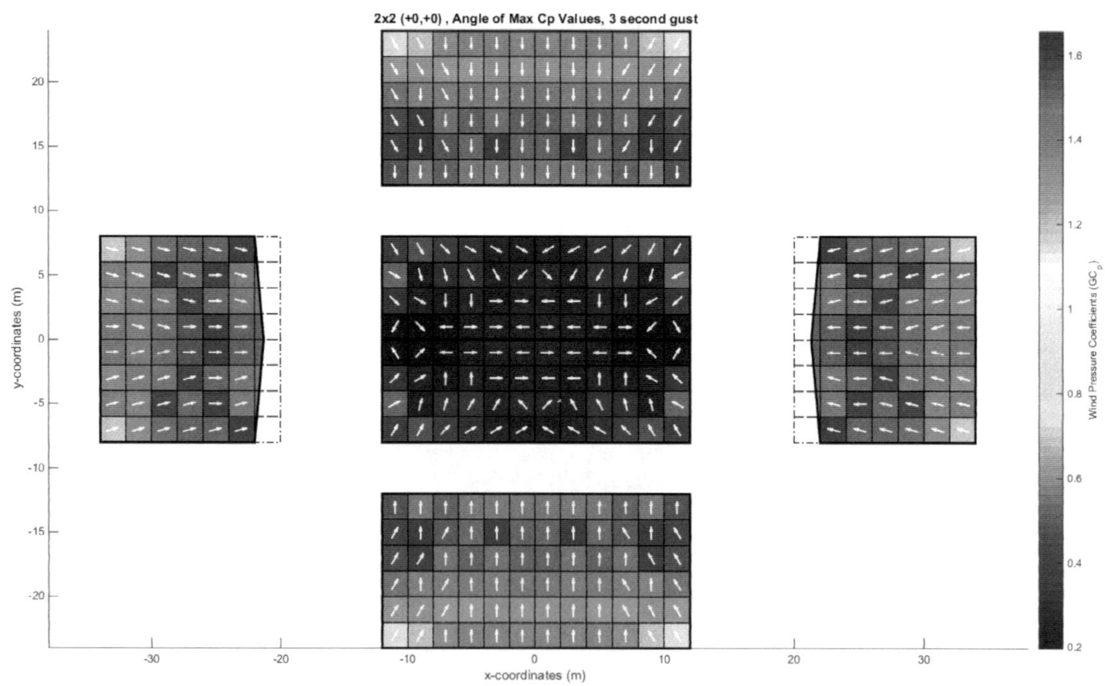

Figure 24. Worst-case wind directions for the positive (max.) pressures of TPU-1

In these plots, the negative data points represent the peak wind suction coefficients, while the positive data points represent the peak coefficients for wind pressure into the building. Current ASCE 7-10 specifications for the positive and negative external pressure coefficients are superimposed to the plots using solid lines. For a given zone and a grid of a given size, say, 50 ft^2 (5 m^2), and all the wind directions available in the measurements, the envelope method explained above produces one peak value of positive pressure and one peak value of negative pressure. The multiple results at, say, 50 ft^2 (5 m^2) come from changing the aspect ratio and the offset of the grid. The wind tunnel tests were only performed every 15° and this sometimes resulted in discontinuities in the pressure results (gaps in vertical spacing between results for a given area). The largest discontinuities appear on the walls, with the leeward walls exhibiting very small pressures compared with the windward walls.

Figures 23 and 24 show the wind direction (arrow) that causes the minimum and the maximum pressures for TPU-1 using a 2 m × 2 m grid. Due to the symmetry of building and terrain, TPU only tested wind directions from 0° and 90°. To extend the results over 0° to 360°, the doubly-symmetric building and terrain are divided into four quadrants. Each grid cell is grouped, through symmetry, with three other grid cells. The peak maximum (or minimum) pressure in each group is assigned to each cell in the group. The wind direction that causes the peak pressure is assigned to each cell in the group, mirrored for symmetry. The colors indicate the magnitude of the wind pressure.

9 DISCUSSION

Figure 13 presented the tributary areas of the irregularly spaced pressure taps, which were automatically calculated over the building's surfaces using the Voronoi diagrams. The Voronoi diagrams (coded in MATLAB) determine the tributary areas of irregularly spaced pressure taps. This ability distinguishes the proposed method as an effective approach in automating the calculation of external pressure coefficients from wind tunnel data.

Figures 14-19 presented the peak external pressure coefficients of TPU-1, TPU-2, and TPU-3 for the minimum and maximum pressure worst-case envelope of all available tested wind directions. The results were shown for a grid area of 2 m × 2 m with no offset, as this was the finest mesh justified by the spacing of the TPU wind tunnel pressure taps. As expected, the highest suction pressures occur in the windward exterior corner of the roof for the three buildings tested.

Figures 20-22 contain the peak external wind pressure coefficients sorted by the ASCE 7-10 zones, and overlaid with ASCE's external pressure coefficients (black lines). Discontinuities (i.e., vertical grouping) in the results for minimum pressure (red) and maximum pressure (blue) are due to the rather large gaps in wind directions tested, since the same surfaces can experience entirely different pressures depending on the wind direction. Discontinuities are also exhibited between peak pressure coefficients for windward and leeward walls (Figs. 20-22 d-e).

It can be observed in these plots that the current ASCE 7-10 specifications for external pressure coefficients on the components and cladding of low-rise buildings are unconservative for all corner, edge, and interior zones of the roof, especially for the negative pressures (suction); similarly, the specifications for the external pressure coefficients on the wall surfaces are unconservative, in both negative (suction) and positive (inward) pressures. The worst-case external pressure coefficient minimum peak value for TPU-1 over all 2 m × 2 m grid areas was -5.54.

Figures 23 and 24 show the critical wind directions that cause the peak minimum or maximum pressure in each grid area. As expected, the peak positive pressure occurs on a wall when the wind is acting perpendicularly to it, and the greatest suction pressure occurs when the wind is acting parallel to it. For the roof, the wind acting at an angle of 60° causes the greatest suction pressures on the roof corner (zone 3). For the horizontal roof edges (zone 2), the suction is strongest for wind directions perpendicular to the edge. For the inclined roof edges (along the gables, also zone 2), the suction is strongest with the wind acting at an angle around 60° for the geometries investigated. For negative wind pressures on the internal regions of the roof (zone 1), the governing wind direction varies widely between 0° and 90°. All angles in this discussion refer to the original TPU data (0° to 90°) and do not include the wind directions (0° to 360°) based on symmetry.

PART C

10 UNCERTAINTIES

The main sources of uncertainty in the pressure coefficients include the modeling of reality with a reduced scale model, the wind tunnel measurements, and the calculation of peak pressures. Long (2004) [28] observes an underestimation of the model-scale pressure coefficients (especially minimum pressure coefficients) compared to corresponding full-scale values. Table 3 compares the peak estimates for the most demanding wind directions and four corresponding pressure taps. Diaz (2006) [29] lists the major factors that contribute to pressure measurement uncertainties in wind tunnels as follows in Table 4 (referenced to $C_p = 1$):

Table 3 Comparison of pressure measurements between model and full scale (adapted from Long 2004) [28]

C_p	Tap 1	Tap 2	Tap 3	Tap 4
Model 1:100	-1.0	-4.4	+2.5	+0.6
Full scale	-1.2	-6.2	+2.5	+0.7

Table 4 Uncertainties in wind tunnel pressure measurements (adapted from Diaz 2006) [29]

Source of uncertainty	Relative uncertainty
Equipment effects	7 % to 13 %
Terrain effects	4 %
Hole dimension	1 %

It is therefore not surprising that different wind tunnels produce different measurements. Hölsher and Niemann (1998) [30] compared measurements from 15 wind tunnels and reported the statistics of maximum and minimum pressures on a floor-mounted cube 50 m high in full scale. For example, for wind directions of 0°, 10° and 40°, the minimum pressure coefficients on the roof from all 15 wind tunnels had a mean of -1.13, -1.43, -1.81 and a standard deviation of 0.12, 0.20 and 0.38, respectively.

Long (2004) [28] compared two different methods to estimate peaks of pressure. The first is the epochal method combined with the Best Linear Unbiased Estimate (Lieblein) [12], whereby a time series is divided into a number of epochs (20 in this case), the peak of each epoch is selected and fit into a Gumbel distribution, and the mean of that distribution (57th percentile) is used to estimate the extreme peak. The second method is the translation method, e.g., Sadek-Simiu (2002) [18], used in this paper. The translation approach depends upon the estimate of the marginal distribution of a non-Gaussian time series, which is typically difficult to perform reliably. According to Long (2004) [28], method 1 provides a better fit to the observed peak and is less conservative than method 2. Method 2 produced 21 estimates of peaks beyond 2 standard deviations of the observed value, compared to 7 for method 1.

Given the unconservatism of the ASCE design pressures, it is worth revisiting the peak estimation procedure. Gavanski, Gurley and Kopp (2016) [31] surveyed the literature and reported that six methods were used historically. According to this survey, the ASCE 7 method is based on "ensemble average of several peaks with some adjustment for intermediate roof slope". It is not clear how the ensemble average method, which averages several maxima or minima taken from equal–length segments of a time series, accounts for the longer duration of actual storms compared to the typical (full scale) duration of wind tunnel tests. In contrast, the translation method used in this paper "uses a full time series rather than observed peaks to obtain a peak cumulative distribution function". The estimated peaks depend on the duration of the windstorm, and this is accounted for by a duration ratio. The peak pressure in this paper is the mean (57th percentile) of the Gumbel distribution, in other words the expectation of the peaks. The survey pointed out that various peak estimation methods used quantiles ranging from 57 % to more than 95 %, with a great many researchers using 78 % or 80 %. This is a complex subject beyond the scope of this paper, but addressed in another paper in this Special Issue by some of the co-authors of this paper. Simiu et al. (2017) [32] argue that, with the assumption that the EV Type I distribution is an appropriate model, the use of the 78th percentile rather than the 57th percentile would be an instance of double counting, by increasing the pressure coefficient above its expected value, while also accounting for the deviation of the pressure from its expected value. A possible counter-argument is that the use of the 78th percentile is consistent with storm durations in excess of one hour (e.g., three hours). Note, however, that if a storm duration longer than one hour were assumed, the expected peak corresponding to it should be estimated directly by using the appropriate duration ratio. Also, the assumption that storm durations are longer than one hour would clearly violate accepted design practice, which follows the convention of a storm duration of one hour (see, e.g., ASCE 7-10, Eq. 26.9-11).

There is good justification to use the 57th percentile here, but the unconservativeness of the ASCE Specifications would be even worse, had the 78th percentile been used. It is not clear what percentile or what probabilistic method forms the basis of the current ASCE pressure coefficient curves.

11 COMPARISON OF THE TWO METHODS

This section compares the results obtained from the two methods explained above applied to data set jp2, Building 7, in suburban terrain exposure (NIST-UWO database). Method B (Voronoi) only used square grid cells of area 4 ft^2 (0.4 m^2) in this partial comparison with Method A, which used all area combinations as explained above. Figures 25-27 show that the results agree reasonably well. Because the grid used in Method B in this limited comparison includes no offset, it does not capture all the possible values of peak pressures. A more complete comparison requires the grid to be shifted in both *x*- and *y*- directions by various offsets to ensure that the peak area-averaged pressure is captured, and the use of rectangles of various dimensions in the grid.

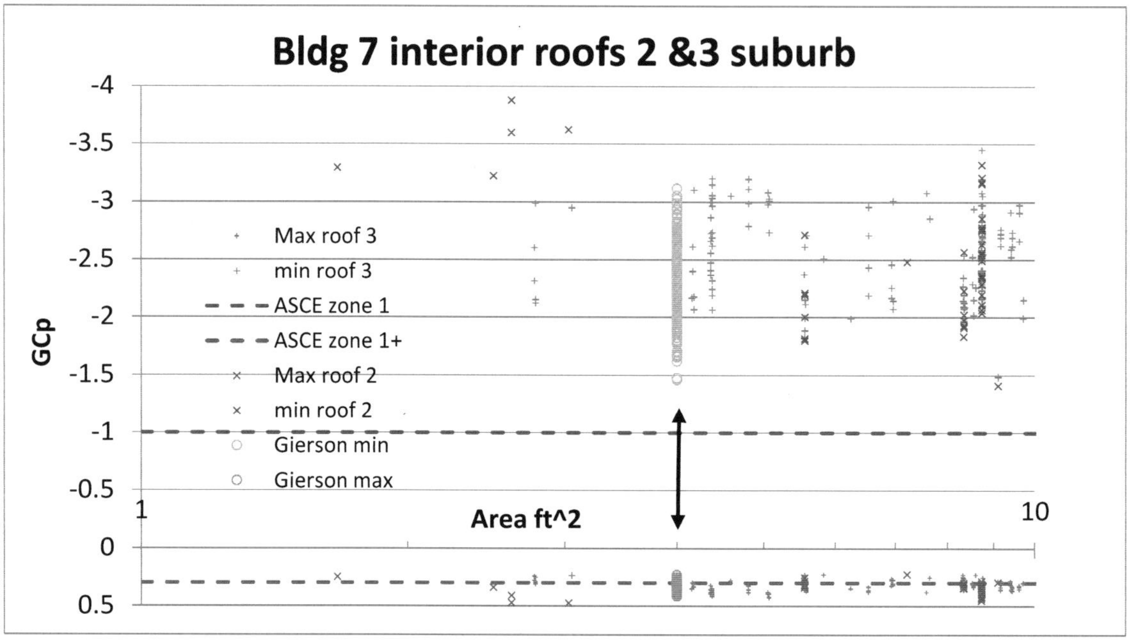

Figure 25. Comparison with Gierson (2015) for Bldg7, roofs 2 & 3 interior, suburban terrain exposure. Gierson's results (circles pointed at by double arrows) are shown for 4 ft^2 (1 ft^2 = 0.0929 m^2).

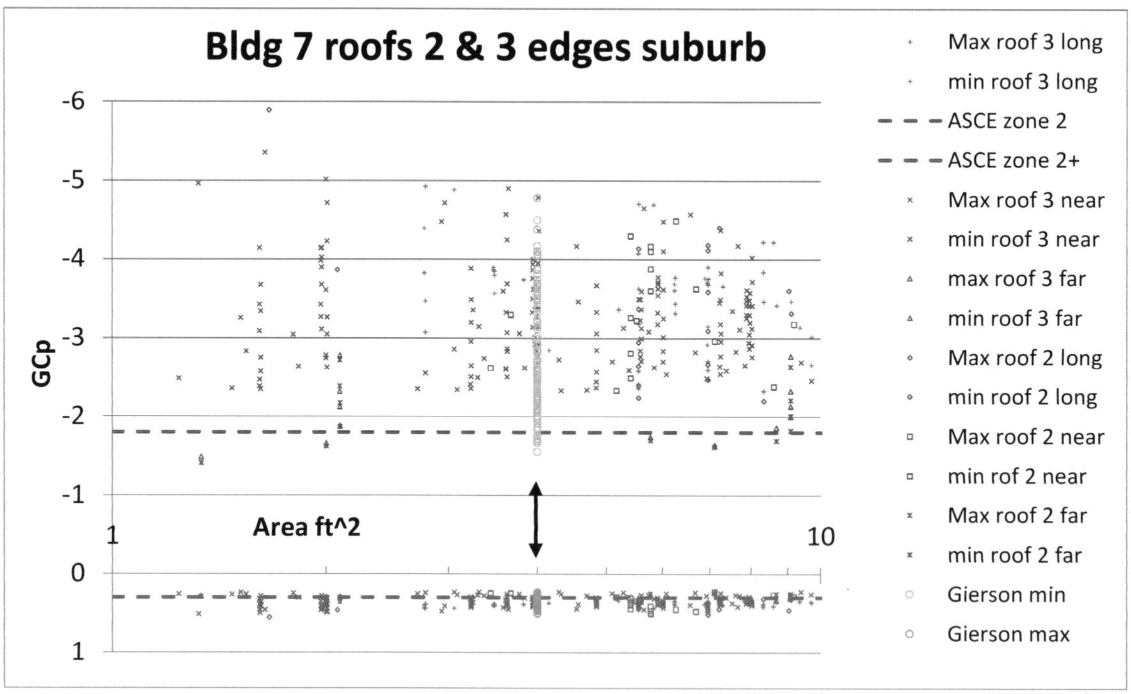

Figure 26. Comparison with Gierson (2015) for Bldg 7, roofs 2 & 3 edges, suburban terrain exposure. Gierson's results (circles pointed at by double arrows) are shown for 4 ft^2 (1 ft2 = 0.0929 m^2).

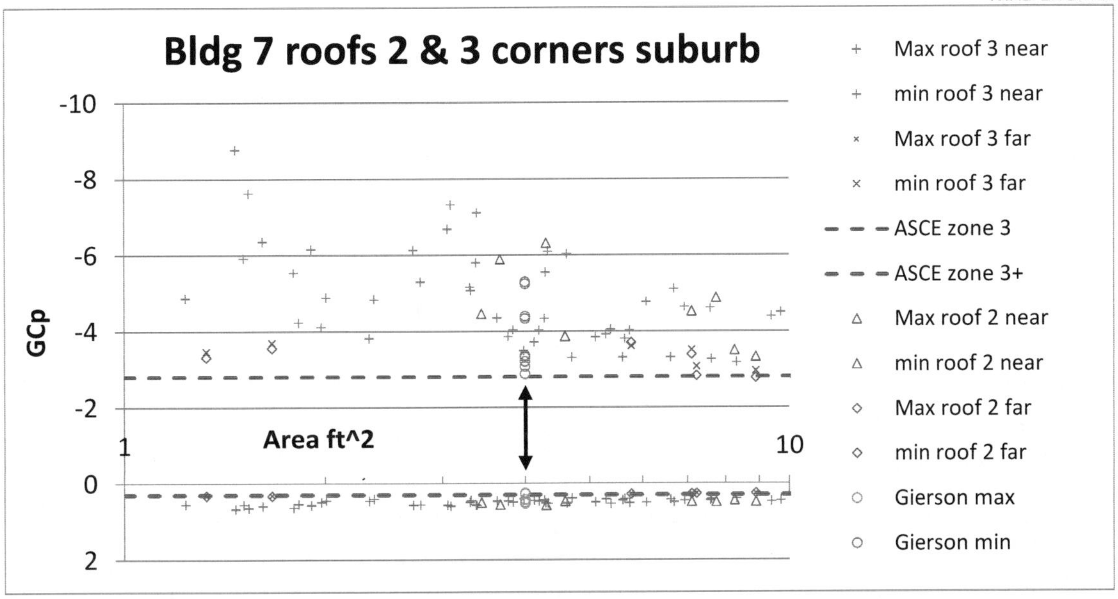

Figure 27. Comparison with Gierson (2015) for Bldg 7, roofs 2 & 3 corners, suburban terrain exposure. Gierson's results (circles pointed at by double arrows) are shown for 4 ft^2 (1 ft^2 = 0.0929 m^2).

12 CONCLUSION

This paper presents two methods used to exploit wind tunnel pressure measurements to arrive at graphs of peak positive and negative pressure coefficients corresponding to various areas over all wind directions. The first method is well suited to pressure taps placed in a rectilinear grid, whereas the second method (Voronoi diagrams) works for a general tap arrangement (but is less efficient than the first method on a rectilinear grid). Such graphs will eventually lead to specifications required for the design of components and cladding by the envelope method. According to the results shown here for one building and one terrain exposure form the NIST-UWO database, the negative wind pressure coefficients (suction) specified by ASCE 7-10 for gable roofs of low-rise buildings are too small, by factors ranging from 1.3 to 2.5, whereas the positive coefficients are adequate. For walls, both positive and negative pressures are underestimated by current ASCE specifications, by factors ranging from 1.6 to 2.0 and 1.2 to 2.1, respectively. Comparable results were obtained from three buildings and one terrain from the TPU database.

Future work will include the validation of these conclusions by alternative methods for the estimation of peaks (e.g., Duthinh, Pintar and Simiu 2016) [33]. All buildings in the NIST-UWO database are being investigated (Duthinh et al. 2015) [17], together with those in the TPU database. Only at the conclusion of the study can more definitive recommendations be made, including possibly changes in the number, size, shape and location of the various wind pressure zones, together with their associated pressure coefficients. This will be the subject of a future paper.

REFERENCES

1. ASCE 7-10 (2010) "Minimum Design Loads for Buildings and Other Structures," American Society of Civil Engineers, Reston, VA

2. NIST (2004) "Extreme Winds and Wind Effects on Structures" (Aerodynamic Database for Rigid Buildings) www.itl.nist.gov/div898/winds/homepage.htm, National Institute of Standards and Technology, Gaithersburg, MD, created 03/05/2004, last updated 05/22/2015, last accessed 09/04/2015

3. T.C.E. Ho, D. Surry, and D. Morrish (2003) "NIST/TTU Cooperative Agreement – Windstorm Mitigation Initiative: Wind Tunnel Experiments on Generic Low Buildings" University of Western Ontario, London, ON, Canada

4. T.C.E. Ho, D. Surry, and M. Nywening (2003) "NIST/TTU Cooperative Agreement – Windstorm Mitigation Initiative: Further Experiments on Generic Low Buildings" Univ. of Western Ontario, London, ON, Canada

5. T.C.E. Ho, D. Surry, D. Morrish, and G.A. Kopp (2005) "The UWO Contribution to the NIST Aerodynamic Database for Wind Loads on Low Buildings Part 1: Archiving Format and Basic Aerodynamic Data" J. Wind Eng. Ind. Aerodyn., **93** (1),1-30

6. Y. Tamura, (2012) "Aerodynamic Database for Low-Rise Buildings," Global Center of Excellence Program, Tokyo Polytechnic University, Tokyo, Japan

7. T. Stathopoulos (1979) "Turbulent Wind Action on Low-Rise Buildings," Ph.D. Thesis, The University of Western Ontario, London, ON, Canada

8. D. Meecham (1988) "Wind Action on Low-Rise Buildings," M.E.Sc. Thesis, The University of Western Ontario, London, ON, Canada

9. T. Stathopoulos, K. Wang, and H. Wu (1999) "Wind Standard Provisions for Low-Building Gable Roof Revisited," 10th Int. Conf. on Wind Eng., Copenhagen, Denmark, 155-162

10. E. Gavanski, B. Kordi, G.A. Kopp, and P.J. Vickery (2013a) "Wind Loads on Roof Sheathing of Houses," J. Wind Eng. & Ind. Aerodyn., **114**, 106-121

11. E. Gavanski, G.A. Kopp, and P.J. Vickery (2013b) "Roof Sheathing Wind Loads on Wood-Frame Residential Houses," 12th Americas Conf. on Wind Eng. (12ACWE), Seattle, WA, June

12. J. Lieblein (1974) "Efficient Methods of Extreme Value Methodology," NBSIR74-602, National Bureau of Standards, Washington, DC

13. P.J. Vickery, G.A. Kopp, and L.A. Twisdale (2013) "Component and Cladding Wind Pressures on Hip and Gable Roofs: Comparisons to the US Wind Loading Provisions," 12th Americas Conf. on Wind Eng. (12ACWE), Seattle, WA, June

14. E. Gavanski, and Y. Uematsu (2014) "Local Wind Pressures Acting on Walls of Low-Rise Buildings and Comparisons to the Japanese and US Wind Loading Provisions," J. Wind Eng. & Ind. Aerodyn.,**132**, 77-91

15. J. Main, and W.P. Fritz (2006) "Database-Assisted Design for Wind: Concepts, Software, and Examples for Rigid and Flexible Buildings," NIST Building Science Series 180, National Institute of Standards and Technology, Gaithersburg, MD

16. F. Habte, A. G. Chowdhury, D. Yeo, and E. Simiu (2016) "Iterative Design of Rigid Structures for Wind Using Time Series of Demand-to-Capacity Indexes," NIST Pub. 919053, ASCE J. Structural Engg. in review

17. D. Duthinh, J.A. Main, and B.M. Phillips (2015) "Methodology to Analyze Wind Pressure Data on Components and Cladding of Low-Rise Buildings," NIST TN 1903, National Institute of Standards and Technology, Gaithersburg, MD, http://nvlpubs.nist.gov/nistpubs/TechnicalNotes/NIST.TN.1903.pdf

18. F. Sadek, and E. Simiu (2002) "Peak Non-Gaussian Wind Effects for Database-Assisted Low-Rise Building Design," ASCE J. Eng. Mech. **128** (5), 530-539, May

19. J.A. Main (2011) "Special-Purpose Software: MATLAB Functions for Estimation of Peaks from Time Series," www.itl.nist.gov/div898/winds/homepage.htm, National Institute of Standards &Technology, Gaithersburg, MD

20. E. Simiu (2011) "Design of Buildings for Wind," 2nd ed. John Wiley, Hoboken, N.J.

21. C.S. Durst (1960) "Wind Speeds Over Short Periods of Time," *Meteor. Mag.* **89**, 181-187

22. E. Simiu and R.H. Scanlan (1996) "Wind Effects on Structures," 3rd ed. Wiley, New York, N.Y.

23. M. Morrison (2015) "Wind Pressure on Cladding of Low-Rise Buildings," private communication

24. G. Voronoi (1908) "New Applications of Continuous Parameters to the Theory of Quadratic Forms," *J. of Pure and Applied Mathematics*, **133,** 97-178. doi:10.1515/crll.1908.133.97

25. B. Delaunay (1934) "On the Empty Sphere." In Memory of Georges Voronoi, Bulletin of the USSR Academy of Sciences, Section: Mathematics and Natural Sciences. 6, 793-800.

26. MATLAB (2014) documentation, The Mathworks, Inc., http://www.mathworks.com/help/matlab/ (last accessed Apr. 29, 2016)

27. A. Hagos, F. Habte, A. Chowdhury, and D. Yeo (2014). "Comparisons of Two Wind Tunnel Pressure Databases and Partial Validation against Full-Scale Measurements." J. Struct. Eng., 10.1061/(ASCE)ST.1943-541X.0001001, 04014065.

28. F. Long (2004) "Uncertainties in pressure coefficients derived from full and model scale data," Master's thesis, Texas Tech University, Lubbock, Tx

29. P.S.Q. Diaz (2006) "Uncertainty analysis of surface pressure measurements on low-rise buildings," Master's thesis, The University of Western Ontario, London, Ontario, Canada

30. N. Hölscher, and H. J. Niemann (1998) "Towards quality assurance for wind tunnel tests: a comparative testing program of the Windtechnologische Gesellschaft," J. Wind Engineering and Industrial Aerodynamic, 74, 599-608, March, DOI: 10.1016/S0167-6105(98)00054-3

31. E. Gavanski, K.R. Gurley, and G.A. Kopp (2016) "Uncertainties in the estimation of local peak pressures on low-rise buildings by using the Gumbel distribution fitting approach," ASCE J. of Structural Engg., vol. 142, no. 11, Nov. paper 04016106.

32. E. Simiu, A.L. Pintar, D. Duthinh, and D. Yeo (2017) "Wind Load Factors for Use in the Wind Tunnel Procedure," ASCE-ASME J. of Risk and Uncertainty in Engineering Systems, Part A: Civil Engineering, Special Issue: Treatment of uncertainties for windstorm risk assessment, in press.

DISCLAIMERS

1. The policy of the National Institute of Standards and Technology is to use the International System of Units (SI) in its technical communications. In this paper, however, building codes and standards are referenced in both customary (as is the practice in US construction industry) and SI units.

2. Some commercial products are identified in this paper for traceability of results. Such identification does not imply recommendation or endorsement by the National Institute of Standards and Technology, nor does it imply that the products identified are necessarily the best available for the purpose.

Comparison of Design Loads for Performance-Based Design of Low-Rise Structures

S.M.A. Hoq[a], J.P. Judd[b]

[a] *Research Assistant, Civil & Arch. Eng., Univ. of Wyoming, 1000 E. Univ. Ave., Laramie, WY, USA*
[b] *Assistant Professor, Civil & Arch. Eng., Univ. of Wyoming, 1000 E. Univ. Ave., Laramie, WY, USA*

ABSTRACT: In this paper, design wind loads for components and cladding based on the envelope and directional procedures are compared with design loads based on wind tunnel tests to determine the suitability of the procedures for performance-based design of low-rise buildings. Wind tunnel loads are extracted from existing databases that containing pressure coefficient histories for low-rise buildings with various aspect ratios, in terms of width, length, and height, roof slopes, subjected to a range of wind speed profiles (terrain exposure categories). Wind loads to be used in the design of the components and cladding are then compared to loads based on wind tunnel tests. An analysis of the temporal and spatial variation in loads is conducted to determine the mean, minimum, and maximum lateral loads on the structure as a function of wind direction. The implications of the procedures for roof sheathing in wood frame construction are briefly discussed through an example examination of roof sheathing panel damage.

KEYWORDS: Performance based design, response history analysis, components and cladding.

1 INTRODUCTION

The current practice in the United States for calculating wind loads on buildings and other structures is given in ASCE 7-10 [1]. The wind load provisions in ASCE 7-10 comprise three procedures: the envelope procedure, the directional procedure, and the wind tunnel procedure. Typically, the envelope procedure is used for low-rise and regularly shaped structures. In ASCE 7-10, buildings with an eave height less than the least horizontal dimension and less than 18.3 m (60 ft) are defined as low rise buildings. The directional procedure is used for tall or irregular structures. Lastly, the wind tunnel procedure is used for tall buildings, flexible structures and irregular structures, as well as for structures with shielding from adjacent buildings.

As depicted in Figure 1, the envelope procedure loads are based on "pseudo" wind pressures that envelope the structural response, whereas the directional loads are based on actual surface pressures as a function of the wind direction. In the wind tunnel procedure, loads are based on surface pressures measured during an atmospheric boundary layer wind tunnel test of a reduced-scale model of the specific structure involved.

Conventionally, wind loads are then employed in either an allowable strength design (ASD) approach, or in a "strength," or load and resistance factor, design (LRFD) approach. Alternatively, the loads can be employed in a performance-based design approach, where the structural system is designed to meet specific performance requirements for a given level of windstorm hazard. Compared to the conventional design approaches, performance-based design has the potential to provide a better understanding of the probably serviceability issues (i.e. damage or loss of functionality) and the potential to better mitigate the impact of extreme windstorms (e.g. hurricanes and thunderstorms).

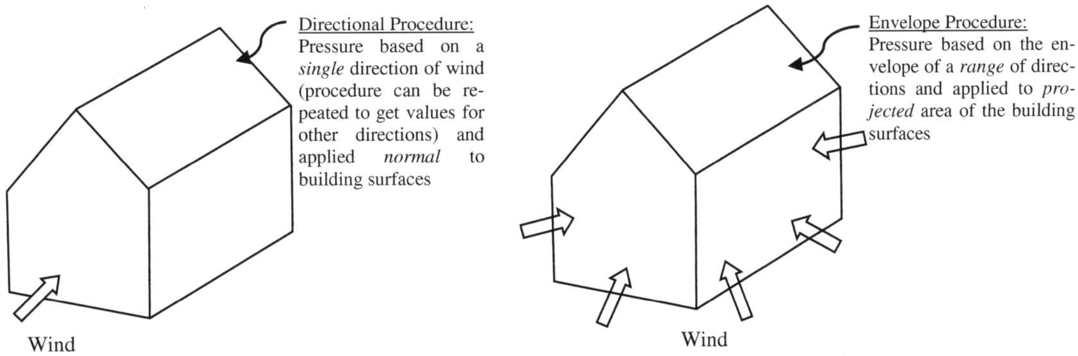

Figure 1. Schematic of directional procedure (left) and envelope procedure (right) to calculate design wind loads.

In this study, loads based on wind tunnel testing are compared with components and cladding loads calculated using the envelope and directional procedures in ASCE 7-10 for a set of low-rise buildings. The study employs data from publicly available wind tunnel test databases for buildings: The National Institute of Standards and Technology (NIST) database [2], and the Tokyo Polytechnic University (TPU) database [3], but in this paper, only the results from the NIST database are reported. Each building in the set of buildings considered has the same plan dimensions, 19.05 m x 12.19 m (62.5 ft x 40 ft), a low-sloped roof (1:12 roof pitch), and has an eave height that varies from 3.66 m (12 ft) to 12.19 m (40 ft).

2 WIND TUNNEL DATA

NIST aerodynamic database contains wind tunnel test data from tests conducted on small-scale models of low-rise buildings at the Boundary Layer Wind Tunnel Lab at the University of Western Ontario (UWO). The database comprises seven plan dimensions, five roof slopes, and eave heights ranging from 3.66 m to 12.19 m (12 ft to 40 ft). Test data is available for both open and suburban terrain conditions for wind angle directions varying from 0° to 360°. In the tests, pressure taps (small holes in the surface of the model used to measure static wind pressure) were installed on different locations in the model. A tube system connected the pressure taps to a solid state high speed pressure scanners to obtain wind pressure measurements. Wind characteristics and simulation of the atmospheric boundary layer were accomplished following the Engineering Sciences Data Unit (ESDU) documents: ESDU 82026[4], ESDU 83045[5] and ESDU 74031[6]. The database is described in detail elsewhere (i.e. [7]) and the simulated wind characteristics compare well with ESDU profiles (see, for example, [7, 8, and 9]).

For this paper, buildings with a plan dimension of 19.05 m x 12.19 m (62.5 ft x 40 ft) and 1:12 roof pitch (4.76° roof slope) were selected in open and suburban terrain (exposure categories "C" and "B," respectively, in ASCE 7-10). The selected scaled models had a length scale of 1:100, relative to the full-scale building dimension. The models had approximately 300 pressure taps on the roof surfaces with densely spaced pressure taps in one of the leading corners (Figure 2). The layout of pressure taps on wall surfaces was similar to the layout on the roof surfaces [7]. Wind tunnel test data is available for both open and suburban exposure for wind angle directions varying from 5° to 90° and 270° to 360° in every 5° interval. A three dimensional model generated by windPRESSURE Database-Assisted Design (DAD) software [2] is presented in Figure 2 (on the following page).

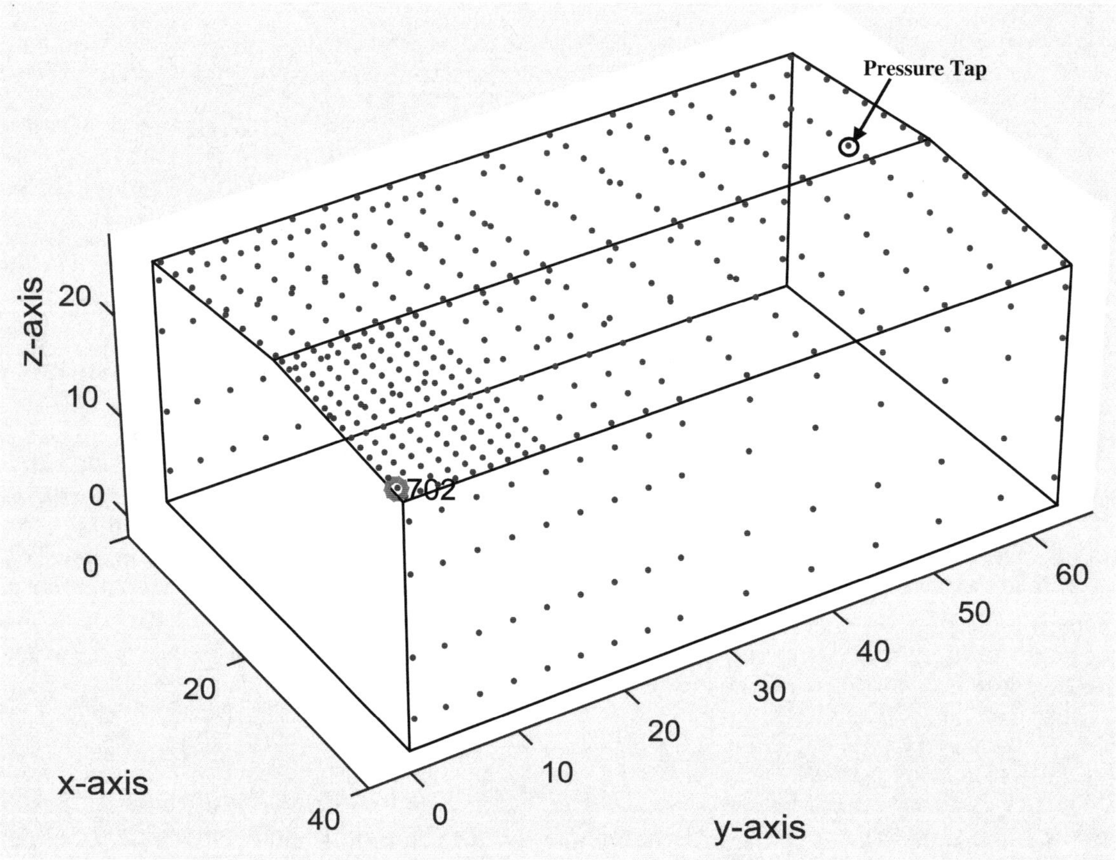

Figure 2. Three-dimensional plot of the building, generated using DAD software [2] (axis dimensions in ft).

All the pressure coefficients in the database were referenced to upper level dynamic pressure in the wind tunnel. The pressure coefficients referenced to upper level of the wind tunnel were converted to the roof (eave) height of the model building by dividing with square of ratio of roof to reference wind speed (V_H/V_{ref}). The wind speed V_{ref} was measured at the upper level of wind tunnel and V_H was measured at the roof height of the building model.

The ratios of V_H/V_{ref} are tabulated in the associated test report [8] and are also provided in the accompanying Hierarchical Data Format (HDF) files [2]. For the plan dimensions considered in this study, velocity ratios for open and suburban exposure are 0.561 and 0.447 for 3.66 m (12 ft) eave height, 0.587 and 0.468 for 5.49 m (18 ft) eave height, 0.620 and 0.487 for 7.32 m (24 ft) eave height, and 0.657 and 0.534 for 12.19 m (40 ft) eave height [8]. The pressure tap layout for the roof zone is shown in Figure 3 (on the following page). The figure is oriented so that the leading corner with closely spaced pressure taps is in the bottom left corner. A close up view of the building model was extracted from the HDF file and is shown in Figure 4 (on the following page).

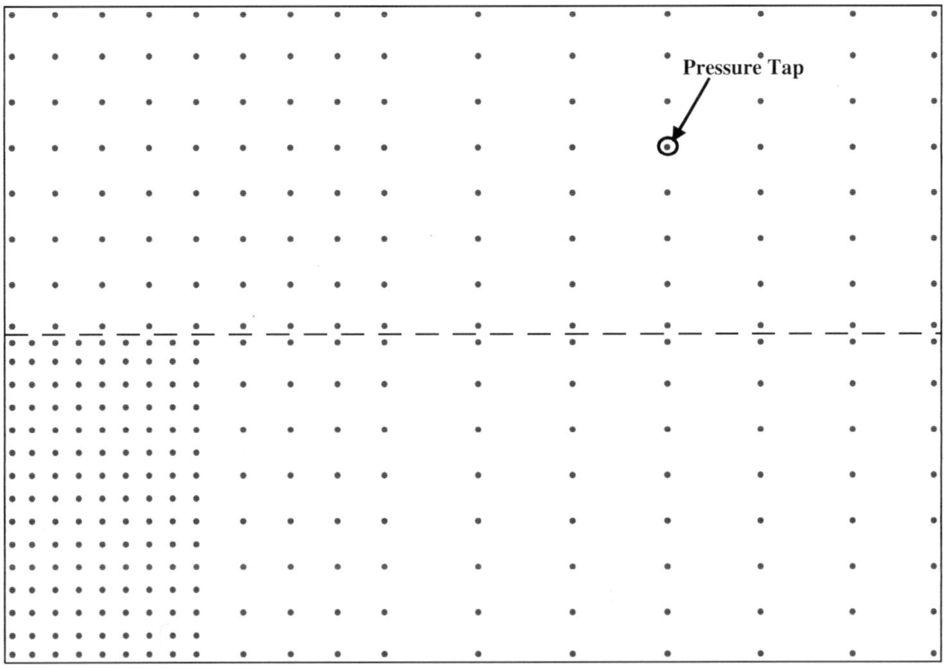

Figure 3. Layout of pressure taps on the roof surface of the building model.

Figure 4. A perspective view of the model building in the wind tunnel [2].

3 PRESSURE COEFFICIENTS

3.1 *Equivalent $GC_{p(eq)}$*

The external pressure coefficient measured in the wind tunnel tests was transformed to an equivalent external pressure coefficient, GC_p corresponding to ASCE 7-10 using the conversion factor equation given in Equation 1 developed by St. Pierre et al. [9].

$$GC_{p(eq)} = \frac{\frac{1}{2}\rho V^2_{h,z_0,mean\ hrly} C_p}{\frac{1}{2}\rho V^2_{10m,OC,3sec\ gust} K_{zt} K_h K_d I} = F_{WT} C_p \quad (1)$$

In Equation 1, $GC_{p(eq)}$ is equal to the ASCE 7-10 equivalent pressure coefficient; ρ is equal to the density of air; $V_{h,z_0,mean\ hrly}$ is equal to the mean hourly velocity, referenced to the mean roof height with the roughness height z_0; $V_{10m,OC,3sec\ gust}$ is equal to the 3-second gust wind speed at a 10-m height in an open terrain; K_{zt} is equal to the topographic factor; K_h is equal to the velocity exposure factor; K_d is equal to the wind directionality factor; F_{WT} is a factor used to convert the wind tunnel pressure coefficient to ASCE 7-10 equivalent pressure coefficient; and C_p is equal to the pressure coefficient measured in the wind tunnel test. The importance factor I, not used in the current ASCE 7 standard-10, is equal to 1.0. The topographic factor K_{zt} was equal to 1.0. Furthermore, the wind tunnel data was not reduced to account for any directionality effect and wind directionality factor K_d was equal to 1.0 in the calculation of the conversion factor F_{WT} [9]. Velocity exposure factors K_h were obtained using ASCE 7-10.

The conversion factor F_{WT} depends on the velocity ratio, expressed in Equation 2 [9].

$$\left(\frac{V_{h,zo,mean\ hr}}{V_{10m,zo=0.03m,3secgust}}\right)^2 = \left(\frac{V_{10m,zo=0.03,mean\ hr}}{V_{10m,zo=0.03m,3secgust}}\right)^2 \times \left(\frac{V_{10m,zo,mean\ hr}}{V_{10m,zo=0.03m,mean\ hr}}\right)^2 \times \left(\frac{V_{h,zo,mean\ hr}}{V_{10m,zo,mean\ hr}}\right)^2 \quad (2)$$

The first term of the right-hand side of Equation 2 is equal to 1.52 based on the durst curve in ASCE 7-10, Fig. C26.5-1. The second term is equal to 1.0 for open exposure, and equal to 0.638 for suburban exposure. The last term was calculated using the associated references and HDF files [2, 7, 8]. The results for the pressure conversion factor F_{WT} between wind tunnel data and ASCE 7-10 are tabulated in Table 1.

Table 1. Pressure conversion factor to convert wind tunnel values to ASCE 7-10 values.

Eave Height	*Open Exposure*		*Suburban Exposure*	
m (ft)	K_h	F_{WT}	K_h	F_{WT}
3.66 (12)	0.85	0.391	0.70	0.300
5.49 (18)	0.88	0.411	0.70	0.329
7.32 (24)	0.94	0.433	0.70	0.357
12.19 (40)	1.04	0.434	0.76	0.395

3.2 *Comparison of pressure coefficients between ASCE 7-10 and Wind Tunnel test data*

In a wind tunnel test, the extreme values of area averaged pressure coefficient are lower than that recorded from a single pressure tap of that area [10]. Thus, the roof of the model building was divided into 0.9 m x 0.9 m (3 ft x 3 ft) areas and a single pressure coefficient was calculated using the area-weighted average of all the pressure tap values bounded by the area. There is 0.3-m (1-ft) strip as offset in X-direction and 0.8-m (2.5-ft) strip as offset in Y-direction, due to the selected size of area and the area of the roof. A comparison between the pressure coefficient time series of a 0.9 m x 0.9 m area with a pressure coefficient time series based on a single pressure tap (Tap No. 702 in Figure 2) is shown in Figure 5 (on the following page), for the 7.32-m building in open exposure with a 330° wind direction for model scale duration of 100 seconds.

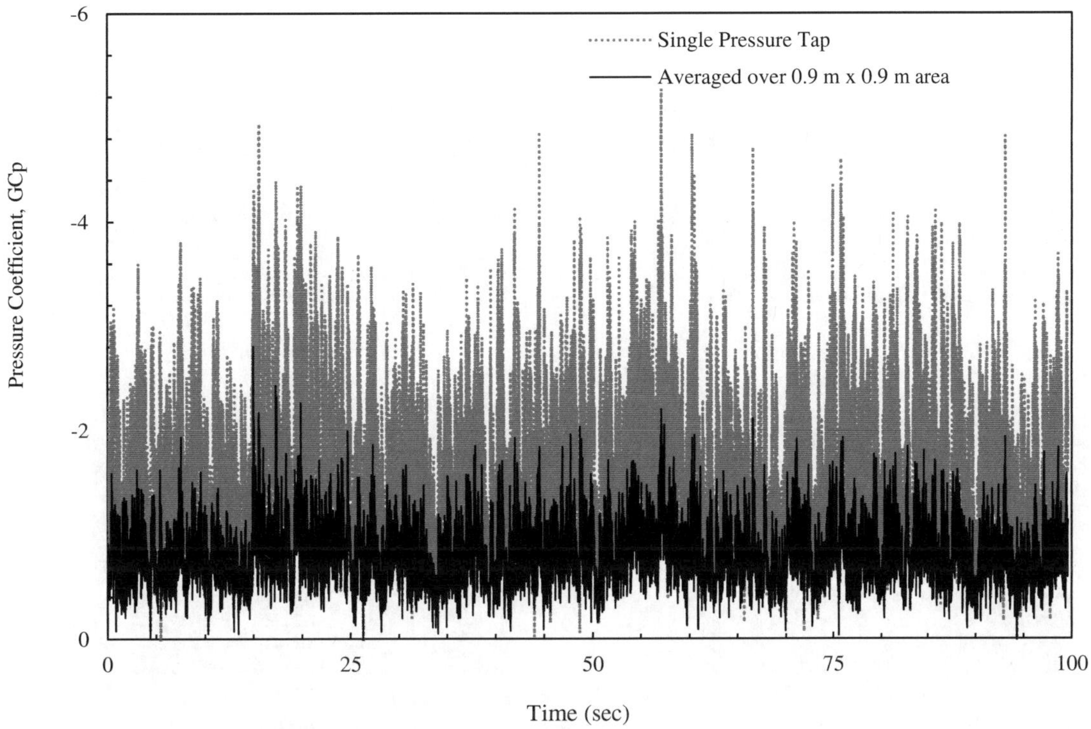

Figure 5. Wind tunnel pressure coefficient time series (model-scale duration).

The maximum, minimum and mean pressure coefficient values for the entire time series were obtained from the HDF files [2] and the area weighted average values were calculated using these values. The extreme pressure coefficient envelopes for wind directions of 270°, 300°, 330° and 360° for a 0.9 m x 0.9 m (3 ft x 3 ft) area are shown in Figure 6 and Figure 7 (on the following page) for the 12.2 m x 19.1 m (40 ft x 62.5 ft) roof. Figure 6 and Figure 7 show plots of the spatially-varying extreme pressure coefficients. The pressure coefficients range from 0.0 to -3.0, where a negative value indicates pressure acting outward (suction). Each plot is scaled to the same range of values. The values were converted from wind tunnel test data to ASCE 7-10 equivalent pressure coefficient using Equation 1. The extreme pressure coefficient values for a single pressure tap along with the extreme value for a 0.9 m x 0.9 m (3 ft x 3 ft) area are tabulated in Table 2. The extreme equivalent ASCE 7-10 pressure coefficients for suburban exposure are larger than the values for open exposure. The difference is due to a smaller values of K_h and V_H/V_{ref} for suburban exposure, compared to open exposure (K_h is used in the calculation of F_{WT}, and V_H/V_{ref} is used to re-reference pressure coefficients to the eave height).

The pressure coefficient values based on the wind tunnel tests are higher than the values given in ASCE 7-10 and depicted in Figure 8. In Figure 8, the design pressure coefficient values from ASCE 7-10 are given for a 11.9 m x 18.3m (39 ft x 60 ft) roof without any offset for a 0.9 m x 0.9 m area. However, instead of considering the extreme values recorded in the HDF files [2], if a moving average is considered, then the pressure coefficient values from the wind tunnel test will be smaller and be in closer agreement with values from ASCE 7-10. For example, for a full-scale mean wind speed of 51.4 m/s (115 mph), the 3-s moving averages for the entire time series were calculated for wind directions of 270°, 300°, 330° and 360°. The envelope of the pressure coefficients for the 3-s moving average are tabulated in Table 2.

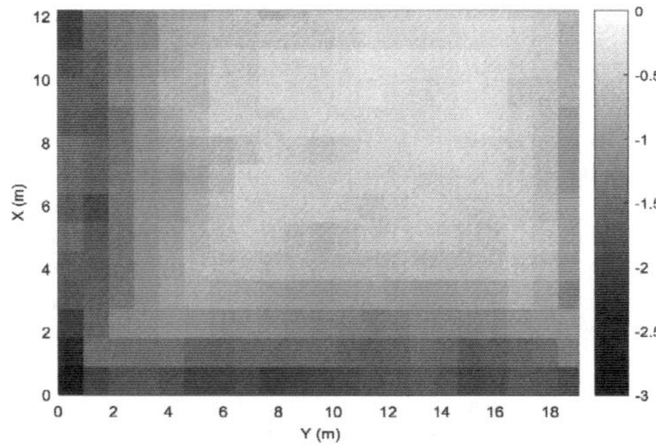

(a) 3.66 m (12 ft) Building.

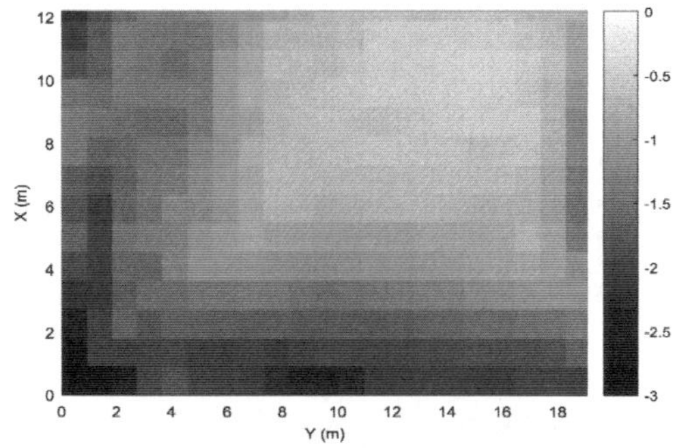

(b) 5.49 m (18 ft) Building.

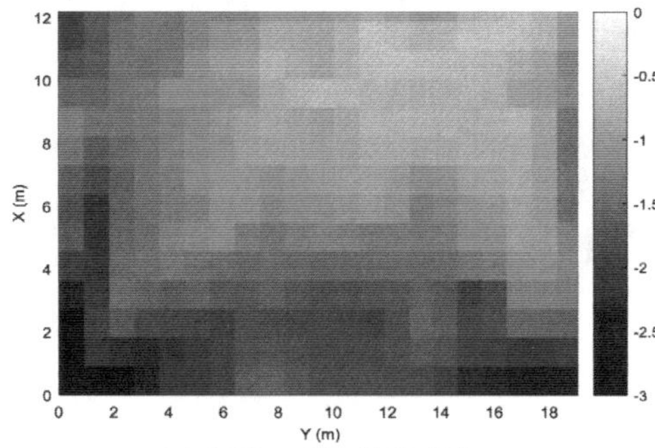

(c) 7.32 m (24 ft) Building.

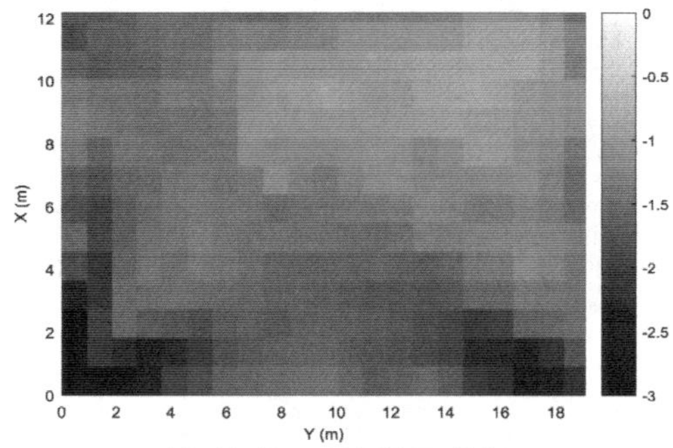

(d) 12.19 m (40 ft) Building.

Figure 6. Spatial variation of extreme equivalent ASCE 7-10 roof C&C pressure coefficients for open exposure.

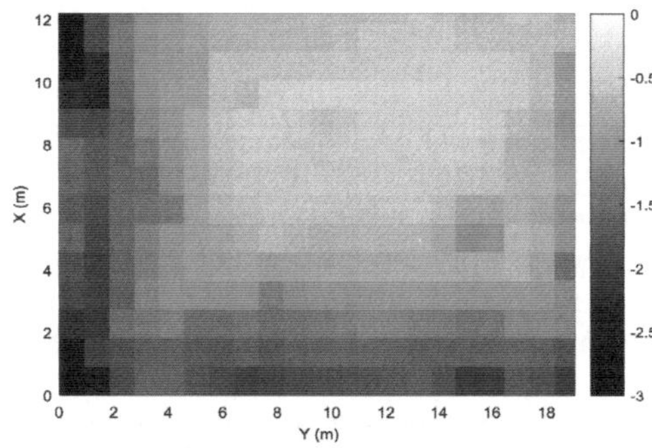

(a) 3.66 m (12 ft) Building.

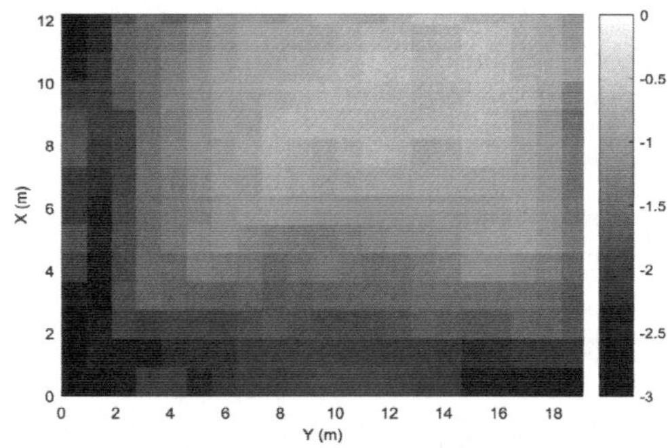

(b) 5.49 m (18 ft) Building.

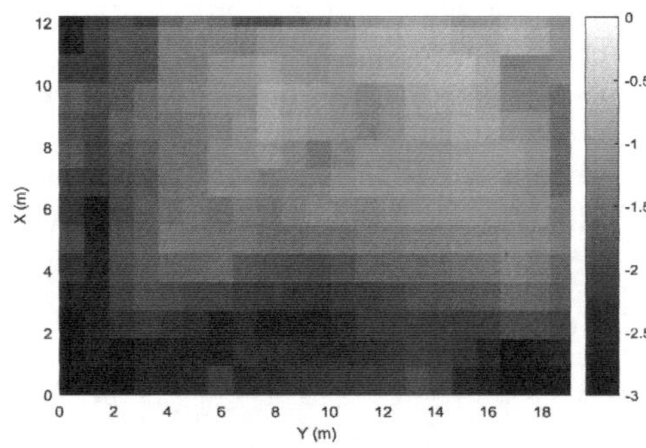

(c) 7.32 m (24 ft) Building.

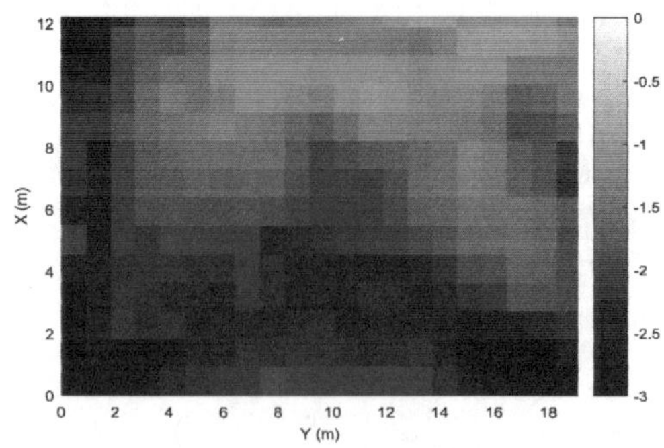

(d) 12.19 m (40 ft) Building.

Figure 7. Spatial variation of extreme equivalent ASCE 7-10 roof C&C pressure coefficients for suburban exposure.

The envelope of pressure coefficient values based on the 3-s moving average were 58% and 48% of the extreme values for open and suburban exposure, respectively. The data indicates that buildings with higher eave heights experienced an ASCE 7-10 equivalent pressure coefficient on the roof zone that was larger compared to buildings with lower eave heights. In making this comparison, note that the equivalent pressure coefficient was intended to account for both the effect of exposure and the effect of elevation by normalizing the value with the velocity exposure factor K_h, using Equation 1. Apparently, for the set of building models considered in this study, the velocity exposure factor K_h in ASCE 7-10 alone did not compensate for the effect of height and exposure on the variation of pressure coefficient.

Table 2. Equivalent ASCE 7-10 pressure coefficient values based on wind tunnel tests.

		Open Exposure		
Eave Height	Extreme $GC_{p(eq)}$		3-s moving $GC_{p(eq)}$ (V = 51.4 m/s)	
m (ft)	Pressure Tap	0.9 m x 0.9 m area	Pressure Tap	0.9 m x 0.9 m area
3.66 (12)	-3.8	-2.8	-2.2	-1.6
5.49 (18)	-5.3	-3.7	-3.0	-2.1
7.32 (24)	-6.0	-4.1	-3.5	-2.6
12.19 (40)	-6.1	-4.3	-3.6	-2.6
		Suburban Exposure		
Eave Height	Extreme $GC_{p(eq)}$		3-s moving $GC_{p(eq)}$ (V = 51.4 m/s)	
m (ft)	Pressure Tap	0.9 m x 0.9 m area	Pressure Tap	0.9 m x 0.9 m area
3.66 (12)	-4.0	-3.3	-1.9	-1.5
5.49 (18)	-5.8	-4.3	-2.9	-2.1
7.32 (24)	-6.3	-4.9	-3.1	-2.2
12.19 (40)	-7.2	-5.4	-3.7	-2.6

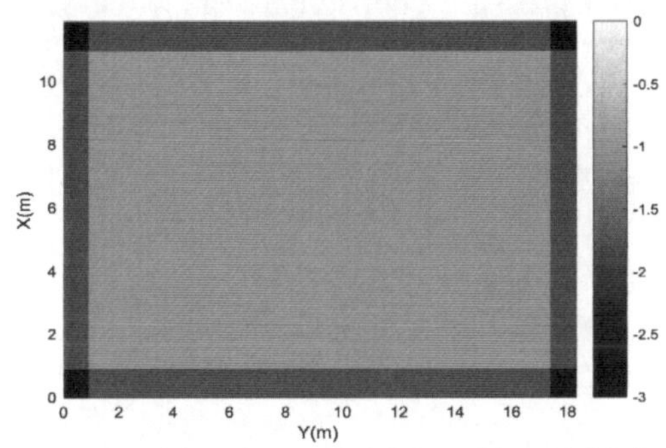

Figure 8. Spatial variation of ASCE 7-10 roof C&C pressure coefficients.

3.3 *Comparison on design wind pressure*

Design wind pressures on components and cladding were calculated following the envelope and directional procedures in ASCE 7-10 and were compared with the wind tunnel test results. The procedure for obtaining wind pressures for components and cladding (C&C) are given in Chapter 30 of ASCE 7-10, comprising Part 1 and Part 2 (simplified method) sections. Part 1 uses Equations 3 and 4, and the simplified method in Part 2 uses tables in ASCE 7-10 with adjustment using Equation 5. Chapter 30 of ASCE 7-10 Part 4 contains a method based on the directional procedure that uses Equation 6 to calculate the design C&C wind pressure. In this study, the design wind pressures were calculated using the directionality factor K_d equal to 0.85, in order to account for the directionality effect on critical load. The internal pressure GC_{pi} was not included (the building was assumed to be sealed).

$$q_z = 0.00256 K_z K_{zt} K_d V^2 \text{ (lb/ft}^2\text{)} \quad (q_z = 0.613 K_z K_{zt} K_d V^2 \text{(N/m}^2\text{)}) \tag{3}$$

$$p = q_h[(GC_p) - (GC_{pi})] \text{ (N/m}^2\text{) (lb/ft}^2\text{)} \tag{4}$$

$$p_{net} = \lambda K_{zt} p_{net30} \quad \text{(lb/ft}^2\text{)} \tag{5}$$

$$p = p_{table}(EAF)(RF)K_{zt} \text{(lb/ft}^2\text{)} \tag{6}$$

In the preceding equations, q_z is equal to the velocity pressure at height z; V is equal to the basic wind speed; p is equal to the design C&C wind pressure; GC_p is equal to the external pressure coefficient; GC_{pi} is equal to the internal pressure coefficient; p_{net} is equal to the net design wind pressure; λ is equal to the adjustment factor for building height and exposure; p_{net30} is equal to the net design wind pressure for exposure category B at a height equal to 9.1 m (30 ft); p_{table} is equal to the design wind pressure on the designated zones given in ASCE 7-10 Table 30.7-2; EAF is equal to the exposure adjustment factor; and RF is equal to the effective area reduction factor.

Design wind pressures for a 0.9 m x 0.9 m (3 ft x 3 ft) roof area based on the ASCE 7-10 procedures are tabulated in Table 3 for a basic wind speed of 51.4 m/s (115 mph). Equations 3 and 4 (ASCE 7-10 Part 1) were used to calculate the design wind pressures from wind tunnel tests.

The design pressure values calculated using the wind tunnel test data were larger than the values calculated using the ASCE 7-10 procedures, except compared to using the envelope procedure in Part 2 and directional procedure in Part 4 for the 3.66-m eave height building in open exposure. Although not shown in Table 3, when the effective wind area is increased from 0.9 m x 0.9 m (3 ft x 3 ft) to a larger area, these differences are lessened.

Figure 9 shows the relationship between the C&C roof pressure coefficient versus the effective area based on the wind tunnel tests, compared the empirical design relationship given in ASCE 7-10. In Figure 9, the effective area varied from 0.9 m x 0.9 m (3 ft x 3 ft) up to 6.1 m x 6.1 m (20 ft x 20 ft), with respect to the leading corner of roof (where the pressure taps are densely spaced). The results are for an area-averaged pressure coefficient for the extreme case and for an open exposure. The figure shows that the difference between the $GC_{p(eq)}$ values for various eave heights were less for larger effective wind areas.

Table 3. Design wind pressure for a 0.9 m x 0.9 m (3 ft x 3 ft) area, for a basic wind speed, V = 51.4 m/s (115 mph).

		Open Exposure		
Eave Height m (ft)	Design Wind Pressure, kN/m² (psf)			
	Wind Tunnel	ASCE Provisions		
		Envelope Procedure		Directional Procedure
		Part 1	Part 2	Part 4
3.66 (12)	-3.3 (-69.7)	-3.3 (-68.6)	-3.5 (-72.7)	-3.5 (-72.8)
5.49 (18)	-4.5 (-94.7)	-3.4 (-70.8)	-3.6 (-75.7)	-3.6 (-75.5)
7.32 (24)	-5.3 (-110.9)	-3.6 (-75.9)	-3.9 (-80.5)	-3.8 (-80.1)
12.19 (40)	-6.2 (-129.2)	-4.0 (-83.7)	-4.3 (-89.5)	-4.3 (-89.5)
		Suburban Exposure		
Eave Height m (ft)	Design Wind Pressure, kN/m² (psf)			
	Wind Tunnel	ASCE Provisions		
		Envelope Procedure		Directional Procedure
		Part 1	Part 2	Part 4
3.66 (12)	-3.2 (-67.1)	-2.7 (-56.3)	-2.9 (-60.1)	-2.4 (-49.3)
5.49 (18)	-4.2 (-86.8)	-2.7 (-56.3)	-2.9 (-60.1)	-2.5 (-51.9)
7.32 (24)	-4.7 (-99.0)	-2.7 (-56.3)	-2.9 (-60.1)	-2.7 (-56.2)
12.19 (40)	-5.7 (-118.2)	-2.9 (-61.3)	-3.1 (-65.5)	-3.1 (-65.2)

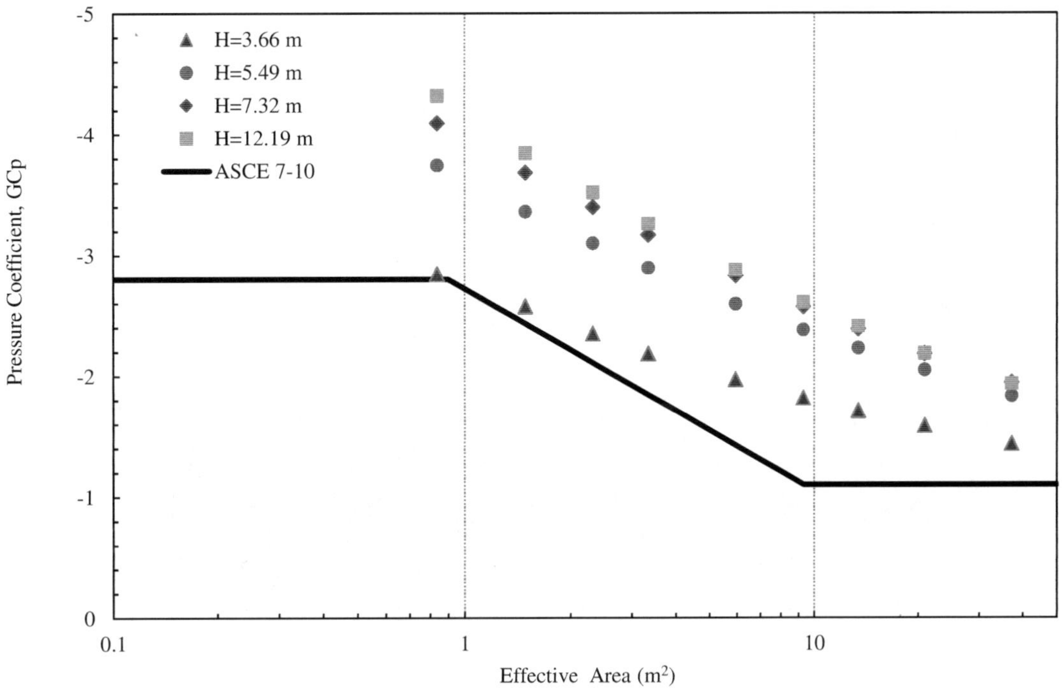

Figure 9. Roof C&C external pressure coefficient (GC_p) for corner zone versus effective wind area.

4 IMPLICATIONS FOR PERFORMANCED-BASED DESIGN OF BUIDLINGS

The implications for performance-based design are illustrated through a simple examination of probable roof sheathing panel damage for the 7.32-m building subjected to a basic wind speed of 51.4 m/s (115 mph) in open country. In this example, only one damage state is considered: panel sheathing-to-framing connection failure due to uplift (suction). The uplift capacity of wood sheathing panels was obtained from previous studies [11, 12] for a standard 2.4 m x 1.2 m (8 ft x 4 ft) panel with 8d (63.5 mm x 3.33 mm) nails with 152 mm edge spacing along panel edges and 305 mm spacing in the panel field. The mean uplift capacity of the panel was 2.8 kN/m^2 (57.7 psf) with a coefficient of variation (COV) equal to 0.2 [13].

Figure 10 shows the spatial variation of extreme equivalent ASCE 7-10 pressure coefficients. The coefficients were based on a horizontal projection of the sheathing panel area 2.4 m x 1.2 m (8 ft x 3.99 ft) mesh. The critical values of pressure coefficients were selected for the three roof zones defined in the ASCE 7-10 envelope procedure (shown in the inset of Figure 11). These critical values were considered as the governing case for the three zones. Assuming a normal distribution, Figure 11 shows fragility curves corresponding to these three roof zones, where the probability of damage (sheathing uplift failure) is expressed as a function of the basic design wind speed. This representation of sheathing fragility may be used in a performance-based design considering a range of basic wind speeds.

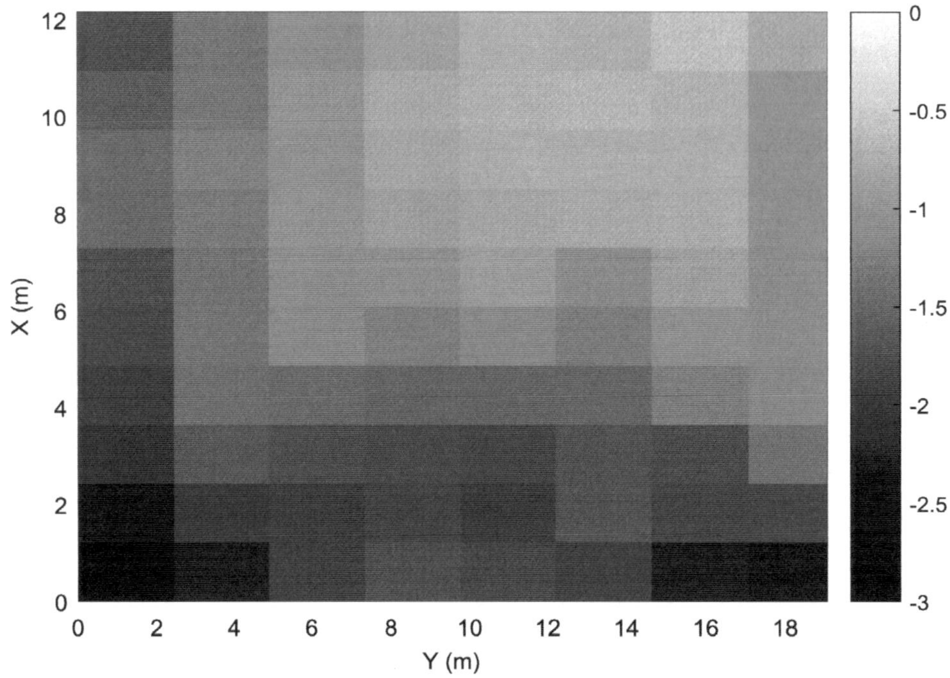

Figure 10. Spatial variation of extreme equivalent ASCE 7-10 roof C&C pressure coefficients for open exposure for a 2.4 m x 1.2 m effective wind area mesh for the 7.32-m height building.

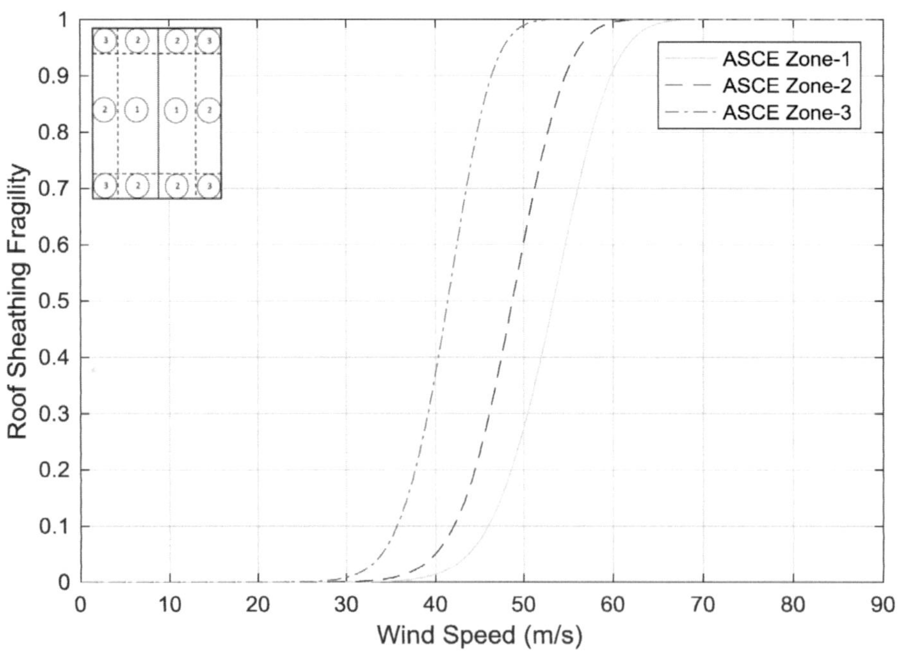

Figure 11. Fragility curve for single roof sheathing panel for three roof zones defined in ASCE 7-10.

The design wind pressure for the bottom-left quarter (the leading corner in the wind tunnel test) of the roof and the associated probability of roof sheathing damage was calculated using the wind tunnel test and compared to using the ASCE 7-10 envelope procedure. The wind tunnel test data leads to a prediction that 28% of the panels will be damaged, based on the extreme C&C roof pressure coefficients. Of course, if the roof pressure was based on a moving average, fewer number of damaged panels would be predicted. The ASCE 7-10 envelope procedure leads to a prediction that only 5% of the panels will be damaged.

5 CONCLUSION

Wind loads for components and cladding of low-rise structures based on wind tunnel tests were compared with design loads calculated using ASCE 7-10 provisions. Wind tunnel test data was obtained from the NIST database for a set of low-slope gable roof low-rise buildings with eave heights of 3.66 m, 5.49 m, 7.32 m and 12.19 m. This study focused on the corner of the roof surface, defined as Zone 3 in ASCE 7-10. The values of pressure coefficients based on wind tunnel tests were larger compared to the values of pressure coefficients calculated using ASCE 7-10. If a moving average for a specified time is considered instead of considering the extreme values from wind tunnel test, the pressure coefficient values will be in closer agreement with the ASCE 7-10 recommended values. Design wind loads calculated from envelope and directional procedures in ASCE 7-10 for components and cladding had a coefficient of variation less than 3.4% and 8.1% for open and suburban exposures, respectively. Design loads for components and cladding based on the wind tunnel tests were 2% to 55% larger than the calculated loads for open country, and 11% to 93% larger than the calculated loads for suburban exposure. The difference between the measured values based on wind tunnel tests and the calculated values increased with increasing building eave height. On the other hand, as the effective wind area increases, the difference between the wind tunnel loads and the loads based on ASCE 7-10 was reduced. The difference between the calculated and the measured pressure coefficients for the range of eave heights considered was also reduced with increasing effective wind areas. For the model buildings considered in this study, the velocity exposure factor K_h in ASCE 7-10 did not seem to fully compensate for the effects of both height and exposure.

This study considered only one set of buildings (e.g. 1:12 roof pitch) and it is recognized that the results may be different for other building types. In fact, this study was part of a wider study that includes both shallow-slope (1:12 to 4:12 roof pitch) and steep-slope (7:12 to 12:12 roof pitch) low-rise buildings. Design wind loads for components and cladding based on the envelope and directional procedures will be compared with design loads based on wind tunnel tests for this wider set of buildings, with the primary objective to compare measured uplift pressures on the steep slope roof with the ASCE 7-16 design C&C loads. The spatial and temporal variation in loads will also be used to evaluate internal forces in roof trusses and to compare those internal forces with conventional truss design methods.

ACKNOWLEDGEMENT

The research was supported in part by the Insurance Institute for Business and Home Safety (IBHS). The authors express appreciation to Dr. Murray Morrison, Senior Research Engineer and Instrumentation Director at the IBHS for his assistance. The views expressed in this paper are those of the authors and do not necessarily reflect the views of those acknowledged.

REFERENCES

1 American Society of Civil Engineers (ASCE) ASCE 7-10. (2010), Minimum Design Loads for Buildings and Other Structures, American Society of Civil Engineers, Reston, VA.
2 http://fris2.nist.gov/winddata/
3 http://www.wind.arch.t-kougei.ac.jp/info_center/windpressure/lowrise/mainpage.html
4 Engineering Science Data Unit (ESDU), Strong winds in the atmosphere boundary layer. Part 1: Mean-hourly wind speeds. Data Item 82026, 1982.
5 Engineering Science Data Unit (ESDU), Strong winds in the atmosphere boundary layer. Part 2: Discrete gust speeds. Data Item 83045, 1983.
6 Engineering Science Data Unit (ESDU), Characteristics of atmospheric turbulence near the ground. Data Item 74031, 1974.
7 Ho, T. C. E., et al., The UWO contribution to the NIST aerodynamic database for wind loads on low buildings: Part 1, Archiving format and basic aerodynamic data, Journal of Wind Engineering and Industrial Aerodynamics 93.1 (2005): 1-30.
8 T.C. E. Ho, D. Surry, and D.P. Morrish (2003), NIST/TTU Cooperative Agreement - Windstorm Mitigation Initiative: Wind Tunnel Experiments on Generic Low Buildings, BLWT-SS20-2003, May 2003.
9 Pierre, LM St., et al., The UWO contribution to the NIST aerodynamic database for wind loads on low buildings: Part 2, Comparison of data with wind load provisions, Journal of Wind Engineering and Industrial Aerodynamics 93.1 (2005): 31-59.
10 Murray J. Morrison, Response of a Two-Story Residential House Under Realistic Fluctuating Wind Loads, Ph.D. Thesis, The University of Western Ontario, London, ONT., Canada, 2010.
11 Rosowsky DV, Cheng N., Reliability of light-frame roofs in high-wind regions, II: Reliability analysis, Journal of Structural Engineering, ASCE 1999;125(7):734–9.
12 Rosowsky DV, Schiff SD., Probabilistic modeling of roof sheathing uplift capacity, In: Proceedings: ASCE specialty conference on probabilistic mechanics and structural reliability, 1996, 334–7.
13 Lee, Kyung Ho, and David V. Rosowsky, Fragility assessment for roof sheathing failure in high wind regions, Engineering Structures 27.6 (2005): 857-868.

Estimation of Probabilistic Wind Load Effects of a Saddle Type Roof Considering Wind Directionality

Min Liu [a], Qingshan Yang [b], Xinzhong Chen [c]

[a] *Beijing's Key Laboratory of Structural Wind Engineering and Urban Wind Environment, School of Civil Engineering, Beijing Jiaotong University, Beijing, China Country, min_liu@bjtu.edu.cn*
[b] *Beijing's Key Laboratory of Structural Wind Engineering and Urban Wind Environment, School of Civil Engineering, Beijing Jiaotong University, Beijing, China Country, qshyang@bjtu.edu.cn*
[c] *National Wind Institute, Department of Civil, Environmental and Construction Engineering, Texas Tech University, Lubbock, TX, United States, xinzhong.chen@ttu.edu*

ABSTRACT

The wind directionality effect is necessary to be considered in the estimation of probabilistic wind load effects for performance-based structural design against wind loading. In this study, the estimation of probabilistic extreme wind load effects of a saddle type large-span roof is presented by integrating the statistics of extreme wind load effects conditional on wind speed and direction and the probability distribution of directional extreme wind speeds. It leads to the estimation of extreme wind load effects for various mean recurrence intervals (MRIs).

Wind tunnel test was carried out at Beijing Jiaotong University to measure the dynamic wind pressures on the hyperbolic paraboloid shape roof under simulated boundary layer flow at different wind directions. Based on the short-term measurement data, the extreme value distributions of Gaussian and non-Gaussian wind load effects are determined. For the non-Gaussian wind load effects, the extreme value distributions are estimated using translation process method through moment-based Hermite model. One of the limitations of the original moment-based Hermite model is that the skewness and kurtosis are statistical moments affected by both positive and negative probability distribution tails, thus are less specific in defining only one of the distribution tails which determines the statistics of maximum or minimum. A new strategy is suggested to improve the accuracy by defining new statistical moments using the distribution greater or lower than the median for the estimation of distribution of maximum or minimum, respectively. The effects of uncertainties of skewness and kurtosis on the estimation of extreme are also studied and the confidence interval of estimation is provided.

The extreme wind speed distribution of each direction is modeled as Gumbel distribution. The joint probability distribution of directional extreme wind speeds is then developed using multivariate extreme value theory with Gaussian copula. The wind climate data in this study were obtained from one of the Xiamen Meteorological Stations recorded from January 1st, 1971 to December 31st, 2007, which is governed by China Meteorological Administration. The distribution of directional wind speed is then combined with the distribution of extreme wind load effect conditional on wind speed and direction for the estimation wind load effects with various MRIs. The influence of correlation of directional wind speeds is examined. The estimation by using simplified sector-by-sector approach is also compared. The effect of structural orientation on the estimation is further examined.

KEYWORDS

Probabilistic, Wind load effects, Directionality effect, Saddle type roof

The 13th Americas Conference on Wind Engineering (13ACWE)
Gainesville, Florida USA, May 21-24, 2017

Assessing the application of influence function approach to evaluate standing seam metal roof response using full scale measurement data

Mahshid Nasiri[a], David Henderson[b], Gregory A. Kopp[c]

[a] *University of Western Ontario, London, Ontario, Canada*
[b] *Cyclone Testing Station, College of Science and Engineering, James Cook University, Townsville, QLD, Australia*
[c] *University of Western Ontario, London, Ontario, Canada*

ABSTRACT: Standards like ASCE-7 calculate wind loads on building components by assigning geometric tributary area to the given element by using half of the distance to an adjacent fixing and assign a uniform influence function over this region between the fixings. This method may not represent the real wind load paths for many cladding systems. This research is focused on understanding load sharing behavior of the cladding (panels) for standing seam metal roofs. A full-scale standing seam metal roof was built in Insurance Research Lab for Better Homes (IRLBH) of University of Western Ontario. By measuring the clip load, produced by single point load at a series of locations on roof, the relative effects of position of an applied load and the load transfer from each position is investigated using the influence function approach.

KEYWORDS: Wind load, Standing seam metal roof, Influence function approach, Full-scale measurements, Load sharing

1 INTRODUCTION

1.1 *Background*

The costliest damages during windstorms do not typically come from the structural frames of industrial and commercial low-rise buildings but from parts of building's envelope like roofing covering. The main cause of roof damage can be identified as the combination of high dynamic wind suctions and positive internal pressure [1]. Standing seam metal roof systems are commonly used in building envelope systems because they are proven to be an excellent watertight, economic and high performance roofing system [2, 3, 4]. The seams, as they stand above the roof waterline, form a watertight seal and they also make a secure connection between adjacent panels which provides structural support to the roof system [5]. Failure of standing seam metal roofs is initiated in two ways; (1) rotation and spreading of the seams as the result of panel deformation, and (2) fracture of roof fasteners components. Due to weak seam-clip connections, standing seam metal roof systems are sensitive to uplift wind pressure [2, 3, 5].

Research with the main topic of responses of the structure under uplifting wind loads were carried out using different methods such as numerical solutions, mathematical formulation, analytical methods, wind-tunnel tests and full-scale experiments [6]. The current method to evaluate the performance of the roof under wind loads is to assess its capability to withstand standard loads using physical tests. Two common physical testing protocols in North America are the Underwriter Laboratories (UL) 580 and American Society for testing and Materials (ASTM) E1592. Neither of these protocols, resemble the real wind load and its temporal and spatial variations [1]. These test methods are structural capacity tests to determine the ability of a panel system to resist uniform static pressure. The roof system capacity is usually controlled by performance of clips where failure initiates during high wind loads [7].

One of the earliest experiments which considered roof subjected to non-static loads, used BRERWULF to simulate uplift pressures. The authors considered temporal variations by applying pre-defined time-varying pressure loads, but, as the result of applying uniform pressure on the roof, the spatial variations were not duplicated [8]. Prevatt et al. used influence surface concept as a representative of changes in the load transfer mechanisms in order to have a better understanding of roof-deformation behavior and load transferring to the clips [5]. They concluded that the majority of the load transfer comes from the two panels directly connected to the clip and while longitudinally, the clip load consists the load effect from at least a complete purlin span away, there is transverse continuity away from the clip, unlike Ho et al. that used a continuous beam model as their theoretical influence surface [9].

Sinno et al. simulated the non-uniform dynamic suctions on standing seam metal roof system using electromagnetic drivers. They tested real roof systems at Mississippi State University (MSU), using electromagnets but had limited resolution of applied forces [10]. Farquhar et al. tested the model-scaled roof of MSU experiments at University of Western Ontario wind tunnel in order to investigate the relationship between uniform uplift failure pressure and true failure wind speed. Despite of using realistic wind loads in UWO experiments, detailed behavior of roof components were not modeled in small-scale. The model consisted of an airtight fabric membrane spanning the test box with piano wires added as standing seams. They concluded that the relationship between effective uniform pressure and responses of roof components under actual wind pressure can be analytically predicted using influence function of clip reactions [11]. Farquhar (2003), calculated the clip reaction influence function along the model seams using equations for the deflection of a beam due to a point load at mid-span. The beam deflection equation is a function of end conditions. The rotational resistance of clips can be assumed between the case of pinned and fixed connections and equivalent to a rotational spring [12].

An estimate of the uplift clip load can be determined by integrating the individual spatial roof loads gathered with their respective influence coefficients for a roof clip;

$$F_U(t) = \int_A P(x, y, t)\, I(x, y)\, dA \tag{1}$$

where F_U = uplift pressure; $P(x ,y, t)$ = spatially and temporally varying pressure; and I(x ,y) = influence function for the given component. By considering uniform pressure, the uplift force on an element in terms of influence function can be expressed as;

$$F_U(t) = P \int_A I(x, y)\, dA \tag{2}$$

thus the effective area of a certain component is;

$$A_{eff} = \int_A I(x, y)\, dA \tag{3}$$

Standards like ASCE-7 use nominal tributary area as effective area for a given component, which can be defined by using half of the distance to an adjacent fixing and assign a uniform influence function over this region between the fixings. The experimental failure mechanisms observed during uniform pressure tests cannot be a representative for non-uniformity of actual wind loads. Also, non-linear structural response and large displacements of standing seam metal roofs under high uplift pressures cause variations in load distribution among roof clips thus the simple tributary area assumptions are not an accurate representative for resultant effective area of components [5].

A wide variety of structural responses ranging from member stresses to bending moments and shears and also deflection can be related to the wind force through the appropriate influence lines. Influence functions can be considered as a direct link between the wind pressure patterns and the responses of the structure [13]. For metal roofs, the influence function is dependent on the panel details and the underlying components system; yet it is not clear for cladding elements mainly because they have relatively small tributary areas and ill-defined load paths, particularly as they deform. The objective of this research is to model the behavior of standing seam metal roofs with a focus on the load sharing among the clips in the undamaged state.

1.2 *Standing Seam Metal Roof (SSMR) System*

A standing seam metal roof can be described as an assembly of metal panels adjacent to each other, joined along their sides by standing seam. Figure 1 shows a typical system of standing seam metal roof. The panels are attached to underlying structural purlins using a clip which is concealed in the seam (between adjacent legs). The hidden clip allows for a joint-free surface, except at transverse ends where panels overlap. The clip also provides a mechanism to accommodate sliding of the panels due to thermal expansion and contraction. Large deformation of seams are expected in high negative pressure region [2]. Also, a high-tension force develops in the clips as they act as supports for the roof. The large deformations combined with the high-tension forces experienced by the clips eventually result in the clip slipping out from the seam causing failure of the roof [2].

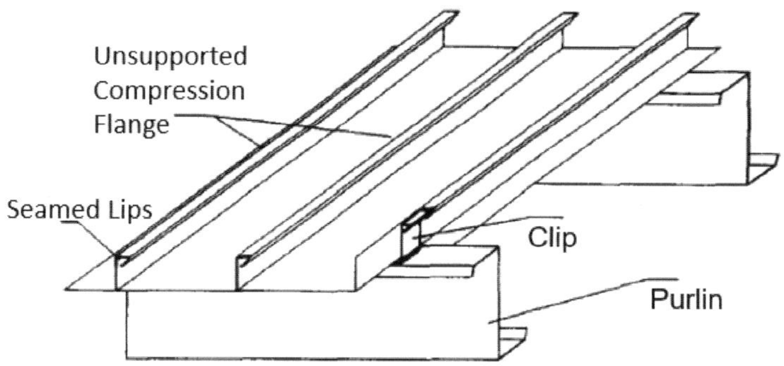

Figure 1. Standing seam roof panel system [2]

Morrison and Reinhold (2015) investigated parameters that affect the design wind loads and standard test methods for low-slope, commercial, metal roof systems. Full-scale experiments in the Insurance Institute for Business and Home Safety's (IBHS) wind tunnel provide the quantification of both the true external wind loads on an SSMR system, and the reaction loads at the clips holding the roof panels in place [14]. The SSMR system which was tested in IBHS experiments, was installed as the roof of a 30ft x 45ft metal building. At each location where two panels meet and cross a purlin, a clip was attached which made the number of installed clips a total number of 158.

2 EXPERIMENTAL SET-UP

2.1 *Roof system for current experiments*

For current experiments, a full-scale replica of IBHS experiments SSMR system was built in Insurance Research Lab for Better Homes (IRLBH) at the University of Western Ontario. Due to smaller available lab space, the replicated roof has the size of 30ft x 37ft, which altered the total number of installed clips to 130. The same trapezoidal standing seam metal roof tested in IBHS, was used in this experiments. 18 metal panels with width of 2ft (0.61 m) and one panel with 1ft width are installed. Figure 2 shows the typical section of tested SSMF system. Locations of 130 clips are shown in figure 3.

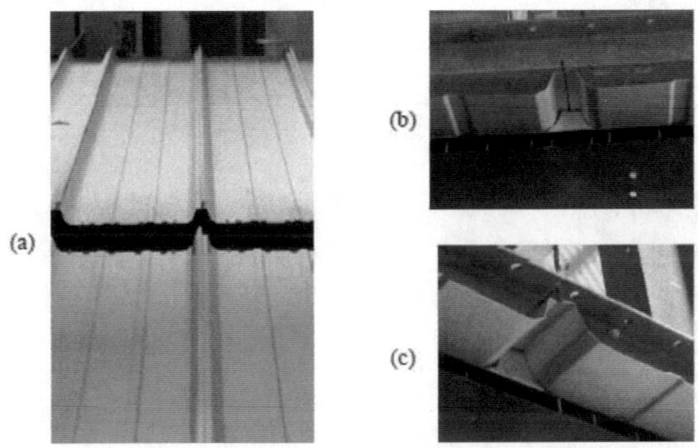

Figure 2. Typical section of tested SSMR system (a) view from above (b) view from below

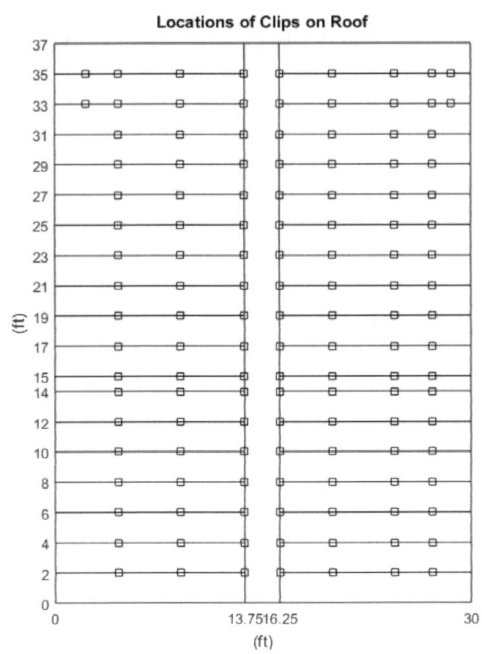

Figure 3. Locations of 130 installed clips and seams

2.2 *Instrumentation*

One of the most common force-measuring devices is the strain-gage load cell. When the load is applied, the strain changes the electrical resistance of the gauges in proportion to the load and the applied load is converted into electrical signals. For current experiment, measurements of the clip reactions were made with S-shape load cells which measure forces in one direction. S shaped load cells can be used in both compression and tension [15]. Figure 4 shows a clip with and without installing a load cell. For any test, only up to 51 load cells can be used simultaneously to measure clip reactions. In order to measure all 130 clip locations, the load cells are needed to be removed depending on which part of the roof is tested.

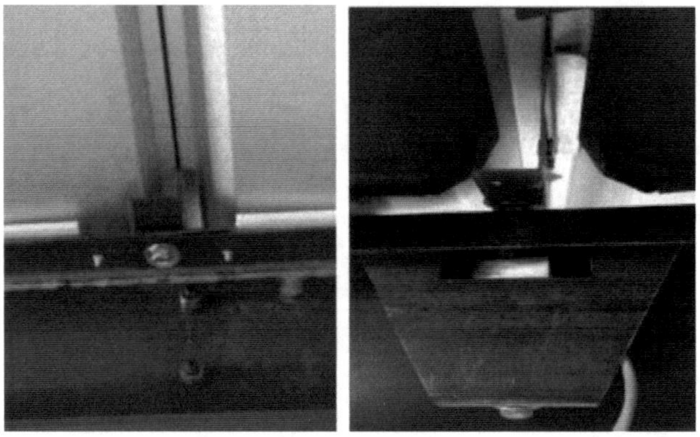

Figure 4. (Left) Clip without installed load cell (Right) Clip with installed load cell

2.3 *Investigating load sharing behavior of SSMR*

The first phase of current experiments is to investigate the load distribution through clips as the roof is subjected to load. As it was mentioned, influence function reflects changes in load transfer mechanisms. By measuring the clip load, produced by single point load at a series of locations on roof, it is possible to determine the relative effects of position of an applied load and the load transfer from each position. The loads transferred to clips were simultaneously recorded and these values are used to compute the influence coefficients. For the first step, load transfer as each clip is subjected to load is investigated and it was followed by two series of tests to investigate the behavior of standing seams. Figure 5 depicts the locations of these tests on SSMR system.

As the first sets of tests, in order to investigate how load transfers as each clip is subjected to load, concentrated load was applied on clip locations using a symmetric frame since it was not possible to apply the load exactly on the clip. Method of load application is shown in figure 6. As it can be seen in figure 5, one of the panels has width of 1 ft. In order to investigate the influence of this difference on load sharing and due to lack of enough space and considering the locations of clips with installed load cells, six clips were chosen to be subjected by load.

Next series of the tests, introduced as series 1 in figure 5, were focused on investigating the load transfer to the edge and the ridge of the SSMR. Concentrated load was applied at different locations on the panels with different distances from the edge while induced clip loads were measured using installed load cells on both sides of the ridge. Schematic plot of locations of clips and load application are shown in figure 7.

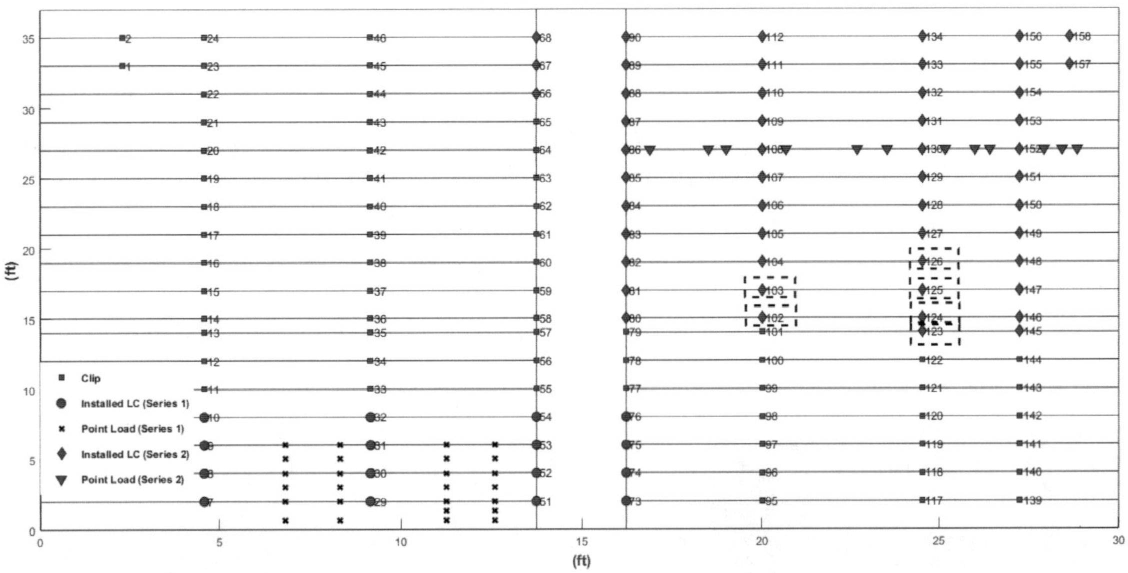

Figure 5. Plot of locations of different sets of tests on the SSMR

Figure 6. Applying load on clip locations using symmetric frame

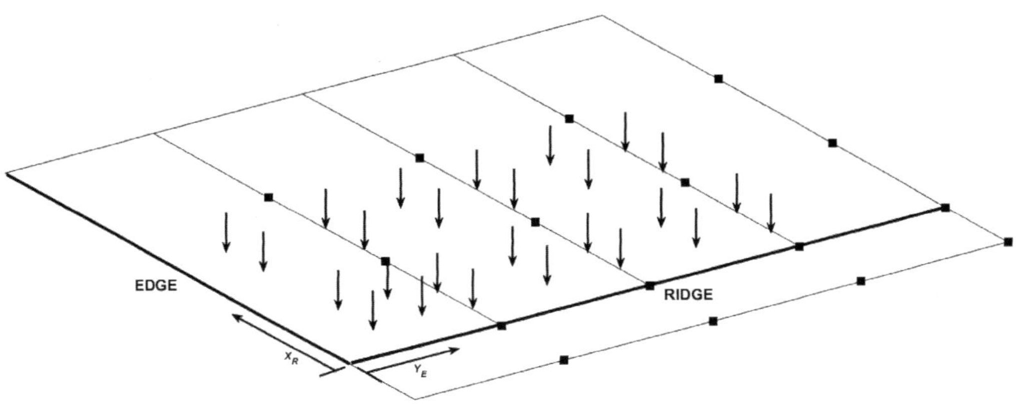

Figure 7. Schematic plot of clips (black rectangles) and load application locations. Y_E = distance from edge and X_R = distance from ridge

The last series of tests, indicated as series 2 in figure 5, were focused on the load sharing along the seams. By applying concentrated load along the standing seam influence functions of clip loads are measured. Considering the standing seam, the points that the seam is connected to the ridge and edge of the roof and the locations of the clips, the seam is idealized as a 4-span beam. Concentrated load is applied at three different locations along each span. Figure 8 shows the idealized beam model of the standing seam and locations of concentrated load and application of concentrated load on roof is shown in figure 9.

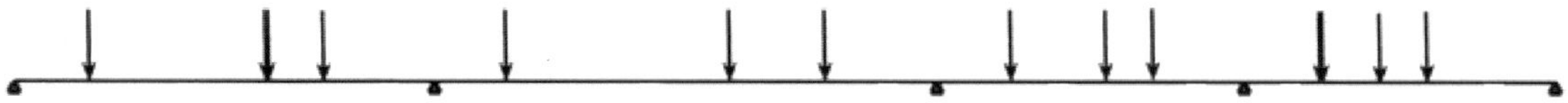

Figure 8. Idealized beam model of the seam and the locations of applied load

Figure 9. Applying concentrated load SSMR system

3 RESULTS

3.1 *Analytical model*

The influence function, g(x, s), represents the deflection of the beam at an observation point, x, associated with the concentrated unit load applied at a source point, s. Thus, for a beam with the length of L which is loaded by distributed load of q(x), the deflection of the beam at point x can be expressed in terms of influence function as;

$$w(x) = \int_{0}^{L} g(x,s)q(s)ds \qquad x \in [0, L] \tag{4}$$

Based on principle of superposition, the deflection at any point on the beam is equal to the resultant of the deflections at that point caused by each of the load acting separately. This method can be used to calculate the influence functions of a continuous beam with several spans. Consider a 4-span beam subjected to a unit point load as it is shown in figure 10. Deflection of determinate beam caused by each of acting loads is shown in figure 11.

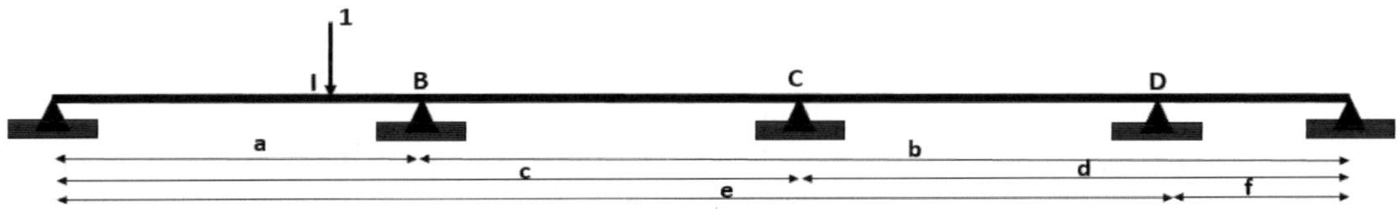

Figure 10. 4-span beam subjected to unit point load

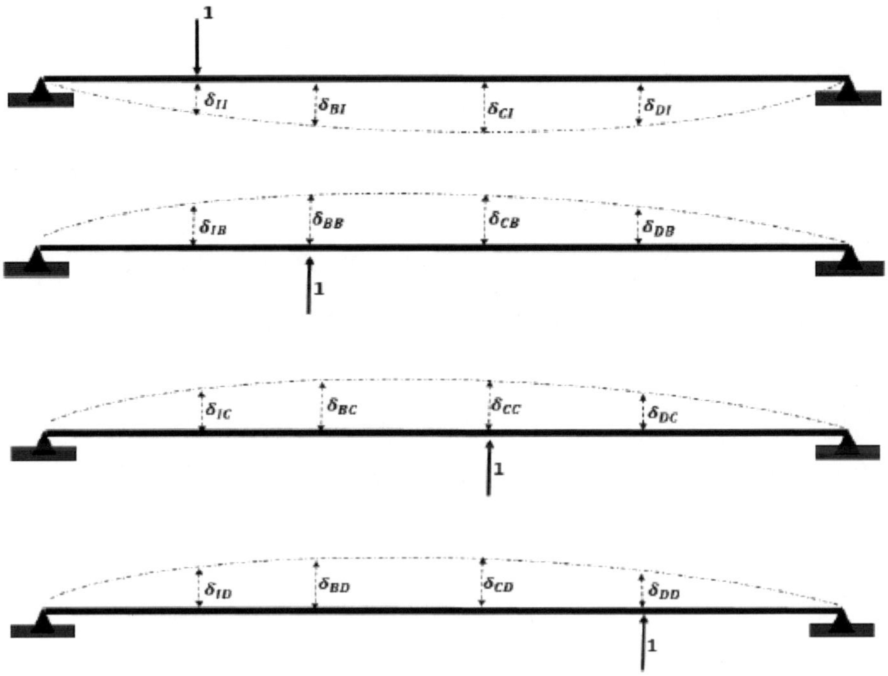

Figure 11. Deflection of determinate beam caused by applying unit load at locations of each support

Considering principle of superposition, the equations of deflections can be expressed as;

$$R_B\delta_{BB} + R_C\delta_{BC} + R_D\delta_{BD} = \delta_{IB} \tag{5}$$

$$R_B\,\delta_{BC} + R_C\,\delta_{CC} + R_D\delta_{CD} = \delta_{IC} \tag{6}$$

$$R_B\,\delta_{BD} + R_C\,\delta_{CD} + R_D\,\delta_{DD} = \delta_{ID} \tag{7}$$

where R_B = reaction at support B; R_C = reaction at support C; R_D = reaction at support D; and δ_{mn}= vertical displacement at m due to unit vertical load at n. It should be noted that δ_{mn} is equal to δ_{nm}. It can be observed from figure 11 that since the influence of each load is considered separately, in order to calculate vertical displacements, multi-span beam can be treated as several double-span beams which makes calculations simpler. By solving equations 5, 6, and 7 the influence functions of reactions at supports are computed.

3.2 *Experimental results*

As it was mentioned, concentrated load was applied on locations of six clips. Typical time series of applied load and induced clip loads for these six cases are shown in figure 12. Since the focus of these experiments is on influence coefficients and amount of transferred load to each clip proportional to amount of applied load, considering the applied load as 100 %, the ratio of induced clip load to applied load is calculated for each clip. Figure 13 shows typical plots of these ratio changes during load application period for these six clips. As it can be seen, the ratio remains constant for most of load application period. Table 1 presents the induced load on clips along the same seams and two adjacent clips parallel to purlins. Figure 14 depicts the changes of induced clip loads around two clips #103 and #125 to have better understanding about effective area of each clip. It can be observed that the applied load on clips is mostly transferred along the seam.

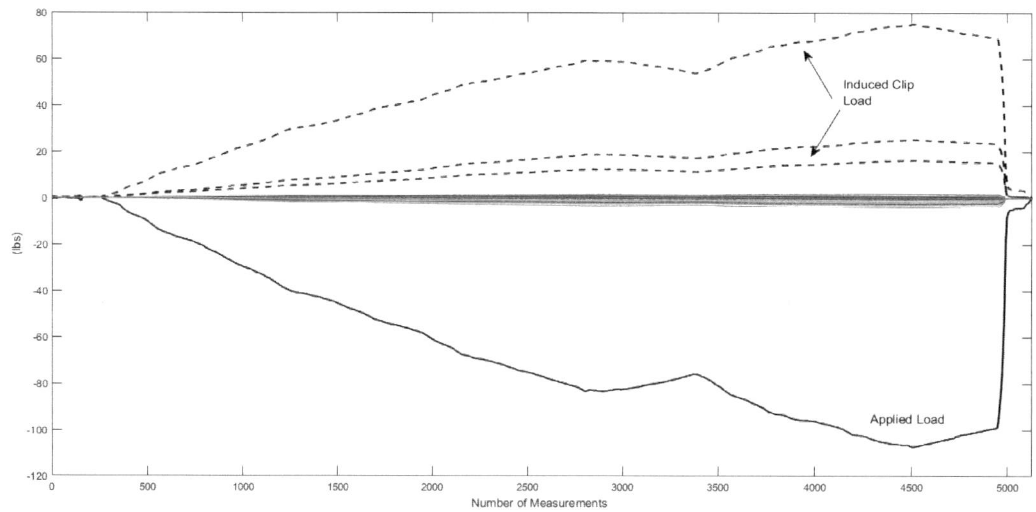

Figure 12. Typical time series of applied load and measured clip loads as the clip is loaded using symmetric frame

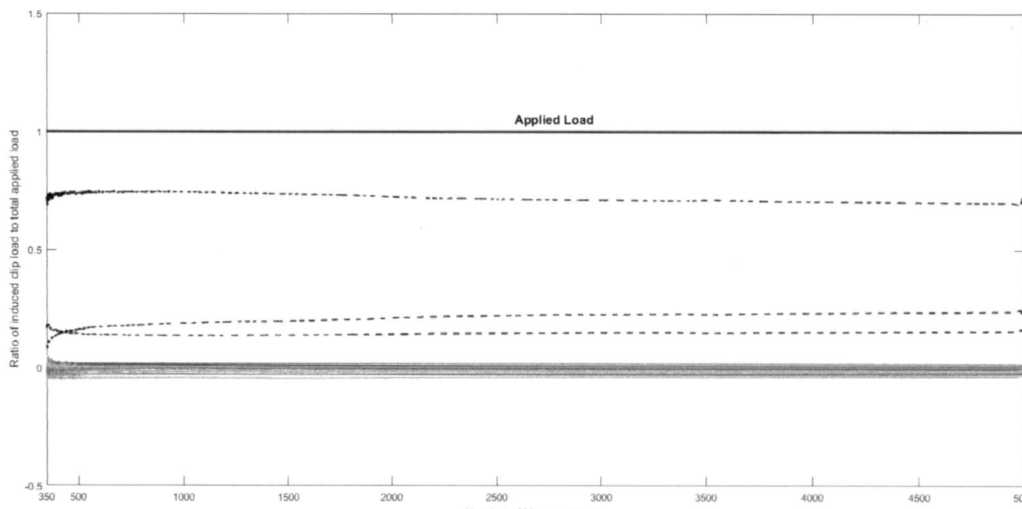

Figure 13. Typical change of induced clip load during load application period

Table 1. Test matrix of concentrated load on clip locations

Location of load application	Induced load ratio on subjected clip	Induced load on clips along the seam		Induced loads on clips along the purlin	
		Clip #	ratio	Clip #	ratio
Clip #102	0.7832	80	0.1245	103	-0.035
		124	0.1385	101	-0.02
		146	-0.01		
Clip #124	0.64	80	-0.028	123	-0.021
		102	0.177	125	-0.019
		146	0.293		
Clip #125	0.725	81	-0.028	124	-0.025
		103	0.15	126	-0.042
		147	0.21		
Clip #103	0.764	81	0.112	102	-0.031
		125	0.165	104	-0.05
		147	-0.0153		
Clip #126	0.67	82	-0.023	125	-0.034
		104	0.165	127	-0.049
		148	0.241		
Clip #123	0.7043	145	0.22		
		104	-0.023	#124	-0.023
		101	0.08		

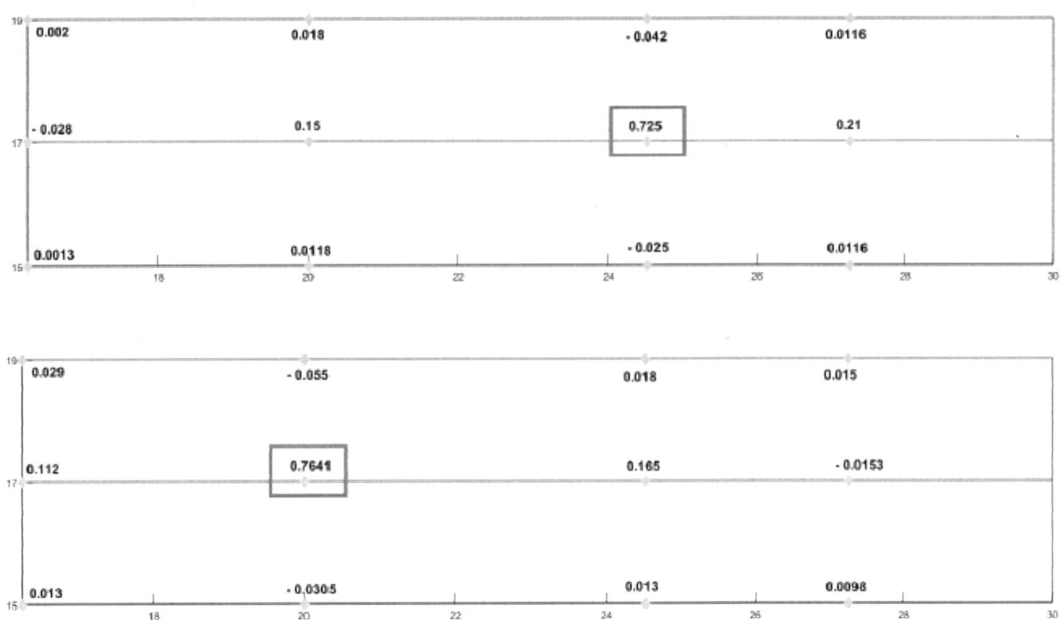

Figure 14. Induced loads on clips surrounding clips #103 and #125

Based on these results, it can be assumed that the behavior of standing seams can be idealized as beams. Several tests conducted measuring induced load on clips as concentrated load was applied on SSMR as it was shown in figures 7 and 8. Considering tested standing seams as 3-span and 4-span beams, influence coefficient of clip reactions were computed by measuring the transferred load ratio. The performance of mentioned analytical model and the application of influence function approach for standing seam is assessed by comparing with measured shared loads. Figures 15 and 16 depict these comparisons for 3-span and 4-span beams respectively.

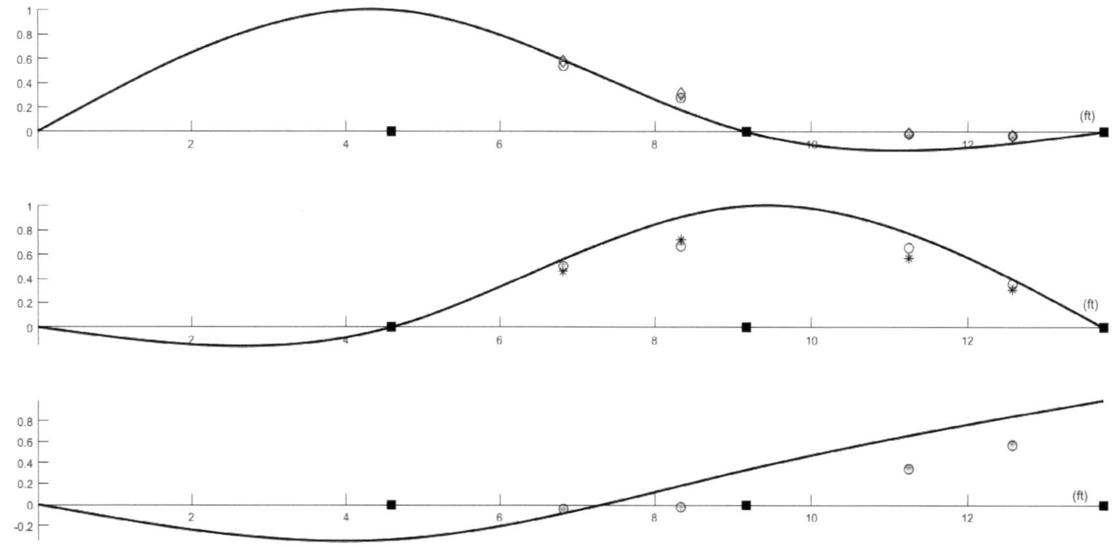

Figure 15. Comparison of measured influence functions of a standing seam with 3 clips and analytical model (-)

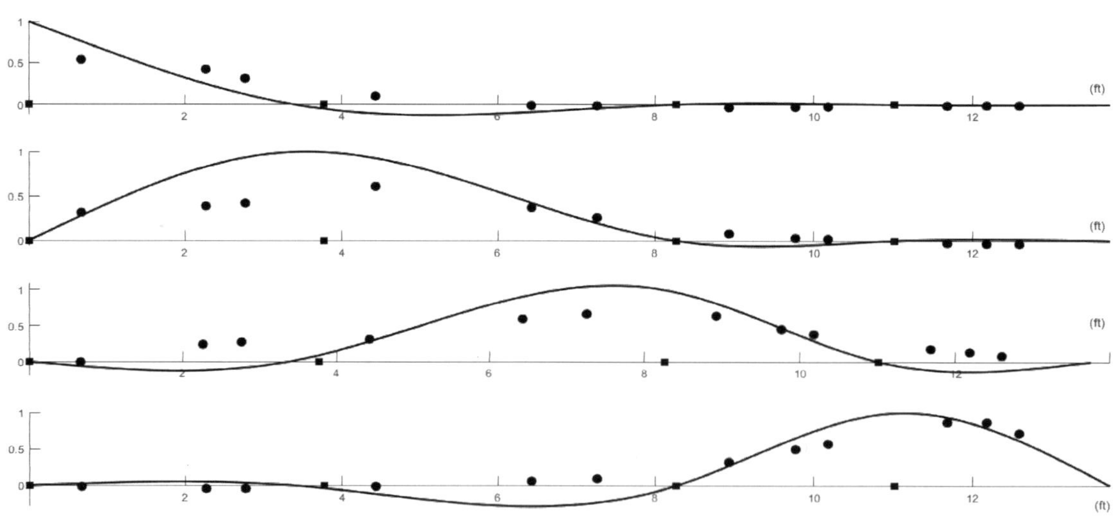

Figure 16. Comparison of measured influence functions of a standing seam with 4 clips and analytical model (-)

4 CONCLUSIONS

Standing seam metal roofs are a common method of cladding commercial and industrial buildings. Loss of this cladding from the building envelope during severe wind storms leads to large losses. Understanding the behavior of standing seam metal roofs with a focus on the load sharing among the clips will lead to better building performance. Standards like ASCE-7, in order to calculate wind loads on roof components like clips, use the simplest approach to determine the effective area of that component. The effective area of clip is calculated regardless of the real load paths on the roof which may be a conservative assumption. A full-scale standing seam metal roof was built in Insurance Research Lab for Better Homes (IRLBH) of the University of Western Ontario. Full-scale measurements improve the determination of aerodynamic loads by eliminating scaling requirements, while not requiring any assumptions about load paths. Influence functions can be considered as a direct link between the wind pressure patterns and the responses of the structure. The application of influence function approach was assessed by comparing with full-scale measurements. It was observed that the behavior of standing seams can be modeled as continuous beams but more measurements are required in order to have better understanding about load sharing and effective area and to be able to consider an influenced surface rather than influence line of an idealized beam.

5 REFERENCES

1 F. Habte, et al., Full-scale testing to evaluate the performance of standing seam metal roofs under simulated wind loading. Eng. Structures, 105 (2015) 231-248.

2 A. A. El Damatty, et al., Component testing and final element modeling of standing seam roofs, J. Thin-walled structures, 41 (2003) 1053-1072.

3 H. M. Ali, and P. E. Sensery, Models for standing seam roofs, J. Wind Eng. Ind. Aerodyn., 91 (2003) 1689-1702.

4 Y. Song, et al, Finite element analysis on deformation characteristics of metal roof panels under negative wind loads, 7th Int. Conf. of measuring technology and mechatronics automation, 2015, China

5 D.O. Prevatt, et al., A technique to assess wind uplift performance of standing seam metal roofs, In: Proc 11th Conf. Roof Technology, 1995.

6 G. Solari, and F. Tubino, Dynamic approach to the wind loading of structures alongwind, crosswind and torsional response, Wind effects on buildings and design of wind-sensitive structures, (2007) 137-166, SpringerWien, New York

7 R. C. Schroter, The load path, A roof's chain of strength is only as good as its weakest link, The Construction Specifier, November 1994

8 N.J. Cook, et al, BRERWULF – The big bad wolf, J. Wind Eng. Ind. Aerodyn. , 29 (1980) 99-107

9 T. C. E. Ho, et al. Characteristic pressure distribution shape and load repetitions for the wind loading of low building roof panels, J. Wind Eng. Ind. Aerodyn., 57 (1995) 261-279.

10 R.R. Sinno, et al., Testing of metal roofing systems under simulated realistic wind loads. In: Proc 11th Conf. Wind Engineering, 2003, 1065-1072.

11 S. Farquhar, et al., Wind tunnel and uniform pressure tests of a standing seam metal roof model, J. Structural Eng., 131(2005) 650-659.

12 S. Farquhar, Wind tunnel and uniform pressure testing of a SSMR model, University of Ontario, London, Canada, 2003

13 A. G. Davenport, How to simplify and generalize wind loads, J. Wind Eng. Ind. Aerodyn., 54/55 (1995) 657-669

14 M.J. Morrison, and T.A. Reinhold, Performance of metal roofing to realistic wind loads and evaluation of current test standards, 14th Int. Conf. on Wind Engineering, 2015, Porto Alegre, Brazil.

15 A. J. Wheeler, and A. R. Ganji, Introduction to engineering experimentation, U. S., Boston: Pearson,2010.

The 13th Americas Conference on Wind Engineering (13ACWE)
Gainesville, Florida USA, May 21-24, 2017

Wind Load Design Recommendations for Separated Sawtooth Roof Structures

Xinlai Peng [a], David O. Prevatt [b]

[a] *Forensic Building Scientist, Unified Investigations & Sciences, Ft Lauderdale, Florida, USA*
[b] *Associate Professor, University of Florida, Gainesville, Florida, USA*

ABSTRACT: Current codes provide specific provisions for design wind loads on monosloped and sawtooth roof structures. Nevertheless, design wind load standards for separated sawtooth roof buildings are still not available in practice. For separated roof structures, the presence of flat roof sections could shield or enhance wind loads on adjacent sloped roofs. In an effort to evaluate the wind effects on separated sawtooth roofs, a wind tunnel study was conducted on a series of 1/100 scale building models. This work compares the single tap peak and area-averaged pressure coefficients on separated sawtooth roofs with those on monosloped roofs and classic sawtooth roofs to assess resemblances or differences among them. Results reveal that there are no significant difference in the maximum peak negative pressure distributions on the monosloped roofs and the first span of two- through five-span sawtooth roofs. Comparison with existing codified standards is also presented in this study and suggests that ASCE 7-10 might underestimate the peak loads on the monosloped roofs, but the codified curves envelop the peak pressures on classic sawtooth roofs. In addition, for the separated sawtooth roofs, the presence of flat roof sections could alter the worst suction in the low edge region of neighboring sloped roof, but shows no influence on the other roof regions. The peak pressures acting on the flat roof sections show some dependence on the separation distance. It is not conservative to use the GC_p values for a single flat roofed building in ASCE 7 for the design of flat roof sections of a separated sawtooth roof.

KEYWORDS: Wind tunnel test; local wind pressure; area-averaged pressure; monosloped roof; classic sawtooth roof; separated sawtooth roof; non-Gaussian natures; GC_p values.

1 INTRODUCTION

Classic sawtooth roof and separated sawtooth roof configurations (Figure 1) are often used in industrial buildings due to their large floor areas as well as advantages in natural lighting and ventilation. Sawtooth roof were far more popular from the mid-twentieth century but many of them are still in existence today. Wind load is usually one of the predominate concerns in the structural design of such buildings due to complex aerodynamic effects and relatively light-weight roofing.

In past four decades, wind tunnel studies have been conducted to better understand the characteristics of wind effects on traditional sawtooth roofs. Holmes (1983; 1984) examined local peak pressures and area-averaged pressures on five-span sawtooth roofs with a pitch of 20° using a 1/200 wind tunnel model. This comprehensive study forms a basis for the codification of design wind loads on sawtooth roofs in the Australian standard (AS/NZS. 2011). Following Holmes' work, Saathoff and Stathopoulos (1992) utilized 1/400 models of a 15° slope to identify local and area-averaged wind loads on monosloped, double-span, and four-span sawtooth roof buildings. The recommended codified values for low-rise buildings with sawtooth roof slopes ranging from 10° to 30° were provided in their study, which have been embodied into ASCE 7 (2010).

However, the conclusions regarding the peak wind uplift loads on sawtooth roof structures from the two studies do not agree, which results in differences in the provisions of the U.S and Australian codes. Prevatt and Cui (2010) revisited this topic of wind loading on monosloped roofs and sawtooth roofs via wind tunnel tests using 1/100 scale building models. In their study, the peak suction at the high corner of a monosloped roof was found to have the same magnitude as on the windward span of a multi-span sawtooth roof. Prevat and Cui (2010) noted that design pressure coefficients in the current ASCE 7 wind load design code for monosloped roof structures are not conservative.

Figure 1: (a) Classic multi-span sawtooth roofs (Sawtooth building studio, CA); and (b) separated sawtooth roofs (Porter 2009).

These previous studies have characterized the wind effects on monosloped roof and sawtooth roof structures, but the provisions associated with the wind loads on separated sawtooth roofs (Figure 1-b) are still not available in current codes and standards. Due to the lack of the relevant provision, engineers often refer to the codified pressure coefficients for classic sawtooth roofs when designing separated sawtooth roofs. However, for such roof geometries, it is unknown whether the presence of flat roof sections may shield or enhance the wind load effects on the adjacent sloped roofs.

As a follow-up to Prevatt and Cui (2010), this present study is intended to identify the similarity and difference in the wind loads on monosloped roof structures, classic and separated sawtooth roofs, and to provide codified pressure coefficients for separated sawtooth roofs. Additionally, the influence of the separation distance and number of roof spans on the roof wind pressures are taken into account.

2 EXPERIMENTAL SETUP

2.1 *Atmospheric boundary layer simulation*

Experiments were conducted in the Clemson University boundary layer wind tunnel, which is an open circuit wind tunnel with an18 m (48 ft) long test section, and 3 m (10 ft) wide by 2.1 m (7 ft) tall cross-section. A full description of the wind tunnel equipment and data acquisition system is included in Prevatt & Cui (2010) and Cui (2007).

An open country terrain and a suburban terrain are considered in this study, having roughness length z_0 of 0.03 m and 0.22 m, respectively. For each exposure, wind velocities and turbulence intensities measured at various heights are converted into the equivalent full-scale values and are compared with the target profiles from Engineering Science Unit Data 86010 (ESDU. 2008) and 82026 (ESDU. 2008). The comparison (Figure 2) shows that except for a negligible deviation for the suburban velocity profiles when the height is less than 6 m, the simulated velocity and turbulence intensity profiles are in a reasonable agreement with the targets.

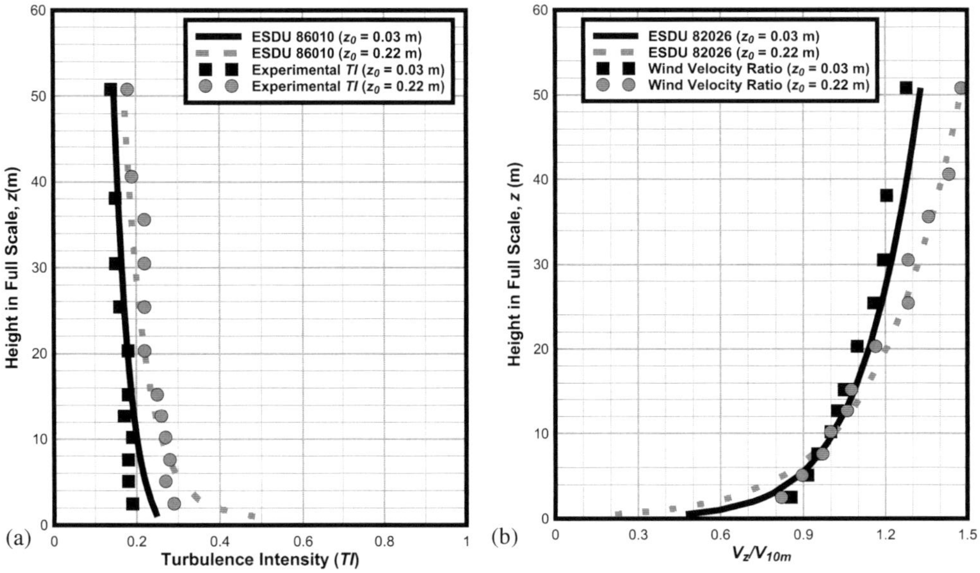

Figure 2: (a) Longitudinal turbulence intensity profiles; and (b) Longitudinal mean hourly velocity profiles.

Figure 3 depicts the power spectra of longitudinal wind velocity measured at a height of 10 m for two exposures. The von Karman and Kamial spectra (Huang et al. 2014) are provided as references. Note that the von Karman spectrum is currently used in the ESDU standard (ESDU 85020, 2008), Japanese Standard (AIJ 2004) and Australian Standard (AS/NZS. 2011). The agreement between the measured and reference spectra is reasonably good.

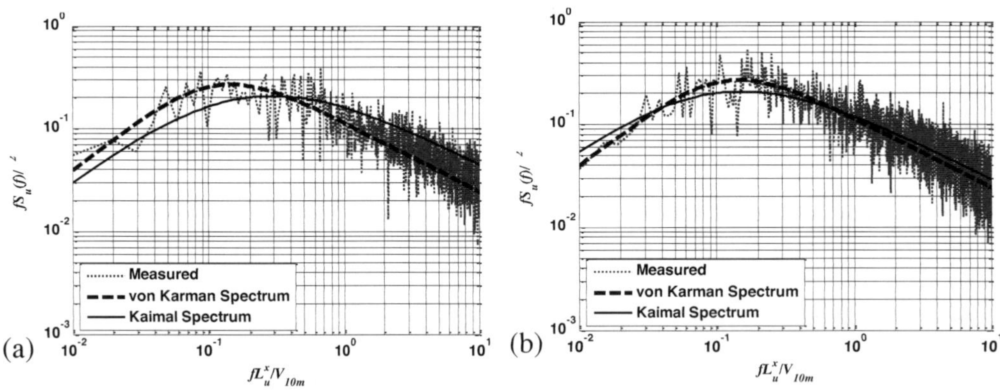

Figure 3: (a) Longitudinal wind spectra at a height of 10 m for open country terrain (z_0 = 0.03 m); and (b) Longitudinal wind spectra at 10 m for suburban terrain (z_0 = 0.22 m);

2.2 *Building model construction*

A length scale of 1/100 is adopted in the present experiments. Various building models with three distinct mean roof heights, three separation distances and one- through five-span roofs were constructed, as listed in

Table 1. In most cases, the wind directions tested for each construction model range from 90° to 270° at 10° intervals. For the sake of consistency with the models in Saathoff and Stathopoulos (1992), the pitch of each sloped roof module is chosen to be 21°. The dimensions of the test models are shown in Figure 4.

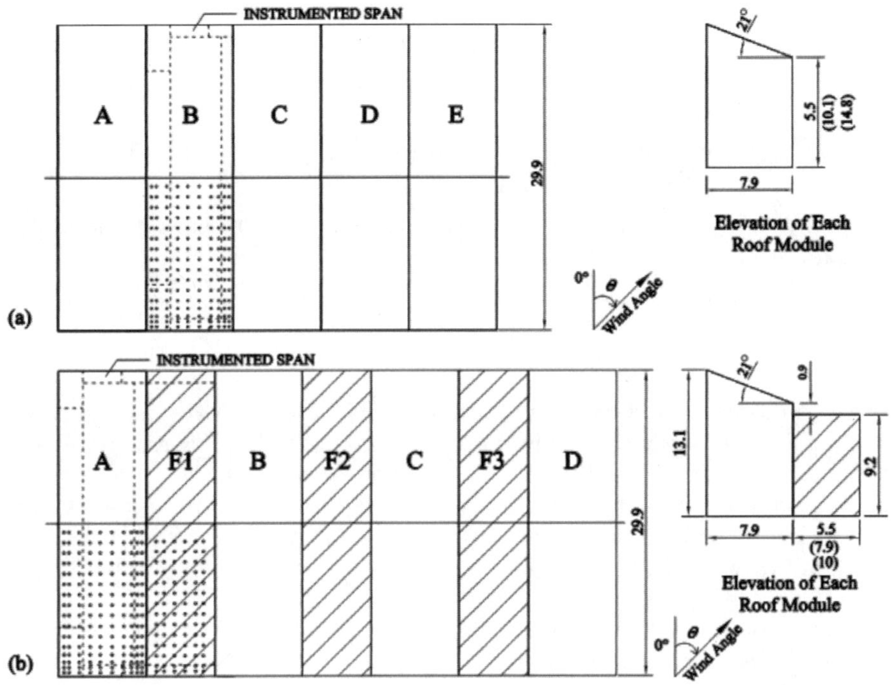

Figure 4: Roof plans and side elevations of: (a) a five-span classic sawtooth roof; and (b) a five-span separated sawtooth roof (Unit: m)

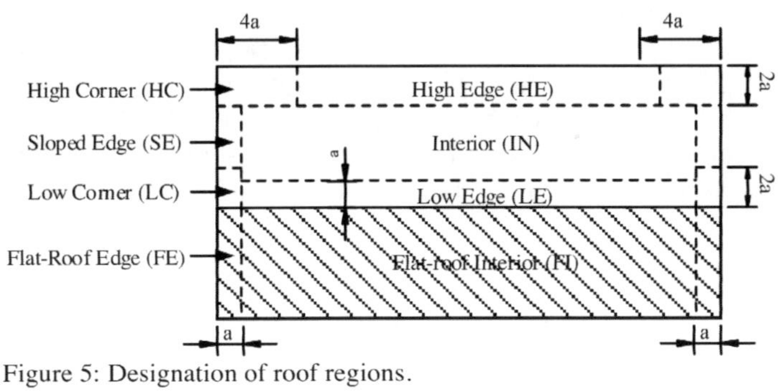

Figure 5: Designation of roof regions.

All of test models are made of Plexiglas sheet. Due to symmetry, the pressure taps on only approximately one-half of the roof surface were used. Accordingly, results for half of roof surface will be presented in the following sections, unless otherwise noted. The location of the instrumented span was changed during the tests in order to collect wind pressure data for the whole sawtooth roof system. The designations of classic sawtooth roof spans and roof zones comply with the ASCE 7 naming convention (see Figures 4 & 5).

Table 1: Summary of building model dimensions and wind tunnel test runs.

Model	Mean Roof Height	Exposure	Wind Direction
	7.0 m	Open Country	90°-270° at 10° intervals
Monosloped Roofs	11.6 m	Open Country	90°-270° at 10° intervals
	16.2 m	Open Country	0°-360° at 10° intervals
2-span Classic Sawtooth Roofs	16.2 m	Open Country	90°-270° at 10° intervals
3-span Classic Sawtooth Roofs	16.2 m	Open Country	90°-270° at 10° intervals
4-span Classic Sawtooth Roofs	16.2 m	Open Country	90°-270° at 10° intervals
	7.0 m	Open Country, Suburban	90°-270° at 10° intervals
5-span Classic Sawtooth Roofs	11.6 m	Open Country, Suburban	90°-270° at 10° intervals
	16.2 m	Open Country	0°-360° at 10° intervals
4-span Separated Sawtooth Roofs with 5.5m, 7.9m and 10m Separations	11.6 m	Suburban	90°-270° at 10° intervals

2.3 *Data processing*

The surface wind pressures on roofs were measured at a referenced experimental wind speed of 13 m/s (at a height of 1.8 m above the ground of wind tunnel) with a sampling frequency of 300 Hz and sampling time duration of 120 s. Based upon a 1/100 length scale and 1/4 wind velocity scale, the 120 s experimental sampling time approximately amounts to 1 hour in full scale, which is sufficient to provide a reliable statistical estimation of peaks. The tubing system and correction methodology can be found in Prevatt and Cui (2010).

All measured surface pressures are normalized by the mean hourly dynamic wind pressures at the mean roof height and converted into the pressure coefficient, $C_{pi}(t)$:

$$C_{pi}(t) = \frac{P_i(t) \quad P_{Static}}{\frac{1}{2} \quad V_{ref}^2} \tag{1}$$

where $C_{pi}(t)$ is the pressure coefficient at the ith tap, $P_i(t)$ represents the total wind pressure measured at the ith tap, and P_{Static} is the static pressure at the reference height. Moreover, ϱ is the air density and V_{ref} is denoted as the measured wind velocity of approach flow at the referenced height, referring to the mean roof height in this study. For the separated sawtooth roofs, the pressures on the sloped and flat roof sections are normalized to the velocity pressures at their mean roof heights, respectively.

Various techniques for the peak estimation have been utilized in the previous investigations. Both Holmes (1983; 1984) and Saathoff & Stathopolous (1992) regarded a single observed peak value recorded during a sampling period as the peak suctions or pressures. Prevatt and Cui (2010) proposed to utilize the regression equation identified from a long time record to extrapolate peak pressures, rather than picking the instantaneous peak. The present work employs the peak estimation procedure developed by Peng et al. (2014; 2013), which has been demonstrated to be simple, robust and reliable via the comparative performance of these available peak estimation methods.

In light of diverse building dimensions and various test conditions, it is worthwhile to re-reference the pressure coefficients to a common standard for comparison. For this reason, the peak wind pressure coefficients from wind tunnel tests are consistently converted into the external pressure coefficients, GC_p, in ASCE 7 (2010). This coefficient is referenced to the 3 sec gust speed at the mean roof height (Vickery et al. 2011). The methodology developed by St. Pierre et al. (2005) is employed for conversion.

3 RESULTS: LOCAL WIND PRESSURE COEFFICIENTS

3.1 *Monosloped roofs and classic sawtooth roofs*

Figure 6 shows the contour plots of peak negative pressure coefficient from all angles for one- through five-span sawtooth roofs. High corners and low corners on Span A of five models sustain larger suctions in magnitude than the rest areas of the roof. However, for the middle and leeward roof slopes, high negative wind pressures occur predominately along the roof ridge and at the low roof corner. Regardless of number of roof spans, the maximum peak pressure coefficients on five models amount to -4.5, which is consistent with the conclusions from Saathoff and Stathopoulos (1992) and Prevatt and Cui (2010).

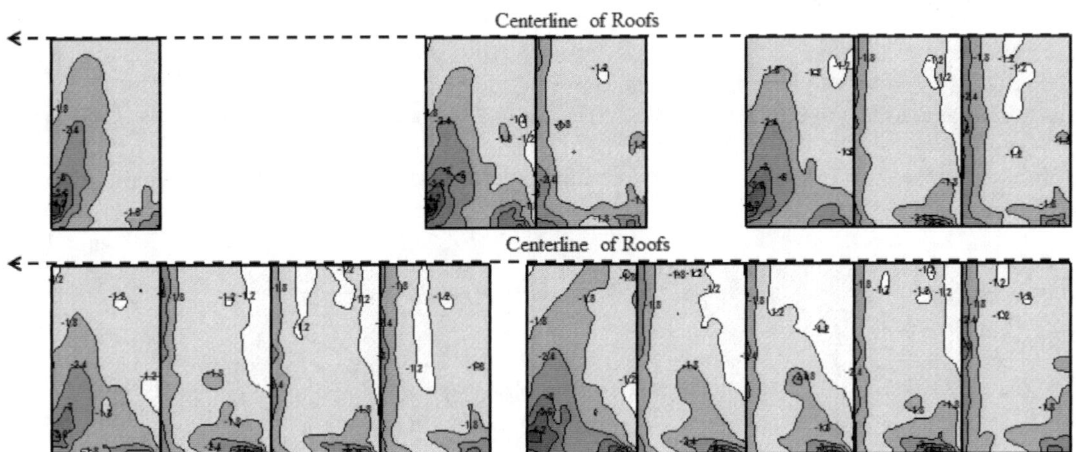

Figure 6: Contour plots of peak pressure coefficients $\hat{C}_p$ for monoslopedd roof and two- through five-span classic sawtooth roofs of 16.2 m mean roof height in open country terrain.

For comparison, the peak pressure coefficients, $\hat{C}_p$, from each region are collected and converted into GC_p values in accordance with the methodology as described in Section 2.3. The GC_p values in six regions of the 16.2m high monosloped roof and first span roof of two- through five-span sawtooth models are presented in Figure 7, along with the codified values for sawtooth structures in ASCE 7-10 (in black dash line). In each plot, the grey dot represents the mean value of all peak negative pressure coefficients in a specific region, whereas the upper and lower whiskers of the range are the maximum and minimum negative peaks, respectively. The comparison reveals that in most cases, the peak pressure coefficients on the mono-sloped roof and first module of the saw-tooth roofs are essentially similar. This indicates that the codified pressure values in ASCE 7-10 may underestimate the high corner pressures on monosloped roofs, and thus it is suggested to design monosloped roofs in the same manner as the windward roof slope of classic sawtooth roofs.

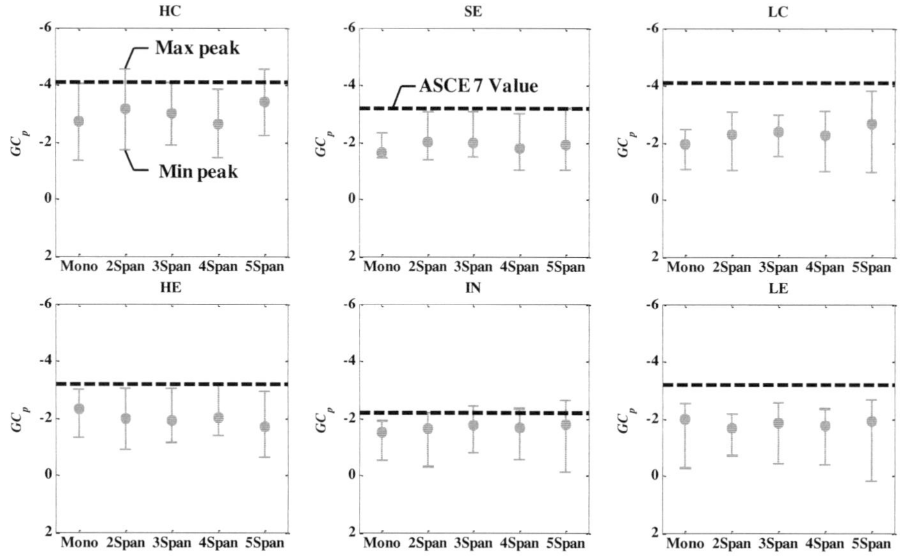

Figure 7: GC_p in six regions of windward roof slope for mono-sloped and 2- through 5-span sawtooth roofs.

3.2 *Separated sawtooth roofs*

Contour plots of peak negative pressure coefficients from all directions for separated sawtooth roofs are illustrated in Figure 8. It can be seen that that the presence of flat roofs alters the pressure magnitude and distribution at the low roof edge of Span A of separated sawtooth roofs. However, there is no significant difference in the rest areas of Span A of classic and separated sawtooth roofs. Additionally, the maximum peak negative pressure coefficients on flat roof sections show some independence of the flat roof width.

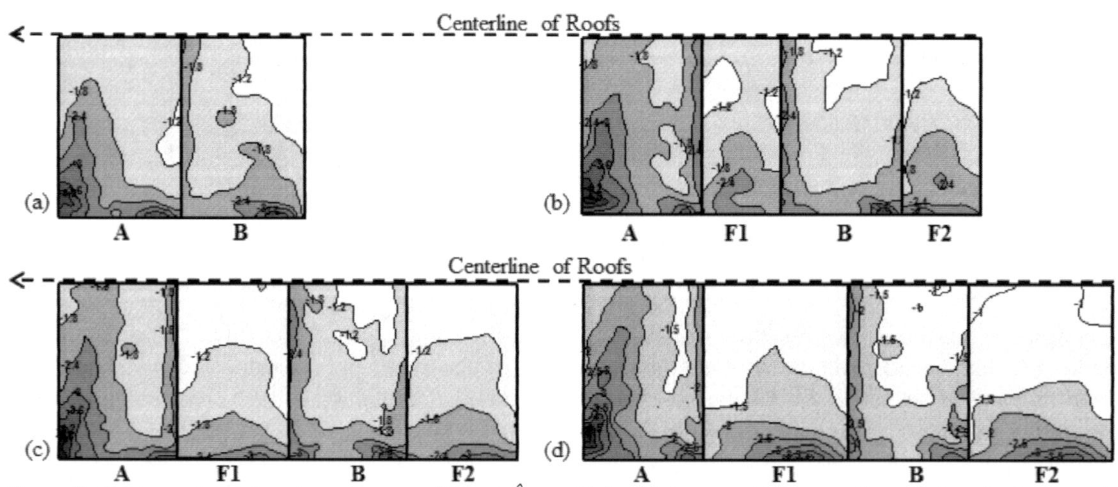

Figure 8: Contour plots of peak pressure coefficient $\hat{Cp}$ on 11.6m high separated sawtooth roofs in suburban terrain: (a) SD = 0 m; (b) SD = 5.5 m; (c) SD = 7.9 m; (d) SD = 10 m.

Similarly, peak negative pressure coefficients on the first and second sloped roof modules of these models are converted into GC_p value and are plotted by roof zones, as shown in Figures 9 and 10.

For simplicity, 'SD0', 'SD5.5, 'SD7.9', and'SD10' are representative of the sawtooth roofs separated by 0m, 5.5m, 7.9m, and 10m wide flat roof sections. Results show that regarding the first roof slope, the maximum negative pressure coefficient at the high corner of the separated sawtooth roofs are comparable to that on the classic sawtooth roof. This implied that the minimum negative peak values on the sloped module are not dependent of the separation portion width. The most significant difference between the classic and separated sawtooth roof is manifested in the lower corner and edge roof areas (see "LC" and "LE" in Figure 9). Nevertheless, there is no statistical difference between the wind pressures on the sloped roof modules of classic and separated sawtooth structures. More importantly, it is observed that the codified values basically envelope the experimental results, indicating that it is feasible to apply the codified values on classic sawtooth roofs to the design of separated sawtooth roofs. The GC_p values on Span B of classic and separated roofs do not show significant difference.

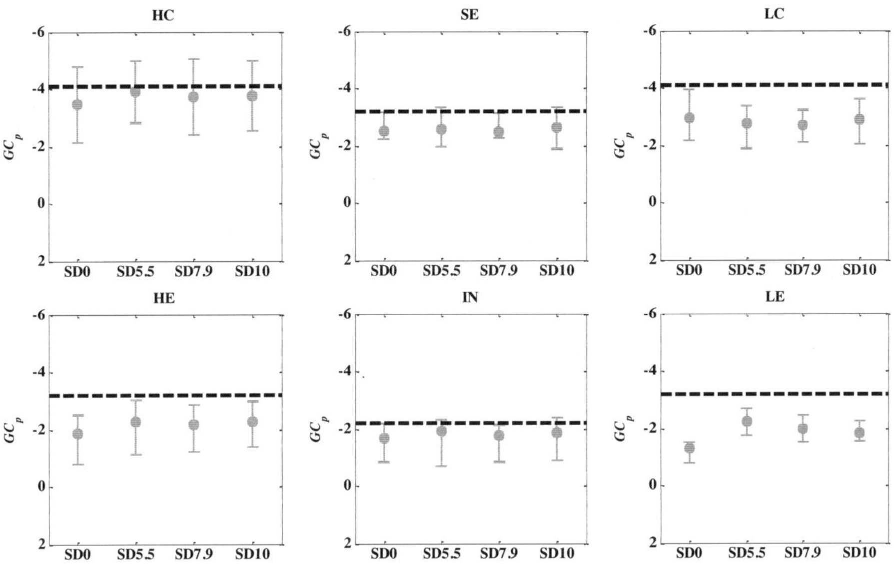

Figure 9: GC_p on Span A of classic and separated sawtooth roofs.

Figure 11 depicts the experimental GC_p values on the flat roof modules of the 11.6m high separated sawtooth models. It is reasonable herein to treat the separation modules of separated sawtooth roof as the lower level flat portion of stepped roof due to the geometric similarity. According to the provision related to the design wind load on stepped roof, the GC_p values in Fig. 30.4-2A of ASCE 7-10 for isolated flat roof of less than 7° roof pitch was plotted as a reference. Apparently, the worst suction in the edge region (FE) of the first flat roof (F1) was similar in magnitude to those of the second flat roof (F2). The highest negative wind pressures at the edge of the separation portion showed somewhat dependence on the separation distance. However, this was not the case for the interior region. More importantly, the ASCE 7-defined values for the flat roofs were not adequate to envelope the wind loads on the flat roof area, which happened to coincide with the finding in Vickery et al. (2011).

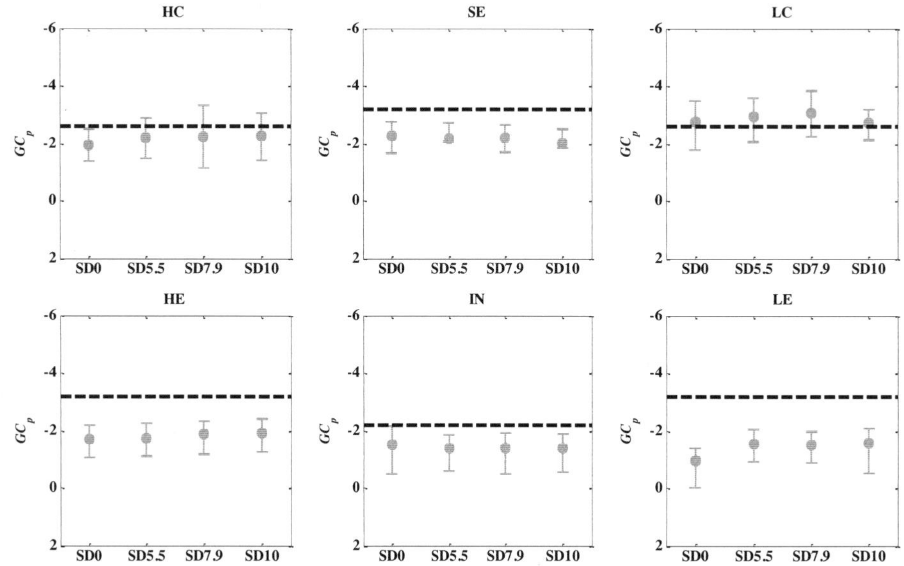

Figure 10: GC_p on Span B of classic and separated sawtooth roofs.

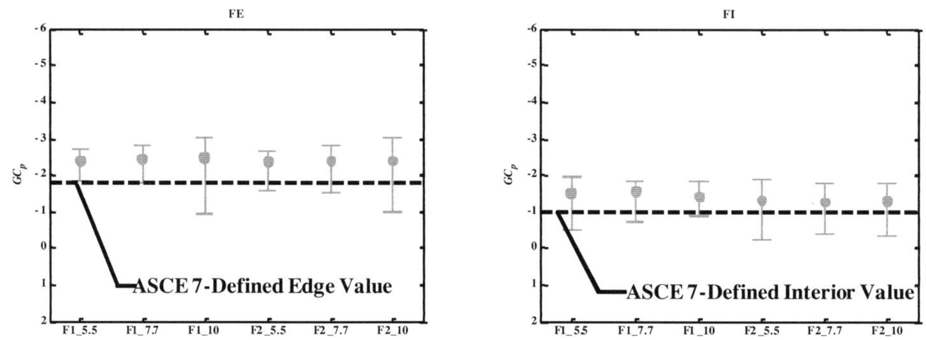

Figure 11: GC_p on the flat roof sections (F1 & F2) of separated sawtooth roofs.

4 AREA-AVERAGED WIND PRESSURE COEFFICIENTS

4.1 *Background*

Single tap peak pressures are useful in an intuitive sense to show the wind load distribution on the roofs. However, the design wind load for a large tributary area might be overestimated if only the tap wind pressure is considered. For this reason, the area-averaged peak pressure coefficients are desirable.

Time series of area-averaged load coefficients over a specific roof area are formed by simultaneously weighted-averaging the pressure histories of all taps within this region. The methodology mentioned in Section 2.3 is then applied to estimate the peaks from the averaged pressure time history. Due to the pressure taps being non-uniformly distributed, a diversity of tap arrangements for spatial averaging and inconsistent representative areas are adopted in the present study, as shown in Table 2.

Table 2: Tap combinations in different regions and corresponding tributary areas

<table>
<tr><td>Location</td><td colspan="7">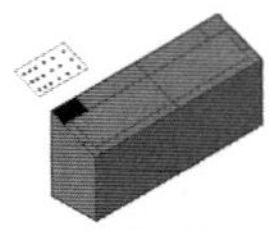</td><td colspan="4">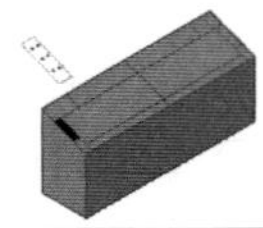</td><td colspan="5"></td></tr>
<tr><td>Tap Arrangement</td><td>1×1</td><td>2×2</td><td>3×2</td><td>3×3</td><td>3×4</td><td>3×5</td><td>3×6</td><td>1×1</td><td>1×2</td><td>2×2</td><td>3×2</td><td>1×1</td><td>2×1</td><td>2×2</td><td>2×3</td><td>2×4</td></tr>
<tr><td>Tributary Area (m^2)</td><td>0.10</td><td>2.56</td><td>3.84</td><td>5.76</td><td>7.68</td><td>9.60</td><td>11.52</td><td>0.10</td><td>1.48</td><td>2.96</td><td>4.44</td><td>0.10</td><td>0.72</td><td>1.44</td><td>2.16</td><td>2.88</td></tr>
</table>

<table>
<tr><td>Location</td><td colspan="6">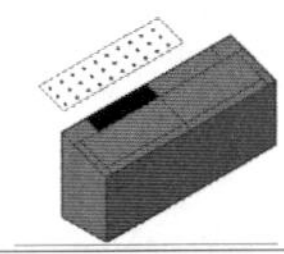</td><td colspan="5">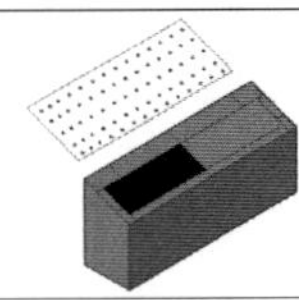</td><td colspan="5"></td></tr>
<tr><td>Tap Arrangement</td><td>1×1</td><td>3×1</td><td>3×2</td><td>3×3</td><td>3×4</td><td>3×5</td><td>1×1</td><td>2×2</td><td>3×3</td><td>4×4</td><td>5×5</td><td>1×1</td><td>2×1</td><td>2×2</td><td>2×3</td><td>2×4</td></tr>
<tr><td>Tributary Area (m^2)</td><td>0.10</td><td>2.45</td><td>4.90</td><td>7.34</td><td>9.79</td><td>12.24</td><td>0.10</td><td>3.86</td><td>8.69</td><td>15.46</td><td>24.15</td><td>0.10</td><td>2.37</td><td>4.73</td><td>7.10</td><td>9.46</td></tr>
</table>

<table>
<tr><td>Location</td><td colspan="5">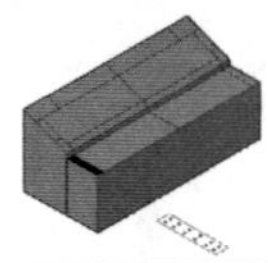</td><td colspan="6">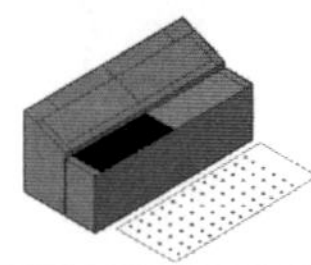</td></tr>
<tr><td>Tap Arrangement</td><td>1×1</td><td>1×2</td><td>2×2</td><td>4×2</td><td>6×2</td><td>1×1</td><td>2×2</td><td>3×3</td><td>4×4</td><td>5×5</td><td>6×6</td></tr>
<tr><td>Tributary Area (m^2)
(Three SDs)</td><td>0.10
(0.10)
(0.10)</td><td>1.10
(1.58)
(2.00)</td><td>2.20
(3.16)
(4.00)</td><td>3.30
(4.74)
(6.00)</td><td>4.40
(6.32)
(8.00)</td><td>0.10
(0.10)
(0.10)</td><td>4.28
(6.14)
(7.78)</td><td>9.62
(13.83)
(17.50)</td><td>17.11
(24.58)
(31.11)</td><td>26.74
(38.40)
(48.60)</td><td>38.50
(55.30)
(70.00)</td></tr>
</table>

4.2 *Separated sawtooth roofs*

Figure 12 presents the variation of worst pressure coefficients with tributary area on Span A & B of separated sawtooth roofs. For the high corner (HC) and low corner (LC) regions, the area-averaged peak pressure coefficients decrease at a larger rate than the other regions. This is expected due to large wind pressure variation and high tap resolutions. The discrepancies of wind loads on Spans A and B are manifested in high corner and edge. As a whole, the codified values in ASCE 7-10 for sawtooth roofs envelop the adverse *GCp* values obtained from the present wind tunnel study with an exception of the low corner of the second roof slope.

The variation of area-averaged pressure coefficients with tributary area on the flat roof sections (F1 and F2) of the separated sawtooth roofs is illustrated in Figure 13. The design values for corner, edge, and interior zones of a single flat roof in ASCE 7-10 are provided for comparison. The worst suction in the edge region of the flat roof shows slight dependence on the separation distance. It is noteworthy that the peak positive pressure coefficients from the experimental study are larger than the codified curves in ASCE 7-10. As such, it is not conservative to utilize the codified values of a flat roof in ASCE 7-10 for the design of separation portions.

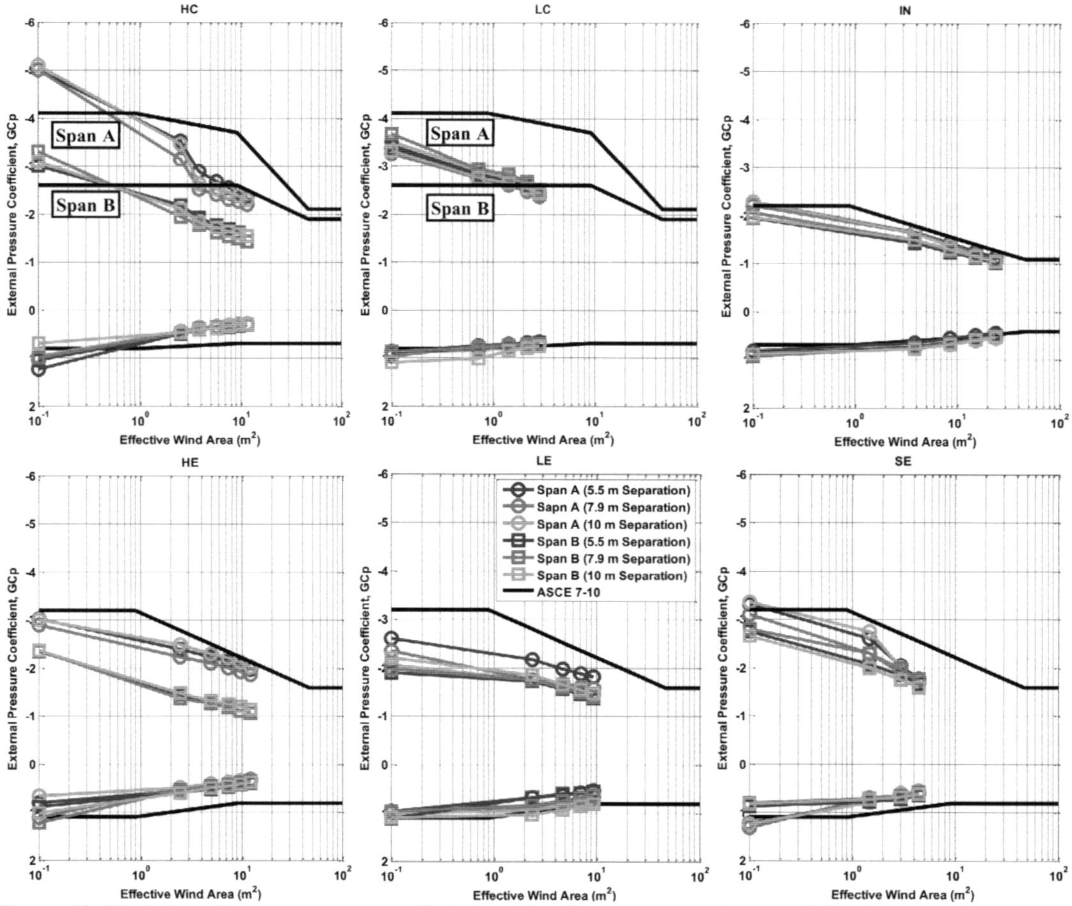

Figure 12: Variation of worst pressure coefficients with tributary area on Span A & B of separated sawtooth roofs from all angles for suburban terrain.

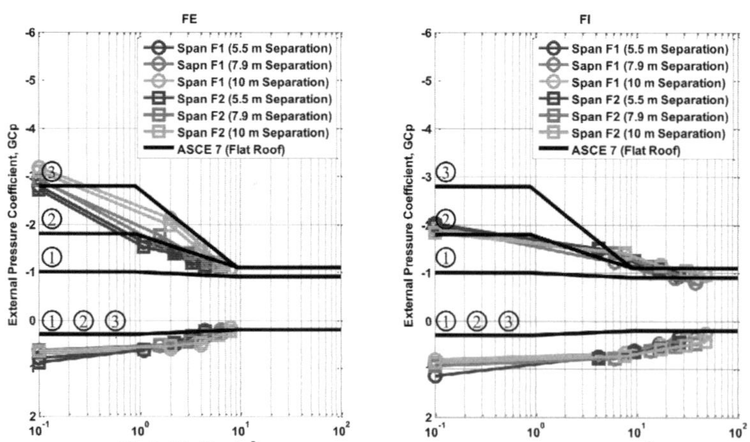

Figure 13: Variation of worst pressure coefficients with tributary area on the flat roof sections of separated sawtooth roofs from all angles for suburban terrain. Note Curve 1, 2 and 3 in the plot represent the codified value for interior, edge and corner regions of a single flat roof.

5 CONCLUSIONS

A comprehensive study of wind effects on monosloped roofs, classis sawtooth roofs, and separated sawtooth roofs was conducted. The following conclusions are drawn:

- The magnitude and pattern of the wind pressures on the monosloped roof are identical to those on the windward roof of classic sawtooth roofs. It is suggested to design monosloped roofs in the same manner as the windward roof slope of classic sawtooth roofs.
- The comparison between the experimental results and the codified GC_p values suggests that current ASCE 7 wind design provision underestimate the peak loads in the low roof corner regions of the middle and leeward roof slopes of classic sawtooth roofs. Similar conclusions are also observed at separated sawtooth roofs.
- Regarding separated sawtooth roofs, the presence of flat roof sections significantly change the wind loads on the low edge region of neighboring roof slope, but shows little influence on the rest roof areas. In most regions, despite the change in the low edge region, the GC_p curves for sawtooth roofs in current ASCE 7 adequately envelop the peak pressures for the separated sawtooth roofs.
- For the flat roof sections of separated sawtooth roofs, the peak wind pressures in the roof edge regions are dependent of the separation distances. It is not conservative to use the GC_p values for a single flat roofed building in ASCE 7 for the design of flat roof sections of a separated sawtooth roof.

6 REFERENCES

AIJ (2004). "Recommendations for loads on Buildings."Architectural Institute of Japan.

AS/NZS. (2011). "Australian/New Zealand Standard: Structural design actions. Part 2, Wind actions." *AS/NZS 1170.2: 2011*, Standards Australia International Ltd., Sydney, Australia.

ASCE 7 (2010). "Minimum design loads for buildings and other structures." *ASCE 7-10*, American Society of Civil Engineers : Structural Engineering Institute, Reston, Va.

Cui, B. (2007). "Wind effects on monosloped and sawtooth roofs." 3274291 Ph.D., Clemson University, United States -- South Carolina.

ESDU. (2008). "Characteristics of atmospheric turbulence near the ground. Part II: single point data for strong winds (neutral atmosphere)." *ESDU 85020*, Engineering Sciences Data Unit, London, UK.

ESDU. (2008). "Characteristics of atmospheric turbulence near the ground. Part III: variations in space and time for strong winds (neutral atmosphere)." *ESDU 86010*, Engineering Sciences Data Unit, London, UK.

ESDU. (2008). "Strong winds in the atmospheric boundary layer. Part I: hourly-mean wind speeds." *ESDU 82026*, Engineering Sciences Data Unit, London, UK.

Holmes, J. D. (1983). "Wind loading of saw-tooth roof buildings, Part I: Point pressures." *Internal Report NO. 83/17.*, Division of Building Research, CSIRO., Melbourne, Australia.

Holmes, J. D. (1984). "Wind loading of saw-tooth roof buildings, Part II: Panel Pressures." *Internal Report NO. 84/1.*, Division of Building Research, CSIRO., Melbourne, Australia.

Huang, P., Peng, X., and Gu, M. (2014). "Aerodynamic devices to mitigate rooftop suctions on a gable roof building." *Journal of Wind Engineering and Industrial Aerodynamics*, 135(0), 90-104.

Kopp, G. A., Mans, C., and Surry, D. (2005a). "Wind effects of parapets on low buildings: Part 1. Basic aerodynamics and local loads." *Journal of Wind Engineering and Industrial Aerodynamics*, 93(11), 817-841.

Peng, X., Yang, L., Gavanski, E., Gurley, K., and Prevatt, D. (2014). "A comparison of methods to estimate peak wind loads on buildings." *Journal of Wind Engineering and Industrial Aerodynamics*, 126(0), 11-23.

Peng, X., Yang, L., Gurley, K. R., Prevatt, D. O., and Gavanski, E. (2013). "Prediction of peak wind loads on a low-rise building." *The 12th Americas Conference in Wind Engineering*Seattle, Washington.

Porter, M. (2009). "New Rice child care center helps keep kids healthy and happy." *Rice Magazine*, Office of Public Affairs, Rice University, Houston, TX.

Prevatt, D. O., and Cui, B. (2010). "Wind tunnel studies on sawtooth and monosloped roofs." *Journal of Structural Engineering*, 136(Compendex), 1161-1171.

Saathoff, P. J., and Stathopouios, T. (1992). "Wind Loads on Buildings with Sawtooth Roofs." *Journal of Structural Engineering*, 118(2), 429-446.

St. Pierre, L. M., Kopp, G. A., Surry, D., and Ho, T. C. E. (2005). "The UWO contribution to the NIST aerodynamic database for wind loads on low buildings: Part 2. Comparison of data with wind load provisions." *Journal of Wind Engineering and Industrial Aerodynamics*, 93(1), 31-59.

Studio, S. B. "http://www.sawtoothbuilding.com/."

Vickery, P. J., Kopp, G. A., and Twisdale, L. A. (2011). "Component and cladding wind pressures on hip and gable roofs: Comparisons to the US wind loading provisions." *13th International Conference on Wind Engineering*Amsterdam, Netheland.

The 13th Americas Conference on Wind Engineering (13ACWE)
Gainesville, Florida USA, May 21-24, 2017

Load Sharing Between Batten to Rafter Connections under Wind Loading

Korah I. Parackal, John D. Ginger, Daniel J. Smith and David J. Henderson

Cyclone Testing Station, James Cook University, Townsville Queensland, Australia

ABSTRACT: Single nailed batten to rafter connections of non-cyclonic region houses were tested under quasi-static ramp loads and under fluctuating dynamic loads determined from a wind tunnel study. Dynamic connection testing showed that nails of connections slip during intermittent 'peak events' in the wind load time history with connection failure occurring after several peak events. Non-linear time-history structural analysis was performed on a system of batten to rafter connections where connection testing data were used to model the nominal force-displacement relationships of the nailed connections. This preliminary computer analysis was able to capture the effects of load sharing and redistribution during nail slips and progressive failures.

KEYWORDS: wind tunnel study, connection testing, time history analysis, progressive failures

1 INTRODUCTION

Batten to rafter connections of light framed timber houses are subject to rapidly fluctuating spatially and temporally varying loads during severe wind events and can fail in a progressive or cascading manner where, upon the failure of one connection, loads are redistributed resulting the overload and failure of neighbouring connections. Recent damage surveys have also indicated that the re-roofing of tile roofs with new corrugated cladding using existing battens can make houses especially vulnerable to wind damage during a storm [1] as original connection uplift requirements are lower due to the tile self-weight. Additionally, newer metal cladding is sometimes fastened to every 2nd or every 3rd existing batten, significantly increasing the tributary area of each batten to rafter connection. This paper presents a study of the performance of aged timber batten to rafter connections under fluctuating wind loads experienced when supporting corrugated metal cladding.

A 1/50 scale wind tunnel test was used to determine fluctuating wind loads on batten to rafter connections which were then applied to individual connections using a servo-hydraulic universal testing machine to determine the connections responses to these loads. Additionally, a structural analysis model of a system of batten to rafter connections and cladding was used to determine the load paths, load sharing and the sequence and direction in which failures are likely to propagate.

2 CONNECTION SPECIMENS

The Cyclone Testing Station (CTS) surveyed a group of 1960's houses in Adelaide to record data for vulnerability modelling. The survey was conducted in the Bedford Park area of Adelaide in collaboration with the Department of Planning, Transport and Infrastructure (DPTI) and the University of Adelaide. Sixteen houses were surveyed and the structural systems and connection details of the roof and walls were recorded.

Most of the houses surveyed were single storeyed with double brick walls and pitched frame hip and valley roofs. Figure 1 shows a typical 1960's single storey double brick house with hardwood pitched roof framing and concrete tile cladding that was surveyed.

During a second trip to Adelaide, batten to rafter connections from two houses were extracted and sent back to the James Cook University materials testing laboratory for testing. One of these was constructed with hardwood rafters and the other with softwood. Timbers were identified by a specialist using appearance and microscopy of the cell structure. Hardwood rafters were identified as Karri (*Eucalyptus diversicolor*), Softwood Rafters as Douglas fir (*Pseudotsuga menziesii*) and battens of both houses as Tasmanian Oak (*Eucalyptus obliqua*).

Connections were securely braced before they were cut away from the roof structure so that they could be interlocked and transported safely via road freight. Rafters and battens were named such that connections on the same rafter and the same batten could be identified.

Moisture contents were determined for the samples using oven drying that showed similar results to an electronic moisture meter that was used during testing conducted later. A comparison to moisture contents taken of samples inside the roof spaces of neighbouring houses with the same material indicated that the levels of moisture had not changed significantly during transit.

Figure 1 An example of the typical double brick pitched framed construction surveyed

Figure 2 Connections being prepared for extraction (left) and being packed for transit (right).

3 WIND TUNNEL TESTS

A wind tunnel study to determine simultaneous loads at batten to rafter connections of a typical roof system was conducted. Tests were carried out in the 2.0m high × 2.5m wide × 22m long boundary layer wind tunnel at the Cyclone Testing Station, James Cook University.

A 1/50 scale model of a rectangular plan, gable roof house, represented Figure 3, was used for this study. Based on survey data from [2] the model is of a 19.8m long by 10m wide house with a 22.5° roof pitch. Rafters and battens are spaced at 900mm and 877mm respectively supporting metal sheet cladding. As shown in Figure 4, rafters and battens are labelled as T1, T2...Tn and B1, B2...Bn, respectively. Batten to rafter connections that were studied are labelled based on the Batten-Rafter intersection T1-B3, T3-B4, etc. within the study area.

Ninety-eight pressure taps were installed on a study area, shown in Figure 3, to capture the spatial and temporally varying pressures near the gable end section of the house. Pressure taps were arranged in a 450 × 439mm (full scale) grid pattern such that the 900 ×8 77mm tributary area of each batten to rafter connection would contain four pressure taps. The model was placed on a turntable in the wind tunnel and pressure time histories $p(t)$ were measured at each tap for wind directions θ in 10° increments.

Pressure taps were connected to Turbulent Flow Instruments pressure transducers and data acquisition system using a tuned tubing system. Pressure signals were low-pass filtered at 500 Hz and sampled at 1000Hz for 24 runs of 30 seconds model scale (10 minutes full scale) for each wind direction.

As described in [2], pressures that influence loads on batten-rafter connections could be determined by taking the average pressure of taps within the tributary area of the batten to rafter connection. For most connections this was four pressure taps, for connections on the roof edges, two taps, and one tap for corner connections. It should be noted that the tributary areas of edge and corner connections are about 1/2 and 1/4th that of typical internal connection.

Batten to rafter connections are subject to fluctuating uplift loads due to turbulence of the atmospheric boundary layer and building induced turbulence from flow separation. For cornering winds, connections near the ridgeline at the gable end experience especially high uplift loads. Fluctuating loads at one of these connection is shown in Figure 5, high loads occur in 'peak events' where the load can be more than 3.5 standard deviations from the mean.

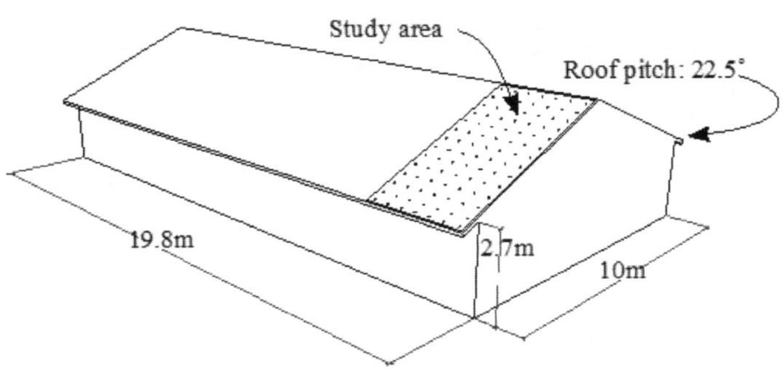

Figure 3 Study area of the wind tunnel model with full-scale dimensions.

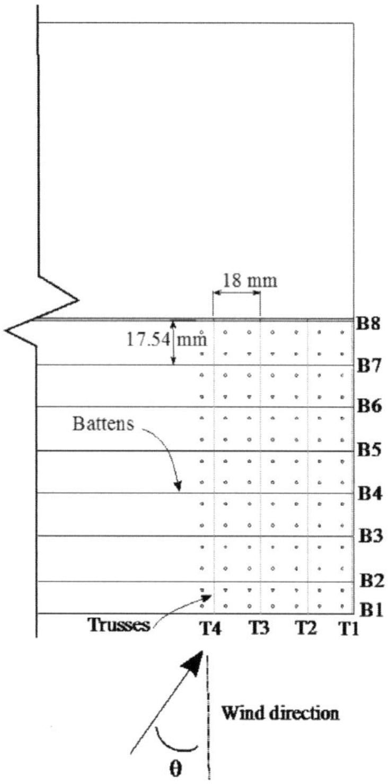

Figure 4 Tap layout and truss/rafter and batten locations in study area

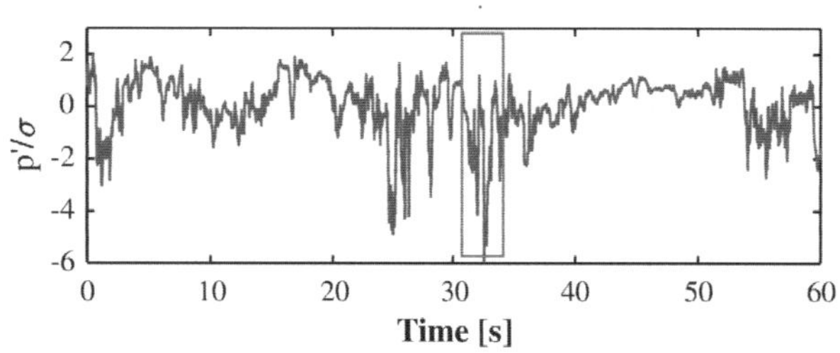

Figure 5 Load time history at a batten to rafter connection showing a intermittent ‘peak event’

4 CORRELATION OF LOADS AMONGST NEIGHBOURING CONNECTIONS

Peak load events experienced at batten to rafter connections occur at different times. The location, size and intensity of these transient high load areas depend on wind direction. Some high load areas occur in approximately the same location for a given wind direction, others move across the roof surface due to the convection of vortices and eddies in the streamflow. The synchrony of peak events occurring at neighbouring connections can be related to the correlation coefficient of the load time-history between two connections.

The level of synchrony between time histories of load at neighbouring connections may influence the likelihood of a progressive failure of batten to rafter connections occurring. If neighbouring connections experience peak events at the same time or near the same time, then a progressive failure is more likely to initiate as connections will also be subject to high uplift loads when loads are redistributed from neighbouring connections that have failed. Correlations of pressure time histories relative to the critical connection T2-B7 are presented in this section to illustrate different patterns of correlation amongst neighbouring connections.

4.1 *Cross Correlation of Load Time-Histories*

The synchrony of the signals can be further related to the cross correlation of pressure time-histories at two connections. The cross correlation coefficient as a function of lag time (τ) of one signal relative to the other is defined as:

$$r_{ij} = \frac{1}{T * \sigma_{pi} * \sigma_{pj}} \int_0^T p_i'(t) * p_j'(t + \tau)\, dt \qquad (1)$$

Where,

- p_i' and p_j' are the fluctuating components of the pressure at locations i and j.
- σ_p is the standard deviation of fluctuating load
- T is the time over which the signal is analysed
- τ is the lag or lead time that the signal at j is shifted relative to i.

Plots of correlation coefficient vs. lag time show patterns of correlation among connections indicating which connections are more correlated to the critical connection and the lag or lead time that causes the maximum correlation. Figure 6 shows such a plot for a 3×3 grid of connections surrounding the critical connection T2-B7 for wind directions 180°, 210° and 270°, where T2-B7 experiences large uplift forces. Additionally, this figure also shows color scale plots of all connections in the study area of the correlation coefficient at zero lag time as well as the lead or lag time that results in the maximum correlation. Pressures at connections are most correlated to that of the critical connection with small lead or lag times, indicating that pressure fluctuations arrive at connections slightly before or after they arrive at the critical connection.

For wind directions 180° to 200°, the connections studied are in the two-dimensional separation region behind the ridgeline. Even though loads are lower than for the critical sector, they are highly correlated along with connections to the left and right, along batten rows. Connections further along the downwind slope experiencing fluctuations at a lag time of about 0 to 0.36 seconds full scale due to the convection of vortices along the leeward roof. However, even though correlations of peak loads at connections for wind directions 180° and 200° are the highest, the magnitude of the peaks are less than 50% at the critical connection T2-B7 for the critical direction 210°.

For the critical sector, 210° to 240°, loads are less correlated among the connections in the study area. Loads at connections diagonally downwind from the critical connection are the most correlated. Lag times of maximum correlation indicate that pressure fluctuations also move diagonally down the roof. Of note is that load time histories at connections immediately to the left and right of the critical connection are less correlated than several connections lower down the roof.

For wind directions 250° to 270°, the correlations increase again. As the wind direction approaches 270°, perpendicular to the gable end, distinct bands of correlation amongst connections along each rafter can be observed. Lag times of maximum correlation indicate that the fluctuations arrive at connections on each rafter slightly after each other.

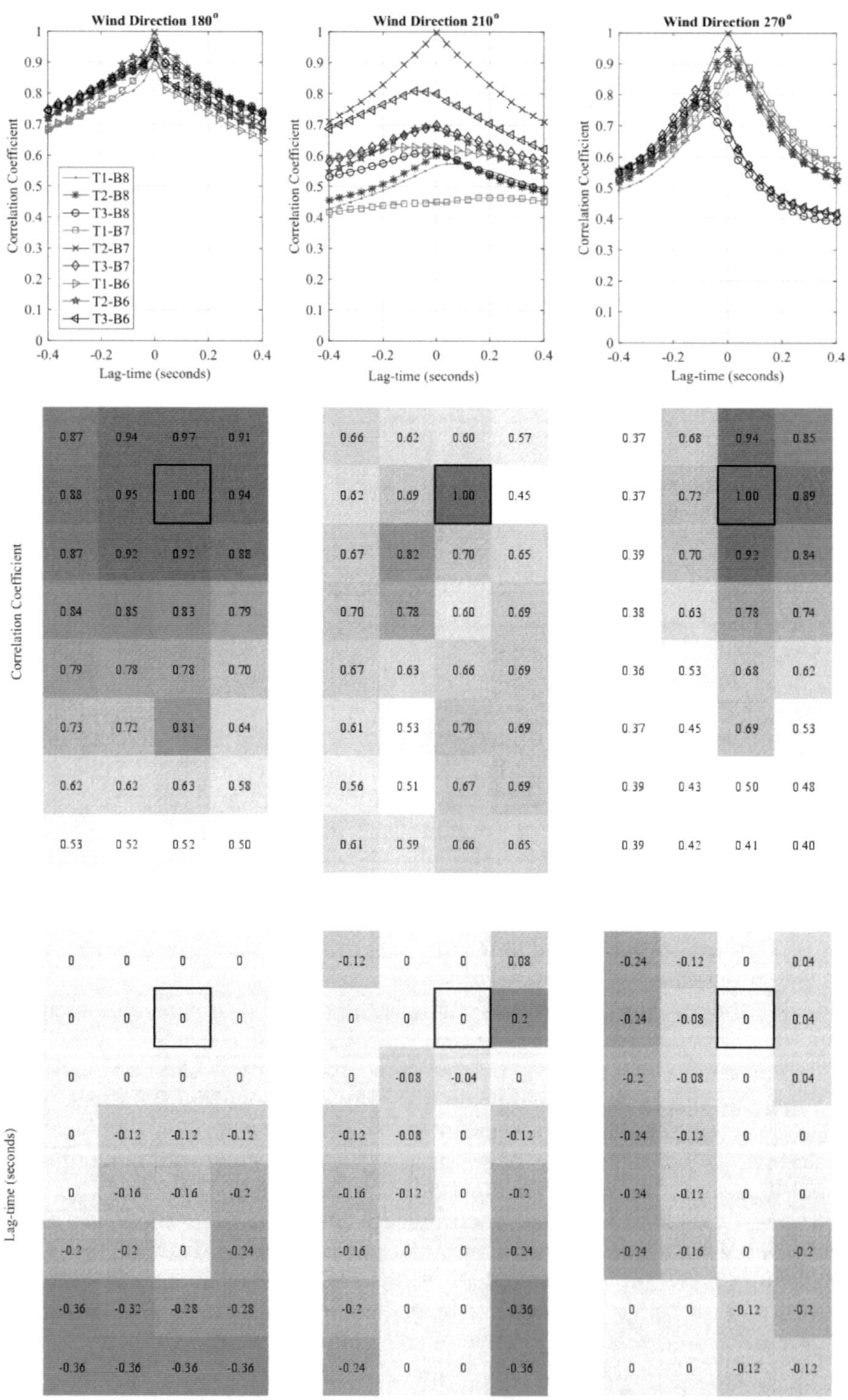

Figure 6: Top row: correlation coefficient with respect to T2-B7 vs. lag time for a 3×3 grid of connections surrounding T2-B7. Middle row: correlation coefficients with respect to T2-B7 at zero lag time for all connections in the study area. Bottom row: lag/lead time in seconds that results in the maximum correlation at T2-B7.

5 CONNECTION TESTING

Laboratory tests were conducted on the hardwood batten to rafter connection specimens. The tests provided data on the performance of in-service nailed connections and enabled quantification of age-related deterioration. Tested samples consisted of:

- Approx. 300mm length Karri Harwood rafter – 120 × 35 mm
- Approx. 300mm length Messmate batten – 25 × 35 mm
- Single flat head plain shank nail – 50 × 2.8 mm

The strength of nailed connections can be influenced by a number of factors; these were recorded for each specimen: 1) moisture content, 2) edge distances, 3) angle of nails, 4) initial gap between batten and rafter, 5) embedment depth of the nail, 6) ring size of rafter timber, 7) orientation of the rings of rafter timber, 8) condition of the nail, 9) any splits in the timber. However, no one factor could be attributed to the performance of the connections upon examination of the testing results.

5.1 *Static testing*

Displacement controlled pull-out tests at a rate of 2.5mm/min were conducted to determine the mean strength of the connections under a slow load rate. A representative load vs. extension plot of a hardwood connection with a strength close to the mean strength is shown in Figure 7.

The connection behaves elastically up to a load of approx. 500N and reaches a maximum load of approx. 600N. Connections then display a plastic region where the nails withdraw while maintaining the same load for 5 to 10mm after which the connection quickly loses its strength with further withdrawal of the nail.

5.2 *Dynamic testing*

Dynamic tests of single batten to rafter connections were undertaken to characterize incremental failure of connections under fluctuating wind loads. Preliminary testing showed that the connection specimens experienced damage only during peak events with low level fluctuations causing only elastic deformation of the connections. To reduce testing time and for added consistency, a 'synthetic' load trace was created consisting of several peak loads repeatedly. Each of these peak loads were the same magnitude and were set to the load in kN developed at a certain wind speed for a connection at an edge region of the gable end.

The results of the dynamic tests still showed large variability in connection performance, with some connections able to survive only 2 or 3 peaks and some able to survive more than 100 peaks. All showed some ductility as they were able to sustain loads at deformations much higher than their elastic limit.

Figure 8 shows an example of connection behaviour under dynamic loading. The vertical bands in this plot indicate the loading and unloading paths during low level load fluctuations between peak events. During peak events, the nail slips causing a permanent withdrawal of the nail. As the pressure fluctuations do not reverse the direction of the load on the nail, the connection deforms elastically resulting in another vertical band in the plot between each peak event. In this case, there was an increase in performance after a slight withdrawal of the nail as the magnitude of each slip decreased near the centre of the plot. After about 7mm of nail withdrawal, the connection rapidly lost strength and failed.

The dynamic tests showed that a connection's elastic stiffness, indicated by the gradient of the loading and unloading paths, does not change with accumulated damage through successive peak events. This indicates that load redistribution to adjacent connections occurs when the load acting on that connection exceeds the yield load of that connection causing nail slip, rather than due to a decrease in elastic stiffness.

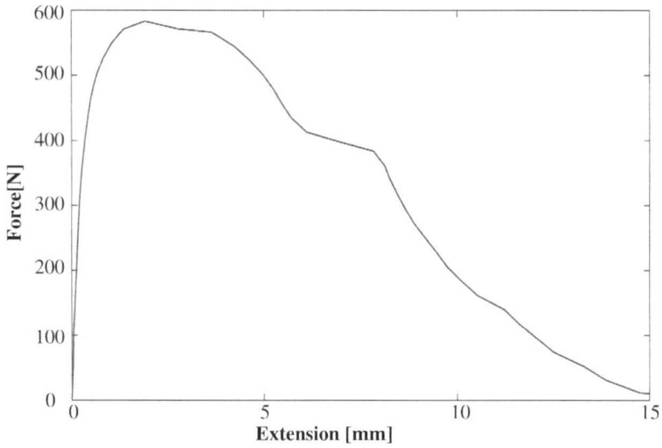

Figure 7 Example of a force vs. displacement curve for static pull-out test of a hardwood batten to rafter connection

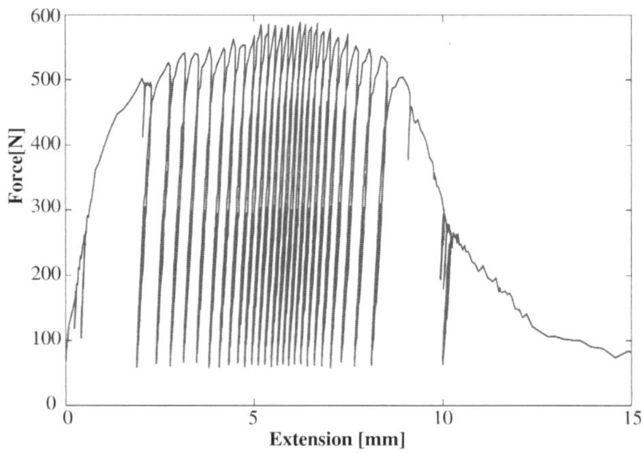

Figure 8 Example of a force vs. displacement behaviour of a hardwood batten to rafter connection under repeated wind uplift peak events

6 STRUCTURAL ANALYSIS MODEL

To determine the characteristics of load redistribution and how a progressive failure may propagate through the roof structure, a non-linear time history analysis using a finite element method structural analysis model was performed. An array of batten to rafter connections representing the connections on the gable roof house described previously were modelled.

Roof cladding was modelled as a thin shell with the thickness selected to give the same flexural rigidity (*EI*) as a 0.42 BMT 'custom orb' cladding profile. A 'stiffness modifier' of 0.1 was applied to *EI* of the sheeting in the direction parallel to the battens to represent the lower bending stiffness in the direction perpendicular to the cladding corrugations. Battens were modelled as frame elements and batten to rafter connections were modelled as multi-linear plastic link elements, with their force-displacement behaviour determined from laboratory tests. These links were assigned a proportional limit of 0.5kN, plastic deformation for 10mm at 0.5kN and complete failure at 15mm extension. The roof structure below is not modelled.

6.1 *Quasi-static pull up analysis*

As described in [3], non-linear time history analysis was performed with a quasi-static ramp load (1kN/min) applied at the location of connection T2-B7. The ramp load is continued as the first connection fails and loads are redistributed to neighbouring connections. Selected time steps of colour scale plots showing the redistribution of load are shown in Figure 9.

As load is applied above T2-B7, the system behaves as a set of springs in parallel, with part of the load being resisted by the connection T2-B7 itself and the remaining resisted by the batten and the cladding in flexure as well as extension of the neighbouring connections. Uplift loads are thus shared among neighbouring connections even when T2-B7 is undamaged and in the elastic range.

Before connection T2-B7 yields as shown in Figure 9(a), a total of 27% of the load is shared among connections immediately to the left and right on the same batten (T3-B7 and T1-B7). Load is also resisted along the corrugations of the cladding, with a total of 21% shared with connections T2-B8 and T2-B6. Finally, 15% is shared among connections diagonally away from the loaded connection. These proportions of load sharing continue to the point where connection T2-B7 yields.

After connection T2-B7 yields, the proportion of load borne by this connection decreases and the amount of load sharing at all neighbouring connections increases. After connections to the left (T3-B7) and right (T1-B7) yield, shown in Figure 9(b), the loads borne by connections on the same batten reduces and load transferred along corrugations increases. Loads are then redistributed along corrugations to connections T2-B8 and T2-B6.

Once all connections adjacent to the loaded connection yield as shown in Figure 9(c), load is shared equally among all the adjacent connections including the loaded connection. However, when the loaded connection fails completely as shown in Figure 9(d), loads are redistributed to diagonal connections.

The next connection to fail completely is T1-B7, to the right of the loaded connection. Loads are again redistributed to connections along the corrugations. At this stage a cascading failure commences with loads being rapidly redistributed as connections fail in succession resulting in the failure of all the connections in the study area.

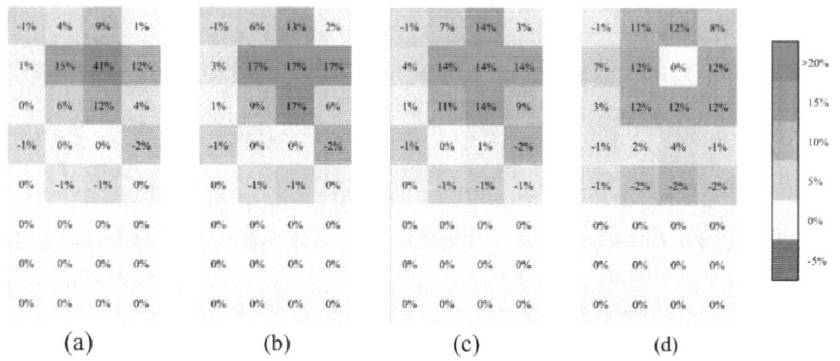

Figure 9 Colour scale plots of batten to rafter connection loads at 5 time steps showing percentage of total load borne by each connection. From Left to right: (a) at and before yield of loaded connection, (b) at yield of connections along battens, (c) at yield of connections along cladding corrugations, (d) at failure of loaded connection. [3]

6.2 *Time history analysis with peak events*

Connection T2-B7 was subjected to a repeated peak event as per the dynamic connection testing. A single connection on its own could not be modelled in the structural analysis program as the structure would become a mechanism when a nail slip occurs, causing the immediate failure of the connection. Multiple nail slips could be observed during laboratory testing as the hydraulic ram of the testing machine would lose pressure and 'drop load' slightly during each nail slip.

Incremental failure of nails was observed similar to those of the individual connection tests. Each nail slip results in loads being redistributed to neighbouring connections. Figure 10 shows the force displacement curves of nine connections including the loaded connection with the nail slip behaviour seen in the repeated diagonal bands. Selected time steps when connection T2-B7 begins to slip are encircled and indicate that nail slips at neighbouring connections occur at different times to that of the loaded connection.

Figure 11 shows load time histories at the nine connections showing the increase in loads at neighbouring connections during nail slips at Connection T2-B7. The increase of loads at neighbouring connections occurs with little time delay, indicating that loads are redistributed effectively instantaneously to neighbouring connections during nail slips. Additionally, load is shared and redistributed in similar directions to the pull up analysis. In this instance the structure becomes unstable after seven peak events, a cascading failure then commences and propagates in a similar manner to that of the pull up analysis.

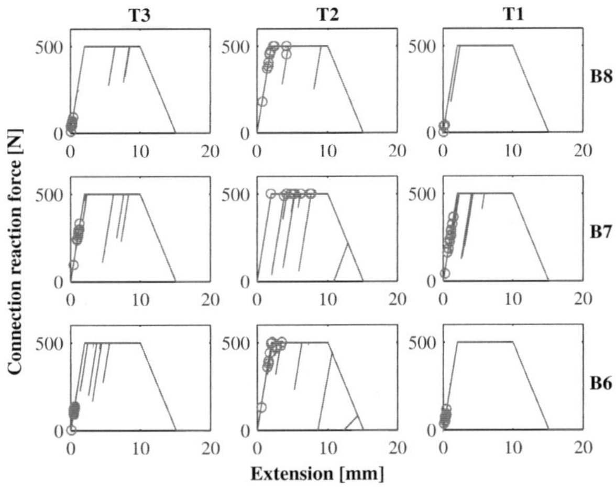

Figure 10 Force-deformation curves of nine connections including the loaded connection showing nail slip behaviour in response to peak events. Time steps encircled indicate instances when connection T2-B7 begins a nail slip. showing that nail slips occur at different times at neighbouring connections.

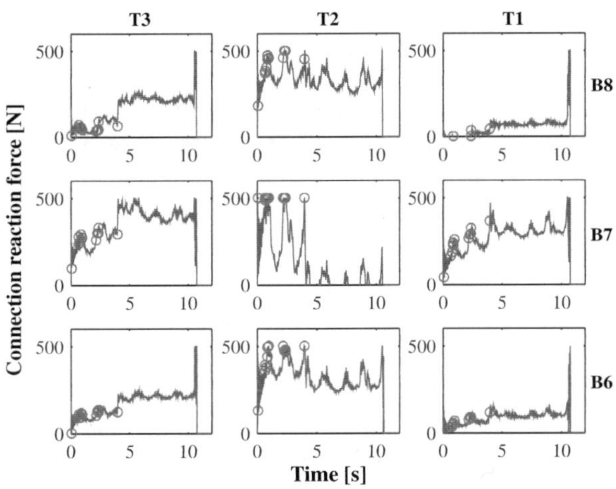

Figure 11 Load time histories at 9 connections showing load redistribution during nails slips at the loaded connection. Time steps encircled indicate instances when connection T2-B7 begins a nail slip.

7 CONCLUSIONS

The structural response of aged batten to rafter connections under fluctuating wind loads was quantified by dynamic connection testing using loads derived from wind tunnel data. Dynamic connection testing of connections showed that connections failed due to incremental nail slip during peak events with their elastic stiffness remaining the same throughout the loading process until failure. Additionally, the connections showed a large variability in performance.

Non-linear structural analysis showed that load sharing and redistribution between batten to rafter connections is a complex process. Loads are shared among neighbouring batten to rafter connections depending on the ratio of stiffness of the battens and cladding as well as the batten and rafter spacing. Additionally, the load path changes whenever a connection yields or fails completely.

Connection failure and load redistribution occurs rapidly but in durations similar to pressure fluctuations experienced on the roof surface. Thus, the correlations of wind pressures across the roof surface may play a significant role in the initiation of a progressive failure. High loads at neighbouring connections were found to be correlated and are sensitive to wind direction. High loads occurred in 'peak events' of about 0.5 to 2.0 seconds duration in localized areas. The location and size of the high load areas are due to different flow separation mechanisms and vary depending on wind direction. It was found that wind directions that cause the highest peak loads are not necessarily those that have the highest correlation of peak events among neighbouring connections. It is possible that the product of the correlation among neighbours with that of the magnitude of load gives an indication of areas that are most vulnerable to a progressive failure initiating

In future work, time history analysis with spatial and temporally varying loads can be used to determine the effects of synchrony of loads at neighbouring connections. Additionally, the effects of variable connection strengths and the presence of defective connections can be examined using the structural analysis model to develop fragility curves for batten to rafter connection failures.

8 ACKNOWLEDGEMENTS

The authors are grateful for the support of the Commonwealth of Australia through the Bushfire and Natural Hazards Cooperative Research Centre. Additionally, the support of The South Australian Department of Planning, Transport and Infrastructure and the University of Adelaide in providing access to the houses for connection sampling.

9 REFERENCES

1 Parackal, K., M. Mason, D. Henderson, et al., Investigation of Damage: Brisbane, 27 November 2014 Severe Storm Event, Cyclone Testing Station, JCU, Report TR60. 2015.
2 Jayasinghe, N.C., The distribution of wind loads and vulnerability of metal clad roofing structures in contemporary Australian houses. 2012, James Cook University: Townsville.
3 Boughton, G., K. Parackal, N. Satheeskumar, et al., Development of a full-scale structural testing program to evaluate the resistance of Australian houses to wind loads. Frontiers in Built Environment. 3,(2017) 21.

The 13th Americas Conference on Wind Engineering (13ACWE)
Gainesville, Florida USA, May 21-24, 2017

Leading Edge Vortex Dynamics on a Flat Roof

Peter Richards [a]

[a] *University of Auckland, Private Bag 92019, Auckland, New Zealand*

ABSTRACT: Roof pressure data from the field testing (Full-Scale) of the Silsoe 6m Cube, for conditions when the wind is nearly perpendicular to one of the vertical faces, has been observed to exhibit occasional, but random in time, high suction pressures on the windward half of the roof. Conditional averaging of data from a number of taps reveals a systematic pattern affecting many points on the roof. Model scale data obtained with a 1:40 scale model in the University of Auckland Boundary Layer Wind Tunnel and with a 1:5 scale model in Florida International University's Wall of Wind (WoW) facility is shown to give very similar event patterns. It is revealed that all of these patterns are essentially the same when expressed in non-dimensional form normalised using parameter related to conditions around the time of the events. Computational Fluid Dynamic (CFD) modelling of the flow over the cube using Large Eddy Simulation also gives a similar pattern of pressures and illustrates the manner in which a strong leading edge vortex is formed and expands to cover most of the roof, displacing the resident vortex associated with the background separating-reattaching flow ahead of it. The data from the Wall of Wind is used to illustrate that the pattern has a band of relatively high pressures spreading across the entire width which moves ahead of the suction wave.

KEYWORDS: Leading edge vortex, cube, flat roof, dynamic behavior.

1 INTRODUCTION

Roof pressure data from the field testing (Full-scale) of the Silsoe 6 m Cube, for conditions when the wind is nearly perpendicular to one of the vertical faces, has been observed to exhibit periodic but random high suction pressures on the windward half of the roof. Conditional averaging of data from a number of taps has revealed a systematic pattern which suggests the high suctions occur when the leading edge vortex strengthens and then grows and weakens. Model scale data obtained with a 1:40 scale model in the University of Auckland boundary layer wind tunnel and with a 1:5 scale model in Florida International University's Wall of Wind (WoW) facility have shown very similar patterns in the event sequence. Computational Fluid Dynamic (CFD) modelling of the flow over the cube using Large Eddy Simulation has also been shown to give a similar pattern of pressures and has illustrated the manner in which a strong leading edge vortex is formed and develops. This paper will focus on situations where the wind is nearly perpendicular to one of the building walls.

2 SOURCES OF DATA

2.1 *The Silsoe 6 m Cube*

The Silsoe 6 m Cube, see Figure 1, was constructed in order to provide a facility for fundamental studies of the interactions between the wind and a structure. This shape was chosen since it represents a simplified building shape, has multiple planes of symmetry and in spite of its simplicity still exhibits many of the complex flow phenomena found on more complex building shapes. Richards et al.[1] was the first of a series of papers that provided full-scale data together with in-depth analysis of the pressure and flow fields. Although the 2001 paper only contained limited mean pressure data it has been used for verification of CFD techniques (Lim et al.[2], Haupt et al.[3]) and for evaluation of experimental facilities such as Florida International University's Wall of Wind (Aly et al.[4]). More recently Richards & Hoxey[5] have provided standard deviation, maximum and minimum pressure coefficients in 15° wind direction increments for vertical ring pressure Taps V2-V9 (V1 was not included in the measurements reported) and horizontal ring Taps H1-H3, where the pressures at other taps in the two rings can be obtained through symmetry.

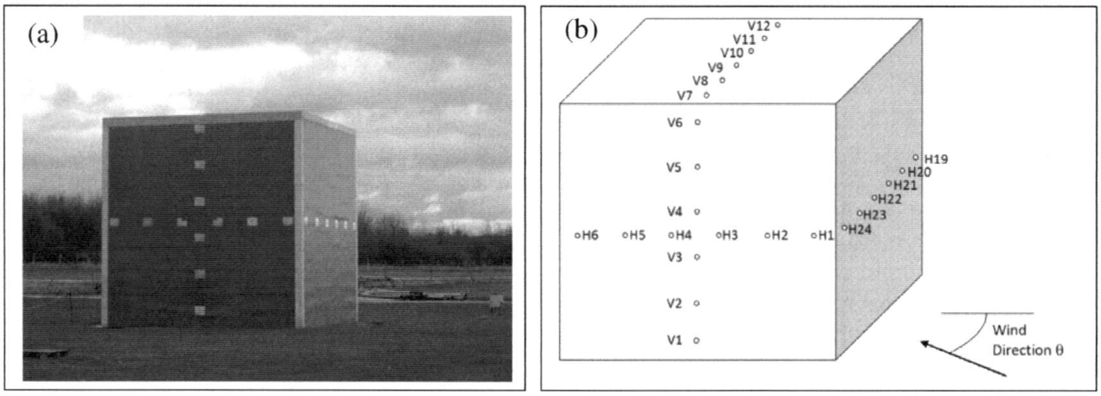

Figure 1. (a) The Silsoe 6 m cube with the pressure taps clearly visible and (b) the numbering system.

For certain tests one quarter of the roof was instrumented so that pressures could be measured over an array of 5 x 6 taps arranged as shown in Figure 2(a). Three of these taps are also part of the vertical ring, for example Column 1, Row 6 (hereafter referred to as C1R6) is Tap V7, C3R6 = V8 and C5R6 = V9. Column 6 was not installed in the full-scale study due to the location of a major structural beam along that line. However this column of taps was included on both the Wall of Wind and Wind Tunnel models discussed below. For a few tests selected roof taps, see Figure 2(b), were measured simultaneously with four sonic anememoters which were laid on the roof alongside Taps V8-V11.

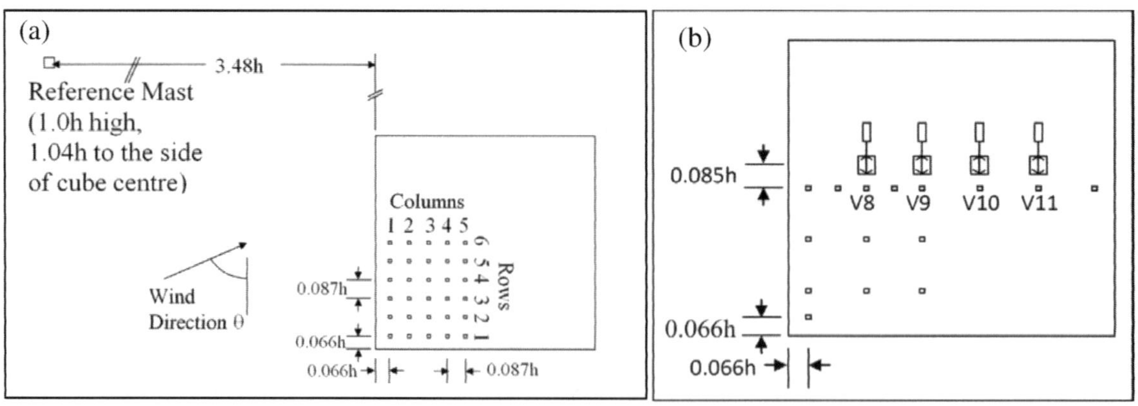

Figure 2. (a) Additional tap locations on one quarter of the cube roof and (b) sonic anemometer locations

2.2 *Wall of Wind*

Florida International University's 12-fan Wall of Wind (WoW), shown in Figure 3, was officially unveiled in August 2012 and is the US's first university research facility capable of simulating Category 5 hurricane winds. With 520 kW behind each 1.8 m tall fan, the Wall of Wind can generate winds of up to 250 km/h. With a test section 4.5 m high by 6 m wide, the Wall of Wind allows researchers, businesses, government agencies and industry to test and analyse how structures and products perform in various hurricane conditions. A 1/5th scale model of the Silsoe Cube has been tested in the FIU Wall of Wind, as shown in Figure 3(b). Pressures were measured at 109 locations corresponding to all taps in the horizontal and vertical rings shown in Figure 1(b), a second vertical ring across the model, 30 additional taps on one corner of the roof replicating those in Figure 2(a) and 4 additional taps in the centre of each quarter of each face (the roof quarter with 30 extra taps already has one in the centre). Each tap was sampled at 512 samples per second for 90 seconds. A transfer function correction was made to compensate for tubing effects. Pressure measurements were made with 37 wind directions from 45° to 225° in 5° increments, with some directions repeated for consistency checks.

Figure 3. The WoW at FIU, (a) the fan inlet and (b) the flow conditioning box and the cube model on the turntable.

2.3 *University of Auckland Wind Tunnel*

The University of Auckland modelling of the Silsoe Cube was conducted at a scale of 1:40, following the approach taken by Richards et al.[6] At this scale the 150 mm edge length made for simple manufacture and manageable location of taps. The tap layout, as illustrated in Figure 4(a) was similar to the Wall of Wind model except for the taps in the centre of each quarter of each face. The 90 taps were connected through 325 mm long, 1.5mm diameter plastic tubes to two 64 channel transducer boxes. The raw pressure data was corrected for tube distortion using a recursive filter.

The old University of Auckland boundary layer wind tunnel was 1.85 m wide and 1.1 m high. Upstream of the model a 7.5 m fetch of roughness blocks, roughness strips and corrugated cardboard mat was used to generate the boundary layer. The boundary layer at low heights was well matched by a log law with a roughness length of 0.47 mm. This corresponds to a full-scale roughness length of 18.8 mm which is slightly rougher than measurements at Silsoe, which were typically 10 mm. The mean wind velocity at cube height was 6.5 m/s, which is similar to the typical full-scale velocity. The pressure measurements were conducted over an angle range of 180° at intervals of 5°. Using symmetry the cases for all angles from 0° to 360° can be constructed. The data was recorded at 600 samples per second for 120 s.

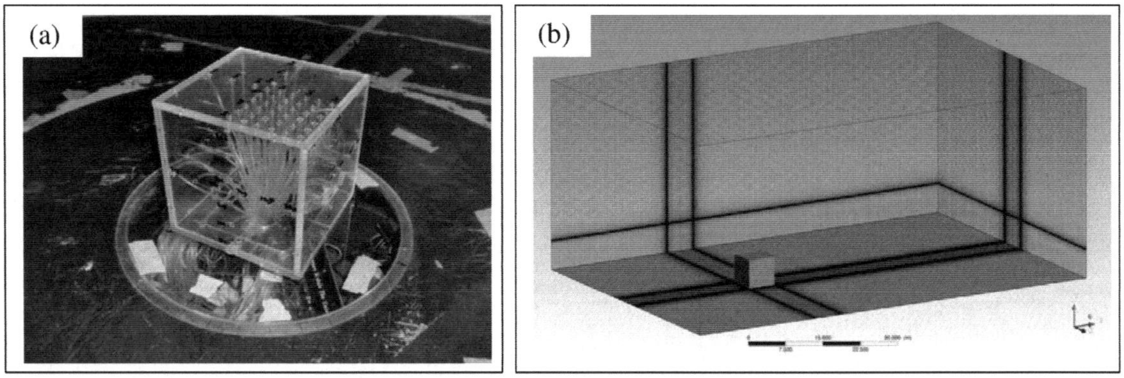

Figure 4. (a) The 1/40th scale wind tunnel model and (b) The grid layout for the Large Eddy Simulation.

2.4 *Computational Fluid Dynamics*

The CFD modelling of the Silsoe Cube used here is described in more detail by Richards & Norris[7]. The simulations were performed using an in-house massively parallel Large Eddy Simulation code, SnS (Armfield et al.[8], Norris[9]), which uses an incompressible non-staggered finite volume formulation based on a structured Cartesian mesh. Second order central differences were employed for approximating the advective and diffusive fluxes in the momentum equations, and an Adams-Bashforth fractional step solver was employed, which gives a solver is second order accurate in both space and time (Armfield & Street[10]). The fractional step method negated the need for iterative coupling at each time step, allowing the efficient calculation of transient flows. Subgrid scale turbulence is modelled using the standard Smagorinsky model (Smagorinsky[11]) with Cs = 0.18 damped at the wall using the method of Mason & Thompson[12], with a rough wall function (Mason and Callen[13]) being applied at the ground boundary.

The majority of calculations were made on a 15h×10h×6.67h computational domain (ie: 90 m long, 60 m wide and 40 m high), using a 246×123×186 mesh with a resolution varying from 0.02 m at the wall, to 0.5 m in the far field (see Figure 4(b)). The cube was modelled as being aerodynamically smooth, while the ground was modelled with a roughness z_0 = 0.01 m. The top of the domain had a free-slip boundary condition applied, and the two side boundaries were periodic. The velocity was prescribed at the upstream boundary, with a mean velocity of 6 m/s at a height of 6 m, and a prescribed pressure outlet boundary condition was applied at the downstream boundary.

The inlet boundary condition at the upstream boundary required the definition of an atmospheric boundary layer, including both the mean profile and the temporal and spatial fluctuations of the flow. This was generated using a precursor calculation of the flow in a 720m×40m×60m empty domain with periodic boundary conditions in the streamwise direction, the flow being driven by a pressure gradient in the streamwise direction. The flow at the low x boundary was sampled at 0.05 s intervals for a period of 75 minutes. This was broken into 12.5 minute blocks of data, which were used for 6 runs modelling the flow around the cube. The flow was allowed to settle for the first 30 s of each run, and then 12 minutes of data was recorded.

The code is parallelised using MPI, and was run on the NeSI Intel Xeon based computational cluster based at the University of Auckland. The momentum equations were solved using a Jacobi solver, with the pressure correction equation being solved using a Bi-Conjugate Gradient Stabilised (BICGSTAB) solver (van der Vorst[14]) preconditioned using block SIP (Stone[15]). When running on 48 cores, 12.5 minutes of simulation time were calculated in approximately 4 days of wallclock time.

3 PEAK SUCTIONS ON THE CUBE ROOF

3.1 *Comparison with quasi-steady predictions*

Full-scale pressure data, as reported by Richards & Hoxey[5], shows that in some cases, such as roof pressures with the flow perpendicular to one face (θ=90°), see Figure 5(a), the minimum pressures coefficient (when defined as the ratio of the minimum surface pressure to the maximum reference dynamic pressure occurring in the same observation period) is more negative than one might expect from a simple quasi-steady prediction. In fact for Tap V8, this situation exists for most wind directions as shown in Figure 5(b). Further, while most data points are only slightly more negative there are occasions when an individual observation is significantly lower.

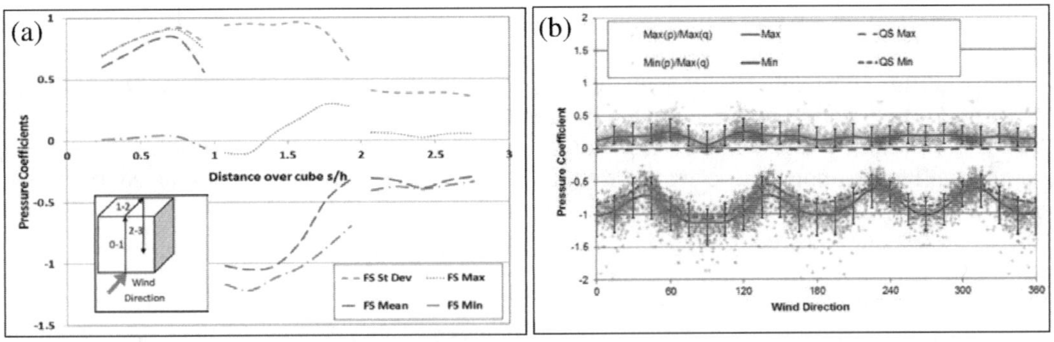

Figure 5. Full-scale pressure coefficients for: (a) The vertical ring of taps for a wind direction of 90° and (b) maximum and minimum pressure coefficients for Tap V8 over a full range of wind directions.

3.2 *Short duration pressure spikes*

Observation of time histories for the taps on the roof, Figure 6(a), shows that the minimum pressure peaks appear to be short duration spikes, primarily affecting Taps V7-9. The form of these spikes can be clarified by expanding the area around the most negative of these spikes (Figure 6(b)). This reveals a sequence of events which affects all of the centreline taps on the roof. This particular event occurs when the reference dynamic pressure, q, is well above the 20 minute average of 22 Pa, however while there is a gust about 2s before the event it isn't reflected in the windward wall pressure at V4 around the time of the event, so doesn't appear to be the cause.

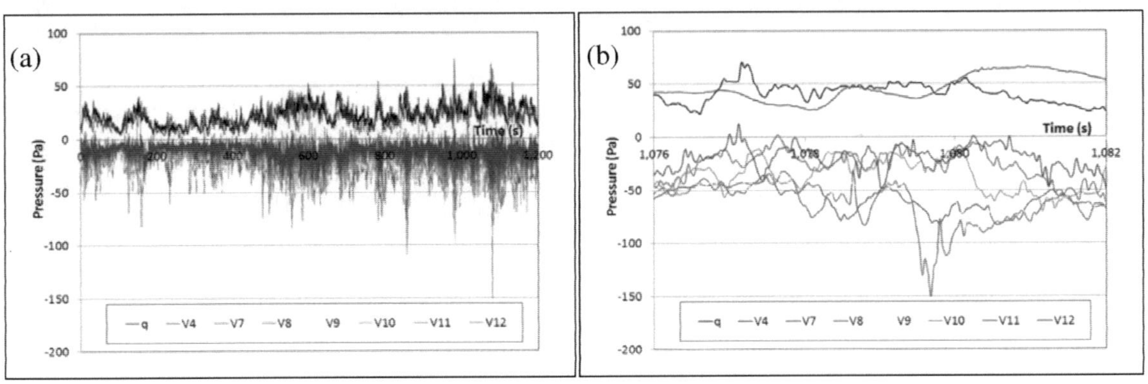

Figure 6. Full-scale pressure data from one 20 minute run during which the wind direction was θ=86.5°; (a) time histories for q, wall Tap V4 and the roof pressures and (b) the same pressures expanded around the time of the lowest pressure which occurred at Tap V8 at a time of 1079.7 s.

3.3 *Conditional averaging*

Where significant events occur on an irregular basis the consistent patterns can be highlighted by using conditional averaging. With this processing technique a number of similar significant events are identified and the time histories around these events averaged. In this way one is attempting to discern what is occurring consistently at times ahead of and after the events. Random fluctuations are hence smoothed out and the repetitive pattern revealed. Figure 7(a) shows the result of applying this analysis to the time history shown in Figure 6(a) and the three subsequent 20 minute records, all of which had similar wind speed ranges and mean wind directions. 33 events were identified, each of which had to satisfy the following criteria:

- The pressure at Tap V8 had to be less than -80 Pa, which is a magnitude more than double the dynamic pressure which was occurring around such events.
- The pressure at Tap V8 had to be the lowest pressure in a period ±5 s of that time. If there was a more negative pressure within that window then that time would be marked instead.

Having identified the 33 events the time series are processed in order to find what the average pressures were at that location and all other pressure taps at the time of the events and at times 1, 2, 3 .... times the sampling interval ahead of and after the identified events. This process reduces, but does not totally eliminate, the effects of random fluctuations in pressures caused by other processes. The exception to this is random effects reducing the pressure at Tap V8 around the time of the events, which are retained and hence somewhat exaggerate the peak for Tap V8 relative to all others.

The pattern in Figure 7(a) begins with a weak suction peak at the most windward of the 8 roof taps (V7 =C1R6), followed by progressively stronger peaks at Tap C2R6 and then the reference Tap V8=C3R6. At times >0 s peaks are observed at Taps C4R6 and V9=C5R6 but these are less significant. The more leeward taps in turn experience a positive going peak followed by a weak suction peak, with these becoming weaker across the roof. Both the reference dynamic pressure (recorded about 4h upstream and 1h to the side at cube height) and the pressure at Tap V4 (near the stagnation point on the windward wall) are 40-50% higher during these events relative to their long term mean but do not show a peak of similar intensity or duration as those experienced by the roof taps. This suggests that such events are associated with periods of stronger winds but are not directly associated with a short duration gust.

Similar patterns have been observed with data from the wind tunnel as shown in Figure 7(b). The pressures here are slightly weaker due to the lower dynamic pressure, but the event duration is much shorter, as might be expected at model scale. In this case the pressure at Wall Tap V4 is similar to its long term average since the wind tunnel boundary layer does not include the very low frequency fluctuations of wind speed which occur in the atmosphere.

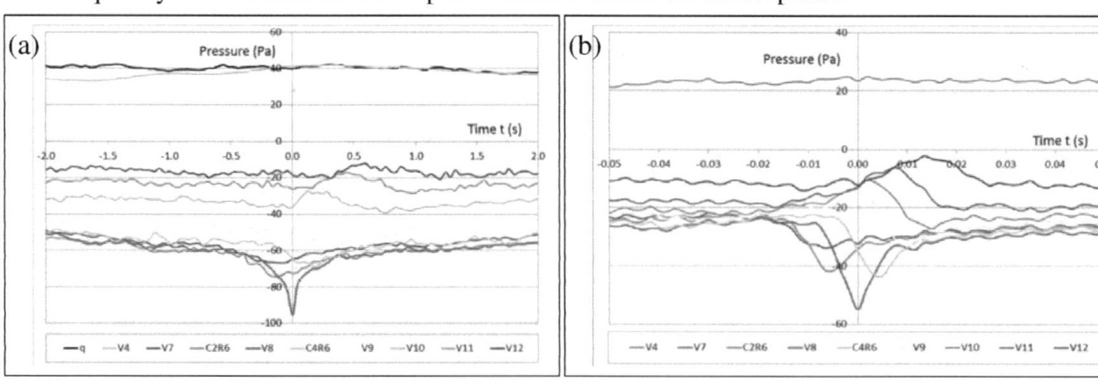

Figure 7. Conditionally averaged result around pressure spikes for selected taps on the vertical ring from (a) Full-scale and (b) Wind tunnel studies. Wind direction ≈90°.

4 CHARACTERISTICS OF PEAK PRESSURE EVENTS

4.1 *Comparison of behaviours from various sources*

Figure 8 is a comparison between the conditionally average events from the four sources of data outlined in Section 2. In order to compensate for the differences in wind speed, dynamic pressure and size, the data has been normalized with the pressures displayed as pressure coefficients. Here the normalizing dynamic pressure is the average of that occurring around the time of the event. For the Full-scale data this was taken as ±5 s either side of the event, with similar ranges being used for the other sources adjusted in relation to the duration of the event in each case. Time is displayed in the non-dimensional form tU/h, where t is the time, U the mean wind speed at cube height and h the height of the cube. These graphs show seven of the taps from the vertical ring, Tap V4 which is near the stagnation point on the windward wall and the six taps V7-12 across the roof centerline. The Tap V4 has been included since it gives an indication of the dynamic pressure that is affecting the cube at the time of the event, whereas the dynamic pressure was either measured some distance away and is hence affected by uncertain time delays, or was measured through systems which have a much lower frequency response.

While there are differences between the four sources of data it is clear that all of them are detecting the same type of phenomenon and that the general pattern occurs similarly in all cases. It also appears that the duration of these events is similar in all cases with the time scaling in proportion to the wind speed and size of the cube. Figure 7 shows that in Full-scale the event occurs between times of -0.5 s to 1.0 s and in the Wind Tunnel between -0.015 s and 0.03 s, however in normalized form these are both from -0.75 to 1.5.

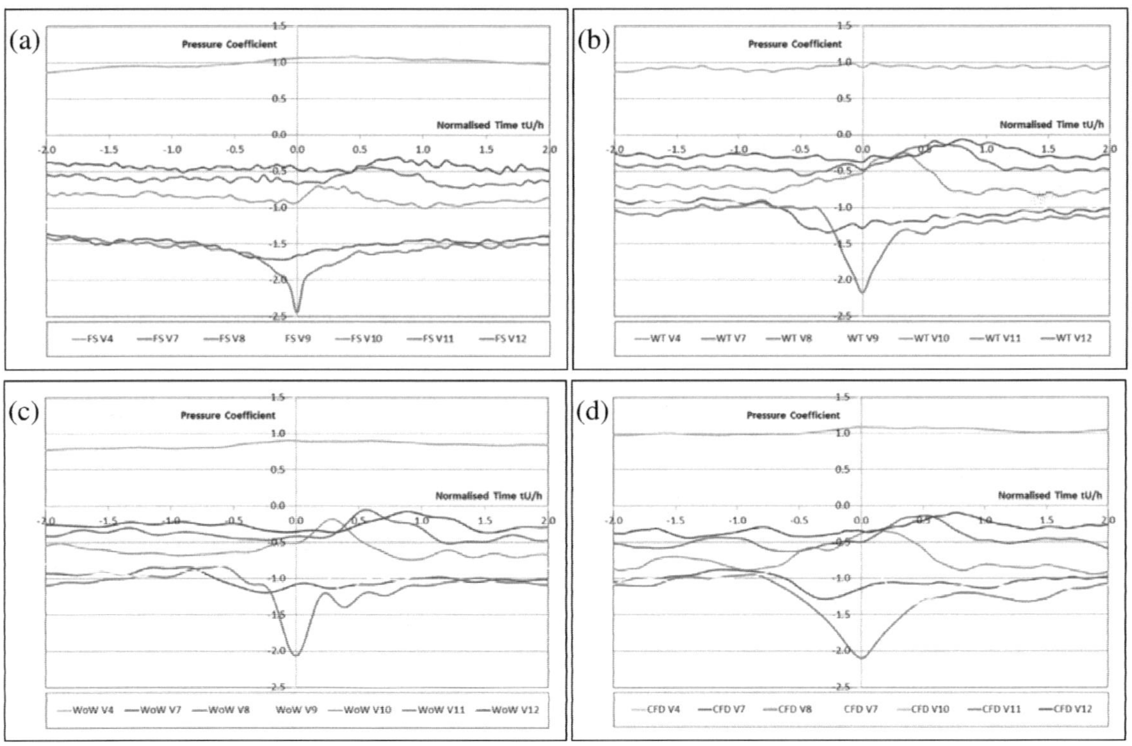

Figure 8. Conditionally averaged windward wall and roof taps in non-dimensional form from (a) Full-scale, (b) Wind tunnel, (c) Wall of Wind and (d) Computational Fluid Dynamics. Wind direction ≈90°.

4.2 *Full-scale results with various wind strengths*

The observations made in the previous section suggest that the strength of the peak event scales with the dynamic pressure prevailing at the time and the duration depends on both the wind speed and the physical scale involved. In addition it may be observed from Figure 6(a) that during the early part of that record the wind is relatively weak but there still seems to be peak events occurring but with lower magnitude. Four 20 minute full-scale records have therefore been conditionally averaged using the following criteria to identify events:

- The pressure at Tap V8 had to be less than the mean pressure coefficient times the maximum dynamic pressure occurring within ±5 s of the time under consideration.
- The pressure at Tap V8 had to be the lowest pressure in a period ±5 s of that time.
- The peak suction pressure must lie within a prescribe range.

This analysis identified a large number of events, as outlined in Table 1, where, as the peak event considered became stronger, the mean dynamic pressure occurring around those events also increased. The ratio of the minimum pressure to the associated dynamic pressure also increase slightly, as illustrated in Figure 9, although a more extensive range is needed to establish this trend with any certainty. It is however clear that the peak pressures are generally between 2 and 3 times the dynamic pressure around the time of the event for a tap where the mean pressure coefficient is only -1.05 for this wind direction.

Table 1. Full-scale peak suction events in prescribed ranges

p_{Min}<	(Pa)	-30	-40	-50	-60	-70	-80	-90	-100
p_{Min}>	(Pa)	-40	-50	-60	-70	-80	-90	-100	-200
Mean p_{Min}	(Pa)	-35.9	-45.6	-55.1	-64.2	-75.6	-83.6	-95.5	-117.6
Mean q	(Pa)	16.3	19.8	23.7	28.8	32.5	36.4	38.5	44.4
Events		29	43	32	47	30	15	10	8
Mean p_{Min}/q		-2.2	-2.3	-2.3	-2.2	-2.3	-2.3	-2.5	-2.7

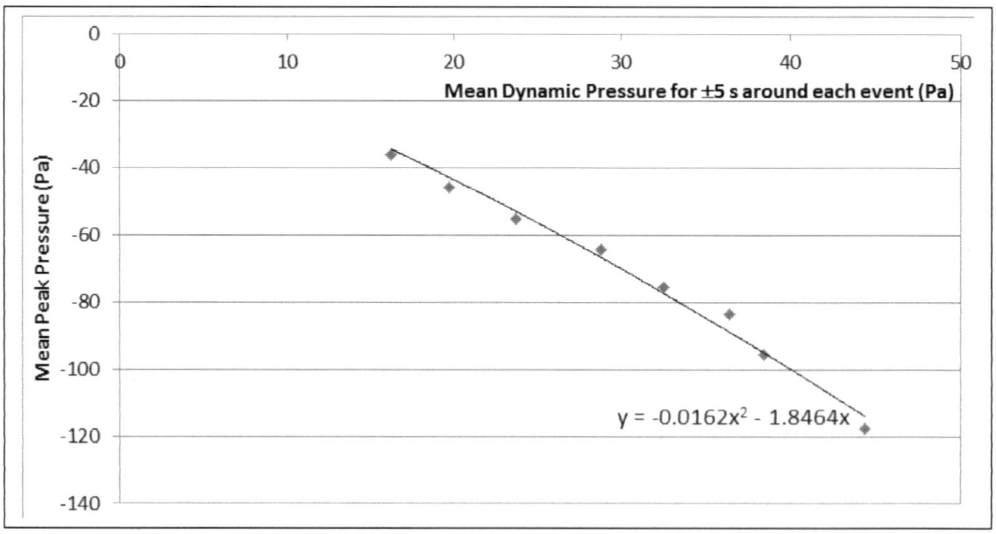

Figure 9. The mean peak pressure occurring within each range in relation to the mean dynamic pressure occurring during periods of ±5 s around each event.

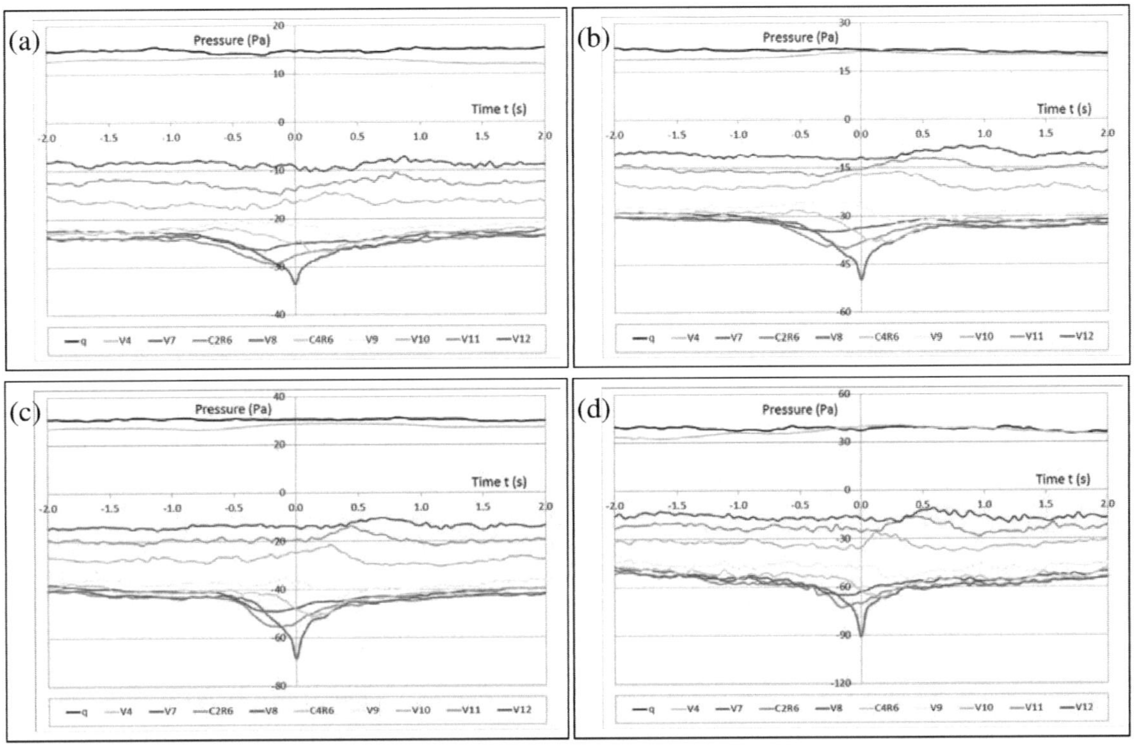

Figure 10. Conditionally averaged Full-scale pressures for peaks at Tap V8 in the ranges (a) -40 Pa< p_{min} < -20 Pa, (b) -60 Pa< p_{min} < -40 Pa, (c) -80 Pa< p_{min} < -60 Pa and (d) -120 Pa< p_{min} < -80 Pa.

It is of course possible that the weaker events are caused by a different phenomenon to that causing the stronger events. However conditional averaging, as illustrated in Figure 10, shows that all of these events have similar characteristics. The range of peak pressures used to create the four parts of Figure 10 are broader than those in Table 1, but they are essentially a combination of two table columns. It is clear that whether the peak at Tap V8 is just over -30 Pa, as in Figure 10(a), or -90 Pa, as in Figure 10(d), the sequence of events is almost exactly the same. Note that the scales of the vertical axes have been changed as the range of pressures considered increased. It can also be seen that the whole sequence occurs over a shorter time period in Figure 10(d) relative to that in Figure 10(a).

4.3 *Simultaneous velocity measurements*

As mentioned in Section 2.1, for some full-scale measurements four sonic anemometers were placed on the roof alongside Taps V8 to V11. These measured the three components of velocity approximately 60 mm above the roof surface and 500 mm to the side of the pressure tap in order to avoid interference. Figure 11(a) shows the conditionally averaged U component velocities (parallel to the roof centerline through V7-V12), while Figure 11(b) shows the vertical component velocities. At times well away from an event the U velocities alongside Taps V8 and V9 show moderately reversed flow, while that alongside Tap V10 weak reversed flow and Tap V11 moderate streamwise flow. This shows that the mean flow field reattaches about 2/3 of the way across the roof. The vertical velocities show a small upward component at Tap V8, horizontal flow at V9 and downwards flow at the other two locations.

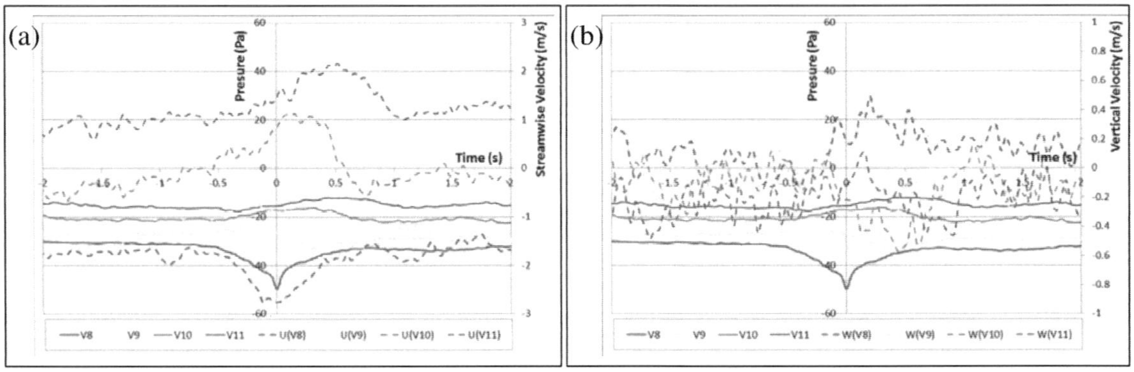

Figure 11. Conditionally averaged Full-scale pressure and velocity data for events where at Tap V8 -60 Pa< p_{min} <-40 Pa, (a) Pressure and streamwise U velocities and (b) pressure and vertical W velocities.

Around the time of the pressure peak at tap V8 the reversed flow in this area becomes more negative, suggesting the formation of a stronger vortex. At the same time the U velocity next to Tap V9 becomes less negative for a period and then changes to being slightly more negative. In a similar manner the U velocities alongside Taps V10 and V11 become more positive for a time before returning towards the general state. With the vertical component the magnitudes are small and so the changes are not as clear. There is a suggestion that for a period after the pressure peak the flow near Tap V8 becomes more upwards, while the other 3 locations show periods when the flow becomes more negative. There was no discernable trend with the transverse component.

4.4 *Large Eddy Simulation of the transient vortex behaviour*

Computational modelling has been used to gain further insight into the flow behavior. Figure 12 is a series of conditionally average flow and pressure fields on the central vertical plane through the cube. The sequence starts 4 s before the highest suction pressure occurs at Tap V8. In the first two images a general pattern of separation and reattachment on the roof exists with a vortex core located near the centre of the roof. This changes at time t=-2 s as a new tight vortex forms near the leading edge, which seems to push the existing vortex downstream. Over the next two images the new vortex grows and strengthens, so that at time t= 0 s the most negative pressures are felt at Tap V8, -1.56 m from the roof centre. At the same time the region just downstream of the roof centre experiences an impinging flow with associated more positive pressures. Over the subsequent couple of seconds the vortex continues to grow but as it does the pressures at its core become less negative and the region of more positive pressures sweeps across the leeward half of the roof. By the time t= 3 s the flow pattern has returned to something similar to that which existed at time t= -4 s.

The LES model isn't perfect in as far as the mean reattachment length is somewhat longer than suggested by the full-scale velocities in Figure 11. Nevertheless reattachment does occur most of the time. Comparison of the images in Figure 12 for times t=-3 s and t= 1 s shows that with the stronger vortex covering the windward half of the roof, at t= 1 s, the pressures are more negative over the windward half and more positive over the leeward half. In addition the reversed flow is strengthened on the windward half of the roof, while most of the leeward half is swept by a moderately strong streamwise flow, exactly as indicated by the anemometers next to Taps V10 & V11 in Figure 11(a). During the later stages as the normal separation-reattachment flow is re-established some of the vorticity from the enhanced vortex is swept downstream.

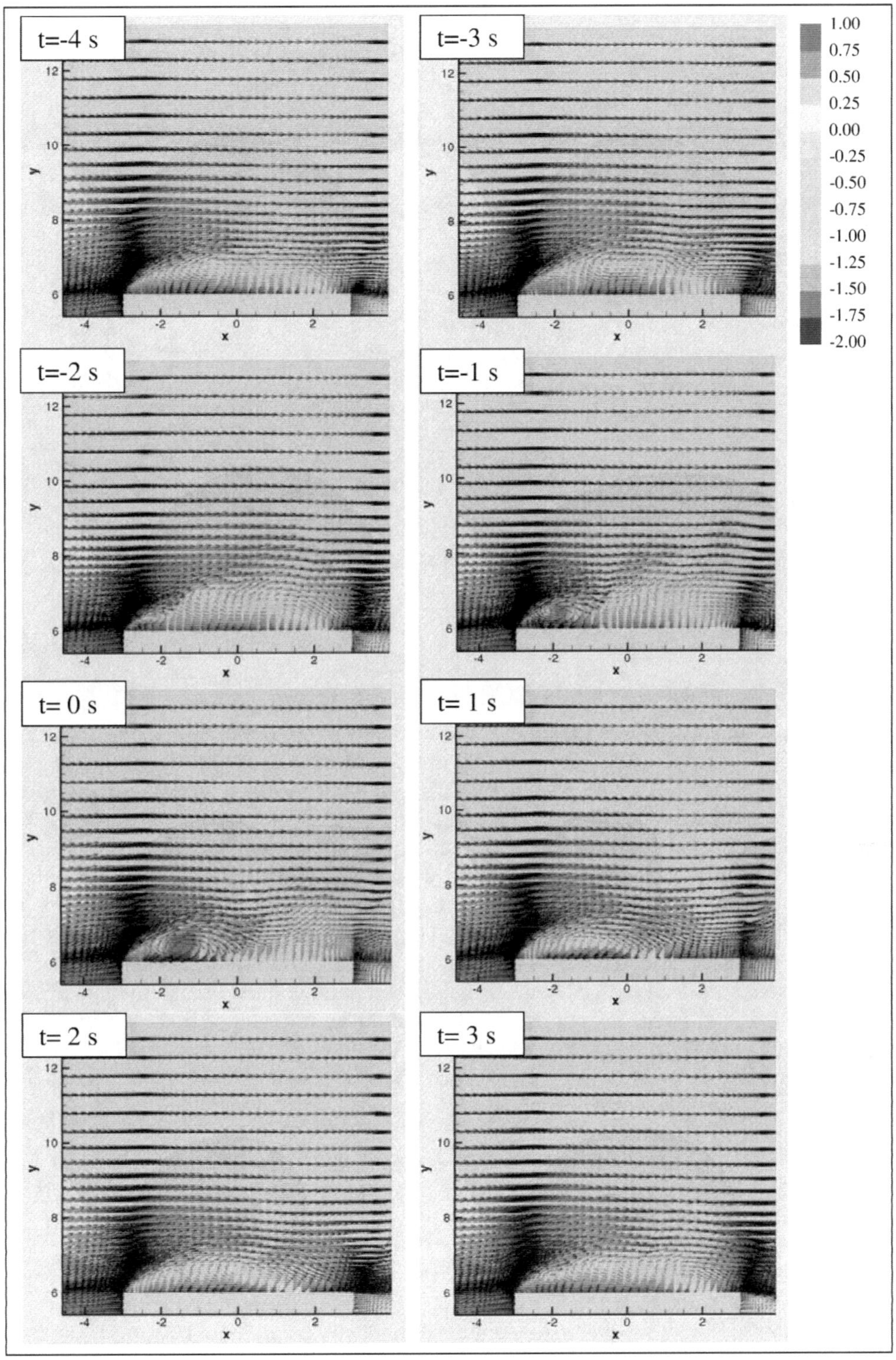

Figure 12. Large Eddy Simulation of the transient formation of a strong vortex on the roof of the 6 m cube. Images are at 1 s intervals. The background contour map shows pressure coefficients, legend top right.

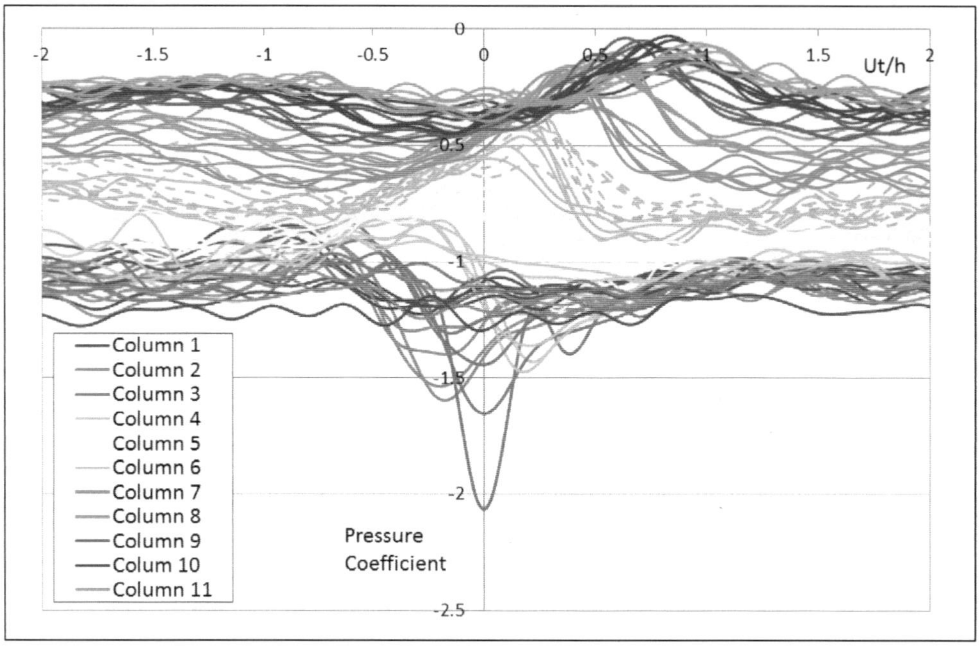

Figure 13. Wall of Wind conditionally averaged pressures for one half of the roof.

4.5 *Wall of Wind pressures for the entire roof*

In previous sections the data presented has focused on pressures occurring along the centerline. However as explained in Section 2, there is additional information available from an array of taps on one quarter of the roof. Analysis of this data shows that particularly with winds nearly normal to the windward face the entire roof is affected during each of these extreme events. Figure 13 displays the pressure coefficient for all taps on one half of the roof from the Wall of Wind. This is the combination of two records, one with a wind direction of 90° where all 36 taps that form columns 1 to 6 are included, and a second data set for wind direction 180° so that the 36 taps are now on a leeward quarter of the roof. The tap used as the event trigger for the 180° case is the one located at the position of V8 for the 90° case, which is a tap approximately ¼ of the way across the roof centreline. The data from taps on the leeward quarter are presented as if columns 7 to 11 actually existed. It may be noted that column 6 exists in both sets and so the full 12 curves are included in Figure 13 with those from the leeward test dashed. The two sets are very similar though not identical. Data sets for the entire roof can be formed for other wind angles by combining measurements from four wind directions.

The data in Figure 13 reveals the following patterns:

- On the windward half of the roof the pressure peak is much stronger in the centre than at the edges,
- On the leeward half of the roof the mainly positive pressure changes are more uniform across the entire width.
- All taps in each column are affected at about the same time, with a clear progression from Column 1 which is affected first to Column 11 which is last.
- At the time that Column 3 is experiencing the highest suctions Columns 5 & 6 are experiencing reduced suction.

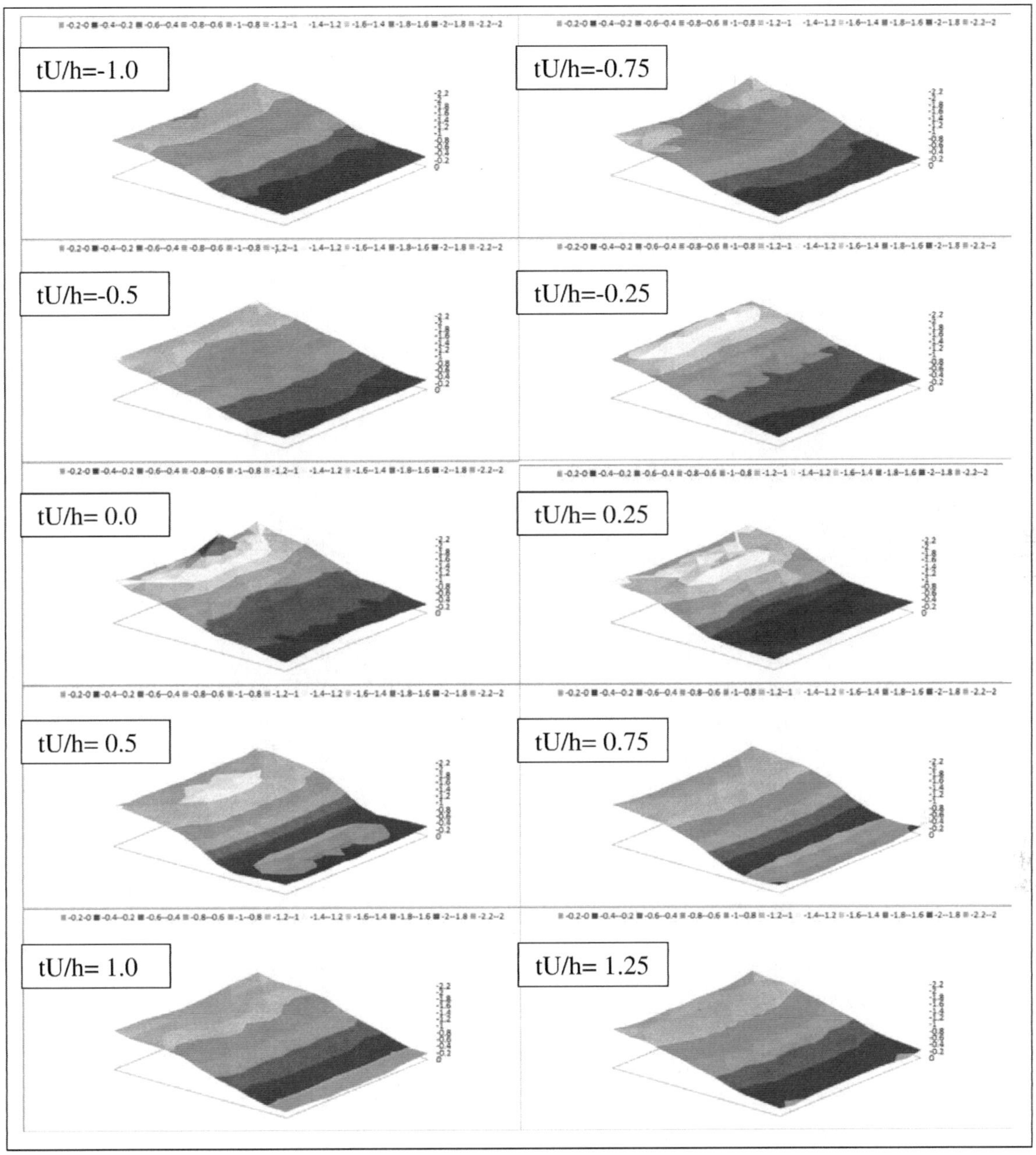

Figure 14. 3D contour maps of the Wall of Wind conditionally averaged pressures for the entire roof as a function of time relative to the peak event at Tap V8. Flow direction top left to bottom right.

Figure 14 presents the same data as a 3D contour map where there is a clear progression of a low pressure region forming near the leading edge and developing downstream. At the same time reduced suctions occur on the leeward half which also moves downwind.

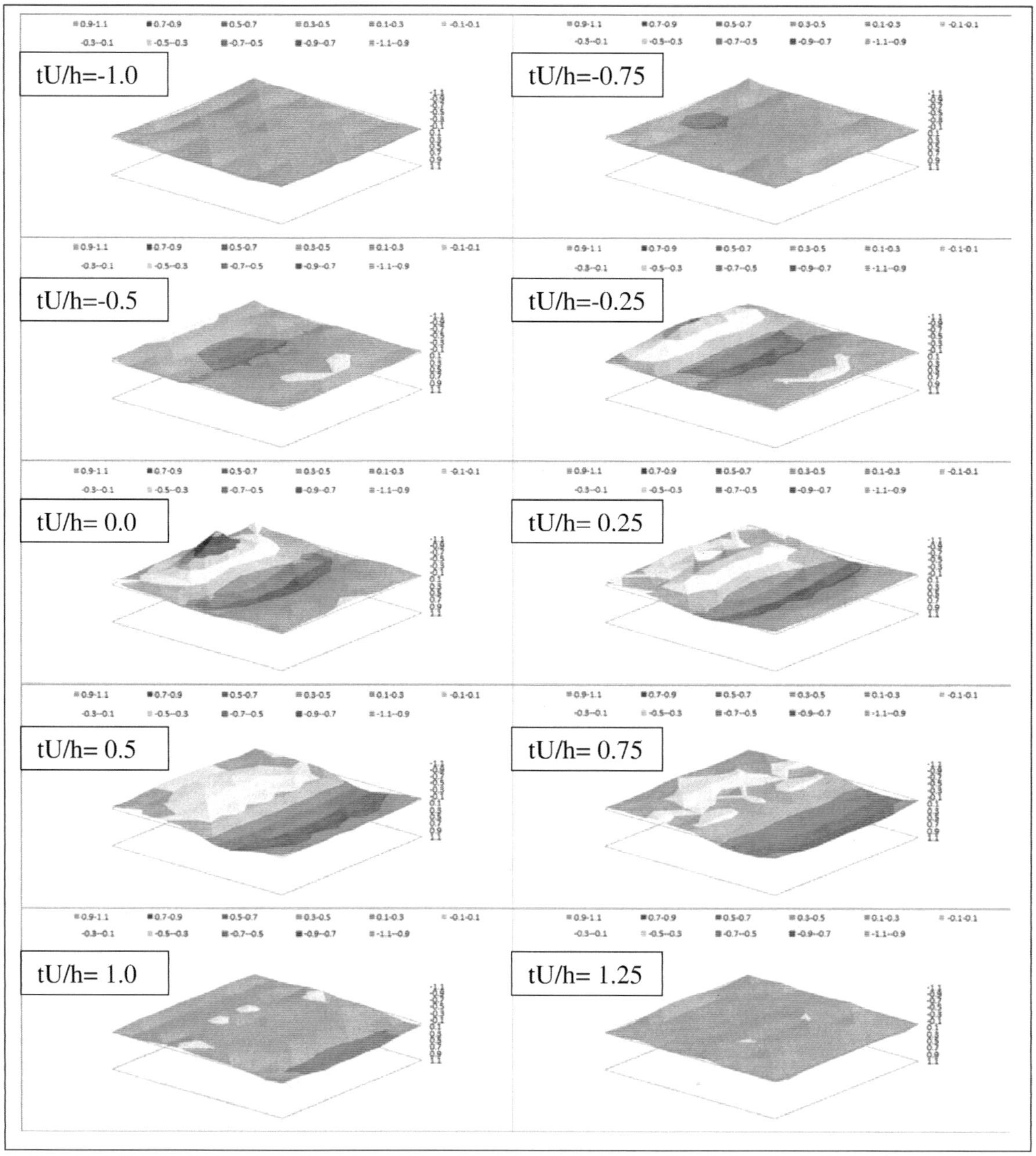

Figure 15. 3D contour maps of the Wall of Wind conditionally averaged pressures relative to the mean pressure at each location.

This progression is even clearer when presented as the difference between the conditionally averaged pressures and those which occur at times remote from an event, as shown in Figure 15. It is perhaps surprising to see that the first indication that an event is beginning is a small positive pressure region at time tU/h=-0.75. This strengthens initially and widens to span the entire width. However as it moves downstream, beyond tU/h= 0.25, it progressively weakens.

At tU/h=-0.5 a suction region develops at the leading edge which over the next two time steps strengthens to become the major suction event. However this is not the only enhanced suction region since a second, but weaker, area can be observed at tU/h=-0.5 and -0.25 ahead of the more positive region. This may be the influence of the vortex which was trapped in the separation-reattachment bubble prior to the formation of the new more intense vortex which has been pushed downstream by the new vortex. As the event develops both the main suction region and relatively positive pressure region spread but also weaken until at about tU/h =1.25 the pressure return to a neutral situation.

5 CONCLUSIONS

Pressure data from field testing of the Silsoe 6 m Cube showed occasional strong suction peaks on the roof when the wind was nearly perpendicular to one of the faces. Conditional averaging of the events showed a pattern of pressure waves affecting the entire roof. Data from wind tunnel testing in the University of Auckland Boundary Layer Wind Tunnel and the University of Florida Wall of Wind (WoW) together with Computational Fluid Dynamic modelling using Large Eddy Simulations (LES) has revealed similar patterns. When expressed in non-dimensional form using an instantaneous pressure coefficient normalized using the average dynamic pressure around the time of the event and normalized time tU/h, where U is the wind speed around the time of the event and h the cube height, the patterns are very similar. Further analysis of the full-scale data, which includes significant low frequency variations in wind speed, has shown that such events occur throughout a record and not just during periods of strong wind. Simultaneous velocity measurements at full-scale and LES modelling show that the pattern is associated with a new intense vortex forming at the leading edge and expanding across the roof, displacing the resident vortex associated with the background separating-reattaching flow pattern. Pressures from FIU's WoW show that the pattern spreads across the entire width of the roof and can be characterized as a relatively positive pressure wave moving ahead of a suction band.

6 ACKNOWLEDGEMENTS

The assistance of Roger Hoxey, Max Haspel, Maryam Asghari Mooneghi and Stuart Norrris with the Full-scale, Wind-tunnel, Wall of Wind and CFD work respectively is gratefully acknowledged.

7 REFERENCES

1 Richards P.J., Hoxey R.P., Short L.J., Wind pressures on a 6m cube, Journal of Wind Engineering and Industrial Aerodynamics 89, 14-15 (2001) 1553-1564.

2 Lim, H.C., Thomas, T.G., Castro, I.P., Flow around a cube in a turbulent boundary layer: LES and experiment, Journal of Wind Engineering and Industrial Aerodynamics, 97, 2 (2009) 96-109.

3 Haupt, S.E., Zajaczkowski, F.J., Peltier, L.J., Detached eddy simulation of atmospheric flow about a surface mounted cube at high Reynolds number. Journal of Fluids Engineering, Transactions of the ASME, 133, 3 (2011), Article number 031002.

4 Aly, A.M., Gan Chowdhury, A. and Bitsuamlak, G., Wind profile management and blockage assessment for a new 12-fan Wall of Wind facility at FIU, Wind and Structures, 14, 4(2011), 285-300.

5 Richards, P.J., Hoxey, R.P., Pressures on a cubic building-Part 1: Full-scale results & Part 2: Quasi-steady and other processes. Journal of Wind Engineering and Industrial Aerodynamics, 102 (2011) 72-86 & 87-96. doi:10.1016/j.jweia.2011.11.004 & 3.

6 Richards, P.J., Hoxey, R.P., Connell, B.D., Lander, D.P., Wind tunnel modelling of the Silsoe Cube. Journal of Wind Engineering and Industrial Aerodynamics, 95 (2007) 1384-1399

7 Richards, P. J., Norris, S., LES modelling of unsteady flow around the Silsoe cube, Journal of Wind Engineering and Industrial Aerodynamics 144 (2015) 70-78, doi:10.1016/j.jweia.2015.03.018 7

8 Armfield, S. W., Norris, S. E., Morgan, P. and Street, R. A, Parallel Non-Staggered Navier-Stokes Solver Implemented on a Workstation Cluster, Proceedings of the Second International Conference on Computational Fluid Dynamics, 30–45, Sydney, Australia, 2002.

9 Norris, S.E., A Parallel Navier-Stokes Solver for Natural Convection and Free Surface Flow. PhD thesis, University of Sydney. 2001.

10 Armfield, S. W. and Street, R., An analysis and comparison of the time accuracy of fractional-step methods for the Navier-Stokes equations on staggered grids, International Journal for Numerical Methods in Fluids, 38 (2002) 255–282.

11 Smagorinsky, J., General circulation experiments with the primitive equations. Mon. Weather Rev., 91 (1963) 99–164.

12 Mason, P.J., Thomson, D.J., Stochastic backscatter in large-eddy simulations of boundary layers. J. Fluid Mech., 242 (1992) 51–78.

13 Mason, P.J., Callen, N.S., On the magnitude of the subgrid-scale eddy coefficient in large-eddy simulations of turbulent channel flow. J. Fluid Mech., 162 (1986) 439–462.

14 van der Vorst, H.A., Iterative solution methods for certain sparse linear systems with a non-symmetric matrix arising from PDE-problems, J. Comput. Phys., 44 (1981) 1-19.

15 Stone, H.L., Iterative solution of implicit approximations of multidimensional partial differential equations, SIAM J. Numer. Anal., 5 (1968) 530-558.

13th Americas Conference on Wind Engineering
Gainesville, Florida, USA
May 21 - 24, 2017

Wind-induced structural forces of L- and T-shaped low-rise buildings with hip roofs

Shuai Shao-1[a,b], Ted Stathopoulos-2[b], Yuji Tian-3[a], Qingshan Yang-4[a]

[a] *Beijing's Key Laboratory of Structural Wind Engineering and Urban Wind Environment, School of Civil Engineering, Beijing Jiaotong University, Beijing, China, 12121097@bjtu.edu.cn, yujitian@bjtu.edu.cn, qshyang@bjtu.edu.cn*

[b] *Department of Building, Civil and Environmental Engineering, Concordia University, Montreal, Canada, statho@bcee.concordia.ca*

ABSTRACT

Previous comprehensive studies have concentrated on wind-induced cladding pressures and structural forces on rectangular low-rise buildings, particularly those with gable roofs. However, low buildings with asymmetric plan views have not yet received the attention they deserve with respect to their interaction with wind. For instance, wind forces on main wind force resisting systems of L-shaped buildings with gable or hip roofs have not yet been examined and wind load provisions for such building configurations are not currently available.

This paper assesses the performance of current wind provisions available in national and international standards for low rectangular buildings with hip roofs. It then focuses on the wind-induced structural loads for the main wind-force resisting system of L- and T-shaped low-rise buildings. Instantaneous vertical total uplifts, horizontal thrusts and several critical bending moments on the tested rigid main frames and bays were estimated based on wind tunnel test results. Additionally, the adequacy of the envelope methodology currently used in North American standards has been evaluated.

In total, 20 rectangular, L- and T-shaped low-rise building models were made at a scale of 1:200 with a 4:12 roof slope, including 5 L-shaped and 10 T-shaped hip-roofed models with different horizontal aspect ratios. Roof and wall pressures were measured at more than 200 locations synchronously in the wind tunnel under simulated open country terrain upstream conditions.

Results show that roof and plan configurations play important roles in varying the frame wind force distributions, due to the considerably influenced wind-induced pressure patterns. Significant discrepancies were observed on wind forces of frames near the interior valley corner, compared with rectangular models around those areas. Moreover, the comparisons between the wind tunnel test results and current rectangular-related provisions in the American as well as the Canadian Standard indicate clear differences. These comparative studies are expected to be instrumental for future development of wind code and standard provisions.

KEYWORDS

Low-rise L- and T-shaped buildings, wind-induced structural forces, wind codes and standards, wind tunnel tests

The 13th Americas Conference on Wind Engineering (13ACWE)
Gainesville, Florida USA, May 21-24, 2017

Estimation of cavity pressures in air-permeable, multi-layer cladding systems using a lumped-leakage approach

Rahul Prabhakaran Nair Sreedevi [a], Gregory A. Kopp [b]

[a]*Department of Civil engineering, Faculty of Engineering, University of Western Ontario, London, Canada*
[b]*Boundary Layer Wind Tunnel Laboratory, University of Western Ontario, London, Canada*

ABSTRACT: Wind loading on the external surface of structures can be estimated by wind tunnel tests. But for air-permeable multi-layers cladding systems like double-skin facades or solar panel arrays on the roof of a building, the estimation of wind loads by conventional wind tunnel tests may not be a feasible solution. This is due to the practical difficulties in exactly simulating the dimensions of the gaps in the physical model with small model scales or other practical issues related to the tubing of the pressure sensors. Pressure equalization plays a major role on the wind loading on individual members of multi-layer systems and this paper discusses the development of an analytical model to estimate the time-varying cavity pressure distributions in a double-layer system with an air-permeable outer layer, given the external pressure on the outer surface. The model uses orifice discharge equation and conservation of mass equation, along with a lumped leakage approach to account the pressure drop due to flow through the gaps. The results from the analytical model are compared with wind-induced loads obtained from the wind tunnel test of roof-mounted photovoltaic solar array system obtained from Stenabaugh (2015).

KEYWORDS: Wind loading, Pressure equalization, Air-permeable cladding.

1 INTRODUCTION

Nowadays, the design community is coming up with wide variety of designs for tall buildings and civil structures, which add value to not only the aesthetic looks but also to the overall performance of the buildings in terms of energy consumption and design wind loading. This urge to come up with creative designs has led to the introduction of unique elements in the outer layer, or façade, of buildings and roofs. But the most important point to be noted here is that the wind-loading mechanism for these new elements are different from those on conventional glass facades, for many reasons. In buildings with double skin façades or multi-layer systems, the outer layer can be porous or have gaps between the layers such that the wind-induced external pressures can be transferred to the cavity between the layers. So, for the design of these systems, it is important to know the mechanisms of overall net wind loading on the external layer, as well as the cavity pressure. The mechanism of transfer of external pressure to the cavity, and, thereby, equalizing the external and cavity pressure, is known as pressure equalization.

The initial attempts to develop an analytical model for pressure equalization were by Killip and Cheetham, (1984); Fazio and Kontopidis, (1988); Baskaran and Brown, (1992); Xie et al, (1992), who used the discharge equation for flow through openings with the assumption of incompressible flow. Burgess (1995) and Van Schijndel and Schols (1998) developed analytical models using the discharge equation for flow through openings and the ideal gas law. Inculet and Davenport (1994), Choi and Wang (1998), and Kumar and Van Schijndel (1999) introduced unsteadiness by using the unsteady discharge equation, i.e., the Helmholtz resonator model of Holmes, (1979) and Vickery, (1986), and predicted the net pressures on the outer layer. To model cavity pressures that are spatially varying because of longitudinal flow in the cavity, Amano et al. (1988), Trung et al. (2010) and Lou et al. (2012) also used Helmholtz resonator model from Holmes, (1979) and Vickery, (1986). Sun and Bienkiewicz (1993) developed an analytical model to estimate the pressure distributions in the cavity underneath loose-laid roof pavers with the assumption of steady flows in the gaps and cavities and, hence, Darcy's law was used to model the mean flow. But this model is applicable only for flow at low Reynolds Number. Bienkiewicz and Sun (1992, 1997) experimentally studied the pressures, especially the cavity pressures beneath the scaled paver systems on the flat roof of a low-rise building for a cornering wind direction. Holmes (2007) also did similar studies for cornering winds and investigated the mechanism behind loading on roof pavers due to flow separation and re-attachment caused due to the rotation of vortices, when the wind is along the corner of the roof. The effect of ratio of gap (G) between the outer layer and the inner layer and the depth of the cavity (H) was examined in large-scale tests conducted at Wall of Wind by Asghari Mooneghi et al. (2014) and concluded that increasing the G/H ratio can lead to reduced net loading on these systems. Ginger et al. (2011), Stenabaugh et al. (2015) and Geurts and Blackmore (2013) studied the wind loading and pressure equalization on solar modules mounted parallel to roofs. Stenabaugh et al. (2015) examined the wind loads on roof-mounted solar arrays and also highlighted the importance of the G/H ratio. Aly and Bitsuamlak (2014) also reported similar results from the studies conducted on solar panel modules mounted parallel to a sloped gable roof system. So from the literature survey, it is clear that there has been a lot of studies done on the topic of pressure equalization; but an accurate, practical, analytical model which can address the issues completely is still lacking. The objective of this paper is to develop of an analytical model which can estimate the time-varying cavity pressure distributions in a double-layer system with an air-permeable outer layer, given the external pressure on the outer surface, using lumped-leakage approach.

2 MATHEMATICAL FORMULATION

The most important equation for fluid flow through an orifice can be derived from the Navier-Stokes equation for incompressible flows,

$$\rho \frac{\partial u_j}{\partial t} + \rho u_k \frac{\partial u_j}{\partial x_k} = -\frac{\partial p}{\partial x_j} + \mu \frac{\partial^2 u_j}{\partial x_j \partial x_j} + \rho f_{bj} \tag{1}$$

where ρ is the density of the fluid, u_j is the velocity component along j direction, u_k is the velocity component along k direction, p is the pressure, μ is the viscosity of the fluid and f_b is the body force.

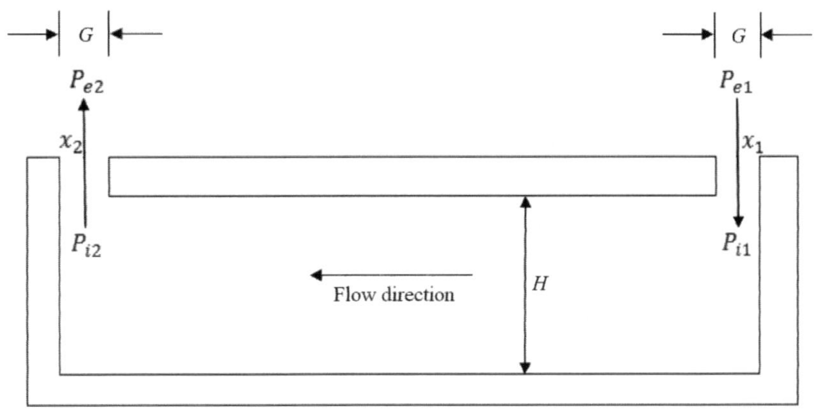

Figure 1. Flow into and out of a double-layer system with two openings (Oh and Kopp, 2014).

Oh and Kopp (2014) and Oh (2014) show that Equation 1, when applied to the system shown in Fig. 1, becomes

$$\rho l_e \dot{U} + C_L \frac{\rho}{2} U|U| + \frac{12\mu l_o}{G^2} U = \Delta P \tag{2}$$

where l_e is the "effective length" of the fluid passing through the orifice, $\dot{U}$ is the average velocity through the area of the opening and is assumed to be constant, C_L is a loss coefficient, l_o is the thickness of the external layer, G is the width of the opening, i.e., the orifice, and ΔP is the differential pressure, which is the difference in external pressures at locations x_1 and x_2, i.e., Pe_1 and Pe_2. The arrows in the figure indicate the direction of flow for a particular pressure distribution. It is also assumed that there is no vertical pressure gradient such that there is a single value for the pressure at each horizontal location.

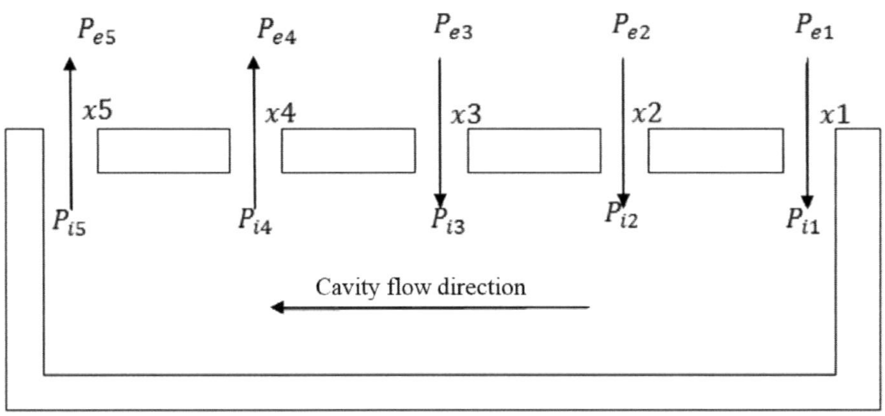

Figure 2. Flow into and out of a double-layer system with multiple openings.

Applying equation 2 for a system with multiple openings, as shown in Fig. 2, will give

$$\rho l_{e1} \dot{U}_{g1}(t) + C_{L1} \frac{\rho}{2} U_{g1}(t)|U_{g1}(t)| + \frac{12\mu l_{o1}}{G_1^2} U_{g1}(t) = P_{e1}(t) - P_{i1}(t) \tag{3}$$

$$\rho l_{e2} \dot{U}_{g2}(t) + C_{L2} \frac{\rho}{2} U_{g2}(t)|U_{g2}(t)| + \frac{12\mu l_{o2}}{G_2^2} U_{g2}(t) = P_{e2}(t) - P_{i2}(t) \tag{4}$$

$$\rho l_{e3} \dot{U}_{g3}(t) + C_{L3} \frac{\rho}{2} U_{g3}(t)|U_{g3}(t)| + \frac{12\mu l_{o3}}{G_3^2} U_{g3}(t) = P_{e3}(t) - P_{i3}(t) \tag{5}$$

$$\rho l_{e4} \dot{U}_{g4}(t) + C_{L4} \frac{\rho}{2} U_{g4}(t)|U_{g4}(t)| + \frac{12\mu l_{o4}}{G_4^2} U_{g4}(t) = P_{e4}(t) - P_{i4}(t) \tag{6}$$

$$\rho l_{e5} \dot{U}_{g5}(t) + C_{L5} \frac{\rho}{2} U_{g5}(t)|U_{g5}(t)| + \frac{12\mu l_{o5}}{G_5^2} U_{g5}(t) = P_{e5}(t) - P_{i5}(t) \tag{7}$$

The velocity at each opening in the outer layer is referred as $U_g(t)$ whereas $P_e(t)$ and $P_i(t)$ refer to the external pressure and cavity pressure, respectively. The numbers in the subscripts (i.e., 1, 2, 3, 4, and 5) denote the location of the gaps, as depicted in Fig. 2. Employing conservation of mass for the system shown in Fig. 2, the flow rate entering the cavity should be equal to the flow rate leaving the cavity, i.e.,

$$Q_1 + Q_2 + Q_3 = Q_4 + Q_5 \tag{8}$$

$$A_1 . U_{g1} + A_2 . U_{g2} + A_3 . U_{g3} = A_4 . U_{g4} + A_5 . U_{g5} \tag{9}$$

where Q is the flow rate through the area of the opening and A is the area of the opening. So there are 6 equations (equations 3 through 7 and 9) and 10 unknowns (U_{g1} to U_{g5} and P_{i1} to P_{i5}).

2.1 *Assumptions considered in the model*

In the current model, it will be assumed that the flow can enter or leave the cavity only through the openings in the external layer such that the remainder of the system is closed. This is to ensure that the cavity volume is well-defined. Moreover, in the current model, the cavity pressure is assumed to be uniform, i.e., the effective G/H ratio is small. With the uniform cavity pressure assumption, the number of unknowns reduced to six (i.e., U_{g1} to U_{g5} and P_i), which means the number of equations and unknowns are same and hence, they can be solved easily (Oh and Kopp, 2014).

3 LUMPED-LEAKAGE APPROACH

It is well-known that when wind reaches a building, the flow is separated at edges, creating large vortices and large pressure gradients on the external surface. Normally, the windward edge of the roof has high suctions, caused by separated and accelerated flow, which reduce as the flow reattaches. Due to the difference in pressure over the external surface, flow may enter the cavity and the gradient in external pressure distribution determines the entry and exit points for the flow through the cavity. For example, as shown in Fig.3 (a), the flow in to the cavity may enter through all the openings in the external surface of region '1' while the flow out of the cavity may come through all the openings in the external surface of the region '2', where the two regions are separated by the red line. The location of the reference (red) line in the figure depends upon the distribution of external pressures, which Oh and Kopp (2014, 2015) called the neutral pressure line.

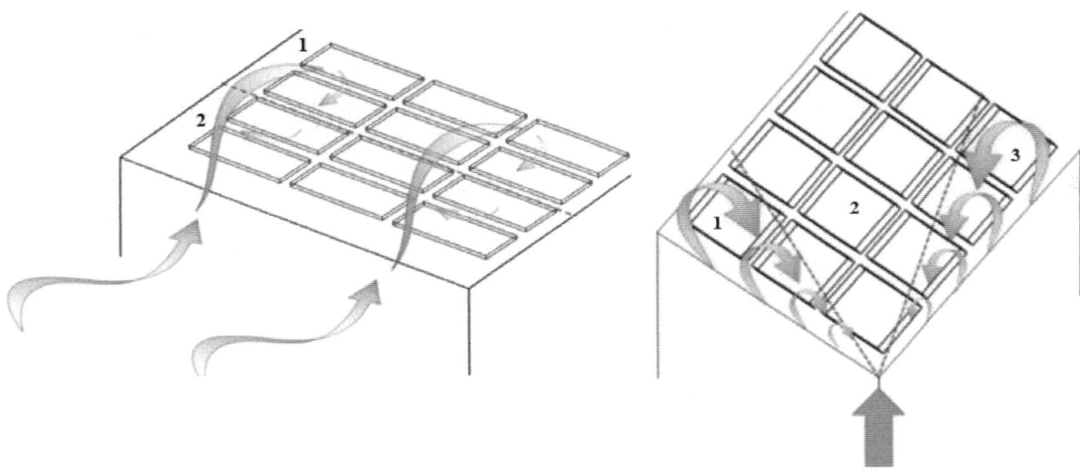

(a) (b)

Figure 3. Schematic representation of flow separation and cavity flow (a) when the wind direction is perpendicular to the face and (b) when the wind direction is coming at an inclined angle to the roof

Moreover, if the flow is entering the cavity through all openings in Region '1', then the external pressure distribution over the entire region '1' have to be accounted. So, instead of using the individual $P_e(t)$ values, an area-averaged external pressure value is given as input to the analytical model. The area-averaged value of the pressure in region 1 will be called P_{AE1} and area-averaged value of the pressure in region 2 will be P_{AE2}.

In addition, the loss coefficient C_L also had to be updated to match with the assumptions stated above. In equation 2, C_L value refers to the loss coefficient through a single opening. But in the current model, the losses due to flow through openings are lumped for a specific area and, hence, instead of loss coefficient for each opening, a single effective loss coefficient, K_L will be used (Guha, 2011). K_L can be estimated using the Bernoulli`s obstruction theory (White, 1999) with a reasonable assumption that the area of the openings are much smaller, when compared to the total area of the panel array.

$$K_L = \frac{1 - (A_o/A)^2}{{C_d}^2} \tag{10}$$

where A_o refers to the area of openings considered in the specific region, A refers to the total area of the specific region and C_d is the discharge coefficient through the lumped openings.

After considering all of the above-mentioned changes to the original discharge equations, the current model as applied to Figure 3(a) where only two areas are considered, which will be evaluated in this work, is:

$$\rho l_{e1}\dot{U}_{g1}(t) + K_{L1}\frac{\rho}{2}U_{g1}(t)|U_{g1}(t)| + \frac{12\mu l_{o1}}{{G_1}^2}U_{g1}(t) = P_{AE1}(t) - P_i(t) \tag{11}$$

$$\rho l_{e2}\dot{U}_{g2}(t) + K_{L2}\frac{\rho}{2}U_{g2}(t)|U_{g2}(t)| + \frac{12\mu l_{o2}}{{G_2}^2}U_{g2}(t) = P_{AE2}(t) - P_i(t) \tag{12}$$

$$A_1U_{g1} + A_2U_{g2} = 0 \tag{13}$$

where $U_{g1}(t)$ and $U_{g2}(t)$ refers to the velocities of flow in to and out of the cavity through the openings in regions 1 and 2, K_{L1} and K_{L2} represent the effective loss coefficients for regions 1 and 2, $P_{AE1}(t)$ and $P_{AE2}(t)$ are the area-averaged external pressures for the regions 1 and 2, respectively, as shown in Figure 3 (a) and $P_i(t)$ represents the uniform cavity pressure.

Now consider the case in which the flow direction is not perpendicular to the face and the wind is coming towards the roof at an angle as shown in Fig. 3 (b). When the wind is coming at such an angle to the roof, the flow is separated at the edge in such a way that corner vortices form. Due to the formation of corner vortices, the distribution of external pressures over the upper surface of the solar panels are different from the pattern discussed in the previous section. In this case it makes sense to think about three distinct pressure regions on the roof, which are labelled in Fig. 3 (b). The estimation of cavity pressure in this scenario can be done using the following four equations

$$\rho l_{e1}\dot{U}_{g1}(t) + K_{L1}\frac{\rho}{2}U_{g1}(t)|U_{g1}(t)| + \frac{12\mu l_{o1}}{{G_1}^2}U_{g1}(t) = P_{AE1}(t) - P_i(t) \tag{14}$$

$$\rho l_{e2}\dot{U}_{g2}(t) + K_{L2}\frac{\rho}{2}U_{g2}(t)|U_{g2}(t)| + \frac{12\mu l_{o2}}{{G_2}^2}U_{g2}(t) = P_{AE2}(t) - P_i(t) \tag{15}$$

$$\rho l_{e3}\dot{U}_{g3}(t) + K_{L3}\frac{\rho}{2}U_{g3}(t)|U_{g3}(t)| + \frac{12\mu l_{o3}}{{G_3}^2}U_{g3}(t) = P_{AE3}(t) - P_i(t) \tag{16}$$

$$A_1U_{g1} + A_2U_{g2} + A_3U_{g3} = 0 \tag{17}$$

4 SOLUTION OF THE ANALYTICAL MODEL

For estimating the (uniform) cavity pressure, the unsteady discharge equations along with conservation of mass equation must be solved and this requires a numerical method. In this study, a second-order, backward-differencing, numerical scheme is used (Chapra and Canale, 2006) where, at each time step, the given time series of the external pressures are used to solve for the (unknown) velocities through the gaps and the cavity pressure.

The area-averaged pressure coefficients are calculated by integrating the pressures of all taps within a specific area simultaneously. The area-averaged pressure coefficient for a given area for the external or upper surface is given by

$$Cp_{AE}(t) = \frac{\sum_{i=1}^{n} Cp_E(t)\ .a_i}{A} \tag{18}$$

where $Cp_E(t)$ are the time history of the pressure coefficients on the upper or external surface of the module, a_i is the tributary area associated with the specific pressure tap 'i' and A is the sum of all tributary areas on one surface of the module. The results obtained from the analytical model are validated by comparing the simulation results with the cavity pressures from the experimental study conducted on a roof-mounted solar panel array by Stenabaugh (2015).

The step-by-step procedure of the working of analytical model are shown in the form of a flowchart in Figure 4. The first step is to examine the pressure distribution on the external surface of the panel array, which is obtained from the wind tunnel tests. The mean external pressure distribution is a crucial factor which decides the average flow through the openings and the location of neutral pressure line. The second step is fixing the neutral pressure line. Initially, since the cavity pressure is unknown, the spatial average of all the external pressures are taken as the cavity pressure. This will help to fix a neutral pressure line and based on the location of that neutral pressure line, the areas of flow entry and exit are decided. In reality, these areas can be quite complicated and hence, the areas are simplified into rectangles or triangles, depending upon the wind direction and pressure distribution pattern. Then the governing equations, mentioned in the previous section, are solved for each of the simplified areas, using the numerical method. Once the equations are solved, we will obtain the cavity pressure and then we have to check whether the flow directions and regions are more precise with the actual distribution. If the predicted cavity pressure are reasonably good, then we can proceed with the calculation of net wind loads on panels and combination of panels for their design. If the predicted cavity pressure is different than the actual pressure distribution, then the model will redo from the second step, i.e., the neutral pressure line fixing. The user then uses the model results to decide the neutral pressure line based on the cavity pressure obtained from the first iteration and this whole process will continue until an accurate cavity pressure is obtained.

5 EXPERIMENTAL SET-UP

The wind tunnel experiments for the solar panel array system were conducted in the Boundary Layer Wind Tunnel II at the University of Western Ontario by Stenabaugh et al. (2015) and Stenabaugh (2015). Figure 5 shows the photo of a sloped roof model in the wind tunnel along and in the top right corner, wind tunnel physical model of the solar panel arrays on the flat roof surrounded by shroud. However the results from the analytical model are compared with the experimental results obtained from a flat roof model as shown in Figure 5.

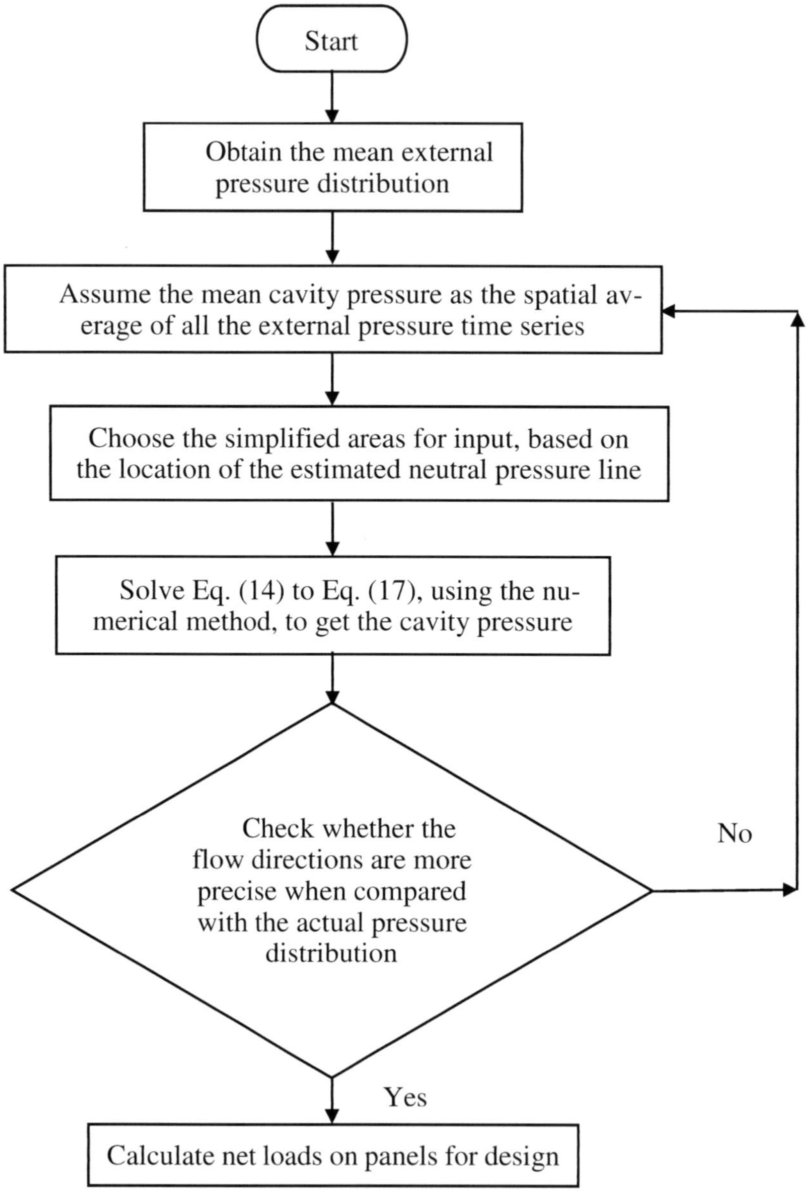

Figure 4. Flowchart showing the algorithm of the analytical model

Figure 5. Photo of the Boundary Layer Wind Tunnel at the University of Western Ontario along with the study model having solar panels on the sloped roof (Stenabaugh, 2015) and (at top right corner) wind tunnel physical model of the solar panel arrays on the flat roof surrounded by shroud

The wind tunnel has a working cross-section of 3.4m (11ft) wide by 2.4m (8ft) high along with an upstream fetch length of 39m (128ft). In the wind tunnel, an open country profile was simulated. The wind tunnel tests were conducted using 1/20 scale physical model. The solar panel modules were modeled as flat panels, made from white Delrin (acetyl) plastic, with equivalent full-scale dimensions of 50cm x 145.5cm, which gives an equivalent module area of 0.73 m^2. The solar panel modules have a thickness of 0.3cm at a model scale of 1:20 and this was maintained to achieve the minimum thickness for the pressure tubes within them. To obtain detailed information about the pressures acting on the external and internal surface of the solar panel array system, a high density pressure tap layout was required. With that intention, 20 pressure taps were placed on each individual solar panel module, i.e., 10 on the upper surface and 10 on the lower surface of each solar panel module.

During the wind tunnel test, experiments were carried out for different configurations and for different roofs. In the current study, we are comparing the results from the analytical model with the experimental results from flat roof configuration having a gap between the panels (G) equivalent to 12cm and the distance between the panel and the roof of 20cm.

The flat roof model used in the wind tunnel test has full-scale dimensions of 15m x 7.5m along with an eave height of 6m. The solar panel array was located at the corner of the flat roof, adjacent to the walls so that it would experience maximum suction due to the flow separations. The results from the analytical model have to be compared with experimental data in which the flow into or out of the cavity can happen only through the gaps between the panel modules. For that purpose, the experimental configuration in which the solar array was surrounded by a shroud was selected. The bottom of the shroud was sealed to the roof surface and the top of the shroud was flush with the top level of solar panels. During the wind tunnel tests, the pressure measurements were made at a reference speed of approximately 14 m/s, which corresponds to a mean wind speed at the mean roof height of 42 m/s in full scale, based on a velocity scale of 1:3. The wind tunnel measurements were collected for 16 wind directions in 22.5° increments. The pressure data samples were collected for 360 seconds at a sampling rate of 400 samples per second, which is equivalent to a full-scale sampling period of approximately 40 minutes. A low-pass filter was used to filter the data at 200 Hz.

6 RESULTS AND DISCUSSION

6.1 *Comparison of time-series and spectra of area-averaged pressure coefficients*

The cavity pressure time histories are estimated by applying the lumped-leakage approach to the orifice flow equations and a segment of the time history is shown for a wind direction of 135 degrees in Figure 6 (a). The time history for the experiments is obtained via an area-average of the measured cavity pressures. There is good agreement between the predicted values and the experimental results and the results showed that the analytical model can capture the variations and distribution of cavity pressure, although there are clearly some differences. Figure 6 (b) shows the comparison of the spectra from the numerical model and the experimental results, which match well over the range of relevant frequencies. Table 1 shows the statistical comparison of the predicted values with the experimental results. From the values in the table, it is clear that there is good agreement between the predicted and experimental results. The similarity between the experimental values and predicted values are not only in mean values. Standard deviation, maximum and minimum values from both time series are compared and the results showed that the predicted cavity pressure time series are similar to the experimental values, with the analytical results within about 5% of the experimental. The model results tend to under-estimate the experimental for the overall fluctuations, but since the model tends to over-estimate at the higher frequencies, the peak values are also over-estimated in magnitude. Similarly, the time series from the predicted results and the experimental results are compared for a few more wind angles and the results are shown in Figure 7 (a) through (f).

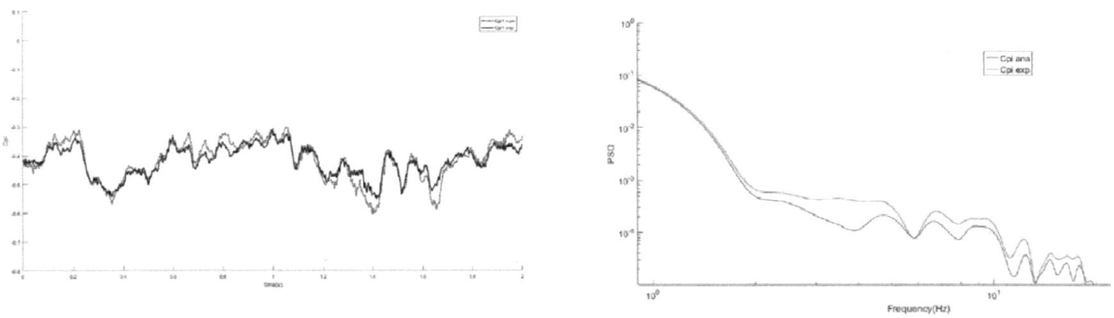

(a) (b)

Figure 6. Comparison of (a) time series and (b) spectra, obtained from analytical model with the experimental results for a wind direction of 135 degrees.

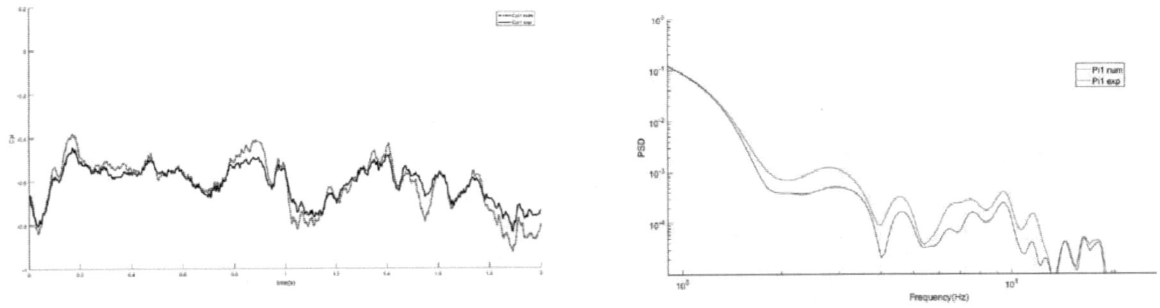

(a)

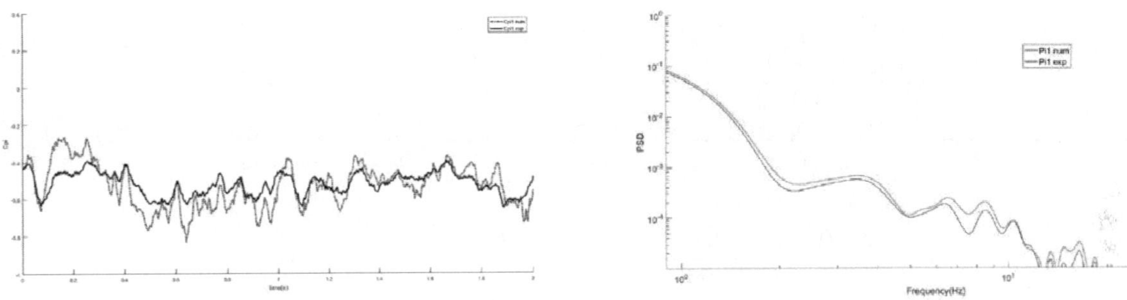

(b)

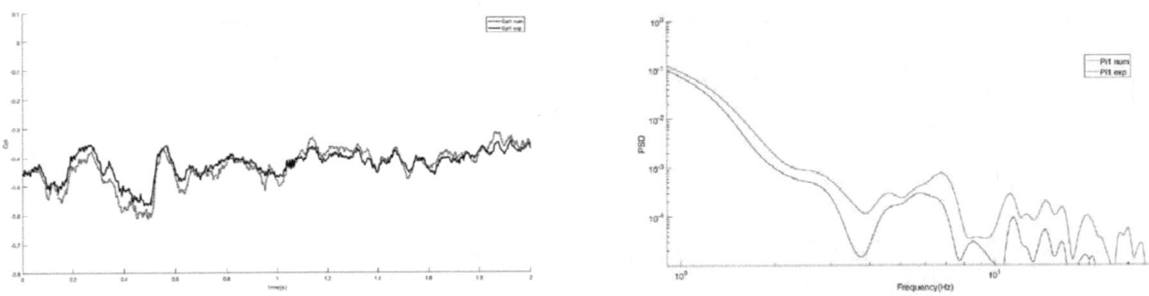

(c)

Figure 7. Comparison of time series and spectra, obtained from analytical model with the experimental results for the wind direction (a) 203 degrees, (b) 225 degrees, (c) 315 degrees.

Table 1. Statistical comparison of results obtained from the analytical model with the spatially averaged time-series obtained from the wind tunnel test.

Wind Direction	Pressure	Mean	RMS	Max	Min
135°	$Cp_{AE}(t)$	-0.561	0.088	-0.256	-1.084
	Cp_i from wind tunnel test	-0.447	0.061	-0.225	-0.806
	Cp_i from analytical model	-0.434	0.059	-0.208	-0.843
	Ratio of Analytical result to wind tunnel result	0.97	0.96	0.92	1.05
203°	$Cp_{AE}(t)$	-0.754	0.114	-0.399	-1.271
	Cp_i from wind tunnel test	-0.610	0.088	-0.287	-1.015
	Cp_i from analytical model	-0.582	0.085	-0.301	-1.071
	Ratio of Analytical result to wind tunnel result	0.95	0.97	1.05	1.06
225°	$Cp_{AE}(t)$	-0.689	0.100	-0.318	-1.212
	Cp_i from wind tunnel test	-0.538	0.074	-0.226	-0.910
	Cp_i from analytical $model_i$	-0.518	0.068	-0.244	-0.919
	Ratio of Analytical result to wind tunnel result	0.96	0.92	1.08	1.01
315°	$Cp_{AE}(t)$	-0.524	0.064	-0.308	-0.833
	Cp_i from wind tunnel test	-0.444	0.053	-0.244	-0.714
	Cp_i from analytical $model_i$	-0.418	0.049	-0.235	-0.747
	Ratio of Analytical result to wind tunnel result	0.94	0.93	0.97	1.05

6.2 *Comparison of panel-averaged cavity pressures with predicted time-series*

The most important parameter considered for the design of solar panel systems, especially in the case of wind loading, are the net loading across the panels as both the external (upper) surface and the cavity (lower) are exposed to wind. Net loading is the difference between the pressures on the upper and lower surfaces and it is what is required for the design of mounting and support systems for the solar panel arrays.

The area-averaged time series of the cavity pressure obtained from the analytical model are also compared with the pressures that are area-averaged over each panel individually. Figure 8 (a) shows the comparison of mean as well as peak pressures, for the135-degree wind direction. The red and blue dots represents the mean and peak results whereas the red and blue lines represent the single values from the analytical model, respectively. The figure shows that the predicted mean area-averaged cavity pressure obtained from the analytical model matches most of the experimental values except at a few locations, where the cavity pressure lack uniformity. In the case of the peak pressures, the predicted values are almost in line with the peak experimental values that are crucial for the design purposes, i.e., the largest magnitude values. However, there are a few hotspots where the peak panel area-averaged pressures are slightly higher than the predicted values. Similarly, the comparison of mean and peak panel-averaged cavity pressures for angles 203, 225 and 315-degrees are shown in Figures 8 (b), (c) and (d), respectively.

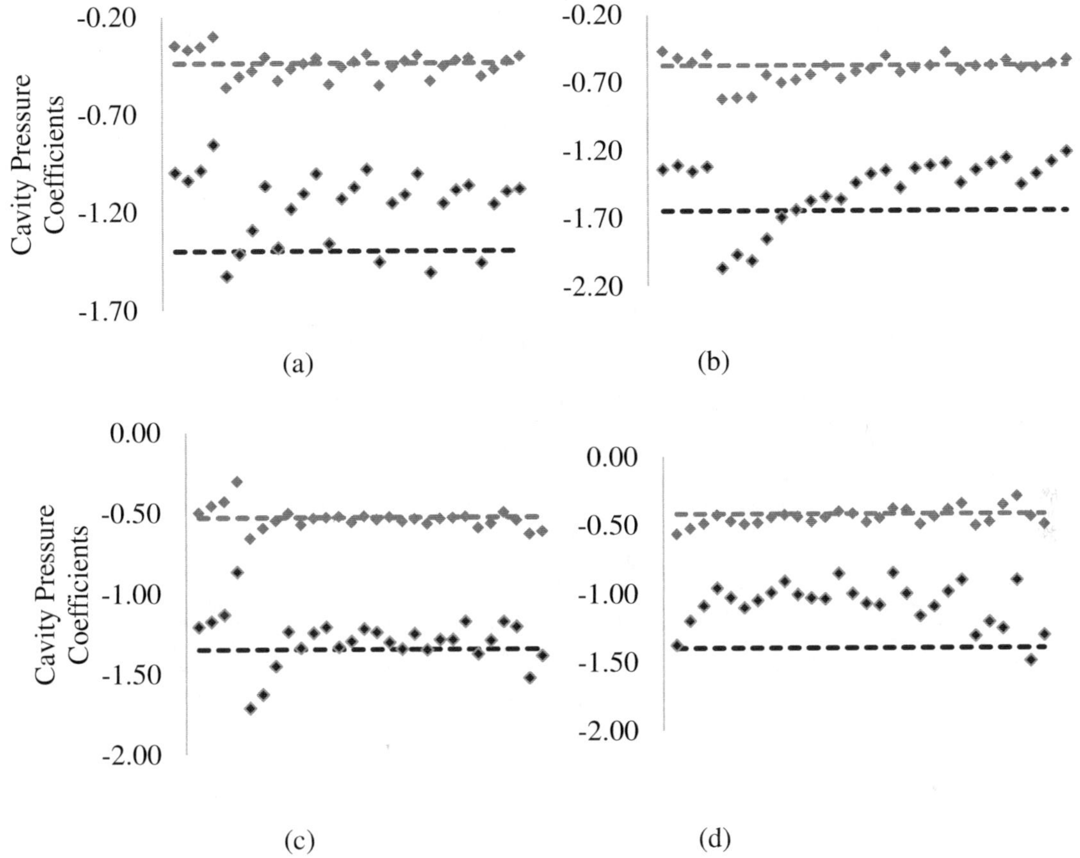

Figure 8. Comparison of panel-averaged cavity pressures from the experimental data with the (single) cavity pressure obtained from the analytical model for (a) 135-degree, (b) 203-degree, (c) 225-degree and (d) 315-degree wind directions.

7 CONCLUSION

In this study, an analytical model was developed, which can estimate the time-varying cavity pressures in a double-layer system with an air-permeable outer layer, given the external pressure on the outer surface as input to the system. The model estimates the cavity pressure based on the location of the neutral pressure line and also uses lumped-leakage approach for accounting the losses through openings. The results show that there is a good statistical agreement between the experimental and analytical results, even though there were some discrepancies which may be due to the assumptions considered in the model. The results obtained in this study are mostly a validation of general framework of the model and as future recommendation, automating the steps in this model has to be done.

8 ACKNOWLEDGEMENTS

This work was funded by the National Science and Engineering Research Council (NSERC) of Canada under the Collaborative Research and Development program and by the Institute for Catastrophic Loss Reduction (ICLR). The ongoing support and encouragement provided by Mr. Paul Kovacs is gratefully acknowledged. Dr. Sarah Stenabaugh was kind enough to provide us with the wind tunnel test data from her doctoral thesis experiments. Her generosity and support is greatly appreciated.

9 REFERENCES

Aly, A., & Bitsuamlak, G. (2014). Wind-induced pressures on solar panels mounted on residential homes. Journal of Architectural Engineering, 20(1): 4013003.

Amano, T., Fujii, K. and Tazaki, S. (1988). Wind loads on permeable roof-blocks in roof insulation systems. Journal of Wind Engineering and Industrial Aerodynamics, 29:39-48.

Asghari Mooneghi, M., Irwin, P., & Gan Chowdhury, A. (2014). Large-scale testing on wind uplift of roof pavers. Journal of Wind Engineering and Industrial Aerodynamics, 128: 22-36.

Baskaran, B, & Brown, W., (1992). Performance of Pressure Equalized Rainscreen Walls under cyclic loading. Journal of Thermal Envelope and Building Science, 16: 183-193.

Bienkiewicz, B., & Sun, Y. (1992). Wind-tunnel study of wind loading on loose-laid roofing systems. Journal of Wind Engineering and Industrial Aerodynamics, 41-44: 1817-1828.

Bienkiewicz, B., & Sun, Y. (1997). Wind loading and resistance of loose-laid roof paver systems. Journal of Wind Engineering and Industrial Aerodynamics, 72(1-3):401-410.

Burgess, J. C., (1995). Air pressure equalization in rainscreened joints by geometric alteration. Building and Environment, 30: 13-18

Chapra, S.C. and Canale, R.P. (2006). Numerical methods for engineers. 5th Edition, McGraw-Hill, pp. 632-634.

Choi, E.C.C. and Wang, Z. (1998). Study on pressure-equalization of curtain wall systems. Journal of Wind Engineering and Industrial Aerodynamics, 73:251-266.

Fazio, P., & Kontopidis, T. (1988). Cavity pressure in rain screen walls. Building and Environment, 23: 137-143.

Geurts, C., & Blackmore, P. (2013). Wind loads on stand-off photovoltaic systems on pitched roofs. Journal of Wind Engineering and Industrial Aerodynamics, 123:239-249.

Ginger, J., Payne, M., Stark, G., Sumant, B., & Leitch, C. (2011). Investigation on Wind Loads Applied to Solar Panels Mounted on Roofs. Townsville, Australia: Cyclone Testing Station.

Guha, T. K, (2011). Internal pressure in low and medium rise buildings subjected to high winds. Doctoral Thesis, The University of Auckland, New Zealand, 2011.

Holmes, J. (2007). Wind Loading of Structures. New York, New York: Taylor & Francis.

Holmes, J.D. (1979). Mean and fluctuating internal pressure induced by wind. Proceedings of the 5th International Conference on Wind Engineering, Colorado State University, Fort Collins, Colorado, USA, July, 1979, pp.435-450.

Inculet, D.R., & Davenport, A.G., (1994). Pressure-equalized rainscreens: A study in the frequency domain. Journal of Wind Engineering and Industrial Aerodynamics, 53: 63-87.

Kilip, I R., & Cheetham D W. (1984). The Prevention of rain Penetration through External Walls and joints by means of Pressure Equalization. Building and Environment, 19: 81-91

Kumar, K.S., and van Schijndel, A.W.M. (1999), "Prediction of pressure equalization performance of rainscreen walls", Wind and Structures, 2(4), 325-345.

Lou, W., Huang, M., Zhang, M. and Lin, N. (2012). Experimental and zonal modeling for wind pressures on double-skin facades of a tall building. Energy and Buildings, 54:179-191.

Oh, J., & Kopp, G. (2014). Modelling of spatially and temporally-varying cavity pressures in air permeable, double-layer roof systems. Building and Environment, 82: 135-150.

Oh, J., & Kopp, G. (2015). An experimental study of pressure distributions within an air permeable, double-layer roof system in regions of separated flow. Journal of Wind Engineering and Industrial Aerodynamics, 138: 1-12.

Stenabaugh, S., (2015). Design Wind Loads for Solar Modules Mounted Parallel to the Roof of a Low-rise Building. Doctoral Thesis, The University of Western Ontario, Canada, 2015.

Stenabaugh, S., Iida, Y., Kopp, G. & Karava, P. (2015). Wind loads on photovoltaic arrays mounted parallel to sloped roofs on low-rise buildings. Journal of Wind Engineering and Industrial Aerodynamics, 139: 16-26.

Stenabaugh, S., Karava, P., & Kopp, G. (2010). Design Wind Loads for Photovoltaic Systems on Sloped Roofs of Residential Buildings. London, Canada: Boundary Layer Wind Tunnel Laboratory.

Stenabaugh, S., Karava, P., & Kopp, G. (2011). Design wind loads for photovoltaic systems on sloped roofs of residential buildings. 13 ICWE. Amsterdam, Netherlands.

Sun, Y. and Bienkiewicz, B. (1993). Numerical simulation of pressure distributions underneath roofing paver systems. Journal of Wind Engineering and Industrial Aerodynamics, 46-47: 517-626.

Trung, V.T., Tamura, Y. and Yoshida, A. (2010). Numerical computation for lower surface pressures on a porous sunshade roof cover sheet. Proceedings of the 5th International Symposium on Computational Wind Engineering, Chapel Hill, North Carolina, USA, May 23-27, 2010.

Van Schijndel, A.W.M. and Schols, S.F.C. (1998). Modeling pressure equalization in cavities. Journal of Wind Engineering and Industrial Aerodynamics, 74-76:641-649.

Vickery, B.J. (1986). Gust factors for internal pressures in low-rise buildings. Journal of Wind Engineering and Industrial Aerodynamics, 23: 259-271.

White, F.M., Fluid Mechanics, 4th edition. 1999, New York, USA: McGraw-Hill.

Xie, J., Schuyler, G.D. and Resar, H.R. (1992). Prediction of net pressure on pressure equalized cavities. Journal of Wind Engineering and Industrial Aerodynamics, 41-44: 2449-2460.

AUTHOR INDEX

AUTHOR INDEX

AUTHOR INDEX

AUTHOR INDEX

American Association for Wind Engineering (AAWE)
c/o Cermak Peterka Peterson, Inc.
1415 Blue Spruce Drive, #3
Fort Collins, Colorado 80524
USA

ISBN 978-1-5108-5482-6

CURRAN ASSOCIATES INC.
proceedings.com

9 781510 854826

13th Americas Conference on Wind Engineering 2017

Gainesville, Florida, USA
21 - 24 May 2017

Volume 2 of 2

ISBN: 978-1-5108-5482-6

Printed from e-media with permission by:

Curran Associates, Inc.
57 Morehouse Lane
Red Hook, NY 12571

Some format issues inherent in the e-media version may also appear in this print version.

Printed by Curran Associates, Inc. (2018)

For permission requests, please contact American Association for Wind Engineering (AAWE) at the address below.

American Association for Wind Engineering (AAWE)
c/o Cermak Peterka Peterson, Inc.
1415 Blue Spruce Drive, #3
Fort Collins, Colorado 80524
USA

Phone: (970) 498-2334
Fax: (970) 221-3124

info@aawe.org

Additional copies of this publication are available from:

Curran Associates, Inc.
57 Morehouse Lane
Red Hook, NY 12571 USA
Phone: 845-758-0400
Fax: 845-758-2633
Email: curran@proceedings.com
Web: www.proceedings.com

TABLE OF CONTENTS

VOLUME 1

CATASTROPHE MODELING II

CATASTROPHE MODELING III

CATASTROPHE MODELING IV

CODES AND STANDARDS I

CODES AND STANDARDS II

CODES AND STANDARDS III

CODES AND STANDARDS IV

COMPUTATIONAL WIND ENGINEERING AND CYBERPHYSICAL MODELING I

COMPUTATIONAL WIND ENGINEERING AND CYBERPHYSICAL MODELING II

COMPUTATIONAL WIND ENGINEERING AND CYBERPHYSICAL MODELING III

DESIGN OPTIMIZATION

HIGH-RISE AND SUPER-TALL BUILDINGS I

HIGH-RISE AND SUPER-TALL BUILDINGS II

LOW-RISE BUILDINGS IV

METEROLOGY I

METEROLOGY II

MULTI-HAZARD ENGINEERING

NHERI I

NHERI II

OTHER STRUCTURES AND WIND ENGINEERING APPLICATIONS I

OTHER STRUCTURES AND WIND ENGINEERING APPLICATIONS II

POSTER SESSION

RESILIENCE

THUNDERSTORMS AND DOWNBURSTS

TORNADOES I

TORNADOES II

TORNADOES III

TORNADOES IV

TORNADOES V

TORNADOES VI

UTILITIES

<u>**WIND DAMAGE ASSESSMENT**</u>

<u>**WIND ENERGY**</u>

<u>**WIND-DRIVEN RAIN, DEBRIS AND EMBERS**</u>

Numerical assessment of partial hip roof failures during tornadoes

Sarah A. Stevenson[a], Gregory A. Kopp[b], and Ayman M. El Ansary[c]

[a] *Department of Civil and Environmental Engineering, University of Western Ontario, London, Canada, ssteve72@uwo.ca*
[b] *Boundary Layer Wind Tunnel Laboratory, University of Western Ontario, London, Canada, gakopp@uwo.ca*
[c] *Department of Civil and Environmental Engineering, University of Western Ontario, London, Canada, aelansa@uwo.ca*

ABSTRACT: This study focuses on investigating the behaviour of wood-frame residential hip roofs during tornadoes. Hip roofs are commonly observed to be more resilient during extreme wind in relation to gable roofs. However, inspection of damage survey data from recent events has revealed a previously unstudied failure mode in which hip roofs suffer partial failure of the framing structure. This paper aims to reveal the conditions under which framing failures may occur. Using the modeling methodology developed herein, a truss tested previously as part of a full-scale hip roof study was modeled. Uplift pressures corresponding to 170 km/h wind speeds were applied to obtain the internal load effects on individual truss members and joints. Using capacity-to-demand ratios possible failure modes were identified. The preliminary results verified that internal truss members and connections are unlikely to fail in trusses where roof-to-wall connections are toe-nailed. In trusses using hurricane straps, however, the truss members and connections along the top chord may become the weak link. Based on the survey data, an additional preliminary observation suggests that stick-frame structures containing few internal members may comprise the most vulnerable hip roofs. Plans for future work to explore this observation will be discussed.

KEYWORDS: Residential Structures, Tornado, Roof Failure, Hip Roof, Enhanced Fujita Scale, Finite Element Modeling

1 INTRODUCTION

Wood-framed houses comprise the highest percentage of housing in North America [1, 2]. The roofs of these structures are highly vulnerable to damage from high wind speeds. Densely populated areas can suffer catastrophic damage due to residential roof failures during relatively weak tornadoes. The resilience of houses during extreme wind events is essential to ensure safety of occupants and minimize damage to internal contents. Significant work has been completed to date to address commonly observed failure modes in residential structures. The current state of knowledge in this area will be introduced prior to defining the present research objectives.

Section 1.1 will introduce the conventional method for assessing tornado intensity through the Enhanced Fujita Scale. This introduction will also discuss the well-understood modes of residential roof failure and provide a basis for subsequent discussions related to failure wind speeds for hip roofs. Section 1.2 will then introduce the impact of residential roof failures on overall insured losses from tornadoes. This information supports the need for further research and development to mitigate roof failures under all critical modes.

Hip roof behaviour will be discussed in Section 1.3 to identify the characteristics of hip roofs that distinguish their performance from that of gable roofs. The available evidence proving that hip roofs are superior under certain failure modes may also point to their susceptibility to other modes. The occurrence of hip roof framing failures will also be introduced through an example failure observed in Angus, Ontario. Section 1.4 will provide additional examples of partial hip roof failure from survey data obtained following recent destructive events in the United States. The failures pictured in this section comprise a small sample of all observed failures. They will assist in characterizing the possible causes of framing failure. The objectives of the present study will then be introduced in Section 1.5 before the research methodology is established in Part 2.

1.1 *The Enhanced Fujita Scale for Residential Structures*

Because it is not practically feasible to assess tornado intensity directly using wind speed measurements, tornadoes are classified according to the levels and types of damage they inflict upon buildings and vegetation. The first standardized method for measuring tornadoes in this way was proposed by Fujita [3] in 1971. The Fujita Scale was used widely for over 30 years before it received significant updates. The Enhanced Fujita (EF) Scale was proposed in 2006 [4] to improve the wind speed estimates from the Fujita Scale and make its widespread use more consistent. In 2013, Environment Canada [5] further modified the EF-Scale to better suit the Canadian context. The present study will refer to the Canadian version of the EF-Scale for assessment of residential structures. More information on the Canadian modifications to the scale can be found in [6].

The EF-Scale provides refined wind speed estimates for 31 categories of common structures and vegetation, referred to as Damage Indicators (DIs). Under each DI, the EF-Scale utilizes the concept of Degrees of Damage (DOD). DODs describe the sequential modes of damage that have been observed to occur in a particular DI. Each DOD is associated with a minimum, maximum, and expected wind speed. These values represent the estimated wind speeds required to cause the specified damage. They can be related back to the EF-Scale wind speeds to rank the intensity of a tornado along the scale, from EF0 to EF5. In the present study, the DI for one- and two-family residences (FR12) is of particular interest. DOD descriptions and wind speed estimates are shown for this DI in Table 1.

Table 1: DOD descriptions and wind speed estimates for the FR12 DI, as adopted by Environment Canada [5].

DOD	Description of Damage	Wind Speed Estimates [km/h]		
		Lower Bound	Expected	Upper Bound
1	Threshold of visible damage	85	105	130
2	Loss of roof covering material (up to 20%), gutters and/or awning; loss of vinyl or metal siding	100	125	155
3	Broken glass in door and windows	125	155	185
4	Uplift of roof deck and loss of significant roof covering material (more than 20%); collapse of chimney; garage doors collapse inward; failure of porch or carport	130	155	185
5	Entire house shifts off foundation	165	195	225
6	Large sections of roof structure removed (more than 50%); most walls remain standing	165	195	230
7	Exterior walls collapsed	180	210	245
8	Most walls collapsed, except small interior rooms	205	245	285
9	All walls collapsed	230	274	320
10	Destruction of engineered and/or well-constructed residence; slab swept clean	265	320	355

The EF-Scale offers a systematic method of assessing tornado intensity, and it helps engineering investigators discern what fails, or what should fail, first so that unexpected failures can be easily identified. The FR12 DI covers typical wood-frame houses constructed in Canada and the United States. In the present study, the DODs of interest are DOD-4 and DOD-6 for residential structures, corresponding to roof sheathing failure and loss of large sections of the roof structure, respectively. Most past research on roof damage focuses on sheathing and roof-to-wall connection (RTWC) failures, and it is apparent that the wind speed estimates for roof damage in the EF-Scale are based heavily on these well-understood failure modes.

As seen in the shaded row of Table 1, DOD-6 describes failure of the RTWC and subsequent loss of large roof sections, which can occur at an expected wind speed of 195 km/h. This wind speed corresponds to relatively weak EF2 tornadoes [5]. DOD-4 occurs at lower wind speeds, however loss of a single sheathing panel can allow for water ingress, which often results in loss of the entire house and its contents. As discussed in the following sections, gable roofs have been observed to perform poorly under these modes, especially DOD-6, relative to neighboring hip roofs of similar construction. In fact, the FR12 listing from the Canadian EF-Scale notes that for hip roof homes, the upper-bound wind speeds for DODs 4 and 6 can be assumed. This is counter to the original EF-Scale documentation [4] which specifies that the lower bound of DOD-6 is due to inadequate construction or large overhangs while the upper bound is due to enhanced construction such as the use of hurricane straps. The difference between the two versions of the EF-Scale is a significant point which warrants further investigation.

1.2 *Residential Roof Failures during Extreme Wind Events*

Residential roof failures are among the most common and expensive types of tornado and hurricane damage and thus have been studied extensively. The density of houses relative to other structures in any populated area leads to high occurrence of residential failures and increased risk to neighboring structures due to blowing debris. As mentioned, the relative vulnerability of residential roofs is additionally important because loss of a single sheathing panel can allow for water ingress. This often leads to loss of the entire structure and its contents due to the heavy rainfall that accompanies extreme wind events. Sparks et al. [7] compared insured losses to observed damage and gradient wind speeds from past hurricanes. Their analysis indicated a sudden rise in insured losses between 70 m/s and 80 m/s gradient wind speeds. Statistics from damage survey observations identified the cause of this rise to be failure of the building envelope through loss of roof sheathing panels and broken windows or doors.

The gradient wind speeds from Sparks et al. [7] can't be directly compared to the EF-Scale wind speed estimates for FR12, however their study showed that over half of insured losses in a major hurricane are due to failure of the building envelopes of houses. Even though the walls don't often collapse, serious damage occurs once the seal provided by the roof or walls has been broken. Repetitive, similar failures are frequently observed across neighborhoods where a common design was implemented or the same party was responsible for construction of many houses. This is often the case in modern subdivisions. Identical failures of similar components suggest that widespread mitigation is possible; improved design approaches and innovative fasteners have been recommended to help mitigate the prevalent sheathing and RTWC failure modes in the work to date.

1.3 *Hip Roof Performance during Extreme Wind Events*

Several past studies have investigated the superior performance of hip roof homes [8, 9], with some more recent works directly investigating hip roof behaviour with regard to roof sheathing (DOD-4) and RTWC (DOD-6) performance [10, 11]. In 1991, Meecham et al. [8] performed wind tunnel testing to further the technical understanding of hip roof performance and found that there is an important relationship between pressure distribution and underlying framing configuration in wood-frame roofs. Despite significant differences between the pressure distributions recorded for the gable and hip roof models, the overall roof uplift and overturning moments were found to be quite similar. This verified that preferable aerodynamic geometry is not the only reason for improved performance of hip roofs.

Meecham's [8] results indicated that the orientation of framing members in a hip roof, relative to the distribution of uplift, provides additional resilience. In contrast, the shape of the gable roof causes higher localized peak pressures and the orientation of framing members result in unfavorable load sharing. In addition to this, hip roofs have RTWCs around their entire perimeters, while gable roofs are only connected to the wall framing along two opposite walls. In combination with the improved load sharing within trussed hip roofs, these factors are generally believed to make hip roofs significantly more resistant to wind uplift.

Following an EF2 tornado that damaged a residential neighborhood in Angus, Ontario on June 17, 2014, Kopp et al. [10] evaluated damage to 101 houses and analyzed the observed failures. Repetitive failures of similar building components were observed across the study area. The statistics of these failures were presented in [10], followed by fragility analyses of comparable gable and hip roofs. The fragility analyses showed that for failure of the roof-to-wall connections in a hip roof, the required wind speeds actually fall above the range specified in the EF-Scale for DOD-6. The failure wind speeds for RTWCs in hip roofs was found to be about 50 km/h higher than the wind speeds required for that in a gable roof, which is greater than one category in the EF-Scale.

Kopp et al. [10] concluded that when toe-nailed RTWCs are used on gable roofs, the RTWCs are more likely to fail than the sheathing panels. In hip roofs, depending on the length of the sheathing fasteners, sheathing loss may be more likely than failure of the RTWC. Both categories of failure occur at much lower wind speeds in gable roofs. Though this study confirmed that hip roofs are more resistant to the common failure modes for residential roofs, failure was still observed in hip roofs or in the hip portions of combination roofs.

An example of hip roof failure from the Angus tornado is shown in Figure 1. This failure does not appear to have occurred simply as sheathing failure or failure of the RTWC. At the left corner of the damaged house there is an opening where it appears that a portion of the roof has been removed. Inside of the opening intact trusses are still visible. This suggests either that hip trusses near the edge of the roof have failed, or that smaller framing members making up the corner of the roof have been removed. In either scenario, partial failure of the roof frame has occurred, though it may have been initiated by failure of the RTWC at the corner.

Figure 1: Partial hip roof failure observed following EF2 tornado in Angus, Ontario, June 2014. (Photo source: authors)

Framing failures in hip roofs may address a gap in the current DOD listing for residential roofs. Although DOD-6 allows for failures other than that of the RTWC, hip framing failures - and the wind speeds at which they are likely to occur - have not been studied in the literature to date. Further evidence for these possible failure modes is examined in the next section.

1.4 *Damage Surveys Showing Hip Roof Failures*

Additional evidence of partial framing failures in hip roofs was obtained following recent destructive tornadoes in the Southern United States, including the Moore, Oklahoma tornado of 2013. This event killed 24 people and was estimated to have caused $3 billion of economic loss [12]. A forensic assessment team of researchers, engineers, and students spent three days surveying the affected areas and documented observed damage.

The Moore, OK tornado was determined to be an EF5 event, with damage ranging from EF0 to EF5 observed across the path of the tornado. EF0 to EF2 winds typically comprise about 85% of the damage area of a strong EF4 or EF5 tornado, and so many stages of damage progression could be identified. The survey performed following this event has informed subsequent research including identification of new methods for improved damage surveys, fragility analyses of house components, and the development of improved laboratory simulations for tornadoes [12]. It also led to changes in the Moore, OK building code such that wood-frame houses have new prescriptive requirements to mitigate DOD-6 damage.

The raw database of photos taken following the Moore tornado, as well as the Tuscaloosa, Alabama and Joplin, Missouri tornadoes of 2011, were used in the present study to examine the nature of hip roof failures. Many instances of partial hip roof failures were identified in the data. As with the findings of the fragility analyses in [10], the observed failures evoked additional questions regarding the likelihood and conditions under which partial hip roof failures may occur. Select examples of the observed failures are shown in Figure 2 and discussed below.

Figure 2a) shows neighboring hip roof homes which exhibit similar failures of the front face of the roof. The RTWC appears to be intact around the remaining perimeter of the roof and it is apparent that several members of the roof frame have failed or been removed, in addition to the sheathing covering this portion. At the right side of the photo the remaining part of the roof is sagging which further indicates that the underlying frame has failed. Several houses along the same street as those in Figure 2a) showed similar failures of hip roof framing. In these cases, it appears that the frame failed at the nailed connections between the members, as no broken lumber is visible.

Figure 2b) shows a similar failure to that in a) but of a much steeper roof. The RTWC appears to be intact, and a large open cavity is visible where both framing members and sheathing have been removed. Similarly to a), it is apparent that this roof didn't exclusively suffer sheathing loss, although the smaller area of sheathing loss at the right side of the photo should be noted. The lack of visible internal members in the cavity, especially those supporting the intact opposite face of the roof, strongly suggest that this roof was built as a stick-frame structure as opposed to one containing prefabricated trusses. Many of the failed hip roofs in the available data appear to have used stick-framing.

Figure 2c) shows a partial failure of a combined hip/gable roof. This failure is unique from those shown in a) and b) because material failure of the wood members is apparent. The RTWC appears to be intact, with the lower part of the roof having lost only sheathing on the right side, and framing members in addition to sheathing on the left. Near the peak of the roof the frame has failed on both faces. This structure appears to contain either trusses or stick-framing with robust connections. As can be seen just above the RTWC, the members were connected or otherwise reinforced using nailed wooden plates.

Upon inspection of the failures shown in Figure 2 and similar damage in the available photos, it becomes apparent that partial framing failures are another possible, repetitive mode of failure occurring in hip roofs. When comparing these hip roof failures to nearby structures from the data, it was determined that framing failures such as those shown in Figure 2 may govern in hip roofs at EF2 wind speeds, rather than RTWC failures or sheathing loss. It was also noted that the construction of the roof may be important. The observed stick-frame failures especially suggest that the performance of stick-frame roofs may need to be distinguished from that of trussed structures in analysis.

1.5 *Objectives*

Based on review of the available literature and damage survey observations, it is clear that modes of roof failure rather than sheathing loss or RTWC failure are possible. This paper is part of a larger study which aims to examine the conditions required for such failure modes to occur. Finite element modeling will be used in lieu of an experimental program so that multiple roof configurations, framing types, and individual trusses can be assessed in a time- and resource-efficient manner. The objective of this paper is to define a method for assessing the wind-induced failure of a single truss, such that it can be replicated to model entire roofs as part of the future work. By performing the proposed analysis on a single truss, preliminary conclusions will also be made regarding the likelihood of roof framing failure within the DOD-4 and DOD-6 range of wind speeds.

a)

b)

c)

Figure 2: Hip roof failures in Moore Oklahoma following the EF5 Tornado of May 21, 2013. a) Framing failure of similar, neighboring hip roofs b) Framing failure of steep, stick-frame hip roof c) Hip roof displaying material failure of the wood framing members (Images provided by Dr. David Prevatt and David Roueche of the University of Florida)

2 METHODOLOGY

2.1 *Overview*

To understand the possibility of member (Figure 2c) or connection (Figure 2a, b) failures within a hip roof frame, the load effects due to wind uplift on the framing elements must be determined and compared to their abilities to withstand those effects. The term "element" in the present study refers to both the lumber framing members and the connections between them. The load effects on an element can include axial, shear, or moment forces that are caused by the transfer of applied uplift from the top face of the structure to the supports at the RTWC. Failure will be predicted by identifying the weakest elements under uplift load effects. This is done through comparison of capacity-to-demand (C/D) ratios across all elements under all relevant load effects.

By applying a uniformly distributed uplift which corresponds to a known wind speed, the finite element model will yield the load effects throughout the truss. These load effects represent the demand on the element. Capacity values for comparison can be determined using code calculations [13, 14] and recommended equations from the literature [15]. The minimum resulting C/D ratios will pinpoint the elements that are prone to failure under the given wind speeds, and the modes under which they are most vulnerable. In addition to analyzing the truss elements, C/D ratios for the RTWCs will be included to assess the likelihood of internal framing failure relative to this expected mode of DOD-6.

Accurate analysis of wood structures must account for the anisotropic properties of wood, the complex behaviour of the connections, and numerous possible failure modes and directions. Due to a lack of published information on the nonlinear behaviour of wood truss members and metal plate connected (MPC) joints, the finite element modeling was limited to the linear range of material behaviour. In this context, an accurate modeling method was selected based on comparison of several common model analogs described in the available literature. The selected method was then validated using experimental data. The model method is described in more detail in Section 2.2, and Section 2.3 shows how the wind loads were determined and applied to the current truss to obtain element-level forces. Following this, element capacities and C/D ratios are presented in Section 2.4.

2.2 *Finite Element Modeling of Single Truss*

Past research involving finite element modeling of wood-framed roofs has included detailed work to represent MPC joint behaviour at the fastener tooth-level [16, 17], as well as larger-scale studies on overall roof behaviour including load sharing and deflection of trusses [18, 19]. Few of the published studies provide an example of modeling to obtain accurate internal member and joint forces for a system of roof framing members. Li [20] published the only relevant model results including member axial forces and moments for a MPC truss. However, Li [20] was unable to obtain satisfactory validation of the moment results. For single trusses, the modeled member moments were close to the experimental data, but other members had significant differences. Due to the apparent gap in research on modeling methodologies for truss member forces, a single model analog could not be deemed most suitable for the present study based on the available literature. Therefore, a detailed validation was carried out in order to determine which of the existing modeling methods would provide the most accurate envelope of internal truss forces.

To construct an analytical model of any structural system, analog decisions must be made carefully to ensure that the numerical model can accurately predict the behaviour under uplift loads. For MPC trusses, it was found that there are two prevailing analog decisions; connection behaviour and member insertion point. Connectivity assumptions are concerned with representing the flexibility behaviour of the MPC joint, and the insertion point affects the concentric or eccentric behaviour of the members.

In the selected modeling software, SAP2000, joints between two or more framing members can be modeled to represent many stiffness conditions, including: fully rigid joints with no internal deformation, pinned joints allowing members to rotate freely, or semi-rigid joints with stiffness parameters that partially limit deformation. The present study considered connectivity analogs where all truss joints were rigid, pinned, or semi-rigid using experimental stiffness parameters from [21]. An additional "combined" case using fully rigid chords and pinned web connections was also considered.

In reality, MPC joints are semi-rigid; they deform under axial loads and moment according to the load-displacement behaviour of the metal plate and its interaction with the wood member. However, without experimental data for joints matching each modeled configuration, it has been deemed unnecessary and inefficient to attempt to model this behaviour accurately [19]. Due to the lack of experimental data for joint and member forces from full-scale trusses, modeled forces could not be validated directly for a single connection analog. Instead, the method yielding the maximum force results for each element was sought to provide a conservative point of comparison to the element capacity estimates. By overestimating the force results, failure modes with high C/D ratios can be deemed improbable and neglected from further study, while those with C/D ratios near or below 1 will be considered critical and explored in greater detail.

The rigid and pinned analogs each represent the two possible extremes with regards to joint rotational stiffness. Accordingly, it was assumed that an envelope of joint and member forces could be obtained by using the rigid and pinned analogs independently, then combining their results by taking the maximum load effects on each element. This assumption was tested and verified by comparing the model results across all four connectivity analogs for 3:12- and 6:12-slope Fink trusses tested in [22, 23]. The method of combining the maximum force results from the pinned and rigid connection analogs was deemed most suitable and efficient for the present study.

Member geometry is typically modelled along the centerlines of the wood members, however this neglects eccentricity that may occur at the joints. An alternative analog was tested where the top and bottom chord members were modeled along their bottom faces, using a "bottom-center" insertion point in SAP2000. Comparing joint forces found that the bottom-of-chord geometric analog consistently provides more conservative joint force and moment values. As previously mentioned, it was decided to use an envelope approach including exaggerated element forces rather than direct validation due to the lack of data and consequent uncertainties.

Model validation used experimental data from a series of test programs carried out at the Forest Products Laboratory in Madison, WI [22, 23, 24]. These experiments tested geometrically-identical Fink trusses under different joint conditions. Wolfe & LaBissoniere [23] tested trusses containing intentionally overdesigned joints to force failure in the wood members, while Wolfe et al. [22] tested trusses of conventional design. A later study [24] provided the only experimental data to include member force and moment measurements, however results from only 3 trusses were applicable to the present validation.

The connectivity and insertion point analogs, tested in combination, resulted in 8 cases which were examined and compared. This work determined that the most practical method for obtaining joint and member internal forces used the bottom-of-chord insertion point and an envelope of results from the rigid and pinned connection analogs. This method was finally validated using the member axial and moment results obtained for the relevant trusses tested in [24]. This step allowed for further refinement of the modeling method to yield more accurate deflection values, and confirmed the approach for estimating the member demand prior to completing the capacity-demand analysis of whole roofs.

2.3 *Uniform Uplift on Hip Roof Truss*

Following definition of the truss modeling method as described in Section 2.2, a model for a realistic hip roof truss experiencing wind uplift was developed. Henderson et al. [11] tested a full-scale, 9 m by 9 m square hip roof in the Insurance Research Lab for Better Homes at the University of Western Ontario. Truss drawings from this project were used in the present analysis. The second full-length truss in the roof was chosen for the present modeling. Figure 3 shows the configuration of the roof tested in [11], and the location of the selected truss, labeled "A2". This truss was selected for the present analysis because it is the first full-length truss from the outer edge of the roof, aside from the 2-ply "A1" truss, and its symmetry simplifies the determination of joint capacities.

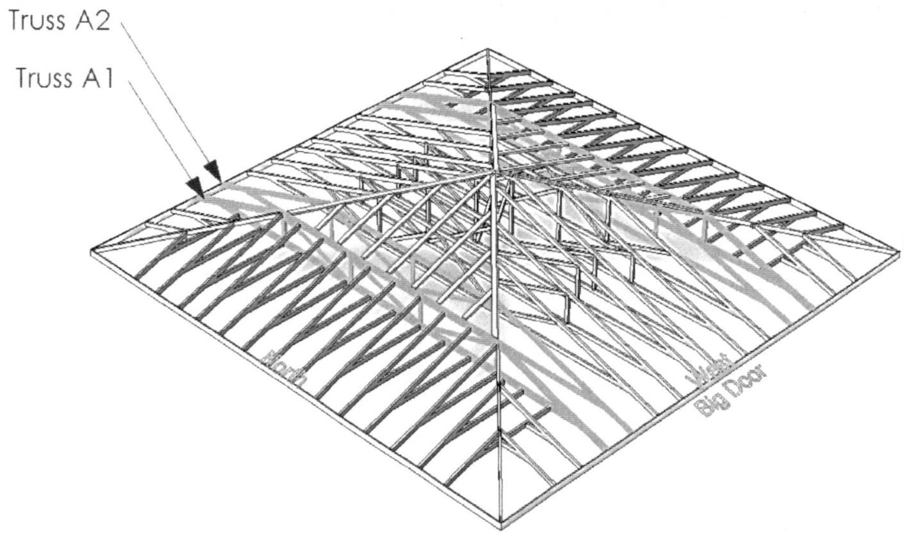

Figure 3: Model rendering of full-scale hip roof tested by Henderson et al. [11]

Figure 4 shows a dimensioned drawing of the truss, which was then modelled using SAP2000 as shown in Figure 5. Metal plate sizes were taken from the drawings prepared by the truss fabricator, but exact locations were estimated by visual inspection. Wind loads were calculated based on the tributary area loading prescribed by the ASCE 7-10 provisions for loads on the main wind force resisting system of a building [25]. The wind speed was taken as 170 km/h (105 mph), which corresponds to the 3-second gust speed specified in design of Occupancy Category 1 buildings in the mainland region of the United States. 170 km/h also corresponds to upper-bound DOD-4 and lower-bound DOD-6 wind speeds for FR12. This relation facilitates assessment of the truss behaviour within the relevant ranges of the EF-scale. The resulting pressure was applied as a uniformly distributed load of 1.22 N/mm acting outwards and normal to the top chord members of the truss. The self-weight of the truss was also included using the mass of SPF No. 2 lumber [26].

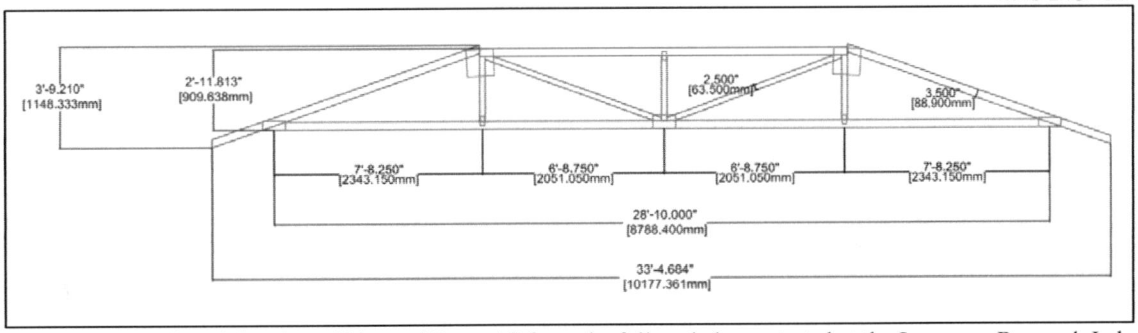

Figure 4: Dimensioned drawing of hip roof Truss A2 from the full-scale house tested at the Insurance Research Lab for Better Homes

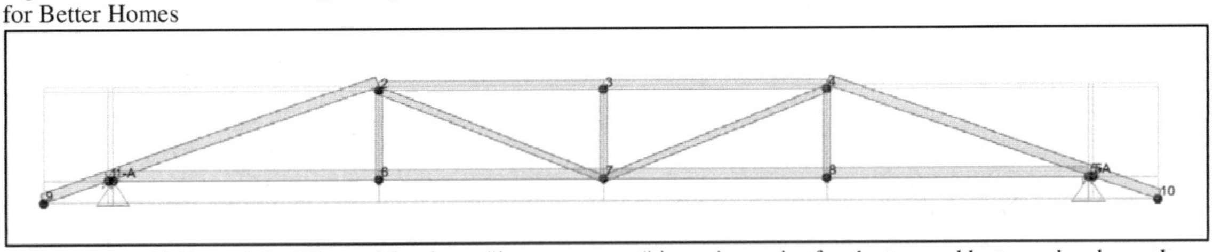

Figure 5: SAP2000 model of Truss A2 using a "bottom-center" insertion point for the top and bottom chord members

The truss was modelled using the rigid and pinned connection analogs in separate models. Following analysis of each model under the described loads, member and joint forces were extracted and compared to yield the maximum values. These results were then compared to the capacity estimates developed in the following section.

2.4 *Member and Joint Capacity*

Member and joint capacities were estimated for Truss "A2" based on material properties and plate strength data matching the lumber and hardware used in Henderson et al.'s [11] hip roof. Truss plate test data for the connections used in this project were obtained from the manufacturer. The mechanical properties of the lumber were obtained from the Canadian Wood Design Manual [26] for SPF 2100Fb-1.8E Mechanically Graded dimensional lumber, which was specified in the truss construction drawings. Member capacities were relatively simple to determine and could be pulled from tabulated data in the manual. Joint capacities, however, required significant effort and research to estimate accurately.

Canadian and American codes prescribe similar procedures for calculating the axial and shear resistance of MPC joints [13, 14]. For joints in tension or shear, three sets of calculations must be carried out to check the material strength of the plate, plus the lateral and slip resistance of the plate-to-wood interface. The equations for these strength estimates typically include material resistance factors, which were neglected in the present analysis, in addition to parameters based on plate strength, wood type, and the direction of loading relative to the plate and wood grain orientation. Details on these equations and parameters can be found in Section 7.5 of [13]. For members joined along two edges using the same plate, there is also a calculation for assessing the combined shear and tension resistance of the joint. An example of such joint is shown in Figure 6. Joints 3 and 9 both include web members that would require this calculation. The top chord members of Joint 3 would also require similar calculations due to the tensile action at the ends of the members in combination with the shear along their lengths. The other joints only include one possible line of action each in shear and tension.

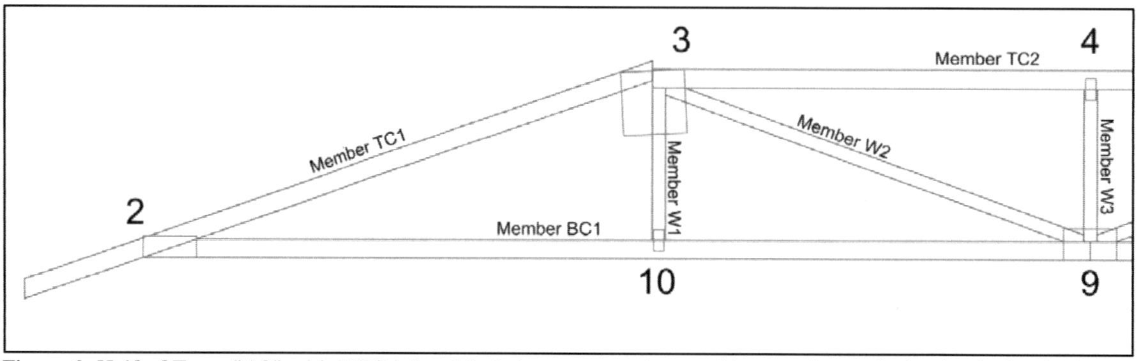

Figure 6: Half of Truss "A2" with MPC joints labeled per convention used in capacity analysis. Analysis of Joints 3 and 9 require additional capacity calculations to assess combined shear and tension as per Section 7.5.4 in [26].

The moment capacities of the joints were estimated based on the equation presented by Lewis et al. [15]. The experimental results presented in [15] indicate that the calculated capacities resulted in a minimum degree of conservatism of 1.5 times (experimental results divided by the calculated strength). The present analysis removed this factor from the calculated results in order to obtain specified strengths for comparison with the unfactored loads from the finite element modeling. The equation from Lewis et al. [15] also accounts for the effects of the axial loads on the members by increasing the moment capacity of joints where the members are in compression, and alternately decreasing the capacity of joints in tension. In compression members, deformation - and therefore failure - of the joint will be limited by the members coming into contact with one another. It is also for this reason that compression members were neglected in the axial strength analysis.

Following calculation of unfactored member and joint capacities, the results from Section 2.3 and Section 2.4 could be combined to determine C/D ratios for each element. Multiple modes of failure were assessed for each joint element according to the number of possible failure modes discussed previously (tension, shear, combined shear/tension, moment, or wood member failure). Table 2 shows the lowest C/D results and identifies the failure mode corresponding to the critical ratio for each element. C/D ratios were also calculated for RTWCs using toenails or the minimum design for hurricane straps. The capacities for these elements were obtained from [27] and [28], respectively. In Table 2, the "failing" elements - those with C/D ratios less than 1 – are indicated by dark shaded cells. Other potentially critical elements with C/D ratios close to 1 are indicated by bold text. The joints with "N/A" C/D ratios either developed compression in the model results or contained members that were continuous and therefore would not inflict the related load effect on the joint.

Table 2: Capacity-Demand results and governing failure mode for Truss "A2" under 1.22 N/mm wind uplift

Truss Element		**Relevant Force Effects and Capacity/Demand Ratios**					
Joints							
Joint	Member	Tension	Critical Mode	Shear	Critical Mode	Moment	Critical Mode
2	*TC1*	2.374	*Plate Slip*	**1.480**	***Member Strength***	2.297	*Member Strength*
2	*BC1*	N/A		10.068	*Lateral Tooth Strength*	5.792	*Lateral Tooth Strength*
2	*TC1 (overhang)*	N/A	*Combined with TC1*	N/A		2.466	*Member Strength*
3	*TC1*	**1.573**	***Member Strength***	**1.954**	***Plate Slip***	**0.651 0.790**	***Member & Plate Fail***
3	*TC2*	**1.273**	***Member Strength***	36.509	*Lateral Tooth Strength*	**0.651**	***Member***
3	*W1*	225.53		13968.4		16.4988	*Plate Slip*
3	*W2*	N/A		N/A		62.8938	
4	*TC2*	**1.273**	***Member Strength***	N/A	*Continuous Member*	N/A	*Continuous Member*
4	*TC2*	**1.273**	***Member Strength***	N/A	*Continuous Member*	N/A	*Continuous Member*
4	*W3*	4.061	*Plate Capacity*	384887.6		4.984	*Lateral Tooth Strength*
9	*BC1*	N/A		1106.3		5.878	*Plate Capacity*
9	*BC2*	N/A		1107.5		2.794	*Member Strength*
9	*W2*	N/A		N/A		28.252	*Lateral Tooth Strength*
9	*W3*	5.364	*Member Strength*	601386.9		37.583	*Lateral Tooth Strength*
9	*W4*	N/A		N/A		29.193	*Lateral Tooth Strength*
10	*BC1*	N/A	*Continuous Member*	457.14		N/A	*Continuous Member*
10	*BC1*	N/A	*Continuous Member*	1539.7		N/A	*Continuous Member*
10	*W1*	261.61		8958.8		10.526	*Lateral Tooth Strength*

Members					
	Tension		Compression	Moment	
Chord Member	**1.277**	***"TC2"***	8.521	**0.666**	***"TC2" around mid-panel***
Web Member	5.347		8.529	15.881	

Roof to Wall Connection			
	Uplift Resistance [N]	Support Reaction [N]	C/D
Toenail [27]	2800	6046.97	**0.463**
H2.5 Hurricane Strap [28]	5840	6046.97	**0.966**

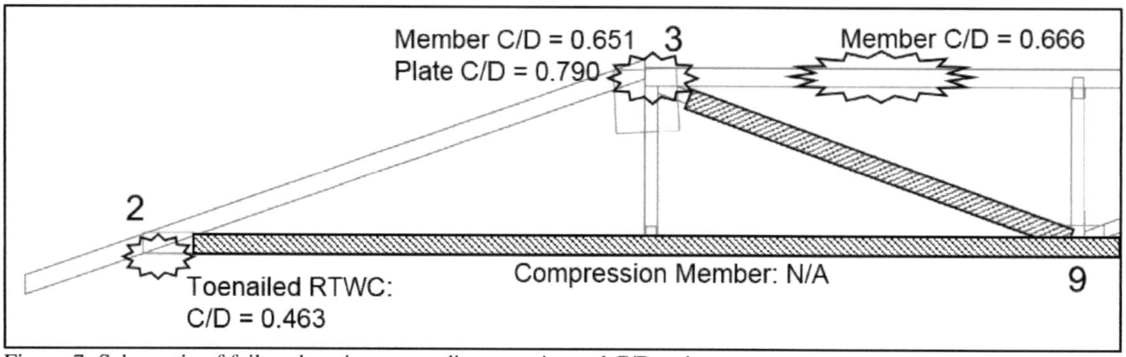

Figure 7: Schematic of failure locations according to estimated C/D ratios.

3 DISCUSSION OF PRELIMINARY RESULTS

3.1 *Critical C/D Ratios in Truss A2*

The results from Table 2 are shown schematically in Figure 7. As can be seen, the C/D ratios for the members and joints vary greatly throughout the truss. The toenailed RTWC has the lowest relative strength by a 40% difference, with a C/D ratio of 0.463 compared to the next-lowest ratio of 0.651 in Joint 3. Because the applied uplift corresponds to a wind speed of 170 km/h, this result agrees with the current understanding of DOD-6 for FR12 by implying that a properly constructed truss will not fail internally, but at the toe-nailed support. As previously mentioned, the original version of the EF-Scale [2] attributes lower-bound DOD-6 to poor construction. Additional work may involve adjusting the uplift load to find the point at which the C/D ratio for toenailed RTWCs first exceeds 1 to verify the wind speed at failure initiation.

In addition to verifying the failure of the toenailed RTWC at lower-bound DOD-6 wind speeds, the present analysis also provides insight into the case where hurricane straps are used. The previous conclusion does not hold in this case. As shown in Table 2, the framing members and truss joints around Joint 3 were found to be more vulnerable than the RTWC, having lower C/D ratios. The predominant failure mode of this joint was determined to be wood member failure, although the possibility of plate failure at this joint should also be noted. The insertion point analog used in this study considers eccentricities that may exist due to poor construction, however including the extreme case may result in exaggerated member forces. Variation in this behaviour could make the C/D ratio for the member exceed that of the plate.

3.2 *Limitations of Current Practice and Challenges*

The complexity of capacity estimation for MPC joints presented an unexpected challenge. Undue effort was required to develop an efficient process for assessing joint capacities for many trusses. Correspondence with practicing engineers to date has suggested that the thorough code-based calculations are not commonly carried out for trusses, and design is either simplified through prescriptive design or use of proprietary "black box" software owned by truss hardware manufacturers.

The objective of this work is to provide a proof-of-concept, using two-dimensional analysis models, regarding which elements or types of construction should be investigated in more detail. Further study will be required to gain a better understanding of the progression of damage in the cases shown in Figures 1 and 2. Variation in the capacity estimates and modeled force results may yield slight changes in the relative C/D results, however the approximations employed presently have provided valuable insight into the possibility of internal failures in trussed hip roofs. Additional work will also provide a better understanding of stick-framed structures, as Figure 2b) suggests that framing failure may especially prevail when too few internal members are included in a structure.

4 FUTURE WORK

4.1 *Capacity Demand Analysis*

Due to the complexity of the joint capacity calculations, the future work will aim to simplify the analysis by identifying patterns in the critical C/D ratios within the first several trusses. It is expected that the C/D ratios at many locations will be sufficiently and consistently high enough to assume that the corresponding elements are "safe" and need not to be assessed further. Further, by comparing the calculated capacities of similar joints across several trusses, it may be possible to determine an appropriate mean capacity for each repetitive joint configuration, eliminating the need to perform intensive capacity calculations for every truss.

4.2 *3-Dimensional Modeling of Hip Roofs*

Following completion of the single truss analysis presented herein, the modeling method will be expanded to an entire hip roof. The future work will include modeling the roof tested by Henderson et al. [9] and validating the 3-dimensional model using experimental reaction forces. Uniform uplift will be applied to compare C/D ratios across all elements of all trusses in the roof. The 3-dimensional case is important because the inclusion of sheathing and transverse members will result in load sharing between trusses and have a significant impact on the load path of applied uplift. The locations of critical C/D ratios may change from the plane truss analysis case. Once the trussed roof is assessed and the probable failure modes are identified, a geometrically-equivalent stick-frame roof will be modelled.

5 CONCLUSIONS

This paper presents the possibility of partial framing failures in hip roof homes under extreme wind loads. This failure mode is contrary to the common understanding of roof failures, upon which the EF-Scale is based, and has not been studied in the published work to date. Observed failures in hip roofs from recent events have sparked inquiry into whether partial failure is probable, and furthermore whether it is more probable than the well-understood RTWC and sheathing failures.

The preliminary results presented in the current study showed that the internal members and connections of a trussed hip roof are unlikely to fail in structures with toenailed RTWCs. Weak links are likely to occur, as predicted in previous research, in the RTWC or sheathing. This conclusion was found not to hold true, however, in the case where hurricane straps are employed at the RTWC. In this case the truss plates and wood members in the top chord were determined to be the most likely point of failure initiation. Further work is required to study the 3-dimensional behaviour of trussed and stick-frame roofs and gain a better understanding of hip roof framing failures as they relate to the EF-Scale.

6 ACKNOWLEDGEMENTS

This work was funded by the Natural Science and Engineering Research Council of Canada under the Collaborative Research and Development program in collaboration with Chaucer Syndicates Ltd. and the Institute for Catastrophic Loss Reduction (ICLR). The ongoing support from Mr. Gero Michel (Chaucer) and Paul Kovacs (ICLR) is gratefully acknowledged. The authors are also grateful to Dr. David Prevatt and David Roueche of the University of Florida for providing damage survey photos and relevant literature.

REFERENCES

[1] M. O. Amini and J. W. van de Lindt, Quantitative insight into rational tornado design wind speeds for residential wood-frame structures using fragility approach, Journal of Structural Engineering, 140 (2014).

[2] C. D. Standohar-Alfano and J. W. van de Lindt, Tornado risk analysis for residential wood-frame roof damage across the United States, Journal of Structural Engineering, 142 (2016).

[3] T. T. Fujita, "Proposed characterization of tornados and hurricanes by area and intensity. Satellite and Mesometeorology Research Project Report 91," University of Chicago, Chicago, IL, 1971.

[4] Wind Science and Engineering Centre, A Recommendation for an Enhanced Fujita Scale, Texas Tech University, Lubbock, TX, 2006.

[5] Environment Canada, Environment and Climate Change Canada: Enhanced Fujita Scale, 6 June 2013. [Online]. Available: https://ec.gc.ca/meteo-weather/default.asp?lang=En&n=41E875DA-1.[Accessed 30 March 2017].

[6] D. M. Sills, P. J. McCarthy and G. A. Kopp, Implementation and application of the EF-scale in Canada, in Extended Abstracts, 27th AMS Conference on Severe Local Storms, Madison, WI, 2014.

[7] P. R. Sparks, S. D. Schiff and T. A. Reinhold, Wind damage to envelopes of houses and consequent insurance losses, Journal of Wind Engineering and Industrial Aerodynamics, 5 (1994) 145-155.

[8] D. Meecham, D. Surry and A. G. Davenport, The magnitude and distribution of wind-induced pressures on hip and gable roofs, Journal of Wind Engineering and Industrial Aerodynamics, 38 (1991) 257-272.

[9] D. Meecham, The improved performance of hip roofs in extreme winds - A case study, Journal of Wind Engineering and Industrial Aerodynamics, 43 (1992) 1717-1726.

[10] G. A. Kopp, E. Hong, E. Gavanski, D. Stedman and D. M. Sills, Assessment of wind speeds based on damage observations from the Angus (Ontario) Tornado of 17 June 2014, Canadian Journal of Civil Envgineering, 44 (2016) 37-47.

[11] D. J. Henderson, M. J. Morrison and G. A. Kopp, Response of toe-nailed, roof-to-wall connections to extreme wind loads in a full-scale, timber-framed, hip roof, Engineering Structures, 56 (2013) 1474-1483.

[12] A. J. Graettinger, C. C. E. Ramseyer, S. Freyne, D. O. Prevatt, L. Myers, T. Dao, R. W. Floyd, L. Holliday, D. Agdas, F. L. Haan, J. Richardson, R. Gupta, R. N. Emerson and C. Alfano, Tornado Damage Assessment in the aftermath of the May 20th 2013 Moore Oklahoma Tornado, The University of Alabama, Tuscaloosa, AL, 2014.

[13] Truss Plate Institute of Canada, Truss Design Procedures and Specifications for Light Metal Plate Connected Wood Trusses, TPIC, Bradford, ON, 2014.

[14] Truss Plate Institute, National Design Standard for Metal Plate Connected Wood Truss Construction, Alexandria, VA: American National Standards Institute (ANSI), 2002.

[15] S. L. Lewis, N. R. Mason, S. M. Cramer, D. C. Wert, P. J. O'Regan, G. Petrov and D. Goclano, Design of Metal Plate Connected Wood Truss Joints for Moment, in 9th World Conference on Timber Engineering, Portland, OR, 2006.

[16] R. O. Foschi, Analysis of wood diaphragms and trusses. Part II: Truss-plate connections, Canadian Journal of Civil Engineering, 4 (1977) 353-362.

[17] X. Liu, Three-Dimensional Modeling of Metal Plate Connected Wood Truss Joints (Doctoral Dissertation), The University of British Columbia, Vancouver, BC, 2013.

[18] K. S. Pfretzschner, Practical Modeling for Load Paths in a Realistic, Light-Frame Wood House (Master's Thesis), Oregon State University, Corvallis, OR, 2012.

[19] G. K. Martin, Evaluation of System Effects and Structural Load Paths in a Wood-Framed Structure (Master's Thesis), Oregon State University, Corvallis, OR, 2010.

[20] Z. Li, A Practical Approach to Model the Behaviour of a Metal-Plate-Connected Wood Truss System (Master's Thesis), Oregon State University, Corvallis, OR, 1996.

[21] M. Vatovec, T. H. Miller, R. Gupta and S. Lewis, Modeling of metal-plate-connected wood truss joints: Part II - Application to overall truss model, Transactions of the American Society of Agricultural Engineers, 40 (1997) 1667-1675.

[22] R. W. Wolfe and T. LaBissoniere, Structural Performance of Light-Frame Roof Assemblies - II: Conventional Truss Assemblies, Forest Products Laboratory, FPL-RP-499, Madison, WI, 1991.

[23] R. W. Wolfe, D. H. Percival and R. C. Moody, Strength and Stiffness of Light-Frame Sloped Trusses, Forest Products Laboratory, FPL-RP-471, Madison, WI, 1986.

[24] R. W. Wolfe, D. Stahl and S. Cramer, Experimental Assessment of Wood Trusses With Square-End Webs, Forest Products Laboratory, FPL-RP-544, Madison, WI, 1996.

[25] Structural Engineering Institute, ASCE 7-10 Minimum Design Loads for Buildings and Other Structures, Reston, VA: American Society of Civil Engineers, 2010.

[26] Canadian Wood Council/Canadian Standards Association, Wood Design Manual: The complete reference for wood design in Canada, Ottawa, ON: Canadian Wood Council, 2015.

[27] M. J. Morrison and G. A. Kopp, Performance of toe-nail connections under realistic wind loading, Engineering Structures, 33 (2011) 69-76.

[28] B. R. Ellingwood, D. V. Rosowsky, Y. Li and J. H. Kim, Fragility assessment of light-frame wood construction subjected to wind and earthquake hazards, Journal of Structural Engineering, 130 (2004) 1921-1930.

[29] D. Rosowsky, T. Reed and K. Tyner, Establishing uplift design values for metal connectors in light-frame construction, Journal of Testing and Evaluation, 26 (1998) 426-433.

Combined Shear and Uplift Testing of Wood Shear Walls

Shideh Shadravan[a], Christine D. Standohar-Alfano[b]

[a] *University of Oklahoma, Norman, OK, U.S., shideh@ou.edu*
[b] *Insurance Institute for Business & Home Safety, Richburg, SC, U.S., calfano@ibhs.org*

ABSTRACT

Each year in the United States, residential, wood-frame construction is damaged due to straight line winds, tornadoes, or hurricanes. While improvements in severe weather warnings has resulted in a reduction in fatalities and injuries, the social and economic impacts from high wind events can result in significant hurdles for communities impacted, especially when families are displaced from their homes. Thus, there is continual interest in cost-effective methodologies to improve wood-frame construction performance in high winds.

The focus of the work presented herein is on the wall-to-foundation connection, which is a critical component in transferring wind-induced lateral and uplift loads from the roof to the foundation. Traditional shear wall construction across most of the United States includes wall studs spaced at 60 cm (24 in) on center where the sheathing is attached with smooth shank nails spaced at 150 mm/300 mm (6 in/12 in) at the edge and field, respectively. Likewise, the walls are commonly attached to the foundation with anchor bolts spaced at 1.8 m (6 ft) on center. By altering the type, number, and spacing of these different connections, wood shear wall behavior under combined shear and uplift loads can be assessed to determine design methodologies which improve performance. Improvements in wall design also considered the cost of improvement as any design changes should also be economically viable.

Testing was performed at the University of Oklahoma FEARS Structural Laboratory. Result indicated that walls with 40 cm (16 in) stud spacing and sheathing fasteners spaced at 100 mm/150 mm (4 in/6 in) had the highest lateral load resistance in comparison to traditionally constructed wood shear walls. Walls constructed with a double bottom plate and anchor bolts nominally spaced at 0.9 m (3 ft) showed high lateral load resistance and deflection about three times greater than traditional shear walls. In tests with anchor bolts nominally spaced at 1.8 m (6 ft) and a single bottom plate, the middle anchor bolt showed the highest load resistance in comparison to the anchor bolts located closest and furthest to the applied load. In addition, using 76 cm x 76 cm (3 in x 3 in) square washer improved the load capacity of the wall in comparison to 10 mm (1.375 in) standard cut washers. Finally, estimates of wall stiffness also indicated that a.) walls with 60 cm (24 in) stud spacing were less stiff in comparison to walls with 40 cm (16 in) stud spacing, even with smaller sheathing fastener spacing; b.) 8d ring shank nails increased the stiffness of the walls in comparison to smooth shank nails; and c.) 76 cm x 76 cm (3 in x 3 in) square washers increased the wall stiffness when compared to the standard cut washers.

KEYWORDS

Shear Walls, Wood, Wind, Lateral Load, Uplift

The 13th Americas Conference on Wind Engineering (13ACWE)
Gainesville, Florida USA, May 21-24, 2017

Structural Response of Lightweight Composite Wall Systems under Spatial Variation of Wind Pressure

Kermelos Woldeyes [a], Yared Shifferaw[a], Tibebu Birhane [b], Girma Bitsuamlak[b]

[a]*Department of Civil, Architectural and Environmental Engineering, Drexel University, Philadelphia, PA, United States*
[b]*Department of Civil and Environmental Engineering, WindEEE Research Institute, University of Western Ontario, London, ON, Canada*

ABSTRACT: This paper augments the preceding work on the numerical evaluation of wind effects on low and midrise hybrid structures by the same authors. The objective of this paper is to examine the structural response of lightweight composite wall systems with various sheathing configurations (gypsum and oriented strand board) under the action of spatial wind loading. Current wind design procedures for lightweight steel wall studs are generally prescriptive in nature and do not consider the spatial and transient variation of wind loading. Furthermore, bracing provided by the sheathing material to a thin-walled member is ignored resulting in overly conservative designs. Finite element analysis of the stability and strength behavior of thin walled roof and wall systems in these hybrid structures under wind loading requires spatial and temporal resolution of wind pressure consistent with both static and dynamic responses. This study will only consider spatial variation of wind pressure. Spatial pressure coefficient readings from the computational fluid dynamics simulation are used to generate the spatially varying wind loading which will be utilized in the finite element modeling of a sheathed single stud wall system configuration. Finite element analysis is undertaken to examine the stability and ultimate strength behavior of the lightweight composite wall system. The results from this research are compared to conventional design approaches for wall studs that ignore sheathing material contribution and will further facilitate in the development of a new design approach that considers the spatiotemporal variation of wind pressure.

KEYWORDS: Thin-Walled Structures, Finite Element Analysis, Computational Fluid Dynamics (CFD), Stability

1 INTRODUCTION

The use of cold formed steel (CFS) in replacing conventional building materials such as concrete, hot-rolled steel and timber has increased due to its advantage in cost-efficiency and ease of construction. Light framed structures using cold-formed steel members are now frequently employed in low and mid-rise residential, commercial and industrial buildings as load-bearing framings for walls, floor and roofs. These members are usually accompanied by cladding and sheathing materials to form building envelopes. During hurricanes and windstorms, these hybrid structural members are acted upon by wind loads that fluctuate both spatially and temporally.

Throughout the world, windstorms have been responsible for causing numerous fatal and property damages. According to Taranath [1], economic and insured losses by hurricane events exceeds losses by other natural disasters such as earthquakes and floods. Between 1986 and 1993, hurricane events in the United States caused a property damage of about 41 billion compared to 6.2 billion for all other natural disasters. In 1992, hurricane Andrew cased about 65 fatalities and insured losses in the region of 16 billion dollars. According to reports by the National Oceanic and Atmospheric Administration (NOAA), the occurrence of hurricanes Katrina (2005) and Sandy (2012) , caused more than 1900 human fatalities and property damages close to 220 billion dollars [2]. Hurricanes usually damage the envelope of a building. Negative pressure zones created by windstorms on roofing systems usually lead to a breach of a building's roof system first. Since hurricanes are associated with heavy rain and storm surges, the breach in the building envelope usually lead to significant interior damages [3]. Therefore, the response of a building component and cladding member under the action of wind loading needs to be investigated.

Light weight wall systems generally include cold formed wall studs connected to sheathing materials. The common types of sheathing materials are oriented strand board (OSB), plywood, and gypsum. Based on the configuration and shape of a building, wall systems can be acted up on by a positive wind pressure or a negative wind suction. Sheathing materials can act as bracings to cold formed steel members providing additional stiffness and coupling for transfer of wind load among adjacent members. The full behavior of sheathed light weight wall systems under wind loading has not been studied well. This is also reflected in current analysis and design approaches used in high wind areas. Most standards and specifications used for wind load determination use a static approach in which wind pressure and suction magnitudes are calculated from mean pressure coefficients that ignore spatial and temporal variation of wind. Furthermore, existing design approaches for high wind areas are limited to prescriptive methods in selection and sizing of wall studs and purlins [4]. In the current study the response of light weight composite wall systems under realistic spatial wind loading conditions is investigated. Single cold formed steel stud members with different sheathing configuration are examined.

2 SPATIAL WIND LOAD DETERMINATION

The spatial wind loading was computed by using numerically computed pressure coefficients on a typical cold-formed light weight steel framed building. The numerical approach is verified with existing wind tunnel pressure coefficients. Unlike wind tunnel experiment, a numerical simulation approach does not put a restriction on the number of pressure taps that can be fit on a building model resulting in a well-defined spatial and temporal variation of wind loading.

The CFS-NEES building is selected for the determination of pressure coefficients on a typical cold-formed steel building. The CFS-NEES building has been used in full scale testing of the performance of multi-story cold formed structures under seismic actions as part of a U.S. National

Science Foundation (NSF) sponsored project [5]. Figure 1 shows the full scale CFS-NEES building. Two regions are considered in the study (1) building wind-ward wall region near the middle of the wind-ward face and (2) Building sidewall region near windward corner. Three wind approaching angles, which are 0^0, 45^0 and 90^0 are considered. Based on the wind directions, the two regions would represent both negative (suction) and positive (pressure) wind loading scenarios. Unique spatial variation is also achieved for each wind approaching angle. Figure 2 shows the resulting spatial variation of wind on a given test panel located at the two regions. Pressure coefficient data for the two wall regions were recorded by dispensing a very fine pressure tap grid layout with a maximum pressure tap tributary area of 1 ft^2. The recorded pressure coefficient data can be converted into wind pressure by using Equation 1.

$$P_{\theta_j} = \frac{1}{2}\rho V_H^2 C_{p,\theta_j} \tag{1}$$

In Equation (1), ρ represent the dry air density measured at sea level. V_H is the mean hourly wind speed at building roof height. $C_{p,\theta j}$ denotes the pressure coefficient reading for a given terrain condition and wind direction θ.

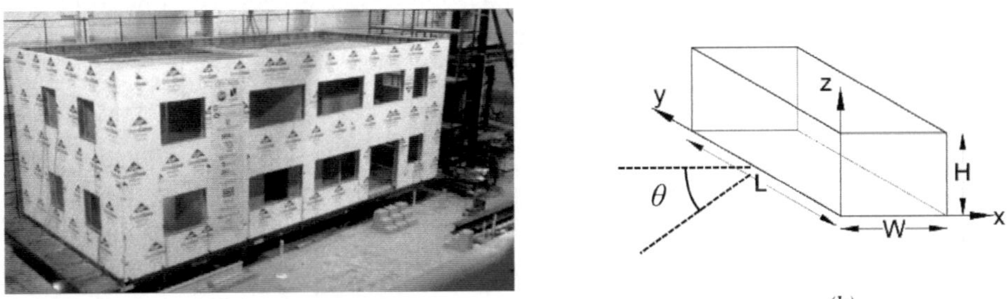

(a) *(b)*

Figure 1. CFS-NEES Cold formed steel building (a) actual full scale building (b) simplified numerical model with wind direction

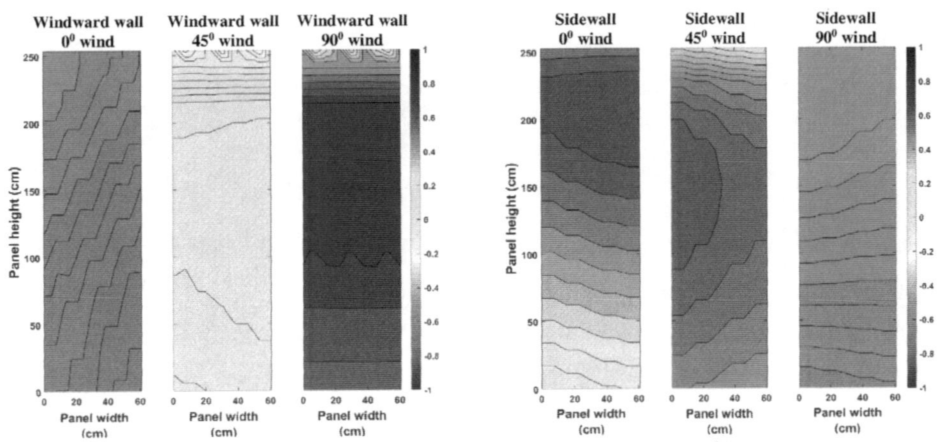

(a) *(b)*

Figure 2. Pressure coefficient distribution over test panel (a) windward wall region (b) sidewall region

3 SHEATHING TYPES AND SHEATHED-WALL STUD SYSTEM CONFIGURATIONS

Wall system and sheathing configuration employed in this study closely followed testing configurations used by Vieira Jr [6] for experimental testing of sheathed cold-formed stud walls under axial compression. Typical sheathed wall stud experiment configurations are shown in Figure 3.

Studs with different sheathing configuration are examined in this research. A sheathed stud wall system of height 244 cm (8 ft.) with a tributary width of 61 cm (2ft) is considered. Two sheathing materials (oriented strand board (OSB) and gypsum board (GYP)) are used where wall sheathed-stud system configurations are made by considering various combinations of exterior and interior wall sheathing material. The configurations include double side sheathed and bare (no sheathing) wall studs. Steel Stud Manufacturers Association (SSMA) sections 362S162-68 and 362T125-68 are used as a stud and track respectively in the configurations. The SSMA naming nomenclature identifies web and flange depth, member thickness and section type [7]. Accordingly, the first three numbers represent web depth in 1/100 of an inch (362 x 1/100 = 3.625 inches). The letter after that represents the type of section (S = Stud and T =Track). Flange width is represented by the next three numbers similar to web depth (162 x 1/100 = 1.625 inches). The last two number represent the thickness of the section in 1/1000 of an inch (68 x 1/1000 = 0.0677 inches). In SI units, the web depth, flange width and member thickness are 9.21 x 4.13 x 0.17 cm and 9.21 x 3.18 x 0.0677 cm for stud and track members respectively.

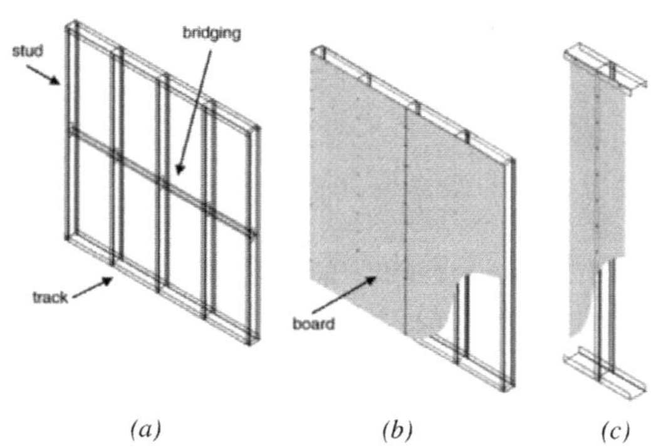

Figure 3 Typical sheathed wall stud testing configuration (a) bare wall system with bridging (b) sheathed wall system with bridging (c) isolated single stud with sheathing [6]

4 FINITE ELEMENT ANALYSIS

A finite element (FE) model is developed for the study of sheathed-stud wall systems under wind loading. Wall systems that consist a single stud are studied. Double sided sheathing with similar interior and exterior sheathing material is considered.

4.1 *Material property and modeling*

For wall stud and track sections, cold-formed steel with elastic modulus and poison's ration of 203 GPa (29500 ksi) and 0.3 respectively is used. Experimentally determined true plastic stress-strain relationship is used to determine the post yield characterization of steel [8]. For post buckling studies, the true stress-strain (σ_t , ε_t) model in Figure 4 is adopted. It is assumed, plastic strain is initiated after yield stress at 0.2% strain offset. The 0.2% offset method is known to give finite element results that agree with experimental results [8-10]. Material properties used for the different sheathing materials are summarized in Table 1.

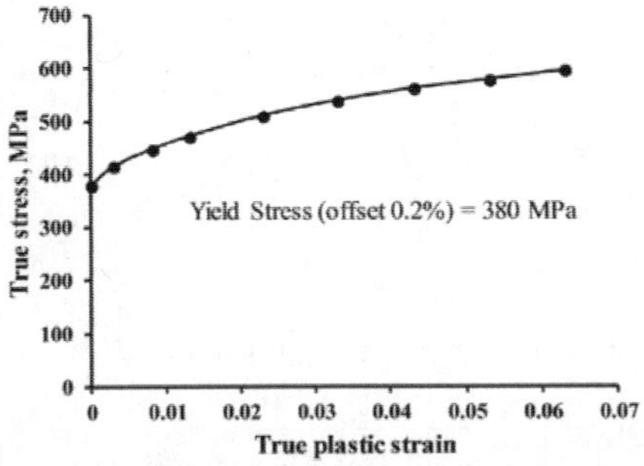

Figure 4. Plastic strain curve with 0.2% yield criteria

Table 1. Sheathing material mechanical properties

Sheathing Material	Modulus of Elasticity - MPa (ksi)	Shear Modulus MPa (ksi)	Poisson's ratio	Sheathing thickness mm (in.)
Gypsum	993 (144)	552 (80)	0.3	11.1 (7/16)
OSB	6,426 (932)	1,310 (190)	0.3	12.7 (1/2)

4.2 *Element type and mesh discretization*

A finite element software ABAQUS is used to carry out buckling and collapse analysis on the different configurations of a sheathed wall stud system [11]. Sheathing, stud and track sections are modeled using shell elements. The S4R shell element, a 4 node, quadrilateral, reduced integration, with large strain deformation formulation is used to define all shell elements [11]. Sheathing, stud and track members are discretized using non-uniform seeding arrangement. Finer discretization is used along the crosswise direction of stud and track members while coarser seeding is employed in the longitudinal direction. Meshing was refined near fastener points on all members.

4.3 *Boundary conditions*

Fasteners are simulated for two types of connections in this research (a) connection between the stud and tracks, and (b) connection between the stud and sheathing. The proper definition of boundary conditions plays an important role in the correct prediction of stability and collapse behavior

According to Vieira Jr [6], the correct idealization of stud to track connection is important in strength and stability determination. The effective length factor of the stud (K), being a function of the end fixity of the member has a significant role in numerical prediction of strength. Test results on the nominal axial strength (P_n) of a bare stud-track configuration under compression had a maximum difference of 3% with theoretical results by both the direct strength method (DSM) and effective width method when considering a fixed-fixed boundary condition (K = 0.5). In addition, first buckling mode shape for a fixed-fixed boundary condition matches experimental results. Therefore, the connection and contact between stud and track defined in the finite element model should simulate a fixed end condition. In this research the methodology employed by Vieira

Jr [6] is adopted to simulate the stud to track connections. Linear multi-point constraint (MPC) equations are used to couple all degrees of freedoms excluding rotation around the axis of the stud to track fastener. This will allow the stud to rotate within the axis of the fastener. The stud is also free to translate (slide) along the longitudinal direction of the track while displacement in the longitudinal axis of the stud is restrained. the general form of the multi point constraint equation and the form implemented in the current research are given in Equation 2 and 3 respectively.

$$A_1 u_i^P + A_2 u_j^Q + \ldots + A_n u_k^R = 0 \tag{2}$$

$$1 u_i^{stud\_node} + 1 u_i^{track\_node} = 0, \quad i = 1, 2, 4 \,\&\, 6 \tag{3}$$

In Equation 2, $A_n u_k^R$ represents a node R, a degree of freedom (*dof*) u_k and a coefficient of A_n which defines any relative displacement or rotation between connected nodes. Using Equation 3, stud to track connections are fully restrained for dof 1, 2, 4 and 6.

In modeling the restriction to the stud due to the sheathing material, two approaches can be employed. One way to provide the restriction by the stud is to include the sheathing in the finite model. However, this could be computationally expensive since it would mean added finite elements. A single stability or collapse finite element analysis could take several hours to prepare, run and post process the output data. More importantly, the sheathing material, being less stiff than the stud, will buckle or reach ultimate collapse strength before the stud in both stability and collapse analysis. Stud members, in this case, would remain unaffected with no significant instability occurring. Similar observation was made in the study of panel-purlin systems under wind loading in which preliminary buckling and collapse of the panel member controlled the stability and collapse behavior [12]. Therefore, this approach would not be appropriate in the study of the sheathed studs under wind loading.

The alternative, which is more computationally economical, is to model restriction due to sheathing material by using linear translational and rotational springs at fastener locations. In this case, traverse variation of wind loading is ignored and is directly applied to the loaded flange of the stud in a form of a distributed load. Values for spring stiffness are taken from experimental tests done by [6]. Table 2 and Figure 5 give the values used in the FE model for spring stiffness and the general spring orientations. The nodes around fastener holes on the stud are connected to a rigid node. Ground springs are then used to connect these rigid nodes to a fixed support. It is assumed that the performance of sheathing material and fasteners will be identical in both wind pressure and suction cases. Finally, tracks at the top and bottom of the stud are restrained both in transition and rotation at the web. Figure 6 portrays FE model for sheathed wall stud, complete with stud, track and sheathing assembly and the simplifications employed.

Table 2. Spring stiffness values used in FE analysis (12-inch (30.5 cm) fastener spacing) [6]

Sheathing Material	Spring	No sheathing, springs idealize both fastener and sheathing
OSB	kx = kz (N/mm)	971
	ky (N/mm)	0.374
	kϕz = kϕx (N.mm /rad)	95309
Gypsum	kx = kz (N/mm)	427
	ky (N/mm)	0.0087
	kϕz = kϕx (N.mm /rad)	95987

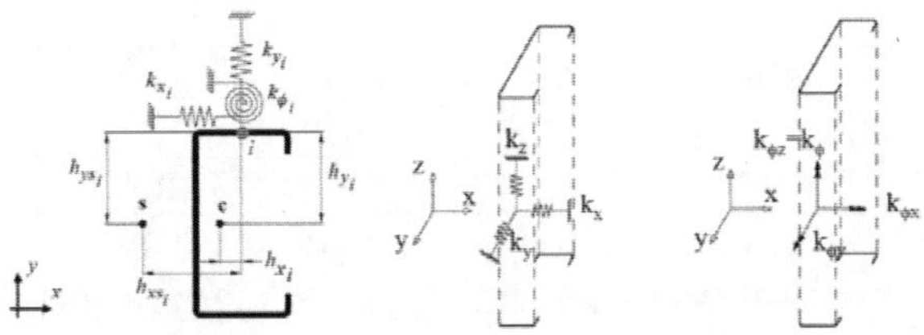

Figure 5. Fastener and sheathing springs (a) spring nomenclatures (b) transitional springs orientation (c) rotational springs orientation [6]

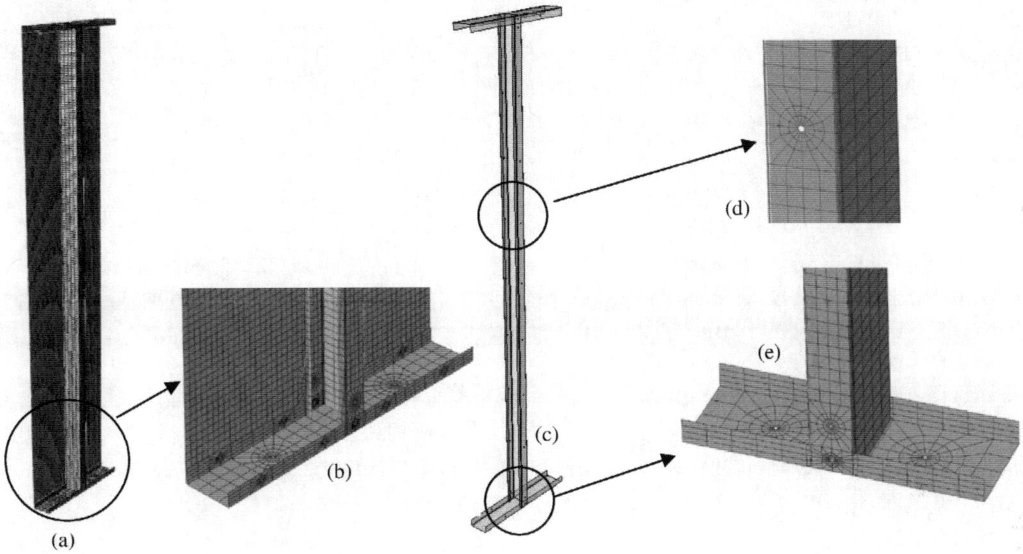

Figure 6. Sheathed stud wall FE models (a) Fully sheathed stud wall system FE model (note: sheathing in front removed for view) (b) sheathing, stud and track connection (c) simplified model with spring (d) fastener hole mesh refinement (e) stud to track connection

5 STABILITY BEHAVIOR OF SHEATHED WALL STUDS UNDER WIND LOADING

In this section, the stability behavior of a sheathed single stud wall system under wind loading is explored. An elastic buckling finite element analysis was conducted on a simplified sheathed stud wall system where springs are used to idealize stiffness of sheathing material and fasteners. The finite element models include both bare and sheathed stud specimens. Sheathing is provided in such a way that similar type of sheathing material is provided on both flanges of the stud. A series of finite element eigen buckling analyses are conducted using ABAQUS and the effect of wind directionality and sheathing restraint on stability are examined. A summary of the results from this study are shown in Table 3.

Table 3. FE eigenvalue analysis result for sheathed and bare wall studs

Specimen ID	Sheathing type	Wind direction	Stud location	Max C_p on stud	q_{app} ($x10^{-3}$)	q_{crd}
					(KN/m)	(KN/m)
GYP_SINGLE_STUD_SW_00	Gypsum	0^0	Side Wall	0.644	0.446	0.850
GYP_SINGLE_STUD_SW_45	Gypsum	45^0	Side Wall	0.566	0.393	0.707
GYP_SINGLE_STUD_SW_90	Gypsum	90^0	Side Wall	-0.589	-0.409	-0.746
GYP_SINGLE_STUD_WW_00	Gypsum	0^0	Wind Ward	-0.629	-0.436	-0.710
GYP_SINGLE_STUD_WW_45	Gypsum	45^0	Wind Ward	0.301	0.209	0.727
GYP_SINGLE_STUD_WW_90	Gypsum	90^0	Wind Ward	0.784	0.544	0.720
GYP_SINGLE_STUD_U_P	Gypsum	U	-	1.000	0.694	0.665
GYP_SINGLE_STUD_U_S	Gypsum	U	-	-1.000	-0.694	-0.681
OSB_SINGLE_STUD_SW_00	OSB	0^0	Side Wall	0.644	0.446	0.867
OSB_SINGLE_STUD_SW_45	OSB	45^0	Side Wall	0.566	0.393	0.722
OSB_SINGLE_STUD_SW_90	OSB	90^0	Side Wall	-0.589	-0.409	-0.764
OSB_SINGLE_STUD_WW_00	OSB	0^0	Wind Ward	-0.629	-0.436	-0.727
OSB_SINGLE_STUD_WW_45	OSB	45^0	Wind Ward	0.301	0.209	0.742
OSB_SINGLE_STUD_WW_90	OSB	90^0	Wind Ward	0.784	0.544	0.736
OSB_SINGLE_STUD_U_P	OSB	U	-	1.000	0.694	0.680
OSB_SINGLE_STUD_U_S	OSB	U	-	-1.000	-0.694	-0.698
BARE_SINGLE_STUD_SW_00	BARE	0^0	Side Wall	0.644	0.446	0.290*
BARE_SINGLE_STUD_SW_45	BARE	45^0	Side Wall	0.566	0.393	0.259*
BARE_SINGLE_STUD_SW_90	BARE	90^0	Side Wall	-0.589	-0.409	-0.722*
BARE_SINGLE_STUD_WW_00	BARE	0^0	Wind Ward	-0.629	-0.436	-0.662*
BARE_SINGLE_STUD_WW_45	BARE	45^0	Wind Ward	0.301	0.209	0.262*
BARE_SINGLE_STUD_WW_90	BARE	90^0	Wind Ward	0.784	0.544	0.259*
BARE_SINGLE_STUD_U_P	BARE	U	-	1.000	0.694	0.247*
BARE_SINGLE_STUD_U_S	BARE	U	-	-1.000	-0.694	-0.648*

1. q_{app} --- applied distributed load at the start of eigen analysis
2. q_{crd} --- critical elastic distortional buckling load of member
3. * --- members that fail in lateral torsional buckling (q_{cre})
4. _P and _S stand for pressure (+) and suction (-) wind loading in case of uniform wind loading case

5.1 *Stud with no sheathing*

As expected for long members with no lateral bracing, in studs with no sheathing (Bare-Bare), the critical mode of buckling was found to be lateral torsional buckling in which the whole member displaces and twists with long buckling half wave lengths (Figure 7a-1 & a-2). Positive wind pressure loading resulted in a lower lateral torsional buckling load when compared with negative wind pressure (suction) loading. This is clearly seen for cases with uniform positive and negative wind pressure loading with the latter being almost twice as high. Similar observations are made for non-uniform wind loading cases with negative wind pressure cases having higher buckling loads. These include 0^0 wind approach angle for the wind ward wall and 90^0 wind approach angle for the side wall. Table 4 gives comparison between uniform and non-uniform wind loading cases for a stud with no sheathing. In all cases, non-uniform wind loading resulted in higher critical elastic buckling load. In addition to lateral torsional buckling, distortional buckling of the compression flange (Figure 7b) and local buckling near the end supports (Figure 7c) are observed.

Table 4. Elastic critical lateral torsional load comparison for wall studs with no sheathing

Specimen ID	Sheathing type	Wind direction	Stud Location	q_{cre} (KN/m)	$\frac{q_{cre\ (non\text{-}uniform)}}{q_{cre\ (uniform)}}$ *
BARE_SINGLE_STUD_SW_00	BARE	0^0	Side Wall	0.290	1.17
BARE_SINGLE_STUD_SW_45	BARE	45^0	Side Wall	0.259	1.05
BARE_SINGLE_STUD_SW_90	BARE	90^0	Side Wall	-0.722	1.11
BARE_SINGLE_STUD_WW_00	BARE	0^0	Wind Ward	-0.662	1.02
BARE_SINGLE_STUD_WW_45	BARE	45^0	Wind Ward	0.262	1.06
BARE_SINGLE_STUD_WW_90	BARE	90^0	Wind Ward	0.259	1.05
BARE_SINGLE_STUD_U_P	BARE	U	-	0.247	-
BARE_SINGLE_STUD_U_S	BARE	U	-	-0.648	-

1. * ... comparison is for non-uniform and uniform wind loading cases with the same line of action

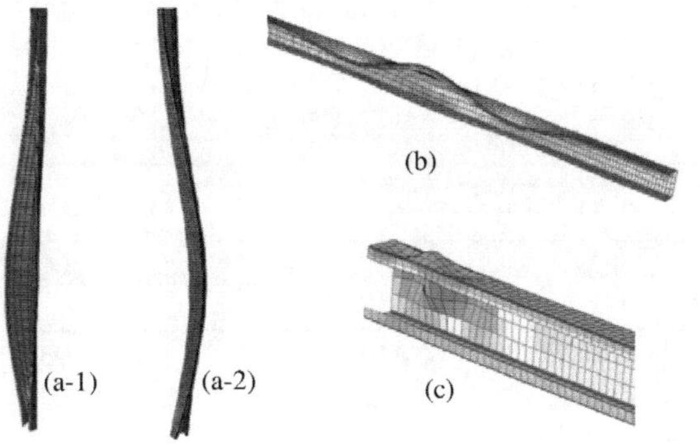

Figure 7. Eigen buckling modes for Bare-Bare studs (a-1) lateral torsional buckling (twist) (a-2) lateral torsional buckling (minor axis bending) (b) compression flange distortional buckling (c) local buckling near support (note: track members removed for better visibility)

5.2 *Sheathed stud wall systems*

Distortional buckling mode of the compression flange was observed to be the critical mode of buckling when sheathing is provided. The sheathing provided a lateral bracing to the stud, hence avoiding lateral torsional buckling. Based on the direction of wind loading, the compression flange could be the exterior loaded flange (positive wind pressure) or interior flange (negative wind suction). Table 5 shows the comparison between uniform and non-uniform wind loading cases and between sheathed and bare specimens under similar wind loading conditions. Uniform wind loading condition was found to be very conservative, with elastic critical distortional buckling load under non-uniform wind loading being on average 11% higher. An increase in the critical elastic buckling load was also observed for sheathed studs when compared with bare specimens. Effect of sheathing in increasing critical elastic buckling load of stud was more significant for members under positive wind loading than members under negative wind loading. However, there is not that much difference in the critical distortional buckling load between gypsum and OSB sheathings. Distortional buckling, characterized by rotation of the flange with respect to the flange/web junction, is mainly affected by the major axis rotation stiffness ($k_{\phi x}$). With $k_{\phi x}$ being similar for the gypsum and OSB sheathings, the difference in distortional stability behavior is negligible. Figure 8 shows typical distortional buckling mode shapes observed.

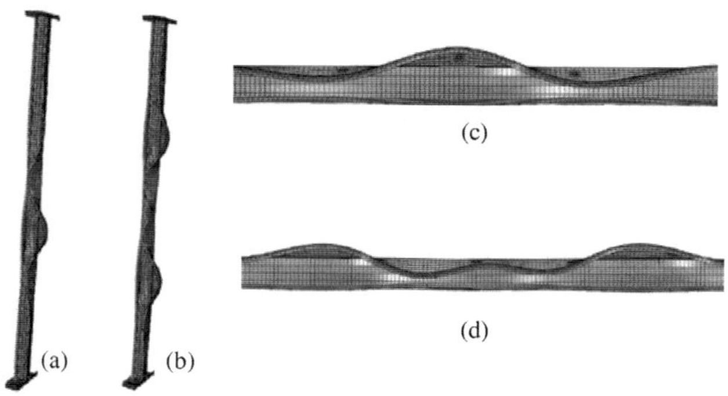

Figure 8. Observed distortional buckling mode shapes for OSB sheathed wall stud (a) first distortional buckling mode (b) higher order distortional buckling mode (c) first distortional buckling mode (zoomed) (d) higher order distortional buckling mode (zoomed)

Table 5. Elastic distortional buckling capacity comparison for sheathed wall studs

Specimen ID	Sheathing type	Wind direction	Stud Location	q_{cre} (KN/m)	$\frac{q_{crd\,(non\text{-}uniform)}}{q_{cre\,(uniform)}}$	$\frac{q_{crd\,(sheathing)}}{q_{cre\,(bare)}}$
GYP_SINGLE_STUD_SW_00	Gypsum	0^0	Side Wall	0.850	1.28	2.93
GYP_SINGLE_STUD_SW_45	Gypsum	45^0	Side Wall	0.707	1.06	2.73
GYP_SINGLE_STUD_SW_90	Gypsum	90^0	Side Wall	-0.746	1.10	1.03
GYP_SINGLE_STUD_WW_00	Gypsum	0^0	Wind Ward	-0.710	1.04	1.07
GYP_SINGLE_STUD_WW_45	Gypsum	45^0	Wind Ward	0.727	1.09	2.77
GYP_SINGLE_STUD_WW_90	Gypsum	90^0	Wind Ward	0.720	1.08	2.78
GYP_SINGLE_STUD_U_P	Gypsum	U	-	0.665	-	2.69
GYP_SINGLE_STUD_U_S	Gypsum	U	-	-0.681	-	1.05
OSB_SINGLE_STUD_SW_00	OSB	0^0	Side Wall	0.867	1.27	2.99
OSB_SINGLE_STUD_SW_45	OSB	45^0	Side Wall	0.722	1.06	2.79
OSB_SINGLE_STUD_SW_90	OSB	90^0	Side Wall	-0.764	1.10	1.06
OSB_SINGLE_STUD_WW_00	OSB	0^0	Wind Ward	-0.727	1.04	1.10
OSB_SINGLE_STUD_WW_45	OSB	45^0	Wind Ward	0.742	1.09	2.83
OSB_SINGLE_STUD_WW_90	OSB	90^0	Wind Ward	0.736	1.08	2.84
OSB_SINGLE_STUD_U_P	OSB	U	-	0.680	-	2.75
OSB_SINGLE_STUD_U_S	OSB	U	-	-0.698	-	1.08

6 COLLAPSE BEHAVIOR OF SHEATHED WALL STUDS UNDER WIND LOADING

In addition to stability, the determination of post buckling collapse behavior of thin-walled members is critical. Non-linear finite element modeling has been successfully used to determine post buckling behavior and collapse strength of thin-walled members in the past [6, 8, 13]. This section describes the procedure adopted in determining post buckling behavior of a sheathed wall stud system under wind loading and discusses the results of the study.

The nonlinear collapse analysis was conducted by using ABAQUS. The ultimate strength of cold-formed steel members are known to be highly affected by material non-linearity and geometric imperfections [8]. Therefore, the finite element modeling for collapse analysis need to be as representative as possible to predict the realistic ultimate strength of the member. Material non-linearity is introduced into the finite element model by defining the true stress-strain relationship that determines the plasticity and post yield behavior of the material. The steel material model given in Figure 4 is adopted for the non-linear material modeling. Eigen buckling mode shapes corresponding to global, distortional and local buckling are used to determine the imperfection shape of the member. Maximum imperfection magnitudes corresponding to each buckling mode shapes are obtained using a the probabilistic approach proposed by Schafer & Peköz [14]. In this approach, maximum imperfection magnitude is obtained from cumulative distribution functions (CDFs) that determine the probability of an imperfection exceeding the maximum imperfection value. Therefore, the nodal coordinates of the finite element model are scaled to the maximum imperfection using the initially defined eigen buckling mode shapes. In this study, maximum imperfection magnitudes corresponding to 25% and 75% CDFs are used. Table 6 gives the maximum imperfection values for 25% and 75% CDFs.

Table 6. CDF values for maximum imperfection [14]

$P(\Delta<d)$	Type-1	Type-2	Global
	d_1/t	d_2/t	
0.25	0.14	0.64	L/2000
0.75	0.66	1.55	L/1000

1. $P(\Delta<d)$---indicates the probability that a random imperfection value, Δ, is less than maximum imperfection, d.
2. d_1/t---maximum local buckling imperfection divided by member thickness(t)
3. d_2/t---maximum distortional buckling imperfection divided by member thickness (t)

ABAQUS provides two options for the determination of the load deformation response and ultimate strength of cold-formed members. These are the modified RIKS method (*STATIC RIKS) and the general static method with artificial damping (*STATIC, STABILIZE) [11]. Both solution methods use the Newton-Raphson algorithm but the latter adds an artificial mass proportional damping to maintain equilibrium in models where local instabilities or rigid body motions are an issue. These issues are common in finite models where there exists contact interaction between different members. The sheathed stud-track arrangement under consideration is such type of finite element model.

Contact modeling in ABAQUS brings about additional complexity to the finite element model and increases the demand for computational time. Constraints enforced due to contact modeling can make it challenging for the Newton-Raphson algorithm iteration to converge and increases equation solver cost [11]. Additional assumptions are made to minimize the issue with contact modeling in which the track sections are removed and replaced with stud end constraints. Figure 9 shows the boundary conditions assumed instead of the restraint provided by the track and the stud-track fasteners. Elastic buckling analysis results using the equivalent boundary conditions show good agreement with corresponding stud-track arrangement.

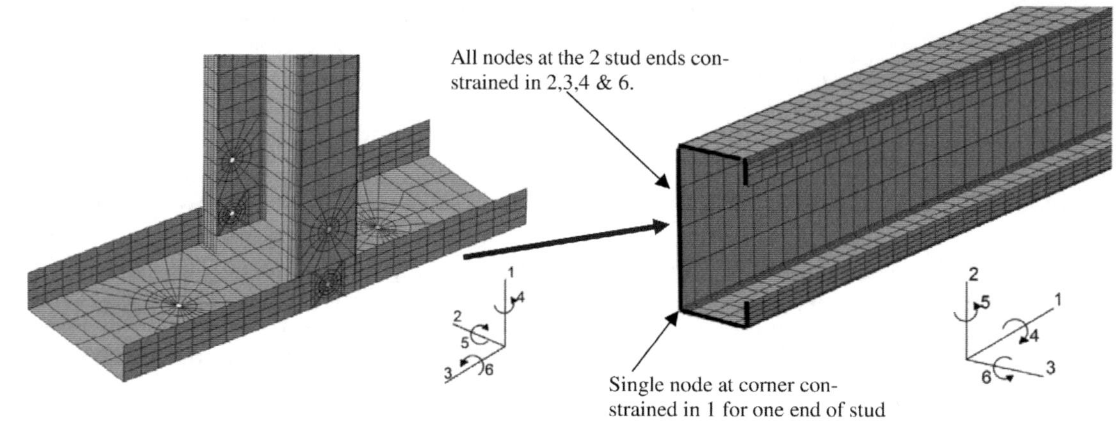

Figure 9. ABAQUS boundary conditions simulating stud-track connection

6.1 *Stud with no sheathing*

In all cases of studs with no sheathing, failure is characterized by a global-distortional interaction. For cases where the wind loading is positive, failure is initiated by a global rotation of the stud along the longitudinal axis and an inward distortional deformation of the loaded compression flange. This is followed by formation of local dents on the web of the member. Plastic failure of material is finally observed at the region of compression flange/web junction near the mid-line of the stud.

When the wind loading is negative (suction), failure is initiated by the outward buckling of the loaded flange which is now in tensile stress. Twisting along the longitudinal axis at the lower end of the stud causes indentations to form near the compression flange/web junction zone. Plastic failure is then observed at the same region of the stud where the indentations are observed. Similar failure mechanisms have been observed in the study of panel-purlin systems under negative wind loading [12].

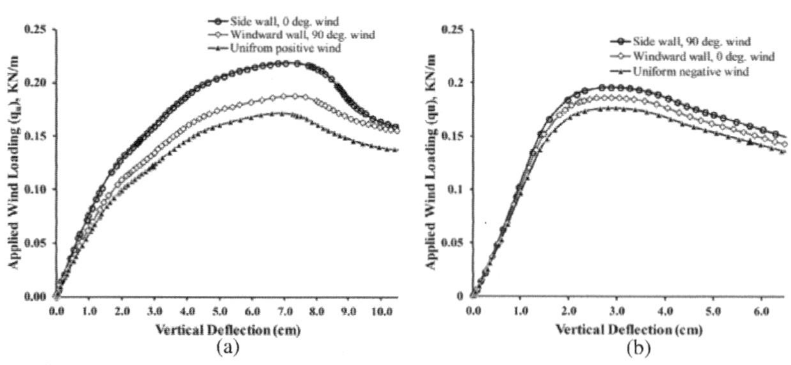

Figure 10. Load-displacement curve for bare stud (75% CDF imperfection) (a) positive wind loading (b) negative wind loading

For both positive and negative wind loading cases, uniform wind loading was found to be highly conservative. Uniform wind loading assumption resulted in a peak wind load that is 25% and 15% less than non-uniform wind loading cases for positive and negative wind loading respectively. Figure 10 shows the load-deformation response comparison between uniform and non-uniform wind loading cases.

6.2 *Sheathed stud wall systems*

Even though failure is still characterized by global-distortional interaction, restriction provided by the sheathing material significantly reduced global twist and translation resulting in higher peak loads. Final failure mechanism is again achieved by plastic failure of material that are localized at fastener locations. For uniform wind loading cases, plastic failure was observed at the fastener located at the midpoint of the compression flange. However, failure location shifts to the point of maximum loading in the case of non-uniform loading showing reliance on the spatial variation of wind. Figure 11 shows typical failure modes observed for uniform and non-uniform wind loading cases.

Sheathed studs were generally found to outperform bare studs (Figure 12-a & b). OSB and gypsum sheathings were found to result in an average of 55% and 45% increase in the ultimate strength of studs. Significant reduction is also observed for deformation magnitudes as indicated by the steep slope of the load displacement curve for sheathed studs. Uniform wind loading assumption still produced conservative results for both negative and positive wind loading but shows a reduced percentage difference when compared with bare stud results (Figure 12-c). Increase in imperfection magnitude from 25% to 75% CDF resulted in a 5% average reduction in ultimate strength of sheathed studs (Figure 12-d). Sheathed studs under negative wind loading were found to be affected more by degree of imperfection with percentage reduction in ultimate strength higher than the average.

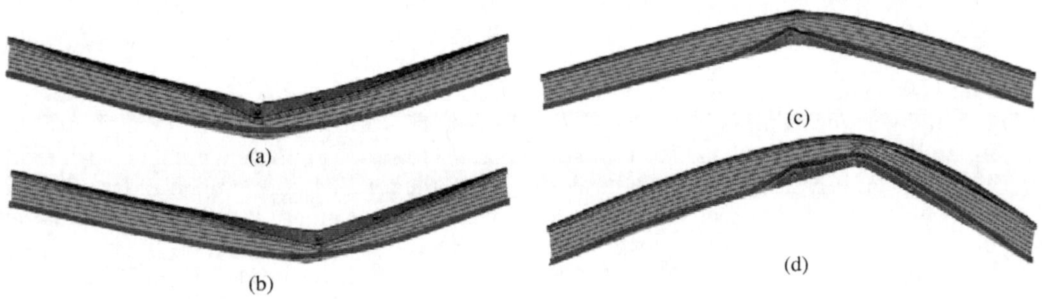

Figure 11. Typical collapse mode for sheathed wall stud (a) uniform (+) (b) non-uniform (+) (c) uniform (-) (d) non-uniform (-)

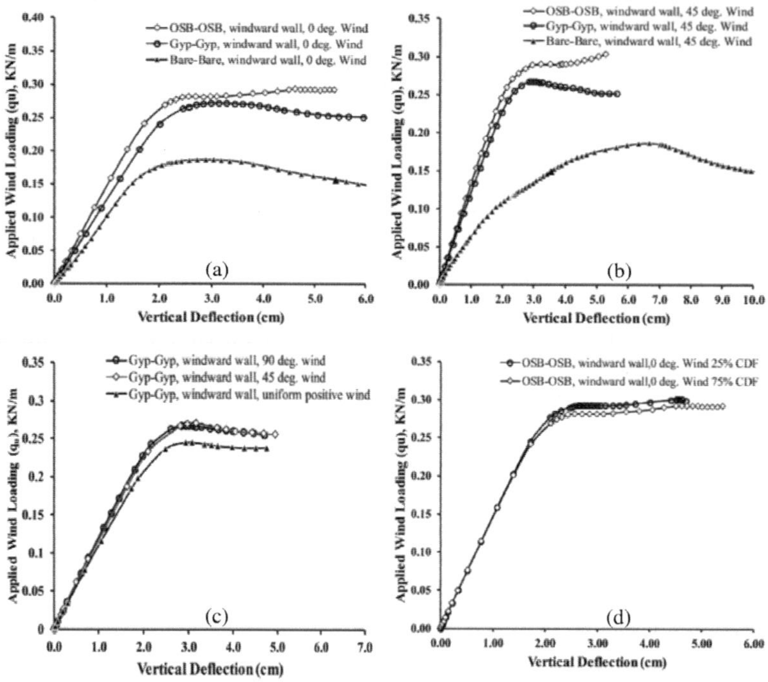

Figure 12. Load-displacement curves for sheathed wall stud (a) sheathed vs bare for 0^0 wind-75% CDF (b) sheathed vs bare for 45^0 wind-75% CDF (c) uniform vs non-uniform for gypsum sheathing-75% CDF (d) imperfection sensitivity for OSB sheathing-0^0 wind

7 CONCLUSIONS

This research is a companion to the preceding work on *numerical evaluation of wind effects on low and midrise hybrid structures* by the authors. The earlier work was done with the objective of determining and creating a database for realistic spatial wind pressure coefficients. A finite element study on the stability and collapse behavior of a sheathed wall stud member under the action of spatially varying wind loading is undertaken in this research. Wall studs with no sheathing and sheathing (oriented-strand board and gypsum) were considered. For studs with no sheathing under wind loading, lateral torsional buckling was observed to be the first mode of buckling. Due to restraint applied by sheathing material, the first buckling mode type changed to distortional buckling of the compression flange for the sheathed wall studs. Sheathing was found to increase the critical elastic buckling load of stud members under uniform and non-uniform wind loading scenarios. An increase by 200-300% and 3-8% over the values for bare stud members under similar wind loading conditions was observed for positive and negative wind loading cases respectively. Uniform wind loading assumption was found to be highly conservative. Non-uniform wind loading assumption for bare and sheathed studs resulted in critical elastic buckling loads 17% and 28% higher than uniform wind loading assumption respectively.

Collapse behavior of bare and sheathed studs under wind loading was characterized by a global-distortional interaction mainly affecting the compression flange/web junction region of the stud. In the case of bare studs, final failure mechanism is achieved by plastic failure of material at the compression flange/web junction region near the mid-length of the stud. Similar failure mechanism was observed for sheathed wall studs. However, plastic failure region was offset to the location of the compression flange fastener that match maximum pressure coefficient on the stud. Sheathed studs showed a 40-60% increase in peak collapse loads when compared with bare studs under similar wind loading conditions. Oriented-strand board, being the stiffer sheathing material, resulted in a 10% average increase in peak collapse load when compared to gypsum. Uniform wind loading assumption was again found to be conservative in predicting peak collapse loads. Peak collapse load values for bare studs under non-uniform wind loading cases were between 5-25% higher than values for uniform wind loading cases, though a reduced percentage difference was observed for sheathed wall studs.

Further work is needed to include other types of wall sheathing materials such as steel sheathing, metallic foam, plywood etc. in the study of stability and collapse behavior of sheathed wall studs. The effect of transient wind loading, separately and in combination with spatial variation of wind loading, on the stability and collapse behavior of thin-walled members should also be investigated. More importantly, experimental testing program to study behavior of thin-walled cladding members under wind loading is required to validate the results of this study.

8 REFERENCES

1 B.S. Taranath, Wind and earthquake resistant buildings: Structural analysis and design, Marcel Dekker, New York, 2005.
2 National Oceanic and Atmospheric Administration (NOAA), Billion-dollar weather and climate disasters, 2017, Available from: https://www.ncdc.noaa.gov/billions/.
3 H.M. Ali and P.E. Senseny, Models for standing seam roofs, Journal of wind engineering and industrial aerodynamics, (2003) 1689-1702.
4 AISI-S230-15, Standard for cold-formed steel framing-prescriptive method for one- and two-family dwellings, American iron and steel institute, Washington D.C., 2015.
5 R.L. Madsen, N. Nakata, and B.W. Schafer, CFS-NEES building structural design narrative, In CFS-NEES - RR01, 2011.
6 L.C.M. Vieira Jr, Behavior and design of sheathed cold-formed steel stud walls under compression, PhD Thesis, Johns Hopkins University, Baltimore, Maryland, 2011
7 Steel Stud Manufacturers Association (SSMA), Product technical guide, 2014.
8 C.D. Moen, Direct strength design of cold - formed steel members with perforations Phd Thesis, Johns Hopkins University, Baltimore, Maryland, 2008
9 C. Yu, Distortional buckling of cold-formed steel members in bending, PhD Thesis, Johns Hopkins University, Baltimore, Maryland, 2005
10 T.O. Rabare and Y. Shifferaw, Stability of perforated cold-formed steel beam-columns, In Structural stability research council annual stability conference, 2015, Orlando, Florida.
11 ABAQUS, ABAQUS/standard version 6.13-1, 2013, Dassault-Systems, Providence, RI.
12 Y. Shifferaw, K. Woldeyes and G. Bitsuamlak, Stability and strength behavior of thin-walled roof-panel-purlin system under wind loading, In Structural stability research council annual stability conference, 2017, San Antonio, Texas.
13 Y. Shifferaw and B. Schafer, Inelastic bending capacity of cold-formed steel members, Journal of structural engineering, (2012) 468-480.
14 B.W. Schafer and T. Peköz, Computational modeling of cold-formed steel: characterizing geometric imperfections and residual stresses, Journal of constructional steel research, (1998) 193-210.

Numerical Evaluation of Wind Effects on Low and Mid Rise Hybrid Structures

Kermelos Woldeyes [a], Yared Shifferaw[a], Tibebu Birhane [b], Girma Bitsuamlak[b]

[a]*Department of Civil, Architectural and Environmental Engineering, Drexel University, Philadelphia, PA, United States*
[b]*Department of Civil and Environmental Engineering, WindEEE Research Institute, University of Western Ontario, London, ON, Canada*

ABSTRACT: This paper is the first part in a two-part study with the objective to numerically examine the aerodynamic response of low and mid-rise hybrid structures (hybrid thin-walled steel and structural steel framing with gypsum and oriented strand board sheathing) where collapse behavior under wind loading is driven by the stability interaction between the thin-walled cladding systems, components, and main wind force resisting framing systems. The oversimplified idealization of low-rise building structural systems, which excludes the coupling between frame elements by cladding material, associated with Database-Assisted Design (DAD) makes the method in its current form unsuitable for thin-walled structures. Furthermore, the current practice of applying uniform static pressure to test thin-walled metal roof panels under wind loading fails to capture the true nature of wind effects on thin-walled structures. To address these shortcomings, a cost-effective approach for estimating wind pressure on the entire building envelope is required. While building specific boundary layer wind tunnel testing is the industry-wide accepted method, it could be costly for low and mid-rise structures. In addition, restrictions on the number of pressure taps that can be fit on a building model in the case of wind tunnel testing reduces spatial wind pressure variation. To circumvent these constraints, the present study establishes pressure coefficient values for a low and midrise thin walled structural system through computational fluid dynamics (CFD). Steady state simulation of wind is done for two building models that are representatives of low and midrise hybrid structure buildings. Spatial pressure coefficient data for critical wind directions is recorded and stored. The numerical model is validated by comparing the results with those obtained from wind tunnel studies conducted at the University of Western Ontario. The building surface pressure from the steady state CFD simulation will be used as an input for the subsequent finite element analysis of components and cladding systems.

KEYWORDS: Computational Fluid Dynamics (CFD), Database-Assisted Design(DAD), Thin Walled Structures, Wind Load

1 INTRODUCTION

In recent years, cold-formed steel members are being widely used in metallic building structures including low and midrise buildings. The ease and economy of fabrication and erection makes them ideal for such type of buildings. Typical structures are made of hybrid cladding systems combining cold-formed steel framing with different types of sheathing material. This includes wall and roof sheathing materials attached to cold-formed stud and purlin members respectively. According to Peköz & Soroushian [1], cold-formed members, which are weak in lateral torsional stability, need to be braced against lateral direction movement and twisting to use their full section capacities. Rigidity and restraint provided by sheathing materials and connections provides this type of bracing and needs to be accounted for in analysis and design of cold-formed members.

Over their life span, cold-formed building systems are acted upon by wind loading and need to fulfill both strength and serviceability requirements. The severe effects that an extreme windstorm event has on human life and property raises the need for accurate determination of wind loading demands and member capacity. One of the main difficulties when dealing with wind is how to accurately represent spatiotemporal variation of wind loading. Current wind loading provisions are based on wind tunnel experiments from the late 70's and are mostly the results of enveloping wind load data sets for a number of parameters including height, roof slope, size and building features such as parapets [2]. Such type of condensed wind load provisions could lead to either underestimation or overestimation of wind loads and building responses. Furthermore, the condensed nature of the data doesn't allow for the post processing of raw wind tunnel data in the study of the aerodynamic properties of low rise buildings. Advancements in data acquisition and storage systems over the years brings about the need for improving condensed wind load data sets. This will help in creating a more elaborated database accessible to the research community in order to further improve wind analysis and design methodologies.

To eliminate the above-mentioned issues, The National Bureau of Standards and Technology in collaboration with the Texas Tech University undertook wind tunnel testing of generic low rise buildings with the objective of developing an aerodynamic database for low rise building design purposes. The tests were conducted at the Boundary Layer Wind Tunnel Laboratory (BLWTL) of the University of Western Ontario (UWO). The database would serve in providing transient and spatial wind load data for a more elaborate static and dynamic structural analysis of low rise buildings and furthermore would be accessible to the public via the internet. The program consists of two sets of tests, (1) on 11 generic building model variations to create an aerodynamic database and (2) on 3 different model variations of the Texas Tech Wind Engineering Research Field Laboratory (WERFL) full scale building for comparison with full scale experiments. Further information regarding the experimental program can be found in [3, 4]. Information about the archival system, testing parameters, comparison of wind pressure data with code provisions and additional information on how to interoperate and use archived data can be found [2, 5, 6]. However, constraints on the number of pressure reading points that can be fit on model scale buildings means that a lot of interpolation is involved to create a more predefined spatiotemporal variation.

Computational fluid dynamics (CFD) provides another alternative in the determination of wind loading demands for buildings. Fielded by recent improvements in computing power, turbulence modeling techniques and availability of CFD software packages, CFD offers a much-needed substitute in the study of wind effects in building structures.

In this paper, CFD is used to conduct steady state simulations of wind on low and midrise buildings and generate spatial wind pressure variation over building cladding surfaces. The results of this study are then used in the study of stability and strength behavior of hybrid sheathed cold-formed steel wall studs under spatial wind loading in part 2 of this study. The UWO wind tunnel data will be used in numerical simulations to define inlet velocity profiles and turbulence intensity.

Numerical wind simulation on a replica of one of the UWO generic models will also be done to validate and increase confidence in the reliability of the numerical simulation of wind approach.

2 COMPUTATIONAL FLUID DYNAMICS WIND SIMULATION

2.1 *Low and midrise building model development*

Two types of building models are considered in the study. The first is a replica of one of the generic low rise building models used for wind tunnel testing at the University of Western Ontario (UWO) in the development of Database Assisted Design (DAD) software (here after called the UWO building) [2, 5-7]. The second building model is aimed to represent low and midrise buildings with light weight cold-formed steel members used as main structural components. This type of building has been used in full scale testing of the performance of multi-story cold-formed structures under seismic actions as part of the U.S. National Science Foundation (NSF) sponsored CFS-NEES project [8]. The CFS-NEES building (here after called the NEES building) has rectangular shape and flat roof with solid parapet along the roof perimeter. The building structure is made of cold-formed steel cee-section as main gravity load carrying members and wood structural panel diaphragms with shear walls as the lateral load resisting system. In both building cases, a 1:100 length scale is used. Figure 1 shows the actual full scale NEES building.

The UWO building shape selected in this study is a low-rise gable roof building with a relatively flat roof slope. Numerical simulation of wind for the UWO building is aimed at validating numerical results with corresponding wind tunnel results. Numerical simulation of wind for the NEES building is aimed at acquiring spatial variation of wind pressure on a typical cold-formed steel building. Figure 2 and Table 1 give shapes and dimensions for the two types of buildings under consideration respectively.

Figure 1. Full scale seismic test of CFS-NEES cold-formed steel building

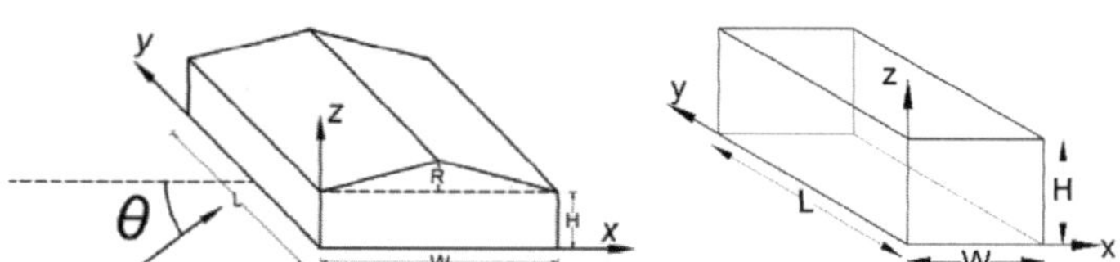

Figure 2. Building shapes and symbol representation for CFD model (a) UWO building (b) NEES building

Table 1. Full scale dimensions of building models

Parameter	UWO building model m (ft)	NEES building model m (ft)
L	76.2 (250)	15.24 (50)
W	48.8 (160)	7 (23)
H	5.5 (18)	5.5 (18)
R	2.03 (6.67)	-

2.2 *Computational domain and boundary conditions*

STARCCM+, a commercially available computational fluid dynamics software, is used to undertake numerical evaluation of wind effects on low and midrise hybrid structures [9]. The preliminary flow domain geometry creation is done by the 3D CAD module of the software.

The two building models vary in terms of building height, width and length ratio. Therefore, based on the building dimension ratios, two numerical flow domains with different sizes are developed. For the UWO building, the inflow boundary condition is set at 12H from the face of the building model. The pressure outlet boundary condition is set at 72H from the velocity inlet, much further away from the flow reattachment point. The width of the flow domain is fixed at 32H and is given a symmetry boundary condition. Another symmetry boundary condition is defined for the upper limit of the flow domain, 10H from the non-slip wall ground boundary condition. The building model surface by itself is also given a non-slip wall boundary condition.

For the NEES building, the velocity inlet boundary condition is defined at 6H from the windward face of the building and the pressure outlet boundary conditions is defined at 25H downstream of the velocity inlet. The flow domain has a width of 10H and a height of 4H. All other boundary condition definitions are kept analogous to the UWO building flow domain. Considering a 1:100 model scale, the flow domain definition results in a computational domain size of 2.96 x 5.76 x 1.8 feet (3.95 x 1.76 x 0.55 meters) and 4.5 x 1.8 x 0.72 feet (1.37 x 0.55 x 0.055 meters) for the UWO and NEES buildings respectively. Figure 3 shows the flow domains defined for each building model and the assigned boundary conditions.

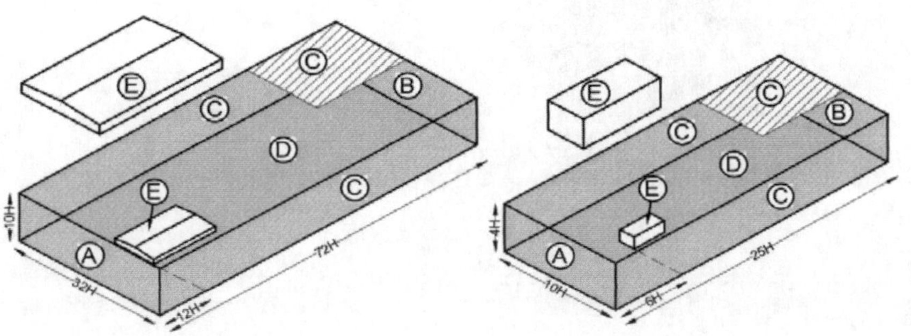

Boundary Conditions
A-velocity inlet
B-pressure outlet
C-symmetry plane
D-no slip wall ground
E-no slip wall building

(a) *(b)*

Figure 3. CFD flow domain and assigned boundary conditions (a) UWO building (b) NEES building

Polyhedral types of meshes are used for the discretization of the flow domain. A non-uniform meshing approach with finer meshing allocated at key locations such as around the building, the wake flow region and the region between the velocity inlet and the building is used. Prism layer meshes with a high aspect ratio are also used at building edges to improve near wall flow results. 5 prism layer meshes with a cell stretching factor of 1.05 is used in this study. The stretching factor controls the growth of prism layers from the wall boundary condition to the core volume mesh. The transition should be smooth with no sudden jump in mesh size to avoid numerical divergence while solving the resolved and sub-grid model scales [10].

For each building mode, two grids are employed to study the grid dependency properties. In this study, for both building models, the designations Grid-1 and Grid-2 represent the finer and coarser mesh domains respectively. For the UWO building, these grids represent a cell count of 1,290,000 and 740,000. Similarly, the grids for the NEES building have a cell count of about 1,500,000 and 880,000. Figure 4 & Figure 5 show the mesh continuums for the NEES and UWO buildings respectively.

(a) *(b)* *(c)*

Figure 4. NEES building computational flow domain views (a) 3d view (b) building close-up view (c) fine-meshing and prism layers around building surface

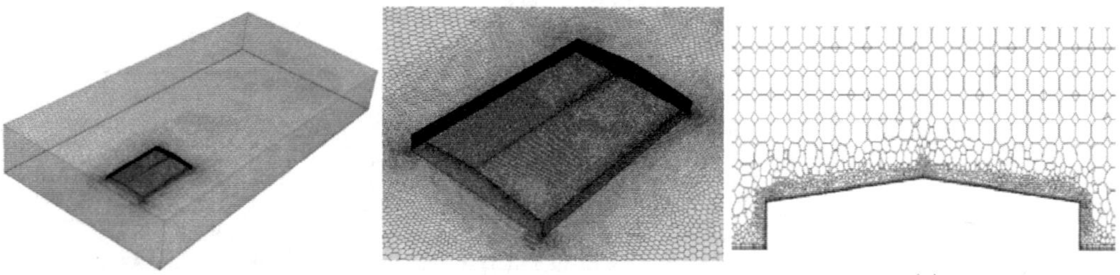

(a) *(b)* *(c)*

Figure 5. UWO building computational flow domain views (a) 3d view (b) building close-up view (c) fine-meshing and prism layers around building surface

2.3 *Inlet velocity definition*

According to Madsen [8], the NEES building is proposed to be built on a site found in Orange County, California. The location can be classified as a suburban area with closely spaced residential buildings in all directions. For a risk category of II, the 3 second gust wind speed in the area is

110 mph [11]. The 3 second gust wind speed must be first converted into hourly mean wind speed before being used to calculate wind pressure associated with wind tunnel or CFD. Such conversion is done by using the Durst Curve [11]. A terrain roughness length (z_o) of 0.3 is used as is the recommendation for large town and city suburbs [12]. Generation of the mean wind velocity, turbulence intensity and turbulence length scale profiles were done per ESDU standards [12-15]. The resulting mean wind velocity and turbulence intensity profiles are compared with wind tunnel recordings of similar exposure category and terrain roughness coefficient as seen in Figure 6. In the case of the UWO building, mean wind velocity and turbulence intensity profiles from the corresponding wind tunnel tests were directly used as an input in the numerical simulation. Like the NEES building, a suburban terrain category is assumed for the UWO building.

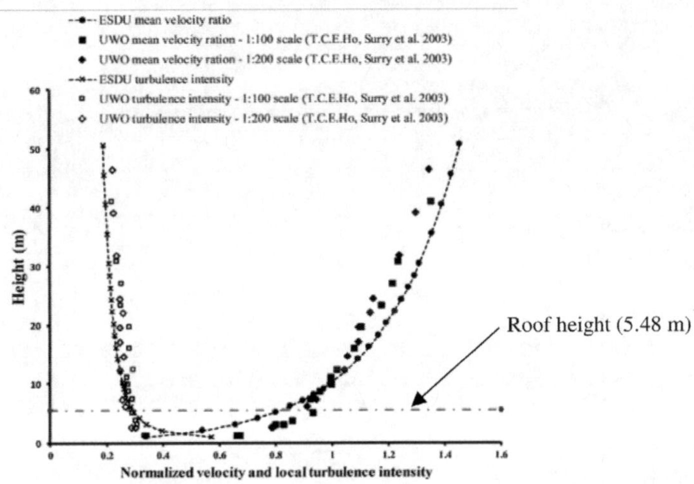

Figure 6. UWO wind tunnel and ESDU velocity and turbulence intensity profiles

3 STEADY STATE SIMULATION

A total of eight steady state simulations were done, four for each building model. Three wind directions and two mesh domains were considered. The three wind directions are 0^0, 45^0 and 90^0. The steady state simulations are used in the study of grid independency and to create spatial pressure coefficient data.

3.1 *Grid dependency study results*

For both building models, a grid independency study was done. Qualitative comparison for the overall spatial pressure distribution is done. Comparison of pressure coefficient results is done by computing the difference (ΔC_p) between finer and coarser grids. A user defined function which maps the difference in pressure coefficient data (ΔC_p) as a contour is used in STARCCM+ for the qualitative comparison. Figure 7 shows the pressure coefficient contour, implementing Grid-1, and the ΔC_p contour of the NEES building model. The ΔC_p plot shows that there is only minimal difference in pressure coefficient over the general surface of the NEES building model. However, there exists a moderate variance at wind-ward edge of the building. In the case of Grid-2, the flow is not well resolved in this region due to the coarseness of the mesh and the moderate difference in pressure coefficient can be attributed to this fact. Similar results were also observed for the UWO building model. In addition to contour plots, the pressure coefficient variation along various perimeter lines of the building are also shown in Figure 8. The contour and line plots illustrate the grid independency of the results by showing comparable pressure coefficient results for both building models.

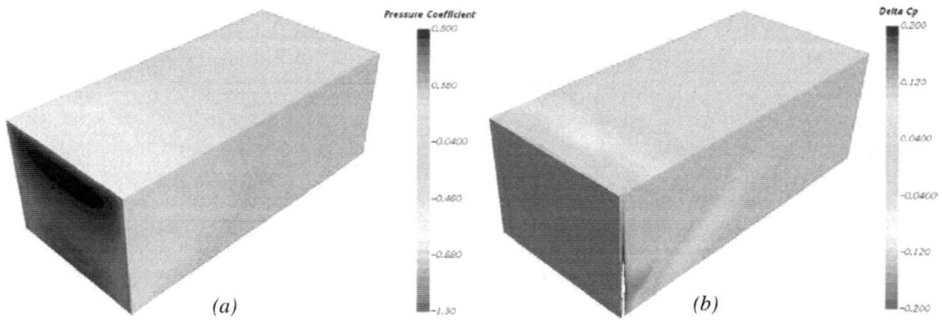

Figure 7. NEES building (a) pressure coefficient (C_p) and (b) ΔC_p plots

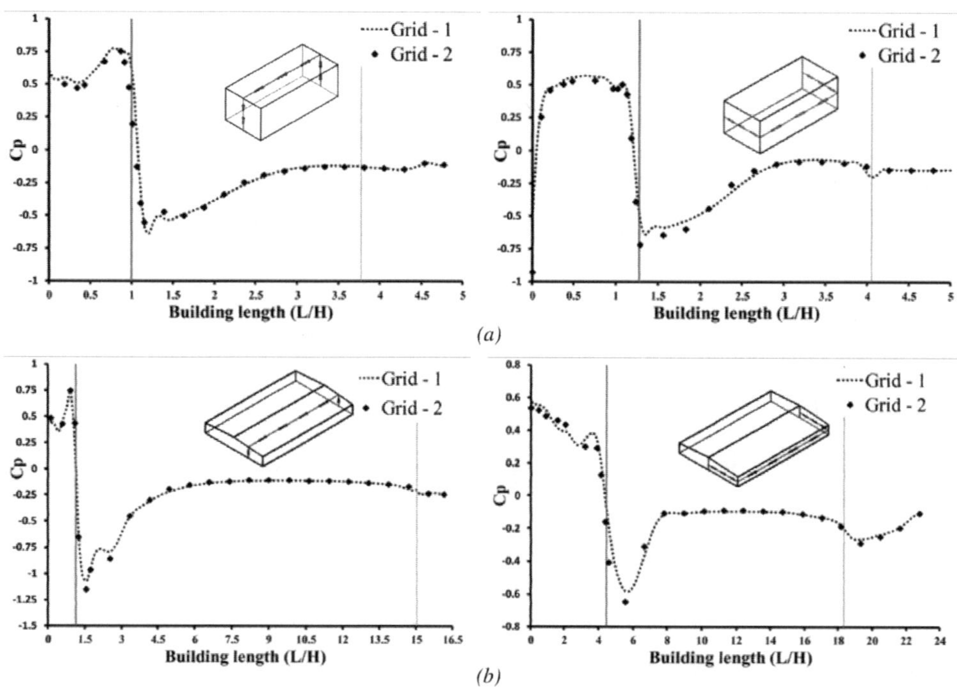

Figure 8. Comparison of pressure coefficient along building perimeter (a) NEES building (b) UWO building

3.2 *Validation with UWO wind tunnel data*

Even though numerical simulation of wind is regarded as an alternative to conventional wind tunnel experiment, the results should always be validated with wind tunnel experiments done on similar or closely matching models. The UWO wind tunnel experiments on low-rise generic buildings will be used to validate the numerical simulations done in this study. In the case of the UWO building model, wind tunnel pressure coefficient recordings on an identical building dimensions exist. However, the NEES building do not match any of the building model dimensions from the UWO wind tunnel experiments. In cases where building model dimensions do not match, scaling of pressure tap coordinates can be used as an alternative approach.

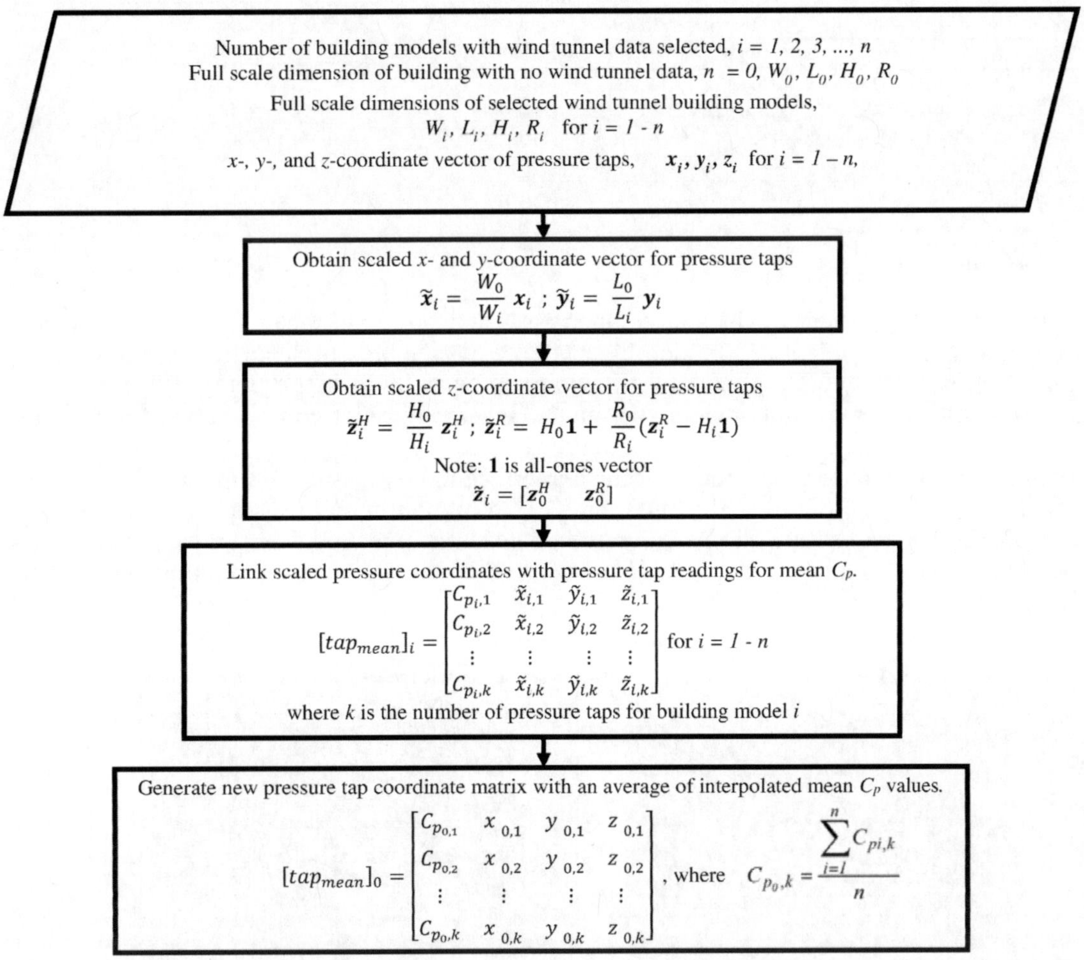

Figure 9. Flow chart for scaled wind pressure generation algorithm

According to Main & Fritz [7], when building model dimensions don't exactly match, interpolated pressure coefficients from several wind tunnel models which closely match and bound the dimensions of a numerical building model, can be used to obtain an enhanced approximation. The DAD approach uses this method in the calculation of pressure coefficients and building responses when no wind tunnel data exists. The flow chart given in Figure 9 shows the procedures that need to be taken to obtain the scaled wind tunnel pressure coefficients.

Four building models from the UWO wind tunnel testing were selected to validate the NEES building numerical wind simulation result. The building dimensions for the selected buildings are given in Table 2. It is assumed that the four building models would bound the dimensions of the NEES building and give a better estimate of wind tunnel pressure coefficients. A MATLAB code is developed to perform the pressure coefficient scaling algorithm shown in Figure 9. The code works for any number of generic low rise gable roof building models.

Table 2. UWO wind tunnel building models used for validation of NEES building numerical C_p results

Sample no.	Plan dimension (m)	(ft)	Height m (ft)	Roof slope	Length scale
1	12.2x19.1	40 x 62.5	5.5 (18)	1:12	1:100
2	15.2x30.5	50 x 100	4.9 (16)	½ : 12	1:100
3	24.4x38.1	80 x 125	5.5 (18)	¼ : 12	1:100
4	30.5x 61	100 x 200	6.1 (20)	½ : 12	1:200

3.2.1 *UWO building*

Prior wind tunnel experiments exist and comparison was made with mean pressure coefficients obtained from an identical building model. Figure 10 shows the pressure coefficient contour plots on the roof of the UWO building model for a wind approach angle of 90^0 obtained from the numerical simulation and the wind tunnel experiment. General distribution of pressure for both numerical and wind tunnel cases show identical trend with peak negative pressure coefficient contour lines converging at opposite windward building corners. Smooth pressure coefficient contour lines are obtained in the case of numerical simulation. Due to limitation on pressure taps and the associated interpolation and extrapolation of pressure tap readings for wind tunnel data, contour lines tend to be irregular. Similar results were also observed on the walls of the building and for the remaining wind directions considered.

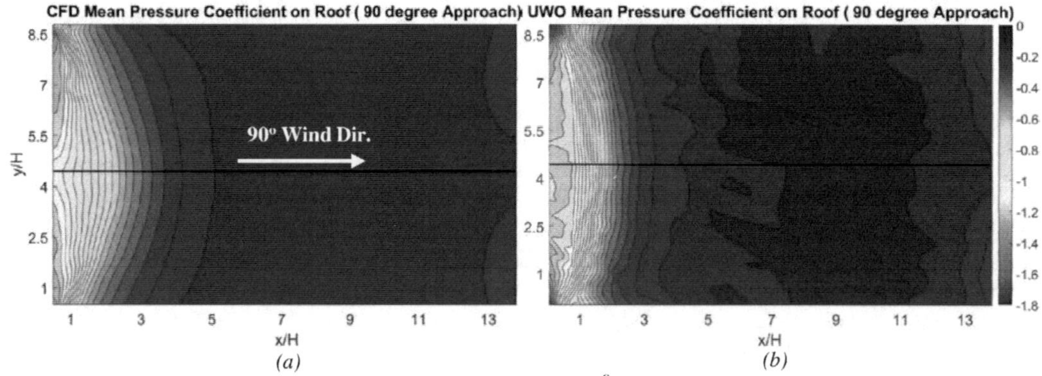

Figure 10. C_p contour plot comparison for UWO building - 90^0 wind direction (a) CFD (b) UWO wind tunnel

3.2.2 *NEES building*

Since no wind tunnel experiments exist that match the NEES building dimensions, numerical wind simulation pressure coefficients for the NEES building are validated using scaled pressure coefficients. Figure 11 shows pressure coefficient contour plot comparisons for windward wall of the NEES building. Identical low and high pressure zones with comparable pressure coefficient magnitudes were observed. The contour lines in the case of numerical simulation are symmetrical along the mid line of the wall. However, there is a loss of symmetry for wind tunnel contour lines. The averaging technique and the scaling implemented on the wind tunnel data clearly has caused this behavior. While being an improved way of checking numerical simulation results, the averaging and scaling approach of wind tunnel pressure coefficients has many uncertainties. These uncertainties arise from the methodology in which wind tunnel recordings are made and from the averaging and scaling techniques implemented.

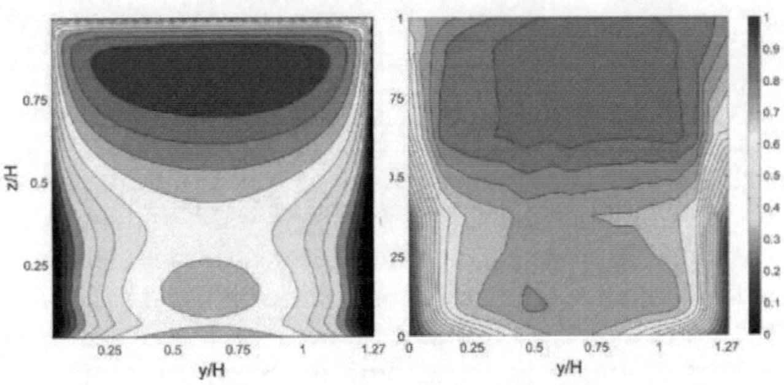

(a) *(b)*

Figure 11. C_p contour plot comparison for NEES building - 90^0 wind direction (a) CFD (b) scaled UWO wind tunnel

On one hand, UWO wind tunnel recordings for pressure coefficient are referenced with a dynamic pressure which is found at a higher height than the roof of the building model. According to Ho [5], high turbulence at roof height level due to building aerodynamics makes measurements at this location highly uncertain. This higher referencing height and dynamic pressure vary for each wind tunnel experiment. Conversion to reference roof height pressure is done by means of a conversion factor which is equal to the ratio of the square of the wind speeds at the two referencing heights. Each building model will also have a different roof height velocity.

On the other hand, the use of linear interpolation and averaging without any probabilistic and statistical input makes the scaling and averaging technique idealistic. Main & Fritz [7] notes the effect of pressure tap separation distance on the correlation between measured pressures at different tap locations. Scaling of pressure taps could either lead to an increase or decrease of separation distance of pressure taps, which affects the quality of the spatial variation of the scaled pressure coefficient readings. Figure 12 illustrates the occurrence of such distortions while scaling pressure tap coordinates. In addition, the method includes data extrapolation for building envelope surfaces outside of the pressure tap grid, such as building edges and corners, which can be very uncertain. Overall the method is a decent way for checking numerical wind simulation results but cannot be assumed to be a reliable way.

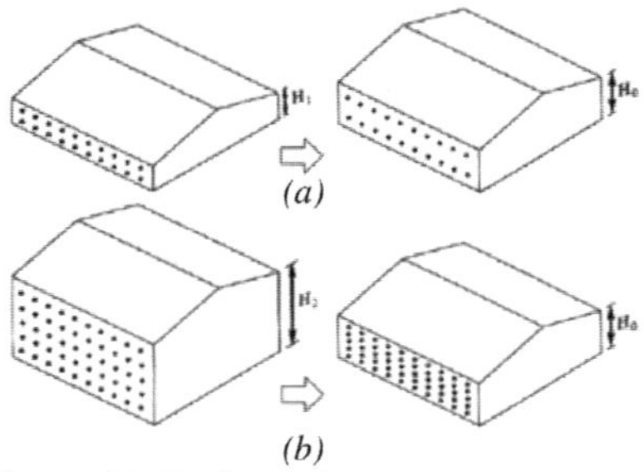

Figure 12. Scaling of pressure tap coordinates (a) Increased separation distance (H_1 to H_0) (b) reduced separation distance (H_2 to H_0) [7]

4 CONCLUSIONS

A computational fluid dynamics wind simulation study was conducted to determine the realistic spatial variation of wind on low and midrise buildings. Focus was given to low and midrise building types that incorporate thin-walled steel structural members. Steady state wind simulations were conducted on two representative building models by considering variable wind approach angles. The CFD results were validated using existing wind tunnel experiments done on identical building models. When wind tunnel experiments that match a given building dimension and shape did not exist, scaled wind tunnel experiment results from closely matching building models were used. An acceptable qualitative agreement was achieved between numerical simulation and wind tunnel experiments. Spatially varying pressure coefficient data was extracted for the various wind approach angles considered which will further be used in the study of the stability and collapse behavior of sheathed wall studs under realistic spatial wind loading. With no restriction imposed on the number of pressure taps that can be fitted on a building model, a well-defined spatial variation of wind loading was achieved.

Future work will be focusing on expanding spatial wind pressure coefficient database by including more wind directional variations and hybrid building types. Research is also being done to add transient pressure coefficient data to the database which will help to examine the full spectra of stability and collapse behavior of thin-walled building cladding systems under spatiotemporal wind loading.

5 REFERENCES

1 T. Peköz and P. Soroushian, Behaviour of C- and Z-purlins under wind uplift, In 6th International Specialty Conference on Cold-Formed Steel Structures, 1982, Missouri University of Science and T echnology.

2 L.M.S. Pierre, G.A. Kopp, D. Surry, and T.C.E. Ho, The UWO contribution to the NIST aerodynamic database for wind loads on low buildings: Part 2. Comparison of data with wind load provisions, Journal of Wind Engineering and Industrial Aerodynamics, (2005) 31-59.

3 T.C.E. Ho, D. Surry, and M. Nywening, NIST/TTU coorperative agreement-windstrom mitigation initiative: Wind tunnel experiments on generic low buildings, 2003, The Boundary Layer Wind Tunnel Laboratory: The University of Western Ontario, London, Ontario, Canada.

4 T.C.E. Ho, D. Surry, and M. Nywening, NIST/TTU cooperative agreement-windstrom mitigation initiative: Further experiments on generic low buildings, 2003, The Boundary Layer Wind Tunnel Laboratory: The University of Western Ontario, London, Ontario, Canada.

5 T.C.E. Ho, D. Surry, D. Morrish, and G.A. Kopp, The UWO contribution to the NIST aerodynamic database for wind loads on low buildings: Part 1. Archiving format and basic aerodynamic data, Journal of Wind Engineering and Industrial Aerodynamics, (2005) 1-30.

6 J.H. Oh, G.A. Kopp, and D.R. Inculet, The UWO contribution to the NIST aerodynamic database for wind loads on low buildings: Part 3. Internal pressures, Journal of Wind Engineering and Industrial Aerodynamics, (2007) 755-779.

7 J.A. Main and W.P. Fritz, Database-assisted design for wind: concepts, software, and examples for rigid and flexible buildings, 2006, U.S. Department of Commerce, Technology Administration, National Institute of Standards and Technology, Gaithersburg, MD.

8 R.L. Madsen, N. Nakata, and B.W. Schafer, CFS-NEES building structural design narrative, In CFS-NEES - RR01, 2011.

9 STARCCM+, STARCCM+ solver, 2015, CD-adapco, Detroit, MI.

10 H. Aboshosha, A. Elshaer, G.T. Bitsuamlak, and A. El Damatty, Consistent inflow turbulence generator for LES evaluation of wind-induced responses for tall buildings, Journal of Wind Engineering and Industrial Aerodynamics, (2015) 198-216.

11 ASCE7-10, Minimum design loads for buildings and other structures: C26 - Wind loads, American Society of Engineers, Reston, Virginia, 2010.

12 Engineering Sceinces Data Unit (ESDU), Strong winds in the atmosphere boundary layer. Part 2: Discrete gust speeds. Data Item 83045, 1983.

13 Engineering Science Data Unit ESDU, Characteristics of atmospheric turbulence near the ground. Data Item 74031, 1974.

14 Engineering Sciences Data Unit (ESDU), Strong winds in the atmosphere boundary layer. Part 1:Mean-hourly wind speeds. Data Item 82026, 1982.

15 Engineering Sceinces Data Unit (ESDU), Characteristics of atmospheric turbulence near the ground. Part 2: Single point data for strong winds (neutral atmosphere). Data item 85020, 1985.

Influence of the vertical angle of attack on building pressures using Quasi-Steady Theory

Antonio Zaldivar de Alba [a], Franklin T. Lombardo [b]

[a]*University of Illinois at Urbana Champaign, 205 North Mathews Ave, Urbana, IL, USA, zldvrdl2@illinois.edu*
[b]*University of Illinois at Urbana Champaign, 205 North Mathews Ave, Urbana, IL, USA, lombaf@illinois.edu*

ABSTRACT: A preliminary quasi-steady model was built to simulate building pressures using quasi-steady theory with the goal of estimating loads during tornadoes and thunderstorms. This model uses a mean pressure coefficient versus horizontal angle of attack relationship obtained by using more than 600 full-scale records of the Wind Engineering Research Field Laboratory (WERFL) at Texas Tech University. After the model was built, it was compared with full-scale WERFL data to test its accuracy. Pressure taps on the roof were found to be the worst performing based on a root mean squared error. Also, a first assessment of the effects of the vertical angle of attack on building pressures during tornadoes was performed using information from previous publications. Finally, a layout of the future development of the quasi-steady model is presented highlighting the importance of the vertical angle of attack on building pressures. It is expected that, after full quasi-steady model is built, it can be used to further understand the effects of the vertical wind component prevalent in tornadoes and thunderstorms.

KEYWORDS: Tornadoes, thunderstorms, quasi-steady theory, low rise, vertical angle of attack.

1 INTRODUCTION

Damage to infrastructure attributable to wind accounts for a large amount of the losses due to natural disasters in U.S. For example, "in terms of property losses covered by insurance, 14 of the 15 costliest natural disasters in U.S. history were due to windstorms, 12 hurricanes and 2 tornado outbreaks" (NIST [1]). Moreover, tornadoes and thunderstorms accounted for 7 out of the 15 natural disasters that surpassed the 1 billion loss threshold in 2016 (NOAA [2]). Furthermore, as Amini and Van de Lindt [3] mentioned, the costs of damage caused by tornadoes and thunderstorms are continuously increasing. Kuligowski et al. [4] reported that critical facilities and high–occupancy buildings were not able to provide life safety protection to occupants during the 2011 Joplin tornado. Furthermore FEMA [5] stated that "well designed, constructed, and maintained critical facilities should experience little if any damage from weak tornadoes, except for window breakage. However, weak tornadoes often cause building envelope damage because many critical facilities have wind resistance deficiencies." This globally escalating loss of property during extreme winds has generated an increased interest in wind engineering (Masters [6]). Also, FEMA [5] reported that tornadoes are responsible for the greatest number of wind-related deaths each year in the United States. Therefore, it is crucial to create, and include in codes and standards, a tornado based design framework providing the tools to engineers to design safe and reliable structures against extreme winds generated by tornadoes and thunderstorms. As an illustration of the nature of damage during tornadoes, a picture taken by the Wind Engineering Group at Illinois tornado damage survey at Naplate, IL is shown in Figure 1.

Figure 1. Tornado Damage at Naplate, IL (Tornado rated EF-3 by NWS [7])

Today, thunderstorms and tornadoes are wind events with the largest uncertainties in wind engineering when designing structures, such as wind speed and wind-building interaction. First, there is limited data near the ground. Second, these phenomena have rapid changes in wind speed and direction, which complicates their simulation in wind tunnels, commonly used to develop codes and standards. These, together with the general belief that designing structures against tornadoes is impractical, has resulted in a lack of codes and standards to design structures against extreme winds generated by tornadoes and thunderstorms. Moreover, Lombardo et al. [8] mentioned that "thunderstorm wind flow also shows physical differences from the stationary boundary layer that may affect wind characteristics and thus wind-induced pressures". This represents a significant issue for codes and standards, designing structures against winds generated by tornadoes and thunderstorms is not as simple as increasing the wind speeds for design, tornado like winds behave in a different way than straight line winds which implies a new philosophy of design.

Moreover, roof pressures are known to be the most critical components in low-rise buildings, where wind induced suction cause roofs to fail under extreme winds and, as Kuligowski et al. [4] and Amini and Van de Lindt [3] mentioned, most of the damage during tornadoes is due to roof loss and failures due to missing vertical load paths are common. Furthermore, Case et al. [9] performed an experiment at the Iowa State University tornado simulator using different models of residential structures and found, without exception, that the vertical force coefficient was greater than either of the horizontal force coefficients and that the peak pressures occurred on the roof for all cases. Therefore, roof pressures are crucial for the estimation of tornado induced loads.

It is important to note that most of tornadoes are rated EF-2 or lower (NOAA [10]), this fact, as Amini and Van de Lindt [3] mentioned, has lead researchers consider that designing structures against extreme winds generated by tornadoes may not be as impractical as originally thought. Regarding this, there are several studies that have investigated tornado induced forces on structures. Selvam and Millett [11] used computer models to conclude that, for the same reference speed, tornadoes result in twice the suction force on the roof than straight line winds. Sengupta et. al. [12] simulated tornado forces on a cubic building using the tornado simulator at Iowa State University and concluded that force coefficients for tornadoes are 1.5 times more than the values specified in ASCE 7-05. Haan et al. [13], using the same simulator, found that the uplift coefficients for tornadoes exceeded the values specified in the ASCE 7-05 provisions by factors between 1.8 and 3.2. Hu et al. [14] found wind loads acting on the gable-roof building model in tornado-like winds were to be at least 3 times higher compared with those in a straight-line. In general, these studies have shown that higher suctions are generated in tornado like winds than in straight line boundary layer winds for the same reference speed.

There are many factors that may be influencing the higher induced forces during tornadoes. One of these factors is the vertical angle of attack which is the focus of this paper and its importance will be further discussed in the next section. Other factors include large pressure deficit, rotational flow and wind speed magnitude, among others.

1.1 *Importance of the vertical angle of attack*

As mentioned earlier, roof pressures are known to be the most critical components in low-rise buildings. More specifically, Banks and Meroney [15], Wu [16] and Marwood and Wood [17] clearly indicate that the highest suctions generated on the roof are beneath the vortex core of conical roof vortices. Marwood and Wood [17] also found that the instantaneous position of a conical roof edge vortex varies significantly and fluctuation in the roof pressure is closely linked to this variation. Wu [16] provided more insight on this variation and showed that the conical vortices are influenced by the wind horizontal and vertical angles of attack, and found that the vertical angle generally modifies the configuration and movement of the vortices. Also, Wu et al. [18] mentioned that sporadically large and fast excursions of the vertical angle of attack can generate extremely high instantaneous pressure coefficients. These large and fast excursions of the vertical angle of attack appeared in the tornado record presented by Lombardo [19].

The vertical angle of attack was often neglected when computing building pressures because high wind speeds are associated with downward wind in the atmospheric boundary layer (Wu and Kopp [20]). Ozmen et al. [21] showed the sensitivity of roof pressure to vertical wind angle of attack by analyzing 3 structures with different pitch angles, finding the largest suctions on the model with smallest pitch angle. In other words, a positive vertical angle of attack (upward wind) will generate higher suctions and a negative one (downward wind) lower. Tornadoes can have significant upwardly directed winds and thus significant increase on pressures as mentioned by Wu and Kopp [20] and observed in the tornado record presented by Lombardo [19].

Selvam and Millett [22] identified the importance of the vertical velocity and mentioned that is necessary to consider a computer model including this effect for a better understanding of tornado induced loads. Moreover, the effect of the wind vertical angle of attack has been recognized as an important one for peak pressures generation on the roof by Richards and Hoxey [23] and Wu [18] therefore, it is necessary to retain its effect for peak pressure determination, however this determination has been an extremely difficult challenge as mentioned by Richards and Hoxey [23]. Although difficult, many researchers have found a better performance of the quasi-steady theory by incorporating in some manner the effect of the vertical angle of attack Letchford and Marwood [24], Wu [16] and Wu and Kopp [20].

Based on the crucial importance of the vertical angle of attack on roof pressures, it is important to investigate its influence as a step forward to create codes and standards that can be used to design safe and reliable structures against extreme winds generated by tornadoes and thunderstorms.

As mentioned earlier, a possible step toward is through quasi-steady theory, an overview of the theory will be presented next in Section 2.

2 OVERVIEW OF THE QUASI-STEADY THEORY

Quasi-steady theory assumes pressure fluctuations follow the variations in velocity. In other words, the highest pressures will occur when the highest speeds are impacting the structure. This simple model has been shown to behave well when predicting average pressures in Letchford et al. [25] and Richards and Hoxey [23]. Also, quasi-steady theory is used in most codes and standards for its simplicity and its ability to relate pressures with peak wind gusts (Holmes [26]).

The full instantaneous equation of the quasi-steady theory is given by:

$$P(t) = \frac{1}{2} pV(t)^2 C_p(\theta, \beta) \qquad (1)$$

where $P(t)$ = pressure at time t; p = air density; $V(t)$ = wind speed at time t; C_p = instantaneous pressure coefficient; θ = horizontal angle of attack; and β = vertical angle of attack.

In Equation (1) the fluctuations in pressure due to building generated turbulence is embedded in the pressure coefficient term that depends on the horizontal and vertical angles of attack.

A challenge when using the quasi-steady theory is its difficulty to predict pressures resulting from wind-structure interaction Gurley et al. [27]. Also, a well-known problem of the quasi-steady theory is the underestimation of peak pressures Letchford et al. [25], Banks and Meroney [15], among others. However, Richards and Hoxey [23] mentioned that the is the simplification of the quasi steady theory the responsible for generating these errors. The commonly made simplifications when using the quasi-steady theory are, the linearization of the horizontal angle of attack-pressure coefficient relationship, the use of the mean pressure coefficient instead of the instantaneous pressure coefficient, and ignoring the effects of the vertical angle of attack Richards and Hoxey [23].

As a first step for generating the full quasi-steady model a simplified equation that ignores the effects of the vertical angle of attack and uses a mean pressure coefficient will be used:

$$P(t) = \frac{1}{2} pV(t)^2 \overline{C_p}(\theta) \qquad (2)$$

where $\overline{C_p}$= mean pressure coefficient; and $P(t)$, p, $V(t)$ and θ where defined in Equation (1).

The development of the quasi-steady model using this equation will be discussed next in Section 3.

3 DEVELOPMENT OF THE QUASI-STEADY MODEL

A model was built to simulate building pressures during extreme winds using quasi-steady theory based on WERFL data. This model is the preliminary version, future development of the model is discussed in Section 5.

3.1 *Overview of the WERFL building at TTU*

The WERFL building is a 30ft wide, 45ft long and 13ft tall building now located at Reese Technology Center at Texas Tech University. The data used to build the model was collected when the building was located at 4th Street. This low-rise building is considered a landmark for wind engineering research throughout the world. The model developed in this work is shown in Figure 2 with the pressure taps in red. A layout of the WERFL building with the identification number for the 204 pressure taps on the four walls and on the roof, is shown in Figure 3.

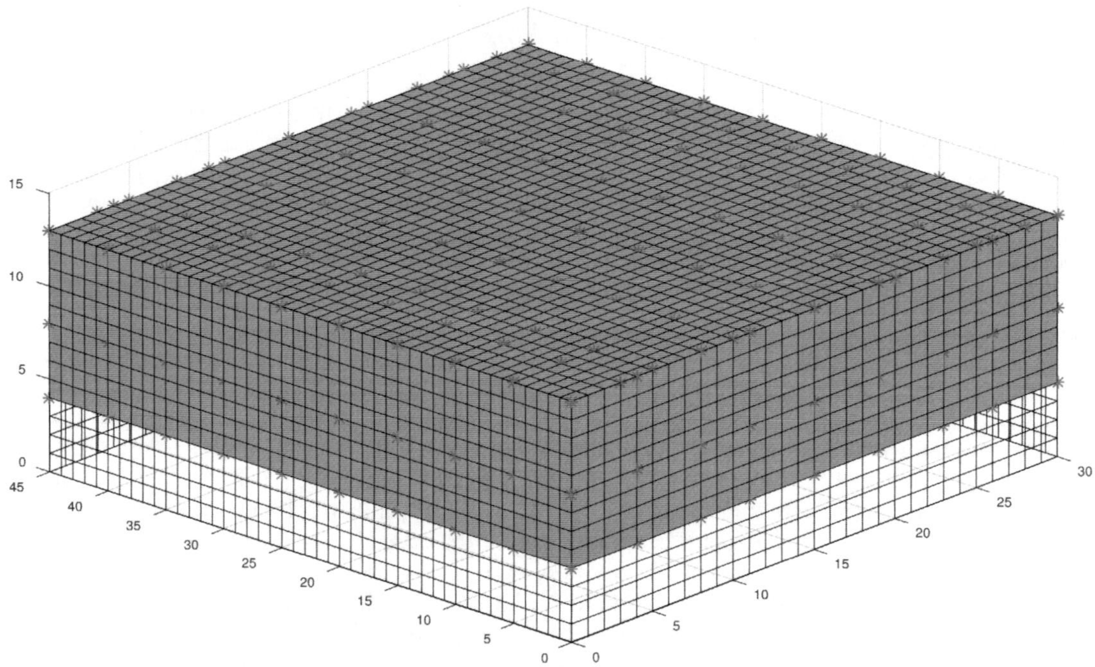

Figure 2. Model of the WERFL building with pressure taps (red)

On the 4th Street site there was also a 160ft tower, measuring wind speeds at several heights (8, 13, 33, 70 and 160 feet), and a sonic anemometer placed above the roof with a height of 30ft. The tower was located approximately 150 feet west of the center of the building (Lombardo [28]). More details of the WERFL building and site can be found in Lombardo [28] and Wu [18].

3.2 *Pressure coefficient vs mean horizontal angle of attack relationship*

To build the pressure coefficient vs mean horizontal angle of attack relationship, data from over 600 15 min records sampled at 30 Hz were used. For every 15 min record a mean pressure coefficient and mean horizontal angle of attack were calculated and, by taking advantage of symmetry, a more robust relationship was built for 55 taps. These was performed previously by Lombardo [28] and the data used here was already arranged in 55 taps mean pressure coefficient vs horizontal angle of attack. Also, the data was smoothed by using a loess smoothing technique in MATLAB as performed by Lombardo [28]. It is important to note that the horizontal angle of attack in the data was previously corrected for building rotation by Lombardo [28]. The data was fitted into a Fourier fit as recommended by Richards and Hoxey [23] to avoid errors generated by the linearization of this relationship. The Fourier fit chosen was the one with the least order having a minimum R^2 value of 0.95. An example for a single tap of the data and the Fourier fitting is shown in Figure 4. The general form of the Fourier fits used is given by:

$$\overline{C_p}(\bar{\theta}) = \sum a_0 + a_n \cos(w\theta) + b_n \sin(w\theta) \quad (3)$$

where $\overline{C_p}$ = mean pressure coefficient; $\bar{\theta}$ = mean horizontal angle of attack; a, b and w = coefficients of the Fourier series; and n = the order of the Fourier series.

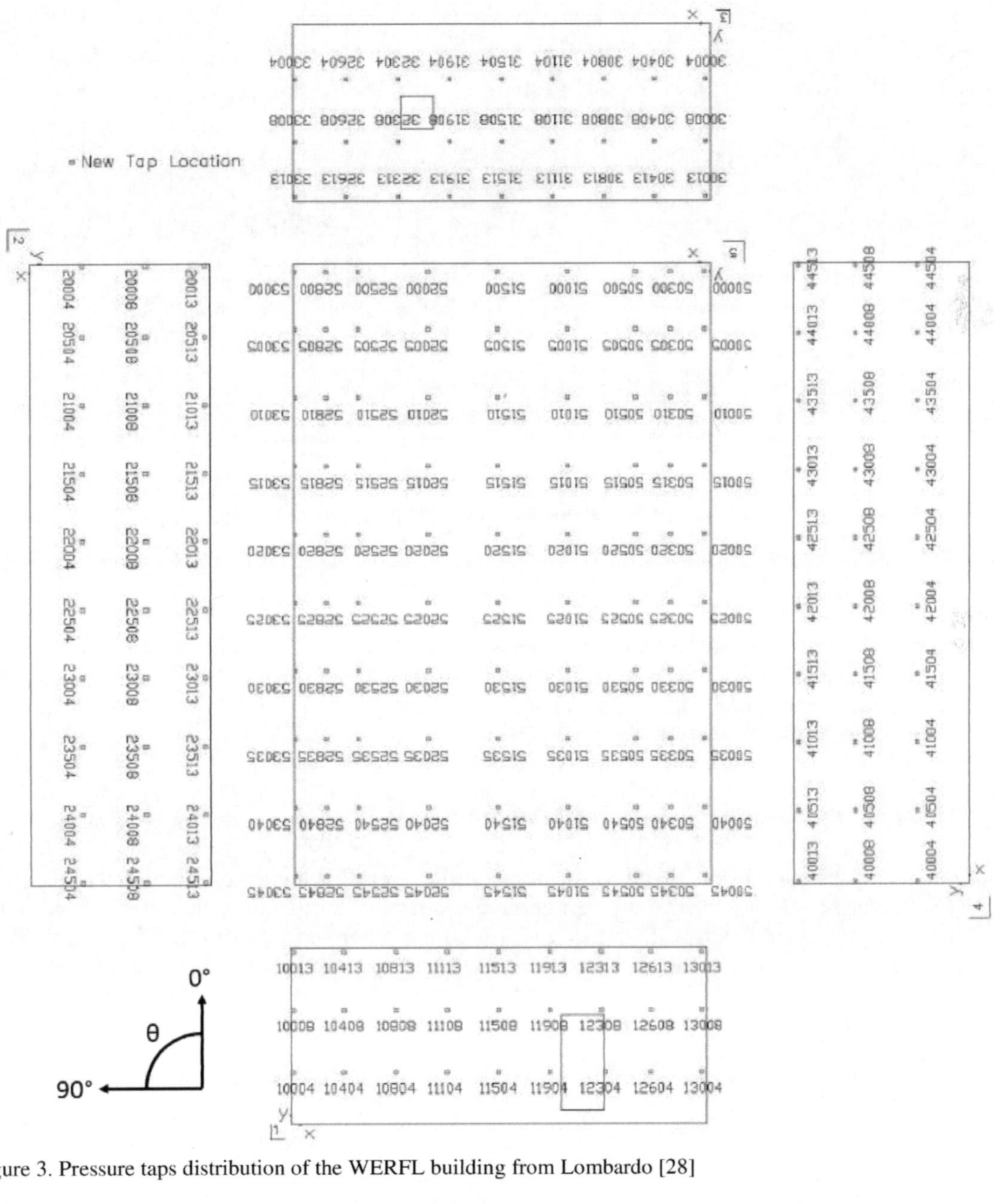

Figure 3. Pressure taps distribution of the WERFL building from Lombardo [28]

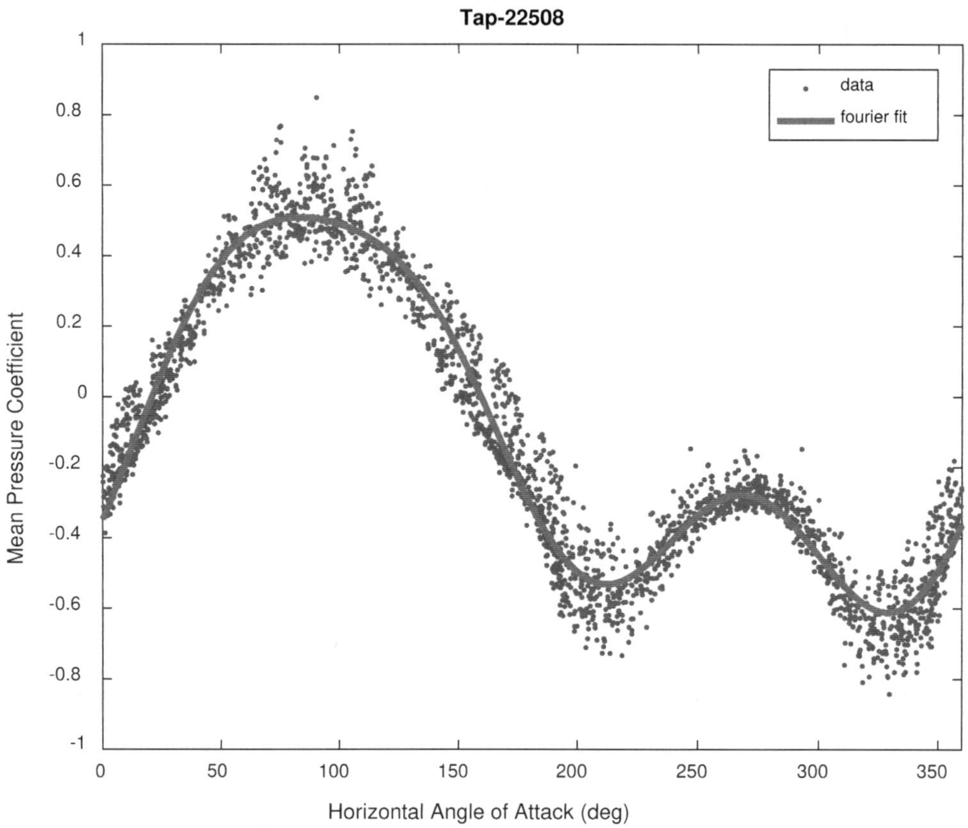

Figure 4. C_p vs Haoa for Tap 22508 (Reflected Taps 42008, 22008, 42508) with 3rd order Fourier Series

The first preliminary model uses the Fourier fits (Eq. 3) of the 55 pressure taps to obtain the pressures using equation 2. Future development of this model will include an instantaneous pressure coefficient vs horizontal angle of attack and the influence of the vertical angle of attack, this will be discussed further in section 5.

4 CALIBRATION OF THE QUASI-STEADY MODEL

The first step of the calibration check was looking at the overall behavior of the model and evaluating if it performed as expected for given angles of attack, this meaning, that positive pressure was obtained on the windward wall and suction on all the other surfaces. For the visualization of pressure coefficients, a cubic interpolation was used and the results of a test run are shown in Figure 5. This basic sanity check was performed for various angles of attack corroborating that the expected behavior was achieved for all cases.

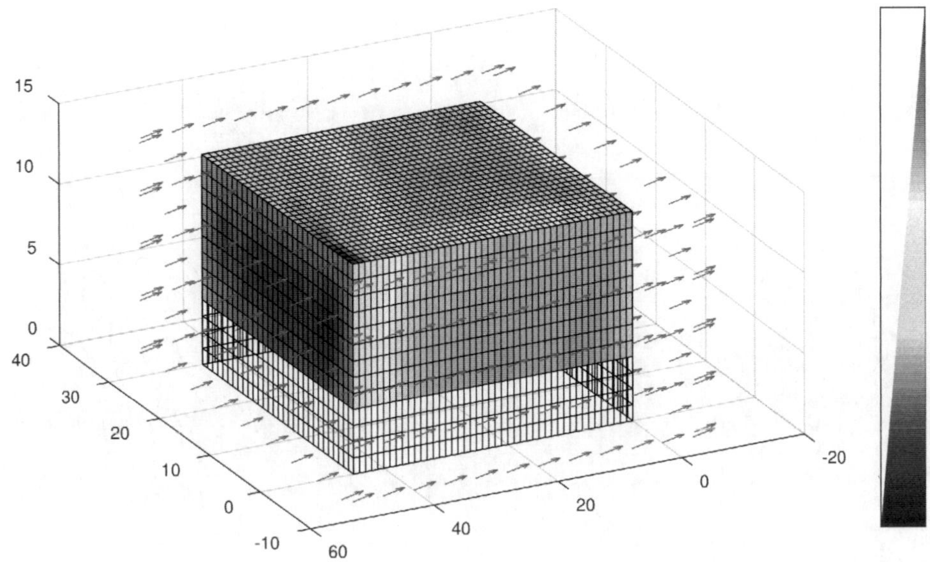

Figure 5. Pressure coefficient results for a run test with 167° angle of attack.

After this basic calibration check a comparison was made with 124 full-scale 15 min records at the WERFL building, the quasi-steady model was run for these records and the root mean squared error $(RMS)_{error}$ was computed for every tap and for each record using the following equation:

$$RMS_{error} = \sqrt{\frac{\sum(P_2 - P_1)^2}{N}} \tag{4}$$

where P_1 = quasi-steady model pressure; P_2 = measured pressure; and N = number of data points.

Once the $(RMS)_{error}$ was computed for the 124 records and the 204 pressure taps, a summation of the errors was performed, the pressure taps with the highest sum were considered the worst performers of the model. The 10 pressure taps with the worst performance are: 50505, 53045, 52845, 52000, 50345, 50045, 53020, 53040, 50040 and 52545. It is important to note that these pressure taps are all located on the roof, where the highest pressures occur. More specifically the worst performing taps are in the roof corners where conical vortex formation is expected to occur for favoring angles of attack. Moreover, it is also on the roof where the influence of the vertical angle of attack on pressures is the greatest.

Another part of the calibration check was comparing the overall behavior of the measured pressures versus the ones obtained by the quasi-steady model. By inspection, it is concluded that the quasi-steady model is performing as expected, it follows the variation in pressure relatively close but it is not able to predict peak pressures. An example of the plots analyzed is shown in Figure 6. In this figure, a good comparison between the quasi-steady model and the measured pressures is observed, with the quasi-steady theory generally underpredicting measured pressures as observed in the residual of Figure 6, but following the patterns observed in the measurement. Furthermore, it is noted that the quasi-steady model is not able to predict peak pressures as shown clearly on the roof pressure tap of Figure 6 were the peak pressure ≈ -700 Pa is under estimated by ≈ 50%.

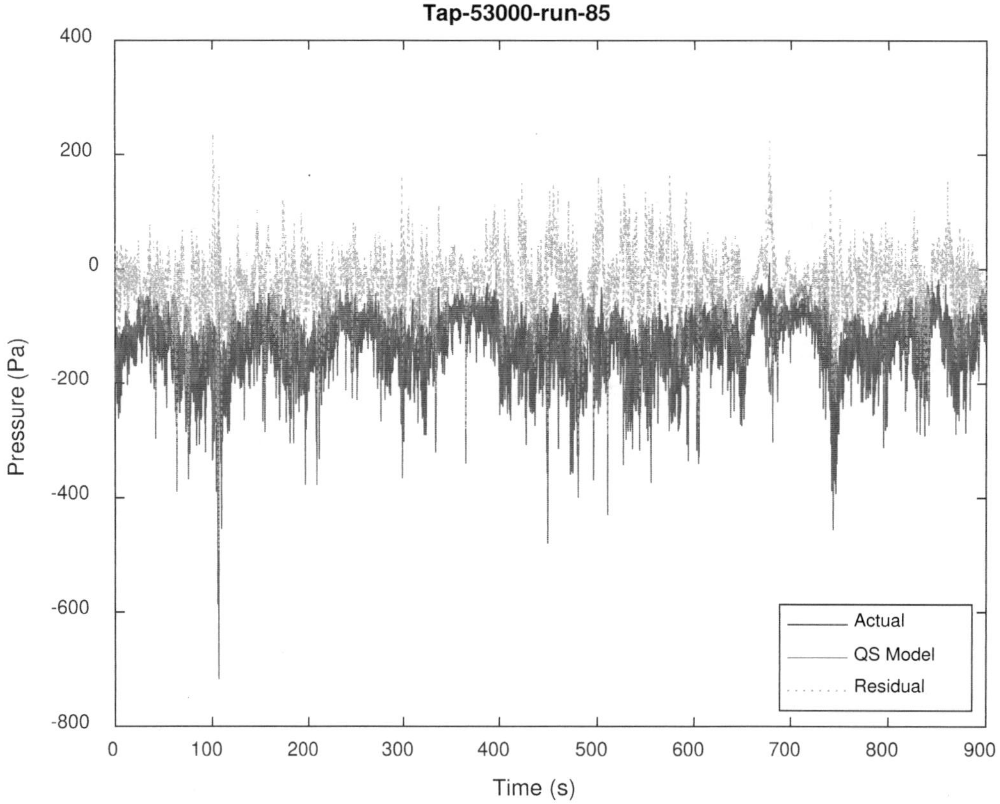

Figure 6. Measured pressure and quasi-steady model pressure for pressure tap 53000

Also, plots of the $(RMS)_{error}$ versus the mean horizontal angle of attack were done for every pressure tap. This will help to analyze the angles of attack for which the pressure tap performs the worst. An example of this type of plot is shown in Figure 7. In this case for the wall pressure tap 31508 (wall of interest shown in red in Figure 7) the angles of attack for which it performs the worst are between 45° to 100°, these angles of attack place the pressure tap on an air flow separation region in a sidewall, which has been recognized as one of the problems of quasi-steady theory when predicting pressures.

The preliminary quasi-steady model showed problems to predict peak pressures and bad performance on roof edges, where the influence of the vertical angle of attack is the greatest. As mentioned previously, an improvement of the quasi-steady theory was obtained by Letchford and Marwood [24], Wu and Kopp [20] and Wu [16] when including, in some manner, the effect of the vertical angle of attack. Moreover, the vertical wind component during tornadoes is significantly greater than in straight line events as shown in the tornado record presented by Lombardo [19], where also high wind speeds are related with high values of the wind vertical component. Thus, adding the effects of the vertical angle of attack to the quasi-steady model is a crucial step forward to obtain an accurate prediction of wind loads generated by tornadoes. Preliminary results of the estimation of tornado pressure coefficients by using the influence of the vertical angle of attack will be discussed next in Section 5.

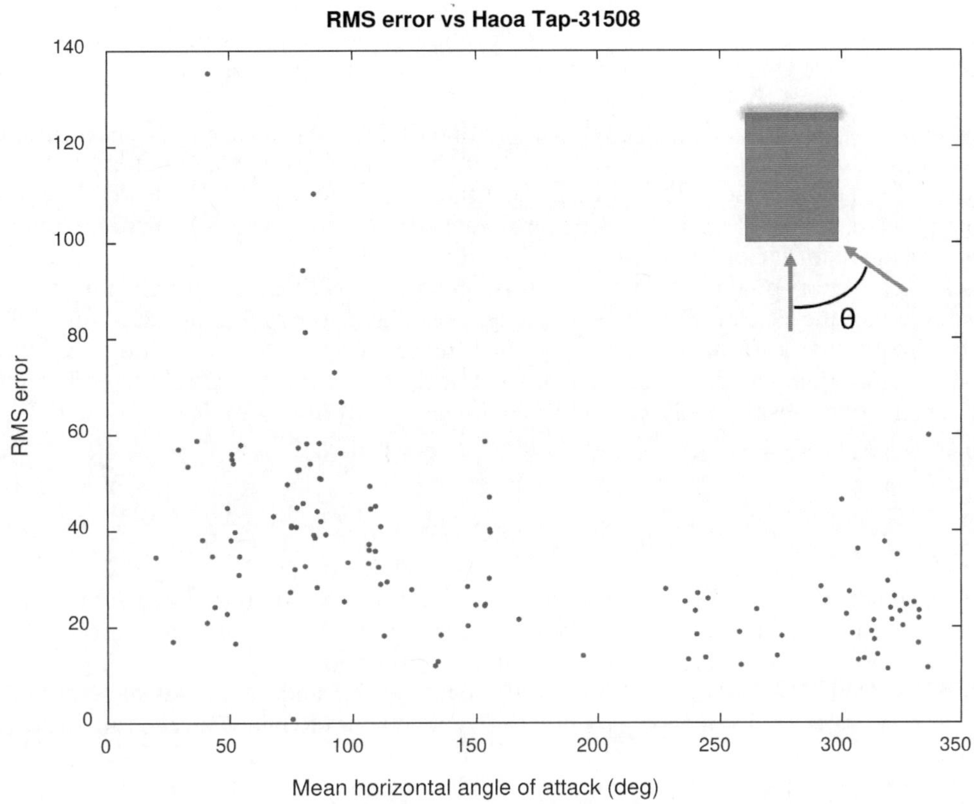

Figure 7. $(RMS)_{error}$ versus mean horizontal angle of attack for wall pressure tap 31508

5 PRELIMINARY RESULTS OF THE INFLUENCE OF THE VERTICAL ANGLE OF ATTACK ON BUILDING PRESSURES

As a first assessment of the influence of the vertical angle of attack on building pressures during tornadoes, the tornado record presented by Lombardo [19] was used. In this tornado record, values of vertical angle of attack as high as 30° related with high wind speeds are encountered. This value of 30° was used to assess the effects of tornado winds on roof pressures by using quasi-steady models that account for the effect of the vertical angle of attack from three previous publications, Letchford and Marwood [24] Richards and Hoxey [23] and Wu and Kopp [20]. These values are compared to pressure coefficients calculated using quasi-steady theory without effects of vertical angle of attack.

Letchford and Marwood [24] found values of mean pressure coefficient for pitch angles ranging from -10 to 10 degrees. Using the pressure tap located closest to the edge and averaging the slope from the lines generated in the positive side of pitch angle a derivative of pressure coefficient to vertical angle of attack was obtained. Using this derivative and the 30° vertical angle of attack from the tornado record presented by Lombardo [28] an approximate increase of 145% on pressure coefficient was found.

Richards and Hoxey [23] obtained a derivative of pressure coefficient to vertical angle of attack of 0.077 times mean pressure coefficient with no elevation angle for a pressure tap located close to the middle of the roof. Using the same procedure as for the previous case an approximate increase of 230% on pressure coefficient was found.

Wu and Kopp [20] used a Fourier fit to build a function of the derivative of pressure coefficient to vertical angle of attack dependent on the horizontal angle of attack using wind tunnel data. Using this information from a tap close to the roof edge and the model presented by Wu and Kopp [20], an approximate increase of 60% on pressure coefficient was found for the 30° vertical angle of attack in the tornado record (Lombardo [19]).

Significant increase on pressure coefficient was obtained for all cases, reaffirming the importance of including the effects of the vertical angle of attack to estimate tornado loads. It is important to mention that a linear relationship of pressure coefficient to vertical angle of attack was used for this preliminary assessment. This may not be the case for high values of vertical angle of attack where the relationship could be highly nonlinear as mentioned by Richards and Hoxey [23].

6 FUTURE IMPROVEMENT OF THE QUASI-STEADY MODEL

The preliminary model developed will serve as a basis for a more robust quasi-steady model of the WERFL building. The future model will use a new pressure coefficient-angle of attack relationship considering both horizontal and vertical angles of attack. This relationship will be built using full-scale data from the WERFL building, future wind tunnel studies and data available from other publications as Wu [16], Letchford and Marwood [24], and Wu and Kopp [20]. The full-scale data available contains information about the vertical angle of attack measured by the sonic anemometer located above the roof, therefore, a correction considering time lag will be needed for the wind speed used as a reference to calculate pressure coefficients. Moreover, the new pressure coefficient vs angle of attack relationship will include the effects of the vertical angle of attack only on the roof, this due to the insensitivity of wall pressures to elevation angles Richards and Hoxey [23].

Another improvement of the quasi-steady model will be the use of instantaneous pressure coefficients. This will be achieved by using WERFL full-scale data making the corresponding corrections due to time lag between the tower measuring wind speed and the WERFL building. To obtain the time lags, two cross correlations will be made; the first one between the wind speed measured by the tower at 13ft, described previously in Section 3.1, and a windward pressure tap; and the second one between the wind speed measured by the sonic anemometer at 30ft and the one measured by the tower at 33ft, both described previously in Section 3.1.

Other important factor considered for future development of the model is the large pressure deficit occurring in tornadoes reported by Karstens [29] and noted in the laboratory by Haan et al. [13]. This large pressure deficit could cause internal pressures to develop and further increase the suction loads on the roof. Furthermore, internal pressures could be also generated due to breach, a model considering these phenomena, large pressure deficit and breach, would be more accurate.

The future improvements of the quasi-steady model will help to reduce the errors caused by the simplification of the quasi-steady theory as mentioned by Richards and Hoxey [23], it is expected that by having a full quasi-steady model the results will significantly improve. Moreover, a full model considering the vertical angle of attack can be used with actual tornado and thunderstorm data to further understand the effects of the vertical wind component on high pressures in tornadoes and thunderstorms.

Furthermore, more data to validate and calibrate the quasi-steady model will be available from the on-development SWERF facility, see Nevill and Lombardo [30] for more details on the capabilities and progress of this full-scale facility at the University of Illinois at Urbana Champaign.

7 CONCLUSIONS

The quasi-steady theory has demonstrated to be a powerful tool to predict building pressures. In this work a preliminary quasi-steady model using a mean pressure coefficient versus horizontal angle of attack relationship was developed. The quasi-steady model performed the worst for pressure taps located on the roof and it underestimated the peak pressures. Furthermore, a preliminary assessment of the effects of the vertical angle of attack on building pressures during tornadoes was performed by using information from previous publications, finding a significant increase on pressure coefficient. Additionally, a proposed future development of the model is outlined, making a special emphasis on the importance of the effects of the vertical angle of attack on building pressures. This future development of the quasi-steady model is a promising path to further understand the effects of the vertical wind component on high pressures in tornadoes and thunderstorms and possibly explain higher pressures obtained by several studies in tornado-like winds.

8 ACKNOWLEDGEMENTS

Wind Engineering Research Field Laboratory at Texas Tech University.

9 REFERENCES

1 National Institute of Standards and Technologies. (NIST). (2017). "Federal Program to Reduce Windstorm Impacts Requests Comments on New Strategic Plan" ⟨https://www.nist.gov/news-events/news/2017/03/federal-program-reduce-windstorm-impacts-requests-comments-new-strategic⟩ (April 23, 2017).

2 National Oceanic and Atmospheric Administration. (NOAA). (2017). "Billion-Dollar Weather and Climate Disasters: Table of Events" ⟨https://www.ncdc.noaa.gov/billions/events/US/1980-2017⟩ (April 23, 2017).

3 M.O. Amini and J.W. Van de Lindt, Quantitative Insight into Rational Tornado Design Wind Speeds for Residential Wood-Frame Structures Using Fragility Approach, Journal of Structural Engineering, 140.7 (2014): 04014033.

4 E.D. Kuligowski, F.T. Lombardo, L.T. Phan, M.L. Levitan, and D.P. Jorgensen. Final Report, National Institute of Standards and Technology (NIST): Technical Investigation of the May 22, 2011, Tornado in Joplin, Missouri. Gaithersburg, MD: U.S. Dept. of Commerce, National Institute of Standards and Technology, 2014. Print.

5 Design Guide for Improving Critical Facility Safety from Flooding and High Winds: Providing Protection to People and Buildings. Washington, D.C.: U.S. Dept. of Homeland Security, FEMA, 2007. Print.

6 F.J. Masters. Grand Challenges in Wind Engineering: Advancing Cyber-Physical Tools to Investigate Structural Performance of Buildings. Frontiers in Built Environment 2 (2016)

7 National Weather Service. (NWS). (2017). "February 28, 2017: Tornado Event (Updated March 8)" ⟨http://www.weather.gov/lot/2017Feb28_tornadoes⟩ (April 23, 2017).

8 F.T. Lombardo, D.A. Smith, J.L. Schroeder, and K.C. Mehta. Thunderstorm characteristics of importance to wind engineering. Journal of Wind Engineering and Industrial Aerodynamics 125 (2014): 121-32.

9 J. Case, P. Sarkar, and S. Sritharan. Effect of low-rise building geometry on tornado-induced loads. Journal of Wind Engineering and Industrial Aerodynamics 133 (2014): 124-34.

10 National Oceanic and Atmospheric Administration. (NOAA). (2017). "U.S. Tornado Climatology." ⟨https://www.ncdc.noaa.gov/climate-information/extreme-events/us-tornado-climatology⟩ (March 3, 2017).

11 R.P. Selvam and P.C. Millett. Computer modeling of tornado forces on buildings. Wind and Structures 6.3 (2003): 209-20. ,

12 A. Sengupta, F.L. Haan, P.P. Sarkar, and V.K. Balaramudu. Transient loads on buildings in microburst and tornado winds. Journal of Wind Engineering and Industrial Aerodynamics 96.10-11 (2008): 2173-187.

13 F.L. Haan, V. Balaramudu, and P.P. Sarkar. Tornado-Induced Wind Loads on a Low-Rise Building. Journal of Structural Engineering J. Struct. Eng. 136.1 (2010): 106-16.
14 H. Hu, Z. Yang, P. Sarkar, and F. Haan. Characterization of the wind loads and flow fields around a gable-roof building model in tornado-like winds. Experiments in Fluids 51.3 (2011): 835-51.
15 D. Banks, and R.N. Meroney. The applicability of quasi-steady theory to pressure statistics beneath roof-top vortices. Journal of Wind Engineering and Industrial Aerodynamics 89.6 (2001): 569-98.
16 F. Wu. Full-scale study of conical vortices and their effects on roof corners, Ph.D. Dissertation, Texas Tech University, 2001.
17 R. Marwood, and C.J. Wood. Conical Vortex Movement and Its Effect on Roof Pressures. Journal of Wind Engineering and Industrial Aerodynamics 69-71 (1997): 589-95.
18 F. Wu, P.P. Sarkar, K.C. Mehta, and Z. Zhao. Influence of Incident Wind Turbulence on Pressure Fluctuations near Flat-roof Corners. Journal of Wind Engineering and Industrial Aerodynamics 89.5 (2001): 403-20.
19 F.T. Lombardo. Engineering Analysis of a Full-Scale, High-Resolution Tornado Wind Record. Journal of Structural Engineering (Submitted).
20 C.H. Wu, and G.A. Kopp. Estimation of Wind-Induced Pressures on a Low-Rise Building Using Quasi-Steady Theory. Frontiers in Built Environment 2 (2016)
21 Y. Ozmen, E. Baydar, and J.P.A.J. Van Beeck. Wind Flow over the Low-rise Building Models with Gabled Roofs Having Different Pitch Angles. Building and Environment 95 (2016): 63-74.
22 R.P. Selvam, and P.C. Millett. Computer modeling of tornado forces on buildings. Wind and Structures 6.3 (2003): 209-20.
23 P.J. Richards, and R.P. Hoxey. Quasi-steady Theory and Point Pressures on a Cubic Building. Journal of Wind Engineering and Industrial Aerodynamics 92.14-15 (2004): 1173-190.
24 C.W. Letchford, and R. Marwood. On the Influence of v and w Component Turbulence on Roof Pressures beneath Conical Vortices. Journal of Wind Engineering and Industrial Aerodynamics 69-71 (1997): 567-77.
25 C.W. Letchford, R.E. Iverson, and J.R. Mcdonald. The Application of the Quasi-Steady Theory to Full Scale Measurements on the Texas Tech Building. Journal of Wind Engineering and Industrial Aerodynamics 48.1 (1993): 111-32.
26 J.D. Holmes, Wind Loading of Structures, CRC Press, Florida, 2001. Print
27 K.R. Gurley., M.A. Tognarelli, and A. Kareem. Analysis and simulation tools for wind engineering. Probabilistic Engineering Mechanics 12.1 (1997): 9-31.
28 F.T. Lombardo. Analysis and Interpretation of Thunderstorm Wind Flow and its Effects on a Bluff Body, Ph.D. Dissertation, Texas Tech University, 2009.
29 C.D. Karstens, T.M. Samaras, B.D. Lee, W.A. Gallus, and C.A. Finley. Near-Ground Pressure and Wind Measurements in Tornadoes. Monthly Weather Review 138.7 (2010): 2570-58.
30 J. Nevill and F.T. Lombardo. Development of a wind engineering field laboratory to characterize transient winds. 13th Americas Conference on Wind Engineering, proceedings (2017).

13th Americas Conference on Wind Engineering
Gainesville, Florida, USA
May 21 - 24, 2017

Atmospheric Boundary Layer Conditions in an Open-Circuit Speed Wind Tunnel in Colombia

Martha Elena Delgado[a], Johannio Marulanda[b], Peter Thomson[c]

[a] *Universidad del Valle, Cali, Colombia, martha.delgado@correounivalle.edu.co*
[b] *Universidad del Valle, Cali, Colombia, johannio.marulanda@correounivalle.edu.co*
[c] *Universidad del Valle, Cali, Colombia, peter.thomson @correounivalle.edu.co*

ABSTRACT

Estimation of wind load is one of the most important requirements for the design of structures. There are several proposed analytical methods to calculate it and experimental methods such as wind tunnel testing using reduced scale models. The Boundary Layer Wind Tunnel of the Universidad del Valle in Colombia is the only one in the country able to reproduce and simulate the atmospheric boundary layer. To achieve the correct conditions, devices designed through empirical formulations must generate turbulence and reach the desired wind profile through a trial and error procedure. In this paper, it was carried out CFD simulations of the boundary layer using as domain geometry the Wind Tunnel of the Universidad del Valle. The wind tunnel, a castellated barrier, spire vortex generators and roughness elements in the test section length were simulate in the domain. Wind profiles were obtained for each arrangement for urban and rural terrain. A comparison between velocity profiles obtained by numerical simulations in the test section, experimental results and the profiles provided by the Colombian Building Code (Reglamento Colombiano de Construcción Sismo Resitente NSR-10) is shown. The methodologies used to reproduce the atmospheric boundary layer agree well with the theoretical profiles and the numerical model. Between the numerical model and Colombian building code, the velocity profile for the exposure C achieved with a setting of 90% and for exposure B 80%.

KEYWORDS

Boundary Layer, Wind Tunnel, Computational Fluids Dynamics, Wink load.

On the use of the weather research and forecasting model to define inflow uncertainty for atmospheric boundary layer simulations

C. García-Sánchez[a,b,c], C. Gorlé[d]
[a] *EMAT, University of Antwerp, Belgium*
[b] *Environmental and Applied Fluid Dynamics, von Karman Institute, Belgium*
[c] *Civil and Environmental Engineering Department, Stanford University, US, gsclara @stanford.edu*
[d] *Civil and Environmental Engineering Department, Stanford University, US*

ABSTRACT

Simulations of Atmospheric Boundary Layer (ABL) flows are widely performed to predict wind flows in diverse environments. The forecast of wind flows over natural terrain and within urban environments has a broad variety of applications, from air quality assessment to pedestrian wind comfort, and wind resource assessment. However, providing accurate predictions of this type of flow remains a challenging task due to the intrinsic variability of the natural system and the wide range of uncertainties involved in the problem.

This research focuses on quantifying the aleatory inflow uncertainty present in any simulation of ABL flows. The uncertainty in the inflow definition is based on the fundamental parameters used to define the standard neutral velocity profile widely used in Reynolds-averaged Navier-Stokes (RANS) simulations for wind engineering purposes: the velocity magnitude and direction at a reference height, and the terrain roughness length. To apply the uncertainty quantification framework, probability density functions (pdfs) of these parameters have to be defined. In this presentation we will present a novel method to define these pdfs using the outcome of ensemble forecasts performed with the Weather Research and Forecasting (WRF) mesoscale model. Once the uncertain variables have been defined in a probabilistic manner, a non-intrusive polynomial chaos method is used to propagate these uncertainties to the quantities of interest, namely the prediction of the velocity magnitude, direction and pollutant concentration at different locations for which measurements are available from field experiments. The proposed approach presents an alternative to the more traditional deterministic coupling between mesoscale weather prediction models and CFD models. It enables accounting for uncertainties in the mesoscale predictions, which would otherwise be directly transferred to the CFD model.

Results will be presented for two very different test cases. The first test case considers the flow over natural terrain as measured during the 1983 field measurement campaign over Askervein Hill (Scotland). In the second case we investigate the flow in an urban environment, considering the Joint Urban 2003 field measurements, completed in Oklahoma City in 2003 (USA). The study focuses primarily on evaluating the potential of the methodology to estimate confidence intervals that encompass the time-averaged experimental measurements, and on identifying the areas of the domain that exhibit a higher sensitivity to the inflow uncertainties. Secondly we will evaluate the influence of the different uncertainties, concluding on their relative effects on the quantities of interest, namely the non-dimensional velocity magnitude, velocity direction and non-dimensional concentration. Finally, general conclusions will be drawn from the differences observed between both test cases, identifying potential areas for future research.

KEYWORDS

Atmospheric boundary layer, dispersion, uncertainty quantification, computational fluid dynamics, real-scale experiments

The 13th Americas Conference on Wind Engineering (13ACWE)
Gainesville, Florida USA, May 21-24, 2017

Wind Data Performance for Estimating Wind Loads on Structures

Matt B. Phelps, P.E. [a], Cliff Fedler, Ph.D., P.E.[b], Don Spradling, EIT [c]

[a]*Texas Tech University, Ph.D. Candidate, Systems Engineering and Management 201 County Road, Hutto, Tx., USA*
[b]*Texas Tech University, Professor Civil Engineering, Department of Civil, Environmental and Construction Engineering, Texas Tech University Box 41023, Lubbock, TX, USA*
[c]*APEC Engineering and Laboratory, 201 County Road 138, Hutto, TX, USA*

Determining the wind load on structures is a fundamental component of structural design and wind damage assessment. Engineers and architects who design with inadequate or incorrect wind data are at risk of designing structures whose wind load capacity may under-perform leading to component fatigue and failure, or over-perform leading to increased and unnecessary cost. Engineers assessing wind loads on existing structures need reliable wind data to compute wind loads. One of the most popular and creditable references for wind loads on structures is the American Society of Civil Engineers, Minimum Design Guide for Buildings and Other Structures (ASCE 7-10, current edition). This document is used by practicing engineers for both design and wind load analysis of a variety of structure types, shapes, and heights. The equation used to calculate the wind velocity pressure from ASCE is based on a force and momentum relationship with the appropriate conversion coefficients combined to use customary units. The standard specifically instructs the engineer, if actual wind anemometer data is to be used, to confirm that the anemometer is correctly cited at 10 meters' height in exposure category C.[6] If the anemometer is not properly cited wind speeds should be corrected to reflect the speeds most likely to occur at the standard height.

According to the World Meteorological Organization (WMO) the standard height for an anemometer is 10 meters above open terrain; however, many anemometers are not located at this reference height.[1] The standard ASCE 7-5, 7-10 and others, cautions the engineer that if actual wind anemometer data is used that it must comply with the reference height of 10 meters. Specifically, ASCE 7-05 6.5.4.2, states that regionally acquired data should be corrected for the basic wind speed. ASCE 7 Minimum Design Guide for Buildings and Other Structures is a design guide; however, it is commonly used for applied load analysis on existing structures and has been upheld as a creditable applied load analysis standard by the US Federal Court System.[9]

Exposure category C describes a condition of roughness, or fetch area, that is similar to smooth grass with few obstructions such as might be found in a large hay field or airport runway system; however, not all anemometers are located in this exposure. Some anemometer locations overlook pine forest that are 20 meters tall, in these cases the tree tops mark the top of the roughness layer and the anemometer should be located 10 meters above the tree canopy The WMO Guide to Meteorological Instruments and Methods of Observation 5.9.1 specifically says corrections can be used to adjust basic wind speeds for height of the reporting anemometer to the referenced height.

Phelps (2013), with the support of the Office of the New Jersey State Climatologist (ONJSC), performed a study of the actual heights of the ONJSC Automated Weather Observation System (AWOS) anemometer heights. There are more than 70 anemometers in the NJ system, 29 anemometers were located and their heights measured and exposure category recorded. Of the 29 anemometers located 20.7% were within +/- 10% of the reference height of 10 meters.[8] Table 1.0 shows the measured heights with negative values being below the referenced height and positive values being above.

Most of the anemometers (58.6%) were located more than 10% below the reference height and 17.2% were above 10% of the same.[8] Based upon these data more than 75% of the measured anemometer heights were not with +/- 10% of the referenced height. In other words, three out of four anemometers were not at the reference height and their data must be corrected to comply with ASCE 7. One anemometer was located on top of a building and others on top of bridges, according to WMO there is little justification to locating an anemometer on top of such structures, and the Guide to Meteorological Instruments and Methods of Observation section 5.9.2 specifically states that anemometers located on top of buildings should be placed at least one building width above the building.[1,4] Wind speed data is not automatically corrected to the reference height. The ONJSC did not know the height of the AWOS anemometers investigated by Phelps; therefore, it is unlikely they could have automatically adjusted to the WMO referenced height of 10 meters. In order to use this data with confidence, it must be collected from the standard reference height and exposure category, or the correction must be applied to the raw velocity data.

The National Centers for Environmental Information (NCEI) maintains a data base of Quality Controlled data that is taken from correctly cited anemometers what are positioned at 10 meters. Not all AWOS are part of this data base, one of the reasons this is so is that their heights have not been confirmed to comply with the referenced height. Those anemometers that are not part of the NCEI Quality Controlled date set and may not be at the correct height or exposure category. ASCE 7 contains a method to change wind data from one exposure category to another, but this conversion is outside of the scope of this paper.

Table 1.0 NJ AWOS data and elevations

APEC
Forensic, Structural & Civil Engineering

APEC, LLC.
PH: 409-766-7500
FAX: 832-369-7664

Island Engineering
1409 39th St.
Galveston, TX 77550

Date	Measured By	Instrument Name	Lat	Long	State	County	Instrument Manufacture	Measured Height(ft)	Exposure Category	Height Variance	%Difference
										(+) is above and (-) is below	
4/28/2013	MBP	Point Pleasant	40.0708	-74.0597	NJ	Ocean	Peet	28.2	C	-4.6	-14.1%
4/28/2013	MBP	Wall Twp.	40.155	-74.0727	NJ	Monmouth	Campbell Scientific CR-10X	10.0	B	-22.8	-69.5%
4/28/2013	MBP	Toms River	39.95547	-74.208	NJ	Ocean	Campbell Scientific CR-10X	11.7	B	-21.1	-64.4%
4/28/2013	MBP	Barnegat	39.76027	74.1105	NJ	Ocean	Unknown	28.6	C	-4.2	-12.9%
4/28/2013	MBP	Harvy Ceder	39.70499	-74.1352	NJ	Ocean	Peet	24.4	C	-8.4	-25.5%
4/28/2013	MBP	Sandy Hook	40.46666	-74.01	NJ	Monmouth	Campbell Scientific CR-10X	19.7	C	-13.1	-40.0%
4/29/2013	MBP	Freehold Twp.	40.25	-74.28	NJ	Monmouth	SSI RWIS	28.5	B	-4.3	-13.1%
5/13/2013	MBP	New Brunswick	40.4727	-74.4225	NJ	Middlesex	Campbell Scientific	36.0	B	3.2	9.8%
5/13/2013	MBP	Keansburg	40.44916	-74.1475	NJ	Monmouth	Unknown	16.5	C	-16.3	-49.7%
5/13/2013	MBP	Teterboro Airport	40.85846	-74.0561	NJ	Bergan	Unknown	8.7	B	-24.1	-73.6%
5/13/2013	MBP	Ocean Grove	40.21017	-74.0032	NJ	Monmouth	Campbell Scientific	34.0	C	1.2	3.7%
5/13/2013	MBP	Sea Girt	40.12044	-74.0328	NJ	Monmouth	Campbell Scientific	28.3	C	-4.6	-13.9%
5/13/2013	MBP	Seaside Heights	39.94383	-74.0738	NJ	Ocean	Peet	39.8	C	7.0	21.2%
5/23/2013	MBP	Atlantic City (NJDOT)	39.37728	-74.4525	NJ	Atlantic	SSI RWIS	49.4	C	16.6	50.7%
5/23/2013	MBP	Atlantic City Marina	39.3785	-74.4241	NJ	Atlantic	Peet	31.7	C	-1.1	-3.4%
5/23/2013	MBP	Egg Harbor (EEH)	39.41947	-74.5378	NJ	Atlantic	CR-10	9.1	B	-23.7	-72.3%
5/23/2013	MBP	West Creek	39.6488	-74.2965	NJ	Ocean	Peet	21.4	C	-11.4	-34.7%
5/25/2013	MBP	Seaside Park	39.91729	-74.07929	NJ	Ocean	Maximum, Inc.	60.8	C	28.0	85.2%
5/26/2013	MBP	Holmdel (Monmouth Co. Park)	40.3776	-74.1809	NJ	Monmouth	Campbell Scientific CR-10X	10.6	B	-22.2	-67.7%
5/26/2013	MBP	Fort Monmouth (Rumson Bridge)	40.3137	-74.049	NJ	Monmouth	Campbell Scientific CR-10X		C		0.0%
5/26/2013	MBP	Gooseneck Pt. Bridge	40.3252	-740196	NJ	Monmouth	Campbell Scientific CR-10X		C		0.0%
5/27/2013	MBP	Howell	40.19204	-74.1962	NJ	Monmouth	Campbell Scientific CR-10X	9.2	B	-23.6	-72.0%
5/27/2013	MBP	Unknown 1	40.16511	-74.236	NJ	Monmouth	Campbell Scientific	32.9	B	0.1	0.4%
5/27/2013	MBP	Berkeley Twp. @ Robert Miller Air Park	39.92578	-74.2981	NJ	Ocean	Campbell Scientific	36.4	C	3.6	11.0%
5/28/2013	MBP	Unknown 2	40.09654	-74.08853	NJ	Ocean	Unknown	40.0	B	7.2	22.0%
5/29/2013	MBP	Jersey City (LSC)	40.70877	-74.053	NJ	Hudson	Campbell Scientific	30.6	B	-2.2	-6.8%
5/23/2013	MBP	Oceanic Bridge	40.38104	-74.0138	NJ	Monmouth	Campbell Scientific	8.5	C	-24.3	-74.1%
7/4/2013	MBP	Sea Bright	40.36567	-73.9747	NJ	Monmouth	Campbell Scientific	10.3	C	-22.5	-68.6%
7/4/2013	MBP	Oswego Lake	39.71539	-74.5142	NJ	Burlington	Campbell Scientific	32.25	C	-0.5	-1.7%
7/4/2013	MBP	Cedar Bridge	39.84	-74.38	NJ	Burlington	Campbell Scientific	28.67	C	-4.1	-12.6%
7/4/2013	MBP	Cream Ridge	40.1185	-74.5256	NJ	Monmouth	Campbell Scientific	33.5	C	0.7	2.1%
							Total	29.0			
						Total within +/- 10% of 32.8 ft.		6.0	20.7%		
							Total below 90%	17.0	58.6%		
							Total above 110%	5.0	17.2%		

In meteorology, stability, or more correctly instability, is measured in Convective Available Potential Energy (CAPE) in units of Joules per kilogram (J/kg). Meteorological measurements of CAPE are made twice daily by weather balloons that are launched around 00:00 and 12:00 UTC; therefor weather balloons are coordinated to launch at the same time. Weather balloons soundings collect temperature, humidity, wind speed and direction, and pressure readings as they ascend and transmit their data. These data are collected and displayed on a skew-t-diagram as shown in Figure 1.0. This convective energy, called CAPE, with higher values of CAPE the greater the atmospheric instability. The higher the instability the more convective energy added to the cloud circulation.

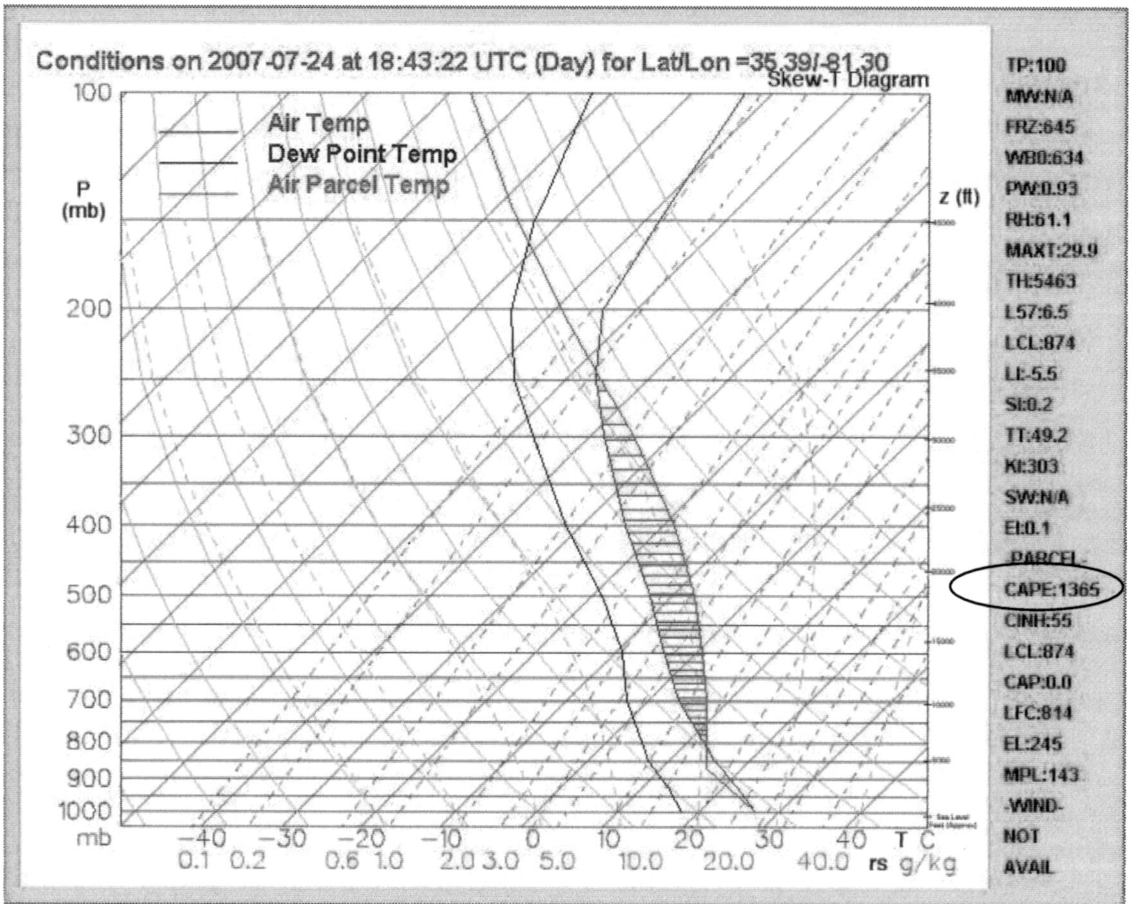

Figure 1.0 Skew-t-diagram with CAPE (shaded area)[2]

CAPE exists within the conditionally unstable layer of the troposphere, the Free Convective Layer (FCL), where an ascending air parcel is warmer than the ambient air. CAPE is measured in joules per kilogram of air (J/kg), see equation 1.0. CAPE values between the FCL and the Level of Free Convection (LFC) can be positive or negative. Any value greater than 0 J/kg indicates instability and an increasing possibility of thunderstorms and hail. Generic CAPE is calculated by integrating vertically the local buoyancy of a parcel from the LFC to the Equilibrium Level (EL).[2]

$$\text{CAPE}=\int_{Z_f}^{Z_n} g \cdot \left(\frac{T_{vparcel}-T_{env}}{T_{env}}\right) dz \qquad 1.0$$

Where:
CAPE - Convective Available Potential Energy (J/kg)
g - gravity (m/s^2)
$T_{vparcel}$ - Virtual Temperature of specific lifted parcel (C)
T_{venv} - Virtual Temperature of the environment at parcel elevation (C)
Z_f - height of the level of free convection (m)
Z_n - height of the equilibrium level (neutral buoyancy) (m)

The WMO provides a specific power law function based upon atmospheric instability and solar radiation, to adjust the wind speed from the measured height to the reference height. The coefficients of the power function are based upon the stability class. The power law equation adopted by WMO was proposed by Stull, 1988, is shown in equation 2.0 with the stability classes and their associated power coefficients.[3]

$$\frac{U_2}{U_1} = \left(\frac{Z_2}{Z_1}\right)^P \qquad 2.0$$

Where:
U_2 – Wind speed at higher elevation (m/s)
U_1 – Wind speed at lower elevation (m/s)
Z_2 – Higher elevation (m)
Z_1 – Lower elevation (m)
P – Power coefficient as function of atmospheric stability (dimensionless)

The power coefficients (P) was determined by using the local net radiation (Rn). Stability class, solar radiation and related coefficients are:

Stability Class		Coefficient Value
Very unstable >= 100	W/m^2	0.1305
Mod. unstable <100 >5	W/m^2	0.1630
Neutral <5 >-5	W/m^2	0.2025
Mod. stable <-5 >-100	W/m^2	0.3570
Very stable <= -100	W/m^2	0.3675

Since few engineers are trained in meteorology they often do not understand the fundamentals of using the WMO power law or where to find the data that the power coefficients are based upon. Without a sufficient understanding of these meteorological principal's uncertainty is increased. Atmospheric stability classes can be found at the National Oceanic and Atmospheric Administration (NOAA) Storm Prediction Center (SPC) web site at http://www.spc.noaa.gov/exper/soundings/. Net solar radiation values can be found at NOAA's National Center for Environmental Information (NCEI) at https://www.ncdc.noaa.gov/data-access/land-based-station-data/land-based-datasets/solar-radiation.

Wind loads can be computed use the equations provide in ASCE 7. The basic equation for velocity pressure is:

$$q_z = 0.00256 K_z K_{zt} K_d V^2 I \qquad 3.0$$

Where:
q_z - velocity pressure at height z (lb/ft^2)
K_z - velocity pressure exposure coefficient as applicable, for each wind direction (dimensionless)
K_{zt} - topography factor, if applicable (dimensionless)
K_d - wind directionality factor (dimensionless)
V - basic wind speed 3-sec. gust (mph) measured at 33 feet in exposure Category C
I - importance factor (dimensionless)
0.00256 is the conversion factor for use with US customary units

Since the velocity pressure is proportionate to the square of the basic wind speed, an increase in the wind speed (V) by 10% will increase the resulting wind load by 21%, whereas an increase in the wind speed (V) by 20% will increase the resulting wind load (q_z) by 42%. The equations for q_z are based upon wind speeds corrected for the standard reference height and exposure category. These results underscore the significance of use these equations with wind speeds measured at or corrected to the referenced height. If the known anemometer height is more, or less, than the reference height, correcting the wind speed with the correct power law coefficient will reduce uncertainty because the anemometer actual height has been adjusted to the referenced height that ASCE 7 equations are based upon.

The typical range of values for the various terms in the base wind load calculation is shown in Table 2.0 for the variables and guidance from ASCE 7-05.

Table 2.0 Wind Load Calculation Information

Term	Notes/Comments	Value Range	ASCE 7-05
K_z	If the site or reporting anemometer location is not exposure Category C, exposure category adjustment factors (λ) are included in Figure 6-2, and 6-3.	1.00 to 1.87*	6.5.6 Exposure
K_{zt}	Wind speed increase due to local topo conditions as specified in 6.5.7.1, otherwise K_{zt} = 1.0	1.00 to 2.96**	6.5.7 Topographic effects
K_d	Table 6-4	0.85 to 0.95	See Section 2.3 and 2.4
V	3-sec. gust wind speed measured at 33 ft. in exposure C***	0 and above	6.5.4.2****
I	Design factor only, based upon class of structure	N/A	6.2

* based upon mean roof height

** based upon H/L_h (K_1), x/L_h (K_2), and z/L_h (K_3)
*** anemometer height and exposure category adjustments must be made in compliance with the World Metrological Organization Guide to Meteorological Instruments and Methods of Observation (WMO-No. 8), 2008.
**** Section 6.5.4.2 specifically addresses anemometer height and exposure siting

There are other equations that can be used to assess wind speeds at various heights. The power law equation shown in Equation 2.0 is the equations adopted by the WMO. The relationship between stability class, net solar radiation and wind speed increases at higher elevations is described mathematically in equation 2.0. In general, the physical relationship of the shape of the wind profile over increased height is shown in Figure 2.0

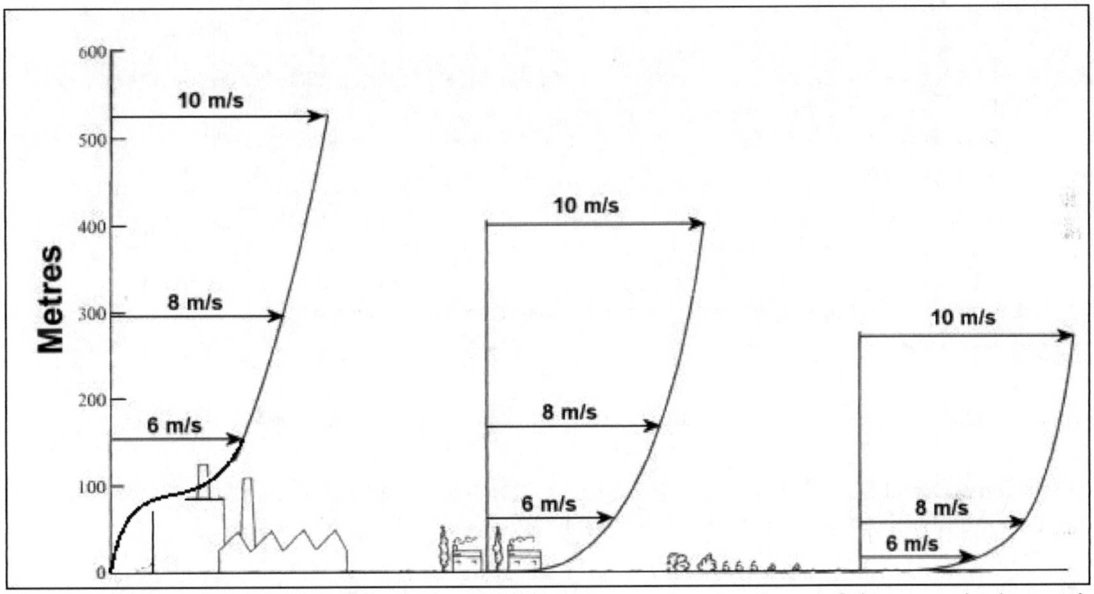

Figure 2.0 Shape of wind profile, category C is on the right, the shape of the curve is that typically associated with a power function

From the provided referenced the user can locate the specific solar radiation, and atmospheric stability for the time and location in question. These data should be used to select the correct power coefficient for equation 2.0. The corrected wind speed should be used in equation 3.0 to compute the most likely wind load occurring at the subject structure.

REFERENCES

1. World Meteorological Organization (WMO) WMO-No.8, WMO-No.8, Seventh Edition, 2008, Guide to Meteorological Instruments and Methods of Observation, ISBN 978-92-63-100008-5

2. National Oceanic and Atmospheric Administration (NOAA), National Weather Service (NWS), JetStream on-line school for weather, http://www.srh.noaa.gov/jetstream/index.html, last accessed on 4/28/2017

3. Stull, Roland B. An Introduction to Boundary Layer Meteorology, Dordrecht, Netherlands: Kluwer Academic Publishers, 1988

4. World Meteorological Organization (WMO) WMO/TD No.1250, 2006, Initial Guidance to Obtain Representative Meteorological Observations at Urban Sites.

5. ONJSC (2014). "Data from Seaside Heights Hook Station," Office of the New Jersey State Climatologist, Rutgers University, Piscataway, NJ. Accessed on March 19, 2014. Retrieved from http://climate.rutgers.edu/njwxnet/dataviewer-stnpt.php?qcarr=&qcreason=&yr=2012&mo=10&dy=30&qc=&element_id[]=1&element_id[]=17&element_id[]=15&element_id[]=20&stn=298&states=NJ&newdc=1

6. ASCE (2005). "ASCE 7-05: Minimum Design Loads for Buildings and Other Structures," American Society of Civil Engineers. Reston, VA. ISBN: 9780784408094.

7. Ahrens, Donald C. Meteorology Today and Introduction to Weather, Climate, and the Environment. 4th edition, St. Paul, MN. West Publishing 1991

8. Phelps, Matt, B., 2013, New Jersey AWOS anemometer height measurements, unpublished

9. Phelps, Matt, B., 2015, Superior Court of New Jersey Law Division- Ocean County Docket No: OCN-L-3378-13 Diann and Christopher Hamilton vs. Chartis Property Casualty

Keywords:
Wind Load, Wind Data, Anemometer, Wind Power Law Equation, Structures, ASCE

The 13th Americas Conference on Wind Engineering (13ACWE)
Gainesville, Florida USA, May 21-24, 2017

Genesis and Track Simulation of Asia Typhoons

Sri Harshitha Polamuri[a], Weichiang Pang[b]

[a] *Clemson University, United states, spolamu@clemson.edu*
[b] *Clemson University, United states, wpang@clemson.edu*

ABSTRACT

Tropical cyclones are among the most common and dangerous natural hazards that effect the coastal regions and islands in Asia. In this study, a stochastic typhoon simulation framework has been developed to simulate the typhoon tracks. The storm genesis and tracks are simulated and validated using historical typhoon database, known as the International Best track data (IBTrACS). The IBTrACS provides the track data of all the tropical storms that occurred in Eastern Pacific Basin. In the simulation framework, the annual spawn rate of typhoons in the East Asia basin is modeled using a negative binomial distribution. The initial or spawn location of individual typhoons is modeled using a Kernal Density Estimate (KDE) model, which specifically considers the spawn density and geographic distribution of historical typhoons. The storm track is modeled using geographically weighted regression (GWR) approach, which explicitly accounts for the movements of historical typhoons (travel speed and heading direction) and the proximity of the observed data to the location of interest. To validate the simulated typhoon tracks, the occurrence rate, storm forward speed, heading direction are compared to the historical records. Kilometer posts along the coast of China, Korea and some key locations in Japan are selected for model validation purpose. The approaching rate and key parameters of typhoons when passing through these key locations are collected. The statistics of these parameters obtained from both simulated and historical storms are examined and compared. Good agreements are observed between the simulated and observed typhoon tracks. The typhoon genesis and track models presented in this paper can be combined with intensity and decay models to simulate the full-track behavior of Asia typhoons.

1. INTRODUCTION

Typhoons are one of the major natural hazards that occur in the western Pacific region. The effects of typhoon include extreme wind and coastal flooding which cause loss of life, damages of buildings and infrastructure, loss of crop and vegetation. In order to minimize the negative impacts of typhoons, preventive measures must be taken to proactively reduce the losses due to typhoons that can occur in the future. Although historical records of typhoons are available for the past events from sources such as: (1) the Japan Meteorological Agency (JMA), where the typhoon history is available from 1951 to 2016, (2) the International Best Track Archive for Climate Stewardship (IBTrACS) data from 1842 to 2015, (3) the China Meteorological Administration (CMA) data from 1949 to 2016. Due to limited observation years, the data are not adequate for estimating the long-term risk of typhoons. Hence, a long-term simulated database of typhoons is essential for implementing measures to reduce the risk and consequences of typhoons. Since the historical catalogues are limited to short-term, it is essential to develop a long-term stochastic typhoon database from which the rate of typhoon occurrence, wind speeds and the amount of damage can be assessed.

The methodology used for simulating synthetic typhoon tracks is based on the probabilistic hurricane simulation approach introduced by Vickery et al. (2000)[1] and later updated in 2009 (Vickery et al. 2009)[2], where hurricanes are simulated using stochastic tracking model approach.

Figure 1 shows the flowchart for the typhoon simulation framework. The typhoon simulation involves several modules that include genesis model, tracking model followed by relative intensity model, central pressure decay model, and wind field model. The typhoon simulation framework begins with genesis model in which the number of storms are randomly generated per year. The storm genesis locations are simulated using the probability density contours developed using historical genesis locations and the kernel density estimation (KDE) method. After genesis, the storm paths are simulated by tracking model using regression equations of change in translational speeds and storm heading direction. Geographically weighted regression (GWR) approach is used to fit the modeling coefficients of the tracking model, which explicitly accounts for the proximity of observed data to the location of interest. Storm track model is followed by relative intensity model where the relative intensity is calculated from the pressure at typhoon eye and sea surface temperature at the location of the typhoon eye. The relative intensity is modeled using a regression equation with the logarithm of previous step relative intensity and sea surface temperature of the storm's previous and current steps as the predictors. Once the storm makes landfall, central pressure deficit decay is modeled as a function of the time after landfall and the central pressure at landfall.

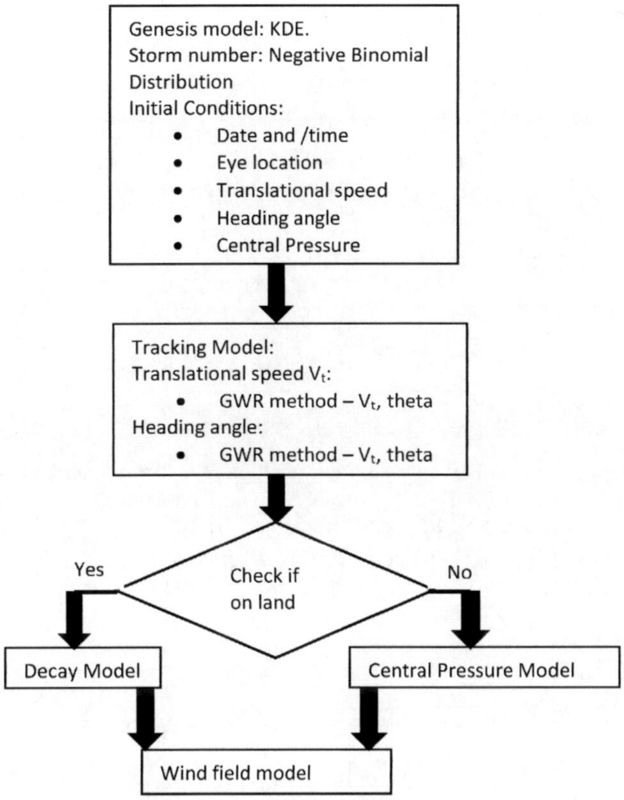

Figure 1: Typhoon simulation model flowchart.

2. METHODOLOGY

2.1. Genesis Model

The number of typhoons per year is generated using a negative binomial distribution. Given the mean μ and standard deviation σ of the observed annual storm frequencies are 36.75 storm/year and 8.18 storm/year, respectively, the negative binomial distribution parameters can be fitted using the method of moments approach:

$$R = \frac{\mu^2}{\sigma^2 - \mu} \tag{1}$$

$$P = \frac{\mu}{\sigma^2} \tag{2}$$

where R and P are estimated as 47.96 and 0.56, respectively using the above equations (i.e. method of moments). The cumulative distribution function (CDF) of the storm frequency using negative binomial fit is plotted against the CDF of observed storm frequencies. From the CDFs shown in Figure 2, it is evident that the fitted negative binomial distribution matches the historical storm rates observed from IBTrACS reasonably well.

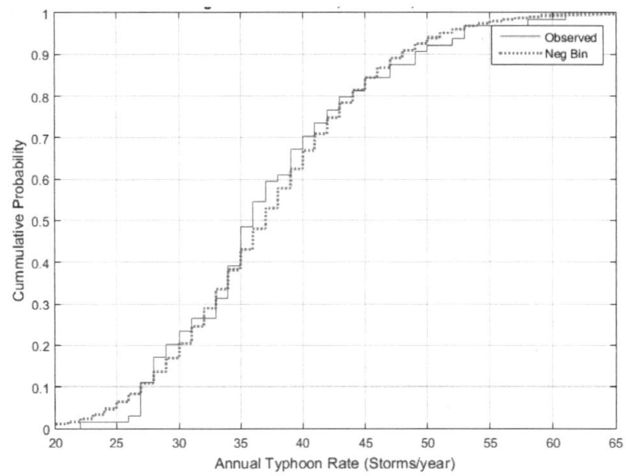

Figure 2: Cumulative distribution function of Typhoon annual frequency.

The initial locations of typhoons are generated based on the Gaussian kernel density estimate (KDE) method by smoothing the observed typhoon spawn locations. The Gaussian kernel density estimator is defined as

$$\hat{f}(X) = \frac{1}{NH}\sum_{i=1}^{N} K\left(\frac{X-X_i}{H}\right) \tag{3}$$

where,

$$K(X) = \frac{1}{\sqrt{2\pi}}e^{-X^2/2} \tag{4}$$

where, N is the total number of spawn locations, H is the bandwidth, X is the generated spawn location, X_i is the historical spawn location, $K(.)$ is the Gaussian probability density function kernel.

In order to estimate the probability density of the initial spawn locations of typhoons using the KDE method, an appropriate bandwidth is required. Choosing a large bandwidth will lead to over smoothing of the data (i.e. spawn probability is approximately the same everywhere). On the other hand, choosing a small bandwidth can create artifacts and sharp variations of probability densities for locations in close proximity. An optimal bandwidth is selected using the rule of thumb method introduced by Bowmen and Azzalini (1997)[3]. The probability density functions of spawn locations were estimated for the overall domain of Asia using the kernel density estimator. The spawning locations are generated for every year based on the estimated probability density functions. The contour plot of simulated spawn locations is compared to that of historical spawn locations in Figure 3. The comparison shows the dispersion of simulated spawn locations is similar to the historical spawn locations.

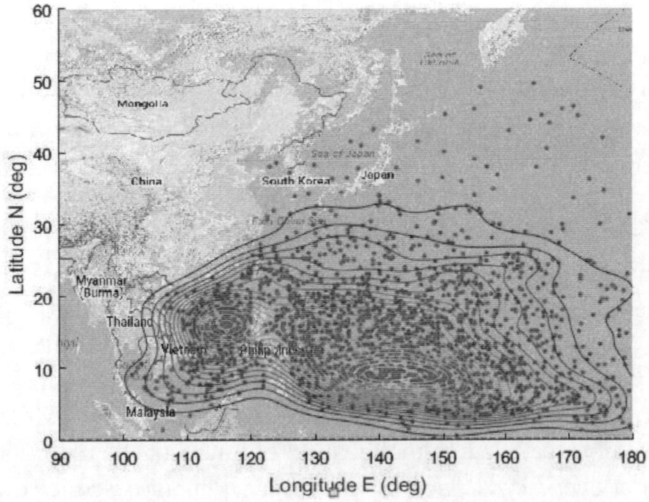

Figure 3a: Real storm spawn locations and the KDE contours.

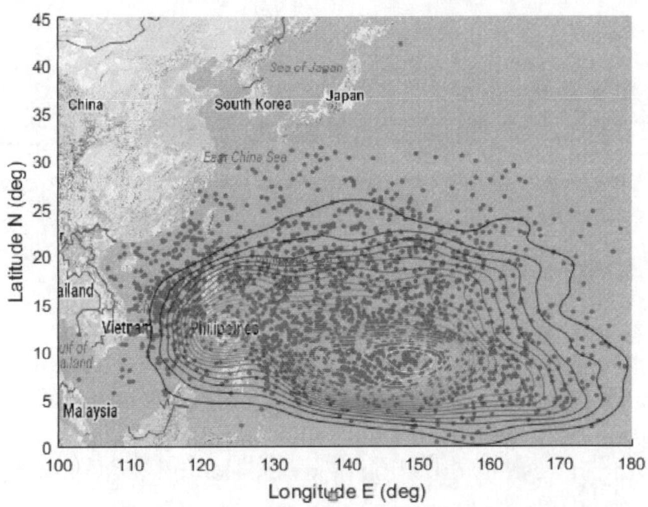

Figure 3b: Simulated spawn locations and the KDE contours.

2.2. Track Model

The methodology for simulating typhoon tracks is based on the empirical track modeling approach introduced by Vickery et al. (2000) for simulation of hurricane risk in the United States (U.S). The storm tracks are modeled using regression equations of change in translational speed (V_t) and heading direction (θ).

$$\Delta lnV_t = a_0 + a_1\psi + a_2\lambda + a_3 lnV_{ti} + a_4\theta_i + \varepsilon \tag{5}$$

$$\Delta\theta = b_0 + b_1\psi + b_2\lambda + b_3 V_{ti} + b_4\theta_i + b_5\theta_{i-1} + \varepsilon \tag{6}$$

V_{ti} and θ_i represent forward speed and angle at time step i. ΔlnV_t is logarithmic difference between V_{ti+1} and V_{ti} *and* $\Delta\theta$ is the difference between θ_{i+1} and θ_i. a_0 to a_4 represent the coefficients for the change in forward speed model. b_0 to b_6 represent the coefficients for the change in heading direction model. Ψ and λ represent the latitude and longitude of the storm eye, respectively. ε is the error term.

2.2.1 GWR Method

Geographically weighted regression (GWR) approach is introduced by Brunsdon et al. (1996; 1997) and Fotheringham et al. (1997a; 1997b)[4] where parameters in an equation are estimated by weighted least square regression approach. GWR method accounts for the location of the data in a geographical space, which allows the estimation and differentiation of the local parameters from the global parameters. GWR approach is used to model the storm tracks, which explicitly accounts for the proximity of the observed data to the location of interest. The regression analysis for track model was performed at every 1^0X1^0 grid for the entire domain based on the observed data that contribute to the location of interest. As the GWR approach already considers the proximity of the data to the location of interest, the location parameters (i.e. latitude and longitude) are removed from the empirical equations 5 and 6 making the new equations as follows,

$$\Delta lnVt = a_0 + a_1 lnVt_i + a_2\theta_i + \varepsilon \quad (7)$$

$$\Delta\theta = b_0 + b_1 Vt_i + b_2\theta_i + b_3\theta_{i-1} + \varepsilon \quad (8)$$

The track coefficients, *a* and *b* are fitted at every 1^0X1^0 using the GWR approach. After obtaining *a* and *b*, the differences between the modeled values and observed values are calculated to quantify the error terms.

3. SAMPLE PRELIMINARY RESULTS

100,000 years of storm tracks were simulated using the forward speed and heading direction models. To assess the simulated tracks, five years of storm tracks are randomly sampled from the simulated database. For comparison, five years of real storm tracks are also plotted from IBTrACS data. The simulated storm tracks are observed to follow similar patterns of the real tracks, in which the storms mainly moving in westerly direction below 30^0 latitude and moving in easterly direction above 30^0 latitude.

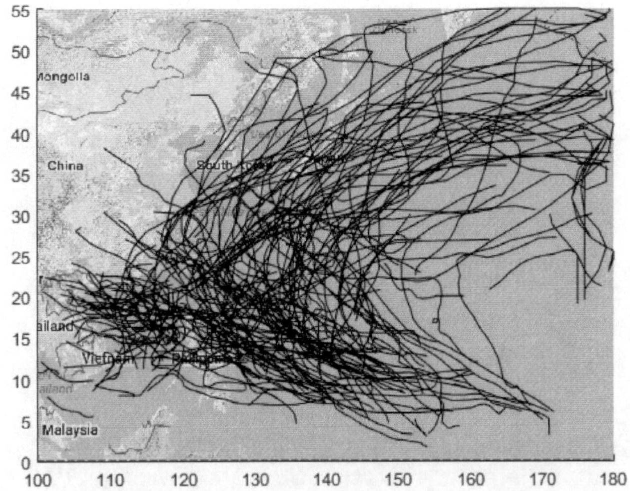

Figure 4a: Real 5 year storm tracks from IBTrACS.

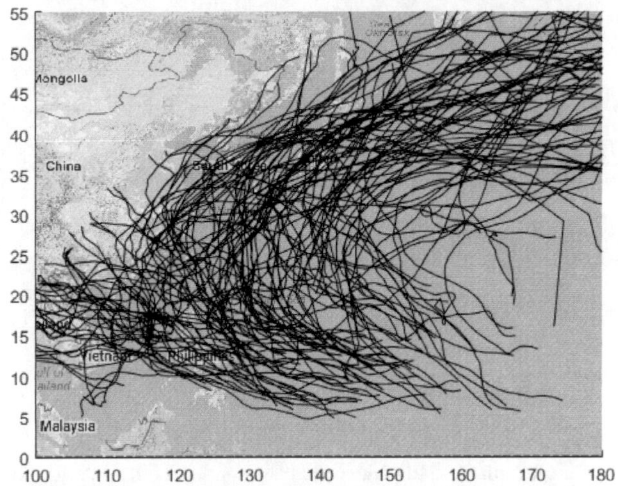

Figure 4b: Simulated 5 year storm tracks.

The key parameters of simulated storms are validated with historical storms at key locations selected in the coastal areas of China, Japan, and Korea. Kilometer posts are assembled at every 100 kilometers along the coastline of the main land of Asia. Storms passing within 250 kilometers radius from each kilometer post are selected. The approaching rate of simulated storms are compared with those past observed storms (Figure 6). The comparison shows the trend of approaching rate of simulated storms along the kilometer posts is similar to that of IBTrACS database. However, discrepancies exist between the simulated and observed approaching rates. The authors are currently investigating the cause of these discrepancies.

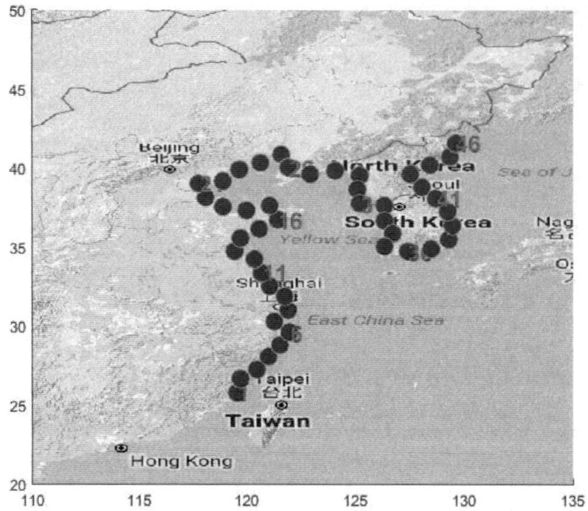

Figure 5: Kilometer posts along the coast of main land Asia.

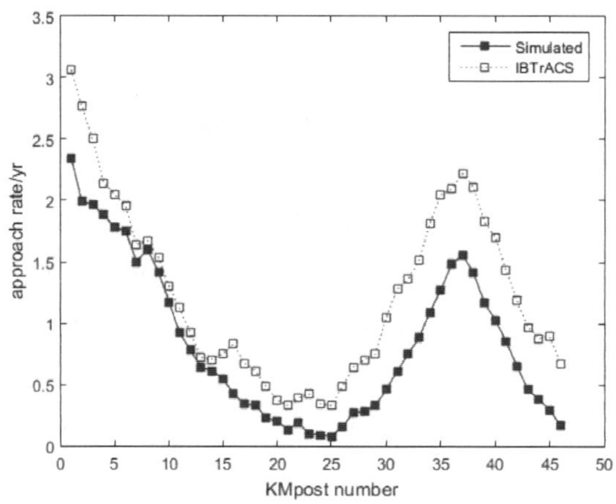

Figure 6: Comparison between the simulated and observed annual approaching rate for Kilometer posts along the main land of Asia.

The distributions of key parameters such as the translational speed and heading angle of selected simulated storms are compared with the observed values for each kilometer post. The statistics of simulated and observed key parameters for each kilometer post are plotted in Figure 7 and Figure 8. The median values vary in the similar trend of observed values along the kilometer posts. The trends of 5 percentile and 95 percentile values match well with the observed values indicating the similarity between both the catalogues.

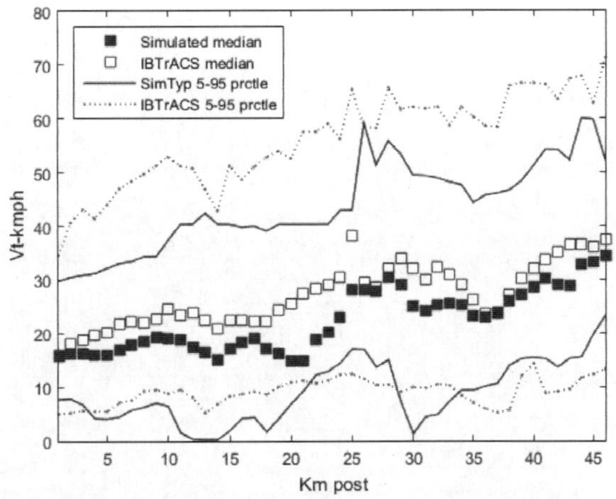

Figure 7: Comparison between simulated and observed translational speeds (median, 5 and 95 percentiles) along the coastline of the main land of Asia.

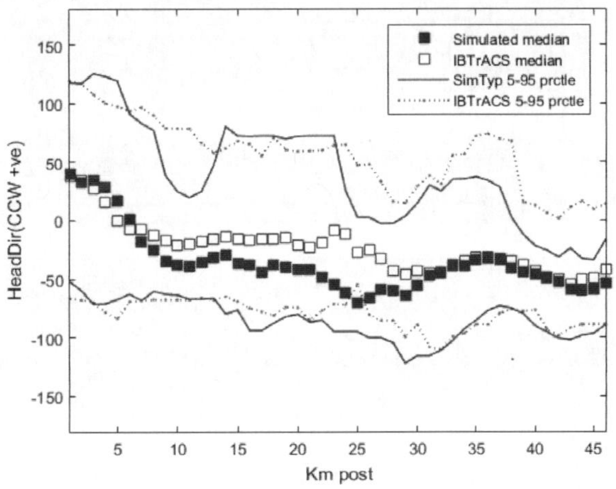

Figure 8: Comparison between simulated and observed heading angles (median, 5 and 95 percentiles) along the coastline of the main land of Asia.

Similar to the kilometer posts selected for the main land, key locations were chosen along the coast of Japan to compare the simulated and observed storm statistics. The storms passing within 250 km radius from each key location are selected to compare the approaching rates, storm translational speeds and heading angles between simulated storms and IBTrACS data.

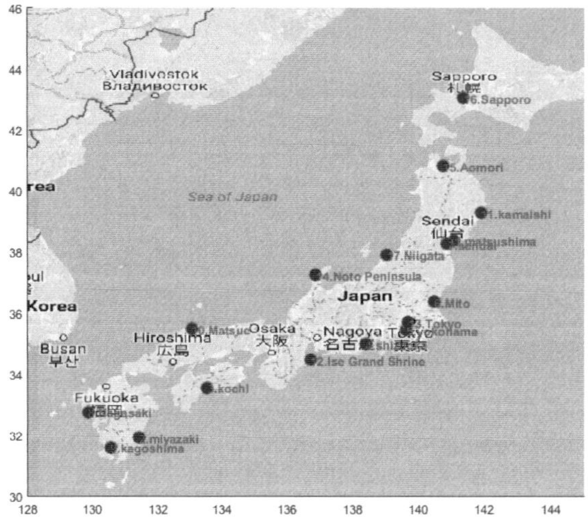

Figure 9: Selected key locations in Japan.

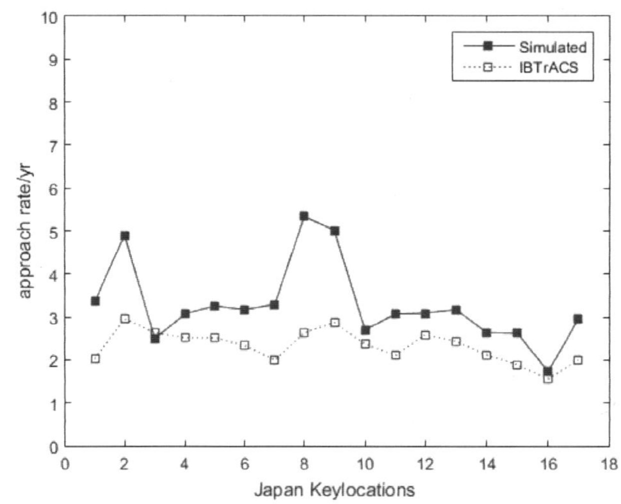

Figure 10: Comparison between the simulated and observed annual approaching rate for key locations in Japan.

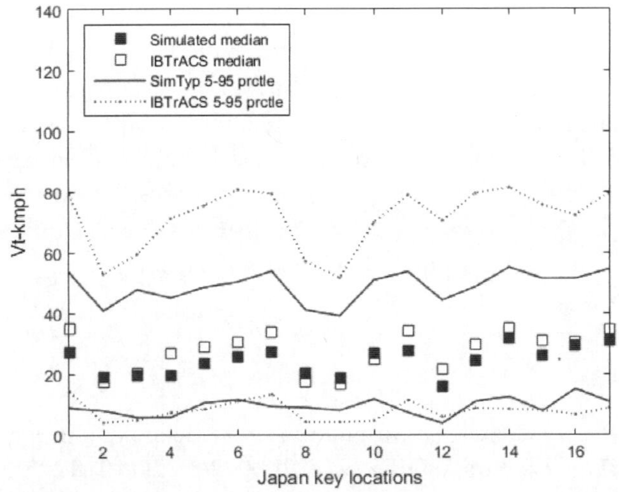

Figure 11: Comparison between simulated and observed translational speeds (median, 5 and 95 percentiles) for selected locations along the coastline of Japan.

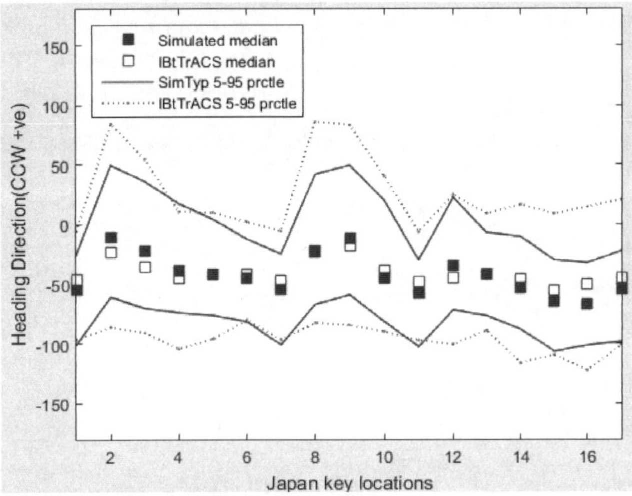

Figure 12: Comparison between simulated and observed heading angles (median, 5 and 95 percentiles) for selected locations along the coastline of Japan.

4. CONCLUSIONS

In this study, a long-term synthetic typhoon track catalogue is generated for the East Asian basin. A stochastic typhoon simulation framework is developed based on the methodology introduced in Vickery et al. (2000) for the U.S hurricanes. The annual storm frequency is estimated using a negative binomial distribution with the distribution parameters fitted using the observed historical storm frequencies in IBTrACS. The genesis or spawn location is generated using geographic dependent probability density functions developed using historical spawn locations and the kernel density estimation (KDE) method. Geographically weighted regression (GWR) approach is used to model the storm tracks, which accounts for proximity of data to location of interest. Since the GWR approach specifically accounts for the locations, the GWR track model does not contain location parameters (i.e. latitude and longitude) in the regression equations. The simulated typhoon track catalogue is validated with historical data by creating kilometer posts along the coastline of China and Korea. The validation was also performed in selected locations along the coast of Japan. The annual approach rates within 250 kilometers of each kilometer post or key location and the storm track parameter statistics were compared to historical data. Comparison results showed reasonable agreement between the simulated and historical catalogues.

5. REFERENCES

1. Vickery PJ, Skerlj PF, Twisdale LA. Simulation of hurricane risk in the US using empirical track model. Journal of structural engineering. 2000 Oct;126(10):1222-37.
2. Vickery PJ, Masters FJ, Powell MD, Wadhera D. Hurricane hazard modeling: The past, present, and future. Journal of Wind Engineering and Industrial Aerodynamics. 2009 Oct 31;97(7):392-405.
3. Bowman, Adrian W., and Adelchi Azzalini. *Applied smoothing techniques for data analysis: the kernel approach with S-Plus illustrations*. Vol. 18. OUP Oxford, 1997.
4. Brunsdon C, Fotheringham AS, Charlton ME. Geographically weighted regression: a method for exploring spatial nonstationarity. Geographical analysis. 1996 Oct 1;28(4):281-98.

State-of-the-art planetary boundary layer modeling and the application of ASCE 7 provisions to super-tall buildings

Emil Simiu[a], Alan Heckert[b], and DongHun Yeo[c]

[a] *National Institute of Standards and Technology, Gaithersburg, MD. 20899, USA, emil.simiu@nist.gov*
[b] *National Institute of Standards and Technology, Gaithersburg, MD. 20899, USA, alan.heckert@nist.gov*
[c] *National Institute of Standards and Technology, Gaithersburg, MD. 20899, USA, donghun.yeo@nist.gov*

ABSTRACT

According to recent Planetary Boundary Layer (PBL) research, neutrally stratified flows can be either of the "truly neutral" or the "conventionally neutral" type. "Truly neutral" flows are characterized by a Kazanski-Monin surface buoyancy flux parameter $\mu = 0$ and a non-dimensional number $\mu_N = N/|f| = 0$, where N is the Brunt-Väisäla frequency and f is the Coriolis parameter. These authors note that that "truly neutral flows are observed during comparatively short transition periods after sunset on a background of residual layers of convective origin," "are often treated as irrelevant because of their transitional nature, and are usually excluded from data analysis;" "neutrally stratified PBLs are almost always 'conventionally neutral,' that is, neutral *and* developing against a background stable stratification. They are characterized by parameters $\mu = 0$, $\mu_N \neq 0$; typically, $50 < \mu_N < 300$. Let u_*, G, and z_0 denote friction velocity, geostrophic speed and roughness length. For numbers μ_N typical of conventionally neutral flows, the dependence upon the Rossby number $Ro = G/(|f|z_0)$ of the geostrophic drag coefficient $C_g = u_*/G$ and the cross-isobaric angle α_0 can be represented by the following expressions:

$$C_g = 0.205/(\log_{10} Ro - 0.556),$$

$$\alpha_0 = (173.58/\log_{10} Ro) - 3.03.$$

The PBL height H can be obtained as follows:

$$\frac{1}{H^2} = \left[\frac{f^2}{(C_R)^2} + \frac{N|f|}{(C_{CN})^2}\right]\frac{1}{u_*^2},$$

where $C_R \approx 0.6$ and $C_{CN} \approx 1.36$.

It has also been recently shown that, at mid-latitudes, for heights of up to the order of a few kilometers, (i) the mean velocities $U(z)$ (parallel to the friction velocity) are monotonically increasing with height, and (ii) the velocities $V(z)$ (normal to the friction velocity) are negligibly small for buildings with height $h < 1$ km. In this paper it is shown that, since PBL heights are considerably greater than the nominal gradient heights z_g specified in the ASCE 7-10 Standard, for $h > z_g$ the Standard's PBL model is not appropriate. To eliminate the possibility that design wind loads meeting the Standard's Section 31.4.3 requirements are underestimated owing to the use of a physically incorrect PBL model, the Standard needs to explicitly provide for an exception to its definition of gradient heights by specifying that, for elevations z such that $z_g < z < h$, the monotonic increase of mean wind speeds with height must be accounted for in the design of buildings with height $h > z_g$. It is also noted in the paper that the mean wind profile can be described by the logarithmic law up to elevations that, at mid-latitudes and for the strong wind speeds of interest in structural design, far exceed those indicated in the ASCE 49-12 Standard.

Therefore, to avoid unconservative designs, an explicit derogation from the ASCE 7 Standard specification of the gradient heights z_g is necessary for buildings with heights greater than z_g.

KEYWORDS

ASCE 7 Standard, ASCE 49-12 Standard, Brunt-Väisäla frequency, conventionally neutral flows, mean wind profiles, Planetary Boundary Layer, super-tall buildings, truly neutral flows.

13th Americas Conference on Wind Engineering
Gainesville, Florida, USA
May 21 - 24, 2017

Consistent analytical urban canopy layer velocity profile for dense building arrangement

Anwar Awol [a]; Girma Bitsuamlak [b]; Fitsum Tariku [c]
a: University of Western Ontario, ON, Canada, ademsis@uwo.ca
b: University of Western Ontario, ON, Canada, gbitsuam@uwo.ca
c: British Colombia Institute of Technology, BC, Canada, fitsum_tariku@bcit.ca

ABSTRACT

Consistent urban canopy layer (UCL) velocity profile is one of the essential methods to characterize local flow around buildings. Based on these local characters; such parameters like pedestrian comfort, pollution transport, and convective heat transfer parameters, urban wind energy potential can be determined in a reliable manner. In the current approach, the velocity is determined using consistent analytical procedure devised by the authors for urban canopy layer of defined morphological state. The derivation process involves the simplification of governing Navier-Stakes equation by employing relevant assumptions for flow in urban roughness settings. First principles from turbulence mixing and eddy viscosity are coupled with linearized frictional drag formulations to come up with the final model differential equation for flow near neighborhoods of various building density. Integration of this differential equation gives rise to a UCL profile that is dependent on the building array density, exposure and climate condition at the site. Results indicate the present UCL profile displays superior traits in terms of the physical essence than conventional relations found in literature that fail to perform well in especially low and high urban density cases. The present profile is able to satisfy the vital conditions such as no-slip near ground, back flow or separation under dense building arrangements and matching the ABL conditions under sparse array scenarios. CFD simulations of several conditions are then utilized to ascertain the consistency of the relation and hence determination of the model coefficients. The results obtained have significant implications in urban design considerations involving pedestrian comfort, pollution transport, convective heat transfer and urban wind energy potential.

KEYWORDS

Urban canopy layer, urban morphology, building density, computational fluid dynamics, array of buildings, velocity profile

The 13th Americas Conference on Wind Engineering (13ACWE)
Gainesville, Florida USA, May 21-24, 2017

Evolutionary Spectra-based Time-varying Coherence Function and Its Engineering Application

Yan Jiang [a], Liuliu Peng [b], Guoqing Huang [c], Xinzhong Chen [d]

[a, b ,c]*Research Center for Wind Engineering, School of Civil Engineering, Southwest Jiaotong University, Chengdu, Sichuan, China*
[d]*National Wind Institute, Department of Civil, Environmental, and Construction Engineering, Texas Tech University, Lubbock, Texas, USA*

ABSTRACT: Priestley's evolutionary spectral theory, due to its physical meaning, is extensively employed in the characterization, modeling and simulation of multivariate nonstationary excitation processes and the structural dynamic analysis under these excitations. When the nonstationary processes are modeled as oscillatory processes, the evolutionary spectra-based coherence function becomes time-invariant. The sigma-oscillatory process theory and Wold-Cramer decomposition model can be used to account for the time-varying coherence function. The Wold-Cramer decomposition model is a special case of the general sigma-oscillatory process theory. Due to the simplicity of Wold-Cramer decomposition model, it is applied to model the nonstationary winds and the corresponding alongwind response analysis framework of tall buildings in engineering field. Taken the measured wind speed data as an numerical example, the time-varying coherence function models of a downburst event are developed. The influence of time-varying coherence of nonstationary winds on the alongwind tall building response is investigated. Results show that use of conventional time-invariant coherence function may underestimate the alongwind tall building response in comparison with the time-varying coherence function.

KEYWORDS: Multivariate nonstationary random process; Evolutionary power spectral density (EPSD); Time-varying coherence function; Downburst; Tall building; Alongwind response.

1 INTRODUCTION

Random excitations such as extreme non-synoptic winds and ground motions are generally modeled as nonstationary processes, which are characterized by time-varying spectra. A variety of methods have been developed to define time-varying spectra, such as evolutionary spectra (Priestley[1] 1965), Wigner-Ville spectra (White and Boashash[2] 1990), short time Fourier transform (Zhan et al[3]. 2006), wavelet transform (Gurley et al[4]. 2003) and empirical mode decomposition (EMD)-based spectra (e.g., Spanos et al[5]. 2007; Huang et al[6]. 2016). Among these methods, the evolutionary power spectral density (EPSD) function characterizes the evolution of the spectral content on the time-frequency domain and can be regarded as the extension of the well-known power spectral density (PSD) to the nonstationary process. Therefore, it has more physical meanings over other spectral models and thus has received wide engineering applications. Especially, it facilitates the simulation of nonstationary processes by the spectral representation approach and the analysis of structural response under nonstationary excitations. EPSD can be estimated by various approaches, such as the classical fixed-window approach (e.g., Priestley[1] 1965; Priestley and Tong[7] 1973), short-time Thomson's multiple-window approach (Conte and Peng[8] 1997), and wavelet transform-based approach (Spanos and Failla[9] 2004;

Huang and Chen[10] 2009). Except the second approach, other two approaches can also be used to estimate the cross EPSD and corresponding coherence function.

Generally, the coherence function of the multivariate nonstationary process has time-varying characteristics. However, due to the difficulty in handling the time-varying coherence, the time-invariant coherence model has been widely used for simplicity. For example, Davenport's coherence model (Davenport[11] 1961) is used to reconstruct nonstationary wind fields (e.g., Peng et al[12]. 2016), while Harichandran and Vanmarcke's coherence model (Harichandran and Vanarcke[13] 1986) is adopted to characterize the nonstationary multiple-point ground motion (e.g., Deodatis[14] 1996). To overcome the limitation of the time-invariant coherence, several attempts have been made to consider the time-varying coherence function. Battaglia[15] (1979) proposed a concept of sigma-oscillatory processes which is the summation of a finite number of oscillatory processes. This concept has been applied to model the univariate nonstationary ground motions and extreme winds (Der Kiureghian and Crempien[16] 1989; Conte and Peng[8] 1997; Huang et al[17]. 2015; Li et al[18]. 2016). In addition, the Wold-Cramer decomposition model has been used to define the time-varying coherence function (Melard and Schutter[19] 1989; Cohen and Walden[20] 2011). Although these theories have been established, less attention has been paid to characterize multivariate nonstationary strong winds with time-varying coherence. More importantly, the structural response features under these excitations are also rarely addressed.

In this paper, the evolutionary spectra-based time-varying coherence function and its application to the analysis of alongwind tall building response under downburst winds are addressed. The main contents are organized as follows. Firstly, the Priestley's classical oscillatory process theory and its limitation on the coherence function are reviewed. Secondly, two evolutionary spectra-based approaches including the sigma-oscillatory process and Wold-Cramer decomposition model are introduced to characterize the processes with time-varying coherence function. Their relationship is also elaborated. Thirdly, the conventional analysis framework of the nonstationary alongwind tall building response is extended to nonstationary winds with time-varying coherence function based on the Wold-Cramer decomposition model. Fourthly, taken measured wind data as an example, the empirical time-varying coherence model is proposed, and the influence of the time-varying coherence function on alongwind tall building responses is highlighted. Finally, concluding remarks are given.

2 CLASSICAL OSCILLATORY PROCESSES

Consider a multivariate zero-mean evolutionary process $\boldsymbol{X}(t)=\left[X_1(t), X_2(t), \cdots, X_n(t)\right]^T$, where T denotes the transpose and n is the number of element processes. Its jth component and corresponding underlying stationary components are expressed as (Priestley[1] 1965; Priestley and Tong[7] 1973)

$$X_j(t)=\int_{-\infty}^{\infty} A_j(\omega,t)e^{i\omega t}dZ_j(\omega) \tag{1}$$

$$X_j^0(t)=\int_{-\infty}^{\infty} e^{i\omega t}dZ_j(\omega) \tag{2}$$

where $A_j(\omega,t),\ j=1,2,\cdots,n$ is a complex-valued slowly-varying modulation function and satisfies $A_j(-\omega,t)=A_j^*(\omega,t)$; ω is circular frequency; $*$ denotes the complex conjugation;

$i=\sqrt{-1}$; $dZ_j(\omega)$ is a complex-valued zero-mean orthogonal incremental process with the following properties

$$dZ_j(-\omega)=dZ_j^*(\omega) \tag{3}$$

$$E\left[dZ_j(\omega)dZ_k^*(\omega')\right]=\delta(\omega-\omega')S_{jk}^0(\omega)d\omega \tag{4}$$

where $\delta(\cdot)$ is the Dirac delta function; $S_{jk}^0(\omega)$ is the cross PSD function between two corresponding stationary element processes $X_j^0(t)$ and $X_k^0(t)$. When $j=k$, $S_{jj}^0(\omega)$ is the auto PSD function of $X_j^0(t)$.

Accordingly, the elements in the spectral matrix $\boldsymbol{S}(\omega,t)$ of the nonstationary process are given as

$$S_{jj}(\omega,t)=\left|A_j(\omega,t)\right|^2 S_{jj}^0(\omega) \tag{5}$$

$$S_{jk}(\omega,t)=A_j(\omega,t)A_k^*(\omega,t)S_{jk}^0(\omega) \tag{6}$$

Similar to the definition of coherence function for the stationary process, the coherence function $\gamma_{jk}(\omega,t)$ of nonstationary processes (the phase is not considered) can be defined as

$$\gamma_{jk}(\omega,t)=\frac{\left|S_{jk}(\omega,t)\right|}{\sqrt{S_{jj}(\omega,t)S_{kk}(\omega,t)}} \tag{7}$$

Substituting Eqs. (5) and (6) into Eq. (7) leads to

$$\gamma_{jk}(\omega,t)=\frac{\left|A_j(\omega,t)A_k^*(\omega,t)S_{jk}^0(\omega)\right|}{\sqrt{\left|A_j(\omega,t)\right|^2 S_{jj}^0(\omega)\left|A_k(\omega,t)\right|^2 S_{kk}^0(\omega)}}=\frac{\left|S_{jk}^0(\omega)\right|}{\sqrt{S_{jj}^0(\omega)S_{kk}^0(\omega)}} \tag{8}$$

Obviously, the coherence function is identical to the time-invariant coherence function of the underlying stationary processes.

In general, the coherence function of nonstationary random processes should be treated as a time-varying function, while the time-invariant coherence function is only a simplified treatment. To overcome this limitation, the evolutionary spectra-based theory with time-varying coherence should be established, which will be discussed in the next section.

3 SIGMA-OSCILLATORY PROCESSES

Consider a multivariate zero-mean evolutionary sigma-oscillatory process $\boldsymbol{Y}(t)=\left[Y_1(t),Y_2(t),\cdots,Y_n(t)\right]^T$. According to Battaglia[15] (1979), the component process $Y_j(t)$ can be defined as the summation of a finite number of oscillatory processes, i.e.,

$$Y_j(t)=\sum_{p=1}^{m}Y_{jp}(t) \tag{9}$$

where $m \le n$ is the number of summation items. When $m=1$, $\boldsymbol{Y}(t)$ becomes an oscillatory process.

Suppose that $Y_{jp}(t)$ and the corresponding underlying stationary components admit the following spectral representation

$$Y_{jp}(t)=\int_{-\infty}^{\infty} A_{jp}(\omega,t)e^{i\omega t}dZ_{jp}(\omega) \tag{10}$$

$$Y_{jp}^{0}(t)=\int_{-\infty}^{\infty} e^{i\omega t}dZ_{jp}(\omega) \tag{11}$$

where $A_{jp}(\omega,t),\ p=1,2,\cdots,m$ is the complex-valued modulation functions of the process $Y_{jp}(t)$ and satisfies $A_{jp}(-\omega,t)=A_{jp}^{*}(\omega,t)$, and $dZ_{jp}(\omega)$ is the complex-valued zero-mean orthogonal incremental process which has

$$dZ_{jp}(-\omega)=dZ_{jp}^{*}(\omega) \tag{12}$$

$$E\left[dZ_{jp}(\omega)dZ_{kq}^{*}(\omega')\right]=\delta(\omega-\omega')S_{jk,pq}^{0}(\omega)d\omega \tag{13}$$

where $S_{jk,pq}^{0}(\omega),\ j,k=1,2,\cdots,n;\ p,q=1,2,\cdots,m$ is the cross PSD between the corresponding underlying stationary processes $Y_{jp}^{0}(t)$ and $Y_{kq}^{0}(t)$. When $j=k$ and $p=q$, $S_{jj,pp}^{0}(\omega)$ is the auto PSD of corresponding underlying stationary process $Y_{jp}^{0}(t)$.

Subsequently, we have

$$S_{jj}(\omega,t)=\sum_{p=1}^{m}\sum_{q=1}^{m} S_{jj,pq}(\omega,t)=\sum_{p=1}^{m}\sum_{q=1}^{m} A_{jp}(\omega,t)A_{jq}^{*}(\omega,t)S_{jj,pq}^{0}(\omega) \tag{14}$$

$$S_{jk}(\omega,t)=\sum_{p=1}^{m}\sum_{q=1}^{m} S_{jk,pq}(\omega,t)=\sum_{p=1}^{m}\sum_{q=1}^{m} A_{jp}(\omega,t)A_{kq}^{*}(\omega,t)S_{jk,pq}^{0}(\omega) \tag{15}$$

Obviously, the coherence function of the sigma-oscillatory process is time-varying.

Consider a special case where

$$dZ_{jp}(\omega)=dZ_{kp}(\omega)=\sqrt{S_{pp}^{0}(\omega)}dZ_{p}(\omega) \tag{16}$$

where $dZ_{p}(\omega),\ p=1,2,\cdots,m$ is the complex-valued zero-mean orthogonal incremental process which satisfies

$$dZ_{p}(-\omega)=dZ_{p}^{*}(\omega) \tag{17}$$

$$E\left[dZ_{p}(\omega)dZ_{q}^{*}(\omega')\right]=\delta(p-q)\delta(\omega-\omega')d\omega \tag{18}$$

Thus, the process component given in Eq. (10) becomes

$$Y_{jp}(t)=\int_{-\infty}^{\infty} A_{jp}(\omega,t)\sqrt{S_{pp}^{0}(\omega)}e^{i\omega t}dZ_{p}(\omega) \tag{19}$$

or can be expressed as

$$Y_{jp}(t) = \int_{-\infty}^{\infty} H_{jp}(\omega,t) e^{i\omega t} dZ_p(\omega) \tag{20}$$

where $H_{jp}(\omega,t) = A_{jp}(\omega,t)\sqrt{S_{pp}^0(\omega)}$.

The corresponding EPSD given by Eqs. (14) and (15) are then expressed as

$$S_{jj}(\omega,t) = \sum_{p=1}^{m} \left| H_{jp}(\omega,t) \right|^2 \tag{21}$$

$$S_{jk}(\omega,t) = \sum_{p=1}^{m} H_{jp}(\omega,t) H_{kp}^*(\omega,t) \tag{22}$$

Obviously, the coherence function is given as

$$\gamma_{jk}(\omega,t) = \frac{\left| \sum_{p=1}^{m} H_{jp}(\omega,t) H_{kp}^*(\omega,t) \right|}{\left[\sum_{p=1}^{m} \left| H_{jp}(\omega,t) \right|^2 \sum_{p=1}^{m} \left| H_{kp}(\omega,t) \right|^2 \right]^{1/2}} \tag{23}$$

which is a time-varying function.

The nonstationary process and the ESPD can be represented in a matrix format as

$$\boldsymbol{Y}(t) = \int_{-\infty}^{\infty} \boldsymbol{H}(\omega,t) e^{i\omega t} d\boldsymbol{Z}(\omega) \tag{24}$$

$$\boldsymbol{S}(\omega,t) = \boldsymbol{H}(\omega,t) \boldsymbol{H}^{T*}(\omega,t) \tag{25}$$

where $d\boldsymbol{Z}(\omega) = \left[dZ_1(\omega), dZ_2(\omega), \cdots, dZ_m(\omega) \right]^T$; $\boldsymbol{H}(\omega,t)$ is a matrix whose elements are $H_{jp}(\omega,t),\ j = 1,2,\cdots,n;\ p = 1,2,\cdots,m$; and $\boldsymbol{S}(\omega,t)$ is a matrix whose elements are $S_{jk}(\omega,t),\ j,k = 1,2,\cdots,n$.

It is noted that the transformation matrix $\boldsymbol{H}(\omega,t)$ can be obtained through Cholesky or eigenvalue decomposition of the spectral matrix $\boldsymbol{S}(\omega,t)$. At this juncture, the decomposition of the nonstationary process is referred to as Wold-Cramer decomposition model (Melard and Schutter[19] 1989; Cohen and Walden[20] 2011). The vector-valued process $\boldsymbol{Y}(t)$ is decomposed into a set of vector-valued subprocesses, such that any two component processes from the same subprocess, say, $Y_{jp}(t)$ and $Y_{kp}(t)$, are statistically fully coherent, while any two component processes from two different subprocesses, say, $Y_{jp}(t)$ and $Y_{kq}(t)$ (where $p \neq q$), are statistically non-coherent. This decomposition is similar to the stochastic decomposition of stationary processes (e.g., Chen and Kareem[21] 2005). This stochastic decomposition model facilitates simulation of nonstationary processes by the classical nonstationary spectral representation method (Deodatis[14] 1996). Based on the above discussions, the Wold-Cramer decomposition model will be applied in the following discussion to describe the nonstationary wind excitations with time-varying coherence.

4 APPLICATION TO ALONGWIND RESPONSE ANALYSES OF TALL BUILDINGS

Consider the alongwind response of a tall building under nonstationary wind excitation. The nonstationary wind speed at the elevation z above the ground is generally described as (e.g., Chen and Letchford[22] 2004; Chen[23] 2008; Solari[24] 2016)

$$U(z,t)=\bar{U}(z,t)+u(z,t) \tag{26}$$

where $\bar{U}(z,t)$ is the time-varying mean wind speed; $u(z,t)$ is a nonstationary random process which can be described by the following Wold-Cramer decomposition model

$$u(z,t)=\sum_{p=1}^{n}\int_{-\infty}^{\infty} H_p(z,\omega,t)e^{i\omega t}dZ_p(\omega) \tag{27}$$

where $H_p(z,\omega,t)$ is a continuous form of $H_{jp}(\omega,t)$ at the elevation z; n is the discretized number of the elevation z.

With the assumption of strip theory, the alongwind force per unit building height at the elevation z above the ground, $P(z,t)$, can be expressed in terms of the oncoming wind speed as (Chen[23] 2008; Huang et al[25]. 2013; Su et al[26]. 2015)

$$P(z,t)=\bar{P}(z,t)+P'(z,t) \tag{28}$$

$$\bar{P}(z,t)=0.5\rho C_D B\bar{U}^2(z,t) \tag{29}$$

$$P'(z,t)=\rho C_D B\bar{U}(z,t)\sum_{p=1}^{n}\int_{-\infty}^{\infty} \chi_D(\omega)H_p(z,\omega,t)e^{i\omega t}dZ_p(\omega) \tag{30}$$

where ρ is the air density and equal to $1.225\ \text{kg/m}^3$; C_D is the drag coefficient and takes 1.2 in this study; B is the building width; $\chi_D(\omega)$ is the aerodynamic admittance function. In this study, Davenport's aerodynamic admittance function is adopted as follows

$$|\chi_D(\omega)|^2=2/\lambda_y(1-1/\lambda_y+1/\lambda_y e^{-\lambda_y}) \tag{31}$$

where $\lambda_y=k_y fD/U_{\max}$; k_y is the decay factor and taken 8 in this study; f is the frequency; $U_{\max}$ is the maximum mean wind speed over the building height.

Subsequently, the time-varying mean and fluctuating components of the generalized force can be given by

$$\bar{Q}(t)=\int_0^H \bar{P}(z,t)\Phi(z)dz \tag{32}$$

$$Q'(t)=\int_0^H P'(z,t)\Phi(z)dz \tag{33}$$

where $\Phi(z)$ is the building fundamental mode shape.

Substituting Eq. (29) into Eq. (32), the following equation can be obtained

$$\bar{Q}(t)=Q_R\frac{1}{H}\int_0^H\frac{\bar{U}^2(z,t)}{U_{\max}^2}\Phi(z)dz \tag{34}$$

where $Q_R = 0.5\rho C_D BHU_{\max}^2$.

The EPSD of $Q'(t)$ can be expressed as

$$S_{Q'}(\omega,t) = 4Q_R^2/(HU_{\max})^2 \int_0^H \int_0^H \frac{\bar{U}(z_1,t)}{U_{\max}} \frac{\bar{U}(z_2,t)}{U_{\max}} \Phi(z_1)\Phi(z_2)\left|\chi_D(\omega)\right|^2 S_u(z_1,z_2,\omega,t)dz_1dz_2 \quad (35)$$

where $S_u(z_1,z_2,\omega,t)$ is the cross EPSD between $u(z_1,t)$ and $u(z_2,t)$, which can be expressed as

$$S_u(z_1,z_2,\omega,t) = \sqrt{S_u(z_1,\omega,t)S_u(z_2,\omega,t)}\gamma_u(z_1,z_2,\omega,t) \quad (36)$$

where $S_u(z_1,\omega,t)$ and $S_u(z_2,\omega,t)$ are the auto EPSDs of $u(z_1,t)$ and $u(z_2,t)$; $\gamma_u(z_1,z_2,\omega,t)$ is the time-varying coherence function between $u(z_1,t)$ and $u(z_2,t)$. It should be noted that if $\gamma_u(z_1,z_2,\omega,t)$ reduce to a time-invariant coherence function $\gamma_u(z_1,z_2,\omega)$, Eq. (35) will reduce to that presented in Chen[23] (2008).

The time-varying mean of the generalized displacement $r(t)$ can be directly evaluated by the quasi-static analysis as $r(t) = \bar{Q}(t)/M_1\omega_1^2$ where M_1 and ω_1 are the generalized mass and structural frequency of the building in fundamental mode, respectively. Meanwhile, the EPSD $S_r(\omega,t)$ of the generalized displacement $r(t)$ can be evaluated in the frequency domain with consideration of transient structural dynamics (e.g., Chen[23][27] 2008, 2016). The time-varying STD of the generalized displacement and those of other responses (e.g., displacement, shear force and bending moment) can be subsequently computed. The time-varying mean and STD can be further used to determine the extreme value distribution based on upcrossing theory (e.g., Huang et al[25]. 2013; Chen[27] 2016).

5 NUMERICAL EXAMPLES

5.1 *Measured downburst winds*

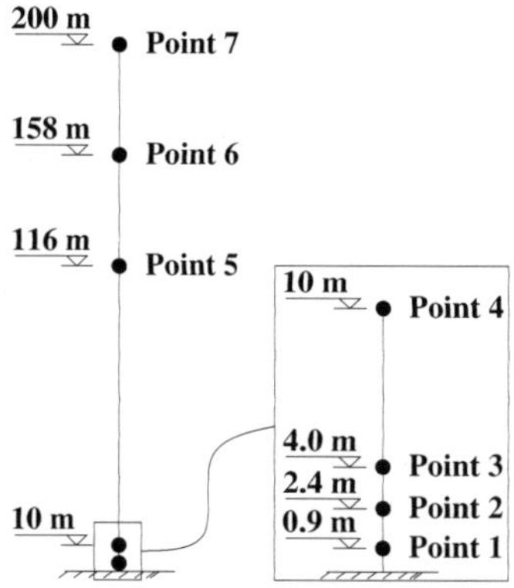

Figure 1. Schematic diagram of the layout of anemometers

In this study, a thunderstorm downburst, which was measured on June 19, 2008 at Texas Tech University, is employed to investigate their effects on structures with consideration of time-varying coherence functions. The anemometers were located on heights of 0.9, 2.4, 4, 10, 116, 158 and 200 m and these measured points are denoted as Points 1-7, as shown in Figure 1. The sampling frequency of both downbursts was 30 Hz. To eliminate the influence of high-frequency noise due to the anemometer response issues, the sampling frequency was downsampled to 1 Hz by averaging the measured wind speed data, which is consistent to the treatment used by Orwig and Schroeder[28] (2007). The detailed information about the field measurement can be referred to Lombardo[29] (2009) and Lombardo et al[30]. (2014).

Since the temporal scales of downburst winds range from 50 to 400 s, the samples with the duration of 1800 s are chosen for the analysis. The wind speed time histories during this event are shown in Figure 2. It can be seen that the event have the "ramp-up" characteristics, i.e., the wind speed increases and then decreases sharply within short time duration. In addition, the wind speed time histories at different heights have similar variation trend.

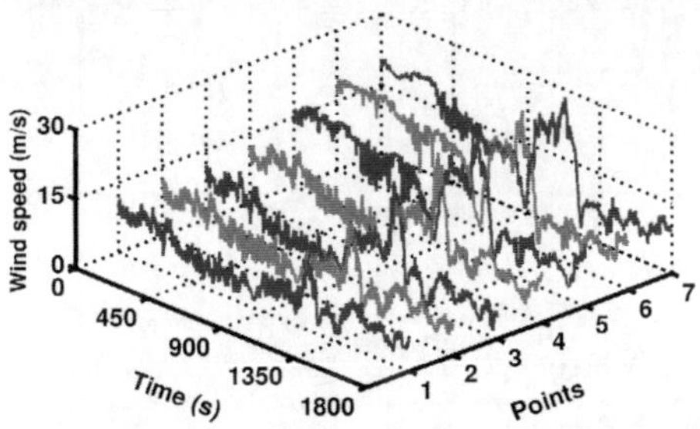

Figure 2. Wind speed time histories of two downburst events

The time-varying mean wind speed is extracted from the measured wind speed records, which can be achieved by using empirical model decomposition (EMD) or discrete wavelet transform (DWT) approach (e.g., Huang and Chen[10] 2009; Su et al[26]. 2015). In this study, the DWT approach is used where Daubechies wavelet of order 10 (Db 10) with decomposition levels of 6 (equivalent to a window size of 64 s) is employed. Accordingly, the maximum frequency embedded in the mean is less than 0.0078 Hz.

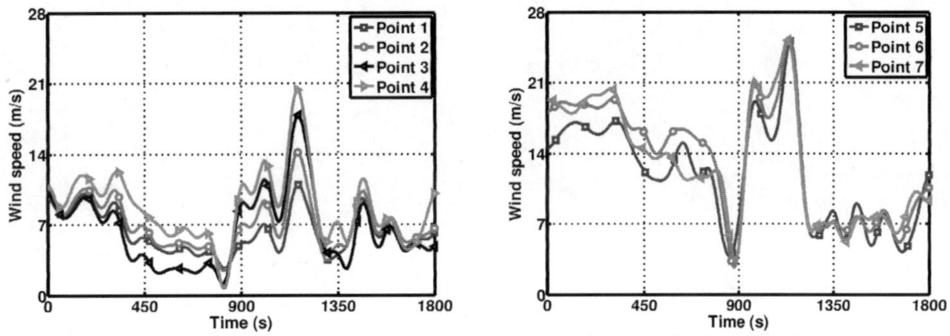

(a) Points 1-4 (b) Points 5-7

Figure 3. Time-varying mean wind speeds

The decomposed time-varying mean wind speeds this event are shown in Figure 3. It can be seen that the time-varying mean wind speeds at different heights have the similar shape. In addition, the wind speeds peak sharply at around 1150 s. Due to the lack of measurements data between height of 10 m and 116 m, this set of measurements are not sufficient to determine the time-varying mean wind speed profile. Hence, only the time variations of mean wind speed at points 4 and 6 are used in the following discussion.

5.2 *Estimation of time-varying coherence functions*

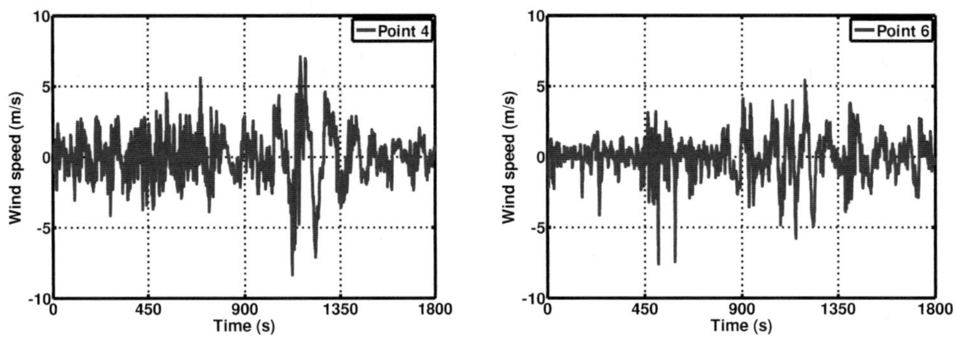

(a) Point 4 (b) Point 6

Figure 4. Wind fluctuations

The wind fluctuations are extracted from the original record by subtracting the time-varying mean component. Figures 4 is the wind fluctuations of Points 4 (10 m high) and 6 (158 m high). The EPSDs of these wind fluctuations can be estimated. For simplicity, the classical Priestley's method is employed here (Priestley and Tong[7] 1973). The essence of this estimation method is to perform a generalized linear transformation on the interested data at a specific frequency by means of a filter function $g(u)$ first. Then, the other weight-function $W_{T'}(t)$ is used to smooth the estimated values for neighboring values of t in order to decrease the estimation error. Clearly, this method is restricted to the uncertainty principle and the selection of filter and weighting functions will have the influence on the estimated results. In this study, these two functions are given by

$$g(u) = \begin{cases} 1/(2\sqrt{h\pi}) & |u| \le h \\ 0 & |u| > h \end{cases} \tag{37}$$

and

$$W_{T'}(t) = \begin{cases} 1/T' & |t| \le \frac{1}{2}T' \\ 0 & \text{otherwise} \end{cases} \tag{38}$$

where the parameters are chosen as $h = 7\Delta t$ and $T' = 200\Delta t$, respectively; and $\Delta t = 1$ s.

The auto and cross EPSDs are estimated as well. For simplicity, only the auto EPSDs are illustrated and analyzed here, as shown in Figure. 5 (where the results in 0.25 to 0.5 Hz are omitted). The time-varying STDs calculated from these auto EPSDs are shown in Figure 6. It is observed that the maximum spectral values all appear at around 1200 s and the overall variation trends are consistent. This observation is also reflected in the comparison of the seven time-

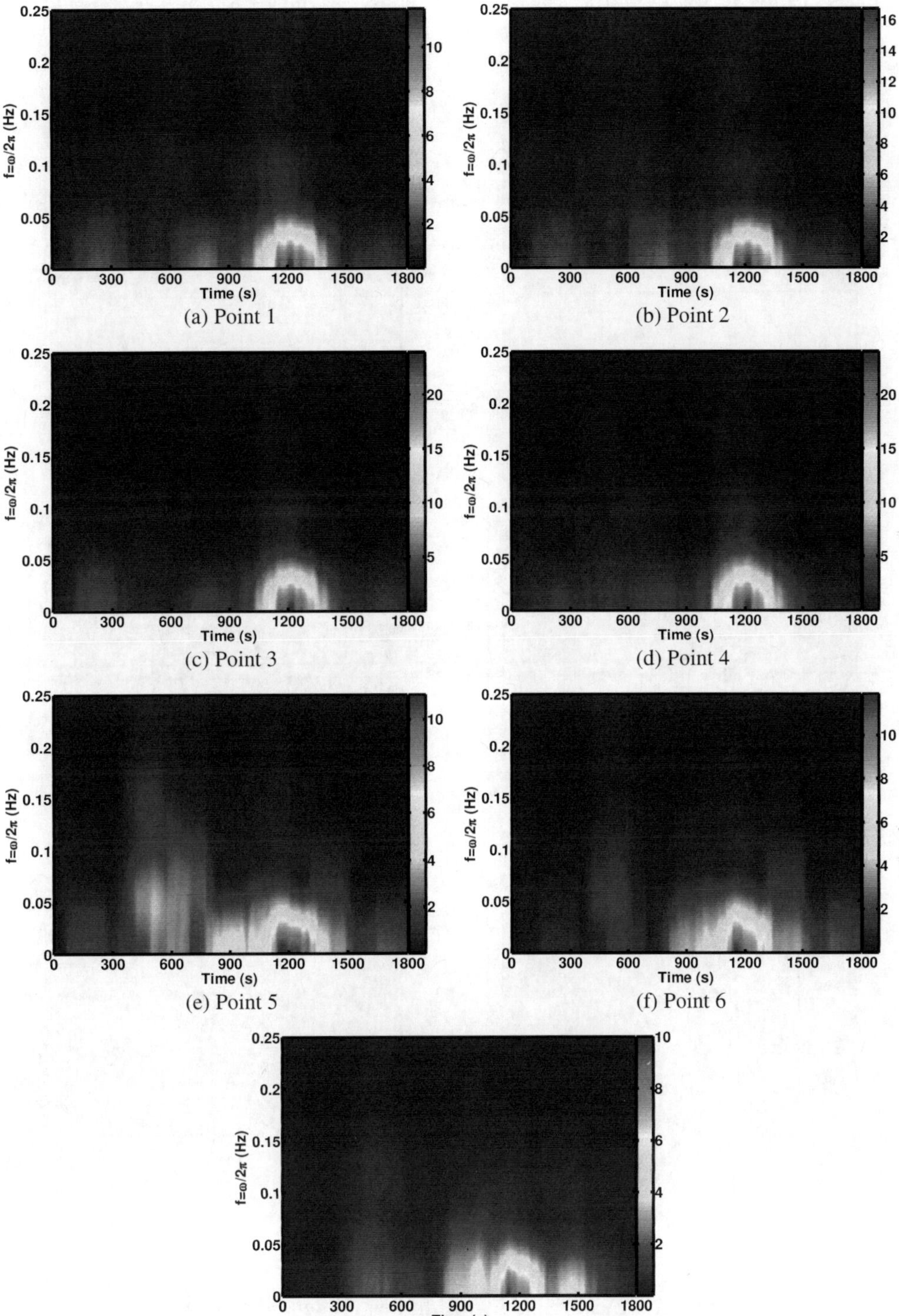

Figure 5. Estimated auto EPSDs

varying STDs. In the following structural response analysis, the EPSD at a representative location will be used to represent the auto EPSDs over the height. Both Points 4 and 6 will be served as the referent point for comparison.

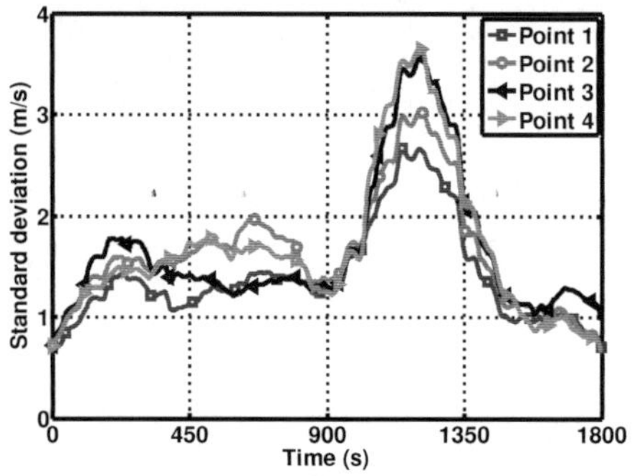

(a) Points 1-4

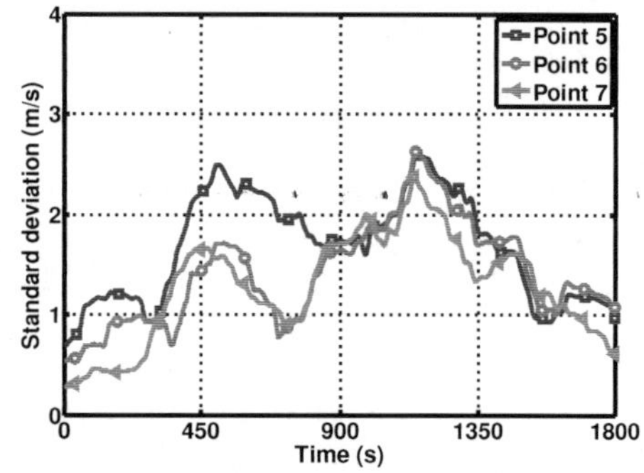

(b) Points 5-7

Figure 6. Estimated time-varying STDs

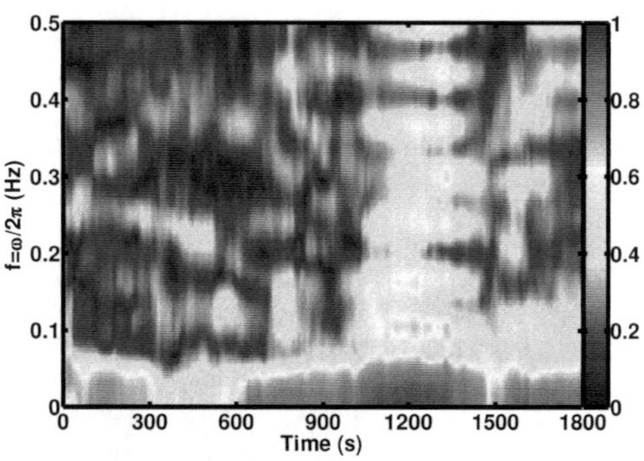

(a) $\gamma_{41}(\omega,t)$

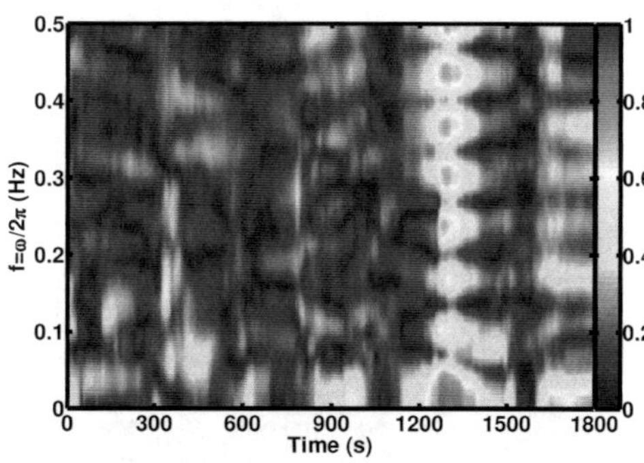

(b) $\gamma_{54}(\omega,t)$

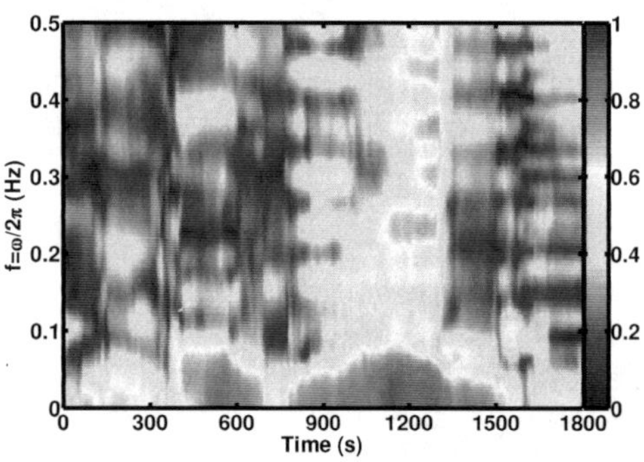

(c) $\gamma_{65}(\omega,t)$

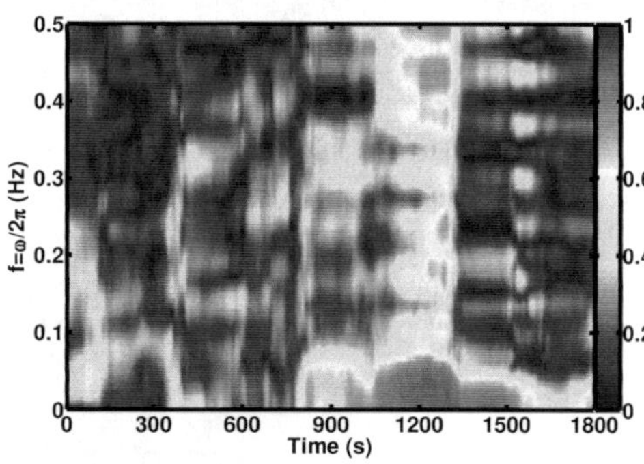

(d) $\gamma_{76}(\omega,t)$

Figure 7. Estimated coherence functions

The coherence functions are calculated from the estimated EPSDs, which are illustrated in Figure 7, where $\gamma_{jk}(\omega,t)$ denotes coherence function between Points j and k. It can be seen that the coherence functions have noticeable time-varying characteristics. Specifically, the coherence functions have much larger value at the duration from 900 to 1350 s, which is consistent with the time variation of STDs, indicating that larger fluctuation has stronger coherence. Similar observations were also obtained by the time-frequency analysis based on the multivariate EMD and wavelet for other measurement data (Huang et al[6]. 2016). It is also observed that the coherence function decreases with the increase of the separation distance. When the separation distance is larger than 100 m, the coherence function becomes very small, as shown in Figure 7b. In the following fitting procedure, this part of measured time-varying coherence functions is neglected. Lastly, the influence of height on the coherence function is insignificant by comparing the coherence functions $\gamma_{65}(\omega,t)$ and $\gamma_{76}(\omega,t)$.

5.3 *Time-varying coherence function models*

The time-varying coherence functions estimated from both events can be fitted into the following two models for the purpose of comparison:

$$\gamma(\Delta z,\omega,t) = \exp\left\{-K_1(\Delta z)\omega\Delta z/[2\pi\sigma_4(t)]\right\} \tag{39}$$

and

$$\gamma(\Delta z,\omega,t) = \exp\left\{-K_2(\Delta z)\omega\Delta z/[2\pi\sigma_6(t)]\right\} \tag{40}$$

where $K_1(\Delta z)$ and $K_2(\Delta z)$ are decay factors and are functions of the distance Δz; $\sigma_4(t)$ and $\sigma_6(t)$ are the time-varying STDs of wind speeds at Points 4 and 6, respectively. For the sake of denotation, these two models are referred to as Models 1 and 2, respectively. Note that the influence of elevation on the coherence function is not considered in this study. For each time-varying coherence function with a specific separation distance, $K_1(\Delta z)$ and $K_2(\Delta z)$ can be obtained through a nonlinear least-square curve fitting. For each event, a total of 10 values of $K_1(\Delta z)$ and $K_2(\Delta z)$ are calculated from the measurements at 7 locations, where the very high values at distance larger than 100 m are neglected here.

Table 1. Decay factors of time-varying coherence functions

Parameters	$K_1(\Delta z)$	$K_2(\Delta z)$
k_1	2.212	1.887
k_2	0.125	0.126
k_3	0.263	0.222

For simplicity, only the fitted results of $K_1(\Delta z)$ are illustrated and analyzed here, as shown in Figure 8. $K_1(\Delta z)$ decreases with the increase of separation distance. Because the coherence function should decrease with the increase of separation distance, $K_1(\Delta z)\Delta z$ will be a positive and monotonically increasing function. The following nonlinear equation is used to fit the discretization value of $K_1(\Delta z)$

$$K_1(\Delta z) = k_1 \exp(-k_2\Delta z) + k_3 \tag{41}$$

where k_1, k_2 and k_3 are constant parameters determined by curve fitting as shown in Figure 8 and Table 1. The comparison of $K_1(\Delta z)\Delta z$ is shown in Figure 9. It is seen that the nonlinear equation can represent the main variation characteristics of $K_1(\Delta z)$ very well. If $K_1(\Delta z)$ is simplified as a constant k_3 which is widely used in the time-invariant coherence function, the estimation error of $K_1(\Delta z)\Delta z$ at the distance less than 50 m is obvious, as shown in Figure 9. The decay factor $K_2(\Delta z)$ is modeled similarly and the results are summarized in Table 1.

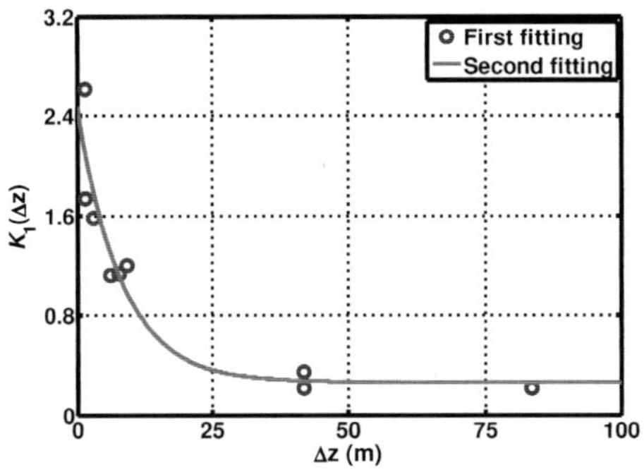

Figure 8. Fitted parameter $K_1(\Delta z)$

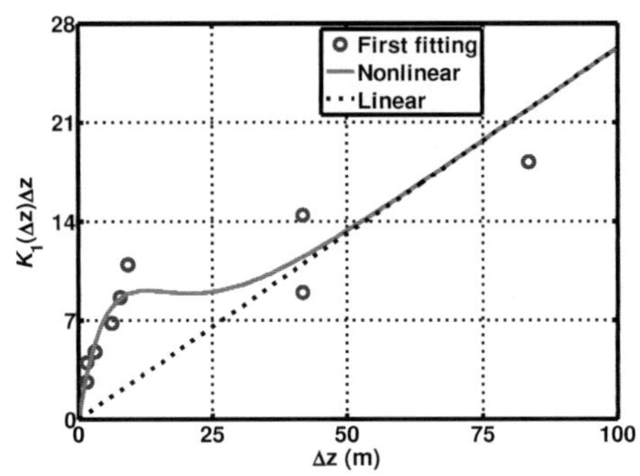

Figure 9. Comparison of different fitting equations

5.4 *Time-invariant coherence function models*

In order to investigate the influence of time-varying coherence function model, the corresponding time-invariant coherence function model is also established. Two time-invariant coherence function models are given by

$$\gamma(\Delta z,\omega)=\exp\left[-K_1(\Delta z)\omega\Delta z/(2\pi\bar{\sigma}_4)\right] \tag{42}$$

and

$$\gamma(\Delta z,\omega)=\exp\left[-K_2(\Delta z)\omega\Delta z/(2\pi\bar{\sigma}_6)\right] \tag{43}$$

where $K_1(\Delta z)$ and $K_2(\Delta z)$ are the same with those in Eqs. (39) and (40); $\bar{\sigma}_4$ and $\bar{\sigma}_6$ are the average values of the $\sigma_4(t)$ and $\sigma_6(t)$ over a selected time duration, respectively. To investigate the influence of the selected duration, two durations including 1-1800 s and 820-1350 s are employed. Note that the second duration is selected according to the "ramp-up" characteristics of the measured downburst events. For better denotation, the models given by Eqs. (42) and (43) associated with the first duration are referred to as Models 3 and 4, respectively, while those using the second duration are referred to as Models 5 and 6, respectively.

The comparison of these four time-invariant coherence function models at the separation distance 5, 50 and 200 m are shown in Figure. 10. As expected, the coherence functions of Models 3 and 4 are obviously smaller than those of Models 5 and 6.

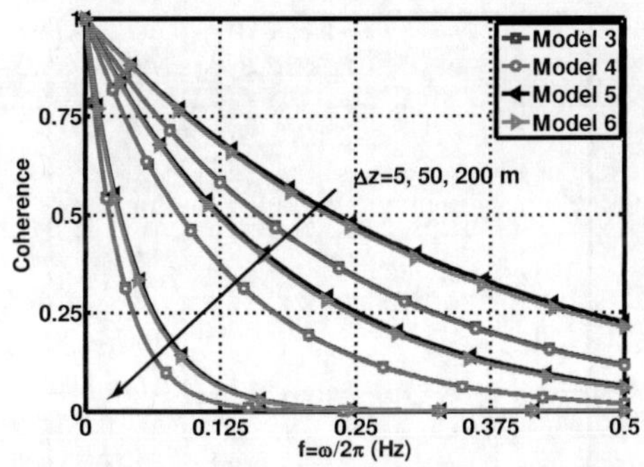

Figure 10. Comparison of time-invariant coherence function Models 3-6

5.5 *Analysis of alongwind tall building response*

In the following discussion, three tall buildings with heights of 100, 200 and 300 m, and square section of $B = D = 40$ m are considered as examples. The fundamental frequency and damping ratio are $f_1 = 46/H$ and $\xi_1 = 0.01$, respectively. In addition, the building density is 192 kg/m^3.

Two simplified time-varying mean wind profiles are considered, which can be expressed as

$$\bar{U}(z,t) = d(t)\bar{U}(z)U_{\max} \tag{44}$$

where the modulation function $d(t)$ is defined as the ratio of mean wind speed of downburst events at the reference height of 10 m (Point 4) or 158 m (Point 6) to its corresponding maximum value U_R; $U_{\max}$ is the maximum wind speed along the vertical profile which is set to the maximum mean wind speed of the measured event, i.e., 25.19 m/s; $\bar{U}(z)$ is the normalized wind profile model. Currently, a number of empirical models have been used to describe the time-invariant mean wind profile of downburst winds. In this study, the Wood model is employed to describe the normalized mean wind profile, which is expressed as (Wood et al. 2001)

$$\bar{U}(z) = 1.55(z/\delta)^{1/6}\left[1 - \text{erf}(0.7z/\delta)\right] \tag{45}$$

where $\text{erf}(z)=2/\sqrt{\pi}\int_0^z e^{-x^2}dx$; δ is the height where the wind speed reaches the half of its maximum value; According to U_R at height of 10 or 158 m and the maximum wind speed $U_{\max}$ of the vertical profile, δ can be calculated from the Wood model.

Table 2. Calculated δ in different cases

	δ (m)	
	Profile 1 (10 m)	Profile 2 (158 m)
Event 1	450	660

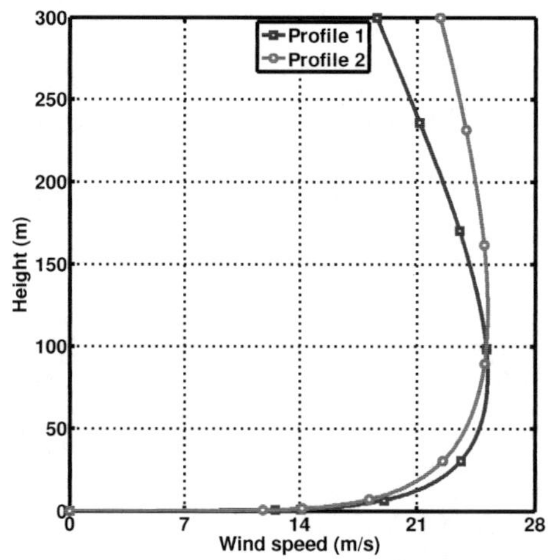

Figure 11. Simplified time-invariant wind profiles

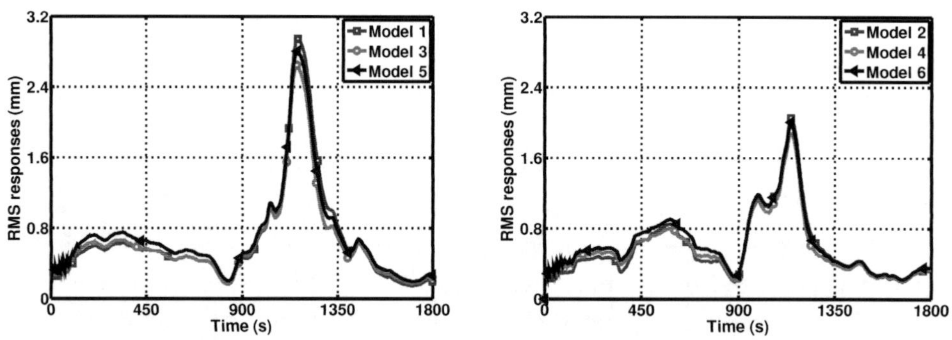

(a) Profile 1 (b) Profile 2

Figure 12. Comparison of RMSs of tip displacement (H =100 m)

For the reference height of 10 and 158 m, the simplified profiles are denoted as Profiles 1 and 2, respectively. The obtained δ are summarized in Table 2 while the corresponding simplified time-invariant profiles are shown in Figure 11. It can be seen that the maximum values of the wind profile for all the cases occur at the height of less than 130 m. In addition, Profile 1 decreases more quickly with the increase of height than that of Profile 2 when the height is larger than that of the maximum wind speed. It implies that the mean response of Profile 1 will be smaller than that of Profile 2 when the building height is larger than some extent.

During the calculation of response analysis, the reference height used in the simplified time-varying mean wind profile, coherence function and auto EPSD will be consistent. For example, when the reference height of 10 m is adopted, the Profile 1, $\sigma_4(t)$ or $\bar{\sigma}_4$ and auto EPSD at Point 4 will be used in the response analysis.

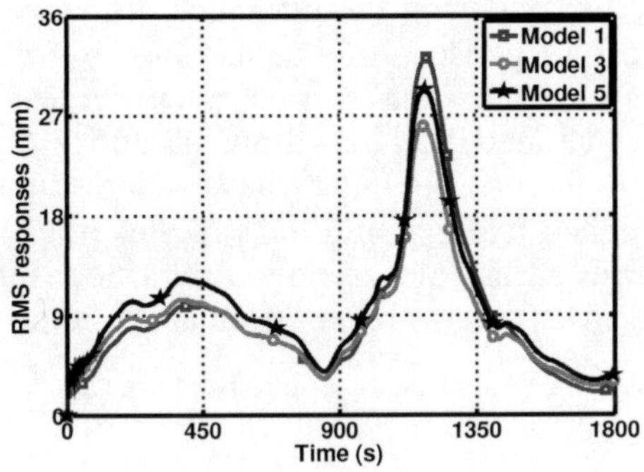

(a) Profile 1

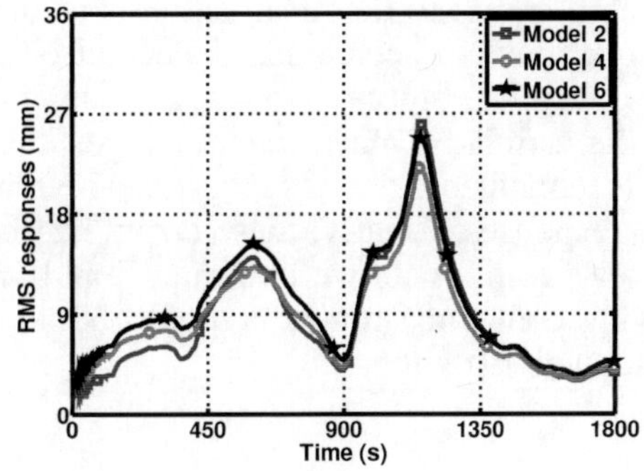

(b) Profile 2

Figure 13. Comparison of RMSs of tip displacement (H =300 m)

Table 3. Comparison of predicted building responses (Profile 1; m)

	$H=100$		$H=200$		$H=300$	
	Mean (0.0105)		Mean (0.0399)		Mean (0.0760)	
Model	RMS	Extreme	RMS	Extreme	RMS	Extreme
1	0.0029	0.0179	0.0138	0.0706	0.0322	0.1421
3	0.0026	0.0171	0.0115	0.0653	0.0260	0.1292
5	0.0028	0.0175	0.0127	0.0681	0.0293	0.1363

Note: Mean, RMS and Extreme denote maximum Mean, RMS and mean extreme

Table 4. Comparison of predicted building responses (Profile 2; m)

	$H=100$		$H=200$		$H=300$	
	Mean (0.0101)		Mean (0.0421)		Mean (0.0892)	
Model	RMS	Extreme	RMS	Extreme	RMS	Extreme
2	0.0021	0.0150	0.0102	0.0638	0.0260	0.1406
4	0.0019	0.0147	0.0089	0.0609	0.0221	0.1327
6	0.0020	0.0149	0.0097	0.0629	0.0247	0.1383

Based on the response analysis framework with the consideration of time-varying coherence function proposed in this study and the conventional counterpart proposed by Chen[23] (2008), the alongwind tall building responses for all cases can be calculated.

Figure 13 shows the comparison of root mean squares (RMSs) of the tip displacement when $H=100$ m is used. Similarly, the building response with $H=300$ m is shown in Figure 14. It can be seen that the difference of RMS response between the cases of the time-varying and time-invariant coherence function is not obvious when $H=100$ m. However, the RMS response under time-varying coherence function is noticeably larger than those under other two time-invariant coherence functions for the building with $H = 300$ m.

Tables 3 and 4 illustrate the estimated response including the maximum mean, maximum RMS and the mean extreme of the tip displacement. It can be found that the maximum mean of the tip displacement among these cases mainly depends on the simplified time-invariant wind profile. Furthermore, the maximum RMSs under the time-varying coherence function are greater than those under the time-invariant coherence functions. With the increase of building height, the discrepancy tends to be obvious. The largest differences may reach to 20% for building with $H = 300$ m.

When it comes to the mean extreme of the tip displacement, the influence of the modeling of coherence function decreases. It is because the mean response component, which is not affected by the coherence function, makes a great contribution to the total response. The largest differences in the estimated mean maximum response are about 10%. Based on the above observations, it is concluded that use of time-invariant coherence function may underestimate the structural responses. Clearly, these errors can be reduced by choosing the time-invariant coherence Model 5 or 6, as shown in Tables 3 and 4. However, in order to obtain the fine results, the time-varying coherence function should be adopted in structural response analysis under thunderstorm downburst winds.

6 CONCLUSIONS

In this paper, an evolutionary spectra-based framework for modeling the nonstationary process with time-varying coherence function had been presented, which could eliminate the limitation of classical modeling of time-invariant coherence function. In addition, the framework was applied to the alongwind response of tall buildings to nonstationary winds in engineering field, which can be seen as the extension of the conventional response analysis framework with time-invariant coherence function of wind fluctuations.

In this framework, the Wold-Cramer decomposition model was derived from the perspective of the sigma-oscillatory process theory and the former could be regarded as the special case of the latter. This nonstationary process model is able to account for the time-varying coherence function, and facilitate the modeling and simulation of nonstationary processes. Based on measurement data, the time-varying coherence function models of a downburst event was derived. The alongwind tall building responses under time-varying coherence and time- invariant coherence models were evaluated. Results showed that time-invariant coherence function may underestimate the alongwind tall building response to nonstationary winds in comparison with the time-varying coherence function.

7 ACKNOWLEDGEMENTS

The support by the Doctoral Innovation Fund of Southwest Jiaotong University and the National Natural Science Foundation of China (Grant No. 51578471, 51408504 and 51478401) are greatly acknowledged. Dr. Lombardo is also greatly acknowledged for providing the measured downburst data.

8 REFERENCES

[1] Priestley, M. B. (1965). "Evolutionary spectra and non-stationary processes." J. Roy. Statist. Soc. Ser. B, 27, 204-237.

[2] White, L. B., and Boashash, B. (1990). "Cross spectral analysis of nonstationary processes." IEEE Trans. Inform. Theory., 36(4), 830-835.

[3] Zhan, Y., Halliday, D., Jiang, P., Liu, X., and Feng, J. (2006). "Detecting time-dependent coherence between non-stationary electrophysiological signals-a combined statistical and time-frequency approach." J. Neurosci. Meth., 156(1), 322-332.

[4] Gurley, K., Kijewski, T., and Kareem, A. (2003). "First-and higher-order correlation detection using wavelet transforms." J. Eng. Mech., 129(2), 188-201.

[5] Spanos, P. D., Giaralis, A., and Politis, N. P. (2007). "Time-frequency representation of earthquake accelerograms and inelastic structural response records using the adaptive chirplet decomposition and empirical mode decomposition." Soil. Dyn. Earthq. Eng., 27(7), 675-689.

[6] Huang, G., Su, Y., Kareem, A., and Liao, H. (2016). "Time-frequency analysis of nonstationary process based on multivariate empirical mode decomposition." J. Eng. Mech., 142(1), 04015065.

[7] Priestley, M. B., and Tong, H. (1973). "On the analysis of bivariate non-stationary processes." J. Roy. Statist. Soc. Ser. B, 35, 153-166.

[8] Conte, J. P., and Peng, B. F. (1997). "Fully nonstationary analytical earthquake ground-motion model." J. Eng. Mech., 123(1), 15-24.

[9] Spanos, P. D., and Failla, G. (2004). "Evolutionary spectra estimation using wavelets." J. Eng. Mech., 130(8), 952-960.

[10] Huang, G., and Chen, X. (2009). "Wavelets-based estimation of multivariate evolutionary spectra and its application to nonstationary downburst winds." Eng. Struct., 31(4), 976-989.

[11] Davenport, A. G. (1961). "The spectrum of horizontal gustiness near the ground in high winds." Q. J. R. Meteorol. Soc., 87(372), 194-211.

[12] Peng, L., Huang, G., Chen, X., and Kareem, A. (2016). "Simulation of multivariate nonstationary random process: A hybrid stochastic wave and proper orthogonal decomposition approach." J. Eng. Mech. (In press).

[13] Harichandran, R., and Vanmarcke, E. (1986). "Stochastic variation of earthquake ground motion in space and time." J. Eng. Mech., 112(2), 154-174.

[14] Deodatis, G. (1996). "Non-stationary stochastic vector processes: seismic ground motion applications." Prob. Eng. Mech., 11, 149-168.

[15] Battaglia, F. (1979). "Some extensions in the evolutionary spectral analysis of a stochastic process." Boll. Un. Mat. Ital. B (5), 16, 1154-1166.

[16] Der Kiureghian, A., and Crempien, J. (1989). "An evolutionary model for earthquake ground motion." Struct. Safety., 6(2), 235-246.

[17] Huang, G., Zheng, H., Xu, Y., and Li, Y. (2015). "Spectrum models for nonstationary extreme winds." J. Struct. Eng., 141(10), 04015010.

[18] Li, Y., Conte, J. P., and Barbato, M. (2016). "Influence of time-varying frequency content in earthquake ground motions on seismic response of linear elastic systems." Earthquake. Eng. Struct. Dyn., 45(8), 1271-1291.

[19] Melard, G., and Schutter, A. H. D. (1989). "Contributions to evolutionary spectral theory." J. Time. Ser. Anal., 10(1), 41-63.

[20] Cohen, E. A., and Walden, A. T. (2011). "Wavelet coherence for certain nonstationary bivariate processes." IEEE Trans. Signal Process., 59(6), 2522-2531.

[21] Chen, X., and Kareem, A. (2005). "Proper orthogonal decomposition-based modeling, analysis, and simulation of dynamic wind load effects on structures." J. Eng. Mech., 131(4), 325-339.

[22] Chen, L., and Letchford, C.W. (2004). "A deterministic-stochastic hybrid model of downbursts and its impact on a cantilevered structure." Eng. Struct., 26, 619-629.

[23] Chen, X. (2008). "Analysis of alongwind tall building response to transient nonstationary winds." J. Struct. Eng., 134(5), 782-791.

[24] Solari, G. (2016). "Thunderstorm response spectrum technique: Theory and applications." Eng. Struct., 108, 28-46.

[25] Huang, G., Chen, X., Liao, H., and Li, M. (2013). "Predicting tall building response to nonstationary winds using multiple wind speed samples." Wind. Struct., 17(2), 227-244.

[26] Su, Y., Huang, G., and Xu, Y. L. (2015). "Derivation of time-varying mean for non-stationary downburst winds." J. Wind. Eng. Ind. Aerodyn., 141, 39-48.

[27] Chen, X. (2016). "Estimation of wind load effects with various mean recurrence intervals with a closed-form formulation." Int. J. Struct. Stab. Dyn., 1550060.

[28] Orwig, K. D., and Schroeder, J. L. (2007). "Near-surface wind characteristics of extreme thunderstorm outflows." J. Wind. Eng. Ind. Aerodyn., 95(7), 565-584.

[29] Lombardo, F. T. (2009). "Analysis and interpretation of thunderstorm wind flow and its effects on a bluff body." Doctoral thesis, Texas Tech University., Lubbock, TX.

[30] Lombardo, F. T., Smith, D. A., Schroeder, J. L., and Mehta, K. C. (2014). "Thunderstorm characteristics of importance to wind engineering." J. Wind. Eng. Ind. Aerodyn., 125, 121-132.

[31] Wood, G. S., Kwok, K. C. S., Motteram, N. A., and Fletcher, D. F. (2001). "Physical and numerical modeling of thunderstorm downbursts." J. Wind. Eng. Ind. Aerodyn., 89(6), 535-552.

Developments in remote sensing of hurricane winds from air and space

Mark Powell [a]

[a]*Risk Management Solutions, Tallahassee, FL, Mark.Powell@rms.com*

ABSTRACT

Observation-based tropical cyclone wind fields are a necessary component of a forecast and warning decision support system and provide the ground truth for developing and validating risk models used by the insurance industry, as well as providing the time-dependent forcing to run storm surge and wave models. While tropical cyclones continue to be monitored by hurricane reconnaissance aircraft in the Atlantic, Eastern Pacific, and Central Pacific basins, the coverage over the rest of the globe is limited. Recent developments in space-based remote sensing, as well as programs utilizing the NOAA research aircraft as a testbed for development of new observing technologies, provide optimism for more complete global coverage of tropical cyclones. NASA's SMAP and ESA's SMOS platforms, originally developed to detect soil moisture and salinity, now show promise in detecting extreme winds. NASA's CYGNSS micro-satellite constellation, launched in December, measures scattering from the specular reflection of GPS signals on the ocean surface to measure high winds from a tropical orbit that provides multiple storm revisits per day. Polarimetric measurements from synthetic aperture radars operated by the Canadian Space Agency have sampled several Atlantic hurricanes, allowing for the development of refined geophysical model functions to retrieve winds from microwave measurements. These promising developments point towards a future in which tropical cyclones are monitored globally. The presentation will review current remote sensing developments under consideration as future data feeds to the RMS-HWind real-time wind analysis system.

KEYWORDS

Tropical cyclone, remote sensing, HWind

Effects of combined surge, wave, and wind loads on the buckling of aboveground storage tanks

Carl Bernier [a], Jamie E Padgett [a]

[a] *Dept. of Civil and Environmental Eng., Rice University, Houston, TX, USA*

ABSTRACT: Aboveground storage tanks (ASTs) have suffered major damage in past hurricanes due to surge, wave, and wind loads. While the vulnerability of ASTs to such loads has been studied in the past, the effects of combined loads have not been fully addressed. This paper explores the effects of combined surge, wave, and wind loads on the buckling behavior of a typical AST. A pressure model for wave impacts is developed using finite element analysis and Artificial Neural Networks. This model is then coupled with an existing wind pressure model to perform buckling analysis for different uncorrelated combinations of wave height and wind speed using an experimental design approach. Results of these analyses are used to estimate the buckling response as a function of tank parameters and load parameters using a stepwise regression analysis. Key insights on the influence of each of the loads are obtained. The results show that wind loads do not significantly affect the buckling behavior of an AST subjected to surge effects and may be reasonably neglected when performing buckling analysis. Lastly, this study highlights the importance of considering wind induced wave loads on ASTs during hurricane events.

KEYWORDS: Aboveground storage tank; storm surge; wave impacts; wind loads; multi-hazard; finite element analysis

1 INTRODUCTION

Aboveground storage tanks (ASTs) are widely used in coastal areas to store large quantities of petrochemical products. Typically, ASTs are constructed of thin-walled steel shells forming a vertical cylinder. The geometry of ASTs allows them to efficiently withstand the pressure generated by liquid inside them. However, this design also leaves these structures sensitive to external loads such as wind, waves, and storm surge. Recent hurricane events in the United States (US) have highlighted this vulnerability. Post-hurricane investigations have shown that the failure modes of ASTs were flotation due to excessive buoyant forces and buckling due to high water or wind pressure [1,2]. Such AST failures result in spillage of hazardous materials in the surrounding environment, costly cleanups, and significant economic impacts, highlighting the importance of evaluating the structural behavior of these storage systems.

Several studies have assessed the behavior and vulnerability of ASTs under hurricane wind [3,4] or storm surge [5]. However, these studies generally focus on a single load, neglecting the possible combined effects of these loads. Post-hurricane investigations have shown that in some cases, the failure of ASTs might have been caused by the combined actions of these loads [1,6]. Thus, it is essential to have a better understanding of the behavior of ASTs under concurrent multi-hazard loads in order to adequately assess the risks associated with these structures during hurricane events. Moreover, very few studies have assessed the behavior of ASTs under hydrodynamic load and wave impacts. Studies focusing on storm surge usually neglect hydrodynamic effects and only consider the hydrostatic effect of storm surge.

Thus, the objective of this paper is to provide a more comprehensive understanding of the behavior of ASTs during hurricanes events by exploring the effects of combined surge, wave, and wind loads on the buckling of a typical AST. The typical AST has a conical roof, a diameter of 15 m, and a height of 10 m corresponding to the median geometry of ASTs located in the Houston Ship Channel (HSC), the largest petrochemical complex in the US located in a hurricane prone area. Buckling is the only failure mode considered here; the AST is anchored to the ground to prevent flotation. A three-dimensional finite element model (FEM) of the AST that takes into account fluid-structure interaction (FSI) is used to derive a model for wave pressure, while an existing model is used for wind pressure. The pressure models are then applied to a separate nonlinear FEM of the storage tank to determine the minimum surge height that causes buckling of the AST. Different combinations of wind speed and wave height are simulated using an uncorrelated statistical experiment design. A regression analysis is then performed to estimate the critical surge height as a function of the AST parameters and load parameters. Finally, the effects of wind and wave load on the critical surge height are discussed and a parameter screening study is performed to further assess the significance of wind and wave loads, providing key insights for future probabilistic studies of tank performance under multi-hazard loads.

2 NUMERICAL MODEL AND BUCKLING ANALYSIS

The buckling analyses are performed using the finite element software LS-Dyna [7]. The numerical model of the case study AST is shown in Figure 1. The AST is designed according to the American Petroleum Institute (API) 650 design standard [8]. Four courses with variable thickness are used to model the shell wall; the thickness varies from 8.0 mm at the bottom course to 6.3 mm at the top course. The AST is designed with a conical roof supported by rafters. A top angle is provided to stiffen the shell while no intermediate wind girders are required. The shell courses are modeled with quadrilateral shell elements. The roof shell is modeled with triangular shell elements while the top angle and rafters are modeled with beam elements. All of the components are modeled using an elastoplastic material model with kinematic hardening and with the physical properties of steel. As the tank is anchored to the ground, the base of the AST is fixed in all directions. Finally, since multiple studies [4] have shown that considering imperfections is critical for an accurate estimation of the buckling behavior of storage tanks, global imperfections of the tank shell are also included in the numerical model using the method proposed by Kameshwar & Padgett [9]. This method uses two dimensional Fourier series with random coefficients to model imperfections.

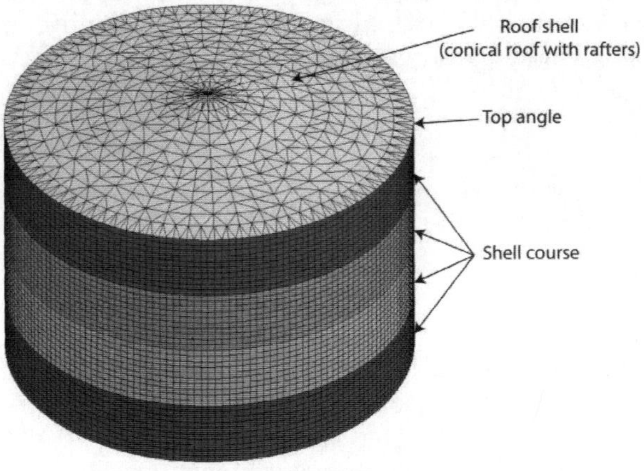

Figure 1: LS-Dyna model of the AST

The internal liquid and external surge loads are modeled using a hydrostatic pressure distribution. In addition, the wind pressure is applied on the dry section of the AST while the hydrodynamic pressure from wave impacts is applied on the wetted section of the AST. The pressure distributions for wind and hydrodynamic loads are detailed in Sections 3 and 4 respectively. The buckling analyses are performed by adapting the procedure proposed by Kameshwar & Padgett [10]. This procedure relies on the critical surge height (S_{cr}) to evaluate the buckling strength of ASTs. The critical surge height corresponds to the minimum water level that causes buckling of the shell. For a given analysis, the wind speed and wave height are fixed while the external surge height is incremented at each time step using a load control scheme. The hydrodynamic pressure and the area where the wind pressure acts are adjusted accordingly for each time step. An equilibrium analysis including material and geometric nonlinearities is performed at each time step. The stability of the tank shell is assessed by monitoring the eigenvalues of the stiffness matrix. Loss of stability (initiation of buckling) occurs when eigenvalues of the stiffness matrix become negative. The time step when eigenvalues become negative provides the critical surge height. The accuracy of this procedure was verified against the arc-length method. Even though wind and waves are transient loads, static buckling analyses are performed neglecting any dynamic effects of the AST. The critical surge height is used as the metric to define the buckling strength rather than the critical wind speed or critical wave height (or combinations of them) since surge is the dominant load acting on the AST; surge pressures are an order of magnitude higher than wind or wave pressures. Moreover, this approach enables an efficient exploration of the effects of each load individually and concurrently on the buckling response of the AST. As detailed in Section 5, the above approach is used to evaluate the critical surge height for multiple combinations of AST parameters, wind speed, and wave height.

3 WIND PRESSURE ON AST

The wind pressure acting on an AST can be described as [11]:

$$p = C_p(\theta)q(z) \tag{1}$$

where $C_p(\theta)$ is the wind pressure coefficient that varies along the circumference of the tank, and $q(z)$ is the vertical velocity profile of the wind pressure. Based on wind tunnel experiments, multiple studies [12,13] or design codes [11,14] proposed wind pressure coefficient distributions. In this study, the model proposed by EN1993-4-1 [11] is adopted since this model takes into account the aspect ratio (D/H = 1.5 for the case study AST) of the AST to determine the wind pressure coefficient. Other available models usually do not distinguish cylindrical structures with different aspect ratios. The EN1993-4-1 model defines the wind pressure coefficient as:

$$\begin{aligned} C_p(\theta) &= a_0 + a_1\cos\theta + a_2\cos 2\theta + a_3\cos 3\theta + a_4\cos 4\theta \\ a_0 &= -0.54 + 0.16(D/H) = -0.30 \\ a_1 &= 0.28 + 0.04(D/H) = 0.34 \\ a_2 &= 1.04 - 0.20(D/H) = 0.74 \\ a_3 &= 0.36 - 0.05(D/H) = 0.28 \\ a_4 &= -0.14 + 0.05(D/H) = -0.07 \end{aligned} \tag{2}$$

The wind pressure coefficient distribution is illustrated in Figure 2 for the case study AST. Typically, the vertical velocity profile is modeled as a power law of the form $q(z) = w_0(h/h_0)^\alpha$ where w_0 is the pressure at the reference height, h is the load height, h_0 is the reference height (10 m for hurricane wind [15]), and α is a coefficient function of the terrain characteristics. However, in this study, the vertical velocity profile is considered to be uniform with $q(z) = w_0$. A uniform profile is used for its simplicity and because preliminary analysis as well as previous studies have shown that the vertical velocity profile shape had limited effects on the buckling behavior of ASTs [3,4]. Finally, no uplift pressure is applied on the tank roof.

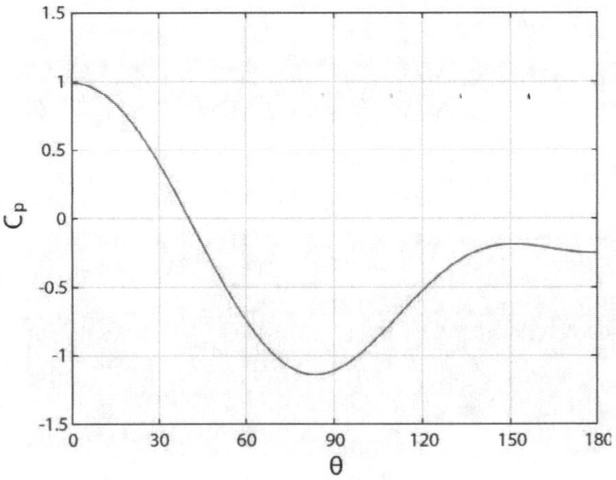

Figure 2: Wind pressure coefficient around the AST

4 HYDRODYNAMIC PRESSURE ON AST

Hydrodynamic pressures caused by wave impacts on circular structures are usually estimated with Morison's equation [16]. However, this equation is only valid when the diameter of the structure is relatively small compared to the wavelength ($D/\lambda < 0.05$). For structures with larger diameters, such as the case study AST, this equation is not valid as diffraction effects become important. To overcome this problem, an FEM model that takes into account fluid-structure interaction (FSI) is developed and used to determine the hydrodynamic pressure acting on the AST.

4.1 *Overview of fluid-structure interaction model*

The FSI simulations are performed using the finite element software LS-Dyna. The LS-Dyna solver is used since it has been validated by previous studies for wave impacts on circular cylinders [17]. The fluids (water and air) are governed by the Navier-Stokes equations. The fluid particles at the air-water interface are tracked using an Arbitrary Lagrangian-Eulerian (ALE) formulation [7]. An overview of the FSI model is presented in Figure 3a). Water and air are modeled with eight-node solid elements with fluid properties. The interface between the fluids and the tank shell is modeled using a multi-physics contact available in LS-Dyna. The tank shell is considered rigid, and taking advantage of its symmetry, only half of the domain is considered. Preliminary analysis has shown that these two assumptions do not significantly affect the pressure distribution on the AST, but greatly reduce the computational cost. The side boundaries of the model are free-slip conditions while the bottom boundary is a non-slip condition. Ambient elements are also added at the downstream face of the model to reduce the reflection of waves at the boundary. The waves are generated at the upstream boundary by using the wavemaker theory derived in Huang & Dong [18]. It takes a minimum of five waves to reach a steady state. An example of wave generation is shown in Figure 3b) for a water depth (S) of 4.0 m and a wave height (H_w) of 1.0 m. The computational time of a FSI analysis is approximately 26 h on a 2.6GHz CPU with 8 GB of RAM.

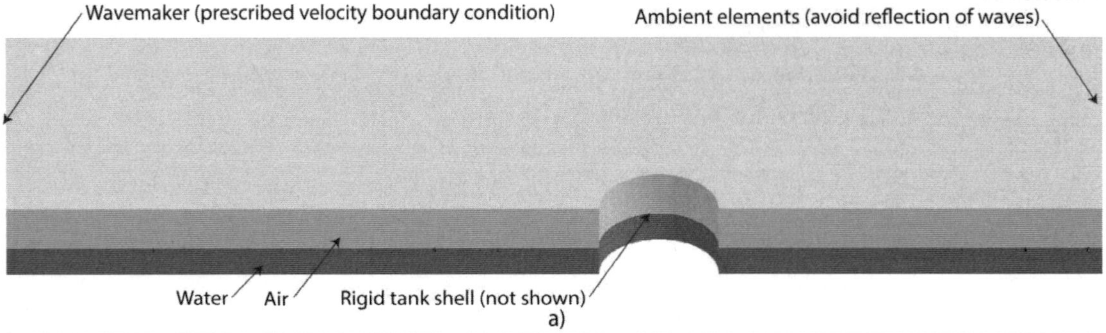

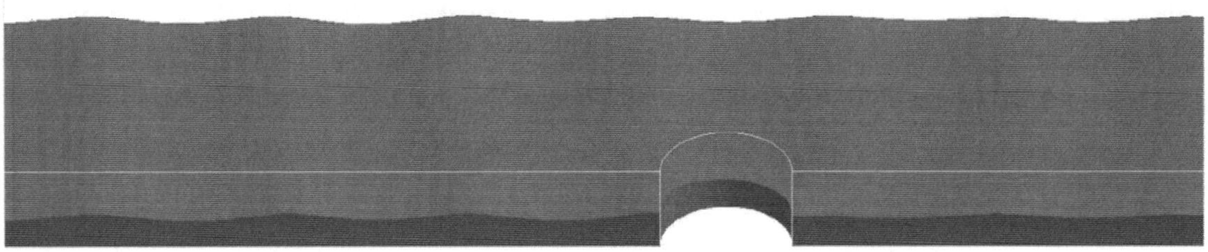

Figure 3: a) Overview of FSI model; b) Example of wave generation (S = 4.0 m and H_w = 1.0 m)

4.2 *Regression model*

Due to the computational cost of a single FSI simulation, it is not practical to simulate all possible combinations of wave heights and surge heights that may be imposed on the tank. Instead, a regression model is developed to estimate the hydrodynamic pressure of wave impacts as a function of the water depth (S) and the wave height (H_w). The development of a regression model requires enough training points to efficiently span the space of the parameters. To do so, Latin Hypercube Sampling (LHS) is used to generate a space filling experimental design [19]. The wave parameters considered in the experimental design are presented in Table 1. The ranges of water depth (S) and wave height (H_w) are obtained from numerical simulations of historic and synthetic hurricanes in the HSC [20]. The wave period (T) is fixed at 4.0 s to correspond to the median wave period in the HSC, and the wavelength (λ) is a function of S and T ($\lambda = 2\pi/k$ where k is obtained from $\omega^2 = gk\tanh(kS)$). Due to the limitations of the wavemaker theory, the minimum water depth is fixed at 2.5 m [18]. Moreover, buckling is not expected to occur below this water depth. Using LHS, a total of 100 samples with different values of S and H_w are generated. For each of the 100 samples, a finite element analysis using the previously described FSI model is performed. For each analysis, the maximum hydrodynamic pressure distribution on the AST is extracted. The maximum hydrodynamic pressure corresponds to the pressure distribution when the wave impact force on the AST is at a maximum minus the hydrostatic pressure. Out of the 100 samples, 95 are used to train the regression model while 5 are used to test it. An example of the hydrodynamic pressure distribution is shown on Figure 4 for a test sample (S = 6.9 m, H_w = 0.4 m).

An Artificial Neural Network (ANN) [21] is employed to obtain a regression model of the hydrodynamic pressure distribution on the AST. Attempts to obtain simpler and closed-form regression models were unsuccessful due to the complexity of the vertical pressure profile. Despite being complex and cumbersome, ANNs are a powerful regression tool. Theoretically, ANNs can approximate any continuous function on a bounded domain given an adequate architecture [21]. However, they are also prone to overfitting. Different ANN architectures were explored and a three-layer ANN with 20 neurons per layer was selected. The inputs of the ANN are the water depth (S), the wave height (H_w), the wavelength (λ), and the position of a point on the AST shell (θ, h). The output of the ANN is the hydrodynamic pressure at a given point. Thus, for a given wave height, it is possible to obtain the hydrodynamic pressure around the case study tank for the whole range of water depths as required for the buckling analysis. The ANN is trained using the Matlab Neural Network Toolbox [22]. With the training samples, the ANN predicts the pressure distribution with a coefficient of determination (R^2) of 0.999 and errors less than 10% on the maximum pressure. With the test sample, the ANN predicts the pressure distribution with a R^2 value of 0.985 and errors less than 15% suggesting that the regression model is not overfitting. Figure 4 also shows the comparison between the LS-Dyna output and the ANN prediction for a test sample. These results highlights an adequate fit of the hydrodynamic pressure distribution, which can be further used for buckling assessment of the AST.

Table 1. Range of wave parameters

Parameter	Parameter name	Range
S (m)	Water depth	2.5 – 9.0
H_w (m)	Wave height	0.0 – 1.5
T (s)	Wave period	4.0
λ (m)	Wavelength	17.4 – 24.2

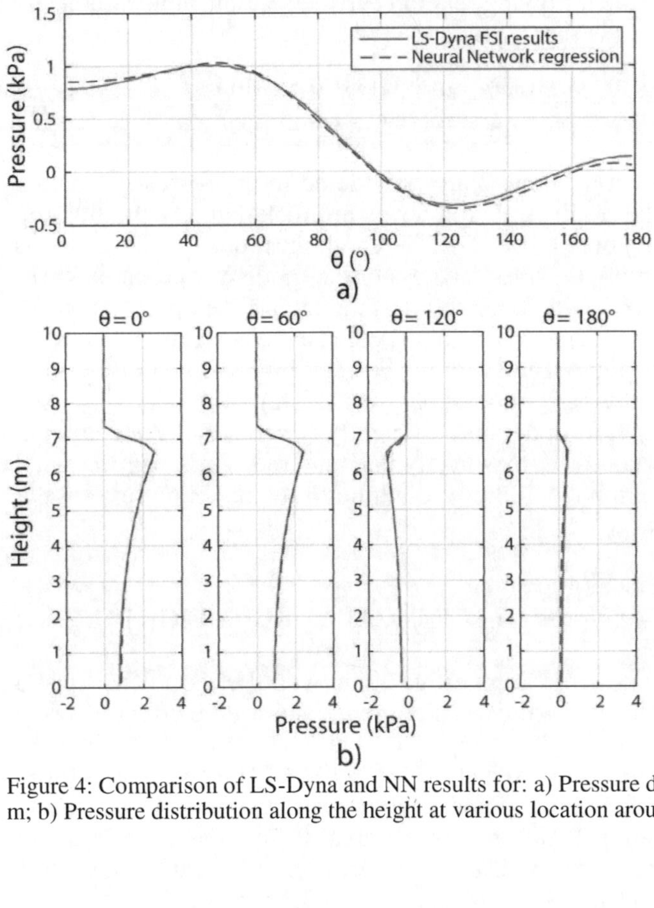

Figure 4: Comparison of LS-Dyna and NN results for: a) Pressure distribution along the AST circumference for h = 1 m; b) Pressure distribution along the height at various location around the AST

5 CRITICAL SURGE HEIGHT ESTIMATION

In order to assess the effect of combined loads on the buckling response of the case study AST, a regression model predicting the critical surge height as a function of the AST parameters, the wind speed, and the wave height is developed. Again, LHS is used to generate training points that span efficiently across the space of the parameters. The parameters considered in the LHS design and their range are summarized in Table 2. In addition to these parameters, the magnitude of the global imperfections of the tank shell is also considered to be variable. A total of 2000 samples are generated; 1800 are training samples while 200 are test samples. For each combination of ρ, L, S_d, U, and H_w, an imperfection model is randomly generated and the critical surge height (S_{cr}) is determined with the buckling analysis procedure outlined in Section 2.

Table 2. Range of parameters in the LHS design

Parameter	Parameter name	Range
ρ	Relative density of internal liquid	0.5 – 1.0
L (m)	Internal liquid height	0.0 – 9.0
S_d (MPa)	Product (steel) design stress	137.0 – 196.0
U (m/s)	Wind speed	0.0 – 100.0
H_w (m)	Wave height	0.0 – 1.5

Using the results of the training samples, the following relation between S_{cr} and tank parameters and load parameters was derived using stepwise regression:

$$S_{cr} = 4.11 - 0.81\rho - 6.21L - 2.69\times10^{-5}S_d - 3.69\times10^{-4}U - 0.33H_w + 8.39L\rho + 0.57\rho^2 + 16.73L^2 - 5.24L\rho^2 + 5.24L^2\rho - 18.41L^3 + 6.19L^4 \quad (3)$$

All first order terms of the parameters shown in Table 2 are considered in the regression model while only the most significant higher order and interaction terms are included. On the training data, the R^2 of this model is 0.978 with a median error of 2.2%. On the test data, the R^2 value is 0.965 and the median error is 2.3%, confirming the adequateness of the model for predicting critical surge buckling heights. It is important to note that the previous model is developed by considering that the loads are uncorrelated. Wave and wind loads are expected to be correlated especially since waves are wind-induced during hurricanes. However, this correlation is a function of other parameters that are site specific (fetch, bathymetry/topography, etc.). Thus, the approach used herein allow the development of a more general model that efficiently covers all of the combinations of wave height and wind speed that could be reasonably encountered. Knowing the wind speed and the associated wave height, Equation (3) allows estimation of the buckling response at any location for a given storm event.

6 EFFECTS OF WIND AND WAVE LOADS ON THE CRITICAL SURGE HEIGHT

With the above regression model, it is now possible to investigate the overall effects of wave and wind loads on the critical surge height of the case study AST. Figure 5 shows the effect of wave height on S_{cr} while Figure 6 shows the effect of wind speed on S_{cr}. In these figures, the internal liquid relative density and product design stress are fixed at 0.75 and 160 MPa respectively. From Figure 5, it is possible to observe that wave height has a non-negligible effect on S_{cr}. For instance, a 1.5 m wave can reduce S_{cr} by approximately 12% for a low internal filling level. For a higher filling level, the reduction of S_{cr} varies between 5 to 10%. However, Figure 6 clearly shows that wind speed has a negligible effect on S_{cr} of the AST. Reductions of less than 0.1% of S_{cr} are observed. This is also true when wind and wave act concurrently on the AST (Fig. 6b)). In this figure, no correlation is assumed between wind and wave loads. Due to the very limited effect of wind loads on S_{cr}, considering this correlation is not expected to significantly affect the results.

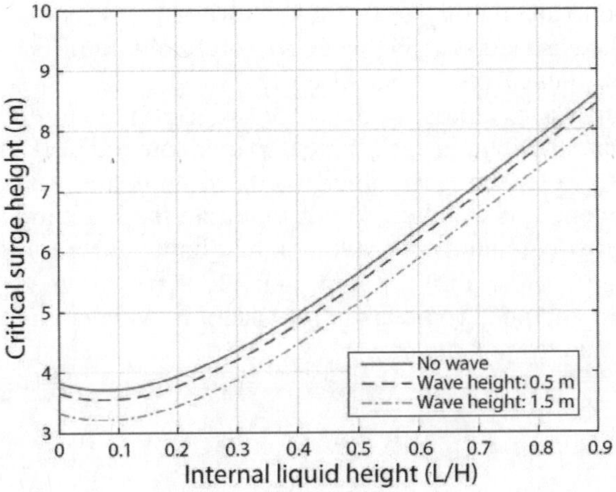

Figure 5: Effect of wave height on the critical surge height

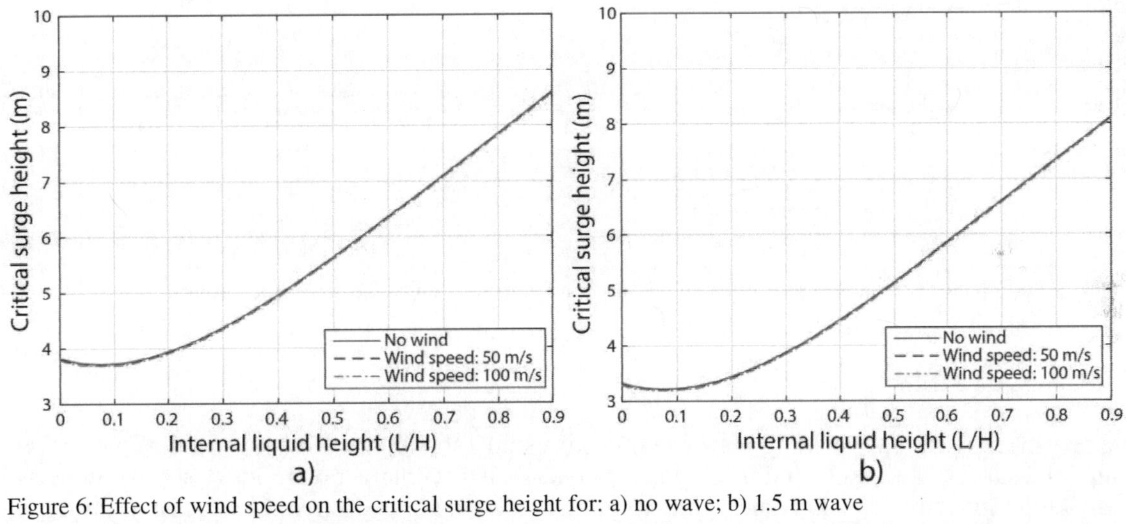

Figure 6: Effect of wind speed on the critical surge height for: a) no wave; b) 1.5 m wave

To further assess the effect of each parameter on the buckling response of the AST, the regression model is used to obtain a perturbation plot (Fig. 7). A perturbation plot shows the global effect of varying each parameter while others are kept constant at their median value. Figure 7 shows that internal liquid height has the strongest influence on S_{cr}. Wind speed and product design stress are not significant parameters and varying these parameters has a very limited influence on S_{cr}. Thus, these two parameters could reasonably be neglected from the regression model (Eq. (3)) without significantly affecting the buckling prediction of the AST. Such results were expected given the regression coefficients obtained in Equation (3). Two potential reasons why wind loads are not significant are: (1) wind pressures are an order of magnitude lower than hydrostatic and hydrodynamic pressures from the surge; (2) due to a vast amount of research [4] and poor performance during past hurricanes, ASTs are now designed to withstand extreme wind pressure [8]. Additional analysis has shown that wind speeds in excess of 110 m/s were required to cause buckling of the case study AST; such high wind speeds are mainly realistic for tornadoes. Product design stress is not significant since buckling of the case study AST occurs before yielding of the tank shell. Wave loads have a more significant influence on the buckling response of the AST compared to wind loads since hydrodynamic pressures are almost of the same order of magnitude as surge pressures. Overall, the above results indicate that wind is not critical to evaluate the buckling response of the AST which is also subjected to storm surge. In other words combined effects of wind and surge loads are minimal. However, even if wind loads do not facilitate the buckling of the AST, wind is still critical to adequately estimate concurrent wave heights that could occur at a specific location during a storm event. In addition, the above results indicate that combined effects between surge and wave loads are important and need to be taken into account. Neglecting wave loads can result in an overestimation of the buckling response.

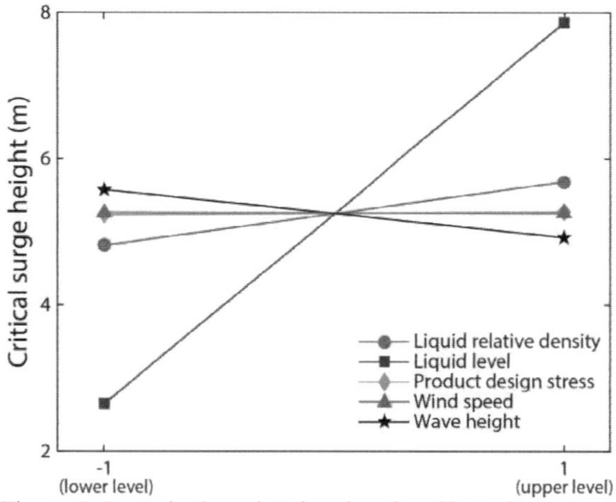

Figure 7: Perturbation plot showing the effect of varying each parameter on the critical surge height

7 CONCLUSIONS

This paper explores the buckling response of a case study aboveground storage tank subjected to combined surge, wave, and wind loads. First, a fluid-structure interaction model for wave impacts is developed given the lack of an available model to estimate hydrodynamic pressure on large diameter structures. This model coupled with an Artificial Neural Network enables the estimation of hydrodynamic pressure for various water levels and wave heights. Hydrodynamic and wind pressures are then applied on a separate numerical model of the AST to perform buckling analyses and determine the critical surge height. Using an experimental design, the critical surge height is expressed as a function of the tank parameters, wind speed, and wave height providing useful insights of the relative effects of each load on the structural behavior of the AST. Results indicate that wind loads are not significant for the representative AST designed to current standards that is also subjected to storm surge, and hence such combined loads may be reasonably neglected when evaluating the buckling capacity. However, wind induced wave loads are important and should be considered to estimate accurately the buckling behavior of ASTs during storm surge. More importantly, these results highlight the need to investigate mitigation measures in order to protect ASTs against surge and wave loads. This study has taken a first step to evaluate the structural behavior of ASTs under multi-hazard loads and provides key insights for future probabilistic studies of tank performance. The results and conclusions presented here are only applicable to the case study AST and hurricane level winds. Future research work will focus on a portfolio of ASTs and on a richer variety of storm surge waves to support fragility and risk assessments of ASTs during hurricane events.

8 ACKNOWLEDGEMENTS

The authors acknowledge the financial support of the National Science Foundation under award #1635784 and the Houston Endowment via the SSPEED Center. The contributions of the first author were also supported in part by the Natural Sciences and Engineering Research Council of Canada. Any opinions, findings, and conclusions or recommendations expressed in this paper are those of the authors and do not necessarily reflect the views of the sponsors.

9 REFERENCES

1 L.A Godoy, Performance of storage tanks in oil facilities damaged by Hurricanes Katrina and Rita, J. Perform. Constr. Facil., 21 (2007) 441-449.
2 V. Cozzani, M. Campedel, E. Renni and E. Krausmann, Industrial accidents triggered by flood events: Analysis of past accidents, J. Hazard. Mater., 175 (2010) 501-509.
3 F.G Flores and L.A. Godoy, Buckling of short tanks due to hurricanes, Eng. Struct., 20 (1998) 752-760.
4 L.A. Godoy, Buckling of vertical oil storage steel tanks: Review of static buckling studies, Thin-Walled Struct., 103 (2016) 1-21.
5 S. Kameshwar and J.E. Padgett, Storm surge fragility assessment of above ground storage tanks, Struct. Saf., (2017 (in review)).
6 H. Sengul, N. Santella, L.J. Steinberg and A.M. Cruz, Analysis of hazardous material releases due to natural hazards in the United States, Disasters, 36 (2012) 723-743.
7 Livermore Software Technology Corporation, LS-Dyna Theory Manual, Livermore, 2017.
8 American Petroleum Institute, Std 650: Welded steel tanks for oil storage, Washington, DC, 2013.
9 S. Kameshwar and J.E. Padgett, Stochastic modeling of geometrici in aboveground storage tanks for probabilistic buckling capacity estimation, ASCE-ASME J. Risk Uncertain. Eng. Syst., Part A: Civ. Eng., 2 (2015) C4015005.
10 S. Kameshwar and J.E. Padgett, Stiffening ring design for prevention of storm surge buckling in above ground storage tanks, Thin-Walled Struct., (2016 (in review)).
11 European Committee for Standardization, EN1993-4-1: Eurocode 3 – Design of steel structures – Part 4-1: Silos, Brussels, 2007.
12 G. Portela and L.A. Godoy, Wind pressures and buckling of cylindrical steel tanks with a conical roof, J. Constr. Steel Res., 61 (2005) 786-807.
13 R. Greiner and P. Derler, Effect of imperfections on wind-loaded cylindrical shells, Thin-Walled Struct., 23 (1995) 271–81.
14 ACI-ASCE Committee 334, ACI334-2R: Reinforced concrete cooling tower – Practice and commentary, 2002.
15 National Oceanic and Atmospheric Administration, H*Wind – Real time hurricane analysis project, 2014.
16 J.R. Morison, M.P. O'Brian, J.W. Johnson and S.A. Schaaf, The force exerted by surface waves on piles, Petroleum Transactions, 2 (1950) 149-154.
17 S.C. Yim and W. Zhang, A multiphysics multiscale 3-D computational wave basin model for wave impact load on a cylindrical structure, Journal of Disaster Research, 4 (2009) 450-461.
18 C.M. Dong and C.J. Huang, On 2-D numerical wave tank in viscous fluid, In Proc. 11th Int. Offshore and Polar Engineering Conf. Stavanger, Norway, 2001, pp. 148-155.
19 M.D. McKay, R.J. Beckman and W.J. Conover, Comparison of three methods for selecting values of input variables in the analysis of output from a computer code, Technometrics, 21 (1979) 239-245.
20 Severe Storm Prediction, Education and Evacuation from Disasters Center, The SSPEED Center 2015 Annual Report, Houston, 2015.
21 K.P. Murphy, Machine Learning: A Probabilistic Perspective, MIT Press, Cambridge, 2012.
22 MathWorks, Matlab – Neural Network Toolbox, 2016.

The 13th Americas Conference on Wind Engineering (13ACWE)
Gainesville, Florida USA, May 21-24, 2017

A Wireless Sensors Network System for Spatially Resolved Storm Surge Measurements

Sunil Chintalapati [a], Chelakara S. Subramanian[b]

[a]*Graduate Student, MAE, Florida Institute of Technology, Melbourne, FL, USA, chintals@fit.edu*
[b]*Professor of MAE, Florida Institute of Technology, Melbourne, FL, USA, subraman@fit.edu,*

ABSTRACT: Preliminary economic damage estimates from Hurricane Matthew indicate near $4-$6 billion out of which 10% is due to storm surge. The need for accurate hydrologic model to assist in flood-damage analysis and design future flood mitigation has prompted FEMA to conduct studies to acquire discrete storm surge measurements via placement of water level sensors prior to a storm event. The crucial aspect of monitoring and transmitting information for the extreme weather condition in nearly real-time is vitally important and requires accurate, robust, and relatively inexpensive sensing systems. The present wireless sensor network (WSN) system for storm surge measurements is an adaptation of the Florida Tech developed hurricane monitoring system which satisfies the above requirements. The WSN system for storm surge application is undergoing trials in lab and field-tests in areas with river and sea water. This paper describes the overview, design, and performance testing of the system.

KEYWORDS: remote sensor network, storm surge measurements, water level sensor, submersible pressure sensor.

1 INTRODUCTION

As areas, proximate to bodies of water (river, lakes, and sea) continue to become densely populated, damage from natural disasters also continue to increase, as demonstrated by hurricanes Katrina, Ike, Sandy, and most recently Matthew [1] [2] [3] [4] [5]. For decades, the U.S. Geological Survey (USGS) has provided critical information on near-shore storm hydrodynamics for decisions regarding emergency response and resource allocation before, during, and immediately after landfall of hurricanes and nor'easter. Much of this information was provided via a few real-time tide gages focused on near-shore waters that supplemented and extended the National Oceanographic and Atmospheric Administration's (NOAA) National Ocean Service (NOS) tide-gage network. In addition, data from temporary sensors and high-water marks (HWM) occasionally obtained by the USGS in cooperation with the Federal Emergency Management Agency (FEMA), the U.S. Army Corps of Engineers (USACE), and various state agencies, provided post-storm documentation of coastal floods; however, neither strategy provided sufficient, timely information for emergency operations or to facilitate improvements in storm-tide and wave modeling and prediction [6].

Preliminary estimates from the most recent Hurricane Matthew indicate economic damage near $4-$6 billion, and does not include insured losses related to additional flooding and business interruption [4] [6]. Of this $4-6 billion, 90% of the insurance claims are expected to be related to wind and 10% is expected to be related to storm surge [5]. Figure 1 shows the model's estimate of the storm surge prior to the arrival of the Hurricane Matthew along east coast of the US. However, storm surge flooding at St. Augustine area was reported to be 2.5 feet above ground level. The measured water levels were at discrete location (active tide gages for Florida is 24, Georgia is 1,

North Carolina is 7, and none for South Carolina [7]) and they do not paint a clear picture of the varying water levels at the entire south east coast.

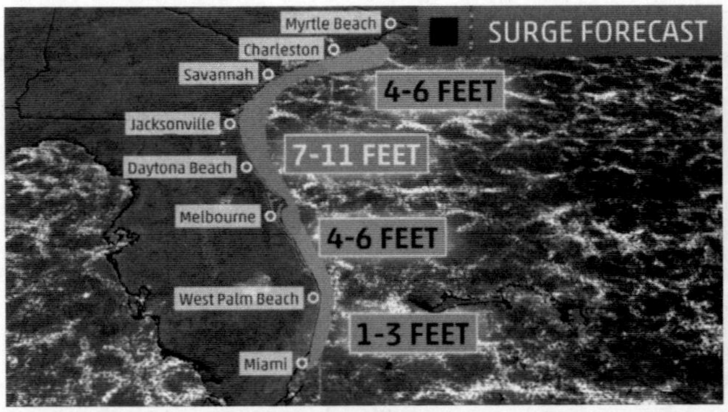

Figure 1. Hurricane Matthew Storm surge forecast [8]

Figure 2. Hurricane Sandy, Storm surge event

The need for hydrologic interpretation of monitoring data to assist in flood-damage analysis and future flood mitigation prompted FEMA to fund sensor placement before Hurricane Sandy to determine how high the water reached after the storm. The results demonstrated how the additional resolution and accuracy of flood depictions due to the data from the sensors resulted in greatly improved damage estimates, as well as vastly improved resiliency planning that would help communities prepare for future storms [9]. Figure 2 shows a before and during image of a surge event off the Melbourne coast during Hurricane Sandy taken by one of the authors of this proposal. Although the water level rise here is approximately a 1-2 ft, the whole beach including the volleyball court is inundated with water.

In order to reduce property damage and prevent loss of life (by evacuation from danger zones), it is essential to better predict the destruction caused by the unexpected events by accurately and preemptively monitoring environmental activity and relaying real-time updates for extreme weather conditions The crucial aspect of monitoring and transmitting information for the extreme weather condition in nearly real-time is vitally important and requires accurate, robust, and relatively inexpensive sensing systems. The sensing systems should also be capable of monitoring a local area of interest while still being relatively easy to deploy. Current damage prediction models are not as robust as they could be due to the existing measurement systems not providing real-time data [10]. Large scale predictive models are available [11] [12] [13] [14] [15], but their resolution (as shown by the predicted hurricane Matthew storm surge forecast and the measured values which are still discrete) to predict small scale local effects are not good enough. Therefore, it is important to develop better prediction models based on actual real-time measurements on a local scale.

2 BACKGROUND WORK

Since 2002, a research team at Florida Institute of Technology (Florida Tech) has been developing a wireless sensor network (WSN) for monitoring the effects of weather conditions on man-made structures during natural disasters, such as a hurricane. The effort was boosted in 2006 when support from NSF (CMMI Award ID # 0625124) allowed for the development of a second generation WSN. The effort was further leveraged in 2009 with complementary support from the Florida Department of Community Affairs Division of Emergency Management. The current result is a 3rd generation WSN, which represents the state of the art in hurricane monitoring at an affordable cost [16] [17] [18] [19] [20] [21] [22] [23]. As of today, it is the only operational and fully wireless local area multi-point sensing network of its kind. Other wireless sensor systems only provide average data over a wide area and, that too, at a much higher cost. For understanding the impact of large-scale natural events on localized communities both global and local scale measurements are necessary, but current systems only provide global measurements, whereas this technology provides local measurements.

The present system (Figure 3)offers at least three fully operational WSN systems each with 32 sensors for hurricane deployment and data collection on pre-instrumented houses that are available through the Florida Coastal Monitoring Program [24]. The network consists of three major sub-systems. The first sub-system includes up to 30 remote sensors units associated base unit and data collection laptop that are installed in the field. This sub-system is referred to as the *field installation* and it is responsible for performing sensing, measurements and data acquisition. Multiple sensor types are integrated within a sensor unit and the data collected by each one of them are sent to the base unit. The sensor elements that are accommodated by the sensor unit are measuring pressure, temperature, wind speed, and wind direction. The second sub-system is the *communication network* that connects individual WSN field installations to a central server. The communication network sends the data from the data collection laptop to a centralized location, which may be located anywhere within the Internet cloud. This approach allows for a real-time availability of the measured data and provides for real time monitoring and control of the WSN field installation. Finally, the third sub-system is the *central server* itself, which processes measured data and presents them to the end user. The processing of the data is performed in an automated fashion and in *near* real time, which no other technologies have been able to reproduce.

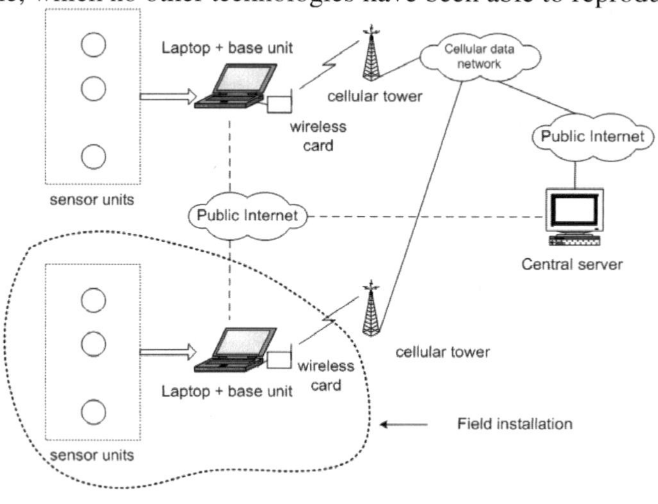

Figure 3. General outline of the Florida Tech's wireless sensor network architecture

In summary, the key features of the wireless multi-sensors network system innovation are, the capability of measuring multi-point with multi-sensors on a structure, collecting data, and sending it to a server to process and publish on the Web in nearly real-time. The other aspect of the system is it has not only the wireless and networking capabilities, but it also has several in-situ controls such as individual sensor remote control, self-negotiating communication, and excellent data throughput, precision, and accuracy; it is robust, reliable, and portable, and allows simultaneous deployment on multiple residences in case of a hurricane landfall within a short time. Additionally, the system has been tested in various wind tunnels and outdoor testing facilities, such as at Florida

International University and at the University of Florida. The objective of the past tests and deployments was collection of real or simulated hurricane wind data.

The assembly exploded view of the pressure/temperature sensor node is shown in Figure 4. The base unit, battery pad, and the field laptop computer are placed in a ventilated water proof box in the vicinity of the house (see Figure 4).

Figure 4. Sensor assembly (on left), Current WSN fully deployed on rooftops of residential houses (on right)

The development of this WSN system is a major achievement from the NSF grant, and has been described in several international journal and conference papers [16] [17] [18] [19] [20] [21] [22] [23]. Three graduate students completed their master thesis on the project [25] [26] [27]. In addition, several undergraduate students have been continuously involved with the research, helping with the design, assembly, testing, and deployment of the sensors.

3 REVIEW OF EXISTING TECHNOLOGIES AND THEIR DEFICIENCIES

The principal agency responsible for monitoring the storm surge levels in the US is the NOAA. The Center for Operational Oceanographic Products and Services (CO-OPS) manages the National Water Level Program (NWLP) and provides operationally sound observations and monitoring capabilities coupled with operational Nowcast/Forecast modeling [28]. These models cater for the insurance company risk assessment modelers, but are currently modeled on historical data, and limited real-time measurements; thus, current models are predictive only, and are only revised after the event, whereas future models based on data provided via the proposed system could be augmented in real-time.

There are several methods used by NOAA, the USGS, and the FEMA to measure storm surge [28] [29]. Prior to the arrival of a hurricane, the USGS team deploys a multitude of instruments to acquire water level and meteorological data. Two of the instruments which are rapidly deployable are [30]:

1. Rapid deployment gages (RDG): Temporary stream gages setup to provide short-term water-level data for critical areas lacking stream gages during an event (see Figure 5). Although the

RDGs are semi-portable, the cost, manpower, and labor hours required to assemble, install, or disassemble and relocate are significantly cost prohibitive and therefore minimize the USGS's ability to rapidly deploy and relocate on-demand (based on the uncertainty of the storm path). Moreover, the overall cost of each RDG, and the associated labor for use per RDG drastically reduces the number of RDGs afforded in any year's budget [31].

Figure 5. USGS Rapid-Deployment Gages a) Overview of the entire system, b) RDG for a bridge installation, c) Closeup view, and d) River bank installation [32]

2. USGS Storm Tide Sensor (STS) Pressure Water Level Sensors are temporary pressure sensors that provide information about storm surge duration, times of surge arrival and retreat, and maximum depths (Figure 6) [28]. In addition to RDGs, the USGS occasionally employs Storm Tide Sensors (STSs), which are more mobile than the RDGs and thus make up for the bulkiness and lack of mobility of the RDGs. However, these mobile storm tide sensors are not capable of transmitting data in real time. Although the data is recorded by the STSs during the storm event, it is only stored locally and collected after the storm has passed, and thus is not provided to any agency in real time.

Figure 6. a) USGS Storm Surge Sensor (non-transmitting) b) close-up view [33]

In summary, the RDGs require significant power provisions, such as a bank of car batteries or solar panels (which can be unreliable given various weather patterns), and the STSs only provide data after the event, in which case neither provide the easy of deployment, reliability, and real-time data-access requisite for accurate and predictive modeling. Notwithstanding the immense strain on manpower to deploy/record/collect storm-tide data locally at various sensors, the lack of real-time data regarding storm surge levels delays the decision-making process regarding critical events, such as evacuation, relief values/controls, diverting water etc. Furthermore, these delays allow for more damage to be done; the prevention of which would more than justify the use of the proposed innovation. Finally, to further enhance the longevity of the proposed system, the WSN can be remotely reconfigured or modified to shut down, reboot, or even change sampling frequency in the absence of any significant weather event so as to save battery life and reduce labor costs needed to maintain the system.

There are several other sensors in the market, none of which provide the level of sophistication, real-time data capture, mobility, or power provisions requisite for current market demands for storm surge measurements, such as those of the USGS [34] [35] [36] [37] [38] [39]. Several of the wireless sensing systems investigated did not provide water level measurements in their sensor suite [40] [41] [42]. The submersible pressure transducer used in our proposed innovation alone is offered by a limited number of companies [43] [44] [45] [46] [47], but are not tied to package system that would address the needs of USGS for storm surge applications. Various wireless sensing systems have been developed by university teams to monitor either bridges or buildings [48] [49] [50] [51] [52] and was never successfully transitioned into a commercial product for storm surge measurements. There are few companies that have a flood monitoring package in list of products, one such company [53] [54] offers an ultrasonic level meter connected to a wireless transmitter and can transmit the data in nearly real-time in a field operation [53]. This product offered is a really competitive option, but relies on ultrasonic level meter, which is ideal for measuring rising or receding water levels in a controlled weather conditions. During conditions such as hurricane or hazardous wind may alter the orientation of the senor placement which would then provide misleading values for the rising water levels. To monitor the storm surge levels, this sensor system would also require additional poles and columns specially constructed so the sensor has a clear unobstructed view of the surface of the water. The additional column, poles or pillar constructed are again prone to deflection from high wind conditions. The proposed sensor system in this proposal is not affected by the deflection or orientation of the sensor because the submersible sensor is still capable of reading the pressure in the fluid column. The orientation or alteration of the sensor head unit does not affect the transmission capability of the sensor head units. Each of the sensor head units will also contain a GPS tag to provide with location information as well as water levels. Some companies also provide a "flood detection and water level monitoring" packages but also clearly indicate the suitable application area are for river level monitoring, highways, railways, and even in cellars [55]. These packages are also not compact and require either a battery or solar power. Some companies also state a flood and water level detection system [56] but clearly state that the usage is limited to families with vacation homes or those who travel frequently, so they can remotely monitor any leaks or water level in the home.

4 WIRELESS SENSOR NETWORK SYSTEM FOR STORM SURGE APPLICATION

This section presents the overview of the WSN concept for storm surge application from the very beginning when the research team found the active need for this application during a NSF I-Corps contract. This section also provides an overview of the feasibility of the WSN concept for storm surge application, as well as development of the prototype.

4.1 *NSF I-Corps*

The wireless sensors network (WSN) system developed with NSF (CMMI 0625124) sponsorship represents the one-of-a-kind for monitoring local environmental (harsh) conditions surrounding a manmade structure exposed to extreme environments (a hurricane or flood). Localized multipoint sensing is vital for accurately modeling the effect of large scale average data (e.g. wind speed) on small scale structures (e.g. residential roof pressure). The WSN system has the capability of measuring pressure and temperature on a roof, collecting data (rate up to 1400 samples/s) and sending it to a server to process and publish on the web in nearly real-time. Also, wind speed and direction are measured by the system with the use of an anemometer. The innovative aspect of the system is it has important wireless and networking capabilities, including individual sensor remote control, self-negotiating communication, and excellent data throughput, precision, and accuracy; it is robust, reliable, and portable, and allows simultaneous deployment on multiple residences in case of a hurricane landfall. Since the land falling hurricanes have random wind fields a large portion of the continuously collected data may be unimportant.

The major goal for the NSF I-Corps contract was to determine if this product was commercially viable. The intent was to adapt the WSN system to other environments like lab settings of full scale or even for normal wind tunnels. The WSN system could also be expanded to several other types of measurements like meteorological, seismic, oceanographic and structure condition monitoring. This goal was to verify whether this technology could make a significant commercial impact.
The preliminary product presented, was funded by the NSF I-Corps contract and a detailed customer discovery was pursued for a period of 7 weeks [57]. The preliminary product consisted of a cluster of 32 wireless pressure sensors in a weather proof enclosures, which can be readily deployed in the field (on top of a house or a building), and be capable of recording pressure measurements over a continuous acquisition period of 12 hours. The technical details and comprehensive capabilities of the product are presented in this reference [16] [17] [18] [19] [20] [21] [22] [23].

The initial value proposition and the associated customer segment was developed using the dynamic Business Model Canvas (see Figure 7) [58] [59] [60] [61].

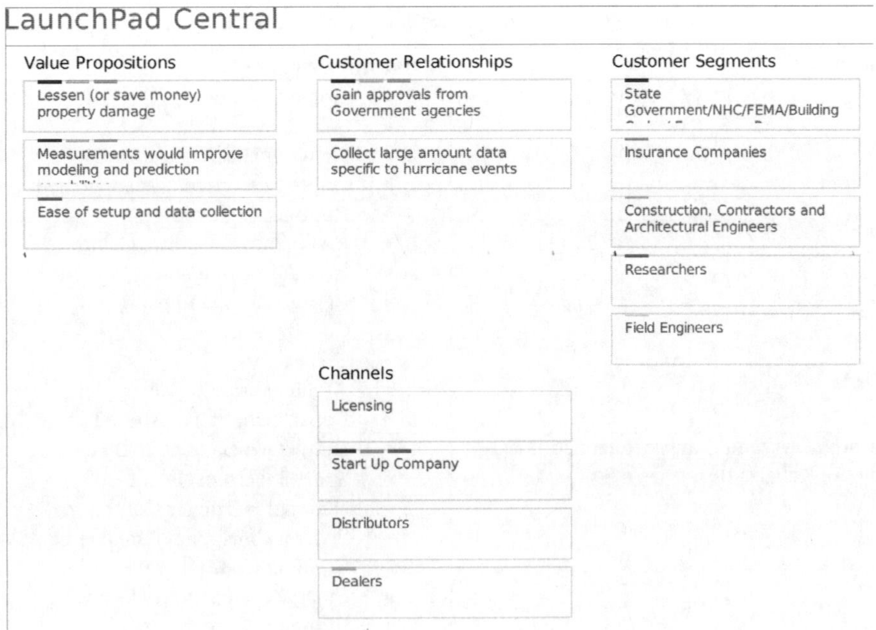

Figure 7. Preliminary Business Model Canvas (CS and VP)

The initial value propositions considered not only the physical component of the product, but also services that would have to be rendered to process the acquired data. This meant that the physical product (cluster of wireless pressure sensors, anemometer, and a laptop with wireless networking enabled) would be deployed by our team. Our team would also collect the data and process the data and if required, perform numerical simulations to identify the potential damage location on the rooftops of a building.

The initial customer segment was really diverse and had specific initial value propositions for each segment.

Table 1 Initial Value Proposition and Customer Segments [57]

Value Proposition	Customer Segments
1. Lessen (or save money) property damage	• Home Owners • State Government • National Hurricane Center • Federal Emergency Management Agency • Building Codes • Emergency Response Center • Insurance Companies • Re-insurance Companies • Construction • Contractors • Architectural Engineers • Department of Transportation (Bridge Safety)
2. Measurements would improve modeling and prediction capabilities	• Home Owners • State Government • National Hurricane Center • Federal Emergency Management Agency • Building Codes • Emergency Response Center • Insurance Companies • Re-insurance Companies • Construction • Contractors • Architectural Engineers • Department of Transportation (Bridge Safety)
3. Ease of setup and data collection	• Researchers • Field Engineers

At the conclusion of the 6 weeks' period (86 interviews), the value proposition for customer segment of home owners was determined to be a passive interest.

Table 2 Customer Feedback [57]

Value Proposition	Must Have	Nice to Have	Don't Care
Lessen (or save money) property damage	15	5	4
Measurements would improve modeling and prediction capabilities	49	19	6
Ease of setup and data collection	5	0	0
Customer Segments			
State Government/NHC/FEMA/Building Codes/ Emergency Response	2	4	4
Insurance Companies	2	3	0
Construction, Contractors and Architectural Engineers	6	4	1
Researchers	15	1	0
Field Engineers	6	1	0
Home Owners	10	9	6
Re-insurance Companies	1	0	0

The cost of acquiring the customer was also prohibitively high and the retention rate was nil. Most of the customer segments deemed pressure measurements as a passive need. A product pivot was initiated after assimilating preliminary data from the interviews. Sensor specific data collection would be desirable to different customer segments.

1. Water level measurements: For example, real time monitoring of the water levels.
2. Structural health monitoring for historical and expensive bridges and structures: Strain measurements
3. Structure or vehicle stability monitoring during operations: Accelerometers
4. Monitor air quality (or any offgassing related issues) in office buildings

4.2 *Modification, Lab Testing, and Data Validation*

The post I-Corps customer discovery determined that there is an active need for a temporary water level sensing system which can provide real-time accurate measurements of rising water levels, and can be easily deployed. To ensure the current WSN system can acquire the water level measurement and continue to operate in real-time, a series of tests were performed at our Hurricane lab at Florida Tech. Figure 8 shows a basic concept of the current sensor node reading the hydrostatic pressure (instead of the ambient pressure reading) at a test location (location 1) and transmitting to the laptop wirelessly located at location 2. A series of tests were also performed to verify the multiple sensors nodes working in tandem to acquire the hydrostatic pressure reading and transmitting it wirelessly.

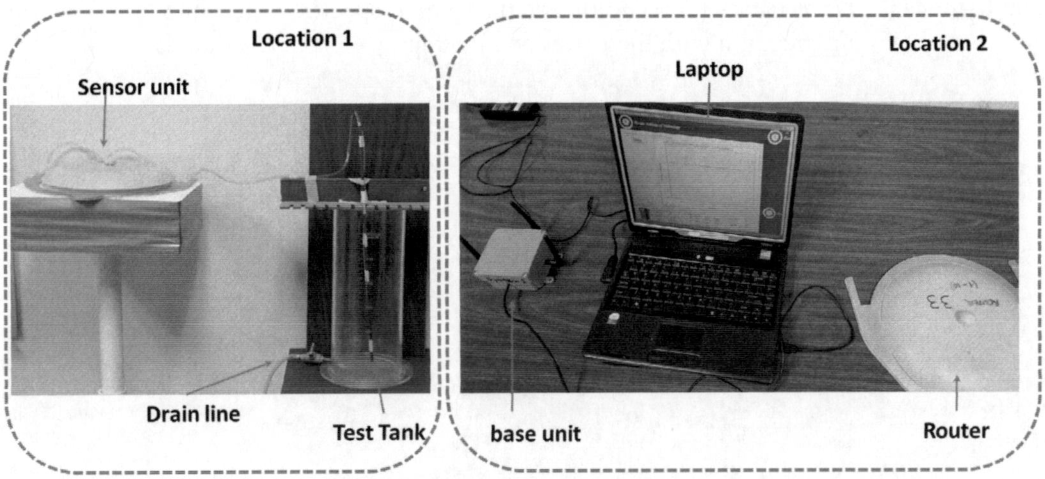

Figure 8. Overview of the setup during Lab Testing

Figure 9 and Figure 10 shows a sample plot of a pressure reading for a receding water level by emptying a test tank. Digital manometer readings corroborated this data as well.

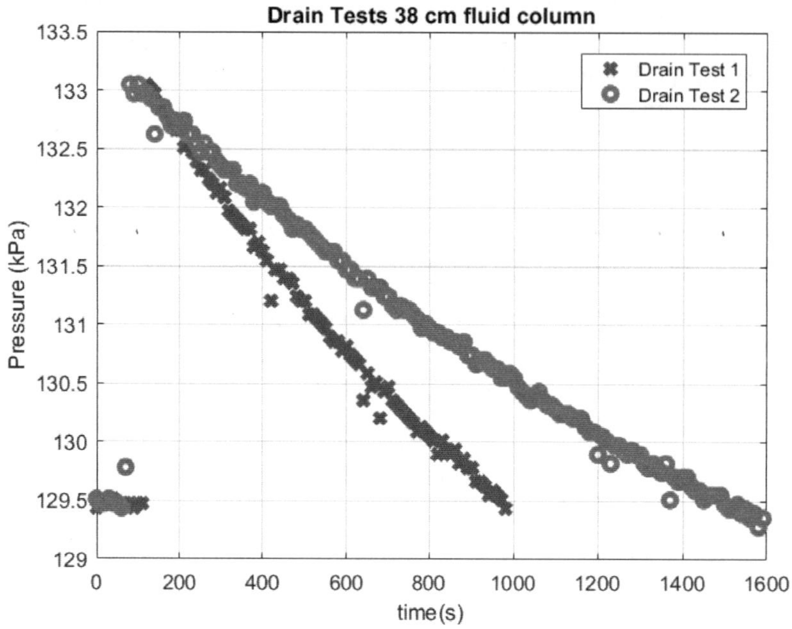

Figure 9. Samples Drain Tests Conducted with a 38 cm Tall Container

Lab Testing also was also conducted with a 71-cm tall container. The sensor (on right of the figure) is adjacent to the tank and measuring the hydrostatic pressure at specified sampling rate. The sensor is on a constant communication with the router and the laptop (left in the figure).

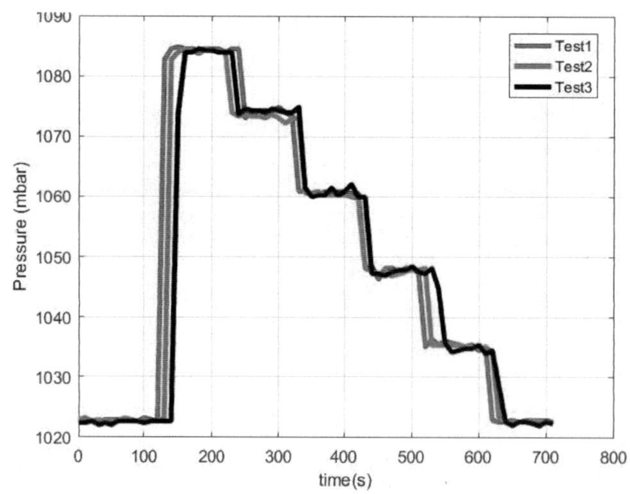

Figure 10. Samples Drain Tests Conducted with a 71 cm Tall Container

4.3 *Storm Surge Prototype Development*

After the completion of the lab testing with the modified sensors, a feature list for a prototype for storm surge measurement was designed. The storm surge prototype would address the need for real-time water level measurements and the network architecture is derived from WSN and consists of three major sub-systems (Figure 11). The first sub-system includes several remote sensors units, associated base unit and data collection laptop that are installed in the field. This sub-system is referred to as the *field installation* and it is responsible for performing sensing, measurements, and data acquisition. The submerged pressure sensor would be connected to a sensor head unit via a long cable (approximately 5-7 m long). The cable length would be adequate and reinforced (to avoid straining the cable as well as protect from debris) to capture rising water level as seen from a strong hurricane such as hurricane Matthew where the recorded peak water levels were at 3.8 m [9].

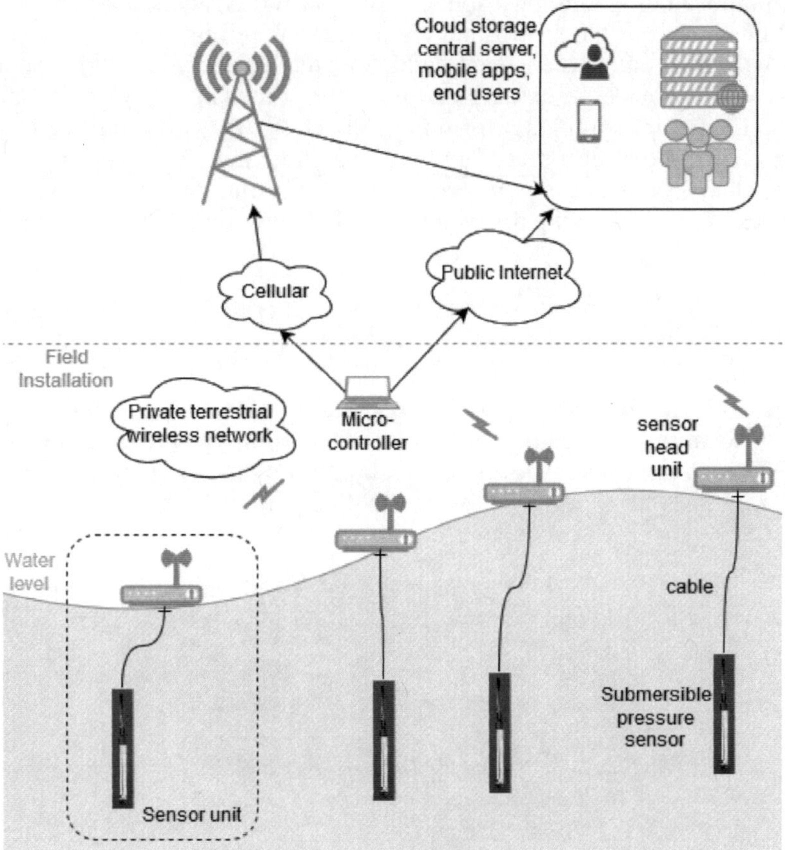

Figure 11. Schematic of the WSN for storm surge application

The second sub-system is the *communication network* that connects individual WSN field installations to a central server. The communication network (private wireless network) sends the data from the data collection device to a centralized server location either through Wi-Fi or cellular data networks (3G – 3rd Generation) using the IEEE 802.15.4 protocol. This approach allows for a real-time availability of the measured data and provides for real time monitoring and control of the WSN field installation. In case the real-time link fails, all the data is stored locally in the memory, which can be retrieved after the storm passes or when the connection is reestablished. Finally, the third sub-system is the *central server (or cloud storage)* itself, which processes measured data and presents them to the end user (via graphs or dedicated specific mobile app). The processing of the data is performed in an automated fashion and in *near* real time, which no other technologies have been able to reproduce. As this is done through the world-wide web, the remote logging- in operation can be performed from a computer anywhere in the world (not restricted to the localized server). A Control Client application is used to monitor status, control individual or group of endpoints (sensor nodes), display sensor data in real-time, and generate logs of sensor data. The design is a user-friendly Graphical User Interface that is meant to be an easy to use method of monitoring and controlling the wireless sensor network. In the WSN system for storm surge, the sensor head unit can be interfaced with a buoy so the sensor head unit is always above water or the sensor head unit can be constrained to a post, pole, or column for a pier. If a buoy is preferred for the purposed of recording sea level rise, it would be anchored to a location to ensure there is minimal movement of the sensor and less strain on the reinforced cable connection. The sensor head unit will also contain a GPS chip which would uniquely identity each sensor units in the network as well as provide an exact location in real time. In an event of failure of one or multiple sensor units, the remaining sensor units in the network will continue to perform flawlessly. The battery pack in the sensor head unit will last more than 5-7 days (on normal operation mode) on a single charge (a metric to validate during field testing of the prototype). Additionally, we could incorporate solar recharging to ensure extended duration for the sensor unit.

4.4 *Overview of the Prototype*

This section consists of making a device capable of storing data from a water pressure transducer at sea (Figure 13). Such data will then be transmitted wirelessly to a remote server to allow its access from any location. The main components of the prototype are discussed below:

Submersible Water Pressure Transducer: This is the sensor that will measure the water pressure. Its output ranges from 4-20 mA, and it is powered with 24V.

Figure 12. Submersible Water Pressure Transducer

Atmeg328P: This 3.3V microcontroller will be in charge of running the WiFi and Cellular modules. It will also collect the sensor's data, save it in an SD card, and retrieve the information when needed to be sent out to the server.

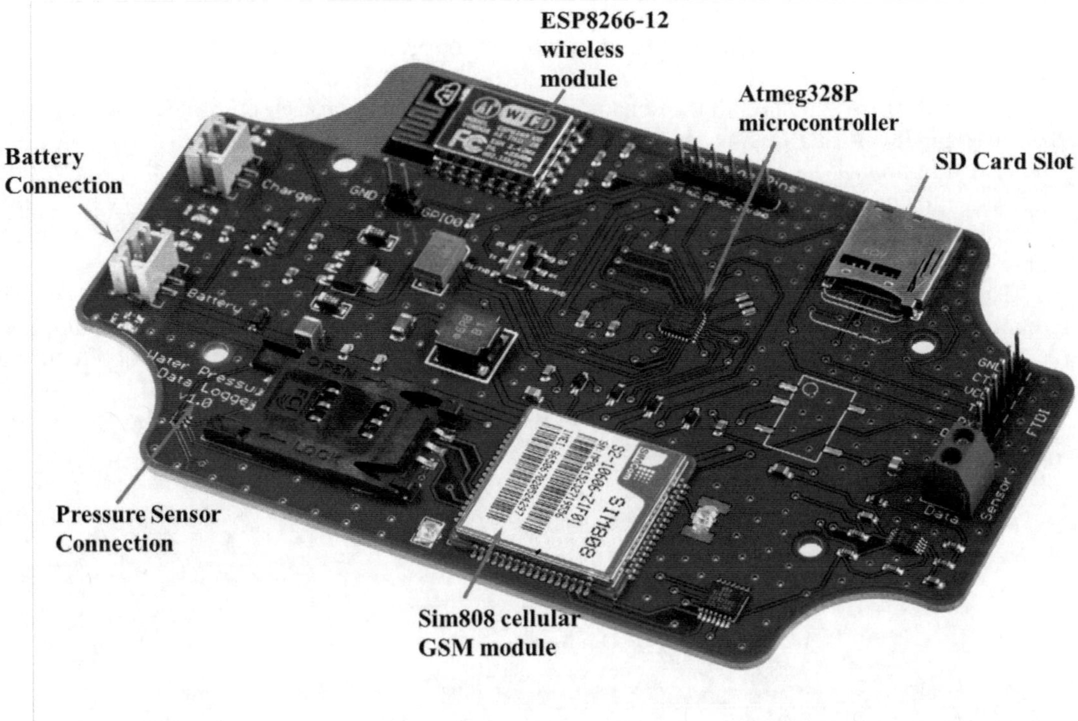

Figure 13. Manufactured prototype with labelled components

Sim808: This cellular GSM GPRS module will be in charge of transferring the data to the server. It supports HTTP and FTP applications. Its quad band allows it to connect to any global GSM network. It communicates directly with the Atmega328P via Software Serial. This module can be powered directly with a 4.2V Lithium-polymer battery.

ESP8266-12: This module has the versatility of granting microcontrollers access to WiFi networks, or acting as a microcontroller with PWM, ADC, and SPI capabilities. It communicates directly with the Atmega328P via Software Serial. This module requires a 3.3V voltage to work properly.

4-20mA ADC: This integrated circuit will convert the sensor's output from milliamps to voltage. The conversion is required so that the Atmega328P can interpret the sensor's output values. It communicates directly with the Atmega328P via I^2C. The chip can be powered with 3.3V or 5V.

3.3V Regulator: This component will regulate the input voltage of the battery to supply 3.3V to the following parts:

- Atmega328P
- ESP8266
- 4-20 mA ADC
- SD Card Reader

Sim Card: A 2G mini SIM card is required for the Cellular module.

DC/DC Booster: This chip will boost the input voltage to 24V. Required for the water sensor to work properly.

Battery Charger: This is a single-cell linear charge management controller suitable for 4.20V, 4.35V, 4.40V, and 4.50V. It features a programmable charge current that ranges from 15mA - 500mA

MicroSD Card: Here is where all the sensor's readings will be stored. It communicates directly with the Atmega328P via SPI using an SD card reader.

Battery: A lithium polymer battery of 4.2V with a capacity of approximately 6000 mAh will be used to power the entire device. Assuming the device is active (transmitting data) every hour, the battery can last up to one day without a recharge. The battery can be recharged through a micro USB cable using a 5V input.

Case: The following case will be used to enclosure and protect the PCB: This case is made of flame retardant polycarbonate. Furthermore, it includes a silicone rubber gasket for dust and moisture resistance. These features make it tough, heat-resistant, water-resistant, and a good electrical insulator. Perfect for electrical hardware used outdoors.

The CAD representation of the completed prototype and the CAD visual of the exploded view is shown in Figure 14.

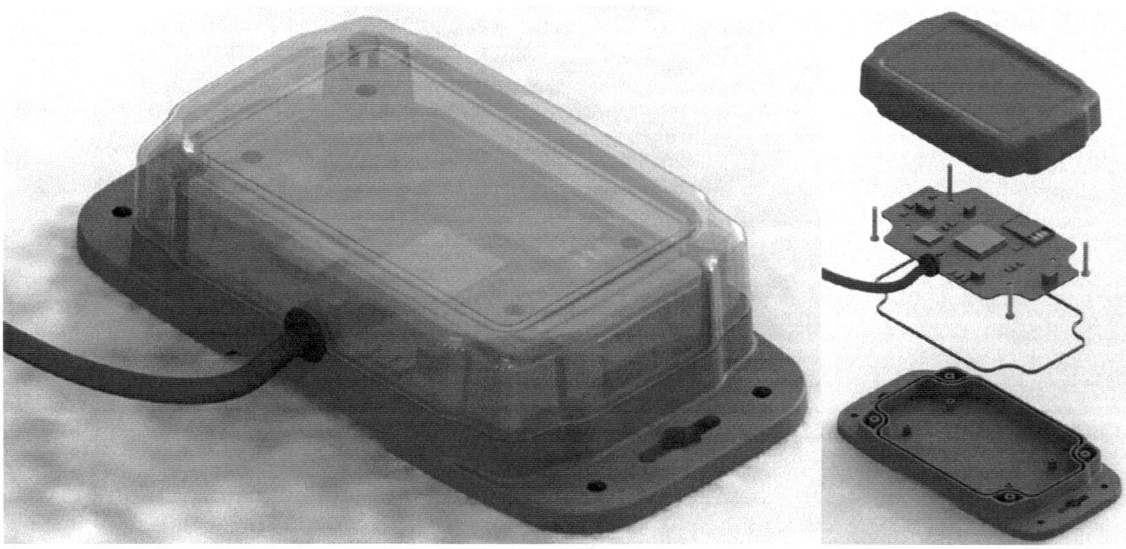

Figure 14. CAD visual of the completed proposed prototype sensor head unit (left), and an exploded view (right)

The workflow needed to acquire detailed storm surge measurements is shown in Figure 15.

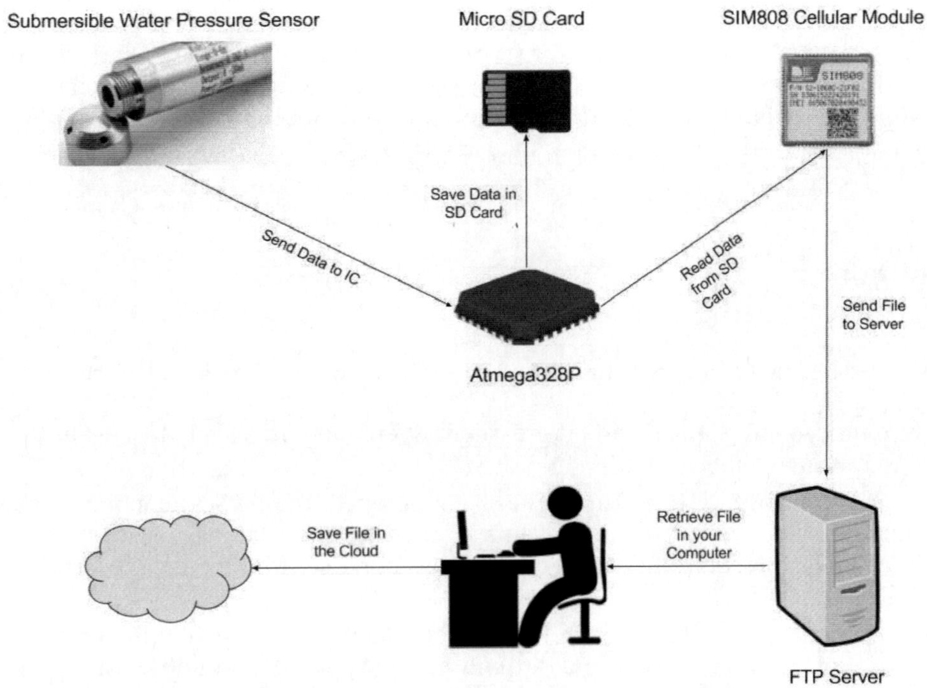

Figure 15. Workflow of the proposed prototype system

5 CONCLUSION AND FUTURE WORK

The key features of this prototype and differentiators from the current state of the art or competing technologies are that there is no product containing a combination of the requisite set of technologies that would meet the needs of the customer; specifically, a light-weight, portable, easy-to-install, wireless, real-time, water propagation enabled device that records and propagates pressure based measurements that are also not limited to specific heights, depths, times, or are subject to damage from wind, or other elements and are contained in a durable, marinized enclosure.

Field testing of the prototype WSN system is underway near the Melbourne beach. Field testing will also be conducted with one of our prospective customers for field performance assessment in collaboration with the local NOAA staff. The system will be deployed and tested at locations specified by NOAA and USGS in addition to other field environments identified during our market and partner validation. The test of success will be the ability to real-time multi-point data acquisition of water level height and remote communication of the data over cellular network.

6 ACKNOWLEDGEMENTS

The development of the WSN system for storm surge application is due in part to NSF I-Corps contract (A Smart Data Acquisition Scheme for Managing Hurricane Measurement Data from a Wireless Multi-sensor Network, NSF IIP- I-Corps 1535760). Special thanks to Robert Mason, and Harry Jenter at USGS. Also, the authors wish to acknowledge the guidance provided from NOAA, FEMA, and USGS.

7 REFERENCES

[1] A. Kimberly, "Hurricane Katrina Facts: Damage and Economic Effects," About.com, [Online]. Available: http://useconomy.about.com/od/grossdomesticproduct/f/katrina_damage.htm. [Accessed 5 October 2015].

[2] A. Kimberly, "How Hurricane Ike Impacted the U.S. Economy," about.com, [Online]. Available: http://useconomy.about.com/od/grossdomesticproduct/f/hurricane_ike.htm. [Accessed 5 October 2015].

[3] A. Kimberly, "What Are the Facts About Sandy's Damage and Economic Impact?," about.com, [Online]. Available: http://useconomy.about.com/od/Disasters/f/Hurricane-Sandy-Economic-Impact.htm. [Accessed 5 October 2015].

[4] T. Worstall , "Hurricane Matthew Damages At $4 to $6 Billion - Yes, This Will Increase GDP," Forbes.com, 9 Oct 2016. [Online]. Available: http://www.forbes.com/sites/timworstall/2016/10/09/hurricane-matthew-damages-at-4-to-6-billion-yes-this-will-increase-gdp/#1335ddf05926. [Accessed 11 Oct 2016].

[5] CoreLogic, "MEDIA ADVISORY: Hurricane Matthew," CoreLogic, 06 10 2016. [Online]. Available: http://www.corelogic.com/about-us/news/media-advisory-hurricane-matthew.aspx. [Accessed 11 10 2016].

[6] D. Rice, "Hurricane Matthew economic damage nears $6 billion," USA TODAY, 8 10 2016. [Online]. Available: http://www.usatoday.com/story/weather/2016/10/08/hurricane-matthew-economic-damage-cost-6-billion/91783304/. [Accessed 11 10 2016].

[7] NOAA Tides and Current, "Water Levels - Station Selection," NOAA, [Online]. Available: https://tidesandcurrents.noaa.gov/stations.html?type=Water+Levels. [Accessed 10 10 2016].

[8] Weather.com, "Hurricane Matthew Recap: Destruction From the Caribbean to the United States," Weather.com, 10 10 2016. [Online]. Available: https://weather.com/storms/hurricane/news/hurricane-matthew-bahamas-florida-georgia-carolinas-forecast. [Accessed 11 10 2016].

[9] C. Kelly and M. Peterson, "Budget Impact: Reducing Exposure to the Risks Posed by Climate Change Saves Money and Lives," 2015.

[10] USGS, "USGS SURGE, WAVE, AND TIDE HYDRODYNAMICS (SWATH) NETWORK," USGS, 10 Feb 2016. [Online]. Available: http://water.usgs.gov/floods/swath/. [Accessed 25 March 2016].

[11] USGS National Research Program, "USGS - Modeling Software," USGS, [Online]. Available: http://water.usgs.gov/nrp/software.php. [Accessed 25 March 2016].

[12] "Catastrophe Models-Mapping the Risk Landscape," Risk Management Solutions, Inc., 2015. [Online]. Available: http://www.rms.com/products/models-cat. [Accessed 7 October 2015].

[13] "About Catastrophe Modeling," AIR Worldwide, 2015. [Online]. Available: http://www.air-worldwide.com/Models/About-Catastrophe-Modeling/. [Accessed 7 October 2015].

[14] "Catastrophe Risk Management," CoreLogic, 2015. [Online]. Available: http://www.corelogic.com/solutions/catastrophe-risk-management.aspx. [Accessed 7 October 2015].

[15] "Catastrophe Management -Aon Benfield Analytics," Aon plc, 2015. [Online]. Available: http://www.aon.com/reinsurance/analytics/catastrophe_management.jsp. [Accessed 7 October 2015].

[16] J.-P. Pinelli, C. S. Subramanian, G. Lapilli and L. Buist, "Application of a Wireless Pressure Sensing System to Coastal Wind Monitoring," *International Journal of Wind and Structures,* vol. 8, no. 3, pp. 179-196, 2005.

[17] C. S. Subramanian, G. Lapilli, F. Kreit, J.-P. Pinelli and I. Kostanic, "Experimental and Computational Performance Analysis of a Multi-Sensor Wireless Network System for HUrricane Monitoring," *Sensors & Transducers Journal,* vol. 10, pp. 206-244, 2011.

[18] C. S. Subramanian, J.-P. Pinelli, I. Kostanic and G. Lapilli, "Analysis and Characterization of HUrricane Winds," *ASCE Journal of Engineering MEchanics,* pp. 1-14, March 2013.

[19] C. S. Subramanian, J.-P. Pinelli, I. Kostanic and G. Lapilli, "Design, Development and Testing of a Wireless Multi-Sensor Network System," *Journal of Mechanics Engineeering and Automation,* vol. 2, pp. 169-183, 2012.

[20] C. S. Subramanian, J.-P. Pinelli, I. Kostanic, L. Buist, B. Van der Veek, A. Velazquez, F. Kreit and C. Otero, "Development and Testing of a Second Generation Wireless Hurricane Wind and Pressure Monitoring System," in *Proceedings. 11th American Conference on Wind Engineering*, San Juan, Puerto Rico, 2009.

[21] C. S. Subramanian, J.-P. Pinelli, G. Lapilli and L. Buist, "A wireless Multi-Point Pressure Sensing System: Design and Operation," *IEEE Sensors Journal,* vol. 5, no. 5, pp. 1068-1074, 2005.

[22] C. Otero, A. Velazquez, I. Kostanic, C. S. Subramanian and J.-P. Pinelli, "Real-Time Monitoring of Hurricane Winds using Wireless Sensor Technology," *Journal of Computers,* vol. 4, no. 12, pp. 1275-1285, 2009.

[23] C. S. Subramanian, G. Lapilli, F. Kreit, J.-P. Pinelli and I. Kostanic, "Experimental and Computational Performance Analysis of a Muti-Sensor Wireless Network System for Hurricane Monitoring," *Sensors & Transducer Journal,* vol. 10, pp. 206-244, 2011.

[24] "Florida Coast Monitoring Program," 5 September 2008. [Online]. Available: http://fcmp.ce.ufl.edu/. [Accessed 8 October 2015].

[25] A. Velazquez, "Wireless Networked Data Acquisition Software for Hurricane Monitoring," in *MS Thesis*, Melbourne, FL, Civil Engineering, Florida Institute of Technology, 2009.

[26] F. Kreit, "Performance Testing of the Wireless Sensor Network System for Hurricane Monitoring," in *MS Thesis*, Melbourne, FL, Department of Mechanical and Aerospace Engineering, Florida Institute of Technology, 2009.

[27] G. Lapilli, "Hurricane Wind Effects: Instrumentation Measurement and Analysis," in *MS Thesis*, Melbourne, FL, Department of Mechanical and Aerospace Engineering, Florida Institute of Technology, 2011.

[28] "Storm Surge and Coastal Inundation - Monitoring Storm Surge," [Online]. Available: http://www.stormsurge.noaa.gov/models_obs_monitoring.html. [Accessed 29 September 2015].

[29] "Hurricane Research Division - NOAA," 14 May 2010. [Online]. Available: ww.aoml.noaa.gov/hrd/tcfaq/F8.html. [Accessed 29 September 2015].

[30] USGS Office of Surface Water, "USGS Storm Tide and Rapid Deployment Gages used during Hurricane Irene," USGS, 5 Feb 2015. [Online]. Available: https://water.usgs.gov/floods/events/2011/irene/SurgeAndRDG.html. [Accessed 29 March 2016].

[31] USGS, "FY 2017 USGS Budget Request," USGS, 2016. [Online]. Available: https://www.usgs.gov/budget/2017/docs/water.pdf. [Accessed 29 March 2016].

[32] USGS, "RDGposter," USGS, 5 October 2015. [Online]. Available: https://water.usgs.gov/floods/resources/rdg/RDGposter.pdf. [Accessed 29 March 2016].

[33] USGS HIF, "USGS Hydrologic Instrumentation Facility," USGS, 29 June 2015. [Online]. Available: http://water.usgs.gov/hif/USGSWATERSENSOR/. [Accessed 25 March 2016].

[34] "Water Level Sensor Liquid Water Droplet Depth Detection Sensor for Arduino hv3n," [Online]. Available: http://www.aliexpress.com/item/Water-Level-Sensor-Liquid-Water-Droplet-Depth-Detection-Sensor-for-Arduino-hv3n/1355075994.html. [Accessed 29 September 2015].

[35] solinst.com, "Water Level Meters," solinst.com, 2016. [Online]. Available: http://www.solinst.com/products/level-measurement-devices/101-water-level-meter/datasheet/. [Accessed 29 March 2016].

[36] "TrueSTEAM Water Level Sensor Assembly - Honeywell," Supplyhouse.com, 2016. [Online]. Available: http://www.supplyhouse.com/Honeywell-50027998-002-TrueSTEAM-Water-Level-Sensor-Assembly?gclid=CJWywfOp5csCFYNkhgod1f8I_g. [Accessed 29 March 2016].

[37] Jet.com, "SONIC H2O LEVEL METER," Jet.com, 2016. [Online]. Available: https://jet.com/product/detail/8b52f0b8345749e8807f61146e747872?jcmp=pla:ggl:gen_hardware_a1:tools_measuring_tools_sensors_a1_other:na:na:na:na:na:2&code=PLA15&ds_c=gen_hardware_a1&ds_cid&ds_ag=tools_measuring_tools_sensors_a1_other&product_id=8b52f0b8345. [Accessed 29 March 2016].

[38] INYOPOOLS.com, "Wtr Water Level Sensor (960090-000)," INYOPOOLS.com, 2016. [Online]. Available: http://www.inyopools.com/Products/07501352029080.htm?gclid=CM76x-ep5csCFddahgoddJwAtw. [Accessed 29 March 2016].

[39] testequipmentdepot.com, "Onset U20L-04, HOBO 13 ft. Freshwater/Saltwater Water Level Data Logger," testequipmentdepot.com, 2016. [Online]. Available: http://www.testequipmentdepot.com/onset/data-loggers/environmental-measurements/freshwater-saltwater-environments/hobo-u20-water-level-data-logger-

u20l04.htm?ref=gbase&gclid=CKbxl7up5csCFcdehgodZbULEw. [Accessed 29 March 2016].

[40] "NI Wireless Sensor Networks," 2015. [Online]. Available: http://www.ni.com/wsn/. [Accessed 29 September 2015].

[41] "WIRELESS Nodes," [Online]. Available: http://www.microstrain.com/wireless/sensors. [Accessed 29 September 2015].

[42] "Wireless Sensor Tags System," [Online]. Available: http://wirelesstag.net/. [Accessed 29 September 2015].

[43] TE.com, "Water Level Sensor," TE.com, 2016. [Online]. Available: http://www.te.com/usa-en/products/sensors/water-level-sensors.html?tab=pgp-story. [Accessed 12 10 2016].

[44] rshydro.co.uk, "Water Level Sensors," rshydro.co.uk, 2016. [Online]. Available: http://www.rshydro.co.uk/water-level-sensors/. [Accessed 12 10 2016].

[45] globalw.com, "WL400 WATER LEVEL SENSOR," globalw.com, 2016. [Online]. Available: http://www.globalw.com/products/levelsensor.html. [Accessed 12 10 2016].

[46] "Water Level, Stage, and Flow Sensors," campbellsci.com, 2016. [Online]. Available: https://www.campbellsci.com/water-level-stage-flow. [Accessed 12 10 2016].

[47] Rickly.com, "Submersible Pressure Transducer, 0-3.4ft/0-1 psi, +/-0.10% Accuracy, 25ft Cable," Rickly.com, 2016. [Online]. Available: http://ricklyhydrologicalcompany.mybigcommerce.com/submersible-pressure-transducer-0-3-4ft-0-1-psi-0-10-accuracy-25ft-cable/. [Accessed 29 March 2016].

[48] R. Allan, "Wireless Sensors Land Anywhere And Everywhere," 21 July 2005. [Online]. Available: http://electronicdesign.com/components/wireless-sensors-land-anywhere-and-everywhere. [Accessed 29 September 2015].

[49] "Better Load Ratings Through Nondestructive Evaluation," 12 June 2015. [Online]. Available: https://www.fhwa.dot.gov/publications/publicroads/98novdec/better.cfm. [Accessed 29 September 2015].

[50] J.-W. Park, S.-H. Sim and H.-J. Jung, "Wireless displacement sensing system for bridges using multi-sensor fusion," vol. 23, no. 4, 2014.

[51] "Permanent Wireless Monitoring System on the Telegraph Road Bridge for Bridge Health Assessment," [Online]. Available: http://www.scsolutions.com/permanent-wireless-monitoring-system-telegraph-road-bridge-bridge-health-assessment. [Accessed 29 September 2015].

[52] M. Ziemba, "Bridge-Saving Sensors Eliminate Infrastructure Deficiency," 29 July 2014. [Online]. Available: http://www.wirelessdesignmag.com/article/2014/07/bridge-saving-sensors-eliminate-infrastructure-deficiency. [Accessed 29 September 2015].

[53] Advantech/B+B, "flood warning and water level monitoring," http://advantech-bb.com/, 2016. [Online]. Available: http://advantech-bb.com/smart-iot-technology-for-flood-and-water-level-monitoring/. [Accessed 10 10 2016].

[54] Senix.com, "General Purpose Ultrasonic Sensors," Senix, 2016. [Online]. Available: https://senix.com/toughsonic-ultrasonic-sensors-overview-old/. [Accessed 12 10 2016].

[55] powelectrics.co.uk, "Water Level Monitoring and Flood Detection," powelectrics.co.uk, 2016. [Online]. Available: http://www.powelectrics.co.uk/remote-

environmental-monitoring/application.asp?page=water_level_monitoring_275. [Accessed 12 10 2016].

[56] lowittalarms.com, "FLOOD AND WATER LEVEL DETECTION," lowittalarms.com, 2016. [Online]. Available: http://lowittalarms.com/home-security/flood-and-water-level-detection/. [Accessed 12 10 2016].

[57] S. Chintalapati, "SUNIL CHINTALAPATI," Florida Institute of Technology, 2015. [Online]. Available: http://my.fit.edu/~chintals/NSF%20I-Corps%20-Team505%20SmartSense/. [Accessed 9 October 2015].

[58] S. Blank, The four steps to the epiphany: Successful strategies for products that win, K&S Ranch..

[59] S. Blank and B. Dorf, The startup owner's manual: The step-by-step guide for building a great company, K&S Ranch, Incorporated., 2012.

[60] A. &. P. Y. Osterwalder, Business model generation: A handbook for visionaries, game changers, and challengers, Hoboken, NJ: Wiley, 2010.

[61] A. P. Y. B. G. &. S. A. Osterwalder, Value Proposition Design: How to Create Products and Services Customers Want, John Wiley & Sons., 2015.

[62] "Water Level Sensors Provide Real-time Flood Warnings in Iowa," Senix, 2016. [Online]. Available: https://senix.com/water-level-sensors-provide-flood-warning/. [Accessed 12 10 2016].

The 13th Americas Conference on Wind Engineering (13ACWE)
Gainesville, Florida USA, May 21-24, 2017

Field Reconnaissance Following the Passage of Hurricane Matthew over Haiti's Tiburon Peninsula

Tracy L. Kijewski-Correa[a], Andrew B. Kennedy[b], David O. Prevatt[c], Alexandros A. Taflanidis[d]

[a] Department of Civil and Environmental Engineering and Earth Sciences & Keough School of Global Affairs, University of Notre Dame, 156 Fitzpatrick Hall, Notre Dame, IN, USA
[b] Department of Civil and Environmental Engineering and Earth Sciences, University of Notre Dame, 156 Fitzpatrick Hall, Notre Dame, IN, USA
[c] School of Sustainable Infrastructure and Environment, University of Florida 365 Weil Hall, Gainesville, FL, USA
[d] Department of Civil and Environmental Engineering and Earth Sciences, University of Notre Dame, 156 Fitzpatrick Hall, Notre Dame, IN, USA

ABSTRACT: Hurricane Matthew made landfall on the southwestern tip of Haiti as a strong Category 4 storm on October 4, 2016. Between the dates of November 18 to 24, 2016, the authors and a team of investigators from the US and Haiti conducted reconnaissance to characterize the hazards and damage extent, using a combination of ground and aerial campaigns as well as eye witness statements. The investigation canvassed the southern coastal communities of Les Cayes, Les Anglais, Port Salut, Chardonnières and La Cahouane, as well as the northern coastal communities of Jérémie, Bonbon and Abricot. This paper summarizes the performance of low-rise masonry construction during this event, while also reflecting on the broader impacts of this disaster on the affected communities and the nation as a whole, which continues to struggle with its recovery from the devastating 2010 earthquake.

KEYWORDS: reconnaissance, hurricane, storm surge, Haiti, Matthew

1 INTRODUCTION

Matthew was the first major hurricane to impact Haiti in more than a half century. It made landfall west of the town of Les Anglais on the southwestern tip of Haiti's Tiburon Peninsula as a strong Category 4 storm at 7 am EDT on October 4, 2016 with estimated sustained wind speeds of 65 m/s (145 mph). The affected region, which is challenging to reach from the capital of Port-au-Prince, sustained extreme damage to its built and natural environment, with at least 1600 fatalities and undoubtedly many since due to injuries, disease and famine. Rainfall in the peninsula was also extreme, estimated by NASA to be 25-50 cm over the course of the storm. Matthew's impacts, much like the 2010 earthquake, were felt most heavily in the private sector, with tens of thousands buildings destroyed, particularly along the southwestern coast, though impacts reached as far as the low-lying areas of the earthquake-affected regions to the east. While this emphasizes again the importance of homes as a foundational element of a community's resilience, the event's strong winds, saltwater inundation and inland flooding due to rainfall, devastated agricultural assets, spiked vector-borne diseases, and impaired the capacity to deliver clean water. These added complexities will present challenges to recovery that were not present in the 2010 earthquake, a disaster from which this nation has yet to recover.

Haiti's ongoing recovery is understandable considering the chronic conditions that have historically impaired the nation's development independent of major disasters. Plagued by war and revolution, deforestation and natural disasters, and fractured political processes born of its heritage as the only successful slave revolt in the Americas, Haiti's informal economy places it among the poorest nation in the western world. Its weak political institutions offer limited support for basic services, including water, sanitation, health care, and education. Widely varying and often informal construction practices have understandably evolved within this context, devoid of consistent code enforcement and credible tenure/land use policies. The implications of this environment and the vulnerabilities it perpetuates have already been demonstrated in the 2010 earthquake [1] and now are only more strongly reiterated with the passage of Hurricane Matthew.

2 METHODOLOGY

The authors assembled a team from the United States with expertise in structural, wind and coastal engineering as well as in remote surveying using unmanned aerial vehicles (UAVs). This team partnered with a Haitian civil engineer and a pair of local community advocates to conduct reconnaissance in Haiti's Tiburon Peninsula (see Acknowledgements for full listing of participants). Reconnaissance began on November 18, 2016 in the southern port city of Les Cayes (Fig. 1 – Pt. 1) and then headed westward from Port Salut and surveying sites along the coastal road to Chardonnières (Fig. 1 – Pt. 2). After assessing Les Anglais and La Cahouane, a pair of communities just east of the landfall region (Fig. 1 – Pt. 3), the team headed eastward assessing additional sites between Les Anglais and Les Cayes (Fig. 1 – Pt. 4). The team then moved (Fig. 1 – Pt. 5) to the northern coastal city of Jérémie, pushing eastward to the communities of Bonbon & Abricot, just east of the hurricane's track (Fig. 1 – Pt. 6). Efforts to reach Dame Marie, west of the hurricane track, were unsuccessful due to recurring landslides. The team concluded its final day of reconnaissance in Tiburon Peninsula on November 24, 2016 in Jérémie (Fig. 1 – Pt. 7).

Three types of assessments were conducted at each site. One team characterized the coastal hazard by surveying high water marks and inundation extent, as inferred from debris lines, damage levels, displaced objects and first-hand accounts. A second team conducted door-to-door detailed assessments of damaged buildings, including classification of primary structural and foundation elements, materials, dimensions, damage mechanisms and damage extent. These were established using direct observation and contact and non-contact measurements accompanied by geotagged photos, videos, and statements from victims to reconstruct the landfall and failure sequence. Wind hazard intensity was mapped using observation-based data products, corroborated with forensic evidence such as overturned rigid bodies like shipping containers. Meanwhile, the UAV collected high-resolution aerial imagery to reconstruct point clouds of the neighborhoods being canvassed on foot for the structural assessments.

3 TYPOLOGIES SURVEYED

This field reconnaissance evaluated the performance of traditional low-rise (1- to 3-story) masonry and reinforced concrete buildings, inclusive of private residences, as well as schools, churches and clinics. The most common typology observed for low-rise construction was load bearing masonry wall systems (classified as unreinforced, reinforced or confined [2]), generally employing concrete masonry units (CMUs). These mechanically-pressed blocks are 15 cm deep (equivalent to the nominal 6 inch block used in the United States) with varying levels of quality based on cement content, aggregate source, production technique and curing conditions. Older buildings like

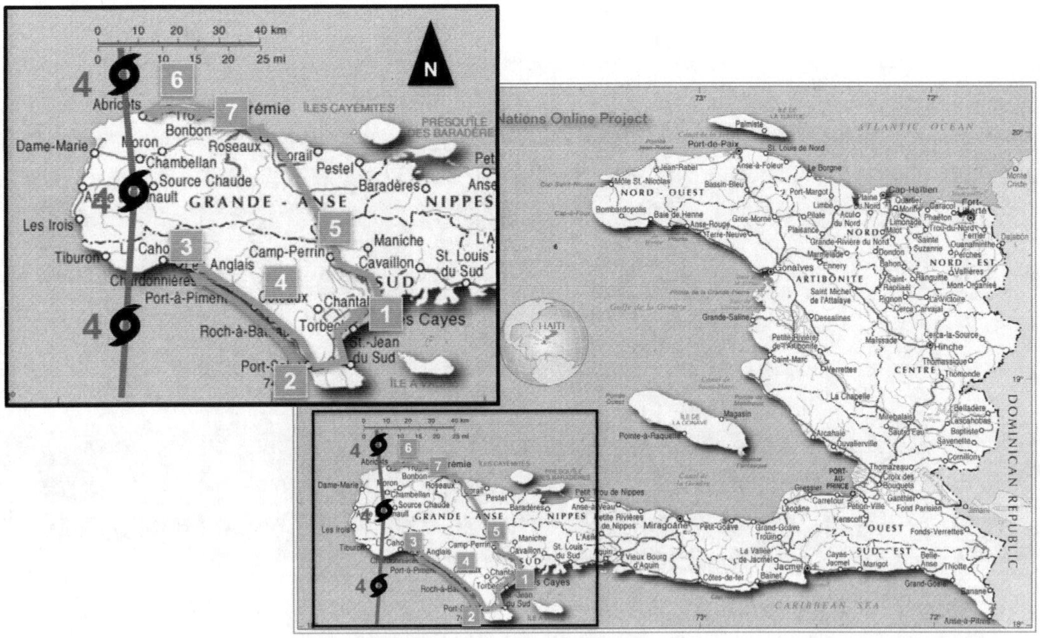

Figure 1. Map of Haiti denoting track of Hurricane Matthew, inset visualizes seven days of reconnaissance efforts in the western Tiburon Peninsula. Background map from the Nations Online Project.

churches and hospitals often employed stone for unreinforced masonry (URM) walls (at times with buttressing). Historical post and beam residential construction was also prevalent throughout the affected region, using native timber to create confining elements infilled by small stones and mortar that is stuccoed to form lightweight wall systems (Fig. 2a-b). Despite their modest size, Haitian cultural practice maintains significant interior partitioning providing out-of-plane stiffness to these masonry wall systems. Home construction is commonly led by informally trained masons, leading in part to the prevalence of load bearing masonry systems that can be built progressively as the homeowner's resources permit.

Low-rise buildings generally employ shallow foundations (slabs with grade beams) to accommodate poor soil conditions and relatively high water tables. Buildings may be slightly elevated (~0.5 m) to accommodate frequent flooding in the rainy season, though most homes are directly on grade. Two major roofing systems are employed in these regions: (i) metal panel roofs framed in wood (see Fig. 2 a-b) or tack welded steel hollow structured sections (HSS) (Fig. 2c) and (ii) concrete slab roofs (Fig. 2d). Wood-framed roofs were far more prevalent in the assessed region due to their low cost and ease of construction. Trusses of varying stiffness have been observed, generally toenailed on-site using either dimensional lumber or machete-milled native woods. Wooden roof framing may be nailed to the top of the URM wall or the wooden ring beam in the post-and-beam construction. When vertical confining elements are used in URM construction (often without horizontal confining elements), the longitudinal reinforcement of these vertical confining elements are wrapped around the roof framing. In rare cases, roof trusses (wood or metal) may have their lower chords cast into reinforced concrete frames or mortared into notches in the top course of CMU walls.

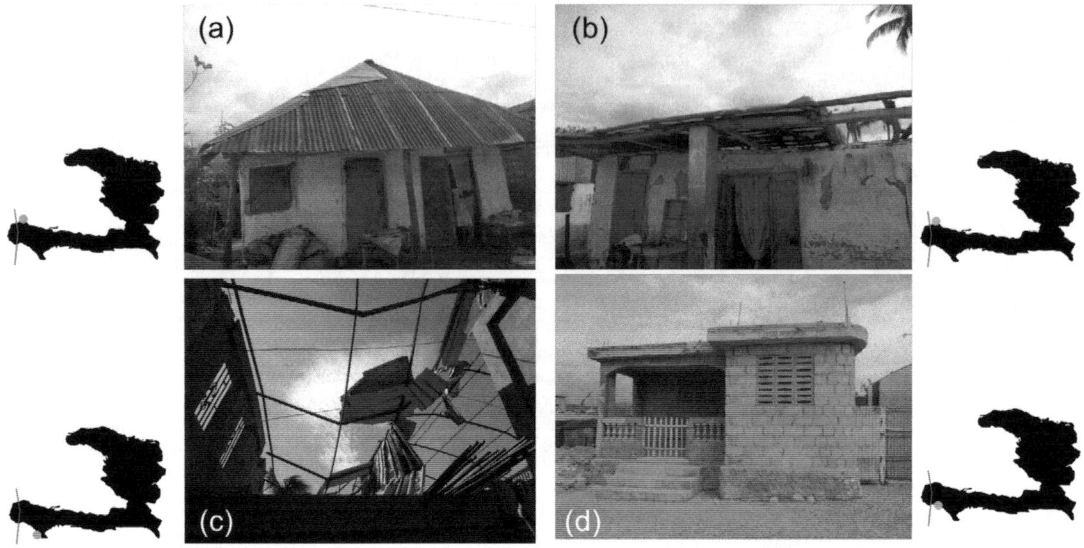

Figure 2. Examples of traditional Haitian post and beam construction in Abricot: (a) historically topped with a hip roof and (b) more recently with low-pitch gable roofs; (c) steel-framed roof in Port Salut; (d) concrete slab roof on unreinforced masonry home in Chardonnières.

4 OBSERVATIONS

4.1 *Performance under Wind Hazards*

Strong winds during Matthew caused damage (often complete) to metal roofs on a wide variety of structures, with wood and HSS framing. Analysis of satellite data surveys affirmed the dramatic extent of roof losses in the affected regions [3]. This included schools, clinics and churches as well as private residences. Wind-induced damage to the primary structure was most prevalent in non-residential construction employing unreinforced masonry walls with elongated aspect ratios and limited out-of-plane support, e.g., churches with open floor plans. In most instances, these buildings had thicker gauge corrugated metal roof panels and higher quality wood trusses that were well-attached to the perimeter walls (often by bending over the longitudinal reinforcement, as described in the previous section). This enabled the winds to generate significant uplift forces to tension the URM walls before the metal panels peeled off or the roof trusses snapped. Though most of these churches remained standing, partially collapsed or bowed masonry walls were commonly observed.

As many residences had weaker roof framing or thinner gauge and even corroded metal panels, complete roof failures preceded the tensioning of the URM walls. The instances where a superior load path was delivered by the roof system resulted regrettably in partial or complete collapse of the URM wall system (Fig. 3). While this may be somewhat expected for URM construction, failures were even observed in wall systems with vertical and horizontal confining elements, as shown previously in Figure 2c. This low-pitch HSS-framed gable roof (note the absence of roof trusses) maintained its load path with the primary structure resulting in a failure of the upper half of the two-story CMU wall (shown leaning inward) before the roof panels began to detach. Ironically, vulnerable URM walls often survived the storm, as the winds exploited the comparatively

Figure 3. Example of unreinforced masonry home in Bonbon constructed with locally-produced CMU weakly confined by only vertical elements. Homes standing in background were constructed by foreign NGO with CMU mechanically-produced off site.

more vulnerable roof system. Exceptions were understandably observed for critical facilities designed and constructed with considerable foreign involvement, such as the pediatric clinic in Port Salut, which employed a high-quality metal roof and non-conventional sandwich panel wall system. As such high-quality metal roofing systems are not accessible to the majority of homeowners, those rebuilding may resort to the use of concrete slab roofs whose sizeable mass maintained sufficient compression in these URM walls to perform uniformly very well in Hurricane Matthew. Regrettably this may lead to further proliferation of a typology that proved deadly in the 2010 Earthquake [4].

Figure 4. Pediatric clinic in Port Salut with (a) loss of metal roof panels on main clinic and (b) minor loss of hip flashing on outdoor triage area.

Figure 5. (a) High elevation wave run-up in a cliffed region near Roche-à-Bateaux, note road is approximately 5 m above sea level; (b) lead author and community advocate L. Presuma interview resident on beachfront in Port Salut, note CMU perimeter of destroyed home with slab foundation and temporary shelter occupying part of home's original footprint; (c) remains of nursing school destroyed by wave runup near Coteaux.

4.2 *Performance under Coastal Hazards*

Shoreline communities in the affected areas were typified by fishermen who require daily access to the sea or those forced to low lying regions as squatters without rights to land. These communities, who had weathered tropical storms and minor hurricanes previously, were largely unprepared for Matthew's storm surge. Waves and storm surge in the assessed regions reached 4.5-7.5 m above sea level along shorelines with steep beaches or cliffs (Fig. 5a) and 1.5-4.5 m along more gentle beaches. Based on interviews with local residents, inundation often took the form of very fast, strong waves similar to those recorded during Super Typhoon Haiyan in 2013 [5]. Close to the ocean in the region of strongest inundation, almost all coastal buildings were partially or completely destroyed. These homes included unreinforced load bearing masonry with only vertical confining elements, as well as traditional post and beam construction. Many occupants remained along the coasts as the hurricane made landfall, either not receiving or disregarding SMS alerts to evacuate until the waves began to physically destroy their homes. At locations like Port Salut, the surge met with inland waters from rivers that overtopped their banks forcing many to swim to higher ground at the storm's peak. Survivors returned after the waters receded to erect temporary structures from storm debris, salvaged materials, and tarps (Fig. 5b). Although official enumerations have not been released, this complete destruction of homes by the action of waves and surge likely numbers in the thousands, particularly along the southwest coast westward from Port Salut. Damage was not as complete, though still significant, on the northern coast. Regrettably, this devastation was not limited solely to residential construction (Fig. 5c).

4.3 *Rainfall and Inland Flooding*

The Matthew-affected regions of Haiti received heavy rainfall during and after the storm. Though not responsible for structural damage, rains damaged the contents of roof-compromised buildings, flooded entire districts in city centers, and hindered recovery efforts by further degrading roads and triggering landslides. The two largest affected cities of Jérémie and Les Cayes both experienced inland flooding of around 1 m in some downtown areas as the steep mountains outside the cities rapidly deposited their water into these and other similarly low-lying coastal areas, overtopping levees along a number of coastal rivers. In the northern city of Jérémie, rainfall continued for weeks after Matthew: according to residents, this post-storm flooding was worse than that experienced during the storm.

5 CONCLUDING REFLECTIONS

Undoubtedly, the Haitian people have shown unparalleled resilience and perseverance in the face of extreme historical, political and natural shocks, though this precious social capital has done little to offset the failures across the range of societal dimensions that underpin a community's resilience. Tangible evidence exists simply by examining the recovery (or lack thereof) of the 2010 earthquake's epicentral regions around Port-au-Prince and Léogâne. Transitory and informal shelters are still prevalent in these communities, many even adapted into what will be permanent shelters for the foreseeable future as residents struggle over issues of land tenure and limited access to financing to rebuild. Those that completely lost their homes in Matthew will now join them in what may be a decades-long recovery. Yet it is likely that Matthew's victims may struggle even more significantly in their process. Although the earthquake instigated catastrophic losses of life and property [6], its recovery did not disrupt agricultural and maritime livelihoods and even permitted salvaging from the rubble. There was little to salvage after Matthew, especially in the low lying coasts where victims often had all that they possessed washed out to sea or completely destroyed by the onslaught of heavy rains once their roofs were lost.

The limited civil infrastructure in these communities was also compromised, with bridges and roads washed-out by the storm or the many landslides that followed. Enumeration and aid delivery over this large and mountainous affected region, with many remote communities far from the capital, was challenging. Without redundant routes and communities dispersed throughout the hillsides, many were completely isolated from aid at the time of this reconnaissance. The few solar powered micro-grids funded by NGOs for communities like Les Anglais were also destroyed. With those heavy rains and latrines overtopped by coastal surge and racing floodwaters came new outbreaks of Cholera. Perhaps even more gravely, Matthew directly struck the nation's "bread basket," uprooting or snapping trees that produce staples like breadfruit, mango, avocado, coconut, bananas and plantains used not just for sustenance in the affected regions, but also sold in the markets of Port-au-Prince to support livelihoods. Meanwhile, saltwater inundation, heavy rains and post-hurricane disease destroyed other produce including the root crops, locally referred to as *vive* or "life." While these staples may recover in months with agricultural aid, the trees will take decades to regrow, producing for now only now a short-term economic boost through the processing of lumber and charcoal (the primary fuel source in Haiti), shipped by the truckloads east toward Port-au-Prince. Sadly, malnutrition was already taking hold in the children of the region at the time of this reconnaissance, while the effects have further rippled through the produce supply chains all the way to the country's capital. Impacts to nutrition and livelihoods has only been compounded as fishing communities struggled to adapt to the dramatic changes in their coastlines, the complete losses of their beaches, and absence of fish and other marine life in once plentiful coves and bays. Coupling the region's geographic disconnect with "Haiti fatigue" among charitable donors, lingering distrust from reports of misappropriated earthquake recovery funds, and the Haitian government's role in controlling the channels for relief after this disaster has resulted in one simple fact: Matthew's victims are struggling and will continue to struggle to meet the basic needs of life for some time to come.

6 ACKNOWLEDGEMENTS

The authors gratefully acknowledge the financial support of the National Science Foundation (CMMI 17-09357) and the guidance offered in this RAPID submission by Dr. Joy Pauschke. Any opinions, findings, and conclusions or recommendations expressed in this material are those of the authors and do not necessarily reflect the views of the National Science Foundation. The authors are also grateful to their collaborators in the field: Scott Schiff of the Applied Technology Council and Kwasi Perry of UAV Survey Inc. The authors further thank Mark Powell of RMS for the permission to use HWind data for this project. This field work would not be possible without the coordination and participation of a number of other individuals, including logistics lead Michael Wilson, Engineering2Empower staff members Lamarre Presuma, Gede Jean Benoit, Edson Jean, and Onel Dossou, drivers Jac Lubin and Brunie St. Fort, and the staff of the Notre Dame Haiti Program. All data and data products will be available in the NHERI Data Depot at www.designsafe-ci.org/.

7 REFERENCES

1 T. Kijewski-Correa, A.A. Taflanidis, D. Mix and R. Kavanagh, An empowerment model for sustainable residential reconstruction in Léogâne, Haiti, after the January 2010 Earthquake, Leadership and Management in Engineering, 12 (2012) 271-287.

2 K. Fink, Contributions to Risk Assessment of Residential Masonry Structures in Developing Nations: Review of Vulnerabilities and Development of a Unified Analytical Modeling Framework, MSCE, University of Notre Dame, 2016.

3 K.K.R. Lai and S. Pecanha, Photos and Detailed Maps Reveal Hurricane Matthew's Brutal Aftermath in Haiti, http://www.nytimes.com/interactive/2016/10/10/world/americas/haiti-hurricane-matthew-aftermath-destruction-maps-photos.html?_r=0.

4 D. Mix, T. Kijewski-Correa and A.A. Taflanidis, Assessment of residential housing in Léogâne, Haiti after the January 2010 earthquake and identification of needs of rebuilding, Earthquake Spectra, 27 (2011) S299-S322.

5 A.B. Kennedy, N. Mori, Y. Zhang, T. Yasuda, S. Chen, Y. Tajima, W. Pecor and K. Toride, Observations and modeling of coastal boulder transport and loading during Super Typhoon Haiyan, Coastal Engineering Journal, 58 (2016) 1640004.

6 E.A Cavallo, A. Powell and O. Becerra, Estimating the Direct Economic Damage of the Earthquake in Haiti, IDB working paper series no IBD-WP-163, Inter-American Development Bank, Washington, DC, 2010.

13th Americas Conference on Wind Engineering
Gainesville, Florida, USA
May 21 - 24, 2017

A Multi-Hazard Tropical Cyclone Disaster Impact Scenario Model for Queensland

Richard J. Krupar III[a] and Matthew S. Mason[a]

[a] *The University of Queensland, St. Lucia, QLD, Australia, r.krupariii@uq.edu.au*
[a] *The University of Queensland, St. Lucia, QLD, Australia, matthew.mason@uq.edu.au*

ABSTRACT

Tropical cyclones pose a significant threat to the built environment and society in Queensland, Australia through the concomitant hazards of wind, rainfall and storm surge. Scenario analysis is one method commonly used to assess potential cyclone wind and flood impacts, where hypothetical cyclone simulations are generated, often based on perturbed historical events, to produce estimates of potential damage to buildings and critical infrastructure should such an event occur. Emergency services managers and insurers use these hypothetical scenarios to effectively manage disaster response resources and risk portfolios for businesses and homeowners. Currently, an openly accessible multi-hazard (i.e. wind, rainfall and storm surge) tropical cyclone disaster impact scenario model does not exist for Queensland. The aim of this project is to build such a model in an effort to improve emergency services and financial risk decision making in the state.

The multi-hazard tropical cyclone disaster impact scenario model is made up of three hazard (wind, rainfall and storm surge) and two vulnerability (wind and flood) models. Wind, rainfall and storm surge hazard footprints, as well as, attendant wind and flood-induced impacts to buildings, critical infrastructure and society, are generated using peer-reviewed models. The wind hazard model is first initialized with key storm input data (e.g. storm track, minimum central pressure, radius of maximum wind), which can be drawn from the Bureau of Meteorology (BoM) historical tropical cyclone database or based on hypothesized storm parameters. The wind model generates gradient level wind speeds that are subsequently adjusted to the surface (i.e. 10-m) through the use of a linear analytic tropical cyclone boundary layer and gust factor models. Maximum surface winds and vertical velocities above the tropical cyclone boundary layer are then used to drive the parametric rainfall and two-dimensional hydrodynamic storm surge models. The rainfall model takes into account orographic and surface terrain interaction, storm asymmetries and rain drift, while the storm surge model solves the non-linear shallow water equations and neglects wave action.

Available in situ wind, rainfall and storm surge measurements collected during Tropical Cyclone Marcia (2015) were obtained to validate the hazard model performance. A Monte Carlo Markov Chain (MCMC) optimization tool is employed to optimize different storm parameters to best replicate the observed tropical cyclone wind speeds, rainfall estimates and storm surge heights. Hazard models are then coupled with Queensland exposure information (i.e. building type, building value, surface terrain conditions, etc.) and both wind and flood vulnerability models to simulate impacts for these events. Validation of these outputs were then based on rapid damage assessment data collected following these events.

Preliminary assessment of model simulations, in particular wind and rainfall, suggests that the uncalibrated model tends to overestimate both observed wind speed and rainfall magnitudes. When wind speed is optimized using the MCMC, the overall root mean square error reduces, however, significant differences still exist at larger storm radii. Wind and flood-related impacts to buildings, infrastructure and society are difficult to benchmark, however, the recently sourced rapid damage assessment impact data will allow such a validation for Cyclone Marcia. These results will be discussed in the full paper.

KEYWORDS

Tropical Cyclones, Scenario Analysis, Vulnerability Modelling, Risk Decision Making, Optimization

Processing Data from Hurricane Matthew in DesignSafe-ci

Harika Gurram[a], Chelakara Subramanian[b], Jean-Paul Pinelli[c], Rituparna Basu[d]

[a,b,d] Mechanical & Aerospace Engineering Department, Florida Tech, Melbourne, Florida, U.S.A.
[c]Civil Engineering Department, Florida Tech, Melbourne, Florida, U.S.A.
[a]hgurram2015@my.fit.edu , [b]subraman@fit.edu , [c]pinelli@fit.edu, [d]rbasu2014@my.fit.edu

ABSTRACT

The DesignSafe.ci is the cyber-infrastructure component of the Natural Hazard Engineering Research Infrastructure (NHERI) funded by the National Science Foundation. It provides a cloud platform to all researchers to archive and share their data on the public domain. It supports research workflows, data analysis and visualization, and the full lifecycle of data required by engineers and scientists, to effectively deal with the threats posed to civil infrastructure by natural hazards. The researchers from Florida Institute of Technology deployed their wireless sensor network (WSN) system on the roof top of a house near the Satellite beach, Melbourne, Florida (longitude and latitude: 28.175944, -80.592710) during the October 6-8, 2017 hurricane Matthew passage to collect real time data. The WSN is an autonomous full scale hurricane data measurement system with 30 pressure and temperature sensors, and an anemometer to measure pressure, temperature, speed and wind direction respectively. All sensors are connected wirelessly through 3 routers to a base modem, which is connected to the laptop for collecting the data. The data from the laptop was uploaded through a cellular network at five minute interval to a Box account, which provides cloud storage and file sharing collaboration. The entire wireless sensor network system was monitored using Team Viewer remote desktop application. All the data collected in the Box was then uploaded into the DesignSafe-ci workspace, which can be shared with researchers. For future deployments, these processes can be performed using the DesignSafe-ci workspace. For post processing and analysis of the uploaded data, DesignSafe-ci provides software tools like Jupyter and MATLAB. To perform the analysis a unique graphical user interface (GUI) application was created using MATLAB, which is capable of analyzing the complete data set on a single run. The authors can run the GUI code instantly on the cloud and provide plots of the physical data, thanks to the virtual desktop connection to the DesignSafe workspace. This paper describes how DesignSafe-ci facilitates the setup, remote monitoring and analysis of the data processing collected during a hurricane. The post processing capabilities for analyzing Hurricane Mathew data are illustrated. A companion paper describes a more detailed analysis of the data.

KEYWORDS: Wireless sensor network, hurricane Matthew, DesignSafe-ci, NHERI, GUI, MATLAB, JUPYTER

1 INTRODUCTION

The effect of hurricane on the residential buildings is estimated using various experimental, computational, and full scale measurements techniques. The Florida Tech Wireless Sensor Networks system is a full scale measurement system that can be used to measure the pressure, temperature wind speed and direction of the hurricane on the roof top of the residential buildings. The real time variations of the properties on the roof of the building due to hurricane is measured and all the data is analyzed and shared in the local computer and on the cloud for monitoring and future post-processing. The Natural Hazard Engineering Research (NHERI) Infrastructure funded by National Science Foundation have created an open access to the workspace on the cloud called DesignSafe-ci. The hurricane full scale measurements can be uploaded on the cloud directly and later post processed and shared. This paper explains the process for collecting, storing and post processing of the hurricane data from the Florida WSN system in DesignSafe-ci workspace, with specific reference to recent hurricane Matthew deployment.

2 WIRELESS NETWORK SYSTEMS

The purpose of Wireless Sensor Network (WSN) system is to measure the pressure, temperature, wind speed and direction on the roof of a house during hurricane to quantify and model the wind loads on full-scale residential building roofs. The WSN consists of three systems and each system has one anemometer, one atmospheric pressure port and three subsystem with total of up to 10 sensors and one router. The sensor pressure transducer measures pressure on the roof top of a residential building, and the propeller vane anemometer, which is deployed on the roof top, measures the wind speed and direction. All these sensor and the anemometer communicate to the base wither directly or through routers. The base connects to laptop, where it stores the raw data as shown in the Figure 1. The hurricane Matthew data is collected on a residential home (longitude and latitude: 28.175944, -80.592710) in Satellite beach, Melbourne, Florida, USA). The full scale digital time histories of the pressure, temperature, wind speed and direction are acquired. This raw data is stored as 15-minute block files in the laptop. The Figure 1: Field Data Collection

shows the setup of the sensors system during hurricane Matthew. The anemometer and sensors are deployed and the data is stored in the laptop connected to the base. This data is synchronized with the Box account to upload all the data instantly on the cloud. In case of any system failure during the data collection, the files are saved on the cloud up to the time of failure. The Figure 3 shows the flow chart of WSN data collection process during a deployment. The system operation can be monitored using a remote desktop control freeware known as TeamViewer. The mobile

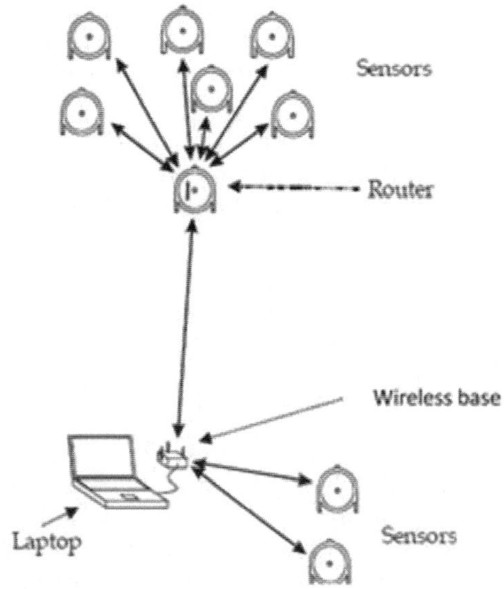

hotspot is serviced by the internet provider which allows internet access anywhere in the field.

This TeamViewer software can be installed on any smart device like smart phone, tablet, IPad, laptop, or, desktop to monitor the WSN system during the hurricane deployment.

Figure 1: Field Data Collection

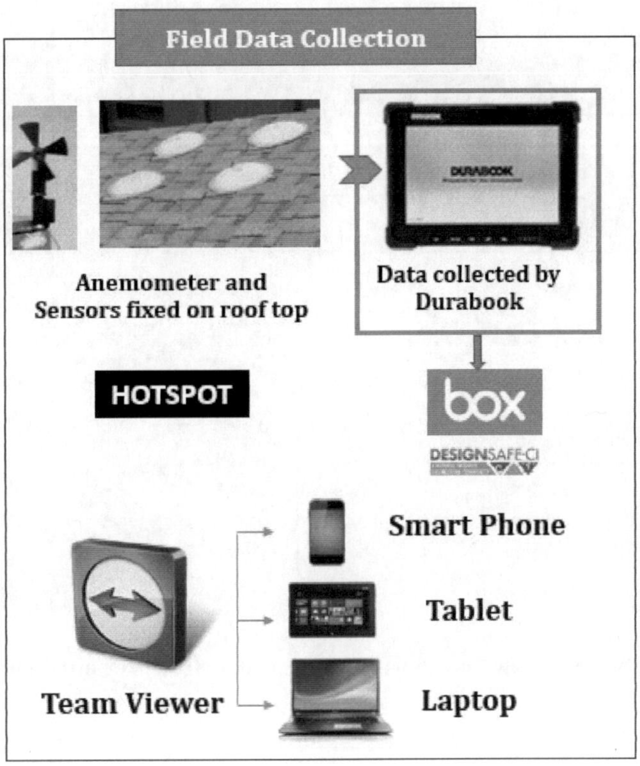

Figure 2: The sensors setup on the roof top during Matthew hurricane

Figure 3: The field data collection

3 DESIGNSAFE WORKSPACE

The DesignSafe-ci is a cyberinfrastructure for the natural hazards engineering community to provide computational tools to manage, analyze, and understand data for natural hazard research. During the event of Hurricane Matthew on October 5th 2016, the WSN data stored in the laptop on the field is directly synchronized to the Box account on the DesignSafe web portal. The data from the Box account is moved to the Project folder. The Project folder on DesignSafe is used to share and process the data. The files are arranged in a systematic manner, the Figure 4 shows the flow chart of the folders arranged in the Project file. The project name is Wireless Sensor Network Hurricane Data, it contains of 3 sub folders which store raw data, documentation and Hurricane Data analysis, respectively. The raw data folder has the raw time histories of pressure, temperature, wind speed and direction. Whereas the documentation folder gives the information about the introduction to WSN deployment details and location of the house, geometry of the roof, user manual for post processing the hurricane data. The Hurricane Data Analysis folder has the post processing code to convert the raw data to physical units and plot the time histories. The processed physical data can be shared with other researchers on cloud. The figure shows the detailed procedure of hurricane data storage and analysis on DesignSafe-ci.

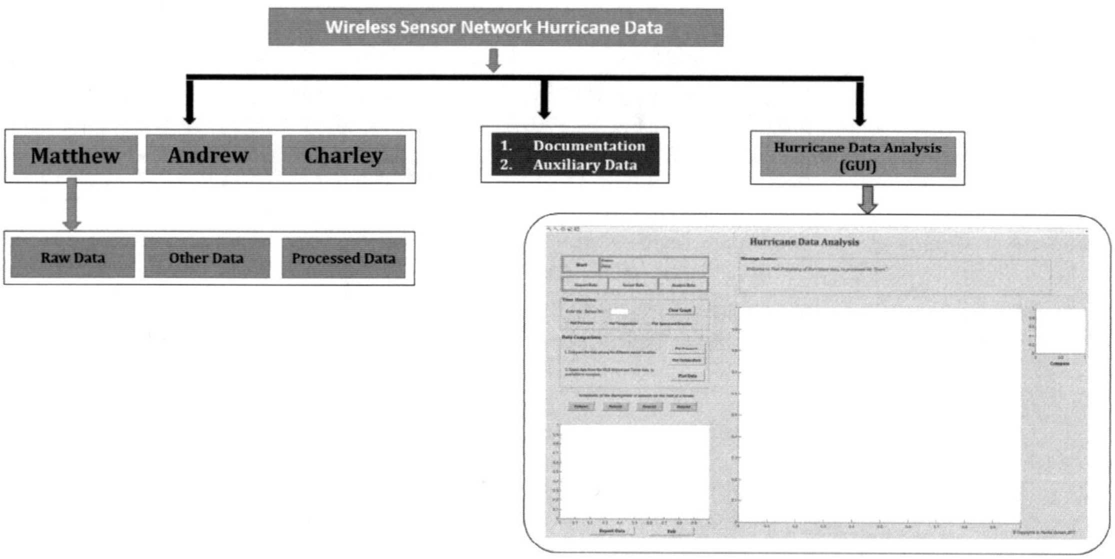

Figure 4: The flow chart of the directory structure on DesignSafe-ci.

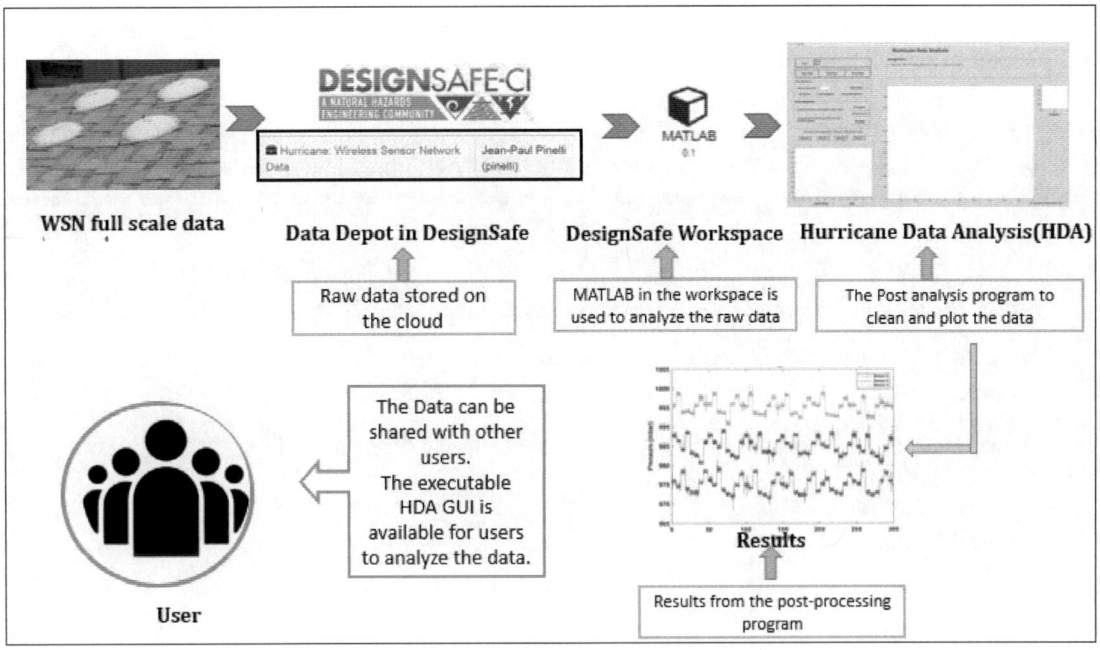

Figure 5: Data Storage and analysis of the hurricane Matthew

4 HURRICANE DATA ANALAYSIS

The WSN system exports the raw data as every 15-minute block file which contains the data from all the sensors and it also has the pressure, temperature, wind speed and direction data all together in the same 15 minute file. The DesignSafe-ci portal provides a compatible workspace to instantly analyze the data on cloud. The Hurricane Data Analysis (HDA) is a graphical user interface (GUI) MATLAB program used to convert the raw data to physical data. The DesignSafe-ci has a work-space which consists of post processing tools like MATLAB, Jupyter, Open Foam, etc. The HDA program is executed using the MATLAB application available in the DesignSafe workspace. The HDA graphical user interface has a step by step procedure to convert the raw data to physical data. The Figure 6 shows the layout of the GUI application. The message center gives the instruction to the user and this interface is also embedded with text to speech add-on to provide user friendly instructions. The 'Start' function starts the program and the path for the raw data should be speci-fied by user. The static text named 'Event' and 'Data' gives the information about the hurricane project. 'Import data' function is used to import all the 15 min raw file and stitches all the data into a matrix with respect to given time period. After importing the data to workspace the data is sorted by sensor ID, this function is performed using the Sensor Data. The function Analyze data is used to convert the raw data into the physical units using calibration coefficients. The program also checks for the packet loss during the communication between the sensors and base. In case of any packet loss the missing data is interpolated. The HDA interface is integrated with picture gallery, plotting workspace and compass. The picture gallery give the details of the sensor location de-ployed on the roof top. The Time histories panel is used to plot the pressure, temperature data, and the wind speed is plotted with respective to the wind direction. The wind direction is shown on the compass. This interface can also be used to compare and correlated the time histories between different sensor locations as well as with external data. External data refers to the other measured or computational data. The time histories are exported using the Export option and the Exit option helps in deleting the intermediate files to save the user memory.

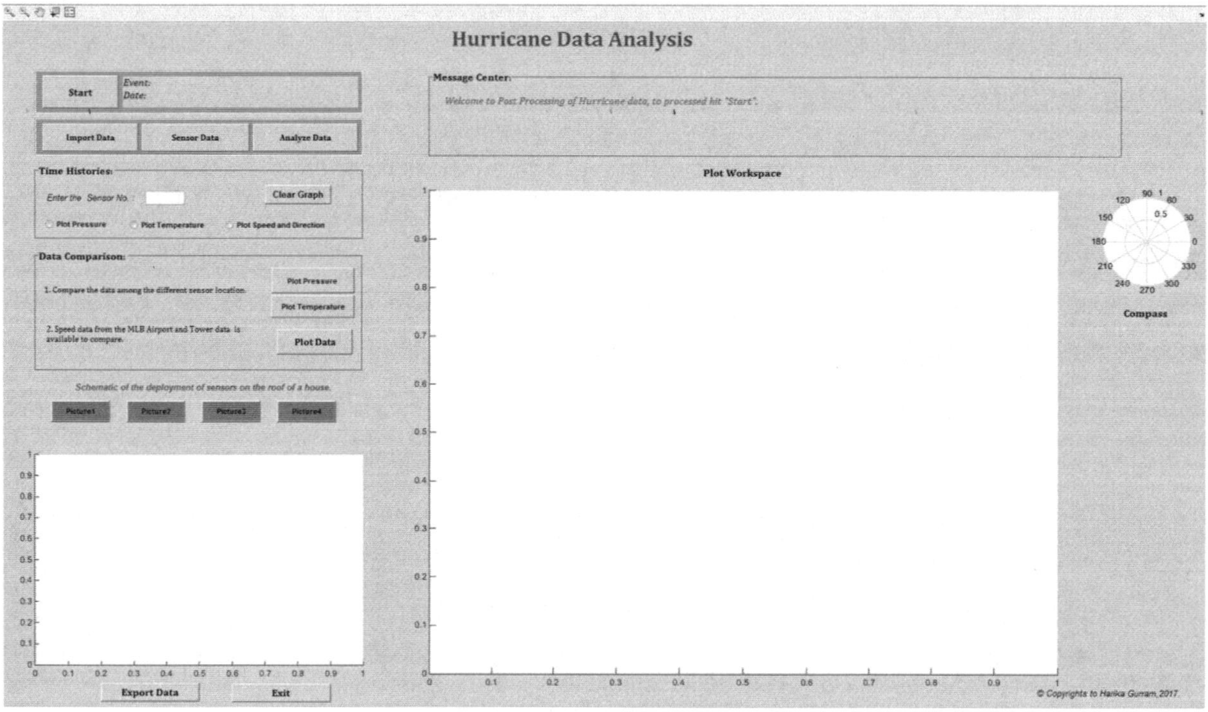

Figure 6: Hurricane data Analysis GUI layout

5 CONCLUSION

The WSN system is used to measure the real time hurricane effect on a full scale residential building during the event. DesignSafe-ci provides a cyber-space to store, analyze and share the data in the cloud. The data collected from the field is synchronized to the cloud. The post processing GUI code is used to convert the raw data to physical data. The time histories with respect to the sensor location can be plotted and compared instantly in the cloud.

6 ACKNOWLEDGEMENT

Authors would like to thank Stephen Mock for helping in deploying the Hurricane Data Analysis graphical user interface on DesignSafe-ci workspace.

7 REFERENCES

[1] "DesignSafe.ci," Natural Hzardus Engineering Research Infrastructure , 2016. [Online]. Available: https://www.designsafe-ci.org/rw/overview/. [Accessed 26 April 2017].

[2] G. Lapilli, "Hurricane Wind Effects: Insturmentation, Measurement and Analysis," Florida Institute of Technology, Melbourne , 2011.

[3] "MATLAB," MathWorks , [Online]. Available: https://www.mathworks.com/. [Accessed 26 April 2017].

[4] "TeamViewer," [Online]. Available: https://www.teamviewer.com/en/. [Accessed 26 April 2017].

[5] Rathje, E., Dawson, C. Padgett, J.E., Pinelli, J.-P., Stanzione, D., Adair, A., Arduino, P., Brandenberg, S.J., Cockerill, T., Dey, C., Esteva, M., Haan,'Jr., F.L., Hanlon, M., Kareem, A., Lowes, L., Mock, S., and Mosqueda, G. 2017. "DesignSafe: A New Cyber-infrastructure for Natural Hazards Engineering," ASCE *Natural Hazards Review*, doi: 10.1061/(ASCE)NH.1527-6996.0000246

13th Americas Conference on Wind Engineering
Gainesville, Florida, USA
May 21 - 24, 2017

Overview of the NHERI Center for Computational Modeling and Simulation of the Effect of Natural Hazards on the Built Environment

Stephen Mahin[a], Camille Crittenton[b], Gregory Deierlein[c], Sanjay Govindjee[d], Ahsan Kareem[e], Laura Lowes[f], Frank McKenna[g] and Matthew Schoettler[h]

[a] *University of California, Berkeley, CA, USA, mahin@berkeley.edu,*
[b] *University of California, Berkeley, CA, USA, ccrittenden@berkeley.edu,*
[c] *Stanford University, Stanford, CA, USA, ggd@stanford.edu,*
[d] *University of California, Berkeley, CA, USA, s_g@berkeley.edu,*
[e] *Notre Dame University, IN, USA, Ahsan.Kareem.1@nd.edu,*
[f] *University of Washington, Seattle, WA, USA, lowes@u.washington.edu,*
[g] *University of California, Berkeley, CA, USA, fmckenna@berkeley.edu,*
[h] *University of California, Berkeley, CA, USA, schoettler@berkeley.edu.*

ABSTRACT

The new NSF-funded Center for Computational Modeling and Simulation provides the natural hazards engineering research and education communities with access to the next generation computational modeling and simulation software tools, user support, and educational materials needed to advance the nation's capability to simulate the impact of natural hazards on structures, lifelines, and communities, and to make informed decisions about the need for and effectiveness of potential mitigation strategies. The new center, known as the SimCenter, is an integral part of NSF's Natural Hazards Engineering Research Infrastructure (NHERI) Program, and is headquartered at the University of California, Berkeley. A team of multidisciplinary investigators from California State University at Long Beach, Georgia Institute of Technology, Georgia State University, Johns Hopkins University, Stanford University, Notre Dame University, University of California campuses at Berkeley, Los Angeles and San Diego, University of Memphis, University of Southern California, and University of Washington will provide core support for the SimCenter. The broad natural hazards engineering community is encouraged to participate in the center's activities.

The various modeling and simulation tools and web-enabled workflow management framework being developed by the SimCenter will be made available through the Discovery Workspace at DesignSafe-ci.org. The SimCenter will provide computational modeling and simulation tools using a new open-source simulation framework that: (1) addresses various natural hazards, such as windstorms, storm surge, tsunamis, and earthquakes; (2) tackles complex, scientific questions of concern to various disciplines involved in natural hazards research, including earth sciences, geotechnical and structural engineering, architecture, urban planning, risk management, social science, public policy, and finance; (3) utilizes machine learning to facilitate and improve modeling and simulation using data obtained from experimental tests, field investigations, and previous simulations; (4) quantifies uncertainties associated with the simulation results obtained; (5) utilizes the high-performance parallel computing, data assimilation, and related capabilities provided by the DesignSafe-ci.org web portal to enable the natural hazards engineering community to easily combine software applications into workflows of unprecedented sophistication and complexity; (6) extends and refines software tools for carrying out performance-based engineering evaluations and supporting decisions that enhance the resilience of communities susceptible to multiple natural hazards; and (7) utilizes existing applications that already provide many of the pieces of desired computational workflows. The resulting cloud-based ecosystem will allow multidisciplinary specialists and various stakeholders to collaborate on solutions to complex natural hazard engineering problems regardless of their local resources and geographic proximity. In tandem with the framework and applications being developed, in-person and online educational programs, including open online courses, will be implemented within a new Virtual Community of Practice (VCoP) that will provide an online meeting place to obtain and exchange community-generated ideas, feedback, best practices for modeling and simulation, insights, and innovations for natural hazards engineering research. The SimCenter will also provide workshops and graduate and undergraduate student research traineeships.

This paper highlights the SimCenter's vision and schedule, and describes specific early deliverables and how to access and use them, as well as opportunities contribute to the development, expansion and implementation of SimCenter tools and framework.

KEYWORDS

Computational simulation, numerical modeling, multi-hazard, multi-scale, high performance computing, risk assessment

Wind Engineering Research at NSF

Kishor C. Mehta, Ph.D.; M of AAWE, Dist. M of ASCE, Fellow SEI, NAE
P. W. Horn Professor of Civil Engineering
Texas Tech University, Lubbock, Texas, USA, kishor.mehta@ttu.edu

Abstract:

The National Science Foundation has reorganized its resilient and sustainable infrastructure (RSI) research program in the last two years. Specifically, fundamental building and structure research is pursued through four programs, namely 1. Engineering for Natural Hazards, 2. Structural and Architectural Engineering and Materials, 3. Natural Hazards Engineering Research Infrastructure, and 4. Geotechnical Engineering and Materials. This reorganization permits more opportunities to propose building research related to wind engineering and wind hazards. In addition, there are many other programs at NSF which can support wind engineering research. Imaginative thinking and incorporation of disciplines other than structural engineering opens a wide variety of programs within NSF that can fund wind engineering research.

Keywords: Research, Wind, NSF, Funding, Fundamental Research

Highlights:

- Wind engineering research funding
- Programs in NSF that fund wind engineering research
- Opportunities for funding for wind engineering research
- Funding programs in NSF

1. [MK1]Introduction

The National Science Foundation is one of the major sources of funding for fundamental research related to wind engineering. The National Earthquake Engineering Simulation (NEES) has funded structural engineering research over a ten year period (2004 – 2014) which has allowed significant advancements in structural engineering. The NEES program was dedicated to improving structural framing to prevent it from collapsing when under earthquake shaking. (Wind engineering research used to primarily be funded under the Hazard Mitigation and Structural Engineering (HMSE) program. This program also entertained proposals related to earthquakes, storm surge, and other hazards in addition to wind hazards.) Since NEES was a long standing dedicated program, many new faculty were educated in earthquake engineering with the result that proposals received in the HMSE program reflected earthquake engineering to a large extent. This phenomena was recognized by the program directors at NSF and remedial actions were pursued when the mandate of the NEES program ended. The reorganization of the Resilient and Sustainable Infrastructure (RSI) now focuses on all natural hazards including wind hazards, and this reorganization now allows funding for wind engineering along with all other natural hazards.

The NEES had established fourteen experimental facilities around the country, and this experimental infrastructure for research was focused on earthquake engineering research. Along with supporting various experimental facilities, the NSF also had NEES-R, a dedicated program that funded research projects to be conducted at NEES experimental facilities. Natural Hazard Experimental Research Infrastructure (NHERI), the new program of experimental facilities, has facilities for experimental research in wind engineering, earthquake engineering, geotechnical engineering, and tsunami and storm surge engineering. Three additional facilities under NHERI (a Simulation Center, the RAPID Research facility, and the Cyberinfrastructure facility) are common facilities for all hazard-related research. The development of NHERI now promotes wind hazard-related research with the same emphasis as all the other natural hazards.

Building-related structural engineering research has focused in the past on the skeleton of the building to make sure that the building survives in the natural hazards of earthquake, hurricane, or tornadoes. Thanks to the excellent work in updating building codes, this focus has been reasonably accomplished. In the future, we will see the definition of structural engineering for buildings change from the design of skeleton to the design of the holistic building that includes the foundation-structure-envelope and the interior system. Vibration and human comfort will continue to be part of the research agenda, with the ultimate goal to develop a building that can be continuously occupied and operated during its useful life. Considerations will be given to the analysis and the design, construction, operation, maintenance, retrofit, and repair of buildings.

The NSF has many different specialized programs where wind engineering research can fit, and it is up to investigators to study the specific focus of each of the different programs and then propose research that advances wind engineering. The new organized programs in the RSI cluster and other NSF programs are delineated here. This paper is based on the experiences of the author while he was the program director at NSF during 2011-2015.

2. Reorganization of Resilient and Sustainable Infrastructure (RSI):

Two years ago, the RSI cluster had five programs. Of these five programs, three programs on physical research on hazards have been reorganized. (Figure 1 shows this reorganization.) The natural hazards of earthquake (and other hazards) under the HMSE program have been merged into one program of Engineering for Natural Hazards (ENH), and the experimental infrastructure NHERI is now part of ENH. A new program of Structural and Architectural Engineering and Materials (SAEM) has been created for fundamental research with a focus on building science. Geotechnical hazard-related research is now part of ENH, while geomechanics, materials and geotechnical engineering are under the reorganized Geotechnical Engineering and Materials (GEM) program. ENH, NHERI and SAEM are discussed in some detail since they involve wind engineering research.

2.1 ENH and NHERI Programs:

NHERI represents a collaboration between eleven facilities that support Engineering for Natural Hazards. These facilities are as follows:

Eight experimental facilities;

- A network coordinating office at Purdue University;
- A computational modeling and simulation center located at the University of California-Berkeley;
- A cyberinfrastructure facility – "DesignSafe" at the University of Texas-Austin.

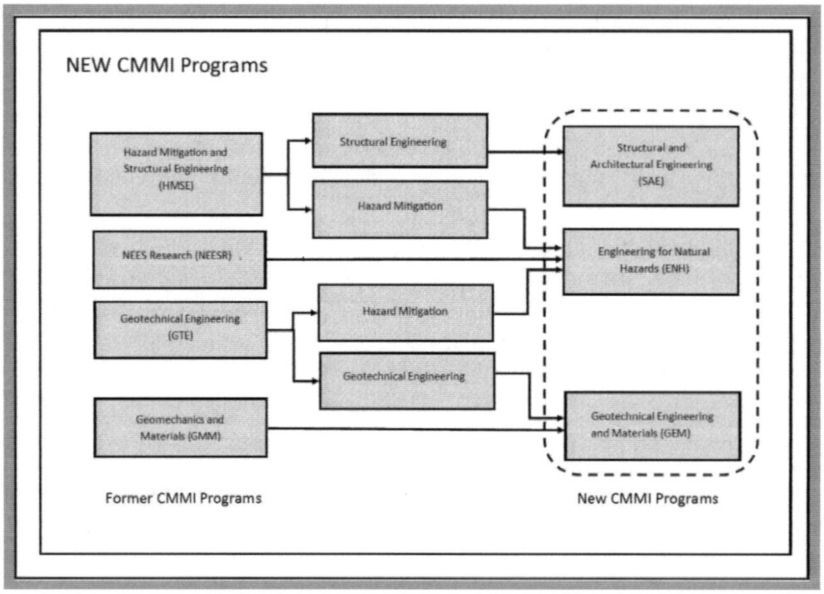

Figure 1. Reorganized organization chart for new CMMI programs.

The experimental facilities are:

- Large-Scale, Multi-Directional, Hybrid Simulation Testing at Lehigh University;
- High-Performance Outdoor Shake Table at the University of California-San Diego;
- Mobile Dynamic Shakers for Field Testing at the University of Texas-Austin;
- Geotechnical Centrifuges at the University of California-Davis;
- Large Wave Flume and Directional Wave Basin at Oregon State University;
- Boundary Layer Wind Tunnel, Wind Load and Dynamic Flow Simulators, and Pressure Loading Actuators at the University of Florida;
- Twelve-Fan Wall of Wind at Florida International University;
- RAPID Research Facility at the University of Washington

The eight experimental facilities include a structural testing, a shake table location, two geotechnical testing locations, a rapid reconnaissance location, a wave research location, and two wind engineering facilities[MK2]. [PE3]The two wind engineering testing facilities are the Wall of Wind (WoW) at Florida International University, and the wind tunnel and full-scale pressure testing facility at the University of Florida. The Cyberinfrastructure facility DesignSafe (located at the University of Texas-Austin) connects to all the facilities (https://www.designsafe-ci.org/facilities/experimental/).

These experimental facilities are a national resource, and are available for researchers to pursue testing at any of each of the facilities. The protocol for investigators interested in pursuing research in natural hazard is to contact the facility of interest prior to preparing the proposal. With agreement of the facility coordinator, a testing research plan can be developed, and the investigator allocates funding for testing in his/her proposal. The facility operators are funded for maintaining the facility for testing, but not for the time and materials required for testing. It should be noted that each of the two experimental facilities for testing for wind engineering are unique and can provide help in planning, testing protocol, and preliminary analysis of results. For further information, each facility site maintains a web site which can provide information on protocols and contacts. Proposals pursuing research at any one of the experimental sites are requested to submit proposals to the NSF ENH program.

The goal of the ENH program is to understand and mitigate the impact of natural hazards of earthquakes, windstorms (such as tornadoes and hurricanes), tsunamis, and landslides on constructed civil infrastructure, including building systems (such as the soil-foundation-structure-envelope-nonstructural system) and other structures (such as tunnels, towers, etc[MK4].). Research may integrate geotechnical, structural, and architectural engineering with advanced discoveries in earth and atmospheric sciences, materials science, dynamical[MK5][PE6] systems and control, systems engineering, decision theory, risk analysis, high performance computational modeling and simulation, and social, behavioral, and economic sciences.

2.2 SAEM Program:

Goal: Buildings that can be continuously occupied and operated

Research topics can include buildings that are cost-effective, that are efficient in resources and energy, that involve the use of sustainable materials, cover the issues of mitigation of fatigue and corrosion, serviceability for deflections and vibrations, and any problem which pursues physics-based computational simulation. Additionally, research in building science that is focused on minimizing fresh water consumption or that can promote self-heating-cooling-ventilating is also encouraged. The integration of the science of architecture, economics and human factors may also be included to pursue research in this program.

2.3 Other Programs at NSF:

- There are many programs at NSF that fund wind engineering research, and it may be necessary to think beyond normal research to fit your research idea in one of the programs. Below is the list of programs, some of which are continuing while other programs are just starting. Each one has specific requirements that the investigators should study on the relevant web site to formulate his/her ideas.
- *Infrastructure Management and Extreme Events (IMEE)*: The IMEE program supports fundamental multidisciplinary research on the impact of hazards and disasters upon civil infrastructure and society.
- *Civil Infrastructure Systems (CIS)*: The Civil Infrastructure Systems (CIS) program supports fundamental and innovative research necessary for designing, constructing, managing, maintaining, operating and protecting efficient, resilient and sustainable civil infrastructure systems.
- *Critical Resilient Interdependent Infrastructure Systems and Processes (CRISP)*: The goals of the CRISP program are to foster an interdisciplinary research community of engineers, computational scientists and social and behavioral scientists, to enhance the understanding of interdependent critical infrastructure systems, and to improve the effectiveness and efficiency with which they deliver services.
- *Prediction of and Resilience against Extreme Events (PREEVENTS)*: Projects that will enhance the understanding of the fundamental processes underlying natural hazards and extreme events, and that will improve our capability to model and forecast such hazards and events.
- *Industry-University Cooperative Research Centers Program (IUCRC)*: This program develops long-term partnerships among industry, academe, and government. The Centers are catalyzed by an investment from the NSF, and are primarily supported by industry Center members with NSF taking a supporting role in the development and evolution of the Center.

- *DCL (17-065) Growing Convergence Research at NSF - Work at the Human-Technology Frontier*: Shaping the Future: (1) research to understand the social and economic consequences of today's emerging technologies and the associated educational needs; and (2) research to develop tools that shape human-technology partnerships, improve worker performance, increase career longevity and job satisfaction, and facilitate the life-long learning of new skills.
- *DCL (17-074) Simulated and Synthetic Data for Infrastructure Modeling (SSDIM)*: NSF (in collaboration with the Department of Homeland Security [DHS]) announces its intention to fund EArly[PE7]-Concept Grants for Exploratory Research (EAGER) proposals in support of research to develop simulated and synthetic data on interdependent critical infrastructures, and thus to improve the understanding and the performance of these systems.
- *Computational and Data-Enabled Science and Engineering (CDS&E)*: The goal of the CDS&E program is to identify and capitalize on new computational and data analysis approaches. The outcome relies on the development, adaptation, and utilization of one or more of the capabilities offered by advancement of both research and infrastructure in computation and data, either through cross-cutting or disciplinary programs.
- *Innovation Corps Program (I-Corps)*: The goals of the I-Corps program are to spur translation of fundamental research, to encourage collaboration between academia and industry, and to train students to understand innovation and entrepreneurship. The purpose of the NSF I-Corps program is to identify NSF-funded researchers who will receive additional support - in the form of mentoring and funding - to accelerate the translation of knowledge derived from fundamental research into emerging products and services that can then attract subsequent third-party funding.
- *Partnerships for International Research and Education (PIRE)*: PIRE is an NSF-wide program that supports international activities across all NSF-supported disciplines. The primary goal of PIRE is to support high quality projects in which advances in research and education could not occur without international collaboration. PIRE seeks to catalyze a higher level of international engagement in the U.S. science and engineering community.
- *Research Coordination Networks (RCN)*: The goal of the RCN program is to advance a field or create new directions in research or education by supporting groups of investigators to communicate and coordinate their research, training and educational activities across disciplinary, organizational, geographic and international boundaries. RCN provides opportunities to foster new collaborations, including international partnerships, and to address interdisciplinary topics.
- *Sustainability Research Network (SRN)*: The goal of the Sustainability Research Networks (SRN) competition is to bring together multidisciplinary teams of researchers, educators, managers, policymakers and other stakeholders to conduct collaborative research that addresses fundamental challenges in urban sustainability.

There are other programs at NSF involving workshops and conferences, but in all cases of proposals, there should be a valid justification for research with a well thought out and logical research plan. The proposals are reviewed by peers, so in writing, the reviewers should be kept in mind. At the same time, the review criteria should be considered.

3. Review Criteria:

Two review criteria, Intellectual Merit and Broader Impact, are the requirements of NSF. There are specific elements that are considered in both criteria. These are:

1) What is the potential for the proposed activity to:
   a. Advance knowledge and understanding within its own field or across different fields (Intellectual Merit); and
   b. Benefit society or advance desired societal outcomes (Broader Impacts)?
2) To what extent do the proposed activities suggest and explore creative, original, or potentially transformative concepts?
3) Is the plan for carrying out the proposed activities well-reasoned, well organized, and based on a sound rationale? Does the plan incorporate a mechanism to assess success?
4) How well qualified is the individual, team, or organization to conduct the proposed activities?
5) Are there adequate resources available to the PI (either at the home organization or through collaborations) to carry out the proposed activities?

It is important to note that the Broader Impact question considers the benefit to society as well as to education and outreach. Most reviewers will look for some indication of reaching out to under-represented groups for students and teachers, or any other outreach aspects.

4. Observations:

Funding from the NSF has become very competitive. The proposals are reviewed by 3 to 5 reviewers, independent of each other, and unless the proposal receives "Good" or better ratings by all the reviewers, the chances of it being funded are small. Reviewers are under pressure to review 8 to 10 proposals on a volunteer basis so if a proposal is difficult to read or a figure is difficult to understand, the proposal is likely to receive a review that is one notch lower than it deserves. It is suggested that innovative and transformative research needs to be done if we are to make an impact on the society with wind engineering, and we have been provided an opportunity to pursue research with good funding; we should take advantage of it.

NHERI Network Coordination Office and Science Plan

Julio Ramirez [a], Thomas L. Smith [b], Billy L. Edge[c], and William Holmes[d]

[a]*Purdue University, West Lafayette, IN 47907, USA*
[b]*TLSmith Consulting Inc., 16681 Boswell Road, Rockton, IL 61072, USA*
[c]*North Carolina State University, Raleigh, NC 27695, USA*
[d] *Rutherford and Chekene, San Francisco, CA 94105, USA*

ABSTRACT: The Natural Hazards Engineering Research Infrastructure (NHERI) is a vibrant NSF distributed, multi-user, national facility that provides the natural hazards engineering community with access to research infrastructure to support the needs of users focused on reducing the impact of earthquakes, tsunamis, windstorms, and storm surge. It consists of eight experimental facilities (EFs), a cyberinfrastructure (DesignSafe-CI) with access to computational resources and simulation tools, a computational modeling and simulation center and a network coordination office. The Network Coordination Office's (NCO's) role is to lead the multi-hazards engineering community and coordinate activities of other NHERI components to better enable research and educational advances that can contribute knowledge and innovation for the nation's civil infrastructure and communities and prevent natural hazard events from becoming societal disasters. The NCO leads network-wide education and community outreach activities and includes several committees such as the central scheduling, education and outreach, and technology transfer committees.

A five-year Science Plan is being developed with support of users and all the NHERI components. The plan is built around the three Grand Challenges:

1. Identify the characteristics of earthquake, windstorm, and associated hazards—including tsunamis, storm surge and waves—that are damaging to civil infrastructure (including lifelines) and disruptive to communities.
2. Evaluate the physical vulnerabilities of civil infrastructure and the population exposure and social vulnerability of communities to earthquake, windstorm, and associated hazards.
3. Create the technologies, and architectural and engineering tools to design, construct, retrofit and operate a multi-hazard resilient and sustainable infrastructure for the Nation.

Each Grand Challenge is supported by five key research questions with research task that should be undertaken to meet the Grand Challenges. The Science Plan will be reviewed and revised each year as needed. This presentation will give an overview of NHERI, the NCO and the science plan. Opportunities for providing input into the science plan will be discussed.

KEYWORDS: Natural hazards, national facility, NHERI, Grand Challenges, earthquakes, windstorms, tsunamis, storm surge and coastal inundation.

13th Americas Conference on Wind Engineering
Gainesville, Florida, USA
May 21 - 24, 2017

DesignSafe: A Cyberinfrastructure for Data Sharing and Analysis in Natural Hazards Engineering

Ellen M. Rathje[a], Clint Dawson[b], Jamie E. Padgett[c], Jean-Paul Pinelli[d], Dan Stanzione[e], Fred L. Haan, Jr.[f], and Ahsan Kareem[g]

[a] *University of Texas, Austin, TX, USA, e.rathje@mail.utexas.edu*
[b] *University of Texas, Austin, TX, USA, clint@ices.utexas.edu*
[c] *Rice University, Houston, TX, USA, jamie.padgett@rice.edu*
[d] *Florida Institute of Technology, Melbourne, FL, USA, pinelli@fit.edu*
[e] *University of Texas, Austin, TX, USA, dan@tacc.utexas.edu*
[f] *Calvin College, Grand Rapids, MI, USA, fhaan@calvin.edu*
[g] *University of Notre Dame, Notre Dame, IN, USA, kareem@nd.edu*

ABSTRACT

The DesignSafe cyberinfrastructure (www.designsafe-ci.org) has been developed to enhance research in natural hazards engineering by providing researchers access to cloud-based tools that support their work to analyze, visualize, and integrate diverse data types. DesignSafe has been developed to support the National Science Foundation-sponsored Natural Hazards Engineering Research Infrastructure (NHERI), but its capabilities are available to the broader natural hazards research community.

The core of the DesignSafe cyberinfrastructure is the Research Workbench, which currently includes the Data Depot and Discovery Workspace. The Data Depot is the central shared data repository that supports the full research lifecycle, from data creation to analysis to curation and publication. Within the Data Depot researchers have access to a private My Data home directory, a semi-private My Projects space for sharing data and results with collaborators, and the Public space that includes Published data sets. The shared-use NHERI experimental facilities use the Data Depot to share data with their users, which facilitates researchers taking advantage of the cyberinfrastructure in their work. With their data safely organized in the Data Depot, researchers can use cloud-based tools in the Discovery Workspace for data analysis, they can share data and results with other members of their research team, and eventually they can publish the data in the Public space for re-use by others.

The Discovery Workspace provides a venue for researchers to analyze, visualize, and transform their data, as well as perform simulations using sophisticated computational tools. The desktop metaphor of the Discovery Workspace includes an Apps window that provides access to a list of available tools and a Data Depot window that provides access to the contents of the Data Depot for use with those tools. Our initial deployment of tools includes open source computational simulation tools, including the computational fluid dynamics (CFD) program OpenFOAM (www.openfoam.org), as well as tools for both data analytics and visualization. The data analytics and visualization tools include commercial programs, such as MATLAB, and open-source programs such as ParaView (www.paraview.org) and Jupyter (jupyter.org). The Jupyter notebook is particularly useful for cloud-based data analysis because it is an electronic notebook that allows users to embed rich text elements, along with computer code, graphs, and visualizations that link to data, all within a single notebook that can be shared through the web. Most of the tools in the Discovery Workspace have access to high performance computing (HPC) resources, making it easy for researchers to employ these resources in their work.

Researchers in the wind engineering community are starting to use the DesignSafe functionalities and resources in their work. At Florida Tech, researchers are currently using MATLAB deployed on DesignSafe to analyze data they recently collected during the passage of Hurricane Matthew and investigate the stationarity of the wind flow field.

Researchers are Calvin College are using MATLAB and Jupyter to analyze peak pressure data sets from low-rise buildings tested in the tornado simulator at Iowa State University for similarities and differences with straight-line boundary layer results. And at the University of Notre Dame, researchers are conducting fundamental studies in the area of aerodynamic shape optimization of tall buildings and long span bridges using CFD simulations in the OpenFOAM environment. The high computational demand for such an optimization couched in a surrogate modeling scheme is being met through DesignSafe, its HPC resources, and the cloud-based Data Depot.

KEYWORDS

Cyberinfrastructure, high performance computing, data analysis, data publishing

NSF Large-Scale, Shared-Use Experimental Facility for Coastal Wave/Surge and Tsunami Hazards

Daniel Cox[a], Pedro Lomonaco[b], and Christopher Higgins[c]

School of Civil and Construction Engineering, Oregon State University, Corvallis, Oregon, USA
[a] dan.cox@oregonstate.edu, [b] Pedro.Lomonaco@oregonstate.edu, [c] Chris.Higgins@oregonstate.edu

ABSTRACT

The US National Science Foundation established the Natural Hazards Engineering Research Infrastructure (NHERI) program to advance the resilience of civil infrastructure and communities to windstorms and seismic hazards. The O.H. Hinsdale Wave Research Laboratory at Oregon State University was funded to host the NHERI Coastal Wave/Surge and Tsunami Experimental Facility (CWST-EF) to serve as a state-of-the-art engineering research, education, and outreach center for coastal wave and surge caused by windstorms and tsunamis caused by earthquakes. CWST-EF includes two unique, large-scale resources for physical model testing of coastal systems: a large wave flume (LWF) and a directional wave basin (DWB). The LWF is the largest of its kind in the US and can generate repeatable regular, irregular, tsunami, and user-defined waves. The LWF is used to induce and measure wave forces on constructed and natural ocean features at near prototype scales, providing high resolution wave structure interactions with minimum distortion due to viscous effects. Studies in the DWB are generally of smaller scale compared to the LWF but have the added advantage of fully 3D capabilities for directional regular, irregular, tsunami, and user-defined waves. These two facilities, along with with state-of-the-art instrumentation and access to computational tools and data archival systems, are available to the research community to encourage collaborative research addressing the grand challenges in community resilience, including the overland flow characteristics of coastal wave/surge and tsunami hazards, induced erosion and scour, hydraulic loads, and the response of coastal infrastructure and natural environments.

KEYWORDS

Resilience, Adaptation, Coastal Hazards, Tsunami, Coastal Engineering, Experiment, Large-scale, Erosion, Scour, Wave Forces, Inundation, Wave runup, Overtopping, Levee

The 13th Americas Conference on Wind Engineering (13ACWE)
Gainesville, Florida USA, May 21-24, 2017

NHERI Lehigh Experimental Facility: Natural Hazards Research Using Real-time Hybrid Simulation

James Ricles [a], Richard Sause [a], Chinmoy Kolay [a], Thomas Marullo [a], Chad Kusko [a]

[a]*Lehigh University, 117 ATLSS Drive, Bethlehem PA, USA*

ABSTRACT: Real-time hybrid simulation (RTHS) is a method that enables a complete system to be considered in a simulation. In an RTHS, a system is divided into two domains: (i) a numerical substructure, where components of the system are computationally modeled; and, (ii) an experimental substructure, where remaining components of the system for which reliable computational models are non-existent are physically modeled in the laboratory. The two domains are coupled through the governing equations of motion, and the response of the system to imposed loading is determined by solving the equations of motion in real time. RTHS is a viable technique to investigate the response of civil infrastructure systems to natural hazards, such as earthquakes, wind storms, and storm surge. To help meet the grand challenge of community resilience to natural hazards, the Natural Hazards Engineering Research Infrastructure (NHERI) Lehigh Experimental Facility (EF) was funded by the National Science Foundation (NSF) to be a world-class, open-access facility that enables researchers to address key research questions associated with the challenge of community resilience. The NHERI Lehigh EF has a unique portfolio of equipment, instrumentation, infrastructure, testbeds, experimental simulation control protocols, large-scale simulation and testing experience along with know-how developed over the past two decades. The unique strength of the NHERI Lehigh EF is accurate, large-scale, multi-degree-of-freedom and multi-directional simulations (RTHS) of the effects of natural hazards on civil infrastructure systems (i.e., buildings, bridges, industrial facilities, etc.) with potential soil-foundation effects. The NHERI Lehigh EF will be presented, reviewing the facility's infrastructure, capabilities, and advanced algorithms developed at the facility for conducting multi-directional RTHS. The discussion will include a brief background presentation of the concept of RTHS. Examples of prior and current projects performed by researchers using the facility will be provided in order generate ideas and motivation among members of the audience to utilize the open-access NEHRI Lehigh EF for their research.

KEYWORDS: Natural Hazards, Dynamic Loading, Structural Systems, Computational Modeling, Experimental Testing, Real-time Hybrid Simulation

The NHERI Wall of Wind Experimental Facility at Florida International University

Arindam Chowdhury [a], Peter Irwin [b], Ioannis Zisis [c], Maryam Refan [d]

[a] *NHERI Wall of Wind Experimental Facility (EF), Department of Civil and Environmental Engineering, Florida International University, Miami, FL, USA, 33174, chowdhur@fiu.edu*

[b] *NHERI Wall of Wind EF, Department of Civil and Environmental Engineering, Florida International University, Miami, FL, USA, 33174, peairwi@fiu.edu*

[c] *NHERI Wall of Wind EF, Department of Civil and Environmental Engineering, Florida International University, Miami, FL, USA, 33174, izisis@fiu.edu*

[d] *NHERI Wall of Wind EF, International Hurricane Research Center, Florida International University, Miami, FL, USA, 33174, mrefan@fiu.edu*

ABSTRACT

Natural hazards engineering research is undergoing far-reaching changes with the introduction of novel and large-scale experimental research facilities that can address the growing economic losses associated with natural disasters. The Wall of Wind (WOW), at Florida International University (FIU), is such an experimental facility, capable of simulating up to Category 5 hurricane winds and wind-driven rain to better understand hurricane effects on buildings and other infrastructures. The WOW has been designated by the National Science Foundation (NSF) as one of the "Experimental Facilities" (EFs) under the Natural Hazards Engineering Research Infrastructure (NHERI) program that aims to allow researchers to enable innovations and prevent natural hazards from becoming societal disasters. The NHERI WOW EF's goal is to facilitate research to improve design practices for structural systems, building envelopes, and lifeline infrastructure. In addition to providing unique experimental resources and testing capabilities to the wind engineering community, the NHERI WOW EF allows for diverse and multi-disciplinary research collaborations that go beyond the traditional structural wind engineering applications.

This paper presents a brief overview of the design and development of the NHERI WOW EF and delineates its ambitious science plan which is designed to assess and mitigate the impacts of hurricane wind, rain, and debris on civil infrastructures. Also, the unique advantages of the NHERI WOW EF, namely multi-scale and high Reynolds number simulations as well as destructive tests, are discussed. To illustrate the NHERI WOW EF's research capabilities, details and results from various case studies, related to large- and full-scale testing of building components and traffic infrastructures, are described.

KEYWORDS

Large-scale testing, infrastructure, hurricane wind and wind-driven rain, Wall of Wind, NHERI

Single-Axis Solar Tracking Systems: State of the Industry

Matthew Browne [a], Scott Gamble [a], Richard Erb [b], Nick McKibben [b]

[a] *RWDI, 600 Southgate Drive, Guelph, ON, Canada, Matthew.Browne@rwdi.com*
[b] *SunPower Corporation, 1414 Harbour Way South, Richmond, CA, USA, Richard.Erb@sunpower.com*

ABSTRACT

Solar energy has seen unprecedented growth across North America and around the world. As generation capacity from solar power plants is becoming a more critical part of the worldwide energy infrastructure, solar power plant reliability becomes increasingly important to protect investment and minimize risk. These large scale solar power plants consist of ground-mounted solar systems of both the fixed-tilt and tracking variety. Single-axis tracking systems relying on a long, flexible central torque tube driven from a single location pose unique challenges to the designer related to physical oscillation of the system. This paper commences with an overview of the design process currently used within the industry for designing these types of structures, including common codes/standards, static model wind tunnel testing, and the treatment of issues related to the dynamic motions induced by wind action. Comparisons of field observation and data to wind tunnel predictions are used to demonstrate the limitations of the rigid model approach and clearly show the presence of an aeroelastic phenomenon.

At present, there are three mechanisms for wind to cause vibration or 'instability' of portions of a solar installation: 1) wake effect resonance, 2) torsional divergence, and 3) a form of flutter. The first has been identified through wind tunnel testing using rigid models and can be described as wake effect resonance caused by turbulence generated from the first row of an array causing resonant vibration of subsequent rows. Although important for design, this mechanism is not indicative of the aeroelastic phenomenon observed in the field. In the case of the second mechanism, the change in torque applied to the row as it rotates can overpower the torque tube's ability to resist, resulting in an effect known as torsional divergence. As this mechanism is also not evident in the field data, we are left with the third mechanism. For systems relying on a central torque tube driven from a central location, a twisting mode of vibration is possible created by a form of flutter generally initiated at the ends of the row.

In classical aeronautics terms, flutter is a self-excited aerodynamic oscillatory instability in which the aerodynamic forces depend on the motion of the structure itself and can lead to very large amplitudes in torsional motion or coupled torsional and vertical motions. Once initiated, flutter causes motions to increase rapidly until failure occurs. The field data presented in this paper clearly shows that the amplitude of rotation increases with increasing wind speed over several cycles, but decreases as quickly when the wind speed is reduced. This phenomenon can be characterized as an 'instability' due to motions of an amplitude that can cause structural failure, and is a real issue for these types of structures. Recommendations to properly investigate this form of instability through sectional and/or full aeroelastic wind tunnel testing are given. Possible mitigation measures, such as changes in aerodynamic shape, structural stiffness or damping, are also examined.

KEYWORDS

Single-axis trackers, solar energy, aeroelastic phenomenon, instability, flutter

13th Americas Conference on Wind Engineering
Gainesville, Florida, USA
May 21 - 24, 2017

Anchorage and Envelope Wind Loads on Roof-Mounted Equipment

Lakshmana S. Doddipatla[a], Gregory A. Kopp[b]

[a]*Center for Property Risk Solutions, FM Global, Research Division, Norwood, MA, USA.*
lakshmana.doddipatla@fmglobal.com
[b]*Boundary Layer Wind Tunnel Laboratory, Western University, London, ON, Canada.*
gakopp@uwo.ca

ABSTRACT

The study investigates anchorage and envelope wind loads on a 1:50 scaled roof-mounted equipment model with equivalent full-scale dimensions of a cube with side lengths of 0.92 m, 3 m, and 4.5 m, and a rectangular prism of 6 x 3 x 1.5 m. The experiments were conducted for four elevations corresponding to 0 m (seated), 1.5 m, 3 m, and 4.5 m, and ten mean wind directions between 0^0 to 90^0 in increments of 10^0. The model scale equipment was placed at eight different locations to represent the corner, perimeter, and field regions of the roof. It is found that the ASCE 7-10 uplift coefficient can be reduced by 40% for some situations, whereas the drag coefficient for anchorage design remains unchanged for elevated equipment compared to seated equipment. It is also found that the envelope pressure coefficients on roof-mounted equipment can be approximately 20% to 25% higher compared to ASCE 7-10 anchorage wind loads for an effective wind area corresponding to 1 m^2, but can be lower by approximately 30 % and 40% for larger effective wind areas corresponding to 10 m^2 and 20 m^2, respectively.

KEYWORDS

Seated Roof-Mounted Equipment, Elevated Roof-Mounted Equipment, Anchorage Wind Loads, Envelope Wind Loads.

Experimental assessment of wind loads on span-wire traffic signals

I. Zisis [a], P. Irwin [b], B. Hajra [c], A.G. Chowdhury [d], M. Matus [e]

[a] *Florida International University, Miami, USA, email: izisis@fiu.edu*
[b] *Florida International University, Miami, USA, email: peairwi@fiu.edu*
[c] *Florida International University, Miami, USA, email: bhajra@fiu.edu*
[d] *Florida International University, Miami, USA, email: chowdhur@fiu.edu*
[e] *Florida International University, Miami, USA, email: mmatu016@fiu.edu*

ABSTRACT: Wind induced damage on span-wire traffic signals have occurred in the past, causing significant financial loses besides disruption of traffic flow. A span-wire system consists of two wires (catenary and messenger wires) connected to posts on either side, with the signals connected to these wires through hangers. One of the objectives of this study was to assess the wind loads on a 22 ft (6.7 m) span-wire system with springs fitted on either side, consisting of two 3-section and one 5-section traffic signals connected through a flexible hanger. This configuration was tested at the Wall of Wind (WOW) experimental facility at Florida International University (FIU), Miami, USA, for, wind directions varying from 0° to 180° and wind speeds ranging from 0 to 150 mph (67 m/s). Based on force-deflection relationships, the 22 ft (6.7 m) span wire system with springs represents a 72 ft (21.5 m) long span-wire system. The advantage of having a 6.7 m span wire system with springs is that it allows the testing of traffic signals for a wide range of wind angles, as compared to the 72 ft (21.5 m) long span wire system that can only be tested for 0°. Therefore, the second objective of this study was to compare results from the 22 ft (6.7 m) span wire system with springs and the 72 ft (21.5 m) long span wire system, in order to assess the correctness of the theoretical considerations. At a given wind speed, the messenger wire experienced higher tensions than the catenary wire. The root mean square (rms) of accelerations increased with increasing wind speeds. Beyond 70 mph (31.2 m/s), the traffic signals experienced an aerodynamic instability. Good comparisons of total drag and lift forces on the traffic signals were obtained from the 22 ft (6.7 m) span wire system with springs and the 72 ft (21.5 m) span wire system at 0°. The results from this study are encouraging and demonstrate the usefulness of the test rig with springs in realistically simulating long span-wire systems.

KEYWORDS: Catenary wire; Messenger Wire; Span-wire traffic signal; Wind load; Wall of Wind

1 INTRODUCTION

Traffic signals are an important part of civil infrastructure that help manage traffic flow within a city. Damage to these signals caused by strong winds can greatly hamper the flow of traffic [1, 2]. Wind induced damage to buildings and traffic signals have occurred in the past in hurricane prone areas of USA [3]. Few studies have been carried out so far on the wind induced damage on traffic signals. For instance, McDonald et al. [4] found that a major cause of structural damage to the traffic signal was the oscillation caused by strong winds. Similarly, wind induced galloping phenomenon was found to be a major cause of damage for traffic signals on cantilevered structures [5]. In the 2003-2004 season, hurricanes comprising of wind speeds exceeding 100 mph (44 m/s) in Florida, USA caused significant structural damage to various civil engineering structures including traffic signals [6]. Span wire traffic signals are increasingly being used in the state of

Florida and other states in USA. A span wire traffic signal generally consists of a catenary wire on top and a pre-tensioned messenger wire at the bottom being supported on either side by masts, while the traffic signals are connected to these two wires through a hanger/connector. Existing standards and specifications such as the ASCE 7-10 [7] and AASHTO, 2009 [8] provide limited information for the safe wind design of span wire traffic signals [6]. For instance, AASHTO [8] provides a constant wind drag coefficient of 1.2 for traffic signals without any consideration for lift. Therefore, this explains the need to assess the wind induced loads on traffic signals so that they may be designed to withstand wind induced forces and remain functional at all times.

One of the main objectives of this paper is to present the development of a 6.7 m long span test rig fitted with coil springs on either side that represents a 21.5 m long span, based on theoretical considerations [9, 10]. The WOW at FIU has a turntable of 4.9 m diameter, which makes it possible to test a 6.7 m span for different wind directions, while the 21.5 m span can only be tested for 0° (wind perpendicular to the signals) due to space restrictions in the WOW. The second objective is to assess the response of the traffic signals and the forces in the wires for the '6.7 m span with springs' and '21.5 m span', for a 'flexible hanger joint' connecting the traffic signals to the wires.

2 DEVELOPMENT OF THE SHORT SPAN RIG WITH COIL SPRINGS

The purpose of constructing a 6.7 m short span with springs on either side is to facilitate testing at different wind directions at the WOW experimental facility at FIU. The 6.7 m test rig with springs on either side represents a 21.5 m long span in full scale, and the development of this test rig was based on theoretical considerations as discussed further.

2.1 *Theoretical Considerations*

For the present study, 3 signals (two 3-section and one 5-section) were placed at 4 ft (1.22 m) spacing, as shown in Figure 1 (a). Detailed derivations for the case of just one signal in the centre of the span are discussed in Irwin et al. [11]. The present study extends this derivation for the case of three signals. Consider the deflections of the messenger wire when three equal forces F_n are applied at the points of attachment of the signal hangers; the effect of differing forces will be explained subsequently. Using similar analysis methods described in Irwin et al. [11], where a single force was applied at mid-span it can be shown that

$$F_n = \frac{2}{3}\frac{T_0}{L_1}x_1 + \frac{EA_1}{3L_1^3}x_1^3 \quad (1)$$

Also it can be shown that

$$F_n = 2\frac{T_0}{L_2}(x_2 - x_1) + \frac{EA_2}{L_2^3}(x_2 - x_1)^3 \quad (2)$$

where

T_0 is initial tension in the wire, EA_1 and EA_2 are the products of effective modulus of elasticity and areas in the wire lengths L_1 and L_2 respectively. It may be noted that these expressions ignore any bending stiffness of the wires and the effect of the clamps at the hanger/wire connections. Normally $EA_2 = EA_1$ but when going to a short span rig with springs it is necessary to keep the distinction.

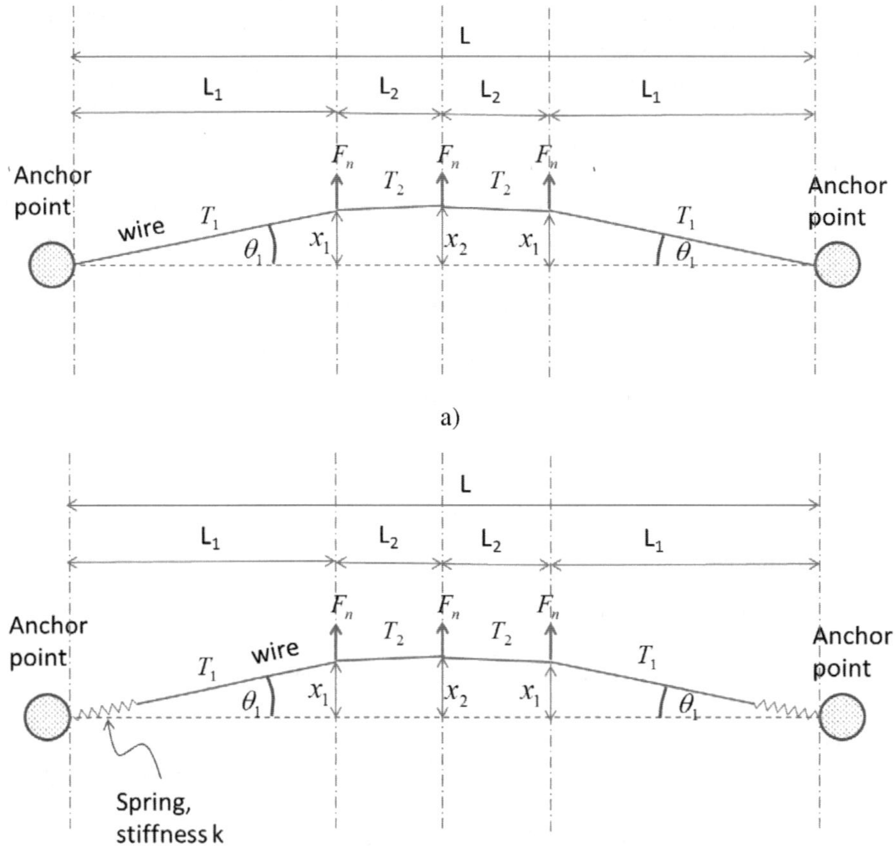

b)

Figure 1. Schematic showing: a) Tensioned wire with three forces symmetrically about mid-span; b) Test rig with springs introduced near the end posts

In the present study, it is assumed that $EA = 1.48 \times 10^6$ lb (6.58 x 10^6 N) which can be derived from stretch tests carried out on the signals for tensions in the range 0 to about 2000 lb (8896 N). For a 21.5 m span we have $L_1 = 9.75$ m and $L_2 = 1.22$ m.
It was found that depending on the assumed value of EA, a spring rate of 100 lb/in (17511 N/m) simulates field spans in the range of about 21.5 to 25.5 m. Additional information on the formulations can be obtained from references [9, 10].

2.2 *Implementation of theory using 6.7 m span rig at FIU*

Figure 2 presents the implementation of the theory in the WOW experimental facility using a 6.7 m span with coil springs on either side of the catenary and messenger wires.

Figure 2. 6.7 m test rig with coil springs on either side at the WOW with two 3-section and one 5-section traffic signals connected to the wires using flexible hanger joint.

In the present study two 3-section signals and one 5-section signal were connected to the catenary and messenger wires using a "flexible hanger joint". The coil springs were attached on either side of the messenger and catenary wires. There were two different hollow structural sections (HSS) that were utilized, HSS 0.25m x 0.15m x 0.009m and HSS 0.15m x 0.15m x 0.009m. The two support columns are 4.72m tall supported on top of HSS 0.25m x 0.15m x 0.076m sections. A sag length of 1.22m on the catenary wire was used to simulate full scale conditions.

2.3 *Development of the 21.5 m long span rig at FIU*

The flexible hanger joint that was used for both the short and long span rig in the present study are shown along with the 21.5 m test rig shown in Figure 3. The same dimensions of HSS were used as in the 6.7 m span rig, although in this case the columns were 21.5 m apart.

a) b)

Figure 3. Span wire traffic signals (21.5 m span): a) flexible joint connection; b) traffic signal set up at the WOW facing the wind

3 INSTRUMENTATION USED FOR THE STUDY

For the present study, 6 degree of freedom load cells, inclinometers and accelerometers were used to measure the wind induced forces, inclinations and accelerations respectively. The load cells can measure forces and moments in 'x', 'y' and 'z' directions [12], while the inclinometers measure the inclinations along two axes in terms of a voltage, which is later converted to degrees [13]. The accelerometers provide accelerations along the three axes [14], and in this study the results are reported as root mean square (rms) values. Figure 4 (a) shows a schematic of the test set up, Figure 4 (b) shows the location of the inclinometers and accelerometers on the traffic signals.

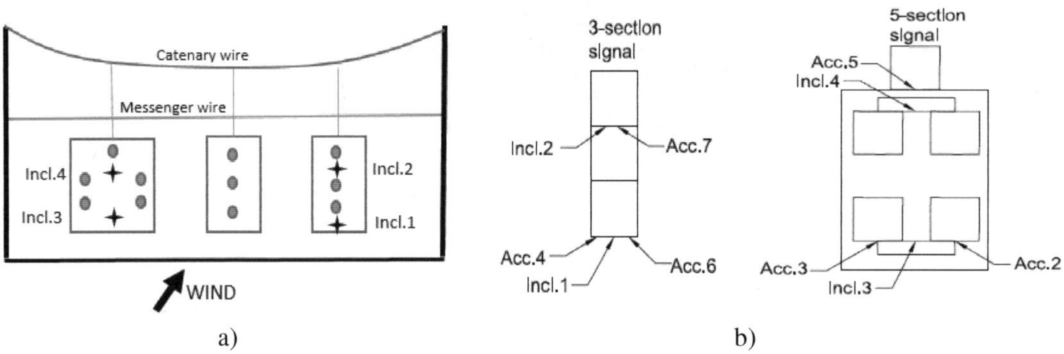

Figure 4. Instrumention used for the study: a) Schematic of the test rig; b) location of accelerometers and inclinometers on the 3-section and 5-section signals

4 RESULTS AND DISCUSSION

Selected results are presented for the total drag and total lift on the traffic signals, rms of accelerations and mean inclinations for the 6.7m span with springs and the 21.5m span.

4.1 *Total drag and total lift on the traffic signals*

Figure 5 (a) presents the total mean drag and total mean lift obtained on the traffic signals on the 6.7 m span with springs and the 21.5 m span rig. Results show that the mean drag increases gradually with increasing wind speed with a maximum value of 2428 N at 67 m/s. The mean drag increases gradually with increasing wind speed of up to 38 m/s, beyond which it is nearly constant. The mean drag force and the mean lift force obtained from the 6.7 m span with springs and the 21.5 m span compare well, thereby illustrating the efficacy of the 6.7 m span with springs to represent a 21.5m long span. The peak drag and peak lift obtained from the 6.7 m span with springs and 21.5m span are shown in Figure 5 (b). Results show that the peak values obtained from the 6.7 m span and 21.5m span compare well at almost points except at wind speed greater than 58 m/s, where the 21.5 m span produces somewhat higher peak drag/lift values than the 6.7 m span. Regardless of these discrepancies, the overall trends of data obtained from the two different spans are very similar which is encouraging.

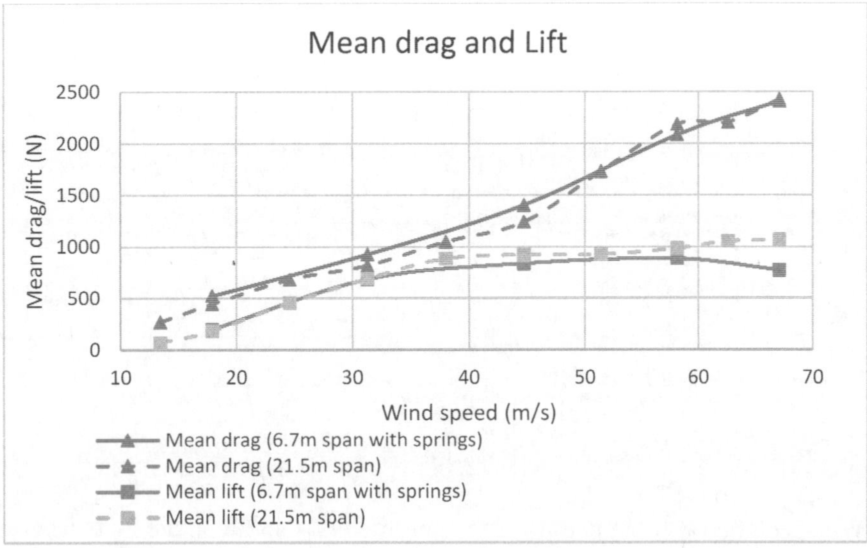

a)

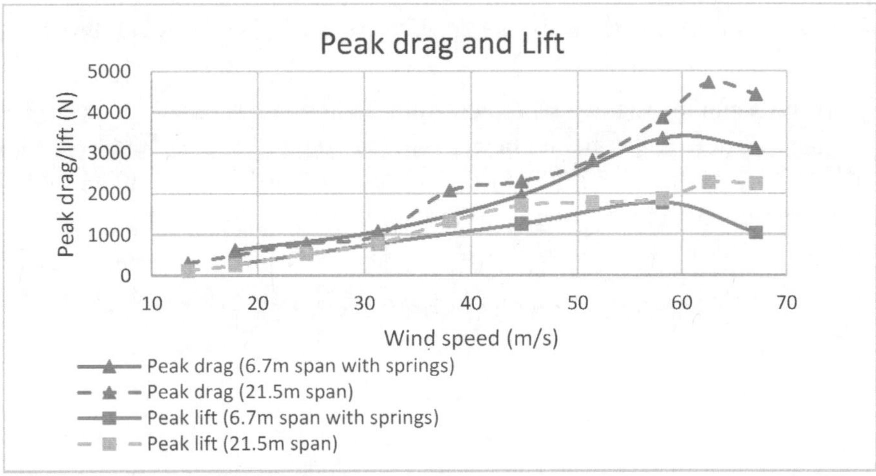

b)

Figure 5. Comparisons of total drag and lift forces on the traffic signals: a) mean; b) peak

4.2 *rms of acceleration on the traffic signal*

Figure 6 presents the rms of accelerations on the 3-section traffic signal obtained from accelerometers 4 and 6 (see Figure 4 (b) for location of accelerometers). Results indicate that the rms values from both accelerometers increase with increasing wind speed. Furthermore, the comparison of rms values on accelerometers 4 between 6.7m span with springs and the 21.5 m span is excellent; similar trends were also observed with accelerometer 6 for both spans. The accelerometers were not used for measurements at wind speeds above 44 m/s to avoid damage to the instruments due to excessive vibration.

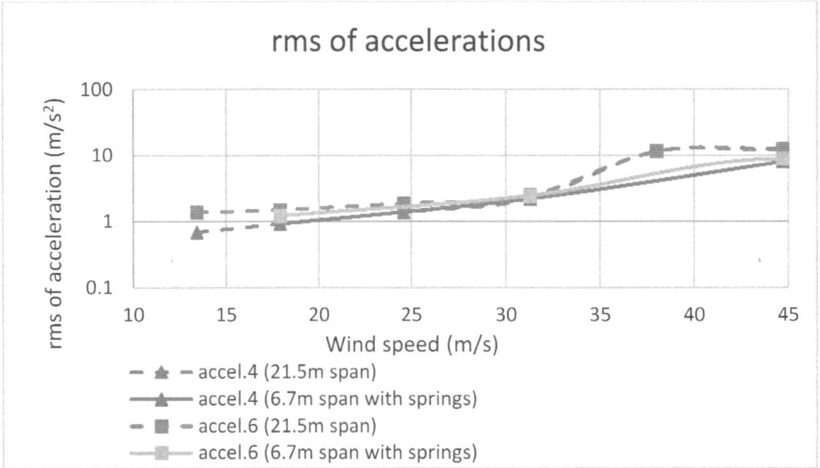

Figure 6. rms of accelerations obtained from accelerometers 4 and 6 on the 3-section signal

4.3 *Mean inclinations of the traffic signal in the along wind and across wind directions*

Figure 7 presents the mean inclinations obtained from inclinometer 2 on the 3-section signal (see Figure 4 (b) for the location of the inclinometer). The mean inclinations increase with increasing wind speed from 0 to 18 m/s. '2-1' refers to the component in the across wind direction, while '2-2' refers to the component in the along wind direction.

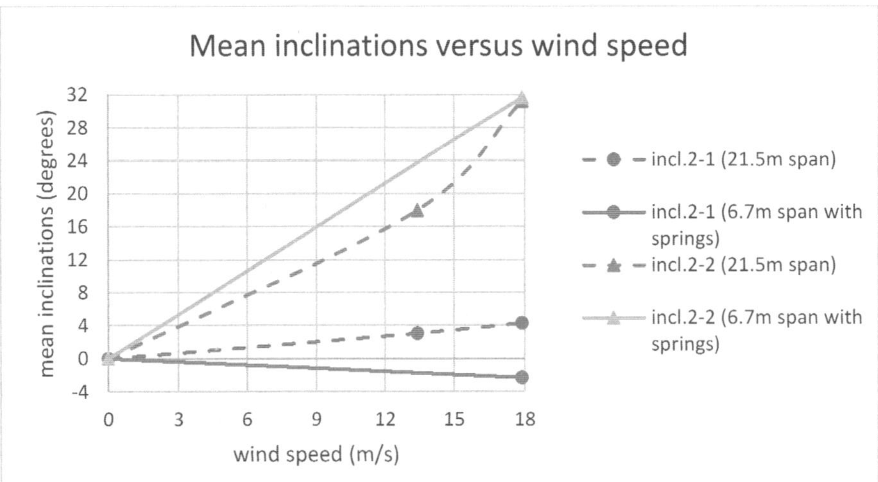

Figure 7. Mean inclinations obtained from inclinometer 2 on the 3-section signal

Beyond 18 m/s the traffic signals had a wide range of inclinations. The maximum mean inclination in the along wind direction was 31.6 degrees which exceeded the acceptable range of 27 degrees. The trends of data obtained from the 6.7 m span and 21.5 m span for the along wind component were quite similar.

At a wind speed of around 53 m/s, a portion of the 3-section polycarbonate signal was damaged, and at a slightly higher speed of 58 m/s the 5-section signal experienced damage. The damage to the signals changed the overall shape of the signals causing a change in the wind flow around them. Future research is needed to analyze span wire traffic signals for different wind directions and different hanger connections.

5 CONCLUSIONS

This paper presented results from an experimental study carried out at the Wall of Wind (WOW) experimental facility at Florida International University (FIU) for two cases: a 6.7 m long span with coil springs on either side and a 21.5 m long span. Both cases consisted of two 3-section and one 5-section traffic signals attached to the assembly through a 'flexible hanger joint'. Based on theoretical considerations, the 6.7 m long span with springs on either end represents a 21.5 m long span. The prime objective of this paper was to validate this theory through experiments at the WOW since the 6.7 m span allows testing in different wind directions, while a 21.5 m span can only be tested for 0° due to the size of the turntable in the WOW. The second objective was to assess the forces in the cables and the traffic signals for the 'flexible hanger joint', for both the 6.7 m span with springs and 21.5 span. Results show good comparison between the 6.7 m span with coil springs and the 21.5 m long span for the mean and peak values of the total drag, total lift, rms of accelerations and mean inclinations experienced by the traffic signals. For both the 6.7 m span with springs and the 21.5 m span, an aerodynamic instability was observed at wind speeds above 31 m/s. The present study clearly shows that a 6.7 m span with springs at either end can successfully be used to replicate a 21.5 m full scale span wire traffic signal system for wind testing. Future research on wind testing of span wire traffic signals must focus on different types of connections, different wind directions and terrain conditions.

6 ACKNOWLEDGEMENTS

The financial contributions of the Florida Department of Transportation (FDOT), Florida, USA is gratefully acknowledged. In particular, the authors would like to thank the WOW technical team for setting up the test rig and instrumentation for carrying out the tests. The authors would also like to thank Horsepower Electric Inc. for their assistance on the setup of the span wire systems.

7 REFERENCES

1. Sivarao SK, Esro M, Anand TJ. Electrical & Mechanical Fault Alert Traffic Light System Using Wireless Technology. International Journal of Mechanical and Mechatronics Engineering. 2010; 10(4):19-22.
2. Zuo, D., Letchford, C. W. Wind-induced vibration of a traffic-signal-support structure with cantilevered tapered circular mast arm. Engineering Structures, 2010; 32(10), 3171-3179.
3. Bigger, J. E., Willingham, M. G., Krimgold, F., Mili, L. Consequences of critical infrastructure interdependencies: lessons from the 2004 hurricane season in Florida. International journal of critical infrastructures, 2009; 5(3): 199-219.
4. McDonnald, J.R., Mehta, K. C., W. Oler, and N. Pulipaka. Wind load effects on signs, luminaires and traffic signal structures. Report 1303-1F, Department of Transportation, Texas, 1995, USA.
5. Kaczinski MR, Dexter RJ, Van Dien JP. Fatigue-resistant design of cantilevered signal, sign and light supports. Transportation Research Board, NCHRP report 412, National Research Council; 1998.

6. Cook, R.A., Masters, F., Rigdon, J.L. Evaluation of Dual Cable Signal Support Systems with Pivotal Hanger Assemblies. FDOT Contract No. BDK75 977-37, University of Florida, Department of Civil and Coastal Engineering, 2012, Gainesville, FL, USA.
7. ASCE 7-10, Minimum Design Loads for Buildings and Other Structures, American Society of Civil Engineers, ASCE, 2010, Virginia, USA.
8. AASHTO, American Association of state Highway and Transportation Officials, Standard specifications for structural supports for highway signs, luminaires and traffic signals, 6th edition, Washington, USA, 2009.
9. Inglis, C., Applied Mechanics for Engineers, Dover Publications, 1963, New York, USA.
10. Irvine, H.M., Studies in the Statics and Dynamics of Simple Cable Systems, California Institute of Technology, Dynamics Laboratory Report, 1974, DYNL-108.
11. Irwin, P, Zisis, I, Berlanga, B, Chowdhury, A, Hajra, B. Wind testing of span-wire traffic signal systems, Proceedings of the CSCE annual conference (Natural Disaster Mitigation Specialty Conference), 2016; June 1-4, London, Ontario, Canada.
12. JR3 Installation manual for Force-Torque sensors, JR3 Inc., Woodland, California, 95776, USA, 2012 (http://www.jr3.com/).
13. Murata Electronics, The SCA 121 T dual axis inclinometer modules data sheet, Murata Electronics, 2016, Japan (http://www.murata.com/en-us).
14. TE Connectivity Sensor solutions, 35203 B Tri axial digital accelerometer specifications, 2015; Akron, 44333, OH, USA (http://www.te.com/usa-en/products/sensors.html).

Wind Loads on a Full-Scale Photovoltaic Array Compared to Those on an Identical Model-Scale Array

Timothy Johnston[a], Murray Morrison[b], Sarah Stenabaugh[c], Mike Gibbons[d], Matthew Browne[e], Scott Gamble[f]

[a] *Insurance Institute for Business & Home Safety, Richburg, SC, USA, tjohnston@ibhs.org*
[b] *Insurance Institute for Business & Home Safety, Richburg, SC, USA, mmorrison@ibhs.org*
[c] *Rowan Williams Davies and Irwin Inc., Guelph, Ontario, Canada, sarah.stenabaugh@rwdi.com*
[d] *Rowan Williams Davies and Irwin Inc., Guelph, Ontario, Canada, mike.gibbons@rwdi.com*
[e] *Rowan Williams Davies and Irwin Inc., Guelph, Ontario, Canada, matthew.browne@rwdi.com*
[f] *Rowan Williams Davies and Irwin Inc., Guelph, Ontario, Canada, scott.gamble@rwdi.com*

ABSTRACT

The increase in market demand and incentive programs for renewable energy sources over the last decade has led to a significant increase in photovoltaic (PV) array installations. Low-slope roofs provide an ideal space to mount solar collectors, making roof-mounted arrays particularly favorable in or near urban centers. This demand has resulted in numerous wind tunnel studies to determine appropriate design loads for PV arrays. However, due to the relative size of the arrays and the complexity of the geometry, model wind tunnel scales of 1:20 and 1:30 have been used in these wind tunnel studies. Typically, geometric scales for low-rise building tests in model-scale boundary layer wind tunnels are 1:50 to 1:100. However, even at these geometric scales it is not possible to fully match the proper flow characteristics. Testing at even larger geometric scales further exacerbates these challenges in terms of matching appropriate flow characteristics, such as turbulence intensities and target energy spectra. To address these problems, several partial turbulence spectrum (PTS) methods have been developed. Nevertheless, these methods rely on quasi-steady assumptions to compensate for those scales of turbulence that are not correctly matched. In addition, due to the small gaps that exist between solar modules and the roof surface, as well as between modules, the results may be subject to Reynolds number effects. In order to quantify the importance of mismatches in turbulence scales and Reynolds number effects, comparisons to full-scale measurements are needed. The present study compares the wind loads on PV arrays from model-scale wind tunnel measurements conducted at RWDI to simulated full-scale measurements taken at IBHS for identical array configurations. The presentation and future publication will present these comparisons in detail and discuss the impact, if any, on design loads of PV arrays.

KEYWORDS

Photovoltaic Array, Wind Loads, Turbulence Scale Effects, Wind Tunnel, Full Scale

The 13th Americas Conference on Wind Engineering (13ACWE)
Gainesville, Florida USA, May 21-24, 2017

Time-domain Analysis of Solar Concentrator Structure under Gust Wind

Bassem Kaabia[a], Sébastien Langlois[a], Sébastien Maheux[a], Frédéric Légeron[b]

[a] *Department of Civil Engineering, Université de Sherbrooke, 2500 boul. de l'Université, Sherbrooke, QC, Canada, J1K 2R1*
[b] *Vice President, Parsons, Abu Dhabi PO5498, UAE*

ABSTRACT: Accurate evaluation of wind load effects on solar structures is important for optimizing their design. Gust wind and its dynamic effect may be significant for this type of structure, however while time-domain numerical simulations are powerful tools to predict the dynamic response, they are rarely used for wind analysis because of the complexity of aerodynamic force models and the difficulty of performing full-scale validations. This paper aims to validate a time-domain dynamic analysis method used to predict the statistical parameters of 10 minute full-scale measurements response of a solar concentrator structure under service wind load conditions. A finite element model of the solar concentrator support was developed and time-domain dynamic (TD) analyses were performed. As a comparison, the acceptability of this method is also revisited in the light of the ASCE 7-10 and the spectral approach prediction. Full-scale measurements and corresponding simulations covered different configurations depending on the pitch angle and loading case. The measured wind speed is converted to drag force, lift force and pitching moment applied to the finite element model. An aerodynamic force model based on the buffeting indicial response function (IRF) associated to the aerodynamic admittance function for flat plate is used for TD analyses of different configurations. Results are compared in terms of statistical parameters, such as the standard deviation and gust effect factors. It was found that, despite the simplified hypotheses taken especially for wind load and structural model, in many cases the proposed approach gives slightly better results for the statistical parameters compared to the spectral approach. Therefore, its overall acceptability is validated. This time-domain approach could then be used to study the linear and nonlinear dynamic structural response of solar concentrator structures under wind load and thus identify optimal design configurations to meet both serviceability and ultimate loading conditions.

KEYWORDS: Finite Element Model, Time-domain Analysis, Wind Load, Solar Concentrator, Spectral Approach, Gust Effect Factor

1 INTRODUCTION

Accurate evaluation of wind loading is a primary concern in optimizing solar structure design. Full-scale measurements are costly solution for the pre-design studies. Wind tunnel investigation, especially conducted in the Sandia Laboratory (Peterka and Derickson, 1992; Hosaya et al., 2008), are well recognized but generalization in quantitative research including dynamic effect is recommended (Banks, 2014) (CPP, Inc). Then, there is a strong need for investigations of numerical tools as a cost-effective alternative. Since its development by Davenport (1961), the stochastic spectral approach is the common method to study the response under turbulent wind. A key concept developed in this approach is the aerodynamic admittance function, which is inspired by thin airfoil theory (Sears, 1941), and adapted by Davenport (1961) to civil engineering structures. The current standard ASCE 7-10 formulation to predict the equivalent static peak response is developed based on the early work of Davenport (1961), and it was improved through research such us the work of Vickery (1968), Kareem (1986) and Solari & Kareem (1998). These methods, based on a frequency domain analysis, require significant simplifications in the structural response modeling to represent the complex dynamic behavior. Time domain methods are better to incorporate the non-linear behavior of the structures, and many studies in time-domain have been conducted for various structure types. In many cases, these studies focus on the modeling of wind force in the time-domain. For example, Chen & Kareem (2002) proposed advanced aerodynamic force models applied for long-span bridges. Letchford et al., (1993) evaluated the acceptability of the quasi-steady approach to predict the peak response and RMS values for full-scale low rise buildings at low frequency range. Other time-domain studies looked at the non-linear behavior of certain types of structures. For example, Wegner (2006) and Gani & Légeron (2010) compared the time-domain nonlinear analysis to the standard equivalent static force method for guyed tower structures. Advanced time-domain methods have rarely been used to study the dynamic response of solar structures. Recently, the research group at the Sandia Laboratory were interested in modeling and simulating the dynamic response of a heliostat structure prototype (Zang et al., 2014; Griffth et al., 2014). However, more studies are needed to evaluate the capacity of different analysis methods to predict the global dynamic response of solar structures to turbulent wind. The

The indicial response is the equivalent time domain representation of the aerodynamic admittance. Wagner in 1925 first used thin airfoil theory to introduce indicial functions to predict self-excited forces on airfoils. Its application is then extended to bluff body sections where thickness effects, unsteady viscous flow effects and flow separation effect could be more important. It could be also used to model the lift, moment and drag force for both the buffeting and the self-excited components. The expressions of the indicial functions can be derived or approximated via various approaches, such as analytical (Chen & Kareem, 2002; Chang, 2010), computational fluid dynamics (Farsani, 2012), or via experimental studies.

The objective of this paper is to present a time-domain method which predicts the dynamic response of solar concentrator prototype under measured gust wind samples and approaches the statistical parameters of the full-scale global response. The acceptability of the method is also reviewed in the light of the results from the equivalent static approach based on the spectral approach.

The present study is composed of two main parts. The first part concerns the description of the numerical model, the time-domain analysis process and the aerodynamic force model. In the second part, the results are presented in its deterministic description by comparing the measured full-scale base moment measurement and the calculated numerical moment in time-domain. Then, the statistical parameters issued from both frequency and time-domain analyses are compared to measurements on a full-scale prototype for over 45 test samples, which were subjected to comparable wind conditions.

2. STUDY DESCRIPTION

2.1. *Concentrated Photovoltaic (CPV) Prototype*

The CPV structure prototype was modeled based on the technical drawings and full 3D model (Fig. 1). For this purpose, Salome-Meca, an open-source simulation platform related to the finite element solver Code Aster, was used. The geometry was composed of three principal parts. First, a double guyed mast is used as a support of the Dense receiver array (DRA). Second, the dish reflector surface is composed of 128 mirrors and a parabolic frame truss supporting structure.

Third, the tower support structure is a truss structure, which is mainly composed of six L203*102*13 section members, and an azimuth motor drive which is fixed to the ground. The HSS 200*200*9.5 torque central beam rests on the tower with two actuators fixed on both sides of the central beam through two roller bearings supported by two vertical rigid plates. The central beam can turn around the transversal X axis by moving the actuators to get the target elevation $\gamma°$. This beam is also considered as the primary load transfer element between the dish frames and the tower supporting structure by principally resisting to induced torque and bending moment. The total weight of the CPV prototype is approximately 135 kN.

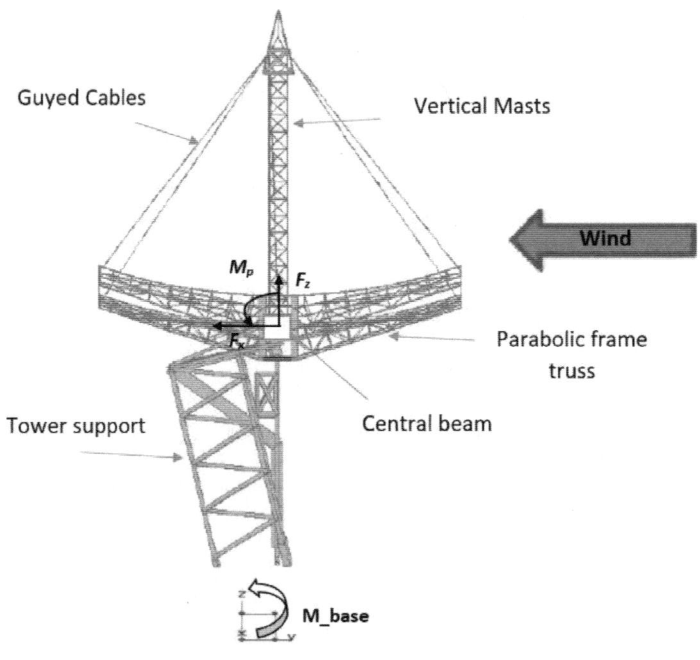

Figure 1. Side view of the CPV structure prototype (3D CAD model).

2.2. Measurement of structure response

Twelve Linear Variable Differential Transformer (LVDT) sensors were installed on the structure. They were attached by pairs to each of the six main tower members. Compression and tension forces are calculated from displacement for each member. The current configuration was used to calculate a global instantaneous base moment in the along wind direction (Fig. 1). The total moment is calculated as a sum of the product of the forces in each member by its distance from the base center. Simultaneously, the wind characteristics were measured using four ultrasonic anemometers. The first and the second anemometers were fixed at 7.5 and 10 m height on the reference tower located at 28 m in the S-W direction from the center point of the prototype. The third and fourth anemometers were fixed at 10 m height on two towers located respectively at 10 m North and 10 m South from the reference tower. The recorded data were treated by 10 min samples with sampling frequency of 20 Hz. Force and moment aerodynamic coefficients are deduced at different prototype configurations. The measurements were taken in different pitch angle γ and different wind characteristics. The detailed description of the experimental program methodology and the main results are given in Kaabia et al., (2017). In this paper, only the configurations in which wind direction is perpendicular to the normal axis to dish collector were analyzed (Fig. 1).

2.2.Numerical Modeling

In this part, two modeling strategies were tested. The first consist of modeling the complete CPV prototype. All the structure components (masts, the parabolic frame truss and the supporting tower truss) were modeled as beam elements. The mirrors were modeled by rigid elements, representing only the part where they are attached to the ribs. The cables were also modeled by beam elements with initial tension applied as a temperature gradient. Fixed boundary conditions were assumed at the tower base. Steel material properties were assigned to all parts of the structure except for the mirrors for which the density of the rigid elements was adjusted to represent the total mass of the mirrors. The supporting tower members were rotated in such a way that they respect the actual orientation of the tower members. The elevation bearing mechanism was modeled as discrete spring element having only one rotational degree of freedom around the central beam (X axis). Because, it was found difficult to quantify the stiffness of the elevation bearing mechanism, spring rigidity was calibrated to obtain the main measured natural frequencies. To do that, the two main natural frequencies at 1.0 and 3.8 Hz, corresponding to the pitching and swaying mode shapes were chosen from the PSD (Power Spectral Density) of the measured base moment for many test samples.

The second strategy consisted in modeling only the supporting tower, the central beam and the drive mechanism. It was assumed that the rest of the structure (parabolic frame and mast) could behave like a rigid body. Then, the entire weight of the parabolic dish collector and masts (~68 kN) was distributed uniformly along the central beam length. Discrete mass elements were affected to the central beam node, where elementary mass matrix including the mass and the mass moment of inertia values with respect to the eccentricity, were assigned.

2.3. Model Simplification and Modal Analysis

Preliminary modal analyses were conducted for the complete structure at different elevation angles. The analysis resulted in a high number of global and local mode shapes. Only a few of them in the low frequency range could be excited and thereby induce an important structural dynamic response at the tower base in the studied along-wind-direction case. For example, based on Davenport (1961), pitching and swaying mode shapes are the most important to be considered for such type of structure. Therefore, it would be of interest to simplify the structure as mentioned in the second modeling strategy. In order to validate this simplified model, a modal analysis was conducted for both the complete and simplified model. It was found that the pitching and swaying mode shapes, corresponding respectively to the rotation of the dish collector (Fig. 2a) and the displacement of the tower in the along wind direction (Fig. 2b) were observed at 1.0 and 3.8 Hz for both models. In order to achieve the objective of studying the global response at the base tower in the along wind direction, it was decided to use only the simplified model for the rest of the study.

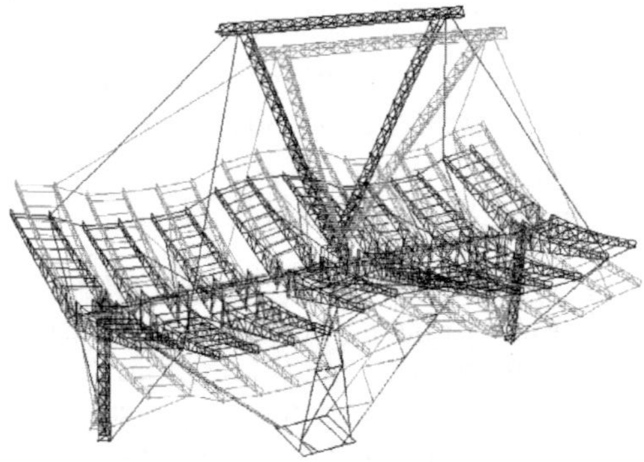

Figure 2a. Pitching mode shape (f1=1.0)

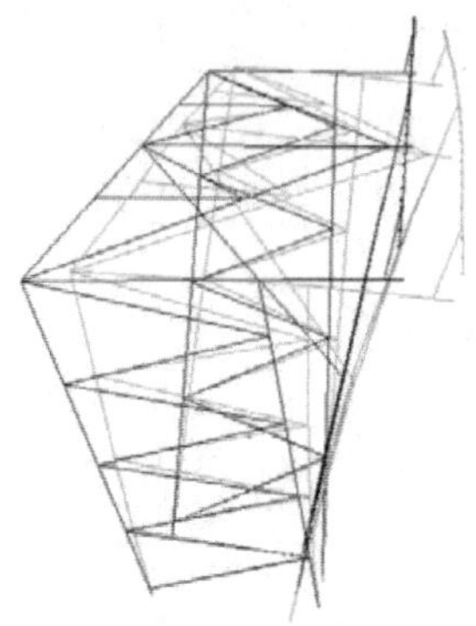

Figure 2b. Swaying mode shape (f2=3.8)

Figure 2. Modal analysis in Code_Aster : principal mode shapes.

2.4. Time Domain Method

Newmark direct integration scheme with coefficients $\beta = 0.25,\ \alpha = 0.5$ was used for the time-domain analysis. This implies that no numerical damping is introduced into the numerical integration. The time step was set to 0.05 s which is equivalent to the experimental sampling period. This gives about 20 calculations per eigen-period for the first mode shape, which ensures that the relative period error is small. The total wind load components including drag and lift forces as well as pitching moment were calculated from the wind speed test samples. Then, they were uniformly distributed through the central beam nodes.

2.4.1. Damping Considerations

The structural damping was included in this study using Rayleigh damping by fixing the structural damping ratio of two chosen natural frequencies 1.0 and 3.8 Hz, corresponding to pitching and swaying mode shapes, to values of ξ=0.25% and ξ =0.1%, respectively. The values were based on the recommendations of Davenport (1964) for a parabolic antenna structure with a similar geometry. It also appears realistic when compared to the experimental study of Griffth et al., (2014). Aerodynamic damping (C) was evaluated according to the quasi-static theory as proceeded in Gani & Légeron (2010) where the aerodynamic damping is evaluated using the simplified formulation. The aerodynamic damping values were introduced as discrete concentrated dampers for drag, lift and moment and distributed equally on the central beam nodes where the wind loads were applied. Eq. 1 present the applied aerodynamic function for the drag force.

$$C = \rho_{air} C_d \overline{U} A \qquad (1)$$

where ρ_{air} is the air density, $\overline{U}$ is the mean wind speed, A is total dish area and C_d is the drag force coefficient used to evaluate the aerodynamic damping for the swaying mode case.

2.4.2. Aerodynamic Force Model

This study is concerned with the prediction of the maximum probable response of the CPV prototype in the along-wind direction. It is conventional to separate the along-wind force into buffeting and self-excited forces. In this paper, it is assumed that the most important component is related to the buffeting forces, and the self-excited force is simply considered by introducing the aerodynamic damping as mentioned in the precedent subsection. The time-domain varying wind load components including drag and lift forces as well as pitching moment were calculated from the wind speed test samples. Then, they were uniformly distributed through the central beam nodes. The lift and moment aerodynamic coefficients used here were evaluated from wind tunnel measurements made by Gani et al., (2011). The drag force coefficients were adjusted according to full-scale measurements on the CPV prototype (Kaabia et al., 2017). The forces in the main supporting tower members are extracted from the numerical analysis and the global base moment is calculated with the same methodology as for the experimental base moment. In addition, it is found that the plate is the form that comes closest to the parabolic dish geometry, for which the experimental aerodynamic admittance was evaluated in the past (Davenport, 1961; Vickery, 1968). The larger the focal ratio f/D, the more solar structures behave like flat plates. Then, the approximate form of the aerodynamic admittance function for buffeting along wind force could be presented as follows:

$$\chi_a{}^2(n) = |\ \Phi(k)|^{\ 2} = \frac{1}{(1+k)^{4/3}} \qquad (2)$$

$$k = \frac{nc}{U} \qquad (3)$$

where $\chi_a{}^2(n)$ is the aerodynamic admittance function, $\Phi(k)$ is Sears' function, k is the reduced frequency, n is the frequency. The variable c is defined as the characteristic scale length taken equal to $\sqrt{A}$, the square root of the projected area of the dish collector depending on its elevation angle. Theoretically, this characteristic length, as compared to the wind wave length, could define how important is the correlation between wind speed and the induced aerodynamic wind force.

The need to use the aerodynamic admittance function concept in time-domain analysis gave rise to the indicial response function (IRF) approach. This time-domain aerodynamic force modeling approach represents a modified form of the quasi-steady nonlinearized approach QS as reviewed in Scanlan (1993) and Chen & Kareem (2002) in the modeling of wind force applied to bridge decks. The non-correlation between turbulent wind speed and the induced force, which is not considered with the quasi-steady model (QS), is mathematically presented with the convolution product where the participation of the precedent turbulent wind speed values in the current wind force are weighted according to the IRF form as shown in equations 4-7. The turbulent lift force and the turbulent pitching moment would be formulated in a similar way.

$$F(t) = \bar{F} + f(t) \qquad (4)$$

$$f(t) = \rho_{air}\bar{U}AC_{fx}\int_0^s [u(s) - \varphi'(s-\sigma)]\, d\sigma \qquad (5)$$

$$\varphi' = 0.075e^{-0.513\,s} + 1.794e^{-2.111s} \qquad (6)$$

$$s = \bar{U}t/\sqrt{A} \qquad (7)$$

As formulated, the total drag force $F(t)$ is linearized to be composed of mean $\bar{F}$ and turbulent parts $f(t)$. φ' is the derivative of the indicial function φ for flat plate, s is the nondimensional time. The derivative of the indicial function φ' for a flat plate is already evaluated in Chang (2010).

2.5. Statistical Dynamic Parameters

The measured moment at the base is the reference parameter to be evaluated in the numerical work. It is considered that the along-wind response is coming from the pitching moment M_p and the drag force F_x. Then the overturning base moment could be deduced from Eq. 8.

$$M = F_x \, . H + M_p \tag{8}$$

Where H is the height of the central beam.

The study focuses especially on comparing the peak response from the experimental and numerical results. Therefore, as is presented in its basic formulation in the literature, the gust effect factor GEF, a key dimensionless parameter to separate the peak response in its background and resonance form from the static gust wind effect, is calculated in Eq. 9.

$$GEF = \frac{M_{max}/\bar{M}}{{G_v}^2} \tag{9}$$

where M_{max} and $\bar{M}$ are respectively the maximum and the 10-min time average moment and G_v is the wind gust factor based on 3-s gust duration in order to fit the ASCE 7-10 current formulation (ASCE 7-10).

In this study, three force modeling methods in time-domain including QS, IRF and a modified QS issued from Lawson's TVL formula (Lawson, 1980) using a 0.3-s moving average wind speed as a size reduction filter were tested. In frequency domain, Davenport's spectral approach in its basic form (Davenport, 1961) and ASCE 7-10 method are implemented and validated. All parameters in ASCE 7-10 as well as Davenport's spectral approach were adapted to the CPV prototype geometry and to the measured wind characteristics, including wind spectrum, turbulence intensity and peak wind factor, for each test sample. The ASCE 7-10 is applied by only including the first mode shape as it is recommended by the code. Then, the peak response factor g_r is calculated using only the first natural frequency. The hypothesis and the procedure are detailed in (Kaabia et al., 2017). However, the Davenport's spectral approach is adapted to calculate the statistical parameters related to the base moment response for 10-min duration by including the two principal mode contributions. It is considered that the along-wind maximum response is coming from the pitching moment M_p and the drag force F_x. The peak value of M_p and F_x do not occur at the same time because they are caused by different flow conditions. Then, it was calculated using the square-root-of-sum-of-squares (SRSS) combination method. The measured wind spectrum is also used for this application. The summary of the mathematical model and the hypothesis is detailed in the following paragraph.

Assuming that wind speed and consequently wind response are stationary random series following the normal stochastic law (Davenport, 1961), the peak moment response and the GEF are then evaluated in Eqs 10-13.

$$M_{max} = \bar{M} + g\sigma_M \tag{10}$$

$$\sigma_M = \sigma_B + \sigma_R \tag{11}$$

$$GLF = \frac{1+\sqrt{g_v{}^2\sigma_B{}^2+g_r{}^2\sigma_R{}^2}}{\bar{M}} \tag{12}$$

$$GEF = \frac{GLF}{G_v{}^2} \tag{13}$$

Where g_v is the quasi-static peak response factor or the wind peak factor. g_r is the resonance peak factor, which is calculated using the extremum statistics law (Davenport, 1961) for a duration T=600 s. σ_B, σ_R are the standard deviation for the quasi-static and the resonance response. σ_B was evaluated as the area under the analytical response spectrum (PSD) by ignoring any dynamic effect. σ_R was evaluated by considering a narrow band response at the natural frequency. It was also more coherent to split the total peak factor g into terms g_v and g_r. This procedure was separately applied for both the drag force and the pitching moment induced response. Then, the combination to get the total response and the statistical parameters was applied.

3. RESULTS AND DISCUSSIONS

3.1.Time domain

This study focuses on determining the maximum base moment response of the CPV prototype subjected to operational wind conditions in a direction coincident with the wind for different configurations. However, it is interesting to have information about the complete time history of the studied response and how often peak values could occur in a time period.

Figs. 3a and 3c present a dynamic time history analysis issued from the QS and IRF approach and compare the results to the measured response for two typical test samples at different configurations. The constant dashed line illustrates the static equivalent peak values issued from ASCE 7-10 prediction using the evaluated 3-s wind gust. The measured response for two samples display average responses of approximately 7 kN.m and 15 kN.m respectively, with some narrow peaks due to significant gusts. It is also shown that the use of IRF method decreases the important fluctuations and the high peaks observed for the QS method, therefore better approaching the measured response.

Figs. 3b and 3d present the PSD calculated from the same test samples. In fig. 3b, it is observed that the graph for the frequency range 0-3.8 Hz is better fitting the measured response for the IRF method. This is consistent with the role played by the indicial response function and the aerodynamic admittance function which attenuates the response above ~0.5 Hz for this case. After this value, the gust becomes ineffective in producing total forces on the structure. The resonant peak at 3.8 Hz also better fits the measured response when using the IRF method. Its frequency peaks are reduced compared to QS method and the standard deviation represented by the surface under the spectrum are better represented. Above ~4 Hz, the simulated response is much lower than the measured response. However, for this frequency the contribution to the global response of the structure is limited. Similar observations can be made from Fig. 3d for the configuration $\gamma = 90°$. The main difference is that the resonant response obtained in the simulations at 3.8 Hz for the swaying mode is much less important in the measurements compared to the pitching mode peak. This may be due to the weak peak contribution of the drag force to the global response in that configuration of the structure. It is concluded that the role played by the IRF in time-domain and by the aerodynamic admittance in frequency domain seems to be essential to predict the response in high frequency range where the natural frequency could exhibit significant energy. On the other hand, the dynamic effect coming from the pitching mode at 1.0 Hz is more important for the configurations γ close to 90°, where the turbulent pitching moment is higher and the wind spectrum energy is more important.

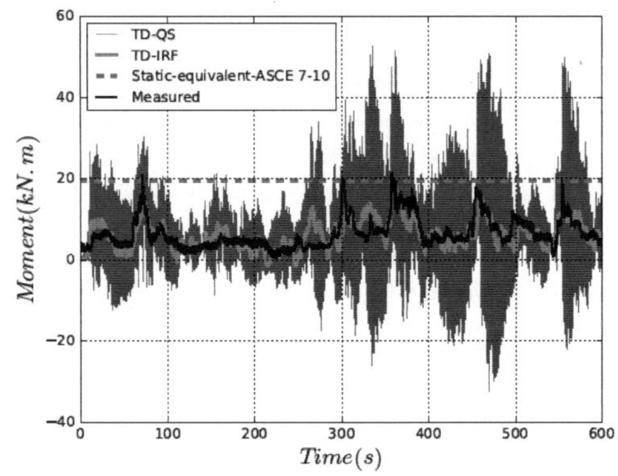

Figure 3a. Instantaneous base moment

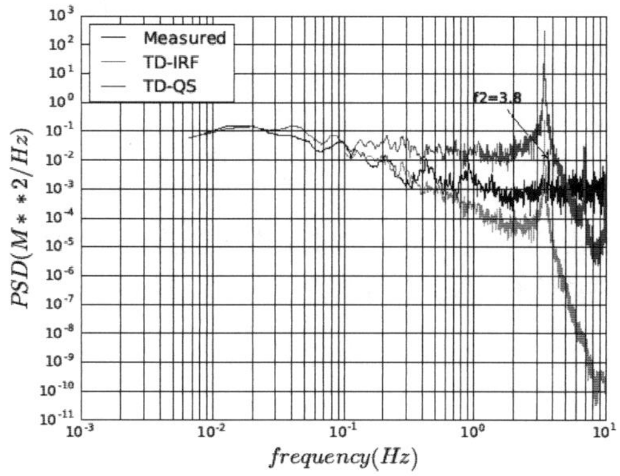

Figure 3a. Normalized PSD

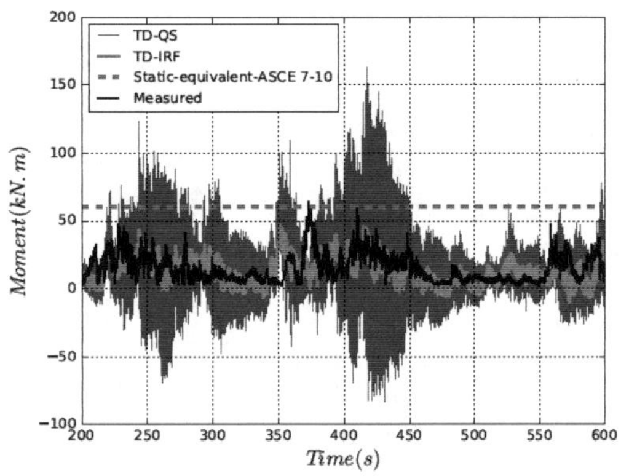

Figure 3c. Instantaneous base moment

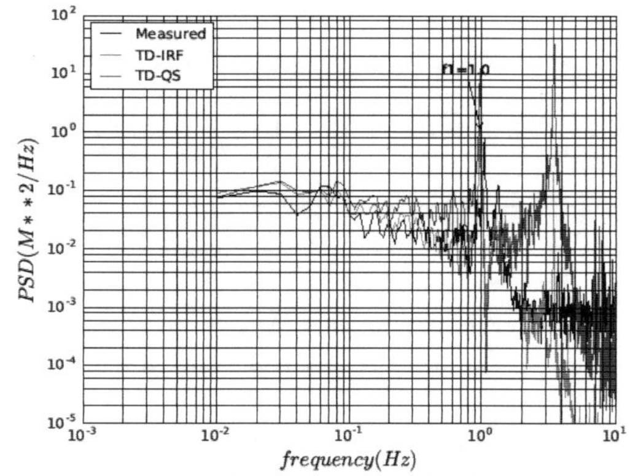

Figure 3d. Normalized PSD

Figure 3. Comparison of measured and simulated moment response for two typical samples at pitch angle $\gamma = 0°$ and $\gamma = 90°$.

3.2. *Statistical parameters comparison*

The statistical study was carried out for 45 test samples in order to calculate the conformity between values derived from the full-scale measured data and the simulated data. The ratios between simulated and measured values were calculated for each test and E represents the average of these ratios over all tests and σ_E is the standard deviation for this ratio. The results are summarized in Table 1. Moreover, five typical test results representing the five tested configurations are shown in Fig. 4. In this figure, the dynamic contribution in the total response is highlighted in lighter hatched color. The dynamic contribution part for each parameter was calculated as a difference between the results from the total dynamic response and the response issued from a time-depending static analysis using the turbulent wind loading.

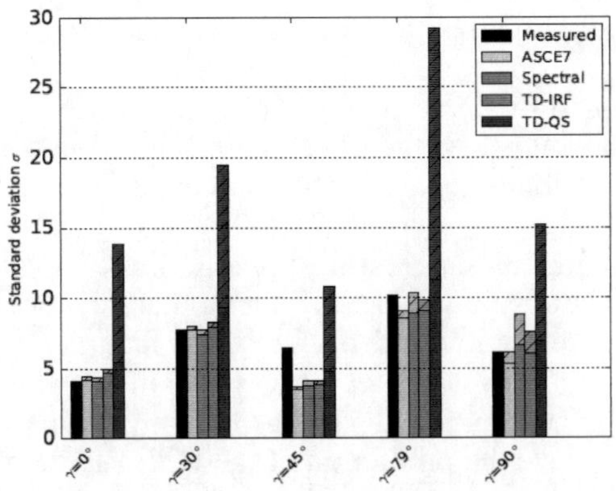

Figure 4a. Standard deviation

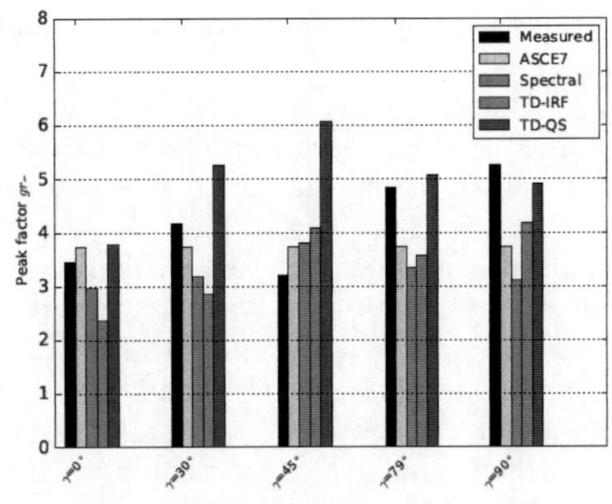

Figure 4b. Peak factor

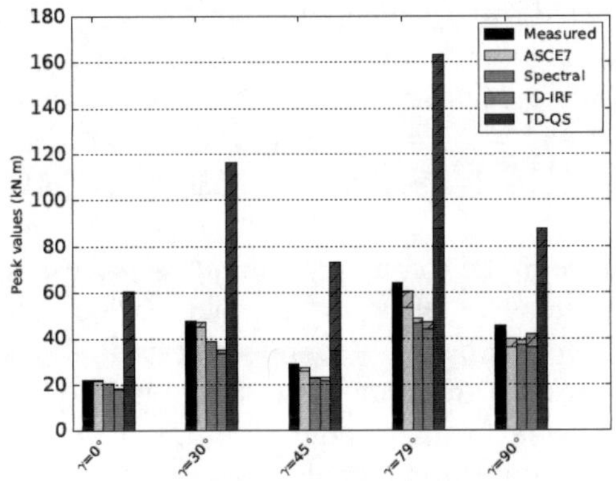

Figure 4c. Peak moment values

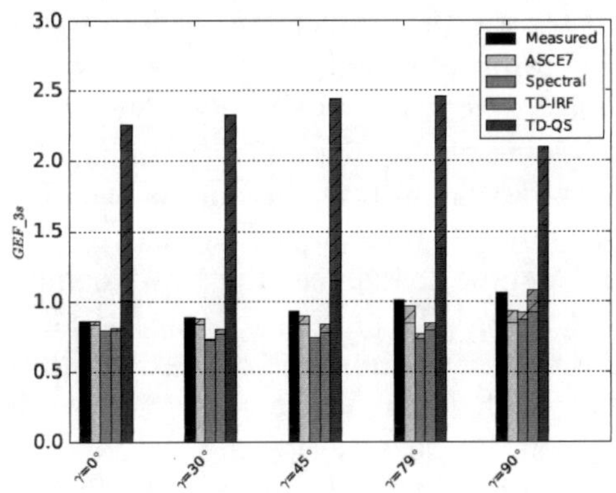

Figure 4d. Gust effect factor

Figure 4. Comparison of different methods in predicting response statistical parameters.

Table 1. Comparison between the calculated statistical parameters from the measured and simulated base moment using different methods

	Spectral stochastic				TD					
Parameters	ASCE7-10		Davenport		QS		Lawson's TVL		IRF	
	E	σ_E	E	σ_E	E	σ_E	E	σ_E	E	σ_E
σ_M	0.85	0.14	0.99	0.16	2.50	0.65	1.81	0.49	0.98	0.24
g	0.86	0.26	0.84	0.22	1.17	0.31	1.15	0.30	0.82	0.20
GEF	0.96	0.18	0.88	0.15	2.38	0.69	1.79	0.43	0.92	0.20

For this final comparison, all the results relating to the base moment, shown in Table 1, were summarized. In general, good agreement and similar results could be observed for the standard deviation, the peak factor and the gust effect factor of the time-domain analyses based on the IRF, the spectral approach and the ASCE 7 formulation. The ASEC 7 and the spectral approach results are always within the limit of $E \pm \sigma_E$ from those of the IRF analyses. Table 1 shows also that, in term of standard deviation values, the background and the resonance response are better represented by the aerodynamic admittance function and the IRF used for the Davenport's spectral approach and the time-domain method, respectively. The ASCE 7 prediction slightly underestimates this value because of its less conservative aerodynamic admittance function compared to flat plate. However, the QS approach highly overestimates the standard deviation for all tests which leads to high discrepancies in predicting the maximum values. In term of peak factor values, more discrepancies are recorded between the investigated methods and the experimental results. It seems that the statistical extrema law introduced to calculate the response factor g_r could contribute in slightly underestimating the peak values for the study of loading cases at service conditions (see Fig .4b).

Overall, it is observed in Table 1 that the IRF method was successful to predict the GEF and the standard deviation of the base moment response. The results of the IRF method are coherent with both Davenport's spectral approach and ASCE 7 prediction and by far better than the QS prediction which is overestimating the peak response for all test samples. It was also investigated that applying the Algebric Combination (AC) instead of the Square-Root-of-Sum-of-Squares (SRSS) for the Davenport's spectral approach, could improve the prediction of the GEF and therefore the peak value by 8% in average. However, the time-domain method is still the most efficient to include correctly the mode shape interaction effect.

It is shown in Figure 4c, that depending on the CPV structure configurations and the wind loading cases, the peak equivalent static response calculated from Davenport's spectral approach or ASCE 7 formulation and the maximum response calculated with the IRF may underestimate or overestimate slightly the possible dynamic response. However, QS approach highly overestimates the total response for all tests. For example, it is shown in Fig. 4a that the dynamic contribution in the standard deviations for QS seems to be out of range and could represent more than about ~50% of the total response for all the selected tests. Therefore, it is confirmed that the filtering role played by the IRF in time domain and the aerodynamic admittance in the frequency domain is successful to evaluate the turbulent response and its dynamic part for this type of structure under service wind conditions. In fact, the unsteady events like flow separation which could be important for this type of structure with a complex aerodynamic profile don't have an instantaneous effect on the body surface. Then, the quasi-steady theory may not be applicable for the prediction of aerodynamic forces in separation regions as investigated in Tieleman (2003). It is also noted that neglecting the second order turbulence effect by linearizing the total force formulation could contribute to reduce the mean value of the calculated time-domain response. This could explain that the spectral approach may give higher peak value compared to the IRF method in Fig. 4c. Then, the GEF is a more consistent parameter to compare the dynamic effect in its normalized form (Fig .4d).

In the same figure, the dynamic part contribution could slightly vary depending on the method for the same test configuration. This point is also confirmed from the previous statistical comparison, and it is found that the contribution of the dynamic part in the peak response for all test samples, in terms of GEF, varies between 0 and 22% for IRF, between 0 and 8% for Davenport's spectral approach, and between 1 and 21% for ASCE 7. It is also observed that dynamic effect is likely to be significant for the configurations around the $\gamma = 90°$ when the pitching moment is important. The non-consideration of this pitching moment with its turbulent component in the numerical analysis could lead to high errors for this type of structure.

It is concluded that the IRF could be confidently used in time-domain analysis instead of the QS as an alternative to the common frequency domain methods. It is also observed that the dynamic effect contribution could be important in the design of such special structures at security position for high wind conditions and the dynamic contribution could be well predicted using the IRF method. The spectral approach give also a coherent result in predicting statistical dynamic parameters. However, it could be sensible to dynamic structural mode interaction and its results alone gives no indication about how often peak values could occur in a period of time.

4. CONCLUDING REMARKS AND FUTURE WORK

The present study is the continuation of previous full-scale work for which the wind speed and the base moment of a CPV prototype were measured instantaneously. This paper shows the results of a deterministic time-domain analysis of the CPV prototype numerical model. It compares this time-domain method to common frequency domain methods as well as full-scale measurements. A statistical comparison study over many test samples has shown that time-domain method, using indicial function derived for a flat plate and applied to the simplified CPV numerical model, can predict with acceptable variation the statistical parameters of the global wind response. The importance of possible dynamic effects caused by the turbulent wind could vary depending on wind cases and structure configurations. It is observed that, even though the drag force is not very important for positions near to the security position at $\gamma = 90°$, the pitching moment could lead to important dynamic effects.

The methodology developed here is a simplified tool to obtain the global response of a solar structure. Important simplifications were made regarding the dynamic behavior of the CPV prototype, for the structural behavior of the cables and dish collector. The across-wind response which could be caused especially by vortex shedding was not considered, owing to the difficulty to perform a validation procedure. Moreover, the proposed method simplifies the aerodynamic force caused by the complex fluid flow around the structure. Nonetheless, the proposed simplification hypotheses could be advantageous for engineering design application for such type of structures. A simplified time-domain method could be effectively used to extend the study for design conditions where high wind speed cases and nonlinearities could be included. It could be also used to conduct parametric study to evaluate the effect of wind and structural parameters to meet optimal design criteria for similar structures under wind load.

5. ACKNOWLEDGEMENTS

The authors would like to thank SpaceWatts division of Gestion TechnoCap for the technical contributions and financial support for the research, along with NSERC, PROMPT and Hydro-Sherbrooke. We are also grateful for the financial contributions of the Canadian Foundation for Innovation (FCI), the Ministère de l'Économie, Science et Innovation of Québec and the Université de Sherbrooke for the unique CPV infrastructures. Finally, the technical contribution of M. Richard Norman for the work related to the CPV prototype is greatly acknowledged.

6. REFERENCES

1 ASCE7-10, Minimum Design Loads for Buildings and other Structures. ASCE, Reston, VA, 2010.

2 Chang, B., Time-Domain Model for Predicting Aerodynamic Loads on a Slender support structure for fatigue design, Journal of Engineering Mechanics, 2010, 136 (6), 736-746.

3 Chen, X., Kareem, A., Advances in Modeling of Aerodynamic Forces on Bridge Decks, Journal of Engineering Mechanics, 2002, 128 (11), 1193-1205.

4 Davenport, A., The Application of Statistical Concept to the Wind Loading of Structures, Journal. I.C.E, 1961.

5 Davenport, A., Buffeting of Large Superficial Structures by Atmospheric Turbulence, Proceedings of the Conference on Large Radio Antennas. Annuals of New York Academy of Sciences, 1964, 116, 11-34.

6 Banks, D., How Wind Load Studies will Impact the Solar Industry, Fort Collins, 2014.

7 Farsani, H. Y., Indicial Function Predictions of Two Dimensional Unsteady Viscous Flows Around Bluff Bodies Based on Discrete Vortex Simulations, 2012.

8 Gani, F., Légeron, F., Dynamic Response of Transmission Lines Guyed Towers Under Wind Loading, Canadian Society of Civil Engineering, 2010, 37, 450-464.

9 Gani, F., Légeron, F., Norman, R., Refinement of Design for Parabolic Truss Structure for Solar Concentrating Photovolataic Tracker System, Annual Conference of the Canadian Society of Civil Engineering, 2011, 788-799.

10 Griffth, T. D., Moya, A. C., Ho, C. K., Hunter, P. S., Structural Dynamics Testing and Analysis for Design Evaluation and Monitoring of Heliostats, Journal of Solar Energy Engineering, 2014, 137 (2), 21010.

11 Hosoya, N., Peterka, J. a., Gee, R. C., Kearney, D., Wind Tunnel Tests of Parabolic Trough Solar Collectors. Tech. rep., National Renewable Energy Laboratory, 2008.

12 Kareem, A., Synthesis of Fluctuating Along-Wind Loads on Tall Buildings, Journal. Mech. Div., ASCE, 1986, 112(1), 121-125.

13 Lawson, T., Wind Effect on Buildings, Applied Science Publishers Ltd.1, 1980.

14 Letchford, C. W., Iverson, R. E., McDonald, J. R., The Application of the Quasi-Steady Theory to Full Scale Measurements on the Texas Tech Building, Journal of Wind Engineering and Industrial Aerodynamics, 1993, 48 (1), 11-132.

15 Peterka, J.A., and Derickson, R.G., Wind Load Design Methods for Ground Based Heliostat and Parabolic Dish Collectors, Sandia National Laboratories, 1992.

16 Scanlan, R. H., Problematics in Formulation of Wind Force Models for Bridge Decks, Journal of Engineering Mechanics, 1993, 119 (7), 1353-1375.

17 Sears, W., Some Aspects of Non-Stationary Airfoil Theory and its Practical Application, Journal. Aero. Sci, 1941, 8(3), 104-108.

18 Solari, G., Kareem, A., On the Formulation of ASCE7-95 Gust Effect Factor, Journal of Wind Engineering and Industrial Aerodynamics, 1998, 77-78, 673-684.

19 Sparling, B. F., Wegner, L. D., Comparison of Frequency and Time-domain Analyses for Guyed Masts in Turbulent Winds, Canadian Society of Civil Engineering, 2006,182, 169-182.

20 Tieleman, H. W., Wind Tunnel Simulation of Wind Loading on Low-rise Structures : A review, Journal of Wind Engineering and Industrial Aerodynamics, 2003, 91(12), 1627-1649.

21 Vickery, B., The Application of Statistical Concept to the Wind Loading of Structures, Journal. Eng. Mech. Div., ASCE, 1968, 94(1), 31-46.

22 Zang, C., Christian, J., Yuan, J., Sment, J., Moya, A., Ho, C., Wang, Z., Numerical Simulation of Wind Loads and Wind Induced Dynamic Response of Heliostats, Energy Procedia 49, 2014, 1582-1591.

Structural Response of Transmission Line Conductors under Thunderstorm Wind

Amal Elawady[1] and Haitham Aboshosha[2]
[1]*Department of Civil and Environmental Engineering, Florida International University, Miami, USA, Email: aelawady@uwo.ca*
[2]*The Boundary Layer Wind Tunnel Laboratory, Western University, London, Ontario, Canada Email: haboshos@uwo.ca*

ABSTRACT

Various studies highlighted long span structures including transmission lines conductors experiencing high stresses under thunderstorm winds. This is due to the localized nature of the thunderstorm downdraft outflow which imposes non-uniform wind loads on the conductor that leads to uneven tension in the spans adjacent to the tower of interest. Previous studies showed that uneven tension can be very significant and in fact was the main reason for a number of transmission line failures. The current study focuses on evaluating conductor response due to thunderstorm wind seen at two selected locations. Thunderstorm wind speeds are modeled using a recently developed and validated framework that couples historical wind records with Monte Carlo simulation. Analysis of the conductors are conducted using two approaches (i) a semi-closed form approach that is able to analyze different variation of the load along the conductor span and (ii) a simplified approach that is suitable for practicing engineers and focuses on modeling an idealized distribution of the wind load. Conductor reactions resulting from the two approaches are compared and are found in a good match. This emphasizes on the applicability of the simplified approach to rapidly evaluate conductor reactions under complex thunderstorm loading.

KEYWORDS

Thunderstorm, Gust Front, Downdraft, Design Speed, Conductors, Reactions, Non-uniform loading.

13th Americas Conference on Wind Engineering
Gainesville, Florida, USA
May 21 - 24, 2017

13th Americas Conference on WIND ENGINEERING

Structural failures due to strong winds in non-wind areas in Mexico

Hernández-Barrios Hugo[a], Huergo Ríos Iván[b]

[a] *Universidad Michoacana de San Nicolás de Hidalgo, Morelia, México, hugohernandezbarrios@yahoo.cm.mx*
[b] *Universidad Nacional Autónoma de México, FESA, UNAM, ing.huergo@hotmail.com*

ABSTRACT

The Mexican Republic is subject to strong winds that appear in the coasts of Pacific Ocean and the Gulf of Mexico, beginning every year in May and extending to the month of December. These areas are classified as hurricane zones and in them there have been a large number of structural failures already reported in the literature. However, in non-wind areas in Mexico there have been recent local phenomena that involve strong wind gusts and have produced a large number of structural failures.

In this paper, a series of structural failures are reported and analyzed in industrial buildings, telecommunication towers, monopoles, building coatings and colonial churches that are located in non-wind areas in Mexico. The above allows to conclude that it is necessary to review and update the wind design considerations including a new zoning of the Mexican Republic.

KEYWORDS

Comfort conditions, high-rise buildings, aerodynamic configuration

The 13th Americas Conference on Wind Engineering (13ACWE)
Gainesville, Florida USA, May 21-24, 2017

Evaluation on Uncertainty of Peak Wind Load Effect with Analytical Approaches

Guoqing Huang [a], Xiaowen Ji [a]

[a] *School of Civil Engineering, Southwest Jiaotong University, Chengdu, China*

ABSTRACT: Estimating the peak value of wind load effects is an important task for engineers in the structural and cladding design. Due to the various sources of uncertainties, the peak wind load effects may have large variations. The current study focuses on the evaluation of the uncertainty for peak value of wind load effects across multiple samples from a wind tunnel. Firstly, the Hermite polynomial model (HPM)-based translation process method is adopted to estimate the peak wind load effect. Secondly, an empirical formula is proposed to estimate the parameters of peak value distribution. Thirdly, two analytical approaches to model the uncertainty of the peak value at any fractile are proposed based on the probability function and the Gaussianity inference on the peak value. Fourthly, numerical examples including the wind pressure coefficient and wind-induce response are provided to illustrate the performances of the proposed approaches.

KEYWORDS: Peak value; Hermite polynomial model; Uncertainty; Analytical approach; Wind load effect

1 INTRODUCTION

In the reliability- and performance-based building structural and cladding design, the determination of the peak value of wind pressure and wind-induced response for a given time interval is essential. In the past decades, various approaches have been developed to determine the peak factor and peak value at a fractile for wind load effect (e.g., Refs. [1-9]). Among these approaches, the Hermite polynomial model (HPM) -based translation process method received significant attention in the estimation of the non-Gaussian wind pressure or response because it has the flexibility of free adjustment on skewness and kurtosis, the straightforwardness and higher accuracy (e.g., Refs. [8,10]). Recently, correlation structures for wind pressures and their peak values have been investigated by Huang et al. [11] and Luo & Huang [12].

It is well known that the peak wind pressure and wind-induced response estimated from short-term samples (T = 10 min or 1 hour) may vary due to many factors such as the randomness of the aerodynamic effect, model selection and calibration errors. Generally, these uncertainties in the peak value estimation may be separated as the aleatory (randomness inherent in the variable) and epistemic (lack of knowledge or data) uncertainties [13]. Minciarelli et al. [14] examined the influences of various knowledge (epistemic) uncertainties such as time interval, terrain roughness and other errors caused by tunnel tests on determination of extreme wind load effects and load factors. Recently, more studies have been carried out based on the long wind pressure data from wind tunnel tests. Yang & Tian [15] modeled the aleatory randomness of the peak factor and associated wind pressure based on the multiple wind pressure samples. Gavanski et al. [16] investigated uncertainties of the sampling duration and peak sampling frequency in peak pressure estimations using Gumbel distribution fitting method. Wu et al. [17] studied influences of epistemic uncertainties due to limited data on wind-induced peak responses with various mean recurrence intervals by the closed-from solution.

Previous uncertainty studies mainly focus on the peak factor (57% fractile for Gumbel distribution). However, the peak values at other fractiles are also important in wind-resistance design. For example, 78% fractile of the peak pressure coefficient is adopted in codes and standards to define the design wind loading in order to consider the uncertainty of both wind speed and pressure coefficient (e.g., Refs. [18,19]). Obviously, it is important to develop an efficient approach to estimate the uncertainties of peak wind load effects associated with various fractiles. In this study, two analytical approaches are proposed to model the uncertainty of aerodynamic randomness on the peak value of non-Gaussian wind load effects at any fractile. First, HPM-based translation process method is introduced to estimate the peak wind pressure and load effect. Second, an empirical formula is proposed to estimate the parameters of peak value distribution based on HPM-based translation process method. Third, the uncertainty of peak value at any fractile can be evaluated from two analytical approaches based on probability distribution function and Gaussianity reference for peak value at a fractile. Fourth, two numerical examples including wind pressure coefficient and wind-induced response are used to evaluate the performances of the proposed approaches. Finally, some conclusions are provided.

2 ESTIMATION OF PEAK WIND LOAD EFFECT

Usually the wind pressure and wind-induced response show mild to strong non-Gaussianity. In this study, HPM-based translation process method is adopted to estimate the peak wind load effect, where the moment-based HPM will be used to estimate the probability density function (PDF) of the parent process and the translation process method can be used to estimate the PDF of the peak value by mapping the non-Gaussian process onto the underlying Gaussian counterpart [20].

Assume $Y(t)$ is a "softening" non-Gaussian process whose mean, standard deviation (STD), skewness and kurtosis are r_1, r_2, r_3 and r_4, respectively. The normalized version of $Y(t)$ is $X(t) = [Y(t) - r_1]/r_2$. It can then be expressed by a standard Gaussian process $U(t)$ via the Hermite polynomial [21]

$$x = k[u + h_3(u^2 - 1) + h_4(u^3 - 3u)] \tag{1}$$

where k ensures that $X(t)$ has unit variance, and h_3 and h_4 control the shape of the distribution. Those parameters are determined by a set of nonlinear equations via Newton-Raphson iteration (e.g., Ref. [22]). The inversion and the corresponding requirement on the monotonicity of the Hermite polynomial (or effective region) can be found in literature (e.g., Refs. [8,23,23]). Then the PDF of $X(t)$ can be determined as

$$f_X(x) = \varphi[u(x)]\frac{du(x)}{dx} \tag{2}$$

where $\varphi(\cdot)$ is standard Gaussian PDF. The distribution of $Y(t)$ is obtained from that of $X(t)$.

For the process with skewness and kurtosis located in the effective region, usually its PDF can be satisfactorily fitted by moment-based HPM. Otherwise, an approximate adjustment for skewness or kurtosis can be carried out [6]. Once the parent distribution is determined, the peak value can be estimated by a translation process method (e.g., Ref. [2]).

For the standard Gaussian process $U(t)$, the cumulative distribution function (CDF) of the peak value U_{pk} during time interval of T is given by

$$F_{U_{pk}}(u_{pk}) = \exp[-\lambda_0 \exp(-u_{pk}^2 / 2)] \tag{3}$$

where $\lambda_0 = \nu_{0,u} T$ is the corresponding mean zero upcrossing number (In following discussions, mean zero is dropped for abbreviation) in T for the Gaussian process $U(t)$ and $\nu_{0,u}$ is the upcrossing rate, which can be computed by

$$\nu_{0,u} = \sqrt{\int_0^\infty f^2 S_U(f) df} \Big/ \sqrt{\int_0^\infty S_U(f) df} \tag{4}$$

where f is natural frequency (Hz); $S_U(f)$ is power spectral density (PSD) function of process $U(t)$. In the preceding equation, $S_U(f)$ can be replaced by PSD function of pressure coefficient time history $Y(t)$, $S_Y(f)$, because the peak is not very sensitive on the variation of the upcrossing rate (e.g., Refs. [2,25]). With CDF of U_{pk} ready, CDF of X_{pk}, peak of $X(t)$ during time interval of T, can be determined by mapping peaks of the Gaussian distribution process onto the non-Gaussian distribution space. The detailed discussion can be found in literature (e.g., Refs. [2,26]). Obviously, the peak value distribution of the normalized non-Gaussian process can be merely derived from skewness, kurtosis and upcrossing number under the framework of HPM-based translation process method.

According to previous studies, the estimated peak value of $X(t)$ will approach Gumbel distribution (e.g., Refs. [8,27]), which is given by

$$F_{X_{pk}}(x_{pk}) = \exp\left[-\exp\left(-\frac{x_{pk} - \delta_x}{\psi_x}\right)\right] \tag{5}$$

where δ_x and ψ_x are location and scale parameters, respectively. Accordingly, the peak value distribution of original non-Gaussian process $Y(t)$ can be directly deduced from that of $X(t)$.

$$\delta_y = r_1 + r_2 \delta_x \tag{6}$$

$$\psi_y = r_2 \psi_x \tag{7}$$

where δ_y and ψ_y are location and scale parameters of peak value distribution of $Y(t)$. Note that additional shape parameter remains unchanged if the generalized extreme value distribution is adopted to fit the peak value.

3 EMPIRICAL FORMULA TO PEAK VALUE DISTRIBUTION OF NORMALIZED PROCESS

Based on the preceding discussion, it is possible to establish the relationship between two sets of parameters where the first set includes the skewness, kurtosis and upcrossing number, and the other one contains the location and scale parameters of the peak value distribution for the normalized process. How to formulate the relationship for these two sets of parameters will be addressed hereafter.

First, the ranges of the concerned skewness, kurtosis and upcrossing rate should be selected. According to previous literature [8,11], the scatter of skewness and kurtosis for wind pressures and panel lift forces on typical building roofs is shown in Figure 1. In this figure, the dashed line denotes the effective region for HPM by Choi & Sweetman [23]. The selected zone for skewness and kurtosis is surrounded by solid lines, which could cover almost all possible pairs of skewness and kurtosis in practice. Based on wind pressure time histories of the two UF building roofs [6], it is found that the upcrossing rate varies from 1.44 to 9.37. In engineering practice, the upcrossing rate can be regarded to be in the range of 1 to 10. Typically, the time interval for determining peak values could be 10 min or 1 hour. Accordingly, the upcrossing number could be from 600 to 36000.

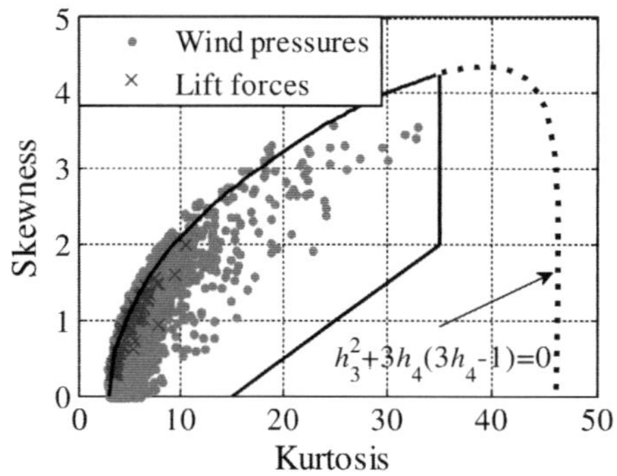

Figure 1. Scatter of skewness and kurtosis for wind load effects

Second, the peak value of the normalized non-Gaussian process is estimated via HPM-based translation process method for the process in the aforementioned ranges of skewness, kurtosis and upcrossing number. The corresponding location and scale parameters are fitted under the assumption that the peak value follows Gumbel distribution. Estimated location and scale parameters under selected upcrossing numbers are shown in Figure 2. Compared with the scale parameter, the location parameter might be sensitive to the smaller upcrossing number.

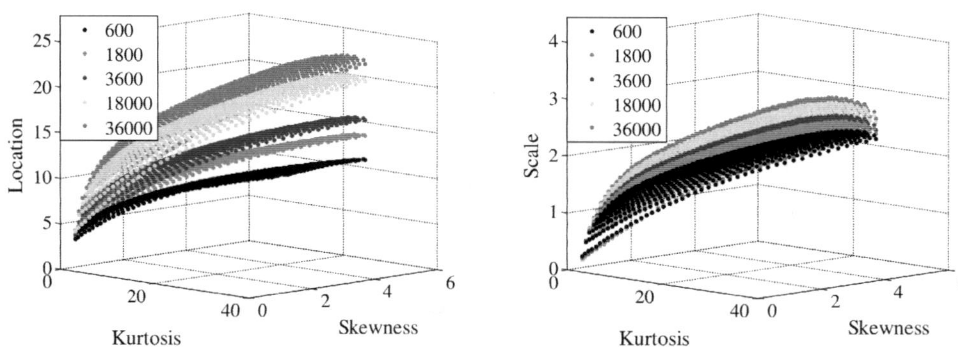

(a) Location parameter (b) Scale parameter

Figure 2. Estimated location and scale parameters under 5 selected mean zero upcrossing numbers

Third, the relationship between these two sets of parameters can be fitted via the multiple linear regression analysis. For location and scale parameters, the empirical formula is obtained as

$$\begin{aligned}\delta_x &= \delta_x(r_3, r_4, \lambda_0) \\ &= a_1 \ln r_4 + a_2 r_3^2 + a_3 \ln^2 r_4 + a_4 \ln^3 r_4 + a_5 \ln \lambda_0 + a_6 r_3 \ln r_4 \\ &\quad + a_7 r_3^2 \ln^2 r_4 + a_8 \ln r_4 \ln \lambda_0 + a_9 r_3^2 \ln \lambda_0 + a_{10} r_3^3 \ln \lambda_0 \\ &\quad + a_{11} r_3 \ln r_4 \ln \lambda_0 + a_{12} r_3^2 \ln r_4 \ln \lambda_0 + a_{13} r_3 \ln^2 r_4 \ln \lambda_0 \\ &\quad + a_{14} r_3^2 \ln^2 r_4 \ln \lambda_0\end{aligned} \tag{8}$$

$$\begin{aligned}\psi_x &= \psi_x(r_3, r_4, \lambda_0) \\ &= b_1 r_3 + b_2 \ln r_4 + b_3 \ln \lambda_0 + b_4 r_3 \ln \lambda_0 + b_5 \ln r_4 \ln \lambda_0 + b_6 r_3^2 \ln \lambda_0 \\ &\quad + b_7 \ln^2 r_4 \ln \lambda_0 + b_8 r_3^3 \ln \lambda_0 + b_9 \ln^3 r_4 \ln \lambda_0 + b_{10} r_3 \ln r_4 \ln \lambda_0 \\ &\quad + b_{11} r_3^2 \ln r_4 \ln \lambda_0 + b_{12} r_3 \ln^2 r_4 \ln \lambda_0 + b_{13} r_3^2 \ln^2 r_4 \ln \lambda_0\end{aligned} \tag{9}$$

The coefficients for both of δ_x and ψ_x are summarized in Table 1. Clearly, the proposed formulae can be directly used to estimate the distribution of the peak value once the skewness, kurtosis and upcrossing number of the wind load effect process are given.

To verify the accuracy of the proposed formulae, the relative error (RE) of the estimated parameter is evaluated. With δ_x as the example, RE for given skewness, kurtosis and upcrossing number is defined as

$$\mathrm{RE} = \frac{\left|\hat{\delta}_x - \delta_x\right|}{\delta_x} \tag{10}$$

where $\hat{\delta}_x$ is the approximately estimated location parameter via Eq. (8), and δ_x is the one directly fitted by the approach in Section 2. For the specified skewness and kurtosis in the aforementioned defined zone, maxima of REs for δ_x across various λ_0 (in the range of 600-36000) can be found in Figure 3a. Similarly, maxima of REs for ψ_x can be found in Figure 3b. It is seen that REs are almost within 3.5% for both δ_x and ψ_x. In addition, RE for the peak value at a fractile can be defined by Eq. (10) in order to further verify the accuracy of the proposed formulae. Similarly to Figure 3a and b, maxima of REs for peak values at fractiles 57% and 78% across various λ_0 are shown in Figure 3c and d. It is observed that the REs are less than 4% for peak values at both fractiles. Hence, the proposed formulae are satisfactorily accurate for estimating the peak value distribution of the normalized process.

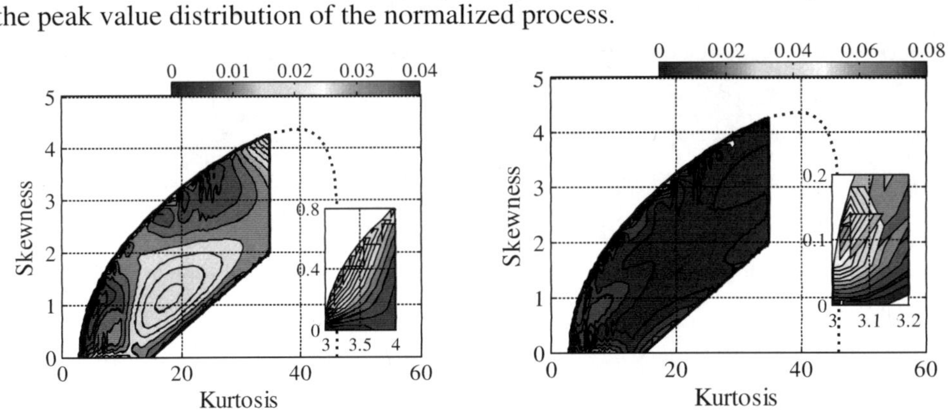

(a) Location parameter

(b) Scale parameter

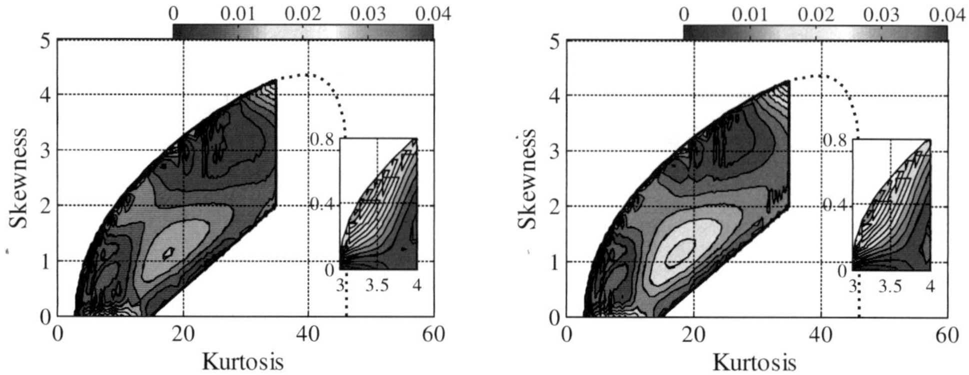

(c) Fractile of 57%

(d) Fractile of 78%

Figure 3. Maxima of relative errors across various λ_0

4 UNCERTAINTY ESTIMATION FOR PEAK VALUES

As discussed in literature (e.g., Ref. [15]), the first four moments (i.e., mean, STD, skewness and kurtosis) of the parent wind pressure process have noticeable variations (or uncertainties) across the samples. Accordingly, the peak value distribution also exhibits the uncertainties. Although the previous study by Yang & Tian [15] showed that the uncertainty of upcrossing number had little effect on peak values, its influence on the peak value needs further investigation according to Figure 2a, where the location parameter is found to be sensitive to the variation of the upcrossing number.

In the following discussion, the moments and upcrossing number will be treated as random variables to account for these uncertainties. To distinguish the notations used for two sets of parameters in previous sections, R_1, R_2, R_3 and R_4 are used to denote random variables for the mean, STD, skewness and kurtosis, respectively; Λ_0 stands for the random variable for the upcrossing number; and Δ and Ψ represent random variables for location and scale parameters, respectively. Two analytical approaches including the approach based on probability distribution function and the approach based on Gaussianity reference for peak value at a fractile (denoted as Approaches 1 and 2) will be introduced to evaluate the peak value uncertainty.

4.1 *Approach based on probability distribution function*

To describe the uncertainties of the first four moments and upcrossing number, the probability distribution is often used. Suppose the marginal CDFs of R_3, R_4 and Λ_0, i.e., $F_{R_3}(r_3)$, $F_{R_4}(r_4)$ and $F_{\Lambda_0}(\lambda_0)$, and the associated correlation coefficient matrix Σ are estimated from the sufficient number of samples or given directly. Then non-Gaussian variables R_3, R_4 and Λ_0 are related to standard Gaussian variables Z_1, Z_2 and Z_3 via isoprobability transformation, respectively, i.e.,

$$z_1 = \Phi^{-1}[F_{R_3}(r_3)];\quad z_2 = \Phi^{-1}[F_{R_4}(r_4)];\quad z_3 = \Phi^{-1}[F_{\Lambda_0}(\lambda_0)] \tag{11}$$

where Φ is the standard Gaussian distribution.

Based on Jacobian transformation, the joint PDF of R_3, R_4 and Λ_0 is obtained as

$$f_{R_3R_4\Lambda_0}(r_3,r_4,\lambda_0)=f_{R_3}(r_3)f_{R_4}(r_4)f_{\Lambda_0}(\lambda_0)\frac{\varphi_3(z_1,z_2,z_3;\Sigma_z)}{\varphi(z_1)\varphi(z_2)\varphi(z_3)} \tag{12}$$

where φ_3 denotes the trivariate joint Gaussian PDF with a correlation coefficient matrix Σ_z, whose element is determined by that of Σ. For example, the correlation coefficient between Z_1 and Z_2 is evaluated by

$$\rho_{R_3R_4}=\int_{-\infty}^{\infty}\int_{-\infty}^{\infty}\left(\frac{r_3-E[R_3]}{\sqrt{D[R_3]}}\right)\left(\frac{r_4-E[R_4]}{\sqrt{D[R_4]}}\right)\varphi_2(z_1,z_2,\rho_{Z_1Z_2})dz_1dz_2 \tag{13}$$

where $\rho_{R_3R_4}$ is the correlation coefficient between R_3 and R_4; $\rho_{Z_1Z_2}$ is the correlation coefficient between corresponding Gaussian variables Z_1 and Z_2; $E[\cdot]$ and $D[\cdot]$ are the mean and variance of the random variable, respectively.

From the joint PDF of R_3, R_4 and Λ_0, it seems straightforward to obtain the joint PDF of Δ_x and Ψ_x via proposed empirical formulae based on Jacobian transformation. However, the inverse of Eq. (8) or (9) is difficult because of no existence of one-on-one mapping between two sets of parameters even when the upcrossing number is regarded to be invariant and only uncertainties of skewness and kurtosis are considered. To obtain the joint PDF of Δ_x and Ψ_x, an efficient approach is to adopt Monte Carlo simulation (MCS), which is given in following four steps. Firstly, a set of samples of R_3, R_4 and Λ_0 can be generated by Nataf transformation, which maintains their correlation structure (e.g., Ref. [11]). Secondly, samples of Δ_x and Ψ_x are computed directly via the empirical formula, i.e., Eqs. (8) and (9). Thirdly, marginal PDFs $f_{\Delta_x}(\delta_x)$ and $f_{\Psi_x}(\psi_x)$, and the correlation coefficient between Δ_x and Ψ_x are then evaluated from these samples. Fourth, the joint PDF of Δ_x and Ψ_x, $f_{\Delta_x\Psi_x}(\delta_x,\psi_x)$ is obtained following Jacobian transformation.

If the randomness of Λ_0 has limited effect on the peak value, the mean of samples of upcrossing number, $\lambda_{0,m}$, can be approximately used. Consequently, Eq. (12) can be simplified as

$$f_{R_3R_4}(r_3,r_4)=f_{R_3}(r_3)f_{R_4}(r_4)\frac{\varphi_3(z_1,z_2;\rho_{Z_1Z_2})}{\varphi(z_1)\varphi(z_2)} \tag{14}$$

Similarly, the joint PDF of Δ_x and Ψ_x can be determined from the joint PDF of R_3 and R_4 by MCS following the aforementioned 4-step procedure.

With the joint PDF of Δ_x and Ψ_x ready, the joint PDF of Δ_y and Ψ_y can be evaluated. Suppose the conditional joint PDF of Δ_y and Ψ_y under conditions of $R_1=r_1$ and $R_2=r_2$ is given as $f_{\Delta_y\Psi_y|R_1=r_1,R_2=r_2}(\delta_y,\psi_y)$. Based on the conditional probability, one has

$$f_{\Delta_y\Psi_yR_1R_2}(\delta_y,\psi_y,r_1,r_2)=f_{\Delta_y\Psi_y|R_1=r_1,R_2=r_2}(\delta_y,\psi_y)f_{R_1R_2}(r_1,r_2) \tag{15}$$

where $f_{\Delta_y\Psi_y R_1 R_2}(\delta_y, \psi_y, r_1, r_2)$ denotes the joint PDF of Δ_y, Ψ_y, R_1 and R_2, and $f_{R_1 R_2}(r_1, r_2)$ is the joint PDF of R_1 and R_2, which can be evaluated following Eq. (12). Based on Jacobian transformation, the following equation can be obtained as

$$f_{\Delta_y\Psi_y|R_1=r_1,R_2=r_2}(\delta_y, \psi_y) = f_{\Delta_x\Psi_x}(\delta_x, \psi_x)|J| \tag{16}$$

where the determinant of Jacobian matrix J is determined by Eqs. (6) and (7), and given by

$$|J| = \frac{1}{r_2^2} \tag{17}$$

Substitute Eqs. (15) to (17) into the following joint PDF of Δ_y and Ψ_y

$$f_{\Delta_y\Psi_y}(\delta_y, \psi_y) = \int_{-\infty}^{\infty}\int_{-\infty}^{\infty} f_{\Delta_y\Psi_y R_1 R_2}(\delta_y, \psi_y, r_1, r_2) dr_1 dr_2 \tag{18}$$

one has

$$f_{\Delta_y\Psi_y}(\delta_y, \psi_y) = \int_{-\infty}^{\infty}\int_{-\infty}^{\infty} f_{R_1 R_2}(r_1, r_2) f_{\Delta_x\Psi_x}(\frac{\delta_y - r_1}{r_2}, \frac{\psi_y}{r_2}) \frac{1}{r_2^2} dr_1 dr_2 \tag{19}$$

The peak value of $Y(t)$ at a fractile q $(0 < q < 1)$, Y_q, is determined by the following equation

$$Y_q = \Delta_y - \Psi_y \ln(-\ln q) \tag{20}$$

With the joint PDF of Δ_y and Ψ_y, the CDF of Y_q can be expressed to be

$$F_{Y_q}(y_q) = \iint_{\delta_y - \psi_y \ln(-\ln q) \le y_q} f_{\Delta_y\Psi_y}(\delta_y, \psi_y) d\delta_y d\psi_y \tag{21}$$

Again based on Jacobian transformation, the PDF of Y_q is obtained as

$$f_{Y_q}(y_q) = \int_{-\infty}^{\infty} f_{\Delta_y\Psi_y}(y_q + \psi_y \ln(-\ln q), \psi_y) d\psi_y \tag{22}$$

4.2 *Approach based on Gaussianity inference for peak value*

Although Approach 1 has the satisfactory accuracy, it may not be applicable when the number of samples is limited. Generally, the parameters in the peak value distribution of the parent process, i.e., Δ_y and Ψ_y, can be regarded to be independent identically distributed. It can be proved that Δ_y and Ψ_y will asymptotically approach joint Gaussian distribution, which is provided in Appendix A. Hence, Y_q given in Eq. (20) will follow Gaussian distribution. How to evaluate the mean and variance of Y_q will be discussed hereafter.

To address Approach 2 concisely, the uncertainty of Λ_0 is not considered firstly, i.e., λ_0 can be replaced by $\lambda_{0,m}$ in Eq. (8). The Taylor expansion of $\delta_x(r_3, r_4, \lambda_{0,m})$ at $r_3 = E[R_3]$ and $r_4 = E[R_4]$ is expressed as

$$\delta_x(r_3,r_4,\lambda_{0,m}) = \delta_x^0 + \left(k\frac{\partial}{\partial r_3} + l\frac{\partial}{\partial r_4}\right)\delta_x^0 + \frac{1}{2!}\left(k\frac{\partial}{\partial r_3} + l\frac{\partial}{\partial r_4}\right)^2 \delta_x^0 + \cdots + \frac{1}{n!}\left(k\frac{\partial}{\partial r_3} + l\frac{\partial}{\partial r_4}\right)^n \delta_x^0 + Rn \tag{23}$$

where $k = r_3 - E[R_3]$ and $l = r_4 - E[R_4]$; $\delta_x^0 = \delta_x(E[R_3], E[R_4], \lambda_{0,m})$ and Rn is the remainder item.

The mean and variance of Δ_x can be approximately estimated by setting n = 2 and 1 in Eq. (23), respectively. These two are given by

$$E[\Delta_x] \approx \delta_x^0 + \frac{1}{2}\frac{\partial^2 \delta_x^0}{\partial r_3^2} D[R_3] + \frac{1}{2}\frac{\partial^2 \delta_x^0}{\partial r_4^2} D[R_4] + \frac{\partial^2 \delta_x^0}{\partial r_3 \partial r_4}\mathrm{Cov}(R_3,R_4) \tag{24}$$

$$D[\Delta_x] \approx \left[\frac{\partial \delta_x^0}{\partial r_3}\right]^2 D[R_3] + \left[\frac{\partial \delta_x^0}{\partial r_4}\right]^2 D[R_4] + 2\frac{\partial \delta_x^0}{\partial r_3}\frac{\partial \delta_x^0}{\partial r_4}\mathrm{Cov}(R_3,R_4) \tag{25}$$

where $\mathrm{Cov}(R_3, R_4)$ is the covariance between R_3 and R_4. Similarly, $E[\Psi_x]$ and $D[\Psi_x]$ can be obtained as well.

Let $\eta(r_3,r_4,\lambda_{0,m}) = \delta_x(r_3,r_4,\lambda_{0,m})\psi_x(r_3,r_4,\lambda_{0,m})$. Its mean can be estimated by replacing δ_x^0 with $\eta^0 = \eta^0(E[R_3], E[R_4], \lambda_{0,m})$ in Eq. (24). Hence, the covariance of Δ_x and Ψ_x is obtained as

$$\mathrm{Cov}(\Delta_x,\Psi_x) \approx \eta^0 + \frac{1}{2}\frac{\partial^2 \eta^0}{\partial r_3^2} D[R_3] + \frac{1}{2}\frac{\partial^2 \eta^0}{\partial r_4^2} D[R_4] + \frac{\partial^2 \eta^0}{\partial r_3 \partial r_4}\mathrm{Cov}(R_3,R_4) - E[\Delta_x]E[\Psi_x] \tag{26}$$

where $\frac{\partial^2 \eta^0}{\partial r_3^2} = \frac{\partial^2 \delta_x^0}{\partial r_3^2}\psi_x^0 + 2\frac{\partial \delta_x^0}{\partial r_3}\frac{\partial \psi_x^0}{\partial r_3} + \frac{\partial^2 \psi_x^0}{\partial r_3^2}\delta_x^0$, $\frac{\partial^2 \eta^0}{\partial r_4^2} = \frac{\partial^2 \delta_x^0}{\partial r_4^2}\psi_x^0 + 2\frac{\partial \delta_x^0}{\partial r_4}\frac{\partial \psi_x^0}{\partial r_4} + \frac{\partial^2 \psi_x^0}{\partial r_4^2}\delta_x^0$ and $\frac{\partial^2 \eta^0}{\partial r_3 \partial r_4} = \frac{\partial^2 \delta_x^0}{\partial r_3 \partial r_4}\psi_x^0 + \frac{\partial \delta_x^0}{\partial r_3}\frac{\partial \psi_x^0}{\partial r_4} + \frac{\partial^2 \psi_x^0}{\partial r_3 \partial r_4}\delta_x^0 + \frac{\partial \delta_x^0}{\partial r_4}\frac{\partial \psi_x^0}{\partial r_3}$.

R_1 and R_2 are associated with the parent process while Δ_x and Ψ_x are related to the peak value of the normalized parent process. Hence, the correlation between these two pairs of parameters is usually weak, and R_1 and R_2 can be regarded to be uncorrelated to Δ_x and Ψ_x (This observation will be further discussed in Section 5.1.1 using wind tunnel data). Then, means of Δ_y and Ψ_y can be directly determined from Eqs. (6) and (7) as

$$E[\Delta_y] = E[R_1] + E[R_2]E[\Delta_x] \tag{27}$$

$$E[\Psi_y] = E[R_2]E[\Psi_x] \tag{28}$$

With Taylor expansion, variances and their covariance can be estimated approximately, and are given by

$$D[\Lambda_y] \approx D[R_1] + D[R_2]E[\Lambda_x]^2 + D[\Lambda_x]E[R_2]^2 \tag{29}$$

$$D[\Psi_y] \approx D[R_2]E[\Psi_x]^2 + D[\Psi_x]E[R_2]^2 \tag{30}$$

$$\mathrm{Cov}(\Lambda_y, \Psi_y) \approx E[\Lambda_x]E[\Psi_x]D[R_2] + E[\Psi_x]\mathrm{Cov}(R_1, R_2) + E[R_2]^2\mathrm{Cov}(\Lambda_x, \Psi_x) \tag{31}$$

With the relation in Eq. (20), the mean and variance of Y_q can be evaluated as

$$E[Y_q] = E[\Lambda_y] - \ln(-\ln q)E[\Psi_y] \tag{32}$$

$$D[Y_q] \approx D[\Lambda_y] + \ln^2(-\ln q)D[\Psi_y] - 2\ln(-\ln q)\mathrm{Cov}(\Lambda_y, \Psi_y) \tag{33}$$

Compared with the Approach 1, Approach 2 is quite convenient and efficient, and suitable for fewer samples. Besides, in this approach, only mean, variance and covariance of first four moments of multiple samples are required. These statistics can be estimated from samples, or derived from their PDFs or given directly. Moreover, the procedure to determine the mean and variance of Y_q can be extended to the case where the uncertainty of Λ_0 is further incorporated. However, it shall be noted that if one or few samples of wind load effects deviate from the population a lot due to other factors (e.g., sudden weather change and undetected instrument fault), the moment estimation in Approach 2 would be influenced severely which leads to inaccuracy in evaluations on peak value uncertainties. In contrast, Approach 1 which models the overall probability of the moments can exclude this effect and remain its robustness in uncertainty evaluations of peak values.

5 NUMERICAL EXAMPLES

The multiple wind pressure samples will be used to evaluate the performance of the proposed methods in modeling the uncertainties of the wind pressure coefficient and wind-induced structural response. The wind pressure data were collected from Boundary Layer Wind Tunnel II at the University of Western Ontario (UWO). One 1:50 scale model (FL30) is adopted (without neighbouring houses and under the suburban exposure). Total 474 taps were distributed on the model roof surface, as shown in Figure 4. The pressures from the wind direction of 120° are used in this study. The sampling frequency for the wind tunnel testing is 400 Hz. The sampling time in the model scale is 3 h. If the full-scale wind speed at 10 m height in suburban exposure is assumed to be 31.7 m/s, the velocity scale will be 1:5. Accordingly, the sampling frequency and time under the full scale are 40 Hz and 30 h, respectively. The very long data (30 h) will be divided into 180 segments of 10 min duration in the following analysis. The details can be found in Refs. [6,10]. In the following discussion, the pressure data are adjusted by multiplying "-1" for better illustration.

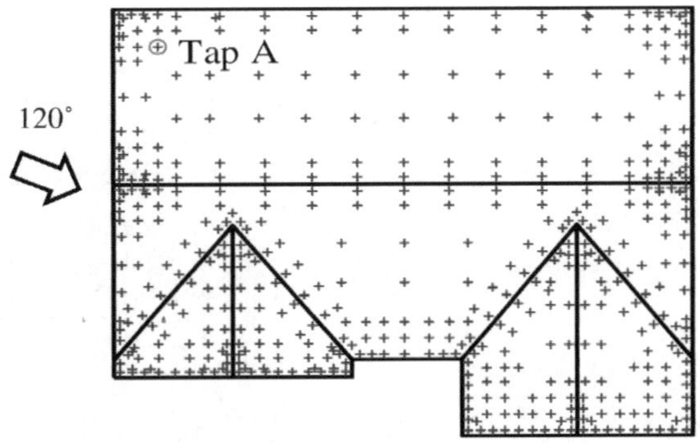

Figure 4. Roof plan and tap arrangement

5.1 *Wind pressure coefficients*

In this section, the correlations among the first four moments and upcrossing number of wind pressure coefficients associated with each tap are examined firstly. Then the effect of the uncertainty of the upcrossing number on the peak value is investigated. Lastly, the performance of two proposed approaches will be evaluated.

5.1.1 Correlations among first four moments and upcrossing number

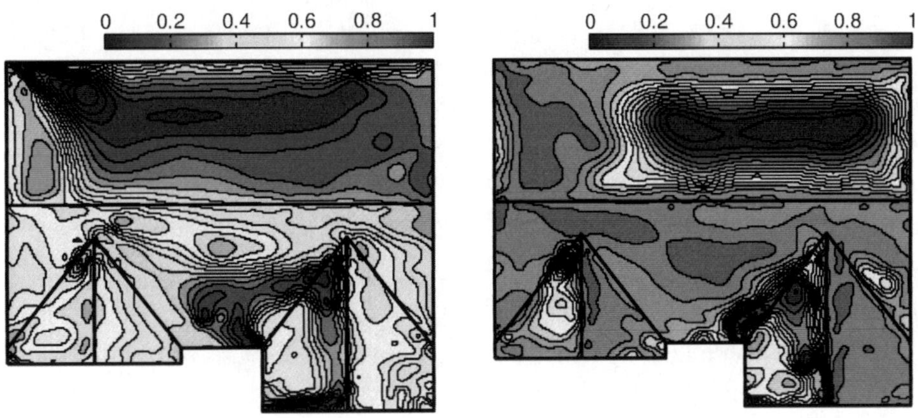

(a) $\rho_{R_1 R_2}$ (b) $\rho_{R_3 R_4}$

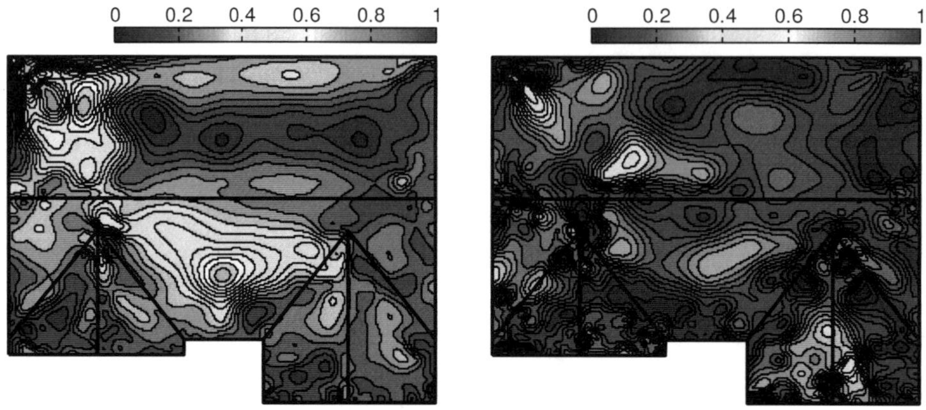

(c) Maxima of $\rho_{R_1R_3}$, $\rho_{R_1R_4}$, $\rho_{R_2R_3}$ and $\rho_{R_2R_4}$ (d) Maxima of $\rho_{R_3\Lambda_0}$ and $\rho_{R_4\Lambda_0}$

Figure 5. Correlation among different pairs of R_1, R_2, R_3, R_4 and Λ_0

The correlation coefficients between two pairs of variables, R_1 and R_2, and R_3 and R_4, for each tap are shown in Figure 5a and b, respectively. It can be seen that the correlation between mean and STD is not very strong, while the correlation between skewness and kurtosis is larger than 0.8 in many areas.

The correlation coefficients among other four pairs of variables, R_1 and R_3, R_1 and R_4, R_2 and R_3, and R_2 and R_4, can be evaluated. The maxima among these coefficients over the roof are shown in Figure 5c. It is observed that the correlation between mean and skewness (or kurtosis) and that between STD and skewness (or kurtosis) are relatively weak, and can be neglected. This observation further validates the inference in Section "Approach based on Gaussianity inference for peak value" that Δ_x and Ψ_x can be regarded to be uncorrelated to R_1 and R_2. This can be explained as follows. Because both of Δ_x and Ψ_x are only functions of R_3 and R_4, and R_1 and R_2 are almost uncorrelated to R_3 and R_4, it can be concluded that Δ_x and Ψ_x are uncorrelated to R_1 and R_2.

Similarly, the correlation coefficients among two pairs of variables, Λ_0 and R_3, and Λ_0 and R_4, can also be computed. The maxima for these coefficients over the roof are plotted in Figure 5d. It can be found that the upcrossing number is almost uncorrelated of skewness and kurtosis.

5.1.2 Effect of upcrossing number uncertainty on peak values

Before the investigation of the effect of the upcrossing number uncertainty on the peak value, the variation characteristics of the upcrossing number is will be discussed. The mean, minimum and maximum of upcrossing number, $\lambda_{0,\mathrm{m}}$, $\lambda_{0,\mathrm{min}}$ and $\lambda_{0,\mathrm{max}}$ are estimated from each set of 180 segments of pressure data. The variation range of upcrossing number can be defined as $\lambda_{0,\mathrm{r}} = \lambda_{0,\mathrm{max}} - \lambda_{0,\mathrm{min}}$. Figure 6 presents the scatter of the mean and range of the upcrossing number for a total of 474 taps. An approximate linear relationship can be found between the mean and range of the upcrossing number. The coefficient of variation (COV) of the upcrossing number for each set of 180 samples can be calculated. Results show that the COV varies from 1.3% to 5.3% for a total of 474 taps. Clearly, the variation of the upcrossing number is relatively weak.

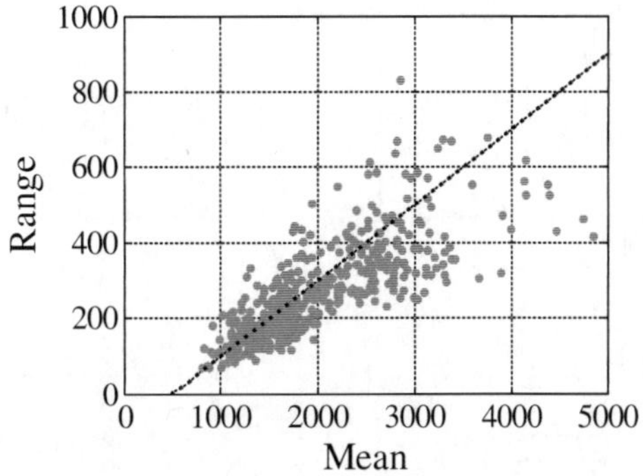

Figure 6. Ranges versus means of upcrossing numbers

To investigate the effect of the upcrossing number uncertainty on the peak value, the peak values of $X(t)$ corresponding $\lambda_{0,\mathrm{m}}$, $\lambda_{0,\min}$ and $\lambda_{0,\max}$ will be evaluated at a fractile. These peak values will be dented as $x_{pk,\mathrm{m}}$, $x_{pk,\min}$ and $x_{pk,\max}$. In the evaluation of peak values, the mean vallues of skewness and kurtosis from 180 samples are used. Results show that the larger of $(x_{pk,\max} - x_{pk,\mathrm{m}})/x_{pk,\mathrm{m}}$ and $(x_{pk,\mathrm{m}} - x_{pk,\min})/x_{pk,\mathrm{m}}$ at fractiles of 57% and 78% for all taps are less than 1.5% . It can be concluded that the uncertainty of upcrossing number has no significant effect on the peak value. Hence, it will not be considered and the mean upcrossing number is adopted hereafter.

5.1.3 Performance evaluation of two analytical approaches

The wind pressure data at Tap A circled in Figure 4 will be used to illustrate the performance of proposed approaches. The mean, STD, skewness and kurtosis are obtained from 180 segments of wind pressures. The histograms of the mean and STD, and the fitted Gaussian PDFs are shown in Figure 7a and b, respectively. It can be seen that the mean and STD almost follow Gaussian distribution. According to the central limit theorem, the mean and the STD will approach Gaussian distribution as the sample size increases. The histograms of the skewness and kurtosis are shown in Figure 7c and d, respectively. Because both of them possess obvious non-Gaussian characteristics, HPM is adopted to model their PDFs, which are also displayed in Figure 7c and d, respectively.

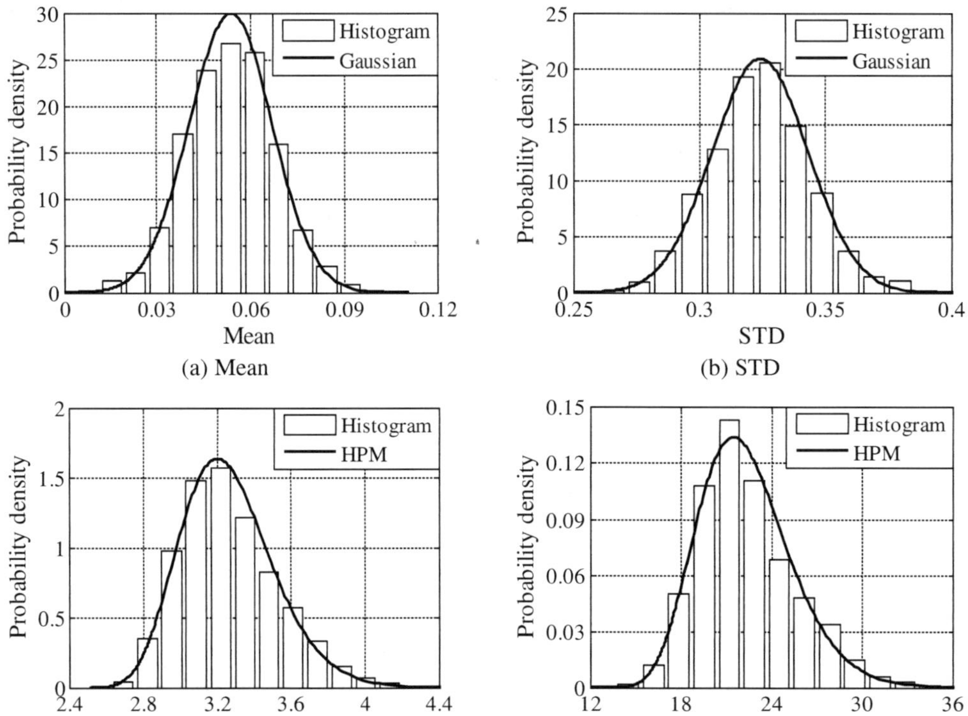

(a) Mean (b) STD

(c)Skewness (d) Kurtosis

Figure 7. PDFs of mean, STD, skewness and kurtosis

To model joint PDFs for mean and STD, and also skewness and kurtosis, the associated correlation coefficients should be directly estimated from 180 segments of pressure data. These correlation coefficients are: $\rho_{R_1R_2} = 0.5487$ and $\rho_{R_3R_4} = 0.9200$. As the PDFs of skewness and kurtosis are modeled by HPM, the corresponding correlation coefficient $\rho_{Z_1Z_2}$ in Eq. (13) can be estimated by a simplified formula, which is [12]

$$\rho_{R_3R_4} = k_3 k_4 (6h_{4,3} h_{4,4} \rho_{Z_1Z_2}^3 + 2h_{3,3} h_{3,4} \rho_{Z_1Z_2}^2 + \rho_{Z_1Z_2}) \tag{34}$$

where k_i, $h_{3,i}$ and $h_{4,i}$ are model coefficients for R_i (i = 3, 4). By this equation, $\rho_{Z_1Z_2} = 0.9213$. Correspondingly, joint PDFs, $f_{R_1R_2}(r_1, r_2)$ and $f_{R_3R_4}(r_3, r_4)$, are obtained, respectively.

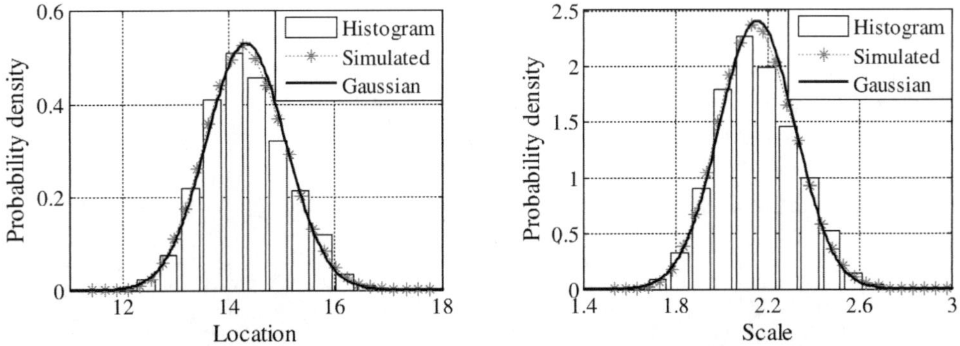

(a) Location parameter (b) Scale parameter

Figure 8. PDFs of location and scale parameters for peak values of normalized process

In following dicussions, the location and scale parameters, and the peak value at a fractile for the normalized or parent wind pressure process can be directly estimated from 180 samples through the method in section 2. These samples will be used to plot the histograms. The histograms of the location and scale parameters associated with the peak value distribution of the normalized process are shown in Figure 8a and b. Following the MCS procedure in section 4.1, simulated samples for these two parameters can be obtained and their probability densities are plotted in Figure 8a and b as well. From the figure, it can be observed that simulated samples match histograms well. Besides, the correlation coefficient between simulated location and scale samples is 0.9932, which meets that estimated from the method in section 2, 0.9955. It is found that these simulated samples approach Gaussian distribution very well. This can be explained as follows. Δ_x and Ψ_x are linear transformation of Δ_y and Ψ_y (The latter pair of variables approach to Gaussian distribution according to Section 4.2). As the variables in the transformation, R_1 and R_2, will converge to fixed values in probability. Hence, Δ_x and Ψ_x will also approach to Gaussian distributions. Based on the fitted Gaussian marginal distributions and the estimated correlation coefficient, the corresponding joint Gaussian PDF of location and scale parameters for peak values of the normalized process can be determined and shown in Figure 9a.

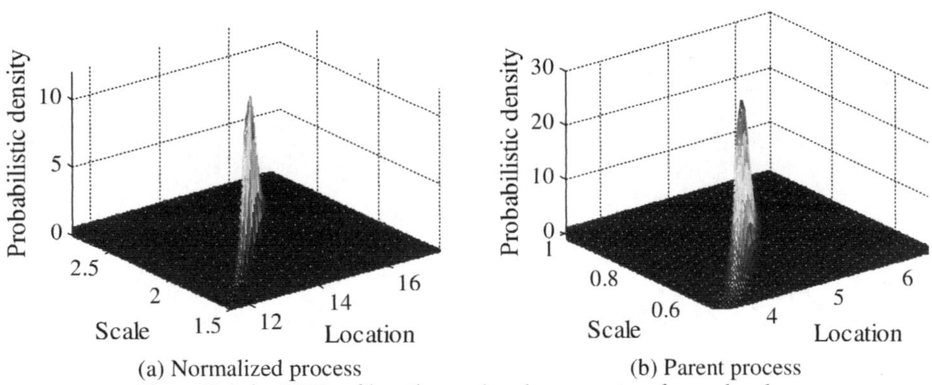

(a) Normalized process (b) Parent process

Figure 9. Joint PDFs of location and scale parameters for peak values

From the joint PDF in Figure 9a, the joint PDF of two parameters of Gumbel distribution for the parent process $Y(t)$ can be evaluated by Eq. (19) and is given in Figure 9b. Marginal

PDFs of both paramters are given in Figure 10. Good fitting between analytical models derived from Approach 1 and corresponding histograms can be found for two parameters. In addition, the Gaussian analytical models for two parameters evaluated by Approach 2 are plotted in Figure 10. Satisfactory fitting exists between the models and their targets as well, which validates the inference derived in section 4.2.

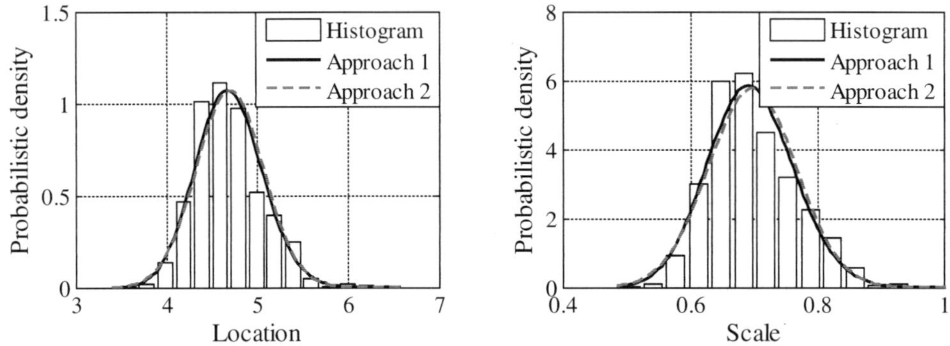

(a) Location parameter (b) Scale parameter

Figure 10. PDFs of location and scale parameters for peak values of parent process

PDFs for peak values at two fractiles of 57% and 78% are evaluated through Eq. (22), which are shown against corresponding histograms in Figure 11a and b. The agreement between them can be observed. Besides, it can be observed that the peak value approximately approaches to Gaussian distribution, which can be validated by the skewness and kurtosis. They are 0.63 (0.64) and 3.48 (3.44) for the peak value at 57% (78%) fractile. From the samples of R_1, R_2, R_3 and R_4, the mean and STD of peak values at two fractiles can be obtained according to Appraoch 2. The corresponding PDFs are plotted in Figure 11a and b as well. Obviously, they can model the uncertainty of the peak value well. The COV of the peak pressure coefficien at Tap A is around 8% for fractiles of 57% and 78%. COVs for all taps are beyond 6.5%, and some of them are beyond 10% and even up to 20%.

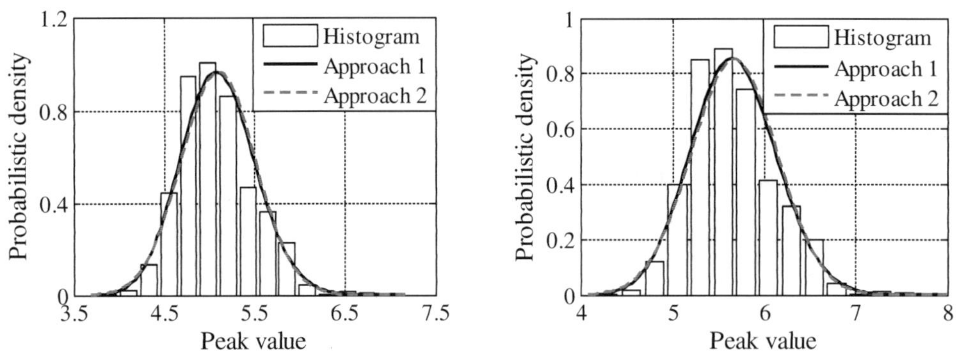

(a) Fractile of 57 (b) Fractile of 78%

Figure 11. PDFs of peak values at fractiles of 57% and 78%

5.2 *Wind-induced response*

Based on FL30 model, the house is designed using the frame structure. Following design codes and specifications for steel structures [28,29], the structual layout, material and sectional dimension for the beam, column and purlin are determined. The finite element (FE) model is build in Sap2000, as plotted in Figure 12. Rigid joints are used to connect columns and the ground. Beams are connectted to inner columns by hinge joints and are rigidly connected to exterior columns. Steel Q345 are used for beams and columns while steel Q235 for purlins. The 30-hour wind pressure (wind speed = 31.7 m/s) time histories are applied to the roof. Time histories of structure responses are obtained via FE analysis and divided into 180 segments as well.

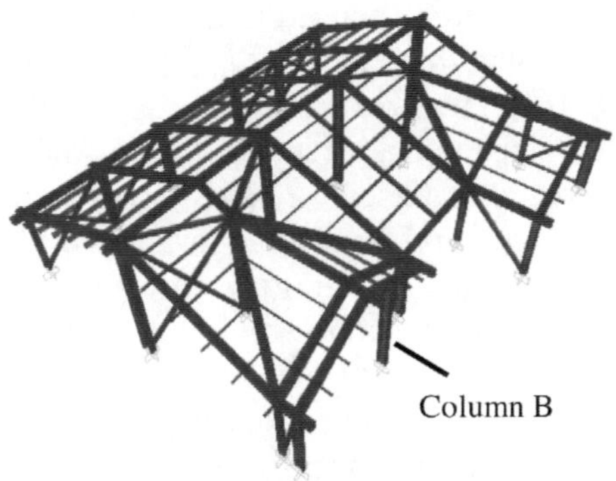

Figure 12. Finite element model of frame house

Because the base bending moment is significant in the structual design, the uncertainty of base bending moments (BBM) in column B will be evaluated for illustration. Similar to wind pressure coefficients, the uncertainty of upcrossing number is neglected. By Approaches 1 and 2, PDFs of peak responses at 57% and 78% fractiles are evaluated and plotted in Figure 13, respectively. It is found that both approaches perform well in uncertainty modeling for peak response. Also that the skewness and kurtosis for the peak value at 57% (78%) fractile are 0.74 (0.73) and 3.74 (3.71), indicating that the distribution of the peak value approximately follows Gaussian distribution. In this study, the COVs of the peak BBM are about 7% for this column under both fractiles.

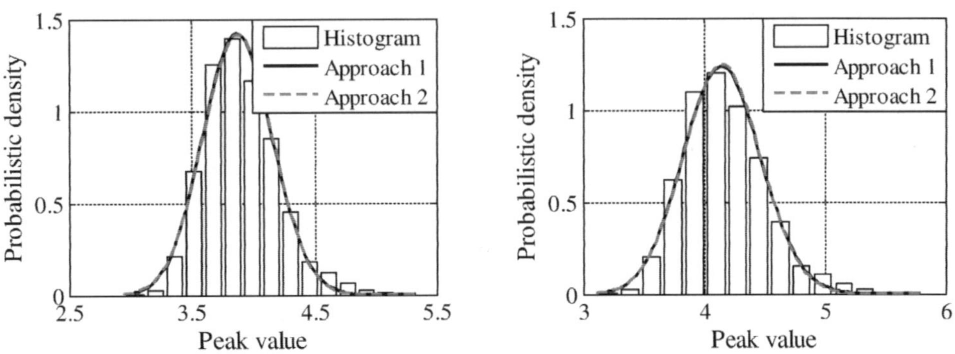

(a) Fractile of 57 (b) Fractile of 78%

Figure 13. PDFs of peak values at 78% fractile for Column B (Unit: kN·m)

6 CONCLUSIONS

The paper addressed how to efficiently model the uncertainty of the peak load effect using very long wind pressure data from a wind tunnel. The major contributions are summarized below.

(1) Based on Hermite polynomial model -based translation process method, the empirical formula to estimate the location and scale parameters of the normalized process was developed for the given skewness, kurtosis and upcrossing number. The evaluation showed this formula was satisfactorily accruate. This formula was very helpful in estimating peak values of wind load effects and modeling their uncertainty.

(2) Two analytical approaches were proposed to model the uncertainty of wind load effects. The first approach is based on probability distribution function where the uncertainties of the first four moments and upcrossing number were modeled by the probability distribution. Although this approach had the satisfactory accuracy, it may not be applicable when samples were limited. The second approach was based on Gaussianity inference for the peak value at a fractile. In this approach, only mean, variance and covariance of first four moments of multiple samples were required, which was suitable for limited samples.

(3) The wind pressure coefficients associated with multiple pressure taps on the roof were used to examine the correlations among first four moments and upcrossing number, and investigate the effect of upcrossing number uncertainty on peak values. Results showed that: i, the correlation between mean and STD was weak while the correlation between skewness and kurtosis was larger than 0.8 in many areas; ii, the correlation between mean and skewness (or kurtosis) and that between STD and skewness (or kurtosis) were relatively weak, and could be neglected; iii, the upcrossing number was almost uncorrelated of skewness and kurtosis; iv, the uncertainty of upcrossing number had no significant effect on the peak value and could be ignored.

(4) The wind pressure coefficients at a typical pressure tap were adopted to illustrate the performance of the proposed approaches. Results showed that, i, the peak value at a fractile approached the Gaussian distribution, which validated the inference derived in section 4.2; ii, both analytical approaches could model the uncertainty of the peak value well. Also it was found that COVs for all taps were beyond 6.5%, and some were beyond 10% and even up to 20%. Furthermore, a frame structure was designed based on the test model. The base bending moment on a column was demonstrated. Results showed that, i, both analytical approaches could model the uncertainty of the peak value well; ii, COV may reach to ahout 7%.

7 REFERENCES

1 Davenport, A. G. (1964). Note on the distribution of the largest value of a random function with application to gust loading. Proc., Instit. Civil Eng., 28, 187-196.

2 Sadek, F., and Simiu, E. (2002). Peak non-Gaussian wind effects for database-assisted low-rise building design. J. Eng. Mech., 2002, 128(5), 530-539.

3 Chen, X., and Huang, G. (2009). Evaluation of peak resultant response for wind-excited tall buildings. Eng. Struct., 31(4), 858-868.

4 Kwon, D.K., and Kareem, A. (2011). Peak factors for non-Gaussian load effects revisited. J. Struct. Eng., 137(12), 1611-1619.

5 Huang, M., Lou, W., Chan, C.M., Lin, N., and Pan, X. (2013). Peak distributions and peak factors of wind-induced pressure processes on tall buildings. J. Eng. Mech., 139(12), 1744-1756.

6 Yang, L., Gurley, K.R., and Prevatt, D.O. (2013). Probabilistic modeling of wind pressure on low-rise buildings. J. Wind Eng. Ind. Aerodyn., 114, 18-26.
7 Ding, J., and Chen, X. (2014). Assessment of methods for extreme value analysis of non-Gaussian wind effects with short-term time history samples. Eng. Struct., 80, 75-88.
8 Huang, G., Luo, Y., Gurley, K. R., and Ding, J. (2016). Revisiting moment-based characterization for wind pressures. J. Wind Eng. Ind. Aerodyn., 151, 158-168.
9 Ma, X., Xu, F., Kareem, A., and Chen, T. (2016). Estimation of Surface Pressure Extremes: Hybrid Data and Simulation-Based Approach. J. Eng. Mech., 04016068.
10 Peng, X., Yang, L., Gavanski, E., Gurley, K., and Prevatt, D. (2014). A comparison of methods to estimate peak wind loads on buildings. J. Wind Eng. Ind. Aerodyn., 126, 11-23.
11 Huang, G., Ji, X., Luo, Y., and Gurley, K.R. (2016) Damage estimation of roof panels considering wind loading correlation. J. Wind Eng. Ind. Aerodyn., 155, 141-148.
12 Luo, Y., and Huang, G. (2016). Characterizing dependence of extreme wind pressures. J. Struct. Eng., 10.1061/(ASCE)ST.1943-541X.0001699, 04016208.
13 Kiureghian, A. D., and Ditlevsen, O. (2009). Aleatory or epistemic? Does it matter? Struct. Safety, 31(2), 105-112.
14 Minciarelli, F., Gioffrè, M., Grigoriu, M., and Simiu, E. (2001). Estimates of extreme wind effects and wind load factors: influence of knowledge uncertainties. Prob. Eng. Mech., 16(4), 331-340.
15 Yang, Q., and Tian, Y. (2015). Model of probability density function of non-Gaussian wind pressure with multiple samples. J. Wind Eng. Ind. Aerodyn., 140, 67-78.
16 Gavanski, E., Gurley, K. R., and Kopp, G. A. (2016). Uncertainties in the Estimation of Local Peak Pressures on Low-Rise Buildings by Using the Gumbel Distribution Fitting Approach. J. Struct. Eng., 142(11).
17 Wu, D., Chen, X., and Yang, Q. (2016). Estimation of wind-induced extreme responses with various mean recurrence intervals considering model uncertainty due to limited data. J. Wind Eng. Ind. Aerodyn., 158, 81-97.
18 Cook, N. J. (1990). The designer's guide to wind loading of building structures, Part 2: static structures. Building Research Establishment Report, Garston, UK.
19 Chen, X., and Huang, G. (2010). "Estimation of Probabilistic Extreme Wind Load Effects: Combination of Aerodynamic and Wind Climate Data." J. Eng. Mech., 136(6), 747-760.
20 Grigoriu, M. (1995). Applied non-Gaussian Processes. Prentice Hall, NJ.
21 Winterstein, S.R. (1988). Nonlinear vibration models for extremes and fatigue. J. Eng. Mech., 114(10), 1772-1790.
22 Ditlevsen, O., Mohr, G., and Hoffmeyer, P. (1996). "Integration of non-Gaussian fields." Prob. Eng. Mech., 11, 15–23.
23 Choi, M., and Sweetman, B. (2010). The Hermite moment model for highly skewed response with application to tension leg platforms. J. Offshore Mech. Arctic Eng., 132(2), 021602.
24 Winterstein, S. R. and MacKenzie, C. A. (2013). Extremes of nonlinear vibration: comparing models based on moments, l-moments, and maximum entropy. J. Offshore Mech. Arctic Eng., 135(2), 021602.
25 Luo, Y., Huang, G., Ding, J., and Gurley, K. R. (2016). Response to "Revisiting moment-based characterization for wind pressures" by G. Huang, Y. Luo, K.R. Gurley and J. Ding. J. Wind Eng. Ind. Aerodyn., 158, 162-163.
26 Huang, G., He, H., Mehta, K.C., and Liu, X. (2015). Data-Based Probabilistic Damage Estimation for Asphalt Shingle Roofing. J. Struct. Eng., 04015065.
27 Holmes, J.D., and Cochran, L.S. (2003). Probability distributions of extreme pressure coefficients. J. Wind Eng. Ind. Aerodyn., 91(7), 893-901.
28 China Association for Engineering Construction Standardization. Technical specification for steel structure of light-weight buildings with gabled frames (CECS102-2002). China Plan Press, 2002.
29 Ministry of Housing and Urban-Rural Development of China. Code for Design of Steel Structures (GB50017-2003). China Plan Press, 2003.

13th Americas Conference on Wind Engineering
Gainesville, Florida, USA
May 21 - 24, 2017

Investigation on wind-induced responses of a 300m high chimney by wind tunnel test

Shuguo Liang[a], Wei YANG[b], Lei WANG[c], Jie Song[d]

[a] *School of Civil Engineering, Wuhan University, Wuhan 430072, China, liangsgwhu@sohu.com*
[b] *School of Civil Engineering, Wuhan University, Wuhan 430072, China, 240682671@qq.com*
[c] *School of Civil Engineering, Henan Polytechnic University, Jiaozuo 454000, China, tumuwanglei@163.com*
[d] *Department of Engineering Mechanics, School of Civil Engineering, Wuhan University, Wuhan, China, jsongaa@connect.ust.hk*

ABSTRACT

In order to acquire the performance of a 300m high chimney which will be built in china under strong wind, a series of wind tunnel tests, including aero-elastic model wind tunnel test, were conducted to investigate the static as well as dynamic responses of the chimney under wind action. The chimney is divided into two parts: the upper part is a cylinder with diameter 22.2m and height 125m; the lower part is a circular truncated cone with top diameter 22.2m, bottom diameter 33.8m and height 175m.

Firstly, Reynolds number effects of surface wind pressure on the chimney were investigated by multi point pressure measurement wind tunnel test of a rigid model. The mean wind pressure distribution on the chimney in practical wind field was simulated by 1/250 geometry scaled wind tunnel model with 40 pasted vertical rough strips equidistributed on the surface of the model. The rough strip is made of sponge, and through a great deal of tests and optimizations, 4mm and 0.5mm are chosen as the width and thickness of the rough strips to obtain the best simulation results. Thus the simulated wind pressure distribution on the wind tunnel model with vertical rough strips is in good agreement with that from the on-site measurements of circular towers.

Secondly, an aero-elastic model of the chimney was designed and produced to simulate the geometry as well as dynam characteristics. The aero-elastic model is a continuous elastomer, and is made of Devcon. The geometry scale, frequency scale and wind speed scale of the model are 1/250, 100 and 1/2.5, respectively. The first mode natural frequency of the model is 25.4 Hz corresponding to 0.251 Hz, which is that of the practical chimney. The damping ratios of the first mode for both the aero-elastic model and practical chimney are 0.02. 40 vertical rough strips as mentioned above were stuck on the surface of the model to simulate the Reynolds number effects of surface wind pressure in actual wind field of the chimney.

Finally, the aero-elastic model wind tunnel test was carried out to investigate wind-induced responses of the chimney under various wind speeds. Both along-wind and across-wind displacements as well as accelerations of the model were measured under wind speeds from 2m/s to 18m/s. The wind tunnel data indicate that the vortex-induced resonance critical wind speed for the first mode of the aero-elastic model is 11.5m/s. Around this wind speed, across-wind dynamic responses of the model appear a sharp peak, which is much bigger than along-wind dynamic responses. Obviously, along-wind and across-wind responses of the model reveal distinctive characters from each other according to wind tunnel data, which are helpful for the wind-resistant design of the chimney.

KEYWORDS

Chimney; wind-induced responses; wind tunnel test; Reynolds number; aero-elastic model

13th Americas Conference on Wind Engineering
Gainesville, Florida, USA
May 21 - 24, 2017

Wind hazard resilient construction mitigation decision-making using a multi-objective optimization algorithm

Fatemeh Orooji[a], Carol Friedland[b]

[a] *Western Kentucky University, Bowling Green, KY, fatemeh.orooji@wku.edu*
[b] *Louisiana State University, Baton Rouge, LA, friedland@lsu.edu*

ABSTRACT

As a result of the increasing trends in wind frequency, intensity, and losses in the United States over the past 50 years, studies have focused on best practices and code-plus practices, and a variety of residential wind hazard mitigation methods have been suggested; however, implementing mitigation requires an initial investment which varies from several hundred to tens of thousands of dollars. Additionally, and more importantly, the economic benefit of mitigation (e.g. avoided loss) may not provide a positive return on investment. The optimal strategy of resilient residential construction should consider multiple objectives such as avoided financial losses and cost of mitigation considering the hazard spatial domain, building configuration, and economic aspects. This study develops a novel method to enhance the mitigation decision-making process. This multi-objective optimization algorithm is developed to maximize wind hazard loss reduction of residential buildings by implementing a variety of mitigation methods that can be customized based on location, years of interest, and building configuration. These conclusions simplify the mitigation decision-making process, ultimately enhancing reliability and reducing risk. To demonstrate the model, a case study of a typical wood frame, one-story, single family house is considered.

KEYWORDS

Multi-objective optimization, Wind resilient construction, Mitigation, Decision-making

The 13th Americas Conference on Wind Engineering (13ACWE)
Gainesville, Florida USA, May 21-24, 2017

13th Americas Conference on WIND ENGINEERING

A Comparative Study of Different Tools for Wind Farm AEP Assessment

Adaiana F. Gomes da Silva[a], Edson L. Zaparoli[b], Cláudia R. Andrade[c]

[a] *Technological Institute of Aeronautics (ITA), Mechanical Engineering Division, São José dos Campos, Brazil, adaiana1@yahoo.com.br*
[b] *Technological Institute of Aeronautics (ITA), Mechanical Engineering Division, São José dos Campos, Brazil, elzaparoli@gmail.com*
[c] *Technological Institute of Aeronautics (ITA), Aeronautical Engineering Division, São José dos Campos, Brazil, claudia@ita.br*

ABSTRACT: In the context of high-resolution local flow characterization for wind energy studies, a variety of microscale numerical approaches is available nowadays. They differ on accuracy according to the application (terrain complexity and hypotheses adopted) and on computational effort. In this study, a mass-conserving and a CFD wind flow models were compared using OpenWind and WindSim tools. The flow over a perfectly idealized flat terrain with uniform surface roughness was simulated with equivalent setup in both models to assure a free tendency analysis. The target is to evaluate the implementation of turbine power curve and a wake model in these two different computational tools, avoiding terrain effects. In this flat terrain, both approaches provided the same (uniform) speed-up field. Wind turbines were distributed over the domain in different arrays and spacing layouts. The power curve implementation, analyzed through gross AEP estimates, showed similar results. A small sensitivity has been verified for net AEP regarding the same wake model implementation (maximum 2.7%), but it was considered negligible. Hence, the techniques may be considered equivalent in terrain-independent cases.

KEYWORDS: Wind energy, wind flow models, CFD, mass-conserving, Annual Energy Production

1 INTRODUCTION

Face to the demand of renewable sources for electric power supply, the wind energy sector has grown enormously worldwide in the past few years, as well as the researches about this topic.

Wind mapping is of utmost importance in several wind energy and wind engineering applications. Wind farms should be installed in high wind potential sites to assure the project profitability. Due to the irregular availability of the wind resources, an accurate comprehension about its distribution over the area of interest is crucial to improve the energy yield prediction. Hence, the motivation of the present study came from the need of improving tools for wind resource assessment.

Currently, numerical modeling of atmospheric flow is the main tool adopted for local flow distribution studies. These mathematical models are designed by integration of appropriate equations with adequate initial and boundary conditions. Limited computer capacity requires that various assumptions are made due to the complex non-linear nature of the equations, that can be simplified according to the problem to be represented, and there is often a compromise between time and computation cost and between physics and geometry complexity [1]. In many cases, steady-state treatments are feasible for solving the time average flow and estimating turbulence statistics. Theoretically, this is valid when the region of the terrain under investigation is small enough so that the time required for an air particle to pass through the region is much smaller than the time over which meteorological synoptic scale phenomena evolve [2]. Furthermore, steady-state solutions are adopted as a good estimate of speed-up average for a long-term velocity conversion.

Depending on the study area extension, on the hypotheses adopted and on which mathematical approach is applied, the stages of wind resources estimation and micrositing can be very time consuming. This is a challenge face to the demand of projects and the huge prosperity of the wind energy sector.

Still, a growing number of wind turbines have been installed in one wind farm to increase the energy yield of the region. Consequently, wake models are necessary to account for the interactions between turbines in operation, caused by this clustered arrangement.

The present study aims to evaluate some current tools for high resolution local flow characterization. Two different microscale numerical approaches are compared. Comparisons of different atmospheric boundary layer (ABL) modeling have been the object of study in relevant previous works, such as [3-8]. Additionally, it is interesting to emphasize Bechmann *et al.* [9], that promoted a large blind comparison of microscale models through the notorious Bolund experiment.

This paper presents the first part of a more general project that relies on investigating how different numerical methods are dependent on the relief. For that, a mass-conserving and a Computational Fluid Dynamics (CFD) wind flow models will be compared by using the softwares OpenWind (WindMap) [10] and WindSim [11]. In this first part, the flow over a perfectly idealized flat terrain with uniform surface roughness was simulated to compare the terrain-independent genuine numerical methods. In this flat terrain, these two approaches provide the same speed-up field. The target of this part of the work is to compare the influence of wake model implementation in each computational tool, as well as to verify their methods of turbine power curve implementation for annual energy production (AEP) computation. The wake effect has been taken into account on the AEP estimations employing a standard analytical wake model, the "Jensen model". Equivalent parameters were used in both models to assure a free tendency analysis, obtaining a validation of their performance to proceed the second part of the ongoing study, where these two tools have been employed to evaluate AEP in a particular complex terrain.

2 WIND FLOW MODELS

In wind industry, microscale models have been increasingly employed to solve steady-state flow in order to obtain 3-dimensional wind resource maps [12-16]. In this context, they are used in projects of new wind farms for micrositing studies, since they can provide high resolution wind maps and describe the wind pattern over the simulation area [7, 13, 14, 17, 18]. Different numerical approaches have been applied, but the most usual are linearized models (including mass-consistent) and CFD models with different ways of turbulence modeling, like RANS (Reynolds Average Navier-Stokes) and LES (Large Eddy Simulation – more restrict application). Their efforts are concentrated on solving at least some governing physical equations of motion, safe the complexity level of each one.

Prospathopoulos *et al.* [2] summarize that steady-state models may be divided into kinematic and dynamic models. Kinematic or mass-consistent models solve only the continuity equation assuming inviscid fluid flow. They are very simple models and require low computational effort, but they cannot take into account the effect of turbulence or predict separation, so they are mainly used in large area applications to derive wind atlases. While dynamic models solve the continuity, the momentum and (possibly) the energy conservation equations and their degree of complexity varies according to the manipulation of the terrain data, the simplifications in the equations, the turbulence model, etc.

Due to its simplicity, mass-conserving models are generally applied for gentle terrain, being usually as capable as others to reproduce neutral ABL wind fields, despite requiring much less computational cost [12, 19]. On the other hand, dynamic models like CFD RANS solvers have shown to be suitable for high resolution mapping of complex terrains, as related in several studies [3, 9, 13, 15] due to the consideration of non-linearity. Nonetheless, Berge *et al.* [4], Neophytou *et al.* [5] and Castellani *et al.* [8], for example, showed that, although in most cases CFD presents better results than other more simplified models, this is not in the totality.

2.1 *Mass-conserving models*

These models solve just one of the physical equations of motion to generate 3D terrain-dependent wind flow: the continuity equation. They do not enclose dynamic equations and the approach is simply to reach a divergence-free velocity field starting from wind observations.

The numerical procedure is based on an initial solution obtained by interpolating/extrapolating wind observations taken in one or more points inside the simulation domain to all the grid points, generating a first guess wind field. In this procedure, the observational data is extrapolated in the vertical direction to the top of the ABL assuming the power law. The horizontal wind speed components at that height are considered to be uniform over the whole domain (geostrophic wind approach), and then they are spread to all other nodes. Next, the friction velocity u^* is defined in each node using the logarithmic wind law. Once the u^* field is known, the wind speed close to the ground can be then calculated (estimated) in all grid points through the power law again. This is the initial wind field, which will be altered according to the interferences caused by the obstacles of the local terrain, giving, in turn, an adjusted wind field, which is subsequently forced to be divergence-free, resulting the mass conservative wind field. [20-23]

The software OpenWind, developed by AWSTruepower, was used in this work in its Basic version as interface to WindMap model, based on NOABL (Numerical Objective Analysis of Boundary Layer), developed in turn by Phillips & Traci [24]. The model was later improved to take into account internal boundary layer growth due to sharp transition in surface roughness [25].

2.2 *CFD models*

CFD consists of a numerical method for solving non-linear partial differential equations (PDE's) that represent a physical flow field. Through a discretization scheme, the continuum solution domain is transformed in a discrete problem with a finite number of nodal points. The PDE's are then integrated over the computational grid and transformed into a system of algebraic equations, that is solved iteratively until a convergent solution is obtained according to pre-established error criteria [26].

CFD solves the RANS equations to simulate the flowfield characteristics (velocity, pressure and temperature). These models generate a basic wind field considering local terrain perturbations and standard inlet boundary conditions. Since they don't take into account the actual local speed values and velocity directions, the CFD results need to be accordingly adjusted to measurements taken in one or more points inside the considered region.

The CFD calculations were performed in this study through the commercial software WindSim. It is a CFD-package based on the more general Phoenics code and was developed in

Norway in the 90's to facilitate the detailed project of wind farms. Considering steady-state, incompressible flow and neutral atmospheric stratification, by default, the RANS equations are discretized and integrated on the computational grid through a finite-volume method approach.

The mean flow is solved through the physical principles of conservation of mass and momentum and the turbulence is parameterized through some closure model, like the standard k-ε, the k-ω model of Wilcox, among others. The lateral inlet boundary condition in WindSim is defined by the logarithmic vertical wind profile, valid from the surface roughness length to the boundary layer height. Above this, the horizontal wind speed is constant, following the geostrophic wind approximation.

3 MODEL SETUP

3.1 *Basic flow calculations*

The input of a terrain description as a digital terrain model (DTM) is necessary to initialize microscale models used for wind energy applications. A wind climatology, that is a statistical description of the wind speed and direction observed in some point inside the domain area, is necessary before starting mass-conserving models (as explained in the previous section), but, for CFD, it is used only to adjust the simulated wind field as a post-processing.

The input used for the simulations performed in both CFD and mass-conserving models were the same. The area of the DTM has covered 20 x 20 km in horizontal dimensions. The terrain is hypothetic, perfectly flat, with an also idealized uniform roughness length, herein defined equal to 0.04 m, typical of cropland/pasture areas.

One wind climatology was positioned in the central point of the domain. The dataset was physically obtained from a meteorological mast ("met mast") at 80 m above ground level (agl) during 3.5 years. This sampling data showed very predominant Easterly winds (see Fig. 1). The geographical identity of this site is indifferent, since the wind data was employed generically only to force the models on adjusting the simulated wind fields to calculate the AEP estimation.

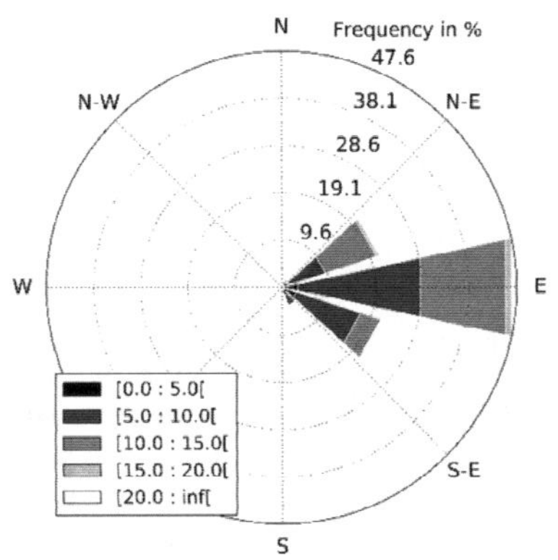

Figure 1. Wind rose of the used wind climatology.

The numerical and physical parameters that can be set by the user in these tools were kept equal. This is necessary to compare the genuine numerical codes. It encloses:

– Boundary layer height: 200 m;
– Number of wind directions: 12 sectors;
– Air density: 1.225 kg/m^3.

Some additional configurations are to be set at the CFD tool, specifically:

– Geostrophic wind speed: 10 m/s;
– Boundary condition at top: no-friction wall;
– Turbulence model: standard k-ε.

3.2 AEP calculations

For AEP estimation, additional inputs of the turbines characteristics are necessary: power and thrust coefficient curves, hub height and rotor diameter.

The following parameters related to AEP calculation and wake effect modeling were used in both models:

– Turbine model: Vestas V-90 [hub height: 80 m; rotor diameter (D): 90 m];
– Wake model: "Jensen model" (Park Model in OpenWind and Model 1 in WindSim);
– Wake influence range: 50D;
– Subsectors for wake calculation: 72 (total).

The objective of a wake model is to assess the energy deficit in the downstream wind field, and not to describe the wind field itself. Jensen model was developed by Jensen [27] and Katíc *et al.* [28]. This model consists of a simple mass conserving application, being the velocity deficit dependent only on the distance behind the rotor, in a linear rate determined by the decay factor.

To test the influence of wake model implementation in each computational tool, different positioning of wind turbines inside the simulation domain have been tested. Configurations with 1, 2 and 4 turbines were used equally spaced around the met mast, being spaced apart one from each other as multiple of rotor diameters: 1.5D, 2D, 3D and 500 m. When using 2 turbines, they were positioned in line with the windflow and also in crosswind (diagonal 45°) array in relation to the predominant wind direction from East (see Figs 1 and 2). The layout with 4 turbines was in square array. When using only 1 turbine, it was located at the position of the climatology, although its position is indifferent, as will be further discussed.

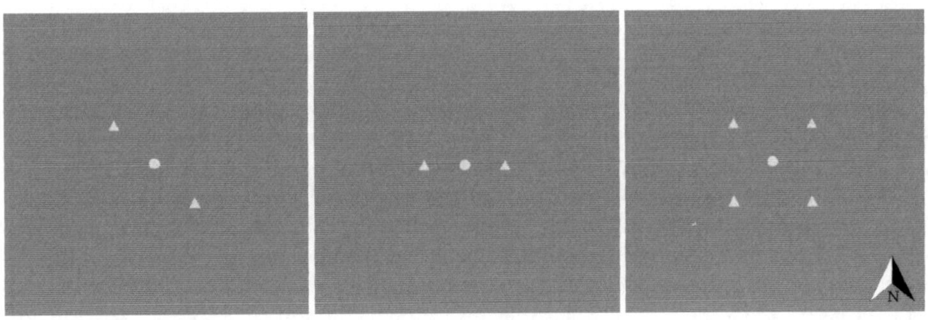

Figure 2. Configurations of objects distribution on the simulation domain, where ● represents the met mast and ▲ represent the wind turbines.

4 RESULTS

The first evaluation to be considered is the wind field generated by each model approach. In this stage, any turbine was included in the wind farm, but only the central met mast provided wind data in order to adjust the basic simulated wind field to the observed wind speed values and generate the so-called wind resource map.

Figure 3 shows the wind resource maps generated by both models. As could be expected for a perfectly flat terrain, both wind fields have uniform wind speed along the whole domain with no speed-ups. This is coherent since any orographic obstacle or roughness changes are present. The constant value around 8.58 m/s agrees with the mean wind speed obtained from the wind dataset. The small difference (from 8.583 to 8.585 m/s) represents 0.07% in terms of energy (the wind power density is proportional to the third power of the wind speed). For terrain-independent cases, CFD and mass-conserving techniques were considered equivalent for simulating the basic flow.

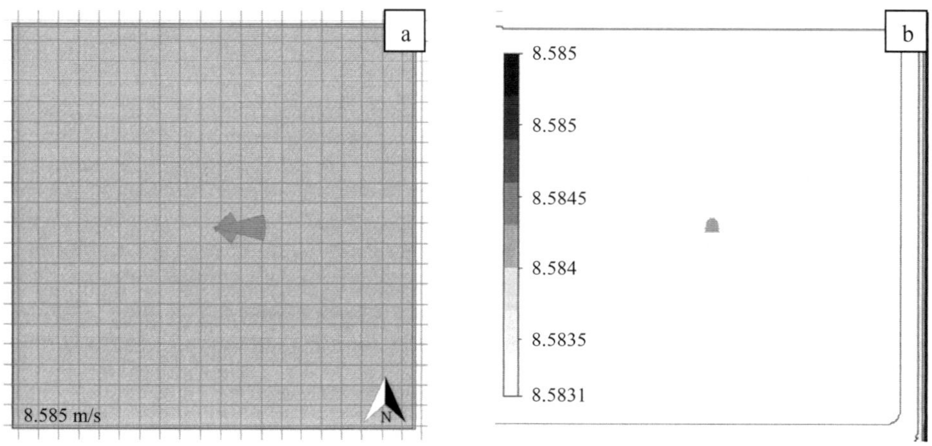

Figure 3. Wind resource maps at 80 m agl: mass-conserving (a) and CFD (b).

In sequence, wind turbines were positioned over the simulation domain as described in the last section to evaluate the AEP assessment.

Table 1 presents the values of gross AEP for all configurations of wind turbines distribution over the domain, as well as the comparison between the estimations obtained with the two wind flow models herein employed. It allows to compare the implementation of AEP calculation in the tools. % diff considers the absolute difference between the models relative to WindSim result. The differences in AEP estimations (0.02 ~ 0.05%) are smaller than the difference due to the wind field properly (0.07%), what means that the implementation of AEP computation in both tools provide the same results. Moreover, it was confirmed that the production is independent of the spacing, since interactions between turbines (wake effects) are not considered yet. And the total AEP is directly proportional to the number of turbines. The configuration with only one turbine is wake-independent and, since the wind field is totally uniform, the position of this turbine is indifferent, resulting always the same AEP.

Table 1. Gross AEP estimated for different layouts of turbines (T) in GWh/year. WS refers to WindSim and OW to OpenWind model.

	Turbine spacing: 500 m			Turbine spacing: 3 D = 270 m		
	2 T in line	2 T diagonal	4 T square	2 T in line	2 T diagonal	4 T square
WindSim	18.781	18.781	37.562	18.781	18.781	37.562
OpenWind	18.790	18.790	37.579	18.790	18.790	37.579
OW - WS	0.009	0.009	0.017	0.009	0.009	0.017
% diff	0.05	0.05	0.05	0.05	0.05	0.05
	Turbine spacing: 2 D = 180 m			Turbine spacing: 1.5 D = 135 m		
	2 T in line	2 T diagonal	4 T square	2 T in line	2 T diagonal	4 T square
WindSim	18.781	18.781	37.562	18.781	18.781	37.562
OpenWind	18.790	18.790	37.579	18.790	18.790	37.579
OW - WS	0.009	0.009	0.017	0.009	0.009	0.017
% diff	0.05	0.05	0.05	0.05	0.05	0.05
	1 T:	WS	OW	% diff		
		9.393	9.395	0.02		

The following analyses are addressed only to the wake model implementation, since the pre-processed wind fields and gross AEP were certified to be equivalent.

The net AEP is presented in Table 2. It discounts from the gross AEP the losses due to the wake effects. All tables are organized similarly, showing the different wind turbines positioning and spacing.

Comparisons for all turbine arrays show that the differences found between the estimations obtained from the two models were small, not higher than 2.7%.

Table 2. Net AEP [GWh/year].

	Turbine spacing: 500 m			Turbine spacing: 3 D = 270 m		
	2 T in line	2 T diagonal	4 T square	2 T in line	2 T diagonal	4 T square
WindSim	18.334	18.664	36.456	17.426	18.364	34.483
OpenWind	18.189	18.666	36.194	17.547	18.445	34.659
OW - WS	-0.145	0.002	-0.262	0.121	0.081	0.176
% diff	0.79	0.01	0.72	0.69	0.44	0.51
	Turbine spacing: 2 D = 180 m			Turbine spacing: 1.5 D = 135 m		
	2 T in line	2 T diagonal	4 T square	2 T in line	2 T diagonal	4 T square
WindSim	16.637	18.123	32.626	16.046	18.040	31.513
OpenWind	16.979	18.232	33.261	16.481	17.962	32.178
OW - WS	0.342	0.109	0.635	0.435	-0.078	0.665
% diff	2.06	0.60	1.95	2.71	0.43	2.11

The smallest differences were observed with the arrangement with 2 turbines in a cross-wind array, not more than 0.6% in net AEP. In this layout, the wake effect between the turbines was very small due to the fact that the most predominant wind direction is from East. Although the flow is disturbed when passing through the upstream turbine rotor, the kinetic energy deficit is weakly felt by the other turbine located more to the North. On the other hand, the highest differences are seen when there are turbines arranged in line along the main wind direction (both with 2 or 4 turbines). The maximum difference of 2.7% between the models has occurred in these cases. In these layouts, the disturbed flow reaches the downstream turbine rotors with a more significant kinetic energy deficit, since they are right behind the shadow of the wakes, affecting their production.

One can also state that the smaller the turbine spacing the larger the wake losses. Since the velocity deficit is inversely proportional to the downstream distance from the rotor, the closest the wind turbines are, the less is the recovery of the kinetic energy lost by the flow (absorbed by the upstream turbine), when reaching the next turbine. The maximum relative difference verified in the wake effects evaluation between the different implementations was 34.45% and the minimum was 5.16%.

From the tables, one can observe still that the smaller the turbine spacing the larger the difference between the estimations from the models. Following the just mentioned theory, this means that the net AEP results were sensible to the wake model implementation in each approach. The wake losses calculated by WindSim were higher than those calculated by OpenWind, except for the largest spacing (500 m). In absolute values, the maximum difference between the models was found in the order of 6.5E-1 GWh/year, what represents approximately 2% in net AEP. This value is higher than 0.07% that was due to the small differences on the wind field values. Then, it may be concluded that there is a difference in the wake model implementation.

5 CONCLUSIONS

In this study, AEP assessments from two different tools for wind resources evaluation were compared to analyze the power curve and a wake model implementation. To avoid the effect of different wind flow modeling in these computational tools, this study was carried for different layouts of wind turbines distributed over a perfectly flat terrain.

The wind fields simulated by both approaches were confirmed to be uniform and similar. When the wake effects were disregarded, the differences between the gross AEP implementation in the tools were negligible (0.05%), smaller than 0.07% due to the small differences on the wind field values. However, when the wake effects were considered, the comparison of the net AEP results showed sensitivity face to the same wake model implementation, with differences up to 2.7% between the results of these two computational tools. The wake losses calculated by WindSim were higher than those calculated by OpenWind, except for the largest spacing (500 m). As the wake kinetic energy deficit is function of the downstream distance from the upstream turbine rotor, the smaller turbines spacing resulted in larger differences between the tools estimations. Considering the predominance of winds from East in the wind climatology used, the highest wake effects (and, consequently, the highest differences between tools) were found in the layouts with in line array of turbines. Further studies in complex terrain are being developed using the present study as validation case.

6 REFERENCES

1 S.J. Wakes, T. Maegli, K.J. Dickinson and M.J. Hilton, Numerical modelling of wind flow over a complex topography, Environ. Modelling & Softw., 25 (2010) 237-247.

2 J.M. Prospathopoulos, E.S. Politis and P.K. Chaviaropoulos, Application of a 3D RANS solver on the complex hill of Bolund and assessment of the wind flow predictions, J. Wind Eng. Ind. Aerodyn. 107-108 (2012) 149-159

3 G.T. Bitsuamlak, T. Stathopoulos, F. Asce and C. Bédard, Numerical evaluation of wind flow over complex terrain: Review, J. Aerosp. Eng., 17 (2004) 135–145.

4 E. Berge, A.R. Gravdahl, J. Schelling, L. Tallhaug and O. Undheim, Wind in complex terrain: a comparison of WAsP and two CFD-models, Proc. European Wind Energy Conference, Athens, Greece, 2006.

5 M. Neophytou, A. Gowardhan and M. Brown, An inter-comparison of three urban wind models using Oklahoma City Joint Urban 2003 wind field measurements, J. Wind Eng. Ind. Aerodyn., 99 (2011) 357-368.

6 P. Beaucage and M.C. Brower, Wind Flow Modeling Performance: Do More Sophisticated Models Produce More Accurate Wind Resource Estimates?, Tech. report, Albany, NY: AWS Truepower, 2012.

7 T. Timander, Reliability of Wind Farm Design Tools in Complex Terrain: A Comparative Study of Commercial Software, Msc. dissertation, Stockholm: KTH School Ind. Eng. Management, 2012.

8 F. Castellani, D. Astolfi, M. Burlando and L. Terzi, Numerical modelling for wind farm operational assessment in complex terrain, J. Wind Eng. Ind. Aerodyn., 147 (2015) 320-329.

9 A. Bechmann, N.N. Sørensen, J. Berg, J. Mann and P.-E. Réthoré, The Bolund Experiment, Part II: blind comparison of microscale flow models, Bound.-Layer Meteorol., 141 (2011) 245–271.

10 AWSTruepower, Openwind: Theoretical Basis and Validation, Tech. report, Albany, NY: AWSTruepower, 2010

11 C. Meissner, WindSim Getting Started: WindSim 7, Tech. report, Tønsberg, Norway: WindSim AS, 2015.

12 K. W. Ayotte, Computational modelling for wind energy assessment, J. Wind Eng. Ind. Aerodyn., 96 (2008) 1571–1590.

13 A. Rasouli and H. Hangan, Microscale Computational Fluid Dynamics simulation for wind mapping over complex topographic terrains, ASME J. Sol. Energy Eng., 135 (2013) 1-18.

14 A. Albani and M.Z. Ibrahim, An assessment of wind energy potential for selected sites in Malaysia using feed-in Tariff Criteria, Wind Eng., 38 (2014) 249–259.

15 B. Blocken, A. Hout, J. Dekker and O. Weiler, CFD simulation of wind flow over natural complex terrain: Case study with validation by field measurements for Ria de Ferrol, Galicia, Spain, J. Wind Eng. Ind. Aerodyn., 147 (2015) 43-57.

16 T. Simões and A. Estanqueiro, A new methodology for urban wind resource assessment, Renew. Energy, 89 (2016) 598-605.

17 J.S. Rodrigo, State-of-the-art of Wind Resource Assessment, Tech. report, Sarriguren: WAUDIT, CENER. Available at: <http://www.waudit-itn.eu/download.php?id=103&parent=79\npapers2://publication/uuid/0E9E3208-4E94-47AA-9C18-36477E8B2FE2>, 2010.

18 S. Jafari, N. Chokani and R.S. Abhari, An immersed boundary method for simulation of wind flow over complex terrain, ASME J. Sol. Energy Eng., 134 (2012) 1-12.

19 N. Gasset, M. Landry and Y. Gagnon, A comparison of wind flow models for wind resource assessment in wind energy applications, Energies, 5 (2012) 4288-4322.

20 C.A. Sherman, A mass-consistent model for wind fields over complex terrain, J. Applied Meteorol., 17 (1978) 312-319.

21 J. Cataldo and M. Zeballos, Roughness terrain consideration in a wind interpolation numerical model, Proc. 11st Americas Conference on Wind Engineering, San Juan, Puerto Rico, 2009.

22 D. Eriksson, A Mass Conserving Wind Model Evaluation with Finite Element, MSc. dissertation, Umeå: Umeå University, 2013.

23 J. Bengtsson, Turbulence Wind Flow Modeling in Complex Terrain, MSc. dissertation, Gothenburg: Chalmers University of Technology, 2015.

24 G. Phillips G and R. Traci, A Preliminary User's Guide for the NOABL Objective Analysis Code, Tech. report, La Jolla, CA: U.S. Department of Energy, 1979.

25 M. Brower (1999) Validation of the Windmap Program and development of MesoMap. Proc. American Wind Energy Association WindPower Conference, Washington, DC, 1999.

26 H.K.E. Versteeg and W. Malalasekera, An Introduction to Computational Fluid Dynamics: the Finite Volume Method, Longman Scientific & Technical, New York, 1995.

27 N. Jensen, A Note on Wind Generator Interaction, Tech. report, Roskilde, Denmark: Risø National Laboratory, 1983.

28 I. Katíc, J. Højstrup and N.O. Jensen, A simple model for cluster efficiency, Proc. European Wind Energy Conference, Rome, Italy, 1986.

13th Americas Conference on Wind Engineering
Gainesville, Florida, USA
May 21 - 24, 2017

Remote Sensing Image Feature Extraction for Automated Wind Damage Assessment

J. Arn Womble[a], Vitor Patriani-Cardoso[a],

[a]*West Texas A&M University, Canyon, Texas, USA,* awomble@wtamu.edu
[a]*West Texas A&M University, Canyon, Texas, USA,* vitorpc93@hotmail.com

ABSTRACT

Remote sensing platforms allow for the rapid collection and preservation of damage data across affected areas. Remote sensing data furthermore provide a basis for the development of automated damage assessments. Automated wind storm damage assessments are needed to assist with rapid allocation of disaster relief resources and for insurance risk modeling and improvement of construction practices. At the present time, damage assessments based on changes in pre- and post-storm remote-sensing images must be made using visual assessments, which are labor-intensive and time-consuming; it is therefore necessary to further automate the process. The accurate automated extraction of pertinent "features" (such as building roofs) from remote-sensing images is a critical limiting factor in making rapid, automated damage assessments. While humans can visually detect changes and assess the associated damage with high accuracy by comparing pre- and post-storm images, it is desired to develop computer algorithms that can also make such damage assessments with high accuracy. This presentation examines issues in a case study of feature extraction from high-resolution before-and-after aerial imagery from the 2013 Moore, OK tornado.

KEYWORDS

Remote sensing, change detection, feature extraction, damage assessment

13th Americas Conference on Wind Engineering
Gainesville, Florida, USA
May 21 - 24, 2017

13th Americas Conference on
WIND ENGINEERING

Photogrammetry for Rapid and Low-Cost Wind Damage Documentation

J. Arn Womble[a], Kenneth R. Leitch[b], Vitor Patriani-Cardoso[c]

[a]*West Texas A&M University, Canyon, Texas, USA, awomble@wtamu.edu*
[b]*West Texas A&M University, Canyon, Texas, USA, kleitch@wtamu.edu*
[c]*West Texas A&M University, Canyon, Texas, USA, vitorpc93@hotmail.com*

ABSTRACT

The ability to produce three-dimensional reality-capture models of wind damaged structures for detailed analysis is enormously helpful for facilitating the study of wind-structure interaction. In the wake of a major wind storm, detailed forensic studies of multiple structures can be prohibitive due to time, access, and safety issues – particularly for investigation teams that must rapidly collect data throughout affected areas before critical evidence is disturbed, demolished, or cleaned up. The rapid collection of detailed damage information for later study using virtual reality models is therefore imperative for thorough analysis of wind damage to structures. Photogrammetric modeling using handheld digital cameras, aerial imaging, and UAV-based imaging provides a low-cost and effective means for rapidly acquiring damage data. Digital images can be obtained quickly in the field, and then subsequently processed to produce 3D virtual reality models useful for detailed forensic analysis. The process relies heavily on acquired skills.

A November 2015 tornado outbreak in the Texas Panhandle provided researchers with the opportunity to explore the application of digital photogrammetric modeling for preserving the damage scene and for obtaining measurements of member sizes for the estimation of the wind resistance of pre-engineered structures. This case study addresses the development and use of photogrammetric modeling techniques for a group of pre-engineered metal buildings subject to this intense tornado.

KEYWORDS

Photogrammetry, remote sensing, forensic engineering, reality capture

13th Americas Conference on Wind Engineering
Gainesville, Florida, USA
May 21 - 24, 2017

Performance-Based Design for Better Resilience

Duzgun Agdas[a], Matthew Mason[b]

[a] *Queensland University of Technology, Brisbane, Australia, duzgun.agdas@qut.edu.au*
[b] *University of Queensland, Brisbane, Australia, matthew.mason@uq.edu.au*

ABSTRACT

Australia has been subjected to many repeating large scale disaster over the last 50 years, and traditional approach to managing these disasters have been in the form of response and recovery. More than 95% of the total federal disaster funding in Australia has been spent on responding to disasters, while mitigation and preparedness received little federal support. This system recently came under scrutiny, possibly due to repeating damage patters in recent floods and cyclones occurred in Queensland. A federal report published on disaster funding clearly articulates the need to move away from response and recovery to better mitigation, and redistribution of responsibility and decision-making for improving overall disaster resilience. There is an initiative to decentralize the decision-making process, in an attempt to motivate investment in increased resilience. What appears to be a part of this initiative, is the adaptation of performance based building codes in addition to more traditional, national prescriptive approach to building design. Building codes are universally accepted to be viable soft mitigation measures; however, their efficiency in mitigation disaster impacts has not been clearly proven effective The added complication to this equation is the potential liabilities introduced by design and verification of building performance. This article is an attempt to analyze the efficacy of the performance based building design guidelines from engineering and governance perspectives, and will conclude with discussions on the implications of this approach for increasing overall disaster resilience.

KEYWORDS

Disaster resilience, performance-based design

13th Americas Conference on Wind Engineering
Gainesville, Florida, USA
May 21 - 24, 2017

Quantifying Socio-Economic Impact of a Tornado by Evaluating Population Dislocation as a Resilience Metric at the Community Level

Hassan Masoomi[a], John W. van de Lindt[a], Lori Peek[b], Bruce Ellingwood[a], Hussam Mahmoud[a], Naiyu Wang[c], Amy Cerato[c], and Kathrina Simonen[d]

[a] *Department of Civil and Env. Engineering, Colorado State University, Fort Collins, CO, USA.*
[b] *Department of Sociology, Colorado State University, Fort Collins, CO, USA.*
[c] *Department of Civil and Env. Science, University of Oklahoma, Norman, OK, USA.*
[d] *Department of Architecture, University of Washington, Seattle, WA, USA.*

ABSTRACT

Policymakers, community leaders, and researchers have been motivated to invest more on understanding of tornado-resilient buildings and communities as a result of lethal and destructive tornadoes over the recent years. In addition to direct losses, indirect losses are among the main concerns in the aftermath of tornadoes since in reality a community is made up of highly coupled networks. Any malfunction in a network may result in disruptions to other networks as a result of cascading failures. The direct and indirect losses can result in socioeconomic crises such as population dislocation which, if significant and long-lasting, can further exacerbate indirect losses over time. In this study, a community was coarsely modeled with its physical-socio-economic attributes to study population dislocation as a socioeconomic resilience metric at the community level. In this regard, recovery of affected physical networks (e.g., electric power network, water network, and buildings) in the wake of a tornado was investigated and linked to students, residences, and employees as socio-economic agents in the community. Then, probability of dislocation for each household was assessed based on the probability that these three agents in the household are affected in a time period from the occurrence of the hazard till the full restoration of the community. Finally, the potential population dislocation for the community was assessed by integrating all the households in the community. This study can be implemented as a decision making tool to prioritize mitigation of infrastructure for a community or for new communities being master planned, and demonstrates the importance of physical-socio-economic interactions in resilience studies.

KEYWORDS

Tornado; Population Dislocation; Resilient Community; Social Disruption; Master-planned Community.

13th Americas Conference on Wind Engineering
Gainesville, Florida, USA
May 21 - 24, 2017

Infrastructure Resilience for Hurricane Hazards

Shuoqi Wang[a], Jose Leon Machado[b], Xian He[c], Eun J. Cha[d] and Dorothy Reed[e]

[a] *University of Washington, Seattle, WA, USA, shuoqw@uw.edu*
[b] *University of Washington, Seattle, WA, USA, jlmaleon@uw.edu*
[c] *University of Illinois, Urbana-Champaign, IL, USA, xianhe2@illinois.edu*
[d] *University of Illinois, Urbana-Champaign, IL, USA, ejcha@illinois.edu*
[e] *University of Washington, Seattle, WA, USA, reed@uw.edu*

ABSTRACT

The proper functioning of society before, during and after hurricanes is dependent upon resilient infrastructure. Although civil infrastructure as defined here is comprised of eleven interdependent systems, our focus is on the modeling of the primary interdependent subsystems of power delivery, water supply and telecommunications. In this paper, we examine resilience through the modeling of inoperability of connected systems post-event using input-output models coupled with a recently derived unified graph approach. Inoperability is directly related to the structural performance of the components of the systems; that is, failure of transmission towers will significantly contribute to the inoperability of the power delivery system. Also, the degree to which inoperabilities in one system such as power delivery will affect others such as telecommunications is a key aspect of capturing the resilience of the entire set of infrastructure systems.

Specifically, our paper will examine models of infrastructure performance using geo-statistical data for past hurricanes. Component and system-level fragilities for past events will be investigated. Interconnected models using in-situ data for hurricanes that made landfall in the USA Gulf of Mexico region will be examined.

KEYWORDS

Resilience, recovery, hurricanes, infrastructure

Resilience Study of Elevated Coastal Residential Buildings subject to Strong Winds

Jeffrey Weston[a], Fang Pan[b], Wei Zhang[c]

[a] *University of Connecticut, Storrs, CT, United States, Jeffrey.weston@uconn.edu*
[b] *Southwest Research Institute, San Antonio, TX, United States, fang.pan@swri.org*
[c] *University of Connecticut, Storrs, CT, United States, wzhang@uconn.edu*

ABSTRACT: Powerful coastal storms have led to a renewed interest in the resiliency of single family residential homes subjected to storm induced winds. Structural components, such as roof framing, connections, sheathing and nailing schedules are all specified in the building codes based on a modular concept. The modular construction techniques in many codes or specifications assume the dependence of the entire structure's safety on the resilience of each structural component, such as a wall or a roof panel. This approach overlooks the interaction of the components that occurs under complex loading. Recently, due to frequent flooding threats to the coastal communities in the northeast regions, many residential buildings have been elevated. Current design codes do not sufficiently specify the wind loading from the wind flow beneath the building structures. In a preliminary effort to address these gaps, a resilience study of elevated coastal residential homes is carried out based on the structural analysis of buildings with different architectural and construction parameters. Specifically, the residential building inventory in a selected coastal community was analyzed and common parameters were used to generate a few analysis models approximating typical homes in that community. Fragility analysis is then carried out based on this representative set of models. The analysis results may give insight to the amount of damage expected at a given wind speed based off construction techniques and foundation type.

KEYWORDS: Elevated coastal residential homes, finite element analysis, fragility analysis, strong winds, building design.

1 INTRODUCTION

The recent increase in the frequency and severity of coastal storms raises concerns over the preparedness of coastal communities to face these events. As sea levels rise, the storm surge impacts from coastal events penetrate further inland and cause more widespread structural damage to coastal communities. In response to this increase, flood insurance premiums have drastically increased for many residential homes in coastal communities. In the future coastal development at the present scale may not be feasible due to the amount of risk it carries. Instead, residential development will be driven by the increasing probabilities of risk associated with current and future estimates of sea-level rise [1]. Although new residential construction in coastal areas may be curtailed, existing homes, especially those in flood zones that have been elevated to mitigate flooding risks account for a large and growing problem in risk management. The Biggert-Waters Flood Insurance Reform Act of 2012 and ultimately the Homeowner Flood Insurance Affordability Act of 2014 lead to a substantial increase in the yearly premium many coastal residents pay to insure their homes. Naturally these modifications to flood insurance premiums were met with resistance from many home owners who felt a mismatch between the insurance companies calculated risk and their perceived level of risk [1].

One common technique to reduce flooding risk and lower yearly insurance premiums is to elevate the foundation of a residential home. While elevating a home may reduce the risk of direct damage from flooding, this modification introduces substantial changes in the load path and stiffness characteristics of the structure. Particularly, the effects of increased exposure during high wind events due to the elevated foundation have not been thoroughly considered or quantified. The use of ASCE7- Minimum Design Loads for Buildings and Other Structures provides the basis for design of the Main Wind-Force Resisting System (MWFRS) for low rise buildings [2]. Unfortunately, direct consideration of the wind loads for column-elevated coastal residential homes is not provided. Clearly by increasing the mean roof height of the structure through the addition of an elevated foundation exposure and thus wind loads are increased. Some interpretations of ASCE7 may justify the usage of this code to calculate the design wind loads for the walls and roof of an elevated residential home. However, by elevating the residential building the exposed sub-structure is now subjected to additional wind loading not currently considered in the design codes. The issue is further complicated by the inclusion of break-away walls which are commonly used to enclose the elevation system. Break-away walls are often included to provide some privacy such that the area beneath the elevated home can be used for storage or as a garage, etc. These walls must be designed such that they fail at a predetermined wave or water induced force allowing flood waters to pass freely below the structure minimizing the loads on the elevated home [3]. In terms of wind loading, under light winds if breakaway walls are present the pressure beneath the floor of the elevated home is likely minimal. For an elevated home without break-away walls, or once the break-away walls have failed, the wind pressure exerted beneath the elevated home is likely significant and needs to be considered when designing an elevation system.

The ideal elevation system will mitigate flood risks without increasing a structure's vulnerability to wind loading. The aim of this paper is to form a comparison between elevated and non-elevated residential structures and thus assess any trade-offs in wind resistance that are sacrificed to achieve increased freeboard. With the collaboration of local communities a database of residential homes has been condensed to only 3 typical home designs to provide a cross section of construction and architectural techniques. Each of these design configurations is analyzed for the regional design wind speed assuming the home is built at grade. Subsequently, a robust elevation system is modeled beneath each home configuration. Loading is adjusted to account for the increased exposure of the buildings then the analysis is re-run at the design wind speed to check for any variations in performance caused by elevating the home.

2 ANALYSIS DETAILS

2.1 *Determination of the Three Representative Analytic Models*

The towns of Fairfield and Milford in Connecticut provided the authors with a GIS dataset containing a variety of information describing the residential building inventory in their communities. Many detailed architectural and structural parameters were provided, however to bound the scope of this work only a select few could be considered. The primary variables of interest are the size of home, the number of stories, age of the home and the roof type. The size of the home and number of stories are expected to influence structural performance because as they increase so do the resultant wind loads. Roof type is significant both from a wind load and structural stiffness perspective. Conveniently, it was observed greater than 97% of the homes sampled had a gable style main roof meaning this is the only roofing design presently considered. The age of a home is another significant design parameter both from a material degradation standpoint as well as the evolution of the construction methods. Material degradation, while very influential, is difficult to characterize across the community, especially given the different level of maintenance performed on individual homes and additions that may have been introduced since the original date of construction. Currently, any reduction in material properties due to age is not considered. In this analysis, age is used to characterize common construction methods in the community. Communicating with town engineers and local builders gave insight to commonly observed framing techniques as well as the types of materials used. Details of the three models included in this study are presented below:

Table 1: Geometric details of the three analysis models

Model:	*1*	*2*	*3*
Floor plan	7.32x14.63m rect.	7.32x14.63m rect.	7.32x14.63m rect.
Stories	1.5	1	2
Roof pitch	9:12	5:12	5:12
Wall framing	3.8x8.9cm stud, 40.6cm o.c.	3.8x8.9cm stud, 40.6cm o.c.	3.8x8.9cm stud, 40.6cm o.c.
Wall sheathing	1.3x25.4cm boards	1.3cm plywood	1.3cm plywood
Roof framing	3.8x14cm rafter, 40.6cm o.c.	3.8x8.9cm truss, 61cm o.c.	3.81x24.13cm rafter, 40.6cm o.c.
Roof sheathing	1.59cm plywood	1.3cm plywood	1.59cm plywood

2.2 *Structural Modeling Considerations*

The framing layout for each of the three models closely follows standard practices detailed by the International Building Code [4]. However in some cases, for instance model 1, the lumber used for roof framing is smaller than what is commonly used in modern construction to capture past construction practices in the community. Model 1 seeks to represent a typical cape-style home in the coastal communities of the northeast region of the United States. These homes typically have steeper roof slopes to allow for additional living space above the first floor which may be considered a half story. Based on the provided community data, homes of this style tended to be older and often used boards for sheathing rather than plywood or oriented strand board (OSB). The roof sheathing for model 1 is modeled as plywood because in older homes from these communities it is likely the original sheathing has been replaced. Model 2 represents a more mass-produced design in which prefabricated engineered trusses are used to frame the roof rather than the stick framing methods used in custom construction. Since the first two models considered only single and 1.5 story homes it is worthwhile to include a model representing the common 2-story configuration. This third model includes heavier lumber in the roof framing as well as thicker roof sheathing because its increased mean roof height necessitates the consideration of higher wind loads.

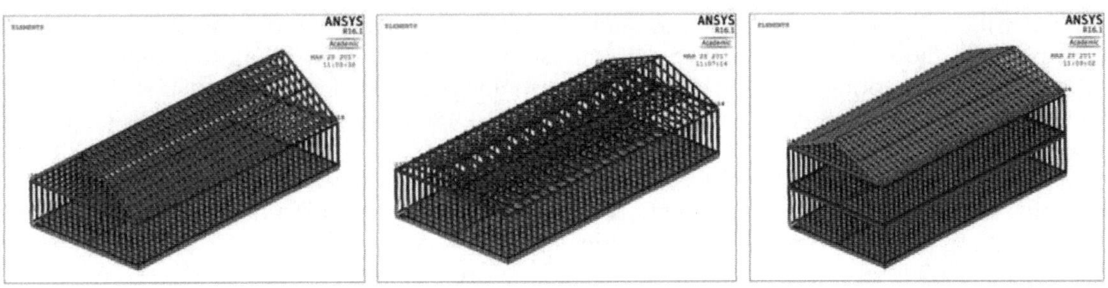

Model 1 Model 2 Model 3

Figure 1: Illustration of framing in analysis models

2.3 *Elevation System*

A variety of acceptable and structurally sound elevation methods exist for residential homes. The basic design consists of a frame to support the floor and rim joists of the home, as well as a series of columns laid out to elevate the structure while providing adequate support. Columns may be timber, continuous concrete, concrete blocks or steel. The support frame is either timber or steel. Many options for foundation design exist and will greatly influence the overall strength of the structure. For instance, columns may be built up on a slab, or more preferably, driven as a pile to an appropriate depth. This analysis utilizes an elevation model consisting of 40.64cm diameter concrete columns, connected to a support frame of W10x12 steel beams. The pile layout is based loosely on Coulbourne's paper [5] along with advice from local builders indicating common practices and finally structural engineering judgement. Elevation height is chosen as 2.74m, which is a common choice in the communities of interest. The elevation system used in all three models is depicted in Fig. 2.

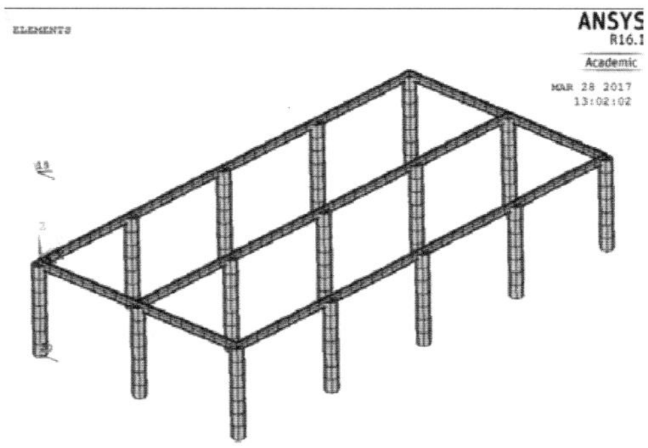

Figure 2: Illustration of elevation system

2.4 *Material Properties and Connection Performance*

The consideration of material properties and nailed connection behavior is critical to the overall performance of the analysis model. The mechanical properties of wood are quite complicated due to varying grain directions, moisture conditions and inclusions such as knots. Even in clear, straight grained samples properties can vary significantly. Due to uncertainty in grain orientation an average isotropic material model representative of properties in the weakest direction is used. All materials are expected to behave in the elastic range with the exception of the nailed connections. Properties for lumber are taken from the Forest Products Laboratories Report: FPL-GTR-190 [6], while sheathing properties are from the American Plywood Association [7]. The W-beam girders are taken to be A992 steel [8] and the concrete columns are assumed to be a 41.4MPa compressive strength mixture [9], [10].

Table 2: Material Properties

Component	Young's Modulus Ex/Ey/Ez (MPa)	Shear Modulus Gxy/Gxz/Gyz (MPa)	Poisson's Ratio
Framing Lumber	8.3e3	-	0.4
Plywood Sheathing	2e3/2e3/13.1e3	1e3/1e3/1e3	0.08/0.08/0.08
Concrete Columns	23.2e3	-	0.3
W10X12 Beams	200e3	-	0.3

Although all structural materials are expected to behave elastically, the connections between them are not. In wooden structures connections provide continuity to the structural members and prove to be, in many cases, more influential in failure progression than the structural members themselves. In fact most failures in wood framed structures are due to connection design, construction or serviceability [11]. In the present study, framing members are connected in the usual way by means of shared nodes, however, sheathing is offset by half the thickness of the framing it joins to and connected with an equivalent parameterized beam connection (EPBC) element that captures connection nonlinearity.

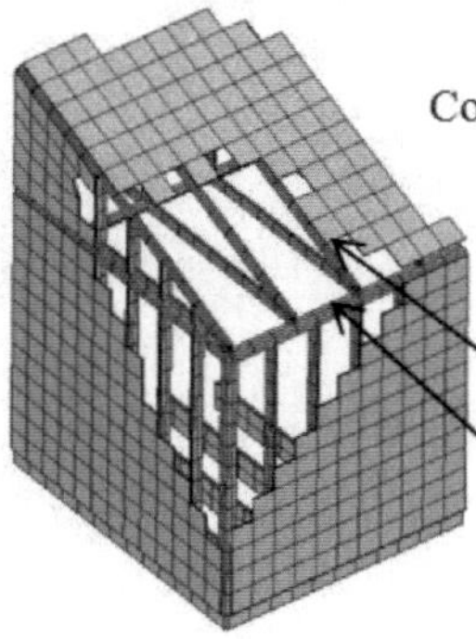

Figure 3: Illustration showing sheathing offset from framing by thickness

The EPBC element is a numerical procedure developed by Weston and Zhang that allows the modeling of nonlinear connection behavior in a similar manner to nonlinear spring elements, with the added benefit of directional load coupling [12]. For this analysis EPBC elements are calibrated to represent the sheathing to framing load-deflection relationship. The actual load deflection curves implemented for the sheathing to frame connection behavior come from the experimental work documented by Dao and Van de Lindt [13] as well as Thampi et al. [14] and reproduced in a similar study of a single residential home on a traditional foundation by Pan et al. [15]. Using experimentally observed load deflection behavior, the system level finite element model can more accurately capture the reduced global structural stiffness caused by connection nonlinearity under large wind loads without incurring the computational costs of detailed connection models.

2.5 *Loading and Solution Procedure*

Wind loading for this analysis was determined per the envelope procedure in ASCE7-10 chapter 28 to simulate wind speeds from 70 to 150mph. The adjustment factors in ASCE7-10 for topology, exposure, etc. are chosen to be representative of a coastal home in Connecticut. Loading is applied as pressures corresponding to 10mph wind speed increments [2]. Wind load pressures for the elevated configurations are recalculated to account for the increase in mean roof height. Additionally, the exposed underside of the first floor is now vulnerable to wind loading. It may be possible to approximate the pressures for this surface as the opposite of those for a flat roof. The true pressure exerted underneath an elevated structure is a topic of discussion and can likely be more accurately characterized by future wind tunnel testing. The envelope procedure specifies four wind angle ranges along with two internal pressure scenarios. Since all the models in consideration have two planes of symmetry the four wind direction ranges specified by the envelope procedure are redundant and only one windward corner of the structure needs to be considered. In the case of negative internal pressure the applied pressures on the structure would cause the sheathing to bear against the framing members putting little to no load on the connections. Since one of the focuses of this analysis is to consider connection failure the positive internal pressure scenario is deemed more appropriate because it leads to higher loads developed in the connections. Furthermore a preliminary analysis of both positive and negative internal pressure coefficient scenarios was conducted to confirm that the positive internal pressure scenario provides greater deflections in the solution.

For the baseline configuration with a traditional foundation, the structure is restrained by fixing nodes along the sill plate where common building code requirements mandate a rigid connection to the foundation [4]. In the elevated configuration the bottoms of the concrete columns are fixed, thus assuming a best case foundation scenario. In the solution process, load application is linear and the analysis is static but includes the effects of geometric nonlinearity. At the end of each load step the analysis model is assessed for failure per the criteria presented in section 3.1 before solving the next incremental load step. The number of failed elements at each solution step is used to construct a vulnerability curve of damage relative to wind speed.

3 RESULTS

3.1 *Sheathing Failure Criteria*

This analysis considers damages in terms of sheathing loss. Sheathing is considered to fail in one of three modes. First, the Von Mises stress in the sheathing may exceed an acceptable material limit. Second, the relative displacement between sheathing and framing may exceed an acceptable value creating a high risk of water intrusion. Third, the nail head may pull through the sheathing constituting a material failure of the sheathing. The failure modes, especially the first and third are undoubtedly coupled. Mesh resolution in the structural model is dictated by the desired nailing schedule so that shell (sheathing) and beam (framing) nodes are readily connected. Due to this restriction on the mesh some accuracy in the solution may be lost and thus considering the axial force in the connection element may be a more accurate means of estimating if pull-through has occurred compared to looking at nodal sheathing stress and comparing them to the material limit. The details of the sheathing failure criteria are presented below:

Table 3: Sheathing Failure Criteria

Failure Criterion	Allowable limit	Comment	Reference
Von Mises stress in sheathing	1.131 MPa	Wet condition and 1/6th safety factor	APA
Nail withdrawal	L/120	Relative displacement, L is spacing of framing	IBC
Nail pull through	316, 409N axial force	Wet condition, 1.3cm and 1.59cm respectively.	[16]

3.2 *Vulnerability and Damage Index*

Evaluation of structural failure must consider economic loss in addition to the remaining stiffness or strength of the structure after an extreme wind event. For instance, damage to a relatively small percentage of structural components may account for a large portion of the insured value of the structure. In an effort to address this complicated problem the HAZUS hurricane damage loss estimation model was constructed by Vickery et al. using historical loss data from previous storms [17]. As part of the HAZUS model, damage states and associated descriptions are provided. Damage is rated both qualitatively from a livability standpoint and quantitatively as a percentage of components lost. Presently, the loss percentages corresponding to each HAZUS defined damage state are converted to an equivalent number of failed sheathing elements based on equivalent area. Impacts from debris as well as window and door failures are not currently considered. The table below presents the usage of HAZUS damage state guidelines in the development of the vulnerability curves constructed in this analysis. Note that the equivalent number of shell elements varies slightly based on which analysis model is under consideration.

Table 4: HAZUS criteria adapted to equivalent number of failed elements

Damage State	Description	Sheathing Damage	#Shell elem. equivalent
0	Very Minor	0	0
1	Minor	0	<66
2	Moderate	1 to 3 panels	<200
3	Severe	3 and <=25%	<1250
4	Destruction	>25%	>=1250

3.3 *General Trends in the Results*

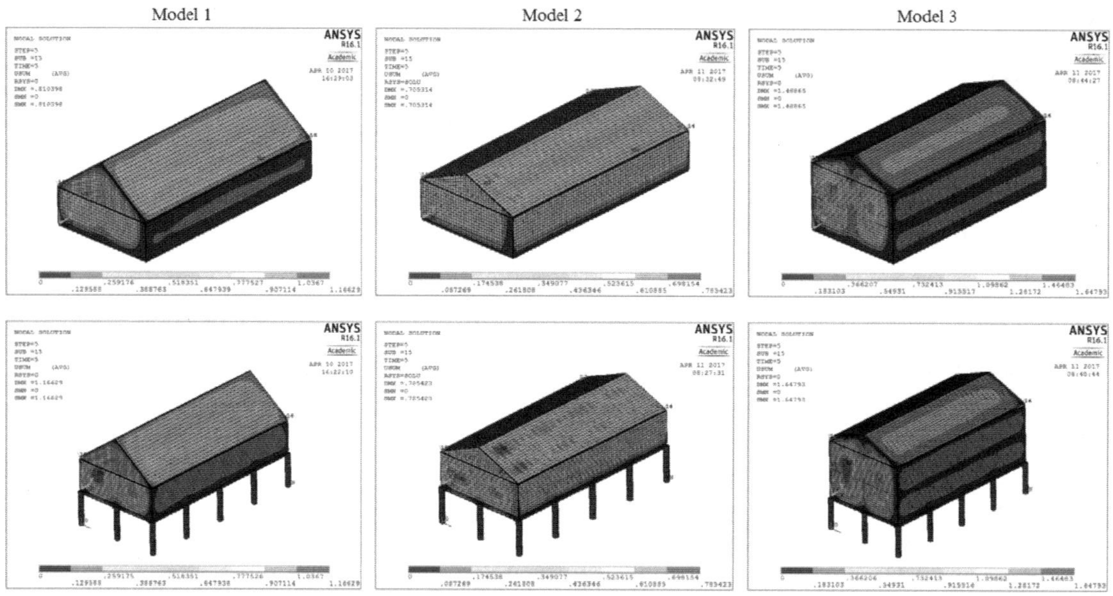

Figure 5: Displacement contours (in) at 110mph wind speed

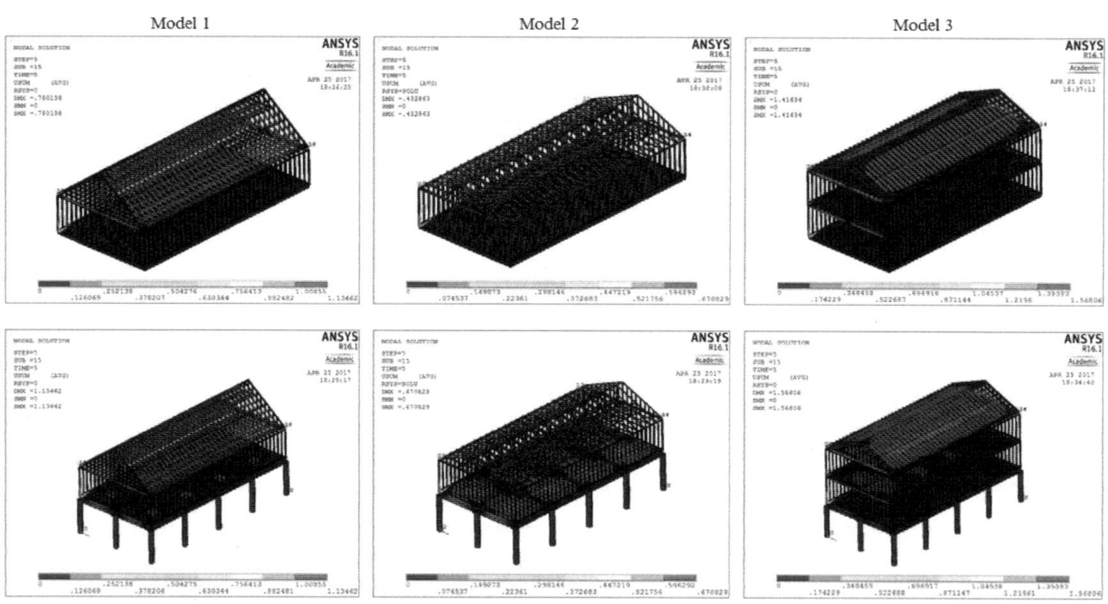

Figure 4: MWFRS Displacement contours (in) at 110mph wind speed

As shown in Figs. 4&5, all elevated configurations of the models experienced larger displacements than their baseline configurations on traditional foundations. Whether or not these larger displacements translated to failure as defined in 3.1 and 3.2 depended on the framing style and sheathing thickness used in that model. It is evident that different styles of roof framing influence the overall structural stiffness. Stick framed models (1 & 3) experienced very non-uniform displacement contours with the largest displacements noted at the ends of the building. Model 2, which used Howe trusses in the roof framing showed far more consistent displacement contours illustrating this design is more effective at transmitting the load throughout the structure. Despite the ability of model 2 to more efficiently transmit forces over the entire structural envelope in most cases the stick framed models (1 & 3) outperformed model 2. There could be a variety of reasons for this observation. First, the stick framed models utilized thicker sheathing to compensate for the lack of engineering in the roof framing design. Second, a Howe truss is designed to efficiently carry weight rather than uplift pressure, however in this analysis uplift proved to be the most damaging to all the models. Finally, since the Howe truss is an engineered structure with well understood performance characteristics smaller lumber can be used. In the stick framed models, uncertainty in the load transfer mechanisms often results in the use of larger lumber for safety and conservatism. It is likely that a Howe truss roofing system designed of larger lumber would outperform the stick framed models. Furthermore, comparing Fig. 5 to Fig. 4 reveals that the displacement of the MWFRS is consistently less than the displacement of the sheathing nodes illustrating that connection deformation is a significant contributor to overall deformation.

3.4 *Vulnerability Curves*

Vulnerability curves for each model were constructed by applying the criteria from Tables 3&4 to each one of the models in both elevated and non-elevated configurations. The total number of sheathing elements that exceeded the pre-established failure criteria are counted and compared to the total sheathing area.

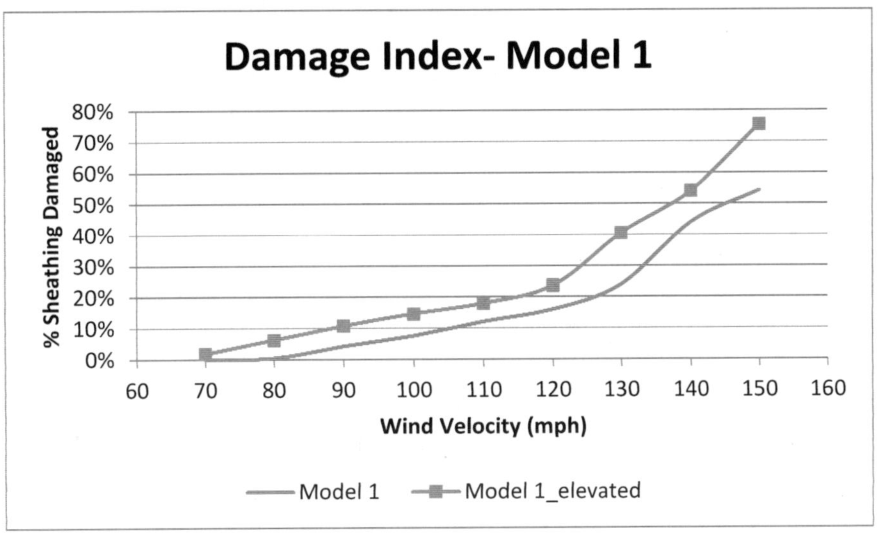

Figure 6: Model 1 Sheathing Damage Index

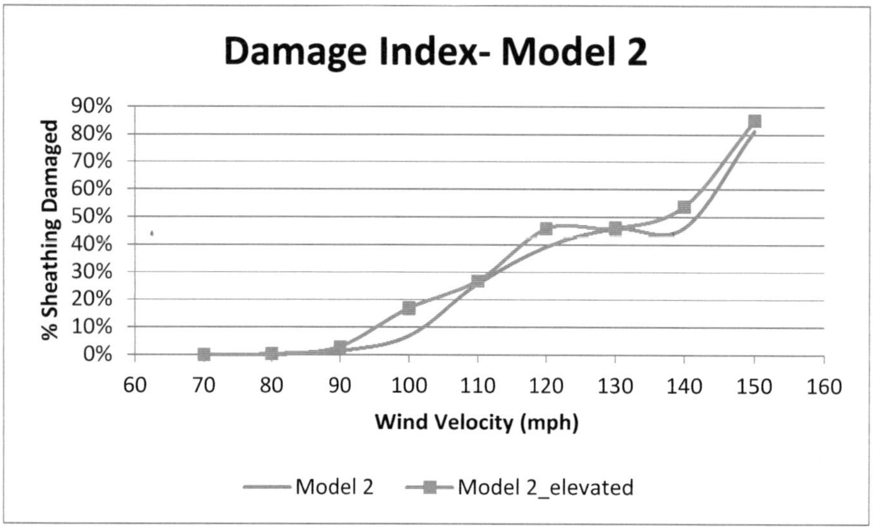

Figure 7: Model 2 Sheathing Damage Index

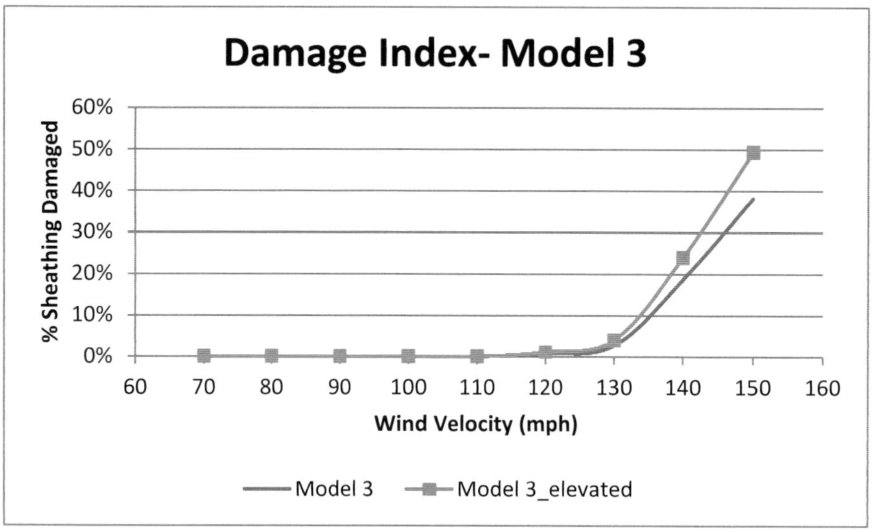

Figure 8: Model 3 Sheathing Damage Index

The vulnerability curves presented in Figs. 6-8 illustrate a penalty in wind resistance for elevated homes. The extent of the penalty depends on the construction techniques of both the home and the elevation system. For model 1 there appears to be a consistent penalty of ~15mph in wind resistance in terms of sheathing damage for the elevated configuration compared to the configuration with a traditional foundation. In model 2 the penalty is less constant across the range of simulated wind speeds but can be as large as ~10mph. The results in Fig.7 from model 3 show the elevation system has an almost negligible effect on sheathing damage at the same wind speed. The less pronounced effects in this model can be attributed to the usage of thicker lumber and sheathing, see Table 1. Although the vulnerability curves for model 3 show little difference between the elevated and non-elevated configurations, the displacement contours as shown in Fig. 4 are significantly different in both shape and magnitude for this, and all models. The contour patterning in Fig.4 helps illustrate the effect of the elevation system on the displacement field in the structure. In the elevated models the contours are clearly shaped by the location of the supports in the elevation system. As previously mentioned the elevation system modeled for this analysis is quite robust. Many elevation systems may be far less stiff and further exaggerate the influence of the support location on structural displacements. Furthermore, the shape of the displacement contours in models 1 and 3 are similar, while the contours for model 2 are quite different. Despite having different roof slopes and using different lumber, models 1 and 3 used the same roof framing configuration. This suggests the roof framing configuration should be considered when designing an elevation system because the roof framing contributes to the overall structural stiffness and influences the dominant load transfer mechanisms in the structure. The usage of the envelope procedure creates some important limitations in the assessment of the analysis results. The envelope procedure was developed to subject the MWFRS with equivalent loading meant to encompass all typical wind loading scenarios for structural design purposes; not necessarily for the response evaluation of existing structures [18]. For this reason the results are best viewed as the elevated configuration model relative to the corresponding non-elevated model rather than as a prediction of the damage a certain wind speed may cause in the absolute sense.

4 CONCLUSIONS

This study explored the effect that elevating a residential home may have on its structural performance during a high wind event. Three separate housing models with different framing techniques and sheathing thicknesses were considered. Each model was analyzed on a traditional foundation and an elevated foundation system for wind speeds ranging from 70-150mph. The model details and the design wind speed (120mph) were chosen to be typical for a shoreline community in Connecticut. The results from the analysis show both the layout and lumber choice for roof framing significantly affect the magnitude and shape of displacement contours developing in the models based on wind pressure loading. Furthermore, displacement solutions are larger and associated element failures are more numerous at a given wind speed for elevated homes compared to their counterparts on traditional non-elevated foundations. Depending on the construction techniques used and the range of wind speeds considered, this analysis suggests the design wind speed penalty caused by elevating a home may be as large as ~15mph. Additionally, a wide variety of substructure designs for elevation exist and will likely play a large role in structural performance.

The findings from this study may help predict the deformation mechanisms resulting from different types of residential construction methods due to high wind loads. Also, the vulnerability curves may give insight to the extent of the design wind speed penalty caused by elevating a home under favorable conditions. It has been demonstrated that elevating a home changes the structural response in addition to increasing wind exposure. For these reasons it is prudent to consider the framing methods used in the original construction when designing the elevation system. It may also be beneficial to explore options that will stiffen the elevated structure and reduce or eliminate the design wind speed penalty caused by elevation. Certain limitations in the usage of design wind pressures as a means of evaluating structural response are well understood and future wind tunnel testing may provide a more realistic pressure loading profile for future analyses.

5 ACKNOWLEDGEMENTS

Funding for this work was partially provided by Connecticut Sea Grant, University of Connecticut through Award No. NA14OAR4170086, Project Number R/CH-1 and the Graduate Assistance in Areas of National Need (GAANN) fellowship, Grant number P200A140212. The support is greatly appreciated. The authors would also like to thank town engineer Laura Pulie of Fairfield, CT and Alex Mora of COWI for their guidance. Any opinions, findings, and conclusions or recommendations expressed in this material are those of the authors and do not necessarily reflect the views of the sponsors.

6 REFERENCES

[1] C. J. McGuire, "U.S. Coastal Flood Insurance, Risk Perception, and Sea-Level Rise: A Perspective," *Coast. Manag.*, vol. 43, no. 5, pp. 459–464, 2015.

[2] ASCE 7-10, *Minimum Design Loads for Buildings and Other Structures*, 7th–10th ed. Reston, VA: American Society of Civil Engineers, 2010.

[3] T. Tomiczek, A. Kennedy, and S. Rogers, "Survival Analysis of Elevated Homes on the Bolivar Peninsula after Hurricane Ike," in *Advances in Hurrican Engineering-Learning from Our Past*, American Society of Civil Engineers, 2013, pp. 108–118.

[4] IBC, "2012 International Building Code," Washington DC, 2011.

[5] W. Coulbourne, "Foundation Design in Coastal Flood Zones," in *Advances in Hurrican Engineering-Learning from Our Past*, American Society of Civil Engineers, 2013, pp. 92–107.

[6] D. E. Kretschmann, "Chapter 5 - Mechanical Properties of Wood," Madison, WI, 2010.

[7] APA, "PDS- Plywood Design Specification," Tacoma, WA, 1997.

[8] American Institute of Steel Construction, *Steel Construction Manual*. Chigago, IL: American Institute of Steel Construction, 2005.

[9] J. McCormac, *Design of Reinforced Concrete*, 10th ed. New York: Harper & Row, 1986.

[10] ACI Committee 318, *Building Code Requirements for Structural Concrete and Commentary*. ACI Committee, 2005.

[11] L. a. Soltis and M. Ritter, "Mechanical Connections in Wood Structures: Chapter 1--Introduction," 1996.

[12] J. Weston and W. Zhang, "Finite Element Modeling of Nailed Connections in Low-rise Residential Home Structures," 2016.

[13] T. N. Dao and J. W. van de Lindt, "New Nonlinear Roof Sheathing Fastener Model for Use in Finite-Element Wind Load Applications," vol. 134, no. October, pp. 1668–1674, 2008.

[14] H. Thampi, V. Dayal, and P. P. Sarkar, "Finite element analysis of interaction of tornados with a low-rise timber building," *J. Wind Eng. Ind. Aerodyn.*, vol. 99, no. 4, pp. 369–377, 2011.

[15] F. Pan, C. S. Cai, W. Zhang, and B. Kong, "Refined damage prediction of low-rise building envelope under high wind load," *Wind Struct.*, vol. 18, no. 6, pp. 669–691, 2014.

[16] B. Herzog and B. Yeh, "Nail Withdrawal and Pull-Through Strength of Structural-Use Panels Nail Withdrawal Strength of Plywood and OSB," *APA – Eng. Wood Assoc.*, 1999.

[17] P. J. Vickery, P. F. Skerlj, J. Lin, L. a. Twisdale, M. a. Young, and F. M. Lavelle, "HAZUS-MH Hurricane Model Methodology. II: Damage and Loss Estimation," *Nat. Hazards Rev.*, vol. 7, no. 2, pp. 94–103, 2006.

[18] J. He, F. Pan, and C. S. Cai, “A review of wood-frame low-rise building performance study under hurricane winds,” *Eng. Struct.*, vol. 141, pp. 512–529, 2017.

13th Americas Conference on Wind Engineering
Gainesville, Florida, USA
May 21 - 24, 2017

Date stamping strategy development for detecting collapse period in Hurricane Resiliency Index Time series

Chengjun Zhu[a,c], Daan Liang[a,b], Bradley T.Ewing[a,c]

[a] *National Wind Institute, Texas Tech University, Lubbock, TX, USA, zhu.chengjun@ttu.edu*
[b] *Department of Construction Engineering & Engineering Technology, Texas Tech University, Lubbock, TX, USA, daan.liang@ttu.edu*
[c] *Rawls College of Business, Texas Tech University, Lubbock, TX, USA, bradley.ewing@ttu.edu*

ABSTRACT

Given that natural disaster cause massive amounts of damage around the world each year, the resiliency of coastal communities has been a major concern of researchers and policymakers. Using the Hurricane Resiliency Index (HRI) developed by Cui, Ewing, Liang & Najat (2015), we applied a series of tests to identify and date-stamp periods of periodically collapsing behavior in what is otherwise a stationary time series. Specifically, we examined the monthly HRI for the Houston MSA (Metropolitan Statistical Area) over the 1990-2015 sample period. Evidence of a collapse signals a break or nonlinearity in the data generating process. Based on the results of test, we obtained the starting and ending dates for multiple periods that exhibited this "bubble" type of behavior. By comparing these periods with historical data on natural disasters (i.e. hurricanes), we then were able to measure the duration of collapse and the time it took for the HRI to return to normal, stationary behavior. We argue that the duration of bubble should be reflective of the resiliency of the study area to hurricane.

Hurricane Resiliency Index (HRI) is a combination of well-chosen indicators to assess and monitor the hurricane resiliency level of coastal communities across geographical boundaries and over time. The HRI is updated monthly and constructed non-parametrically by assigning fixed standardization factors as weights to each of the indicators, commonly referred to as the National Bureau of Economic Research(NBER) nonparametric approach (Conference Board 2001). As a coincident index, it reflects the current state (or "health") of a local community while taking into account the condition of the built environment, labor situation, and energy distribution condition and represents the disaster resiliency of a given community (Cui, Ewing, Liang & Najat 2015).

Based on the HRI, we performed the Augmented Dickey-Fuller (ADF) test to detect the presence of a unit root in the HRI time series. The rolling ADF (RADF) test is a rolling version of the test in which the statistic is calculated over a rolling window of fixed size. Furthermore, we performed the more recent test, supremum ADF (SADF) test to calculate the relevant test statistic and the derivation of its corresponding critical values by Monte Carlo simulations.

Previous research has shown that when the sample period includes multiple episodes of heightened activity and collapse, the SADF test may suffer from reduced power and can be inconsistent, failing to reveal the existence of multiple bubbles. In order to detect multiple periods of collapse, we applied an enhanced approach to overcome the difficulty arising from the complex nonlinear structure involved in multiple bubble phenomena, which is the generalized SADF (GSADF) test. Based on results of the GSADF test, we obtained the date and duration of multiple periods of collapsing episodes.

Examining the history of Hurricane Ike, Hurricane Rita, and other major economics events (e.g., financial crisis/housing bubble) we identified multiple periodically collapsing episodes which accord well with the landfall of hurricanes and tropical storms. The result provides information as to the resiliency of the greater Houston community to hurricanes and, perhaps more importantly, how long the effects of hurricane disaster may last.

In future research, we plan to apply GSADF test on HRI time series in different MSAs and compare the durations of different areas.

KEYWORDS

Hurricane Resiliency Index, Rolling Window ADF Test, SADF Test, GSADF Test

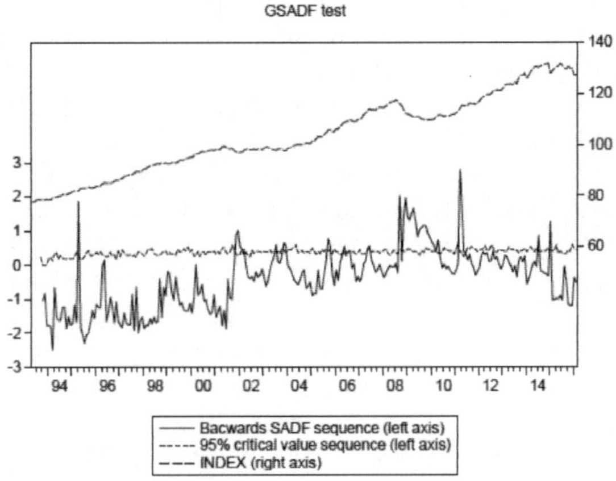

REFERENCES

Yuepeng Cui, Daan Liang, Bradley T. Ewing, Ali Nejat (2015). "Development, Specification and Validation of Hurricane Resiliency Index". *Natural Hazards,* 2149–2165.

Itamar Caspi (2014). "rtadf: Testing for Bubbles with EViews".

Peter C. B. Phillips, Shu-Ping Shi, Jun Yu (2015). "Testing for Multiple Bubbles". *International Economic Review,* 1042-1078.

The 13th Americas Conference on Wind Engineering (13ACWE)
Gainesville, Florida USA, May 21-24, 2017

13th Americas Conference on
WIND ENGINEERING

New Framework for Estimating Thunderstorm Design Speed

Haitham Aboshosha, Thomas Mara, Peter Case

The Boundary Layer Wind Tunnel Laboratory, Western University, London, Ontario, Canada N6A 5B9

ABSTRACT: The current study presents a framework to obtain design wind speeds associated with thunderstorm gust fronts using historical data and Monte Carlo simulations. This framework is analogous to hurricane modeling where (i) a numerical model capable of describing thunderstorm gust front as a function of a number model parameters (called thunderstorm parameters) is developed and validated, (ii) the model is used to describe previous gust front records seen at a meteorological station and identify the model parameters, then (iii) Monte Carlo simulation is utilized to simulate thunderstorm gust front events happening in thousands of simulation years employing the probability distribution of the model parameters. Then, the simulated events are analyzed to obtained speeds suitable for design. The framework is utilized to study thunderstorm wind speeds at two meteorological stations (Lubbock Preston Smith International Airport & Seattle-Tacoma International Airport). The results indicate for Lubbock (located in a thunderstorm active region) that thunderstorm wind speeds can be higher than synoptic wind speeds (15% at 700 mean recurrence interval MRI) and speeds defined in the ASCE 7-10 (10% for 700 MRI). The results for Seattle-Tacoma (not an active thunderstorm region) indicate that synoptic wind speeds are always higher. The model is validated throughout successful prediction of the speed-return period relationship for events seen at both stations. The proposed framework represents a useful validated tool to quantify thunderstorm gust front speeds suitable for design.

KEYWORDS: Thunderstorm, Gust Front, Downburst, Design Speed, Historical Records, Numerical Simulations, Localized Nature.

1 INTRODUCTION

Thunderstorm is a very frequent event in mid latitudes that can be described as convective process where warm air at the earth's surface is exchanged with cool air higher in the atmosphere. Warm air enters the cloud base from underneath forming updrafts while cold air exits from underneath forming downdrafts. These downdrafts when hit the ground convect radially causing outflows of high wind. The leading edge of the outflow is typically referred to as the gust front. Strong downdrafts are typically referred to as downbursts or microbursts. Medium and Intense downdrafts can lead to strong horizontal winds near the ground.

There are numerous failures of structures due to gust fronts/downbursts, and despite that most design codes or standards do not include provisions for those events. One of the main challenges hindering the inclusion of gust front and downburst winds in design codes is the lack of comprehensive full scale measurements primarily due to the localized nature of those downdrafts which have a typical size of 500-4000 m (Hjelmfelt 1988).

Many of the initial numerical simulations focused on the steady-state downdrafts (Oseguera and Bowles, 1988; Vicroy, 1991 and 1992; Selvam and Holmes, 1992; Holmes and Oliver, 2000; Wood et al., 2001; Chay et al., 2006; Xu and Hangan, 2008), while more recent numerical simulations focused on the unsteady characteristics (Kim and Hangan, 2007; Mason et al., 2009, 2010; Vermeire et al. 2011; Orf et al., 2012, Aboshosha et al. 2015). These studies provide very good insights regarding the outflow wind field formed from occurs strong downdraft (downburst) and its differences to synoptic wind. However, these simulations do not simulate either the translation of the storm or dissipation of the outflow or both. Abd-Elaal et al. (2013) proposed a numerical thunderstorm model that accounts for storm translation and outflow dissipation. They utilized the model to fit a number of thunderstorm records successfully. However, such a model is based on Reynolds Averaged Navier Stoke (RANS) turbulence modeling, which inherently can't predict the peak thunderstorm speeds due to the way the turbulence was modeled.

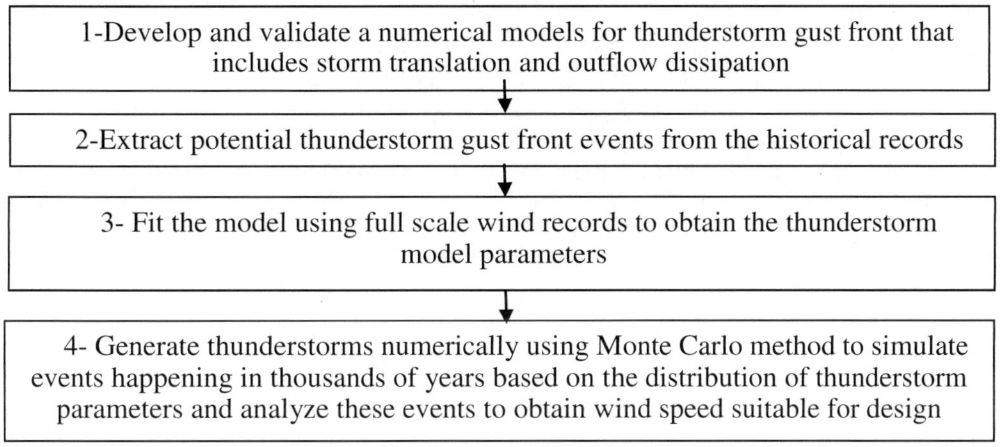

Fig. 1 Suggested Thunderstorm Framework

A number of studies in the literature documented and investigated full scale thunderstorm events. These studies can be classified into two groups, with the first group is aimed at characterizing a specific event identified based on the meteorological conditions (Charba 1974, Goff 1976, Wakimoto 1982, Hjelmfelt 1988, Fujita 1990, Gast and Schroeder 2003), Holmes et al. (2008) and Gunter and Schroeder (2013). The second group is focused on automatic separation and classification of large wind datasets into thunderstorm events and synoptic events (Gomes and Vickery 1976, Riera and Nanni 1989, Twisdale and Vickery 1992, Choi 1999, Choi and Hidayat 2002, Choi and Tanurdjaja 2002, Cook et al. 2003, Duranona et al. 2006, Lombardo et al. 2009, Rowcroft 2011, De Gaetano et al. 2014). This group aims at developing statistical models for gust fronts and downburst speeds suitable for the design. The recent studies conducted by Solari et al. (2012) and De Gaetano et al. (2014) showed that thunderstorm gust fronts are non-stationary and non-Gaussian with a low mean speed and a gust factor G_v greater than 1.25 times theoretical gust factor for synoptic wind G_0. This criterion of having a relatively high gust factor is very useful in distinguishing between thunderstorm and synoptic wind records and is used in the current study.

The study presented in this paper belongs to the second grouping of research and aims at estimating design wind speeds by analyzing historical wind records already available at meteorological stations. Although, these records are typically available at a coarse temporal resolution (i.e., every 1 hour), when wind exceeds a certain threshold, or when there is a drastic change in the wind characteristics, measurements are taken at an increased frequency.

There is an analogy between historical wind records for thunderstorm gust fronts (resulting from medium and intense downdrafts) and hurricanes as both are somehow localized (regardless of the difference in their sizes) and are not of sufficient length or completeness to estimate potential speeds suitable for design. Inspired by previous research on hurricanes (Russell and Schueller

1974, Tryggvason et al. 1976, Georgiou et al. 1983, Vickery et al. 2000a, b, Vickery et al. 2009, Li and Hong 2016), this study presents a new framework to estimate wind speeds of thunderstorm gust front events suitable for design. The flow chart presented in Figure 1 describes the framework. Firstly, a thunderstorm gust front model is developed and validated. This model is based on the CFD simulations by Aboshosha et al. (2015), but accounts for the storm translation and downdraft outflow dissipation. Description and validation of that model is provided in Section 2.1. Secondly, as will be discussed in Section 2.2, a technique is suggested to distinguish between historical records for thunderstorm gust front and records for synoptic wind events. The technique is employed to distinguish events from two sample wind stations (one is located in a region known to be affected by thunderstorm events while the other is not) are selected and wind records are analyzed to identify past thunderstorm events. These events are then used to calibrate the model by obtaining the model parameters and their probability distributions. Thirdly, as provided in Section 3, thunderstorm gust fronts are simulated using Monte Carlo method for five thousands of years and analyzed to estimate the speeds suitable for the design. The main conclusions obtained from the study are summarized in Section 4.

2 THUNDERSTORM GUST FRONT MODEL

2.1 *Model Description*

The model is based on the downburst wind field obtained from the Computational Fluid Dynamic (CFD) Large Eddy Simulations (LES) developed by Aboshosha et al. (2015) and modified to account for storm translation and outflow dissipation. Aboshosha et al. (2015) conducted CFD simulations for strong thunderstorm downdraft (downburst) events with no translation speed. The simulations are based on the impinging jet method (Fujita 1985) with a jet of 1000 m diameter located at 2000 m elevation that impinges towards the ground forming the outflow gust front. Figure 2 shows the computational domain and grid used for the simulations, spatial variations of the radial velocity at the time instance of peak wind speed and temporal variation of radial speed at the location of the maximum peak. It should be noted that radial wind speeds obtained from the model were spatially averaged circumferentially and then temporally to obtain the mean component V_{rm}(r, z, t), which is a function of the radial distance r, height z and time t, as shown Figure 2d.

Resulting wind field from the simulations is a function of the downdraft jet diameter, D_j, and the jet speed, V_j, and can be scaled using the approach developed by Hangan et al. (2003) and Shehata et al. (2005) to obtain the wind field for other D_j and V_j.

A modification of the CFD model is developed by Aboshosha and Mara (2017) to account for storm translation and the outflow dissipation. The modified model contain 10 parameters required to describe a thunderstorm downdraft and its outflow gust front. These parameters are shown in Figure 2e and are summarized as follows: initial downburst location (R_0, φ_0) with respect to the anemometer; translation speed and direction (V_t and θ); ratio between ambient speed and the translation speed (n_{ST}); jet diameter (D_j); jet speed (V_j); initial time (t_0) and time instances controlling the dissipation (t_1, t_2).

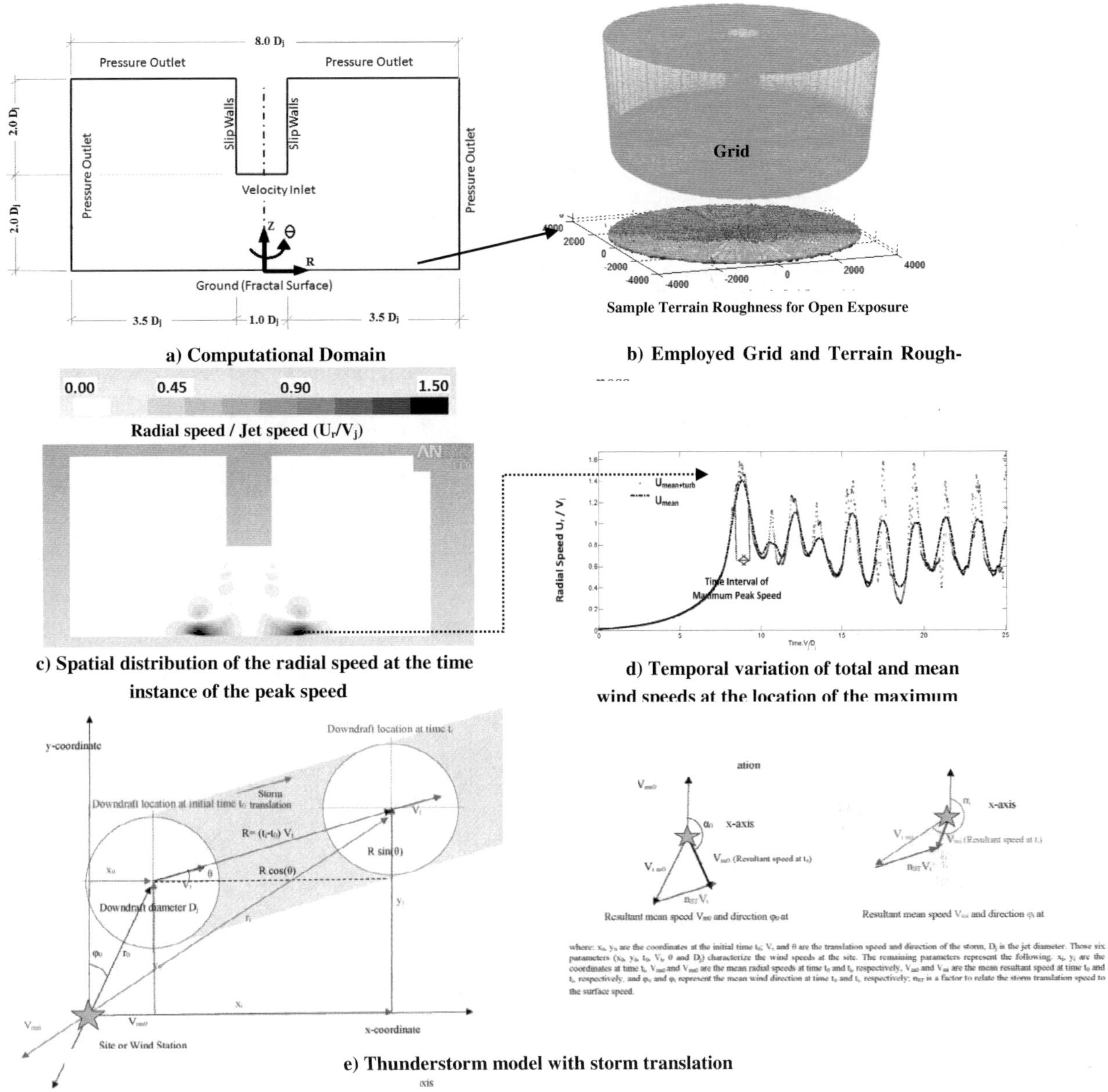

a) Computational Domain

b) Employed Grid and Terrain Roughness

c) Spatial distribution of the radial speed at the time instance of the peak speed

d) Temporal variation of total and mean wind speeds at the location of the maximum

e) Thunderstorm model with storm translation

Fig. 2 Thunderstorm Gust Front Model

2.2 *Model Validation*

The model is validated throughout successful description of the wind speed and direction seen in previous thunderstorm events happened in a) Andrews Air Force (AAFB), near Washington, D.C., U.S.A in 1983 (Fujita 1985), with the data collected at 5.0 m height, and b) Reese Air Force Base (RAFB), near Lubbock, Texas, U.S.A in 2002 from tower 4 at 10 m height (Gast and

Schroeder, 2005). In the validation process, each thunderstorm records is split into a number *ns* of segments (each having a 10 min duration or less) and statistics of the peak speed, mean speed and mean direction are used as the fitting targets of the thunderstorm model. The fitting process is conducted using the Genetic algorithm by minimizing the objective function expressed in Equation 1. Such a function represents the error between statistics from observed data and statics from the model and equals to 0 at an exact match between the modeled and recorded speeds/ directions.

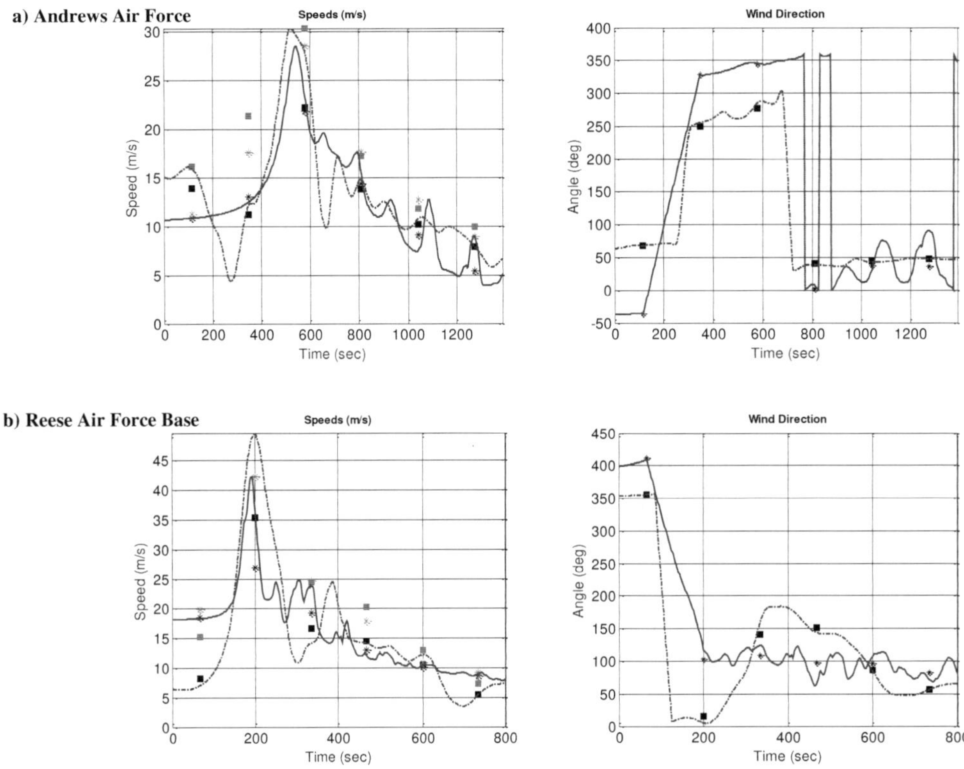

Fig. 3 Validation of the Gust Front Model

$$F_{obj} = \sum_{s=1}^{ns} \left(\left(1 - \frac{V_{mis}}{V_{ms}}\right)^2 + \left(1 - \frac{V_{pis}}{V_{ps}}\right)^2 + \left(\cos(\alpha_i) - \cos(\alpha_{is})\right)^2 + \left(\sin(\alpha_i) - \sin(\alpha_{is})\right)^2 \right) \qquad \text{Equation 1}$$

where V_{mi}, V_{pis}, α_i are the mean wind speed, peak wind speed and mean wind direction obtained from the model; V_{ms}, V_{ps} and α_s are the mean wind speed, peak wind speed and mean wind direction obtained from the observation records, s is the number of the segment.

In the genetic algorithm, a number of 100 sets (i.e. population) is used to generate the 10 parameters (i.e. 100 sets with 10 parameters each) and the objective function is evaluated to identify best few sets that may lead to the global minimum (i.e. referred to as the parent sets). Then another 100 sets are generated based on the characteristics (i.e. genes) taken from the parent sets identified from the earlier generation. The process is repeated for a number of 40 generations until the parameter set leading to the global minimum is identified.

The model is validated by assessing the goodness of the fit, quantified by the R-squared value, resulting from multiple thunderstorm gust front events. Table 1 summarizes thunderstorm parameters resulting from the fitting process and the R^2 value while Figure 3 shows the sample thunderstorm records and their model fit. As indicated from the figure and inferred from the high R^2 value close to 1, the model can sufficiently describe the evolution of the wind speed and direction along the time.

Table 1. Thunderstorm Parameters for sample two events

Param./ Event	r_0 (m)	φ_0 (deg)	V_t (m/s)	n_{ST}	θ (deg)	D_j (m)	V_j (m/s)	t_0 (sec)	t_1 (sec)	t_2 (sec)	$max(V_g)$ (m/s)	R^2
a	1858	156	5.1	3.51	58	1520	45.8	109	813	630	43	0.99
b	2844	111	2.6	0.75	304	2140	17.1	895	2561	2357	13	0.97

*Anemometer is assumed at 10 m height, and an open terrain condition is considered for all downburst cases; V_g is the gust speed

3 APPLICATION OF THE FRAMEWORK

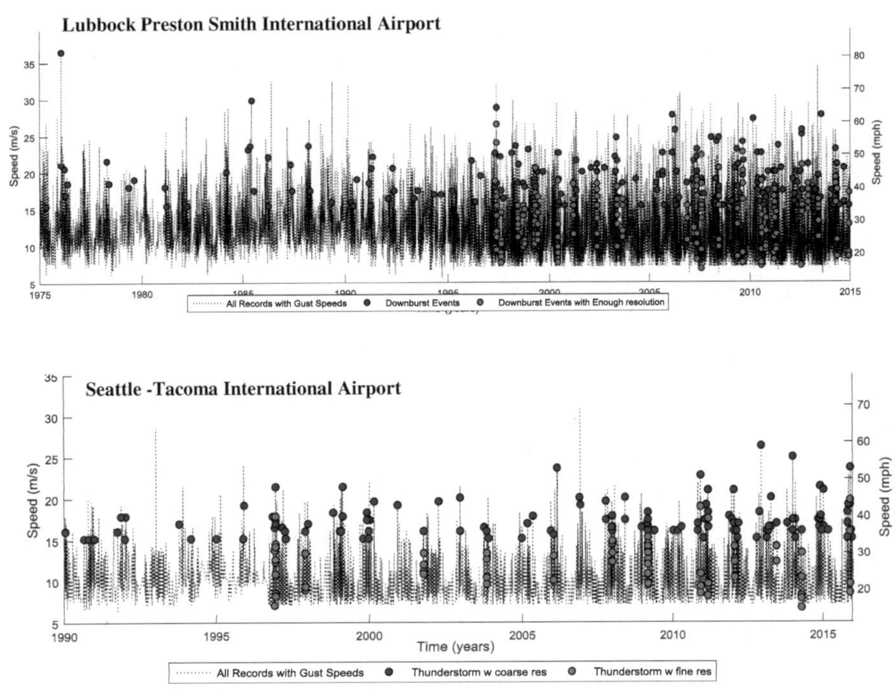

Fig. 4 Extraction of Thunderstorm Gust Front Events

After the model was validated, the framework is applied to study thunderstorm gust speeds for sample 2 meteorological stations: (i) Lubbock Preston Smith International Airport, which is located in a thunderstorm active region and (ii) Seattle -Tacoma International Airport, which is located is a thunderstorm inactive region. This is to show the behavior of the model is these distinct conditions.

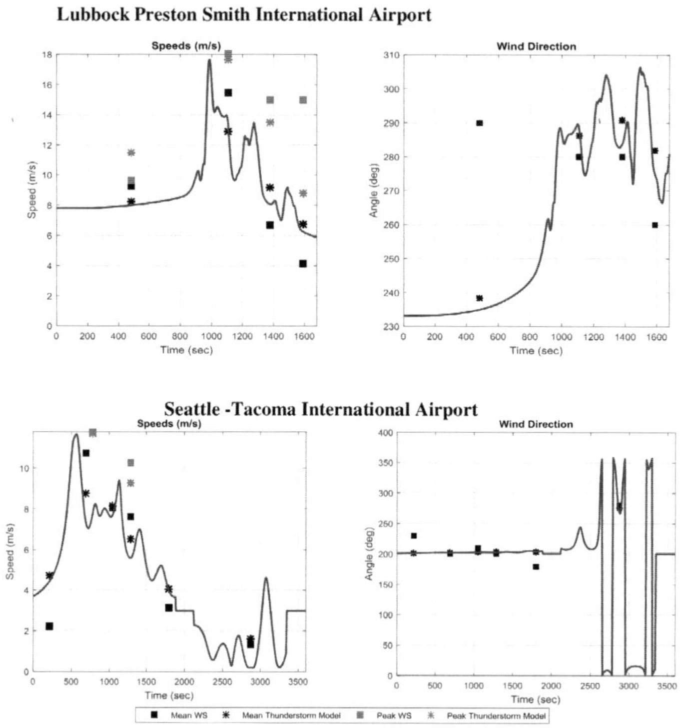

Fig. 5 Sample Fit of the Historical Records

Following to framework steps summarized in Figure 1, after proving the validity of the model, thunderstorm gust events need to be extracted from the historical records. This is conducted as using a 3 step filter as follows. Firstly, wind records are typically provided at meteorological stations in terms of mean and gust speeds. Since gust speeds are provided when wind is intense (which is the case with thunderstorm gust front), only records with available gust speeds are considered as the candidates for thunderstorms. Secondly, theoretical Gust factor assuming a synoptic wind is estimated G_{v0} for wind approaching from various angles, then is compared with the actual gust factor G_v. Records with $G_v > 1.25\ G_0$ are taken as thunderstorm gust front events. Thirdly, identified records from the earlier step are further filtered based on the temporal resolution. Records with high temporal resolution (provided each 10-15 min) are considered good candidates for the fitting with the thunderstorm model to quantify the thunderstorm parameters. Records with coarse resolution and those with high resolution are used to quantify the average number of events per year.

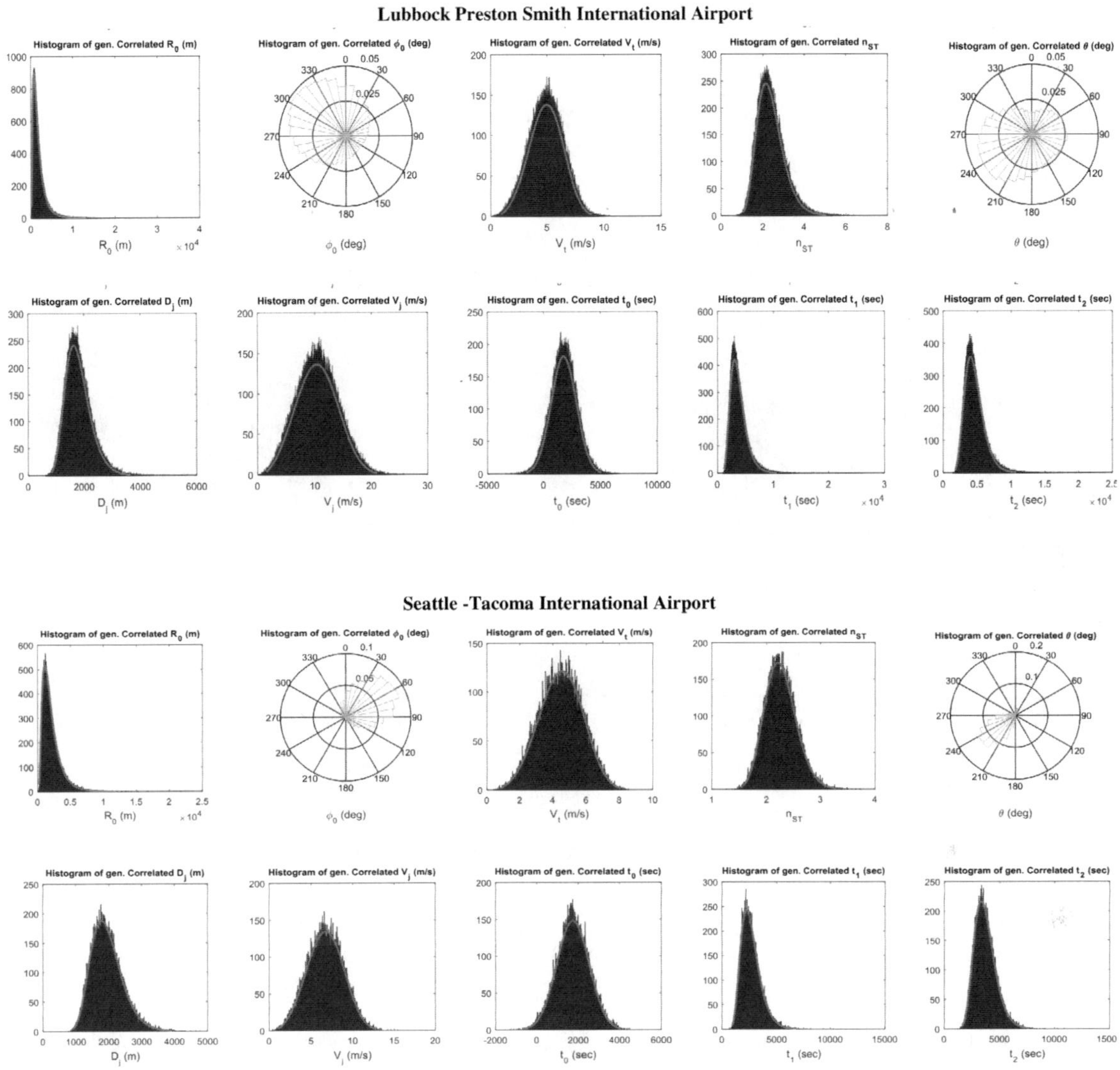

Fig. 6 Probability Distribution of the Thunderstorm Parameters

Records having gust speeds (first step filter) are shown by the dotted lines in Figure 4, while those resulting from the second and third filters are shown by the blue circles for records with coarse resolution and by the red circles for records with the fine resolution.

The third step of the framework (see figure 1) involves fitting the thunderstorm model with the records and estimate the thunderstorm parameters. A number of 40 and 32 records with fine time resolution (red circles in Figure 4) are fit to obtain the model parameters. Figure 5 shows sample of the fit records from both meteorological stations.

The fourth step in the framework represents finding the probability distributions of the model parameters (see Figure 1) and generating parameters for thousands of years based on their probability distribution. Figure 6 shows probability distribution of the parameters. Thunderstorm parameters happening in 5000 years are generated using the probability distribution shown in Figure 6 and correlation among them is introduced according to Equation 2.

$$X_{cor} = LX_{un} \qquad \text{Equation 2}$$

Where X_{cor} is the correlated parameters, X_{un} is the uncorrelated parameters, L is the lower triangle matrix which satisfies $L.L^{T} = C$ where C is the correlation matrix

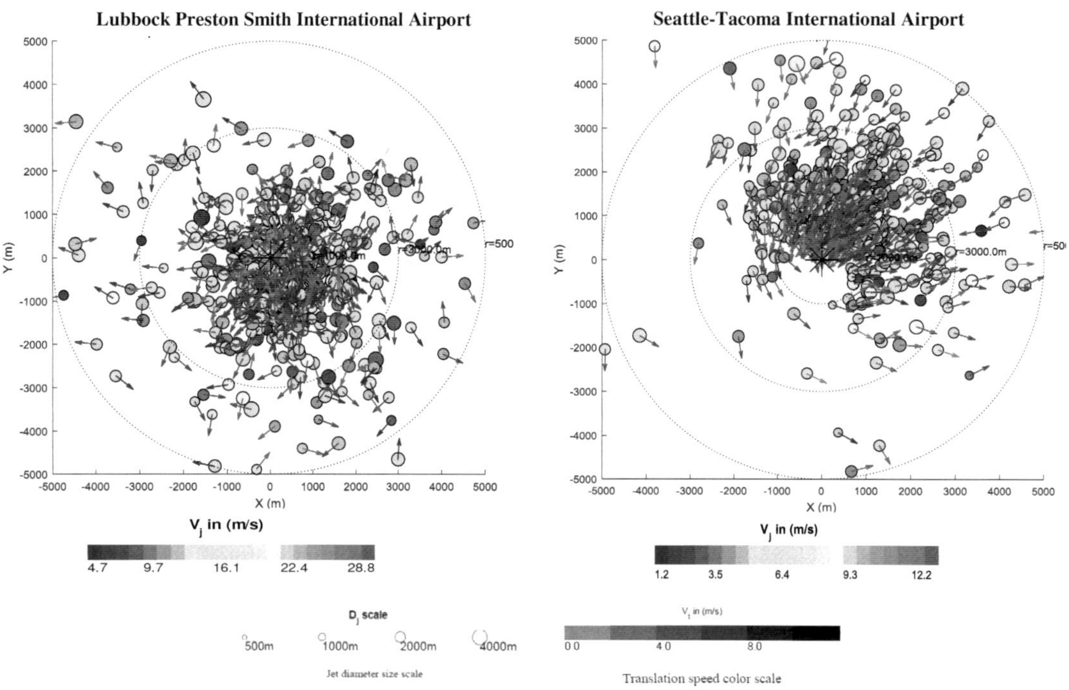

Fig. 7 Sample Downdrafts in a 100-year period

Table 2 Distribution of the thunderstorm parameters

	Param / Dist.	R_0 (m)	ϕ_0 (deg)	V_t (m/s)	n_{ST}	θ (deg)	D_j (m)	V_j (m/s)	T_0 (sec)	T_1 (sec)	T_2 (sec)
Lubbock	Type	Log	von	Log	Log	von	Log	Weib	Nor-	Log	Log
	Const	7.25	-0.88	5.43	0.83	1.39	7.55	11.64	1763.	7.94	7.95
	Const	0.75	0.62	3.77	0.26	0.64	0.25	3.20	1105.	0.51	0.51
Seattle	Type	Log	von	Log	Log	von	Log	Weib	Nor-	Log	Log
	Const	7.44	1.00	4.95	0.81	-2.24	7.55	7.44	1753.	7.83	7.88
	Const	0.65	1.84	3.86	0.13	8.93	0.24	3.47	771.1	0.34	0.39

The constants a, b represent the mean and standard deviation for normal and lognormal distributions and represent the scale and the shape parameters for Weibull distribution, and represent the location and concentration parameters for von Mises distribution, respectively.

Figure 7 shows sample parameters (location R_0, ϕ_0, jet speed V_j, translation speed $n_{ST}.V_j$, and direction θ) for a number of 100 years at both sites. The generated parameters are fed back to the thunderstorm model described in section 2 to identify thunderstorm speeds. Variation of the wind speeds over simulated 5000 years are shown in Figure 8.

Probability of exceedance P_{exc} of a certain speed V is evaluated and used to generate wind speed- return period relationship using Equation 3.

$$R(V) = \frac{1}{\lambda P(v > V)} \qquad \text{Equation 3}$$

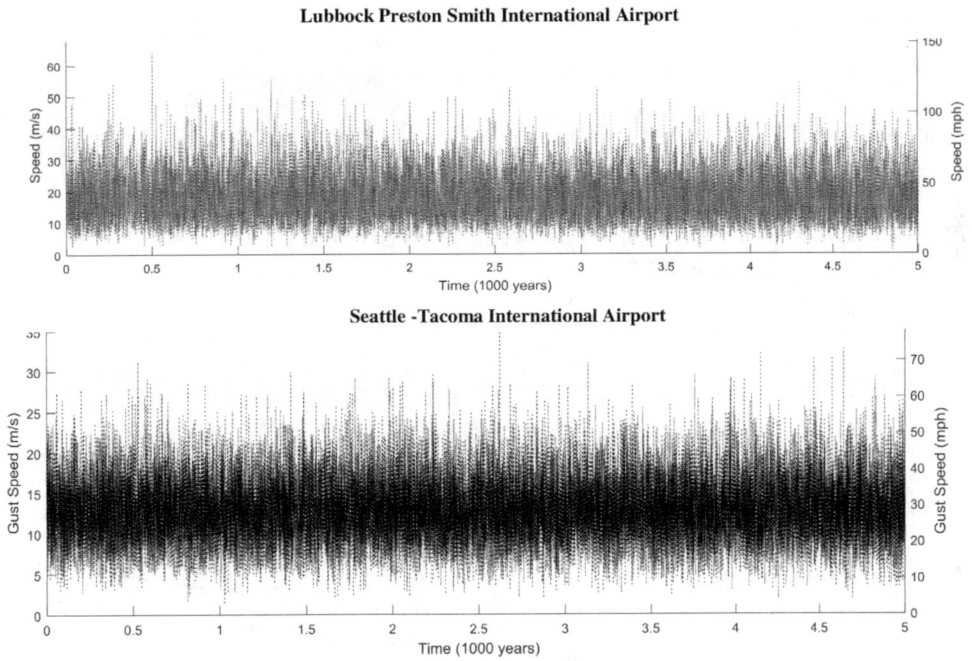

Fig. 8 Simulated Thunderstorm Gust Speeds

Figure 9 shows thunderstorm speeds versus return periods for the two stations given by the black "+" symbol. Gumbel fit of those thunderstorm speeds is provided as well. As indicated from the figures, thunderstorm speeds at Lubbock are much higher than thunderstorm speeds at Seattle-Tacoma. Wind speeds seen at both stations (including synoptic and thunderstorm winds) are treated as synoptic wind, which is a common practice when analyzing meteorological records, and the speed versus the return period and its Gumbel fit are also shown in Figure 8 (blue "+ " and continuous line). Design wind speeds provided in the ASCE 7-10 (2010) are also included in the figure for the comparison purpose. As it can be seen, speeds predicted by the ASCE 7-10 (2010) match or exceed the speeds seen by analyzing the records as synoptic wind. That is expected since the code is based on treating all records as synoptic wind and the code in general tends to be slightly conservative to account for other uncertainty factors.

Comparison between thunderstorm speeds and synoptic wind speeds or ASCE speeds indicate that for Lubbock, which is a thunderstorm active area, thunderstorm speeds are higher than synoptic wind and ASCE speeds for return periods greater than 15 year. For a 700 mean recurrence interval (MRI), which is used by the ASCE for Category II risk, thunderstorm speed is 32% higher than synoptic wind speed (56.8 m/s or 127mph compared with 43.0 m/s or 96 mph) and 10% higher than ASCE wind speed (56.8 m/s or 127mph compared with 51.4 m/s or 115 mph). This is not the case for Seattle-Tacoma station, where thunderstorm speeds are much lower than synoptic wind speeds. The trend of having thunderstorm wind speed higher than synoptic wind speed for thunderstorm active region at long return period (i.e. 15 year or more) is attributed to the fact that thunderstorms are localized events with a maximum speed that can be easily missed from the records taken at a meteorological station especially when the available records are not long enough.

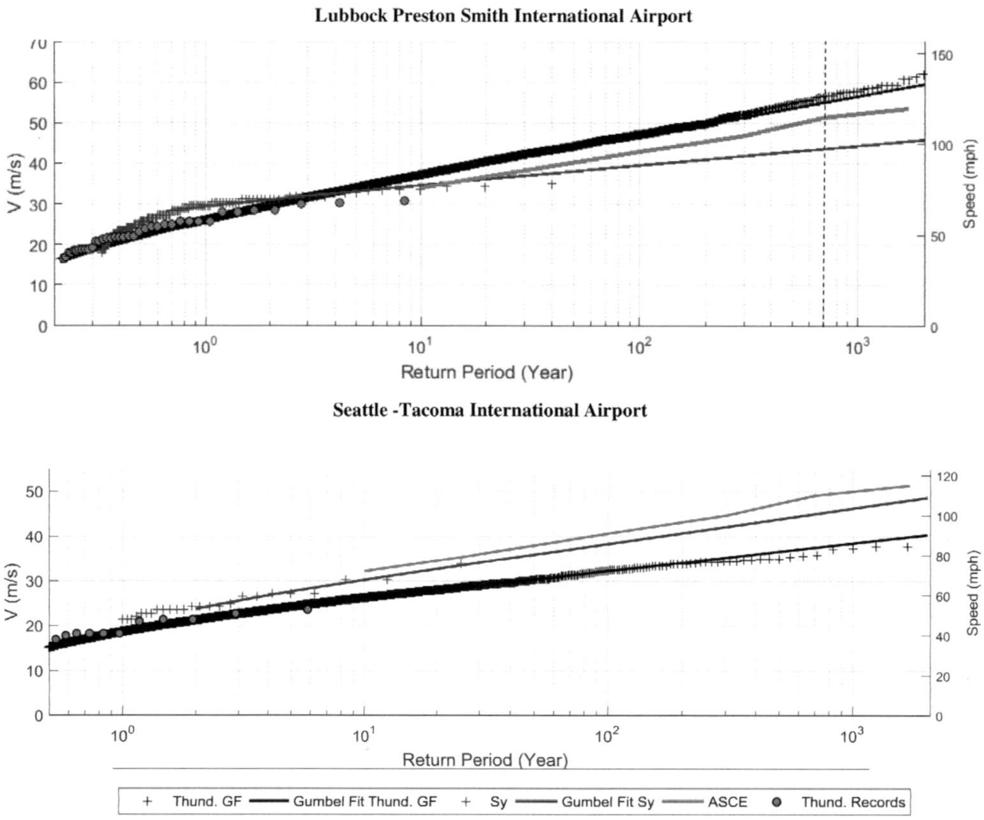

Fig. 9 Thunderstorm Gust Front and Synoptic Wind Speeds

Validation of the framework is assessed by comparing simulated thunderstorm speeds with speeds seen in the records (from step 2 of the framework – see figure 1) as shown in Figures 8. The comparison indicates that the framework successfully describe the wind speed- return period relationship matching what was seen at the sites, which validates the accuracy of the model.

4 CONCLUSIONS

A new framework to quantify thunderstorm speeds is developed and validated. This framework is analogous with hurricane modeling where some records available at wind stations are selected and used to fit a numerical model to obtain the model parameters, then those model parameters are used to generate events for thousands of simulation years to obtain design speeds.

The numerical model is validated throughout successful description of wind speeds and direction seen at some previous thunderstorm events. Then, the model is implemented in the suggested framework to study thunderstorm speeds happening at Lubbock and Seattle-Tacoma metrological stations.

The results indicate that, for Lubbock, which is a thunderstorm active region, thunderstorm wind speeds is higher than synoptic wind speeds (reaches 32 % at 700 mean recurrence interval) and speeds defined in the ASCE 7-10 (reaches 10% at 700 mean recurrence interval). The results also indicate for Seattle-Tacoma, which is a thunderstorm inactive region, that synoptic wind speeds are always higher than thunderstorm speeds. The model is validated throughout successful description of the speed-return period relationship for events seen at the stations.

The proposed framework represents a useful validated tool to quantify thunderstorm gust front speeds suitable for design while accounting for localized nature of the events both in space and time.

5 REFERENCES

1 Abd-Elaal, E.-S., Mills, J.E., Ma, X., 2013. A coupled parametric-CFD study for determining ages of downbursts through investigation of different field parameters. J. Wind Eng. Ind. Aerodyn. 123, 30–42.

2 Aboshosha, H., Bitsuamlak, G., El Damatty A. 2015. Turbulence characterization of downbursts using LES. Journal of Wind Engineering and Industrial Aerodynamics. 136, 44–61.

3 ASCE 7-10. 2010. Minimum Design Loads for Buildings and Other Structures. Reston, VA: ASCE.

4 Boundary Layer Wind Tunnel Laboratoy (BLWTL). 2007. Wind Tunnel Testing: A General Outline. The University of Western Ontario.

5 Chay, M.T., Albermani, F., Wilson, R., 2006. Numerical and analytical simulation of downburst wind loads. Engineering Structures 28(2), 240-254.

6 Choi, E.C.C., 1999. Extreme wind characteristics over Singapore – an area in the equatorial belt. J. Wind Eng. Ind. Aerodyn. 83, 61–69.

7 Choi, E.C.C., Hidayat, F.A., 2002. Gust factors for thunderstorm and nonthunderstorm winds. J. Wind Eng. Ind. Aerodyn. 90, 1683–1696.

8 Choi, E.C.C., Tanurdjaja, A., 2002. Extreme wind studies in Singapore. An area with mixed weather system. J. Wind Eng. Ind. Aerodyn. 90, 1611–1630.

9 Charba, J., 1974. Application of gravity current model to analysis of squall-line gust front. Mon. Weather Rev. 102, 140–156.

10 Cook, N.J., 1982. Towards better estimation of extreme winds. J. Wind Eng. Ind. Aerodyn. 9, 295–323.

11 Duranona, V., Sterling, M., Baker, C.J., 2006. An analysis of extreme non-synoptic winds. J. Wind Eng. Ind. Aerodyn. 95, 1007–1027.

12 Fujita, T.T., 1985. The downburst Report of Projects NIMROD and JAWS. The University of Chicago.

13 Fujita, T.T. 1990. Downbursts: meteorological features and wind field characteristics. J. Wind Eng. Ind. Aerodyn. 36(1), 75-86.

14 De Gaetano P, Repetto MP, Repetto T, Solari G. Separation and classification of extreme wind events from anemometric records. J Wind Eng Ind Aerodyn 2014;126:132–43.

15 Gast, K.D., Schroeder, J.L., 2003. Supercell rear-flank downdraft as sampled in the 2002 thunderstorm outflow experiment. In: Proceedings of the 11th International Conference on Wind Engineering. Lubbock, Texas.

16 Georgiou PN, Davenport AG, Vickery BJ (1983). Design wind speeds in regions dominated by tropical cyclones. J Wind Eng Ind Aerod, 13, 139-152.

17 Gunter, W.S., Schroeder, J.L., 2013. High-resolution full-scale measurements of thunderstorm outflow winds. In: Proceedings of the 12th Americas Conference on Wind Engineering. Seattle, Washington

18 Gast, K. D. & Schroeder, J. L., 2005. Extreme wind events observed in the 2002 thunderstorm outflow experiment. In: Proceedings of the 10th Americas Conference on Wind Engineering (10 ACWE), Baton Rouge, LA, USA.
19 Goff, R.G., 1976. Vertical structure of thunderstorm outflows. Mon. Weather Rev. 104, 1429–1440.
20 Gomes, L., Vickery, B.J., 1976. On thunderstorm wind gusts in Australia. Civil Eng.Trans. Inst. Eng. Aust. 18, 33–39.
21 Hangan, H., Roberts, D., Xu, Z. Kim, J., 2003. Downburst simulation, Experimental and numerical challenges, in: Proceedings of the 11th International Conference on Wind Engineering, Lubbock, Texas, Electronic Version.
22 Hjelmfelt, M.R., 1988. Structure and Life Cycle of Microburst Outflows Observed in Colorado. Journal of Applied Meteorology 27(8), 900-927.
23 Holmes, J.D. & Oliver, S.E., 2000. An empirical model of a downburst. Engineering Structures 22(9), 1167-1172.
24 Holmes, J.D., Hangan, H.M., Schroeder, J.L., Letchford, C.W., Orwig, K.D., 2008. A forensic study of the Lubbock-Reese downdraft of 2002. Wind Struct. 11 (2), 19–39.
25 Oseguera, R. & Bowles, R., 1988. A simple, analytic 3-dimensional downburst model based on boundary layer stagnation flow. NASA Technical Memorandum, 100632.
26 Orf, L.G., Kantor, E., Savory, E., 2012. Simulation of a downburst-producing thunderstorm using a very high resolution three-dimensional cloud model. Journal of Wind Engineering and Industrial Aerodynamics 104, 547-557.
27 Kim, J. & Hangan, H., 2007. Numerical simulations of impinging jets with application to downbursts. Journal of Wind Engineering and Industrial Aerodynamics 95(4), 279-298.
28 Li, S., Hong, H. 2-16. Typhoon wind hazard estimation for China using an empirical track model. Nat Hazards (2016) 82:1009–1029
29 Lombardo, F.T., Main, J.A., Simiu, E., 2009. Automated extraction and classification of thunderstorm and non-thunderstorm wind data for extreme-value analysis. J.Wind Eng. Ind. Aerodyn. 97, 120–131.
30 Mason, M.S., Wood, G.S., Fletcher, D.F., 2009. Numerical simulation of downburst winds. Journal of Wind Engineering and Industrial Aerodynamics 97(11-12), 523-539.
31 Mason, M.S., Fletcher, D.F., Wood, G.S., 2010. Numerical simulation of idealised three-dimensional downburst wind fields. Engineering Structures 32(11), 3558-3570.
32 Russell, L.R. and Schueller, G.F., "Probabilistic Models for Texas Gulf Coast Hurricane Occurrences", J. Pet. Tech., March 1974.
33 Riera, J.D., Nanni, L.F., 1989. Pilot study of extreme wind velocities in a mixed climate considering wind orientation. J. Wind Eng. Ind. Aerodyn. 32, 11–20.
34 Rowcroft, J., 2011. Vertical wind shear profiles in downburst events and the insufficiency of wind turbine design codes. In: Proceedings of the 13th International Conference on Wind Engineering. Amsterdam, Netherlands.
35 Selvam, R. & Holmes, J.D., 1992. Numerical simulation of thunderstorm downdrafts. Journal of Wind Engineering and Industrial Aerodynamics 44 (1-3), 2817-2825.
36 Shehata, A.Y., El Damatty, A.A., Savory, E. 2005. Finite element modeling of transmission line under downburst wind loading. Finite Elements in Analysis and Design, 42, 71-89.
37 Solari, G., Repetto, M.P., Burlando, M., De Gaetano, P., Pizzo, M., Tizzi, M., Parodi, M., 2012. The wind forecast for safety management of port areas. J. Wind Eng. Ind.Aerodyn. 104–106, 266–277.
38 Tryggvason, B.V., Surry, D. and Davenport, A.G. "Predicting Wind-Induced Response in Hurricane Zones", Journal of the Structural Division, ASCE, Vol. 102, No. ST12, Proc. Paper 12630, December 1976, pp 2333-2350.
39 Twisdale, L.A., Vickery, P.J., 1992. Research on thunderstorm wind design parameters. J. Wind Eng. Ind. Aerodyn. 41, 545–556.
40 Vermeire, B.C., Orf, L.G. and Savory, E., 2011. Improved modelling of downburst outflows for wind engineering applications using a cooling source approach. Journal of Wind Engineering and Industrial Aerodynamics 99(8), 801-814.
41 Wakimoto, R.M., 1982. The life cycle of thunderstorm gust fronts as viewed with Doppler radar and rains data. Weather Rev. 110, 1060–1082.
42 Vicroy, D., 1991. A simple, analytical, axisymmetric microburst model for downdraft estimation. NASA Technical Memorandum, 10405.
43 Vicroy, D., 1992. Assessment of microburst models for downdraft estimation. Journal of Aircraft 29 (6), 1043–1048.
44 Vickery, P.J., Skerlj, P.F. and L.A. Twisdale (2000a), "Simulation of Hurricane Risk in the U.S. Using and Empirical Track Model, Journal of Structural Division, ASCE, October.
45 Vickery, P.J., Skerlj, P.F. Steckley, A.C. and L.A. Twisdale (2000b) "A Hurricane Wind Field Model for Use in Hurricane Wind Speed Simulations", Journal of the Structural Division, ASCE, October.

46 Vickery PJ, Masters FJ, Powell MD, Wadhera D (2009). Hurricane hazard modeling: The past, present and future. J Wind Eng Ind Aerod, 97, 392-405.
47 Wood, G.S., Kwok, K.C., Motteram, N.A., Fletcher, D.F., 2001. Physical and numerical modelling of thunderstorm downbursts. Journal of Wind Engineering and Industrial Aerodynamics 89 (6), 535–552.
48 Xu, Z. & Hangan, H., 2008. Scale, boundary and inlet condition effects on impinging jets. Journal of Wind Engineering and Industrial Aerodynamics 96 (12), 2383–2402.

13th Americas Conference on Wind Engineering
Gainesville, Florida, USA
May 21 - 24, 2017

Thunderstorm wind directionality

David Banks[a], Dejiang Chen[b], Anke Beyer-Lout[c]

[a] *CPP Inc, Fort Collins, CO, USA, dbanks@cppwind.com*
[b] *CPP Inc, Fort Collins, CO, USA, dchen@cppwind.com*
[c] *CPP Inc, Fort Collins, CO, USA, abeyer-lout@cppwind.com*

ABSTRACT

One of the underlying premises of the new ASCE 7-16 design wind speed map is the need to separately analyze extreme value wind speed distributions for different storm types, particularly thunderstorms. Over much of the USA, Mexico and Canada, thunderstorms dictate the design wind speed for recurrence intervals of 50 years or more. Our investigation of the wind-time-histories of these storms indicates that all wind directions are equally likely to be associated with the peak wind. The storms tend to have a distinct signature, with a very sudden rise in wind speed accompanied by a sudden significant shift in wind direction. These characteristics are consistent with what we expect from a thunderstorm downburst. We demonstrate this pattern using data from several airports, with an in-depth examination of two sites, one in Texas and one in Arizona.

Not all downbursts are accompanied by thunderstorm flags in the raw weather data, so some judgment is required when identifying the downbursts. In addition, outliers (erroneous data) can look like brief downbursts. A method to characterize the uncertainty associated with these decisions is described, along with a means to incorporate it into a probabilistic design wind load calculation.

For structural design, it is often critical to understand how extreme wind speeds vary with wind direction at a given site. We show how removing thunderstorms significantly improves Gumbel fits to the synoptic winds that otherwise dictate design wind speeds, and sets a floor for the design wind speed for wind directions where other storms produce low wind speeds for high mean recurrence intervals.

With the extreme wind speeds evenly distributed across all wind directions, the appropriate design wind speed in thunderstorm-dominated climates can be estimated from the overall thunderstorm Gumbel fit using the directionality of structural vulnerability; the more narrow the band of high wind load coefficients, the lower the effective design wind speed. Preferably, though, a storm-passage or Monte Carlo analysis can be implemented, and the results of this analysis are compared to the standard ASCE 7 Kd assumption for a typical utility scale solar structure.

KEYWORDS

Extreme value wind speed, Solar wind loads, Thunderstorm downburst

13th Americas Conference on Wind Engineering
Gainesville, Florida, USA
May 21 - 24, 2017

A Simple Vortex Model of a Thunderstorm Downburst – a Parametric Study

Michael Jesson[a], Mark Sterling[b]

[a] *Department of Civil Engineering, University of Birmingham, Birmingham, UK. m.a.jesson@bham.ac.uk*
[b] *Department of Civil Engineering, University of Birmingham, Birmingham, UK.*

ABSTRACT

With the recognition that they cause the design wind speeds for structures in many parts of the world, the study of wind loading due to thunderstorm downbursts has increased over the last decade. Downbursts are short-lived, small-scale events with wind-speeds reaching approximately 150mph, similar to a category EF3 tornado. These winds last for only a few minutes, and are centred on a downdraft region with a diameter of only 1 - 2km. The impingement of this downdraft on the ground creates a high-pressure region and radial outflow. A primary, ring vortex also forms at the shear layer between the downdraft and the surrounding air, and this vortex is transported to ground level and then outwards with the radial flow. It has also been shown that a secondary, counter-rotating vortex forms due to the interaction of the primary vortex and ground roughness. Unlike atmospheric boundary layer (ABL) flows, which increase monotonically with altitude in the familiar logarithmic profile, this combination of flow components results in a high-speed region close to the ground (between 30m and 100m altitude), with a decrease in wind-speed above this point. The wind loading due to downbursts has been shown to differ from that of ABL flows, adding to the importance of understanding their interaction with man-made structures.

Due to the practical difficulties of measuring downburst winds at full-scale, numerical and physical simulations have been employed to further understand the mechanisms which cause the high wind-speeds, and their implications for structures. Apparatus for the physical simulation of downbursts is complex and generally too large to permit easy modification to vary the simulated downburst which they produce. Numerical simulation techniques including Large Eddy Simulation and cloud models have been used, but these are computationally expensive and time consuming to run. This paper introduces a vortex model of a downburst, which includes the three major flow components (outflow, primary vortex, secondary vortex) and models the vortices as simple Rankine type vortices. The evolution of these vortices, in terms of their position, shape and strength, is configurable, and the model takes only a few seconds to run.

Output from the model shows that, from the edge of the downdraft to past the radial distance at which the maximum radial velocity occurs, it is able to match the vertical profiles of velocity as measured in the University of Birmingham Transient Wind Simulator. At distances significantly greater than the position at which the maximum radial velocity occurs, the model deviates from the experimental data, possibly due to more complex breakdown of the vortices than is captured by the model. The speed and simplicity of the model facilitates multiple runs with modified vortex evolution, and details of such a parametric study will be given in the paper presentation.

KEYWORDS

Downbursts, thunderstorms, numerical modelling, vortices.

The 13th Americas Conference on Wind Engineering (13ACWE)
Gainesville, Florida USA, May 21-24, 2017

A Preliminary Examination of Structural Fragility for a Cantilever Structure Subjected to a Thunderstorm Downburst Loading

Viet Le [a], Luca Caracoglia [a]

[a]*Northeastern University, Boston, Massachusetts, USA*

ABSTRACT: Non-stationary downburst winds can induce large-amplitude vibration on vertical slender structures. Since the physical structure of a thunderstorm is currently not fully predictable, several assumptions (modeling simplifications) and uncertainty sources (modeling errors) are present. In this paper, the propagation of errors (due to random variability) through the dynamical system and through limited knowledge of the downburst structure and loading is considered by utilizing a Monte Carlo based quasi-static simulation method for random drag coefficient. To examine the along-wind dynamic response of a simplified vertical cantilever structure, numerical integration of the linear and nonlinear dynamics is considered. The concept of structural fragility curves associated with the maximum lateral vibration of a vertical slender structure is extended to non-stationary wind loads due to thunderstorm downbursts. Furthermore, a preliminary parametric study is conducted to investigate the influence of changing factors in the downburst aerodynamic loading.

KEYWORDS: Downburst Model, Non-stationary Loads, Fragility Analysis, Performance-based Engineering, Transient Response

1 INTRODUCTION

1.1 *The downburst*

Downbursts are meteorological wind phenomena typically occurring during thunderstorms. They were first recognized by Fujita and Byers in 1977 after the investigation of several deadly aircraft accidents [1-5]. In the following decade, a significant number of aircraft accidents caused by downburst-driven wind shear were reported. Due to difficulty in physically identifying downbursts (unlike tornadoes), several on-site systematic recordings during thunderstorms were obtained by Doppler radar stations as part of projects such as the Northern Illinois Meteorological Research in Downbursts (NIMROD) in 1978 and the Joint Airport Weather Studies (JAWS) in 1982 [1, 2]. Using full-scale observations and records, meteorologists and researchers have formulated analytical models to characterize the main features of thunderstorm downburst winds [6-15]. These include a touchdown point, about which the storm is centered, a high-velocity non-stationary wind field, and a boundary layer that differs greatly from that of synoptic winds. The downburst first evolves from an intense vertical downdraft of wind that lands at a touchdown point from which it then radially diverges and decays over a short period of time. As this outburst of wind flows out from the touchdown point, the downburst travels along a path with translational velocity producing transient and non-synoptic wind fields for a short duration of time between a few to dozens of

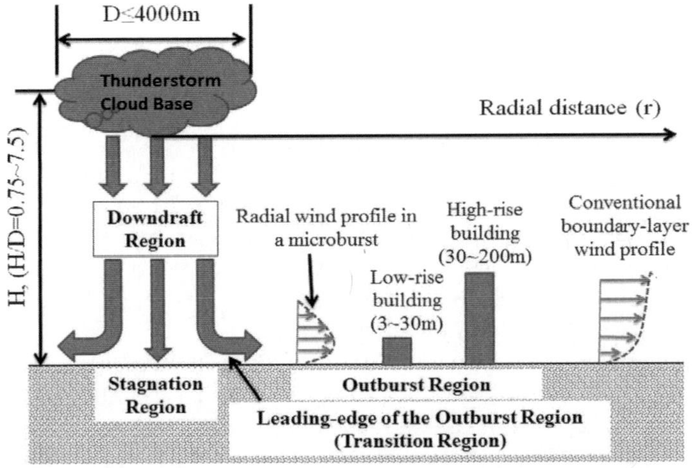

Figure 1. Schematic of a microburst – reproduced from Zhang, Sarkar *et al.* [10]

minutes. Its velocity consists of several main components: radial, vertical, horizontal, and translational [8, 9]. In addition, the velocity can be defined as a combination of a slowly varying mean wind speed and rapidly varying non-stationary turbulence. A schematic of a microburst (a type of downburst) along with a comparison between an example radial wind profile and a conventional boundary-layer wind profile is shown in Figure 1, reproduced from Ref. [10].

From a civil engineering perspective, downbursts present challenging problems as the combination of a rapidly-varying non-stationary turbulence field with a slowly-varying mean wind speed can generate complex aerodynamic loadings that lead to large amplitude vibrations. In addition, the structural response of a system under non-stationary wind loading is difficult to analyze in the frequency domain due to coupled and nonlinear equations of motions [15]. Analysis is necessary in time domain which increases computational demand or in wavelet domain which has been recently proposed. Despite these issues, research work, both through numerical analysis and physical simulation, provide insight into the importance of understanding the downburst for civil engineering applications. For example, using a transient wind simulator consisting of a 1 m diameter impinging jet with aperture control, Jesson *et al.* [16] subjected low-, mid-, and high-rise building models to conditions similar to those present in downbursts and discovered that certain building types would experience high levels of negative pressure coefficients. Le and Caracoglia [17-19] utilized the wavelet-Galerkin method to obtain the stochastic structural dynamic response of tall buildings from digitally simulated non-stationary, transient wind loadings. They selected compactly supported Daubechies wavelets and employed the Galerkin approach to transform the time-varying differential equations of motion into a discrete system of algebraic linear equations in the wavelet domain. Additional developments in techniques coupled with robust physical experimentation can aid researchers in understanding the effects that transient wind loads impose on civil engineering structures.

1.2 *Performance-based engineering*

Numerical analysis and physical simulation of non-stationary wind velocity fields typically involve various assumptions (e.g. axisymmetry of the downburst, nonevolutionary spectra of turbulence, spatial-temporal intensification functions, or pressure and load aerodynamics) due to limited knowledge of the downburst structure and loading. Thus, significant errors from random variability accompany predictions or measurements of structural response and must be considered using probabilistic models.

Owners or engineers can use probability to establish the likelihood that a target structure may exceed a pre-defined limit state or hazard intensity. Limit states are thresholds of structural performance specified by the design criteria or intended objective [20]. Modelling this likelihood of exceedance of a limit state is an integral aspect of performance based engineering [21]. Advancements in computational structural analysis have spurred interest in the study of performance based engineering especially for the unique excitation mechanisms found in the field of wind engineering. Recent research endeavors have utilized methods such as Monte Carlo based simulations to generate fragility curves for structures under stationary wind loads [22-26].

This paper capitalizes from recent studies on the estimation of downburst-induced load and response by using Monte Carlo simulation to account for error propagation and variability in a fragility approach. A simplified vertical cantilever "point-like structure" is subjected to digitally simulated lateral downburst wind loads using a recently developed formulation for the mean wind velocity field [17-19]. In addition, turbulence is simulated with an evolutionary spectral representation methodology utilizing digital modulation to describe the time-varying downburst intensity. The linear and nonlinear dynamics of the structure are examined and fragility curves accounting for wind load variability through random drag coefficient are calculated. Furthermore, a parametric study is conducted to investigate the effects of maximum mean wind speed and distance between the downburst touchdown point and the building structure's location. The preliminary results discussed in this paper are part of an ongoing research activity oriented towards the development of a performance-based methodology for fragility analysis of the structural response due to non-stationary downburst wind loads.

2 SIMULATION OF WIND VELOCITY COMPONENTS OF DOWNBURST

2.1 *Basic structure of a downburst*

Outside the core of the downdraft, velocity features of a downburst can be described by a slowly time varying mean wind velocity (non-turbulent) and rapidly, temporally varying wind fluctuations (turbulent velocity). A simulation of the downburst wind fields typically reproduces these properties through a "deterministic-stochastic" approach, in which the turbulent and non-turbulent components are separately generated and subsequently combined through vector summation [12, 27-29]. The deterministic mean characterizes the time-varying mean wind speed while the partially coherent stochastic high-frequency fluctuation wind field describes the turbulence. This study applies a similar deterministic-stochastic approach to model the wind field. The transient horizontal downburst velocities are defined in three-dimensional space:

$$\boldsymbol{U}(x,y,z,y) = \bar{\boldsymbol{U}}(x,y,z,y) + \boldsymbol{u}'(x,y,z,t) \tag{1}$$

where the transient horizontal downburst velocity, $\boldsymbol{U}(x,y,z,t)$, is the vector sum of mean wind speed, $\bar{\boldsymbol{U}}(x,y,z,t)$, and amplitude modulated turbulence, $\boldsymbol{u}'(x,y,z,t)$.

2.2 *Slowly, time-varying mean wind velocity*

In this study, the downburst mean wind velocity, $\bar{U}_z$, profile as a function of height z is calculated with Vicroy's model [8] as shown in Equation 2.

$$\bar{U}_z(z) = 1.22\left[e^{-0.15z/z_{max}} - e^{-3.2175z/z_{max}}\right]\bar{U}_{max} \quad (2)$$

where z_{max} is the elevation from the ground at which the maximum mean wind velocity in the wind profile $\bar{U}_{max}$ occurs. Time-dependent intensification functions proposed by Chay, Albermani *et al.* [12] and Chen and Letchford [28] can then be used to generate the downburst mean wind field.

$$\bar{\boldsymbol{U}}(x, y, z, t) = \boldsymbol{U}_r(x, y, z, t) + \boldsymbol{U}_{tran}(x, y, z) \quad (3a)$$

$$U_r(x, y, z, t) = \Pi(t)\bar{U}_z(z)g(r) \quad (3b)$$

$$\Pi(t) = \begin{cases} t/t_0 & t<t_0 \\ exp(-(t-t_0)/T) & t>t_0 \end{cases} \quad (3c)$$

$$g(r) = \begin{cases} r/r_{max} & r<r_{max} \\ exp(-(r-r_{max})/R)^2) & r>r_{max} \end{cases} \quad (3d)$$

The resultant downburst mean wind velocity (slowly varying and time dependent), $\bar{\boldsymbol{U}}$, is the vector summation of the radial wind velocity, $\boldsymbol{U_r}$, and the translational velocity, $\boldsymbol{U_{tran}}$. For this study, translational wind speed is a constant 10 m/s. Radial wind velocity, U_r, is calculated using Equation 3b along with the intensification functions of Equations 3c and 3d. Equation 3c describes the evolution and intensity of the downburst through its lifecycle (Fig. 2) where t is the time since the downburst has landed at its touchdown point, t_0 is the time at which the non-turbulent velocity reaches the maximum intensity, and T is the duration of the downburst (total time of the simulation). Equation 3d represents the intensification of the downburst as a function of the horizontal radial distance, r, between the center of the downburst and the point of interest (e.g. geometric center of a building structure). At a certain location r_{max} from the core of the downburst, the intensity of the wind velocity reaches a maximum. R is the radial length scale. For the simulations in this study, various maximum mean wind velocities are used at z_{max} = 80 m while the travelling downburst has parameters t_0 = 6 minutes, r_{max} = 1000 m, and R = 700 m [30]. The duration period of the downburst, T, is 12 minutes.

The magnitude from the vector summation of Equation 3a is shown below in Equation 4.

$$\bar{U}(x, y, z, t)^2 = \bar{U}_r(x, y, z, y)^2 + \bar{U}_{tran}(x, y, z, y)^2 + 2 \times \bar{U}_r(x, y, z, y) \times \bar{U}_r(x, y, z, y) \times \cos(\beta(x, y, t)) \quad (4)$$

$$\mathbf{cos(\theta(x, y, t)) = [\bar{U}(x, y, z, y)^2 + \bar{U}_{tran}(x, y, z, y)^2 - \bar{U}_r(x, y, z, y)^2]/[2 \times \bar{U}(x, y, z, t) \times \bar{U}_{tran}(x, y, z, y)]} \quad (5)$$

Here, β is the angle between the radial and translational velocity vectors. Equation 5 calculates the angle θ which is the deviation from the prevailing travelling direction of the storm (based on $\boldsymbol{U_{tran}}$) due to downburst mean wind speed. In this study, it coincides with the deviation from the along-wind axis due to the downburst mean wind speed.

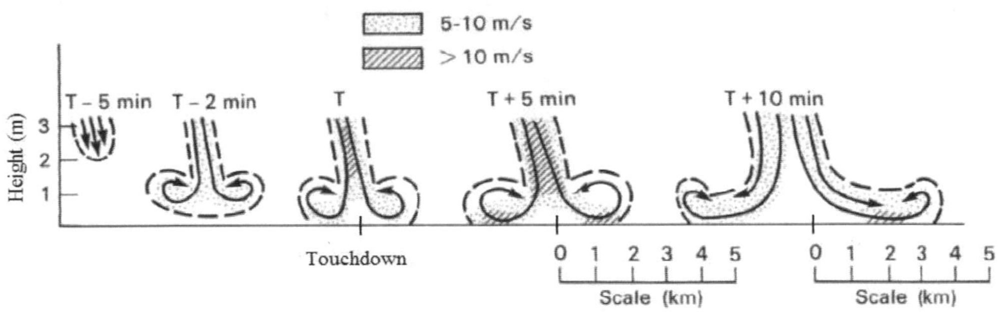

Figure 1. Schematic of the lifecycle of a downburst from touchdown to divergence – reproduced from Hjelmfelt [7].

2.3 *Rapidly, temporally varying wind fluctuations*

The partially-correlated non-stationary turbulent wind field is simulated by superposition of partially-correlated stationary processes with amplitude modulated by a slowly time-varying deterministic function [17-19]. The technique behind this approach stems from evolutionary power spectral density functions (EPSD). To simulate the one-dimensional multi-variate stationary turbulence field, Di Paola [31] has suggested the discretization of the stochastic turbulence into the superposition of single-variate, single-dimensional processes obtained through the proper orthogonal decomposition of the cross power spectral density (PSD) matrix. Using this method and the Kaimal spectrum for longitudinal wind fluctuations [32], the discretized time-dependent stochastic stationary turbulence, $u_p(x_p,y_p,z_p,t)$ at discrete locations or nodes $p = 1, 2, \ldots, M$ can be written as the following summation of M-independent processes each with frequency ω_l:

$$u_p(x_p, y_p, z_p, t) = 2\sqrt{\Delta\omega}\sum_{m=1}^{M}\sum_{l=1}^{n}\psi_{ml}\sqrt{\Lambda_m(\omega_l)}g_{ml}(t) \tag{6}$$

in which $\Delta\omega = \omega_{up}/n$ is a circular frequency interval or "step" with ω_{up} as the upper cut-off circular frequency and n as the number of circular frequencies used by the wave-superposition method. ω_l is a generic circular frequency ($\omega_l = l\Delta\omega$). ψ_{ml} is the l-th eigenvector of node p normalized with respect to the identity matrix and Λ_m represents the eigenvalues of the PSD matrix for node p. Equation 7 for $g_{ml}(t)$ represents harmonic waves of the zero-mean normal single-variate, single-dimension, processes.

$$g_{ml}(t) = R_{ml}\cos(\omega_l t) + I_{ml}\sin(\omega_l t) \tag{7}$$

R_{ml} and I_{ml} are pairs of independent, standard, normally distributed (zero expectation, unit variance) random numbers generated by the Box-Muller algorithm [33].

Turbulence amplitudes in a thunderstorm are uniformly modulated using a deterministic cosine function [34]. The modulation depends on time t but is independent of ω and the position [18]:

$$\hat{A}_p(x, y, z, t) = ((1 - \cos(2\pi t/T_0)/2)^{\eta} \tag{8}$$

T_0 is a reference duration which controls the time of maximum amplitude and η is a positive integer even number for the width of the cosine window. In this investigation, T_0 is half the time duration of the simulated downburst and η is 2. For node p, the stationary signal is at last combined with the amplitude modulation function as follows:

$$u_p'(x_p, y_p, z_p, t) = \hat{A}_p(\omega, x_p, y_p, z_p, t)u_p(x_p, y_p, z_p, t) \tag{9}$$

$u_p'(x_p,y_p,z_p,t)$ is the spatially-correlated zero-mean stationary turbulence process in the same direction as the resultant downburst mean wind velocity, $\bar{\boldsymbol{U}}$. This process is repeated for turbulence $v_p'(x_p,y_p,z_p,t)$ in the lateral direction perpendicular to $\bar{\boldsymbol{U}}$. This component uses the Kaimal spectrum for winds in the lateral direction instead [32]. Both turbulence components, thus, are generated using spectrum models developed by Kaimal.

3 DYNAMIC RESPONSE DUE TO DOWNBURST WINDS

3.1 *Simplified, vertical cantilever, "point-like structure"*

The system is modelled after a simplified, slender, vertical cantilever structure. The "point-like structure" resembles a thin plate with an area A of 2 m^2, mass m of 6000 kg, stiffness constant k_x of 3840 N/m in the along-wind axis and k_y of 4646.4 N/m in the across-wind axis, and damping ratio ζ of 0.01. The height h of the structure is at an elevation of z = 30 m.

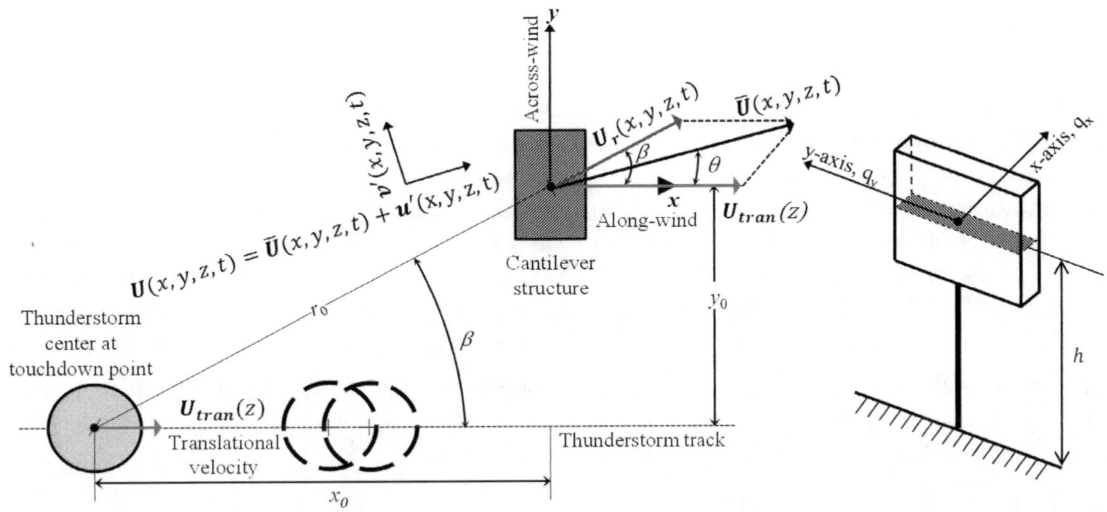

Figure 2. Schematic showing the placement of the structure with respect to the thunderstorm's path of travel. A three-dimensional view of the structure is also provided.

Two scenarios are considered for the location of the structure with respect to the storm's touchdown point. The first involves placing the structure in the path of the travelling downburst. This lends to a single degree-of-freedom (1DOF) analysis in the along-wind direction for the structural motion. The second scenario imposes an offset to the travelling downburst. The analysis is two-dimensional (2DOF) and includes turbulence and structural motion in the across-wind direction.

Figure 2 provides a schematic of the described system for the 2DOF scenario. It is assumed that the along-wind (*"x-"*) axis corresponds to the prevailing travelling direction of the storm (based on $\boldsymbol{U_{tran}}$). The local coordinate axis, perpendicular to it, is the across-wind axis (*"y-"*).

3.2 *Governing dynamic equations*

The equation dictating the time-dependent motion of the structure, $q(t)$, in the 1DOF scenario is derived from the dynamic equation of motion for a "point-like structure" subjected to a drag force (other aerodynamic forces such as transverse or lift force, moments and torque have been neglected for this study):

$$m\ddot{q}(t) + c\dot{q}(t) + k_x q(t) = \frac{1}{2}\rho U^2 A C_D \tag{10}$$

m, c, k_x are the known mass, damping, and stiffness coefficients with values given in the previous section, ρ is the air density taken as 1.25 kg/m^3, A is the area of the point (representative of a plate), C_D is the drag coefficient, and U includes the total downburst wind velocity component at time t and the effects of fluid-structure interaction.

Only for the 1DOF scenario, a nonlinear damping of the Van-der-Pol type (μ = 0.01) is introduced to study the effects of non-linearity in the structural dynamics [18]. The wind velocity, U, is calculated by the simple addition of mean wind velocity, $\bar{U}$, turbulence u', and opposing structural motion, $\dot{q}'$. After linearization to remove second-order effects of turbulence and structural motion, the differential equation of the response is given in Equation (11).

$$m\ddot{q}(t) + c(1 - \mu r(t)^2)\dot{q}(t) + k_x r(t) = \frac{1}{2}\rho A C_D \left(\overline{U}^2 + 2\overline{U}u' - 2\overline{U}_r(t)\right) \tag{11}$$

The addition of turbulence and structural motion in the across-wind direction adds complexity to the motion equation in the 2DOF scenario. Hence, nonlinear damping has been later excluded from the formulation when a more accurate 2DOF model is considered.

The wind velocity components in the 2DOF scenario cannot simply be summed together for vector addition as is done in the simpler 1DOF case. The angle, β, between the radial velocity, U_r, and the translational velocity, U_{tran}, must be considered as shown in Equation 4 to calculate the resultant mean wind velocity, $\bar{U}$, with angle θ measured from the along-wind axis. Velocity of structural motion is first designated as follows: $\dot{q}_x{}'$ which is parallel to $\bar{U}$ (i.e. direction denoted by angle θ) and $\dot{q}_y{}'$ which is perpendicular to $\bar{U}$. Turbulence components, u' and as v', are also projected along the same pair of directions. The square of the magnitude of the effective velocity vector, after accounting for turbulence and structural motion (aerodynamic damping), is considered and leads to U_{eff}^2:

$$U_{eff}^2 = (U + u - \dot{q}_x{}'(t))^2 + (v - \dot{q}_y{}'(t))^2 \tag{12}$$

The terms $\dot{q}_x{}'$ and $\dot{q}_y{}'$ are subsequently projected onto the along-wind and across-wind axes using angle θ:

$$\dot{q}_x{}'(t) = \dot{q}_x(t)\cos(\theta) + \dot{q}_y(t)\sin(\theta) \tag{13}$$

$$\dot{q}_y{}'(t) = -\dot{q}_x(t)\sin(\theta) + \dot{q}_y(t)\cos(\theta) \tag{14}$$

A change in angle, γ, is introduced to model the fluctuation from the originally defined angle θ due to the turbulence and structural motion. It is calculated as follows:

$$\gamma = \tan^{-1}\left(\left(v - \dot{q}_y{}'(t)\right) \Big/ \left(\bar{U} + u - \dot{q}_x{}'(t)\right) \right) \tag{15}$$

After assuming small-angle deviations and $\bar{U}$ being significantly greater than u - $\dot{q}_x{}'$, γ becomes:

$$\gamma = \left(v - \dot{q}_y{}'(t)\right) \Big/ U \tag{16}$$

The angle from the along-wind axis, at which the effective wind speed is directed, presently considers both mean wind (slowly varying) speed and turbulence fluctuations; is denoted by φ, the summation of θ and γ. The structural response in the along-wind and across-wind directions can finally be formulated by projecting the resulting drag force in the along-wind and across-wind axes shown in Figure 3. Eliminating second-order effects from turbulence and structural motion and assuming small angles, the responses of the structure in both directions due to the travelling downburst can be found:

$$m\ddot{q}_x(t) + c_x\dot{q}_x(t) + k_x q_x(t) = \frac{1}{2}\rho A C_D \left(\overline{U}^2 + 2\overline{U}u' - 2\overline{U}\left(\dot{q}_x(t)\cos(\theta) + \dot{q}_y(t)\sin(\theta)\right)\right)\cos(\varphi) \tag{17}$$

$$m\ddot{q}_y(t) + c_y\dot{q}_y(t) + k_y q_y(t) = \frac{1}{2}\rho A C_D \left(\overline{U}^2 + 2\overline{U}u' - 2\overline{U}\left(\dot{q}_x(t)\cos(\theta) + \dot{q}_y(t)\sin(\theta)\right)\right)\sin(\varphi) \tag{18}$$

3.3 *Random simulation of drag coefficients*

Random perturbation, thus far, has only been introduced in the generation of the standard Gaussian amplitudes of the harmonic waves superimposed for turbulence. To account for additional error propagation or uncertainty in the system, random simulation utilizing the Monte Carlo method is performed to produce a sample of reference drag coefficients. As summarized by Di Paola [31], "[the Monte Carlo method] mainly consists in generating sample functions having prescribed probabilistic characteristics, then for each sample function the linear or non-linear dynamic structural analysis is performed. From the time histories of the response the probabilistic analysis is made on the sample functions of the response for ensuring overall safety of the various elements of the structure."

Monte Carlo methods have been previously employed in fragility and expectancy costs studies to examine wind hazards on vulnerable structures such as tall buildings and long-span bridges [23-26]. They statistically assess the dynamic response due to "uncertain loading scenarios" in which modelling simplifications and measurement errors are considered. The Monte Carlo method is used here to generate stochastic occurrences of the reference drag coefficient, C_{D0}, at the beginning of the simulation. This variable is assumed to follow a standard normal probability distribution with a mean of 1.2 and coefficient of variation of 0.10. These values are compatible with a large sign or support structure such as a highway sign or luminaire. If quasi-steady wind load hypothesis is used but a direction-dependent drag coefficient is introduced to simulate a non-symmetric shape in the source of aerodynamic load, this drag coefficient must be recalculated at each time instant using Equation (19) where φ is the sum of deviation from the along-wind axis due to downburst mean wind speed and turbulence fluctuations.

$$C_D = C_{D0} - 0.20(1 - cos(2\varphi)) \quad (19)$$

To linearize the wind loads, subsequent instances of drag force in time are calculated from a process utilizing the derivative of the drag coefficient with respect to angle φ. This is described below by Equation (20).

$$C_{D-lin} = C_{D0} + \left({dC_D}/{d\varphi}\Big|_{\theta} \right) \varphi \quad (20)$$

where $dC_D/d\alpha|_\theta$ is the linear rate change of the drag coefficient with respect to the angle φ, approximately evaluated at an "average" angle of attack θ (roughly coincident with the relative direction between the slowly-varying downburst velocity and the local direction x). It must be noted that Equation 20 approximates the exact relationship between relative wind angle and the local axis (orientation) of the structure. Therefore, it is only adequate for small angles only. Preliminary investigation through inspection of Equation 20 has confirmed that the approximation is acceptable in most cases and, therefore, it is used in the subsequent sections of this study. Additional investigation, beyond the scope of this paper, is under way to further examine the implications of the approximate Equation 20 on the predictions of dynamic response for point-like structures.

3.4 *Evaluating the differential equations of motion*

Approximate solutions to the differential equations can be obtained through a variety of approaches both in time and wavelet domains. This study utilizes the classic fourth order Runge-Kutta-Fehlberg (RKF) numerical integration method to approximate solutions in the time domain. For each time step that constitutes the temporal discretization, the RKF method requires information of the variables for multiple instances in the future to calculate an approximation of the integration for a differential equation. In this study, the integration result is the structural response, q, while the variables needed in the future include mean wind speed, turbulence, and deviation angle φ. The assumption of constant loads (and accelerations) during each time step of the RKF integration has been made to reduce computing time. A time step of 0.05s is used.

The time step is calibrated to minimize the influence of any modeling simplifications and to allow the elimination of second-order effects without compromising accuracy. The effect of linearization of the drag force coefficient is not fully investigated herein but may possibly be considered in future studies. Figure 3 compares the along-wind response in the 2DOF scenario produced with and without the above-described assumptions. From Equation 19, variation of the drag coefficient (0.2/1.2) may be large with respect to the angle of attack and may not be possible on a realistic structure. Responses are thus normalized by the area of the cantilever system. $\bar{U}_{max}$ is 60 m/s, x_0 is 1500 m, and y_0 is 1000 m. Drag coefficients and turbulence realizations are identical in

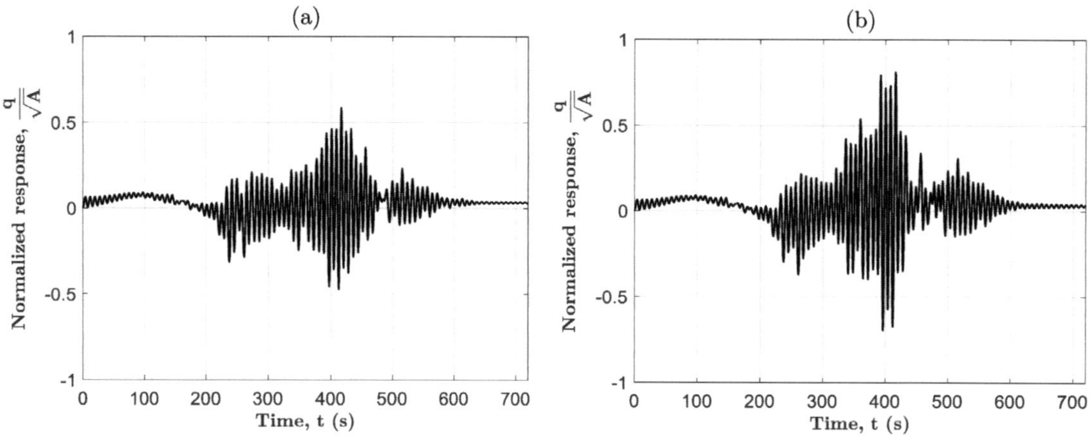

Figure 3. Example of along-wind response (a) with simplifying assumptions and (b) without any assumptions.

both cases. The results show that assumptions made for enabling numerical integration may underestimate deflections of the cantilever structure. Differences from this solution technique require further examination; this is beyond the scope of this study but must be accounted for in future investigations.

4 RESULTS AND ANALYSIS

4.1 *Parametric study*

Using the 1DOF scenario, a parametric study of x_0 is conducted for ten values: x_0 ={250 m, 500 m, 1000 m, 1500 m, 2000 m, 2500 m, 3000 m, 4000 m, 6000 m, 8000 m}; y_0 is always equal to zero (y_0 = 0 m). A sample random population of 5000 C_{D0} was generated. The quantity C_{D0} is the reference drag coefficient of the structure (zero angle of attack), which initializes the time-domain numerical integration of the differential equation of motion, Equation (11). Therefore, 5000 along-wind response sets are obtained for each value of x_0 by Monte Carlo sampling. Figure 9 shows the results for two separate downbursts with maximum mean velocities $\bar{U}_{max}$ = {40 m/s, 65 m/s} and a translational downburst velocity of U_{tran} = 10 m/s. For each distance x_0, the maximum value of the dynamic response time history is extracted. The collection of the maxima allows examination of the statistics of the response. The median, $q^m{}_{MED}$, of the sample is determined and normalized with respect to the square root of the plate's projected area, $\sqrt{A}$. The normalized median response exhibits absolute maximum at a "critical" touchdown position of x_0 = 4000 m for both velocities (Fig. 5). With x_0 = 2500 m, the structural response in Figure 5 shows a second local peak. Special considerations must be taken to properly assess the deflection of the cantilever system at this touchdown position. A higher resolution of x_0 may perhaps reveal more cases exhibiting similar local features.

An investigation is also conducted for $\bar{U}_{max}$ using variables between 40 m/s and 65 m/s. Even though the upper limit has a less plausible physical meaning, it is selected to enable further investigation of the results. It is expected that allowing the wind field to achieve higher speeds may result in strong responses and large structural motion. Figure 5 verifies this trend. The normalized median of the maximum $q^m{}_{MED}/\sqrt{A}$ is plotted against the maximum mean wind speed of the downburst, $\bar{U}_{max}$. An exponential function is used for the empirical curve fitting of the numerical data where a and b are suitable constants.

$$q^m_{MED}\Big/\sqrt{A} = aU_{max}{}^{b} \tag{21}$$

At the critical touchdown point of 4000 m relative to the building center position, a downburst achieving $\bar{U}_{max}$ of 65 m/s has the potential to impose more than double the amount of structural motion in comparison with a storm that only reaches a maximum wind speed of 40 m/s. Performance-based analysis in the following section must include the increased potential for exceedance of limit states when downbursts achieve higher maximum mean wind speeds and when they land at certain critical locations.

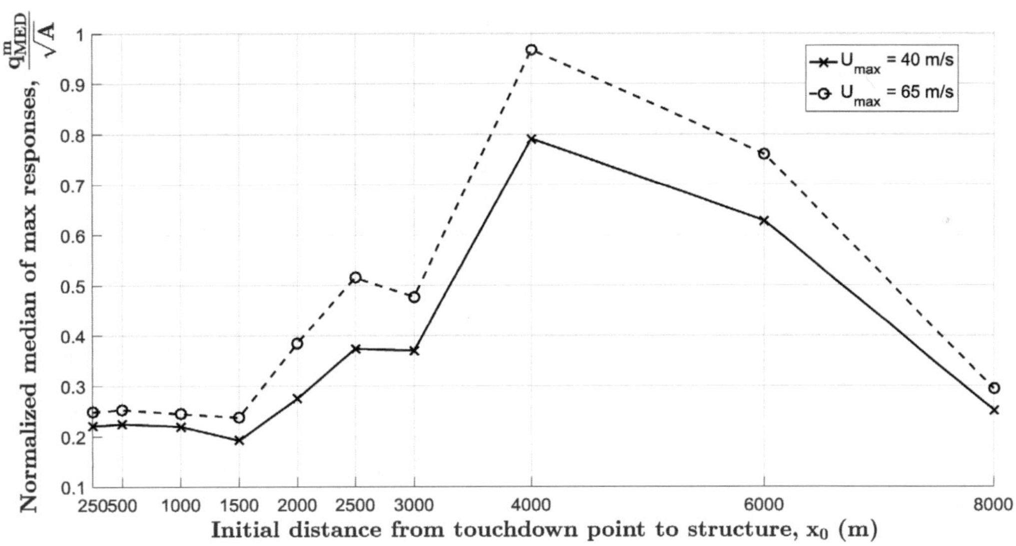

Figure 4. Along-wind response of the benchmark cantilever structure as a function of downburst's touchdown point on the ground.

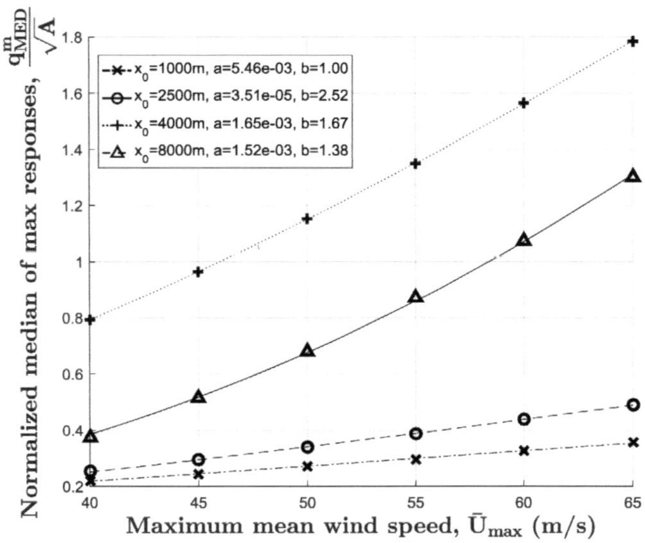

Figure 5. Examination of the empirical relationship between $\bar{U}_{max}$ and the median value of the non-stationary response maxima.

4.2 *Fragility analysis*

A parametric study is conducted to evaluate the performance of the cantilever structure against different scenarios of travelling downburst wind loads. Additionally, structural response must be examined in conjunction with the influence from a random variability or uncertainty in the wind loading [35]. This study constitutes a fragility analysis where randomization of parameters in the aerodynamic loading, such as drag coefficient and turbulence, reflects the presence of measurement or modelling errors in both in-situ wind properties and wind tunnel data variability, possible among various laboratories [36]. A typical fragility analysis includes fragility curves, estimated by determining the complementary cumulative distribution function (CCDF) that a structure will exceed a pre-defined limit state (i.e. hazard level or threshold) conditional on the reference downburst velocity (hazard intensity measure). These threshold levels can be used as indicators of structural performance, for example, by the owner or the engineer. For this study, three limit states, T_i, are defined as fractions of the height, h, of the cantilever structure. In the absence of either a specific standard or design recommendations for downburst winds, the proposed T_i values are a preliminary attempt at defining realistic thresholds for structural motion that apply to a full-scale "point-like structure" (or building), subjected to winds induced by a downburst.

$$\{T_1, T_2, T_3\} = h * \left\{\frac{1}{50}, \frac{1}{100}, \frac{1}{200}\right\} = \{0.6 \text{ m}, 0.3 \text{ m}, 0.15 \text{ m}\} \tag{22}$$

The CCDF, F_T, of the maximum structural motion, q_{max}, exceeding these limit states is represented as a function of the maximum downburst "mean" wind speed, $\bar{U}_{max}$. It is numerically determined by finding the probability, P, that a generic random variable $X = q_{max}$ exceeds the thresholds set by Equation (22), conditional on the presence of a downburst with $V_z = \bar{U}_{max}$.

$$F_T(\bar{U}_{max}) = P[X > T_i | V_z = \bar{U}_{max}] \tag{23}$$

The calculated fragility curves are approximations of the probability that the cantilever system's structural motion will exceed a limit state (i.e. threshold) when exposed to an incoming, travelling system with maximum mean wind speed, $\bar{U}_{max}$.

Figure 6 illustrates two examples of fragility curves approximated from the maximum structural responses selected from simulations where U_{tran} = 10 m/s. The interval of maximum mean wind speeds between 40 m/s and 65 m/s coincides with medium to extremely severe wind velocities for a downburst. Additional simulation sets are generated at lower velocities to ensure proper curve fitting. The log-normal cumulative distribution function (CDF) is selected to model the fragility curves due to a relative simplicity, in a linear least-squares regression, to determine the appropriate model parameters. The fragility curves approximated from Equation (23) are modelled as follows using the log-normal CDF, Φ, with *erf* as the error function, μ as the location parameter, and σ as the scale parameter [37]:

$$\Phi(\bar{U}_{max}) = {}^{1}/_{2}\left[1 + erf\left({}^{ln(\bar{U}_{max}) - \mu}/_{\sqrt{2}\sigma}\right)\right] \tag{24}$$

Additional simulations with lower maximum mean wind speeds between 5 and 35 m/s are needed for proper fitting.

As the limit states become more stringent (i.e. lower in magnitude), the probability of exceedance increases for the same $\bar{U}_{max}$ value. In addition, the downburst, when landing at x_0 = 4000 m from the building center position as opposed to x_0 = 1500 m, will have a much higher likelihood of exceeding a limit state at the same $\bar{U}_{max}$. This finding is corroborated by the results of the previously conducted parametric study, in which the median of the normalized maximum responses is greatest at x_0 = 4000 m for a downburst translating with speed U_{tran} = 10 m/s. In a practical application, an engineer or an owner may gauge the likelihood that a structural design exceeds a given threshold, established to meet requirements such as human comfort or secondary system sensitivity. Additional parameters may be incorporated in the fragility analysis and more simulations sets will increase the resolution of the curves. Furthermore, this example of performance-based thunderstorm load and response analysis can be extended to other structural models subjected to a wider range of downburst cases.

For the 2DOF scenario, the touchdown point of the downburst is offset in the across-wind direction by 1000 m (y_0 = 1000 m). The same six maximum mean wind speeds are used for the Monte Carlo simulations with random reference drag coefficient. Additional simulation sets are generated to develop the fragility curves as a function of a wider range of maximum mean wind speeds from 5 m/s to 100 m/s. The curves are once again fitted with a log-normal CDF. The results in both along-wind and across-wind directions are shown in Figure 7. In the along-wind direction, the more stringent limit state (T_3) has a higher likelihood of being exceeded than in the across-wind direction. For limit states T_1 and T_2, the opposite is true since the probabilities of exceedance in the along-wind direction more gradually increase than the one in the across-wind direction. In addition, the slopes of these two curves can be associated with a measure of the uncertainty in the structural behavior for a given threshold [38]. The steeper increments (greater slope) in probability of exceedance for the curves in the across-wind direction indicate more structural uncertainty for the cantilever system. Compared to the structural motion in the along-wind direction, a smaller variation in $\bar{U}_{max}$ can cause the structure to deflect beyond any threshold in the across-wind direction. The tighter cluster of the curves demonstrates that the system has a diminished aptitude in controlling structural motion beyond an initial sign of deflection exceedance. Further investigation is needed to evaluate fragility curves under a wider variety of conditions (e.g. varying U_{tran}, x_0, y_0, or even building type). A more efficient means for calculating structural responses while maintaining a sufficiently large sample size thus becomes a crucial task, if fragility analysis is to be implemented for broader cases.

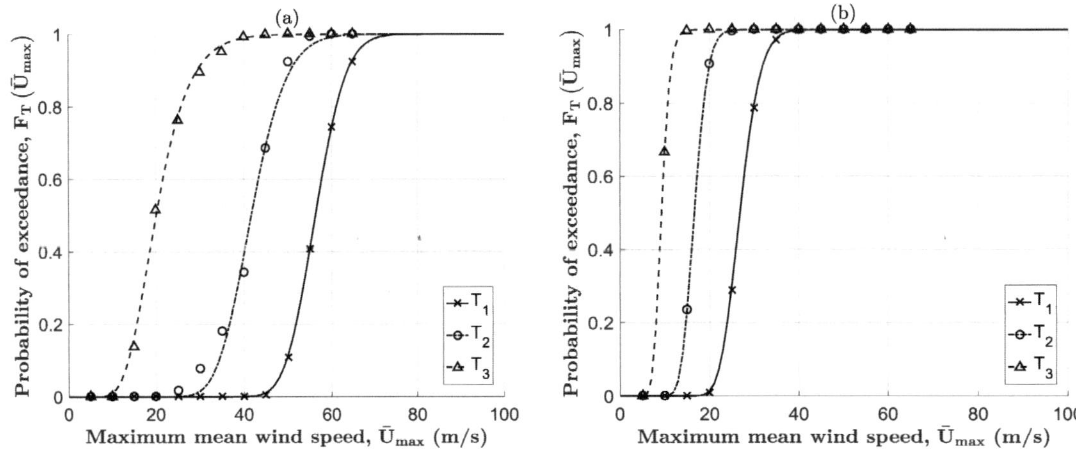

Figure 6. Fragility curves for the 1DOF scenario for (a) x_0 = 1500 m and for (b) x_0 = 4000 m.

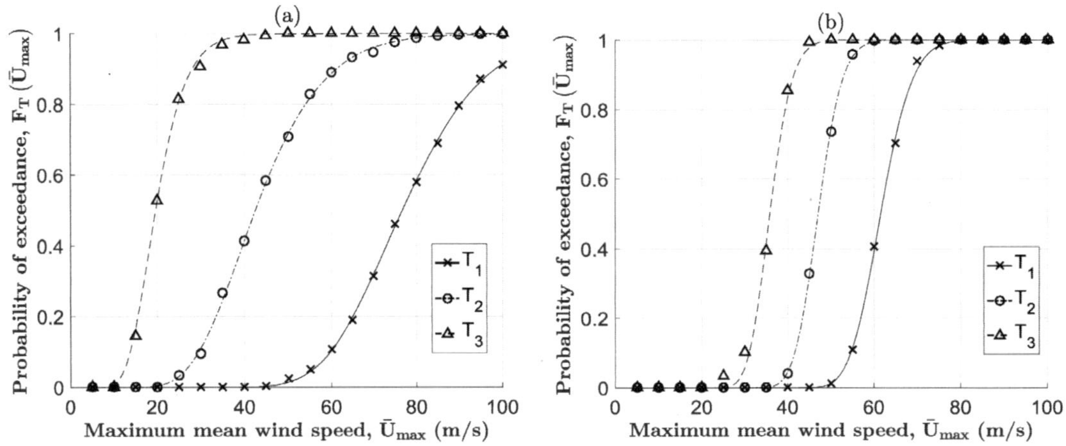

Figure 7. Fragility curves for the 2DOF scenario for (a) along-wind response and (b) across-wind response (x_0 = 1500 m and y_0 = 1000 m).

5 CONCLUSIONS AND FUTURE STUDIES

Recent implementation of fragility analysis for non-stationary wind response of a "point-like structure", subject to thunderstorm downburst winds is presented. A benchmark vertical cantilever "point-like structure" at an elevation of 30 m is digitally subjected to simulated transient wind velocity fields. The structural responses are determined by Runge-Kutta-Fehlberg (RKF) 4th order method and numerical integration of the differential equations of motion for single degree-of-freedom and two degree-of-freedom systems. Error propagation and random variability are simulated through a Monte Carlo sampling approach for the reference drag coefficient and through random Gaussian variables in the synthetic generation of non-stationary turbulence. The main findings from this study are as follows: (i) Utilizing time-dependent intensity functions, the mean wind velocity field of a travelling downburst can be digitally simulated. Turbulent wind fluctuations are replicated with an evolutionary power spectral density approach, involving proper orthogonal decomposition of the power spectral density matrix and amplitude modulation using a cosine function. (ii) Simplifications made to enable RKF numerical integration of a large population of synthetically-generated downburst events, along with linearization of second-order terms in the wind loads, is found to underestimate results for structural response. Other methods for the solution of the differential equations of motion, which are less computationally demanding, must be investigated to avoid these assumptions. Additional investigation for direction-dependent drag forces is required since wind direction may significantly vary due to the downburst radial profile. (iii) A parametric study demonstrates the influence of relative distance between downburst and structure and maximum wind velocity on the stochastic structural response. Certain combinations between these two parameters result in considerable increments in the structural dynamic motion. (iv) Lastly, fragility analysis in both single and two degree-of-freedom scenarios shows promising potential as a feasible approach in the design of a simplified vertical structure subjected to downburst wind loading to meet selected limit states. Probabilities of exceedance of limit states are determined as a function of maximum "mean" wind speed of the downburst, used as intensity measure.

Future investigations may include more rigorous structural analysis modeling that considers across-wind and torsional responses. In addition, the fragility analysis can be expanded to more complex loadings and building models. This may require implementation of innovative techniques to determine accurate values of the structural response while utilizing less computing resource. Finally, additional parametric studies are still needed to clarify the effects of other factors such as the continuous variation in the wind direction with time, downburst translational velocity, or variable turbulence intensity of the wind field.

6 ACKNOWLEDGEMENTS

This material is based upon work supported by the National Science Foundation of the United States (NSF), under CMMI-1434880 award. Any opinions, findings and conclusions or recommendations are those of the authors and do not necessarily reflect the views of the NSF.

7 REFERENCES

1 T.T. Fujita, The downburst: microburst and macroburst, (1985).

2 T.T. Fujita, DFW microburst on August 2, 1985. 1986: University of Chicago.

3 National Transportations Safety Board (NTSB). Aircraft Accident Report--Delta Air Lines, Inc., Lockheed L-1011-385-1, N726DA, Dallas/Fort Worth International Airport, Texas, August 2, 1985, 1985: Washington, D. C. NTSB-AAR-76-8. PB86-910406.

4 National Transportation Safety Board (NTSB). Aircraft Accident Report -- Eastern Air Lines, Inc., Boeing 727-225 John F. Kennedy International Airport Jaimaica, New York. June 24, 1975, 1975. NTSB-AAR-76-8.

5 National Transportations Safety Board (NTSB). Aircraft Accident Report--Pan American Wrold Airways, Clipper 759, N4737, Boeing 727-235, New Orleans International Airport, Kenner, Louisiana, July 9, 1982, 1982: Washington D.C. NTSB/AAR-83/02. PB83-910402.

6 J.W. Wilson, et al., Microburst wind structure and evaluation of Doppler radar for airport wind shear detection, Journal of Climate and Applied Meteorology, 23 (6) 898-915 (1984).

7 M.R. Hjelmfelt, Structure and life cycle of microburst outflows observed in Colorado, Journal of Applied Meteorology, 27 (8) 900-927 (1988).

8 D.D. Vicroy, A simple, analytical, axisymmetric microburst model for downdraft estimation, (1991).

9 R.M. Oseguera and R.L. Bowles, A simple, analytic 3-dimensional downburst model based on boundary layer stagnation flow, NASA Technical Memorandum, (1988).

10 Y. Zhang, P. Sarkar, and H. Hu, An experimental study on wind loads acting on a high-rise building model induced by microburst-like winds, Journal of Fluids and Structures, 50 547-564 (2014).
11 E.-S. Abd-Elaal, J.E. Mills, and X. Ma, Empirical models for predicting unsteady-state downburst wind speeds, Journal of Wind Engineering and Industrial Aerodynamics, 129 49-63 (2014).
12 M.T. Chay, F. Albermani, and R. Wilson, Numerical and analytical simulation of downburst wind loads, Engineering Structures, 28 (2) 240-254 (2006).
13 H. Hu, Z. Yang, and P. Sarkar, Dynamic wind loads and wake characteristics of a wind turbine model in an atmospheric boundary layer wind, Experiments in Fluids, 52 (5) 1277-1294 (2012).
14 D.K. Kwon and A. Kareem, Gust - Front Factor: A New Framework for Wind Load Effects on Structures, Journal of Structural Engineering, 5 (2009).
15 A. Kareem and T. Wu, Wind-induced effects on bluff bodies in turbulent flows: Nonstationary, non-Gaussian and nonlinear features, Journal of Wind Engineering and Industrial Aerodynamics, 122 21-37 (2013).
16 M. Jesson, et al., Aerodynamic forces on the roofs of low-, mid- and high-rise buildings subject to transient winds, Journal of Wind Engineering and Industrial Aerodynamics, 143 42-49 (2015).
17 T.-H. Le and L. Caracoglia, Coupled dynamic response of a tall building in a simulated thunderstorm downburst. in Proceedings of the 14th International Conference on Wind Engineering (ICWE14). 2015. p. 21-26.
18 T.-H. Le and L. Caracoglia, Reduced-order wavelet-Galerkin solution for the coupled, nonlinear stochastic response of slender buildings in transient winds, Journal of Sound and Vibration, 344 179-208 (2015).
19 T.-H. Le and L. Caracoglia, Wavelet-Galerkin analysis to study the coupled dynamic response of a tall building against transient wind loads, Engineering Structures, 100 763-778 (2015).
20 Eurocode-Basis of structural Design. 2002, European Committee for Standardization.
21 B.R. Ellingwood, et al., Fragility Assessment of Light-Frame Wood Construction Subjected to Wind and Earthquake Hazards, Journal of Structural Engineering, 130 (12) 1921-1930 (2004).
22 B.R. Ellingwood and P.B. Tekie, Wind load statistics for probability-based structural design, Journal of Structural Engineering, 125 (4) 453-463 (1999).
23 W. Cui and L. Caracoglia, Simulation and analysis of intervention costs due to wind-induced damage on tall buildings, Engineering Structures, 87 183-197 (2015).
24 M.A. Smith and L. Caracoglia, A Monte Carlo based method for the dynamic "fragility analysis" of tall buildings under turbulent wind loading, Engineering Structures, 33 (2) 410-420 (2011).
25 R. Bashor and A. Kareem, Probabilistic Performance Evaluation of Buildings: An Occupant Comfort Perspective. in 12th International Conference of Wind Engineering. 2007. Cairns, Australia. p. 1-6.
26 D.-W. Seo and L. Caracoglia, Estimating life-cycle monetary losses due to wind hazards: Fragility analysis of long-span bridges, Engineering Structures, 56 1593-1606 (2013).
27 M.T. Chay, R. Wilson, and F. Albermani, Gust occurrence in simulated non-stationary winds, Journal of Wind Engineering and Industrial Aerodynamics, 96 (10-11) 2161-2172 (2008).
28 L. Chen and C.W. Letchford, A deterministic–stochastic hybrid model of downbursts and its impact on a cantilevered structure, Engineering Structures, 26 (5) 619-629 (2004).
29 Y.L. Xu and J. Chen, Characterizing Nonstationary Wind Speed Using Empirical Mode Decomposition, Journal of Structural Engineering, 130 (6) 912-920 (2004).
30 J.D. Holmes and S.E. Oliver, An empirical model of a downburst, Engineering Structures, 22 (9) 1167-1172 (2000).
31 M. Di Paola, Digital simulation of wind field velocity, Journal of Wind Engineering and Industrial Aerodynamics, 74-76 91-109 (1998).
32 J. Kaimal, et al., Spectral characteristics of surface - layer turbulence, Quarterly Journal of the Royal Meteorological Society, 98 (417) 563-589 (1972).
33 C.P. Robert, Monte carlo methods. 2004: Wiley Online Library.
34 G. Solari, Thunderstorm response spectrum technique: Theory and applications, Engineering Structures, 108 28-46 (2016).
35 L. Caracoglia, Comparison of Reduced-Order Models to Analyze the Dynamics of a Tall Building under the Effects of Along-Wind Loading Variability, ASCE-ASME Journal of Risk and Uncertainty in Engineering Systems, Part A: Civil Engineering, 2 (2) C4015002 (2016).
36 W.H. Melbourne, Comparison of measurements on the CAARC standard tall building model in simulated model wind flows, Journal of Wind Engineering and Industrial Aerodynamics, 6 (1-2) 73-88 (1980).

37 E. Weisstein. *Log Normal Distribution*. [Webpage]; Available from: http://mathworld.wolfram.com/LogNormalDistribution.html.
38 B.R. Ellingwood, LRFD: implementing structural reliability in professional practice, Engineering Structures, 22 (2) 106-115 (2000).

The 13th Americas Conference on Wind Engineering (13ACWE)
Gainesville, Florida USA, May 21-24, 2017

A simple vortex model of a thunderstorm downburst – a parametric evaluation

Mike Jesson and Mark Sterling

University of Birmingham, Edgbaston, Birmingham, UK

ABSTRACT: Thunderstorm downbursts are short-lived, transient extreme wind events which can cause wind speeds equivalent to a category EF3 tornado (~150mph). The complex flow field which they produce has previously been the subject of time-expensive numerical modelling. It has been demonstrated, however, that there is a large, random variation in full-scale downbursts and so a quick, easily varied model would be of benefit to engineers calculating dynamic loading on structures. This paper introduces a simple and computationally inexpensive vortex model of a downburst, which is shown accurately models the main features of the flow field in a physically simulated thunderstorm downburst.

KEYWORDS: Transient winds, numerical model, vortex, thunderstorm, downburst.

1 INTRODUCTION

A thunderstorm downburst is a transient, highly localised extreme wind event which can cause wind speeds of 150mph, equivalent to a category EF3 tornado ([1]). These events are created by the cooling of (and precipitation within) warm, moist, rising air in a convective thunderstorm cell, which then reverses direction to form a downdraft which impinges on the ground. A primary ring vortex forms around the downdraft, and is carried radially outwards with the outflow from the impingement point. The superposition of the outflow and vortex flow fields creates a region of very high wind speed. Numerical simulations also indicate the development of a smaller vortex at the base of the primary ([2], [3]), caused by the interaction of the flow and ground roughness. This secondary vortex further accelerates the flow. The combination of these flow elements results in a flow field which is very different from that seen in atmospheric boundary layer (ABL) flows. Unlike ABL winds, which for the purposes of design are regarded as statistically stationary over a number of hours and uniform over tens of kilometres, downbursts typically have a lifetime of only a few minutes (Figure 1) and a downdraft radius of approximately 1 – 2km. The vertical distribution of radial wind-speed typically has a peak maximum (u_m, the spatio-temporal maximum over the whole flow field) close to the ground, at a height (z_m) of 30 – 100m (Figure 2; [4], [5]). One feature of downbursts is the variability of such events, with no two recorded downbursts producing precisely the same flow fields ([6]–[8]), although there are clear similarities between the large-scale characteristics.

There is growing consensus that extreme weather events such as downbursts may become more frequent due to climate change. Consequently, efforts have been made to understand the wind loading which they exert. Due to the difficulty in predicting where and when a downburst will occur, along with the usual issues of variability, the use of full-scale measurements for the determination of downburst wind loading is problematic (though the work of Lombardo ([7]) provides a very useful data set for validation). For this reason, simulations are used to model downbursts and (in some cases) their effects on structures, both physically (e.g. [9]–[14]) and numerically ([3], [15], [16]). The more advanced physical simulators, such as that used by McConville et al. ([17]) and Jesson et al. ([12], [13]), model the transient nature of a downburst event, and exhibit the run-to-run variation which has been seen as with full-scale events ([17]).

This variation limits the insight which can be gained, although general loading patterns may be quantified (e.g. [12], [13]). On the numerical side, techniques such as Large Eddy Simulation and cloud models have been used, with the lifecycle of the downburst being simulated from the initial downdraft to the formation and motion of the ring vortex. Although they are of importance in elucidating the mechanisms which drive a downburst and lead to their high wind speeds, these techniques are computational expensive.

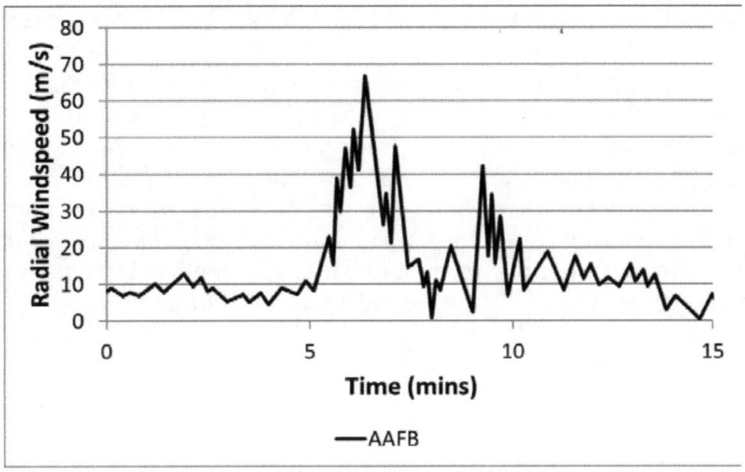

Figure 1: Radial wind-speed time-series from the Andrews Air Force Base downburst event, 5m above ground (adapted from [1]).

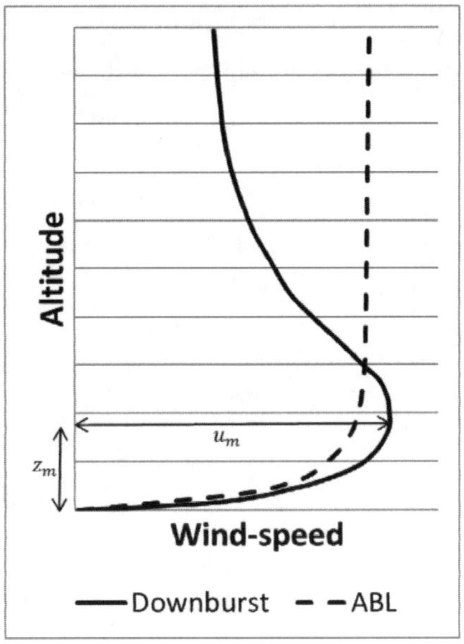

Figure 2: Schematic vertical distribution of velocity for a downburst and ABL flow. u_m is the maximum radial wind-speed of the downburst, z_m is the altitude at which this wind-speed occurs (adapted from [5]).

A simple numerical model of a downburst is presented in this paper. Before presenting the model definition, a brief description of the University of Birmingham Transient Wind Simulator (UoB-TWS) is given in Section 3; experimental data from the UoB-TWS provides the reference data for model validation. The model, which is described fully in Section 4, calculates the velocity field as a superposition of a primary vortex, secondary vortex and linear outflow velocity. Section 5 compares model output with the UoB-TWS data, and includes a parametric study which identifies the important parameters in defining and creating a downburst flow field. Finally, important conclusions from this work are presented.

2 UNIVERSITY OF BIRMINGHAM TRANSIENT WIND SIMULATOR (UOB-TWS)

The UoB-TWS is a vertical impinging jet downburst simulator with a length scale estimated as 1:1600, and is described fully by Jesson et al. ([12], [13]). Aperture control is used to simulate the rapid flow accelerations which occur in full-scale downbursts, and the simulator has been shown to simulate the transient aspects of a downburst flow ([12], [13], [17]). Run-to-run variation is seen in the simulations, as has been noted in full-scale events and mentioned in the introduction. In order to investigate the generic aspects of downbursts, while minimising the effects of such variation, an ensemble-mean approach has been used in analysing the UoB-TWS data. Thus, time-series from multiple runs are averaged according to:

$$u(t) = \frac{1}{N}\sum_{n=1}^{N} u_n(t) \tag{1}$$

where n is the run index, N is the total number of runs in the ensemble, $u(t)$ is the ensemble-mean velocity time-series and $u_n(t)$ is the velocity time-series from the nth experimental run. Ensemble-mean values are used in this paper.

The aim of the original UoB experiments was to measure the wind loading on building models in a simulated downburst. The velocity measurements had two purposes: Firstly, to identify the position, (x_m, z_m), of the peak maximum outflow velocity (found to be $\frac{x_m}{D} = 1.50$, $\frac{z_m}{D} = 0.02$), where D is the diameter of the simulated downdraft), and secondly to ensure that the vertical profile of radial velocity at this point was consistent with full-scale data (which was demonstrated by comparison with the work of Hjelmfelt ([5]; see [13])). "Complete" profiles were measured at $\frac{x}{D} = 1.00,\ 1.50,\ 2.00$ and 2.50, with partial profiles at $\frac{x}{D} = 1.25$ and 1.75 to verify that the $\frac{x}{D} = 1.50$ profile included the maximum velocity point.

3 THE VORTEX MODEL

3.1 *Model Development*

An early version of the vortex model has been presented by Jesson and Sterling ([18]) and this description is expanded and updated here. A non-translating downburst is simulated, i.e., the downburst is not part of a larger storm which carries the downburst with it (although incorporating the translation of the storm would be relatively straightforward. This permits the assumption that the downdraft creates an axially symmetric outflow around the impingement point, meaning that model is 2-D within a cylindrical polar coordinate system; variation occurs along the radial (x) and vertical (z) directions only. The respective velocities are u and w, and the velocity field is assumed to be the superposition of three, independent velocity fields, one from each of the main flow structures:

- The main outflow from the downdraft impingement point.

- The primary ring vortex.
- The secondary vortex.

This superposition is a technique applied in (inviscid) Potential Flow models, as is the use of mirrored vortices (Figure 3) to ensure that the condition of zero flow across the ground plane is met. The mirroring of the vortices also accelerates the radial flow close to the boundary, as required by continuity to reflect the contraction of the flow field by the ground plane. In standard Potential Flow theory, the flow is assumed to be inviscid, leading to vortices with a singularity at the centre. In this model, each vortex is an independent (viscous) Rankine-type vortex. For a circular Rankine vortex with a core of radius R and circulation Γ, the tangential wind speed at a radial distance r from the centre, $V_\theta(r)$, is given by:

$$V_\theta(r) = \begin{cases} \dfrac{\Gamma r}{2\pi R^2} = \dfrac{\Gamma\sqrt{X^2+Z^2}}{2\pi(X_R^2+Z_R^2)} for\ r \leq R \\ \dfrac{\Gamma}{2\pi r} = \dfrac{\Gamma}{2\pi\sqrt{X^2+Z^2}} for\ r > R \end{cases} \quad (2)$$

where, in the Cartesian form, (X, Z) is a general point in the domain, (X_R, Z_R) lies on the core boundary, and capitals indicate a local coordinate system with its origin at the vortex centre (Figure 4).

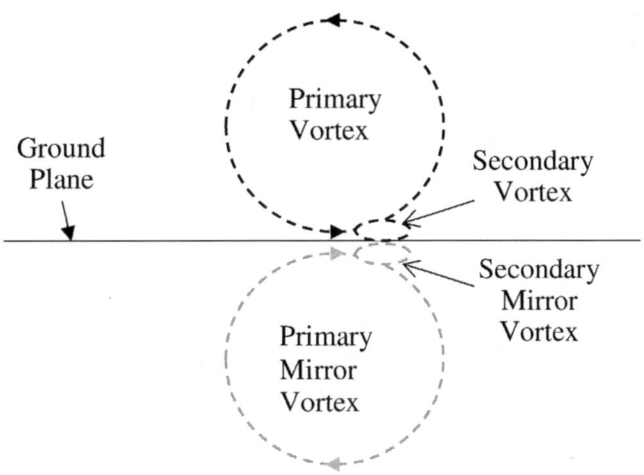

Figure 3: Schematic application of mirrored vortices.

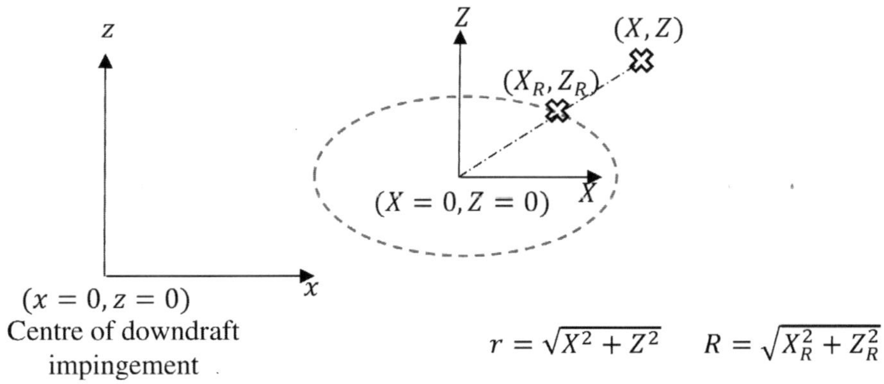

Figure 4: Definition of the coordinate system.

In order to estimate the relative positions and speeds of the three elements of the flow field, reference was made to rudimentary flow visualisation work at the University of Birmingham ([17], [19]) and the numerical work of Mason ([3]) and Kim & Hangan ([2]). From these it was observed that:

- The primary vortex changes shape (becoming elliptical) and weakens (falling circulation) with time, i.e. as it spreads out from the centre of the impingment zone. In addition to the reduction in circulation, it also lifts from the ground.
- The secondary vortex is initially not present, but strengthens with time before weakening. It forms at the leading edge of the primary vortex, at ground level but, according to Mason's CFD simulations, ([3]) is lifted by the rotation of the primary vortex.

The vortex model includes parameters to govern the initial size, shape, position and strength of each vortex, and also the rate of change of these parameters. For the secondary vortex, a "half-life" can be configured, such that the magnitude of the vortex increases up to the half-life, following which the vortex weakens. The full list of parameters is given in Table 1.

The evolution of the vortices from circular to elliptical has been incorporated in the model through specifying the rate of change of the length of the ellipse axes, one of which is parallel to the x-axis and the other parallel to the z-axis. The circumference of the vortex core is thus defined by:

$$\frac{X_R^2}{a(t)^2} + \frac{Z_R^2}{b(t)^2} = 1 \tag{3}$$

where $a(t)$ and $b(t)$ are the lengths of the principal axes, which vary with time, t. $V_\theta(r)$ is then calculated using a modified version of (2) such that r is the distance along the line joining the centre of the ellipse and the point at which the velocity is calculated, and R is the ellipse radius along that line (Figure 4).

Table 1 Vortex model parameters.

Parameter	"Best Fit"	
	Primary Vortex	Secondary Vortex
Initial circulation (m^2/s)	16	0
Rate of change of circulation (m^2/s^2)	-28.00	-2.00
Initial x, z radii (m)	0.30, 0.30	0.04, Calc.
Rate of change of x, z radii (m/s)	-0.20, Calc.	0.00, Calc.
Initial centre position, x, z (m)	Calc., Calc.	Calc. ,Calc.
Translation speed, x, z-directions (m/s)	Calc.,-0.25	Calc.,Calc.
Half-life (s)	N/A	0.20
Final Outflow Velocity (m/s)	10.00	
Time Step (s)	0.01	
Grid Spacing* (m)	0.25	

*Grid spacing does not affect the values calculated for each grid node.

"Calc." Indicates a parameter which is calculated from other parameter values.

3.2 *Parameter Quantification*

Values for the model parameters were initially estimated from the references stated above, and then refined to improve the fit of the model to the UoB-TWS data while ensuring that the values remained representative of the reference data. The refinement process involved incrementing model parameters over multiple runs and then identifying the "best fit" configuration. In order to make the number of runs manageable (by reducing the number of varied parameters and hence parameter combinations) it was assumed that:

- The primary vortex is initially circular.
- The primary vortex always starts with its bottom edge touching the ground plane and its rear edge touching the edge of the downdraft region (i.e., the initial position is a function of the initial size, and does not vary independently; the vortex "edge" is at radius R).
- The combination of the rate of change of z-radius and z-translation speed is such that $R = 0.02m = z_{max}$ when the secondary vortex has its greatest circulation.
- The secondary vortex starts with its rear edge at the centreline of the primary vortex.
- At its vertical centre, the secondary vortex spans the distance between the ground plane and the primary vortex.
- The translation speed is the outflow speed. This is initially zero, but increases to the final outflow velocity in a time of 1.6 times the secondary vortex half-life.
- The secondary vortex moves horizontally at the same speed as the primary vortex. No relative motion has been included in this version of the model.

Configurations for which the peak maximum velocity did not occur at $\frac{x}{D} = 1.50$ and $0.01 \leq \frac{z}{D} \leq 0.04$ (as seen in the UoB data), or for which the peak maximum velocity was more than $\pm 0.1m/s$ from the experimental value were discarded. For the acceptable configurations, "Best Fit" was evaluated using the root mean square (RMS) difference between the UoB experimental data and the model output. The data used for this calculation included the vertical profiles of local maximum radial velocity at $\frac{x}{D} = 1.00$, 1.25, 1.50 and 1.75, and also the velocity time-series at the position of maximum radial velocity. The vertical profile at $\frac{x}{D} = 2.00$ and 2.50 were not used for reasons discussed below.

The Best Fit values for the model parameters are given in Table 1.

4 RESULTS AND DISCUSSION

Using the Best Fit model parameters shown in Table 1, the vertical profiles of radial velocity have been calculated for the same radial positions as were used for the UoB-TWS experiments (Figure 5), and the time-series calculated at the position of global maximum radial velocity (Figure 6). For $\frac{x}{D} \leq 1.75$ the experimental data are modelled well by the Best Fit profiles. As the secondary vortex strengthens the "nose" of the typical downburst vertical profile becomes evident. Beyond this position, the vortices decay rapidly in the physical simulations, a decay not accurately modelled by the vortex model (for this reason, these vertical sections were not used for the best fit calculation outlined in Section 4.2). Another difference is in the vertical spread of the "nose", which spans the lowest three data points in the UoB-TWS data. The vortex model creates a sharp nose where the outflow velocities from the two Rankine-type vortices combine; each vortex will have a clearly defined radius of maximum tangential velocity, resulting in this sharply defined, high-speed region. It is possible that the interaction of the primary and secondary vortices is distorting the flow field in this region, making the Rankine-type model inappropriate. Notwithstanding these differences, the vortex model is shown to give a good representation of the flow field over the main region of interest, from the edge of the downdraft to the radius of maximum velocity and slightly beyond.

At the point of global maximum radial velocity, the model also captures the main features of the velocity time-series (Figure 6). The initial acceleration is slightly lower (by about 5%) in the vortex model, with a narrower peak (by about 30%). The deceleration phase from the model also matches the experimental data, with the final outflow velocity matched. The narrower peak may be due to the simplifications and assumptions made in the position and motion of the secondary vortex (which is not modelled as moving up and around the perimeter of the primary vortex, indicated by Mason et al. ([3])). Incorporation of these more complex relationships may be included in future model development.

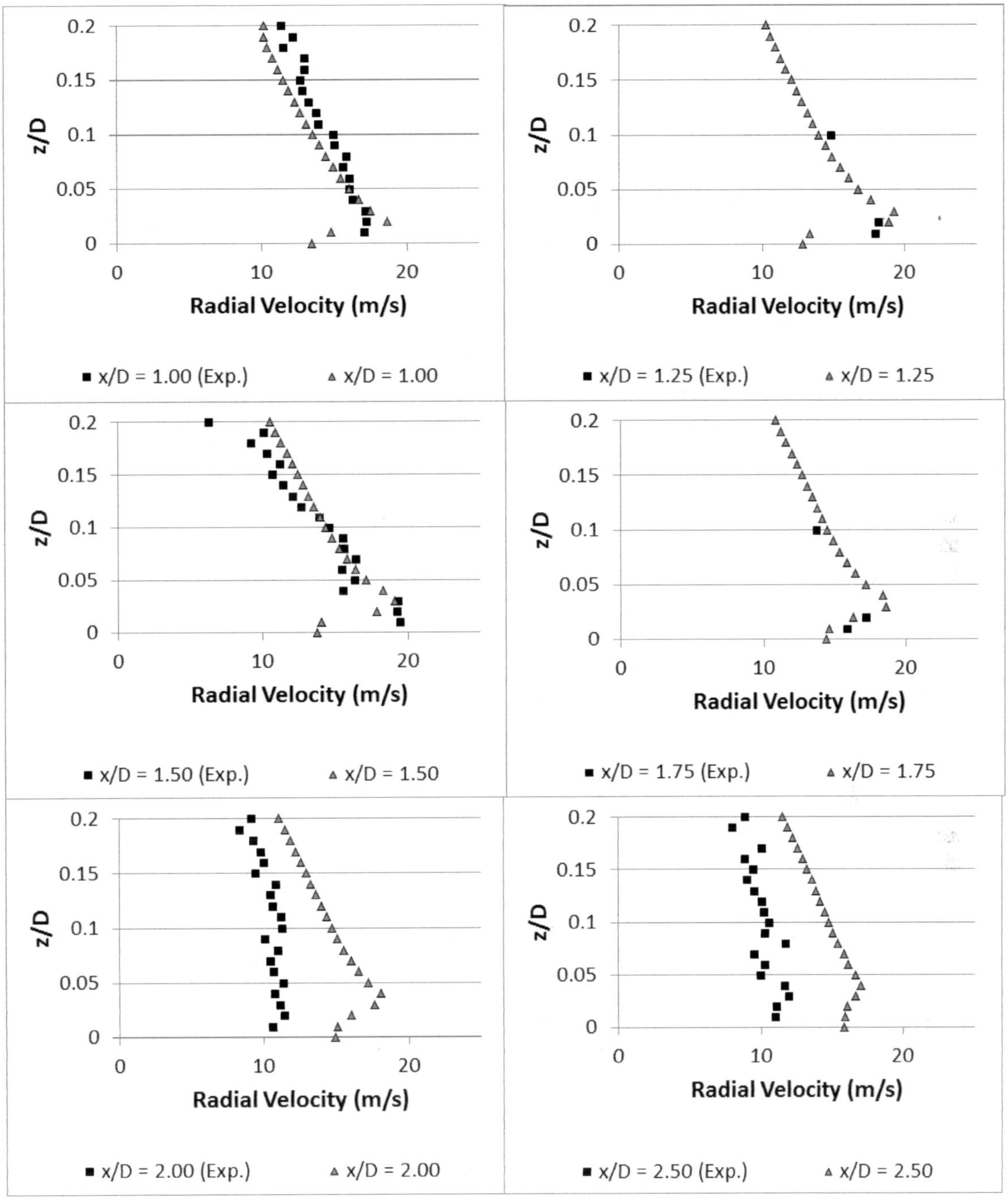

Figure 5: Comparison of vortex model Best Fit output with UoB-TWS experimental data.

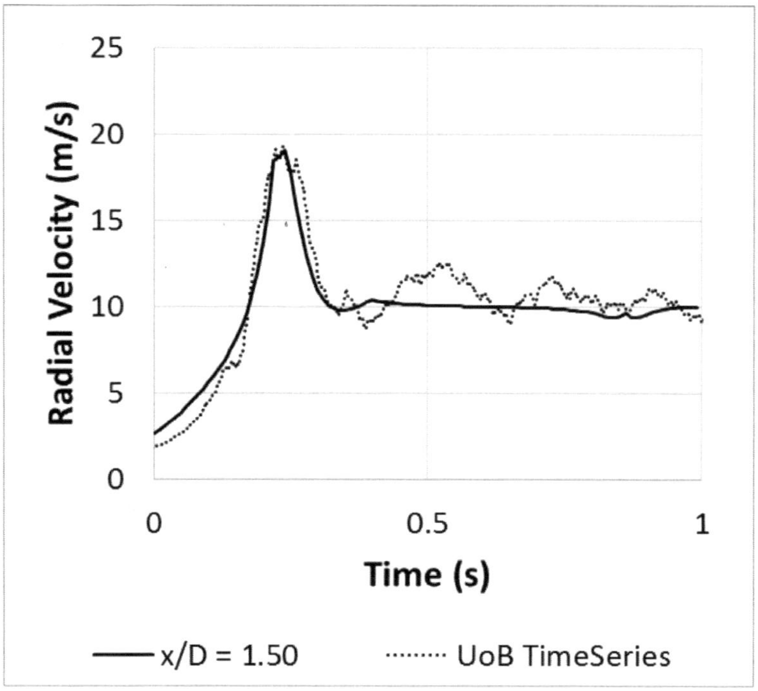

Figure 6: Model and experimental radial velocity time-series at the point of the global maximum velocity in the UoB experiments.

5 CONCLUSIONS

In this paper, we have outlined a simple, analytical model, which is shown to appropriately capture the main features of a thunderstorm downburst. It is acknowledged that a number of simplifying assumptions have been made along the way and that a Rankine-type model, originally conceived for laminar flow, has been used to model a highly turbulent event. Nevertheless, the results are remarkable and offer, for the first time, a computationally inexpensive tool which could be used for design purposes.

6 ACKNOWLEDGEMENTS

The authors would like to thank the UK Engineering and Physical Science Research Council, which funded the original research on which this model is based under grant number EP/J008281/1.

7 REFERENCES

1 T. T. Fujita, 'Downburst: Microburst and macroburst', *Univ. Chic. Press IL*, p. pp.–pp. 122, 1985.

2 J. Kim and H. Hangan, 'Numerical simulations of impinging jets with application to downbursts', *J. Wind Eng. Ind. Aerodyn.*, vol. 95, pp. 279–298, 2007.

3 M. S. Mason, G. S. Wood, and D. F. Fletcher, 'Numerical simulation of downburst winds', *J. Wind Eng. Ind. Aerodyn.*, vol. 97, pp. 523–539, 2009.

4 T. T. Fujita and R. M. Wakimoto, 'Five scales of airflow associated with a series of downbursts on 16 July 1980', *Mon. Weather Rev.*, vol. 109, pp. 1439–1456, 1981.

5 M. R. Hjelmfelt, 'Structure and life cycle of microburst outflows observed in Colorado', *J. Appl. Meteorol.*, vol. 27, no. August, pp. 900–927, 1988.

6 E. C. C. Choi, 'Field measurement and experimental study of wind speed profile during thunderstorms', *J. Wind Eng. Ind. Aerodyn.*, vol. 92, pp. 275–290, 2004.

7 F. T. Lombardo, 'Thunderstorm characteristics of wind engineering importance', presented at the In Proceedings of the 13th International Conference on Wind Engineering, 2011.

8 F. T. Lombardo, D. A. Smith, J. L. Schroeder, and K. C. Mehta, 'Thunderstorm characteristics of importance to wind engineering', *J. Wind Eng. Ind. Aerodyn.*, vol. 125, no. 0, pp. 121–132, 2014.

9 K. Butler and A. Kareem, 'Physical and numerical modeling of downburst generated gust fronts', presented at the Proceedings of the 12th International Conference on Wind Engineering, Cairns, Australia, 2007, pp. 791–798.

10 M. T. Chay and C. W. Letchford, 'Pressure distributions on a cube in a simulated thunderstorm downburst - Part A: stationary downburst observations', *J. Wind Eng. Ind. Aerodyn.*, vol. 90, no. 7, pp. 711–732, 2002.

11 M. T. Chay and C. W. Letchford, 'Pressure distributions on a cube in a simulated thunderstorm downburst - Part B: moving downburst observations', *J. Wind Eng. Ind. Aerodyn.*, vol. 90, no. 7, pp. 733–753, 2002.

12 M. Jesson, M. Sterling, C. Letchford, and C. Baker, 'Aerodynamic forces on the roofs of low-, mid- and high-rise buildings subject to transient winds', *J. Wind Eng. Ind. Aerodyn.*, vol. 143, no. 0, pp. 42–49, 2015.

13 M. Jesson, M. Sterling, C. Letchford, and M. Haines, 'Aerodynamic forces on generic buildings subject to transient, downburst-type winds', *J. Wind Eng. Ind. Aerodyn.*, vol. 137, no. 0, pp. 58–68, 2015.

14 T. S. Lundgren, J. Yao, and N. N. Masour, 'Microburst modelling and scaling', *J. Fluid Mech.*, vol. 239, pp. 461–488, 1992.

15 L. G. Orf, C. Oreskovic, E. Savory, and E. Kantor, 'Circumferential analysis of a simulated three-dimensional downburst-producing thunderstorm outflow', *J. Wind Eng. Ind. Aerodyn.*, vol. 135, no. 0, pp. 182–190, 2014.

16 B. C. Vermeire, L. G. Orf, and E. Savory, 'Improved modelling of downburst outflows for wind engineering applications using a cooling source approach', *J. Wind Eng. Ind. Aerodyn.*, vol. 99, no. 8, pp. 801–814, 2011.

17 A. C. McConville, M. Sterling, and C. J. Baker, 'The physical simulation of thunderstorm downbursts using an impinging jet', *Wind Struct.*, vol. 12, no. 2, pp. 133–149, 2009.

18 M. Jesson and M. Sterling, 'A Simple Vortex Model of a Thunderstorm Downburst', presented at the 12th UK Conference on Wind Engineering, 2016.

19 M. R. Haines, 'The simulation of non-synoptic effects and their implications for engineering structures', University of Birmingham, UK, Birmingham, 2015.

Tornado Damage Analysis of Residential Buildings Coupled with Electrical Power Network for the 2011 Joplin Tornado

N. Attary[a], J.W. van de Lindt[a], M. Koliou[a], S. Smith[b] and V.U. Unnikrishnan[a]

[a] *Colorado State University, Fort Collins, CO, U.S.A., email: nattary@colostate.edu, jwv@engr.colostate.edu, Maria.Koliou@colostate.edu,vipin@colostate.edu*
[b] *Missouri Southern State University, Joplin, MO, U.S.A., smith-st@mssu.edu*

ABSTRACT

Many modeling techniques are being developed and used to assess the resilience of communities to natural and human-made hazards around the world. These models require systematic validation using the results of real events to ensure that influencing factors not modeled are accounted for when needed. An EF5 tornado struck Joplin MO on May 22nd, 2011, and became the costliest and deadliest single tornado in U.S. history. In this study, the residential buildings and electrical power network of the city of Joplin MO were modeled to partially hindcast the 2011 tornado event and to estimate damage, loss, and restoration of functionality for the community. To model the electrical power network, detailed data of the network, including electrical substations, transmission lines and distribution lines were considered. A weighted cellular automata technique was used to estimate the service area of each of the substations based on their capacities and demand of residential structures. Residential buildings were categorized into five archetypes and their tornado fragility curves were developed using an Assembly-based Vulnerability (ABV) approach. The dependency of the electrical power network with the residential buildings is an important issue that affects the functionality of the community and was considered in this study. Both intrinsic and extrinsic functionality curves were developed for residential buildings to evaluate their recovery response. The accuracy of the results of the modeling simulations was validated through comparison with available data from building inspections immediately following the tornado.

KEYWORDS

Tornado, Interdependency, Hindcast, Joplin, Electrical Power Network, Residential Buildings, Functionality

13th Americas Conference on Wind Engineering
Gainesville, Florida, USA
May 21 - 24, 2017

Effect of laboratory simulated dynamic and translating downburst on the pressure distributions of a low-rise building

Jubayer Chowdhury[a], Horia Hangan[b]

[a] *The Wind Engineering, Energy and Environment (WindEEE) Research Institute, London, Ontario, Canada, cjubaye@uwo.ca*
[b] *The Wind Engineering, Energy and Environment (WindEEE) Research Institute, London, Ontario, Canada, hmhangan@uwo.ca*

ABSTRACT

Downburst is a strong downdraft of wind which is developed at the upper levels of a thunderstorm. During a thunderstorm, when hails melt or raindrops evaporate, latent heat is extracted from the surrounding air which significantly cools down the air. The cold heavy air descends from the cloud base causing the downdraft of wind. As the downdraft hits the ground, it spreads out radially and produces extremely powerful and damaging winds close to the ground surface in all directions. According to Munich Re, about 60% of the insured losses in 2015 were caused due to the extreme winds during severe thunderstorms. For synoptic winds, significant advancements have been achieved to estimate design wind loads on structures from various wind tunnels tests and has been a basis for the building codes around the world (USA: ASCE 7-10, 2013; Australia/New Zealand: AS / NZS 1170.2:2011; Canada: NBCC 2010, 2012). In contrast, the loading due to extreme thunderstorm events, some of which produce downbursts, is not yet well understood.

In this study, wind loads on a typical low-rise building are investigated under a large laboratory simulated downburst flow produced in the WindEEE Dome at Western University. The WindEEE Dome has the capability to produce both synoptic and non-synoptic wind events. The test chamber at the WindEEE Dome has a 25 meters hexagonal footprint in the horizontal plane and has a total of 106 fans. Among which, 100 fans are along the perimeter walls of the test chamber and 6 fans are in a plenum above the test chamber. The upper plenum is connected to the test chamber through a 5 m diameter bellmouth opening with mechanical louvers. To create a dynamic downburst flow, the louvers are closed with the six fans running to build up pressure in the upper plenum and once the pressure is built, the louvers are opened suddenly. To create a translating downburst, the bellmouth opening with louvers are translated at a maximum of 2m/s over a translating path of 5 meters with the fans running in the upper plenum.

A 1:100 scaled model of one of the NIST database generic low-rise buildings was selected for this study. Previously the aerodynamic loading on the same building due to a steady state stationary downburst has been analysed by the authors of this abstract. For the stationary dynamic downburst (includes sudden opening of the louver), the relative locations and orientations of the building with respect to the downburst were varied. For the translating case, a steady downburst (continuous radial impinging jet) was chosen to translate over the building. The wind profiles in the empty test chamber were analyzed first to investigate the flow characteristics for dynamic and translating downbursts. Pressure distributions on the surfaces of the building, in terms of pressure coefficients, were evaluated to examine the effect of downburst on the aerodynamic loading of a low-rise building and compared with Atmospheric Boundary Layer loading cases.

KEYWORDS

Downburst, low-rise building, wind loads, WindEEE

13th Americas Conference on Wind Engineering
Gainesville, Florida, USA
May 21 - 24, 2017

Methodology to Estimate Tornado Wind Speeds Based on Site-Specific Damage Observations

Peter L. Datin

Risk Management Solutions, Newark, CA, USA, peter.datin@rms.com

ABSTRACT

As part of the ASCE Wind Speed Estimation in Tornadoes Standards Committee, the Forensics sub-committee is working on developing a method to estimate, with some level of confidence, a tornado wind speed range for site-specific investigations. Currently, there is no standardized method for estimating tornado wind speeds outside of using the degrees of damage with the Enhanced Fujita scale. However, by using site-specific information about failed and non-failed components, a reasonable tornado wind speed range can be established. The proposed method accounts for the uncertainty of wind speed estimates and building component resistances by using probability distributions for both component resistances and wind loading factors. This paper examines the proposed methodology for developing a best estimate of the damaging wind speed at a given building location, which is based on a reliability index and accounts for the uncertainty in both the resistance and the load. The selection of the reliability index values used is discussed. Resistance distributions for common building components can be found using laboratory testing or in the literature. Wind load distributions can also be found in the literature for various building surfaces. The relationship with the ASCE 7 standard design wind pressure equations is also presented including the factors that need to be modified to properly account for tornado wind loads.

KEYWORDS

Tornado wind speed, tornado damage, reliability index, wind resistance distribution, wind load distribution

13th Americas Conference on Wind Engineering
Gainesville, Florida, USA
May 21 - 24, 2017

Reported wind damage in Uruguay and registered wind speeds during 2014

Valeria Durañona [a], Sofía Orteli [b] and Andrés Guggeri [c]

[a, b, c] *Institute of Fluid Mechanics and Environmental Engineering, School of Engineering, University of the Republic, Julio Herrera y Reissig 565, Montevideo, Uruguay,* [a] *valeriad@fing.edu.uy,* [b] *sofiao@fing.edu.uy and* [c] *aguggeri@fing.edu.uy*

ABSTRACT

The installation of a network of automatic weather stations in Uruguay since 2008 has allowed different analyses of the characteristics of high wind speeds events that occur in the country. This network is composed of around 25-30 stations, in a country with an area of approximately 175,000 km^2.

In particular, a comparison between reported wind damage and high wind speeds registered by this network of stations along 2014 is presented in this paper.

Wind damage reports were obtained carrying out a digital search in one of the main Uruguayan newspapers, El País, looking for news that included the words in Spanish meaning 'wind', 'storm' and 'tornado'. For each wind damage report found, relevant information such as date of occurrence of the reported high wind event, affected locations, measured wind speeds, duration of the event, description of wind damage and of other meteorological events that had taken place during its occurrence (as intense rain, hail, etc.) was extracted. In this way, many high wind events were found and characterized.

This information could sometimes be complemented with reports from different institutions and/or meteorologists, information from local or regional web pages, YouTube videos of severe weather episodes, etc. These other sources of information contained in many cases pictures taken, witnesses description of the characteristics of the events and of the damage produced, which could also help characterizing the events and estimating the highest wind speeds that had occurred.

On the other hand, a search for wind gust speeds greater than 80km/h (22.2m/s) registered by the mentioned network of automatic weather stations during 2014 was performed. These stations were distributed across the country and placed in open and level rural terrain, without significant obstacles in their surroundings. They measured, with a sampling frequency of 1Hz, the mean, standard deviation and maximum and minimum values of wind speed, wind direction and temperature every 10 minutes at heights between 10 and 100m.

After obtaining the dates when wind damage had been reported and the dates when high wind speeds had been registered by this network, i.e., wind gusts greater than 22.2m/s, a comparison between both sets of dates were carried out, finding that the intersection of both sets was relatively small.

This result would mean that there is an important number of high wind events that produce damage in Uruguay, but are not usually registered by the network of stations, and that many high wind events with the potential of causing wind damage are not reported in newspapers. The first conclusion would point out that an important number of high wind events produce damage in areas of length scales of 200km or less, which is the maximum separation between stations. In some cases, it was found that wind gusts measured at stations 30 to 50km apart were very different. The second conclusion would mean that many more high wind events occur in Uruguay than those usually reported in newspapers.

Information on the types of damage, typical weather conditions, and monthly and geographical distributions are also indicated.

KEYWORDS

High wind events, Uruguay, wind damage, non-synoptic high winds, severe convective activity

13th Americas Conference on Wind Engineering
Gainesville, Florida, USA
May 21 - 24, 2017

A Predictive Model for Structural Tornado Damage to Residential Structures Using Housing Data from the 2011 Joplin, MO Tornado

Alyssa C. Egnew[a], Dr. David O. Prevatt[b], David Roueche[c]

[a] *University of Florida, Gainesville, FL, USA, aegnew@ufl.edu*
[b] *University of Florida, Gainesville, FL, USA, dprev@ce.ufl.edu*
[c] *University of Florida, Gainesville, FL, USA, david.roueche@ufl.edu*

ABSTRACT

Improvement upon existing building codes helps new houses perform better in extreme wind events, such as tornadoes, but there are still millions of existing houses that remain vulnerable. Hence, there is a need to know which residential structures are most at risk for tornado damage, so that city officials can allocate mitigation resources accordingly. The information on house characteristics in a given community varies, but typically includes year built, appraised value, number of stories, and square footage of the houses. Thus, city officials stand to benefit from a model based on these readily available input characteristics that can assess how likely a house is to incur specific tornado damage levels. To see the effect of each characteristic on the expected tornado damage levels, a multinomial regression model can be built through combining damage observations and realtor or tax assessor databases in the aftermath of a large tornado event. However, damage observations after a tornado are typically classified into discrete damage states having a natural progressive order, i.e., ordinal response data. Standard multinomial linear regression models are inappropriate for this data type. Hence, a proportional odds model is used to predict the likelihood of being in a given damage state given predictor variables of year built, appraised value, number of stories, square footage of the house, and estimated wind speed at the house location. The model output is the odds of the expected level of damage being less than or equal to a reference damage level, as compared to greater than that reference level. This paper demonstrates the model using a database of house locations and damage observations for 1,349 houses collected after the 2011, EF-5 Joplin, Missouri tornado. The database of damaged homes was first augmented, house by house, to contain street addresses and parcel IDs that could be cross referenced with the Jasper County Tax Assessor's database of house characteristics. In this way, the year built, appraised value, square footage, and number of stories for each home was obtained for a final sample size of 488 homes. The maximum wind speed encountered by each home was assigned using each home's location with respect to the tornado centerline and the near-surface wind model of the Joplin tornado developed by Lombardo et al. (2015).

The model rejected the number of stories as being a significant predictor of damage level. However, an increase in year built or square footage all resulted in greater odds of having a higher state of damage. Conversely, an increase in appraised value resulted in lower odds of having a higher state of damage. The most influential parameter on damage was the appraised value. These results can be further investigated regarding the underlying correlations to other parameters; perhaps the appraised value of the house is significant because of better maintenance, which allows it to perform better in a tornado than a house with decay. This proportional odds model, when coupled with other related studies, would be useful for city officials to evaluate the most vulnerable residential areas.

KEYWORDS

Tornado, residential structures, Joplin, degree of damage, proportional odds

The 13th Americas Conference on Wind Engineering (13ACWE)
Gainesville, Florida USA, May 21-24, 2017

13th Americas Conference on
WIND ENGINEERING

Close to Ground Wind Field in a Vortex Chamber using CFD

Damaso Dominguez [a,b], R. Panneer Selvam [c]

[a]*University of Arkansas, Graduate Student, Fayetteville, 4190 Bell Engineering, U.S.A*
[b]*Universidad Tecnológica de Panamá, Estudiante de Doctorado, Panamá*
[c]*University of Arkansas, Womble Professor, Fayetteville, 4190 Bell Engineering, U.S.A*

ABSTRACT: Measurements of wind velocities close to the ground are difficult to perform due to risky conditions or limitations regarding the available technology, such as radar observation restrictions. A numerical tornado simulator has been designed to model different tornado-like vortices, and measure those parameters such as pressure and force coefficients, which normally are not available at low elevations. The analytical model uses Rankine Combined Vortex Model (RCVM) and hence the profile is the same at various heights, while in the vortex chamber the wind field varies at different heights from the ground for different swirl ratios. In this work, the detailed wind field at various heights for different swirl ratios are computed using CFD. The investigation shows that, at various heights close to the ground, the wind field is different, and it implies that for different building heights, different wind fields are interacting due to a tornado. For lower swirl ratios such as 0.2, the tornado does not touch down and for swirl ratios such as 0.4 or higher, the tornado touches the ground and becomes a multiple cell tornado beyond S>0.95. The variation of velocity with respect to radius is clearly observed. The effects of grid resolution at different parts of the chamber on wind field are considered. The computed tangential wind field at various heights and vector plots are presented in order to provide more details of the wind field for various swirl ratios. For various building heights and widths, different wind fields are interacting, and hence the force coefficients will be different.

KEYWORDS: Tornado simulator, chamber, Tornado wind field, swirl ratio, CFD.

1 INTRODUCTION

Tornadoes generate the strongest winds above ground level, and they cause severe damage over vegetation and structures on their paths. Nowadays, there are several ways to obtain parameters from tornadoes to study, them such as measurements in the field using Doppler radar, experimental simulations, and Computational Fluid Dynamics (CFD) simulations. The current technology available with radar does not allow researchers to investigate the wind field close to ground, where the majority of low-rise buildings (≈3.0 m) are present in tornado-prone areas. Normally for Doppler radar, the conditions of the moving tornado can be investigated for levels above 20m or more [1,2,3,4] due to earth curvature and hills on the field. Doppler radar data is retrieved at inappropriate heights for our comparisons. There are only a few measurements taken at our interest elevation, but so far there is not an analysis of tangential velocity profile, peak velocities, etc. coming from radar measurements at elevations less than 20m.

From the experimental point of view, there are some factors that affect the tornado-like–vortex phenomena, such as the existence of boundary of the simulation equipment and the scales used [5]. In an experimental tornado simulator, the boundary conditions will change the environment of the field. In other words, the surrounding wall imposes different conditions from the real tornado ambient or field conditions. Sometimes the hardware used to measure velocities (Probe) cannot be located at the desired elevation for the study due to scale. One more thing that cannot allow researchers to properly simulate the real condition, as it happens with the majority of experimental works, is the reduction of the updraft flow radius of the system, and it can be seen in the work done by [6]. Furthermore, variations of swirl ratios cannot be appropriate [7] because they have limited values for the inlet region and the updraft opening.

In addition, for experimental simulations, the scale of the simulations affects greatly the results. In an experimental tornado simulator, the boundary conditions will change the conditions of the field. In other words, the surrounding wall imposes different conditions from the real tornado ambient or field conditions.

Close to the ground, the terrain roughness directly affects the wind field, and the radial and vertical velocities are not reported so far. Capturing properly the boundary layer is vital to obtain reasonable results on computer tornado simulations. According to [8], there is a fundamental difficulty when trying to capture turbulent eddies in the tornado core close to the ground, and this difficulty is the mesh spacing close to the ground. The aforementioned condition is found in the work done by [6], where they do not have proper refinement to capture the parameters that we will present here such as different peak velocities in tangential velocity profile. They used a mesh with minimum refinement of $0.01H_o$ (H_o is the height of inflow opening, an illustration is presented below) in a domain of 2km by 2km. Therefore, some grids will be developed to study the effect of grid refinement over the flow.

Wilson and Rotunno [9] also worked with computer simulations and a domain with a total opened updraft similar to [7]. One of the differences between the last two works is the height of inlet flow region, which is very small for [9] when compared it with the radius of the outflow. Although the work of [9] has a proper refinement in the horizontal direction, the mesh grid size in the vertical direction is coarse as it is for [6], affecting the quality of velocity results close to the ground.

Liu and Ishihara [10] used radial refinement of $0.005H_o$, and $0.0005H_o$ in the vertical direction; this is even more in the vertical direction, but they do not have peak velocities for the tangential velocity profile, which can be due to the fact that their minimum elevation, when reporting results, is over 50 m above ground level (AGL).

Some recent works such as [11], about measurements close to the ground, used an armored vehicle to collect data at lower levels as low as 3.5 m AGL. Although this vehicle has an anemometer, which can record wind velocities at the inner core at such low elevations, the report of Wurman does not include a velocity profile picture or graph at the previously mentioned elevation. He also stated that there is not significant change in velocities at the lower levels (3.5m-30m AGL), and this shows a considerable disagreement with previous works such as [7, 12, 13], in which references stated that a peak tangential velocity is found close to the ground.

1.1 *Objective*

Due to the limitations stated above, for our area of study, a computer tornado simulator will be used here to study the wind field close to the ground. To do that, several swirl ratios will be used to analyze the wind field and tangential velocities for different tornado strengths in a 2D model (axisymmetric), and those results will be compared with 3D results that come from computer simulations as well. Moreover, some results from experimental work such as [4] will be considered for comparison.

2 MATHEMATICAL MODEL AND COMPUTER MODELLING

For this study, our research group develop a computer program that will be used to examine the characteristics of the flow of tornado-like vortices. Some important considerations about this 2D program are as follows: the flow is symmetric in the vertical direction, unsteady and axisymmetric. According to [8], data published by [14] about Fine-scale Doppler observations, the tornado flow within 0.5 km of tornado center is almost axisymmetric, which follows the consideration previously stated here. Figure 1a shows the computational domain, whose dimensions are: $r=r_o=1.0$ (updraft radius), $H_o=1.0$ (inlet region height), and $h=2^*\ H_o=2.0$ (total height of the domain), all dimensionless. The domain dimensions considered here are similar to the work done by [8], in which they limited their domain to 1km x 1km x 2km. Likewise, we are focused on the interactions of the tornado-like vortex with structures close to the ground, and so our domain has similar scope. The flow field is computed in an axisymmetric grid system. Three grids will be used here to discuss the effects of mesh refinement and for general comparison. These meshes have minimum spacing of $0.01H_o$, $0.005H_o$, and $0.001h_o$ with 999, 1419, and 3000 nodes, respectively. Moreover, the minimum spacing used by [8], which has similar domain, is $0.0015H_o$. This minimum spacing is similar to our smaller minimum refinement, which allow us to have external values for comparison. The maximum spacing is $0.1H_o$ in both directions. The growing factor for the grid is 1.1 and the computer model was run for 5 time units with a time step of 0.1 to satisfy the Courant-Friedrich-Lewy (CFL) condition. The Reynolds number is $Re=1x10^8$ [34] and the reference velocities are: V_r: 1.0 and $V_\theta = SC_1$. Here S is the swirl ratio from the input data and $C_1=0.0975$. The boundary effect is considered using the formula:

$$V_z = C_1 ln\frac{z+z_o}{z_o} \tag{1}$$

where $z_o = 0.035$m (open country, Exposure C), and the C_1 is calculated as follows:

$$C_1 = \frac{V_{ref}}{ln\frac{H_o+z_o}{z_o}} = \frac{1.0}{ln\frac{1000+0.035}{0.035}} = 0.0975$$

Here $V_{ref}=1.0$, $H_o=1000$ m (inlet height). More details about Law of the Wall are presented in the next section.

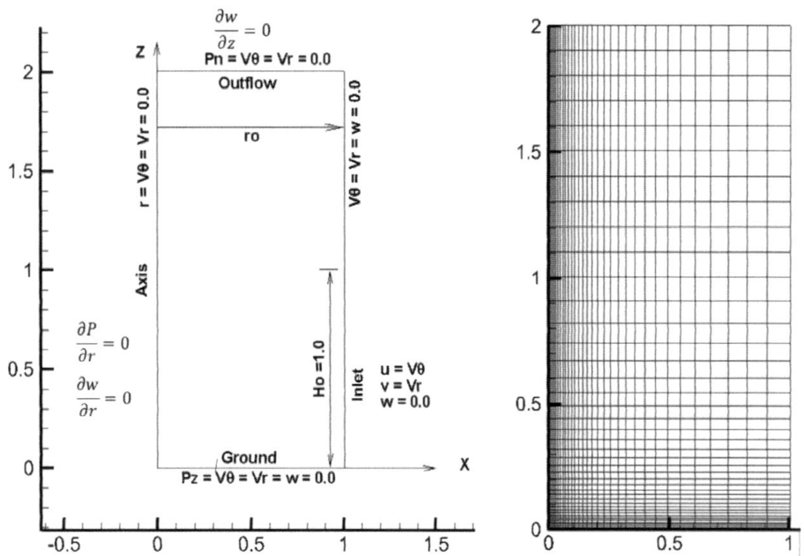

Figure 1. a) Computational domain, b) Sample Grid.

Figure 1b shows a sample grid, in which we are keeping more refinement near the surface and at the tornado inner core in order to properly capture the boundary layer and some other properties of the vortex, such as critical radius (r_c), and the horizontal location of maximum tangential velocity ($V_{\theta, max}$).

The Navier-Stokes equations are solved for different Swirl ratios between 0.20 and 0.95. From the experimental point of view, [4] developed a series of trials using S ranging from 0.12 to 1.29 with the tornado simulator at the University of London, Canada, results which are taken into account for comparison.

2.1 *Solution Method and Boundary Conditions*

The differential equations used in the computer program are shown below in terms of Cartesian coordinates (x,y). As it was stated before, the flow is considered axisymmetric, and then x is the radial coordinate, y is the vertical coordinate (axial), and some other terms such as ξ and g (gravity) [30]. The values of ξ are as follows: for plane geometry ξ=0, and for cylindrical geometry ξ=1.0. For this study, we have cylindrical geometry, then ξ=1.

The mass continuity equation is:

$$\frac{\partial u}{\partial x} + \frac{\partial v}{\partial y} + \xi\frac{u}{x} = 0 \tag{3}$$

The differential Navier-Stokes equations in Cartesian coordinates are:

$$\frac{\partial u}{\partial t} + \frac{\partial u^2}{\partial x} + \frac{\partial uv}{\partial y} + \xi\frac{u^2}{x} = -\frac{\partial p}{\partial x} + g_x + \nu\left(\frac{\partial^2 u}{\partial x^2} + \frac{\partial^2 u}{\partial y^2} + \xi\left(\frac{1}{x}\frac{\partial u}{\partial x} - \frac{u}{x^2}\right)\right) \tag{4}$$

$$\frac{\partial v}{\partial t} + \frac{\partial v^2}{\partial y} + \frac{\partial uv}{\partial x} + \xi\frac{uv}{x} = -\frac{\partial p}{\partial y} + g_Y + \nu\left(\frac{\partial^2 v}{\partial x^2} + \frac{\partial^2 v}{\partial y^2} + \xi\left(\frac{1}{x}\frac{\partial v}{\partial x}\right)\right) \tag{5}$$

The velocities (u,v) go along the coordinate directions (x,y), the ratio of pressure to constant density is p (normalized with density), g_x and g_y are the body accelerations, and the kinematic viscosity is ν. For our model, u=V_θ and v=V_r; w is the vertical velocity.

The CFD model uses Semi-implicit Pressure Linked Equations (SIMPLE) algorithm to solve the incompressible Navier-Stokes equations (from the momentum equations, u and v are implicitly solved), the convection term is approximated using Upwind (UW) for space approximation, and Euler Forward (Explicit) is the method used for time approximation. The model has a rectangular grid system and the variables are storage in a staggered (Yaqui-type) configuration [31, 33] to avoid the problem of pressure-velocity decoupling seen on the non-staggered configuration.

The Control Volume Method (CVM) with Preconditioned Conjugate Gradient (PCG) is used to solve the Navier-Stokes Equations. To account for the turbulent nature of the system, Large Eddy Simulation (LES) was implemented in our model.

The boundary conditions used for the model can be seen in the sketch presented in Figure 1a. The no-slip condition is applied at the ground surface and the vertical wall above the inlet region. The inlet vertical velocity was considered as zero, and the outlet radial and tangential velocities were assumed zero. A special consideration is presented here due to the application of the Law of the Wall in the computer model. The Law of the Wall is applied to the ground surface of the chamber and the vertical wall above the inlet region. Below, there is summary of the steps used to derive this equation considering the distance from the wall (y), the mean velocity (ū), the shear stress (τ_w), the fluid density (ρ), and the fluid kinematic viscosity (ν). The dimensions of the quantities are listed next: y[L], ū(y)[L/T], τ_w [M/LT^2], ρ[M/L^3], ν[L^2/T]. The friction velocity can be defined as:

$$u* = \sqrt{\tau w/\rho} \tag{6}$$

The wall shear stress is based on the velocity gradient in the direction normal to the surface of the wall [32]:

$$\tau_W = \mu \frac{\partial u}{\partial y}\Big|_{y=0} \tag{7}$$

where U is the fluid velocity along the wall, μ is the dynamic viscosity. Using dimensional analysis, the velocity can be described as a function of the fluid velocity and the friction velocity:

$$u^{+} = \frac{U}{u*} \tag{8}$$

3 EFFECTS OF SWIRL RATIO FOR THE TORNADO SIMULATOR

The swirl ratio is a fundamental parameter for the tornado simulator because it controls the size and the strength of the tornado. The swirl ratio is the ratio of vortex circulation to the accompanying inflow rate [15]. Other authors such as [4, 9, 16, 17], have defined the swirl ratio as the ratio of the angular momentum to radial momentum in the vortex, which can be expressed as:

$$S = \frac{\Gamma_\infty}{\frac{2QH_O}{r_O}} = \frac{tan\theta}{2a} \tag{9}$$

where $\Gamma_\infty = 2\,\pi\, r_c, V_{\theta,max}$, the aspect ratio is $a=H_o/r_o$; h is the height of the inlet (space between the floor and the vortex chamber wall); r_o is the top opening radius; and Q is the outlet volume, r_c is the radius corresponding to $V_{\theta,max}$.

For our numerical simulator, we will use the next formula, which also was used by [4, 7, 23]:

$$S = \frac{V_\theta}{2aV_r} = \frac{tan\theta}{2a} = \frac{V_\theta}{2V_r\frac{H_O}{r_O}} \tag{10}$$

where V_θ is the tangential velocity, V_r is the radial velocity, $a=H_o/r_o$ is the aspect ratio.

The swirl ratio is a parameter so important for the simulation that the structure of the vortex can be considered as controlled by this factor. In other words, the swirl ratio can lead to a single-cell vortex or two-cell vortex depending upon if it is low (S=0.31) or high (S=0.65), respectively [17].

Liu and Ishihara [10] performed several simulations using different swirl ratios over smooth and rough terrain. These authors compared their results with experimental data provided by [18, 19, 20, 21], which gives us valuable swirl values or sample range for our computer simulations.

According to [10], the flow patterns can change easily due to changes of the swirl ratio; for example, the flow pattern can vary quickly from a single-cell vortex (S=0.4) to a vortex touch-down (S=0.6). This variability of flow patterns finishes after the tornado touch-down, and the flow becomes similar for larger S like 3.8. Because of this tornado behavior due to S, the S-values used by [10] are very close for small S (variations of 0.2 units) and the time step is higher when using S close to 3.8 (variations of ≈0.5 units).

Some important conclusions derived by [10] regarding the influence of the swirl ratio over the tornado behavior are:

- When the swirl ratio is high (3.8), the tornado becomes less sensitive to motion. In other words, the tangential velocity profile is almost identical with the profile of a stationary tornado.
- For low swirl ratio (0.6), the tornado becomes weak and the tangential velocity decreases slightly, but the vertical velocity changes considerably when giving translation to the vortex. At low elevation, the highest vertical velocity decreases from $2.8W_o$ to $0.5W_o$ where W_o is upward velocity at outlet of simulator. Moreover, the effect of swirl ratio on vertical velocity becomes weak for higher swirl ratios such as 2.0 or 3.8.

- A significant change on radial profile of pressure coefficient was noticed by [10] when comparing translational and stationary tornadoes for different swirl ratios. Here the pressure increases drastically in the center.
- For tornadoes with S=0.6, the influence of translation is significant. The vortex is tilted just close to the ground.

4 EVALUATION OF GRID SIZE RESOLUTION AND COMPARISON OF WIND FIELD FOR DIFFERENT SWIRL RATIOS USING NUMERICAL ANALYSIS.

As it was stated before, swirl ratio (S) has significant influence over several properties of tornadoes such as pressure, tangential and vertical velocity. Furthermore, grid refinement is another factor that can affect the output considerably.

4.1 *Effects of Grid Resolution on the Accuracy of Capturing Tangential Velocities*

One of the preliminary parameters to be assessed when working with numerical modeling is how proper the grid refinement is for the desired study. We will present a comparison of effects of different S for three different grid sizes (minimum dimensionless spacing of $0.01H_o$, $0.005H_o$ and $0.001H_o$) using Control Volume Procedures. The grid size is determinant to capture parameters like the downdraft, the touchdown and peak velocities in the tangential velocity profile. Having an appropriate grid spacing close to the investigated structure or site is fundamental to achieve proper results. As it was stated by [15], the grid size can easily affect the outputs of the numerical model. Although the mesh refinement close to ground is fundamental to obtain results according to reality, it is necessary to consider the computational limitations. Hence, a graphic comparison of V_θ for three different grid sizes is to be presented below.

Figure 2 shows a comparison of tangential velocity (V_θ) for a swirl ratio of S=0.4 and minimum grid spacing (MGS) of $0.01H_o$, $0.005H_o$ and $0.001H_o$. The left side of Figure 2 shows the whole domain with tangential velocity distribution, and the right side shows a close-up of the picture on the left. At approximately the same elevation and horizontal distance from the tornado center, Figures 2b, 2d and 2f show that V_θ close to the ground changes when varying the grid size. The magnitude of V_θ changes from $3.74V_{ref}$ to $4.78V_{ref}$ when varying the grid size from $0.01H_o$ to $0.001H_o$. Another important parameter that can be noticed in Figure 2 is the elevation of the inner downdraft. For the coarse grid (MGS=0.01 H_o), the elevation of the inner downdraft is about 50m AGL; for the mesh with MGS=0.005 H_o, the elevation of the inner downdraft is 20m, but for the fine mesh (MGS=0.001 H_o), we have touchdown stage.

Due to the details discussed above about V_θ regarding the effects of mesh refinement, the discussions herein will be based on results obtained from the mesh with MGS=$0.001H_o$. Additional features supporting the selection of this mesh will be presented in the next section.

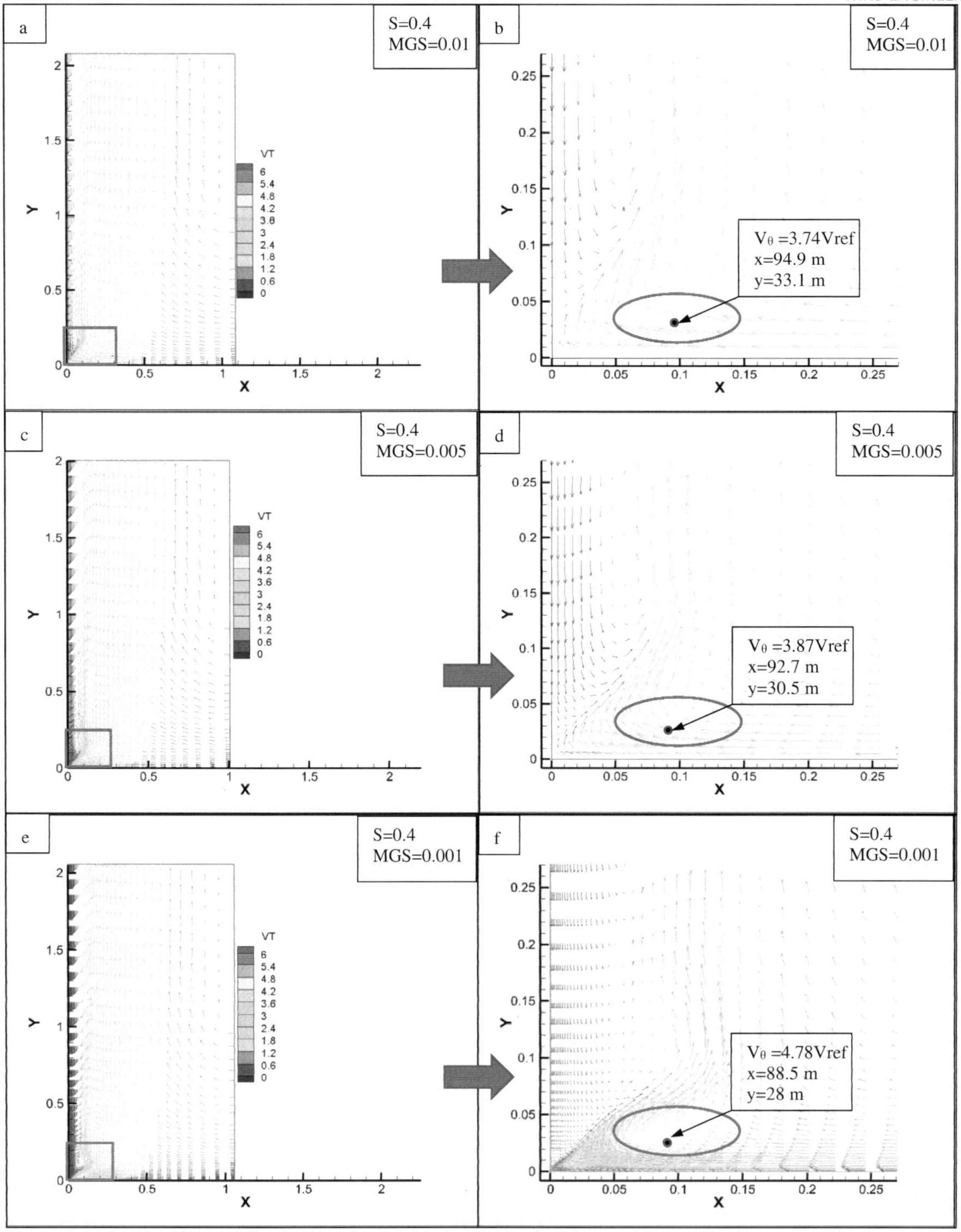

Figure 2. Left side: Tangential velocities for swirl ratio =0.4 and grid sizes. Right side: Close-up of the bottom left corner of each picture on the left side.

4.2 *Effects of Swirl Ratio on Tangential Velocity Close to the Ground*

The wind field close to the ground has significant variations for several S. Figure 3 shows the wind field for S from 0.2 to 0.95 and grid size is MGS=0.01H_o. Here, the touchdown happens for S $\geq$ 0.60, and for all the used S the wind field follows the RCVM profile. In addition, Figure 4 shows the results for the same range of S but using MGS=0.005H_o. Likewise, the touchdown happens for S $\geq$ 0.60 and the wind field profiles have the same shape as the previous grid.

On the contrary, when using MGS=0.001H_o in Figure 5, the touchdown can be captured for S$\geq$0.4 and the wind field is different from the RCVM profile. Here we have two peak velocities (except for S=0.20) instead of one as it happens in Figure 3 and Figure 4. In other words, the velocity distribution close to the ground for several S is different from wind field above 40m. Beyond 40m-AGL, the wind profile starts following the same shape as the RCVM profile. Moreover, this situation (wind field with two peak velocities) was reported by [22] at height of 87.5m (0.0875 dimensionless), but it is necessary to clarify that the boundary conditions in an experimental simulator are greatly affected by scale, reduction of the updraft opening, availability of space close to the ground surface to perform measurements, etc. The reason for those two peak velocities at elevations close to the ground can be explained based on the components of velocity for a tornado. The three components of tornado velocity are: the radial (V_r), tangential (V_θ) and vertical (V_z) velocities, but several previous works consider RCVM, where V_r and V_z are assumed equal to zero. Because a real tornado is characterized by a non-zero radial component, especially close to ground, the wind field plotted in Figure 5 is no longer following completely the RCVM profile, and different peak V_θ are found in the vortex profile. In other words, the results of our model are comparable to real tornadoes.

Figure 6a (left side) shows the V_θ contour plot for the entire domain, and a close-up of the region (Figure 6a, right side) close to the ground, where the different peak velocities are located for the wind profile; this is presented together with the wind field profile for S=0.6 (Figure 6b). Regarding the touchdown phenomenon, [7, 10], used Control Volume Method (CVM), and reported that the touchdown happens for S > 0.4. Furthermore, [23] argued that the vortex touchdown happens at S=0.5 when using CVM procedure.

Previous investigations, experimental and numerical, have not captured these peaks velocities close to the ground because they do not have enough resolution (grid spacing) or the data was collected at heights in which the effects of the boundary layer are not significant.

For all these reasons, and the ones established in the previous section, MGS=0.001H_o is an adequate grid spacing to properly capture the effects on the boundary layer and the real velocity distribution close to the ground.

Tangential velocity profiles at different heights are shown in Figures 3, 4 and 5. The profiles shown in Figure 5 are in agreement with the conditions (results) found by [17]. Ishihara et al. pointed out that there is a proportional relationship between V_θ and S. This means that the maximum V_θ for a small S (S=0.3) is little compared with the maximum V_θ for a higher S (S=0.75). It is important to underline, (based on Figure 5), that the wind field is very similar for several S beyond S=0.30, but the wind field for S=0.2 is quite different. This can be explained from the point of view that the flow is laminar for small S. In general, with high S, the vortex broadens, having larger radius. Moreover, Figure 5b shows that when the elation is increasing, the wind profile starts becoming RCVM. This happens at about 10m AGL.

b

Z≈21m
MGS=0.01

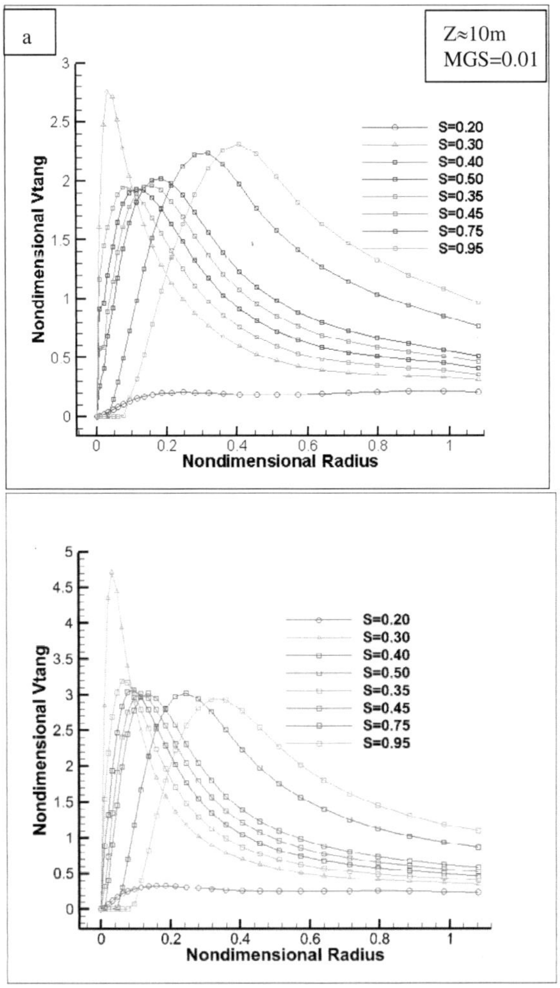

Figure 3. Wind field for different swirl ratios and heights (z). a) MGS=0.01H$_o$, z=10m; b) MGS=0.01H$_o$, z=21m

b

Z≈10.5m
MGS=0.005

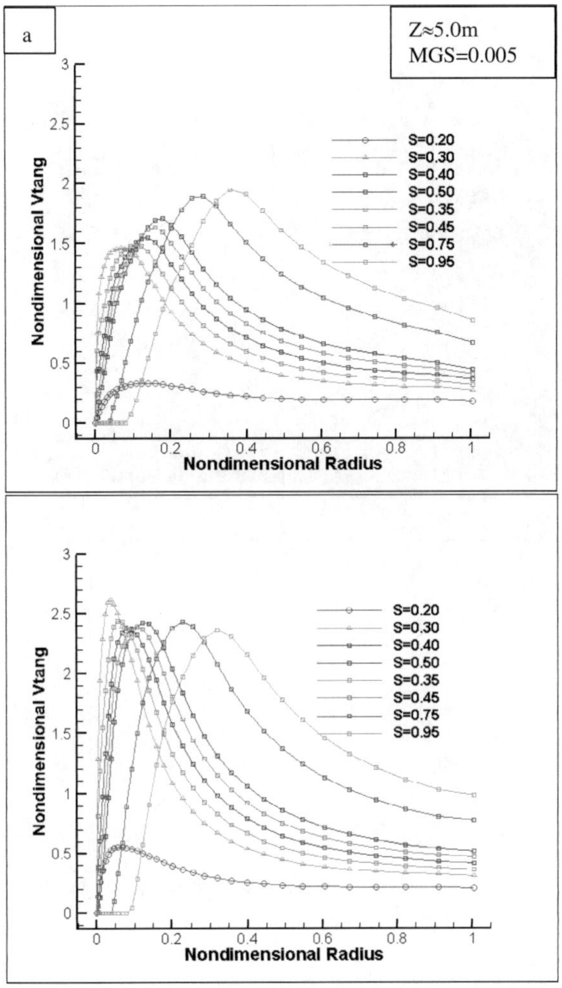

Figure 4. Wind field for different swirl ratios and heights (z). a) MGS=0.005H$_o$, z=5m; b) MGS=0.005H$_o$, z=10.5m

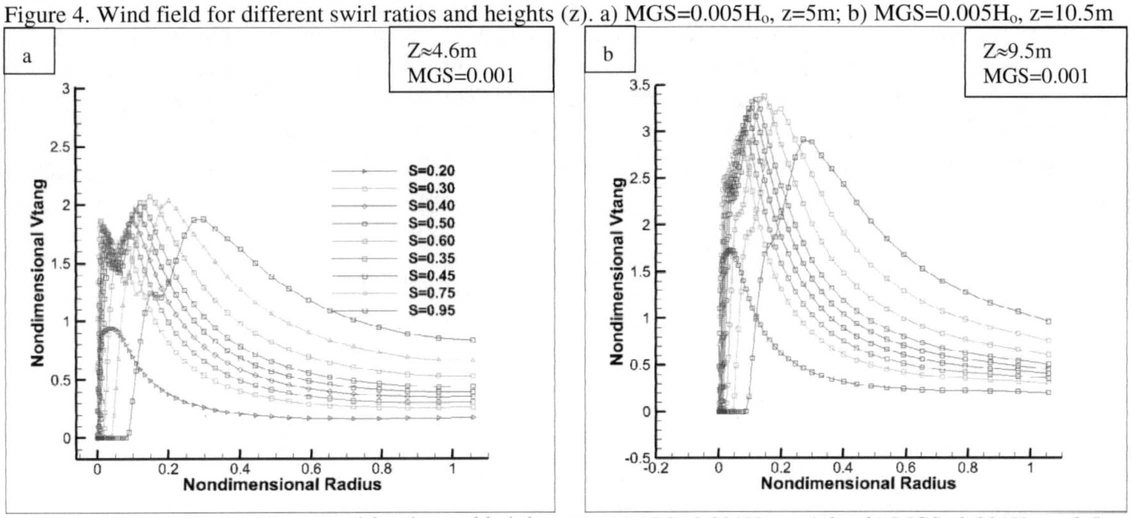

Figure 5. Wind field for different swirl ratios and heights (z). a) MGS=0.001H$_o$, z=4.6m; b) MGS=0.001H$_o$, z=9.5m

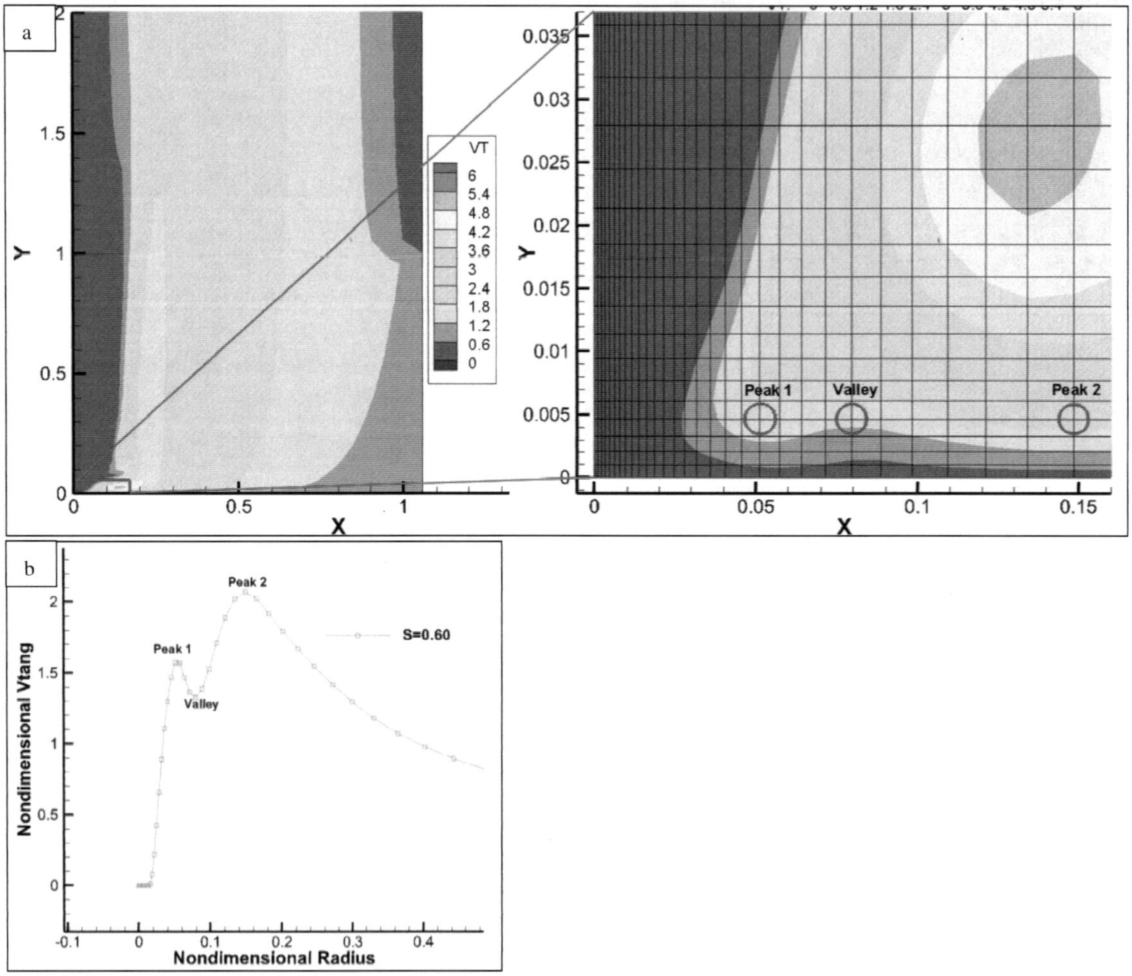

Figure 6. Zoomed region close to the ground where the wind profile was extracted for S=0.6 at elevation of 4.6m.

In addition, to improve the comparison between our results and previous numerical results, we can quote the work done by [8] because it has similar minimum grid refinement close to the ground surface and domain. [8], using Finite Difference method (FDM), stated that the horizontal velocity at 27 m AGL is 110 m/s, which normalized regarding to the inlet velocity (16.6 m/s) is equal to 6.6 for S=0.94. Likewise, at 28 m AGL, we found that the non-dimensional tangential velocity is equal to 4.5 when S=0.95.

5 COMPARISON OF WIND FIELD CLOSE TO THE GROUND OF NUMERICAL TORNADO SIMULATORS AND EXPERIMENTAL TORNADO SIMULATORS

Keeping in mind that experimental research results can be affected by the parameters aforementioned, we will present a comparison based on the available data.

It has been previously stated by several authors [4, 18, 21, 25, 26, 27, 29] that the tornado inner core expands when increasing S. In fact, our numerical tornado simulator shows that the core radius enlarges if we consider increments of S for several simulations. That situation is presented in Figures 3 and 7, in which the inner blue core is widening while increasing S.

Refan [4] reported that the touchdown occurs when $S \approx 0.57$. In order to provide proofs about the agreement between her work and our simulations, Figure 7 shows that effectively the touchdown happens for $S \geq 0.5$. An accurate description of how the touchdown occurs is provided by [18] in which they also agreed that the vortex breakdown reaches the ground surface when $S>0.5$.

Our major objective in this paper is the study of tornado effects over civil structures on the ground as it was stated previously. Experimentally, it has been demonstrated [4, 12, 21, 28] that close to the ground there is an intensification of the V_θ, which causes destruction or intense damages over low-rise building along the tornado path. V_θ changes very fast from zero to a maximum V_θ at some height, which depends on S and the horizontal distance from the tornado center (radius). Normally, the maximum V_θ for different S is found at heights less than 50m, and this is the zone where due to beam angle and obstructions, the measurements of Doppler on Wheels (DOW) radar are not trustworthy, as it is well explained by [28].

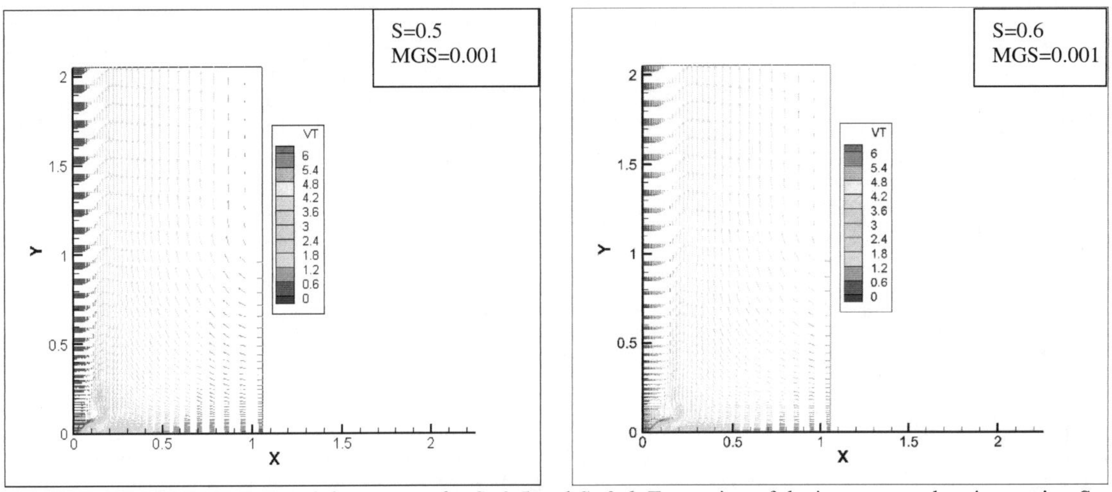

Figure 7. Illustration of the touchdown stage for S=0.5 and S=0.6. Expansion of the inner core when increasing S.

Figure 8 clearly depicts that the maximum V_θ happens close to the ground, which is the most destructive factor for low-rise buildings facing tornadoes [17, 24]. A rapid increment on V_θ occurs at the elevation where low-rise buildings are located, and this is the reason for large failures in civil structures, as it was stated by [12, 13]. Figure 9 shows that the same situation (intensified flow on the ground) happens for several $S \geq 0.3$, which is in good agreement with the experimental work reported by [4]. According to Figures 8 and 9 and Table 1, the maximum V_θ for all S considered is located in a range of $4.0V_{ref} \leq V_\theta \leq 4.9V_{ref}$ (Recall $V_{ref} = SC_1$). Due to the low tangential velocity in the inner core, for S=0.95 in Figures 9a-c, V_θ is almost zero.

Having this information, it is possible to measure the maximum V_θ and the height at which it happens for different tornado sizes, and we can obtain appropriate values for designing low-rise buildings in tornado-prone areas.

Another point that needs to be mentioned is the magnitude of V_θ above approximately 100m. The profile (magnitude of V_θ) is very sensitive depending upon the horizontal distance from the tornado center, and it can be seen in Figure 9a-b, in which for all S values (except for S=0.20), all the profiles of V_θ are almost constant (vertical) with values between 0 and 1.0, but in Figure 9c, the majority of the profiles have an unsteady shape with a more ample range of values in the considered domain. Because the inner core for the majority of the swirl ratios ($S \leq 0.6$) is far from a vertical section at 245.5 from the center, Figure 9d shows that the profiles for V_θ are constant for heights above 100m, but for S=0.75 and S=0.95 there is still a peak velocity close to the ground.

These two peak velocities (for S=0.75 and S=0.95) in Figure 9d happen because the vertical section (at 245m approximately, vertically) was taken out of the inner core.

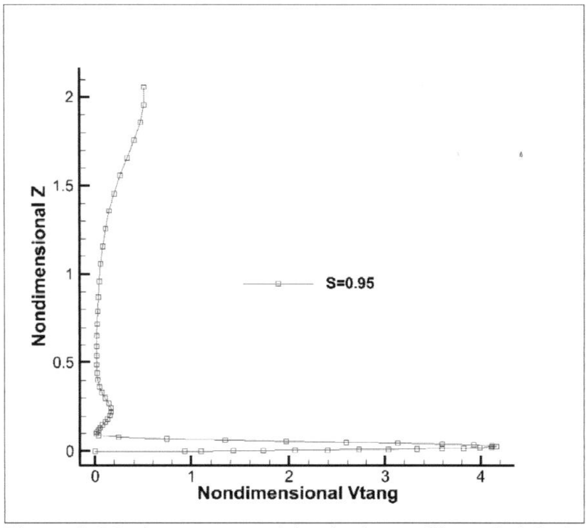

Figure 8. Maximum tangential velocity profile vs. height for S=0.95 at 245.5 m from the tornado center.

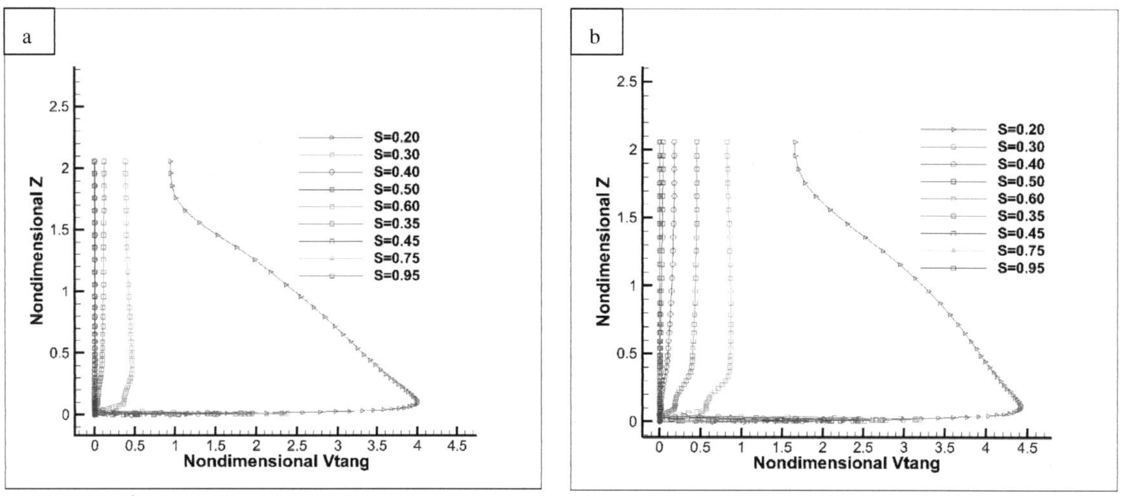

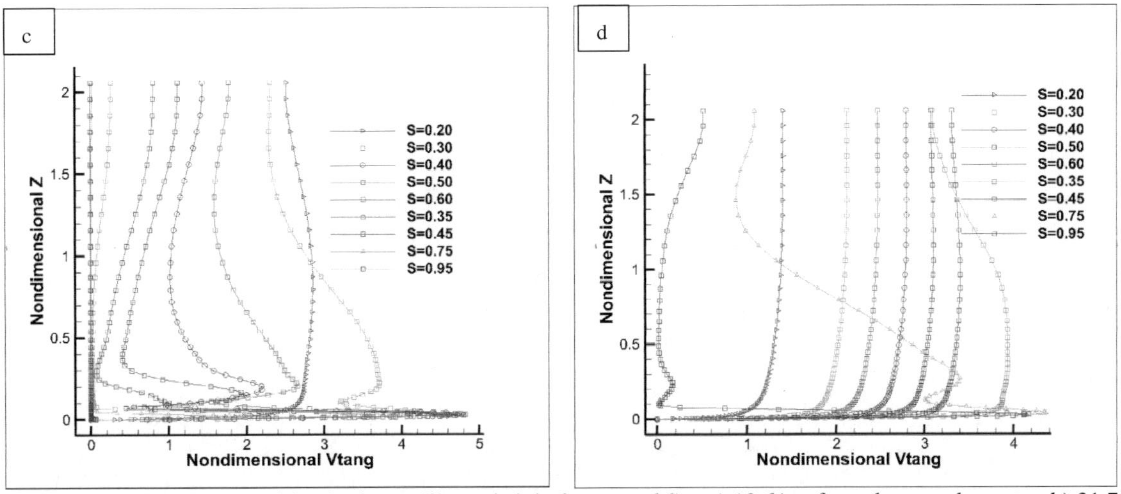

Figure 9. Maximum tangential velocity profile vs. height for several S at a) 13.61 m from the tornado center, b) 31.76 m from the tornado center, c) 88.36 m from the tornado center, d) 245.51 m from the tornado center. MGS=0.001 h_o

6 MAXIMUM V_Θ AND V_Θ' VARIATION AT 3.3 M-HEIGHT.

Because low-rise buildings are greatly affected by peak velocities close to the ground, details of maximum V_θ and velocities at 3.3 m-elevation will be presented below.

Table 1 displays the maximum values of V_θ, the corresponding elevations (Z), and the horizontal location from the tornado center (X), for different S values. This table also shows that the peak V_θ for the majority of the tested S, is located at ≈ 30m AGL. Moreover, as it happens with the critical radius (r_c), the horizontal distance from the tornado center to the peak V_θ increases when increasing S. The higher peak tangential velocity ($4.98V_{ref}$) was found for $0.5 \leq S \leq 0.6$, which is in agreement with the range of S when the touchdown happens. In addition, this table shows that in the majority of the cases, Doppler radar measurements are not capturing the peak V_θ because mostly these measurements are performed at above 50 m AGL.

Table 1. Maximum V_θ for several S at different locations from the tornado center (X).

Swirl Ratio	$V_{\theta,max}/V_{ref}$	Z (m)	X (m)
S=0.20	4.4192	109.2	28.0
S=0.30	4.6180	31.8	64.0
S=0.35	4.6384	31.8	79.5
S=0.40	4.7778	28.0	88.5
S=0.45	4.9026	28.0	98.3
S=0.50	4.9814	28.0	109.2
S=0.60	4.9897	28.0	134.2
S=0.75	4.7714	28.0	181.9
S=0.95	4.2745	28.0	271.0
S=1.29	4.6873	31.8	364.0

Table 2 shows the values of V_θ at 3.3 m heights, which is the common elevation of many of structures (low-rise buildings) on tornado-prone areas. These values can be used to analyze the real magnitude of V_θ that affects buildings. With these values, a better design can be performed because V_θ is the larger velocity term (component) close to the ground [12], and therefore, the most necessary to consider. At height ≈ 3m AGL tangential velocities could be 47-80% greater than V_{ref} for some swirl ratios which is very important for design considerations. In parallel, experimental data reported by [29] shows that the highest velocities occur close to the ground, having peak values some 30% bigger than velocities at upper levels.

Table 2. V_θ at 3.3 m height for several S.

Swirl Ratio	V_θ/V_{ref}	X (m)
S=0.20	0.7018	40.5
S=0.30	1.5253	9.5
S=0.35	1.4793	15.9
S=0.40	1.4691	21.4
S=0.45	1.5126	109.2
S=0.50	1.5748	121.1
S=0.60	1.6393	148.6
S=0.75	1.6482	201.1
S=0.95	1.5692	299.1
S=1.29	1.8007	364.0

7 CONCLUSIONS

A numerical tornado simulator has been built to investigate the wind field close to the ground for several swirl ratios. There are some conclusions that arise from the current work:

a) The wind field varies when changing the swirl ratio and keeping the same domain and mesh refinement.
b) Mesh refinement has a great influence over the results of the wind field. Just a proper mesh resolution is capable of presenting that close to the ground wind field does not follow Rankine Combine Vortex Model.
c) Analysis of velocity distribution shows that V_θ is not constant, having two peak velocities in the region close by the maximum V_θ.
d) The tornado keeps a structure of one-cell vortex for S≤0.4 without touchdown, but when S≥0.5, the downdraft at the inner core touchdown.
e) The maximum tangential velocity occurs close to the ground, commonly at the first 50 m above ground level, except for small tornadoes (S=0.2).
f) V_θ at lower elevations (close to the ground) levels could reach so higher values over the critical velocity (V_{ref}) such as 4.99 V_{ref}, which affects directly civil structures.
g) Because the wind profile close to the ground no longer follows the Rankine Combine Vortex Model profile when working with a vortex chamber simulator, different wind fields are interacting with low-rise buildings.
h) Close to the tornado center (radius<90m), Figures 9a-c, the vertical profile of V_θ has a peak near the ground, but above heights of around 100 m, the profile becomes constant (vertically) having $V_\theta \leq 2.0\, V_{ref}$ (except for S=0.2). Moreover, V_θ is very small when using values nearby the center of the inner core for S≥0.45.
i) Far from the tornado center (≥90 m) and heights above 100 m, Figure 9d, the V_θ profiles become constant (vertically) for heights above the peak velocity location as it happens in the previous point, but their values now are on the range of $0 \leq V_\theta \leq 4.0\, V_{ref}$. Here, V_θ values are small or zero for S≥0.6.

8 REFERENCES

1 W. C. Lee and J. Wurman, Diagnosed three-dimensional axisymmetric structure of the Mulhall Tornado on 3 May 1999. Bulletin of the American Meteorological Society, 62(2005) 2373-2393.

2 R. M. Wakimoto, P. Stauffer, W. C. Lee, N. T. Atkins, and J. Wurman, Fine-scale structure of the LaGrange, Wyoming Tornado during VORTEX2: GBVTD and photogrammetric analyses. Monthly Weather Review, 140(2012) 3397-3418.

3 J. Wurman, J. Straka, E. Rasmussen, M. Randall, and A. Zahrai, Design and deployment of a portable, pencil-beam, pulsed, 3-cm Doppler radar. Journal of Atmospheric and Oceanic Technology, 14(1997) 1502-1512.

4 M. Refan, Physical simulation of tornado-like vortices. Doctoral Dissertation. The University of Western Ontario, London, Ontario, Canada, 2014.

5 D. R. Smith, Effect of boundary conditions on numerically simulated tornado-like vortices. Journal of the Atmospheric Sciences, 44(1987) 648-656.

6 W. A. Gallus Jr, F. L. Haan Jr, P. P. Sarkar, K. Le, and J. Wurman, Comparison of numerical model and laboratory simulator tornado wind fields with radar observations of the Spencer, South Dakota tornado. In Symposium on the Challenges of Severe Convective Storms, 86th AMS Annual Meeting, Atlanta, GA, 2006. American Meteorological Society.

7 H. Hangan, and J. D. Kim, Swirl ratio effects on tornado vortices in relation to the Fujita scale. Wind and Structures, 11(2008) 291-302.

8 W. S. Lewellen, D. C. Lewellen, and R. I. Sykes, Large-eddy simulation of a tornado's interaction with the surface. Journal of the Atmospheric Sciences, 54(1997) 581-605.

9 T. Wilson and R. Rotunno, Numerical simulation of a laminar end-wall vortex and boundary layer. The Physics of Fluids, 29(1986) 3993-4005.

10 Z. Liu and T. Ishihara, Study of the effects of translation and roughness on tornado-like vortices by large-eddy simulations. Journal of Wind Engineering and Industrial Aerodynamics, 151(2016) 1-24.

11 J. Wurman, K. Kosiba, and P. Robinson, In situ, Doppler radar, and video observations of the interior structure of a tornado and the wind damage relationship. Bulletin of the American Meteorological Society, 94(2013) 835-846.

12 W. Zhang, and P. P. Sarkar, Near-ground tornado-like vortex structure resolved by particle image velocimetry (PIV). Experiments in Fluids, 52(2012) 479-493.

13 T. Ishihara, and Z. Liu, Numerical study on dynamics of a tornado-like vortex with touching down by using the LES turbulent model. Wind and Structures, 19(2014) 89-111.

14 J. Wurman, J. M. Straka, and E. N. Rasmussen, Fine-scale Doppler radar observations of tornadoes. Science, 272(1996) 1774-1777.

15 L. Kuai, F. L. Haan Jr, W. A. Gallus Jr, and P. P. Sarkar, CFD simulations of the flow field of a laboratory-simulated tornado for parameter sensitivity studies and comparison with field measurements. Wind and Structures, 11(2008) 75-96.

16 R. Rotunno, Numerical simulation of a laboratory vortex. Journal of the Atmospheric Sciences, 34(1977), 1942-1956.

17 T. Ishihara, S. Oh, and Y. Tokuyama, Numerical study on flow fields of tornado-like vortices using the LES turbulence model. Journal of Wind Engineering and Industrial Aerodynamics, 99(2011) 239-248.

18 C. Church, J. T. Snow, G. L. Baker, and E. M. Agee, Characteristics of tornado-like vortices as a function of swirl ratio: A laboratory investigation. Journal of the Atmospheric Sciences, 36(1979) 1755-1776.

19 N. Monji, A laboratory investigation of the structure of multiple vortices. Meteorological Society of Japan Journal, 63(1985) 703-713.

20 A. R. Mishra, D. L. James, and C. W. Letchford, Physical simulation of a single-celled tornado-like vortex, part A: flow field characterization. Journal of Wind Engineering and Industrial Aerodynamics, 96(2008) 1243-1257.

21 P. H. Tari, R. Gurka, and H. Hangan, Experimental investigation of tornado-like vortex dynamics with swirl ratio: the mean and turbulent flow fields. Journal of Wind Engineering and Industrial Aerodynamics, 98(2010), 936-944.

22 M. Refan, and H. Hangan, Characterization of tornado-like flow fields in a new model scale wind testing chamber. Journal of Wind Engineering and Industrial Aerodynamics, 151(2016) 107-121.

23 D. Natarajan, and H. Hangan, Large eddy simulations of translation and surface roughness effects on tornado-like vortices. Journal of Wind Engineering and Industrial Aerodynamics, 104(2012) 577-584.

24 T. Maruyama, A numerically generated tornado-like vortex by large eddy simulation. In Proceedings of 7th Asia Pacific Conference on Wind Engineering, Taipei, Taiwan, 2009.
25 N. B. Ward, The exploration of certain features of tornado dynamics using a laboratory model. Journal of the Atmospheric Sciences, 29(1972) 1194-1204.
26 R. P. Davies-Jones, The dependence of core radius on swirl ratio in a tornado simulator. Journal of the Atmospheric Sciences, 30(1973) 1427-1430
27 C. R. Church, and J. T. Snow, Laboratory models of tornadoes. The Tornado: its structure, dynamics, prediction, and hazards. Geophysical Monograph, (1993) 277-295.
28 P. Sarkar, F. Haan, W. Gallus Jr, K. Le, and J. Wurman, Velocity measurements in a laboratory tornado simulator and their comparison with numerical and full-scale data. In Proceedings of the 37th US-Japan panel on wind and seismic effects joint meeting, Tsukuba, Japan, 2005.
29 G. L. Baker, and C. R. Church, Measurements of core radii and peak velocities in modeled atmospheric vortices. Journal of the Atmospheric Sciences, 36(1979) 2413-2424.
30 C. W. Hirt, B. D. Nichols, and N. C. Romero, SOLA: A numerical solution algorithm for transient fluid flows. NASA STI/Recon Technical Report N, 75, 32418, 1975.
31 F. H. Harlow, and J. E. Welch, Numerical calculation of time-dependent viscous incompressible flow of fluid with free surface. The physics of fluids, 8(1965), 2182-2189.
32 A. Neale, D. Derome, B. Blocken, and J. Carmeliet, CFD calculation of convective heat transfer coefficients and validation–Part 2: Turbulent flow, 2006.
33 Selvam, R. P. Computation of pressures on Texas Tech building. Journal of Wind Engineering and Industrial Aerodynamics, 43(1992), 1619-1627.
34 B. H. Fiedler, and G. S. Garfield. Axisymmetric vortex simulations with various turbulence models. CFD Letters, 2(2010), 112-122.

13th Americas Conference on Wind Engineering
Gainesville, Florida, USA
May 21 - 24, 2017

Stochastic Simulation of Tornado Tracks for the Continental United States

Fanfu Fan[a], Weichiang Pang[b]

[a] *Clemson University, Clemson, U.S., ffan@clemson.edu*
[b] *Clemson University, Clemson, U.S., wpang@clemson.edu*

ABSTRACT

According the records of Storm Prediction Center (SPC) of the National Oceanic Atmospheric Administration (NOAA), approximately more than 1000 tornadoes spawned every year in the continental United States. Tornadoes often cause significant damages, injuries and economic losses to the affected regions. Compared to hurricanes, the damage footprint of a tornado is relatively small. While the damage potential of a tornado is high, the occurrence of a tornado at a given geographic location can be characterized as a low probability but high consequent event. The SPC maintains a database of historical tornado events with approximately 65 years of data. Since the historical data is limited and the impact of a tornado is very localized, it is not feasible to use solely the raw historical data or tracks to properly quantify the tornado risk for a given structure or a community that has not been affected by historical tornadoes. This paper presents a stochastic simulation framework to generate synthetic tornado tracks for the Continental United States. The annual spawn rate is modeled using a negative binomial distribution. The spawn or touchdown locations are simulated using geographic dependent kernel density estimation (KDE) GIS maps, which specifically account for the variability of tornado properties at different geographic regions (e.g. tornado alley, Dixie alley and etc.). A database of one million years of simulated tornado tracks have been simulated using the Clemson high performance computing facility, the Palmetto Cluster. The track parameters include the tornado occurrence rate, intensity (EF-scale), touchdown location, touchdown time, path length, maximum path width, and path direction. All these parameters are geographic dependent, meaning the properties vary depending on the geographic locations. The simulated spawn rates and other key parameters for select major cities in the United States were compared to the observations. Good agreements were observed between the simulation and observation. Using the simulated track database and a 2D tornado wind field model, the wind speed distributions for different return periods were computed for select cities. As an illustration for the application of the tornado database, the computed wind speed distributions were used in a fragility framework to quantify the damage risk and loss potential of tornado hazards to a low-rise building.

KEYWORDS

Tornado hazards, Tornado database, Stochastic simulation, Risk analysis, Wind speed, Tornado wind field

13th Americas Conference on Wind Engineering
Gainesville, Florida, USA
May 21 - 24, 2017

Effect of a 2016 tornado in rural areas of Jiangsu province in China.

Weihua Fang [a], Donghua Pan [b], Jean-Paul Pinelli [c]
Jingwei Hou[d]
[a] *Beijing Normal University, Beijing, China, weihua.fang@bnu.edu.cn*
[b] *National Disaster Reduction Center of China, Beijing, China, pandonghua@ndrcc.gov.cn*
[c] *Florida Institute of Technology, Melbourne, Florida, USA, pinelli@fit.edu*
[d] *Beijing Normal University, Beijing, China, jingwei.hou @mail.bnu.edu.cn*

ABSTRACT

Tornadoes are rare events in China. However, in the afternoon hours of June 23, 2016, a tornado ravaged Funing and Sheyang counties, in the Jiangsu Province of China. A total of 98 people were killed and severe economic losses were experienced. Several industrial structures, including large factories, were destroyed, communications towers, and power infrastructures were damaged, and rural residential areas in 21 villages suffered extensive damage. Some of the event was captured on video cameras. The paper will describe the field observations, including UAV remote sensing imagery and surviving videos of the event. Preliminary analysis of the causes and sequence of structural failure for some of these structures and infrastructure will be provided, as well as preliminary wind speed estimates based on the observed damage. Recommendations will be discussed to prevent or reduce similar failures in the future, and to promote wind mitigation in that region of China.

KEYWORDS

Tornado, wind damage, field survey, UAV remote sensing, mitigation.

Development of a wind engineering field laboratory to characterize transient winds

Justin Nevill[a], Franklin T. Lombardo[b]

[a]*University of Illinois at Urbana Champaign, 205 North Mathews Avenue, Urbana, Illinois, USA*
[b]*University of Illinois at Urbana Champaign, 205 North Mathews Avenue, Urbana, Illinois, USA*

ABSTRACT: The University of Illinois at Urbana-Champaign has established the Smart Wind Engineering Research Facility (SWERF) to address the demonstrated need for a full-scale wind engineering facility. Through stationary and mobile initiatives SWERF will aim to provide observation of tornadoes and thunderstorm winds, full-scale structural and aerodynamic testing, and validation for wind-tunnel and numerical simulations. This report provides an overview of the facility and results of a preliminary data collection period. The facility overview includes site characteristics, infrastructure, and instrumentation. Preliminary results include site wind climate characterization and description of three thunderstorm events.

KEYWORDS: thunderstorm, downburst, tornado, full-scale, validation

1 INTRODUCTION

The high cost of damage from transient wind events, such as tornadoes and thunderstorms, is well documented. Fourteen of the fifteen costliest natural disasters in US history can be categorized as windstorms. These events have caused over 4,500 deaths and $70 billion in losses since 1980. Transient wind events other than hurricanes represent only two of the fifteen costliest events, but caused over $22.6 billion in damage during the five-years spanning 2011 to 2015. [1, 2]

Despite the cost and visibility of windstorm damage, the ground-level characteristics of transient events continue to be poorly understood [1]. Wind tunnel evidence showing that the gust structure of the wind has a significant effect on structural loads demonstrates the need for field observations during transient events [3]. Few quantitative field observations of transient winds and the loads induced on structures by these winds have been conducted. The majority of available field data captures the interaction of simple geometries and non-transient winds in open terrain. The characteristics of transient winds are not easily generalized into a single behavior. Additional observations and analysis may allow classification of thunderstorm gusts into multiple types with distinct behavior [4].

Standard engineering and construction practice continues to be based solely on surface boundary layer characteristics, excluding the effects of transient winds, despite the plurality of evidence that this assumption is unconservative [4]. Low-rise design and construction often relies on the minimum requirements of ASCE 7-10, the provisions of which can be more than 50% lower than values obtained through wind tunnel testing [5, 6]. Wind tunnel simulations underestimate peak loads on low-rise structures, and design values obtained from wind tunnel testing can vary significantly between wind tunnels. Wind tunnel and numerical simulations continue to need correction and validation through comparison with full-scale, natural-wind experiments and observation. [5, 7, 8, 9, 10, 11]

A broad range of well-designed wind engineering facilities have been constructed to explore the relationship between extreme winds and structural failure – addressing many of the disadvantages of traditional boundary-layer wind tunnel testing. Full-scale buildings and components can be tested to destruction at Florida International University's "Wall of Wind" [12] and the large test chamber at the Insurance Institute for Business and Home Safety (IBHS) Research Center [13]. Tornado-like vortices can be generated at a model scale in chambers at Texas Tech, Iowa State University, and University of Western Ontario [14, 15, 16]. The dynamic effects of wind loading on building components can be measured using facilities such as University of Florida's "MAWLS" [17] and the "Three Little Pigs" project at UWO [18]. High-quality data from modern, full-scale field experimentation is needed to validate the results produced by these facilities and provide accurate representations of natural wind behavior in transient events [19].

1.1 *The Smart Wind Engineering Research Facility*

The University of Illinois at Urbana-Champaign is developing the Smart Wind Engineering Research Facility (SWERF) to meet the identified needs of full-scale wind engineering observation, testing, and validation [20]. The facility is designed to specifically address transient wind characterization, bluff-body aerodynamics, and structure-obstruction interactions. The guiding principle of the facility is to become a community asset by responding to the full-scale wind engineering needs of the academic and non-academic communities, hosting extra-disciplinary research and testing where appropriate, and actively seeking collaborative, symbiotic projects.

1.2 *Physical site*

SWERF is located in Rantoul, IL on property that was formerly part of the decommissioned Chanute Air Force Base. The site is adjacent to the Rantoul National Aviation Center and the University of Illinois's Advanced Transportation Research and Engineering Laboratory (ATREL). SWERF is 15 miles north of the University of Illinois at Urbana-Champaign campus. The site elevation ranges from 220 to 223 meters (720 to 732 feet) above sea level. The immediate area is largely agricultural with corn and soybean fields dominating the landscape.

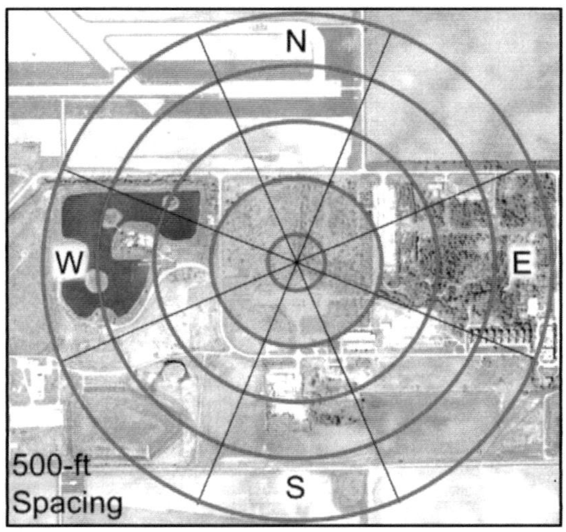

Figure 1 - SWERF, development within 1500 feet. The SWERF property lies mostly within the central shaded region. One of the runways of Rantoul National Aviation Center, a public use airport, is visible (top left).

The region within a fetch of 2,100 meters (7,000 feet) north, west, and south of the property is largely open with scattered trees and low-rise buildings. The region within a fetch of 900 meters (3,000 feet) to the east is wooded and developed with scattered low-rise buildings. The suburban development of Rantoul lies within a fetch of one-half to three miles north to west of the property. Beyond the regions mentioned previously, the surrounding terrain is predominantly open farmland. The closest heavily-developed areas are 10 miles south at Urbana-Champaign and 25 miles southeast at Danville, IL.

Figure 1 shows the near-field site characteristics around SWERF. The region beyond the property in most directions is expected to be characterized by scattered obstacles, except for the wooded region east of the property where wooded/suburban roughness is expected. Qualitative analysis of the surrounding terrain suggests a surface roughness between 0.2 meters and 0.4 meters in most directions, assuming obstructions within the SWERF site are limited. Surface roughness for winds from the east is anticipated near 0.5 meters. [21]

Trees and brushy undergrowth may be removed from portions of SWERF to decrease the natural roughness within the property boundary, allowing greater control of the roughness length (z_0) through the addition of designed roughness elements. This development may be coupled with a limited study of near-field roughness effects.

Figure 2 shows the distribution of wind speed and direction as reported by the CMI station at Willard Airport in Champaign, IL, approximately 20 miles from SWERF. The wind climate at SWERF is expected to be very similar to this observation. The mean 2-minute wind speed is 5.2 m/s (11.6 mph) and the speed-weighted mean wind direction is 232.7°.

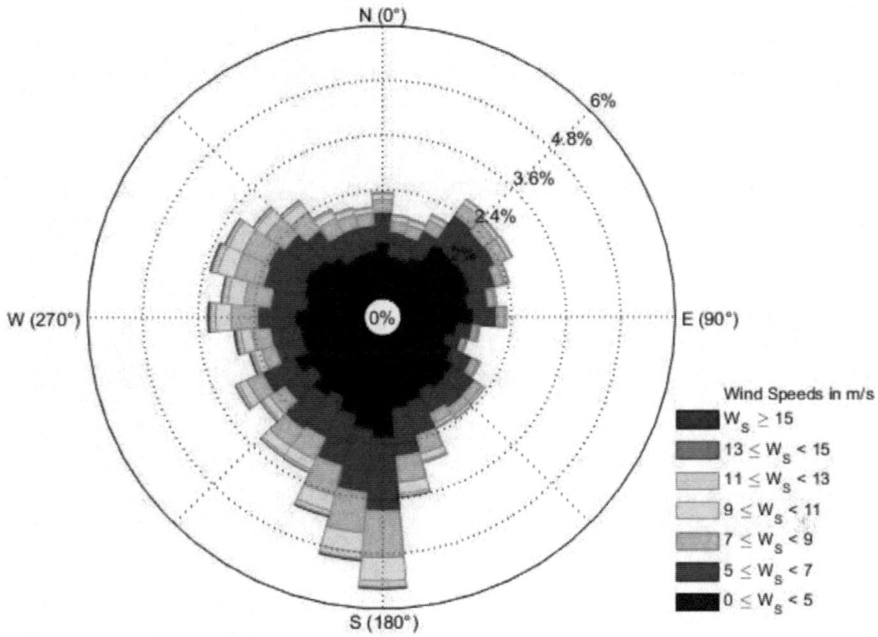

Figure 2 - Distribution of 2-minute mean wind speed and direction at CMI station at Willard Airport in Champaign, IL collected January 1, 1973 to December 8, 2010.

1.3 *Initial Instrumentation*

The first set of stationary sensors will be installed at the SWERF facility starting May 2017. This section provides a brief overview of the infrastructure and instrumentation that will be operable in 2017. Additional infrastructure and instrumentation will be selected based on project requirements and community demand.

1.3.1 *45.72-meter (150-ft) wind tower*

Baseline wind measurements for the SWERF site will be collected with instrumentation mounted on a 45.72-meter tower located near the centroid of the property. The tower is a guyed steel lattice tower with a 0.9-meter (3-ft) triangular cross section. Figure 1 provides a map of SWERF facility locations.

The initial instrument package was selected for all-season reliability with the goal of calculating basic wind engineering metrics throughout the year. Barometric pressure, temperature, and 3-axis wind velocity will be measured at 2, 4, 10, 22, and 45.72-meter elevations. Relative humidity will be measured at the 4-meter elevation only. Wind velocity will be measured with a triplet of Gill propeller anemometers supported by custom orthogonal mounts. Table 1 summarizes the sensor models selected and the heights at which they will be mounted.

Table 1 - Wind climate observation sensors

Observation height (ft)	Observation height (m)	Anemometer Gill 27106R triplet (UVW)	Pressure RM Young 61302V	Temperature RM Young 41342VC	Temperature & Relative Humidity RM Young 41382VC
6.56	2	X	X	X	
13.12	4	X	X		X
32.81	10	X	X	X	
72.18	22	X	X	X	
150	45.72	X	X	X	

Data collection will be accomplished at the tower with a National Instruments 9135 compact-DAQ running LabVIEW RT with all data transmitted on an IP broadcast. Continuous data collected will initially be set at a 10-hertz sampling frequency. All aspects of the data collection system will accommodate additional sensors and higher sampling frequencies when required by project specifications.

1.3.2 *SWERF HQ*

An existing low-rise building on the western edge of the property will serve as the central hub of the facility (SWERF HQ). It will house initial data processing, data storage, office space, a conference room, and a sensor laboratory. This structure will also serve as a foundation for full-scale wind engineering experiments.

SWERF HQ is an existing 10.9x13.4x4-meter (36x44x13-ft) CMU structure with steel joists supporting a metal deck. The dimensions of this structure are well-suited to comparisons with the 9.1x13.7x4-meter (30x45x13-ft) WERFL building located at Texas Tech University in Lubbock, Texas that has served as a foundation for full-scale observations since 1989 and is the basis for numerous wind tunnel and numerical simulations. [22, 10]

Instrumentation for this structure has not been selected and will conform to specific project requirements.

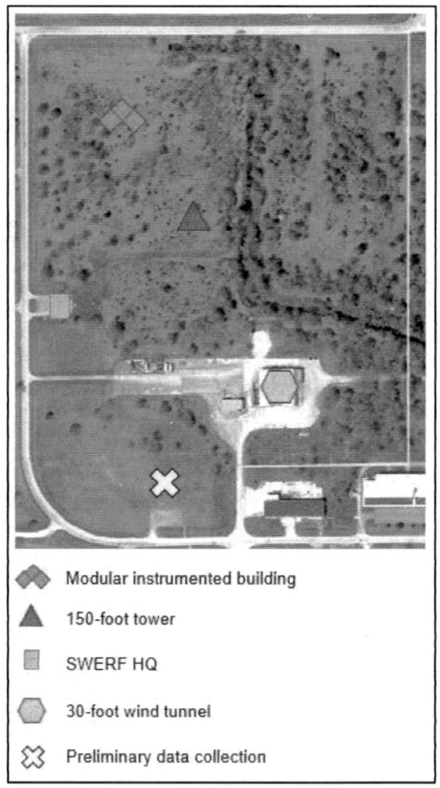

Figure 3 - SWERF facility map

1.3.3 *9.14-meter (30-ft) wind tunnel*

A 9.14-meter, open-return wind tunnel is being assembled within an existing building on the southern edge of the SWERF property. This sub-sonic wind tunnel was previously used by the Aerospace Engineering Department at the University of Illinois and is being rehabilitated after multiple years in storage. Significant modifications are expected during the rehabilitation process. The specifications of the wind tunnel and related data collection have yet to be determined.

Wind-tunnel modifications and data collection will aim to provide a facility that will meet the local demand for boundary layer testing and provide calibration for sensors used and developed at SWERF.

1.4 *Mobile instrumentation*

The observation and characterization of transient wind events are among the initial motivations for developing SWERF. An improved understanding of the ground-level behavior of thunderstorm gusts and tornadic winds would provide a basis for extreme-wind structural and façade design. The development of mobile instrumentation to measure surface-level wind during tornadoes and thunderstorm events where they are likely to occur is key to achieving this goal.

Mobile instrumentation will also supplement stationary equipment at SWERF, facilitating full-scale experimentation where it is desirable to measure the wind field at specific locations around and between structures and obstacles.

The mobile sensor currently under development at SWERF is a pressure-sphere anemometer based on the "Extreme Turbulence Probe" developed by NOAA. The sensor measures pressure at 30 points distributed near the horizontal axis of the sphere. Wind velocity is calculated from the gradient of pressures around the stagnation point. [23]

While the NOAA probe relied on independent signal processing and data collection, the current prototype is being designed with a circuit board that incorporates all necessary functions: analog pressure sensors, signal processing, and data collection. This simple modification will allow for a sensor that costs less to produce and consumes less electricity. The current design is housed in an arbitrarily-large 18"-diameter sphere, but future iterations will be housed in a smaller spherical chassis, with the size determined by the board design.

Future off-site projects will aim to combine in-situ measurements from mobile instrumentation with post-storm damage surveys, damage-pattern vortex modeling, and radar-based wind field measurements. [24, 25, 26]

1.5 *Modular instrumented building*

The second major phase of development at SWERF will aim to offer versatility in experimentation unprecedented in full-scale, natural-wind experimentation. The current concept is to construct a modular, rapidly reconfigurable, full-scale instrumented building near the center of the northern half of the SWERF property. Independent rapidly reconfigurable full-scale obstructions and roughness elements would supplement the test environment. The combination of mobile instrumentation with stationary instrumentation would provide conventional measurements and allow characterization of the wind field at any location.

2 PRELIMINARY RESULTS

The motivation for preliminary data collection was to begin characterization of the SWERF site and provide a foundation of experience for members of the SWERF team.

2.1 *Preliminary data collection*

Preliminary data collection was conducted at SWERF between June 2, 2016 and October 3, 2016 using a 10-meter mobile tower on loan from the University of Florida where it serves as part of the Florida Coastal Monitoring Program. The UF tower collects 3-axis wind velocity at 5-meter and 10-meter heights and temperature, pressure, and relative humidity near the 5-meter height. [27, 28]

The UF tower remained stationary through the duration of preliminary data collection near the south edge of SWERF, as shown in Figure 3. The location was selected because this portion of the property was clear of near-field obstructions. Immediate plans for SWERF do not include resuming data collection at this location, but an instrumented tower or mobile sensor at this location may be added to future development plans.

During preliminary data collection, 82 non-consecutive days of data were collected. Including all data collecting in this period, the mean 2-minute wind speed measured at the 10-meter height was 2.9 m/s (6.5 mph) and the speed weighted mean direction was 185°. Note that the June through September period of preliminary data collection only provides an observation of meteorological summer. Among these records, 131 ten-minute "stationary" segments and three thunderstorm records were identified for further analysis.

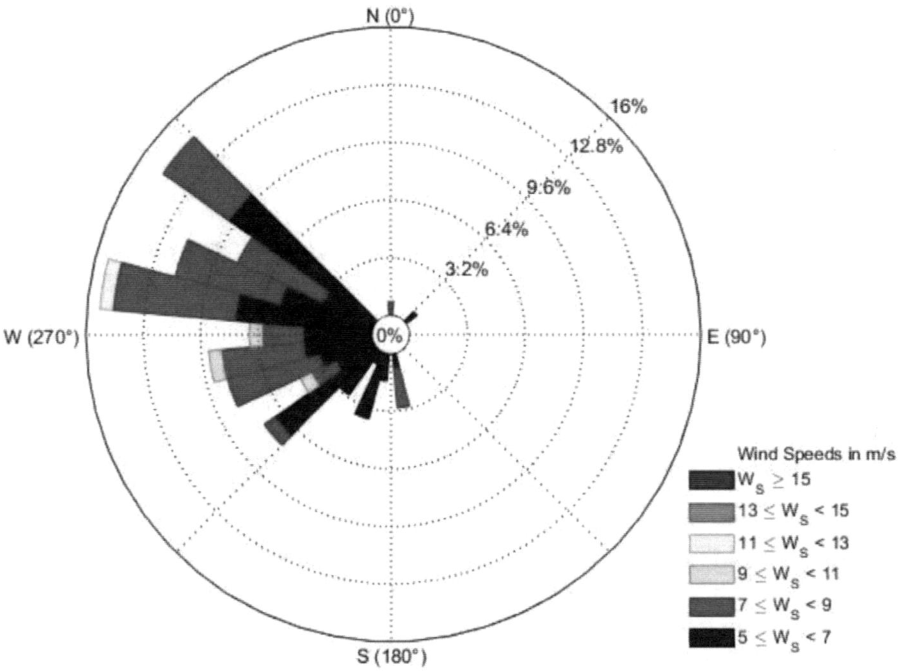

Figure 4 - Distribution of 10-minute mean wind speeds and direction measured at 10-meter height for stationary segments

2.2 *Stationary data*

Stationary segments were identified for use in analytic methods where stationary ergodic behavior is assumed.

2.2.1 *Methodology*

Each record was divided into ten-minutes segments. Segments were identified as "stationary" if the primary and crosswind components passed a runs test for stationarity and the mean wind speed was at least 5 m/s (11.2 mph). 131 ten-minute segments met the criteria. Alternate tests for stationarity, segments of length other than ten minutes, and varying segment initial times to maximize the number of stationary segments were all considered but were not implemented.

2.2.2 *Overview of stationary segments*

As shown in Figure 4, the stationary segments originated primarily from the west. The mean 10-minute wind speed measured at 10 meters above grade for the stationary segments is 7.0 m/s (15.7 mph) and the speed-weighted average of the wind direction is 269°. Note that the selection process excludes wind speeds below 5 m/s (11.2 mph) for consideration as "stationary".

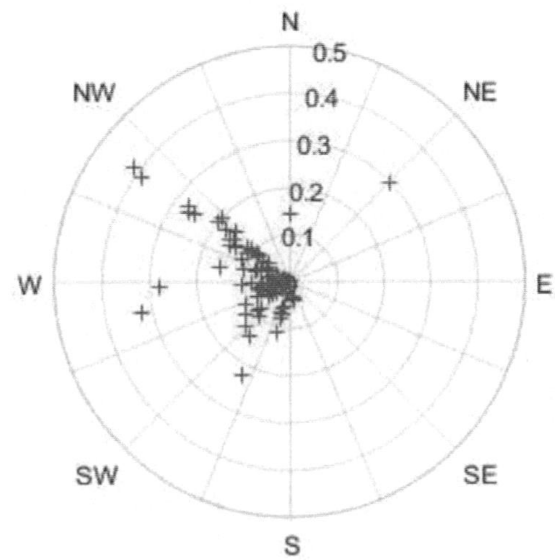

Figure 5 – Roughness length in meters calculated from preliminary data. Limited data prevents meaningful representation of central tendency.

2.2.3 *Roughness length*

Roughness length (z_0) was calculated using the turbulence intensity method. Other methods of calculations were considered, but the short duration of data collection and limitation of two observation heights prevented a meaningful comparison. [29] [30]

As shown in Figure 5 the calculated roughness length varies with median values around 0.2 meters to 0.3 meters for winds from the west. Reliable measures of roughness length from all directions will require additional measurements of stationary segments. The calculated roughness lengths are within the range of the expected values.

2.2.4 *Spectral density*

Figure 6 shows the mean spectral density calculated from the preliminary collection of 10-minute stationary records.

As anticipated by previous observations, a "micro-meteorological" peak in the power spectrum is visible at a period of approximately 1-minute. [31]

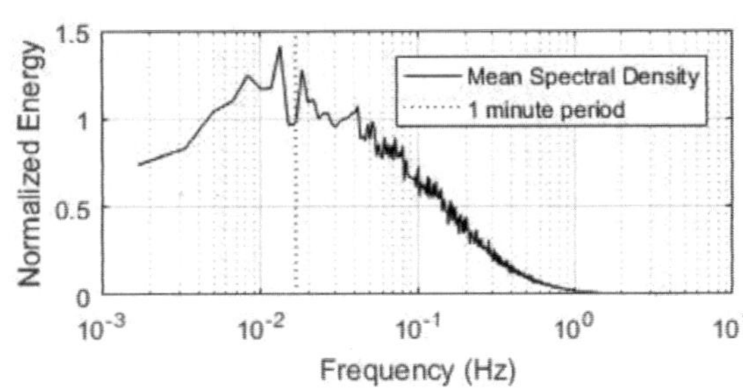

Figure 6 - Mean spectral density of 10-minute stationary records. SWERF, 2016.

2.3 *Transient events*

The current study focused on passive observation of the wind climate at SWERF – as such, active pursuit of transient events was not included. Comparative analysis of transient wind meteorological origin and gust structure are beyond the scope of this report.

2.3.1 *Methodology*

Few significant transient wind events were expected during preliminary data collection due to the season of observation. Criteria for identification of transient events within the current study were selected to provide an inclusive representation of the dataset. Future studies will likely focus on events with higher minimum wind speeds.

Events with a peak 3-second mean wind speed above 20 m/s (44.7 mph) were selected for consideration as transient events in the current study. Events that met the wind speed threshold were then compared with National Weather Service records to ensure they coincided with reported thunderstorm activity. Three events met the wind speed threshold and each coincided with reported thunderstorms.

2.3.2 *June 21, 2016 thunderstorm event*

The first transient event recorded at SWERF occurred on June 21, 2016. The peak 3-second gust was 35.8 m/s (80.1 mph). The NWS reported thunderstorm activity near Rantoul, IL. The NWS Storm Prediction Center received reports of trees and power-lines blown down, damage to an aircraft hangar at the Rantoul Aviation Center, and damage to a storage unit in the Rantoul area. The same weather system caused an outbreak of tornadoes near Chicago, IL on June 22. [32]

UIUC deployed a small team to survey damage near the aviation center, allowing the correlation between the transient event recorded at SWERF and damage to adjacent properties.

This wind event is completely contained within a thirty-minute interval. Within a five-minute period, the wind speed at five and ten meters rapidly increases from less than 10 m/s (22.4 mph) to greater than 35 m/s (78.3 mph) then decreases to below 10 m/s. The vertical component of the wind velocity at 10 meters shows a significant upward component. Within the primary five-minute gust four distinct pulses are observed. A second minor spike in wind speed follows the primary event.

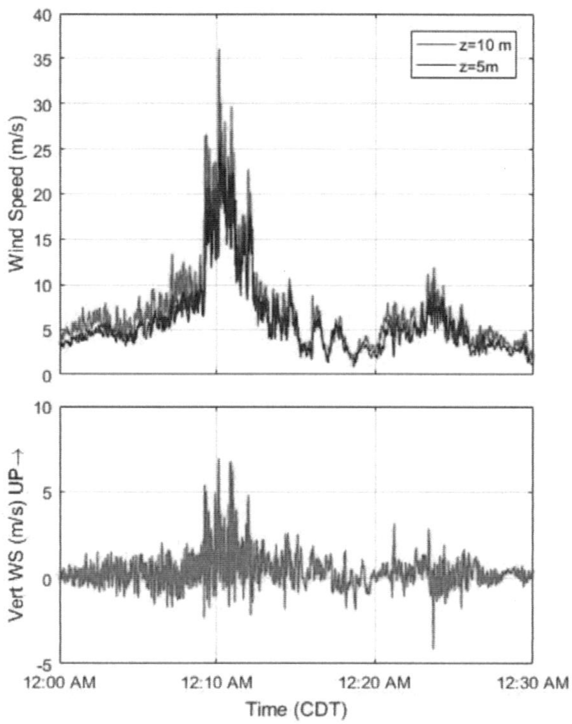

Figure 7 - June 21, 2016 transient wind event, wind speed and vertical component. SWERF, Rantoul, IL.

The team from UIUC noted damage on multiple structures during their damage survey following this event. The most extensive damage was noted at the aviation center, immediately north of SWERF, where large sections of steel wall panel and roofing were dislocated from the structure. There was no observable structural damage to the aviation center. To the west of SWERF, tearing was noted on a fabric enclosure maintained as part of the UIUC Advanced Transportation Research and Engineering Laboratory. Multiple uprooted or snapped trees were also observed, but tree damage was limited to smaller trees and unhealthy mature trees.

2.3.3 *August 25, 2016 thunderstorm event*

On August 25, 2016, the 10-meter mobile tower at SWERF recorded the event shown in Figure 8. The maximum 3-second gust of this event was 22.3 m/s (49.9 mph).

This record coincides with NWS reports of thunderstorms near SWERF, including reports of fallen trees in central Illinois. [33]

The wind speed increases from under 5 m/s (11.2 mph) to over 20 m/s (44.7 mph) in less than 5 minutes, then gradually decays. The vertical component of the wind speed is nonzero, but maintains a nearly non-zero mean throughout the event.

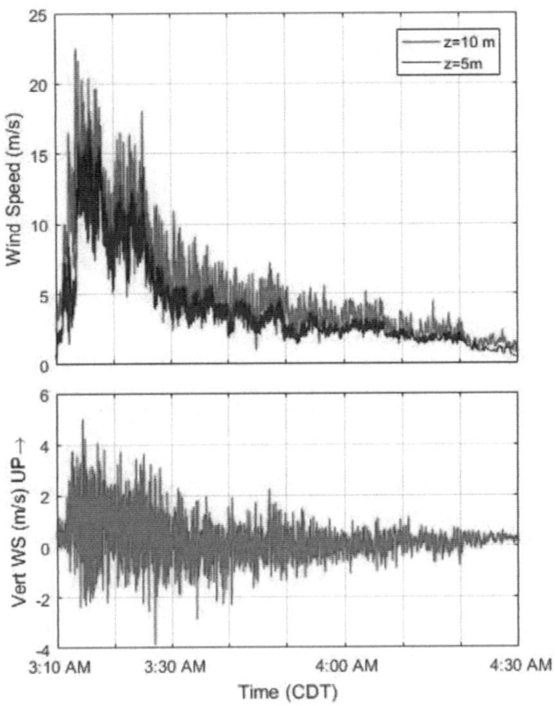

Figure 8 - August 25, 2016 transient wind event, wind speed and vertical component. SWERF, Rantoul, IL.

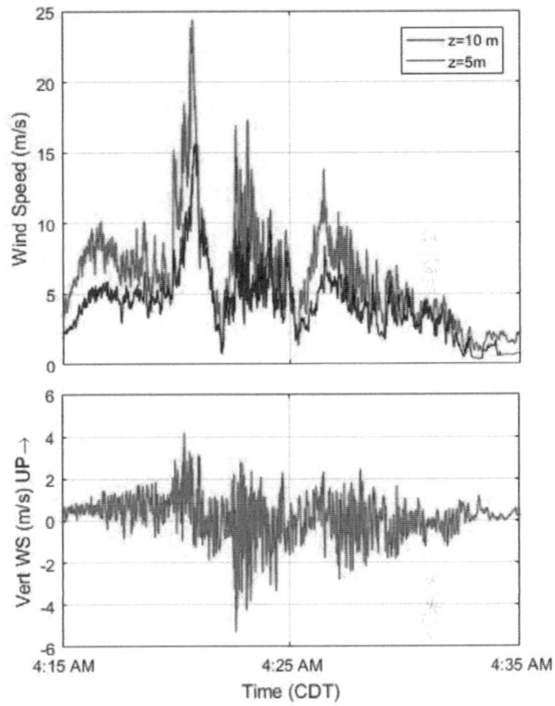

Figure 9 - August 26, 2016 transient wind event, wind speed and vertical component. SWERF, Rantoul, IL

2.3.4 *August 26, 2016 thunderstorm event*

The mobile tower at SWERF recorded the event shown in Figure 9 on August 26, 2016. The maximum 3-second gust was recorded at 24.0 m/s (53.7 mph). This event includes a small increase in wind speed for a duration of five minutes followed by three distinct pulses.

The vertical component of the wind speed records a slight upward flow during the first pulse and a primarily downward flow during the second pulse. The vertical component of the third pulse is less extreme and primarily downward.

3 CONCLUSIONS

The results of preliminary data collection and analysis at SWERF support previous assumptions on roughness length and provide a preview of the data that will be collected with the 47.6m instrumented tower.

A larger volume of data, collected across multiple years and all seasons, is necessary before wind climate characterization of the site can be generalized as indicative of the site.

The three thunderstorm events collected during preliminary data collection hints that SWERF is likely to capture multiple transient events with gusts above the NWS "severe" classification threshold of 26 m/s (58 mph) in any single year.

A wider range of results are anticipated within 2017, after the initial instrumentation is operating and the first round of full-scale experimentation begins at SWERF.

Immediate plans for the SWERF facility include two equally important tasks: implementing data collection with the initial instrumentation and incorporating community feedback into the evolving design of SWERF.

4 ACKNOWLEDGEMENTS

Dr. Forrest Masters, Pedro Fernandez Caban, and the Department of Civil & Coastal Engineering at University of Florida are recognized for their loan and technical support of the 10-meter mobile tower used during preliminary data collection.

5 REFERENCES

[1] NIST, "Federal program to reduce windstrorm impacts," NIST, 14 March 2017. [Online]. Available: https://www.nist.gov/news-events/news/2017/03/federal-program-reduce-windstorm-impacts-requests-comments-new-strategic. [Accessed 11 April 2017].

[2] NWS, "Natural hazard statistics," NOAA, 6 April 2016. [Online]. Available: www.nws.noaa.gov/os/hazstats.shtml. [Accessed 11 April 2017].

[3] Commitee to Review the Need fo a Large-scale Test Facility for Research on the Effects of Extreme Winds on Structures, "Review of the need for a large-scale test facility for research on the effects of extreme winds on structures," National Academy Press, Washington, DC, 1999.

[4] F. T. Lombardo, D. A. Smith, J. L. Schroeder and K. C. Mehta, "Thunderstorm characteristics of importance to wind engineering," *Journal of Wind Engineering and Industrial Aerodynamics,* vol. 125, pp. 121-132, 2014.

[5] A. Hagos, F. Habte, A. G. Chowdhury and D. Yeo, "Comparisons of two wind tunnel pressure databases and partial validation against full-scale measurements," *Journal of structural engineering,* vol. 140, no. 10, 2014.

[6] ASCE, Minimum design loads for buildings and other structures, Reston, VA: American Society of Civil Engineers, 2010.

[7] T. Ho, D. Surry and D. Morrish, "NIST/TTU Cooperative Agreement - windstorm mitigation initiative: wind tunnel experiments on generic low buildings," Univeristy of Western Ontario, London, Ontario, 2003.

[8] W. P. Fritz, B. Bienkiewicz, B. Cui, O. Flamand, T. Ho, H. Kikitsu, C. W. Letchford and E. Simiu, "International comparison of wind tunnel estimates of wind effects on low-rise buildings: test-related uncertainties," *Journal of structural engineering,* vol. 134, no. 12, pp. 1887-1890, 2008.

[9] H. W. Tieleman, M. A. Elsayed and M. R. Hajj, "Peak wind load comparison: theoretical estimates and ASCE 7," *Journal of Structural Engineering,* vol. 132, no. 7, pp. 1150-1157, 2006.

[10] D. A. Smith, S. M. Morse and K. C. Mehta, "Wind Engineering Research Field Laboratory selected data sets for comparison to model-scale, full-scale and computation fluid dynamics," Texas Tech University, Lubbock, TX, 2017.

[11] B. L. Sill, N. J. Cook and C. Fang, "The Aylesbury comparative experiment: a final report," *Journal of Wind Engineering and Industrial Aerodynamics,* vol. 41, no. 44, pp. 1553-1564, 1992.

[12] Florida International University, "Wall of Wind," FIU, 2017. [Online]. Available: https://cee.fiu.edu/research/facilities/wall-of-wind/. [Accessed 18 April 2017].

[13] Insurance Institute for Business & Home Safety, "The IBHS research facility," IBHS, 2017. [Online]. Available: http://disastersafety.org/ibhs-research-center/research-center-main/. [Accessed 18 April 2017].

[14] Texas Tech University, "VorTECH," TTU, 7 December 2016. [Online]. Available: https://www.depts.ttu.edu/nwi/research/facilities/vortech.php. [Accessed 18 April 2017].

[15] Iowa State University, "Tornado microburst simulator," ISU, 2017. [Online]. Available: https://weea.aere.iastate.edu/sample-page/tornado-micro-burst-simulator/. [Accessed 18 April 2017].

[16] University of Western Ontario, "WindEEE Research Center," UWO, 2013. [Online]. Available: http://www.eng.uwo.ca/windeee/. [Accessed 18 April 2017].

[17] Florida State University, "Multi-Axis Wind Load Simulator," FSU, 19 August 2015. [Online]. Available: http://multihazard.eng.ufl.edu/experimentation/testing-apparatuses/wind-engineering/multi-axis-wind-load-simulator/. [Accessed 18 April 2017].

[18] Univeersity of Western Ontario, "IRLBH," UWO, 22 June 2010. [Online]. Available: http://www.eng.uwo.ca/irlbh/. [Accessed 18 April 2017].

[19] G. A. Kopp, M. J. Morrison and D. J. Henderson, "Full-scale testing of low-rise, residential buildings with realistic wind loads," *Journal of Wind Engineering and Industrial Aerodynamics,* vol. 104, no. 106, pp. 25-39, 2012.

[20] Wind Engineering Research Laboratory, "SWERF," The University of Illinois at Urbana-Champaign, 17 March 2017. [Online]. Available: https://publish.illinois.edu/ftlombardo/swerf/. [Accessed 24 April 2017].

[21] Y. Tamura and A. Kareem, Advanced Structural Wind Engineering, Japan: Springer, 2013.

[22] M. L. Levitan, K. C. Mehta, C. V. Chok and D. L. Millsaps, "An overview of Texas Tech's Wind Engineering Field Reserch Laboratory," *Journal of Wind Engineering and Industrial Aerodynamics,* vol. 36, pp. 1037-1046, 1990.

[23] R. M. Eckman, R. J. Dobosy, D. L. Auble, T. W. Strong and T. L. Crawford, "A pressure-sphere anemometer for measuring turbulance in hurricanes," *Journal of Atmospheric and Oceanic Technology,* vol. 24, pp. 994-1007, 2006.

[24] F. T. Lombardo, D. B. Rouche and D. O. Prevatt, "Comparison of two methods of near-surface wind speed estimation in the 22 May, 2011 Joplin, Missouri tornado," *Journal of Wind Engineering and Aerodynamics,* vol. 138, pp. 87-97, 2015.

[25] National Severe Storms Laboratory, "VORTEX Southeast," NOAA, 2017. [Online]. Available: http://www.nssl.noaa.gov/projects/vortexse/. [Accessed 18 April 2017].

[26] J. Wurman and S. Gill, "Finescale radar observations of the Dimmit, Texas (2 June 1995), tornado," *Monthly Weather Review,* vol. 128, pp. 2135-2164, 2000.

[27] University of Florida Civil and Coastal Engineering, "Florida Coastal Monitoring Program," 5 September 2008. [Online]. Available: http://fcmp.ce.ufl.edu. [Accessed 16 April 2017].

[28] J. Balderrama, F. J. Masters, K. R. Gurley, D. O. Prevatt, L. D. Aponte-Bermudez, T. A. Reinhold, J. P. Pinelli, C. S. Subramanian, S. D. Schiff and A. G. Chowdhury, "The Florida Coastal Monitoring Program (FCMP): a review," *Journal of wind engineering and industrial aerodynamics,* vol. 99, pp. 979 995, 2011.

[29] M. R. Conder, R. E. Peterson and D. A. Smith, "Site characterization of the Texas Tech Wind Engineering Research Field Laboratory," in *14th Symposium on boudary lary turbulance*, Aspen, CO, 2000.

[30] J. A. Campbell, "Wind Engineering Research Field Laboratory Characterization," Texas Tech University, Lubbock, TX, 1995.

[31] I. Van der Hoven, "Power spectrum of horizontal wind speed in the frequecy range from 0.0007 to 900 cycles per hour," *Journal of Meteorology,* vol. 14, pp. 160-164, 1957.

[32] National Weather Service, "Storm Prediction Center," NOAA, 21 June 2016. [Online]. Available: http://www.spc.noaa.gov/climo/reports/160620_rpts.html. [Accessed 11 April 2017].

[33] National Weather Service, "Strorm Prediction Center," NOAA, 25 August 2016. [Online]. Available: http://www.spc.noaa.gov/climo/reports/160824_rpts.html. [Accessed 11 April 2017].

13th Americas Conference on Wind Engineering
Gainesville, Florida, USA
May 21 - 24, 2017

Effect of Topography on Tornado Flow Field

Alireza Razavi-1[a], Partha P. Sarkar-2[b]

[a] *Iowa State University, Ames, USA, razavi@iastate.edu*
[b] *Iowa State University, Ames, USA, ppsarkar@iastate.edu*

ABSTRACT

While influence of tornado parameters such as swirl ratio, Reynolds number and aspect ratio on its flow field have been studied for decades, one parameter which is almost missing in the literature is topography. Observing the topography profiles of the paths of some destructive tornadoes in recent history, i.e., Joplin, MO of 2011, Tuscaloosa, AL of 2011 and Moore, OK of 2013, it is clear that neither of them occurred on a flat terrain. It suggests the importance of investigation on the effects of topography on tornado flow field.

In the current study, the effects of three different topographies 2D-ridge, 2D-escarpment and 2D-valley, on near-ground tornado flow field were investigated using the ISU Tornado Simulator. In this paper, the detailed results from the 2D-ridge which is representative of this study will be presented for brevity. The dimensions of the 2D-ridge were extracted from the ASCE7-5 standard. Two tornado translation speeds of 0.15 m/s and 0.61 m/s were considered to see their effect on the wind speed profiles. It showed that the lower translation speed resulted in larger peak horizontal velocities. To find the wind speed-up or amplification factor for the tornado translating over the 2D-ridge, time histories of wind speed profiles with elevation at different locations on the topography were compared to the ones on a flat terrain. For straight line winds, it is known that the speed-up factor depends on location and elevation only. For tornado flow, however, it was found that it depends on the location lateral to the mean translation path and gust averaging time as well. To eliminate dependency of amplification factor on location lateral to the mean translation path, the wind profile along any lateral location which produces the largest load on a specific class of structure such a low-rise building was selected. The reasoning behind this is the intrinsic randomness in the path of tornadoes which urges us to consider the worst scenario. The transient effect was considered using two approaches. The first one is to determine maximum horizontal winds occurring at every location along the topography for different elevations from the ground, ignoring the time of occurrence. The second one is related to loading on a structure which is exposed to only a part of the horizontal wind profile that will vary with the height of the structure. To account for this effect, three categories of structure based on height were considered. For each category, profile which produced the largest drag was selected. Finally, selected velocity profiles at each location on topography were compared to the one producing the largest load on flat terrain. This resulted in tornado speed-up ratio (TSR) for each structure category at each point along the tornado's mean path along the topography. Results show the largest TSR of about 14% close to the windward base of the 2D-ridge. Results how that the best location for low-rise buildings to experience the least effect of the wind speed up in a tornado is at the leeward base of the 2D-ridge.

KEYWORDS

Tornado Simulator, Tornado Flow Field, Topography Effects, 2D-ridge

13th Americas Conference on Wind Engineering
Gainesville, Florida, USA
May 21 - 24, 2017

13th Americas Conference on
WIND ENGINEERING

Tornado Damage Modeling of Single Family Wood Frame House

S. S. Banik[a], P. J. Vickery[a], and L. A. Twisdale[a]

[a] *Applied Research Associates, Inc. Raleigh, N. C., USA, sbanik@ara.com*
[a] *Applied Research Associates, Inc. Raleigh, N. C., USA, pvickery@ara.com*
[a] *Applied Research Associates, Inc. Raleigh, N. C., USA, ltwisdale@ara.com*

ABSTRACT

Single family wood frame house is the most common damage indicator (DI) used by meteorologist to rate tornadoes with intensities greater than EF3. Based upon the observed degree of damage (DoD), tornado windspeeds are subjectively inferred and an EF rating is assigned to the tornado. This paper presents a computerized engineering modeling approach to produce the DoD damage probability matrices for single family houses subjected to simulated tornado strikes.

The modeling approach is a tornado enhanced version of the simulation methods used in FEMA's HAZUS Hurricane Model and ARA's HurLoss model. The damage model uses time stepping of a three dimensional tornado windfield to compute component and system wind loads on a 3D model of the structure. The external pressure load is modeled using a combination of atmospheric pressure change (APC) load and static wind pressure load while the internal pressure load is modeled using the flow continuity of the air entering and exiting the building through openings. Component resistances are based on experimental results and finite element modeling. Progressive failure modeling includes effects of internal pressure changes due to failed components and building leakage, and strength degradation of walls once the roof is failed. In the current version of the model, the density of wind borne debris in the hurricane environment has been enhanced for tornado wind-borne debris using engineering judgment.

This paper discusses the details of the modeling of tornado loads on the components of the house and damage/failure of the house frame. The modeled failure modes include: pressure failure of roof cover and roof deck, uplift failure of the whole roof, pressure failure of the windows and doors, bending, shear and torsional failure of walls, overturning and sliding failure of the whole building, and wall failure of an internal room. A number of house cases with varied construction characteristics have been analyzed for a set of simulated tornadoes. The results show the inherent complexities of the tornado structure interaction and broad DoD probabilities within each tornado windspeed bin $P(d \mid v)$. The effects of building location within vs. outside the tornado core, small vs large tornadoes, and low vs high translational speed are illustrated. The results are also compared with those for simulated hurricanes and straight wind fields.

KEYWORDS

Tornadoes; wind damage; wood frame house; APC

13th Americas Conference on Wind Engineering
Gainesville, Florida, USA
May 21 - 24, 2017

A Methodology for Estimating Relative Safety of Tornado Refuge Areas

L. A. Twisdale[a], L. A. Mudd[b], and M. Levitan[c]

[a] *Applied Research Associates, Inc. Raleigh, N. C., USA, ltwisdale@ara.com*
[b] *Applied Research Associates, Inc. Raleigh, N. C., USA, lmudd@ara.com*
[c]*Nat. Inst. of Standards and Technology, Gaithersburg, MD, USA, marc.levitan@nist.gov*

ABSTRACT

Tornadoes often occur with little warning and produce more deaths than any other natural hazard in the US. As most buildings do not currently have shelters or safe rooms specifically designed to resist tornadic loads, other approaches to reducing risks to life safety are needed. Identification of the best available tornado refuge areas within an existing building or group of buildings (such as a campus) is an option, but one that presents a significant challenge to building design professionals. Qualitative guidance and checklists for design professionals have been available in the form of the FEMA P-431series of publications and associated web pages. This paper presents a new concept for evaluating possible refuge area spaces within a building and across different structural types of commercial and institutional buildings. The approach uses probabilistic structural reliability methods coupled with tornado fragility functions to quantitatively rank the best available refuge areas. The potential refuge area is identified and geometrically located within the structure according to its floor location and wall locations relative to the exterior walls of the structure. Pre-calculated basic fragilities are converted to refuge area specific fragilities based on checklist inputs, logic, and calculations within a macro-enabled excel workbook. The emphasis of the work thus far has been on wall and roof collapse fragilities. The wall fragilities are based on wall type, roof type, span lengths, wall heights, and top support. An aggregate refuge area fragility is developed from the refuge area wall and roof fragilities, considering its position within the structure and the presence of walls between the refuge area and exterior walls. The aggregate refuge area fragility is then integrated with the region's tornado hazard curve to produce a failure frequency, which is converted to a relative score. The score reflects the relative safety of the space from wall and roof collapse. Frame failures and missile fragilities can be addressed in a similar manner. Several examples are presented for the following cases: one story reinforced masonry (RM), unreinforced masonry (URM), Tilt-up Buildings, and Frame Buildings (currently, the frame is assumed to not collapse). The examples include cases for refuge areas with exterior walls and interior walls and with all interior walls. Interior walls types considered thus far include URM, RM and partition walls. The analysis method includes fragilities for glazed openings and doors. The results show strong scoring differentiation based on wall strength, effect of glazing and doors, room size, location within the building, and the presence of overhead protection.

KEYWORDS

Tornadoes; best available refuge area; relative safety; structures

Y

13th Americas Conference on Wind Engineering
Gainesville, Florida, USA
May 21 - 24, 2017

13th Americas Conference on WIND ENGINEERING

Probabilistic Analysis of EF Scale Windspeeds for Tornado Hazard Analysis

L. A. Twisdale[a] , S. Banik[a], P. J. Vickery[a], S. Quayyum[a], M. Levitan[b], L. Phan[b]

[a] *Applied Research Associates, Inc. Raleigh, N. C., USA, ltwisdale@ara.com*
[a] *Applied Research Associates, Inc. Raleigh, N. C., USA, sbanik@ara.com*
[a] *Applied Research Associates, Inc. Raleigh, N. C., USA, pvickery@ara.com*
[a] *Applied Research Associates, Inc. Raleigh, N. C., USA, squayyum@ara.com*
[b] *Nat. Inst. for Standards and Technology, Gaithersburg, MD, USA, marc.levitan@nist.gov*
[b] *Nat. Inst. for Standards and Technology, Gaithersburg, MD, USA, long.phan@nist.gov*

ABSTRACT

Fujita introduced the concept of rating tornadoes based on observed damage in 1971. Fujita's method of damage-based intensity classification used photos and brief word descriptions to describe typical damage for each F scale intensity level (F0-F5). His method of rating tornadoes based on the maximum observed damage within the tornado path continues today with the Enhanced Fujita (EF) scale rating system. The EF scale was adopted by the NWS in 2007 and implemented into the NWS's EF Scale Toolkit. The EF scale includes 28 damage indicators (DIs), each with various degrees of damage (DoD), and subjectively-estimated windspeed ranges for each DoD.

The process of transforming tornado damage intensity ratings into windspeeds is a critical step in the development of tornado windspeeds for engineering design. In order to achieve acceptable performance for tornado-resistant designs, the engineering framework for the analysis and design of structures (forward process) for specified tornado windspeeds and associated load effects should be consistent with the methods used for the estimation of windspeeds from damage (reverse process).

This paper illustrates how the results of engineering modeling/analysis of tornado damage to the most commonly used DIs to rate tornadoes can be used to estimate windspeed probability distributions from EF scale tornado classifications. The inputs to this analysis are the DoD damage probability matrices given tornado windspeeds $P(d \mid v)$. For simple DIs, in which the damage is a single outcome such as failure, the damage probability matrix reduces to an engineering fragility. We use Bayesian methods to develop probability distributions of windspeed given damage $P(v \mid d)$ and, then, the probability distribution of windspeed given an EF scale rating, $P(v \mid EF)$. To reflect variations in important structural characteristics for a given DI, we use region-specific building stock weighting methods and tornado damage investigation data. For example, single family residential homes are the most common DI for tornadoes rated ≥ EF 3. For intense tornadoes, high/max DOD rating for multiple single family homes are required for the storm to receive a rating of EF4 or EF5. For these cases, sequential Bayesian updating is used to estimate the windspeed probability distributions where multiple rated DIs are used in the rating process. Preliminary engineering estimates of the resulting windspeed distributions for the EF scale era

(2007-current) are illustrated using the methodology described herein. These engineering-based windspeed distributions can be used in tornado hazard models to produce windspeed vs. return period.

KEYWORDS

Tornado windspeeds; EF Scale; Bayesian updating;

13th Americas Conference on Wind Engineering
Gainesville, Florida, USA
May 21 - 24, 2017

Development and Validation of a Tornado Damage Model

Peter J. Vickery[a], Sudhan Banik[b], Lawrence A. Twisdale Jr.[c], and Long Phan[d]

[a]Principal Engineer, IntraRisk Division, Applied Research Associates, Raleigh, NC, United States, pvickery@ara.com
[b]Senior Engineer, IntraRisk Division, Applied Research Associates, Raleigh, NC, United States, sbanik@ara.com
[c]Executive Vice-President, IntraRisk Division, Applied Research Associates, Raleigh, NC, United States, ltwisdale@ara.com
[c]Leader, Structures Group, National Institute of Standards and Technology, Gaithersburg, MD, United States, long.phan@nist.gov

ABSTRACT

Modeling loads and damage to buildings in tornado winds has received increased attention of the last decade. A number of experimental studies performed in tornado simulators have suggested that wind pressures on buildings in tornadoes are much different than those produced in atmospheric boundary layer (ABL) flow conditions. However, direct comparisons of exterior pressure coefficients derived in tornado simulators to those developed using ABL wind tunnels have been clouded through differences in normalization of wind speeds and though the contribution of the atmospheric pressure change (APC) to the exterior pressures. In this study we make the assumption that wind loads produced by tornadoes can be approximated using a combination of pressure coefficients from ABL wind tunnels and the proper consideration of APC. The assumptions are validated through comparisons of modeled and observed damage to buildings from a historical tornado.

The load model uses a combination of tornado wind field model including both wind speeds and APC, and a model for the directionally dependent external pressure coefficient, GCp(θ), developed using ABL wind tunnel data. During a tornado simulation, pressures on the exterior of the building are computed using the directional pressure coefficients multiplied by the dynamic pressure obtained using the undisturbed tornado wind speed at mean roof height directly above the location for which loads are being computed. This approach results in wind loads that vary over the structure due to variations of GCp(θ) and variations in both the local tornado wind speeds and directions.

The tornado load model is coupled with a time-stepping damage model and a wind speed dependent wind borne debris impact model. At each time step, external pressures are computed using the local tornado wind speeds and APC. While the building is undamaged, the internal pressure is computed assuming uniform leakage over the exterior of the building and a spatially invariant internal pressure. When the building is breached due to failure of a fenestration, roof deck or wall section, a new value of the internal pressure is computed, taking into account the size of the breach and the existing background leakage. Again, the resulting internal pressure is assumed to be spatially invariant. The model considers failures to roof deck, roof trusses, fenestrations, and walls. Walls treated include reinforced and unreinforced masonry and tilt-up construction.

The tornado damage model was validated through comparisons of modeled and observed damage to two big box stores that occurred during the May 22, 2011 Joplin tornado. Structural models of the two big box stores were developed using the design drawings acquired by members of the NIST Joplin tornado investigation team. The tornado damage model was able to reproduce the observations that one of the buildings was destroyed and that the damage on the other building varied from complete at one end of the building and undamaged at the other end.

The results of the two validation cases demonstrated that the observed damage patterns could not be duplicated without the modeling of the APC and its spatial and temporal variation. The results suggest that ignoring the effects of APC on large buildings will lead to erroneous estimates of tornado wind loads on building, associated damage, and the wind speed required to produce a given damage state.

KEYWORDS

Tornado, Damage, Model, APC, Joplin

The 13th Americas Conference on Wind Engineering (13ACWE)
Gainesville, Florida USA, May 21-24, 2017

13th Americas Conference on
WIND ENGINEERING

Effects of Aspect Ratio on Tornado-Like Vortices Simulated in a Large-Scale Tornado Simulator

Zhuo Tang[a], Liang Wu[b], Changda Feng[c], Delong Zuo[d], Darryl James[e]

[a] *National Wind Institute, Texas Tech University, Lubbock, TX, United States, zhuo.tang@ttu.edu*
[b] *National Wind Institute, Texas Tech University, Lubbock, TX, United States, liang.wu@ttu.edu*
[c] *National Wind Institute, Texas Tech University, Lubbock, TX, United States, changda.feng@ttu.edu*
[d] *National Wind Institute, Texas Tech University, Lubbock, TX, United States, delong.zuo@ttu.edu*
[e] *National Wind Institute, Texas Tech University, Lubbock, TX, United States, carryl.james@ttu.edu*

ABSTRACT: Experiments were conducted in a large-scale Ward-type tornado simulator to study tornado-like vortices. Both flow velocities and the pressures at the surface beneath the vortices were measured. An interpretation of these measurements enabled an assessment of the mean flow field as well as the mean and fluctuating characteristics of the surface pressure deficit, which is a manifestation of the flow fluctuation aloft. An emphasis was placed on the effect of the aspect ratio of the tornado simulator on the characteristics of the simulated flow and the corresponding surface pressure deficit, especially the evolution of these characteristics due to the transition of the flow from a single-celled vortex to a two-celled vortex with increasing swirl ratio.

KEYWORDS: Tornado-like vortex, surface pressure deficit, aspect ratio, swirl ratio.

1 INTRODUCTION

Tornadoes are among the most devastating natural hazards. This has remained true despite the continued improvement in building materials and construction technology. For example, many recent major tornadoes, including the Joplin and Tuscaloosa–Birmingham tornadoes in 2011, the Moore tornado in 2013 and the Dallas area tornado in 2015, have all caused more than one billion dollars in losses and tens of or, in the case of the Joplin tornado, more than 100 fatalities.

A major reason for the continued devastations by tornados is that the understanding of this type of wind storm and their loading on structures remains inadequate. To date, high-fidelity, high frequency measurement of full-scale tornados is still challenging due to the destructive nature of tornados and the difficulty in predicting tornado formation [1]. Also, despite the advancement in numerical simulation of tornado winds based on computational fluid dynamics [e.g., 2, 3, 4], the capability of numerical models to capture the turbulence in tornadic flow and the interaction between the flow and structures remains limited. As a result, most studies of tornado effects on structures have been based on laboratory simulation of tornado-like flow and measurements of loading by the simulated flow on structure models.

Laboratory simulation of tornado-like flow started four decades ago when Ying and Chang [5] built the first tornado simulator. Since then, many simulators of various types and sizes have been constructed. The so-called Ward-type simulator [6], which uses a fan or multiple fans at the top to generate updraft, a mechanism such as rotating screens or turning vanes at the bottom to supply the circulation in the inflow and a baffle to decouple the exhaust flow and the simulated vortex, has perhaps been the most prominent. It was used in the early experiments that revealed many major characteristics of tornadic flow and the dependence of these characteristics on the parameters that govern the dynamics and geometry of tornado vortices [e.g., 6, 7, 8, 9]. These experiments suggested that the swirl ratio controls the transition of the flow from a laminar state to a turbulent state and subsequently from a single-celled vortex to a two-celled vortex and eventually to multiple vortices. By contrast, the radial Reynolds number and the aspect ratio were observed to have secondary effects. Recently, the concepts formulated by Ward have been used in the development of the WindEEE dome at the University of Western Ontario, which can generate translating vortices. Experiments in a scaled-down model of this simulator further illustrated the evolution of the flow with increasing swirl ratio [10].

In addition to the Ward-type simulators, other facilities have also been developed to simulate tornado-like flows. The most notable one is perhaps the simulator at Iowa State University, which differs from Ward-type simulators in that it is of a closed-circuit type instead of an open type, that it uses turning vanes at its top instead of bottom to provide the circulation in the flow, and that it can simulate translating instead of stationary vortices. This simulator has been used in many comprehensive studies of tornado loading on buildings [e.g., 11].

Despite the continued advancement in laboratory simulation of tornado-like vortices, most of the previous studies have focused on the mean characteristics of the simulated flow and those of the resultant surface pressure deficit. In a recent study, both the mean characteristics of flows simulated in a large-scale Ward-type simulator and the characteristics of both the mean and the fluctuating components of the surface pressure were investigated [12]. It was revealed that the swirl ratio controls not only the mean structure of the flow and the surface pressure, but also the fluctuation of the surface pressure, which reflects the fluctuation of the flow aloft. It was also found that the radial Reynolds number does not meaningfully affect either the mean or the fluctuation characteristics of the flow if this number is sufficiently large. This paper presents the outcomes of a follow-up study that further investigate the effects of the aspect ratio on the flows generated in the same simulator.

2 EXPERIMENTAL CONFIGURATIONS

The experiments described herein are conducted in a large-scale Ward-type simulator at Texas Tech University known as VorTECH. As schematically depicted in Figure 1, this simulator has a cylindrical testing chamber of 10.2 m in diameter, an updraft hole of octagon cross-section that is 4 m in diameter, 64 turning vanes at the periphery of the chamber that can be aligned at various orientations to control the circulation in the inflow, 8 fans at the top, the speed of which can be varied to control the amount of updraft, and a honeycomb system that functions as the baffle in a Ward-type simulator. While the diameter of the updraft hole is fixed, the heights of the chamber and the turning vanes can be adjusted between 1 m and 2 m to control the internal aspect ratio of the apparatus.

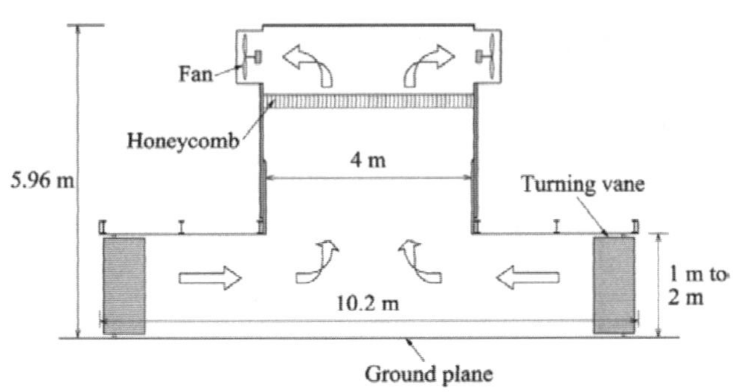

Figure 1. A schematic illustration of VorTECH

A Cobra probe (Turbulent Flow Instrumentation Pty Ltd) with a dedicated data acquisition system and a 12-hole Omniprobe (Aeroprobe Corporation) sampled by a Scanivalve pressure scanner were used for flow velocity measurements at grid points in a plane through the simulator axis. The Cobra probe can measure flows at frequencies of more than 2000 Hz. However, it can only accept flows within a ±45° cone. The Omniprobe accepts flow angles within ±150°, but cannot accurately measure high frequency turbulence. In this study the Cobra probe was used whenever at least 98% percent of the flow was reported by the manufacturer-supplied software to be within a ±45° cone; otherwise the Omniprobe was used. The same Scanivalve system that was used to sample the Omniprobe was also used to measure pressure on the surface of the simulator floor. The Cobra probe was sampled at 600 Hz; the Omniprobe and the pressure taps for surface pressure measurements were sampled at 300 Hz. Each individual velocity and pressure measurement had a duration of 2 minutes. Velocity measurements were conducted once, but surface pressure measurements were repeated 10 times to provide ensembles.First line of text or heading

3 ESTIMATION OF CONTROLLING PARAMETERS

According to many previous studies [e.g., 7, 13], the dynamics of tornado-like vortices is controlled by the swirl ratio and the radial Reynolds number and the geometry of such vortices are governed by the aspect ratios. The definitions of the swirl ratio, the radial Reynolds number and the internal aspect ratio, which is often the controlling aspect ratio with fixed external aspect ratios and will simply be referred to as the aspect ratio subsequently, are:

$$S = r_0\Gamma / (2Qh) \tag{1}$$

$$Re_r = Q / (2\pi v) \tag{2}$$

$$a = h / r_0 \tag{3}$$

where h and r_0 are the depth and radius of the convergence region, which are the height of the turning vanes and the radius of the updraft hole, respectively, for VorTECH, Q is the volume flow rate per unit axial length, Γ is the circulation, and v is the kinematic viscosity of air. The swirl ratio and the radial Reynolds number have been calculated in various manners [e.g., 1, 7]. In this study, the velocities of the flow at discrete locations along a vertical line at the edge of the convergence region are measured and used as the basis for the calculation of the volume flow rate and the circulation as

$$Q = \frac{2\pi r_0}{h}\sum_{n=1}^{N}\left(V_{r,n} \times \Delta h_n\right); \quad \Gamma = \frac{2\pi r_0}{h}\sum_{n=1}^{N}\left(V_{\theta,n} \times \Delta h_n\right) \tag{4}$$

in which $V_{r,n}$ and $V_{\theta,n}$ are the mean radial and tangential velocities of the flow at the nth point of measurement along the vertical line, Δh_n, is half the distance between the two measurement points surrounding this point, and N is the total number of measurement points. Δh_n increased from 0.5 cm over the heights near the simulator floor, where the flow velocity varies rapidly along the height, to 10 cm over the upper heights, where the variation of the flow velocity is less significant.

4 CHARACTERISTICS OF SIMULATED FLOW AND THE RESULTANT SURFACE PRESSURE DEFICIT

In the In the subsequent presentations, axial and radial components of the flow that point upward and toward the simulator axis, respectively, are deemed positive. With the assumption that the mean flow characteristics is axisymmetric, the maximum mean tangential velocity at a height of z above the floor is defined as the local maximum mean tangential velocity, $\overline{V}_{\theta z\max}$, and the maximum of the local maximum mean tangential velocities are defined as the global maximum mean tangential velocity, $\overline{V}_{\theta\max}$. Correspondingly, the local core radius at every height z, r_{cz}, is defined as the distance between the point of maximum mean tangential velocity and the simulator axis, and the representative core radius, r_c, is defined as the distance between the global maximum mean tangential velocity and the simulator axis. The height at which the global maximum mean tangential velocity is achieved is designated z_c

4.1 *Mean component of the flow*

Figure 2 shows the mean flow fields of four representative vortices generated in VorTECH, two at an aspect ratio of unity and the other at an aspect ratio of 0.5. The radial Reynolds numbers for all four simulations are between 3.11×10^5 and 7.21×10^5, which are larger than the value beyond which the characteristics of the simulated flow are essentially independent of this parameter [7, 10, 12]. The arrows in the graphs represent the mean radial ($\overline{V}_r$) and axial ($\overline{V}_z$) components of the flow; the background colors represent the mean tangential velocities ($\overline{V}_\theta$) resulted from linear interpolations of the measured mean tangential velocities. The effects of the swirl ratio are clearly shown in the mean flow fields. At each aspect ratio, the core of the simulated vortex is larger at the larger swirl ratio. More importantly, the two vortices at the smaller swirl ratios both have a single cell and the vortices at the higher swirl ratios both have two cells, which are signified by the apparent downdraft in the center of these vortices. This is consistent with observations in previous studies [7] that tornado-like flow transitions from a single-celled vortex to a double-celled vortex with increasing swirl ratio. Figure 2 also shows an effect of the aspect ratio on the simulated vortices. For the single-celled vortex, the one simulated at the smaller aspect ratio has a larger core although the corresponding swirl ratios are similar. This reflects the effect of the aspect ratio on the geometry of simulated vortices.

The effects of the swirl ratio and aspect ratio on the evolution of the core of the simulated vortex can also be observed in Figure 3. It can be seen that at each aspect ratio, the local core radius of the two-celled vortex at each height is much larger than that of the single-celled vortex at the same height. Although the two swirl ratios of the single-celled and two-celled vortices at the aspect ratio of 0.5 are slightly larger than the swirl ratios of the corresponding types of the vortices at the aspect ratio of unity, the local core radius of the vortices at the aspect ratio of unity is larger than the core radius of the vortices at the aspect ratio of 0.5.

In addition, Figure 3 (b) suggest that for the two single-celled vortices, the local maximum mean tangential velocity is almost constant over the height of measurement. For the two-celled vortices, however, the local maximum mean tangential velocity increases with height until reaching the global maximum mean tangential velocity at a low level and subsequently decreases with increasing height.

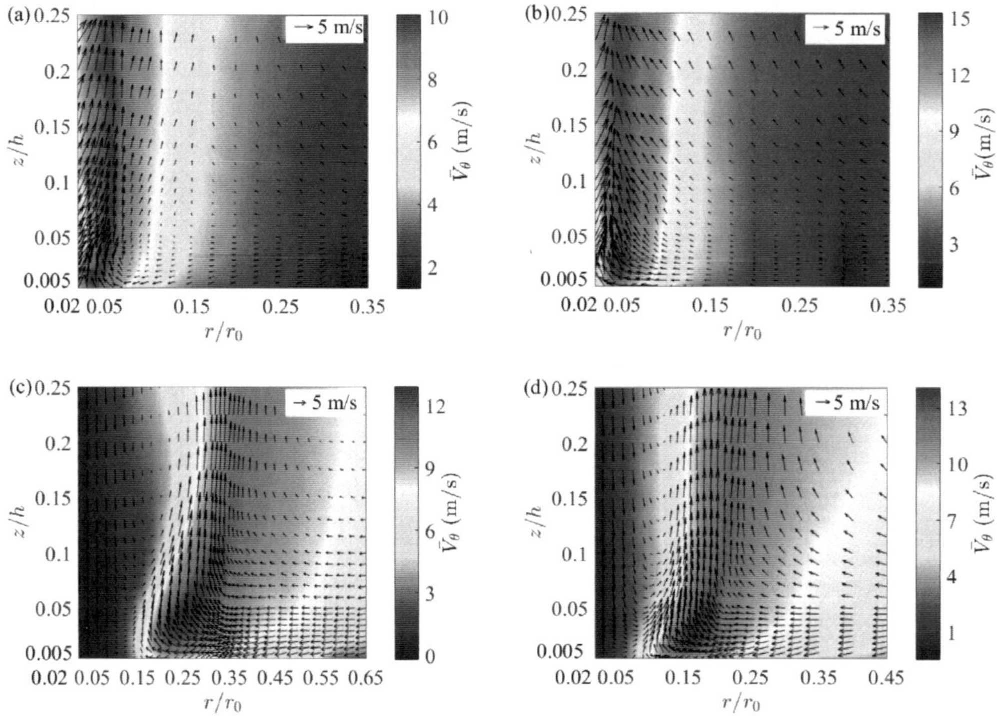

Figure 2. Mean flow fields at aspect ratios and swirl ratios of (a) S=0.17, a=1.0, (b) S=0.24, a=0.5, (c) S=0.84, a=1.0 and (d) S=0.78, a=0.5

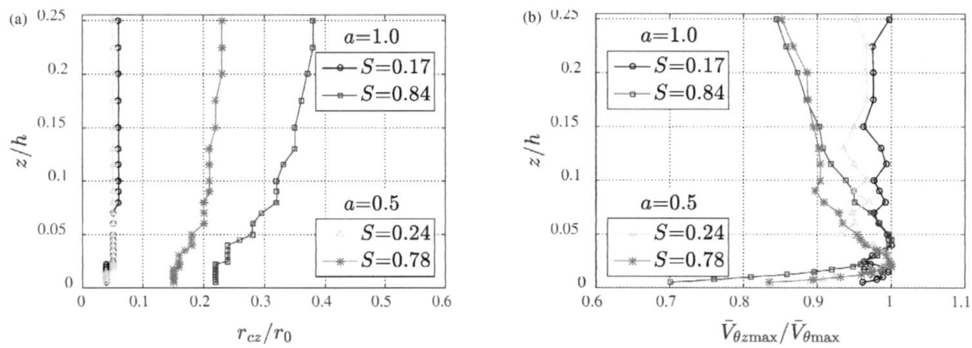

Figure 3. Profiles of (a) local core radius and (b) maximum mean tangential velocity at four swirl ratios

Figure 4 shows the radial profiles of the mean tangential velocity at the height of the maximum tangential velocity, z_c, and those at two heights below and two heights above z_c. Also included in this figure are the radial mean tangential velocity profiles prescribed by Burgers-Rott vortex model [14, 15] for the single-celled vortices and the Sullivan vortex model [16] for the two-celled vortices, which are represented by equations (5) and (6), respectively:

$$\bar{V}_\theta = 1.4\bar{V}_{\theta\max}\left(r/r_c\right)^{-1}\left\{1-\exp\left[-1.2564\left(r/r_c\right)^2\right]\right\} \tag{5}$$

$$\bar{V}_\theta = \bar{V}_{\theta\max} \left(r/r_c\right)^{2.4} \left[0.3 + 0.7\left(r/r_c\right)^{7.89}\right]^{-0.435} \tag{6}$$

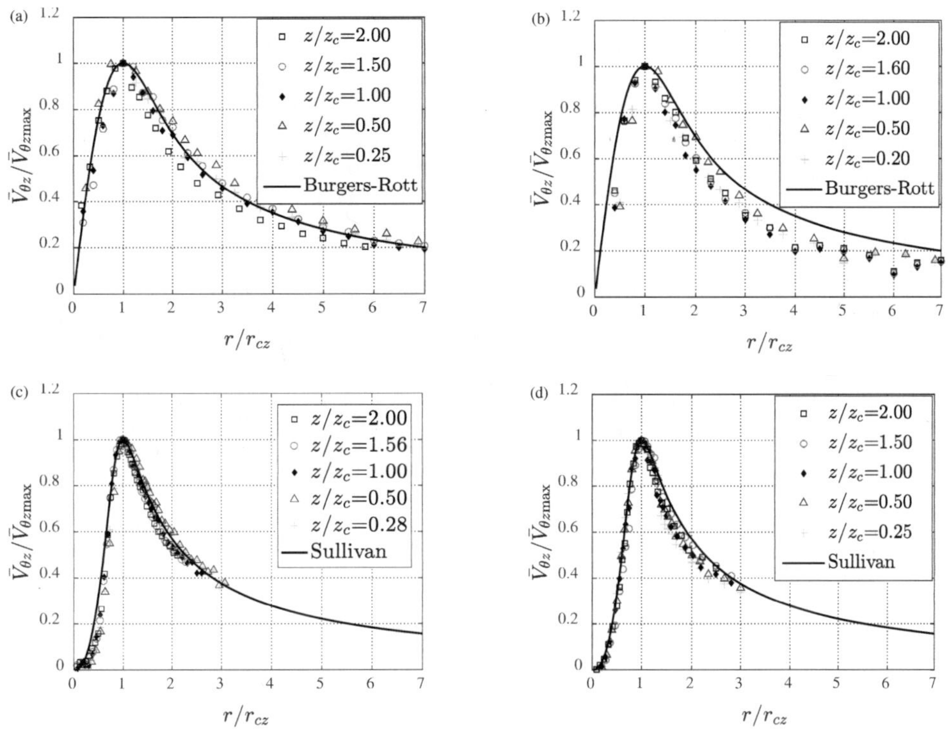

Figure 4. Radial profiles of mean tangential velocity of vortices simulated at (a) S=0.17, a=1.0, (b) S=0.24, a=0.5, (c) S=0.84, a=1.0 and (d) S=0.78, a=0.5

It is seen that the data closely follow the prescriptions by the two models regardless of the swirl ratio or aspect ratio, or the height of measurement. By contrast, a previous study [12] has suggested that the combined Rankine vortex model cannot closely represent the radial variation of the mean tangential velocity in the core of the vortices generated in this same simulator. This observation agrees with a conclusion from recent radar observations that the Burgers–Rott vortex model is a better fit to the tangential velocity profile in one-celled tornadoes than the Rankine vortex model [17, 18]. The difference between the performances of the models is believed to be primarily because the Burgers-Rott and Sullivan models are directly derived from the Navier–Stokes equations, while the empirical combined Rankine vortex model does not consider the radial and axial components of the flow and treats the flow inside the core as a rotating rigid body.

However, although the Burgers-Rott and Sullivan models can closely fit the radial profile of the mean tangential velocities, their performances in fitting the profiles of the other mean flow components are not satisfactory. As an illustration, Figure 5 and Figure 6 show representative radial profiles of the mean radial velocity and axial profiles of the mean axial velocity of the two simulated single-celled vortices and the corresponding profiles prescribed the Burgers-Rott vortex model as

$$\bar{V}_r = ar; \quad \bar{V}_z = 2az \tag{7}$$

where $a = Q/(2\pi h^3)$.

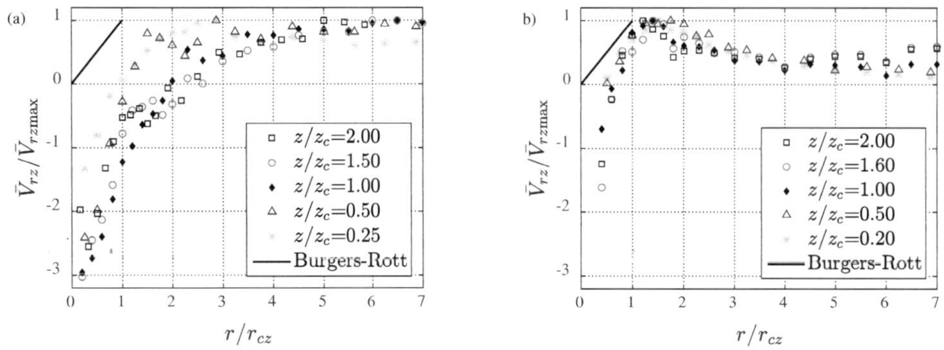

Figure 5. Radial profiles of mean radial velocity of single-celled vortices simulated as (a) S=0.17, a=1.0 and (b) S=0.24, a=0.5

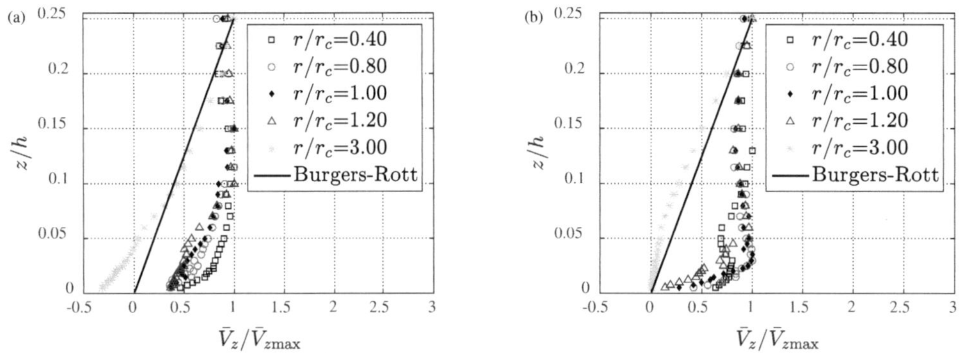

Figure 6. Axial profiles of mean axial velocity of single-celled vortices simulated as (a) S=0.17, a=1.0 and (b) S=0.24, a=0.5

The mean velocity profiles apparently depend on the axial and radial position, respectively. It is also clear that the profile of the mean radial velocity prescribed by the model is not close to the corresponding profiles based on experimental data for either vortex, and that the axial profile of the mean axial velocity prescribed by the model is not close to the corresponding experimental data either, especially for the axial component of the flow around or inside the core of the vortices.

4.2 *Mean surface pressure deficit*

The effects of the swirl ratio and aspect ratio on the vortices are also reflected in the characteristics of the mean surface pressure deficit caused by the flow aloft. Figure 7 (a) and (b) show the dependence of the radial profiles of the mean surface pressure deficit ($\bar{P}$) on the swirl ratio for aspect ratios of unity and 0.5, respectively. The transition of the mean surface pressure profile due to increasing swirl ratio can be clearly observed. At both aspect ratios, the mean surface pressure profiles at the smaller swirl ratios have a single valley and the magnitude of the mean pressure deficit increases with increasing swirl ratio. As the swirl ratio further increases, the profile of the mean surface pressure deficit around the center of the vortex flattens and develops into a shape with two valleys, indicating a corresponding transition of the flow. Figure 7 also clearly shows the effect of the aspect ratio on the mean surface pressure. At similar aspect ratios but different aspect ratios, the magnitude of the mean surface pressure at the same radial positions can be very different. It is of interest to notice that the flattening of the profile, which signifies the breakdown of the vortex reaching the floor, starts at a swirl ratio around 0.22 for the simulation at an aspect ratio of unity, but the same starts at a swirl ratio of 0.36 for simulation at an aspect ratio of 0.5. Such influence of the aspect ratio on the mean surface pressure deficit can also be clearly observed in Figure 8, which shows the correlation between the maximum mean surface pressure deficit ($\overline{P}_{\min}$) and the swirl ratio.

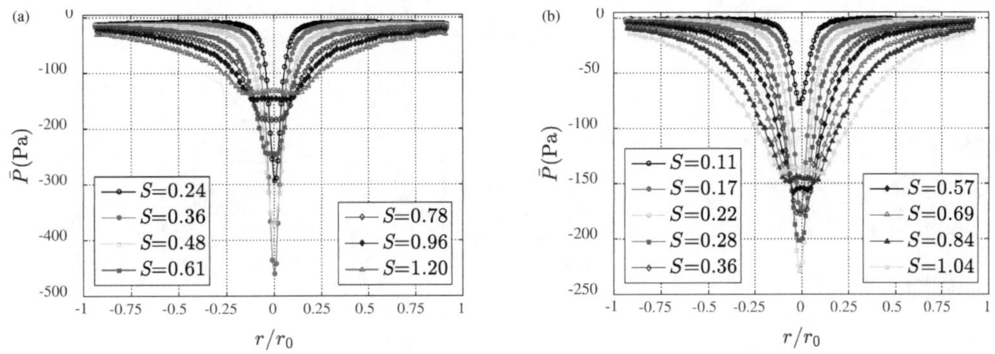

Figure 7. Dependence of radial profile of mean surface pressure deficit on swirl ratio for (a) a=0.5 and (b) a=1.0

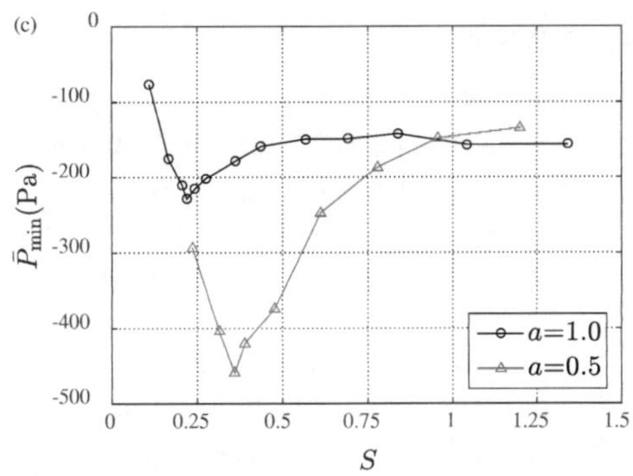

Figure 8. Minimum mean pressure deficit versus swirl ratio

5 FLUCTUATING SURFACE PRESSURE DEFICIT

Figure 9 shows the radial profiles of the mean, standard deviation, skewness and kurtosis of the coefficient of the surface pressure deficit calculated as $C_P = P/(0.5\rho\overline{V}_{\theta\max}^2)$, where P is the surface pressure deficit and ρ is air density.

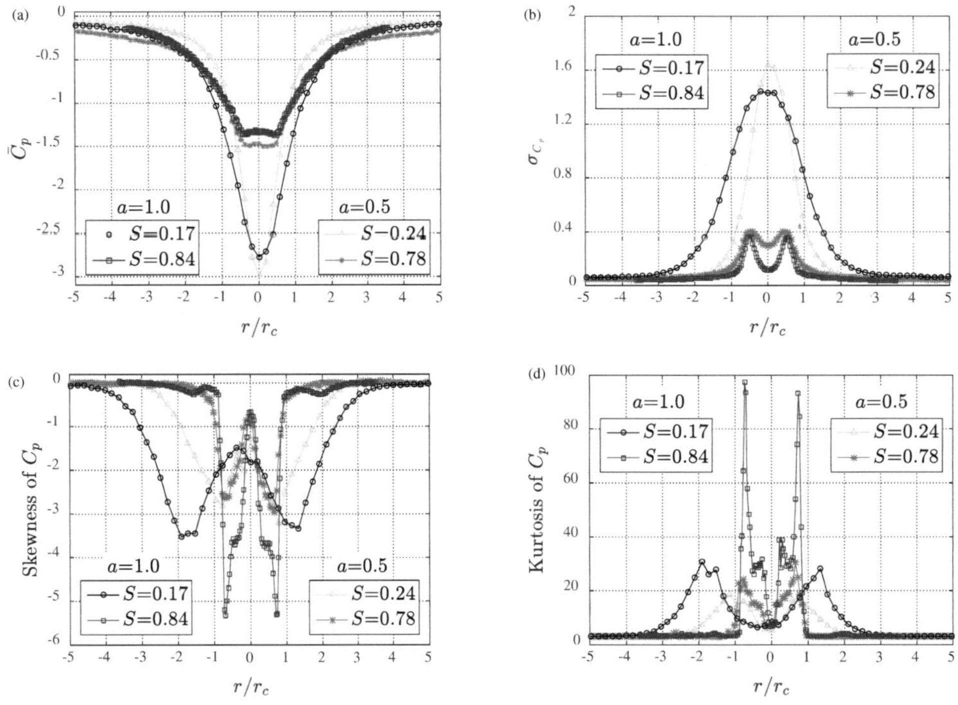

Figure 9. Dependence of surface pressure fluctuation on swirl ratio and aspect ratio

The effect of swirl ratio on the surface pressure fluctuation that can be observed in these graphs is remarkable. For example, the radial profiles of the variance of the pressure fluctuation beneath the single-celled vortices have a single peak, but the corresponding profiles for the two-celled vortices have a bi-modal distribution. Also, the profiles of the skewness and kurtosis clearly suggest that the surface pressure fluctuation inside and surrounding the core of the vortex is highly non-Gaussian regardless of the vortex type. However, the exact probability distribution of the pressure is highly dependent on the swirl ratio. Figure 9 also suggests that the surface pressure fluctuation is influenced by the aspect ratio. For example, for vortices of the same type at similar swirl ratios, while the shapes of radial profiles of the variance of the surface pressure deficit are of the same type, the magnitudes of the variance at the same radial location can be very different. The same can be concluded regarding the skewness and kurtosis of the surface pressure fluctuation. This is an indication that the aspect ratio can also affect the fluctuation of the flow above the surface.

6 CONCLUSIONS

A large-scale Ward-type simulator is used to simulate tornado-like vortices. It is found that the mean characteristics of the flow and those of the surface pressure deficit depend critically on the swirl ratio and that the aspect ratio also affects these characteristics. The transition of the flow from a single-celled vortex to a two-celled vortex and the corresponding transition of the mean flow and surface pressure deficit characteristics with increasing swirl ratio is clearly observed, and the aspect ratio affects not only the geometry of both single-celled and two-celled vortices but also the swirl ratio at which the critical transition occurs. Despite the effect of the aspect ratio, however, the Burgers-Rott and Sullivan vortex models can closely fit the radial profiles of the mean tangential velocity of single-celled and two-celled vortices, respectively. The experimental data also suggests that the surface pressure fluctuation can be highly non-Gaussian and that the swirl ratio critically affects the pressure fluctuation while the aspect ratio has a secondary effect.

REFERENCES

1 Haan, F.L., Jr, P.P. Sarkar, and W.A. Gallus, *Design, construction and performance of a large tornado simulator for wind engineering applications.* Engineering Structures, 2008. 30(4): p. 1146-1159.

2 Lewellen, W.S., D.C. Lewellen, and R.I. Sykes, *Large-Eddy Simulation of a Tornado's Interaction with the Surface.* Journal of the Atmospheric Sciences, 1997. 54(5): p. 581-605.

3 Natarajan, D. and H. Hangan, *Large eddy simulations of translation and surface roughness effects on tornado-like vortices.* Journal of Wind Engineering and Industrial Aerodynamics, 2012. 104–106: p. 577-584.

4 Ishihara, T., S. Oh, and Y. Tokuyama, *Numerical study on flow fields of tornado-like vortices using the LES turbulence model.* Journal of Wind Engineering and Industrial Aerodynamics, 2011. 99(4): p. 239-248.

5 Ying, S.J. and C.C. Chang, *Exploratory Model Study of Tornado-Like Vortex Dynamics.* Journal of the Atmospheric Sciences, 1970. 27(1): p. 3-14.

6 Ward, N.B., *The Exploration of Certain Features of Tornado Dynamics Using a Laboratory Model.* Journal of the Atmospheric Sciences, 1972. 29(6): p. 1194-1204.

7 Church, C.R., et al., *Characteristics of Tornado-Like Vortices as a Function of Swirl Ratio: A Laboratory Investigation.* Journal of the Atmospheric Sciences, 1979. 36(9): p. 1755-1776.

8 Church, C.R., J.T. Snow, and E.M. Agee, *Tornado Vortex Simulation at Purdue University.* Bulletin of the American Meteorological Society, 1977. 58(9): p. 900-908.

9 Lund, D.E. and J.T. Snow, *Laser Doppler Velocimeter Measurements in Tornadolike Vortices*, in *The Tornado: Its Structure, Dynamics, Prediction, and Hazards.* 1993, American Geophysical Union. p. 297-306.

10 Refan, M. and H. Hangan, *Characterization of tornado-like flow fields in a new model scale wind testing chamber.* Journal of Wind Engineering and Industrial Aerodynamics, 2016. 151: p. 107-121.

11 Haan, F.L., Jr, V. Balaramudu, and P.P. Sarkar, *Tornado-Induced Wind Loads on a Low-Rise Building.* Journal of Structural Engineering, 2010. 136(1): p. 106-116.

12 Tang, Z., et al. *Simulations of tornado-like vortices in a large-scale ward-type tornado simulator.* in *8th International Colloquium on Bluff Body Aerodynamics and Applications.* 2016. Boston, Massachusetts.

13 Davies-Jones, R.P., *The Dependence of Core Radius on Swirl Ratio in a Tornado Simulator.* Journal of the Atmospheric Sciences, 1973. 30(7): p. 1427-1430.

14 Burgers, J.M., *A mathematical model illustrating the theory of turbulence.* Advances in applied mechanics, 1948. 1: p. 171-199.

15 Rott, N., *On the viscous core of a line vortex.* Zeitschrift für angewandte Mathematik und Physik ZAMP, 1958. 9(5): p. 543-553.

16 Sullivan, R.D., *A two-cell vortex solution of the Navier-Stokes equations.* Journal of the Aerospace Sciences, 1959. 26(11): p. 767-768.

17 Bluestein, H.B., et al., *The Structure of Tornadoes near Attica, Kansas, on 12 May 2004: High-Resolution, Mobile, Doppler Radar Observations.* Monthly Weather Review, 2007. 135(2): p. 475-506.

18 Kosiba, K. and J. Wurman, *The Three-Dimensional Axisymmetric Wind Field Structure of the Spencer, South Dakota, 1998 Tornado.* Journal of the Atmospheric Sciences, 2010. 67(9): p. 3074-3083.

13th Americas Conference on Wind Engineering
Gainesville, Florida, USA
May 21 - 24, 2017

Surface Pressures Dependency on Reynolds Number and Swirl Ratio in Tornado Vortices

Maryam Refan[a,*], Horia Hangan[a]
[a] *WindEEE Research Institute, University of Western Ontario, Institution, London, ON, Canada, hmhangan@uwo.ca*

ABSTRACT

Despite all the achievements in the field of laboratory simulation of tornado vortices, these simulations have always faced a number of major limitations including the limited size of simulators and the Reynolds number (*Re*) similarity. The relatively small size of simulators results in large geometric scaling ratios which, in return, reduces the resolution of measurements and makes the wind-induced load measurements impractical. On the other hand, the dependency of simulated tornado vortex characteristics on *Re* has not been thoroughly investigated, mainly due to the inability of the tornado simulators to cover a wide range of *Re*.

A series of measurements were performed to characterize the surface static pressure distribution of stationary and translating tornado-like vortices and its variation with the swirl ratio (S) and the *Re*. The WindEEE Dome at the University of Western Ontario was used to produce up to 4.5 m wide and 4 m tall tornado vortices. A pressure system along with a 1.1m in diameter floor panel equipped with four hundred pressure taps were used to measure surface pressures. The measurements were carried out at four swirl ratios (S= 0.2-0.8) and at Re=1.6×10^4 and 2×10^6.

A substantial vortex wandering was observed for low swirl stationary vortices. An estimation of the vortex wandering over time was obtained by calculating the RMS of the radius at which the minimum pressure was recorded. The footprint of vortex structure evolution with S, as reported in the literature, was apparent in the center pressure drops. Negative peak pressure coefficients, calculated using a reference maximum tangential velocity, for the surface panel exposed to a translating vortex suggested a slightly lower suction at S=0.76 when compared to S=0.48. This is mainly due to a larger core size and less localized pressure drop for vortices with higher S but also due to the opposed tendencies of pressure coefficient and the reference tangential velocity variations with S.

The availability of tornado flow characteristics data for WindEEE Dome and its prototype, Model WindEEE (1:11 scale), provided a unique opportunity to explore the effect of *Re* on vortex characteristics. Radial profiles of averaged pressure deficits for seven radial Reynolds numbers ($1.6\times10^4 \le Re \le 2\times10^6$) obtained from the 1:11 Model (at S=0.73) and the WindEEE Dome (at S=0.76) were plotted. It was observed that surface pressure deficits are independent of the radial Reynolds number for $Re \ge 4.5\times10^4$, yet highly dependent on the swirl ratio.

The surface pressure distribution of tornado vortices provides useful insights into the vortex dynamics in the near surface region where velocity measurements are difficult. This simulator, for the first time, provides the foundation for tornado wind-induced load studies of building and structures with a relatively large geometric scale and consequently a high spatial resolution. The outcomes of this research will impact both future analytical modeling of tornado vortices as well as the resulting aerodynamic loading on various building shapes.

KEYWORDS

Tornado vortex, laboratory simulations, surface pressure, Reynolds number dependency

Improved Characterization of Damage Patterns from Idealized Tornado Models

Daniel M. Rhee [1], Franklin T. Lombardo [1]

[1]*University of Illinois at Urbana-Champaign, Urbana-Champaign, IL, USA*

ABSTRACT: Tornadoes have caused significant damage and casualties in the past decades. These losses have spurred efforts toward tornado-based design in codes and standards, which requires rigorous estimates of wind characteristics and wind-induced loading of tornadoes. Due to the difficulty of obtaining in-situ measurements of tornadic near-surface wind speeds, a promising method of estimating near-surface wind speed based upon the damage inflicted is developed. The method includes fall directions of trees and crops in residential and agricultural areas that exhibit distinct fall patterns. The observed fall patterns are used to estimate parameters of the Rankine vortex model and reproduce the near-surface wind field. The wind field then can compared with wind speed based on structural damage.

KEYWORDS: Tornado, Near-surface, Wind speeds, Tree-fall, Soybean

1 INTRODUCTION

The total number of tornadoes recorded in the United States has been increasing over the past decades [1]. Annually, there are over 1,200 tornadoes on average, resulting in nearly $1 billion total loss each year [2]. Thus, tornado-based design has become prominent. In order to implement tornado-based design, a robust method of estimating the near-surface wind speed of the tornado is critical for estimating the wind-induced loading by a tornado. However, obtaining in-situ measurements of tornado wind speed has been a challenge. A majority of the world's tornadoes occur in relatively low population areas such as agricultural areas. As Fujita stresses the importance of estimating the wind-speed in rural areas, tornadoes are often time underrated in agricultural areas due to the lack of number of structures in the vicinity [3]. The current method of estimating near-surface wind speed using the structural damage, such as the EF-scale, has been favored though there are issues regarding inconsistency and subjectivity [4]. Therefore, a method for estimating near-surface of tornadic wind speed without structures is necessitous. A more rigorous method of using the tree-fall patterns from tornadoes was developed. This study uses the Rankine vortex model to simulate the tree-fall patterns and compare to the observed pattern generated by tornadoes. The tree-fall model can be applied to different circumstances and can be also used as an independent comparison to the structural damage.

2 GENERALIZED TREE-FALL MODEL

2.1 *Methodology*

The tree-fall model produces wind-field and tree-fall patterns based on Rankine Vortex model. The Rankine vortex model is a simple vortex model that is widely used to describe wind distribution of hurricanes and tornadoes [5]. The model has two regions: 1) the rotational velocity increases until the radius of maximum wind speed and then 2) the velocity decays exponentially afterwards as shown in Figure 1. The equation of Rankine vortex model is as follows:

$$V_{rot}(r) = V_{max}\left(\frac{r}{RMW}\right)^{\varphi}, \quad (r \leq RMW)$$

$$V_{rot}(r) = V_{max}\left(\frac{RMW}{r}\right)^{\varphi}, \quad (r > RMW) \tag{1}$$

where V_{rot} is the rotational speed at radius r, V_{max} is the maximum wind speed generated by the vortex model, RMW is the radius at which V_{max} is reached, and φ is the decay exponent. The typical φ value used for tree-fall model has been 1.0 [6]. However, studies have shown that the exponent value generally ranges from 0.5 to 0.7 based on multiple tornado cases [7, 8].

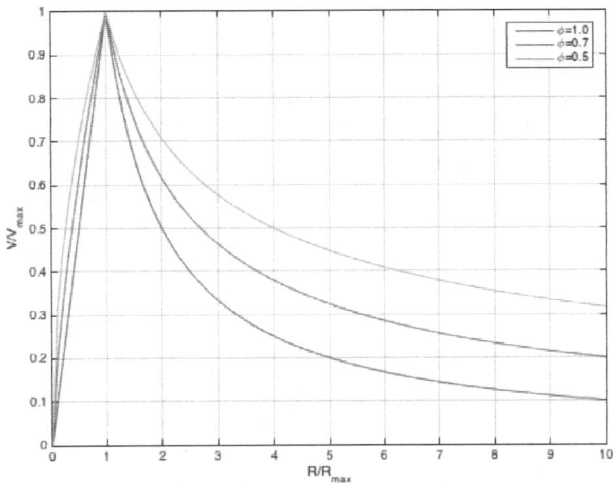

Figure 1. Normalized Rankine vortex model with different decay exponents

The rotational wind speed (V_{rot}) of tornado has two components: the radial component (V_r) and tangential component (V_θ). Depending on the angle (α), the angle between the rotational wind speed and the radial wind component, the rotational wind speed changes its direction. The rotational speed will be purely radial at $\alpha = 0°$ and purely tangential at $\alpha = 90°$. Adding the translation speed of the tornado (V_T) to the rotational speed produces the resulting wind speed (V) in the wind field vortex at any r. Figure 2 illustrates the wind components at a specific location.

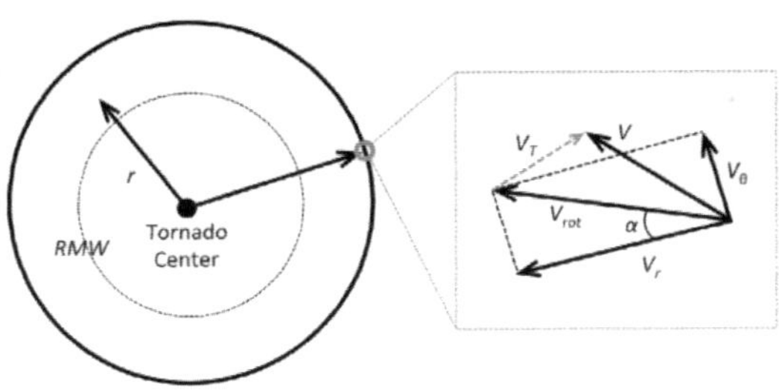

Figure 2. Illustration of the Rankine vortex model (left) and the wind components and overall wind speed (right) at a specific point

Two additional parameters are used in the tree-fall model that determine the tree-fall pattern [9]. The first parameter is G_{max}, which is the ratio between maximum rotational wind speed (V_{max}) and translation speed of the tornado (V_T). The equation for G_{max} is provided in Eqn. (2) and the maximum resultant wind speed ($\hat{V}$) in the tornado can be derived as shown in Eqn. (3).

$$G_{max} = \frac{V_{max}}{V_T} \quad (2)$$

$$\hat{V} = V_T(G_{max} + 1) \quad (3)$$

Second parameter is the average critical wind speed at which a tree will fall (V_c). It is assumed the tree will fall when the resultant wind speed (V) exceeds the critical wind speed.

All of these parameters, including the parameters in the Rankine vortex model, are used in a factorial design to find the "best-matched" combination of parameters. Different combinations of these parameters produce different outcomes as the interaction between these parameters varies. First, the initial range of these parameters is roughly estimated based on the tree-fall pattern and interaction plots. Then, iteration process is done comparing the outcome of the model to the observed data until the "best-matched" parameters are found. The tree-fall model produces three outputs. First is the damage width (DW), which is the total width of the tree damage. Second is the damage ratio (DR), which is the ratio of DW on either side (i.e., south and north) of the convergence line. Convergence line is the estimated location where the tree-fall pattern converges. Depending on the tree-fall pattern, the damage ratio may not exist and thus not be used. Examples of these properties are illustrated in Figures 3 and 4. Third is the tree-fall direction, which is the angle the tree falls measured clock-wise with respect to true north. The collection of individual tree-fall directions forms the overall tree-fall pattern of the tornado. Any out these outputs of the simulated model can be used compared to the observed pattern along the tornado track [10]. Once the "best-matched" parameters are found, the model can generate a wind field that "best" represents the actual tornado wind field.

2.2 *Pattern Tendency with Different Combination of Parameters*

The tree-fall pattern changes significantly with its parameters. In case of high value of G_{max}, (generally 4.0-6.0) the rotational speed of the tornado is much greater than its translation speed and the trees will more likely to fall inward toward the center of rotation vortex and the opposite direction of translation. A low G_{max} value (generally 1.0-3.0) suggests that the translational speed dominates and the tree will more likely orient in the direction of translation. Typically, the tree-fall orientation changes with G_{max} value between 3.0 and 4.0. With all the other parameters kept the same, the damage width (DW) significantly increases as G_{max} increases because the maximum rotational wind speed ($\hat{V}$) is stronger, which suggests that significant tornadoes (EF-4 and EF-5) typically result in high G_{max} value, ranging from 4.0 to 6.0. On the other hand, the damage ratio (DR) is not affected much. Examples of tree-fall pattern with two different G_{max} values are shown in Figure 3. The damage width (DW) increases from 0.6 miles to 1.6 miles as the G_{max} value increases from 3.0 to 4.5. The tree-fall converges to the center of the tornado towards the tornado motion for G_{max} of 3.0 and towards the center of rotation vortex for G_{max} of 4.5. With the convergence line being in the center, the damage width in the south side and the north side are the same for both cases. Thus, the damage ratio (DR) is both 1.0.

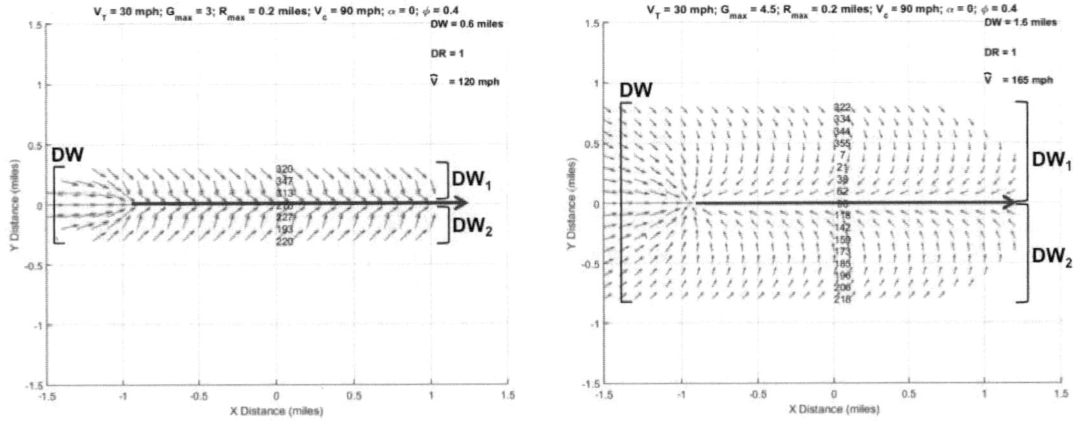

Figure 3. Tree-fall patterns produced with G_{max} = 3.0 (left) and G_{max} = 4.5 (right). Note that tornado center and its motion is from left to right (blue arrow) and all other parameters were kept the same for both simulations. DW is the total damage width, DW_1 is the damage width on the north side of convergence line, and DW_2 is the damage width on the south side of convergence line. DR is the ratio between DW_2 and DW_1.

Alpha (α) value also contributes significantly to the tree-fall pattern. For alpha value greater than 0° and less than 90°, the tree-fall model will create a cyclonic swirling pattern. The damage ratio (DR) is greatly affected by alpha. As alpha value increases, the damage ratio increases significantly. However, the damage width (DW) is not affected by alpha. Examples of tree-fall pattern with two different alpha values are shown in Figure 4. The DWs are 1.5 miles and 1.6 miles for α value of 45° and 90°, respectively. However, the convergence line is formed at the one seventh of the total damage width from the top for α value of 45°, resulting in a DR of 6.5. On the other hand, no convergence line is found and the DR increases to infinity as α increases to 90°. A statistical relationship can be developed to closely analyze the relationship between the location of convergence line and DR.

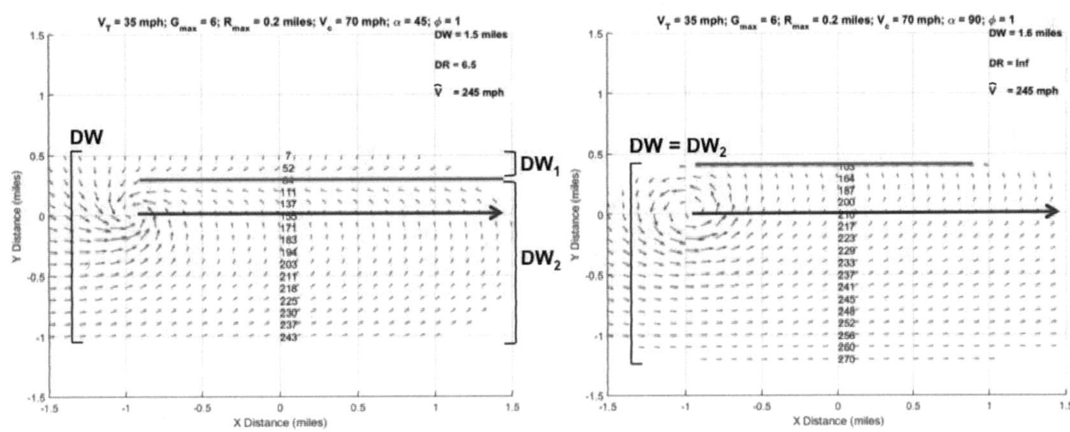

Figure 4. Tree-fall patterns produced with α = 45° (left) and α = 90° (right). Note that tornado motion is from left to right (blue arrow) and all other parameters were kept the same for both simulations. The convergence line is indicated in red line.

These relationships can be shown through the individual parameter and their interaction effects in Figure 5. As shown in the top, the damage width (DW) is affected the most by the translational speed (V_T). It has a positive correlation with G_{max}, RMW, and V_T and a negative correlation with V_c and phi regardless of interaction with other inputs parameters in the Rankine vortex. Considering the interaction between G_{max} and RMW, higher G_{max} and RMW lead to large DW, whereas the interaction is reversed with V_c as lower V_c leads to higher DW because lower critical tree fall wind speed would cause more trees to fall. On the other hand, damage ratio (DR) is almost solely affected by α parameter. Increase in α results in increase in DR regardless of other parameters. However, the slopes are relatively flat for all the other parameters with little interaction with each other. The interaction plots can be used for general interpretation of the Rankine vortex parameters [11]. With different combination of parameters, the tree-fall model can produce numerous tree-fall patterns that be used to match real-life tornado tree-fall patterns. Applications of the tree-fall model are shown in the next section.

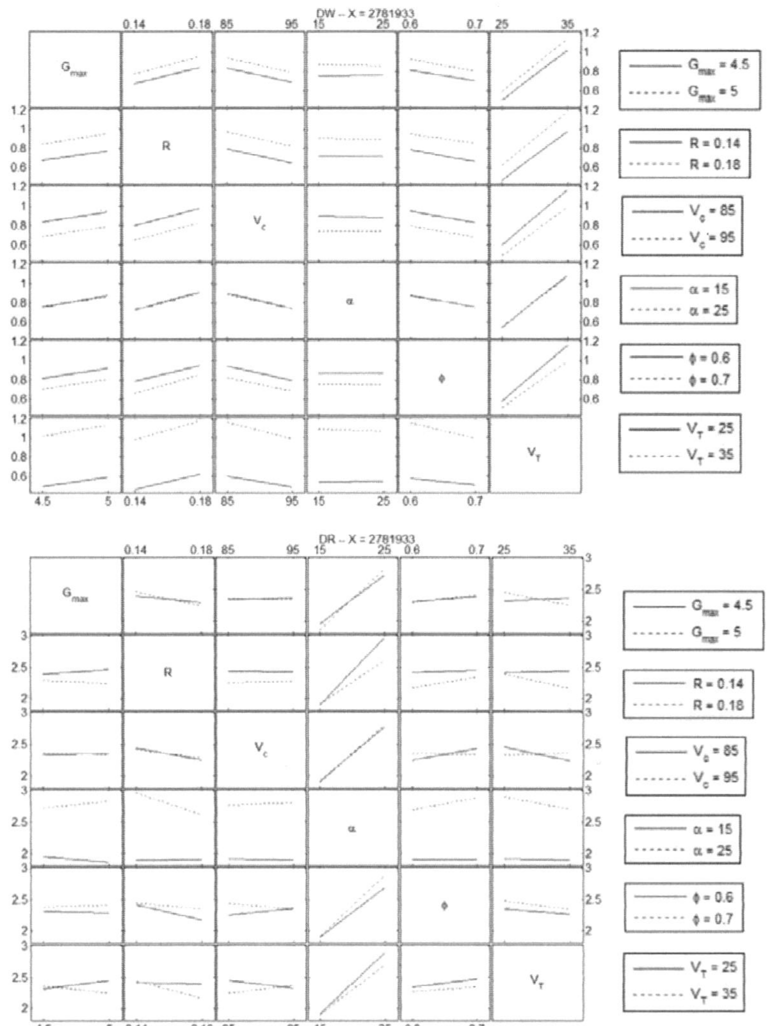

Figure 5. Illustration of interaction effect with different parameters on damage width (top) and damage ratio (bottom). Values of the parameters are shown in box on the right side [11]

3 APPLICATIONS

3.1 *May 22, 2011 Joplin, MO tornado*

On the 22 May 2011, a devastating EF-5 rated tornado struck the city of Joplin, Missouri and caused 116 fatalities and nearly $2 billion total losses. The damage and location of nearly 1200 residential houses and the tree-fall direction and location of nearly 5000 trees were documented by ASCE and NIST, respectively, as part of their post-storm reconnaissance [11, 12]. The tree-fall model has been applied to the May 22, 2011 Joplin, Missouri tornado and the simulated wind field has been created with "best-matched" parameters shown in Table 1.

Table 1. Best-estimated values and ranges of parameters for Joplin tornado

Parameter	Best-estimated value	Range
Alpha (α)	20	15–25
Phi (φ)	0.65	0.60–0.70
RMW (m)	257	225–290
V_c (m/s)	40	38–42
G_{max}	4.8	4.5–5.0

Figure 6 shows the map of city of Joplin with the maximum wind speed from the tree-fall model converted to EF-scale overlaid with differences between the EF-scale from tree-fall model and EF-scale using the structural damage of residential houses. Approximately 34% of residential houses showed no difference in EF rating, 76% of residential houses showed either no difference or ±1 EF rating, and the remaining 24% showed more than ±1 EF rating difference near the center of the tornado. One possible discrepancy is caused by the subjectivity of rating EF-scale of residential houses [10].

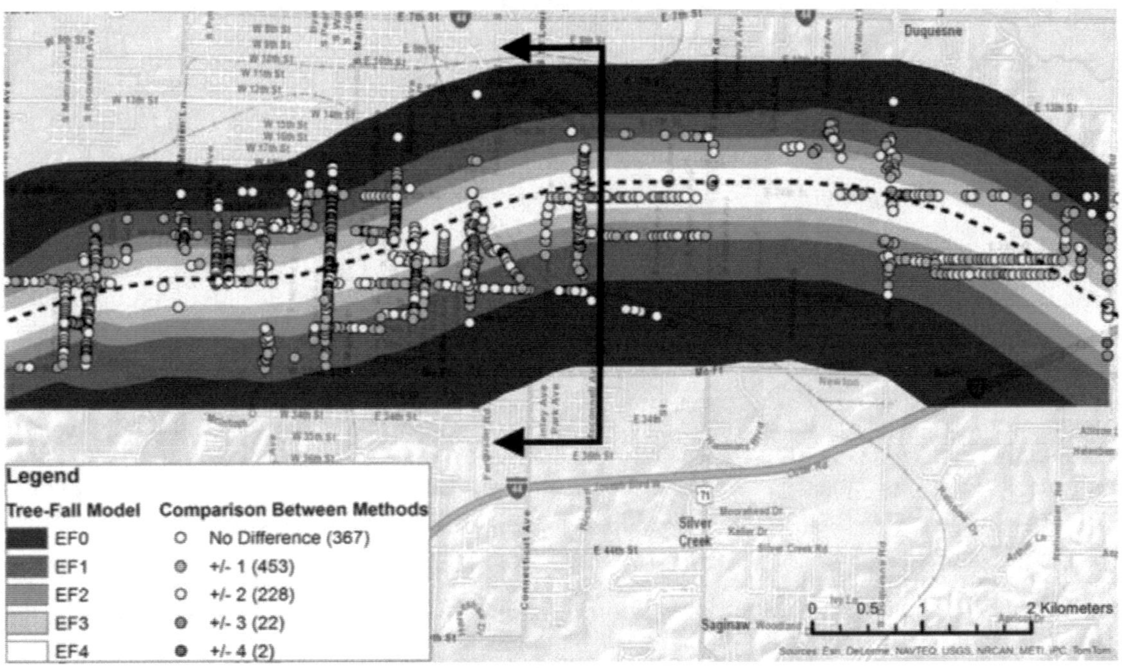

Figure 6. EF-scale wind field from tree-fall model (contour) and comparison to EF-scale from structural damage (point). The difference in EF-scale is color coded with white being no difference and red being +/- 4 difference [10]

With the maximum wind speed from the tree-fall model and the damage state (DS) known, a fragility curve can then be developed. The DS was classified, using the DODs for FR12 (One- and Two-Family Residences) of the EF-scale. Then, the probability of structures meeting or exceeding a DOD at a given wind speed was determined through fragility functions which are shown in Figure 7. A progressive nature of the fragilities is evident, with the median failure wind speed increasing with DOD [13]. Inclusion of many different uncertainties, such as roof type, orientation of the structures with respect to the tornado path, in the fragility curves can be used to further define the vulnerability of structures in different aspect.

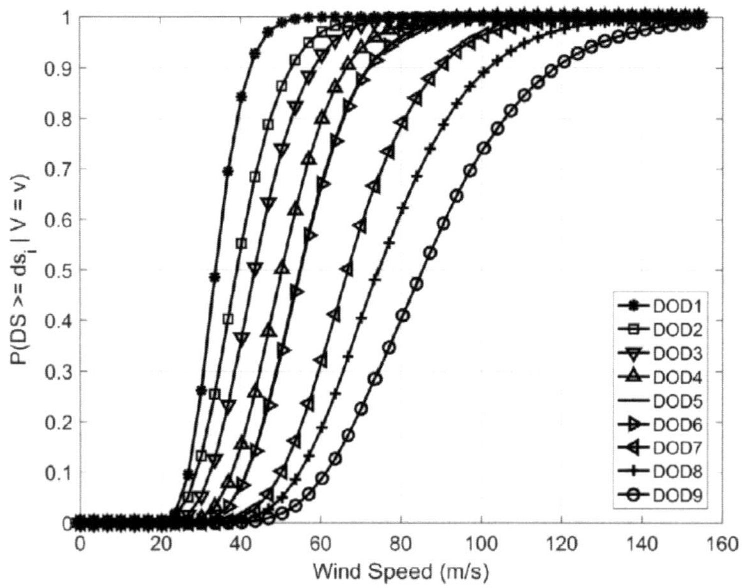

Figure 7. Empirical fragilities for damage measures corresponding to the nine Degrees of Damage for FR12 (One- and Two-Family Residences) of the EF-Scale [13].

3.2 *February 28, 2017 Naplate, IL tornado*

On 28 February 2017, multiple tornadoes touched down in the state of Illinois and the eastern part of Missouri. This prompted the Wind Engineering Group at University of Illinois at Champaign-Urbana to conduct a damage survey on one of the tornadoes. The location and the DODs of residential houses and the location and fall direction of trees, poles, and signs were recorded and mapped into ArcGIS shown in Figure 8. The fallen trees are pointing toward the direction of the tornado translation, which resembles the tree fall pattern of the left figure in Figure 3.

From the DOD of the residential structures, an estimated wind speed can be extracted based on structural damage. Similar to the May 22, 2011 Joplin tornado, a near-surface wind field created from the tree-fall model can be compared with the wind speed from structural damage. Furthermore, laboratory testing on distribution poles and street signs can be incorporated to further improve the model. With a simple laboratory testing of finding the plastic moment, the critical wind speed of these structures can be back-calculated, which provides the lower bound wind speed of the tornado [14]. The fall direction of cantilever structures also provides valuable information of the wind direction of the tornado.

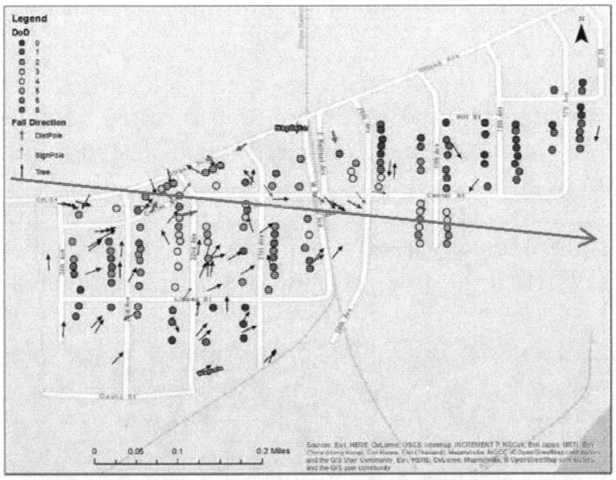

Figure 8. Location of damaged residential homes with its associated DOD and fall direction of fallen trees, signs, and poles for the February 28, 2017 Naplate, IL tornado survey. The red arrow indicates the estimated tornado center and translation direction

3.3 *September 9, 2016 Sidney, IL tornado*

Assessing tornadic damages and estimating near-surface wind speed in agricultural areas has been a challenge due to lack of structures. Thus, the assessment of EF ratings is often biased and underrated in agricultural areas [3]. On the 9 September 2016, an EF-2 rated tornado touched down near Sidney, IL [15]. Interesting soybean patterns were noticed, prompting the Wind Engineering Group to conduct a ground survey in the soybean field and apply the tree-fall model to this case. The location and direction of the soybean fall were recorded and displayed with red arrows in Figure 9. Interestingly, convergent and divergent patterns of soybeans were also found in this survey, shown in black and blue arrow in Figure 9 respectively. These patterns illustrate possible indications of multiple vortices although they were not considered in this study. A low G_{max} value (1.0-3.0) was presumed initially as the soybean-fall pattern pointed toward the direction of the tornado translation. A translational speed of 30 knots, averaged along the entire tornado path, was used in the simulated model since the exact translational speed at the survey area was unknown. An assumed value of critical wind speed of soybean was used as there has been little research in the critical wind speed of soybeans. Applying the methodology described in Section 2.1, the simulated model produced a G_{max} value of 2.2 and alpha value of 5°. The "best-matched" parameters are shown in Table 2.

Table 2. Best-estimated values and ranges of parameters for sidney tornado

Parameter	Best-estimated value	Range
Alpha (α)	5	0–10
Phi (φ)	0.65	0.60–0.70
RMW (m)	12.5	10-15
V_c (m/s)	36	34–48
G_{max}	2.2	2.1–2.3

The maximum wind speed was calculated as 110 mph, which is associated with EF-1 in EF-scale, whereas the National Weather Service rated the tornado an EF-0 in the surveyed area [15]. The simulated pattern, which is shown in yellow arrow, shows good agreement with the observed pattern. The damage width and damage ratio comparison between simulated model and observed pattern at different sections are shown in Figure 10. Both outputs (DW and DR) show moderately good agreement, resulting 1.5% to 40% difference between the simulated and observed. Some possible discrepancies could be the incorrect choices of translation speed (V_T) and critical wind speed of soybean fall (V_c). It is possible that the translational speed is significantly different in the surveyed section than the averaged translational speed and the assumed critical wind speed of soybean fall is incorrect. Although they were not considered in this study, the types of vegetation and growth stage of crops can be also incorporated, allowing tree-fall model to be apply to soybeans and other different vegetation fields. Investigation of correlation of the critical wind speed at different growth stage for different types of vegetation in a wind tunnel testing is the focus of future research as further refinement of tree-fall model in agricultural area is necessary.

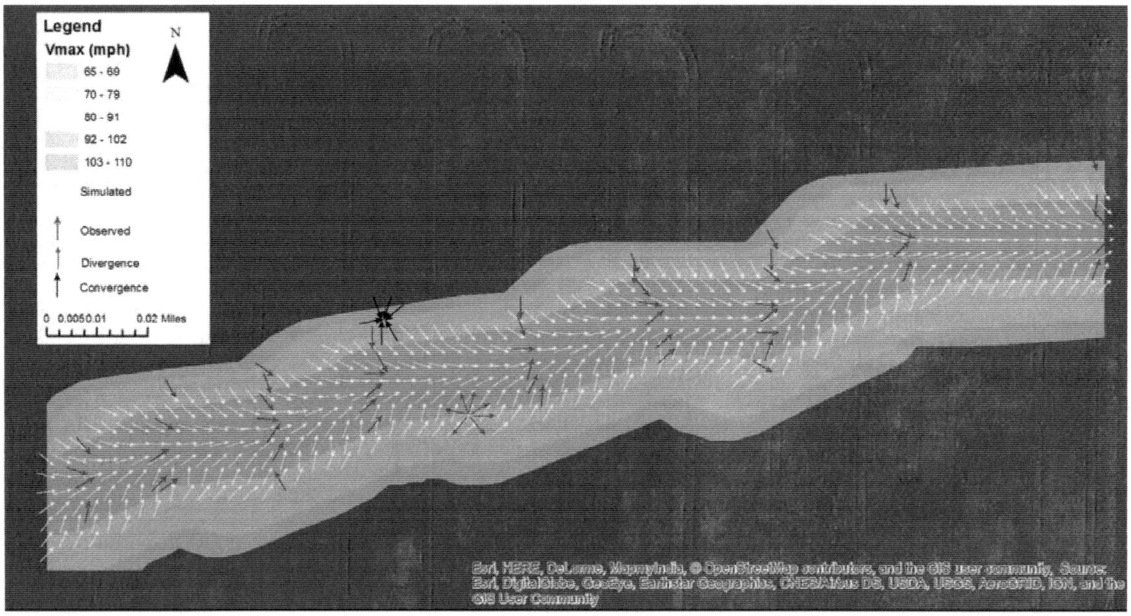

Figure 9. Observed (red arrow) and simulated (yellow arrow) soybean-fall pattern with wind field (contour) overlaid, including convergent (black arrow) and divergent (divergence) patterns. Note that the tornado is moving from west to east

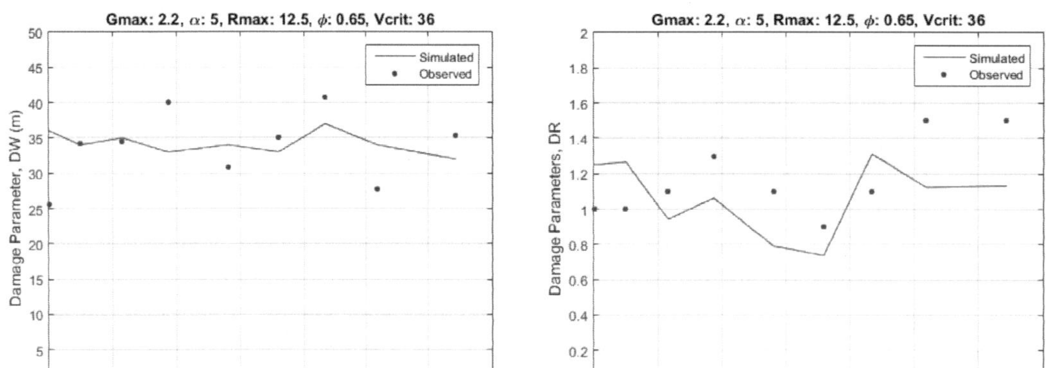

Figure 10. Damage width (left) and damage ratio (right) comparison between observed (point) and simulated (line) soybean-fall pattern. Note that the tornado path is taken from the starting point of simulation.

3.4 *April 27, 2011 Tuscaloosa, AL tornado*

Tornadoes with very long tracks often change their characteristics along the path. The 27 April 2011, Tuscaloosa-Birmingham, AL tornado translated through forest and urban area with a total path length of 80 miles [16]. In mountainous terrain, tornadic damage pattern may change because the topographic variation can influence the damage severity [17]. In the Tuscaloosa tornado case, the tornado exited heavily forested area in mountainous terrain and entered the urban area. An abrupt change of terrain can influence the traits of the tornado, so the parameters of the tree-fall model will likely change accordingly. Figure 11 shows a segment of the tornado track with tree-falls in the city of Tuscaloosa. There are three sections with distinctly different tree-fall patterns and damage width and ratio, which indicate a change of tornado characteristics. Thus, by changing the parameters along the path, the tree-fall model can produce a wind field where the characteristics of the tornado change.

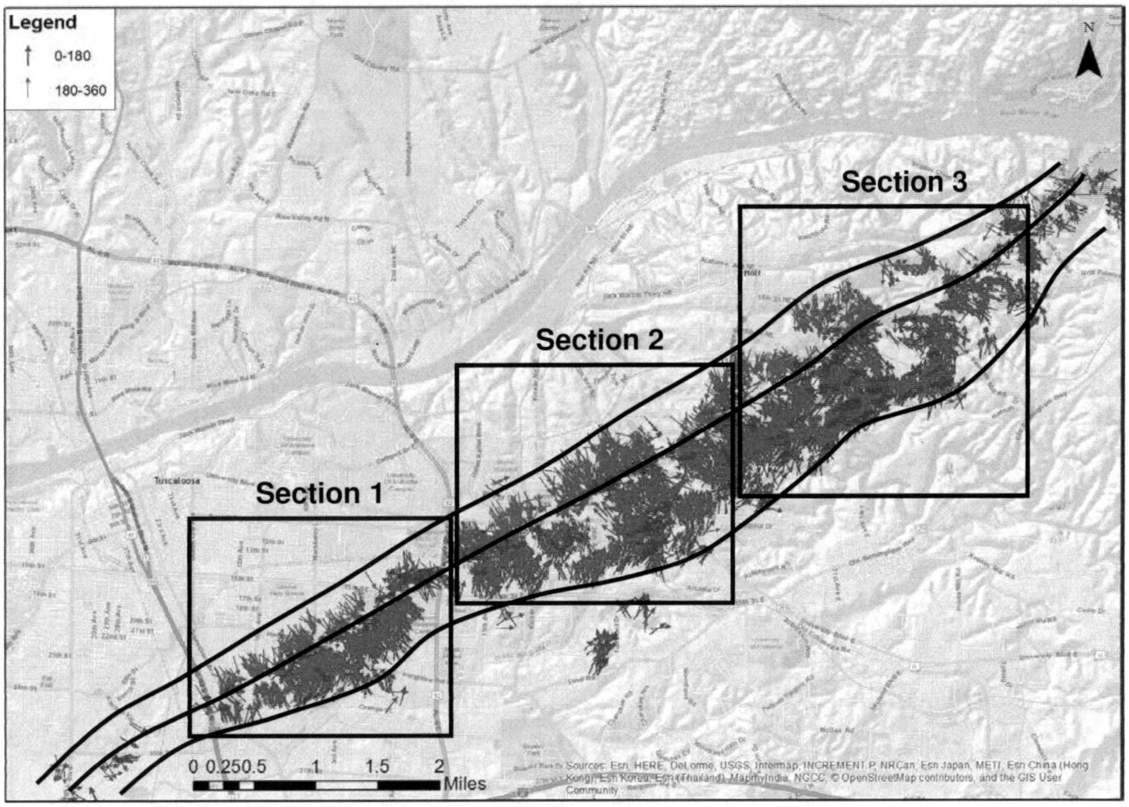

Figure 11. Partial segments of tree-fall pattern of Tuscaloosa–Birmingham, AL tornado in city of Tuscaloosa. Note that the tornado moves from south-west to north-east. Black lines indicate the approximated damage area and the center of the tornado. Tree-fall angle from 0° to 180° and from 180° to 360° with respect to the convergence line are colored in red and blue, respectively

4 CONCLUSION

The tree-fall model uses the Rankine vortex and its parameters to simulate the observed tree-fall pattern and produces an estimated near-surface wind speed of the tornado. The model can be validated through comparison with the wind speed estimated from structural damage in the damaged area. The tree-fall model was applied to the May 22, 2011 Joplin, MO tornado and September 9, 2016 Sidney, IL tornado. The wind speed estimated from tree-fall model for the Joplin tornado was compared to wind speed from the residential houses, which 76% of residential houses showed either no difference or ±1 EF rating. The tree-fall pattern simulated for the Sidney tornado showed good agreement. However, the model outputs had moderate differences with the observed pattern. This suggests that the tree-fall model is accurate using tree-fall pattern, but may need to be refined more to accurately predict the wind speed in agricultural areas.

With further validation and incorporating other factors, the tree-fall model shows a lot of potential for application to many different actual tornadoes. Application to vegetation other than trees in agricultural areas with further research is also viable. The tree-fall model can be a robust tool to estimate near surface wind speed of tornado independent of structural damage and investigate details of reconstructed wind field. With a reliable estimate of the tornado wind field, fragility curves can also be eventually developed for tornado damage assessment using nearby structures. These estimated wind fields and fragility curves can be used in tornado-based design and further to interpret the vulnerability to tornadoes.

5 ACKNOWLEDGEMENTS

The authors gratefully acknowledge the financial support of the VORTEX-SE program by NOAA (Award No: NA150AR4590228) to conduct post-damage surveys of multiple tornadoes.

6 REFERENCES

1 NOAA National Centers for Environmental Information, U.S. Tornado Climatology: Historical Records and Trends, <https://www.ncdc.noaa.gov/climate-information/extreme-events/us-tornado-climatology/trends>.
2 S.A. Changnon, Tornado losses in the United States, Natural Hazards Review, 10.4 (2009) 145-150.
3 J.L. Guyer and M.L. Moritz, On issues of tornado damage assessment and F-scale assignment in agricultural areas, In: Preprints: 83rd AMS Annual Meeting Symposium on F-scale and Severe-Weather Damage Assessment, 2003.
4 R. Edwards, *et al.*, Tornado intensity estimation: past, present, and future. Bull. Am. Meteorol. Soc., 94.5 (2013) 641–653.
5 D.B. Giaiotti and F. Stel, The Rankine vortex model, University of Trieste-International Centre for Theoretical Physics, Visco(UD), Italy, 2006.
6 V. Beck and N. Dotzek, Reconstruction of near-surface tornado wind fields from forest damage, J. Appl. Meteorol. Climatol., 49.7 (2010) 1517-1537.
7 K. Kosiba and J. Wurman, The three-dimensional axisymmetric wind field structure of the Spencer, South Dakota, 1998 tornado, J. Atmos. Sci., 67.9 (2010) 3074–3083.
8 J. Wurman and R.A. Curtis, The 30 May 1998 Spencer, South Dakota, Storm. Part II: Comparison of observed damage and radar-derived winds in the tornadoes, Mon. Weather Rev., 133.1 (2005) 97-119.
9 A.P. Holland, A.J. Riordan, E.C. Franklin, A simple model for simulating tornado damage in forests, J. Appl. Meteorol. Climatol., 45.12 (2006) 1597-1611.
10 F.T. Lombardo, *et al.*, Comparison of two methods of near-surface wind speed estimation in the 22 May, 2011 Joplin, Missouri Tornado, J. Ind. Aerodyn., 2015, Vol. 138, pp. 87-97.
11 E.D. Kuligowski *et al.*, Technical investigation of the May 22, 2011, tornado in Joplin, Missouri, Natl. Inst. Stand. Technol., 2014, 428.
12 D.O. Prevatt *et al.*, Joplin, Missouri, Tornado of May 22, 2011: Structural Damage Survey and Case for Tornado-resilient Building Codes. ASCE Publications, 2013.
13 D. Roueche *et al.*, An Empirical Approach to Evaluating the Tornado Fragility of Residential Structures., J. Struct. Eng.

14. G.N. Boughton, *et al.*, Tropical Cyclone Yasi Structural Damage to Buildings-Technical Report, Cyclone Testing Station, James Cook University, Australia, 2011, No. 57.
15 NWS, 2016. Summary of September 9th Tornadoes in East Central Illinois. In: National Weather Service Report. National Weather Service Weather Forecast Office.
16 NWS, 2011. Tuscaloosa-Birmingham tornado summary. In: National Weather Service Report. National Weather Service Weather Forecast Office.
17 J.B. Cannon *et al.*, Landscape-scale characteristics of forest tornado damage in mountainous terrain, Landscape Ecology, 31.9 (2016) 2097-2114.

13th Americas Conference on Wind Engineering
Gainesville, Florida, USA
May 21 - 24, 2017

Cost-Effectiveness of Tornado Resistant Construction

David B. Roueche[a], Bryan Wood[b], David O. Prevatt[c]

[a] *University of Florida, Gainesville, FL, USA, aegnew@ufl.edu*
[b] *Assurant, Inc, Springfield, OH, USA, bryan.wood@assurant.com*
[c] *University of Florida, Gainesville, FL, USA, david.roueche@ufl.edu*

ABSTRACT

The number of homes in tornado-prone regions of the United States is expected to nearly double by 2100. This vast increase in human exposure to tornadoes will lead to higher losses and human impacts independent of any increases in tornado activity due to climate change. Building codes and wind design standards do not currently consider tornadoes, as tornado-resistant construction has traditionally been considered cost-prohibitive given the extreme wind loads they induce and the low probability of an individual structure being impacted by a tornado.

This paper evaluates the cost-effectiveness of strengthening a wood-frame residential structure to a level where it is repairable after a tornado event. The annual expected losses are determined by convolving tornado hazard, fragility and loss models for a prototype home located in Tuscaloosa, AL and Moore, OK. The prototype home is a rectangular, wood-frame home with a gable roof. The tornado hazard models are developed from the Storm Prediction Center database of reported tornadoes in the United States. The tornado fragility models are engineering-based fragility functions for roof and wall damage calibrated against damage observations to 1,291 structures from the 2011 Joplin, MO tornado and 718 structures from the 2015 Dallas, TX tornado. The loss model is developed from claims data for 350 homes that were damaged in the 2015 Dallas, TX tornado and assessed by a field survey team from the University of Florida and a local engineering firm. The 2015 Dallas, TX tornadoes had peak wind speeds estimated at over 170 mph, providing a full range of structural damage and insured loss ratios that were used in this study.

The paper demonstrates the critical point at which a structure is damaged to the extent that it is deemed a total loss. The cost of preventing total loss is compared against the expected annual loss for the building location to evaluate the cost effectiveness of strengthened construction. Discussion is given to the need for assessing cost-effectiveness of tornado design on a community level rather than solely for an individual structure. The results have important implications for the future of tornado design policies, particularly as it relates to future building codes in tornado-prone regions.

KEYWORDS

Tornado, residential structures, loss ratios, cost-benefit

The 13th Americas Conference on Wind Engineering (13ACWE)
Gainesville, Florida USA, May21-24, 2017

WRF Simulation of Strong Wind Gust Induced by Tornado-genesis Storm

Tao Tao [a], Tetsuro Tamura [a], Hidenori Kawai [a]

[a] *Environmental Science and Technology, Tokyo Institute of Technology, 4259 Nagatsuta, Midori, Yokohama, kanagawa, Japan*

ABSTRACT: On May 6th 2012, a strong storm process attacked Kanto Plain in which three tornadoes was observed and caused serious damages. To understand this strong convective process and the behavior of local strong gust, authors have simulated this process by mesoscale meteorological model WRF (Weather Research and Forecast). Two nested domains have been employed downscaling the horizontal grid resolution from 1.25 km to 0.25 km. A convective storm has been generated and local gust phenomenon occurs with extremely high horizontal wind velocity. Meanwhile, to weaken the influence of "gray zone", nested domains largely shifted from 1.25 km to 0.125 km and further 42 m grid resolution are employed where WRFLES (WRF Large Eddy Simulation) is applied to the inner domains. Tornado-like structure appears while the rotation and upward wind speed are not strong enough comparing to a typical tornado case. The complexity of realistic tornado structure shows much asymmetric characteristics in wind distribution than ideal tornado experiments or simulations.

KEYWORDS: Tornado, WRF, Severe Local Storm, Wind Gust.

1 INTRODUCTION

From wind engineering point of view, besides field observations, experimental or numerical techniques can also be utilized for understanding detailed structures of tornadoes. However, both techniques, including CFD (Computation Fluid Dynamic), are normally applied by making models under idealized conditions. Accordingly, CFD does not deal with the complex atmospheric behaviors existing in realistic tornadoes. To overcome this drawback, an alternative approach such as mesoscale meteorological model can be employed for generation of tornado process under the drive of realistic meteorological condition. Meanwhile in recent meteorological analysis, the focusing points are mainly located on the mechanism of tornado-related surroundings such as mesocyclone or atmospheric unstable convection in the upper atmosphere. The near-ground wind gust in tornadic flows has not been taken into consideration in these analyses. This study tries to investigate the characteristics of near-ground wind gust for tornado with complicated wind distribution using the mesoscale meteorological model.

Severe local meteorological phenomena, such as storm and local gust front are attracting increasing attention because of its high unpredictability. Together with the local gust, thunderstorm also appears at the same time. In particular, a supercell storm is indicated by the existence of a persistent mesocyclone with strong updraft in a convective system (Klemp 1987). The overwhelming majority of devastating tornadoes are mentioned as a connection with the supercell storms (Markowski et.al 2014). To discuss the tornado generation and behaviors, the generation of supercell mesocyclone should be firstly focused on. The vertical updraft tendency has been discussed from theoretical, numerical and experimental methods. Cintineo et al (2013) described the prediction of supercell thunderstorm evolution by idealized simulation of WRF

(Weather Research and Forecast). In his research, quantity analysis of several tornadogensis sensitive variables are utilized, such as CAPE (convective available potential energy), UH (up helicity) and reflection strength. The updraft explicitly embodies the sucking strength of air aloft which is in fact a release of the CAPE. The mesocyclone strength is typically the order of 10^{-2}/s. The tilted vorticity comes from near ground friction needs several kilometers ascending to reach the strength of an apparent mesocyclone structure, so that the mesocyclone is always formed on a mid-level height. When a low-level mesocyclone is once formed, it implies that a tornadogensis may occur.

The utilization of meteorological model on real tornado simulation is very limited so far. It should be mentioned that a "scale gap" is located between the meteorological data and specified local phenomena like mesocyclone or tornado. The temporal and spatial scale of the former data is about the order of hours and kilometers, while the latter one is less than an hour and hundreds of meters. It becomes more serious that to explicitly capture a typical supercell related tornado, resolution of tens of meters and minutes is acquired. The large "scale gap" and complex generation dynamic of local strong convective system would cause great difficulty in successfully reproducing a tornado process, not to mention prediction. To reduce this problem, sometimes assimilation method are be employed (Xue 2014) which relies on a high resolution observation data covering a large 3d-area. Yokota (2016) successfully predicted low-level mesocyclone by assimilation of surface meteorological data and radar data. Without assimilation method, Mashiko (2009, 2016) reproduced the tornado case in an outer-rainband minisupercell of typhoon and the case of Tsukuba city tornado. In his papers, the sources for tornado vorticity are discussed.

The "gray zone" problem has been encountered in downscaling meteorology research recent years due to the increasing higher resolution simulations. In conventional mesoscale model, horizontal grid resolution is about tens kilometers, so that it can be assumed that the turbulence energy scale is much smaller than horizontal grid size. For the vertical grid whose resolution commonly varies from tens to hundreds meters, one-dimensional assumption based turbulence models are set up which deal with the vertical turbulent flux effect whose quantity is much larger than horizontal turbulent flux. Meanwhile, when horizontal resolution approaches about one hundred meters or less, LES (Large Eddy Simulation) method is naturally considered because a large part of turbulence energy is located between integral turbulence scale and grid resolution. However, between the LES limit (slightly less than integral turbulence scale) and mesoscale limit (assuming non turbulent effect), both mesoscale model and LES model are less reliable. This intermediate scale is named as gray zone or "Terra Incognita" (Wyngaard, 2004).

In this paper, focusing on a severe storm process with tornado and extreme local gust, a numerical simulation by WRF model is conducted. After introducing the meteorological case (section 2) and numerical design (section 3), the results from WRF model with a 5:1 nesting ratio (Case A) and WRFLES model with a 10:1 nesting ratio (Case B) will be discussed on section 4. Based on the results from fine grid resolution, the complexity of real time tornado structure will be also discussed.

2 CASE OVERVIEW

On 6 May 2012, an F3 tornado attacked the Tsukuba city, Ibaraki-ken in Japan, which is one of the most devastative tornadoes recorded since 1960 in Japan. The tornado caused 1 fatality, 37 injures and severe property damage. The tornado is generated at about 03:35 (UTC+00) and moves from southwest to northeast direction with a track of about 17 km long. Figure 1b shows the tornado path on the map. A hook-shaped echo pattern and a strong updraft contained mesocyclone are detected by an observation Doppler radar which is located about 15 km south from the storm (Yamauchi et al, 2014). The supercell related reflection pattern is shown on Figure 1a.

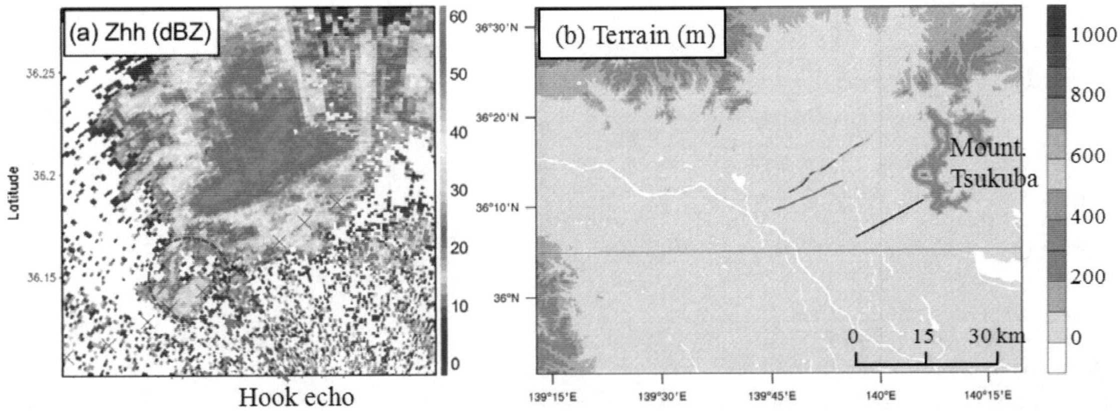

Figure 1. (a): The reflectivity of the supercell over Tsukuba observed by radar from MRI-C radar. The figure is extracted from yamauchi (2014) paper. × refers to the tornado track. (b): Topographic map of the area around Tsukuba city, and the black, red, blue, green lines refer to observed tornado path, WRF250m, WRFLES125m, WRFLES42m simulated tornado-like rotation paths respectively.

3 NUMERICAL SETTING

Numerical simulation of this severe local storm case is conducted by WRF-ARW (Advance Research WRF) model with nested domains. For both Case A and B, the outermost domain calculation condition is the same whose resolution is 1250 m by conventional PBL schemes. Here we name it as WRF1250m and so as other domains. This domain covers the majority part of Honshou Island of Japan centering on Tsukuba city. Domain arrangement is shown on Table 1.

The basic equation of WRF is set up based on compressible Euler equation. Advection terms are calculated by 5/3 order upwind schemes for horizontal/vertical component. For turbulence model, the horizontal component is not taken into account which means no diffusion is imposed; vertical component is calculated in PBL scheme. Time-split RK3 scheme is utilized for time integration to get rid of acoustic model that filtering effects are included.

The topography and land cover data is interpolated from USGS (United States Geological Survey) whose resolution is 30". GPV (Grid Points Values) data from JMA (Japan Meteorological Agency) is utilized to provide the atmospheric conditions with grid size of 5 km and update frequency of every three hours. ECMWF (European Centre of Medium-Range Weather Forecasts) Era-interim dataset provides four layers of soil moisture and temperature whose resolution is 0.75°. The simulation period is 6 hours long which starts from 00:00, 6 May.

From the lowest level to the top level (100mb), the level height interval is increasing steady from about 16 meters to about 310 meters. Near ground ten levels depths increase smoothly from 16 m to 74 m with largest adjacent level depth ratio of 1.29.

In mesoscale meteorological model, parameterization scheme is one of the key points that influence the atmosphere movement and meteorological performance. Considering the smaller grid size comparing to typical mesoscale cases (resolution of tens kilometers), in this study cumulus cloud scheme is shut down. Other physical schemes are shown as Table 2. Important physical schemes that sensitively influence convection are PBL (Planet Boundary Layer) schemes and microphysics schemes. In fact, we have taken several combinations of these two schemes to test their performance in reproducing this strong convection phenomenon (results not shown). The best combination is taking MYJ (Mellow-Yamada-Janjic) for PBL and Purdue Lin (Lin et al) for microphysics which generates a 50 minutes duration mesocyclone structure with larger than 30 m/s updraft.

Table 1. Design of the domains

	WRF1250m	Case A	Case B	
		WRF250m	WRFLES125m	WRFLES42m
Grids dimensions	721 × 721	901 × 701	901 × 701	1501 × 1201
Horizontal grid size (m)	1250	250	125	41.6667
Large time step (s)	3	0.6	0.3	0.1
Intergration time (UTC+00)	0000-0600	0200-0400	0200-0400	0250-0320

Table 2. Physics schemes

Long/short Radiation	RRTMG
Surface Layer	Monin-Obukhov
Land surface	Unified Noah

In Case A simulation, domain arrangement is shown on Table 1. The second domain resolution of 0.25 km occupies the most part of the Kanto Plain (eastern Japan). One way nesting is adopted which means there is no feedback from the inner domain to the parent one. Vertical nesting is not available in recent WRF model during concurrent run, so all domains have same vertical levels arrangement which is 68 levels on vertical direction. Physical schemes are the same as domain WRF1250m.

Considering the influence of gray zone, 250 m resolution is sensitive to this range. So that naturally we adopt the idea using a larger ratio to avoid this possible influence. In Case B simulation, WRF1250m is downscaled directly to WRFLES125m by ratio 10:1. The WRFLES model is employed to calculated turbulence effect instead of PBL schemes in WRFLES125m. In WRFLES model, a 1.5 order turbulence closure based on prognostic TKE (Turbulent Kinetic Energy) is utilized in which TKE is calculated from the effect of shear production, buoyancy and dissipation. Other physical schemes and numerical methods are the same as domain WRF1250m and WRF250m. A domain of 41.67 meters horizontal grid resolution, named as WRFLES42m is further nested on WRFLES125m in order to capture more detailed structures.

4 RESULTS AND DISCUSSION

4.1 *Results of Case A*

Figure 2 shows the mesocyclone structure calculated in WRF250m on different heights at 03:03, 6 May. Hydrometeors zone been sucked approaching a low pressure area which forms a hook-like shape can be observed. The maximum vertical velocity exceeds 32 m/s at 3 km high within a rotation area. The rotation area can also be clearly shown by vertical vorticity pattern. The pressure on the centre of the mesocyclone is about 4 hPa lower than the surrounding. Obviously the rotation occurs around the pressure deficit centre from the pattern of horizontal speed vectors. The rotation zone stretches throughout three heights (2 km, 3 km, 4 km as shown) which could be regarded as not only a mid-level mesocyclone but also a low-level mesocyclone. The stretching of rotating towards ground implies the possibility of tornadogensis (Mashiko 2016). The mesocyclone diameter is about 4 km, which is slightly smaller than a typical supercell mesocyclone happened in United States (Xue et al 2014).

A weak rotation zone near ground also appears and forms a tube-like vertical vorticity which implies the possible generation of tornado as shown on Figure 3. Considering the 250 m horizontal resolution, it would be difficult to directly resolve a tornado structure because a finer grid resolution is required. The track of the rotation center (defined as largest vertical vorticity location on 100 m high) is plotted on Figure 1b. The weak rotation near ground lasts from 03:03 to 03:09.

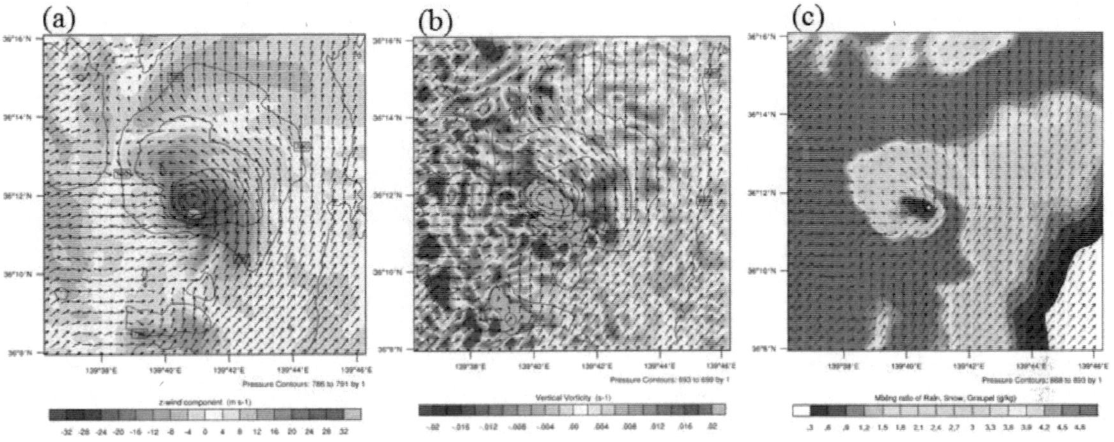

Figure 2. The mesocyclone structure simulated on WRF250m. The heights of figure a, b and c are 1 km, 2 km, 3 km above the sea respectively, and contours refer to hydrometer (g/kg), vertical velocity (m/s), vertical vorticity (s^{-1}) respectively. The black lines show isobar with 1 hPa interval and vectors refer to horizontal velocity direction.

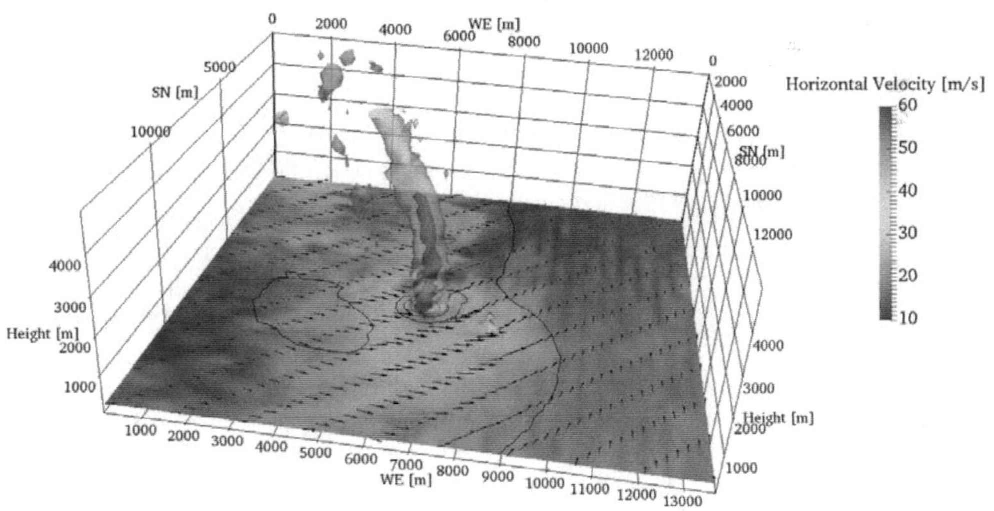

Figure 3. The 3D visualization of weak rotation structure from WRF250m. The vertical vorticity isosurfaces of 0.05 s^{-1} and 0.1 s^{-1} are plotted by 3D contours. Plane contour and vectors show horizontal velocity magnitude and direction near ground (200 m above the ground). Line contours denote isobars at interval of 1 hPa.

4.2 *Results of Case B*

4.2.1 *Verification of tornado-like structure genesis*

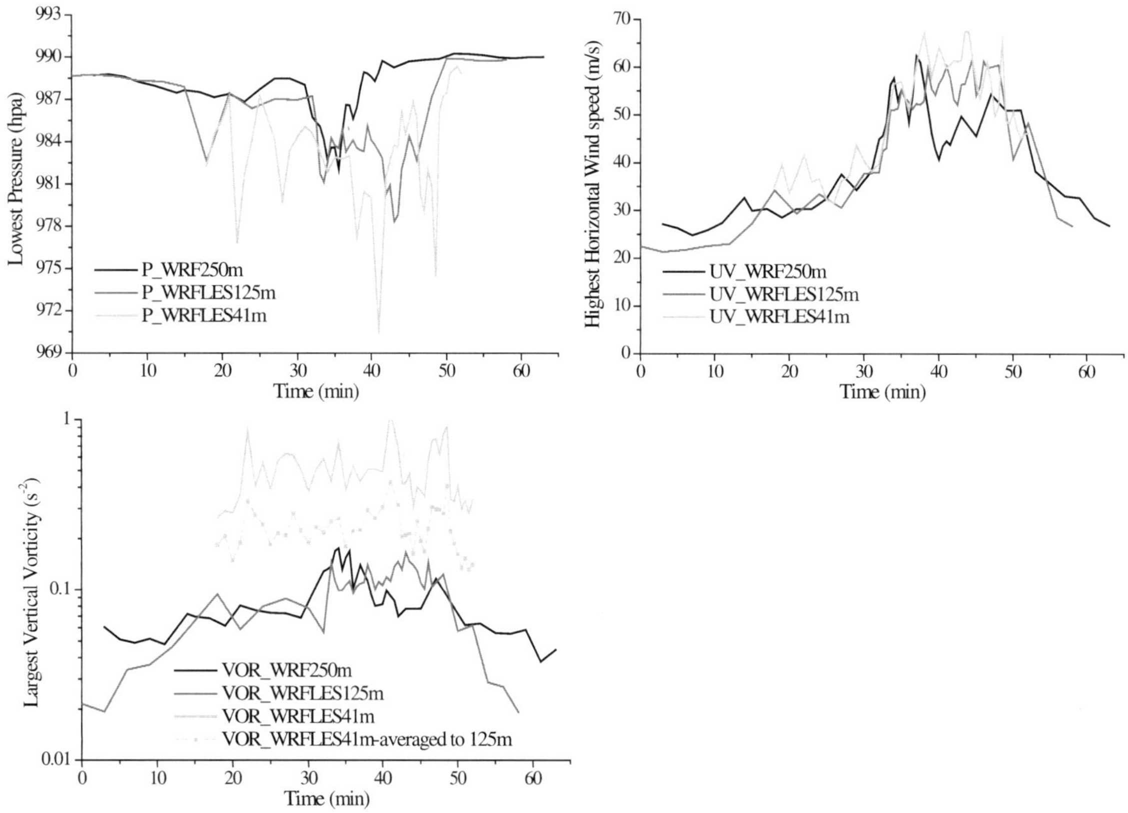

Figure 4. Time series of lowest pressure, maximum horizontal wind velocity, maximum vertical vorticity. The values are calculated at the height of 100 m above the ground, the area is an about 5 km square whose center is determined by largest vertical vorticity. The dotted green line means that the WRFLES41m wind field is averaged to 125 m horizontal resolution grids and vorticity is calculated based on the averaged data. The start time (Time = 0) refers to 02:46.

A simple verification of tornado-like structure genesis can be found from several meteorological variables histories as shown on Figure 4. The maximum wind speed exceeds 65 m/s which reaches the level of EF2 (Enhanced Fujita) level. Maximum vorticity is about 1.0 s^{-1} which is larger than a typical tornado criterion 0.65 s^{-1}. The 125 m grid averaged vorticity from WRF42m is larger than WRFLES125m result that implies more intensified rotation is generated in WRFLES42m. The duration of large wind speed over 60 m/s is extended in WRFLES cases than WRF case which implies the possibility of resolving more strong gust front near the ground. For WRF model, because of greater diffusion and rougher grid size, it is less likely to keep an "outstanding" wind speed for a long time whose sharp gradient will be easily smoothed.

To be mentioned, in sigma coordinate which is used in WRF model, a large skewness of vertical grid may cause truncation error, which easily occurs with large grid aspect ratio and steep terrain (Haney, 1991). This problem can be relieved by removing a reference state from governing equations. In this study, considering the strong convective system with tornadogensis existence, the sharp horizontal pressure gradient around the tornado center may lead to the skewness of vertical level. Considering this possibility, we have checked the grid skewness situation near the "sharp" pressure gradient location (pressure deficit zone), with skewness ratio defined as Eq. 1.

$$R_{skew} = \left| \frac{\sigma}{H} \frac{\partial H}{\partial x} \right| \delta x / \delta\sigma \tag{1}$$

where, the sigma coordinate is defined as $\sigma = z/H$. z, H refer to height above ground and model top height. $\delta x, \delta\sigma$ mean the horizontal grid and vertical grid spacing.

The lowest ten levels around the pressure deficit zone with 100×100 horizontal grids are testified. The maximum R_{skew} is only 3.84%, which is much less than the threshold value of 1.0 (Haney, 1991) that would lead to truncation error. It implies that the possible influence of pressure gradient induced vertical grid skewness is negative in this study.

4.2.2 *The complex 3D rotation structure*

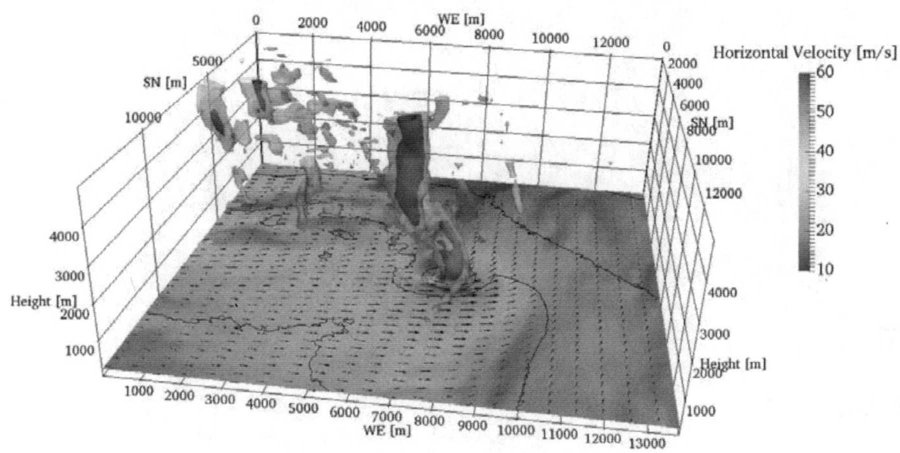

Figure 5. 3D visualization of weak tornado-like structure on WRFLES125m. Contours' descriptions are same as Figure 3.

Figures 5-6 show the vorticity structure of mid-level mesocyclone and lower-level near ground rotation from two WRFLES domains at 03:09. Actually a near ground strong rotation in WRFLES42m is regarded as tornado-like structure because of its vorticity strength (see 4.2.1). The rotation center tracks are also plotted on Figure 1b following the same definition method of WRF250m which last from about 03:03 to about 03:15. The path of WRFLES is deviated to a northern position than WRF250m, possibly because of large interpolation ratio between WRFLES125m and WRF1250m which causes some inaccuracy on WRFLES125m boundary.

Comparing to the rotation structure on WRF250m, the rotation on WRFLES125m is intensified especially at mid-level mesocyclone region on about 3 km high. A low-level mesocyclone-like structure below 2 km high is observed which also stretches to the ground. However, the connection between the mid-level mesocyclone and low-level mesocyclone seems weak. It indicates that this mid-level mesocyclone is not the dominant "pump" for accelerating and maintaining near ground rotation.

From WRFLES42m results shown on Figure 6, more detailed and sophisticated structure can be observed compared to WRFLES125m described previously. Both mid-level (about 3 km high) and low-level (from about 0.5 km to about 2 km high) mesocyclone are clearly resolved. Meanwhile beneath the low-level mesocyclone, a strong rotation structure near ground is observed which is regarded as tornado-like structure.

A complex and weak connection between mid-level mesocyclone, low-level mesocyclone and near ground tornado-like structure is shown. The mid-level mesocyclone seems weakly stretching to the tornado-like structure without directly connected to the low-level mesocyclone. Vertical vorticity of low-level mesocyclone is actually extending to near ground level, while this extended zone is horizontally large and much exceeds the area of tornado-like structure. These structure relationships can be illustrated by the near ground rotation pattern showed on Figures 8-10 on which there is as strong rotation center (tornado-like structure) located in whole near ground rotation (extension from low-level mesocyclone)

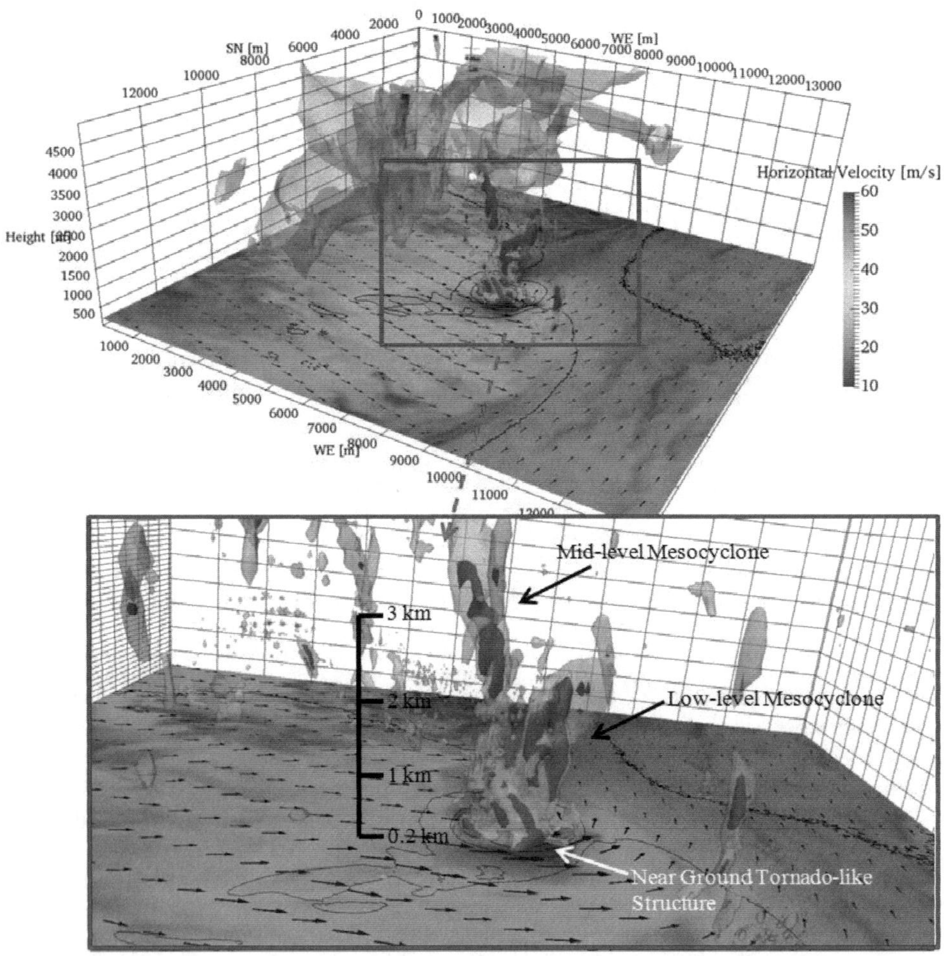

Figure 6. 3D visualization of weak tornado-like structure on WRFLES42m. The vertical vorticity isosurfaces of 0.14 s^{-1} and 0.2 s^{-1} are plotted by 3D contours. Green contour refers to hydrometeors. Other contours' descriptions are same as Figure 3.

4.2.3 *The Asymmetric of real tornado structure*

In this part, the results from WRFLES42m will be used to discuss the tornado structure. Firstly the axisymmetric characteristics are shown on Figure 7. To be mentioned, the wind field rotation center defined by the maximum vertical vorticity on different heights is shifted to same location so that the centers are stacked vertically.

At the radius about 200 m and 450 m, two peak values are shown which are different from common tornado generator experiments. It indicates the complexity of tornado structure whose detail will be described in the following part.

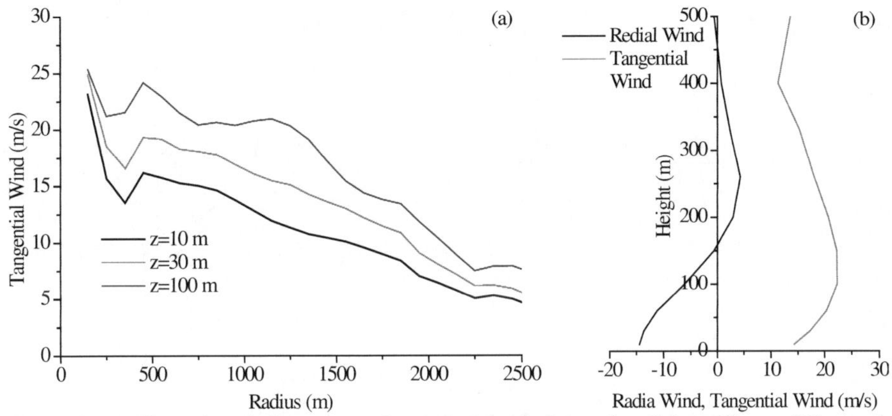

Figure 7. (a) The azimuthally averaged tangential wind speed radial profile at different heights of WRFLES42m. The center is determined by the maximum vertical vorticity location. (b) The vertical profile the radial and tangential winds of the simulated tornado-like structure in WRFLES42m. The wind are azimuthally averaged for 400 m < r < 500 m, where r refers the distance to center determined by the maximum vertical vorticity location at each height.

Figures 8-10 show the horizontal velocity distributions on different heights. The moving speed of rotation center is removed from the horizontal velocity field. The rotation center is determined as the location of maximum vertical vorticity at 100 m high.

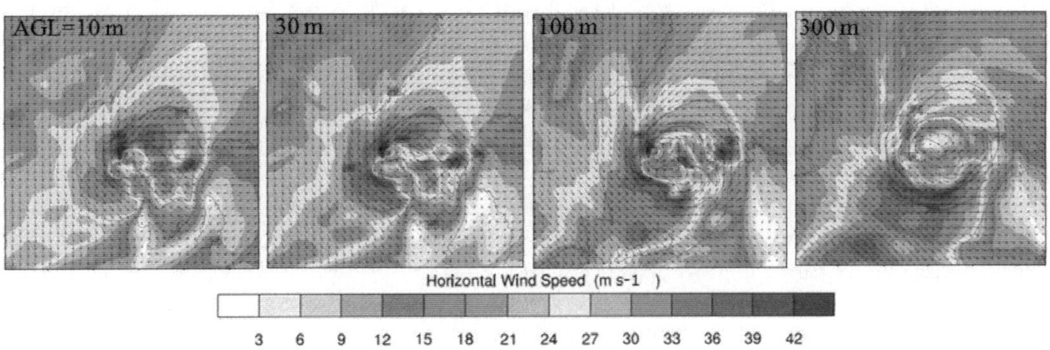

Figure 8. Horizontal wind velocity pattern on different heights above the ground

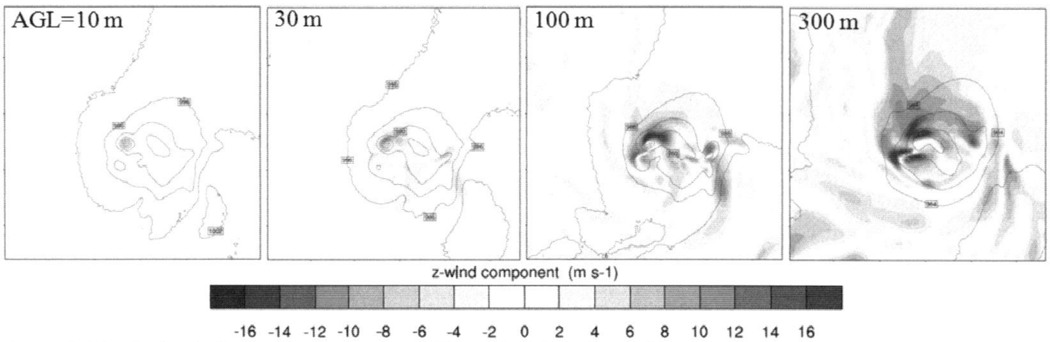

Figure 9. Vertical wind velocity pattern on different heights above the ground

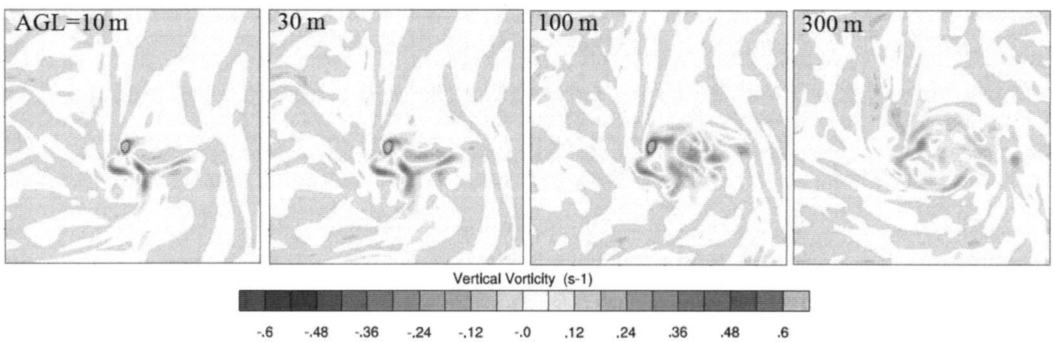

Figure 10. Vertical vorticity pattern on different heights above the ground

Corresponding to the indication mentioned from the discussion of Figure 6, a strong pressure deficit zone concentrated in a small area can be found inside a large but weak deficit zone shown on Figure 8-9. These two pressure deficits zones are actually corresponding the tornado-like structure and low-level mesocyclone extension to ground. In the strong deficit zone both local horizontal velocity maximum and RFD (Rear Front Downburst) structures can also be clearly observed. This feature can also be depicted from the vertical vorticity field on Figure 10 that a strong rotation center near the ground whose maximum vertical vorticity exceeds 1.0 s^{-1} appears corresponding to the strong pressure deficit location. The strong rotation zone is smoothened at about 300 m high which turns into a rotation zone with larger radius corresponding to the extension of low-level mesocyclone. The weak connection between mesocyclone and tornado-like structure may cause the hardness to maintain a strong and concentrated rotation zone.

The asymmetrical characteristics can be obviously observed from both horizontal and vertical wind components. From meteorology point of view, the complex influence from such as RFD and temperature gradient near the ground is very sensitive to the real case tornado generation. However these influences are commonly neglected in ideal tornado wind experiment. This real time tornado simulation provides an example emphasizing these influences.

5 CONCLUSION

Mesoscale meteorological model is employed to simulate a severe storm process with tornadogenesis on Tsukuba city to explore the local wind gust characteristics in a realistic tornado condition. Conclusions can be drawn as below.

By applying nesting method, the grid resolution can be downscaled from kilometers to tens of meters to capture the behavior of local storm and tornado process while also keep the influence from synoptic scale. A tornado process is reproduced in 40 meters resolution by a triple nesting simulation while its rotation concentration needs improvement. A relatively weak mesocyclone with no sufficient updraft force may cause the weak connection between the near ground rotation and mesocyclone.

WRFLES model whose diffusion effect is smaller than PBL tends to result in more complex structures under that circumstance of strong friction and convection. The grid skewness induced by pressure deficit is moderate in this convective storm case.

The complexity of a real tornado structure is described with obvious asymmetry characteristics comparing to typical tornado generator experiments or ideal numerical simulation.

6 ACKNOWLEDGEMENTS

This study was funded by the MEXT project: "Advancement of meteorological and global environmental predictions utilizing observational 'Big Data'" of the social and scientific priority issues (Theme 4) to be tackled by using post K computer of the FLAGSHIP2020 Project (hp160229). The authors' research is also supported by scholarship from China Scholarship Council. The authors acknowledge support and simulation instructions from Wataru Mashiko in Japan Meteorology Agency and helpful suggestions from scholars in post-K projection group.

7 REFERENCES

1 J.B. Klemp, Dynamics of tornadic thunderstorms. Annual review of fluid mechanics, 19(1) (1987) 369-402.
2 P.M. Markowski and Y.P. Richardson, The influence of environmental low-level shear and cold pools on tornadogenesis: Insights from idealized simulations. Journal of the Atmospheric Sciences, 71(1) (2014) 243-275.
3 R.M. Cintineo and D.J. Stensrud, On the predictability of supercell thunderstorm evolution. Journal of the Atmospheric Sciences, 70(7) (2013) 1993-2011.
4 M. Xue, M. Hu and A.D. Schenkman, Numerical prediction of the 8 May 2003 Oklahoma City tornadic supercell and embedded tornado using ARPS with the assimilation of WSR-88D data, Weather and Forecasting, 29(1) (2014) 39-62.
5 S. Yokota, H. Seko, M. Kunii, et al., The Tornadic Supercell on the Kanto Plain on 6 May 2012, Polarimetric Radar and Surface Data Assimilation with EnKF and Ensemble-Based Sensitivity Analysis, Monthly Weather Review, 144(9) (2016) 3133-3157.
6 W. Mashiko, H. Niino and T. Kato, Numerical simulation of tornadogenesis in an outer-rainband minisupercell of Typhoon Shanshan on 17 September 2006, Monthly Weather Review, 137(12) (2009) 4238-4260.
7 W. Mashiko, A Numerical Study of the 6 May 2012 Tsukuba City Supercell Tornado. Part I: Vorticity Sources of Low-Level and Midlevel Mesocyclones, Monthly Weather Review, 144(3) (2009) 1069-1092.
8 J. C. Wyngaard, Toward numerical modeling in the "Terra Incognita", Journal of the atmospheric sciences, 61(14) (2004) 1816-1826.
9 H. Yamauchi, H. Niino, O. Suzuki, Y. Shoji, E. Sato, A. Adachi and W. Mashiko, Vertical structure of the Tsukuba F3 tornado on 6 May 2012 as revealed by a polarimetric radar, (2013) In Proceedings of 36th AMS Radar Conference.
10 R. L. Haney, On the pressure gradient force over steep topography in sigma coordinate ocean models, Journal of physical Oceanography, 21(4) (1991) 610-619.

13th Americas Conference on Wind Engineering
Gainesville, Florida, USA
May 21 - 24, 2017

Study on Most Unfavorable Translating Paths of Tornadoes for Tornado-resistant Design of Civil Structures

Jiahao Zu[a], Guirong Yan[b], Kakkattukuzhy Mathai Isaac[c]

[a]*Department of Civil, Architectural and Environmental Engineering, Missouri University of Science and Technology, Rolla, MO, US, jzfx7@mst.edu*
[b]*Department of Civil, Architectural and Environmental Engineering, Missouri University of Science and Technology, Rolla, MO, US, yang@mst.edu*
[c]*Department of Mechanical and Aerospace Engineering, Missouri University of Science and Technology, Rolla, MO, US, isaac@mst.edu*

ABSTRACT

Tornadoes are one of the most violent natural hazards. They happen very frequently in the United States, with about 1200 tornadoes per year based on the NOAA data. They result in severe property damage and fatalities each year. The devastation from recent tornadoes has left no doubt as to the vulnerability of the Central United States to tornadoes, and prompted a renewed interest to better understand tornadoes, their interactions with civil engineering structures and the performance of structures that lie in the paths of tornadoes. Although the wind effects of tornadoes on cubic low-rise or high-rise buildings have been widely studied, the tornadic wind effects on dome structures have not been sufficiently explored.

Long-span spherical dome structures are usually built for public assembly venues, such as sports stadiums and auditoriums. Because of the large surface area, this type of structure suffers from large wind loads. In order to study the tornado effect on the dome structure which can be used in the structure design, a wind field of Tornado Spencer without a structure in it is simulated first numerically (CFD) and compared with radar-observed data for verification. Then, the translating tornados with a dome structure in it that translate along two different translating paths are studied to find the most unfavorable scenario for structural design. By comparison of pressure, force and moment coefficients between the two paths, different design schemes are proposed for different parts of the dome structure subjected to tornadic wind loads. When designing the dome structure, wind pressure in Path 2 for designing envelops, tornado induced force on dome structure in Path 1, as well as tornado induced moment on dome structure in Path 2 for designing foundations, should be considered, in order to establish an economical and safe structure.

KEYWORDS

CFD, Dome structure, Translation speed, Tornadic effect.

Tornado Rates in China and Post-Disaster Investigation of Tornado in China on June 23, 2016

Rong Gao[a], Qingshan Yang[a], Tian Li[a], Yukio Tamura[a]
[a] Beijing's Key Laboratory of Structural Wind Engineering and Urban Wind Environment, School of Civil Engineering, Beijing Jiaotong University, Beijing, China
e-mail: ronggao@bjtu.edu.cn, qshyang@bjtu.edu.cn, 10231113@bjtu.edu.cn, yukio@arch.t-kougei.ac.jp

ABSTRACT

Tornadoes are the most violent storms that can destroy or seriously damage civil structures and cause huge losses to human societies. Statistical studies and post-disaster investigations of tornadoes have been done in many countries such as America and Japan. In China, both event frequency and tornado strengths are much lower than in America. Moreover, tornadoes occurred mostly in open terrain, which caused neither large losses nor complete records. Thus, less attention was paid to them in early years. However, with increasing urbanization, rural population density is growing, and tornadoes are causing more damage and are experienced by more people than before. Furthermore, as more and more people are using cell phones, witnesses can take photos of tornadoes and post them online, thus focusing increasing attention to them. This paper consists of two parts. The first part gives an overview of tornado occurrences in China in the past decade, and studies their times and spatial distribution. The second part presents a post-disaster investigation of a tornado that occurred on June 23th, 2016, in Yancheng, Jiangsu, China, which was ranked EF4 by the China Meteorological Administration and resulted in the deaths of 99 people and injuries to 846 others. Damage ratings were assigned using both the Enhanced Fujita Scale and the Japanese Enhanced Fujita Scale because some Chinese building forms are excluded from EF-Scale but are included in the JEF-Scale. Failure mechanisms of low-rise brick buildings, which are common in Chinese rural areas, and industrial buildings are studied. Analyses of damage to a road sign, a utility pole and a low-rise building were made to estimate peak wind gust velocities, which were calculated as 51m/s, 47m/s and 73m/s in different studied places. These statistical researches and field studies suggest that the current building standards may need improvement, and structural retrofits in areas with high tornado frequency are also necessary. Meanwhile, a Chinese Enhanced Fujita Scale should be developed to fit Chinese building forms.

KEYWORDS

Tornado, post-disaster investigation, tornado wind speed, F-scale, EF-scale, JEF-scale

The 13th Americas Conference on Wind Engineering (13ACWE)
Gainesville, Florida USA, May 21-24, 2017

An investigation of translation and static pressure effects on tornado-induced peak pressures on low-rise buildings using the DesignSafe cyberinfrastructure

Fred L. Haan, Jr.[a], Partha P. Sarkar[b]

[a] *Calvin College, Grand Rapids, MI, USA, fhaan@calvin.edu*
[b] *Iowa State University, Ames, IA, USA, ppsarkar@iastate.edu*

ABSTRACT: This study used a laboratory tornado simulator to investigate tornado-induced peak pressures on a low-rise building. Specifically, the role of the vortex static pressure and of vortex translation speed was investigated. One primary question being considered was whether or not tornado-induced loading can be explained mainly with straight-line wind aerodynamics with vortex-induced static pressure superposed. By subtracting static pressure from the pressure signals and normalizing all the peak pressures with a dynamic pressure produced from local velocities, this study found that for some cases, straight-line wind aerodynamics might be sufficient. For several building zones, however, the tornado-induced pressures still exceeded what might be expected from straight-line winds. Tornado event duration was found to increase pressure peaks significantly. Increasing duration times from 10 sec to 100 sec, could increase the pressure peaks by 40%. The DesignSafe cyberinfrastructure helped facilitate the analysis phase of this project.

KEYWORDS: tornado, low-rise buildings, static pressure, cyberinfrastructure, DesignSafe

1 INTRODUCTION

This study sought to use a laboratory tornado simulator to investigate how tornado-induced loads on a low-rise building behave. Specifically, behavior with respect to the static pressure generated by the vortex and with respect to the translation of the vortex were investigated. Given the nature of tornado-induced wind loads, one might ask, "Are these loads the same as straight-line winds except for the static pressure induced by the vortex?" Or one might ask, "If I normalize the tornado-induced pressure by a local wind velocity, will I get pressure coefficients similar to that of straight-line winds?" These questions, along with how they are affected by vortex translation speed, were addressed in this study. To conduct this study, the DesignSafe cyberinfrastructure was employed [1]. This cyberinfrastructure (CI) has the capability to analyze and share all data sets in the cloud (www.designsafe-ci.org).

The work reported here was conducted with the Iowa State University tornado simulator (Haan et al. [2]). Numerous studies have been conducted over the past decades on tornado-induced loads on buildings. Chang [3] and Jischke [4] were among the first to use laboratory tornado simulators specifically to quantify loads on buildings. Mishra [5] reported significant differences in loading between straight-line boundary layer tests and tornado simulator tests. Haan et al. [6] found tornado-induced roof pressure coefficients that were 50-60% larger than ASCE 7 provisions. Integrated uplift loads were found to be 2 to 3 times that predicted with ASCE 7 provisions. The large differences between those vertical loads and ASCE 7 are part of the motivation for the present work. Kikitsu et al. [7] examined the role of internal pressure along with external pressure and found that total uplift loading on a building can reduce significantly depending on the size and orientation of holes in the building envelope. Sabareesh et al. [8,9] used a tornado simulator to examine pressures on low-rise buildings with respect to ground roughness. Ground roughness was reported to increase internal and external pressures. Although internal pressure plays an important role in tornado-induced loading, this study focused on identifying the behavior of external pressures induced by tornado vortices. The ISU simulator was the largest of its kind when built, but the VorTECH facility at Texas Tech and the WindEEE Dome at the University of Western Ontario (Refan and Hangan [10]) are significantly larger and hold the promise of testing higher Reynolds numbers and larger surface fetch than has been possible before.

2 TORNADO SIMULATOR TEST CONDITIONS

As stated earlier, this project was conducted in the Iowa State University tornado simulator. The building model was made of acrylic and had the plan dimensions (W) of 91mm by 91mm. The gable roof of the building had an angle of 35°, a maximum height of 66mm and an eave height of 36mm. The model was fabricated with 89 pressure taps arranged and numbered as shown in Figure 1. An electronic pressure scanner was used to acquire data from the taps. All vortices translated in the positive x-direction and rotated counter-clockwise (also indicated in Figure 1).

The vortex used for this project is a low-swirl, single-celled vortex (referred to as the Vane1 vortex in Haan et al. [2]. Velocity measurements were conducted with a TFI Cobra probe sampling at 2500 Hz. All pressure measurements were conducted with Scanivalve electronic pressure scanners sampling at 430 Hz. Figure 2 illustrates the experimental setup and shows the coordinate system definition and the velocity vector definitions for tangential (V_θ) and radial (V_r) components (V_z is the vertical component and is not illustrated). The next section of this paper presents the velocity field measurements.

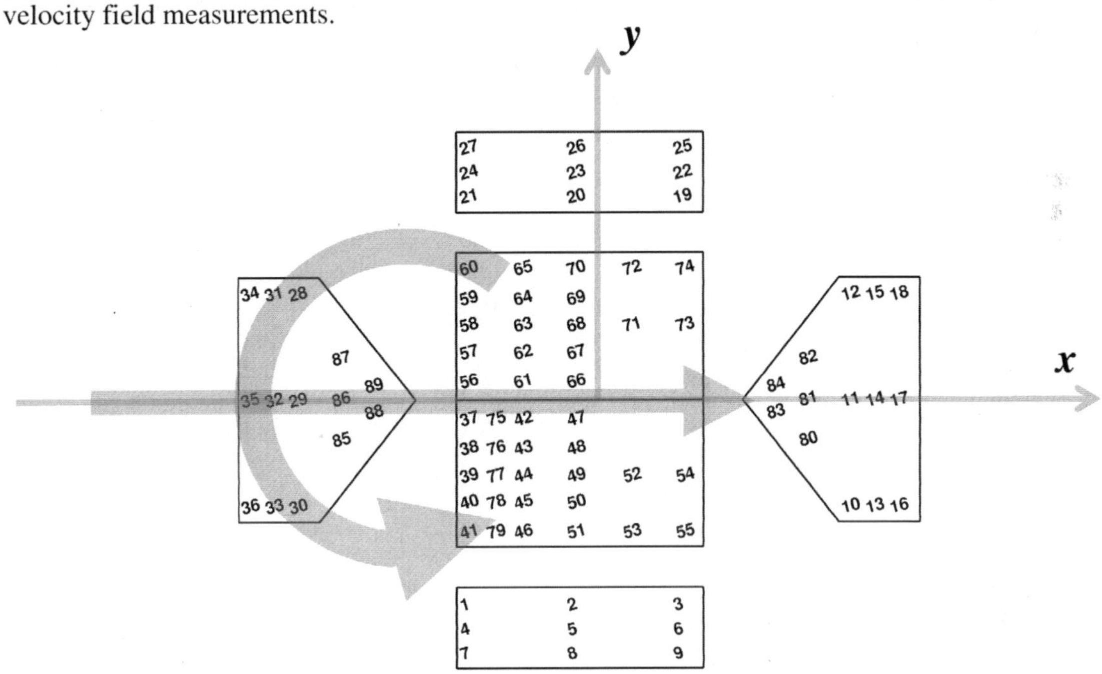

Figure 1. Pressure tap numbering for the low-rise building model including the tornado rotation and translation direction.

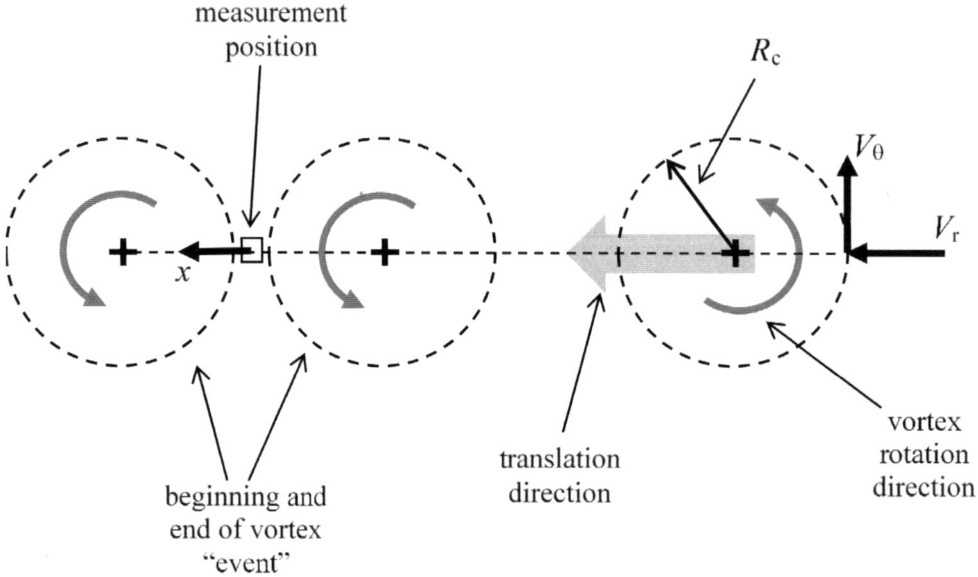

Figure 2. Schematic diagram of the test setup defining the coordinate system with respect to the measurement position. The measurement position was used for the both Cobra probe velocity measurements and building model pressure measurements. Also illustrated is the nomenclature for velocity measurement and the definition of a vortex "event," that is, the time it takes for a vortex to traverse the measurement position.

3 TORNADO-INDUCED VELOCITY AND STATIC PRESSURE

3.1 *Velocity measurements*

Velocity measurements were conducted as the vortex translated past the measurement position. The velocity profiles presented come from measurements with vortex translating at 0.61 m/s (although the building pressure tests in this project considered translation tests from 0.15 m/s to 0.61 m/s). The velocity signals were broken into 0.1 sec time segments. Within each segment a mean and peak value was found. These values were averaged over five vortex translation trials to produce the plots in Figures 3 and 4. Figure 3 is the horizontal velocity (V_H) computed as the vector sum of the radial and horizontal components. Figure 4 shows the averaged and peak values of the vertical velocity (V_Z). It should be noted how large these vertical velocities can be. Inside the vortex core, peak values can be 60-70% of the mean horizontal velocity. All positions in this paper are reported at x/R_c positions where x is defined from the measurement position as shown in Figure 2 and R_c is the vortex core radius (0.22m for this study).

The Reynolds numbers formed using the horizontal velocity and the building model height ranged from from 1.8×10^4 away from the core to 4.4×10^4 at the core radius.

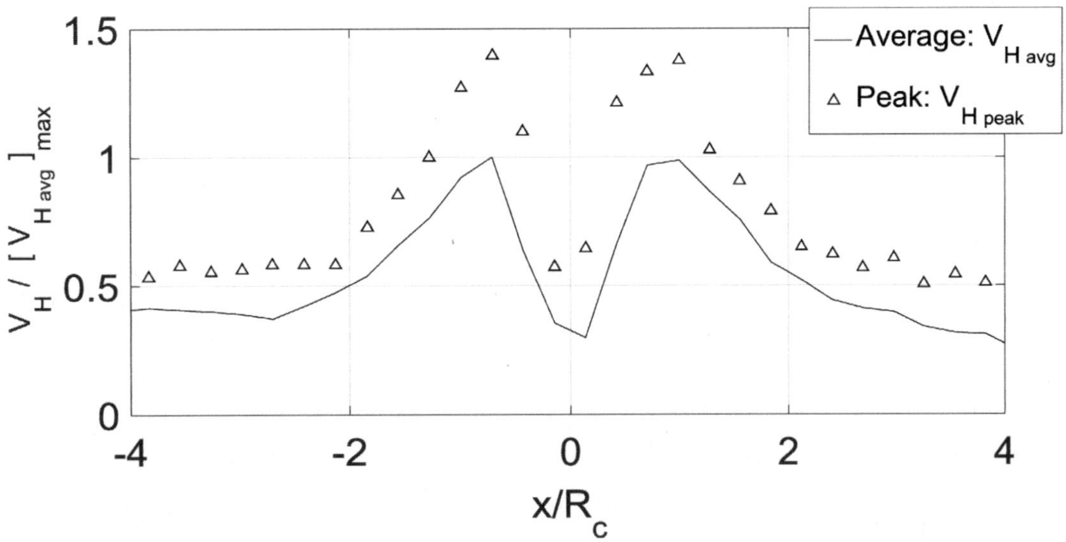

Figure 3. Horizontal velocity profile at building eave height as the vortex translates by. Each average and peak value corresponds to a local 0.1 sec bin averaged over five separate vortex translations. The horizontal velocity is the vector sum of the radial (V_r) and tangential (V_θ) components of velocity. Vortex translation speed was 0. 61 m/s.

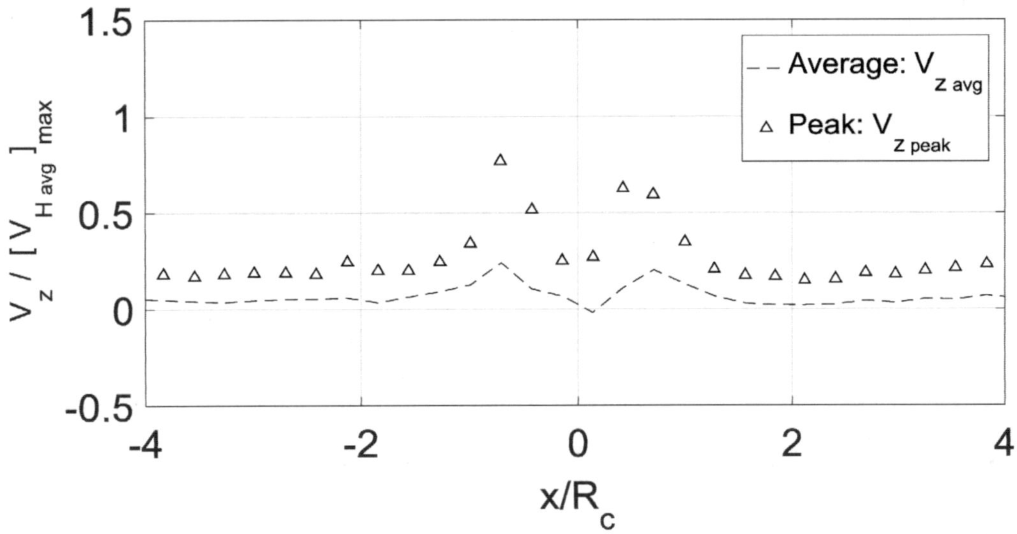

Figure 4. Vertical velocity profile at building eave height as the vortex translates by. Each average and peak value corresponds to a local bin averaged over five separate vortex translations. Vortex translation speed was 0. 61 m/s.

3.2 *Static pressure measurements*

Figure 5 presents the static pressure generated by the vortex. The static pressure presented here is the average of all the building model pressure taps. This estimate of the static pressure was compared to that measured on the ground plane of the simulator and to that measured with the Cobra probe. They were all consistent. Like the velocity measurements of the previous section, this static pressure was divided into 0.1 sec bins to form average and peak values as well. The static pressure peaks near the core could be 25% higher than the average values.

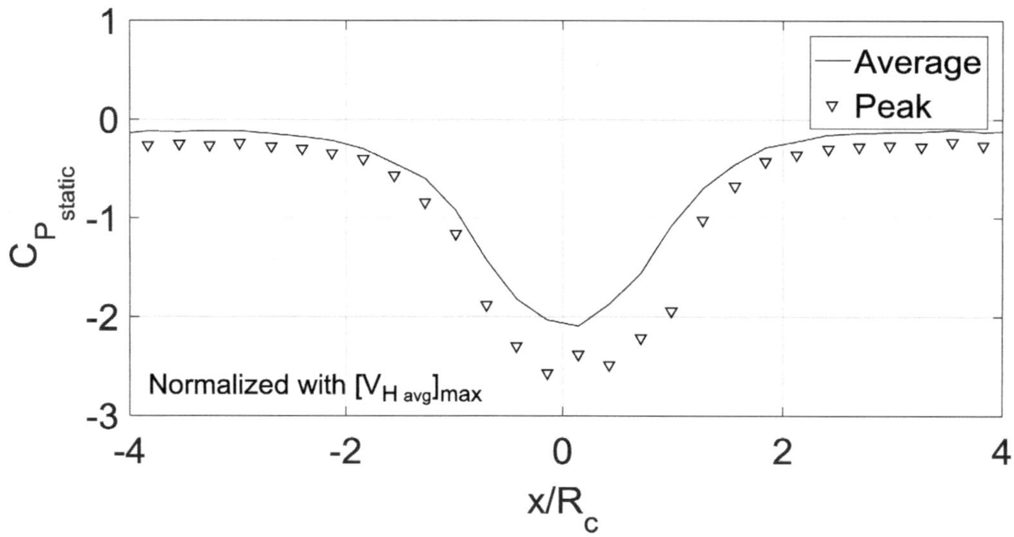

Figure 5. Static pressure induced by the vortex. Average and peak values were found from dividing the original signal into 0.1 sec segments. Vortex translation speed was 0. 61 m/s.

4 TORNADO-INDUCED PRESSURES ON A LOW-RISE BUILDING

4.1 *Pressure analysis*

To address the question of how similar vortex-induced pressures are to those induced by straight-line wind, the vortex static pressure was subtracted from all building model pressures, and those building model pressures were normalized by a local dynamic pressure to account for the transient nature of the vortex translation. Analytically this can be described as shown in Equation (1) below. Figure 6 illustrates the subtraction of the static pressure by showing a plot of $C_p(x)$ for pressure tap #63 both with and without the static pressure. All pressures in this figure are normalized with the maximum value of $V_{Hpeak}(x)$. Figure 6 also includes a comparison with ASCE 7 provisions [11] for this location on the roof.

$$C_p(x) = \frac{(p(x)-p_{ref})_{peak}}{\frac{1}{2}\rho\left[V_{H_{peak}}(x)\right]^2} - \frac{p_{static}(x)-p_{ref}}{\frac{1}{2}\rho\left[V_{H_{peak}}(x)\right]^2} = \frac{(p(x)-p_{ref})_{peak}}{\frac{1}{2}\rho\left[V_{H_{peak}}(x)\right]^2} - C_{p_{static}}(x) \quad (1)$$

where $p(x)$ is the pressure signal, p_{ref} is the reference pressure for the measurements (in this case, this was a location in the lab outside the tornado simulator), $p_{static}(x)$ was the static pressure induced by the vortex and estimated as the average of all the pressure taps on the building.

Figure 7 illustrates the normalization of the pressures with local values of $V_{Hpeak}(x)$. In this figure, the pressure signals (one of which is shown in Figure 6) were divided in 0.1-sec segments and normalized as shown in Equation (1). Median values and confidence intervals are shown in the figure. The statistics are based on 10 runs of the same configuration. From this figure, one can see that the worst negative peak pressure occur before the center of the vortex core hits the structure, that is, for an x value around $-1.3R_c$. The worst value of peak pressure occurs at different tornado vortex positions for different locations on the structure. Figure 8 summarizes the x/R_c values at which the worst case pressure occurs for the entire structure.

Some but not all of the patterns of Figure 8 are somewhat intuitive. Many of the worst case pressures on the roof or about the corners occur when the angle of the incident wind would be expected to create separated regions about those building locations. Figure 9 shows zones that have been delineated to identify areas where pressure behavior is somewhat consistent. For example, pressures at tap #63 (shown in Figures 6 and 7) show the largest negative values when the vortex center has not yet reached the model and the roof region of tap #63 would be expected to be leeward to the incident wind.

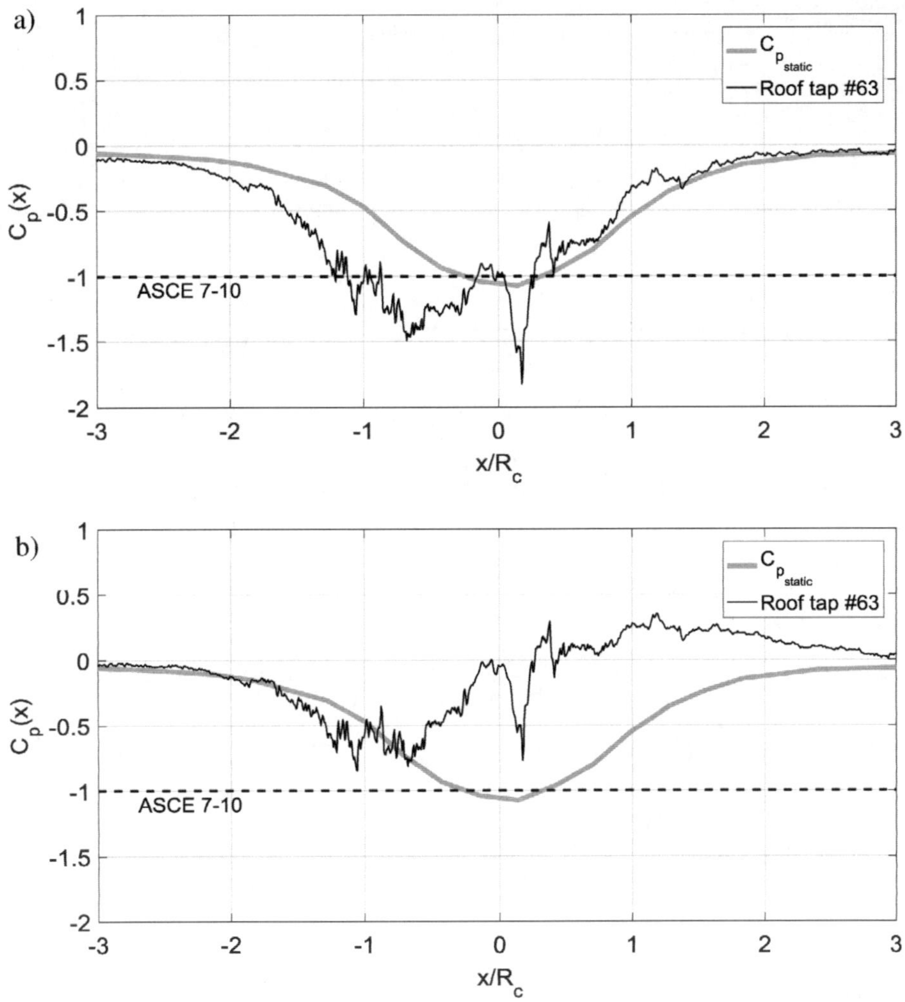

Figure 6. Pressure coefficient for roof tap #63. (a) Original signal with the static pressure included. (b) Static pressure removed according to Equation 1. Both of these signals came from tests at the maximum vortex translation speed of 0.61 m/s. The static pressure and the roof pressure were both normalized using the maximum value of $V_{Hpeak}(x)$.

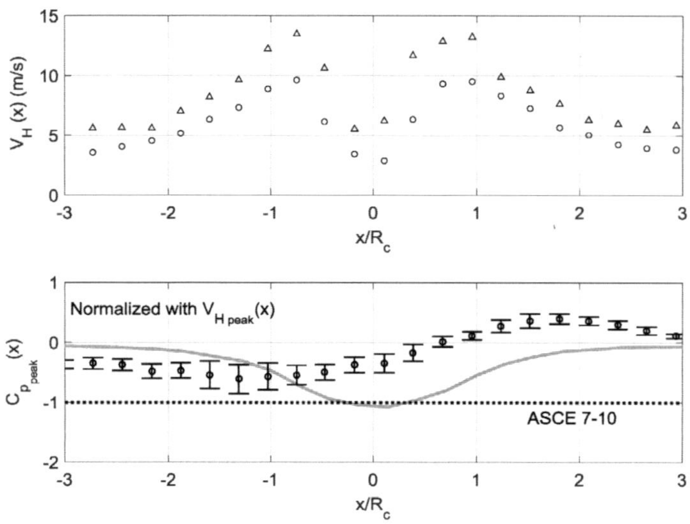

Figure 7. (a) Local horizontal velocity and (b) local peak pressures (from tap #63) normalized with that horizontal velocity. Pressure coefficient for roof tap #63. In the pressure plot, the circles represent the median peak value and the error bars represent 95% confidence intervals.

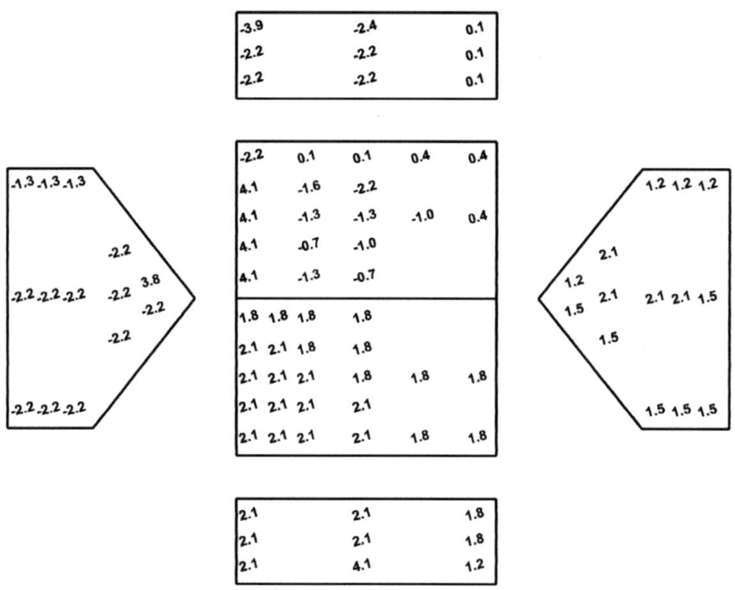

Figure 8. Each number on this plot represents the position of the tornado vortex (represented at x/R_c values where R_c is the vortex core radius) when the worst case negative pressure occurs on the building (after subtracting the static pressure from each pressure). This is for maximum vortex translation speed (0.61 m/s).

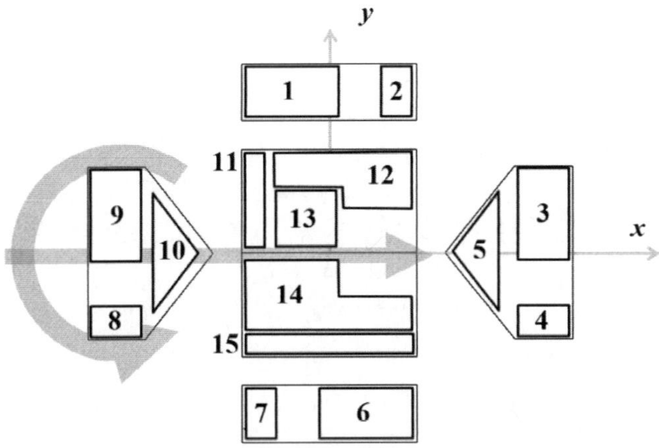

Figure 9. Zones where pressure peak behavior is somewhat consistent in terms of when peaks occur.

4.2 *Pressure analysis*

This section presents data from various zones in Figure 9 to illustrate the different behavior identified for different parts of the building. Figure 10 shows data from pressure tap #77 in zone 14. This might be considered the opposite of tap #63. The behavior shows the most negative peaks coming when the vortex has passed the building the zone 14 would be expected to experience flow separation. The behavior of pressures at tap #84 in zone 5 is a little harder to interpret (see Figure 11). One might expect flow from a vortex at either $+1.0R_c$ or $-1.0\ R_c$ to create a separated region in zone 5, the worst pressures occur beyond $+1.0\ R_c$. This may be caused by a separated region induced by the static pressure. It is not clear without more study.

Data from tap #73 in zone 12 is shown in Figure 12 and shows a different location for the worst pressure peaks than previous zones. This pressure tap experienced its worst pressure when the vortex core was on top of the building. While this might be expected for any pressure tap if the static pressure was still part of the signal, it is not clear why this should happen at this location. This tap location is not clearly in a separated region and yet its pressure peaks exceed what we might expect from ASCE 7.

Clearly there are some aerodynamic mechanisms that must still be investigated. The pressures do not appear to be merely due to straight-line wind with a static pressure superposed.

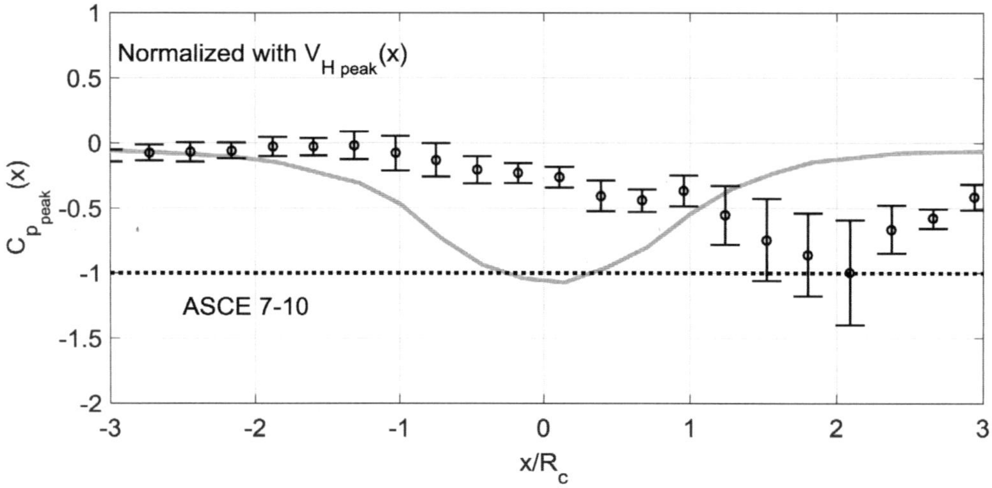

Figure 10. Local peak pressures from tap #77 normalized with local horizontal velocity. In the pressure plot, the circles represent the median peak value and the error bars represent 95% confidence intervals.

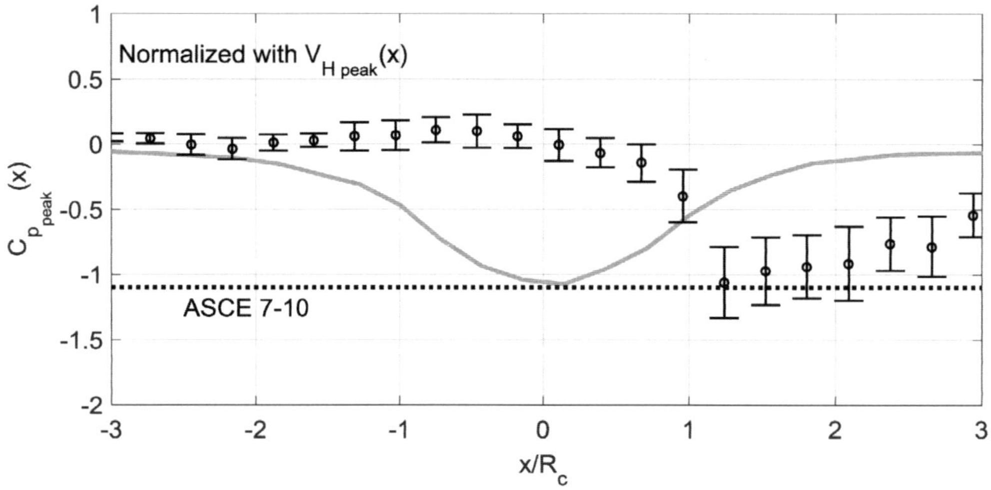

Figure 11. Local peak pressures from tap #84 normalized with local horizontal velocity. In the pressure plot, the circles represent the median peak value and the error bars represent 95% confidence intervals.

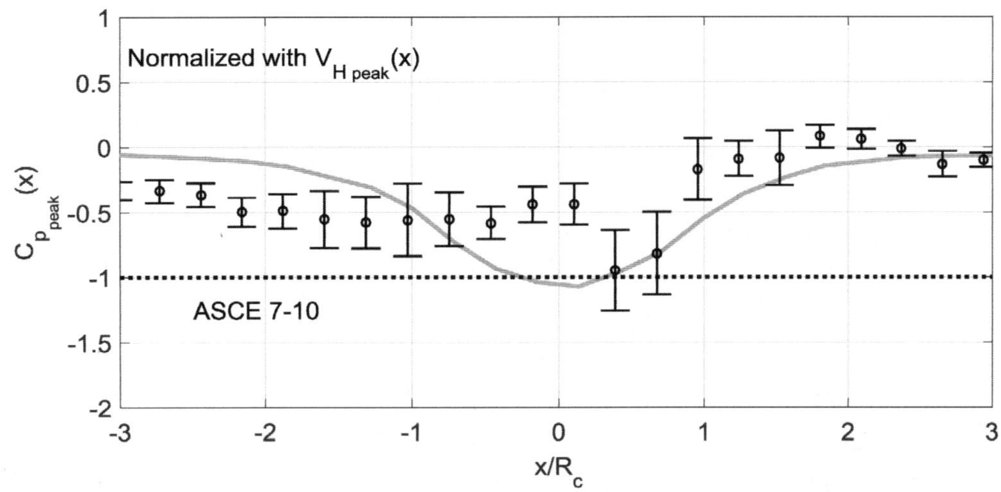

Figure 12. Local peak pressures from tap #73 normalized with local horizontal velocity. In the pressure plot, the circles represent the median peak value and the error bars represent 95% confidence intervals.

5 TRANSLATION EFFECTS ON TORNADO-INDUCED PRESSURES

Two different approaches were used to study duration effects. One method considered pressure peaks observed for different vortex translation speeds. A second method considered a stationary vortex and pressure peaks observed over different amounts of time of exposure to that stationary vortex.

The method of observing peaks for different vortex translation speeds is similar to what was done for Haan et al. [6], but here the analysis was conducted differently. In this analysis, the duration of the vortex event was estimated as the time that a point on the building is exposed to the vortex core, that is, the time it takes the vortex to translate past a single point (as illustrated in Figure 2). The duration, τ, then was estimated as follows in equation (2):

$$\tau = \frac{2R_c}{V_t} \tag{2}$$

where V_t is the vortex translation speed. With this study's translation speeds of 0.15 m/s to 0.61 m/s, this resulted in duration values from 0.7 sec. to 2.9 sec. Each translation speed was tested 10 times, peak C_p values were found for each of these 10 trials. A median peak value was found for the 10 trials for each translation speed.

For the stationary vortex cases, data were divided in a manner similar to that employed by Kopp and Morrison [12]. A time series of 48 seconds was acquired for each stationary vortex. This time series was divided into segments as long as 10 sec. and as short as 0.1 sec. Given the nature of turbulent flow pressure fluctuations, the longer time segments should result in larger peak pressures. The peak pressure was obtained for each segment and the median peak was found for the entire 48 sec. time series.

All the peak values calculated as described above (for translating and stationary vortices) were normalized with the median pressure coefficient for the entire stationary vortex time series. These values are all presented in Figure 13 for pressure tap #63. As the duration increases, the peak pressure coefficient increases. The rate of increase is somewhat steeper when the vortex is stationary. For this particular pressure tap, an increase in duration from 10 sec to 100 sec increase the pressure peak by more than 40%. More details of this analysis along with a discussion of this behavior over the entire structure can be found in Haan [15].

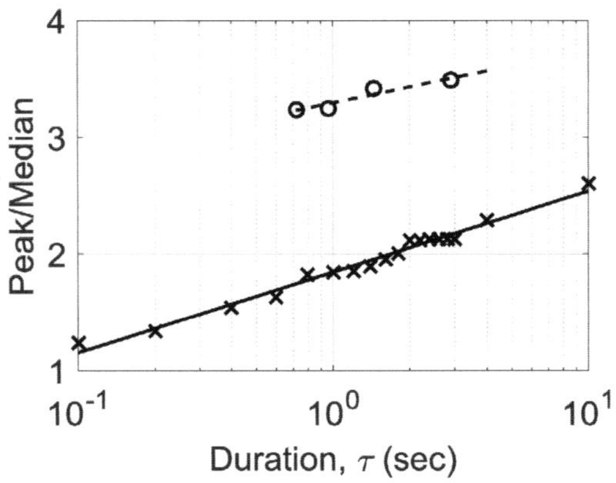

Figure 13. Peak Cp to median Cp ratios as a function of tornado event duration for pressure tap #63.

6 CONCLUSIONS

This study investigated the role of tornado-induced static pressure and duration on peak pressures on a low-rise building. Studies such as Haan et al. [6] have suggested that peak pressures on buildings in tornadoes were up to 50% (or more) higher than straight-line ABL values, and this study showed that much but not all of this increase can be explained by the static pressure of the vortex. Subtracting the static pressure from pressure time series and normalizing by a local horizontal peak velocity brought peak pressures closer to what one would expect from straight-line ABL flows. However, the data showed that peaks in some portions of the building could still be larger the ASCE 7-10 provisions.

To consider duration effects, pressure peak results from both translating vortex and stationary vortex tests were considered. Increases in duration were found to increase peak pressure coefficient magnitudes. Depending on the zone of the building, changing duration from 10 sec. to 100 sec. was found to increase peak magnitudes significantly from 10% to 40%. Work such as this might lead to factors that could adjust tornado pressure coefficients for the effect of event duration.

This project only considered results from a single-celled, low-swirl vortex. Future work must investigate the effects of other tornado parameters such as tornado diameter relative to building size and tornado structure. There are several aerodynamic mechanisms that still need to be investigated before tornado-induced loading can be understood relative to loading from straight-line winds.

7 ACKNOWLEDGEMENTS

The authors gratefully acknowledge the support of the U.S. National Science Foundation under grants 0220006 and 0239070, the work of Bill Rickard, and the work of numerous undergraduate assistants at Iowa State University.

8 REFERENCES

1 E.M. Rathje, C. Dawson, J.E. Padgett, J.P. Pinelli, D. Stanzione, A. Adair, P. Arduino, S.J. Brandenburg, T. Cockerill, C. Dey, M. Esteva, F.L. Haan, Jr., M. Hanlon, A. Kareem, L. Lowes, S. Mock, G. Mosqueda, DesignSafe: New Cyberinfrastructure for Natural Hazards Engineering, Natural Hazards Review, v. 18, n. 3 (2017).
2 F.L. Haan, Jr., P.P. Sarkar, W.A. Gallus, Design, construction and performance of a large tornado simulator for wind engineering applications, Engineering Structures, 30, (2008) 1146-1159.
3 C.C. Chang, Tornado wind effects on buildings and structures with laboratory simulation. In: Proceedings of the 3rd International Conference on Wind Effects on Buildings and Structures (1971) 231–240.
4 M.C. Jischke, B.D. Light, Laboratory simulation of tornadic wind loads on a rectangular model structure. Journal of Wind Engineering and Industrial Aerodynamics, 13 (1983) 371-382.
5 A.R. Mishra, D.L. James, C.W. Letchford, Physical simulation of a single-celled tornado-like vortex, Part B: Wind loading on a cubic model. Journal of Wind Engineering and Industrial Aerodynamics, 96 (2008) 1258-1273.
6 F.L. Haan, Jr., V.K. Balaramudu, P.P. Sarkar, Tornado-Induced Wind Loads on a Low-Rise Building. ASCE J. Struct. Eng. v. 136, n. 1 (2010) 106-116.
7 H. Kikitsu, P.P. Sarkar, F.L. Haan, Experimental study on tornado-induced loads of low-rise buildings using a large tornado simulator. 13th International Conference on Wind Engineering Amsterdam, Netherlands (2011).
8 G.R. Sabareesh, M. Matsui, Y. Tamura, Characteristics of internal pressure and resulting roof wind force in tornado-like flow, Journal of Wind Engineering and Industrial Aerodynamics, 112 (2013) 52-57.
9 G.R. Sabareesh, M. Matsui, Y. Tamura., Ground roughness effects on internal pressure characteristics for buildings exposed to tornado-like flow", Journal of Wind Engineering and Industrial aerodynamics, 122 (2013) 113-117.
10 M. Refan, H. Hangan, Characterization of tornado-like flow fields in a new model scale wind testing chamber. Journal of Wind Engineering and Industrial Aerodynamics, v. 151, April 2016 (2016) 107-121.
11 ASCE (2010). Minimum Design Loads for Buildings and Other Structures. American Society for Civil Engineers, Reston, VA, USA.
12 G.A. Kopp, M.J. Morrison, Discussion of "Tornado-Induced Wind Loads on a Low-Rise Building" by F.L. Haan Jr., V.K. Balaramudu, and P.P. Sarkar. ASCE Journal of Structural Engineering, v. 137, n. 12 (2011) 1620-1624.
13 F.L. Haan, Jr., An examination of static pressure and duration effects on tornado-induced peak pressures on a low-rise building, Frontiers in Built Environment, v. 3, April (2017) doi.org/10.3389/fbuil.2017.00020.

13th Americas Conference on Wind Engineering
Gainesville, Florida, USA
May 21 - 24, 2017

Empirical Validation of Engineering-based Tornado Fragility Functions

David B. Roueche[a], David O. Prevatt[b], Frank T. Lombardo[c], Fred L. Haan[d]

[a] *University of Florida, Gainesville, FL, USA, david.roueche@ufl.edu*
[b] *University of Florida, Gainesville, FL, USA, dprev@ce.ufl.edu*
[c] *University of Illinois at Urbana-Champaign, Champaign, IL, USA, lombaf@illinois.edu*
[d] *Calvin College, Grand Rapids, MI, USA, flh2@calvin.edu*

ABSTRACT

Research on tornado-induced wind loads on low-rise structures has advanced considerably with the advent of larger laboratory tornado simulators, improved computational fluid dynamics methods and resources, and more detailed post-tornado damage investigations. However, to date there has been little interaction or comparison between these various research efforts.

In this paper we compare tornado-induced wind loads obtained using a laboratory tornado simulator at Iowa State University to empirical observations of the response of wood frame residential structures to the 2011 Joplin, MO tornado. This comparison is made by means of tornado fragility curves, which define the conditional probability of a damage state being met or exceeded for a given tornado wind speed intensity. Fragility curves are compared for limit states corresponding to the loss of roof sheathing panels and the failure of roof-to-wall connections.

The empirical tornado fragility curves are derived using observed damage levels to 1,241 residential structures in the 2011 Joplin, MO tornado. Damage levels correspond to the Degrees of Damage for 1- and 2-story residential structures in the Enhanced Fujita (EF) Scale. Peak wind speeds are estimated at each location of a damaged structure using a tornado wind field model conditioned on tree-fall patterns also observed in the 2011 Joplin, MO tornado.

Analytical tornado fragilities are developed through a Monte Carlo simulation framework. Tornado-induced loads are taken from the laboratory tornado simulator at Iowa State University and resistances for wood roof sheathing and roof-to-wall connections are taken to represent typical structures damaged in the Joplin tornado. Ten thousand simulations are performed, randomly varying the location of the building model from the center of the tornado, the orientation of the building model, the tornado-induced wind loads for a given location and building orientation, and the resistance of the structural components of the model. Analytical fragilities are developed from the output of the simulations and compared with the empirical fragilities from the Joplin, MO tornado.

The results of this paper demonstrate a methodology for validating tornado-induced wind loads from experimental simulations against post-tornado damage data. Agreement between analytical and empirical models of tornado-induced wind loads is evaluated. The research highlights the benefits of additional validation studies as tornado research advances.

KEYWORDS

Tornado, residential structures, fragility functions, empirical models

The Influence of Tangential to Translational Velocity Ratio on Tornado Force Coefficients on Building Using CFD

Majdi A. Yousef[a], R. Panneer, Selvam[b]

[a]*University of Arkansas, Fayetteville, Arkansas, USA, myousef@uark.edu*
[b]*University of Arkansas, Fayetteville, Arkansas, USA, rps@uark.edu*

ABSTRACT: The ratio (V_θ/V_t) of tangential (V_θ) to translational (V_t) velocity reported in recent studies is 10 or greater, which is larger than the field observation ratios. In this work the effect of V_θ/V_t ratio on tornado force coefficients on building will be investigated with systematic study. The V_θ/V_t ratios are considered to be 1, 3, 6 and 8 for comparison. These ratios are very much in agreement with field observation ratios. The Computational Fluids dynamics (CFD) Model will be used to calculate the effect V_θ/V_t ratio on tornado force coefficient for a dome, cubic and prism building. The effect of grid resolution in the domain is considered. The magnitudes of the forces were found to be larger for slower translation speed or higher V_θ/V_t ratios. For faster translation speeds or, lower V_θ/V_t ratio, the maximum force coefficients shifted to the left of the time history.

KEYWORDS: CFD, Tornado Simulator, tornado force coefficients, tangential and translational velocity, dome and prism building

1 INTRODUCTION

Every year in the United States, about 1,200 tornadoes cause 60-65 fatalities, 1,500 injuries and at least 400 million dollars in economic damage, as reported by the American National Weather Service [1]. The U.S. Natural Hazard Statistics considers tornado losses as the second largest loss next to flood [2]. In order to mitigate this damage, it is necessary to design buildings that are more resistant to tornadoes. Tornadoes produce different type of wind forces than a Straight-Line (SL) wind. The first requirement for accomplishing this goal is a better understanding of tornado-structure interaction and tornado-induced loads on buildings. The development in tornado wind modeling can lead to a better prediction of tornado maximum forces. Then, the outcome can be implemented for improving the design standards.

The challenges to understand the tornado-structure interaction date back to 1970. In-site measurements of tornadic winds around a structure were valuable to assess the actual wind effects [3]. It was reported that it difficult to acquire in site measurements [4]. Thus, researchers have started studying the tornadic wind fields on structures in laboratory tornado simulators or using CFD. Several studies utilize laboratory and computer tornado simulators to study tornado force and pressure on building as summarized in Table 1. Most of the work was on one or two tornado translation speed of tornado effect on building forces. In addition, the V_θ/V_t ratios were reported in recent studies that have measured the wind loads on low-rise buildings in simulated tornadoes is 10 or greater, which is larger than the field observation ratios. In this work, the effect of V_θ/V_t ratio on tornado force coefficients over buildings will be investigated with systematic study.

In this work, the effect of V_θ/V_t ratio on tornado force coefficients over buildings will also be investigated with systematic study. The V_θ/V_t ratios are considered to be 1, 3, 6 and 8 for comparison. These ratios are very much in agreement with field observation ratios. The V_θ/V_t ratio average from real tornadoes has been reported to be from 1.0 to 8.0 [16]. The CFD computer model based on Rankine Combined Vortex Model will be used to calculate the effect V_θ/V_t ratio on tornado force coefficient for dome (DM1), cubic (CM1) and prism (PM1) building.

Table 1. Influence of the different V_θ/V_t ratios on tornadic wind fields on structures

Reference	$V\theta/Vt$	Building Shape	Model	Cx	Cy	Cz
[5]	35	Tall Cube	Exp	2.01	2.01	1.77
	18			1.78	1.78	1.66
[6]	78	Gable roof	Exp.	0.75	1.20	2.4
	26			0.70	1.00	2.0
[7]	40	Cube	Exp.	1.97	1.97	1.24
	20			1.82	1.82	1.22
[7]	40	Tall Cube	Exp.	2.17	2.17	1.54
	20			1.75	1.75	1.78
[7]	40	Tall Cube	Num.	2.01	2.01	1.77
	20			1.78	1.78	1.66
[7]	40	Cube	Num.	1.57	1.57	1.09
	20			1.4	1.4	0.98
[8]	80	Gable roof	Exp.	1.1	1.2	3
[9]	18	Gable roof	Exp.	0.9	0.7	2.8
[10]	24	Tall cube	Exp.	2.0	0.4	0.7
[11]	2	Cube	Num.	0.82	1.36	1.81

2 COMPUTER MODEL

The Navier-Stokes (NS) equations are solved to compute the flow around the structure. The flow equations are approximated by either Finite Different Method (FDM) or Finite Element Method (FEM). The FDM code to study flow over cuboids, rectangular prisms and cubes has been used [12], [13] and [14]. This is based on orthogonal grid system, and it is computationally very efficient. The FDM code is used in this work to study flow over prisms. However, the FDM code had more error in transporting the tornado like vortex when it used to study flow around a dome. Therefore, the FEM code based on body fitted grid was developed to study flow around a dome. The superiority of FEM to FDM in transporting vortices is reported [15]. The FEM code to study the tornado-terrain interaction is used [16]. They validated this numerical model by comparing the results with wind tunnel measurement. The FEM code is used in this work to study flow around a dome. The FEM code takes more computer time and hence parallel computing is utilized. More detail about parallel computing and the methods of solving the NS equations are reported [16].

3 TORNADO NUMERICAL MODEL

Tornado wind field models provide the velocities at different times as a tornado move. This model needs to satisfy the Navier-Stokes equations. The Rankine Combined Vortex Model (RCVM) is the simplest model that can satisfy the Navier Stokes (NS) equations [17].This model consists of two different flow fields, which are the force vortex region and free regions vortex as shown in Figure 1. In the force vortex region, the tangential velocity of tornado, V_θ, increases linearly with the distance from the tornado center, r, where $r \leq r_t$. In the free vortex region, the tangential velocity is decreasing inversely to the radius the region where $r > r_t$. Equations representing the two fields are shown in Figure 1. The turbulence is modeled using Large Eddy Simulation (LES). Then the approximated equations are solved and explained using a semi-implicit method [16].

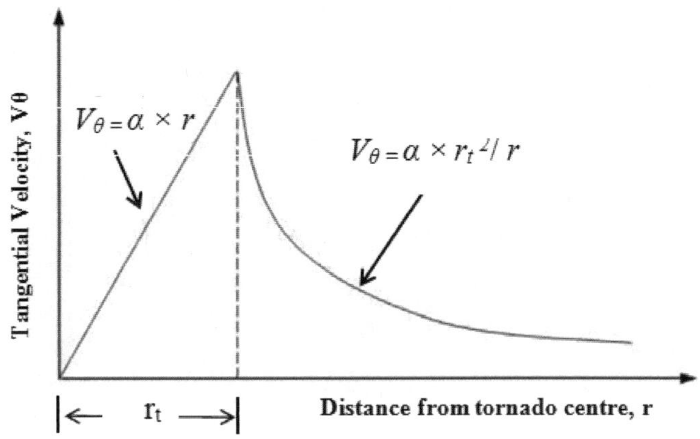

Figure 1. Rankine combined vortex model

4 PROBLEM GEOMETRY AND BOUNDARY CONDITIONS

In this study, the physical modeled is the parallel interaction of a vortex with a long, rigid, structure (e.g. dome, cube and prism) that is immersed in a fee steam. The geometry of the problem is illustrated in Figure 2. The counterclockwise rotating vortex travels along the x-axis with a constant velocity, V_t and here it is taken as translational velocity. The vortex flow and free stream of a constant velocity is smoothly introduced into the computational domain. The numerical calculations are conducted based on the dimensionless values. The fluid density ($\rho = 1$), fluid kinematic viscosity ($v = 1/150$). The dimensions of dome and prism are listed in Table 2. The simulated vortex travels with a translational velocity (V_t) of 3.0, 1.0, 0.5 and 0.375 unit (30, 10, 5, 3.75 m/s). The vortex tangential flow velocity (V_θ) is equal to 3.0 units (30 m/s). Thus, the maximum horizontal flow velocity (V_{max}) (a sum of the free stream flow and the maximum tangential velocity) is 6.0, 4.0, 3.5 and 3.375 units (60, 40, 35, 33.75 m/s). The reference length is 10 m. The total simulation time is 120 units. The simulation starts at t = − 60 units and ends at t = 60 units. However, the time axis is translated forward with 60 units for preventing the use of negative time values, during post-processing.

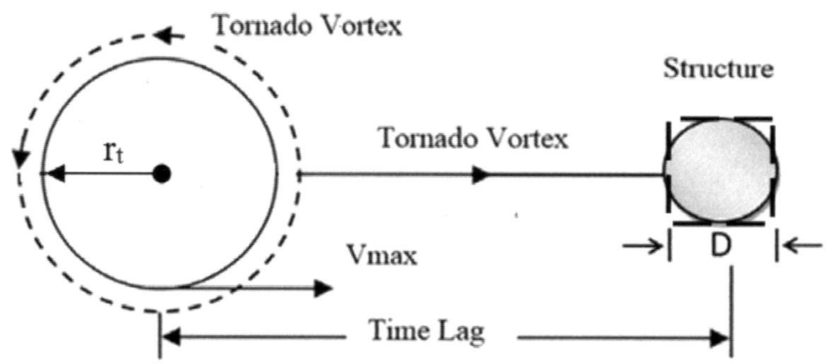

Figure 2. Tornado like vortex approaching a building

Table 2. Dimensions of dome,Cubic and prism models

Name	Shape	Units	Length=Width *(L=D)*	Height *(H)*	Area *(Ax=Ay)*	Area *(Az)*	Volume *(V)*
DM1	Dome	m	20.0	10.0	157	314	2,094
CM2	Cube	m	10.0	10.0	100	100	1,000
PM3	prism	m	17.72	10.0	177.2	314	3140

The building (e.g. dome, cube and prism) is located at reasonable distance from the boundary of the computational domain. The velocities are considered to be zero on the surface of the rectangular prism and hemispherical dome (no-slip condition). The logarithmic law is used to model the bottom boundary layer. The reference velocity is taken as the velocity at the building top. The boundary condition for vortex-structure interaction was described [11]. The NS equations are used for solving the interior velocities and pressures at each time step. The Force coefficients are computed by integrating the pressure and shear force acting on the structure surface. The force coefficients are calculated using the following equations:

$$C_X = F_x/(0.5\rho V^2 A) \quad (1)$$
$$Cy = Fy/(0.5\rho V^2 A) \quad (2)$$
$$Cz = Fz/(0.5\rho V^2 A) \quad (3)$$
$$Cp = \Delta p/(0.5\rho V^2) \quad (4)$$

Where, *Cx*, *Cy* and *Cz* are the computed force coefficients in the *x*, *y*, and *z* respectively. *Fx*, *Fy* and *Fz* are respective forces in *x*, *y*, and *z* directions, ρ is the density of air, *V* is the reference velocity and *v* is the kinematic viscosity of air. *Cp* is the mean pressure coefficient, Δp is the pressure difference, and *P - Pref* (*Pref* is equal to 0.0). The reference velocity in the tornado wind field is the maximum velocity, which is equal to $V\theta + Vt$. By integrating the pressure in each direction on the surface, the forces are computed.

5 COMPUTATIONAL DOMAIN AND GRID GENERATION

The dissipative and the convective effects modify the vortex structure and strength over the simulation. Those effects are strongly dependent on the computational grid quality and computational domain dimensions. Computational domain is a rectangle block with dimensions (L_D=60) × (D_D=60) × (H_D=45) units. In this section, the dome model and prism model boundary layer is resolved by fine grid refinement. The grid spacing used in the x and y-plane, around the dome and prism, are shown in Figure 3. The grid spacing from the building is increased by 1.2 times the previous spacing until it reaches 0.25H spacing. The 0.25H spacing is applied only on the 6 × r_t wide lane on the vortex path and around the dome and prism. Outside the path, the grid spacing is equal to 0.5H units[14].The first grid spacing next to the building is equal to 0.005H, where H is the height of the building [11].

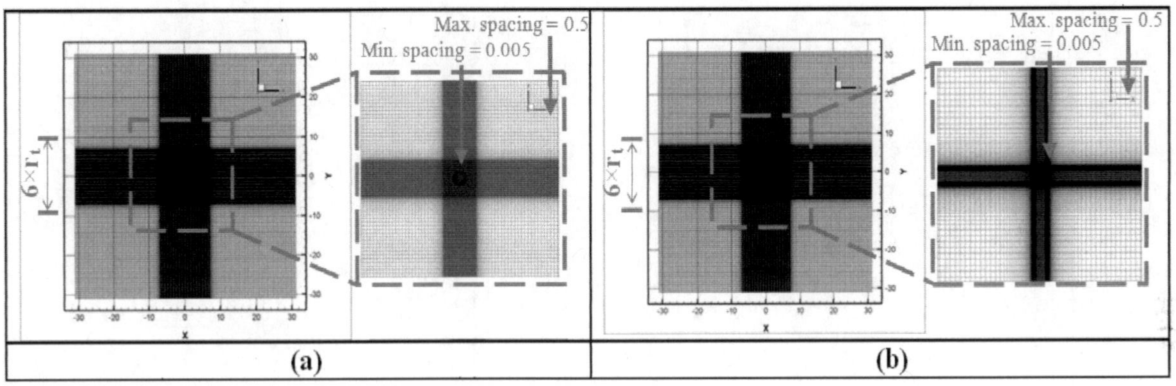

Figure 3. Generated grid (a) dome overall view (b) prism overall view

6 RESULTS AND DISCUSSION

6.1 *Tornado vortex bending and displacement during the travel*

The Figure 4 illustrates the pressure field of the different tornado speed. It was noticed that as the tornado speed increases, the upper portion of the tornado inclines forward as shown in Figure 4. That may affect the position and value of tornado forces. The effect of different V_Θ/V_t ratio of tornado force coefficients on building will be disused V_Θ/V_t in the next section.

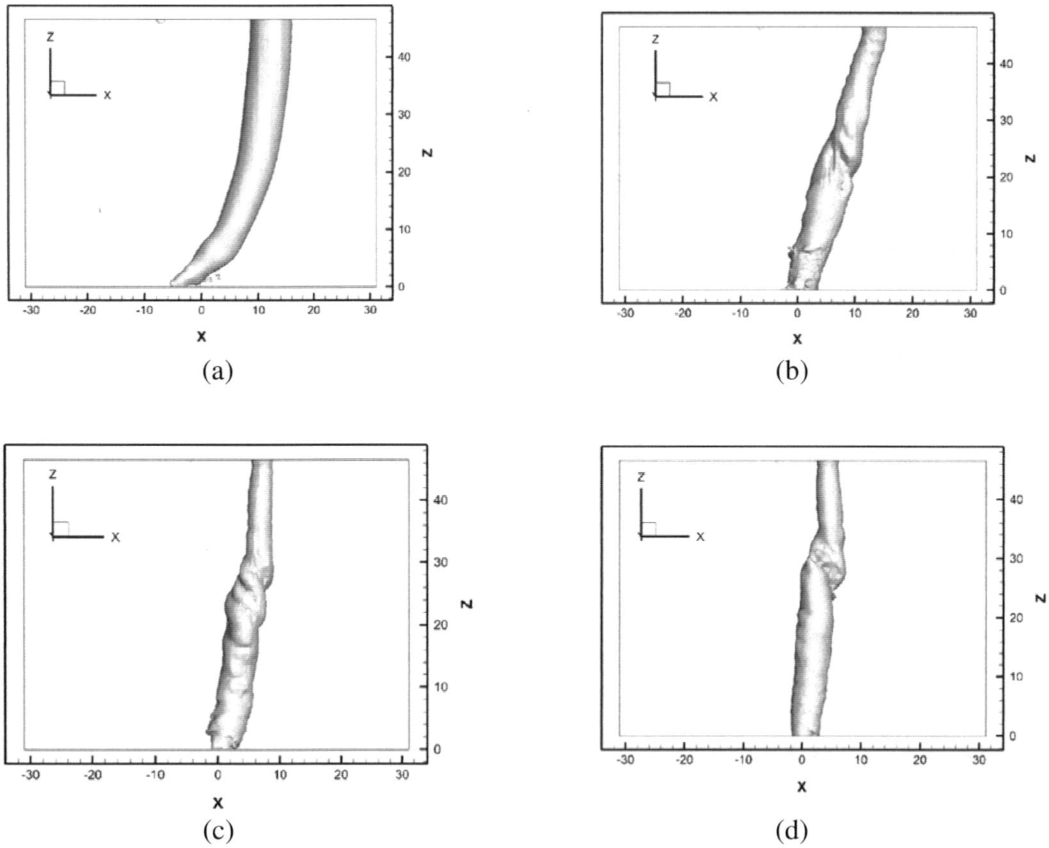

Figure 4. *xz*-plane of tornado vortex at 24 units for (a) $V\theta/Vt$ = 1.0, (b) $V\theta/Vt$ = 3.0, (c) $V\theta/Vt$ = 6.0 and (d) $V\theta/Vt$ = 8.0.

6.2 *Effect of the ratio of the tangential to translational velocity on tornado force coefficients*

At each time step, large quantities of data were produced. At this time, the forces produced in the x, y and z-directions were calculated. The computed tornado force coefficients C_x, C_y, C_z were plotted against time. The forces were computed by integrating the pressure on a cubic and prism and dome building in the x, y and z directions, respectively.

6.2.1 *The x-direction force coefficients*

The x direction force coefficient time histories of each model for V_Θ/V_t = 1, 3, 6, 8 are given in Figure 5. The C_x time histories of the three models (*CM, PM, DM*) were composed of several distinct parts. The time histories begin at a value close to zero and continue almost constantly until the tornado is close enough to begin affecting the pressure on the surface of the model. The second part causes a force in the negative x direction (opposite of the direction of translation) as the tornado reaches the model and the model begins to have negative surface pressures caused by the pressure drop due to the swirling winds. As the tornado passes over the building model, the force coefficient in the direction of translation returns to zero and then becomes increasingly positive until it reaches a peak and then returns to zero as the tornado continues past it.

For the V_Θ/V_t =1 case, the C_x changes from zero and then back to zero between the time values of 23 and 28 indicating that the loading of the building in the x direction begins when the tornado is about a distance equal to one core diameter from the center of the building. For the ratio (V_Θ/V_t = 3, 6 and 8), the tornado loading mainly occurs between the non-dimensional time of time from 18, 15, 12 units to 28, 30, 35 units. This was caused by the lagging behind of the lower portion of the tornado due to the faster translation speed. The low V_θ/V_t ratio had faster translations speeds which shifted the entire time history with respect to the *x*-axis that measured the distance of the center of the vortex to the center of the building.

6.2.2 *The y-direction force coefficients*
The y-direction (perpendicular to the direction of translation) force coefficient time histories of each model for V_θ/V_t = 1, 3, 6, 8 are given in Figure 6. The C_y time histories also followed a distinct pattern. As the tornado approaches the building model, the tornado tangential velocity component is parallel to the positive y- axis. The positive pressures on the windward side of the building caused by the strong tangential velocity overcome the negative pressures due to the vortex causing the force coefficient to reach a positive peak. As the tornado moved over the center of the building, the tangential velocity component comes from the opposite direction causing the y- direction force coefficient to peak in the opposite direction.

The non-dimensional time histories for cubic building considering all ratios (V_θ/V_t = 1, 3, 6, 8) reach their positive peak at a time t = 27, 24, 22, 20 units, respectively. Similar trends could be seen for prism and dome model as shown in Figure 6.5. For faster translation speeds or, lower V_θ/V_t ratio, the maximum force coefficients shifted to the left of the time history. For all models considering all V_θ/V_t ratios the positive and negative peaks were symmetric around C_y = 0. The envelopes of the peak values are symmetric.

6.2.3 *The z-direction force coefficients*
The vertical (z) force coefficient time histories of the three models for the V_Θ/V_t = 1, 3, 6, 8 were shown in Figures 7. There were two important observations that should be made concerning the C_z coefficients. The first was that the C_z time histories peak at a much higher value than C_x, C_y. Where C_x and C_y never reached or exceeded C_z value. The second observation was that the pressures on the roofs of the models contribute to the tornado force in the vertical direction. In fact, only the vertical components of the pressures that act normal to the surface of the roof contribute to the vertical force. The time histories above clearly show two distinct peaks. This is most likely due to the high swirl ratio vortex simulated for this study. The valley between the two peaks does not drop to zero because even though the tangential velocity component of the swirling wind approaches zero at the center of the vortex, the pressures on the surface of the building are still affected by the pressure drop at the center of the tornado.

The duration of the loading on the roofs of the three models (CM, PM, DM) between the peaks for the ratios (V_Θ/Vt = 1, 3, 6, 8) are about 1, 3, 7, 10 units, respectively. The cubic model has much higher peaks than prism model that is geometrically the same except the dimensions. The prism has larger width and length than the cubic. The reason for the considerable difference between the tornado force coefficients on the cubic and prism model is that the relation between the tornado's diameter (*td*) and cubic and prism width (*D*) is not the same. The tornado's diameter is about 6 and 3.3 times larger than the width of the cubic and prism building, respectively. The maximum force coefficient on building increases when ratio between tornado and building' width increases. A comparison of forces coefficients shows that for all models (*CM, PM, DM*) the slower moving tornado produces higher force coefficients.

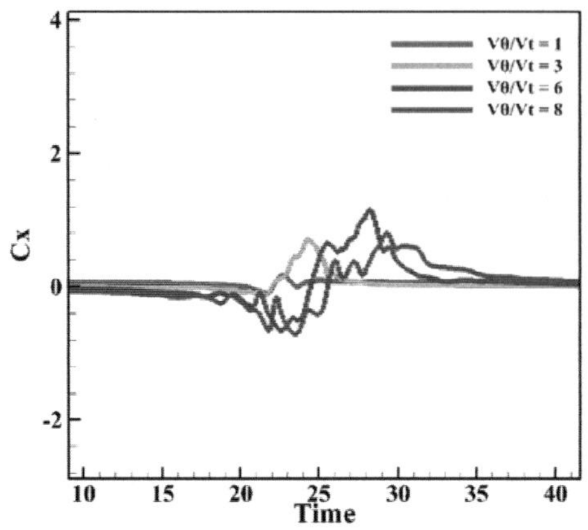

(a)

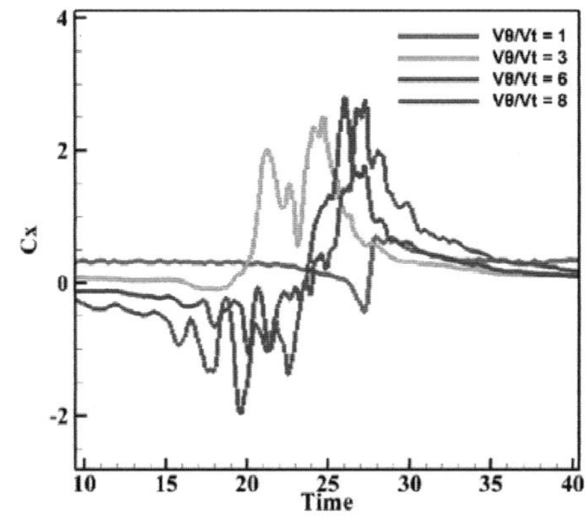

(b)

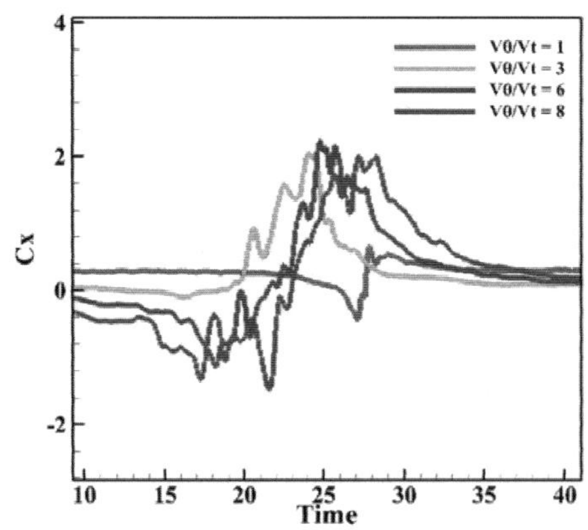

(c)

Figure 5. Tornado force coefficients in x-direction due to different V_θ/V_t (1, 3, 6 and 8) ratios on: (a) dome, (b) cubic and (c) prism

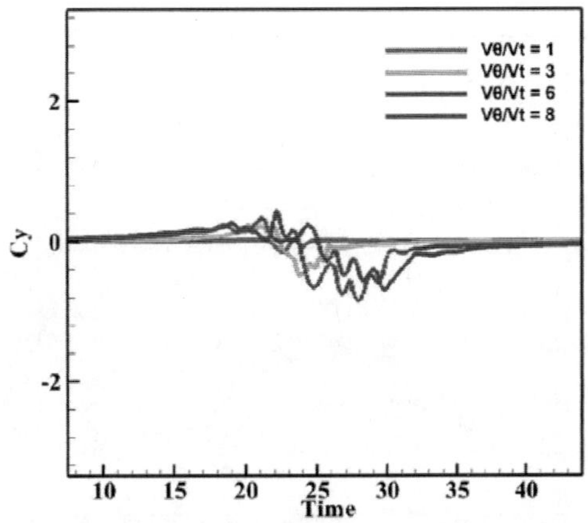

(a)

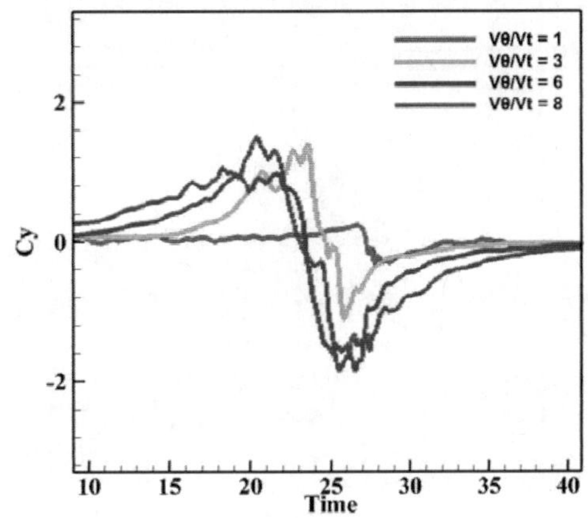

(b)

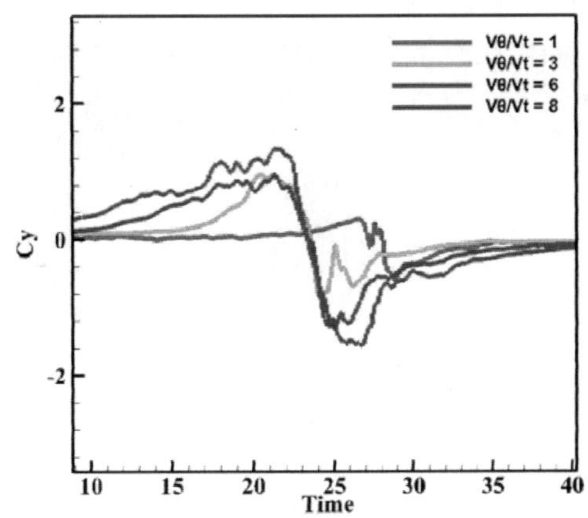

(c)

Figure 6. Tornado force coefficients in *y*-direction due to different V_θ/V_t (1, 3, 6 and 8) ratios on: (a) dome, (b) cubic and (c) prism

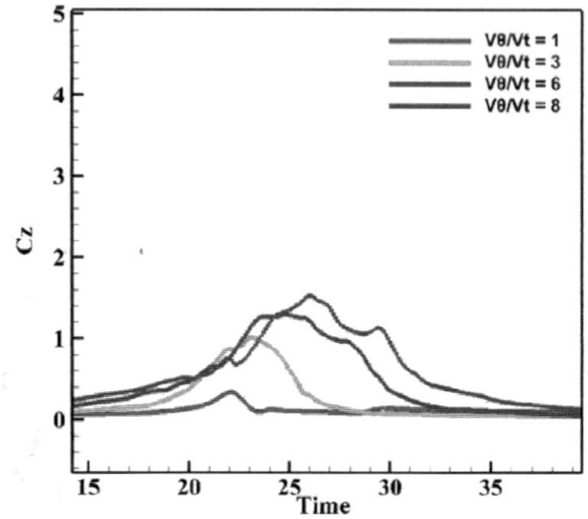

(a)

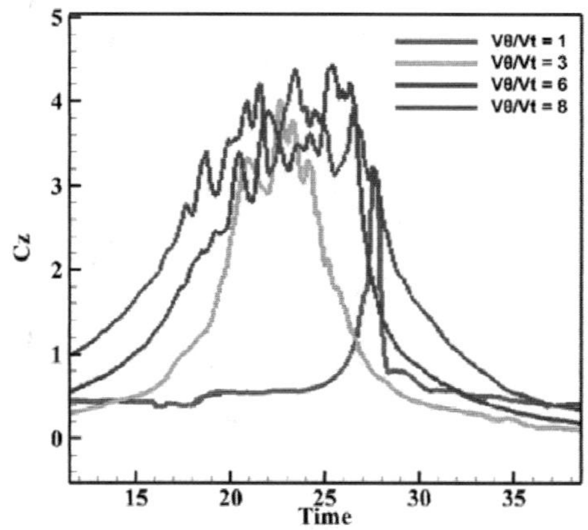

(b)

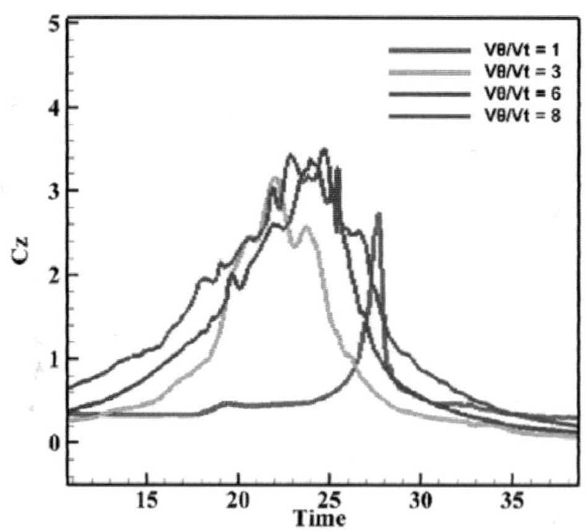

(c)

Figure 7. Tornado force coefficients in *x*-direction due to different V_θ/V_t (1, 3, 6 and 8) ratios on: (a) dome, (b) cubic and (c) prism.

7 RESULTS AND DISCUSSION

The ratio of tangential (V_θ) to translational (V_t) velocity reported in recent studies is 10 or greater, which is larger than the field observation ratios. The influence of V_θ/V_t ratios on the tornado force coefficient for a cubic, prism and dome building were compared using CFD Model. The V_θ/V_t ratios were considered to be 1, 3, 6, and 8 for comparison. These ratios were very much in agreement with field observation ratios. The conclusions arrived from this comparison are:

When the tornado translation speed increases, the upper portion of tornado moved forward more than bottom part due to boundary layer interaction at the bottom part.

The magnitudes of the forces were found to be larger for slower translation speeds, high V_θ/V_t ratio.

For faster translation speeds or, lower V_θ/V_t ratio, the maximum values shifted to the left of the time history

8 REFERENCES

1 NOAA, Enhanced F Scale for Tornado Damage, 2012, Retrieved from http://www.spc.noaa.gov/faq/tornado/ef-scale.html
2 NWS, Tuscaloosa-Birmingham Tornado, 2011, Retrieved from http://www.srh.noaa.gov/bmx/?n=event_04272011tuscbirm
3 K.C. Mehta, J.E. Minor and J.R. McDonald, Wind speed analysis of April 3-4 tornadoes, J. Struct. Div., 102(1976), 1709-1724.
4 J. Wurman, K. Kosiba and P. Robinson, In situ, Doppler radar and video observations of the interior structure of a tornado and the wind-damage relationship, Bull. Amer. Meteor. Society, 94 (2013), 835-846.
5 P.P. Sarkar, F.L. Haan, V. Balaramudu and A. Sengupta, Laboratory simulation of tornado and microburst to assess wind loads on buildings, Proc., ASCE Structures Congress, ASCE, Reston,Va, 2006
6 J.M. Case, Effect of building geometry on wind loads on low rise buildings in a laboratory-simulated tornado with a high swirl ratio, Master thesis, Iowa State University, Ames, Iowa, 2011.
7 A. Sengupta, F.L. Haan, , P.P Sarkar and V. Balaramudu, Investigation of tornado transient loads on buildings in microburst and tornado winds, J. Wind Eng. Indus. Aerodyn., 96 (2008) 2173-2187.
8 F.L. Haan, V.K. Balaramudu and P.P. Sarkar, Tornado-induced wind loads on a low-rise building, J. Struc. Eng. 136 (2010) 106-116.
9 H. Hu, Z. Yang, P. Sarkar and F. Haan, Characterization of the wind loads and flow fields around a gable roof building model in tornado-like winds, Exp. fluids, 51(2011) 835-851.
10 Z. Yang, P. Sarkar and H. Hu,Visualization of the flow structures around a gable-roofed building model in tornado-like winds, J. Vis., 13 (2011), 285-288.
11 P. Selvam and, P. Millett, Large eddy simulation of the tornado-structure interaction to determine structural loadings, Wind Stru., 8(2005) 49-60.
12 R.P. Selvam and, P. Millett, Computer modeling of tornado forces on buildings, Wind and Struc., 6(2003) 209 - 220.
13 N.H. Alrasheedi and R.P. Selvam, Computing tornado forces on different building sizes. In Proc.13th Inter. Confere on Wind Eng. (ICWE13), Amsterdam, Netherlands, 2011.
14 M.P. Gorecki and R.P. Selvam,Rankine combined vortex interaction with a rectangular prism, int. J. Comput. Fluid Dyn., (2015)1-13.
15 R.P. Selvam, Computational procedures in grid based computational bridge aerodynamics, In Bridge Aerodynamics, Larsen, A. and Esdahl (eds), Balkema, Rotterdam, 1998, pp. 327-336.
16 S.N. Ahmed and R.P. Selvam, Ridge effects on tornado path deviation. Int. J. Civil and Struc. Eng. Research, 3(2015) 273-294.
17 W.S. Lewellen, Tornado Vortex Theory. In: The Tornado: Its Structure, Dynamics, Prediction, and Hazards, Church C, Burgess D, Doswell C, Davies-Jones R, Geophysical Monograph 79, American Geophysical Union,1993, pp19-39.

The 13th Americas Conference on Wind Engineering (13ACWE)
Gainesville, Florida USA, May 21-24, 2017

Modelling translating tornado-like wind flow using CFD simulations

Fangping Yuan [a], Guirong Yan [b], Ryan Honerkamp [b], Kakkattukuzhy M. Isaac[a]

[a]*Department of Mechanical and Aerospace Engineering, Missouri University of Science and Technology, 1201 N. State St., Rolla, MO, USA*
[b]*Department of Civil and Architectural and Environmental Engineering, Missouri University of Science and Technology, 1201 N. State St., Rolla, MO, USA*

ABSTRACT: This study investigates and compares two different strategies to simulate the translating motion of tornadoes, which are to establish the relative motion between the tornado and the ground, or to move the swirling wind flow directly. To be specific, two approaches based on the two different strategies are applied to model the translating motion of the tornado-like wind flow. By comparing the generated overall and near-ground wind flow with the real-world tornadoes, the proper way to simulate the translation of tornadoes is determined.

KEYWORDS: Tornado Simulation, Translating tornado, Tornado-like vortices, CFD simulations

1 INTRODUCTION

The recent devastating tornado events in Tuscaloosa, Birmingham and Huntsville, Alabama as well as Chattanooga, Tennessee have underscored the importance to conduct in-depth research on tornado dynamics and wind effects of tornadoes on civil engineering structures[1-3]. Considering that it is challenging to obtain the in situ measurements during real-world tornado events[4,5], researchers started to study the tornadic wind fields and wind effects on civil structures in laboratory tornado simulators.

However, the testing facility cannot replicate the full intensity and scale of real-world tornadoes in terms of wind speed and impact. Therefore, the current authors developed an approach to numerically simulate a type of laboratory tornado simulator that can produce translating tornado-like wind flow[6]. This study will be to further discuss how to properly simulate the translation of tornadic wind flow by applying proper boundary conditions on the computational domain.

Two different strategies have been applied to simulate the translating motion of tornadoes. The first one is to simulate the tornado translation by moving the ground to establish the relative motion between the tornadic wind field and the ground. Related research can be referred to the following work[7-13]. The other strategy to simulate the translating motion of tornadoes is achieved by directly moving the generated swirling wind flow while keeping the ground fixed[14-17], which is same as the real situation[9,10].

This study is to investigate how to numerically simulate the translating motion of swirling wind flow properly for the type of laboratory tornado simulator that can produce translating swirling wind flow. In this study, two different approaches based on the above two simulation strategies are applied and compared. The remainders of this paper are organized as follows. First, the detailed information on the modelling of the laboratory tornado simulator is provided. Second, four different approaches to numerically simulate the translating motion of the tornado simulator are described. Third, the overall and near-ground wind flow fields obtained by each approach are presented and compared. Finally, conclusions are drawn.

2 NUMERICAL MODELLING OF LABORATORY TORNADO SIMULATOR

2.1 *Description of the tornado simulator*

The tornado simulator to be numerically simulated in this study follows the same mechanism to generate tornado-like vortices as the one at Iowa State University (ISU)[14]. The dimension of the simulator in this study is slightly smaller than the one at ISU. Different from the tornado simulator at ISU, the chamber/duct designed here has a curved shape, and the guide vanes follow the contour curved duct. The schematic diagram and the corresponding solid CAD model of this tornado simulator are presented in Figure 1(A) and 1(B), respectively. Table 1 presents the dimensions of the physical simulator. The proposed flow circulation in a radial-vertical plane is illustrated in Figure 1(B). The detailed information on the simulated tornado simulator can be referred to our previous paper[6].

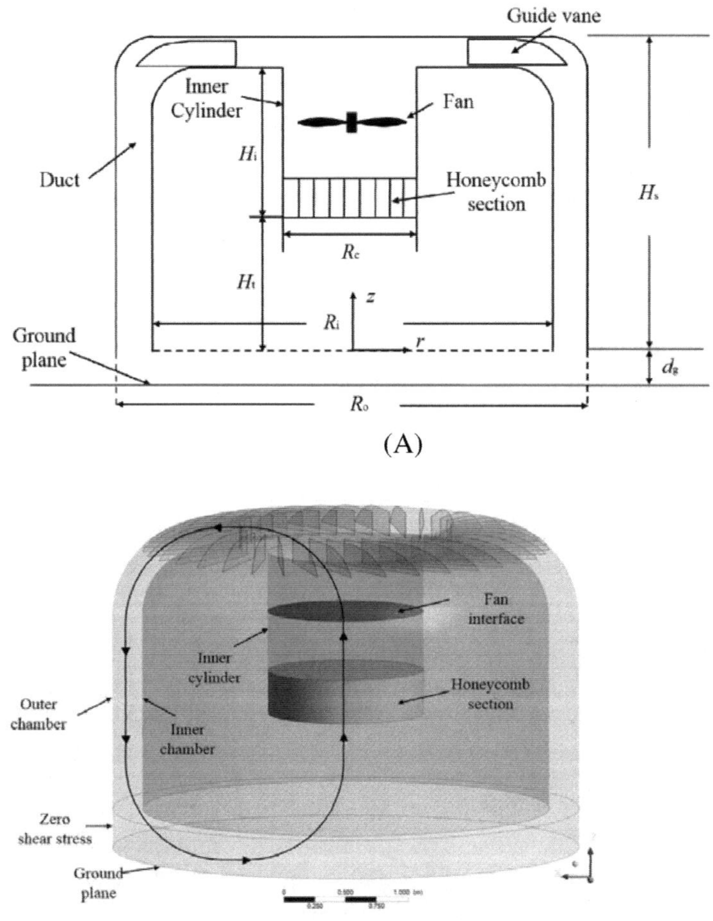

Figure 1. Tornado simulator with a curved chamber. (A) Schematic diagram; and (B) 3D CAD model.

Table 1. Dimensions of the tornado simulator

Height of the chamber, H_s	91.13 in. /2.315m
Diameter of the outer chamber, R_o	148.16 in. /3.763m
Diameter of the inner chamber, R_i	131.10 in. /3.330m
Diameter of the inner cylinder, R_c	49.25 in. /1.251m
Height of the inner cylinder, H_i	51.90 in. /1.318m
Thickness of the honeycomb section, d_h	13.90 in. /0.353m
Distance between the chamber bottom and ground plane, d_g	12.40 in. /0.315m
Length of the vane, L_v	29.42 in. /0.747m
Height of the vane, H_v	8.03 in. /0.204m
Distance between the bottom of inner cylinder and the bottom of chamber, H_t	30.63 in. /0.778m
Vane setting angle (from radial direction)	30°

2.2 *Modelling the fan and the honeycomb section and simulation setup*

In this study, the fan is modeled using the FLUENT fan boundary condition[18]. Herein the fan is treated as an interface, across which a static pressure jump is applied to drive the air flow. A pressure jump ΔP=373 Pa is applied at the fan interface in this study. The honeycomb is modeled using the porous media model[19]. In FLUENT, the porous media is modeled by adding additional momentum source term to the standard Navier-Stokes equations. Users can specify the porosity, viscous and inertial resistant factors to determine the porous media. More details of the porous media modelling can be found in our previous work.

The tornado simulator is designed as a flow generator with the air circulating inside, driven by the fan. Therefore, no inflow and outflow boundary conditions are applied to the outer chamber. For the simulator itself, the guide vanes, inner cylinder, outer and inner chambers (duct walls) are all set up as no-slip boundary conditions, which indicates zero velocity at the walls. Initially, the velocity is zero everywhere, and the standard sea-level atmospheric pressure is applied to the whole computational domain except at the fan interface.

A transient, incompressible three-dimensional CFD simulation is conducted using the commercial software package, ANSYS FLUENT. Large eddy simulation (LES) is used to model the turbulence. The finite volume-based SIMPLEC scheme is used with the second-order bounded implicit scheme for the time discretization. The time step size is set up as Δt=0.01 s. A total of 6 seconds of simulation is run. For the first 2 seconds, assume that the simulator does not move, this is to allow the tornado-like vortices to be established. Then another 4 seconds of simulation is run to obtain the translating wind field.

3 APPROACHES FOR MODELLING THE TRANSLATING MOTION OF TORNADIC WIND FIELD

3.1 *Approach 1: moving wall boundary with a velocity inlet boundary*

The translation of a tornado is quite similar to a moving car. Being inspired by the approach to simulate a moving car or aircraft, we propose this approach to numerically simulate the translating tornado. In this approach, the tornado simulator is covered by a quarter ellipsoid computational domain, as shown in Figure 2. In this computational domain, the curved surface is served as a velocity inlet boundary, and the vertical surface on the right side is served as an outflow boundary. The bottom of the enclosure is served as the ground plane. To simulate the translation of tornadoes, a moving wall boundary condition is applied to the ground plane. If the tornado simulator translates in the positive x direction at a speed of u_t, then the directions of the moving wall should be in the negative x direction and the speed should be u_t as well, as shown in Figure 2(A). This simulation approach not only can simulate the relative motion between the simulator and the ground, but also can include the interaction between the simulator and surrounding air in the simulation.

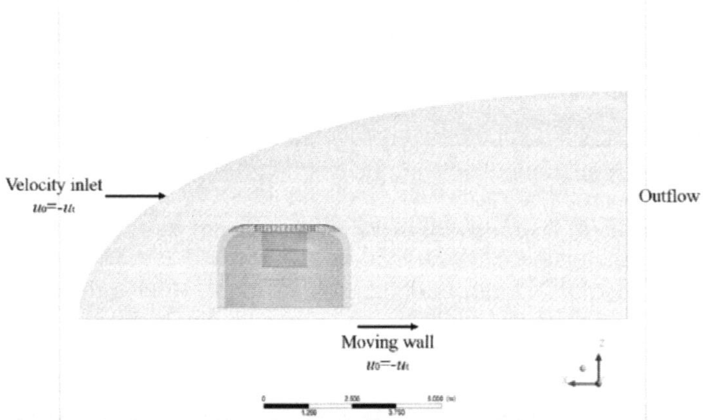

(A)

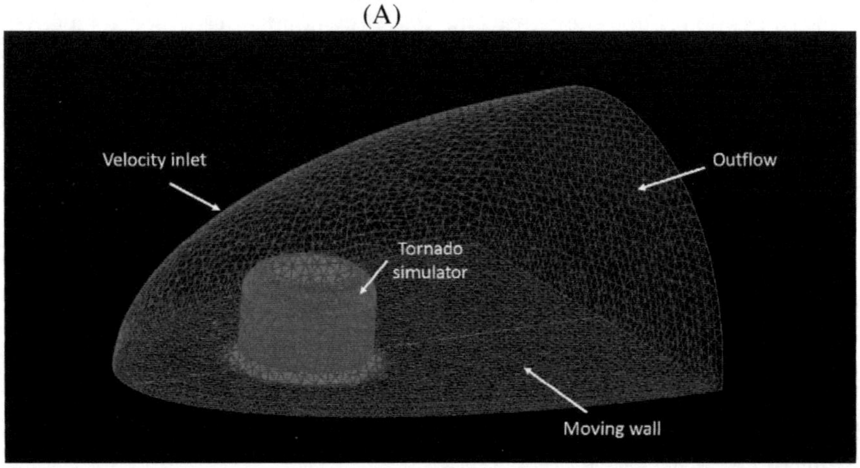

(B)

Figure 2. Schematic diagram of approach 1 (A), and mesh of approach 1 (B).

3.2 *Approach 2: moving simulator directly with the sliding mesh technique*

The translation of tornadoes can be simulated by directly moving the chamber of the simulator while keeping the ground plane fixed. The laboratory tornado simulator at ISU applied this approach to simulate the translation of tornadoes[14-16]. To be specific, the moving of the simulator is induced by an overhead crane. Numerically, this can be achieved by applying the sliding mesh technique in FLUENT. Figure 3 presents the mesh created when using this approach. In this numerical model, a box with a thickness of 12.4 inches is placed below the chamber to simulate the "near-ground domain" beneath the chamber with the bottom of this box representing the ground boundary. The length and width of the box are 279.08 inches and 164.17 inches, respectively. Since different shapes of cells are used for the chamber and "near-ground domain", a mesh interface has to be created between the bottom surface of the chamber and the top surface of the "near-ground domain". This approach assumes that the air flow outside the chamber does not affect the flow field generated inside the chamber and underneath the chamber. Therefore, this approach only considers the fluid dynamics inside the chamber and the near-ground domain under the chamber, while neglecting the flow field outside the chamber, which significantly reduces the computational cost. In this study, the chamber translates at a speed of u_t =0.5 m/s in the positive x direction. More details of this approach can be found in our previous work[6].

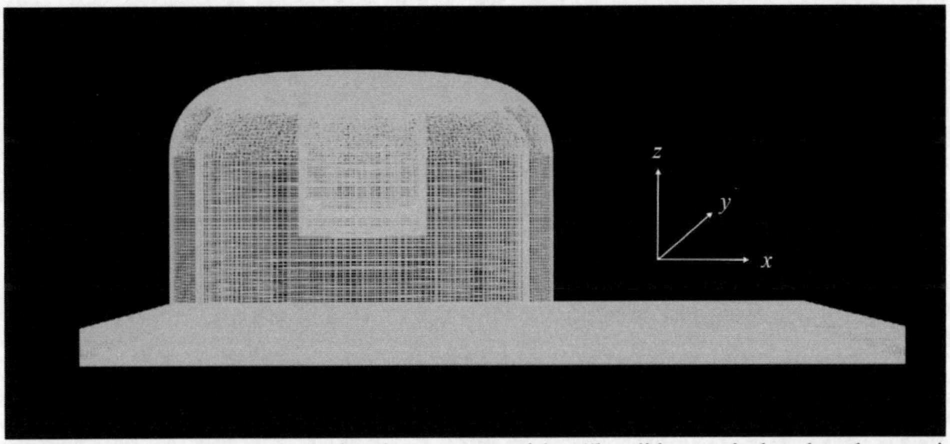

Figure 3. Mesh when translating is present, with a "box" beneath the chamber to simulate the near-ground domain. Approach 2.

4 RESULTS AND DISCUSSIONS

This section presents the generated translating tornadic wind field using two different approaches described in Section 3. In all cases, the simulation is first run for 2 seconds as the stationary case until a developed tornadic wind field is established and then another 4 seconds of simulation is run for the translating motion. For example, for Approach 1, the simulation is run for 2 seconds with the bottom wall (ground plane) fixed as a no-slip wall, and then the bottom wall is switched to the moving wall boundary for the following 4 seconds. The mean flow properties presented in the following sections are time-averaged over the 4 seconds from t=2 s-6 s. The translating speed of the simulator is 0.5 m/s in the positive x direction in both cases.

4.1 *Approach 1, moving wall boundary with a velocity inlet boundary*

In approach 1, both a moving wall boundary and a velocity inlet are implemented to simulate the relative motions between the simulator and ground plane, and the interaction between the simulator and the surrounding air, respectively. Figure 4 shows the overall mean static pressure and plot of mean velocity vectors on radial-vertical plane at y=0.

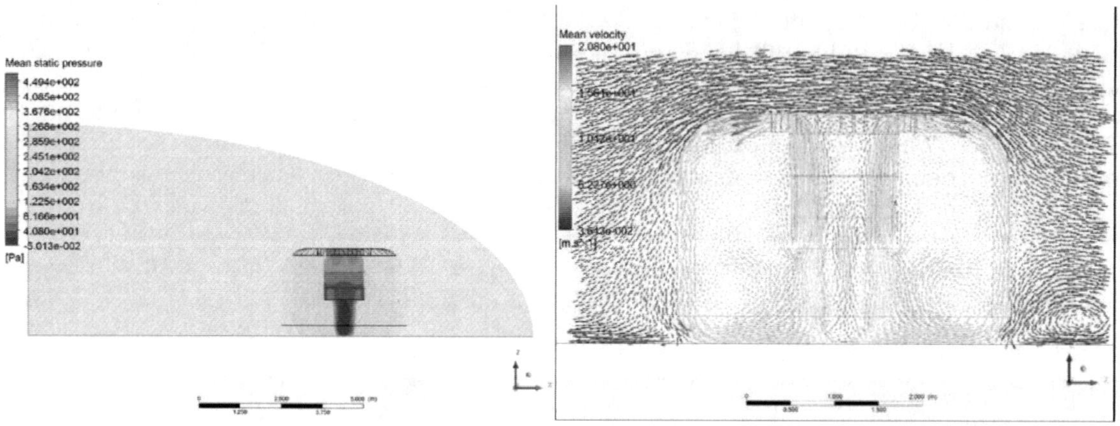

(A) (B)

Figure 4. Time-averaged static pressure contours (A) and velocity vectors (B) on radial-vertical plane at y=0. Translating case, approach 1.

In this case, the maximum mean tangential velocity is 15.35 m/s at r_c =0.303m and z=-0.254 m, and the volume flow rate is Q=8.92 m^3/s. The corresponding swirl ratio according to the definition from ISU[14] in this case is computed as S=0.496. Figure 5 presents the time-averaged static pressure, tangential, radial and axial velocity contours on radial-vertical plane at y=0 between the ground plane and the bottom of the inner cylinder. From the results, we can find that the induced wind structure is axisymmetric and vertical to the ground plane.

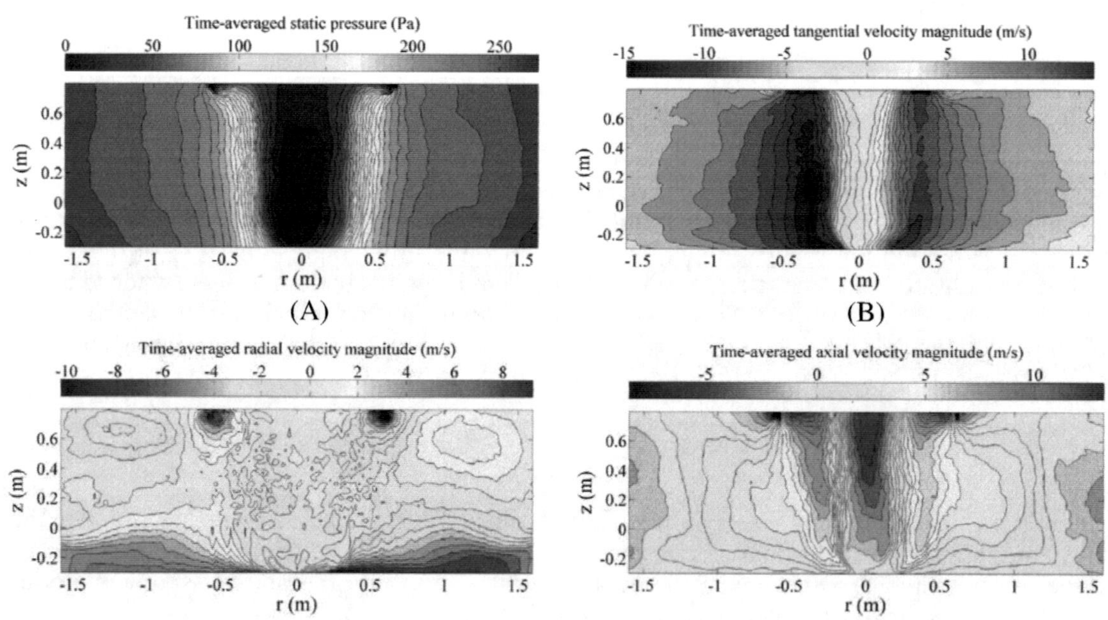

(C) (D)

Figure 5. Mean static pressure (A), tangential velocity (B), radial velocity (C) and axial velocity contours at radial-vertical plane *y*=0. Translating case, approach 1.

4.2 *Approach 2, moving simulator directly with sliding mesh technique*

Approach 2 relies on the sliding mesh technique to directly move the simulator along the positive *x* direction with a speed of 0.5 m/s. Figure 6 presents the mean static pressure contours and the vectors of the mean resultant velocity of the radial and vertical (axial) components on the radial-vertical plane at *y*=0 between the ground plane and the bottom of the inner cylinder. The axisymmetric flow field is now skewed due to the translating motion, which is quite similar to a real-world tornado[9,10].

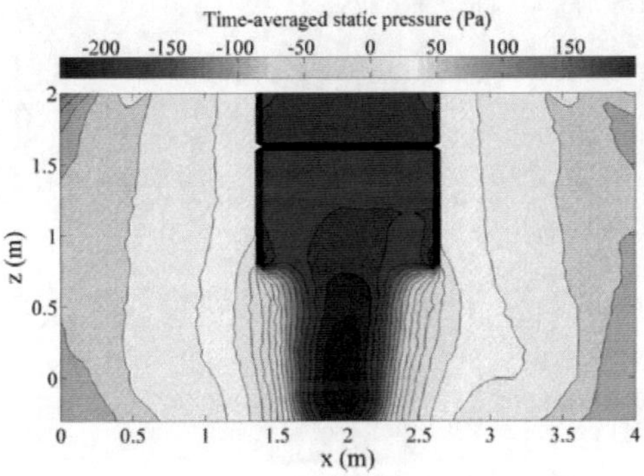

(A)

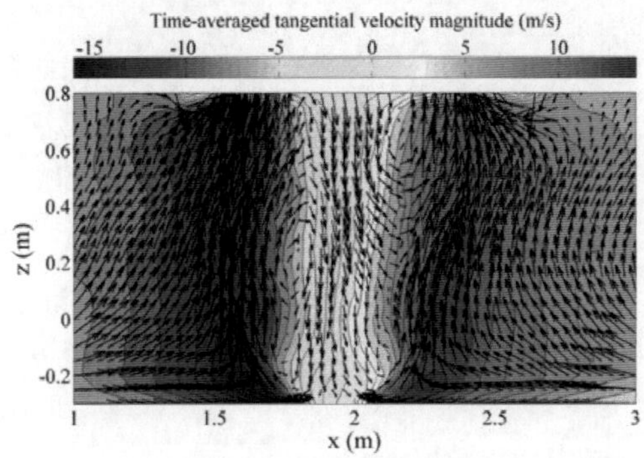

(B)

Figure 6. Time-averaged static pressure (A), and velocity vectors (B) on radial-vertical plane at *y*=0. Translating case, approach 2.

The maximum tangential velocity now occurs at r_c=0.275 m at a height z=-0.275 m with a magnitude of $V_{t,max}$=16.395 m/s, which is about 6.8% larger than the maximum tangential velocity in approach 1 (15.35 m/s). Moreover, the core radius is approximately 9.2% smaller than that in approach 1. This results in a reduction of swirl ratio from 0.496 for approach 1 to 0.437 for approach 2. Figure 7 presents the near-ground time-averaged static pressure, tangential, radial and axial velocity contours on the radial-vertical plane at *y*=0 between the ground plane and the bottom of the inner cylinder. More details including the velocity profiles at different elevations can be found in our work[6].

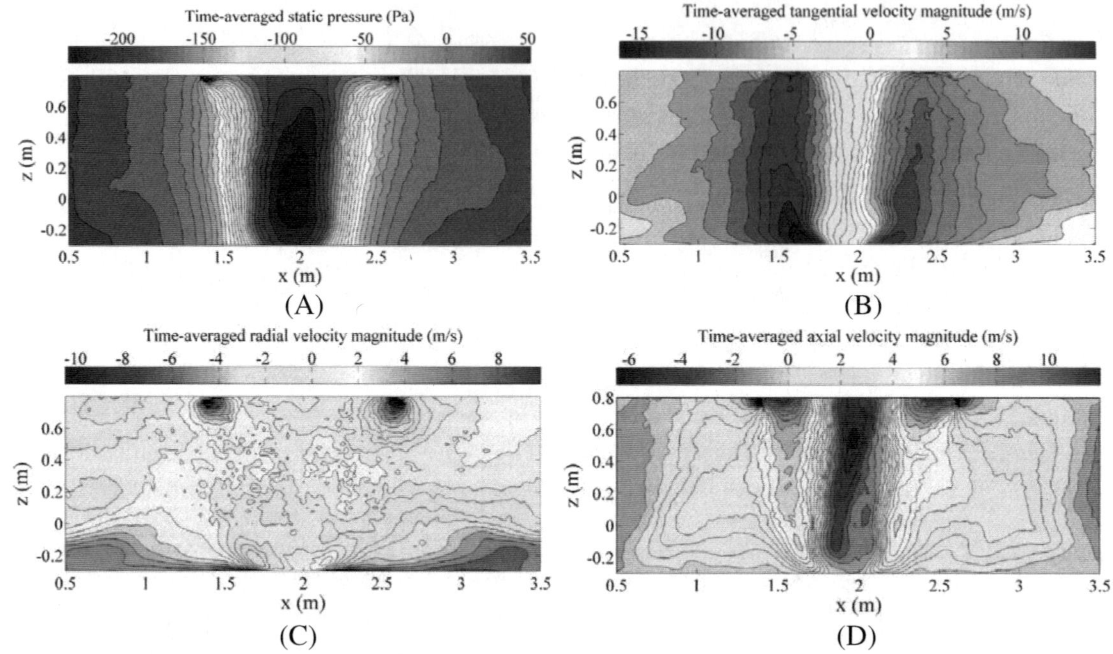

Figure 7. Mean static pressure (A), tangential velocity (B), radial velocity (C) and axial velocity contours at radial-vertical plane y=0. Translating case, method 3.

4.3 *Comparison and discussion*

From the generated overall and near-ground flow fields, we find that from approach 1 to approach 2, the maximum mean tangential velocity is increased, but the core radius and the corresponding swirl ratio is decreased. Furthermore, in approach 2, a skewed wind structure can be generated which is not found in approach 1. Considering that real tornado is translating and the induced near-ground wind structure is also skewed[9,10], we may conclude that to move the tornado simulator directly (approach 2) might be a more appropriate way to simulate the translating tornadoes.

5 CONCLUSION AND FUTURE WORK

In this paper, a numerical model of a type of laboratory tornado simulator that can generate translating tornadoes is established. Two approaches based on different strategies are applied to simulate the translating motion of the induced tornadic wind field. Overall and near-ground wind fields are extracted and investigated. Results from different approaches are compared with each other and with real-world tornadoes. Results show that to move the tornado simulator directly (approach 2) is a better way to simulate the translating tornadoes than to construct the relative motion (approach 1). In the future, different translating speed can be tested in such a model with approach 2 to better understand the impact on flow structures due to the translating effect.

6 ACKNOWLEDGEMENTS

The authors greatly appreciate the financial support from National Science Foundation, the Hazard Mitigation and Structural Engineering program, through the project, “Damage and Instability Detection of Civil Large-scale Space Structures under Operational and Multi-hazard Environments” (Award NO.: 1455709)

7 REFERENCES

1 T.W. Moore, On the temporal and spatial characteristics of tornado days in the United States, Atmospheric Research, 184(2017) 56-65.

2 K.M. Simmons, D. Sutter and R. Pielke, Normalized tornado damage in the United States, Environmental Hazards, 12(2013) 132-147.

3 A.B. Smith and J.L. Matthews, Quantifying uncertainty and variable sensitivity within the US billion-dollar weather and climate disaster cost estimate, Natural Hazards, 77(2015) 1829-1851.

4 E. Savory, Parke, A.R. Gerard, M. Zeinoddini, N. Toy and P. Disney, Modeling of tornado and microburst-induced wind loading and failure of a lattice transmission tower, Engineering Structures, 23(2001) 265-375.

5 K.M. Simmons and D. Sutter, 2005: WSR-88D Radar, tornado warning, and tornado casualties, Weather Forecasting, 20(2005) 301-310.

6 F Yuan, G. Yan, R. Honorkamp and K.M. Isaac, Numerical simulation of Laboratory tornado simulator that can produce translating tornadic wind flow, submitted to Journal of wind engineering and industrial aerodynamics (under review).

7 C.C. Chang, First real man made tornado is generated in laboratory cage by space scientists at the catholic University of America. *News item of PI of CUA,* 1966.

8 N.B. Ward, The exploration of certain features of tornado dynamics using a laboratory model. Journal of the Atmospheric Sciences, 29(6)(1972) 1194-1204.

9 C.R. Alexander and J. Wurman, The 30 May 1998 Spencer, South Dakota, Strom. Prat I: The Structure Evolution and Environment of The Tornadoes, Monthly Weather Review, 133(2004) 72-96.

10 J. Wurman and C.R. Alexander, The 30 May 1998 Spencer, South Dakota, storm. Part II: Comparison of observed damage and radar-derived winds in the tornados, AMS monthly weather reviews, 133(2005) 97-118.

11 C.J. Diamond and E.M. Wilkins, Translation Effects on Simulated Tornadoes, Journal of the Atmosphere Science, 41(1984) 2574-2580.

12 D. Natarajan and H. Hangan, Large eddy simulations of translation and surface roughness effects on tornado-like vortices, Journal of wind engineering and industrial aerodynamics, 104(2012) 577-584.

13 Z Liu, T. Ishihara, Study of the effects of translation and roughness on tornado-like vortices by large-eddy simulations, Journal of Wind Engineering and Industrial Aerodynamics, 151(2016) 1-24.

14 F.L. Haan, P.P. Sarkar and W.A. Gallus, Design, construction and performance of a large tornado simulator for wind engineering applications, Engineering Structures, 30(2008) 1146-1159.

15 W. Zhang and P.P. Sarkar, Near-ground tornado-like vortex structure resolved by particle image velocimetry (PIV), Exp Fluid, 52(2012) 479-493.

16 Z. Yang, P.P. Sarkar and H. Hu, An experimental study of a high-rise building model in tornado-like winds, Journal of Fluid and Structures, 27(2011) 471-486.

17 V.P. Pham, N. Tsuyoshi, N. Kojiro and K. Hirotoshi, A numerical study of the effects of moving tornado-like vortex on a cube, The Seventh International Colloquium on Bluff Body Aerodynamics and Applications (BBAA7), Shanghai, China, Sept -6, 2012.

18 ANSYS FLUENT fan boundary condition user guide.

19 ANSYS FLUENT porous media user guide.

Tornado Hazard Assessment for the Continental of United States

Fanfu Fan[a], Weichiang Pang[b], Jin Wang[c]

[a]*Clemson University, Clemson, SC, U.S.*
[b]*Clemson University, Clemson, SC, U.S.*
[c]*Clemson University, Clemson, SC, U.S.*

ABSTRACT: According the records of Storm Prediction Center (SPC) of the National Oceanic Atmospheric Administration (NOAA), approximately more than 1000 tornadoes spawned every year in the continental United States. Tornadoes often cause significant damages, injuries and economic losses to the affected regions. Compared to hurricanes, the damage footprint of a tornado is relatively small. While the damage potential of a tornado is high, the occurrence of a tornado at a given geographic location can be characterized as a low probability but high consequent event. The SPC maintains a database of historical tornado events with approximately 65 years of data. Since the historical data is limited and the impact of a tornado is very localized, it is not feasible to use solely the raw historical data or tracks to properly quantify the tornado risk for a given structure or a community that has not been affected by a past tornado event. This paper presents a stochastic simulation framework to generate synthetic tornado tracks for the Continental United States. The annual spawn rate was modeled using a negative binomial distribution. The spawn or touchdown locations were simulated using spawn probability density GIS maps, which specifically account for the variability of tornado properties at different geographic regions (e.g. tornado alley, Dixie alley and etc.), developed using the kernel density estimation (KDE). A database of 500,000 years of simulated tornado tracks were simulated using the Clemson high performance computing facility, the Palmetto Cluster. The track parameters included the tornado occurrence rate, intensity (EF-scale), touchdown location, touchdown time, and path direction. All these parameters are geographic dependent, meaning the properties vary based on the geographic locations. The simulated spawn rates and other key parameters for selected major cities in the United States were compared to the observations. Good agreements were observed between the simulations and observations. Using the simulated track database and a 2D tornado wind field model, the wind speed distributions for different return periods were computed for selected cities. As an illustration for the application of the tornado database, the computed wind speed distributions were used in a fragility framework to quantify the damage risk and loss potential of tornado hazards to a low-rise building.

KEYWORDS: Tornado hazards, Risk analysis, Simulation, Wind speed, Tornado wind field

1 INTRODUCTION

On average, the continental of the United States is subjected to more than one thousand tornadoes every year, causing significant financial losses and social disruption. The Storm Prediction Center (SPC), a division of the National Oceanic Atmospheric Administration (NOAA), maintains a database of more than 60,000 tornado events recorded since 1953. Compared to a hurricane, the influence area of a tornado is relatively small. Even with over 60,000 of known historical tornado events, many places in tornado prone region have not been hit by a tornado. Therefore, it is not feasible to directly determine the risk due to tornadoes for a location or small region using solely the observed tornado events. In this study, a stochastic simulation program for simulating tornadoes based on the statistics of historical data was developed.

The Monte Carlo Simulation (MCS) technique was employed to develop the tornado simulation program. MCS is a computational method that utilizes repeated sampling of random numbers from a sequence of probability distributions to obtain the behaviors or responses of a relatively complex system or phenomena with random outcomes. The MCS technique has been widely used in assessing the risk of natural hazards with relatively rare (i.e. low probability) occurrences but high consequences. Meyer[1] employed the MCS approach to study significant tornado occurrence distribution in the continental of the United States. Daneshvaran[2] evaluated the spatial frequency of occurrence of tornadoes in the United States and estimated the losses of tornado and hail outbreaks. Banik[3] used a stochastic model for assessing the probability of exceedance for the maximum tornado wind speed in Southern Ontario, Canada.

2 METHOD OF ANALYSIS

2.1 *Data sources description and processing*

The intensity of a tornado is rated based on as a classification called, the Enhanced Fujita scale (EF scale). It should be noted that the original tornado classification method, Fujita scale (F scale), was originally proposed by Dr. Theodore Fujita in 1971 and it was not incorporated into the tornado rating until 1973. The Enhanced Fujita (EF) scale was later introduced in 2007. There are two historical archives that have been widely used in tornado related research. The Grazulis database collected over 10,000 tornadoes for the period of 1921 to 1995. The SPC database contains over 60,000 tornadoes covering events from 1953 to present. Both of these databases contain tornado information such as, occurrence location, starting time, and damage classification. However, the Grazulis database only includes F2 and stronger tornadoes and it does not cover the tornadoes in recent years. Instead of conducting field measurement of tornado intensity based on degree of damage, recorded tornado intensities before 1973 were assigned purely according to the newspaper reports or photographs.

The SPC tornado database from 1973 to 2015 were used in this study. Early years data were excluded because the tornado intensity rating may be inaccurate and the tornado records may be incomplete due to the lack of reporting for non-newsworthy tornadoes. Another reason to use the tornado data set after 1973 is because a clear difference in trends between the annual occurrence rates for pre and post 1973 data has been observed. On average, the SPC tornado database shows a steady increase of an approximately 55 tornadoes per year from 1950 to 1973 resulting in the annual occurrence rate for 1973 to be almost four times higher than that of the 1950. In other words, the records prior to 1973 may be incomplete; hence, they were excluded from this study.

After excluding the data prior to 1973, the SPC database has 46,227 tornado records for the period of 1973 to 2015. The weak tornadoes (EF 0 and EF 1) comprise about 85% of all tornadoes. The strong tornadoes (EF 2 and EF 3) comprise about 14% of all tornadoes while the violent tornadoes (EF 4 and EF 5) only account for about 1% of all tornadoes. Even though many areas with high tornado occurrences are at risk from violent tornadoes, many of these regions have not been hit by EF 4 and EF 5 tornadoes over the relatively short observation period of the SPC database. Thus, a stochastic simulation method was developed to generate more tornado tracks and to quantify the long-term tornado hazards. During the preparation of this paper, a database of 500,000 years of simulated tornadoes was generated. More simulations may be conducted if necessary for quantifying the tornado hazards of areas with low tornado activities.

2.2 *Methodology*

The main tornado simulation framework includes three sub-models, namely, the genesis, track and wind field models. The genesis model was used to simulate the tornado annual occurrence rate and spawn locations. The track model was utilized to simulate the tornado parameters (such as intensity, heading direction, width etc.) according to its spawn location. The wind field model was used to compute the tornado wind speed along the tornado track. The parameters of each simulated tornado include the intensity (EF scale), touchdown location in terms of the latitude and longitude, touchdown date and time, path length, path width and heading direction.

The parameters of the simulated tornadoes in this study are geographic dependent. In other words, the tornado parameters (e.g. EF scale) were sampled from probability distributions that vary based on geographic location. For example, the likelihood of a violent tornado (EF 4 or EF 5) spawns in Kansas, a tornado prone area, is expected to be significantly higher than that a location along the eastern coast of the United States. To achieve a geographic dependent simulation, the Kernel Density Estimation method (KDE) was applied in both the genesis model and the track model. The KDE method is one of the most commonly used spatial analytical techniques, which is used to model the spatial variation of the probability density of a random variable. A bivariate KDE algorithm was utilized to estimate an optimal bandwidth for each tornado parameter. A large bandwidth will causes "oversmoothed" estimation, which obscures much of the data variation details and makes the estimated probability densities similar everywhere. In comparison, if a small bandwidth is applied, the estimated density function will contain too many spurious data artifacts and sharp changes in probability density values between close proximity locations. The bandwidth selection technique proposed by Botev[4] was applied to this study. This method has superior computation efficiency and better performance in the estimation of multimodal density function when compared to other plug-in method or 'rules of thumb' method. The algorithm evaluates the best-fit bandwidth according to the spatial distribution and size of the sample space.

2.3 *Genesis Model*

According to the previous research by Standohar[5], negative binominal distribution was determined to be a distribution suitable for describing the tornado annual frequency. The probability density function of the negative binomial distribution is given by:

$$\Pr(X = N_{Tor}) = \binom{N_{Tor} + r - 1}{N_{Tor}} p^{N_{Tor}} (1-p)^r \tag{1}$$

where N_{Tor} is the number of tornados in a year (annual spawn frequency), p=0.0165 and r=18.02 are the parameters of the negative binomial distribution fitted using the annual spawn frequency obtained from the SPC database.

The two annual spawn frequency parameters were estimated through the maximum likelihood method using the SPC tornado database from 1973 to 2015. Figure 1 shows the quantile-to-quantile (Q-Q) plot of observed versus modeled annual spawn frequencies. The Q-Q plot is a graphical method for comparing the empirical (observed) and modeled probability distributions by plotting their quantiles against each other. If the empirical and modeled probabilities have identical distributions, the points on the Q-Q plot will lie along a 45-degree straight line. In Figure 1, the modeled tornado annual occurrence rates are plotted on the x-axis, and the corresponding quantile values from the tornado historical database are plotted on the y-axis. The data points correspond to 10th,

20th, 30th, etc. percentiles of the two probability distributions. From Figure 1, the majority of the points are close to the 45-degree line. The modeled probability distribution underestimates the empirical dataset for high annual occurrence rates (>1500 tornadoes/year). The slight difference between the modeled and observed data only occurs at the 93th percentile and beyond, so the fitted negative normal distribution is deemed an acceptable distribution for modeling the annual spawn rate.

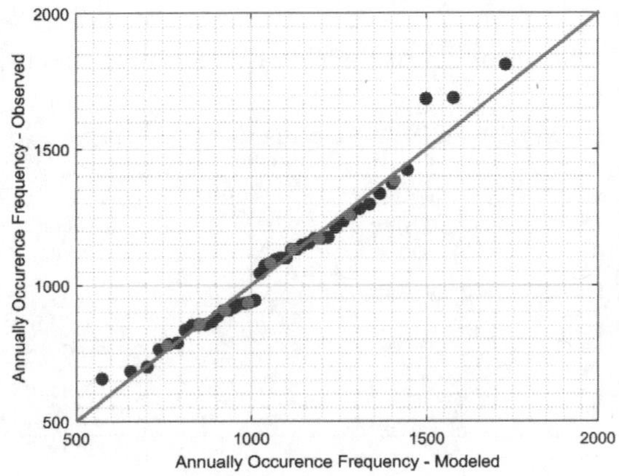

Figure 1. Q-Q plot of observed and modeled tornado annually spawn frequencies.

To consider the potential variation in tornado occurrences due to climatological differences in the U.S. (Coleman[6], Farney[7]), this study modeled the tornado touchdown location as a geographic dependent parameter, which considers higher occurrence probabilities for areas where relatively large numbers of tornadoes have been observed. Each modeled tornado spawn location was randomly generated around a known historical spawn location by using a bivariate normal random number generator.

The simulation procedure for the tornado genesis model is as follows:

(1) Randomly sample a total number of tornadoes (N_{Tor}) to be generated for a simulation year using the fitted negative binominal distribution (p=0.0165, r=18.02);

(2) Randomly pick a past recording year (1973-2015) and select N_{Tor} observed tornado spawn locations in that year as the seeds for spawning tornadoes;

(3) For each of the N_{Tor} tornadoes, use a bivariate normal distribution generator with the mean (μ) equal to the latitude and longitude of the selected seed tornado spawn point and the covariance matrix (σ) equal to the optimal bandwidth, determined using the KDE method, to generate a random spawn location.

(4) If any of the modeled tornado spawn locations are outside of the land boundary, regenerate the spawn locations of those tornadoes until they are all on the land.

To prevent the genesis model from being trapped in an infinite loop, a threshold for regeneration of the spawn locations was set. For each simulation, if the re-sampling threshold was reached in step (4), those re-sampled tornadoes with their spawn locations outside of the land boundary were removed and the N_{Tor} was reduced accordingly for the simulation year.

2.4 *Track Model*

For each tornado spawned using the genesis model, six additional tornado parameters, namely (1) EF-scale, (2) spawn hour, (3) spawn date/month, (4) heading angle (measured clockwise from the true North), (5) length, and (6) width, were simulated using the track model. To ensure the simulated tornado parameters follow the geographic patterns of historical tornado records, four of the six tornado parameters were analyzed separately using GIS KDE maps developed for the parameter of interest. Table 1 shows the grouping of the four tornado parameters.

Table 1: Grouping of tornado parameters

Group B_i	Parameter			
	EF-scale	Heading Angle (deg°)	Spawn Hour	Spawn Date/Month
B_1	EF 0	0 – 22.5	1	Early Jan.
B_2	EF 1	22.5 – 45	2	Late Jan.
B_3	EF 2	45 – 67.5	3	Early Feb.
...	...	...	...	...
B_{n-1}	EF 4	315 – 337.5	23	Early Dec.
B_n	EF 5	337.5 – 360	24	Late Dec.

For a specific tornado parameter, for instance the EF-scale, the historical tornadoes were categorized into different groups (B_i) and each group contained tornadoes with the same characteristic (e.g. EF-scale equal to 2). The tornado spawn records from these grouped datasets were used to generate the probability density contour maps using the KDE method. The KDE contours reflect the spatial distribution and concentration of tornadoes with the same characteristic. The developed probability density contour maps were used to determine the point estimate for probability density of tornadoes at a given location with the specified group of the parameter of interest, Figure 2.

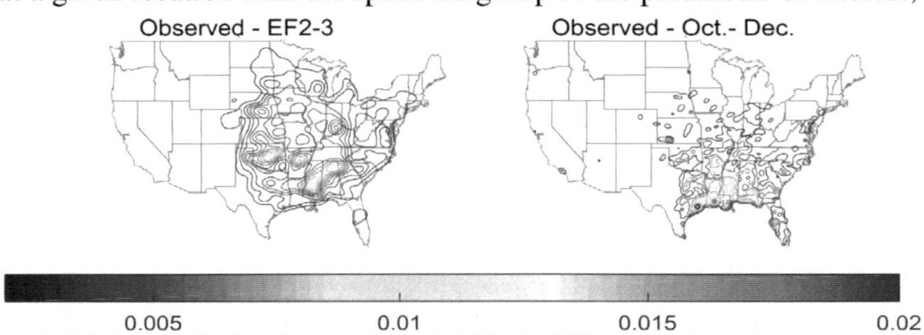

Figure 2: Spatial distribution for tornado probability by EF-scale (left) and spawn month (right).

For example, consider the determination of the EF-scale for a tornado j at location $(Lat, Lon)_j$. According to the Bayes' theorem, the probability of a tornado occurs at location $(Lat, Lon)_j$ and its EF-scale is equal to B_i is:

$$P(B_i \cap (Lat, Lon)_j) = P((Lat, Lon)_j \cap B_i) = P((Lat, Lon)_j \mid B_i) \times P(B_i) \tag{2}$$

where $P(B_i)$ is the probability of the EF-scale of the tornado is equal to B_i. $P\big((Lat, Lon)_j | B_i\big)$ denotes the point estimate for the probability of a tornado spawned at location $(Lat, Lon)_j$ given that the EF-scale of the tornado is B_i. To obtain the point estimate probability based on EF-scale, a set of probability density maps were developed by grouping the historical tornadoes into groups B_1 to B_6 for tornadoes with EF-scales equal to 0 to 5, respectively. These probability density maps were used to determine the probability mass function for EF-scale B_i given a tornado spawned at location $(Lat, Lon)_j$:

$$P(B_i \mid (Lat, Lon)_j) = \frac{\sum_{m=1}^{i} P((Lat, Lon)_j \mid B_m)) \times P(B_m)}{\sum_{k=1}^{6} P((Lat, Lon)_j \mid B_k)) \times P(B_k)} \quad (3)$$

The location specific probability mass function was used to simulate the EF-scale of the tornado. Similar procedures were applied to simulate the tornado spawn hour, spawn date/month and heading angle. For determining the tornado spawn hour, B_i represents the tornado spawn hour in a day (1 to 24). For the heading angle, the data were grouped into eight equal bins with a 22.5-degree increment. For the spawn date and month, the data were divided into 24 groups with each month split into two segments, first half and second half of the month. It should be noted that the first half of each month always contains 15 days and the second half of the month contains the remaining days of that month. After the determination of the tornado spawned month segment (i.e. early January, late January etc.), the probability mass function for the days in that month segment was used to determine the actual day in a year (1 to 365).

Tornado path length and maximum width were simulated according to the EF-scale. Based on the study of Brooks[8], tornado path length (PL) and maximum path width (PW) were determined to correlate to the tornado intensity. A stronger tornado tends to have a longer path length and wider path width than that of the weaker ones. Brooks also modeled the tornado path lengths and maximum widths using the two-parameter Weibull distribution. The cumulative distribution function of the Weibull distribution is:

$$F(x) = 1 - \exp(-(x/c)^d) \quad (4)$$

where c and d are the scale and shape parameters of the Weibull distribution. The distribution parameters for the length and width conditioned on the EF-scale were fitted using the maximum likelihood method and are shown in Table 2.

Table 2. Fitted Weibull distribution parameters for tornado length and width.

	Par.	Enhanced Fujita Scale					
		EF 0	EF 1	EF 2	EF 3	EF 4	EF 5
Length (km)	c	1.16	4.28	10.3	25.4	44.7	67.7
	d	0.677	0.729	0.796	1.04	1.13	1.55
Width (m)	c	41.3	93.6	187	415	702	952
	d	1.01	0.943	0.908	0.994	1.16	1.38

2.5 *Wind Field Model*

2.5.1 *Wind field along the tornado path*

The intensity of a tornado along the track usually degrades as the friction with the ground dissipates the energy of the tornado. After examined 150 tornado tracks, Twisdale[9] determined the intensity variation along the track of tornadoes. The fractions of the tornado strength for each of the highest EF-scales attained by a tornado are shown in Figure 2. In this study, it was assumed that the maximum intensity occurs at the middle of the track and the lowest intensity occurs at the beginning and the end of the tornado track.

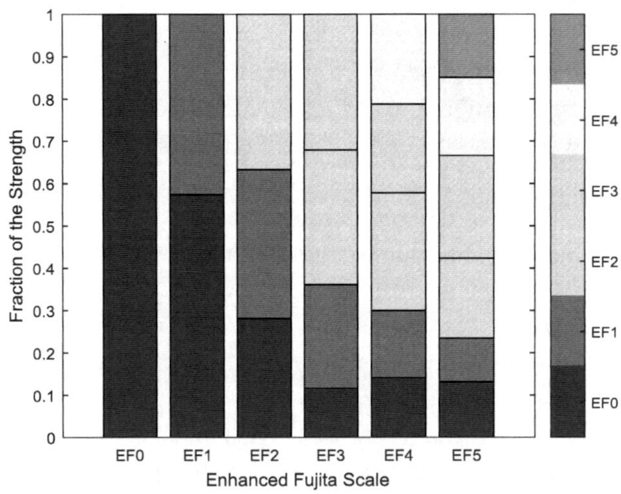

Figure 3. Tornado intensity variation along the track (adapted from Twisdale[9]).

Table 3. Tornado EF-scale and wind speed

EF-scale	Wind Speed	
	(mph.)	(km/h)
EF 0	65-85	105-137
EF 1	86-110	138-177
EF 2	111-135	178-218
EF 3	136-165	219-266
EF 4	166-200	267-322
EF 5	Over 200	Over 322

2.5.2 *Wind field across the tornado path*

The wind speed variation across the tornado vortex was modeled using the modified Rankine vortex model in which the tangential velocity can be computed as:

$$V_{\tan}(r) = \frac{r\Gamma_{\infty}}{\pi(r^2 + r_c^2)} \tag{5}$$

where r is the radial coordinate with $r = 0$ at the center of the tornado vortex and r_c is the core radius where the maximum tangential velocity ($V_{tan,max}$) occurs, $V_{tan,max}$ is assumed to be uniformly distributed over the range of corresponding EF-scale (Table 3). $\Gamma_\infty = 2\pi r_c V_{tan,max}$ is the maximum vortex strength. Substitute $\Gamma_\infty = 2\pi r_c V_{tan,max}$ into equation (5) yields:

$$V_{\tan}(r) = \frac{r\Gamma_{\infty}}{\pi(r^2 + r_c^2)} = \frac{2r(r_c V_{\tan,\max})}{r^2 + r_c^2} \tag{6}$$

collect the r_c terms in equation (6) gives the following expression:

$$r_c^2 - (\frac{2rV_{\tan,\max}}{V_{\tan}(r)})r_c + r^2 = 0 \tag{7}$$

Note that r_c is not equal to the maximum path width (PW). According to the SPC database, the maximum path width is a damage based value and it represents the maximum damage width caused by a tornado. Damage to a building typically occurs when the tangential wind speed reaches the minimum wind speed of an EF 0 tornado (105 km/h). Thus, it is reasonable to assume the tangential wind speed $V_{tan}(r)$ is equal to the lower bound wind speed of EF 0 (105 km/h) at $r = PW/2$. set $r = PW/2$ and $V_{tan}(r) = 105\ km/h$ in equation (7),

$$r_c^2 - (\frac{PW}{105} V_{\tan,\max})r_c + (\frac{PW}{2})^2 = 0 \tag{8}$$

solve for r_c and $V_{\tan}(r)$ is determined.

To account for the asymmetry in wind speeds caused by the tornado translation speed (V_T), truncated normal distributions describing the tornado translation speed developed by Twisdale[10] were used in this study to construct the asymmetric wind field (Table 4).

Table 4. Tornado translation speed distribution parameters

EF-scale	Tornado Translation Speed			
	Lower bound (km/h)	Upper bound (km/h)	Mean (km/h)	Standard Deviation (km/h)
EF 0	8	56	32	12
EF 1	8	72	40	15.4
EF 2	8	80	48	16.6
EF 3	8	88	56	17.6
EF 4	8	104	72	20.2
EF 5	8	120	88	22.4

3 RESULT AND DISCUSSION

The tornado simulation framework discussed in the previous sections was used to generate 500,000 years of tornadoes, which includes approximately half a billion individual tornadoes. A simple comparison of simulated and observed tornado tracks across or within 64.4 km (40 mi) from the Oklahoma City is shown in Figure 5 and the tornado count for each EF-scale is shown in Table 5.

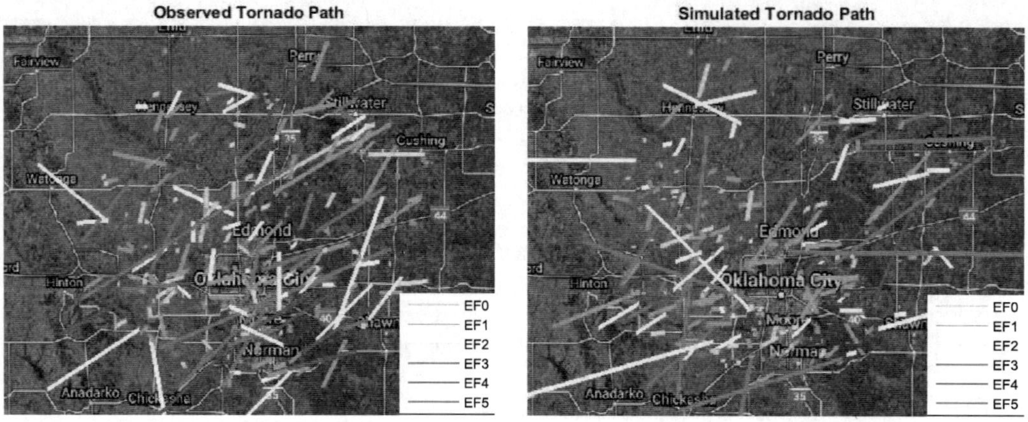

Figure 4: Observed (left) and simulated (right) tornado tracks near Oklahoma City

Table 5: Numbers of simulated and observed tornado tracks near Oklahoma City

EF-scale	Tornado in Oklahoma City	
	Observed (1973-2015)	Simulated (43 years)
EF 0	120	129
EF 1	124	115
EF 2	41	39
EF 3	20	19
EF 4	10	3
EF 5	3	2

Comparisons were further made between the simulated and observed statistics for tornado intensity, spawn month and spawn hour for selected cities. Three cities were selected according to the degrees of tornado activity in the regions. These cities were: (1) Des Moines (41.577, -93.617) located in High Plains, (2) Oklahoma City (35.457, -97.514) located in tornado alley, and (3) Indianapolis (39.777, -86.148) located in Midwest region. A search radius of 40km (25 mi) from the city center was used to identify the tornadoes that affected the city of interest. These tornadoes were used to compare the statistics of the simulated and actual observed tornadoes.

3.1 *Tornado Spawn Month/Time*

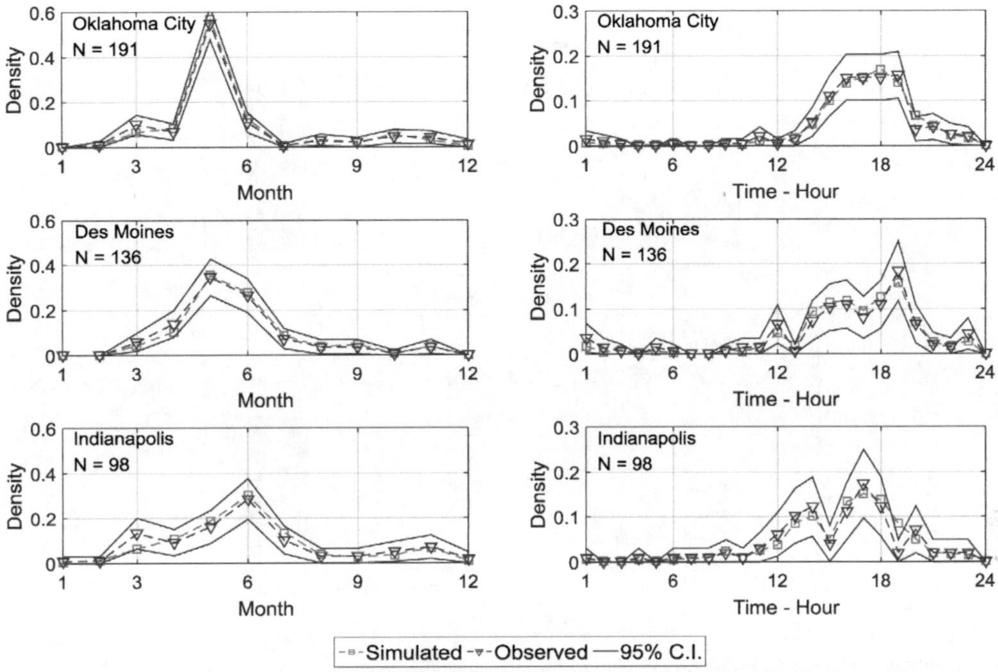

Figure 5. Probability mass functions of tornado spawn hour (left) and spawn month (right) for selected cities.

The probability mass functions of tornado spawn hour and month for the three selected locations are shown in Figure 5. The red dash lines are the means of the simulated PMFs and the blue dash lines are the observed PMFs. The 95% confidence intervals are plotted as blue solid lines. Also shown in Figure 5 are the actual numbers of historical tornadoes observed (N) for the three cities. It can be seen from Figure 5 that the PMFs of the simulated tornado spawn time and month match the observed data. All three cities experience high tornado activities from March to June. Some study have shown that, nocturnal tornado is the main reason that causes high fatality rate in the South-eastern region of the United States (Ashley[11,12]). Unlike other cities which tornado often observed during the afternoon, Figure 5 shows Des Moines has a high probability of observing tornadoes at night.

3.2 *Tornado intensity*

The annual tornado spawn frequency within a radius of 40km (25mi) of the selected cities for different EF-scales are plotted in Figure 6. The histograms show the simulated tornado frequencies and the x markers are the means annual frequencies of historical events. Also shown in Figure 6 are the 95% confidence intervals. The model shows a very good simulation result in those selected cities, all test values are within 95% confidence interval except those have no tornado observed in the recorded period. Note that there was no EF 5 tornado reported in Indianapolis and Des Moines over the record period (1973 to 2015); however, this does not mean the occurrence probability of EF 5 tornadoes in these cities are zero. Based on 500,000 years of simulated tornadoes, the model predicted annual spawn frequencies for EF 5 tornadoes for Indianapolis and Des Moines are 0.13 and 0.14, respectively.

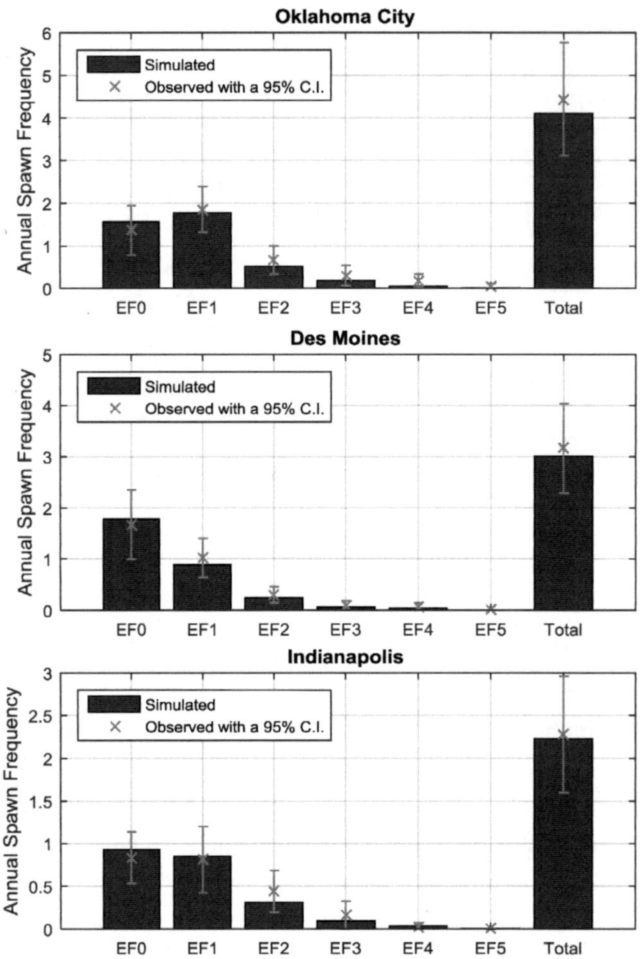

Figure 6. Annual spawn frequencies by EF-scales for Oklahoma City (top), Des Moines (middle), and Indianapolis (bottom).

3.3 *Mean recurrence interval*

The mean recurrence interval (MRI) for tornado induced wind speed at a given site can be determined using the following equation:

$$MRI(v_i > V) = \frac{1}{\lambda P(v_i > V)} = \frac{Y}{n} \tag{11}$$

where n is the total number of tornadoes with wind speeds v_i at the site higher than the threshold wind speed V, and Y is the number of simulation years (i.e., 500,000 years in this study). The MRI curves for tornado induced wind speeds for the three cities considered in this study are plotted in Figure 7. Every point on the MRI curve of a site represents the peak wind speed observed at the site for a particular simulated tornado event. These wind speeds when coupled with a given target hazard level can be used as reference design wind speeds for structural design purpose.

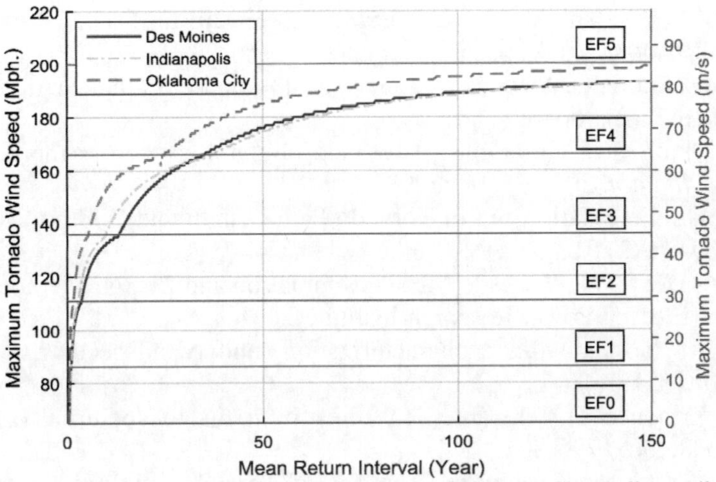

Figure 7: Mean recurrence interval plots of tornado wind speed for Oklahoma City, Indianapolis and Des Moines.

4 CONCLUSION

In this study, a stochastic tornado track simulation program was used along with a wind field model to simulate 500,000 years of synthetic tornadoes and the corresponding wind speeds. In the simulation program, the annual spawn rate was modeled using a negative binomial distribution. The spawn or touchdown locations were simulated using geographic dependent kernel density estimation (KDE) GIS maps, which specifically account for the variability of tornado properties at different geographic regions (e.g. tornado alley, Dixie alley and etc.). In the wind field model, tornado wind speed was determined using degradation model along the path and modified Rankine vortex model across the path. The track parameters include the tornado occurrence rate, intensity (EF-scale), touchdown location, touchdown time. All these parameters are geographic dependent, meaning the properties vary depend on the geographic locations. The simulated spawn rates and other parameters for three selected cities were compared to observations and the modeled results compared well with the observed tornado records. As an illustrative example for design application, wind speed distributions were generated with respect to the mean recurrence intervals at selected cities. Even though this study was only performed for several selected cities, the procedure presented here can be applied to the whole United States. The simulated tornado database can be used for performance-based design, risk assessment and loss estimation for tornado hazards.

5 REFERENCES

1. Meyer CL, Brooks H, Kay M. A Hazard Model For Tornado Occurrence In The United States. 2002.
2. Daneshvaran S, Morden RE. Tornado risk analysis in the United States. *J Risk Financ.* 2007;8(2):97-111.
3. Banik SS, Hong HP, Kopp GA. Assessment of tornado hazard for spatially distributed systems in southern Ontario. *J Wind Eng Ind Aerodyn.* 2008;96(8-9):1376-1389.

4. Botev ZI, Grotowski JF, Kroese DP. Kernel density estimation via diffusion. *Ann Stat.* 2010;38(5):2916-2957.
5. Standohar-Alfano CD, van de Lindt JW. Empirically Based Probabilistic Tornado Hazard Analysis of the United States Using 1973–2011 Data. *Nat Hazards Rev.* 2014;16(1):4014013.
6. Coleman T a., Dixon PG. An objective analysis of tornado risk in the United States. *Weather Forecast.* 2013;29(2):366–376.
7. Farney TJ, Dixon PG. Variability of tornado climatology across the continental United States. *Int J Climatol.* 2014;35(10):2993-3006.
8. Brooks HE. On the Relationship of Tornado Path Length and Width to Intensity. *Weather Forecast.* 2004;19(2):310-319.
9. Twisdale LA, Dunn WL. Tornado Missile Simulation and Design Methodology. *ERPI NP-2005.* 1981;2(Electric Power Research Institute).
10. Twisdale LA. Tornado data characterization and windspeed risk. *J Struct Div.* 1978;104(10):1611-1630.
11. Ashley WS, Krmenec AJ, Schwantes R. Vulnerability due to Nocturnal Tornadoes. *Weather Forecast.* 2008;23(5):795-807.
12. Ashley WS. Spatial and Temporal Analysis of Tornado Fatalities in the United States: 1880–2005. *Weather Forecast.* 2007;22(6):1214-1228.

The 13th Americas Conference on Wind Engineering (13ACWE)
Gainesville, Florida USA, May 21-24, 2017

Dual-Doppler Radar, In Situ Anemometric and Ground Damage Observations of the 27 November 2014 Brisbane Supercell

Richard J. Krupar III [a], Matthew S. Mason [a], Daniel J. Smith [b], Joshua Soderholm [a], Alain Protat [c], W. Scott Gunter [d]

[a]*The School of Civil Engineering, The University of Queensland, St. Lucia, Queensland, Australia*
[b]*Cyclone Testing Station, James Cook University, Townsville, Queensland, Australia*
[c]*Centre for Australian Weather and Climate Research, Bureau of Meteorology-CSIRO, Melbourne, Victoria, Australia*
[d]*Earth and Space Sciences, Columbus State University, Columbus, Georgia, USA*

ABSTRACT: The primary aims of this study were to characterize the three-dimensional wind structure of the supercell that impacted Brisbane on 27 November 2014 and compare available Doppler radar data with near-surface in situ wind histories and ground-based damage survey data. Single- and dual-Doppler radar data show that the storm downdraft impacted the Archerfield Airport automatic weather station (AWS), which recorded peak one-minute mean and maximum three-second gust wind speeds of 31.4 ms^{-1} and 39.2 ms^{-1} respectively. After passing through Archerfield, the storm evolved into a supercell, but produced weaker wind speeds at the Brisbane AWS. Dual-Doppler radar horizontal wind profiles above the AWSs evolved from synoptic boundary layer wind profiles into thunderstorm outflow wind profiles that varied spatially and temporally. Most of the wind damage observed after the event was around Archerfield, while hail damage was observed across the entire study region.

KEYWORDS: Supercell, Thunderstorm Outflow, Doppler Radar, Wind Profiles, Automatic Weather Station, Damage Survey.

1 INTRODUCTION

Operational fixed-site ground-based Doppler radar studies have provided pertinent information regarding the kinematic and thermodynamic evolution of convective thunderstorms [1-6]. This early radar work coupled with aircraft and ground-based in situ observations contributed to the development of the long-standing conceptual model of the near-surface structure of supercell thunderstorms [7]. Whilst operational fixed-site ground-based Doppler radars have contributed significant knowledge to our understanding of both kinematic and thermodynamic convective thunderstorm structure, they are not resolute enough, both horizontally or temporally, to resolve storm-scale mechanisms that drive convective thunderstorm features. This known limitation has led to the development of ground-based high-resolution mobile research Doppler radars [8-9]. The advent and deployment of these systems in field campaigns such as Severe Convective Outflow in Thunderstorms [10] has increased the body of knowledge on the three-dimensional kinematic evolution of convective thunderstorms. However, more Doppler radar-derived thunderstorm wind profiles need to be examined to better understand how they vary from storm-to-storm and how they can be compared with traditional in situ wind measurements to include them in current building design standards.

On 27 November 2014, an intense supercell impacted South East Queensland with maximum three-second gust wind speeds of 39.2 ms^{-1} and tennis ball size hailstones [11]. While damaging winds did not reach design limits specified by [12], the 39.2 ms^{-1} three-second gust wind speed recorded at the Archerfield Airport Bureau of Meteorology (BoM) Automatic Weather Station

(AWS) was the second highest gust wind speed ever observed at that station. Furthermore, the supercell caused approximately $1.4B (AUD) in total insured loss, making it the costliest Australian natural catastrophe in 2014 [11].

At the time of the event, both a BoM S-band Doppler radar (Mt. Stapylton) and dual-wavelength research Doppler radar (CP2) were simultaneously scanning the storm. Beneath the Doppler radar sweeps, two BoM AWSs (Archerfield Airport and Brisbane) were also recording near-surface (i.e. 10-m) in situ kinematic and thermodynamic data. In the immediate aftermath of the storm, the Queensland Fire and Emergency Services (QFES) conducted a ground-based rapid damage assessment (RDA) in and around Brisbane to gather information about the extent of damage caused by the storm. Given the unique juxtaposition of Doppler radar, AWS, and ground-based QFES RDA data in Brisbane during and after the passage of the storm, this study aims to couple the available observations to (1) characterize the three-dimensional wind structure of the storm as it passed over Brisbane and (2) compare the low-level (i.e. 200 m) Doppler radar data with the BoM AWS in situ wind histories and QFES RDA damage observations.

A description of the data and methods employed in this study are presented in section 2, followed by a discussion of the storm environment in section 3. Doppler radar observations (wind and reflectivity), BoM AWS wind records and QFES RDA data are discussed and compared in section 4. Conclusions and suggestions for future work follow in section 5.

2 DATA AND METHODS

2.1 *Doppler radar data and analysis*

Doppler radar data used in this study were collected by CP2 and Mt. Stapylton Doppler radars, which are both S-band (10.9 cm and 10.0 cm respectively) wavelength systems with half-power beam widths of 0.96° and 0.90° respectively. CP2 is a dual-wavelength research Doppler radar with X-band (3.2 cm) capability, however, for the purposes of maintaining wavelength continuity in the dual-Doppler synthesis, X-band data was not considered. Mt. Stapylton has a Nyquist velocity almost double that of CP2 (Table 1), making CP2 more susceptible to aliasing in severe convective thunderstorms. During the event, CP2 used eleven elevation tilts to collect a full volume of radar data in five-minutes, while Mt. Stapylton used fourteen elevation tilts to collect data in five-minutes and thirty seconds (Table 2).

Table 1. CP2 and Mt. Stapylton Doppler radar specifications

Doppler Radar	Location (°)	PRF (Hz)	Beam width (°)	Range resolution (m)	Nyquist velocity (ms^{-1})	Antenna height (m)
CP2	152.863, -27.669	< 1,700	0.96	150	27	187
Mt. Stapylton	153.240, -27.718	666 – 1,000	0.90	250	52	174

Table 2. CP2 and Mt. Stapylton Doppler radar scanning strategies during data collection

Doppler Radar	Scanning strategies	Azimuth range (°)	Elevation sequence (°)	Data collection time (sec)
CP2	PPI	360	0.9, 1.7, 2.4, 3.2, 4.7, 6.5, 9.1, 12.8, 17.8	300
Mt. Stapylton	PPI	360	0.5, 0.9, 1.3, 1.8, 2.4, 3.0, 4.2, 5.6, 7.4, 10.0, 13.3, 17.9, 23.9, 32.0	330

Mt. Stapylton is located on an isolated hill around 150 m above mean sea level to the east of Beenleigh, Queensland [13]. The radar coverage extends well out to about 200 km in most directions except when the radar look angle is south and veering to the east. Here the low-level coverage reduces to approximately 150 km due to terrain in the nearby Great Dividing Range and Lamington Plateau [13]. Before it was decommissioned, CP2 was located at Redbank Plains, Queensland, where the radar could capture the influence of the nearby coast and complex terrain on South East Queensland weather activity. Despite complex terrain beam blockage issues, CP2 could provide reliable single Doppler operational coverage over Brisbane and was located close enough to Mt. Stapylton (approximately 37.8 km to the west-northwest of Mt. Stapylton) to create a dual-Doppler network (Figure 1).

Before a dual-Doppler synthesis can be performed, radial velocities that exceed the Nyquist velocity must be "unfolded" from the measured radial velocities within the Nyquist interval. This process of unfolding is called "dealiasing" and was conducted using a region-based dealiasing method [14]. The method identifies regions of similar Doppler radial velocities as a dynamic network graph and unfolds them against one another by minimizing the radial velocity differences between neighbouring range gates. Regions with the largest number of shared range gate edges are unfolded first. Given that Mt. Stapylton has a Nyquist velocity of 52 ms^{-1}, only the 0.9° and 1.7° radial velocity sweeps from CP2 required dealiasing.

Once dealiased and ground clutter removed, single-Doppler radar radial velocities from each radar were gridded to a common Cartesian grid with 1-km x 1-km x 200-m dimensions. In general, the dual-Doppler wind retrieval technique employed in this study [15] consists in minimizing a cost function including the two non-colinear radial velocities from CP2 and Mt-Stapylton as weak constraints, a terminal fall velocity-reflectivity relationship, and the airmass continuity equation as a strong constraint. Using a radius of influence of approximately 1 km that increased with range, a weighted combination of upward and downward intergrations of the airmass continuity equation were performed between the ground and cloud top to minimize errors associated with just integrating upward only. Linear time interpolation of the CP2 and Mt Stapylton radial velocities to a single time of reference was performed using two consecutive volumetric radar scans to account for storm evolution between radar volumes. The time of reference is the start time of the second scan.

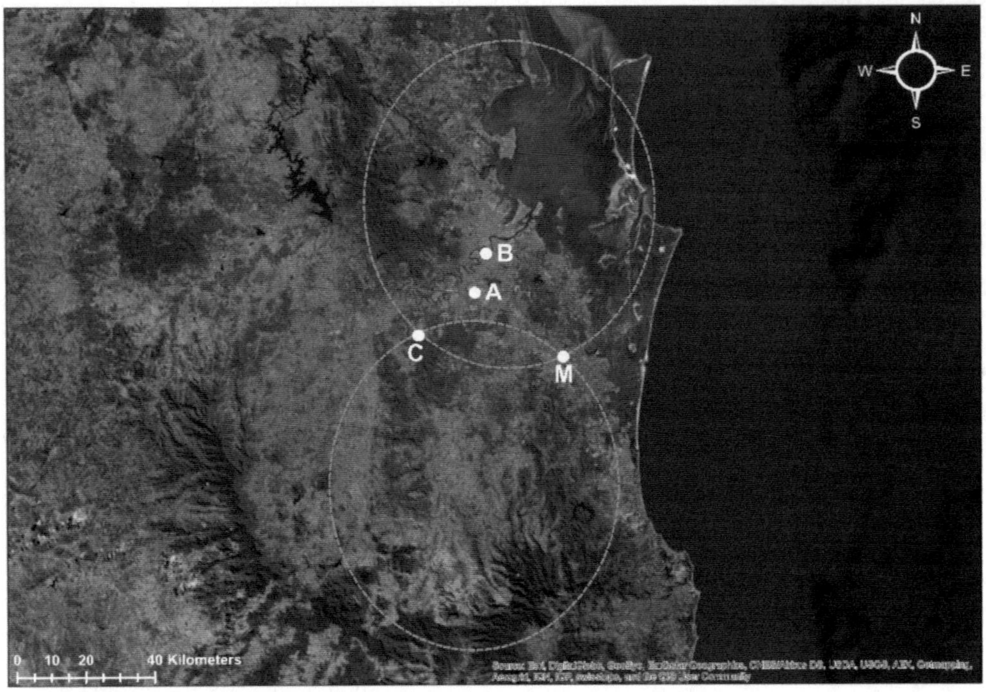

Figure 1. Dual-Doppler radar lobes for CP-2 (C) and Mt. Stapylton (M) Doppler radars over the Archerfield Airport (A) and Brisbane (B) BoM AWSs during the Brisbane supercell on 27 November 2014. A crossing angle of 30° is assumed to generate the theoretical lobes using [16].

2.2 *Automatic Weather Stations*

The Archerfield Airport and Brisbane BoM AWSs provided continuous observations of air temperature, relative humidity, barometric pressure, wind speed and direction through the duration of the supercell event (Figure 2). Each AWS uses a Synchrotac 706 three-cup anemometer that has a distance constant of 13-m [17] and is processed using a three-second moving average procedure [17 – 18]. This 'double filter' attenuates the wind gust spectra at frequencies above 0.2 Hz [17] and will therefore modulate measured three-second gust values but will have little influence on one-minute averages. For this study, however, no correction for this filtering process has been attempted. One-minute barometric pressure, air temperature, relative humidity and rainfall observations were also sourced to characterize the time of arrival of the thunderstorm downdraft and other pertinent storm features.

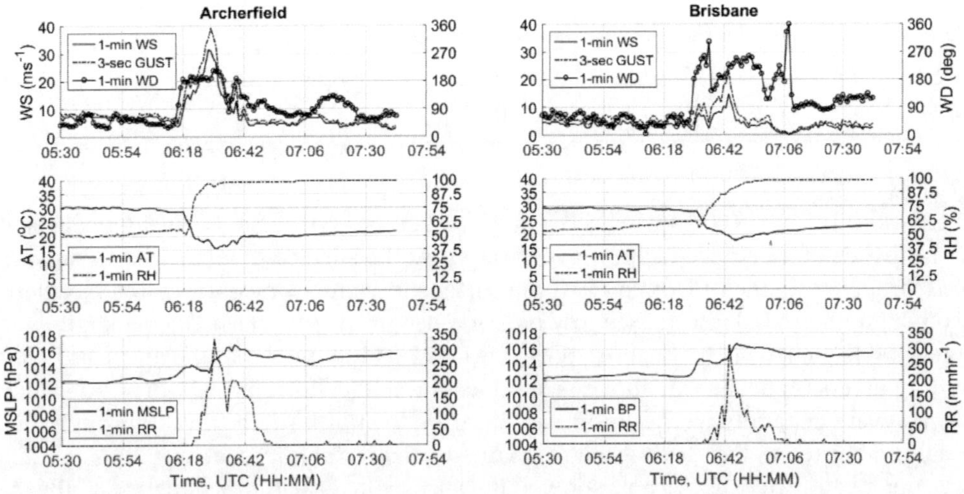

Figure 2. (*Top*) Archerfield and Brisbane AWS one-minute mean wind speed (WS in ms^{-1}) and direction (WD in deg) and maximum three-second gust (GUST in ms^{-1}) wind histories. (*Middle*) Archerfield and Brisbane AWS one-minute mean air temperature (AT in °C) and relative humidity (RH in %) histories. (*Bottom*) Archerfield and Brisbane AWS mean sea level pressure (MSLP in hPa) and rain rate (RR in $mmhr^{-1}$) histories. Measurements were collected on 27 November 2014 and time is shown in UTC on the x-axes.

2.3 *Queensland Fire and Emergency Services (QFES) Rapid Damage Assessment (RDA)*

Rapid damage assessment of residential structures was performed by specially trained Urban Search and Rescue personnel using hand-held electronic devices in the hours following the storm. Due to the nature of QFES objectives, RDA surveys tend to emphasize areas where more severe damages pose a potential threat to life safety, so are not a uniform assessment of damage over all structures. That said, a total of 3,343 RDA data points were included in the analyzed event set. The relevant observations collected for each point included: location, damage state (undamaged, minor, moderate, severe), presence of tree damage, presence of debris, and a brief description of the property damage (e.g., "window damage from hail"). Photographs of damage were also collected in many instances.

To enable damage footprint comparisons with the gridded maximum dual-Doppler radar horizontal velocity and single-Doppler reflectivity swaths, RDA data points were assigned either hail or wind damage modes. The assignments were applied using engineering judgement, term searches within the description of property damage (e.g. search for the word "hail") and assumptions based on the presence of tree damage or debris (i.e. both would yield an assignment of wind damage). In selected areas, the resulting damage mode assignments were validated by comparison with the supplied photographs.

Summary statistics for the hail and wind damage were computed and are shown in Table 3. Most damage (2,810 records) was minor and was attributed to hail. However, 90 and 25 instances of 'moderate' (e.g., failure of fascia, awnings and smaller sections of roofing) and 'severe' (e.g. complete roof loss, etc.) wind damage respectively were observed from Brisbane central business district (CBD) through to Forest Lake (Southeast of Brisbane), particularly near the suburbs of West End and Yeronga.

Table 3. Summary of hail and wind damage observed during the QFES RDA.

Damage Mode	Minor	Moderate	Severe	Count
Hail	97%	3%	0%	2425
Wind	61%	31%	8%	295

3 STORM ENVIRONMENT

Beginning at 00:00 UTC (10:00 am AEST) On 27 November 2014, South East Queensland was under the influence of a marginal upper-level trough. At the surface, a relatively broad surface trough axis stretched southeast from a 1010 mb surface low pressure center in the Northern Territory through another 1010 mb surface low pressure center over western Queensland. An atmospheric sounding launched at Brisbane Airport at 00:00 UTC showed that the air mass below the weak capping inversion near 1 km above ground level (AGL) was relatively dry and stable (Figure 3). Winds were generally weak, unidirectional from the north to northwest through the lower 1 km AGL, and backed to the south through 2 km AGL. Several hours later at 05 UTC, another sounding launched at Brisbane Airport showed the surface layer had moistened (e.g. the temperature and dewpoint profiles were closer together). A shallow pocket of dry air was present below 1 km and had likely been advected eastward throughout the late morning/early afternoon. Winds over the lower 1 km had veered from the east-northeast to the south-southwest as evidenced by the hodograph further supporting low-level moisture advection. Marginal wind shear existed across Brisbane to promote mixing and a source of vorticity for thunderstorms early in the analysis period. Moreover, the advection of dry air over the moist inland air mass in the low-levels helped destabilize the atmosphere over Brisbane.

Convective initiation commenced over the Border Ranges, Lamington, and Mount Barney National Parks between 03:00 – 04:00 UTC as the surface pressure fell across South East Queensland in response to the eastward propagation of the upper-level trough and surface trough axis (not shown). With ample inland low-level moisture in place as evidenced by the presence of a broad inland dryline, the southern range multicell storms began to intensify and slowly move off the high terrain to the north-northeast as a cluster through 06:00 UTC. At 06:12 UTC, a weak southeasterly change in wind direction in association with a sea breeze front promoted a north-northeasterly trajectory for the far northern multicell cluster, specifically the far northeastern storm. The surface winds near the multicell continued to veer and possibly provided low-level streamwise vorticity for the multicell thunderstorm as it approached Brisbane.

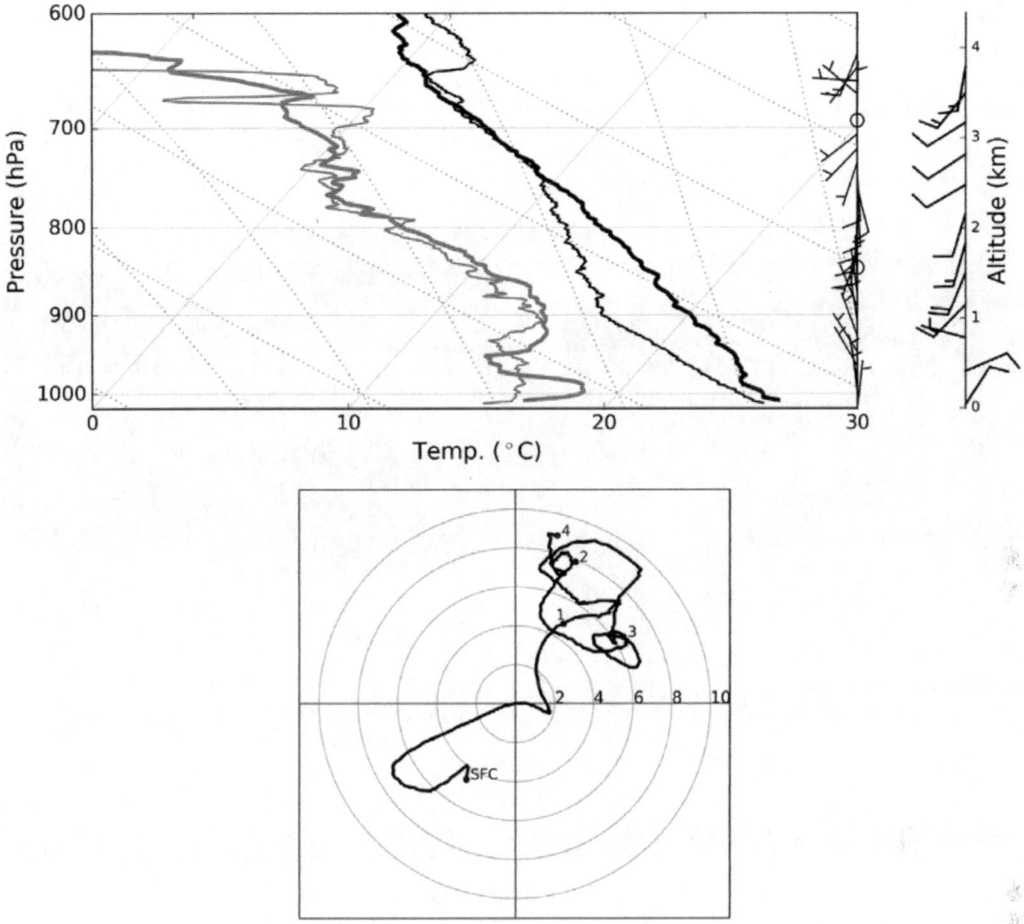

Figure 3. Atmospheric temperature and dewpoint profiles taken inland at 00:00 UTC (thin solid lines and wind barbs) and at Brisbane Airport at 05:00 UTC (thick solid lines and wind barbs). The hodograph for the 05:00 UTC sounding is also displayed with range rings indicating wind speeds in knots.

4 ANALYSIS OF STORM EVOLUTION

4.1 *Doppler radar observations*

An individual multicell thunderstorm began to emerge from a cluster of thunderstorms south of the dual-Doppler radar baseline around 05:48 UTC (not shown). This multicell thunderstorm continued to intensify as it moved through the baseline between 06:06 – 06:18 UTC (not shown). The storm evolved into a high-end multicell thunderstorm before the downdraft passed over the Archerfield Airport AWS between 06:24 – 06:30 UTC (Figure 4a – 4b). Dual-Doppler radar horizontal wind speeds at 200 m were between 25 – 30 ms^{-1} during this time frame and blanketed most of Archerfield. At 06:36 UTC, the reflectivity field at 200 m began to resemble a more classic Southern Hemisphere supercell (Figure 4c) pattern as evidenced by the hook signature. Also, a mesocyclone became more prominent in the 1,500 – 3,000 m dual-Doppler radar wind fields as the supercell approached Brisbane (not shown). Between 06:36 – 06:42 UTC, the downdraft and maximum reflectivity core passed over Brisbane and the radar hook signature began to occlude. This occlusion continued through 06:54 UTC as the downdraft weakened and the storm moved away from Brisbane.

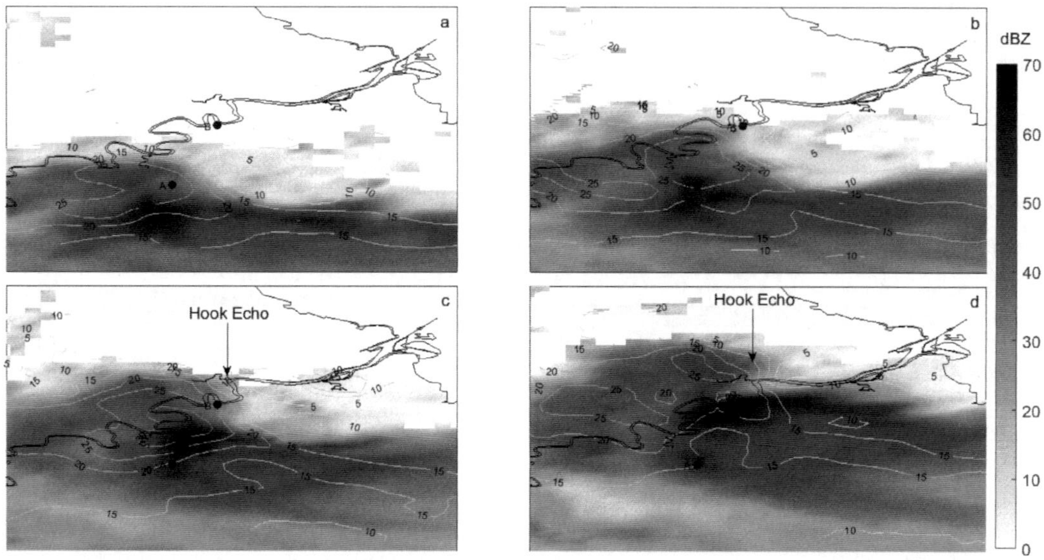

Figure 4. Four panel time history of Mt. Stapylton objectively analyzed 200-m radar reflectivity (dBZ) and dual-Doppler radar horizontal velocity from 06:24 – 06:42 UTC (a – h) on 27 November 2014 overlaid on the study region. Circles A and B represent the location of the Archerfield and Brisbane AWSs respectively. Panels are displayed in six-minute time intervals.

Dual-Doppler radar horizontal velocities were plotted between 200 m – 4,000 m at the grid points closest to the Archerfield Airport and Brisbane AWSs in Figures 5 and 6, respectively. At Archerfield Airport (Figure 5), the 06:12 UTC and 06:18 UTC dual-Doppler horizontal wind speeds remained relatively uniform with height over the lowest 2 km and remained below 10 ms^{-1}. Between 06:18 – 06:24 UTC, the dual-Doppler radar wind profile rapidly evolved to one with a wind speed maximum below 1,500 m with a rapid drop in wind speeds above this height. The same profile shape was observed at 06:30 UTC, however, wind speeds decreased as the storm moved away from the site as too did the height of observed maximum wind speed.

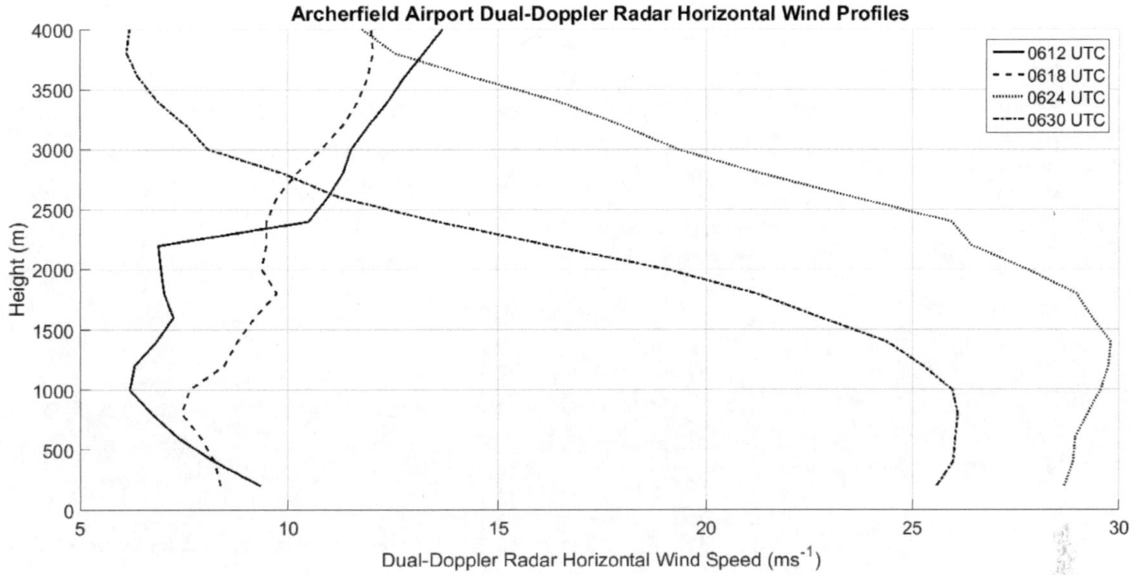

Figure 5. Dual-Doppler radar horizontal wind profiles generated at the grid point closest to the location of the Archerfield Airport AWS.

Beginning at 06:30 UTC over Brisbane (Figure 6), the dual-Doppler wind profile evolved from a pre-thunderstorm outflow wind regime to one more representative of a thunderstorm outflow. wind profile with a wind speed maximum evident below 1,000 m like at Archerfield Airport from 06:30 UTC to 06:36 UTC. In contrast to Archerfield Airport, the dual-Doppler wind profile quickly lost its thunderstorm outflow shape at 06:42 UTC. This is explained by the fact that the Brisbane AWS was clipped by the eastern edge of the outflow so did not experience a prolonged thunderstorm outflow regime. Whilst information in the lowest 200 m of the atmosphere was not readily available, it is clearly seen that the shape and intensity of the wind profile does not stay the same across the entire thunderstorm outflow feature and the height of the observable wind maximum varies both spatially and temporally. These wind profile observations validate the need for collection of more wind profiles in thunderstorms so they can be characterized properly for inclusion in current building design standards around the globe.

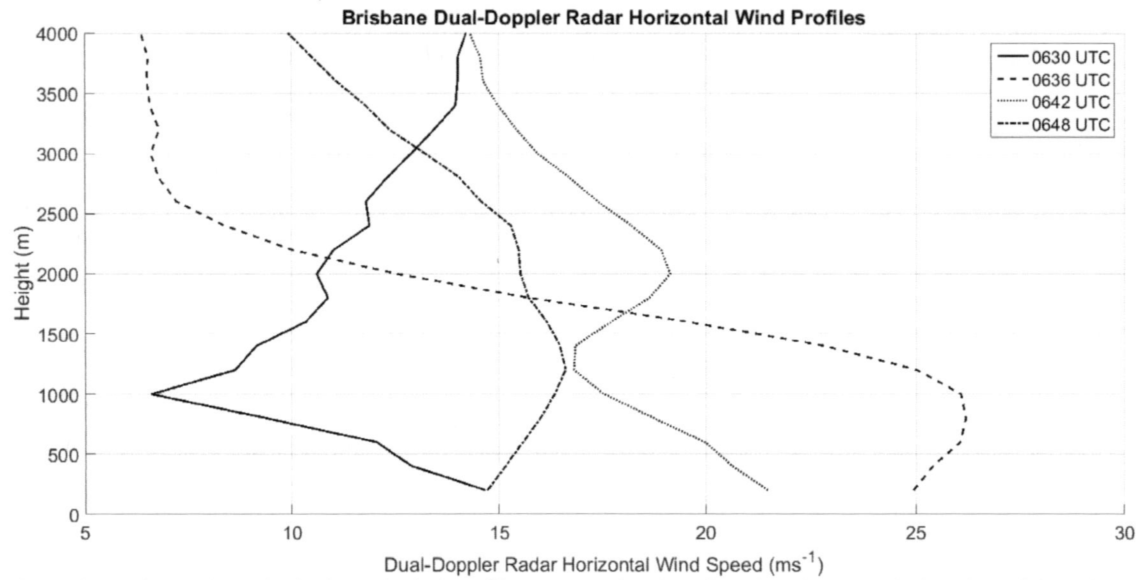

Figure 6. Dual-Doppler radar horizontal wind profiles generated at the grid point closest to the location of the Brisbane AWS.

4.2 *AWS and Dual-Doppler radar horizontal velocity comparisons*

Doppler radar systems have a unique spatial advantage over conventional point-based anemometry because they can measure the wind over a larger area and vertical depth. However, few comparisons have been made between the two measurement platforms to determine an appropriate method for relating one to the other. Issues around spatio-temporal averaging of single- and dual-Doppler radar velocity measurements during sampling and synthesis, for example, mean that direct comparison to anemometer wind measurements are not straightforward. Understanding this relationship, however, will be crucial for the acceptance of Doppler derived wind velocities as a reliable alternative to the traditional point-based anemometry currently used.

Until such time that a suitable comparison methodology is developed, simplified methods must be used for integrating data from these two platforms. Here, an approach of simply viewing both three-second gust and one-minute mean anemometer wind speed data against dual-Doppler velocities is adopted. The lowest dual-Doppler radar analysis level, i.e. 200 m, horizontal wind speeds at the grid point closest to the Archerfield Airport and Brisbane AWS stations were plot alongside one-minute mean and maximum three-second gust wind speeds recorded at 10 m (Figure 7). The distances between the Archerfield Airport and Brisbane AWS stations and the closest grid point are 193 m and 187 m respectively. Inferences about the outflow profile shape relative to the in situ anemometric wind histories below were made and discussion around these observations is supplemented by the thermodynamic time histories shown in Figure 3.

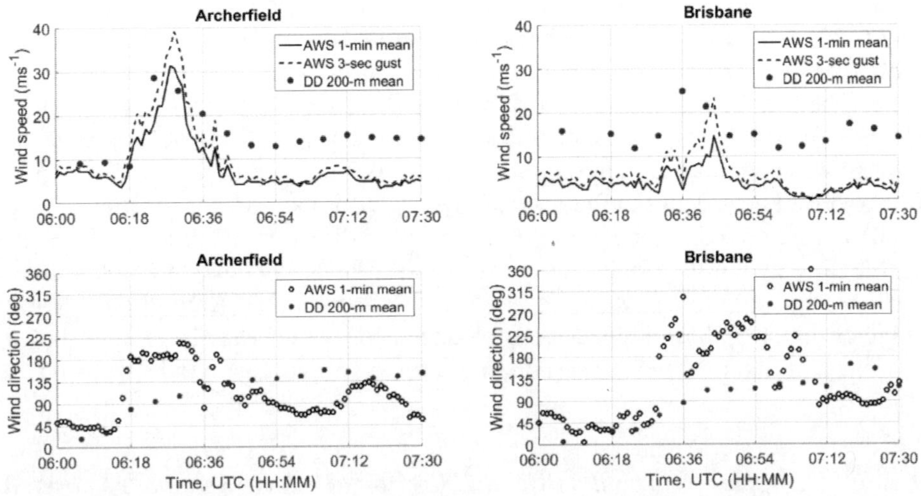

Figure 7. Wind history comparisons between dual-Doppler radar 200 m horizontal wind speeds and both Archerfield (*top*) and Brisbane (*bottom*) one-minute mean and maximum three-second gust wind speeds.

Beginning after 06:18 UTC, the air temperature decreased from 28.7°C to 15.4 °C in roughly thirteen minutes at Archerfield Airport and was accompanied by a relative humidity increase from 53% to 95%. The station mean sea level pressure fluctuated between 1013 – 1015 mb and exhibited a double peak structure. A subtle decrease in mean sea level pressure was observed at 06:28 UTC in association with passage of the leading edge of the downdraft. The subtle decrease in mean sea level pressure has been speculated by [19] to be dynamically generated by converging and rising air along the leading edge of the gust front. The second peak is associated with the passage of the downdraft and is as observed in experimental impinging jet tests as the jet itself moves over a point [20]. The double peak occurs due to the re-intensification process, and the passage of a secondary surge shortly following the initial gust front.

During the passage of the downdraft, the one-minute mean and maximum three-second gust wind speeds accelerated from below 10 ms^{-1} before 06:18 UTC to 31.4 ms^{-1} and 39.2 ms^{-1} at 06:28 UTC and 06:29 UTC and the wind direction veered to the south. The dual-Doppler radar horizontal wind speed also accelerated, however, when the peak one-minute mean and three-second gust wind speeds occurred, there was no dual-Doppler radar horizontal wind speed to compare with the AWS since the temporal resolution is much coarser. However, it is likely that at 10 m winds were just as strong, if not stronger than at 200 m during the peak of the outflow. The dual-Doppler wind direction histories veered but did not capture the approximate 135° wind direction change as the outflow passed over Archerfield Airport. The lack of temporal resolution and reduced influence of surface roughness on the wind at 200 m versus 10 m are believed to help explain the difference in wind direction between the AWS and radar, but further examination is necessary to fully address the differences observed in the wind direction histories.

The air temperature began decreasing at the Brisbane AWS after 06:31 UTC from 28.2°C to 17.8 °C in about sixteen minutes. The relative humidity increased from 61% to 92%, while the station mean sea level pressure fluctuated between 1014 – 1017 mb and exhibited a double peak structure. The one-minute mean and maximum three-second gust wind speeds did not accelerate as quickly as at Archerfield Airport and only reached 14.7 ms^{-1} and 23.3 ms^{-1} respectively at

06:44 UTC. The one-minute mean wind direction veered from the northeast to the west-north-west between 06:28 – 06:36 UTC, sharply backed to south-southeast before veering again to west-southwest. The dual-Doppler radar horizontal wind speed also did not accelerate and peaked above and before the Brisbane AWS. When the Brisbane AWS peak one-minute mean and maximum three-second gust wind speeds occurred, the 0642 UTC dual-Doppler radar horizontal wind speed was comparable in magnitude to the AWS maximum three-second gust wind speed despite the differences between the measurement platforms. The dual-Doppler wind direction histories also veered but did not capture the same wind direction detail as the Brisbane AWS. Local shielding by multi-story buildings and vegetation are believed to have influenced the wind speed and direction histories at this location. Also, unlike the Archerfield Airport AWS, the eastern edge of the downdraft clipped the Brisbane AWS, which may have contributed to the weaker near-surface one-minute mean and maximum three-second gust wind speeds that were observed.

4.3 *Comparisons of Doppler radar reflectivity and horizontal velocity to damage observations*

Hail and wind damage modes derived from the QFES RDA data collected in the storm aftermath were compared with maximum 200 m radar reflectivity and dual-Doppler radar velocity data (Figure 8). The maximum observed radar reflectivity and dual-Doppler velocity values were calculated over the duration of the storm analysis period and were contoured on top of point density plots of the hail and wind damage to examine qualitative spatial relationships between the datasets.

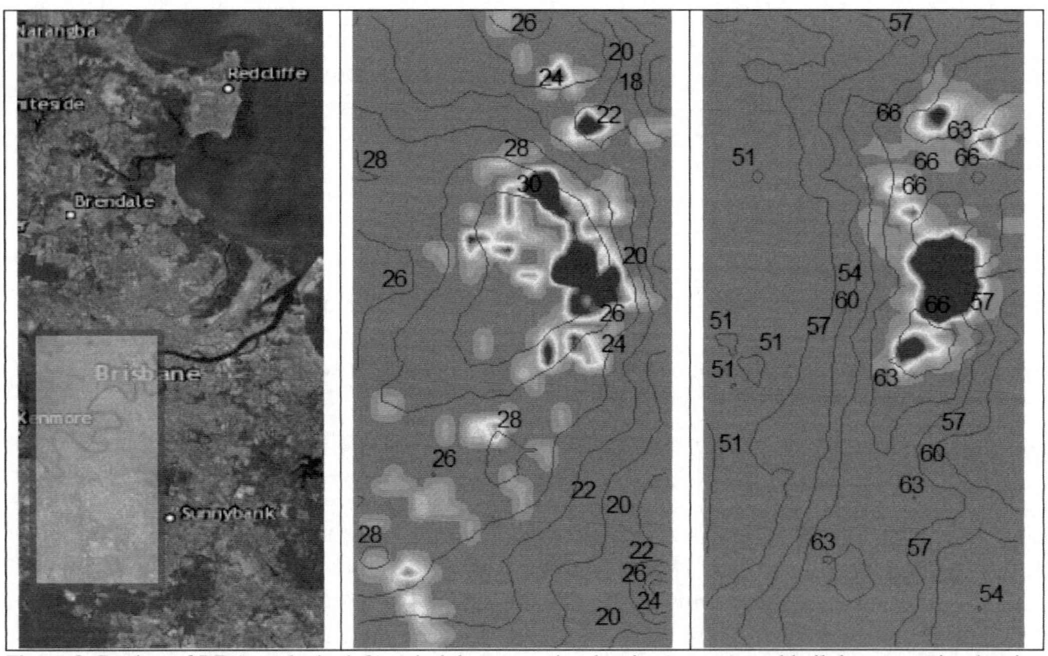

Figure 8. Region of RDA analysis (left), wind damage point density (center), and hail damage point density (right) based on RDA analysis region. Wind speed (m/s) contours (center) and reflectivity contours (right) based on radar data analysis. Note: refer to Table 3 for relative count of hail and wind damage mode points.

As the storm propagated to the north-northeast, impacted suburbs included, but were not limited to: Archerfield, Moorooka, Annerley, West End, Brisbane CBD, Spring Hill and Herston. Focusing primarily on the wind damage mode, the areas of highest wind damage were noted along the east and northeast leading edge of the downdraft (i.e. north of Archerfield Airport) with some isolated pockets north of the observed downdraft. Over Brisbane, the density of wind damage significantly decreased and this can partially be explained by the fact that Brisbane was clipped by the eastern edge of a weakening downdraft. This observation supports the claim by [21] that structures may experience different degrees of damage depending on whether the structure is in the core portion of the downdraft or not. Regarding the hail damage experienced during the event, most of it was collocated or just west of the maximum 200 m radar reflectivity values above 63 dBZ. Looking back at the radar reflectivity panel time histories in Figure 4, the hail damage appears to scale to the horizontal extent of the more intensity precipitation echoes (i.e. 63 dBZ or greater values), which is in general agreement with the observations of [22 – 24] in severe convective windstorms.

5 SUMMARY AND CONCLUSIONS

The unique positioning of the CP2 and Mt. Stapylton Doppler radars with the Archerfield Airport and Brisbane AWSs allowed for a preliminary characterization of the three-dimensional wind structure of the 27 November, Brisbane Supercell outflow. Also, unique and preliminary comparisons between the single-Doppler radar reflectivity and dual-Doppler radar horizontal wind speed with ground-based RDA data collected by QFES were made possible, and relationships between the radar and ground-based damage survey data were examined.

Dual-Doppler radar horizontal wind profiles were inspected over the lower 4,000 m of the boundary layer over both Archerfield Airport and Brisbane AWSs. These wind profiles revealed that the structure and intensity of the wind profiles did not remain constant over the duration of the event, but rather varied spatially and temporally. Dual-Doppler radar horizontal velocity data was limited to heights above 200 m due to surrounding topography and the scanning strategies employed, thus, making it difficult to make direct comparisons at the surface (i.e. 10 m AGL) and inferences about the wind profile structure in and above the surface layer. However, it was shown that the wind profile during this outflow event do not follow a typical synoptic boundary layer wind profile, as used in [2] and other building design standards. More convective thunderstorm wind profiles need to be collected with mobile research radars to better understand how the wind profile evolves below 200 m.

Finally, comparisons of the maximum 200 m dual-Doppler radar horizontal wind speeds and radar reflectivity with QFES RDA data validated the observations of [21 – 24] who showed that wind damage depends on whether a structure is in the downdraft core or not and overall ground damage due to hail coincides with radar reflectivity values about 60 dBZ. Area around and north of Archerfield Airport experienced the most wind damage of all the regions examined since it was directly impacted by the core of the downdraft near its peak intensity. In contrast to Archerfield Airport, whilst the storm acquired supercellular characteristics and the reflectivity increased in magnitude, the downdraft weakened and the eastern edge clipped Brisbane, resulting in far less wind damage. This coupled with local shielding effects is potentially limited the wind speed experienced in and around Brisbane. However, a large area of minor hail damaged was still experienced and was the primary reason for the significant insured losses during the event.

ACKNOWLEDGEMENTS

This research was supported in part by the Australian Research Council (DE150101347), an Advance Queensland Fellowship, and Suncorp Group. The authors would like to extend a special thank you to the Bureau of Meteorology for providing the Doppler radar and automatic weather station data used in this study, as well as, the Queensland Fire and Emergency Services for collecting and sharing their rapid damage assessment data.

REFERENCES

1 P.S. Ray, R.J. Doviak, G.B. Walker, D. Sirmans, J. Carter, and B. Bumgarner, Dual-Doppler observation of a tornadic storm, J. Appl. Meteor., 14 (1975) 1521–1530.

2 E.A. Brandes, Flow in severe thunderstorms observed by dual-Doppler radar, Mon. Wea. Rev., 105 (1977) 113–120.

3 E.A. Brandes, Gust front evolution and tornado genesis as viewed by Doppler radar, J. Appl. Meteor., 16 (1977) 333–338.

4 R.A. Kropfli and L.S. Miller, Kinematic structure and flux quantities in a convective storm from dual-Doppler radar observations, J. Atmos. Sci., 33 (1976) 520–529.

5 G.M. Heymsfield, Kinematic and dynamic aspects of the Harrah tornadic storm analyzed from dual-Doppler radar data, Mon. Wea. Rev., 106 (1978) 233–254.

6 P.S. Ray, K.K. Wagner, K.W. Johnson, J.J. Stephens, W.C. Bumgarner, and E.A. Mueller, Triple-Doppler observations of a convective storm, J. Appl. Meteor., 17 (1978)1201–1212.

7 L.R. Lemon and C.A. Doswell, Severe thunderstorm evolution and mesocyclone structure as related to tornadogenesis, Mon. Wea. Rev., 107 (1979) 1184–1197.

8 J. Wurman, J. Straka, E. Rasmussen, M. Randall, and A. Zahrai, Design and deployment of a portable, pencil-beam, pulsed, 3-cm Doppler radar, J. Atmos. Oceanic Technol., 14 (1997) 1502–1512.

9 M.I. Biggerstaff, L.J. Wicker, J. Guynes, C. Ziegler, J.M. Straka, E.N. Rasmussen, A. Doggett, L.D. Carey, J.L. Schroeder, and C. Weiss, The Shared Mobile Atmospheric Research and Teaching radar: A collaboration to enhance research and teaching, Bull. Amer. Meteor. Soc., 86 (2005) 1263–1274.

10 W.S. Gunter and J.L. Schroeder, High-resolution full-scale measurements of thunderstorm outflow winds, J. Ind. Aerodyn., 138 (2015) 13–26.

11 Insurance Council of Australia, ICA Dataglobe, (2017) http://www.icadataglobe.com/access-catastrophe-data/.

12 Standards Australia, AS/NZS 1170.2:2011 structural design actions. Part II: wind actions, (2011).

13 Commonwealth of Australia, Bureau of Meteorology, Queensland radar site information, (2017) http://www.bom.gov.au/australia/radar/info/qld_info.shtml.

14 J.J. Helmus and S.M. Collis, The Python ARM Radar Toolkit (Py-ART), a library for working with weather radar data in the Python programming language, J. Open Research Software. 4 2016, 25.

15 A. Protat and I. Zawadzki, A variational method for real-time retrieval of three-dimensional wind field from multiple-Doppler bistatic radar network data. J. Atmos. Oceanic Technol., 16 (1999), 432–449.

16 R.P. Davies-Jones, Dual-Doppler radar coverage area as a function of measurement accuracy and spatial resolution, J. Appl. Meteor., 18 (1979) 1229–1233.

17 C. Miller, J. Holmes, D. Henderson, J. Ginger, and M. Morrison, The response of the dines anemometer to gusts and comparisons with cup anemometers, J. Atmos. Oceanic Technol., 30 (2013) 1320–1336.

18 J.D. Holmes and J.D. Ginger, The gust wind speed duration in AS/NZS 1170.2, Australian J. of Struct. Eng., 13 (2012) 207–218.

19 V.N. Mahale, G. Zhang, and M. Xue, Characterization of the 14 June 2011 Norman, Oklahoma, downburst through dual-polarization radar observations and hydrometeor classification. J. Appl. Meteor. Climatol., 55 (2016) 2635–2655.

20 C.W. Letchford and M.T. Chay, Pressure distributions on a cube in a simulated thunderstorm downburst. Part B: moving downburst observations, J. Ind. Aerodyn., 90 (2002) 733–753.

21 J. Wurman and C.R. Alexander, The 30 May 1998 Spencer, South Dakota, storm. Part II: comparison of observed damage and radar-derived winds in the tornadoes, Mon. Wea. Rev., 133 (2005) 97–119.

22 N.T., Atkins, J.M. Arnott, R.W. Przybylinski, R.A. Wolf, and B.D. Ketcham, Vortex structure and evolution within bow echoes. Part I: single-Doppler and damage analysis of the 29 June 1998 derecho, Mon. Wea. Rev., 132 (2004) 2224–2242.

23 N.T., Atkins, C.S. Bouchard, R.W. Przybylinski, R.J. Trapp, and G. Schmocker, Damaging surface wind mechanisms within the 10 June 2003 saint louis bow echo during BAMEX, Mon. Wea. Rev., 133 (2005), 2275–2296.

24 D.M. Wheatley, R.J. Trapp, and N.T. Atkins, 2006: Radar and damage analysis of severe bow echoes observed during BAMEX. Mon. Wea. Rev., 134 (2006) 791–806.

13th Americas Conference on Wind Engineering
Gainesville, Florida, USA
May 21 - 24, 2017

Tornado Hazard Mapping for Tornado-Resistant Design Buildings and Infrastructure

L. Phan[a], M. Levitan[b], L. Twisdale[c]

[a] *National Institute of Standards and Technology, Gaithersburg, Maryland, USA, long.phan@nist.gov*
[b] *National Institute of Standards and Technology, Gaithersburg, Maryland, USA, marc.levitan@nist.gov*
[c] *Applied Research Associates, Inc., Raleigh, North Carolina, USA, ltwisdale@ara.com*

ABSTRACT

Tornadoes typically affect much smaller geographic areas compared with other natural hazards like earthquakes and hurricanes, but occur at a much higher frequency and cause more deaths than those two hazards combined. During the period from 1950 through 2011, U.S. tornadoes caused about 5,600 fatalities. This number well exceeds the toll for U.S. hurricanes and earthquakes over the same period (3,102 and 459, respectively). U.S. average annual property losses due to tornadoes are approximately the same as losses due to hurricanes. Inflation adjusted insured catastrophe losses for the two decades between 1995 and 2014 were $155B for tornadoes and tornadic storms, compared to $161B for hurricanes and tropical storms (in 2014 dollars).

The National Institute of Standards and Technology (NIST) conducted a detailed investigation of the deadliest tornado since official record-keeping began in 1950, the May 22, 2011 EF-5 Joplin tornado, which caused 161 fatalities and more than 1,000 injuries. NIST found that the majority of the fatalities, 135 of 161 (more than 84%), occurred inside buildings. This, despite an official tornado warning time of about 17 minutes, greater than the National Weather Service (NWS) national average warning time of approximately 14 minutes. This particular finding gives rise to an important question: can building safety in tornado be enhanced through tornado-resistant design procedure and practice?

The NIST investigation report also outlined 16 specific recommendations, among which was recommendation calling for the development of nationally accepted Performance-Based Design (PBD) standards for tornado-resistant design of buildings and infrastructure. The conventional wisdom is that designing for tornadoes is not warranted since the probability of a tornado strike on any particular building is considered too low. However, research underway towards implementation of the NIST tornado recommendations appears to indicate that tornado strike probabilities may be an order of magnitude greater than previously assumed. Additionally, considering the lopsided death tolls caused by tornadoes and earthquakes, one can justifiably argue that tornado resistant design should have the level of importance as seismic design. This is not the case currently, as tornadoes remain the only major natural hazard not considered in the design of buildings and other structures in modern building codes and standards.

Designing for tornadoes requires a set of tornado hazard maps that accurately reflect the tornado risks for different regions of the U.S. and can be used to derive appropriate tornado design criteria for structural design purposes. Such maps need to be developed using up-to-date tornado database, tornado parameters, and science-based tornado risk assessment methodology, with appropriate consideration of tornado climatology. This paper describes a major, on-going research and development effort to create these new tornado hazard maps.

KEYWORDS

Buildings, design criteria, tornado-resistant design, tornado hazard maps, tornado risk assessment methodology.

The 13th Americas Conference on Wind Engineering (13ACWE)
Gainesville, Florida USA, May 21-24, 2017

A 3D Tornado Wind Field Model for Multiple-Cell Tornado Vortex

Jin Wang [a], Weichiang Pang [b], Shuyang Cao [c], Fanfu Fan [b]

[a]*Clemson University, Clemson, SC, USA*
Tongji University, Shanghai, China
[b]*Clemson University, Clemson, SC, USA*
[c]*Tongji University, Shanghai, China*

ABSTRACT: Tornadoes can cause severe damages to buildings and other structures. Understanding the tornado wind field near the ground is important for tornado risk assessment, emergency planning and hazard mitigation purposes. This paper presents the development of a new 3D tornado wind field model for multiple-cell tornado vortex based on experimental data. A parametric tornado wind field model consisting of three equations for describing the three wind velocity components was developed. One of the key features of this new model is that it is able to characterize the vertical profile of tangential wind speed at different radial locations relative to the core radius, including the shearing profile inside the core radius and boundary-layer-like profile outside the core radius.

KEYWORDS: Multiple-cell tornado vortex, Wind field model, Three dimensional wind velocities, Parametric model

1 INTRODUCTION

Tornadoes with high wind speeds can cause severe damages to buildings and other structures. As such, tornadoes can have significant impacts on the safety and economy of a society. The mean annual occurrence rate of tornadoes in the United States is about 1000 (events/year). Accurate assessment of tornado risk is essential for emergency planning and hazard mitigation purposes. The tornado wind field near the ground is an important element for tornado risk assessment.

Several theoretical models were developed based on fluid dynamics theory and other relevant principles of physics such as the Navier-Stokes equations of motions (e.g. Kuo, 1970; Wen, 1975). Since the pioneer work of Chang (1971) on tornado simulator, several studies on experimental simulation of tornado vortices were conducted and parametric tornado wind field models were proposed (Ward 1972; Church et al., 1979; Davies-Jones 1979; Haan et al., 2008; Wang et al., 2016). The one dimensional (1D) Rankine model (Rankine, 1872) and its modified version (Refan 2014) are among the most commonly used parametric models for simulating the tornado wind field. While a 1D or 2D wind field model is simple, the actual tornado vortex is three dimensional (3D). Hence, a 1D or 2D wind field model may not be adequate for certain applications such as studying the flight of debris in a tornado vortex. Many previously discussed existing tornado wind field models were developed for modelling single-cell vortex. These models are not suitable for modelling multiple-cell vortices. More intense tornadoes are typically associated with a multiple-cell vortex and have been observed to cause more severe damages than tornadoes with a single-cell vortex.

In this study, a 3D tornado wind field model for multiple-cell vortex was developed based on the measured wind velocities of a tornado-like flow simulated using a tornado vortex simulator in Tongji University, China. The 3D wind velocities were measured using the TFI (Turbulent Flow

Instrumentation) Cobra Probe. This new 3D tornado wind field model consists of a set of three empirical equations in a cylindrical coordinate system, which can be used to describe the tangential, radial and axial wind velocities of both single-cell and multiple-cell tornado vortices.

2 EXPERIMENTAL SETUP

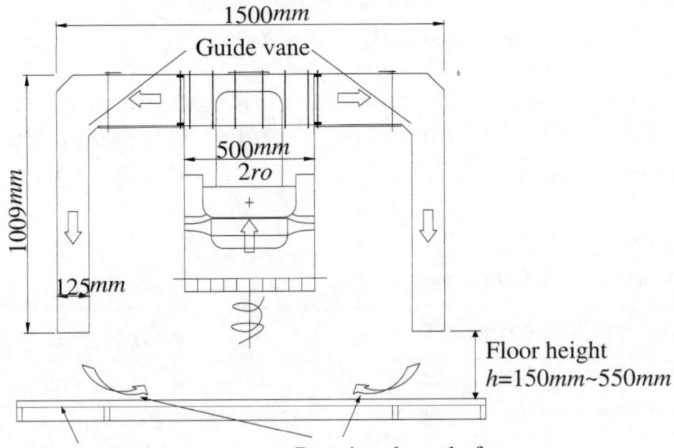

Figure 1. Schematic diagram of tornado vortex simulator at Tongji University

The tornado vortex simulator as shown in Figure 1 at Tongji University was utilized to obtain the tornado wind flow data used in this study (Wang et al., 2016). A suspended overhead circular duct of 1.5 m in diameter and 1.009 m in height with a 0.5 m-diameter updraft hole (r_o = 250 mm) which housed the controlling fan (maximum flow rate is 4.8 m^3s^{-1}, and maximum rotational speed is 3500 rpm) used to generate the strong updraft needed for creating the tornado vortex. The simulator floor is adjustable in the vertical direction, enabling the simulation of a range of heights for the inflow layer (h = 150 mm ~ 550 mm). Both the fan and guide vanes are located at the top of the simulator, which allowed more space on the simulator floor for placement of the scaled structural models and sensors for measuring wind speeds and pressures. In addition, the suspended overhead tornado vortex simulator can translate along the longitudinal direction of the simulator floor at a user prescribed speed (maximum speed is 0.4*m/s*) to simulate the translational effect of tornadoes.

The tornado simulator experiments used for the development of the 3D tornado wind field model were conducted for stationary tornadoes without considering the translation speed. The flow structure of a tornado vortex is controlled by a parameter known as the swirl ratio, *S*, which describes the momentum exchange between the rotational and radial flows. In this study, the swirl ratio is defined as $S = \tan\theta / (2a)$, where θ and a are guide vane angle and the inflow height to fan radius aspect ratio, respectively (Mitsuta and Monji, 1984; Matsui and Tamura, 2009). The aspect ratio is taken as the ratio of the inflow height to the radius of the simulator fan, $a = h / r_o$. In this study, the inflow height h was fixed at 400mm above the ground. The wind field of tornadoes with five different swirl ratios were measured. The value of the swirl ratio was controlled by adjusting the guide vane angle from 20 to 60 degrees at an increment of 10 degrees, corresponding to S = 0.15 to 0.72.

3 WIND FIELD MODEL

In order to allow the scaling of the model vortex into full-scale tornadoes of different widths and sizes, all length scales in the 3D tornado wind field model are non-dimensionalized by the vortex core radius r_{c0}, which corresponds to the maximum tangential velocity u_{max0} of the whole wind field as shown in Table 1. In addition, all the three velocity components, tangential velocity u, radial velocity v and axial velocity w are normalized by u_{max0}. z_0 is the elevation corresponding to u_{max0}. r_c denotes the core radius corresponding to the maximum tangential velocity u_{max} at each elevation. It should be noted that r_{ca} is the radius corresponding to the maximum axial velocity w_{max} at each elevation, while r_{cr} is the radius corresponding to the transition position from inward flow to outward flow with zero radial velocity. Note that all the uppercase variables are normalized dimensionless quantities and the lowercase variables are dimensional variables of original scales.

Table 1. A list of variables used in the 3D tornado wind field model.

r_o	Updraft hole radius	a	Aspect ratio (h/r_o)
h	Inflow height	S	Swirl ratio
θ	Guide vane angle	z	Measurement height or elevation
r	Radial distance from the vortex center		
z_0	Elevation of the maximum tangential velocity		
u	Tangential velocity	v	Radial velocity
w	Axial velocity	w_{max}	Maximum axial velocity at individual elevation
u_{max}	Maximum tangential velocity at individual elevation		
u_{max0}	Maximum tangential velocity of the entire wind field		
r_c	Core radius corresponding to u_{max} at individual elevation		
r_{c0}	Core radius corresponding to u_{max0}		
r_{cr}	Radial distance of the transition from radial inflow to radial outflow		
r_{ca}	Radial distance corresponding to w_{max}		

$$R=\frac{r}{r_{c0}},\ R_t=\frac{r}{r_c},\ R_{ct}=\frac{r_c}{r_{c0}},\ R_{cr}=\frac{r_{cr}}{r_{c0}},\ R_{ca}=\frac{r_{ca}}{r_{c0}},\ Z=\frac{z}{r_{c0}},\ Z_0=\frac{z_0}{r_{c0}},$$
$$U=\frac{u}{u_{max0}},\ V=\frac{v}{u_{max0}},\ W=\frac{w}{u_{max0}}, \tag{1}$$
$$U_t=\frac{u}{u_{max}},\ U_{max}=\frac{u_{max}}{u_{max0}},\ W_a=\frac{w}{w_{max}},\ W_{max}=\frac{w_{max}}{u_{max0}}$$

3.1 *Tangential Velocity Model*

Figure 2 (left) depicts the shape of the vertical profile for maximum tangential velocity of a tornado vertex. The maximum tangential velocity of the entire wind field for a tornado vortex has been observed to occur at an elevation above the ground (denoted as Z_0 in Figure 2). For a no-slip boundary condition, the tangential velocity at the ground (i.e. $Z = 0$) can be assumed to be zero.

The following equation is utilized to characterize the vertical profile for maximum tangential velocity.

$$U_{\max}(Z) = \frac{\alpha \times Z}{Z^{a_1} + \beta} = \frac{a_1 \times Z_0^{a_1 - 1} \times Z}{Z^{a_1} + (a_1 - 1) \times Z_0^{a_1}} \quad (2)$$

$$\alpha = a_1 \times Z_0^{a_1 - 1}$$

$$\beta = (a_1 - 1) \times Z_0^{a_1}$$

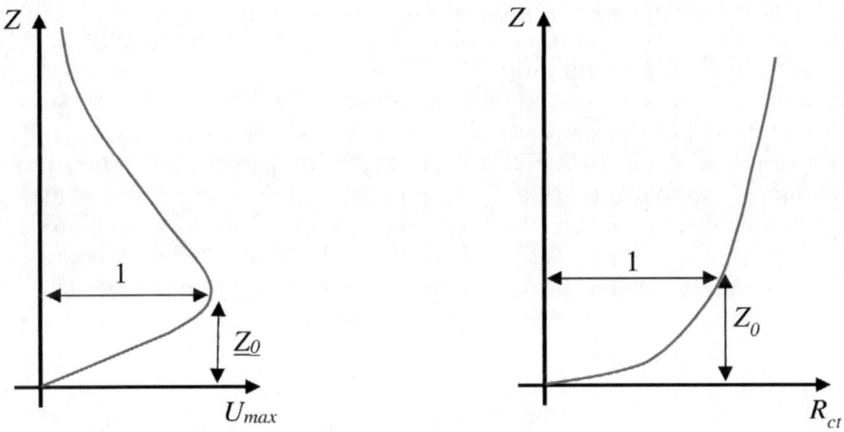

Figure 2. Vertical profiles of maximum tangential wind speed (left) and core radius (right).

Where, parameter a_1 controls the shape of the vertical profile for maximum tangential wind speed. Parameter Z_0 is a location parameter that marks the elevation of the global maximum tangential velocity (i.e. maximum of the entire wind field).

The vertical profile of core radius is given in Equation 3.

$$R_{ct}(Z) = \frac{a_2 \times Z}{Z + (a_2 - 1) \times Z_0} \quad (3)$$

Where, a_2 is the shape parameter for the vertical profile of core radius.

The Rankine model (Rankine, 1872) is one of the most commonly used models for describing the horizontal profile of tangential wind velocity. Refan (2014) modified the Rankine model to account for the swirl motion of the vortex. Both the original and modified Rankine models do not consider the radial and axial velocity components. In addition, the vertical variation or profile of tangential velocity is not considered in the original and modified Rankine models. In this study, a new tangential velocity model is proposed (Equation. 4).

$$U(R, Z) = \frac{a_3 \left(\frac{R}{R_{ct}(Z)} \right)}{\left(\frac{R}{R_{ct}(Z)} \right)^{a_3} + a_3 - 1} \times U_{\max}(Z) \quad (4)$$

Note that the horizontal profile model for tangential velocity is a function of the vertical profile models for maximum tangential velocity and core radius (i.e. Equations 2 and 3, respectively). a_1 and a_2 are the model parameters for the vertical profiles of maximum tangential velocity and core radius. a_3 is the shape parameter for the horizontal profile for tangential velocity which describes the decay rate of the tangential velocity outside of the core radius. The proposed new tangential velocity model expressed by Equation 4 satisfies three boundary conditions: (1) the far-field tangential velocity approaches zero, $U \to 0$ as $R \to \infty$; (2) the maximum tangential velocity occurs at the core radius, $U = U_{max}$ at $R = R_{ct}$; and (3) the slope or change in tangential velocity is zero at the core radius, $U' = dU/dR_{ct} = 0$ at $R = R_{ct}$. Once given the dimensionless radius R and elevation Z, the tangential velocity can be determined using Equation 4. In this study, the values for parameters a_1, a_2 and a_3 were fitted using the least squares error fitting method and the experimentally measured tornado vortices for different swirl ratios.

Figure 3 shows the comparisons between the horizontal velocity profiles obtained from the tornado simulator and the proposed model for $S = 0.72$. It can be seen that the proposed model fits the tangential velocity profiles reasonably well. Figure 4 shows the comparisons of vertical profiles of tangential velocity at various normalized distances ($0 \leq R \leq 5$) between the experimental data and proposed model. Due to difficulty in placing the wind speed sensor close to the simulator floor, the wind field below an elevation of 10 mm was not measured. The model captures both the velocity shearing or double curvature profile inside the vortex core ($R < 1$) and single curvature profile similar to a typical boundary-layer-type profile outside the vortex core ($R > 1$).

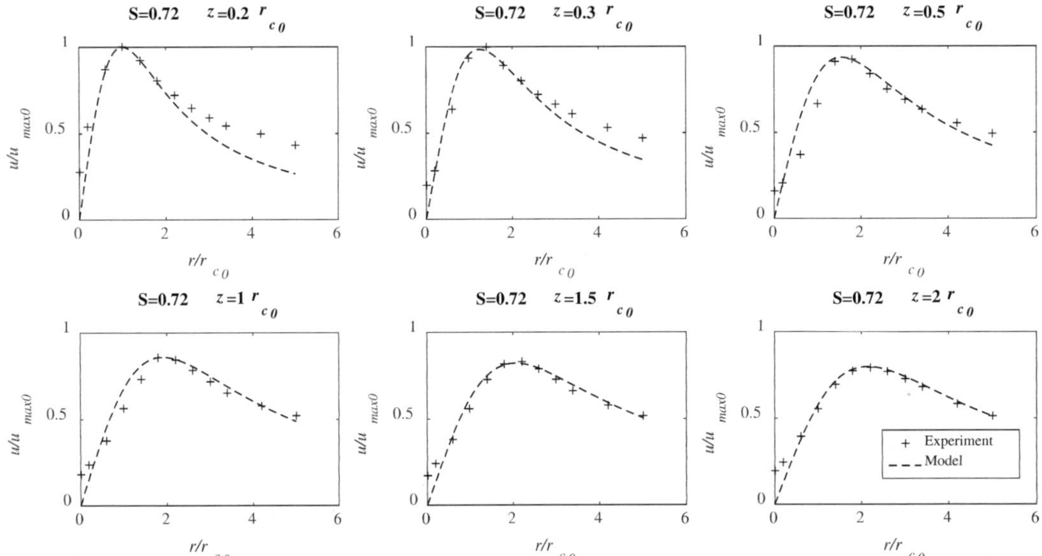

Figure 3. Horizontal profiles of tangential velocity for swirl ratio S=0.72 at selected elevations.

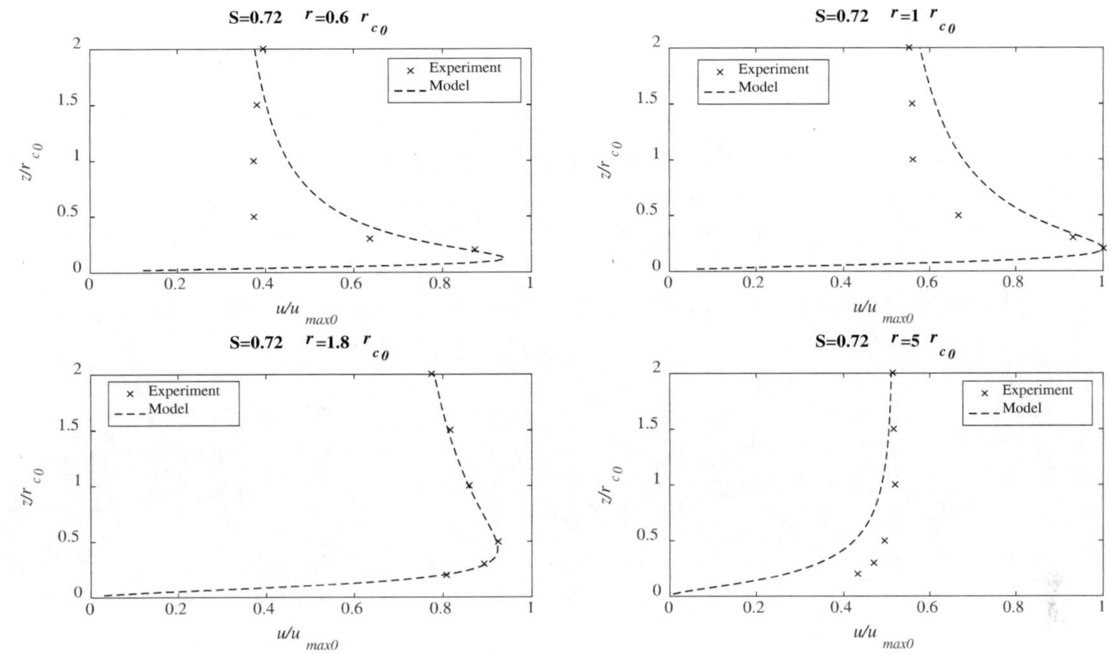

Figure 4. Vertical profiles of tangential velocity for swirl ratio $S = 0.72$ at selected radial distances.

3.2 *Radial Velocity Model*

Unlike the single-cell tornado vortex, in which all the radial velocities flow toward the center, only the radial velocities outside of certain radial distances in a multiple-cell vortex (or a vortex in a transitional state between single-cell and multiple cell) flow inward and the radial velocities inside the core radius flow outward. The normalized radial distance where reversal of radial velocity occurs as a function of elevation is modeled using the following equation.

$$R_{cr}(Z) = \frac{b_1 Z}{Z + b_2} \quad (5)$$

Where, b_1 and b_2 are parameters that control the flow reversal location. While Equation 5 is developed to characterize the radial velocity profile of multiple-cell tornadoes, the radial velocity profiles of single-cell tornadoes can also be represented by setting $b_1 = 0$ (i.e. all radial flows are inward). The complete radial velocity profile is given by the following equation.

$$V(R,Z) = \frac{\gamma \times R \times (R_{cr}(Z) - R)}{R^3 + \eta}$$

$$\gamma = \frac{b_3 Z}{Z^{b_4} + b_5} \quad (6)$$

$$\eta = b_6 Z + b_7$$

In which, γ and η are dependent on the independent parameters b_3, b_4, b_5, b_6 and b_7. Note that the radial profiles for multiple-cell vortices of different swirl ratios can be modeled by adjusting the seven modeling parameters (i.e. b_1, b_2, … b_7). Parameters γ and η control the magnitude of the radial velocity for different elevations.

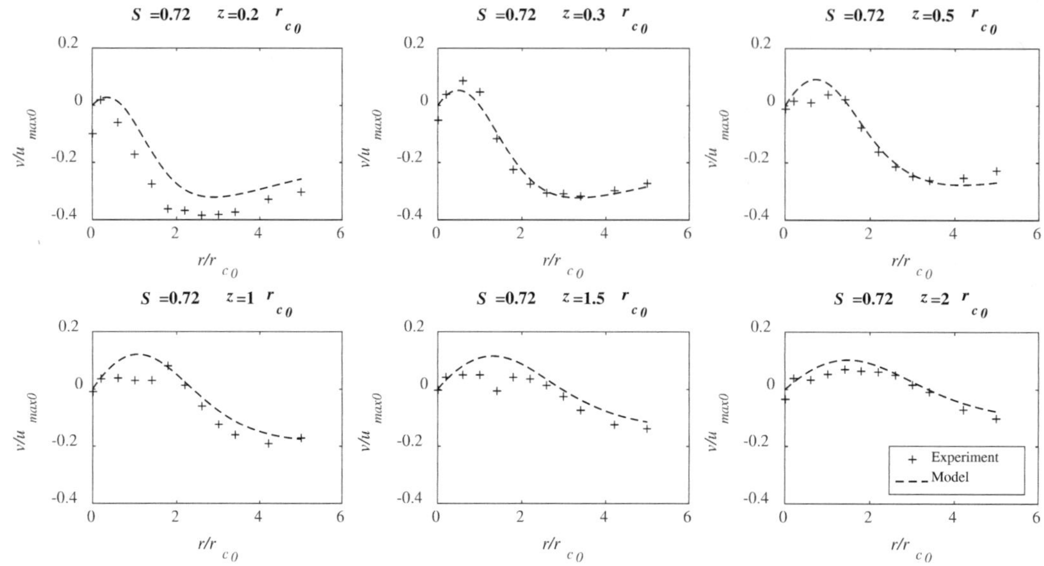

Figure 5. Horizontal profiles of radial velocity for swirl ratio S=0.72 at selected elevations.

Figure 5 shows the horizontal distribution of the measured and the fitted normalized radial velocities (Equation 6) for S= 0.72 at six different elevations. It should be noted that the radial velocities are normalized by the overall maximum tangential velocity u_{max0} of the whole wind field. Similarly, the radial distance is normalized by the corresponding core radius r_{c0} of the entire wind field. A negative radial velocity means the flow is inward (toward the center of the tornado vortex) and a positive radial velocity indicates the flow is outward.

3.3 *Axial Velocity Model*

In a single-cell tornado vortex, the axial velocities in the core of the tornado flow upward. On the other hand, the axial velocities near the center of a multiple-cell tornado typically flow downward. Similar to the tangential velocity profile, the maximum axial flow occurs at a radial distance away from the center of the multiple-cell vortex. The following equation, similar to that of the radial velocity core radius equation, is used to model the core radius of axial velocity.

$$R_{ca}(Z) = \frac{c_1 Z}{Z + c_2} \qquad (7)$$

Where, c_1 and c_2 are the modeling parameters for the core radius of axial velocity. The vertical profile for maximum axial velocity is given by the following equation:

$$W_{max}(Z) = \frac{c_3 Z}{Z + c_4} \qquad (8)$$

Where, c_3 and c_4 are the modeling parameters for the vertical profile of the maximum axial velocity. It should be noted that the axial velocities are normalized by the overall maximum tangential velocity u_{max0} of the whole wind field. This is the same normalization approach used for the radial velocity model. Similarly, the radial distance is normalized by the corresponding core radius r_{c0} of the entire wind field.

The complete axial velocity profile model is given below.

$$W(R,Z)=\frac{5\times R_{cw}(Z)^4\times R+\tau-4\times R_{cw}(Z)^5}{R^5+\tau}\times W_{max}(Z)$$
$$\tau=\frac{4\times W_{max}(Z)\times[R_{cw}(Z)]^5\times(Z+c_5)}{W_{max}(Z)\times(Z+c_5)-c_6\times Z} \quad (9)$$

Where τ is dependent on independent parameters c_5 and c_6. It is assumption that the axial velocity is zero at the ground (i.e. Z=0) and the magnitude of the axial velocity increases with elevation and approaches an asymptotic constant at a high elevation. Additionally, three boundary conditions are assumed: (1) the far-field axial velocity approaches zero, $w\rightarrow 0$ as $R\rightarrow\infty$; (2) the maximum axial velocity occurs at the core radius, $W= W_{max}$ at R_{ca}; and (3) the slope or change in axial velocity is zero at core radius of R_{cw}, $W' = dW/dR = 0$ at $R=R_{ca}$. The modeling equation for axial velocity profile, shown in Equation 9, was developed based on the above assumption and boundary conditions.

Figure 6 shows the horizontal distributions of the normalized axial velocities for $S = 0.72$. The sign convention used in the axial velocity profile model is similar to that of the radial profile model, where a negative sign means the flow is downward and a positive sign indicates the axial flow is upward.

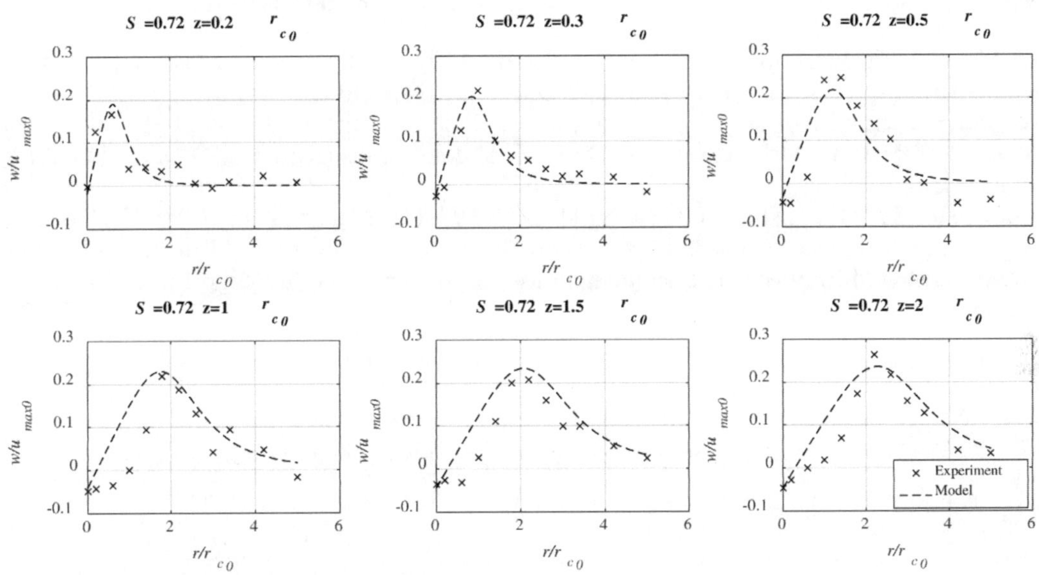

Figure 6. Horizontal profiles of axial velocity for swirl ration S=0.72 at different elevations.

4 EXAMPLE WIND FIELDS

The modelled flow vectors in the RZ-plane for the vortices of two swirl ratios ($S = 0.15$ and 0.72) are shown in Figure 7. Both the radial distance r and elevation z are normalized by r_{c0}. For the low swirl ratio vortex ($S = 0.15$), the flow vectors in the RZ-plane shows that outside the core, the air is moving radially toward the vortex center and then turning upward near the core radius. The flows around the vortex center move downward to the ground and the flow direction is reversed (i.e. upward) near the core radius. At high swirl ratio ($S = 0.72$), the flow structure is similar to the phenomenon shown in Figure 7(a), for the low swirl ratio, except that a more obvious structure of downward axial flow can be observed around the vortex center. The modelled velocity fields are consistent with the observations in present experimental results and other research (e.g. Church et al. 1979; and Haan et al., 2008).

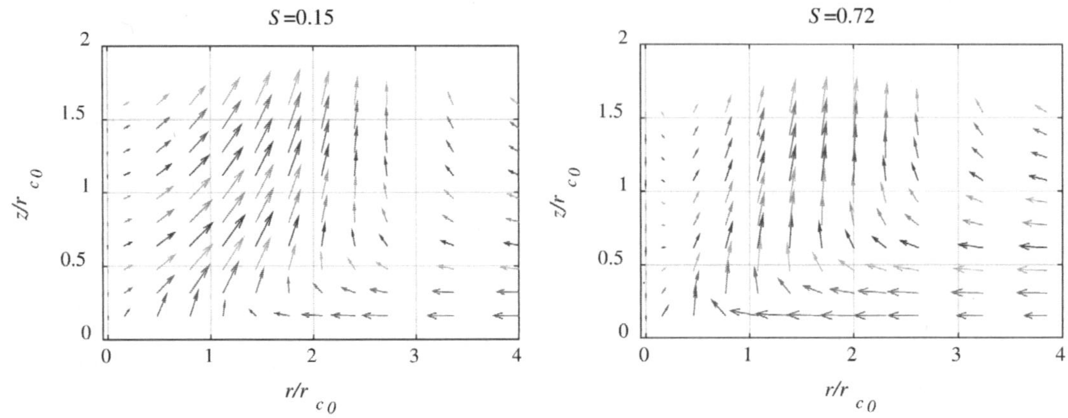

(a) $S = 0.15$ (b) $S = 0.72$
Figure 7. Vertical cross-sectional vector plots of multiple-cell vortices for (a) S=0.15 and (b) S=0.72.

The horizontal cross-sectional views of the model multiple-cell vortices are shown in Figure 8 for two swirl ratios, a low $S = 0.15$ and high $S = 0.72$, respectively. From Figure 8, it can be observed that the vortex with high swirl ratio ($S = 0.72$) shows higher ratio of tangential-to-radial flow than that of the low swirl ratio vortex ($S = 0.15$) at core radius ($r/r_{c0} = 1$). These flow patterns, which are typical for multiple-cell vortices, are consistent with the experimental observations and findings of several previous studies (e.g. Church et al. 1979; and Haan et al., 2008). For visualization purpose, the 3D tornado wind fields that correspond to the 2D wind field shown in Figures 7 and 8 are presented in Figure 9. The parameters used to generate the 3D wind fields for $S = 0.15$ and 0.72 are presented in Table 2.

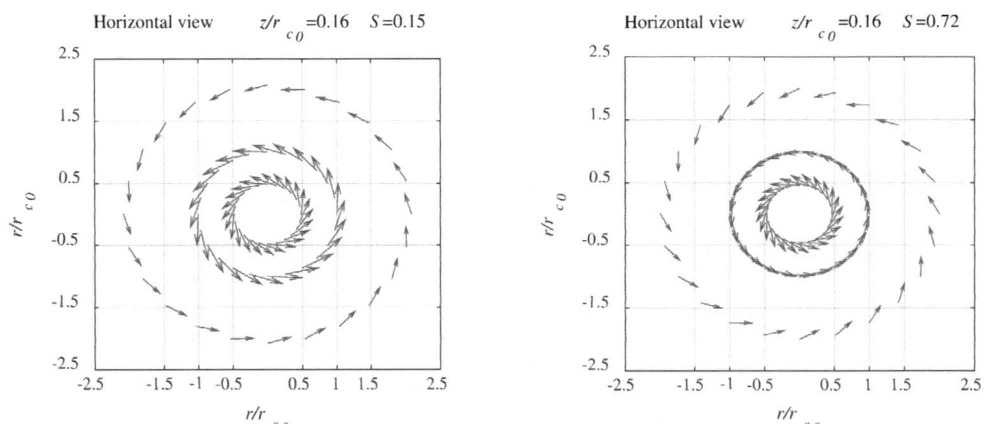

(a) $S = 0.15$ (b) $S = 0.72$
Figure 8. Horizontal cross-sectional vector plots of multiple-cell vortices for (a) S=0.15 and (b) S=0.72.

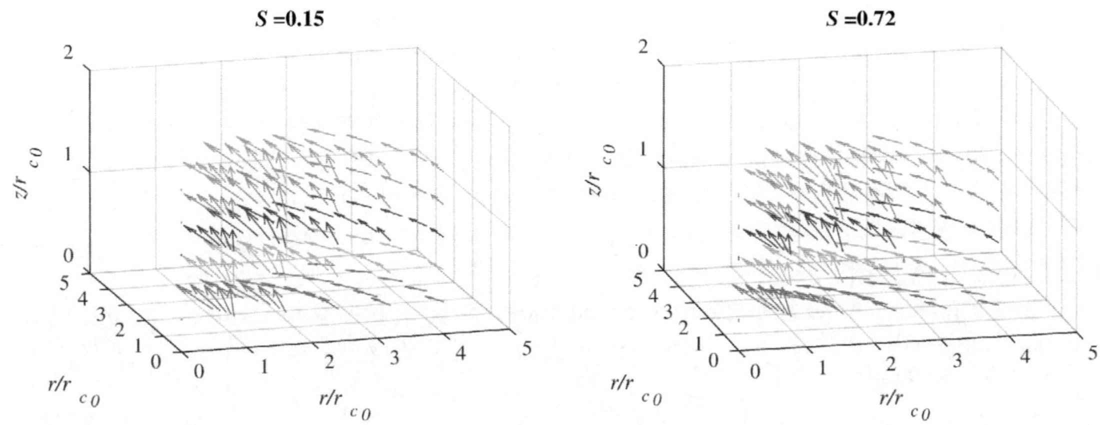

(a) S=0.15 (b) S=0.72

Figure 9. 3D vector plots of multiple-cell vortices for (a) S=0.15 and (b) S=0.72.

Table 2. Fitted modeling parameters of selected tornado wind field models.

Parameters		S=0.15	S=0.72
Tangential velocity	a_1	1.188	1.166
	a_2	1.723	2.460
	a_3	2.625	2.301
Radial velocity	b_1	3.163	5.565
	b_2	0.266	1.457
	b_3	1.256	4.415
	b_4	1.500	1.500
	b_5	0.069	0.537
	b_6	2.513	46.922
	b_7	5.195	-7.741
Axial velocity	c_1	2.135	3.195
	c_2	0.334	0.814
	c_3	0.231	0.252
	c_4	0.054	0.052
	c_5	0.086	0.402
	c_6	-0.012	-0.056

5 CONCLUSIONS

A new three dimensional tornado wind field model for multiple-cell vortices is presented in this paper. The new three dimensional tornado wind field model is developed based on the results of tornado simulator experiments. It has been shown that the proposed model provides more accurate descriptions for the three velocity components of wind flows measured in a tornado simulator than other single-cell models (e.g. Rankine model). One of the key features of this new model is that it is able to characterize the vertical distribution of tangential wind speed at different radial locations relative to the core radius, including the shearing profile or double curvature profile inside the core radius and boundary-layer profile outside the core radius. In addition, the new model can consider the downward flow of axial velocity in multiple-cell tornadoes near the vortex center. It should be noted that the proposed model is suitable for the stationary tornado vortex.

This three dimensional wind field model can be utilized to develop a risk assessment framework for tornado hazards. Specifically, the proposed three dimensional tornado wind field model can be used in conjunction with a simulated tornado tracks to generate tornado wind hazard maps for structural design or building codes. Finally, the three dimensional tornado wind field may also be used in a debris flight model to simulate the trajectories of tornado-borne debris.

6 ACKNOWLEDGEMENTS

The first author greatly appreciates the financial supports provided by Tongji University in China and Clemson University in the USA. Additional supports were provided by the China Scholarship Council (CSC Scholarship).

7 REFERENCES

1 Chang, C.C., 1971. Tornado wind effects on buildings and structures with laboratory simulation. In: Committee, J.O. (Ed.), Proceedings of the Third International Conference on Wind Effects on Buildings and Structures, pp. 231–240.

2 Church, C.R., Snow, J.T., Baker, G.L. and Agee, E.M., 1979. Characteristics of tornado-like vortices as a function of swirl ratio: a laboratory investigation. Journal of the Atmospheric Sciences, 36, 1755–1766.

3 Davies-Jones, R.P., 1973. The dependence of core radius on swirl ratio in a tornado simulator. Journal of the Atmospheric Sciences, 30(7), pp.1427-1430.

4 Haan, F. L., Sarkar, P. P., Gallus, W. A., 2008. Design, Construction and performance of a large tornado simulator for wind engineering applications. Eng. Struct. J. Earthquake Wind Ocean Eng. 30, 1146-1159.

5 Kuo, H.L., 1971. Axisymmetric flows in the boundary layer of a maintained vortex. Journal of the Atmospheric sciences, 28, pp.20-41.

6 Matsui, M., Tamura, Y., 2009. Influence of incident flow conditions on generation of tornado-like flow. Proceedings of the 11th American Conference on Wind Engineering, Puerto Rico, USA.

7 Mitsuta, Y., Monji, N., 1984. Development of a laboratory simulator for small scale atmospheric vortices. Natural Disaster Science, 6(1):43-53.

8 Rankine, W.J.M., 1872. A manual of applied mechanics. Charles Griffin and Company.

9 Refan, M., 2014. Physical simulation of tornado-like vortices. PhD Dissertation, The University of Western Ontario, London, Ontario, Canada.

10 Wang, J., Cao, S., Pang, W. and Cao, J., 2016. Experimental Study on Effects of Ground Roughness on Flow Characteristics of Tornado-Like Vortices. Boundary-Layer Meteorology, pp.1-21.

11 Ward, N.B., 1972. The exploration of certain features of tornado dynamics using a laboratory model. Journal of the Atmospheric Sciences, 29, 1194–1204.

12 Wen, Y. K., 1975. Dynamic Wind Loads on Tall Buildings. Journal of the Structural Division, 101, pp. 169–185.

13th Americas Conference on Wind Engineering
Gainesville, Florida, USA
May 21 - 24, 2017

Tornado Damage Preservation via 3D Reality Capture

M. Ebrahim Mohammadi[a], Richard L. Wood[a], J. Arn Womble[b], Douglas A. Smith[c], Elizabeth I. Louden[c]

[a]*University of Nebraska-Lincoln, Lincoln, NE, rwood@unl.edu; me.m@huskers.unl.edu*
[b]*West Texas A&M University, Canyon, TX, awomble@wtamu.edu*
[c]*Texas Tech University, Lubbock, TX, doug.smith@ttu.edu; elizabeth.louden@ttu.edu*

ABSTRACT

Structural design for the resistance of tornado effects is complicated by a lack of knowledge about the intensity and structure of tornado winds. Since direct measurements of tornado wind speeds are difficult, dangerous, and rare, it is necessary for wind engineers, structural engineers, atmospheric scientists, and catastrophe modelers to utilize proxy methods (such as damage to structures) to better estimate tornado wind speeds. Damage to simple engineered structures is of particular interest, since the lateral resistance of such structures can often be readily estimated. Detailed forensic investigation of wind-induced failures requires detailed data collection and analysis, which can be prohibitive due to time, access, and safety issues – particularly for investigation teams who must rapidly collect data throughout affected areas before critical evidence is disturbed, demolished, or cleaned up. Newly available remote-sensing technologies, including laser (lidar) scanning, digital photogrammetry, and UAV-based imaging have provided revolutionary new digital possibilities for rapidly preserving highly-detailed and comprehensive 3D damage scenes for subsequent forensic analysis.

A recent case study has highlighted the use of such remote-sensing technologies. An intense tornado outbreak severely damaged a group of engineered structures at an industrial facility near Pampa, TX in November 2015. This facility contained multiple engineered structures for which structural resistances could be estimated, thereby enabling the estimation of tornado wind speeds required to cause the observed damage. Site access was prohibited due to safety and security liability concerns; however, investigators were able to acquire remote-sensing data from multiple platforms to preserve the damage scene for subsequent detailed analyses, as part of an NSF-sponsored RAPID grant. Ground-based lidar scanning and terrestrial digital imaging from the property line as well as aerial and satellite imaging provided effective solutions for rapidly, safely, and accurately capturing and sharing the damage scene via the NHERI DesignSafe-ci community.

KEYWORDS

Remote sensing, lidar, UAV, virtual reality, tornado damage

Non-Synoptic Wind-Induced Response of Long-Span Bridges:

A Hybrid CFD-CSD-based Approach

Jianming Hao[a], Teng Wu[b]

[a] *University at Buffalo, Buffalo, NY 14260, USA, jhao2@buffalo.edu*
[b] *University at Buffalo, Buffalo, NY 14260, USA, tengwu@buffalo.edu*

ABSTRACT

The non-synoptic wind events such as downbursts and tornadoes present violent impacts on the civil infrastructures, resulting in significant human fatalities and economic losses. The long-span bridges located in the coastline and mountainous areas are prone to suffer from the thunderstorms, associated with transient downburst/tornado winds. Therefore, it is important to examine the behaviors of long-span bridges under such wind events.

In this study, the time history of non-turbulent downburst wind field is modeled based on the computational fluid dynamics (CFD) approach. Several CFD-based schemes have been introduced to generate winds in downbursts such as the impinging jet, cooling source model, and the approach based on a very high-resolution three-dimensional cloud model. Among these, the impinging jet model, with a reasonable computational cost, will be employed here since it has demonstrated high-fidelity simulation especially for the near-surface radial winds, which is particularly dominant downburst wind field for present investigation of bridge aerodynamics and aeroelasticity. The simulated time-varying mean wind velocity field of a typical downburst event is validated using available full-scale measurement and laboratory test data. The correlated nonstationary fluctuations of the downburst are stochastically simulated by the Hilbert-wavelet scheme, and then superimposed onto the CFD-based transient mean wind field.

The obtained turbulent downburst wind field is employed as the dynamic inputs of a long-span suspension bridge modeled based on the computational structural dynamics (CSD) approach. Specifically, the line-element-based 3-D finite element model with equivalent sectional properties, which could well capture the global dynamic behaviors of bridges, is utilized. The aerodynamic and aeroelastic couplings in this hybrid CFD-CSD-based scheme are modeled using the 2-D indicial response functions that could accurately represent the transient bridge aerodynamics under non-synoptic winds. A time-domain buffeting analysis of the long-span bridge based on abovementioned simulation framework is carried out to investigate the transient and nonstationary effects on the non-synoptic wind-induced structural response.

KEYWORDS

Non-synoptic winds, long-span bridges, computational fluid dynamics, computational structural dynamics, transient aerodynamics

The 13th Americas Conference on Wind Engineering (13ACWE)
Gainesville, Florida USA, May 21-24, 2017

Hurricane Performance Assessment of Power Distribution Lines Using Multi-scale Matrix-based System Reliability Analysis Method

Yousef Mohammadi Darestani [a], Abdollah Shafieezadeh [b], Reginald DesRoches[c]

[a]PhD Candidate, The Ohio State University, Columbus, Ohio, United States,
mohammadidarestani.1@osu.edu
[b]Assisstance Professor, The Ohio State University, Columbus, Ohio, United States,
shafieezadeh.1@osu.edu
[c]Karen and John Huff School Chair and Professor, Georgia Institute of Technology, Atlanta, Georgia, United States
reginald.desroches@ce.gatech.edu

ABSTRACT: During hurricanes, a considerable portion of damage to electric power systems occurs in distribution networks. In order to minimize induced power outages and associated economic loss, network reliability analysis can significantly help by providing information on the likely performance of distribution systems and inform strengthening decisions. This paper expands upon and improves the system level reliability analysis of distribution networks by incorporating a Multi-scale Matrix-based System Reliability method. This approach accounts for correlations in the failure of poles. The results indicate that correlated failure events have significant impact on the system level reliability of distribution lines.

KEYWORDS: System-level Reliability Analysis, Hurricanes, Power Distribution Lines, Multi-scale Matrix-based System Reliability.

1 INTRODUCTION

Electric power is one of the most important components of societies as it provides the supply of energy for households, industries and organizations, among others. Any disruption in the constant flow of electricity results in considerable economic losses as well as societal and organizational disruptions. Electric power systems consist of three main subsystems including generation, transmission, and distribution. During hurricanes, distribution systems supported by wood poles are one of the most vulnerable components of electric power systems [1, 2]. This is partly due to the high rate of decay in the wood material that results in considerable reduction in the strength of wood poles.

A distribution network can be considered as a graph [3] where substations, customers and poles constitute the nodes and the conductors (cables) are the edges. In terms of the physical connectivity of the nodes, in a distribution network there are several cut-sets (paths) between the customers and the substations. However, in terms of the electric network, under normal operating conditions, the distribution line operates as a radial tree structure [3, 4]. In these configurations, the system is composed of a number of switches whose status can be closed or open at any time. In radial systems, the status of the switches are chosen such that each customer is connected to exactly one substation [3, 4]. When a fault occurs in the system, utility companies modify the status of switches in order to find an operating cut-set that is able to provide electricity for individual customers and accordingly, minimize the number of customers without power. This procedure is called "reconfiguration" [3, 4]. For network reliability purposes, in order to have an accurate estimation of the performance of the system, the reconfiguration procedure should be considered. Accounting for the reconfiguration procedure requires system level reliability analysis of different cut-sets of a distribution network. These cut-sets can be in order to of a few miles with a few hundred poles.

Previous studies on reliability and resilience assessment of distribution systems considered statistically independent failure events for utility poles [5-7]. However, this assumption may not be accurate because it is expected that the failure events of adjacent poles in distribution networks are statistically dependent. This is due to the similarities in the wind loadings, environmental conditions that impact the long-term decay in wood, and structural characteristics, among other factors. Another limitation of current reliability studies of distribution systems is that in the performance assessment of the utility poles, the impact of adjacent spans on the pole of interest is neglected [1, 8-10]. In existing distribution lines, due to the difference in the age of adjacent poles and the span length of neighboring spans, this impact is considerable [2].

This paper presents a systematic procedure for the assessment of the probability of failure of individual cut-sets of distribution networks. This procedure considers the impacts of structural couplings as well as effects of statistical dependence in the hurricane reliability analysis of distribution lines. The effect of adjacent spans is reflected by adopting an equivalent boundary model developed by Mohammadi-Darestani et al. [2]. In addition, to account for the correlated failure events, a Multi-scale Matrix-based System Reliability (Multi-scale MSR) method developed by Song and Kang [11] is adopted. This method is an extension of Matrix-based System Reliability (MSR) method [12]. Multi-scale MSR estimates the reliability of systems with large number of components by decomposing the system to a set of links where each link consists of a few components (utility poles). Subsequently, the reliability of the system is obtained in two levels. The first level is the link failure analysis and the second level is the link connectivity analysis [11]. The proposed procedure can be integrated with reconfiguration processes to obtain the reliability of distribution networks.

2 PROBABILITY OF FAILURE OF WOOD POLES IN OVERHEAD DISTRIBUTION LINES

In reliability analysis, failure is defined as the event where the limit state function ($G(X)$) is smaller than zero. Therefore, the probability of failure of a system with limit state function $G(X)$ can be obtained as

$$P_f = P[G(X) < 0] = \int_y P[G(X) < 0 | IM = y]\, f_{IM}(y) dy \tag{1}$$

where IM is the intensity measure for the hazard, $f_{IM}(y)$ is the probability density function of the intensity measure, and $P[G(X) < 0 | IM = y]$ is the fragility function which defines the conditional probability of failure given $IM = y$.

For utility poles, the primary mode of failure occurs when the wind induced moment demand at the ground-line is greater than the corresponding moment capacity [1-2, 8-10]. Assuming a lognormal distribution for the random variables representing the moment capacity and demand, the limit state function for poles can be defined as the natural logarithm of the safety factor as follows

$$G(X) = ln(R) - \ln(S) \tag{2}$$

where R is the modulus of rupture of wood poles and S is the maximum wind-induced bending stress in the ground-line section of the poles.

2.1 *Capacity Model for Wood Poles*

According to Shafieezadeh et al. [1], a lognormal distribution can be defined for the capacity of wood poles. The mean value of this lognormal distribution can be found as:

$$E(\Sigma_R) = K_s K_c K_h K_L A C_{GL}^B \times \exp[k_{0.05,0.5,\infty} \times \sqrt{\ln(\rho_R^2 + 1)}] \quad (3)$$

where C_{GL} is the circumference of the pole at the ground-line, and A and B are regression parameters. For Southern Pine wood poles, these parameters were derived as 3.482×10^7 and 0.320, respectively [13]. In addition, K_s is the adjustment factor for the size effect. K_c is the adjustment factor for conditioning. K_h is the calibration factor for the historic precedent in conventional design. K_L is the adjustment factor for the effect of load sharing [13]. Σ_R is a random variable representing the modulus of rupture of wood poles and ρ_R is the coefficient of variation (COV) of the modulus of rupture. The recommended value of COV for Southern Pines is 0.169 by [14]. k is the tolerance factor which can be obtained from [1].

Knowing the mean value of the lognormal distribution from Eq. (3) and the coefficient of variation, which is assumed to be 0.169 [14], the random variable representing the capacity of new wood poles can be defined.

2.2 *Effects of In-Service Decay on the Capacity of Poles*

Shafieezadeh et al. [1] proposed a time-dependent model for the decay in wood poles. Based on this model, the expected value and variance of the capacity of wood poles are

$$E[R|T=t] = E[R_0][1 - \min(\max(a_1 t - a_2, 0), 1) \times \min(\max(b_1 t - b_2, 0), 1)] \quad (4.a)$$

$$\begin{aligned} Var[R|T=t] = {} & (Var[R_0] + E[R_0]^2)\,(1 - \min(\max(b_1 t - b_2, 0), 1)) \\ & + \{(Var[R_0] + E[R_0]^2)[Var[L|T] \\ & + (1 - \min(\max(a_1 t - a_2, 0), 1))^2]\} \times (\min(\max(b_1 t - b_2, 0), 1)) \\ & - E[R_0]^2 \times [1 - \min(\max(a_1 t - a_2, 0), 1) \\ & \times \min(\max(b_1 t - b_2, 0), 1)]^2 \end{aligned} \quad (4.b)$$

where R is the capacity of the wood pole, R_0 is the capacity of the new pole, t is the age of the wood pole in terms of years, and $Var[L|T]$ is the time-dependent variance of the loss of the capacity of wood poles. The variance of the loss of capacity of wood poles is assumed as 0.11 based on the data reported in [15]. Parameters a_1 and a_2 are found as 0.014418 and 0.10683, respectively [15]. Parameters b_1 and b_2 are obtained as 0.004 and 0.04, respectively [1, 15].

2.3 *Demand Model for Wood Poles*

To estimate wind loads on utility poles, the static gust wind load model for non-building structures in ASCE07 [16] is adopted here. This model defines the wind force per unit length as:

$$f_w = q_z G C_f D \quad (5)$$

where q_z is the velocity pressure at height z on the pole, G is the gust-effect factor, C_f is the force coefficient, and D is the diameter perpendicular to the wind direction. The wind pressure is obtained from:

$$q_z = 0.613 K_z K_d K_{zt} V^2 \quad (6)$$

where K_z is the velocity pressure exposure coefficient, K_d is the wind directionality factor, K_{zt} is the topographic factor, V is the 3-second gust wind speed at 10 m above the ground.

The mean values of these factors are provided in ASCE07 [16]. Although some of the factors presented in Eqs. (5-6) are assumed deterministic, some others are considered as random variables. For the random variables, Table 1 presents the probability distribution model and the coefficient of variation. These models can be used to perform Monte Carlo simulations in order to obtain the probability distribution of demand on the wood poles.

Table 1

Parameter	Distribution	COV
G	Normal	0.11
C_f	Normal	0.12
K_z	Normal	0.16
K_d	Normal	0.08

2.4 *Boundary model for the effect of Adjacent Spans on the Wind Performance of Utility Poles*

The coupling effects among adjacent spans can be considered through an equivalent boundary model proposed by Mohammadi-Darestani et al. [2]. The proposed boundary model accounts for the impact of adjacent spans on the pole of interest via a set of equivalent springs and equivalent forces that represent the stiffness and the wind-induced forces of adjacent spans on the pole of interest, respectively. In order to obtain equivalent boundary model parameters, the stiffness and equivalent wind induced forces for each component of the distribution line should be obtained. Mohammadi-Darestani et al. [2] obtained the stiffness of the poles through a flexibility method. In addition, the stiffness of conductors were determined by considering a parabolic shape for the conductors under wind loads. Subsequently, a stiffness matrix and a force vector were obtained for the segment of the distribution line next to the poles of interest. Finally, the equivalent stiffness and equivalent force representing the impact of adjacent spans on the pole of interest were obtained through a superposition procedure. Details of the derivation of the boundary model are provided in details in [2].

3 MULTI-SCALE MATRIX-BASED SYSTEM RELIABILITY ASSESSMENT

In system level reliability of distribution networks, conventional Monte Carlo simulation methods are not capable of considering correlation in the failure events of the components of the system. In order to overcome this issue, more advanced methods such as Dichotomized Gaussian Method (DGM) have been proposed in the literature [17]. However, these methods become computationally inefficient for systems with small probabilities of failure, as they need a considerably large number of samples in order to produce reliable estimations. An alternative for simulation based techniques is the Matrix-based system reliability method (MSR), which obtain the probability of failure of the system by simple matrix procedures and is also capable of considering correlation in the failure events of the components of the system [12]. However, this method becomes computationally inefficient for systems with large number of components. In order to overcome this issue, Multi-scale Matrix-based system reliability method (Multi-scale MSR) has been proposed to perform reliability analysis in multiple levels [11]. This method decomposes the system into a set of subsystems called links. Subsequently, the MSR method is employed in two levels. The first level is the link failure analysis and the second level is the link connectivity analysis. At both levels, the order of the system is reduced and therefore, MSR method can be used efficiently. The details of MSR method can be found in [11-12]. The Multi-scale MSR method is adopted in this study to analyze the hurricane system level reliability of distribution lines. The level of correlation between the components of each link is predefined. Therefore, this correlation level can be considered in the link failure analysis. However, the level of correlation between the links are not known. In order to estimate this correlation, the probability of the union of the failure events of links i and j can be defined assuming the failure event of these two links are represented by a bivariate standard normal variable with unknown correlation coefficient ρ_{ij}.

$$P(F_i \cup F_j) = P(F_i) + P(F_j) - P(F_i)P(F_j) \; - \int_0^{\rho_{ij}} \varphi_2(\beta_i, \; \beta_j, \rho)d\rho \tag{7}$$

where β_i is the reliability index for link i and is calculated by Eq. (8) and $\varphi_2(x, y, \rho)$ is the PDF of a bivariate standard normal distribution with correlation coefficient ρ and is calculated by Eq. (9)

$$\beta_i = -\Phi^{-1}(P(F_i)) \tag{8}$$

where Φ^{-1} is the inverse CDF of a standard normal distribution

$$\varphi_2(x, y, \rho) = \frac{1}{2\pi\sqrt{1-\rho^2}} \exp\left(-\frac{x^2 - 2\rho xy + y^2}{2(1-\rho^2)}\right) \tag{9}$$

The probability of failure of links i and j and the probability of the union of their failure events can be obtained through an MSR method. The remaining unknown in Eq. (7) is ρ_{ij} which can be found through a numerical process. When ρ_{ij} is estimated, the link connectivity analysis can be performed efficiently by an MSR method.

4 NUMERICAL STUDY

In this study, in order to investigate the effect of correlation on the reliability of cut-sets of distribution lines, a cut-set of a distribution line with 200 poles is considered. The poles are assumed to be class 5 with 11 m height. Moreover, four lines of ASCR Penguin conductors are considered. Penguin conductors have a diameter of 14.3 mm and an ultimate strength of 37.1 kN. The initial horizontal component of tension in the conductors is assumed to be 25% of their ultimate strength.

Multi-scale MSR method is employed to obtain the probability of failure of the distribution line at three correlation levels and six different mean ages of the poles. In addition, the probability of failure of the system for the uncorrelated and fully correlated cases of the failure events of the system are provided (Figure 1.). The level of correlation is defined by a length scale that defines the spatial correlation for the failure events of the adjacent poles through the following equation

$$\rho_{ij} = \exp\left(-\frac{|x_i - x_j|}{L_{scale}}\right) \tag{10}$$

where x_i and x_j are the location of poles i and j, respectively and L_{scale} is the length scale.

Results indicate that the level of correlation has a considerable impact on the probability of failure of distribution lines. For example in Figure. 1.a, the probability of failure of a distribution line where all poles are new is provided. As expected, increase in the level of correlation reduces the probability of failure of the system. This is due to the fact that when individual components are correlated, the likelihood of joint failure of individual poles increases. The increase in the probability of the intersection of the failure events of the poles leads to a decrease in the probability of the union of the failure events of the poles. For the new poles (Figure 1.a), the probability of failure of the system for a wind speed equal to 44.7 m/s in the case of no correlation is 0.75. For length scales of 100 m, 200 m, and 400 m, the probabilities of failure are found as 0.71, 0.66, and 0.58, respectively. For the case of fully correlated failure events, the probability of failure is 0.41. This indicates a 5.3%, 12.0%, 22.7 %, and 45.3% reduction in the probability of failure of the system, respectively, compared to when pole failures are assumed to be uncorrelated. Similar trends can be observed for lines with other mean ages. As it is evident from these results, the uncorrelated case represents the worst case scenario for the probability of failure of the distribution system. However, in reality due to the similarities in the wind history, environmental conditions and structural properties of adjacent poles, it is very likely that utility poles exhibit a non-negligible level of correlation in their failure events. The assumption of uncorrelated failure events therefore results in an overestimation of the probability of failure of the distribution system; the extent of the difference depends on the level of spatial correlation.

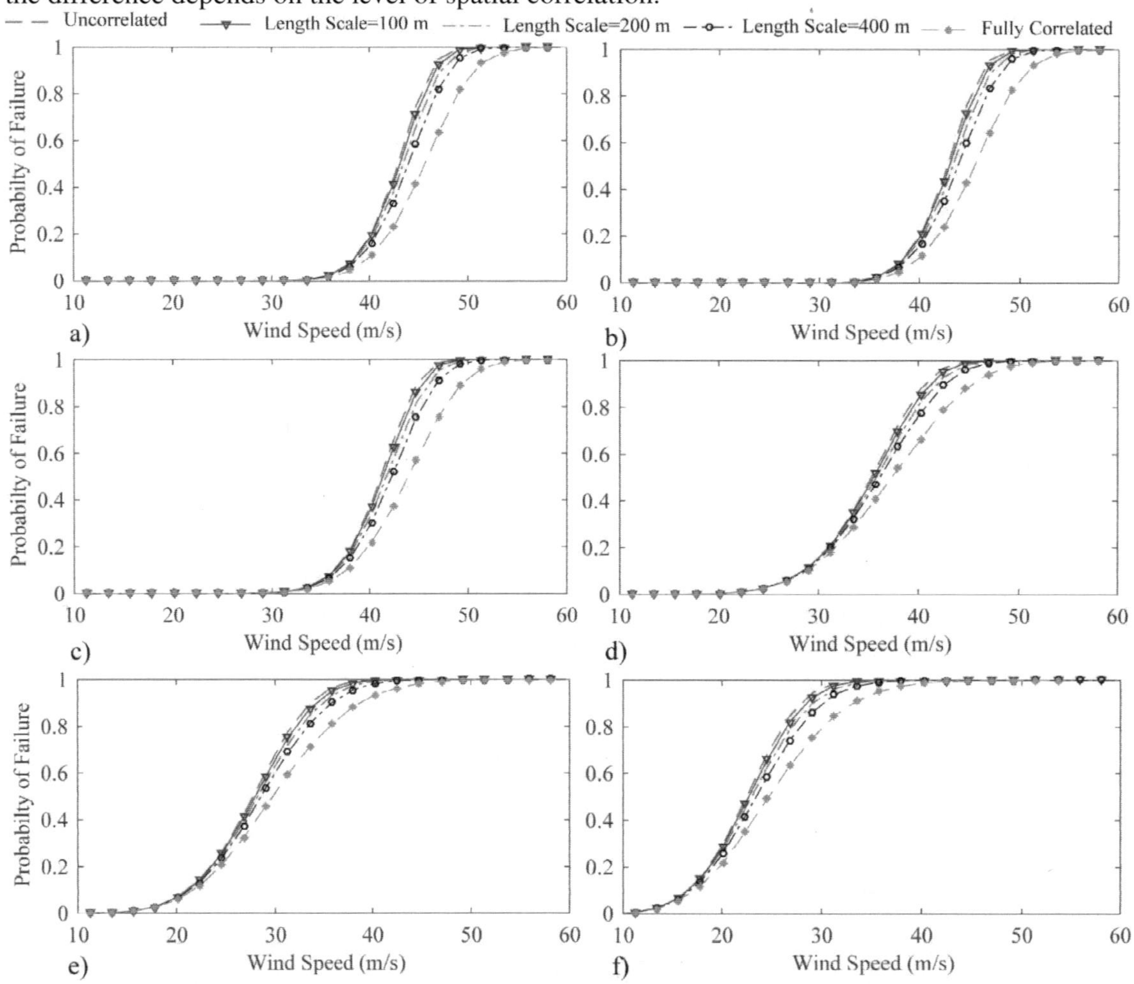

Figure 1. Probability of failure of the 200 poles distribution line for a) new poles, b) mean age of 10 years old, c) mean age of 20 years old, d) mean age of 30 years old, e) mean age of 40 years old, and f) mean age of 50 years old

5 SUMMARY AND CONCLUSIONS

The hurricane vulnerability of distribution lines highlights the need for accurate system reliability methods. This study expands upon and improves the hurricane reliability analysis of distribution lines by using a Multi-scale Matrix-based System Reliability method for estimating the probability of failure of cut-sets of distribution systems with statistically dependent components. In this approach, the system is divided into a set of links, where each link consists of a set of components. Subsequently, the reliability of the system is estimated at two levels. The first level is the link failure analysis and the second level is the link connectivity analysis.

The above procedure is applied to analyze the hurricane reliability of a cut-set of a distribution line with 200 poles and investigate the impact of correlation among the failure events of utility poles. Numerical results for distribution lines with 200 poles and different mean ages of poles in the lines indicated the significant impacts of correlated failure events on the system reliability of distribution lines.

The common assumption of uncorrelated failure events for utility poles in reliability analysis of distribution lines appears to yield the worst-case scenario for the failure of these systems. Considering that adjacent poles have similarities in wind loading histories, rate of decay, and structural characteristics, among other factors, it is expected that a non-negligible correlation exists among the failure events of poles. Therefore, the assumption of uncorrelated events may lead to considerable overestimations of the probability of failure of distribution networks. This may adversely impact management plans for inspection, maintenance, repair and replacement of poles. Therefore, for risk-informed decision purposes, a reasonable level of correlation should be considered for the failure events of adjacent poles in distribution lines.

6 ACKNOWLEDGEMENT

This paper is based upon work supported by the National Science Foundation under Grant No. CMMI-1333943 and 1334733. This support is greatly appreciated.

7 REFERENCES

1 Shafieezadeh, Abdollah, Urenna P. Onyewuchi, Miroslav M. Begovic, and Reginald DesRoches. "Age-dependent fragility models of utility wood poles in power distribution networks against extreme wind hazards." IEEE Transactions on Power Delivery 29, no. 1 (2014): 131-139.
2 Mohammadi-Darestani, Yousef, Abdollah Shafieezadeh, and Reginald DesRoches. "An equivalent boundary model for effects of adjacent spans on wind reliability of wood utility poles in overhead distribution lines." Engineering Structures 128 (2016): 441-452.
3 Shirmohammadi, Dariush, and H. Wayne Hong. "Reconfiguration of electric distribution networks for resistive line losses reduction." IEEE Transactions on Power Delivery 4, no. 2 (1989): 1492-1498.
4 Zhu, Ji Zhong. "Optimal reconfiguration of electrical distribution network using the refined genetic algorithm." Electric Power Systems Research 62, no. 1 (2002): 37-42.
5 Onyewuchi, Urenna P., Abdollah Shafieezadeh, Miroslav M. Begovic, and Reginald DesRoches. "A probabilistic framework for prioritizing wood pole inspections given pole geospatial data." IEEE Transactions on Smart Grid 6, no. 2 (2015): 973-979.
6 Salman, Abdullahi M., Yue Li, and Emilio Bastidas-Arteaga. "Maintenance optimization for power distribution systems subjected to hurricane hazard, timber decay and climate change." Reliability Engineering & System Safety (2017).
7 Ouyang, M., and L. Dueñas-Osorio. "Multi-dimensional hurricane resilience assessment of electric power systems." Structural Safety 48 (2014): 15-24.
8 Ryan, Paraic C., Mark G. Stewart, Nathan Spencer, and Yue Li. "Reliability assessment of power pole infrastructure incorporating deterioration and network maintenance." Reliability Engineering & System Safety 132 (2014): 261-273.
9 Salman, Abdullahi M., and Yue Li. "Age-dependent fragility and life-cycle cost analysis of wood and steel power distribution poles subjected to hurricanes." Structure and Infrastructure Engineering 12.8 (2016): 890-903.

10 Salman, Abdullahi M., Yue Li, and Mark G. Stewart. "Evaluating system reliability and targeted hardening strategies of power distribution systems subjected to hurricanes." Reliability Engineering & System Safety 144 (2015): 319-333.
11 Song, Junho, and Seung - Yong Ok. "Multi - scale system reliability analysis of lifeline networks under earthquake hazards." Earthquake Engineering & Structural Dynamics 39, no. 3 (2010): 259-279.
12 Song, Junho, and Won-Hee Kang. "System reliability and sensitivity under statistical dependence by matrix-based system reliability method." Structural Safety 31, no. 2 (2009): 148-156.
13 Wolfe, Ronald W., Jozef Bodig, and Patricia K. Lebow. Derivation of nominal strength for wood utility poles. US Department of Agriculture, Forest Service, Forest Products Laboratory, (2001).
14 ANSI O5.1, Specifications and Dimensions for Wood Poles, Amer. Nat. Std. Inst. (ANSI), New York (2008).
15 Li, Y., S. Yeddanapudi, J. D. McCalley, A. A. Chowdhury, and M. Moorehead. "Degradation-path model for wood pole asset management." In Power Symposium, Proceedings of the 37th Annual North American, pp. 275-280. IEEE (2005).
16 Minimum design loads for buildings and other structures. American Society Civil Engineers (ASCE). Reston, VA, USA. ASCE 07-10; 2010.
17 Emrich, Lawrence J., and Marion R. Piedmonte. "A method for generating high-dimensional multivariate binary variates." The American Statistician 45, no. 4 (1991): 302-304.

The 13th Americas Conference on Wind Engineering (13ACWE)
Gainesville, Florida USA, May 21-24, 2017

Limit State Evaluation for Transmission Tower-Line System

Xuan Li [a], Huawei Niu [b], Fang Pan [c], Wei Zhang[d]

[a] *University of Connecticut, Storrs, Connecticut, USA*
[b] *Wind Engineering Research Center, Hunan University, Changsha, China*
[c] *Southwest Research Institute, San Antonio, Texas, USA*
[d] *University of Connecticut, Storrs, Connecticut, USA*

ABSTRACT: The physical infrastructure of the power systems, including the high-voltage transmission towers and lines as well as the poles and wires for power distribution at a lower voltage level, is critical for the resilience of the community to avoid power outages under the extreme weather related loads. In the current engineering practices, single circuit lattice steel towers linked by transmission lines are widely used to form power transmission systems. This study is to develop a limit state evaluation method for typical transmission tower-line system subjected to strong wind loadings. Wind tunnel experiments were carried out to obtain the drag coefficients for different panels of the transmission towers as well as for the transmission lines. Incremental dynamic analysis (IDA) is carried out to find the capacity curve for the transmission tower-line system. The probabilities of failure are evaluated based on two predefined limit states. The results show that the uncertainties from meteorology have considerable influence on the reliability of the transmission tower-line system.

KEYWORDS: Transmission tower-line system, wind tunnel experiments, dynamic analysis, capacity curves, limit state evaluation.

1 INTRODUCTION

The transmission tower-line system, consisting of lattice steel support towers and conductors, plays an important role in the electrical transmission and distribution network systems. Under the threats from natural hazards and man-made disasters, the failure of individual structural component and/or system and the possible cascading failures could cause large area blackouts with a huge economic loss, which are of great concern to the stakeholders, decision makers, and the community residents. In the Americas, Australia and South Africa, about 80% of the transmission tower failures are due to the strong wind loadings from the extreme weather events and such failures could lead to large area power outages [1]–[5]. Therefore, understanding the structural capacity of the transmission tower-line system and the characteristics of the interactions between structures and strong winds are essential to avoid future failure of the transmission tower-line system and achieve a quick recovery of regional areas from future extreme natural hazards.

Currently, many codes, specifications or manuals are available for the transmission tower-line system design, such as the ANSI National Electric Safety Code (NESC) (ANSI, 2006), the International Electrotechnical Commission (IEC) Standard 60826:2003 (IEC, 2003), and the ASCE Manual 74 (ASCE, 2010), etc. In these codes or design manuals, only linear elastic behavior is considered. However, under the strong winds, the nonlinear inelastic behavior could become dominated, which could be crucial for evaluating the structural safety and reliability under strong winds. Meanwhile, the drag coefficient of the lattice frames is based on the solidity ratio in the design codes, which could not effectively represent the load distributions of the wind drag forces on the transmission towers with different structural geometries. To better understand wind-structure interactions, wind tunnel experiments are usually carried out to obtain the drag coefficient of the towers and conductors in different wind directions considering their real geometry [6], [7].

Due to the localized and unpredictable characteristics of the wind environment at different sites, the wind time histories and therefore the drag forces acting on the transmission tower-line system could be site specific. The transmission tower-line system could have different responses comparing with the responses of the structures under the specified loads defined in the codes or specifications. In addition, with a more dynamic climate environment, more extreme weather events are expected leading to a more adverse service condition with an increased failure probability of structures [8]. Therefore, it is essential to evaluate structural performance systematically in different limit states for various failure modes to meet the arising challenges of infrastructure management and the associated cost-benefit optimization demand [9].

After the introduction section, the paper is organized as the following five sections. Section 2 describes the numerical model for the transmission tower-line system and the simulated wind loading. Section 3 introduces the wind tunnel experiments and the drag coefficients for both the tower and the conductor are obtained. Section 4 briefly introduces IDA analysis and capacity curves for different wind angles are presented. Section 5 evaluates the probability of failure according to different limit states. Section 6 delivers some brief conclusions from this study.

2 MODELING THE TOWER-LINE SYSTEM AND WIND LOADING

2.1 *Finite element modeling of the transmission tower-line system*

The transmission tower line system is modeled in ANSYS as three same single circuit lattice steel towers being connected by three four-bundled conductors of four spans. The prototype of the transmission tower-line system is shown in Figure 1. The 550-kV transmission tower is 68.6 m height and the side length of the square base is 14.094 m. The structural components of the tower are built with steel angle sections of a 210 GPa elastic modulus. The nonlinear behavior of the tower is modeled by considering both the material nonlinearity and the geometric nonlinearity. More specifically, these two nonlinearities are included by using bilinear elastoplastic material property with 235 MPa yield strength and by implementing large deformation analysis, respectively. The prototype of a single transmission tower is same as that used by Zhang et al. (2015) [9].

Figure 1. Model of the transmission tower-line system

The distance between adjacent towers is 450 m and the conductor sag is 18 m based on the catenary equation. Because the two ground lines on the top have very small diameters compared with the conductors, the wind drag forces can be neglected and the two ground lines are not modeled in the structural model. Detailed properties of a single cable among the four-bundled conductor are listed in Table 1. In the transmission tower-line model, the conductors are connected to the transmission towers using suspension type insulator strings. The insulator string is composed of 34 insulators each with the mass of 13 kg and the equivalent characteristic parameters of the insulator string used in the numerical model are listed in Table 2. The finite element model built in the commercial software ANSYS uses 3D beam element with a total of 2221 elements and 878 nodes for each transmission tower and tension-only link element with the length of 5 m for the conductors and 0.195 m for the insulator strings. The four nodes of each transmission tower at the ground level are fixed and the leftmost and rightmost endpoints of the conductors are also fixed.

Table 1. Conductor properties

Item	Value
Transmission line type	JL/G1A-630/45 Aluminum conductor steel reinforced
Cross-section area (mm^2)	673.6
Outside diameter (mm)	33.8
Elastic modulus (GPa)	63
Mass per unit length (kg/m)	2.0792
Calculated tensile force (kN)	150450

Table 2. Insulator string parameters

Item	Value
Cross-section area (m^2)	0.07065
Density (kg/m^3)	920
Elastic modulus (GPa)	280
Poisson ratio	0.28
Initial force (kN)	36.7

2.2 *Modeling of wind loads*

The wind loads on a single transmission tower are applied in the same way as Zhang et al. (2015) did [9]. For the sake of brevity, only the calculation method for the wind drag forces on the conductors is given and defined as:

$$F_c = 0.5\rho V^2 \sin^2 \beta C_{dc} dL \tag{1}$$

where F_c is the total wind loading on the conductor; ρ is the density of air; V is the total wind velocity at the height of the conductors consisting of the mean wind speed and the fluctuating component of the wind speed; β is the wind yaw angle between the wind direction and the conductors; C_{dc} is the drag coefficient for the four-bundled conductor obtained through wind tunnel experiments; d is the diameter of a single cable; and L is the span length of the conductors.

In the present study, the mean wind speed is calculated based on the power law. The fluctuating wind speed is assumed to be a Gaussian stochastic process and the spectral representation method is used to generate the fluctuating component of wind speed based on spectral density functions [10]. The Kaimal's spectrum [11] is implemented to simulate the longitudinal and lateral fluctuating wind speed, and the Lumley-Panofsky's spectrum is used for simulating the vertical fluctuating wind speed.

3 WIND TUNNEL EXPERIMENTS

3.1 *Drag coefficients of transmission towers*

The drag coefficients for the transmission tower can be obtained through wind tunnel experiments. The wind tunnel experiments for the transmission tower model were carried out in HD-2 wind tunnel at Hunan University, China. The test section of the wind tunnel is 2.5 m high, 3.0 m wide and 17.0 m long. Considering both the size limitation of the wind tunnel and the fabrication accuracy of the transmission tower model, a rigid model with 1/40 scale is used to achieve a better measurement of the drag coefficients. The rigid model in the wind tunnel is shown in Figure 2. Detailed information about the experimental measuring procedures and the tested drag coefficients for each subsection in different wind directions can be found in [9]. For a demonstration purpose, only four wind yaw angles of 0°, 30°, 60° and 90° are taken into account in the present study.

Figure 2. Rigid tower model in wind tunnel

Figure 3. Conductor model in wind tunnel

3.2 *Drag coefficients of conductors*

In the transmission tower-line system, the four-bundled ACSR (aluminum conductor steel reinforced) conductors with a distance of 400 mm between adjacent cables are used, which are supported by insulator strings on the transmission towers. The characteristic parameters of the four-bundled conductors can be found in Table 1. To measure the drag coefficients for the conductor, wind tunnel experiments are also carried out for the real conductor as shown in Figure 3. The four-bundled conductor model is connected to a six-component force balance on a turntable and the drag coefficients are tested in the smooth flow. Considering the effect of Reynold number and wind attack angle, two wind speeds are chosen to measure the drag coefficients for the four-bundled conductor model at different wind attack angles (α) ranging from 0° to 45° with an interval of 7.5°. The illustration of the wind attack angle is shown in Figure 4. Then the drag coefficients for the four-bundled conductor are calculated in the same way as for the transmission tower. It should also be noted that the projected area is calculated from one single cable instead of the entire four cables.

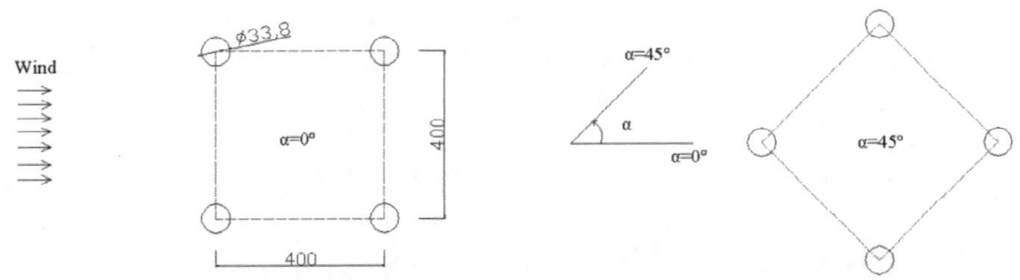

Figure 4. Shape of the 4-boundled conductor and attack angle in wind tunnel tests

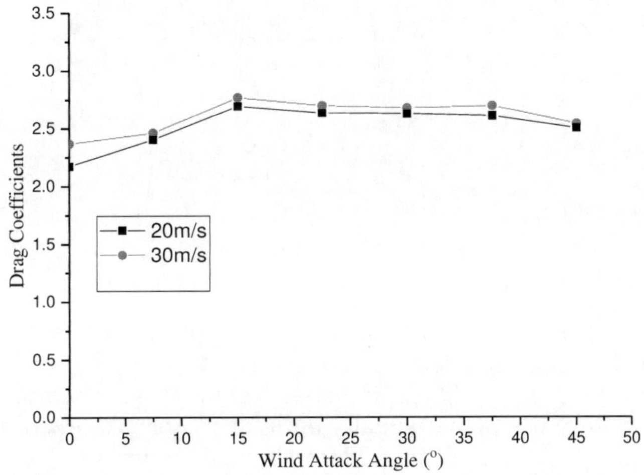

Figure 5. Drag coefficients for 4-bundled conductor

The tested drag coefficients of the four-bundled conductor model for each wind speed and wind attack angle are shown in Figure 5. As shown in Figure 5, the difference in drag coefficients between these two wind speeds are limited and the drag coefficients vary mildly with wind attack angles due to the gradually varying projected area of the four-bundled conductor. For both wind speeds, the maximum drag coefficients occur at the attack angle of 15°.

4 INCREMENTAL DYNAMIC ANALYSIS

4.1 *Dynamic analysis*

To obtain the capacity curve and perform the limit state evaluation of the entire transmission tower-line system, the dynamic analysis is carried out in time domain. Based on the wind loads defined for the towers and conductors and the drag coefficients obtained from wind tunnel experiments, the dynamic response of the transmission tower-line system can be obtained. Figure 6 shows the time history of the tower top tip displacement and the base shear force at the wind speed of 30 m/s and the wind yaw angle of 90°. Based on Figure 6, a gust response factor of 1.285 and 1.375, which is defined as the ratio of the maximum response to the mean response, can be obtained for the top tip displacement and the base shear force, respectively.

Meanwhile, it should be mentioned that the dynamic analysis is carried out using 80 s wind load time history and a ramp load of 3 s is applied at the beginning of the simulation to reduce the dynamic instability caused by the suddenly applied wind loading. Also, a sensitivity analysis is carried out for 80 s and 150 s simulations and there is limited difference between these two cases for the maximum top tip displacement and base shear force. Therefore the simulation time is chosen as 80 s to save computational time.

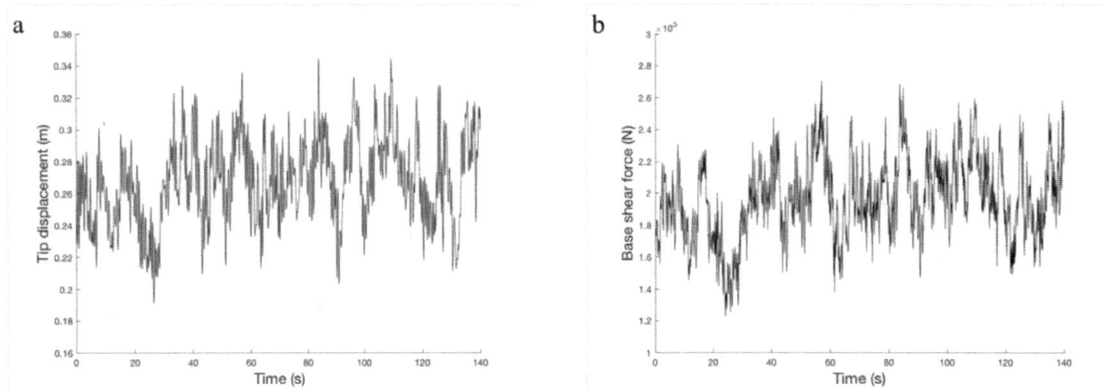

Figure 6. Time history for: (a) top tip displacement and (b) base shear force

4.2 *Incremental dynamic analysis*

IDA is originally used in earthquake engineering to evaluate the structure capacity and performance under seismic loading. Recently, some researchers applied IDA method to assess structural capacity of transmission towers under wind loading [12]–[14]. Similar to scaling the given ground motion time history, the wind speed time history is scaled gradually until the incipience of structural collapse, at which the numerical solution does not converge. Through the IDA approach, time histories of structural dynamic response at different wind loading levels can be obtained and therefore the structure capacity can be assessed.

Some damage measures (DM) need to be defined in order to implement IDA method for the evaluation of structural capacity. In this study, the maximum top tip displacement and the maximum base shear force are selected as DMs. Through the implementation of the IDA method, time histories of the top tip displacement and the base shear force corresponding to each wind yaw angle and wind speed are obtained and the dynamic response are then used to establish the capacity curve.

4.3 *Capacity curve*

The capacity curves for different wind yaw angles are obtained based on the maximum top tip displacement and the maximum base shear force. Figure 7 shows the relationship between the top tip displacement and the base shear force for the four wind directions considered in this study. All capacity curves exhibit a similar behavior of moving from elastic range to plastic range as the top tip displacement increases and only limited differences can be found for the force-displacement curves in the lower elastic range. As shown in Figure 7, the structure is in an elastic range when the top tip displacement is below 0.75 m or 0.5 m and the base shear force is below 550 kN or 400 kN for 0° wind direction and the other three wind directions, respectively. As yielding begins to occur in the structural model, the force-displacement curve for 0° deviates from that for the other three wind directions and is parallel to the other three curves in the plastic range. A 24% difference in the base shear force at the same top tip displacement are kept in the plastic range. It is noteworthy that the force-displacement curves for the wind direction of 30°, 60° and 90° are almost the same which implies that the transmission tower exhibits a similar elastic-plastic behavior under the skew wind conditions.

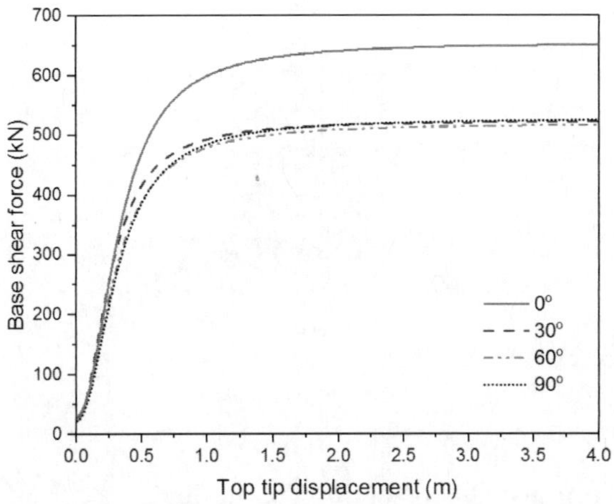

Figure 7. Capacity curves for the four wind directions

5 LIMIT STATE EVALUATION

5.1 *Probability density function*

To evaluate the capacity of the tower-line system from the viewpoint of probability, the distribution types for the peak values of certain performance parameters that we are interested in need to be formed for each wind direction and wind speed. Considering the DMs we used to evaluate the capacity curves, the top tip displacement is chosen as one of the performance parameters. Meanwhile, the swing angle of insulator strings is also chosen as another performance parameter since many codes have a requirement for the swing angle of insulator strings under strong wind loading, such as ASCE Manual 74 [15] and Design Manual for High Voltage Transmission Lines [16]. Based on the assumption of Gaussian stochastic process for the fluctuating wind speed and the sensitivity analysis for the length of simulation time, the response of the transmission tower-line system is assumed ergodic and the simulation results could well represent the structural behavior during high intensity wind events, such as tornados, hurricanes, and downbursts. The peak values of the top tip displacement and the swing angle of insulator strings are then fitted to certain probability distribution type based on Chi-Square test shown below:

$$\sum_{i=1}^{m}\frac{\left(n_i-e_i\right)^2}{e_i}\le C \tag{2}$$

where m is the number of bins in a histogram; n_i is the observed frequency within the ith bin; e_i is the theoretical frequency for a given distribution within the ith bin; and C is used as a threshold value determined by the significance level and the degree of freedom.

For the sake of brevity, only the plots of probability density function fitting for the top tip displacement and the swing angle of insulator strings at the wind yaw angle of 90° and the wind speed of 15 m/s are shown in Figure 8. Based on the Chi-Square test, the Lognormal and Gamma distribution type are used as the distribution type for the top tip displacement and the swing angle of insulator strings, respectively. As shown in Figure 8, these two kinds of distribution type fit the histogram very well with a deviation of 2.57 and 1.05, which are less than their respective threshold values of 15.09 and 13.28 in the Chi-Square test.

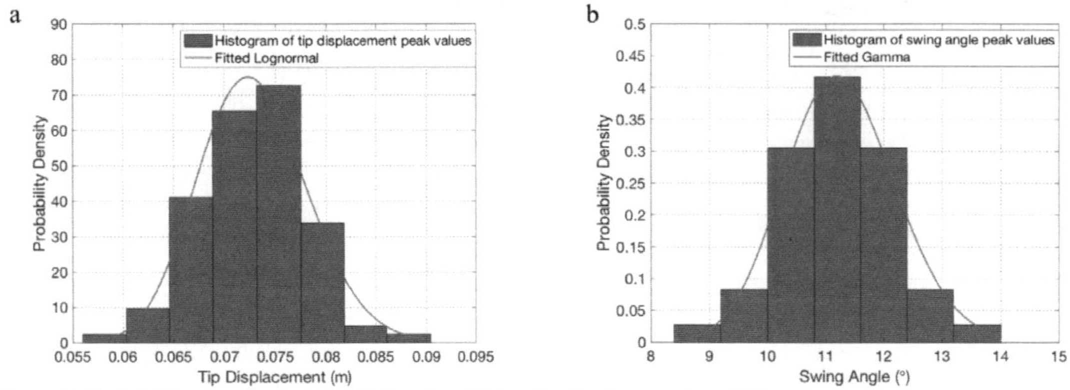

Figure 8. Probability density function fitting for: (a) top tip displacement and (b) swing angle

5.2 *Limit state formulation*

In the present study, two limit state functions are formulated considering the motion of the tower top tip and the swing angle of the insulator strings and the failure criteria are chosen based on the guidelines and design manuals. In FEMA guidelines, the maximum drift to height ratio $\theta_{max} = 2\%$ is defined as the limit state for steel moment-resisting frames with type-1 connections for the immediate occupancy structural performance level [17]. Since no design codes for the transmission tower have defined such a limit state, $\theta_{max} = 2\%$ is used as the limit state to evaluate the probability of failure ($P_{\theta i}$) in the ith wind direction due to large top tip displacements which could be obtained by evaluating the limit state function shown below:

$$g_\theta = \theta - \theta_{max} \tag{3}$$

where g_θ is a failure function such that $g_\theta < 0$ implies a structural failure due to large top tip displacement; θ is the drift (top tip displacement) to height ratio and its probability density function could be obtained from the distribution type fitting method mentioned in Section 5.1; and $\theta_{max} = 2\%$ is the limit state defined for structural capacity.

Suspension insulator strings supporting transmission conductors are usually free to swing about their upper connection point to the crossarms of transmission towers, and therefore it is necessary to limit swing angles to ensure enough clearances between conductors and structures under various weather conditions [16]. In Design Manual for High Voltage Transmission Lines [16], the minimum allowable clearance is defined for no wind, moderate wind and high wind conditions to prevent flashovers between conductors and towers. Since the goal is to evaluate the structure capacity in extreme weather events, the high wind clearance (1.215 m) is used to calculate the maximum allowable swing angle (ϕ_{max}). Based on the geometry of our transmission tower model, $\phi_{max} = 50°$ is used as the limit state to evaluate the probability of failure ($P_{\phi i}$) due to large swing angle of insulator strings. Similar as the limit state function defined for θ, the limit state function for ϕ is defined as:

$$g_\phi = \phi - \phi_{max} \tag{4}$$

where g_ϕ is the failure function for the swing angle, ϕ is the swing angle of insulator strings of which the probability density function could also be obtained from distribution type fitting method, and $\phi_{max} = 50°$ is another limit state defined for structural capacity.

5.3 *Limit state evaluation*

After defining the limit state functions, the probability of failure for each defined failure mode in each wind direction can be evaluated using Eq. 3 and Eq. 4, respectively, based on the meteorological data from a given site. In this study, three groups of wind speeds are selected according to 50-year, 100-year and 200-year return period from the wind historical data at the Xiaoyangshan Meteorological Observatory [18], [19]. For a 50-year return period wind, the probability of failure due to large top tip displacement is very close to 0 for all wind direction except for $\beta = 300°$, which has a value of 1, and the probability of failure due to large swing angle behaves in a similar way with non-zero values of 0.771 at $\beta = 270°$ and 0.047 at $\beta = 300°$. It is noteworthy that the value of 1 for the probability of failure of θ implies that the drift to height ratio has exceeded our defined limit state (θ_{max} = 2%) under the wind speed in that wind direction. Figure 9 shows these two kinds of probability of failure in each wind direction under 50-year return period wind. Similar plots can also be obtained for the 100-year and 200-year return period wind as shown in Figure 10 and Figure 11.

As shown in Figure9-11, the range of wind yaw angles related to the large top tip displacement failures keeps increasing with the rising of the wind speed and the towers are more vulnerable at $\beta = 270° - 330°$ due to strong wind speed in these directions. However, the wind directions with larger probability of failure for the swing angle is mainly at $\beta = 270°$ where the wind with higher speed blows on the conductors perpendicularly and result in a larger swing angle of insulator strings.

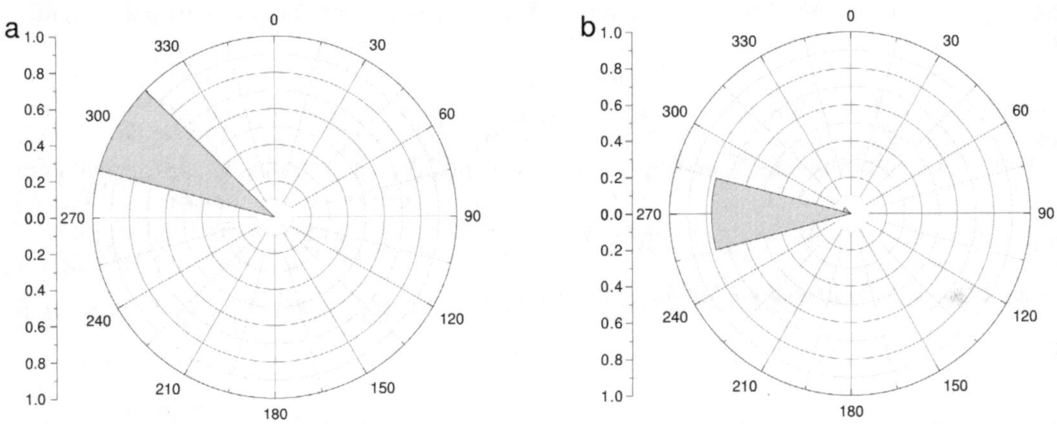

Figure 9. Probability of failure at 50-year return period for: (a) top tip displacement and (b) swing angle

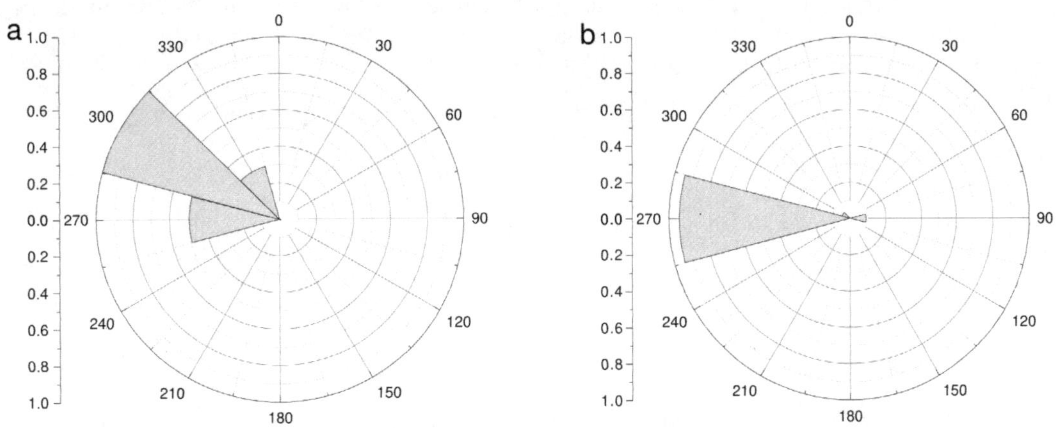

Figure 10. Probability of failure at 100-year return period for: (a) top tip displacement and (b) swing angle

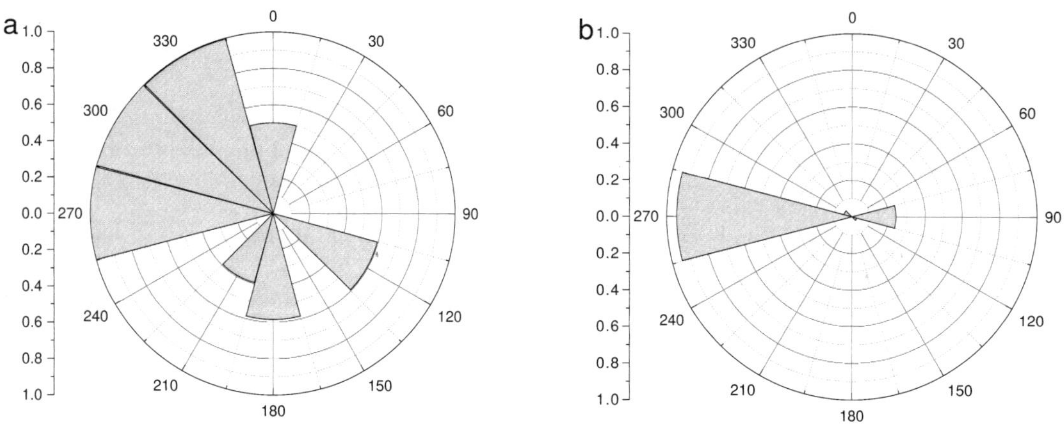

Figure 11. Probability of failure at 200-year return period for: (a) top tip displacement and (b) swing angle

6 CONCLUDING REMARKS

In the present study, a limit state evaluation approach is carried out for evaluating the performance of the transmission tower-line system by considering the uncertainties from the environment. Two limit states for the top tip displacement and the swing angle of insulator strings are defined based on the guidelines and design manuals. Probability of failure of these two failure modes are evaluated based on the meteorological data from one meteorological observatory.

From the present study, the following conclusions are drawn: (1) elastic-plastic behavior could occur in the transmission towers when wind speed is large enough; (2) the structural capacities under skew wind and lateral wind conditions are similar and are lower than the longitudinal wind condition, which implies the tower-line system is more robust when the wind direction is along the conductors; (3) probability of failure for the large drift to height ratio and swing angle both vary with the wind yaw angles, whereas the variability of the range of wind yaw angles leading to appreciable probability of failure is different for these two failure modes at different return periods.

It is noteworthy that other failure modes, such as fatigue damages and large tensile forces in conductors, could also be integrated into the limit state evaluation approach developed in this study. In addition, system reliability analysis can also be carried out by combining the results from several single limit state evaluations to have a deeper and comprehensive understanding of the global performance of the transmission tower-line system.

7 REFERENCES

[1] D. Dempsey and H. B. White, “Winds wreak havoc on lines,” *Transm. Distrib. World*, vol. 48, no. 6, pp. 32–42, 1996.

[2] P. McCarthy and M. Melsness, “Severe weather elements associated with September 5, 1996 hydro tower failures near Grosse Isle, Manitoba, Canada,” *Manitoba Environ. Serv. Centre, Environ. Canada*, p. 21, 1996.

[3] E. Savory, G. A. R. Parke, M. Zeinoddini, N. Toy, and P. Disney, “Modelling of tornado and microburst-induced wind loading and failure of a lattice transmission tower,” *Eng. Struct.*, vol. 23, no. 4, pp. 365–375, 2001.

[4] A. De Oliveira, E. Silva, J. C. P. De Medeiros, A. M. Loredo-Souza, M. M. Rocha, L. I. Rippel, E. A. Carpeggiani, and G. J. Z. Núñez, “Wind loads on metallic latticed transmission line towers,” *Proc. CIGRE Sess. 2006*, 2006.

[5] M. Jacobs, "13 of the Largest Power Outages in History — and What They Tell Us About the 2003 Northeast Blackout," *Union Concerned Sci. Sci. a Heal. planet safer world*, no. August 8, 2013.

[6] T. G. Mara and T. C. E. Ho, "Design loads for transmission towers under skewed wind loading," in *Structures Congress 2011*, 2011, pp. 1246–1257.

[7] A. M. Loredo-Souza, "The aeroelastic wind tunnel modeling of transmission lines and their behaviour under severe boundary layer winds," no. July, pp. 101–110, 2014.

[8] G. A. Meehl, F. Zwiers, J. Evans, T. Knutson, L. Mearns, and P. Whetton, "Trends in extreme weather and climate events: issues related to modeling extremes in projections of future climate change," *Bull. Am. Meteorol. Soc.*, vol. 81, no. 3, pp. 427–436, 2000.

[9] W. Zhang, J. Zhu, H. Liu, and H. Niu, "Probabilistic Capacity Assessment of Lattice Transmission Towers Under Strong Wind," *Front. Built Environ.*, vol. 1, no. October, pp. 1–12, 2015.

[10] G. Deodatis, "Simulation of ergodic multivariate stochastic processes," *J. Eng. Mech.*, vol. 122(8), no. 8, pp. 778–787, 1996.

[11] J. C. Kaimal, J. C. Wyngaard, Y. Izumi, and O. R. Cote, "Spectral characteristics of surface layer turbulence," *Quart. J. Roy. Meteorol. Soc.*, vol. 98, no. 453, pp. 563–589, 1972.

[12] S. Banik, H. Hong, and G. A. Kopp, "Assessment of Structural Capacity of an Overhead Power Transmission Line Tower under Wind Loading," *BBAA VI Int. Colloq. Bluff Bodies Aerodyn. Appl.*, pp. 20–24, 2008.

[13] S. S. Banik, H. P. Hong, and G. A. Kopp, "Assessment of capacity curves for transmission line towers under wind loading," *Wind Struct. An Int. J.*, vol. 13, no. 1, pp. 1–20, 2010.

[14] T. G. Mara, "Capacity Assessment of a Transmission Tower Under Wind Loading," 2013.

[15] ASCE, "Guidelines for electrical transmission line structural loading," *ASCE Manuals Reports Eng. Pract. No. 74, Ed. by C. Jerry Wong Michael D. Miller, Prep. by Task Comm. Struct. Load. Comm. Electr. Transm. Struct. Struct. Eng. Inst.*, 2010.

[16] USDA, "Design Manual for High Voltage Transmission Lines," *Bull. 1724E-200*, no. May, 2015.

[17] FEMA, "Recommended Seismic Evaluation and Upgrade Criteria for Existing Welded Steel Moment-Frame Buildings, prepared for the Federal Emergency Management Agency," *FEMA-351, Washington, DC*, p. 256, 2000.

[18] Y. Ge, Y. Yang, and Z. Zhang, "Report on the wind tunnel experiments of Donghai cable-stayed bridge," *Dep. Bridg. Eng. Tongji Univ. Shanghai*, 2003.

[19] W. Zhang, C. S. Cai, and F. Pan, "Fatigue reliability assessment for long-span bridges under combined dynamic loads from winds and vehicles," *J. Bridg. Eng.*, vol. 18(8), pp. 735–747, 2013.

The 13th Americas Conference on Wind Engineering (13ACWE)
Gainesville, Florida USA, May 21-24, 2017

Assessing the Wind-Induced Response of High-Mast Illumination Poles (HMIPs) using Field Data

Y.-C. Chen[a], L. Manuel[b], M.A. Morovat[c], D. Zuo[d]

[a]*University of Texas, Dept. of Civil, Arch. & Env. Engr., Austin, TX, USA, ycchen@utexas.edu*
[b]*University of Texas, Dept. of Civil, Arch. & Env. Engr., Austin, TX, USA, lmanuel@mail.utexas.edu*
[c]*University of Texas, Dept. of Civil, Arch. & Env. Engr., Austin, TX, USA, morovatma@utexas.edu*
[d]*Texas Tech University, Dept. of Civil, Env. & Const. Engr., Lubbock, TX, USA, delong.zuo@ttu.edu*

ABSTRACT: High-Mast Illumination Poles (HMIPs) are commonly used to provide lighting along highways and at interchanges. HMIPs are tapered steel poles fabricated in sections with heights sometimes exceeding 50 meters. In recent years, collapses of HMIPs have been reported in the United States. These collapses have been attributed, in part, to fatigue-induced damage at the HMIP shaft-to-base plate connection. In order to better understand the fatigue loading environment at the base of HMIPs, we discuss a field measurement campaign underway that is seeking to characterize associated wind loading regimes and ultimately employ the data in fatigue reliability studies. We are primarily focused on loading that is associated with buffeting and vortex shedding. Various parameters such as the mean wind speed, vertical wind profile, wind direction, and turbulence intensity estimated from the data are used to characterize the wind environment and how it results in different types of HMIP response. Strain gauges installed at the bottom of the HMIP are also used to identify vibration modes and relate response amplitudes to the wind. Stress range distributions at a detail of interest—conditional on 5-minute averaged wind field parameters—are estimated from the field data. Together with uncertainty assessments of fatigue capacity, to be established separately from ongoing laboratory measurements, a framework to assess the reliability against fatigue damage of HMIPs under different wind regimes is developed. While measurements from four HMIPs are being gathered, only data from a site in Austin, Texas are presented here. The presence of pre-existing cracks from a galvanizing process often used with HMIPs and how they influence fatigue life is a separate subject of investigation for which the present study is relevant.

KEYWORDS: High-mast illumination poles; vortex shedding; buffeting; wind-induced vibration

1 INTRODUCTION

High-Mast Illumination Poles (HMIPs) are typically tapered steel poles fabricated in sections with heights ranging from 20 to over 50 meters. Figure 1 shows a typical HMIP structure located in Texas. The advantage of HMIPs over standard luminaire posts is a reduction in the number of luminaire structures that are needed to illuminate large areas. A few collapses of HMIPs have been reported, however, in the U.S. lately. These collapses have been attributed, in part, to fatigue failures at the HMIP shaft-to-base plate connection. Cracks resulting from the galvanizing process have been reported in past studies [1, 2]; these, in turn, are believed to be responsible for a reduction in fatigue resistance. On the load (demand) side, vortex-induced vibration at low wind speeds is also believed to be responsible for higher induced stresses; VIV has also been reported for HMIPs in past studies [3]. Both the reduced capacity and the increased demand from the wind loading need to be evaluated in order to assess the remaining life of HMIPs in the field.

According to past studies [3], the two types of aerodynamic response of HMIPs of greatest concern are vortex-induced vibration (VIV) and buffeting. VIV leads to large deflections in the cross-wind direction when the vortex shedding frequency locks in with the natural frequency of the HMIP. Vortex shedding for HMIPs is more common at low wind speeds. Buffeting response of HMIPs is enhanced by natural wind gusts or turbulence. Buffeting affects both along-wind and cross-wind response; its influence increases with increasing wind speed and turbulence intensity.

Figure 1. Typical HMIP in Texas

The response of an HMIP at a long-term field monitoring at a site in Austin, TX is the subject of this study; we discuss the observed response of the HMIP under different wind conditions. A procedure to evaluate the reliability of HMIPs based on Miner's damage accumulation model [4] using collected field data is outlined.

2 FIELD MONITORING PROGRAM

HMIPs in Texas consist of a variety of designs with varying parameters such as the height, number of sides, design wind speed, and whether or not a ground sleeve is used at the bottom of the pole. Based on previous studies [1], a specific HMIP design consisting of a 12-sided, 45.7 m high pole, with a 35.8 m/s design wind speed and without a ground sleeve has a higher chance of experiencing cracks in the shaft-to-base region during the galvanization process. This particular design is being studied in laboratory tests (not discussed here) in order to assess the fatigue capacity of the pre-cracked detail. The present study is related to field monitoring intended to assess the fatigue demand for various wind conditions. In order to consider representative and different wind conditions across the state, five HMIPs are selected at different locations in Texas for the field monitoring study. The monitoring locations and the HMIP designs there are 12-sided, 45.7 m high poles without ground sleeves and with two different design wind speeds as listed in Table 1. The design wind speed in Galveston is higher than at the other locations since that region generally experiences higher winds. The present study only discusses the monitored site in Austin but a procedure is outlined here for the use of field data in fatigue reliability assessments that can be applied to other sites such as those with contrasting wind characteristics and the data from the five sites in Table 1 can all be used then.

Table 1. Monitored HMIPs in Texas

Location	Design Speed m/s	Yearly Wind Information: Average Wind Speed m/s	Yearly Wind Information: Max Gust m/s
Austin	35.8	3.5	33.5
El Paso	35.8	3.5	27.3
Dallas	35.8	4.1	30.4
Laredo	35.8	4.2	25.5
Galveston	44.7	5.0	26.8

The selected HMIPs for the 35.8 m/s design wind speed consist of three segments with a taper slope of 14.58 mm/m in diameter along their heights. The top and bottom diameters (measured flat to flat) are 19.69 cm and 82.87 cm, respectively. The three segments have equal lengths of 15.7 meter each; the two lower segments have a wall thickness of 8 mm whereas the top segment has a wall thickness of 6.4 mm. The 76.2-mm thick base plate is welded to the tube and connected to a concrete base using 10 threaded anchor rods. The lighting apparatus with 12 luminaire lights at the top of the pole weighs about 816 kilograms.

Figure 2 shows a systematic diagram of the field monitoring setup. A sonic anemometer and a propeller anemometer are installed at 9.1 m and 24.4 m above the base plate, respectively. An ultrasonic anemometer (R.M. Young, Model 81000) is used to record turbulence as well as wind speed and direction. The additional propeller anemometer (R.M. Young, Model 05103V) at 24.4 m above the base records wind speed at a different height in order to estimate the wind profile (shear). Both anemometers are installed on a 1.6 meter aluminum cantilever arm that is attached to the pole with steel straps. Three tri-axial accelerometers (PCB, Model 3713E1110G and 3713E112G) are attached to the HMIP to monitor motions of the pole; a ±2G accelerometer is installed 9.1 m above the base plate whereas two ±10G accelerometers are installed 15.2 m and 24.4 m above the base plate. A total of 4 strain gauges are installed 1.9 m above the base plate in the four cardinal directions to record induced strains resulting from motions of the pole. The wind, pole motion, and strain data are continuously recorded by means of a National Instruments data acquisition system consisting of a cRIO-9024 controller, a cRIO-9114 chassis, and three different modules: NI-9239 (for accelerometers and ultrasonic anemometer), NI-9237 (for strain gauges), and NI-9219 (for propeller anemometer and power supply monitoring). The accelerometers, strain gauges, and ultrasonic anemometers collect data sampled at 50 Hz while the propeller anemometer data is sampled at 1 Hz. The data are frequently downloaded using a wireless modem (AirLink Raven RV50) to free up space on the acquisition system. The sensors, acquisition system, and modem are all powered by a system consisted of a 12V battery (DURHR12-540), two 90-watt solar panels (Ameresco 90J) and a charge controller (MorningStar, Model SS-MPPT-15L) that controls charging and power supply to the system.

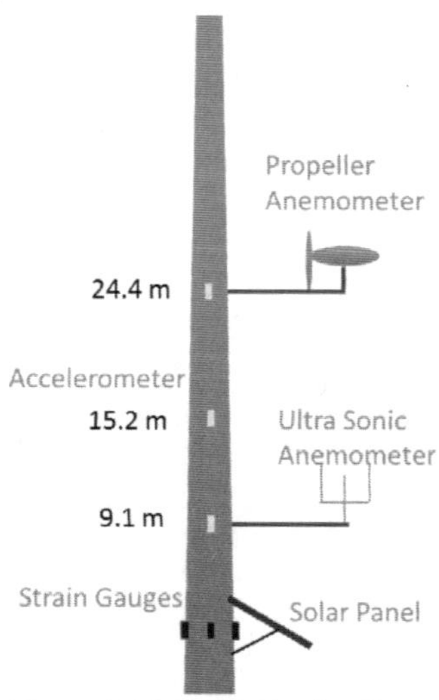

Figure 2. Field Monitoring Setup

3 DATA PROCESSING

The continuous data streams consisting of wind conditions, motion response, and induced stress response are broken down into 5-minute segments and then analyzed. The wind data consist of the average wind speed and direction from both anemometers, an estimated power law exponent (α) of an assumed mean wind speed profile, and turbulence intensity at the ultrasonic anemometer level.

The motion response of the pole is investigated using similar methods employed in a research study of a cantilevered mast arm for a traffic signal structure [5]. According to the method employed, it is necessary to extract displacements contributed to by different participating modes of vibration. The data processing flowchart is presented in Fig. 3.

First, the acceleration time series recorded in the N-S and E-W directions are transformed into along-wind and cross-wind directions as follows:

$$a_\theta(t) = a_N(t)\cdot\cos\theta + a_E(t)\cdot\sin\theta \tag{1}$$

where θ is the clockwise angle from North to the along-wind direction. Second, both along-wind and cross-wind acceleration time series are numerically integrated twice to obtain displacement time series. A detrend process using a db10-level 7 wavelet is implemented after each integration to remove any resulting offset and slowly varying trend. Third, the displacement time series data are band-passed using a 0.3 Hz frequency band with a 5th-order Butterworth filter applied around the first three natural frequencies (Table 2) that are identified by peak picking from the power spectra of the acceleration data. Fourth, the Hilbert transform [6] of the displacement time series is obtained so as to yield the envelope of the peak displacements and the instantaneous frequencies for each mode. The average displacement of the envelope process and the average frequency for each mode in the along-wind and cross-wind directions at all three locations of the accelerometers are obtained in this manner. Note that the model-based natural frequencies in Table 2 are obtained using the commercial software, SAP2000 while the field-based frequencies are from averaged power spectra based on measured accelerations.

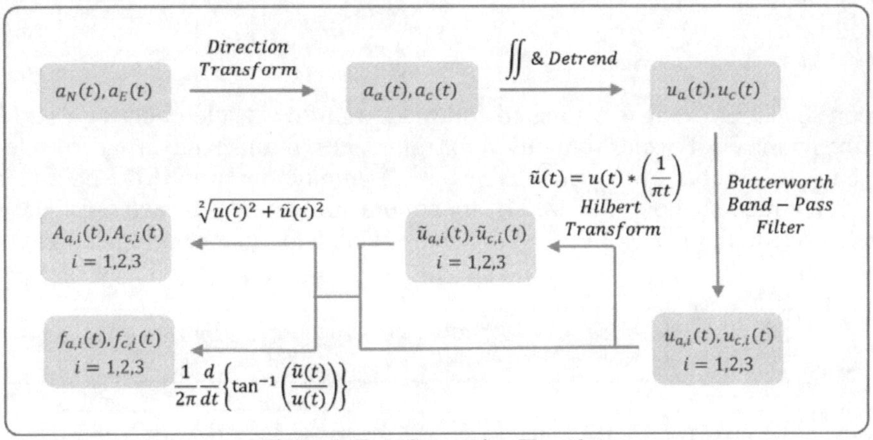

Figure 3. Data Processing Flowchart

Table 2. Natural Frequencies of the Monitored HMIP in Austin

Modes	Field	Model	Difference
	Hz	Hz	%
1	0.28	0.29	3.6
2	1.20	1.23	2.5
3	3.14	3.37	7.3

Figure 4 shows an example 5-minute acceleration time series that is processed so as to yield displacement time series for the three primary modes. Identified frequencies for the three modes (Fig. 4(g)) are consistent with the natural frequencies from the power spectral density (PSD) plots of the transformed displacement record as can be confirmed from studying Fig. 5.

In addition to estimation of the displacement amplitudes of the important response modes, the contribution of each mode is estimated using Spectral Proper Transformation (SPT) or Proper Orthogonal Decomposition (POD) implemented in the frequency domain [7]. The cross-power spectral density function matrix for frequencies of interest is estimated using time series at three heights. Eigenvalues of these matrices represent the variance contributed to by the mode described by the corresponding eigenvectors at that frequency. Since the second and third eigenvalues are relatively small compared to the first (largest) eigenvalue, the contributions at each frequency are represented by the largest eigenvalue. The contribution of each mode is estimated based on the sum of the largest eigenvalues in a 0.3 Hz frequency band centered on the three modal frequencies. The eigenvector corresponding to the largest eigenvalue at the three modal frequencies represents the vibrating mode shape of the structure. Figure 6 shows that the estimated eigenvectors from data match the mode shapes derived using the commercial software, SAP2000.

The induced stress response as derived from the strain gauge data are also rotated into along-wind and cross-wind directions in a manner as was indicated for accelerations using Eq. (3). The orthogonal strain gauges provide two pairs of along-wind and cross-wind strain time series that are averaged. Stress time series are derived from the strain time series and are subsequently used to obtain stress range histograms using the rainflow cycle counting algorithm [8].

The concept of an "equivalent" fatigue load (*EFL*) is introduced here to quantify the fatigue damage accumulated in a 5-minute segment. The N_0-cycle *EFL* is the constant-amplitude stress range which over N_0 cycles causes the same amount of fatigue damage as the variable-amplitude stress ranges, $S_{R,i}$, cause in each 5-minute segment. This concept of *EFL* is frequently used in the fatigue analysis of wind turbines [9] and its value easily distinguishes the level of damage from two different 5-minute segments. The *EFL* is defined as follows:

$$EFL = \sqrt[3]{\frac{\sum S_{R,i}^3}{N_0}} \tag{2}$$

where $S_{R,i}$ represent stress ranges estimated following rainflow cycle counting and N_0 is a pre-determined fixed number of cycles (only used for reference). In this study, N_0 is selected to be 84, which represents the number of response cycles in a 5-minute duration if the pole vibrates in its first mode with its frequency of 0.28 Hz. The maximum stress range in each 5-minute segment is also computed along with the *EFL*s in both the along-wind and cross-wind directions.

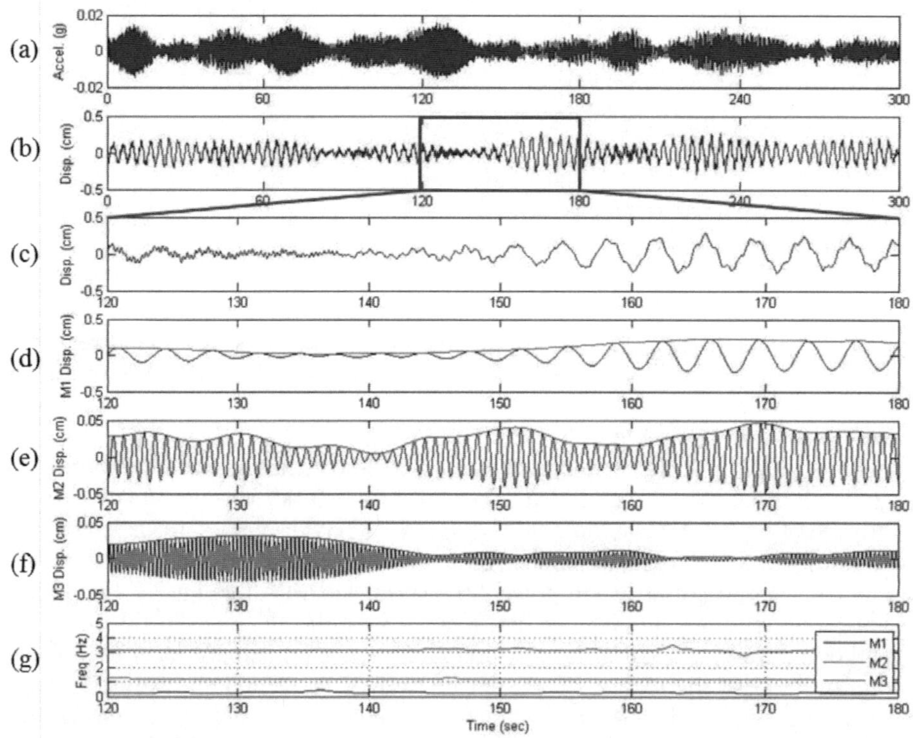

Figure 4. 5-Minute Displacement Time Series in Three Primary Modes Derived from Acceleration Data

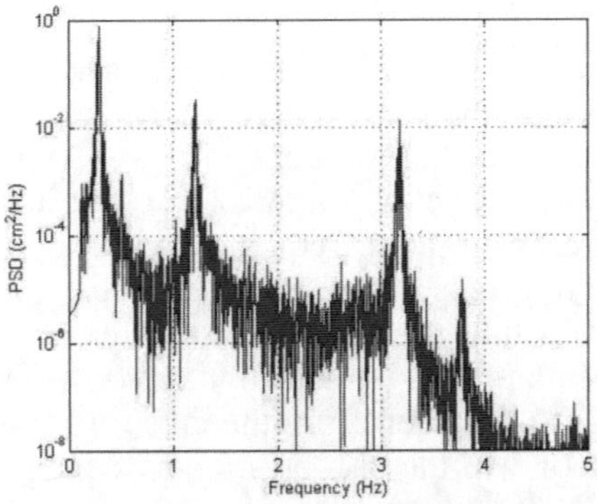

Figure 5. Power Spectral Density of the Selected Displacement Time Series

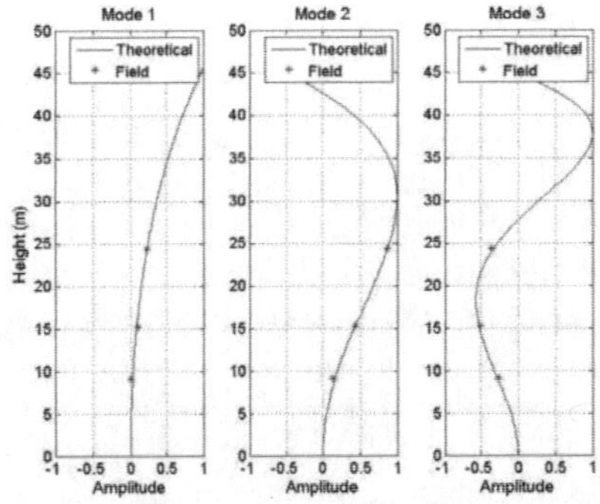

Figure 6. Mode Shapes - Calculated versus Measured

4 DYNAMIC RESPONSE USING FIELD DATA

Data collected from approximately four months in Austin, TX are used in this study to illustrate the effects of different wind parameters. Figure 7 shows characteristics of the wind environment (at 9.1 m above the HMIP base plate) during the monitoring period. Figure 7a shows the relative likelihood of different wind speeds and wind directions during the period. Figure 7b shows the number of occurrences of different 5-min average wind speed bins. Hourly wind speed and wind direction time series from the first 12 weeks of the monitoring period are presented in Fig. 7c.

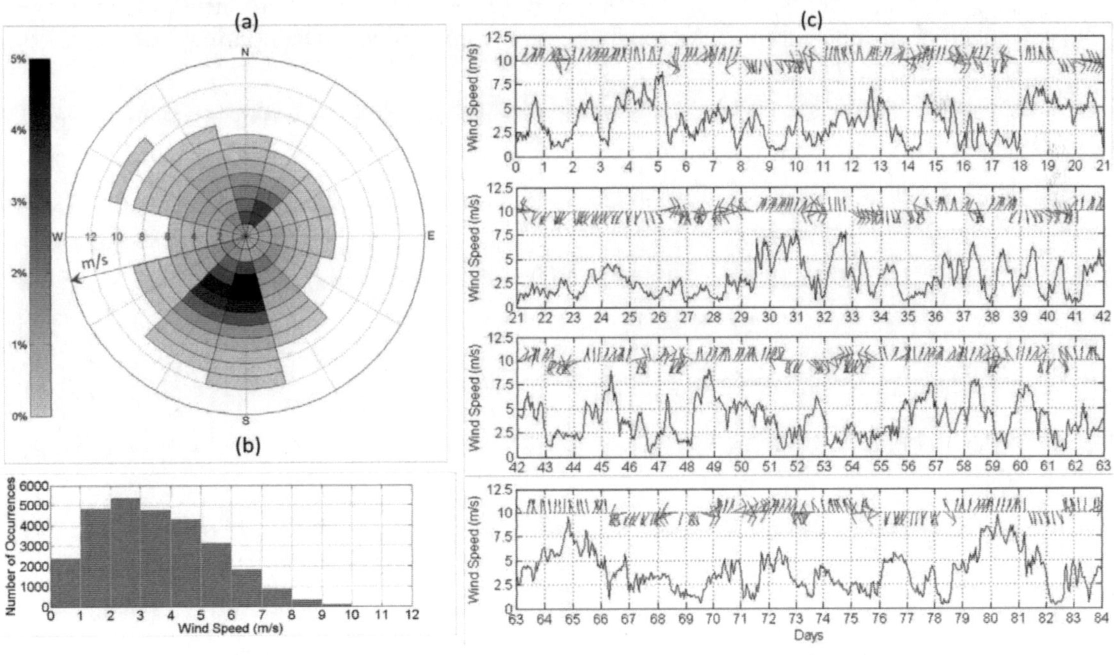

Figure 7. The Wind Environment in Austin as Recorded over the Four-Month Monitoring Period

4.1 *Dominant Response*

Figures 8a-8c show a sample first-mode dominant strain time series obtained from strain gauges attached to the South and West faces of the HMIP. The average wind speed at 9.1 meters above the base plate is 3.7 m/s from the North (0°) and is aligned with the higher response levels in the North-South direction (Fig. 8c). As seen in the close-up of a few cycles (Fig. 8b), the dominant frequency in the along-wind direction (South) is close to the first modal frequency whereas the cross-wind direction (West) exhibits some small-amplitude vibrations at higher frequencies.

Figures 8d-8f show a sample second-mode dominant strain time series from the same strain gauges. The average wind speed at 9.1 meters above the base plate is 1.4 m/s from the North-East (45°) which is perpendicular to the vibrating direction (Fig. 8f). As seen from the close-up (Fig. 8e), the vibrating frequencies for both strain gauges correspond to the second mode frequency, which is presumably related to vortex shedding.

Figures 8g-8i show a sample third-mode dominant strain time series from the same strain gauges. The average wind speed at 9.1 meters above the base plate is 1.4 m/s from the South-West (225°) which is perpendicular to the vibrating direction (Fig. 8i). As seen from the close-up (Fig. 8h), the vibrating frequency from both strain gauges is close to the third mode frequency.

Figure 9 shows stress range histograms that result from rainflow cycle counting based on the strain time series presented in Fig. 8. The stress range histograms for first-mode dominant time series indicate a greater number of occurrences of higher stress range bins for the South direction which is consistent with direction of the larger vibrating amplitude indicated in Figure 8c. The second- and third-mode dominant time series also bring about larger vibrating amplitudes (and stress ranges) in the South direction which can be confirmed by studying Fig. 9. The number of stress cycles generally increases with higher mode; this is a result of the higher vibrating frequencies. However, as seen from Fig. 9, stress range amplitudes for the third mode are relatively small; most of the cycles are confined to low stress range bins and thus relatively little fatigue damage will result from third-mode dominant stress data than from second-mode dominant data.

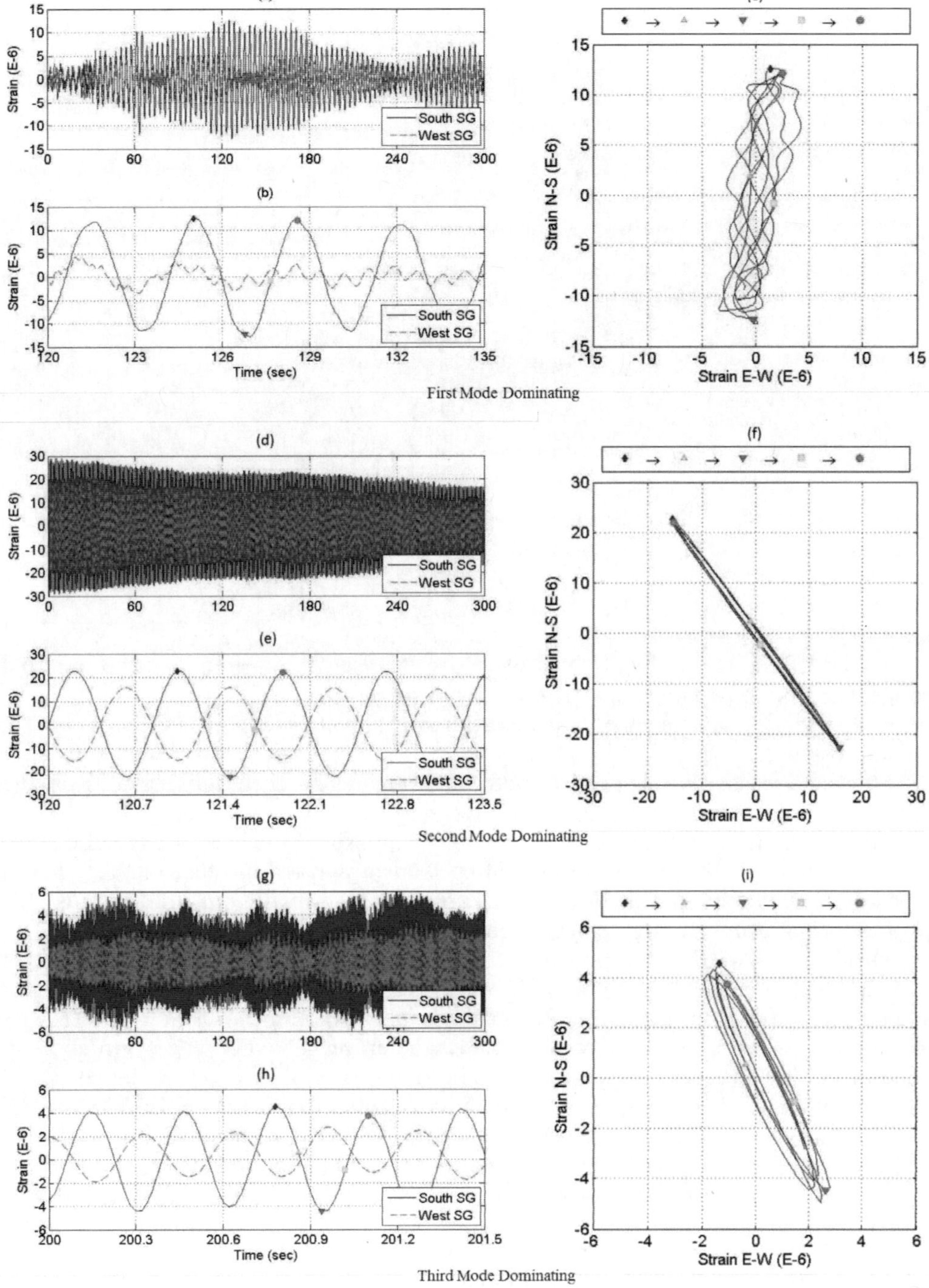

Figure 8. Strain Time Series Corresponding to Different Dominating Modes

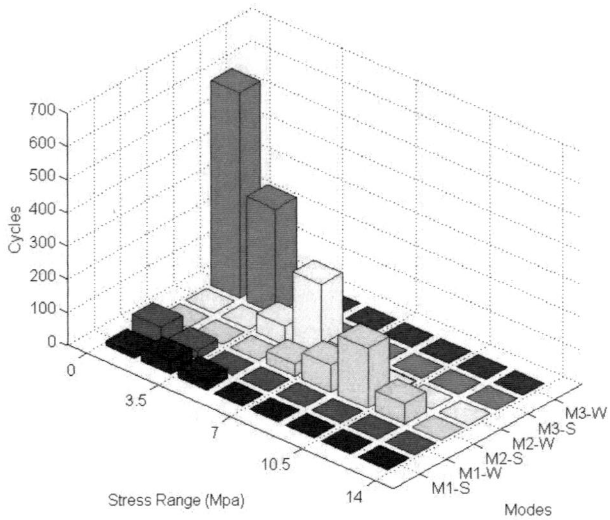

Figure 9. Stress Range Histograms from Strain Gauges from Time Series with Different Dominating Modes (5-min duration)

4.2 *Effect of Wind Speed*

Figure 10 shows HMIP along-wind and cross-wind displacements 24.4 m above the base plate decomposed into the first three modes for different wind speeds recorded at 9.1 m above the base plate. It is seen that Mode 1 displacements generally increase with increasing speed in both the along-wind and cross-wind directions. This response is attributed mostly to turbulence that typically increases with increasing wind speed in both directions as has been observed in previous studies [3]. Mode 2 displacements indicate a clear peak in the low wind speed range that is thought to be due to vortex shedding since the amplified dynamic response only occurs in a specific narrow wind speed range. Although Mode 3 vortex shedding is also reported in other studies [11], the observed Mode 3 contributions indicated in Fig. 10 are relatively small compared to those from the two other modes and can therefore be neglected.

In order to clearly identify HMIP response that is associated with vortex shedding, mean displacements for dominating modes among the collected data are plotted against the reduced velocity (U_r) (Fig. 11), which is the reciprocal of the Strouhal number (St); i.e., $U_r = 1/\mathrm{St} = U/f_s \cdot D$, where U and D are taken to be the wind velocity and pole diameter, respectively, at 30.5 m above the base (f_s is the vortex shedding frequency). Since both the wind velocity and the diameter of the pole vary with height, the 30.5 m location was selected to represent the whole pole since the maximum displacement in the second mode occurs at that height. The wind speed at 30.5 m is estimated using a power law with the wind speed measurements at 9.1 m and 24.4 m. Following the spectral proper transformation carried out, the mode with the highest contribution to the displacement variance is regarded as the dominating mode. The average frequency (f_s) of the dominating mode is estimated using the Hilbert transform.

The significant peak over a relatively narrow band that corresponds to a reduced speed of around 5, as seen in Fig. 11, is close to the vortex-shedding lock-in reduced speed reported in the literature [3]. Higher response levels in the cross-wind direction over the lock-in speed range are also an indication of vortex shedding and vortex-induced vibration (VIV). From the figure, we can see that the lock-in response is only coming from the second mode. Mode 1 is influenced by turbulence and greater response levels result only at higher wind speeds; Mode 3 does not contribute significantly to either cross-wind or along-wind response.

Equivalent fatigue load (*EFL*) values computed for both the along- and cross-wind directions are plotted versus wind speed at 9.1 m above the base plate and shown in Figure 14. Higher *EFL* levels in the low wind speed range are due to the significant VIV response in the cross-wind direction.

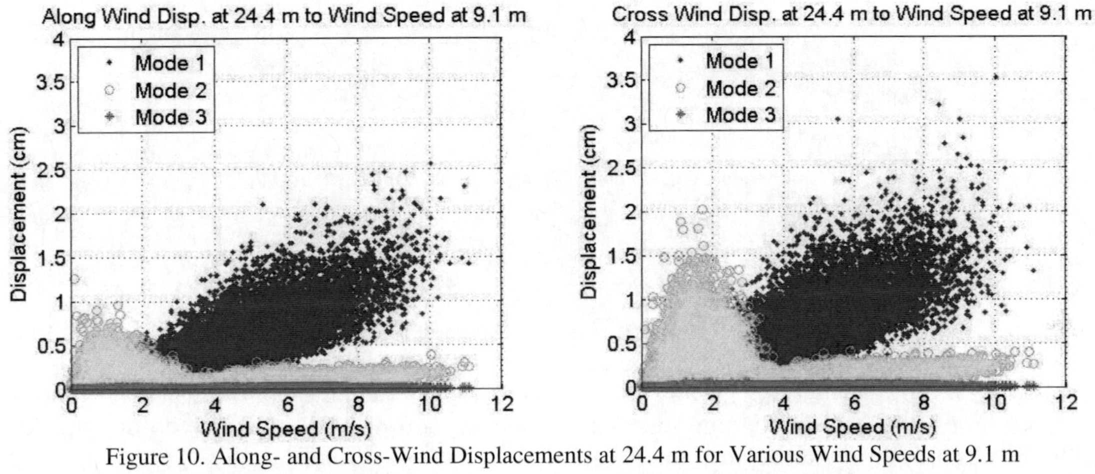

Figure 10. Along- and Cross-Wind Displacements at 24.4 m for Various Wind Speeds at 9.1 m

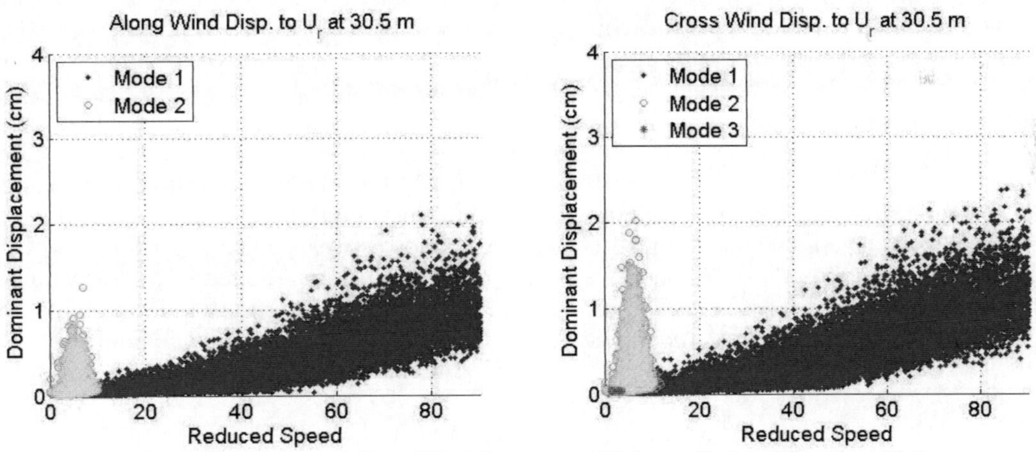

Figure 11. Along- and Cross-Wind Response at 24.4 m to Reduced Speeds at 30.5 m

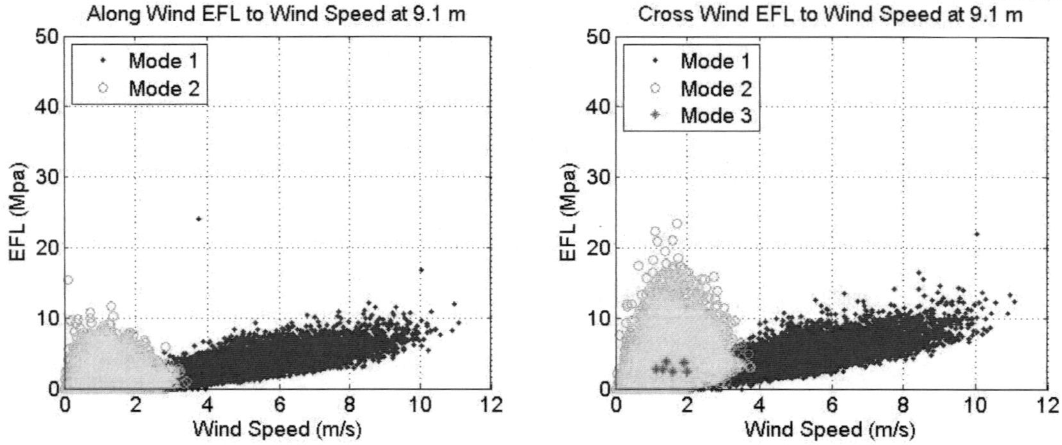

Figure 12. Along- and Cross-Wind *EFL* for Various Wind Speeds at 9.1 m

4.3 *Effect of Turbulence*

Smooth winds with low turbulence intensity provide favorable conditions for vortex-induced vibration of cantilevered mast arms for traffic signals [5]. Upon examining turbulence data at the monitored site for this HMIP study, it was found that due to the orientation of the arm that supports the ultrasonic anemometer, a shielding or blockage effect from the pole itself is observed. This leads to the absence of low turbulence intensity levels in the 270° direction (West) as can be seen in Fig. 13. A highway bridge is in close proximity in direction due North of the pole; this also leads to additional blockage for some directions. In order to study the effect of atmospheric turbulence on HMIP motions and to avoid drawing incorrect conclusions, winds coming from behind the pole and in the direction of the bridge are neglected leaving only data with wind directions between 90° and 240° for analysis.

Figure 14 shows plots of the along-wind maximum stress ranges versus the cross-wind maximum stress ranges for low wind speeds (1 to 2 m/s) and higher wind speed (8 to 9 m/s). At the low wind speeds (Fig. 14a), data for the time series with higher turbulence intensity levels are seen to result in low induced stress ranges in both the along- and cross-wind directions while with lower turbulence intensity levels, higher induced stress ranges are seen particularly in the cross-wind direction due to vortex shedding. At the higher wind speeds (Fig. 14b), the recorded stress ranges show have no preferential magnitude variation in either the along- or cross-wind directions; the response in both directions show a tendency to increase with increasing turbulence intensity.

5 WIND-INDUCED FATIGUE LOADING AND RELIABILITY ANALYSIS

Fatigue loading and damage on the monitored HMIP can be estimated using the stress time series derived from the strain gauge measurements. From measurements at a single site, one can learn about the likelihood of greater fatigue, say, from low wind speeds and low turbulence levels as well as from some other distinct wind conditions. It is expected that wind regimes may differ at different locations resulting in different fatigue demands on the structure. Characteristics of the wind environment may be described in terms of the occurrence likelihood of different wind speeds and directions. An example of such a characterization for the monitored site in Austin is presented in Fig. 7a.

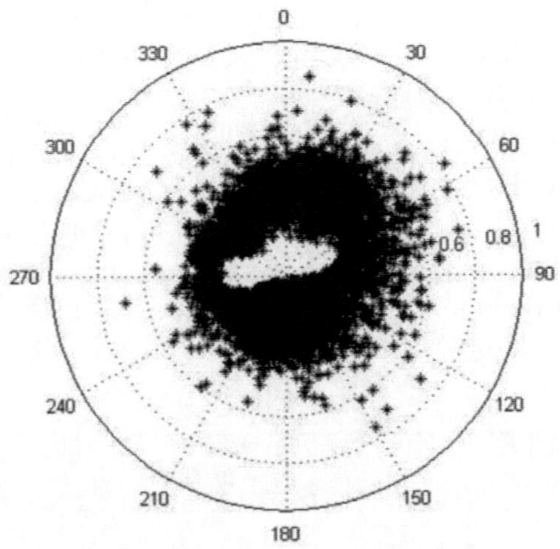

Figure 13. Measured Turbulence Intensity Levels for Different Wind Directions

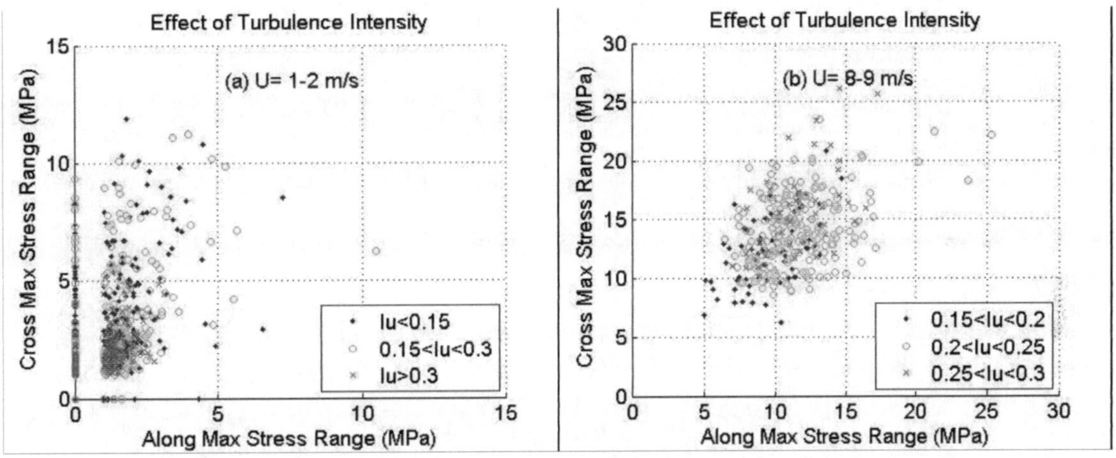

Figure 14. Along- and Cross-wind Maximum Stress Ranges for Different Turbulence Intensities

5.1 *EFL Variation with Wind Speed and Direction*

With knowledge of the wind direction in each 5-minute data segment, one can recover, using Eq. (1) applied to strains, strain time series at any location around the perimeter of the 12-sided HMIP. Accordingly, for 12 different angles (relative to the wind), θ_r, at 30° intervals, *EFL* values are computed at the Austin site. The database then conveniently yields an *EFL*(U, θ_r) matrix or table of fatigue damage information conditional on a wind speed bin, U, and for a direction of the wind relative to the detail of interest, θ_r. Only six different θ_r values need to be included in the important *EFL*(U, θ_r) table since *EFL*$(U, \pi + \theta_r)$ = *EFL*(U, θ_r). Figure 15 shows the computed *EFL*(U, θ_r) information derived using all the data from the Austin site. Higher *EFL* values implying greater fatigue demand in the cross-wind direction (θ_r=90°) are clearly seen at lower wind speeds indicating the influence of VIV in fatigue damage there.

At any site of interest where no measurements are available but where an HMIP is to be evaluated to assess its probability of failure due to fatigue within a specified period or time or where its remaining life is to be estimated, it would be important to have some information on the wind regime summarized, say, by the relative likelihood, α_{ij}, of winds of different speeds, i, and from different directions, j. Note that $EFL(U, \theta_r)$ effectively provides information on EFL_{ij} that can be used in a fatigue reliability analysis when combined with uncertainty information on fatigue capacity from laboratory testing of HMIP details. In Table 1, we mentioned other sites where HMIPs are monitored; EFL_{ij} data from those sites can also be used to supplement the EFL_{ij} data from the Austin site. Finally, we note that the different sites might have other contrasting characteristics such as wind shear and turbulence that are not accounted for with the simple (i, j) binning; by using more sites to define EFL_{ij} for the reliability model, we would achieve a more widely applicable approach for fatigue life assessment.

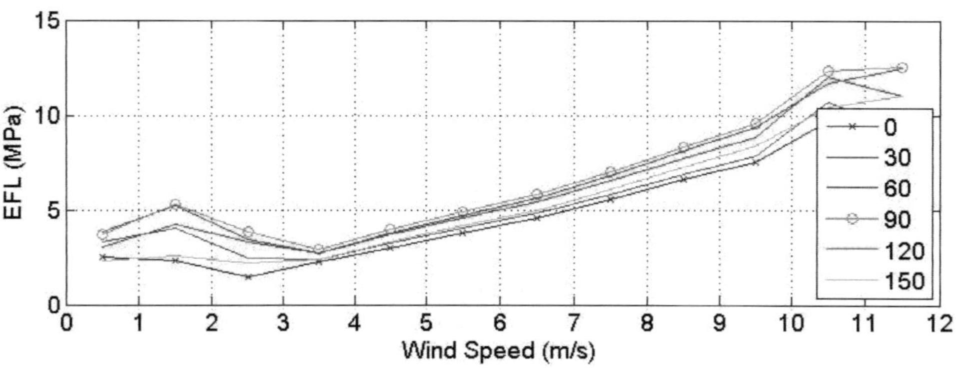

Figure 15. Representative *EFL* Values for Different Wind Speeds (at Different Relative Angles to the Wind)

5.2 *Fatigue Reliability Analysis*

Assuming that Miner's fatigue damage accumulation rule may be applied, a limit state function defining fatigue failure can be expressed as follows:

$$g\,(T;\, \Delta,\, A,\, \alpha_{ij},\, EFL_{ij}) = \Delta \cdot A - \sum_{i=1}^{12} \sum_{j=1}^{f} \alpha_{ij} \cdot \frac{T}{5} \cdot EFL_{ij}^{3} \cdot N_0 < 0 \tag{3}$$

where Δ is a critical damage accumulation index accounting for uncertainty in Miner's rule and often modeled as a lognormal random variable with a median value of unity and a coefficient of variation of 30%. The parameter, A, in Eq. (3) must be estimated from experiments or from knowledge of the endurance and fatigue capacity of the HMIP detail under evaluation (uncertainty may be accounted for in modeling A). Equation (3) allows the use of EFL_{ij} based on field measurements and α_{ij} from site-specific wind data to evaluate the probability of failure due to fatigue within time T expressed in minutes. Note that N_0 appears in Eq. (3) because the *EFL* definition implied equivalent fatigue load assuming N_0 constant-amplitude cycles in 5 minutes.

6 CONCLUSIONS

The dynamic response of a monitored HMIP in Austin, Texas was analyzed and presented in this study. Natural frequencies and mode shapes extracted from the field data were found to match corresponding results quite well that were based on a model using commercial structural analysis software. Large-amplitude vortex-induced vibration response in the cross-wind direction occurred at low wind speeds as was seen in previous studies. VIV mainly accompanied vibration in the second mode while the third mode was rarely excited significantly and could even be neglected. Turbulence was found to be more important at higher wind speeds where both the along-wind and cross-wind response in the first mode were significant. Damage expressed in terms of EFL for low wind speeds may exceed that for high wind speeds due to the higher vibration frequencies and higher motion amplitudes resulting from VIV. Since the wind speed and direction relative to a target location around the pole greatly affects the likely induced fatigue damage there, a comprehensive framework has been proposed to incorporate and account for damage contribution for different wind regimes and distributions (specified, say, by the probability of occurrence of different wind speed and wind direction bins) at any HMIP site of interest. The accumulated fatigue damage demand may be combined with the fatigue capacity information on HMIP details to permit a fatigue reliability assessment or an assessment of the remaining life of in-service HMIPs.

7 ACKNOWLEDGEMENTS

The authors gratefully acknowledge the funding received from the Texas Department of Transportation (TxDOT) for Project 0-6829: Fatigue Resistance and Reliability of High Mast Illumination Poles (HMIPs) with Pre-Existing Cracks. They also acknowledge discussions with other researchers (Drs. Michael Engelhardt, Todd Helwig, Patricia Clayton, and Eric Williamson) at Ferguson Structural Engineering Laboratory (FSEL) who are involved in the fatigue capacity studies of HMIPs that complement the wind load-related studies reported herein.

8 REFERENCES

1 Pool, CS 2010, Effect of Galvanization on the Fatigue Strength of High Mast Illumination Poles, M.S. Thesis, University of Texas at Austin, Austin, Texas
2 Kleineck, J 2011, Galvanizing Crack Formation at Base Plate to Shaft Welds of High Mast Illumination Poles, M.S. Thesis, University of Texas at Austin, Austin, Texas
3 Chang, B, Phares, BM, Sarkar, PP & Wipf, TJ 2009, 'Development of a Procedure for Fatigue Design of Slender Support Structures Subjected to Wind-Induced Vibration,' Journal of the Transportation Research Board, no. 2131, pp. 23-33.
4 Miner, MA 1945, 'Cumulative Damage in Fatigue,' Journal of Applied Mechanics, vol. 12, no. 3, pp. 159-164.
5 Zuo, D, Letchford, CW 2010, 'Wind-induced vibration of a traffic-signal-support structure with cantilevered tapered circular mast arm,' Eng. Struct., 32(10), 3171-3179
6 Bendat, JS, Piersol, AG 2002, Random data analysis and measurement procedures, 3rd edition, John Wiley and Sons, New York, NY
7 Srivilairit, T, Manuel, L 2009, 'Vortex-Induced Vibration and Coincident Current Velocity Profiles for a Deepwater Drilling Riser,' Journal of Offshore Mechanics and Arctic Engineering, 131(2), 021101
8 Downing, S, Socie, D 1982, 'Simple Rainflow Counting Algorithm,' International Journal of Fatigue, 4(1), 31-40
9 Ragan P and Manuel, L 2007, "Comparing Estimates of Wind Turbine Fatigue Loads using Time-Domain and Spectral Methods," Wind Engineering, Vol. 31, No. 2, pp. 83-99.
10 National Cooperative Highway Research Program (NCHRP) 2012, Fatigue Loading and Design Methodology for High-Mast Light Towers, Research report no. 718 prepared by RJ Connor, SH Collicott, AM DeSchepper, RJ Sherman, JA Ocampo, Transportation Research Board, Washington, DC.

11 Magenes, L 2011, Fatigue Assessment of High Mast Illumination Poles Using Field Measurements, M.S. Thesis, University of Texas at Austin, Austin, Texas
12 AASHTO 2015, LRFD Specifications for Structural Supports for Highway Signs, Luminaires and Traffic Signals, 1st edition, American Association of State Highway and Transportation Officials, Washington, DC

An Equivalent Dynamic Boundary Model for Effects of Adjacent Spans in Overhead Distribution Lines during Hurricanes

Yousef Mohammadi Darestani[a], Abdollah Shafieezadeh[b]

[a] *The Ohio State University, Columbus, Ohio, United States, mohammadidarestani.1@osu.edu*
[b] *The Ohio State University, Columbus, Ohio, United States, shafieezadeh.1@osu.edu*

ABSTRACT

Electricity supports various activities in a society and any disruption in its constant flow may result in a large economic loss. In the United States, hurricanes are responsible for a considerable portion of power outages. During these events, distribution systems are one of the most vulnerable parts of the grid, which highlights the need for developing techniques to reliably assess their wind performance. Given high fluctuations in wind pressure fields during hurricanes and the large amplitude vibrations induced in conductors, static analyses may not be adequate for accurate assessment of wind performance of overhead distribution lines; this can be addressed via dynamic analysis. Since distribution lines are very long and consist of a large number of spans, performing dynamic analysis on the entire distribution line is not practically feasible due to significantly high computational demands. In order to overcome this issue, past studies have either neglected boundary effects and modeled only the pole of interest or considered only a few spans on each side of the pole of interest and performed convergence studies to determine how many spans need to be considered. The latter approach is more accurate than the former method, but it is still very time consuming due to the fact that geometric nonlinearities in conductors and the need to model them using tension-only materials increase the computational time considerably with each added span in the dynamic analysis. Alternatively, this study proposes a more practical yet accurate approach to consider the impact of adjacent spans in distribution lines through an equivalent dynamic boundary model. For this purpose, a dynamic model representing the steady-state wind response of the adjacent spans is developed. The input and output of the model are the wind force history and the force transferred from adjacent spans to the pole of interest, respectively. Subsequently, model reduction techniques are used to reduce the order of the produced model, while preserving its critical dynamic characteristics. Among various model reduction techniques, Rational Krylov method is adopted here due to its stability in dealing with very large dynamic systems. The generated reduced order model cannot be directly converted to a physical dynamic model with stiffness, damping, and mass matrices; therefore, it cannot be directly represented in Finite Element platforms. In order to address this issue, an equivalent physical dynamic model with the same number of degrees of freedom as the order of the reduced system is considered. Subsequently, the mass, stiffness, and damping of the physical system is obtained through matching the transfer functions of the two systems using optimization techniques. The proposed equivalent dynamic boundary model for the effects of adjacent spans in a distribution line on the pole or line segment of interest is validated using a series of dynamic analyses of detailed Finite Element models of distribution lines. Results indicate that the proposed equivalent dynamic boundary model can accurately capture boundary effects with significantly reduced computational demand.

KEYWORDS

Reliability Analysis, Hurricanes, Overhead Distribution Lines, Equivalent Dynamic Boundary Model, Model Reduction Techniques

The 13th Americas Conference on Wind Engineering (13ACWE)
Gainesville, Florida USA, May 21-24, 2017

Structural Resiliency Assessment of a Single-circuit Overhead Distribution System under Extreme Wind Hazard

Hao Yuan [a], Wei Zhang [a], Jin Zhu [a], Amvrossios Bagtzoglou [a]

[a]*University of Connecticut, Storrs, CT, USA*

ABSTRACT: A reliable and resilient overhead distribution system is crucial for end customers in the regions that suffered from extreme weather conditions. From the perspective of structural engineering peer, the resiliency of the physical distribution line infrastructure, i.e. a pole-wire system, is discussed in the paper. The structural resiliency of the power system is decomposed into structural reliability and hardening/inspection priority, while the corresponding approach is proposed. To illustrate the reliability assessment, fragility analysis is performed for a three-span pole-wire system under both extreme wind loads and tree/brank impact load. Finally, the whole approach is applied to a realistic single-circuit overhead distribution system. The system reliability, critical pole location, and critical wind direction is reported. The hardening priority is also suggested.

KEYWORDS: Wind, Reliability, Resilience, Distribution system, Single-circuit

1 INTRODUCTION

A reliable and resilient power grid is crucial for the public safety and economy, especially in regions that suffer from frequent extreme weather conditions. The challenge of asset management and associated cost-benefit optimization prior, during and/or shortly after extreme weather and security events are critical to enable the resiliency of the power grid and the community under various perturbations or interruptions with constraints such as limited availability of utility crews or equipment. As one of the major units of the power system, overhead distribution lines are the closest related to the end customers and act as "lifeline" but often underrated compared with transmission line. Existing distribution lines suffered extensive damage and the associated failures induced most of the outages during various hazards like hurricanes or winter storms in decades. The failure of the physical distribution infrastructure can directly induce power service interruption and potential life-threatening accidents, such as dropping conductors, short-circuit induced fires, or fallen poles. Therefore, a robust distribution infrastructure system is an essential part of the entire power system. However, current system resilience analyses on power system mainly focus on power devices and power delivery after the several literature review [1–5]. The resiliency of physical infrastructure, such as poles and wires for overhead power distribution system, is usually ignored in the system network evaluation process, letting alone the interdependency and interactions between the physical infrastructure resiliency, power delivery, and vegetation management plans and practices.

For the physical infrastructure, the resiliency study of pole-wire system is usually focused on a single pole failure, or on a serial pole-conductor structural system in a straight line. However, a typical overhead distribution structural system could be much more complicated including poles with various top configurations, primary conductors, secondary/neutral lines, telecom lines, guy wires, and various large attachments like transformers. Also as the most commonly used configuration in the US, a radial distribution system has a geometric network with one main feeder and multiple lateral lines while the wind speed and attack angle can be different for each segment. For resiliency evaluation and risk assessment of power distribution system in a community, modeling of the physical infrastructure including thousands of poles and even more related attached devices could be very complicated if not impossible. Therefore, there is a necessity to model the system automatically with existing databases or related design manuals or specifications.

Historically, the design and application of the pole-wire system were de-veloped in an empirical way from lab/field tests and experience of success and failure, and with these, utility companies built up their own enterprise standards [6]. Certainly, these practicing standards are governed by the National Electrical Safety Code [7] but only on safety issues. Under normal operation and historical employment, the design margin is sufficient or even conservative. However, since pole-wire systems are not designed like trans-mission systems for high reliability to sustain extreme loads but allowed for failures under storms, the climate change and frequent extreme weathers in recent years brought more often failure events of them, inducing large area outage, long restoration time and high interruption costs [8]. Therefor more attention was paid on the reliability and resiliency of the power-wire systems. Reliability-based design methods for distribution poles have been integrated into design standards [6]. Sa et al. [9], Li et al. [10], Bjarnadottir et al. [8], Shafieezadeh et al. [11], and Ryan et al. [12,13], gradually established the reliability assessment framework of individual pole against the climate change and extreme weather conditions like hurricanes, while more associated factors were incorporated, such as age deterioration and network maintenance. However for the poles and wires as a mechanical system, the research is still limited. Darestani et al. [14] developed a serial pole-wire system in a straight line and then condensed the effects of adjacent spans to the equivalent boundary condition of a single pole model, hence it is still an individual pole evaluation essentially. Salman et al. [15] did perform a reliability evaluation including both pole and line failures, but based on an ideal three span submodel. With the fragility function and importance index of each component, the targeted hardening strategies for resiliency were recommended to decision makers. Potvin and Short [16] also did work on the resiliency testing of pole-wire systems and tried to develop strategies to mechanically coordinate overhead lines in a way that reduces frequency and/or duration of repair during/after storms, considering the power component mostly like to fail should also be the one that can be repaired with minimum cost and time. For power network level assessment, Park et al. [17] assessed of the Florida electric-power network system against hurricanes; Reed et al. [18] advocate a systems-based life-line performance model to interdependent electric power delivery infrastructure resilience subject to weather-related hazards; Bollinger and Dijkema [19] did a sophisticated network theory-based resilience evaluation on Dutch electricity transmission network under two types of extreme events – floods and heat waves.

Thus, the high vulnerability of overhead distribution system against such extreme wind hazards highlights the need for reliable hazard induced damage assessment to provide an effective decision-making strategy of pre-storm reinforcement and post-storm restoration. From the above discussion, the existing methods for distribution system reliability assessment mostly focus on the electric system, or only on pole failure, or on a serial pole-conductor structural system in a straight line. However, a typical overhead distribution structural system should include poles with various top configurations, primary conductors, secondary/neutral lines, telecom lines, guy wires and large attachments like transformers. Also, a radial distribution system has a geometric network with one main feeder and multiple lateral lines while the wind speed and attack angle can be different for each segment. Therefore in the present paper, a more realistic structural resiliency assessment framework is proposed for a single-circuit overhead distribution system under strong wind. The framework is based on Monte Carlo Simulation (MCS) on finite element model of a single-circuit overhead distribution system. The parametric finite element model (FEM) integrating with utility infrastructure database can physically reduce the uncertainty by directly modelling a realistic network and apply to community-based power grid assessment [20].

The paper is organized in following parts: Section 2 discusses the general concept of resilience from structural perspective and proposes a structural resilience assessment approach; Section 3 provides a fragility analysis of a three-span pole-wire system by proposed approach; Section 4 presents a case study on a realistic single-circuit distribution system; Section 5 does some conclusions and outlook.

2 STRUCTURAL RESILIENCY ASSESSMENT OF POLE-WIRE SYSTEM

2.1 *Infrastructure resilience, power system resilience and structural resilience of powerline*

Resilience has broad features and its definitions and metrics are various and still in evolution [21]. Here for infrastructure, the definition by the National Infrastructure Advisory Council (NIAC, USA) which adopted by several researchers [22–24] is also taken here as:

"*...the ability to reduce the magnitude and/or duration of disruptive events. The effectiveness of a resilient infrastructure or enterprise depends upon its ability to anticipate, absorb, adapt to, and/or rapidly recover from a potentially disruptive event.*"

The main feature of this resilience concept is shown in Fig. 1 and some projections to power system resilience is also presented. New technologies such as smart grid, micro grid, and wide area monitoring applications could increase situational awareness as well as enable faster restoration of the system. Since the resilience of power system is focused on the power delivery instead of structural response, the physical infrastructure such as pole-wire system only act as a support system and may not be evaluated alone for resilience. Therefore, the structural resilience is defined as structural reliability and hardening/inspection priority through structural analysis and damage assessment. The damage assessment of physical infrastructure system acts as a link between weather forecast/vegetation condition and power system performance. Together with power resilience module, forecasting natural disaster related power system disturbances, hardening and pre-storm operations, and restoration models can be developed.

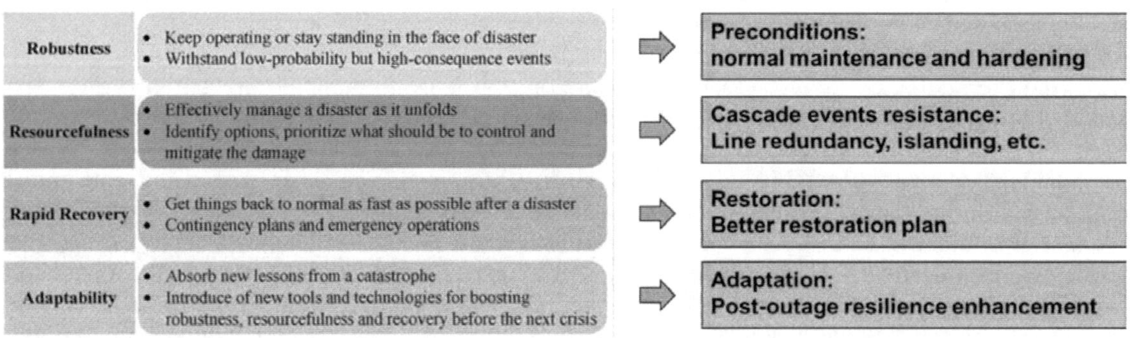

Figure 1. Resilience main features by NIAC and projection to power system resilience in extreme event cycles.

2.2 *Input and outcome*

The damage assessment of physical infrastructure system acts as a link between weather forecast/vegetation condition and power system performance. Structural models are suitable to link the weather forecast and the hardening investment guidance by component failure rate evaluation. The input and output of the process is shown in Fig. 2. The weather and vegetation information can change the load, boundary conditions and material property of structural model in both time and space. The structural assessment of powerline infrastructure under extreme weather conditions can provide the fragility curves to be used in outage prediction model and total system performance model. It can also locate and rank critical components for hardening and restoration plan, serving a resilient purpose.

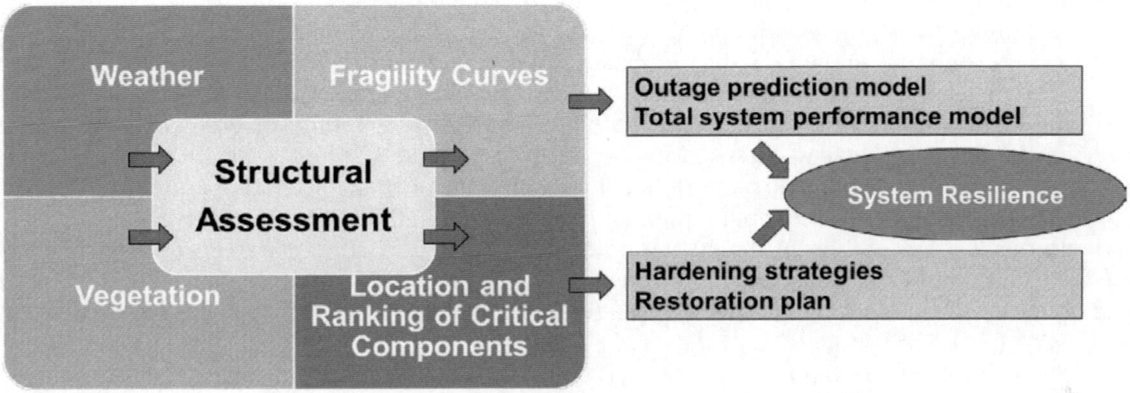

Figure 2. Input and output of structural assessment module and relation to system resilience.

2.3 *Procedure for structural resiliency assessment of distribution line under extreme wind*

The framework is carried out in four steps. Firstly, a parametric modeling script automatically constructs the finite element model in ANSYS [25] for a radial distribution network between two reclosers. The network can have a realistic vein-like geometric configuration with poles (beam elements) and wires (link elements). Secondly, the wind load can be generated randomly based on codes or numerical simulation [26] and applied to poles and wires. The effect of large attachments such as transformers can be simulated as increased wind loads and gravity loads on poles. Thirdly, for various material properties and wind scenarios, MCS will be carried out on the instantiated model to get the failure rate of each component (such as in ANSYS Probabilistic Design System (PDS)). Finally, the structural failure associated distribution system reliability can be assessed using topological power-flow tree with the main line as the trunk and the tap lines as the branches. The fragility curve for the system can be obtained, and the reliability index such as storm-based system average interruption frequency index (STAIFI) can also be calculated as the customer percentage affected by interruptions. With reliability importance measure, the components can be ranked according to the influence to the system reliability. Based on all these results, the decision maker can develop hardening strategies for the overhead distribution system.

3 FRAGILITY STUDY OF A THREE-SPAN POLE-WIRE SYSTEM

3.1 *Fragility of pole-wire system with age-degradation of pole under extreme wind load*

In an extreme event such as hurricane or wind storm, extreme wind can induce considerable damage to a pole-wire system by direct wind blowing, falling tree and flying debris. To evaluate the damage from wind, the fragility analysis is performed. Two load cases are considered in present study. The first one is conventional wind application on pole and wire, while the second load case is the falling tree loads. Wind fragility is the conditional probability of failure of pole-wire system subjected to a given wind load/speed. The wind load is calculated based on extreme wind loading (Grade B construction) of NESC 2017 [7] for current study with Equation 1:

$$\text{Wind Pressure in N/m}^2 = 0.613 \cdot (V_{\text{m/s}})^2 k_z G C_f \qquad (1)$$

where V is 3 s gust wind speed in m/s at 10 m aboveground, G is gust response facto, and C_f is the force coefficient (shape factor).

In falling tree load case, the point loads are estimated from typical street-side tree weight and can be evenly put in the mid-span of each primary conductor [27]. The falling tree load can also be taken as cascading pole failure impact load which has a similar load configuration. MCS on finite element model was used to calculate the probability of failure of pole and wire while varying the basic 3-s gust wind speed or the falling tree load. In limit state function, demand terms are ground-line moment of pole and maximum stress in wire; correspondingly, the resistance terms are ground-line rupture moment and yield strength of wire, respectively.

In our analysis, a typical three-span pole-wire system with fiberglass (FG) crossarm (9.2 cm ×11.7 cm × 244 cm) construction is constructed. 13.7 m tall (45 ft) pole bottom is fixed and end poles are supported by guy wire. The phase conductors are 336 AAC (TW) and neutral lines are made of 1/0 ACSR if any.

Two classes of wood poles, i.e. Class 3 (C3) and Class 2 (C2), are evaluated for resiliency-purpose alternative. In the US, wood such as southern yellow pine (SP) is the primary pole material which has relative large uncertainty in material properties such as strength, elastic modulus or density. These properties can vary with time because such natural material is sensitive to time-accumulative environmental influence from temperature, humidity, fungi, etc. The age-degradation effect is modeled as time-variation in strength distribution by [11,12]. The model is taken here (refer to [11] for equations) with the assumption that the CCA treated pole in our study are also suitable to apply. The degraded ratio of mean pole strength is calculated as 0.9774, 0.8111, and 0.4734 for poles at 30 years, 60 years and 90 years, respectively.

3.2 *Model parameters*

The model parameters used in this section and Section 4 are presented in Table 1. The wood elastic modulus and density of SP are correlated with its fiber strength by 0.58 and 0.79, respectively [28]. In fact, wind parameters need corrections considering vegetation. However, this leaves for future study.

Table 1 Model parameters and their statistics

Parameters	Probability Distribution	Mean	COV	Source
Fiber strength (SP) (kPa)	Lognormal	55158 as new, varies with time	0.15 as new, varies with time	[6][11]
Wood density (kg/m^3)	Normal	500	0.04	[28]
Wood elastic modulus (GPa)	Lognormal	11.59		[28]
Pole diameter (mm)	Uniform between bounds of each class			[29]
C_f	Normal	1	0.12	[7][11]
G (pole)	Normal	0.97	0.11	[7][11]
G (wire)	Normal	0.88	0.11	[7][11]
k_z (pole)	Normal	1	0.06	[7][6]
k_z (wire)	Normal	1.1	0.06	[7][6]
Falling tree load (kN)	Normal	Varies	0.1	Assumed
Elastic modulus (FG) (GPa)	Deterministic	75.84		Utility database
Elastic modulus (1/0	Deterministic	158.56		Utility data-

ACSR) (GPa)			base
Elastic modulus (336 AAC) (GPa)	Deterministic	68.94	Utility data-base
Density (FG) (kg/m^3)	Deterministic	2768	Utility data-base
Density (1/0 ACSR) (kg/m^3)	Deterministic	6643	Utility data-base
Density (336 AAC) (kg/m^3)	Deterministic	2768	Utility data-bas
Diameter (1/0 ACSR) (mm)	Deterministic	18.3	Utility data-base
Diameter (336 AAC) (mm)	Deterministic	24.4	Utility data-base
Yield strength (1/0 ACSR) (mm)	Deterministic	635	Utility data-base
Yield strength (336 AAC) (MPa)	Deterministic	276	Utility data-base

3.3 *Finite element analysis results and fragility curves*

From FEA results from nonlinear analysis based on large deflection small strain theory, the wire stress is far below its strength in all simulations under both load cases. This means the second failure mode of wire break is impossible in current ideal model. However in reality, the wire can have weak connections, ties or grips/clamps to induce the wire failure modes. Therefore, only results and fragility of poles are presented. Moment, deformation and load plots as the main FEA results are shown in Figure 3. As the results of 10000 MCS, the final fragility curves for different class poles under each load case are shown in Figure 4. It can be concluded that the Class 2 pole is a little better than Class 3 pole. In wind load case, the probability of failure is very low until the wind speed increases to 22-45 m/s (50-100 mph) for pole with different ages. The fragility curves are sensitive to the age-degradation model.

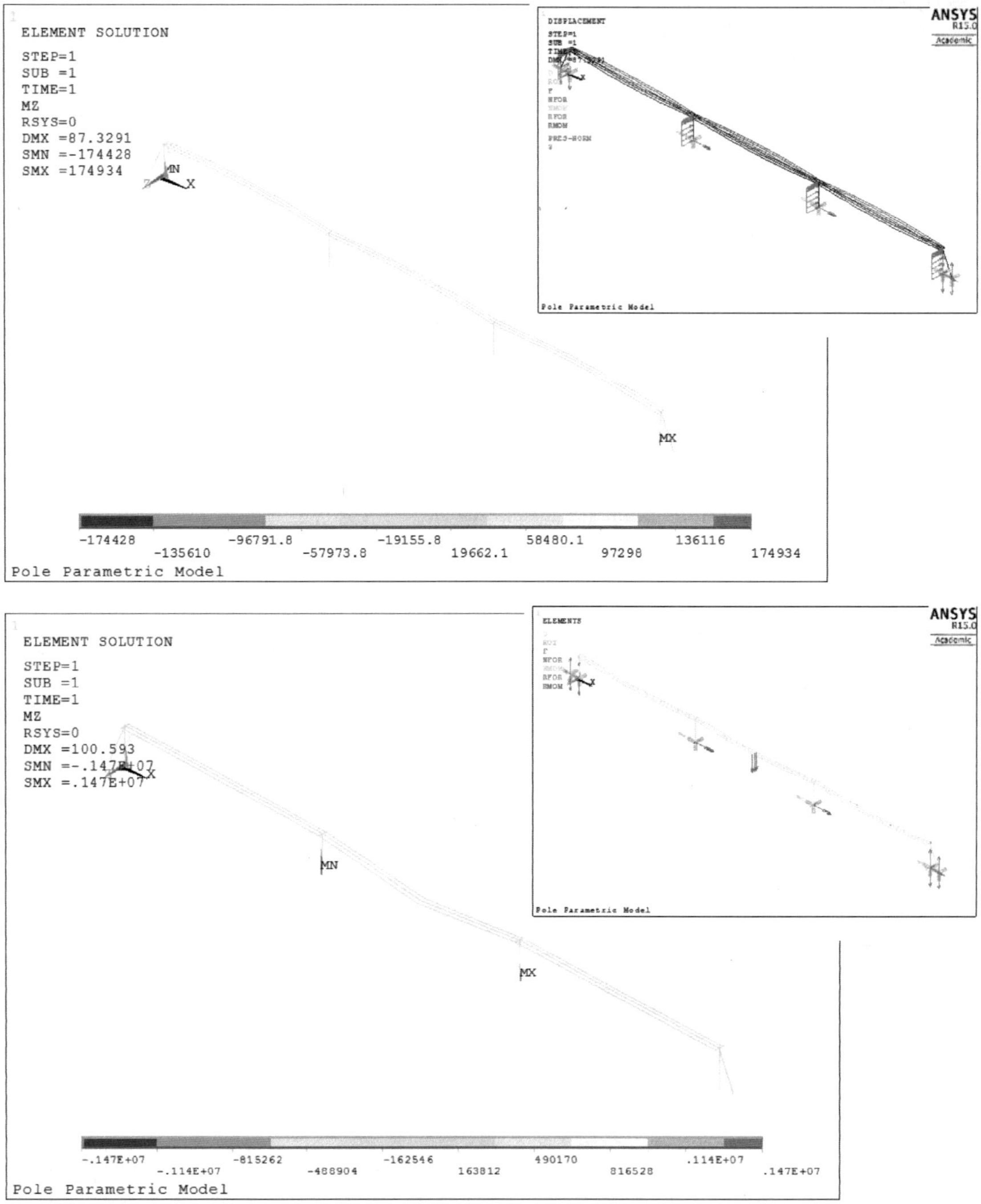

Figure 3. FEA of a three-span pole-wire system under direct wind load (upper) and falling tree load (lower)

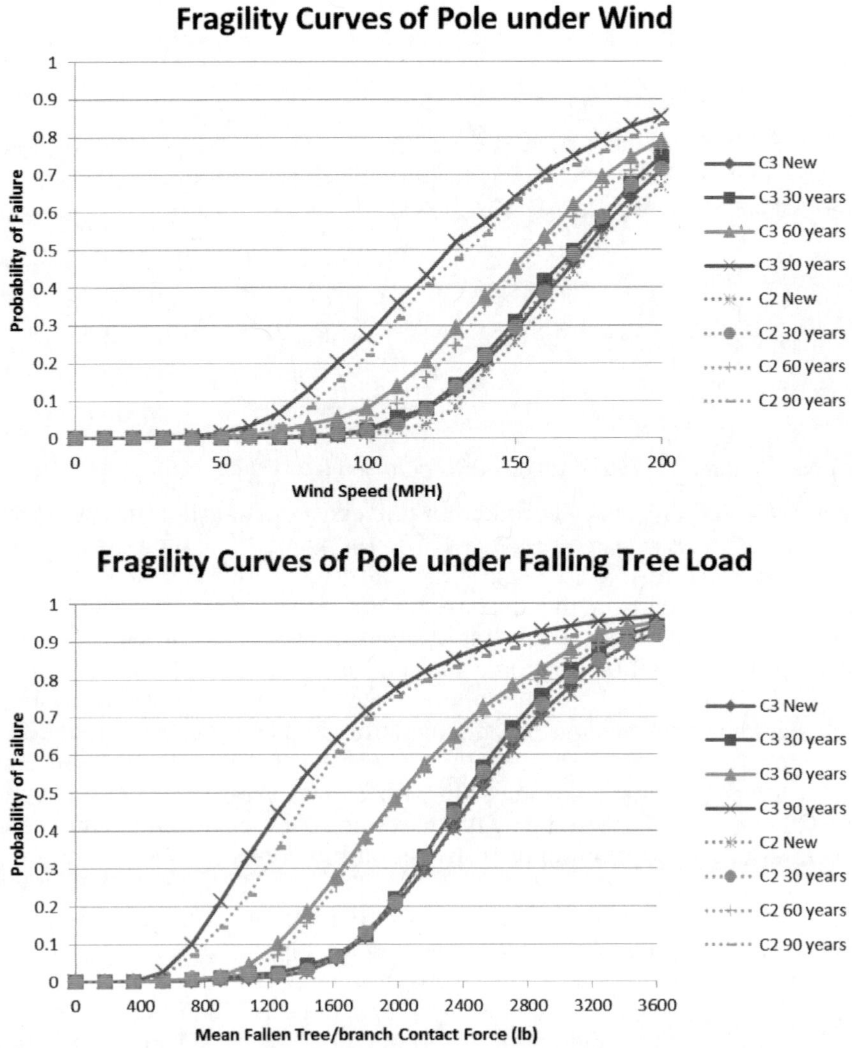

Figure 4. Fragility curves of age-degraded poles under direct wind load (upper) and falling tree load (lower).

4 A CASE STUDY ON A REALISTIC SINGLE-CIRCUIT OVERHEAD DISTRIBUTION SYSTEM

4.1 *Overview of the single-circuit distribution line*

A powerline sample for demonstration is taken along: Richmond Hill Rd., N Porchuck Rd., Old Mill Rd. in Greenwich, CT. The layout of the pole-wire system is shown in Fig. 4. From the figure, the sample powerline is single-circuit with one mainline and several lateral lines. ML switch is used to isolate powerline downstream and keep the upstream line in normal operation when outage happens in downstream mainlines. The fuses or switches between lateral line and mainline are not shown but assumed existing. This means the failure of lateral line cannot affect the mainline power delivery.

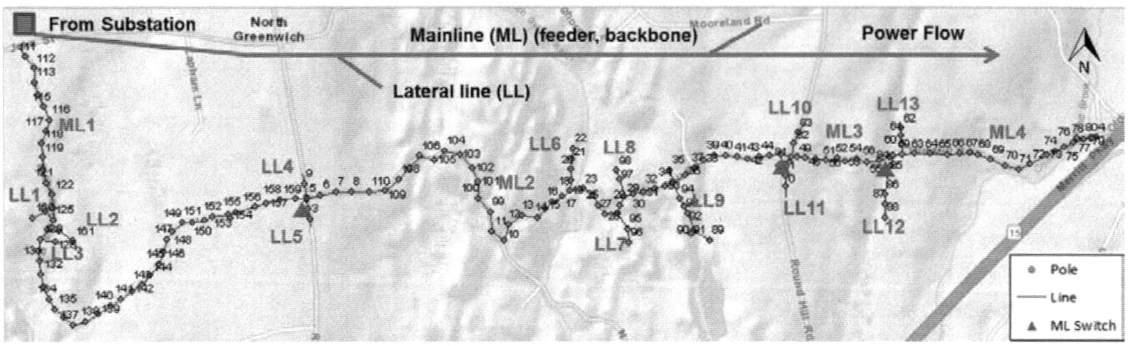

Figure 4. Layout of the overhead distribution system for case study.

4.2 *Parametric finite element modelling and analysis of the pole-wire system*

With the Eversource Energy Center, an energy corporation-university partnership, the utility infrastructure database and LiDAR datasets can be accessed. The following finite element model is constructed through parametric FEM in ANSYS integrating with ArcGIS & MS Excel based parameter databases. Based on information from the database for this powerline sample, the Class 2 and Class 3 poles are the majority with a few Class 4 pole, and the ages are all around 30 years. Since the wire is very strong itself based on Section 3 and the falling tree load statistics is not disclosure, the following analysis is focused on pole failure when direct wind blows the system. Performing with the same parameters in Section 3.2, the modeling and analysis are summarized in Figure 5 for concise. Figure 5a is the element mesh for the whole system, while different line colors indicate different materials (green–wood pole, blue–fiber glass crossarm, deep blue–1/0 ACSR wire, red–336 AAC wire, same for b and d). Guy wires are not modeled since most guy wires cannot be installed in effective position reported by the utility company. The transformer load is not considered in this simulation since researchers goes into two opposite directions. Some researchers think it is important such as [30] while others like [15] says the effect is small. The lateral view of the powerline in Figure 5b indicates the model is built with consideration of the terrain. The sample powerline climb up and down hills, turning a lot in horizontal plane as well. The importance of case-sensitive analysis can be observed from this kind of configuration. Figure 5d shows two different line configurations in this simulation, i.e. crossarm construction and armless vertical construction, which can provide very different structural response but is often ignored by most researchers. Even though some old constructions are in replacement, mass of them are still in operation and prone to extreme wind. Figure 5c is the structural moment plot which indicates the maximum moment location of pole is at the ground line. In the finite element analysis, since the cable is too flexible, a weak spring is implemented and prestress stiffen for easier convergence in nonlinear analysis with larger deflection and small strain.

Since the wind in an extreme weather event such as hurricane does not have a fixed direction, the simulation is carried out for different direction (0 ~ 360 degree by 45 degree increase) of 40 m/s (90 mph) wind which is assumed for mean wind speed for next extreme event. A higher wind can also be applied based on the location of the community. If preferred, the fragility curve of each pole in the system can be obtained as well in a similar manner shown in Section 3. Without the wind direction distribution in an event, conditional probability of failure under different wind direction is obtained and shown in Figure 6 as an example. The wind-angle unsymmetry of failure rate is due to unsymmetrical boundary conditions which is also studied in [14] but in a symmetric manner.

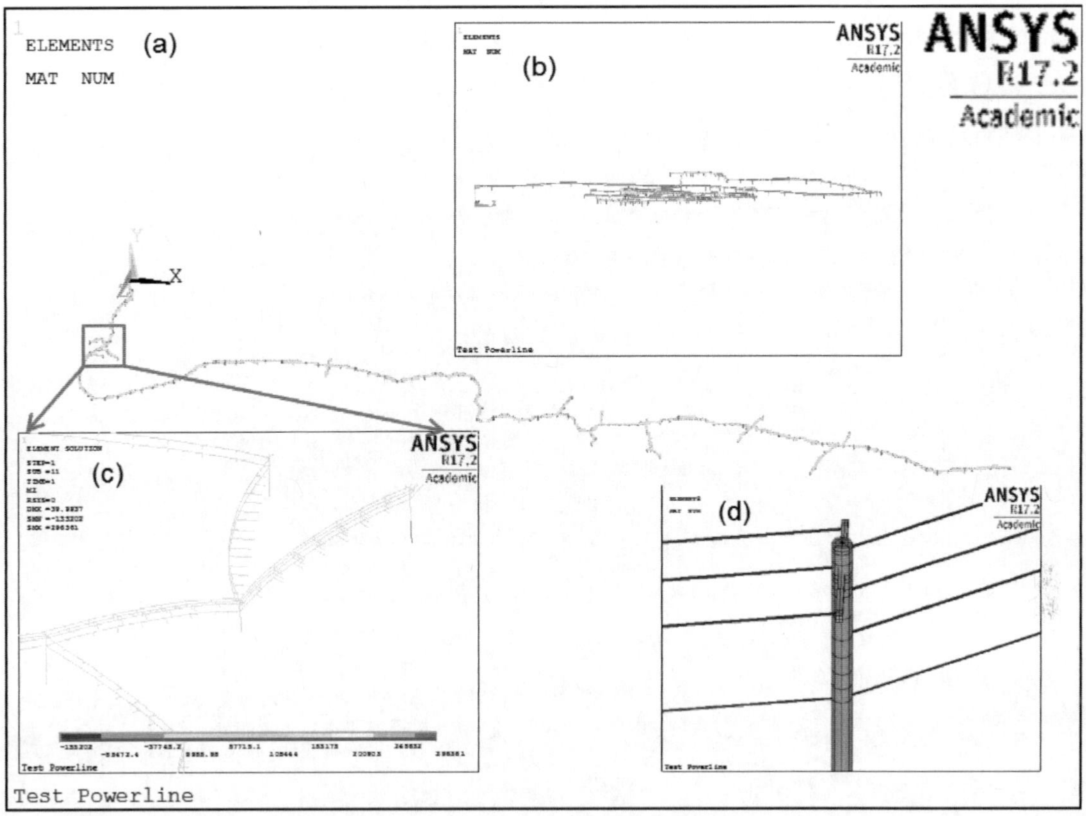

Figure 5. Finite element modeling and analysis of pole-wire system under extreme wind

4.3 *Critical poles and critical wind direction for the region*

Based on the analysis from Section 4.2, critical poles are located as shown in Figure 7 according to the failure rate of each pole such as Pole 8 shown in Figure 6. And the critical wind direction for the region can also be identified for this sample powerline segment, which is northeast wind based on all simulations. From Figure 7, all critical poles are belong to mainline which indicates that mainline are more vulnerable to wind events, possibly due to higher loading from conductors and pole-top configuration. With the observation of topography along the powerline, the critical poles prefer the hill ridge, sharp turning point, and plane area with larger span which is equivalent to larger wind forces. This finding also proves the necessity of such detailed finite element analysis when evaluating the community-level powerline resilience.

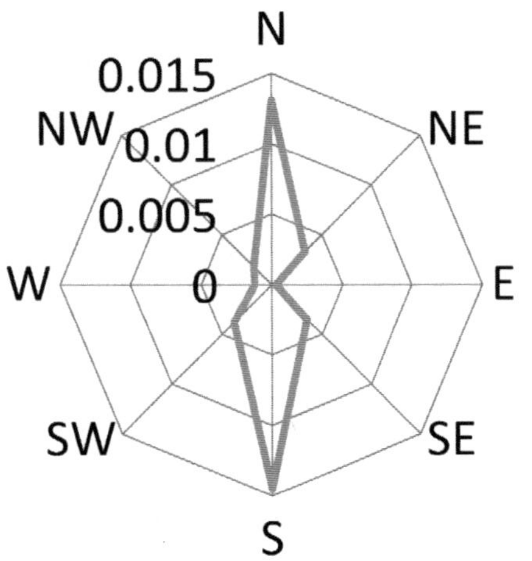

Figure 6. Probability of failure of Pole 8 under 40 m/s winds with different directions.

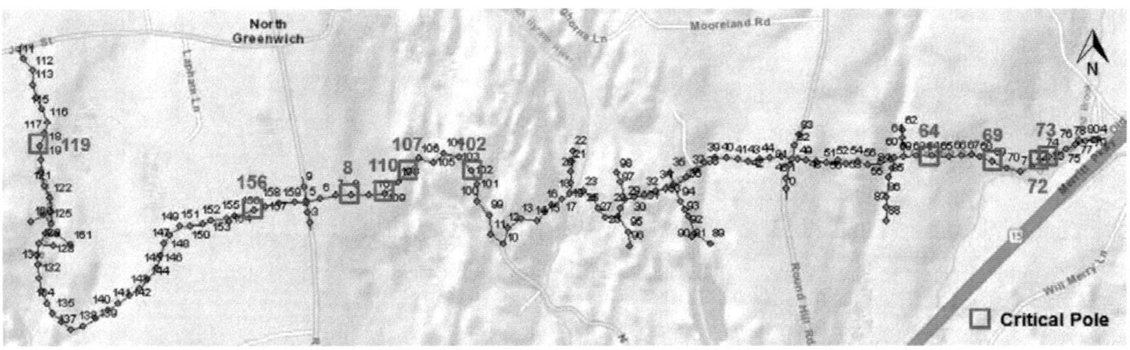

Figure 7. Locations of top 10 identified critical poles (rank by decreasing max failure rate: 107, 102, 73, 69, 110, 8, 156, 72, 64, 119).

4.4 *System reliability and hardening priority*

4.4.1 *System reliability index*

The structural part of system resilience assessment is finished and the workflow with the failure rate data and critical pole location data should go to the power flow topology network for further analysis and system resilience evaluation. In present paper, for completeness, a simple system reliability index is taken as Equation 3 similar with [15], assuming all customers are along lateral lines,

$$R_S = 1 - \sum_{i=1}^{N} Q_{L_i} \frac{C_i}{C} \tag{2}$$

where, Q_{Li} is the probability that power is not delivery to *i* th lateral line and can be calculated by serial and parallel probability of line failure in the radial system, N is the number of lateral lines in the system, C_i/C is the weighting factor for *i* th lateral line while C is the representative of power load. Assume each customer has the same weight and each pole serve the same number of customers, C_i/C can be replaced by pole amount percentage of i th lateral line in the whole system.

The failure of lines such ML1 or LL1 is described as pole failure in current study even though other failure modes can happen in real scenario. The upper bound value of line failure rate is used assuming the poles along a line are fully dependent as a serial system. Finally, the system reliability index $R_S = 0.3342$.

4.4.2 *Hardening priority and lateral line importance measure*

The mainline should be hardened by the order ML2, ML1, ML4, ML3 based on calculated probability of failure where the length or pole amount along can be concluded as a primary influence factor from Figure 4. The increasing line length or more poles can induce higher uncertainty and larger probability to get a weak pole. For lateral line, risk achievement worth (RAW) is adopted for measurement of importance measure similar as [15]. RAW is defined as the ratio of the conditional system unreliability if component i has failed to actual system unreliability, i.e.

$$RAW(i) = \frac{1 - R_S(Q_{L_i} = 1)}{1 - R_S} \tag{3}$$

The rank of RAW shown in Figure 8 can be taken as the priority of hardening or post-event restoration. RAW of LL9 is relative high due to its length or pole amount as well.

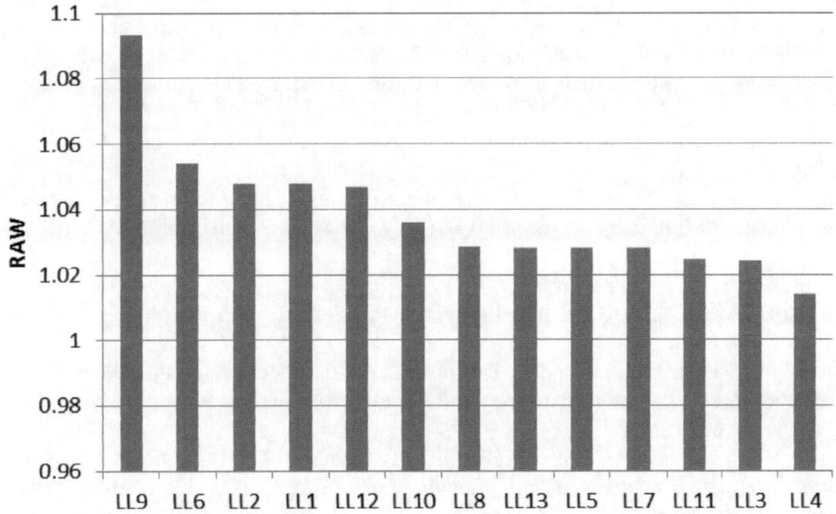

Figure 8. Risk achievement worth of lateral lines.

5 CONCLUSIONS

A structural resiliency approach is proposed for pole-wire system. By the approach, the system reliability can be obtained and hardening priority can be provided. A demonstration on simple model and a realistic implementation is performed to illustrate the structural resiliency assessment approach. The random wind load used in present study is based on codes [7]. With the in-house computer program [26] to simulate time-varied event wind field with certain vegetation correction, time history of wind from a hurricane event can be applied to the community-level powerline in the future. The effect from transformer, guy wire, foundation and surface treatment will also be considered. If the falling tree/flying debris–wind relationship is disclosed, effect from falling trees or flying debris can also be included for system simulation in future studies.

6 ACKNOWLEDGEMENTS

This work is partially supported by the Eversource Energy Center pilot projects. This support is greatly appreciated. The contents of the paper reflect only the views of the authors.

7 REFERENCES

1 M.N. Albasrawi, N. Jarus, K.A. Joshi and S.S. Sarvestani, Analysis of reliability and resilience for smart grids, Proc. - Int. Comput. Softw. Appl. Conf. (2014) 529–534. doi:10.1109/COMPSAC.2014.75.
2 D. Ahmad and C.K. Chanda, A framework for resilience performance analysis of an electrical grid, 2016 2nd Int. Conf. Control. Instrumentation, Energy Commun. CIEC 2016. (2016) 392–396. doi:10.1109/CIEC.2016.7513735.
3 A. Kwasinski, Quantitative model and metrics of electrical grids' resilience evaluated at a power distribution level, Energies. 9 (2016) 1–27. doi:10.3390/en9020093.
4 Y. Wang and C. Guo, The study on effect of random maintenance quality of components on composite power system reliability, Asia-Pacific Power Energy Eng. Conf. APPEEC. (2013). doi:10.1109/APPEEC.2013.6837263.
5 K. Barker, J.E. Ramirez-Marquez and C.M. Rocco, Resilience-based network component importance measures, Reliab. Eng. Syst. Saf. 117 (2013) 89–97. doi:10.1016/j.ress.2013.03.012.
6 American Society of Civil Engineers (ASCE), Reliability-Based Design of Utility Pole Structures (No. 111), 2006.
7 A. Standards and N. Standards, 2012 National Electrical Safety Code (NESC), 2011. doi:10.1109/ieeestd.2011.5967877.
8 S. Bjarnadottir, Y. Li and M.G. Stewart, Hurricane risk assessment of power distribution poles considering impacts of a changing climate, J. Infrastruct. Syst. 19 (2013) 12–24. doi:10.1061/(ASCE)IS.1943-555X.0000108.
9 Y. Sa, Reliability Analysis of Electric Distribution Lines, 2002.
10 L. Li and L. Shen, A centroidal voronoi tessellation based approach for creating grain morphology for crystal plasticity finite element simulations, (2006).
11 A. Shafieezadeh, U.P. Onyewuchi, M.M. Begovic and R. Desroches, Age-dependent fragility models of utility wood poles in power distribution networks against extreme wind hazards, IEEE Trans. Power Deliv. 29 (2014) 131–139. doi:10.1109/TPWRD.2013.2281265.
12 P.C. Ryan, M.G. Stewart, N. Spencer and Y. Li, Reliability assessment of power pole infrastructure incorporating deterioration and network maintenance, Reliab. Eng. Syst. Saf. 132 (2014) 261–273. doi:10.1016/j.ress.2014.07.019.
13 P.C. Ryan, M.G. Stewart, N. Spencer and Y. Li, Probabilistic analysis of climate change impacts on timber power pole networks, Int. J. Electr. Power Energy Syst. 78 (2016) 513–523. doi:10.1016/j.ijepes.2015.11.061.
14 Y. Mohammadi Darestani, A. Shafieezadeh and R. DesRoches, An equivalent boundary model for effects of adjacent spans on wind reliability of wood utility poles in overhead distribution lines, Eng. Struct. 128 (2016) 441–452. doi:10.1016/j.engstruct.2016.09.052.
15 A.M. Salman, Y. Li and M.G. Stewart, Evaluating system reliability and targeted hardening strategies of power distribution systems subjected to hurricanes, Reliab. Eng. Syst. Saf. 144 (2015) 319–333. doi:10.1016/j.ress.2015.07.028.
16 J. Potvin and T. Short, Resiliency Testing of Overhead Distribution Components and Systems, (n.d.) 2–6.
17 Y.J. Park, C.R. Glagola, K.R. Gurley and K. Son, Performance assessment of the Florida electric-power network system against hurricanes, Nat. Hazards Rev. 15 (2014) 1–7. doi:10.1061/(ASCE)NH.1527-6996.0000132.

18 D. Reed, S. Wang, K. Kapur and C. Zheng, Systems-Based Approach to Interdependent Electric Power Delivery and Telecommunications Infrastructure Resilience Subject to Weather-Related Hazards, J. Struct. Eng. 142 (2015) C4015011. doi:10.1061/(ASCE)ST.1943-541X.0001395.
19 L.A. Bollinger and G.P.J. Dijkema, Evaluating infrastructure resilience to extreme weather – The case of the Dutch electricity transmission network, Eur. J. Transp. Infrastruct. Res. 16 (2016) 214–239.
20 B. Wareing, Total environmental protection of woodpole overhead lines, In IEE Semin. Improv. Reliab. Woodpole Overhead Lines (Ref. No. 2000/031), IET, 2000, pp. 6–1.
21 H.H. Willis and K. Loa, Measuring the Resilience of Energy Distribution Systems, (2015) 1–25.
22 Y. Wang, C. Chen, J. Wang and R. Baldick, Research on Resilience of Power Systems Under Natural Disasters A Review, IEEE Trans. Power Syst. 31 (2015) 1–10. doi:10.1109/TPWRS.2015.2429656.
23 M. Panteli and P. Mancarella, Modeling and evaluating the resilience of critical electrical power infrastructure to extreme weather events, IEEE Syst. J. PP (2015) 1–10. doi:10.1109/JSYST.2015.2389272.
24 R.J. Campbell, Weather-Related Power Outages and Electric System Resiliency, Congr. Res. Serv. Rep. August 28 (2012) 1–15. doi:R42696.
25 ANSYS.Inc, ANSYS® Academic Research, Release 17.2, (2016).
26 J. Zhu and W. Zhang, Numerical Simulation of Wind and Wave Fields for Coastal Slender Bridges, J. Bridg. Eng. 22 (2016). http://ascelibrary.org/doi/abs/10.1061/(ASCE)BE.1943-5592.0001002.
27 D. World, Design modifications lessen outage threat, (1996).
28 D. V Doyle and L.J. Markwardt, Properties of southern pine in relation to strength grading of dimension lumber. Res. Pap. FPL 64. U.S., Dep. Agric. For. Serv. For. Prod. Lab. Madison, WI. (1966).
29 P. Edition, 2017 NESC Handbook, 2017.
30 R.E. Brown, Electric Power Distribution Reliability, Second Edition, 2008. https://books.google.com.mx/books?id=CVNW8qW3ggwC.

The 13th Americas Conference on Wind Engineering (13ACWE)
Gainesville, Florida USA, May 21-24, 2017

Effects of Hurricane Matthew on a Florida residence roof

Harika Gurram[a], Chelakara Subramanian[b], Jean-Paul Pinelli[c], Noel Hadj Mbarek[d]

[a,b] *Mechanical & Aerospace Engineering Department, Florida Tech, Melbourne, Florida, U.S.A.*
[c] *Civil Engineering Department, Florida Tech, Melbourne, Florida, U.S.A.*
[d] *Mechanical Engineering, ENISE, Saint-Etienne, France.*

[a]hgurram2015@my.fit.edu , [b]subraman@fit.edu , [c]pinelli@fit.edu, [d]nhadjmbarek@my.fit.edu

ABSTRACT

The wind pressure field was measured on the roof of a Satellite Beach, Florida residence (longitude and latitude: 28.175944, -80.592710) using the Florida Tech's wireless sensor network system (WSN), during the passage of Hurricane Matthew. This paper describes the measurement setup and analysis of the data. Twenty absolute pressure sensor were strategically placed on the roof top of single story home, which is a block away from the North Atlantic Ocean. A reference static pressure port was placed at the ground level under a covered porch. A tower installed in the vicinity by the Florida Coastal Monitoring (FCMP) provided the wind data. The data was collected at a sampling rate of 10 samples per second. The data was continuously collected for seventy two hours starting at 17h00 (UTC) on October 5, 2016. The raw output form WSN system was collected and stored on the commercial cloud storage 'Box', which is linked to the DesignSafe.ci workspace. The analysis investigates if the hurricane data represents a stationary or non-stationary flow field. The variable interval time averaging (VITA) technique is used to determine the stationarity of the signal.

KEYWORDS: Wireless sensor network, Hurricane Matthew, VITA analysis, Non-stationary signal analysis, DesignSafe,

1 INTRODUCTION

The Hurricane Matthew was a Category 5 tropical cyclone started in the Atlantic Ocean, it hit the east coast of the Florida on 5th October 2016 as a Category 3 hurricane. To estimate the impact of the real time hurricane on the full scale residential building a wireless sensor network system is deployed on the roof top of a Satellite Beach, Florida, building. The pressure, temperature wind speed and direction data are continuously collected for 3 days. Previous full-scale hurricane measurements (Subramanian, et al, 2013, Tao, et al, 2016) suggest that the highly intense period of hurricane wind field is non-stationary. This means the traditional Gaussian statistical analysis for stationary signals cannot be expected to work well for the hurricane process. There are different ways of analyzing a non-stationary signal (Bendat and Piersol, 2010). The approach used here, called pseudo–stationary analysis, is to find a typical averaging time of a continuous time varying signal which will yield a converged average value. Subramanian, et al, 2013 found such pseudo averaging time exists for 3 hurricanes and it is about 300 seconds. Then the higher order statistics are based on this pseudo average value. This paper presents the analysis (variable interval time average, VITA) and results of the hurricane Matthew impact on the roof pressures of a U.S. coastal home.

2 WIRELESS SENSOR NETWORK DEPLOYMENT

The sensor system is deployed on the residential building close to the Atlantic Ocean (longitude and latitude: 28.175944, -80.592710). The house roof is preinstalled with the brackets to hold the sensors. To measure the local critical pressure affecting the roof top, the sensors were distributed on the roof as shown in Figure 1. These sensors are deployed for 3 days, starting on October 5, 2016 to collect the pressure data 'Before', 'After' and 'During' the hurricane. The Wireless Networks System, comprising of 22 pressure and temperature sensors, a vane anemometer, a base unit, and 3 routers, is easy and quick to deploy, it takes only an hour for five people to deploy the system. All the sensor systems are turned on prior to the deployment. The base unit was installed alongside the laptop inside a waterproof and ventilated box, which was secured under a porch with sand bags. The computer with backup battery power pad (the battery unit which supplies power for 10 hours in case of electricity outage) was powered by the house power supply and an internet mobile modem was also connected to the laptop. And the hotspot was powered using a USB cable attached to the laptop. The data acquisition was monitored by the remote desktop (or smart phones) connection using Team View Application. The raw data output in 15 minute block files are synchronized to the cloud storage, in addition to being stored in the laptop. The data collected is analyzed and the following section gives the details.

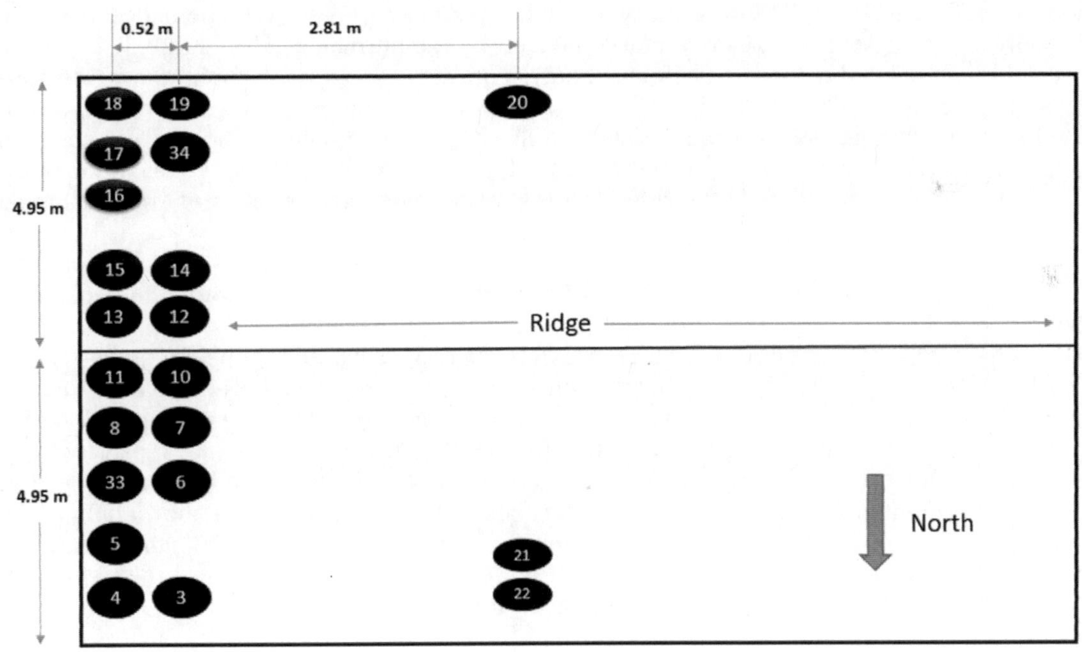

Figure 1: Sensor location on the roof

3 HURRICANE MATTHEW ANALYSIS

The hurricane Matthew reached the test location around 4:00 am of 6th October 2017, Figure 2 shows the track of the hurricane as provided by the National Hurricane Center. The entire system was deployed for 3 complete days which captured the pressure variation before, after and during the hurricane. The raw data is converted to the physical data using the individual sensor calibration coefficients. The calibration is performed by establishing the linear relationship between the known applied pressure values and the binary data output from the sensor. The time histories of pressure and temperature are analyzed, where temperature variations were minimal throughout the deployment but the pressure series varied based on the time and location of the sensor. The Figure 3 shows the pressure time history of the sensor 3. To describe the pressure variations, firstly the optimum averaging period is determined so as to represent entire data set variation. By choosing too small sample time, the mean of the data appears as a fluctuating component, whereas by choosing a large sample time the variation will be smoothed out. The – 6 show the normalized pressure time history for Sensor 7 for sensors 14 and 20 respectively where the instantaneous pressure data is normalized with the peak suction value, i.e. the minimum value,. These plots show that the average peak suction pressure depends on the averaging period and before reaches the steady-state (constant) values after 150 to 300 seconds. Therefore in order to compute the statistics of pressure data, the 300 seconds averaging is used, for the entire pressure dataset of the hurricane. Incidentally, Gabriel [1] reported the same optimum averaging period for hurricanes Gordon (2000), Isidore (2002), and Rita (2005) wind speed variations.

The overall roof pressure variations is analyzed by diving the entire hurricane periods into 3 subperiods as 'before hurricane', 'during hurricane' and 'after hurricane'. A mathematical equation is developed to represent the mean variation induced by the hurricane. The Figure 7 shows the pressure variation with the trend line before the hurricane, the Figure 8 shows the pressure variation during the hurricane and, Figure 9 shows the pressure variation after the hurricane. The pressure variation is also dependent on the location of the sensor on the roof, the pressure sensors installed on the edges of the roof are subjected to the maximum gust, and whereas the sensors closer to the ridge of the roof are protected because of some damping effect due to the roof boundary layer flow.

The averages pressure and root mean square of the pressure signal is analyzed simultaneously. The

Figure **10** -13show the root mean square (RMS) pressure variation plots for sensors 14, 4, 6, and 19, respectively. It is observed that the sensor locations near the edges have large fluctuations about the mean. The difference between the ambient pressure (reference) port and the sensor pressures on the roof top is not large, but significant enough to cause some impact. Apart from the pressure distribution the local wind speed and direction variation during the hurricane was also analyzed. The (Florida Coastal Monitoring Project) FCMP tower is deployed in the vicinity of the test house location and the wind speed and direction data collected. The data from the Melbourne airport weather station are used for the analysis. Figure 14 shows the locations of the test house, FCMP tower and Melbourne airport weather station. Figure **145** shows the wind speed variations at Melbourne airport as a function of time. The FCMP tower data is 3 second segmented average of wind speed and direction at 10 meters for the Satellite Beach, FL and Melbourne airport data is averaged for every 5 minutes. The variation of pressure data from the wireless sensor network is compared to the wind speed variations during the hurricane event. The cross correlation function at a lag 0, fluctuating between -1 and 1 are used to quantify this interdependency. The curves closer to 1 or -1 shows the maximum correlation. For the present case the correlation coefficient is -0.92,

which indicated that both the signal are highly correlated and the negative sign confirms that their magnitudes are opposite to each other as shown in
Figure **16**. The Figure 17 shows the corresponding wind direction variation at the same location, mainly from the North-east, then West and South-West directions. The wind speed and direction data is also represented into 3 different sections i.e. 'before hurricane', 'during hurricane' and 'after hurricane'. A mathematical equation is fitted with respect to direction as shown in the Figure **18**-20 for wind speed in North- East, South-West and West direction, respectively. The variation of the mean for the wind speed of the entire dataset is also modeled.

4 CONCLUSION

The hurricane Matthew data was collected by deploying the Wireless Sensor Network system on the roof of Satellite Beach house, the data was also uploaded to the cloud during the hurricane event for online monitoring. The wireless system measured the pressure distribution over the roof top. The wind speed and wind direction data were obtained from the FCMP tower and Melbourne airport weather station. The pressure data is analyzed by selecting optimum sample period which yielded a steady value for the peak suction pressure. This averaging time was found to be around 300 seconds. The RMS pressure fluctuations are quantified with respect to the 300-second mean pressure. The mean pressure data correlated well with the wind speed data and the mathematical expressions are developed for different segments, before hurricane, during hurricane and after hurricane. During this measurements the wind direction were mostly from North- East, South-West and West on the house.

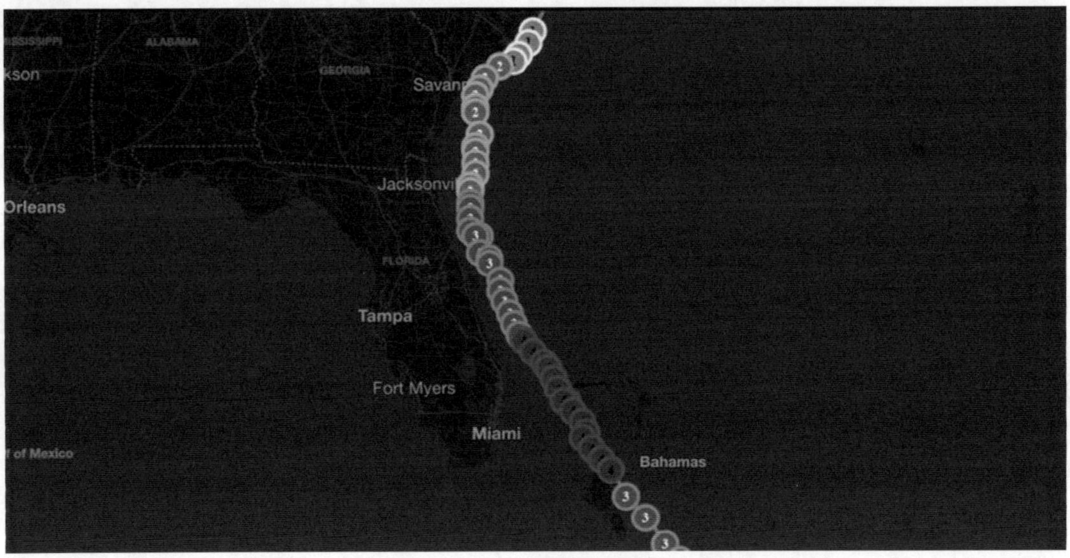

Figure 2: Track of hurricane Matthew

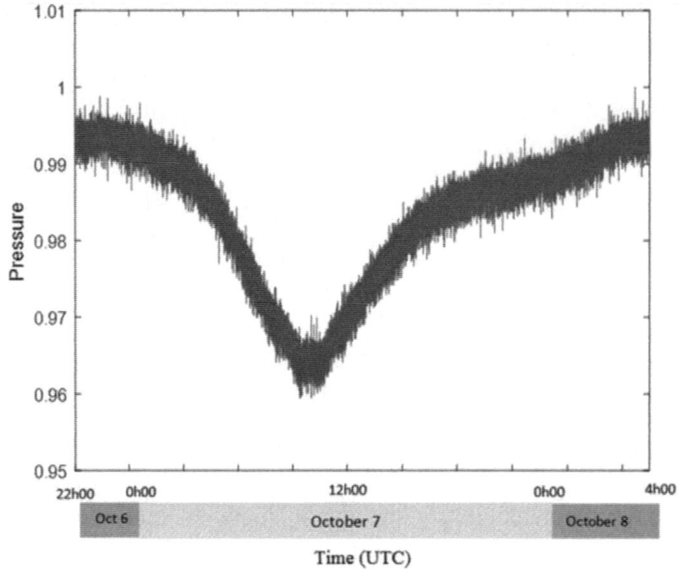

Figure 3: Pressure time history for sensor 3

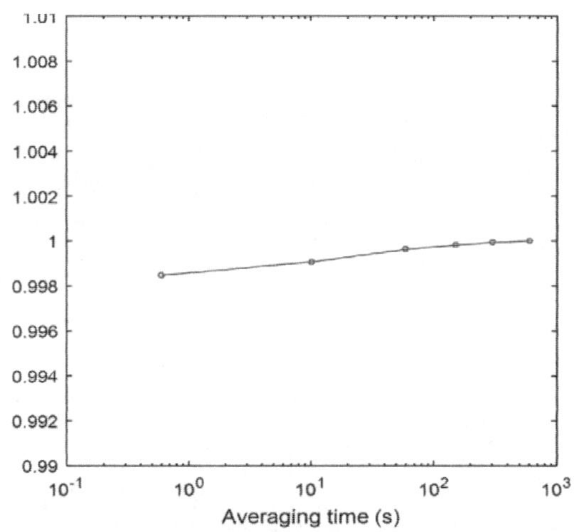

Figure 4: Normalized pressure for sensor 7

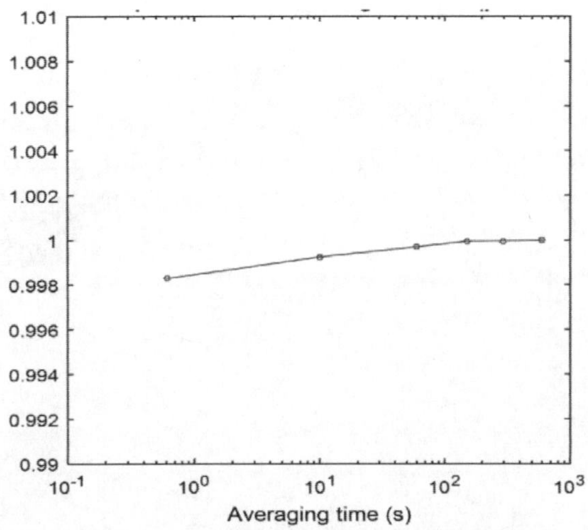

Figure 5: Normalized pressure for sensor 14

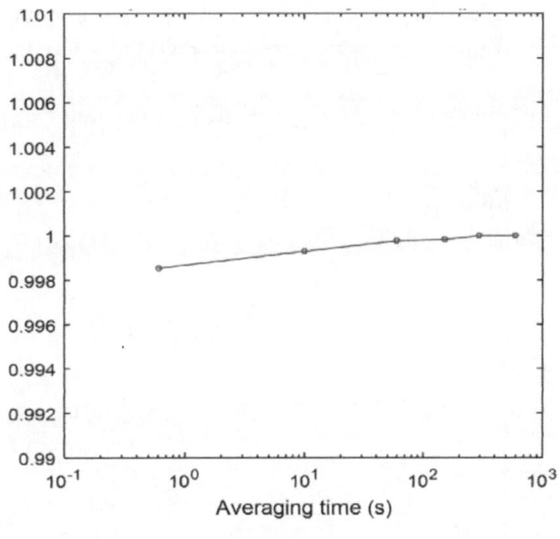

Figure 6: Normalized pressure for sensor 20

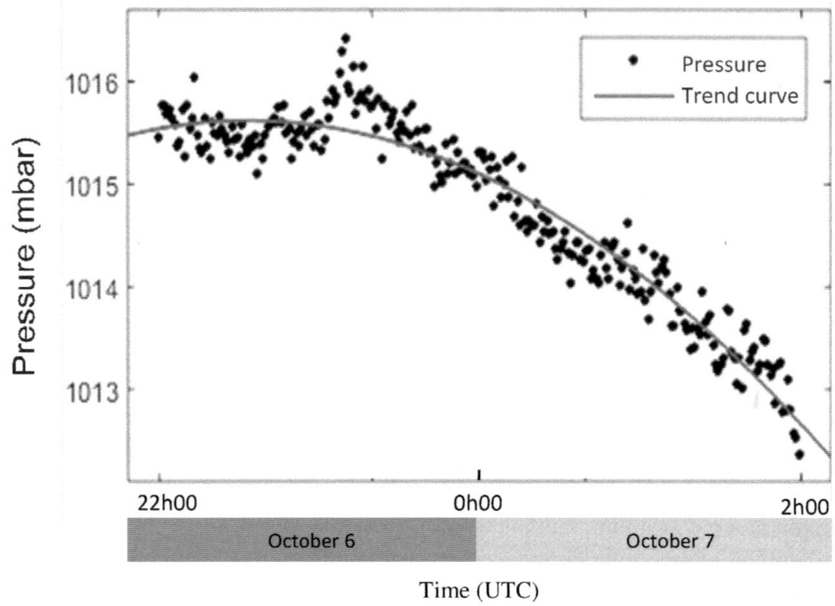

f(x) = (-1.781e-10)*x^2 + (7.465e-06)*x + 1016

Figure 7: Pressure variation trend for before hurricane

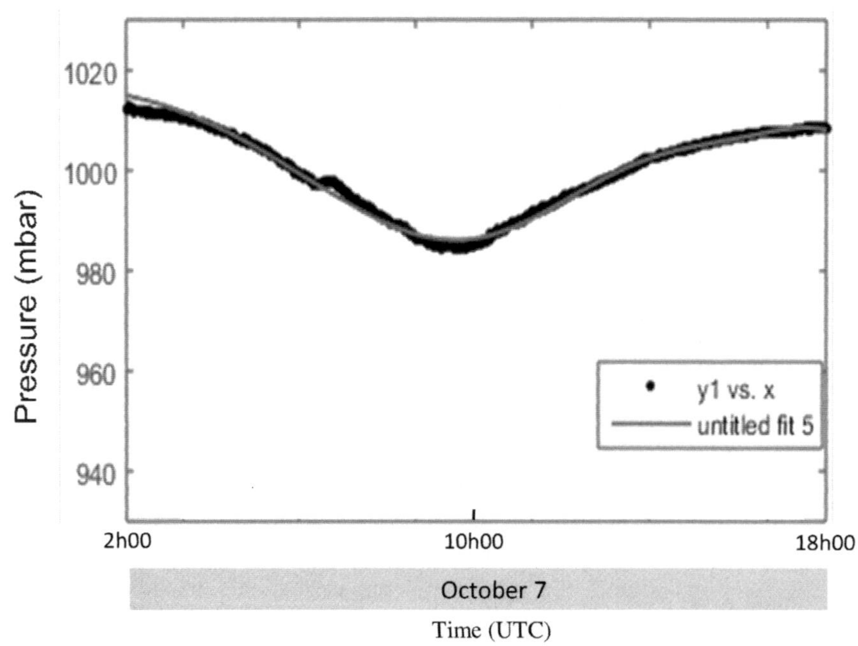

f(x) = 1016*exp(-((x+2)/8.414)^2) + 158.3*exp(-((x-2.224)/0.8462)^2) + 90.8*exp(-((x-1.104)/0.9891)^2)

Figure 8: Pressure variation for the entire duration of the hurricane

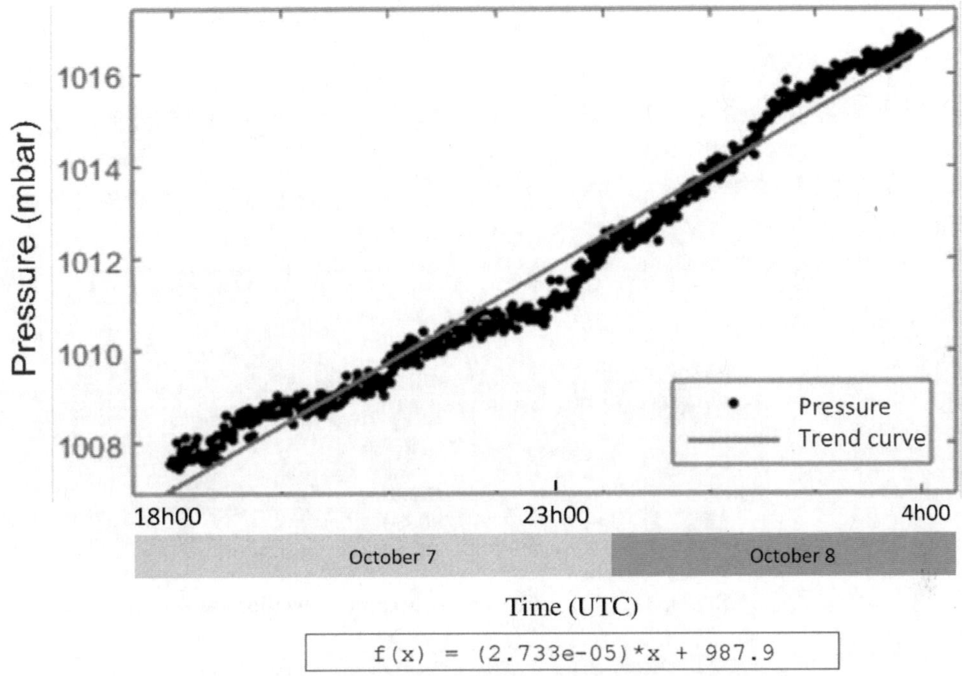

Figure 9: Pressure variation trend for after the hurricane

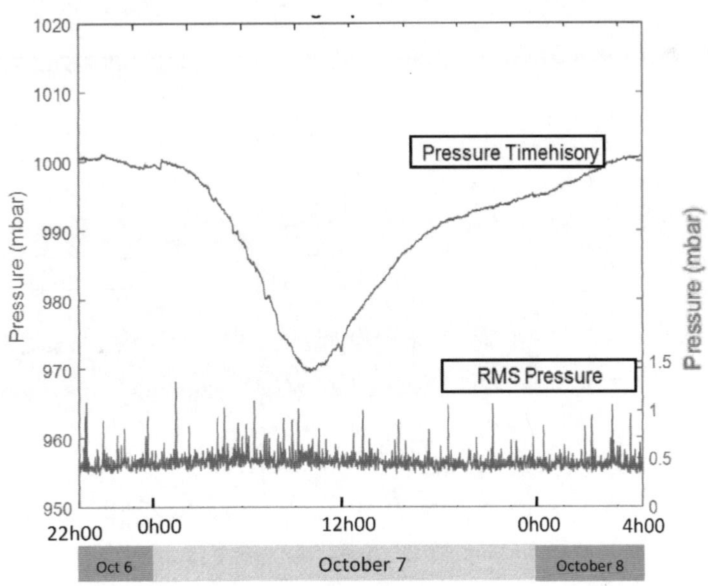

Figure 10: RMS and Average pressure variation for sensor 14

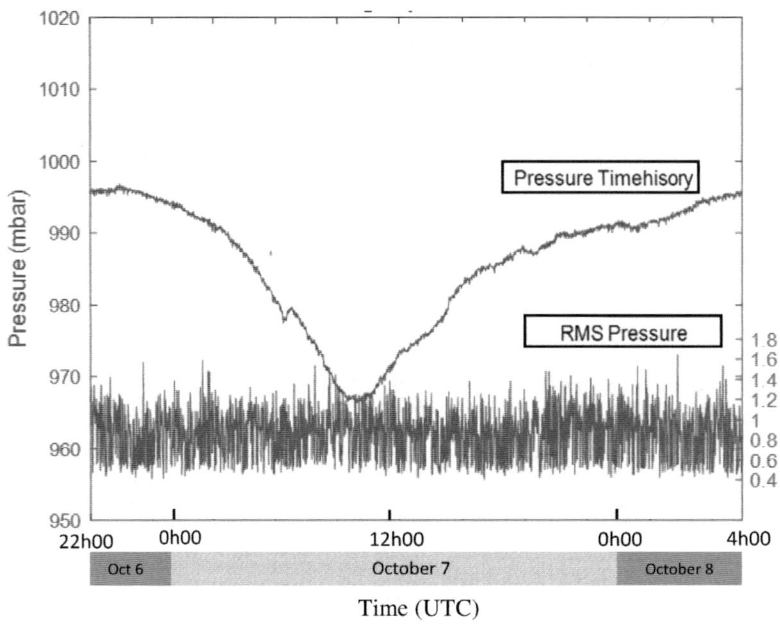

Figure 11: RMS and Average pressure variation for sensor 4

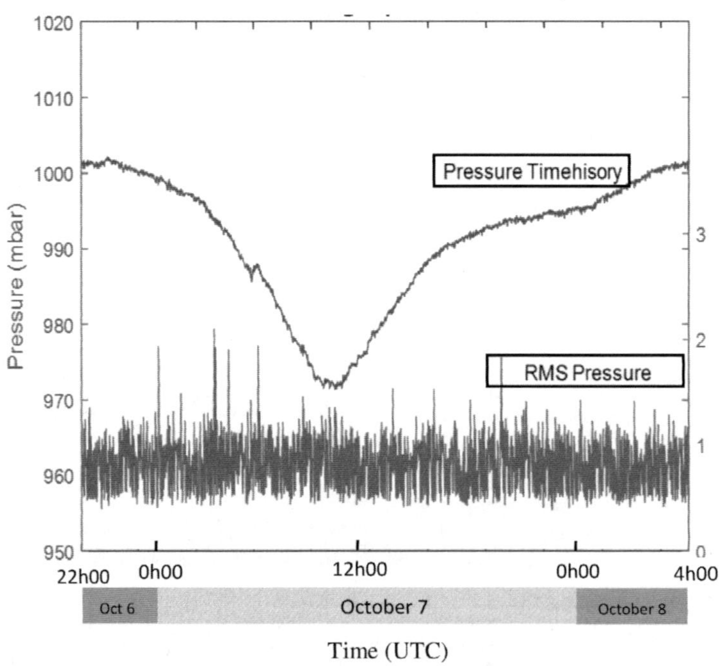

Figure 12: RMS and Average pressure variation for sensor 6

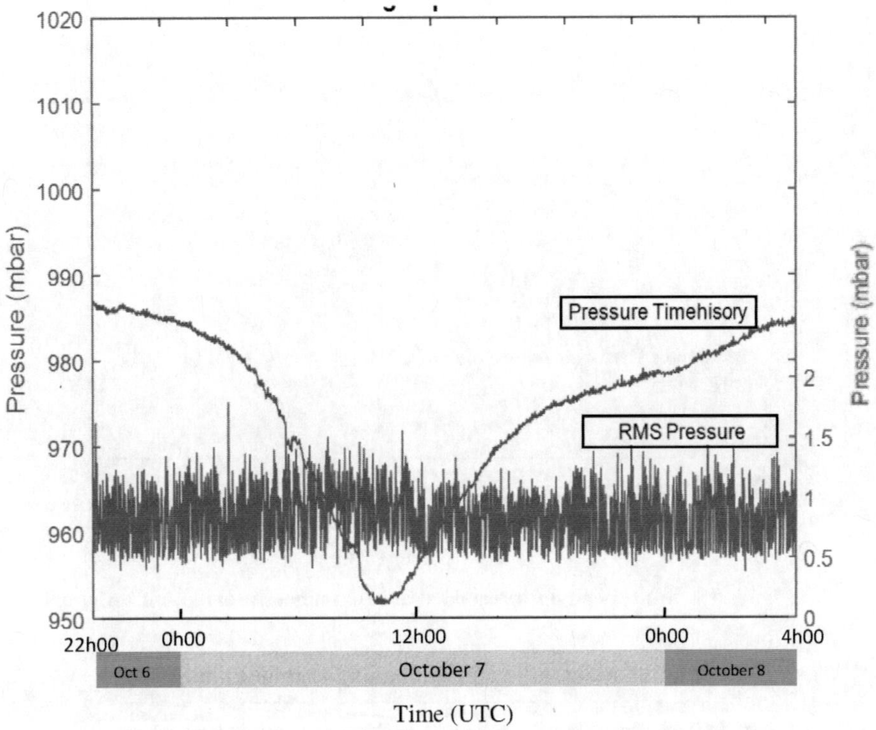

Figure 13: RMS and Average pressure variation for sensor 19

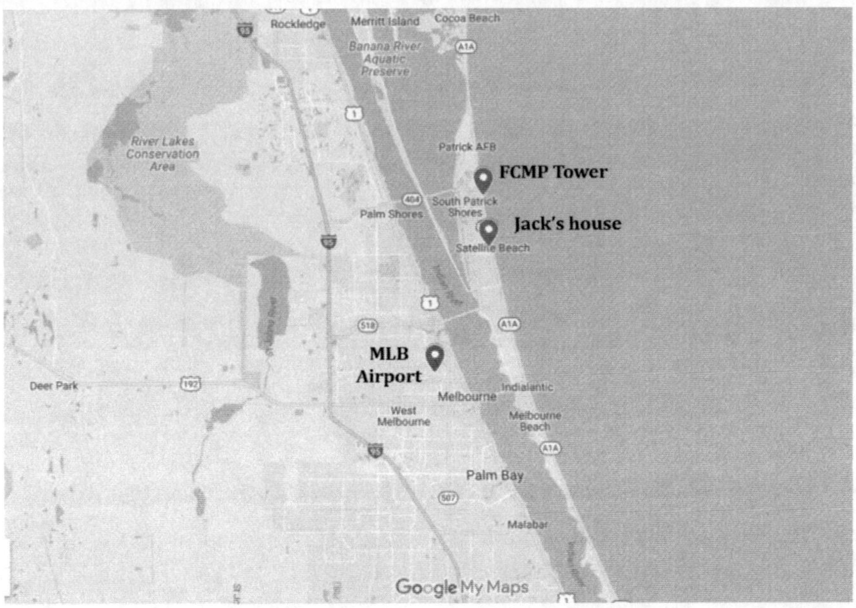

Figure 14: locations of the test house, FCMP tower and Melbourne airport weather station

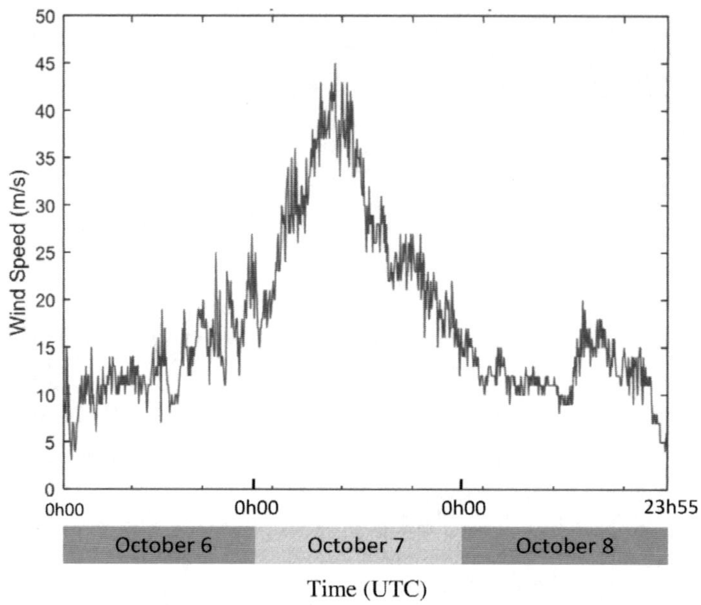

Figure 15: wind speed data recorded at Melbourne International Airport

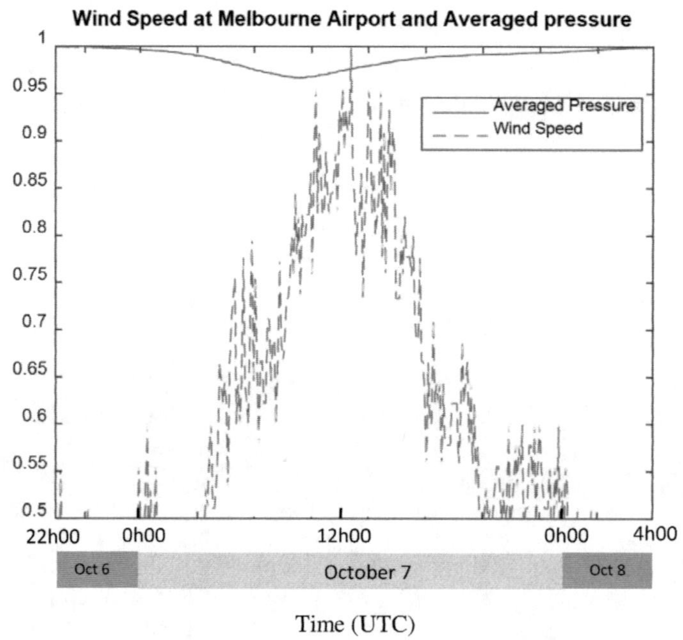

Figure 16: Comparing the wind speed data from Melbourne Airport data to Averaged pressure data

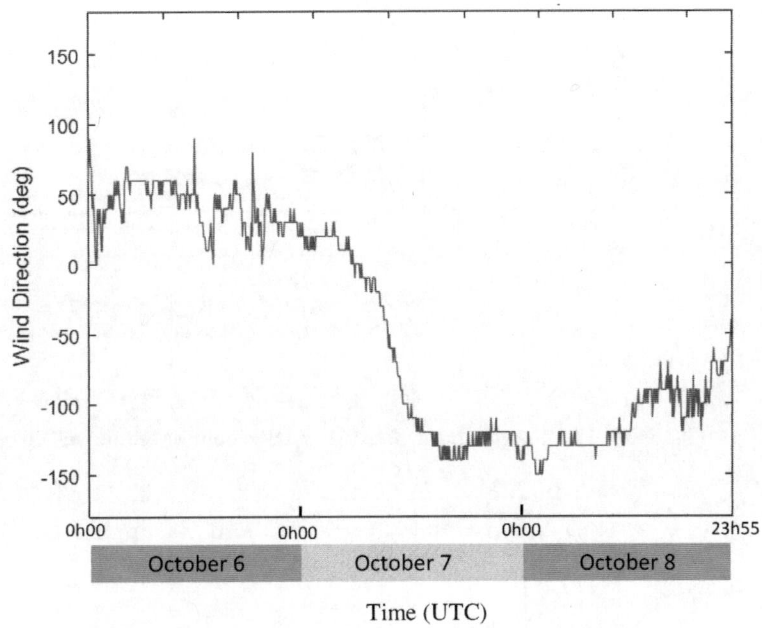

Figure 17: Wind direction data from Melbourne Airport data

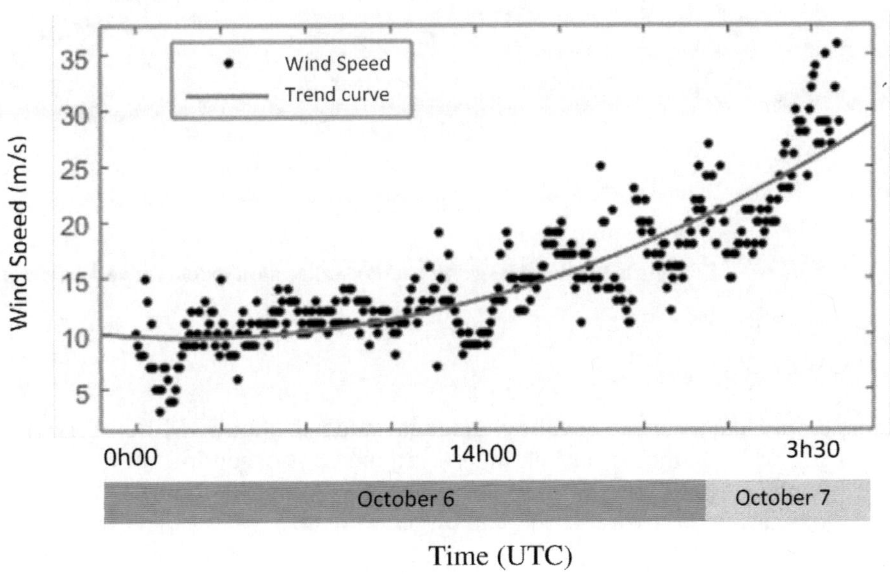

f(x) = 1,708*x^2 + 4,967*x + 13,28

Figure 18: Wind speed in the North-East direction (Melbourne Airport data)

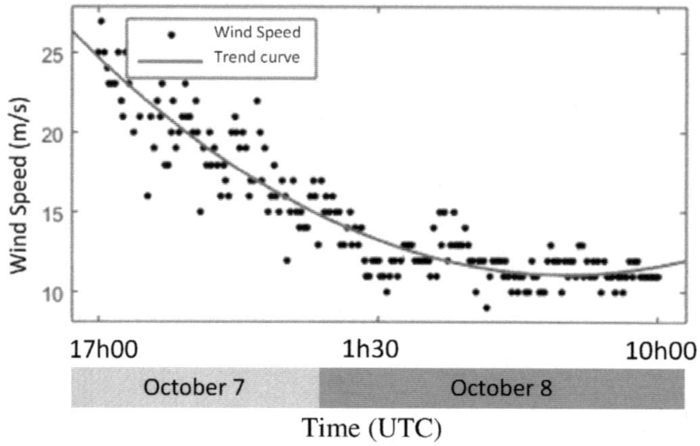

f(x) = 1,653*x^2 – 3,796*x + 13,27

Figure 19: Wind speed in the South-West direction (Melbourne Airport data)

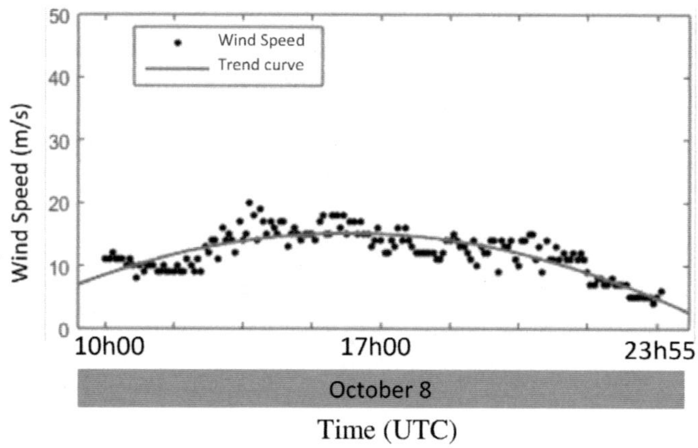

f(x) = -2,854*x^2 – 1,169*x + 14,98

Figure 20: Wind speed in the West direction (Melbourne Airport data)

5 ACKNOWLEDGEMENT

Authors would like to thank the fellow graduate students Nicolas Miller, Priyanka L Kanherkar, Dhruv Sangal and Mohammad for helping during the deployment of the wireless sensor network system. The authors are also very thankful to the FCMP and Melbourne Airport weather station authorities for sharing their wind speed and direction data.

6 REFERENCES

[1] G. Lapilli, "Hurricane Wind Effects: Insturmentation, Measurement and Analysis," Florida Institute of Technology, Melbourne , 2011.

[2] C. Subramanian, J.-P. Pineli, C. D. Lapilli and L. Buist, "A Wireless Multipoint Pressure Sensing System: Design and Operation," *IEEE sensors journal,* vol. 5, p. 5, 2005.

[3] T. Tao, , H. Wang, , and T. Wu, "Comparative Study of the Wind Characteristics of a Strong Wind Event Based on Stationary and Nonstationary Models," ASCE *J. Struc. Eng. 2,* 2016.

[4] J.S. Bendat and A.G. Piersol, A.G., Random Data Analysis and Measurement Procedures, 4th Ed. John Wiley and Sons, Inc., Hoboken N.J., 2010

[5] C. Subramanian, G. Lapilli and J-P Pinelli, "Analysis and Characterization of Hurricane Winds' ASCE *J. Engr Mechanics*, vol. 139, (3), 2013

13th Americas Conference on Wind Engineering
Gainesville, Florida, USA
May 21 - 24, 2017

Hurricane Matthew Multi-Hazard Damage Survey

Filmon Habte[a], Nozar Kishi[b], and Marc Simon[c]

[a] *Karen Clark and Company (KCC), Boston, Massachusetts, USA (fhabte@karenclarkandco.com)*
[b] *Karen Clark and Company (KCC), Boston, Massachusetts, USA (nkishi@karenclarkandco.com)*
[c] *Karen Clark and Company (KCC), Boston, Massachusetts, USA (msimon@karenclarkandco.com)*

ABSTRACT

Hurricane Matthew was the 13th storm, and the second major hurricane of the 2016 Atlantic tropical season. Matthew maintained category 3 - category 4 intensity through most of its track, causing significant damages to Haiti and Bahamas. It then became a category 2 hurricane west of St. Augustine, FL, and made landfall as a category 1 west of Mc Clellanville, SC, on October 8, 2016. It travelled on land up to Myrtle Beach, SC before heading east to the Atlantic and further weakening into a tropical storm. Although Hurricane Matthew made a brief landfall in South Carolina, its track in the Northern Atlantic region was mostly parallel to the US coastline, enabling it to affect multiple states. Post disaster surveys help in understanding the extent of damage caused, they have also been very important in identifying building structures and components most susceptible to hazards, their damage patterns, and improving construction practices. For example, most of the available prescriptive attachment details for roof tiles in codes and standards were developed / modified based on damage experienced during previous hurricanes. Karen Clark and Company (KCC) engineers conducted a survey of the damage in some of the highly impacted areas. Ground personnel were deployed to specific zones that included a mixture of residential and commercial properties and field observations were collected and analyzed. Hurricane Matthew was a very good demonstration of the multi-hazardous nature of tropical cyclones as damages due to high wind speeds, storm surge and inland flooding were observed. Some of the noted observations include; progressiveness of metal roof failures, shingle roof failures, importance dune elevation in storm surge damage, importance of roof age, vulnerability of gas station canopies, etc…. Herein, we discuss the detailed results of our post disaster damage survey.

KEYWORDS

Hurricane Matthew, Damage Survey, Damage patters, Multi hazard, Field observations

13th Americas Conference on Wind Engineering
Gainesville, Florida, USA
May 21 - 24, 2017

Analysis of Claims Data and Damage Investigations from Three Tropical Cyclones in Queensland

Daniel J. Smith[a], David J. Henderson[b], Richard J. Krupar III[c] and Matthew S. Mason[d]

[a] *Cyclone Testing Station, James Cook University, Townsville, Australia, daniel.smith8@jcu.edu.au*
[b] *Cyclone Testing Station, James Cook University, Townsville, Australia, david.henderson@jcu.edu.au*
[c] *The University of Queensland, St. Lucia, Australia, r.krupariii@uq.edu.au*
[d] *The University of Queensland, St. Lucia, Australia, matthew.mason@uq.edu.au*

ABSTRACT

Tropical cyclone loss estimation models (i.e. catastrophe models) are used by insurers, government and emergency responders to plan for and manage the potential losses from extreme events. The hazard component of a tropical cyclone loss model is generally based on tropical cyclone boundary layer theory with adjustments for local topography, shielding and terrain roughness to estimate the hazard parameters (i.e. wind, rain and/or wind-borne debris) for a particular site or region. The vulnerability component of the model comprises a set of functions that estimate the relationship between hazard loading intensity and the expected losses for various types of structures. These relationships can be validated and enhanced using insurance claims data and damage investigations for tropical cyclones. In this paper, claims data and field investigations from three tropical cyclones (Tropical Cyclones Larry, Yasi and Marcia occurring in 2006, 2011 and 2015 respectively) in Queensland, Australia are used to review the relationships between tropical cyclone hazard parameters and residential building damage level for various building attributes.

To estimate maximum gust wind speeds at individual claim locations, parametric wind field models for the three events were used with local multipliers (estimated via Landsat imagery) for shielding, topography and terrain roughness. The data were binned based on the ratio of claim value to property value (i.e. loss ratio) as an indicator of damage intensity. Using sampling theory, a subset of claims within each loss ratio bin were analyzed in detail to identify typical modes of failure. The detailed analysis included assessor's reports, photos and repair quotes for each property in addition to damage investigation data compiled by the research and emergency response communities.

The analysis suggests that at lower wind speeds, non-structural attributes of the building and surrounding area (e.g. permeability of building openings, fencing, trees, outdoor furniture, etc.) have stronger influence on losses than structural attributes (e.g., construction type, roof type, etc.). In the absence of detailed information about building features, age of construction has been found to be a useful proxy for some building characteristics unknown in a dataset. Through detailed analysis of the more comprehensive 2015 event dataset, the research will attempt to delineate between damages to different components of the buildings (e.g., roof cover, interior, appurtenances, etc.) to provide more refined validation data for component-based vulnerability model simulations.

KEYWORDS

Tropical Cyclone, Hurricane, Insurance, Risk, Vulnerability

Recent Advances in Remote-Sensing Assessments of Wind Damage

J. Arn Womble[a], Richard L. Wood[b], M. Ebrahim Mohammadi[b], Ronald T. Eguchi[c], Shubharoop Ghosh[c]

[a]*West Texas A&M University, Canyon, TX, awomble@wtamu.edu*
[b]*University of Nebraska-Lincoln, Lincoln, NE, rwood@unl.edu; me.m@huskers.unl.edu*
[c]*ImageCat, Inc., Long Beach, CA, rte@imagecatinc.com; sg@imagecatinc.com*

ABSTRACT

Remote-sensing technologies provide an effective basis for the rapid and objective assessment of wind damage. While the development of remote-sensing based assessments has progressed markedly over the past decade, automated algorithms have not yet achieved the speed, objectivity, and reliability desired for practical implementation in time-critical damage assessments. Current standard practice for making swift, objective, and widespread assessments of wind damage currently consists of rapid visual interpretation of first-available imagery. Techniques for rapidly and practically accomplishing widespread damage assessments by visual inspection have been implemented in recent major tornado outbreaks in Birmingham-Tuscaloosa, Alabama and Joplin, Missouri.

Meanwhile, rapidly emerging technologies, such as unmanned aerial systems (UAS), laser (lidar) scanners, and digital photogrammetric modeling are helping to improve both the speed and the accuracy of damage assessments, in particular for rapid and target-specific data collection at very high spatial resolutions. Applications of these emerging technologies following recent severe tornadoes in Joplin (2011), at Pilger, Nebraska (2014) and Pampa, Texas (2015) have demonstrated their rapidly emerging roles in helping to refine strategies for making rapid semi-automated damage assessments. The paper discusses the significant guidance provided by these recent case studies in the ongoing development of new technologies for rapidly preserving and analyzing wind damage evidence.

KEYWORDS

Remote sensing, lidar, laser scanning, UAV, photogrammetry, tornado damage

13th Americas Conference on Wind Engineering
Gainesville, Florida, USA
May 21 - 24, 2017

Natural Hazard and Damage Assessment of Building Infrastructures in a Coastal Community

Jin Zhu[a], Di Wu[b], Jeffery Weston[c], Wei Zhang[d]

[a] *Graduate Student, Dept. of Civil and Environmental Engineering, Univ. of Connecticut, Storrs, CT 06269, jin.zhu@uconn.edu*
[b] *Graduate Student, Dept. of Civil and Environmental Engineering, Univ. of Connecticut, Storrs, CT 06269, di.4.wu@uconn.edu*
[c] *Graduate Student, Dept. of Civil and Environmental Engineering, Univ. of Connecticut, Storrs, CT 06269, Jeffery.Weston@uconn.edu*
[d] *Assistant professor, Dept. of Civil and Environmental Engineering, Univ. of Connecticut, Storrs, CT 06269, wzhang@engr.uconn.edu*

ABSTRACT

The hurricane associated high wind and coastal flooding can cause damages on infrastructures, including buildings, bridges, etc. in a coastal community with potential large economic and social impacts. A probabilistic natural hazard and damage assessment approach is presented to evaluate the coastal storm related natural hazards and associated damages on building infrastructures. A probabilistic model of hurricane induced wind, waves and surge is first developed for the hurricane risk assessment. The hurricane occurrence and intensity are modeled through Poisson process and Weibull distribution, and the wave is modeled through empirical model, respectively. The parameters of hurricane wind and wave models are derived from a statistical analysis of storms affecting the Northeastern United States and include radius of maximum winds, central pressure difference, translation speed, hurricane categories, etc. Meanwhile, the coastal storm surge is modeled using the generalized extreme value (GEV) model with the parameters estimated from the long term water level observation record along the Long Island Sound coastline in the state of Connecticut. With the predicted wave and storm surge, combined with deterministic astronomical tides, the extreme water level is estimated. Compared with the existing GIS flooding maps in the coastal community, the resultant extra water level will be used to generate the coastal flood inundation map. Finally, the generated natural hazard maps including the coastal wind and coastal flood inundation map with various mean recurrence interval (MRI) will be used together with empirical damage prediction models for the wind induced damage and the flood induced damage, to estimate the expected damage losses for the building infrastructures in the coastal community. A case study is presented based on the data from a Connecticut coastal community.

KEYWORDS

Hurricane, damage assessment, statistics, wind speed, wave, storm surge, coastal flooding, risk assessment.

Optimization of Alternative Towers for Wind Turbines in Low Wind Resource Regions

Anwer Al-Kaimakchi [a], Sungmoon Jung [a], Michelle Rambo-Roddenberry [a], Gholamreza Amirinia [a]

[a] *Department of Civil and Environmental Engineering, Florida A&M University – Florida State University College of Engineering, 2525 Pottsdamer street, Tallahassee, FL 32310, USA,*

ABSTRACT

Taller wind turbines have been proposed for low wind speed sites, where conventional 80 m tower cannot produce enough electricity. Gaining 20-45% more power are possible by increasing the height of tower from 80 m to 140 m in moderate wind velocity regions as shown by other researchers. The aim of this study is to come up with optimum designs under two conflicting objectives. The conflicting objectives were, minimizing the initial wind turbine cost and maximizing power production, considered in determining best design. Different design variables and constraints were defined for each tower design. Due to the highly constrained, non-convex and non-linear nature of the design problem, Genetic algorithm has been chosen as a solver for the problem. The towers were analyzed for load conditions according to IEC 61400-1.

Keywords: Optimization, wind turbine tower, Genetic algorithm, power, payback period.

1 INTRODUCTION

Wind turbine towers have been through many developments. Two decades ago, conventional steel towers were dominating the industry. Industries started to increase height of the towers to reach higher wind speed in order to decrease cost of energy, which makes wind energy a better resource among others. As height of the tower increases, diameter at the base should increase too to make the tower able to resist wind loading and be stable. Due to transportation constraints, diameter of the steel tower cannot be greater than 4.5 m [1]. In order to overcome this constraint, manufacturers came up with precast concrete towers. They can be divided into vertical rings and longitudinal segments, transported and erected at the construction site. To go even higher with another tower option, wind turbine industries started manufacturing hybrid towers which are combination of steel and concrete towers. Industries tackle the obstacle of steel tower diameter constraint with hybrid towers. Hybrid towers now can reach height of 159m, new generation tower from Enercon [2]. With these new options, towers started to reach higher wind speeds and as a result cost of energy goes down.

The main goal of the industries is to minimize the cost of energy in order to compete with other sources, such as fossil fuel and solar. Historically, people used wind power to sail a long time ago. However, using the wind as a source to generate electricity started in the late 19th century [3]. Wind energy industry is fairly new when compared to other sources of energy. During the last 25 years, wind energy is recorded as the fastest growing source of energy. An extreme effort has been made in the past 20 years from industries to make wind energy a competitive source of energy. At least 20% of the energy in the United States is expected to be generated from onshore and offshore wind farms by 2030 [3]. Also by 2030, a cost reductions of energy of 20%-30% from wind turbines are expected by many studies [4].

The objective of this paper is to find an optimum design under conflicting objectives, minimizing wind turbine cost and maximizing power production, for three alternative towers. The three options were steel, concrete and hybrid towers. Design variables were chosen to be diameter, thickness and height of tower and number of strands for concrete tower. Design constraints were buckling, yielding, shear stresses for steel tower and ultimate and service limit states for concrete tower. Those constraints have been used to control the stability of the tower. Design problems were conceptual design, so design of the flange for steel tower, connection between concrete parts and between steel and concrete towers for the hybrid tower were not considered in the optimization process. Nonetheless, cost of these parts were taken from previous studies and added as a cost to the design problem. Genetic algorithm was chosen as a solver for the optimization problem. Multiple optimal solutions were obtained which are called Pareto-optimal solutions.

2 LITRATURE REVIEW

A multi-objective optimization problem (MOOP) works with more than one objective function [5]. The best common uses of multi-objective optimization is by combining conflicting objective functions. For example in this paper, one objective function was minimizing cost of the wind turbine and another objective function was maximizing power production. In the past 20 years, wind turbines have improved and scattered worldwide. Number of designs and optimizations of wind turbine towers have associated this improvement as shown in figure 1. To solve a more sophisticated optimization problem, multi-objective algorithm was used in about less than 25% of wind turbine optimization publications [6].

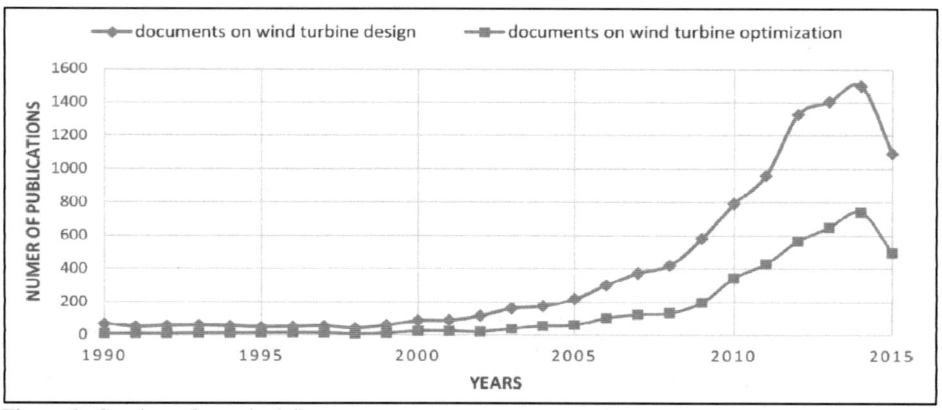

Figure 2. Caption of a typical figure.

Researchers have used multi-objective optimization for wind engineering problem or wind turbine blade optimization [7-9]. However, it was not yet used for optimization of wind turbine tower.

Wind industries tend toward higher tower to reduce cost of energy by catching more energy from higher attitude. Steel tower has limitation in diameter of the tower due to transportation difficulties [1]. Thus, turbine manufactures in the last decade have been seeking new feasible and cost-effective solutions of wind turbine towers. Concrete towers seemed to be the best solution as there is no limitation for the transportation part. Also, concrete has been chosen to build tower structures similar in lay out to the wind turbine tower such as: chimneys, poles and bridge piers [10, 11]. To take most advantages out of those materials, steel and concrete; industries have built hybrid towers to reach higher wind speed. Lower part is constructed from precast concrete, tower is erected in the field. Upper part is fabricated from steel, this part is light in weight.

2.1 *Steel tower*

Genetic algorithm was used by Yoshida [12] to optimize steel wind turbine tower. Main single objective was to minimize mass of the tower under different constraints and load conditions. Uys et al. [13] presented a design procedure to minimize cost of steel tower. Design variables were average thickness of shell, and height, thickness and number of ring stiffeners. The steel tower was slightly conical. Cost function included material and manufacturing costs. Shell buckling and local buckling of flat ring stiffeners were used as design constraints. Results showed that objective function was controlled by the minimum number of ring stiffeners. Karpat F. [14] used a particle swarm optimization algorithm to minimize cost of steel wind turbine towers. Design variables were wall thickness, diameter, and dimensions of the flat ring stiffeners. Design constraints were natural frequency, shell buckling and local buckling of flat ring stiffeners. Perelmuter and Yurchenko [15] used an improved gradient method to minimize weight, not cost, of steel tower. Design variable were height, diameter, shell thickness and wheel diameter.

Enercon has provided a new method to assemble longitudinal sections of tubular steel tower using grout joint [16]. The new Enercon design's divides the tubular steel towers into longitudinal sections. Those sections are smaller and lighter, which make them easy to transport. Before the erection on construction site, they are assembled together into complete tubular sections. There is a unique procedure for this method. Also, there is an alternative method for cold places, low temperature, where concrete cannot be used.

2.2 *Concrete tower*

Concrete is a versatile material which has been used structurally in many different ways. Structural concrete can be used as reinforced, prestressed or normal (mass). Reinforced concrete has been in industry for at least 100 year and prestressed concrete for over 70 years. Yet, mass concrete has been used in construction for a very long history [17].

Based on Grünberg and Göhlmann [18], the Enercon E-126/6 MW is the most powerful wind turbine in the world. The hub height is 135m and rotor diameter is 126m. Tower is divided vertically and horizontally. Enercon and Acciona wind power industries have made a greater improvement in the design of precast concrete tower. Enercon produces precast concrete towers with different hub heights. Cylindrical segments fit together to form the tower, diameter of the segment decreases with height.

Ma and Meng [19] designed a prestressed concrete tower with a regular octagon cross-section. The tower was tapered and had interior ribs on each side. The objective function was minimum cost of concrete tower and tendon. The constraints were stress, displacements and natural frequency of the tower. Finite element method and optimization were incorporating in the procedure to get the best design.

Lofty [11] designed a new type of tower that has a triangular cross section and there was a column at each corner of the triangle. Precast panels were oriented between the columns. Lofty designed 75m and 100m towers for his new proposed design and steel. The concrete and steel towers' performance were compared; results showed that the proposed design had a lower initial cost than regular steel tower.

Serna J. and Jimeno J. [20] provided outcomes of the Spanish company, Inneo Torres, for concrete towers. 80m, 100m and 120m concrete towers were the products of the company. The rated power was ranging from 1.5 MW to 4.5 MW. The tower is divided into several segments and assembled at the construction site. A rate of 2.5 towers in a week can be achieved by using fewer and longer segments with simple and rapid systems for the use of the joints.

2.3 *Hybrid tower*

A hybrid tower has a hub height of 133m and machine rating of 2.3MW was erected by the Dutch company, Mecal [18]. Concrete tower has five rings, each ring was divided into four quarter-circle elements; and four straight elements that fitted between the quarters for ease transportation. Vertically, each segments has a height of 15m. One of the advantages of the hybrid tower is steel part would be replaced after 20 years as the design life for the concrete tower is 40 years.

Enercon proposed a new generation of hybrid tower [2]. This new proposed design improves the efficiency of production, accelerates assembly times on construction site. The newest hybrid tower has hub height of 159 m. The hybrid tower has a combination of conical and cylindrical segments. Early 2017, the first new generation hybrid tower was constructed. Based on the author knowledge, literature is so limited for design or optimization of the hybrid tower.

3 ILLUSTRATION OF MULTI-OBJECTIVE OPTIMIZATION

The multi-objective optimization minimizes multiple objective functions as shown in figure 2. Feasible objective space is the area to the right and above curve. It is important to mention that the feasible objective space contains optimal and non-optimal solutions.

Any solution from the feasible objective space can be chosen and compared. Solutions A to H are non-dominated solutions, as none of them has lower cost and 1/power than the rest. For some pairs of solutions, it can be noticed that solution A is better than D in cost objective but is worse than in 1/power objective. For other sets of solutions, it can be seen that one solution E is better than another solution 3 in both objectives. Solutions 1 to 4 are dominated solutions. No single solution can be selected as best solution, in both objectives, due to conflicting nature of the objectives.

Pareto-optimal solutions are those which are lying on the curve, A to H, and they are special in the term of multi-objective optimization. Pareto-optimal front is the curve which is formed by joining of the optimal solutions. None of the solutions lying on the Pareto front can be selected as the optimum solution with respect to both objectives; as discussed before. In our example, solution D is better than all dominated solutions, 1 to 4, in both objectives.

In this research, the two objective functions were minimization of cost of wind turbine and 1/power, it is the same to say maximization of power production. As multi-objective optimization uses to minimize the objectives.

By comparing solution E with 4, it can be observed that solution E is better than solution 4 in both objectives. Since solution E has lower cost and 1/power than solution 4, this means that solution 4 is dominated by solution E. However, by comparing solution H and 4, it can be noticed that solution H is better in the 1/power objective but is worse in the cost objective. The presence of solution E and others lying on the Pareto-optimal front make 4 a dominated solution, as solution 4 can be dominated by solution E. Thus, the non-dominated solutions should be compared with any solution to see if they belong to dominated solutions or not. The idea of figure 2 was inspired by [5, 7, 21].

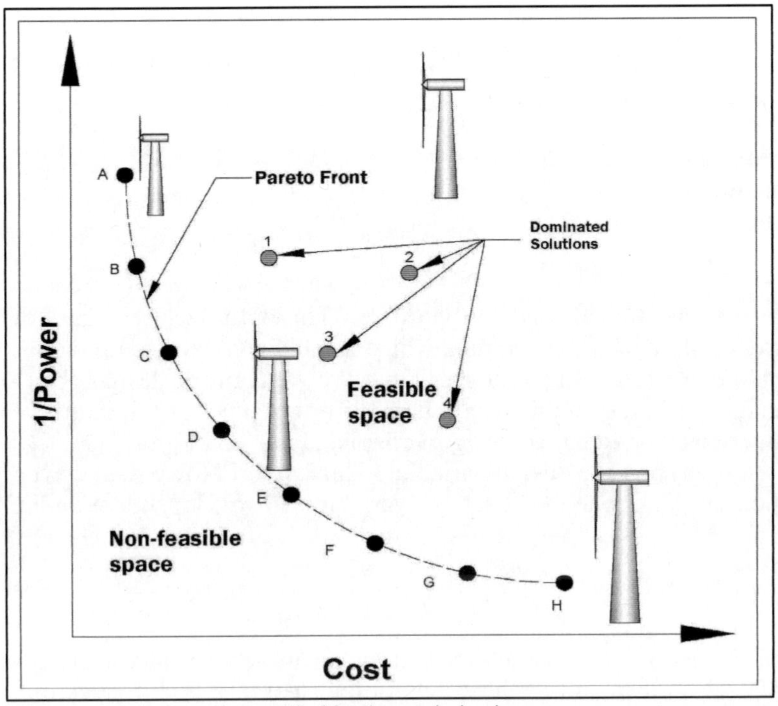

Figure 1 Pareto Front for multi objective optimization.

4 LOADING

Throughout service life of wind turbine towers, they are subjected to many different load conditions. These loads can be classified into two groups: turbine loads and tower loads. Turbine loads are loads caused from the operation of turbine, and are applied to the tower at hub height. In this paper, thrust load is the turbine load. Thrust coefficient was taken from [22] as 0.3. Tower loads are loads originated from self-weight of tower, dead weight of wind turbine, direct wind force on the tower and earthquake force if exist. This direct wind force was simplified to a static equivalent force. There is no single design document for wind turbine in the United States, thus several standards were used. IEC 61400-1 [23] categorizes turbines into three classes (I, II, III) based on typical wind speeds within their projected operating environment. To achieve a more accurate wind turbine operating environment, each class has sub-category (A, B or C) for wind turbulence characteristics. Gust design wind speeds was used in ASCE 7-10 to calculate the direct wind force on the tower [ASCE 7-10]. Dead weight of the wind turbine was calculated based NREL report [24]. Mass estimates of the wind turbine components were a function of machine rating and rotor diameter.

5 OPTIMIZATION PROBLEM FORMULATION

The design problem is formulated as an optimization problem. In order to solve this optimization problem, an appropriate design objectives, design variables and design constraints must be defined. Due to the highly constrained, non-convex and non-linear nature of the design problem, Genetic algorithm has been chosen as a solver.

5.1 *Design objective*

Two conflicting design objectives were minimizing cost of tower and wind turbine and reciprocal of power equation, which also means maximizing power.
minF(X) = min $[f_1(X) \quad f_2(X)]$
$f_1(X)$ is cost of supported structure and wind turbine
$f_2(X)$ is reciprocal of power equation

5.2 *Design variables*

A design variable is a parameter which is controlled by optimizer. The optimizer searches for values of design variables, should satisfy design constraints, which minimize objective functions. Also, those design variables should be within defined ranges. Thus, the value of the design objective is evaluated based on the design variables. In this paper, three different types of wind turbine towers were optimized. Each tower has its number of design variables. First, steel tower had five design variable. Second, concrete tower had nine design variables. Third, hybrid tower had eleven design variables. The rotor diameter was not one of the design variables; it was dependent on the height of the tower [25].

5.3 *Design constraints*

Constraint is divided into two parts, inequality and equality. All the design constraints were formulated as an inequality constraints. The inequality constraints for the steel tower were buckling, bending and shear stresses and ultimate and service limit states for the concrete tower. For the hybrid tower, both inequality constraints for steel and concrete towers were applied. All the design constraints for the three types of towers were continuous, smooth and differentiable.

6 COST BREAKDOWN

The cost calculation by LaNier [10] covered all wind tower elements, sections are shown in figure 3, except turbine hub and rotor. Cost of shop fabrication of steel tower, section separated in figure 3 left, was $0.9 per Lb. This represents a percentage from total cost of steel tower. Erection cost of concrete tower was based on the concrete unit volume. $336 was cost of cubic yard for concrete; and section separated in figure 4 middle is percentile of erection for concrete tower. For hybrid tower, cost of shop fabrication of steel tower was the same, $0.9 per Lb. However, erection of concrete tower cost was assumed as $320 per cubic yard. Both sections were separated in figure 4 right. Then, tower's whole cost for each type was back calculated from the separated section percentile. Unit cost value of any material and elements were calculated at the time report was written and it was not projected or compared with today's value. For more details about cost calculation and what it included, the authors recommend to check report [10].
Cost estimation of tower is very difficult as there is no direct data available. High level of accuracy for the cost estimations requires construction and logistic information. Wind turbine companies and wind farm developers have a detailed construction cost data. Yet they don't provide it to public as it is their right to keep guarded the business [17]. Some cost information are available but they are very general and dependent on other components. In this study, judgment and some assumptions have been made for cost estimations.

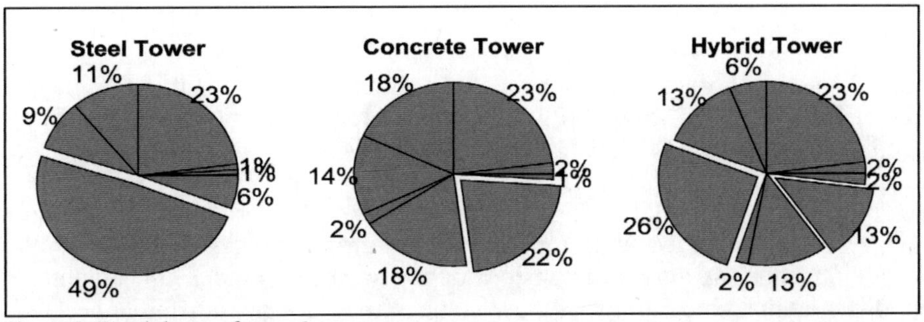

Figure 3. Breakdown of tower's cost

7 WIND RESOURCE

The main focus of this paper was low wind resource regions. Wind speed used here was 5.43m/s at 10m height. Figure 4 shows power production and rotor diameter change with height. As mentioned before blade length was dependent on height of tower. The maximum rotor diameter was 126m [26]. The trend relationship between tower height and blade length was taken from study done by [25]. Ashuri et al. [27] also provided a trend between rotor diameter and rated power output of 27 wind turbines. At height 100m, rotor diameter reached its maximum value 126m. After this height, rotor diameter was constant. The minimum preferred clearance distance, defined by [26], was 15m between blade tip projected down and ground level. However, [28] used a distance of 13.5m. Clearance distance here was higher than both previous values as this study followed [26] trends.

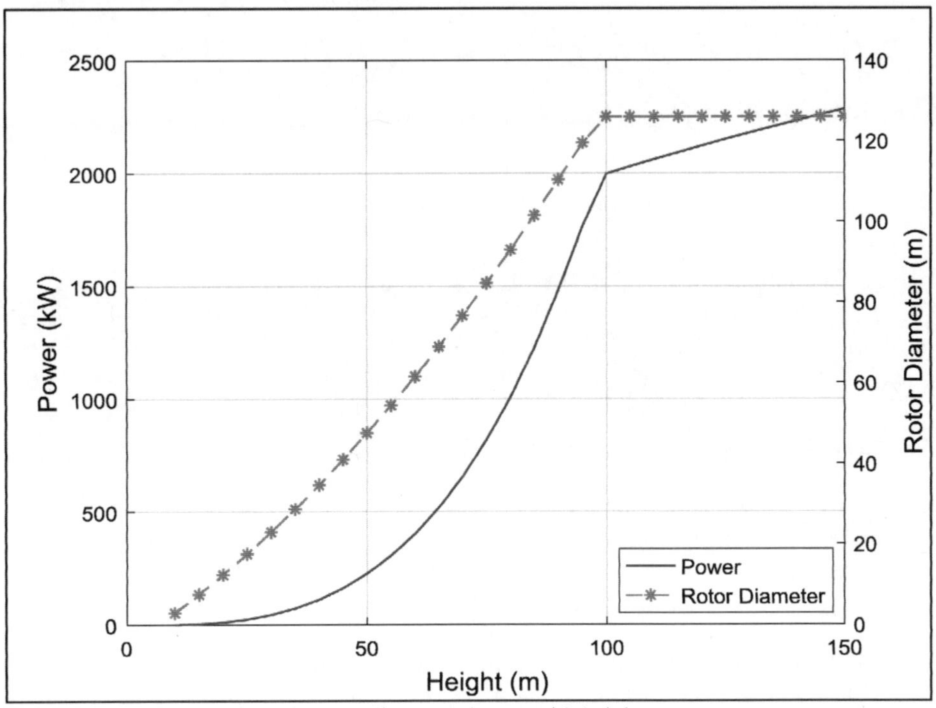

Figure 4. Power production and rotor diameter change with height.

8 RESULTS

The National Renewable Energy Laboratory has done a study about cost and construction methodology of three different types of wind towers [29]. The study was done only for a particular height, 100m, and machine rating of 1.5MW, 3.6MW and 5MW. All the costs were related to 100m height and those machine ratings. However, unit cost of material, such as steel and concrete, is the same for any height but equipment, labor and constructability costs were changing for different heights. Shop fabrication of tower, including material, had a certain portion from the whole cost of tower. This portion depends mainly on dimensions of the tower, volume. This percentage was relatively same for different machine ratings; as the volume of the tower was different because of turbine mass, for the same height.

Figure 5 shows pareto front generated by genetic algorithm for the steel wind turbine. Solver selected each point in figure 5 with optimum design variables. Cost of wind turbine increased with height as shown in figure 6. More power can be generated from higher altitudes as presented in figure 4. Thus, reciprocal of power equation deceased as the cost increased.

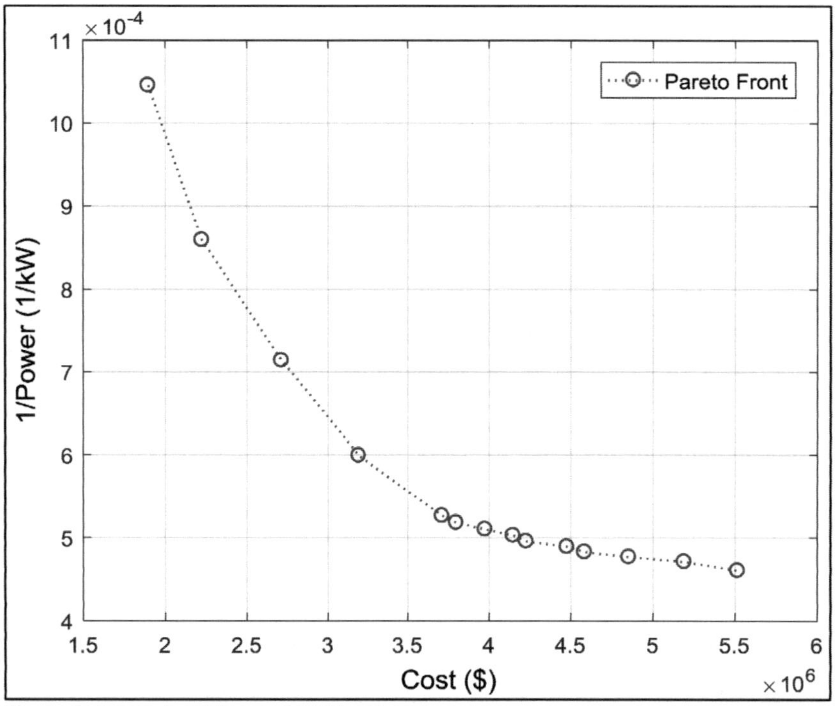

Figure 5. Pareto front for steel turbine tower.

Figure 6 shows the cost of wind turbine for the three options. Despite of the steel tower's base diameter exceeded 4.5m, transportation constraint, for 85 m height and above, it was not a competitive solution. This was one of the reasons that made industries think about alternative support structures, concrete and hybrid. For the concrete and hybrid towers, costs were relatively close in most height. Hybrid tower dominated at height greater than 100 m.

The concrete center [17] concluded that concrete taller towers, supporting large turbines, can save up 30% compared to their steel counterpart. An increase in the cost gap between steel and concrete towers with height as shown in figure 6. 150 m concrete tower height can save 30% compared to its steel counterpart. The percentage of saving was quite less for lower height.

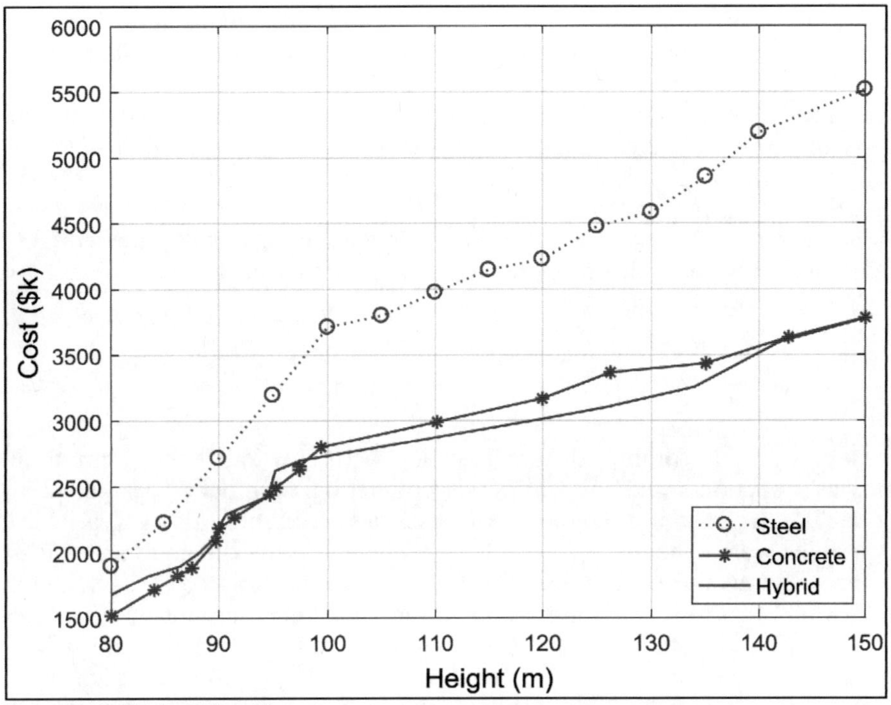

Figure 6. Cost of wind turbine.

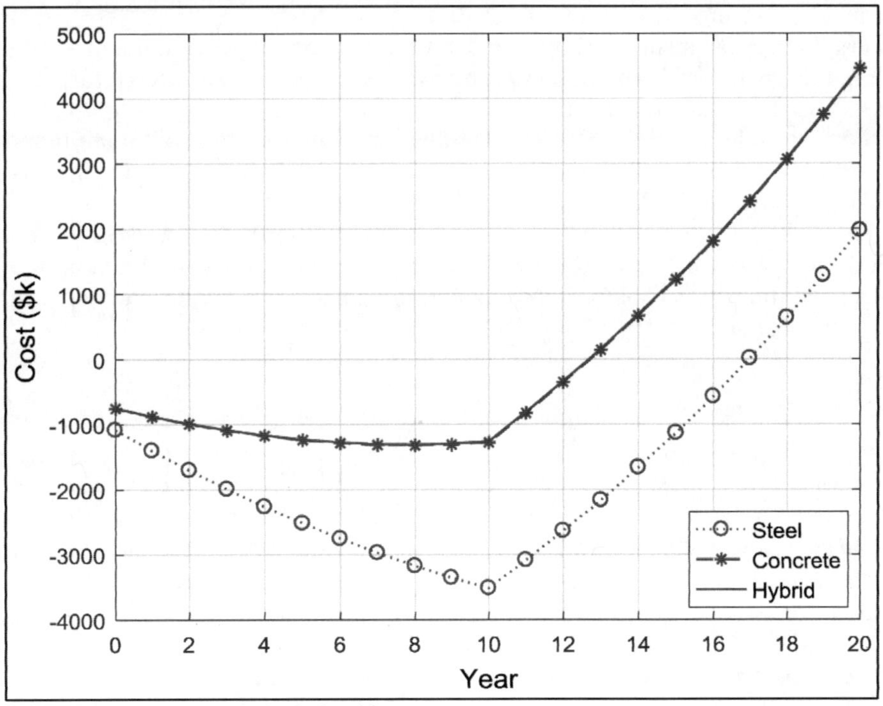

Figure 7. Payback period of wind turbines.

Those numbers proved a strong matching between the concrete center conclusion and this paper. As mentioned before, the cost estimation of the towers were based mainly on their weights. Thus, this indicates cost assumptions made here were comparatively realistic.

Annual energy produced is used to pay back total capital cost of wind system, this is called payback period. As shown in figure 6, concrete and hybrid towers were competitive and steel tower was more expensive. Figure 7 shows payback period for every tower at height 150 m. Annual energy production was calculated using swept area method. However, swept area method gives high AEP compared to power curve method. The AEP, swept area method, was divided by 2.38. 2.38 was average of AEP of 30 turbines calculated by swept area method to AEP calculated by power curve method.

9 CONCLUSION

Multi-objective optimization was done for three different tower structures using genetic algorithm. First objective function was cost and second one was power production. Number of design variables were different for each tower structure. Hybrid tower had eleven, highest, design variables among others. Those variables were concrete and steel heights, base and top diameter and thickness for both concrete and steel towers, and number of strands. Design constraints were buckling, bending and shear stresses for steel tower and ultimate and service limit states for the concrete tower. Low wind resource region was selected for this design problem. IEC 64100-1 and ASCE 7-10 were used for wind conditions design. The direct wind force was simplified to a static equivalent force. Due to limitation in the cost estimation data, cost estimation formulas was based mainly on NREL reports [24, 29]. Relationship between global wind turbine components, height and rotor diameter and machine rating, were based on [25].

The concrete center concluded that steel wind turbine can be 30% more expensive than its concrete counterpart for height higher than 70m and supporting large turbine. Even though turbines were not big, low speed, difference in cost between steel and concrete turbines reached up to 30% at height 150 m. The cost gap increased with height. Moreover, concrete and hybrid turbines seemed to be competitive in most heights. At some heights, hybrid turbine was better which clarified why turbines industries started to build those types of tower structure.

For future studies, authors are currently planning to add blade length as a design variable and also conducting the study for high wind resource zones. Investigate the results of changing the wind resource region and see impact of it on the optimization results. Moreover, the authors are planning to conduct the study for another cost assumption.

10 REFERENCE

[1] Nicholson JC. Design of wind turbine tower and foundation systems: optimization approach. 2011.

[2] Enercon. Series production od EP4 components has begun. Windblatt. Germany: Enercon; 2016.

[3] Lindenberg S. 20% Wind Energy By 2030: Increasing Wind Energy¿ s Contribution to US Electricity Supply: Diane Publishing; 2009.

[4] Lantz E, Wiser R, Hand M. The past and future cost of wind energy. National Renewable Energy Laboratory, Golden, CO, Report No NREL/TP-6A20-53510. 2012.

[5] Deb K. Multi-objective optimization using evolutionary algorithms: Wiley; 2005.

[6] Chehouri A, Younes R, Ilinca A, Perron J. Wind Turbine Design: Multi-Objective Optimization. arXiv preprint arXiv:161000975. 2016.

[7] Patil A, Jung S, Lee S, Kwon S-D. Mitigation of vortex-induced vibrations in bridges under conflicting objectives. Journal of Wind Engineering and Industrial Aerodynamics. 2011;99:1243-52.
[8] Wang L, Wang T-g, Luo Y. Improved non-dominated sorting genetic algorithm (NSGA)-II in multi-objective optimization studies of wind turbine blades. Applied Mathematics and Mechanics. 2011;32:739-48.
[9] Sessarego M, Dixon K, Rival D, Wood D. A hybrid multi-objective evolutionary algorithm for wind-turbine blade optimization. Engineering Optimization. 2015;47:1043-62.
[10] Lewin TJ. An investigation of design alternatives for 328-ft (100-m) tall wind turbine towers. 2010.
[11] Lotfy I. Prestressed concrete wind turbine supporting system. 2012.
[12] Yoshida S. Wind turbine tower optimization method using a genetic algorithm. Wind Engineering. 2006;30:453-69.
[13] Uys P, Farkas J, Jarmai K, Van Tonder F. Optimisation of a steel tower for a wind turbine structure. Engineering structures. 2007;29:1337-42.
[14] Karpat F. A virtual tool for minimum cost design of a wind turbine tower with ring stiffeners. Energies. 2013;6:3822-40.
[15] Perelmuter A, Yurchenko V. Parametric optimization of steel shell towers of high-power wind turbines. Procedia Engineering. 2013;57:895-905.
[16] Enercon. Second EP4 prototype installed. windblatt. Germany: Enercon; 2016.
[17] Tricklebank A, Magee B, Halberstadt P. Concrete towers for onshore and offshore wind farms. The Concrete Centre: Surrey. 2007.
[18] Grünberg J, Göhlmann J. Concrete Structures for Wind Turbines: John Wiley & Sons; 2013.
[19] Ma H, Meng R. Optimization design of prestressed concrete wind-turbine tower. Science China Technological Sciences. 2014;57:414-22.
[20] Serna JaJ, J. Precast Concrete Wind Tower - The Rise to the Next Level in Hub Height and Support Capacity for Large Turbines. Modern Energy Review. 2010;2:54-9.
[21] Dufo-López R, Bernal-Agustín JL, Yusta-Loyo JM, Domínguez-Navarro JA, Ramírez-Rosado IJ, Lujano J et al. Multi-objective optimization minimizing cost and life cycle emissions of stand-alone PV–wind–diesel systems with batteries storage. Applied Energy. 2011;88:4033-41.
[22] Chantharasenawong C, Jongpradist P, Laoharatchapruek S. Preliminary design of 1.5-MW modular wind turbine tower. The 2nd TSME International Conference on Me-chanical Engineering2011.
[23] Commission IE. IEC 61400-1: Wind turbines part 1: Design requirements. International Electrotechnical Commission. 2005.
[24] Fingersh LJ, Hand MM, Laxson AS. Wind turbine design cost and scaling model. 2006.
[25] de Lara Garcıa JS. Wind turbine database: Modelling and analysis with focus on upscaling: Master's thesis, Department of Applied Mechanics, Chalmers University of Technology; 2013.
[26] Jonkman J, Butterfield S, Musial W, Scott G. Definition of a 5-MW reference wind turbine for offshore system development. National Renewable Energy Laboratory, Golden, CO, Technical Report No NREL/TP-500-38060. 2009.
[27] Ashuri T. Beyond classical upscaling: integrated aeroservoelastic design and optimization of large offshore wind turbines: TU Delft, Delft University of Technology; 2012.
[28] Ashuri T, Zaaijer MB, Martins JR, Van Bussel GJ, Van Kuik GA. Multidisciplinary design optimization of offshore wind turbines for minimum levelized cost of energy. Renewable Energy. 2014;68:893-905.
[29] LaNier MW. LWST Phase I Project Conceptual Design Study: Evaluation of Design and Construction Approaches for Economical Hybrid Steel/Concrete Wind Turbine Towers; June 28, 2002--July 31, 2004. National Renewable Energy Lab., Golden, CO (US); 2005.

13th Americas Conference on Wind Engineering
Gainesville, Florida, USA
May 21 - 24, 2017

Utilizing Radar-Measured Velocity Fields to Provide Extended Inflow Forecasts to Individual Turbines Within a Wind Plant

James B. Duncan Jr.[a], Brian D. Hirth[b], John L. Schroeder[c]

[a] *National Wind Institute, Texas Tech University, Lubbock, TX, USA, james.b.duncan@ttu.edu*
[b] *National Wind Institute, Texas Tech University, Lubbock, TX, USA, brian.hirth@ttu.edu*
[c] *Geosciences Department, Texas Tech University, Lubbock, TX, USA, john.schroeder@ttu.edu*

ABSTRACT

The control system of a modern wind turbine typically possesses no knowledge of the upstream flow. Variabilities embedded within the inflow can only be accounted for once the disturbance has passed through the rotor sweep and manifested itself in the dynamics of the wind turbine system. As a result, a modern wind turbine almost exclusively operates in a reactive state. Research has shown that a snapshot of the approaching wind field can provide for improved turbine performance through proactive wind turbine control. The integration of proactive control has the ability to greatly increase the performance characteristics of the wind turbine. These improvements include enhancements to power production as well as reductions in generator fatigue and structural loads across the blade, tower, and gearbox. Observations by nacelle-mounted LIDAR (Light Detection and Ranging) systems currently constitute the industry standard for providing measurements of the upstream flow for proactive control. Whereas nacelle-mounted LIDARs primarily sample the near-upstream region of the wind turbine, the use of specialized Doppler radar technology has the ability to drastically expand the upstream extent of the wind field that can be measured and fed into the control system of a turbine for proactive control. This broadened look at the upstream boundary layer flow allows for turbine inflow to be forecasted in excess of 60 seconds in advance.

Data collected by two Texas Tech University Ka-band (TTUKa) Doppler radars will be used in this study to demonstrate the ability of radar-retrieved velocity fields in providing extended forecasts of turbine inflow. A dual-Doppler (DD) deployment was executed in December 2014 yielding three-dimensional maps of the horizontal boundary layer flow surrounding a subset of turbines within a wind plant. Typically, Taylor's hypothesis would be employed to advect velocity measurements recorded upstream of a turbine to its location. Research indicates, however, that the propagation of upstream measurements based solely upon the governing wind speed and direction is not sufficient for extended inflow forecasts. An innovative spatial mapping technique, which analyzes the translation of boundary layer features across an isolated portion of the DD domain, has been developed to account for the actual advection of momentum in the derivation of the extended inflow forecasts. To establish the merit of the derived predictions, the radar-forecasted inflow will be coupled with the Supervisory Control and Data Acquisition (SCADA) data of seven operational wind turbines located across the leading and trailing row of the wind plant. Through analysis of the turbine's pitch schedule, the radar-forecasted inflow can additionally be expressed as a pitch angle denoting optimal turbine performance. Discrepancies between the radar-forecasted and actual blade pitch angle results in undesired fluctuations in the outputted generator speed and power time histories. Alternatively, during periods when the turbine's pitch operates in consonance with the radar-forecasted pitch schedule, the level of variability in the outputted generator speed and power time histories was greatly reduced. These results assert that turbine performance can be enhanced through the development of radar-derived inflow forecasts to allow for advanced proactive wind turbine control.

KEYWORDS

TTUKa Doppler Radar, Dual-Doppler, Spatial Mapping of Boundary Layer, Proactive Wind Turbine Control

13th Americas Conference on Wind Engineering
Gainesville, Florida, USA
May 21 - 24, 2017

Optimize the Wind Farm Layout Using a Modified Genetic Algorithm

Jianxiao Hu[a], Qingshan Yang[a], Jian Zhang[a]

[a] Institution, Beijing's Key Laboratory of Structural Wind Engineering and Urban Wind Environment,

School of Civil Engineering, Beijing Jiaotong University, Beijing, China. 10274063@bjtu.edu.cn (J.H.),

qshyang@bjtu.edu.cn(Q.Y.), zhangjian@bjtu.edu.cn(J.Z)

ABSTRACT

As the most important method to harvest wind energy, wind farm has been installed worldwide in various scales and forms. However, the cluster of wind turbines causes the problem that the upstream turbines extract the kinetic energy from the coming flow while diminish the power generation of the downstream turbines, which is also known as the wake effect. Existing studies showed that the wake effect causes a 10%-20% loss of the total power generation in normal wind farms. To reduce the loss caused by the wake effects, the number and the position of the turbine should be carefully studied. More turbine means higher total power generation with lower marginal cost, nevertheless, it also leads reduction in efficiency of the wind farm as a whole. Sophisticated arrangement of the wind turbine should balance the power output and the cost and yield the highest profit. And this procedure is known as wind farm layout optimization.

Genetic algorithm (GA) is the most commonly used method to deal with the wind farm layout optimization problem. In GA, the coding and programming method heavily influence the optimizing procedure and results. In this study, a modified genetic algorithm is proposed. The initial population and the mutation operator are modified to account for the Boolean property of the code. With the proposed method, a 2km×2km wind farm with flat terrain is optimized under three different wind scenarios and the optimization results are compared with previous studies. The comparisons show that the proposed method is of high accuracy and can reduce the cost per power production efficiently.

KEYWORDS

Wind farm layout, Genetic algorithm, Boolean code, Initial population, Mutation

The 13th Americas Conference on Wind Engineering (13ACWE)
Gainesville, Florida USA, May 21-24, 2017

Wind Fields during the Evening Transition Period and their Influence on Wind Turbine Loads

N.-Y. Lu[a], S. Basu[b], L. Manuel[c]

[a]*University of Texas at Austin, Dept. of Civil, Arch. and Env. Eng., Austin, TX, USA,* nylu@utexas.edu
[b]*Delft University of Tech., Fac. of Civil Eng. and Geosc., Delft, The Netherlands,* s.basu@tudelft.nl
[c]*University of Texas at Austin, Dept. of Civil, Arch. and Env. Eng., Austin, TX, USA,* lmanuel@utexas.edu

ABSTRACT: The late afternoon hours in the diurnal cycle precede the development of the nocturnal stable boundary layer. This so-called evening transition (ET) period is often when energy demands peak. To capture physical characteristics of the wind fields in the atmospheric boundary layer (ABL)—particularly the interplay of shear and turbulence—during this evening transition period, stochastic simulation approaches though inadequate are commonly used to assess wind turbine loads because they are more tractable. Large-eddy simulation (LES), on the other hand, may be used to generate high-resolution ABL turbulent flow fields. We present a suite of idealized LES 4-dimensional flow fields that define a database representing different combinations of large-scale atmospheric conditions (characterized by, for instance, geostrophic wind) and surface boundary conditions (characterized by surface heat flux). Our ultimate objective is to evaluate the performance of wind turbines during the ET period. Accordingly, we conduct a statistical analysis of turbine-scale variables (including hub-height mean wind speed, standard deviation of wind speed in the streamwise direction, and wind speed shear) and discuss patterns of inter-relationships among them. Surface heat fluxes are seen to greatly influence wind shear. Finally, we apply the database of LES-based inflow wind fields in aeroelastic simulation of a 5-MW wind turbine model. We discuss how turbine loads change as the wind field evolves during the evening transition. We also discuss extreme and fatigue loads on the rotor and tower that result from different ABL conditions.

KEYWORDS: Atmospheric boundary layer; evening transition; inflow; large-eddy simulation; turbulence; wind turbine loads

1 INTRODUCTION

In recent years, wind energy has emerged as one of the fastest growing renewables; with this growth, some challenges for further development have become evident. For example, contemporary design standards—from, for example, the International Electrotechnical Commission (IEC) [1]—for turbine inflow generation do not cover the realistic dynamics and physics in the atmospheric boundary layer (ABL). Moreover, the failures of turbines in service have drawn the attention of engineers especially due to the fatigue-sensitive composite materials used for blades. Such challenges served as motivation for us to propose generating a suite of LES-ABL simulations and associated turbine loads evaluations that take into consideration inflow wind field spatial structure and characteristics that reflect a wide range of atmospheric conditions.

The evening transition, defined as a change over time in atmospheric stability from a convective boundary layer to a neutral boundary layer, has a direct influence on the structure of the nocturnal boundary layer and the strength of low-level jets (LLJs). This transitional boundary layer has drawn the interest of meteorologists for many years [2-5]; it has been found that large-scale forcings (e.g., geostrophic winds [2]) as well as surface boundary conditions (e.g., surface heat fluxes [3]) play important roles in the evolution of these transitional wind fields. Additionally, power production would be impacted in these periods of enhanced electricity demands common during the evening transition [6]. As such, there is interest in an investigation of the characteristics of these transitional wind fields and of associated assessment of wind turbine loads. Large-eddy simulation (LES), because of its relatively low computational costs for ABL simulations, has been successfully applied in the study of transitional wind fields in previous studies [7-12]. It is widely accepted that LES can describe physical details of atmospheric turbulence more realistically than conventional stochastic simulation approaches (such as TurbSim [13]) and can thus be expected to simulate evening transition wind fields also more accurately, thus allowing LES-generated inflow wind fields to be used in the estimation of turbine loads during the ET period.

In the following paragraphs of Section 1, we introduce the framework for the LES-ABL computations and the process of inflow field generation for the assessment of turbine loads. Next, Section 2 summarizes the statistical analyses of ET wind fields with emphasis on the relations among turbine-scale wind field variables and the influence of larger scale "external" variables on them. Turbine load statistics are discussed in Section 3 with an explanation of how the external variables influence both fatigue and extreme turbine loads. Finally, we present the most significant findings and conclusions from this study and discuss how this work lays the foundation for prospective research, including the simulation of downbursts which are common in the late-afternoon period during or after the ET period.

1.1 *LES-ABL computations*

LES is employed to generate wind velocity fields in the evening transitional boundary layer. We select several combinations of large-scale environmental conditions and boundary conditions as input parameters in order to generate the wind fields for the convective boundary layer (CBL). An in-house simulation code is utilized to generate the transitional wind fields with the following initialization settings:

$$
\begin{aligned}
U(z) &= G && \text{for } z_0 < z \\
V(z) &= 0 && \text{for } z_0 < z \\
W(z) &= 0 && \text{for } z_0 < z \\
\Theta(z) &= 300 && \text{for } z_0 < z \leq H_i \\
\Theta(z) &= 300 + N(z - H_i) && \text{for } z > H_i
\end{aligned}
\tag{1}
$$

where $U(z)$, $V(z)$, and $W(z)$ are velocity (m/s) components in the longitudinal, lateral and vertical directions, respectively, at height z (m) above ground; G is the geostrophic wind speed (m/s); $\Theta(z)$ is potential temperature (K); z_0 is the aerodynamic roughness length (m); H_i is the initial boundary layer height (m); and N is the inversion strength (K/m). To generate turbulence, Gaussian perturbations with standard deviations of 0.1 m/s and 0.1 K are added to the vertical velocity and potential temperature fields, respectively. The initialized fields are governed by Navier-Stokes equations, with the constraints of the large-scale forcing terms and boundary conditions, to generate a 2-hour LES-based wind velocity field. The first hour of spin-up is discarded to guarantee a well-developed inertial range in the spectra of the simulated flow fields.

Among all the LES inputs, we focus on two external variables—geostrophic wind speed G (m/s) and surface heat flux Hp (K-m/s)—as we investigate characteristics of the ET wind fields and associated turbine loads. Table 1 shows 12 combinations of different G and Hp values used for the LES settings in this study. Run No. 4 (denoted as "control" in the table) is selected as a baseline for comparison against other cases in subsequent discussion. The geostrophic wind speed

G is a constant forcing value for each LES run, while the surface heat flux Hp serves to establish a lower boundary condition for the CBL and has a constant value for the first hour and decays to zero using a cosine function over the second hour. Other external variables are also required for the LES model but these are held constant (e.g., the aerodynamic roughness length z_0 is set as 0.1 m; the initial boundary layer height H_i is 500 m; the inversion strength N is 0.003 K/m; and the Coriolis frequency f is 10^{-4} sec^{-1}) thus ruling out their effects in this study.

In the LES computations, the domain size is set as 10,000 m × 10,000 m × 2,000 m using a 360 × 360 × 72 grid. A time step of 0.25 sec is selected; thus, there are a total of 28,800 time steps for each 2-hour run. In order to generate inflow fields for the wind turbine model, a 2D spatio-temporal field with dimensions 360 (y) × 72 (z) × 240 (t) is extracted from the middle cross-section (y-z plane) of the full 3D domain output. These 2D data serve as the inflow input for the aeroelastic simulation that yield turbine loads.

Table 1. List of LES runs with different combinations of external variables

Run No.	1	2	3	4 (**control**)	5	6	7	8	9	10	11	12
G (m/s)	10	10	10	12	12	12	15	15	15	18	18	18
Hp (K-m/s)	0.00	0.01	0.05	0.00	0.01	0.05	0.00	0.01	0.05	0.00	0.01	0.05

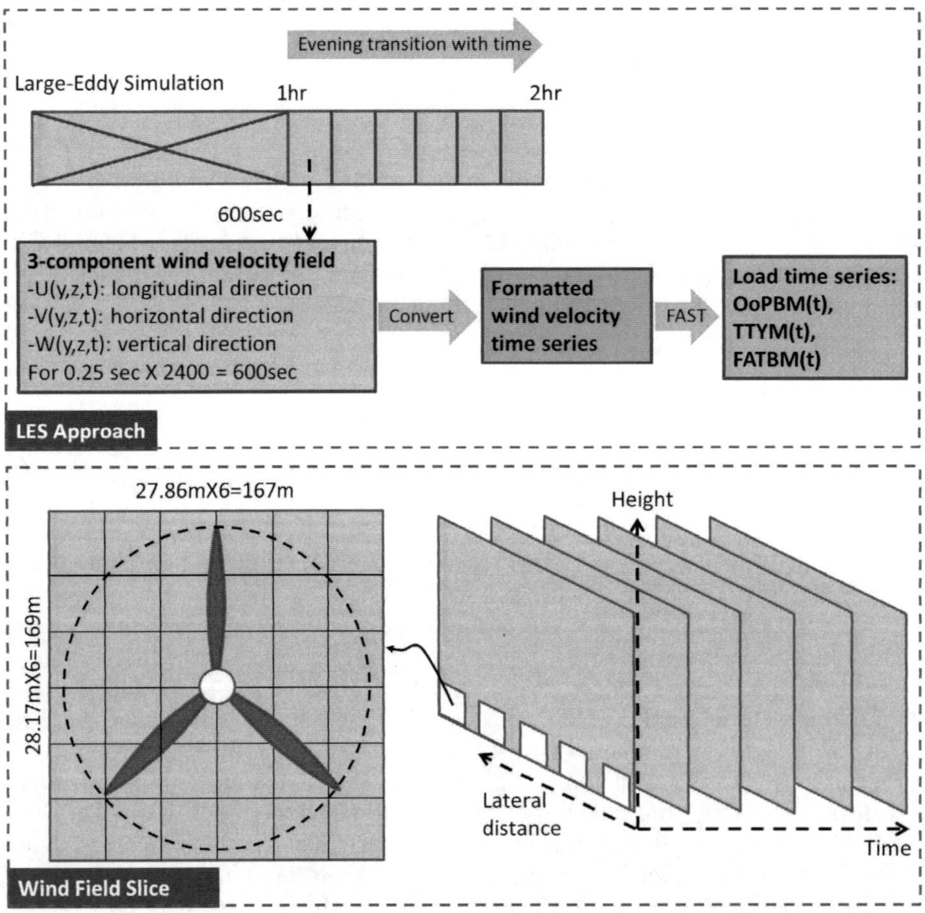

Figure 1. Temporal (top) and spatial (bottom) segmentations of LES wind velocity fields.

1.2 *Turbine-scale inflow generation*

Figure 1 schematically represents the framework of the LES-ABL computations and describes details related to turbine-scale inflow generation for loads assessment. After obtaining the 2D LES-based wind velocity fields, we extract 5 non-overlapping flow field slices from the bottom of the 2D domain along the lateral direction. These field slices are selected so as to cover a vertical ranging from z = 14 m to z = 183 m because the blade tips of the selected 5-MW turbine model span a distance ranging from 27 m to 153 m above ground level. Adjacent 2D field slices are separated by a distance equal to at least five times the rotor radius so as to limit correlation in the wind field characteristics of the individual slices. As stated, the first hour during spin-up in the LES computation is discarded; the time series representing the second hour are separated into six 10-minute segments that serve as distinct flow fields for subsequent analyses. As such, 30 sets of flow fields are generated for each of the 12 LES cases (Table 1), yielding a total of 360 flow fields that form the ET database for wind field statistical analysis and inflow for turbine loads assessment.

1.3 *Turbine model for load simulations*

In this study, the NREL 5-MW Onshore Baseline Wind Turbine model [14] is used to carry out all the turbine load response simulations. The model captures the characteristics of a conventional three-bladed upwind turbine with a variable-speed and collective pitch-control system. Selected properties and dimensions of the turbine model are shown in Table 2. The open-source program, FAST [15], for aeroelastic simulation is utilized to obtain turbine load time series resulting from the selected inflow fields. These time series of simulated loads on various components of the turbine are used in the analysis of both fatigue and extremes.

Table 2. Selected wind turbine model properties and dimensions

Properties and dimensions	Values
Power rating	5 MW
Rotor orientation	Upwind
Rotor configuration	3 blades
Rotor diameter	126 m
Hub height	90 m
Cut-in, rated, and cut-out wind speeds	3 m/s, 11.4 m/s, and 25 m/s
Cut-in and rated rotor speeds	6.9 rpm and 12.1 rpm
Rotor mass	110,000 kg
Nacelle mass	240,000 kg
Tower mass	347460 kg

1.4 *Fatigue and extreme loads*

We are interested in both extreme and fatigue loads during the ET period and accordingly undertake a statistical analysis of these loads using all the 360 FAST simulations. Extreme loads (ExL) refer to the largest value of the selected load measure that occurs in a 10-minute time series while the notion of a 1,000-cycle "equivalent fatigue load" (EFL) is used when considering fatigue loads. The EFL is defined as follows:

$$\text{EFL} = \sqrt[m]{\sum_{i=1}^{N} \frac{S_i^m}{1000}} \qquad (2)$$

By this definition, the implication is that the accumulated damage due to 1,000 cycles of stress cycles of constant amplitude, EFL, is the same as that due to the N cycles of variable amplitude, S_i, that are simulated in each 10-minute time series (also, m is the Wöhler exponent taken to be 10 for carbon-fiber composite). This convenient definition for fatigue loads is such that a 10-minute load time series that places more fatigue demand on a turbine component due to simulated S_i levels will lead to higher EFL values; also, the choice of 1,000 cycles is arbitrary and does not affect conclusions drawn. The particular choice of 1,000 refers to stress amplitude cycles with an average rate of about 5 cycles every 3 seconds.

2 CHARACTERISTICS OF LES-BASED WIND FIELDS

2.1 *Turbine-scale variables*

We compute turbine-scale flow field statistics from the 360 LES wind velocity time series to investigate characteristics of the ET. Several turbine-scale statistics during the ET period are considered—of greatest interest for wind turbine loads assessment are mean wind velocity at hub height in the longitudinal (streamwise) direction, U_h, the standard deviation of the wind speed components at hub height in the longitudinal direction, σ_U, and the estimated wind shear exponent, α, assuming a power law wind shear profile. We are interested in the inter-relationships among these parameters. In Fig. 2, statistics extracted from the 360 flow fields are summarized so that all three of the important turbine-scale statistical parameters may be studied together. The parameters, α and σ_U, are normalized to have zero mean and unit standard deviation by using the following derived variables:

$$\mathrm{z}_\alpha = \frac{\alpha - m_\alpha}{s_\alpha} \qquad (3)$$

$$\mathrm{z}_{\sigma U} = \frac{\sigma_U - m_{\sigma U}}{s_{\sigma U}} \qquad (4)$$

where m_α and $m_{\sigma U}$ represent the mean values of the wind shear exponent and the turbulence standard deviation, respectively, while s_α and $s_{\sigma U}$ represent corresponding sample standard deviations of the same parameters. The data in Fig. 2 are separated based on the values of U_h indicated by different scales of gray and different symbols. It is clear that the data are mostly in the upper-right and lower-left corners which suggests that, in the ET flow fields, wind shear and turbulence are positive correlated with each other. It is clear from the figure that there is also positive correlation between mean wind speed and turbulence level.

2.2 *Influence of external flow field variables on turbine-scale variables*

We discuss the influence of large-scale "external" flow field variables, geostrophic wind speed and surface heat flux, on various turbine-scale wind field variables. Figure 3 shows scatter plots of wind field statistics (wind shear and hub-height mean wind speed) for three different surface heat flux values; the data are separated out according to geostrophic wind speed (in m/s). In the plots, *R* indicates the sample correlation coefficient between the two variables, α and U_h. First, it is quite clear that higher geostrophic wind speeds leads to larger hub-height mean wind speeds, and since we have seen that U_h is highly correlated with hub-height turbulence, σ_U, we find that higher geostrophic wind speeds also result in greater turbulence. Second, the data suggest that larger heat flux levels result in stronger correlation between U_h and α. Importantly, then, the extent of the correlation of wind shear with geostrophic wind speed depends on the value of the surface heat flux.

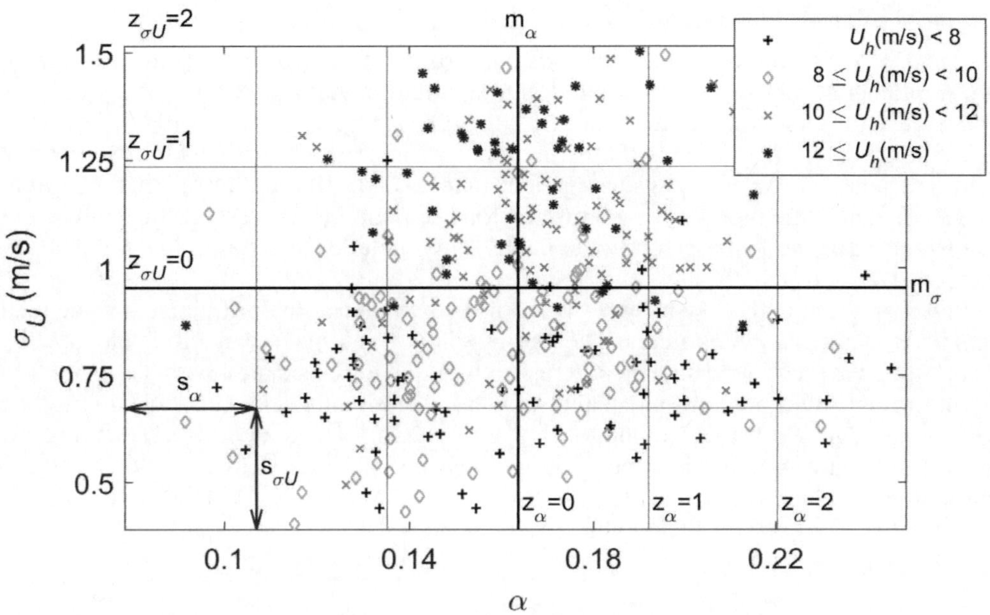

Figure 2. Relationships between wind shear and turbulence classified by U_h based on 360 LES-generated wind fields.

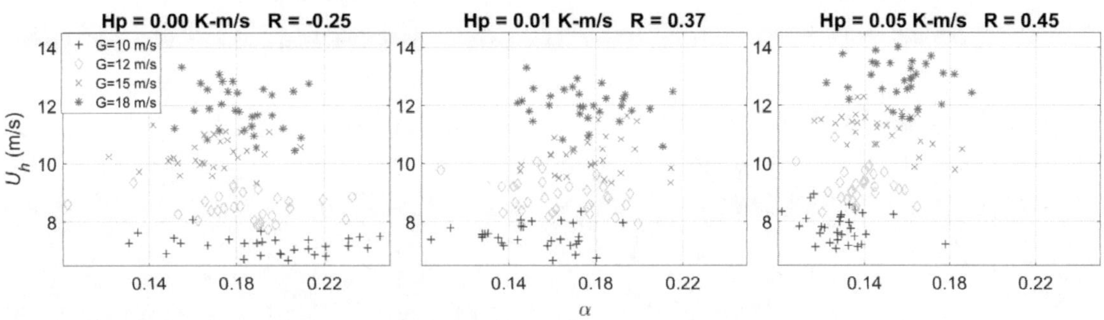

Figure 3. Variation of turbine-scale wind field statistics extracted from 10-minute time series, for different geostrophic wind speeds, with changes in surface heat flux (R denotes the correlation coefficient between the wind shear exponent and hub-height mean wind speed).

3 TURBINE LOADS ASSESSMENT

In the statistical analysis of wind turbine loads, we focus on the following load variables: the blade root out-of-plane bending moment (OoPBM), the fore-aft tower base moment (FATBM), and the tower-top yaw moment (TTYM). These three turbine loads have contrasting characteristics and are often of interest when studying turbine performance under various external conditions such as the evening transition period.

3.1 *Influence of turbine-scale inflow field variables on turbine loads*

We consider a linear regression model relating each turbine extreme or fatigue load statistic to any turbine-scale inflow parameter, x_i, as follows:

$$\text{Turbine Load Statistic} = C_0 + \sum_i C_i x_i \tag{5}$$

where C_0 and C_i are estimated using the loads data from the 360 10-minute simulations All variables are normalized to have zero mean and unit standard deviation. A total of six turbine load statistics (representing fatigue and extreme loads, i.e., ExL and EFL, for OoPBM, FATBM, and TTYM) are studied. Figure 4 shows the estimated coefficients, C_i, for seven different turbine-scale inflow parameters. We use p-values greater than 0.05 to filter out terms that are not appropriate to be included in the model for each turbine load statistic; the corresponding coefficient, C_i, is then set to zero. The model based on Eq. (5), thus, only includes turbine-scale parameters whose p-values are less than 0.05.

From Figure 4, we note that the turbine-scale inflow parameters that influence turbine loads the most are U_h, σ_U, and α. In particular, the hub-height longitudinal mean wind velocity, U_h, has a dominant influence on all load statistics except extreme loads associated with TTYM, where the vertical wind velocity component's mean and standard deviation have large contributions. We also note that σ_U has a greater influence on OoPBM and TTYM loads than on FATBM; this is likely because the tower base moment is affected mostly by the longitudinal wind velocity, U_h. Wind shear (through the parameter, α) mostly affects only OoPBM and TTYM loads and fatigue loads are more directly influenced by shear than extreme loads.

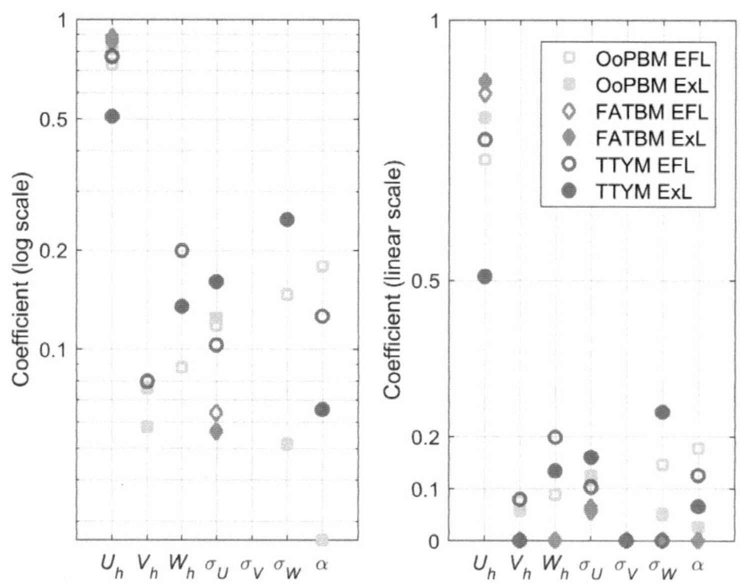

Figure 4. Model coefficients for turbine extreme load (ExL) and fatigue load (EFL) statistics based on regression on normalized turbine-scale inflow parameters.

Results derived from analyses of the turbine loads may be further examined by studying scatter plots of the load statistics and their variation with different turbine-scale inflow field parameters. Figure 5 shows TTYM load statistics (EFL and ExL in MN-m) for different U_h, σ_U, and α values. The gray scale used for the load statistics are an indication of magnitude; the load statistics from all the 360 simulations are separated by groups based on the hub-height mean wind speed, U_h. Generally, for all wind speeds, the data follow a trend going from the lower-left corner (low α, low σ_U) to the upper-right corner (high α, high σ_U) as the load level increases. This indicates that the turbine loads are positively correlated with turbulence and wind shear; however, the influence of wind shear is lower than that of turbulence.

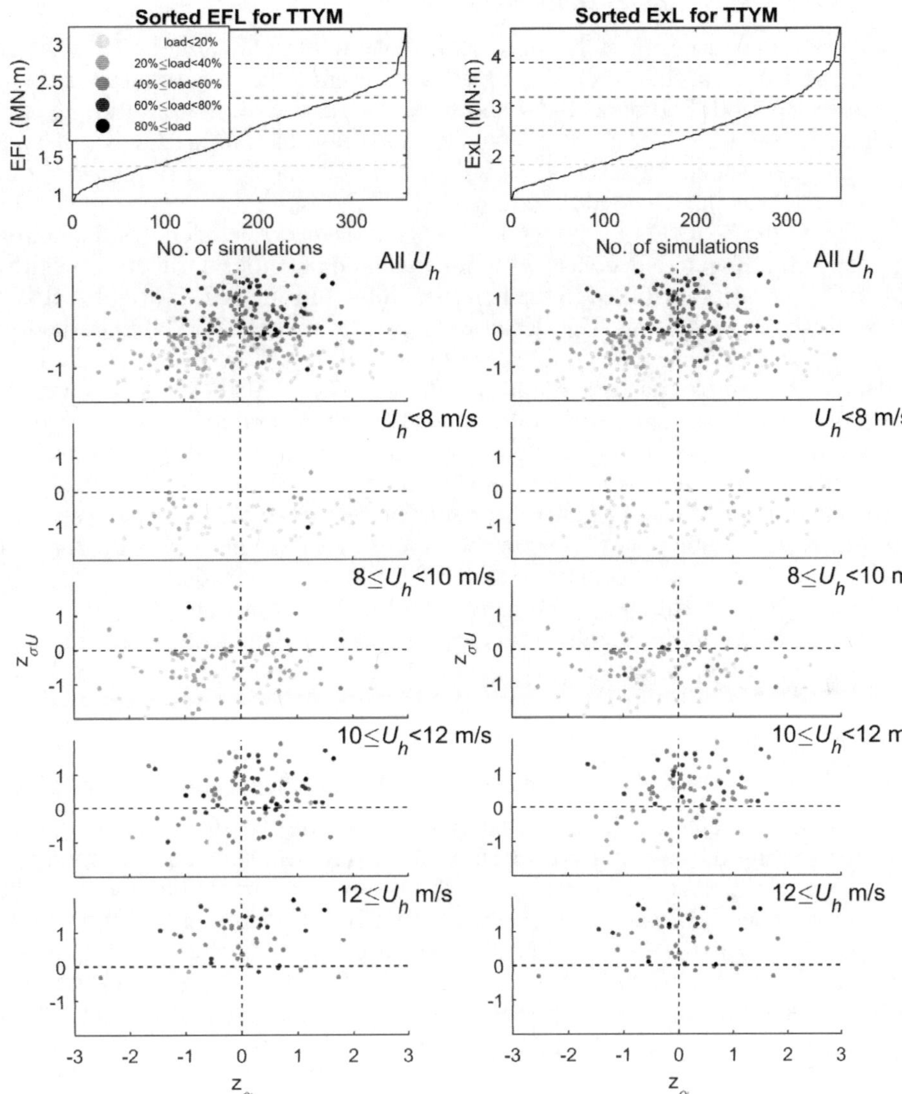

Figure 5. Fatigue and extreme load statistics of TTYM (in MN-m) versus turbine-scale inflow field parameters.

Turbine loads are seen to increase with U_h. Compared with OoPBM and FATBM (not shown), TTYM load are less sensitive to changes in U_h; there are only a few simulations with load values above the 80% level (deep black dots) that results when the mean wind speed, U_h, is greater than 12 m/s. This is because TTYM has a greater influence from the localized vertical wind velocity mean and turbulence (Fig. 4) near the hub and is relatively less affected by U_h. This finding is consistent with conclusions drawn from the regression model where the smallest coefficient of U_h

was estimated for the TTYM extreme load. When U_h exceeds the turbine's rated wind speed of 11.4 m/s, the pitch control system would limit load levels. Since the longitudinal mean wind velocity rarely exceeds the rated wind speed in this study, the effect of pitch control effect does not affect our statistical analysis of loads.

3.2 *Influence of external variables on turbine loads*

We investigate next the influence of the large-scale "external" field variables (i.e., the geostrophic wind speed and surface heat flux) on turbine loads by studying the variation of load statistics with time during the ET period. Figures 6 and 7 show the variation with time of turbine load statistics every 10 minutes during the 1 hour of the evening transition for fatigue loads (left) and extreme loads (right) for four selected LES cases. In Figure 6, we consider different geostrophic wind speeds and hold the heat flux fixed at 0.00 K-m/s, while in Figure 7 the geostrophic wind speed is fixed at 12 m/s and the heat flux is changed. The black dashed lines denotes the control case. In general, all the loads are seen to decrease with time over the one hour during the evening transition period. This finding can be attributed to the fact that all the turbine-scale variables also decay with time over the one-hour simulation. The decay with time is least for TTYM extreme loads that do not change significantly over the evening transition period.

It is evident from Figure 6 that larger loads result with increased geostrophic wind speeds. Considering both fatigue and extremes, the OoPBM loads appear to be most strongly correlated with geostrophic wind speed whereas TTYM loads indicate greater variability and obscure the trend. The influence of heat flux on turbine load statistics (as seen in Figure 7) is relatively smaller than that of geostrophic wind speed. In general, greater heat flux levels may be expected to lead to larger values of load statistics. However, in the case of TTYM and with EFL for OoPBM, this trend is not as clear. There may be an explanation for this if we consider wind shear. In Figure 4, we saw that wind shear contributes to the EFL of OoPBM the most and then to TTYM loads. In addition, increased heat flux levels also reduce wind shear; hence, the two competing effects of flux and shear lead to a less influence of flux levels on turbine load statistics.

4 CONCLUSIONS

In this study, we have created an extensive database of high-resolution (~28 m), four-dimensional flow fields representing the evening transition period using an idealized LES model. The large-scale atmospheric conditions (e.g., geostrophic wind speed) and surface boundary conditions (e.g., heat flux) were systematically varied to establish a database for further statistical analyses. Turbine-scale wind field statistics (such as the hub-height mean wind speed, hub-height turbulence standard deviation, and wind shear) were studied and their mutual correlations are influenced by variations in the external large-scale variables. In addition, we carried out statistical studies of loads on a wind turbine by using the LES-generated flow fields to create the inflow fields for turbine aeroelastic response simulations. A database of both fatigue and extreme load statistics was developed from the output load time series for the selected 5-MW turbine model. A multivariate regression model was developed in order to assess contributions from different turbine-scale wind field variables to the load statistics; these in turn can be used to assess the effects of external variables. It is concluded that the geostrophic wind speed and heat flux both influence turbine loads to different degrees.

Based on the results of this study, the LES-ABL simulation framework is seen to successfully capture characteristics of the evening transition from convective to neutral boundary layer. The wind field and turbine load statistics suggest that the mean wind speed and turbulence as well as associated turbine loads decay over the one-hour period representing the evening transition. These investigations on the wind fields during the evening transition are considered to be fundamental in precursor downburst scenarios that often occur in the late afternoon and early evening. We expect to refine the evaluation of downburst wind field characteristics and associated turbine loads by incorporating appropriate conditions during the evening transition and, thus, further improve the design for wind turbines under these extreme loading scenarios.

5 ACKNOWLEDGEMENTS

The authors gratefully acknowledge the financial support received from the National Science Foundation (NSF) under Grant Nos. 1336304 and 1336760.

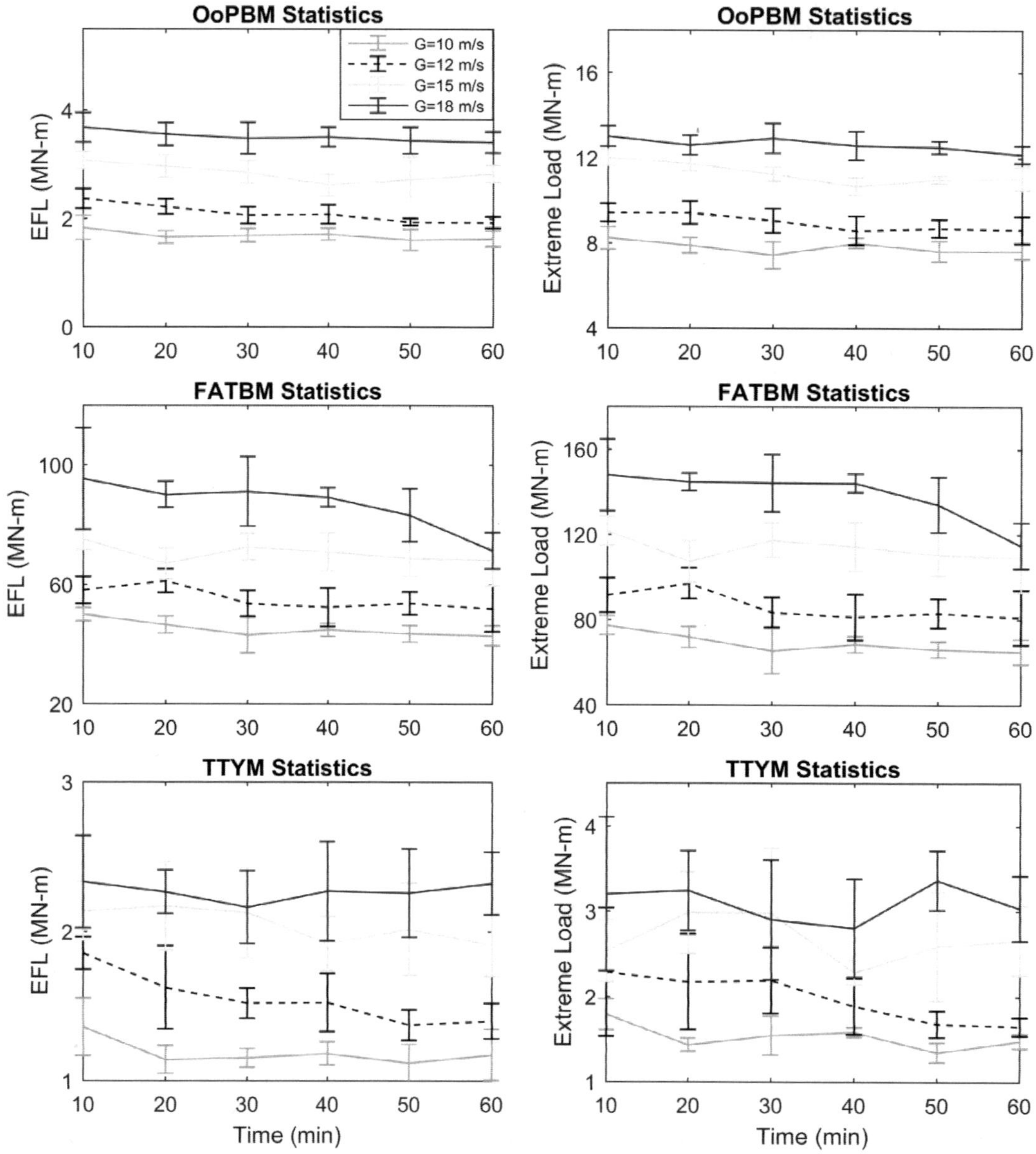

Figure 6. Variation of turbine load statistics with time for different geostrophic wind speeds (*Hp*=0.00 K-m/s).

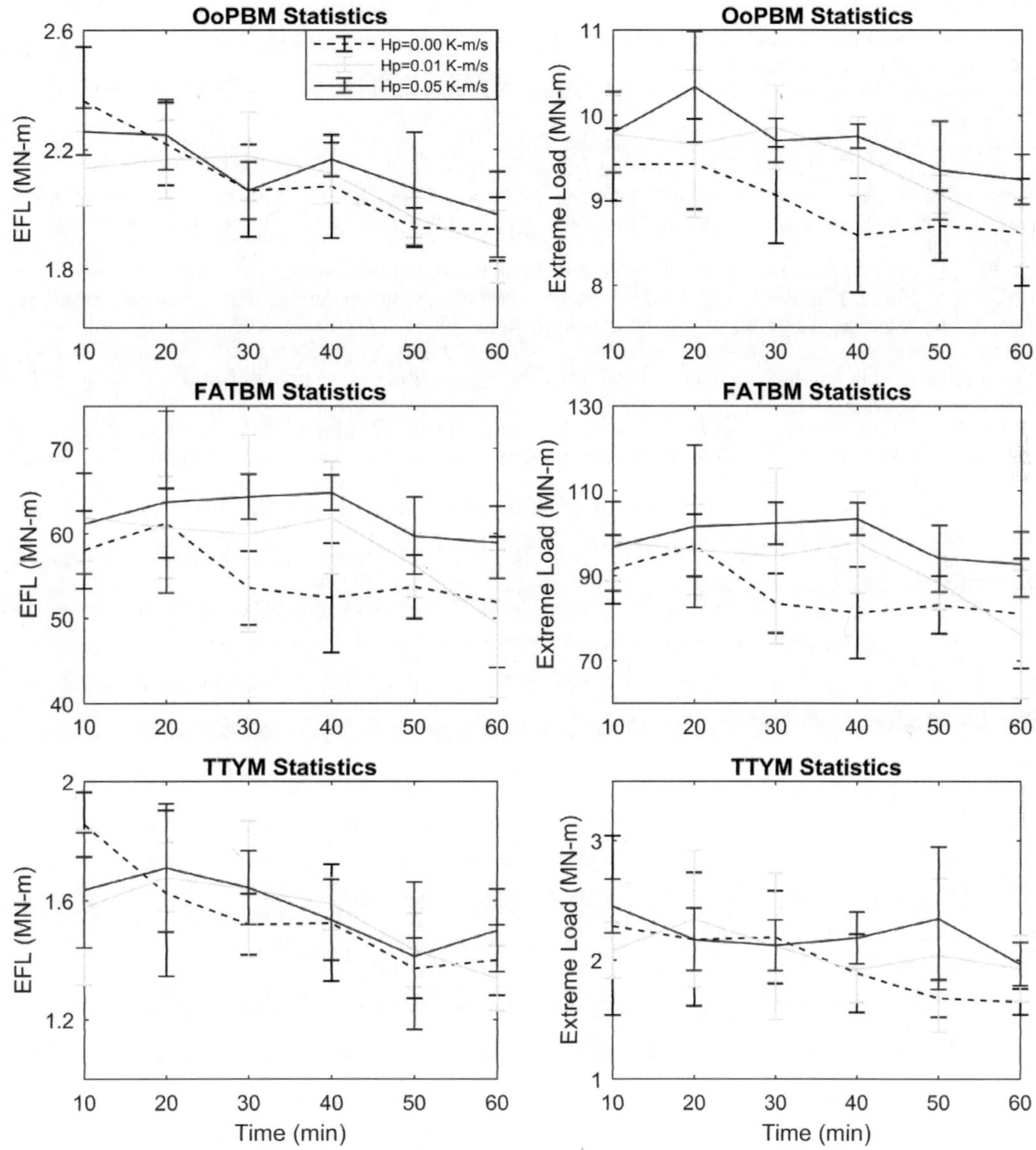

Figure 7. Variation of turbine load statistics with time for different heat fluxes (G=12 m/s).

6 REFERENCES

1 Wind Turbine Generator Systems - Part 1: Design Requirements, 3rd Edition, International Electrotechnical Commission, IEC 61400-1, Geneva, Switzerland, 2005.
2 L. Mahrt, The early evening boundary layer transition, Quarterly Journal of the Royal Meteorological Society, 107, pp. 329-343, 1981.
3 A. Grant, An observational study of the evening transition boundary-layer, Quarterly Journal of the Royal Meteorological Society, 123, pp. 657-677, 1997.
4 A.R. Columbié, Atmospheric Boundary Layer Evening Transitions over West Texas, Ph.D. Dissertation, Texas Tech University, 2008.
5 M. Sastre, C. Yagüe, C. Román-Cascón, Atmospheric Boundary-Layer Evening Transitions: A Comparison Between Two Different Experimental Sites, Boundary-Layer Meteorology, 157, pp. 375-399, 2015.
6 CAISO, What the duck curve tells us about managing a green grid, http://www.caiso.com/, California ISO.
7 F. T. M. Nieuwstadt, R. A. Brost, The decay of convective turbulence. Journal of the Atmospheric Sciences 43 : 532-546, 1986.
8 Z. Sorbjan, Decay of convective turbulence revisited. Boundary-Layer Meteorology 82: 503-517, 1997.
9 D. Pino, H.J. Jonker, J.V.Arellano, A.Dosio, Role of shear and the inversion strength during sunset turbulence over land: characteristic length scales, Boundary-Layer Meteorology, 121, pp. 537-556, 2006.
10 U. Rizza, M. Miglietta, G. Degrazia, O. Acevedo, E. Marques Filho, Sunset decay of the convective turbulence with Large-Eddy Simulation under realistic conditions, Physica A: Statistical Mechanics and its Applications, 392, pp. 4481-4490, 2013.
11 U. Rizza, M. Miglietta, O. Acevedo, V. Anabor, G. Degrazia, A. Goulart, Zimmerman, H., Large-eddy simulation of the planetary boundary layer under baroclinic conditions during daytime and sunset turbulence, Meteorological Applications, 20, pp. 56-71, 2013.
12 B. Jayaraman, J. Brasseur, Transition in Atmospheric Turbulence Structure from Neutral to Convective Stability States, 32nd ASME Wind Energy Symposium, pp. 0868, 2014.
13 B.J. Jonkman, TurbSim user's guide: version 1.50, National Renewable Energy Laboratory, Technical Report NREL/TP-500-46198, Golden, CO, 2009.
14 J.M. Jonkman, S. Butterfield, W. Musical, G. Scott, Definition of a 5MW reference wind turbine for an offshore system development, National Renewable Energy Laboratory, Technical Report NREL/TP-500-38060, Golden, CO, 2007.
15 J.M. Jonkman, M.L. Buhl, FAST User's Guide, National Renewable Energy Laboratory, Technical Report NREL/EL-500-38230, Golden, CO, 2005.

The 13th Americas Conference on Wind Engineering (13ACWE)
Gainesville, Florida USA, May 21-24, 2017

Mitigation of Monopile Offshore Wind Turbines under Wind and Wave Loading

Chao Sun [a], Vahid Jahangiri [a]

[a]*Louisiana State University, Baton Rouge, USA*

ABSTRACT: Mitigation of fixed-bottom monopile offshore wind turbines subjected to wind and wave loading is studied in this paper. A semi-active tuned mass damper (STMD) with tunable natural frequency and damping ratio is utilized to mitigate the tower top dynamic response. Soil effects and damage are considered. Wind turbulence is generated via mapping a three dimensional wind field profile onto the rotating blades. Aerodynamic loading is computed using the Blade Element Momentum method where the Prandtl's tip loss factor and the Glauert correction are considered. Wave loading is computed using Morison's equation together with the strip theory. The National Renewable Energy Lab monopile 5 MW baseline wind turbine model is employed to examine the performance of the STMD. Results indicate that the STMD has significant effectiveness improvement over the passive TMD in mitigating the tower and foundation response under wind and wave loadings when soil effects and damage are considered.

KEYWORDS: Offshore wind turbines; mitigation; soil effects (SE); damage; semi-active control.

1 INTRODUCTION

Due to environmental concerns, wind energy production and consumption are growing remarkably worldwide. Because of the steadier and higher wind speed in the marine area, offshore wind plants are becoming an essential sector for wind energy production. However, due to the combined wind and wave loading and other marine environmental effects, excessive vibration and fatigue load will result, adversely influencing the structural integrity and service. To address this issue, increasing research effort is being exerted to mitigate the response and fatigue load. Structural vibration control, which has been successfully employed in civil structures, is being studied in recent years to protect the offshore wind turbines.

Structural control on wind turbines, especially the offshore wind turbines, is a relatively young field. There are three basic control strategies: passive, semi-active and active [1-3]. Passive device include the tuned mass damper (TMD), the tuned liquid damper (TLD) and the tuned liquid column damper (TLCD). Murtagh *et al.* [4] used a passive TMD to mitigate the along-wind vibration of a wind turbine. Findings indicated that response reduction could be achieved when the TMD was tuned to the predominant frequency. To control both the fore-aft and side-side vibration, Lackner *et al.* [5] used dual passive TMDs placed in the nacelle. Performance of the dual TMDs model were examined through controlling a barge-type floating offshore wind turbine and a fixed-bottom monopile offshore wind turbine. It was found that the passive TMDs can improve the structural response of offshore wind turbines. Colwell *et al.* [6] studied the vibration control of an offshore wind turbine using the TLCD where a joint wind wave model was used to generate correlated wind wave loadings. The authors concluded that implementing TLCDs could decrease the structural costs and prolong the tower life. Although passive control techniques provide reduction when the parameters are finely tuned, they might lose their effectiveness as a result of environmental/system variations.

In comparison, semi-active techniques can track the operational/environmental variations and thus are more applicable to systems with time-variant parameters. Semi-active TMDs (STMDs) have been studied and demonstrated effective in mitigating vibrations of linear and nonlinear systems subjected to stationary and non-stationary excitations [7, 8, 9, 10, 11]. Huang and Arrigan *et al.* [12, 13] explored the mitigation of wind turbine blades using the STMDs retuned in real-time via an STFT based control algorithm. The authors found that the STMDs could significantly mitigate the blade responses under varying operational/environmental conditions. While in Refs [7, 12, 13], only the frequency of the STMDs was retuned yet the damping ratio remained constant. Recently, Sun *et al.* [14] further advanced the STFT-based control algorithm by incorporating retuning of the damping ratio of the STMD in addition to the natural frequency retuning. By doing so, the kinetic energy flow and dissipation rate can be maximized and additional mitigation effect can be achieved. The authors examined the performance of the STMD and the control algorithm for seismic protection and observed effective response reduction.

In addition to semi-active control, active control of wind turbines have been studied and demonstrated effective to overcome the issues caused by system/operational variations. Staino *et al.* [15, 16] studied edgewise vibration control of wind turbine blades using active tendons mounted inside each blade. The authors found that the proposed active control scheme can significantly improve the edgewise response of the blades under aerodynamic loading or with variable rotational speed. Fitzgerald *et al.* [17, 18] utilized active tuned mass dampers (ATMDs) to control the in-plane vibration of wind turbine blades. Results indicated that the ATMDs can achieve greater response reduction than the passive TMDs.

As one of the primary offshore wind turbine options, the monopile fixed-bottom offshore wind turbines are widely employed and under active investigation. In most existing literatures, the monopile offshore wind turbines are assumed to be completely fixed in the sea bed. This assumption, however, is inappropriate when soil is relatively soft. Veletsos *et al.* [19] showed that flexible soil underneath the foundation increases the damping property and reduces the structural natural frequency. Bush *et al.* concluded in [20] that inclusion of the flexible-based foundation can affect the dynamic behavior of the wind turbines. Furthermore, cyclic wind and wave loading can cause damage (stiffness degradation) to the foundation [21] and the structure. The accumulated damage will potentially change the structural natural frequency. As a result, variation of the structural natural frequency due to soil effects (SE) and/or damage will cause detuning of a passive TMD which is tuned to the natural frequency of the original structure. To address this issue, Fitzgerald *et al.* [22] used active TMD (ATMD) to control the response of the wind turbine subjected to wind loading with the SE considered. The authors found that the proposed active control scheme was promising when the SE needs to be included. However, mitigation of the monopile offshore wind turbines under wind, wave loading and the combined SE and damage presence was not studied in Ref. [22].

The present study proposes to use the STMD to mitigate the structural response of the monopile fixed-bottom offshore wind turbines subjected to wind and wave loadings with the SE and damage presence considered. It essentially advances the research reported in Ref. [13]. On one hand, real-time tuning of the STMD includes tuning of both the natural frequency and the damping ratio via a modified control algorithm. On the other hand, the effectiveness of the STMD is examined under combined and more complicated environmental effects, including wind and wave loading, the SE and damage presence. The focus of the present study is to examine the performance of the proposed STMD when the SE and damage presence cause the potential variation of the system dominant response frequency. A mathematical model of the monopile wind turbine coupled with the STMD is established. Dynamic interaction between the blades and the tower and the STMD is modeled. Aerodynamic loading, wave loading and gravity loading are incorporated in the model. To illustrate the effectiveness of the STMD, a passive TMD is used for comparison. Structural responses are calculated via solving the established coupled dynamic equations. Results show that the STFT based tracking algorithm can rapidly capture the variation of structural natural frequency caused by the SE and/or damage presence in the foundation and the tower. As a result, the STMD exhibits excellent effectiveness in mitigating the response while its passive counterpart becomes ineffective when the SE and damage are considered.

2 ESTABLISHMENT OF THE ANALYTICAL MODEL

A fully coupled three dimensional dynamic model with and without the STMD is established using the Euler-Lagrangian equation. The model contains 10 degree-of-freedom (DOF) for the uncontrolled system and 11 DOF for the controlled system. Since the fundamental vibration dominates the dynamic response, the present study establishes the equation of motion using the fundamental mode of the blade and the tower.

2.1 *Model Description*

A general monopile offshore wind turbine model subjected to the combination of wind and wave loading is illustrated in Fig. 1. As shown in Fig. 1, the global coordinate system originates at the intersection of the tower center line and the mean sea level (MSL). The left plot of Fig. 1 shows the original model and right plot shows the deformed geometry of the wind turbine under wind and wave loading. As described in the Introduction section, the SE are considered and represented by a translational spring k_x and a rotational spring k_ϕ. The damping property of the soil are also considered by introducing two dash-pot parameters c_x and c_ϕ. Parameters q_9 and q_{10} in Fig. 1 denote the translation and rotation coordinates of the foundation.

In present study, parameters $q_1 \sim q_3$ are defined to denote the edgewise coordinates of the three blades and $q_4 \sim q_6$ denote the flapwise coordinates. Parameters q_7 and q_8 denote the relative coordinates of the nacelle in the fore-aft and side-side directions with respect to the foundation. Parameter q_{11} denotes the relative coordinate of the STMD with respect to the nacelle. The blades rotating speed is Ω and the azimuthal angle $\psi_j(t)$ (see Fig. 2) of the jth blade can be expressed as:

$$\psi_j = \Omega t + \frac{2\pi}{3}(j-1),\ j = 1,2,3 \tag{1}$$

It is worthy to note that the edgewise and flapwise vibration are essentially coupled due to the geometry of the blade. This coupling effect can be modeled based on the pre-twist angle when analyzing the dynamics of the blade [23]. In the present study, the coupling effect is not included because the focus of this study is to evaluate the performance of the STMD in mitigating the response of the nacelle and the tower where the coupling effect of edgewise and flapwise vibration has minimal effect. Hence, they are assumed to be independent for simplification.

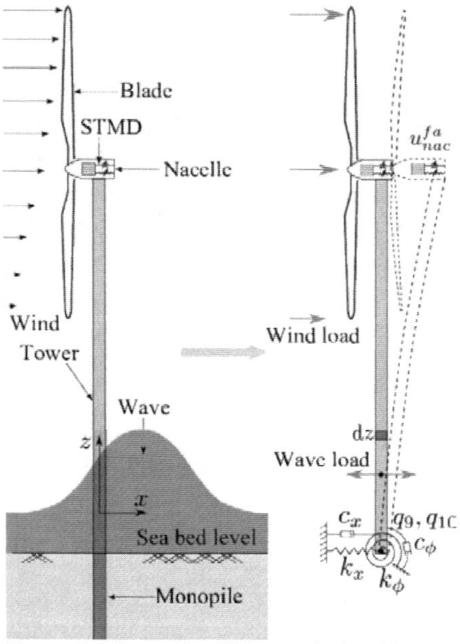

Figure 1. Monopile offshore wind turbine coupled with an STMD under wind and wave loading. Left: original schematic model; right: simplified deformed model for analysis.

2.2 *Euler-Lagrangian Equation*

The Euler-Lagrangian equation is used to derive the analytical model of the monopile offshore wind turbine system. The Euler-Lagrangian equation is listed as follows:

$$\frac{d}{dt}\frac{\partial T(t,\tilde{q}(t),\dot{\tilde{q}}(t))}{\partial \dot{q}_i(t)}-\frac{\partial T(t,\tilde{q}(t),\dot{\tilde{q}}(t))}{\partial q_i(t)}+\frac{\partial V(t,\tilde{q}(t))}{\partial q_i(t)}=Q_i(t) \tag{2}$$

where T and V are the system kinetic and potential energy, $\tilde{q}(t)$ is the generalized coordinates vector, $Q_i(t)$ is the generalized force corresponding to the ith component of $\tilde{q}(t)$. Sign $(\dot{\ })$ denotes the first derivative with respect to time.

2.3 *Kinetic Energy*

In order to represent the motion of the blade, a local coordinate system $x'y'z'-o'$ originating at the center of the hub front surface (static state) is established as shown in Fig. 2 where the nacelle has a side-side displacement u_{nac}^{ss} and a fore-aft displacement u_{nac}^{fa}. Displacement of the jth blade is illustrated in Fig. 2. Variables $u_{je}(r,t)$ and $u_{jf}(r,t)$ represent the edgewise and flapwise displacements of the point at a distance r from the blade root; variables f_{1e} and f_{1f} denote the edgewise and flapwise fundamental mode shape. The motion of the nacelle in z direction is minimal and ignored.

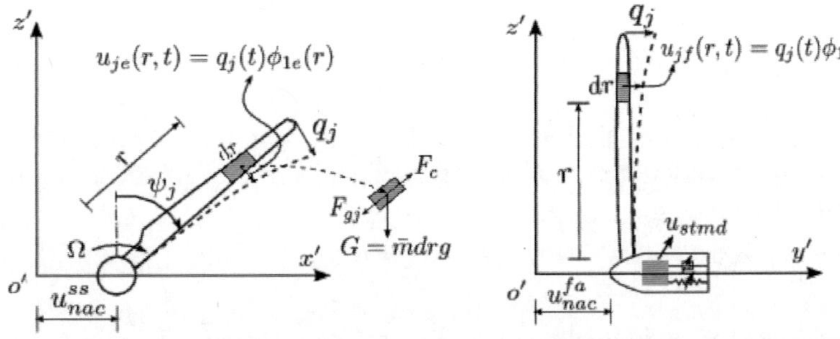

Figure 2. Displacements of the nacelle and the jth blade.

The absolute velocity of the nacelle in the fore-aft direction v_{nac}^{fa} and the side-side direction v_{nac}^{ss} can be expressed as:

$$v_{nac}^{fa} = \dot{q}_7 + \dot{q}_9 + h\dot{q}_{10} \qquad v_{nac}^{ss} = \dot{q}_8 \tag{4}$$

The resultant absolute velocity of the nacelle v_{nac} can be written as:

$$v_{nac} = \sqrt{(v_{nac}^{fa})^2 + (v_{nac}^{ss})^2} \tag{5}$$

Similarly, the absolute velocity of the STMD v_{stmd} can be represented as:

$$v_{stmd} = v_{nac}^{fa} + \dot{q}_{11} = \dot{q}_7 + \dot{q}_9 + h\dot{q}_{10} + \dot{q}_{11} \tag{6}$$

Consider an infinitesimal unit dr of the jth blade in Fig. 2, the coordinate of the unit in the $x'y'z'$- o' system can be represented as:

$$\begin{aligned} x'_r &= u_{nac}^{ss} + r\sin\psi_j + u_{je}\cos\psi_j \\ y'_r &= u_{nac}^{fa} + u_{jf} \\ z'_r &= r\cos\psi_j - u_{je}\sin\psi_j \end{aligned} \tag{7}$$

Taking the first derivative of the coordinates gives the velocity component as:

$$\begin{aligned} \dot{x}'_r &= v_{nac}^{ss} + \Omega r\cos\psi_j + \dot{q}_j\phi_{1e}\cos\psi_j - \Omega q_j\phi_{1e}\sin\psi_j \\ \dot{y}'_r &= v_{nac}^{fa} + \dot{q}_{j+3}\phi_{1f} \\ \dot{z}'_r &= -\Omega r\sin\psi_j - \dot{q}_j\phi_{1e}\sin\psi_j - \Omega q_j\phi_{1e}\cos\psi_j \end{aligned} \tag{8}$$

The absolute velocity magnitude of the unit dr is:

$$v_{bj}(r,t) = \sqrt{\dot{x}'^2_r + \dot{y}'^2_r + \dot{z}'^2_r} \tag{9}$$

As shown in Fig. 1, the absolute velocity $v_{tow}(z,t)$ of an infinitesimal unit dz of the tower can be expressed as:

$$v_{tow} = \sqrt{\left[\dot{q}_8\phi_{1t}\right]^2 + \left[\dot{q}_7\phi_{1t} + \dot{q}_9 + z\dot{q}_{10}\right]^2} \tag{10}$$

where ϕ_{1t} denotes the fundamental mode shape (side-side and fore-aft) of the tower in the side-side and fore-aft directions.

Therefore, the kinetic energy of the entire wind turbine system can be expressed as:

$$T = \frac{1}{2}\sum_{j=1}^{3}\int_0^R \bar{m} v_{bj}^2(r,t)dr + \frac{1}{2}M_{nac}v_{nac}^2 + \frac{1}{2}\int_0^h \bar{M} v_{tow}^2 dz + \frac{1}{2}M_f \dot{q}_9^2(t) + \frac{1}{2}I_f \dot{q}_{10}^2(t) + \frac{1}{2}m_s v_{stmd}^2(t) \tag{11}$$

where M_{nac} is the mass of the nacelle (including the hub mass); M_f and I_f denote the mass and moment of inertial of the foundation; $\bar{m}$ and $\bar{M}$ denote the mass density per length of the blade and the tower; m_s denotes the mass of the STMD.

2.4 *Potential Energy*

Strain energy due to shear deformation is neglected in this study. The total potential energy of the blades is calculated considering the strain energy of the blades due to bending, the centrifugal stiffening effect and the gravity related potential energy.

As shown in Fig. 2(a), the centrifugal force F_c and the gravity force component F_{gj} acting along the jth blade at a distance r from the hub can be expressed as:

$$F_c(r) = \Omega^2 \int_r^R \bar{m}(x) x dx \tag{12}$$

where x denotes the axial distance of the current considered element from the hub.

$$F_{gj}(r) = -\int_r^R \bar{m}(x) g \cos\psi dx = -g\cos\psi \int_r^R \bar{m}(x) dx \tag{13}$$

Therefore, the total potential energy of the blades can be written as:

$$\begin{aligned} V_b &= \frac{1}{2}\sum_{j=1}^{3}\int_0^R \left[EI_{eg}(r)(u''_{je})^2 + EI_{fp}(r)(u''_{jf})^2 + (F_c + F_{gj})(u'_{jf})^2\right]dr \\ &= \frac{1}{2}\sum_{j=1}^{3}\left[(k_{eg} + k_{ge,eg} - k_{gr,eg}\cos\psi_j)q_j^2 + (k_{fp} + k_{ge,fp} - k_{gr,fp}\cos\psi_j)q_{j+3}^2\right] \end{aligned} \tag{14}$$

where

$$\begin{aligned} &k_{eg} = \int_0^R EI_{eg}(r)(\varphi''_{1e})^2 dr,\ k_{fp} = \int_0^R EI_{fp}(r)(\varphi''_{1f})^2 dr, \\ &k_{ge,eg} = \Omega^2 \int_0^R \left[\int_r^R \bar{m}(x) x dx\right](\varphi'_{1e})^2 dr,\ k_{ge,fp} = \Omega^2 \int_0^R \left[\int_r^R \bar{m}(x) x dx\right](\varphi'_{1f})^2 dr, \\ &k_{gr,eg} = g\int_0^R \left[\int_r^R \bar{m}(x) x dx\right](\varphi'_{1e})^2 dr,\ k_{gr,fp} = g\int_0^R \left[\int_r^R \bar{m}(x) x dx\right](\varphi'_{1f})^2 dr \end{aligned} \tag{15}$$

I_{eg} and I_{fp} are the moment of inertia in the edgewise and flapwise direction, Ω is the turbine rotating speed, g is the acceleration due to gravity. Sign $()'$ and $()''$ denote the first and second derivatives with respect to length r.

Hence, the total potential energy V of the wind turbine system is:

$$V = V_b + \frac{1}{2}k_t^{fa} q_7^2(t) + \frac{1}{2}k_t^{ss} q_8^2(t) + \frac{1}{2}k_x q_9^2(t) + \frac{1}{2}k_\varphi q_{10}^2(t) + \frac{1}{2}k_s q_{11}^2(t) \tag{16}$$

where k_t^{fa} and k_t^{ss} denote the fore-aft and side-side stiffness of the tower, k_x and k_φ denote the translational and rotational stiffness of the foundation, k_s denotes the stiffness of the STMD.

2.5 *System Equation of Motion*

Substituting Eqn.11 and Eqn.16 into Eqn.2 yields the equation of motion of the wind turbine, which can be written in a matrix form as:

$$\tilde{M}\ddot{\tilde{q}} + \tilde{C}\dot{\tilde{q}} + \tilde{K}\tilde{q} = \tilde{Q}_{wind} + \tilde{Q}_{wv} \tag{17}$$

where $\tilde{M}$, $\tilde{C}$ and $\tilde{K}$ are the system mass, damping and stiffness matrices. Two sets of system matrices for the uncontrolled and controlled systems are derived. Variables $\tilde{Q}_{wind}$ and $\tilde{Q}_{wv}$ are the generalized force vectors corresponding to the wind and wave loading, which are derived in the following section.

2.6 *Validation of the Model*

The established model is benchmarked using the open source program FAST developed by Jonkman et al. [24]. The NREL 5MW baseline OC3 monopile offshore wind turbine whose properties are presented in Section 5.1 is utilized as a reference model. Natural frequencies of the reference wind turbine are calculated using FAST and the established model separately and compared. It is noted that linearization is implemented using FAST and then the natural frequencies are obtained via eigen analysis with respect to the linearized state space matrices. In comparison, the natural frequencies obtained from the established model are directly computed through eigen analysis with respect to the matrices in Eqn. (17). The obtained natural frequencies from the two approaches are compared in Table 1 where one can find that the established model predicts the natural frequencies with high accuracy. Therefore, the established model in the present study is valid for further application.

Table 1. Comparison between the established model and FAST

	Tower fore-aft (Hz)	Tower side-side (Hz)	Blade flapwise (Hz)	Blade flapwise (Hz)
Established model	0.302	0.305	0.747	1.100
FAST	0.318	0.320	0.744	1.103

3 LOADING

This section presents the derivation of the generalized force vectors $\tilde{Q}_{wind}$ and $\tilde{Q}_{wv}$ corresponding to the wind and wave loading based on the Principles of Virtual Work. Wind speed time-history is generated for each of the blade element via a three dimensional wind field model. The aerodynamic loading is calculated based on the BEM method. Linear wave and the strip theory are used to calculate the wave loading. Details are illustrated in the following subsections.

3.1 *Wind Turbulence Model*

Wind velocity $v(z,t)$ can be represented by the summation of a constant mean velocity $\overline{v}(z)$ and a turbulent component $\tilde{v}(z,t)$, i.e.

$$v(z,t) = \overline{v}(z) + \tilde{v}(z,t) \tag{18}$$

In the present study, the logarithmic wind profile is adopted to calculate the mean velocity $\overline{v}(z)$, i.e.

$$\overline{v}(z) = V_{ref}\frac{\log(z / z_0)}{\log(H_{ref} / z_0)} \tag{19}$$

where V_{ref} is the mean velocity at the reference height $H_{ref} = 90m$. Parameter z_0 is the length of roughness and its value is $z_0 = 0.03$.

The turbulent component of wind velocity $\tilde{v}(t)$ is computed using the IEC Kaimal spectral model which is described by the following equations:

$$S_v(f) = \frac{4I^2 L_c}{(1 + 6fL_c / \bar{v})^{5/3}} \tag{20}$$

where $S_v(f)$ is the power spectral density function, f is the wind frequency in Hz, I is the wind turbulence intensity and L_c is an integral scale parameter.

To account for the spatial dependency of wind velocity $\tilde{v}(z,t)$ at different points, the cross spectra between two points i and j are defined as:

$$S_{ij}(f) = Coh(i,j;f)\sqrt{S_{ii}(f)S_{jj}(f)} \tag{21}$$

where S_{ij} is the cross spectra, S_{ii} and S_{jj} are the auto spectra at points i and j, respectively. Referring to the IEC spectral mode [25], the spatial coherence function is given as:

$$Coh(i,j;f) = \exp\left(-a\sqrt{\left(\frac{fL}{\bar{v}_{hub}}\right)^2 + \left(\frac{0.12L}{L_c}\right)^2}\right) \tag{22}$$

where L is the distance between points i and j on the grid, a is the coherence decrement, and L_c is a coherence scale parameter, $\bar{v}_{hub}$ is the mean hub-height wind speed. In terms of IEC-64000-1 3rd, $a = 12$ and $L_c = 340.2m$ are adopted in the present study.

Based on Eqns. (19~22), a three dimensional wind field profile covering the domain of the rotor disk is generated using the TurbSim program [25]. Matlab code has been developed to map the full wind field profile onto each span station of the rotating blades.

In the present study, mean velocity is $\bar{v} = 20m/s$ at the hub height and turbulence intensity $TI = 0.18$ is used. As an illustration, the wind velocity time history at the center of the tip element of blade 1 is shown in Fig. 3.

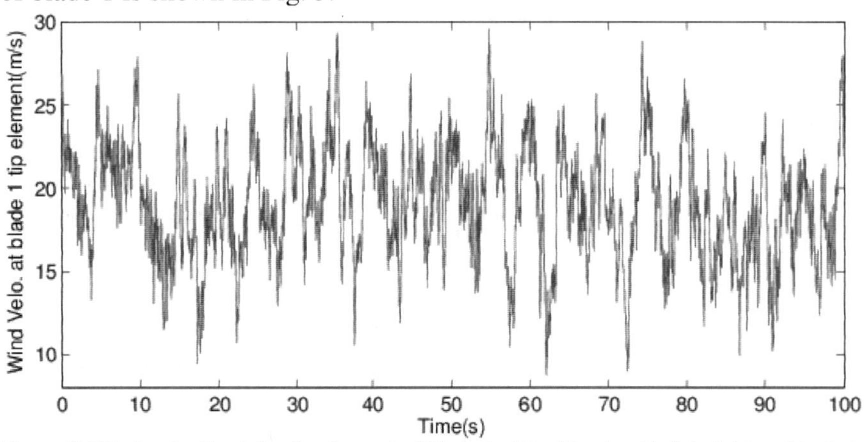

Figure 3.Wind velocity at the tip element of blade 1. ($\bar{v} = 20m/s$ at hub height and turbulence intensity $TI = 18\%$)

3.2 *Aerodynamic Loading*

In the present study, the Blade Element Momentum (BEM) theory is used to estimate the aerodynamic loading acting on the rotating blades. The BEM method is a combination of the momentum theory and the blade element theory. Time series of the aerodynamic loading are computed based on the momentum theory, the blade characteristics and the operational conditions. The input parameters include the rotor geometry (number of blades, twist, chord distribution, and the airfoils), wind speed and the blade rotational velocity [27].

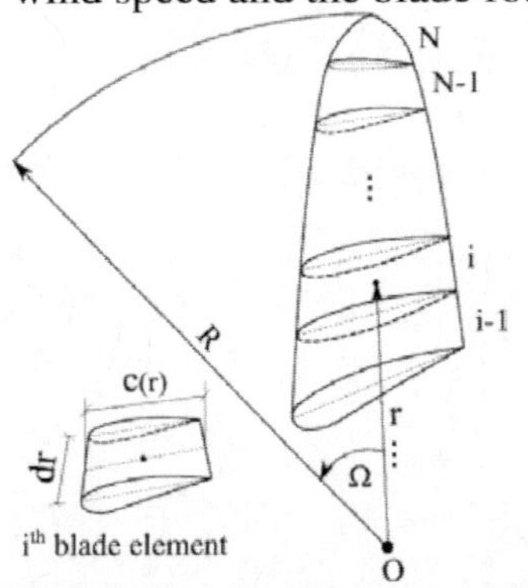

Figure 4. Turbine blade discretized into N blade elements for BEM analysis.

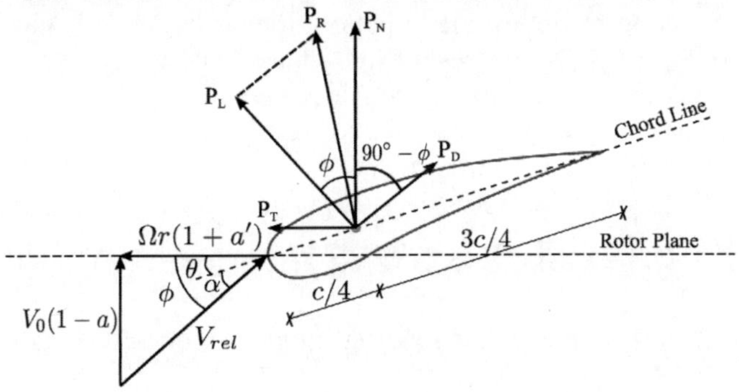

Figure 5. Blade element section subjected to local velocity and aerodynamic loading.

Fig. 4 illustrates a general turbine blade discretized into N elements for BEM analysis where R is the rotor radius and Ω is the rotation velocity. It is assumed in the BEM theory that no radial dependency exists along the blade span and thus the elements can be analyzed independently via performing the momentum theory. In Fig. 4, the ith blade element at a distance r is detailed where dr is the element span length and $c(r)$ is the chord length at the element mid-span.

An arbitrary blade element experiencing local velocities and aerodynamic loading is illustrated in Fig. 5. The relative wind velocity V_{rel} with reference to the element shown in Fig. 5 can be expressed by combining the axial velocity $V_0(1-a)$ and the tangential velocity $\Omega r(1+a')$ as:

$$V_{rel} = \sqrt{\left[V_0(1-a)\right]^2 + \left[\Omega r(1+a')\right]^2} \tag{23}$$

where a and a' are the axial velocity and tangential velocity induction factors.

The flow angle ϕ can be calculated by

$$\phi = \tan^{-1}\frac{V_0(1-a)}{\Omega r(1+a')} \tag{24}$$

The attack angle α between the relative velocity V_{rel} and chord line is obtained as:

$$\alpha = \phi - \theta \tag{25}$$

where θ is the summation of the pitch angle and the twist which is predetermined by the airfoil.

In terms of the attack angle α, the lift and drag coefficients C_l and C_d can be determined from the airfoil data. The lift force P_L which is perpendicular and the drag force which is parallel to the relative velocity can be computed as:

$$P_L = \frac{1}{2}\rho V_{rel}^2 c C_l, \quad P_D = \frac{1}{2}\rho V_{rel}^2 c C_d \tag{26}$$

where ρ is the density of air and c is the chord length.

The normal and the tangential coefficients C_N and C_T are defined as:

$$C_N = \cos\phi C_l + \sin\phi C_d, \quad C_T = \sin\phi C_l - \cos\phi C_d \tag{27}$$

The normal and tangential forces P_N and P_T can be calculated as:

$$P_N = \frac{1}{2}\rho V_{rel}^2 c C_N, \quad P_T = \frac{1}{2}\rho V_{rel}^2 c C_T \tag{28}$$

Eqns. (23) through (28) demonstrate the primary procedure to calculate the aerodynamic load. In real application, a and a' are unknown and need to be determined via iterations. Matlab code has been developed to calculate the time series of P_N and P_T in the present study based on the algorithm proposed in [27]. Prandtl's tip loss factor and Glauert correction are considered in the Matlab code. Next the Principle of Virtual Work is applied to calculate the generalized aerodynamic load.

Under P_N and P_T, the virtual work δW_{wl} done by external wind load is:

$$\delta W_{wl} = \sum_{j=1}^{3}\left\{\int_0^R P_{Tj}(r,t)\left[\phi_{1e}\delta q_j + \delta u_{nac}^{ss}\cos(\psi_j)\right]dr + \int_0^R P_{Nj}(r,t)\left[\phi_{1f}\delta q_{j+3} + \delta u_{nac}^{fa}\right]dr\right\} \tag{29}$$

where $P_{Tj}(r,t)$ and $P_{Nj}(r,t)$ denote the tangential and normal wind loading intensity per unit length on the jth blade.

In terms of the principles of work and energy, the generalized force Q_j can be determined as:

$$Q_j = \frac{\partial(\delta W_{wl})}{\partial(\delta q_j)} \tag{30}$$

Substituting Eqn. (29) into Eqn. (30) yields the wind induced generalized forces:

$$\begin{aligned}
&Q_{j,wind} = \int_0^R P_{Tj}(r,t)\phi_{1e}dr, \quad Q_{j+3,wind} = \int_0^R P_{Nj}(r,t)\phi_{1f}dr, \quad j = 1,2,3 \\
&Q_{7,wind} = \sum_{j=1}^{3}\int_0^R P_{Nj}(r,t)dr, \quad Q_{8,wind} = \sum_{j=1}^{3}\int_0^R P_{Tj}(r,t)dr\cos(\psi_j) \\
&Q_{9,wind} = Q_{7,wind}, \quad Q_{10,wind} = hQ_{7,wind}, \quad Q_{11,wind} = 0
\end{aligned} \tag{31}$$

It is noted that the wind loading acting on the tower is relatively small as compared to that on the blade and thus is ignored. Based on Eqn. (32), the aerodynamic loading corresponding to an average wind speed of $v_0 = 20m/s$ at the hub height and a turbulence intensity of $TI = 18\%$ are computed. As an illustration, the first 60 second data of $Q_{7,wind}$ is shown in Fig. 6.

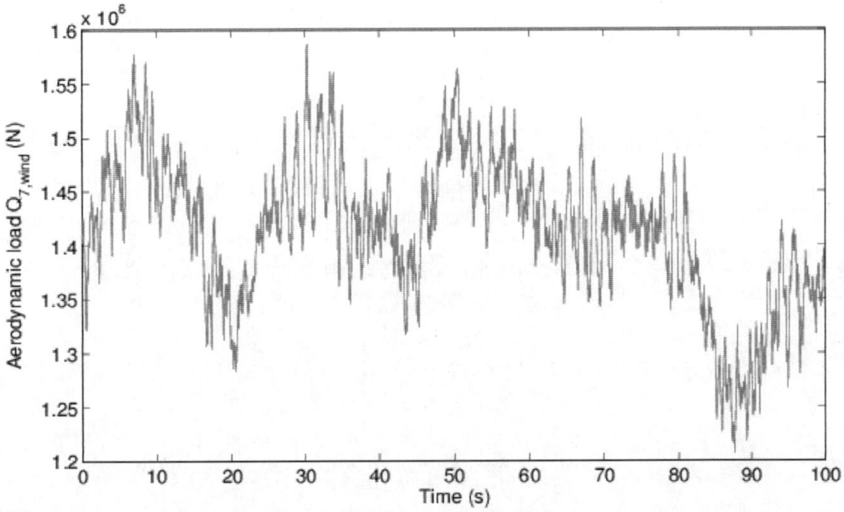

Figure 6. Aerodynamic loading applied on the nacelle in fore-aft direction. ($\overline{v} = 20m/s$ at hub height and turbulence intensity $TI = 18\%$)

3.3 *Wave Loading*

Wave loading on circular cylindrical structural members of fixed offshore structures can be estimated using Morison's equation [28]. For the monopile offshore wind turbine, the horizontal force dF acting on a strip of length dz can be written as:

$$dF = \frac{\pi D^2}{4} C_M \rho \ddot{u} dz + \frac{\rho}{2} C_D D \dot{u} \left| \dot{u} \right| dz \tag{32}$$

where C_M and C_D are the mass and drag coefficients ($C_M = 1.0$ and $C_D = 1.2$ are adopted in the present paper); ρ is water density ($1025 kg/m^3$), D is the diameter of the tower and the monopile; $\ddot{u}$ and $\dot{u}$ are the wave induced horizontal acceleration and velocity of fluid particles.

In terms of linear wave theory (Airy theory) [28], the wave surface elevation $\eta(t)$, the fluid particles velocity $\dot{u}$, and acceleration $\ddot{u}$ can be expressed as:

$$\eta(t) = \frac{1}{2} H_s \cos(kx - \omega t) \tag{33}$$

$$\dot{u} = \frac{H_s \pi \cosh\left[k(z + d_w)\right]}{T_w \sinh(kd_w)} \cos(kx - \omega t) \tag{34}$$

$$\ddot{u} = \frac{2 H_s \pi^2 \cosh\left[k(z + d_w)\right]}{T_w^2 \sinh(kd_w)} \sin(kx - \omega t) \tag{35}$$

where H_s is wave height in meter, k is wave number in m^{-1}, ω is wave frequency in rad/s, T_w is the wave period in s, z is the vertical ordinate from mean water level.

Given water depth z, the parameters ω and k are related by the dispersion equation [28]:

$$k \tanh kz = \omega^2 / g \tag{36}$$

Hence, for any given water depth z, the wave number k can be determined via solving Eqn. (36). Virtual work δW_{wv} done by the wave load along virtual displacement δu_{tow} of the tower can be written as:

$$\delta W_{wv} = \int_0^h dF\,\delta u_{tow} = \int_0^h dF\left(f_{1t}\delta q_7 + \delta q_9 + z\delta q_{10}\right) \tag{37}$$

Substituting Eqn. (37) into Eqn. (30) yields the generalized forces corresponding to wave:

$$Q_{7,wave} = \int_{-d_w}^{\eta(t)} f_{1t}(z)dF = \sum_{i=1}^{N_z} f_{1t}(z_i)\left[\frac{\rho\pi D^2(z_i)}{4}C_M\ddot{u}(z_i,t)\Delta z + \frac{\rho}{2}C_D D(z_i)\dot{u}(z_i,t)\left|\dot{u}(z_i,t)\right|\Delta z\right] \tag{38}$$

$$Q_{9,wave} = \int_{-d_w}^{\eta(t)} dF = \sum_{i=1}^{N_z}\left[\frac{\rho\pi D^2(z_i)}{4}C_M\ddot{u}(z_i,t)\Delta z + \frac{\rho}{2}C_D D(z_i)\dot{u}(z_i,t)\left|\dot{u}(z_i,t)\right|\Delta z\right] \tag{39}$$

$$Q_{10,wave} = \int_{-d_w}^{\eta(t)} z\,dF = \sum_{i=1}^{N_z}\left[\frac{\rho\pi D^2(z_i)}{4}C_M\ddot{u}(z_i,t)\Delta z + \frac{\rho}{2}C_D D(z_i)\dot{u}(z_i,t)\left|\dot{u}(z_i,t)\right|\Delta z\right] z_i \tag{40}$$

where d_w denotes the water depth and N_z denotes the number of segments that the wetted portion of the tower is divided and Δz is the segment length.

Operational wave condition, $H_s = 6m$ and $T = 10.4s$ are adopted. Based on Eqns. (32 ~ 40), the wave induced modal force $Q_{7,wv}$, the resultant force $Q_{9,wv}$ and the bending moment $Q_{10,wv}$ are determined. Other wave induced generalized forces are zero.

4 CONTROL ALGORITHM

A short-time Fourier transformation (STFT) based control algorithm is presented in this section.

4.1 *Short-time Fourier Transformation*

STFT is a widely used tool to extract the time-frequency characteristics of non-stationary signals. In addition to its application in vibration control, the STFT can be used for damage detection. Fitzgerald et al. [29] developed a STFT-based algorithm to detect damage of wind turbine blades. The research findings indicated that the frequency-tracking technique can identify the stiffness loss and damage.

The basic idea of STFT is to divide the time series of a signal into a number of time segments and to analyze each of the segments using Fourier transform to determine the frequencies existing in it. The spectrum is obtained for each different time, and the totality of the spectra is the distribution of the signal in time-frequency domain [30]. Principles and the associated equations are illustrated as follows.

Consider a signal $x(\tau)$, multiplying the signal by a window function $h(\tau - t)$ yields:

$$\hat{x}(\tau) = x(\tau)h(\tau - t) \tag{41}$$

where $\hat{x}(\tau)$ is a weighted signal; t is the fixed time and τ is the running time.
Applying Fourier transformation to $\hat{x}(\tau)$ gives the spectrum $S(t,\omega)$ at the fixed time t:

$$S(t,\omega) = \frac{1}{2\pi}\int e^{-j\omega\tau}\hat{x} = \frac{1}{2\pi}\int e^{-j\omega\tau}x(\tau)h(\tau - t) \tag{42}$$

The power spectral density $P(t, w)$ at time t is obtained as:

$$P(t,w) = |S(t,w)|^2 = S(t,w).\overline{S(t,w)} \tag{43}$$

At the fixed time t_i, the dominant frequency w_{id} is calculated using Eqn.(44) and Eqn. (45).

$$w_{inst} = \{ w | P(t_i, w) = \max\{P(t_i, w)\}\} \tag{44}$$

$$w_{id} = \frac{\sum_{k=\max\{1,i-m+1\}}^{i} w_{inst}(t_k)\max\{P(t_k,w)\}}{\sum_{k=\max\{1,i-m+1\}}^{i} \max\{P(t_k,w)\}} \tag{45}$$

where w_{inst} is the instantaneous frequency and w_{id} is the dominant frequency at t_i calculated through averaging the values of the instantaneous frequencies over m time steps, $m = 3$ in the present study.

4.2 *Real-time tuning of the STMD*

In the resent study, the nacelle fore-aft displacement is tracked and the power spectrum density $P(t_i, w)$ is computed via Eqns. (41~43) continuously. Then the current frequency of the STMD is retuned based on the control algorithm described as follows. In the beginning at $t = 0$, the STMD circular natural frequency is $w_s = f_{opt} w_n$ and damping ratio is set to $z_s = 0$ where f_{opt} is the optimal frequency ratio to be determined next and w_n is the original circular natural frequency of the nacelle/tower top.

Control algorithm

1. A moving window of n time points ($n = 500$ in the present study) of signal at time t_i is convolved with a Hamming window to compute the power spectral density $P(t,w)$ (Eqns. (41)~(43)) of the nacelle displacement, producing a vector P_i. The length L_h of the Hamming window is set to $L_h = 1024$ and the vector P_i is of size $NN \times 1$, where $NN = L_h / 2 + 1$.
2. The dominant frequency w_{id} at time t_i is calculated using Eqns. (44)~(45)
3. If $|w_{id} - w_n| / w_n <= d$, the stiffness of the STMD is retuned such that $w_s = w_{id}$; else, it is set to $w_s = f_{opt} w_n$ where $d = 0.1$.
4. The relative displacement of the STMD q_{11} is recorded in a vector D_s. If the current magnitude is less than that of the previous amplitude, i.e. $|D_s(k)| \le |D_s(k-1)|$, the damping ratio of the STMD is set to be $z_s = 2z_{opt}$; else it is set to $z_s = 0$.
   Parameters f_{opt} and z_{opt} are determined as $f_{opt} = 0.99$ and $z_{opt} = 11\%$ via using the modal analysis approach illustrated in [14].

5 RESULTS

Performance of the proposed STMD to mitigate wind and wave induced response of the monopole offshore wind turbines is evaluated and the results are presented. Matlab code has been developed to model the 10-DOF uncontrolled system and the 11-DOF controlled system. The NREL 5 MW baseline wind turbine model is used.

5.1 *Model Parameters*

Blade utilized in the model is the $LM61.5P2$ (manufactured by LM Wind power) with a mass of 17740 kg. The total rotor radius is 63 m and the blade length is 61.5 m because the hub radius is 1.5m. The fundamental mode shape (edgewise and flapwise) of the blade and the tower are represented using six-order polynomials (see Eqn. (47)) and illustrated in Fig. 7. More details of the wind turbine can be found in Ref. [31].

$$\begin{aligned} f_{1e}(\bar{r}) &= -0.6952\bar{r}^6 + 2.3760\bar{r}^5 - 3.5772\bar{r}^4 + 2.5337\bar{r}^3 + 0.3627\bar{r}^2 \\ f_{1f}(\bar{r}) &= -2.2555\bar{r}^6 + 4.7131\bar{r}^5 - 3.2452\bar{r}^4 + 1.7254\bar{r}^3 + 0.0622\bar{r}^2 \\ f_{1t}(\bar{h}) &= -0.6952\bar{h}^6 + 2.3760\bar{h}^5 - 3.5772\bar{h}^4 + 2.5337\bar{h}^3 + 0.3627\bar{h}^2 \end{aligned} \quad (46)$$

where f_{1e} and f_{1f} denote the edgewise and flapwise fundamental mode shape of the blade; parameter f_{1t} denotes the tower fore-aft and side-side fundamental mode shape. Variables $\bar{r} = r / 61.5$ and $\bar{h} = h / 87.6$ denote the normalized blade radius and tower height.

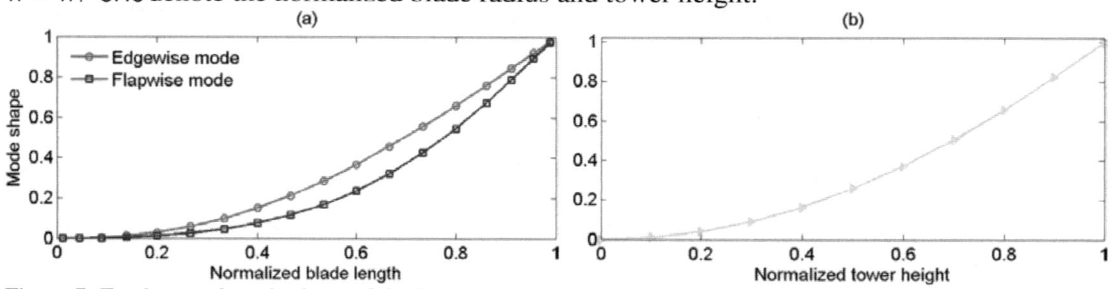

Figure 7. Fundamental mode shape of the blade and the tower. (a): mode shape of the blade (b): mode shape of the tower.

5.2 *Soil Effects and Damage Parameters*

As described in Section 2.1, the SE are considered via using a linear spring and a dash-pot as illustrated in Fig. 1. With reference to [32], value of parameters k_x and k_f are obtained as: $k_x = 3.89E9N / m, k_f = 1.14E11Nm / rad$ to represent clay soil condition. Soil damping properties c_x and c_f are selected such that the corresponding damping ratio are $z_x = z_f = 0.6\%$.

Due to cyclic wind and wave loading, damage is likely to happen in the foundation and the tower [21]. It is worthy to note that the mechanism of damage as well as the damage development is complex, which is not the focus of the present study. Here it is assumed that the damage occurs at a given time and accumulates linearly with respect to time. Three damage cases referred to as DC I, DC II and DC III are considered in the present study. In DC I, damage occurs to the foundation; in DC II, damage occurs to the tower; in DC III, DC I and DC II occur one after another. Fig. 8 (a) and (b) show the damage development corresponding to DC I and DC II where d denotes the damage coefficient. In Fig. 8(a) and (b), it is assumed that the damage development has a duration of 50s. It is worthy to note that this assumption is merely a conceptual reflection of real damage (which in reality might develop for months or years) to demonstrate the performance of the STMD

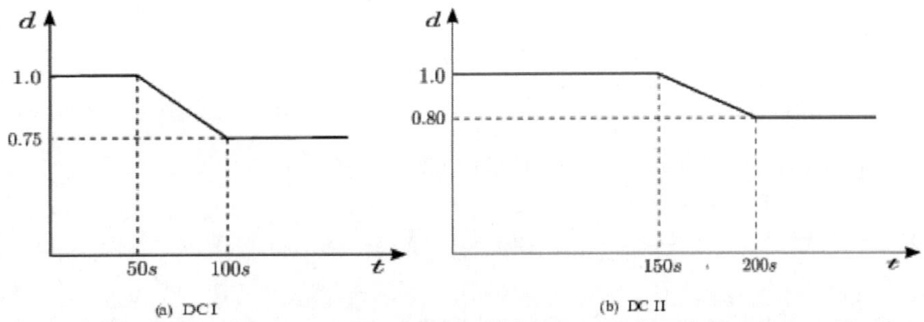

Figure 8. Illustration of the gradually developed damage. (a) foundation damage development referred to as DC I; (b) tower structural damage development referred to as DC II.

5.3 *Responses of Uncontrolled Wind Turbine*

With reference to [33], a representative wind and wave loading case DLC 1.3_3 is used.

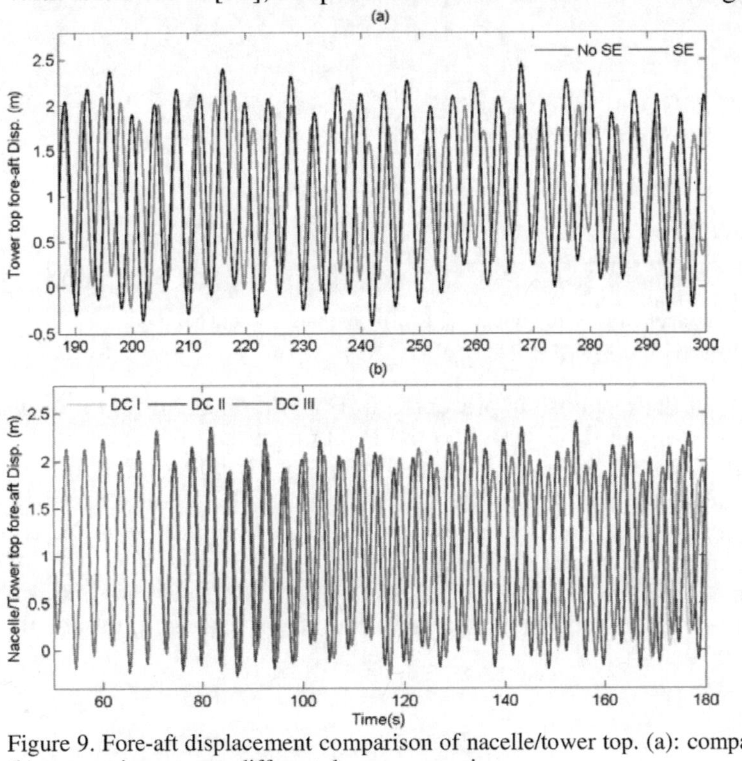

Figure 9. Fore-aft displacement comparison of nacelle/tower top. (a): comparison between with SE and without SE; (b): comparison among different damage scenarios.

Fig. 9(a) shows the nacelle displacement with and without SE. It is found that the nacelle vibrates with a lower frequency in the case of SE than in that without SE. However, the response amplitude in the case of SE is relatively larger. Fig. 9(b) shows the nacelle response in three damage cases: DC I, DC II and DC III. It can be observed that after $t = 50\ s$, the response frequency in DC I and DC III (the red and green lines actually coincide) gradually becomes slower than that in DC II due to the foundation damage initiation at $t = 50\ s$. After $t = 150\ s$, response in the case of DC III becomes slower than that in DC I due to the tower damage presence.

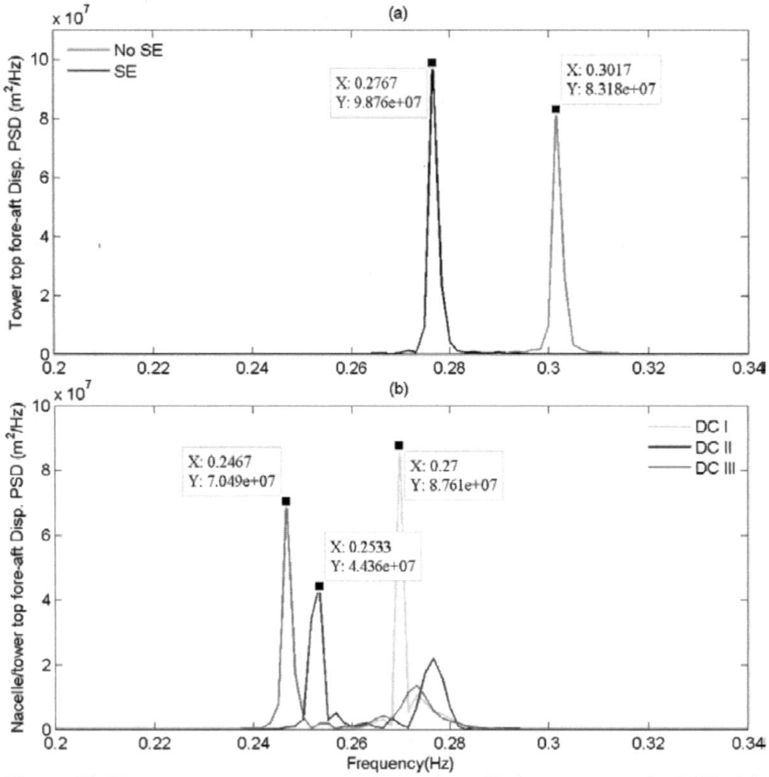

Figure 10. Response spectrum comparison of nacelle/tower top under LC1. (a): comparison between with SE and without SE; (b): comparison among different damage scenarios.

The power spectra density (PSD) of the response is presented in Fig. 10. One can find in Fig. 10 (a) that the dominant frequency is decreased by 10% and the PSD magnitude is increased by around 19% when SE is considered. Through comparing the result in Fig. 10 (a) and that in Fig. 10 (b), it can be found that the dominant frequency in the cases of DC I, DC II and DC III is decreased gradually. It is also found that close damage extent (20% stiffness reduction in DC II) in the tower causes a larger frequency variation (8.5%) than damage in the foundation (25% stiffness reduction in DC I) (2.4%). Moreover, it can be found in Fig. 10(b) that a second remarkable frequency appears in DC II and DC III, which is essentially caused by damage presence in the tower.

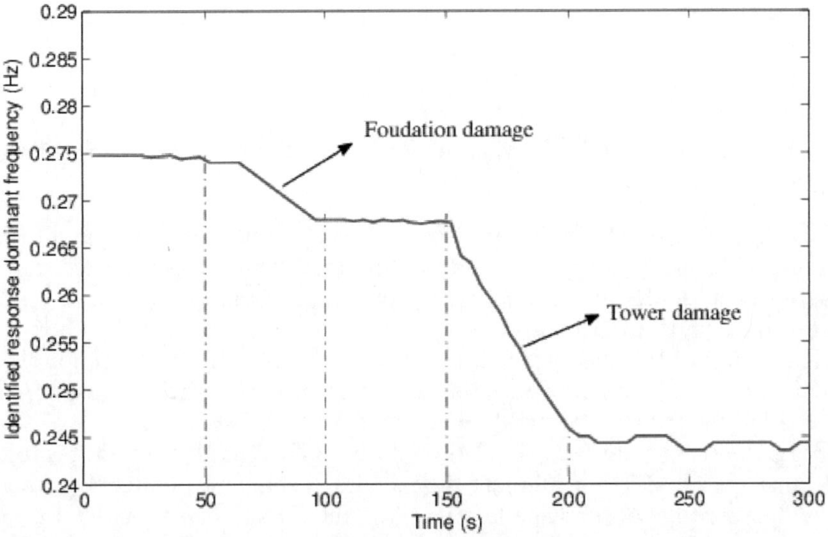

Figure 11. Identified time-varying dominant frequency of the nacelle response under damage case DC III.

5.4 *Responses of Controlled Wind Turbine*

This section presents the mitigation of the nacelle/tower top response controlled by the STMD. Loading, operational conditions and damage scenarios are the same as that defined in Section 5.3. To show the advantage of the STMD, a passive TMD designed using Warburton's formula [26] shown in Eqn. (47) is utilized for comparison. Different mass ratios from 1% to 5% have been tested in the present study and they all provide considerable mitigation. Herein the result in the case of 1% mass ratio is presented, that is, the mass ratio of the STMD μs and the passive TMD μt are set to $\mu_s = \mu_t = 1\%$. Based on Eqn. (47), frequency ratio and damping ratio of the TMD is determined as $f_t = 0.993$, $\zeta_t = 5\%$. Frequency and damping ratio of the STMD is retuned in real-time based on the control algorithm described in Section 4.2.

$$f_t = \sqrt{\frac{1-\mu/2}{1+\mu}},\quad \zeta_t = \sqrt{\frac{\mu(1-\mu/4)}{4(1+\mu)(1-\mu/2)}} \qquad (47)$$

Key advantage of the STMD and the associated control algorithm is that the response frequency variation caused by SE and/or damage can be tracked and identified in real-time. Fig. 11 illustrates the identified time-varying dominant response frequency of the nacelle under damage case DC III (damage occurs to the foundation and the tower). Two obvious frequency drops can be observed in Fig. 11. One can find that the initiation and duration of the two frequency drops agree well with that of the damage development illustrated in Fig. 8 (a) and (b). The two frequency drops occurring from 50 s to 100 s and from 150 s to 200 s essentially signal the foundation and tower damage. Therefore, the proposed real-time control algorithm can track the time-varying dominant response frequency accurately and timely.

Fig. 12 (a)-(d) show the nacelle displacement time-history comparison between the passive TMD and the STMD in four scenarios: SE, SE with DC I, SE with DC II and SE with DC III. In Fig. 12 (a), it is found that the STMD outperforms the passive TMD in mitigating structural response when SE is considered. By comparing Fig. 12 (b) and Fig. 12 (a), one can find that from t = 276.7 s to t = 284.1 s, the response amplitude in the case of TMD has been increased by around 30% while that with STMD remains a low level when damage is present in the foundation after $t = 50\ s$.

This is because that the passive TMD becomes partially off-tuned and less effective due to the foundation damage. In addition, one can further find in 12 (c) that the response amplitude from $t = 268.9\ s$ to $t = 283.1\ s$ in the case of TMD is doubled as compared to Fig. 12 (a) while that with the STMD remains a low level. Furthermore, a highest response amplitude increase from $t = 278\ s$ to $t = 283.1\ s$ can be observed in 12 (d)(damage is present in foundation and tower), signaling that the TMD becomes severely off-tuned and ineffective as damage accumulates in both the foundation and the tower.

Fig. 13 (a) and (b)show the response spectrum comparison between the passive TMD and the STMD in the case of DC I and DC III, separately. It is worth noting that the response spectrum in the case of DC II is similar to that of DC III in Fig. 13 (b) and thus not shown. In Fig. 13(a), one can find that the STMD can significantly improve the mitigation of the PSD magnitude when 7 compared to the passive TMD. In Fig. 13 (b), the response spectrum in the case of TMD has double peaks because of damage in the tower (discussed in Section 5.3). In comparison, the two response spectrum peaks are significantly mitigated to a minimal level when the STMD is used.

Acceleration time history comparison between the TMD and the STMD is illustrated in Fig. 14. It is indicated in Fig. 14 that the STMD can mitigate the nacelle acceleration better than the TMD. In addition to mitigate the response of the nacelle/tower top, the STMD can provide effective mitigation to the foundation. Fig. 15 show the foundation in-plane rotation under damage case III. One can find that the foundation response in the case of STMD is mitigated more effectively than that with TMD. This will help mitigate the fatigue load of the foundation so as to prolong its service life and reduce the maintenance cost.

TMD stroke is a critical index when evaluating the performance because of the limited space in the nacelle. Fig. 16 compares the stroke between the STMD and the TMD in the case of DC II. It is found that the STMD has a much lower stroke than the TMD.

To sum up, the monopile offshore wind turbines under wind and wave loading are inevitably subjected to SE and damage, which will result in variation of the system response characteristics, including the amplitude and the response frequency contents. This variation can render the passive TMD off-tuned and ineffective. To the contrary, application of the STMD can track the frequency variation and retune in real-time so as to prevent the detuning and to provide significant mitigation for the nacelle, the tower and the foundation.

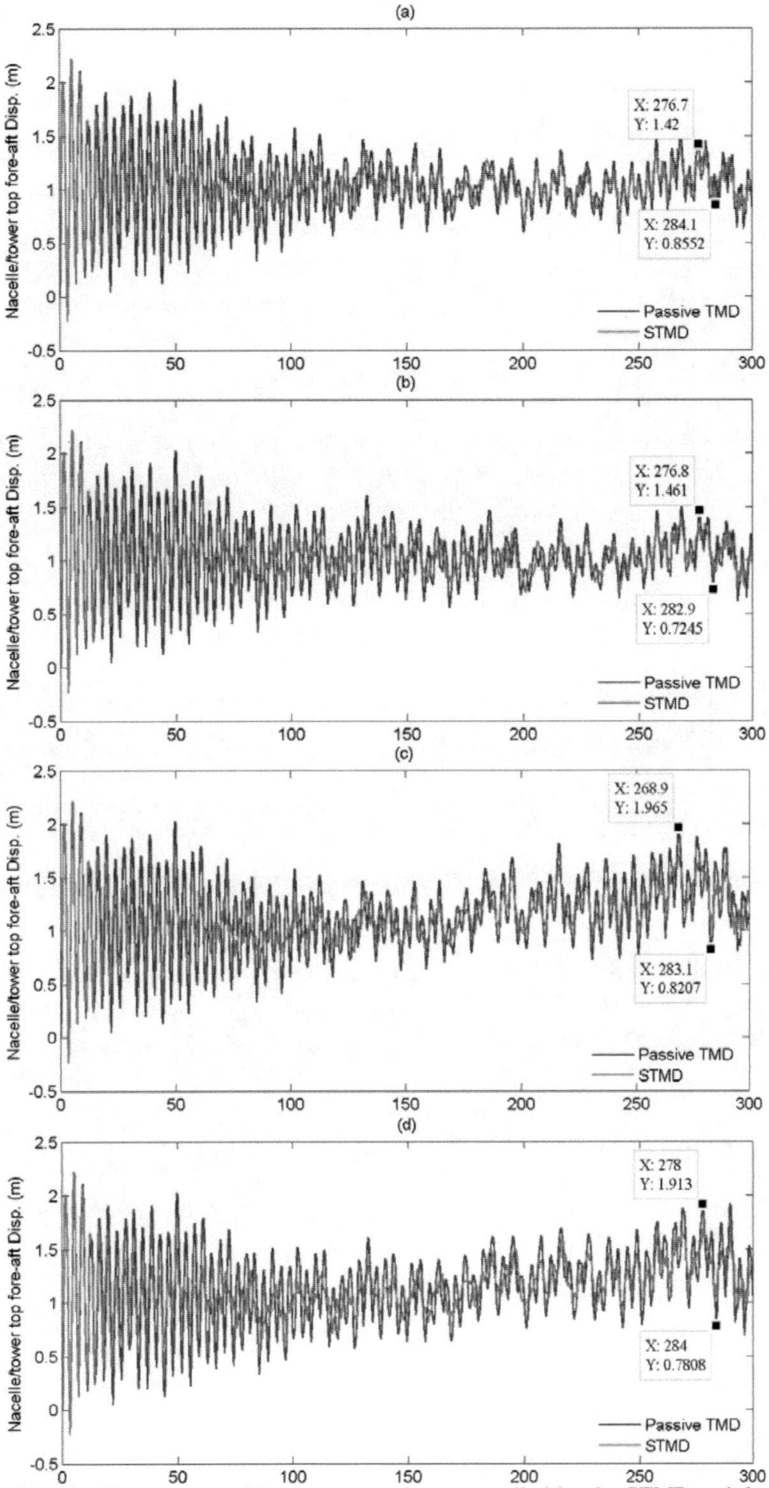

Figure 12. Nacelle response mitigation comparison controlled by the STMD and the TMD. (a): SE without damage (b): under DC I, (b): under DC II, (c): under DC III.

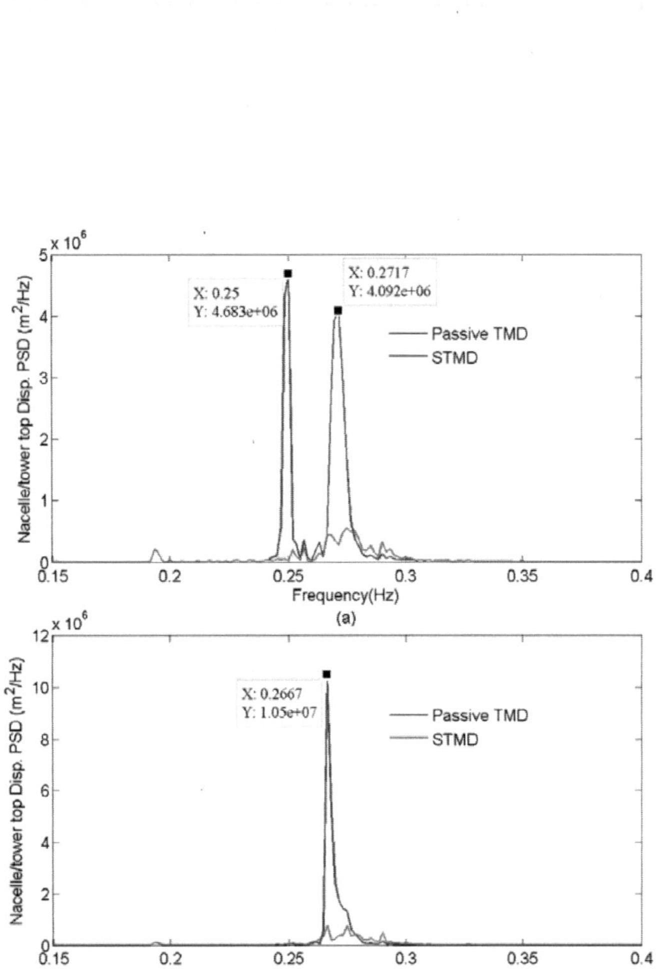

Figure 13. Nacelle response mitigation comparison of the STMD and the TMD. Upper: under damage case DC I; lower DC III.

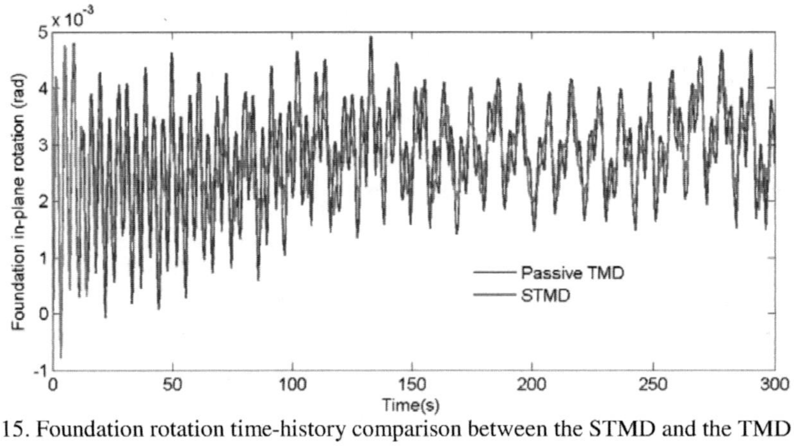

Figure 15. Foundation rotation time-history comparison between the STMD and the TMD under damage case DC III.

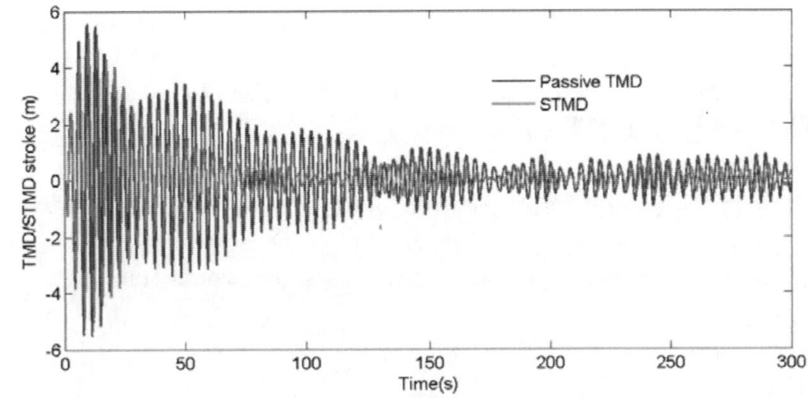

Figure 16. Stroke comparison between the STMD and TMD.

6 CONCLUSIONS

The present study establishes an analytical model of the monopile offshore wind turbine controlled by an STMD and subjected to wind and wave loading. Soil effects and damage presence in the foundation and/or the tower are considered. Complete Matlab code has been developed to compute the aerodynamic/hydrodynamic loading and to solve the established equations. The wind turbine responses are simulated via solving the established equations. Performance of the STMD is evaluated and compared with a passive TMD. Based on the results and discussions presented, the following four key conclusions can be obtained:

1. Consideration of the SE in the model influences the wind turbine response characteristics, including the response magnitude as well as the dominant frequency. The dominant frequency is decreased and the response PSD magnitude is increased. The SE influences the response of the nacelle/tower to a more extent than influencing response of the blades. Similar conclusions can be drawn when damage is present in the foundation and the tower.
2. When damage is present in the foundation or/and the tower, the dominant frequency of the nacelle response spectrum will be decreased further. Given similar level of damage, DC II (damage in the tower) results in more decrease of the dominant frequency than DC I (damage in the foundation) does. A combination of the two (DC III) causes the most significant frequency variation.
3. Because of the dominant frequency variation, the traditional passive TMD which is tuned to the original structural natural frequency becomes off-tuned. As damage continues to accumulate in the foundation and the tower, this detuning effect will be escalated and the passive TMD becomes ineffective.
4. The proposed STFT-based control algorithm can capture the SE and damage presence in the foundation and tower. Real-time tuning of the STMD natural frequency and damping property enables the STMD to be highly effective in mitigating the response of the nacelle/tower and the foundation, thereby protecting the wind turbines and prolonging its fatigue life. In addition, the STMD experiences smaller stroke than the passive TMD, which is beneficial to the nacelle from the perspective of practical application. Overall, the presented results indicate that the STMD can provide significant advantage over the passive TMD when considering the SE and damage presence.

ACKNOWLEDGMENT

This work was supported by Louisiana State University Start-up Fund (fund number is 127150013), the Faculty Research Grant (fund number is 127159132) and the Innovation in Engineering Research (FIRE) Grant provided by the College of Engineering at Louisiana State University. The author is grateful for all the support.

APPENDIX A

System matrices involved in Eqn. (17) for the uncontrolled 10-DOF system are:

$$\tilde{M} = \begin{bmatrix} m_1 & 0 & 0 & 0 & 0 & 0 & 0 & m_{18} & 0 & 0 \\ 0 & m_1 & 0 & 0 & 0 & 0 & 0 & m_{28} & 0 & 0 \\ 0 & 0 & m_1 & 0 & 0 & 0 & 0 & m_{38} & 0 & 0 \\ 0 & 0 & 0 & m_2 & 0 & 0 & m_{47} & 0 & m_{47} & hm_{47} \\ 0 & 0 & 0 & 0 & m_2 & 0 & m_{47} & 0 & m_{47} & hm_{47} \\ 0 & 0 & 0 & 0 & 0 & m_2 & m_{47} & 0 & m_{47} & hm_{47} \\ 0 & 0 & 0 & m_{47} & m_{47} & m_{47} & m_7 & 0 & m_7 & hm_7 \\ m_{18} & m_{28} & m_{38} & 0 & 0 & 0 & 0 & m_8 & 0 & 0 \\ 0 & 0 & 0 & m_{47} & m_{47} & m_{47} & m_7 & 0 & m_9 & hm_7 \\ 0 & 0 & 0 & hm_{47} & hm_{47} & hm_{47} & hm_7 & 0 & hm_7 & I_f + h^2 m_7 \end{bmatrix} \quad (48)$$

where

$$m_1 = \int_0^R \bar{m}(r)\phi_{1e}^2 dr,\ m_2 = \int_0^R \bar{m}(r)\phi_{1f}^2 dr,\ m_{j8} = \int_0^R \bar{m}(r)\phi_{1e}^2 dr \cos\psi_j,\ j = 1,2,3$$

$$m_{47} = \int_0^R \bar{m}(r)\phi_{1f} dr,\ M_{1t} = \int_0^h \bar{M}(z)\phi_{1t}^2 dz,\ m_8 = m_7 = 3m_0 + M_{nac} + M_{hub} + M_{1t}$$

$$m_9 = 3m_0 + M_{nac} + M_{hub} + M_t + M_f,\ M_t = \int_0^h \bar{M}(z) dz$$

Parameter m_0 denotes the mass of a single blade, M_{nac} and M_{hub} denote the mass of the nacelle and the hub, $\bar{M}$ denote the mass density (kg/m) of the tower, M_t denotes the mass of the tower, h_0 denotes the tower height, M_f and I_f denote the mass and moment of inertia of the foundation.

$$\tilde{K} = \begin{bmatrix} k_{b1,eg} & 0 & 0 & 0 & 0 & 0 & 0 & 0 & 0 & 0 \\ 0 & k_{b2,eg} & 0 & 0 & 0 & 0 & 0 & 0 & 0 & 0 \\ 0 & 0 & k_{b3,eg} & 0 & 0 & 0 & 0 & 0 & 0 & 0 \\ 0 & 0 & 0 & k_{b1,fp} & 0 & 0 & 0 & 0 & 0 & 0 \\ 0 & 0 & 0 & 0 & k_{b2,fp} & 0 & 0 & 0 & 0 & 0 \\ 0 & 0 & 0 & 0 & 0 & k_{b3,fp} & 0 & 0 & 0 & 0 \\ 0 & 0 & 0 & 0 & 0 & 0 & k_7 & 0 & 0 & 0 \\ \Omega^2 m_{18} & -\Omega^2 m_{28} & -\Omega^2 m_{38} & 0 & 0 & 0 & 0 & k_8 & 0 & 0 \\ 0 & 0 & 0 & 0 & 0 & 0 & 0 & 0 & k_x & 0 \\ 0 & 0 & 0 & 0 & 0 & 0 & 0 & 0 & 0 & k_f \end{bmatrix} \quad (49)$$

where

$$k_{bj,eg} = k_{eg} + k_{ge,eg} - k_{gr,eg}\cos\psi_j - \Omega^2 m_1,\quad k_{bj,eg} = k_{fp} + k_{ge,fp} - k_{gr,fp}\cos\psi_j,$$

$$k_7 = k_8 = \int_0^h EI_t\left(\phi_{11}'\right)^2 dz$$

$$\tilde{C} = \begin{bmatrix} c_{b,eg} & 0 & 0 & 0 & 0 & 0 & 0 & 0 & 0 & 0 \\ 0 & c_{b,eg} & 0 & 0 & 0 & 0 & 0 & 0 & 0 & 0 \\ 0 & 0 & c_{b,eg} & 0 & 0 & 0 & 0 & 0 & 0 & 0 \\ 0 & 0 & 0 & c_{b,fp} & 0 & 0 & 0 & 0 & 0 & 0 \\ 0 & 0 & 0 & 0 & c_{b,fp} & 0 & 0 & 0 & 0 & 0 \\ 0 & 0 & 0 & 0 & 0 & c_{b,fp} & 0 & 0 & 0 & 0 \\ 0 & 0 & 0 & 0 & 0 & 0 & c_7 & 0 & 0 & 0 \\ 2\Omega\bar{m}_{18} & -2\Omega\bar{m}_{28} & -2\Omega\bar{m}_{38} & 0 & 0 & 0 & 0 & c_8 & 0 & 0 \\ 0 & 0 & 0 & 0 & 0 & 0 & 0 & 0 & c_x & 0 \\ 0 & 0 & 0 & 0 & 0 & 0 & 0 & 0 & 0 & c_f \end{bmatrix} \tag{50}$$

where

$$\bar{m}_{8j} = \int_0^R \bar{m}(r)\phi_{1e}\,dr\,\sin\psi_j,\ j = 1,2,3$$

$c_{bj,eg}$ and $c_{bj,fp}$ denote the edgewise and flapwise structural and aerodynamic damping of the blade, c_7 and c_8 denote the damping in the fore-aft and side-side directions, including the structural and aerodynamic damping property of the nacelle and the tower, c_x and c_f denote the in-plane translational and rotational damping of the foundation.

APPENDIX B

System matrices involved in Eqn. (17) for the controlled 11-DOF system are:

$$\tilde{M} = \begin{bmatrix} m_1 & 0 & 0 & 0 & 0 & 0 & 0 & m_{18} & 0 & 0 & 0 \\ 0 & m_1 & 0 & 0 & 0 & 0 & 0 & m_{28} & 0 & 0 & 0 \\ 0 & 0 & m_1 & 0 & 0 & 0 & 0 & m_{38} & 0 & 0 & 0 \\ 0 & 0 & 0 & m_2 & 0 & 0 & m_{47} & 0 & m_{47} & hm_{47} & 0 \\ 0 & 0 & 0 & 0 & m_2 & 0 & m_{47} & 0 & m_{47} & hm_{47} & 0 \\ 0 & 0 & 0 & 0 & 0 & m_2 & m_{47} & 0 & m_{47} & hm_{47} & 0 \\ 0 & 0 & 0 & m_{47} & m_{47} & m_{47} & m_7 + m_s & 0 & m_7 & hm_7 & m_s \\ m_{18} & m_{28} & m_{38} & 0 & 0 & 0 & 0 & m_8 & 0 & 0 & 0 \\ 0 & 0 & 0 & m_{47} & m_{47} & m_{47} & m_7 & 0 & m_9 + m_s & hm_7 & m_s \\ 0 & 0 & 0 & hm_{47} & hm_{47} & hm_{47} & hm_7 & 0 & hm_7 & I_1 + h^2 m_s & hm_s \\ 0 & 0 & 0 & 0 & 0 & 0 & m_s & 0 & m_s & hm_s & m_s \end{bmatrix} \tag{51}$$

where $I_1 = I_f + h^2 m_7$, m_s denotes the mass of the STMD

$$
\tilde{K}=\begin{bmatrix}
k_{b1,eg} & 0 & 0 & 0 & 0 & 0 & 0 & 0 & 0 & 0 & 0 \\
0 & k_{b2,eg} & 0 & 0 & 0 & 0 & 0 & 0 & 0 & 0 & 0 \\
0 & 0 & k_{b3,eg} & 0 & 0 & 0 & 0 & 0 & 0 & 0 & 0 \\
0 & 0 & 0 & k_{b1,fp} & 0 & 0 & 0 & 0 & 0 & 0 & 0 \\
0 & 0 & 0 & 0 & k_{b2,fp} & 0 & 0 & 0 & 0 & 0 & 0 \\
0 & 0 & 0 & 0 & 0 & k_{b3,fp} & 0 & 0 & 0 & 0 & 0 \\
0 & 0 & 0 & 0 & 0 & 0 & k_7 & 0 & 0 & 0 & 0 \\
\Omega^2 m_{18} & -\Omega^2 m_{28} & -\Omega^2 m_{38} & 0 & 0 & 0 & 0 & k_8 & 0 & 0 & 0 \\
0 & 0 & 0 & 0 & 0 & 0 & 0 & 0 & k_x & 0 & 0 \\
0 & 0 & 0 & 0 & 0 & 0 & 0 & 0 & 0 & k_f & 0 \\
0 & 0 & 0 & 0 & 0 & 0 & 0 & 0 & 0 & 0 & k_s
\end{bmatrix} \tag{52}
$$

$$
\tilde{C}=\begin{bmatrix}
c_{b,eg} & 0 & 0 & 0 & 0 & 0 & 0 & 0 & 0 & 0 & 0 \\
0 & c_{b,eg} & 0 & 0 & 0 & 0 & 0 & 0 & 0 & 0 & 0 \\
0 & 0 & c_{b,eg} & 0 & 0 & 0 & 0 & 0 & 0 & 0 & 0 \\
0 & 0 & 0 & c_{b,fp} & 0 & 0 & 0 & 0 & 0 & 0 & 0 \\
0 & 0 & 0 & 0 & c_{b,fp} & 0 & 0 & 0 & 0 & 0 & 0 \\
0 & 0 & 0 & 0 & 0 & c_{b,fp} & 0 & 0 & 0 & 0 & 0 \\
0 & 0 & 0 & 0 & 0 & 0 & c_7 & 0 & 0 & 0 & -c_s \\
2\Omega\bar{m}_{81} & -2\Omega\bar{m}_{82} & -2\Omega\bar{m}_{83} & 0 & 0 & 0 & 0 & c_8 & 0 & 0 & 0 \\
0 & 0 & 0 & 0 & 0 & 0 & 0 & 0 & c_x & 0 & 0 \\
0 & 0 & 0 & 0 & 0 & 0 & 0 & 0 & 0 & c_f & 0 \\
0 & 0 & 0 & 0 & 0 & 0 & 0 & 0 & 0 & 0 & c_s
\end{bmatrix} \tag{53}
$$

REFERENCES

1. Housner G., Bergman L.A., Caughey T.K., Chassiakos A.G. etc. Structural control: past, present, and future. Journal of Engineering Mechanics 1997; 123(9): 897–971.
2. Spencer B., Nagarajaiah S. State of the art of structural control Journal of Structural Engineering 2003; 129(7): 845–856.
3. Soong T, Spencer B. Supplemental energy dissipation: state-of-the-art and state-of-the practice. Engineering Structures 2002; 24(7): 243–259.
4. Murtagh P, Ghosh A, Basu B, Broderick B. Passive control of wind turbine vibrations including blade/tower interaction and rotationally sampled turbulence. Wind Energy 2007; 11: 305–317.
5. Lackner M., Rotea M. Passive structural control of offshore wind turbines. Wind Energy 2011; 14: 373–388.
6. Colwell S., Basu B. Tuned liquid column dampers in offshore wind turbines for structural control. Engineering Structures 2009; 31: 358–368.
7. Nagarajaiah S., Sonmez E. Structures with semiactive variable stiffness single/multiple tuned mass dampers. Journal of Structural Engineering 2007; 133(1): 67-77.
8. Nagarajaiah S. Adaptive passive, semiactive, smart tuned mass dampers: identification and control using empirical mode decomposition, hilbert transform, and short-term fourier transform. Structural Control and Health Monitoring 2009; 16(7-8): 800-841.
9. Eason R.P., Sun C., Nagarajaiah S., Dick A.J. Attenuation of a linear oscillator using a nonlinear and a semi-active tuned mass damper in series. Journal of Sound and Vibration 2013; 332(1): 154-166.
10. Sun C., Eason R.P., Nagarajaiah S., Dick A.J. Hardening Düffing Oscillator Attenuation Using a Nonlinear TMD, a Semi-active TMD and a Multiple TMD. Journal of Sound and Vibration 2013; 332(4): 674-686.
11. Sun C., Nagarajaiah S., Dick A. J. Experimental Investigation of Vibration Attenuation Using Nonlinear Tuned Mass Damper and Pendulum Tuned Mass Damper in Parallel. Nonlinear Dynamics 2014; 78(4): 2699–2715.

12. Huang C., Arrigan J., Nagarajaiah S., Basu B. Semi-Active Algorithm for Edgewise Vibration Control in Floating Wind Turbine Blades. Conference: 12th Biennial International Conference on Engineering, Construction, and Operations in Challenging Environments 2010, p. 2097–2110.
13. Arrigan J., Pakrashi V., Basu B., Nagarajaiah S. Control of flapwise vibrations in wind turbine blades using semiactive tuned mass dampers. Struct. Control Health Monit 2011; 18: 840–851.
14. Sun C., Nagarajaiah S. Study on Semi-active Tuned Mass Damper with Variable Damping and Stiffness under Seismic Excitations. Structural Control and Health Monitoring 2014; 21(6): 890-906.
15. Staino A., Basu B., Nielsen S. Actuator control of edgewise vibrations in wind turbine blades. J. Sound Vib. 2012; 331: 1233–1256.
16. Staino A., Basu B. Dynamics and control of vibrations in wind turbines with variable rotor speed. Engineering Structures 2013; 56: 58–67
17. Fitzgerald B., Basu B., Nielsen S. Active tuned mass dampers for control of in-plane vibrations of wind turbine blades Structural Control and Health Monitoring 2013; 20: 1377–1396.
18. Fitzgerald B., Basu B. Cable connected active tuned mass dampers for control of in-plane vibrations of wind turbine blades J. Sound Vib. 2014; 333: 5980–6004.
19. Veletsos A.S., Verbic B. Vibration of viscoelastic foundations. Earthquake Engineering and Structural Dynamics 1973; 2(1): 87–102.
20. Bush E., Manuel L. Foundation models for offshore wind turbines. 47th AIAA Aerospace Sciences Meeting Including The New Horizons Forum and Aerospace Exposition; 2009, p. 1–7.
21. Achmus M., Kuo Y., Abdel-Rahman K. Behavior of monopile foundations under cyclic lateral load. Computers and Geotechnics 2009; 36(5): 725–735.
22. Fitzgerald B., Basu B. Structural control of wind turbines with soil structure interaction included. Engineering Structures 2016; 111: 131–151.
23. Zhu T. The vibrations of pre-twisted rotating Timoshenko beams by the Rayleigh–Ritz method. Computational Mechanics 2011; 47: 395–408.
24. Jonkman J.M., Buhl M. FAST User's Guide. National Renewable Energy Laboratory, Technical Report, NREL/EL- 500-38230, Golden, CO; 2005.
25. Jonkman B.J., Kilcher L. TurbSim User's Guide: Version 1.06.00, National Renewable Energy Laboratory, Technical Report,Golden, CO; 2012.
26. Warburton G.B. Optimum absorber parameters for minimizing vibration response. Earthquake Eng. Structural Dynamics 1981, 9(3): 251–262.
27. Hansen M. Aerodynamics of wind turbines (Second Edition). Earthscan, London, UK, 2000.
28. Faltinsen O. Sea Loads on Ships and Offshore Structures. Cambridge University Press: Cambridge, UK, 1990.
29. Fitzgerald B., Arrigan J., Basu B. Damage detection in wind turbine blades using time-frequency analysis of vibration signals. The 2010 International Joint Conference on Neural Networks (IJCNN). IEEE 2010 : p.1–5.
30. Cohen L. Time-frequency analysis: theory and applications. Prentice Hall, 1995.
31. Jonkman J., Butterfield S., Musial W., Scott G. Definition of a 5-MW Reference Wind Turbine for Offshore Systems Development, National Renewable Energy Laboratory, Technical Report, Golden, CO; 2009.
32. Carswell W., Johansson J., Lovholt F., Arwade S.R., Madshus C., DeGroot D.J., Myers A.T. Foundation damping and the dynamics of offshore wind turbine monopiles. Renewable Energy 2015, 80: 724–736.
33. IEC. Wind turbines. Part 3: design requirements for offshore wind turbines. IEC 61400-3 (ed. 1), Geneva, Switzerland: International Electrotechnical Commission, 2009.

13th Americas Conference on Wind Engineering
Gainesville, Florida, USA
May 21 - 24, 2017

Critical Scouring Velocity of Loosely Laid Roof Aggregate

Lakshmana S. Doddipatla[a]

[a] *Center for Property Risk Solutions, FM Global, Research Division, Norwood, MA, USA, lakshmana.doddipatla@fmglobal.com*

ABSTRACT

Windborne debris are one of the primary causes of building damage during severe wind storms. Windborne debris arise from a variety of sources with a prominent one being the roof aggregate. The debris can accelerate and travel long distances, and have significant momentum to induce substantial damage on downstream buildings. Although most studies in this area are focused on debris flight, very few studies have addressed the initiation mechanism and the building parameters that influence the initial movement of roof aggregate. This paper provides a summary of previous studies on how various building parameters affect critical scouring velocity of roof aggregate, and how these data can be used from a risk mitigation point of view. The review suggests that the available data can be used for only up to a building height of 45 m, with and without parapets, and with no roof-mounted equipment. However, roof aggregate including loosely laid roof ballast is currently used up to an approximate building height of 150 m in some cities and these buildings usually have substantial roof-mounted equipment and parapets that are typically less than 1.8 m high. The available data may not necessarily provide reliable recommendations when extrapolated to these high-rise buildings as the aerodynamics of low-rise to high-rise buildings are different, and can be influenced by the presence of roof-mounted equipment. This study identifies areas that need further research to address the critical scouring velocity of roof aggregate for low-rise and high-rise buildings.

KEYWORDS

Roof aggregate, Roof ballast, Critical scouring velocity, Effect of parapet height, Effect of roof-mounted equipment.

13th Americas Conference on Wind Engineering
Gainesville, Florida, USA
May 21 - 24, 2017

Ember flight modeling and experiments

Ali Tohidi[a], N. B. Kaye[b]

[a] *University of Maryland, College Park, MD, USA, atohidi@umd.edu*
[b] *Clemson University, Clemson, SC, USA, nbkaye@clemson.edu*

ABSTRACT

Ember storms, and the resulting spot fires, are a leading cause of home destruction at the wildland urban interface (WUI). Embers form from burning vegetation and structures, are lofted into the atmospheric boundary layer by the fire plume, and are then transported downwind where they can start secondary fires. Standard ember flight models build on the classic debris flight equations though almost exclusively use simplified versions that have the effect of treating embers as compact debris. This eliminates lift forces and converts a complex six degree of freedom problem in to a two degree of freedom problem with a two dimensional trajectory. However, experimental and field evidence shows that embers are not compact but rather cylindrical and have complex three dimensional trajectories. This paper presents results of the first combined experimental and computational study of ember flight that fully accounts for drag, lift, and rotational forces applied to cylindrical debris in a model fire wind field.

A series of boundary layer wind tunnel experiments were conducted in which model embers were lofted into the wind field using a vertical jet. An air blower was installed below the wind tunnel floor and connected to a 15.2 cm diameter nozzle. Non-combusting square cylindrical model embers made of rigid insulation foam were then released one at a time just above the nozzle. The air jet blew the embers up into the wind field where they were subsequently transported downwind. Tests were run for combinations of three wind speeds, three jet speeds, and three different aspect ratio embers with 200 ember releases per parameter combination. Each ember's trajectory was recorded from the side using an HD video camera. Particle tracking software was developed in MATLAB to follow each ember's path. In parallel to these experiments a full scale LES simulation of the jet and wind tunnel was developed in OpenFOAM. The unsteady LES velocity field was then coupled with the six degree of freedom rod-like debris flight equations to computationally replicate the wind tunnel experiments. Due to the enormous computational cost of running the LES only two of the nine wind fields were simulated.

Both the experimental and computational results show that ember flight is highly variable. The coefficient of variation of both the maximum rise height and the downwind flight distance were of the order of 50%. In general the computational and experimental results gave similar values for the mean of the maximum rise height and downwind flight distance though the computational results had higher standard deviations about that mean. A sensitivity analysis conducted using the LES wind field indicated that the trajectory statistics are quite sensitive to the release location of the embers across the diameter of the jet. The results show that ember flight models require a full six degree of freedom debris flight model coupled with an unsteady wind field to fully capture the statistics of ember flight trajectories and the resulting risk of spot fires.

KEYWORDS

Wildfire spread, windborne debris, Embers, Spot fires, Wind tunnel testing, Computational Wind Engineering.

13th Americas Conference on Wind Engineering
Gainesville, Florida, USA
May 21 - 24, 2017

Interior Damage of Residential Buildings due to Wind-Driven Rain Intrusion

Farzaneh Raji [a], Nicholas Miller [b], Ioannis Zisis [c], Jean-Paul Pinelli [d], Arindam Gan Chowdhury [e]

[a] *Florida International University, Miami, Florida, fraji001@fiu.edu*
[b] *Florida Institute of Technology, Melbourne, Florida, nmiller2012@my.fit.edu*
[c] *Florida International University, Miami, Florida, izisis@fiu.edu @fiu.edu*
[d] *Florida Institute of Technology, Melbourne, Florida, pinelli@fit.edu*
[e] *Florida International University, Miami, Florida, chowdhur@fiu.edu*

ABSTRACT

The amount of economic losses caused by hurricanes has increased dramatically in past few decades. In order to establish and refine hurricane risk mitigation strategies, it is necessary to develop accurate hurricane loss estimation models. Understanding the mechanism of the hurricane induced damages and the influence of these damages on the sustained losses is fundamental for developing such models.

One of the major causes of residential building losses during a hurricane event is associated with the interior damages. Many of the post hurricane field surveys have proven that rain water intrusion is the main reason for damage to the building interior, which typically represents the highest potential insurance losses. Despite the importance of rain induced damages, the challenges in simulating Wind Driven Rain (WDR) in traditional structural laboratories and wind tunnels have hindered the progress of adequate research on water intrusion effects on the building interior. Among the few available wind tunnel based research studies on WDR, one recent study focused on deposition of WDR on residential building surfaces and estimated direct impinging rain and surface runoff coefficients. However, the study didn't delve into WDR propagation inside the building which is an important aspect to be considered for modeling of interior losses. This research aims to propose a method for the experimental evaluation of WDR propagation and deposition on interior elements of buildings during hurricane events.

To experimentally investigate the WDR propagation and consequent interior damages, the WDR is simulated at Wall of Wind (WOW) experimental laboratory at Florida International University (FIU). The target characteristics of WDR associated with hurricanes are obtained from data acquired during three past hurricane events. A 1:4 scale model of a typical low-rise residential building of Florida is tested with several exterior damage scenarios to investigate the effect of location and area of defects and breaches on the rain propagation. The building interior elements incorporated into the model include ceiling, partitions and flooring. The water intrusion path and the amount of water deposited on the building interior elements are investigated and the results are presented as a function of the percentage of intruded water that has penetrated into the building. The tests are performed for different wind angles of attack to account for the wind direction effect. The sustained damage by each interior element can later be evaluated based on the total amount of water that has reached that element over a duration of time. The damage will also be dependent on other parameters including the drying capacity for each element.

The resulted data can be applied in hurricane loss projection models in order to improve their loss estimation accuracy. Specifically, the Florida Public Hurricane Loss Model (FPHLM), which has already used the recent data on direct impinging rain and surface runoff coefficients, will attempt to directly use the rain propagation data to further improve its interior damage estimation methodology. Currently, the FPHLM uses a linear relationship to relate the volume of water ingress to the percentage of the interior that is damage. The results of this research could refine this relationship by allowing for distinct modeling of the propagation of water through the interior parts of a building and refining the relationship between the exposure of an interior element to water and the damage incurred.

KEYWORDS

Wind Driven Rain, Rain Propagation, Interior Damages, Hurricane Loss, Residential Buildings

A Theoretical Model for Rapid Estimates of Rainfall during Tropical Cyclones

Reda Snaiki, Teng Wu
University at Buffalo, State University of New York, Buffalo, NY, USA

ABSTRACT: A new, physical-based rainfall model for translating tropical cyclones was proposed in this study. This theoretical model was developed essentially based on the surface boundary condition where the total momentum flux (total surface stress) was decomposed into two contributions from wind and rain, respectively. An analytical wind field was utilized during the model construction leading to a linear analytical solution of the surface rain rate. The proposed rainfall model explicitly depends on parameters for a typical tropical cyclone wind field simulation, namely storm location, approach angle, translation speed, radius of maximum wind, pressure profile, surface drag coefficient, and turbulent diffusivity. Hence, it could be readily implemented into state-of-the-art tropical cyclone risk assessment using the Monte Carlo sampling methodology. Sensitivity analysis of the rainfall field to the abovementioned parameters was comprehensively conducted. The results generated by the present model are consistent with field measurements.

KEYWORDS: Tropical cyclone, rain rate, wind field simulation, rain-induced stress.

1 INTRODUCTION

Tropical cyclones are responsible for the substantial part of natural hazard-induced economic and life losses through high winds, torrential rain and wind-driven storm surge. Among these, the rainfall-induced inland-flooding contributes to the most significant portion of the tropical cyclone related damages (e.g., Landsea 2000; Rappaport 2000). Therefore, rainfall hazards simulation has attracted interest of a number of researchers to better assess the rain field inside the tropical cyclone. While there have been considerable advances in improving the simulation accuracy of tropical cyclone rain field based on advanced numerical weather prediction models, they are not practical for risk assessment due to their high computational demands.

In this study, a new, physical-based rainfall model for tropical cyclones was derived for high-efficiency simulations. It is primarily based on the surface boundary condition in which the total momentum flux or the total surface stress was decomposed into two contributions from wind and rain, respectively. Actually, the rain-induced stress was demonstrated to be as important as the wind-induced one since the raindrops interact with the near-surface wind and transfer momentum to the surface (Caldwell and Elliot 1971; 1972; Zhao et al. 2013). Moreover, raindrops produce both vertical and horizontal stresses on the surface. The vertical impact velocity of raindrop generates the vertical stress. Whereas the horizontal wind speed that accelerates the raindrops is responsible for the horizontal stress. Several previous studies (e.g., Trenberth et al. 2007) suggested that the rain rate was related to the horizontal wind speed. Hence, the low-level horizontal wind speed may be the dominant factor for a reasonable simulation of the tropical cyclone rainfall. The proposed model offers the possibility to use any reliable wind field model for the rain intensity estimation, however, in this study an analytical wind field model recently developed by Snaiki and Wu (2016) was utilized leading to a linear analytical solution of the surface rain rate. The model explicitly depends on various parameters for a typical tropical cyclone wind field simulation, namely storm location, approach angle, translation speed, radius of maximum wind, pressure profile, surface drag coefficient, and turbulent diffusivity. Hence, it could be readily implemented into state-of-the-art tropical cyclone risk assessment using the Monte Carlo sampling methodology. Sensitivity analysis of the rainfall field to several essential parameters in the rain rate simulation was comprehensively conducted. The new analytical rain rate model was validated using observation data obtained during hurricane Dennis (1999).

2 THEORETICAL MODEL FOR RAINFALL ESTIMATION

The rain intensity formula has been derived in this study based on the surface boundary condition in which the total momentum flux could be decomposed into various surface stresses. Generally, several factors can contribute to the total surface stress, namely the turbulent fluxes, rain effects, spray, airborne sediment if it exists (Saltation theory) and convection-induced stress. In this study it will be assumed that no airborne sediment exists therefore their momentum flux contribution will be disregarded as well as the convection-induced stress (Huang 2012). Also, the spray does not add extra momentum to the system but only redistribute the wind's momentum near the surface (Andreas 2004). Moreover, spray can evaporate quickly (Emanuel 1995) and therefore its contribution becomes negligible (e.g., Wu 1973; Fairall et al. 1994). On the other hand, the rain induced momentum flux can be significant since the raindrops interact with the near-surface wind and transfer momentum to the surface (Caldwell and Elliot 1971; 1972), which could be as important as the wind-induced stress. For example, Zhao et al. (2013) compared the horizontal rain-induced stress at various rain rates and the wind stress and concluded that, depending on the horizontal wind speed and the rain rate, the horizontal stress by rain can have the same order of magnitude with that by wind stress. Therefore, it is important to take the rain-induced stress contribution to the total stress near the surface.

The total horizontal shear stress or momentum flux density τ is expressed as:

$$\tau = \rho_a u_{*\,effective}^{2} \tag{1}$$

where $u_{*effective}$ = effective frictional velocity during rain; and ρ_a = air density. The total stress τ is then partitioned between two momentum contributions such that:

$$\tau = \tau_a + \tau_r \tag{2}$$

where τ_a = momentum flux contribution from wind; and τ_r = momentum flux contribution from rain. The stress τ_a can be estimated based on the first-order closure commonly used for the atmospheric surface layer as follows:

$$\tau_a = \rho_a k_m \left.\frac{dV_{wind}}{dz}\right|_{z=0} \tag{3}$$

where k_m = eddy viscosity; and V_{wind} = wind speed. A widely used parametrization of the rain stress τ_r that has been incorporated in several models, e.g., the one-dimensional mixed-layer model developed by Clayson and Kantha (1999), the bulk parametrization outlined by Fairall et al. (1996), and the Regional Ocean Modelling System (ROMS), relates the surface rain-induced stress to the rain rate and wind speed (Caldwell and Elliott 1971) as:

$$\tau_r = \gamma \rho_r V_{wind} R \tag{4}$$

where ρ_r = density of rainwater; γ = empirically determined factor varying between 0.8 and 0.9 in most cases except for situation where only very small drops are involved; and R = rain rate. In this study $\gamma = 0.85$ will be adopted (Caldwell and Elliott 1971; Wong and Toumi 2016).

On the other hand, the effective frictional velocity $u_{*effective}$ can be related to the effective roughness length $z_{0,effective}$ based on the general equation of the Charnock relation (Charnock 1955; Huang 2012):

$$u_{*\,effective}^{2} = g \cdot z_{0,effective} / \alpha \tag{5}$$

where g = gravitational acceleration constant; and α = empirical constant denoted as the Charnock constant. Then, Eqs. (1), (2), (3), (4), and (5) lead to the following formula of the rain rate:

$$R = \frac{\rho_a g z_{0,effective} / A_0 - \rho_a k_m \left. dV_{wind} / dz \right|_{z=0}}{0.85 \rho_r V} \tag{6}$$

The model could be extended from 1D to 2D for tropical cyclones with a total wind velocity vector $\mathbf{V}_{wind}$ expressed in the cylindrical coordinates as:

$$\mathbf{V}_{wind} = \mathbf{v}_r + \mathbf{v}_\theta \tag{7}$$

where $\mathbf{v}_r$ = radial component of the wind velocity; $\mathbf{v}_\theta$ = azimuthal component of the wind velocity. Furthermore the roughness length can be related to the bulk transfer coefficient of momentum (drag coefficient) $C_{d,effective}$ based on the following relationship (Paulson, 1970; Huang 1979):

$$C_{d,effective} = \frac{\kappa^2}{\ln\left(\frac{z}{z_{0,effective}}\right)^2} \tag{8}$$

where κ = Von Karman coefficient; z = height above the ground. It should be stressed out that the values of $z_{0,effective}$ or $C_{d,effective}$ are obtained from field measurements such us the data provided from the GPS dropsondes (e.g., Powell et al. 2003; Moon et al. 2007; Andreas et al. 2012; Zijlema et al. 2012). Therefore, the rain rate intensity of the tropical cyclone becomes:

$$R = \frac{\rho_a C_{d,effective}}{0.85 \rho_r} \sqrt{\left(v_r - \frac{\partial v_r}{\partial z} \frac{k_m}{C_{d,effective} |\mathbf{V}_{wind}|}\right)^2 + \left(v_\theta - \frac{\partial v_\theta}{\partial z} \frac{k_m}{C_{d,effective} |\mathbf{V}_{wind}|}\right)^2} \tag{9}$$

The evaluation of the rain rate as suggested here requires the horizontal wind speed components as input. Therefore, any reliable linear, non-linear or empirical wind field model can be used in this regards. In this study, the recently developed linear analytical model by Snaiki and Wu (2016) will be utilized which is convenient for the risk assessment.

3 WIND FIELD MODEL

The wind field used for the rain rate estimation is based on the analytical wind field model proposed by Snaiki and Wu (2016). A brief discussion of this model will be presented in this section. The governing equation of the wind field is described as follows:

$$\frac{\partial \boldsymbol{v}}{\partial t}+\boldsymbol{v}.\nabla \boldsymbol{v}=-\frac{1}{\rho}\nabla p-f\boldsymbol{k}\times \boldsymbol{v}+\boldsymbol{F} \tag{10}$$

where f = Coriolis parameter; $\boldsymbol{k}$ = unit vector in the vertical direction; and $\boldsymbol{F}$ = frictional force. In order to solve Eq. (10), the decomposition method is used where the wind velocity $\boldsymbol{v}$ is expressed as:

$$\boldsymbol{v}=\boldsymbol{v}_g+\boldsymbol{v}' \tag{11}$$

where $\boldsymbol{v}_g$ = gradient wind in the free atmosphere; and $\boldsymbol{v}'$ = frictional component near the ground surface. Therefore two equations can be derived from Eq. (10):

$$\frac{\partial \boldsymbol{v}_g}{\partial t}+\boldsymbol{v}_g.\nabla \boldsymbol{v}_g=-\frac{1}{\rho}\nabla p-f\boldsymbol{k}\times \boldsymbol{v}_g \tag{12a}$$

$$\frac{\partial \boldsymbol{v}'}{\partial t}+\boldsymbol{v}'.\nabla \boldsymbol{v}'+\boldsymbol{v}'.\nabla \boldsymbol{v}_g+\boldsymbol{v}_g.\nabla \boldsymbol{v}'=-f\boldsymbol{k}\times \boldsymbol{v}'+\boldsymbol{F} \tag{12b}$$

The solution of the gradient wind speed could be solved straightforwardly in the cylindrical coordinate system (Georgiou 1985). Hence:

$$v_{\theta g}=\frac{\left(-c\sin\left(\theta-\beta\right)-fr\right)}{2}+\left[\frac{\left(-c\sin\left(\theta-\beta\right)-fr\right)^2}{4}+\frac{r}{\rho}\frac{\partial p}{\partial r}\right]^{1/2} \tag{13}$$

The radial velocity is then obtained from the continuity equation which is usually disregarded as suggested by Meng et al. (1995) due to its insignificant effects.

To obtain the frictional wind speed component, the nonlinear Eq. (12b) is first simplified using the scale analysis approach, and then linearized leading to the frictional wind speed formulas as:

$$u'(\theta,z')=\left(\alpha/\beta\right)^{1/2}\times Real\left\{A_0\times e^{(q_0 z')}+A_1\times e^{(q_1 z'+i\theta)}+A_{-1}\times e^{(q_{-1}z'-i\theta)}\right\} \tag{14a}$$

$$v'(\theta,z')=Imag\left\{A_0\times e^{(q_0 z')}+A_1\times e^{(q_1 z'+i\theta)}+A_{-1}\times e^{(q_{-1}z'-i\theta)}\right\} \tag{14b}$$

where (u',v') = frictional components of the wind velocity; $q_1=-(1+i)(\gamma+\sqrt{\alpha\beta}-\phi)^{1/2}$; $q_{-1}=-(1+i)(-\gamma+\sqrt{\alpha\beta}-\phi)^{1/2}$; $q_0=-(\alpha\beta)^{1/4}$; $\alpha=\xi_g/2K$; $\beta=\xi_{ag}/2K$; $\xi_g=2v_{\theta g}/r+f$ is the absolute angular velocity; $\xi_{ag}=\partial v_{\theta g}/\partial r+v_{\theta g}/r+f$ is the vertical component of absolute vorticity of the gradient wind; K = the turbulent diffusivity which is assumed to be constant in this study; $\gamma=v_{\theta g}/2Kr$; $\phi=(1/2Kr)\partial v_{\theta g}/\partial\theta$; and z'= new vertical coordinate used as the base of the computation scheme where z'=0 is located above z_{10} (i.e., the 10 m height above the mean height of roughness elements) (Meng et al. 1995). The other necessary parameters needed for the simulation are presented as follows:

$$A_0 = -X_3/(X_1 + X_2 X_4) \tag{15a}$$

$$A_1 = icC_d e^{-i\beta}\left(A_0 + A_0^*\right)/\left[4K\left(q_1 - q_{-1}^*\right)\right] \tag{15b}$$

$$A_{-1} = icC_d e^{i\beta}\left(A_0 + A_0^*\right)/\left[4K\left(q_1^* - q_{-1}\right)\right] \tag{15c}$$

$$X_1 = \left[q_0 + \frac{fr}{K}C_d - \frac{2\eta C_d}{K} - \frac{c^2 C_d^2}{4K^2\left(q_1 - q_{-1}^*\right)} + \frac{c^2 C_d^2}{4K^2\left(q_1^* - q_{-1}\right)}\right] \tag{15d}$$

$$X_2 = \left[-q_0^* - \frac{fr}{K}C_d + \frac{2\eta C_d}{K} - \frac{c^2 C_d^2}{4K^2\left(q_1 - q_{-1}^*\right)} + \frac{c^2 C_d^2}{4K^2\left(q_1^* - q_{-1}\right)}\right] \tag{15e}$$

$$X_3 = -2iC_d\left(\eta - fr/2\right)^2/K \tag{15f}$$

$$X_4 = -\left[-q_0 - \frac{fr}{2K}C_d + \frac{\eta C_d}{K}\right]/\left[-q_0^* - \frac{fr}{2K}C_d + \frac{\eta C_d}{K}\right] \tag{15g}$$

$$\eta = \left[\frac{\left(-c\sin(\theta - \beta) - fr\right)^2}{4} + \frac{r}{\rho}\frac{\partial p}{\partial r}\right]^{1/2} \tag{15h}$$

4 MODEL VALIDATION

Hurricane Dennis affected large parts of North Carolina coast areas with strong winds and heavy rains (Lonfat et al. 2004). To simulate the rain rate profile of hurricane Dennis, the wind parameters were obtained from the National Hurricane Center's North Atlantic Hurricane Database (HURDAT). The necessary parameters needed for the tropical cyclone wind field simulation are: β approach angle; c translation velocity of the hurricane; p_c central pressure; Δp central pressure difference; R_{max} radius of maximum winds; B Holland's parameter; ψ latitude; and λ longitude. On the other hand, the tropical cyclone rainfall data can be obtained from the Tropical Rainfall Measuring Mission (TRMM) database.

The rain rate was simulated using the proposed model and is depicted in Figure 1. Good agreement between the simulations and observations is highlighted except for some localized regions which are mostly due to local intensifications caused by rainbands.

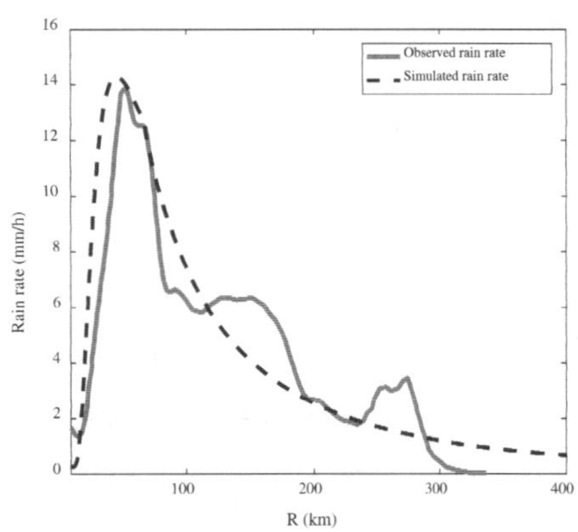

Figure 1. Comparison of the radial profile of the azimuthally averaged rain rates of Hurricane Dennis on 28 Aug 1999

5 SENSITIVITY ANALYSIS

Numerous studies have demonstrated based on field measurement and numerical simulations that the rainfall distribution depends on environmental factors (e.g., roughness) and inherent tropical cyclone factors (e.g., intensity, translation speed) (Lonfat et al, 2004). The proposed rainfall model explicitly depends on these parameters. Figure 2 illustrates the sensitivity of the rainfall intensity to several selected parameters, namely B Holland parameter, r_m radius of maximum wind, c translation speed, Δp central pressure difference, k_m turbulent diffusivity, and z_0 equivalent roughness length. The baseline scenario is taken as: $B=1$, $r_m = 30km$, $c = 10m/s$, $\Delta p = 70hpa$, $k_m = 50m^2/s$ and $z_0 = 0.001m$. The radial profile of the rain rate was taken at $\theta = 0°$ (counterclockwise positive from the East) for all simulations.

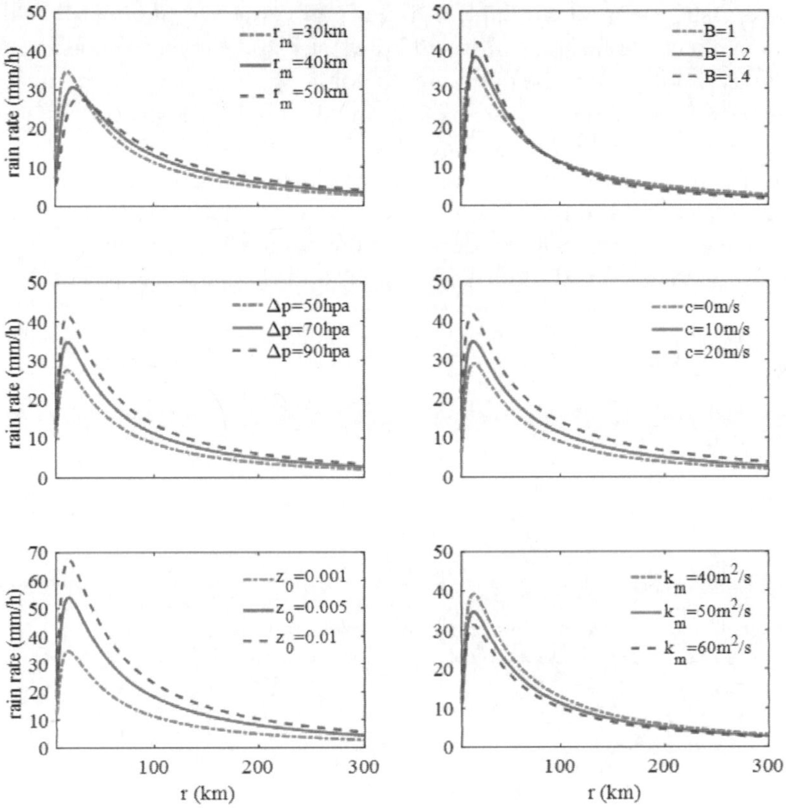

Figure 2. Sensitivity analysis of the rain rate based on the proposed model at $\theta = 0°$.

As it can be concluded from Figure 2, the central pressure difference Δp, the translational tropical cyclone speed c and the surface roughness z_0 have significant effects on the rain rate. For example, higher value of Δp, which indicates larger maximum wind v_m (Holland et al. 2010), leads to higher values of the rain intensity. This is in accordance to the observations and findings of the literature (e.g., Lonfat et al., 2004; Tuleya et al., 2007; Langousis and Veneziano 2009). Similarly, an increase in the translational tropical cyclone speed results in the enhancement of the total precipitation. In addition, the surface roughness can alter substantially the rain intensity which has been reported by several studies (e.g., Trenberth et al., 2007; Langousis and Veneziano 2009; and Lin et al., 2010). On the other hand, lower values of the radius of maximum wind leads to more peaked profiles concentrated towards the tropical cyclone center. Low sensitivity of the rain rate to the vertical turbulent diffusivity is noted as well. The Holland's parameter *B* mainly modify the decay rate of the rainfall intensity profile and has small effects on the maximum rain rate.

6 CONCLUDING REMARKS

A new, theoretical rain field model was proposed in this study for rapid rain intensity estimation. The rain rate formula was essentially developed based on the surface boundary condition where the total momentum is decomposed into several contributions from different sources. The proposed formula indicates that the spatial distribution of the rainfall field is governed by the wind field inside the tropical cyclone. This confirms the findings of several previous studies where it has been shown that the low-level horizontal wind speed is the dominant factor for tropical cyclone rainfall. A recently developed analytical wind field model was utilized during the model construction leading to a linear analytical solution of the surface rain rate. The sensitivity analysis was comprehensively carried out to investigate the rain rate dependence on several physical tropical cyclone parameters. The proposed model offers good simulation results and is consistent with tropical cyclone observations. It can be readily used in conjunction with the Monte Carlo techniques to perform risk analysis of tropical cyclone hazards.

7 ACKNOWLEDGEMENTS

The support for this project provided by the NSF Grant # CMMI 15-37431 is gratefully acknowledged.

8 REFERENCES

Andreas, E.L., 2004. Spray stress revisited. Journal of Physical Oceanography, 34(6), pp.1429-1440.

Andreas, E.L., Mahrt, L. and Vickers, D., 2012. A new drag relation for aerodynamically rough flow over the ocean. Journal of the Atmospheric Sciences, 69(8), pp.2520-2537.

Caldwell, D.R. and Elliott, W.P., 1971. Surface stresses produced by rainfall. Journal of Physical Oceanography, 1(2), pp.145-148.

Caldwell, D.R. and Elliott, W.P., 1972. The effect of rainfall on the wind in the surface layer. Boundary-Layer Meteorology, 3(2), pp.146-151.

Charnock, H., 1955. Wind stress on a water surface. Quarterly Journal of the Royal Meteorological Society, 81(350), pp.639-640.

Clayson, C.A. and Kantha, L.H., 1999. Turbulent kinetic energy and its dissipation rate in the equatorial mixed layer. Journal of physical oceanography, 29(9), pp.2146-2166.

Emanuel, K.A., 1995. Sensitivity of tropical cyclones to surface exchange coefficients and a revised steady-state model incorporating eye dynamics. Journal of the Atmospheric Sciences, 52(22), pp.3969-3976.

Fairall, C.W., Kepert, J.D. and Holland, G.J., 1994. The effect of sea spray on surface energy transports over the ocean. Global Atmos. Ocean Syst, 2(2-3), pp.121-142.

Fairall, C.W., Bradley, E.F., Rogers, D.P., Edson, J.B. and Young, G.S., 1996. Bulk parameterization of air-sea fluxes for tropical ocean-global atmosphere coupled-ocean atmosphere response experiment. Journal of Geophysical Research: Oceans, 101(C2), pp.3747-3764.

Georgiou, P.N., 1986. Design wind speeds in tropical cyclone-prone regions.

Holland, G.J., Belanger, J.I. and Fritz, A., 2010. A revised model for radial profiles of hurricane winds. Monthly Weather Review, 138(12), pp.4393-4401.

Holton, J.R. and Hakim, G.J., 2012. An introduction to dynamic meteorology (Vol. 88). Academic press.

Huang, C.H., 2012. Modification of the Charnock wind stress formula to include the effects of free convection and swell. Advanced Methods for Practical Applications in Fluid Mechanics, p.47.

Kummerow, C., Barnes, W., Kozu, T., Shiue, J. and Simpson, J., 1998. The tropical rainfall measuring mission (TRMM) sensor package. Journal of atmospheric and oceanic technology, 15(3), pp.809-817.

Landsea, C.W., 2000. Climate variability of tropical cyclones: past, present and future. Storms. Routledge, New York, pp.220-241.

Langousis, A. and Veneziano, D., 2009. Theoretical model of rainfall in tropical cyclones for the assessment of long-term risk. Journal of Geophysical Research: Atmospheres, 114(D2).

Lin, N., Smith, J.A., Villarini, G., Marchok, T.P. and Baeck, M.L., 2010. Modeling extreme rainfall, winds, and surge from Hurricane Isabel (2003). Weather and Forecasting, 25(5), pp.1342-1361.

Lonfat, M., Marks Jr, F.D. and Chen, S.S., 2004. Precipitation distribution in tropical cyclones using the Tropical Rainfall Measuring Mission (TRMM) microwave imager: A global perspective. Monthly Weather Review, 132(7), pp.1645-1660.

Meng, Y., Matsui, M. and Hibi, K., 1995. An analytical model for simulation of the wind field in a typhoon boundary layer. Journal of Wind Engineering and Industrial Aerodynamics, 56(2-3), pp.291-310.

Moon, I.J., Ginis, I., Hara, T. and Thomas, B., 2007. A physics-based parameterization of air–sea momentum flux at high wind speeds and its impact on hurricane intensity predictions. Monthly weather review, 135(8), pp.2869-2878.

Pomeroy, J.W. and Male, D.H., 1987. Wind transport of seasonal snowcovers. In Seasonal Snowcovers: Physics, Chemistry, Hydrology (pp. 119-140). Springer Netherlands.

Powell, M.D., Vickery, P.J. and Reinhold, T.A., 2003. Reduced drag coefficient for high wind speeds in tropical cyclones. Nature, 422(6929), pp.279-283.

Rappaport, E.N., 2000. Loss of life in the United States associated with recent Atlantic tropical cyclones. Bulletin of the American Meteorological Society, 81(9), pp.2065-2073.

Raupach, M.R., 1991. Saltation layers, vegetation canopies and roughness lengths. In Aeolian Grain Transport 1 (pp. 83-96). Springer Vienna.

Simpson, J., Adler, R.F. and North, G.R., 1988. A proposed tropical rainfall measuring mission (TRMM) satellite. Bulletin of the American meteorological Society, 69(3), pp.278-295.

Snaiki, R., Wu, T., 2016. Temperature and Moisture Effects on the Tropical Cyclone Boundary Layer: Pressure and Wind. Proceedings of 8th International Colloquium on Bluff-Body Aerodynamics and its Application (BBAA-VIII), June, 2016, Boston, USA.

Trenberth, K.E. and Fasullo, J., 2007. Water and energy budgets of hurricanes and implications for climate change. Journal of Geophysical Research: Atmospheres, 112(D23).

Tuleya, R.E., DeMaria, M. and Kuligowski, R.J., 2007. Evaluation of GFDL and simple statistical model rainfall forecasts for US landfalling tropical storms. Weather and forecasting, 22(1), pp.56-70.

Wong, B. and Toumi, R., 2016. Effect of extreme ocean precipitation on sea surface elevation and storm surges. Quarterly Journal of the Royal Meteorological Society, 142(699), pp.2541-2550.

Wu, J., 1973. Spray in the atmospheric surface layer: Laboratory study. Journal of Geophysical Research, 78(3), pp.511-519.

Zhao, D., Ma, X., Liu, B. and Xie, L., 2013. Rainfall effect on wind waves and the turbulence beneath air-sea interface. Acta Oceanologica Sinica, 32(11), pp.10-20.

Zijlema, M., Van Vledder, G.P. and Holthuijsen, L.H., 2012. Bottom friction and wind drag for wave models. Coastal Engineering, 65, pp.19-26.

AUTHOR INDEX

AUTHOR INDEX

AUTHOR INDEX

AUTHOR INDEX

American Association for Wind Engineering (AAWE)
c/o Cermak Peterka Peterson, Inc.
1415 Blue Spruce Drive, #3
Fort Collins, Colorado 80524
USA

ISBN 978-1-5108-5482-6

9 781510 854826

CURRAN ASSOCIATES INC.
proceedings.com